Social Sciences

Sports and Recreation

A Library of Functions

Identity function

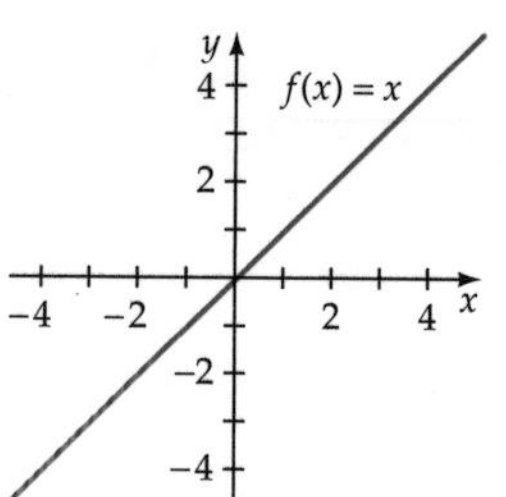

Linear function

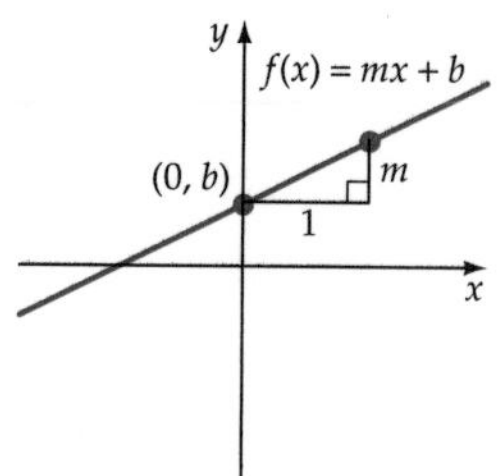

Constant function

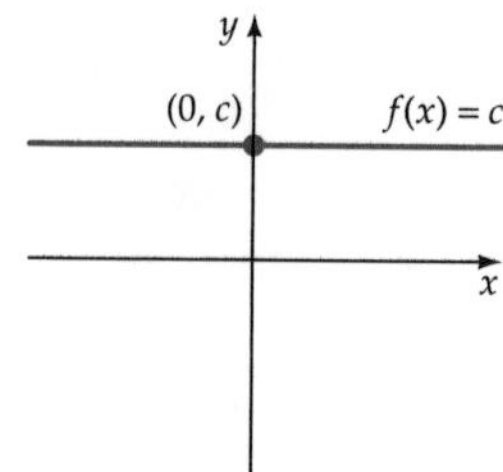

Absolute value function

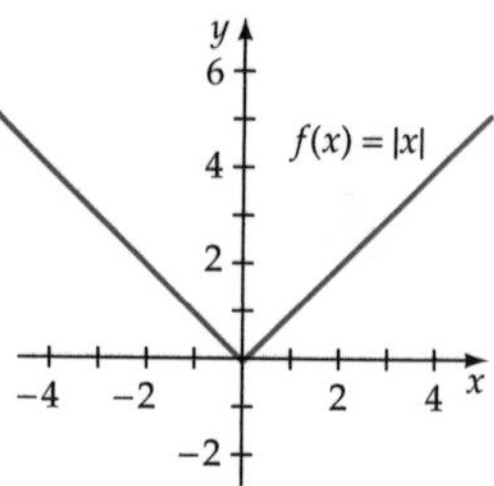

Squaring function

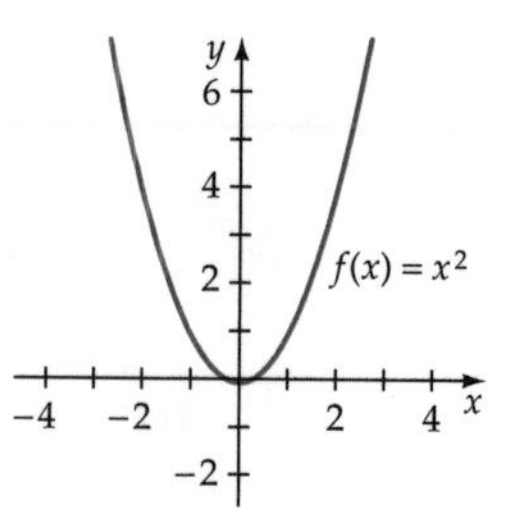

Cubing function

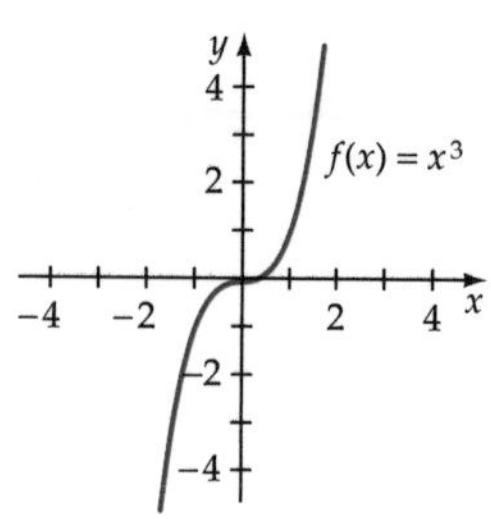

Square root function

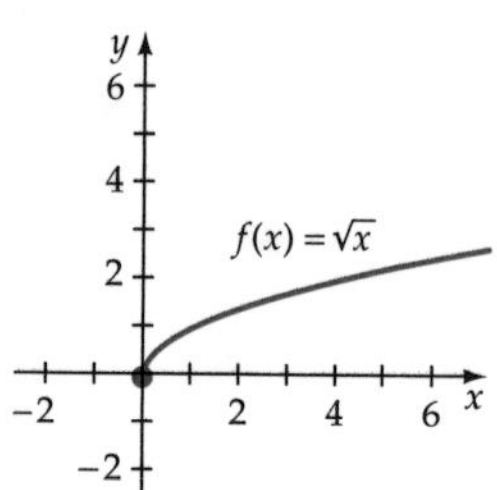

Cube root function

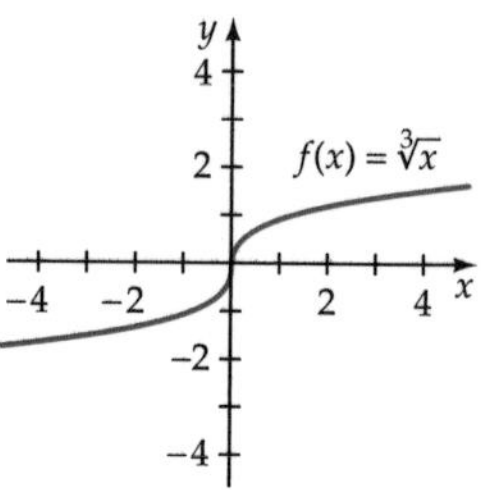

Exponential function

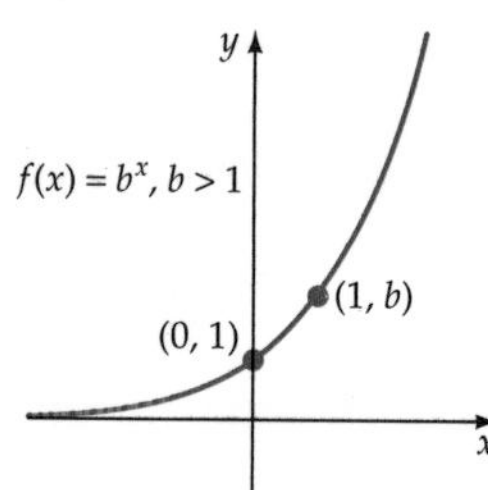

Exponential function

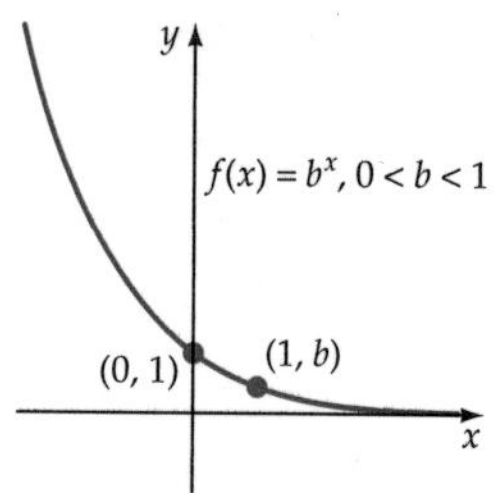

Logarithmic function

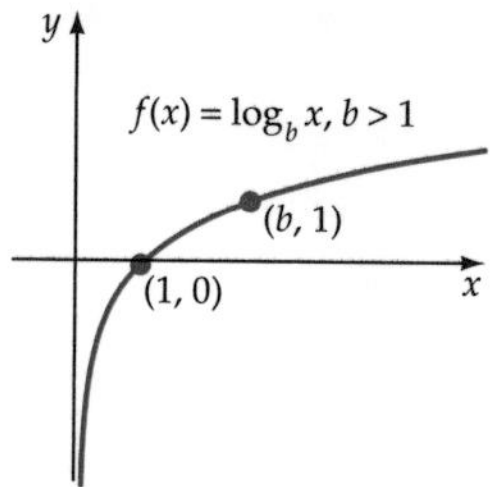

Logarithmic function

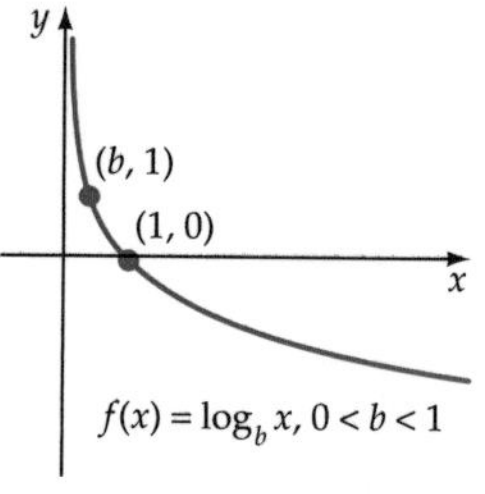

Logistic function

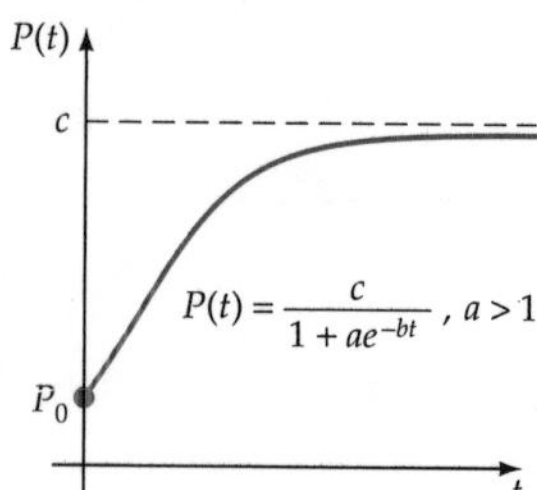

Logistic function

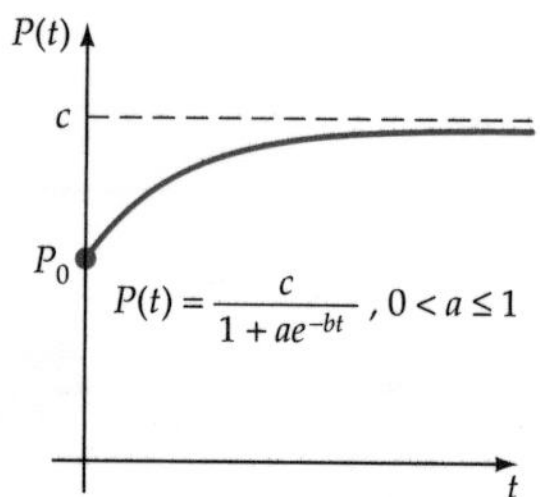

Reciprocal function

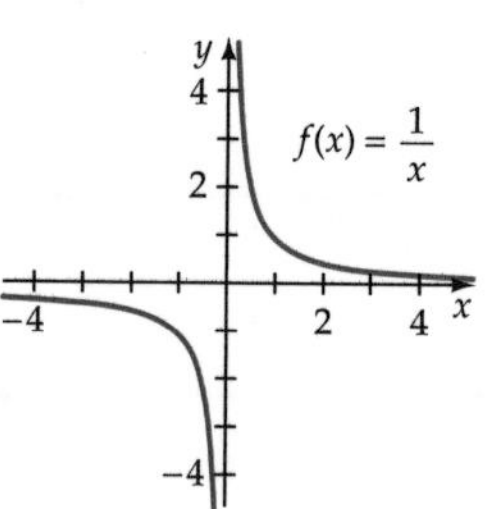

A rational function

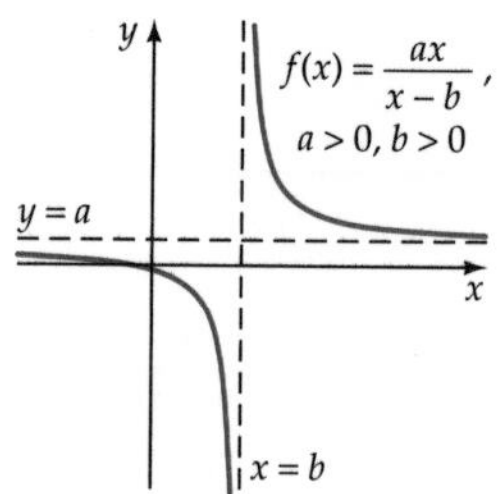

Instructor's Annotated Edition

Applied College Algebra

Richard N. Aufmann

Richard D. Nation

Daniel K. Clegg

Palomar College

HOUGHTON MIFFLIN COMPANY

Boston New York

Publisher: Jack Shira
Senior Sponsoring Editor: Lynn Cox
Senior Development Editor: Dawn Nuttall
Assistant Editor: Jennifer King
Project Editor: Kathleen Deselle
Senior Production/Design Coordinator: Carol Merrigan
Manufacturing Manager: Florence Cadran
Senior Marketing Manager: Danielle Potvin
Marketing Associate: Nicole Mollica

Cover photos: Whale leaping from sea, © Veer/Digital Vision; Wall Street, New York, USA, © Veer/Digital Vision; Cycling Race, © Veer/Digital Vision; Space Shuttle, © PunchStockStockbyte; Graduates portrait, © John Henley/CORBIS

Photo credits: **Chapter P:** p. 1, John Anderson Photography; p. 2, AP/Wide World Photos; p. 56, Jose Luis Pelaez, Inc./CORBIS. **Chapter 1:** p. 81, AP/Wide World Photos; p. 103, The Granger Collection; p. 108, PhotoDisc/Getty Images; p. 111, AP/Wide World Photos; p. 119, Courtesy of NASA and STSci; p. 136, David Young-Wolff/PhotoEdit, Inc.; p. 139, Reuters NewMedia Inc./CORBIS; p. 146, Brian Bahr/Getty Images. **Chapter 2:** p. 161, Myrleen Ferguson Cate/PhotoEdit, Inc.; p. 222, Bob David/Golden Gate Bridge; p. 223, ML Sinibaldi/CORBIS; p. 229, Stuart Hannagan/Allsport/Getty Images; p. 230, Sandy Felsenthal/CORBIS; p. 231, Roger Ressmeyer/CORBIS. **Chapter 3:** p. 251, Jonathan Nourak/PhotoEdit, Inc. **Chapter 4:** p. 311, Strauss/Curtis/CORBIS; p. 321, Topham/The Image Works; p. 340, The Granger Collection; p. 352, CORBIS; p. 373, Richard T. Nowitz/CORBIS. **Chapter 5:** p. 383, Getty Images; p. 391, Bettmann/CORBIS; p. 397, Charles O'Rear/CORBIS; p. 401, Bettmann/CORBIS; p. 419, AFP/CORBIS; p. 457, AP/Wide World. **Chapter 6:** p. 577, Tom Nebbia/CORBIS. **Chapter 7:** p. 579, Deborah Davis/PhotoEdit, Inc.; p. 582, Bettmann/CORBIS; p. 591, ARPL/Topham/The Image Works. **Chapter 8:** p. 645, Cindy Charles/PhotoEdit, Inc.; p. 664, Bettmann/CORBIS.

Printed in the U.S.A.

Library of Congress Control Number: 2002109361

Student Text ISBN: 0-618-07363-9
Instructor's Annotated Edition ISBN: 0-618-07364-7

123456789-VH-07 06 05 04 03

Contents

CHAPTER 1 Equations and Inequalities 81

APPLICATIONS

CHAPTER 2 Introduction to Functions 161

APPLICATIONS

CHAPTER 3 Properties of Functions 251

CHAPTER 4 Polynomial and Rational Functions 311

CHAPTER 5 Exponential and Logarithmic Functions 383

APPLICATIONS

Additional Topics on the Internet*

(available only at math.college.hmco.com)

Measures of Central Tendency and Dispersion
Mean
Median
Mode
Range
Standard Deviation
Variance

Distribution of Data and the Empirical Rule
Stem-and-Leaf Diagrams
Frequency Distributions and Histograms
Normal Distributions and the Empirical Rule
z-scores

Mathematical Induction
Principle of Mathematical Induction
Extended Principle of Mathematical Induction

Additional Matrix Topics
Input-Output Analysis
Cramer's Rule

*These topics are fully covered with Examples, Check Your Progress (with solutions), and exercises (with answers).

Preface

Applied College Algebra is a new text designed to assist students in making connections between mathematics and its applications. Our goal is to develop a student's mathematical skills through appropriate use of applications and to establish links between abstract mathematical concepts and visual representations or concrete applications.

Our hallmark *interactive approach,* which encourages students to practice a skill or concept as it is presented and get immediate feedback, is also highlighted in this text. For each numbered Example within a section, there is a similar *Check Your Progress* problem for the student to try. The numbered example is worked out; the *Check Your Progress* is for the student to work. By solving this problem, the student actively practices concepts as they are presented in the text. There are *complete worked-out* solutions to the *Check Your Progress* problems in an appendix. Students can compare their solution to the solution in the appendix and thereby obtain immediate feedback on the concept. In addition, by providing complete worked-out solutions to the *Check Your Progress,* it significantly increases the number of examples available for students to refer to when doing homework or studying for a test.

The application of algebra is a central theme of this text. We have not only provided applications from traditional disciplines such as the physical sciences and engineering but have incorporated, among others, applications from business, economics, social science, life science, health science, and sports. The application topics provide instructors with numerous options for engaging students with diverse interests and allow instructors to customize the course to address the needs of students with a variety of career goals.

Through the use of applications, we demonstrate to students that mathematics has a vast array of tools that can be used to solve relevant, meaningful problems. Modeling, analytic representation, and verbal representations of problems and their solutions are encouraged. We have also integrated numerous data analysis exercises throughout the text and encourage students to derive meaningful conclusions about the data.

In some cases, we have incorporated a writing component to an exercise that asks the student to write a few sentences explaining the meaning of an answer in the context of the problem. Additional writing exercises are integrated throughout most exercise sets. These exercises ask students to make a conjecture based on some given facts, restate a concept in their own words, provide a written answer to a question, or research a topic and write a short report.

We have paid special attention to the standards suggested by AMATYC and have made a serious attempt to incorporate these standards in the text. Problem solving, critical analysis, function concept, connecting mathematics to other disciplines through applications, multiple representations of concepts, and the appropriate use of technology are all integrated within this text. Our goal is to provide students with a variety of analytical tools that will make them more effective quantitative thinkers and problem solvers.

Chapter Opening Features

Chapter Opener

Each chapter begins with a **Chapter Opener** that illustrates a specific application of a concept from the chapter. This application references an exercise in the chapter where students solve a problem related to the chapter opener topic.

57. *Wedding Expenses* The function $C(t) = 17t^2 + 128t + 5900$ models the average cost of a wedding reception and the function $W(t) = 38t^2 + 291t + 15{,}208$ models the average cost of a wedding, where $t = 0$ represents the year 1990 and $0 \le t \le 12$. The rational function

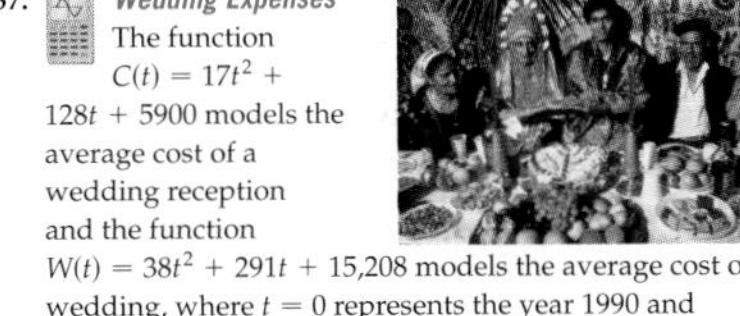

$$R(t) = \frac{C(t)}{W(t)} = \frac{17t^2 + 128t + 5900}{38t^2 + 291t + 15{,}208}$$

gives the relative cost of the reception compared to the cost of a wedding.

a. Use $R(t)$ to estimate the relative cost of the reception compared to the cost of a wedding for the years $t = 0$, $t = 7$, and $t = 12$. Round your results to the nearest tenth of a percent.

page 373

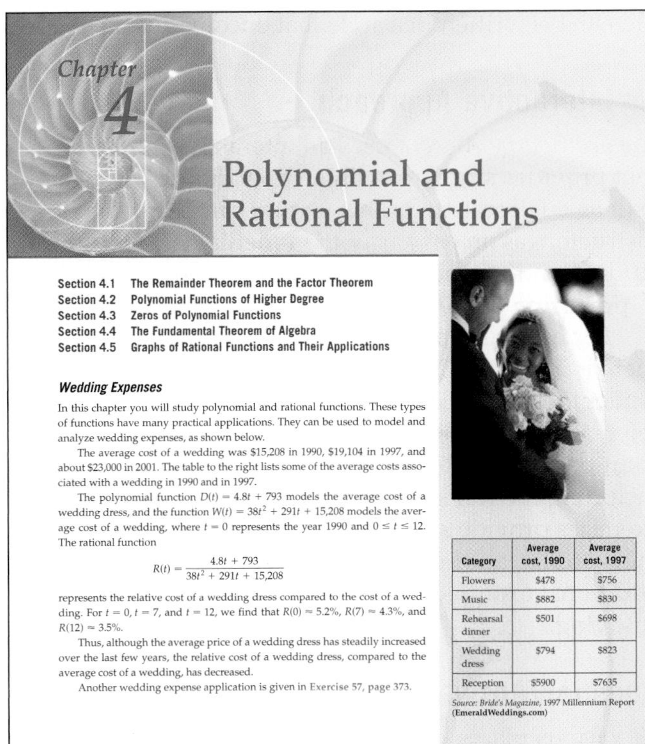

Chapter 4

Polynomial and Rational Functions

Wedding Expenses

In this chapter you will study polynomial and rational functions. These types of functions have many practical applications. They can be used to model and analyze wedding expenses, as shown below.

The average cost of a wedding was \$15,208 in 1990, \$19,104 in 1997, and about \$23,000 in 2001. The table to the right lists some of the average costs associated with a wedding in 1990 and in 1997.

The polynomial function $D(t) = 4.8t + 793$ models the average cost of a wedding dress, and the function $W(t) = 38t^2 + 291t + 15{,}208$ models the average cost of a wedding, where $t = 0$ represents the year 1990 and $0 \le t \le 12$. The rational function

$$R(t) = \frac{4.8t + 793}{38t^2 + 291t + 15{,}208}$$

represents the relative cost of a wedding dress compared to the cost of a wedding. For $t = 0$, $t = 7$, and $t = 12$, we find that $R(0) \approx 5.2\%$, $R(7) \approx 4.3\%$, and $R(12) \approx 3.5\%$.

Thus, although the average price of a wedding dress has steadily increased over the last few years, the relative cost of a wedding dress, compared to the average cost of a wedding, has decreased.

Another wedding expense application is given in Exercise 57, page 373.

Category	Average cost, 1990	Average cost, 1997
Flowers	\$478	\$756
Music	\$882	\$830
Rehearsal dinner	\$501	\$698
Wedding dress	\$794	\$823
Reception	\$5900	\$7635

Source: Bride's Magazine, 1997 Millennium Report (EmeraldWeddings.com)

page 311

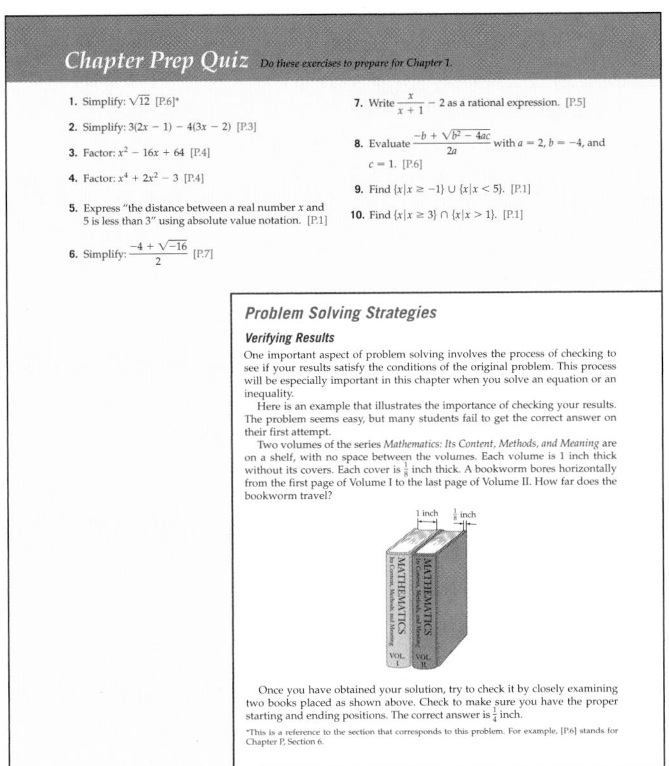

Chapter Prep Quiz *Do these exercises to prepare for Chapter 1.*

1. Simplify: $\sqrt{12}$ [P.6]*
2. Simplify: $3(2x - 1) - 4(3x - 2)$ [P.3]
3. Factor: $x^2 - 16x + 64$ [P.4]
4. Factor: $x^4 + 2x^2 - 3$ [P.4]
5. Express "the distance between a real number x and 5 is less than 3" using absolute value notation. [P.1]
6. Simplify: $\frac{-4 + \sqrt{-16}}{2}$ [P.7]
7. Write $\frac{x}{x+1} - 2$ as a rational expression. [P.5]
8. Evaluate $\frac{-b + \sqrt{b^2 - 4ac}}{2a}$ with $a = 2$, $b = -4$, and $c = 1$. [P.6]
9. Find $\{x \mid x \ge -1\} \cup \{x \mid x < 5\}$. [P.1]
10. Find $\{x \mid x \ge 3\} \cap \{x \mid x > 1\}$. [P.1]

Problem Solving Strategies

Verifying Results

One important aspect of problem solving involves the process of checking to see if your results satisfy the conditions of the original problem. This process will be especially important in this chapter when you solve an equation or an inequality.

Here is an example that illustrates the importance of checking your results. The problem seems easy, but many students fail to get the correct answer on their first attempt.

Two volumes of the series *Mathematics: Its Content, Methods, and Meaning* are on a shelf, with no space between the volumes. Each volume is 1 inch thick without its covers. Each cover is $\frac{1}{8}$ inch thick. A bookworm bores horizontally from the first page of Volume I to the last page of Volume II. How far does the bookworm travel?

Once you have obtained your solution, try to check it by closely examining two books placed as shown above. Check to make sure you have the proper starting and ending positions. The correct answer is $\frac{1}{4}$ inch.

*This is a reference to the section that corresponds to this problem. For example, [P.6] stands for Chapter P, Section 6.

page 82

Prep Quiz and Problem Solving Strategies

Chapter Prep Quizzes occur at the beginning of each chapter and test students on previously covered concepts that are required in order to succeed in the upcoming chapter. Next to each question, in brackets, is a reference to the section of the text that contains the concepts related to the question to allow students to refer back for help. All answers are provided in Answers to Selected Exercises.

Problem Solving Strategies give students insight into successful problem-solving strategies and help students better understand how they are used.

Aufmann Interactive Method (AIM)

This text is written in a style that encourages the student to interact with the textbook.

An Interactive Approach

Applied College Algebra uses an interactive approach that provides students with an opportunity to try a skill as it is presented. This feature can be used by instructors as an easy way to immediately check for student understanding and to actively engage students in practicing concepts as they are presented.

For each numbered Example within a section, there is a similar *Check Your Progress* problem for the student to work. Each *Check Your Progress* problem references a page in the back of the text where the full solution is presented—rather than just the answer. By including the full solution, the *Check Your Progress* exercises provide students immediate feedback on their understanding of the concepts and serve as additional examples for students to refer to while studying.

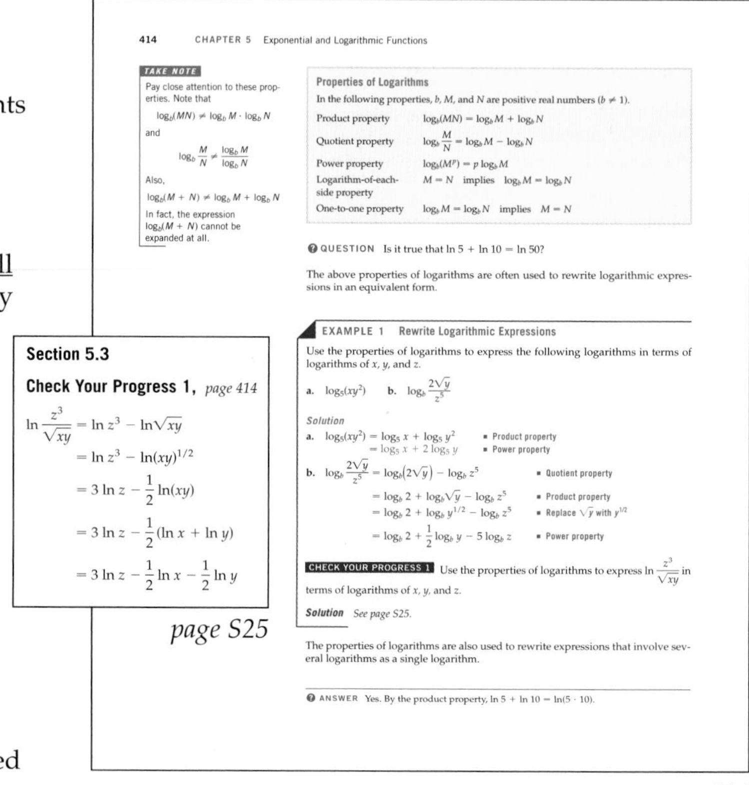

414 CHAPTER 5 Exponential and Logarithmic Functions

TAKE NOTE
Pay close attention to these properties. Note that

$\log_b(MN) \neq \log_b M \cdot \log_b N$

and

$\log_b \frac{M}{N} \neq \frac{\log_b M}{\log_b N}$

Also,

$\log_b(M+N) \neq \log_b M + \log_b N$

In fact, the expression $\log_b(M+N)$ cannot be expanded at all.

Properties of Logarithms

In the following properties, b, M, and N are positive real numbers ($b \neq 1$).

Product property	$\log_b(MN) = \log_b M + \log_b N$
Quotient property	$\log_b \frac{M}{N} = \log_b M - \log_b N$
Power property	$\log_b(M^p) = p \log_b M$
Logarithm-of-each-side property	$M = N$ implies $\log_b M = \log_b N$
One-to-one property	$\log_b M = \log_b N$ implies $M = N$

QUESTION Is it true that $\ln 5 + \ln 10 = \ln 50$?

The above properties of logarithms are often used to rewrite logarithmic expressions in an equivalent form.

EXAMPLE 1 Rewrite Logarithmic Expressions

Use the properties of logarithms to express the following logarithms in terms of logarithms of x, y, and z.

a. $\log_5(xy^2)$ b. $\log_b \frac{2\sqrt{y}}{z^5}$

Solution

a. $\log_5(xy^2) = \log_5 x + \log_5 y^2$ • Product property
$= \log_5 x + 2\log_5 y$ • Power property

b. $\log_b \frac{2\sqrt{y}}{z^5} = \log_b(2\sqrt{y}) - \log_b z^5$ • Quotient property
$= \log_b 2 + \log_b \sqrt{y} - \log_b z^5$ • Product property
$= \log_b 2 + \log_b y^{1/2} - \log_b z^5$ • Replace $\sqrt{y}$ with $y^{1/2}$
$= \log_b 2 + \frac{1}{2}\log_b y - 5\log_b z$ • Power property

CHECK YOUR PROGRESS 1 Use the properties of logarithms to express $\ln \frac{z^3}{\sqrt{xy}}$ in terms of logarithms of x, y, and z.

Solution See page S25.

The properties of logarithms are also used to rewrite expressions that involve several logarithms as a single logarithm.

ANSWER Yes. By the product property, $\ln 5 + \ln 10 = \ln(5 \cdot 10)$.

Section 5.3

Check Your Progress 1, *page 414*

$$\ln \frac{z^3}{\sqrt{xy}} = \ln z^3 - \ln\sqrt{xy}$$
$$= \ln z^3 - \ln(xy)^{1/2}$$
$$= 3\ln z - \frac{1}{2}\ln(xy)$$
$$= 3\ln z - \frac{1}{2}(\ln x + \ln y)$$
$$= 3\ln z - \frac{1}{2}\ln x - \frac{1}{2}\ln y$$

page S25

page 414

Question/Answer

At various places during a discussion, we ask students to respond to a **Question** about the material being presented. This question encourages students to pause and think about the mathematics. To make sure students do not miss important information, and to help those students studying independently, the **Answer** to the question is provided as a footnote at the bottom of the page.

AIM for Success

Welcome to *Applied College Algebra*. As you begin this course, we know two important facts: (1) We want you to succeed. (2) You want to succeed. To do that requires an effort from each of us. For the next few pages, we are going to show you what is required of you to achieve that success and how you can use the features of this text to be successful.

Motivation One of the most important keys to success is motivation. We can try to motivate you by offering interesting or important ways mathematics can benefit you. But, in the end, the motivation must come from you. On the first day of class, it is easy to be motivated. Eight weeks into the term, it is harder to keep that motivation.

page xxiii

AIM for Success Student Preface

This "how to use this book" student preface explains what is required of a student to be successful in mathematics and how this text has been designed to foster student success through the Aufmann Interactive Method (AIM). *AIM for Success* can be used as a lesson on the first day of class or as a project for students to complete to strengthen their study skills. There are suggestions for teaching this lesson in the *Instructor's Resource Manual* and on the *Class Prep CD*.

Real Data and Applications

Applications

One way to motivate an interest in mathematics is through applications. This carefully integrated, applied approach generates student awareness of the value of algebra as a relevant real-life tool.

Applications in this text are taken from many disciplines to address the diverse interests and backgrounds of students. Topics include agriculture, business, carpentry, chemistry, construction, Earth science, health science, education, manufacturing, nutrition, real estate, sports, and sociology.

Wherever appropriate, applications use problem-solving strategies to solve practical problems.

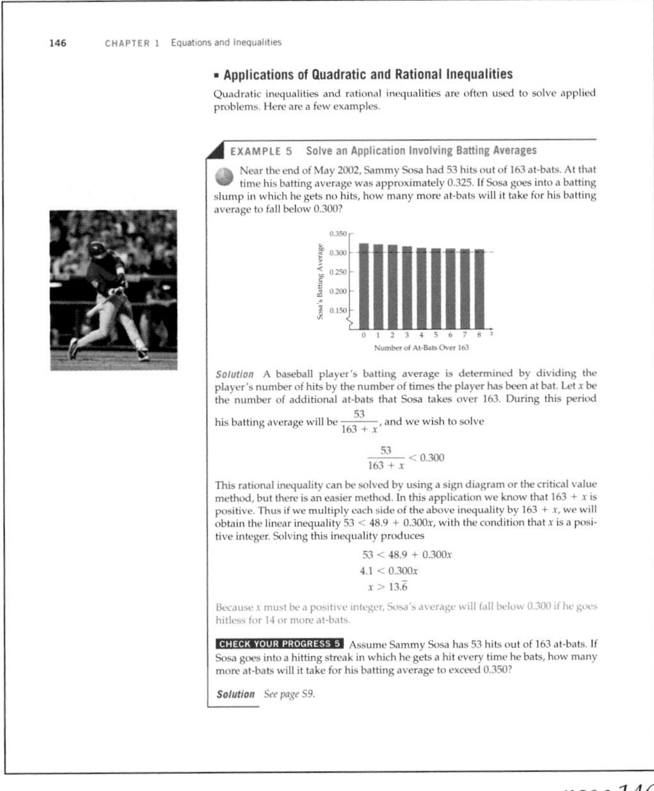

146 CHAPTER 1 Equations and Inequalities

Applications of Quadratic and Rational Inequalities

Quadratic inequalities and rational inequalities are often used to solve applied problems. Here are a few examples.

EXAMPLE 5 Solve an Application Involving Batting Averages

Near the end of May 2002, Sammy Sosa had 53 hits out of 163 at-bats. At that time his batting average was approximately 0.325. If Sosa goes into a batting slump in which he gets no hits, how many more at-bats will it take for his batting average to fall below 0.300?

Solution A baseball player's batting average is determined by dividing the player's number of hits by the number of times the player has been at bat. Let x be the number of additional at-bats that Sosa takes over 163. During this period his batting average will be $\frac{53}{163+x}$, and we wish to solve

$$\frac{53}{163+x} < 0.300$$

This rational inequality can be solved by using a sign diagram or the critical value method, but there is an easier method. In this application we know that $163 + x$ is positive. Thus if we multiply each side of the above inequality by $163 + x$, we will obtain the linear inequality $53 < 48.9 + 0.300x$, with the condition that x is a positive integer. Solving this inequality produces

$$53 < 48.9 + 0.300x$$
$$4.1 < 0.300x$$
$$x > 13.\overline{6}$$

Because x must be a positive integer, Sosa's average will fall below 0.300 if he goes hitless for 14 or more at-bats.

CHECK YOUR PROGRESS 5 Assume Sammy Sosa has 53 hits out of 163 at-bats. If Sosa goes into a hitting streak in which he gets a hit every time he bats, how many more at-bats will it take for his batting average to exceed 0.350?

Solution See page S9.

page 146

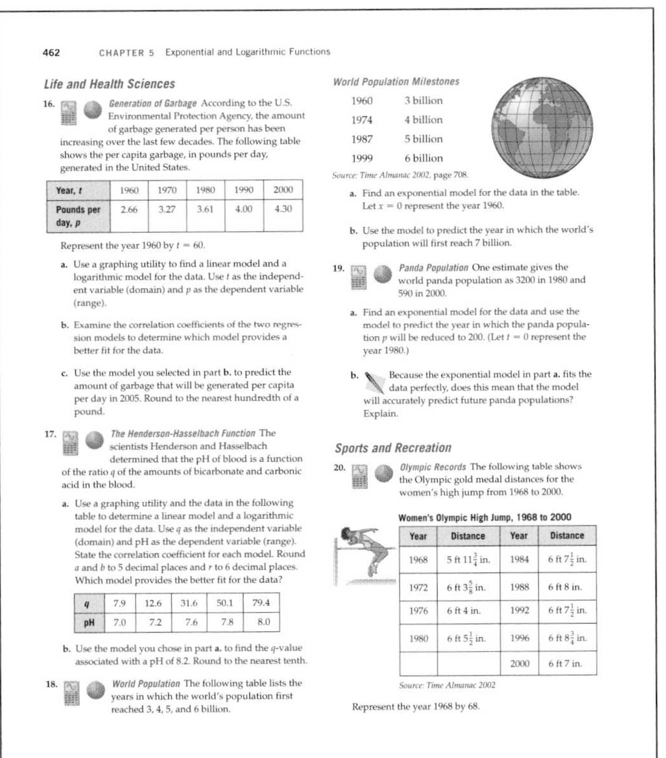

462 CHAPTER 5 Exponential and Logarithmic Functions

Life and Health Sciences

16. *Generation of Garbage* According to the U.S. Environmental Protection Agency, the amount of garbage generated per person has been increasing over the last few decades. The following table shows the per capita garbage, in pounds per day, generated in the United States.

Year, t	1960	1970	1980	1990	2000
Pounds per day, p	2.66	3.27	3.61	4.00	4.30

Represent the year 1960 by $t = 60$.

a. Use a graphing utility to find a linear model and a logarithmic model for the data. Use t as the independent variable (domain) and p as the dependent variable (range).

b. Examine the correlation coefficients of the two regression models to determine which model provides a better fit for the data.

c. Use the model you selected in part **b.** to predict the amount of garbage that will be generated per capita per day in 2005. Round to the nearest hundredth of a pound.

17. *The Henderson-Hasselbach Function* The scientists Henderson and Hasselbach determined that the pH of blood is a function of the ratio q of the amounts of bicarbonate and carbonic acid in the blood.

a. Use a graphing utility and the data in the following table to determine a linear model and a logarithmic model for the data. Use q as the independent variable (domain) and pH as the dependent variable (range). State the correlation coefficient for each model. Round a and b to 5 decimal places and r to 6 decimal places. Which model provides the better fit for the data?

q	7.9	12.6	31.6	50.1	79.4
pH	7.0	7.2	7.6	7.8	8.0

b. Use the model you chose in part **a.** to find the q-value associated with a pH of 8.2. Round to the nearest tenth.

18. *World Population* The following table lists the years in which the world's population first reached 3, 4, 5, and 6 billion.

World Population Milestones

1960	3 billion
1974	4 billion
1987	5 billion
1999	6 billion

Source: Time Almanac 2002, page 708.

a. Find an exponential model for the data in the table. Let $x = 0$ represent the year 1960.

b. Use the model to predict the year in which the world's population will first reach 7 billion.

19. *Panda Population* One estimate gives the world panda population as 3200 in 1980 and 590 in 2000.

a. Find an exponential model for the data and use the model to predict the year in which the panda population p will be reduced to 200. (Let $t = 0$ represent the year 1980.)

b. Because the exponential model in part **a.** fits the data perfectly, does this mean that the model will accurately predict future panda populations? Explain.

Sports and Recreation

20. *Olympic Records* The following table shows the Olympic gold medal distances for the women's high jump from 1968 to 2000.

Women's Olympic High Jump, 1968 to 2000

Year	Distance	Year	Distance
1968	5 ft $11\frac{3}{4}$ in.	1984	6 ft $7\frac{1}{2}$ in.
1972	6 ft $3\frac{5}{8}$ in.	1988	6 ft 8 in.
1976	6 ft 4 in.	1992	6 ft $7\frac{1}{2}$ in.
1980	6 ft $5\frac{1}{2}$ in.	1996	6 ft $8\frac{3}{4}$ in.
		2000	6 ft 7 in.

Source: Time Almanac 2002

Represent the year 1968 by 68.

page 462

Real Data

Real data examples and exercises, identified by [real data icon], ask students to analyze and solve problems taken from actual situations. Students often work with tables, graphs, and charts drawn from a variety of disciplines.

Technology

Integrating Technology

The Integrating Technology feature contains discussions that can be used to further explore a concept using technology. Some introduce technology as an alternative way to solve certain problems and others provide suggestions for using a calculator to solve certain problems and applications.

Additionally, optional graphing calculator examples and exercises (identified by [calculator icon]) are presented throughout the text.

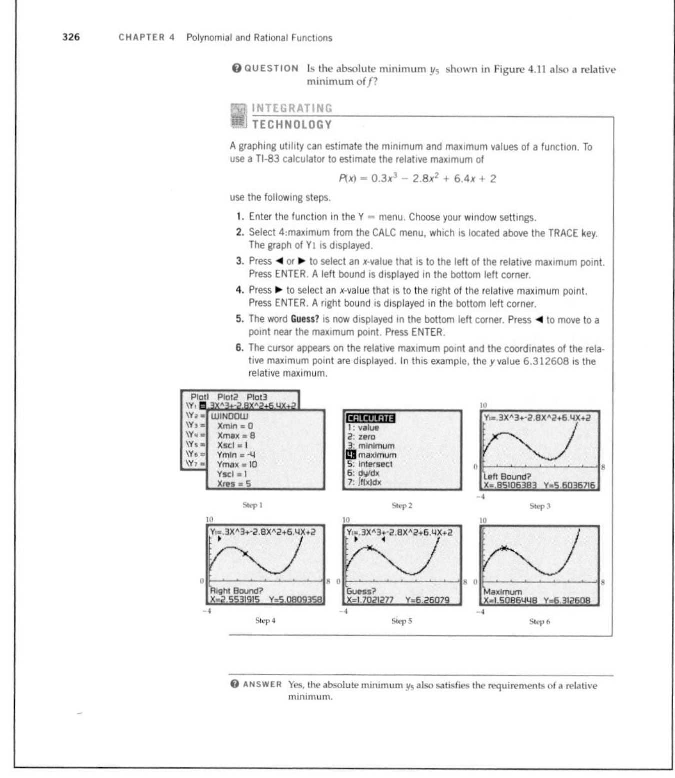
326 CHAPTER 4 Polynomial and Rational Functions

? QUESTION Is the absolute minimum y_5 shown in Figure 4.11 also a relative minimum of f?

INTEGRATING TECHNOLOGY

A graphing utility can estimate the minimum and maximum values of a function. To use a TI-83 calculator to estimate the relative maximum of

$$P(x) = 0.3x^3 - 2.8x^2 + 6.4x + 2$$

use the following steps.

1. Enter the function in the Y = menu. Choose your window settings.
2. Select 4:maximum from the CALC menu, which is located above the TRACE key. The graph of Y1 is displayed.
3. Press ◀ or ▶ to select an x-value that is to the left of the relative maximum point. Press ENTER. A left bound is displayed in the bottom left corner.
4. Press ▶ to select an x-value that is to the right of the relative maximum point. Press ENTER. A right bound is displayed in the bottom left corner.
5. The word **Guess?** is now displayed in the bottom left corner. Press ◀ to move to a point near the maximum point. Press ENTER.
6. The cursor appears on the relative maximum point and the coordinates of the relative maximum point are displayed. In this example, the y value 6.312608 is the relative maximum.

Step 1 Step 2 Step 3

Step 4 Step 5 Step 6

? ANSWER Yes, the absolute minimum y_5 also satisfies the requirements of a relative minimum.

page 326

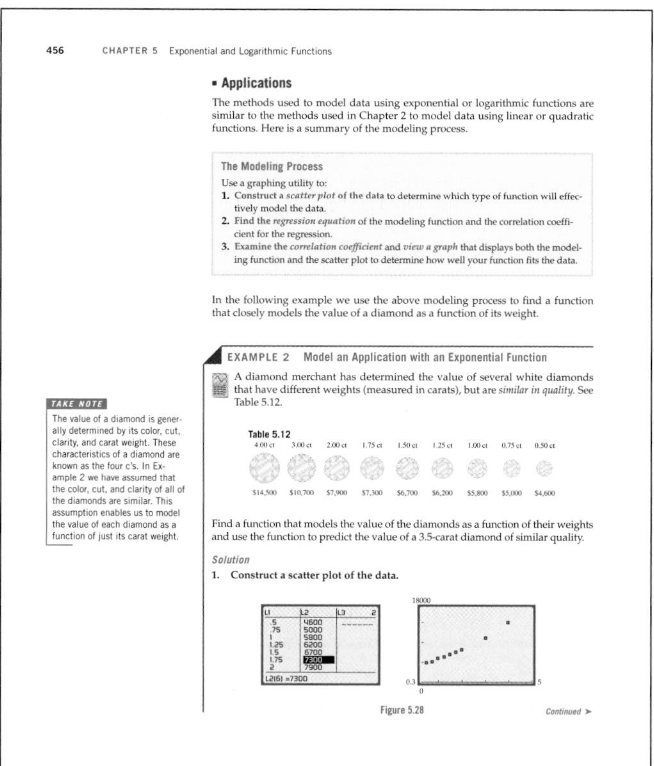
456 CHAPTER 5 Exponential and Logarithmic Functions

Applications

The methods used to model data using exponential or logarithmic functions are similar to the methods used in Chapter 2 to model data using linear or quadratic functions. Here is a summary of the modeling process.

The Modeling Process

Use a graphing utility to:

1. Construct a *scatter plot* of the data to determine which type of function will effectively model the data.
2. Find the *regression equation* of the modeling function and the correlation coefficient for the regression.
3. Examine the *correlation coefficient* and *view a graph* that displays both the modeling function and the scatter plot to determine how well your function fits the data.

In the following example we use the above modeling process to find a function that closely models the value of a diamond as a function of its weight.

EXAMPLE 2 Model an Application with an Exponential Function

A diamond merchant has determined the value of several white diamonds that have different weights (measured in carats), but are *similar in quality*. See Table 5.12.

TAKE NOTE

The value of a diamond is generally determined by its color, cut, clarity, and carat weight. These characteristics of a diamond are known as the four c's. In Example 2 we have assumed that the color, cut, and clarity of all of the diamonds are similar. This assumption enables us to model the value of each diamond as a function of just its carat weight.

Table 5.12

4.00 ct	3.00 ct	2.00 ct	1.75 ct	1.50 ct	1.25 ct	1.00 ct	0.75 ct	0.50 ct
\$14,500	\$10,700	\$7,900	\$7,300	\$6,700	\$6,200	\$5,800	\$5,000	\$4,600

Find a function that models the value of the diamonds as a function of their weights and use the function to predict the value of a 3.5-carat diamond of similar quality.

Solution

1. **Construct a scatter plot of the data.**

Figure 5.28

Continued ➤

page 456

Modeling

Special modeling sections, which rely heavily on the use of a graphing calculator, are incorporated within the text. These optional sections introduce the idea of a mathematical model using various real-world data sets, which further motivate students and help them see the relevance of mathematics.

Student Pedagogy

This text was designed to be an understandable resource for students. Special emphasis was given to readability and effective pedagogical use of color to highlight important words and concepts.

Icons

The     icons at each objective head remind students of the many and varied additional resources available for each objective.

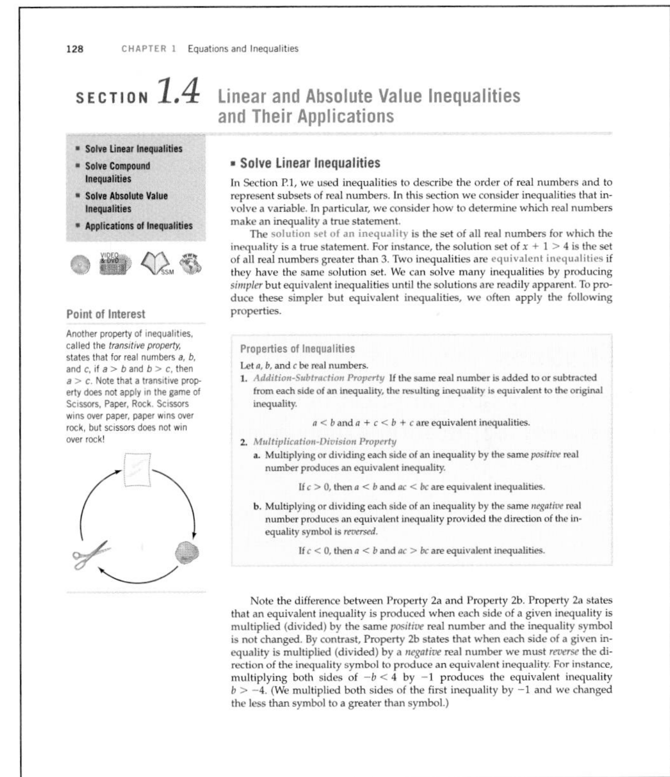

128 CHAPTER 1 Equations and Inequalities

SECTION 1.4 Linear and Absolute Value Inequalities and Their Applications

- Solve Linear Inequalities
- Solve Compound Inequalities
- Solve Absolute Value Inequalities
- Applications of Inequalities

Point of Interest

Another property of inequalities, called the *transitive property,* states that for real numbers a, b, and c, if $a > b$ and $b > c$, then $a > c$. Note that a transitive property does not apply in the game of Scissors, Paper, Rock. Scissors wins over paper, paper wins over rock, but scissors does not win over rock!

Solve Linear Inequalities

In Section P.1, we used inequalities to describe the order of real numbers and to represent subsets of real numbers. In this section we consider inequalities that involve a variable. In particular, we consider how to determine which real numbers make an inequality a true statement.

The solution set of an inequality is the set of all real numbers for which the inequality is a true statement. For instance, the solution set of $x + 1 > 4$ is the set of all real numbers greater than 3. Two inequalities are equivalent inequalities if they have the same solution set. We can solve many inequalities by producing *simpler* but equivalent inequalities until the solutions are readily apparent. To produce these simpler but equivalent inequalities, we often apply the following properties.

Properties of Inequalities

Let a, b, and c be real numbers.

1. *Addition-Subtraction Property* If the same real number is added to or subtracted from each side of an inequality, the resulting inequality is equivalent to the original inequality.

   $a < b$ and $a + c < b + c$ are equivalent inequalities.

2. *Multiplication-Division Property*
   a. Multiplying or dividing each side of an inequality by the same *positive* real number produces an equivalent inequality.

      If $c > 0$, then $a < b$ and $ac < bc$ are equivalent inequalities.

   b. Multiplying or dividing each side of an inequality by the same *negative* real number produces an equivalent inequality provided the direction of the inequality symbol is *reversed.*

      If $c < 0$, then $a < b$ and $ac > bc$ are equivalent inequalities.

Note the difference between Property 2a and Property 2b. Property 2a states that an equivalent inequality is produced when each side of a given inequality is multiplied (divided) by the same *positive* real number and the inequality symbol is not changed. By contrast, Property 2b states that when each side of a given inequality is multiplied (divided) by a *negative* real number we must *reverse* the direction of the inequality symbol to produce an equivalent inequality. For instance, multiplying both sides of $-b < 4$ by -1 produces the equivalent inequality $b > -4$. (We multiplied both sides of the first inequality by -1 and we changed the less than symbol to a greater than symbol.)

page 128

Key Terms and Important Concepts

A blue bold font is used whenever a **key term** is first introduced.

Important Concepts are presented in yellow boxes in order to highlight these concepts and serve as an easy-to-find reference.

Point of Interest

These margin notes contain interesting comments about mathematics, its history, or its application.

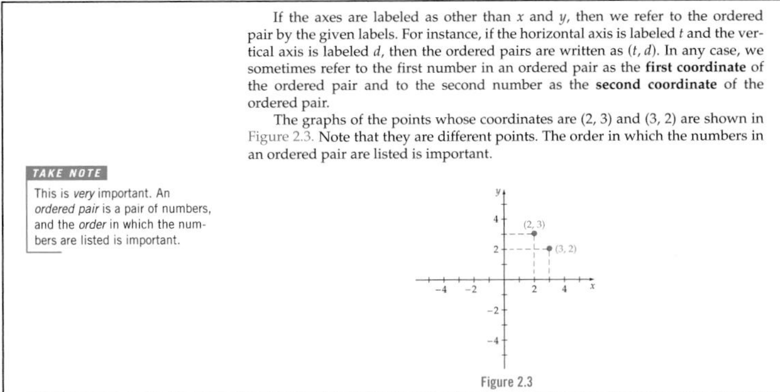

If the axes are labeled as other than x and y, then we refer to the ordered pair by the given labels. For instance, if the horizontal axis is labeled t and the vertical axis is labeled d, then the ordered pairs are written as (t, d). In any case, we sometimes refer to the first number in an ordered pair as the **first coordinate** of the ordered pair and to the second number as the **second coordinate** of the ordered pair.

The graphs of the points whose coordinates are (2, 3) and (3, 2) are shown in Figure 2.3. Note that they are different points. The order in which the numbers in an ordered pair are listed is important.

TAKE NOTE

This is *very* important. An *ordered pair* is a pair of numbers, and the *order* in which the numbers are listed is important.

Figure 2.3

page 164

Take Note

These margin notes alert students to a point requiring special attention or are used to highlight the concept under discussion.

Exercises

Topics for Discussion

Topics for Discussion provide questions related to key concepts of the section. Instructors can use these to initiate class discussions or to ask students to write about concepts presented in the section.

Exercises

The exercise sets of *Applied College Algebra* emphasize skill building, skill maintenance, and applications. Concept-based writing or developmental exercises have also been integrated within the exercise sets.

Icons identify appropriate writing , group, data analysis , and graphing calculator 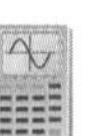 exercises.

Applications

Whenever possible, applications of mathematics are emphasized. Application exercises are grouped under one of five categories:

Business and Economics ***Life and Health Sciences***
Social Sciences ***Sports and Recreation***
Physical Sciences and Engineering

Each application exercise has a title that further describes the particular application.

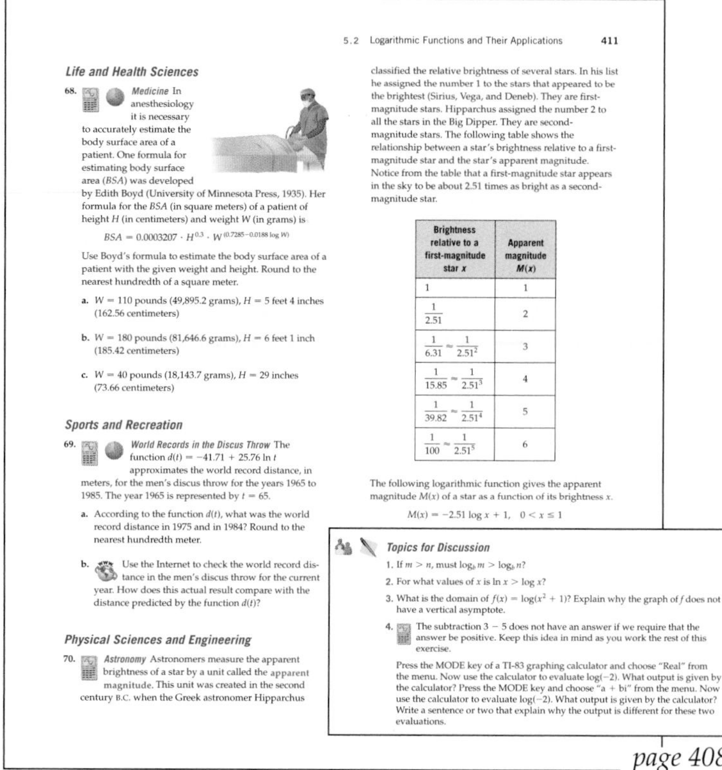

5.2 Logarithmic Functions and Their Applications 411

Life and Health Sciences

68. ***Medicine*** In anesthesiology it is necessary to accurately estimate the body surface area of a patient. One formula for estimating body surface area (BSA) was developed by Edith Boyd (University of Minnesota Press, 1935). Her formula for the BSA (in square meters) of a patient of height H (in centimeters) and weight W (in grams) is

$$BSA = 0.0003207 \cdot H^{0.3} \cdot W^{(0.7285-0.0188 \log W)}$$

Use Boyd's formula to estimate the body surface area of a patient with the given weight and height. Round to the nearest hundredth of a square meter.

a. $W = 110$ pounds (49,895.2 grams), $H = 5$ feet 4 inches (162.56 centimeters)

b. $W = 180$ pounds (81,646.6 grams), $H = 6$ feet 1 inch (185.42 centimeters)

c. $W = 40$ pounds (18,143.7 grams), $H = 29$ inches (73.66 centimeters)

Sports and Recreation

69. ***World Records in the Discus Throw*** The function $d(t) = -41.71 + 25.76 \ln t$ approximates the world record distance, in meters, for the men's discus throw for the years 1965 to 1985. The year 1965 is represented by $t = 65$.

a. According to the function $d(t)$, what was the world record distance in 1975 and in 1984? Round to the nearest hundredth meter.

b. Use the Internet to check the world record distance in the men's discus throw for the current year. How does this actual result compare with the distance predicted by the function $d(t)$?

Physical Sciences and Engineering

70. ***Astronomy*** Astronomers measure the apparent brightness of a star by a unit called the apparent magnitude. This unit was created in the second century B.C. when the Greek astronomer Hipparchus classified the relative brightness of several stars. In his list he assigned the number 1 to the stars that appeared to be the brightest (Sirius, Vega, and Deneb). They are first-magnitude stars. Hipparchus assigned the number 2 to all the stars in the Big Dipper. They are second-magnitude stars. The following table shows the relationship between a star's brightness relative to a first-magnitude star and the star's apparent magnitude. Notice from the table that a first-magnitude star appears in the sky to be about 2.51 times as bright as a second-magnitude star.

Brightness relative to a first-magnitude star x	Apparent magnitude $M(x)$
1	1
$\frac{1}{2.51}$	2
$\frac{1}{6.31} \approx \frac{1}{2.51^2}$	3
$\frac{1}{15.85} \approx \frac{1}{2.51^3}$	4
$\frac{1}{39.82} \approx \frac{1}{2.51^4}$	5
$\frac{1}{100} \approx \frac{1}{2.51^5}$	6

The following logarithmic function gives the apparent magnitude $M(x)$ of a star as a function of its brightness x.

$$M(x) = -2.51 \log x + 1, \quad 0 < x \le 1$$

Topics for Discussion

1. If $m > n$, must $\log_b m > \log_b n$?
2. For what values of x is $\ln x > \log x$?
3. What is the domain of $f(x) = \log(x^2 + 1)$? Explain why the graph of f does not have a vertical asymptote.
4. The subtraction $3 - 5$ does not have an answer if we require that the answer be positive. Keep this idea in mind as you work the rest of this exercise.

Press the MODE key of a TI-83 graphing calculator and choose "Real" from the menu. Now use the calculator to evaluate $\log(-2)$. What output is given by the calculator? Press the MODE key and choose "a + bi" from the menu. Now use the calculator to evaluate $\log(-2)$. What output is given by the calculator? Write a sentence or two that explain why the output is different for these two evaluations.

page 411

page 408

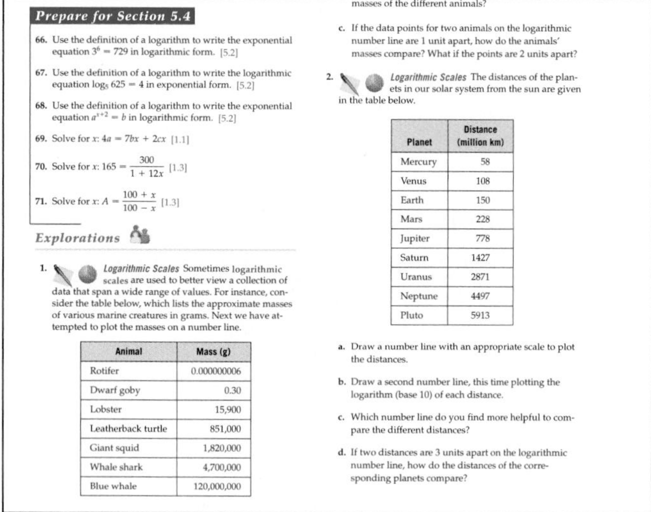

Prepare for Section 5.4

66. Use the definition of a logarithm to write the exponential equation $3^6 = 729$ in logarithmic form. [5.2]

67. Use the definition of a logarithm to write the logarithmic equation $\log_5 625 = 4$ in exponential form. [5.2]

68. Use the definition of a logarithm to write the exponential equation $a^{x+2} = b$ in logarithmic form. [5.2]

69. Solve for x: $4a = 7bx + 2cx$ [1.1]

70. Solve for x: $165 = \frac{300}{1 + 12x}$ [1.3]

71. Solve for x: $A = \frac{100 + x}{100 - x}$ [1.3]

Explorations

1. ***Logarithmic Scales*** Sometimes logarithmic scales are used to better view a collection of data that span a wide range of values. For instance, consider the table below, which lists the approximate masses of various marine creatures in grams. Next we have attempted to plot the masses on a number line.

Animal	Mass (g)
Rotifer	0.000000006
Dwarf goby	0.30
Lobster	15,900
Leatherback turtle	851,000
Giant squid	1,820,000
Whale shark	4,700,000
Blue whale	120,000,000

masses of the different animals?

c. If the data points for two animals on the logarithmic number line are 1 unit apart, how do the animals' masses compare? What if the points are 2 units apart?

2. ***Logarithmic Scales*** The distances of the planets in our solar system from the sun are given in the table below.

Planet	Distance (million km)
Mercury	58
Venus	108
Earth	150
Mars	228
Jupiter	778
Saturn	1427
Uranus	2871
Neptune	4497
Pluto	5913

a. Draw a number line with an appropriate scale to plot the distances.

b. Draw a second number line, this time plotting the logarithm (base 10) of each distance.

c. Which number line do you find more helpful to compare the different distances?

d. If two distances are 3 units apart on the logarithmic number line, how do the distances of the corresponding planets compare?

page 426

Exercises to Prepare for the Next Section

Every section's exercise set (except for the last section of a chapter) contain exercises that allow students to practice the previously learned skills and concepts students will need to be successful in the next section. Next to each question, in brackets, is a reference to the section of the text that contains the concepts related to the question for students to easily review. All answers are provided in Answers to Selected Exercises.

Explorations are provided at the end of each exercise set and are designed to encourage students to do research and write about what they have learned. These Explorations generally emphasize critical thinking skills and can be used as collaborative learning exercises or as extra credit assignments.

End of Chapter

Chapter Summary

At the end of each chapter there is a Chapter Summary that includes **Key Terms** and **Essential Concepts and Formulas** that were covered in the chapter. These chapter summaries provide a single point of reference as the student prepares for a test. Each key term and concept references the page number from the lesson where the term or concept was first introduced.

Chapter True/False Exercises

Following each chapter summary are true/false exercises. These exercises are intended to help students understand concepts and can be used to initiate class discussions.

Chapter Review Exercises

Review exercises are found at the end of each chapter. These exercises are selected to help the student integrate all of the topics presented in the chapter.

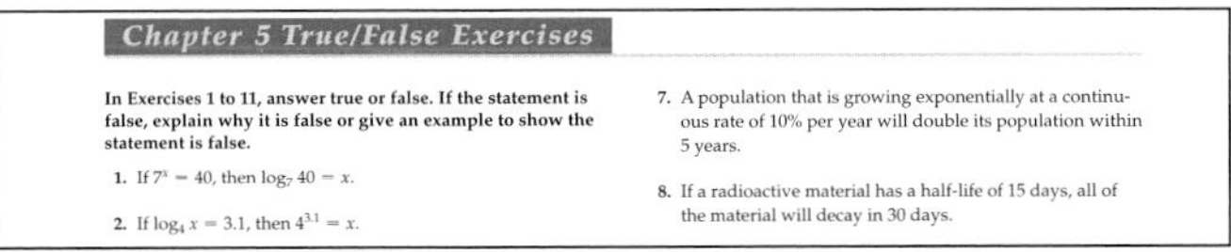

Chapter 5 True/False Exercises

In Exercises 1 to 11, answer true or false. If the statement is false, explain why it is false or give an example to show the statement is false.

1. If $7^x = 40$, then $\log_7 40 = x$.
2. If $\log_4 x = 3.1$, then $4^{3.1} = x$.

7. A population that is growing exponentially at a continuous rate of 10% per year will double its population within 5 years.
8. If a radioactive material has a half-life of 15 days, all of the material will decay in 30 days.

page 479

Chapter 5 Review Exercises

In Exercises 1 to 12, solve each equation. Do not use a calculator.

1. $\log_5 25 = x$
2. $\log_3 81 = x$
3. $\ln e^3 = x$
4. $\ln e^{\pi} = x$

17. $f(x) = 2^x - 3$
18. $f(x) = 2^{(x-3)}$
19. $f(x) = \frac{1}{3} \log x$
20. $f(x) = 3 \log x^{1/3}$

page 479

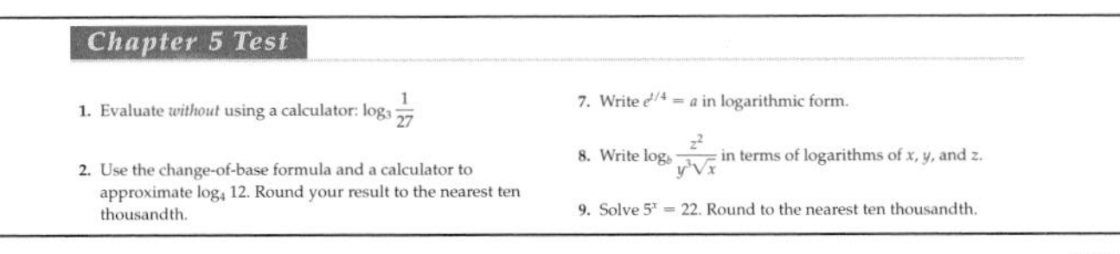

Chapter 5 Test

1. Evaluate *without* using a calculator: $\log_3 \frac{1}{27}$
2. Use the change-of-base formula and a calculator to approximate $\log_4 12$. Round your result to the nearest ten thousandth.

7. Write $e^{1/4} = a$ in logarithmic form.
8. Write $\log_b \frac{z^2}{y^3\sqrt{x}}$ in terms of logarithms of x, y, and z.
9. Solve $5^x = 22$. Round to the nearest ten thousandth.

page 482

Cumulative Review Exercises

1. Solve $|x - 4| \le 2$. Write the solution set using interval notation.
2. Solve $\frac{x}{2x - 6} \ge 1$. Write the solution set using set-builder notation.

4. The height, in feet, of a ball released with an initial upward velocity of 44 feet per second and at an initial height of 8 feet is given by $h(t) = -16t^2 + 44t + 8$, where t is the time in seconds after the ball is released. Find the maximum height the ball will reach.

page 483

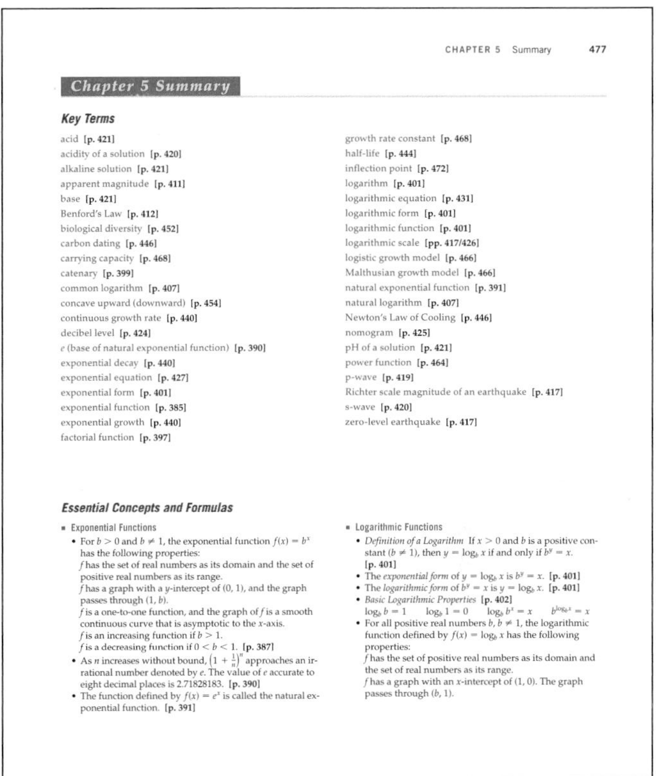

CHAPTER 5 Summary 477

Chapter 5 Summary

Key Terms

acid [p. 421]
acidity of a solution [p. 420]
alkaline solution [p. 421]
apparent magnitude [p. 411]
base [p. 421]
Benford's Law [p. 412]
biological diversity [p. 452]
carbon dating [p. 446]
carrying capacity [p. 468]
catenary [p. 399]
common logarithm [p. 407]
concave upward (downward) [p. 454]
continuous growth rate [p. 440]
decibel level [p. 424]
e (base of natural exponential function) [p. 390]
exponential decay [p. 440]
exponential equation [p. 427]
exponential form [p. 401]
exponential function [p. 385]
exponential growth [p. 440]
factorial function [p. 397]
growth rate constant [p. 468]
half-life [p. 444]
inflection point [p. 472]
logarithm [p. 401]
logarithmic equation [p. 431]
logarithmic form [p. 401]
logarithmic function [p. 401]
logarithmic scale [pp. 417/426]
logistic growth model [p. 466]
Malthusian growth model [p. 466]
natural exponential function [p. 391]
natural logarithm [p. 407]
Newton's Law of Cooling [p. 446]
nomogram [p. 425]
pH of a solution [p. 421]
power function [p. 464]
p-wave [p. 419]
Richter scale magnitude of an earthquake [p. 417]
s-wave [p. 420]
zero-level earthquake [p. 417]

Essential Concepts and Formulas

- Exponential Functions
  - For $b > 0$ and $b \neq 1$, the exponential function $f(x) = b^x$ has the following properties:
    f has the set of real numbers as its domain and the set of positive real numbers as its range.
    f has a graph with a y-intercept of $(0, 1)$, and the graph passes through $(1, b)$.
    f is a one-to-one function, and the graph of f is a smooth continuous curve that is asymptotic to the x-axis.
    f is an increasing function if $b > 1$.
    f is a decreasing function if $0 < b < 1$. [p. 387]
  - As n increases without bound, $\left(1 + \frac{1}{n}\right)^n$ approaches an irrational number denoted by e. The value of e accurate to eight decimal places is 2.71828183. [p. 390]
  - The function defined by $f(x) = e^x$ is called the natural exponential function. [p. 391]
- Logarithmic Functions
  - *Definition of a Logarithm* If $x > 0$ and b is a positive constant ($b \neq 1$), then $y = \log_b x$ if and only if $b^y = x$. [p. 401]
  - The *exponential form* of $y = \log_b x$ is $b^y = x$. [p. 401]
  - The *logarithmic form* of $b^y = x$ is $y = \log_b x$. [p. 401]
  - *Basic Logarithmic Properties* [p. 402]
    $\log_b b = 1$ $\quad \log_b 1 = 0$ $\quad \log_b b^x = x$ $\quad b^{\log_b x} = x$
  - For all positive real numbers b, $b \neq 1$, the logarithmic function defined by $f(x) = \log_b x$ has the following properties:
    f has the set of positive real numbers as its domain and the set of real numbers as its range.
    f has a graph with an x-intercept of $(1, 0)$. The graph passes through $(b, 1)$.

page 477

Chapter Test

The Chapter Test exercises are designed to simulate a possible test of the material in the chapter.

Cumulative Review Exercises

Cumulative Review Exercises, which appear at the end of each chapter (beginning with Chapter 1), allow students to refresh previously developed skills and concepts.

The answers to all Chapter Review Exercises, all Chapter Test exercises, and all Cumulative Review Exercises are given in Answers to Selected Exercises. Along with the answer, there is a reference to the section that pertains to each exercise. This further illustrates how the text supports students while they are studying and preparing for exams.

Instructor's Annotated Edition

The Instructor's Annotated Edition includes the following features:

Instructor Notes give suggestions for teaching concepts, warnings about common student errors, or historical notes.

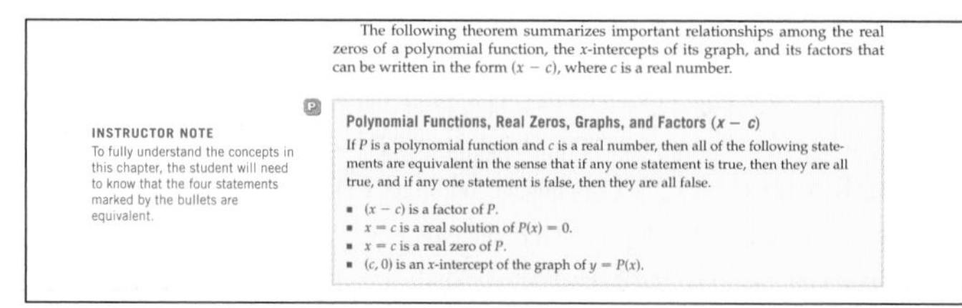
The following theorem summarizes important relationships among the real zeros of a polynomial function, the x-intercepts of its graph, and its factors that can be written in the form $(x - c)$, where c is a real number.

INSTRUCTOR NOTE
To fully understand the concepts in this chapter, the student will need to know that the four statements marked by the bullets are equivalent.

Polynomial Functions, Real Zeros, Graphs, and Factors $(x - c)$
If P is a polynomial function and c is a real number, then all of the following statements are equivalent in the sense that if any one statement is true, then they are all true, and if any one statement is false, then they are all false.

- $(x - c)$ is a factor of P.
- $x = c$ is a real solution of $P(x) = 0$.
- $x = c$ is a real zero of P.
- $(c, 0)$ is an x-intercept of the graph of $y = P(x)$.

page 330

Next to many of the graphs or tables in the text, there is a P icon that indicates that a Microsoft **PowerPoint® slide** of that figure is available. These slides (along with PowerPoint Viewer) are available on the *Class Prep CD* and can also be downloaded from our web site at **math.college.hmco.com/instructors**. These slides can also be printed as transparency masters.

An **Alternative to Example** note accompanies every example and offers an additional example for an instructor to use in class.

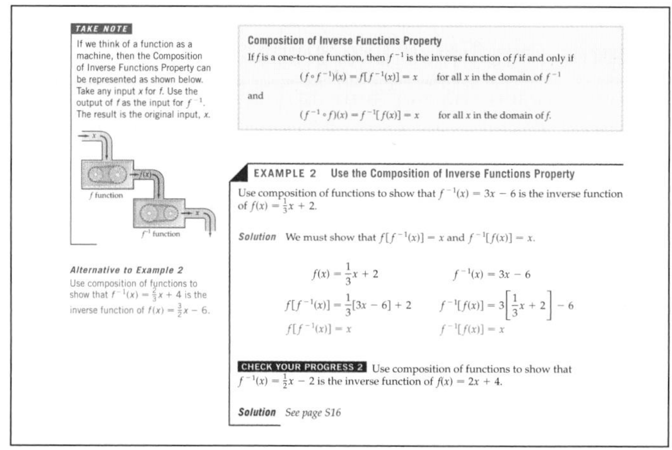
TAKE NOTE
If we think of a function as a machine, then the Composition of Inverse Functions Property can be represented as shown below. Take any input x for f. Use the output of f as the input for f^{-1}. The result is the original input, x.

Composition of Inverse Functions Property
If f is a one-to-one function, then f^{-1} is the inverse function of f if and only if

$$(f \circ f^{-1})(x) = f[f^{-1}(x)] = x \quad \text{for all } x \text{ in the domain of } f^{-1}$$

and

$$(f^{-1} \circ f)(x) = f^{-1}[f(x)] = x \quad \text{for all } x \text{ in the domain of } f.$$

EXAMPLE 2 Use the Composition of Inverse Functions Property

Use composition of functions to show that $f^{-1}(x) = 3x - 6$ is the inverse function of $f(x) = \frac{1}{3}x + 2$.

Solution We must show that $f[f^{-1}(x)] = x$ and $f^{-1}[f(x)] = x$.

$$f(x) = \frac{1}{3}x + 2 \qquad f^{-1}(x) = 3x - 6$$

$$f[f^{-1}(x)] = \frac{1}{3}[3x - 6] + 2 \qquad f^{-1}[f(x)] = 3\left[\frac{1}{3}x + 2\right] - 6$$

$$f[f^{-1}(x)] = x \qquad f^{-1}[f(x)] = x$$

Alternative to Example 2
Use composition of functions to show that $f^{-1}(x) = \frac{2}{3}x + 4$ is the inverse function of $f(x) = \frac{3}{2}x - 6$.

CHECK YOUR PROGRESS 2 Use composition of functions to show that $f^{-1}(x) = \frac{1}{2}x - 2$ is the inverse function of $f(x) = 2x + 4$.

Solution *See page S16*

page 265

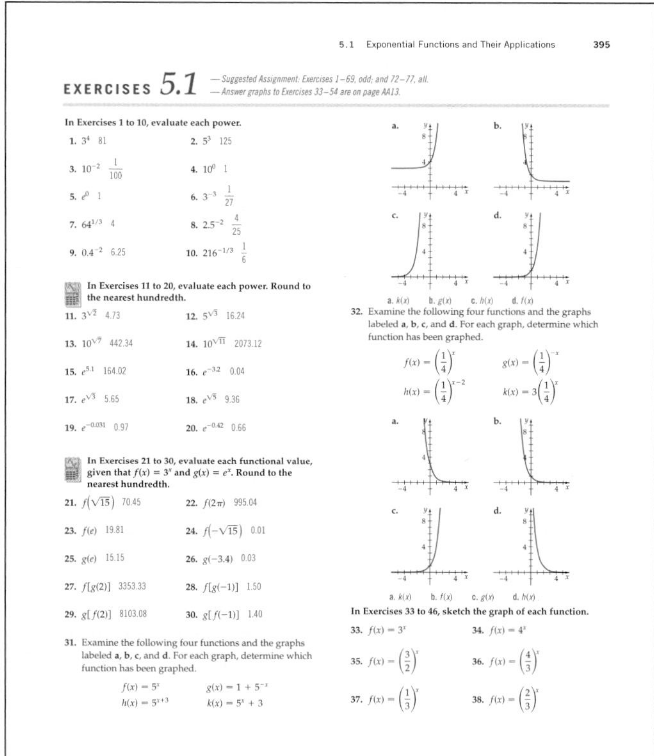
5.1 Exponential Functions and Their Applications 395

EXERCISES 5.1 — *Suggested Assignment: Exercises 1–69, odd; and 72–77, all.*
— *Answer graphs to Exercises 33–54 are on page AA13.*

In Exercises 1 to 10, evaluate each power.

1. 3^4 81 **2.** 5^3 125

3. 10^{-2} $\frac{1}{100}$ **4.** 10^0 1

5. e^0 1 **6.** 3^{-3} $\frac{1}{27}$

7. $64^{1/3}$ 4 **8.** 2.5^{-2} $\frac{4}{25}$

9. 0.4^{-2} 6.25 **10.** $216^{-1/3}$ $\frac{1}{6}$

In Exercises 11 to 20, evaluate each power. Round to the nearest hundredth.

11. $3^{\sqrt{2}}$ 4.73 **12.** $5^{\sqrt{3}}$ 16.24

13. $10^{\sqrt{7}}$ 442.34 **14.** $10^{\sqrt{11}}$ 2073.12

15. $e^{5.1}$ 164.02 **16.** $e^{-3.2}$ 0.04

17. $e^{\sqrt{3}}$ 5.65 **18.** $e^{\sqrt{5}}$ 9.36

19. $e^{-0.031}$ 0.97 **20.** $e^{-0.42}$ 0.66

In Exercises 21 to 30, evaluate each functional value, given that $f(x) = 3^x$ and $g(x) = e^x$. Round to the nearest hundredth.

21. $f(\sqrt{15})$ 70.45 **22.** $f(2\pi)$ 995.04

23. $f(e)$ 19.81 **24.** $f(-\sqrt{15})$ 0.01

25. $g(e)$ 15.15 **26.** $g(-3.4)$ 0.03

27. $f[g(2)]$ 3353.33 **28.** $f[g(-1)]$ 1.50

29. $g[f(2)]$ 8103.08 **30.** $g[f(-1)]$ 1.40

31. Examine the following four functions and the graphs labeled **a**, **b**, **c**, and **d**. For each graph, determine which function has been graphed.

$$f(x) = 5^x \qquad g(x) = 1 + 5^{-x}$$
$$h(x) = 5^{x+3} \qquad k(x) = 5^x + 3$$

a. b. c. d.

a. $k(x)$ b. $g(x)$ c. $h(x)$ d. $f(x)$

32. Examine the following four functions and the graphs labeled **a**, **b**, **c**, and **d**. For each graph, determine which function has been graphed.

$$f(x) = \left(\frac{1}{4}\right)^x \qquad g(x) = \left(\frac{1}{4}\right)^{-x}$$
$$h(x) = \left(\frac{1}{4}\right)^{x-2} \qquad k(x) = 3\left(\frac{1}{4}\right)^x$$

a. b. c. d.

a. $k(x)$ b. $f(x)$ c. $g(x)$ d. $h(x)$

In Exercises 33 to 46, sketch the graph of each function.

33. $f(x) = 3^x$ **34.** $f(x) = 4^x$

35. $f(x) = \left(\frac{3}{2}\right)^x$ **36.** $f(x) = \left(\frac{4}{3}\right)^x$

37. $f(x) = \left(\frac{1}{3}\right)^x$ **38.** $f(x) = \left(\frac{2}{3}\right)^x$

page 395

A **Suggested Assignment** is provided for each section. Answers for all exercises are provided.

INSTRUCTOR RESOURCES

Applied College Algebra has a complete set of support materials for the instructor.

Instructor's Annotated Edition This edition contains a replica of the student text and additional items just for the instructor. These include *Instructor Notes, Alternative to Example* notes, *PowerPoint transparency icons, Suggested Assignments,* and *Answers to all exercises.*

Instructor's Solutions Manual The *Instructor's Solutions Manual* contains worked-out solutions for all exercises in the text.

Instructor's Resource Manual with Testing This resource includes a lesson plan for the *AIM for Success* student preface, four ready-to-use printed *Chapter Tests* per chapter, and a *Printed Test Bank* providing a printout of one example of each of the algorithmic items on the *HM Testing* CD-ROM program.

HM ClassPrep with HM Testing CD-ROM *HM ClassPrep* contains a multitude of text-specific resources for instructors to use to enhance the classroom experience. These resources can be easily accessed by chapter or resource type and can also link you to the text's web site. *HM Testing* is our computerized test generator and contains a database of algorithmic test items as well as providing **online testing** and **gradebook** functions.

Instructor Text-Specific Web Site The resources available on the *Class Prep CD* are also available on the instructor web site at **math.college.hmco.com/instructors**. Appropriate items are password protected. Instructors also have access to the student part of the text's web site.

STUDENT RESOURCES

Student Solutions Manual The *Student Solutions Manual* contains complete solutions to all odd-numbered exercises in the text.

Math Study Skills Workbook by Paul D. Nolting This workbook is designed to reinforce skills and minimize frustration for students in any math class, lab, or study skills course. It offers a wealth of study tips and sound advice on note taking, time management, and reducing math anxiety. In addition, numerous opportunities for self-assessment enable students to track their own progress.

HM eduSpace® Online Learning Environment *eduSpace®* is a text-specific online learning environment that combines an algorithmic tutorial program with homework capabilities. Specific content is available 24 hours a day to help you further understand your textbook.

HM mathSpace™ Tutorial CD-ROM This tutorial CD-ROM allows students to practice skills and review concepts as many times as necessary by providing algorithmically generated exercises and step-by-step solutions for practice.

SMARTHINKING™ live, online tutoring Houghton Mifflin has partnered with SMARTHINKING to provide an easy-to-use and effective online tutorial service. **Whiteboard Simulations** and **Practice Area** promote real-time visual interaction. Three levels of service are offered:

- **Text-Specific Tutoring** provides real-time, one-on-one instruction with a specially qualified "e-structor."
- **Questions Any Time** allows students to submit questions to the tutor outside the scheduled hours and receive a reply within 24 hours.
- **Independent Study Resources** connect students with around-the-clock access to additional educational services, including interactive web sites, diagnostic tests and Frequently Asked Questions posed to SMARTHINKING e-structors.

Houghton Mifflin Instructional Videos and DVDs Text-specific videos and DVDs, hosted by Dana Mosely, cover all sections of the text and provide a valuable resource for further instruction and review. Next to every objective head, the VIDEO & DVD icon serves as a reminder that the objective is covered in a video/DVD lesson.

Student Text-Specific Web Site Online student resources can be found at this text's web site at **math.college.hmco.com/students**.

ACKNOWLEDGMENTS

The authors would like to thank the people who have reviewed this manuscript and provided many valuable suggestions.

Randall Allbritton, *Daytona Beach Community College, FL*
Judy Barclay, *Cuesta College, CA*
Heidi Barrett, *Arapahoe Community College, CO*
Jesse W. Byrne, *University of Central Oklahoma, OK*
RoseMarie Castner, *Canisius College, NY*
Douglas M. Colbert, *University of Nevada–Reno, NV*
Jacqueline Donofrio, *Monroe Community College, NY*
Richard T. Driver, *Washburn University, KS*
Michael W. Ecker, *Pennsylvania State University–Wilkes-Barre, PA*
Susan Haller, *St. Cloud State University, MN*
Thomas P. Kline, *University of Northern Iowa, IA*
Susann Kyriazopoulos, *DeVry University–Chicago, IL*
Zongzhu Lin, *Kansas State University, KS*
Phyllis Meckstroth, *San Diego Mesa College, CA*
Debbie K. Millard, *Florida Community College, FL*
Lauri Semarne
David S. Tucker, *Midwestern State University, TX*

Special thanks to Sandy Doerfel, *Palomar College*, for her preparation of the solutions manuals and for her contribution to the accuracy of the textbook.

AIM for Success

INSTRUCTOR NOTE

See the *Instructor's Resource Manual* or *Class Prep CD* for suggestions on how to teach this lesson.

Welcome to *Applied College Algebra*. As you begin this course, we know two important facts: (1) We want you to succeed. (2) You want to succeed. To do that requires an effort from each of us. For the next few pages, we are going to show you what is required of you to achieve that success and how you can use the features of this text to be successful.

Motivation

One of the most important keys to success is motivation. We can try to motivate you by offering interesting or important ways mathematics can benefit you. But, in the end, the motivation must come from you. On the first day of class, it is easy to be motivated. Eight weeks into the term, it is harder to keep that motivation.

To stay motivated, there must be outcomes from this course that are worth your time, money, and energy. List some reasons you are taking this course. Do not make a mental list—actually write them out.

TAKE NOTE

Motivation alone will not lead to success. For instance, suppose a person who cannot swim is placed in a boat, taken out to the middle of a lake, and then thrown overboard. That person has a lot of motivation to swim but there is a high likelihood the person will drown without some help. Motivation gives us the desire to learn but is not the same as learning.

____

____

____

Although we hope that one of the reasons you listed was an interest in mathematics, we know that many of you are taking this course because it is required to graduate, it is a prerequisite for a course you must take, or because it is required for your major. Although you may not agree that this course is necessary, it is! If you are motivated to graduate or complete the requirements for your major, then use that motivation to succeed in this course. Do not become distracted from your goal to complete your education!

Commitment

To be successful, you must make a commitment to succeed. This means devoting time to math so that you achieve a better understanding of the subject.

List some activities (sports, hobbies, talents such as dance, art, or music) that you enjoy and at which you would like to become better.

Activity	Time Spent	Time Wished Spent

Thinking about these activities, put the number of hours that you spend each week practicing these activities next to the activity. Next to that number, indicate the number of hours per week you would like to spend on these activities.

Whether you listed surfing or sailing, aerobics or restoring cars, or any other activity you enjoy, note how many hours a week you spend doing it. To succeed in math, you must be willing to commit the same amount of time. Success requires some sacrifice.

The "I Can't Do Math" Syndrome

There may be things you cannot do, such as lift a two-ton boulder. You can, however, do math. It is much easier than lifting the two-ton boulder. When you first learned the activities you listed above, you probably could not do them well. With practice, you got better. With practice, you will be better at math. Stay focused, motivated, and committed to success.

It is difficult for us to emphasize how important it is to overcome the "I Can't Do Math" Syndrome. If you listen to interviews of very successful athletes after a particularly bad performance, you will note that they focus on the positive aspect of what they did, not the negative. Sports psychologists encourage athletes to always be positive—to have a "Can Do" attitude. Develop this attitude toward math.

Strategies for Success

Textbook Reconnaissance Right now, do a 15-minute "textbook reconnaissance" of this book. Here's how:

First, read the table of contents. Do it in three minutes or less. Next, look through the entire book, page by page. Move quickly. Scan titles, look at pictures, notice diagrams.

A textbook reconnaissance shows you where a course is going. It gives you the big picture. That's useful because brains work best when going from the general to the specific. Getting the big picture before you start makes details easier to recall and understand later on.

Your textbook reconnaissance will work even better if, as you scan, you look for ideas or topics that are interesting to you. List three facts, topics, or problems that you found interesting during your textbook reconnaissance.

____

____

____

The idea behind this technique is simple: It's easier to work at learning material if you know it's going to be useful to you.

Not all the topics in this book will be "interesting" to you. But that is true of any subject. Surfers find that on some days the waves are better than others, musicians find some music more appealing than other music, computer gamers find some computer games more interesting than others, car enthusiasts find some cars more exciting than others. Some car enthusiasts would rather have a completely restored 1957 Chevrolet than a new Ferrari.

Know the Course Requirements To do your best in this course, you must know exactly what your instructor requires. Course requirements may be stated in a *syllabus*, which is a printed outline of the main topics of the course, or they may be presented orally. When they are listed in a syllabus or on other printed pages, keep them in a safe place. When they are presented orally, make sure to take complete notes. In either case, it is important that you understand them completely and follow them exactly. Be sure you know the answer to the following questions.

1. What is your instructor's name?
2. Where is your instructor's office?
3. At what times does your instructor hold office hours?

4. Besides the textbook, what other materials does your instructor require?
5. What is your instructor's attendance policy?
6. If you must be absent from a class meeting, what should you do before returning to class? What should you do when you return to class?
7. What is the instructor's policy regarding collection or grading of homework assignments?
8. What options are available if you are having difficulty with an assignment? Is there a math tutoring center?
9. Is there a math lab at your school? Where is it located? What hours is it open?
10. What is the instructor's policy if you miss a quiz?
11. What is the instructor's policy if you miss an exam?
12. Where can you get help when studying for an exam?

Remember: Your instructor wants to see you succeed. If you need help, ask! Do not fall behind. If you are running a race and fall behind by 100 yards, you may be able to catch up but it will require more effort than had you not fallen behind.

TAKE NOTE

Besides time management, there must be realistic ideas of how much time is available. There are very few people who can *successfully* work full-time and go to school full-time. If you work 40 hours a week, take 15 units, spend the recommended study time given at the right, and sleep 8 hours a day, you will use over 80% of the available hours in a week. That leaves less than 20% of the hours in a week for family, friends, eating, recreation, and other activities.

Time Management We know that there are demands on your time. Family, work, friends, and entertainment all compete for your time. We do not want to see you receive poor job evaluations because you are studying math. However, it is also true that we do not want to see you receive poor math test scores because you devoted too much time to work. When several competing and important tasks require your time and energy, the only way to manage the stress of being successful at both is to manage your time efficiently.

Instructors often advise students to spend twice the amount of time outside of class studying as they spend in the classroom. Time management is important if you are to accomplish this goal and succeed in school. The following activity is intended to help you structure your time more efficiently.

List the name of each course you are taking this term, the number of class hours each course meets, and the number of hours you should spend studying each subject outside of class. Then fill in a weekly schedule like the one on the following page. Begin by writing in the hours spent in your classes, the hours spent at work (if you have a job), and any other commitments that are not flexible with respect to the time that you do them. Then begin to write down commitments that are more flexible, including hours spent studying. Remember to reserve time for activities such as meals and exercise. You should also schedule free time.

	Monday	Tuesday	Wednesday	Thursday	Friday	Saturday	Sunday
7–8 a.m.							
8–9 a.m.							
9–10 a.m.							
10–11 a.m.							
11–12 p.m.							
12–1 p.m.							
1–2 p.m.							
2–3 p.m.							
3–4 p.m.							
4–5 p.m.							
5–6 p.m.							
6–7 p.m.							
7–8 p.m.							
8–9 p.m.							
9–10 p.m.							
10–11 p.m.							
11–12 a.m.							

We know that many of you must work. If that is the case, realize that working 10 hours a week at a part-time job is equivalent to taking a three-unit class. If you must work, consider letting your education progress at a slower rate to allow you to be successful at both work and school. There is no rule that says you must finish school in a certain time frame.

Schedule Study Time As we encouraged you to do by filling out the time management form above, schedule a certain time to study. You should think of this time the way you would the time for work or class—that is, reasons for missing study time should be as compelling as reasons for missing work or class. "I just didn't feel like it" is not a good reason to miss your scheduled study time.

Although this may seem like an obvious exercise, list a few reasons you might want to study.

____

____

____

Of course we have no way of knowing the reasons you listed, but from our experience one reason given quite frequently is "To pass the course." There is nothing wrong with that reason. If that is the most important reason for you to study, then use it to stay focused.

One method of keeping to a study schedule is to form a ***study group***. Look for people who are committed to learning, who pay attention in class, and who are punctual. Ask them to join your group. Choose people with similar educational goals but different methods of learning. You can gain insight from seeing the material from a new perspective. Limit groups to four or five people; larger groups are unwieldy.

There are many ways to conduct a study group. Begin with the following suggestions and see what works best for your group.

1. Test each other by asking questions. Each group member might bring two or three sample test questions to each meeting.
2. Practice teaching each other. Many of us who are teachers learned a lot about our subject when we had to explain it to someone else.
3. Compare class notes. You might ask other students about material in your notes that is difficult for you to understand.
4. Brainstorm test questions.
5. Set an agenda for each meeting. Set approximate time limits for each agenda item and determine a quitting time.

And finally, probably the most important aspect of studying is that it should be done in relatively small chunks. If you can only study three hours a week for this course (probably not enough for most people), do it in blocks of one hour on three separate days, preferably after class. Three hours of studying on a Sunday is not as productive as three hours of paced study.

Text Features That Promote Success

Preparing for a Chapter Before you begin a new chapter, you should take some time to review previously learned skills. There are two ways to do this. The first is to complete the ***Cumulative Review Exercises***, which occurs after every chapter (except Chapter P). For instance, turn to page 483. The questions in this review are taken from the previous chapters. The answers for all these exercises can be found on page A29. Turn to that page now and locate the answers for the Chapter 5 Cumulative Review. After the answer to the first exercise, which is [2, 6], you will see the section reference [1.4]. This means that this question was taken from Chapter 1, Section 4. If you missed this question, you should return to that section and restudy the material.

A second way of preparing for a new chapter is to complete the ***Chapter Prep Quiz***. This quiz focuses on the particular skills that will be required for the new chapter. Turn to page 384 to see a Prep Quiz. Note that a section reference is given for each question. The answers for the Prep Quiz are the first set of answers in the answer section for a chapter. Turn to page A23 to see the answers for the Chapter 5 Prep Quiz. If you answer a question incorrectly, restudy the section for which the question was taken.

Before the class meeting in which your professor begins a new section, you should browse through the material, being sure to note each word in bold type. These words indicate important concepts that you must know in order to learn the material. Do not worry about trying to understand all the material. Your professor is there to assist you with that endeavor. The purpose of browsing through the material is so that your brain will be prepared to accept and organize the new information when it is presented to you.

Turn to page 3. Write down the title of Section P.1. Then write down the words in the section that are in bold print. It is not necessary for you to understand the meaning of these words. You are in this class to learn their meaning.

______	______	______	______
______	______	______	______
______	______	______	______
______	______	______	______
______	______	______	______

Math is Not a Spectator Sport To learn mathematics you must be an active participant. Listening and watching your professor do mathematics is not enough. Mathematics requires that you interact with the lesson you are studying. If you filled in the blanks above, you were being interactive. There are other ways this textbook has been designed to help you be an active learner.

Example/Check Your Progress Pairs One of the key instructional features of this text is Example/Check Your Progress pairs. Note that each Example is completely worked out and the Check Your Progress problem following the example is not. Study the worked-out example carefully by working through each step. Then work the Check Your Progress. If you get stuck, refer to the page number following the Check Your Progress, which directs you to the page on which the Check Your Progress is solved—a complete worked-out solution is provided. Try to use the given solution to get a hint for the step you are stuck on.

When you have completed your solution, check your work against the solution we provided. (Turn to page S26 to see the solution of Check Your Progress 1). Be aware that frequently there is more than one way to solve a problem. Your answer, however, should be the same as the given answer. If you have any question as to whether your method will "always work," check with your instructor or with someone in the math center.

EXAMPLE 1 Solve an Exponential Equation

Use the Equality of Exponents Theorem to solve $2^{x+1} = 32$.

Solution

$$2^{x+1} = 32$$

$$2^{x+1} = 2^5 \quad \text{▪ Write each side as a power of 2.}$$

$$x + 1 = 5 \quad \text{▪ Equate the exponents.}$$

$$x = 4 \quad \text{▪ Solve the resulting equation.}$$

▪ CHECK

$$2^{x+1} \stackrel{?}{=} 32$$

$$2^{4+1} \stackrel{?}{=} 32 \quad \text{▪ Let } x = 4.$$

$$2^5 \stackrel{?}{=} 32$$

$$32 = 32$$

CHECK YOUR PROGRESS 1 Solve the exponential equation $3^{5-2x} = \frac{1}{9}$.

Solution *See page S26.*

page 427

Check Your Progress 1, *page 427*

$$3^{5-2x} = \frac{1}{9}$$

$$3^{5-2x} = 3^{-2}$$

$$5 - 2x = -2$$

$$-2x = -7$$

$$\frac{-2x}{-2} = \frac{-7}{-2}$$

$$x = \frac{7}{2}$$

page S26

Browse through the textbook and write down the page numbers where two other Example/Check Your Progress pairs occur.

Remember: Be an active participant in your learning process. When you are sitting in class watching and listening to an explanation, you may think that you understand. However, until you actually try to do it, you will have no confirmation of the new knowledge or skill. Most of us have had the experience of sitting in class thinking we knew how to do something only to get home and realize that we didn't.

TAKE NOTE

There is a strong connection between reading and being a successful student in math or in any other subject. If you have difficulty reading, consider taking a reading course. Reading is much like other skills. There are certain things you can learn that will make you a better reader.

Word Problems Word problems are difficult because we must read the problem, determine the quantity we must find, think of a method to do that, and then actually solve the problem. In short, we must formulate a *strategy* to solve the problem and then devise a *solution*. If you have difficulty with a word problem, write down the known information. Be very specific. Write out a phrase or sentence that states what you are trying to find. Ask yourself whether there are known formulas that relate the known and unknown quantities. Do not ignore the word problems. They are an important part of mathematics.

TAKE NOTE

If a rule has more than one part, be sure to make a notation to that effect.

Rule Boxes Pay special attention to rules placed in boxes. These rules give you the reasons certain types of problems are solved the way they are. When you see a rule, try to rewrite the rule in your own words.

Equality of Exponents Theorem

If $b^x = b^y$, then $x = y$, provided $b > 0$ and $b \neq 1$.

page 427

Find and write down two page numbers on which there are examples of rule boxes.

Chapter Exercises When you have completed studying a section, do the exercises in the exercise set that correspond with that section. Math is a subject that needs to be learned in small sections and practiced continually in order to be mastered. Doing all of the exercises in each exercise set will help you master the problem-solving techniques necessary for success. As you work through the exercises for a section, check your answers to the odd-numbered exercises with those in the back of the book.

Preparing for a Test There are important features of this text that can be used to prepare for a test.

- Chapter Summary
- Chapter Review Exercises
- Chapter Test

After completing a chapter, read the Chapter Summary. This summary is divided into two sections: *Key Terms* and *Essential Concepts and Formulas*. (See page 477 for the Chapter 5 Summary.) This summary highlights the important topics covered in the chapter. The page number following each topic refers you to the page in the text on which you can find more information about the concept.

Following the Chapter Summary are Chapter Review Exercises (see page 479) and a Chapter Test (see page 482). Doing the review exercises is an important way of testing your understanding of the chapter. The answer to each review exercise is given at the back of the book, along with its section reference. After checking your answers, restudy any section from which a question you missed was taken. It may be helpful to retry some of the exercises for that section to reinforce your problem-solving techniques.

The Chapter Test should be used to prepare for an exam. We suggest that you try the Chapter Test a few days before your actual exam. Take the test in a quiet place and try to complete the test in the same amount of time you will be allowed for your exam. When taking the Chapter Test, practice the strategies of successful test takers: (1) Scan the entire test to get a feel for the questions; (2) Read the directions carefully; (3) Work the problems that are easiest for you first; and perhaps most importantly, (4) Try to stay calm.

When you have completed the Chapter Test, check your answers. If you missed a question, review the material in that section and rework some of the exercises from that section. This will strengthen your ability to perform the skills in that section.

Your career goal goes here. →

Is it difficult to be successful? YES! Successful music groups, artists, professional athletes, chefs, and ______________ have to work very hard to achieve their goals. They focus on their goals and ignore distractions. The things we ask you to do to achieve success take time and commitment. We are confident that if you follow our suggestions, you will succeed.

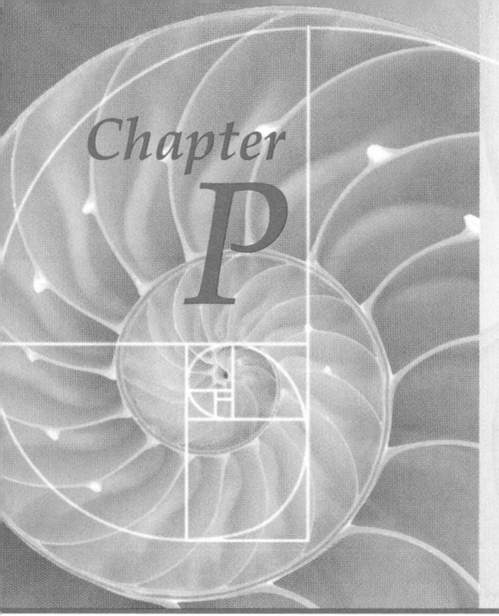

Chapter P Preliminary Concepts

Which Came First: Writing or Counting?

Archaeologist Denise Schmandt-Besserat of the University of Texas, Austin has presented convincing evidence that writing was an outgrowth of counting. As hunters and gatherers banded together to form agricultural communities (around 7500 B.C.), the need to count goods belonging to a member of the community became important. For instance, if a farmer wanted to store three bushels of wheat in a community silo, there needed to be a way to record that fact.

Professor Schmandt-Besserat realized that little geometric objects called *tokens*, which earlier were thought to be ornamental jewerly, were the symbols used to represent various goods. The farmer's wheat might be represented by a cone and a sphere might represent a certain animal. A record of ownership was made by pressing a token in wet clay. As tokens became more complex, it became easier to use a sharp stick to draw an imprint on the wet clay. These clay tablets show the earliest attempts at writing. And all of this just because someone needed to count.

Besides the way we write, math symbolism has changed over time. For instance, $2 + 3 = 5$ was not so simple before 1557 because the equal symbol had not been invented. Furthermore, the plus sign did not first occur in print until 1489. Exercises 1 and 2 on page 12 use other math symbols that have evolved over time.

Denise Schmandt-Besserat

Geometric tokens

Chapter Prep Quiz *Do these exercises to prepare for Chapter P.*

1. Add: $48 + (-53)$ -5
2. Multiply: $(-5)(-12)$ 60
3. Evaluate: $(-5)^3$ -125
4. What is the reciprocal of $-\frac{4}{3}$? $-\frac{3}{4}$
5. Divide: $\frac{3}{4} \div \frac{5}{8}$ $\frac{6}{5}$
6. Simplify: $\frac{24}{36}$ $\frac{2}{3}$
7. Add: $-\frac{5}{8} + \frac{3}{8}$ $-\frac{1}{4}$
8. What is the least common multiple of 6 and 8? 24
9. Evaluate $3ab^2$ when $a = 4$ and $b = -2$. 48
10. Simplify: $4x - 9x$ $-5x$

Problem Solving Strategies

Polya's Four-Step Process

Your success in mathematics and your success in the workplace are heavily dependent on your ability to solve problems. One of the foremost mathematicians to study problem solving was George Polya (1887–1985). The basic structure that Polya advocated for problem solving has four steps, as outlined below.

1. Understand the Problem
   - Can you restate the problem in your own words?
   - Can you determine what is known about this type of problem?
   - Is there missing information that you need in order to solve the problem?
   - Is there information given that is not needed?
   - What is the goal?
2. Devise a Plan
   - Make a list of the known information.
   - Make a list of information that is needed to solve the problem.
   - Make a table or draw a diagram.
   - Work backwards.
   - Try to solve a similar but simpler problem.
   - Research the problem to determine whether there are known techniques for solving problems of its kind.
   - Try to determine whether some pattern exists.
   - Write an equation.
3. Carry Out the Plan
   - Work carefully.
   - Keep an accurate and neat record of all your attempts.
   - Realize that some of your initial plans will not work and that you may have to return to Step 2 and devise another plan or modify your existing plan.
4. Review Your Solution
   - Make sure that the solution is consistent with the facts of the problem.
   - Interpret the solution in the context of the problem.
   - Ask yourself whether there are generalizations of the solution that could apply to other problems.
   - Determine the strengths and weaknesses of your solution. For instance, is your solution only an approximation to the actual solution?
   - Consider the possibility of alternative solutions.

Point of Interest

George Polya was born in Hungary and moved to the United States in 1940. In 1942, he moved to California and began teaching at Stanford University, where he remained until his retirement. While at Stanford, he published 10 books and a number of articles for mathematics journals. Of the books Polya published, *How To Solve It* (1945) is one of his best known. In this book Polya outlines a strategy for solving problems. This strategy is frequently applied to problems in mathematics, but it can be used to solve problems from virtually any discipline.

INSTRUCTOR NOTE

The *Chapter Prep Quiz* is a means to test your students' mastery of *prerequisite* material. It is assumed in the coming chapter that students are familiar with this material. Each question identifies the section to review (if necessary) and all answers are provided in the Answers to Selected Exercises.

SECTION P.1 Real Numbers

- Sets of Numbers
- Union and Intersection of Sets
- Absolute Value and Distance
- Interval Notation
- Order of Operations Agreement
- Simplifying Variable Expressions

Point of Interest

Georg Cantor (1845–1918) was a German mathematician who developed many new concepts that dealt with the theory of sets. Much of Cantor's work was controversial. One of the simplest of the controversial concepts concerned points on a line segment. For instance, consider the line segment *AB* and the line segment *CD* shown below.

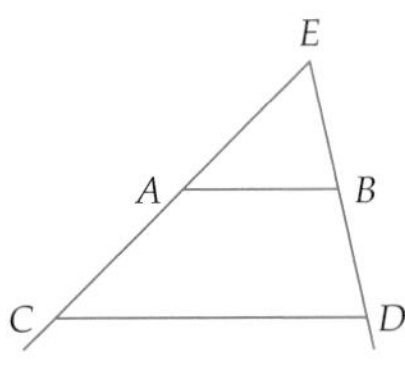

Which line segment, *AB* or *CD*, do you think contains the most points? Cantor was able to show that they both contain the same number of points. In fact, he was able to show that any line segment—no matter how short—contains the same number of points as a line, or as a plane, or as all of three-dimensional space.

Sets of Numbers

It seems to be a human characteristic to group similar items. For instance, a biologist groups animals with similar characteristics into groups called *phyla*. Humans belong to the phylum Chordata. Astronomers classify groups of stars into galaxies. The solar system containing Earth belongs to the Milky Way galaxy, which is a spiral galaxy.

Mathematicians place objects with similar properties into sets. A **set** is a collection of objects. The objects in a set are called the **elements** of the set.

The **roster method** of writing sets encloses a list of the elements in braces. The set of sections within an orchestra is written {brass, percussion, string, woodwind}.

The numbers that we use to count objects, such as the number of students enrolled in a university are the *natural numbers*.

$$\textbf{Natural numbers} = \{1, 2, 3, 4, 5, 6, 7, 8, 9, 10, \ldots\}$$

The three dots mean that the list of natural numbers continues on and on and that there is no highest natural number.

Each natural number greater than 1 is a prime number or a composite number. A **prime number** is a natural number greater than 1 that is evenly divisible only by itself and 1. The first six prime numbers are 2, 3, 5, 7, 11, and 13. A natural number greater than 1 that is not a prime number is a **composite number**. The numbers 4, 6, 8, 9, and 10 are the first five composite numbers.

QUESTION What is the seventh prime number? What is the sixth composite number?

That natural numbers do not have a symbol to denote the concept of none—for instance, the number of trees on the moon. The whole numbers include zero and the natural numbers.

$$\textbf{Whole numbers} = \{0, 1, 2, 3, 4, 5, 6, 7, 8, 9, 10, \ldots\}$$

The whole numbers do not provide all the numbers that are necessary in applications. For example, a chemist needs numbers to describe temperatures below zero.

$$\textbf{Integers} = \{\ldots, -5, -4, -3, -2, -1, 0, 1, 2, 3, 4, 5, \ldots\}$$

The integers $\ldots, -5, -4, -3, -2, -1$ are negative integers. The integers 1, 2, 3, 4, 5, ... are positive integers. Note that the natural numbers and the positive integers are the same set of numbers. The integer zero is neither positive nor negative.

ANSWER The seventh prime number is 17. The sixth composite number is 12.

Still other numbers are necessary to solve the variety of application problems that exist. For instance, a carpenter may need to cut a piece of wood $6\frac{3}{4}$ inches long. The numbers that include fractions are called *rational numbers.*

$$\textbf{Rational numbers} = \left\{\frac{p}{q}, \text{ where } p \text{ and } q \text{ are integers and } q \neq 0\right\}$$

Examples of rational numbers include $\frac{2}{3}$, $-\frac{9}{2}$, and $\frac{5}{1}$. Note that $\frac{5}{1} = 5$; all integers are rational numbers. The number $\frac{4}{\pi}$ is not a rational number because π is not an integer.

A fraction can be written in decimal notation by dividing the numerator by the denominator. For example, $\frac{7}{20} = 7 \div 20 = 0.35$ and $\frac{5}{11} = 5 \div 11 = 0.\overline{45}$, where the bar over 45 indicates that those digits repeat without end.

Some numbers cannot be written as terminating or repeating decimals—for example, 0.13113111311113..., $\sqrt{13} = 3.6055513\ldots$, and $\pi = 3.1415926\ldots$. These numbers have decimal representations that neither terminate nor repeat. They are called **irrational numbers.** The rational numbers and the irrational numbers taken together are the **real numbers.**

TAKE NOTE

Recall that a perfect square integer is one that is the square of an integer. For instance, $9 = 3^2$, $49 = 7^2$, and $121 = 11^2$ are examples of perfect squares. The square root of any number that is *not* a perfect square is an irrational number. Thus $\sqrt{6}$, $\sqrt{13}$, and $\sqrt{125}$ are examples of irrational numbers.

EXAMPLE 1 Classify Real Numbers

Determine which of the following numbers are

a. integers. **b.** rational numbers. **c.** irrational numbers.
d. real numbers. **e.** prime numbers.

$$0 \quad -3.4 \quad \frac{5}{2} \quad 47 \quad 15 \quad -12 \quad 4.212121\ldots \quad \sqrt{37} \quad -1.10110111011110\ldots$$

Alternative to Example 1
Exercise 2, page 12

Solution

a. Integers: 0, 47, 15, −12
b. Rational numbers: 0, −3.4, $\frac{5}{2}$, 47, 15, −12, 4.212121...
c. Irrational numbers: $\sqrt{37}$, −1.10110111011110...
d. Real numbers: 0, −3.4, $\frac{5}{2}$, 47, 15, −12, 4.212121..., $\sqrt{37}$, −1.10110111011110...
e. Prime numbers: 47

CHECK YOUR PROGRESS 1 Determine which of the following numbers are

a. integers. **b.** rational numbers. **c.** irrational numbers.
d. real numbers. **e.** prime numbers.

$$-13 \quad 4.\overline{142} \quad \frac{5}{2} \quad 29 \quad -12 \quad \frac{2}{\sqrt{7}} \quad \pi \quad 4.32789123409$$

Solution *See page S0.* **a.** −13, 29, −12 **b.** −13, $4.\overline{142}$, $\frac{5}{2}$, 29, −12, 4.32789123409 **c.** $\frac{2}{\sqrt{7}}$, π **d.** All **e.** 29

Point of Interest

Archimedes (c. 287–212 B.C.) was the first to calculate π with any degree of precision. He was able to show that

$$3\frac{10}{71} < \pi < 3\frac{1}{7}$$

from which we get the approximation $3\frac{1}{7} = \frac{22}{7} \approx \pi$. The use of the symbol π for this quantity was introduced by Leonhard Euler (1707–1783) in 1739, approximately 2000 years after Archimedes.

The set of natural numbers is an example of an **infinite set;** the pattern of numbers continues without end. It is impossible to list all the elements of an infinite set. The set of prime numbers less than 15 is written $\{2, 3, 5, 7, 11, 13\}$. This is an example of a **finite set;** all the elements of the set can be listed.

It is common to designate a set by a capital letter. For instance, if A is the set of the natural numbers less than 6, then $A = \{1, 2, 3, 4, 5\}$.

The symbol $\in$ is used to refer to the elements of a set. This symbol is read "is an element of." The symbol $\notin$ means "is not an element of."

$$\text{Given } B = \{1, 3, 5\}, \text{ then } 1 \in B \text{ and } 3 \in B \text{ but } 6 \notin B.$$

The **empty set,** or **null set,** is the set that contains no elements. The symbol $\varnothing$ is used to represent the empty set. The set of people who have run a two-minute mile is the empty set.

A second method of representing a set is **set-builder notation.** Set-builder notation can be used to describe almost any set, but it is especially useful when writing infinite sets. For instance, the set

$$\{2n \mid n \in \text{natural numbers}\}$$

is read as "the set of all numbers $2n$ such that n is a natural number." By replacing n by each of the natural numbers, this is the set of positive even numbers: $\{2, 4, 6, 8, \ldots\}$.

The set of real numbers greater than 2 is written

$$\{x \mid x > 2, x \in \text{real numbers}\}$$

and is read "the set of all x such that x is greater than 2 and x is an element of the real numbers."

Much of the work we do in this text uses the real numbers. With this in mind, we will frequently write, for instance, $\{x \mid x > 2, x \in \text{real numbers}\}$ as $\{x \mid x > 2\}$, where we assume that x is a real number.

TAKE NOTE

The order in which the elements of a set are listed is not important. For instance, the set of prime numbers less than 15 given at the right could have been written $\{11, 3, 5, 2, 13, 7\}$. It is customary, however, to list the elements of a set in numeric order.

Point of Interest

A **fuzzy set** is one in which each element is given a "degree" of membership. An element that has degree 0 does not belong to the set. An element of degree 1 belongs to the set. An element that partially belongs to a set has a degree between 0 and 1. The concepts behind fuzzy sets are used in a wide variety of applications such as programming traffic lights and washing machines, and computer speech recognition programs.

Point of Interest

Sophie Germain (1776–1831) was born in Paris, France. Because enrollment in the university she wanted to attend was available only to men, Germain attended under the name of Antoine-August Le Blanc. Eventually her ruse was discovered, but not before she came to the attention of Pierre Lagrange, one of the best mathematicians of the time. He encouraged her work and became a mentor to her. A certain type of prime number, called a Germain prime number, is named after her. It is one such that p and $2p + 1$ are both prime. For instance, 23 is a Germain prime because $23 = 2(11) + 1$ and 23 and 11 are prime numbers. Germain primes are used in public key cryptography, a method used to send secure communications over the Internet.

▪ Union and Intersection of Sets

Just as operations such as addition and multiplication are performed on real numbers, operations are performed on sets. Two operations performed on sets are union and intersection.

The **union of two sets,** written $A \cup B$, is the set of all elements that belong to either A or B. In set-builder notation, this is written

$$A \cup B = \{x \mid x \in A \text{ or } x \in B\}$$

For instance, given $A = \{2, 3, 4\}$ and $B = \{0, 1, 2, 3\}$, $A \cup B = \{0, 1, 2, 3, 4\}$. Note that an element that belongs to both sets is listed only once.

The **intersection of two sets,** written $A \cap B$, is the set of all elements that are common to both A and B. In set-builder notation, this is written

$$A \cap B = \{x \mid x \in A \text{ and } x \in B\}$$

For instance, given $A = \{2, 3, 4\}$ and $B = \{0, 1, 2, 3\}$, $A \cap B = \{2, 3\}$.

Alternative to Example 2
Exercises 4, 6, and 8, page 12

EXAMPLE 2 Find the Union and Intersection of Sets

Find each intersection or union given $A = \{0, 2, 4, 6, 10, 12\}$, $B = \{0, 3, 6, 12, 15\}$, and $C = \{1, 2, 3, 4, 5, 6, 7\}$.

a. $A \cup C$ **b.** $B \cap C$

c. $A \cap (B \cup C)$ **d.** $B \cup (A \cap C)$

Solution

a. $A \cup C = \{0, 1, 2, 3, 4, 5, 6, 7, 10, 12\}$ ▪ The elements that belong to A or C

b. $B \cap C = \{3, 6\}$ ▪ The elements that belong to B and C

c. First determine $B \cup C = \{0, 1, 2, 3, 4, 5, 6, 7, 12, 15\}$. Then
$A \cap (B \cup C) = \{0, 2, 4, 6, 12\}$ ▪ The elements that belong to A and $(B \cup C)$

d. First determine $A \cap C = \{2, 4, 6\}$. Then $B \cup (A \cap C) = \{0, 2, 3, 4, 6, 12, 15\}$
▪ The elements that belong to B or $(A \cap C)$

CHECK YOUR PROGRESS 2 Find each intersection or union given $A = \{-2, -1, 0, 1, 2\}$, $B = \{-4, -2, 0, 2, 4, 6\}$, and $C = \{3, 4, 5, 6, 7\}$.

a. $A \cap C$ **b.** $B \cup C$

c. $A \cup (B \cap C)$ **d.** $B \cap (A \cup C)$

Solution *See page S0.* **a.** $\varnothing$ **b.** $\{-4, -2, 0, 2, 3, 4, 5, 6, 7\}$ **c.** $\{-2, -1, 0, 1, 2, 4, 6\}$ **d.** $\{-2, 0, 2, 4, 6\}$

If the intersection of two sets A and B is the empty set, then A and B are said to be **disjoint** sets. For instance, suppose $A = \{2, 4, 6, 8\}$ and $B = \{1, 3, 5, 7, 9\}$. Then $A \cap B = \varnothing$ and A and B are disjoint.

▪ Absolute Value and Distance

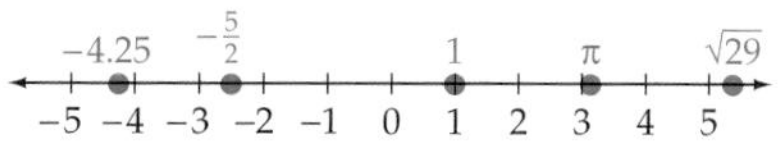

Figure P.1

The real numbers can be represented geometrically by a **coordinate axis** called a **real number line. Figure P.1** shows a portion of a real number line. The number associated with a point on a real number line is called the **coordinate** of the point. The point corresponding to zero is called the **origin.** Every real number corresponds to a point on the number line and every point on the number line corresponds to a real number.

The *absolute value* of a real number a, denoted $|a|$, is the distance between a and 0 on the number line. For instance, $|3| = 3$ and $|-3| = 3$ because both 3 and -3 are 3 units from zero. See **Figure P.2** below.

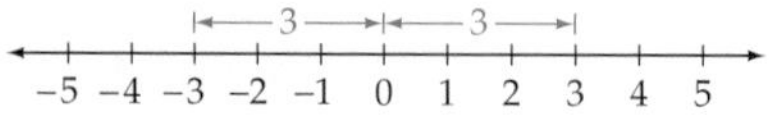

Figure P.2

In general, if $a \geq 0$, then $|a| = a$. However, if $a < 0$, then $|a| = -a$ because $-a$ is positive when $a < 0$. This leads to the following definition.

TAKE NOTE

The second part of the definition of absolute value states that if $a < 0$, then $|a| = -a$. For instance, if $a = -4$, then

$$|a| = |-4| = -(-4) = 4$$

Definition of Absolute Value

The **absolute value** of the real number a is defined by

$$|a| = \begin{cases} a & \text{if } a \geq 0 \\ -a & \text{if } a < 0 \end{cases}$$

The definition of the *distance* between two points on a real number line makes use of absolute value.

Distance Between Points on a Real Number Line

If a and b are the coordinates of two points on a number line, the **distance** between the graph of a and the graph of b, denoted by $d(a, b)$, is given by $d(a, b) = |a - b|$.

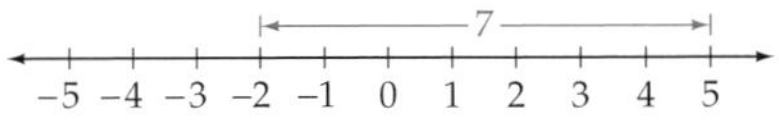

Figure P.3

As an example of this definition, the distance between the point whose coordinate is -2 and the point whose coordinate is 5 is given by

$$d(-2, 5) = |-2 - 5| = |-7| = 7$$

Note from **Figure P.3** that there are 7 units between -2 and 5 on the number line. Also note that the order of the coordinates does not matter.

$$d(5, -2) = |5 - (-2)| = |7| = 7$$

Alternative to Example 3

Express the distance between x and 4 is 5 using absolute value.

■ $|x - 4| = 5$

EXAMPLE 3 Use Absolute Value to Express the Distance Between Two Points

Express the distance between a and -3 on the number line using absolute value.

Solution $d(a, -3) = |a - (-3)| = |a + 3|$

CHECK YOUR PROGRESS 3 Express the distance between x and 4 on the number line using absolute value.

Solution *See page S0.* $|x - 4|$

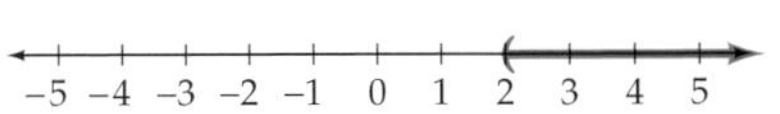

Figure P.4

■ Interval Notation

The graph of $\{x \mid x > 2\}$ is shown in **Figure P.4**. The set is the real numbers greater than 2. The parenthesis at 2 indicates that 2 is not included in the set. Rather than write this set of real numbers using set-builder notation, we frequently write the set in **interval notation** as $(2, \infty)$, where the infinity symbol, ∞, is used to indicate the numbers in the interval do not end. See the Take Note on page 8.

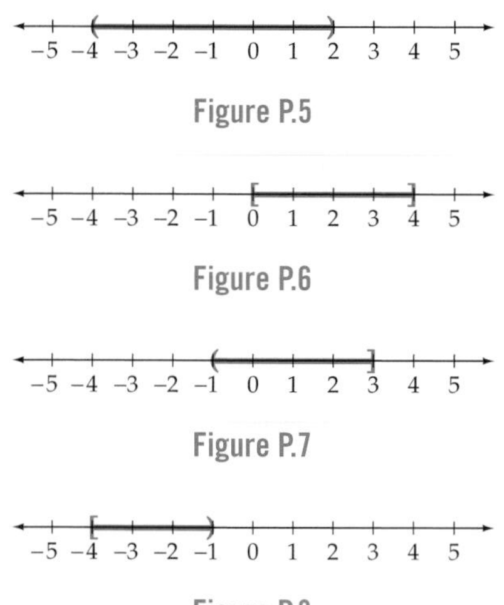

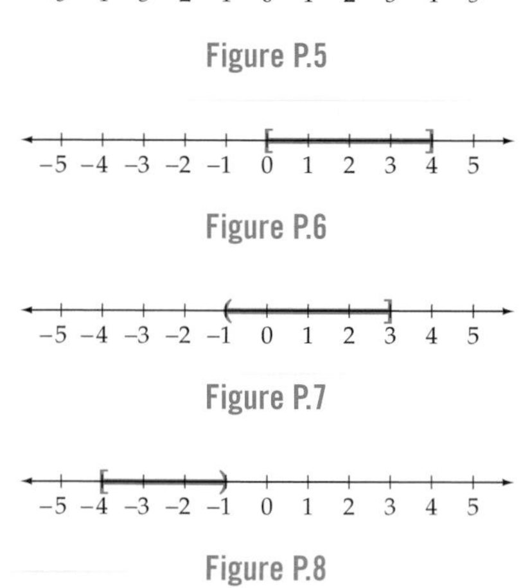

Figure P.5

Figure P.6

Figure P.7

Figure P.8

In general, the interval notation

(a, b) represents all real numbers between a and b, not including a and b. This is an **open interval.** In set-builder notation, we write $\{x \mid a < x < b\}$. For instance, the graph of $(-4, 2)$ is shown in Figure P.5.

$[a, b]$ represents all real numbers between a and b, including a and b. This is a **closed interval.** In set-builder notation, we write $\{x \mid a \le x \le b\}$. For instance, the graph of $[0, 4]$ is shown in Figure P.6. The brackets at 0 and 4 indicate that those numbers are included in the graph.

$(a, b]$ represents all real numbers between a and b, not including a but including b. This is a **half-open interval.** In set-builder notation, we write $\{x \mid a < x \le b\}$. For instance, the graph of $(-1, 3]$ is shown in Figure P.7.

$[a, b)$ represents all real numbers between a and b, including a but not including b. This is a **half-open interval.** In set-builder notation, we write $\{x \mid a \le x < b\}$. For instance, the graph of $[-4, -1)$ is shown in Figure P.8.

TAKE NOTE

It is *never* correct to use a bracket when using the infinity symbol. For instance, $[-\infty, 3]$ is not correct and neither is $[2, \infty]$. Neither negative infinity nor positive infinity is a real number and therefore cannot be contained in an interval.

Alternative to Example 4
Exercise 30, page 13

EXAMPLE 4 Graph a Set in Interval Notation

Graph $(-\infty, 3]$. Write the interval in set-builder notation.

Solution The set is the real numbers less than or equal to 3. In set-builder notation, this is the set $\{x \mid x \le 3\}$. Draw a bracket at 3, and darken the number line to the left of 3 as shown in Figure P.9.

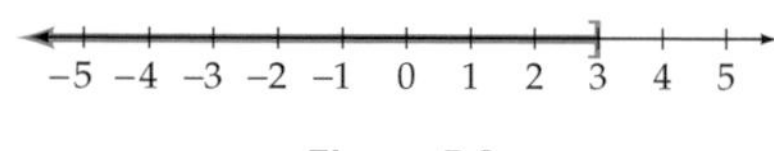

Figure P.9

CHECK YOUR PROGRESS 4 Graph the interval $(-4, 3)$. Write the interval in set-builder notation.

Solution *See page S0.* $\{x \mid -4 < x < 3\}$

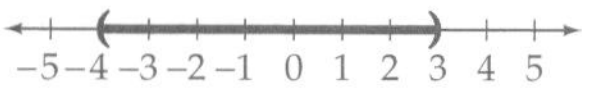

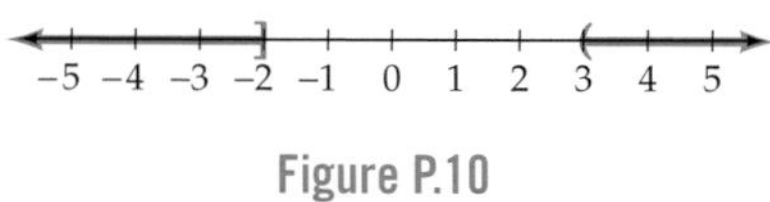

Figure P.10

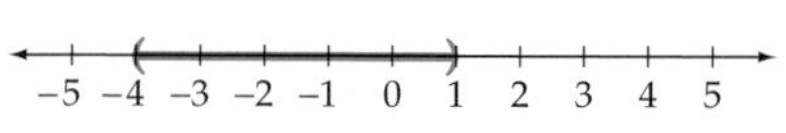

Figure P.11

Alternative to Example 5
Exercises 36, 40, and 48, page 13

The set $\{x \mid x \le -2\} \cup \{x \mid x > 3\}$ is the set of real numbers that are either less than or equal to -2 or greater than 3. We could also write this in interval notation as $(-\infty, -2] \cup (3, \infty)$. The graph of the set is shown in Figure P.10.

The set $\{x \mid x > -4\} \cap \{x \mid x < 1\}$ is the set of real numbers that are greater than -4 *and* less than 1. Note from Figure P.11 that this set is the interval $(-4, 1)$, which can be written in set-builder notation as $\{x \mid -4 < x < 1\}$.

EXAMPLE 5 Graph Intervals

Graph the following. Express **a.** and **b.** in interval notation. Express **c.** and **d.** in set-builder notation.

a. $\{x \mid x \le -1\} \cup \{x \mid x \ge 2\}$ **b.** $\{x \mid x \ge -1\} \cap \{x \mid x < 5\}$

c. $(-\infty, 0) \cup [1, 3]$ **d.** $[-1, 3] \cap (1, 5)$

Continued ➤

TAKE NOTE

For part **d.**, the graphs of $[-1, 3]$ (in red) and $(1, 5)$ (in blue) are shown below.

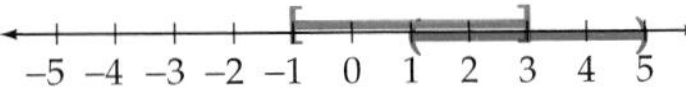

Note that the intersection of the sets occurs where the graphs intersect. Although $1 \in [-1, 3]$, $1 \notin (1, 5)$. Therefore, 1 does not belong to the intersection of the sets. On the other hand, $3 \in [-1, 3]$ and $3 \in (1, 5)$ and therefore 3 belongs to the intersection of the sets. Thus the intersection is $\{x \mid 1 < x \leq 3\}$.

Solution

a. $-5\ -4\ -3\ -2\ -1\ \ 0\ \ 1\ \ 2\ \ 3\ \ 4\ \ 5$ $\quad (-\infty, -1] \cup [2, \infty)$

b. $-5\ -4\ -3\ -2\ -1\ \ 0\ \ 1\ \ 2\ \ 3\ \ 4\ \ 5$ $\quad [-1, 5)$

c. $-5\ -4\ -3\ -2\ -1\ \ 0\ \ 1\ \ 2\ \ 3\ \ 4\ \ 5$ $\quad \{x \mid x < 0\} \cup \{x \mid 1 \leq x \leq 3\}$

d. $-5\ -4\ -3\ -2\ -1\ \ 0\ \ 1\ \ 2\ \ 3\ \ 4\ \ 5$ $\quad \{x \mid 1 < x \leq 3\}$

CHECK YOUR PROGRESS 5 Graph the following. Express **a.** in interval notation and express **b.** in set-builder notation.

a. $\{x \mid x \geq -1\} \cup \{x \mid x \geq 3\}$ **b.** $(-\infty, 1) \cap [-1, 3]$

Solution *See page S0.* See graphs at the left.
a. $[-1, \infty)$ b. $\{x \mid -1 \leq x < 1\}$

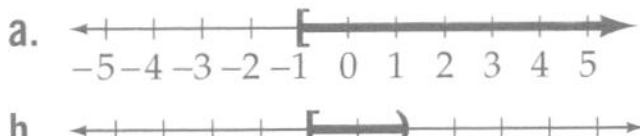

Order of Operations Agreement

The value V of an investment of \$5000 after t years at an annual simple interest rate of 6% is given by the equation

$$V = 5000 + 300t$$

Then the value of the investment

after 1 year is	$5000 + 300(1) = 5000 + 300 = 5300$
after 2 years is	$5000 + 300(2) = 5000 + 600 = 5600$
after 3 years is	$5000 + 300(3) = 5000 + 900 = 5900$
after 4 years is	$5000 + 300(4) = 5000 + 1200 = 6200$

Notice that the expression $5000 + 300(4)$ has two operations, addition and multiplication. When an expression contains more than one operation, the operations must be performed in a specified order, as listed below in the Order of Operations Agreement.

The Order of Operations Agreement

Step 1 Perform operations inside grouping symbols. Grouping symbols include parentheses, brackets, fraction bars, absolute value symbols, and radical symbols.
Step 2 Evaluate exponential expressions.
Step 3 Do multiplication and division as they occur from left to right.
Step 4 Do addition and subtraction as they occur from left to right.

Therefore, the expression $5000 + 300(4)$ is simplified by first performing the multiplication $300(4)$ and then performing the addition, as we did above.

TAKE NOTE

Recall that $a - b$ can be rewritten as $a + (-b)$. Therefore,

$3x^2 - 4xy + 5x - y - 7$

$= 3x^2 + (-4xy) + 5x + (-y) + (-7)$

In this form, we can see that the terms (addends) are $3x^2$, $-4xy$, $5x$, $-y$, and -7.

One of the ways in which the Order of Operations Agreement is used is to evaluate **variable expressions.** The addends of a variable expression are called terms. The terms for the expression at the right are $3x^2$, $-4xy$, $5x$, $-y$, and -7. Observe that the sign of a term is the sign that immediate precedes it.

$$3x^2 - 4xy + 5x - y - 7$$

The terms $3x^2$, $-4xy$, $5x$, and $-y$ are variable terms. The term -7 is a constant term. Each variable term has a numerical coefficient and a variable part. The numerical coefficient for the term $3x^2$ is 3; the numerical coefficient for the term $-4xy$ is -4; the numerical coefficient for the term for $-y$ is -1. When the numerical coefficient is 1 or -1 (as in x and $-x$), the 1 is usually not written.

To evaluate a variable expression, replace the variables by their given values and then use the Order of Operations Agreement to simplify the result.

Alternative to Example 6

a. Evaluate $-x^2 - x(3x + 2)$ when $x = -3$.

b. Evaluate $2xy^2 - 4(x^2 - y)$ when $x = 2$ and $y = -4$.

▪ **a.** -30

▪ **b.** 32

EXAMPLE 6 Evaluate a Variable Expression

a. Evaluate $\dfrac{x^3 - y^3}{x - y}$ for $x = 2$ and $y = -3$.

b. Evaluate $(x + 2y)^2 - 4z$ when $x = 3$, $y = -2$, and $z = -4$.

Solution

a. $\dfrac{x^3 - y^3}{x - y}$

$$\frac{2^3 - (-3)^3}{2 - (-3)} = \frac{8 - (-27)}{2 - (-3)} = \frac{35}{5} = 7$$

b. $(x + 2y)^2 - 4z$

$$[3 + 2(-2)]^2 - 4(-4) = [3 + (-4)]^2 - 4(-4) = (-1)^2 - 4(-4) = 1 - 4(-4) = 1 - (-16) = 1 + 16 = 17$$

CHECK YOUR PROGRESS 6 Evaluate $3ab - 4(2a - 3b)$ when $a = 4$ and $b = -3$.

Solution *See page S0.* -104

▪ Simplifying Variable Expressions

Variable expressions are simplified by using the properties of real numbers.

TAKE NOTE

The additive inverse of a number is also called the opposite of the number. For instance, -3 is the additive inverse, or opposite, of 3. The multiplicative inverse of a number is also called the **reciprocal** of the number. For instance, the multiplicative inverse, or reciprocal, of $\frac{2}{3}$ is $\frac{3}{2}$.

Properties of Real Numbers

If a, b, and c are real numbers, then the following properties hold true.

Commutative Property of Addition
$a + b = b + a$

Commutative Property of Multiplication
$ab = ba$

Associative Property of Addition
$(a + b) + c = a + (b + c)$

Associative Property of Multiplication
$(ab)c = a(bc)$

Additive Identity Property
There exists a unique number 0 such that
$a + 0 = 0 + a = a$

Multiplicative Identity Property
There exists a unique number 1 such that
$a \cdot 1 = 1 \cdot a = a$

Inverse Property of Addition
For each number a, there exists a unique real number $-a$, called the *additive inverse* of a, such that

$a + (-a) = (-a) + a = 0$

Inverse Property of Multiplication
For each *nonzero* number a, there exists a unique real number $1/a$, called the *multiplicative inverse* of a, such that

$a \cdot \frac{1}{a} = \frac{1}{a} \cdot a = 1, a \neq 0$

Distributive Property
$a(b + c) = ab + ac$

For instance, to simplify $(6x)2$, both the commutative and associative properties of multiplication are used.

$$\begin{aligned}(6x)2 &= 2(6x) && \text{▪ Commutative property of multiplication}\\ &= (2 \cdot 6)x && \text{▪ Associative property of multiplication}\\ &= 12x\end{aligned}$$

TAKE NOTE

Normally we will not show all the steps involved in the simplification of a variable expression. For instance, we will just write

$(6x)2 = 12x$,

$3(4p + 5) = 12p + 15$, and

$3x^2 + 9x^2 = 12x^2$

It is important to know, however, that every step in the simplification depends on one of the properties of real numbers.

To simplify $3(4p + 5)$, use the distributive property.

$$\begin{aligned}3(4p + 5) &= 3(4p) + 3(5) && \text{▪ Distributive property}\\ &= 12p + 15\end{aligned}$$

Terms that have the same variable part are called like terms. The distributive property is also used to simplify an expression with like terms, such as $3x^2 + 9x^2$.

$$\begin{aligned}3x^2 + 9x^2 &= (3 + 9)x^2 && \text{▪ Distributive property}\\ &= 12x^2\end{aligned}$$

Note from this example that like terms are combined by adding the coefficients of the like terms.

QUESTION Are the terms $2x^2$ and $3x$ like terms?

ANSWER No. The variable parts are not the same. The variable part of $2x^2$ is $x \cdot x$. The variable part of $3x$ is x.

Alternative to Example 7

Simplify: $6 - 4[2x - 3(2x + 1)]$

■ $16x + 18$

INSTRUCTOR NOTE

Some students may have difficulty understanding that, for example, $2x^2$ and $3x$ are not like terms. One analogy that may help these students is to explain that if they wanted to make the classroom larger, it would be necessary to add *square* feet (area) to the room, not just feet (length).

EXAMPLE 7 Simplify Variable Expressions

a. Simplify $5 + 3(2x - 6)$.

b. Simplify $2(x^2 - 3x + 4) - 4(2x^2 + 5x - 2)$.

Solution

a. $5 + 3(2x - 6) = 5 + 6x - 18$ ■ Use the distributive property.

$= 6x - 13$ ■ Add the constant terms.

b. $2(x^2 - 3x + 4) - 4(2x^2 + 5x - 2) = 2x^2 - 6x + 8 - 8x^2 - 20x + 8$

$= -6x^2 - 26x + 16$

CHECK YOUR PROGRESS 7 Simplify the following.

a. $6 - 3(5a - 4)$

b. $-3(2a - 5b + 1) + 5(3a - b + 2)$

Solution *See page S0.* a. $-15a + 18$ b. $9a + 10b + 7$

Topics for Discussion

1. Give an example of **a.** an integer that is not a natural number, **b.** a rational number that is not an integer, and **c.** a real number that is not a rational number.
2. If $K = \{\text{irrational numbers}\}$ and $Q = \{\text{rational numbers}\}$, find **a.** $K \cup Q$ and **b.** $K \cap Q$.
3. If the proposed simplification shown at the right is correct, so state. If it is incorrect, show a correct simplification.
4. Are there any even prime numbers? If so, name them.
5. Does every real number have an additive inverse? Does every real number have a multiplicative inverse?

$2 \cdot 3^2 = 6^2 = 36$

1. $-\frac{1}{5}$: rational, real; 0: integer, rational, real; -44: integer, rational, real; π: irrational, real; 3.14: rational, real; 5.05005000500005...: irrational, real; $\sqrt{49}$: integer, rational, real, prime; 53: integer, rational, real, prime

2. $\frac{5}{\sqrt{7}}$: irrational, real; $\frac{5}{7}$: rational, real; 31: integer, rational, real, prime; $-2\frac{1}{2}$: rational, real; 4.235653907493: rational, real; 51: integer, rational, real; 0.888...: rational, real

EXERCISES P.1

—*Suggested Assignment: Exercises 1–77, every other odd; 79–86, all; 87–103, odd; and 104–109, all.*

—*Answer graphs to Exercises 23–34 and 35–50 are on page AA1.*

In Exercises 1 and 2, determine whether each number is an integer, a rational number, an irrational number, a prime number, or a real number.

1. $-\frac{1}{5}, 0, -44, \pi, 3.14, 5.05005000500005\ldots, \sqrt{49}, 53$

2. $\frac{5}{\sqrt{7}}, \frac{5}{7}, 31, -2\frac{1}{2}, 4.235653907493, 51, 0.888\ldots$

In Exercises 3 to 12, perform the operations given $A = \{-3, -2, -1, 0, 1, 2, 3\}$, $B = \{-2, 0, 2, 4, 6\}$, $C = \{0, 1, 2, 3, 4, 5, 6\}$, and $D = \{-3, -1, 1, 3\}$.

3. $A \cup B$ $\{-3, -2, -1, 0, 1, 2, 3, 4, 6\}$
4. $C \cup D$ $\{-3, -1, 0, 1, 2, 3, 4, 5, 6\}$
5. $A \cap C$ $\{0, 1, 2, 3\}$
6. $C \cap D$ $\{1, 3\}$
7. $B \cap D$ $\emptyset$
8. $B \cup (A \cap C)$ $\{-2, 0, 1, 2, 3, 4, 6\}$

9. $D \cap (B \cup C)$ $\{1, 3\}$

10. $(A \cap B) \cup (A \cap C)$ $\{-2, 0, 1, 2, 3\}$

11. $(B \cup C) \cap (B \cup D)$ $\{-2, 0, 1, 2, 3, 4, 6\}$

12. $(A \cap C) \cup (B \cap D)$ $\{0, 1, 2, 3\}$

In Exercises 13 to 18, write each expression without absolute value symbols.

13. $|0|$ 0

14. $\left|-\frac{1}{2}\right|$ $\frac{1}{2}$

15. $|3| - |-7|$ -4

16. $|3| \cdot |-4|$ 12

17. $|a^2 + 7|$ $a^2 + 7$

18. $|2x^2 + 1|$ $2x^2 + 1$

In Exercises 19 to 22, use absolute value notation to describe the given situation.

19. The distance between x and 3 $|x - 3|$

20. The distance between a and -2 $|a + 2|$

21. The distance between x and -2 is 4. $|x + 2| = 4$

22. The distance between z and 5 is 1. $|z - 5| = 1$

In Exercises 23 to 34, graph each set. Write sets given in interval notation in set-builder notation and write sets given in set-builder notation in interval notation.

23. $(-2, 3)$ $\{x \mid -2 < x < 3\}$

24. $[1, 5]$ $\{x \mid 1 \le x \le 5\}$

25. $[-5, -1]$ $\{x \mid -5 \le x \le -1\}$

26. $(-3, 3)$ $\{x \mid -3 < x < 3\}$

27. $[2, \infty)$ $\{x \mid x \ge 2\}$

28. $(-\infty, 4)$ $\{x \mid x < 4\}$

29. $\{x \mid 3 < x < 5\}$ $(3, 5)$

30. $\{x \mid x < -1\}$ $(-\infty, -1)$

31. $\{x \mid x \ge -2\}$ $[-2, \infty)$

32. $\{x \mid -1 \le x < 5\}$ $[-1, 5)$

33. $\{x \mid 0 \le x \le 1\}$ $[0, 1]$

34. $\{x \mid -4 < x \le 5\}$ $(-4, 5]$

In Exercises 35 to 50, graph each set.

35. $(-\infty, 0) \cup [2, 4]$

36. $(-3, 1) \cup (3, 5)$

37. $(-4, 0) \cap [-2, 5]$

38. $(-\infty, 3] \cap (2, 6)$

39. $(1, \infty) \cup (-2, \infty)$

40. $(-4, \infty) \cup (0, \infty)$

41. $(1, \infty) \cap (-2, \infty)$

42. $(-4, \infty) \cap (0, \infty)$

43. $[-2, 4] \cap [4, 5]$

44. $(-\infty, 1] \cap [1, \infty)$

45. $(-2, 4) \cap (4, 5)$

46. $(-\infty, 1) \cap (1, \infty)$

47. $\{x \mid x < -3\} \cup \{x \mid 1 < x < 2\}$

48. $\{x \mid -3 \le x < 0\} \cup \{x \mid x \ge 2\}$

49. $\{x \mid x < -3\} \cup \{x \mid x < 2\}$

50. $\{x \mid x < -3\} \cap \{x \mid x < 2\}$

In Exercises 51 and 52, a. state the number of terms in the variable expression, b. name the numerical coefficient of each variable term, and c. name the constant term.

51. $5 - 3x^2 + 4xy - z + 4y^2$ **a.** 5 **b.** $-3, 4, -1, 4$ **c.** 5

52. $x^3y^3 - 2xy + 3y^2 - 9$ **a.** 4 **b.** $1, -2, 3$ **c.** -9

In Exercises 53 to 64, evaluate the variable expression for $x = 3$, $y = -2$, and $z = -1$.

53. $-y^3$ 8

54. $-y^2$ -4

55. $2xyz$ 12

56. $-3xz$ 9

57. $-2x^2y^2$ -72

58. $2y^3z^2$ -16

59. $xy - z(x - y)^2$ 19

60. $(z - 2y)^2 - 3z^3$ 12

61. $\dfrac{x^2 + y^2}{x + y}$ 13

62. $\dfrac{2xy^2z^4}{(y - z)^4}$ 24

63. $\dfrac{3y}{x} - \dfrac{2z}{y}$ -3

64. $(x - z)^2(x + z)^2$ 64

In Exercises 65 to 78, simplify the variable expression.

65. $3(2x)$ $6x$

66. $-2(4y)$ $-8y$

67. $3(2 + x)$ $6 + 3x$

68. $-2(4 + y)$ $-8 - 2y$

69. $\frac{2}{3}a + \frac{5}{6}a$ $\frac{3}{2}a$

70. $\frac{3}{4}x - \frac{1}{2}x$ $\frac{1}{4}x$

71. $2 + 3(2x - 5)$ $6x - 13$

72. $4 + 2(2a - 3)$ $4a - 2$

73. $5 - 3(4x - 2y)$ $5 - 12x + 6y$

74. $7 - 2(5n - 8m)$ $7 - 10n + 16m$

75. $3(2a - 4b) - 4(a - 3b)$ $2a$

76. $5(4r - 7t) - 2(10r + 3t)$ $-41t$

77. $5a - 2[3 - 2(4a + 3)]$ $21a + 6$

78. $6 + 3[2x - 4(3x - 2)]$ $30 - 30x$

79. ***Area of a Triangle*** The area of a triangle is given by Area $= \frac{1}{2}bh$, where b is the length of the base of the

triangle and h is its height. Find the area of a triangle whose base is 3 inches and whose height is 4 inches. 6 in^2.

80. ***Volume of a Box*** The volume of a rectangular box is given by Volume $= lwh$, where l is the length, w is the width, and h is the height of the box. Find the volume of a classroom that is 40 feet long, 30 feet wide, and 12 feet high. 14,400 ft^3

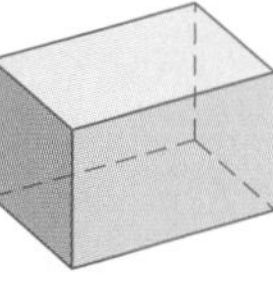

Business and Economics

81. ***Profit from Sales*** The profit, in dollars, a company earns from the sale of x bicycles is given by

$$\text{Profit} = -0.5x^2 + 120x - 2000$$

Find the profit the company earns from selling 110 bicycles. $5150

82. ***Magazine Circulation*** The circulation, in thousands of subscriptions, of a new magazine n months after its introduction can be approximated by

$$\text{Circulation} = \sqrt{n^2 - n + 1}$$

Find, to the nearest hundredth, the circulation of the magazine after 12 months. 11.53 thousand

Life and Health Sciences

83. ***Heart Rate*** The heart rate, in beats per minute, of a certain runner during a cool-down period can be approximated by

$$\text{Heart rate} = 65 + \frac{53}{4t + 1}$$

where t is the number of minutes of cool down. Find the runner's heart rate after 10 minutes. Round to the nearest whole number. 66 beats/min

84.  ***Body Mass Index*** According to the National Institute of Health, body mass index (BMI) is a measure of body fat based on height and weight that applies to adult men and women. Values between 18.5 and 24.9 are considered healthy. BMI is calculated by

$$\text{BMI} = \frac{705w}{h^2}$$

where w is the weight of the person in pounds and h is the person's height in inches. Find the BMI for a person who weighs 160 pounds and is 5 feet 10 inches tall. Round to the nearest tenth. 23.0

Physical Sciences and Engineering

85. ***Physics*** The height, in feet, of a ball t seconds after it is thrown upward is given by Height $= -16t^2 + 80t + 4$. Find the height of the ball 2 seconds after it has been released. 100 ft

86. ***Chemistry*** Salt is being added to water in such a way that the concentration, in grams per liter, is given by Concentration $= \dfrac{50t}{t + 1}$, where t is the time in minutes after the introduction of the salt. Find the concentration of salt after 24 minutes. 48 g/L

In Exercises 87 to 94, state the property of real numbers that is used.

87. $(ab^2)c = a(b^2c)$ Associative property of multiplication

88. $2x - 3y = -3y + 2x$ Commutative property of addition

89. $4(2a - b) = 8a - 4b$ Distributive property

90. $(3x)y = y(3x)$ Commutative property of multiplication

91. $4ab + 0 = 4ab$ Additive identity property

92. $1 \cdot (4x) = 4x$ Multiplicative identity property

93. $4 \cdot \dfrac{1}{4} = 1$ Inverse property of multiplication

94. $ab + (-ab) = 0$ Inverse property of addition

In Exercises 95 to 98, perform the operation given A is any set.

95. $A \cup A$ A

96. $A \cap A$ A

97. $A \cap \emptyset$ $\emptyset$

98. $A \cup \emptyset$ A

99. If A and B are two sets and $A \cup B = A$, what can be said about B? All elements of B are contained in A.

100. If A and B are two sets and $A \cap B = B$, what can be said about B? All elements of B are contained in A.

101. What set is represented by $(-\infty, \infty)$? All real numbers

102. Write $|2 - \pi|$ without absolute value symbols. $\pi - 2$

103. Why are 2 and 5 the only prime numbers whose difference is 3?
Adding 3 to an odd prime number would result in an even number greater than 2. All even numbers greater than 2 are not prime.

Prepare for Section P.2

104. Simplify: $2^2 \cdot 2^3$ 32

105. Simplify: $\dfrac{4^3}{4^5}$ $\dfrac{1}{16}$

106. Simplify: $(2^3)^2$ 64

107. Simplify: $3.14(10^5)$ 314,000

108. True or false: $3^4 \cdot 3^2 = 9^6$ False

109. True or false: $(3 + 4)^2 = 3^2 + 4^2$ False

Explorations 

1. *Search Engines* Some Internet search engines use the operators OR and AND when searching for a topic. For instance, using the search engine AltaVista, a recent search for "music" produced about 22.2 million sites that contained the word *music*. A search for "MP3" produced 4.9 million sites that mentioned MP3. However, when we searched for "music AND MP3," there were 2.6 million sites that mentioned the words *music* and *MP3*.

a. Explain the search described above in terms of intersections of sets. Answers will vary.

b. Explain how a search engine might respond to a search for "music OR MP3." Answers will vary.

c. Try some Internet searches using the operators OR and AND. A good place to start is **www.altavista.com/sites/search/adv.** Answers will vary.

2. *Number Theory* A number n has the following properties.

When n is divided by 6, the remainder is 5.

When n is divided by 5, the remainder is 4.

When n is divided by 4, the remainder is 3.

When n is divided by 3, the remainder is 2.

When n is divided by 2, the remainder is 1.

What is the smallest possible value of n? 59

3. *Intervals* In addition to finding unions and intersections of intervals, it is possible to apply other operations to intervals. For instance, $(-1, 2)^2$ is the interval that results from squaring every number in the interval $(-1, 2)$. This gives $[1, 4)$. Thus $(-1, 2)^2 = [1, 4)$.

a. Find $(-4, 2)^2$. $[0, 16)$

b. Find $\dfrac{1}{[-2, 3]}$, the reciprocal of every number in $[-2, 3]$.

c. Find ABS$(-4, 5)$, the absolute value of every number in $(-4, 5)$. $[0, 5)$

3b. $\left(-\infty, -\dfrac{1}{2}\right] \cup \left[\dfrac{1}{3}, \infty\right)$

SECTION P.2 Integer Exponents and Scientific Notation

- Multiplication of Exponential Expressions
- Powers of Exponential Expressions
- Division of Exponential Expressions
- Scientific Notation

Multiplication of Exponential Expressions

Recall that the exponential expression 3^4 means to use the base, 3, as a factor four times. Therefore, $3^4 = 3 \cdot 3 \cdot 3 \cdot 3 = 81$. For the variable exponential expression x^6, x is the base and 6 is the exponent. The exponent indicates the number of times the base occurs as a factor. Therefore,

$$x^6 = \overbrace{x \cdot x \cdot x \cdot x \cdot x \cdot x}^{x \text{ is a factor six times.}}$$

The product of exponential expressions with the *same* base can be simplified by writing each expression in factored form and then writing the result with an exponent.

$$x^3 \cdot x^4 = \underbrace{\overbrace{(x \cdot x \cdot x)}^{3 \text{ factors}} \cdot \overbrace{(x \cdot x \cdot x \cdot x)}^{4 \text{ factors}}}_{7 \text{ factors}}$$
$$= x \cdot x \cdot x \cdot x \cdot x \cdot x \cdot x$$
$$= x^7$$

Note that adding the exponents results in the same product.

$$x^3 \cdot x^4 = x^{3+4} = x^7$$

Point of Interest

One billion, which is 10^9, is too large a number for most of us to comprehend. If a computer were to start counting from 1 to 1 billion, writing to the screen one number every second of every day, it would take over 31 years for the computer to complete the task.

And if you think a billion is a large number, consider a googol. A googol is 1 with 100 zeros after it, or 10^{100}. Edward Kasner is the mathematician credited with thinking up this number, and his nine-year-old nephew is said to have thought up the name. The two then coined the word *googolplex,* which is 10^{googol}.

As a final point—the search engine **Google.com** is a take off on the word *googol.*

Alternative to Example 1

Simplify: $x^4 \cdot x^7 \cdot x$

■ x^{12}

Alternative to Example 2

Simplify: $(-2x^2y)(-3x^3y^6)$

■ $6x^5y^7$

This suggests the following rule for multiplying exponential expressions.

Rule for Multiplying Exponential Expressions

If m and n are integers, then $x^m \cdot x^n = x^{m+n}$.

As an example of this rule,

$$c^5 \cdot c^7 = c^{5+7} = c^{12}$$

■ The bases are the same. Add the exponents.

EXAMPLE 1 Multiply Exponential Expressions

Simplify: $a^3 \cdot a \cdot a^5$

Solution $a^3 \cdot a \cdot a^5 = a^{3+1+5} = a^9$ ■ Recall that $a = a^1$.

CHECK YOUR PROGRESS 1 Simplify: $m \cdot m^2 \cdot m^3$

Solution *See page S0.* m^6

QUESTION Why can't the exponential expression x^5y^3 be simplified?

EXAMPLE 2 Multiply Exponential Expressions

Simplify: $(4ab^4)(-3a^2b^3)$

Solution

$$(4ab^4)(-3a^2b^3) = [4(-3)]a^{1+2}b^{4+3}$$

■ Multiply the coefficients. Multiply the variables by adding the exponents on the like bases.

$$= -12a^3b^7$$

CHECK YOUR PROGRESS 2 Simplify: $(12p^4q^3)(-3p^5q^2)$

Solution *See page S0.* $-36p^9q^5$

■ Powers of Exponential Expressions

The expression $(x^4)^3$ is an example of a *power of an exponential expression;* the exponential expression x^4 is raised to the third (3) power.

ANSWER The bases are not the same.

The power of an exponential expression can be simplified by writing the power in factored form and then using the rule for multiplying exponential expressions.

$$(x^4)^3 = x^4 \cdot x^4 \cdot x^4 = x^{4+4+4} = x^{12}$$

Note that multiplying the exponent inside the parentheses by the exponent outside the parentheses results in the same product.

$$(x^4)^3 = x^{4\cdot 3} = x^{12}$$

This suggests the following rule for simplifying powers of exponential expressions.

Rule for Simplifying the Power of an Exponential Expression

If m and n are integers, then $(x^m)^n = x^{m\cdot n}$.

As an example of this rule, $(z^2)^5 = z^{2\cdot 5} = z^{10}$.

QUESTION Which of the following expressions is the multiplication of two exponential expressions and which is the power of an exponential expression?

a. $q^4 \cdot q^{10}$ **b.** $(q^4)^{10}$

The expression $(a^2b^3)^2$ is the *power of the product* of two exponential expressions, a^2 and b^3. The power of the product of exponential expressions can be simplified by writing the product in factored form and then using the rule for multiplying exponential expressions.

Write the exponential expression in factored form. Use the rule for multiplying exponential expressions.

$$(a^2b^3)^2 = (a^2b^3)(a^2b^3) = a^{2+2}b^{3+3} = a^4b^6$$

Note that multiplying each exponent inside the parentheses by the exponent outside the parentheses results in the same product.

$$(a^2b^3)^2 = a^{2\cdot 2}b^{3\cdot 2} = a^4b^6$$

Rule for Simplifying Powers of Products

If m, n, and p are integers, then $(x^my^n)^p = x^{m\cdot p}y^{n\cdot p}$.

Using this rule, we would simplify $(5z^3)^2$ as follows.

$(5z^3)^2 = 5^{1\cdot 2}z^{3\cdot 2}$ ▪ Multiplying each exponent inside the parentheses by the exponent outside the parentheses. Note that $5 = 5^1$.

$= 5^2z^6$

$= 25z^6$ ▪ $5^2 = 25$

ANSWER **a.** This is the multiplication of two exponential expressions. q^4 is multiplied by q^{10}. The result is q^{14}. **b.** This is the power of an exponential expression. q^4 is raised to the 10th power. The result is q^{40}.

Alternative to Example 3

Simplify: $(-3a^3b^2)^3$

▪ $-27a^9b^6$

TAKE NOTE

If Example 3 were changed to $-(u^3v^8)^6$ (the negative sign has been placed outside the parentheses), then

$$-(u^3v^8)^6 = -(u^{3\cdot 6}v^{8\cdot 6}) = -u^{18}v^{48}$$

In this case, the answer is negative.

EXAMPLE 3 Simplify a Power of an Exponential Expression

Simplify: $(-u^3v^8)^6$

Solution

$(-u^3v^8)^6 = (-1)^{1\cdot 6}u^{3\cdot 6}v^{8\cdot 6}$ ▪ Multiply each exponent inside the parentheses by the exponent outside the parentheses. Note that $(-u^3v^8)^6 = (-1u^3v^8)^6$.

$= u^{18}v^{48}$ ▪ $(-1)^6 = 1$

CHECK YOUR PROGRESS 3 Simplify: $(-2x^3y^7)^3$

Solution *See page S0.* $-8x^9y^{21}$

For some products, it is necessary to use the rule for simplifying powers of products and the rule for multiplying exponential expressions.

Alternative to Example 4

Simplify: $(-4x^2y^3)^2(2xy^4)^3$

▪ $128x^7y^{18}$

EXAMPLE 4 Simplify a Product of Exponential Expressions

Simplify: $(2a^2b)(2a^3b^2)^3$

Solution

$(2a^2b)(2a^3b^2)^3 = (2a^2b)(8a^9b^6)$ ▪ $(2a^3b^2)^3 = 2^{1\cdot 3}a^{3\cdot 3}b^{2\cdot 3} = 8a^9b^6$

$= 16a^{11}b^7$ ▪ $(2a^2b)(8a^9b^6) = (2\cdot 8)a^{2+9}b^{1+6} = 16a^{11}b^7$

CHECK YOUR PROGRESS 4 Simplify: $(-xy^4)(-2x^3y^2)^2$

Solution *See page S0.* $-4x^7y^8$

▪ Division of Exponential Expressions

The quotient of two exponential expressions with the *same* base can be simplified by writing each expression in factored form, dividing by the common factors, and then writing the result with an exponent.

$$\frac{x^6}{x^2} = \frac{\overset{1}{\cancel{x}} \cdot \overset{1}{\cancel{x}} \cdot x \cdot x \cdot x \cdot x}{\underset{1}{\cancel{x}} \cdot \underset{1}{\cancel{x}}} = x^4$$

Note that subtracting the exponents results in the same quotient.

$$\frac{x^6}{x^2} = x^{6-2} = x^4$$

This example suggests that to divide exponential expressions with like bases, subtract the exponents.

Rule for Dividing Exponential Expressions

If m and n are integers and $x \neq 0$, then $\dfrac{x^m}{x^n} = x^{m-n}$.

As an example of this rule, $\dfrac{c^8}{c^5} = c^{8-5} = c^3$.

Alternative to Example 5

Simplify: $\frac{x^6y}{x^6y^2}$

■ $\frac{1}{y}$

EXAMPLE 5 Simplify the Quotient of Exponential Expressions

Simplify: $\frac{x^5y^7}{x^4y^2}$

Solution $\frac{x^5y^7}{x^4y^2} = x^{5-4}y^{7-2}$ ■ Subtract the exponents on the like bases.

$= xy^5$ ■ Note that $x^{5-4} = x^1 = x$. The exponent 1 is not written.

CHECK YOUR PROGRESS 5 Simplify: $\frac{a^7b^6}{ab^3}$

Solution *See page S0.* a^6b^3

The expression below has been simplified in two ways: by dividing by common factors, and by using the rule for dividing exponential expressions.

$$\frac{x^3}{x^3} = \frac{\overset{1}{\cancel{x}} \cdot \overset{1}{\cancel{x}} \cdot \overset{1}{\cancel{x}}}{\underset{1}{\cancel{x}} \cdot \underset{1}{\cancel{x}} \cdot \underset{1}{\cancel{x}}} = 1 \qquad \frac{x^3}{x^3} = x^{3-3} = x^0$$

Because $\frac{x^3}{x^3} = 1$ and $\frac{x^3}{x^3} = x^0$, we use the following definition of zero as an exponent.

Definition of x^0

If $x \neq 0$, then $x^0 = 1$. The expression 0^0 is undefined.

TAKE NOTE

Notice at the right that $(-28)^0 = 1$ but $-(2x)^0 = -1$. You can think of $-(2x)^0$ as $-1(2x)^0 = -1(1) = -1$.

This definition applies to any nonzero expression raised to the zero power. For instance,

$$(-28)^0 = 1 \qquad (2a^2 + 2a - 1)^0 = 1 \qquad (4x^2y^3)^0 = 1 \qquad -(2x)^0 = -1$$

The expression below has been simplified in two ways: by dividing by common factors, and by using the rule for dividing exponential expressions.

$$\frac{x^3}{x^5} = \frac{\overset{1}{\cancel{x}} \cdot \overset{1}{\cancel{x}} \cdot \overset{1}{\cancel{x}}}{\underset{1}{\cancel{x}} \cdot \underset{1}{\cancel{x}} \cdot \underset{1}{\cancel{x}} \cdot x \cdot x} = \frac{1}{x^2} \qquad \frac{x^3}{x^5} = x^{3-5} = x^{-2}$$

Because $\frac{x^3}{x^5} = \frac{1}{x^3}$ and $\frac{x^3}{x^5} = x^{-2}$, $\frac{1}{x^2}$ must equal x^{-2}. Therefore, the following definition of a negative exponent is used.

Point of Interest

In the fifteenth century, the expression $12^{2\overline{m}}$ was used to mean $12x^{-2}$. The use of $\overline{m}$ reflected an Italian influence. In Italy, m was used for minus and p was used for plus. It was understood that $2\overline{m}$ referred to an unnamed variable. Isaac Newton, in the seventeenth century, advocated the use of a negative exponent.

Definition of Negative Exponents

If n is a positive integer and $x \neq 0$, then $x^{-n} = \frac{1}{x^n}$ and $\frac{1}{x^{-n}} = x^n$.

An exponential expression is in simplest form when there are no negative exponents in the expression. For instance, n^{-3} is not in simplest form. To write this expression in simplest form, we use the definition of negative exponents.

$$n^{-3} = \frac{1}{n^3}$$

Similarly, $\frac{1}{c^{-4}}$ is not in simplest form. To write this expression in simplest form, use the definition of negative exponents.

$$\frac{1}{c^{-4}} = c^4$$

Observe that $3^{-4} = \frac{1}{3^4} = \frac{1}{81}$ and that $\frac{1}{81}$ is a *positive* number. A negative exponent does not affect whether a number is positive or negative.

QUESTION How are **a.** b^{-8} and **b.** $\frac{1}{2^{-3}}$ rewritten in simplest form?

Alternative to Example 6

Simplify: $\frac{12x^4y^{-8}}{18x^{-2}y^2}$

■ $\frac{2x^6}{3y^{10}}$

EXAMPLE 6 Simplify a Quotient

Simplify: $\frac{48x^2y^6}{36x^5y^{-2}}$

Solution

$$\frac{48x^2y^6}{36x^5y^{-2}} = \frac{4x^{2-5}y^{6-(-2)}}{3}$$

■ Subtract the exponents on the like bases. Note that $\frac{48}{36} = \frac{12 \cdot 4}{12 \cdot 3} = \frac{4}{3}$.

$$= \frac{4x^{-3}y^8}{3} = \frac{4y^8}{3x^3}$$

■ Write the answer in simplest form.

CHECK YOUR PROGRESS 6 Simplify: $\frac{-35a^6b^{-2}}{25a^{-3}b^5}$

Solution *See page S0.* $-\frac{7a^9}{5b^7}$

Powers of the quotient of two exponential expressions are simplified using the following rule.

Rule for Simplifying Powers of Quotients

If m, n, and p are integers and $y \neq 0$, then $\left(\frac{x^m}{y^n}\right)^p = \frac{x^{m \cdot p}}{y^{n \cdot p}}$.

ANSWER **a.** $b^{-8} = \frac{1}{b^8}$ **b.** $\frac{1}{2^{-3}} = 2^3 = 8$

Alternative to Example 7

Simplify: $\left(\dfrac{-2x^{-1}y^2}{4x^2y^2}\right)^{-3}$

■ $-8x^9$

EXAMPLE 7 Simplify a Power of a Quotient

Simplify: $\left(\dfrac{3xy^5}{6x^2y^{-3}}\right)^{-2}$

Solution

$$\left(\frac{3xy^5}{6x^2y^{-3}}\right)^{-2} = \left(\frac{x^{1-2}y^{5-(-3)}}{2}\right)^{-2} = \left(\frac{x^{-1}y^8}{2}\right)^{-2}$$

■ Simplify $\dfrac{3xy^5}{6x^2y^{-3}}$.

$$= \frac{x^{-1(-2)}y^{8(-2)}}{2^{1(-2)}} = \frac{x^2y^{-16}}{2^{-2}}$$

■ Use the rule for simplifying powers of quotients.

$$= \frac{4x^2}{y^{16}}$$

■ Write the answer in simplest form.

CHECK YOUR PROGRESS 7 Simplify: $\left(\dfrac{4a^3b^2}{2a^4b^2}\right)^{-3}$

Solution *See page S0.* $\dfrac{a^3}{8}$

The rules of exponents are true for all integers. In Example 8 and Check Your Progress 8, the rule for multiplying exponential expressions and the definition of negative exponents are used.

Alternative to Example 8

Simplify: $(-4x^3)^{-2}(8x^3)^2$

■ 4

EXAMPLE 8 Simplify an Expression with Negative Exponents

Simplify: $(-3a)(4a^{-3})^{-2}$

Solution

$$(-3a)(4a^{-3})^{-2} = (-3a)(4^{1(-2)}a^{-3(-2)})$$

■ Simplify $(4a^{-3})^{-2}$.

$$= (-3a)(4^{-2}a^6) = \frac{-3a \cdot a^6}{4^2}$$

■ Use the definition of negative exponents.

$$= -\frac{3a^7}{16}$$

■ Write the answer in simplest form.

CHECK YOUR PROGRESS 8 Simplify: $(-3ab)(2a^3b^{-2})^{-3}$

Solution *See page S1.* $-\dfrac{3b^7}{8a^8}$

■ Scientific Notation

Very large and very small numbers are encountered in the fields of science and engineering. For example, each gram of oxygen contains approximately 602,300,000,000,000,000,000,000 atoms and the charge of an electron is 0.000000000000000000160 coulomb. These numbers can be written and read more easily in scientific notation. In **scientific notation,** a number is expressed as a product of two factors, one a number between 1 and 10 and the other a power of 10.

TAKE NOTE

Look at the example at the right: $0.00000612 = 6.12 \times 10^{-6}$. Using the definition of negative exponents,

$$10^{-6} = \frac{1}{10^6} = \frac{1}{1{,}000{,}000} = 0.000001$$

Because $10^{-6} = 0.000001$, we can write

$$6.12 \times 10^{-6} = 6.12 \times 0.000001 = 0.00000612$$

which is the number we started with. We have not changed the value of the number; we have just written it in another form.

To change a number written in decimal notation to scientific notation, write it in the form $a \times 10^n$, where $1 \leq a < 10$ and n is an integer.

For numbers greater than 10, move the decimal point to the right of the first digit. The exponent n is positive and equal to the number of places the decimal point has been moved.

For numbers less than 1, move the decimal point to the right of the first nonzero digit. The exponent n is negative. The absolute value of the exponent is equal to the number of places the decimal point has been moved.

$240{,}000 = 2.4 \times 10^5$

$93{,}000{,}000 = 9.3 \times 10^7$

$0.00301 = 3.01 \times 10^{-3}$

$0.00000612 = 6.12 \times 10^{-6}$

Here are some additional examples of writing a number in scientific notation.

$$824{,}300{,}000{,}000 = 8.243 \times 10^{11}$$
$$57{,}000{,}000{,}000 = 5.7 \times 10^{10}$$
$$0.000000961 = 9.61 \times 10^{-7}$$
$$0.00000001703 = 1.703 \times 10^{-8}$$

? QUESTION Write the number of atoms in a gram of oxygen (602,300,000,000,000,000,000,000) and the charge of an electron (0.000000000000000000160) in scientific notation.

Point of Interest

- Approximately 3.1×10^7 orchid seeds weigh 1 ounce.
- Computer scientists measure an operation in nanoseconds. 1 nanosecond = 1×10^{-9} seconds
- If a spaceship traveled 25,000 miles per hour, it would require approximately 2.7×10^9 years to travel from one end of our universe to the other.

Changing a number written in scientific notation to decimal notation also requires moving the decimal point.

When the exponent on 10 is positive, move the decimal point to the right the same number of places as the exponent.

$7.83 \times 10^6 = 7{,}830{,}000$

$5.093 \times 10^8 = 509{,}300{,}000$

$4.93 \times 10^{-5} = 0.0000493$

$9.1 \times 10^{-6} = 0.0000091$

When the exponent on 10 is negative, move the decimal point to the left the same number of places as the absolute value of the exponent.

Here are some additional examples of writing a number given in scientific notation in standard form.

$$8.98 \times 10^7 = 89{,}800{,}000$$
$$1.073 \times 10^{11} = 107{,}300{,}000{,}000$$
$$2.9 \times 10^{-8} = 0.000000029$$
$$4.45 \times 10^{-3} = 0.00445$$

Scientists and engineers perform multiplication and division on numbers written in scientific notation in a manner similar to that used for multiplying and dividing exponential expressions. The power of 10 corresponds to the variable and the number between 1 and 10 corresponds to the coefficient of the variable.

? ANSWER 6.023×10^{23}; 1.60×10^{-19}

Alternative to Example 9

a. Simplify: $(6.4 \times 10^{-8})(5.7 \times 10^{-6})$

b. Simplify: $\dfrac{1.3 \times 10^{-10}}{5.2 \times 10^{5}}$

- **a. 3.648×10^{-13}**
- **b. 2.5×10^{-16}**

TAKE NOTE

In part **a.** of Example 9, observe that 12.402×10^4 is not in scientific notation because 12 is not a number between 1 and 10. Moving the decimal point one place to the left means we must add 1 to the exponent 4, producing 1.2402×10^5.

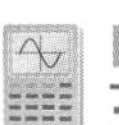

INTEGRATING TECHNOLOGY

The EE key (above the comma key) on a TI-83 calculator is used to enter numbers in scientific notation. For instance, to calculate part **a.** in Check Your Progress 9, enter 2.4 [2nd] [EE] [(-)] 9 [×] 1.6 [2nd] [EE] 3 [ENTER]

A typical screen is shown below.

EXAMPLE 9 Operations on Numbers in Scientific Notation

Simplify: **a.** $(2.34 \times 10^{6})(5.3 \times 10^{-2})$ **b.** $\dfrac{4.9 \times 10^{5}}{1.4 \times 10^{12}}$

Solution

a. $(2.34 \times 10^{6})(5.3 \times 10^{-2}) = [2.34(5.3)] \times 10^{6+(-2)}$ ▪ Multiply 2.34 times 5.3. Add the exponents on 10.

$= 12.402 \times 10^{4}$ ▪ Simplify.

$= 1.2402 \times 10^{5}$ ▪ Write the number in scientific notation.

b. $\dfrac{4.9 \times 10^{5}}{1.4 \times 10^{12}} = \dfrac{4.9}{1.4} \times 10^{5-12}$ ▪ Divide $\dfrac{4.9}{1.4}$. Subtract the exponents on 10.

$= 3.5 \times 10^{-7}$

CHECK YOUR PROGRESS 9 Simplify.

a. $(2.4 \times 10^{-9})(1.6 \times 10^{3})$ **b.** $\dfrac{5.6 \times 10^{-10}}{8.0 \times 10^{-4}}$

Solution *See page S1.* a. 3.84×10^{-6} b. 7.0×10^{-7}

Topics for Discussion

1. Explain the difference between simplifying $(-3)^8$ and simplifying -3^8.
2. Does $(2x - 4)^0 = 1$ for all values of x? If not, what value(s) of x must be excluded?
3. Each of the simplifications below is incorrect. Explain why and correct the error.

   a. $x^2 \cdot x^4 = x^8$ **b.** $(3a^5)^2 = 3a^{10}$
4. If $x = -2$, is x^{-4} a positive or a negative number?
5. Which of the following numbers are not in scientific notation?

   a. 3.0×10^5 **b.** 30.0×10^6 **c.** 0.30×10^{-5}

   Rewrite those numbers so that they are in scientific notation.

EXERCISES P.2

— *Suggested Assignment: Exercises 1–63, odd; 64, 65, 66, 67, 68, 71; and 77–82, all.*

In Exercises 1 to 14, evaluate each expression.

1. -5^3 -125

2. $(-5)^3$ -125

3. $\left(\dfrac{2}{3}\right)^0$ 1

4. -6^0 -1

5. 4^{-2} $\frac{1}{16}$

6. 3^{-4} $\frac{1}{81}$

7. $\frac{1}{2^{-5}}$ 32

8. $\frac{1}{3^{-3}}$ 27

9. $\left(\frac{2}{3}\right)^{-2}$ $\frac{9}{4}$

10. $\left(-\frac{4}{3}\right)^{-3}$ $-\frac{27}{64}$

11. $\frac{2^{-3}}{6^{-3}}$ 27

12. $\frac{4^{-2}}{2^{-3}}$ $\frac{1}{2}$

13. $(3^2 - 4^5)^0$ 1

14. $(8^3 - 9^6)^0$ 1

In Exercises 15 to 44, write the exponential expression in simplest form.

15. x^{-5} $\frac{1}{x^5}$

16. $\frac{1}{y^{-4}}$ y^4

17. $2x^{-4}$ $\frac{2}{x^4}$

18. $3y^{-2}$ $\frac{3}{y^2}$

19. $(-2ab^4)(-3a^2b^4)$ $6a^3b^8$

20. $(9xy^2)(-2x^2y^2)$ $-18x^3y^4$

21. $\frac{6a^8}{3a^4}$ $2a^4$

22. $\frac{12x^5}{6x^4}$ $2x$

23. $\frac{3x^3}{x^5}$ $\frac{3}{x^2}$

24. $\frac{5v^4}{v^8}$ $\frac{5}{v^4}$

25. $2xy(-3x^2)(4x^3y^4)$ $-24x^6y^5$

26. $(-3b^5)(2ab^2)(-2ab^2c^2)$ $12a^2b^9c^2$

27. $\frac{36a^{-2}b^3}{3ab^4}$ $\frac{12}{a^3b}$

28. $\frac{-48ab^{10}}{-32a^4b^3}$ $\frac{3b^7}{2a^3}$

29. $(-2m^3n^2)(-3mn^2)^2$ $-18m^5n^6$

30. $(2a^3b^2)^3(-4a^4b^2)$ $-32a^{13}b^8$

31. $(x^{-2}y)^2(xy)^{-2}$ $\frac{1}{x^6}$

32. $(x^{-1}y^2)^{-3}(x^2y^4)^{-3}$ $\frac{1}{x^3y^{18}}$

33. $\frac{(-4x^2y^3)^2}{(2xy^2)^3}$ $2x$

34. $\frac{(-3a^2b^3)^2}{(-2ab^4)^3}$ $-\frac{9a}{8b^6}$

35. $\left(\frac{a^{-2}b}{a^3b^{-4}}\right)^2$ $\frac{b^{10}}{a^{10}}$

36. $\left(\frac{x^{-3}y^{-4}}{x^{-2}y}\right)^{-2}$ x^2y^{10}

37. $x^{3n} \cdot x^{4n}$ x^{7n}

38. $2a^{2n+1}(3a^{n-1})$ $6a^{3n}$

39. $(a^n)^{2n}$ a^{2n^2}

40. $(x^2)^n$ x^{2n}

41. $\frac{x^{2n-3}}{x^{n-4}}$ x^{n+1}

42. $\frac{a^{5n}}{a^{8n}}$ $\frac{1}{a^{3n}}$

43. $(2ab^2)^4 + (3a^2b^4)^2$ $25a^4b^8$

44. $\frac{4x^4}{y^{-2}} + \left(\frac{y^{-1}}{x^2}\right)^{-2}$ $5x^4y^2$

In Exercises 45 to 48, write the number in scientific notation.

45. 2,011,000,000,000 2.011×10^{12}

46. 49,100,000,000 4.91×10^{10}

47. 0.000000000562 5.62×10^{-10}

48. 0.000000402 4.02×10^{-7}

In Exercises 49 to 52, change the number from scientific notation to decimal notation.

49. 3.14×10^7 31,400,000

50. 4.03×10^9 4,030,000,000

51. -2.3×10^{-6} −0.0000023

52. 6.14×10^{-8} 0.0000000614

In Exercises 53 to 60, perform the indicated operation and write the answer in scientific notation.

53. $(3 \times 10^{12})(9 \times 10^{-5})$ 2.7×10^8

54. $(8.9 \times 10^{-5})(3.4 \times 10^{-6})$ 3.026×10^{-10}

55. $\frac{9 \times 10^{-3}}{6 \times 10^8}$ 1.5×10^{-11}

56. $\frac{2.5 \times 10^8}{5 \times 10^{10}}$ 5×10^{-3}

57. $\frac{(3.2 \times 10^{-11})(2.7 \times 10^{18})}{1.2 \times 10^{-5}}$ 7.2×10^{12}

58. $\frac{(6.9 \times 10^{27})(8.2 \times 10^{-13})}{4.1 \times 10^{15}}$ 1.38×10^0

59. $\frac{(4.0 \times 10^{-9})(8.4 \times 10^5)}{(3.0 \times 10^{-6})(1.4 \times 10^{18})}$ 8×10^{-16}

60. $\frac{(7.2 \times 10^8)(3.9 \times 10^{-7})}{(2.6 \times 10^{-10})(1.8 \times 10^{-8})}$ 6×10^{19}

61. If $2^x = y$, then find 2^{x-4} in terms of y. $\frac{y}{16}$

62. Use the expressions $(2 + 3)^{-2}$ and $2^{-2} + 3^{-2}$ to show that $(x + y)^{-2} \neq x^{-2} + y^{-2}$.

63. If a and b are nonzero numbers and $a < b$, is the statement $a^{-1} < b^{-1}$ a true statement? Give a reason for your answer. No

In Exercises 64 to 74 write the answer in scientific notation.

Business and Economics

64. ***National Debt*** In May of 2002, the U.S. national debt was approximately 5.97×10^{12} dollars. At that time, the population of the U.S. was 2.87×10^8. In May of 2002, what was the average U.S. debt per person? $\approx 2.08 \times 10^4$ dollars/person

65. ***Chocolate Chip Cookie Consumption*** In 1930, at the Toll House Restaurant, Ruth Wakefield invented the modern-

day version of the chocolate chip cookie. Assume that approximately 2.9×10^9 chocolate chip cookies are consumed in the U.S. each year. If the U.S. population is 2.87×10^8, what is the average annual U.S. chocolate chip cookie consumption per person?
1.01×10 cookies per person

Life and Health Sciences

66. *Color Monitors* A color monitor for a computer can display 2^{32} colors. A physiologist estimates that the human eye can detect approximately 36,000 different colors. How many colors, to the nearest thousand, would go undetected by a human using this monitor?
4,294,931,000 colors

67. *Weight of an Orchid Seed* An orchid seed weighs approximately 3.2×10^{-8} ounce. If a package of seeds contains 1 ounce of orchid seeds, how many seeds are in the package? 3.125×10^7 seeds

68. *Volume of a Cell* The radius of a cell is 1.5×10^{-4} millimeter. Assuming the cell is a sphere, find the volume of the cell. (*Hint:* Volume of a sphere $= \frac{4}{3}\pi r^3$, where r is the radius of the sphere.)
$4.5\pi \times 10^{-12}$ mm^3

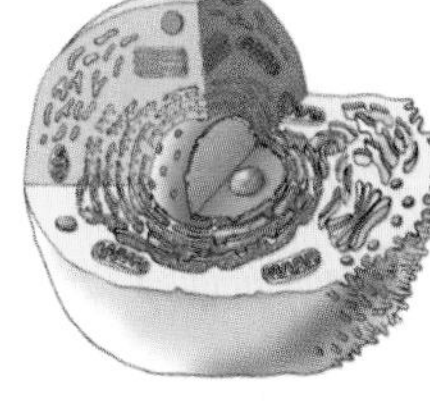

Physical Sciences and Engineering

69. *Computer Calculations* A computer can do one addition in 4×10^{-9} second. How many additions can the computer do in 1 minute? 1.5×10^{10} additions

70. *Speed of Light* Light travels approximately 3×10^8 meters per second. Using this number, how many meters does light travel in 24 hours? 2.592×10^{13} m

71. *Astronomical Unit* Earth's mean distance from the sun is 9.3×10^7 miles. This distance is called an *astronomical unit* (AU). Jupiter is 5.2 AU from the sun. Find the distance in miles from Jupiter to the sun. 4.836×10^8 mi

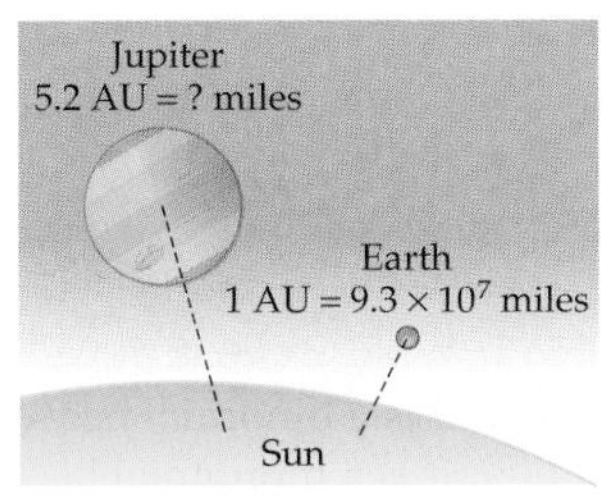

72. *Astronomy* The sun is approximately 1.5×10^{11} meters from Earth. If light travels 3×10^8 meters per second, how long (in minutes) does it take light from the sun to reach Earth? 8 min

73. *Mass of an Atom* One gram of hydrogen contains 6.023×10^{23} atoms. Find the mass of one hydrogen atom.
1.66×10^{-24} g

74. *Travel the Milky Way* The Milky Way is estimated to be 5.6×10^{19} miles in diameter. If a spaceship could travel 1,000,000 miles per hour, how long would it take (in hours) for the spaceship to travel the diameter of the Milky Way? 5.6×10^{13} h

75. Which is larger, $(10^{10})^{10}$ or $10^{(10^{10})}$? $10^{(10^{10})}$

76. How many digits are in the product $4^{50} \cdot 5^{100}$? 101 digits

Prepare for Section P.3

77. Simplify: $-3(2a - 4b)$ [P.1] $-6a + 12b$

78. Simplify: $5 - 2(2x - 7)$ [P.1] $19 - 4x$

79. Simplify: $2x^2 + 3x - 5 + x^2 - 6x - 1$ [P.1] $3x^2 - 3x - 6$

80. Simplify: $4x^2 - 6x - 1 - 5x^2 + x$ [P.1] $-x^2 - 5x - 1$

81. True or false: $4 - 3x - 2x^2 = -2x^2 - 4x + 4$ [P.1] False

82. True or false: $\frac{12 + 15}{4} = \frac{\overset{3}{\cancel{12}} + 15}{\cancel{4}} = 18$ [P.1] False

Explorations 

1. *Finance* You plan to save 1¢ the first day of the month, 2¢ the second day, 4¢ the third day, and to continue this pattern of saving twice what you saved on the preceding day for every day in a month that has 30 days.

a. How much money, in dollars, will you need to save on the 30th day? \$5,368,709.12

b. How much money, in dollars, will you have saved after 30 days? (*Suggestion:* Note that after 2 days, you have saved $2^2 - 1 = 3$¢; after 3 days, you have saved $2^3 - 1 = 7$¢; and after 4 days, you have saved $2^4 - 1 = 15$¢.) \$10,737,418.23

SECTION P.3 Operations on Polynomials

- Introduction to Polynomials
- Addition and Subtraction of Polynomials
- Multiplication of Polynomials
- Division of Polynomials

INSTRUCTOR NOTE

An analogy may help students understand these terms. Dogs, lions, and monkeys are specific types of animals, just as monomials, binomials, and trinomials are specific types of polynomials.

TAKE NOTE

Note that the standard form of a polynomial is written in order of decreasing powers of the variable. The polynomial $3x^2 - 4x^3 + 2 - 3x$ is not in standard form, but can be rewritten as $-4x^3 + 3x^2 - 3x + 2$.

Also note that the terms of a polynomial take the signs immediately preceding them. For example, the terms of $4x^3 - 5x^2 - x + 8$ are $4x^3$, $-5x^2$, $-x$, and 8.

Introduction to Polynomials

Research into automobile tire performance has shown that under- or overinflated tires are not only a safety hazard, but contribute to tire wear. For one such tire, the number of safe driving miles is given by

$$\text{Safe driving miles} = -642x^2 + 42{,}100x - 649{,}000$$

where x is the air pressure in pounds per square inch (psi). For instance, for an air pressure of 28 psi, we have

$$\begin{aligned}\text{Safe driving miles} &= -642x^2 + 42{,}100x - 649{,}000\\ &= -642(28)^2 + 42{,}100(28) - 649{,}000\\ &= 26{,}472\end{aligned}$$

The number of safe driving miles for this tire at a tire pressure of 28 psi is approximately 26,000 miles.

The variable expression $-642x^2 + 42{,}100x - 649{,}000$ is a **polynomial.**

Standard Form of a Polynomial

The **standard form of a polynomial** of degree n in the variable x is

$$a_nx^n + a_{n-1}x^{n-1} + \cdots + a_2x^2 + a_1x + a_0$$

where $a_n \neq 0$ and n is a nonnegative integer. The coefficient a_n is called the leading coefficient, and a_0 is the **constant term.**

The degree of the polynomial $-642x^2 + 42{,}100x - 649{,}000$ is 2, the leading coefficient is -642, and the constant term is $-649{,}000$.

A polynomial of *one* term is a **monomial.** $-5z^2$ is a monomial.
A polynomial of *two* terms is a **binomial.** $7x + 1$ is binomial.
A polynomial of *three* terms is a **trinomial.** $3a^2 + 5a - 6$ is a trinomial.

QUESTION Is each of the following expressions a polynomial? If so, is the expression a monomial, a binomial, or a trinomial?

a. $4x^2 + 9$ **b.** $-\frac{2}{5}ab$ **c.** $x^2 - 4xy + 9$ **d.** $\frac{2}{z} + 1$

ANSWER **a.** It is a polynomial. It is a binomial because it has two terms, $4x^2$ and 9. **b.** It is a polynomial. It is a monomial because it has one term. **c.** It is a polynomial. It is a trinomial because it has three terms, x^2, $-4xy$, and 9. **d.** $\frac{2}{z} = 2z^{-1}$ and therefore the exponent on the variable is a negative integer. (See the definition above.) Thus $\frac{2}{z} + 1$ is not a polynomial.

Note from the definition of a polynomial that a constant is a polynomial. For instance, 3 is a polynomial (it is a monomial). The degree of a *nonzero* constant polynomial is 0. Thus the degree of 3 is 0. The constant polynomial 0 is a special case and has no degree.

The **degree of a polynomial** in one variable is its largest exponent. For instance,

- The degree of $3x + 4$ is 1. It is a first-degree, or **linear,** polynomial.
- The degree of $4y^2 - 5y + 3$ is 2. It is a second-degree, or **quadratic,** polynomial.
- The degree of $5x^3 + 4x^2 - 6x - 7$ is 3. It is a third-degree, or **cubic,** polynomial.
- The degree of $7z^4 + 8z^3 - z^2 - 7z + 8$ is 4. It is a fourth-degree polynomial.

Addition and Subtraction of Polynomials

Polynomials are added by combining like terms.

Alternative to Example 1
Add: $(2x^2 - 14x - 8) + (-3x^2 + 7)$
■ $-x^2 - 14x - 1$

EXAMPLE 1 Add Polynomials

Add: $(4x^2 + 8x - 9) + (3x^2 - 12x - 7)$

Solution

$$\begin{aligned}(4x^2 + 8x - 9) + (3x^2 - 12x - 7) &= (4x^2 + 3x^2) + (8x - 12x) + (-9 - 7) \quad \text{■ Combine like terms.}\\ &= 7x^2 - 4x - 16\end{aligned}$$

CHECK YOUR PROGRESS 1 Add: $(-4x^2 - 3xy + 2y^2) + (3x^2 - 4y^2)$

Solution *See page S1.* $-x^2 - 3xy - 2y^2$

Recall that the definition of subtraction is addition of the opposite.

$$a - b = a + (-b)$$

This definition holds true for polynomials. Polynomials can be subtracted by adding the opposite of the second polynomial to the first. The opposite of a polynomial is the polynomial with the sign of every term changed.

The opposite of the polynomial $x^2 + 5x - 7$ is $-x^2 - 5x + 7$.

The opposite of the polynomial $-3a^3 - 5a + 2$ is $3a^3 + 5a - 2$.

QUESTION What is the opposite of the polynomial $5d^4 - 6d^2 + 9$?

ANSWER The opposite of $5d^4 - 6d^2 + 9$ is $-5d^4 + 6d^2 - 9$.

Alternative to Example 2
Subtract:
$(5x - 7) - (2x^2 - x - 7)$
■ $-2x^2 + 6x$

EXAMPLE 2 Subtract Polynomials

Subtract: $(4z^2 - 7) - (6z^2 - 3z - 4)$

Solution

$$\begin{aligned}(4z^2 - 7) - (6z^2 - 3z - 4) \\ &= (4z^2 - 7) + (-6z^2 + 3z + 4) && \text{■ Add the opposite polynomial.} \\ &= -2z^2 + 3z - 3 && \text{■ Combine like terms.}\end{aligned}$$

CHECK YOUR PROGRESS 2 Subtract: $(5x^2 - 3x + 4) - (-6x^3 - 2x + 8)$

Solution *See page S1.* $6x^3 + 5x^2 - x - 4$

A company's revenue is the money the company earns by selling its products. A company's cost is the money it spends to manufacture and sell its products. A company's profit is the difference between its revenue and its cost. This relationship is expressed by the formula Profit = revenue − cost. This formula is used in Example 3 and Check Your Progress 3.

Alternative to Example 3
Exercise 58, page 35

EXAMPLE 3 Determine a Monthly Profit

A company manufactures and sells computer desks. The total monthly cost, in dollars, to produce n desks is $25n + 3500$. The company's revenue, in dollars, from selling all n desks is $-0.35n^2 + 140n$. Express, in terms of n, the company's monthly profit. Use this expression to find the profit for a month in which the company sells 250 desks.

Solution The monthly profit is the difference between revenue and cost: revenue $= -0.35n^2 + 140n$ and cost $= 25n + 3500$. Therefore,

$$\begin{aligned}\text{Profit} &= \text{revenue} - \text{cost} \\ &= (-0.35n^2 + 140n) - (25n + 3500) \\ &= -0.35n^2 + 140n - 25n - 3500 \\ &= -0.35n^2 + 115n - 3500\end{aligned}$$

The monthly profit is $-0.35n^2 + 115n - 3500$.
To find the monthly profit for selling 250 desks, replace n by 250 and simplify.

$$\begin{aligned}\text{Monthly profit} &= -0.35n^2 + 115n - 3500 \\ &= -0.35(250)^2 + 115(250) - 3500 \\ &= 3375\end{aligned}$$

The monthly profit for selling 250 desks is \$3375.

CHECK YOUR PROGRESS 3 A company's total monthly cost, in dollars, to manufacture and sell n DVD burners is $45n + 4500$. The company's revenue, in dollars, from selling all n DVD burners is $-0.25n^2 + 180n$. Express, in terms of n, the company's monthly profit. Use this expression to find the profit for a month in which the company sells 310 DVD burners.

Solution *See page S1.* \$13,325

■ Multiplication of Polynomials

To multiply a polynomial by a monomial, use the distributive property and the rule for multiplying exponential expressions.

Alternative to Example 4

a. Multiply: $(5 - 4x)(-3x)$

b. Multiply: $-4x^3(3x^2 - 5x - 7)$

■ a. $12x^2 - 15x$

■ b. $-12x^5 + 20x^4 + 28x^3$

EXAMPLE 4 Multiply a Polynomial by a Monomial

Multiply. **a.** $(4x - 3)(-6x)$ **b.** $2x^2(3x^2 - 4x - 8)$

Solution

a. $(4x - 3)(-6x) = 4x(-6x) - 3(-6x)$ ■ Use the distributive property.

$= -24x^2 + 18x$ ■ Simplify.

b. $2x^2(3x^2 - 4x - 8) = (2x^2)3x^2 - (2x^2)4x - (2x^2)8$ ■ Use the distributive property.

$= 6x^4 - 8x^3 - 16x^2$ ■ Simplify.

CHECK YOUR PROGRESS 4 Multiply. **a.** $(5x^2 - 4x)(-3x^2)$ **b.** $-y^3(4y^3 - 2y - 7)$

Solution *See page S1.* a. $-15x^4 + 12x^3$ b. $-4y^6 + 2y^4 + 7y^3$

INSTRUCTOR NOTE

For the example at the right, ask students whether the distributive property could have been applied as $(2x + 3)3x^2 - (2x + 3)4x - (2x + 3)6$. Have them verify that the same answer is obtained in both cases.

We choose to show this method because it corresponds to the vertical format shown below.

Multiplication of two polynomials requires repeated application of the distributive property. Study the product of $(2x + 3)(3x^2 - 4x - 6)$ shown below.

$(2x + 3)(3x^2 - 4x - 6)$

$= 2x(3x^2 - 4x - 6) + 3(3x^2 - 4x - 6)$ ■ Use the distributive property to multiply each term of $2x + 3$ by $3x^2 - 4x - 6$.

$= (6x^3 - 8x^2 - 12x) + (9x^2 - 12x - 18)$ ■ Use the distributive property to multiply $3x^2 - 4x - 6$ by $2x$ and by 3.

$= 6x^3 + x^2 - 24x - 18$ ■ Simplify.

INSTRUCTOR NOTE

Before doing an example similar to the one at the right, provide an illustration of multiplying two whole numbers—for example, 473×28. Compare the procedure for multiplying two polynomials to this example.

Two polynomials can also be multiplied using a vertical format similar to that used for multiplication of whole numbers. Using the polynomials given above, we have

$$\begin{array}{rrrrl} & 3x^2 & - 4x & - 6 & \\ & & 2x & + 3 & \\ \hline & 9x^2 & - 12x & - 18 & \leftarrow 3(3x^2 - 4x - 6) \\ 6x^3 & - 8x^2 & - 12x & & \leftarrow 2x(3x^2 - 4x - 6) \\ \hline 6x^3 & + x^2 & - 24x & - 18 & \text{■ Add the terms in each column.} \end{array}$$

Notice that the same product is obtained in each case. The vertical format is just a convenient method of using the distributive property. You may, however, use either method.

Alternative to Example 5

Multiply:
$(3x^2 + 4x^2 - 2)(2x^2 - 3)$
▪ $6x^5 + 8x^4 - 9x^3 - 16x^2 + 6$

EXAMPLE 5 Multiply Polynomials

Multiply: $(2v^3 - 3v + 1)(4v - 2)$

Solution

$$\begin{array}{rrrrrl} & & 2v^3 & - 3v & + 1 & \\ & & & 4v & - 2 & \\ \hline & -4v^3 & & + 6v & - 2 & = -2(2v^3 - 3v + 1) \\ 8v^4 & & - 12v^2 & + 4v & & = 4v(2v^3 - 3v + 1) \\ \hline 8v^4 & - 4v^3 & - 12v^2 & + 10v & - 2 & \end{array}$$

▪ Add the terms in each column.

CHECK YOUR PROGRESS 5 Multiply: $(4x^3 - 3x^2 + x - 4)(2x - 5)$

Solution *See page S1.* $8x^4 - 26x^3 + 17x^2 - 13x + 20$

TAKE NOTE

The FOIL method is not really a different way of multiplying two polynomials. It is based on the distributive property.

$(2x + 3)(x + 5)$
$= (2x + 3)x + (2x + 3)5$
$= 2x^2 + 3x + 10x + 15$
$= 2x^2 + 13x + 15$

Note that the terms of

$2x^2 + 3x + 10x + 15$

are the same as the products found using the FOIL method.

The product of two binomials can be found by using a shortcut called the **FOIL** method, which is based on the distributive property. The letters of FOIL stand for **F**irst, **O**uter, **I**nner, and **L**ast.

To multiply $(2x + 3)(x + 5)$:

Multiply the **F**irst terms.	$(2x + 3)(x + 5)$	$2x \cdot x = 2x^2$
Multiply the **O**uter terms.	$(2x + 3)(x + 5)$	$2x \cdot 5 = 10x$
Multiply the **I**nner terms.	$(2x + 3)(x + 5)$	$3 \cdot x = 3x$
Multiply the **L**ast terms.	$(2x + 3)(x + 5)$	$3 \cdot 5 = 15$

F O I L

Add the products.	$(2x + 3)(x + 5) = 2x^2 + 10x + 3x + 15$
Combine like terms.	$= 2x^2 + 13x + 15$

Alternative to Example 6

a. $(4x - 3)(2x + 7)$
b. $(2a - 5b)(3a + 4b)$
▪ a. $8x^2 + 22x - 21$
▪ b. $6a^2 - 7ab - 20b^2$

EXAMPLE 6 Multiply Using the FOIL Method

Find the products. **a.** $(4x + 5)(3x - 2)$ **b.** $(x - 3y)(2x + 5y)$

Solution

a. $(4x + 5)(3x - 2) = 4x(3x) + 4x(-2) + 5(3x) + 5(-2)$
$= 12x^2 - 8x + 15x - 10$
$= 12x^2 + 7x - 10$

b. $(x - 3y)(2x + 5y) = x(2x) + x(5y) + (-3y)(2x) + (-3y)(5y)$
$= 2x^2 + 5xy - 6xy - 15y^2$
$= 2x^2 - xy - 15y^2$

Continued ➤

CHECK YOUR PROGRESS 6 Multiply. **a.** $(3z - 4)(2z - 5)$ **b.** $(4a + 5b)(4a - 5b)$

Solution *See page S1.* a. $6z^2 - 23z + 20$ b. $16a^2 - 25b^2$

Powers of a binomial are expanded by repeated multiplication. For instance,

$$\begin{aligned}(3x + 5)^2 &= (3x + 5)(3x + 5) = 3x(3x) + 3x(5) + 5(3x) + 5(5) && \text{▪ Use FOIL.}\\ &= 9x^2 + 30x + 25\end{aligned}$$

$$\begin{aligned}(2x - 3)^3 &= \underbrace{(2x - 3)(2x - 3)}(2x - 3)\\ &= (4x^2 - 12x + 9)(2x - 3) && \text{▪ Multiply } (2x - 3)(2x - 3).\\ &= 8x^3 - 36x^2 + 54x - 27 && \text{▪ Multiply } 4x^2 - 12x + 9 \text{ by } 2x - 3.\end{aligned}$$

There are many applications in which it is necessary to multiply two polynomials.

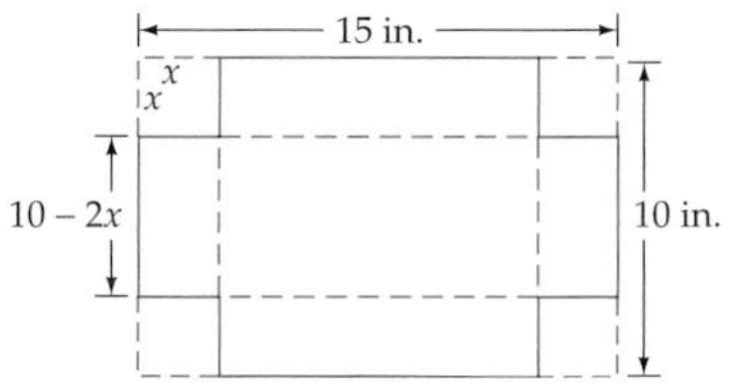

Alternative to Example 7
Exercise 64, page 36

EXAMPLE 7 Determine the Volume of a Box

A rectangular piece of cardboard measures 15 inches by 10 inches. An open box is formed by cutting four squares that measure x inches on a side from the corners of the cardboard and then folding up the sides, as shown in the figure at the left. Determine the volume of the box in terms of x. Use your expression to find the volume of the box when x is 1.5 inches.

Solution To determine the volume of the box in terms of x, use the formula for the volume of a box, $V = LWH$. Substitute variable expressions for L, W, and H. Then multiply.

$$\begin{aligned}V &= LWH\\ &= (15 - 2x)(10 - 2x)x && \text{▪ } L = 15 - 2x,\ W = 10 - 2x,\ H = x\\ &= (4x^2 - 50x + 150)x && \text{▪ Multiply } (15 - 2x)(10 - 2x)\\ &= 4x^3 - 50x^2 + 150x && \text{▪ Multiply by } x.\end{aligned}$$

The volume of the box in terms of x is $(4x^3 - 50x^2 + 150x)$ cubic inches.
To determine the volume when $x = 1.5$, substitute 1.5 for x in the above expression. Then simplify the numerical expression.

$$\begin{aligned}V &= 4x^3 - 50x^2 + 150x\\ &= 4(1.5)^3 - 50(1.5)^2 + 150(1.5) && \text{▪ Replace } x \text{ by 1.5.}\\ &= 126\end{aligned}$$

When x is 1.5 inches, the volume of the box is 126 cubic inches.

CHECK YOUR PROGRESS 7 The base and height of a triangle are each $(a + 2)$ feet. Find the area of the triangle in terms of a. What is the area of the triangle when a is 4?

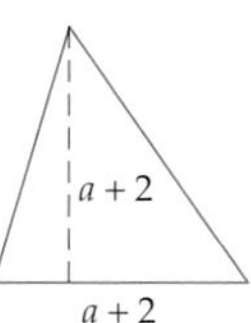

Solution *See page S1.* $\frac{1}{2}a^2 + 2a + 2$ 18 square feet

▪ Division of Polynomials

TAKE NOTE

Recall that a fraction bar acts as a grouping symbol. Division of a polynomial by a monomial is an application of the distributive property.

To divide a polynomial by a monomial, divide each term of the polynomial by the monomial. For instance,

$$\frac{16x^3 - 8x^2 + 12x}{4x} = \frac{16x^3}{4x} - \frac{8x^2}{4x} + \frac{12x}{4x}$$
▪ Divide each term in the numerator by the denominator.

$$= 4x^2 - 2x + 3$$
▪ Simplify.

To find the quotient of two polynomials, we use a method similar to that used to divide whole numbers. Consider $(6x^3 - 16x^2 + 23x - 5) \div (3x - 2)$.

$$3x - 2\overline{)6x^3 - 16x^2 + 23x - 5}$$

INSTRUCTOR NOTE

It may help some students understand the division of polynomial process if you first divide two whole numbers, such as $7345 \div 3$.

Think: $\dfrac{6x^3}{3x} = 2x^2$

Multiply: $2x^2(3x - 2) = 6x^3 - 4x^2$

Subtract.

$$\begin{array}{r} 2x^2 \phantom{{}- 16x^2 + 23x - 5} \\ 3x - 2\overline{)6x^3 - 16x^2 + 23x - 5} \\ \underline{6x^3 - 4x^2 \phantom{{}+ 23x - 5}} \\ -12x^2 + 23x - 5 \end{array}$$

Think: $\dfrac{-12x^2}{3x} = -4x$

Multiply: $-4x(3x - 2) = -12x^2 + 8x$

Subtract.

$$\begin{array}{r} 2x^2 - 4x \phantom{{}+ 23x - 5} \\ 3x - 2\overline{)6x^3 - 16x^2 + 23x - 5} \\ \underline{6x^3 - 4x^2 \phantom{{}+ 23x - 5}} \\ -12x^2 + 23x - 5 \\ \underline{-12x^2 + 8x \phantom{{}- 5}} \\ 15x - 5 \end{array}$$

Think: $\dfrac{15x}{3x} = 5$

Multiply: $5(3x - 2) = 15x - 10$

Subtract.

$$\begin{array}{r} 2x^2 - 4x + 5 \phantom{{}+ 23x - 5} \\ 3x - 2\overline{)6x^3 - 16x^2 + 23x - 5} \\ \underline{6x^3 - 4x^2 \phantom{{}+ 23x - 5}} \\ -12x^2 + 23x - 5 \\ \underline{-12x^2 + 8x \phantom{{}- 5}} \\ 15x - 5 \\ \underline{15x - 10} \\ 5 \end{array}$$

TAKE NOTE

$\frac{20}{3}$ written as a mixed number is $6\frac{2}{3}$. Recall, however, that $6\frac{2}{3}$ means $6 + \frac{2}{3}$, which is in the form quotient $+ \frac{\text{remainder}}{\text{divisor}}$.

Thus $(6x^3 - 16x^2 + 23x - 5) \div (3x - 2) = 2x^2 - 4x + 5$ with a remainder of 5.

Although there is nothing wrong with writing the answer as we did above, it is more common to write the remainder as a fraction, as we do when dividing whole numbers. (See the Take Note at the left.) Using this method, we write

$$\overbrace{\frac{6x^3 - 16x^2 + 23x - 5}{\underbrace{3x - 2}_{\text{Divisor}}}}^{\text{Dividend}} = \overbrace{2x^2 - 4x + 5}^{\text{Quotient}} + \frac{5 \leftarrow \text{Remainder}}{3x - 2 \leftarrow \text{Divisor}}$$

In this example, $6x^3 - 16x^2 + 23x - 5$ is called the dividend, $3x - 2$ is the divisor, $2x^2 - 4x + 5$ is the quotient, and 5 is the remainder. The dividend is equal to the product of the divisor and quotient, plus the remainder. That is,

$$\underbrace{6x^3 - 16x^2 + 23x - 5}_{\text{Dividend}} = \underbrace{(2x^2 - 4x + 5)}_{\text{quotient}} \cdot \underbrace{(3x - 2)}_{\text{divisor}} + \underbrace{5}_{\text{remainder}}$$

Point of Interest

Digital signals consist of a sequence of zeros and ones. These signals are used in applications such as communicating with satellites, in high-definition television (HDTV), in music or movie DVDs, and in many other instances. Any time a digital signal is sent, there is the possibility that a 0 is changed to a 1 or that a 1 is changed to a 0. If this happens, the quality of, say, the music on a DVD is not what it should be. To check whether a signal has been distorted, *error-correcting codes* are used. These codes can determine the digit that was changed and correct the error. One method of doing this relies on the division of polynomials that have coefficients of 0 and 1.

Before dividing polynomials, make sure that the polynomials are written in descending order. In some cases, it is helpful to insert a 0 in the dividend for a missing term (one whose coefficient is 0) so that like terms align in the same column. This is shown in Example 8.

QUESTION What is the first step you should perform to find the quotient of $(2x + 1 + x^2) \div (x - 1)$?

EXAMPLE 8 Divide Polynomials

Divide: $\dfrac{3x^2 + x^4 - 10 - 6x}{x^2 + 3x - 5}$

Solution Write the numerator in descending order. Then divide.

$$\frac{3x^2 + x^4 - 10 - 6x}{x^2 + 3x - 5} = \frac{x^4 + 3x^2 - 6x - 10}{x^2 + 3x - 5}$$

$$\begin{array}{r} x^2 - 3x + 17 \\ x^2 + 3x - 5\overline{)x^4 + 0x^3 + 3x^2 - 6x - 10} \\ \underline{x^4 + 3x^3 - 5x^2 } \\ -3x^3 + 8x^2 - 6x - 10 \\ \underline{-3x^3 - 9x^2 + 15x } \\ 17x^2 - 21x - 10 \\ \underline{17x^2 + 51x - 85} \\ -72x + 75 \end{array}$$

■ Writing $0x^3$ for the missing term helps align like terms in the same column.

Thus $\dfrac{3x^2 + x^4 - 10 - 6x}{x^2 + 3x - 5} = x^2 - 3x + 17 + \dfrac{-72x + 75}{x^2 + 3x - 5}$.

CHECK YOUR PROGRESS 8 Divide: $\dfrac{2x^3 - x^2 + 5}{x - 3}$

Solution *See page S1.* $2x^2 + 5x + 15 + \dfrac{50}{x - 3}$

Alternative to Example 8

Divide:
$\dfrac{12 + x - 19x^2 + x^3 + 6x^4}{2x^2 - x - 3}$

■ $3x^2 + 2x - 4 + \dfrac{3x}{2x^2 - x - 3}$

ANSWER Write the dividend in descending order as $x^2 + 2x + 1$.

Topics for Discussion

1. Give an example of a third-degree polynomial.
2. Give an example of **a.** a monomial, **b.** a binomial, and **c.** a trinomial.
3. Give an example of an expression that is not a polynomial.
4. State the distributive property and explain how it is used to multiply two binomials.
5. If a polynomial of degree 3 is added to a polynomial of degree 3, is the resulting polynomial of degree 3?
6. If a polynomial of degree 3 is multiplied by a polynomial of degree 2, what is the degree of the resulting polynomial?

EXERCISES P.3

—Suggested Assignment: Exercises 1–65, odd; and 67–72, all.

In Exercises 1 to 6, for each polynomial, determine its a. standard form, b. degree, c. leading coefficient, and d. number of terms.

1. $2x + x^2 - 7$
a. $x^2 + 2x - 7$ b. 2 c. 1 d. 3

2. $-3x^2 - 11 - 12x^4$
2a. $-12x^4 - 3x^2 - 11$ b. 4 c. -12 d. 3

3. $x^3 - 1$
a. $x^3 - 1$ b. 3 c. 1 d. 2

4. $4x^2 - 2x + 7$
a. $4x^2 - 2x + 7$ b. 2 c. 4 d. 3

5. $2x^4 + 3x^2 - 5 - 4x^2$
a. $2x^4 - x^2 - 5$ b. 4 c. 2 d. 3

6. $3x^2 - 5x^3 + 7x - 1$
a. $-5x^3 + 3x^2 + 7x - 1$ b. 3 c. -5 d. 4

In Exercises 7 to 20, perform the indicated operations and write the result in standard form.

7. $(3x^2 + 4x + 5) + (2x^2 + 7x - 2)$ $5x^2 + 11x + 3$

8. $(5y^2 - 7y + 3) + (2y^2 + 8y + 1)$ $7y^2 + y + 4$

9. $(4x^3 - 2x + 7) + (5x^3 + 8x^2 - 1)$ $9x^3 + 8x^2 - 2x + 6$

10. $(5x^4 - 3x^2 + 9) + (3x^3 - 2x^2 - 7x + 3)$ $5x^4 + 3x^3 - 5x^2 - 7x + 12$

11. $(r^2 - 2r - 5) - (3r^2 - 5r + 7)$ $-2r^2 + 3r - 12$

12. $(7a^2 - 4a + 11) - (-2a^2 + 11a - 9)$ $9a^2 - 15a + 20$

13. $(u^3 - 3u^2 - 4u + 8) - (u^3 - 2u + 4)$ $-3u^2 - 2u + 4$

14. $(4a^3 - 6a^2 + 5a - 7) - (5a^3 - 6a^2 - 4)$ $-a^3 + 5a - 3$

15. $(4x - 5)(2x^2 + 7x - 8)$ $8x^3 + 18x^2 - 67x + 40$

16. $(2x + 1)(3x^2 - 4x - 5)$ $6x^3 - 5x^2 - 14x - 5$

17. $(x - 4)(x^3 - 2x^2 + 5x - 1)$ $x^4 - 6x^3 + 13x^2 - 21x + 4$

18. $(x + 3)(x^3 + 4x^2 - 5x - 7)$ $x^4 + 7x^3 + 7x^2 - 22x - 21$

19. $(2x - 3)(2x^3 - x - 3)$ $4x^4 - 6x^3 - 2x^2 - 3x + 9$

20. $(3x + 2)(4x^3 - 6x^2 - 5)$ $12x^4 - 10x^3 - 12x^2 - 15x - 10$

In Exercises 21 to 32, use the FOIL method to find the indicated product.

21. $(2x + 4)(5x + 1)$
$10x^2 + 22x + 4$

22. $(5x - 3)(2x + 7)$
$10x^2 + 29x - 21$

23. $(4z - 3)(z + 4)$
$4z^2 + 13z - 12$

24. $(y - 5)(2y + 7)$
$2y^2 - 3y - 35$

25. $(2x - 3)^2$
$4x^2 - 12x + 9$

26. $(2a + 7)^2$
$4a^2 + 28a + 49$

27. $(4z - 7)(4z + 7)$
$16z^2 - 49$

28. $(2x + 5)(2x - 5)$
$4x^2 - 25$

29. $(5x - 4y)(2x + 3y)$
$10x^2 + 7xy - 12y^2$

30. $(3m - 5n)(3m + 4n)$
$9m^2 - 3mn - 20n^2$

31. $(2x + 5y)^2$
$4x^2 + 20xy + 25y^2$

32. $(4x - 5y)^2$
$16x^2 - 40xy + 25y^2$

In Exercises 33 to 40, perform the indicated operations and simplify.

33. $(4d - 1)^2 - (2d - 3)^2$ $12d^2 + 4d - 8$

34. $(5x - 3)^2 - (2x + 3)^2$ $21x^2 - 42x$

35. $(x + y)(x^2 - xy + y^2)$ $x^3 + y^3$

36. $(x - y)(x^2 + xy + y^2)$ $x^3 - y^3$

37. $(3c - 2)(4c + 1)(5c - 2)$ $60c^3 - 49c^2 + 4$

38. $(4d - 5)(2d - 1)(3d - 4)$ $24d^3 - 74d^2 + 71d - 20$

39. $(x + 4)^3$ $x^3 + 12x^2 + 48x + 64$

40. $(2x - 3)^3$ $8x^3 - 36x^2 + 54x - 27$

In Exercises 41 to 56, find the indicated quotient.

41. $(x^2 + 5x + 4) \div (x + 4)$ $x + 1$

42. $(x^2 - x - 12) \div (x - 4)$ $x + 3$

43. $(2x^2 - 5x + 3) \div (2x - 3)$ $x - 1$

44. $(6x^2 + 5x - 6) \div (3x - 2)$ $2x + 3$

45. $(x^2 - 9) \div (x - 3)$ $x + 3$

46. $(4x^2 - 25) \div (2x + 5)$ $2x - 5$

47. $\dfrac{3x^2 - x - 1}{x + 2}$ $3x - 7 + \dfrac{13}{x + 2}$

48. $\dfrac{5x^2 + 4x + 3}{x - 4}$ $5x + 24 + \dfrac{99}{x - 4}$

49. $(x^3 + 2x^2 - 4x - 8) \div (x - 2)$ $x^2 + 4x + 4$

50. $(x^3 + 2x^2 - 7x - 12) \div (x + 3)$ $x^2 - x - 4$

51. $\dfrac{x^3 - 4x^2 + 3x - 5}{x - 4}$ $x^2 + 3 + \dfrac{7}{x - 4}$

52. $\dfrac{x^3 - 6x^2 + 15x + 24}{x + 4}$ $x^2 - 10x + 55 - \dfrac{196}{x + 4}$

53. $(x^3 + 4x + 16) \div (x + 4)$ $x^2 - 4x + 20 - \dfrac{64}{x + 4}$

54. $(x^3 - 2x^2 + 10) \div (x - 2)$ $x^2 + \dfrac{10}{x - 2}$

55. $(2x^3 + 4x - 5) \div (x + 3)$ $2x^2 - 6x + 22 - \dfrac{71}{x + 3}$

56. $(3x^3 - 6x + 15) \div (x - 2)$ $3x^2 + 6x + 6 + \dfrac{27}{x - 2}$

Business and Economics

57. ***Gross Domestic Product*** The gross domestic product (GDP), in trillions of dollars, for the U.S. can be approximated by $-0.027x^2 + 0.67x + 7.11$, where $x = 0$ corresponds to the year 1996. According to this model, what will be the U.S. GDP in 2010? $11.198 trillion

58. ***Sales of Inkjet Printers*** A manufacturer has determined that the revenue from selling x inkjet printers is given by Revenue $= -0.1x^2 + 130x$ and the cost to produce x inkjet printers is given by Cost $= 40x + 8000$. Express the company's profit in terms of x. Use this expression to find the profit from selling 400 inkjet printers.

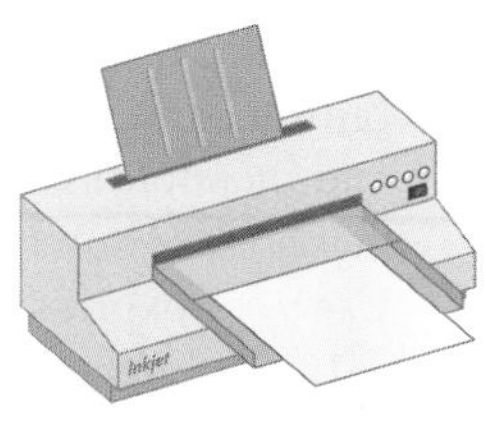

$-0.1x^2 + 90x - 8000$; $12,000

59. ***Travel Agency Sales*** The owners of a travel agency have determined that their revenue, in dollars, from selling x tickets for a tour is given by Revenue $= -0.25x^2 + 48x$. The cost for offering the tour is given by Cost $= 8x$. Express the travel agency's profit in terms of x. Use this expression to find the profit from selling 95 tickets. $-0.25x^2 + 40x$; $1543.75

Life and Health Sciences

60. ***Reaction Times*** Based on data from one experiment, the reaction time, in hundredths of a second, of a person to a visual stimulus varies according to age and is given by $0.005x^2 - 0.32x + 12$, where x is the age of the person in years. Find the reaction time to the stimulus for a person who is 25 years old. 7.125 hundredths of a second

61. ***Air Velocity of a Cough*** The velocity, in meters per second, of the air that is expelled during a cough is given by Velocity $= 6r^2 - 10r^3$, where r is the radius of the trachea in centimeters. **a.** $-10r^3 + 6r^2$ **b.** 0.30625 m/s

    a. Find the velocity as a polynomial in standard form.

    b. Find the velocity of the air expelled during a cough when the radius of the trachea is 0.35 centimeter.

Sports and Recreation

62. ***Cycling*** The air resistance (in pounds) on a cyclist riding a bicycle in an upright position can be approximated by $0.015v^2 + 0.1v$, where v is the velocity of the cyclist in miles per hour. Find the resistance on a cyclist traveling at 15 miles per hour. 4.875 lb

Physical Sciences and Engineering

63. ***Constructing a Box*** A rectangular piece of cardboard measures 40 centimeters by 30 centimeters. An open box

is formed by cutting four squares that measure x centimeters on a side from the corners of the cardboard and then folding up the sides, as shown in the figure below. Determine the volume of the box in terms of x. Use your expression to find the volume of the box when x is 10 centimeters. $4x^3 - 140x^2 + 1200x$; 2000 cm^3

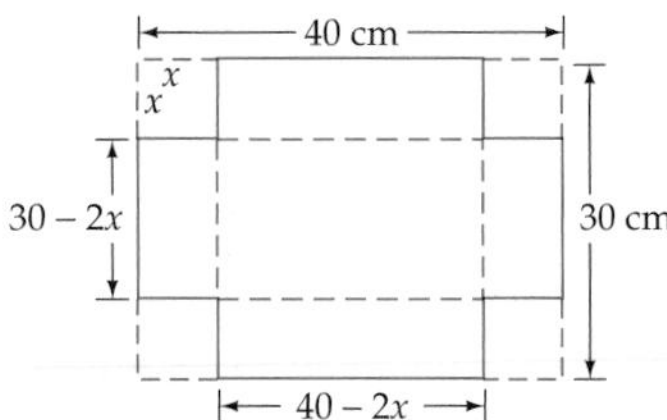

64. ***Constructing a Trough*** A long, thin sheet of tin 30 inches wide is to be made into a trough by bending up two sides until they are perpendicular to the bottom, as shown in the figure below. What is the cross-sectional area? $-2x^2 + 30x$ in.^{2}

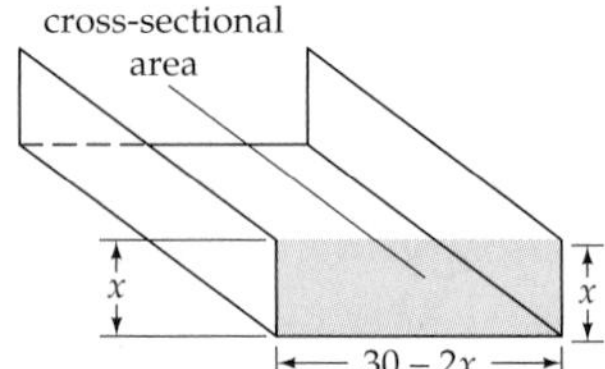

65. ***Highway Engineering*** On an expressway, the recommended *safe distance* between cars, in feet, is given by Safe distance $= 0.015v^2 + v + 10$, where v is the velocity of the car in miles per hour. Find the safe distance between two cars traveling at 50 miles per hour. 97.5 ft

66. ***Meteorology*** The Fahrenheit temperature in a city during a particular 12-hour period is given by

$$\text{Temperature} = 0.05(t - 1)(t - 4)(t - 8) + 44$$

where t is the time in hours after 9:00 A.M.

a. Find the temperature as a polynomial in standard form. a. $0.05t^3 - 0.65t^2 + 2.2t + 42.4$

b. Find the temperature at 1:00 P.M. b. 44°F

Prepare for Section P.4

In Exercises 67 to 72, replace the question mark to make a true statement.

67. $3x^3 \cdot ? = 12x^5$ $4x^2$

68. $4x^2y^3 \cdot ? = 8x^2y^4$ $2y$

69. $4(3a + ?) = 12a + 20$ 5

70. $2x(3x - ?) = 6x^2 - 2x$ 1

71. $(x - 2)(?) = x^2 + 3x - 10$ $x + 5$

72. Express x^6 as a power of **a.** x^2 and **b.** x^3. a. $(x^2)^3$ b. $(x^3)^2$

Explorations 

1. ***Prime Numbers*** Fermat's Little Theorem states, "If n is a prime number and a is any natural number, then $a^n - a$ is divisible by n." For instance, if $n = 11$ and $a = 14$, $\frac{14^{11} - 14}{11} = 368{,}142{,}288{,}150$. The important aspect of this theorem is that no matter what natural number is chosen for a, $a^n - a$ is evenly divisible by n. Knowing whether a number is prime plays a central role in the security of computer systems. A restatement (called the *contrapositive*) of Fermat's Little Theorem is "If n is a number and a is some number for which $a^n - a$ is not divisible by n, then n is not a prime number." This restatement is often used to show that a number is not prime. For example, if $n = 14$, then $\frac{2^{14} - 2}{14} = \frac{8191}{7}$, and thus there is some number ($a = 2$) for which $2^{14} - 2$ is not evenly divisible by 14. This tells us that 14 is not prime.

a. Explain the meaning of the contrapositive (used above) of a theorem. Use your explanation to write the contrapositive of "If two triangles are congruent, then they are similar."
If two triangles are not similar, then they are not congruent.

b. $7^{14} - 7$ is divisible by 14. Explain why this does not contradict the fact that 14 is not a prime.
Fermat's Little Theorem does not apply when n is not a prime.

c. Explain the meaning of the converse of a theorem. State the converse of Fermat's Little Theorem.

d. The number 561 has the property that a $a^{561} - a$ is divisible by 561 for all natural numbers a. Can you use Fermat's Little Theorem to conclude that 561 is a prime number? Explain.
No. It is not given that the converse of Fermat's Little Theorem is true.

e. Suppose that $a^n - a$ is divisible by n for all values of a. Can you conclude that n is a prime number? Explain.
No. It is not given that the converse of Fermat's Little Theorem is true.

f. Find a definition of a *Carmichael number*. What do Carmichael numbers have to do with the information in Exercise e?
A Carmichael number is a number that is not prime but satisfies Fermat's Little Theorem.

1c. If n is a natural number and $a^n - a$ is divisible by n for all natural numbers a, then n is a prime number.

SECTION P.4 Factoring

- Greatest Common Factor
- Factoring Trinomials
- Special Factoring
- Factoring by Grouping
- General Factoring

Writing a polynomial as a product of polynomials of lower degree is called **factoring**. Factoring is an important procedure that is often used to simplify fractional expressions and to solve equations.

In this section we consider only the factorization of polynomials that have integer coefficients. Also, we are concerned only with factoring over the integers. That is, we search only for polynomial factors that have integer coefficients.

Greatest Common Factor

The first step in any factorization of a polynomial is to use the distributive property to factor out the **greatest common factor (GCF)** of the terms of the polynomial. Given two or more exponential expressions with the same prime number base or the same variable base, the GCF is the exponential expression with the smallest exponent. For example,

$$2^3 \text{ is the GCF of } 2^3, 2^5, \text{ and } 2^8 \quad \text{and} \quad a \text{ is the GCF of } a^4 \text{ and } a.$$

The GCF of two or more monomials is the product of the GCFs of all of the *common* bases. For example, to find the GCF of $27a^3b^4$ and $18b^3c$, factor the coefficients into prime factors and then write each common base with its smallest exponent.

$$27a^3b^4 = 3^3 \cdot a^3 \cdot b^4 \qquad 18b^3c = 2 \cdot 3^2 \cdot b^3 \cdot c$$

The only common bases are 3 and b. The product of these common bases with their smallest exponents is 3^2b^3. The GCF of $27a^3b^4$ and $18b^3c$ is $9b^3$.

The expressions $3x(2x + 5)$ and $4(2x + 5)$ have a common *binomial* factor, which is $2x + 5$. Thus, the GCF of $3x(2x + 5)$ and $4(2x + 5)$ is $2x + 5$.

Alternative to Example 1

Factor out the GCF.

a. $15a^2 - 5a$
b. $6x^2y^2 - 12xy^2 - 9y^2$
c. $(2a - b)(x + y) - (2a - b)(x - y)$

- a. $5a(3a - 1)$
- b. $3y^2(2x^2 - 4x - 3)$
- c. $2y(2a - b)$

EXAMPLE 1 Factor Out the Greatest Common Factor

Factor out the GCF.

a. $10x^3 + 6x$ **b.** $15x^{2n} + 9x^{n+1} - 3x^n$ (where n is a positive integer)

c. $(m + 5)(x + 3) + (m + 5)(x - 10)$

Solution

a. $10x^3 + 6x = (2x)(5x^2) + (2x)(3)$ ▪ The GCF is $2x$.

$= 2x(5x^2 + 3)$ ▪ Factor out the GCF.

b. $15x^{2n} + 9x^{n+1} - 3x^n$

$= (3x^n)(5x^n) + (3x^n)(3x) - (3x^n)(1)$ ▪ The GCF is $3x^n$.

$= 3x^n(5x^n + 3x - 1)$ ▪ Factor out the GCF.

Continued ➤

c. $(m + 5)(x + 3) + (m + 5)(x - 10)$
$= (m + 5)[(x + 3) + (x - 10)]$ ▪ Use the distributive property to factor out $(m + 5)$.
$= (m + 5)(2x - 7)$ ▪ Simplify.

CHECK YOUR PROGRESS 1 Factor: $6a^3b^2 - 12a^2b + 72ab^3$

Solution *See page S1.* $6ab(a^2b - 2a + 12b^2)$

▪ Factoring Trinomials

Some trinomials of the form $x^2 + bx + c$ can be factored by a trial procedure. This method makes use of the FOIL method in reverse. For example, consider the following products:

$$(x + 3)(x + 5) = x^2 + 5x + 3x + (3)(5) = x^2 + 8x + 15$$
$$(x - 2)(x - 7) = x^2 - 7x - 2x + (-2)(-7) = x^2 - 9x + 14$$
$$(x + 4)(x - 9) = x^2 - 9x + 4x + (4)(-9) = x^2 - 5x - 36$$

The coefficient of x is the sum of the constant terms of the binomials.

The constant term of the trinomial is the product of the constant terms of the binomials.

Points to Remember to Factor $x^2 + bx + c$

1. The constant term c of the trinomial is the product of the constant terms of the binomials.
2. The coefficient b in the trinomial is the sum of the constant terms of the binomials.
3. If the constant term c of the trinomial is positive, the constant terms of the binomials have the same sign as the coefficient b of the trinomial.
4. If the constant term c of the trinomial is negative, the constant terms of the binomials have opposite signs.

Alternative to Example 2
Factor: $x^2 - 5x - 6$
▪ $(x - 6)(x + 1)$

EXAMPLE 2 Factor a Trinomial of the Form $x^2 + bx + c$

Factor: $x^2 + 7x - 18$

Solution We must find two binomials whose first terms have a product of x^2 and whose last terms have a product of -18; also, the sum of the product of the outer terms and the product of the inner terms must be $7x$. Begin by listing the possible integer factorizations of -18 and the sum of those factors.

Continued ➤

Factors of −18	Sum of the Factors
$1 \cdot (-18)$	$1 + (-18) = -17$
$(-1) \cdot 18$	$(-1) + 18 = 17$
$2 \cdot (-9)$	$2 + (-9) = -7$
$(-2) \cdot 9$	$(-2) + 9 = 7$

■ Stop. This is the desired sum.

Thus −2 and 9 are the numbers whose sum is 7 and whose product is −18. Therefore,

$$x^2 + 7x - 18 = (x - 2)(x + 9)$$

The FOIL method can be used to verify that the factorization is correct.

CHECK YOUR PROGRESS 2 Factor: $b^2 + 12b - 28$

Solution *See page S1.* $(b - 2)(b + 14)$

The trial method can sometimes be used to factor trinomials of the form $ax^2 + bx + c$, which do not have a leading coefficient of 1. We use the factors of a and c to form trial binomial factors. Factoring trinomials of this type may require testing many factors. To reduce the number of trial factors, make use of the following points.

P

Points to Remember to Factor $ax^2 + bx + c$, $a > 0$

1. If the constant term of the trinomial is positive, the constant terms of the binomials have the same sign as the coefficient b in the trinomial. For instance, $2x^2 - 7x + 3 = (2x - 1)(x - 3)$ and $6x^2 + 17x + 12 = (3x + 4)(2x + 3)$.
2. If the constant term of the trinomial is negative, the constant terms of the binomials have opposite signs. For instance, $4x^2 + 13x - 12 = (4x - 3)(x + 4)$.
3. If the terms of the trinomial do not have a common factor, then neither binomial will have a common factor. For instance, $8x^2 + 2x - 15$ does not have a common factor and neither do the two factors $2x + 3$ and $4x - 5$.

? **QUESTION** Is $(2x - 6)(3x + 1)$ a possible factorization of $6x^2 + 5x - 6$?

? **ANSWER** No. $(2x - 6)$ has a common factor of 2, but 2 is not a common factor of $6x^2 - 5x - 6$.

Alternative to Example 3

Factor: $6x^2 - 23x - 18$

■ $(2x - 9)(3x + 2)$

EXAMPLE 3 Factor a Trinomial of the Form $ax^2 + bx + c$

Factor: $6x^2 - 11x + 4$

Solution Because the constant term of the trinomial is positive and the coefficient of the x term is negative, the constant terms of the binomials will both be negative. This time we find factors of the first term as well as factors of the constant term.

Factors of $6x^2$	Factors of 4 (both negative)
$x, 6x$	$-1, -4$
$2x, 3x$	$-2, -2$

Use these factors to write trial factors. Use the FOIL method to see whether any of the trial factors produce the correct middle term. If the terms of a trinomial do not have a common factor, then a binomial factor cannot have a common factor (point 3). Such trial factors need not be checked.

Trial Factors	Middle Term	
$(x - 1)(6x - 4)$	Common factor	■ $6x$ and 4 have a common factor.
$(x - 4)(6x - 1)$	$-1x - 24x = -25x$	
$(x - 2)(6x - 2)$	Common factor	■ $6x$ and 2 have a common factor.
$(2x - 1)(3x - 4)$	$-8x - 3x = -11x$	■ This is the correct middle term.

Thus $6x^2 - 11x + 4 = (2x - 1)(3x - 4)$.

TAKE NOTE

Check the factorization by multiplying:

$(2x - 1)(3x - 4) = 6x^2 - 11x + 4$

This is the original polynomial, so the factorization is correct.

CHECK YOUR PROGRESS 3 Factor: $57y^2 + 4y - 28$

Solution *See page S2.* $(3y - 2)(19y + 14)$

Sometimes it is impossible to factor a polynomial into the product of two polynomials having integer coefficients. Such polynomials are said to be nonfactorable over the integers. For example, $x^2 + 3x + 7$ is nonfactorable over the integers because there are no integers whose product is 7 and whose sum or difference is 3.

If you have difficulty factoring a trinomial, you may wish to use the following theorem. It will indicate whether the trinomial is factorable over the integers.

Factorization Theorem

The trinomial $ax^2 + bx + c$, with integer coefficients a, b, and c, can be factored as the product of two binomials with integer coefficients if and only if $b^2 - 4ac$ is a perfect square.

Alternative to Example 4

a. Determine whether $2x^2 + 4x + 5$ is factorable over the integers.
b. Determine whether $5x^2 - 11x - 6$ is factorable over the integers.

- **a. No**
- **b. Yes. $(5x + 2)(x - 3)$**

EXAMPLE 4 Apply the Factorization Theorem

Determine whether each trinomial is factorable over the integers.

a. $4x^2 + 8x - 7$ **b.** $6x^2 - 5x - 4$

Solution

a. The coefficients of $4x^2 + 8x - 7$ are $a = 4$, $b = 8$, and $c = -7$. Applying the factorization theorem yields

$$b^2 - 4ac = 8^2 - 4(4)(-7) = 176$$

Because 176 is not a perfect square, the trinomials is nonfactorable over the integers.

b. The coefficients of $6x^2 - 5x - 4$ are $a = 6$, $b = -5$, and $c = -4$. Thus

$$b^2 - 4ac = (-5)^2 - 4(6)(-4) = 121$$

Because 121 is a perfect square $(121 = 11^2)$, the trinomial is factorable over the integers. Using the methods we have developed, we find

$$6x^2 - 5x - 4 = (3x - 4)(2x + 1)$$

CHECK YOUR PROGRESS 4 Determine whether $16a^2 - 8a - 35$ is factorable over the integers.

Solution *See page S2.* Yes. $b^2 - 4ac = 2304 = 48^2$

▪ Special Factoring

Some polynomials of degree greater than 2 can be factored by the trial procedure. Consider $2x^6 + 9x^3 + 9$. Because all the signs of the trinomial are positive, the coefficients of all the terms in the binomial factors must be positive.

Factors of $2x^6$	Factors of 9 (both positive)
	1, 9
$x^3, 2x^3$	3, 3

The factors $(x^3 + 3)$ and $(2x^3 + 3)$ are the only trial factors whose product has the correct middle term, $9x^3$. Thus $2x^6 + 9x^3 + 9 = (x^3 + 3)(2x^3 + 3)$.

Some polynomials can be factored by making use of the following factoring formulas.

Factoring Formulas

Special Form	Formula(s)
Difference of two squares	$x^2 - y^2 = (x + y)(x - y)$
Perfect-square trinomials	$x^2 + 2xy + y^2 = (x + y)^2$ $x^2 - 2xy + y^2 = (x - y)^2$
Sum of cubes	$x^3 + y^3 = (x + y)(x^2 - xy + y^2)$
Difference of cubes	$x^3 - y^3 = (x - y)(x^2 + xy + y^2)$

TAKE NOTE

The polynomial $x^2 + y^2$ is the sum of two squares. You may be tempted to factor it in a manner similar to the method used on the difference of two squares; however, $x^2 + y^2$ is nonfactorable over the integers.

The monomial a^2 is the square of a, and a is called a square root of a^2. The factoring formula

$$x^2 - y^2 = (x + y)(x - y)$$

indicates that the difference of two squares can be written as the product of the sum and the difference of the square roots of the squares.

Alternative to Example 5

Factor: $27x^2 - 48$

■ $\mathbf{3(3x - 4)(3x + 4)}$

EXAMPLE 5 Factor the Difference of Squares

Factor: $49x^2 - 144$

Solution

$$49x^2 - 144 = (7x)^2 - (12)^2$$ ■ Recognize the difference-of-squares form.

$$= (7x + 12)(7x - 12)$$ ■ The binomial factors are the sum and the difference of the square roots of the squares.

CHECK YOUR PROGRESS 5 Factor: $81b^2 - 16c^2$

Solution *See page S2.* $(9b - 4c)(9b + 4c)$

A perfect-square trinomial is a trinomial that is the square of a binomial. For example, $x^2 + 6x + 9$ is a perfect-square trinomial because

$$(x + 3)^2 = x^2 + 6x + 9$$

Every perfect-square trinomial can be factored by the trial method, but it generally is faster to factor perfect-square trinomials by using the factoring formulas.

Alternative to Example 6

Factor: $8x^2 - 40xy + 50y^2$

■ $\mathbf{2(2x - 5y)^2}$

EXAMPLE 6 Factor a Perfect-Square Trinomial

Factor: $16m^2 - 40mn + 25n^2$

Solution

$$16m^2 - 40mn + 25n^2$$

$$= (4m)^2 - 2(4m)(5n) + (5n)^2$$ ■ Recognize the perfect-square trinomial form.

$$= (4m - 5n)^2$$

Continued ➤

CHECK YOUR PROGRESS 6 Factor: $b^2 - 24b + 144$

Solution *See page S2.* $(b - 12)^2$

The product of the same three factors is called a cube. For example, $8a^3$ is a cube because $8a^3 = (2a)^3$. The cube root of a cube is one of the three equal factors. To factor the sum or the difference of two cubes, use the appropriate factoring formula. It helps to use the following patterns, which involve the signs of the terms.

Same sign; Opposite signs

$$x^3 + y^3 = (x + y)(x^2 - xy + y^2)$$

Same sign; Opposite signs

$$x^3 - y^3 = (x - y)(x^2 + xy + y^2)$$

In the factorization of the sum or difference of two cubes, the terms of the binomial factor are the cube roots of the cubes. For example,

$$8a^3 - 27b^3 = (2a)^3 - (3b)^3 = (2a - 3b)(4a^2 + 6ab + 9b^2)$$

Alternative to Example 7

Factor:

a. $27x^3 + 1$

b. $27x^3 - 125y^3$

▪ **a.** $(3x + 1)(9x^2 - 3x + 1)$

▪ **b.** $(3x - 5y)(9x^2 + 15xy + 25y^2)$

EXAMPLE 7 Factor the Sum or Difference of Cubes

Factor: **a.** $8a^3 + b^3$ **b.** $a^3 - 64$

Solution

a. $8a^3 + b^3 = (2a)^3 + b^3$ ▪ Recognize the sum-of-cubes form.

$= (2a + b)(4a^2 - 2ab + b^2)$ ▪ Factor.

b. $a^3 - 64 = a^3 - 4^3$ ▪ Recognize the difference-of-cubes form.

$= (a - 4)(a^2 + 4a + 16)$ ▪ Factor.

CHECK YOUR PROGRESS 7 Factor: $b^3 + 64$

Solution *See page S2.* $(b + 4)(b^2 - 4b + 16)$

▪ Factoring by Grouping

Some polynomials can be factored by grouping. Pairs of terms that have a common factor are first grouped together. The process makes repeated use of the Distributive Property, as shown in the following factorization of $6y^3 - 21y^2 - 4y + 14$.

$6y^3 - 21y^2 - 4y + 14$

$= (6y^3 - 21y^2) - (4y - 14)$ ▪ Group the first two terms and the last two terms.

$= 3y^2(2y - 7) - 2(2y - 7)$ ▪ Factor out the GCF from each of the groups.

$= (2y - 7)(3y^2 - 2)$ ▪ Factor out the common binomial factor.

TAKE NOTE

$-a + b = -(a - b)$. Thus $-4y + 14 = -(4y - 14)$.

When you factor by grouping, some experimentation may be necessary to find a grouping that is of the form of one of the special factoring formulas.

Alternative to Example 8

Factor:

a. $9x^2 - 6xy + y^2 - 1$

b. $4a^2 + 6a - b^2 - 3b$

■ a. $(3x - y - 1)(3x - y + 1)$

■ b. $(2a - b)(2a + b + 3)$

EXAMPLE 8 Factor by Grouping

Factor by grouping. **a.** $a^2 + 10ab + 25b^2 - c^2$ **b.** $p^2 + p - q - q^2$

Solution

a. $a^2 + 10ab + 25b^2 - c^2$

$= (a^2 + 10ab + 25b^2) - c^2$ ■ Group the terms of the perfect-square trinomial.

$= (a + 5b)^2 - c^2$ ■ Factor the trinomial.

$= [(a + 5b) + c][(a + 5b) - c]$ ■ Factor the difference of squares.

$= (a + 5b + c)(a + 5b - c)$ ■ Simplify.

b. $p^2 + p - q - q^2$

$= p^2 - q^2 + p - q$ ■ Rearrange the terms.

$= (p^2 - q^2) + (p - q)$ ■ Regroup.

$= (p + q)(p - q) + (p - q)$ ■ Factor the difference of squares.

$= (p - q)(p + q + 1)$ ■ Factor out the common factor $(p - q)$.

CHECK YOUR PROGRESS 8 Factor: $a^2y^2 - ay^3 + ac - cy$

Solution *See page S2.* $(ay^2 + c)(a - y)$

■ General Factoring

Here is a general factoring strategy for polynomials.

General Factoring Strategy

1. Factor out the GCF of all terms.
2. Try to factor a binomial as
   a. the difference of two squares.
   b. the sum or difference of two cubes.
3. Try to factor a trinomial
   a. as a perfect-square trinomial.
   b. using the trial method.
4. Try to factor a polynomial with more than three terms by grouping.
5. After each factorization, examine the new factors to see whether they can be factored.

Alternative to Example 9

Factor: $x^5 - 4x^3 + 2x^2 - 8$

■ $(x - 2)(x + 2)(x^3 + 2)$

EXAMPLE 9 Factor Using the General Factoring Strategy

Completely factor: $x^6 + 7x^3 - 8$

Solution

Factor $x^6 + 7x^3 - 8$ as the product of two binomials.

$$x^6 + 7x^3 - 8 = (x^3 + 8)(x^3 - 1)$$

Now factor $x^3 + 8$, which is the sum of two cubes, and factor $x^3 - 1$, which is the difference of two cubes.

$$x^6 + 7x^3 - 8 = (x + 2)(x^2 - 2x + 4)(x - 1)(x^2 + x + 1)$$

CHECK YOUR PROGRESS 9 Factor: $81y^4 - 16$

Solution *See page S2.* $(3y - 2)(3y + 2)(9y^2 + 4)$

Topics for Discussion

1. Discuss the meaning of the phrase *nonfactorable over the integers.*

2. You know that if $ab = 0$, then $a = 0$ or $b = 0$. Suppose a polynomial is written in factored form and then set equal to zero. For instance, suppose

$$x^2 - 2x - 15 = (x - 5)(x + 3) = 0$$

Discuss what implications this has for the values of x. Do not answer this question only for the polynomial above, but also discuss the implications for any polynomial written as a product of linear factors and then set equal to zero.

3. Let P be a polynomial of degree n. Discuss the number of possible distinct linear polynomials that can be a factor of P.

4. If n is a natural number, ***n* factorial** is given by $n! = n(n - 1)(n - 2) \cdots 3 \cdot 2 \cdot 1$. Explain why none of the following consecutive integers is a prime number.

$$5! + 2 \qquad 5! + 3 \qquad 5! + 4 \qquad 5! + 5$$

How many numbers are in the following list of consecutive integers? How many of those numbers are prime numbers?

$$k! + 2, k! + 3, k! + 4, k! + 5, \ldots, k! + k$$

Explain why this result means that there are arbitrarily long sequences of consecutive natural numbers that do not contain a prime number.

TAKE NOTE

Here are some examples of factorials.

$5! = 5 \cdot 4 \cdot 3 \cdot 2 \cdot 1 = 120$

$7! = 7 \cdot 6 \cdot 5 \cdot 4 \cdot 3 \cdot 2 \cdot 1 = 5040$

EXERCISES P.4

— Suggested Assignment: Exercises 1–83, odd; and 85–90, all.

In Exercises 1 to 8, factor the expression.

1. $5x + 20$ $5(x+4)$
2. $8x^2 + 12x - 40$ $4(2x^2 + 3x - 10)$
3. $-15x^2 - 12x$ $-3x(5x+4)$
4. $-6y^2 - 54y$ $-6y(y+9)$
5. $10x^2y + 6xy - 14xy^2$ $2xy(5x + 3 - 7y)$
6. $18a^2b^3 - 24ab^2 + 6a$ $6a(3ab^3 - 4b^2 + 1)$
7. $(x-3)(a+b) + (x-3)(a+2b)$ $(x-3)(2a+3b)$
8. $(x-4)(2a-b) + (x+4)(2a-b)$ $(2a-b)(2x)$

In Exercises 9 to 22, factor each trinomial.

9. $x^2 + 7x + 12$ $(x+3)(x+4)$
10. $x^2 + 9x + 20$ $(x+5)(x+4)$
11. $a^2 - 10a - 24$ $(a-12)(a+2)$
12. $y^2 - 5y - 36$ $(y-9)(y+4)$
13. $6x^2 + 25x + 4$ $(6x+1)(x+4)$
14. $8a^2 - 26a + 15$ $(4a-3)(2a-5)$
15. $51x^2 - 5x - 4$ $(17x+4)(3x-1)$
16. $21v^2 + 26v - 15$ $(7v-3)(3v+5)$
17. $6x^2 + xy - 40y^2$ $(3x+8y)(2x-5y)$
18. $8x^2 + 10xy - 25y^2$ $(2x+5y)(4x-5y)$
19. $x^4 + 6x^2 + 5$ $(x^2+5)(x^2+1)$
20. $x^4 + 11x^2 + 18$ $(x^2+2)(x^2+9)$
21. $6x^4 + 23x^2 + 15$ $(6x^2+5)(x^2+3)$
22. $9x^4 + 10x^2 + 1$ $(9x^2+1)(x^2+1)$

In Exercises 23 to 28, use the factorization theorem to determine whether each trinomial is factorable over the integers.

23. $8x^2 + 26x + 15$ Factorable over the integers
24. $4a^2 - 7a + 2$ Nonfactorable over the integers
25. $15a^2 + 19a + 6$ Factorable over the integers
26. $5v^2 + 33v - 14$ Factorable over the integers
27. $6x^2 - 14x + 5$ Nonfactorable over the integers
28. $12c^2 - 11c - 15$ Factorable over the integers

In Exercises 29 to 38, factor each difference of squares.

29. $x^2 - 9$ $(x-3)(x+3)$
30. $x^2 - 64$ $(x-8)(x+8)$
31. $4a^2 - 49$ $(2a-7)(2a+7)$
32. $9a^2 - 64b^2$ $(3a-8b)(3a+8b)$
33. $1 - 100x^2$ $(1-10x)(1+10x)$
34. $1 - 121y^2$ $(1+11y)(1-11y)$
35. $x^4 - 9$ $(x^2+3)(x^2-3)$
36. $y^4 - 196$ $(y^2+14)(y^2-14)$
37. $(x+5)^2 - 4$ $(x+3)(x+7)$
38. $(x-3)^2 - 16$ $(x+1)(x-7)$

In Exercises 39 to 46, factor each perfect-square trinomial.

39. $x^2 + 10x + 25$ $(x+5)^2$
40. $y^2 + 6y + 9$ $(y+3)^2$
41. $a^2 - 14a + 49$ $(a-7)^2$
42. $c^2 + 20c + 100$ $(c+10)^2$
43. $4x^2 + 12x + 9$ $(2x+3)^2$
44. $25y^2 + 40y + 16$ $(5y+4)^2$
45. $z^4 + 4z^2w^2 + 4w^4$ $(z^2+2w^2)^2$
46. $9x^4 - 30x^2y^2 + 25y^4$ $(3x^2-5y^2)^2$

In Exercises 47 to 54, factor each sum or difference of cubes.

47. $x^3 - 8$ $(x-2)(x^2+2x+4)$
48. $w^3 + 125$ $(w+5)(w^2-5w+25)$
49. $8x^3 - 27y^3$ $(2x-3y)(4x^2+6xy+9y^2)$
50. $64u^3 - 27v^3$ $(4u-3v)(16u^2+12uv+9v^2)$
51. $8 - x^6$ $(2-x^2)(4+2x^2+x^4)$
52. $1 + y^{12}$ $(y^4+1)(y^8-y^4+1)$
53. $(x-2)^3 - 1$ $(x-3)(x^2-3x+3)$
54. $(y+3)^3 + 8$ $(y+5)(y^2+4y+7)$

In Exercises 55 to 60, factor by grouping in pairs.

55. $3x^3 + x^2 + 6x + 2$ $(3x+1)(x^2+2)$
56. $18w^3 + 15w^2 + 12w + 10$ $(6w+5)(3w^2+2)$
57. $ax^2 - ax + bx - b$ $(x-1)(ax+b)$
58. $6a^2 + 3ab - 2ac - bc$ $(2a+b)(3a-c)$
59. $6w^3 + 4w^2 - 15w - 10$ $(3w+2)(2w^2-5)$
60. $10z^3 - 15z^2 - 4z + 6$ $(2z-3)(5z^2-2)$

In Exercises 61 to 80, use the general factoring strategy to completely factor each polynomial. If the polynomial does not factor, then state that it is nonfactorable over the integers.

61. $18x^2 - 2$ $2(3x-1)(3x+1)$
62. $4bx^3 + 32b$ $4b(x+2)(x^2-2x+4)$
63. $16x^4 - 1$ $(2x-1)(2x+1)(4x^2+1)$
64. $81v^4 - 256$ $(3v-4)(3v+4)(9v^2+16)$
65. $12ax^2 - 23axy + 10ay^2$ $a(3x-2y)(4x-5y)$
66. $6ax^2 - 19axy - 20ay^2$ $a(6x+5y)(x-4y)$
67. $3bx^3 + 4bx^2 - 3bx - 4b$ $b(3x+4)(x-1)(x+1)$
68. $2x^6 - 2$ $2(x+1)(x-1)(x^2+x+1)(x^2-x+1)$
69. $72bx^2 + 24bxy + 2by^2$ $2b(6x+y)^2$
70. $64y^3 - 16y^2z + yz^2$ $y(8y-z)^2$
71. $(w-5)^3 + 8$ $(w-3)(w^2-12w+39)$
72. $5xy + 20y - 15x - 60$ $5(y-3)(x+4)$

73. $x^2 + 6xy + 9y^2 - 1$
$(x + 3y - 1)(x + 3y + 1)$

74. $4y^2 - 4yz + z^2 - 9$
$(2y - z - 3)(2y - z + 3)$

75. $8x^2 + 3x - 4$
Nonfactorable over the integers

76. $16x^2 + 81$
Nonfactorable over the integers

77. $5x(2x - 5)^2 - (2x - 5)^3$
$(2x - 5)^2(3x + 5)$

78. $6x(3x + 1)^3 - (3x + 1)^4$
$(3x - 1)(3x + 1)^3$

79. $4x^2 + 2x - y - y^2$
$(2x - y)(2x + y + 1)$

80. $a^2 + a + b - b^2$
$(a + b)(a + 1 - b)$

In Exercises 81 and 82, find all positive values of k such that the trinomial is a perfect-square trinomial.

81. $x^2 + kx + 16$ 8

82. $36x^2 + kxy + 100y^2$ 120

In Exercises 83 and 84, find k such that the trinomial is a perfect-square trinomial.

83. $x^2 + 16x + k$ 64

84. $x^2 - 14xy + ky^2$ 49

Prepare for Section P.5

85. What is the smallest natural number divisible by 6 and 8?
24

86. Add: $\frac{2}{3} + \frac{1}{4}$ $\frac{11}{12}$

87. Subtract: $\frac{2}{5} - \frac{1}{2}$ $-\frac{1}{10}$

88. Simplify: $1 + \cfrac{1}{2 + \cfrac{1}{3}}$
$\frac{10}{7}$

89. Simplify: $\left(\frac{w}{x}\right)^{-1}\left(\frac{y}{z}\right)^{-1}$
$\frac{xz}{wy}$

90. Simplify: $\frac{x^3 - 4x^2 + 2x - 8}{x - 4}$ $x^2 + 2$

Explorations

1. ***Geometry*** The ancient Greeks used geometric figures and the concept of area to illustrate many algebraic concepts. The factoring formula $x^2 - y^2 = (x + y)(x - y)$ can be illustrated by the following figure.

   a. Which regions are represented by $(x + y)(x - y)$?
   Regions I + II + III

   b. Which regions are represented by $x^2 - y^2$?
   Regions II + III + V

   c. Explain why the area of the regions listed in **a.** must equal the area of the regions listed in **b.**
   Regions I + II + III = Regions II + III + V or $(x + y)(x - y) = x^2 - y^2$.

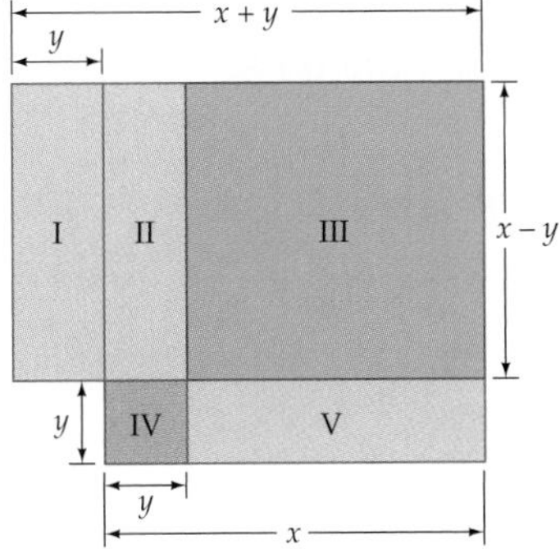

2. ***Geometry*** What algebraic formula does the following geometric figure illustrate? $(x + y)^2 = x^2 + 2xy + y^2$

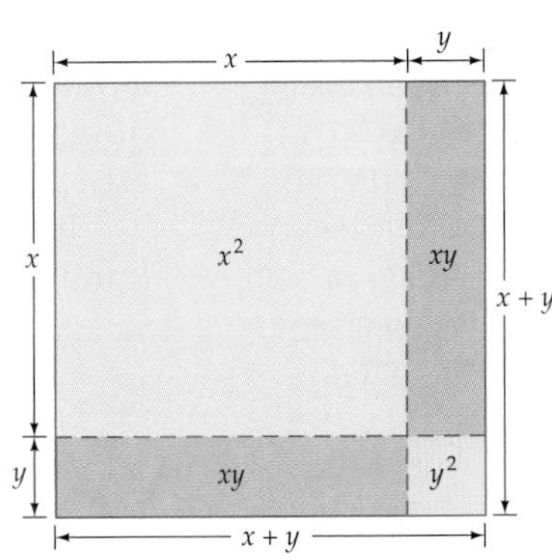

3. ***Geometry*** Show how the figure below can be used to illustrate the factoring formula for the difference of two cubes.
Answers will vary.

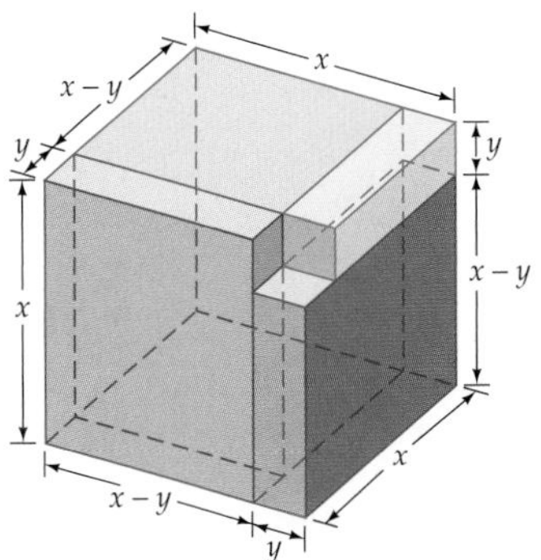

SECTION P.5 Rational Expressions

- Simplify a Rational Expression
- Operations on Rational Expressions
- Determining the LCD of Rational Expressions
- Complex Fractions
- Applications of Rational Expressions

Point of Interest

Evidence left by early Egyptians more than 3600 years ago shows that they used, with one exception, unit fractions—that is, fractions whose numerators are 1. The one exception was $\frac{2}{3}$. A unit fraction was represented by placing an oval over the symbol for the number in the denominator. For instance, $\frac{1}{4}$ = 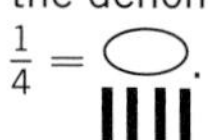.

A **rational expression** is a fraction in which the numerator and denominator are polynomials. For example,

$$\frac{3}{x+1} \quad \text{and} \quad \frac{x^2 - 4x - 21}{x^2 - 9}$$

are rational expressions.

The domain of a rational expression is the set of all real numbers that can be used as replacements for the variable. Any value of the variable that causes division by zero is excluded from the domain of the rational expression. For example, the domain of

$$\frac{7x}{x^2 - 5x} \quad x \neq 0, x \neq 5$$

is the set of all real numbers except 0 and 5. Both 0 and 5 are excluded values because the denominator $x^2 - 5x$ equals zero when $x = 0$ and also when $x = 5$. Sometimes the excluded values are specified to the right of a rational expression, as shown here. However, a rational expression is meaningful only for those real numbers that are not excluded values, regardless of whether the excluded values are specifically stated.

Rational expression have properties similar to the properties of rational numbers.

Properties of Rational Expressions

For all rational expressions P/Q and R/S where $Q \neq 0$ and $S \neq 0$,

Equality $\quad \frac{P}{Q} = \frac{R}{S}$ if and only if $PS = QR$

Equivalent expressions $\quad \frac{P}{Q} = \frac{PR}{QR}, R \neq 0$

Sign $\quad -\frac{P}{Q} = \frac{-P}{Q} = \frac{P}{-Q}$

Simplify a Rational Expression

To simplify a rational expression, factor the numerator and the denominator. Then use the equivalent expressions property to eliminate factors common to both the numerator and the denominator. A rational expression is *simplified* when 1 is the only common factor of both the numerator and the denominator.

Alternative to Example 1

Simplify: $\dfrac{20 - 9x + x^2}{x^2 - x - 20}$

■ $\dfrac{x - 4}{x + 4}$

TAKE NOTE

A rational expression such as $(x + 3)/3$ does not simplify to $x + 1$ because

$$\frac{x+3}{3} = \frac{x}{3} + \frac{3}{3} = \frac{x}{3} + 1$$

Rational expressions can be simplified by dividing factors common to the numerator and the denominator, but not terms.

EXAMPLE 1 Simplify a Rational Expression

Simplify: $\dfrac{7 + 20x - 3x^2}{2x^2 - 11x - 21}$

Solution

$$\begin{aligned}\frac{7 + 20x - 3x^2}{2x^2 - 11x - 21} &= \frac{(7 - x)(1 + 3x)}{(x - 7)(2x + 3)} && \text{■ Factor.}\\ &= \frac{-(x - 7)(1 + 3x)}{(x - 7)(2x + 3)} && \text{■ Use } (7 - x) = -(x - 7).\\ &= \frac{-\cancel{(x - 7)}(1 + 3x)}{\cancel{(x - 7)}(2x + 3)}\\ &= \frac{-(1 + 3x)}{2x + 3} = -\frac{3x + 1}{2x + 3} \qquad x \neq 7, x \neq -\frac{3}{2}\end{aligned}$$

CHECK YOUR PROGRESS 1 Simplify: $\dfrac{2x^2 - 5x - 12}{2x^2 + 5x + 3}$

Solution *See page S2.* $\dfrac{x - 4}{x + 1}$

■ Operations on Rational Expressions

Arithmetic operations are defined on rational expressions just as they are on rational numbers.

P

Arithmetic Operations Defined on Rational Expressions

For all rational expressions P/Q, R/Q, and R/S where $Q \neq 0$ and $S \neq 0$,

Addition	$\dfrac{P}{Q} + \dfrac{R}{Q} = \dfrac{P + R}{Q}$
Subtraction	$\dfrac{P}{Q} - \dfrac{R}{Q} = \dfrac{P - R}{Q}$
Multiplication	$\dfrac{P}{Q} \cdot \dfrac{R}{S} = \dfrac{PR}{QS}$
Division	$\dfrac{P}{Q} \div \dfrac{R}{S} = \dfrac{P}{Q} \cdot \dfrac{S}{R} = \dfrac{PS}{QR}, \quad R \neq 0$

Factoring and the equivalent expressions property of rational expressions are used in the multiplication and division of rational expressions.

Alternative to Example 2

Simplify: $\frac{15x^2 + 11x - 12}{25x^2 - 9} \div \frac{3x^2 + 13x + 12}{10x^2 + 11x + 3}$

■ $\frac{2x + 1}{x + 3}$

EXAMPLE 2 Divide a Rational Expression

Simplify: $\frac{x^2 + 6x + 9}{x^3 + 27} \div \frac{x^2 + 7x + 12}{x^3 - 3x^2 + 9x}$

Solution

$$\frac{x^2 + 6x + 9}{x^3 + 27} \div \frac{x^2 + 7x + 12}{x^3 - 3x^2 + 9x}$$

$$= \frac{(x + 3)^2}{(x + 3)(x^2 - 3x + 9)} \div \frac{(x + 4)(x + 3)}{x(x^2 - 3x + 9)}$$ ■ Factor.

$$= \frac{(x + 3)^2}{(x + 3)(x^2 - 3x + 9)} \cdot \frac{x(x^2 - 3x + 9)}{(x + 4)(x + 3)}$$ ■ Multiply by the reciprocal.

$$= \frac{\cancel{(x + 3)^2}x\cancel{(x^2 - 3x + 9)}}{\cancel{(x + 3)}\cancel{(x^2 - 3x + 9)}(x + 4)\cancel{(x + 3)}}$$ ■ Simplify.

$$= \frac{x}{x + 4}$$

CHECK YOUR PROGRESS 2 Simplify: $\frac{x^2 - 16}{x^2 + 7x + 12} \cdot \frac{x^2 - 4x - 21}{x^2 - 4x}$

Solution *See page S2.* $\frac{x - 7}{x}$

Addition of rational expressions with a common denominator is accomplished by writing the sum of the numerators over the common denominator. For example,

$$\frac{5x}{18} + \frac{x}{18} = \frac{5x + x}{18} = \frac{6x}{18} = \frac{x}{3}$$

If the rational expressions do not have a common denominator, then they can be written as equivalent rational expressions that have a common denominator by multiplying the numerator and denominator of each rational expression by the required polynomial. The following procedure can be used to determine the least common denominator (LCD) of rational expressions. It is similar to the process used to find the LCD of rational numbers.

■ Determining the LCD of Rational Expressions

1. Factor each denominator completely and express repeated factors using exponential notation.
2. Identify the largest power of each factor in any single factorization. The LCD is the product of each factor raised to its largest power.

For example,

$$\frac{1}{x + 3} \quad \text{and} \quad \frac{5}{2x - 1}$$

have an LCD of $(x + 3)(2x - 1)$. The rational expressions

$$\frac{5x}{(x+5)(x-7)^3} \quad \text{and} \quad \frac{7}{x(x+5)^2(x-7)}$$

have an LCD of $x(x + 5)^2(x - 7)^3$.

Alternative to Example 3

a. $\dfrac{x}{12} + \dfrac{7x}{12}$

b. $\dfrac{2x}{x-3} - \dfrac{x-2}{x+3}$

c. $\dfrac{x}{x^2-16} - \dfrac{2x+1}{x^2-5x+4}$

■ a. $\dfrac{2x}{3}$

■ b. $\dfrac{x^2+11x-6}{(x-3)(x+3)}$

■ c. $\dfrac{-x^2-10x-4}{(x-4)(x+4)(x-1)}$

EXAMPLE 3 Add and Subtract Rational Expressions

Perform the indicated operation and then simplify if possible.

a. $\dfrac{5x}{48} + \dfrac{x}{15}$ **b.** $\dfrac{x}{x^2-4} - \dfrac{2x-1}{x^2-3x-10}$

Solution

a. Determine the prime factorization of the denominators.

$$48 = 2^4 \cdot 3 \quad \text{and} \quad 15 = 3 \cdot 5$$

The desired common denominator is the product of each of the prime factors raised to its largest power. Thus the common denominator is $2^4 \cdot 3 \cdot 5 = 240$. Write each rational expression as an equivalent rational expression with a denominator of 240.

$$\frac{5x}{48} + \frac{x}{15} = \frac{5x \cdot 5}{48 \cdot 5} + \frac{x \cdot 16}{15 \cdot 16} = \frac{25x}{240} + \frac{16x}{240} = \frac{41x}{240}$$

b. Factor each denominator to determine the LCD of the rational expressions.

$$x^2 - 4 = (x+2)(x-2)$$
$$x^2 - 3x - 10 = (x+2)(x-5)$$

The LCD is $(x + 2)(x - 2)(x - 5)$. Forming equivalent rational expressions that have the LCD, we have

$$\begin{aligned}
&\frac{x}{x^2-4} - \frac{2x-1}{x^2-3x-10} \\
&= \frac{x(x-5)}{(x+2)(x-2)(x-5)} - \frac{(2x-1)(x-2)}{(x+2)(x-5)(x-2)} \\
&= \frac{x^2-5x-(2x^2-5x+2)}{(x+2)(x-2)(x-5)} = \frac{x^2-5x-2x^2+5x-2}{(x+2)(x-2)(x-5)} \\
&= \frac{-x^2-2}{(x+2)(x-2)(x-5)} = -\frac{x^2+2}{(x+2)(x-2)(x-5)}
\end{aligned}$$

CHECK YOUR PROGRESS 3 Simplify: $\dfrac{3y-1}{3y+1} - \dfrac{2y-5}{y-3}$

Solution *See page S2.* $\dfrac{-3y^2+3y+8}{(3y+1)(y-3)}$

Complex Fractions

A **complex fraction** is a fraction whose numerator or denominator or both contain one or more fractions. Complex fractions can be simplified by using one of the following two methods.

Methods for Simplifying Complex Fractions

Method 1: Multiply by 1 in the form of $\frac{\text{LCD}}{\text{LCD}}$

1. Determine the LCD of all the fractions in the complex fraction.
2. Multiply both the numerator and the denominator of the complex fraction by the LCD.
3. If possible, simplify the resulting rational expression.

Method 2: Multiply the numerator by the reciprocal of the denominator.

1. Simplify the numerator to a single fraction and the denominator to a single fraction.
2. Using the definition for dividing fractions, multiply the numerator by the reciprocal of the denominator.
3. If possible, simplify the resulting rational expression.

Alternative to Example 4

Simplify: $\dfrac{\dfrac{5x+3}{x^2+3} - \dfrac{2}{x}}{\dfrac{4}{x+3} - \dfrac{1}{x}}$

■ $\dfrac{(x+2)(x+3)}{x^2+3}$

TAKE NOTE

For Example 4 we have used Method 2 to simplify the complex fractions. For Example 5 on the next page, we have used Method 1.

EXAMPLE 4 Simplify Complex Fractions

Simplify: $\dfrac{\dfrac{2}{x-2} + \dfrac{1}{x}}{\dfrac{3x}{x-5} - \dfrac{2}{x-5}}$

Solution First simplify the numerator to a single fraction and then simplify the denominator to a single fraction.

$$\frac{\dfrac{2}{x-2} + \dfrac{1}{x}}{\dfrac{3x}{x-5} - \dfrac{2}{x-5}} = \frac{\dfrac{2 \cdot x}{(x-2) \cdot x} + \dfrac{1 \cdot (x-2)}{x \cdot (x-2)}}{\dfrac{3x-2}{x-5}}$$

■ Simplify numerator and denominator.

$$= \frac{\dfrac{2x + (x-2)}{x(x-2)}}{\dfrac{3x-2}{x-5}} = \frac{\dfrac{3x-2}{x(x-2)}}{\dfrac{3x-2}{x-5}}$$

$$= \frac{\cancel{3x-2}}{x(x-2)} \cdot \frac{x-5}{\cancel{3x-2}}$$

■ Multiply the numerator by the reciprocal of the denominator.

$$= \frac{x-5}{x(x-2)}$$

CHECK YOUR PROGRESS 4 Simplify: $\dfrac{3 - \dfrac{2}{a}}{5 + \dfrac{3}{a}}$

Solution *See page S2.* $\dfrac{3a-2}{5a+3}$

Alternative to Example 5
Simplify: $(4a^{-1} - b^{-1})^{-1}$
▪ $\frac{ab}{4b - a}$

TAKE NOTE

It is a mistake to write

$$\frac{c^{-1}}{a^{-1} + b^{-1}} \quad \text{as} \quad \frac{a + b}{c}$$

because a^{-1} and b^{-1} are *terms* and cannot be treated as factors.

EXAMPLE 5 Simplify a Fraction

Simplify the fraction $\frac{c^{-1}}{a^{-1} + b^{-1}}$.

Solution The fraction written without negative exponents becomes

$$\frac{c^{-1}}{a^{-1} + b^{-1}} = \frac{\frac{1}{c}}{\frac{1}{a} + \frac{1}{b}}$$

▪ Use $x^{-n} = \frac{1}{x^n}$.

$$= \frac{\frac{1}{c} \cdot abc}{\left(\frac{1}{a} + \frac{1}{b}\right)abc}$$

▪ Multiply the numerator and the denominator by *abc*, which is the LCD of the fraction in the numerator and the fraction in the denominator.

$$= \frac{ab}{bc + ac}$$

CHECK YOUR PROGRESS 5 Simplify: $\frac{e^{-2} - f^{-1}}{ef}$

Solution *See page S2.* $\frac{f - e^2}{e^3 f^2}$

▪ Applications of Rational Expressions

Alternative to Example 6
Exercise 64, page 56

EXAMPLE 6 Solve an Application

The *average speed* for a round trip is given by the complex fraction

$$\frac{2}{\frac{1}{v_1} + \frac{1}{v_2}}$$

where v_1 is the average speed on the way to the destination and v_2 is the average speed on the return trip. Find the average speed for a round trip if $v_1 = 50$ miles per hour and $v_2 = 40$ miles per hour.

Solution Evaluate the complex fraction with $v_1 = 50$ and $v_2 = 40$.

$$\frac{2}{\frac{1}{v_1} + \frac{1}{v_2}} = \frac{2}{\frac{1}{50} + \frac{1}{40}} = \frac{2}{\frac{1 \cdot 4}{50 \cdot 4} + \frac{1 \cdot 5}{40 \cdot 5}}$$

▪ Substitute and simplify the denominator.

$$= \frac{2}{\frac{4}{200} + \frac{5}{200}} = \frac{2}{\frac{9}{200}}$$

$$= 2 \cdot \frac{200}{9} = \frac{400}{9} = 44\frac{4}{9}$$

The average speed of the round trip is $44\frac{4}{9}$ miles per hour.

Continued ➤

CHECK YOUR PROGRESS 6 According to Einstein's Theory of Relativity, the sum of the speeds of objects approaching each other is given by the complex fraction

$$\frac{v_1 + v_2}{1 + \frac{v_1 v_2}{c^2}}$$

where v_1 and v_2 are the speeds of the objects and c is the speed of light.

a. Evaluate this expression with $v_1 = 1.2 \times 10^8$ miles per hour, $v_2 = 2.4 \times 10^8$ miles per hour, and $c = 6.7 \times 10^8$ miles per hour.

b. Simplify the complex fraction.

Solution *See page S2.* **a.** 3.38×10^8 mph **b.** $\frac{c^2(v_1 + v_2)}{c^2 + v_1 v_2}$

QUESTION In Example 6, why is the speed of the round trip *not* the average of v_1 and v_2?

Topics for Discussion

1. Discuss the meaning of the phrase *rational expression.* Is a rational expression the same as a fraction? If not, give some examples of fractions that are not rational expressions.

2. What is the domain of a rational expression?

3. Explain why the following is *not* correct.

$$\frac{2\cancel{x^2} + 5}{\cancel{x^2}} = 2 + 5 = 7$$

4. Consider the rational expression $\frac{x^2 - 3x - 10}{x^2 + x - 30}$. By simplifying this expression, we have

$$\frac{x^2 - 3x - 10}{x^2 + x - 30} = \frac{(x - 5)(x + 2)}{(x - 5)(x + 6)} = \frac{x + 2}{x + 6}$$

Does this really mean that $\frac{x^2 - 3x - 10}{x^2 + x - 30} = \frac{x + 2}{x + 6}$ for every value of x? If not, for what values of x are the two expressions equal?

ANSWER Because you were traveling slower on the return trip, the return trip took longer than the time spent going to your destination. More time was spent traveling at the slower speed. Thus the average speed for the round trip is less than the average of v_1 and v_2.

EXERCISES P.5

— *Suggested Assignment: Exercises 1–63, odd; and 64–72, all.*

In Exercises 1 to 10, simplify each rational expression.

1. $\dfrac{x^2 - x - 20}{3x - 15}$ $\quad \dfrac{x+4}{3}$

2. $\dfrac{2x^2 + 7x + 6}{2x^2 + 9x + 10}$ $\quad \dfrac{2x+3}{2x+5}$

3. $\dfrac{x^3 - 9x}{x^3 + x^2 - 6x}$ $\quad \dfrac{x-3}{x-2}$

4. $\dfrac{x^3 + 125}{2x^3 - 50x}$ $\quad \dfrac{x^2 - 5x + 25}{2x(x-5)}$

5. $\dfrac{a^3 + 8}{a^2 - 4}$ $\quad \dfrac{a^2 - 2a + 4}{a-2}$

6. $\dfrac{y^3 - 27}{-y^2 + 11y - 24}$ $\quad -\dfrac{y^2 + 3y + 9}{y-8}$

7. $\dfrac{x^2 + 3x - 40}{-x^2 + 3x + 10}$ $\quad -\dfrac{x+8}{x+2}$

8. $\dfrac{2x^3 - 6x^2 + 5x - 15}{9 - x^2}$ $\quad -\dfrac{2x^2 + 5}{x+3}$

9. $\dfrac{4y^3 - 8y^2 + 7y - 14}{-y^2 - 5y + 14}$ $\quad -\dfrac{4y^2 + 7}{y+7}$

10. $\dfrac{x^3 - x^2 + x}{x^3 + 1}$ $\quad \dfrac{x}{x+1}$

In Exercises 11 to 40, simplify each expression.

11. $\left(-\dfrac{4a}{3b^2}\right)\left(\dfrac{6b}{a^4}\right)$ $\quad -\dfrac{8}{a^3 b}$

12. $\left(\dfrac{12x^2y}{5z^4}\right)\left(-\dfrac{25x^2z^3}{15y^2}\right)$ $\quad -\dfrac{4x^4}{yz}$

13. $\left(\dfrac{6p^2}{5q^2}\right)^{-1}\left(\dfrac{2p}{3q^2}\right)^2$ $\quad \dfrac{10}{27q^2}$

14. $\left(\dfrac{4r^2s}{3t^3}\right)^{-1}\left(\dfrac{6rs^3}{5t^2}\right)$ $\quad \dfrac{9s^2t}{10r}$

15. $\dfrac{x^2 + x}{2x + 3} \cdot \dfrac{3x^2 + 19x + 28}{x^2 + 5x + 4}$ $\quad \dfrac{x(3x+7)}{2x+3}$

16. $\dfrac{x^2 - 9}{x^2 - 5x + 6} \cdot \dfrac{x^2 + 2x - 8}{x^2 + 7x + 12}$ $\quad 1$

17. $\dfrac{3x - 15}{2x^2 - 50} \cdot \dfrac{2x^2 + 16x + 30}{6x + 9}$ $\quad \dfrac{x+3}{2x+3}$

18. $\dfrac{y^3 - 8}{y^2 + y - 6} \cdot \dfrac{y^2 + 3y}{y^3 + 2y^2 + 4y}$ $\quad 1$

19. $\dfrac{12y^2 + 28y + 15}{6y^2 + 35y + 25} \div \dfrac{2y^2 - y - 3}{3y^2 + 11y - 20}$ $\quad \dfrac{(2y+3)(3y-4)}{(2y-3)(y+1)}$

20. $\dfrac{z^2 - 81}{z^2 - 16} \div \dfrac{z^2 - z - 20}{z^2 + 5z - 36}$ $\quad \dfrac{(z+9)^2(z-9)}{(z+4)^2(z-5)}$

21. $\dfrac{a^2 + 9}{a^2 - 64} \div \dfrac{a^3 - 3a^2 + 9a - 27}{a^2 + 5a - 24}$ $\quad \dfrac{1}{a-8}$

22. $\dfrac{6x^2 + 13xy + 6y^2}{4x^2 - 9y^2} \div \dfrac{3x^2 - xy - 2y^2}{2x^2 + xy - 3y^2}$ $\quad \dfrac{2x+3y}{2x-3y}$

23. $\dfrac{p+5}{r} + \dfrac{2p-7}{r}$ $\quad \dfrac{3p-2}{r}$

24. $\dfrac{2s + 5t}{4t} + \dfrac{-2s + 3t}{4t}$ $\quad 2$

25. $\dfrac{x}{x-5} + \dfrac{7x}{x+3}$ $\quad \dfrac{8x(x-4)}{(x-5)(x+3)}$

26. $\dfrac{2x}{3x+1} + \dfrac{5x}{x-7}$ $\quad \dfrac{x(17x-9)}{(3x+1)(x-7)}$

27. $\dfrac{5y-7}{y+4} - \dfrac{2y-3}{y+4}$ $\quad \dfrac{3y-4}{y+4}$

28. $\dfrac{6x-5}{x-3} - \dfrac{3x-8}{x-3}$ $\quad \dfrac{3(x+1)}{x-3}$

29. $\dfrac{4z}{2z-3} + \dfrac{5z}{z-5}$ $\quad \dfrac{7z(2z-5)}{(2z-3)(z-5)}$

30. $\dfrac{2a-1}{a+3} - \dfrac{3a-7}{2a+1}$ $\quad \dfrac{a^2 - 2a + 20}{(2a+1)(a+3)}$

31. $\dfrac{x}{x^2 - 9} - \dfrac{3x - 1}{x^2 + 7x + 12}$ $\quad \dfrac{-2x^2 + 14x - 3}{(x-3)(x+3)(x+4)}$

32. $\dfrac{m - n}{m^2 - mn - 6n^2} + \dfrac{3m - 5n}{m^2 + mn - 2n^2}$ $\quad \dfrac{4(m-2n)^2}{(m-3n)(m+2n)(m-n)}$

33. $\dfrac{1}{x} + \dfrac{2}{3x-1} \cdot \dfrac{3x^2 + 11x - 4}{x - 5}$ $\quad \dfrac{(2x-1)(x+5)}{x(x-5)}$

34. $\dfrac{2}{y} - \dfrac{3}{y+1} \cdot \dfrac{y^2 - 1}{y + 4}$ $\quad -\dfrac{(3y-8)(y+1)}{y(y+4)}$

35. $\dfrac{q+1}{q-3} - \dfrac{2q}{q-3} \div \dfrac{q+5}{q-3}$ $\quad \dfrac{-q^2 + 12q + 5}{(q-3)(q+5)}$

36. $\dfrac{p}{p+5} + \dfrac{p}{p-4} \div \dfrac{p+2}{p^2 - p - 12}$ $\quad \dfrac{p(p^2 + 9p + 17)}{(p+5)(p+2)}$

37. $\dfrac{1}{x^2 + 7x + 12} + \dfrac{1}{x^2 - 9} + \dfrac{1}{x^2 - 16}$ $\quad \dfrac{3x^2 - 7x - 13}{(x+3)(x+4)(x-3)(x-4)}$

38. $\dfrac{2}{a^2 - 3a + 2} + \dfrac{3}{a^2 - 1} - \dfrac{5}{a^2 + 3a - 10}$ $\quad \dfrac{3(7a-5)}{(a-2)(a-1)(a+1)(a+5)}$

39. $\left(1 + \dfrac{2}{x}\right)\left(3 - \dfrac{1}{x}\right)$ $\quad \dfrac{(x+2)(3x-1)}{x^2}$

40. $\left(4 - \dfrac{1}{z}\right)\left(4 + \dfrac{2}{z}\right)$ $\quad \dfrac{2(2z+1)(4z-1)}{z^2}$

In Exercises 41 to 58, simplify each complex fraction.

41. $\dfrac{4 + \dfrac{1}{x}}{1 - \dfrac{1}{x}}$ $\quad \dfrac{4x+1}{x-1}$

42. $\dfrac{4 + \dfrac{1}{a}}{2 - \dfrac{3}{a}}$ $\quad \dfrac{4a+1}{2a-3}$

43. $\dfrac{\dfrac{x}{y}-2}{y-x}$ $\quad \dfrac{x-2y}{y(y-x)}$

44. $\dfrac{3+\dfrac{2}{x-3}}{4+\dfrac{1}{2+\dfrac{1}{x}}}$ $\quad \dfrac{(3x-7)(2x+1)}{(x-3)(9x+4)}$

45. $\dfrac{5-\dfrac{1}{x+2}}{1+\dfrac{3}{1+\dfrac{3}{x}}}$ $\quad \dfrac{(5x+9)(x+3)}{(x+2)(4x+3)}$

46. $\dfrac{\dfrac{1}{(x+h)^2}-1}{h}$ $\quad -\dfrac{(x+h+1)(x+h-1)}{h(x+h)^2}$

47. $\dfrac{1+\dfrac{1}{b-2}}{1-\dfrac{1}{b+3}}$ $\quad \dfrac{(b+3)(b-1)}{(b-2)(b+2)}$

48. $r-\dfrac{r}{r+\dfrac{1}{3}}$ $\quad \dfrac{r(3r-2)}{3r+1}$

49. $\dfrac{1-\dfrac{1}{x^2}}{1+\dfrac{1}{x}}$ $\quad \dfrac{x-1}{x}$

50. $\dfrac{1}{\dfrac{1}{a}+\dfrac{1}{b}}$ $\quad \dfrac{ab}{a+b}$

51. $2-\dfrac{m}{1-\dfrac{1-m}{-m}}$ $\quad 2-m^2$

52. $\dfrac{\dfrac{x+h+1}{x+h}-\dfrac{x}{x+1}}{h}$ $\quad \dfrac{2x+h+1}{h(x+h)(x+1)}$

53. $\dfrac{\dfrac{1}{x}-\dfrac{x-4}{x+1}}{\dfrac{x}{x+1}}$ $\quad \dfrac{-x^2+5x+1}{x^2}$

54. $\dfrac{\dfrac{2}{y}-\dfrac{3y-2}{y-1}}{\dfrac{y}{y-1}}$ $\quad -\dfrac{3y^2-4y+2}{y^2}$

55. $\dfrac{\dfrac{1}{x+3}-\dfrac{2}{x-1}}{\dfrac{x}{x-1}+\dfrac{3}{x+3}}$ $\quad \dfrac{-x-7}{x^2+6x-3}$

56. $\dfrac{\dfrac{x+2}{x^2-1}+\dfrac{1}{x+1}}{\dfrac{x}{2x^2-x-1}+\dfrac{1}{x-1}}$ $\quad \dfrac{(2x+1)^2}{(x+1)(3x+1)}$

57. $\dfrac{\dfrac{x^2+3x-10}{x^2+x-6}}{\dfrac{x^2-x-30}{2x^2-15x+18}}$ $\quad \dfrac{2x-3}{x+3}$

58. $\dfrac{\dfrac{2y^2+11y+15}{y^2-4y-21}}{\dfrac{6y^2+11y-10}{3y^2-23y+14}}$ $\quad 1$

In Exercises 59 to 62, simplify each algebraic fraction. Write all answers with positive exponents.

59. $\dfrac{a^{-1}+b^{-1}}{a-b}$ $\quad \dfrac{a+b}{ab(a-b)}$

60. $(1+a^{-1})^{-1}$ $\quad \dfrac{a}{a+1}$

61. $\dfrac{a^{-1}b-ab^{-1}}{a^2+b^2}$ $\quad \dfrac{(b-a)(b+a)}{ab(a^2+b^2)}$

62. $(a+b^{-2})^{-1}$ $\quad \dfrac{b^2}{ab^2+1}$

Business and Economics

63. ***Annuity*** Equal payments made at equal time intervals (such as a monthly car payment) are called *annuities*. The present value of an ordinary annuity is given by

$$R\left[\frac{1-\dfrac{1}{(1+i)^n}}{i}\right]$$

where n is the number of payments of R dollars, each incurring an interest rate of i% (as a decimal). Simplify the complex fraction. $R\left[\dfrac{(1+i)^n-1}{i(1+i)^n}\right]$

64. ***Average Cost*** A manufacturer of flat-screen computer monitors has determined that the change in the average cost per monitor to produce one additional monitor is given by

$$\frac{\left(250+\dfrac{20}{n}\right)-\left(250+\dfrac{20}{n+1}\right)}{n}$$

where n is the number of monitors already produced. Simplify the complex fraction. $\dfrac{20}{n^2(n+1)}$

Physical Sciences and Engineering

65. ***Average Speed*** According to Example 6, the average speed for a round trip for which the average speed on the way to the destination is v_1 and the average speed on the return trip is v_2 is given by the complex fraction.

$$\frac{2}{\dfrac{1}{v_1}+\dfrac{1}{v_2}}$$

a. Find the average speed, to the nearest hundredth, for a round trip by helicopter with $v_1 = 180$ miles per hour and $v_2 = 110$ miles per hour. 136.55 mph

b. Simplify the complex fraction. $\dfrac{2v_1v_2}{v_1+v_2}$

66. ***Relativity Theory*** According to Einstein's Theory of Relativity, the speed at which two objects are moving toward each other is given by

$$\frac{v_1+v_2}{1+\dfrac{v_1v_2}{c^2}}$$

where v_1 and v_2 are the speeds of the two objects and c is the speed of light. Evaluate this expression with $v_1 = 1.5 \times 10^8$ meters per second, $v_2 = 1.5 \times 10^8$ meters per second, and $c = 3.0 \times 10^8$ meters per second.
2.4×10^8 m/s

Prepare for Section P.6

In Exercises 67 to 69, simplify the expression.

67. $(4x^3y)(-2x^2y^5)$ [P.2] $-8x^5y^6$

68. $(2x^3)^4$ [P.2] $16x^{12}$

69. $(2x - 3y)(2x + 3y)$ [P.3] $4x^2 - 9y^2$

70. $(a + 2c)(?) = a^2 - 4c^2$ [P.4] $a - 2c$

71. If $x^n \cdot x^n = x^6$, find n. [P.2] 3

72. $x^3 + 8 = (x + 2)(?)$ [P.4] $x^2 - 2x + 4$

Explorations 

1. ***Continued Fractions*** The complex fraction shown at the right is called a continued fraction. The three dots in $\dfrac{1}{1 + \cdots}$ indicate that the pattern continues in the same manner. A convergent of a complex fraction is an approximation of the continued fraction that is found by stopping the process at some point.

$$\cfrac{1}{1 + \cfrac{1}{1 + \cfrac{1}{1 + \cfrac{1}{1 + \cdots}}}}$$

a. Calculate the convergent $C_2 = \cfrac{1}{1 + \cfrac{1}{1 + 1}}$. 0.6667

b. Calculate the convergent $C_3 = \cfrac{1}{1 + \cfrac{1}{1 + \cfrac{1}{1 + 1}}}$. 0.6

c. Calculate the convergent 0.6154

$$C_5 = \cfrac{1}{1 + \cfrac{1}{1 + \cfrac{1}{1 + \cfrac{1}{1 + \cfrac{1}{1 + 1}}}}}$$

d. Show that $C_5 \approx \dfrac{-1 + \sqrt{5}}{2}$. Using some techniques from more advanced math courses, it can be shown that the convergents of the continued fraction become closer and closer to $\dfrac{-1 + \sqrt{5}}{2}$.

2. ***Representation of π*** There are a few continued-fraction representations for π. Find two of these representations. Compute the value of π accurate to four decimal places using a convergent from each of the continued fractions you found.

SECTION P.6 Rational Exponents and Radicals

- Rational Exponents and Radicals
- Simplify Radical Expressions

Rational Exponents and Radicals

To this point, the expression b^n has been defined for real numbers b and integers n. Now we wish to extend the definition of exponents to include rational numbers so that expressions such as $2^{1/2}$ will be meaningful. Not just any definition will do. We want a definition of rational exponents for which the properties of integer exponents are true. The following example shows the direction we can take to accomplish our goal.

If the rule for multiplying exponential expressions is to hold true for rational exponents, then for the rational numbers p and q, $b^p b^q = b^{p+q}$. For example,

$$9^{1/2} \cdot 9^{1/2} \quad \text{must equal} \quad 9^{1/2+1/2} = 9^1 = 9$$

Thus $9^{1/2}$ must be a square root of 9. That is, $9^{1/2} = 3$.

The example suggests that $b^{1/n}$ can be defined in terms of roots according to the following definition.

Definition of $b^{1/n}$

If n is an even positive integer and $b \geq 0$, then $b^{1/n}$ is the nonnegative real number such that $(b^{1/n})^n = b$.

If n is an odd positive integer, then $b^{1/n}$ is the real number such that $(b^{1/n})^n = b$.

As examples,

- $25^{1/2} = 5$ because $5^2 = 25$.
- $(-64)^{1/3} = -4$ because $(-4)^3 = -64$.
- $16^{1/2} = 4$ because $4^2 = 16$.
- $-16^{1/2} = -(16^{1/2}) = -4$.
- $(-16)^{1/2}$ is not a real number.
- $(-32)^{1/5} = -2$ because $(-2)^5 = -32$.

If n is an even positive integer and $b < 0$, then $b^{1/n}$ is a *complex number.* Complex numbers are discussed in the next section.

To define expressions such as $8^{2/3}$, we will extend our definition of exponents even further. Because we want the power property $(b^p)^q = b^{pq}$ to be true for rational exponents also, we must have $(b^{1/n})^m = b^{m/n}$. With this in mind, we make the following definition.

Definition of $b^{m/n}$

For all positive integers m and n such that m/n is in simplest form, and for all real numbers b for which $b^{1/n}$ is a real number,

$$b^{m/n} = (b^{1/n})^m = (b^m)^{1/n}$$

INTEGRATING TECHNOLOGY

Some graphing calculators do not evaluate $b^{m/n}$ when $b < 0$. Try entering (-8)^(2/3). The answer should be 4, but some calculators display an error message for this expression. You can still use your calculator to evaluate this expression, but you must use parentheses. You can enter ((-8)^(1/3))^2 to evaluate $(-8)^{2/3}$.

Because $b^{m/n}$ is defined as $(b^{1/n})^m$ and also as $(b^m)^{1/n}$, we can evaluate expressions such as $8^{4/3}$ in more than one way. For example, because $8^{1/3}$ is a real number, $8^{4/3}$ can be evaluated in either of the following ways:

$$8^{4/3} = (8^{1/3})^4 = 2^4 = 16$$
$$8^{4/3} = (8^4)^{1/3} = 4096^{1/3} = 16$$

Of the two methods, the $b^{m/n} = (b^{1/n})^m$ method is usually easier to apply, provided you can evaluate $b^{1/n}$.

The following exponent properties were stated earlier, but they are restated here to remind you that they have now been extended to apply to rational exponents.

Properties of Rational Exponents

If p, q, and r represent rational numbers and a and b are positive real numbers, then

Product $\quad b^p \cdot b^q = b^{p+q}$

Quotient $\quad \dfrac{b^p}{b^q} = b^{p-q}$

Power $\quad (b^p)^q = b^{pq} \qquad (a^p b^q)^r = a^{pr} b^{qr}$

$$\left(\frac{a^p}{b^q}\right)^r = \frac{a^{pr}}{b^{qr}} \qquad b^{-p} = \frac{1}{b^p}$$

Recall that an exponential expression is in simplest form when no powers of powers or negative exponents appear and each base occurs at most once.

Alternative to Example 1

Simplify: $\left(\dfrac{4x^{-2}y^4}{32xy^{-2}}\right)^{-1/3}$

■ $\dfrac{2x}{y^2}$

EXAMPLE 1 Simplify Exponential Expressions

Simplify: $\left(\dfrac{x^2y^3}{x^{-3}y^5}\right)^{1/2}$ (Assume $x > 0$, $y > 0$.)

Solution

$$\left(\frac{x^2y^3}{x^{-3}y^5}\right)^{1/2} = (x^{2-(-3)}y^{3-5})^{1/2} = (x^5y^{-2})^{1/2} = x^{5/2}y^{-1} = \frac{x^{5/2}}{y}$$

CHECK YOUR PROGRESS 1 Simplify: $\left(\dfrac{16x^4y^2}{2x^{-2}y^5}\right)^{-1/3}$ (Assume $x > 0$, $y > 0$)

Solution *See page S2.* $\dfrac{y}{2x^2}$

▪ Simplify Radical Expressions

Radicals, expressed by the notation $\sqrt[n]{b}$, are also used to denote roots. The number b is the radicand, and the positive integer n is the index of the radical.

Point of Interest

The formula for kinetic energy (energy of motion) that is used in Einstein's Theory of Relativity involves a radical.

$$\text{K.E.} = mc^2\left[\frac{1}{\sqrt{1 - \dfrac{v^2}{c^2}}} - 1\right]$$

where m is the mass of the object at rest, v is the speed of the object, and c is the speed of light.

Definition of $\sqrt[n]{b}$

If n is a positive integer and b is a real number such that $b^{1/n}$ is a real number, then $\sqrt[n]{b} = b^{1/n}$.

If the index n equals 2, then the radical $\sqrt[2]{b}$ is written as simply $\sqrt{b}$, and it is referred to as the principal square root of b or simply the square root of b.

The symbol $\sqrt{b}$ is reserved to represent the nonnegative square root of b. To represent the negative square root of b, write $-\sqrt{b}$. For example, $\sqrt{25} = 5$, whereas $-\sqrt{25} = -5$.

Definition of $\left(\sqrt[n]{b}\right)^m$

For all positive integers n, all integers m, and all real numbers b such that $\sqrt[n]{b}$ is a real number, $\left(\sqrt[n]{b}\right)^m = \sqrt[n]{b^m} = b^{m/n}$.

When $\sqrt[n]{b}$ is a real number, the equations

$$b^{m/n} = \sqrt[n]{b^m} \quad \text{and} \quad b^{m/n} = \left(\sqrt[n]{b}\right)^m$$

can be used to write exponential expressions such as $b^{m/n}$ in radical form. Use the denominator n as the index of the radical and the numerator m as the power of the radicand or as the power of the radical. For example,

$$(5xy)^{2/3} = \left(\sqrt[3]{5xy}\right)^2 = \sqrt[3]{25x^2y^2}$$

▪ Use the denominator 3 as the index of the radical and the numerator 2 as the power of the radical.

The equations

$$b^{m/n} = \sqrt[n]{b^m} \quad \text{and} \quad b^{m/n} = \left(\sqrt[n]{b}\right)^m$$

can also be used to write radical expressions in exponential form. For example,

$$\sqrt{(2ab)^3} = (2ab)^{3/2}$$

▪ Use the index 2 as the denominator of the power and the exponent 3 as the numerator of the power.

The definition of $\sqrt[n]{b^m}$ can often be used to evaluate radical expressions. For instance,

$$\sqrt[3]{8^4} = 8^{4/3} = \left(8^{1/3}\right)^4 = 2^4 = 16$$

Care must be exercised when simplifying even roots (square roots, fourth roots, sixth roots, . . .) of variable expressions. Consider $\sqrt{x^2}$ when $x = 5$ and when $x = -5$.

Case 1 If $x = 5$, then $\sqrt{x^2} = \sqrt{5^2} = \sqrt{25} = 5 = x$.

Case 2 If $x = -5$, then $\sqrt{x^2} = \sqrt{(-5)^2} = \sqrt{25} = 5 = -x$.

These two cases suggest that

$$\sqrt{x^2} = \begin{cases} x, & \text{if } x \geq 0 \\ -x, & \text{if } x < 0 \end{cases}$$

Recalling the definition of absolute value, we can write this more compactly as $\sqrt{x^2} = |x|$.

Simplifying odd roots of a variable expression does not require using the absolute value symbol. Consider $\sqrt[3]{x^3}$ when $x = 5$ and when $x = -5$.

Case 1 If $x = 5$, then $\sqrt[3]{x^3} = \sqrt[3]{5^3} = \sqrt[3]{125} = 5 = x$.

Case 2 If $x = -5$, then $\sqrt[3]{x^3} = \sqrt[3]{(-5)^3} = \sqrt[3]{-125} = -5 = x$.

Thus $\sqrt[3]{x^3} = x$.

Although we have illustrated this principle only for square roots and cube roots, the same reasoning can be applied to other cases. The general result is given below.

Definition of $\sqrt[n]{b^n}$

If n is an even natural number and b is a real number, then

$$\sqrt[n]{b^n} = |b|$$

If n is an odd natural number and b is a real number, then

$$\sqrt[n]{b^n} = b$$

Here are some examples of these properties.

$$\sqrt[4]{16z^4} = 2|z| \qquad \sqrt[5]{32a^5} = 2a$$

❓ **QUESTION** What is the correct simplification of

a. $(x^6)^{1/6}$ **b.** $(x^5)^{1/5}$

Because radicals are defined in terms of rational powers, the properties of radicals are similar to those of exponential expressions.

P

Properties of Radicals

If m and n are natural numbers and a and b are nonnegative real numbers, then

Product $\sqrt[n]{a} \cdot \sqrt[n]{b} = \sqrt[n]{ab}$

Quotient $\dfrac{\sqrt[n]{a}}{\sqrt[n]{b}} = \sqrt[n]{\dfrac{a}{b}}$

Index $\sqrt[m]{\sqrt[n]{a}} = \sqrt[mn]{a}$

A radical is in simplest form if it meets all of the following criteria.

1. The radicand contains only powers less than the index. ($\sqrt{x^5}$ does not satisfy this requirement because 5, the exponent, is greater than 2, the index.)
2. The index of the radical is as small as possible. ($\sqrt[9]{x^3}$ does not satisfy this requirement because $\sqrt[9]{x^3} = x^{3/9} = x^{1/3} = \sqrt[3]{x}$.)
3. The denominator has been rationalized. That is, no radicals appear in the denominator. ($1/\sqrt{2}$ does not satisfy this requirement.)
4. No fractions appear under the radical sign. ($\sqrt[4]{2/x^3}$ does not satisfy this requirement.)

❓ **ANSWER** **a.** $|x|$ **b.** x

Radical expressions are simplified by using the properties of radicals. Here are some examples.

Alternative to Example 2

Simplify:

a. $\sqrt{27x^3y^5}$

b. $\sqrt[5]{32x^6y^{12}z^{10}}$

■ a. $3|x|y^2\sqrt{3xy}$

■ b. $2xy^2z^2\sqrt[5]{xy^2}$

EXAMPLE 2 Simplify Radical Expressions

Simplify.

a. $\sqrt[4]{32x^3y^4}$ b. $\sqrt[3]{162x^4y^6}$

Solution

a. $\sqrt[4]{32x^3y^4} = \sqrt[4]{2^5x^3y^4} = \sqrt[4]{(2^4y^4)\cdot(2x^3)}$ ■ Factor and group factors that can be written as a power of the index.

$= \sqrt[4]{2^4y^4}\cdot\sqrt[4]{2x^3}$ ■ Use the product property of radicals.

$= 2|y|\sqrt[4]{2x^3}$ ■ Recall that for n even, $\sqrt[n]{b^n} = |b|$.

b. $\sqrt[3]{162x^4y^6} = \sqrt[3]{(2\cdot 3^4)x^4y^6}$ ■ Factor and group factors that can be written as a power of the index.

$= \sqrt[3]{(3xy^2)^3\cdot(2\cdot 3x)}$

$= \sqrt[3]{(3xy^2)^3}\cdot\sqrt[3]{6x}$ ■ Use the product property of radicals.

$= 3xy^2\sqrt[3]{6x}$ ■ Recall that for n odd, $\sqrt[n]{b^n} = b$.

CHECK YOUR PROGRESS 2 Simplify: $\sqrt[3]{-81x^6y^4}$

Solution *See page S2.* $-3x^2y\sqrt[3]{3y}$

Like radicals have the same radicand and the same index. For instance,

$$3\sqrt[3]{5xy^2} \quad \text{and} \quad -4\sqrt[3]{5xy^2}$$

are like radicals. Addition and subtraction of like radicals are accomplished by using the distributive property. For example,

$$4\sqrt{3x} - 9\sqrt{3x} = (4-9)\sqrt{3x} = -5\sqrt{3x}$$
$$2\sqrt[3]{y^2} + 4\sqrt[3]{y^2} - \sqrt[3]{y^2} = (2+4-1)\sqrt[3]{y^2} = 5\sqrt[3]{y^2}$$

TAKE NOTE

Using rational exponents, the expression $2\sqrt{3} + 6\sqrt{5}$ can be written

$$2(3)^{1/2} + 6(5)^{1/2}$$

The bases of the exponential expressions are different and therefore the terms are not like terms. Thus the expression cannot be simplified.

The sum $2\sqrt{3} + 6\sqrt{5}$ cannot be simplified further because the radicands are not the same. The sum $3\sqrt[3]{x} + 5\sqrt[4]{x}$ cannot be simplified because the indices are not the same.

Sometimes it is possible to simplify a radical expression that does not appear to contain like radicals by first simplifying each term.

Alternative to Example 3
Simplify:
$6a^2\sqrt{12ab^5} - 3b^2\sqrt{27a^5b}$
■ $3a^2b^2\sqrt{3ab}$

EXAMPLE 3 Combine Radical Expressions

Simplify: $5x\sqrt[3]{16x^4} - \sqrt[3]{128x^7}$

Solution

$$\begin{aligned} 5x\sqrt[3]{16x^4} - \sqrt[3]{128x^7} &= 5x\sqrt[3]{2^4x^4} - \sqrt[3]{2^7x^7} && \text{▪ Factor.} \\ &= 5x\sqrt[3]{2^3x^3} \cdot \sqrt[3]{2x} - \sqrt[3]{2^6x^6} \cdot \sqrt[3]{2x} && \text{▪ Group factors that can be written as a power of the index.} \\ &= 5x\left(2x\sqrt[3]{2x}\right) - 2^2x^2 \cdot \sqrt[3]{2x} && \text{▪ Use the product property of radicals.} \\ &= 10x^2\sqrt[3]{2x} - 4x^2\sqrt[3]{2x} && \text{▪ Simplify.} \\ &= 6x^2\sqrt[3]{2x} \end{aligned}$$

CHECK YOUR PROGRESS 3 Simplify: $\sqrt[3]{54xy^3} + \sqrt[3]{128xy^3}$

Solution *See page S3.* $7y\sqrt[3]{2x}$

Multiplication of radical expressions is accomplished by using the distributive property. For instance,

$$\begin{aligned} \sqrt{5}\left(\sqrt{20} - 3\sqrt{15}\right) &= \sqrt{5}\left(\sqrt{20}\right) - \sqrt{5}\left(3\sqrt{15}\right) && \text{▪ Use the distributive property.} \\ &= \sqrt{100} - 3\sqrt{75} && \text{▪ Multiply the radicals.} \\ &= 10 - 3 \cdot 5\sqrt{3} && \text{▪ Simplify.} \\ &= 10 - 15\sqrt{3} \end{aligned}$$

Finding the product of more complicated radical expressions may require repeated use of the distributive property.

Alternative to Example 4
Expand: $\left(2\sqrt{5} - 3\right)^2$
■ $29 - 12\sqrt{5}$

EXAMPLE 4 Multiply Radical Expressions

Multiply: $\left(\sqrt{3} + 5\right)\left(\sqrt{3} - 2\right)$

Solution

$$\begin{aligned} \left(\sqrt{3} + 5\right)\left(\sqrt{3} - 2\right) &= \left(\sqrt{3} + 5\right)\sqrt{3} - \left(\sqrt{3} + 5\right)2 && \text{▪ Use the distributive property.} \\ &= \left(\sqrt{3}\sqrt{3} + 5\sqrt{3}\right) - \left(2\sqrt{3} + 2 \cdot 5\right) && \text{▪ Use the distributive property.} \\ &= 3 + 5\sqrt{3} - 2\sqrt{3} - 10 \\ &= -7 + 3\sqrt{3} \end{aligned}$$

CHECK YOUR PROGRESS 4 Multiply: $\left(3\sqrt{7} - 5\right)\left(2\sqrt{7} + 3\right)$

Solution *See page S3.* $27 - \sqrt{7}$

To rationalize the denominator of a fraction means to write the fraction in an equivalent form that does not involve any radicals in its denominator.

Alternative to Example 5

a. $\dfrac{2x}{\sqrt[3]{x^2}}$ b. $\dfrac{5}{\sqrt{2x}}$

■ a. $2\sqrt[3]{x}$ ■ b. $\dfrac{5\sqrt{2x}}{2|x|}$

EXAMPLE 5 Rationalize the Denominator

Rationalize the denominator. **a.** $\dfrac{5}{\sqrt[3]{a}}$ **b.** $\sqrt{\dfrac{3}{32y}}$

Solution

a. $\dfrac{5}{\sqrt[3]{a}} = \dfrac{5}{\sqrt[3]{a}} \cdot \dfrac{\sqrt[3]{a^2}}{\sqrt[3]{a^2}} = \dfrac{5\sqrt[3]{a^2}}{\sqrt[3]{a^3}} = \dfrac{5\sqrt[3]{a^2}}{a}$ ■ Use $\sqrt[3]{a} \cdot \sqrt[3]{a^2} = \sqrt[3]{a^3} = a$.

b. $\sqrt{\dfrac{3}{32y}} = \dfrac{\sqrt{3}}{\sqrt{32y}} = \dfrac{\sqrt{3}}{4\sqrt{2y}} = \dfrac{\sqrt{3}}{4\sqrt{2y}} \cdot \dfrac{\sqrt{2y}}{\sqrt{2y}} = \dfrac{\sqrt{6y}}{8|y|}$

TAKE NOTE

For the solution of Example 5b,
$4\sqrt{2y} \cdot \sqrt{2y} = 4\sqrt{4y^2}$
$= 4 \cdot 2|y| = 8|y|$

CHECK YOUR PROGRESS 5 Rationalize the denominator for each of the following.

a. $\dfrac{3}{\sqrt[3]{9}}$ **b.** $\sqrt{\dfrac{5}{18x}}$

Solution *See page S3.* a. $\sqrt[3]{3}$ b. $\dfrac{\sqrt{10x}}{6|x|}$

To rationalize the denominator of a fractional expression such as

$$\frac{1}{\sqrt{m} + \sqrt{n}}$$

we make use of the **conjugate** of $\sqrt{m} + \sqrt{n}$, which is $\sqrt{m} - \sqrt{n}$. The product of these conjugate pairs does not involve a radical.

$$\left(\sqrt{m} + \sqrt{n}\right)\left(\sqrt{m} - \sqrt{n}\right) = m - n$$

In Example 6 we use the conjugate of the denominator to rationalize the denominator.

Alternative to Example 6

a. $\dfrac{4}{3 - \sqrt{5}}$ b. $\dfrac{x - 2\sqrt{7}}{x + 3\sqrt{7}}$

■ a. $3 + \sqrt{5}$

■ b. $\dfrac{x^2 - 5x\sqrt{7} + 42}{x^2 - 63}$

EXAMPLE 6 Rationalize the Denominator

Rationalize the denominator.

a. $\dfrac{2}{\sqrt{3} + \sqrt{2}}$ **b.** $\dfrac{a + \sqrt{5}}{a - \sqrt{5}}$

Solution

a. $\dfrac{2}{\sqrt{3} + \sqrt{2}} = \dfrac{2}{\sqrt{3} + \sqrt{2}} \cdot \dfrac{\sqrt{3} - \sqrt{2}}{\sqrt{3} - \sqrt{2}} = \dfrac{2\sqrt{3} - 2\sqrt{2}}{3 - 2} = 2\sqrt{3} - 2\sqrt{2}$

b. $\dfrac{a + \sqrt{5}}{a - \sqrt{5}} = \dfrac{a + \sqrt{5}}{a - \sqrt{5}} \cdot \dfrac{a + \sqrt{5}}{a + \sqrt{5}} = \dfrac{a^2 + 2a\sqrt{5} + 5}{a^2 - 5}$

CHECK YOUR PROGRESS 6 Rationalize the denominator: $\dfrac{5}{2\sqrt{3} - 3}$

Solution *See page S3.* $\dfrac{10\sqrt{3} + 15}{3}$

Topics for Discussion

1. Given that a is a real number, discuss when the expression $a^{p/q}$ represents a real number.
2. The expressions $-a^n$ and $(-a)^n$ do not always represent the same number. Discuss the situations in which the two expressions are equal and those in which they are not equal.
3. If you enter the expression for $\sqrt{5}$ on your calculator, the calculator will respond with 2.236067977 or some number close to this number. Is this the exact value of $\sqrt{5}$? Is it possible to find the exact decimal value of $\sqrt{5}$ with a calculator? with a computer?

EXERCISES P.6

— Suggested Assignment: Exercises 1–85, every other odd; and 87–98, all.

In Exercises 1 to 16, evaluate each expression.

1. $4^{3/2}$ 8
2. $16^{3/2}$ 64
3. $-9^{1/2}$ -3
4. $-25^{3/2}$ -125
5. $-64^{2/3}$ -16
6. $125^{4/3}$ 625
7. $(-64)^{2/3}$ 16
8. $(-27)^{4/3}$ 81
9. $9^{-3/2}$ $\frac{1}{27}$
10. $32^{-3/5}$ $\frac{1}{8}$
11. $625^{-1/4}$ $\frac{1}{5}$
12. $-64^{-4/3}$ $-\frac{1}{256}$
13. $\left(\frac{4}{9}\right)^{1/2}$ $\frac{2}{3}$
14. $\left(\frac{16}{25}\right)^{3/2}$ $\frac{64}{125}$
15. $\left(\frac{1}{8}\right)^{-4/3}$ 16
16. $\left(\frac{8}{27}\right)^{-2/3}$ $\frac{9}{4}$

In Exercises 17 to 28, simplify each exponential expression.

17. $(4a^{2/3}b^{1/2})(2a^{1/3}b^{3/2})$ $8ab^2$
18. $(6a^{3/5}b^{1/4})(-3a^{1/5}b^{3/4})$ $-18a^{4/5}b$
19. $(-3x^{2/3})(4x^{1/4})$ $-12x^{11/12}$
20. $(-5x^{1/3})(-4x^{1/2})$ $20x^{5/6}$
21. $(81x^4y^{12})^{1/4}$ $3xy^3$
22. $(27x^3y^6)^{2/3}$ $9x^2y^4$
23. $\dfrac{16z^{3/4}}{12z^{1/4}}$ $\dfrac{4z^{1/2}}{3}$
24. $\dfrac{6a^{2/5}}{9a^{3/5}}$ $\dfrac{2}{3a^{1/5}}$
25. $\dfrac{9a^{3/4}b}{3a^{2/3}b^2}$ $\dfrac{3a^{1/12}}{b}$
26. $\dfrac{12x^{1/6}y^{1/4}}{16x^{3/4}y^{1/2}}$ $\dfrac{3}{4x^{7/12}y^{1/4}}$
27. $\left(\dfrac{m^2}{n^4}\right)^{-1/2}$ $\dfrac{n^2}{|m|}$
28. $\left(\dfrac{r^3s^{-2}}{rs^4}\right)^{1/4}$ $\dfrac{r^{1/2}}{s^{3/2}}$

In Exercises 29 to 34, write each exponential expression in radical form.

29. $(3x)^{1/2}$ $\sqrt{3x}$
30. $3x^{1/2}$ $3\sqrt{x}$
31. $(5x)^{2/3}$ $\sqrt[3]{25x^2}$
32. $(3a)^{3/4}$ $\sqrt[4]{27a^3}$
33. $(a^2 + b^2)^{1/2}$ $\sqrt{a^2 + b^2}$
34. $(x^3 + y^3)^{1/3}$ $\sqrt[3]{x^3 + y^3}$

In Exercises 35 to 40, write each radical expression in exponential form.

35. $\sqrt[3]{4z}$ $(4z)^{1/3}$
36. $\sqrt[4]{5x}$ $(5x)^{1/4}$
37. $\sqrt[5]{a^3}$ $a^{3/5}$
38. $\sqrt[4]{z^5}$ $z^{5/4}$
39. $4\sqrt{x^3}$ $4x^{2/3}$
40. $-5\sqrt[3]{z^2}$ $-5z^{2/3}$

In Exercises 41 to 50, simplify each radical expression.

41. $\sqrt{45}$ $3\sqrt{5}$
42. $\sqrt{75}$ $5\sqrt{3}$
43. $\sqrt[3]{24}$ $2\sqrt[3]{3}$
44. $\sqrt[3]{135}$ $3\sqrt[3]{5}$
45. $\sqrt[3]{-135}$ $-3\sqrt[3]{5}$
46. $\sqrt[3]{-250}$ $-5\sqrt[3]{2}$

47. $\sqrt{24x^2y^3}$ $2|xy|\sqrt{6y}$

48. $\sqrt{18x^5y^4}$ $3x^2y^2\sqrt{2x}$

49. $\sqrt[3]{16a^3y^7}$ $2ay^2\sqrt[3]{2y}$

50. $\sqrt[3]{54m^2n^7}$ $3n^2\sqrt[3]{2m^2n}$

In Exercises 51 to 58, simplify each radical and then combine like radicals.

51. $2\sqrt{32} - 3\sqrt{98}$ $-13\sqrt{2}$

52. $5\sqrt[3]{32} + 2\sqrt[3]{108}$ $16\sqrt[3]{4}$

53. $-8\sqrt[4]{48} + 2\sqrt[4]{243}$ $-10\sqrt[4]{3}$

54. $2\sqrt[3]{40} - 3\sqrt[3]{135}$ $-5\sqrt[3]{5}$

55. $4\sqrt[3]{32y^4} + 3y\sqrt[3]{108y}$ $17y\sqrt[3]{4y}$

56. $-3x\sqrt[3]{54x^4} + 2\sqrt[3]{16x^7}$ $-5x^2\sqrt[3]{2x}$

57. $x\sqrt[3]{8x^3y^4} - 4y\sqrt[3]{64x^6y}$ $-14x^2y\sqrt[3]{y}$

58. $4\sqrt{a^5b} - a^2\sqrt{ab}$ $3a^2\sqrt{ab}$

In Exercises 59 to 68, find the indicated products and express each result in simplest form.

59. $(\sqrt{5} + 3)(\sqrt{5} + 4)$ $17 + 7\sqrt{5}$

60. $(\sqrt{7} + 2)(\sqrt{7} - 5)$ $-3\sqrt{7} - 3$

61. $(\sqrt{2} - 3)(\sqrt{2} + 3)$ -7

62. $(2\sqrt{7} + 3)(2\sqrt{7} - 3)$ 19

63. $(3\sqrt{z} - 2)(4\sqrt{z} + 3)$ $12z + \sqrt{z} - 6$

64. $(4\sqrt{a} - \sqrt{b})(3\sqrt{a} + 2\sqrt{b})$ $12a + 5\sqrt{ab} - 2b$

65. $(\sqrt{x} + 2)^2$ $x + 4\sqrt{x} + 4$

66. $(\sqrt{x} - 3)^2$ $x - 6\sqrt{x} + 9$

67. $(\sqrt{x - 3} + 2)^2$ $x + 1 + 4\sqrt{x - 3}$

68. $(\sqrt{2x + 1} - 3)^2$ $2x + 10 - 6\sqrt{2x + 1}$

In Exercises 69 to 80, simplify each expression by rationalizing the denominator. Write the result in simplest form.

69. $\dfrac{2}{\sqrt{2}}$ $\sqrt{2}$

70. $\dfrac{3x}{\sqrt{3}}$ $x\sqrt{3}$

71. $\sqrt{\dfrac{5}{18}}$ $\dfrac{\sqrt{10}}{6}$

72. $\sqrt{\dfrac{7}{40}}$ $\dfrac{\sqrt{70}}{20}$

73. $\dfrac{3}{\sqrt[3]{2}}$ $\dfrac{3\sqrt[3]{4}}{2}$

74. $\dfrac{2}{\sqrt[3]{4}}$ $\sqrt[3]{2}$

75. $\dfrac{4}{\sqrt[3]{8x^2}}$ $\dfrac{2\sqrt[3]{x}}{x}$

76. $\dfrac{2}{\sqrt[4]{4y}}$ $\dfrac{\sqrt[4]{4y^3}}{y}$

77. $\dfrac{3}{\sqrt{3} + 4}$ $-\dfrac{3\sqrt{3} - 12}{13}$

78. $\dfrac{2}{\sqrt{5} - 2}$ $2\sqrt{5} + 4$

79. $\dfrac{6}{2\sqrt{5} + 2}$ $\dfrac{3\sqrt{5} - 3}{4}$

80. $\dfrac{-7}{3\sqrt{2} - 5}$ $3\sqrt{2} + 5$

In Exercises 81 and 82, rationalize the numerator.

81. $\dfrac{\sqrt{4 + h} - 2}{h}$ $\dfrac{1}{\sqrt{4 + h} + 2}$

82. $\dfrac{\sqrt{9 + h} - 3}{h}$ $\dfrac{1}{\sqrt{9 + h} + 3}$

83. Evaluate $\left(\sqrt{2}^{\sqrt{2}}\right)^{\sqrt{2}}$. 2

84. Does $\sqrt{a^2 + b^2} = a + b$ for all values of a and b? If not, find values of a and b for which the statement is not true. No. Not true for $a \neq 0$; $b \neq 0$.

85. For what values of x does $\sqrt{x^2} = -x$? $x \leq 0$

86. For what values of x does $\sqrt[3]{x^3} = -x$? 0

Business and Economics

87. ***Cellular Phone Production*** An electronics firm estimates that the revenue R it will receive from the sale of x cell phones (in thousands) can be approximated by $R = 450x(2^{-0.007x})$. What is the estimated revenue when the company sells 20,000 cell phones? \$8167.67

Life and Health Sciences

88. ***Drug Potency*** The amount A (in milligrams) of Digoxin, a drug taken by cardiac patients, remaining in the blood t hours after a patient takes a 2-milligram dose is given by $A = 2(10^{-0.0078t})$.

 a. How much Digoxin remains in the blood of a patient 4 hours after taking a 2-milligram dose? 1.86 mg

 b. Suppose a patient takes a 2-milligram dose of Digoxin at 1:00 P.M. and another 2-milligram dose at 5:00 P.M. How much Digoxin remains in the patient's blood at 6:00 P.M.? 3.79 mg

Social Sciences

89. ***World Population*** An estimate of the world's future population is given by $P = 5.9(2^{0.025n})$, when n is the number of years after 2000 and P is the world population in billions. Using this estimate, what will the world's population be in 2020? 8.34 billion

90. ***Learning Theory*** In a psychology experiment, students were given a nine-digit number to memorize. The percent P of students who remembered the number t minutes after it was read to them can be given by $P = 90 - 3t^{2/3}$. What percent of the students remembered the number after 1 hour? 44%

Physical Sciences and Engineering

91. *Oceanography* The percent P of light that will pass to a depth d, in meters, at a certain place in the ocean is given by $P = 10^{2-d/40}$. Find, to the nearest percent, the amount of light that will pass to a depth of **a.** 10 meters and **b.** 25 meters below the surface of the ocean.
a. 56% b. 24%

92. *Food Science* The number of hours h needed to cook a pot roast that weighs p pounds can be approximated by $h = 0.9p^{3/5}$.

a. Find the time, to the nearest tenth of an hour, required to cook a 12-pound pot roast. 4.0 h

b. If one pot roast is twice the weight of a second pot roast, is the cooking time for the heavier pot roast twice the cooking time for the lighter one? No

Prepare for Section P.7

In Exercises 93 to 98, simplify the expression.

93. $(2 - 3x) - (4 - 5x)$ [P.1] $2x - 2$

94. $(4 + 3x)(2 - 5x)$ [P.3] $-15x^2 - 14x + 8$

95. $\dfrac{x^2 - x - 6}{9 - x^2}$ [P.3] $-\dfrac{x+2}{x+3}$

96. $(2 - 5x)^2$ [P.3] $25x^2 - 20x + 4$

97. $\dfrac{2x}{x+1} - \dfrac{3}{x-1}$ [P.5] $\dfrac{2x^2 - 5x - 3}{x^2 - 1}$

98. Which of the following polynomials, if any, is not factorable over the integers? [P.4] b

a. $81 - x^2$ **b.** $9 + z^2$

Explorations 

1. *Relativity Theory* A moving object has energy, called kinetic energy, by virtue of its motion. As mentioned earlier in this chapter, the theory of relativity uses

$$K.E_r = mc^2\left[\frac{1}{\sqrt{1 - \dfrac{v^2}{c^2}}} - 1\right]$$

as the formula for kinetic energy. When the speed of an object is much less than the speed of light (3.0×10^8 meters per second), the formula

$$K.E_n = \frac{1}{2}mv^2$$

is used. In each formula, v is the velocity of the object in meters per second, m is its rest mass in kilograms, and c is the speed of light given above.

In **a.** through **e.**, calculate the percent error for each of the given velocities. The formula for percent error is

$$\%\text{ error} = \frac{|K.E_r - K.E_n|}{K.E_r} \times 100$$

a. $v = 30$ meters per second (speeding car on an expressway) ≈0%

b. $v = 240$ meters per second (speed of a commercial jet) ≈0%

c. $v = 3.0 \times 10^7$ meters per second (10% of the speed of light) 0.75%

d. $v = 1.5 \times 10^8$ meters per second (50% of the speed of light) 19.2%

e. $v = 2.7 \times 10^8$ meters per second (90% of the speed of light) 68.7%

f. Use your answers from **a.** through **e.** to give a reason why the formula for kinetic energy given by $K.E_n$ is adequate for most of our common experiences involving motion (walking, running, bicycle, car, plane).

g. According to relativity theory, the mass m of an object changes as its velocity according to

$$m = \frac{m_0}{\sqrt{1 - \dfrac{v^2}{c_2}}}$$

where m_0 is the rest mass of the object. The approximate rest mass of an electron is 9.11×10^{-31} kilogram. What is the percent change, from its rest mass, in the mass of an electron that is traveling at $0.99c$ (99% of the speed of light)? 608.9%

h. According to the theory of relativity, a particle (such as an electron or a spacecraft) cannot exceed the speed of light. Explain why the equation for $K.E_r$ suggests such a conclusion.
If $v = c$, then the denominator is 0 and division by 0 is undefined.

SECTION P.7 Complex Numbers

- Complex Numbers
- Addition and Subtraction of Complex Numbers
- Multiplication of Complex Numbers
- Division of Complex Numbers
- Powers of i

Point of Interest

It may seem strange to just invent new numbers, but that is how mathematics evolves. For instance, negative numbers were not an accepted part of mathematics until well into the thirteenth century. In fact, these numbers were often referred to as "fictitious numbers."

In the seventeenth century, René Descartes called the square roots of negative numbers *imaginary numbers,* an unfortunate choice of words, and started using the letter i to denote these numbers. These numbers were subjected to the same skepticism as negative numbers.

It is important to understand that these numbers are not *imaginary* in the dictionary sense of the word, just as negative numbers are not really "fictitious."

If you think of a number line, then the numbers to the right of zero are positive numbers and the numbers to the left of zero are negative numbers. One way to think of an imaginary number is *up* or *down* from zero.

Complex Numbers

Recall that $\sqrt{9} = 3$ because $3^2 = 9$. Now consider the expression $\sqrt{-9}$. To find $\sqrt{-9}$, we need to find a number c such that $c^2 = -9$. However, the square of any real number c (except zero) is a *positive* number. Consequently, we must expand our concept of numbers to include numbers whose squares are negative numbers.

Around the seventeenth century, a new number, called an *imaginary number,* was defined so that a negative number would have a square root. The letter i was chosen to represent the number whose square is -1.

Definition of i

The number i, called the **imaginary unit,** is the number such that $i^2 = -1$.

The principal square root of a negative number is defined in terms of i.

Principal Square Root of a Negative Number

If a is a positive real number, then $\sqrt{-a} = i\sqrt{a}$. The number $i\sqrt{a}$ is called an **imaginary number.**

Here are some examples of imaginary numbers.

$$\sqrt{-36} = i\sqrt{36} = 6i \qquad \sqrt{-18} = i\sqrt{18} = 3i\sqrt{2}$$
$$\sqrt{-23} = i\sqrt{23} \qquad \sqrt{-1} = i\sqrt{1} = i$$

It is customary to write i in front of the radical sign, as we did for $i\sqrt{23}$, to avoid confusing $\sqrt{a}\,i$ with $\sqrt{ai}$.

The real numbers and the imaginary numbers make up the *complex numbers.*

Complex Numbers

A **complex number** is a number of the form $a + bi$, where a and b are real numbers and $i = \sqrt{-1}$. The number a is the **real part** of $a + bi$, and b is the **imaginary part.**

Here are some examples of complex numbers.

$-3 + 5i$	Real part: -3; imaginary part: 5
$2 - 6i$	Real part: 2; imaginary part: -6
5	Real part: 5; imaginary part: 0
$7i$	Real part: 0; imaginary part: 7

Note from these examples that a real number is a complex number whose imaginary part is zero and that an imaginary number is a complex number whose real part is zero.

Point of Interest

The imaginary unit i is important in the field of electrical engineering. However, because the letter i is used by engineers as the symbol for electrical current, these engineers use j for the imaginary unit.

? QUESTION What is the real part and the imaginary part of $3 - 5i$?

Note from the following diagram that the real numbers are a subset of the complex numbers and the imaginary numbers are a subset of the complex numbers. The set of real numbers and the set of imaginary numbers do not intersect.

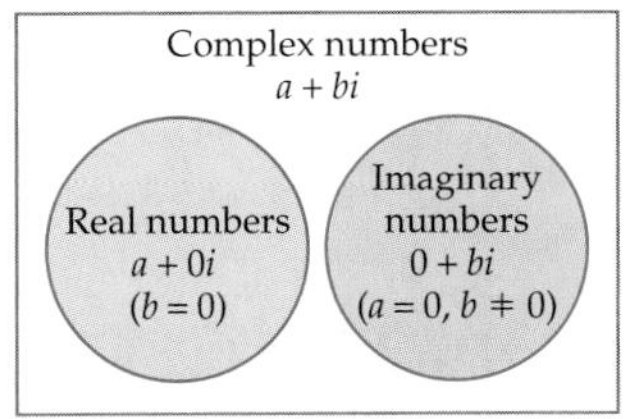

Example 1 illustrates writing a complex number in its standard form of $a + bi$.

Alternative to Example 1
Write $\sqrt{9} - \sqrt{-9}$ in standard form.
■ $3 - 3i$

EXAMPLE 1 Write a Complex Number in Standard Form

Write $7 + \sqrt{-45}$ in the form $a + bi$.

Solution

$$\begin{aligned} 7 + \sqrt{-45} &= 7 + i\sqrt{45} \\ &= 7 + i\sqrt{9} \cdot \sqrt{5} \\ &= 7 + 3i\sqrt{5} \end{aligned}$$

CHECK YOUR PROGRESS 1 Write $4 - \sqrt{-72}$ in the form $a + bi$.

Solution *See page S3.* $4 - 6i\sqrt{2}$

■ Addition and Subtraction of Complex Numbers

All of the standard arithmetic operations that are applied to real numbers can be applied to complex numbers.

Definition of Addition and Subtraction of Complex Numbers

If $a + bi$ and $c + di$ are complex numbers, then

Addition $(a + bi) + (c + di) = (a + c) + (b + d)i$

Subtraction $(a + bi) - (c + di) = (a - c) + (b - d)i$

? ANSWER Real part: 3; imaginary part: −5

Basically these rules state that to add two complex numbers, add the real parts and add the imaginary parts. To subtract two complex numbers, subtract the real parts and subtract the imaginary parts.

Alternative to Example 2

Simplify:

a. $3i - (4 + 2i)$
b. $5 - (6 - 3i)$
c. $(3 + 7i) - (6 + 7i)$

■ a. $-4 + i$ ■ b. $-1 + 3i$
■ c. -3

EXAMPLE 2 Add or Subtract Complex Numbers

Simplify.

a. $(7 - 2i) + (-2 + 4i)$ **b.** $(-8 + 5i) - (7 - i)$

Solution

a. $(7 - 2i) + (-2 + 4i) = [7 + (-2)] + (-2 + 4)i = 5 + 2i$

b. $(-8 + 5i) - (7 - i) = (-8 - 7) + [5 - (-1)]i = -15 + 6i$

CHECK YOUR PROGRESS 2 Simplify.

a. $(-3 + 5i) + (-7 - 5i)$ **b.** $(6 - 3i) - (6 - 4i)$

Solution *See page S3.* a. -10 b. i

■ Multiplication of Complex Numbers

When multiplying complex numbers, the term i^2 is frequently a part of the product. Recall that $i^2 = -1$. Therefore,

$$3i(5i) = 15i^2 = 15(-1) = -15$$
$$-2i(6i) = -12i^2 = -12(-1) = 12$$
$$4i(3 - 2i) = 12i - 8i^2 = 12i - 8(-1) = 8 + 12i$$

TAKE NOTE

Recall that the definition of the product of radical expressions requires that the radicand be a *positive* number. Therefore, when multiplying expressions containing *negative* radicands, we must first rewrite the expression using i and a positive radicand.

When multiplying square roots of negative numbers, first rewrite the radical expressions using i. For instance,

$$\begin{aligned}\sqrt{-6} \cdot \sqrt{-24} &= i\sqrt{6} \cdot i\sqrt{24} \\ &= i^2\sqrt{144} = -1 \cdot 12 \\ &= -12\end{aligned}$$

■ $\sqrt{-6} = i\sqrt{6}$, $\sqrt{-24} = i\sqrt{24}$

Note from this example that it would have been incorrect to multiply the radicands of the two radical expressions. To illustrate:

$$\sqrt{-6} \cdot \sqrt{-24} = \sqrt{(-6)(-24)} = \sqrt{144} = 12, \textit{ not } -12$$

? QUESTION What is the product of $\sqrt{-2}$ and $\sqrt{-8}$?

To multiply two complex numbers, we use the following definition.

? ANSWER $\sqrt{-2} \cdot \sqrt{-8} = i\sqrt{2} \cdot i\sqrt{8} = i^2\sqrt{16} = -1 \cdot 4 = -4$

Definition of Multiplication of Complex Numbers

If $a + bi$ and $c + di$ are complex numbers, then

$$(a + bi)(c + di) = (ac - bd) + (ad + bc)i$$

Because every complex number can be written as a sum of two terms, it is natural to perform multiplication on complex numbers in a manner similar to that used to multiply binomials. We will also make use of the definition $i^2 = -1$. By using this analogy, you can multiply complex numbers without memorizing the above definition. The procedure is similar for the FOIL method.

Alternative to Example 3

Simplify:

a. $(5 + 3i)(2 - 5i)$

b. $\left(2 - 3\sqrt{-5}\right)\left(7 + 4\sqrt{-5}\right)$

▪ a. $25 - 19i$

▪ b. $74 - 13i\sqrt{5}$

EXAMPLE 3 Multiply Complex Numbers

Simplify.

a. $(3 - 4i)(2 + 5i)$ b. $\left(2 + \sqrt{-3}\right)\left(4 - 5\sqrt{-3}\right)$

Solution

a.
$$\begin{aligned}(3 - 4i)(2 + 5i) &= 6 + 15i - 8i - 20i^2 \\ &= 6 + 15i - 8i - 20(-1) && \text{▪ Replace } i^2 \text{ by } -1. \\ &= 6 + 15i - 8i + 20 && \text{▪ Simplify.} \\ &= 26 + 7i\end{aligned}$$

b.
$$\begin{aligned}\left(2 + \sqrt{-3}\right)\left(4 - 5\sqrt{-3}\right) &= \left(2 + i\sqrt{3}\right)\left(4 - 5i\sqrt{3}\right) \\ &= 8 - 10i\sqrt{3} + 4i\sqrt{3} - 5i^2\sqrt{9} \\ &= 8 - 10i\sqrt{3} + 4i\sqrt{3} - 5(-1)(3) \\ &= 8 - 10i\sqrt{3} + 4i\sqrt{3} + 15 = 23 - 6i\sqrt{3}\end{aligned}$$

CHECK YOUR PROGRESS 3 Simplify.

a. $(4 + 3i)(3 + 5i)$ b. $\left(5 - \sqrt{-5}\right)\left(2 - 3\sqrt{-5}\right)$

Solution *See page S3.* a. $-3 + 29i$ b. $-5 - 17i\sqrt{5}$

INTEGRATING TECHNOLOGY

Some graphing calculators can be used to perform operations on complex numbers. Here are some typical screens for a TI-83 Plus.

Press MODE. Use the down arrow key to highlight $a + bi$.

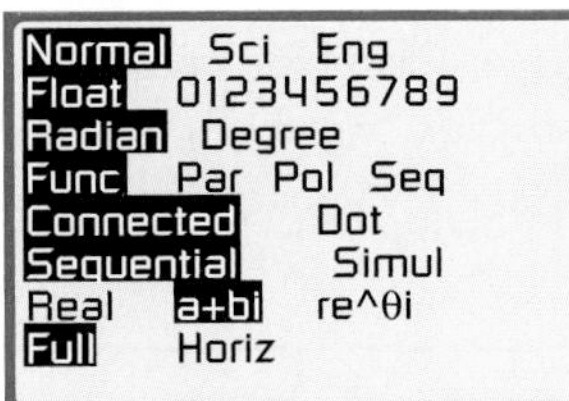

Press ENTER 2nd [QUIT].

Here are two examples of computations on complex numbers. To enter an i, use 2nd [i].

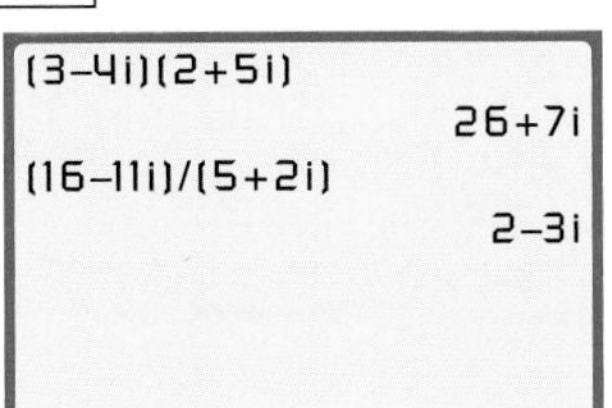

▪ Division of Complex Numbers

Recall that the number $\frac{3}{\sqrt{2}}$ is not in simplest form because there is a radical expression in the denominator. Similarly, $\frac{3}{i}$ is not in simplest form because $i = \sqrt{-1}$. To write this expression in simplest form, multiply the numerator and denominator by i.

$$\frac{3}{i} \cdot \frac{i}{i} = \frac{3i}{i^2} = \frac{3i}{-1} = -3i$$

Here is another example.

$$\frac{3-6i}{2i} = \frac{3-6i}{2i} \cdot \frac{i}{i} = \frac{3i-6i^2}{2i^2} = \frac{3i-6(-1)}{2(-1)}$$
$$= \frac{3i+6}{-2} = -3 - \frac{3}{2}i$$

Recall that to simplify the quotient $\frac{2+\sqrt{3}}{5+2\sqrt{3}}$, we multiply the numerator and denominator by the conjugate of $5 + 2\sqrt{3}$, which is $5 - 2\sqrt{3}$. In a similar manner, to find the quotient of two complex numbers, we multiply the numerator and denominator by the conjugate of the denominator.

The complex numbers $a + bi$ and $a - bi$ are called **complex conjugates** or conjugates of each other. The conjugate of the complex number z is denoted by $\overline{z}$. For instance,

$$\overline{2+5i} = 2-5i \quad \text{and} \quad \overline{3-4i} = 3+4i$$

Consider the product of a complex number and its conjugate. For instance,

$$(2+5i)(2-5i) = 4 - 10i + 10i - 25i^2$$
$$= 4 - 25(-1) = 4 + 25$$
$$= 29$$

Note that the product is a *real* number. This is always true.

Product of Complex Conjugates

The product of a complex number and its conjugate is a real number. That is, $(a + bi)(a - bi) = a^2 + b^2$.

INSTRUCTOR NOTE

Some students do not see that multiplying the numerator and denominator by the conjugate of the denominator is really division. It may help to remind these students that $\frac{8}{2} = 4$ and $2 \cdot 4 = 8$.

Similarly, $\frac{16-11i}{5+2i} = 2 - 3i$ and $(5 + 2i)(2 - 3i) = 16 - 11i$.

For instance, $(5 + 3i)(5 - 3i) = 5^2 + 3^2 = 25 + 9 = 34$.

The next example shows how the quotient of two complex numbers is determined by using conjugates.

Alternative to Example 4

Simplify: $\frac{10-5i}{2+i}$

■ $3 - 4i$

EXAMPLE 4 Divide Complex Numbers

Simplify: $\frac{16-11i}{5+2i}$

Solution

$$\frac{16-11i}{5+2i} = \frac{16-11i}{5+2i} \cdot \frac{5-2i}{5-2i}$$

■ Multiply the numerator and denominator by the conjugate of the denominator.

$$= \frac{80-32i-55i+22i^2}{5^2+2^2}$$
$$= \frac{80-32i-55i+22(-1)}{25+4}$$
$$= \frac{80-87i-22}{29}$$
$$= \frac{58-87i}{29} = 2 - 3i$$

Continued ➤

CHECK YOUR PROGRESS 4 Simplify: $\dfrac{3 + 2i}{5 - i}$

Solution *See page S3.* $\frac{1}{2} + \frac{1}{2}i$

Powers of i

The following powers of i illustrate a pattern:

$$\begin{aligned} i^1 &= i & i^5 &= i^4 \cdot i = 1 \cdot i = i \\ i^2 &= -1 & i^6 &= i^4 \cdot i^2 = 1(-1) = -1 \\ i^3 &= i^2 \cdot i = (-1)i = -i & i^7 &= i^4 \cdot i^3 = 1(-i) = -i \\ i^4 &= i^2 \cdot i^2 = (-1)(-1) = 1 & i^8 &= (i^4)^2 = 1^2 = 1 \end{aligned}$$

Because $i^4 = 1$, $(i^4)^n = 1^n = 1$ for any integer n. Thus it is possible to evaluate powers of i by factoring out powers of i^4, as shown in the following example.

$$i^{27} = (i^4)^6 \cdot i^3 = 1^6 \cdot i^3 = 1 \cdot (-i) = -i$$

The following theorem can be used to evaluate powers of i. Essentially, it makes use of division to eliminate powers of i^4.

Powers of i

If n is a positive integer, then $i^n = i^r$, where r is the remainder of the division of n by 4.

Alternative to Example 5
Evaluate: i^{146}
■ -1

EXAMPLE 5 Evaluate a Power of i

Evaluate: i^{153}

Solution Use the powers of i theorem.

$$i^{153} = i^1 = i$$ ■ Remainder of $153 \div 4$ is 1.

CHECK YOUR PROGRESS 5 Evaluate: i^{214}

Solution *See page S3.* -1

Topics for Discussion

1. What is an imaginary number? What is a complex number?
2. How are the real numbers related to the complex numbers?
3. Is zero a complex number?
4. What is the conjugate of a complex number?
5. If a and b are real numbers and $ab = 0$, then $a = 0$ or $b = 0$. Is the same true for complex numbers? That is, if u and v are complex numbers and $uv = 0$, is one of the numbers u or v equal to zero?

EXERCISES P.7

— *Suggested Assignment: Exercises 1–79, odd.*

In Exercises 1 to 10, write the complex number in standard form.

1. $\sqrt{-81}$ $9i$
2. $\sqrt{-64}$ $8i$
3. $\sqrt{-98}$ $7i\sqrt{2}$
4. $\sqrt{-27}$ $3i\sqrt{3}$
5. $\sqrt{16} + \sqrt{-81}$ $4 + 9i$
6. $\sqrt{25} + \sqrt{-9}$ $5 + 3i$
7. $5 + \sqrt{-49}$ $5 + 7i$
8. $6 - \sqrt{-1}$ $6 - i$
9. $8 - \sqrt{-18}$ $8 - 3i\sqrt{2}$
10. $11 + \sqrt{-48}$ $11 + 4i\sqrt{3}$

In Exercises 11 to 36, simplify and write the complex number in standard form.

11. $(5 + 2i) + (6 - 7i)$ $11 - 5i$
12. $(4 - 8i) + (5 + 3i)$ $9 - 5i$
13. $(-2 - 4i) - (5 - 8i)$ $-7 + 4i$
14. $(3 - 5i) - (8 - 2i)$ $-5 - 3i$
15. $(1 - 3i) + (7 - 2i)$ $8 - 5i$
16. $(2 - 6i) + (4 - 7i)$ $6 - 13i$
17. $(-3 - 5i) - (7 - 5i)$ -10
18. $(5 - 3i) - (2 + 9i)$ $3 - 12i$
19. $8i - (2 - 8i)$ $-2 + 16i$
20. $3 - (4 - 5i)$ $-1 + 5i$
21. $5i \cdot 8i$ -40
22. $(-3i)(2i)$ 6
23. $\sqrt{-50} \cdot \sqrt{-2}$ -10
24. $\sqrt{-12} \cdot \sqrt{-27}$ -18
25. $3(2 + 5i) - 2(3 - 2i)$ $19i$
26. $3i(2 + 5i) + 2i(3 - 4i)$ $-7 + 12i$
27. $(4 + 2i)(3 - 4i)$ $20 - 10i$
28. $(6 + 5i)(2 - 5i)$ $37 - 20i$
29. $(-3 - 4i)(2 + 7i)$ $22 - 29i$
30. $(-5 - i)(2 + 3i)$ $-7 - 17i$
31. $(4 - 5i)(4 + 5i)$ 41
32. $(3 + 7i)(3 - 7i)$ 58
33. $\left(3 + \sqrt{-4}\right)\left(2 - \sqrt{-9}\right)$ $12 - 5i$
34. $\left(5 + 2\sqrt{-16}\right)\left(1 - \sqrt{-25}\right)$ $45 - 17i$
35. $\left(3 + 2\sqrt{-18}\right)\left(2 + 2\sqrt{-50}\right)$ $-114 + 42i\sqrt{2}$
36. $\left(5 - 3\sqrt{-48}\right)\left(2 - 4\sqrt{-27}\right)$ $-422 - 84i\sqrt{3}$

In Exercises 37 to 54, write each expression as a complex number in standard form.

37. $\dfrac{6}{i}$ $-6i$
38. $\dfrac{-8}{2i}$ $4i$
39. $\dfrac{6 + 3i}{i}$ $3 - 6i$
40. $\dfrac{4 - 8i}{4i}$ $-2 - i$
41. $\dfrac{1}{7 + 2i}$ $\dfrac{7}{53} - \dfrac{2}{53}i$
42. $\dfrac{5}{3 + 4i}$ $\dfrac{3}{5} - \dfrac{4}{5}i$
43. $\dfrac{2i}{1 + i}$ $1 + i$
44. $\dfrac{5i}{2 - 3i}$ $-\dfrac{15}{13} + \dfrac{10}{13}i$
45. $\dfrac{5 - i}{4 + 5i}$ $\dfrac{15}{41} - \dfrac{29}{41}i$
46. $\dfrac{4 + i}{3 + 5i}$ $\dfrac{1}{2} - \dfrac{1}{2}i$
47. $\dfrac{3 + 2i}{3 - 2i}$ $\dfrac{5}{13} + \dfrac{12}{13}i$
48. $\dfrac{8 - i}{2 + 3i}$ $1 - 2i$
49. $\dfrac{-7 + 26i}{4 + 3i}$ $2 + 5i$
50. $\dfrac{-4 - 39i}{5 - 2i}$ $2 - 7i$
51. $(3 - 5i)^2$ $-16 - 30i$
52. $(2 + 4i)^2$ $-12 + 16i$
53. $(1 + 2i)^3$ $-11 - 2i$
54. $(2 - i)^3$ $2 - 11i$

In Exercises 55 to 62, evaluate the power of i.

55. i^{15} $-i$
56. i^{66} -1
57. $-i^{40}$ -1
58. $-i^{51}$ i
59. $\dfrac{1}{i^{25}}$ $-i$
60. $\dfrac{1}{i^{83}}$ i
61. i^{-34} -1
62. i^{-52} 1

In Exercises 63 to 68, evaluate $\dfrac{-b + \sqrt{b^2 - 4ac}}{2a}$ for the given values of a, b, and c. Write your answer as a complex number in standard form.

63. $a = 3$, $b = -3$, $c = 3$ $\dfrac{1}{2} + \dfrac{\sqrt{3}}{2}i$
64. $a = 2$, $b = 4$, $c = 4$ $-1 + i$
65. $a = 2$, $b = 6$, $c = 6$ $-\dfrac{3}{2} + \dfrac{\sqrt{3}}{2}i$
66. $a = 2$, $b = 1$, $c = 3$ $-\dfrac{1}{4} + \dfrac{\sqrt{23}}{4}i$

67. $a = 4, b = -4, c = 2$ $\frac{1}{2} + \frac{1}{2}i$

68. $a = 3, b = -2, c = 4$ $\frac{1}{3} + \frac{\sqrt{11}}{3}i$

The property that states that the product of conjugates of the form $(a + bi)(a - bi)$ is equal to $a^2 + b^2$ can be used to factor the sum of two perfect squares over the set of complex numbers. For example, $x^2 + y^2 = (x + yi)(x - yi)$. In Exercises 69 to 74, factor the binomial over the set of complex numbers.

69. $x^2 + 16$ $(x + 4i)(x - 4i)$

70. $x^2 + 9$ $(x + 3i)(x - 3i)$

71. $z^2 + 25$ $(z + 5i)(z - 5i)$

72. $z^2 + 64$ $(z + 8i)(z - 8i)$

73. $4x^2 + 81$ $(2x + 9i)(2x - 9i)$

74. $9x^2 + 1$ $(3x + i)(3x - i)$

75. Show that if $x = 1 + 2i$, then $x^2 - 2x + 5 = 0$.

76. Show that if $x = 1 - 2i$, then $x^2 - 2x + 5 = 0$.

77. When we think of the cube root of 8, $\sqrt[3]{8}$, we normally mean the *real* cube root of 8 and write $\sqrt[3]{8} = 2$. However, there are two other cube roots of 8 that are complex numbers. Verify that $-1 + i\sqrt{3}$ and $-1 - i\sqrt{3}$ are cube roots of 8 by showing that $\left(-1 + i\sqrt{3}\right)^3 = 8$ and $\left(-1 - i\sqrt{3}\right)^3 = 8$.

78. It is possible to find the square root of a complex number. Verify $\sqrt{i} = \frac{\sqrt{2}}{2}(1 + i)$ by showing that $\left[\frac{\sqrt{2}}{2}(1 + i)\right]^2 = i$.

79. Simplify $i + i^2 + i^3 + i^4 + \cdots + i^{28}$. 0

80. Simplify $i + i^2 + i^3 + i^4 + \cdots + i^{100}$. 0

Explorations

— Answer graph to Exercises 1–8 is on page AA1.

Argand Diagram Just as we can graph a real number on a real number line, we can graph a complex number. This is accomplished by using one number line for the real part of the complex number and another number line for the imaginary part of the complex number. These two number lines are drawn perpendicular to each other and intersect at their respective origins, as shown at the right.

The result is called the *complex plane* or an *Argand diagram,* after Jean-Robert Argand (1768–1822), an accountant and amateur mathematician. Although Argand is given credit for this representation of complex numbers, Caspar Wessel actually conceived the idea before Argand.

To graph the complex number $3 + 4i$, start at 3 on the real axis. Now move 4 units up (for positive numbers move up; for negative numbers move down) and place a dot at that point, as shown in the diagram. Graphs of several other points are also shown.

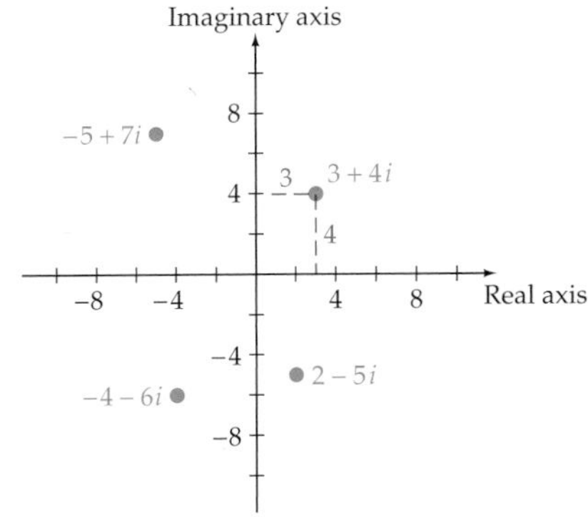

In Exercises 1 to 8, graph the complex number.

1. $2 + 5i$ **2.** $4 - 3i$ **3.** $-2 + 6i$ **4.** $-3 - 5i$

5. 4 **6.** $-2i$ **7.** $3i$ **8.** -5

The absolute value of a complex number is given by $|a + bi| = \sqrt{a^2 + b^2}$. In Exercises 9 to 12, find the absolute value of the complex number.

9. $2 + 5i$ $\sqrt{29}$ **10.** $4 - 3i$ 5 **11.** $-2 + 6i$ $2\sqrt{10}$ **12.** $-3 - 5i$ $\sqrt{34}$

13. The additive inverse of $a + bi$ is $-a - bi$. Show that the absolute value of a complex number and the absolute value of its additive inverse are equal.

14. A *real* number and its additive inverse are the same distance from zero but on opposite sides of zero on a real number line. Describe the relationship between the graphs of a complex number and its additive inverse. They are the same distance from the origin in the complex plane.

Chapter P Summary

Key Terms

absolute value **[p. 7]**
binomial **[p. 26]**
closed interval **[p. 8]**
complex conjugates **[p. 72]**
complex fraction **[p. 52]**
complex numbers **[p. 68]**
composite number **[p. 3]**
conjugate **[p. 64]**
coordinate **[p. 6]**
coordinate axis **[p. 6]**
cubic polynomial **[p. 27]**
degree of a polynomial **[p. 27]**
distance **[p. 7]**
elements of a set **[p. 3]**
empty set **[p. 5]**
factoring a polynomial **[p. 37]**
finite set **[p. 5]**
greatest common factor (GCF) **[p. 37]**
half-open interval **[p. 8]**
imaginary number **[p. 68]**
imaginary part of a complex number **[p. 68]**
infinite set **[p. 5]**
integers **[p. 3]**
intersection of two sets **[p. 5]**
interval notation **[p. 7]**
irrational numbers **[p. 4]**
linear polynomial **[p. 27]**
monomial **[p. 26]**
natural numbers **[p. 3]**
null set **[p. 5]**
open interval **[p. 8]**
origin **[p. 6]**
polynomial **[p. 26]**
prime number **[p. 3]**
quadratic polynomial **[p. 27]**
rational expression **[p. 48]**
rational numbers **[p. 4]**
real number line **[p. 6]**
real numbers **[p. 4]**
real part of a complex number **[p. 68]**
reciprocal **[p. 11]**
roster method **[p. 3]**
set **[p. 3]**
set-builder notation **[p. 5]**
scientific notation **[p. 21]**
standard form of a polynomial **[p. 26]**
trinomial **[p. 26]**
union of two sets **[p. 5]**
variable expression **[p. 10]**
whole numbers **[p. 3]**

Essential Concepts and Formulas

- **Union of Two Sets**

$$A \cup B = \{x \mid x \in A \text{ or } x \in B\}$$ **[p. 5]**

- **Intersection of Two Sets**

$$A \cap B = \{x \mid x \in A \text{ and } x \in B\}$$ **[p. 5]**

- **Properties of Real Numbers**

P

	Addition	Multiplication
Commutative	$a + b = b + a$	$ab = ba$
Associative	$(a + b) + c = a + (b + c)$	$(ab)c = a(bc)$
Identity	$a + 0 = a$	$a(1) = a$
Inverse	$a + (-a) = 0$	$a\left(\frac{1}{a}\right) = 1$
Distributive	$a(b + c) = ab + ac$	

[p. 11]

Distance Between Two Points on a Real Number Line

If a and b are the coordinates of two points on a real number line, then the distance between the two points is given by $d(a, b) = |a - b|$. **[p. 6]**

Properties of Exponents

(Assume all denominators are nonzero.)

$$x^m \cdot x^n = x^{m+n} \qquad (x^m)^n = x^{m \cdot n}$$

$$(x^m y^n)^p = x^{m \cdot p} y^{n \cdot p} \qquad \frac{x^m}{x^n} = x^{m-n}$$

If $x \neq 0$, then $x^0 = 1$. $\quad x^{-n} = \dfrac{1}{x^n}$

$$\left(\frac{x^m}{y^n}\right)^p = \frac{x^{m \cdot p}}{y^{n \cdot p}}$$ **[p. 15]**

Order of Operations Agreement

- Perform operations inside grouping symbols. Grouping symbols include parentheses, brackets, fraction bars, absolute value symbols, and radical symbols.
- Evaluate exponential expressions.
- Do multiplication and division as they occur from left to right.
- Do addition and subtraction as they occur from left to right. **[p. 9]**

Properties of Radicals

Let a and b be positive real numbers and n be a positive integer.

- If n is a positive integer and $b^{1/n}$ is a real number, then $\sqrt[n]{b} = b^{1/n}$.
- If n is an even integer, then $\sqrt[n]{b^n} = |b|$. If n is an odd integer, then $\sqrt[n]{b^n} = b$.
- $\sqrt[n]{a} \cdot \sqrt[n]{b} = \sqrt[n]{ab}$
- $\dfrac{\sqrt[n]{a}}{\sqrt[n]{b}} = \sqrt[n]{\dfrac{a}{b}}$
- $\sqrt[m]{\sqrt[n]{a}} = \sqrt[mn]{a}$
- $\sqrt{-a} = i\sqrt{a}$ **[p. 61]**

Some factoring formulas are:

- $a^2 - b^2 = (a - b)(a + b)$
- $a^2 - 2ab + b^2 = (a - b)^2$
- $a^2 + 2ab + b^2 = (a + b)^2$
- $a^3 - b^3 = (a - b)(a^2 + ab + b^2)$
- $a^3 + b^3 = (a + b)(a^2 - ab + b^2)$ **[p. 42]**

Operations on Complex Numbers

- $(a + bi) + (c + di) = (a + c) + (b + d)i$
- $(a + bi) - (c + di) = (a - c) + (b - d)i$
- $(a + bi)(c + di) = (ac - bd) + (ad + bc)i$
- $\dfrac{a + bi}{c + di} = \dfrac{a + bi}{c + di} \cdot \dfrac{c - di}{c - di}$ **[p. 70]**

Multiply numerator and denominator by the conjugate of the denominator.

Chapter P True/False Exercises

In Exercises 1 to 10, answer true or false. If the answer is false, give an example to show that the statement is false.

1. If a and b are real numbers, then $|a - b| = |b - a|$. True
2. The number 2.134894023 is a rational number. True
3. If a is a real number, then $a^2 \geq a$. False. Let $a = \frac{1}{2}$.
4. The sum of two irrational numbers is an irrational number. False. $\pi + (-\pi) = 0$
5. The product of two irrational numbers is an irrational number. False. $\sqrt{2} \cdot \sqrt{8} = \sqrt{16} = 4$
6. The sum or product of two real numbers is a real number. True
7. The quotient of any two real numbers is a real number. False. $a \div b$ is not a real number when $b = 0$.
8. Using set-builder notation, we write $[2, \infty)$ as $\{x|x \leq 2\}$. False $\{x|x \geq 2\}$
9. $\sqrt{-5} \cdot \sqrt{-5} = \sqrt{25} = 5$ False. $\sqrt{-5} \cdot \sqrt{-5} = i\sqrt{5} \cdot i\sqrt{5} = i^2 \cdot 5 = -5$
10. $\sqrt{a^2 + b^2} = a + b$ False. Let $a = 3$ and $b = 4$.

Chapter P Review Exercises

—Answer graphs to Exercises 7–10 are on page AA1.

In Exercises 1 to 4, classify each number as one or more of the following: integer, rational number, irrational number, real number, prime number, composite number.

1. 3 Integer, rational, real, prime [P.1]

2. $\sqrt{7}$ Irrational, real [P.1]

3. $-\frac{1}{2}$ Rational, real [P.1]

4. $0.\overline{5}$ Rational, real [P.1]

In Exercises 5 and 6, use $A = \{1, 5, 7\}$ and $B = \{2, 3, 5, 11\}$ to find the indicated intersection or union.

5. $A \cup B$ $\{1, 2, 3, 5, 7, 11\}$ [P.1]

6. $A \cap B$ $\{5\}$ [P.1]

In Exercises 7 and 8, graph each set and write the set using interval notation.

7. $\{x \mid -4 < x \le 2\}$ $(-4, 2]$ [P.1]

8. $\{x \mid x \le -1\} \cup \{x \mid x > 3\}$ $(-\infty, 1] \cup (3, \infty)$ [P.1]

In Exercises 9 and 10, graph each interval and write the interval in set-builder notation.

9. $[-3, 2)$ $\{x \mid -3 \le x < 2\}$ [P.1]

10. $(-1, \infty)$ $\{x \mid x > -1\}$ [P.1]

In Exercises 11 and 12, find the distance on the real number line between the points whose coordinates are given. Write the answer without using absolute value symbols.

11. $-3, 14$ 17 [P.1]

12. $\pi, 5$ $5 - \pi$ [P.1]

In Exercises 13 and 14, evaluate each expression.

13. $-5^2 - 11$ -36 [P.1]

14. $\dfrac{(2 \cdot 3^{-2})^2}{3^{-3} \cdot 2^3}$ $\dfrac{1}{6}$ [P.2]

In Exercises 15 and 16, simplify each expression.

15. $(3x^2y)(2x^3y)^2$ $12x^8y^3$ [P.2]

16. $\left(\dfrac{2a^2b^3c^{-2}}{3ab^{-1}}\right)^2$ $\dfrac{4a^2b^8}{9c^4}$ [P.2]

In Exercises 17 and 18, write each number in scientific notation.

17. 620,000 6.2×10^5 [P.2]

18. 0.0000017 1.7×10^{-6} [P.2]

In Exercises 19 and 20, change each number from scientific notation to decimal form.

19. 3.5×10^4 35,000 [P.2]

20. 4.31×10^{-7} 0.000000431 [P.2]

In Exercises 21 to 26, perform the indicated operation and express the result as a polynomial in standard form.

21. $(2a^2 + 3a - 7) + (-3a^2 - 5a + 6)$ $-a^2 - 2a - 1$ [P.3]

22. $(5b^2 - 11) - (3b^2 - 8b - 3)$ $2b^2 + 8b - 8$ [P.3]

23. $(2x - 3)(5x + 2)$ $10x^2 - 11x - 6$ [P.3]

24. $(2x^2 + 3x - 4)(2x - 3)$ $4x^3 - 17x + 12$ [P.3]

25. $(3y - 5)^2$ $9y^2 - 30y + 25$ [P.3]

26. $(3x^3 - 4x^2 + x - 4) \div (x - 3)$ $3x^2 + 5x + 16 + \dfrac{44}{x - 3}$ [P.3]

In Exercises 27 to 30, completely factor each polynomial.

27. $3x^2 + 30x + 75$ $3(x + 5)^2$ [P.4]

28. $25x^2 - 30xy + 9y^2$ $(5x - 3y)^2$ [P.4]

29. $100a^2 - 4b^2$ $4(5a + b)(5a - b)$ [P.4]

30. $16a^3 + 250$ $2(2a + 5)(4a^2 - 10a + 25)$ [P.4]

In Exercises 31 and 32, simplify each rational expression.

31. $\dfrac{6x^2 - 19x + 10}{2x^2 + 3x - 20}$ $\dfrac{3x - 2}{x + 4}$ [P.5]

32. $\dfrac{4x^3 - 25x}{8x^4 + 125x}$ $\dfrac{2x - 5}{4x^2 - 10x + 25}$ [P.5]

In Exercises 33 to 36, perform the indicated operation and simplify if possible.

33. $\dfrac{10x^2 + 13x - 3}{6x^2 - 13x - 5} \cdot \dfrac{6x^2 + 5x + 1}{10x^2 + 3x - 1}$ $\dfrac{2x + 3}{2x - 5}$ [P.5]

34. $\dfrac{15x^2 + 11x - 12}{25x^2 - 9} \div \dfrac{3x^2 + 13x + 12}{10x^2 + 11x + 3}$ $\dfrac{2x + 1}{x + 3}$ [P.5]

35. $\dfrac{x}{x^2 - 9} + \dfrac{2x}{x^2 + x - 12}$ $\dfrac{x(3x + 10)}{(x + 3)(x - 3)(x + 4)}$ [P.5]

36. $\dfrac{3x}{3x - 1} - \dfrac{x + 1}{x - 1}$ $\dfrac{-5x + 1}{(3x - 1)(x - 1)}$ [P.5]

In Exercises 37 and 38, simplify each complex fraction.

37. $\dfrac{2 + \dfrac{1}{x - 5}}{3 - \dfrac{2}{x - 5}}$ $\dfrac{2x - 9}{3x - 17}$ [P.5]

38. $\dfrac{1}{2 + \dfrac{3}{1 + \dfrac{4}{x}}}$ $\dfrac{x + 4}{5x + 8}$ [P.5]

In Exercises 39 and 40, evaluate each exponential expression.

39. $25^{-1/2}$ $\frac{1}{5}$ [P.6]

40. $-27^{2/3}$ -9 [P.6]

In Exercises 41 to 44, simplify each expression.

41. $x^{2/3} \cdot x^{3/4}$ $x^{17/12}$ [P.6]

42. $\left(\frac{8x^{5/4}}{x^{1/2}}\right)^{2/3}$ $4x^{1/2}$ [P.6]

43. $\left(\frac{x^2y}{x^{1/2}y^{-3}}\right)^{1/2}$ $x^{3/4}y^2$ [P.6]

44. $(x^{1/2} - y^{1/2})(x^{1/2} + y^{1/2})$ $x - y$ [P.6]

In Exercises 45 to 58, simplify each radical expression.

45. $\sqrt{48a^2b^7}$ $4|a|b^3\sqrt{3b}$ [P.6]

46. $\sqrt{12a^3b}$ $2|a|\sqrt{3ab}$ [P.6]

47. $\sqrt{\frac{54xy^2}{10x}}$ $\frac{3|y|\sqrt{15}}{5}$ [P.6]

48. $\sqrt{\frac{24xyz^3}{15z^6}}$ $\frac{2\sqrt{10xyz}}{5z^2}$ [P.6]

49. $\sqrt[3]{135x^2y^7}$ $3y^2\sqrt[3]{5x^2y}$ [P.6]

50. $\sqrt[3]{-250x^4y^6}$ $-5xy^2\sqrt[3]{2x}$ [P.6]

51. $4\sqrt{8x^5} - 2x\sqrt{32x^3}$ 0 [P.6]

52. $(2\sqrt{3} + 1)(4\sqrt{3} - 5)$ $19 - 6\sqrt{3}$ [P.6]

53. $\frac{2}{\sqrt{5}}$ $\frac{2\sqrt{5}}{5}$ [P.6]

54. $\frac{x}{\sqrt{x^3}}$ $\frac{\sqrt{x}}{x}$ [P.6]

55. $\frac{7x}{\sqrt[3]{2x^2}}$ $\frac{7\sqrt[3]{4x}}{2}$ [P.6]

56. $\frac{5y}{\sqrt[3]{9y}}$ $\frac{5\sqrt[3]{3y^2}}{3}$ [P.6]

57. $\frac{5}{2 - \sqrt{3}}$ $10 + 5\sqrt{3}$ [P.6]

58. $\frac{\sqrt{x}}{\sqrt{x} + 2}$ $\frac{x - 2\sqrt{x}}{x - 4}$ [P.6]

In Exercises 59 and 60, write the number in the form $a + bi$.

59. $5 - \sqrt{-7}$ $5 - i\sqrt{7}$ [P.7]

60. $2 + \sqrt{-18}$ $2 + 3i\sqrt{2}$ [P.7]

In Exercises 61 to 68, perform the indicated operation and write the answer in simplest form.

61. $(2 - 3i) + (4 + 2i)$ $6 - i$ [P.7]

62. $(4 + 7i) - (6 - 3i)$ $-2 + 10i$ [P.7]

63. $2i(3 - 4i)$ $8 + 6i$ [P.7]

64. $(4 - 3i)(2 + 7i)$ $29 + 22i$ [P.7]

65. $(3 + i)^2$ $8 + 6i$ [P.7]

66. i^{345} i [P.7]

67. $\frac{4 - 6i}{2i}$ $-3 - 2i$ [P.7]

68. $\frac{2 - 5i}{3 + 4i}$ $-\frac{14}{25} - \frac{23}{25}i$ [P.7]

Business and Economics

69. ***Sale of Golf Balls*** The number of golf balls, in millions, a manufacturer can sell each year at a price of p dollars per golf ball is given by $5 - 0.2\sqrt{p + 21}$. How many golf balls can the manufacturer sell at a price of \$3 per ball? 4.02 million [P.6]

70. ***Total Revenue*** The revenue, in thousands of dollars, a company earns for the sale of softball bats is given by $-0.05x^2 + 15x$, where x is the number of bats in thousands. Find the revenue this company earns for selling 25,000 bats. \$343,750 [P.3]

Life and Health Sciences

71. ***Flu Epidemic*** The percent of people in an office who will get the flu t days after the first person contracts the flu is given by $15 + 0.5t^{3/4}$. What percent of the office will contract the flu after 2 weeks? 18.6% [P.6]

72. ***Physician's Salary*** The salary, in thousands of dollars, of a physician specializing in internal medicine is given by Salary $= -0.15x^2 + 6.3x + 70$, where x is the number of years practicing medicine. Using this model, how much will this physician earn after practicing for 10 years? \$118,000 [P.3]

Social Sciences

73. ***Employment*** As a result of a new manufacturing plant relocating to a small city, the unemployment rate in the city began to decrease. The percent of people unemployed n months after the plant opened is given by $\frac{0.5 + \frac{6.5}{n}}{0.2 + \frac{1}{n}}$. Write the complex fraction in simplest form. $\frac{5n + 65}{2n + 10}$ [P.5]

74. ***Tuition*** The annual tuition, in thousands of dollars, at a certain university is given by Tuition $= 4\sqrt{2n + 1}$, where n is the number of years after 2000. Find the tuition, to the nearest hundred, for this university in 2010. \$18,300 [P.6]

75. ***Labor Contract*** A new labor contract for nurses at a hospital states that the hourly wage of a nurse is given by Hourly wage $= 17.45(1.055)^n$, where n is the number of years after the contract goes into effect. Find the hourly wage of a nurse at this hospital 3 years after the contract goes into effect. \$20.49/per hour [P.6]

Physical Sciences and Engineering

76. ***Grand Canyon*** The height, in feet, of a rock above the Colorado River that is thrown from the north rim of the Grand Canyon is given by Height $= 5700 - 16t^2$, where t is the number of seconds after the rock is released. Find the height of the rock above the river 5 seconds after it is released. 5300 ft [P.3]

77. ***Drag Forces on a Car*** The wind resistance force, in pounds, on a certain car traveling at x miles per hour is given by Resistance $= 0.05x^2 - 0.01x$. Find the wind resistance when the car is traveling at 25 miles per hour. 31 lb [P.3]

Chapter P Test

1. Given $A = \{3, 4, 5, 6, 7\}$ and $B = \{4, 6, 8, 10\}$, find $A \cap B$. $\{4, 6\}$ [P.1]

2. Write 0.0000341 in scientific notation. 3.41×10^{-5} [P.2]

3. Write 2.03×10^8 in standard form. 203,000,000 [P.2]

4. Simplify: $(-3x^2y^3)^3(2x^4)$ $-54x^{10}y^9$ [P.2]

5. Simplify: $\dfrac{18x^2y^7}{24x^5y^5}$ $\dfrac{3y^2}{4x^3}$ [P.2]

6. Evaluate: $-27^{-2/3}$ $-\dfrac{1}{9}$ [P.6]

7. Subtract: $(2x^2 - 3x - 7) - (x^3 + 2x^2 - 4x + 1)$ $-x^3 + x - 8$ [P.3]

8. Multiply: $(3x^2 + 4x - 2)(2x - 3)$ $6x^3 - x^2 - 16x + 6$ [P.3]

9. Divide: $(3x^2 + 5x - 7x + 4) \div (x - 2)$ $3x + 4 + \dfrac{12}{x - 2}$ [P.3]

In Exercises 10 and 11, factor completely over the integers.

10. $2x^2 - 5x - 3$ $(2x + 1)(x - 3)$ [P.4]

11. $16x^4 - 2xy^3$ $2x(2x - y)(4x^2 + 2xy + y^2)$ [P.4]

12. Simplify: $\dfrac{x^2 - 2x - 15}{25 - x^2}$ $-\dfrac{x + 3}{x + 5}$ [P.5]

13. Simplify: $\dfrac{x^2 - x - 6}{x^2 - 9} \cdot \dfrac{x^2 + 7x + 12}{x^2 + 4x - 5}$ $\dfrac{(x + 2)(x + 4)}{(x + 5)(x - 1)}$ [P.5]

14. Simplify: $\dfrac{2x^2 + 3x - 2}{x^2 - 3x} \div \dfrac{2x^2 - 7x + 3}{x^3 - 3x^2}$ $\dfrac{x(x + 2)}{x - 3}$ [P.5]

15. Simplify: $x - \dfrac{x}{x + \dfrac{1}{2}}$ $\dfrac{x(2x - 1)}{2x + 1}$ [P.5]

16. Simplify: $3\sqrt[3]{16a^6b^7} - 4ab\sqrt[3]{54a^3b^4}$ $-6a^2b^2\sqrt[3]{2b}$ [P.6]

17. Simplify: $\dfrac{5}{\sqrt[3]{25}}$ $\sqrt[3]{5}$ [P.6]

18. Simplify: $\dfrac{5 - \sqrt{3}}{2 + 3\sqrt{3}}$ $-\dfrac{19 - 17\sqrt{3}}{23}$ [P.6]

In Exercises 19 to 22, perform the indicated operation.

19. $(4 - 3i) - (2 - 5i)$ $2 + 2i$ [P.7]

20. $(2 + 5i)(1 - 4i)$ $22 - 3i$ [P.7]

21. $\dfrac{3 + 4i}{5 - i}$ $\dfrac{11}{26} + \dfrac{23}{26}i$ [P.7]

22. i^{97} i [P.7]

23. A developer of computer games estimates that the revenue, in thousands of dollars, from selling a new game can be given by Revenue $= -0.008x^2 + 50x$, where x is the number of games sold. The cost to produce x games is given by Cost $= 10x + 3000$. Write the profit in terms of x. What is the profit for selling 1000 of these games? Profit $= -0.008x^2 + 40x - 3000$; \$29,000 [P.3]

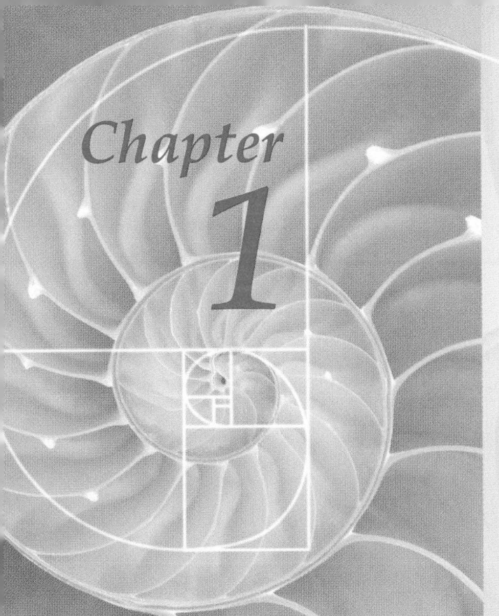

Chapter 1

Equations and Inequalities

Forensic Science and Mathematics

Forensic scientists often use equations and inequalities when they attempt to identify a deceased person, and to determine the cause and the manner of death. In some cases they may have only a few bones with which to gather their clues. In situations such as these, equations and inequalities called *regression formulas* are used to estimate the height of the deceased person by measuring the lengths of some of the person's bones. This procedure is explained in Exercise 75, page 137.

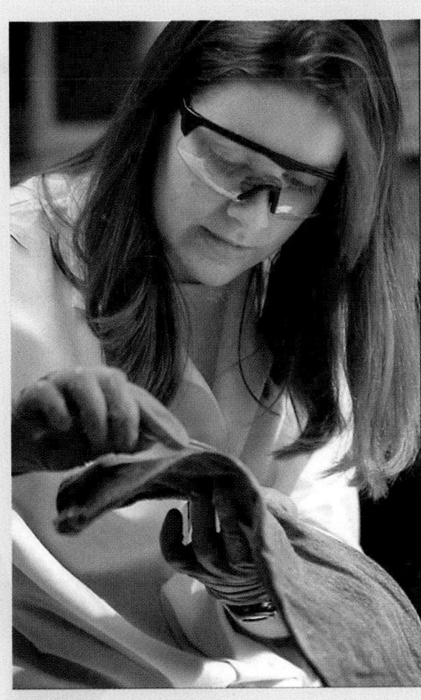

Chapter Prep Quiz

Do these exercises to prepare for Chapter 1.

1. Simplify: $\sqrt{12}$ [P.6]* $2\sqrt{3}$

2. Simplify: $3(2x - 1) - 4(3x - 2)$ [P.3] $-6x + 5$

3. Factor: $x^2 - 16x + 64$ [P.4] $(x - 8)^2$

4. Factor: $x^4 + 2x^2 - 3$ [P.4] $(x + 1)(x - 1)(x^2 + 3)$

5. Express "the distance between a real number x and 5 is less than 3" using absolute value notation. [P.1] $|x - 5| < 3$

6. Simplify: $\dfrac{-4 + \sqrt{-16}}{2}$ [P.7] $-2 + 2i$

7. Write $\dfrac{x}{x + 1} - 2$ as a rational expression. [P.5] $-\dfrac{x + 2}{x + 1}$

8. Evaluate $\dfrac{-b + \sqrt{b^2 - 4ac}}{2a}$ with $a = 2$, $b = -4$, and $c = 1$. [P.6] $\dfrac{2 + \sqrt{2}}{2}$

9. Find $\{x | x \geq -1\} \cup \{x | x < 5\}$. [P.1] $(-\infty, \infty)$

10. Find $\{x | x \geq 3\} \cap \{x | x > 1\}$. [P.1] $\{x | x \geq 3\}$

Problem Solving Strategies

Verifying Results

One important aspect of problem solving involves the process of checking to see if your results satisfy the conditions of the original problem. This process will be especially important in this chapter when you solve an equation or an inequality.

Here is an example that illustrates the importance of checking your results. The problem seems easy, but many students fail to get the correct answer on their first attempt.

Two volumes of the series *Mathematics: Its Content, Methods, and Meaning* are on a shelf, with no space between the volumes. Each volume is 1 inch thick without its covers. Each cover is $\frac{1}{8}$ inch thick. A bookworm bores horizontally from the first page of Volume I to the last page of Volume II. How far does the bookworm travel?

Once you have obtained your solution, try to check it by closely examining two books placed as shown above. Check to make sure you have the proper starting and ending positions. The correct answer is $\frac{1}{4}$ inch.

INSTRUCTOR NOTE

The *Chapter Prep Quiz* is a means to test your students' mastery of *prerequisite* material. It is assumed in the coming chapter that students are familiar with this material. Each question identifies the section to review (if necessary) and all answers are provided in the Answers to Selected Exercises.

*This is a reference to the section that corresponds to this problem. For example, [P.6] stands for Chapter P, Section 6.

SECTION 1.1 Linear Equations and Their Applications

- Introduction to Equations
- Linear Equations
- Strategies for Solving Application Problems

Introduction to Equations

A formula that is sometimes used to compute the ideal weight for an adult male in the U.S., based on averages, is Ideal weight $= 6 \cdot h - 254$, where h is the man's height in inches. For instance, a man who is 6 feet (72 inches) tall would have an ideal weight of $6 \cdot 72 - 254 = 178$ pounds. If, on the other hand, we would like to know what height corresponds to an ideal weight of 145 pounds, we need to determine the number of inches h such that $6h - 254 = 145$. This means we must *solve an equation* rather than evaluate a formula.

An **equation** is a statement about the equality of two expressions. If either of the expressions contains a variable, the equation may be a true statement for some values of the variable and a false statement for other values. For example, the equation $2x + 1 = 7$ is a true statement for $x = 3$, but it is false for any number other than 3. The number 3 is used to **satisfy** the equation $2x + 1 = 7$ because substituting 3 for x produces $2(3) + 1 = 7$, which is a true statement.

To solve an equation means to find all values of the variable that satisfy the equation. The values that satisfy an equation are called **solutions** or **roots** of the equation. Equations can have exactly one solution, more than one solution, or no solutions at all. For instance, 2 and 3 are both solutions of the equation $5x - x^2 = 6$ (you should verify this), whereas the equation $\sqrt{x} + 2 = 0$ does not have any solutions.

It can be difficult to find solutions of equations by trial and error. Instead, we normally work backwards by converting the equation to a simpler one. **Equivalent equations** are equations that have exactly the same solution(s). We can solve some equations involving the variable x by producing a sequence of equivalent equations until we produce an equation or equations of the form

$$x = \text{constant}$$

To produce these equivalent equations that lead us to the solution(s), we often perform one or more of the following procedures.

Procedures That Produce Equivalent Equations

1. Simplify an expression on either side of the equation by procedures such as (i) combining like terms or (ii) applying the properties explained in Chapter P, such as the commutative, associative, and distributive properties.

   $2x + 3 + 5x = -11$ and $7x + 3 = -11$ are equivalent equations.

2. Add or subtract the same quantity to (from) both sides of the equation.

   $3x - 7 = 2$ and $3x = 9$ are equivalent equations.
   (Add 7 to both sides.)

3. Multiply or divide both sides of the equation by the same nonzero quantity.

   $5x = 10$ and $x = 2$ are equivalent equations.
   (Divide each side by 5.)

Alternative to Example 1

A retailer uses the formula Price $= 1.8c + 6$ to determine the price a sweater will sell for, where c is the wholesale cost of the sweater in dollars. What is the wholesale cost of a sweater priced at $71.70?

▪ **$36.50**

EXAMPLE 1 Ideal Weight

Use the formula Ideal weight $= 6 \cdot h - 254$ to compute the ideal weight for an adult male in the U.S., based on averages, where h is the man's height in inches. What height corresponds to an ideal weight of 145 pounds?

Solution We need to solve the equation $6h - 254 = 145$.

$$6h - 254 = 145$$

$$6h - 254 + 254 = 145 + 254 \qquad \text{▪ Add 254 to each side.}$$

$$6h = 399$$

$$\frac{6h}{6} = \frac{399}{6} \qquad \text{▪ Divide each side by 6.}$$

$$h = 66.5$$

An adult male 66.5 inches tall has an ideal weight of 145 pounds.

CHECK YOUR PROGRESS 1 A carpet installer replaces carpet in offices and uses the formula Cost $= 350 + 1.6A$, where A is the number of square feet of carpet to be replaced, to determine the cost. In how large an office can the carpet be replaced for $3000?

Solution *See page S3.* 1656.25 ft^2

▪ Linear Equations

The equation that was solved in Example 1 is an example of a *linear equation*.

Definition of a Linear Equation

A **linear equation** in the single variable x is an equation that can be written in the form

$$ax + b = 0$$

where a and b are real numbers, with $a \neq 0$.

Linear equations are generally solved by applying the procedures that produce equivalent equations.

Alternative to Example 2

Solve: $\frac{2}{5}x + 3 = -5$

▪ **−20**

EXAMPLE 2 Solve a Linear Equation

Solve: $\frac{3}{4}x - 6 = 0$

Continued ➤

TAKE NOTE

Check your solution by substituting 8 for x in the original equation.

$$\frac{3}{4}x - 6 = 0$$
$$\frac{3}{4}(8) - 6 \stackrel{?}{=} 0$$
$$0 = 0 \quad \text{True}$$

Solution

$$\frac{3}{4}x - 6 = 0$$
$$\frac{3}{4}x - 6 + 6 = 0 + 6 \qquad \text{▪ Add 6 to each side.}$$
$$\frac{3}{4}x = 6$$
$$\left(\frac{4}{3}\right)\left(\frac{3}{4}x\right) = \left(\frac{4}{3}\right)6 \qquad \text{▪ Multiply each side by } \frac{4}{3}.$$
$$x = 8$$

Because 8 satisfies the original equation (see the Take Note), 8 is the solution.

CHECK YOUR PROGRESS 2 Solve: $\frac{2}{3}x + 9 = 1$

Solution *See page S3.* -12

If an equation involves several fractions, it is helpful to first multiply each side of the equation by the LCD of all the fractions to produce an equivalent equation that does not contain fractions.

Alternative to Example 3

Solve: $\frac{1}{2}x - \frac{7}{3} = \frac{2}{3}x - 2$

▪ -2

EXAMPLE 3 Solve a Linear Equation Containing Fractions

Solve: $\frac{2}{3}x + 10 - \frac{x}{5} = \frac{36}{5}$

Solution

$$\frac{2}{3}x + 10 - \frac{x}{5} = \frac{36}{5}$$
$$15\left(\frac{2}{3}x + 10 - \frac{x}{5}\right) = 15\left(\frac{36}{5}\right) \qquad \text{▪ Multiply each side of the equation by 15, the LCD of the fractions.}$$
$$10x + 150 - 3x = 108 \qquad \text{▪ Simplify.}$$
$$7x + 150 = 108$$
$$7x + 150 - 150 = 108 - 150 \qquad \text{▪ Subtract 150 from each side.}$$
$$7x = -42$$
$$\frac{7x}{7} = \frac{-42}{7} \qquad \text{▪ Divide each side by 7.}$$
$$x = -6 \qquad \text{▪ Check as before.}$$

CHECK YOUR PROGRESS 3 Solve: $\frac{7x}{3} - \frac{2}{3} = \frac{3}{4} - \frac{5x}{3}$

Solution *See page S3.* $x = \frac{17}{48}$

An equation may not appear to be a linear equation at first, but may simplify to the form of a linear equation. When simplified, linear equations will not include expressions such as x^2, x^3, $\sqrt{x}$, or $\frac{1}{x}$.

QUESTION Is $\sqrt{2} - 5^2x = \frac{1}{4}$ a linear equation?

Alternative to Example 4
Solve:
$(3x - 4)(x + 2) = x(3x + 5)$
■ $-\frac{8}{3}$

EXAMPLE 4 Simplify and Solve an Equation

Solve: $(x + 2)(5x + 1) = 5x(x + 1)$

Solution

$$(x + 2)(5x + 1) = 5x(x + 1)$$

$$5x^2 + 11x + 2 = 5x^2 + 5x \quad \text{■ Simplify each product.}$$

$$11x + 2 = 5x \quad \text{■ Subtract } 5x^2 \text{ from each side.}$$

$$6x + 2 = 0 \quad \text{■ Subtract } 5x \text{ from each side.}$$

The result is a linear equation. Solve as before.

$$6x = -2 \quad \text{■ Subtract 2 from each side.}$$

$$x = -\frac{1}{3} \quad \text{■ Divide each side by 6.}$$

CHECK YOUR PROGRESS 4 Solve: $4x(x + 2) - 1 = (2x - 3)(2x + 1)$

Solution *See page S4.* $x = -\frac{1}{6}$

As in Example 1, real-world situations can often be described using linear equations.

Alternative to Example 5
Exercise 68, page 96

EXAMPLE 5 An Application to Amusement Park Attendance

According to the International Association of Amusement Parks and Attractions, attendance at amusement parks has been increasing, on average, over the last decade. The annual attendance can be approximated by the equation

$$\text{Attendance} = 6.2x + 252$$

where attendance is measured in millions of people and x is the number of years *after* 1990 ($x = 0$ corresponds to 1990). If the increase in attendance continues to follow the same trend, use the equation to predict when the attendance will first reach 500 million.

ANSWER Yes. The equation can be written as $-5^2x + \sqrt{2} - \frac{1}{4} = 0$, which is in the form $ax + b = 0$, where $a = -5^2$ and $b = \sqrt{2} - \frac{1}{4}$.

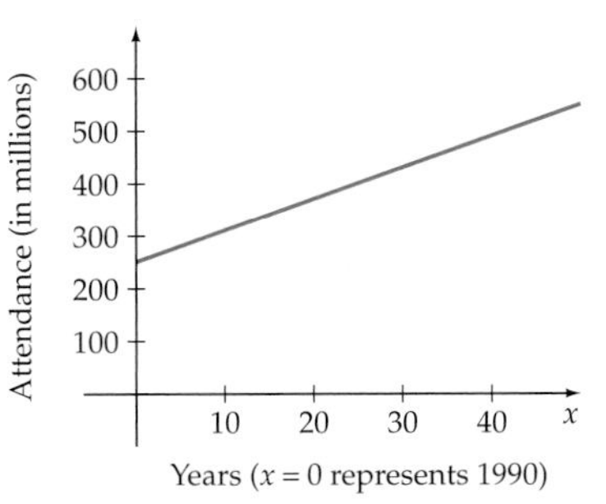

Solution Replace Attendance by 500 and solve for x.

$$\text{Attendance} = 6.2x + 252$$
$$500 = 6.2x + 252$$
$$248 = 6.2x$$
$$\frac{248}{6.2} = \frac{6.2x}{6.2}$$
$$40 = x$$

The year 2030 is 40 years after 1990, so we can predict that the attendance will first reach 500 million people in 2030.

CHECK YOUR PROGRESS 5 Data from the U.S. Patent and Trademark Office suggest that the number of patents that have been issued each year in the U.S. since 1993 can be approximated by the equation

$$\text{Number of patents (in thousands)} = 5.4x + 110$$

where x is the number of years after 1993. Using this equation, determine in what year the number of patents first exceeded 150,000.

Solution *See page S4.* 2000

Many known formulas lead to linear equations when we attempt to determine an unknown quantity. For instance, the perimeter of a rectangle is given by $P = 2l + 2w$, where l is the length of the rectangle and w is the width. (See the inside back book cover.) If the perimeter of a rectangle measures 76 feet and the width is 14 feet, the length can be determined.

$$76 = 2l + 2 \cdot 14 \qquad \blacksquare\ P = 76, w = 14$$
$$76 = 2l + 28$$
$$48 = 2l$$
$$24 = l$$

Formulas with several unknowns, also called **literal equations,** can be solved (or rearranged) for a specified variable, as shown in the next example.

Alternative to Example 6
Solve: $2m + n = mp$ for m.

■ $m = \dfrac{n}{2 - p}$

EXAMPLE 6 Solve a Literal Equation

a. Solve the literal equation $P = 2l + 2w$ for l.

b. Solve $xy - z = yz$ for y.

Solution

a.

$$P = 2l + 2w$$
$$P - 2w = 2l \qquad \blacksquare\ \text{Subtract } 2w \text{ from each side to isolate the } 2l \text{ term.}$$
$$\frac{P - 2w}{2} = l \qquad \blacksquare\ \text{Divide each side by 2.}$$
$$l = \frac{P - 2w}{2}$$

TAKE NOTE

In Example **6a.**, the solution $l = \dfrac{P - 2w}{2}$ can also be written as $l = \dfrac{P}{2} - w$.

Continued ➤

TAKE NOTE

In Example **6b.**, the restriction $x - z \neq 0$ is necessary to ensure that each side is divided by a *nonzero* expression.

b. To solve for y, first isolate the terms that involve the variable y on the left side of the equation.

$$xy - z = yz$$

$$xy - yz - z = 0$$ ▪ Subtract yz from each side so that all terms that contain y are on the same side of the equation.

$$xy - yz = z$$ ▪ Add z to each side to isolate the terms that contain y.

$$y(x - z) = z$$ ▪ Factor y from each term on the left side of the equation.

$$y = \frac{z}{x - z}$$ ▪ Divide each side of the equation by $x - z$, $x - z \neq 0$.

CHECK YOUR PROGRESS 6 Solve the following equation for c: $ac = 2b + 3c$

Solution *See page S4.* $c = \frac{2b}{a - 3}$

▪ Strategies for Solving Application Problems

INSTRUCTOR NOTE

Encourage your students to use the guidelines even on "simple" problems. For example, have your students solve the following problem.

A bottle and a cork together cost $1.10. If the bottle costs $1.00 more than the cork, what is the cost of the bottle and what is the cost of the cork?

Some students may respond that the bottle costs $1.00 and the cork costs $0.10; however, those students who use the problem-solving guidelines will find that the bottle costs $1.05 and the cork costs $0.05.

Alternative to Example 7

The width of a rectangle is 3 feet less than half the length. If the perimeter of the rectangle is 60 feet, find the length and width of the rectangle.

▪ **length: 22 ft; width: 8 ft.**

Linear equations emerge in a variety of application problems. In solving such problems, it helps to apply specific techniques in a series of small steps. The following general guidelines will help with the remaining examples in this section.

Guidelines for Solving Application Problems

1. Read the problem carefully. If necessary, reread the problem several times.
2. When appropriate, draw a sketch and label parts of the drawing with the specific information given in the problem.
3. Determine the unknown quantities and label them with variables. Write down any equation that relates the variables.
4. Use the information from Step 3, along with a known formula or some additional information given in the problem, to write an equation.
5. Solve the equation obtained in Step 4. Check to see whether the result satisfies all the conditions of the original problem.

EXAMPLE 7 Solve an Application

The length of a rectangular garden is 2 feet greater than three times the width. If the perimeter of the garden is 92 feet, find the width and the length of the garden.

Solution Follow the guidelines given above.

1. Read the problem carefully.
2. Draw a rectangle as shown in Figure 1.1.

Continued ➤

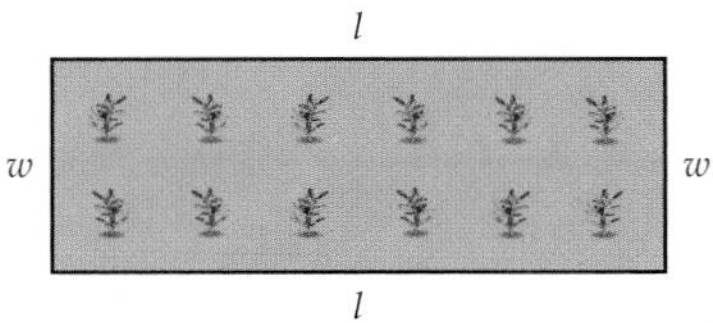

Figure 1.1

3. Label the length of the rectangle l and the width of the rectangle w. The problem states that the length l is 2 feet greater than three times the width w. Thus l and w are related by the equation

$$l = 3w + 2$$

4. Because the problem involves the length, width, and perimeter of a rectangle, we use the geometric formula $2l + 2w = P$ (see the inside back cover). To write an equation that involves only constants and a single variable (say, w), substitute 92 for P and $3w + 2$ for l.

$$2l + 2w = P$$
$$2(3w + 2) + 2w = 92$$

5. Solve for the unknown w.

$$6w + 4 + 2w = 92$$
$$8w + 4 = 92$$
$$8w = 88$$
$$w = 11$$

Because the length l is 2 more than three times the width,

$$l = 3(11) + 2 = 35$$

A check verifies that 35 is 2 more than three times 11. Also, twice the length (70) plus twice the width (22) gives the perimeter (92). The width of the rectangle is 11 feet and the length is 35 feet.

CHECK YOUR PROGRESS 7 The width of a rectangle is 1 meter more than half the length of the rectangle. If the perimeter of the rectangle is 110 meters, find the width and the length.

Solution *See page S4.* Length: 36 m; width: 19 m

Alternative to Example 8
Exercise 70, page 96

EXAMPLE 8 Solve a Uniform Motion Problem

A runner runs a course at a constant speed of 6 miles per hour. One hour after the runner begins, a cyclist starts on the same course at a constant speed of 15 miles per hour. How long after the runner starts does the cyclist overtake the runner?

Solution This is an example of a *uniform motion problem.* For objects moving at a constant speed, the distance traveled can be computed by using the formula $d = rt$, where d is the distance traveled, r is the rate of speed, and t is the time.

If we represent the time (in hours) the runner has spent on the course when the cyclist overtakes the runner by t, then the time the cyclist is on the course is $t - 1$. We can use a table to organize the information and help determine an expression for the distance each person travels.

Continued ➤

	Rate r	·	time t	=	distance d
Runner	6	·	t	=	$6t$
Cyclist	15	·	$t - 1$	=	$15(t - 1)$

Figure 1.2

Figure 1.2 indicates that the runner and the cyclist cover the same distance. Thus

$$\begin{aligned} \text{Runner distance} &= \text{cyclist distance} \\ 6t &= 15(t - 1) \\ 6t &= 15t - 15 \\ -9t &= -15 \\ t &= 1\frac{2}{3} \end{aligned}$$

A check will verify that the cyclist overtakes the runner $1\frac{2}{3}$ hours after the runner starts.

INSTRUCTOR NOTE
Encourage your students to make a sketch for all uniform motion application problems. Without a sketch, some students will not be able to write the equation that relates the given information and the unknown.

CHECK YOUR PROGRESS 8 A motorboat left a harbor and traveled to an island at an average rate of 15 knots (1 knot = 1 nautical mile per hour; a nautical mile is approximately 1.15 statute miles). The average speed on the return trip was 10 knots. If the total trip took 7.5 hours, how far is the harbor from the island?

Solution *See page S4.* 45 nautical miles

Alternative to Example 9
Exercise 48, page 95

EXAMPLE 9 Solve an Investment Problem

An accountant invests part of a \$6000 bonus in a 5% simple interest account and the remainder of the money at 8.5% simple interest. Together the investments earn \$370 per year. Find the amount invested at each rate.

Solution Simple interest (which is computed and paid only once per year) is computed using the formula $I = Prt$, where I is the interest earned, P is the principal (the initial amount of money), r is the simple interest rate per year, and t is the number of years.

The \$6000 has been split between two different accounts. Let x be the amount invested at 5%. The remainder of the money is given by $\$6000 - x$, which is the amount invested at 8.5%. Using the simple interest formula with $t = 1$ year gives

$$\begin{aligned} \text{Interest earned at } 5\% &= x \cdot 0.05 = 0.05x \\ \text{Interest earned at } 8.5\% &= (6000 - x) \cdot (0.085) = 510 - 0.085x \end{aligned}$$

The total interest earned on the two accounts equals \$370.

$$\begin{aligned} \text{Interest earned at } 5\% + \text{interest earned at } 8.5\% &= \$370 \\ 0.05x + (510 - 0.085x) &= 370 \\ -0.035x + 510 &= 370 \\ -0.035x &= -140 \\ x &= 4000 \end{aligned}$$

Continued ➤

Therefore, the accountant invested \$4000 at 5% and the remaining \$2000 at 8.5%. You should check these values to be sure no errors were made.

CHECK YOUR PROGRESS 9 A total of \$7500 is deposited into two simple interest accounts. On one account the annual simple interest rate is 5% and on the second account the annual simple interest rate is 7%. The amount of interest earned for 1 year was \$405. How much was invested in each account?

Solution *See page S4.* \$6000 invested at 5%; \$1500 invested at 7%

The next example involves the mixture of two solutions that have different concentrations of a common substance. The formula $pA = Q$, where p is the percent of concentration, A is the amount of the solution, and Q is the quantity of a substance in the solution, describes the relationship between the amounts of substance and solution. For instance, in 4 liters of a 25% acid solution, p is the percent of acid (so $p = 0.25$), A is the total amount of solution (4 liters), and Q is the amount of acid in the solution, which equals $0.25 \cdot 4$ liters $= 1$ liter.

Alternative to Example 10
Exercise 66, page 96

INSTRUCTOR NOTE

Encourage students to ask, "Does the answer seem reasonable?" For instance, in Example 10, if 300 milliliters of the 11% solution are mixed with 300 milliliters of the 6% solution, the result is 600 milliliters of $8\frac{1}{2}$% solution (the average of 11% and 6%). A weaker 8% solution requires slightly more of the 6% solution and less of the 11% solution.

EXAMPLE 10 An Application to Chemical Mixtures

A chemist mixes an 11% hydrochloric acid solution with a 6% hydrochloric acid solution. How many milliliters (ml) of each solution should the chemist use to make a 600-milliliter solution that is 8% hydrochloric acid?

Solution Let x be the number of milliliters of the 11% solution. Because the final solution will have a total of 600 milliliters of fluid, $600 - x$ is the number of milliliters of the 6% solution. See the figure below.

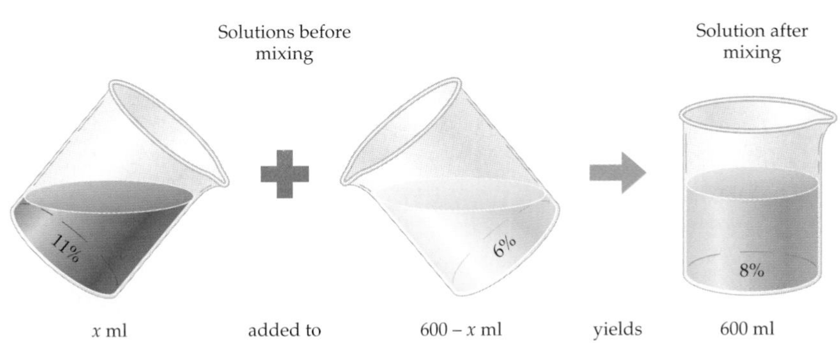

Because all of the hydrochloric acid in the final solution comes from either the 11% solution or the 6% solution, the number of milliliters of hydrochloric acid in the

Continued ➤

11% solution added to the number of milliliters of hydrochloric acid in the 6% solution must equal the number of milliliters of hydrochloric acid in the 8% solution.

$$\begin{pmatrix}\text{ml of acid in} \\ \text{11\% solution}\end{pmatrix} + \begin{pmatrix}\text{ml of acid in} \\ \text{6\% solution}\end{pmatrix} = \begin{pmatrix}\text{ml of acid in} \\ \text{8\% solution}\end{pmatrix}$$

$$0.11x \quad + \quad 0.06(600 - x) \quad = \quad 0.08(600)$$

$$0.11x + 36 - 0.06x = 48$$

$$0.05x + 36 = 48$$

$$0.05x = 12$$

$$x = 240$$

Therefore, the chemist should use 240 milliliters of the 11% solution and 360 milliliters of the 6% solution to make a 600-milliliter solution that is 8% hydrochloric acid.

CHECK YOUR PROGRESS 10 How many liters of a 40% sulfuric acid solution should be mixed with 4 liters of a 24% sulfuric acid solution to produce a 30% solution?

Solution *See page S4.* 2.4 L

Finally, we look at an example in which more than one person or machine is working on a task and we want to determine how long it will take to complete the task. The equation

$$\text{Rate of work} \times \text{time worked} = \text{part of task completed}$$

computes the portion of a task finished by each person or machine. For example, if a painter can paint a wall in 15 minutes, then the painter can paint $\frac{1}{15}$ of the wall in 1 minute. The painter's *rate of work* is $\frac{1}{15}$ of the wall each minute. In general, if a task can be completed in x minutes, then the rate of work is $\frac{1}{x}$ of the task each minute.

Alternative to Example 11
Exercise 56, page 95

EXAMPLE 11 Solve a Work Problem

Pump A can fill a pool in 6 hours, and pump B can fill the same pool in 3 hours. How long will it take to fill the pool if both pumps are used?

Solution Because pump A fills the pool in 6 hours, $\frac{1}{6}$ represents the part of the pool filled by pump A in 1 hour. Because pump B fills the pool in 3 hours, $\frac{1}{3}$ represents the part of the pool filled by pump B in 1 hour.

Continued ➤

Business and Economics

45. ***Retail*** It costs a manufacturer of sunglasses \$8.95 to produce a pair of sunglasses that sells for \$29.99. How many pairs of sunglasses must the manufacturer sell to make a profit of \$17,884? 850

46. ***Retail*** It costs a restaurant owner 18 cents per glass for orange juice, which is sold for 75 cents per glass. How many glasses of orange juice must the restaurant owner sell to make a profit of \$2337? 4100

47. ***Investment*** An investment advisor invested \$14,000 in two accounts. One investment earned 8% annual simple interest and the other investment earned 6.5% annual simple interest. The total amount of interest earned for 1 year was \$1024. How much was invested in each account? \$7600 invested at 8%, \$6400 invested at 6.5%

48. ***Investment*** A total of \$12,000 is invested in two accounts earning simple interest. On one account the annual interest rate is 6%, and on the second account the rate is 4%. After 1 year, the total amount of interest earned was \$580. How much was invested in each account? \$5000 invested at 6%, \$7000 invested at 4%

49. ***Investment*** An investment of \$2500 is made at an annual simple interest rate of 5.5%. How much additional money must be invested at an annual simple interest rate of 8% so that the total interest earned is 7% of the total investment? \$3750

50. ***Investment*** An investment of \$4600 is made at an annual simple interest rate of 6.8%. How much additional money must be invested at an annual simple interest rate of 9% so that the total interest earned is 8% of the total investment? \$5520

51. ***Commerce*** A ballet performance brought in \$61,800 on the sale of 3000 tickets. If the tickets sold for \$14 and \$25, how many of each type of ticket were sold? 1200 at \$14 and 1800 at \$25

52. ***Commerce*** A coffee shop decides to blend a coffee that sells for \$12 per pound with a coffee that sells for \$9 per pound to produce a blend that will sell for \$10 per pound. How much of each blend should be used to yield 20 pounds of the new blend? $6\frac{2}{3}$ lb of the \$12 coffee and $13\frac{1}{3}$ lb of the \$9 coffee

53. ***Share an Expense*** Three people decide to share the cost of a yacht equally. By bringing in an additional partner, they can reduce the cost to each person by \$4000. What is the total cost of the yacht? \$48,000

54. ***Share an Expense*** A group of five private pilots would like to purchase a used Cessna airplane and split the cost equally. If they can find a sixth person to share the costs, they can each contribute \$2000 less. What is the cost of the airplane? \$60,000

55. ***Install Electrical Wiring*** An electrician can install the electric wires in a house in 14 hours. A second electrician requires 18 hours. How long would it take both electricians, working together, to install the wires? 7.875 h

56. ***Interior Painting*** A painter can repaint a room of a house in 4 hours. His assistant takes 6 hours to paint the same size room. How long would it take the two painters, working together, to paint the room? 2.4 h

Life and Health Sciences

To benefit from an aerobic exercise program, many experts recommend that you exercise three to five times a week for 20 minutes to an hour. It is also important that your heart rate be in the *training zone*, which is defined by the following linear equations, where a is your age in years and your heart rate is in beats per minute.*

Maximum exercise heart rate = $0.85(220 - a)$

Minimum exercise heart rate = $0.65(220 - a)$

57. ***Maximum Exercise Heart Rate*** Find the maximum exercise heart rate and the minimum exercise heart rate for a person who is 25 years old. (Round to the nearest beat per minute.) Maximum: 166 beats/min; minimum: 127 beats/min

58. ***Maximum Exercise Heart Rate*** How old is a person who has a maximum exercise heart rate of 153 beats per minute? 40 yrs old

59. ***Nursing*** How many liters of water should be evaporated from 160 liters of a 12% saline solution so that the solution that remains is a 20% saline solution? 64 L

*"The Heart of the Matter," *American Health*, September 1995.

60. ***Finding an Average*** A biology student has test scores of 80, 82, 94, and 71. What score does the student need on the next test to produce an average score of 85? 98

61. ***Finding an Average*** In a botany class, a student has test scores of 90, 74, 82, and 90. Only the final exam remains, which counts as two regular tests. What score does the student need on the final exam to produce an average score of 85? 87

Sports and Recreation

62. ***Amusement Parks*** The revenues of all the amusement and theme parks in the United States have been increasing since 1990. An equation that approximates the total revenues of all parks is given by

$$\text{Revenues (in billions)} = 0.35x + 5.7$$

where x is the number of years after 1990. Using this equation, determine between what two years the revenues for all amusement and theme parks in the U.S. first exceeded \$10 billion. (*Source: Amusement Business* magazine as reported in the San Diego *Union-Tribune,* March 19, 2000) Between 2002 and 2003

63. ***Uniform Motion*** Running at an average rate of 6 meters per second, a sprinter ran to the end of a track. The sprinter then jogged back to the starting point at an average rate of 2 meters per second. The total time for the sprint and the jog back was 2 minutes 40 seconds. Find the length of the track. 240 m

64. ***Uniform Motion*** Marlene rides her bicycle to her friend Jon's house and returns home by the same route. Marlene rides her bike at constant speeds of 6 miles per hour on level ground, 4 miles per hour when going uphill, and 12 miles per hour when going downhill. If her total time riding was 1 hour, how far is it to Jon's house? 3 mi

65. ***Automotive*** A radiator contains 6 liters of a 25% antifreeze solution. How much should be drained and replaced with pure antifreeze to produce a 33% antifreeze solution? 0.64 L

66. ***Motorcycle Fuel*** A motorcyclist poured too much of a fuel additive into the gas tank, so that the tank currently contains 3 gallons of gasoline with a 2% additive mixture. The manufacturer recommends only a $\frac{1}{2}$% mixture. How much gasoline mix should be drained and replaced with pure gasoline to achieve a $\frac{1}{2}$% mixture? 2.25 gal

Physical Sciences and Engineering

67. ***Computer Science*** The percent of a file that remains to be downloaded using a dialup Internet connection for a certain modem is given by the equation

$$\text{Percent remaining} = 100 - \frac{42{,}000}{N}t$$

where N is the size of the file in bytes and t is the number of seconds since the download began. In how many minutes will 25% of a 500,000-byte file remain to be downloaded? Round to the nearest minute. 15 min

68. ***Aviation*** The number of miles that remain to be flown by a commercial jet traveling from Boston to Los Angeles can be approximated by the equation

$$\text{Miles remaining} = 2650 - 475t$$

where t is the number of hours since leaving Boston. In how many hours will the plane be 1000 miles from Los Angeles? Round to the nearest tenth. 3.5 h

69. ***Uniform Motion*** A plane leaves an airport traveling at an average speed of 240 kilometers per hour. How long will it take a second plane traveling the same route at an average speed of 600 kilometers per hour to catch up with the first plane if it leave 3 hours later? 2 h

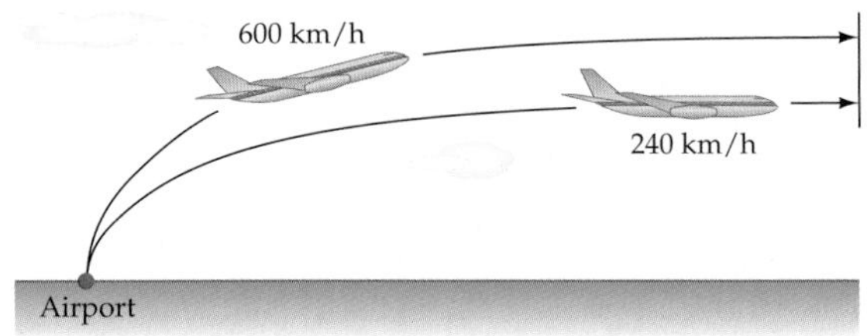

70. ***Uniform Motion*** A plane leaves Chicago headed for Los Angeles traveling at 540 miles per hour. One hour later, a second plane leaves Los Angeles headed for Chicago traveling at 660 miles per hour. If the air route from Chicago to Los Angeles is 1800 miles, how long will it take for the first plane to pass the second plane? How far from Chicago will the planes be at that time? 2.05 h; 1107 mi

71. ***Metallurgy*** How many grams of pure silver must a silversmith mix with a 45% silver alloy to produce 200 grams of a 50% alloy? $18\frac{2}{11}$ g

72. ***Metallurgy*** How much pure gold should be melted with 15 grams of 14-karat gold to produce 18-karat gold? *Hint:*

A karat is a measure of the purity of gold in an alloy. Pure gold measures 24 karats. An alloy that measures x karats is $\frac{x}{24}$ gold. For example, 18-karat gold is $\frac{18}{24} = \frac{3}{4}$ gold. 10 g

73. ***Chemistry*** A chemist starts with 5 quarts of a 48% acid solution. How much 32% acid solution should she add to produce a 37% solution? 11 qt

74. ***Temperature*** The relationship between the Fahrenheit temperature (F) and the Celsius temperature (C) is given by the formula

$$F = \frac{9}{5}C + 32$$

a. At what temperature will a Fahrenheit thermometer and a Celsius thermometer read the same? $-40°$

b. A materials engineer heats a metal bar until it begins to melt, and he notices that the Fahrenheit temperature is exactly twice the Celsius temperature. At what Fahrenheit temperature did the metal melt? 320°F

The Archimedean law of the lever states that for a lever to be in a state of balance with respect to a point called the fulcrum, the sum of the downward forces times their respective distances from the fulcrum on one side of the fulcrum must equal the sum of the downward forces times their respective distances from the fulcrum on the other side of the fulcrum. The accompanying figure shows this relationship.

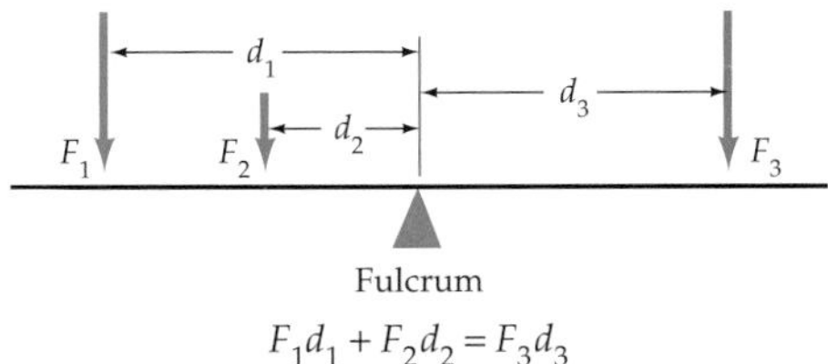

$$F_1d_1 + F_2d_2 = F_3d_3$$

75. ***Locate the Fulcrum*** A 100-pound person 8 feet to the left of the fulcrum and a 40-pound person 5 feet to the left of the fulcrum balance with a 160-pound person on a teeter-totter. How far from the fulcrum is the 160-pound person? 6.25 ft

76. ***Locate the Fulcrum*** A lever 21 feet long has a force of 117 pounds applied to one end and a force of 156 pounds applied to the other end. Where should the fulcrum be located to produce a state of balance? 12 ft from the 117-lb force

77. ***Determine a Force*** How much force applied 5 feet from the fulcrum is needed to lift a 400-pound weight that is located on the other side of the fulcrum, 0.5 foot from the fulcrum? 40 lb

78. ***Determine a Force*** Two workers need to lift a 1440-pound rock. They use a 6-foot steel bar with the fulcrum 1 foot from the rock, as the accompanying figure shows. One worker applies 180 pounds of force to the other end of the lever. How much force will the second worker need to apply 1 foot from that end to lift the rock? 135 lb

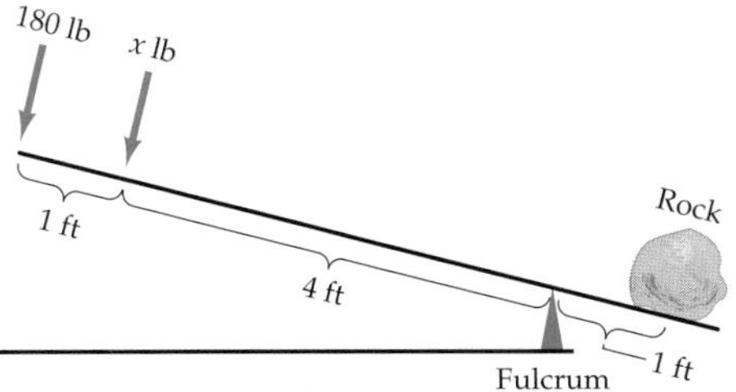

79. ***Speed of Sound in Air*** Two seconds after firing a rifle at a target, the shooter hears the impact of the bullet. Sound travels at 1100 feet per second and the bullet at 1865 feet per second. Determine the distance to the target (to the nearest foot). 1384 ft

80. ***Speed of Sound in Water*** Sound travels through sea water 4.62 times faster than through air. The sound of an exploding mine on the surface of the water and partially submerged reaches a ship through the water 4 seconds before it reaches the ship through the air. How far is the ship from the explosion? Round to the nearest foot. Use 1100 feet per second as the speed of sound through the air. 5615 ft

Prepare for Section 1.2

81. Expand: $(x + 4)^2$ [P.3] $x^2 + 8x + 16$

82. Factor: $x^2 - x - 42$ [P.4] $(x + 6)(x - 7)$

83. Factor: $2x^2 + 3x - 9$ [P.4] $(2x - 3)(x + 3)$

84. Write $\sqrt{-9}$ as a complex number. [P.7] $3i$

85. If $a = -2$, $b = -3$, and $c = 5$, evaluate [P.1] 1

$$\frac{-b - \sqrt{b^2 - 4ac}}{2a}$$

86. If $x = 5 - 2i$, evaluate $x^2 - 2x$. [P.7] $11 - 16i$

Explorations 

1. **Baseball** In baseball, a perfect game is a game in which one of the teams gives up no hits, no walks, and no errors. Statistics show that a batter will get on base roughly 30% of the time. Thus the probability that a pitcher will retire two batters in a row is $0.7^2 = 0.49$. The probability that a pitcher will retire 27 batters in succession and thus pitch a perfect game is 0.7^{27}.*

   a. Explain why the linear equation

   $$p = 2(0.7^{27})x$$

   provides a good estimate of the number of perfect games p we can expect after x games are completed. See the Instructor's Resource Manual.

   b. Check a major league baseball almanac to determine how many perfect games have been played in the last 45 years and how many games have been played in the last 45 years. See the Instructor's Resource Manual.

   c. Use the linear equation in **a.** to estimate how many perfect games we should expect to have been pitched over the last 45 years of major league baseball. How does this result compare with the actual result found in **b.**? See the Instructor's Resource Manual.

*_A Mathematician Reads the Newspaper_ by John Allen Paulos (New York: BasicBooks, A Division of Harper-Collins Publishers, Inc., 1995).

SECTION 1.2 Quadratic Equations and Their Applications

- Solve Quadratic Equations by Factoring
- Solve Quadratic Equations by Taking Square Roots
- Solve Quadratic Equations by Completing the Square
- Solve Quadratic Equations by Using the Quadratic Formula
- The Discriminant of a Quadratic Equation
- Applications of Quadratic Equations

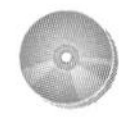

Point of Interest

The term *quadratic* is derived from the Latin word *quadrare*, which means "to make square." Because the area of a square that measures x units on each side is x^2, we refer to equations that can be written in the form $ax^2 + bx + c = 0$ as equations that are quadratic in x.

Solve Quadratic Equations by Factoring

In Section 1.1 you solved linear equations. In this section you will learn to solve a type of equation that is referred to as a *quadratic equation.*

Definition of a Quadratic Equation

A **quadratic equation** in x is an equation that can be written in the **standard quadratic form**

$$ax^2 + bx + c = 0$$

where a, b, and c are real numbers and $a \neq 0$.

Several methods can be used to solve a quadratic equation. For instance, if you can factor $ax^2 + bx + c$ into linear factors, then $ax^2 + bx + c = 0$ can be solved by applying the following property.

The Zero Product Principle

If A and B are algebraic expressions such that $AB = 0$, then $A = 0$ or $B = 0$.

The **zero product principle** states that if the product of two factors is zero, then at least one of the factors must be zero. In Example 1, the zero product principle is used to solve a quadratic equation.

Alternative to Example 1

Solve each equation by factoring.

a. $4x^2 - 2 = 7x$

b. $36x^2 - 12x + 1 = 0$

▪ a. $-\frac{1}{4}, 2$

▪ b. $\frac{1}{6}$

INSTRUCTOR NOTE

Encourage students to check possible solutions in the original equation rather than checking by using an equation they produced from the original equation.

EXAMPLE 1 Solve by Factoring

Solve each quadratic equation by factoring.

a. $x^2 + 2x - 15 = 0$ **b.** $2x^2 - 5x = 12$

Solution

a.

$$x^2 + 2x - 15 = 0$$

$$(x - 3)(x + 5) = 0 \quad \text{▪ Factor.}$$

$$x - 3 = 0 \qquad x + 5 = 0 \quad \text{▪ Set each factor equal to zero.}$$

$$x = 3 \qquad x = -5 \quad \text{▪ Solve each linear equation.}$$

A check shows that 3 and −5 are both solutions of $x^2 + 2x - 15 = 0$.

b.

$$2x^2 - 5x = 12$$

$$2x - 5x - 12 = 0 \quad \text{▪ Write in standard quadratic form.}$$

$$(x - 4)(2x + 3) = 0 \quad \text{▪ Factor.}$$

$$x - 4 = 0 \qquad 2x + 3 = 0 \quad \text{▪ Set each factor equal to zero.}$$

$$x = 4 \qquad 2x = -3 \quad \text{▪ Solve each linear equation.}$$

$$x = -\frac{3}{2}$$

A check shows that 4 and $-\frac{3}{2}$ are both solutions of $2x^2 - 5x = 12$.

CHECK YOUR PROGRESS 1 Solve each quadratic equation by factoring.

a. $6x^2 - x - 12 = 0$ **b.** $6x^2 + x = 15$

Solution *See page S5.* a. $-\frac{4}{3}, \frac{3}{2}$ b. $-\frac{5}{3}, \frac{3}{2}$

In the following example we solve $x^2 - 8x + 16 = 0$ by factoring.

$$x^2 - 8x + 16 = 0$$

$$(x - 4)(x - 4) = 0 \quad \text{▪ Factor.}$$

$$x - 4 = 0 \qquad x - 4 = 0 \quad \text{▪ Set each factor equal to zero.}$$

$$x = 4 \qquad x = 4 \quad \text{▪ Solve each linear equation.}$$

The only solution of $x^2 - 8x + 16 = 0$ is 4. In this situation the single solution 4 is called a **double solution** or **double root** because it was produced by solving the two identical equations $x - 4 = 0$, both of which have 4 as a solution.

▪ Solve Quadratic Equations by Taking Square Roots

In the following example we solve $x^2 = 25$ by factoring.

$$x^2 = 25$$

$$x^2 - 25 = 0$$

$$(x - 5)(x + 5) = 0 \quad \text{▪ Factor.}$$

$$x - 5 = 0 \qquad x + 5 = 0 \quad \text{▪ Set each factor equal to zero.}$$

$$x = 5 \qquad x = -5 \quad \text{▪ Solve each linear equation.}$$

The solutions of $x^2 = 25$ can also be found by taking the square root of each side of the equation. In the following work, a plus or minus sign is placed in front of the square root of 25 to produce both solutions. The notation $x = \pm 5$ means $x = 5$ or $x = -5$.

$$x^2 = 25$$

$$x = \pm\sqrt{25}$$ ■ Take the square root of each side of the equation. Insert a plus or minus sign in front of the radical on the right.

$$x = \pm 5$$

$$x = 5 \quad \text{or} \quad x = -5$$

We will refer to the above method of solving a quadratic equation as the **square root procedure.**

> **The Square Root Procedure**
>
> If $x^2 = c$, then $x = \sqrt{c}$ or $x = -\sqrt{c}$, which can also be written as $x = \pm\sqrt{c}$.

Alternative to Example 2

Use the square root procedure to solve each equation.

a. $2x^2 = 72$

b. $(x + 8)^2 = 81$

■ **a.** ± 6

■ **b.** $-17, 1$

TAKE NOTE

For a review of complex numbers, see Section P.7.

EXAMPLE 2 Solve by Using the Square Root Procedure

Use the square root procedure to solve each equation.

a. $3x^2 + 12 = 0$ **b.** $(x + 1)^2 = 49$

Solution

a. $3x^2 + 12 = 0$

$$3x^2 = -12$$ ■ Solve for x^2.

$$x^2 = -4$$

$$x = \pm\sqrt{-4}$$ ■ Take the square root of each side of the equation. Insert a plus or minus sign in front of the radical on the right.

$$x = \pm 2i$$

b. $(x + 1)^2 = 49$

$$x + 1 = \pm\sqrt{49}$$ ■ Take the square root of each side of the equation.

$$x = -1 \pm 7$$ ■ Simplify.

$$x = 6 \quad \text{or} \quad -8$$

CHECK YOUR PROGRESS 2 Use the square root procedure to solve each equation.

a. $2x^2 - 128 = 0$ **b.** $(x - 3)^2 = 81$

Solution *See page S5.* **a.** ± 8 **b.** -6 or 12

Point of Interest

Mathematicians have studied quadratic equations for centuries. Many of the initial quadratic equations they studied were a result of trying to solve a geometry problem. One of the most famous, which dates from around 500 B.C., concerns "squaring a circle." The question was, "Is it possible to construct a square whose area is the same as the area of a given circle?" For these early mathematicians, to construct meant to draw with only a straightedge and a compass. It was approximately 2300 years later that mathematicians were able to prove that such a construction is impossible.

▪ Solve Quadratic Equations by Completing the Square

Consider the following binomial squares and their perfect-square trinomial products.

Square of a Binomial	=	Perfect-Square Trinomial
$(x+5)^2$	=	$x^2+10x+25$
$(x-3)^2$	=	x^2-6x+9

In each of the above perfect-square trinomials, the coefficient of x^2 is 1 and the constant term is the square of half of the coefficient of the x term.

$$x^2+10x+25, \qquad \left(\frac{1}{2}\cdot 10\right)^2=25$$

$$x^2-6x+9, \qquad \left(\frac{1}{2}\cdot(-6)\right)^2=9$$

Adding to a binomial of the form x^2+bx the constant term that makes the binomial a perfect-square trinomial is called **completing the square.** For example, to complete the square of x^2+8x, add

$$\left(\frac{1}{2}\cdot 8\right)^2=16$$

to produce the perfect-square trinomial $x^2+8x+16$.

Completing the square is a powerful procedure because it can be used to solve *any* quadratic equation. For instance, to solve $x^2-6x+13=0$, begin by writing the variable terms on one side of the equation and the constant term on the other side.

$$x^2-6x+13=0$$

$$x^2-6x=-13$$ ▪ Subtract 13 from each side of the equation.

$$x^2-6x+9=-13+9$$ ▪ Complete the square by adding $\left[\frac{1}{2}(-6)\right]^{-2}=9$ to each side of the equation.

$$(x-3)^2=-4$$ ▪ Factor and solve by the square root procedure.

$$x-3=\pm\sqrt{-4}$$

$$x-3=\pm 2i$$

$$x=3\pm 2i$$

The solutions of $x^2-6x+13=0$ are $3-2i$ and $3+2i$. You can check these solutions by substituting each solution into the original equation. For instance, the following check shows that $3-2i$ does satisfy the original equation.

$$x^2-6x+13=0$$

$$(3-2i)^2-6(3-2i)+13\stackrel{?}{=}0$$ ▪ Substitute $3-2i$ for x.

$$9-12i+4i^2-18+12i+13\stackrel{?}{=}0$$ ▪ Simplify.

$$4+4(-1)\stackrel{?}{=}0$$

$$0=0$$ ▪ The left side equals the right side, so $3-2i$ checks.

Alternative to Example 3

Solve $x^2 + 3 = x$ by completing the square.

■ $\frac{1}{2} \pm \frac{i\sqrt{11}}{2}$

Point of Interest

Ancient mathematicians thought of "completing the square" in a geometric manner. For instance, to complete the square of $x^2 + 8x$, draw a square that measures x units on each side and add four rectangles that measure 1 unit by x units to the right side and the bottom of the square.

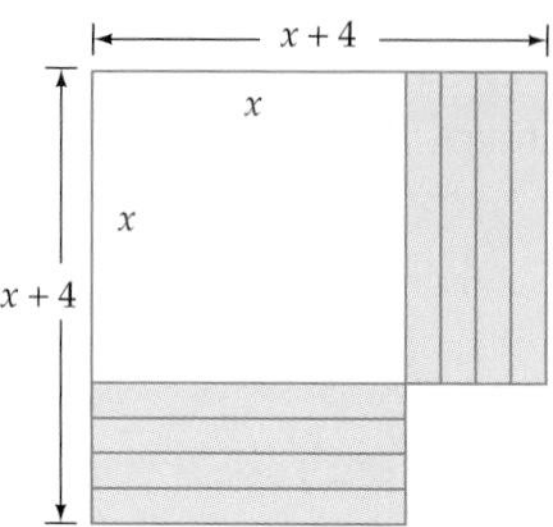

Each of the rectangles has an area of x square units, so the total area of the figure is $x^2 + 8x$. To make this figure a complete square, we must add 16 squares that measure 1 unit by 1 unit, as shown below.

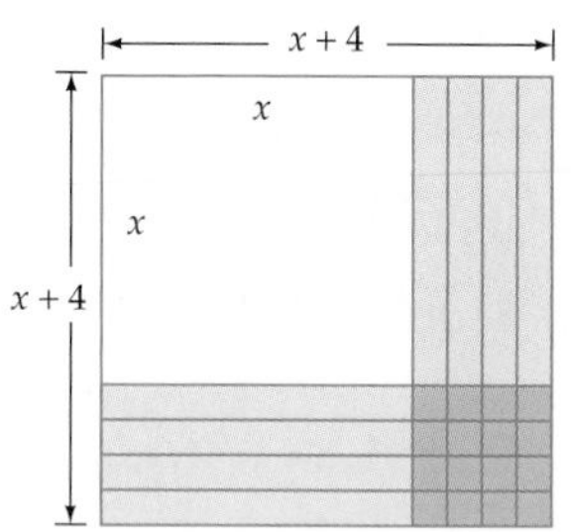

This figure is a *complete square* whose area is

$$(x + 4)^2 = x^2 + 8x + 16$$

Alternative to Example 4

Solve $3x^2 + 9x - 1 = 0$ by completing the square.

■ $\frac{-9 \pm \sqrt{93}}{6}$

EXAMPLE 3 Solve by Completing the Square

Solve $x^2 = 2x + 6$ by completing the square.

Solution

$$x^2 = 2x + 6$$

$$x^2 - 2x = 6$$ ■ Isolate the constant term.

$$x^2 - 2x + 1 = 6 + 1$$ ■ Complete the square.

$$(x - 1)^2 = 7$$ ■ Factor and simplify.

$$x - 1 = \pm\sqrt{7}$$ ■ Apply the square root procedure.

$$x = 1 \pm \sqrt{7}$$ ■ Solve for x.

The exact solutions of $x^2 = 2x + 6$ are $x = 1 - \sqrt{7}$ and $x = 1 + \sqrt{7}$. A calculator can be used to show that $1 - \sqrt{7} \approx -1.646$ and $1 + \sqrt{7} \approx 3.646$. The decimals -1.646 and 3.646 are approximate solutions of $x^2 = 2x + 6$.

CHECK YOUR PROGRESS 3 Solve $x^2 + 6x - 5 = 0$ by completing the square.

Solution *See page S5.* $-3 \pm \sqrt{14}$

Completing the square by adding the square of half of the coefficient of the x term requires that the coefficient of the x^2 term be 1. If the coefficient of the x^2 term is not 1, then first multiply each term on each side of the equation by the reciprocal of the coefficient of x^2 to produce a coefficient of 1 for the x^2 term.

EXAMPLE 4 Solve by Completing the Square

Solve $2x^2 + 8x - 1 = 0$ by completing the square.

Solution

$$2x^2 + 8x - 1 = 0$$

$$2x^2 + 8x = 1$$ ■ Isolate the constant term.

$$\frac{1}{2}(2x^2 + 8x) = \frac{1}{2}(1)$$ ■ Multiply both sides of the equation by the reciprocal of the leading coefficient.

$$x^2 + 4x = \frac{1}{2}$$

$$x^2 + 4x + 4 = \frac{1}{2} + 4$$ ■ Complete the square.

$$(x + 2)^2 = \frac{9}{2}$$ ■ Factor and simplify.

$$x + 2 = \pm\sqrt{\frac{9}{2}}$$ ■ Apply the square root procedure.

$$x = -2 \pm 3\sqrt{\frac{1}{2}}$$ ■ Solve for x.

$$x = -2 \pm 3\frac{\sqrt{2}}{2}$$ ■ Simplify.

$$x = \frac{-4 \pm 3\sqrt{2}}{2}$$

Continued ➤

The solutions of $2x^2 + 8x - 1 = 0$ are $x = \frac{-4 + 3\sqrt{2}}{2}$ and $x = \frac{-4 - 3\sqrt{2}}{2}$.

CHECK YOUR PROGRESS 4 Solve $4x^2 + 32x - 3 = 0$ by completing the square.

Solution *See page S5.* $\frac{-8 \pm \sqrt{67}}{2}$

Point of Interest

Evariste Galois (1811–1832)

The quadratic formula provides the solutions to the general quadratic equation

$$ax^2 + bx + c = 0$$

Formulas have also been developed to solve the general cubic

$$ax^3 + bx^2 + cx + d = 0$$

and the general quartic

$$ax^4 + bx^3 + cx^2 + dx + e = 0$$

However, the French mathematician Evariste Galois, shown above, was able to prove that there are no formulas that can be used to solve "by radicals" general equations of degree 5 or larger.

Shortly after the completion of his remarkable proof, Galois was shot in a duel. It has been reported that as Galois lay dying he said to his brother, Alfred, "Take care of my work. Make it known. Important." When Alfred broke into tears, Evariste told him, "Don't cry, Alfred. I need all my courage to die at twenty."

(*Source: Whom the Gods Love* by Leopold Infeld. The National Council of Teachers of Mathematics, 1978, page 299.)

▪ Solve Quadratic Equations by Using the Quadratic Formula

Completing the square on $ax^2 + bx + c = 0$ $(a \neq 0)$ produces a formula for x in terms of the coefficients a, b, and c. The formula is known as the **quadratic formula,** and it can be used to solve *any* quadratic equation.

The Quadratic Formula

If $ax^2 + bx + c = 0, a \neq 0$, then

$$x = \frac{-b \pm \sqrt{b^2 - 4ac}}{2a}$$

PROOF We assume a is a positive real number. If a were a negative real number, then we could multiply each side of the equation by -1 to make it positive.

$$ax^2 + bx + c = 0 \quad (a \neq 0)$$ ▪ Given.

$$ax^2 + bx = -c$$ ▪ Isolate the constant term.

$$x^2 + \frac{b}{a}x = -\frac{c}{a}$$ ▪ Multiply each term on each side of the equation by $1/a$.

$$x^2 + \frac{b}{a}x + \left(\frac{b}{2a}\right)^2 = \left(\frac{b}{2a}\right)^2 - \frac{c}{a}$$ ▪ Complete the square.

$$\left(x + \frac{b}{2a}\right)^2 = \frac{b^2}{4a^2} - \frac{c}{a}$$ ▪ Factor the left side. Simplify the power on the right side.

$$\left(x + \frac{b}{2a}\right)^2 = \frac{b^2}{4a^2} - \frac{4a}{4a} \cdot \frac{c}{a}$$ ▪ Use a common denominator to simplify the right side.

$$x + \frac{b}{2a} = \pm\sqrt{\frac{b^2 - 4ac}{4a^2}}$$ ▪ Apply the square root theorem.

$$x + \frac{b}{2a} = \pm\frac{\sqrt{b^2 - 4ac}}{2a}$$ ▪ Because $a > 0$, $\sqrt{4a^2} = 2a$.

$$x = -\frac{b}{2a} \pm \frac{\sqrt{b^2 - 4ac}}{2a}$$ ▪ Add $-\frac{b}{2a}$ to each side.

$$x = \frac{-b \pm \sqrt{b^2 - 4ac}}{2a}$$ ■

As a general rule, you should first try to solve quadratic equations by factoring. If the factoring process proves difficult, then solve either by using the quadratic formula or by completing the square.

Alternative to Example 5

Solve each equation by using the quadratic formula.

a. $2x^2 + 3x = 5$

b. $x^2 - 5x = -7$

■ a. $1, -\frac{5}{2}$

■ b. $\frac{5}{2} \pm \frac{i\sqrt{3}}{2}$

TAKE NOTE

Although the equation in Example 5**a.** can be solved by factoring, we have solved it by using the quadratic formula to illustrate the procedures involved in applying the quadratic formula.

EXAMPLE 5 Solve by Using the Quadratic Formula

Use the quadratic formula to solve each of the following.

a. $4x^2 - 4x - 3 = 0$ **b.** $x^2 = 3x + 5$

Solution

a. For the equation $4x^2 - 4x - 3 = 0$, we have $a = 4$, $b = -4$, and $c = -3$. Substituting in the quadratic formula produces

$$x = \frac{-b \pm \sqrt{b^2 - 4ac}}{2a}$$

$$= \frac{-(-4) \pm \sqrt{(-4)^2 - 4(4)(-3)}}{2(4)}$$

$$= \frac{4 \pm \sqrt{64}}{8} = \frac{4 \pm 8}{8} = \frac{1 \pm 2}{2} = \frac{3}{2} \quad \text{or} \quad -\frac{1}{2}$$

The solutions of $4x^2 - 4x - 3 = 0$ are $x = \frac{3}{2}$ and $x = -\frac{1}{2}$.

b. The standard form of $x^2 = 3x + 5$ is $x^2 - 3x - 5 = 0$. Substituting $a = 1$, $b = -3$, and $c = -5$ in the quadratic formula produces

$$x = \frac{-(-3) \pm \sqrt{(-3)^2 - 4(1)(-5)}}{2(1)}$$

$$= \frac{3 \pm \sqrt{29}}{2}$$

The solutions of $x^2 = 3x + 5$ are $x = \frac{3 + \sqrt{29}}{2}$ and $x = \frac{3 - \sqrt{29}}{2}$.

CHECK YOUR PROGRESS 5 Use the quadratic formula to solve each of the following

a. $12x^2 - x - 6 = 0$ **b.** $x^2 = 2x - 2$

Solution *See page S5.* a. $-\frac{2}{3}, \frac{3}{4}$ b. $1 \pm i$

? QUESTION Can the quadratic formula be used to solve any quadratic equation $ax^2 + bx + c = 0$ with real coefficients and $a \neq 0$?

■ The Discriminant of a Quadratic Equation

The solutions of $ax^2 + bx + c = 0$, $a \neq 0$ are given by

$$x = \frac{-b \pm \sqrt{b^2 - 4ac}}{2a}$$

? ANSWER Yes. However, it is sometimes easier to find the solutions by factoring, by the square root procedure, or by completing the square.

The expression under the radical, $b^2 - 4ac$, is called the **discriminant** of the equation $ax^2 + bx + c = 0$. If $b^2 - 4ac \geq 0$, then $\sqrt{b^2 - 4ac}$ is a real number. If $b^2 - 4ac < 0$, then $\sqrt{b^2 - 4ac}$ is not a real number. Thus the sign of the discriminant can be used to determine whether the solutions of a quadratic equation are real numbers.

TAKE NOTE

For a review of complex conjugates, see Section P.7.

The Discriminant and the Solutions of a Quadratic Equation

The equation $ax^2 + bx + c = 0$, with real coefficients and $a \neq 0$, has as its discriminant $b^2 - 4ac$.

- If $b^2 - 4ac > 0$, then $ax^2 + bx + c = 0$ has *two distinct real solutions.*
- If $b^2 - 4ac = 0$, then $ax^2 + bx + c = 0$ has one *real solution.* The solution is a double solution.
- If $b^2 - 4ac < 0$, then $ax^2 + bx + c = 0$ has *two distinct nonreal complex solutions.* The solutions are conjugates of each other.

Alternative to Example 6

For each equation, determine the discriminant and state the number of real solutions.

a. $4x^2 - 5x = -1$
b. $9x^2 - 30x + 25 = 0$
c. $2x^2 - x + 5 = 0$

- **a. 9; two real solutions**
- **b. 0; one real solution**
- **c. −39; no real solutions**

EXAMPLE 6 Use the Discriminant to Determine the Number of Real Solutions

For each equation, determine the discriminant and state the number of real solutions.

a. $2x^2 - 5x + 1 = 0$ **b.** $3x^2 + 6x + 7 = 0$ **c.** $x^2 + 6x + 9 = 0$

Solution

a. The discriminant of $2x^2 - 5x + 1 = 0$ is $b^2 - 4ac = (-5)^2 - 4(2)(1) = 17$. Because the discriminant is positive, $2x^2 - 5x + 1 = 0$ has two distinct real solutions.

b. The discriminant of $3x^2 + 6x + 7 = 0$ is $b^2 - 4ac = 6^2 - 4(3)(7) = -48$. Because the discriminant is negative, $3x^2 + 6x + 7 = 0$ has no real solutions.

c. The discriminant of $x^2 + 6x + 9 = 0$ is $b^2 - 4ac = 6^2 - 4(1)(9) = 0$. Because the discriminant is 0, $x^2 + 6x + 9 = 0$ has a single real solution.

CHECK YOUR PROGRESS 6 For each equation, determine the discriminant and state the number of real solutions.

a. $4x^2 - 5x + 3 = 0$ **b.** $6x^2 - 23x + 20 = 0$ **c.** $4x^2 + 12x + 9 = 0$

Solution *See page S5.* **a.** −23; no real solutions **b.** 49; two real solutions **c.** 0; one real solution

Applications of Quadratic Equations

A **right triangle** contains one 90° angle. The side opposite the 90° angle is called the **hypotenuse.** The other two sides are called **legs.** The lengths of the sides of a right triangle are related by a theorem known as the Pythagorean Theorem.

The Pythagorean Theorem

If a and b denote the lengths of the legs of a right triangle and c the length of the hypotenuse, then $c^2 = a^2 + b^2$.

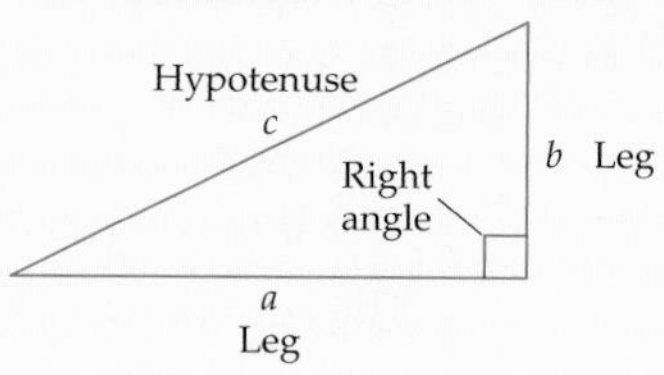

The **Pythagorean Theorem** states that the square of the length of the hypotenuse of a right triangle is equal to the sum of the squares of the lengths of the legs. This theorem is often used to solve applications that involve right triangles.

Alternative to Example 7

A modern computer screen measures 17 inches diagonally and its aspect ratio is 5 to 3. Find the width and the height of the screen. Round to the nearest tenth of an inch.

- **width 14.6 in.; height 8.7 in.**

EXAMPLE 7 Determine the Dimensions of a Television Screen

A television screen measures 60 inches diagonally and its *aspect ratio* is 16 to 9. This means that the ratio of the width of the screen to the height of the screen is 16 to 9. Find the width and height of the screen. Round to the nearest tenth of an inch.

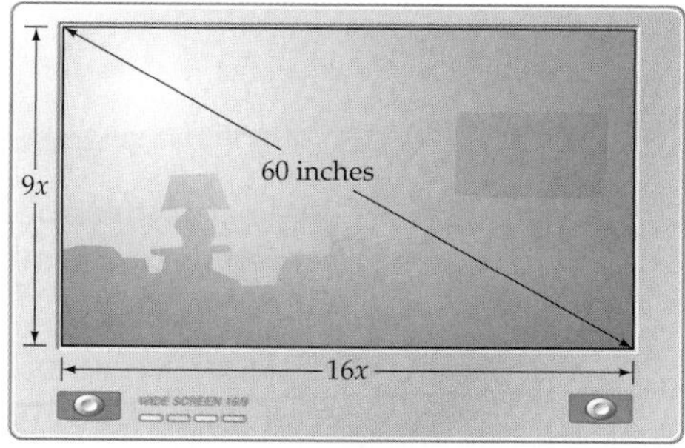

A 60-inch television screen with a 16:9 aspect ratio

Solution Let $16x$ represent the width of the screen and let $9x$ represent the height of the screen. Applying the Pythagorean Theorem gives us

$$(16x)^2 + (9x)^2 = 60^2$$

$$256x^2 + 81x^2 = 3600$$ ■ **Solve for x^2.**

$$337x^2 = 3600$$

$$x^2 = \frac{3600}{337}$$ ■ **Apply the square root procedure.**

$$x = \sqrt{\frac{3600}{337}} \approx 3.268 \text{ inches}$$ ■ **The plus or minus sign is not used in this application because we know x is positive.**

Thus the height of the screen is about $9(3.268) \approx 29.4$ inches and the width of the screen is about $16(3.268) \approx 52.3$ inches.

Continued ➤

CHECK YOUR PROGRESS 7 A television screen measures 54 inches diagonally and its aspect ratio is 4 to 3. Find the width and the height of the screen.

Solution *See page S5.* 43.2 in. by 32.4 in.

Alternative to Example 8
Exercise 60, page 110

EXAMPLE 8 Determine the Dimensions of a Candy Bar

At the present time a company makes rectangular solid candy bars that measure 5 inches by 2 inches by 0.5 inch. Due to difficult financial times, the company has decided to keep the price of the candy bar fixed and reduce the volume of the bar by 20%. What should be the dimensions of the new candy bar if it is decided to keep the height at 0.5 inch and to keep the length of the candy bar 3 inches longer than the width?

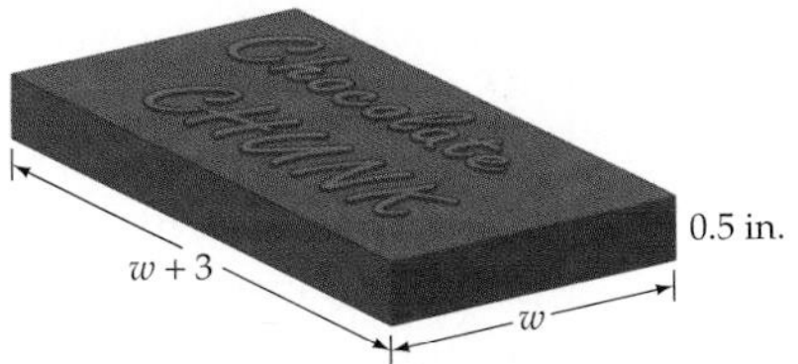

Solution The volume of a rectangular solid is given by $V = lwh$. The original candy bar had a volume of $5 \cdot 2 \cdot 0.5 = 5$ cubic inches. The new candy bar will have a volume of $80\%(5) = 0.80(5) = 4$ cubic inches.

Let w represent the width and $w + 3$ represent the length of the new candy bar. For the new candy bar we have:

$$lwh = V$$

$$(w + 3)(w)(0.5) = 4 \quad \text{• Substitute in the volume formula.}$$

$$(w + 3)(w) = 8 \quad \text{• Multiply each side by 2.}$$

$$w^2 + 3w = 8$$

$$w^2 + 3w - 8 = 0 \quad \text{• Write in } ax^2 + bx + c = 0 \text{ form.}$$

$$w = \frac{-(3) \pm \sqrt{(3)^2 - 4(1)(-8)}}{2(1)} \quad \text{• Use the quadratic formula.}$$

$$= \frac{-3 \pm \sqrt{41}}{2}$$

$$\approx 1.7 \quad \text{or} \quad -4.7$$

We can disregard the negative value because the width must be positive. The width of the new candy bar should be 1.7 inches, to the nearest 0.1 inch. The length should be 3 inches longer, which is 4.7 inches.

CHECK YOUR PROGRESS 8 What should be the dimensions, to the nearest 0.1 inch, of the new candy bar in Example 8 if it is decided to keep the height at 0.5 inch and to keep the length of the candy bar 2.5 times longer than the width?

Solution *See page S6.* Width 1.8 in.; length 4.5 in.

INTEGRATING TECHNOLOGY

In many application problems it is helpful to use a calculator to estimate the solutions of a quadratic equation by applying the quadratic formula. For instance, the following figure shows the use of a graphing calculator to estimate the solutions of $w^2 + 3w - 8 = 0$.

```
(-3+√(3²-4*1*(-8)))/2
            1.701562119
(-3-√(3²-4*1*(-8)))/2
           -4.701562119
```

INSTRUCTOR NOTE

Inform your students that many calculators have a "recall procedure" that allows a user to quickly recall and edit a previous entry. For instance, in the above TI-83 calculator display, the second entry was produced by recalling the first entry and changing the plus sign in front of the radical to a subtraction sign. On a TI-83 calculator, this recall procedure is accomplished by pressing 2nd ENTRY. ENTRY is located above the ENTER key.

Quadratic equations are often used to determine the height (position) of an object that has been dropped or projected. For instance, the *position equation* $s = -16t^2 + v_0t + s_0$ can be used to estimate the height of a projected object near the surface of the earth at a given time t, in seconds. In this equation, v_0 is the initial velocity of the object in feet per second and s_0 is the initial height of the object in feet.

Alternative to Example 9
Exercise 70, page 111

EXAMPLE 9 Determine the Time of Descent

A ball is thrown *downward* with an initial velocity of 5 feet per second from the Golden Gate Bridge, which is 220 feet above the water. How long will it take for the ball to hit the water? Round your answer to the nearest 0.01 second.

Solution The distance s, in feet, of the ball above the water after t seconds is given by $s = -16t^2 - 5t + 220$. We have replaced v_0 with -5 because the ball is thrown downward. (If the ball had been thrown upward, we would use $v_0 = 5$.) To determine the time it takes the ball to hit the water, substitute 0 for s in the equation $s = -16t^2 - 5t + 220$ and solve for t. In the following work, we have solved by using the quadratic formula.

$$0 = -16t^2 - 5t + 220$$

$$t = \frac{-(-5) \pm \sqrt{(-5)^2 - 4(-16)(220)}}{2(-16)}$$ ▪ Use the quadratic formula.

$$= \frac{5 \pm \sqrt{14{,}105}}{-32}$$ ▪ Use a calculator to simplify and solve for t.

$$\approx -3.87 \quad \text{or} \quad 3.56$$

Because the time must be positive, we disregard the negative value. The ball will hit the water in about 3.56 seconds.

CHECK YOUR PROGRESS 9 A baseball player hits a baseball upward with an initial vertical velocity of 88 feet per second. The distance s, in feet, of the ball above the ground after t seconds is given by $s = -16t^2 + 88t + 5$. How long will it take for the ball to hit the ground? Round your answer to the nearest 0.1 second.

Solution *See page S6.* 5.6 s

Topics for Discussion

1. Name the four methods of solving a quadratic equation that have been discussed in this section. What are the advantages and disadvantages of each?
2. If x and y are real numbers and $xy = 0$, then $x = 0$ or $y = 0$. Do you agree with this statement? Explain.

3. If x and y are real numbers and $xy = 1$, then $x = 1$ or $y = 1$. Do you agree with this statement? Explain.
4. Explain how to complete the square on $x^2 + bx$.
5. If the discriminant of the quadratic $ax^2 + bx + c = 0$ with real coefficients and $a \neq 0$ is negative, then what can be said concerning the solutions of the equation?

EXERCISES 1.2

— Suggested Assignment: Exercises 1–77, odd; 68; and 79–84.

In Exercises 1 to 10, solve each quadratic equation by factoring and applying the zero product principle.

1. $x^2 - 2x - 15 = 0$ $-3, 5$
2. $x^2 + 3x - 10 = 0$ $-5, 2$
3. $2x^2 - x = 1$ $-\frac{1}{2}, 1$
4. $2x^2 + 5x = 3$ $-3, \frac{1}{2}$
5. $8x^2 + 189x - 72 = 0$ $-24, \frac{3}{8}$
6. $12x^2 - 41x + 24 = 0$ $\frac{8}{3}, \frac{3}{4}$
7. $3x^2 - 7x = 0$ $0, \frac{7}{3}$
8. $5x^2 = -8x$ $0, -\frac{8}{5}$
9. $(x - 5)^2 - 9 = 0$ $2, 8$
10. $(3x + 4)^2 - 16 = 0$ $-\frac{8}{3}, 0$

In Exercises 11 to 20, use the square root procedure to solve each quadratic equation.

11. $x^2 = 81$ ± 9
12. $x^2 = 225$ ± 15
13. $2x^2 = 48$ $\pm 2\sqrt{6}$
14. $3x^2 = 144$ $\pm 4\sqrt{3}$
15. $3x^2 + 12 = 0$ $\pm 2i$
16. $4x^2 + 20 = 0$ $\pm i\sqrt{5}$
17. $(x - 5)^2 = 36$ $-1, 11$
18. $(x + 4)^2 = 121$ $-15, 7$
19. $(x - 3)^2 + 16 = 0$ $3 \pm 4i$
20. $(x + 2)^2 + 28 = 0$ $-2 \pm 2i\sqrt{7}$

In Exercises 21 to 32, solve each quadratic equation by completing the square.

21. $x^2 + 6x + 1 = 0$ $-3 \pm 2\sqrt{2}$
22. $x^2 + 8x - 10 = 0$ $-4 \pm \sqrt{26}$
23. $x^2 - 2x - 15 = 0$ $-3, 5$
24. $x^2 + 2x - 8 = 0$ $-4, 2$
25. $x^2 + 4x + 5 = 0$ $-2 \pm i$
26. $x^2 - 6x + 10 = 0$ $3 \pm i$
27. $x^2 + 3x - 1 = 0$ $\frac{-3 \pm \sqrt{13}}{2}$
28. $x^2 + 7x - 2 = 0$ $\frac{-7 \pm \sqrt{57}}{2}$
29. $2x^2 + 4x - 1 = 0$ $\frac{-2 \pm \sqrt{6}}{2}$
30. $2x^2 + 10x - 3 = 0$ $\frac{-5 \pm \sqrt{31}}{2}$
31. $3x^2 - 8x = -1$ $\frac{4 \pm \sqrt{13}}{3}$
32. $4x^2 - 4x = -15$ $\frac{1}{2} \pm \frac{\sqrt{14}}{2}i$

In Exercises 33 to 46, solve each quadratic equation by using the quadratic formula.

33. $x^2 - 2x = 15$ $-3, 5$
34. $x^2 - 5x = 24$ $-3, 8$
35. $x^2 = -x + 1$ $\frac{-1 \pm \sqrt{5}}{2}$
36. $x^2 = -x - 1$ $-\frac{1}{2} \pm \frac{\sqrt{3}}{2}i$
37. $2x^2 + 4x = -1$ $\frac{-2 \pm \sqrt{2}}{2}$
38. $2x^2 + 4x = 1$ $\frac{-2 \pm \sqrt{6}}{2}$
39. $3x^2 - 5x + 3 = 0$ $\frac{5}{6} \pm \frac{\sqrt{11}}{6}i$
40. $3x^2 - 5x + 4 = 0$ $\frac{5}{6} \pm \frac{\sqrt{23}}{6}i$
41. $\frac{1}{2}x^2 + \frac{3}{4}x - 1 = 0$ $\frac{-3 \pm \sqrt{41}}{4}$
42. $\frac{2}{3}x^2 - 5x + \frac{1}{2} = 0$ $\frac{15 \pm \sqrt{213}}{4}$
43. $24x^2 = 22x + 35$ $-\frac{5}{6}, \frac{7}{4}$
44. $72x^2 + 13x = 15$ $-\frac{5}{9}, \frac{3}{8}$
45. $0.5x^2 + 0.6x = 0.8$ $-2, \frac{4}{5}$
46. $1.2x^2 + 0.4x - 0.5 = 0$ $-\frac{5}{6}, \frac{1}{2}$

In Exercises 47 to 56, determine the discriminant of the quadratic equation and then state the number of real solutions of the equation. Do not solve the equation.

47. $2x^2 - 5x - 7 = 0$ 81; two real solutions
48. $x^2 + 3x - 11 = 0$ 53; two real solutions
49. $3x^2 - 2x + 10 = 0$ -116; no real solutions
50. $x^2 + 3x + 3 = 0$ -3; no real solutions
51. $x^2 - 20x + 100 = 0$ 0; one real solution
52. $4x^2 + 12x + 9 = 0$ 0; one real solution
53. $24x^2 = -10x + 21$ 2116; two real solutions
54. $32x^2 - 44x = -15$ 16; two real solutions
55. $12x^2 + 15x = -7$ -111; no real solutions
56. $8x^2 = 5x - 3$ -71; no real solutions
57. For what values of k does $x^2 - 6x + k = 0$ have two distinct real number solutions? $k < 9$
58. Show that the equation $x^2 + bx - 4 = 0$ always has two distinct real number solutions regardless of the value of b. The discriminant is $b^2 + 16$, which is positive for any value of b.

59. ***Animal Enclosure Dimensions*** A veterinarian wishes to use 132 feet of chain-link fencing to enclose a rectangular region and subdivide the region into two smaller rectangular regions, as shown in the following figure. If the total enclosed area is 576 square feet, find the dimensions of the enclosed region. 12 ft by 48 ft or 32 ft by 18 ft

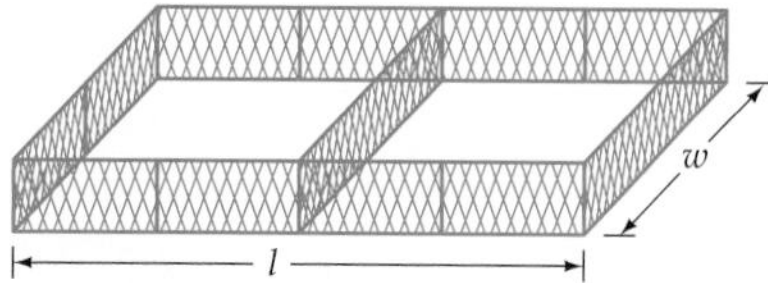

60. ***Construction of a Box*** A square piece of cardboard is formed into a box by cutting out 3-inch squares from each of the corners and folding up the sides, as shown in the figure. If the volume of the box needs to be 126.75 cubic inches, what size square piece of cardboard is needed? 12.5 in. by 12.5 in.

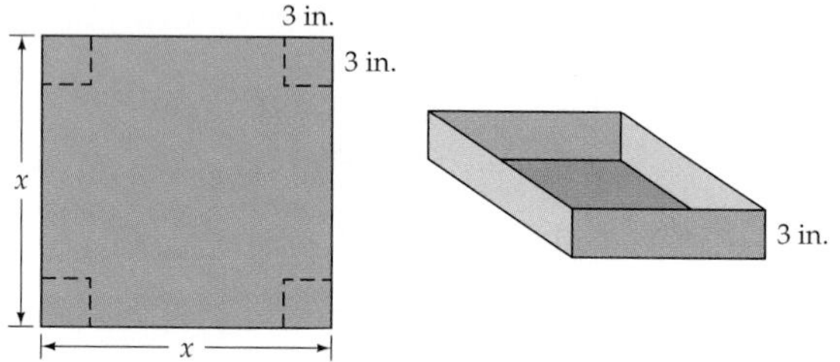

Business and Economics

61. ***Publishing Costs*** The cost, in dollars, of publishing x books is $C(x) = 40{,}000 + 20x + 0.0001x^2$. How many books can be published for \$250,000? 10,000 books

62. ***Cost of a Wedding*** The average cost of a wedding, in dollars, is modeled by

$$C(t) = 38t^2 + 291t + 15{,}208$$

where $t = 0$ represents the year 1990 and $0 \leq t \leq 14$. Use the model to determine the year during which the average cost of a wedding first reached \$19,000. 1996

63. ***Revenue*** The price, in dollars, for a certain product is given by $p = 26 - 0.01x$, where x is the number of units sold per month. The monthly revenue is given by $R = xp$. What number of items sold produces a monthly revenue of \$16,500? 1100 or 1500 items

64. ***Profit*** A company has determined that the profit, in dollars, it can expect from the manufacture and sale of x tennis racquets is given by

$$P = -0.01x^2 + 168x - 120{,}000$$

How many racquets should the company manufacture and sell to earn a profit of \$518,000? 5800 or 11,000 racquets

Life and Health Sciences

65. ***Quadratic Growth*** A plant's ability to create food through the process of photosynthesis is dependent on the surface area of its leaves. A biologist has determined that the surface area A of a maple leaf can be closely approximated by the formula $A = 0.72(1.28)h^2$, where h is the height of the leaf in inches.

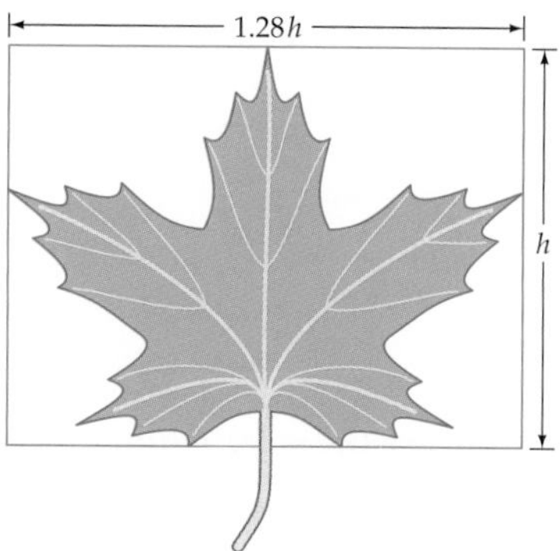

Find

a. the surface area of a maple leaf with a height of 7 inches. Round to the nearest 0.1 square inch. 45.2 in.^{2}

b. the height of a maple leaf with an area of 92 square inches. Round to the nearest 0.1 inch. 10.0 in.

66. ***Population Density of a City*** A city health official has determined that the population density D (in people per square mile) of a city is related to the distance x, in miles, from the center of the city by $D = -45x^2 + 190x + 200$, $0 < x < 5$. At what distances from the center of the city does the population density equal 250 people per square mile? Round each result to the nearest 0.1 mile. 0.3 mi and 3.9 mi

Sports and Recreation

67. ***Baseball Diamond Dimensions*** How far, to the nearest tenth of a foot, is it from home plate to second base on a baseball diamond? *Hint:* The bases in a baseball diamond form a square that measures 90 feet on each side. See the following figure. 127.3 ft

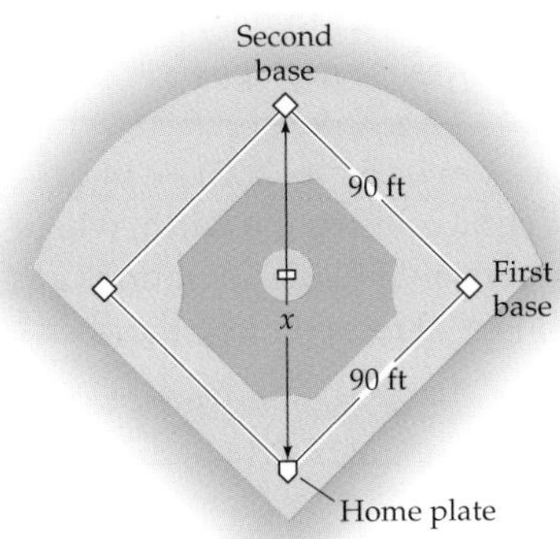

68. *Daredevil Motorcycle Jump* In March of 2000, Doug Danger made a successful motorcycle jump over an L-1011 jumbo jet. The horizontal distance of his jump was 160 feet and his height, in feet, during the jump was approximated by $h = -16t^2 + 25.3t + 20$, for $t \geq 0$. He left the take-off ramp at a height of 20 feet and he landed on the landing ramp at a height of about 17 feet. How long, to the nearest tenth of a second, was he in the air? 1.7 s

69. *Height of a Rocket* A model rocket is launched upward with an initial velocity of 220 feet per second. The height, in feet, of the rocket t seconds after the launch is given by $h = -16t^2 + 220t$. How many seconds after the launch will the rocket be 350 feet above the ground? Round to the nearest 0.1 second. 1.8 s and 11.9 s

70. *Baseball* The height h, in feet, of a baseball above the ground t seconds after it is hit is given by $h = -16t^2 + 52t + 4.5$. Use this equation to determine the number of seconds, to the nearest 0.1 second, from the time the ball is hit until the ball hits the ground. 3.3 s

71. *Baseball* Two equations can be used to track the position of the baseball t seconds after it is hit. For instance, $h = -16t^2 + 50t + 4.5$ gives the height, in feet, of a baseball t seconds after it is hit and $s = 103.9t$ gives the horizontal distance, in feet, the ball is from home plate t seconds after it is hit. Use these equations to determine whether this particular baseball will clear a 10-foot fence positioned 360 feet from home plate. No

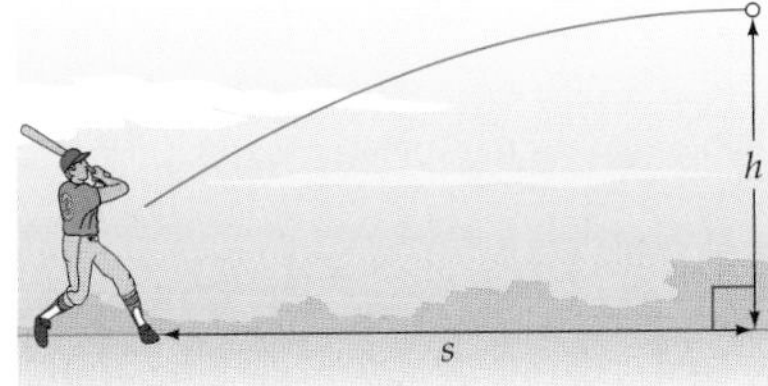

72. *Basketball* Michael Jordan was known for his "hang time," which is the amount of time a player is in the air when making a jump toward the basket. An equation that approximates the height s, in inches, of one of Jordan's jumps is given by $s = -16t^2 + 26.6t$, where t is time in seconds. Use this equation to determine Michael Jordan's hang time, to the nearest 0.1 second, for this jump. 1.7 s

Social Sciences

73. *Number of Handshakes* If everyone in a group of n people shakes hands with everyone other than themselves, then the total number of handshakes h is given by

$$h = \frac{1}{2}n(n - 1)$$

The total number of handshakes that are exchanged by a group of people is 36. How many people are in the group? 9 people

74. *Median Age at First Marriage* During the first 60 years of the twentieth century, couples tended to marry at younger and younger ages. During the last 40 years, that trend was reversed. (*Source:* U.S. Census Bureau, **www.Census.gov**)

Median Age at First Marriage, for Women

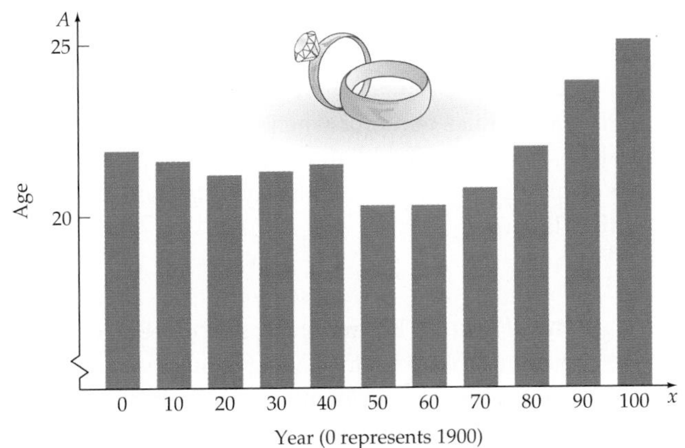

The median age A, in years, at first marriage for women can be modeled by

$$A = 0.0013x^2 - 0.1048x + 22.5256$$

where $x = 0$ represents the year 1900 and $x = 100$ represents the year 2000. Use the model to predict in what year in the future the median age at first marriage for women will first reach 26 years. 2005

75. ***Percent of Divorced Citizens*** The percent P of U.S. citizens who are divorced can be closely approximated by $P = -0.0016t^2 + 0.225t + 6.201$, where t is time in years, with $t = 0$ representing 1980. Use this model to predict in what year the percent of U.S. citizens who are divorced will first reach 11.0%. (*Source:* U.S. Census Bureau, **www.Census.gov**) 2006

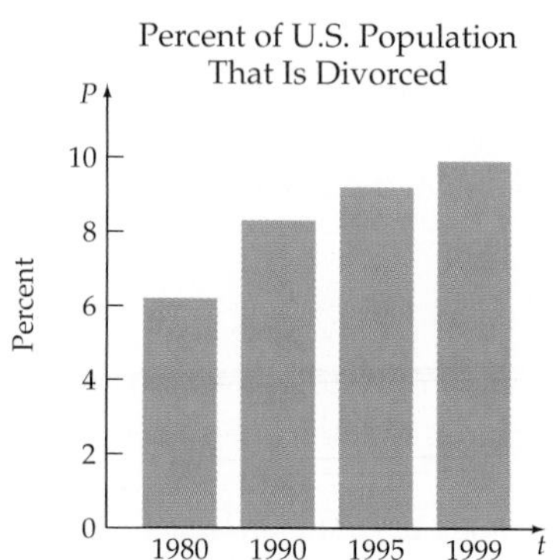

Physical Sciences and Engineering

76. ***Automotive Engineering*** The number of feet N that a car needs to stop on a certain road surface is given by $N = -0.015v^2 + 3v$, $0 \le v \le 90$, where v is the speed of the car in miles per hour when the driver applies the brakes. What is the maximum speed, to the nearest mile per hour, that a motorist can be traveling and stop the car within 100 feet? 42 mph

77. ***Orbital Debris*** The amount of space debris orbiting the Earth has been increasing at an alarming rate. In 1995, there were about 14 million pounds of debris orbiting Earth, and by the year 2000 the amount of debris had increased to over 25 million pounds. (*Source:* **http://orbitaldebris.jsc.nasa.gov/**)

The equation $A = 0.05t^2 + 2.25t + 14$ closely models the amount of debris orbiting the Earth, where A is the amount of debris in millions of pounds and t is the time in years, with $t = 0$ representing the year 1995. Use the equation to

a. estimate the amount of orbital debris we can expect in the year 2006. 44.8 million pounds

b. estimate in what year the amount of orbital debris will first reach 50 million pounds. 2007

78. ***Traffic Control*** Traffic engineers install "flow lights" at the entrances of freeways to control the number of cars entering the freeway during times of heavy traffic. For a particular freeway entrance, the number of cars N waiting to enter the freeway during the morning hours can be approximated by $N = -5t^2 + 80t - 280$, where t is the time of the day and $6 \le t \le 10.5$. According to this model, when will there be 35 cars waiting to enter the freeway? 7 A.M. and 9 A.M.

Prepare for Section 1.3

79. Factor: $x^3 - 16x$ [P.4] $x(x + 4)(x - 4)$

80. Factor: $x^4 - 36x^2$ [P.4] $x^2(x + 6)(x - 6)$

81. Evaluate: $8^{2/3}$ [P.6] 4

82. Evaluate: $16^{3/2}$ [P.6] 64

83. Find $\left(1 + \sqrt{x - 5}\right)^2$, $x > 5$. [P.3] $x + 2\sqrt{x - 5} - 4$

84. Find $\left(2 - \sqrt{x + 3}\right)^2$, $x > -3$. [P.3] $x - 4\sqrt{x + 3} + 7$

Explorations 

1. *Golden Rectangles* A rectangle is a *golden rectangle* provided its length l and its width w satisfy the equation

$$\frac{l}{w} = \frac{l + w}{l}$$

 a. Solve this formula for l. *Hint:* Multiply both sides of the equation by wl and then use the quadratic formula to solve for l in terms of w. $l = \left(\frac{1 + \sqrt{5}}{2}\right)w$

 b. If the width of a golden rectangle measures 101 feet, what is the length of the rectangle? Round to the nearest 0.1 foot. 163.4 ft

 c. Measure the width and the length of a credit card. Would you say that the credit card closely approximates a golden rectangle? Explain.

 Yes. The length of a credit card is about $\left(\frac{1 + \sqrt{5}}{2}\right)$ times the width.

2. *Garfield's Proof of the Pythagorean Theorem* President James A. Garfield is credited with an original proof of the Pythagorean Theorem. Use the Internet or a library to find Garfield's proof. Write the steps in Garfield's proof with the supporting reason(s) for each step. Answers will vary.

SECTION 1.3 Other Types of Equations

- Absolute Value Equations
- Polynomial Equations
- Rational Equations
- Radical Equations
- Equations That Are Quadratic in Form

Absolute Value Equations

Recall that the absolute value of a real number x is the distance between the number x and 0 on the real number line. For instance, $|5| = 5$ and $|-5| = 5$. Equations involving absolute value often have more than one solution. For example, the solution set of the equation $|x| = 3$ is the set of all real numbers that are 3 units from 0, namely 3 and -3. See Figure 1.3.

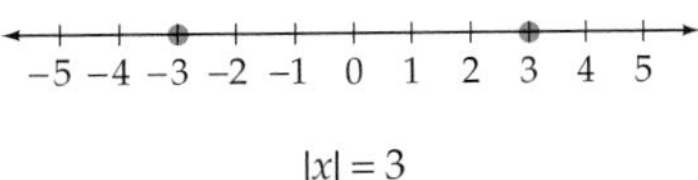

$|x| = 3$

Figure 1.3

The following property is used to solve **absolute value equations.**

Property of Absolute Value Equations

For any variable expression E and any nonnegative real number k,

$$|E| = k \quad \text{if and only if} \quad E = k \quad \text{or} \quad E = -k$$

Alternative to Example 1
Solve: $|4x + 3| = 15$
■ $3, -\frac{9}{2}$

TAKE NOTE

Some absolute value equations do not have any real number solutions. For example, $|x + 2| = -5$ is false for all values of x. Because an absolute value is always nonnegative, the equation is never true. When solving an absolute value equation such as $|3x - 7| = 5$, remember to keep the $3x - 7$ the same when writing the two resulting equations $3x - 7 = 5$ and $3x - 7 = -5$. *Do not* change the sign inside the absolute value bars.

EXAMPLE 1 Solve an Absolute Value Equation

Solve: $|2x - 5| = 21$

Solution $|2x - 5| = 21$ is equivalent to saying $2x - 5 = 21$ or $2x - 5 = -21$. Solving each of these equations produces

$$2x - 5 = 21 \quad \text{or} \quad 2x - 5 = -21$$
$$2x = 26 \qquad\qquad 2x = -16$$
$$x = 13 \qquad\qquad x = -8$$

Therefore, the solutions of $|2x - 5| = 21$ are -8 and 13.
As a check, if $x = -8$, then $|2(-8) - 5| = |-16 - 5| = |-21| = 21$, and if $x = 13$, then $|2x - 5| = |2(13) - 5| = |26 - 5| = |21| = 21$.

CHECK YOUR PROGRESS 1 Solve: $|3x + 4| = 16$

Solution *See page S6.* $x = 4, -\frac{20}{3}$

? QUESTION Is it true that $|-x| = x$?

■ Polynomial Equations

A **polynomial equation** is any equation that can be expressed as $P = 0$, where P is a polynomial expression. Linear equations and quadratic equations are examples of polynomial equations. While many higher degree polynomial equations are difficult or even impossible to solve, some polynomial equations of higher degree can be solved by factoring and using the zero product property that we used in solving quadratic equations.

Alternative to Example 2
Solve: $x^3 - 2x^2 = 35x$
■ $-5, 0, 7$

TAKE NOTE

If you attempt to solve Example 2 by dividing each side of the equation by x, you will produce the equation $x^2 - 16 = 0$, which has only -4 and 4 as solutions. In this case, the solution $x = 0$ has been lost. Thus dividing each side of the equation by the variable x does not produce an equivalent equation. To avoid this common mistake, factor out any variable factors that are common to each term instead of dividing each side of the equation by the variable factor.

EXAMPLE 2 Solve a Polynomial Equation by Factoring

Solve: $x^3 - 16x = 0$

Solution

$$x^3 - 16x = 0$$
$$x(x^2 - 16) = 0 \quad \text{■ Factor out the GCF, } x.$$
$$x(x + 4)(x - 4) = 0 \quad \text{■ Factor the difference of squares.}$$

Set each factor equal to zero.

$$x = 0 \quad \text{or} \quad x + 4 = 0 \quad \text{or} \quad x - 4 = 0$$
$$x = -4 \qquad\qquad x = 4$$

A check will show that -4, 0, and 4 are solutions of the original equation.

Continued ➤

? ANSWER No. For example, if $x = -3$, then this equation states that $|-(-3)| = -3$ or $3 = -3$, which is false.

CHECK YOUR PROGRESS 2 Solve: $x^4 - 36x^2 = 0$

Solution *See page S6.* $x = 0, \pm 6$

▪ Rational Equations

A **rational equation** is an equation involving fractional expressions in which the numerator and denominator of each fraction is a linear, quadratic, or other polynomial expression. (See Section P.5.) For example, $\dfrac{3}{x+5}$ and $\dfrac{x+4}{x^2-4}$ are rational expressions.

Normally a rational equation can be simplified by first multiplying both sides of the equation by a variable expression that allows the equation to be written in a form that does not involve fractions. Multiplying each side of an equation by the same nonzero number always produces an equivalent equation. If, however, each side of an equation is multiplied by an expression that involves a variable, then we must restrict the variable so that the expression is not equal to zero. Otherwise, incorrect solutions can result. Example 3b illustrates this pitfall. To be safe, always check your solutions.

Alternative to Example 3

Solve: $\dfrac{x}{x+4} - 3 = \dfrac{2}{x+4}$

▪ -7

EXAMPLE 3 Solve a Rational Equation

Solve each equation.

a. $\dfrac{x}{x-3} = \dfrac{9}{x-3} - 5$ **b.** $1 + \dfrac{x}{x-5} = \dfrac{5}{x-5}$

Solution

a. To produce a simpler equivalent equation, multiply each side of the equation by $x - 3$ so that the denominators cancel. However, note that the expression $x - 3$ would equal zero if x were 3, so we must include the restriction $x \neq 3$.

$$\frac{x}{x-3} = \frac{9}{x-3} - 5$$
$$(x-3)\left(\frac{x}{x-3}\right) = (x-3)\left(\frac{9}{x-3} - 5\right)$$
$$x = (x-3)\left(\frac{9}{x-3}\right) - (x-3)5$$
$$x = 9 - 5x + 15$$
$$6x = 24$$
$$x = 4$$

To verify the solution, substitute 4 for x in the *original* equation:

$$\frac{4}{4-3} = \frac{9}{4-3} - 5$$
$$4 = 9 - 5$$
$$4 = 4$$

Thus the solution is $x = 4$.

TAKE NOTE

When we multiply both sides of an equation by $x - a$, we assume that $x \neq a$.

Continued ➤

INSTRUCTOR NOTE

To demonstrate to students why the restriction on the variable is needed, show the following "proof" that 2 = 1. See if the students can identify the incorrect step.

$a = b$	Given
$a^2 = ab$	Multiply by a.
$a^2 - b^2 = ab - b^2$	Subtract b^2.
$(a + b)(a - b) = b(a - b)$	Factor.
$\frac{(a + b)(a - b)}{(a - b)} = \frac{b(a - b)}{(a - b)}$	Divide by $a - b$.
$a + b = b$	Simplify.
$b + b = b$	Substitute b for a ($a = b$).
$2b = b$	Simplify.
$2 = 1$	Divide by b.

b. To produce a simpler equivalent equation, multiply each side of the equation by $x - 5$, with the restriction that $x \neq 5$.

$$1 + \frac{x}{x - 5} = \frac{5}{x - 5}$$

$$(x - 5)\left(1 + \frac{x}{x - 5}\right) = (x - 5)\left(\frac{5}{x - 5}\right)$$

$$(x - 5)1 + (x - 5)\left(\frac{x}{x - 5}\right) = 5$$

$$x - 5 + x = 5$$

$$2x = 10$$

$$x = 5$$

Although we have obtained 5 as a proposed solution, 5 is *not* a solution of the original equation because it contradicts our restriction $x \neq 5$. Substituting 5 for x in the original equation results in denominators of 0. In this case the original equation has no solution.

CHECK YOUR PROGRESS 3 Solve: $\frac{2x}{x - 2} + 3 = \frac{4}{x - 2}$

Solution *See page S6.* No solution

Alternative to Example 4

Exercise 114, page 125

EXAMPLE 4 A Medical Application of Rational Equations

Young's rule is a rule-of-thumb method used to determine the portion of an adult dose of medication to administer to a child. If A represents the age of the child, the rule states

$$\text{Fraction of adult dose} = \frac{A}{A + 12}$$

If, according to Young's rule, a child should receive $\frac{1}{3}$ of an adult dose, what is the age of the child?

Solution If the fraction of an adult dose is $\frac{1}{3}$, then according to Young's rule,

$$\frac{1}{3} = \frac{A}{A + 12}$$

$$3(A + 12) \cdot \frac{1}{3} = 3(A + 12) \cdot \left(\frac{A}{A + 12}\right)$$ ▪ **Multiply each side by 3(A + 12).**

$$A + 12 = 3A$$ ▪ **Simplify.**

$$12 = 2A$$

$$6 = A$$

Thus a 6-year-old child would require $\frac{1}{3}$ of an adult dose.

Continued ➤

CHECK YOUR PROGRESS 4 Using Young's rule, what age child would require 52% of an adult dose?

Solution *See page S6.* A 13-year-old child

▪ Radical Equations

Radical equations involve variable expressions contained in a root, such as $\sqrt[3]{x}$ or $\sqrt{2x+3}$. To solve a radical equation containing a square root, the root must be "undone" by squaring.

To simplify such an equation, both sides of the equation can be squared, although the result may not be equivalent to the original equation. We will explain this shortly.

Alternative to Example 5

Solve: $\sqrt{x-3} = 5$

▪ 28

EXAMPLE 5 Solve a Radical Equation

Solve: $\sqrt{x+4} = 3$

Solution

$$\sqrt{x+4} = 3$$
$$\left(\sqrt{x+4}\right)^2 = 3^2 \qquad \text{▪ Square each side.}$$
$$x + 4 = 9$$
$$x = 5$$

Check the solution:

$$\sqrt{x+4} = 3$$
$$\sqrt{5+4} \stackrel{?}{=} 3 \qquad \text{▪ Substitute 5 for } x.$$
$$\sqrt{9} \stackrel{?}{=} 3$$
$$3 = 3 \qquad \text{▪ 5 checks.}$$

The only solution is 5.

CHECK YOUR PROGRESS 5 Solve: $\sqrt{x-5} = 6$

Solution *See page S6.* $x = 41$

Squaring both sides of an equation as we did in the previous example is an example of using the **power principle.**

The Power Principle

If P and Q are algebraic expressions and n is a positive integer, then every solution of $P = Q$ is also a solution of $P^n = Q^n$.

Some care must be taken when using the power principle because the equation $P^n = Q^n$ may have more solutions than the original equation $P = Q$. As an example, consider the simple equation $x = 3$. The only solution is the real number 3. If we square each side of the equation to produce $x^2 = 9$, we get both 3 and -3 as solutions. The -3 is called an *extraneous solution* because it is not a solution of the original equation $x = 3$.

Extraneous Solutions

Any solution of $P^n = Q^n$ that is not a solution of $P = Q$ is called an **extraneous solution.** Extraneous solutions *may* be introduced whenever we raise each side of an equation to an *even* power.

When solving radical equations, always check your answers for extraneous solutions.

Alternative to Example 6

Solve: $x - \sqrt{5x - 11} = 5$

▪ 12

EXAMPLE 6 A Radical Equation with an Extraneous Solution

Solve: $x = 2 + \sqrt{2 - x}$

Solution Remember that individual terms cannot be squared, only entire sides of an equation. Therefore, we first isolate the radical expression.

$$x = 2 + \sqrt{2 - x}$$

$$x - 2 = \sqrt{2 - x} \qquad \text{▪ Isolate the radical.}$$

$$(x - 2)^2 = \left(\sqrt{2 - x}\right)^2 \qquad \text{▪ Square each side of the equation.}$$

$$x^2 - 4x + 4 = 2 - x$$

$$x^2 - 3x + 2 = 0 \qquad \text{▪ Collect and combine like terms.}$$

$$(x - 2)(x - 1) = 0 \qquad \text{▪ Factor.}$$

$$x - 2 = 0 \quad \text{or} \quad x - 1 = 0$$

$$x = 2 \qquad\qquad x = 1 \qquad \text{▪ Proposed solutions}$$

Check for $x = 2$: $x = 2 + \sqrt{2 - x}$

$$2 \stackrel{?}{=} 2 + \sqrt{2 - (2)} \qquad \text{▪ Substitute 2 for } x.$$

$$2 \stackrel{?}{=} 2 + \sqrt{0}$$

$$2 = 2 \qquad \text{▪ 2 is a solution.}$$

Check for $x = 1$: $x = 2 + \sqrt{2 - x}$

$$1 \stackrel{?}{=} 2 + \sqrt{2 - (1)} \qquad \text{▪ Substitute 1 for } x.$$

$$1 \stackrel{?}{=} 2 + \sqrt{1}$$

$$1 \neq 3 \qquad \text{▪ 1 is not a solution.}$$

The check shows that 1 is not a solution; it is an extraneous solution that we created by squaring each side of the equation. The only solution is 2.

CHECK YOUR PROGRESS 6 Solve: $x = \sqrt{5 - x} + 5$

Solution *See page S6.* $x = 5$

Alternative to Example 7
Exercise 122, page 126

EXAMPLE 7 An Application to Orbiting Satellites

Satellites orbiting planets (or moons) obey the equation

$$v = \sqrt{\frac{GM}{r}}$$

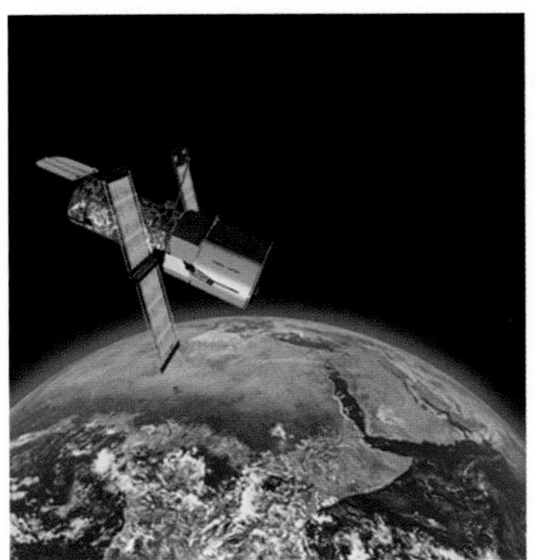

The Hubble space telescope

where v is the velocity of the satellite in meters per second, M is the mass of the planet in kilograms, r is the radius from the center of the planet to the orbit path (measured in meters), and G is a constant equal to approximately $6.67 \times 10^{-11}\ \text{m}^3/(\text{s}^2 \cdot \text{kg})$. The Hubble space telescope completes one orbit in less than 2 hours and must travel at approximately 7550 meters per second (about 17,000 miles per hour). How far above the earth's surface does the Hubble telescope orbit? (The mass of the earth is 5.98×10^{24} kilograms and its radius is 6.37×10^6 meters.)

Solution The given values are in the correct units, so we substitute the values into the given equation:

$$7550 = \sqrt{\frac{(6.67 \times 10^{-11})(5.98 \times 10^{24})}{r}}$$

$$57{,}002{,}500 = \frac{(6.67 \times 10^{-11})(5.98 \times 10^{24})}{r}$$ ▪ Square both sides.

$$57{,}002{,}500r = (6.67 \times 10^{-11})(5.98 \times 10^{24})$$ ▪ Multiply each side by r.

$$r = \frac{(6.67 \times 10^{-11})(5.98 \times 10^{24})}{57{,}002{,}500}$$ ▪ Divide each side by 57,002,500.

$$r \approx 6{,}997{,}342$$

Rounding, the radius from the center of the earth to the orbit path is about 7.00×10^6 meters. Because the earth's radius is 6.37×10^6 meters, the telescope is orbiting $7.00 \times 10^6 - 6.37 \times 10^6 = 0.63 \times 10^6$ meters, or about 630,000 meters (about 390 miles) above the earth's surface.

CHECK YOUR PROGRESS 7 Compute the orbit radius (measured from the center of the earth) of a satellite that is orbiting the earth at 5000 meters per second. Round to the nearest million meters.

Solution *See page S7.* 16,000,000 m

Some equations that involve fractional exponents can be solved by raising each side of the equation to a reciprocal power. For example, to solve $x^{1/3} = 4$, raise each side of the equation to the third power to find that $x = 64$. Be sure to check all proposed solutions to determine whether they are actual solutions or extraneous solutions.

Alternative to Example 8

Solve: $(x^2 - 8)^{2/3} = 4$

■ ± 4

EXAMPLE 8 Solve an Equation That Involves Fractional Exponents

Solve: $(x + 4)^{3/2} = 27$

Solution Because the equation involves a $\frac{3}{2}$ power, start by raising each side of the equation to the $\frac{2}{3}$ power $\left(\frac{2}{3}\text{ is the reciprocal of }\frac{3}{2}\right)$.

$$[(x + 4)^{3/2}]^{2/3} = 27^{2/3}$$

$$x + 4 = 9 \qquad \text{■ Think: } 27^{2/3} = (\sqrt[3]{27})^2 = 3^2 = 9.$$

$$x = 5 \qquad \text{■ Solve for } x.$$

Check to verify that $x = 5$ is a solution of the original equation.

CHECK YOUR PROGRESS 8 Solve: $(4z + 7)^{1/3} = 2$

Solution *See page S7.* $z = \frac{1}{4}$

■ Equations That Are Quadratic in Form

Some polynomial equations of higher degree can be written in the form

$$au^2 + bu + c = 0, \qquad a \neq 0$$

where u is an algebraic expression involving x. For example, the equation

$$4x^4 - 25x^2 + 36 = 0$$

can be written as

$$4(x^2)^2 - 25x^2 + 36 = 0$$

If we substitute u for x^2, then our original equation becomes

$$4u^2 - 25u + 36 = 0$$

Such equations are called **quadratic in form.** The equation can be solved for u using the methods of Section 1.2, and then the solutions of the original equation can be found from the relationship between u and x.

? QUESTION Is the equation $x^9 - 5x^3 + 1 = 0$ quadratic in form?

? ANSWER No. If we substitute $u = x^3$ in the middle term, then the first term is u^3, not u^2.

Alternative to Example 9

Solve: $x^4 - 8x^2 - 9 = 0$

■ ± 3

EXAMPLE 9 Solve an Equation That Is Quadratic in Form

Solve: $4x^4 - 25x^2 + 36 = 0$

Solution As mentioned on the preceding page, the equation can be written as $4(x^2)^2 - 25x^2 + 36 = 0$. The substitution $u = x^2$ (and consequently $u^2 = x^4$) gives $4u^2 - 25u + 36 = 0$. Factor the quadratic polynomial on the left side of the equation.

$$4u^2 - 25u + 36 = 0$$
$$(4u - 9)(u - 4) = 0$$
$$4u - 9 = 0 \quad \text{or} \quad u - 4 = 0$$
$$u = \frac{9}{4} \qquad\qquad u = 4$$

Substitute x^2 for u to produce

$$x^2 = \frac{9}{4} \quad \text{or} \quad x^2 = 4$$
$$x = \pm\sqrt{\frac{9}{4}} \qquad x = \pm\sqrt{4}$$
$$x = \pm\frac{3}{2} \qquad x = \pm 2$$

■ Check the solutions in the original equation.

The solutions are $-2, -\frac{3}{2}, \frac{3}{2}$, and 2.

CHECK YOUR PROGRESS 9 Solve: $x^4 - 10x^2 + 9 = 0$

Solution *See page S7.* $x = \pm 3, \pm 1$

Polynomials of degree greater than 4, as well as expressions with fractional or negative exponents, can also be quadratic in form. The following table lists examples of equations that are quadratic in form, along with an appropriate substitution that will enable you to write the equation in the form $au^2 + bu + c = 0$.

Equations That Are Quadratic in Form

Original equation	Substitution	$au^2 + bu + c$ form
$x^4 - 8x^2 + 15 = 0$	$u = x^2$	$u^2 - 8u + 15 = 0$
$x^6 + x^3 - 12 = 0$	$u = x^3$	$u^2 + u - 12 = 0$
$x^{1/2} - 9x^{1/4} + 20 = 0$	$u = x^{1/4}$	$u^2 - 9u + 20 = 0$
$2x^{2/3} + 7x^{1/3} - 4 = 0$	$u = x^{1/3}$	$2u^2 + 7u - 4 = 0$
$15x^{-2} + 7x^{-1} - 2 = 0$	$u = x^{-1}$	$15u^2 + 7u - 2 = 0$

Alternative to Example 10
Solve: $2x^{1/2} - 5x^{1/4} + 3 = 0$
■ $1, \frac{81}{16}$

EXAMPLE 10 Solve an Equation That Is Quadratic in Form

Solve: $3x^{2/3} - 5x^{1/3} - 2 = 0$

Solution Making the substitution $u = x^{1/3}$ (and so $u^2 = x^{2/3}$) gives $3u^2 - 5u - 2 = 0$.

$$3u^2 - 5u - 2 = 0$$

$$(3u + 1)(u - 2) = 0 \quad \text{■ Factor.}$$

$$3u + 1 = 0 \quad \text{or} \quad u - 2 = 0$$

$$u = -\frac{1}{3} \qquad u = 2$$

$$x^{1/3} = -\frac{1}{3} \qquad x^{1/3} = 2 \quad \text{■ Replace } u \text{ with } x^{1/3}.$$

$$x = -\frac{1}{27} \qquad x = 8 \quad \text{■ Cube each side.}$$

A check will verify that both $-\frac{1}{27}$ and 8 are solutions.

CHECK YOUR PROGRESS 10 Solve: $6x^{2/3} - 7x^{1/3} - 20 = 0$

Solution *See page S7.* $x = -\frac{64}{27}, \frac{125}{8}$

Topics for Discussion

1. Is the statement $|x - 3| = x + 3$ true? Explain.
2. Consider the equation $(x^2 - 1)(x - 2) = 3(x - 2)$. Dividing each side of the equation by $x - 2$ gives $x^2 - 1 = 3$. Is this second equation equivalent to the first equation?
3. In the equation $\frac{1}{x} - \frac{x}{x-2} = \frac{3}{x-2}$, what restrictions are placed on the variable x?
4. In order to solve the equation $\frac{1}{a} = \frac{1}{b} + \frac{1}{c}$ for the variable a, a student takes the reciprocal of each term to arrive at $a = b + c$. Did this procedure produce an equivalent equation? Explain.
5. If P and Q are algebraic expressions and n is a positive integer, then the equation $P^n = Q^n$ is equivalent to the equation $P = Q$. Do you agree? Explain.
6. A tutor claims that the equation $x^{-2} - \frac{2}{x} = 15$ is quadratic in form. Do you agree? If so, what would be an appropriate substitution that would enable you to write the equation as a quadratic equation?

EXERCISES 1.3 — *Suggested Assignment: Exercises 1–125, every other odd; 127–132.*

In Exercises 1 to 14, solve each absolute value equation for x.

1. $|x| = 4$ $4, -4$
2. $|x| = 7$ $7, -7$
3. $|x - 5| = 2$ $7, 3$
4. $|x - 8| = 3$ $11, 5$
5. $|2x - 5| = 11$ $8, -3$
6. $|2x - 3| = 21$ $12, -9$
7. $|2x + 6| = 10$ $2, -8$
8. $|2x + 14| = 60$ $23, -37$
9. $|4x + 1| = -5$ No solution
10. $|7x - 2| = -10$ No solution
11. $\left|\frac{x - 4}{2}\right| = 8$ $20, -12$
12. $\left|\frac{x + 3}{4}\right| = 6$ $21, -27$
13. $2|x + 3| + 4 = 34$ $12, -18$
14. $3|x - 5| - 16 = 2$ $11, -1$

In Exercises 15 to 22, find all real solutions of each polynomial equation by factoring and using the zero product principle.

15. $x^3 - 25x = 0$ $0, \pm 5$
16. $x^3 - x = 0$ $0, \pm 1$
17. $2x^5 - 18x^3 = 0$ $0, \pm 3$
18. $2x^4 - 98x^2 = 0$ $0, \pm 7$
19. $x^4 - 3x^3 - 40x^2 = 0$ $0, -5, 8$
20. $x^4 + 3x^3 = 18x^2$ $0, 3, -6$
21. $x^4 - 16x^2 = 0$ $0, \pm 4$
22. $x^4 - 16 = 0$ ± 2

In Exercises 23 and 24, find all complex solutions by factoring and using the quadratic formula.

23. $x^3 - 8 = 0$ $2, -1 \pm i\sqrt{3}$
24. $x^3 + 8 = 0$ $-2, 1 \pm i\sqrt{3}$

In Exercises 25 to 38, solve each rational equation and check your solution(s).

25. $\frac{3}{x + 2} = \frac{5}{2x - 7}$ 31
26. $\frac{4}{y + 2} = \frac{7}{y - 4}$ -10
27. $\frac{30}{10 + x} = \frac{20}{10 - x}$ 2
28. $\frac{6}{8 + x} = \frac{4}{8 - x}$ $\frac{8}{5}$
29. $\frac{3x}{x + 4} = 2 - \frac{12}{x + 4}$ No solution
30. $\frac{t}{t - 4} + 3 = \frac{4}{t - 4}$ No solution
31. $2 + \frac{9}{r - 3} = \frac{3r}{r - 3}$ No solution
32. $\frac{8}{2m + 1} - \frac{1}{m - 2} = \frac{5}{2m + 1}$ 7
33. $\frac{5}{x - 3} - \frac{3}{x - 2} = \frac{4}{x - 3}$ $\frac{7}{2}$
34. $\frac{4}{x - 1} + \frac{7}{x + 7} = \frac{5}{x - 1}$ $\frac{7}{3}$
35. $\frac{x}{x - 3} = \frac{x + 4}{x + 2}$ -12
36. $\frac{x}{x - 5} = \frac{x + 7}{x + 1}$ 35
37. $\frac{x + 3}{x + 5} = \frac{x - 3}{x - 4}$ 1
38. $\frac{x - 6}{x + 4} = \frac{x - 1}{x + 2}$ $-\frac{8}{7}$

In Exercises 39 to 52, use the power principle to solve each radical equation. Check all proposed solutions.

39. $\sqrt{x - 1} = 4$ 17
40. $\sqrt{x + 8} = 5$ 17
41. $\sqrt{x - 4} - 6 = 0$ 40
42. $\sqrt{10 - x} = 4$ -6
43. $x = 3 + \sqrt{3 - x}$ 3
44. $x = 1 + \sqrt{11 + x}$ 5
45. $2x = \sqrt{4x + 15}$ $\frac{5}{2}$
46. $x = \sqrt{12x - 35}$ $5, 7$
47. $\sqrt[3]{x - 1} = -3$ -26
48. $\sqrt[3]{4 - x} = 4$ -60
49. $\sqrt[4]{5 - 2x} = \sqrt[4]{7x - 4}$ 1
50. $\sqrt[3]{7x - 3} = \sqrt[3]{2x - 7}$ $-\frac{4}{5}$
51. $\sqrt[3]{2x^2 + 5x - 3} = \sqrt[3]{x^2 + 3}$ $1, -6$
52. $\sqrt[4]{x^2 + 20} = \sqrt[4]{9x}$ $4, 5$

In Exercises 53 to 62, use the power principle to solve each equation containing fractional exponents. Check all proposed solutions.

53. $(3x + 5)^{1/3} = (-2x + 15)^{1/3}$ 2
54. $(7 - 3t)^{1/3} = 3$ $-\frac{20}{3}$
55. $(x + 4)^{2/3} = 9$ $23, -31$
56. $(x - 5)^{3/2} = 125$ 30
57. $(2x - 6)^{3/5} = 8$ 19
58. $(12 - 3x)^{4/5} = 81$ -77
59. $(4x)^{2/3} = (30x + 4)^{1/3}$ $2, -\frac{1}{8}$
60. $z^{2/3} = (3z - 2)^{1/3}$ $2, 1$
61. $4x^{3/4} = x^{1/2}$ $0, \frac{1}{256}$
62. $x^{3/5} = 2x^{1/5}$ $0, \pm 4\sqrt{2}$

In Exercises 63 to 82, find all real solutions of the equation by first rewriting the equation in quadratic form.

63. $x^4 - 9x^2 + 14 = 0$ $\pm\sqrt{7}, \pm\sqrt{2}$

64. $x^4 - 12x^2 + 32 = 0$ $\pm 2, \pm 2\sqrt{2}$

65. $2x^4 - 11x^2 + 12 = 0$ $\pm 2, \pm\frac{\sqrt{6}}{2}$

66. $6x^4 - 7x^2 + 2 = 0$ $\pm\frac{\sqrt{2}}{2}, \pm\frac{\sqrt{6}}{3}$

67. $x^6 + x^3 - 6 = 0$ $\sqrt[3]{2}, -\sqrt[3]{3}$

68. $6x^6 + x^3 - 15 = 0$ $\frac{\sqrt[3]{12}}{2}, -\frac{\sqrt[3]{45}}{3}$

69. $21x^6 + 22x^3 = 8$ $-\frac{\sqrt[3]{36}}{3}, \frac{\sqrt[3]{98}}{7}$

70. $-3x^6 + 377x^3 = 250$ $\frac{\sqrt[3]{18}}{3}, 5$

71. $x^{1/2} - 3x^{1/4} + 2 = 0$ $1, 16$

72. $2x^{1/2} - 5x^{1/4} - 3 = 0$ 81

73. $3x^{2/3} - 11x^{1/3} - 4 = 0$ $-\frac{1}{27}, 64$

74. $5b^{2/3} + 6b^{1/3} - 8 = 0$ $\frac{64}{125}, -8$

75. $9x^4 = 30x^2 - 25$ $\pm\frac{\sqrt{15}}{3}$

76. $4x^4 - 28x^2 = -49$ $\pm\frac{\sqrt{14}}{2}$

77. $x^{2/5} - 1 = 0$ ± 1

78. $2x^{2/5} - x^{1/5} = 6$ $-\frac{243}{32}, 32$

79. $x^{-2} - 9x^{-1} = 22$ $\frac{1}{11}, -\frac{1}{2}$

80. $\frac{1}{x^2} + \frac{3}{x} - 10 = 0$ $\frac{1}{2}, -\frac{1}{5}$

81. $9x - 52\sqrt{x} + 64 = 0$
Hint: Write x as $\left(\sqrt{x}\right)^2$. $\frac{256}{81}, 16$

82. $8x - 38\sqrt{x} + 9 = 0$
Hint: Write x as $\left(\sqrt{x}\right)^2$. $\frac{1}{16}, \frac{81}{4}$

In Exercises 83 and 84, solve each equation. Round your solution(s) to the nearest hundredth.

83. $x^4 - 3x^2 + 1 = 0$ $\pm 0.62, \pm 1.62$

84. $2x^2 = \sqrt{10x^2 - 3}$ $\pm 0.59, \pm 1.47$

In Exercises 85 to 94, solve each literal equation for the indicated variable.

85. $|2x - a| = b \ (b > 0)$; x $x = \frac{a \pm b}{2}$

86. $3|x - d| = c \ (c > 0)$; x $x = d \pm \frac{c}{3}$

87. $a^2 + b^2 = 9$; b $b = \pm\sqrt{9 - a^2}$

88. $\frac{x^2}{a^2} + \frac{y^2}{b^2} = 1$; x $x = \pm\sqrt{a^2 - \frac{a^2y^2}{b^2}}$ or $\pm\frac{a}{b}\sqrt{b^2 - y^2}$

89. $\frac{1}{f} = \frac{1}{d_0} + \frac{1}{d_1}$; f (astronomy) $f = \frac{d_0 d_1}{d_1 + d_0}$

90. $S = \frac{a_1}{1 - r}$; r (mathematics) $r = \frac{S - a_1}{S}$

91. $\frac{P_1V_1}{T_1} = \frac{P_2V_2}{T_2}$; V_2 (chemistry) $V_2 = \frac{P_1V_1T_2}{P_2T_1}$

92. $\frac{w_1}{w_2} = \frac{f_2 - f}{f - f_1}$; f_1 (hydrostatics) $f_1 = \frac{w_1 f - w_2 f_2 + w_2 f}{w_1}$

93. $\sqrt{x} - \sqrt{y} = \sqrt{z}$; x $x = \left(\sqrt{y} + \sqrt{z}\right)^2$

94. $x - y = \sqrt{x^2 + y^2 + 5}$; x $x = -\frac{5}{2y}$

In Exercises 95 to 98, determine whether the given equations are equivalent.

95. $3x - 11 = -5$, $\frac{3x - 11}{x - 2} = \frac{-5}{x - 2}$ Not equivalent

96. $3x - 9 = x - 3$, $\frac{3x - 9}{x - 3} = \frac{x - 3}{x - 3}$ Not equivalent

97. $\frac{1}{t} = \frac{1}{a} + \frac{1}{b}$, $t = \frac{ab}{a + b}$, where t is a variable and a and b are nonzero constants, $a \neq -b$. Equivalent

98. $\frac{2}{x} = \frac{1}{x - 1}$, $2(x - 1) = x$ Equivalent

In Exercises 99 to 104, determine the values of x that make each equation true.

99. $|x + 4| = x + 4$ $\{x \mid x \geq -4\}$

100. $|x - 1| = x - 1$ $\{x \mid x \geq 1\}$

101. $|x + 7| = -(x + 7)$ $\{x \mid x \leq -7\}$

102. $|x - 3| = -(x - 3)$ $\{x \mid x \leq 3\}$

103. $|x - 2| + |x + 4| = 8$ $3, -5$

104. $|x + 1| - |x - 3| = 4$ $\{x \mid x \geq 3\}$

105. Solve $\left(\sqrt{x} - 2\right)^2 - 5\sqrt{x} + 14 = 0$ for x. *Hint:* Use the substitution $u = \sqrt{x} - 2$, and then rewrite the equation so that it is quadratic in terms of the variable u. $9, 36$

106. Solve $\left(\sqrt[3]{x} + 3\right)^2 - 8\sqrt[3]{x} = 12$ for x. *Hint:* Use the substitution $u = \sqrt[3]{x} + 3$, and then rewrite the equation so that it is quadratic in terms of the variable u. $-1, 27$

107. ***Radius of a Circle*** The radius r of a circle inscribed in a triangle with sides of lengths a, b, and c is given by

$$r = \sqrt{\frac{(s - a)(s - b)(s - c)}{s}}$$

where $s = \frac{1}{2}(a + b + c)$.

a. Find the length of the radius of a circle inscribed in a triangle with sides of lengths 5 inches, 6 inches, and 7 inches. See the following figure. Round to the nearest hundredth of an inch. $r \approx 1.63$ in.

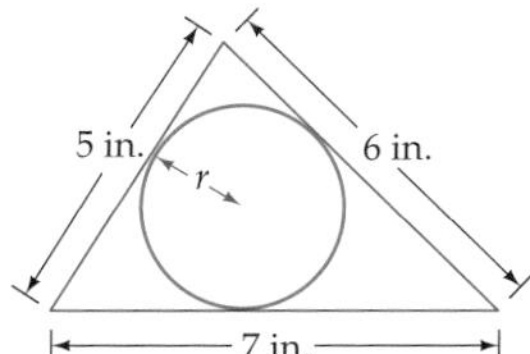

b. The radius of a circle inscribed in an equilateral triangle measures 2 inches. What is the exact length of each side of the equilateral triangle? $4\sqrt{3}$ in.

108. *Radius of a Circle* The radius r of a circle that is circumscribed about a triangle with sides of lengths a, b, and c is given by

$$r = \frac{abc}{4\sqrt{s(s-a)(s-b)(s-c)}}$$

where $s = \frac{1}{2}(a + b + c)$.

a. Find the radius of a circle that is circumscribed about a triangle with sides of lengths 7 inches, 10 inches, and 15 inches. Round to the nearest hundredth of an inch. 8.93 in.

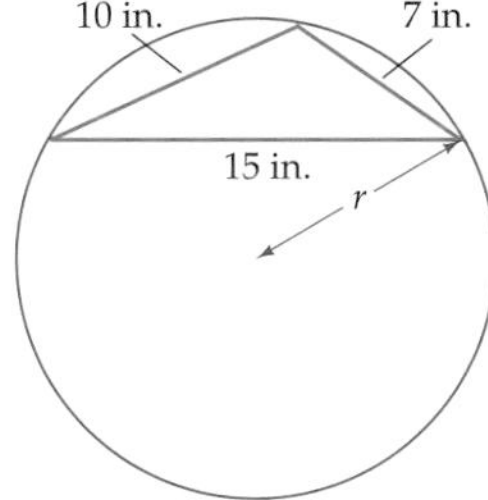

b. A circle with radius 5 inches is circumscribed about an equilateral triangle. What is the exact length of each side of the equilateral triangle? $5\sqrt{3}$ in.

Business and Economics

109. *Construction Materials* A construction company ordered 4000 cubic feet of sand to mix into cement. When it was delivered, it was poured from a dump truck and formed a right circular cone whose height was one-third the diameter of the base. What was the diameter of the base of the pile? (Round to the nearest tenth of a foot.) 35.8 ft

110. *Precious Metals* A jeweler has two small, solid spheres made of gold. One is 8 millimeters in diameter and the other is 12 millimeters in diameter. She will melt the spheres and recast the gold into a single cube. What is the length of each edge of the cube? Round your answer to the nearest tenth of a millimeter. 10.5 mm

111. *Build a Fence* A worker can build a fence in 8 hours. With the help of an assistant, the fence can be built in 5 hours. How long would it take the assistant to build the fence working alone? $13\frac{1}{3}$ h

112. *Roof Repair* A roofer and an assistant can repair a roof in 6 hours. The assistant can complete the repair working alone in 14 hours. If both the roofer and the assistant work together for 2 hours and then the assistant is left to finish the job, how much longer will the assistant need to finish the repairs? $9\frac{1}{3}$ h

Life and Health Sciences

113. *Medical Dosage* If Young's rule (see Example 4, page 116) is used to determine that a child's dosage of a medication should be $\frac{3}{7}$ of an adult dose, how old is the child? 9 years old

114. *Medical Dosage* If Young's rule (see Example 4) is used to determine that a child's dosage of a medication should be 40% of an adult dose, how old is the child? 8 years old

Sports and Recreation

115. *Golf Score* A golfer has played six 18-hole games so far this year, and his score is averaging 88. If he can score 80 on each game he plays in the future, how many more games will he need to play to bring his average down to 83? 10 games

116. *Spectator Crowd* The crowd watching a high school basketball game currently includes 60 women. Soon 100 more women arrive, which doubles the percentage of women in the crowd. How large was the crowd before the 100 women arrived? 300 people

Physical Sciences and Engineering

117. *Chemistry* A conically shaped chemical flask has a height h of 4 inches and a lateral surface area L of 15π square inches. What is the radius r of the flask? (The formula for the lateral surface area of a cone is $L = \pi r\sqrt{r^2 + h^2}$.) 3 in.

118. ***Geology*** Many volcanoes are approximately conical in shape. One volcano is an estimated 1200 feet high and has an approximate lateral surface area of 1.1 square miles. Assuming the volcano is shaped like a cone, compute the diameter of the volcano. Round your answer to the nearest 10 feet. (Use the formula for the lateral surface area of a cone given in Exercise 117.) 6020 ft

119. ***Pendulum*** The period T of a pendulum is the time it takes the pendulum to complete one swing from left to right and back. For a pendulum near the surface of the earth,

$$T = 2\pi\sqrt{\frac{L}{32}}$$

where T is measured in seconds and L is the length of the pendulum in feet. Find the length of a pendulum that has a period of 4 seconds. Round to the nearest tenth of a foot. 13.0 ft

120. ***Distance to the Horizon*** On a ship, the distance d that you can see to the horizon is given by $d = 1.5\sqrt{h}$, where h is the height of your eye measured in feet above sea level and d is measured in miles. How high is the eye level of a navigator who can see 14 miles to the horizon? Round to the nearest foot. 87 ft

121. ***Electronic Circuits*** Capacitors are used in electronic circuits to store electrical charges. The capacitance of a capacitor, measured in microfarads (μF), is a measure of its capacity to store electrical charge. When two capacitors are connected in series (one follows the other along the same path), their respective capacitances C_1 and C_2 are related to the equivalent capacitance C of the circuit by $C = \frac{C_1C_2}{C_1 + C_2}$. If a circuit requires an equivalent capacitance of 18 μF and one capacitor has a capacitance of 24 μF, what capacitance will be required for the second capacitor? 72 μF

122. ***Satellite Orbit*** Some satellites are put into geosynchronous orbit, which means they are above the same location on the earth's surface at all times. To maintain this orbit, the satellites travel at 3070 meters per second. How far above the earth's surface are the satellites orbiting? (See Example 7, page 119.) Approximately 36 million meters

123. ***Satellite Orbit*** A satellite is moving at 4600 meters per second as it orbits around the earth. Determine the distance from the center of the earth to the satellite's orbit. (See Example 7, page 119.) Approximately 19 million meters

In Exercises 124 and 125, consider that the depth s from the opening of a well to the water in the well can be determined by measuring the total time between the instant you drop a stone and the time you hear it hit the water. The time (in seconds) it takes the stone to hit the water is given by Time $= \sqrt{s}/4$, where s is measured in feet. The time (also in seconds) required for the sound of the impact to travel up to your ears is given by Time $= s/1100$. Thus the total time T (in seconds) between the instant you drop a stone and the moment you hear its impact is

$$T = \frac{\sqrt{s}}{4} + \frac{s}{1100}$$

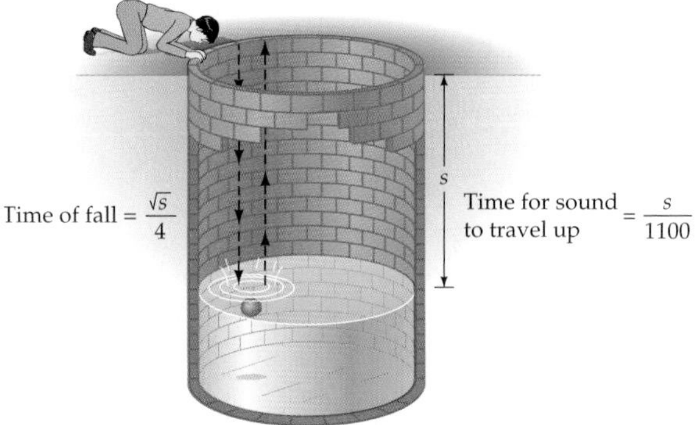

124. ***Time of Fall*** One of the world's deepest water wells is 7320 feet deep. Find the time between the instant you drop a stone and the time you hear it hit the water if the surface of the water is 7100 feet below the opening of the well. Round your answer to the nearest tenth of a second. 27.5 s

125. Solve $T = \frac{\sqrt{s}}{4} + \frac{s}{1100}$ for s. $s = \left(\frac{-275 + 5\sqrt{3025 + 176T}}{2}\right)^2$

126. ***Depth of a Well*** Use the result of Exercise 125 to determine the depth from the opening of a well to the water level if the time between the instant you drop a stone and the moment you hear its impact is 3 seconds. Round your answer to the nearest foot. 133 ft

Prepare for Section 1.4

127. Which is correct, $-16 < -18$ or $-16 > -18$? [P.1] $-16 > -18$

128. Find the intersection of the sets $\{x|x > 3\}$ and $\{x|x < 8\}$. [P.1] $\{x|3 < x < 8\}$

129. Write the union of the sets $\{x|x \le -2\}$ and $\{x|x > 5\}$ in interval notation. [P.1] $(-\infty, -2] \cup (5, \infty)$

130. Solve for x: $-3x + 8 = 7 + 2x$ [1.1] $\frac{1}{5}$

131. What values of x make the equation $|2x - 5| = 7$ true? [1.3] $6, -1$

132. If a taxi ride costs \$2.25 plus \$0.85 for each mile traveled, write a formula that computes the total cost of the taxi ride after traveling m miles. [1.1] Cost $= 2.25 + 0.85m$

Explorations

1. ***A Work Problem and Its Extensions*** If a pump can fill a pool in A hours, and a second pump can fill the same pool in B hours, then the total time T, in hours, to fill the pool with both pumps working together is given by

$$T = \frac{AB}{A + B}$$

a. Verify this formula See the Instructor's Resource Manual.

b. Consider the case of a pool that is to be filled by three pumps. One pump can fill the pool in A hours, the second in B hours, and the third in C hours. Derive a formula in terms of A, B, and C for the total time T needed to fill the pool. $T = \dfrac{ABC}{AB + AC + BC}$

c. Consider the case of a pool that is to be filled by n pumps. One pump can fill the pool in A_1 hours, the second in A_2 hours, the third in A_3 hours, . . . , and the nth pump can fill the pool in A_n hours. Write a formula in terms of $A_1, A_2, A_3, \ldots, A_n$ for the total time T needed to fill the pool.

$$T = \frac{A_1A_2A_3 \cdots A_n}{(A_2A_3A_4 \cdots A_n) + (A_1A_3A_4 \cdots A_n) + (A_1A_2A_4 \cdots A_n) + \cdots + (A_1A_2A_3 \cdots A_{n-1})}$$

That is, T is given by the product of the A's divided by the sum of products of the A's taken $(n - 1)$ at a time.

The chart below is called an *alignment chart* or a *nomogram* for the formula

$$T = \frac{AB}{A + B}$$

If you know any two of the values of A, B, and T, then you can use the alignment chart to determine the unknown value. For example, the straight line segment that connects 3 on the A-axis with 6 on the B-axis crosses the T-axis at 2. Thus the total time required to fill the pool if a pump that takes 3 hours to fill the pool and a pump that takes 6 hours to fill the pool work together is 2 hours.

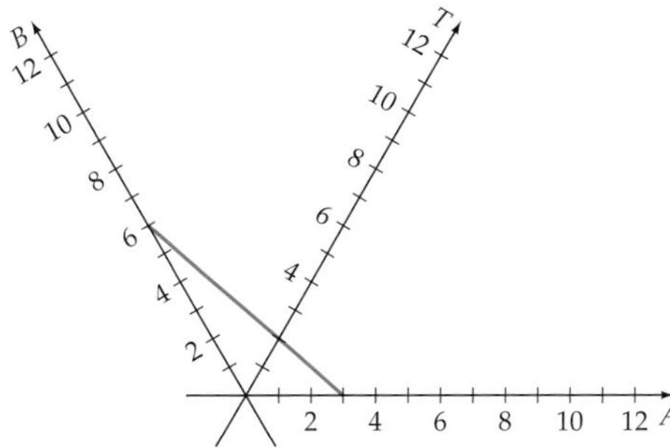

Alignment Chart for $T = \dfrac{AB}{A + B}$

d. Consider the case of a pool that is to be filled by three pumps. One pump can fill the pool in $A = 6$ hours, the second in $B = 8$ hours, and the third in $C = 12$ hours. Write a few sentences explaining how you could make use of the alignment chart above to show that it takes about 2.7 hours for the three pumps to fill the pool when they work together. See the Instructor's Resource Manual.

2. ***The Reduced Cubic*** The mathematician Francois Vieta knew a method of solving the "reduced cubic" $x^3 + mx + n = 0$ by using the substitution $x = m/(3z) - z$.

a. Show that this substitution results in the equation $z^6 - nz^3 - \dfrac{m^3}{27} = 0.$ See the Instructor's Resource Manual.

b. Show that the equation in part **a.** is quadratic in form. See the Instructor's Resource Manual.

c. Solve the equation in part **a.** for z. $z = \sqrt[3]{\dfrac{n}{2} + \sqrt{\dfrac{n^2}{4} + \dfrac{m^3}{27}}}$

d. Use your solution from part **c.** to find the real solution(s) of the equation $x^3 + 3x = 14$. $x = 2$

SECTION 1.4 Linear and Absolute Value Inequalities and Their Applications

- Solve Linear Inequalities
- Solve Compound Inequalities
- Solve Absolute Value Inequalities
- Applications of Inequalities

Point of Interest

Another property of inequalities, called the *transitive property*, states that for real numbers a, b, and c, if $a > b$ and $b > c$, then $a > c$. Note that a transitive property does not apply in the game of Scissors, Paper, Rock. Scissors wins over paper, paper wins over rock, but scissors does not win over rock!

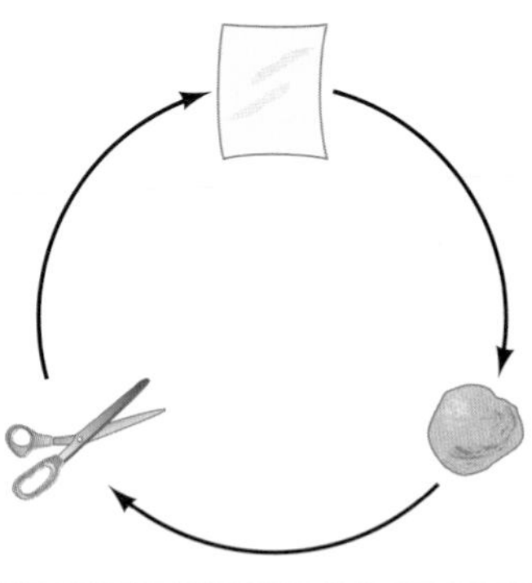

▪ Solve Linear Inequalities

In Section P.1, we used inequalities to describe the order of real numbers and to represent subsets of real numbers. In this section we consider inequalities that involve a variable. In particular, we consider how to determine which real numbers make an inequality a true statement.

The **solution set of an inequality** is the set of all real numbers for which the inequality is a true statement. For instance, the solution set of $x + 1 > 4$ is the set of all real numbers greater than 3. Two inequalities are **equivalent inequalities** if they have the same solution set. We can solve many inequalities by producing *simpler* but equivalent inequalities until the solutions are readily apparent. To produce these simpler but equivalent inequalities, we often apply the following properties.

Properties of Inequalities

Let a, b, and c be real numbers.

1. *Addition-Subtraction Property* If the same real number is added to or subtracted from each side of an inequality, the resulting inequality is equivalent to the original inequality.

$$a < b \text{ and } a + c < b + c \text{ are equivalent inequalities.}$$

2. *Multiplication-Division Property*
   a. Multiplying or dividing each side of an inequality by the same *positive* real number produces an equivalent inequality.

$$\text{If } c > 0, \text{ then } a < b \text{ and } ac < bc \text{ are equivalent inequalities.}$$

   b. Multiplying or dividing each side of an inequality by the same *negative* real number produces an equivalent inequality provided the direction of the inequality symbol is *reversed*.

$$\text{If } c < 0, \text{ then } a < b \text{ and } ac > bc \text{ are equivalent inequalities.}$$

Note the difference between Property 2a and Property 2b. Property 2a states that an equivalent inequality is produced when each side of a given inequality is multiplied (divided) by the same *positive* real number and the inequality symbol is not changed. By contrast, Property 2b states that when each side of a given inequality is multiplied (divided) by a *negative* real number we must *reverse* the direction of the inequality symbol to produce an equivalent inequality. For instance, multiplying both sides of $-b < 4$ by -1 produces the equivalent inequality $b > -4$. (We multiplied both sides of the first inequality by -1 and we changed the less than symbol to a greater than symbol.)

Alternative to Example 1

Solve each of the following inequalities.

a. $3x - 5 < 7$

b. $-2x + 8 \geq 14$

- a. $x < 4$
- b. $x \leq 3$

TAKE NOTE

Solutions of inequalities are often stated using set-builder notation or interval notation. For instance, the solutions of $2x + 1 < 7$ can be written in set-builder notation as $\{x \mid x < 3\}$ or in interval notation as $(-\infty, 3)$. For a review of set-builder notation, see Section P.1.

EXAMPLE 1 Solve Linear Inequalities

Solve each of the following inequalities.

a. $2x + 1 < 7$ **b.** $-3x - 2 \leq 10$

Solution

a. $2x + 1 < 7$

$2x < 6$ ▪ Add −1 to each side and keep the inequality symbol as is.

$x < 3$ ▪ Divide each side by 2 and keep the inequality symbol as is.

The inequality $2x + 1 < 7$ is true for all real numbers less than 3. In set-builder notation the solution set is given by $\{x \mid x < 3\}$. In interval notation the solution set is $(-\infty, 3)$. See the following figure.

b. $-3x - 2 \leq 10$

$-3x \leq 12$ ▪ Add 2 to each side and keep the inequality symbol as is.

$x \geq -4$ ▪ Divide each side by −3 and reverse the direction of the inequality symbol.

The inequality $-3x - 2 \leq 10$ is true for all real numbers greater than or equal to −4. In set-builder notation the solution set is given by $\{x \mid x \geq -4\}$. In interval notation the solution set is $[-4, \infty)$. See the following figure.

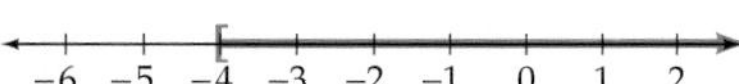

CHECK YOUR PROGRESS 1 Solve each of the following inequalities. Write each solution set using set-builder notation and interval notation.

a. $3x - 2 > 4$ **b.** $-2x + 7 \leq 5$

Solution *See page S7.* a. $\{x \mid x > 2\}$; $(2, \infty)$ b. $\{x \mid x \geq 1\}$; $[1, \infty)$

▪ Solve Compound Inequalities

A **compound inequality** is formed by joining two or more inequalities with the connective word *and* or *or*. The solution set of a compound inequality formed by using the *and* connective is the *intersection* of the solution sets of the individual inequalities. The solution set of a compound inequality formed by using the *or* connective is the *union* of the solution sets of the individual inequalities.

Alternative to Example 2

Solve each compound inequality.

a. $x + 4 > 3$ and $3x + 2 \geq 14$

b. $x + 3 < 7$ or $x - 5 > 6$

▪ **a.** $x \geq 4$

▪ **b.** $x < 4$ or $x > 11$

TAKE NOTE

For a review of the set operations of intersection and union, see Section P.1.

EXAMPLE 2 Solve Compound Inequalities

Solve each compound inequality.

a. $x + 3 > 4$ and $2x + 1 > 15$ **b.** $2x < 10$ or $x + 1 > 9$

Solution

a. $x + 3 > 4$ and $2x + 1 > 15$

$x > 1$ $\quad 2x > 14$ ▪ Solve each individual inequality.

$x > 7$

The solution set is the intersection of $\{x|x > 1\}$ and $\{x|x > 7\}$. Thus the solution set is $\{x|x > 1\} \cap \{x|x > 7\} = \{x|x > 7\}$. In interval notation the solution set is $(7, \infty)$. See the following figure.

b. $2x < 10$ or $x + 1 > 9$

$x < 5$ $\quad x > 8$ ▪ Solve each individual inequality.

The solution set is the union of $\{x|x < 5\}$ and $\{x|x > 8\}$. Thus the solution set is $\{x|x < 5\} \cup \{x|x > 8\} = \{x|x < 5 \text{ or } x > 8\}$. In interval notation the solution set is $(-\infty, 5) \cup (8, \infty)$. See the following figure.

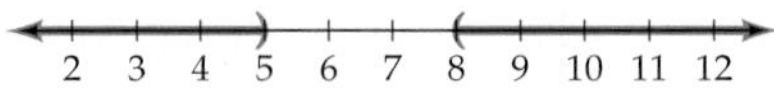

CHECK YOUR PROGRESS 2 Solve each compound inequality. Write each solution set using set-builder notation and interval notation.

a. $x + 5 > 11$ and $2x + 3 > 7$ **b.** $3x < 12$ or $x + 4 > 7$

Solution *See page S7.* **a.** $\{x|x > 6\}$; $(6, \infty)$ **b.** $\{x|x \in \text{real numbers}\}$; $(-\infty, \infty)$

INSTRUCTOR NOTE

The compound inequality $a < b$ *and* $b < c$ can be written in the compact form $a < b < c$; however, the compound inequality $a < b$ *or* $b < c$ cannot be expressed in a compact form. Explain to your students why writing $x < 2$ *or* $x > 3$ as $2 > x > 3$ is incorrect.

The inequality $12 < x + 5 < 19$ is equivalent to the compound inequality $12 < x + 5$ *and* $x + 5 < 19$. You can solve $12 < x + 5 < 19$ by finding the intersection of the solution sets of $12 < x + 5$ and $x + 5 < 19$, but it is easier to just subtract 5 from each of the *three* parts of the inequality, as shown below.

$$12 < x + 5 < 19$$

$$12 - 5 < x + 5 - 5 < 19 - 5$$ ▪ Subtract 5 from each of the three parts of the inequality.

$$7 < x < 14$$

The solution set of $12 < x + 5 < 19$ is $\{x|7 < x < 14\}$. See the following figure.

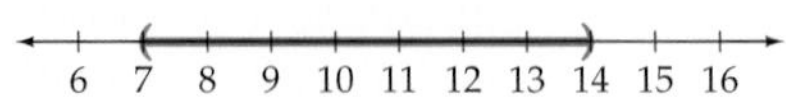

■ Solve Absolute Value Inequalities

An inequality that involves absolute value symbols, such as $|x - 1| < 3$, is called an **absolute value inequality.** The solution set of $|x - 1| < 3$ is the set of all real numbers that lie within 3 units of 1. Thus the solution set consists of all numbers between -2 and 4. See the following figure. In interval notation, the solution set is $(-2, 4)$.

$|x - 1| < 3$

The solution set of the absolute value inequality $|x - 1| > 3$ is the set of all real numbers whose distance from 1 is *greater than* 3. Thus the solution set consists of all numbers less than -2 *or* greater than 4. In interval notation, the solution set is $(-\infty, -2) \cup (4, \infty)$. See the following figure.

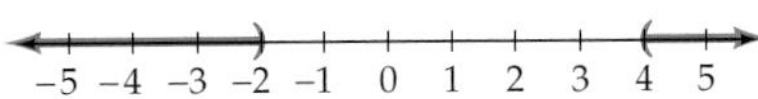

TAKE NOTE

In geometric terms, the solutions of $|x| \le k$, $k > 0$ are represented by all x that lie between k and $-k$, inclusive.

$|x| \le k$

The solutions of $|x| \ge k$ are represented by all x that are less than or equal to $-k$ or greater than or equal to k.

$|x| \ge k$

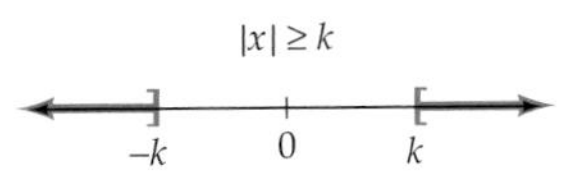

The following properties are often used to solve absolute value inequalities.

Properties of Absolute Value Inequalities

For any variable or variable expression E and any nonnegative real number k:

1. $|E| \le k$ if and only if $-k \le E \le k$

2. $|E| \ge k$ if and only if $E \le -k$ or $E \ge k$

These properties also hold when the $<$ symbol is substituted for the $\le$ symbol and when the $>$ symbol is substituted for the $\ge$ symbol.

In Example 3 we make use of the above properties to solve absolute value inequalities.

Alternative to Example 3

Solve each absolute value inequality.

a. $|2x - 1| < 9$

b. $|x - 2| > 7$

■ **a.** $-4 < x < 5$

■ **b.** $x < -5$ or $x > 9$

EXAMPLE 3 Solve Absolute Value Inequalities

Solve each absolute value inequality.

a. $|3x - 1| < 7$ **b.** $|4x - 3| \ge 5$

Solution

a. $|3x - 1| < 7$ if and only if $-7 < 3x - 1 < 7$.

$$-7 < 3x - 1 < 7$$

$$-6 < 3x < 8 \quad \text{■ Add 1 to each of the three parts of the inequality.}$$

$$-2 < x < \frac{8}{3} \quad \text{■ Multiply each part of the inequality by } \frac{1}{3}.$$

In interval notation the solution set is $\left(-2, \frac{8}{3}\right)$. See the following figure.

TAKE NOTE

Some absolute value inequalities have a solution set that consists of all real numbers. For instance, it is easy to see that $|x + 2| \ge 0$ is true for all values of x, because every absolute value expression is nonnegative.

Continued ➤

b. $|4x - 3| \geq 5$ if and only if $4x - 3 \leq -5$ or $4x - 3 \geq 5$. Solving each of these inequalities produces

$$\begin{aligned} 4x - 3 &\leq -5 & \text{or} \quad 4x - 3 &\geq 5 \\ 4x &\leq -2 & 4x &\geq 8 \\ x &\leq -\frac{1}{2} & x &\geq 2 \end{aligned}$$

Therefore, the solution set is $\left(-\infty, -\frac{1}{2}\right] \cup [2, \infty)$. See the following figure.

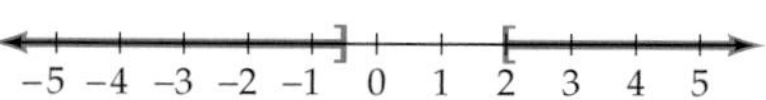

CHECK YOUR PROGRESS 3 Solve each absolute value inequality.

a. $|2x - 1| \leq 6$ **b.** $|3x + 2| > 10$

Solution *See page S7.* a. $\left[-\frac{5}{2}, \frac{7}{2}\right]$ b. $(-\infty, -4) \cup \left(\frac{8}{3}, \infty\right)$

? QUESTION What is the solution set of $|x - k| \geq 0$, regardless of the value of the constant k?

▪ Applications of Inequalities

Many applied problems can be solved by using inequalities.

Alternative to Example 4
Exercise 64, page 135

EXAMPLE 4 Solve an Application Concerning Leases

A real estate company needs a new copy machine. The company has decided to lease either the model ABC machine for \$75 a month plus 5 cents per copy or the model XYZ machine for \$210 a month and 2 cents per copy. Under what conditions is it less expensive to lease the XYZ machine?

Solution Let x represent the number of copies the company produces per month. The dollar costs per month are $75 + 0.05x$ for model ABC and $210 + 0.02x$ for model XYZ. It will be less expensive to lease model XYZ provided:

$$210 + 0.02x < 75 + 0.05x$$

$$210 - 0.03x < 75 \qquad \text{▪ Subtract } 0.05x \text{ from each side.}$$

$$-0.03x < -135 \qquad \text{▪ Subtract 210 from each side.}$$

$$x > 4500 \qquad \text{▪ Divide each side by } -0.03\text{. Reverse the inequality.}$$

The company will find it less expensive to lease model XYZ if it produces over 4500 copies per month.

Continued ➤

? ANSWER The solution set is the set of all real numbers. This is easy to see because, by definition, an absolute value is nonnegative.

CHECK YOUR PROGRESS 4 A school must decide whether to lease a copy machine for 3 years or purchase the machine. If the school leases the machine, it will pay \$150 a month plus \$.015 per copy. If it purchases the machine, it will pay \$12,400 (which includes the purchase price and the maintenance for 3 years) and \$.005 per copy. The school expects that after 3 years the copy machine will have a value of \$0 and it will need to lease or purchase a new machine. Under what conditions will the school find that leasing the copy machine for a 3-year period is less expensive than purchasing the machine?

Solution *See page S8.* Leasing will be less expensive provided the school produces fewer than 700,000 copies during the 3-year period.

Alternative to Example 5
Exercise 76, page 137

EXAMPLE 5 Solve an Application Concerning Test Scores

Tyra has test scores of 70 and 81 in her biology class. To receive a C grade, she must obtain an average greater than or equal to 72 but less than 82. What range of test scores on the one remaining test will enable Tyra to get a C for the course?

Solution The average of three test scores is the sum of the scores divided by 3. Let x represent Tyra's next test score. The requirements for a C grade produce the following inequality.

$$72 \le \frac{70 + 81 + x}{3} < 82$$

$$216 \le 70 + 81 + x < 246$$ ▪ Multiply each part of the inequality by 3.

$$216 \le 151 + x < 246$$ ▪ Simplify.

$$65 \le x < 95$$ ▪ Solve for x by subtracting 151 from each part of the inequality.

To get a C in the course, Tyra's remaining test score must be in the interval $[65, 95)$.

CHECK YOUR PROGRESS 5 Natasha has test scores of 85, 64, and 92 in her algebra class. What is the range of test scores she can receive on her fourth test to produce an average that is at least 84? Assume that the highest possible test score is 100.

Solution *See page S8.* $95 \le x \le 100$

Alternative to Example 6
Exercise 78, page 137

EXAMPLE 6 Solve an Application Concerning Temperatures

A chemical solution for developing photographs needs to satisfy the condition $|C - 20| < 5$, where C is temperature measured in degrees Celsius.

a. What is an acceptable temperature range, expressed in degrees Celsius, for the solution?

b. Express the temperature range in part **a.** in degrees Fahrenheit.

Continued ➤

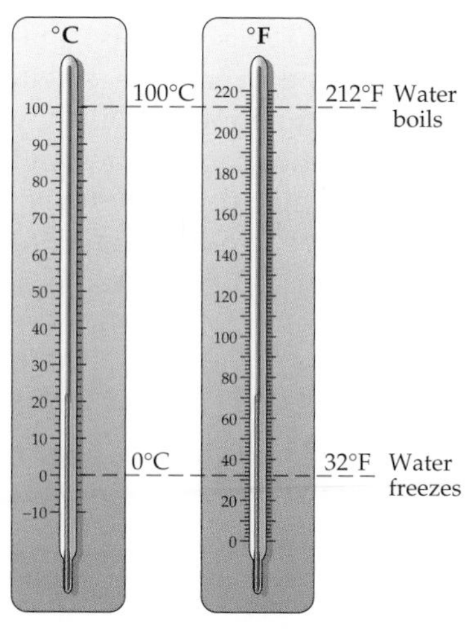

Solution

a. The inequality $|C - 20| < 5$ can be written as $-5 < C - 20 < 5$. Adding 20 to each part of this inequality yields $15 < C < 25$. The chemical solution needs to be kept between 15° and 25° Celsius.

b. A formula that relates a Celsius temperature C to its equivalent Fahrenheit temperature F is $C = \frac{5}{9}(F - 32)$. From part **a.** we know that $15 < C < 25$, where C is measured in degrees Celsius. Substituting $\frac{5}{9}(F - 32)$ for C yields

$$15 < \frac{5}{9}(F - 32) < 25$$

$$27 < F - 32 < 45 \quad \text{■ Multiply each part of the inequality by } \tfrac{9}{5}.$$

$$59 < F < 77 \quad \text{■ Add 32 to each part of the inequality.}$$

The chemical solution needs to be kept between 59° and 77° Fahrenheit.

CHECK YOUR PROGRESS 6 The average daily minimum-to-maximum temperature range for the city of Palm Springs during the month of January is 41 to 68 degrees Fahrenheit. What is the corresponding temperature range measured on the Celsius temperature scale?

Solution *See page S8.* 5° to 20° Celsius

Topics for Discussion

1. If $x < y$, then $xz < yz$. Do you agree? Explain.
2. Can the solution set of the compound inequality $x < -3$ or $x > 5$ be expressed as $-3 > x > 5$? Explain.
3. If $-a < b$, then it must be true that $a > -b$. Do you agree? Explain.
4. Do the inequalities $x < 4$ and $x^2 < 4^2$ both have the same solution set? Explain.
5. If $k < 0$, then $|k| = -k$. Explain why this is a true statement.

EXERCISES 1.4

— Suggested Assignment: Exercises 1–83, odd; and 84–89.

In Exercises 1 to 10, use the properties of inequalities to solve each inequality. Write each solution set in set-builder notation.

1. $2x + 3 < 11$ $\{x \mid x < 4\}$

2. $3x - 5 > 16$ $\{x \mid x > 7\}$

3. $x + 4 > 3x + 16$ $\{x \mid x < -6\}$

4. $5x + 6 < 2x + 1$ $\left\{x \mid x < -\frac{5}{3}\right\}$

5. $-6x + 1 \geq 19$ $\{x \mid x \leq -3\}$

6. $-5x + 2 \leq 37$ $\{x \mid x \geq -7\}$

7. $-3(x + 2) \leq 5x + 7$ $\left\{x \mid x \geq -\frac{13}{8}\right\}$

8. $-4(x - 5) \geq 2x + 15$ $\left\{x \mid x \leq \frac{5}{6}\right\}$

9. $-4(3x - 5) > 2(x - 4)$ $\{x \mid x < 2\}$

10. $3(x + 7) \leq 5(2x - 8)$ $\left\{x \mid x \geq \frac{61}{7}\right\}$

In Exercises 11 to 24, solve each compound inequality. Write the solution set using set-builder notation.

11. $4x + 1 > -2$ and $4x + 1 \leq 17$ $\left\{x \mid -\frac{3}{4} < x \leq 4\right\}$

12. $2x + 5 > -16$ and $2x + 5 < 9$ $\left\{x \mid -\frac{21}{2} < x < 2\right\}$

15. $\left\{x \middle| -\frac{3}{8} \le x < \frac{11}{4}\right\}$

13. $10 \ge 3x - 1 \ge 0$ $\left\{x \middle| \frac{1}{3} \le x \le \frac{11}{3}\right\}$

14. $0 \le 2x + 6 \le 54$ $\{x | -3 \le x \le 24\}$

15. $20 > 8x - 2 \ge -5$

16. $4 \le 10x + 1 \le 51$ $\left\{x \middle| \frac{3}{10} \le x \le 5\right\}$

17. $-4x + 5 > 9$ or $4x + 1 < 5$ $\{x | x < 1\}$

18. $2x - 7 \le 15$ or $3x - 1 \le 5$ $\{x | x \le 11\}$

19. $3x - 3 > 9$ or $-x + 2 < 3$ $\{x | x > -1\}$

20. $4x - 3 \le -5$ or $4x - 3 \ge 5$ $\left\{x \middle| x \le -\frac{1}{2} \text{ or } x \ge 2\right\}$

21. $-2x + 6 > -2$ and $3x + 1 \le 16$ $\{x | x < 4\}$

22. $\frac{1}{2}x - 5 \ge 2$ and $-\frac{1}{3}x + 3 < 10$ $\{x | x \ge 14\}$

23. $\frac{3}{4}x - 6 \ge 8$ or $-\frac{1}{2}x + 6 < 20$ $\{x | x > -28\}$

24. $0.2x - 3 \ge 5$ and $-0.75x - 6 < 2$ $\{x | x \ge 40\}$

In Exercises 25 to 46, solve each absolute value inequality and write the solution set using interval notation.

25. $|2x - 1| > 4$ $\left(-\infty, -\frac{3}{2}\right) \cup \left(\frac{5}{2}, \infty\right)$

26. $|2x - 9| < 7$ $(1, 8)$

27. $|x + 3| \ge 5$ $(-\infty, -8] \cup [2, \infty)$

28. $|x - 10| \ge 2$ $(-\infty, 8] \cup [12, \infty)$

29. $|3x - 10| \le 14$ $\left[-\frac{4}{3}, 8\right]$

30. $|2x - 5| \ge 1$ $(-\infty, 2] \cup [3, \infty)$

31. $|4 - 5x| \ge 24$ $(-\infty, -4] \cup \left[\frac{28}{5}, \infty\right)$

32. $|3 - 2x| \le 5$ $[-1, 4]$

33. $|x - 5| \ge 0$ $(-\infty, \infty)$

34. $|x - 7| \ge 0$ $(-\infty, \infty)$

35. $|x - 4| \le 0$ $\{4\}$

36. $|2x + 7| \le 0$ $\left\{-\frac{7}{2}\right\}$

37. $|5x - 1| < -4$ $\varnothing$

38. $|2x - 1| < -9$ $\varnothing$

39. $|2x + 7| \ge -5$ $(-\infty, \infty)$

40. $|3x + 11| \ge -20$ $(-\infty, \infty)$

41. $2|x + 4| + 3 \ge 9$ $(-\infty, -7] \cup [-1, \infty)$

42. $5|2x - 3| + 1 < 31$ $\left(-\frac{3}{2}, \frac{9}{2}\right)$

43. $6 \ge |3 - x|$ $[-3, 9]$

44. $9 < |6 + 2x|$ $\left(-\infty, -\frac{15}{2}\right) \cup \left(\frac{3}{2}, \infty\right)$

45. $\left|\frac{x - 3}{2.4}\right| \ge 5$ $(-\infty, -9] \cup [15, \infty)$

46. $\left|\frac{x - 40}{1.5}\right| \le 2$ $[37, 43]$

In Exercises 47 to 54, write an absolute value inequality that has the indicated solution.

47. $(1, 5)$ $|x - 3| < 2$

48. $(10, 18)$ $|x - 14| < 4$

49. $(-\infty, 4) \cup (8, \infty)$ $|x - 6| > 2$

50. $(-\infty, 12) \cup (30, \infty)$ $|x - 21| > 9$

51. $[7, 11]$ $|x - 9| \le 2$

52. $[50, 80]$ $|x - 65| \le 15$

53. $(-\infty, 8] \cup [12, \infty)$ $|x - 10| \ge 2$

54. $(-\infty, 22] \cup [64, \infty)$ $|x - 43| \ge 21$

In Exercises 55 to 60, write an absolute value inequality that has the solution set shown by the graph.

55. 

$|x| \le 4$

56. 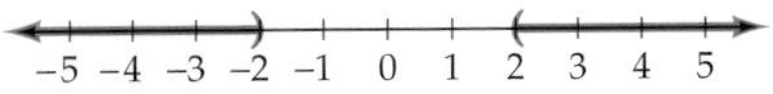

$|x| > 2$

57. 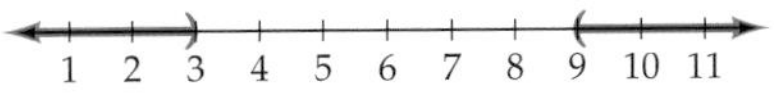

$|x - 6| > 3$

58. 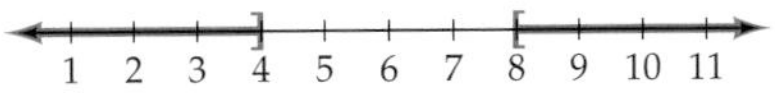

$|x - 6| \ge 2$

59. 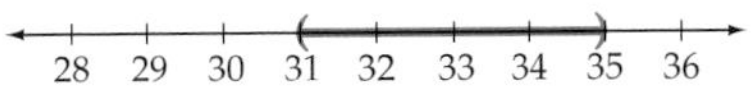

$|x - 33| < 2$

60. 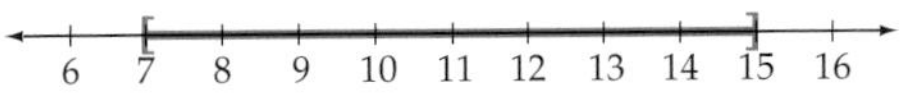

$|x - 11| \le 4$

Business and Economics

61. ***Sum of Consecutive Odd Integers*** The sum of three consecutive odd integers is between 63 and 81. Find all sets of integers that satisfy this condition. {21, 23, 25}, {23, 25, 27}

62. ***Sum of Consecutive Even Integers*** The sum of three consecutive even integers is between 36 and 54. Find all sets of integers that satisfy this condition. {12, 14, 16}, {14, 16, 18}

63. ***Personal Finance*** A bank offers two checking account plans. Under plan A you pay a monthly fee of \$3.00 and \$.06 per check. Under plan B you pay a monthly fee of \$8.00 and \$.01 per check. Under what conditions is it less expensive to use plan B? More than 100 checks per month are written.

64. ***Consumer Spending*** You can rent a car for the day from Company A for \$42.00 plus \$.05 per mile. Company B charges \$34.00 plus \$.14 per mile. Under what conditions is it cheaper to rent a car for one day from company A? You drive more than $88.\overline{8}$ miles per day.

65. ***Consumer Spending*** Herman wishes to spend a maximum of \$1800 for the purchase of a computer. On his purchase he must pay a state tax of 8%, \$110 for an extended warranty, and shipping charges of \$65. What is the price range of the computers he should consider? $\le$\$1504

66. ***Production and Profit*** A company that manufactures calculators has fixed costs of \$18,500 per month. The cost of manufacturing each calculator is \$12. The company sells each calculator for \$39. How many calculators should be manufactured and sold each month to make a profit? At least 686

67. ***Production and Profit*** A company that manufactures cellular telephones has fixed costs of \$27,800 per month. The cost of manufacturing each telephone is \$37. The company sells each telephone for \$149. How many telephones need to be manufactured and sold each month to make a profit? At least 249

68. ***Consumer Spending*** You purchased 2.5 pounds of salmon at \$9.00 per pound. The scale used to weigh the salmon was accurate to within 0.1 pound. What is the range of possible amounts, in cents, that you may have been over- or undercharged? (*Hint:* A scale that measures to the nearest 0.1 pound may be off by as much as 0.05 pound.) −45¢ to +45¢

69. ***Consumer Spending*** Eugene purchased 2 ounces of Russian Beluga caviar. The price was \$85 per ounce. The scale used to weigh the caviar was accurate to within 0.1 ounce. What is the range of possible dollar amounts that Eugene may have been over- or undercharged? (*Hint:* A scale that measures to the nearest 0.1 ounce may be off by as much as 0.05 ounce.) −\$4.25 to +\$4.25

70. ***Consumer Spending*** Ronda wants to rent one of two apartments. There is less than \$150 difference between the monthly rental fees of the apartments. One of the apartments rents for \$575 per month. What is the range of possible monthly rental fees for the other apartment?
More than \$425 but less than \$725

71. ***Consumer Spending*** The price of a pair of Revo sunglasses and the price of a pair of Bolle sunglasses differ by more than \$48. The price of the Revo sunglasses is \$218.

   a. Write an absolute value inequality that expresses the relationship between the price B, in dollars, of the Bolle sunglasses and the price, in dollars, of the Revo sunglasses. $|B - 218| > 48$

   b. Use interval notation to describe the price range, in dollars, that is possible for the Bolle sunglasses.
   $(0, 170) \cup (266, \infty)$

72. ***Shipping Requirements*** Federal Express (FedEx) will only ship packages for which the length is less than or equal to 119 inches and the length plus the girth is less than or equal to 165 inches. The length of a package is defined as the length of the longest side and the girth is defined as twice the width plus twice the height of the package. If a box has a length of 42 inches and a width of 38 inches, determine the possible range of heights for this package if you wish to ship it by FedEx. (*Source:* **http://www.iship.com**) ≤ 23.5 in.

73. ***Shipping Requirements*** United Parcel Service (UPS) will only ship packages for which the length is less than or equal to 108 inches and the length plus the girth (see Exercise 72) is less than or equal to 130 inches. If a box has a length of 34 inches and width of 22 inches, then determine the possible range of heights for this package if you wish to ship it by UPS. (*Source:* **http://www.iship.com**) ≤ 26 in.

74. ***Movie Ticket Prices*** The average U.S. movie ticket price P, in dollars, can be modeled by

$$P = 0.218t + 4.02, \quad t \ge 0$$

where $t = 0$ represents the year 1994. According to this model, in what year will the average price of a movie ticket first exceed \$6.50? (*Source:* National Association of Theatre Owners, **http://www.natoonline.org/statisticstickets.htm**) 2005

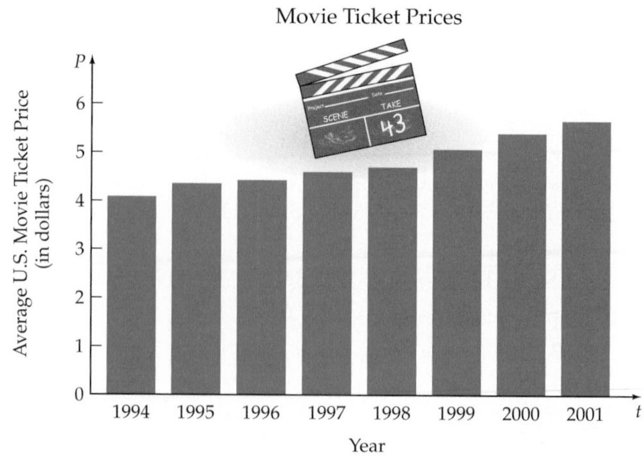

Life and Health Sciences

75. **Forensic Science** Forensic specialists can estimate the height of a deceased person from the lengths of the person's bones. These lengths are substituted into mathematical inequalities. For instance, an inequality that relates the height h, in centimeters, of an adult female and the length f, in centimeters, of her femur is

$$|h - (2.47f + 54.10)| \leq 3.72$$

a. Use the above inequality to estimate the possible range of heights, rounded to the nearest 0.1 centimeter, for an adult female whose femur measures 32.24 centimeters.

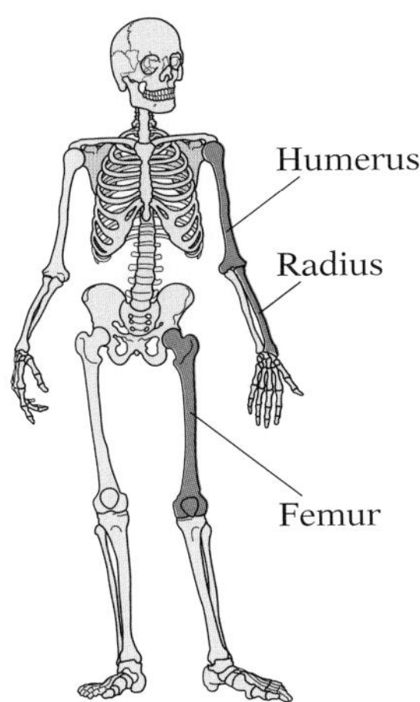

Forensic specialists use the terms *potential stature* and *living stature*. Potential stature is the height of a person whose height has not been affected by the aging process. Living stature is the height of a person whose height has been affected by the aging process. If a person is over age 30, a formula is used to adjust for the loss of height due to aging. One adjustment formula that is used subtracts $0.06(a - 30)$ centimeters from a person's potential stature, where a is the age of the person in years and $a > 30$. For example, a 65-year-old person will have lost about $0.06(65 - 30) = 2.1$ centimeters due to the aging process.

130.0 to 137.5 cm

b. An inequality that is used to calculate the potential stature of an adult male from the length r of his radius is

$$|h - (3.32r + 85.43)| \leq 4.57$$

where h and r are both in centimeters. Use this inequality to estimate the range of potential statures for an adult male whose radius measures 26.36 centimeters. If it has been determined that this man died at age 55, then determine the range of heights for his living stature.

Potential stature: 168.4 to 177.5 cm; living stature: 166.9 to 176.0 cm

c. A formula that is used to calculate the potential stature of an adult female from the length r of her radius is $h = (4.74r + 54.93) \pm 4.24$, where h and r are both in centimeters. Write this formula as an inequality involving

$$|h - (4.74r + 54.93)|.$$

(*Source:* Based on data from Western Kentucky University, **http://www.wku.edu/**)

$|h - (4.74r + 54.93)| \leq 4.24$

76. **Course Grade** An average of 68 to 79 in a biology class receives a C grade. A student has test scores of 82, 72, 64, and 95 on four tests. Find the range of scores on the fifth test that will give the student a C grade for the course. [27, 82]

Sports and Recreation

77. **Basketball Dimensions** A basketball is to have a circumference of 29.5 inches to 30.0 inches. Find the acceptable range of diameters for the basketball. Round results to the nearest hundredth of an inch. 9.39 to 9.55 in.

78. **Football Dimensions** According to the National Collegiate Athletic Association (NCAA), the length x of a football, in inches, must satisfy the following inequality. (*Source:* **http://www.infoplease.com**)

$$\left|x - 11\frac{5}{32}\right| \leq \frac{9}{32}$$

Find the acceptable range of lengths for an NCAA football. $10\frac{7}{8}$ to $11\frac{7}{16}$ in.

Social Sciences

79. **Beverage Consumption Trends** The following chart shows that in recent years, the U.S. per capita consumption of coffee has been decreasing while the per capita consumption of bottled water has been increasing. The per capita consumption of coffee can be modeled by

$$C = -0.940t + 21.2, \quad t \geq 0$$

where $t = 0$ represents the year 1995. The per capita consumption of bottled water can be modeled by

$$B = 0.594t + 10.2, \quad t \geq 0$$

where $t = 0$ represents the year 1995. According to these models, in what year will the per capita consumption of coffee first fall below the per capita consumption of bottled water? 2002

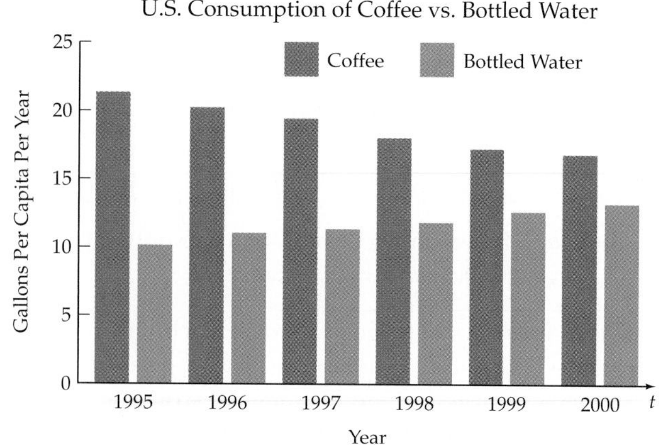

Source: **http://www.fas.usda.gov**

80. ***Cost of Tuition*** The yearly cost C, in dollars, of tuition and fees for public U.S. institutions of higher education can be modeled by

$$C = 82.29t + 2185, \quad t \geq 0$$

where $t = 0$ represents the school year 1995–1996. According to this model, in what school year will the cost of tuition and fees first exceed \$3000? (*Source:* National Center for Education Statistics, U.S. Dept. of Education)

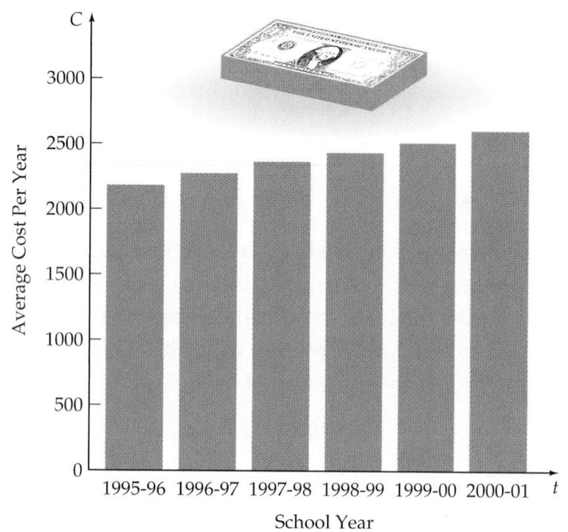

2004–2005

81. ***Number of Degrees Conferred*** The following bar graph shows the number of bachelor's degrees conferred in the U.S. for selected years. (*Source:* National Center for Education Statistics, U.S. Dept. of Education)

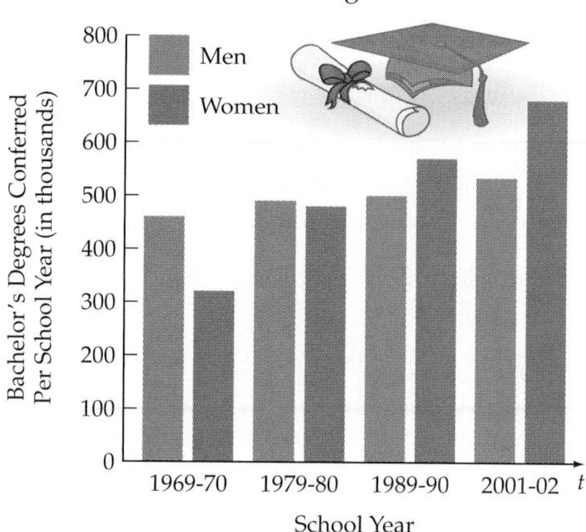

The number of men M receiving bachelor's degrees can be modeled by $M = 1.59t + 463, t \geq 0$, where $t = 0$ represents the school year 1969–1970 and M is measured in thousands.

a. According to this model, in what school year will the number of men receiving bachelor's degrees first exceed 530,000? 2011–2012

The number of women W receiving bachelor's degrees can be modeled by $W = 10.8t + 338, t \geq 0$, where $t = 0$ represents the school year 1969–1970 and W is measured in thousands.

b. According to this model, in what school year will the number of women receiving bachelor's degrees first exceed 750,000? 2007–2008

Physical Sciences and Engineering

82. ***Fluid Measurement*** A beaker has an inner radius of 2 centimeters. To what height h (to the nearest 0.1 centimeter) should you fill the beaker if you need to measure 750 cubic centimeters of a solution with an error of 15 cubic centimeters or less? 58.5 to 60.9 cm

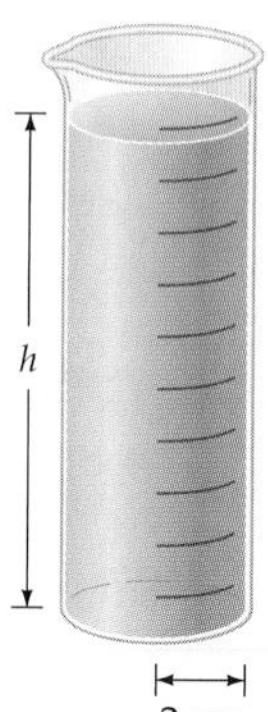

83. *Tolerance* A machinist is producing a circular cylinder on a lathe. The circumference of the cylinder must be 28 inches, with a tolerance of 0.15 inch. The **tolerance** of a part is the acceptable amount by which the part may vary from a given measurement. What is the range of radii, to the nearest 0.01 inch, that will produce acceptable results? 4.43 to 4.48 in.

Prepare for Section 1.5

84. Simplify: $[7 - (-3)]^2$ [P.1] 100

85. Simplify: $\sqrt{(4-1)^2 + (5-1)^2}$ [P.6] 5

86. Simplify: $\sqrt{96}$ [P.6] $4\sqrt{6}$

87. Factor: $x^2 - 14x + 49$ [P.4] $(x-7)^2$

88. Evaluate $3x^2 - 2x + 5$ for $x = -4$. [P.3] 61

89. Solve: $2x^2 - 18 = 0$ [1.2] ± 3

Explorations 

1. A coin is considered a fair coin if it has an equal chance of landing heads up or tails up. To decide whether a coin is a fair coin, a statistician tosses the coin 100 times and records the number of tails, t. The statistician is prepared to state that the coin is a fair coin provided

$$\left|\frac{t-50}{1.58}\right| \le 2.33$$

a. Determine what values of t will cause the statistician to state that the coin is a fair coin. 47, 48, 49, 50, 51, 52, or 53

b. Pick a coin and test it to see whether it is a fair coin according to the criterion stated above. Answers will vary.

SECTION 1.5 Quadratic and Rational Inequalities and Their Applications

- Use a Sign Diagram to Solve Quadratic Inequalities
- Use the Critical Value Method to Solve Quadratic Inequalities
- Use a Sign Diagram to Solve Rational Inequalities
- Use the Critical Value Method to Solve Rational Inequalities
- Applications of Quadratic and Rational Inequalities

Use a Sign Diagram to Solve Quadratic Inequalities

A **sign diagram** indicates over what intervals an algebraic expression is negative, zero, or positive. For instance, the algebraic expression $x - 1$ is

- negative if $x < 1$,
- zero if $x = 1$, and
- positive if $x > 1$.

The following sign diagram conveys this information in a visual manner.

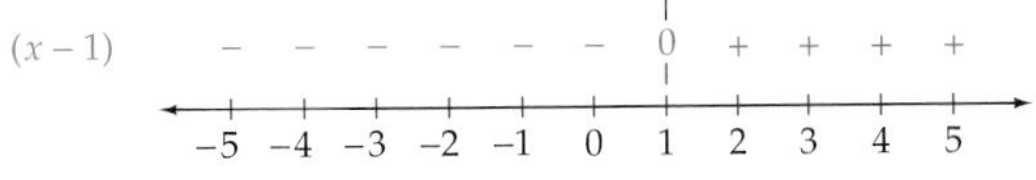

A sign diagram for $(x - 1)$

Some quadratic inequalities can be solved by constructing a sign diagram. The following procedure explains the necessary steps.

Solve a Quadratic Inequality by Using a Sign Diagram

1. Write the inequality so that one side of the inequality is a quadratic polynomial and the other side is 0.
2. Factor the quadratic polynomial into linear factors.
3. Draw a sign diagram for each of the linear factors and use the diagram to determine where the quadratic polynomial is positive and where it is negative.

Alternative to Example 1

Use a sign diagram to solve each inequality.

a. $x^2 + 2x - 8 < 0$

b. $x^2 + x \geq 6$

- **a.** $-4 < x < 2$
- **b.** $x \leq -3$ or $x \geq 2$

TAKE NOTE

If the product of two factors is negative, then one factor must be positive and the other negative.

EXAMPLE 1 Use a Sign Diagram to Solve Quadratic Inequality

Use a sign diagram to solve each inequality.

a. $x^2 + x - 2 < 0$ **b.** $x^2 \geq -x + 12$

Solution

a. Factor $x^2 + x - 2$ to produce $(x - 1)(x + 2) < 0$. Draw a sign diagram for each of the linear factors $(x - 1)$ and $(x + 2)$.

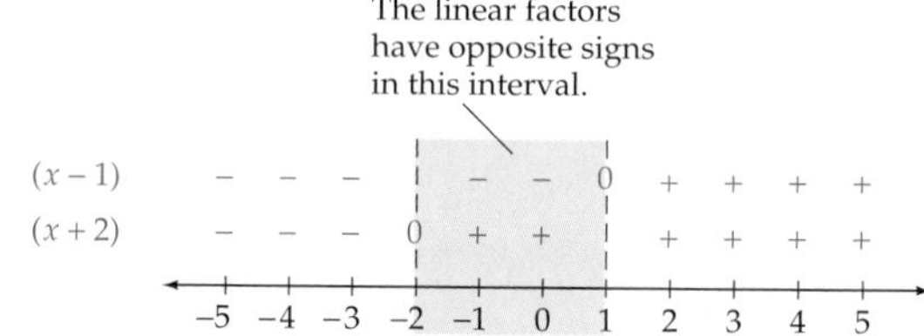

A sign diagram for $(x - 1)$ and $(x + 2)$

To determine where $(x - 1)(x + 2) < 0$, examine the sign diagram to see where the factors $(x - 1)$ and $(x + 2)$ have *opposite* signs. The above sign diagram shows that the only interval on which $(x - 1)$ and $(x + 2)$ have opposite signs is $(-2, 1)$. Thus the solution set of $x^2 + x - 2 < 0$ is $(-2, 1)$. See the following figure.

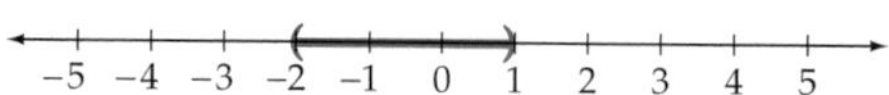

b. Rewrite $x^2 \geq -x + 12$ as $x^2 + x - 12 \geq 0$. Factor to produce $(x + 4)(x - 3) \geq 0$. Draw a sign diagram for each of the linear factors.

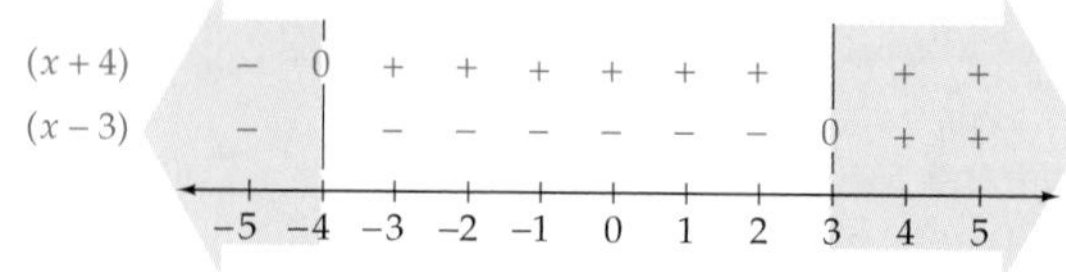

A sign diagram for $(x + 4)$ and $(x - 3)$

Continued ➤

TAKE NOTE

If the product of two factors is positive or zero, then either both factors are positive, both factors are negative, or at least one factor is zero.

To determine where $(x + 4)(x - 3) \geq 0$, examine the sign diagram to determine where $(x + 4)$ and $(x - 3)$ have the *same* sign or where they equal 0. The sign diagram shows that $(x + 4)$ and $(x - 3)$ are both positive on $(3, \infty)$ and are both negative on $(-\infty, -4)$. We also know that $(x + 4) = 0$ at $x = -4$ and $(x - 3) = 0$ at $x = 3$. Thus the solution set of $(x + 4)(x - 3) \geq 0$, which is equivalent to the original inequality, is $(-\infty, -4] \cup [3, \infty)$. See the following figure.

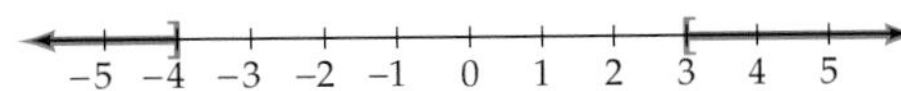

CHECK YOUR PROGRESS 1 Use a sign diagram to solve each inequality.

a. $x^2 - 4x + 3 \leq 0$ **b.** $2x^2 - x > 3$

Solution *See page S8.* a. $[1, 3]$ b. $(-\infty, -1) \cup \left(\frac{3}{2}, \infty\right)$

▪ Use the Critical Value Method to Solve Quadratic Inequalities

Using a sign diagram to solve a quadratic inequality requires that you first write the quadratic polynomial as a product of linear factors. If you cannot factor a quadratic over the integers, then you should try the critical value method, which makes use of the zeros of a quadratic polynomial.

Zeros of a Polynomial

If $P(x)$ is a polynomial, then the values of x for which $P(x) = 0$ are called the **zeros** of $P(x)$.

The critical value method relies on the following sign property of polynomials.

Sign Property of Polynomials

Between any two consecutive intervals formed by the real zeros of a polynomial the polynomial will be positive or negative, but not both.

The **sign property of polynomials** indicates that a polynomial will not change sign on any of the consecutive intervals formed by its real zeros. Here are the steps needed to solve a quadratic inequality by the critical value method.

TAKE NOTE

A critical value should not be used as a test value.

For a review of open intervals, see Section P.1.

Solve a Quadratic Inequality by the Critical Value Method

1. Write the inequality so that one side of the inequality is a quadratic polynomial and the other side is 0.
2. Find the real zeros of the quadratic polynomial. They are the **critical values of the quadratic inequality.** The consecutive open intervals formed by these values are the **test intervals.**
3. Pick one number x, called a **test value,** from each of the test intervals and evaluate the quadratic polynomial at each test value.
   - If the polynomial is positive for a particular test value, then the polynomial is positive for all x-values in that test interval.
   - If the polynomial is negative for a particular test value, then the polynomial is negative for all x-values in that test interval.

Alternative to Example 2

Use the critical value method to solve $x^2 - 2x - 5 < 0$.

- $1 - \sqrt{6} < x < 1 + \sqrt{6}$

EXAMPLE 2 Use the Critical Value Method to Solve a Quadratic Inequality

Use the critical value method to solve $x^2 - 2x - 4 < 0$.

Solution Use the quadratic formula to find the real zeros of $x^2 - 2x - 4$. The real zeros are $1 - \sqrt{5} \approx -1.2$ and $1 + \sqrt{5} \approx 3.2$. Sketch a number line that shows the test intervals formed by these real zeros.

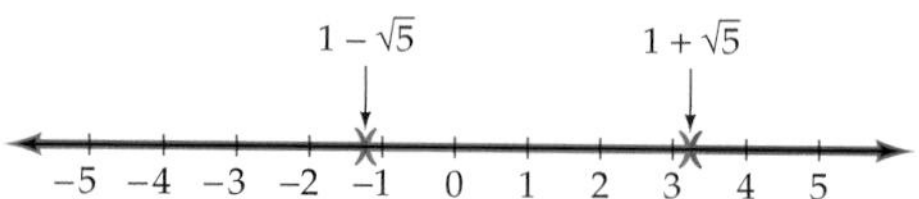

The test intervals are shown in red, blue, and green.

Pick a convenient test value from each of the test intervals. For instance, we picked -3, 0, and 4 as our test values. Evaluate $x^2 - 2x - 4$ for each test value.

- For $x = -3$, $x^2 - 2x - 4 = (-3)^2 - 2(-3) - 4 = 11$, which is *positive.*
- For $x = 0$, $x^2 - 2x - 4 = (0)^2 - 2(0) - 4 = -4$, which is *negative.*
- For $x = 4$, $x^2 - 2x - 4 = (4)^2 - 2(4) - 4 = 4$, which is *positive.*

These results show that $x^2 - 2x - 4$ is negative for all values of x between $1 - \sqrt{5}$ and $1 + \sqrt{5}$, positive for $x > 1 + \sqrt{5}$, and positive for $x < 1 - \sqrt{5}$. Thus the solution set of $x^2 - 2x - 4 < 0$ is $\left(1 - \sqrt{5}, 1 + \sqrt{5}\right)$. See the following figure.

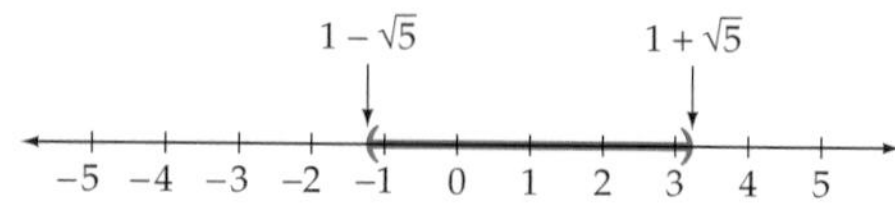

CHECK YOUR PROGRESS 2 Use the critical value method to solve $x^2 - 4x \leq 1$.

Solution *See page S8.* $\left[2 - \sqrt{5}, 2 + \sqrt{5}\right]$

Use a Sign Diagram to Solve Rational Inequalities

A **rational expression** is the quotient of two polynomials. An inequality that involves a rational expression is called a **rational inequality.** For instance,

$$\frac{x-3}{x+1} > 0$$

is a rational inequality. Many rational inequalities can be solved by using a sign diagram, as illustrated in the following example.

Alternative to Example 3

Use a sign diagram to solve $\frac{2x+1}{x-3} < 1$.

▪ $-4 < x < 3$

TAKE NOTE

Solving

$$\frac{3x+4}{x+1} \le 2$$

by multiplying both sides of the inequality by $x + 1$ does not yield the correct solution set. One problem with this approach is that we do not know whether $x + 1$ is positive or negative. Thus we do not know whether we should reverse the inequality symbol. Another problem is the fact that $x + 1$ has a zero of -1. If we multiply both sides by $x + 1$, we will not have -1 as a critical value of the inequality.

EXAMPLE 3 Use a Sign Diagram to Solve a Rational Inequality

Use a sign diagram to solve $\frac{3x+4}{x+1} \le 2$.

Solution Write the inequality so that 0 appears on one side of the inequality.

$$\frac{3x+4}{x+1} \le 2$$

$$\frac{3x+4}{x+1} - 2 \le 0$$

Write the left side as a single rational expression.

$$\frac{3x+4}{x+1} - \frac{2(x+1)}{x+1} \le 0 \qquad \text{▪ Rewrite with } (x+1) \text{ as the common denominator.}$$

$$\frac{3x+4-2x-2}{x+1} \le 0 \qquad \text{▪ Simplify.}$$

$$\frac{x+2}{x+1} \le 0$$

The following sign diagram shows that for all values of x on the interval $(-2, -1)$ the quotient $(x + 2)/(x + 1)$ is negative. On the other intervals the quotient $(x + 2)/(x + 1)$ is positive.

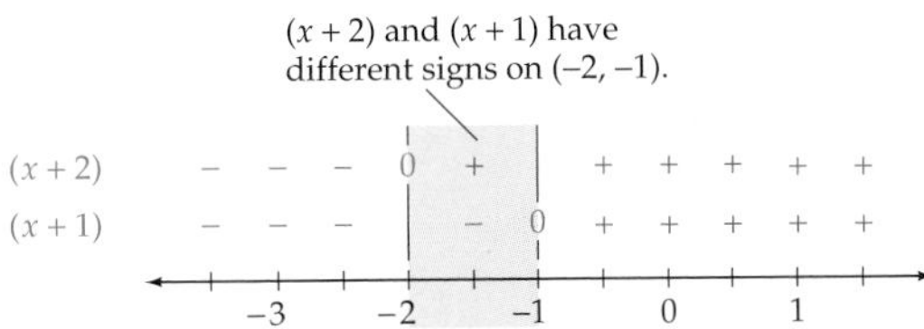

Thus the solution set is $[-2, -1)$. Note that -2 is included in the solution set because $(x + 2)/(x + 1) = 0$ for $x = -2$. However, -1 is not included in the solution set because the denominator $(x + 1)$ is zero for $x = -1$. See the following figure.

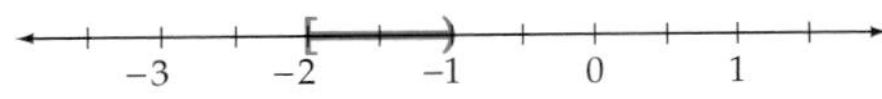

Continued ➤

CHECK YOUR PROGRESS 3 Use a diagram to solve $\dfrac{3x + 1}{x - 2} \geq 4$.

Solution *See page S9.* $(2, 9]$

▪ Use the Critical Value Method to Solve Rational Inequalities

If a rational inequality is written such that one side of the inequality is a rational expression and the other side is 0, then the **critical values of the rational inequality** are the real zeros of the numerator and the real zeros of the denominator of the rational expression. Rational inequalities can be solved by using their critical values.

Solve a Rational Inequality by the Critical Value Method

1. Write the inequality so that one side of the inequality is a rational expression and the other side is 0.
2. Find the critical values of the rational expression. The consecutive open intervals formed by the critical values are the test intervals.
3. Pick one number x, called a test value, from each of the test intervals and evaluate the rational expression at each test value.
   - If the rational expression is positive for a particular test value, then the rational expression is positive for all x-values in that test interval.
   - If the rational expression is negative for a particular test value, then the rational expression is negative for all x-values in that test interval.

Alternative to Example 4

Use the critical value method to solve $\dfrac{x^2 - 2x - 5}{x - 4} \geq 2$.

▪ $1 \leq x \leq 3$ or $x > 4$

INSTRUCTOR NOTE

Remind students that when they solve a rational inequality by the critical value method, the inequality should first be written so that a single rational expression appears on one side and 0 appears on the other side of the inequality.

EXAMPLE 4 Use the Critical Value Method to Solve a Rational Inequality

Use the critical value method to solve $\dfrac{x^2 - x - 11}{x - 5} \geq 1$.

Solution Write the inequality so that a single rational expression appears on one side and 0 appears on the other side of the inequality.

$$\frac{x^2 - x - 11}{x - 5} \geq 1$$

$$\frac{x^2 - x - 11}{x - 5} - 1 \geq 0$$

$$\frac{x^2 - x - 11}{x - 5} - \frac{x - 5}{x - 5} \geq 0$$

$$\frac{x^2 - 2x - 6}{x - 5} \geq 0$$

Continued ➤

The real zero of the denominator is 5. Using the quadratic formula, we find the real zeros of the numerator to be $1 - \sqrt{7} \approx -1.6$ and $1 + \sqrt{7} \approx 3.6$. The critical values of the inequality are 5, $1 - \sqrt{7}$, and $1 + \sqrt{7}$. Sketch the number line that shows the test intervals formed by these critical values.

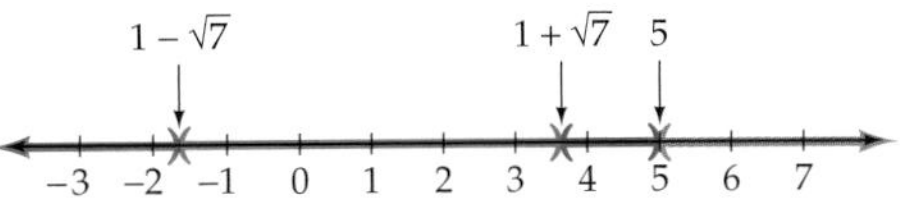

The four test intervals

Pick a test value from each of the test intervals. For instance, we picked −2, 0, 4, and 6 as our test values. Evaluate $\dfrac{x^2 - 2x - 6}{x - 5}$ for each value.

- For $x = -2$, $\dfrac{x^2 - 2x - 6}{x - 5} = -\dfrac{2}{7}$, which is *negative*. Thus the interval $\left(-\infty, 1 - \sqrt{7}\right)$ is not part of the solution set.
- For $x = 0$, $\dfrac{x^2 - 2x - 6}{x - 5} = \dfrac{6}{5}$, which is *positive*. Thus the interval $\left(1 - \sqrt{7}, 1 + \sqrt{7}\right)$ is part of the solution set.
- For $x = 4$, $\dfrac{x^2 - 2x - 6}{x - 5} = -2$, which is *negative*. Thus the interval $\left(1 + \sqrt{7}, 5\right)$ is not part of the solution set.
- For $x = 6$, $\dfrac{x^2 - 2x - 6}{x - 5} = 18$, which is *positive*. Thus the interval $(5, \infty)$ is part of the solution set.

The numbers $1 - \sqrt{7}$ and $1 - \sqrt{7}$ are included in the solution set because they make $(x^2 - 2x - 6)/(x - 5)$ equal to zero. The number 5 is not included in the solution set because $(x^2 - 2x - 6)/(x - 5)$ is undefined for $x = 5$. Thus the solution set is $\left[1 - \sqrt{7}, 1 + \sqrt{7}\right] \cup (5, \infty)$. See the following figure.

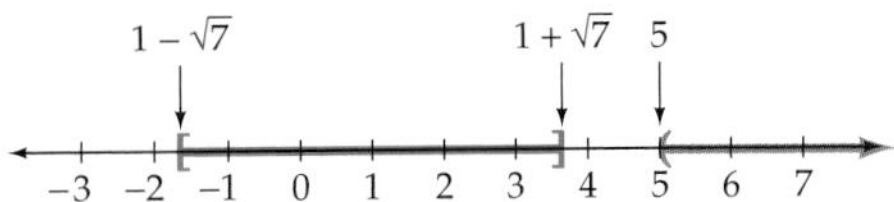

CHECK YOUR PROGRESS 4 Use the critical value method to solve $\dfrac{x^2 - 6x + 2}{x - 3} > 2$.

Solution *See page S9.* $\left(4 - 2\sqrt{2}, 3\right) \cup \left(4 + 2\sqrt{2}, \infty\right)$

QUESTION Do $\dfrac{x + 4}{x - 1} < 2$ and $x + 4 < 2x - 2$ have the same critical value(s)?

ANSWER No. The inequality $(x + 4)/(x - 1) < 2$ has both 1 and 6 as critical values. The inequality $x + 4 < 2x - 2$ has only 6 as a critical value.

Applications of Quadratic and Rational Inequalities

Quadratic inequalities and rational inequalities are often used to solve applied problems. Here are a few examples.

Alternative to Example 5
Exercise 42, page 150

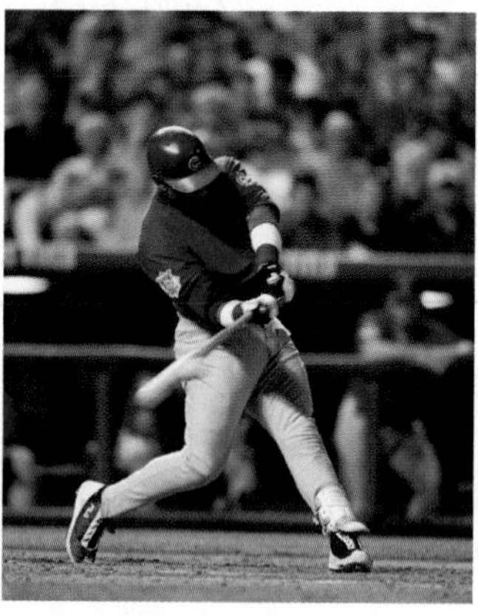

EXAMPLE 5 Solve an Application Involving Batting Averages

Near the end of May 2002, Sammy Sosa had 53 hits out of 163 at-bats. At that time his batting average was approximately 0.325. If Sosa goes into a batting slump in which he gets no hits, how many more at-bats will it take for his batting average to fall below 0.300?

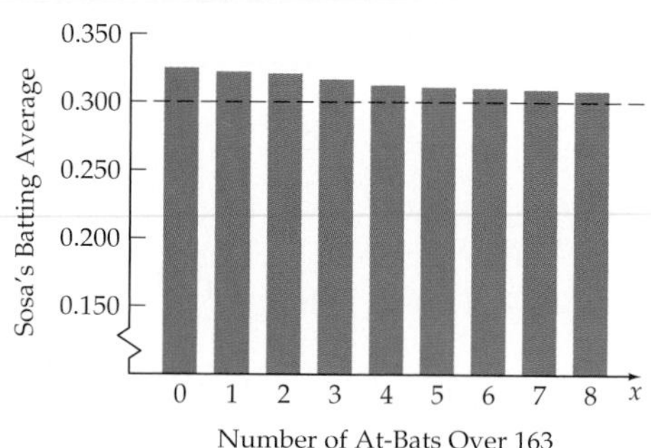

Solution A baseball player's batting average is determined by dividing the player's number of hits by the number of times the player has been at bat. Let x be the number of additional at-bats that Sosa takes over 163. During this period his batting average will be $\frac{53}{163 + x}$, and we wish to solve

$$\frac{53}{163 + x} < 0.300$$

This rational inequality can be solved by using a sign diagram or the critical value method, but there is an easier method. In this application we know that $163 + x$ is positive. Thus if we multiply each side of the above inequality by $163 + x$, we will obtain the linear inequality $53 < 48.9 + 0.300x$, with the condition that x is a positive integer. Solving this inequality produces

$$53 < 48.9 + 0.300x$$
$$4.1 < 0.300x$$
$$x > 13.\overline{6}$$

Because x must be a positive integer, Sosa's average will fall below 0.300 if he goes hitless for 14 or more at-bats.

CHECK YOUR PROGRESS 5 Assume Sammy Sosa has 53 hits out of 163 at-bats. If Sosa goes into a hitting streak in which he gets a hit every time he bats, how many more at-bats will it take for his batting average to exceed 0.350?

Solution *See page S9.* 7 or more at-bats

Alternative to Example 6
Exercise 48, page 151

EXAMPLE 6 Solve an Application Concerning Population Densities

The population density D (in people per square mile) of a city is related to the distance x, in miles, from the center of the city by $D = \dfrac{4500x}{2x^2 + 25}$, $0 < x < 12$. Describe the region of the city in which the population density exceeds 200 people per square mile.

Solution We need to solve $\dfrac{4500x}{2x^2 + 25} > 200$ for x. We could solve this rational inequality by the critical value method, but we know that $2x^2 + 25 > 0$, so it is easier to just produce an equivalent inequality by multiplying both sides of the inequality by $2x^2 + 25$.

$$\frac{4500x}{2x^2 + 25} > 200$$
$$4500x > 400x^2 + 5000$$
$$0 > 400x^2 - 4500x + 5000 \quad \text{■ Write the inequality with 0 on one side.}$$
$$0 > 4x^2 - 45x + 50 \quad \text{■ Divide both sides by 100.}$$
$$0 > (4x - 5)(x - 10) \quad \text{■ Factor.}$$

A sign diagram shows that $0 > (4x - 5)(x - 10)$ for all x in the interval $(1.25, 10)$. The region of the city in which the population density exceeds 200 people per square mile is the region that is more than 1.25 miles but less than 10 miles from the center of the city. See the following figure.

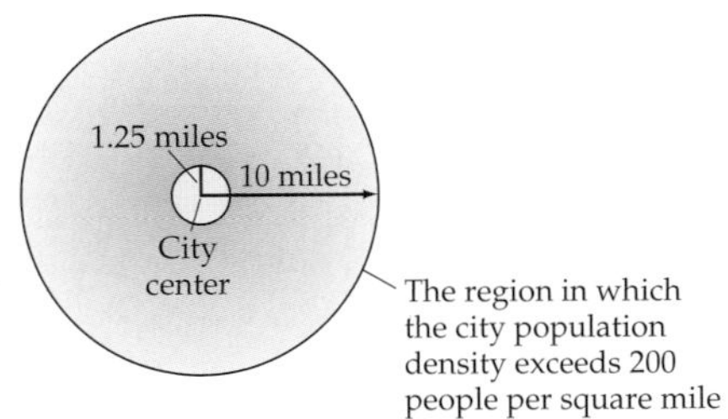

CHECK YOUR PROGRESS 6 The population density D (in people per square mile) of a city is related to the distance x, in miles, from the center of the city by $D = \dfrac{2200x}{5x^2 + 16}$, $0 < x < 6$. Describe the region of the city in which the population density exceeds 100 people per square mile. Round to the nearest tenth of a mile.

Solution *See page S9.* The region that is more than 0.9 mi but less than 3.5 mi from the center of the city

In many business applications a company is interested in the **cost** C of manufacturing x items, the **revenue** R generated by selling all of the items, and the **profit** P made by selling the items.

In the next example, the cost of manufacturing x tennis racquets is given by $C = 32x + 120{,}000$ dollars. The 120,000 represents the **fixed cost** because it remains constant regardless of how many racquets are manufactured. The $32x$ represents the **variable cost** because this term varies depending on how many racquets are manufactured. Each additional racquet costs the company an additional $32.

The revenue received from the sale of x tennis racquets is given by $R = x(200 - 0.01x)$ dollars. The quantity $(200 - 0.01x)$ is the price the company charges for each tennis racquet. The price varies depending on the number of racquets that are manufactured. For instance, if the number of racquets x that the company manufactures is small, the company will be able to demand almost $200 for each racquet. As the number of racquets that the company manufactures increases (approaches 20,000), the company will only be able to sell *all* of the racquets if it decreases the price of each racquet.

The following **profit formula** shows the relationship between profit P, revenue R, and cost C.

$$P = R - C$$

Alternative to Example 7
Exercise 46, page 151

EXAMPLE 7 Solve a Business Application

A company determines that the cost C, in dollars, of producing x tennis racquets is $C = 32x + 120{,}000$. The revenue R, in dollars, from selling all of the tennis racquets is $R = x(200 - 0.01x)$.

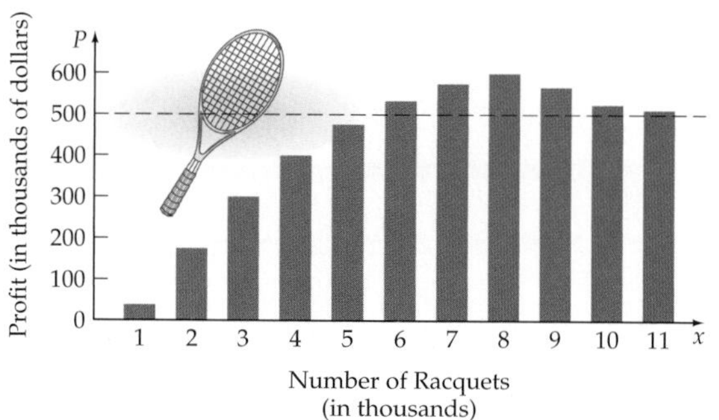

How many racquets should the company manufacture and sell if the company wishes to earn a profit of at least $500,000?

Solution

The profit is given by

$$\begin{aligned} P &= R - C \\ &= x(200 - 0.01x) - (32x + 120{,}000) \\ &= 200x - 0.01x^2 - 32x - 120{,}000 \\ &= -0.01x^2 + 168x - 120{,}000 \end{aligned}$$

The profit will be at least $500,000 provided

$$\begin{aligned} -0.01x^2 + 168x - 120{,}000 &\geq 500{,}000 \\ -0.01x^2 + 168x - 620{,}000 &\geq 0 \end{aligned}$$

Continued ➤

Using the quadratic formula, we find that the approximate critical values of this last inequality are 5474.3 and 11,325.7. Test values show that the inequality is positive only on the interval (5474.3, 11,325.7). Thus the company should manufacture at least 5475 tennis racquets but not more than 11,325 tennis racquets to produce the desired profit.

CHECK YOUR PROGRESS 7 Using the information in Example 7, determine how many racquets the company should manufacture and sell if the company wishes to earn a profit of at least $550,000.

Solution *See page S10.* At least 6514 but not more than 10,286

Topics for Discussion

1. How does the method used to find the critical values of a quadratic inequality differ from the method used to find the critical values of a rational inequality?
2. Explain why the quadratic inequality $x^2 + 9 > 0$ has no critical numbers.
3. What are the only possible solution sets of a quadratic inequality that has no critical numbers?
4. Are the inequalities $\dfrac{x-4}{x^2-9} > 1$ and $x - 4 > x^2 - 9$ equivalent? Explain.
5. Explain how you could easily determine, without using a sign diagram or the critical value method, that the solution set of $x^2 + 5 > 0$ is the set of all real numbers.

EXERCISES 1.5

— *Suggested Assignment: Exercises 1–49, odd.*

In Exercises 1 to 24, use a sign diagram to solve each inequality. Write each solution set using interval notation.

1. $x^2 + 7x > 0$ $(-\infty, -7) \cup (0, \infty)$

2. $x^2 - 5x \le 0$ $[0, 5]$

3. $x^2 - 16 \le 0$ $[-4, 4]$

4. $x^2 - 49 > 0$ $(-\infty, -7) \cup (7, \infty)$

5. $x^2 + 7x + 10 < 0$ $(-5, -2)$

6. $x^2 + 5x + 6 < 0$ $(-3, -2)$

7. $x^2 - 3x \ge 28$ $(-\infty, -4] \cup [7, \infty)$

8. $x^2 < -x + 30$ $(-6, 5)$

9. $2x^2 - x \le 1$ $\left[-\frac{1}{2}, 1\right]$

10. $2x^2 \ge x + 6$ $\left(-\infty, -\frac{3}{2}\right] \cup [2, \infty)$

11. $6x^2 - 19x > -15$ $\left(-\infty, \frac{3}{2}\right) \cup \left(\frac{5}{3}, \infty\right)$

12. $8x^2 - 29x < 12$ $\left(-\frac{3}{8}, 4\right)$

13. $6x^2 \le -23x + 4$ $\left[-4, \frac{1}{6}\right]$

14. $4x^2 \ge -5x + 9$ $\left(-\infty, -\frac{9}{4}\right] \cup [1, \infty)$

15. $\dfrac{x-5}{x-8} > 0$ $(-\infty, 5) \cup (8, \infty)$

16. $\dfrac{x+3}{x-2} < 0$ $(-3, 2)$

17. $\dfrac{2x+1}{x-6} \le 0$ $\left[-\frac{1}{2}, 6\right)$

18. $\dfrac{3x+4}{x-5} \ge 0$ $\left(-\infty, -\frac{4}{3}\right] \cup (5, \infty)$

19. $\dfrac{x-6}{x-5} \le 3$ $\left(-\infty, \frac{9}{2}\right] \cup (5, \infty)$

20. $\dfrac{x-5}{2x+3} \ge 1$ $\left[-8, -\frac{3}{2}\right)$

21. $\dfrac{x^2-9}{x-6} > 0$ $(-3, 3) \cup (6, \infty)$

22. $\dfrac{x^2-16}{2x+5} \ge 0$ $\left[-4, -\frac{5}{2}\right) \cup [4, \infty)$

23. $\dfrac{x^2 + 4}{3x - 5} < 0$ $\left(-\infty, \dfrac{5}{3}\right)$ **24.** $\dfrac{x^2 + 1}{2x - 7} > 0$ $\left(\dfrac{7}{2}, \infty\right)$

In Exercises 25 to 40, use the critical value method to solve each inequality. Write each solution set using interval notation.

25. $x^2 - 4x + 1 > 0$ $\left(-\infty, 2 - \sqrt{3}\right) \cup \left(2 + \sqrt{3}, \infty\right)$

26. $x^2 - 6x + 7 \le 0$ $\left[3 - \sqrt{2}, 3 + \sqrt{2}\right]$

27. $12x^2 - 52x - 9 \ge 0$

28. $36x^2 + 19x - 6 < 0$ $\left(-\dfrac{3}{4}, \dfrac{2}{9}\right)$

29. $2x^2 + 5x - 1 \le 0$

30. $3x^2 + 4x + 1 \le 0$ $\left[-1, -\dfrac{1}{3}\right]$

31. $8x^2 > 45x + 18$ $\left(-\infty, -\dfrac{3}{8}\right) \cup (6, \infty)$

32. $6x^2 < 43x - 40$ $\left(\dfrac{43 - \sqrt{889}}{12}, \dfrac{43 + \sqrt{889}}{12}\right)$

33. $\dfrac{x^2 - 6x + 4}{x - 3} \le 0$ $\left(-\infty, 3 - \sqrt{5}\right] \cup \left(3, 3 + \sqrt{5}\right]$

34. $\dfrac{x^2 - 8x + 14}{x - 4} \ge 0$ $\left[4 - \sqrt{2}, 4\right) \cup \left[4 + \sqrt{2}, \infty\right)$

35. $\dfrac{x^2 - 3x - 4}{x - 2} \ge 0$ $[-1, 2) \cup [4, \infty)$

36. $\dfrac{x^2 + 6x + 9}{x - 5} \le 0$ $(-\infty, 5)$

37. $\dfrac{x^2 - 2}{x + 3} > 1$

38. $\dfrac{x^2 - 5}{x - 6} > 2$ $(6, \infty)$

39. $\dfrac{x^2}{x - 4} \le 3$ $(-\infty, 4)$

40. $\dfrac{x}{x - 4} < 5$ $(-\infty, 4) \cup (5, \infty)$

Business and Economics

27. $\left(-\infty, -\dfrac{1}{6}\right] \cup \left[\dfrac{9}{2}, \infty\right)$

41. ***Publishing*** A publisher has determined that if x books are published, the average cost per book is given by

$$\overline{C} = \frac{14.25x + 350{,}000}{x}$$

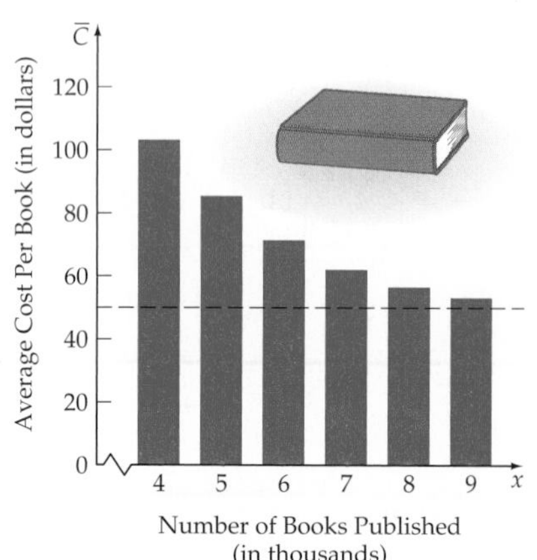

How many books should to be published if the company wants to bring the average cost per book below \$50? At least 9791

42. ***Manufacturing*** A company manufactures running shoes. The company has determined that if it manufactures x pairs of shoes, the average cost, in dollars, per pair is

$$\overline{C} = \frac{0.00014x^2 + 12x + 400{,}000}{x}$$

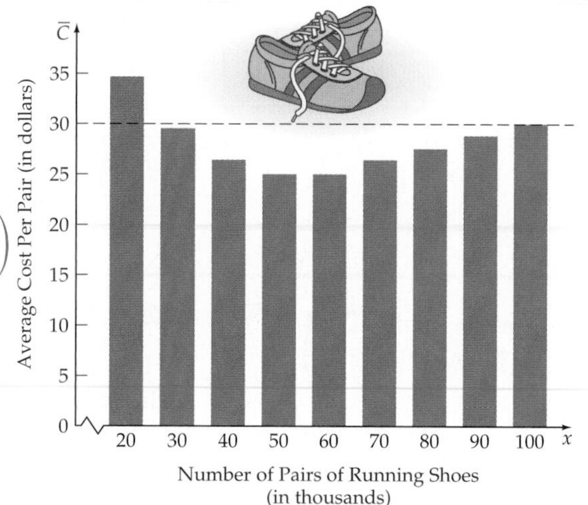

How many pairs of running shoes should the company manufacture if it wishes to bring the average cost below \$30 per pair? From 28,572 to 99,999

Life and Health Sciences

43. ***Medical Dosage for Children*** A formula that is used to determine the dosage d of a pain relieving drug for children is

$$d = \frac{500t}{t + 12}, \quad 2 \le t \le 16$$

where t is the age of the child in years and d is in milligrams. The patients in the pediatrics wing of a hospital range in age from 3 to 8 years. What is the range of dosages of the drug that can be prescribed for these patients? 100 to 200 mg

44. ***Pest Infestation*** A botanist has determined that the population density d of the European corn borer can be approximated by

$$d = 0.01t^2 + 0.05t$$

where t is the number of days since the start of an infestation and d is the number of corn borers per plant. During what time period will the density of the corn borers grow from 3 corn borers per plant to 5 corn borers per plant? 15 to 20 days

29. $\left[\dfrac{-5 - \sqrt{33}}{4}, \dfrac{-5 + \sqrt{33}}{4}\right]$ 37. $\left(-3, \dfrac{1 - \sqrt{21}}{2}\right) \cup \left(\dfrac{1 + \sqrt{21}}{2}, \infty\right)$

Sports and Recreation

45. ***Contour of a Football Field*** Some football fields are built in a parabolic mound shape so that water will drain off the field. A model for the parabolic contour of a particular football field is

$$h = -0.0002348x^2 + 0.0375x$$

where h is the height of the field, in feet, at a distance of x feet from one sideline. Describe the portion of the field for which $h > 6$ inches. Round your results to the nearest 0.1 foot. More than 14.7 ft but less than 145.0 ft from a sideline

Social Sciences

46. ***Median Age at First Marriage*** The median age A, in years, at first marriage for women can be modeled by

$$A = 0.00118t^2 - 0.095t + 22.503$$

where $t = 0$ represents the year 1900 and $t = 100$ represents the year 2000. Use this model to determine during what years from 1900 to 2000 the median age at first marriage for women fell below 21 years. (*Source:* U.S. Census Bureau, **www.Census.gov**) 1922–1958

47. ***Median Age at First Marriage*** The median age A, in years, at first marriage for men can be modeled by

$$A = 0.00136x^2 - 0.130x + 26.353$$

where $x = 0$ represents the year 1900 and $x = 100$ represents the year 2000.

Median Age at First Marriage, for Men

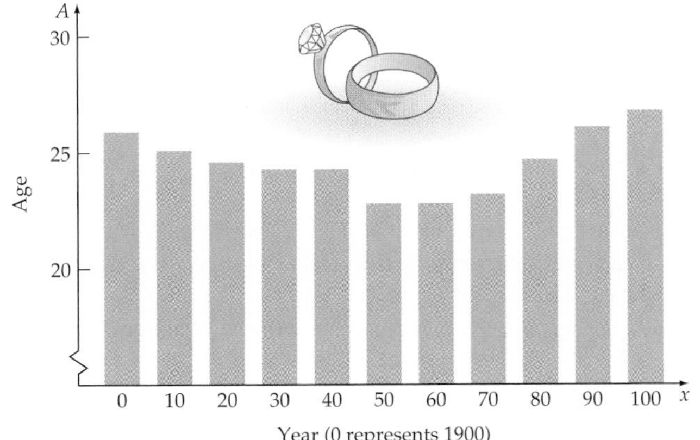

Use the model to determine during what years from 1900 to 2000 the median age at first marriage for men was above 25 years. (*Source:* U.S. Census Bureau, **www.Census.gov**) 1900–1911 and 1984–2000

48. ***Population Density of a City*** The population density D (in people per square mile) of a city is related to the distance x, in miles, from the center of the city by $D = -45x^2 + 190x + 200, 0 < x < 5$. Describe the region of the city in which the population density exceeds 300 people per square mile. Round critical values to the nearest tenth of a mile.
More than 0.6 mi but less than 3.6 mi from the city center

Physical Sciences and Engineering

49. ***Height of a Projectile*** The equation $s = -16t^2 + v_0t + s_0$ gives the height s, in feet, of an object t seconds after the object is thrown directly upward from a height of s_0 feet above the ground with an initial velocity of v_0 feet per second. A ball is thrown directly upward from ground level with an initial velocity of 64 feet per second. Find the time interval during which the ball attains a height of more than 48 feet. More than 1 s but less than 3 s

Explorations 

1. ***Design of a Container*** A cylindrical soft drink can is to be constructed such that it will have a volume of 355 milliliters.

   a. Write an equation for the height h of the can in terms of its radius r. $h = \dfrac{355}{\pi r^2}$

   b. Write an equation for the surface area A of the can in terms of its radius r. $A = \dfrac{2\pi r^3 + 710}{r}$

   c. An engineer has determined that a radius of 3.84 centimeters minimizes the surface area of the can (and thus minimizes the cost of the can), but the managers of the company have decided to construct the can using a radius of 3.1 centimeters. Explain some of the reasons the managers might give to justify their decision.
   Answers will vary. For example, the can with the smaller radius of 3.1 cm is easier to hold with one hand.

Chapter 1 Summary

Key Terms

absolute value equation **[p. 113]**
absolute value inequality **[p. 131]**
addition-subtraction property **[p. 128]**
completing the square **[p. 101]**
compound inequality **[p. 129]**
cost **[p. 147]**
critical values of a quadratic inequality **[p. 142]**
critical values of a rational inequality **[p. 144]**
critical value method **[p. 142]**
discriminant **[p. 105]**
double solution or double root **[p. 99]**
equation **[p. 83]**
equivalent equations **[p. 83]**
equivalent inequalities **[p. 128]**
extraneous solution **[p. 118]**
fixed cost **[p. 147]**
hypotenuse **[p. 105]**
leg **[p. 105]**
linear equation **[p. 84]**
literal equation **[p. 87]**
multiplication-division property **[p. 128]**
polynomial equation **[p. 114]**
power principle **[p. 117]**
profit **[p. 147]**
profit formula **[p. 148]**
properties of absolute value inequalities **[p. 131]**
properties of inequalities **[p. 128]**
Pythagorean Theorem **[p. 106]**
quadratic equation **[p. 98]**
quadratic formula **[p. 103]**
quadratic in form **[p. 120]**
radical equation **[p. 117]**
rational equation **[p. 115]**
rational expression **[p. 143]**
rational inequality **[p. 143]**
revenue **[p. 147]**
right triangle **[p. 105]**
roots of an equation **[p. 83]**
satisfy an equation **[p. 83]**
sign diagram **[p. 139]**
sign property of polynomials **[p. 141]**
solution set of an inequality **[p. 128]**
solutions of an equation **[p. 83]**
square root procedure **[p. 100]**
standard quadratic form **[p. 98]**
test interval **[p. 142]**
test value **[p. 142]**
tolerance **[p. 139]**
variable cost **[p. 148]**
zero **[p. 141]**
zero product principle **[p. 98]**

Essential Concepts and Formulas

■ Solving an Equation

Solving an equation can be accomplished by producing a sequence of equivalent equations until an equation of the form variable = constant is produced. The following procedures can be used.

1. Simplify either side of the equation by combining like terms and applying the commutative, associative, and distributive properties.
2. Add or subtract the same quantity to (from) both sides of the equation.
3. Multiply or divide both sides of the equation by the same nonzero quantity. **[p. 83]**

■ Guidelines for Solving Application Problems

1. Read the problem carefully. If necessary, reread the problem several times.
2. When appropriate, draw a sketch and label parts of the drawing with the specific information given in the problem.
3. Determine the unknown quantities and label them with variables. Write down any equation that relates the variables.
4. Use the information from Step 3, along with a known formula or some additional information given in the problem, to write an equation.

5. Solve the equation obtained in Step 4 and check to see whether your result satisfies all the conditions of the original problem. **[p. 88]**

Quadratic Equation

A quadratic equation in x is an equation that can be written in the standard quadratic form $ax^2 + bx + c = 0$, where a, b, and c are real numbers and $a \neq 0$. **[p. 98]**

Zero Product Principle

If A and B are algebraic expressions such that $AB = 0$, then $A = 0$ or $B = 0$. **[p. 98]**

Square Root Procedure

If $x^2 = c$, then $x = c$ or $x = -c$. This can also be written as $x = \pm\sqrt{c}$. **[p. 100]**

Quadratic Formula

If $ax^2 + bx + c = 0, a \neq 0$, then $x = \dfrac{-b \pm \sqrt{b^2 - 4ac}}{2a}$.

[p. 103]

Profit Formula

$P = R - C$, where P is the profit, R is the revenue, and C is the cost. **[p. 148]**

Discriminant of a Quadratic Equation

The equation $ax^2 + bx + c = 0$, with real coefficients and $a \neq 0$, has as its discriminant $b^2 - 4ac$.
If $b^2 - 4ac > 0$, then $ax^2 + bx + c = 0$ has two distinct real solutions.
If $b^2 - 4ac = 0$, then $ax^2 + bx + c = 0$ has one real solution. The solution is a double solution.
If $b^2 - 4ac < 0$, then $ax^2 + bx + c = 0$ has two distinct nonreal complex solutions. **[p. 105]**

Pythagorean Theorem

If a and b are the lengths of the legs of a right triangle and c is the length of the hypotenuse, then $c^2 = a^2 + b^2$. **[p. 106]**

Absolute Value Equations

For any variable expression E and any nonnegative real number k, $|E| = k$ if and only if $E = k$ or $E = -k$. **[p. 113]**

Solving Polynomial Equations

Some polynomial equations can be solved by factoring and using the zero product principle. For equations that are quadratic in form, a substitution of u for a variable expression can be made so that the equation can be written in the form $au^2 + bu + c = 0$. The quadratic equation can be solved for u. The solution of the original equation is then determined from the relationship between the original variable and u. **[p. 121]**

Solving Rational Equations

Rational equations can be simplified by multiplying both sides of the equation by a variable expression such that the denominators cancel. The variable must be restricted so that the variable expression is not equal to zero. **[p. 115]**

The Power Principle

If P and Q are algebraic expressions and n is a positive integer, then every solution of $P = Q$ is also a solution of $P^n = Q^n$. **[p. 117]**

Solving Radical Equations

Radical equations can be solved by isolating the radical expression and then using the power principle. To solve equations in which the radical expression contains a fractional exponent, raise both sides of the equation to a reciprocal power. Solution candidates must be checked, as the power principle can introduce extraneous solutions. **[p. 118]**

Properties of Inequalities

1. *Addition-Subtraction Property*
   If the same real number is added to or subtracted from each side of an inequality, the resulting inequality is equivalent to the original inequality.
2. *Multiplication-Division Property*
   a. Multiplying or dividing each side of an inequality by the same positive real number produces an equivalent inequality.
   b. Multiplying or dividing each side of an inequality by the same negative real number produces an equivalent inequality provided the direction of the inequality symbol is reversed. **[p. 128]**

Absolute Value Inequalities

For any variable or variable expression E and any nonnegative real number k:

1. $|E| \leq k$ if and only if $-k \leq E \leq k$
2. $|E| \geq k$ if and only if $E \leq -k$ or $E \geq k$ **[p. 131]**

Quadratic and Rational Inequalities

Quadratic inequalities and many rational inequalities can be solved by using a sign diagram or the critical value method. **[p. 140 and p. 142]**

Chapter 1 True/False Exercises

In Exercises 1 to 10, answer true or false. If the answer is false, then explain why or give a counterexample to show that the statement is false.

1. Adding the same constant to each side of a given equation produces an equation that is equivalent to the original equation. True
2. The solution set of $x^2 = 9$ is $\{3\}$. False; $\{-3, 3\}$
3. If $a > b$, then $-a < -b$. True
4. The discriminant of $ax^2 + bx + c = 0$ is $\sqrt{b^2 - 4ac}$. False; $b^2 - 4ac$
5. The absolute value equation $|-3x - 4| = 8$ is equivalent to $3x + 4 = 8$. False; $3x + 4 = 8$ or $3x + 4 = -8$
6. The equations

$$x = \sqrt{12 - x} \quad \text{and} \quad x^2 = 12 - x$$

are equivalent equations.
False. The first equation has a solution of 3, whereas the second equation has both 3 and -4 as solutions.

7. The equation

$$x^{4/3} - 3x^{2/3} + 9 = 0$$

is quadratic in form. True

8. The solution set of $|x - a| < b$, $b > 0$, is the interval $(a - b, a + b)$. True
9. Every quadratic equation $ax^2 + bx + c = 0$ with real coefficients such that $ac < 0$ has two distinct real solutions. True
10. The only critical values of $(x^2 - 4)/(x - 3) > 0$ are -2 and 2. False; 3 is also a critical value.

Chapter 1 Review Exercises

In Exercises 1 to 30, solve each equation for the unknown.

1. $x - 2(5x - 3) = -3(-x + 4)$ $\frac{3}{2}$ [1.1]
2. $3x - 5(2x - 7) = -4(5 - 2x)$ $\frac{11}{3}$ [1.1]
3. $\frac{4x}{3} - \frac{4x - 1}{6} = \frac{1}{2}$ $\frac{1}{2}$ [1.1]
4. $\frac{3x}{4} - \frac{2x - 1}{8} = \frac{3}{2}$ $\frac{11}{4}$ [1.1]
5. $\frac{x}{x + 2} + \frac{1}{4} = 5$ $-\frac{38}{15}$ [1.3]
6. $\frac{y - 1}{y + 1} - 1 = \frac{2}{y}$ $-\frac{1}{2}$ [1.3]
7. $\frac{8}{y + 3} = 4 - \frac{2y}{y + 3}$ -2 [1.3]
8. $\frac{z - 3}{z + 4} = \frac{z + 6}{z - 1}$ $-\frac{3}{2}$ [1.3]
9. $x^2 - 5x + 6 = 0$ 3, 2 [1.2]
10. $6x^2 + x - 12 = 0$ $\frac{4}{3}, -\frac{3}{2}$ [1.2]
11. $3x^2 - x - 1 = 0$ $\frac{1 \pm \sqrt{13}}{6}$ [1.2]
12. $x^2 - x + 1 = 0$ $\frac{1}{2} \pm \frac{\sqrt{3}}{2}i$ [1.2]
13. $3x^3 - 5x^2 = 0$ $0, \frac{5}{3}$ [1.3]
14. $2x^3 - 8x = 0$ $0, \pm 2$ [1.3]
15. $6x^4 - 23x^2 + 20 = 0$ $\pm\frac{2\sqrt{3}}{3}, \pm\frac{\sqrt{10}}{2}$ [1.3]
16. $3x + 16\sqrt{x} - 12 = 0$ $\frac{4}{9}$ [1.3]
17. $\sqrt{x^2 - 15} = \sqrt{-2x}$ $-5, 3$ [1.3]
18. $\sqrt{x^2 - 24} = \sqrt{2x}$ $-4, 6$ [1.3]
19. $\sqrt[3]{2x^2 - 2x} = \sqrt[3]{x^2 + 3x + 14}$ $7, -2$ [1.3]
20. $(6 - 3x)^{3/5} = 27$ -79 [1.3]
21. $\frac{1}{(y + 3)^2} = 1$ $-4, -2$ [1.3]
22. $\frac{1}{(2s - 5)^2} = 4$ $\frac{9}{4}, \frac{11}{4}$ [1.3]
23. $|x - 3| = 2$ 1, 5 [1.3]
24. $|x + 5| = 4$ $-9, -1$ [1.3]
25. $|2x + 1| = 5$ $-3, 2$ [1.3]
26. $|3x - 7| = 8$ $-\frac{1}{3}, 5$ [1.3]
27. $(x + 2)^{1/2} + x(x + 2)^{3/2} = 0$ $-2, -1$ [1.3]
28. $x^2(3x - 4)^{1/4} + (3x - 4)^{5/4} = 0$ $-4, 1, \frac{4}{3}$ [1.3]
29. $(2x - 1)^{2/3} + (2x - 1)^{1/3} = 12$ $-\frac{63}{2}, 14$ [1.3]
30. $6(x + 1)^{1/2} - 7(x + 1)^{1/4} - 3 = 0$ $\frac{65}{16}$ [1.3]

In Exercises 31 to 36, solve the literal equation for the indicated variable.

31. $e = mc^2$; m $m = \frac{e}{c^2}$ [1.1]

32. $V = \pi r^2 h$; h $h = \frac{V}{\pi r^2}$ [1.1]

33. $A = \frac{h}{2}(b_1 + b_2)$; b_1 $b_1 = \frac{2A - hb_2}{h}$ [1.1]

34. $P = 2(l + w)$; w $w = \frac{P - 2l}{2}$ [1.1]

35. $P = \frac{A}{1 + rt}$; t $t = \frac{A - P}{Pr}$ [1.1]

36. $F = G\frac{m_1 m_2}{s^2}$; m_1 $m_1 = \frac{Fs^2}{Gm_2}$ [1.1]

In Exercises 37 to 58, solve each inequality. Write the solution set using interval notation.

37. $-3x + 4 \geq -2$ $(-\infty, 2]$ [1.4]

38. $-2x + 7 \leq 5x + 1$ $\left[\frac{6}{7}, \infty\right)$ [1.4]

39. $-2x + 1 > 9$ or $3x + 1 < 10$ $(-\infty, 3)$ [1.4]

40. $2x - 3 \leq 15$ or $x - 1 \leq 7$ $(-\infty, 9]$ [1.4]

41. $x - 2 > 3$ and $-x + 4 < 1$ $(5, \infty)$ [1.4]

42. $2x - 5 \leq -9$ and $4x - 1 \geq 9$ $\varnothing$ [1.4]

43. $-x + 5 > -1$ and $3x + 2 \leq 14$ $(-\infty, 4]$ [1.4]

44. $\frac{1}{3}x - 2 \geq 3$ or $-\frac{1}{2}x + 1 < 6$ $(-10, \infty)$ [1.4]

45. $x^2 + 3x - 10 \leq 0$ $[-5, 2]$ [1.5]

46. $x^2 - 2x - 3 > 0$ $(-\infty, -1) \cup (3, \infty)$ [1.5]

47. $61 \leq \frac{9}{5}C + 32 \leq 95$ $\left[\frac{145}{9}, 35\right]$ [1.4]

48. $30 < \frac{5}{9}(F - 32) < 65$ $(86, 149)$ [1.4]

49. $\frac{x + 3}{x - 4} > 0$ $(-\infty, -3) \cup (4, \infty)$ [1.5]

50. $\frac{x(x - 5)}{x - 7} \leq 0$ $(-\infty, 0] \cup [5, 7)$ [1.5]

51. $\frac{2x}{3 - x} \leq 10$ $\left(-\infty, \frac{5}{2}\right] \cup (3, \infty)$ [1.5]

52. $\frac{x}{5 - x} \geq 1$ $\left[\frac{5}{2}, 5\right)$ [1.5]

53. $\frac{x^2 - 1}{x - 4} \leq 2$ $(-\infty, 4)$ [1.5]

54. $\frac{x^2 - 9}{x + 2} > 3$

55. $|x - 3| \leq 5$ $[-2, 8]$ [1.4]

56. $|x + 2| > 3$ $(-\infty, -5) \cup (1, \infty)$ [1.4]

57. $|3x - 4| < 2$ $\left(\frac{2}{3}, 2\right)$ [1.4]

58. $|2x - 3| \geq 1$ $(-\infty, 1] \cup [2, \infty)$ [1.4]

54. $\left(\frac{3 - \sqrt{69}}{2}, -2\right) \cup \left(\frac{3 + \sqrt{69}}{2}, \infty\right)$ [1.5]

In Exercises 59 to 62, write an absolute value inequality that has the given solution.

59. $(4, 10)$ One possible answer: $|x - 7| < 3$ [1.4]

60. $(-\infty, 3) \cup (5, \infty)$ One possible answer: $|x - 4| > 1$ [1.4]

61. $(-\infty, 1] \cup [6, \infty)$ One possible answer: $\left|x - \frac{7}{2}\right| \geq \frac{5}{2}$ [1.4]

62. $[44, 60]$ One possible answer: $|x - 52| \leq 8$ [1.4]

In Exercises 63 to 66, determine the discriminant of the quadratic equation and state the number of real solutions of the equation.

63. $2x^2 - 5x - 7 = 0$ 81; two real solutions [1.2]

64. $x^2 + 3x - 11 = 0$ 53; two real solutions [1.2]

65. $3x^2 - 2x + 10 = 0$ −116; no real solutions [1.2]

66. $x^2 + 3x + 3 = 0$ −3; no real solutions [1.2]

67. ***Sum of Natural Numbers*** The sum S of the first n natural numbers 1, 2, 3, 4, . . . , n is given by

$$S = \frac{1}{2}n(n + 1)$$

How many consecutive natural numbers, starting with 1, must be added to produce a sum of 820? 40 [1.2]

68. ***Number of Diagonals*** The number of diagonals D of a polygon with n sides is given by the quadratic equation $D = \frac{1}{2}n(n - 3)$.

a. How many diagonals does a polygon with 10 sides have? 35

b. How many sides does a polygon have if it has 90 diagonals? 15

c. Explain how you know there is no polygon with exactly 100 diagonals.

The equation $100 = \frac{1}{2}n(n - 3)$ has no natural number solutions. [1.2]

Business and Economics

69. ***Investment*** A total of \$5500 was deposited into two simple interest accounts. On one account the annual simple interest rate is 4%, and on the second account the annual simple interest rate is 6%. The amount of interest earned for 1 year was \$295. How much was invested in each account? \$1750 at 4%, \$3750 at 6% [1.1]

70. ***Construction*** A mason can build a wall in 9 hours less than an apprentice. Together they can build the wall in 6 hours. How long would it take the apprentice, working alone, to build the wall? 18 h [1.3]

71. ***Ticket Sales*** An art show brought in \$33,196 on the sale of 4526 tickets. The adult tickets sold for \$8 and the student tickets sold for \$2. How many of each type of ticket were sold? 4024 adult tickets and 502 student tickets [1.1]

72. ***Maintenance Cost*** Eighteen owners share the maintenance cost of a condominium complex. If six more units are sold, the maintenance cost will be reduced by \$12 per month for each of the present owners. What is the total monthly maintenance cost for the complex? \$864 [1.1]

73. ***Average Cost of Watches*** A manufacturer of watches has determined that if x watches are manufactured, the average cost per watch is given by

$$\overline{C} = \frac{8.50x + 112{,}000}{x}$$

How many watches should be manufactured if the company wants to bring the average cost per watch below \$16? At least 14,934 watches [1.5]

74. ***Revenue*** A company that manufacturers calculators finds that the monthly revenue from a particular style of calculator is $R = 72x - 2x^2$, where x is the price in dollars of each calculator. Find the range of prices for which the monthly revenue is greater than \$576. Between \$12 and \$24 [1.5]

75. ***Price*** A company finds that the price it can command for its DVDs is given by

$$p = \frac{22{,}000 - x}{x}, 0 < x \leq 22{,}000$$

where x is the number of DVDs that consumers are willing to purchase per month at a price of p dollars per DVD. How many DVDs should the company produce per month if it wishes the price to exceed \$4.00 per DVD? 1 to 4399 DVDs [1.5]

76. ***Cost*** A motorcycle dealership finds that the cost C of ordering, preparing, and storing x motorcycles is

$$C = 900x + \frac{255x + 4000}{x}, 1 \leq x \leq 60$$

How many motorcycles can the dealership order if it needs to keep the cost under \$30,000? 1 to 32 motorcycles [1.5]

77. ***Revenue*** The demand for a certain product is given by $p = 26 - 0.01x$, where x is the number of units sold per month and p is the price, in dollars, at which each item is sold. The monthly revenue is given by $R = xp$. What number of items sold produces a monthly revenue of at least \$15,000? 865 to 1735 items [1.5]

Life and Health Sciences

78. ***Pain Relief*** The pain relieving effect of the drug naproxen sodium is approximated by

$$p = \frac{100x^2}{x^2 + 0.03}$$

where p is the percent of the pain that is relieved from x grams of the drug. If 80% to 90% pain relief is to be obtained, how many grams of naproxen sodium should be administered? Round your critical values to the nearest thousandth of a gram. 0.346 g to 0.520 g [1.5]

79. ***Dimensions of a House*** The square footage of a one-story rectangular house is 1920 square feet and the length of the house is 34 feet longer than the width. Find the length and width of the house. 64 ft by 30 ft [1.2]

80. ***Window Dimensions*** An architect designs a rectangular window for which the length of the glass is 4 feet more than the height. If the area of the glass is 28 square feet, determine the height and length of the glass to the nearest tenth of a foot. 3.7 ft by 7.7 ft [1.2]

Sports and Recreation

81. ***Rectangular Court*** A group of children are playing dodge ball in a rectangular court whose length is 9 feet less than twice the width. If the perimeter of the court is 54 feet, find the width and length. Width 12 ft, length 15 ft [1.1]

82. ***Boating to an Island*** A motorboat left a harbor and traveled to an island at an average rate of 8 knots (1 knot = 1 nautical mile per hour). The average speed on the return trip was 6 knots. If the total trip took 7 hours, how far is it from the harbor to the island?
24 nautical miles [1.1]

Social Sciences

83. ***Illegal Drug Traffic*** A state government estimates that the cost of intercepting p percent of the illegal drugs that flow through a border checkpoint is given by

$$C = \frac{55p}{100 - p}, \quad 0 < p < 100$$

where C is the cost in millions of dollars. If the state government plans to fund the border checkpoint by allocating 10 to 14 million dollars, what is the range of the percent of illegal drugs the checkpoint can expect to intercept? Round your percents to the nearest 0.1%.
Between 15.4% and 20.3% [1.5]

Physical Sciences and Engineering

84. ***Diameter of a Cone*** As sand is poured from a chute, it forms a right circular cone whose height is one-fourth the diameter of the base. What is the diameter of the base of the cone when the cone has a volume of 144 cubic feet? Round to the nearest foot. 13 ft [1.3]

85. ***Pendulum*** The period T of a pendulum is the time it takes the pendulum to complete one swing from left to right and back. For a pendulum near the surface of the earth,

$$T = 2\pi\sqrt{\frac{L}{32}}$$

where T is measured in seconds and L is the length of the pendulum in feet. Find the length of a pendulum that has a period of 15 seconds. Round to the nearest foot.
182 ft [1.3]

Chapter 1 Test

1. Solve: $2(3x - 5) + 7 = -3(x - 2)$ 1 [1.1]

2. Solve: $\left|\frac{3}{4}x - \frac{1}{2}\right| = 5$ $-6, \frac{22}{3}$ [1.3]

3. Solve $6x^2 - 13x - 8 = 0$ by factoring and applying the zero product property. $-\frac{1}{2}, \frac{8}{3}$ [1.2]

4. Solve $2x^2 - 8x + 1 = 0$ by completing the square. $\frac{4 \pm \sqrt{14}}{2}$ [1.2]

5. Solve $3x^2 - 5x - 1 = 0$ by using the quadratic formula.

6. Determine the discriminant of $2x^2 - 5x + 2 = 0$ and state the number of real solutions of the equation.
Discriminant: 9; two real solutions [1.2]

7. Solve the compound inequality

$$2x - 5 \le 11 \text{ or } -3x + 2 > 14$$

Write the solution set using set-builder notation.
$\{x \mid x \le 8\}$ [1.4]

8. Solve the compound inequality

$$2x - 3 < 7 \text{ and } -5x + 1 \le 11$$

Write the solution set using interval notation. $[-2, 5)$ [1.4]

9. Solve $\left|\frac{2}{3}x - 4\right| < 2$. Write the solution set using interval notation. $(3, 9)$ [1.4]

10. Solve: $x - \sqrt{x - 7} = 19$ 23 [1.3]

11. Solve: $x^4 - 2x^2 = 15$ $\pm\sqrt{5}, \pm i\sqrt{3}$ [1.3]

12. Solve $\frac{x^2 - 5x + 4}{x - 2} \ge 0$. Write the solution set using interval notation. $[1, 2) \cup [4, \infty)$ [1.5]

13. Solve $\frac{x - 3}{2x - 5} \ge 1$. Write the solution set using set-builder notation. $\{x \mid 2 \le x < 2.5\}$ [1.5]

14. Solve the literal equation $ab - 4w = bw$ for w.

15. A company invests a total of $50,000 in two different accounts, one that earns 5% annual simple interest and a second that earns 7% annual simple interest. After 1 year the combined interest earned was $2940. How much money was invested in each account?
$28,000 at 5%, $22,000 at 7% [1.1]

5. $\frac{5 \pm \sqrt{37}}{6}$ [1.2]

14. $w = \frac{ab}{b + 4}$ [1.1]

16. A commercial pilot needs to start the landing procedure when the airplane is 180 miles from the airport. If the plane is flying from Los Angeles to Seattle, a distance of 1150 miles, at an average rate of 460 miles per hour, how long after takeoff should the pilot start the landing procedure? State your answer in hours and minutes, with the minutes rounded to the nearest minute. 2 h 7 min [1.1]

17. The revenue, in dollars, earned by selling x inkjet printers is given by $R = 200x - 0.004x^2$. The cost, in dollars, of manufacturing x inkjet printers is $C = 65x + 320{,}000$. How many printers should be manufactured and sold to earn a profit of at least $600,000? 9475 to 24,275 printers [1.5]

18. The price of a digital camera and the price of a 35mm camera differ by more than $210. The price of the 35mm camera is $485.

    a. Write an absolute value inequality that expresses the relationship between the price of the digital camera d and the price of the 35mm camera. $|d - 485| > 210$

    b. It is known that the price of the digital camera is less than $900. Use interval notation to describe the possible prices, in dollars, for the digital camera. $(0, 275) \cup (695, 900)$ [1.4]

19. An inequality that is used by forensic specialists to calculate the potential stature h of an adult female from the length t of her tibia bone is

$$|h - (2.90t + 61.53)| \leq 3.66$$

where h and t are both in centimeters. Use this inequality to estimate the range of potential statures for an adult female whose tibia measures 28.40 centimeters. (*Source:* Western Kentucky University, **http://www.wku.edu/**) 140.23 cm to 147.55 cm [1.4]

20. A company manufactures fax machines. The company has determined that if it manufacturers x fax machines, the average cost, in dollars, per machine is

$$\overline{C} = \frac{0.0028x^2 + 52x + 205{,}000}{x}$$

How many fax machines should the company manufacture if it wishes to bring the average cost per machine below $100? 8067 to 9076 fax machines [1.5]

Cumulative Review Exercises

1. Evaluate: $4 + 3(-5)$ -11 [P.1]

2. Write 0.00017 in scientific notation. 1.7×10^{-4} [P.2]

3. Perform the indicated operations and simplify: $(3x - 5)^2 - (x + 4)(x - 4)$ $8x^2 - 30x + 41$ [P.3]

4. Factor: $8x^2 + 19x - 15$ $(8x - 5)(x + 3)$ [P.4]

5. Simplify: $\dfrac{7x - 3}{x - 4} - 5$ $\dfrac{2x + 17}{x - 4}$ [P.5]

6. Simplify: $a^{2/3} \cdot a^{1/4}$ $a^{11/12}$ [P.6]

7. Find: $(2 + 5i)(2 - 5i)$ 29 [P.7]

8. Solve: $2(3x - 4) + 5 = 17$ $\dfrac{10}{3}$ [1.1]

9. Solve $2x^2 - 4x = 3$ by using the quadratic formula. $\dfrac{2 \pm \sqrt{10}}{2}$ [1.2]

10. Solve: $|2x - 6| = 4$ 1, 5 [1.3]

11. Solve: $x = 3 + \sqrt{9 - x}$ 5 [1.3]

12. Factor to solve: $x^3 - 36x = 0$ $-6, 0, 6$ [1.3]

13. Solve: $2x^4 - 11x^2 + 15 = 0$ $\pm\sqrt{3}, \pm\dfrac{\sqrt{10}}{2}$ [1.3]

14. Solve the compound inequality

$$3x - 1 > 2 \text{ or } -3x + 5 \geq 8$$

Write the solution set using set-building notation. $\{x \mid x \leq -1 \text{ or } x > 1\}$ [1.4]

15. Solve $|x - 6| \geq 2$. Write the solution set using interval notation. $(-\infty, 4] \cup [8, \infty)$ [1.4]

16. Solve $\dfrac{x - 2}{2x - 3} \geq 4$. Write the solution set using set-builder notation. $\left\{x \,\middle|\, \dfrac{10}{7} \leq x < \dfrac{3}{2}\right\}$ [1.5]

17. A fence built around the border of a rectangular field measures a total of 200 feet. If the length of the field is 16 feet longer than the width, find the dimensions of the field. Length 58 ft, width 42 ft [1.1]

18. A worker can cover a parking lot with asphalt in 10 hours. With the help of an assistant, the work can be done in 6 hours. How long would it take the assistant, working alone, to cover the parking lot with asphalt? 15 h [1.3]

19. An average score of 80 or above but less than 90, in a history class receives a B grade. Rebecca has scores of 86, 72, and 94 on three tests. Find the range of scores she could receive on the fourth test that would give her a B grade for the course. Assume that the highest test score she can receive is 100. 68 to 100 [1.4]

20. A highway patrol department estimates that the cost of ticketing p percent of the speeders who travel on a freeway is given by

$$C = \frac{600p}{100 - p}, 0 < p < 100$$

where C is in thousands of dollars.

a. If the highway patrol department plans to fund their program to ticket speeding drivers with \$100,000 to \$180,000, what is the range of the percent of speeders the department can expect to ticket? Round your percents to the nearest 0.1%. Between 14.3% and 23.1%

b. Do you think that the additional \$80,000 above the lower cost of \$100,000 is money well spent? Explain. Answers will vary. [1.5]

Introduction to Functions

Telematics: The Wired Car

Telematics is an automotive communications technology that uses global positioning satellite systems and cellular phone technology to locate a vehicle's position and offer the driver a wide range of safety and entertainment features. The car is wirelessly connected to a service center that can unlock the car when the driver has locked the keys inside, send medical or police assistance for a driver in distress, tell the driver where the next gas station is located, deliver navigation assistance, and provide a host of other services.

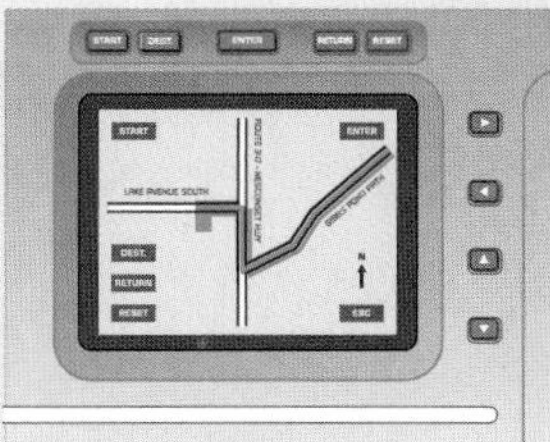

The graph at the right shows the projected number of telematics-enabled vehicles through 2006. From this graph, observe that there is a relationship between the year and the number of telematics-enabled cars. For instance, we can see that in 2002 there were approximately 4,410,000 telematics-enabled cars.

This relationship between the year and the number of telematics-enabled cars is an example of a *function,* one of the most important concepts in mathematics and its applications. Exercise 58 on page 190 is another example of a graphical description of a function.

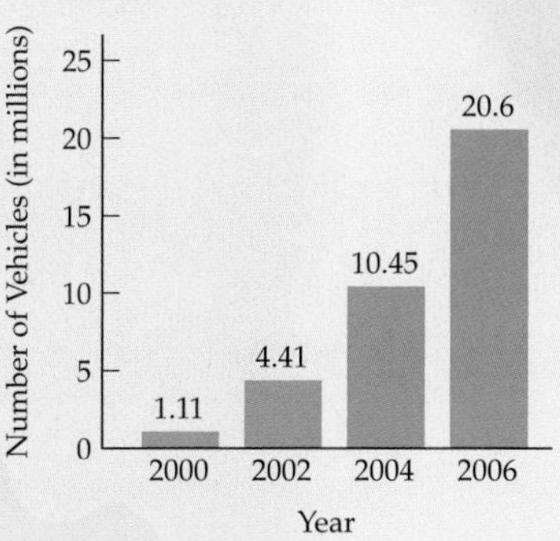

Source: Telematics Research Group

Chapter Prep Quiz

Do these exercises to prepare for Chapter 2.

1. Simplify: $\sqrt{75}$ [P.6]* $5\sqrt{3}$

2. Evaluate $x^2 - 3x - 2$ when $x = -3$. [P.1] 16

3. What term must be added to $x^2 + 6x$ so that the resulting polynomial is a perfect square? [1.2] 9

4. Solve $3x + 4y = 8$ for y. [1.1] $y = -\frac{3}{4}x + 2$

5. Solve: $x^2 - 2x - 2 = 0$ [1.2] $1 \pm \sqrt{3}$

6. Solve $\frac{y - 3}{x + 1} = 2$ for y. [1.3]
   $y = 2x + 5$

7. What is the distance between the two points whose coordinates on the number line are -2 and 4? [P.1]
   6

8. For what value of x is it impossible to evaluate the expression $\frac{x}{x + 3}$? [P.5] -3

9. Evaluate $\sqrt{(a - b)^2 + (c - d)^2}$ when $a = 1, b = -2, c = -3$, and $d = 4$. [P.1] $\sqrt{58}$

10. If $a = 3x - 2$ and $a = -x + 6$, what is the value of a? [1.1] 4

INSTRUCTOR NOTE

The *Chapter Prep Quiz* is a means to test your students' mastery of prerequisite material that is assumed in the coming chapter. Each question identifies the section to review (if necessary) and all answers are provided in the Answers to Selected Exercises.

Problem Solving Strategies

Making a Table

Suppose a game is played in such a way that a player starts at O in the diagram below and tosses a coin.

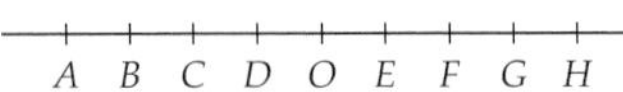

H	T	Result
4	0	*H*
3	1	*F*

If the result is heads, the player moves one place to the right; if the result is tails, the player moves one place to the left. At the new point, the process is repeated. Is it possible, after four tosses, for the player to rest on D or E?

To solve this problem, begin by thinking about the outcomes of various tosses of a coin. If four heads are tossed, the player will arrive at H; if four tails are tossed, the player will arrive at A.

If three heads and one tail are tossed, the situation is a little more complicated. Using the diagram below, we can determine that the resulting position is always F.

The results so far are shown in the table. Complete the table to answer the question posed above.

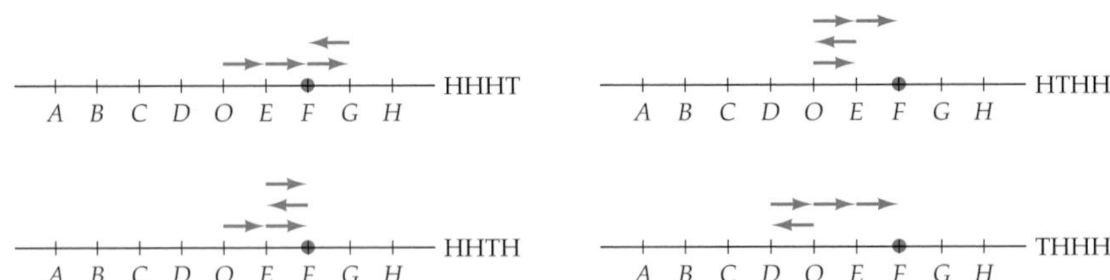

*This is a reference to the section that corresponds to this problem. For example, [P.6] stands for Chapter P, Section 6.

SECTION 2.1 Rectangular Coordinates and Graphs

- Introduction to a Rectangular Coordinate System
- Distance Formula
- Midpoint Formula
- Graphs of Equations in Two Variables
- Equation of a Circle

Point of Interest

The concept of a coordinate system developed over time, culminating in 1637 with the publication of *Discourse On the Method for Rightly Directing One's Reason and Searching for Truth in the Sciences* by René Descartes (1596–1650) and *Introduction to Plane and Solid Loci* by Pierre de Fermat (1601–1665). Of the two mathematicians, Descartes is usually given more credit for developing the concept of a coordinate system. In fact, he became so famous in Le Haye, the town in which he was born, that the town was renamed Le Haye-Descartes.

▪ Introduction to a Rectangular Coordinate System

A cartographer (a person who makes a map) divides a city into little squares as shown on the map of Washington, D.C. below. Telling a visitor that the White House is located in square A3 enables the visitor to locate the White House within a small area of the map.

(1) Department of State
(2) FBI Building
(3) Lincoln Memorial
(4) National Air and Space Museum
(5) National Gallery of Art
(6) Vietnam Veterans Memorial
(7) Washington Monument
(8) White House

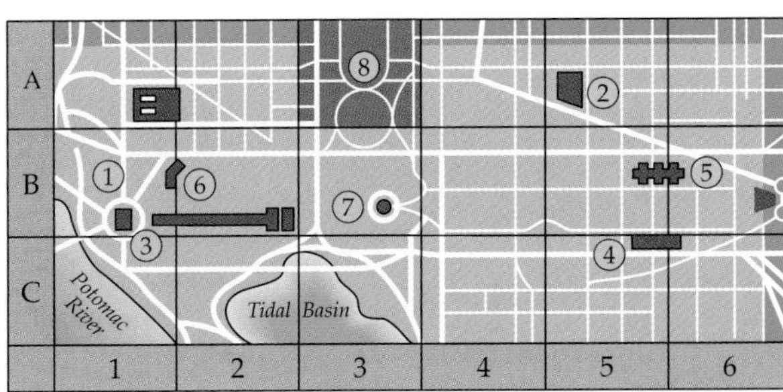

In mathematics we encounter a similar problem, that of locating a point in a plane. One way to solve the problem is to use a *rectangular coordinate system.* A **rectangular coordinate system,** shown in Figure 2.1, is formed by two number lines, one horizontal and one vertical, that intersect at the zero point of each line. The point of intersection is called the **origin.** The two lines are called the **coordinate axes** or simply the axes. Frequently, the horizontal axis is labeled the x-axis and the vertical axis is labeled the y-axis. In this case, the axes form what is called the **xy-plane.**

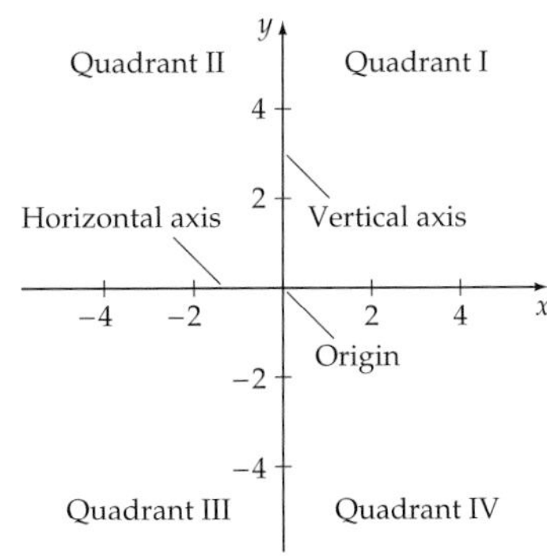

Figure 2.1

The two axes divide the plane into four regions called **quadrants** that are numbered counterclockwise, using Roman numerals, from I to IV starting at the upper right.

Each point in the plane can be identified by a pair of numbers called an **ordered pair.** The first number of the ordered pair measures a horizontal change from the y-axis and is called the **abscissa,** or **x-coordinate.** The second number of the ordered pair measures a vertical change from the x-axis and is called the **ordinate,** or **y-coordinate.** The ordered pair (x, y) associated with a point is also called the **coordinates of the point.**

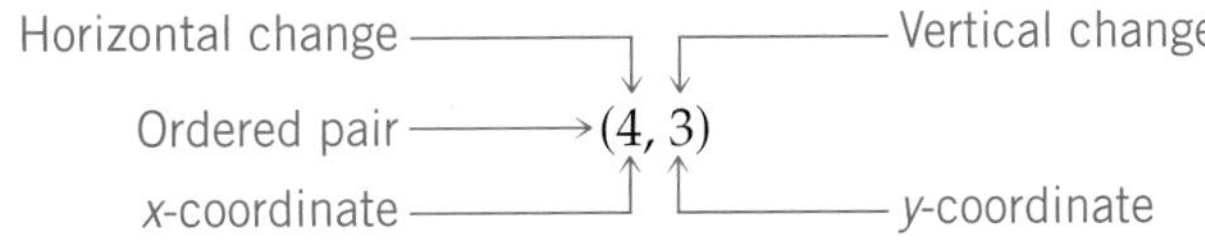

To **graph,** or **plot,** a point means to place a dot at the coordinates of the point. For example, in Figure 2.2, to graph the ordered pair (4, 3), start at the origin. Move 4 units to the right and then 3 units up. Draw a dot. To graph (−3, −4), start at the origin. Move 3 units to the left and then 4 units down. Draw a dot.

The **graph of an ordered pair** is the dot drawn at the coordinates of the point in the plane. The graphs of the ordered pairs (4, 3) and (−3, −4) are shown at the right.

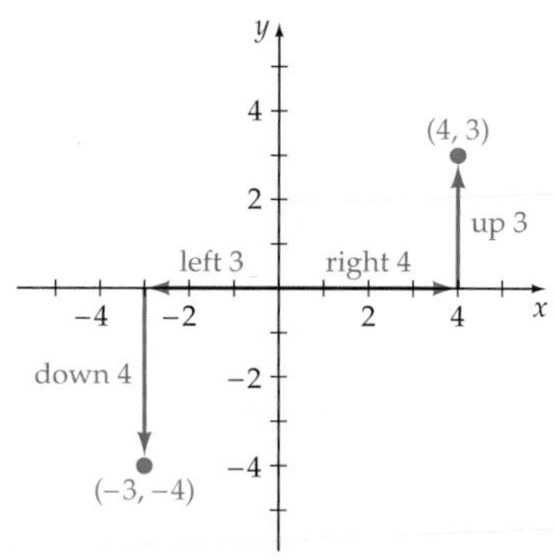

Figure 2.2

If the axes are labeled as other than x and y, then we refer to the ordered pair by the given labels. For instance, if the horizontal axis is labeled t and the vertical axis is labeled d, then the ordered pairs are written as (t, d). In any case, we sometimes refer to the first number in an ordered pair as the **first coordinate** of the ordered pair and to the second number as the **second coordinate** of the ordered pair.

The graphs of the points whose coordinates are (2, 3) and (3, 2) are shown in Figure 2.3. Note that they are different points. The order in which the numbers in an ordered pair are listed is important.

TAKE NOTE

This is *very* important. An *ordered pair* is a pair of numbers, and the *order* in which the numbers are listed is important.

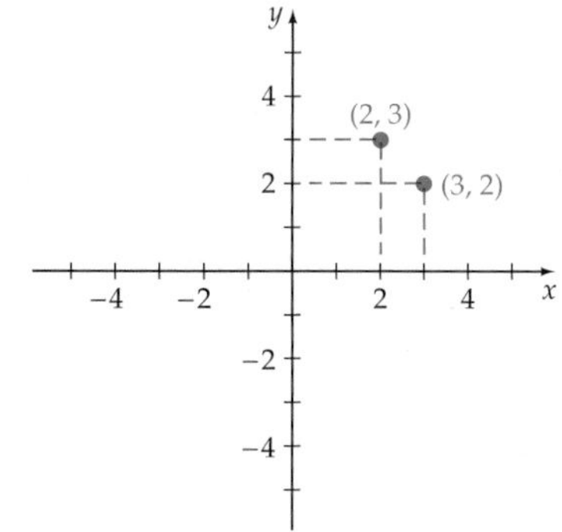

Figure 2.3

▪ Distance Formula

Using the Pythagorean Theorem, we can derive a formula for the distance between any two points in the plane. Consider the right triangle shown in Figure 2.4. The *vertical* distance between P_1 and P_2 is $|y_1 - y_2|$. The *horizontal* distance between P_1 and P_2 is $|x_1 - x_2|$. The absolute value symbol is needed to ensure that the distance is a nonnegative number.

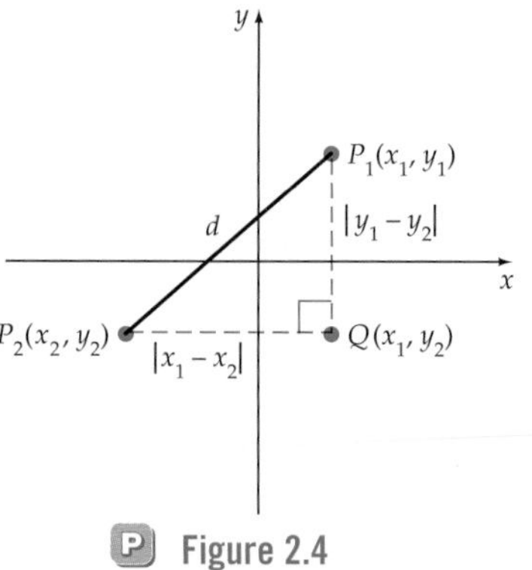

Figure 2.4

The quantity d^2 is calculated by applying the Pythagorean Theorem to the right triangle P_1P_2Q.

$$d^2 = |x_1 - x_2|^2 + |y_1 - y_2|^2$$
$$= (x_1 - x_2)^2 + (y_1 - y_2)^2$$

The distance, d, is the square root of d^2.

$$d = \sqrt{(x_1 - x_2)^2 + (y_1 - y_2)^2}$$

Distance Formula

If $P_1(x_1, y_1)$ and $P_2(x_2, y_2)$ are two points in the plane, then the distance $d(P_1, P_2)$ between the two points is given by

$$d(P_1, P_2) = \sqrt{(x_1 - x_2)^2 + (y_1 - y_2)^2}$$

Alternative to Example 1

Find the distance between $P_1(-3, 5)$ and $P_2(3, -5)$. Give an exact answer and an answer rounded to the nearest hundredth.

▪ **$2\sqrt{34}$, 11.66**

EXAMPLE 1 Find the Distance Between Two Points

Find the distance between $P_1(2, -3)$ and $P_2(-3, 1)$. Give an exact answer and an answer rounded to the nearest hundredth.

Solution

$$d(P_1, P_2) = \sqrt{(x_1 - x_2)^2 + (y_1 - y_2)^2}$$ ▪ Use the distance formula.

$$= \sqrt{[2 - (-3)]^2 + (-3 - 1)^2}$$

$$= \sqrt{5^2 + (-4)^2} = \sqrt{25 + 16}$$

$$= \sqrt{41}$$ ▪ Exact answer

$$\approx 6.40$$ ▪ Approximate answer

CHECK YOUR PROGRESS 1 Find the distance between $P_1(-4, 0)$ and $P_2(-2, 5)$. Give an exact answer and an answer rounded to the nearest hundredth.

Solution *See page S10.* $\sqrt{29}$, 5.39

▪ Midpoint Formula

If you make arrangements to meet a friend or associate at a place that is halfway between the two of you, the point you have selected is called the *midpoint*. The **midpoint** of a line segment is the point that is equidistant from the endpoints. On the line segment in Figure 2.5, point C is equidistant from point A and point B.

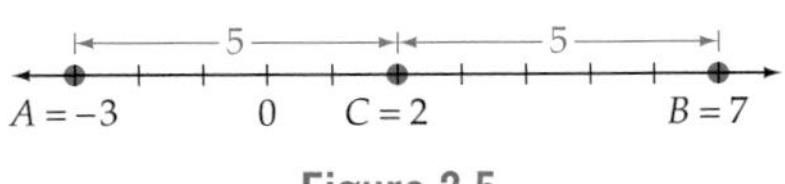

Figure 2.5

The coordinate of C is the average of the coordinates of A and B.

$$C = \frac{-3 + 7}{2} = \frac{4}{2} = 2$$

For a line segment in the plane, the coordinates of the midpoint are the averages of the x-coordinates of the endpoints and the y-coordinates of the endpoints.

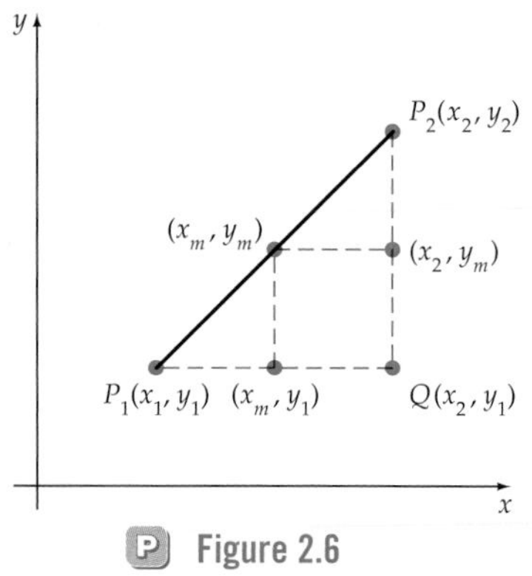

Figure 2.6

The coordinates of the midpoint of line segment P_1P_2 in Figure 2.6 are (x_m, y_m). The intersection of the horizontal line segment through P_1 and the vertical line segment through P_2 is Q, with coordinates (x_2, y_1).

The x-coordinate x_m of the midpoint of line segment P_1P_2 is the same as the x-coordinate of the midpoint of line segment P_1Q. It is the average of the x-coordinates of the points P_1 and P_2.

$$x_m = \frac{x_1 + x_2}{2}$$

Similarly, the y-coordinate y_m of the midpoint of line segment P_1P_2 is the same as the y-coordinate of the midpoint of line segment P_2Q. It is the average of the y-coordinates of the points P_1 and P_2.

$$y_m = \frac{y_1 + y_2}{2}$$

Midpoint Formula

If $P_1(x_1, y_1)$ and $P_2(x_2, y_2)$ are two points in the plane, then the coordinates of the midpoint, (x_m, y_m), of the line segment between the two points are given by

$$x_m = \frac{x_1 + x_2}{2} \quad \text{and} \quad y_m = \frac{y_1 + y_2}{2}$$

Alternative to Example 2

Find the midpoint of the line segment between $P_1(-2, 4)$ and $P_2(6, -5)$.

▪ $\left(4, -\frac{1}{2}\right)$

EXAMPLE 2 Find the Midpoint of a Line Segment

Find the midpoint of the line segment between $P_1(3, -2)$ and $P_2(-5, -3)$.

Solution

$$x_m = \frac{x_1 + x_2}{2} = \frac{3 + (-5)}{2} = \frac{-2}{2} = -1 \qquad y_m = \frac{y_1 + y_2}{2} = \frac{-2 + (-3)}{2} = \frac{-5}{2} = -\frac{5}{2}$$

The coordinates of the midpoint are $\left(-1, -\frac{5}{2}\right)$.

CHECK YOUR PROGRESS 2 Find the midpoint of the line segment between $P_1(-3, 4)$ and $P_2(4, -4)$.

Solution *See page S10.* $\left(\frac{1}{2}, 0\right)$

▪ Graphs of Equations in Two Variables

One purpose of a coordinate system is to draw a picture of the solutions of an **equation in two variables.** Examples of equations in two variables are shown at the right.

$$y = 3x - 2$$
$$x^2 + y^2 = 25$$
$$s = t^2 - 4t + 1$$

A **solution of an equation in two variables** is an ordered pair that makes the equation a true statement. For instance, as shown below, (2, 4) is a solution of $y = 3x - 2$ but $(3, -1)$ is not a solution of the equation.

$y = 3x - 2$			$y = 3x - 2$		
4	$3(2) - 2$	■ $x = 2, y = 4$	-1	$3(3) - 2$	■ $x = 3, y = -1$
4	$6 - 2$		-1	$9 - 2$	
$4 = 4$		■ Checks.	$-1 \neq 7$		■ Does not check.

In addition to (2, 4), there are many other solutions of the equation $y = 3x - 2$. For instance, $(-1, -5)$, $(0, -2)$, $(1, 1)$, and $\left(\frac{2}{3}, 0\right)$ are all solutions of the equation. In fact, for every real number x, we can find a corresponding value of y. Therefore, there are an infinite number of solutions of the equation.

? QUESTION Is $(-2, -8)$ a solution of $y = 3x - 2$?

The **graph of an equation in two variables** is a drawing of the ordered pair solutions of the equation. To create a graph of an equation, find some of the ordered pair solutions of the equation, plot the corresponding points, and then connect the points with a smooth curve.

Alternative to Example 3
Exercise 32, page 174

TAKE NOTE

In Example 3, we have shown all the calculations for the ordered pairs. Normally an input/output table would show only the results of the calculations. The table can be displayed vertically or horizontally.

x	y
−2	−8
−1	−5
0	−2
1	1
2	4
3	7

x	−2	−1	0	1	2	3
y	−8	−5	−2	1	4	7

EXAMPLE 3 Graph an Equation in Two Variables

Graph $y = 3x - 2$.

Solution To find ordered pair solutions, select various values of x and calculate the corresponding values of y. Plot the ordered pairs. After the ordered pairs have been graphed, draw a smooth curve through the points.

It is convenient to keep track of the solutions in a table. This table is sometimes referred to as an **input/output table.** The values of x are in the *inputs*; the values of y are the *outputs*. Once the value of x (the input) is chosen, the value of y (the output) depends on x. Therefore, we say that y is the **dependent variable** and x is the **independent variable.**

x	3x − 2 = y	(x, y)
−2	$3(-2) - 2 = -8$	(−2, −8)
−1	$3(-1) - 2 = -5$	(−1, −5)
0	$3(0) - 2 = -2$	(0, −2)
1	$3(1) - 2 = 1$	(1, 1)
2	$3(2) - 2 = 4$	(2, 4)
3	$3(3) - 2 = 7$	(3, 7)

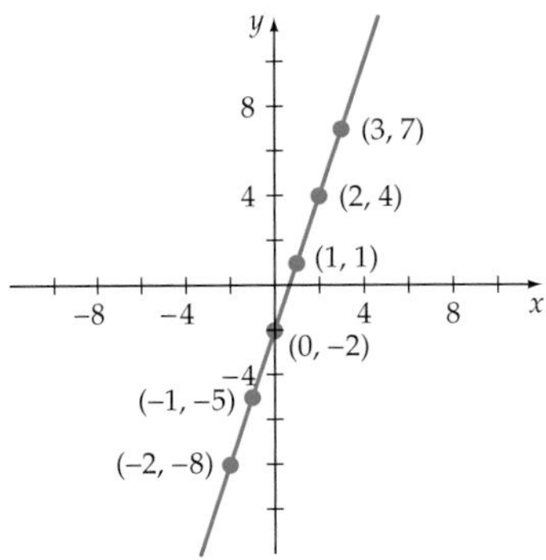

Continued ➤

? ANSWER Yes, $-8 = 3(-2) - 2$.

CHECK YOUR PROGRESS 3 Graph $y = -2x + 3$.

Solution *See page S10.*

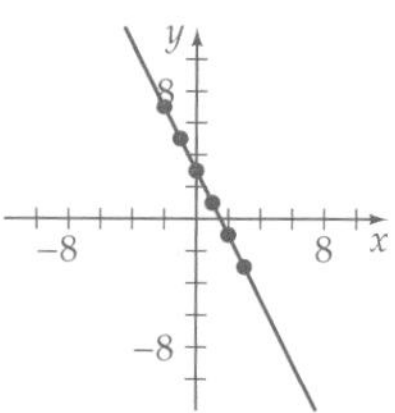

Point of Interest

Maria Agnesi (1718–1799) wrote *Foundations of Analysis for the Use of Italian Youth*, one of the most successful textbooks of the eighteenth century. The French Academy authorized a translation into French in 1749, noting that "there is no other book, in any language, which would enable a reader to penetrate as deeply, or as rapidly, into the fundamental concepts of analysis." A curve that Agnesi discusses in her text is given by the equation $y = \dfrac{a^3}{a^2 + x^2}$. Unfortunately, due to a translation error from Italian to English, the curve became known as the "witch of Agnesi."

The graph of $y = 3x - 2$ is shown again at the right. Note that the ordered pair $\left(\frac{4}{3}, 2\right)$ is a solution of the equation and is a point on the graph. The ordered pair (4, 8) is *not* a solution of the equation and is *not* a point on the graph. Every ordered pair solution of the equation is a point on the graph and every point on the graph is an ordered pair solution of the equation.

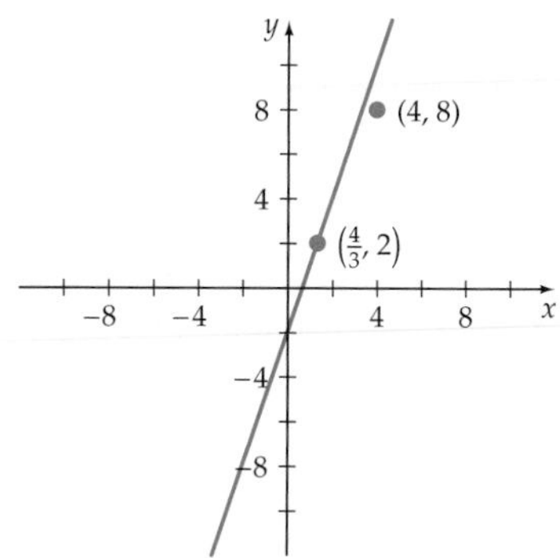

Graphing a more complicated equation such as $y = -2x^3 + 6x + 2$ may require finding many solutions of the equation before a graph can be drawn. For instance, if we calculate ordered pair solutions when $x = -2, -1, 0, 1,$ and 2, we have Figure 2.7A. It is difficult to see the shape of the graph from just these ordered pairs. Adding ordered pair solutions when $x = -1.5, -0.5, 0.5,$ and 1.5 produces Figure 2.7B. This is a little more helpful, but we still need more points. Plotting points whose x-coordinates vary from -2 to 2 using an increment of 0.25 gives Figure 2.7C. With this many points, we can draw the graph in Figure 2.7D.

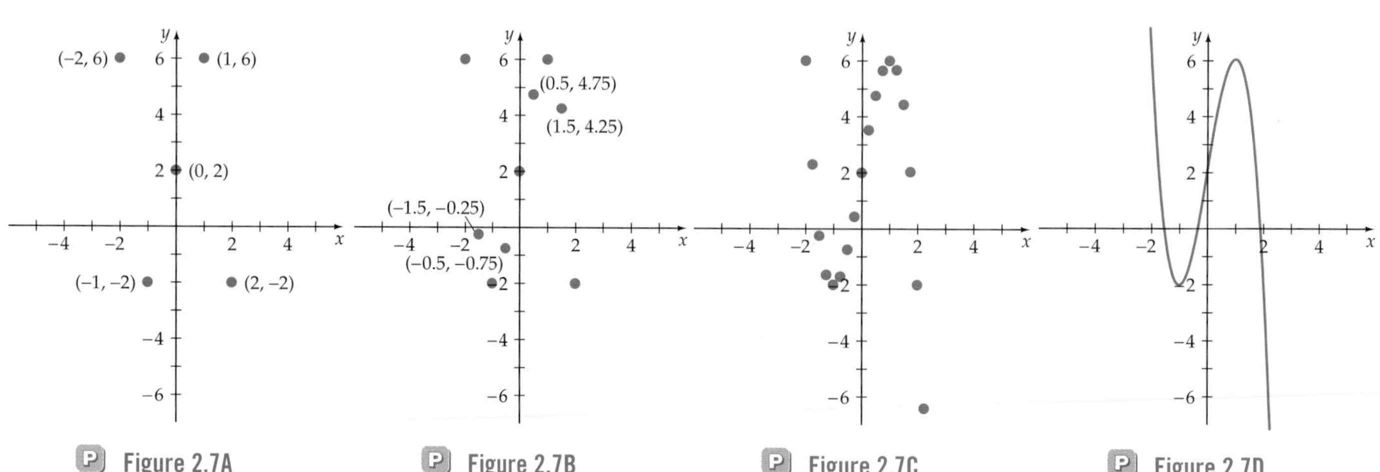

Figure 2.7A Figure 2.7B Figure 2.7C Figure 2.7D

Alternative to Example 4
Exercise 38, page 174

TAKE NOTE

As this example shows, it may be necessary to graph quite a number of points before a reasonably accurate graph can be drawn.

EXAMPLE 4 Graph an Equation in Two Variables

Graph $y = x^2 + 4x$.

Solution Select various values of x and calculate the corresponding values of y. Plot the ordered pairs. After the ordered pairs have been graphed, draw a smooth curve through the points. Here is a table showing some possible ordered pairs.

x	$x^2 + 4x = y$	(x, y)
-5	$(-5)^2 + 4(-5) = 5$	$(-5, 5)$
-4	$(-4)^2 + 4(-4) = 0$	$(-4, 0)$
-3	$(-3)^2 + 4(-3) = -3$	$(-3, -3)$
-2	$(-2)^2 + 4(-2) = -4$	$(-2, -4)$
-1	$(-1)^2 + 4(-1) = -3$	$(-1, -3)$
0	$(0)^2 + 4(0) = 0$	$(0, 0)$
1	$(1)^2 + 4(1) = 5$	$(1, 5)$

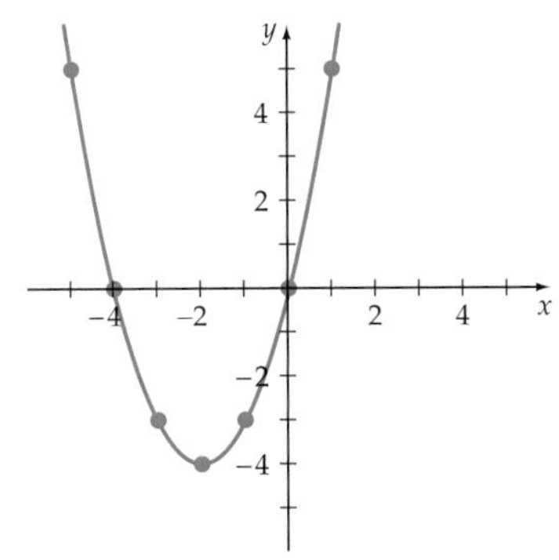

CHECK YOUR PROGRESS 4 Graph $y = x^2 - 1$.

Solution *See page S10.*

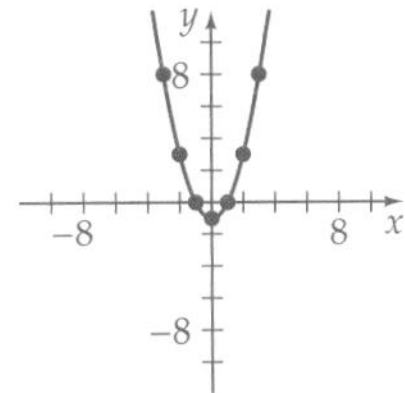

INTEGRATING TECHNOLOGY

Graphing more complicated graphs such as the one in Figure 2.7D is frequently accomplished with a graphing utility. Here are some typical screens from a TI-83 graphing calculator that was used to create the graph of $y = -2x^3 + 6x + 2$.

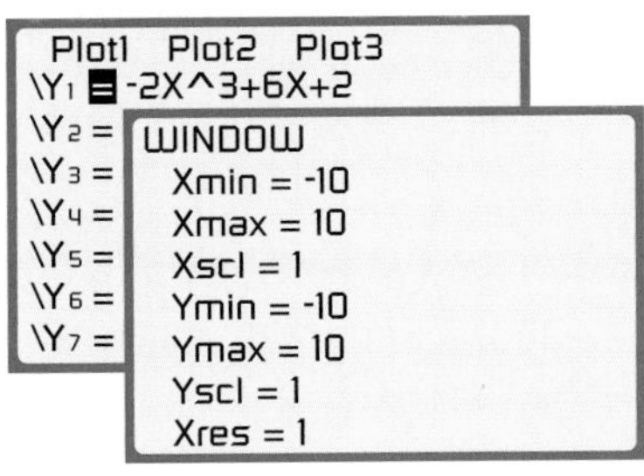

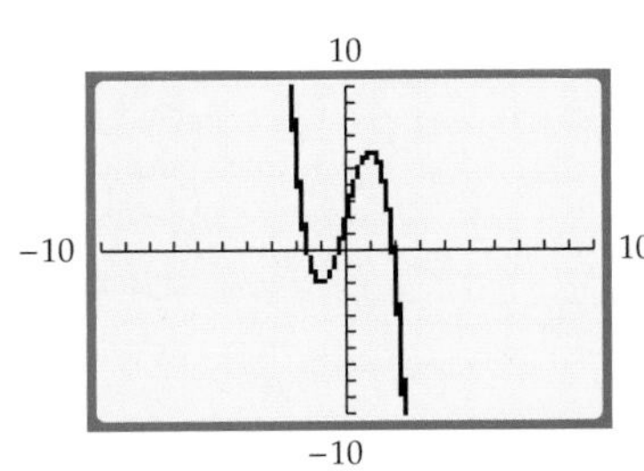

You may want to try creating the graph in Example 4 with a graphing calculator. Some typical screens are shown below.

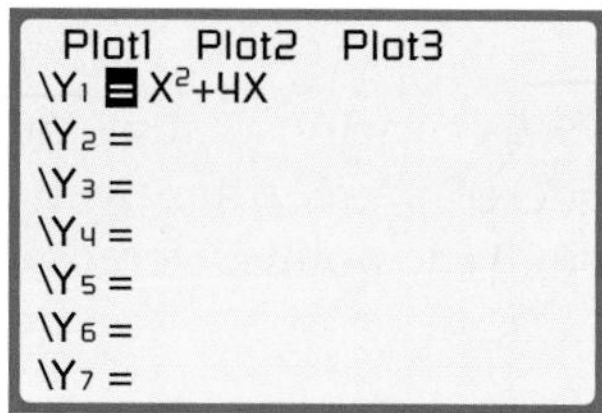

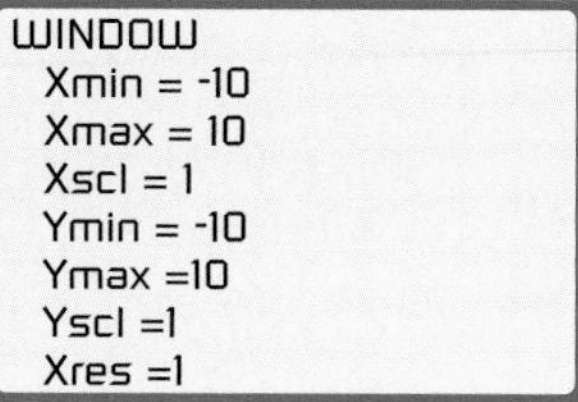

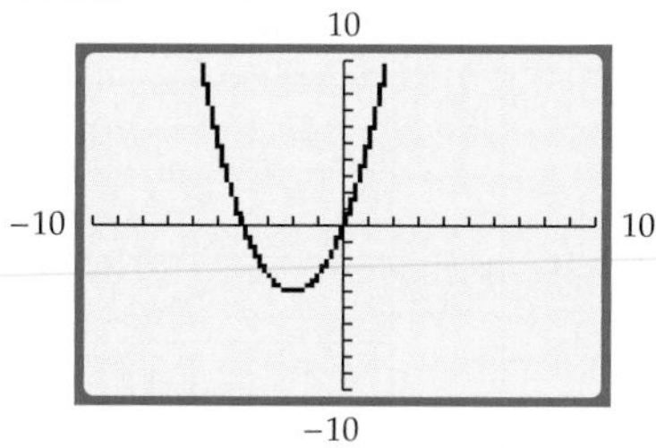

▪ Equation of a Circle

Some graphs can be sketched by recognizing the form of the equation. For instance, as you proceed through this text, you will learn that an equation of the form $y = mx + b$ is always a straight line. A circle is another graph that can be drawn from its equation without plotting points.

Definition of a Circle

A **circle** is the set of points in the plane that are a fixed distance from a specified point. The distance is the *radius* of the circle, and the specified point is the *center* of the circle.

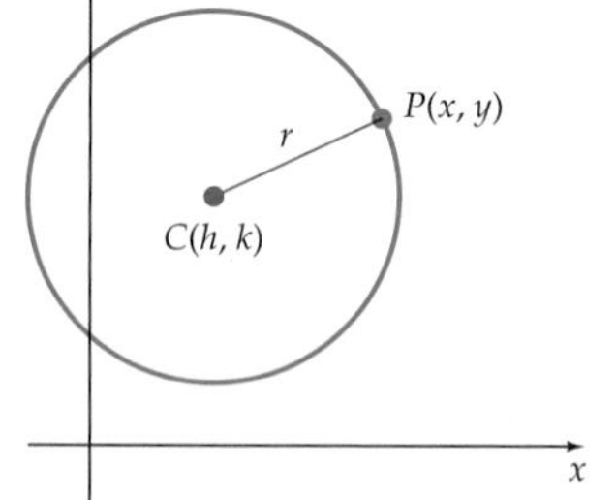

Figure 2.8

The graph in Figure 2.8 shows a circle with center $C(h, k)$ and an arbitrary point $P(x, y)$ on the circle. The radius of the circle is r. Using the distance formula, we can find what is called the *standard form* of the equation of the circle.

$$r = \sqrt{(x - h)^2 + (y - k)^2}$$

$$r^2 = (x - h)^2 + (y - k)^2$$ ▪ Square each side of the equation.

Standard Form of the Equation of a Circle

The **standard form of the equation of a circle** with center $C(h, k)$ and radius r is $(x - h)^2 + (y - k)^2 = r^2$.

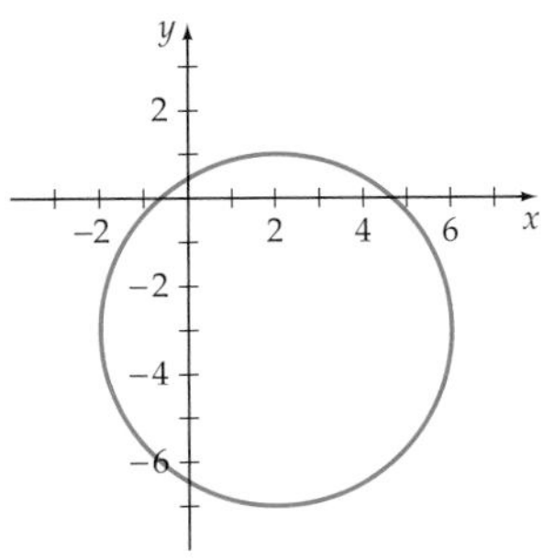

Figure 2.9

For instance, the equation $(x - 2)^2 + (y + 3)^2 = 16$ is the equation of a circle. This is equivalent to the standard form of the equation of the circle.

$$(x - 2)^2 + [y - (-3)]^2 = 4^2$$

from which we can determine that $h = 2$, $k = -3$, and $r = 4$. Thus the graph of the equation is a circle with center at $(2, -3)$ and a radius of 4, as shown in Figure 2.9.

If a circle is centered at the origin (0, 0), then $h = 0$ and $k = 0$ and the standard form of the equation of a circle simplifies to

$$x^2 + y^2 = r^2$$

For example, the graph of $x^2 + y^2 = 9$ is a circle centered at the origin with a radius of 3.

Alternative to Example 5

The endpoints of the diameter of a circle are $P_1(5, -1)$ and $P_2(3, 7)$. Find the standard form of the equation of the circle.

■ $(x - 4)^2 + (y - 3)^2 = 68$

EXAMPLE 5 Find an Equation of a Circle Given a Diameter

The endpoints of the diameter of a circle are $P_1(-3, 2)$ and $P_2(1, 4)$. (See Figure 2.10.) Find the standard form of the equation of the circle.

Solution The center of the circle is the midpoint of the diameter. Thus

$$x_m = \frac{x_1 + x_2}{2} = \frac{-3 + 1}{2} = \frac{-2}{2} = -1 \qquad y_m = \frac{y_1 + y_2}{2} = \frac{2 + 4}{2} = \frac{6}{2} = 3$$

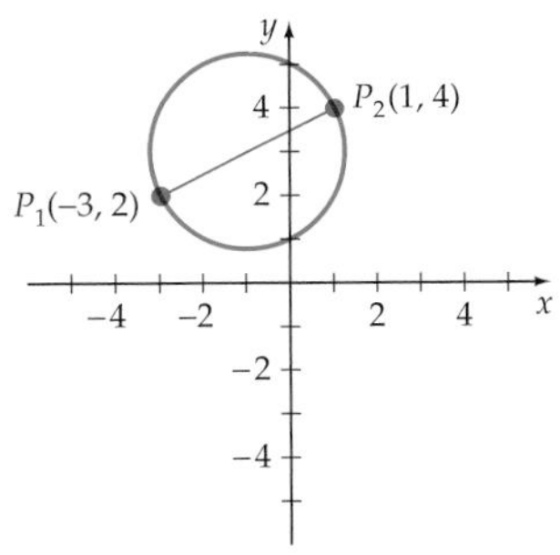

Figure 2.10

The center of the circle is $C(-1, 3)$. Use the coordinates of the center of the circle, the coordinates of P_1 or P_2, and the distance formula to find the radius of the circle. We use P_1 here.

$$r = \sqrt{[-1 - (-3)]^2 + (3 - 2)^2} = \sqrt{2^2 + 1^2} = \sqrt{5}$$

Because $r = \sqrt{5}$, $r^2 = 5$. The standard form of the equation of the circle is $(x + 1)^2 + (y - 3)^2 = 5$.

CHECK YOUR PROGRESS 5 Find the standard form of the equation of the circle that has center $C(-4, -2)$ and contains the point $P(-1, 2)$.

Solution *See page S10.* $(x + 4)^2 + (y + 2)^2 = 25$

If we rewrite the solution to Example 5 by squaring and combining like terms, we produce

$$(x + 1)^2 + (y - 3)^2 = 5$$
$$(x^2 + 2x + 1) + (y^2 - 6y + 9) = 5$$
$$x^2 + y^2 + 2x - 6y + 5 = 0$$

This form of the equation of a circle is known as the **general form of the equation of a circle.** By completing the square, it is always possible to write the general form $x^2 + y^2 + Ax + By + C = 0$ in the standard form $(x - h)^2 + (y - k)^2 = r^2$.

Alternative to Example 6

Find the center and radius of the circle whose equation is $x^2 + y^2 + 8x - 6y + 4 = 0$

▪ $C(-4, 3)$; $r = \sqrt{21}$

EXAMPLE 6 Find the Center and Radius of a Circle by Completing the Square

Find the center and radius of the circle whose equation is

$$x^2 + y^2 - 6x + 4y - 3 = 0$$

Solution First rearrange and group the terms as shown below.

$$(x^2 - 6x) + (y^2 + 4y) = 3$$

Now complete the squares of $x^2 - 6x$ and $y^2 + 4y$.

$$(x^2 - 6x) + (y^2 + 4y) = 3$$
$$(x^2 - 6x + 9) + (y^2 + 4y + 4) = 3 + 9 + 4$$
▪ Add 9 and 4 to each side of the equation.
$$(x - 3)^2 + (y + 2)^2 = 16$$
$$(x - 3)^2 + [y - (-2)]^2 = 4^2$$

The last equation is in standard form. From this equation, the center is $C(3, -2)$ and the radius is 4. See Figure 2.11.

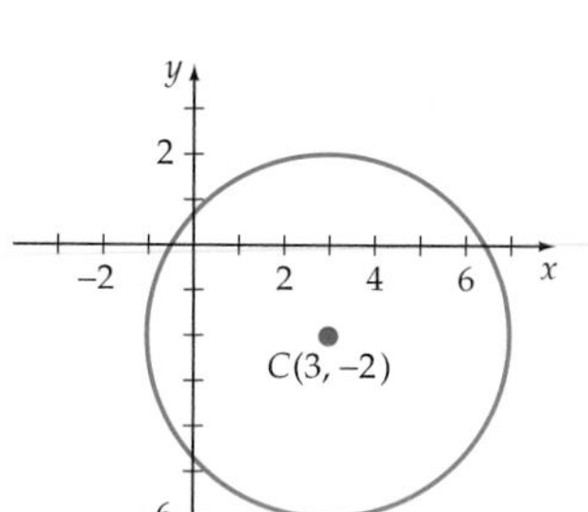

Figure 2.11

CHECK YOUR PROGRESS 6 Find the center and radius of the circle given by $x^2 + y^2 + 8x - 5y + 12 = 0$.

Solution *See page S11.* $C\left(-4, \frac{5}{2}\right)$; $r = \frac{\sqrt{41}}{2}$

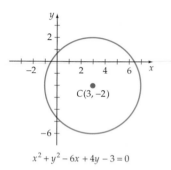

Topics for Discussion

1. Describe a rectangular coordinate system. Include in your description the concepts of axes, origin, ordered pair, and quadrant.
2. What is the graph of an ordered pair?
3. What is the graph of an equation?
4. How can the distance between two points in the plane be found?
5. What is the midpoint of a line segment and how are its coordinates calculated?
6. What is the difference between the standard form of the equation of a circle and the general form of the equation of a circle?

EXERCISES 2.1

— *Suggested Assignment: Exercises 1–77, odd; and 73–78, all.*
— *Answer graphs to Exercises 1–6, 23–40, and 59–72 are on pages AA2–AA3.*

In Exercises 1 to 6, answer each question on a separate graph.

1. Graph the ordered pairs (0, −1), (2, 0), (3, 2), and (−1, 4).

2. Graph the ordered pairs (−1, −3), (0, −4), (0, 4), and (3, −2).

3. Draw a line through all points with an x-coordinate of 2.

4. Draw a line through all points with an x-coordinate of −3.

5. Draw a line through all points with a y-coordinate of −3.

6. Draw a line through all points with a y-coordinate of 4.

In Exercises 7 to 12, find the length and midpoint of the line segment between the two points.

7. $P_1(2, -4)$, $P_2(3, 1)$
$\sqrt{26}$, (2.5, −1.5)

8. $P_1(4, -1)$, $P_2(-2, -5)$
$2\sqrt{13}$, (1, −3)

9. $P_1(-1, 5)$, $P_2(-2, -3)$
$\sqrt{65}$, (−1.5, 1)

10. $P_1(-6, 2)$, $P_2(-4, -2)$
$2\sqrt{5}$, (−5, 0)

11. $P_1(0, -3)$, $P_2(-2, -1)$
$2\sqrt{2}$, (−1, −2)

12. $P_1(4, -4)$, $P_2(0, -1)$
5, (2, −2.5)

The median of a triangle is a line segment from a vertex of the triangle to the midpoint of the side opposite that vertex. Use this definition for Exercises 13 and 14.

13. The coordinates of a triangle are $A(-5, -3)$, $B(2, 7)$, and $C(5, -1)$. Consider the median from A to line segment BC. Find the coordinates of the endpoint on the line segment BC of this median. (3.5, 3)

14. The coordinates of a triangle are $A(-7, 8)$, $B(5, -1)$, and $C(0, -6)$. Consider the median from B to line segment AC. Find the coordinates of the endpoint on the line segment AC of this median. (−3.5, 1)

15. Complete the input/output table.

Input, x	−3	−2	−1	0	1	2	3
Output, $2x - 3$							

−9, −7, −5, −3, −1, 1, 3

16. Complete the input/output table.

Input, x	−3	−2	−1	0	1	2	3
Output, $1 - 3x$							

10, 7, 4, 1, −2, −5, −8

17. Complete the input/output table.

Input, x	−3	−2	−1	0	1	2	3
Output, $x^2 + 2x - 4$							

−1, −4, −5, −4, −1, 4, 11

18. Complete the input/output table.

Input, x	−3	−2	−1	0	1	2	3
Output, $x^2 - 3x - 5$							

13, 5, −1, −5, −7, −7, −5

Business and Economics

19. ***Motorhome Production*** The weekly revenue, in dollars, received by a company that customizes motorhomes is given by $2880x - 1.25x^2$, where x is the number of motorhomes produced each week.

    a. Complete an input/output table for $x = 10, 20, 30,$ and 40. 28,675; 57,100; 85,275; 113,200

    b. What is the meaning of the output when $x = 30$? The revenue for customizing 30 motorhomes is $85,275.

20. ***Farming*** The value of a farmer's crop, in hundreds of dollars, is given by $1000\sqrt{x} + 120x$, where x is the number of acres planted.

    a. Complete an input/output table for $x = 16, 25, 36,$ and 49. 5920; 8000; 10,320; 12,880

    b. What is the meaning of the output when $x = 49$? The revenue from planting 49 acres is $1,280,000.

Life and Health Sciences

21. ***Medication Effectiveness*** The concentration, in milligrams per liter, t hours after a medication has been taken by a patient is given by $\frac{t}{t^2 + 5}$.

    a. Complete an input/output table for $t = 1, 2, 3,$ and 4. Round to the nearest hundredth. 0.17, 0.22, 0.21, 0.19

    b. What is the meaning of the output when $t = 3$? After 3 hours, the concentration of the medication is about 0.21 mg/L.

22. ***Flu Epidemic*** The rate at which a flu epidemic spreads through a college campus is given by $0.000025x(5000 - x)$, where x is the number of people infected and the rate is the number of people that will be infected the next day.

    a. Complete an input/output table for $x = 200, 400, 600,$ and 800. 24, 46, 66, 84

    b. What is the meaning of the output when $x = 600$? When 600 people are infected, an additional 66 people will be infected the next day.

In Exercises 23 to 30, graph the ordered-pair solutions of the given equation for the indicated values of x.

23. $y = x^2$ when $x = -2, -1, 0, 1,$ and 2.

24. $y = -x^2 + 1$ when $x = -2, -1, 0, 1,$ and 2.

25. $y = |x + 1|$ when $x = -5, -3, 0, 3,$ and 5.

26. $y = -2|x|$ when $x = -3, -1, 0, 1,$ and 3.

27. $y = -x^2 + 2$ when $x = -2, -1, 0, 1,$ and 2.

28. $y = -x^2 + 4$ when $x = -3, -1, 0, 1,$ and 3.

29. $y = x^3 - 2$ when $x = -1, 0, 1,$ and 2.

30. $y = -x^3 + 1$ when $x = -1, 0, 1,$ and 2.

In Exercises 31 to 40, graph each equation.

31. $y = 2x - 1$

32. $y = -3x + 2$

33. $y = \frac{2}{3}x + 1$

34. $y = -\frac{x}{2} - 3$

35. $y = \frac{1}{2}x^2$

36. $y = \frac{1}{3}x^2$

37. $y = 2x^2 - 1$

38. $y = -3x^2 + 2$

39. $y = |x - 1|$

40. $y = |x - 3|$

In Exercises 41 to 52, find the standard form of the equation of the circle from the given conditions.

41. Center $(1, -5)$; radius 2 $(x - 1)^2 + (y + 5)^2 = 4$

42. Center $(-4, 3)$; radius 4 $(x + 4)^2 + (y - 3)^2 = 16$

43. Center $(0, 1)$; radius $\sqrt{7}$ $x^2 + (y - 1)^2 = 7$

44. Center $\left(\frac{2}{3}, -\frac{3}{4}\right)$; radius 5 $\left(x - \frac{2}{3}\right)^2 + \left(y + \frac{3}{4}\right)^2 = 25$

45. Center $(3, 4)$, passing through $P(-1, 2)$ $(x - 3)^2 + (y - 4)^2 = 20$

46. Center $(-2, -1)$, passing through $P(1, -3)$ $(x + 2)^2 + (y + 1)^2 = 13$

47. Center $(0, 0)$, passing through $P(3, 4)$ $x^2 + y^2 = 25$

48. Center $(1, -2)$, passing through $P(2, -5)$ $(x - 1)^2 + (y + 2)^2 = 10$

49. Endpoints of a diameter are $P_1(1, 4)$ and $P_2(5, -2)$. $(x - 3)^2 + (y - 1)^2 = 13$

50. Endpoints of a diameter are $P_1(2, -3)$ and $P_2(0, 1)$. $(x - 1)^2 + (y + 1)^2 = 5$

51. Endpoints of a diameter are $P_1(-1, 4)$ and $P_2(-5, 3)$. 51. $(x + 3)^2 + \left(y - \frac{7}{2}\right)^2 = \frac{17}{4}$

52. Endpoints of a diameter are $P_1(5, -2)$ and $P_2(2, 1)$. $\left(x - \frac{7}{2}\right)^2 + \left(y + \frac{1}{2}\right)^2 = \frac{9}{2}$

In Exercises 53 to 58, find the center and radius of the circle given by each equation.

53. $x^2 + y^2 - 6x + 5 = 0$ $C(3, 0); r = 2$

54. $x^2 + y^2 - 6x - 4y + 12 = 0$ $C(3, 2); r = 1$

55. $x^2 + y^2 - 10x + 4y + 20 = 0$ $C(5, -2); r = 3$

56. $x^2 + y^2 - 2x + 8y - 2 = 0$ $C(1, -4); r = \sqrt{19}$

57. $x^2 + y^2 - 5y - 3 = 0$ $C\left(0, \frac{5}{2}\right); r = \frac{\sqrt{37}}{2}$

58. $x^2 + y^2 - x + 3y - 1 = 0$ $C\left(\frac{1}{2}, -\frac{3}{2}\right); r = \frac{\sqrt{14}}{2}$

In Exercises 59 to 72, use a graphing calculator to graph the equation.

59. $y = -\frac{1}{2}x - 2$

60. $y = 2x - 4$

61. $y = x^2 + 2x - 4$

62. $y = -x^2 + 2x - 1$

63. $y = 2|x| - 1$

64. $y = 2|x| + 2$

65. $y = \sqrt{1 + x}$

66. $y = \sqrt{4 - x}$

67. $y = |x| + 1$

68. $y = |x| - 1$

69. $y = x^3 + 2x^2$

70. $y = x^3 - 3x^2$

71. $y = x^3 - x^2 + x - 1$

72. $y = x^3 + x^2 - x + 1$

Prepare for Section 2.2

73. Simplify: $(3x^2 - x + 1) - (x^2 - 3)$ [P.3] $2x^2 - x + 4$

74. Simplify: $(2x - 1)(3x + 2)$ [P.3] $6x^2 + x - 2$

75. At what numbers is it impossible to evaluate the rational expression $\frac{x - 2}{x^2 - 1}$? [P.6] -1 and 1

76. Solve: $x^2 + 2x = 5$ [1.2] $-1 \pm \sqrt{6}$

77. Suppose $a < b$. What inequality symbol should replace the question mark in $3a - 2 ? 3b - 2$? [1.4] $<$

78. Suppose $a < b$. What inequality symbol should replace the question mark in $2 - 3a ? 2 - 3b$? [1.4] $>$

Explorations 

1. ***Verify a Geometry Theorem*** Use the midpoint formula and the distance formula to prove that the midpoint M of the hypotenuse of a right triangle is equidistant from each of the vertices of the triangle. (*Suggestion:* Label the vertices of the triangle as shown in the figure at the right.)

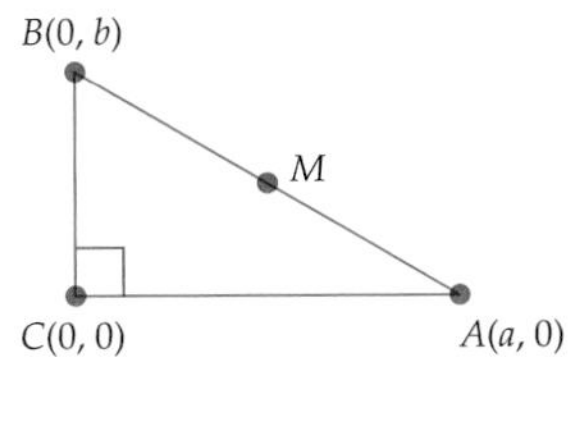

2. ***Solve a Quadratic Equation Geometrically*** In the seventeenth century, Descartes (and others) solved equations by using both algebra and geometry. This project outlines the method used by Descartes to solve certain quadratic equations using the figure below.

   a. Show that $d(Q, B)$ is a solution of $x^2 = 2ax + b^2$.

   b. Show that $d(P, B)$ is a solution of $x^2 = -2ax + b^2$.

   c. Show that $d(S, B)$ and $d(T, B)$ are solutions of $x^2 = 2ax - b^2$.

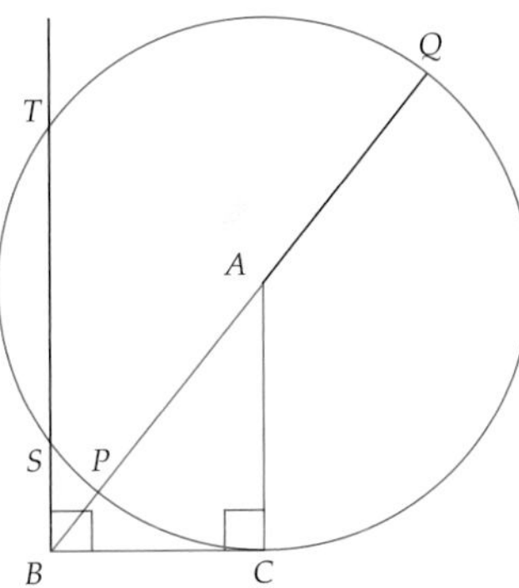

SECTION 2.2 Relations and Functions

- Introduction to Relations and Functions
- Evaluating Functions
- Graphs of Functions
- Properties of Functions
- Increasing and Decreasing Functions
- Applications of Functions

INSTRUCTOR NOTE

For the equation $d = 0.05s^2$ at the right, have students find a few more ordered pairs—say, for 20 mph and 50 mph.

▪ Introduction to Relations and Functions

Exploring relationships between known quantities frequently results in sets of ordered pairs. For instance, on a certain type of road, the distance a car will skid with its brakes locked depends on the speed of the car before the brakes are applied. The equation $d = 0.05s^2$ gives the distance d, in feet, the car will skid when the brakes are applied if the car is traveling at a speed of s miles per hour. The table below shows how far the car skids depending on its speed for various values of s.

Speed, s (in miles per hour)	10	25	30	45	60	70
Distance car skids, d (in feet)	5	31.25	45	101.25	180	245

The numbers in the table can also be written as ordered pairs in which the first coordinate of the ordered pair is the speed of the car when the brakes are applied and the second coordinate is the distance the car skids. The ordered pairs are (10, 5), (25, 31.25), (30, 45), (45, 101.25), (60, 180), and (70, 245). The ordered pairs from the table above are only some of the possible ordered pairs. Other possibilities are (42, 88.2), (55, 151.25), (65, 211.25), and many more.

? QUESTION Concerning the above example, what is the meaning of the ordered pair (55, 151.25)?

The table at the right shows a grading scale for an exam. Some possible ordered pairs for the relationship between exam score and letter grade are (95, A), (72, C), (84, B), (92, A), (61, D), and (88, B).

Test score	Grade
90–100	A
80–89	B
70–79	C
60–69	D
0–59	F

INSTRUCTOR NOTE

After you have introduced the four ways of describing a relationship between two quantities, ask students to name them. (Ans: Equation, table, list of ordered pairs, and graph)

A third way of describing a relationship between two quantities is a graph. The graph in Figure 2.12, based on data from The Strategis Group, shows the projected increase in broadband (high speed) wireless subscribers.

The ordered pairs from the graph are (2001, 410), (2002, 1185), (2003, 2385), (2004, 3875), and (2005, 5405).

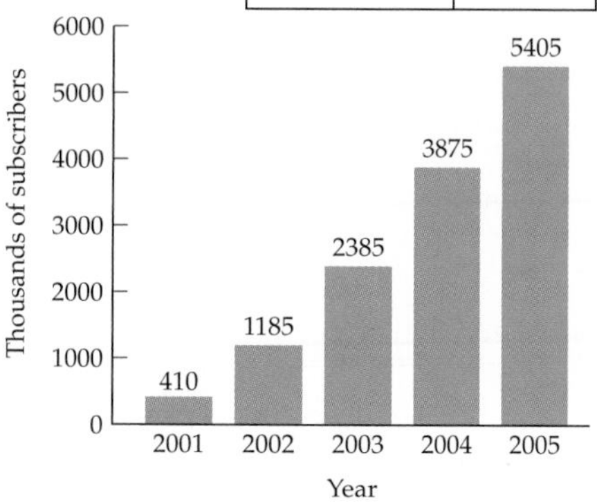

Figure 2.12

? ANSWER A car traveling 55 miles per hour will skid 151.25 feet after the brakes are applied.

For each of the situations above, ordered pairs were used to show the relationship between two quantities. In mathematics, a set of ordered pairs is called a *relation.*

Definition of Relation

A **relation** is any set of ordered pairs. The **domain** of the relation is the set of first coordinates of the ordered pairs. The **range** of the relation is the set of second coordinates of the ordered pairs.

For the relation between exam score and letter grade, we have

$$\text{Domain} = \{0, 1, 2, 3, 4, \ldots, 97, 98, 99, 100\} \quad \text{and}$$
$$\text{Range} = \{\text{A, B, C, D, F}\}$$

Here is one more example of a relation. Consider a meteorologist who is taking temperature and percent humidity readings at a weather station at a certain time each day. Here are the data for 10 consecutive days.

Temperature (°F)	72	75	67	71	70	73	68	69	75	68
Percent humidity	47	50	39	41	36	50	37	36	44	48

This relation is the set of ordered pairs

$$\{(72, 47), (75, 50), (67, 39), (71, 41), (70, 36),$$
$$(73, 50), (68, 37), (69, 36), (75, 44), (68, 48)\}$$

TAKE NOTE

The idea of *function* is one of the most important concepts in mathematics. It is a concept you will encounter throughout this text.

Although relations are important in mathematics, the concept of *function* is especially useful in applications.

Definition of a Function

A **function** is a relation in which no two ordered pairs have the same first coordinate and different second coordinates.

Point of Interest

Functions are used in many applications. For instance, there are functions to calculate π to millions of digits, functions to predict your chances of winning the lottery, functions to estimate the age of a fossil, and functions to calculate your monthly car payment. These are only some of the ways in which functions are used.

The temperature-humidity relation given earlier is *not* a function because the ordered pairs (75, 50) and (75, 44) have the same first coordinate and different second coordinates. The ordered pairs (68, 37) and (68, 48) also have the same first coordinate and different second coordinates.

Now consider the exam score and letter grade relation discussed earlier. If there were two ordered pairs with the same first coordinate and different second coordinates, it would mean that two students with the same exam score (first coordinate) would receive different letter grades (second coordinate). But this does not happen with a grading scale. Thus there are no two ordered pairs with the same first coordinate and different second coordinates. The exam score and grade relation is a function.

Alternative to Example 1

What are the domain and range of the following relation? Is the relation a function?

$\{(1, 4), (3, 7), (2, 1), (4, 5), (6, 6), (5, 1)\}$

- **Domain: {1, 2, 3, 4, 5, 6}**
  Range: {1, 4, 5, 6, 7}
  Yes

TAKE NOTE

Recall that when listing elements of a set, duplicate elements are listed only once.

EXAMPLE 1 Determine the Domain and Range of a Relation

What are the domain and range of the following relation? Is the relation a function?

$$\{(2, 4), (3, 6), (4, 7), (5, 4), (3, 2), (6, 8)\}$$

Solution Domain $= \{2, 3, 4, 5, 6\}$ Range $= \{2, 4, 6, 7, 8\}$

The relation is not a function because there are two ordered pairs, (3, 6) and (3, 2), with the same first coordinate and different second coordinates.

CHECK YOUR PROGRESS 1 What are the domain and range of the following relation? Is the relation a function?

$$\{(1, 1), (2, 1), (3, 1), (4, 1), (5, 1), (6, 1), (7, 1)\}$$

Solution *See page S11.* Domain: {1, 2, 3, 4, 5, 6, 7}; Range: {1}; Yes

Although a function can be described in terms of ordered pairs, in a table, or by a graph, a major focus of this text will be functions defined by equations in two variables. For instance, when gravity is the only force acting on a falling body, a function that describes the distance s, in feet, the object will fall in t seconds can be given by the equation $s = 16t^2$.

Given a value of t (time), the value of s (the distance the object falls) can be found. For instance, if $t = 3$, then $s = 144$. Because the distance the object falls depends on how long it has been falling, s is the *dependent variable* and t is the *independent variable.* Some of the ordered pairs of this function are (3, 144), (1, 16), (0, 0), and $\left(\frac{1}{4}, 1\right)$. The set of ordered pairs can be written as (t, s), where $s = 16t^2$. By substituting $16t^2$ for s, we can also write the set of ordered pairs as $(t, 16t^2)$. For the equation $s = 16t^2$, we say that "distance is a function of time." Not all equation in two variables define a function. For instance,

$$y^2 = x^2 + 9$$

is not an equation that defines a function. As shown below, the ordered pairs (4, 5) and (4, −5) belong to the solution set of the equation.

y^2	$= x^2 + 9$
5^2	$4^2 + 9$
25	$16 + 9$
25	$= 25$

- **Let $(x, y) = (4, 5)$. Replace x by 4 and y by 5.**
- **(4, 5) checks.**

y^2	$= x^2 + 9$
$(-5)^2$	$4^2 + 9$
25	$16 + 9$
25	$= 25$

- **Let $(x, y) = (4, -5)$. Replace x by 4 and y by −5.**
- **(4, −5) checks.**

Consequently, there are two ordered pairs with the same first coordinate, 4, but *different* second coordinates, 5 and −5; the equation does not define a function. The phrase "y is a function of x," or a similar phrase with different variables, is used to describe those equations in two variables that define functions.

TAKE NOTE

The notation $f(x)$ does *not* mean f *times* x. The letter f stands for the name of the function, and $f(x)$ is the value of the function at x.

▪ Evaluating Functions

Functional notation is frequently used to describe those equations that define functions. Just as x is commonly used as a variable, the letter f is commonly used to name a function.

To describe the relationship between a number and its square using functional notation, we can write $f(x) = x^2$. The symbol $f(x)$ is read "the value of f at x" or "f of x." The symbol $f(x)$ is the **value of the function** and represents the value of the dependent variable for a given value of the independent variable. We will often write $y = f(x)$ to emphasize the relationship between the independent variable, x, and the dependent variable, y. **Remember: y and $f(x)$ are different symbols for the same number.** Also, the *name* of the function is f; the *value* of the function at x is $f(x)$. For instance, the equation $d = 0.05s^2$ discussed at the beginning of this section could be written as $d(s) = 0.05s^2$. The name of the function is d.

The letters used to represent a function are somewhat arbitrary. All of the following equations represent the same function.

$f(x) = x^2$ $g(t) = t^2$ $P(v) = v^2$ ▪ These equations represent the square function.

The process of finding $f(x)$ for a given value of x is called **evaluating a function.** For instance, to evaluate $f(x) = x^2$ when x is 4, replace x by 4 and simplify.

$$f(x) = x^2$$
$$f(4) = 4^2 = 16$$
▪ Replace x by 4. Then simplify.

The *value* of the function is 16 when $x = 4$. An ordered pair of the function is (4, 16).

You may think of a function as a machine that turns one number into another number. For instance, you can think of the "square" function machine at the right as taking an input (a number from the domain) and creating an output (a number in the range) that is the square of the input.

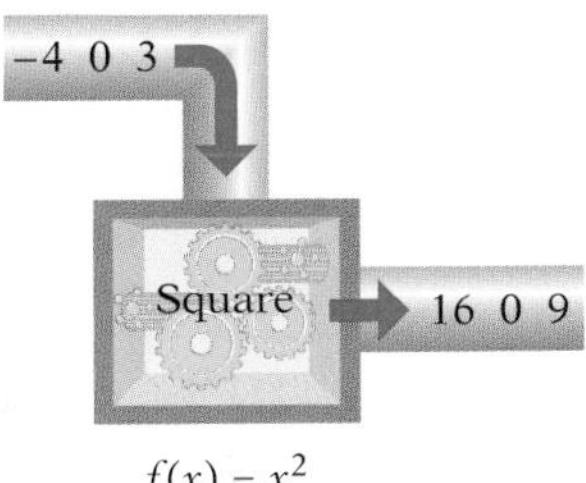

$f(x) = x^2$

Alternative to Example 2

Evaluate $g(t) = 2t^2 - 5t + 3$ when $t = -1$.

▪ 10

TAKE NOTE

To evaluate a function, you can use open parentheses in place of the variable. For instance,

$$p(s) = s^2 - 4s - 1$$
$$p(\;) = (\;)^2 - 4(\;) - 1$$

To evaluate the function, fill each of the parentheses with the same number and then simplify.

EXAMPLE 2 Evaluate a Function

Evaluate $p(s) = s^2 - 4s - 1$ when $s = -2$.

Solution

$$p(s) = s^2 - 4s - 1$$
$$p(-2) = (-2)^2 - 4(-2) - 1$$
$$= 11$$
▪ Replaces s by −2. Then simplify.

The value of the function is 11 when $s = -2$.

CHECK YOUR PROGRESS 2 Given $f(z) = 2z^3 - 4z$, find $f(-4)$.

Solution *See page S11.* −112

It is also possible to evaluate a function at a variable expression. For instance, to evaluate $N(v) = v^2 + v$ when $v = 3 - h$, proceed as follows.

$$N(v) = v^2 + v$$
$$N(3 - h) = (3 - h)^2 + (3 - h)$$ ▪ Replace v by $3 - h$.
$$= (9 - 6h + h^2) + (3 - h)$$ ▪ Simplify.
$$= h^2 - 7h + 12$$

The value of N when $v = 3 - h$ is $h^2 - 7h + 12$.

▪ Graphs of Functions

The **graph of a function** is the graph of the ordered pairs that belong to the function. The graph of the speed-distance function given earlier is shown in Figure 2.13. The horizontal axis represents the domain of the function (speed of the car); the vertical axis represents the range of the function (distance traveled after the brakes are applied).

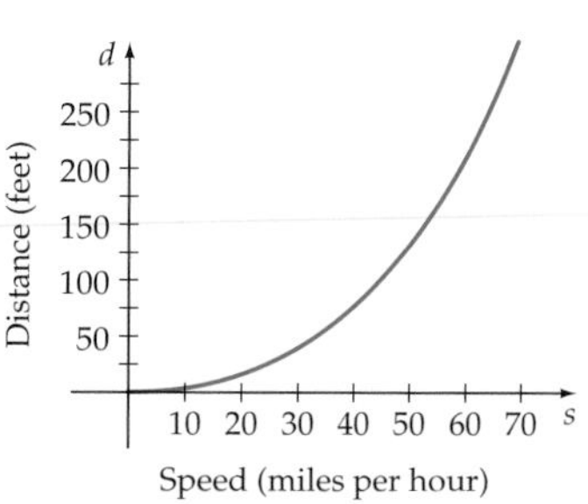

Figure 2.13

The graph of a function can be drawn by finding ordered pairs of the function, plotting the points corresponding to the ordered pairs, and then connecting the points with a smooth curve. For example, to graph $f(x) = x^3 + 1$, select several values of x and evaluate the function at those values. Recall that $f(x)$ and y are different symbols for the same quantity.

TAKE NOTE

We are creating the graph of an equation in two variables as we did earlier, this time using functional notation. That is, instead of writing $y = x^3 + 1$, we write $f(x) = x^3 + 1$.

x	$y = f(x) = x^3 + 1$	(x, y)
-2	$y = f(-2) = (-2)^3 + 1 = -7$	$(-2, -7)$
-1	$y = f(-1) = (-1)^3 + 1 = 0$	$(-1, 0)$
0	$y = f(0) = (0)^3 + 1 = 1$	$(0, 1)$
1	$y = f(1) = (1)^3 + 1 = 2$	$(1, 2)$
2	$y = f(2) = (2)^3 + 1 = 9$	$(2, 9)$

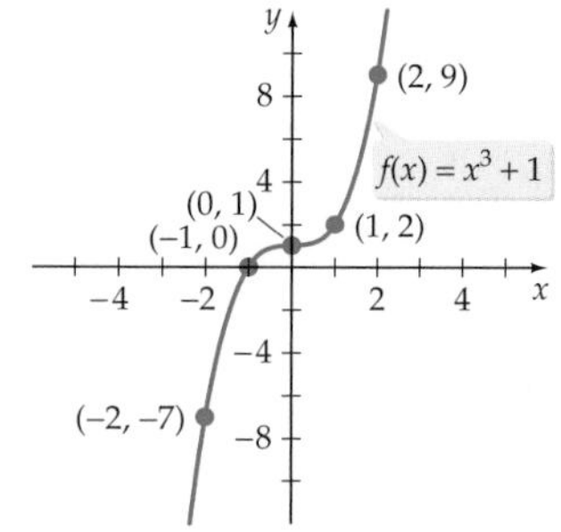

Plot the ordered pairs and draw a smooth curve through the points.

Alternative to Example 3
Exercise 28, page 188

EXAMPLE 3 Graph a Function

Graph $h(x) = x^2 - 3$.

Solution Find some ordered pairs of the function by evaluating the function at various value of x. The results can be recorded in a table.

Continued ➤

TAKE NOTE

Recall that a polynomial is an expression of the form

$$a_nx^n + a_{n-1}x^{n-1} + \cdots + a_1x + a_0$$

where each exponent on a variable is a non-negative integer. Thus $x^2 - 3$ and $-x^2 + 2$ are polynomials. $2x^{-2} + 3x^2 - 1$ is not a polynomial because one of the variables has a negative exponent.

Alternative to Example 4
Exercise 30, page 188

INTEGRATING TECHNOLOGY

If we create a graph by hand, the process of evaluating the function will give us some idea of how small or large the y-values will be. When a graphing calculator is used to draw the graph of a function, it may be necessary to experiment with a *viewing window* in which to see the graph. The standard viewing window is a good place to begin, but may not be appropriate for all graphs. For the graph of the polynomial function given by $f(x) = x^4 + x^3 - 18x^2 - 16x + 32$, shown below, we used a viewing window of Xmin = −6, Xmax = 6, Xscl = 1, Ymin = −100, Ymax = 100, Yscl = 10.

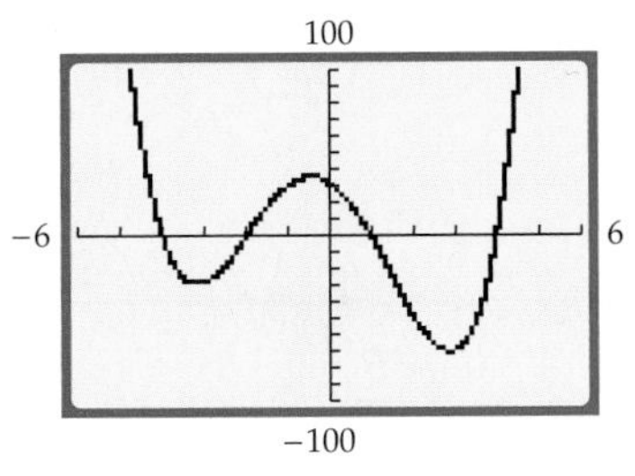

x	$h(x) = x^2 - 3$	(x, y)
−3	$h(-3) = (-3)^2 - 3 = 6$	$(-3, 6)$
−2	$h(-2) = (-2)^2 - 3 = 1$	$(-2, 1)$
−1	$h(-1) = (-1)^2 - 3 = -2$	$(-1, -2)$
0	$h(0) = (0)^2 - 3 = -3$	$(0, -3)$
1	$h(1) = (1)^2 - 3 = -2$	$(1, -2)$
2	$h(2) = (2)^2 - 3 = 1$	$(2, 1)$
3	$h(3) = (3)^2 - 3 = 6$	$(3, 6)$

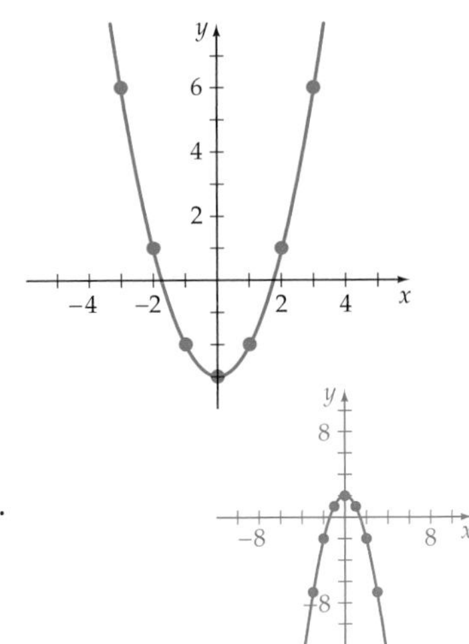

CHECK YOUR PROGRESS 3 Graph $f(x) = -x^2 + 2$.

Solution *See page S11.*

The functions in Example 3 and Check Your Progress 3 are called *polynomial functions* because $x^2 - 3$ and $-x^2 + 2$ are polynomials. We can also produce the graphs of other types of functions. In Example 4, we graph a function that involves absolute value.

EXAMPLE 4 Graph a Function

Graph $f(x) = |2x + 4|$.

Solution Find some ordered pairs of the function by evaluating the function at various values of x. The results can be recorded in a table.

x	$f(x) = \lvert 2x + 4 \rvert$	(x, y)
−4	$f(-4) = \lvert 2(-4) + 4 \rvert = 4$	$(-4, 4)$
−3	$f(-3) = \lvert 2(-3) + 4 \rvert = 2$	$(-3, 2)$
−2	$f(-2) = \lvert 2(-2) + 4 \rvert = 0$	$(-2, 0)$
−1	$f(-1) = \lvert 2(-1) + 4 \rvert = 2$	$(-1, 2)$
0	$f(0) = \lvert 2(0) + 4 \rvert = 4$	$(0, 4)$
1	$f(1) = \lvert 2(1) + 4 \rvert = 6$	$(1, 6)$
2	$f(2) = \lvert 2(2) + 4 \rvert = 8$	$(2, 8)$

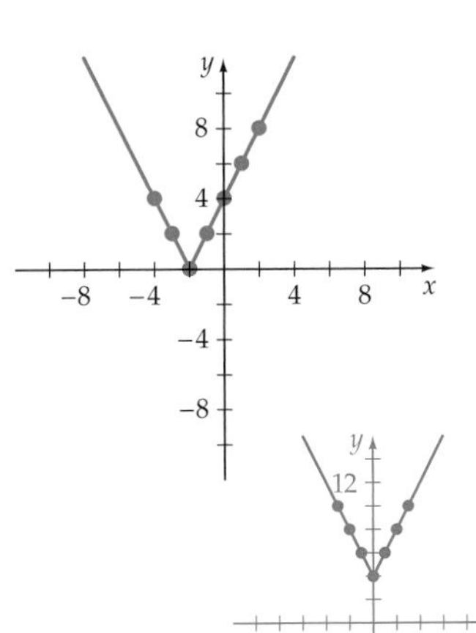

CHECK YOUR PROGRESS 4 Graph $g(x) = 2|x| + 4$.

Solution *See page S11.*

Alternative to Example 5
Exercise 86, page 193

INTEGRATING TECHNOLOGY

The *standard viewing window* for a graphing calculator is usually defined as
Xmin = −10, Xmax = 10,
Ymin = −10, and
Ymax = 10.

INTEGRATING TECHNOLOGY

These screens are from a TI-83 calculator. Note that although the domain is $\{x|x \geq -3\}$, we chose Xmin to be −5 so that the beginning part of the graph would be visible. We could have chosen any number less than −3.

EXAMPLE 5 Graph a Function Using a Graphing Calculator

Use a graphing calculator to produce a graph of $R(x) = \sqrt{x+3}$.

Solution The viewing window and the graph are shown below. Note that when $x < -3$, $\sqrt{x+3}$ is not a real number. (For instance, if $x = -4$, then $\sqrt{-4+3} = \sqrt{-1}$, which is not a real number.) Therefore, the domain of R is $\{x|x \geq -3\}$. Using interval notation, this domain is $[-3, \infty)$.

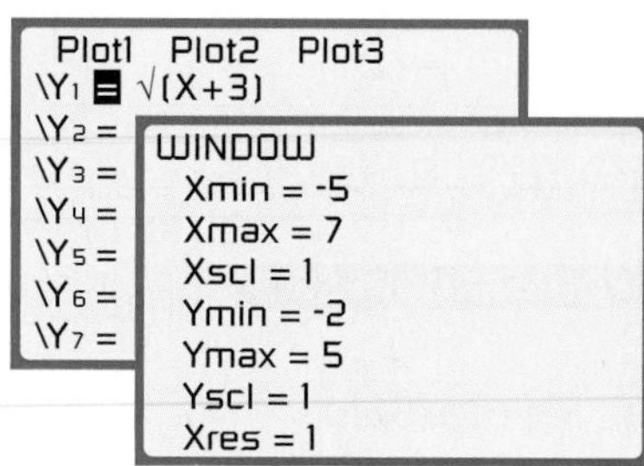

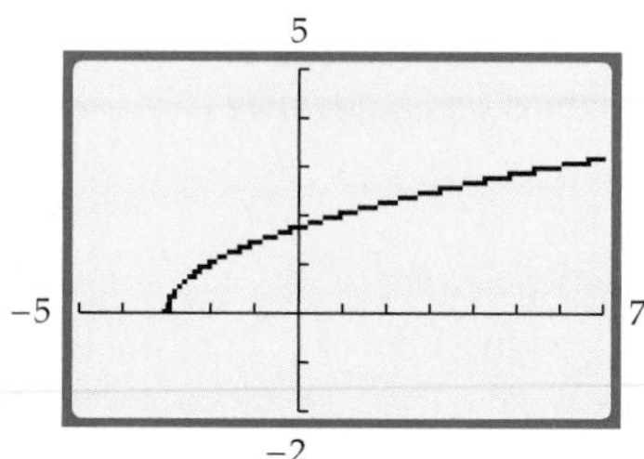

CHECK YOUR PROGRESS 5 Graph $f(x) = \sqrt{1-x}$ by using a graphing calculator.

Solution *See page S11.*

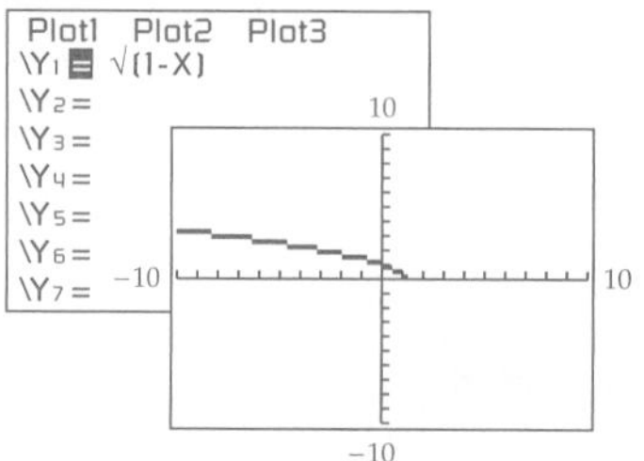

When we create the graph of a function, we generally assume that the domain of the function is all the real numbers for which the value of the function is a real number. For instance:

- The domain of $f(x) = x^3$ is the set of real numbers because every real number can be raised to the third power.
- The domain of $h(x) = \dfrac{2x}{x-3}$ is the set of all real numbers except 3, because when $x = 3$, $h(3) = \dfrac{6}{3-3} = \dfrac{6}{0}$, which is undefined.
- The domain of $g(x) = \sqrt{x}$ is $\{x|x \geq 0\}$, because when x is a negative number, $\sqrt{x}$ is not a real number. For instance, if $x = -4$, then $g(-4) = \sqrt{-4} = 2i$, which is not a real number. (Recall that $\sqrt{-1} = i$.)

Alternative to Example 6
Find the domain of
$$g(x) = \frac{x-2}{x^2 - 5x + 6}$$
- **Real numbers except $x \neq 2, 3$**

EXAMPLE 6 Find the Domain of a Function

Find the domain of $f(x) = \dfrac{x}{x^2 - 4}$.

Continued ➤

Solution We must exclude from the domain any number that makes the denominator equal to zero. To find the numbers to exclude, solve $x^2 - 4 = 0$ for x.

$$x^2 - 4 = 0$$

$$(x - 2)(x + 2) = 0 \qquad \text{▪ Solve the quadratic equation for } x \text{ by factoring.}$$

$$x - 2 = 0 \qquad x + 2 = 0$$

$$x = 2 \qquad x = -2$$

The numbers -2 and 2 must be excluded from the domain of f. The domain includes all real numbers except -2 and 2.

CHECK YOUR PROGRESS 6 Find the domain of $g(x) = \sqrt{2x - 6}$.

Solution *See page S12.* $[3, \infty)$

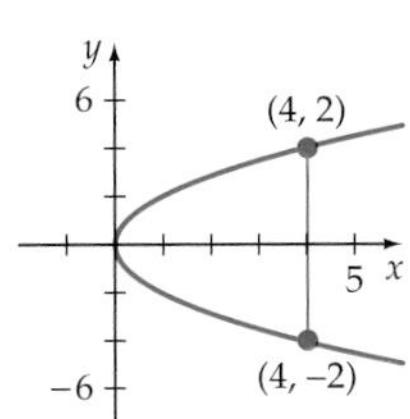

Figure 2.14

▪ Properties of Functions

Consider the graph in Figure 2.14. The two ordered pairs (4, 2) and (4, −2) belong to the graph, and these points lie on a vertical line. These two ordered pairs have the same first coordinates but different second coordinates and therefore the graph is not the graph of a function. With this observation in mind, we can give a visual method to determine whether a graph is the graph of a function.

Vertical Line Test for the Graph of a Function

A graph is the graph of a function if and only if every vertical line intersects the graph at most once.

This graphical interpretation of a function is often described by saying that each value in the domain of the function is paired with *exactly one* value in the range of the function.

TAKE NOTE

In the second graph at the right, there are values of x for which there is only one value of y. For instance, when $x = -5$, $y = 4$. To be a function, however, *every* value of x in the domain of the function must have exactly one value of y. If there is even one value of x that is paired with two or more values of y, the condition for a function is not met.

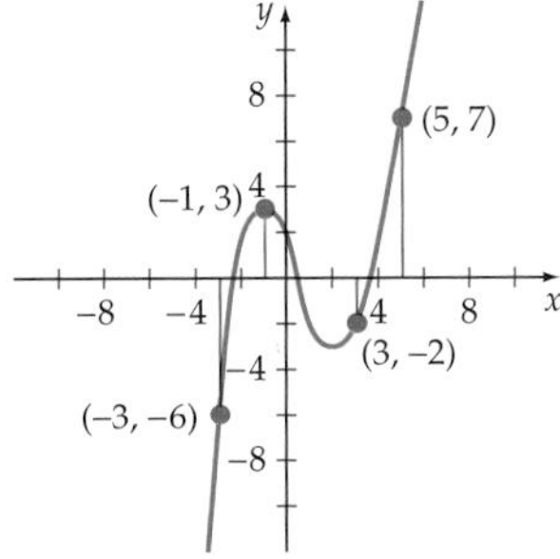

For each x there is *exactly one* value of y. For instance, when $x = -3$, $y = -6$. This is the graph of a function.

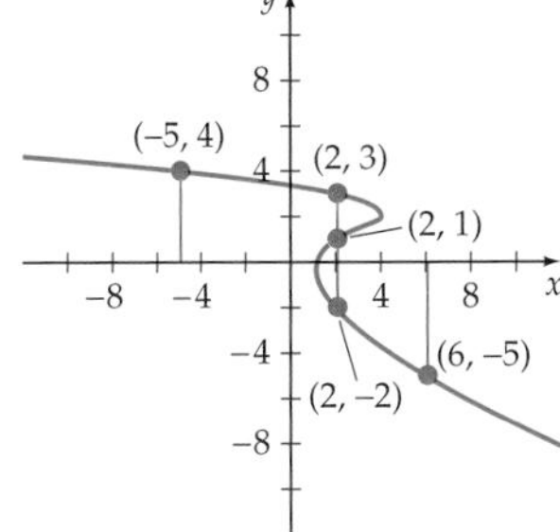

Some values of x can be paired with more than one value of y. For instance, 2 can be paired with -2, 1, and 3. This is not the graph of a function.

Alternative to Example 7
Exercise 42, page 189

EXAMPLE 7 Use the Vertical Line Test

Use the vertical line test to determine whether the graph is the graph of a function.

a.

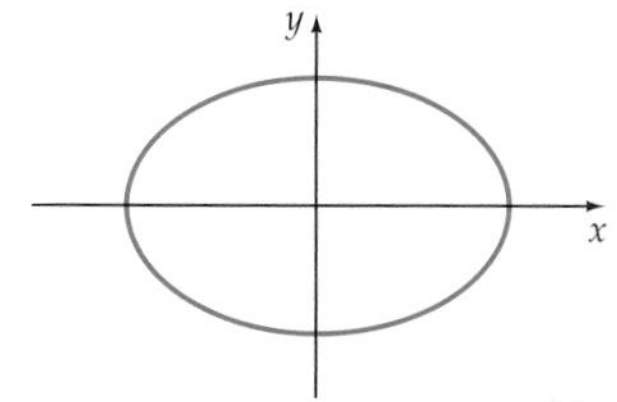

b.

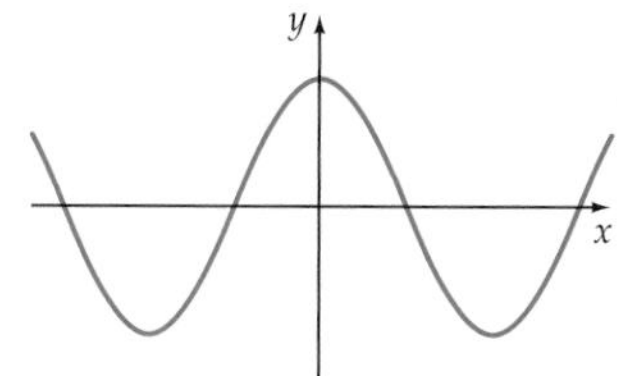

Solution

a. As shown at the right, there are vertical lines that intersect the graph at more than one point. Therefore, the graph is not the graph of a function.

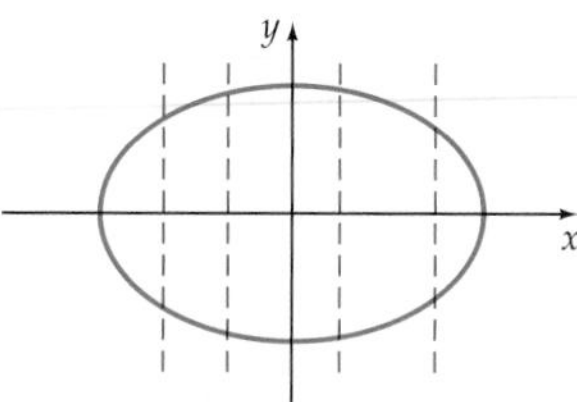

b. For the graph at the right, every vertical line intersects the graph at most once. Therefore, the graph is the graph of a function.

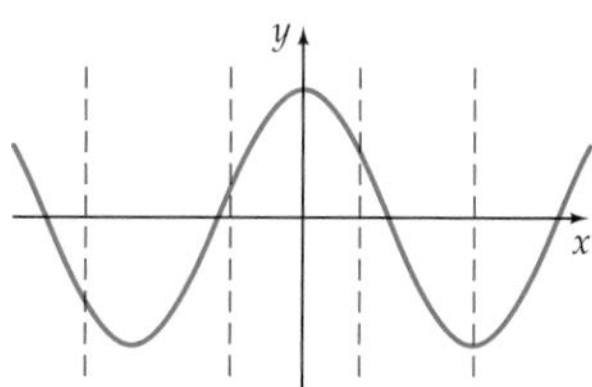

CHECK YOUR PROGRESS 7 For each of the graphs below, determine whether the graph defines a function.

a.

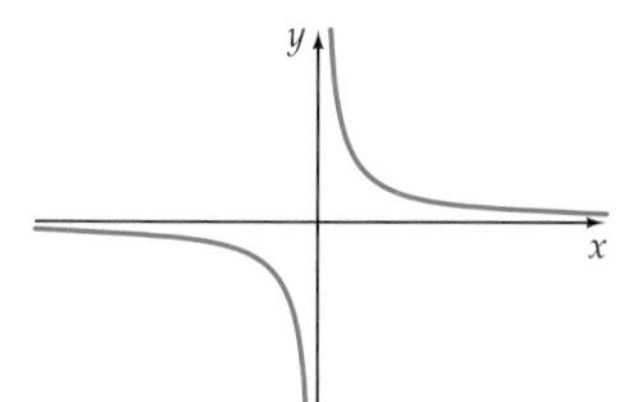

b.

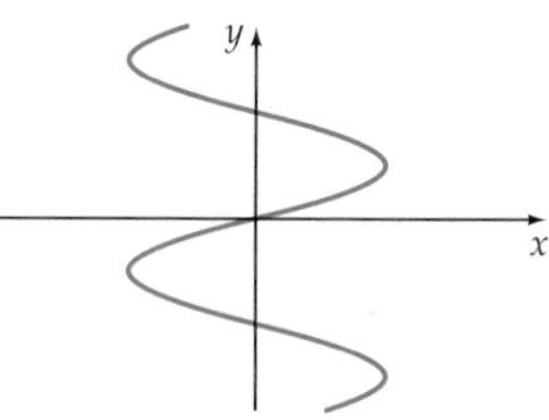

Solution *See page S12.* a. Function b. Not a function

▪ Increasing and Decreasing Functions

Consider the graph in Figure 2.15 on the next page. As we move from left to right, the graph rises on the interval $(-\infty, -5]$, remains at the same height on the interval $[-5, 2]$, and falls on the interval $[2, \infty)$. The function is said to be *increasing* on the interval $(-\infty, -5]$, *constant* on the interval $[-5, 2]$, and *decreasing* on the interval $[2, \infty)$.

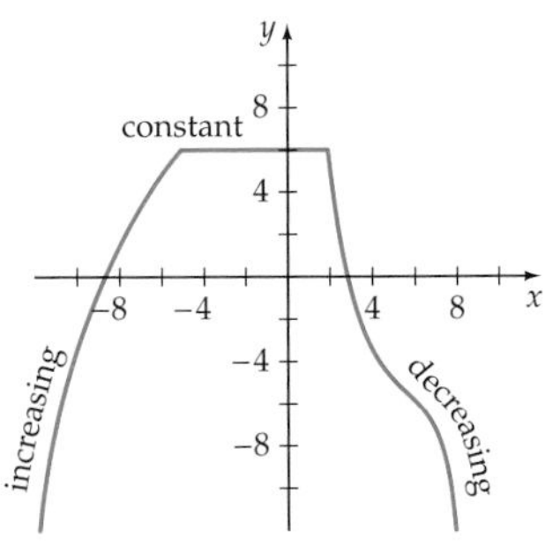

Figure 2.15

Increasing, Decreasing, and Constant Intervals of Functions

If a and b are elements of an interval I that is a subset of the domain of a function f, then

- f is **increasing** on I if $f(a) < f(b)$ whenever $a < b$.
- f is **decreasing** on I if $f(a) > f(b)$ whenever $a < b$.
- f is **constant** on I if $f(a) = f(b)$ for all a and b.

Recall that a function is a relation in which no two ordered pairs have the same first coordinate and different second coordinates. Consider the graphs below.

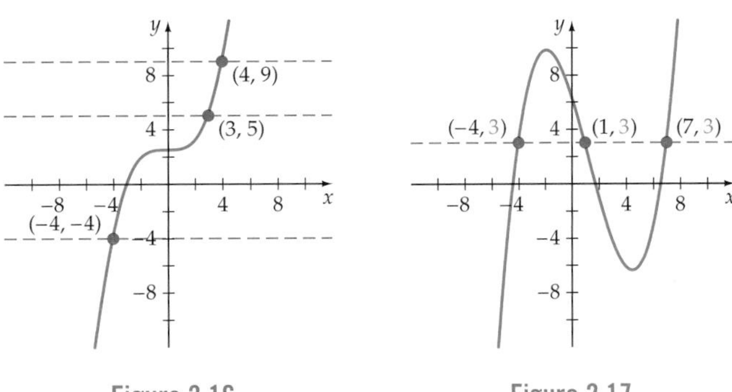

Figure 2.16 Figure 2.17

The graphs satisfy the vertical line test and therefore are the graphs of functions. Thus for each value of x there is exactly one value of y. In Figure 2.16, note that the reverse is true. That is, for each value of y there is exactly one value of x. This is not the case in Figure 2.17. There are some values of y (for instance, $y = 3$) for which there are different values of x. Thus $(-4, 3)$, $(1, 3)$, and $(7, 3)$ are all points on the graph that have the same y-coordinate but different x-coordinates.

A **one-to-one function** satisfies the additional requirement that no two ordered pairs have the same second coordinate and different first coordinates. The graph in Figure 2.17 is not a one-to-one function because $(-4, 3)$, $(1, 3)$, and $(7, 3)$ all belong to the function.

In a manner similar to applying the vertical line test, we can apply a **horizontal line test** to identify one-to-one functions.

Horizontal Line Test for a One-to-One Function

If every horizontal line intersects the graph of a function at most once, then the graph is the graph of a one-to-one function.

For instance, every horizontal line intersects the graph in Figure 2.18 at most once. It is the graph of a one-to-one function. Some horizontal lines intersect the graph in Figure 2.19 at more than one point. It is not the graph of a one-to-one function.

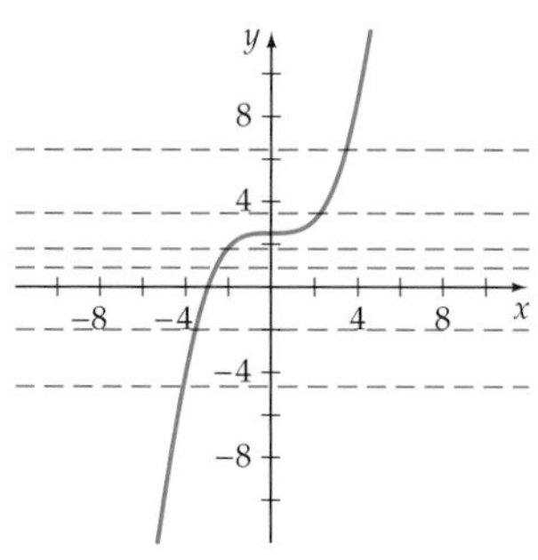

Every horizontal line intersects this graph at most once. It is the graph of a one-to-one function.

Figure 2.18

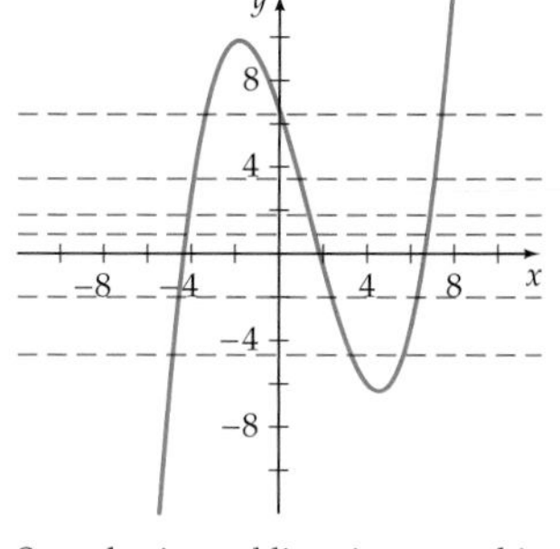

Some horizontal lines intersect this graph at more than one point. It is *not* the graph of a one-to-one function.

Figure 2.19

▪ Applications of Functions

Any letter or combination of letters can be used to name a function. In the next example, the letters *SA* are used to name a *S*urface *A*rea function. In this case, *SA* is the name of the function and does not mean *S* times *A*.

Alternative to Example 8
Exercise 55, page 190

EXAMPLE 8 Solve an Application to Geometry

The surface area of a cube (the sum of the areas of the six faces of the cube) is given by $SA(s) = 6s^2$, where $SA(s)$ is the surface area of the cube and s is the length of one side of the cube. Find the surface area of a cube that has a side of length 10 centimeters.

Solution

$$\begin{aligned} SA(s) &= 6s^2 \\ SA(10) &= 6(10)^2 && \text{▪ Replace } s \text{ by 10.} \\ &= 6(100) && \text{▪ Simplify.} \\ &= 600 \end{aligned}$$

The surface area is 600 square centimeters.

CHECK YOUR PROGRESS 8 If m points are placed in the plane in such a way that no three points lie on the same line, then the number of different line segments that can be drawn between the points, $N(m)$, is given by $N(m) = \dfrac{m(m-1)}{2}$. Find the total number of line segments that can be drawn between 12 different points in the plane.

5 points
10 line segments

Solution *See page S12.* 66 line segments

Graphing calculators frequently use combinations of letters for various functions. The screen below from a TI-83 calculator is accessed by pressing MATH. Here abs denotes the absolute value function.

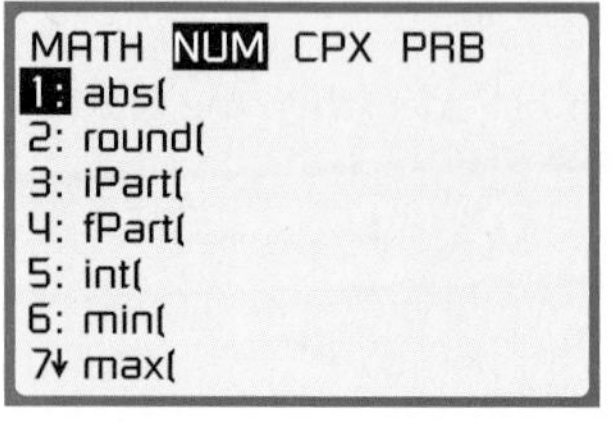

Alternative to Example 9
Exercise 62, page 191

EXAMPLE 9 Solve an Application to Economics

Generally, as manufacturers produce more and more of a certain commodity (for instance, a computer chip), the price per unit of the commodity decreases. The revenue a manufacturer receives for selling the commodity is the product of the price per unit and the number of units sold. Suppose the price per unit (in dollars) of a computer chip is given by $p(x) = 450 - \frac{x}{5000}$, where x represents the number of chips sold. Find the revenue function for the manufacturer of this chip. Find the manufacturer's revenue for selling 25,000 chips.

Solution

$$\text{Revenue} = (\text{price per unit}) \cdot (\text{number of units})$$
$$R(x) = p(x) \cdot x$$
$$= \left(450 - \frac{x}{5000}\right)x = 450x - \frac{x^2}{5000}$$

The revenue function is $R(x) = 450x - \frac{x^2}{5000}$.

To find the manufacturer's revenue for selling 25,000 chips, evaluate the revenue function for $x = 25{,}000$.

$$R(x) = 450x - \frac{x^2}{5000}$$
$$R(25{,}000) = 450(25{,}000) - \frac{25{,}000^2}{5000} = 11{,}125{,}000$$

The manufacturer's revenue for selling 25,000 chips is \$11,125,000.

CHECK YOUR PROGRESS 9 The profit function for manufacturing a product is the difference between the revenue received from selling the product and the cost to produce the product. Suppose that the revenue function for the sale of x ceramic cups is given by $R(x) = 13x - 0.015x^2$ and cost function to produce x ceramic cups is given by $C(x) = 5x + 3$. Find the profit function as a polynomial in standard form.

Solution *See page S12.* $P(x) = -0.015x^2 + 8x - 3$

Topics for Discussion

1. What is the difference between a relation and a function?
2. What is the domain of a function? What is the range of a function?
3. Explain how to evaluate a function.
4. If f is a function, how does f differ from $f(x)$?
5. If $f(a) = b$, is $P(a, b)$ a point on the graph of f?
6. How can you visually determine whether a graph is the graph of a function?
7. What is a one-to-one function, and how can you graphically determine whether a function is one-to-one?

EXERCISES 2.2

— *Suggested Assignment: Exercises 1–77, odd; 81, 85, 89, 93; and 94–99, all.*
— *Answer graphs to Exercises 23–32 and 86–93 are on page AA3.*

In Exercises 1 to 6, determine whether the set of ordered pairs is a function. Determine the domain and range of the relation.

1. $\{(-3, 1), (-2, 2), (1, 5), (4, -7)\}$
   Yes; domain $= \{-3, -2, 1, 4\}$, range $= \{-7, 1, 2, 5\}$
2. $\{(-5, 4), (-2, 3), (0, 1), (3, 2), (7, 11)\}$
   Yes; domain $= \{-5, -2, 0, 3, 7\}$, range $= \{1, 2, 3, 4, 11\}$
3. $\{(1, 5), (2, 5), (3, 5), (4, 5), (5, 5)\}$
   Yes; domain $= \{1, 2, 3, 4, 5\}$, range $= \{5\}$
4. $\{(1, 0), (10, 1), (100, 2), (1000, 3), (10{,}000, 4)\}$
   Yes; domain $= \{1, 10, 100, 1000, 10{,}000\}$, range $= \{0, 1, 2, 3, 4\}$
5. $\{(2, 3), (4, 5), (6, 7), (8, 9), (6, 8)\}$
   No; domain $= \{2, 4, 6, 8\}$, range $= \{3, 5, 7, 8, 9\}$
6. $\{(-1, 2), (-2, 4), (2, 4), (-2, 5), (4, 9)\}$
   No; domain $= \{-1, -2, 2, 4\}$, range $= \{2, 4, 5, 9\}$

In Exercises 7 to 16, determine whether the equation defines y as a function of x.

7. $y = 3x - 1$ Yes
8. $y = x$ Yes
9. $y = x^2 - x$ Yes
10. $y = x^3 - 2x + 1$ Yes
11. $x + y = 5$ Yes
12. $2x - y = 7$ Yes
13. $y^2 = x - 1$ No
14. $|y| = x$ No
15. $x^2 + y^2 = 25$ No
16. $|x| + |y| = 10$ No

In Exercises 17 to 22, evaluate the function.

17. Given $f(x) = 4x + 5$, find:
   a. $f(2)$ 13 b. $f(-2)$ -3 c. $f(0)$ 5
   d. $f\left(\frac{3}{4}\right)$ 8 e. $f(a)$ $4a + 5$ f. $f(2a)$ $8a + 5$
18. Given $f(x) = 2x^2 - 1$, find:
   a. $f(3)$ 17 b. $f(-2)$ 7 c. $f(0)$ -1
   d. $f\left(\frac{1}{2}\right)$ $-\frac{1}{2}$ e. $f(-a)$ $2a^2 - 1$ f. $f(a + 1)$ $2a^2 + 4a + 1$
19. Given $g(x) = x^2 + 2x - 1$, find:
   a. $g(1)$ 2 b. $g(-3)$ 2 c. $g(-a)$ $a^2 - 2a - 1$
   d. $g(0)$ -1 e. $g(a + 1)$ $a^2 + 4a + 2$ f. $g(3c)$ $9c^2 + 6c - 1$
20. Given $g(x) = \frac{3}{x}, x \neq 0$, find:
   a. $g(2)$ $\frac{3}{2}$ b. $g(-2)$ $-\frac{3}{2}$ c. $g(10{,}000)$ 0.0003
   d. $g\left(\frac{3}{4}\right)$ 4 e. $g(0.001)$ 3000 f. $g(z + 5)$ $\frac{3}{z + 5}$
21. Given $f(x) = \frac{2x}{x - 1}, x \neq 1$, find:
   a. $f(4)$ $\frac{8}{3}$ b. $f(-1)$ 1 c. $f(0)$ 0
   d. $f\left(\frac{2}{3}\right)$ -4 e. $f(1.0001)$ 20,002 f. $f(0.9999)$ $-19{,}998$
22. Given $P(x) = \frac{|x|}{x}, x \neq 0$, find:
   a. $P(5)$ 1 b. $P(-3)$ -1 c. $P(1000)$ 1
   d. $P\left(-\frac{1}{2}\right)$ -1 e. $P(-0.001)$ -1 f. $P\left(\sqrt{3}\right)$ 1

In Exercises 23 to 32, draw a graph of the function.

23. $f(x) = 2x - 3$
24. $f(x) = -x + 2$
25. $f(x) = -\frac{2}{3}x + 1$
26. $f(x) = \frac{5}{3}x - 2$
27. $f(x) = x^2 + 3x$
28. $f(x) = x^2 - 4x$
29. $f(x) = |x - 3|$
30. $f(x) = |2x + 2|$
31. $f(x) = 2|x| + 2$
32. $f(x) = x^2 + 2x - 3$

In Exercises 33 to 40, determine the domain of the function.

33. $f(x) = x^2 + 4x - 1$
   All real numbers
34. $f(x) = x^3 + x - 1$
   All real numbers
35. $f(x) = \frac{x + 1}{x - 1}$
   All real numbers except $x = 1$
36. $f(x) = \frac{x^2}{x^2 - 9}$
   All real numbers except $x = -3$ and $x = 3$
37. $f(x) = \sqrt{4 - x}$
   $(-\infty, 4]$
38. $f(x) = \sqrt{2x + 6}$ $[-3, \infty)$
39. $f(x) = \frac{x - 1}{x^2 + 4}$
   All real numbers
40. $f(x) = \frac{2x}{x^2 + 2x + 2}$
   All real numbers

41. Use the vertical line test to determine whether the graph is the graph of a function.

a.

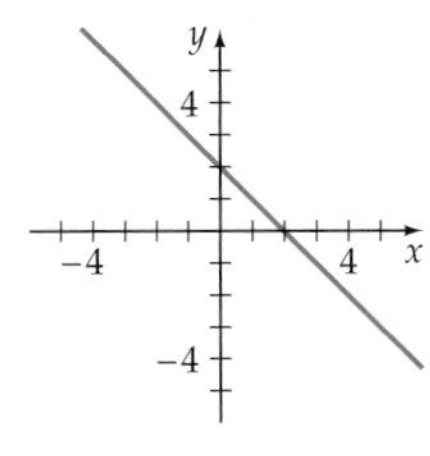

Yes

b.

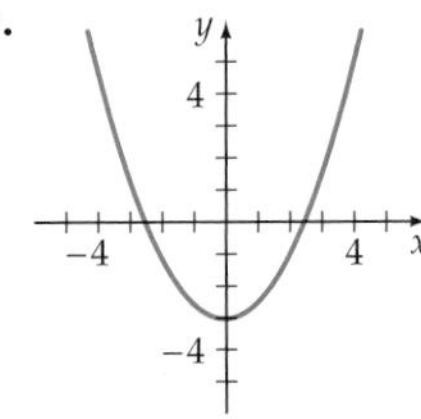

Yes

c.

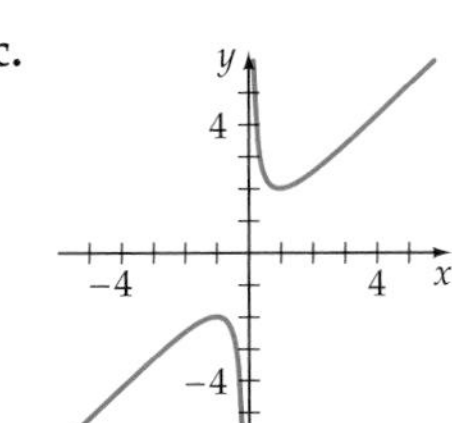

Yes

d.

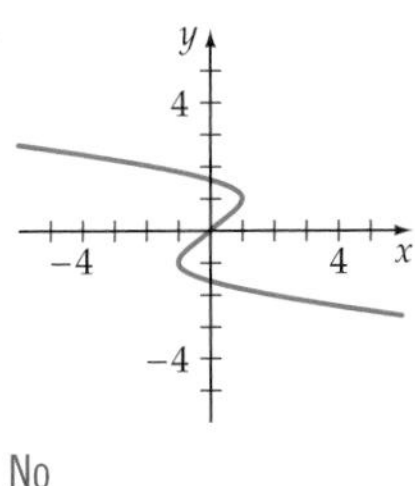

No

42. Use the vertical line test to determine whether the graph is the graph of a function.

a.

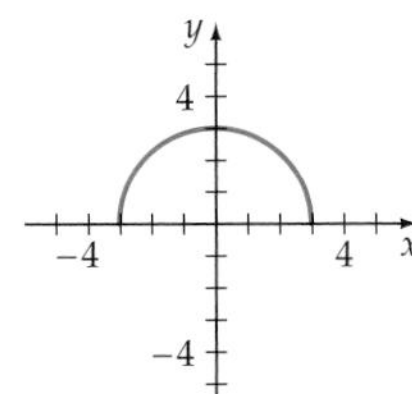

Yes

b.

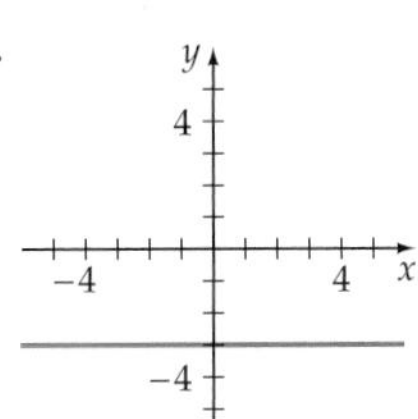

Yes

c.

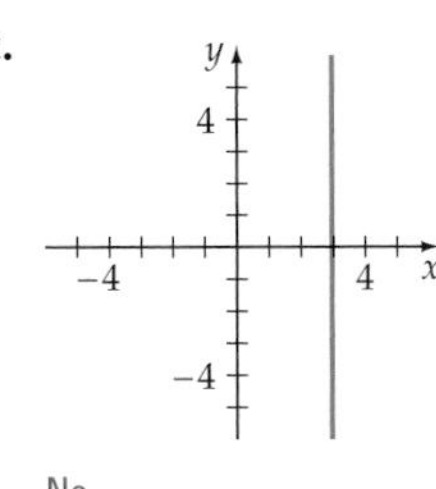

No

d.

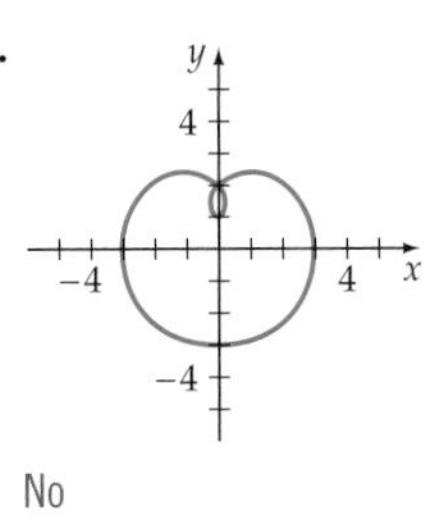

No

In Exercises 43 to 52, use the given graph to identify the intervals over which the function is increasing, constant, or decreasing.

43.

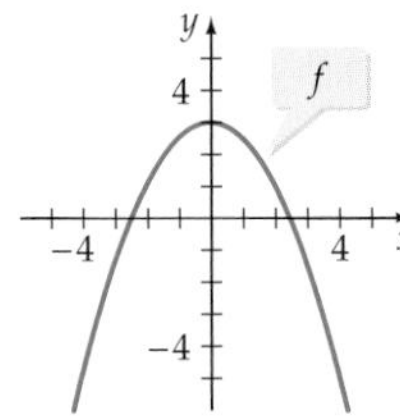

Increasing on $(-\infty, 0]$, decreasing on $[0, \infty)$

44.

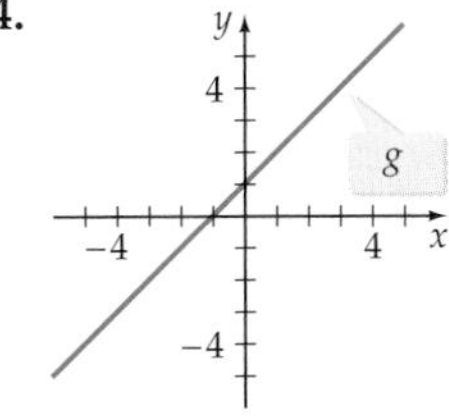

Increasing on $(-\infty, \infty)$

45.

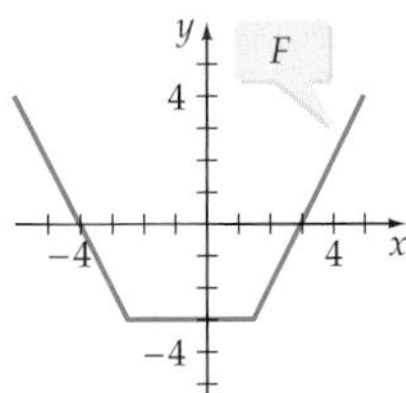

Decreasing on $(-\infty, -2]$, constant on $[-2, 1]$, increasing on $[1, \infty)$

46.

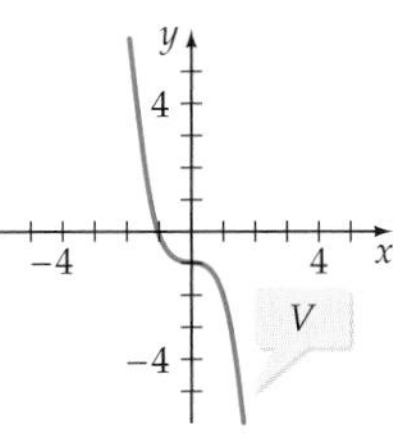

Decreasing on $(-\infty, \infty)$

47.

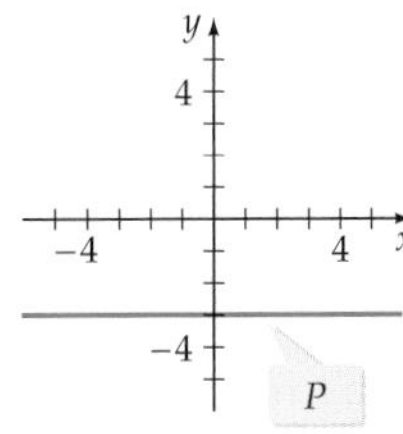

Constant on $(-\infty, \infty)$

48.

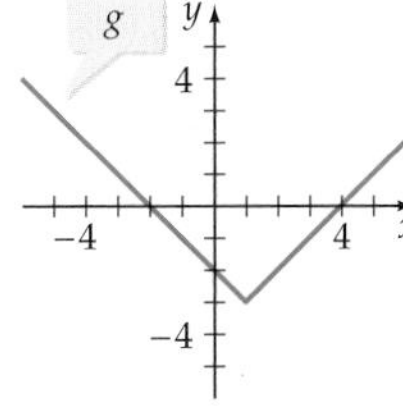

Decreasing on $(-\infty, 1]$, increasing on $[1, \infty)$

49.

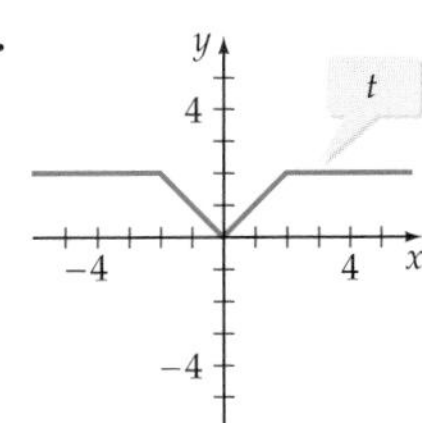

Constant on $(-\infty, -2]$, $[2, \infty)$, decreasing on $[-2, 0]$, increasing on $[0, 2]$

50.

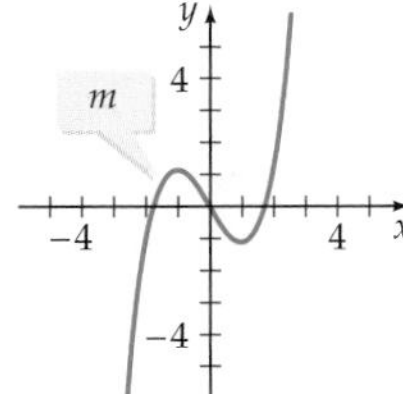

Increasing on $(-\infty, -1]$, $[1, \infty)$, decreasing on $[-1, 1]$

51.

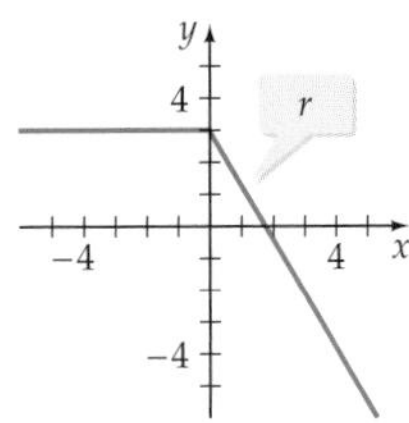

Constant on $(-\infty, 0]$, decreasing on $[0, \infty)$

52.

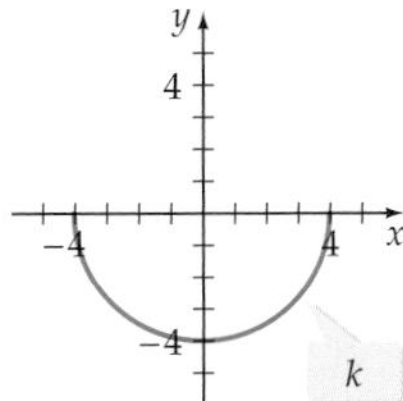

Decreasing on $[-4, 0]$, increasing on $[0, 4]$

53. Use the horizontal line test to determine which of the following functions are one-to-one.

a. f as shown in Exercise 43 No

b. g as shown in Exercise 44 Yes

c. F as shown in Exercise 45 No

d. V as shown in Exercise 46 Yes

e. P as shown in Exercise 47 No

54. Use the horizontal line test to determine which of the following functions are one-to-one.

a. g as shown in Exercise 48 No

b. t as shown in Exercise 49 No

c. m as shown in Exercise 50 No

d. r as shown in Exercise 51 No

e. k as shown in Exercise 52 No

55. ***Geometry*** A rectangle has a length of l feet and a perimeter of 50 feet.

a. Write the width w of the rectangle as a function of its length. $w(l) = 25 - l$

b. Write the area A of the rectangle as a function of its length. $A(l) = 25l - l^2$

56. ***Geometry*** A cone has an altitude of 15 centimeters and a radius of 3 centimeters. A right circular cylinder of radius r and height h is inscribed in the cone as shown in the figure. Use similar triangles to write h as a function of r. $h(r) = 5r$

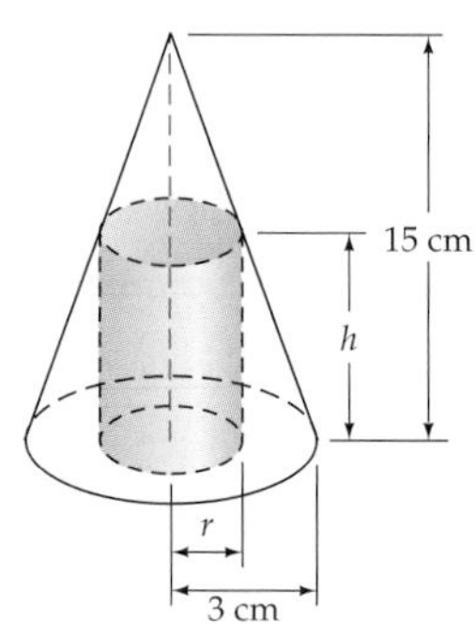

57. ***Number Sense*** The sum of two numbers is 20. Let x represent one of the numbers.

a. Write the second number y as a function of x. $y(x) = 20 - x$

b. Write the product P of the two numbers as a function of x. $P(x) = 20x - x^2$

Business and Economics

58. ***Credit Card Debt*** The following graph shows the amount of interest a person with credit card debt of \$1000 will pay and the time it will take to repay the debt if that person makes only the minimum monthly payment (and makes no additional purchases).

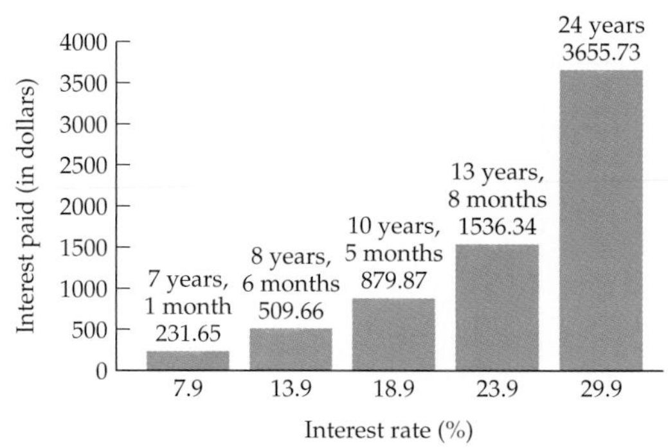

Source: **CardWeb.com**

a. How much interest will this person pay if the credit card interest rate is 18.9% (a fairly typical rate)? \$879.87

b. How many years and months will it take to repay the debt if the interest rate is 23.9%? 13 years 8 months

59. ***Pipe Manufacturing*** A business finds that the number of feet f of pipe it can sell per week is a function of the price p, in cents per foot, as given by

$$f(p) = \frac{320{,}000}{p + 25}$$

Complete the following table by evaluating f (to the nearest 10 feet) for the indicated values of p.

p	40	50	60	75	90
$f(p)$					

4920, 4270, 3760, 3200, 2780

60. ***Cost of MP3 Players*** A manufacturer finds that the cost in dollars to produce x portable MP3 players is given by $C(x) = \left(225 + 1.4\sqrt{x}\right)^2$, $100 \le x \le 1000$. Complete the following table by evaluating $C(x)$ (to the nearest dollar) for the indicated numbers of MP3 players.

x	100	250	500	750	1000
$C(x)$					

57,121, 61,076, 65,692, 69,348, 72,507

61. ***Bus Depreciation*** A bus was purchased for \$80,000. Assuming the bus depreciates at a rate of \$6500 per year (straight-line depreciation) for the first 10 years, write the value v of the bus as a function of the time t (measured in years) for $0 \le t \le 10$. $v(t) = 80{,}000 - 6500t$

62. ***Profit for Keyboards*** The selling price for one computer keyboard is \$17. The total cost (in dollars) to produce x keyboards is given by $C(x) = 6x + 40$. Assuming the keyboard manufacturer can sell all units produced, find

a. the revenue R as a function of x. $R(x) = 17x$

b. the profit P as a function of x. $P(x) = 11x - 40$

c. What is the profit for selling 300 keyboards? \$3260

63. ***Profit for Photographic Paper*** A manufacturer produces a photographic paper for which the cost function is given by $C(x) = 28x + 400$, where x is the number of reams of paper produced. The manufacturer sells each ream for \$37. Assuming the manufacturer sells all reams produced,

a. write the revenue R as a function of x. $R(x) = 37x$

b. write the profit P as a function of x. $P(x) = 9x - 400$

c. find the profit for selling 250 reams. \$1850

64. ***Profit for PDAs*** A manufacturer's selling price for a personal digit assistant (PDA) is given by

$p(x) = 125 - \dfrac{x}{500}$, where x is the number of PDAs sold and $p(x)$ is the price per PDA in dollars.

a. Write the revenue R as a function of x. $R(x) = 125x - \dfrac{1}{500}x^2$

b. If the cost to produce x PDAs is given by $C(x) = 300 + 50x - 0.001x^2$, find the profit function.

c. What is the profit for selling 10,000 PDAs? \$649,700

64b. $P(x) = -\dfrac{1}{1000}x^2 + 75x - 300$

65. ***Profit for Memory Chips*** A computer business analyst has determined that the selling price p, in dollars, of a memory chip is given by $p(x) = 225 - \dfrac{x}{300}$, where x is the number of memory chips sold in thousands. The cost, in dollars, to produce x thousand memory chips is given by $C(x) = -0.025x^2 + 40x + 215$.

a. Write the revenue R as a function of x. $R(x) = 225x - \dfrac{1}{300}x^2$

b. Find the profit function. $P(x) = \dfrac{13}{600}x^2 + 185x - 215$

c. What is the profit for selling 10,000 memory chips? \$1637.17

66. ***Boat Depreciation*** A sailboat was purchased for \$44,000. Assuming the boat depreciates at a rate of \$4200 per year (straight-line depreciation) for the first 8 years, write the value v of the boat as a function of the time t (measured in years) for $0 \le t \le 8$. $v(t) = 44{,}000 - 4200t$

Life and Health Sciences

67. ***Conditioning*** An athlete swims from point A to point B at a rate of 2 miles per hour and runs from point B to point C at the rate of 8 miles per hour. Use the dimensions in the figure to write the time t required to reach point C as a function of x. $t(x) = \dfrac{\sqrt{x^2+1}}{2} + \dfrac{3-x}{8}$

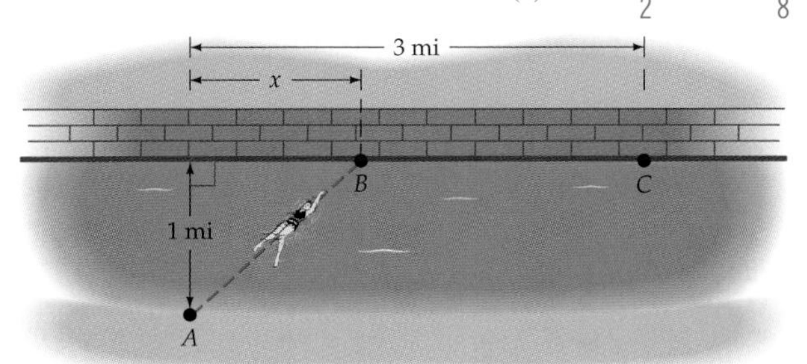

68. ***Apple Yields*** The yield Y of apples per tree is related to the amount x (in pounds per year) of a particular type of fertilizer applied by

$$Y(x) = 400 - \frac{2000}{(x-1)^2}$$

Complete the following table by evaluating Y (to the nearest apple) for the indicated fertilizer applications.

x	5	10	12.5	15	20
Y(x)					

275, 375, 385, 390, 394

69. ***Grape Yields*** The yield of grapes for a certain plot of land depends on the amount of water the grapes receive. The yield can be modeled by $Y(x) = -0.4x^3 + 3.25x^2 - 5.8x$, where x is the number of gallons of water (in thousands) per acre per week and $Y(x)$ is the grape yield in tons.

a. Using this model, determine the yield of grapes when 3500 gallons of water per acre per week are used. 2.3625 tons

b. Using this model, determine the decrease in yield if watering is increased from 4500 gallons per acre per week to 5000 gallons per acre per week. 1.0125 tons

Social Sciences

70. **Centenarians** Based on projections by the U.S. Census Bureau, a model that approximates the number of centenarians (people one hundred years old and older) living in the U.S. is given by

$$H(x) = 0.3214x^2 - 2.1629x + 96.4286$$

where $0 \leq x \leq 50$, $x = 0$ corresponds to the year 2000, and $H(x)$ is the number of centenarians in thousands.

a. Using this model, predict the number of centenarians in 2010. 106,940 people

b. Using this model, predict the increase in the number of centenarians from 2005 to 2015. 42,651 people

71. **Tuition and Income Increases** The graph below shows the percent increase in tuition at U.S. public colleges and universities and the percent increase in median family income for selected years.

a. In what years was the percent increase in median income at least as large as the percent increase in tuition? 1981, 1996, 1999, 2000

b. In what year was the difference between the percent increase in median income and the percent increase in tuition more than 6%? 1982, 1983, 1991, 1992

Percentage Change in Income and Public College Tuition from Previous Year

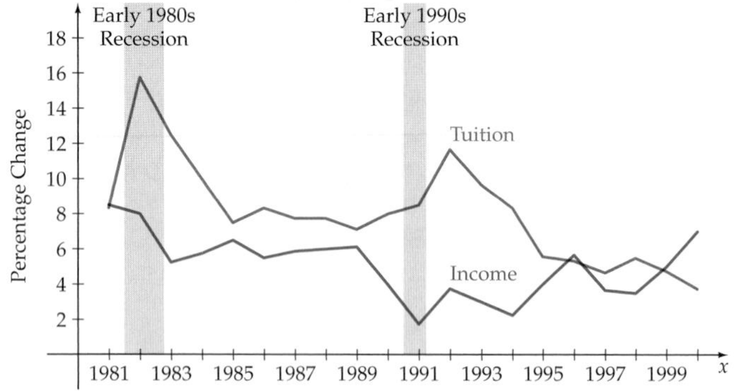

Sources: Income: U.S. Bureau of the Census. "Median Income for Four Person Families, By State" (2001).
Tuition: Washington State Higher Education Coordinating Board, *Tuition and Fee Rates: A National Comparison*, volumes 1985, 1990, 1995, 2000.

Physical Sciences and Engineering

72. **Fluid Dynamics** Water is running out of a conical funnel that has an altitude of 20 inches and a radius of 10 inches, as shown in the following figure.

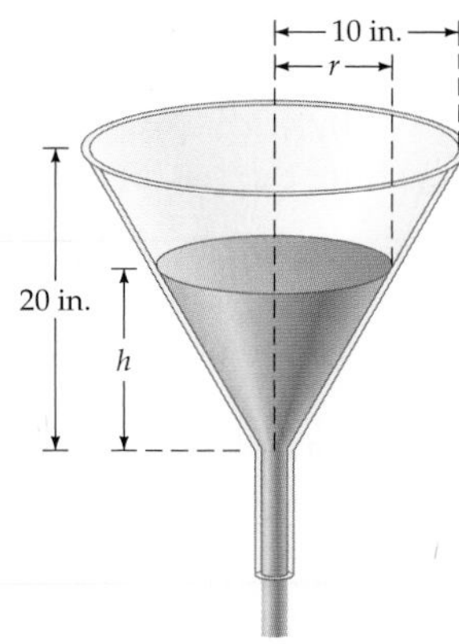

a. Write the radius r of the water as a function of its depth h. $r(h) = \frac{1}{2}h$

b. Write the volume V of the water as a function of its depth h. $V(h) = \frac{1}{12}\pi h^3$

73. **Aeronautics** For the first minute of flight, a hot air balloon rises vertically at a rate of 3 meters per second. If t is the time in seconds that a balloon has been airborne, write the distance d between the balloon and a point on the ground 50 meters from the point of lift-off as a function of t. $d(t) = \sqrt{9t^2 + 2500}$

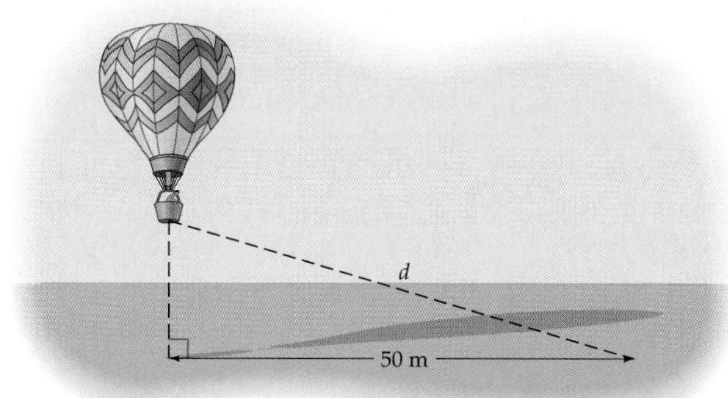

74. **Box Construction** An open box is to be made from a square piece of cardboard of dimensions 30 inches by 30 inches by cutting out squares of area x^2 from each corner, as shown in the figure.

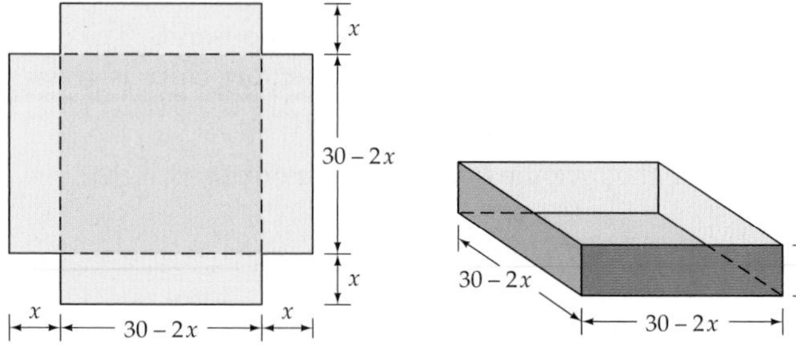

a. Express the volume V of the box as a function of x.
$V(x) = 4x^3 - 120x^2 + 900x$

b. State the domain of V. $0 < x < 15$

75. ***Navigation*** At 12:00 noon ship A is 45 miles due south of ship B and is sailing north at a rate of 8 miles per hour. Ship B is sailing east at a rate of 6 miles per hour. Write the distance d between the ships as a function of the time t, where t is in hours and $t = 0$ represents 12:00 noon.
$d(t) = \sqrt{100t^2 - 720t + 2025}$

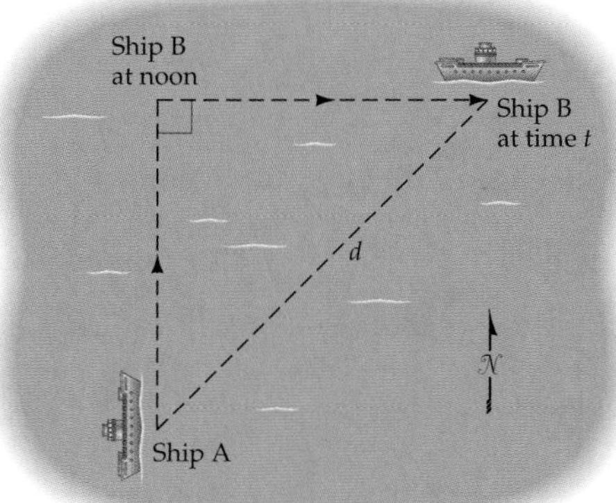

In Exercises 76 to 85, determine whether the given number b is in the range of the function. [*Suggestion:* If b is in the range of a function f then there must exist a number a in the domain of f for which $f(a) = b$. For instance, to determine whether 5 is in the range of $f(x) = 2x + 1$, there must be a solution of the equation $f(x) = 2x + 1 = 5$. Because 2 is a solution of this equation, there is a number (2) in the domain of f for which $f(2) = 2(2) + 1 = 5$. See the graph below.]

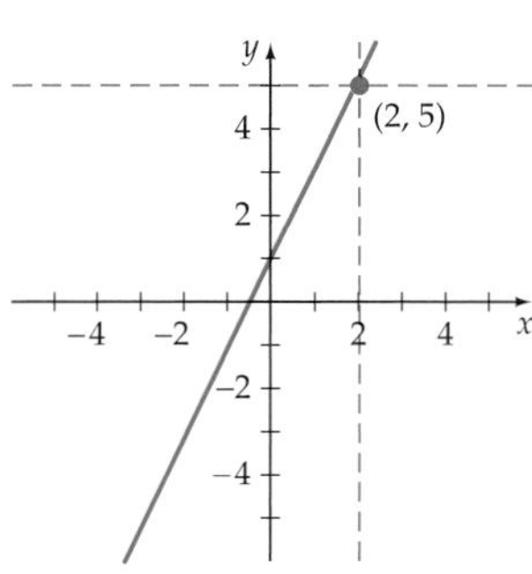

A number b will be in the range of f if a horizontal line through $y = b$ intersects the graph of f. The numbers (there may be more than one) in the domain are the x-coordinates of the points of intersection and of the horizontal line and the graph.

76. $b = 1; f(x) = 2x - 4$ Yes

77. $b = -2; f(x) = 3x + 1$ Yes

78. $b = 0; f(x) = x^2 - 1$ Yes

79. $b = 2; f(x) = x^2 + 1$ Yes

80. $b = 4; f(x) = x^2 + 5$ No

81. $b = -3; f(x) = x^2 + x$ No

82. $b = 1; f(x) = \dfrac{x}{x + 1}$ No

83. $b = 2; f(x) = \dfrac{2x}{x - 3}$ No

84. $b = 2; f(x) = \dfrac{x}{x + 1}$ Yes

85. $b = 1; f(x) = \dfrac{2x}{x - 3}$ Yes

In Exercises 86 to 93, use a graphing calculator to draw the graph of the equation.

86. $f(x) = x|x|$

87. $f(x) = |x + 1| + |x - 2|$

88. $f(x) = \dfrac{10}{x^2 + 1}$

89. $f(x) = \dfrac{3x}{x^2 + 1}$

90. $f(x) = x^3 - 3x + 1$

91. $f(x) = x^3 + 6x - 1$

92. $f(x) = x^4 - 4x^3 + 2x^2 - x - 1$

93. $f(x) = x^4 + 2x^3 - x^2 + x - 4$

Prepare for Section 2.3

94. If the graph of an equation passes through the x-axis, what is the y-coordinate of the point of intersection? [2.1]
0

95. If the graph of an equation passes through the y-axis, what is the x-coordinate of the point of intersection? [2.1]
0

96. Let $P(2, -1)$ and $Q(-3, 5)$ be two points in a plane.

a. What is the *horizontal* distance between the two points? [2.1] 5

b. What is the *vertical* distance between the two points? [2.1] 6

97. What are the coordinates of the midpoint of the line segment connecting the points $P(3, -2)$ and $Q(-1, -4)$? [2.1]
$(1, -3)$

98. What is the length of the line segment connecting the points $P(-3, 4)$ and $Q(-2, -3)$? [2.1] $5\sqrt{2}$

99. Describe the graph of all ordered pairs of the form $(x, 1)$, where x is a real number. [2.2]
A horizontal line passing through (0, 1)

Explorations 

1. ***Zeller's Congruence*** A formula known as *Zeller's Congruence* uses a function called the **greatest integer function** or **floor function.** The value of this function at x is the greatest integer less than or equal to x. The greatest integer function is symbolized by $\lfloor x \rfloor$. Here are some examples.

$$\lfloor 2.9 \rfloor = 2 \qquad \lfloor 7 \rfloor = 7$$

$$\left\lfloor \frac{1}{2} \right\rfloor = 0 \qquad \lfloor \pi \rfloor = 3$$

$$\lfloor -3.1 \rfloor = -4 \qquad \lfloor -2 \rfloor = -2$$

   a. Evaluate $\lfloor 4.7 \rfloor$. 4 **b.** Evaluate $\lfloor 6.1 \rfloor$. 6

   c. Evaluate $\lfloor 8 \rfloor$. 8 **d.** Evaluate $\lfloor -2.1 \rfloor$. −3

   Zeller's Congruence can be used to determine the day of the week on which a given date fell or will fall. The formula is

$$z = \left\lfloor \frac{13m - 1}{5} \right\rfloor + \left\lfloor \frac{y}{4} \right\rfloor + \left\lfloor \frac{c}{4} \right\rfloor + d + y - 2c$$

   where c is the first two digits of the year, y is the last two digits of the year, d is the day of the month, and m is the month. (For the month, March = 1, April = 2, . . . , December = 10. The months January and February are assigned the values 11 and 12.) Once z is calculated, that number is divided by 7. The remainder gives the day of the week, with 0 = Sunday, 1 = Monday, . . . , 6 = Saturday.

   For instance, to find the day of the week for July 4, 2009, use $c = 20$, $y = 9$, $d = 4$, and $m = 5$.

$$z = \left\lfloor \frac{13(5) - 1}{5} \right\rfloor + \left\lfloor \frac{9}{4} \right\rfloor + \left\lfloor \frac{20}{4} \right\rfloor + 4 + 9 - 2(20)$$

$$= \lfloor 12.8 \rfloor + \lfloor 2.25 \rfloor + \lfloor 5 \rfloor + 4 + 9 - 2(20)$$

$$= 12 + 2 + 5 + 4 + 9 - 40 = -8$$

   If we divide the result by 7, the remainder is −1. When a remainder is negative, add 7 to find the day of the week. In this case $-1 + 7 = 6$. July 4, 2009 falls on a Saturday.

   e. Find the day of the week for July 4, 1776. Thursday

   f. Find the day of the week for January 1, 2010. Friday

   g. Find the day of the week on which you were born. Answers will vary.

SECTION 2.3 Properties of Linear Functions

- Intercepts
- Slope of a Line
- Slope-Intercept Form of a Straight Line
- Parallel and Perpendicular Lines
- Equations of the Form $Ax + By = C$

▪ Intercepts

The graph in Figure 2.20 shows the pressure on a diver as the diver descends into the ocean. The equation of this graph can be represented by $P(d) = 64d + 2100$, where $P(d)$ is the pressure in pounds per square foot on a diver d feet below the surface of the ocean.

For instance, when $d = 2$, we have

$$P(d) = 64d + 2100$$

$$P(2) = 64(2) + 2100 \qquad \text{▪ Replace } d \text{ by 2.}$$

$$= 128 + 2100$$

$$= 2228$$

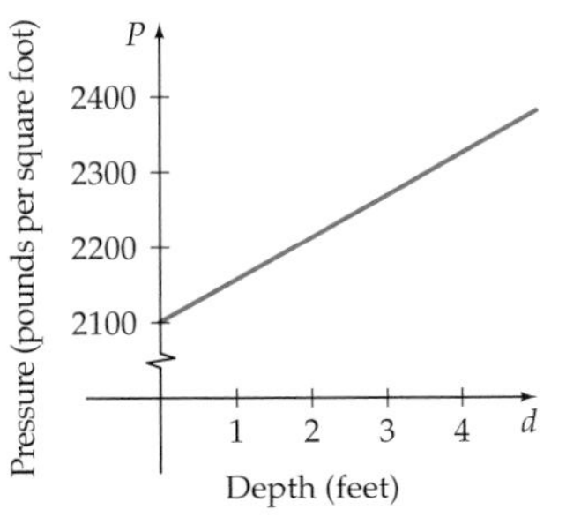

Figure 2.20

The pressure on a diver 2 feet below the ocean's surface is 2228 pounds per square foot.

The equation $P(d) = 64d + 2100$ is an example of a *linear function.*

TAKE NOTE

The jagged line along the vertical axis in the graph above indicates that a portion of the axis is not displayed.

Linear Function

A **linear function** is one that can be written in the form $f(x) = mx + b$ or $y = mx + b$, where m and b are constants.

Here are some examples.

$$y = 2x + 5 \qquad \blacksquare\ m = 2, b = 5$$

$$s(t) = \frac{2}{3}t - 1 \qquad \blacksquare\ m = \frac{2}{3}, b = -1$$

$$v = -2s \qquad \blacksquare\ m = -2, b = 0$$

$$y = 3 \qquad \blacksquare\ m = 0, b = 3$$

$$f(x) = 2 - 4x \qquad \blacksquare\ m = -4, b = 2$$

Note that different variables can be used to designate a linear function.

? QUESTION Which of the following are linear function?

a. $f(x) = 2x^2 + 5$ **b.** $f(x) = 1 - 3x$

Consider the equation $y = 2x + 4$. The graph of the equation is shown in Figure 2.21. From the graph we can see that when $x = -2$, $y = 0$, and the graph crosses the x-axis at $(-2, 0)$. The point $(-2, 0)$ is called the **x-intercept** of the graph.

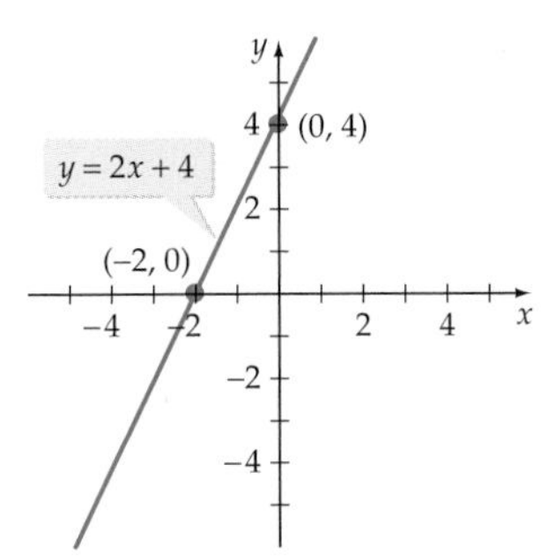

Figure 2.21

When $x = 0$, $y = 4$, and the graph crosses the y-axis at $(0, 4)$. The point $(0, 4)$ is called the **y-intercept** of the graph.

? ANSWER **a.** Because $2x^2 + 5$ is not a first-degree polynomial, $y = 2x^2 + 5$ is not a linear function. **b.** Because $1 - 3x$ is a first-degree polynomial, $y = 1 - 3x$ is a linear function.

Alternative to Example 1

Find the x- and y-intercepts of the graph of $y = 5x - 6$.

■ $\left(\frac{6}{5}, 0\right)$; $(0, -6)$

EXAMPLE 1 Find x- and y-intercepts

Find the x- and y-intercepts of the graph of $y = -3x + 2$.

Solution When a graph crosses the x-axis, the y-coordinate of the point is 0. Therefore, to find the x-intercept, replace y by 0 and solve the equation for x.

$$y = -3x + 2$$

$$0 = -3x + 2 \quad \text{■ Replace } y \text{ by 0.}$$

$$-2 = -3x \quad \text{■ Solve for } x.$$

$$\frac{2}{3} = x$$

The x-intercept is $\left(\frac{2}{3}, 0\right)$.

When a graph crosses the y-axis, the x-coordinate of the point is 0. Therefore, to find the y-intercept, replace x by 0 and solve for y.

$$y = -3x + 2$$

$$y = -3(0) + 2 \quad \text{■ Replace } x \text{ by 0. Then simplify.}$$

$$= 2$$

The y-intercept is $(0, 2)$.

CHECK YOUR PROGRESS 1 Find the x- and y-intercepts of the graph of $y = \frac{1}{2}x + 3$.

Solution *See page S12.* $(-6, 0)$; $(0, 3)$

TAKE NOTE

To find the y-intercept of $y = mx + b$, let $x = 0$. Then

$$y = mx + b$$

$$y = m(0) + b$$

$$= b$$

The y-intercept is $(0, b)$.

In Example 1, note that the y-coordinate of the y-intercept of $y = -3x + 2$ has the same value as b in the equation $y = mx + b$. This is always true. The y-intercept of the graph of $y = mx + b$ is $(0, b)$. (See the Take Note at the left.)

If we replace d by 0 in the linear equation that modeled the pressure on a diver, we have

$$P(d) = 64d + 2100$$

$$P(0) = 64(0) + 2100 = 2100$$

In this case, the P-intercept (the intercept on the vertical axis) is $(0, 2100)$. In the context of this application, this means that the pressure on a diver 0 feet below the ocean's surface is 2100 pounds per square foot. Another way of saying 0 feet below the ocean's surface is "at sea level." Thus, the pressure on a diver, or anyone else for that matter, is 2100 pounds per square foot at sea level.

Both the x- and y-intercepts can have meaning in application problems. This is demonstrated in the next example.

Alternative to Example 2
Exercise 60, page 207

EXAMPLE 2 Solve an Application Involving Intercepts

After a parachute is deployed, an equation that models the height of the parachutist above the ground is $s = -10t + 2800$, where s is the height (in feet) of the parachutist t seconds after the chute is deployed. Find the intercepts on the vertical and horizontal axes and explain what they mean in the context of the problem.

Solution To find the intercept on the vertical axis, replace t by 0 and solve for s.

$$s = -10t + 2800$$
$$s = -10(0) + 2800 = 2800$$

The intercept on the vertical axis is (0, 2800). This means that the parachutist was 2800 feet above the ground when the parachute was deployed.

To find the intercept on the horizontal axis, set $s = 0$ and solve for t.

$$\begin{aligned} s &= -10t + 2800 \\ 0 &= -10t + 2800 \\ -2800 &= -10t \\ 280 &= t \end{aligned}$$

The intercept on the horizontal axis is (280, 0). This means that the parachutist reached the ground ($s = 0$) 280 seconds after the parachute was deployed.

CHECK YOUR PROGRESS 2 An equation that models the altitude of a certain small plane as it descends to an airport is given by $s = -20t + 8000$, where s is the height (in feet) of the plane above the airport t seconds after it begins its descent. Find the intercepts on the vertical and horizontal axes and explain what they mean in the context of the problem.

Solution *See page S12.* t-intercept: (400, 0). This means that the plane will land in 400 s. s-intercept: (0, 8000). This means that the plane was at an altitude of 8000 ft before starting to descend.

■ Slope of a Line

Consider again the equation $P(d) = 64d + 2100$ that models the pressure on a diver as the diver descends below the ocean's surface. From Figure 2.22, note that as the depth of the diver increases by 1 foot, the pressure on the diver increases by 64 pounds per square foot. This can be verified algebraically.

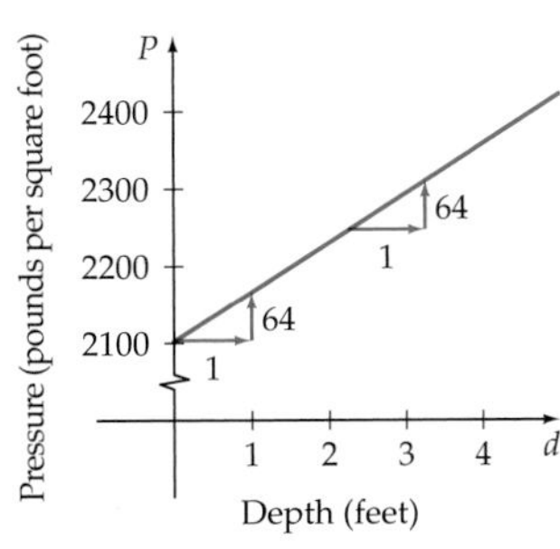

Figure 2.22

$P(0) = 64(0) + 2100 = 2100$ ■ Pressure at sea level

$P(1) = 64(1) + 2100 = 2164$ ■ Pressure after descending 1 foot

$2164 - 2100 = 64$ ■ Change in pressure

If we choose two other depths that differ by 1 foot, such as 2.5 and 3.5, the change in pressure is still 64 pounds per square foot.

$P(2.5) = 64(2.5) + 2100 = 2260$ ■ Pressure at 2.5 feet below surface

$P(3.5) = 64(3.5) + 2100 = 2324$ ■ Pressure at 3.5 feet below surface

$2324 - 2260 = 64$ ■ Change in pressure

The value of the *slope* of a line is the change in the vertical direction caused by a 1-unit increase in the horizontal direction. For $P(d) = 64d + 2100$, the slope is 64. In the context of this problem, the slope means that the pressure on the diver increases by 64 pounds per square foot for each additional 1 foot the diver descends.

The slope of a line can be calculated by using the coordinates of any two points on the line, not just points whose x-coordinates differ by 1.

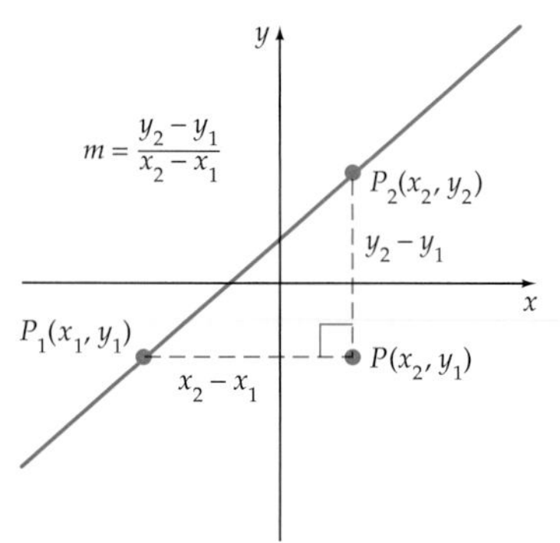

Figure 2.23

Slope Formula

Let $P_1(x_1, y_1)$ and $P_2(x_2, y_2)$ be two points on a line, as in Figure 2.23. Then the **slope**, m, of the line through the two points is the ratio of the change in the y-coordinates to the change in the x-coordinates.

$$\text{Slope} = m = \frac{\text{change in } y}{\text{change in } x} = \frac{y_2 - y_1}{x_2 - x_1}, \; x_1 \neq x_2$$

QUESTION Why is the restriction $x_1 \neq x_2$ required in the definition of slope?

Alternative to Example 3

Find the slope between the two points $(-1, -3)$ and $(4, -1)$.

■ $\frac{2}{5}$

EXAMPLE 3 Find the Slope of a Line

Find the slope of the line between the two points.

a. $P_1(-4, -3)$ and $P_2(-1, 1)$ **b.** $P_1(-2, 3)$ and $P_2(1, -3)$

c. $P_1(-1, -3)$ and $P_2(4, -3)$ **d.** $P_1(4, 3)$ and $P_2(4, -1)$

Solution

a. $(x_1, y_1) = (-4, -3), (x_2, y_2) = (-1, 1)$

$$m = \frac{y_2 - y_1}{x_2 - x_1} = \frac{1 - (-3)}{-1 - (-4)} = \frac{4}{3}$$

The slope is $\frac{4}{3}$. A *positive* slope indicates that the line slants *upward* to the right. For this particular line, the value of y *increases* by $\frac{4}{3}$ when x increases by 1.

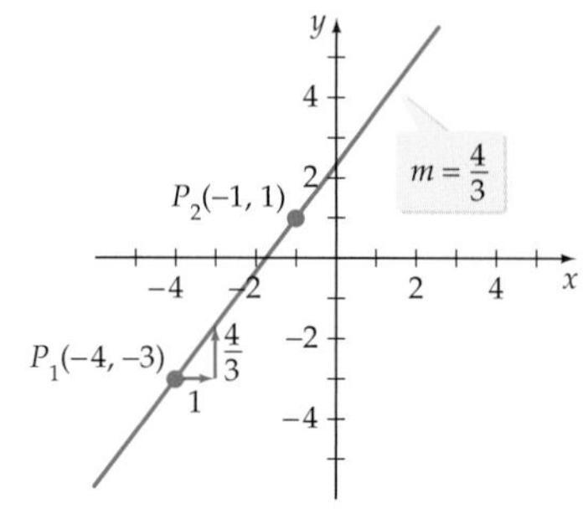

b. $(x_1, y_1) = (-2, 3), (x_2, y_2) = (1, -3)$

$$m = \frac{y_2 - y_1}{x_2 - x_1} = \frac{-3 - 3}{1 - (-2)} = \frac{-6}{3} = -2$$

The slope is -2. A *negative* slope indicates that the line slants *downward* to the right. For this particular line, the value of y *decreases* by 2 when x increases by 1.

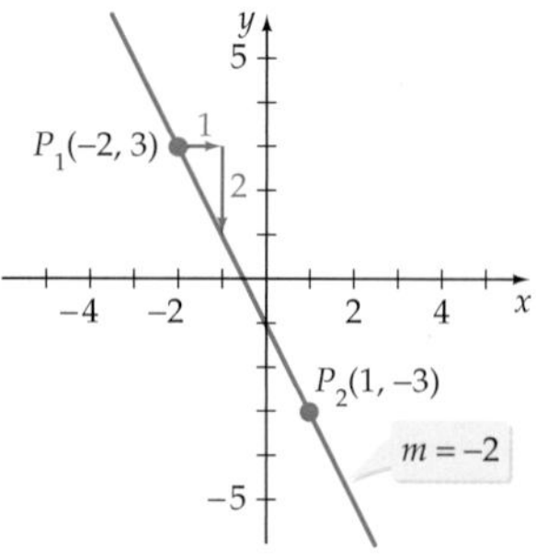

Continued ➤

ANSWER If $x_1 = x_2$, then the difference $x_2 - x_1 = 0$. This would make the denominator 0, and division by 0 is undefined.

c. $(x_1, y_1) = (-1, -3), (x_2, y_2) = (4, -3)$

$$m = \frac{y_2 - y_1}{x_2 - x_1} = \frac{-3 - (-3)}{4 - (-1)} = \frac{0}{5} = 0$$

The slope is 0. A *zero* slope indicates that the line is *horizontal*. For this particular line, the value of y stays the same when x increases by 1.

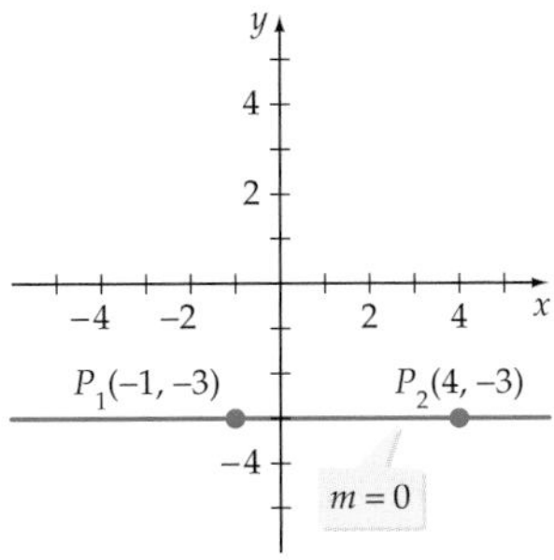

TAKE NOTE

Here is a summary of the results in Example 3.
A line with *positive* slope slants upward to the right.
A line with *negative* slope slants downward to the right.
A horizontal line has 0 slope.
A vertical line has no slope.

d. $(x_1, y_1) = (4, 3), (x_2, y_2) = (4, -1)$

$$m = \frac{y_2 - y_1}{x_2 - x_1} = \frac{-1 - 3}{4 - 4} = \frac{-4}{0}$$ ▪ Undefined

There is no slope. If the denominator of the slope formula is zero, the line has *no slope*. Sometimes we say that the slope of the line is *undefined*.

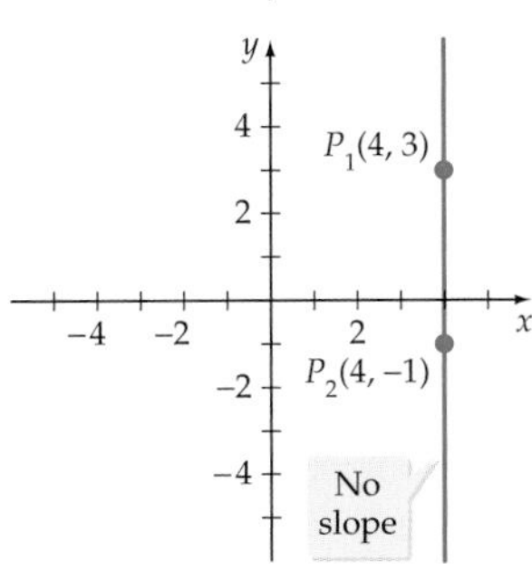

CHECK YOUR PROGRESS 3 Find the slope of the line between the two points.

a. $P_1(-6, 5)$ and $P_2(4, -5)$ b. $P_1(-5, 0)$ and $P_2(-5, 7)$
c. $P_1(-7, -2)$ and $P_2(8, 8)$ d. $P_1(-6, 7)$ and $P_2(1, 7)$

Solution *See page S12.* a. -1 b. Undefined c. $\frac{2}{3}$ d. 0

▪ Slope-Intercept Form of a Straight Line

Let $y = 2x - 3$ and choose any two values of x. We will use -1 and 4, but any two different values of x could be used. Using these two values of x, we can find two points on the graph of $y = 2x - 3$.

$$\begin{aligned} y &= 2x - 3 \\ &= 2(-1) - 3 \quad \text{▪ Replace } x \text{ by } -1. \\ &= -5 \end{aligned}$$

One point is $(-1, -5)$.

$$\begin{aligned} y &= 2x - 3 \\ &= 2(4) - 3 \quad \text{▪ Replace } x \text{ by } 4. \\ &= 5 \end{aligned}$$

One point is $(4, 5)$.

The graph of the line and the two points on the line are shown in Figure 2.24. If we use these two points to calculate the slope, we have

$$m = \frac{y_2 - y_1}{x_2 - x_1} = \frac{5 - (-5)}{4 - (-1)} = \frac{10}{5} = 2$$

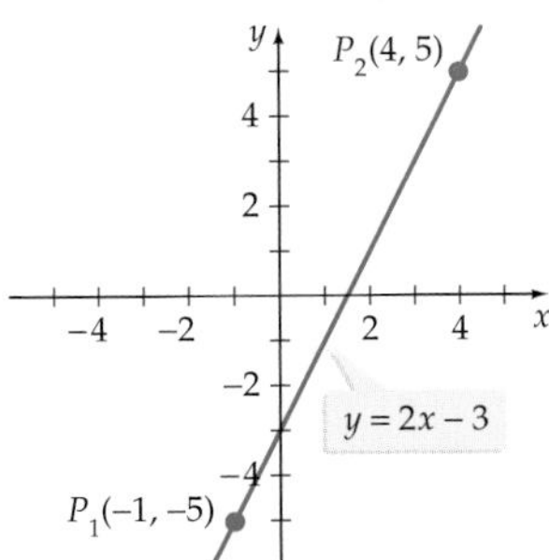

Figure 2.24

Note that the value of the slope is 2 and the coefficient of x in the equation $y = 2x - 3$ is 2. This is not a coincidence. The slope of the graph of $y = mx + b$ is m, the coefficient of x. Recall from an earlier discussion that the y-intercept of $y = mx + b$ is $(0, b)$. Combining the last two sentences gives us the following.

Slope-Intercept Form of the Equation of a Line

The equation $y = mx + b$ is called the **slope-intercept form** of a straight line. The graph of $y = mx + b$ is a line with slope m and y-intercept $(0, b)$.

QUESTION What is the slope and y-intercept of each of the following?

a. $y = -2x + 3$ **b.** $y = x + 4$ **c.** $y = 3 - 4x$

Alternative to Example 4
Exercise 40, page 207

EXAMPLE 4 Graph a Line Using the Slope and y-Intercept

Graph $y = -\frac{2}{3}x + 4$ by using the slope and y-intercept.

Solution From the equation, the slope is $-\frac{2}{3}$ and the y-intercept is $(0, 4)$. Place a dot at the y-intercept. We can write the slope as

$$m = -\frac{2}{3} = \frac{-2}{3} = \frac{\text{change in } y}{\text{change in } x}$$

Starting from the y-intercept, move 2 units down and 3 units to the right and place another point. Now draw a line through the two points.

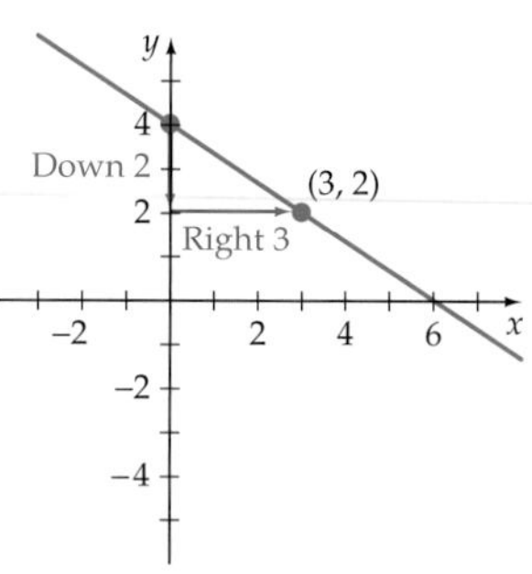

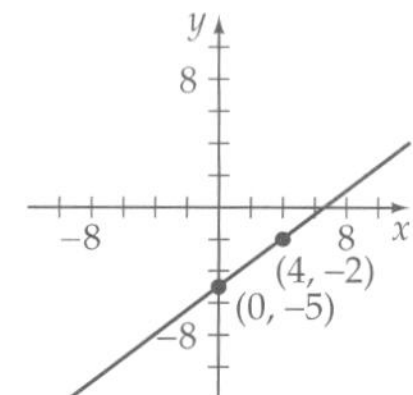

CHECK YOUR PROGRESS 4 Graph $y = \frac{3}{4}x - 5$ by using the slope and y-intercept.

Solution *See page S12.*

In Example 4, we used the slope and y-intercept to graph a line. We can also graph a line given the slope and any point on the line.

Alternative to Example 5
Exercise 74, page 209

EXAMPLE 5 Graph a Line Given Its Slope and a Point

Draw the line that passes through $P(-2, 4)$ and has slope $-\frac{3}{4}$.

Solution Place a dot at $(-2, 4)$ and then rewrite $-\frac{3}{4}$ as $\frac{-3}{4}$. Starting from $(-2, 4)$, move 3 units down (the change in y) and then 4 units to the right (the change in x). Place a dot at that location and then draw a line through the two points.

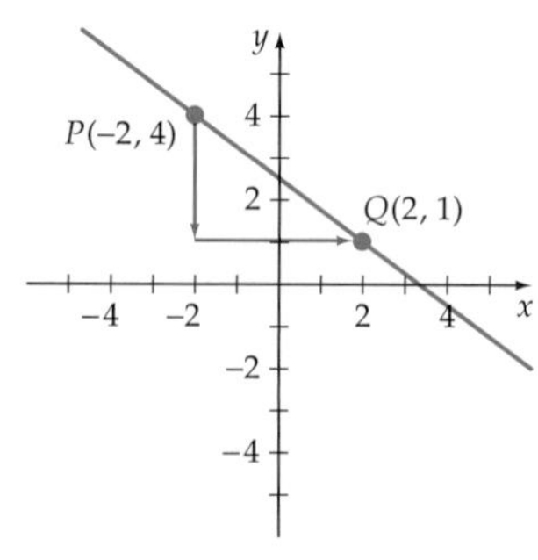

Continued ➤

ANSWER **a.** Slope: −2, y-intercept: (0, 3) **b.** Slope: 1, y-intercept: (0, 4) **c.** Slope: −4, y-intercept: (0, 3)

CHECK YOUR PROGRESS 5 Draw the line that passes through $P(2, 4)$ and has slope -1.

Solution *See page S12.*

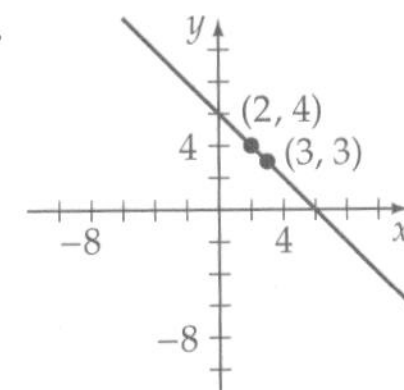

Recall that the value of the slope is the change in y for a *1-unit* change in x. For instance, a slope of -3 means that y decreases by 3 as x increases by 1; a slope of 2 means that y increases by 2 as x increases by 1; and a slope of $\frac{4}{3}$ means that y increases by $\frac{4}{3}$ as x increases by 1.

Because y is changing as x changes, we frequently refer to slope as a measure of the *rate of change* of y with respect to x. For instance, if a line has a slope of 2, then the rate of change of y with respect to x is 2. This means that y increases by 2 as x increases by 1. If the slope is $-\frac{1}{2}$, then the rate of change of y with respect to x is $-\frac{1}{2}$. This means that y decreases by $\frac{1}{2}$ as x increases by 1. It is because slope measures a *rate of change* that it is useful in many applications.

TAKE NOTE

Speed is a measure of the change in distance for a change in time. Six miles per hour means the distance changes by 6 miles for a 1-hour change in time. This is the same as saying that y increases by 6 as x increases by 1, or that the slope is $\frac{6}{1} = 6$. The *rate of change* of distance with respect to time is 6 miles per hour. The slope of the graph is 6.

Suppose a jogger is running at a constant speed of 6 miles per hour. Then the equation $d = 6t$ relates the time running t to the distance traveled d. An input/output table is shown below.

***t*, in hours**	0	0.5	1	1.5	2	2.5
***d*, in miles**	0	3	6	9	12	15

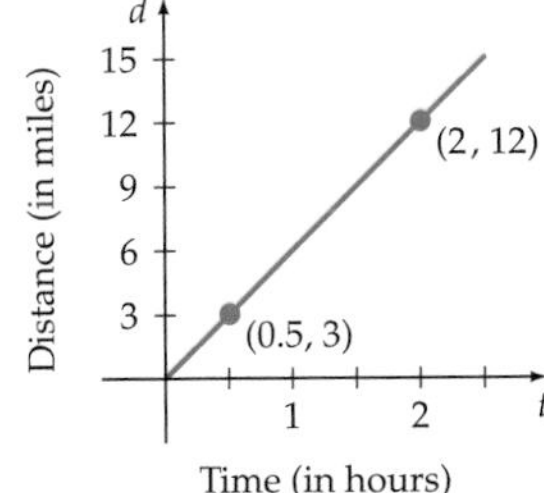

Because $d = 6t$ is a linear equation in two variables, the slope of the graph is 6. This can be confirmed by choosing any two points on the graph above and finding the slope of the line between the two points. The points (0.5, 3) and (2, 12) are used here.

$$m = \frac{\text{change in } d}{\text{change in } t} = \frac{12 \text{ miles} - 3 \text{ miles}}{2 \text{ hours} - 0.5 \text{ hours}} = \frac{9 \text{ miles}}{1.5 \text{ hours}} = 6 \text{ miles per hour}$$

This example demonstrates that the slope of the graph of an object in uniform motion is equal to the speed of the object. In a more general sense, any time we discuss the speed of an object, we are discussing the slope of the graph that describes the relationship between the distance the object travels and the time it travels.

Alternative to Example 6
Exercise 58, page 207

EXAMPLE 6 Solve an Application Involving Slope

The equation $T = -6.5x + 20$ approximates the temperature T, in degrees Celsius, x kilometers above sea level. What is the slope of the graph of this equation? Write a sentence that explains the meaning of the slope in the context of the problem.

Solution For the graph of $T = -6.5x + 20$, the slope is the coefficient of x. Therefore, the slope is -6.5. The value of the slope means that the temperature is decreasing (because the slope is negative) 6.5°C for each 1 kilometer increase in height above sea level.

CHECK YOUR PROGRESS 6 The distance d, in miles, a homing pigeon can fly in t hours can be approximated by $d = 50t$. Find the slope of the graph of this equation. What is the meaning of the slope in the context of the problem?

Solution *See page S13.* Slope: 50; The speed of the homing pigeon is 50 mph.

▪ Parallel and Perpendicular Lines

The graphs of $y = 2x - 3$ and $y = 2x + 4$ are shown in Figure 2.25. Note from the equations that the lines have the same slope and different y-intercepts. Different lines that have the same slope are called **parallel lines;** the graphs of the lines never intersect.

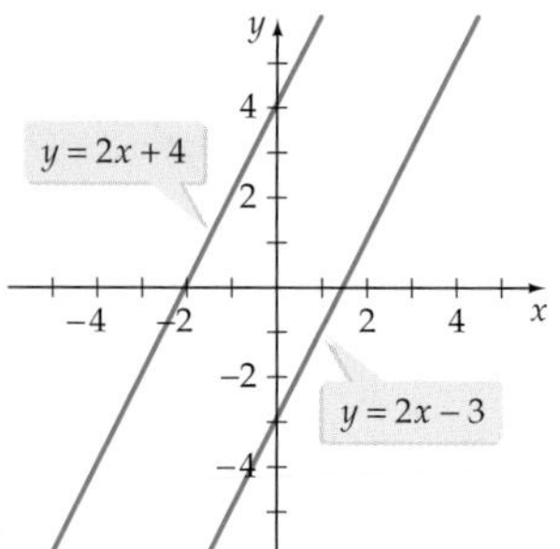

Figure 2.25

Parallel Lines

Two nonvertical lines with slopes m_1 and m_2 are parallel lines if and only if $m_1 = m_2$. Vertical lines are also parallel lines, but they have no slope.

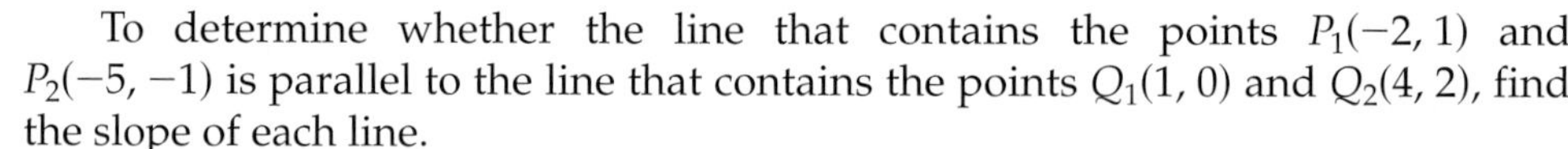

To determine whether the line that contains the points $P_1(-2, 1)$ and $P_2(-5, -1)$ is parallel to the line that contains the points $Q_1(1, 0)$ and $Q_2(4, 2)$, find the slope of each line.

$$\text{Slope between } P_1 \text{ and } P_2\text{: } m_1 = \frac{y_2 - y_1}{x_2 - x_1} = \frac{-1 - 1}{-5 - (-2)} = \frac{-2}{-3} = \frac{2}{3}$$

$$\text{Slope between } Q_1 \text{ and } Q_2\text{: } m_2 = \frac{y_2 - y_1}{x_2 - x_1} = \frac{2 - 0}{4 - 1} = \frac{2}{3}$$

The slopes are equal ($m_1 = m_2$). Therefore, the lines are parallel.

QUESTION Is the graph of $y = -\frac{1}{2}x + 2$ parallel to the graph of $y = -x + 2$?

ANSWER No. The slope of one line is $-\frac{1}{2}$ and the slope of the other line is -1. The slopes are not equal, so the graphs are not parallel.

Two lines that intersect at right angles are called **perpendicular lines.** The following theorem enables us to determine whether the graphs of two lines are perpendicular.

> **Slopes of Perpendicular Lines**
>
> If m_1 and m_2 are the slopes of two lines, neither of which is vertical, then the lines are perpendicular if and only if $m_1m_2 = -1$. A vertical line is perpendicular to a horizontal line.

Solving $m_1m_2 = -1$ for m_1 gives $m_1 = -\frac{1}{m_2}$. This last equation states that the slopes of perpendicular lines are negative reciprocals of each other.

To determine whether the line through $P_1(4, 2)$ and $P_2(-2, 5)$ is perpendicular to the line containing $Q_1(-4, 3)$ and $Q_2(-3, 5)$, find the slope of each line.

INSTRUCTOR NOTE

Ask students to determine whether the line through $P_1(3, -4)$ and $P_2(6, -2)$ is perpendicular to the line containing $Q_1(-4, 1)$ and $Q_2(-1, -1)$.
[Ans: The slopes are $\frac{2}{3}$ and $-\frac{2}{3}$ and therefore the lines are not perpendicular.]

$$\text{Slope between } P_1 \text{ and } P_2\text{: } m_1 = \frac{y_2 - y_1}{x_2 - x_1} = \frac{5 - 2}{-2 - 4} = \frac{3}{-6} = -\frac{1}{2}$$

$$\text{Slope between } Q_1 \text{ and } Q_2\text{: } m_2 = \frac{y_2 - y_1}{x_2 - x_1} = \frac{5 - 3}{-3 - (-4)} = \frac{2}{1} = 2$$

Multiply the slopes. $m_1m_2 = \left(-\frac{1}{2}\right)2 = -1$. The product of the slopes is -1. The graphs of the lines are perpendicular. Note also that $-\frac{1}{2}$ and 2 are negative reciprocals of each other.

INTEGRATING TECHNOLOGY

When using a graphing calculator to graph lines, the resulting graph may give the impression that two lines are not perpendicular when in fact they are. This is due to the size of pixels (picture elements). For instance, the graphs of $y = -\frac{1}{2}x + 4$ and $y = 2x - 3$ are shown at the right in the standard viewing window. Although the graphs should be perpendicular, they do not appear to be from the graph. This distortion can be fixed by using the SQUARE viewing window on your calculator. A sample setting of this window on a TI-83 is shown at the left.

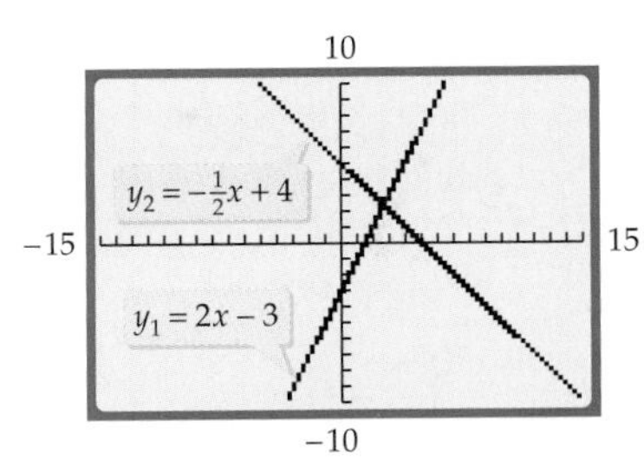

ZOOM MEMORY
1: ZBox
2: Zoom In
3: Zoom Out
4: ZDecimal
5: ZSquare
6: ZStandard
7↓ ZTrig

WINDOW
Xmin = -15.16129...
Xmax = 15.16129...
Xscl = 1
Ymin = -10
Ymax = 10
Yscl = 1
Xres = 1

The graphs of $y = -\frac{1}{2}x + 4$ and $y = 2x - 3$ are shown at the right in the SQUARE viewing window. Note that the graphs now appear to be perpendicular. The appearance of a graph is influenced by the values of Xmin, Xmax, Ymin, and Ymax.

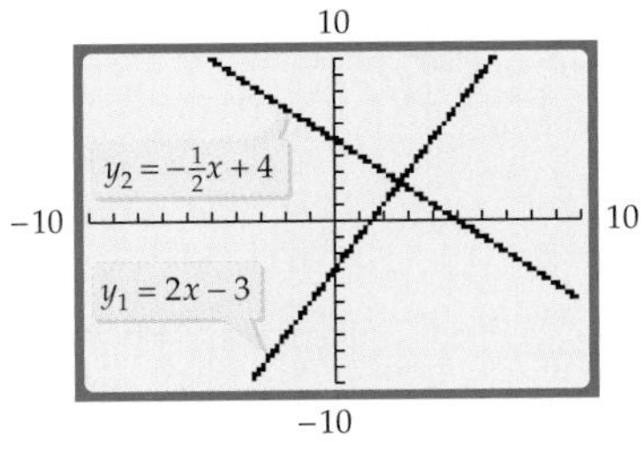

▪ Equations of the Form $Ax + By = C$

Sometimes the equation of a line is written in the form $Ax + By = C$. This is called the **standard form of the equation of a line.** For instance, $3x + 4y = 12$ is in standard form. If an equation is in standard form, we can solve for y and write the equation in slope-intercept form.

$$
\begin{aligned}
3x + 4y &= 12 && \text{▪ Standard form}\\
3x - 3x + 4y &= -3x + 12 && \text{▪ Subtract } 3x \text{ from each side.}\\
4y &= -3x + 12 &&\\
\frac{4y}{4} &= \frac{-3x + 12}{4} && \text{▪ Divide each side by 4.}\\
y &= -\frac{3}{4}x + 3 && \text{▪ Slope-intercept form}
\end{aligned}
$$

Once the equation is written in slope-intercept form, we can use the technique of Example 4 to graph the equation. The graph is shown in Figure 2.26.

Another way to create the graph of a line when the equation is in standard form is to find the x- and y-intercepts. This method is shown below.

Figure 2.26

To find the x-intercept, let $y = 0$ and then solve for x.

$$
\begin{aligned}
3x + 4y &= 12\\
3x + 4(0) &= 12\\
3x &= 12\\
x &= 4
\end{aligned}
$$

The x-intercept is $(4, 0)$.

To find the y-intercept, let $x = 0$ and then solve for y.

$$
\begin{aligned}
3x + 4y &= 12\\
3(0) + 4y &= 12\\
4y &= 12\\
y &= 3
\end{aligned}
$$

The y-intercept is $(0, 3)$.

Plot the intercepts and draw a line through the points, as shown in Figure 2.26.

Alternative to Example 7
Exercise 48, page 207 with the additional text ". . . by finding the x- and y-intercepts."

EXAMPLE 7 Graph a Line by Using Its Intercepts

Graph $2x - 5y = 10$ by finding the x- and y-intercepts.

Solution

To find the x-intercept, let $y = 0$ and solve for x.

$$
\begin{aligned}
2x - 5y &= 10\\
2x - 5(0) &= 10\\
2x &= 10\\
x &= 5
\end{aligned}
$$

The x-intercept is $(5, 0)$.

To find the y-intercept, let $x = 0$ and solve for y.

$$
\begin{aligned}
2x - 5y &= 10\\
2(0) - 5y &= 10\\
-5y &= 10\\
y &= -2
\end{aligned}
$$

The y-intercept is $(0, -2)$.

Plot the intercepts and draw a line through the two points.

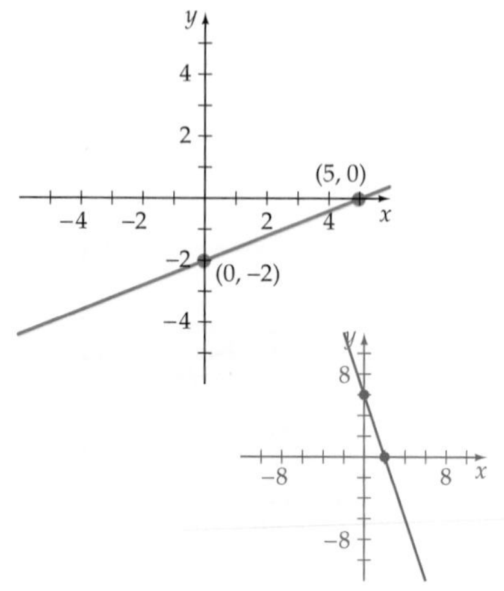

CHECK YOUR PROGRESS 7 Graph $3x + y = 6$ by finding the x- and y-intercepts.

Solution *See page S13.* x-intercept $(2, 0)$; y-intercept $(0, 6)$

A linear equation in which one of the variables is missing has a graph that is either a horizontal or a vertical line. The equation $y = -2$ can be written in standard form as

$$0x + y = -2 \qquad \blacksquare\ A = 0, B = 1, \text{ and } C = -2$$

Because $0x = 0$ for all values of x, the value of y is -2 for all values of x.

Some of the possible ordered-pair solutions of $y = -2$ are given in the table below. The graph is the horizontal line shown at the right.

x	−4	−1.5	0	1	3
y	−2	−2	−2	−2	−2

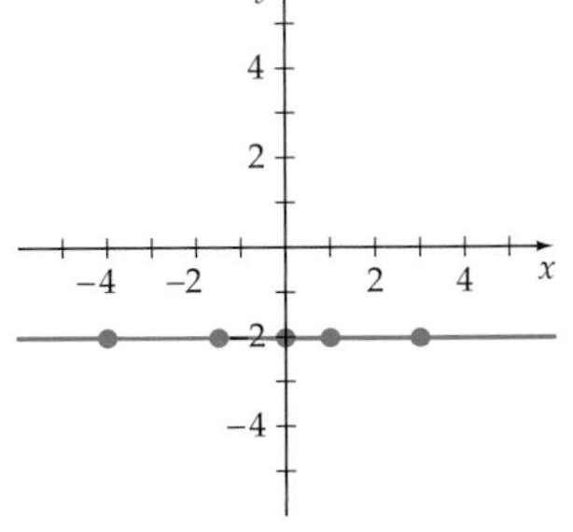

For the equation $x = 3$, the coefficient of y is zero. The equation $x = 3$ can be written in standard form as

$$x + 0y = 3 \qquad \blacksquare\ A = 1, B = 0, \text{ and } C = 3.$$

No matter what value of y is chosen, $0y = 0$ and therefore x is always 3.

Some of the possible ordered-pair solutions of $x = 3$ are given in the table below. The graph is the vertical line shown at the right.

x	3	3	3	3	3
y	−3	−2	0	1	4

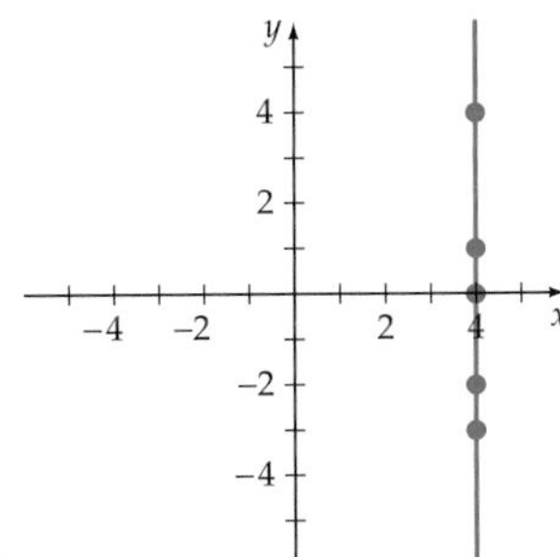

Alternative to Example 8
Exercise 52, page 207

y
4
2
−4 −2
2 4 x
−2
−4

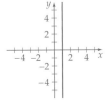

EXAMPLE 8 Graph a Horizontal or Vertical Line

Graph: $y + 1 = 0$

Solution Solve for y.

$$y + 1 = 0$$
$$y = -1$$

The graph is a horizontal line through $(0, -1)$.

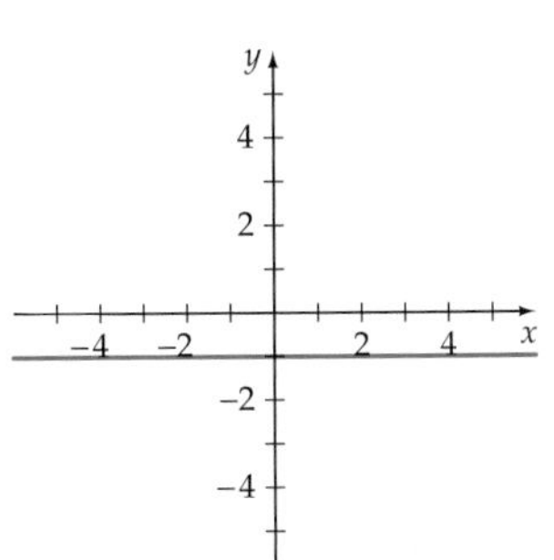

CHECK YOUR PROGRESS 8 Graph: $x - 1 = 0$

Solution *See page S13.*

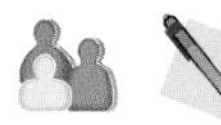

Topics for Discussion

1. Give an example of a linear equation in two variables that
   a. is in slope-intercept form.
   b. is in standard form.
2. What is the x-coordinate of the y-intercept? What is the y-coordinate of the x-intercept?
3. What is the slope of the line whose equation is $y = -2x + 3$?
4. Explain how to find the slope of a line given two points on the line.
5. If two lines are parallel, what can be said about their slopes?
6. If two nonvertical lines are perpendicular, what can be said about their slopes?
7. Give two examples of real-world quantities that are mathematically treated as slopes.

EXERCISES 2.3

— *Suggested Assignment: Exercises 1–69, odd; 71–83, every other odd; and 84–89, all.*
— *Answer graphs to Exercises 35–54 and 74–77 are on page AA4.*

In Exercises 1 to 14, find the x- and y-intercepts of the graph of the equation.

1. $y = 3x - 6$ (2, 0), (0, −6)
2. $y = 2x + 8$ (−4, 0), (0, 8)
3. $y = \frac{2}{3}x - 4$ (6, 0), (0, −4)
4. $y = -\frac{3}{4}x + 6$ (8, 0), (0, 6)
5. $y = -x - 4$ (−4, 0), (0, −4)
6. $y = -\frac{x}{2} + 1$ (2, 0), (0, 1)
7. $3x + 4y = 12$ (4, 0), (0, 3)
8. $5x - 2y = 10$ (2, 0), (0, −5)
9. $2x - 3y = 9$ (4.5, 0), (0, −3)
10. $4x + 3y = 8$ (2, 0), $\left(0, \frac{8}{3}\right)$
11. $\frac{x}{2} + \frac{y}{3} = 1$ (2, 0), (0, 3)
12. $\frac{x}{3} - \frac{y}{2} = 1$ (3, 0), (0, −2)
13. $x - \frac{y}{2} = 1$ (1, 0), (0, −2)
14. $-\frac{x}{4} + \frac{y}{3} = 1$ (−4, 0), (0, 3)

In Exercises 15 to 26, find the slope of the line containing the points P_1 and P_2.

15. $P_1(1, 3), P_2(3, 1)$ −1
16. $P_1(2, 3), P_2(5, 1)$ $-\frac{2}{3}$
17. $P_1(-1, 4), P_2(2, 5)$ $\frac{1}{3}$
18. $P_1(3, -2), P_2(1, 4)$ −3
19. $P_1(-1, 3), P_2(-4, 5)$ $-\frac{2}{3}$
20. $P_1(-1, -2), P_2(-3, 2)$ −2
21. $P_1(0, 3), P_2(4, 0)$ $-\frac{3}{4}$
22. $P_1(-2, 0), P_2(0, 3)$ $\frac{3}{2}$
23. $P_1(2, 4), P_2(2, -2)$ No slope
24. $P_1(4, 1), P_2(4, -3)$ No slope
25. $P_1(2, 3), P_2(-1, 3)$ 0
26. $P_1(3, 4), P_2(0, 4)$ 0

In Exercises 27 to 34, determine whether the line through points P_1 and P_2 is parallel to, perpendicular to, or neither parallel nor perpendicular to the line through points Q_1 and Q_2.

27. $P_1(1, 3), P_2(3, 1); Q_1(2, -1), Q_2(4, 1)$ Perpendicular
28. $P_1(-1, 4), P_2(2, 5); Q_1(2, 4), Q_2(3, 1)$ Perpendicular
29. $P_1(-1, 3), P_2(-4, 5); Q_1(1, 4), Q_2(-1, 7)$ Neither
30. $P_1(0, 3), P_2(4, 0); Q_1(2, -3), Q_2(-2, -6)$ Neither
31. $P_1(-3, -4), P_2(2, 2); Q_1(-4, 1), Q_2(-9, 7)$ Neither
32. $P_1(2, 3), P_2(-1, 5); Q_1(5, -2), Q_2(2, 0)$ Parallel
33. $P_1(-1, 0), P_2(5, -1); Q_1(3, 5), Q_2(-3, 6)$ Parallel
34. $P_1(7, -4), P_2(-2, -1); Q_1(6, -3), Q_2(4, -9)$ Perpendicular

In Exercises 35 to 44, graph the equation using the slope and y-intercept.

35. $y = 2x - 4$ **36.** $y = -3x + 4$

37. $y = -x + 3$ **38.** $y = x - 2$

39. $y = \dfrac{2x}{3} - 3$ **40.** $y = \dfrac{1}{2}x + 2$

41. $y = -\dfrac{3}{2}x$ **42.** $y = \dfrac{3}{4}x$

43. $y = \dfrac{x}{3} - 1$ **44.** $y = -\dfrac{3}{2}x + 6$

In Exercises 45 to 54, graph the equation.

45. $2x + 3y = -6$ **46.** $3x - 4y = 12$

47. $2x - 5y = 10$ **48.** $x + 3y = 6$

49. $x + 4 = 0$ **50.** $y - 3 = 0$

51. $y + 2 = 0$ **52.** $x - 2 = 0$

53. $3x + y = 9$ **54.** $2x - 3y = 12$

55. Vertical intercept: (0, 100,000), horizontal intercept: (40, 0). The vertical intercept represents the amount in the account ($100,000) when withdrawals begin. The horizontal intercept means that after 40 months, there is $0 remaining in the account.

Business and Economics

55. ***Retirement Planning*** A retired biologist begins withdrawing money from a retirement account according to the linear function $A(t) = 100{,}000 - 2500t$, where A is the amount remaining in the account t months after withdrawals begin. Find and discuss the meaning of the intercepts on the vertical and horizontal axes.

56. ***Fuel Prices*** The increase in fuel prices for the first 6 months of a recent year are shown in the graph below. Find the slope of the line between the two points shown on the following graph. Write a sentence that states the meaning of the slope in the context of this problem.

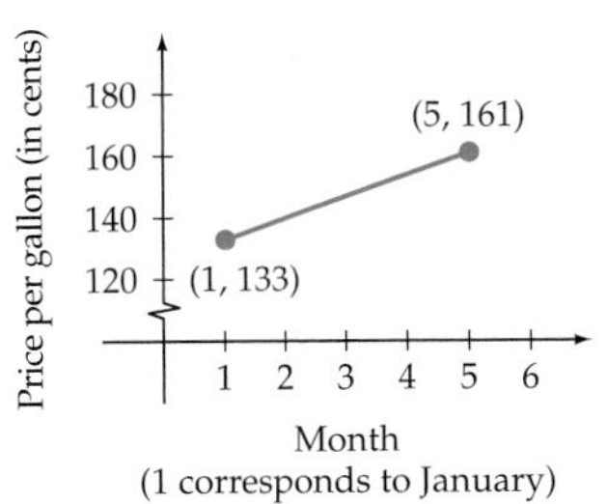

Slope = 7. The slope means that fuel prices are increasing 7 cents per month.

57. ***Business Travel*** The function $s(t) = 6000 - 500t$ gives the remaining distance a business executive must travel during a flight from Los Angeles to Paris, where s is the remaining distance, in miles, t hours after the flight begins. Find and discuss the meaning of the intercepts on the vertical and horizontal axes. Vertical intercept: (0, 6000), horizontal intercept: (12, 0). The vertical intercept means that Paris is 6000 mi from Los Angeles. The horizontal intercept means that after 12 h, the jet will arrive in Paris.

58. ***Graphic Arts*** The graph below shows the relationship between the size of an advertising brochure and the time required to print the brochure using a color laser printer. Find the slope of the line between the two points shown on the graph. Write a sentence that states the meaning of the slope in the context of this problem.

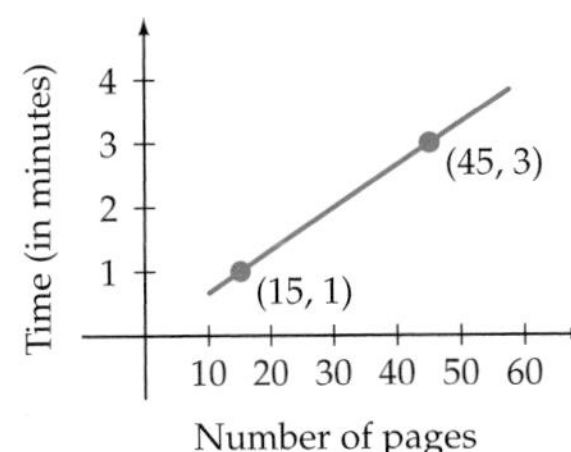

Slope $= \dfrac{1}{15}$. The slope means that printing takes $\dfrac{1}{15}$ minute per page.

Life and Health Sciences

59. ***Cricket Chirps*** There is a relationship between the number of times a cricket chirps per minute and the air temperature. A linear equation that models this relationship is given by $y = 7x - 30$, where x is the temperature in degrees Celsius and y is the number of chirps per minute. Find and discuss the meaning of the x-intercept. x-intercept $= \left(\dfrac{30}{7}, 0\right)$. The x-intercept represents the temp. at which chirping stops. Below this temp. chirping is negative and meaningless.

60. ***Food Science*** A can of frozen orange juice is taken from a freezer and gradually warmed. The equation $T = 3x - 6$ gives the Celsius temperature T of the juice x minutes after it is removed from the freezer. Find and discuss the meaning of the intercepts on the vertical and horizontal axes.

60. T-intercept: (0, −6); x-intercept: (2, 0). The T-intercept represents the temperature of the juice in the freezer before it is taken out. The x-intercept represents the point when the juice reaches 0°C.

61. ***Food Science*** The graph below shows the relationship between the temperature inside an oven and the time elapsed since the oven was turned off. Find the slope of the line between the two points shown on the graph. Write a sentence that states the meaning of the slope in the context of this problem.

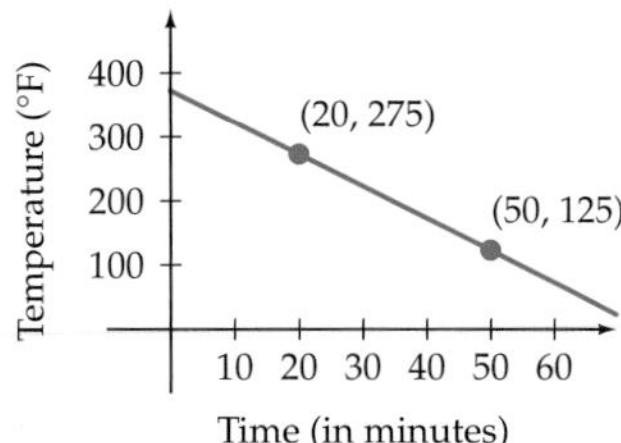

Slope $= -5$. The slope means that the oven temperature is decreasing 5°F/min.

62. ***Heart Rate*** The graph below shows the relationship between the age of a person and the maximum recommended exercise heart rate for a person who exercises regularly. Find the slope of the line between the two points shown on the graph. Write a sentence that states the meaning of the slope in the context of this problem.

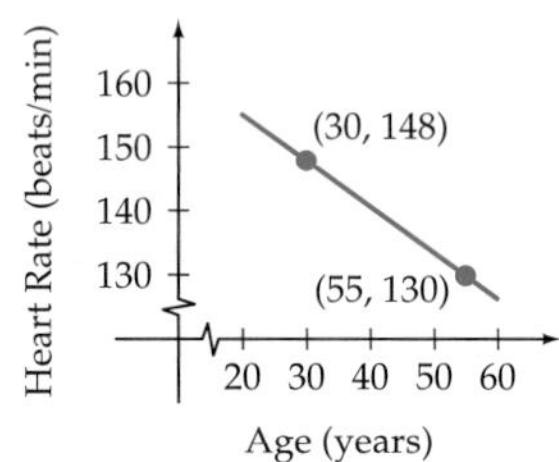

Slope $= -0.72$. The slope means that the recommended exercise heart rate decreases 0.72 beats/min/year.

63. ***Jogging Speeds*** Lois and Terri start from the same place on a jogging course. Lois is jogging at 9 kilometers per hour and Terri is jogging at 6 kilometers per hour. The graphs below show the total distances traveled by each jogger and the total distance between Lois and Terri. Which lines represent which distances?

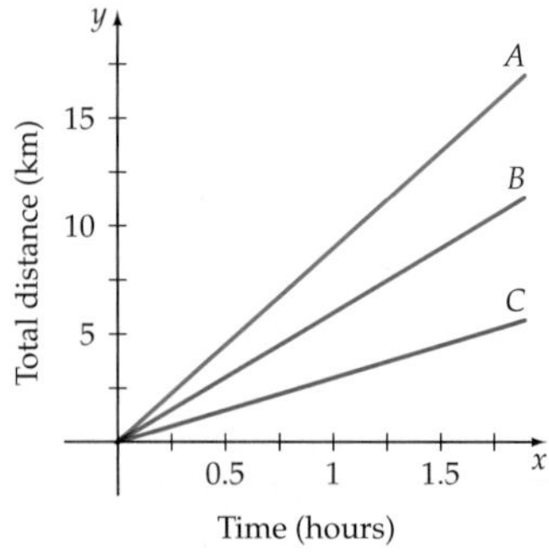

Lois is line A. Terri is line B. The distance between them is line C.

Physical Sciences and Engineering

64. ***Oceanography*** The graph below shows the relationship between the speed of sound in water and the temperature of the water. Find the slope of this line and write a sentence that explains the meaning of the slope in the context of this problem. Slope $= 2.875$. The slope means that the speed of sound in water increases 2.875 m/s for each 1 degree increase in temperature.

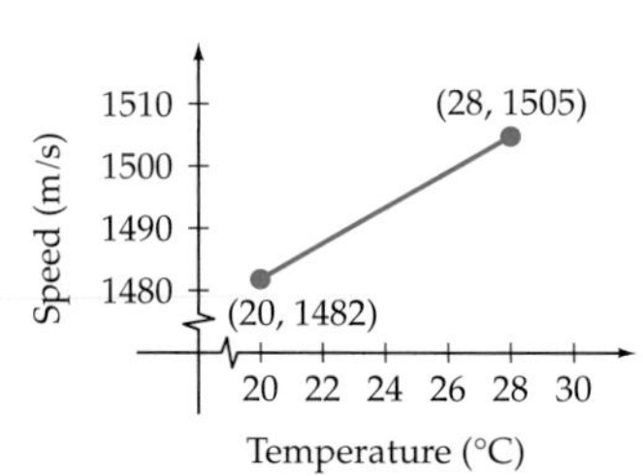

65. ***Computer Science*** The graph below shows the relationship between the time, in seconds, it takes to download a file and the size of the file in megabytes. Find the slope of the line between the two points shown on the graph. Write a sentence that states the meaning of the slope in the context of this problem.

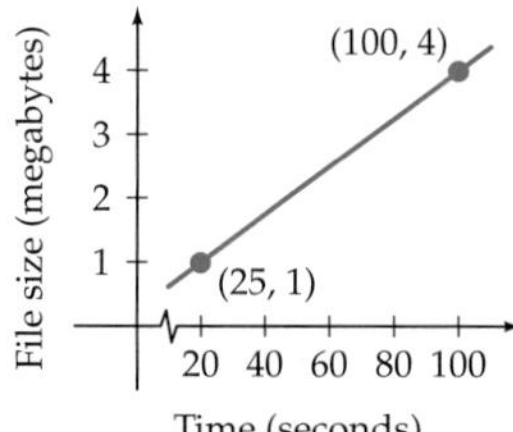

Slope $= 0.04$. The slope means that 0.04 megabytes are downloaded each second.

66. ***Wheelchair Ramps*** The American National Standards Institute (ANSI) states that the slope for a wheelchair ramp must not exceed $\frac{1}{12}$.

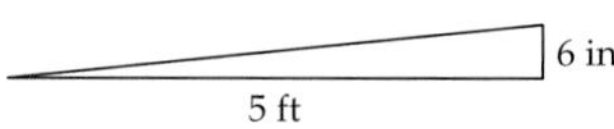

a. Does a ramp that is 6 inches high and 5 feet long meet the requirements of ANSI? No

b. Does a ramp that is 12 inches high and 170 inches long meet the requirements of ANSI? Yes

67. ***Wheelchair Ramps*** A ramp for a wheelchair must be 14 inches high. What is the minimum length of this ramp so that it meets the ANSI requirements in Exercise 66? 168 in.

68. ***Chemistry*** A chemist is filling two cans from a faucet that releases water at a constant rate. Can 1 has a diameter of 20 millimeters and can 2 has a diameter of 30 millimeters. The depth of the water in each can is measured at 5-second intervals. The graph of the results is shown below. On the graph, which line represents which can?

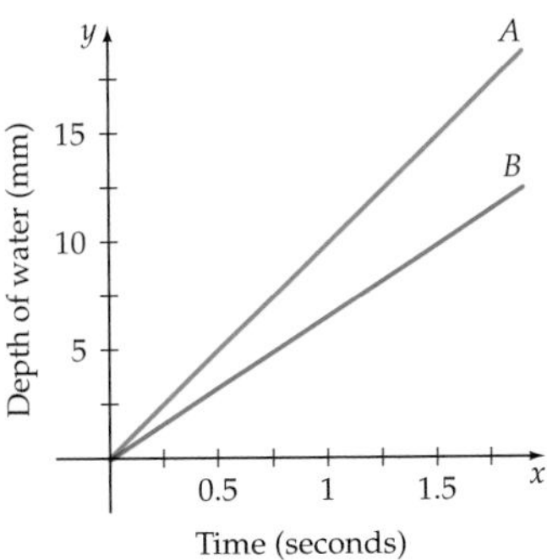

Can 1 is A. Can 2 is B.

Sports and Recreation

69. ***Swimming Pools*** The graph below shows the number of gallons of water remaining in a pool t hours after a valve is opened to drain the pool. Find the slope of the line between the two points shown on the graph. Write a sentence that states the meaning of the slope in the context of this problem. Slope $= -1.08$. The slope means that the water is flowing out of the pool at a rate of 1080 gal/h.

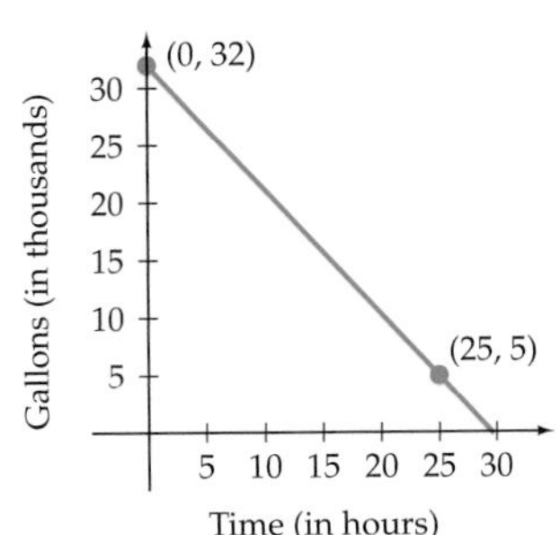

70. If (2, 3) are the coordinates of a point on a line that has slope 2, what is the y-coordinate of the point on the line whose x-coordinate is 4? 7

71. If (−1, 2) are the coordinates of a point on a line that has slope −3, what is the y-coordinate of the point on the line whose x-coordinate is 1? −4

72. If (1, 4) are the coordinates of a point on a line that has slope $\frac{2}{3}$, what is the y-coordinate of the point on the line whose x-coordinate is −2? 2

73. If (−2, −1) are the coordinates of a point on a line that has slope $\frac{3}{2}$, what is the y-coordinate of the point on the line whose x-coordinate is −6? −7

74. Graph the line that passes through the point (−1, −3) and has slope $\frac{4}{3}$.

75. Graph the line that passes through the point (−2, −3) and has slope $\frac{5}{4}$.

76. Graph the line that passes through the point (−3, 0) and has slope −3.

77. Graph the line that passes through the point (2, 0) and has slope −1.

78. What effect does increasing the coefficient of x have on the graph of $y = mx + b$? It increases the slope.

79. What effect does decreasing the coefficient of x have on the graph of $y = mx + b$? It decreases the slope.

80. What effect does increasing the constant term have on the graph of $y = mx + b$? It increases the y-intercept.

81. What effect does decreasing the constant term have on the graph of $y = mx + b$? It decreases the y-intercept.

82. Do the graphs of all straight lines have a y-intercept? If not, give an example of one that does not.
No. The graph of $x - 2 = 0$ has no y-intercept.

83. If two lines have the same slope and the same y-intercept, must the graphs of the lines be the same? If not, give an example. Yes

Prepare for Section 2.4

In Exercises 84 and 85, solve for x.

84. $x = 2 + 3(2x + 1)$ [1.1] −1

85. $(x + 1)(x - 2) = 4$ [1.2] −2, 3

In Exercises 86 and 87, solve for y.

86. $y + 3 = -2(x - 1)$ [1.1] $y = -2x - 1$

87. $\frac{y + 1}{x - 2} = -\frac{1}{2}$ [1.3] $y = -\frac{1}{2}x$

88. Describe the graph of the ordered pairs $(x, -1)$, where x is any real number. [2.2]
A horizontal line passing through $(0, -1)$

89. Describe the graph of the ordered pairs (x, x), where x is any real number. [2.2]
A diagonal line passing through the origin and bisecting quadrants I and III

Explorations 

1. ***Prove That the Product of the Slopes of Non-vertical Perpendicular Lines is −1*** In the diagram below, lines l_1 and l_2 are perpendicular and line segment AC has length l.

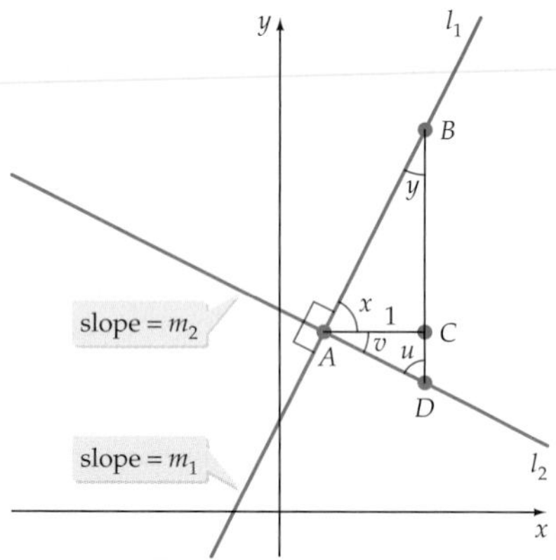

a. Show that the length of BC is m_1.

b. Show that the length of CD is $-m_2$. Note that because m_2 is a negative number, $-m_2$ is a positive number.

c. Show that triangles ACB and ACD are similar right triangles. (*Suggestion:* Show that the measure of angle ADC equals the measure of angle BAC.)

d. Because the ratios of corresponding sides of similar triangles are equal, show that $\frac{m_1}{1} = \frac{1}{-m_2}$.

e. Use the equation in part **d.** to show that $m_1 m_2 = -1$.

SECTION 2.4 Linear Models

- Find the Equation of a Line Given a Point and the Slope
- Find the Equation of a Line Given Two Points
- Find Equations of Parallel and Perpendicular Lines

▪ Find the Equation of a Line Given a Point and the Slope

Suppose that a car uses 0.05 gallon of gas per mile driven and that the fuel tank, which holds 18 gallons of gas, is full. Using this information, we can determine a linear model for the amount of fuel remaining in the gas tank.

Recall that a linear equation in two variables is one that can be written in the form $y = mx + b$, where m is the slope of the line and b is the y-intercept. The slope is the rate at which the car is burning fuel. Because the car is consuming the fuel, the amount of fuel in the tank is decreasing. Therefore, the slope is negative and we have $m = -0.05$.

The amount of fuel in the tank depends on the number of miles x the car is driven. Before the car starts (that is, when $x = 0$), there are 18 gallons of gas in the tank. The y-intercept is $(0, 18)$.

TAKE NOTE

When trying to create a linear model, the slope is the quantity that is expressed by using the word *per*. The car discussed at the right uses 0.05 gallons *per* mile. The slope is negative because the amount of fuel in the tank is decreasing.

Using this information, we can create the linear equation.

$$y = mx + b$$

$$y = -0.05x + 18$$ ■ Replace m by -0.05; replace b by 18.

The linear equation that models the amount of fuel remaining in the tank is $y = -0.05x + 18$, where y is the amount of fuel remaining, in gallons, after driving x miles. An equation such as this is sometimes referred to as a **linear model** because it is a mathematical model of the situation. The graph of the equation is shown in Figure 2.27.

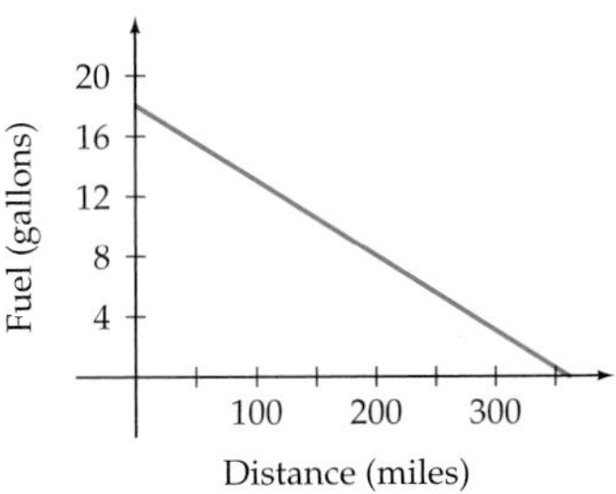

Figure 2.27

The x-intercept of this graph is the point at which $y = 0$. For this application, this means the point at which there are 0 gallons of fuel remaining in the tank. Replacing y by 0 in $y = -0.05x + 18$ and solving for x will give the number of miles that can be driven before running out of gas.

$$y = -0.05x + 18$$

$$0 = -0.05x + 18$$ ■ Replace y by 0.

$$-18 = -0.05x$$

$$360 = x$$

The car can travel 360 miles before running out of gas.

QUESTION Why does it make no sense to allow the values of x for $y = -0.05x + 18$ to exceed 360?

Alternative to Example 1

A rental company purchases a truck for \$18,600. The truck requires an average of \$7.50 per day in maintenance. Find a linear function that expresses the cost y of owning the truck after x days.

■ $y = 7.5x + 18,600$

EXAMPLE 1 Find a Linear Model

Suppose a 20-gallon gas tank contains 2 gallons of gas when a motorist decides to fill up the tank. If the gas pump fills the tank at a rate of 0.08 gallon per second, find a linear model that gives the amount of fuel in the tank t seconds after fueling begins.

Solution Because there are 2 gallons of gas in the tank when fueling begins (when $t = 0$), the y-intercept is $(0, 2)$. The slope is the rate at which fuel is being added to the tank. Because the amount of fuel in the tank is increasing, the slope is positive and we have $m = 0.08$. To find the linear equation, replace m and b by their values.

$$y = mt + b$$

$$y = 0.08t + 2$$ ■ Replace m by 0.08; replace b by 2.

The linear equation is $y = 0.08t + 2$, where y is the number of gallons of fuel in the tank t seconds after fueling begins.

Continued ➤

ANSWER If $x > 360$, then $y < 0$. This means that the tank has negative gallons of gas. For instance, if $x = 400$, $y = -2$.

CHECK YOUR PROGRESS 1 The boiling point of water at sea level is 100°C. The boiling point decreases 3.5°C per 1 kilometer increase in altitude. Find a linear equation that gives the boiling point of water in terms of altitude.

Solution *See page S13.* $y = -3.5x + 100$

In the previous example, the known point on the graph of the linear equation was the y-intercept. This information enabled us to determine b in $y = mx + b$. In some instances, a point other than the y-intercept is given. In such cases, the following *point-slope formula* is used to find the equation of the line.

TAKE NOTE

Using parentheses may help when substituting into the point-slope formula.

$$y - y_1 = m(x - x_1)$$
$$\downarrow \quad \downarrow \quad \downarrow$$
$$y - (\ ) = (\ )[x - (\ )]$$

Point-Slope Formula

Let $P_1(x_1, y_1)$ be a point on a line and let m be the slope of the line. Then the equation of the line can be found using the point-slope formula

$$y - y_1 = m(x - x_1)$$

Alternative to Example 2

Find the equation of the line that passes through $P(-2, 5)$ and has slope 4.

■ $y = 4x + 13$

EXAMPLE 2 Find the Equation of a Line Given a Point and the Slope

Find the equation of the line that passes through $P(1, -3)$ and has slope -2.

Solution

$$y - y_1 = m(x - x_1) \quad \text{■ Use the point-slope formula.}$$
$$y - (-3) = -2(x - 1)$$
$$y + 3 = -2x + 2 \quad \text{■ } m = -2, (x_1, y_1) = (1, -3)$$
$$y = -2x - 1$$

CHECK YOUR PROGRESS 2 Find the equation of the line that passes through $P(-2, 2)$ and has slope $-\frac{1}{2}$.

Solution *See page S13.* $y = -\frac{1}{2}x + 1$

The next example shows how linear models can be used to solve real-life applications.

Alternative to Example 3

A cab charges $1.50 for the first mile and $1.20 for each mile after the first. Find a linear model for the price of a cab ride after x miles and then use the model to find the price of a cab ride after 9 miles.

■ $P = 1.2x + 0.3$; $11.10

EXAMPLE 3 Find a Linear Equation That Models the Value of a Car

Suppose that data from the Kelley Blue Book show that the value of a certain car decreases approximately $250 per month. If the value of the car 2 years after it was purchased was $14,000, find a linear equation that models the value of the car after x months of ownership. Use this equation to find the value of the car after 3 years of ownership.

Solution Let V represent the value of the car after x months. Then $V = 14{,}000$ when $x = 24$ (2 years is 24 months). The car is decreasing $250 per month in

Continued ➤

value. Therefore, the slope is −250. Now use the point-slope formula to find the linear model.

$$V - V_1 = m(x - x_1)$$
$$V - 14{,}000 = -250(x - 24)$$
$$V - 14{,}000 = -250x + 6000$$
$$V = -250x + 20{,}000$$

A linear equation that models the value of the car is $V = -250x + 20{,}000$.
To find the value of the car after 3 years (36 months), replace x by 36 and solve for V.

$$V = -250x + 20{,}000$$
$$V = -250(36) + 20{,}000 \quad \text{▪ Replace } x \text{ by 36. Then solve for } V.$$
$$= -9000 + 20{,}000 = 11{,}000$$

The value of the car is \$11,000 after 36 months of ownership.

CHECK YOUR PROGRESS 3 A vitamin manufacturer has determined that the cost to produce a 250-tablet bottle of a certain vitamin is \$2.50 per bottle. If the total cost to produce 1000 of these bottles is \$3500, find a linear model that gives the cost to produce n bottles of the vitamin.

Solution *See page S13.* $C = 2.50n + 1000$

▪ Find the Equation of a Line Given Two Points

In the previous examples, we found the equation of a line given its slope and a point on the line. However, if the slope is not given but two points are known, these points can be used to find the slope of the line. We can then find the equation of the line through the two points using the calculated slope and one of the two points.

Alternative to Example 4
Find the equation of the line that passes through $P_1(-1, -7)$ and $P_2(3, 5)$.
▪ $y = 3x - 4$

TAKE NOTE

In Example 4, we could have used the point $P_2(3, 2)$ instead of $P_1(6, -4)$. In this case, we get

$$y - 2 = -2(x - 3)$$
$$y - 2 = -2x + 6$$
$$y = -2x + 8$$

This is the same equation as the one at the right.

EXAMPLE 4 Find the Equation of a Line Given Two Points

Find the equation of the line that passes through $P_1(6, -4)$ and $P_2(3, 2)$.

Solution Find the slope of the line between the two points.

$$m = \frac{y_2 - y_1}{x_2 - x_1} = \frac{2 - (-4)}{3 - 6} = \frac{6}{-3} = -2$$

Use the point-slope formula to find the equation of the line.

$$y - y_1 = m(x - x_1)$$
$$y - (-4) = -2(x - 6) \quad \text{▪ } m = -2,\ x_1 = 6,\ y_1 = -4.$$
$$y + 4 = -2x + 12$$
$$y = -2x + 8$$

Continued ➤

CHECK YOUR PROGRESS 4 Find the equation of the line that passes through $P_1(-2, 3)$ and $P_2(4, 1)$.

Solution *See page S13.* $y = -\frac{1}{3}x + \frac{7}{3}$

Alternative to Example 5
Exercise 48, page 219

EXAMPLE 5 Solve an Application

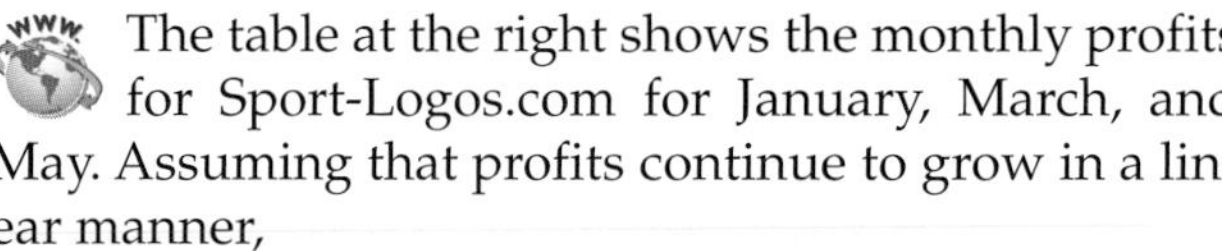

The table at the right shows the monthly profits for Sport-Logos.com for January, March, and May. Assuming that profits continue to grow in a linear manner,

Month	Profit ($)
January	20,000
March	23,000
May	26,000

a. find a linear model for these data.

b. write a sentence that explains the meaning of the slope of this line in the context of the problem.

c. use the model to predict the profit for the following December.

Solution Let the month of the year be represented by x, where $x = 1$ corresponds to January, and let the monthly profit be represented by P. The month and profit can be represented by the points $P_1(1, 20{,}000)$, $P_2(3, 23{,}000)$, and $P_3(5, 26{,}000)$.

a. To find a linear model, choose any two points and find the slope of the line between the two points. We will use P_1 and P_3, but any two points could be used.

$$m = \frac{26{,}000 - 20{,}000}{5 - 1} = \frac{6000}{4} = 1500$$

Now use the point-slope formula to find the equation of the line.

$$P - P_1 = m(x - x_1)$$
$$P - 20{,}000 = 1500(x - 1) \qquad \blacksquare\ m = 1500,\ x_1 = 1,\ P_1 = 20{,}000$$
$$P - 20{,}000 = 1500x - 1500$$
$$P = 1500x + 18{,}500$$

The linear model is $P = 1500x + 18{,}500$.

b. The slope of the line is 1500. This means that the profit is increasing by $1500 per month.

c. To find the profit in December, replace x by 12 and solve for P.

$$P = 1500x + 18{,}500$$
$$= 1500(12) + 18{,}500$$
$$= 36{,}500$$

The profit in December will be $36,500.

CHECK YOUR PROGRESS 5 The table at the right shows the change in the trade-in value of a sports car for various months. Assuming that the value of the sports car continues to decrease in the same manner,

Month	Value ($)
February	27,000
March	26,700
April	26,400
May	26,100

Continued ➤

a. find a linear model for these data.

b. write a sentence that explains the meaning of the slope of this line in the context of the problem.

c. use the model to predict the value of the car in January of the next year.

Solution *See page S13.* a. $y = -300x + 27{,}000$ b. The value of the car is decreasing by \$300 each month. c. \$23,700

Find Equations of Parallel and Perpendicular Lines

Recall from the last section that if two lines have slopes m_1 and m_2, then the graphs of these lines are *parallel* if and only if $m_1 = m_2$ and the graphs of the lines are perpendicular if and only if $m_1 m_2 = -1$ or, equivalently, $m_1 = -\frac{1}{m_2}$.

Using this information, we can find the equations of lines parallel or perpendicular to a given line.

Alternative to Example 6

Find the equation of the line that is parallel to the graph of $3x + 4y = 8$ and passes through the point $(3, -4)$

■ $y = -\frac{3}{4}x - \frac{7}{4}$

Point of Interest

When Euclid (circa 300 B.C.) stated the principles of geometry, the fifth axiom was called the *parallel postulate.* It stated that given a line and a point not on the line, only one line parallel to the given line could be drawn.

In the early 1800s, three mathematicians (Carl Gauss, János Bolyai, and Nikolai Lobachevsky), independent of one another, challenged this axiom and theorized that more than one parallel line could be drawn. Thus began what is now called non-Euclidean geometry. Their work eventually helped Albert Einstein to formulate the general theory of relativity.

EXAMPLE 6 Find the Equation of a Line That Is Parallel to a Given Line

Find the equation of the line that is parallel to the graph of $2x - 3y = 12$ and passes through the point $P(1, -1)$.

Solution Because the lines are parallel, the slope of the unknown line is the same as the slope of the given line. Write $2x - 3y = 12$ in slope-intercept form by solving for y.

$$\begin{aligned} 2x - 3y &= 12 \\ -3y &= -2x + 12 \\ y &= \frac{2}{3}x - 4 \end{aligned}$$

The slope of the given line is $\frac{2}{3}$.

Because the lines are parallel, the slope of the line through $P(1, -1)$ is also $\frac{2}{3}$. Use the point-slope formula to find the equation of this line.

$$\begin{aligned} y - y_1 &= m(x - x_1) \\ y - (-1) &= \frac{2}{3}(x - 1) \qquad \blacksquare\ m = \tfrac{2}{3},\ (x_1, y_1) = (1, -1) \\ y + 1 &= \frac{2}{3}x - \frac{2}{3} \\ y &= \frac{2}{3}x - \frac{5}{3} \end{aligned}$$

CHECK YOUR PROGRESS 6 Find the equation of the line that is parallel to the graph of $3x + 5y = 15$ and passes through the point $P(-2, 3)$.

Solution *See page S14.* $y = -\frac{3}{5}x + \frac{9}{5}$

In Example 6, we left the answer in slope-intercept form. However, we could have written the equation in standard form.

$$y = \frac{2}{3}x - \frac{5}{3}$$

$$3y = 3\left(\frac{2}{3}x - \frac{5}{3}\right)$$ ▪ Multiply each side by 3.

$$3y = 2x - 5$$

$$-2x + 3y = -5$$ ▪ Subtract $2x$ from each side. The equation is in standard form. $A = -2$, $B = 3$, and $C = -5$.

Alternative to Example 7

Find the equation of the line that is perpendicular to the graph of $y = -\frac{3}{4}x - 3$ and passes through the point $(-2, 5)$.

▪ $y = \frac{4}{3}x + \frac{23}{3}$

EXAMPLE 7 Find the Equation of a Line That Is Perpendicular to a Given Line

Find the equation of the line that is perpendicular to the graph of $y = \frac{2}{5}x + 1$ and passes through the point (5, 3).

Solution Because the lines are perpendicular, the value of the slope of the unknown line is the negative reciprocal of the slope of the given line. The slope of the given line is $\frac{2}{5}$. Therefore, the slope of the perpendicular line is $-\frac{5}{2}$. Now use the point-slope formula to find the equation of the line.

$$y - y_1 = m(x - x_1)$$

$$y - 3 = -\frac{5}{2}(x - 5)$$ ▪ $m = -\frac{5}{2}$, $(x_1, y_1) = (5, 3)$

$$y - 3 = -\frac{5}{2}x + \frac{25}{2}$$

$$y = -\frac{5}{2}x + \frac{31}{2}$$

CHECK YOUR PROGRESS 7 Find the equation of the line that is perpendicular to the graph of $5x - 3y = 15$ and passes through the point $P(-2, -3)$.

Solution *See page S14.* $y = -\frac{3}{5}x - \frac{21}{5}$

There are many applications of the concept of perpendicular lines. For example, suppose a ball is being whirled on the end of a string. If the string breaks, the initial path of the ball is on a line that is perpendicular to the radius of the circle. This situation is discussed in the next example.

Alternative to Example 8
Exercise 56, page 220

EXAMPLE 8 Solve an Application

Suppose a ball is being whirled on the end of a string and that the center of rotation is the origin of a coordinate system. See the figure at the right. If the string breaks when the ball is at the point whose coordinates are $P(6, 3)$, find the initial path of the ball.

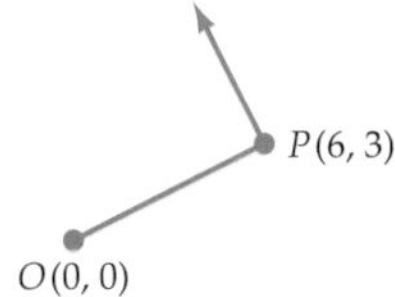

Solution The initial path of the ball is perpendicular to the line through OP. Therefore, the slope of the path of the ball is the negative reciprocal of the slope of the line between O and P.

The slope between O and P is $m = \dfrac{y_2 - y_1}{x_2 - x_1} = \dfrac{3 - 0}{6 - 0} = \dfrac{1}{2}$. The slope of the line that models the initial path of the ball is the negative reciprocal of $\frac{1}{2}$. Therefore, the slope of the initial path is -2. To find the equation of the path, use the point-slope formula.

$$\begin{aligned} y - y_1 &= m(x - x_1) \\ y - 3 &= -2(x - 6) \\ y - 3 &= -2x + 12 \\ y &= -2x + 15 \end{aligned}$$

The initial path of the ball is along the line whose equation is $y = -2x + 15$.

CHECK YOUR PROGRESS 8 Suppose a ball is being whirled on the end of a string and that the center of rotation is the origin of a coordinate system. See the figure at the right. If the string breaks when the ball is at the point whose coordinates are (2, 8), find the initial path of the ball.

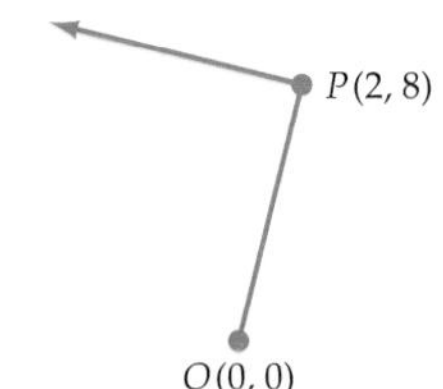

Solution *See page S14.* $y = -\frac{1}{4}x + \frac{17}{2}$

Topics for Discussion

1. What is a linear model?
2. What is the point-slope formula and how is it used?
3. If the coordinates of two points on a line are known, explain how the equation of the line through those points can be found.
4. Suppose a biologist has determined that there is a linear model that relates the height of a tree y to the age x. In the equation $y = mx + b$, would you expect m to be positive or negative? Why?
5. Suppose a baked potato is taken from a hot oven and placed on a plate to cool. If y represents the temperature of the potato and x represents the time the potato has been out of the oven, would you expect m to be positive or negative in the equation $y = mx + b$? Why?

EXERCISES 2.4

— *Suggested Assignment: Exercises 1–55, odd; and 57–62, all.*

In Exercises 1 to 16, find the equation of the line that contains the given point and has the given slope.

1. $P(0, 5), m = 2$
$y = 2x + 5$

2. $P(2, 3), m = \frac{1}{2}$ $y = \frac{1}{2}x + 2$

3. $P(-1, 7), m = -3$
$y = -3x + 4$

4. $P(0, 0), m = \frac{1}{2}$ $y = \frac{1}{2}x$

5. $P(3, 5), m = -\frac{2}{3}$
$y = -\frac{2}{3}x + 7$

6. $P(0, -3), m = -1$
$y = -x - 3$

7. $P(-2, -3), m = 0$
$y = -3$

8. $P(4, -5), m = -2$
$y = -2x + 3$

9. $P(-2, 0), m = \frac{3}{2}$
$y = \frac{3}{2}x + 3$

10. $P(-2, -4), m = \frac{1}{4}$ $y = \frac{1}{4}x - \frac{7}{2}$

11. $P(4, -5), m = 2$
$y = 2x - 13$

12. $P(-3, -2), m = 0$
$y = -2$

13. $P(-3, 5), m = 3$
$y = 3x + 14$

14. $P(2, 0), m = \frac{5}{6}$ $y = \frac{5}{6}x - \frac{5}{3}$

15. $P(5, 1), m = -\frac{4}{5}$
$y = -\frac{4}{5}x + 5$

16. $P(-3, 5), m = -\frac{1}{4}$
$y = -\frac{1}{4}x + \frac{17}{4}$

In Exercises 17 to 32, find the equation of a line between the given points.

17. $P_1(0, 2), P_2(3, 5)$
$y = x + 2$

18. $P_1(0, -3), P_2(-4, 5)$
$y = -2x - 3$

19. $P_1(0, 3), P_2(2, 0)$
$y = -\frac{3}{2}x + 3$

20. $P_1(-2, -3), P_2(-1, -2)$
$y = x - 1$

21. $P_1(2, 0), P_2(0, -1)$
$y = \frac{1}{2}x - 1$

22. $P_1(3, -4), P_2(-2, -4)$
$y = -4$

23. $P_1(-2, 5), P_2(-2, -5)$
$x = -2$

24. $P_1(2, 1), P_2(-2, -3)$
$y = x - 1$

25. $P_1(-3, -1), P_2(2, 4)$
$y = x + 2$

26. $P_1(0, 4), P_2(2, 0)$
$y = -2x + 4$

27. $P_1(-3, -5), P_2(4, -5)$
$y = -5$

28. $P_1(4, 1), P_2(3, -2)$
$y = 3x - 11$

29. $P_1(3, 1), P_2(-3, -2)$
$y = \frac{1}{2}x - \frac{1}{2}$

30. $P_1(-3, 3), P_2(-2, 3)$
$y = 3$

31. $P_1(3, 2), P_2(3, -4)$
$x = 3$

32. $P_1(1, -3), P_2(-2, 4)$
$y = -\frac{7}{3}x - \frac{2}{3}$

33. Find the equation of the line that is parallel to $y = -3x - 1$ and passes through $P(1, 4)$. $y = -3x + 7$

34. Find the equation of the line that is parallel to $y = \frac{2}{3}x + 2$ and passes through $P(-3, 1)$. $y = \frac{2}{3}x + 3$

35. Find the equation of the line that contains the point $(-2, -4)$ and is parallel to the graph of $2x - 3y = 2$.
$y = \frac{2}{3}x - \frac{8}{3}$

36. Find the equation of the line that contains the point $(3, 2)$ and is parallel to the graph of $3x + y = -3$. $y = -3x + 11$

37. Find the equation of the line that contains the point $(4, 1)$ and is perpendicular to the graph of $y = -3x + 4$.
$y = \frac{1}{3}x - \frac{1}{3}$

38. Find the equation of the line that contains the point $(2, -5)$ and is perpendicular to the graph of $y = \frac{5}{2}x - 4$.
$y = -\frac{2}{5}x - \frac{21}{5}$

39. Find the equation of the line that contains the point $(-1, -3)$ and is perpendicular to the graph of $3x - 5y = 2$.
$y = -\frac{5}{3}x - \frac{14}{3}$

40. Find the equation of the line that contains the point $(-1, 3)$ and is perpendicular to the graph of $2x + 4y = -1$.
$y = 2x + 5$

Business and Economics

41. ***Sales Commission*** An account executive receives a base salary plus a commission. On \$20,000 in monthly sales, the account executive would receive \$1800. On \$50,000 in monthly sales, the account executive would receive \$3000. Determine a linear function that yields the compensation of the account executive for a given amount of monthly sales. Use this model to determine the compensation of the account executive for \$85,000 in monthly sales.
$c = \frac{1}{25}s + 1000$; \$4400

42. ***Truck Manufacturing*** A manufacturer of pickup trucks has determined that 50,000 trucks per month can be sold at a price of \$12,000 per truck. At a price of \$11,000, the number of trucks sold per month increases to 55,000. Determine a linear function that predicts the number of trucks that will be sold at a given price. Use this model to predict the number of trucks that will be sold at a price of \$12,500.
$n = -5p + 110{,}000$; 47,500 trucks

43. ***Calculator Manufacturing*** A manufacturer of graphing calculators has determined that 10,000 calculators per week will be sold at a price of \$95 per calculator. At a price of \$90, it is estimated that 12,000 calculators will be sold. Determine a linear function that predicts the number of calculators that will be sold at a given price. Use this model to predict the number of calculators per week that will be sold at a price of \$75.
$n = -400p + 48{,}000$; 18,000 calculators

44. ***Forecasting*** An egg rancher has determined that the weekly sales of eggs (in dozens) depends on the price (in dollars) of the eggs. During one week the rancher sold 2500 one-dozen cartons of eggs at \$.77 per dozen. During another week, the rancher sold 3000 one-dozen cartons of eggs at \$.72 per dozen. Find a linear function that predicts the number of dozens of eggs sold in terms of the price per dozen. If the rancher lowers the price of a dozen eggs by 1 cent, how many more dozen eggs will the rancher expect to sell?
$n = -10{,}000p + 10{,}200$; 100 more dozen eggs

45. ***Hospitality Industry*** The operator of a hotel estimates that 500 rooms per night will be rented if the room rate per night is \$75. For each \$10 increase in the price of a room, six fewer rooms will be rented. Determine a linear function that predicts the number of rooms that will be rented for a given price per room. Use this model to predict the number of rooms that will be rented if the room rate is \$100 per night. $r = -\frac{3}{5}p + 545$; 485 rooms

46. ***Income Distribution*** Based on data from the U.S. Census Bureau, 8.4% of U.S. workers in 1992 earned over \$100,000. In 2000, 13.4% of U.S. workers earned over \$100,000.

  a. Find a linear function that gives the percent of workers earning in excess of \$100,000 for a given year.
  $y = 0.625x - 1236.6$

  b. In 1998, there were approximately 104,000,000 workers in the U.S. According to your model, how many of these workers earned over \$100,000? 12,636,000 people

47. ***New Home Construction*** A general building contractor estimates that the cost to build a new home is \$30,000 plus \$85 for each square foot of floor space in the house. Determine a linear function that gives the cost of building a house that contains a given number of square feet of floor space. Use this model to determine the cost to build a house that contains 1800 square feet of floor space.
$c = 85f + 30{,}000$; \$183,000

48. ***Automobile Manufacturing*** A molten piece of metal for a car frame is allowed to cool in a controlled environment. The temperature (in degrees Fahrenheit) of the metal after it is removed from a smelter for various times (in minutes) is shown in the table.

Time (min)	Temperature (°F)
15	2500
20	2450
30	2350
60	2050

  a. Find a linear model for the temperature of the metal after t minutes. $T = -10t + 2650$

  b. Explain the meaning of the slope of this line in the context of the problem.
  The slope means that the temperature is decreasing 10° per minute.

  c. Assuming temperature continues to decrease at the same rate, what will be the temperature in 3 hours?
  850°F

Life and Health Sciences

49. ***Blood Pressure*** A reduction in a patient's blood pressure depends on the number of milligrams of medication the patient receives. The table at the right shows the blood pressures of patients for various doses.

Dose (mg)	Blood pressure
5	200
9	188
12	179
15	170
20	155

  a. Find a linear model for the blood pressure of a patient after taking x milligrams of this medication. $p = -3x + 215$

  b. Write a sentence explaining the meaning of the slope of this line in the context of the problem.
  The slope means that blood pressure is decreasing at a rate of 3 units per milligram.

  c. Using this model, what will be the blood pressure of a patient who takes 25 milligrams of this medication?
  140

50. ***Ecology*** Water is being added to a pond in a bird sanctuary that already contains 1000 gallons of water. The table at the right shows the number of gallons of water in the pond after selected times in minutes.

Time (min)	Gallons
30	1750
60	2500
80	3000
120	4000

  a. Find a linear model for the number of gallons of water in the pond after t minutes. $g = 25t + 1000$

b. Write a sentence that explains the meaning of the slope of this line in the context of this problem.
The slope means that 25 gal of water per minute are entering the pond.

c. Assuming water continues to flow into the pond at the same rate, how many gallons of water will be in the pond after 6 hours? 10,000 gallons

Physical Sciences and Engineering

51. *Aeronautics* A plane travels 830 miles in 2 hours at a constant speed. Determine a linear model that predicts the number of miles the plane can travel in a given amount of time. Use this model to predict the distance the plane will travel in $4\frac{1}{2}$ hours. $d = 415t$; 1867.5 mi

52. *Oceanography* Seawater contains salt that decreases the temperature at which the seawater will freeze. If pure water freezes at 32°F and seawater containing 35 parts per million (ppm) of salt freezes at 28.6°F, find a linear model that gives the Fahrenheit temperature at which seawater with x ppm of salt will freeze. $t = -\frac{17}{175}x + 32$

53. *Boiling Point of Water* When sugar is added to water, the solution has a higher boiling point than pure water. The table at the right gives the approximate boiling points (in degrees Celsius) of water as various amounts of sugar are added to the water.

Sugar (grams)	Boiling point (°C)
20	100.104
30	100.156
40	100.208
60	100.312
80	100.416

a. Find a linear model for the boiling point of the solution in terms of the number of grams of sugar added. $b = 0.0052s + 100$

b. Write a sentence that explains the meaning of the slope of this line in the context of the problem. The slope means that the boiling point increases 0.0052°C for each additional gram of sugar added.

c. Using this model, what is the boiling point of a pure water solution to which 50 grams of sugar have been added? 100.26°C

54. *Freezing Point of Water* When sugar is added to water, the solution has a lower freezing point than pure water. The table at the right gives the approximate freezing point (in degrees Celsius) of water as various amounts of sugar are added to the water.

Sugar (grams)	Freezing point (°C)
20	−0.372
30	−0.558
40	−0.744
60	−1.116
80	−1.488

a. Find a linear model for the freezing point of the solution in terms of the number of grams of sugar added. $F = -0.0186s$

b. Write a sentence explaining the meaning of the slope of this line in the context of the problem. The slope means that the freezing point decreases −0.0186°C for each additional gram of sugar added.

c. Using this model, what is the freezing point of a pure water solution to which 25 grams of sugar have been added? −0.465°C

55. *Rotation of a Ball* Suppose a ball is being whirled on the end of a string and that the center of rotation is the origin of a coordinate system. If the string breaks when the ball is at the point whose coordinates are $P(-2, 5)$, find the equation of the line on which the initial path of the ball will follow. $y = \frac{2}{5}x + \frac{29}{5}$

56. *Rotation of a Ball* Suppose a ball is being whirled on the end of a string and that the center of rotation is the origin of a coordinate system. If the string breaks when the ball is at the point whose coordinates are $P(-1, -3)$, find the equation of the line on which the initial path of the ball will follow. $y = -\frac{1}{3}x - \frac{10}{3}$

Prepare for Section 2.5

57. Factor: $2x^2 - x - 3$ [P.4] $(2x - 3)(x + 1)$

58. Complete the square of $x^2 - 6x$ and write the result as the square of a binomial. [1.2] $x^2 - 6x + 9 = (x - 3)^2$

59. Find the x- and y-intercepts for $y = 6 - 2x$. [2.3]
(3, 0), (0, 6)

60. Does the graph of $y = x^2 + 1$ pass through the x-axis? Explain your answer. [P.2]
No. $x^2 + 1 > 0$ for all real numbers x.

In Exercises 61 and 62, solve for x.

61. $x^2 + 3x = 4$ [1.2]
$-4, 1$

62. $x^2 + 2x - 4 = 0$ [1.2]
$-1 \pm \sqrt{5}$

Explorations 

1. ***Distance from a Point to a Line*** In Section 2.1, we discussed the formula used to find the distance between two points. In this Exploration, we will derive the formula for the distance between a point and a line. By definition, the distance between a point and a line is the length of the line segment from the point perpendicular to the line.

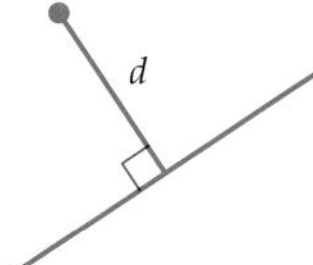

To complete the Exploration, recall the following properties of a transversal intersecting two parallel lines.

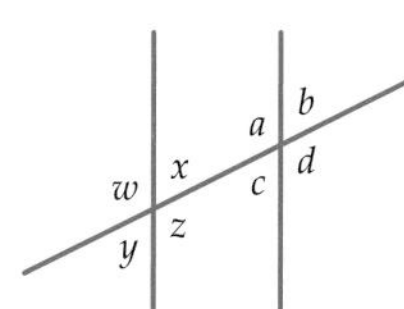

The measures of angles *a*, *d*, *w*, and *z* are equal.

The measures of angles *b*, *c*, *x*, and *y* are equal.

Also, recall that the ratios of corresponding sides of similar triangles are equal.

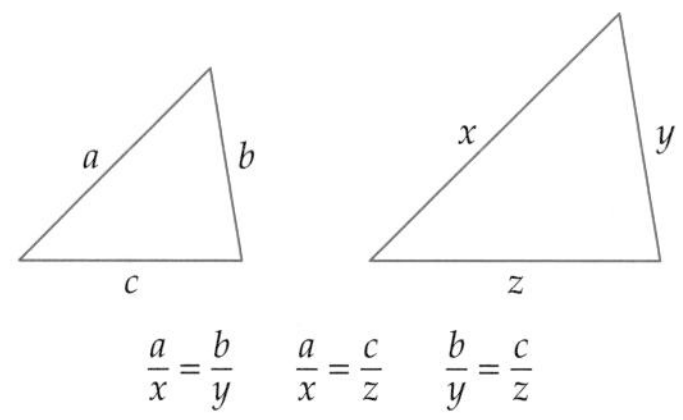

For the remaining portion of this Exploration, use the figure below.

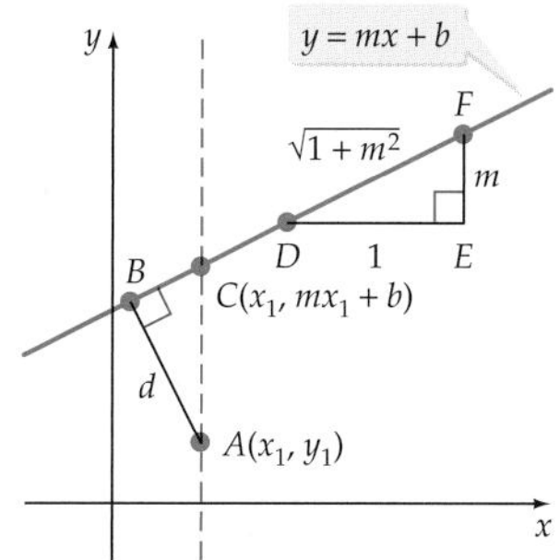

a. Using the properties of parallel lines cut by a transversal, explain why $\angle ACB = \angle DFE$.

b. Explain why triangles *ABC* and *DEF* are similar triangles. (*Hint:* Recall that if corresponding angles of two triangles are equal, the triangles are similar.)

c. What is the length of line segment *CA*? Remember to include absolute value signs to ensure the distance is positive.

d. Using the properties of similar triangles, find a formula for the length of *d*.

SECTION 2.5 Quadratic Functions

- Properties of Quadratic Functions
- Intercepts of Quadratic Functions
- Minimum and Maximum Values of Quadratic Functions
- Applications

▪ Properties of Quadratic Functions

Recall that a linear function is one of the form $f(x) = mx + b$. The graph of a linear function has certain characteristics. It is a straight line with slope m and y-intercept $(0, b)$.

A **quadratic function** is a function of the form $f(x) = ax^2 + bx + c$, $a \neq 0$. Examples of quadratic functions are given below.

$$f(x) = x^2 - 3x + 1 \qquad a = 1,\ b = -3,\ c = 1$$
$$g(t) = -2t^2 - 4 \qquad a = -2,\ b = 0,\ c = -4$$
$$h(p) = 4 - 2p - p^2 \qquad a = -1,\ b = -2,\ c = 4$$
$$f(x) = 2x^2 + 6x \qquad a = 2,\ b = 6,\ c = 0$$

The graph of a quadratic function, which is called a **parabola,** also has certain characteristics. The graphs of two quadratic functions are shown in Figure 2.28.

Point of Interest

The suspension cables for some bridges, such as the Golden Gate bridge, have the shape of a parabola.

The picture above shows the roadway of the bridge being assembled in sections and attached to the suspender ropes. The bridge was opened to vehicles on May 28, 1937.

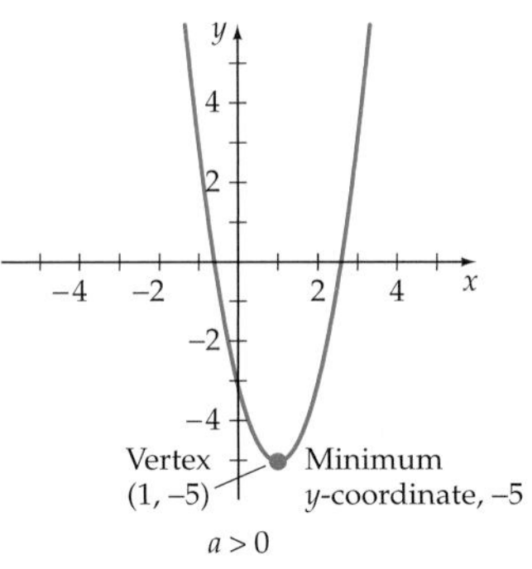

Figure 2.28A

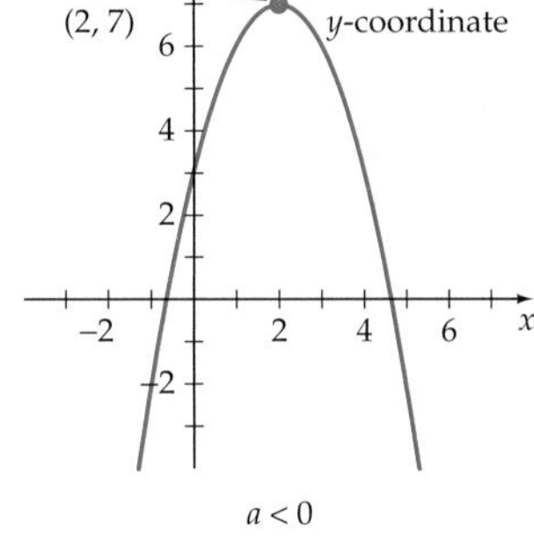

Figure 2.28B

For Figure 2.28A, $f(x) = 2x^2 - 4x - 3$. The value of a is *positive* ($a = 2$) and the graph opens up. For Figure 2.28B, $f(x) = -x^2 + 4x + 3$. The value of a is *negative* ($a = -1$) and the graph opens down. The point at which the graph of a parabola has a minimum or a maximum is called the vertex of the parabola. The **vertex of a parabola** is the point with the smallest y-coordinate when $a > 0$ and the point with the largest y-coordinate when $a < 0$.

TAKE NOTE

The axis of symmetry is a vertical line. The vertex of the parabola lies on the axis of symmetry.

The **axis of symmetry** of the graph of a quadratic function is a line that passes through the vertex of the parabola and is parallel to the y-axis. To understand the axis of symmetry, think of folding the graph along that line. The two portions of the graph will match up. Because the axis of symmetry is a vertical line and passes through the vertex, its equation can be determined once the vertex is determined. Its equation is $x =$ constant, where the constant is the x-coordinate of the vertex.

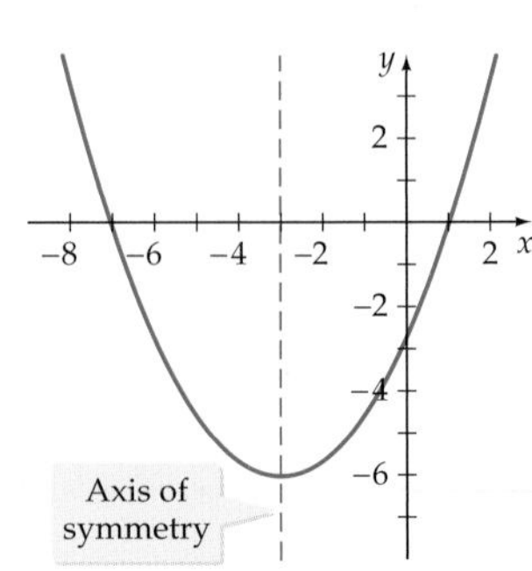

Point of Interest

The movie *Contact* is based on a novel by astronomer Carl Sagan. In the movie, Jodie Foster plays an astronomer who is searching for extraterrestrial intelligence. One scene from the movie takes place at the Very Large Array (VLA) in New Mexico. The VLA consists of 27 large radio telescopes whose dishes are paraboloids, the three-dimensional version of a parabola. A parabolic shape is used because of its unique reflective property: parallel rays of light or radio waves that strike the surface of the parabola are reflected to the same point. This point is called the *focus* of the parabola.

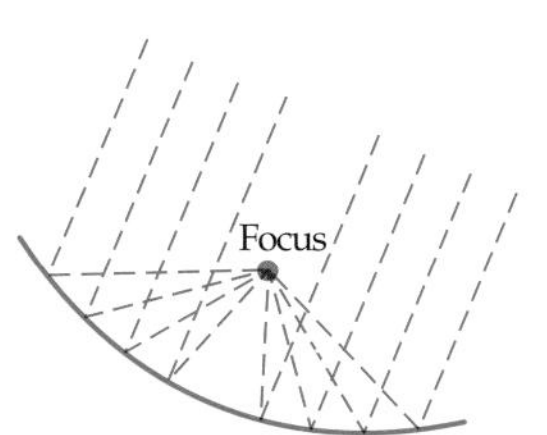

The following formula will enable us to determine the vertex and axis of symmetry without having to graph the function.

Vertex of a Parabola

Let $f(x) = ax^2 + bx + c$ be the equation of a parabola. The coordinates of the vertex are $\left(-\frac{b}{2a}, f\left(-\frac{b}{2a}\right)\right)$. The equation of the axis of symmetry is $x = -\frac{b}{2a}$.

QUESTION **a.** The axis of symmetry of a parabola is the line $x = -6$. What is the x-coordinate of the vertex of the parabola? **b.** The vertex of a parabola is $(3, -2)$. What is the equation of the axis of symmetry of the parabola? ***Alternative to Example 1***

Find the vertex and axis of symmetry of the parabola whose equation is $f(x) = 2x^2 - 12x - 3$

▪ **(3, −21), $x = 3$**

EXAMPLE 1 Find the Vertex and Axis of Symmetry of a Parabola

Find the vertex and axis of symmetry of the parabola whose equation is $f(x) = -3x^2 + 6x + 1$.

Solution

$$-\frac{b}{2a} = -\frac{6}{2(-3)} = 1$$
▪ **Find the x-coordinate of the vertex. $a = -3$, $b = 6$**

$$f(x) = -3x^2 + 6x + 1$$
$$f(1) = -3(1)^2 + 6(1) + 1$$
$$y = 4$$
▪ **Find the y-coordinate of the vertex by replacing x by 1 and solving for y.**

The vertex is $(1, 4)$. The equation of the axis of symmetry is $x = 1$, the x-coordinate of the vertex.

CHECK YOUR PROGRESS 1 Find the vertex and axis of symmetry of the parabola whose equation is $y = x^2 - 2$.

Solution *See page S14.* $V(0, -2)$; $x = 0$

▪ Intercepts of Quadratic Functions

Recall that a point at which a graph crosses the x- or y-axis is called an *intercept* of the graph. The x-intercepts of the graph of an equation occur when $y = 0$; the y-intercepts occur when $x = 0$.

The graph of $f(x) = x^2 + 3x - 4$ is shown in Figure 2.29. The points whose coordinates are $(-4, 0)$ and $(1, 0)$ are x-intercepts of the graph. The point whose coordinates are $(0, -4)$ is the y-intercept. We can algebraically determine the x- and y-intercepts by solving an equation.

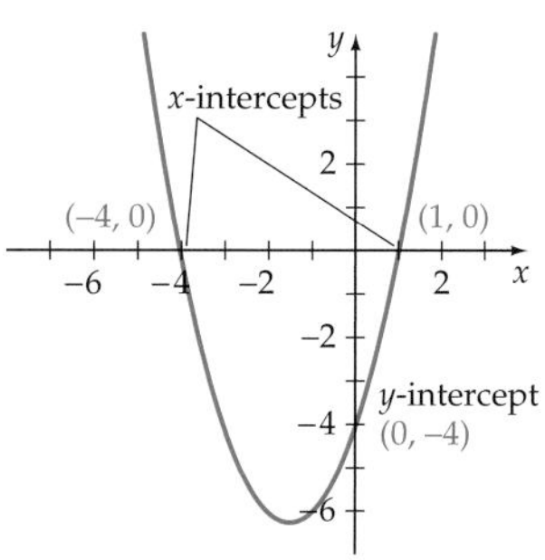

Figure 2.29

ANSWER **a.** The x-coordinate of the vertex is -6. **b.** The equation of the axis of symmetry is $x = 3$.

Alternative to Example 2

Find the x- and y-intercepts of the parabola given by the equation

$$y = x^2 + 6x + 2$$

■ $(-3 - \sqrt{7}, 0), (-3 + \sqrt{7}, 0)$; $(0, -2)$

TAKE NOTE

When the quadratic equation has a double root as in part **a.**, the graph of the quadratic equation is said to be *tangent* to the x-axis, meaning that it touches the x-axis in exactly one point. See the graph at the right.

TAKE NOTE

Try to solve the quadratic equation by factoring (as we did in part **a.**). If that cannot be done easily, then use the quadratic formula, as we did in part **b.** Note that if a graphing calculator is used to check an algebraic solution, the answers are given as decimals.

$$-1 - \sqrt{3} \approx -2.7320508$$

$$-1 + \sqrt{3} \approx 0.73205081$$

EXAMPLE 2 Find the x- and y-Intercepts of a Parabola

Find the x- and y-intercepts of the parabola given by the equation.

a. $y = 4x^2 + 4x + 1$ **b.** $y = x^2 + 2x - 2$

Solution

a. $y = 4x^2 + 4x + 1$

$0 = 4x^2 + 4x + 1$ ■ Let $y = 0$.

$0 = (2x + 1)(2x + 1)$ ■ Solve for x by factoring.

$2x + 1 = 0 \qquad 2x + 1 = 0$

$x = -\dfrac{1}{2} \qquad x = -\dfrac{1}{2}$

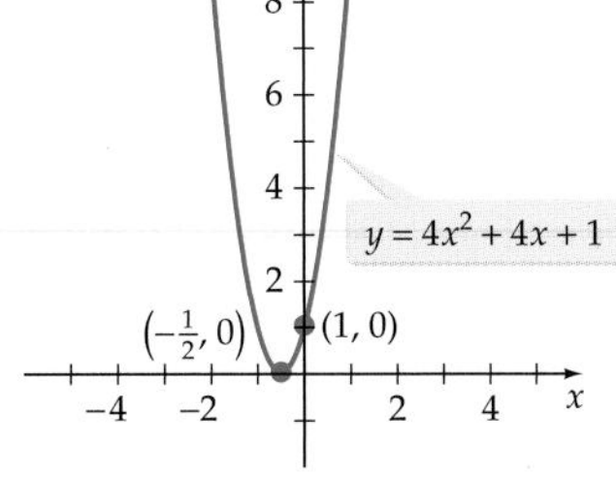

The x-intercept is $\left(-\frac{1}{2}, 0\right)$.

$y = 4x^2 + 4x + 1$

$y = 4(0)^2 + 4(0) + 1$ ■ Let $x = 0$.

$y = 1$

The y-intercept is $(0, 1)$.

b. $y = x^2 + 2x - 2$

$0 = x^2 + 2x - 2$ ■ Let $y = 0$. The trinomial $x^2 + 2x - 2$ is nonfactorable over the integers.

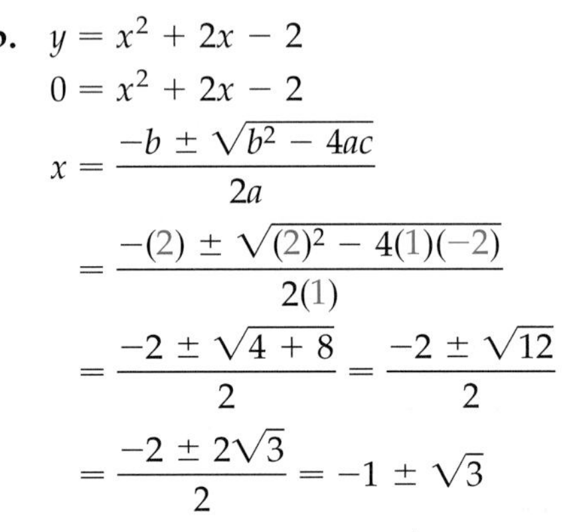

$$x = \frac{-b \pm \sqrt{b^2 - 4ac}}{2a}$$ ■ Use the quadratic formula to solve for x.

$$= \frac{-(2) \pm \sqrt{(2)^2 - 4(1)(-2)}}{2(1)}$$ ■ $a = 1, b = 2, c = -2$

$$= \frac{-2 \pm \sqrt{4 + 8}}{2} = \frac{-2 \pm \sqrt{12}}{2}$$

$$= \frac{-2 \pm 2\sqrt{3}}{2} = -1 \pm \sqrt{3}$$

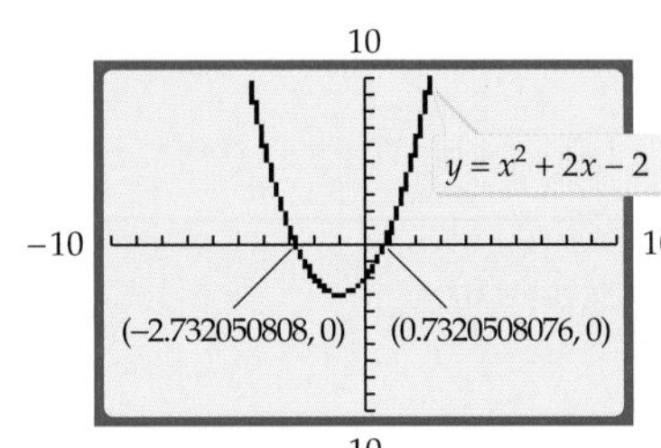

The x-intercepts are $\left(-1 - \sqrt{3}, 0\right)$ and $\left(-1 + \sqrt{3}, 0\right)$.

$y = x^2 + 2x - 2$

$y = 0^2 + 2(0) - 2$ ■ Let $x = 0$.

$y = -2$

The y-intercept is $(0, -2)$.

CHECK YOUR PROGRESS 2 Find the x- and y-intercepts of the parabola given by the equation.

a. $y = 2x^2 - 5x + 2$ **b.** $y = x^2 + 4x + 4$

Solution *See page S14.* **a.** $\left(\frac{1}{2}, 0\right)$ and $(2, 0)$; $(0, 2)$ **b.** $(-2, 0)$; $(0, 4)$

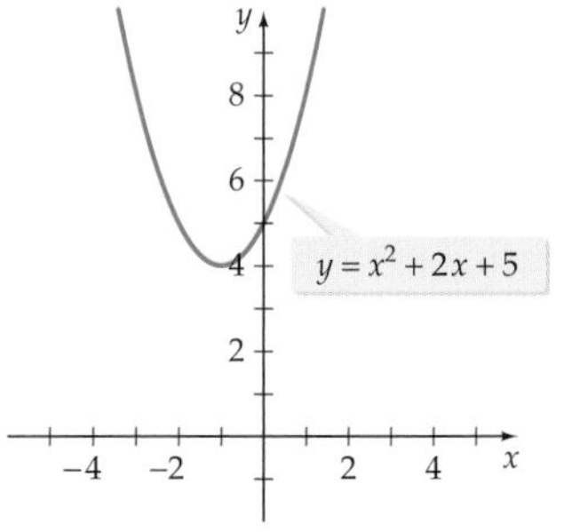

Figure 2.30

The graph of $y = x^2 + 2x + 5$ is shown in Figure 2.30. Note that the graph has a y-intercept but no x-intercepts. If we attempt to find x-intercepts as we did in Example 2, we will find that the equation has nonreal complex number solutions.

$$y = x^2 + 2x + 5$$
$$0 = x^2 + 2x + 5$$

- Let $y = 0$. The trinomial $x^2 + 2x + 5$ is nonfactorable over the integers.

$$x = \frac{-b \pm \sqrt{b^2 - 4ac}}{2a}$$

- Use the quadratic formula to solve for x. $a = 1$, $b = 2$, $c = 5$

$$= \frac{-(2) \pm \sqrt{(2)^2 - 4(1)(5)}}{2(1)} = \frac{-2 \pm \sqrt{4 - 20}}{2}$$
$$= \frac{-2 \pm \sqrt{-16}}{2} = \frac{-2 \pm 4i}{2} = -1 \pm 2i$$

Because intercepts on the axes are real numbers and this equation has nonreal complex number solutions, there are no x-intercepts.

TAKE NOTE

Calculus is a branch of mathematics that demonstrates, among other things, how to find the maximum or minimum values of functions other than quadratic functions. These are very important problems in applied mathematics. For instance, an automotive engineer wants to design a car whose shape will *minimize* the effect of air flow. Another engineer tries to *maximize* the efficiency of a car's engine. Similarly, an economist may try to determine what business practices will *minimize* cost and *maximize* profit.

▪ Minimum and Maximum Values of Quadratic Functions

Note that for the graphs below, when $a > 0$, the vertex of the parabola is the point with the minimum y-coordinate. When $a < 0$, the vertex of the parabola is the point with the maximum y-coordinate.

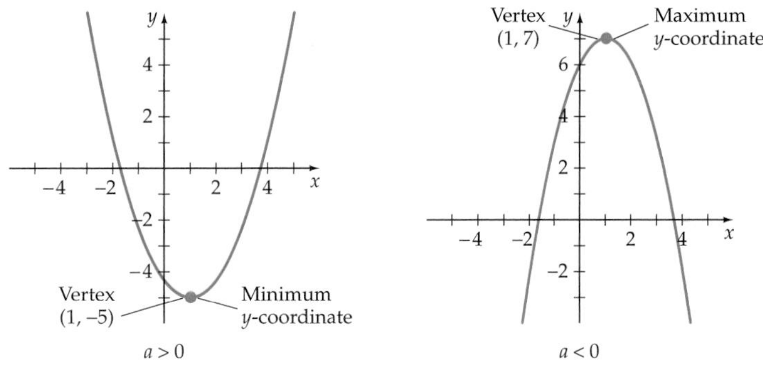

$f(x) = ax^2 + bx + c$

Finding the minimum or maximum value of a quadratic function is a matter of finding the y-coordinate of the vertex of the graph of the function.

QUESTION The vertex of a parabola that opens up is $(-4, 7)$. What is the minimum value of the function?

ANSWER The minimum value of the function is 7, the y-coordinate of the vertex.

Alternative to Example 3
Find the minimum value of

$f(x) = -3x^2 - 2x + 5$

■ $\frac{16}{3}$

EXAMPLE 3 Find the Maximum Value of a Quadratic Function

Find the maximum value of $f(x) = -2x^2 + 4x + 3$.

Solution

$$-\frac{b}{2a} = -\frac{4}{2(-2)} = 1$$

■ Find the x-coordinate of the vertex. $a = -2$, $b = 4$

$$f(x) = -2x^2 + 4x + 3$$
$$f(1) = -2(1)^2 + 4(1) + 3$$
$$f(1) = 5$$

■ Find the y-coordinate of the vertex by evaluating the function at 1.

The vertex is (1, 5). The maximum value of the function is 5, the y-coordinate of the vertex.

CHECK YOUR PROGRESS 3 Find the minimum value of $f(x) = 2x^2 - 3x + 1$.

Solution *See page S15.* $-\frac{1}{8}$

INTEGRATING TECHNOLOGY

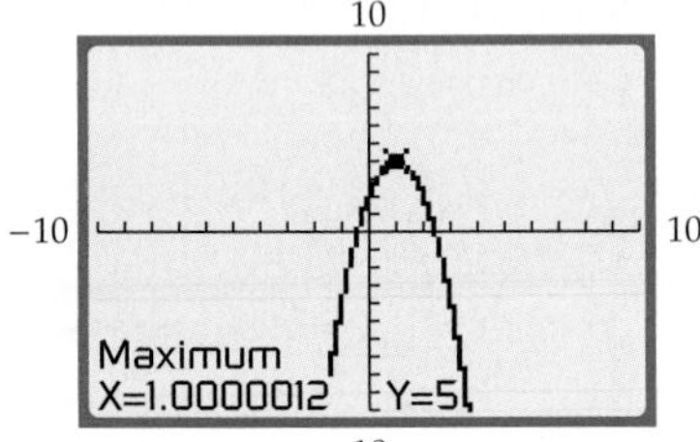

Graphing calculators are programmed to estimate the minimum and maximum values of functions. By performing this calculation for a quadratic function, the vertex of the parabola can be determined.

The determination of the vertex for the equation in Example 3 is shown at the left. For the keystrokes needed to accomplish this, see your owner's manual or visit our web site at **math.college.hmco.com**. Because of rounding errors that occur during the calculations, the exact values of the coordinates of the vertex may not be displayed. This is the case for the graphing calculator screen at the left. The exact coordinates of the vertex are (1, 5).

■ Applications

Alternative to Example 4
Exercise 54, page 230

EXAMPLE 4 Solve a Minimization Application

A mining company has determined that the cost c, in dollars per ton, of mining a mineral is given by $c(x) = 0.2x^2 - 2x + 12$, where x is the number of tons of the mineral mined. Find the number of tons of the mineral that should be mined to minimize the cost. What is the minimum cost?

Solution To find the number of tons of the mineral that should be mined to minimize the cost, and to find the minimum cost, find the x-coordinate and y-coordinate of the vertex of the graph of $c(x) = 0.2x^2 - 2x + 12$.

$$-\frac{b}{2a} = -\frac{-2}{2(0.2)} = 5$$

■ Find the x-coordinate of the vertex. $a = 0.2$, $b = -2$

Continued ➤

To minimize the cost, 5 tons of the mineral should be mined.

$$c(x) = 0.2x^2 - 2x + 12$$
$$c(5) = 0.2(5)^2 - 2(5) + 12$$
$$c(5) = 7$$

▪ Find the y-coordinate of the vertex by replacing x by 5 and solving for y. Recall that the y-coordinate is the value of the function at 5, or $c(5)$.

The minimum cost per ton is \$7.

CHECK YOUR PROGRESS 4 The height s, in feet, of a ball thrown straight up is given by $s(t) = -16t^2 + 64t + 4$, where t is the time in seconds. Find the time it takes the ball to reach its maximum height. What is the maximum height?

Solution *See page S15.* 2 s; 68 ft

Alternative to Example 5
Exercise 48, page 229

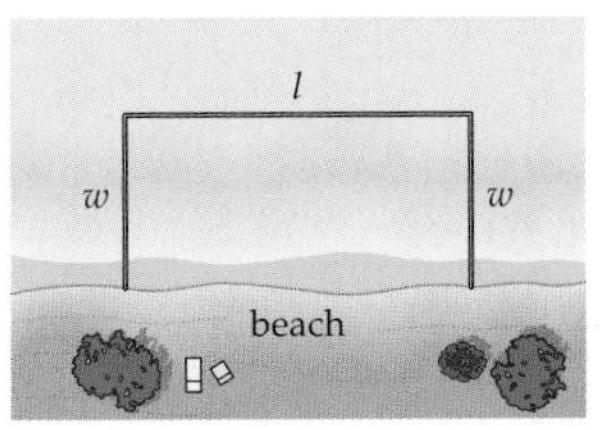

EXAMPLE 5 Solve a Maximization Application

A lifeguard has 600 feet of rope with buoys attached to lay out a rectangular swimming area on a lake. If the beach forms one side of the rectangle, find the dimensions of the rectangle that will create the greatest swimming area.

Solution Let l represent the length of the rectangle, let w represent the width of the rectangle, and let A (which we want to maximize) represent the area of the rectangle. See the figure at the left. Use these variables to write expressions for the length of the rope and area of the rectangle.

Length of the rope: $w + l + w = 600$ ▪ There is 600 feet of rope.

$$2w + l = 600$$

Area: $A = lw$

Our goal is to maximize A. To do this, we first write A in terms of a single variable. This can be accomplished by solving $2w + l = 600$ for l and then substituting into $A = lw$.

$$2w + l = 600$$
$$l = -2w + 600$$ ▪ Solve for l.

$$A = lw$$
$$= (-2w + 600)w$$ ▪ Substitute $-2w + 600$ for l.
$$A = -2w^2 + 600w$$ ▪ Simplify. This is now a quadratic equation.

Find the w-coordinate of the vertex.

$$w = -\frac{b}{2a} = -\frac{600}{2(-2)} = 150$$ ▪ $a = -2$, $b = 600$

The width is 150 feet. To find l, replace w by 150 in $l = -2w + 600$ and solve for l.

$$l = -2w + 600$$
$$l = -2(150) + 600 = -300 + 600 = 300$$

The length is 300 feet. The dimensions of the rectangle are 150 feet by 300 feet.

Continued ➤

INTEGRATING TECHNOLOGY

A graphing calculator check for Example 5 is shown below. It may be necessary to experiment with different viewing windows so that the graph will be displayed correctly.

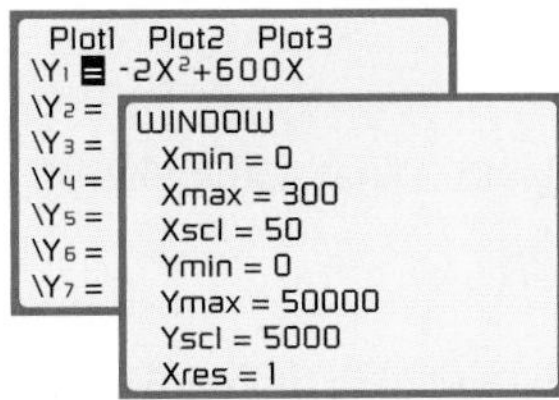

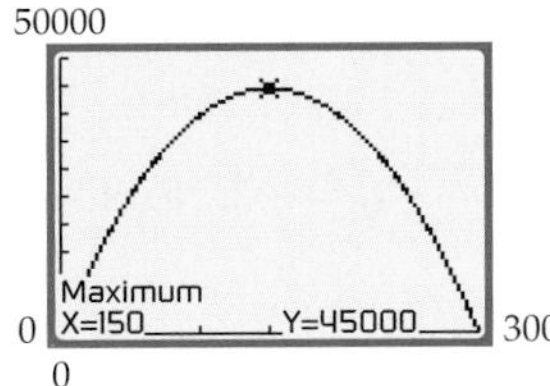

This last screen shows that the x-coordinate of the maximum is 150, as we calculated in Example 5.

CHECK YOUR PROGRESS 5 A gardener has 44 feet of fencing to enclose a rectangular vegetable garden. What dimensions of the rectangular garden will give the gardener the maximum area in which to plant?

Solution *See page S15.* 11 ft by 11 ft

Topics for Discussion

1. For the quadratic function given by $f(x) = ax^2 + bx + c$, explain why we state that $a \neq 0$.
2. Does $f(x) = -x^2 + 4x - 5$ have a maximum value or a minimum value? Explain.
3. How many x-intercepts can the graph of a quadratic function have? How many y-intercepts can the graph of a quadratic function have?
4. What is the axis of symmetry for the graph of a parabola?
5. What is the difference between the graph of $f(x) = x^2$ and the graph of $g(x) = \frac{x^3}{x}$?

EXERCISES 2.5

— Suggested Assignment: Exercises 1–61, odd; and 62–67, all.

17. $\left(\frac{1}{2}, -\frac{9}{4}\right)$; $x = \frac{1}{2}$

1. The axis of symmetry of a parabola is the line $x = 0$. The point $(-2, -3)$ lies on the parabola. Use the symmetry of a parabola to find a second point on the graph. $(2, -3)$
2. The axis of symmetry of a parabola is the line $x = 1$. The point $(3, 0)$ lies on the parabola. Use the symmetry of a parabola to find a second point on the graph. $(-1, 0)$
3. The axis of symmetry of a parabola is the line $x = 2$. The point $(4, -4)$ lies on the parabola. Use the symmetry of a parabola to find a second point on the graph. $(0, -4)$
4. The axis of symmetry of a parabola is the line $x = -1$. The point $(1, -1)$ lies on the parabola. Use the symmetry of a parabola to find a second point on the graph. $(-3, -1)$
5. The axis of symmetry of a parabola is the line $x = -5$. What is the x-coordinate of the vertex of the parabola? -5
6. The axis of symmetry of a parabola is the line $x = 8$. What is the x-coordinate of the vertex of the parabola? 8
7. The vertex of a parabola is $(7, -9)$. What is the axis of symmetry of the parabola? $x = 7$
8. The vertex of a parabola is $(-4, 10)$. What is the axis of symmetry of the parabola? $x = -4$

In Exercises 9 to 20, find the vertex and axis of symmetry of the graph of the equation.

9. $y = x^2 - 2$ $(0, -2)$; $x = 0$
10. $y = x^2 + 2$ $(0, 2)$; $x = 0$
11. $f(x) = -x^2 - 1$ $(0, -1)$; $x = 0$
12. $f(x) = -2x^2 + 3$ $(0, 3)$; $x = 0$
13. $y = -\frac{1}{2}x^2 + 2$ $(0, 2)$; $x = 0$
14. $y = \frac{1}{2}x^2$ $(0, 0)$; $x = 0$
15. $f(x) = 2x^2 + 4x$ $(-1, -2)$; $x = -1$
16. $f(x) = x^2 - 2x$ $(1, -1)$; $x = 1$
17. $f(x) = x^2 - x - 2$
18. $f(x) = x^2 - 3x + 2$ $\left(\frac{3}{2}, -\frac{1}{4}\right)$; $x = \frac{3}{2}$
19. $y = 2x^2 - x - 5$ $\left(\frac{1}{4}, -\frac{41}{8}\right)$; $x = \frac{1}{4}$
20. $y = 2x^2 - x - 3$ $\left(\frac{1}{4}, -\frac{25}{8}\right)$; $x = \frac{1}{4}$

In Exercises 21 to 32, find the x-intercepts and y-intercepts of the parabola given by the equation. 23. $(-2, 0), \left(-\frac{3}{4}, 0\right); (0, 6)$

21. $y = 2x^2 - 4x$ $(0, 0), (2, 0); (0,0)$

22. $y = 3x^2 + 6x$ $(0, 0), (-2, 0); (0, 0)$

23. $f(x) = 4x^2 + 11x + 6$

24. $f(x) = x^2 - 2$ $(\sqrt{2}, 0), (-\sqrt{2}, 0); (0, -2)$

25. $y = x^2 + 2x - 1$ $(-1 - \sqrt{2}, 0), (-1 + \sqrt{2}, 0); (0, -1)$

26. $y = x^2 + 4x - 3$ $(-2 + \sqrt{7}, 0), (-2 - \sqrt{7}, 0); (0, -3)$

27. $y = -x^2 - 4x - 5$ None; $(0, -5)$

28. $y = x^2 - 2x + 5$ None; $(0, 5)$

29. $y = 2x^2 - 8x + 1$

30. $y = 3x^2 + 9x - 3$

31. $y = 3x^2 + 6x - 2$

32. $y = 4x^2 - 2x + 2$ None; $(0, 2)$

29. $\left(\frac{4 - \sqrt{14}}{2}, 0\right), \left(\frac{4 + \sqrt{14}}{2}, 0\right); (0, 1)$

In Exercises 33 to 40, find the minimum or maximum value of the quadratic function.

33. $f(x) = x^2 - 2x + 3$ 2

34. $f(x) = 2x^2 + 4x$ -2

35. $f(x) = -2x^2 + 4x - 5$ -3

36. $f(x) = 3x^2 + 3x - 2$ $-\frac{11}{4}$

37. $f(x) = x^2 + 2x - 4$ -5

38. $f(x) = 2x^2 + 6x - 3$ -7.5

39. $f(x) = -x^2 - x + 2$ $\frac{9}{4}$

40. $f(x) = -3x^2 + 4x - 2$ $-\frac{2}{3}$

41. Which of the following equations has the largest minimum value? c

a. $y = x^2 - 2x - 3$ **b.** $y = x^2 - 10x + 20$

c. $y = 3x^2 - 3$

30. $\left(\frac{-3 - \sqrt{13}}{2}, 0\right), \left(\frac{-3 + \sqrt{13}}{2}, 0\right); (0, -3)$

42. Which of the following equations has the smallest maximum value? a

a. $y = -x^2 + 4x - 1$ **b.** $y = 3 - 2x - 2x^2$

c. $y = -x^2 - 6x - 1$

31. $\left(\frac{-3 - \sqrt{15}}{3}, 0\right), \left(\frac{-3 + \sqrt{15}}{3}, 0\right); (0, -2)$

Business and Economics

43. ***Cost of Producing a Lens*** A camera lens manufacturer estimates that the average monthly cost C, in dollars, of producing a lens is given by the function $C(x) = 0.1x^2 - 20x + 2000$, where x is the number of lenses produced each month. Find the number of lenses the company should produce per month in order to minimize the average cost. 100 lenses

44. ***Physician Income*** The net annual income I, in dollars, of a family physician can be modeled by the function $I(x) = -290(x - 48)^2 + 148{,}000$, where x is the age of the physician in years and $27 \le x \le 70$. Find **a.** the age at which the physician's income will be a maximum and **b.** the maximum income. **a.** 48 years **b.** \$148,000

45. ***Microwave Oven Revenue*** A manufacturer of microwave ovens believes that the revenue R, in dollars, the company receives is related to the price P, in dollars, of an oven by the function $R(P) = 125P - 0.25P^2$. What price of an oven will give the maximum revenue? \$250

46. ***Fuel Efficiency*** The fuel efficiency of an average car is given by the equation $E(v) = -0.018v^2 + 1.476v + 3.4$, where E is the fuel efficiency in miles per gallon and v is the speed of the car in miles per hour. What speed will yield the maximum fuel efficiency? What is the maximum fuel efficiency? 41 mph; 33.658 mpg

Life and Health Sciences

47. ***Algae Growth*** A pool is treated with a chemical to reduce the amount of algae. The amount of algae in the pool t days after the treatment can be approximated by $A(t) = 12t^2 - 168t + 1150$. How many days after treatment will the pool have the least amount of algae? 7 days

48. ***Agriculture*** A rancher has 200 feet of fencing to build a rectangular corral alongside an existing fence. Determine the dimensions of the corral that will maximize the enclosed area. 100 ft by 50 ft

Sports and Recreation

49. ***Pole Vaulting*** At the 2000 Olympic games, Stacy Dragila of the U.S. set an Olympic pole vault record. The path y, in feet, of her jump can be approximated by $y = -6.7037x^2 + 20.1111x$. What was the maximum height of her jump? Round to the nearest hundredth. 15.08 ft

50. ***Stadium Design*** Some football fields are built in a parabolic mound shape so that water will drain off the field. A model for the shape of the field is given by $h(x) = -0.00023475x^2 + 0.0375x$, where h is the height of the field in feet at a distance of x feet from the sideline. What is the maximum height of the field? Round to the nearest tenth. 1.5 ft

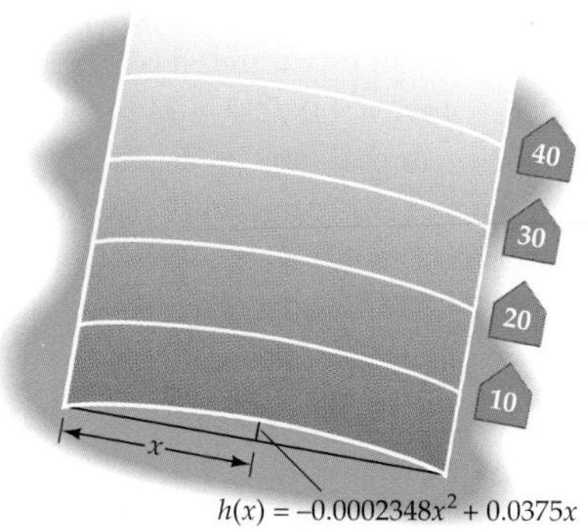

51. ***Path of a Punt*** The path of a punt from a professional football player can be approximated by $s(x) = -0.01432x^2 + 1.1918x + 4.5$, where $s(x)$ is the height of the football, in feet, x feet from the punter. What is the maximum height of the football during the punt? Round to the nearest foot. 29 ft

Physical Science and Engineering

52. ***Projectile Motion*** The height s, in feet, of a rock thrown upward and over a cliff 50 feet above the surface of the ocean at an initial speed of 64 feet per second is given by $s(t) = -16t^2 + 64t + 50$, where t is the time in seconds.
    a. Find the maximum height above the ocean that the rock will attain. 114 ft
    b. What are the t-intercepts and what are their significance in this problem? Round to the nearest tenth.

53. ***Projectile Motion*** The height s, in feet, of a ball thrown upward at an initial speed of 80 feet per second from a platform 50 feet high is given by $s(t) = -16t^2 + 80t + 50$, where t is the time in seconds. Find the maximum height above the ground that the ball will attain. 150 ft

54. ***Suspension Bridges*** The suspension cable that supports a small footbridge hangs in the shape of a parabola. The height h, in feet, of the cable above the bridge is given by $h(x) = 0.25x^2 - 0.8x + 25$, where x is the distance in feet from one end of the bridge. What is the minimum height of the cable above the bridge? 24.36 ft

55. ***Physics*** Karen, who is standing on the ground, is throwing an apple to her brother Saul, who is standing on the balcony of their home. The height h, in feet, of the apple above the ground t seconds after it is thrown is given by $h(t) = -16t^2 + 32t + 4$. If Saul's outstretched arms are 18 feet above the ground, will the apple ever be high enough so that he can catch it? Yes

56. ***Fountain Construction*** The Buckingham Fountain in Chicago shoots water from a nozzle at the base of the fountain. The height h, in feet, of the water above the ground t seconds after it leaves the nozzle is approximated by

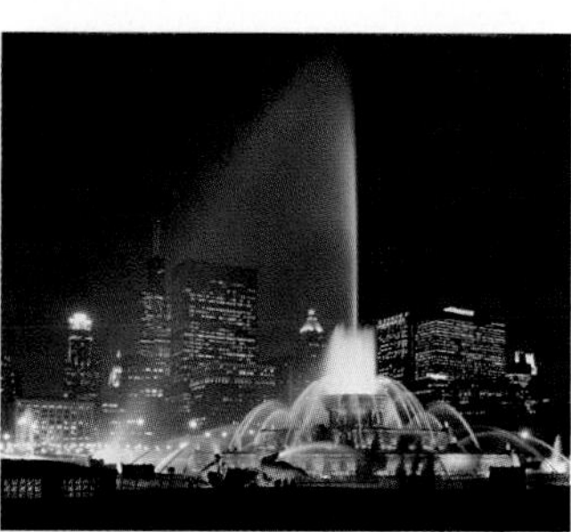

$$h(t) = -16t^2 + 90t + 15.$$

What is the maximum height of the water spout to the nearest tenth of a foot? 141.6 ft

57. ***Stopping Distance*** On wet concrete, the stopping distance s, in feet, of a car traveling v miles per hour is given by $s(v) = 0.55v^2 + 1.1v$. At what speed could a car be traveling and still stop at a stop sign 44 feet away? 8 mph

In Exercises 58 and 59, find the value of k such that the graph of the function contains the given point.

58. $f(x) = x^2 - 3x + k; (2, 5)$ $k = 7$

59. $f(x) = 2x^2 + kx - 3; (4, -3)$ $k = -8$

60. For $f(x) = 2x^2 - 5x + k$, we have $f\left(-\frac{3}{2}\right) = 0$. Find another value of x for which $f(x) = 0$. $x = 4$

61. For what values of k does the graph of

$$f(x) = 2x^2 - kx + 8$$

just touch the x-axis without crossing it? $k = 8, -8$

Prepare for Section 2.6

62. Find the slope of the line between $P_1(3, -4)$ and $P_2(-1, -2)$. [2.3] $-\frac{1}{2}$

63. Find the x- and y-intercepts of the graph of $3x + 4y = 12$. [2.3] (4, 0), (0, 3)

64. What is the slope and y-intercept of $y = 2 - \frac{1}{2}x$? [2.3]

64. Slope: $-\frac{1}{2}$, y-intercept: (0, 2)

52. b. The t-intercepts, rounded to the nearest tenth, are (−0.7, 0) and (4.7, 0). Because −0.7 represents negative time, it has no significance in this problem. The positive t-intercept means that the rock will hit the ocean approximately 4.7 s after it is released.

65. Find the equation of the line passing through $P(2, 3)$ with slope -2. [2.4] $y = -2x + 7$

66. Find the equation of the line passing through $P_1(-2, 1)$ and $P_2(1, -3)$. [2.4] $y = -\frac{4}{3}x - \frac{5}{3}$

67. Water is being poured into a tank such that the height h, in feet, of the water can be approximated by $h = 3t + 5$, where t is the number of minutes the water has been flowing into the tank.

   a. At what rate, in feet per minute, is the water rising in the tank? 3 ft/min

   b. At what height was the water in the tank when the valve was turned on? [2.4] 5 ft

Explorations 

Reflective Properties of a Parabola The fact that the graph of a quadratic equation in two variables is a parabola is based on the following geometric definition of a parabola.

> **Definition of a Parabola**
>
> A **parabola** is the set of points in the plane that are equidistant from a fixed line (the **directrix**) and a fixed point (the **focus**) not on the directrix.

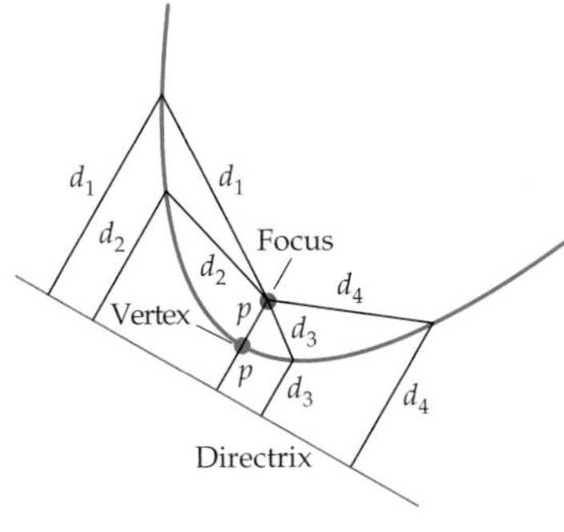

This geometric definition of a parabola is illustrated in the figure at the left. Basically, for a point to be on a parabola, the distance from the point to the focus must equal the distance from the point to the directrix. Note also that the vertex is halfway between the focus and the directrix. This distance is traditionally labeled p.

The equation of a parabola that opens up with vertex at the origin can be written in terms of the distance p between the vertex and the focus as $y = \frac{1}{4p}x^2$. For this equation, the coordinates of the focus are $(0, p)$. For instance, to find the coordinates of the focus for $y = 2x^2$, let $\frac{1}{4p} = 2$ and solve for p.

$$\frac{1}{4p} = 2$$

$$\frac{1}{4} = 2p \qquad \text{▪ Multiply each side of the equation by } p.$$

$$\frac{1}{8} = p \qquad \text{▪ Divide each side of the equation by 2.}$$

The coordinates of the focus are $\left(0, \frac{1}{8}\right)$.

1. Find the coordinates of the focus for the parabola whose equation is $y = 0.4x^2$. $\left(0, \frac{5}{8}\right)$

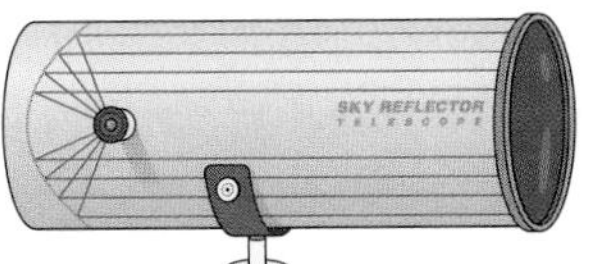

Optical telescopes operate on the same principle as radio telescopes, except that light hits a mirror that has been shaped into a paraboloid. The light is reflected to the focus, where another mirror reflects it through a lens to the observer. See the diagram above.

2. 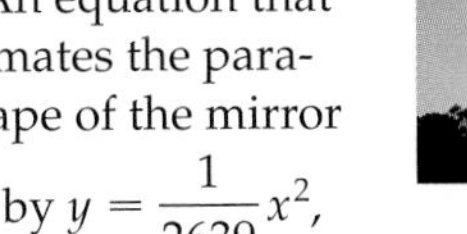 The telescope at the Palomar Observatory in California has a mirror that is 200 inches across. An equation that approximates the parabolic shape of the mirror is given by $y = \frac{1}{2639}x^2$, where x and y are measured in inches. How far is the focus from the vertex of the mirror? 659.75 in.

Palomar Observatory with the shutters open

If a point on a parabola whose vertex is at the origin is known, then the equation of the parabola can be found. For instance, if (4, 1) is a point on a parabola with vertex at the origin, then we find the equation of the parabola as follows.

$$y = \frac{1}{4p}x^2 \qquad \text{▪ Begin with the equation of the parabola.}$$

$$1 = \frac{1}{4p}(4)^2 \qquad \text{▪ The known point is (4, 1). Replace } x \text{ by 4 and } y \text{ by 1.}$$

$$1 = \frac{4}{p} \qquad \text{▪ Solve for } p.$$

$$p = 4$$

The equation of the parabola is $y = \frac{1}{16}x^2$.

3. Find a flashlight and measure the diameter of the top and the height of the reflective surface. See the diagram at the right. If a coordinate axis is set up as shown, find the equation of the parabola. Answers will vary.

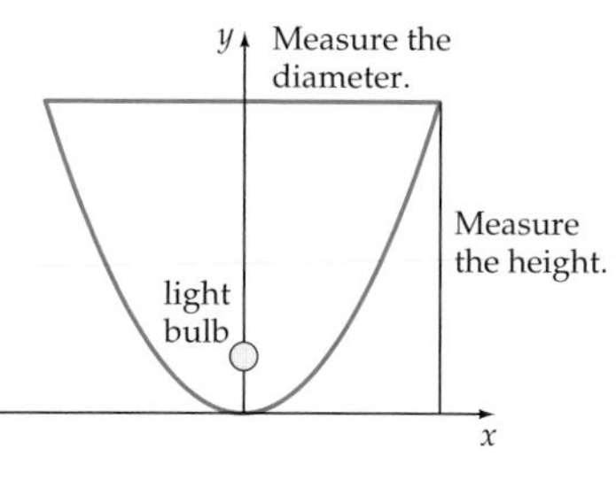

4. Find the location of the focus. Explain why the light bulb should be placed at this point. Answers will vary.

5. $\left(0, \frac{g}{2\omega^2}\right)$

A new type of mirror for a telescope is called a liquid mirror and is created by placing mercury in a container and then rotating the container. The shape of the mercury forms a paraboloid. The equation of the parabolic cross-section is given by $y = \frac{\omega^2}{2g}x$, where $\omega = 2\pi t$ (t is the number of revolutions per second) is the angular speed of the container and g is the acceleration due to gravity.

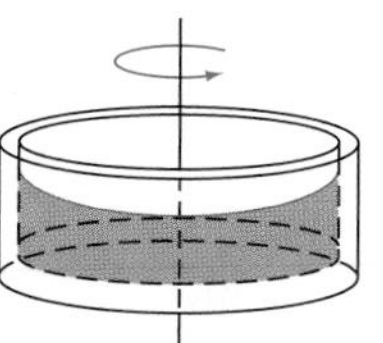

5. Express the coordinates of the focus in terms of ω^2 and g.

6. If a container makes 1 revolution in 5 seconds and $g = 32$ feet per second squared, find the distance from the vertex of the mirror to the focus. 10.13 ft

7. How many revolutions per second are necessary to have a focus 4 feet above the vertex of the mirror? 0.32 rps

SECTION 2.6 Modeling Data Using Regression

- Linear Regression
- Correlation Coefficient

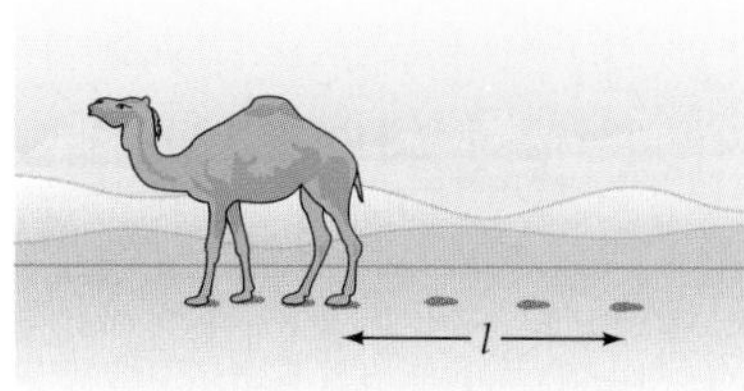

Figure 2.31

▪ Linear Regression

In many applications, scientists try to determine whether two variables are related. If they are related, they then try to find an equation that can be used to *model* the relationship. For instance, zoology professor R. McNeill Alexander wanted to determine whether the *stride length* of a dinosaur, as shown by its fossilized footprints, could be used to estimate the speed of the dinosaur. *Stride length* for an animal is defined as the distance l from a particular point on a footprint to that same point on the next footprint of the same foot. See Figure 2.31.

Because no dinosaurs were available, Alexander and fellow scientist A. S. Jayes carried out experiments with many types of animals, including adult men, dogs, camels, ostriches, and elephants. The results of these experiments tended to support the idea that the speed s of an animal is related to the animal's stride length l. To better understand this relationship, examine the data in Table 2.1, which is similar to, but less extensive than, the data collected by Alexander and Jayes.

Table 2.1 Speed for Selected Stride Lengths

a. Adult men

Stride length (meters)	2.5	3.0	3.3	3.5	3.8	4.0	4.2	4.5
Speed (meters per second)	3.4	4.9	5.5	6.6	7.0	7.7	8.3	8.7

b. Dogs

Stride length (meters)	1.5	1.7	2.0	2.4	2.7	3.0	3.2	3.5
Speed (meters per second)	3.7	4.4	4.8	7.1	7.7	9.1	8.8	9.9

c. Camels

Stride length (meters)	2.5	3.0	3.2	3.4	3.5	3.8	4.0	4.2
Speed (meters per second)	2.3	3.9	4.4	5.0	5.5	6.2	7.1	7.6

A graph of the ordered pairs in Table 2.1 is shown in Figure 2.32. In this graph, which is called a **scatter diagram** or **scatter plot,** the horizontal axis represents the stride lengths in meters and the vertical axis represents the average speeds in meters per second. The scatter diagram seems to indicate that for each of the three species, a larger stride length generally produces a faster speed. Also notice that for each species, a line can be drawn such that all of the points representing that species are on or very close to the line. Thus the relationship between speed and stride length appears to be a linear relationship.

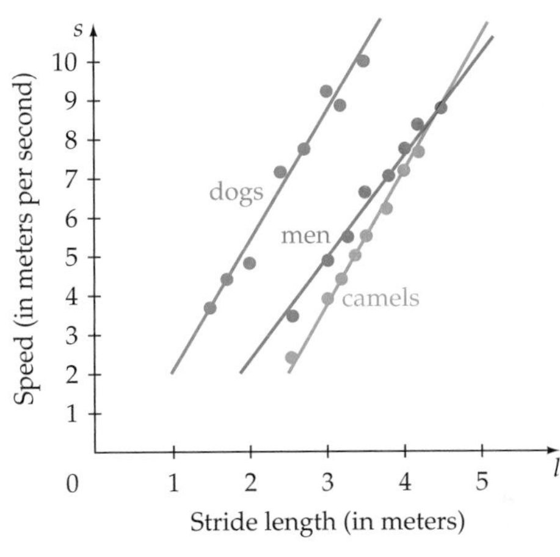

Figure 2.32

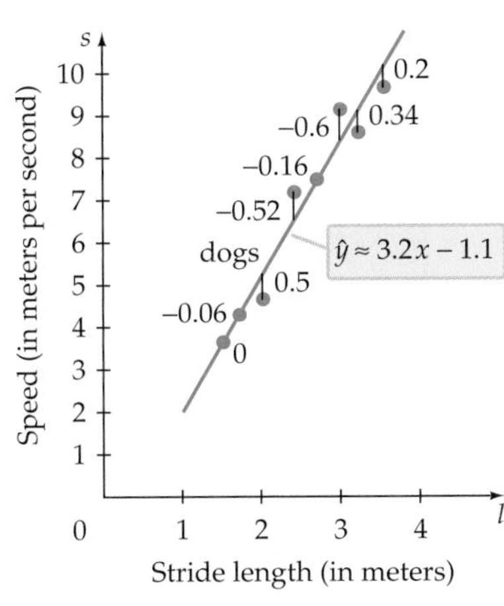

Figure 2.33

To find an equation that models that relationship, we could use the techniques of the previous section—select two of the data points and then find the equation of the line between those two points. However, scientists are generally interested in the line called the *line of best fit* or the *least-squares regression line.*

Definition of the Least-Squares Regression Line

The **least-squares regression line** of a set of data points is the line that minimizes the sum of the squares of the vertical deviations from each data point to the regression line.

The least-squares regression line is also called the least-squares line. The equation, with values rounded to the nearest tenth, of the least-squares line for the data that concerns the dogs is $\hat{y} = 3.2x - 1.1$. Figure 2.33 shows the graph of these data and the graph of the least-squares line.

In Figure 2.33, the vertical deviations from the ordered pairs to the least-squares line are 0, −0.06, 0.5, −0.52, −0.16, −0.6, 0.34, and 0.2. The sum of the squares of these vertical deviations is

$$0^2 + (-0.06)^2 + 0.5^2 + (-0.52)^2 + (-0.16)^2 + (-0.6)^2 + 0.34^2 + 0.2^2 = 1.0652$$

The graph of $\hat{y} = 3.2x - 1.1$ is the least-squares line for the data because for all other lines, the sum of the squares of the vertical deviations is greater than 1.0625. It is traditional to use the symbol $\hat{y}$ (pronounced y-hat) in place of y in the equation of a least-squares line. This also helps us differentiate its y-values from the y-values of the given ordered pairs.

The equations used to calculate a regression line are somewhat cumbersome. Fortunately, these equations are preprogrammed into most graphing calculators. We will now illustrate the technique for a TI-83 calculator using the data set for the stride length and speed of a dog. On our web site at **math.college.hmco.com,** there are instructions for some other calculators.

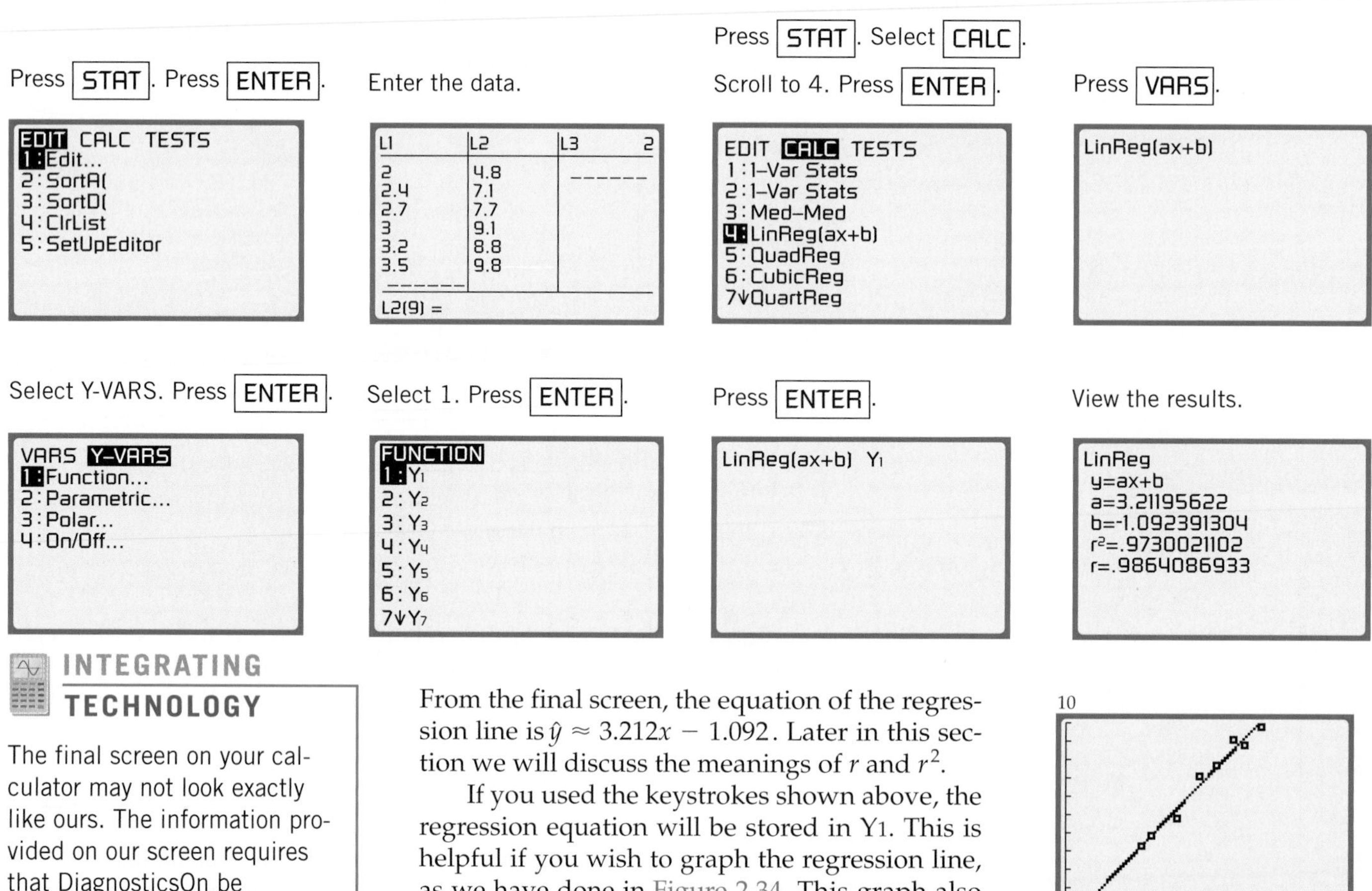

INTEGRATING TECHNOLOGY

The final screen on your calculator may not look exactly like ours. The information provided on our screen requires that DiagnosticsOn be enabled. This is accomplished by using the following keystrokes: 2nd CATALOG D (Scroll to DiagnosticsOn) ENTER ENTER.

From the final screen, the equation of the regression line is $\hat{y} \approx 3.212x - 1.092$. Later in this section we will discuss the meanings of r and r^2.

If you used the keystrokes shown above, the regression equation will be stored in Y1. This is helpful if you wish to graph the regression line, as we have done in Figure 2.34. This graph also shows the data points. These are displayed by using STATPLOT, located above the Y= key.

Figure 2.34

The data on the next page show the horsepower and EPA mileage estimates for the 10 best cars of 2002, as ranked by *Car and Driver* magazine. We will use this data in Example 1. A scatter plot of the data is shown in Figure 2.35.

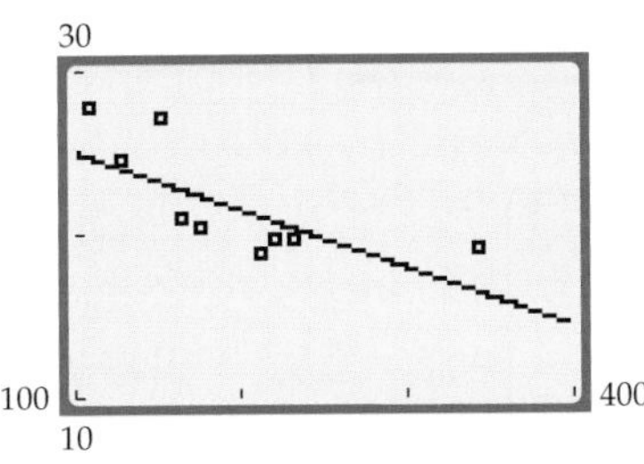

Figure 2.35

Horsepower and EPA Mileage Estimates for 10 Selected Cars

Car model	Horse-power	Mileage (mpg)	Car model	Horse-power	Mileage (mpg)
Acura RSX	160	27	Ford Focus	110	28
Audi A4	170	22	Honda Accord	135	25
BMW 3 Series	184	21	Honda S2000	240	20
BMW 5 Series	184	21	Porsche Boxster	217	19
Corvette	350	19	Subaru Impreza	227	20

Source: **www.caranddriver.com**

Alternative to Example 1
Exercise 14, page 240

```
LinReg
 y=ax+b
 a=-.0379534727
 b=29.70340156
 r²=.5962894986
 r=-.7721978364
```

EXAMPLE 1 Find a Linear Regression Equation

Find the **linear regression equation** that predicts EPA mileage for a given horsepower engine. Using this model, what is the EPA mileage estimate for a 250-horsepower car? Round to the nearest whole number.

Solution Using your calculator, enter the data from the table. Then have the calculator produce the values for the regression equation. Your results should be similar to those shown at the left. The equation of the regression line, with constants rounded to the nearest thousandth, is $y = -0.038x + 29.703$.

To find the EPA mileage estimate for a 250-horsepower car, evaluate the regression equation for $x = 250$.

$$\begin{aligned} y &= -0.038x + 29.703 \\ &= -0.038(250) + 29.703 \\ &\approx 20 \end{aligned}$$

The equation predicts that the EPA mileage estimate for a 250-horsepower car is approximately 20 miles per gallon.

INTEGRATING TECHNOLOGY

If you followed the steps we gave earlier and stored the regression equation in Y1, you can evaluate the regression equation using the following keystrokes.

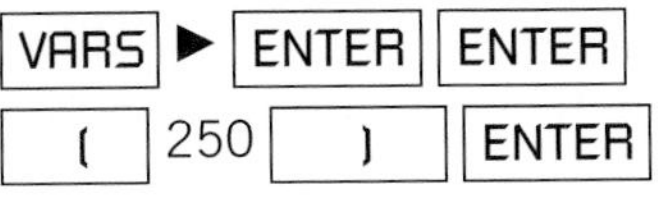

CHECK YOUR PROGRESS 1 Using the data at the beginning of this section, find the linear regression equation, with constants rounded to the nearest thousandth, that predicts the speed of a camel given its stride length. What speed does the regression equation predict for a camel with a stride length of 3 meters? Round to the nearest tenth.

Solution *See page S15.* $y = 3.130x - 5.547$; 3.8 m/s

▪ Correlation Coefficient

If the *linear correlation coefficient* r of a linear equation is positive, then the relationship between the variables is a **positive correlation.** In this case if one variable increases, then the other variable also tends to increase. If r is negative, then the linear relationship between the variables is a **negative correlation.** In this case if one variable increases, the other variable tends to decrease. Figure 2.36 shows some scatter diagrams, along with the type of *linear correlation* that exists between the x and y variables. If r is positive, then the closer r is to 1, the stronger the linear relationship between the variables. If r is negative, then the closer r is to -1, the stronger the linear relationship between the variables.

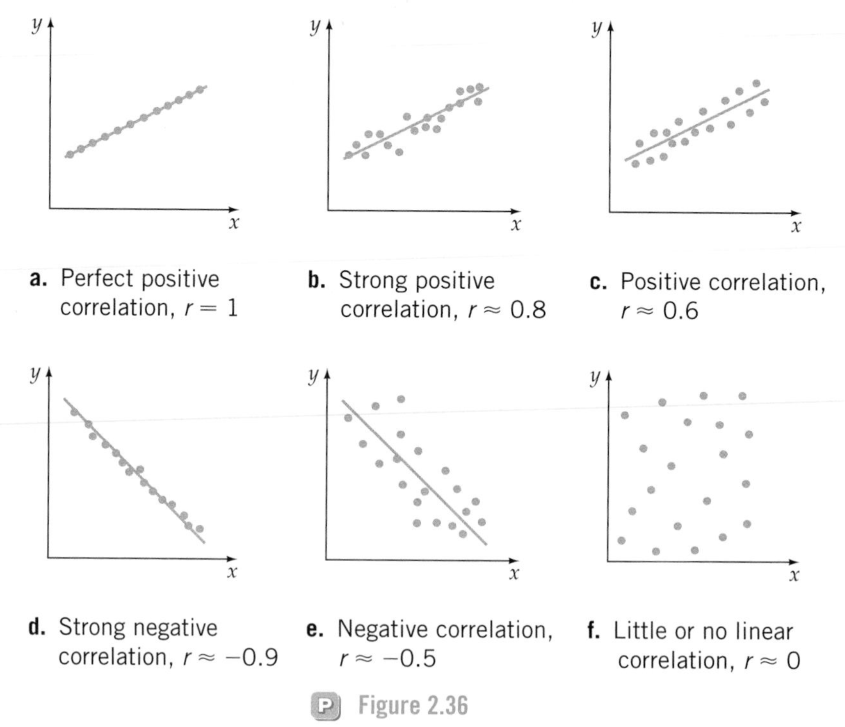

a. Perfect positive correlation, $r = 1$

b. Strong positive correlation, $r \approx 0.8$

c. Positive correlation, $r \approx 0.6$

d. Strong negative correlation, $r \approx -0.9$

e. Negative correlation, $r \approx -0.5$

f. Little or no linear correlation, $r \approx 0$

Figure 2.36

Point of Interest

Karl Pearson (1857–1936) spent most of his career as a mathematics professor at University College, London. Some of his major contributions concerned the development of statistical procedures such as regression analysis and correlation. He was particularly interested in applying these statistical concepts to the study of heredity. The term standard deviation was invented by Pearson. Because of his work in the area of correlation, the formal name given to the linear correlation coefficient is the Pearson product moment coefficient of correlation. Pearson was a co-founder of the statistical journal Biometrika.

Linear Correlation Coefficient

The **linear correlation coefficient** r is a measure of how closely the points of a data set can be modeled by a straight line. If $r = -1$, the data set can be modeled *exactly* by a straight line with a negative slope. If $r = 1$, the data set can be modeled *exactly* by a straight line with a positive slope. For all data sets, $-1 \leq r \leq 1$.

A scatter plot of the horsepower rating of a car and its EPA mileage estimate is shown in Figure 2.37, along with the graph of the regression line. Note that the slope of the regression line is negative. This indicates that as the horsepower of a car increases, its mileage estimate decreases. Note also that for these data, the value of r on the calculator screen is negative, $r \approx -0.772$.

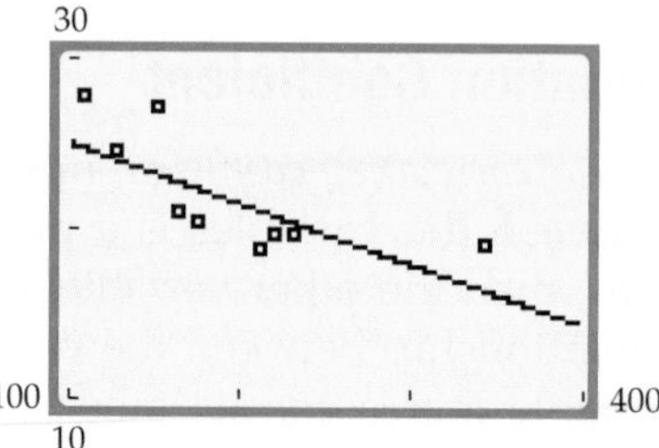


```
LinReg
 y=ax+b
 a=-.0379534727
 b=29.70340156
 r²=.5962894986
 r=-.7721978364
```

Figure 2.37

Alternative to Example 2
Exercise 22, page 242

TAKE NOTE

The data for the Porsche Boxster assumes that the condition of the car is excellent. The only variable that changes is the odometer reading.

EXAMPLE 2 Find a Linear Correlation Coefficient

  The data in the table below show the trade-in values of a 1998 Porsche Boxster for various odometer readings.

Trade-in Value of a 1998 Porsche Boxster, June 2002

Odometer reading (in thousands)	Trade-in value (in dollars)
30	27,075
35	26,775
40	26,275
45	25,950
55	24,975

Source: Kelley Blue Book web site, **www.kbb.com**

Find the linear correlation coefficient for a regression model that predicts trade-in value on the basis of odometer reading.

```
LinReg
 y=ax+b
 a=-84.59459459
 b=29678.37838
 r²=.9937364649
 r=-.9968633131
```

Solution Enter the data from the table and then use the calculator keystrokes for finding a regression equation. From the screen at the left, the correlation coefficients is approximately −0.997.

CHECK YOUR PROGRESS 2   The number of runs scored and the number of games won for selected Major League Baseball teams in the 2001 season are shown in the table below. Find the correlation coefficient for a linear regression model that predicts the number of games won on the basis of the number of runs scored.

Team	Runs scored	games won
Seattle Mariners	927	116
Oakland Athletics	884	102
New York Yankees	804	95
Boston Red Sox	772	82
Anaheim Angels	691	75
Toronto Blue Jays	767	80
Minnesota Twins	771	85
Chicago White Sox	798	83
Cleveland Indians	897	91

Source: **espn.com**

Solution *See page S15.* 0.876

The scatter plots for the stride length data for dogs and the Porsche Boxster data are shown in Figure 2.38, along with the graphs of their respective regression lines.

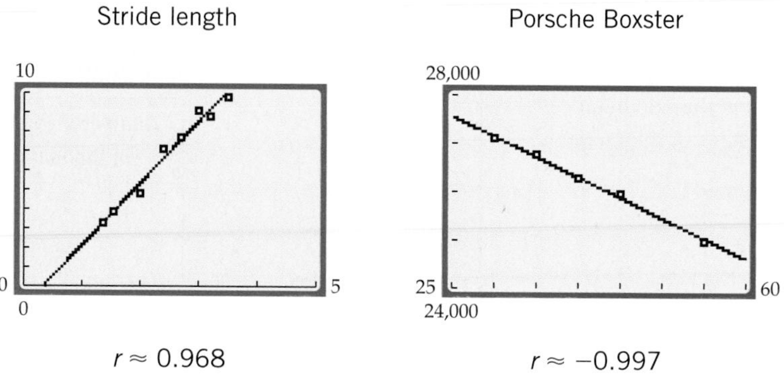

Figure 2.38

For the stride length data, the slope of the regression line is positive and r is positive. For the Porsche data, the slope of the regression line is negative and r is negative. The sign of r and the sign of the slope are always the same.

Researchers calculate a regression equation to determine a relationship between two variables. The researcher wants to know whether a change in one variable produces a predictable change in the second variable. The value of r^2 tells the researcher the extent of that relationship.

Coefficient of Determination

The **coefficient of determination** is r^2. It measures the percent of the total variation in the dependent variable that is explained by the regression equation.

For the horsepower/EPA mileage data that precedes Example 1, $r^2 \approx 0.596$. This means that approximately 60% of the total variation in the dependent variable (horsepower) can be attributed to the independent variable (EPA mileage estimate). This also means that car horsepower alone does not predict with certainty the EPA mileage estimate. Other factors, such as aerodynamic design, are also involved in the mileage rating of a car.

? QUESTION What is the coefficient of determination for the odometer reading/trade-in value data for the Porsche Boxster, given in Example 2, and what is its significance?

? ANSWER $r^2 \approx 0.994$. This means that for a Porsche Boxster in excellent shape, 99.4% of the total variation in trade-in value can be attributed to the odometer reading.

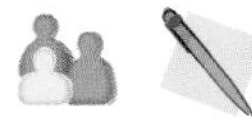

Topics of Discussion

1. What is the purpose of calculating the equation of a regression line?
2. Discuss the implications of the following correlation coefficients: $r = -1$, $r = 0$, and $r = 1$.
3. Discuss the coefficient of determination and what its value says about a data set.
4. Suppose an infant's weight is measured once a month for 2 years. If a regression equation were calculated for the weight of the infant in terms of the age of the infant in months, would r be positive, zero, or negative? Why?
5. Suppose a person purchases a new car. If data were collected giving the value of the car in terms of its age, would r be positive, zero, or negative? Why?
6. Suppose that in a college history class, data are collected giving the heights of students and the students' scores on an exam. Would you expect r to be close to 1, close to 0, or close to -1? Why?

EXERCISES 2.6

— Suggested Assignment: Exercises 1–25, odd; and 26.

In Exercises 1 to 4, use the scatter diagram to determine whether the correlation coefficient is negative, close to zero, or positive.

1. 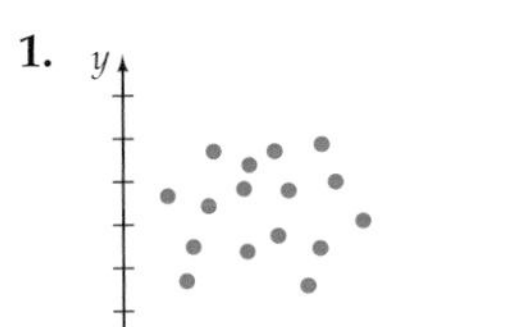

Zero

2. 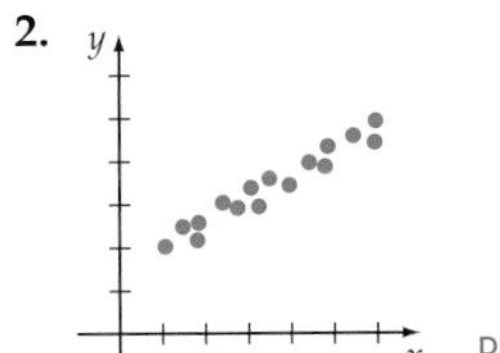

Positive

3. Negative

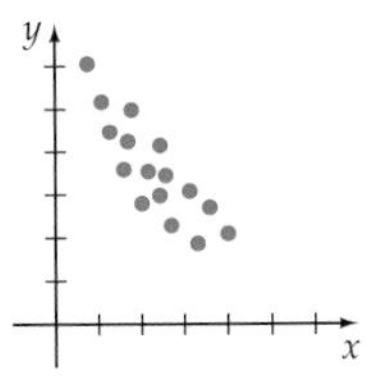

4. Positive

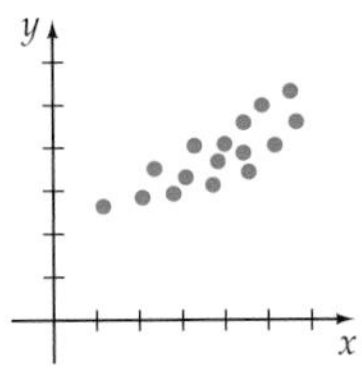

In Exercises 5 and 6, determine for which scatter diagram, A or B, the coefficient of determination is closer to 1.

5. 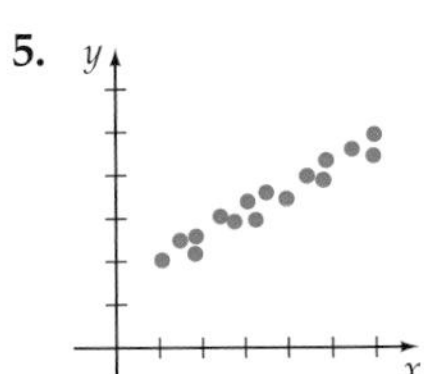

Figure A

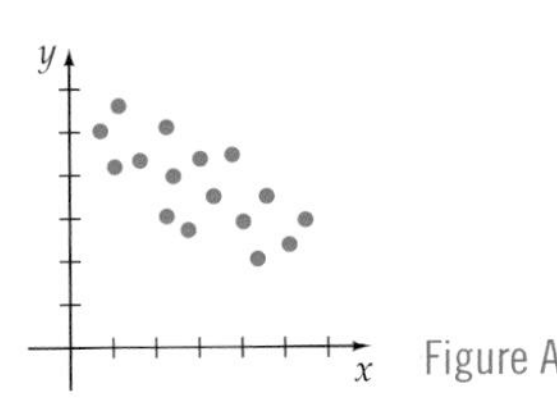

Figure B

Figure A

6. 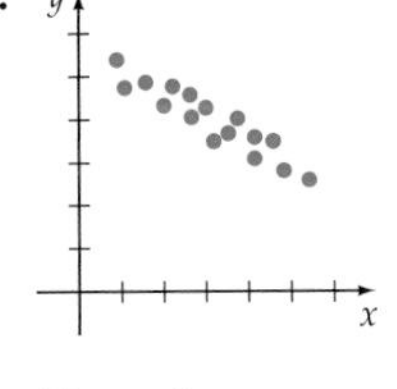

Figure A

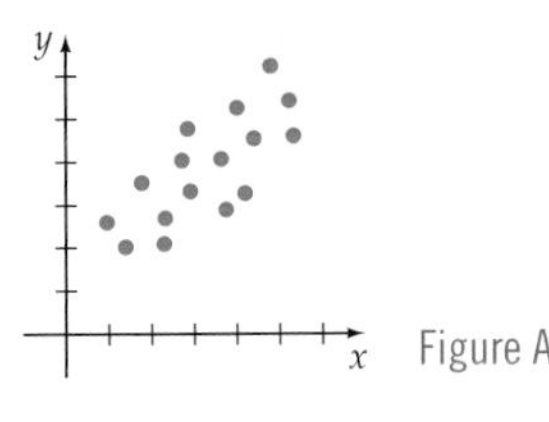

Figure B

Figure A

In Exercises 7 to 13, find the linear regression equation for the given set. Round the slope and y-intercept to the nearest thousandth.

7. $\{(2, 6), (3, 6), (4, 8), (6, 11), (8, 18)\}$ $y = 2.009x + 0.560$
8. $\{(2, -3), (3, -4), (4, -9), (5, -10), (7, -12)\}$
$y = -1.919x + 0.459$
9. $\{(-3, 11.8), (-1, 9.5), (0, 8.6), (2, 8.7), (5, 5.4)\}$
$y = -0.723x + 9.234$
10. $\{(-7, -11.7), (-5, -9.8), (-3, -8.1), (1, -5.9), (2, -5.7)\}$
$y = 0.659x - 6.658$
11. $\{(1.3, -4.1), (2.6, -0.9), (5.4, 1.2), (6.2, 7.6), (7.5, 10.5)\}$
$y = 2.223x - 7.364$
12. $\{(-1.5, 8.1), (-0.5, 6.2), (3.0, -2.3), (5.4, -7.1), (6.1, -9.6)\}$
$y = -2.302x + 4.814$
13. $\{(-1, -3.1), (0, -2.9), (1, 0.8), (2, 6.8), (3, 15.9)\}$
$y = 4.77x - 1.27$

Business and Economics

When computing the linear regression equation in the following exercises, round the constants to the nearest ten-thousandths.

14. *Trade-in Value of a Car* The data in the table below show the trade-in values for a 2-door, 1998 Ford Explorer utility vehicle in excellent condition for various odometer readings (in thousands of miles).

Trade-in Value of a 1998 Ford Explorer, June 2002

Odometer reading	Trade-in value	Odometer reading	Trade-in value
45	$9560	65	$8660
50	$9435	75	$8110
55	$9285	90	$7410
60	$8960		

Source: Kelley Blue Book web site, **www.kbb.com**

a. Compute the linear regression equation for these data. $y = -50.1856x + 11928.8119$

b. On the basis of this model, what is the expected trade-in value of a similar Ford Explorer with 80,000 miles on the odometer? Round to the nearest whole number. $7914

15. *EPA Estimates* The table below shows the EPA mileage estimates for city and highway driving for 10 selected 2002 cars.

EPA mileage estimates for selected cars

Car	City mpg	Highway mpg
Toyota MR2	25	30
Mazda MX5	23	28
Audi TT	22	31
BMW Z3	21	28
Mitsubishi Eclipse	22	30
Honda Civic HX	36	44
Suzuki Swift	36	42
Toyota Echo	34	41
Chevrolet Prizm	32	42
Mitsubishi Mirage	32	39

Source: **money.cnn.com**

a. Compute the linear regression equation for these data. $y = 1.0268x + 6.4402$

b. On the basis of this model, estimate the highway miles per gallon for a car whose EPA city estimate is 24 miles per gallon. Round to the nearest whole number. 31 mpg

16. *Manufacturing a Motor* Permanent magnet direct current motors are used in a variety of industrial applications. For these motors to be effective, there must be a strong linear relationship between the current (in amps, A) supplied to the motor and the resulting torque (in Newton-centimeters, N-cm) produced by the motor. A randomly selected motor is chosen from a production line and tested with the following results.

Direct Current Motor Data at 12 Volts

A	N-cm	A	N-cm
7.3	9.4	8.5	8.6
11.9	2.8	7.9	4.3
5.6	5.6	14.5	9.5
14.2	4.9	12.7	8.3
7.9	7.0	10.6	4.7

Based on the data in this table, is the chosen motor effective? Explain. No. $r \approx 0.001$, which is close to 0.

Life and Health Sciences

17. *Dinosaur Wingspan* Using data from fossils, archeologists are able to estimate various characteristics of dinosaurs. Suppose an archeologist compiles the data in the table below, which shows the length, in centimeters, of the humerus and the total wingspan, in centimeters, of several pterosaur, which are extinct flying reptiles of the order Pterosauria.

Pterosaur Data

Humerus	Wingspan	Humerus	Wingspan
24	548	20	490
32	679	27	591
22	437	15	300
17	379	15	285
13	277	9	185
4.4	83	4.4	76
3.2	43	3.2	49
1.5	8	5.6	110

a. Compute the linear regression equation for these data. $y = 22.6029x - 21.8128$

b. On the basis of this model, what is the projected wingspan of the pterosaur Quetzalcoatlus northropi, which is thought to be the largest of the prehistoric birds, if its humerus is 54 centimeters? Round to the nearest whole number. 1199 cm

18. ***Botany*** The data in the table are based on a study by R. A. Fisher of various flowers of the Iris family. The width, in centimeters, and length, in centimeters, of the petal for selected Versicolor species are shown in the table.

Iris Data

Width	Length	Width	Length
13	45	16	47
14	47	12	40
10	33	10	41
15	45	10	33
14	39	12	39

Source: Fisher, R. A. The Use of Multiple Measurements in Axonomic Problems. *Annals of Eugenics 7*, 1936, 179–188.

a. Compute the linear regression equation for these data. $y = 1.9009x + 16.9481$

b. On the basis of the model, what is the estimated petal length of an iris that has a petal width of 18 centimeters? Round to the nearest whole number. 51 cm

19. ***Botany*** The study by R. A. Fisher (see Exercise 18) also measured the width, in centimeters, and length, in centimeters, of the sepal for these flowers. Some of the data are shown below.

Iris Data

Width	Length	Width	Length
28	57	33	63
32	70	26	58
23	50	27	58
29	60	24	49
27	52	27	58

a. Compute the linear regression equation for these data. $1.7534x + 9.1063$

b. On the basis of the equation, what is the estimated sepal length of an Iris that has a sepal width of 25 centimeters? Round to the nearest whole number. 53 cm

20. ***Health and Fitness*** The body mass index (BMI) of a person is a measure of the person's ideal body weight. The data in the table below show the change in BMI for different weights for a person 5 feet 6 inches tall.

BMI Data for Person 5′6″ Tall

Weight (pounds)	BMI	Weight (pounds)	BMI
110	17.8	160	25.6
120	19.4	170	27.4
125	20.2	180	29.0
135	21.8	190	30.7
140	22.6	200	32.3
145	23.4	205	33.1
150	24.2	215	34.7

Source: National Center for Chronic Disease Prevention

a. Find the linear regression equation for the data. $y = 0.1611x + 0.0299$

b. On the basis of this model, what is the estimated BMI for a person 5 feet 6 inches tall whose weight is 158 pounds? Round to the nearest tenth. 25.5

21. ***Health and Fitness*** The BMI (see Exercise 20) of a person depends on height as well as weight. The table below shows the change in BMI for a 150-pound person as height (in inches) changes.

BMI Data for 150-pound Person

Height (inches)	BMI	Height (inches)	BMI
60	29.3	71	20.9
62	27.4	72	20.3
64	25.7	73	19.8
66	24.2	74	19.3
67	23.4	75	18.7
68	22.8	76	18.3
70	21.5		

Source: National Center for Chronic Disease Prevention

a. Find the linear regression equation for the data. $y = -0.6747x + 69.0371$

b. On the basis of the model, what is the estimated BMI for a 150-pound person who is 5 feet 8 inches tall? Round to the nearest tenth. 23.2

22. ***Life Expectancy for Men*** The average remaining lifetimes for men of various ages in the U.S. are given in the table below.

Average Remaining Lifetime for Men

Age	Years	Age	Years
0	73.9	65	16.1
15	59.8	75	10.0
35	41.1		

Source: National Institute of Health

Based on the data in this table, is there a strong correlation between a man's age and the average remaining lifetime for that man? Explain.
Yes. $r \approx -0.999$, which is very close to -1.

23. ***Life Expectancy for Women*** The average remaining lifetimes for women of various ages in the U.S. are given in the table below.

Average Remaining Lifetime for Women

Age	Years	Age	Years
0	79.4	65	19.1
15	65.1	75	12.1
35	45.7		

Source: National Institute of Health

a. Based on the data in this table, is there a strong correlation between a woman's age and the average remaining lifetime for that woman? Yes

b. Find the linear regression equation for the data.
$y = -0.9040x + 78.6309$

c. On the basis of this model, what is the estimated remaining lifetime of a woman of age 25? Round to the nearest tenth. 56.0 yr

24. ***Infant Development*** The table below gives the typical weight, in pounds, of a normally developing infant boy at various ages in months.

Infant Development

Months	Weight	Months	Weight
12	22.8	24	27.7
15	24.5	27	28.9
18	25.5	30	30.0
21	26.8	33	30.6

Source: National Institute of Health, Centers for Disease Control

Based on these data, what is the expected weight of a child who is 3 years old? 32.1 lb

25. ***Infant Development*** The table below shows the typical head circumference, in centimeters, of a normally developing infant girl at various ages in months.

Infant Development

Months	Head circumference	Months	Head circumference
12	17.7	24	18.7
15	18.2	27	18.9
18	18.3	30	19.1
21	18.5	33	19.2

Source: National Institute of Health, Centers for Disease Control

Is there a strong linear relationship between age and head circumference? Explain.
Yes. $r \approx 0.984$, which is very close to 1.

Physical Sciences and Engineering

26. ***Astronomy*** In 1929, Edwin Hubble published a paper that revolutionized astronomy. His paper discussed the distance from extra-galactic nebula to the Milky Way galaxy and the nebula's velocity with respect to the Milky Way. The data are given in the table on the following page. Distance is measured in megaparsecs (1 megaparsec equals 1.918×10^{19} miles) and velocity (called *recession velocity*) in kilometers per second. A negative velocity means the nebula is moving toward the Milky Way; a positive velocity means the nebula is moving away from the Milky Way.

Recession Velocities

Distance	Velocity	Distance	Velocity
0.032	170	0.9	650
0.034	290	0.9	150
0.214	−130	0.9	500
0.263	−70	1.0	920
0.275	−185	1.1	450
0.275	−220	1.1	500
0.45	200	1.4	500
0.5	290	1.7	960
0.5	270	2.0	500
0.63	200	2.0	850
0.8	300	2.0	800
0.9	−30	2.0	1090

Source: Hubble, E. "A Relationship Between Distance and Radial Velocity Among Extra-Galactic Nebulae," Proceedings of the National Academy of Science, 1929, 168.

a. Find the linear regression model for these data.
$y = 454.1584x - 40.7836$

b. On the basis of this model, what is the recession velocity of a nebula that is 1.5 megaparsecs from the Milky Way? 640 km/s

Explorations  

Median-Median Lines **Another linear equation used to model data is called the median-median line. This line uses *summary points*, which are calculated using the medians of subsets of the independent and dependent variables. The median of a data set is the middle number, or the average of the two middle numbers, when the data set is listed in numerical order. For instance, to find the median of {8, 12, 6, 7, 9}, first arrange the data in numerical order.**

6, 7, 8, 9, 12

The median is 8, the number in the middle. To find the median of {15, 12, 20, 9, 13, 10}, arrange the numbers in numerical order.

9, 10, 12, 13, 15, 20

The median is 12.5, the average of the two middle numbers.

$$\text{Median} = \frac{12 + 13}{2} = 12.5$$

The median-median line is determined by dividing a data set into three equal groups. (If the set cannot be divided into three equal groups, the first and third groups should be equal. For instance, if there are 11 data points, divide the set into groups of four, three, four.) The slope of the median-median line is the slope of the line through the x-medians and y-medians of the first and third sets of points. The median-median line passes through the average of the x- and y-medians of all three sets.

A graphing calculator can be used to find the median-median line. This line, along with the linear regression line, is shown below for the data in the given table.

Table A

x	y
2	3
3	5
4	4
5	7
6	8
7	9
8	12
9	12
10	14
11	15
12	14

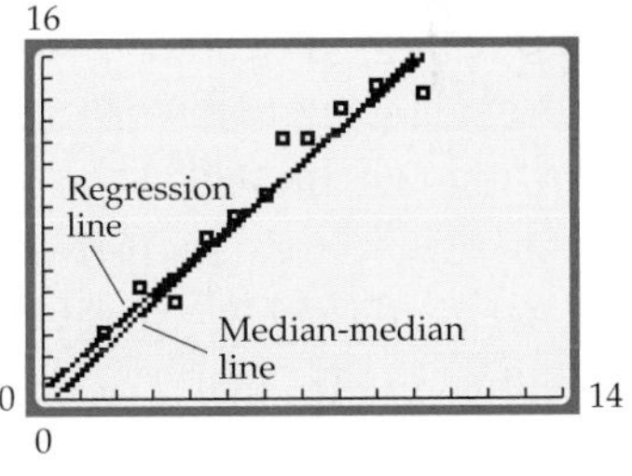

1. Find the median-median line for the data in Exercise 17. $y = 23.9904x - 36.8013$

2. Find the median-median line for the data in Exercise 18. $y = 2.8889x + 4.5185$

3. Consider the data set {(1, 3), (2, 5), (3, 7), (4, 9), (5, 11), (6, 13), (7, 15), (8, 17)}.

a. Find the linear regression line for these data.
$y = 2x + 1$

b. Find the median-median line for these data.
$y = 2x + 1$

c. What conclusion might you draw from the answers to parts **a.** and **b.**?
When the data are an exact linear fit, the linear regression equation and the median-median equation are exactly the same.

Chapter 2 Summary

Key Terms

abscissa (x-coordinate) **[p. 163]**
axis of symmetry **[p. 222]**
circle **[p. 170]**
coefficient of determination **[p. 238]**
constant function **[p. 185]**
coordinate axes **[p. 163]**
coordinates of a point **[p. 163]**
correlation coefficient **[p. 235]**
decreasing function **[p. 185]**
dependent variable **[p. 167]**
directrix **[p. 231]**
domain **[p. 177]**
equation in two variables **[p. 166]**
focus **[p. 231]**
function **[p. 177]**
functional notation **[p. 179]**
general form of the equation of a circle **[p. 172]**
graph of a function **[p. 180]**
graph of an ordered pair **[p. 164]**
graph of an equation in two variables **[p. 167]**
horizontal line test **[p. 185]**
increasing function **[p. 185]**
independent variable **[p. 167]**
input/output table **[p. 167]**
least-squares regression line **[p. 233]**
linear correlation coefficient **[p. 236]**
linear function **[p. 195]**
linear model **[p. 211]**
linear regression equation **[p. 235]**
median of a triangle **[p. 173]**
midpoint **[p. 165]**
one-to-one function **[p. 185]**
ordered pair **[p. 163]**
ordinate (y-coordinate) **[p. 163]**
origin **[p. 163]**
parabola **[p. 222]**
parallel lines **[p. 202]**
perpendicular lines **[p. 203]**
quadratic function **[p. 222]**
quadrants **[p. 163]**
range **[p. 177]**
rectangular coordinate system **[p. 163]**
relation **[p. 177]**
scatter diagram (scatter plot) **[p. 233]**
slope **[p. 198]**
slope-intercept form **[p. 200]**
solution of an equation in two variables **[p. 167]**
standard form of the equation of a circle **[p. 170]**
value of a function **[p. 179]**
vertex of a parabola **[p. 222]**
vertical line test **[p. 183]**
x-intercept **[p. 195]**
y-intercept **[p. 195]**
xy-plane **[p. 163]**

Essential Concepts and Formulas

- **Distance Formula**
  If $P_1(x_1, y_1)$ and $P_2(x_2, y_2)$ are two points in the plane, then the distance $d(P_1, P_2)$ between the two points is given by

$$d(P_1, P_2) = \sqrt{(x_1 - x_2)^2 + (y_1 - y_2)^2} \quad \textbf{[p. 165]}$$

- **Midpoint Formula**
  If $P_1(x_1, y_1)$ and $P_2(x_2, y_2)$ are two points in the plane, then the midpoint of the line segment between the two points is given by (x_m, y_m), where

$$x_m = \frac{x_1 + x_2}{2} \quad \text{and} \quad y_m = \frac{y_1 + y_2}{2} \quad \textbf{[p. 166]}$$

- **Standard Form of the Equation of a Circle**
  Let $C(h, k)$ be the coordinates of the center of a circle of radius r. Then the equation of the circle is

$$(x - h)^2 + (y - k)^2 = r^2 \quad \textbf{[p. 170]}$$

- **General Form of the Equation of a Circle**
  The general form of the equation of a circle is

$$x^2 + y^2 + Ax + By + C = 0 \quad \textbf{[p.172]}$$

- **Ways to Describe a Function**
  - a set of ordered pairs
  - a table
  - an equation
  - a graph **[p. 179]**

- **Slope Formula**
  Let $P_1(x_1, y_1)$ and $P_2(x_2, y_2)$ be two points on a line.

$$m = \frac{\text{change in } y}{\text{change in } x} = \frac{y_2 - y_1}{x_2 - x_1}, x_1 \neq x_2 \quad \textbf{[p. 198]}$$

- **Slopes of Parallel Lines**
  Two nonvertical lines with slopes m_1 and m_2 are parallel lines if and only if $m_1 = m_2$. Vertical lines are also parallel lines. **[p. 202]**

- **Slopes of Perpendicular Lines**
  If m_1 and m_2 are the slopes of two lines, neither of which is vertical, then the lines are perpendicular if and only if $m_1 m_2 = -1$.
  A vertical line is perpendicular to a horizontal line. **[p. 203]**

- **Slope-Intercept Form of the Equation of a Line**
  The equation $y = mx + b$ is called the slope-intercept form of a straight line. The graph of $y = mx + b$ is a straight line with slope m and y-intercept $(0, b)$. **[p. 200]**

- **Standard Form of a Straight Line**
  The equation $Ax + By = C$ is called the standard form of the equation of a line. **[p. 204]**

- **Point-Slope Formula of a Straight Line**
  Let $P_1(x_1, y_1)$ be a point on a line and let m be the slope of the line. Then the equation of the line can be found using the point-slope formula

$$y - y_1 = m(x - x_1) \quad \textbf{[p. 212]}$$

- **Linear Function**
  A linear function is a function of the form

$$f(x) = ax + b \quad \textbf{[p. 195]}$$

- **Quadratic Function**
  A quadratic function is a function of the form

$$f(x) = ax^2 + bx + c, a \neq 0 \quad \textbf{[p. 222]}$$

Chapter 2 True/False Exercises

In Exercises 1 to 10, answer true or false. If the answer is false, explain why it is false or give an example to show the statement is false.

1. If the slope of a line is positive, then as x decreases y decreases. True

2. A possible coordinate for a y-intercept is $(0, 3)$. True

3. The equation of *any* line can be written in the form $y = mx + b$.
False. The vertical line $x = 2$ cannot be written in the form $y = mx + b$.

4. The equation of *any* line can be written in the form $Ax + By = C$. True

5. Consider the graph of an equation of the form $y = mx + 2$. As the slope increases, the y-intercept increases.
False. The slope m and y-intercept are independent of each other.

6. If m is the slope of a line and (x_1, y_1) is a point on the line, then the point-slope formula is

$$y - y_1 = m(x - x_1). \quad \text{True}$$

7. The distance between $P(0, 0)$ and $Q(3, 4)$ is 5. True

8. If the graphs of two nonvertical lines are perpendicular, then the product of the slopes of the lines is 1.
False. The product of the slopes of nonvertical perpendicular lines is -1.

9. If two lines have the same slope and different y-intercepts, then the graphs of the lines will intersect at some point.
False. Two lines with the same slope but different y-intercepts are parallel and will never meet.

10. If the correlation coefficient is positive, the slope of the regression line is positive. True

Chapter 2 Review Exercises

—Answer graphs to Exercises 11–14 and 23–36 are on pages AA4–AA5.

1. Does the set of ordered pairs

$$\{(1, 20), (2, 40), (-1, -10), (3, 15), (2, 30), (4, 30)\}$$

define a function? No [2.2]

2. Does the set of ordered pairs

$$\{(-2, 3), (0, 4), (-1, -5), (1, 4), (2, 3), (6, -5)\}$$

define a function? Yes [2.2]

In Exercises 3 to 8, does the equation define y as a function of x?

3. $y = 3x - 5$ Yes [2.2]
4. $2x + 3y = 7$ Yes [2.2]
5. $x^2 = y$ Yes [2.2]
6. $y = \frac{1}{x}$ Yes [2.2]
7. $x^2 - y^2 = 9$ No [2.2]
8. $y^2 = x^3$ No [2.2]

9. If $f(x) = x^2 - 3x - 5$, find:

   a. $f(-2)$ 5
   b. $f(1)$ -7
   c. $f(c)$ $c^2 - 3c - 5$
   d. $f(2 + h)$ $h^2 + h - 7$
   e. $f(2a)$ $4a^2 - 6a - 5$
   f. $2f(a)$ $2a^2 - 6a - 10$ [2.2]

10. If $R(x) = \frac{x}{x - 5}$, find:

   a. $R(4)$ -4
   b. $R(-1)$ $\frac{1}{6}$
   c. $R(0)$ 0
   d. $R(a + 2)$ $\frac{a+2}{a-3}$
   e. $R(5z)$ $\frac{z}{z-1}$
   f. $5R(z)$ $\frac{5z}{z-5}$ [2.2]

11. Draw a line through all points with a y-coordinate of 1.

12. Draw a line through all points with an x-coordinate of -2.

13. Draw a line through all points whose y-coordinate is twice the x-coordinate.

14. Draw a line through all points whose y-coordinates has the opposite sign of the x-coordinate.

In Exercises 15 to 18, find the coordinates of the midpoint and the length of the line segment between the two points. Round the lengths of the line segments to the nearest hundredth.

15. $P_1(3, 5), P_2(-7, 1)$ Midpoint $(-2, 3)$, distance 10.77 [2.1]
16. $P_1(-2, -1), P_2(2, -5)$ Midpoint $(0, -3)$, distance 5.66 [2.1]
17. $P_1(5, -3), P_2(-2, -1)$ Midpoint $(1.5, -2)$, distance 7.28 [2.1]
18. $P_1(6, -3), P_2(5, -2)$ Midpoint $(5.5, -2.5)$, distance 1.41 [2.1]

In Exercises 19 and 20, find the standard form of the equation of the circle from the given conditions.

19. Center is $C(-3, 5)$, radius is 3.
   $(x + 3)^2 + (y - 5)^2 = 9$ [2.1]
20. Endpoints of a diameter are $P(3, 1)$ and $Q(-1, 5)$.
   $(x - 1)^2 + (y - 3)^2 = 8$ [2.1]

In Exercises 21 and 22, find the center and radius of the circle from the given equation.

21. $x^2 + y^2 + 6x - 2y - 6 = 0$ $C(-3, 1)$; $r = 4$ [2.1]
22. $x^2 + y^2 - 4x + 8y + 1 = 0$ $C(2, -4)$; $r = \sqrt{19}$ [2.1]

In Exercises 23 to 32, graph the equation.

23. $y = 2 - 3x$
24. $y = x + 2$
25. $2x - 5y = 10$
26. $3x + 5y = 15$
27. $y = \frac{1}{2}x^2 + 1$
28. $y = 1 - x^2$
29. $f(x) = -2x^2 + 4x$
30. $f(x) = x^2 + 2x + 1$
31. $f(x) = |2x - 4|$
32. $f(x) = |x| - 3$

33. Graph the line that passes through the point $(-1, -3)$ and has slope $\frac{4}{3}$.

34. Graph the line that passes through the point $(-2, -3)$ and has slope 2.

35. Graph the line that passes through the point $(2, -4)$ and has slope -1.

36. Graph the line that passes through the point $(-4, 5)$ and has slope $-\frac{2}{3}$.

In Exercises 37 to 42, find the slope of the line between the two points.

37. $P_1(-1, 3), P_2(3, -1)$ -1 [2.3]
38. $P_1(4, -5), P_2(-2, 3)$ $-\frac{4}{3}$ [2.3]
39. $P_1(-2, 4), P_2(-2, -5)$ No slope [2.3]
40. $P_1(4, -1), P_2(-3, -1)$ 0 [2.3]
41. $P_1(6, 1), P_2(-3, 7)$ $-\frac{2}{3}$ [2.3]
42. $P_1(4, -3), P_2(8, 3)$ $\frac{3}{2}$ [2.3]

43. If (3, 4) are the coordinates of a point on a line that has slope $\frac{2}{3}$, what is the y-coordinate of the point on the line whose x-coordinate is -3? 0 [2.3/2.4]

44. If (1, 0) are the coordinates of a point on a line that has slope $-\frac{3}{4}$, what is the y-coordinate of the point on the line whose x-coordinate is -7? 6 [2.4]

45. If (1, 3) are the coordinates of a point on a line that has slope -3, what is the y-coordinate of the point on the line whose x-coordinate is 3? -3 [2.4]

46. If $(2, -1)$ are the coordinates of a point on a line that has slope $\frac{1}{2}$, what is the y-coordinate of the point on the line whose x-coordinate is 3? $-\frac{1}{2}$ [2.3/2.4]

In Exercises 47 to 50, determine whether the line between points P_1 and P_2 is parallel or perpendicular to the line through points Q_1 and Q_2.

47. $P_1(5, 2), P_2(-1, 6); Q_1(5, -1), Q_2(7, 2)$ Perpendicular [2.4]

48. $P_1(8, -3), P_2(4, 1); Q_1(5, -5), Q_2(3, -3)$ Parallel [2.4]

49. $P_1(2, 5), P_2(-1, 3); Q_1(2, 1), Q_2(8, 5)$ Parallel [2.4]

50. $P_1(-1, 5), P_2(-1, 2); Q_1(4, 1), Q_2(-4, 1)$ Perpendicular [2.4]

In Exercises 51 to 56, find the equation of the line passing through the given point with the given slope.

51. $P(6, -2), m = -2$
$y = -2x + 10$ [2.4]

52. $P(-3, 1), m = 1$
$y = x + 4$ [2.4]

53. $P(-4, -3), m = \frac{3}{4}$
$y = \frac{3}{4}x$ [2.4]

54. $P(3, -2), m = -\frac{2}{3}$
$y = -\frac{2}{3}x$ [2.4]

55. $P(3, 2), m = -\frac{1}{3}$
$y = -\frac{1}{3}x + 3$ [2.4]

56. $P(5, -3), m = \frac{6}{5}$
$y = \frac{6}{5}x - 9$ [2.4]

57. Find the equation of the line that is parallel to $y = \frac{2}{3}x + 1$ and passes through $P(-1, 4)$. $y = \frac{2}{3}x + \frac{14}{3}$ [2.4]

58. Find the equation of the line that is parallel to $y = -2x - 3$ and passes through $P(4, -2)$.
$y = -2x + 6$ [2.4]

59. Find the equation of the line that contains the point $(-2, -4)$ and is parallel to the graph of $2x + 3y = 6$.
59. $y = -\frac{2}{3}x - \frac{16}{3}$ [2.4]

60. Find the equation of the line that is perpendicular to $y = 2x - 3$ and passes through $P(5, -3)$.
$y = -\frac{1}{2}x - \frac{1}{2}$ [2.4]

61. Find the equation of the line that is perpendicular to $y = -\frac{3}{4}x + 1$ and passes through $P(3, -6)$. $y = \frac{4}{3}x - 10$ [2.4]

62. Find the equation of the line that contains the point $P(3, -1)$ and is parallel to the graph of $5x + 3y = 15$.
$y = -\frac{5}{3}x + 4$ [2.4]

63. Find the equation of the line that passes through $P_1(4, -5)$ and $P_2(2, -4)$. $y = -\frac{1}{2}x - 3$ [2.4]

64. Find the equation of the line that passes through $P_1(2, 0)$ and $P_2(-3, 2)$. $y = -\frac{2}{5}x + \frac{4}{5}$ [2.4]

In Exercises 65 to 68, find the x- and y-intercepts of the graph of the function.

65. $f(x) = x^2 + 4x - 21$
$(-7, 0), (3, 0); (0, -21)$ [2.5]

66. $f(x) = 4x^2 + 4x + 1$

67. $f(x) = x^2 + 3x - 1$

68. $f(x) = x^2 - 2x + 3$
No x-intercepts; y-intercept is $(0, 3)$. [2.5]

69. Find the vertex and axis of symmetry for $f(x) = 2x^2 + 6x - 1$.
Vertex: $(-1.5, -5.5)$; axis of symmetry: $x = -1.5$ [2.5]

In Exercises 70 and 71, find the maximum or minimum value of the function.

70. $f(x) = -x^2 + 4x - 2$
Maximum is 2. [2.5]

71. $f(x) = 2x^2 + 4x - 5$
Minimum is -7. [2.5]

Business and Economics

66. $\left(-\frac{1}{2}, 0\right)$; $(0, 1)$ [2.5]

72. ***Computer Game Profit*** The graph below shows the profit a manufacturer of computer games earns for sales of one of its games. Find the slope of the line between the two points shown on the graph. Write a sentence that states the meaning of the slope in the context of this problem.
Slope = 2. The slope means that the profit is increasing at a rate of \$2 per game sold. [2.3]

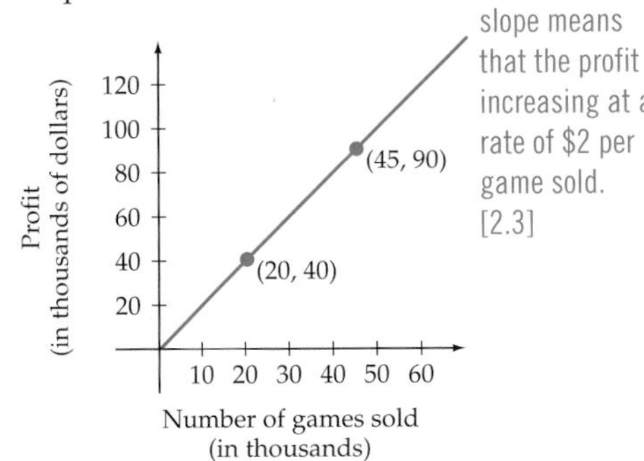

73. ***Value of Jewelry*** The weights, in grams, of various 14-karat gold bracelets and their prices in dollars are given in the following table.

67. $\left(\frac{-3 - \sqrt{13}}{2}, 0\right), \left(\frac{-3 + \sqrt{13}}{2}, 0\right)$; $(0, -1)$ [2.5]

Gold Bracelets

Weight	Cost	Weight	Cost
10	$225	30	$800
25	$500	50	$1300
34	$800	45	$1000
66	$1500	15	$250

a. Find the linear regression equation, with coefficients of the constants rounded to the nearest ten-thousandth, for these data. $y = 24.4710x - 44.3159$

b. Based on the regression equation, what is the predicted price of a gold bracelet that weighs 54 grams? Round to the nearest dollar. $1277 [2.6]

74. ***Parking Lot Design*** A company wants to fence in a parking lot using one side of its headquarters building as one side of the parking lot. If the company has 1000 feet of fencing, what are the dimensions of the rectangular parking lot of maximum area? 250 ft by 500 ft [2.5]

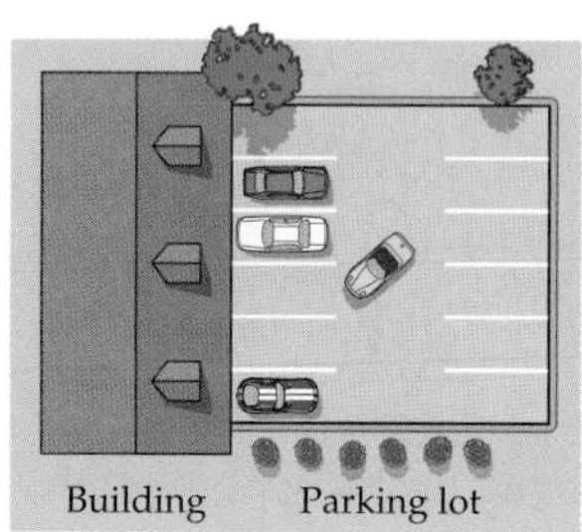

Life and Health Sciences

75. ***Heart Rate*** The heart rate (in beats per minute) of an athlete is measured at various time intervals (in minutes) after terminating an exercise program. The data are shown in the table at the right.

Time	Heart rate
3	135
5	115
8	85
10	65

a. Find a linear model for the athlete's heart rate t minutes after terminating the exercises. $y = -10t + 165$

b. Write a sentence that explains the meaning of the slope of this line in the context of the problem. The slope means that the athlete's heart rate is decreasing at a rate of 10 beats/min.

c. Use the model to determine the athlete's heart rate 6 minutes after terminating the exercise program. 105 beats/min [2.3/2.6]

76. ***Lake Pollution*** The amount of pollution, in milligrams of pollutants per liter of water, in a lake t years after 1995 is given by $p(t) = t^2 - 10t + 25.5$. In what year were the pollutants in this lake at a minimum? What was the minimum amount of pollutants per liter of water? 2000; 0.5 mg/L [2.5]

Social Sciences

77. ***Education*** The math and verbal SAT scores for eight selected students are given in the table below.

SAT Scores

Math	Verbal	Math	Verbal
450	500	650	625
525	560	500	560
600	660	560	620
490	525	480	510

a. Find the linear regression equation, with coefficients rounded to the nearest ten-thousandth, for these data. $y = 0.796x + 146.646$

b. Based on the regression equation, what is the predicted verbal score of a student who scores 550 on the math portion of the test? 584 [2.6]

Physical Sciences and Engineering

78. ***Braking Distances*** The graph below shows how the speed of a car changes as its brakes are applied. Find the slope of the line between the two points shown on the graph. Write a sentence that states the meaning of the slope in the context of this problem. Slope $= -10$. The slope means that the speed of the car is decreasing at a rate of 10 mph per second. [2.3]

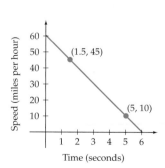

79. *Meteorology* The height (in feet) of a weather balloon above Earth's surface t seconds after it is released is given in the table at the right.

Time	Height
30	90
60	180
90	270
120	360

a. Find a linear model for the height of the balloon t seconds after it is released. $y = 3t$

b. Write a sentence that explains the meaning of the slope of this line in the context of the problem.
The slope means that the height of the balloon is increasing at a rate of 3 ft/s.

c. Use the model to determine the height of the balloon 200 seconds after it is released. 600 ft [2.3/2.6]

Chapter 2 Test

—Answer graphs to Exercises 1–5 are on page AA5.

Graph each of the following equations.

1. $y = 3 - \frac{1}{2}x$ [2.2]

2. $f(x) = x^2 - 2x - 3$ [2.5]

3. $4x + 3y = 12$ [2.3]

4. $x = 3$ [2.3]

5. Graph the line that passes through $P_1(-3, 2)$ and has slope $-\frac{3}{2}$. [2.3]

6. Find the coordinates of the midpoint and the length of the line segment between $P_1(3, -2)$ and $P_2(5, 4)$. Round the length to the nearest hundredth.
Midpoint (4, 1), distance 6.32 [2.1]

7. Find the slope of the line that passes through

a. $P_1(2, 1)$ and $P_2(-1, 5)$. $-\frac{4}{3}$

b. $P_1(-5, 3)$ and $P_2(-5, -1)$. No slope [2.3]

8. A diameter of a circle has endpoints $P(4, -1)$ and $Q(-2, 5)$. Find the standard form of the equation of the circle. $(x - 1)^2 + (y - 2)^2 = 18$ [2.1]

9. Find the equation of the line that contains the point $P(6, -1)$ and is parallel to the graph of $x + 3y = 15$.
$y = -\frac{1}{3}x + 1$ [2.4]

10. Normally, as the price of a commodity increases, the number of units of the commodity that can be sold at that price decreases. For instance, as the price of strawberries increases, the number of pounds of strawberries a grocer can sell decreases. The following graph shows this situation for a certain grocer.

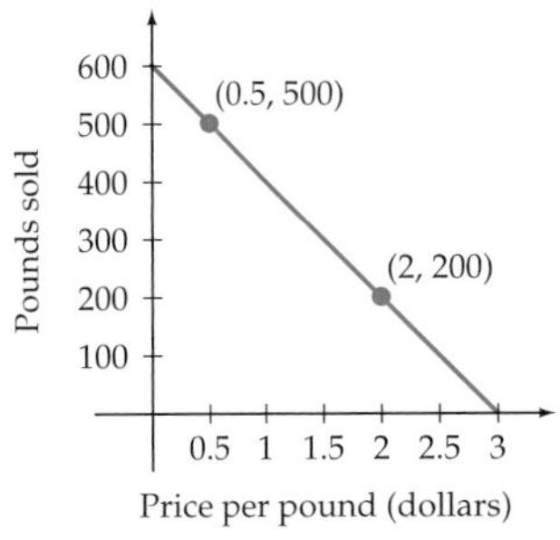

a. Find the slope of the line. Slope = −200

b. Write a sentence that explains the meaning of the slope in the context of this problem.
The slope means that for each $1 increase in the price of a pound of strawberries, the grocer sells 200 fewer pounds.

c. What is the intercept on the horizontal axis? (3,0)

d. Write a sentence that explains the meaning of the intercept in part **c.** in the context of the problem.
If the grocer charges $3 per pound, the grocer will not sell any strawberries. [2.3/2.4]

11. The total cost, in dollars, to produce x 50-gigabyte hard drives for a personal computer is shown in the table at the right.

Number of hard drives	Total cost
2500	307,500
3000	365,000
3500	422,500
4000	480,000

a. Find a linear model for the total cost to produce x 50-gigabyte hard drives. $y = 115x + 20{,}000$

b. Use the model to determine the total cost to produce 2750 50-gigabyte hard drives. $336,250 [2.6]

12. Let $f(x) = 2x^2 - x - 6$. Find

a. the y-intercept. $(0, -6)$

b. the x-intercepts. $\left(-\frac{3}{2}, 0\right)$ and $(2, 0)$

c. the vertex of the graph of f. $\left(\frac{1}{4}, -\frac{49}{8}\right)$

d. the axis of symmetry of the graph of f. $x = \frac{1}{4}$ [2.5]

13. The height h, in feet, of a stream of water x feet from a fire hose held by a firefighter can be approximated by $h = -0.01x^2 + x + 5$. What is the highest point on a building at which the firefighter can spray water? 30 ft [2.5]

14. The following table shows the weights, in ounces, of various oranges before and after they are peeled.

Orange Weights

Before	After	Before	After
6.5	5.5	7.2	6.5
8.1	7.5	6.8	6.0
7.6	7.0	7.4	6.3
8.1	6.9	8.3	7.3

a. Find the linear regression equation, with the constants rounded to the nearest ten-thousandth, for these data. $y = 0.980x - 0.723$

b. Based on the regression equation, what is the predicted peeled weight of an orange that weighs 7.0 ounces before it is peeled? 6.1 oz [2.6]

Cumulative Review Exercises

1. What property of real numbers is demonstrated by $2(3x - 5) = 6x - 10$? Distributive property [P.1]

2. Which of the numbers $-3, -\frac{2}{3}, 0, 3$, and $\sqrt{2}$ are not integers? $-\frac{2}{3}, \sqrt{2}$ [P.1]

In Exercises 3 to 8, simplify the expression.

3. $3 - 5(2x - 1)$
$-10x + 8$ [P.1]

4. $(2x^3y^2)(-3xy^4)$
$-6x^4y^6$ [P.2]

5. $\dfrac{8a^4b^3}{12ab^5}$ $\dfrac{2a^3}{3b^2}$ [P.2]

6. $(3x - 1)(2x + 5)$
$6x^2 + 13x - 5$ [P.3]

7. $\dfrac{x^2 + 4x - 5}{x^2 + 7x + 10}$
$\dfrac{x - 1}{x + 2}$ [P.6]

8. $\dfrac{2}{x + 3} - \dfrac{4}{x - 1}$
$\dfrac{-2x - 14}{(x + 3)(x - 1)}$ [P.6]

In Exercises 9 to 14, solve for x.

9. $5 - 2(3x - 4) = 13$
0 [1.1]

10. $x^2 - 3x - 2 = 0$ $\dfrac{3 \pm \sqrt{17}}{2}$ [1.2]

11. $(2x - 3)(x + 1) = 3$
$-\frac{3}{2}, 2$ [1.2]

12. $5x - 2y = 14$ $x = \dfrac{2y + 14}{5}$ [1.3]

13. $x^4 - x^2 - 2 = 0$
$\pm\sqrt{2}, \pm i$ [1.3]

14. $3x - 1 < 5x + 7$
$x > -4$ [1.4]

15. Find the distance between the points $P_1(5, -2)$ and $P_2(-1, -3)$. $\sqrt{37}$ [2.1]

16. Given $F(x) = 3x^2 + x - 5$, find $F(-2)$. 5 [2.2]

17. Find the equation of the line between the points $P_1(1, -3)$ and $P_2(2, -1)$. $y = 2x - 5$ [2.4]

18. A long-distance runner started on a course running at a constant speed of 8 miles per hour. One hour later, a cyclist traveled the same course at a constant speed of 12 miles per hour. How long after the runner started did the cyclist overtake the runner? 3 h [1.1]

19. The temperature t hours after 7:00 A.M. at a lifeguard stand on a beach is given by $T = -t^2 + 11t + 50$, where T is the temperature in degrees Fahrenheit. At what times was the temperature 78°F? 11:00 A.M., 2:00 P.M. [1.2]

20. The temperature T, in degrees Fahrenheit, inside an oven can be modeled by $T = 55t + 75$, where t is the time in minutes after the oven has been turned on. What is the rate, in degrees per minute, at which the temperature inside the oven is increasing? 55°F/min [2.3]

Chapter 3

Properties of Functions

Credit Card Numbers

Credit card numbers are created in a very special way using what is called a *modulo* function. Knowing how the credit card number is created allows companies to accept credit cards over the Internet. For instance, 5234 8213 3410 1298 is a valid credit card number but 5234 3128 3410 1298 is not.

When credit card information is sent over the Internet, it must be encrypted, that is, the numbers scrambled in such a way that it would be impossible for an unauthorized person to determine the credit card number. The way a number is encrypted is also based on a modulo function. When the credit card number is received by a merchant, the number is decrypted or unscrambled to its original form using another function, called an *inverse* function. Finding inverse functions is one of the topics of this chapter. Exercises 53 and 54 on page 272 demonstrate one way in which a function and its inverse can be used to encode and decode messages.

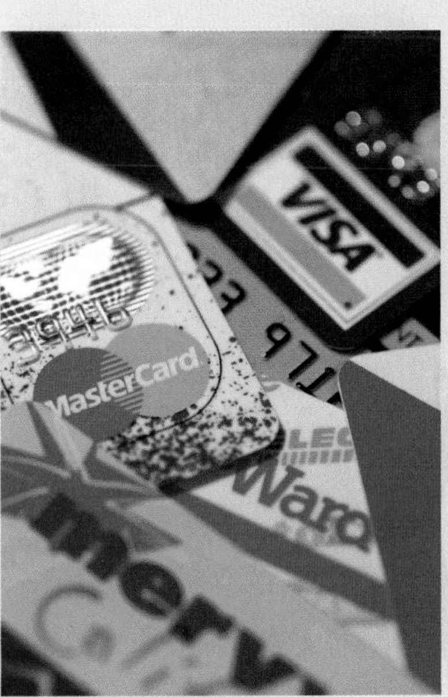

Chapter Prep Quiz

Do these exercises to prepare for Chapter 3.

1. If $f(x) = x^2 - x$, find $f(-3)$. [2.2] 12

2. If $g(x) = 2x^2 - x - 1$, find $g(-3 + h)$. [2.2] $2h^2 - 13h + 20$

3. If $F(x) = 2x^2 + 1$, find $F(-x)$ in simplest form. [2.2] $2x^2 + 1$

4. If $h(x) = x^3 - x$, what is the relationship between $h(-3)$ and $-h(-3)$? [2.2] One is the negative of the other.

5. Subtract: $(2x^3 - x^2 + 1) - (4x^3 + 3x - 5)$ [P.3] $-2x^3 - x^2 - 3x + 6$

6. Multiply: $(2x + 4)(x^2 + 1)$ [P.3] $2x^3 + 4x^2 + 2x + 4$

7. Solve for y: $3x - 5y = 15$ [1.1] $y = \frac{3}{5}x - 3$

8. Solve for y: $x = \dfrac{y}{y + 1}$ [1.3] $y = \dfrac{x}{1 - x}$

9. If $y = 3u + 2$ and $u = 2x - 1$, find y in terms of x. [P.3] $y = 6x - 1$

10. If $y = u^2 - 1$ and $u = x + 4$, find y in terms of x. [P.3] $y = x^2 + 8x + 15$

Problem Solving Strategies

Counterexamples

A **counterexample** is an example that shows that a statement is not always true. For instance, consider the statement "Every number has a reciprocal." A counterexample is the number 0, because 0 has no reciprocal. Because there is a counterexample, the given statement is not true. If we modify the statement to read "Every *nonzero* number has a reciprocal," then we have a statement that is always true. In mathematics, a theorem has no counterexamples.

Which of the following are theorems? If the statement is not a theorem, give a counterexample.

1. The product of two real numbers is always greater than either of the factors.
2. For all real numbers x and y, if $x > y$, then $x^2 > y^2$.
3. The square root of a positive number is always smaller than the number.
4. The sum of two prime numbers is an even number.
5. For all real numbers x and y, $|x + y| = |x| + |y|$.
6. For all real numbers x, $\sqrt{x^2} = |x|$.
7. For all real numbers x and y, if $x < y$, then $\frac{1}{x} > \frac{1}{y}$.
8. If x is any number, then $x^2 \geq 0$.

For all true/false questions, the answer is false if there is a counterexample to the statement. Exercise 61 in Section 3.2 is such a question.

INSTRUCTOR NOTE

The *Chapter Prep Quiz* is a means to test your students' mastery of prerequisite material that is assumed in the coming chapter. Each question identifies the section to review (if necessary) and all answers are provided in the Answers to Selected Exercises.

SECTION 3.1 Algebra of Functions

- Basic Operations on Functions
- Difference Quotient
- Composition of Functions

Basic Operations on Functions

Just as we can add, subtract, multiply, and divide numbers, we can perform similar operations on functions. For instance, consider the functions $f(x) = x^2$ and $g(x) = 2x + 1$. We can create a new function h as follows.

$$h(x) = f(x) + g(x) = x^2 + 2x + 1$$

In a similar manner, we can create the difference, product, or quotient of two functions.

Operations on Functions

For all values of x for which both $f(x)$ and $g(x)$ are defined, we define the following functions.

Sum	$(f + g)(x) = f(x) + g(x)$
Difference	$(f - g)(x) = f(x) - g(x)$
Product	$(f \cdot g)(x) = f(x) \cdot g(x)$
Quotient	$\left(\frac{f}{g}\right)(x) = \frac{f(x)}{g(x)}$ provided $g(x) \neq 0$

Alternative to Example 1

Let $f(x) = x^2 - 2$ and $g(x) = 4x - 3$. Find
a. $(f \cdot g)(3)$
b. $(f - g)x$
c. $\left(\frac{f}{g}\right)(-3)$
d. $(f + g)x$

- a. 63
- b. $x^2 - 4x + 1$
- c. $-\frac{7}{15}$
- d. $x^2 + 4x - 5$

EXAMPLE 1 Operations on Functions

Let $f(x) = 2x + 3$ and $g(x) = x^2 + 1$. Find:

a. $(f - g)(4)$ **b.** $(f \cdot g)(-2)$ **c.** $\left(\frac{f}{g}\right)(x)$ **d.** $(f + g)(x)$

Solution

a. $(f - g)(4) = f(4) - g(4) = [2(4) + 3] - [4^2 + 1] = 11 - 17 = -6$

b. $(f \cdot g)(-2) = f(-2) \cdot g(-2) = [2(-2) + 3] \cdot [(-2)^2 + 1] = -1 \cdot 5 = -5$

c. $\left(\frac{f}{g}\right)(x) = \frac{f(x)}{g(x)} = \frac{2x + 3}{x^2 + 1}$

d. $(f + g)(x) = f(x) + g(x) = (2x + 3) + (x^2 + 1) = x^2 + 2x + 4$

CHECK YOUR PROGRESS 1 Let $f(x) = 3x - 2$ and $g(x) = x^2 - 1$. Find:

a. $(f + g)(3)$ **b.** $\left(\frac{f}{g}\right)(4)$ **c.** $(f \cdot g)(x)$ **d.** $(f - g)(x)$

Solution *See page S15.* a. 15 b. $\frac{2}{3}$ c. $3x^3 - 2x^2 - 3x + 2$ d. $-x^2 + 3x - 1$

INSTRUCTOR NOTE

Before proceeding with Example 2, you might give students a simple example of the concept that profit equals revenue minus cost. For instance, if a company earned \$5 on the sale of a computer game and it cost the company \$2 to produce the game, what is the profit on the sale of one game? What is the profit on the sale of *n* games?

Alternative to Example 2

Exercise 38, page 308

In economics, profit is equal to revenue minus cost. We can represent this symbolically by letting P, R, and C represent, respectively, the profit, revenue, and cost to produce x products. Then $P(x) = R(x) - C(x)$. Thus, to find a profit function, a business analyst must find the difference between the revenue and cost functions.

EXAMPLE 2 Find a Profit Function

A business analyst has determined that the revenue, in thousands of dollars, from the sale of x boats is given by $R(x) = 23x - 0.1x^2$. The cost, in thousands of dollars, to produce x boats is given by $C(x) = 12x + 25$. Find the profit function for this product. Determine the profit for selling 50 boats.

Solution To find the profit function, find the difference between the revenue and cost functions.

$$\begin{aligned} \text{Profit} &= \text{revenue} - \text{cost} \\ P(x) &= R(x) - C(x) \\ &= (23x - 0.1x^2) - (12x + 25) \\ &= -0.1x^2 + 11x - 25 \end{aligned}$$

The profit function is given by $P(x) = -0.1x^2 + 11x - 25$. To determine the profit for selling 50 boats, evaluate the profit function when $x = 50$.

$$\begin{aligned} P(x) &= -0.1x^2 + 11x - 25 \\ P(50) &= -0.1(50)^2 + 11(50) - 25 = 275 \end{aligned}$$

The profit is \$275,000.

CHECK YOUR PROGRESS 2 The revenue, in thousands of dollars, from the sale of x computers is given by $R(x) = 33x - 0.05x^2$. The cost, in thousands of dollars, to produce x computers is given by $C(x) = 24x + 103$. Find the profit function for selling x computers. Determine the profit for selling 50 computers.

Solution *See page S16.* $P(x) = -0.05x^2 + 9x - 103$; \$222,000

Difference Quotient

Have you ever been standing in a line and heard someone say, "At this *rate*, we'll be here forever"? It is not usually the case that the person objects to the length of the line (although that could be true) but to the fact that the line is moving very slowly. It is the rate of movement rather than the length that is of concern.

There are many instances in applied mathematics where the rate at which something is happening is important. For instance, consider two companies, Verona Software and Prime Software, whose profits increase from \$1 million to \$5 million. The amount of increase is the same for both companies (\$4 million), but suppose Verona Software increased its profits over a 2-year period, whereas Prime Software increased its profits over a 4-year period. Then the average rate of increase of Verona's profits is \$2 million per year and the average rate of increase

TAKE NOTE

It is important to note that we are discussing the *rate* of change, not the amount of change. The amount of change is the same for both companies. The rate of change is different.

for Prime's profits is \$1 million per year. Because Verona's profits are increasing more rapidly, Verona may have more appeal to an investor than Prime Software.

One way to measure the average rate of change of a function is to use the *difference quotient*.

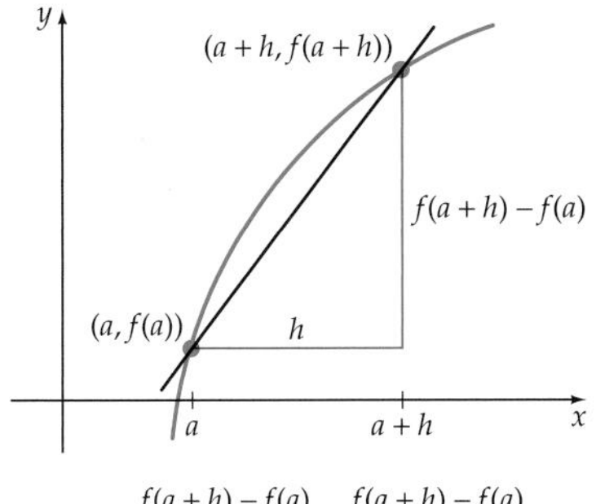

$$m = \frac{f(a+h)-f(a)}{a+h-a} = \frac{f(a+h)-f(a)}{h}$$

Figure 3.1

Difference Quotient

The expression

$$\frac{f(a+h)-f(a)}{h}, h \neq 0$$

is called the **difference quotient** of f.

Notice from Figure 3.1 that the difference quotient is the slope of the line through $(a, f(a))$ and $(a + h, f(a + h))$.

Alternative to Example 3

Determine the difference quotient $\frac{f(a+h)-f(a)}{h}$ for $x^2 + x$.

■ $2a + h + 1$

TAKE NOTE

Because $f(x) = x^2 - 2x$, $f(a + h) = (a + h)^2 - 2(a + h)$. Always include parentheses around the expression for $f(a)$, as we did here for $(a^2 - 2a)$. This will reduce the chances of making a subtraction error.

EXAMPLE 3 Find a Difference Quotient

Determine the difference quotient $\frac{f(a+h)-f(a)}{h}$ for $f(x) = x^2 - 2x$.

Solution

$$\frac{f(a+h)-f(a)}{h} = \frac{(a+h)^2 - 2(a+h) - (a^2 - 2a)}{h}$$

$$= \frac{a^2 + 2ah + h^2 - 2a - 2h - a^2 + 2a}{h}$$

$$= \frac{2ah + h^2 - 2h}{h} = \frac{h(2a + h - 2)}{h}$$

$$= 2a + h - 2$$

CHECK YOUR PROGRESS 3 Determine the difference quotient $\frac{f(3+h)-f(3)}{h}$ for $f(x) = \frac{1}{x}$.

Solution *See page S16.* $-\frac{1}{9 + 3h}$

The difference quotient $2a + h - 2$ from Example 3 is the slope of the secant line (the line through two points on a graph) through the points $(a, f(a))$ and $(a + h, f(a + h))$. For instance, if $a = 3$ and $h = 2$, then

$$(a, f(a)) = (3, f(3)) = (3, 3)$$

$$(a + h, f(a + h)) = (3 + 2, f(3 + 2)) = (5, f(5)) = (5, 15)$$

The difference quotient is

$$2a + h - 2$$

$$2(3) + 2 - 2 = 6 \qquad \text{■ } a = 3, h = 2$$

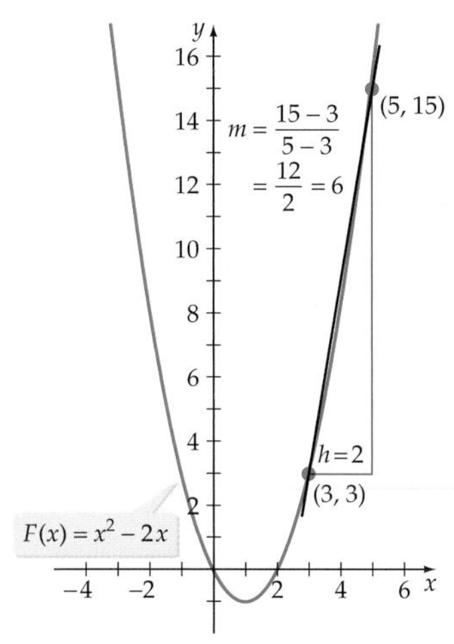

Figure 3.2

Thus the slope of the secant line through (3, 3) and (5, 15), as shown in Figure 3.2, is 6. This is the average rate of change in the value of the function on the interval $3 \le x \le 5$.

The difference quotient can be used to calculate the average rate of change for many different situations. Suppose, for instance, that the revenue function for an electronic organizer is given by $R(x) = 45x - 0.07x^2$. A financial analyst could use the difference quotient to find the average rate of change of the revenue for the sale of between 100 and 150 units. To do this, the analyst would use the difference quotient for the revenue function with $a = 100$ and $h = 50$. Then $a + h = 100 + 50 = 150$.

$$\frac{R(a+h) - R(a)}{h}$$

$$\frac{R(100+50) - R(100)}{50} = \frac{R(150) - R(100)}{50}$$

$$= \frac{[45(150) - 0.07(150)^2] - [45(100) - 0.07(100)^2]}{50}$$

$$= \frac{5175 - 3800}{50} = \frac{1375}{50} = 27.5$$

This means that for sales of between 100 and 150 organizers, the average rate of change of revenue is increasing by \$27.50 per organizer.

If we looked at a different interval, say sales of between 400 and 425 organizers ($a = 400, h = 25$), the average rate of change of the revenue would be different.

$$\frac{R(a+h) - R(a)}{h}$$

$$\frac{R(400+25) - R(400)}{25} = \frac{R(425) - R(400)}{25}$$

$$= \frac{[45(425) - 0.07(425)^2] - [45(400) - 0.07(400)^2]}{25}$$

$$= \frac{6481.25 - 6800}{25} = \frac{-318.75}{25} = -12.75$$

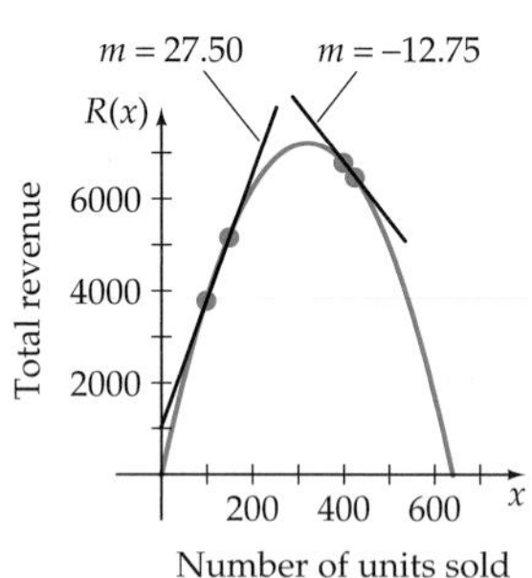

Figure 3.3

This means that for sales of between 400 and 425 organizers, the average rate of change of revenue is decreasing by \$12.75 per organizer. In other words, the company is receiving less revenue per organizer. It is important to note that the revenue is not negative; the *average rate of change* of the revenue is negative. See Figure 3.3.

Average rate of change is also used in the study of objects in motion. The average velocity of an object is given by the difference quotient

$$\frac{s(t + \Delta t) - s(t)}{\Delta t}$$

where $s(t)$ is the distance the object travels in time t. We have used the symbol Δt for h. The symbol Δt is typical notation for engineers and physicists who study motion. We read Δt as "delta t," and it represents "change of time."

INSTRUCTOR NOTE

Some students have difficulty with the idea of decreasing revenue per unit. One explanation that may help these students is to talk about volume discounts. In this case, the manufacturer is receiving less revenue per unit and hoping to make up for this decrease with increased sales.

Alternative to Example 4

The distance, in feet, traveled by a ball rolling down a ramp after t seconds is given by $s(t) = 4t^2$. Evaluate the average velocity of the ball between $t = 3$ seconds and $t = 6$ seconds.

■ 36 ft/s

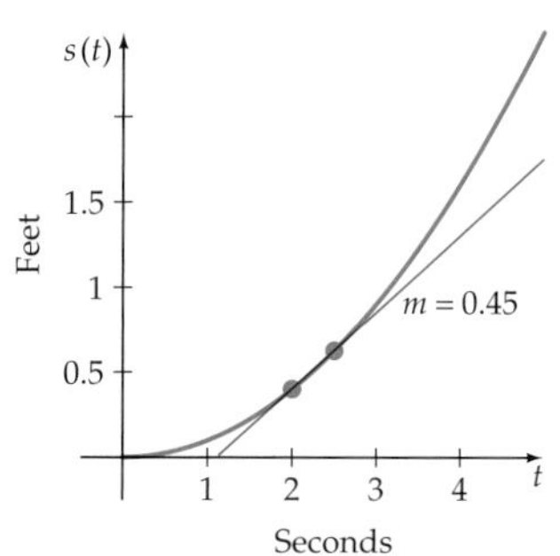

Figure 3.4

EXAMPLE 4 Find Average Velocity

The distance, in feet, a ball will roll down a ramp t seconds after it is released is given by $s(t) = 0.1t^2$. Find the average velocity of the ball between $t = 2$ seconds and $t = 2.5$ seconds.

Solution Use the difference quotient with $t = 2$ and $\Delta t = 0.5$. (The time has changed from 2 seconds to 2.5 seconds, so the "change in time" is $2.5 - 2 = 0.5$ second.)

$$\frac{s(t + \Delta t) - s(t)}{\Delta t}$$

$$\frac{s(2 + 0.5) - s(2)}{0.5} = \frac{s(2.5) - s(2)}{0.5}$$

$$= \frac{0.1(2.5)^2 - 0.1(2)^2}{0.5} = \frac{0.1(6.25) - 0.1(4)}{0.5}$$

$$= 0.45$$

The average velocity of the ball between 2 seconds and 2.5 seconds is 0.45 foot per second. Note from Figure 3.4 that the slope of the line through the two points is 0.45. This graph illustrates again that average velocity is mathematically described by slope.

CHECK YOUR PROGRESS 4 The distance, in feet, a car has traveled t seconds after it begins accelerating on a freeway on-ramp is given by $s(t) = 1.5t^2$. Find the average velocity of the car between $t = 10$ seconds and $t = 12$ seconds.

Solution *See page S16.* 33 ft/s

■ Composition of Functions

Composition of functions is another way in which functions can be combined. This method of combining functions uses the output of one function as the input for a second function.

Suppose the spread of oil from a leak in a tanker can be approximated by a circle with the tanker at its center. The radius r (in feet) of the spill t hours after the leak begins is given by $r(t) = 150\sqrt{t}$. The area of the spill is the area of a circle and is given by the formula $A(r) = \pi r^2$. To find the area of the spill 4 hours after the leak begins, we first find the radius of the spill and then use that number to find the area of the spill.

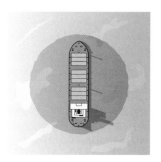

$$\begin{aligned} r(t) &= 150\sqrt{t} \\ r(4) &= 150\sqrt{4} && \text{■ } t = 4 \text{ hours} \\ &= 150(2) \\ &= 300 \end{aligned}$$

$$\begin{aligned} A(r) &= \pi r^2 \\ A(300) &= \pi(300^2) && \text{■ } r = 300 \text{ feet} \\ &= 90{,}000\pi \\ &\approx 283{,}000 \end{aligned}$$

The area of the spill after 4 hours is approximately 283,000 square feet.

There is an alternative way to solve this problem. Because the area of the spill depends on the radius and the radius depends on the time, there is a relationship between area and time. We can determine this relationship by evaluating the formula for the area of a circle using $r(t) = 150\sqrt{t}$. This will give the area of the spill as a function of time.

$$\begin{aligned} A(r) &= \pi r^2 \\ A[r(t)] &= \pi[r(t)]^2 && \text{▪ Replace } r \text{ by } r(t). \\ &= \pi\left[150\sqrt{t}\right]^2 && \text{▪ } r(t) = 150\sqrt{t} \\ A(t) &= 22{,}500\pi t && \text{▪ Simplify.} \end{aligned}$$

The area of the spill as a function of time is $A(t) = 22{,}500\pi t$. To find the area of the oil spill after 4 hours, evaluate this function at $t = 4$.

$$\begin{aligned} A(t) &= 22{,}500\pi t \\ A(4) &= 22{,}500\pi(4) && \text{▪ } t = 4 \text{ hours} \\ &= 90{,}000\pi \\ &\approx 283{,}000 \end{aligned}$$

This is exactly the same result we calculated earlier.

The function $A(t) = 22{,}500\pi t$ is referred to as the *composition* of A with r. The notation $A \circ r$ is used to denote this composition of functions. That is,

$$(A \circ r)(t) = 22{,}500\pi t$$

Definition of the Composition of Two Functions

Let f and g be two functions such that $g(x)$ is in the domain of f for all x in the domain of g. Then the **composition** of the two functions, denoted by $f \circ g$, is the function whose value at x is given by $(f \circ g)(x) = f[g(x)]$.

The function defined by $(f \circ g)(x)$ is also called the **composite** of f and g. We read $(f \circ g)(x)$ as "f circle g of x" and $f[g(x)]$ as "f of g of x."

Consider the functions $f(x) = 2x - 1$ and $g(x) = x^2 - 3$. The expression $(f \circ g)(-1)$ (or, equivalently, $f[g(-1)]$) means to evaluate the function f at $g(-1)$.

$$\begin{aligned} g(x) &= x^2 - 3 \\ g(-1) &= (-1)^2 - 3 && \text{▪ Evaluate } g \text{ at } -1. \\ &= -2 \end{aligned}$$

$$\begin{aligned} f(x) &= 2x - 1 \\ f(-2) &= 2(-2) - 1 = -5 && \text{▪ Evaluate } f \text{ at } g(-1) = -2. \end{aligned}$$

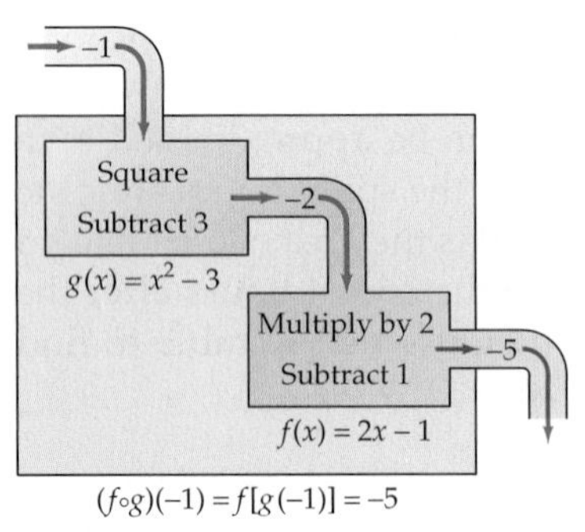

Figure 3.5

If we applied the function machine analogy discussed earlier, the composition $(f \circ g)(-1)$ would look something like Figure 3.5.

The requirement in the definition of the composition of two functions that $g(x)$ be in the domain of f for all x in the domain of g is important. For instance, let

$$f(x) = \frac{1}{x - 1} \quad \text{and} \quad g(x) = 3x - 5$$

When $x = 2$,

$$g(2) = 3(2) - 5 = 1$$

$$f[g(2)] = f(1) = \frac{1}{1-1} = \frac{1}{0}$$ ▪ Undefined

In this case, $g(2)$ is not in the domain of f. Thus the composition $(f \circ g)(x)$ is not defined at 2.

We can find a general expression for $f[g(x)]$ by evaluating f at $g(x)$. For instance, using $f(x) = 2x - 1$ and $g(x) = x^2 - 3$ as in Figure 3.5, we have

$$f(x) = 2x - 1$$

$$f[g(x)] = 2[g(x)] - 1$$ ▪ Replace x by $g(x)$.

$$= 2[x^2 - 3] - 1$$ ▪ Replace $g(x)$ by $x^2 - 3$.

$$= 2x^2 - 7$$ ▪ Simplify.

In general, the composition of functions is not a commutative operation. That is, $(f \circ g)(x) \neq (g \circ f)(x)$. To verify this, we will compute the composition $(g \circ f)(x) = g[f(x)]$, again using the functions $f(x) = 2x - 1$ and $g(x) = x^2 - 3$.

$$g(x) = x^2 - 3$$

$$g[f(x)] = [f(x)]^2 - 3$$ ▪ Replace x by $f(x)$.

$$= [2x - 1]^2 - 3$$ ▪ Replace $f(x)$ by $2x - 1$.

$$= 4x^2 - 4x - 2$$ ▪ Simplify.

Thus $f[g(x)] = 2x^2 - 7$, which is not equal to $g[f(x)] = 4x^2 - 4x - 2$. Therefore, $(f \circ g)(x) \neq (g \circ f)(x)$ and composition is not a commutative operation.

? QUESTION Let $f(x) = x - 1$ and $g(x) = x + 1$. Then $f[g(x)] = g[f(x)]$. (You should verify this statement.) Does this contradict the statement we made that composition is not a commutative operation?

Alternative to Example 5

Given $f(x) = x^2 - x + 3$ and $g(x) = 3x + 2$, evaluate each composite function.

a. $f[g(x)]$ **b.** $g[f(-2)]$

▪ **a.** $9x^2 + 9x + 5$

▪ **b.** 29

INSTRUCTOR NOTE

The graph below can be used to illustrate Example 5**a.**

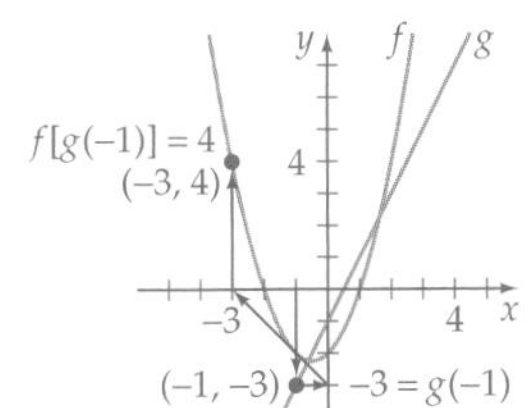

EXAMPLE 5 Find the Composition of Two Functions

Given $f(x) = x^2 + x - 2$ and $g(x) = 2x - 1$, evaluate each composite function.

a. $f[g(-1)]$ **b.** $g[f(x)]$

Solution

a. $g(x) = 2x - 1$

$$g(-1) = 2(-1) - 1 = -3$$ ▪ To evaluate $f[g(-1)]$, first evaluate $g(-1)$.

$$f(x) = x^2 + x - 2$$

$$f[g(-1)] = f(-3) = (-3)^2 + (-3) - 2$$ ▪ Substitute the value of $g(-1)$ for x in $f(x)$.

$$= 4$$ ▪ Simplify.

Continued ➤

? ANSWER No. When we say that composition is not a commutative operation, we mean that generally, given any two functions, $(f \circ g)(x) \neq (g \circ f)(x)$. However, there may be particular instances in which $(f \circ g)(x) = (g \circ f)(x)$. It turns out that these particular instances are quite important, as we shall see later.

b.

$$g(x) = 2x - 1$$

$$g[f(x)] = 2[f(x)] - 1$$ ▪ To find $g[f(x)]$, replace x in $g(x)$ by $f(x)$.

$$= 2[x^2 + x - 2] - 1$$ ▪ $f(x) = x^2 + x - 2$

$$= 2x^2 + 2x - 5$$ ▪ Simplify.

CHECK YOUR PROGRESS 5 Given $g(x) = 4x + 1$ and $h(x) = x^2 + 1$, evaluate each composite function.

a. $g[h(-2)]$ **b.** $h[g(x)]$

Solution *See page S16.* a. 21 b. $16x^2 + 8x + 2$

Topics for Discussion

1. What does it mean to add, subtract, multiply, or divide two functions?
2. If $f(x) = 2x + 4$ and $g(x) = 3x - 1$, write an email to a friend explaining how to evaluate $(f \circ g)(4)$.
3. If $f(x) = 2x + 1$, find $(f \cdot f)(x)$ and $(f \circ f)(x)$. Do these operations yield the same function?
4. If f is a linear function, what is the difference quotient of f?

EXERCISES 3.1

— *Suggested Assignment: Exercises 1–47, odd; 48, and 49–54.*

22. $-\dfrac{2}{a^2 + ah + 2a + h + 1}$

In Exercises 1 to 12, let $f(x) = x^2 - x - 2$ and $g(x) = 2x - 7$. Evaluate each of the following.

1. $(f + g)(3)$ 3
2. $(f - g)(-1)$ 9
3. $(f - g)\left(\frac{1}{2}\right)$ $\frac{15}{4}$
4. $(f + g)\left(-\frac{1}{2}\right)$ $-\frac{37}{4}$
5. $(fg)(2)$ 0
6. $(fg)(-3)$ -130
7. $(gf)\left(\frac{3}{2}\right)$ 5
8. $(gf)\left(\frac{1}{2}\right)$ $\frac{27}{2}$
9. $\left(\frac{f}{g}\right)(4)$ 10
10. $\left(\frac{f}{g}\right)(1)$ $\frac{2}{5}$
11. $\left(\frac{f}{g}\right)\left(\frac{1}{2}\right)$ $\frac{3}{8}$
12. $\left(\frac{g}{f}\right)(3)$ $-\frac{1}{4}$

In Exercises 13 to 22, find the difference quotient for the given function.

13. $f(x) = 2x - 6$ 2
14. $f(x) = 3x + 1$ 3
15. $f(x) = x^2 - 5$ $2a + h$
16. $f(x) = x^2 + 10$ $2a + h$
17. $f(x) = x^2 - 3x$ $2a + h - 3$
18. $f(x) = x^2 + 5x - 1$ $2a + h + 5$
19. $f(x) = 2x^2 + 2x - 5$ $4a + 2h + 2$
20. $f(x) = 2x^2 - 3$ $4a + 2h$
21. $f(x) = \dfrac{1}{x}$ $-\dfrac{1}{a^2 + ah}$
22. $f(x) = \dfrac{2}{x + 1}$

In Exercises 23 to 28, evaluate the composite function given $f(x) = 2x + 2$ and $g(x) = 3x - 1$.

23. $(f \circ g)(2)$ 12
24. $g[f(0)]$ 5
25. $f[g(2)]$ 12
26. $(f \circ g)(-2)$ -12
27. $g[f(x)]$ $6x + 5$
28. $(f \circ g)(x)$ $6x$

In Exercises 29 to 34, evaluate the composite function given $g(x) = x^2 - 1$ and $h(x) = x + 3$.

29. $g[h(2)]$ 24
30. $(g \circ h)(-1)$ 3

31. $h[g(-2)]$ 6

32. $g[h(-2)]$ 0

33. $(h \circ g)(x)$ $x^2 + 2$

34. $g[h(x)]$ $x^2 + 6x + 8$

In Exercises 35 to 40, evaluate the composite function given $f(x) = x^2 + 2x$ and $h(x) = 2x - 1$.

35. $(f \circ h)(-1)$ 3

36. $h[f(-1)]$ -3

37. $(h \circ f)(2)$ 15

38. $(f \circ h)(2)$ 15

39. $f[h(x)]$ $4x^2 - 1$

40. $(h \circ f)(x)$ $2x^2 + 4x - 1$

In Exercises 41 to 46, evaluate the composite function given $f(x) = x + 2$ and $g(x) = x^3$.

41. $(f \circ g)(3)$ 29

42. $g[f(-4)]$ -8

43. $f[g(2)]$ 10

44. $(g \circ f)(-1)$ 1

45. $(f \circ g)(x)$ $x^3 + 2$

46. $(g \circ f)(x)$ $x^3 + 6x^2 + 12x + 8$

Business and Economics

47. ***Manufacturing Cameras*** Suppose the manufacturing cost $M(x)$, in dollars, to manufacture a digital camera is given by the function $M(x) = \dfrac{50x + 10{,}000}{x}$, where x is the number of digital cameras sold. A camera store will sell the cameras by marking up the manufacturing cost per camera by 60%.

   a. Express the selling price of a camera as a function of the number of cameras manufactured. That is, find $(S \circ M)(x)$. $\dfrac{80x + 16{,}000}{x}$

   b. Find $(S \circ M)(5000)$. 83.2

   c. Explain the meaning of the answer in part **b.** The selling price of a camera is \$83.20 when 5000 cameras are sold.

48. ***Manufacturing Fax Machines*** The number of fax machines m that a factory can produce per day is a function of the number of hours h it operates.

$$m(h) = 250h, 0 \le h \le 10$$

The daily cost c to manufacture m fax machines is given by the function

$$c(m) = 0.025m^2 + 10m + 1000$$

   a. Find $(c \circ m)(h)$. $1562.5h^2 + 2500h + 1000$

   b. Evaluate $(c \circ m)(10)$. \$182,250

   c. Write a sentence that explains the meaning of the answer in part **b.**
   When the factory operates 10 hours per day, the cost of manufacturing the fax machines is \$182,250 per day.

Prepare for Section 3.2

49. Solve $2x + 5y = 15$ for y. [1.1] $y = -\dfrac{2}{5}x + 3$

50. Solve $x = \dfrac{y+1}{y}$ for y. [1.1] $y = \dfrac{1}{x-1}$

51. Given $f(x) = \dfrac{2x^2}{x-1}$, find $f(-1)$. [2.2] -1

52. Suppose $a < b$. Replace the question mark in $5 - 4a \;?\; 5 - 4b$ to make a true statement. [1.4] $>$

53. Suppose $a < b$. What other condition on b must be true so that $a^2 > b^2$? [1.4] $b \le 0$

54. What is the domain of $f(x) = \sqrt{x+2}$? [2.2] $\{x \mid x \ge -2\}$

Explorations 

1. A computer outlet store is offering a 15% discount on the purchase of any computer. The total cost T to a customer is the sale price S of the computer plus a sales tax that is 5.5% of the sale price. Let r be the regular price of a computer.

   a. Explain the meaning of $S(r) = 0.85r$. It is the sale price of a computer before sales tax.

   b. Explain the meaning of $T(S) = 1.055S$. It is the price of the computer after sales tax.

   c. Find $(T \circ S)(r)$. $(T \circ S)(r) = 0.89675r$

   d. What is the meaning of the value of $(T \circ S)(r)$? It is the price of the computer after sales tax.

   e. Does the composition of S with T, $S \circ T$, make any sense in the context of this problem? Why or why not? No. The sale price is unknown, so the tax cannot be found.

2. The function $F(x) = \frac{9}{5}x + 32$ converts x degrees Celsius into degrees Fahrenheit. The function $C(k) = k - 273.16$ converts k degrees Kelvin into degrees Celsius.

   a. Explain the meaning of $(F \circ C)(k)$. It is the Fahrenheit temperature given the Kelvin temperature.

   b. Does $(C \circ F)(x)$ make any sense in the context of this exercise? No. $F(x)$ is the Fahrenheit temperature given a Celsius temperature. The domain of C is Kelvin temperatures.

   c. The answer to part **b.** gives a real-world application of a property that composition of functions does not have. What is that property? Commutative property

SECTION 3.2 Inverse Functions

- Introduction to Inverse Functions
- Graphs of Inverse Functions
- Composition of a Function and Its Inverse
- Applications

Introduction to Inverse Functions

Consider the "doubling" function $f(x) = 2x$ that doubles every input. Some of the ordered pairs of this function are

$$\left\{(-4, -8), (-1.5, -3), (1, 2), \left(\frac{5}{3}, \frac{10}{3}\right), (7, 14)\right\}$$

Now consider the "halving" function $g(x) = \frac{1}{2}x$ that takes one-half of every input. Some of the ordered pairs of this function are

$$\left\{(-8, -4), (-3, -1.5), (2, 1), \left(\frac{10}{3}, \frac{5}{3}\right), (14, 7)\right\}$$

Observe that the coordinates of the ordered pairs of g are the reverse of the coordinates of the ordered pairs of f. This is always the case for f and g. Here are two more examples.

$$f(5) = 2(5) = 10 \qquad g(10) = \frac{1}{2}(10) = 5$$

Ordered pair: $(5, 10)$ Ordered pair: $(10, 5)$

$$f(a) = 2(a) = 2a \qquad g(2a) = \frac{1}{2}(2a) = a$$

Ordered pair: $(a, 2a)$ Ordered pair: $(2a, a)$

For these functions, f and g are called *inverse functions* of one another.

Inverse Function

If the coordinates of the ordered pairs of a function g are the reverse of the coordinates of the ordered pairs of a function f, then g is said to be the **inverse function** of f.

TAKE NOTE

It is important to remember the information in the paragraph at the right. If f is a function and g is the inverse of f, then

Domain of g = range of f

and

Range of g = domain of f.

Because the coordinates of the ordered pairs of the inverse function g are the reverse of the coordinates of the ordered pairs of the function f, the domain of g is the range of f and the range of g is the domain of f.

Not all functions have an inverse that is a function. Consider, for instance, the "square" function $S(x) = x^2$. Some of the ordered pairs of S are

$$\{(-3, 9), (-1, 1), (0, 0), (1, 1), (3, 9), (5, 25)\}$$

If we reverse the coordinates of the ordered pairs, we have

$$\{(9, -3), (1, -1), (0, 0), (1, 1), (9, 3), (25, 5)\}$$

This set of ordered pairs is not a function because there are ordered pairs, for instance $(9, -3)$ and $(9, 3)$, with the same first coordinate and different second coordinates. In this case, S has an inverse *relation* but not an inverse *function*.

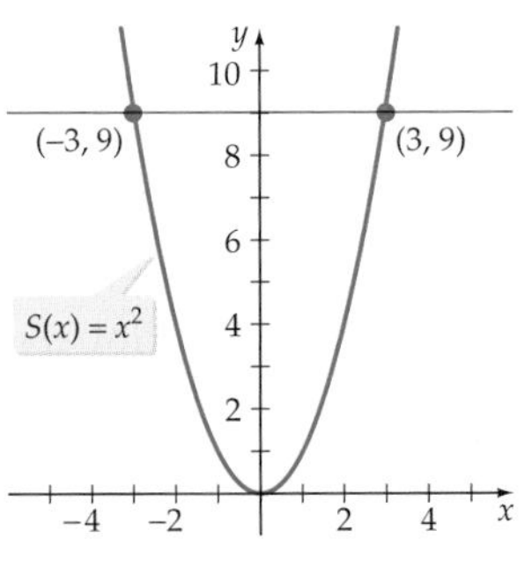

Figure 3.6

A graph of S is shown in Figure 3.6. Note that $x = -3$ and $x = 3$ produce the same value of y. Thus the graph of S fails the horizontal line test and therefore S is not a one-to-one function. This observation is used in the following theorem.

Condition for an Inverse Function

A function f has an inverse function if and only if f is a one-to-one function.

Recall that increasing functions or decreasing functions are one-to-one functions. Thus we can state the following alternative condition for a function to have an inverse.

Alternative Condition for an Inverse Function

If f is an increasing function or a decreasing function, then f has an inverse function.

INSTRUCTOR NOTE

After introducing the notation for inverse functions, give students an example such as $f(2) = 12$ and ask them to find $f^{-1}(12)$.

QUESTION Which of the functions graphed below has an inverse function?

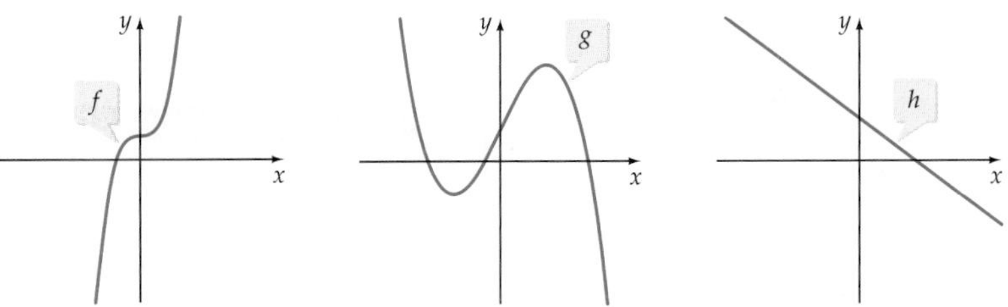

If a function g is the inverse of a function f, we usually denote the inverse function by f^{-1} rather than g. For the doubling and halving functions f and g discussed earlier, we write

$$f(x) = 2x \qquad f^{-1}(x) = \frac{1}{2}x$$

TAKE NOTE

$f^{-1}(x)$ does not mean $\frac{1}{f(x)}$. For $f(x) = 2x$, $f^{-1}(x) = \frac{1}{2}x$ but $\frac{1}{f(x)} = \frac{1}{2x}$.

Graphs of Inverse Functions

Because the coordinates of the ordered pairs of the inverse of a function f are the reverse of the coordinates of f, we can use them to create a graph of f^{-1}.

ANSWER The graph of f is the graph of an increasing function. Therefore, f is a one-to-one function and has an inverse function. The graph of h is the graph of a decreasing function. Therefore, h is a one-to-one function and has an inverse function. The graph of g is not the graph of a one-to-one function. g does not have an inverse function.

Alternative to Example 1

Sketch the graph of f^{-1} given that f is the function shown in

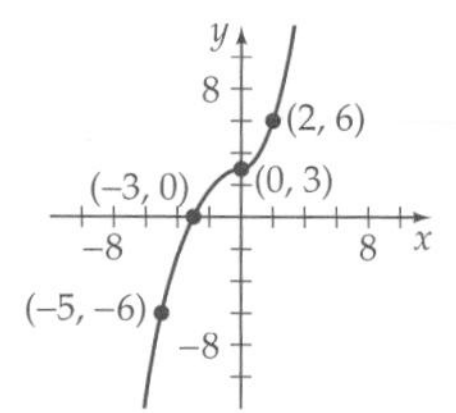

EXAMPLE 1 Sketch the Graph of the Inverse of a Function

Sketch the graph of f^{-1} given that f is the function shown in Figure 3.7.

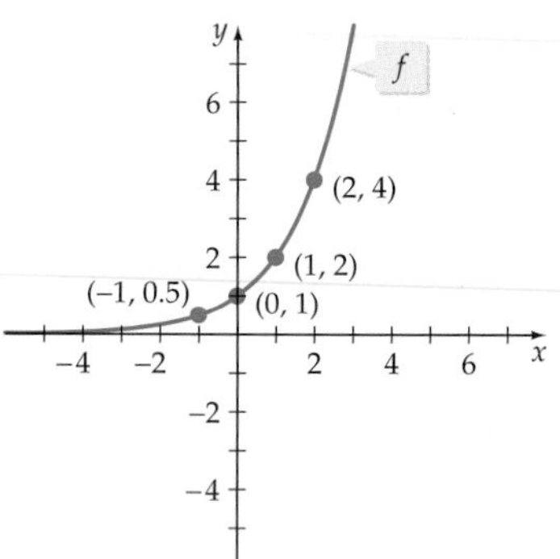

Figure 3.7

Solution Because the graph of f passes through $(-1, 0.5)$, $(0, 1)$, $(1, 2)$, and $(2, 4)$, the graph of f^{-1} must pass through $(0.5, -1)$, $(1, 0)$, $(2, 1)$, and $(4, 2)$. Plot the points and then draw a smooth graph through the points, as shown in Figure 3.8.

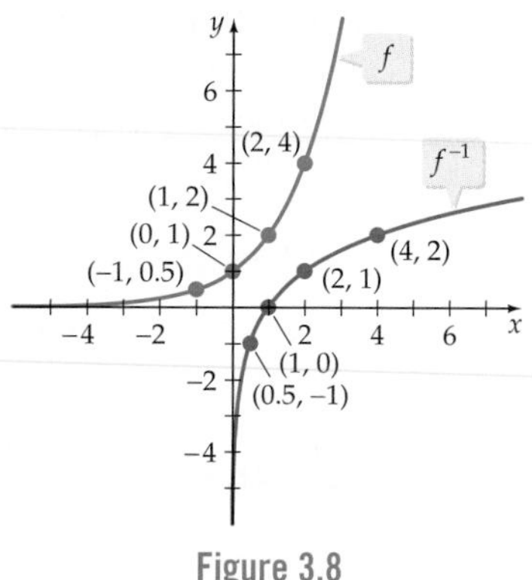

Figure 3.8

CHECK YOUR PROGRESS 1 Sketch the graph of f^{-1} given that f is the function shown in Figure 3.9.

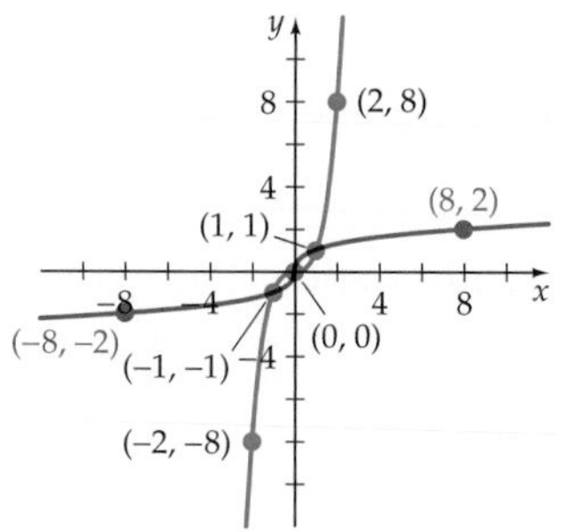

Figure 3.9

Solution *See page S16.* See Figure 3.9.

The graph from the solution to Example 1 is shown again in Figure 3.10. Note that the graph of f^{-1} is symmetric to the graph of f with respect to the line $y = x$. If the graph were folded along the dashed line, the graph of f would lie on top of the graph of f^{-1}. This is a characteristic of all graphs of functions and their inverses. In Figure 3.11, although S does not have an inverse that is a function, the graph of the inverse relation S^{-1} is symmetric to S with respect to the line $y = x$.

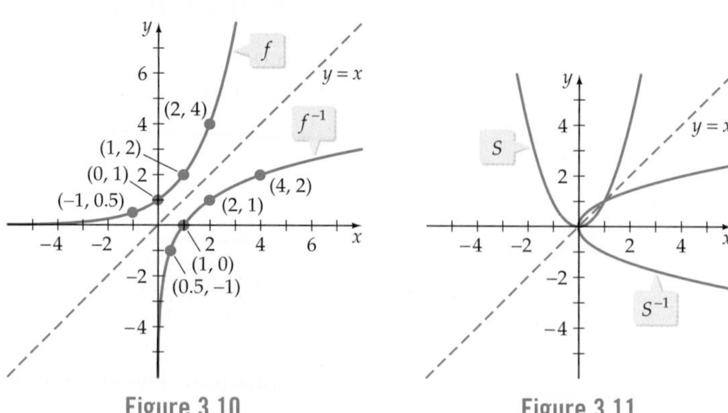

Figure 3.10

Figure 3.11

INSTRUCTOR NOTE

It may help some students if you show them other instances of reversing operations. For example, if you take the number 7, add 5, and then subtract 5, you are back to 7.

▪ Composition of a Function and Its Inverse

Observe the effect, as shown below, of taking the composition of functions that are inverses of one another.

$$f(x) = 2x \qquad\qquad g(x) = \frac{1}{2}x$$

$$f[g(x)] = 2\left[\frac{1}{2}x\right] \quad \text{▪ Replace } x \text{ by } g(x). \qquad g[f(x)] = \frac{1}{2}[2x] \quad \text{▪ Replace } x \text{ by } f(x).$$

$$f[g(x)] = x \qquad\qquad g[f(x)] = x$$

This property of the composition of inverse functions always holds true. When taking the composition of inverse functions, the inverse function reverses the effect of the original function. For the two functions above, f doubles a number and g halves a number. If you double a number and then take one-half of the result, you are back to the original number.

TAKE NOTE

If we think of a function as a machine, then the Composition of Inverse Functions Property can be represented as shown below. Take any input x for f. Use the output of f as the input for f^{-1}. The result is the original input, x.

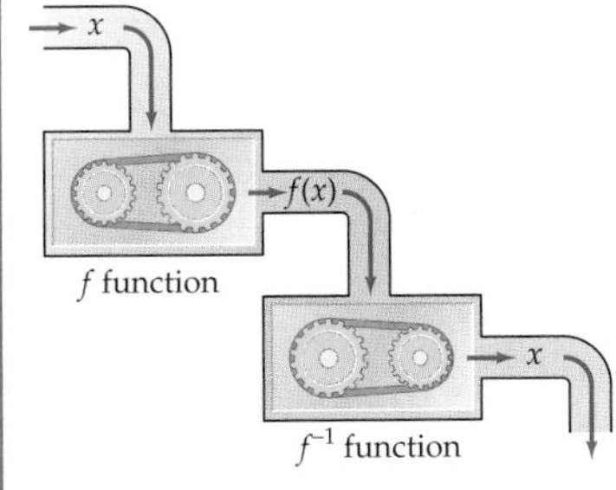

Composition of Inverse Functions Property

If f is a one-to-one function, then f^{-1} is the inverse function of f if and only if

$$(f \circ f^{-1})(x) = f[f^{-1}(x)] = x \qquad \text{for all } x \text{ in the domain of } f^{-1}$$

and

$$(f^{-1} \circ f)(x) = f^{-1}[f(x)] = x \qquad \text{for all } x \text{ in the domain of } f.$$

EXAMPLE 2 Use the Composition of Inverse Functions Property

Use composition of functions to show that $f^{-1}(x) = 3x - 6$ is the inverse function of $f(x) = \frac{1}{3}x + 2$.

Solution We must show that $f[f^{-1}(x)] = x$ and $f^{-1}[f(x)] = x$.

$$f(x) = \frac{1}{3}x + 2 \qquad\qquad f^{-1}(x) = 3x - 6$$

$$f[f^{-1}(x)] = \frac{1}{3}[3x - 6] + 2 \qquad\qquad f^{-1}[f(x)] = 3\left[\frac{1}{3}x + 2\right] - 6$$

$$f[f^{-1}(x)] = x \qquad\qquad f^{-1}[f(x)] = x$$

Alternative to Example 2

Use composition of functions to show that $f^{-1}(x) = \frac{2}{3}x + 4$ is the inverse function of $f(x) = \frac{3}{2}x - 6$.

CHECK YOUR PROGRESS 2 Use composition of functions to show that $f^{-1}(x) = \frac{1}{2}x - 2$ is the inverse function of $f(x) = 2x + 4$.

Solution *See page S16.*

INTEGRATING TECHNOLOGY

We mentioned the importance of the SQUARE viewing window when we discussed the graphs of perpendicular lines earlier in the text.

INTEGRATING TECHNOLOGY

In the standard viewing window of a calculator, the distance between two tic marks on the x-axis is not equal to the distance between two tic marks on the y-axis. As a result, the graph of $y = x$ does not appear to bisect the first and third quadrants. See Figure 3.12. This anomaly is important if a graphing calculator is being used to check whether two functions are inverses of one another. Because the graph of $y = x$ does not appear to bisect the first and third quadrants, the graphs of f and f^{-1} will not appear to be symmetric about the graph of $y = x$. The graphs of $f(x) = \frac{1}{3}x + 2$ and $f^{-1}(x) = 3x - 6$ from Example 2 are shown in Figure 3.13. Notice that the graphs do not appear to be quite symmetric about the graph of $y = x$.

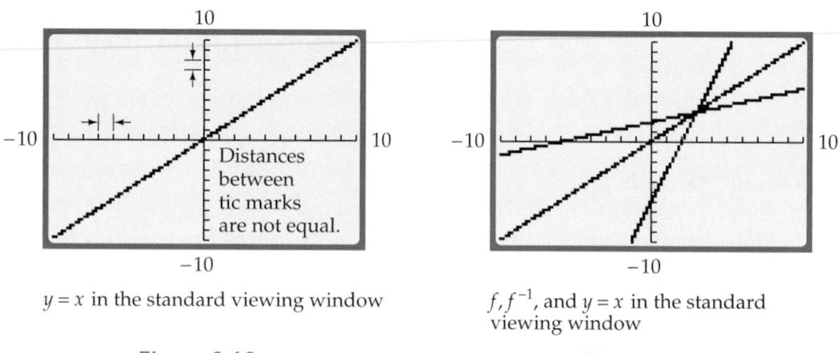

$y = x$ in the standard viewing window

Figure 3.12

f, f^{-1}, and $y = x$ in the standard viewing window

Figure 3.13

To get a better view of a function and its inverse, it is necessary to use the SQUARE viewing window, as in Figure 3.14. In this window, the distance between two tic marks on the x-axis is equal to the distance between two tic marks on the y-axis.

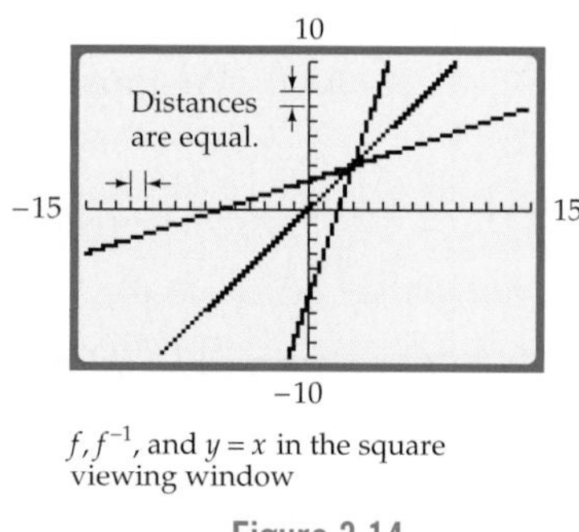

f, f^{-1}, and $y = x$ in the square viewing window

Figure 3.14

If a one-to-one function f is defined by an equation, then we can use the following method to find the equation for f^{-1}.

Steps for Finding the Inverse of a Function

1. Substitute y for $f(x)$.
2. Interchange x and y.
3. Solve, if possible, for y in terms of x.
4. Substitute $f^{-1}(x)$ for y.

Alternative to Example 3
Find the inverse of
$f(x) = \frac{1}{3}x - 6.$
▪ $f^{-1}(x) = 3x + 18$

EXAMPLE 3 Find the Inverse of a Function

Find the inverse of $f(x) = 3x + 8$.

Solution

$$f(x) = 3x + 8$$
$$y = 3x + 8 \quad \text{▪ Replace } f(x) \text{ by } y.$$
$$x = 3y + 8 \quad \text{▪ Interchange } x \text{ and } y.$$
$$x - 8 = 3y \quad \text{▪ Solve for } y.$$
$$\frac{x-8}{3} = y$$
$$\frac{1}{3}x - \frac{8}{3} = f^{-1}(x) \quad \text{▪ Replace } y \text{ by } f^{-1}(x).$$

The inverse function is given by $f^{-1}(x) = \frac{1}{3}x - \frac{8}{3}$.

TAKE NOTE

If the ordered pairs of f are given by (x, y), then the ordered pairs of f^{-1} are given by (y, x). That is, x and y are interchanged. This is the reason for Step 2 at the right.

CHECK YOUR PROGRESS 3 Find the inverse of $f(x) = 4x - 8$.

Solution *See page S16.* $f^{-1}(x) = \frac{1}{4}x + 2$

Alternative to Example 4
Find the inverse of
$f(x) = \frac{x-1}{x}, x \neq 0.$
▪ $f^{-1}(x) = -\frac{1}{x-1}, x \neq 1$

EXAMPLE 4 Find the Inverse of a Function

Find the inverse of $f(x) = \frac{2x+1}{x}, x \neq 0$.

Solution

$$f(x) = \frac{2x+1}{x}$$
$$y = \frac{2x+1}{x} \quad \text{▪ Replace } f(x) \text{ by } y.$$
$$x = \frac{2y+1}{y} \quad \text{▪ Interchange } x \text{ and } y.$$
$$xy = 2y + 1 \quad \text{▪ Solve for } y.$$
$$xy - 2y = 1$$
$$y(x - 2) = 1 \quad \text{▪ Factor the left side.}$$
$$y = \frac{1}{x-2}$$
$$f^{-1}(x) = \frac{1}{x-2}, x \neq 2 \quad \text{▪ Replace } y \text{ by } f^{-1}(x).$$

Continued ➤

CHECK YOUR PROGRESS 4 Find the inverse of $f(x) = \dfrac{x}{x-2}, x \neq 2.$

Solution *See page S17.* $f^{-1}(x) = \dfrac{2x}{x-1}, x \neq 1$

? QUESTION If f is a one-to-one function and $f(4) = 5$, what is $f^{-1}(5)$?

The graph of $f(x) = x^2 + 4x + 3$ is shown in Figure 3.15A. The function f is not a one-to-one function and therefore does not have an inverse function. However, the function given by $G(x) = x^2 + 4x + 3$ shown in Figure 3.15B, for which the domain is restricted to $\{x|x \geq -2\}$, is a one-to-one function and has an inverse function G^{-1}. This is shown in Example 5.

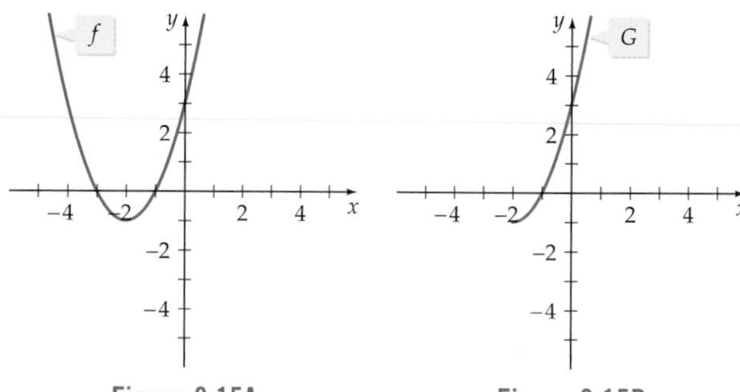

Figure 3.15A Figure 3.15B

Alternative to Example 5

Find the inverse of $H(x) = x^2 - 6x + 2$, where the domain of H is $\{x|x \geq 3\}$.

- $H^{-1}(x) = \sqrt{x+7} + 3, x \geq -7$

EXAMPLE 5 Find the Inverse of a Function with a Restricted Domain

Find the inverse of $G(x) = x^2 + 4x + 3$, where the domain of G is $\{x|x \geq -2\}$.

Solution

$$G(x) = x^2 + 4x + 3$$

$$y = x^2 + 4x + 3$$
- Replace $G(x)$ by y.

$$x = y^2 + 4y + 3$$
- Interchange x and y.

$$x = (y^2 + 4y + 4) - 4 + 3$$
- Solve for y by completing the square of $y^2 + 4y$.

$$x = (y+2)^2 - 1$$
- Factor.

$$x + 1 = (y+2)^2$$
- Add 1 to each side of the equation.

$$\sqrt{x+1} = \sqrt{(y+2)^2}$$
- Take the square root of each side of the equation.

$$\pm\sqrt{x+1} = y + 2$$
- Recall that if $a^2 = b$, then $a = \pm\sqrt{b}$.

$$\pm\sqrt{x+1} - 2 = y$$

Because the domain of G is $\{x|x \geq -2\}$, the range of G^{-1} is $\{y|y \geq -2\}$. This means that we must choose the positive value of $\pm\sqrt{x+1}$. Thus $G^{-1}(x) = \sqrt{x+1} - 2$. See Figure 3.16.

Continued ➤

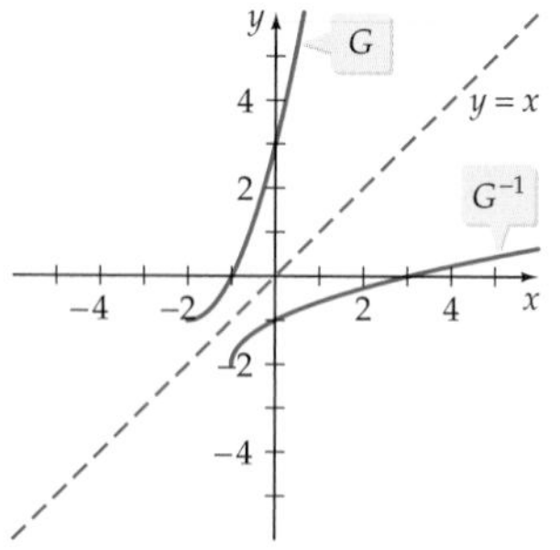

Figure 3.16

TAKE NOTE

Recall that the range of a function f is the domain of f^{-1}, and the domain of f is the range of f^{-1}.

? ANSWER Because f^{-1} is the inverse function of f, the coordinates of the ordered pairs of f^{-1} are the reverse of the coordinates of the ordered pairs of f. Therefore, $f^{-1}(5) = 4$.

CHECK YOUR PROGRESS 5 Find the inverse of $f(x) = x^2 - 6x$, where the domain of f is $\{x|x \le 3\}$.

Solution *See page S17.* $f^{-1}(x) = -\sqrt{x+9} + 3$

▪ Applications

There are practical applications of finding the inverse of a function. Here is one in which a shirt size in the U.S. is converted to a shirt size in Italy. Finding the inverse function gives the function that converts a shirt size in Italy to a shirt size in the U.S.

Alternative to Example 6
Exercise 50, page 272

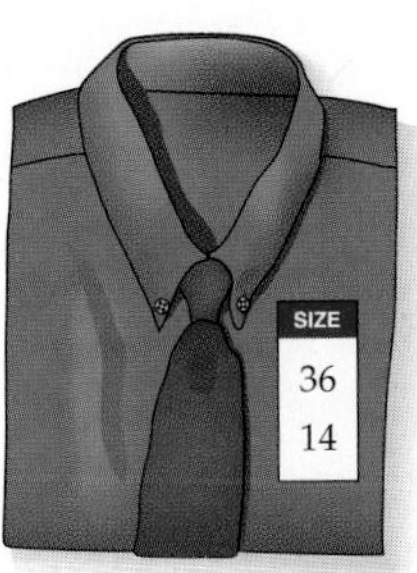

EXAMPLE 6 Solve an Application

The function $IT(x) = 2x + 8$ converts a man's shirt size x in the United States to the equivalent shirt size in Italy.

a. Use IT to determine the equivalent Italian shirt size for a size 16.5 U.S. shirt.

b. Find IT^{-1} and use IT^{-1} to determine the U.S. men's shirt size that is equivalent to an Italian shirt size of 36.

Solution

a. $IT(16.5) = 2(16.5) + 8 = 33 + 8 = 41$
A size 16.5 U.S. shirt is equivalent to a size 41 Italian shirt.

b. To find the inverse function, begin by substituting y for $IT(x)$.

$$IT(x) = 2x + 8$$
$$y = 2x + 8$$
$$x = 2y + 8 \quad \text{▪ Interchange } x \text{ and } y.$$
$$x - 8 = 2y \quad \text{▪ Solve for } y.$$
$$\frac{x-8}{2} = y$$

In inverse notation, the above equation can be written as

$$IT^{-1}(x) = \frac{x-8}{2} \quad \text{or} \quad IT^{-1}(x) = \frac{1}{2}x - 4$$

Substitute 36 for x to find the equivalent U.S. shirt size.

$$IT^{-1}(36) = \frac{1}{2}(36) - 4 = 18 - 4 = 14$$

Thus a size 36 Italian shirt is equivalent to a size 14 U.S. shirt.

CHECK YOUR PROGRESS 6 The function $h(x) = 3x - 10$ converts a woman's hat size x in the United States to the equivalent hat size in France. Find h^{-1} and use h^{-1} to determine the hat size in the U.S. of a French hat size of 62.

Solution *See page S17.* $h^{-1}(x) = \frac{1}{3}x + \frac{10}{3}$; 24

TAKE NOTE

The graphing calculator will draw the inverse even if the function is not one-to-one.

INTEGRATING TECHNOLOGY

Some graphing utilities can be used to draw the graph of the inverse of a function without the user having to find the inverse function. For instance, Figure 3.17 shows the graph of $f(x) = 0.1x^3 - 4$. The graphs of f and f^{-1} are both shown in Figure 3.18, along with the graph of $y = x$. Note that the graph of f^{-1} is the reflection of the graph of f with respect to the graph of $y = x$. The display shown in Figure 3.18 was produced on a TI-83 graphing calculator by using the DrawInv command, which is in the DRAW menu.

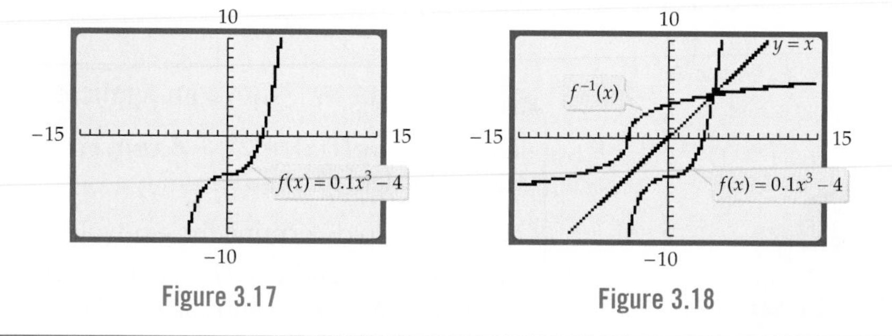

Figure 3.17 Figure 3.18

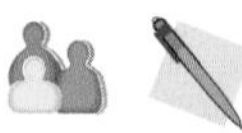

Topics for Discussion

1. If $f(x) = 3x + 1$, what are the values of $f^{-1}(2)$ and $[f(2)]^{-1}$?
2. How are the domain and range of a one-to-one function f related to the domain and range of the inverse function of f?
3. How is the graph of the inverse of a function f related to the graph of f?
4. The function $f(x) = -x$ is its own inverse. Find at least two other functions that are their own inverses.
5. What are the steps in finding the inverse of a one-to-one function?

EXERCISES 3.2

— *Suggested Assignment: Exercises 1–8; 9–53, odd; and 57–61.*
— *Answer graphs to Exercises 37–44 are on page AA6.*

In Exercises 1 to 4, assume that the given function has an inverse function.

1. Given $f(3) = 7$, find $f^{-1}(7)$. 3
2. Given $g(-3) = 5$, find $g^{-1}(5)$. −3
3. Given $h^{-1}(-3) = -4$, find $h(-4)$. −3
4. Given $f^{-1}(7) = 0$, find $f(0)$. 7
5. If 3 is in the domain of f^{-1}, find $f[f^{-1}(3)]$. 3
6. If f is a one-to-one function and $f(0) = 5$, $f(1) = 2$, and $f(2) = 7$, find:

   a. $f^{-1}(5)$ 0 **b.** $f^{-1}(2)$ 1
7. The domain of the inverse function f^{-1} is the _____ of f. range
8. The range of the inverse function f^{-1} is the _____ of f. domain

In Exercises 9 to 12, find the inverse of the function. If the function does not have an inverse function, write "no inverse function."

9. $\{(-3, 1), (-2, 2), (1, 5), (4, -7)\}$ $\{(1, -3), (2, -2), (5, 1), (-7, 4)\}$

10. $\{(-5, 4), (-2, 3), (0, 1), (3, 2), (7, 11)\}$
$\{(4, -5), (3, -2), (1, 0), (2, 3), (11, 7)\}$

11. $\{(0, 1), (1, 2), (2, 4), (3, 8), (4, 16)\}$
$\{(1, 0), (2, 1), (4, 2), (8, 3), (16, 4)\}$

12. $\{(1, 0), (10, 1), (100, 2), (1000, 3), (10{,}000, 4)\}$
$\{(0, 1), (1, 10), (2, 100), (3, 1000), (4, 10{,}000)\}$

In Exercises 13 to 30, find $f^{-1}(x)$. State any restriction on the domain of f^{-1}.

13. $f(x) = 2x + 4$ $f^{-1}(x) = \frac{1}{2}x - 2$

14. $f(x) = x - 1$ $f^{-1}(x) = x + 1$

15. $f(x) = 3x - 7$ $f^{-1}(x) = \frac{1}{3}x + \frac{7}{3}$

16. $f(x) = -3x - 8$ $f^{-1}(x) = -\frac{1}{3}x - \frac{8}{3}$

17. $f(x) = -2x + 5$ $f^{-1}(x) = -\frac{1}{2}x + \frac{5}{2}$

18. $f(x) = -x + 3$ $f^{-1}(x) = -x + 3$

19. $f(x) = \dfrac{2x}{x - 1}, x \neq 1$ $f^{-1}(x) = \dfrac{x}{x - 2}, x \neq 2$

20. $f(x) = \dfrac{x + 2}{2x}, x \neq 0$ $f^{-1}(x) = \dfrac{2}{2x - 1}, x \neq \frac{1}{2}$

21. $f(x) = \dfrac{x - 1}{x + 1}, x \neq -1$

22. $f(x) = \dfrac{2x - 1}{x + 3}, x \neq -3$

23. $f(x) = x^2 + 1, x \geq 0$ $f^{-1}(x) = \sqrt{x - 1}, x \geq 1$

24. $f(x) = x^2 - 4, x \geq 0$ $f^{-1}(x) = \sqrt{x + 4}, x \geq -4$

25. $f(x) = \sqrt{x - 2}, x \geq 2$ $f^{-1}(x) = x^2 + 2, x \geq 0$

26. $f(x) = \sqrt{4 - x}, x \leq 4$ $f^{-1}(x) = -x^2 + 4, x \geq 0$

27. $f(x) = x^2 + 4x, x \geq -2$ $f^{-1}(x) = \sqrt{x + 4} - 2, x \geq -4$

28. $f(x) = x^2 - 6x, x \leq 3$ $f^{-1}(x) = -\sqrt{x + 9} + 3, x \geq -9$

29. $f(x) = x^2 + 4x - 1, x \leq -2$ $f^{-1}(x) = -\sqrt{x + 5} - 2, x \geq -5$

30. $f(x) = x^2 - 6x + 1, x \geq 3$ $f^{-1}(x) = \sqrt{x + 8} + 3, x \geq -8$

In Exercises 31 to 36, use composition of functions to determine whether *f* and *g* are inverses of one another.

31. $f(x) = 4x; g(x) = \dfrac{x}{4}$ Yes

32. $f(x) = 3x; g(x) = \dfrac{1}{3x}$ No

33. $f(x) = 4x - 1; g(x) = \dfrac{1}{4}x + \dfrac{1}{4}$ Yes

34. $f(x) = \dfrac{1}{2}x - \dfrac{3}{2}; g(x) = 2x + 3$ Yes

35. $f(x) = -\dfrac{1}{2}x - \dfrac{1}{2}; g(x) = -2x + 1$ No

36. $f(x) = 3x + 2; g(x) = \dfrac{1}{3}x - \dfrac{2}{3}$ Yes

In Exercises 37 to 44, draw the graph of the inverse relation. Is the inverse relation a function?

37.

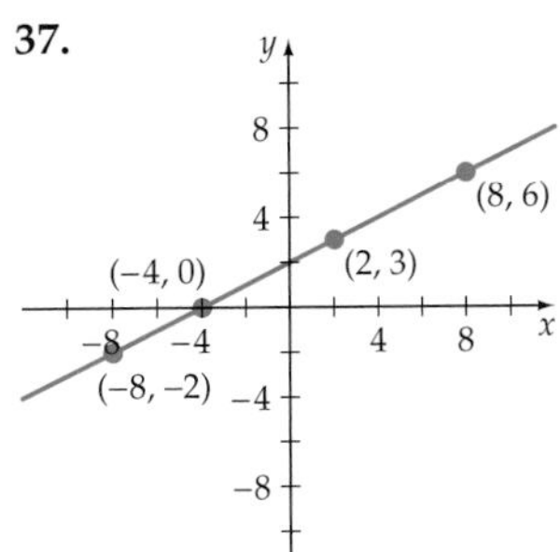

Yes

38.

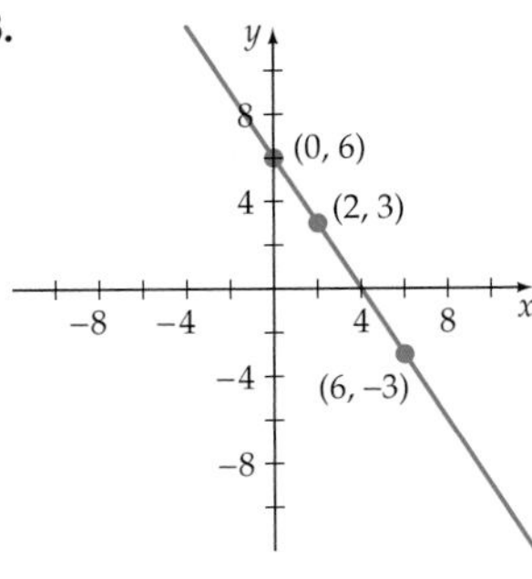

Yes

39.

Yes

40.

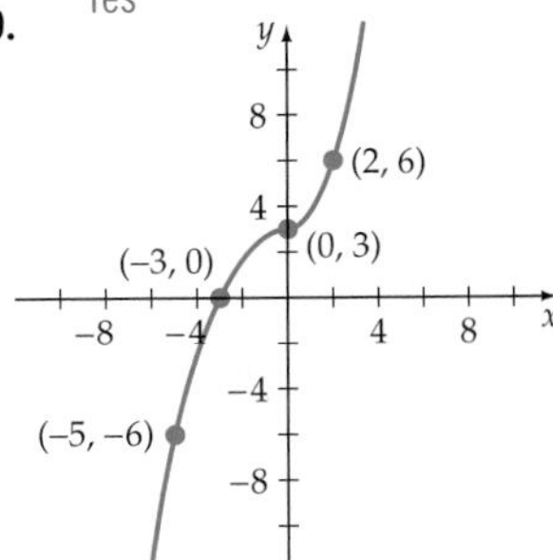

Yes

41.

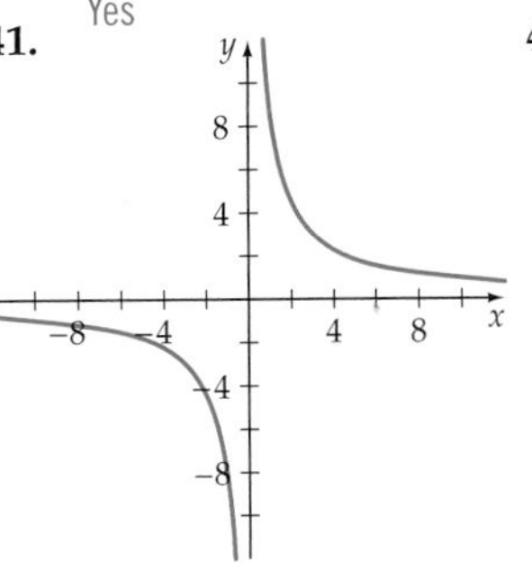

Yes

42.

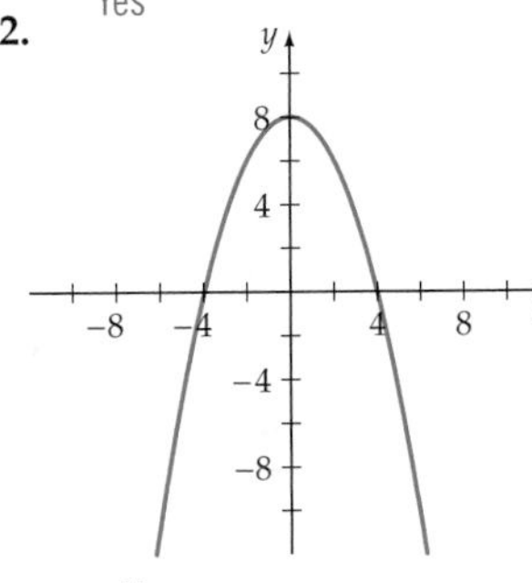

No

43.

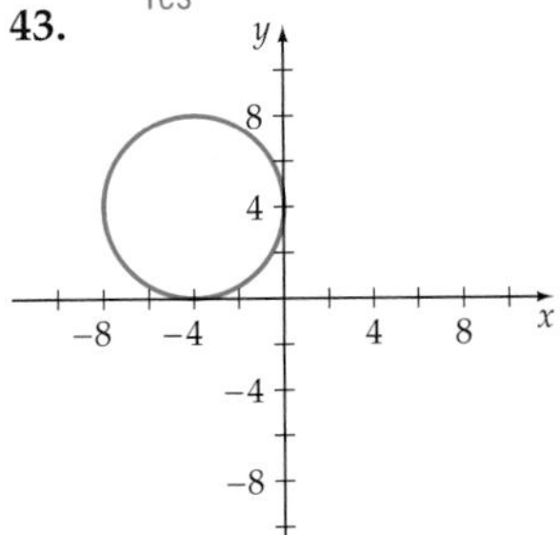

No

44.

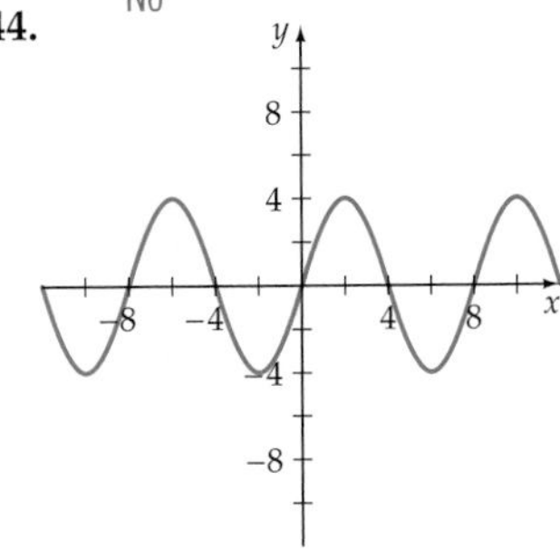

No

45. *Geometry* The volume of a cube is given by $V(x) = x^3$, where x is the measure of the length of a side of the cube. Find $V^{-1}(x)$ and explain what it represents.
$V^{-1}(x) = \sqrt[3]{x}$. V^{-1} finds the length of a side of a cube given the volume.

21. $f^{-1}(x) = \dfrac{x + 1}{1 - x}, x \neq 1$ 22. $f^{-1}(x) = \dfrac{3x + 1}{2 - x}, x \neq 2$

46. ***Unit Conversions*** The function $f(x) = 12x$ converts feet, x, into inches, $f(x)$. Find $f^{-1}(x)$ and explain what it determines. $f^{-1}(x) = \frac{x}{12}$. f^{-1} converts x inches into feet.

47. ***Unit Conversions*** A conversion function such as the one in Exercise 46 converts a measurement in one unit into another unit. Is a conversion function always a one-to-one function? Does a conversion function always have an inverse function? Explain your answer.

48. ***Grading Scale*** Does the grading scale function given below have an inverse function? Explain your answer.

No. It is not a one-to-one function. For a given grade, there is more than one score that can be associated with that grade.

Score	Grade
90–100	A
80–89	B
70–79	C
60–69	D
0–59	F

47. Yes. Yes. A conversion function is a nonconstant linear function. All nonconstant linear functions have inverses that are also functions.

Business and Economics

49. ***Fashion*** The function $s(x) = 2x + 24$ can be used to convert a U.S. woman's shoe size to an Italian woman's shoe size.

Determine the function $s^{-1}(x)$ that can be used to convert an Italian woman's shoe size to its equivalent U.S. shoe size. $s^{-1}(x) = \frac{1}{2}x - 12$

50. ***Fashion*** The function $K(x) = 1.3x - 4.7$ converts a man's shoe size in the United States to the equivalent shoe size in the United Kingdom. Determine the function $K^{-1}(x)$ that can be used to convert a United Kingdom man's shoe size to its equivalent U.S. shoe size. $K^{-1}(x) = \frac{x + 4.7}{1.3}$

51. ***Compensation*** The monthly earnings $E(s)$, in dollars, of a software sales executive is given by $E(s) = 0.05s + 2500$, where s is the value, in dollars, of the software sold by the executive during the month. Find $E^{-1}(s)$ and explain how the executive could use this function. $E^{-1}(x) = 20s - 50{,}000$. From the monthly earnings (s), the executive can find the value of the software sold.

52. ***Postage*** Does the first-class postage rate function in the next column have an inverse function? Explain your answer. No. It is not a one-to-one function. For a given cost, there is more than one weight that can be associated with that cost.

Weight (in ounces)	Cost
$0 < w \leq 1$	\$.37
$1 < w \leq 2$	\$.60
$2 < w \leq 3$	\$.83
$3 < w \leq 4$	\$1.06

53. ***Internet Commerce*** Functions and their inverses can be used to create secret codes that are used to secure business transactions made over the Internet. (See the chapter opener.) Let A = 10, B = 11, . . . , and Z = 35. Let $f(x) = 2x - 1$ define a coding function. Code the word MATH (M–22, A–10, T–29, H–17), which is 22102917, by finding $f(22102917)$. Now find the inverse of f and show that applying f^{-1} to the output of f returns the original word.
44205833; $f^{-1}(x) = \frac{1}{2}x + \frac{1}{2}$; $f^{-1}(44205833) = 22102917$

54. ***Cryptography*** A friend is using the letter-number correspondence in Exercise 53 and the coding function $f(x) = 2x + 3$. Suppose this friend sends you the coded message 5658602671. Decode this message. STUDY

In Exercises 55 to 60, answer the question without finding the equation of the linear function.

55. Suppose f is a linear function, $f(2) = 7$, and $f(5) = 12$. If $f(4) = c$, then is c less than 7, between 7 and 12, or greater than 12? Explain your answer.
Because the function is increasing and 4 is between 2 and 5, c must be between 7 and 12.

56. Suppose f is a linear function, $f(1) = 13$, and $f(4) = 9$. If $f(3) = c$, then c is less than 9, between 9 and 13, or greater than 13? Explain your answer.
Because the function is decreasing and 3 is between 1 and 4, c must be between 9 and 13.

57. Suppose f is a linear function, $f(2) = 3$, and $f(5) = 9$. Between which two numbers is $f^{-1}(6)$? Between 2 and 5

58. Suppose f is a linear function, $f(5) = -1$, and $f(9) = -3$. Between which two numbers is $f^{-1}(-2)$? Between 5 and 9

59. Suppose g is a linear function, $g^{-1}(3) = 4$, and $g^{-1}(7) = 8$. Between which two numbers is $g(5)$? Between 3 and 7

60. Suppose g is a linear function, $g^{-1}(-2) = 5$, and $g^{-1}(0) = -3$. Between which two numbers is $g(0)$? Between -2 and 0

61. True or false: If f is a function and $f(a) = f(b)$, then $a = b$. False

Prepare for Section 3.3

62. Let $f(x) = x^2 - x$. Write $f(x - 2)$ in simplest form. [2.2] $x^2 - 5x + 6$

63. Let $g(x) = \dfrac{x}{x + 2}$. Write $g(2x)$ in simplest form. [2.2] $\dfrac{x}{x + 1}$

64. If $f(x) = x^2 - 3$, does $f(-x) = f(x)$? [2.2] Yes

65. If $g(x) = x^3 - 2x$, does $g(-x) = -g(x)$? [2.2] Yes

66. Let $f(x) = 4x - 5$. Does $f(3x) = 3f(x)$? [2.2] No

67. Let $f(x) = 3x + 4$. Does $f(x + 2) = f(x) + 2$? [2.2] No

Explorations 

1. ***Intersection Points of the Graphs of f and f^{-1}*** For each of the following, graph f and its inverse function.

i. $f(x) = 2x - 4$ **ii.** $f(x) = -x + 2$

iii. $f(x) = x^3 + 1$ **iv.** $f(x) = x - 3$

v. $f(x) = -3x + 2$ **vi.** $f(x) = \dfrac{1}{x}$

a. Do the graphs of a function and its inverse always intersect? No

b. If the graphs of a function and its inverse intersect at one point, what is true about the coordinates of the point of intersection? They are equal.

c. Can the graphs of a function and its inverse intersect at more than one point? Yes. Consider $y = x$.

SECTION 3.3 Transformations of Graphs

- Translations of Graphs
- Reflections of Graphs
- Stretching and Compressing Graphs
- Symmetry
- Even and Odd Functions

▪ Translations of Graphs

Figure 3.19 shows the graphs of $y_1 = |x|$, $y_2 = |x| + 2$, and $y_3 = |x| - 3$. Notice that the shapes of the three graphs are exactly the same; only their positions in the xy-plane differ. These graphs are examples of *transformations* of a graph. A **transformation** of a graph is a change in its position, shape, or size.

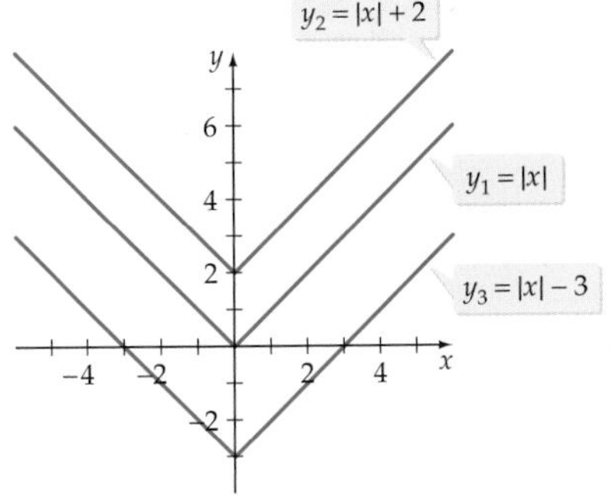

Figure 3.19

The graph of $y_2 = |x| + 2$ is the same as the graph of $y_1 = |x|$, but shifted up 2 units. In function notation, if $f(x) = |x|$, then we can write $y_2 = |x| + 2$ as $y_2 = f(x) + 2$. [Replace $|x|$ by $f(x)$.] Because, for the same input value, each output value of $y_2 = f(x) + 2$ is 2 more than the output value of $y_1 = f(x)$, each point on the graph of $y_2 = f(x) + 2$ is 2 units higher than the corresponding point on the graph of $y_1 = f(x)$. We say that the graph of $y_2 = f(x) + 2$ is a vertical translation of the graph of $y_1 = f(x)$.

Similarly, the graph of $y_3 = |x| - 3$ is the same as the graph of $y_1 = |x|$, but shifted down 3 units. Because each output value of $y_3 = f(x) - 3$ is 3 less than the corresponding output value of $y_1 = f(x)$, each point on the graph of $y_3 = f(x) - 3$ is 3 units lower than the corresponding point on the graph of $y_1 = f(x)$.

Vertical Translations

If f is a function and c is a positive constant, then the graph of

- $y = f(x) + c$ is the graph of $y = f(x)$ shifted vertically *up* c units.
- $y = f(x) - c$ is the graph of $y = f(x)$ shifted vertically *down* c units.

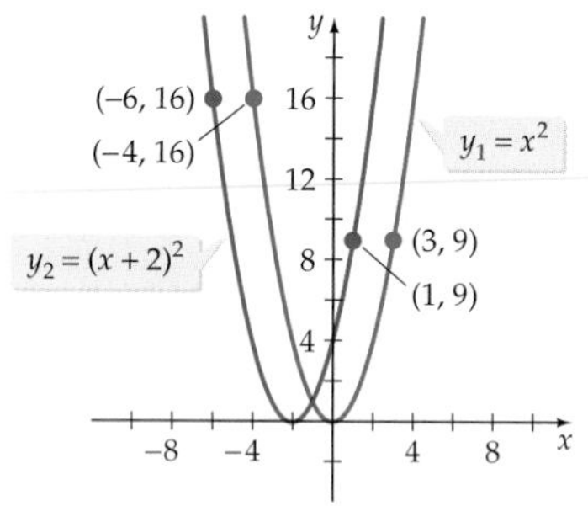

Figure 3.20

Now consider the graphs of $y_1 = f(x) = x^2$ and $y_2 = f(x + 2) = (x + 2)^2$, shown in Figure 3.20. The graph of $y_2 = f(x + 2) = (x + 2)^2$ is the same as the graph of $y_1 = f(x) = x^2$, but shifted to the left 2 units. To understand this, observe in Figure 3.20 that when $x = 3$, $f(3) = 3^2 = 9$, so $(3, 9)$ is a point on the graph. If we input $x = 1$, which is 2 units to the left of the original value of $x = 3$, into $y_2 = f(x + 2) = (x + 2)^2$, we obtain $y_2 = f(1 + 2) = (1 + 2)^2 = 3^2 = 9$. Hence $(1, 9)$ is on the graph of $y_2 = f(x + 2) = (x + 2)^2$. That is, when the input value for y_2 is 2 less than the input value for y_1, $y_2 = y_1$. Here is another example.

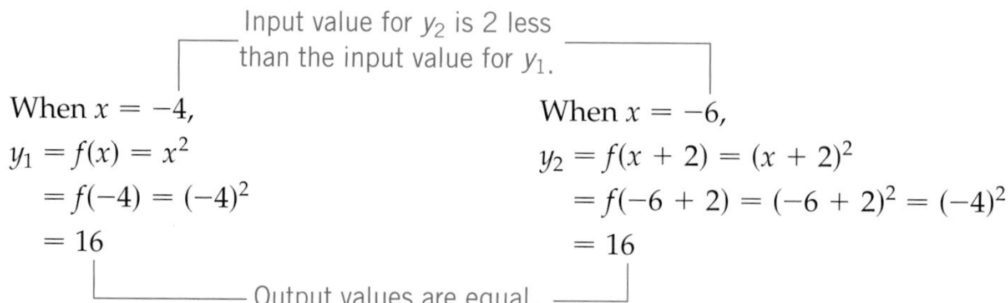

Input value for y_2 is 2 less than the input value for y_1.
When $x = -4$,
$y_1 = f(x) = x^2$
$= f(-4) = (-4)^2$
$= 16$
When $x = -6$,
$y_2 = f(x + 2) = (x + 2)^2$
$= f(-6 + 2) = (-6 + 2)^2 = (-4)^2$
$= 16$
Output values are equal.

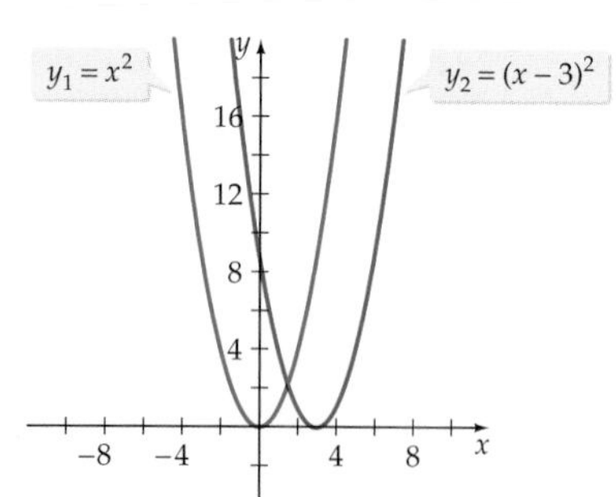

Figure 3.21

Thus the graph of $f(x) = (x + 2)^2$ is the graph of $f(x) = x^2$ shifted to the left 2 units. See Figure 3.20. This is an example of a **horizontal translation.**

Note in Figure 3.21 that the graph of $f(x) = (x - 3)^2$ is the graph of $f(x) = x^2$ shifted 3 units to the right.

Horizontal Translations

If f is a function and c is a positive constant, then the graph of

- $y = f(x + c)$ is the graph of $y = f(x)$ shifted horizontally *left* c units.
- $y = f(x - c)$ is the graph of $y = f(x)$ shifted horizontally *right* c units.

Vertical and horizontal translations are illustrated in Figure 3.22 and Figure 3.23.

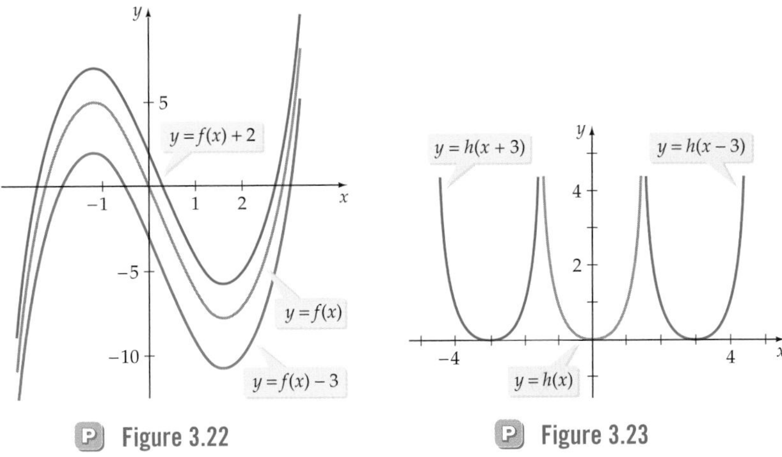

Figure 3.22

Figure 3.23

Alternative to Example 1
Exercise 22, page 290

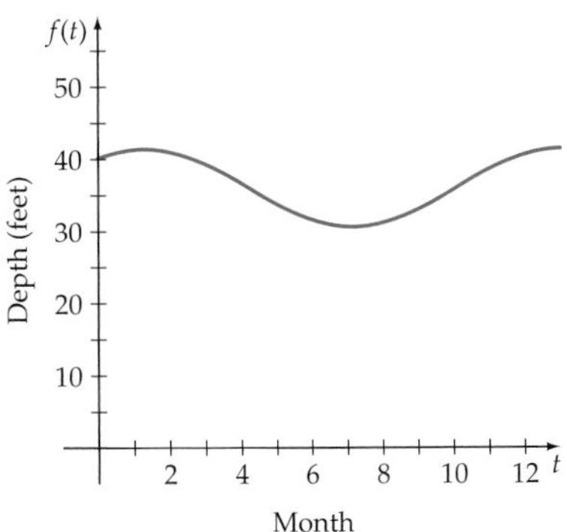

Figure 3.24

EXAMPLE 1 An Application of Translation to Lake Depth

Figure 3.24 shows the graph of a function f that gives the depth, in feet, of a lake at a particular location over the last year (t represents the number of months after January 1).

a. Sketch a graph of $g(t) = f(t) + 5$.

b. If the depth of the lake at the same location next year follows the function $g(t) = f(t) + 5$, describe how the depth of the lake compares to its depth the previous year.

c. Sketch a graph of $h(t) = f(t + 5)$.

d. If the depth of the lake during the first half of next year (at the same location) follows the function $h(t) = f(t + 5)$, describe how the depth of the lake compares to its depth the previous year.

Solution

a. The graph of $g(t) = f(t) + 5$ has the same shape as the graph of $f(t)$, but is shifted vertically upward 5 units. Note that the graph starts at the point $(0, 45)$ rather than $(0, 40)$.

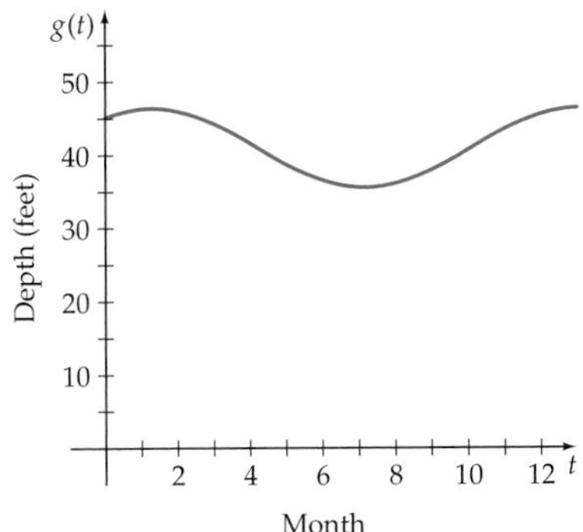

Continued ➤

b. At any given time next year, the depth of the lake will be 5 feet higher than it was the same time the previous year.

c. The graph of $h(t) = f(t + 5)$ has the same shape as the graph of $f(t)$, but is shifted horizontally to the left 5 units. The graph starts at the point $(-5, 40)$ rather than $(0, 40)$.

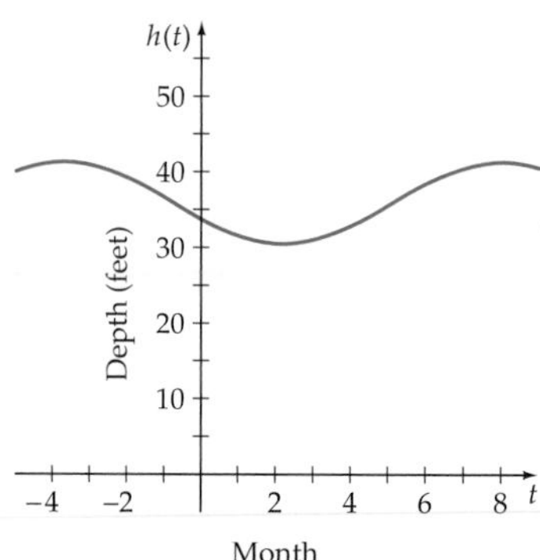

d. On each day during the first half of next year, the depth of the lake will be the same as it was 5 months *after* that day the previous year. Notice that the smallest depth last year occurred approximately during the seventh month, whereas the smallest depth next year will occur approximately during the second month.

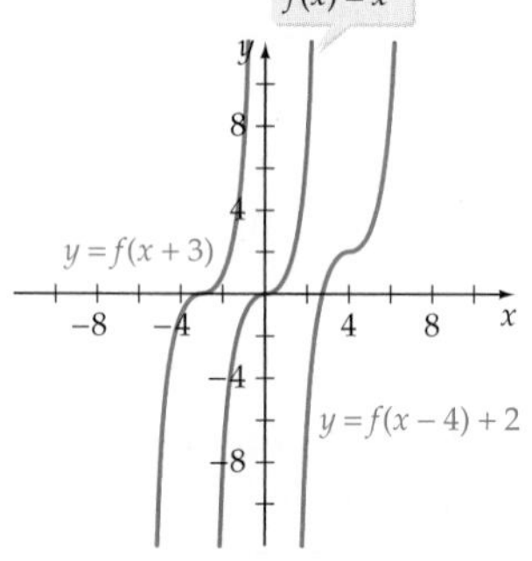

Figure 3.25

CHECK YOUR PROGRESS 1 Given the graph of $y = f(x)$ shown in Figure 3.25, graph

a. $y = f(x + 3)$. **b.** $y = f(x - 4) + 3$.

Solution *See page S17.* See the graph at the left.

QUESTION If $(1, 5)$ is on the graph of $y = f(x)$, find a point on the graph of $y = f(x - 3) + 2$.

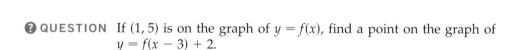

INTEGRATING TECHNOLOGY

A graphing calculator can be used to display translations of graphs. For the graphs in Check Your Progress 1, enter x^3 into Y1. For Y2 enter Y1$(x + 3)$ and for Y3 enter Y1$(x - 4) + 3$. Now graph the functions in the standard viewing window. Some typical screens are shown on the following page.

ANSWER $y = f(x - 3) + 2$ moves the point $(1, 5)$ 3 units to the right and up 2 units. The new point is $(4, 7)$.

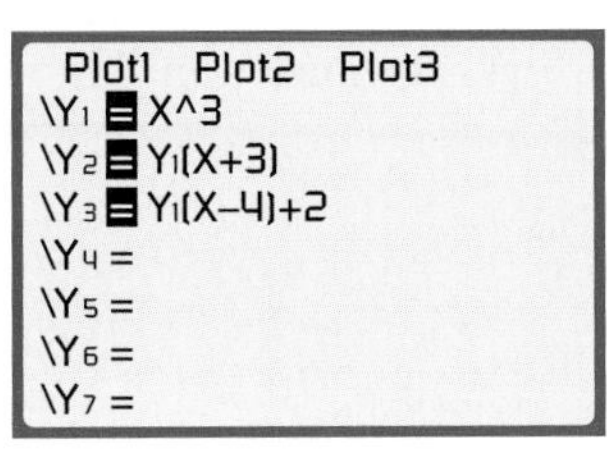

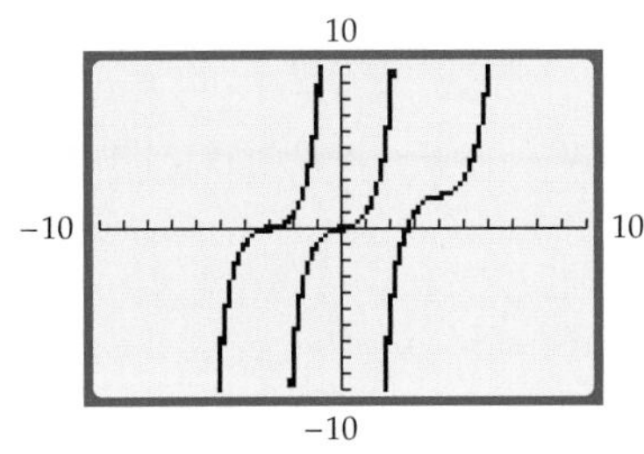

▪ Reflections of Graphs

The graph of $f(x) = \sqrt{x}$ is shown in Figure 3.26. How would the graph of $y = -f(x)$ compare? In effect, we are changing the sign of every output value of the function. So, positive y-values become negative, and negative y-values become positive. The graph of $y = -f(x) = -\sqrt{x}$ is shown in Figure 3.27. It appears that the graph of $y = -f(x)$ is the reflection across the x-axis of the graph of $y = f(x)$.

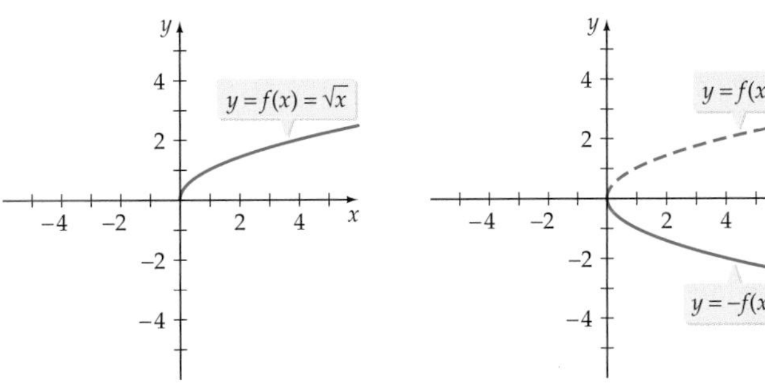

Figure 3.26

Figure 3.27

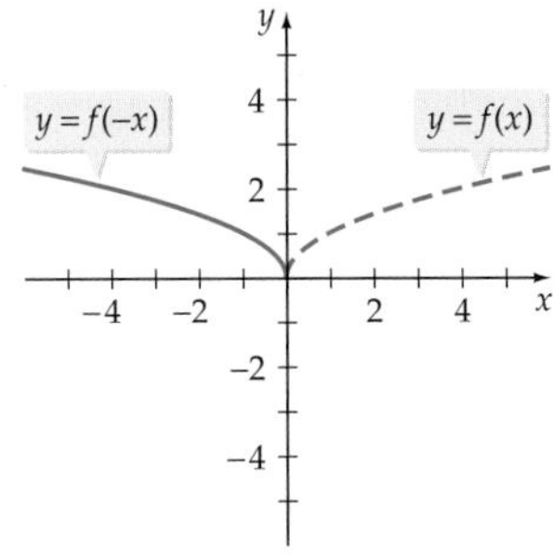

Figure 3.28

$f(x) = \sqrt{x},\ f(-x) = \sqrt{-x}$

Now consider the graph of $y = f(-x)$. If we again use $f(x) = \sqrt{x}$, then $f(-x) = \sqrt{-x}$. Notice that if we input a value for x, the sign is changed *before* the root is taken. From the graph in Figure 3.28, it appears that this time the graph is reflected across the y-axis.

P

Reflections

The graph of

- $y = -f(x)$ is the graph of $y = f(x)$ reflected across the x-axis.
- $y = f(-x)$ is the graph of $y = f(x)$ reflected across the y-axis.

As an example of these reflections, consider the graph in Figure 3.29. The graph of $y = -f(x)$ is shown in Figure 3.30 and the graph of $y = f(-x)$ is shown in Figure 3.31.

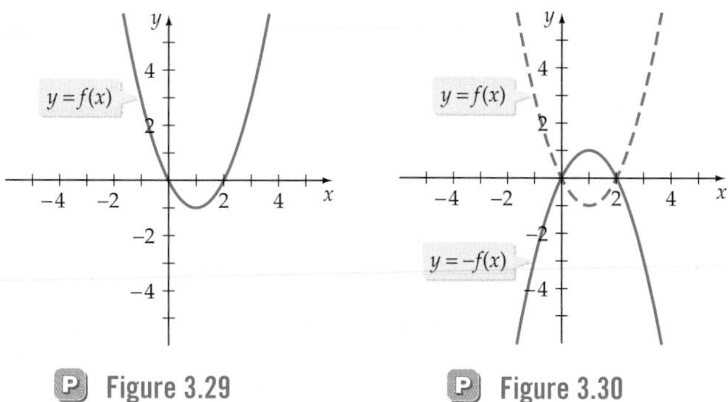

Figure 3.29 Figure 3.30

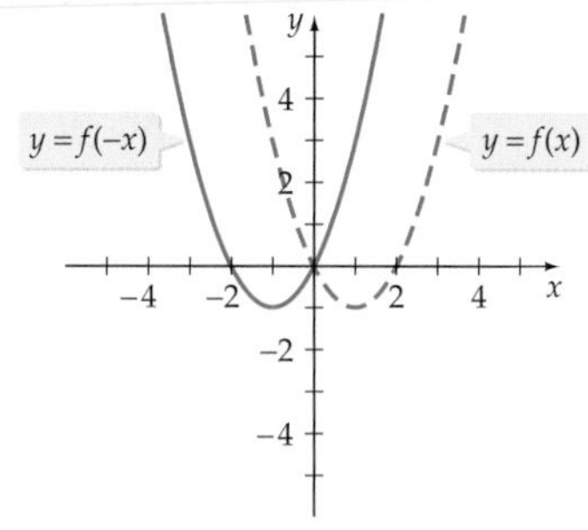

Figure 3.31

Alternative to Example 2
Exercise 14, page 289

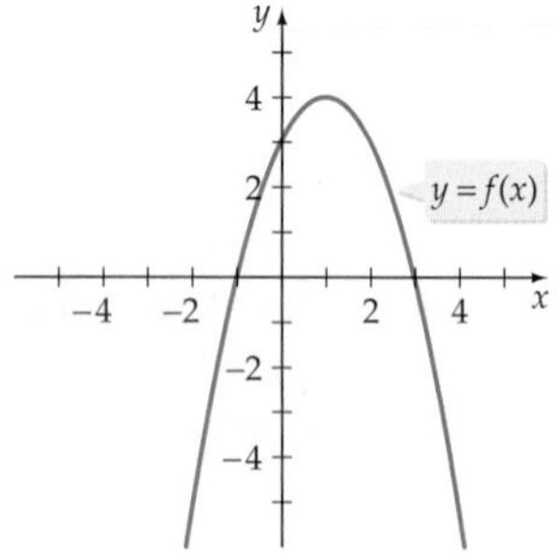

Figure 3.32

EXAMPLE 2 Draw a Graph Using Translation and Reflection

Use the graph of $y = f(x)$ in Figure 3.32 to graph

a. $y = -f(x + 1) + 3$

b. $y = f(-x - 2) - 3$.

Solution

a. The graph is produced in steps. First move the graph 1 unit to the left $[y = f(x + 1)]$. Reflect the graph through the x-axis $[y = -f(x + 1)]$. Next move the graph 3 units up $[y = -f(x + 1) + 3]$, as shown on the next page.

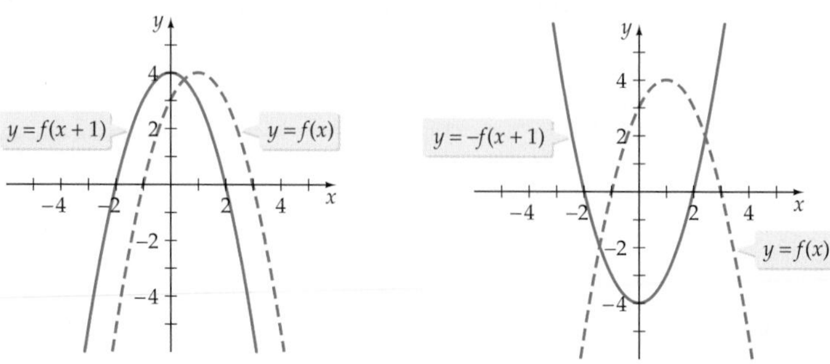

Continued ➤

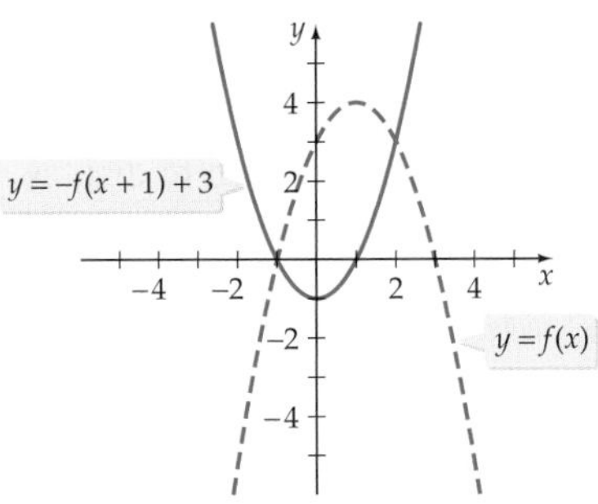

b. Produce the graph in steps. First move the graph 2 units to the right $[y = f(x - 2)]$. Reflect the graph through the y-axis $[y = f(-x - 2)]$. Move the graph 3 units down $[y = f(-x - 2) - 3]$.

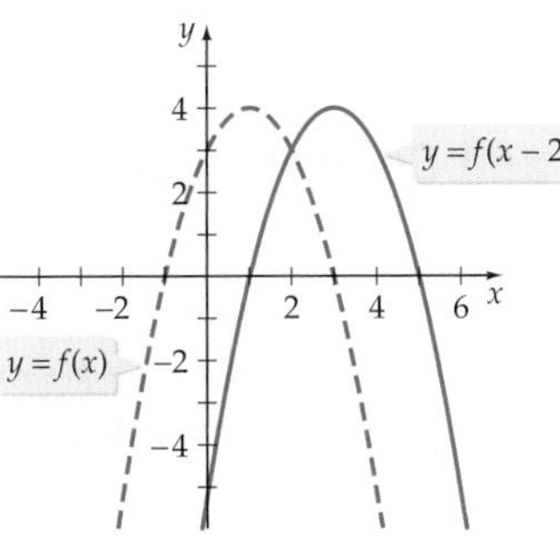

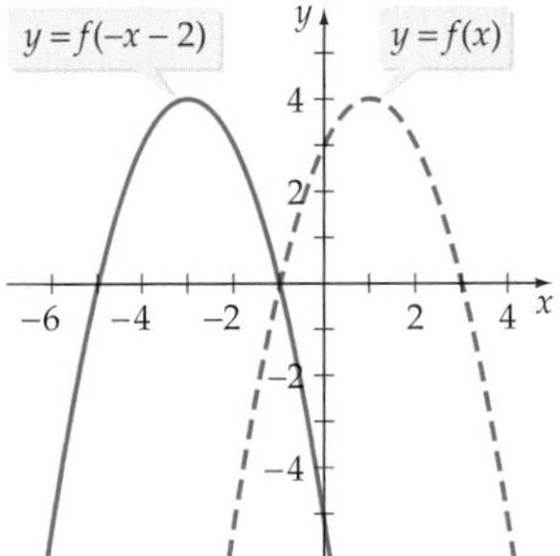

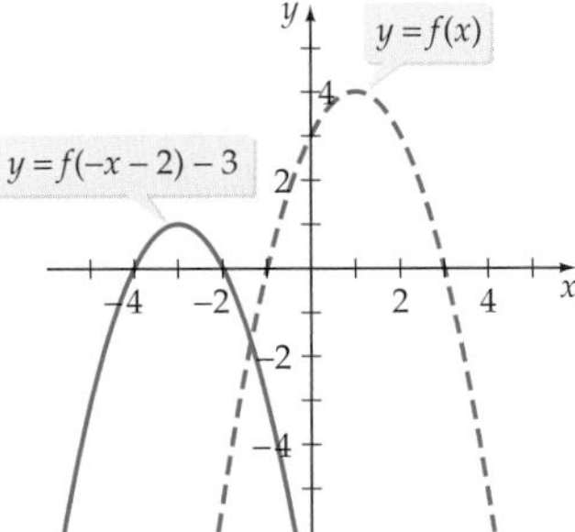

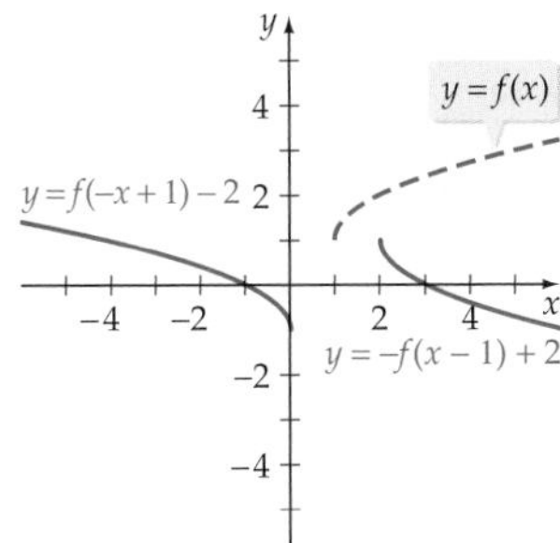

Figure 3.33

CHECK YOUR PROGRESS 2 Use the graph of $y = f(x)$ shown in Figure 3.33 to graph

a. $y = -f(x - 1) + 2.$

b. $y = f(-x + 1) - 2.$

Solution *See page S17.* See Figure 3.33.

INTEGRATING TECHNOLOGY

A graphing calculator can be used to display reflections of graphs. For the graphs in Check Your Progress 2, enter $\sqrt{x-1}+1$ into Y1. For Y2 enter $-$Y1$(x-1)+2$ and for Y3 enter Y1$(-x+1)-2$. Now graph the functions. Some typical screens are shown below.

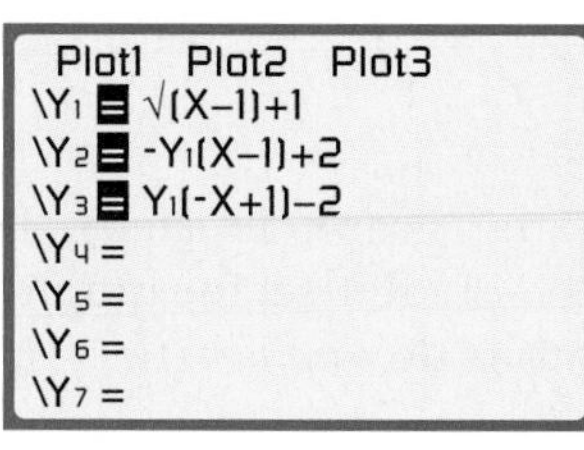

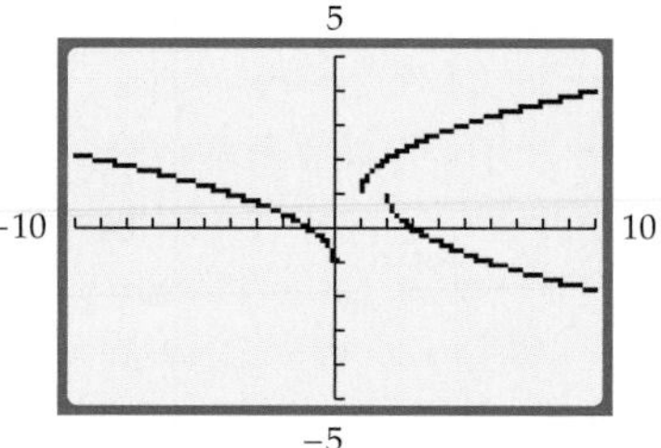

▪ Stretching and Compressing Graphs

From Figure 3.34, observe that each point on the graph of $y = 2f(x)$ has a y-value that is double the y-value of the corresponding point on the graph of $y = f(x)$. Figure 3.34 shows the graph of $f(x) = \sqrt{9 - x^2}$ along with the graph of $g(x) = 2f(x) = 2\sqrt{9 - x^2}$. The graph of g appears to be similar to the graph of f, but stretched vertically. Similarly, in Figure 3.35, we have the graphs of $f(x) = \sqrt{9 - x^2}$ and $h(x) = \frac{1}{2}f(x) = \frac{1}{2}\sqrt{9 - x^2}$; notice that the graph of h is similar to the graph of f but is compressed vertically by a factor of 2.

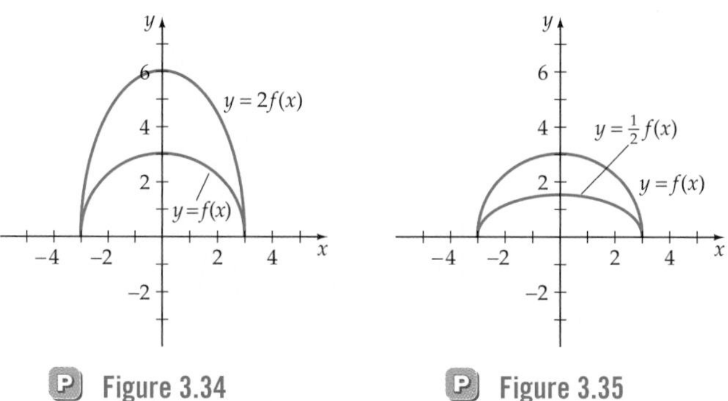

Figure 3.34

Figure 3.35

TAKE NOTE

The graph of $y = c \cdot f(x)$ $(c > 1)$ is the graph of $y = f(x)$ stretched away from the x-axis. Points on the x-axis do not move. Points above the x-axis move higher, and points below the x-axis move lower.

The graph of $y = c \cdot f(x)$ $(0 < c < 1)$ is the graph of $y = f(x)$ compressed toward the x-axis. Points on the x-axis do not move. Points above the x-axis move lower, and points below the x-axis move higher.

Vertical Stretching and Compressing of Graphs

If f is a function and c is a positive constant, then

- if $c > 1$, the graph of $y = c \cdot f(x)$ is the graph of $y = f(x)$ *stretched* vertically by a factor of c away from the x-axis.
- if $0 < c < 1$, the graph of $y = c \cdot f(x)$ is the graph of $y = f(x)$ *compressed* vertically by a factor of c toward the x-axis.

? QUESTION If c is a negative constant, how does the graph of $y = c \cdot f(x)$ compare to the graph of $y = f(x)$?

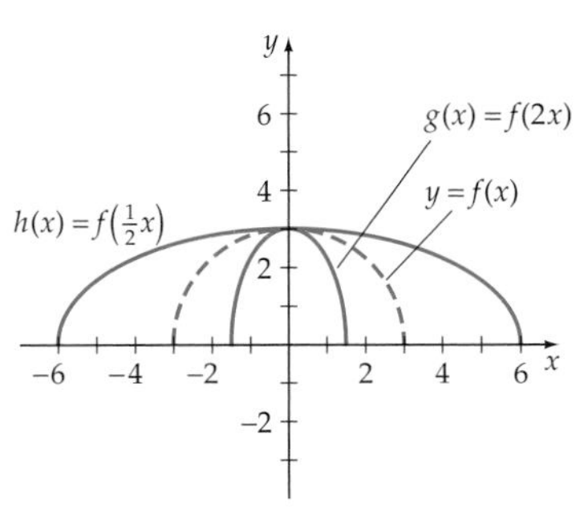

Figure 3.36

If $y = c \cdot f(x)$ is the graph of $y = f(x)$ stretched or compressed vertically, then you might guess that $y = f(c \cdot x)$ is the graph of $y = f(x)$ stretched or compressed horizontally. This guess is correct, but be watchful of how the value of the constant c affects the graph. Figure 3.36 shows the graphs of the following functions.

$$f(x) = \sqrt{9 - x^2}$$
$$g(x) = f(2x) = \sqrt{9 - (2x)^2} = \sqrt{9 - 4x^2}$$
$$h(x) = f\left(\frac{1}{2}x\right) = \sqrt{9 - \left(\frac{1}{2}x\right)^2} = \sqrt{9 - \frac{1}{4}x^2}$$

It appears that $y = f(2x)$ is the graph of $y = f(x)$ compressed horizontally, and $y = f\left(\frac{1}{2}x\right)$ is the graph of $y = f(x)$ stretched horizontally.

Horizontal Compressing and Stretching of Graphs

If f is a function and c is a positive constant, then

- if $c > 1$, the graph of $y = f(c \cdot x)$ is the graph of $y = f(x)$ *compressed* horizontally by a factor of $\frac{1}{c}$ toward the y-axis.
- if $0 < c < 1$, the graph of $y = f(c \cdot x)$ is the graph of $y = f(x)$ *stretched* horizontally by a factor of $\frac{1}{c}$ away from the y-axis.

If the point (x, y) is on the graph of $y = f(x)$, then the graph of $y = f(cx)$ will contain the point $\left(\frac{1}{c}x, y\right)$.

Alternative to Example 3
Exercise 36, page 292

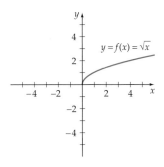

EXAMPLE 3 Use Stretching and Compressing to Draw a Graph

Graph the following by stretching or compressing the graph of $y = \sqrt{x}$ shown in Figure 3.37.

a. $y = 3\sqrt{x}$ **b.** $y = \sqrt{2x}$ **c.** $y = 2\sqrt{\frac{1}{2}x}$

Figure 3.37

Solution

a. The graph of $y = 3\sqrt{x}$ is a vertical stretch, by a factor of 3, of the graph of $y = \sqrt{x}$. For instance, the point (1, 1) is on the graph of $y = \sqrt{x}$. Multiplying the y-value by 3, we get the point (1, 3), which is on the graph of $y = 3\sqrt{x}$. Similarly, the point (4, 2) on the graph of $y = \sqrt{x}$ corresponds to (4, 6) on the graph of $y = 3\sqrt{x}$ on the following page.

Continued ➤

? ANSWER In addition to being stretched or compressed, the graph is reflected across the x-axis.

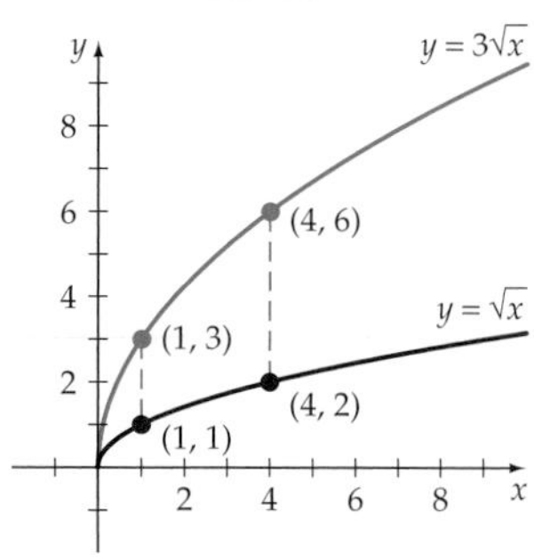

b. The graph of $y = \sqrt{2x}$ is a horizontal compression, by a factor of $\frac{1}{2}$, of the graph of $y = \sqrt{x}$. The point (1, 1) is on the graph of $y = \sqrt{x}$; the corresponding point on the graph of $y = \sqrt{2x}$ has an x-coordinate with half the value, so the graph will include the point $\left(\frac{1}{2}, 1\right)$. Similarly, the point (4, 2) on the graph of $y = \sqrt{x}$ corresponds to (2, 2) on the graph of $y = \sqrt{2x}$.

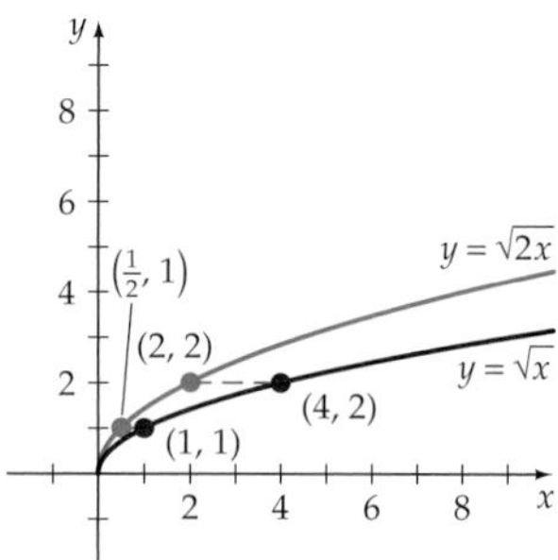

c. The graph of $y = 2\sqrt{\frac{1}{2}x}$ is both a vertical stretch (by a factor of 2) and a horizontal stretch $\left(\text{by a factor of } \frac{1}{\frac{1}{2}} = 2\right)$ of the graph of $y = \sqrt{x}$.

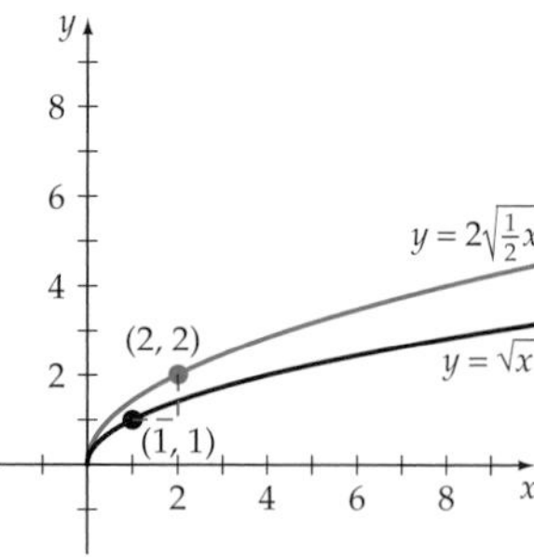

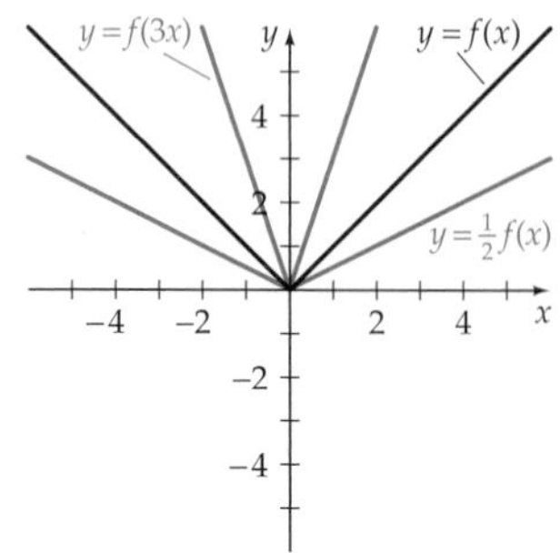

Figure 3.38

CHECK YOUR PROGRESS 3 Graph the following by stretching or compressing the graph of $y = f(x)$ shown in Figure 3.38.

a. $y = \frac{1}{2}f(x)$ **b.** $y = f(3x)$

Solution *See page S18.* See the graph at the left.

Graphs of some functions can be constructed by using a combination of translations, reflections, and stretching or compressing. When using a combination of transformations, try to use the same sequence of steps. Here is one suggestion.

Steps for Transforming a Graph

1. Do the horizontal translation.
2. Do reflections.
3. Do vertical or horizontal stretching/compressing.
4. Do the vertical translation.

For instance, consider the graph of $y = f(x)$ shown below. To create the graph of $y = -2f(x - 1) + 3$, we followed the steps outlined above.

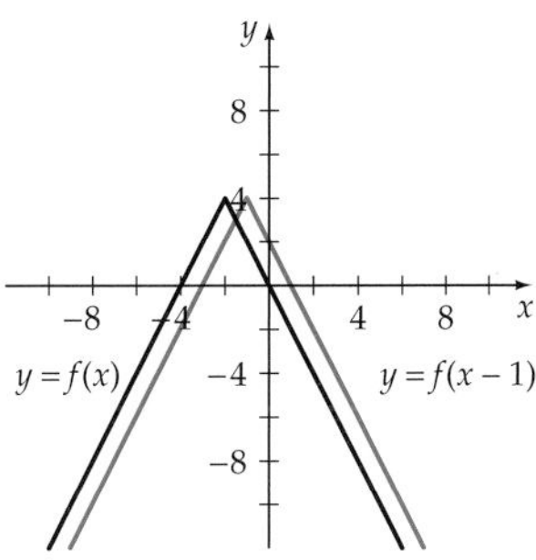

Horizontal translation

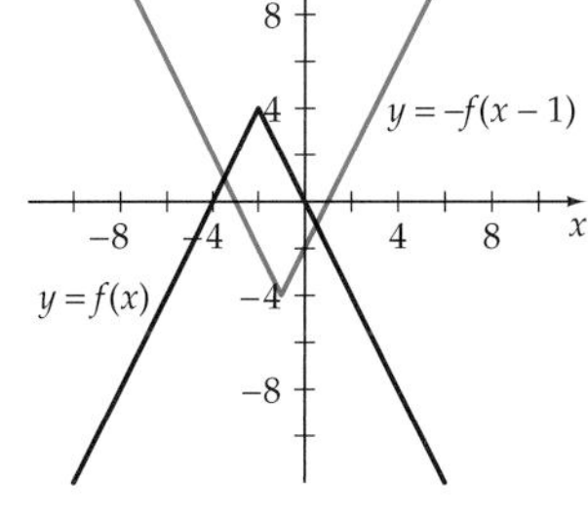

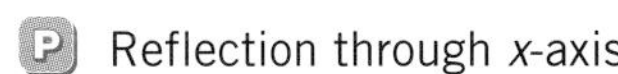
Reflection through x-axis

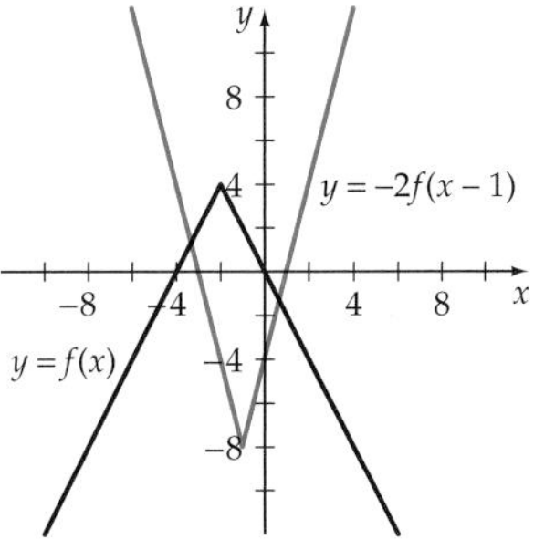

Vertical stretching

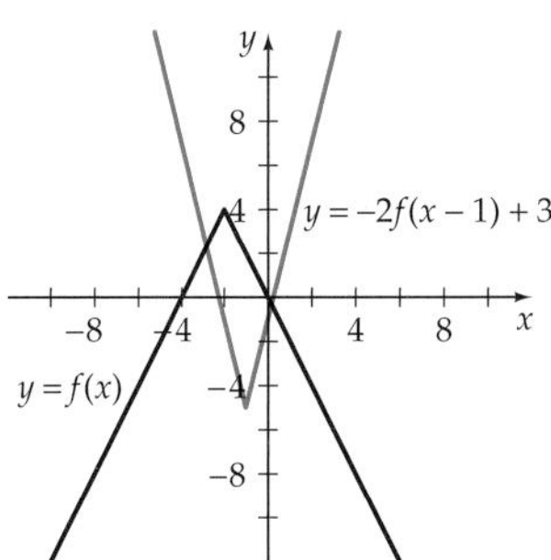

Vertical translation

Alternative to Example 4
Exercise 30, page 291

EXAMPLE 4 Solve an Application

A researcher believes that the shape of the graph of $f(x) = \sqrt{x}$ in Figure 3.39 would fit the points in her scatter plot shown in Figure 3.40. Use transformations to create a function whose graph fits the scatter plot.

Continued ➤

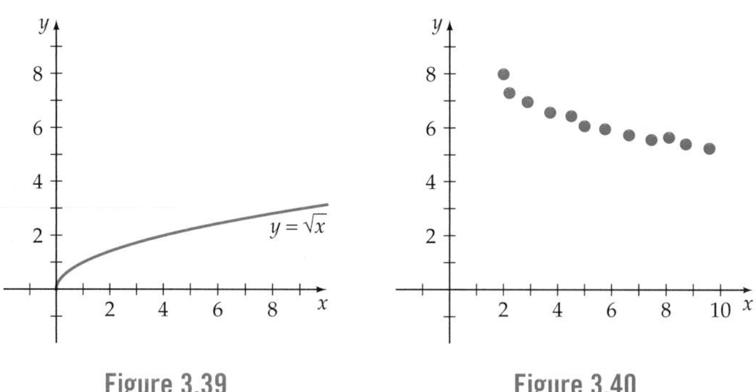

Figure 3.39

Figure 3.40

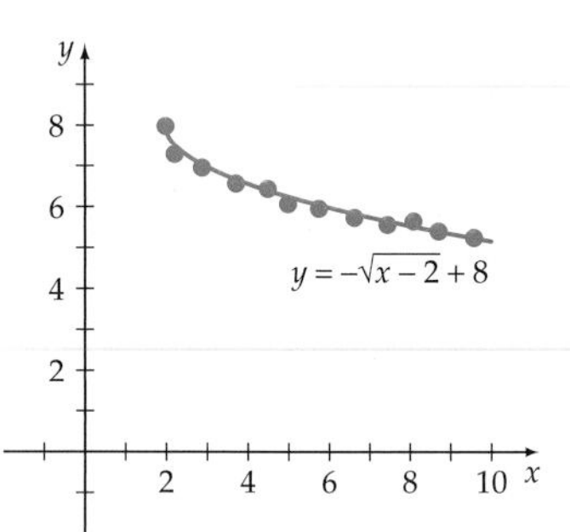

Figure 3.41

Solution The function f in Figure 3.39 has the right shape, but needs to be reflected and moved in order to match the pattern of the data points plotted in Figure 3.40. First the graph of $y = f(x)$ must be reflected across the x-axis by using the equation $y = -f(x)$. We then need to move the graph so that it starts at the point (2, 8) rather than at the origin. To shift the graph right 2 units and up 8 units, we use $y = -f(x - 2) + 8$. The corresponding equation is

$$y = -f(x - 2) + 8 = -\sqrt{x - 2} + 8$$

See Figure 3.41.

CHECK YOUR PROGRESS 4 Use transformations of the graph of $f(x) = \sqrt{2x - x^2}$ to find an equation for g, whose graph is shown at the right.

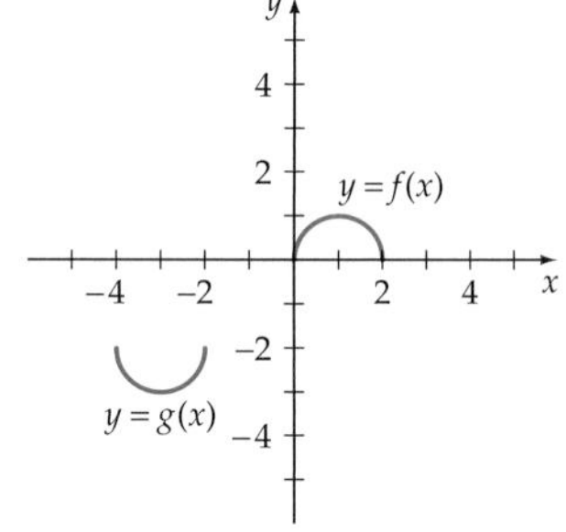

Solution *See page S18.* $y = -\sqrt{-x^2 - 6x - 8} - 2$

▪ Symmetry

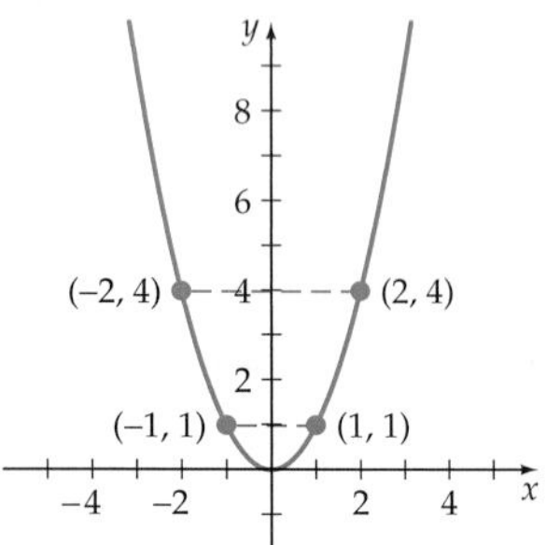

Figure 3.42

The parabola whose equation is $f(x) = x^2$ has a special property. If we reflect its graph across the y-axis [that is, if we graph $y = f(-x)$], the graph looks exactly the same. (See Figure 3.42.) The graph is said to be **symmetric with respect to the y-axis.** The graph has the property that if we were to fold the page along the y-axis, the point $(-1, 1)$ would coincide with the point (1, 1), the point $(-2, 4)$ would coincide with the point (2, 4), and so on. The left half of the graph is a *mirror image* of the right half of the graph.

In general, if any point (x, y) is on a graph that is symmetric with respect to the y-axis, $(-x, y)$ must also be on the graph (see Figure 3.43). This suggests an algebraic way to test the graph of an equation for symmetry with respect to the

y-axis: replace x by $-x$ in the equation and check whether the resulting equation is the same as the original. For instance,

$$y = f(x) = x^2$$
$$y = f(-x) = (-x)^2 = x^2$$ ■ Replace x by $-x$ and then simplify.

After simplifying, both equations are $y = x^2$. Replacing x by $-x$ did not change the equation. The graph of $f(x) = x^2$ is symmetric with respect to the y-axis.

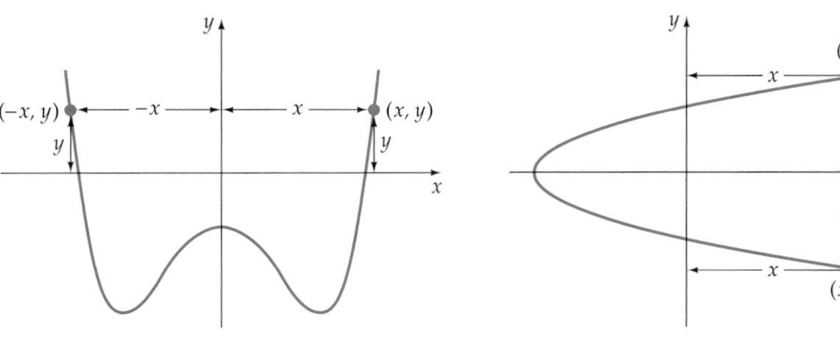

Figure 3.43

Symmetry with respect to the y-axis

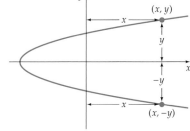

Figure 3.44

Symmetry with respect to the x-axis

Figure 3.44 shows a graph that is **symmetric with respect to the x-axis.** The top half is a mirror reflection of the bottom half, with the x-axis acting as the mirror. Notice that if a point (x, y) is on a graph that is symmetric with respect to the x-axis, then the point $(x, -y)$ must also be on the graph.

TAKE NOTE

After you apply one of the tests for symmetry, some algebraic simplification may be required to determine whether the resulting equation is unaltered.

Tests for Symmetry with Respect to the x-axis or y-axis

The graph of an equation is symmetric with respect to

- the y-axis if the replacement of x with $-x$ leaves the equation unaltered.
- the x-axis if the replacement of y with $-y$ leaves the equation unaltered.

Alternative to Example 5

Exercise 62, page 294

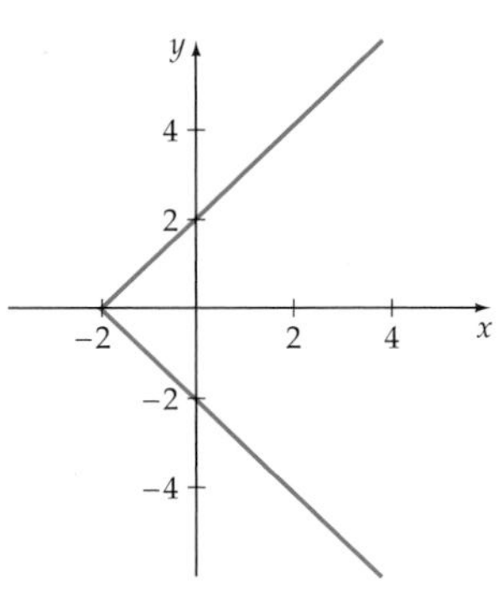

Figure 3.45

EXAMPLE 5 Determine Coordinate Axis Symmetries of a Graph

Determine whether the graph of the equation $x = |y| - 2$ has symmetry with respect to either the x- or y-axis.

Solution The equation $x = |y| - 2$ is altered by the replacement of x with $-x$. That is, if we simplify $-x = |y| - 2$, which yields $x = -|y| + 2$, we do not get the original equation $x = |y| - 2$. This implies that the graph of $x = |y| - 2$ is not symmetric with respect to the y-axis. However, the equation $x = |y| - 2$ is left unaltered by the replacement of y with $-y$. If we simplify the equation $x = |-y| - 2$, we arrive at the original equation $x = |y| - 2$. Thus the graph is symmetric with respect to the x-axis. The graph of $x = |y| - 2$, shown in Figure 3.45, confirms this result.

Continued ➤

CHECK YOUR PROGRESS 5 Determine whether the graph of the equation $y = x^2 + 2$ has symmetry with respect to either the x- or the y-axis.

Solution *See page S18.* Symmetry with respect to the y-axis

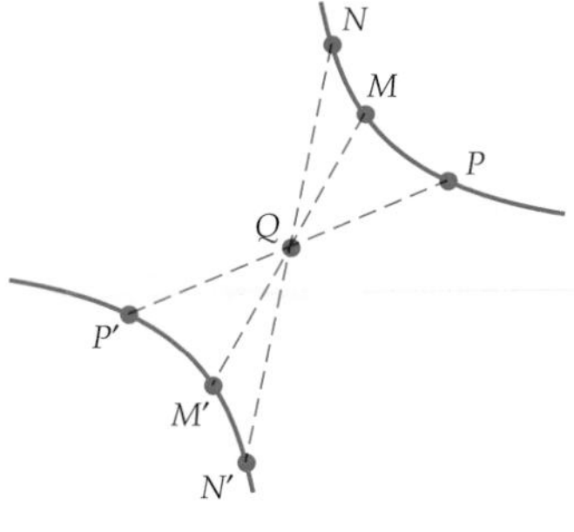

Figure 3.46

A graph can be **symmetric with respect to a point.** Figure 3.46 shows a graph that is symmetric with respect to the point Q. Notice that for each point P on the graph, there is a point P' on the graph for which Q is the midpoint of the line segment joining P and P'.

When we discuss symmetry with respect to a point, we frequently use the origin. The graph in Figure 3.47 is symmetric with respect to the origin; notice that if any point (x, y) is on the graph, the point $(-x, -y)$ must also be on the graph. This suggests a method to test an equation for symmetry with respect to the origin.

TAKE NOTE

Another way to visualize symmetry with respect to the origin is to imagine rotating the graph 180° about the origin; the graph should look the same. You can also visualize symmetry with respect to the origin as reflecting the graph across the x-axis and then across the y-axis, and arriving at the same graph.

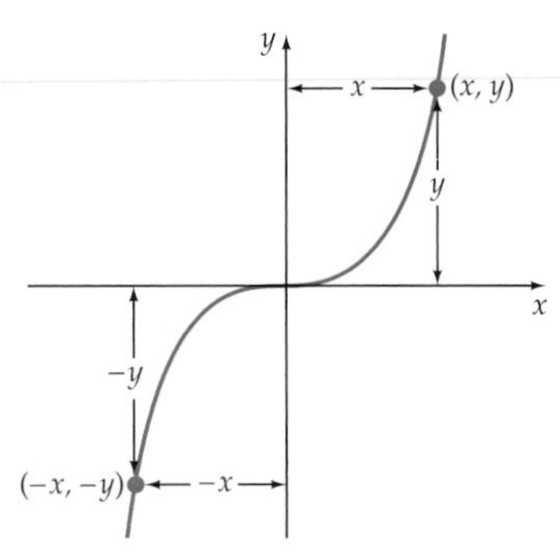

Figure 3.47

Symmetry with respect to the origin

Test for Symmetry with Respect to the Origin

The graph of an equation is symmetric with respect to the origin if the replacement of x with $-x$ and y with $-y$ leaves the equation unaltered.

Alternative to Example 6
Exercise 70, page 294

EXAMPLE 6 Determine Origin Symmetry

Determine whether the graph of the equation $xy = 4$ is symmetric with respect to the origin.

Solution If we replace x with $-x$ and y with $-y$, the equation becomes $(-x)(-y) = 4$, which simplifies to $xy = 4$. Thus the equation is left unaltered, and the graph of $xy = 4$ is symmetric with respect to the origin. The graph is shown in Figure 3.48.

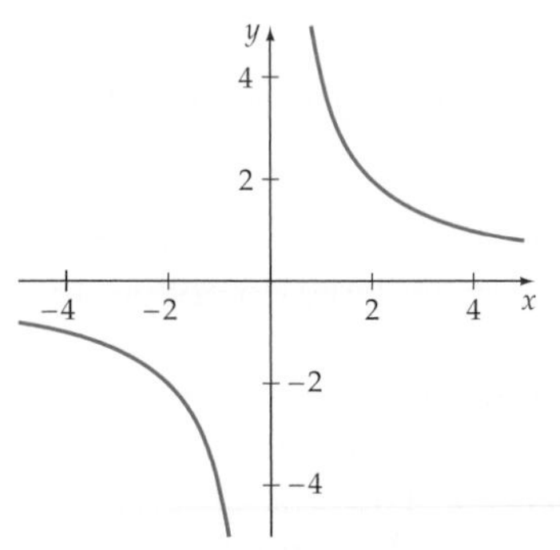

Figure 3.48

CHECK YOUR PROGRESS 6 Determine whether the graph of the equation $y = x^3 + 1$ is symmetric with respect to the origin.

Solution *See page S18.* Not symmetric with respect to the origin

Some graphs possess more than one type of symmetry. For example, the graph of the equation $|x| + |y| = 2$ has symmetry with respect to the x-axis, the y-axis, and the origin. (See Figure 3.49.)

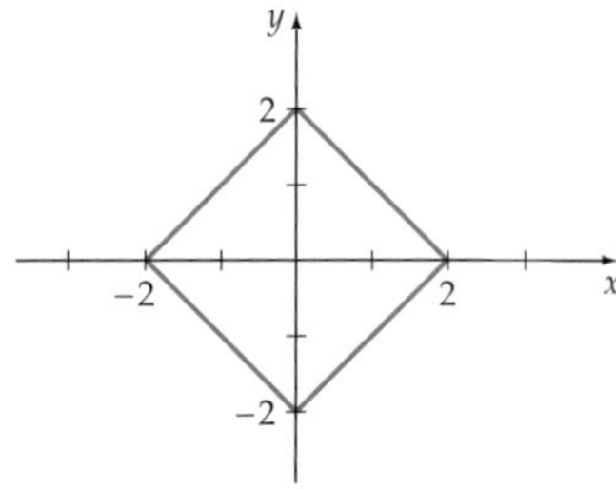

Figure 3.49

▪ Even and Odd Functions

Functions whose graphs have certain symmetries are given special names. If a function has a graph that is symmetric with respect to the y-axis, the function is called *even*. Recall that for the graph of an equation to be symmetric with respect to the y-axis, we must be able to replace x with $-x$ while leaving the equation unaltered. In function notation, we simply state that $f(-x) = f(x)$. See Figure 3.50.

A function whose graph is symmetric with respect to the origin is called an *odd* function. For an equation to be symmetric with respect to the origin, we must be able to replace x with $-x$ and y with $-y$ without altering the equation. In terms of functions, this means $-[f(-x)] = f(x)$ or $f(-x) = -f(x)$. See Figure 3.51.

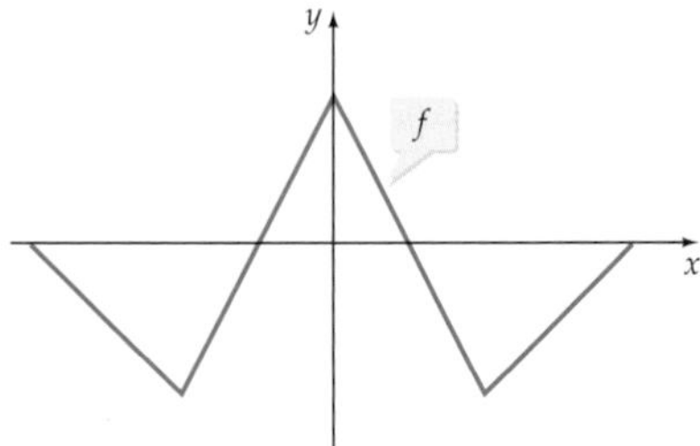

Figure 3.50

The graph of an even function is symmetric with respect to the y-axis.

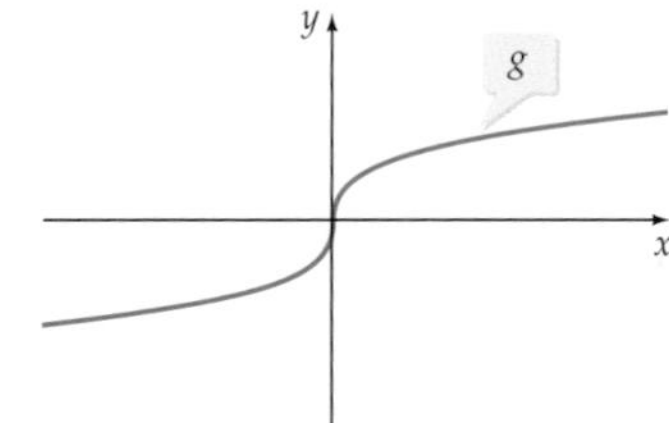

Figure 3.51

The graph of an odd function is symmetric with respect to the origin.

Definition of Even and Odd Functions

The function f is an **even function** if

$$f(-x) = f(x) \qquad \text{for all } x \text{ in the domain of } f$$

The function f is an **odd function** if

$$f(-x) = -f(x) \qquad \text{for all } x \text{ in the domain of } f$$

Alternative to Example 7

Determine whether each function is even, odd, or neither.

a. $f(x) = x^3 + 1$

b. $f(x) = x^3 + x$

■ **a. Neither**

■ **b. Odd**

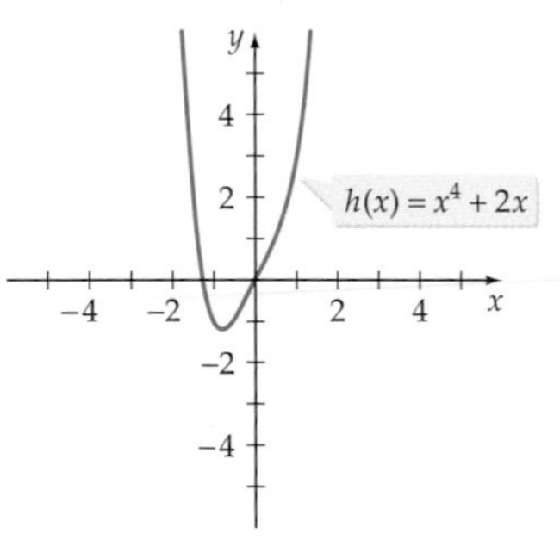

Figure 3.52

EXAMPLE 7 Characterize Functions as Even or Odd

Determine whether each function is even, odd, or neither.

a. $f(x) = x^3$

b. $h(x) = x^4 + 2x$

Solution Replace x with $-x$ and simplify.

a. $f(-x) = (-x)^3 = -x^3 = -(x^3) = -f(x)$

This function is an odd function because $f(-x) = -f(x)$.

b. $h(-x) = (-x)^4 + 2(-x) = x^4 - 2x$

Because $h(-x) = x^4 - 2x$, which is not equal to either $h(x)$ or $-h(x)$, this function is neither an even nor an odd function. Figure 3.52 shows that its graph is not symmetric with respect to the y-axis or the origin.

CHECK YOUR PROGRESS 7 Determine whether each function is even, odd, or neither.

a. $F(x) = |x|$

b. $g(x) = x^3 - x^2$

Solution *See page S18.* F is even. g is neither even nor odd.

Topics for Discussion

1. What does it mean to reflect a graph across the x-axis or across the y-axis?
2. Explain how the graphs of the equations $y_1 = 2x^3 - x^2$ and $y_2 = 2(-x)^3 - (-x)^2$ are related.
3. Given the graph of $y = f(x)$, explain how to obtain the graph of $y = f(x - 3) + 1$.
4. Using the graph in Figure 3.53, graph $y = f(2x)$ and $y = 2f(x)$. Are the graphs of $y = f(2x)$ and $y = 2f(x)$ the same? If f is some function, does $f(ax) = af(x)$?
5. What does it mean for a graph to be symmetric with respect to the x-axis? How do you determine algebraically whether a graph is symmetric with respect to the x-axis? with respect to the y-axis?
6. How can you visually determine whether a graph is symmetric with respect to the origin? How can you algebraically determine whether a graph has this symmetry?
7. What does it mean for a function to be an even function? an odd function?

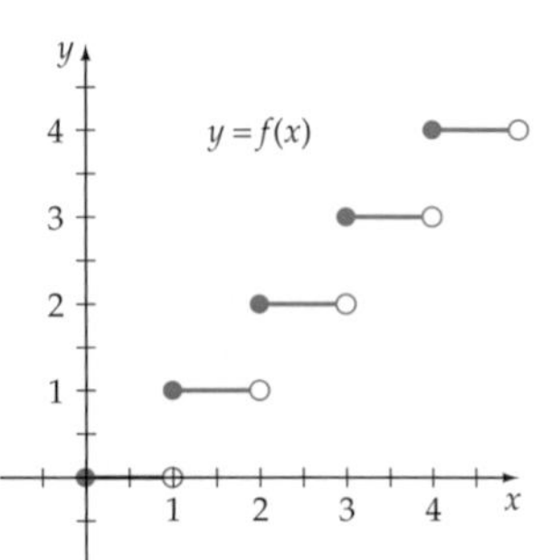

Figure 3.53

EXERCISES 3.3

— *Suggested assignment: 1–59, odd; 61–97, every other odd; and 98–106.*
— *Answer graphs to Exercises 11–18, 23–28, 31–40, 55–60, and 91–98 are on pages AA6–AA9.*

1. The point (2, 5) is on the graph of $y = f(x)$. Find the coordinates of a point on the graph of $y = f(x) - 3$. (2, 2)

2. The point (1, −2) is on the graph of $y = f(x)$. Find the coordinates of a point on the graph of $y = f(x) + 1$. (1, −1)

3. The point (−2, 3) is on the graph of $y = f(x)$. Find the coordinates of a point on the graph of $y = f(x + 1)$. (−3, 3)

4. The point (2, −5) is on the graph of $y = f(x)$. Find the coordinates of a point on the graph of $y = f(x - 4)$. (6, −5)

5. The point (5, −2) is on the graph of $y = f(x)$. Find the coordinates of a point on the graph of $y = f(x + 1) - 2$. (4, −4)

6. The point (−3, 4) is on the graph of $y = f(x)$. Find the coordinates of a point on the graph of $y = f(x - 3) + 4$. (0, 8)

7. The point (6, −3) is on the graph of $y = f(x)$. Find the coordinates of a point on the graph of $y = 2f(x - 1) + 1$. (7, −5)

8. The point (−4, 1) is on the graph of $y = f(x)$. Find the coordinates of a point on the graph of $y = -2f(x + 3) + 2$. (−7, 0)

9. The point (−2, 0) is on the graph of $y = f(x)$. Find the coordinates of a point on the graph of $y = -3f(2x) + 4$. (−1, 4)

10. The point (3, 5) is on the graph of $y = f(x)$. Find the coordinates of a point on the graph of $y = 2f(6x) - 4$. $\left(\frac{1}{2}, 6\right)$

11. The graph of a function f is shown. Draw a graph of the following.

a. $y = f(x) + 4$ **b.** $y = f(x) - 2$

c. $y = f(x + 1)$ **d.** $y = f(x - 3)$

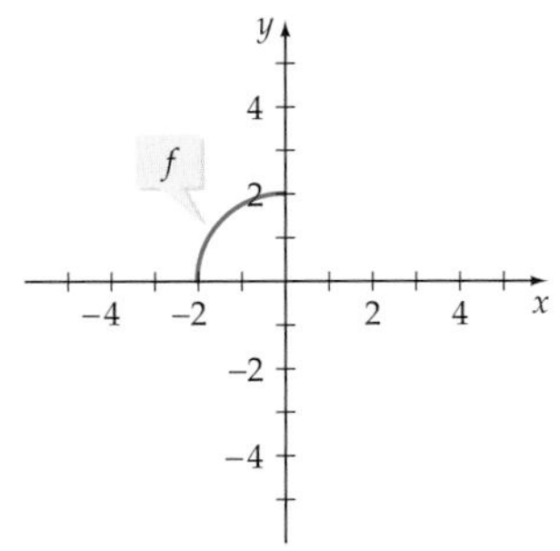

12. The graph of a function g is shown. Draw a graph of the following.

a. $y = g(x) - 1$ **b.** $y = g(x) + 3$

c. $y = g(x + 5)$ **d.** $y = g(x - 4)$

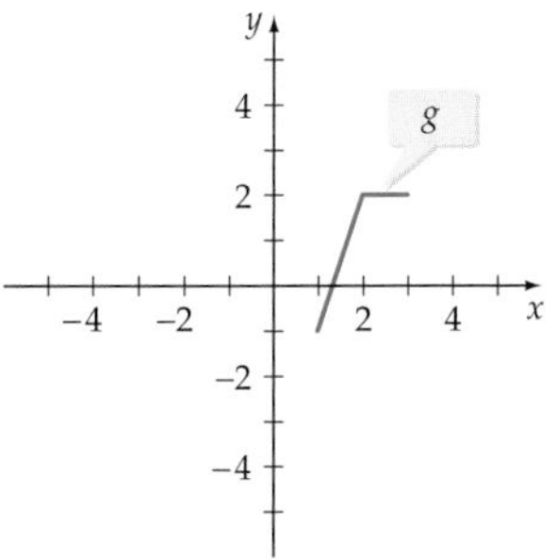

13. Use the graph of the function f in Exercise 11 to draw a graph of:

a. $y = f(x - 2) + 3$

b. $y = f(x + 3) + 1$

14. Use the graph of the function g in Exercise 12 to draw a graph of:

a. $y = g(x + 4) - 5$

b. $y = g(x - 1) - 1$

15. Use the graph of $y = x^3$ (Figure 3.25, page 276) to draw a graph of:

a. $y = x^3 + 3$ **b.** $y = (x - 2)^3$

c. $y = (x + 2)^3 - 1$ **d.** $y = -x^3 + 1$

e. $y = -(x - 1)^3 - 2$

16. Use the graph of $y = \sqrt{x}$ (Figure 3.26, page 277) to draw a graph of:

a. $y = \sqrt{x} - 4$ **b.** $y = \sqrt{x + 3}$

c. $y = \sqrt{x - 1} + 2$ **d.** $y = -\sqrt{x - 2}$

e. $y = -\sqrt{x} - 2$

17. Use the graph of $f(x) = |x|$ shown below to draw a graph of:

a. $y = f(x - 3)$

b. $y = -f(x - 3)$

c. $y = f(x - 1) + 1$

d. $y = -f(x + 2) - 3$

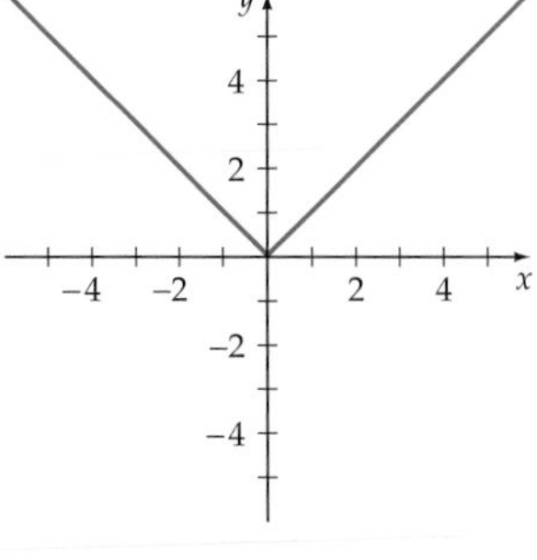

18. Use the graph of $f(x) = \sqrt{16 - x^2}$ shown below to draw a graph of:

a. $y = f(x + 1) - 2$

b. $y = -f(x + 2) - 1$

c. $y = f(x - 1) + 3$

d. $y = -f(x - 2) + 2$

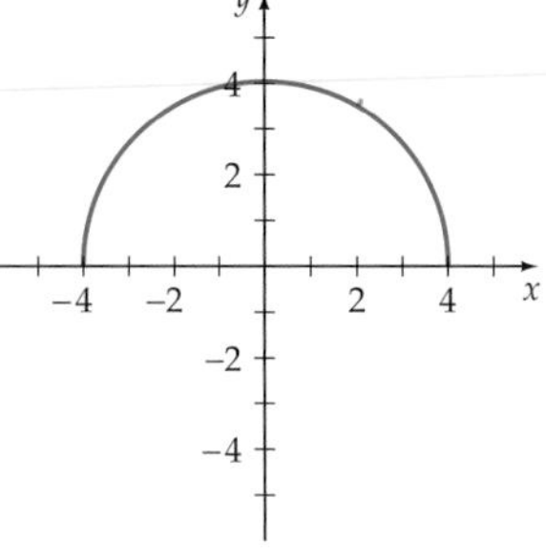

Business and Economics

19. ***Office Temperatures*** Suppose $T(t)$ is the function that gives the temperature T at time t in the CEO's office at a particular company. (t is the number of hours after midnight.)

a. The vice president's office is always kept 4 degrees cooler than the CEO's office. Write a function, in terms of $T(t)$, for the temperature at time t in the vice president's office. $f(t) = T(t) - 4$

b. The reception area has the temperature control set to the same temperature as the CEO's office, but changes in temperature always seem to lag a half hour behind temperature changes in the CEO's office. Write a function, in terms of $T(t)$, for the temperature at time t in the reception area. $g(t) = T\left(t - \frac{1}{2}\right)$

c. The temperature in the mailroom can be described by $T(t - 1) - 2$. Describe how the temperature in the mailroom compares with the temperature in the CEO's office. The temperature is 2 degrees cooler and the change in temperature lags behind by 1 hr.

20. ***Compensation*** Ken has worked at the same company for many years. The function $A(t)$ gives his annual salary A at any time t, where t is the number of years after 1980.

a. Ken's friend Gabriel works at the same company, but she always earns one-and-a-half times as much per year as he does. Write a function, in terms of $A(t)$, that gives her annual salary. $f(t) = 1.5A(t)$

b. Another coworker, Jean, is always paid at an annual salary rate that is 15% less than Ken's rate. Write a function, in terms of $A(t)$, that gives her annual salary. $g(t) = 0.85A(t)$

Social Sciences

21. ***Demographics*** The population of Springfield is given by $P(t)$, where t is the number of years since 1950.

a. During any given year, the population of Sunnyville is about 8000 less than the population of Springfield. Write a function, in terms of $P(t)$, for the population of Sunnyville. $f(t) = P(t) - 8000$

b. Woodside is a neighboring city whose population during any given year matches the population of Springfield 10 years prior. Write a function, in terms of $P(t)$, for the population of Woodside. $g(t) = P(t - 10)$

c. The population of another nearby city, River Glen, can be described by $P(t - 5) + 2000$. Describe how the population of River Glen compares to the population of Springfield.
The population is 2000 people greater than the population of Springfield 5 years ago.

22. ***Emergency Services*** The sound intensity (loudness, given in decibels) $I(t)$ of a police siren after t seconds is shown in the graph below.

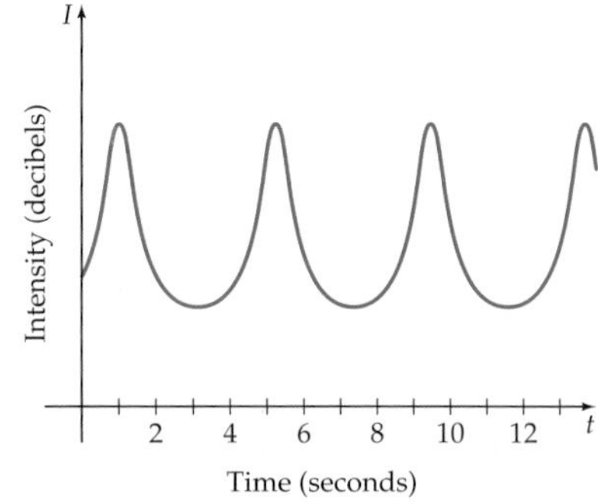

a. If an ambulance's siren is given by $I(2t)$, describe how its siren compares to the police siren. It has a higher pitch.

b. A fire engine's siren follows the same pattern as the police siren, but is always twice as loud. Write a function, in terms of $I(t)$, for the fire engine's siren. $f(t) = 2I(t)$

c. A paramedic's siren is given by $0.5I\left(\frac{t}{2}\right)$. Describe how the siren compares to the police siren.
The siren's loudness is reduced by a factor of 2 and the pitch is lower.

23. Use the graph of the function f in Exercise 11 to draw a graph of:

a. $y = f(-x)$ **b.** $y = -f(x)$

24. Use the graph of the function g in Exercise 12 to draw a graph of:

a. $y = -g(x)$ **b.** $y = g(-x)$

25. Use the graph of the function f in Exercise 11 to draw a graph of:

a. $y = f(-x) + 3$ **b.** $y = -f(x - 1) - 1$

26. Use the graph of the function g in Exercise 12 to draw a graph of:

a. $y = -g(-x)$

b. $y = -g(x + 3) + 4$

27. The graph of $F(x) = (x - 1)^{2/3}$ is shown below. Find a formula and sketch a graph for:

a. $y = F(x + 3) - 1$
$y = (x + 2)^{2/3} - 1$

b. $y = -F(x)$
$y = -(x - 1)^{2/3}$

c. $y = F(-x)$
$y = (-x - 1)^{2/3}$

d. $y = F(-x) + 2$
$y = (-x - 1)^{2/3} + 2$

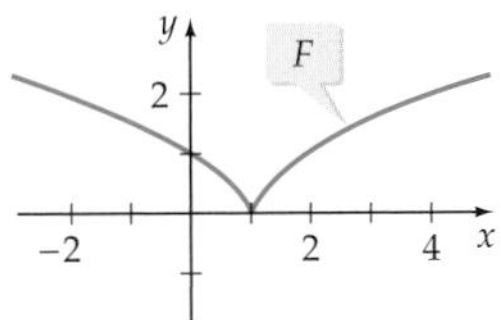

28. The graph of $E(x) = |x - 1| + 1$ is shown below. Find a formula and sketch a graph for:

a. $y = E(x - 2)$
$y = |x - 3| + 1$

b. $y = -E(x)$
$y = -|x - 1| - 1$

c. $y = E(-x) - 3$
$y = |x + 1| - 2$

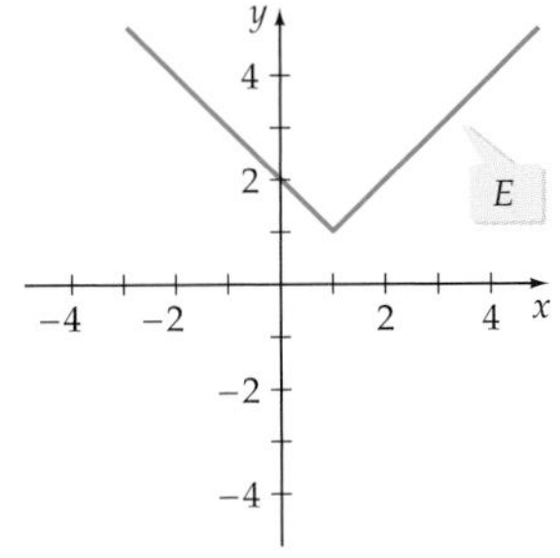

29. The data points in the scatter plot below seem to fit the shape of the graph of $y = \sqrt{x}$. Use transformations to create a function whose graph provides a close fit to the points in the scatter plot. $y = \sqrt{x - 4} + 2$

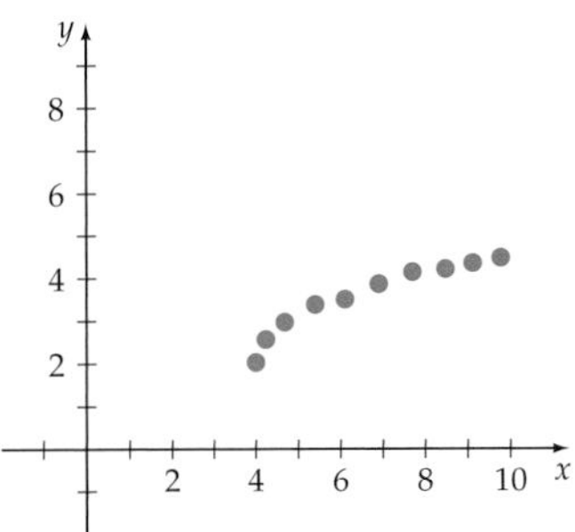

30. The data points in the scatter plot below seem to fit the shape of the graph of $y = \sqrt{3x}$. Use transformations to create a function whose graph provides a close fit to the points in the scatter plot. $y = -\sqrt{3x - 12} + 12$

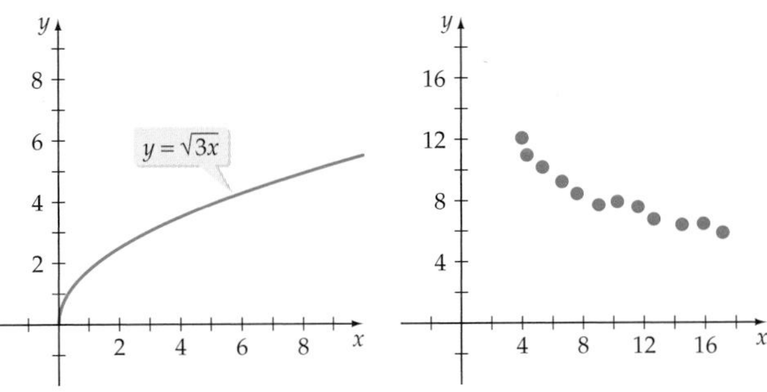

31. The graph of a function f is shown below. Draw a graph of **a.** $y = 2f(x)$, **b.** $y = 3f(x)$, and **c.** $y = \frac{1}{2}f(x)$.

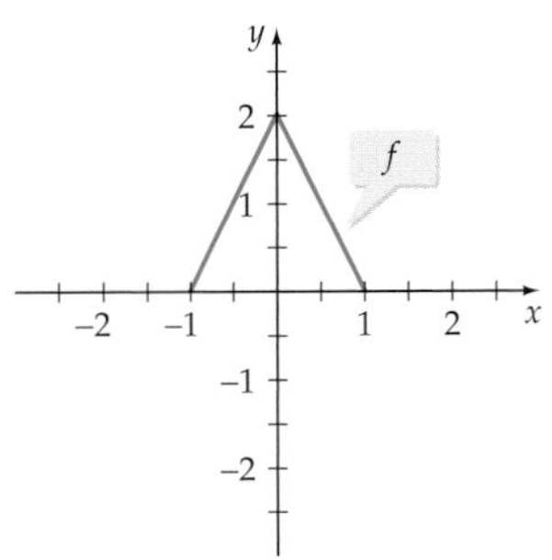

32. The graph of a function g is shown on the following page. Draw a graph of **a.** $y = 2g(x)$, **b.** $y = 4g(x)$, and **c.** $y = \frac{1}{2}g(x)$.

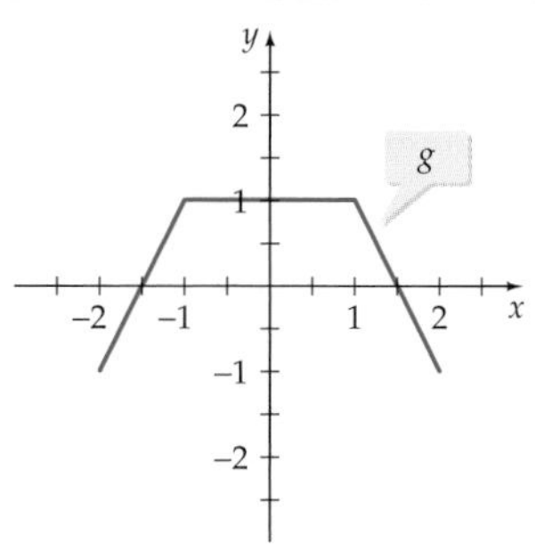

33. Use the graph of $y = f(x)$ to sketch the graphs of **a.** $y = f(2x)$ and **b.** $y = f\left(\frac{1}{2}x\right)$.

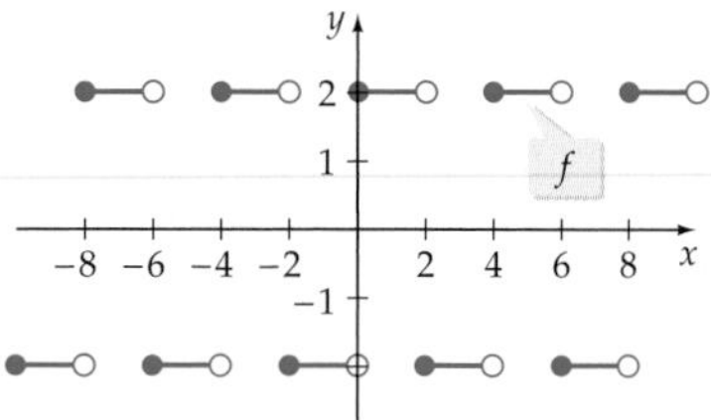

34. Use the graph of $y = g(x)$ to sketch the graphs of **a.** $y = g(2x)$ and **b.** $y = g\left(\frac{1}{2}x\right)$.

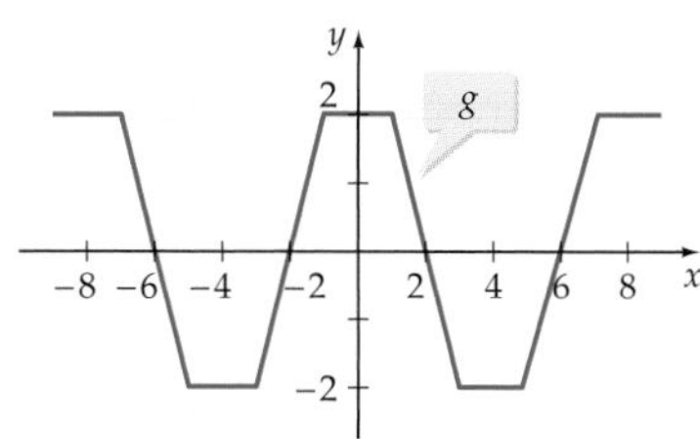

35. Use the graph of $y = \sqrt{x}$ (Figure 3.26, page 277) to draw a graph of:

a. $y = 2\sqrt{x}$ **b.** $y = \sqrt{\dfrac{x}{3}}$

c. $y = -\sqrt{2x}$ **d.** $y = -\dfrac{\sqrt{x}}{2}$

36. Use the graph of $y = |x|$ (Figure 3.19, page 273) to draw a graph of:

a. $y = 2|x|$ **b.** $y = \left|\dfrac{1}{4}x\right|$

c. $y = -|3x|$ **d.** $y = -2|x + 1|$

37. Use the graph of $f(x) = x^2$ to draw the graph of:

a. $y = -2f(x)$ **b.** $y = f(2x + 4)$

c. $y = \dfrac{1}{2}f(x - 1)$ **d.** $y = -2f(x + 1) - 1$

38. Use the graph of $f(x) = x^3$ to draw the graph of:

a. $y = -\dfrac{1}{2}f(x)$ **b.** $y = f\left(-\dfrac{1}{2}x\right) - 1$

c. $y = \dfrac{1}{2}f(x - 1) + 1$ **d.** $y = -\dfrac{1}{2}f(-x)$

39. The graph of $f(x) = \sqrt{4 - x^2}$ is shown below. Find a formula and sketch a graph for:

a. $y = 2f(x)$ **b.** $y = f(2x)$

c. $y = \dfrac{1}{2}f(x)$ **d.** $y = f\left(\dfrac{1}{2}x\right)$

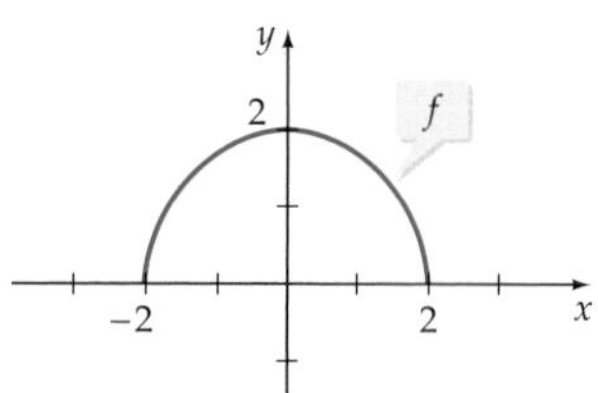

40. The graph of $m(x) = x^2 - 2x - 3$ is shown below. Find a formula and sketch a graph for:

a. $y = \dfrac{1}{2}m(x)$ **b.** $y = m\left(\dfrac{1}{3}x\right)$

c. $y = -\dfrac{1}{2}m(x) + 3$

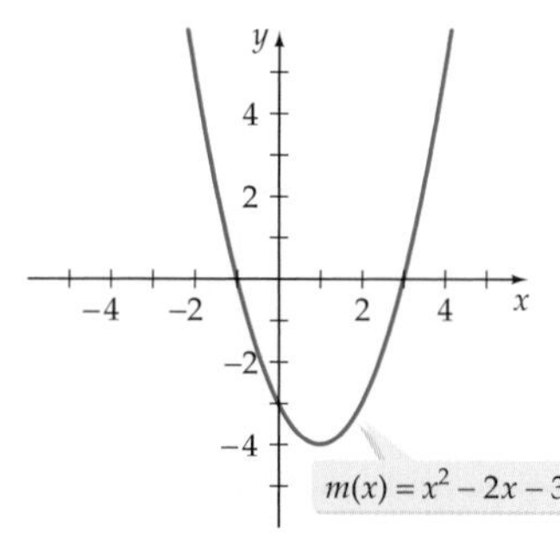

In Exercises 41 to 48, the graph of the function can be obtained by transforming the graph of the function f shown below. Match each graph to one of the following equations.

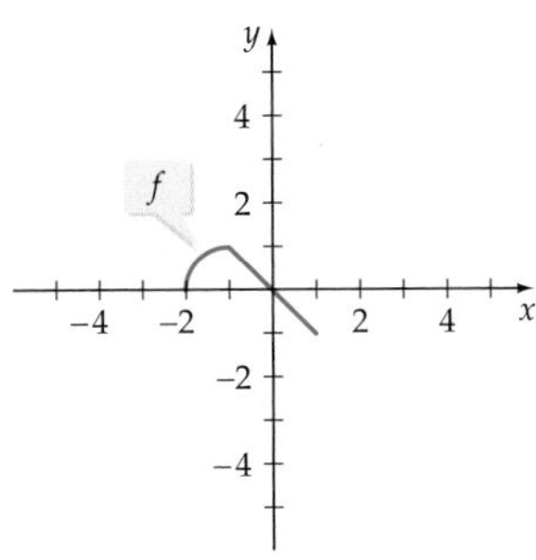

I. $y = 2f(x-2)$ **II.** $y = f(-x) - 3$

III. $y = f(x+2) - 1$ **IV.** $y = f\left(\dfrac{x}{2}\right) + 2$

V. $y = 2f(x-3) + 1$ **VI.** $y = f(x-3) + 2$

VII. $y = f\left[\dfrac{1}{2}(x-2)\right] + 1$ **VIII.** $y = -f(x-1) + 2$

41. **42.**

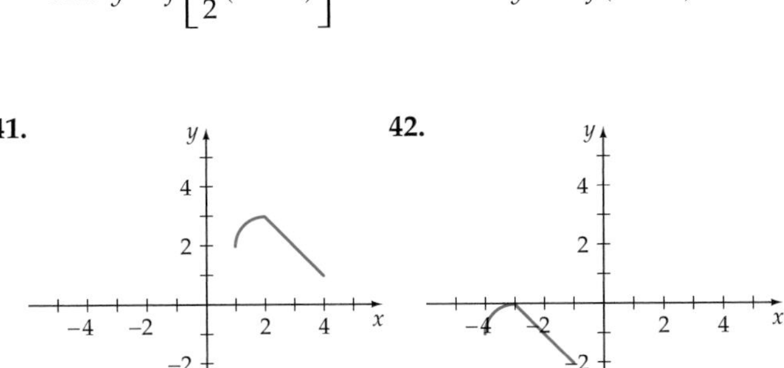

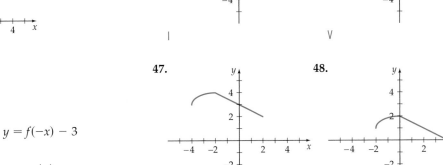

43. **44.**

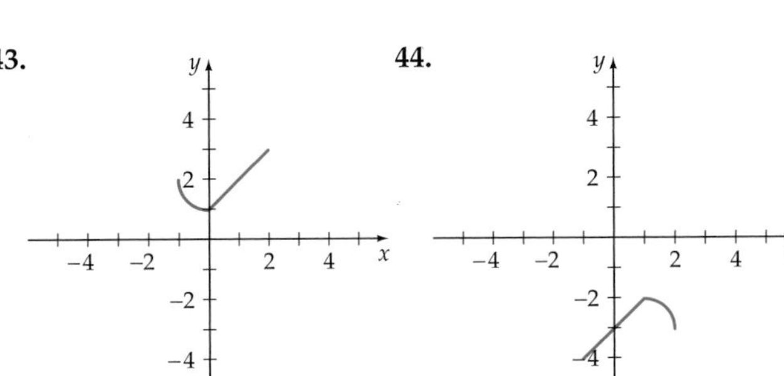

45. **46.** **47.** **48.**

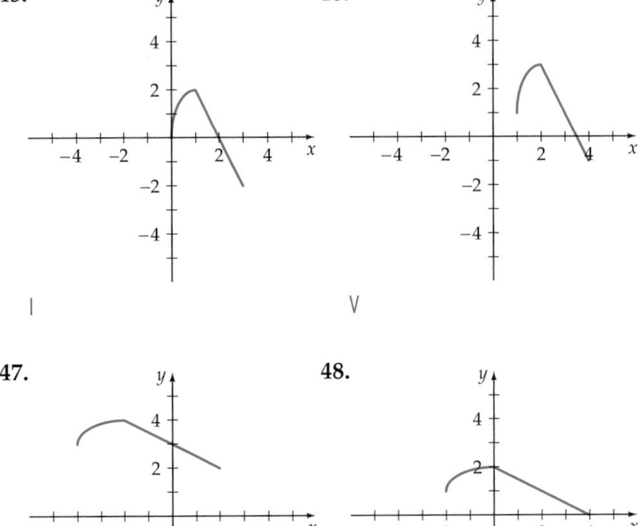

In Exercises 49 to 54, the given graph can be obtained by applying transformations of the graph of $g(x) = \sqrt{8x - x^4}$ shown below. Write an equation for the given graph.

$y = g(x)$

50. $y = g(x-3) + 2 = \sqrt{8(x-3) - (x-3)^4} + 2$

49. **50.**

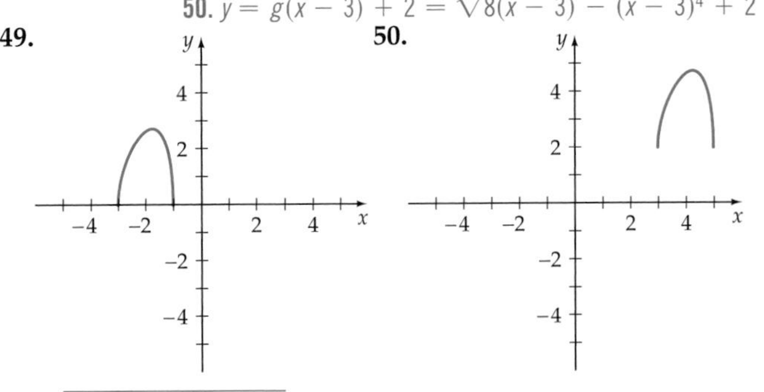

51.

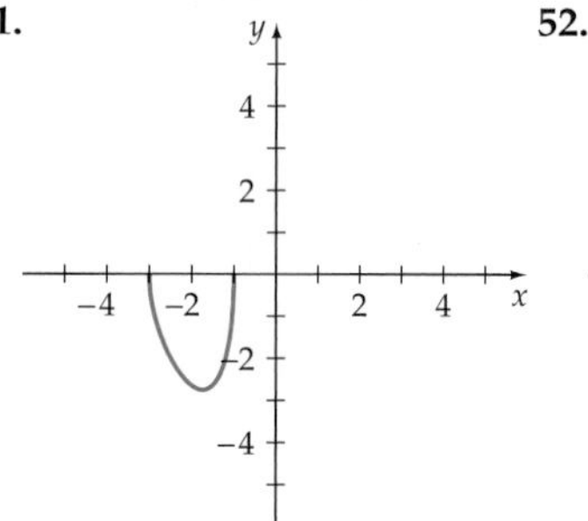

52.

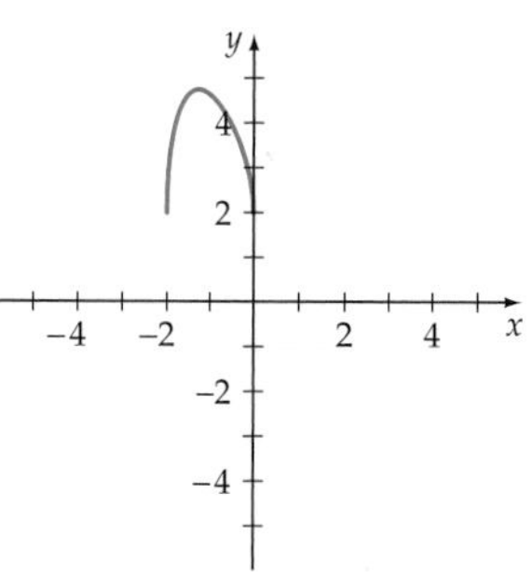

$y = -g(x+3) = -\sqrt{8(x+3) - (x+3)^4}$ $y = g(-x) + 2 = \sqrt{-8x - x^4} + 2$

53. 54.

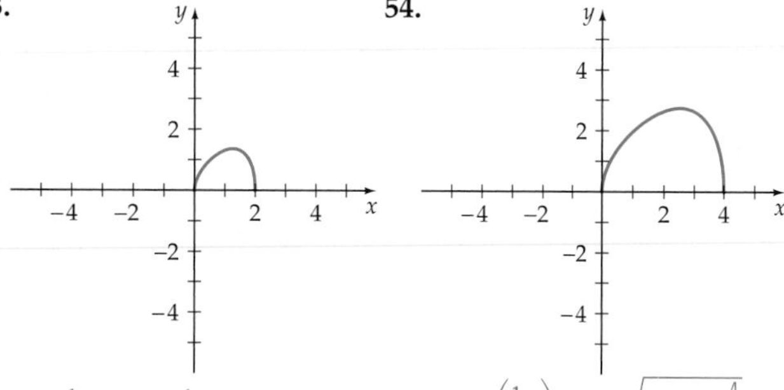

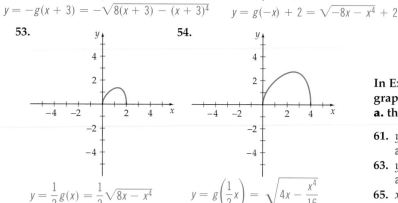

$y = \frac{1}{2}g(x) = \frac{1}{2}\sqrt{8x - x^4}$ $y = g\left(\frac{1}{2}x\right) = \sqrt{4x - \frac{x^4}{16}}$

In Exercises 55 and 56, a portion of a graph that is symmetric with respect to the x-axis is shown. Sketch the remaining portion of the graph.

55. 56.

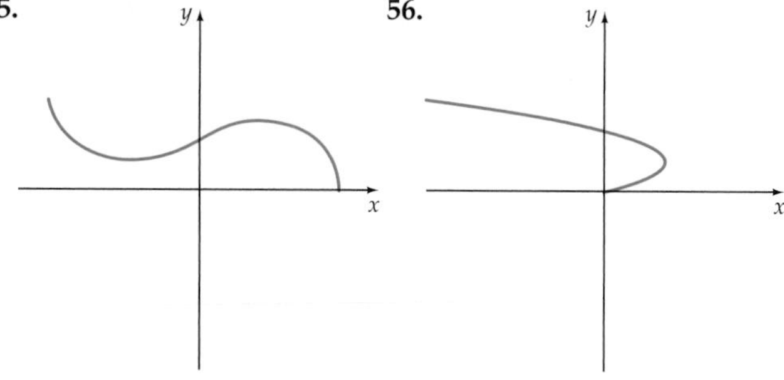

In Exercises 57 and 58, a portion of a graph that is symmetric with respect to the y-axis is shown. Sketch the remaining portion of the graph.

57. 58.

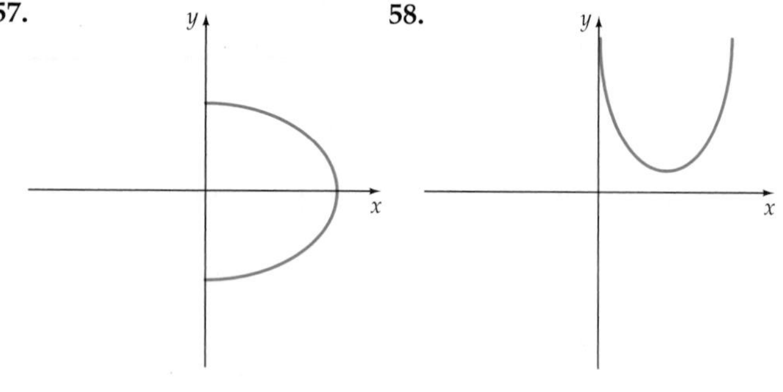

In Exercises 59 and 60, a portion of a graph that is symmetric with respect to the origin is shown. Sketch the remaining portion of the graph.

59. 60.

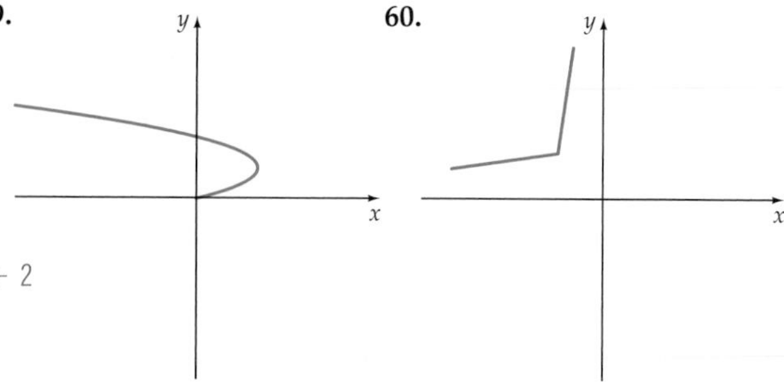

In Exercises 61 to 68, determine algebraically whether the graph of the given equation is symmetric with respect to a. the x-axis and b. the y-axis.

61. $y = 2x^2 - 5$
 a. No b. Yes

62. $x = 3y^2 - 7$
 a. Yes b. No

63. $y = x^3 + 2$
 a. No b. No

64. $y = x^5 - 3x$
 a. No b. No

65. $x^2 + y^2 = 9$
 a. Yes b. Yes

66. $x^2 - y^2 = 10$
 a. Yes b. Yes

67. $x^2 = y^4$
 a. Yes b. Yes

68. $|x| - |y| = 6$
 a. Yes b. Yes

In Exercises 69 to 76, determine algebraically whether the graph of the given equation is symmetric with respect to the origin.

69. $y = x + 1$ No

70. $y = x^3 - x$ Yes

71. $xy = 5$ Yes

72. $y = \frac{9}{x}$ Yes

73. $x^2 + y^2 = 10$ Yes

74. $x^2 - y^2 = 4$ Yes

75. $x^2 = y^3$ No

76. $|y| = |x|$ Yes

In Exercises 77 and 82, the graph of a function is given. State whether the function appears to be even, odd, or neither.

77. 78.

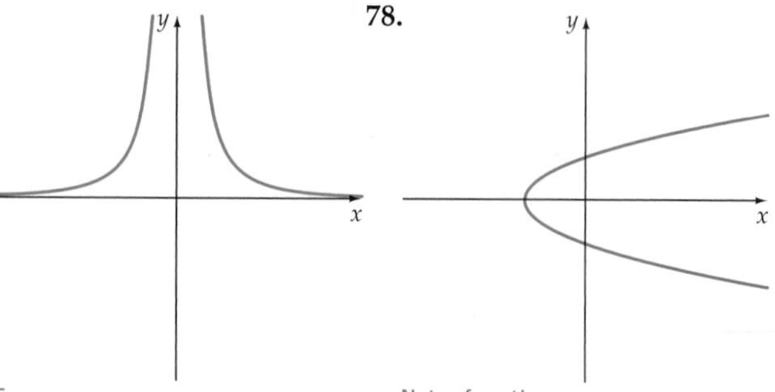

Even Not a function

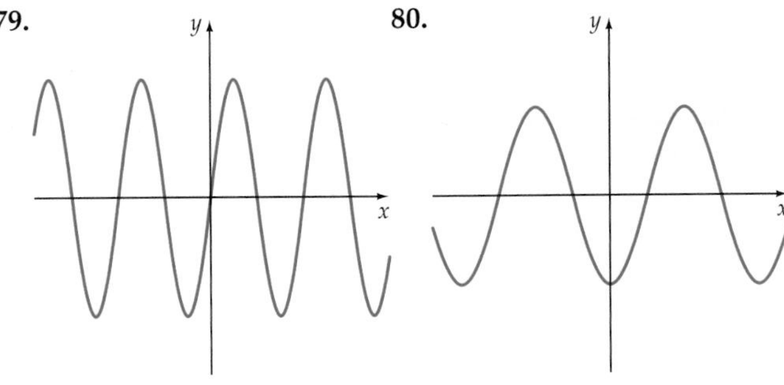

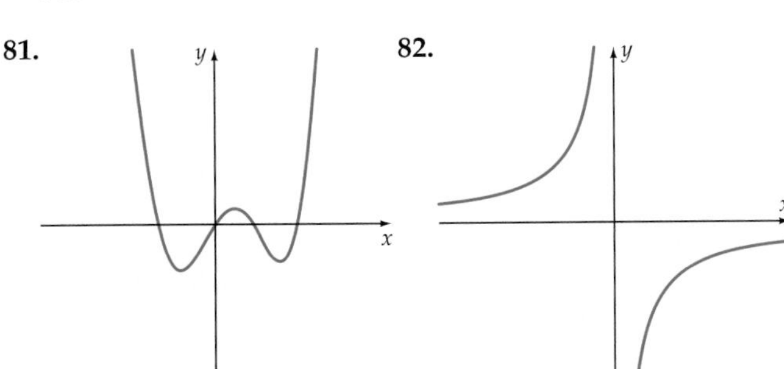

In Exercises 83 to 90, determine algebraically whether the given function is even, odd, or neither.

83. $g(x) = x^2 - 7$ Even

84. $h(x) = x^2 + 1$ Even

85. $F(x) = x^5 + x^3$ Odd

86. $G(x) = 2x^5 - 10$ Neither

87. $v(x) = 2x + x^2$ Neither

88. $H(x) = 3|x|$ Even

89. $r(x) = \sqrt{x^2 + 4}$ Even

90. $u(x) = \sqrt{3 - x^2}$ Even

91. A portion of the graph of a function *f* is given. Sketch the remaining portion of the graph if *f* is **a.** an even function or **b.** an odd function.

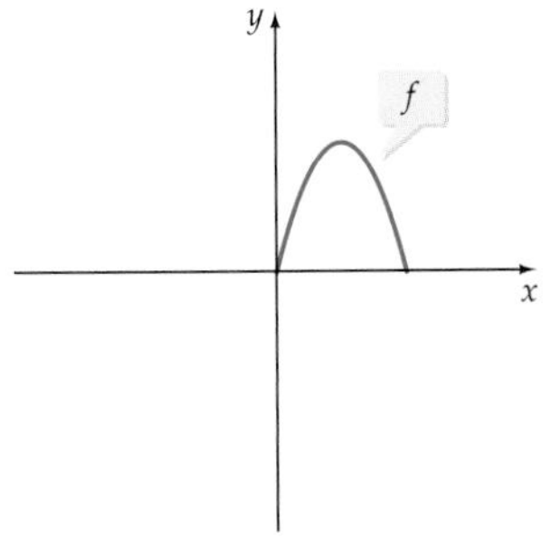

92. A portion of the graph of a function g is given. Sketch the remaining portion of the graph if *g* is **a.** an even function or **b.** an odd function.

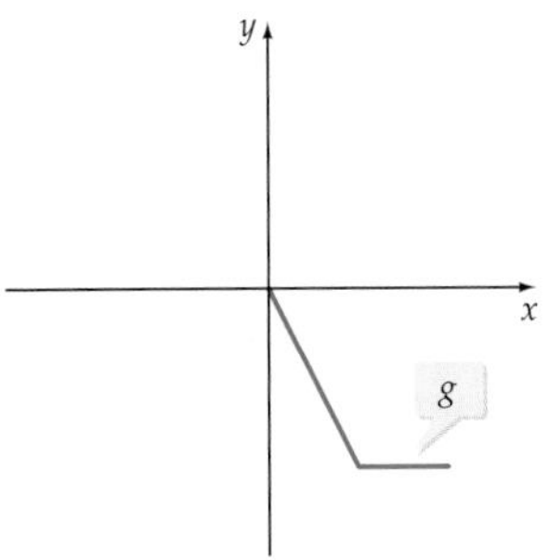

In Exercises 93 to 98, graph the function on a graphing calculator and use the graph to visually determine whether the function is even, odd, or neither.

93. $f(x) = \dfrac{x^3}{x^2 - 1}$ Odd

94. $f(x) = \dfrac{x^3}{x^2 + x}$ Neither

95. $f(x) = \dfrac{5x^3}{x^5 + x}$ Even

96. $f(x) = \dfrac{x^3 - 4x}{x^2 + 1}$ Odd

97. $f(x) = |x^2 + x| + |x^2 - x|$ Even

98. $f(x) = \dfrac{|x^3 - x|}{1 - |x|}$ Even

99. Use the graph of $f(x) = 2/(x^2 + 1)$ below to determine equations for the graphs shown in **a.** and **b.**

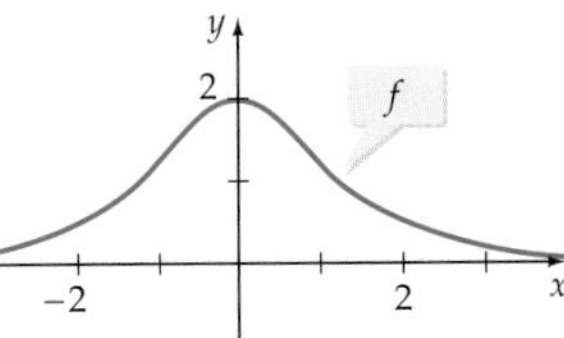

a.

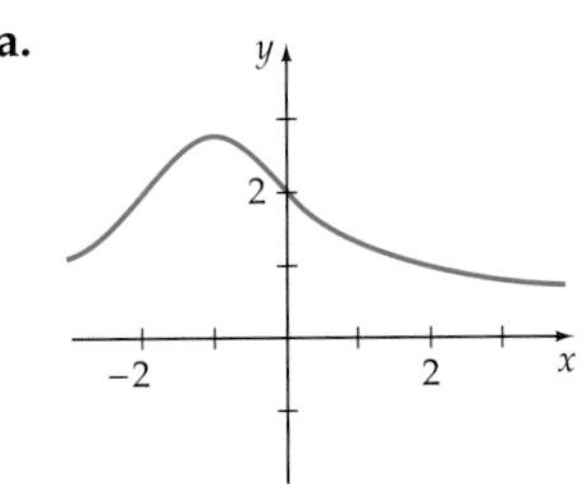

$y = \dfrac{2}{(x + 1)^2 + 1} + 1$

b.

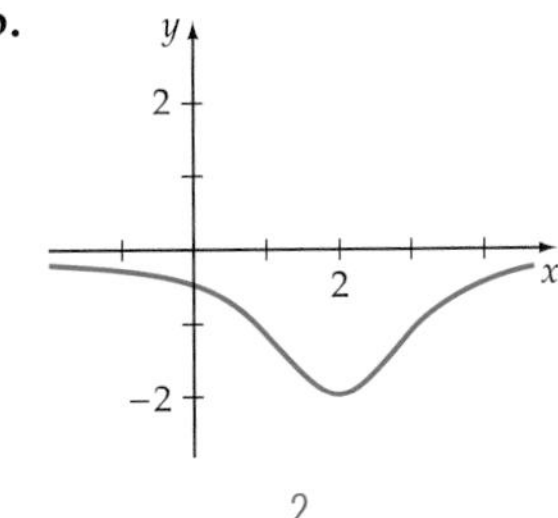

$y = -\dfrac{2}{(x - 2)^2 + 1}$

100. Use the graph of $f(x) = x\sqrt{2 + x}$ below to determine equations for the graphs shown in **a.** and **b.**

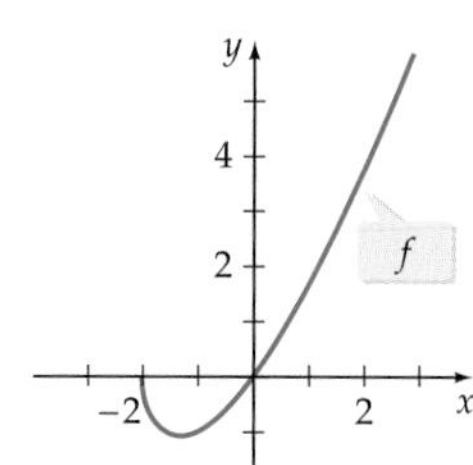

a.

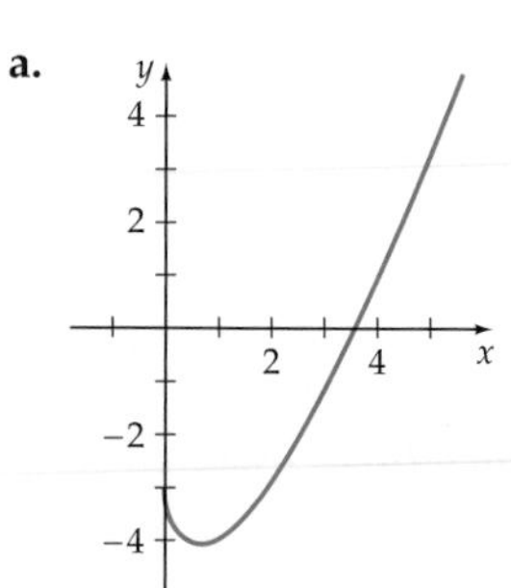

$y = (x - 2)\sqrt{x} - 3$

b.

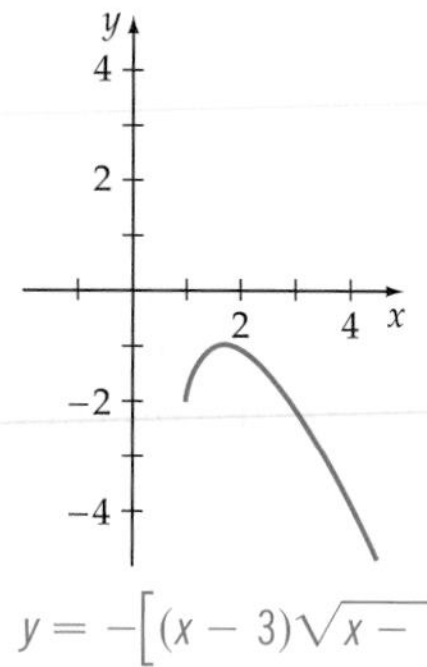

$y = -\left[(x - 3)\sqrt{x - 1} + 2\right]$

Prepare for Section 3.4

In Exercises 101 to 103, solve for x.

101. $\dfrac{2}{5} = \dfrac{10}{x}$ [1.3] 25

102. $\sqrt{2x + 1} = 3$ [1.3] 4

103. $\dfrac{6x}{x + 1} = 4$ [1.3] 2

104. If $y = kx + 2$ and $y = 3$ when $x = 2$, find k. [1.1] $\dfrac{1}{2}$

105. If $F = \dfrac{k}{d^2}$ and $F = 4$ when $d = 3$, find k. [1.1] 36

106. Solve $F = \dfrac{km}{r^2}$ for r. [1.2] $r = \pm\sqrt{\dfrac{km}{F}}$

Explorations 

1. ***Families of Functions*** A family of functions is a group of functions that are related in some way. For instance, the formula $f(x) = |x + c| + 1$, where c can be any value, gives a family of functions. Each value of c gives a different function, and the graphs of the functions look similar to each other.

a. Use a graphing calculator to graph $f(x) = |x + c| + 1$ for $c = -4, -2, 0, 2$, and 4. Sketch the five curves on the same axes and describe how the graphs are related.
The graphs are horizontal translations of $y = |x| + 1$.

b. Sketch the graphs of $f(x) = \dfrac{c^2x^2 + 1}{\sqrt{cx}}$ for $c = -2, 2, -1$, and 1 describe how the graphs are related.
The graphs of f when $c < 0$ are reflections through the y-axis of the graphs of f when $c > 0$.

c. Sketch the graphs of $h(x) = c + \sqrt{1 + 2cx - c^2 - x^2}$ for $c = -4, -2, 0, 2$, and 4 and describe how the graphs are related.
The graphs of h when $c < 0$ are reflections through the y-axis of the graphs of h when $c > 0$.

2. ***Symmetry with Respect to the Line $y = x$*** In this section we investigated symmetry with respect to a vertical line (the y-axis) and a horizontal line (the x-axis). Graphs can also be symmetric with respect to a diagonal line. Figure A shows a graph that is symmetric with respect to the diagonal line $y = x$. Notice that if a point (a, b) is on the graph, then (b, a) must also be on the graph. This leads to a method of testing an equation for symmetry about the line $y = x$: Switch x and y; if the equation is unaltered, its graph will be symmetric with respect to the line $y = x$. Determine algebraically whether the graphs of the following equations are symmetric with respect to the line $y = x$.

a. $y = \dfrac{1}{x}$ Yes

b. $x^2 - y^2 = 3$ No

c. $x^3 = y^3 + 1$ No

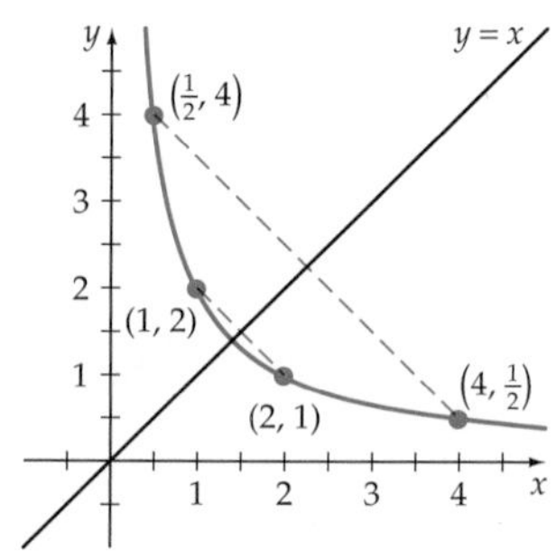

Figure A

SECTION 3.4 Variation

- Direct Variation
- Inverse Variation
- Joint Variation and Combined Variation

Direct Variation

Many real-life situations can be modeled using variables that are related by a type of function called a **variation.** For instance, a stone dropped into a pond generates circular ripples whose circumferences and diameters are increasing. The equation $C = \pi d$ expresses the relationship between the circumference C of a circle and its diameter d. If d increases, then C increases. In fact, if d doubles in size, C also doubles in size. The circumference C is said to *vary directly* as the diameter d.

Definition of Direct Variation

The variable y **varies directly** as the variable x, or y is **directly proportional** to x, if and only if

$$y = kx$$

where k is a constant called the **constant of proportionality** or the **variation constant.**

Direct variation occurs in many daily applications. For example, if the cost of a pound of potatoes is \$.52, then the cost C to purchase x pounds of potatoes is *directly proportional* to the number of pounds x. That is, $C = 0.52x$. In this case, the constant of proportionality is 0.52.

QUESTION Suppose a jogger is running at 8 miles per hour. Then the distance d (in miles) the jogger runs is directly proportional to the time t (in hours) spent running. What is the direct variation equation for this situation?

To solve a problem that involves a variation, we typically write a general equation that relates the variables and then use the given information to solve for the variation constant. Once the variation constant is known, the variation equation can be written.

ANSWER $d = 8t$

Alternative to Example 1
Exercise 22, page 303

EXAMPLE 1 Direct Variation Applied to Sales Tax

The amount of state sales tax on a purchase is directly proportional to the cost of the item. Suppose the state sales tax on a shirt that costs \$32.00 is \$2.48. Find the variation equation for this situation. Use the equation to find the sales tax on a sweater that costs \$40.

Solution Write an equation that relates the amount of sales tax A to the cost of the purchase c. Because A varies directly as c, we have $A = kc$. We are given that $A = 2.48$ when $c = 32$.

$$A = kc$$
$$2.48 = k(32) \quad \text{▪ } A = 2.48,\ c = 32$$
$$0.0775 = k$$

Replace k by 0.0775. The variation equation is $A = 0.0775c$.

To find the sales tax on the sweater, use the variation equation.

$$A = 0.0775c$$
$$= 0.0775(40) = 3.1 \quad \text{▪ Replace } c \text{ by 40.}$$

The sales tax is \$3.10.

CHECK YOUR PROGRESS 1 The distance sound travels varies directly as the time it travels. If sound travels 4020 meters in 12 seconds, find the distance sound will travel in 15 seconds.

Solution *See page S18.* 5025 m

Direct Variation as the *n*th Power

If y varies as the nth power of x, then

$$y = kx^n$$

where k is a constant.

Alternative to Example 2
Exercise 30, page 304

EXAMPLE 2 Determine the Speed of a Near-Earth Satellite

For a near-Earth circular orbit, the speed of a satellite varies directly as the square root of the distance from the satellite to the center of Earth. Suppose that a satellite is traveling at 18,200 miles per hour at an altitude of 265 miles above Earth. Find the variation equation. Use the variation equation to determine the speed required for a satellite to stay in a near-Earth circular orbit 1665 miles above the surface of Earth. Use 3960 miles as the radius of Earth.

Solution Write an equation that relates the speed of the satellite v to its distance from the center of Earth r. Because v varies directly as the square root of r, we

Continued ➤

have $v = k\sqrt{r}$. We are given that $v = 18{,}200$ when $r = 4225$. (Note that r is measured from the center of Earth, so $r = 3960 + 265 = 4225$.)

$$v = k\sqrt{r}$$
$$18{,}200 = k\left(\sqrt{4225}\right) \qquad \text{▪ } v = 18{,}200,\ r = 4225$$
$$\frac{18{,}200}{\sqrt{4225}} = k$$
$$280 = k$$

Replace k by 280. The variation equation is $v = 280\sqrt{r}$.

To determine the speed required for a satellite to stay in a near-Earth circular orbit 1665 miles above the surface of Earth, use the variation equation.

$$v = 280\sqrt{r}$$
$$= 280\sqrt{5625} \qquad \text{▪ Replace } r \text{ by 5625. Here } r = 1665 + 3960 \text{ because } r \text{ is measured from Earth's center.}$$
$$= 21{,}000$$

The satellite's required speed is 21,000 miles per hour.

CHECK YOUR PROGRESS 2 The distance s that an object falls from rest (neglecting air resistance) varies directly as the square of the time t that it has been falling. If an object falls 144 feet in 3 seconds, how far will it fall in 7 seconds?

Solution *See page S19.* 784 ft

▪ Inverse Variation

Two variables can also vary *inversely*.

Definition of Inverse Variation

The variable y **varies inversely** as the nth power of x, or y is **inversely proportional** to the nth power of x, if and only if

$$y = \frac{k}{x^n}$$

where k is the variation constant.

In the definition of inverse variation, when $n = 1$, we usually say "y varies inversely as x" or "y is inversely proportional to x."

Alternative to Example 3
Exercise 36, page 304

EXAMPLE 3 Use Inverse Variation to Determine Light Intensity

The illumination a source of light provides is inversely proportional to the square of the distance from the light source. If the illumination at a distance of 10 feet from the source is 50 foot-candles, what is the illumination at a distance of 15 feet from the source?

Continued ➤

Solution Write an equation that relates the illumination of the light source I to its distance from the source r. Because I varies inversely as the square of d, we have $I = \frac{k}{d^2}$. We are given that $I = 50$ when $d = 10$.

$$I = \frac{k}{d^2}$$

$$50 = \frac{k}{10^2} \qquad \text{▪ } I = 50, d = 10$$

$$5000 = k$$

Replace k by 5000. The variation equation is $I = \frac{5000}{d^2}$.

To determine the illumination 15 feet from the light source, use the variation equation.

$$I = \frac{5000}{d^2}$$

$$= \frac{5000}{15^2} \qquad \text{▪ Replace } d \text{ by 15.}$$

$$\approx 22.2$$

The illumination is approximately 22.2 foot-candles.

CHECK YOUR PROGRESS 3 Boyle's Law states that the volume V of a sample of gas (at a constant temperature) varies inversely as the pressure P. The volume of a gas in a J-shaped tube is 75 milliliters when the pressure is 1.5 atmospheres. Find the volume of the gas when the pressure is increased to 2.5 atmospheres.

Solution *See page S19.* 45 ml

▪ Joint Variation and Combined Variation

Some variations involve more than two variables.

Definition of Joint Variation

The variable z **varies jointly** as the variables x and y if and only if

$$z = kxy$$

where k is a constant.

Alternative to Example 4
Exercise 38, page 304

EXAMPLE 4 Joint Variation Applied to Kinetic Energy

The kinetic energy K of a moving body varies jointly as its mass and the square of its velocity. If the kinetic energy of a ball whose mass is 2 kilograms and that is traveling at 20 meters per second is 400 newton-meters (a measure of energy in the metric system), find the kinetic energy of the ball when it is moving at 30 meters per second.

Continued ➤

Solution Write an equation that relates the kinetic energy K of the ball to its mass m and velocity v. Because K varies jointly as the mass and the square of the velocity, we have $K = kmv^2$. We are given that $K = 400$ when $m = 2$ and $v = 20$.

$$K = kmv^2$$
$$400 = k(2)(20)^2 \qquad \blacksquare\ K = 400,\ m = 2,\ v = 20$$
$$400 = 800k$$
$$\frac{1}{2} = k$$

Replace k by $\frac{1}{2}$. The variation equation is $K = \frac{1}{2}mv^2$.

To determine the kinetic energy when the ball is traveling at 30 meters per second, use the variation equation.

$$K = \frac{1}{2}mv^2$$
$$= \frac{1}{2}(2)(30)^2 \qquad \blacksquare\ \text{Replace } m \text{ by 2 and } v \text{ by 30.}$$
$$= 900$$

The kinetic energy is 900 newton-meters.

CHECK YOUR PROGRESS 4 The cost of insulating the ceiling of a house varies jointly as the thickness, in inches, of the insulation and the area, in square feet, of the ceiling. It costs $175 to insulate a 2100-square-foot ceiling with insulation that is 4 inches thick. Find the cost of insulating a 2400-square-foot ceiling with insulation that is 6 inches thick.

Solution *See page S19.* $300

Combined variation involves more than one type of variation.

Alternative to Example 5
Exercise 34, page 304

EXAMPLE 5 Combined Variation Applied to Electrical Resistance

The electrical resistance R of a wire varies directly as the length l of the wire and inversely as the square of the radius r. If a wire with a radius of 4 millimeters and a length of 50 meters has a resistance of 3 ohms, find the resistance of a similar wire that is 75 meters long with a radius of 5 millimeters.

Solution Write an equation that relates the resistance R of the wire to its length l and radius r. Because R varies directly as the length and inversely as the square of the radius, we have $R = \frac{kl}{r^2}$. We are given that $R = 3$ when $l = 50$ and $r = 4$.

$$R = \frac{kl}{r^2}$$
$$3 = \frac{k(50)}{4^2} \qquad \blacksquare\ R = 3,\ l = 50,\ r = 4$$
$$48 = 50k \qquad \blacksquare\ \text{Solve for } k.$$
$$0.96 = k$$

Continued ➤

Replace k by 0.96. The variation equation is $R = \dfrac{0.96l}{r^2}$.

To determine the resistance of a similar wire 75 meters long with a radius of 5 millimeters, use the variation equation.

$$R = \frac{0.96l}{r^2}$$

$$= \frac{0.96(75)}{5^2} \quad \text{■ Replace } l \text{ by 75 and } r \text{ by 5.}$$

$$= 2.88$$

The resistance is 2.88 ohms.

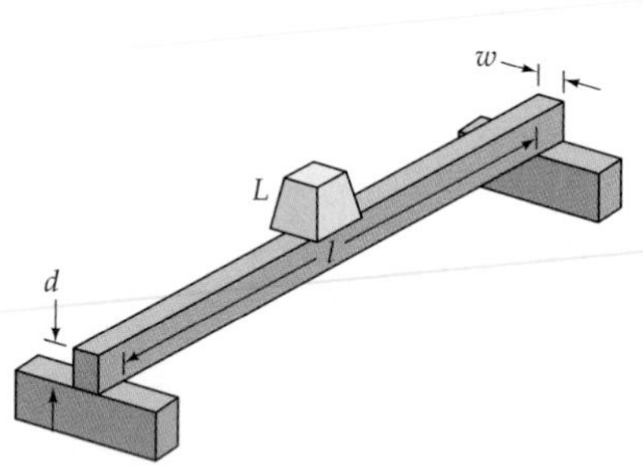

Figure 3.54

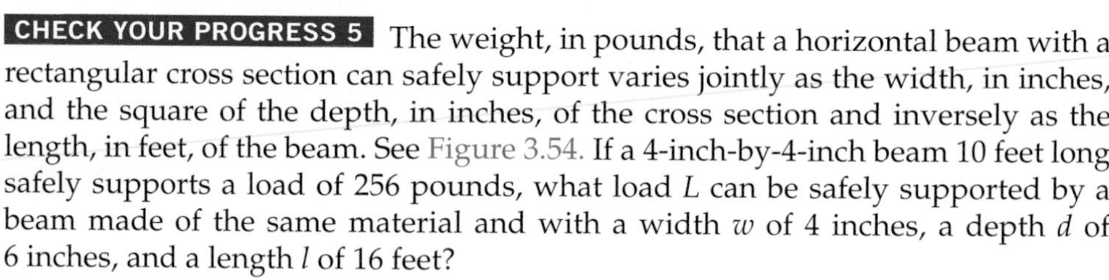

CHECK YOUR PROGRESS 5 The weight, in pounds, that a horizontal beam with a rectangular cross section can safely support varies jointly as the width, in inches, and the square of the depth, in inches, of the cross section and inversely as the length, in feet, of the beam. See Figure 3.54. If a 4-inch-by-4-inch beam 10 feet long safely supports a load of 256 pounds, what load L can be safely supported by a beam made of the same material and with a width w of 4 inches, a depth d of 6 inches, and a length l of 16 feet?

Solution *See page S19.* 360 lb

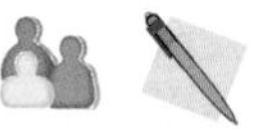

Topics for Discussion

1. Given that the variation constant $k > 0$ and that A varies directly as b, then A _____ when b increases and A _____ when b decreases.
2. Given that the variation constant $k > 0$ and that A varies inversely as b, then A _____ when b increases and A _____ when b decreases.
3. All direct variations can be written in form $y =$ _____ and all inverse variations can be written in the form $y =$ _____.
4. What does it mean to say that V varies jointly as x and y?
5. The volume V of a right circular cylinder varies jointly as the square of the radius of the base and the height. Explain how the volume changes as **a.** h is tripled, **b.** r is tripled, and **c.** h is doubled and r is doubled.

EXERCISES 3.4

— Suggested Assignment: Exercises 1–37, odd; and 38–40.

In Exercises 1 to 12, write an equation that represents the relationship between the given variables. Use k as the variation constant.

1. s varies directly as t. $s = kt$
2. v varies directly as the square of s. $v = ks^2$
3. y is inversely proportional to x. $y = \dfrac{k}{x}$
4. z varies inversely to the square of x. $z = \dfrac{k}{x^2}$
5. S varies jointly as n and p. $S = knp$
6. w varies jointly as r and the square root of t. $w = kr\sqrt{t}$

7. P varies jointly as n, R, and t. $P = knRt$

8. y varies directly as x and inversely as the square of p. $y = \frac{kx}{p^2}$

9. A is directly proportional to the square of s and inversely proportional to the square root of r. $A = \frac{ks^2}{\sqrt{r}}$

10. A varies jointly as v and the cube of z. $A = kvz^3$

11. F varies jointly as m_1 and m_2 and inversely as the square of r. $F = \frac{km_1m_2}{r^2}$

12. T varies jointly as t and r and the square of q. $T = ktrq^2$

In Exercises 13 to 20, write the equation that expresses the relationship among the variables, and then use the given data to solve for the variation constant.

13. s varies directly as t, and $s = 40$ when $t = 20$. $s = kt, k = 2$

14. y is directly proportional to x, and $y = 38$ when $x = 114$. $y = kx, k = \frac{1}{3}$

15. v is directly proportional to the square of r, and $v = 64$ when $r = 4$. $v = kr^2, k = 4$

16. A varies directly as r, and $A = 44$ when $r = 24$. $A = kr, k = \frac{11}{6}$

17. s varies jointly as v and the square of t, and $s = 10$ when $v = 20$ and $t = 5$. $s = kvt^2, k = \frac{1}{50}$

18. p varies directly as v and inversely as the square root of w, and $p = 0.04$ when $v = 8$ and $w = 0.04$. $p = \frac{kv}{\sqrt{w}}, k = 0.001$

19. V varies jointly as l, w, and h, and $V = 300$ when $l = 5$, $w = 6$, and $h = 10$. $V = klwh, k = 1$

20. t varies directly as the square of r and inversely as the cube of s, and $t = 10$ when $r = 5$ and $s = 0.2$. $t = \frac{kr^2}{s^3}, k = 0.0032$

Business and Economics

21. ***Revenue*** The revenue a company earns is directly proportional to the number of articles sold. If the company has a revenue of \$50,000 on the sale of 2500 cell phone headsets, what revenue can the company expect on the sale of 3000 headsets? \$60,000

22. ***Property Tax*** The property tax on a home is directly proportional to the value of the home. If the property tax on a \$500,000 home is \$6000, what is the property tax on home with a value of \$300,000? \$3600

23. ***Audio Technology*** The manufacturer of stereo speakers knows that the loudness, in decibels, of a stereo speaker is inversely proportional to the square of the distance of the listener from the speaker. If the loudness is 28 decibels at a distance of 8 feet, what is the loudness when the listener is 4 feet from the speaker? 112 decibels

24. ***Civil Engineering*** The owner of a highway construction company knows that the maximum load a cylindrical column of circular cross section can support varies directly as the fourth power of the diameter and inversely as the square of the height. If a column 2 feet in diameter and 10 feet high supports up to 6 tons, how much of a load can a column 3 feet in diameter and 14 feet high support? 15.5 tons

Life and Health Sciences

25. ***Light Intensity*** The intensity of a light source on the retina is inversely proportional to the square of the distance of the retina from the source of light. If the illumination at a distance of 20 feet from the light source is 100 foot-candles, what is the illumination at a distance of 40 feet from the source? 25 foot-candles

26. ***Pressure on a Whale*** The pressure on a whale is directly proportional to the depth of the whale below the surface of the ocean. If the pressure at a depth of 1000 feet is 8900 pounds per square inch, find the pressure at a depth of 2500 feet. 22,250 lb/in^2

Physical Sciences and Engineering

27. ***Potential Energy*** The potential energy of an object varies jointly as the mass of the object and its height above the ground. If the potential energy of an object with a mass of 2 kilograms and a height of 10 meters above the ground is 196 newton-meters, find the potential energy of the same mass at a height of 5 meters above the ground. 98 newton-meters

28. ***Charles's Law*** Charles's Law states that the volume V occupied by a gas (at a constant pressure) is directly proportional to its absolute temperature T. An experiment with a balloon shows that the volume of the

balloon is 0.85 liter at 270°K (absolute temperature). What will the volume of the balloon be when its temperature is 324°K? (*Note:* Absolute temperature is measured on the Kelvin scale. A difference of 1 unit on the Kelvin scale is the same as a difference of 1 degree on the Celsius scale; however, 0° on the Kelvin scale corresponds to −273.18° on the Celsius scale.) 1.02 L

29. ***Hooke's Law*** Hooke's Law states that the distance a spring stretches varies directly as the weight on the spring. If a weight of 80 pounds stretches a spring 5 inches, how far will a weight of 90 pounds stretch the spring? 5.625 in.

30. ***Projectile Motion*** The range of a projectile is directly proportional to the square of its velocity. If a motorcyclist can make a jump of 140 feet by coming off a ramp at 60 miles per hour, find the distance the motorcyclist could expect to jump if the speed coming off the ramp were increased to 65 miles per hour. Round to the nearest tenth. 164.3 ft

31. ***Pendulum Motion*** The period T of a certain pendulum (the time it takes the pendulum to make one complete oscillation) varies directly as the square root of its length L. A pendulum 3 feet long has a period of 1.9 seconds.

    a. Find the period of a pendulum 10 feet long. 3.5 s

    b. What is the length of a pendulum that has a 2-second period? Round to the nearest tenth. 3.3 ft

32. ***Construction*** The volume V of a right circular cone varies jointly as the square of the radius r and the height h. State what happens to V when

    a. r is doubled. V is four times larger.

    b. h is doubled. V is doubled.

    c. both r and h are doubled. V is eight times larger.

33. ***Civil Engineering*** The load that a horizontal beam can safely support varies jointly as the width and the square of the depth d of the beam. If a beam with a width of 2 inches and a depth of 6 inches safely supports up to 200 pounds, how many pounds can a beam of the same length that has a width of 4 inches and a depth of 4 inches be expected to support? Round to the nearest tenth. 177.8 lb

34. ***Ideal Gas Law*** The ideal gas law states that the volume V of a gas varies jointly as the number n of moles of gas and the absolute temperature T and inversely as the pressure P. What happens to V when n is tripled and P is reduced by a factor of $\frac{1}{2}$? V is six times as large.

35. ***Electronics Technology*** The electrical resistance R of a wire varies directly as the length l of the wire and inversely as the square of its radius r. If a wire with a radius of 0.5 millimeter and a length of 25 meters has a resistance of 2 ohms, find the resistance of a similar wire that is 50 meters long and has a radius of 2 millimeters. 0.25 ohm

36. ***Meteorite Speed*** A meteorite approaching the earth has a velocity that varies inversely as the square root of its distance from the center of the earth. The meteorite has a velocity of 3 miles per second at a distance of 4900 miles from the center of the earth. Find the velocity of the meteorite when it is 4225 miles from the center of the earth. Round to the nearest hundredth. 3.23 mi/s

37. ***Civil Engineering*** The load L that a horizontal beam can safely support varies jointly as the width v and the square of the depth d of the beam and inversely as its length l. If a 12-foot beam with a width of 4 inches and a depth of 8 inches safely supports 800 pounds, how many pounds can a 16-foot beam that has a width of 3.5 inches and a depth of 6 inches be expected to support? Round to the nearest tenth. 295.3 lb

38. ***Force on an Airplane Wing*** The force exerted by air moving over a surface (such as plane's wing) varies jointly as the area of the surface and the square of the velocity of the air. If the force on a surface whose area is 25 square feet is 20 pounds when the air is moving at 15 miles per hour, find the force on a surface whose area is 30 square feet when the air is moving at 20 miles per hour. Round to the nearest tenth. 42.7 lb

39. ***Automotive Technology*** The force needed to keep a car from skidding on a curve varies jointly as the weight of the car and the square of its speed and inversely as the radius of the curve. It takes 2800 pounds of force to keep an 1800-pound car from skidding on a curve with a radius of 425 feet and traveling at 45 miles per hour. What force is needed to keep the same car from skidding when it takes a similar curve with a radius of 450 feet at 55 miles per hour? Round to the nearest pound. 3950 lb

40. ***Construction*** A cylindrical log is to be cut so that it will yield a beam that has a rectangular cross section of depth d and width w. The stiffness of a beam of given length is jointly proportional to its width and the cube of its depth. The diameter of the log is 18 inches. What depth will yield the stiffest beam: $d = 10$ inches, $d = 12$ inches, $d = 14$ inches, or $d = 16$ inches? $d = 16$ in.

Explorations

1. ***Light Intensity*** The illumination a light source provides is directly proportional to the strength of the source and inversely proportional to the square of the distance from the source. Two light sources are 10 feet apart. The strength of the light source at point B is 4 times the strength of the light source at point A.

   a. Let C be a point x units from A on the line segment $\overline{AB}$. Write a formula for the amount of illumination at C, assuming that the strength of the source is 10 and the constant of variation is one. $I = 10\left[\frac{1}{x^2} + \frac{4}{(10 - x)^2}\right]$

   b. Use the table feature of a graphing utility to evaluate your formula for $x = 2$ to $x = 8$ in increments of 0.1.
   Use the table feature of a graphing calculator to produce a table of values for the equation in part **a**. Use TblStart $= 2$ and ΔTbl $= 0.1$.

   c. Explain how the intensity changes as x changes.
   Intensity decreases until x is approximately 3.9, and then starts to increase.

   d. Based on values from the table, approximately how far from A is the intensity a minimum? 3.9 ft

2. ***Planetary Motion*** Kepler's Third Law states that the time needed for a planet to make one complete revolution about the sun is directly proportional to the $\frac{3}{2}$ power of the average distance between the planet and the sun. Earth, which averages 93 million miles from the sun, completes one revolution in 365 days. Find the average distance from the sun to Mars if Mars completes one revolution about the sun in 686 days. $\approx$142,000,000 mi

Chapter 3 Summary

Key Terms

combined variation **[p. 301]**
composition (composite) of two functions **[p. 258]**
compressing a graph **[p. 280]**
constant of proportionality **[p. 297]**
counterexample **[p. 252]**
difference quotient **[p. 255]**
directly proportional to **[p. 297]**
direct variation **[p. 297]**
even function **[p. 287]**
horizontal translation of a graph **[p. 274]**
inverse function **[p. 262]**
inverse variation **[p. 299]**
inversely proportional **[p. 299]**
joint variation **[p. 300]**
odd function **[p. 287]**
origin symmetry **[p. 286]**
point symmetry **[p. 286]**
reflection of a graph **[p. 277]**
stretching a graph **[p. 280]**
variation **[p. 297]**
variation constant **[p. 297]**
vertical translation of a graph **[p. 273]**
x-axis symmetry **[p. 285]**
y-axis symmetry **[p. 284]**

Essential Concepts and Formulas

- **Operations on Functions**

For all values of x for which both $f(x)$ and $g(x)$ are defined, we define the following functions.

Sum $(f + g)(x) = f(x) + g(x)$

Difference $(f - g)(x) = f(x) - g(x)$

Product $(f \cdot g)(x) = f(x) \cdot g(x)$

Quotient $\left(\frac{f}{g}\right)(x) = \frac{f(x)}{g(x)}, g(x) \neq 0$ **[p. 253]**

- **Difference Quotient**

$\frac{f(a + h) - f(a)}{h}, h \neq 0$ **[p. 255]**

- **Composition of Inverse Functions Property**

If f is a one-to-one function, then f^{-1} is the inverse function of f if and only if $(f \circ f^{-1})(x) = f[f^{-1}(x)] = x$ for all x in the domain of f^{-1} and $(f^{-1} \circ f)(x) = f^{-1}[f(x)] = x$ for all x in the domain of f. **[p. 265]**

- **Steps for Finding the Inverse of a Function**

1. Substitute y for $f(x)$.
2. Interchange x and y.
3. Solve, if possible, for y in terms of x.
4. Substitute $f^{-1}(x)$ for y. **[p. 267]**

- **Transformations**

If f is a function and c is a positive constant, then

- the graph of $y = f(x) + c$ is the graph of $y = f(x)$ shifted vertically up c units.
- the graph of $y = f(x) - c$ is the graph of $y = f(x)$ shifted vertically down c units.
- the graph of $y = f(x + c)$ is the graph of $y = f(x)$ shifted horizontally left c units.
- the graph of $y = f(x - c)$ is the graph of $y = f(x)$ shifted horizontally right c units.
- the graph of $y = -f(x)$ is the graph of $y = f(x)$ reflected across the x-axis.
- the graph of $y = f(-x)$ is the graph of $y = f(x)$ reflected across the y-axis.
- for $c > 1$, the graph of $y = c \cdot f(x)$ is the graph of $y = f(x)$ stretched vertically by a factor of c away from the x-axis.
- for $0 < c < 1$, the graph of $y = c \cdot f(x)$ is the graph of $y = f(x)$ compressed vertically by a factor of c toward the x-axis.
- for $c > 1$, the graph of $y = f(c \cdot x)$ is the graph of $y = f(x)$ compressed horizontally by a factor of $\frac{1}{c}$ toward the y-axis.
- for $0 < c < 1$, the graph of $y = f(c \cdot x)$ is the graph of $y = f(x)$ stretched horizontally by a factor of $\frac{1}{c}$ away from the y-axis. **[pp. 274–284]**

- **Steps for Transforming a Graph**

1. Do the horizontal translation.
2. Do reflections.
3. Do vertical or horizontal stretching/compressing.
4. Do the vertical translation. **[p. 283]**

- **Symmetry**

The graph of an equation has symmetry with respect to

- the y-axis if the replacement of x with $-x$ leaves the equation unaltered.
- the x-axis if the replacement of y with $-y$ leaves the equation unaltered.
- the origin if the replacement of x with $-x$ and y with $-y$ leaves the equation unaltered. **[p. 285]**

- **Even and Odd Functions**

The function f is

- an *even* function if $f(-x) = f(x)$ for all x in the domain of f. Its graph is symmetric with respect to the y-axis.
- an *odd* function if $f(-x) = -f(x)$ for all x in the domain of f. Its graph is symmetric with respect to the origin. **[p. 287]**

- **Variation**

- y varies directly as the nth power of x if and only if $y = kx^n$ for some nonzero constant k.
- y varies inversely as the nth power of x if and only if $y = \frac{k}{x^n}$ for some nonzero constant k.
- z varies jointly with x and y if and only if $z = kxy$ for some nonzero constant k. **[pp. 298–300]**

Chapter 3 True/False Exercises

In Exercises 1 to 12, answer true or false. If the statement is false, give an example that shows the statement is false.

1. Every function has an inverse relation. True

2. Every function has an inverse function.
False. $\{(2, 4), (3, 4)\}$ is a function whose inverse is not a function.

3. Let f be any function. Then $f(a) = f(b)$ implies $a = b$.
False. $f(x) = |x|$ is a function and $f(-2) = f(2)$, but $-2 \neq 2$.

4. Let f be any function. Then $f(a + b) = f(a) + f(b)$.

5. If $c > 0$, then the graph of $y = f(x - c)$ is the graph of $y = f(x)$ shifted to the right c units. True

6. If (a, b) are the coordinates of a point on the graph of $y = f(x)$, then (a, kb) are the coordinates of a point on the graph of $y = kf(x)$. True

7. If y varies inversely as x, then if x is doubled, y is doubled. False. y is halved.

4. False. Let $f(x) = x^2$. Then $f(2 + 3) = f(5) = 5^2 = 25$. However, $f(2) + f(3) = 2^2 + 3^2 = 4 + 9 = 13$.

8. If z varies jointly as x and y, then z equals a nonzero constant times the product of x and y. True

9. If f is a function and $a < b$, then $f(a) < f(b)$.

10. If f is a one-to-one function and $a > b$, then $f(a) \neq f(b)$. True

11. If f is a function and $a > b$, then $f(a) \neq f(b)$. False. Let $f(x) = x^2$, $a = 2$, and $b = -2$. Then $a > b$, but $f(a) = f(b)$.

12. For a one-to-one function f whose inverse function is f^{-1}, the domain of f^{-1} is the range of f. True

9. False. Let $f(x) = -x$, $a = 2$, and $b = 3$. Then $f(2) = -2$ and $f(3) = -3$. Thus $a < b$ but $f(a) > f(b)$.

Chapter 3 Review Exercises

—Answer graphs to Exercises 7–14 are on page AA9.

1. Let $f(x) = 2x^2 + 1$ and $g(x) = x - 3$. Find:

   a. $(f + g)(2)$ 8 b. $(g - f)(x)$ $-2x^2 + x - 4$

   c. $(f \cdot g)(x)$ $2x^3 - 6x^2 + x - 3$ d. $\left(\dfrac{f}{g}\right)(5)$ $\dfrac{51}{2}$ [3.1]

2. Let $f(x) = x^2 - x$ and $g(x) = 2x + 7$. Find:

   a. $(f - g)(3)$ -7 b. $(g + f)(x)$ $x^2 + x + 7$

   c. $(f \cdot g)(x)$ $2x^3 + 5x^2 - 7x$ d. $\left(\dfrac{f}{g}\right)(-1)$ $\dfrac{2}{5}$ [3.1]

3. Let $f(x) = x^2 - x - 1$ and $g(x) = 2x - 1$. Find:

   a. $(f \circ g)(-1)$ 11 b. $(f \circ g)(2)$ 5 c. $(g \circ f)(2)$ 1

   d. $(f \circ g)(x)$ $4x^2 - 6x + 1$ e. $(g \circ f)(x)$ $2x^2 - 2x - 3$ f. $(g \circ g)(3)$ 9 [3.1]

4. Let $f(x) = 2x + 5$ and $g(x) = 3x^2 - 1$. Find:

   a. $(g \circ f)(0)$ 74 b. $(g \circ f)(-1)$ 26 c. $(f \circ g)(-1)$ 9

   d. $(f \circ g)(x)$ $6x^2 + 3$ e. $(g \circ f)(x)$ $12x^2 + 60x + 74$ f. $(f \circ f)(2)$ 23 [3.1]

In Exercises 5 and 6, find the difference quotient $\dfrac{f(a + h) - f(a)}{h}$.

5. $f(x) = 5x - 1$ 5 [3.1]

6. $f(x) = 3x^2 - 1$ $6a + 3h$ [3.1]

In Exercises 7 to 10, use the graph of g to sketch a graph of each function.

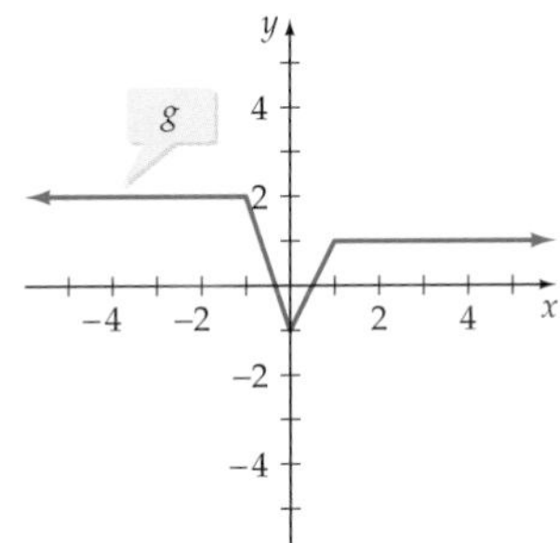

7. $y = g(x - 1) + 1$

8. $y = g(x + 2) - 3$

9. $y = g(-x + 2)$

10. $y = -g(x + 2)$

In Exercises 11 to 14, use the graph of f to sketch a graph of each function.

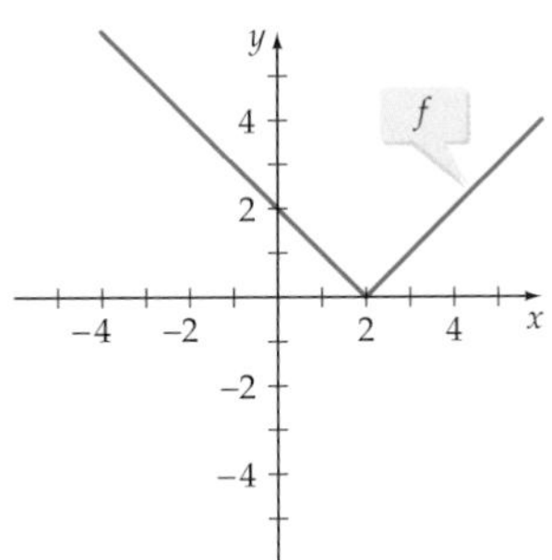

11. $y = f(x + 3) - 2$

12. $y = f(-x) + 1$

13. $y = -f(2x)$

14. $y = 2f(x - 1)$

In Exercises 15 to 20, determine whether the graph of the equation is symmetric with respect to a. the x-axis, b. the y-axis, or c. the origin.

15. $f(x) = x^2 - 7$ y-axis [3.3]

16. $x = y^2 + 3$ x-axis [3.3]

17. $g(x) = x^3 - 4x$ Origin [3.3]

18. $y^2 = x^2 + 4$ x-axis, y-axis, origin [3.3]

19. $|y| = |x|$ x-axis, y-axis, origin [3.3]

20. $xy = 8$ Origin [3.3]

In Exercises 21 to 24, determine whether f is an even function, an odd function, or neither.

21. $f(x) = x^3 + x$ Odd [3.3]

22. $f(x) = x^4 + x$ Neither [3.3]

23. $f(x) = x^2 - |x|$ Even [3.3]

24. Odd [3.3]

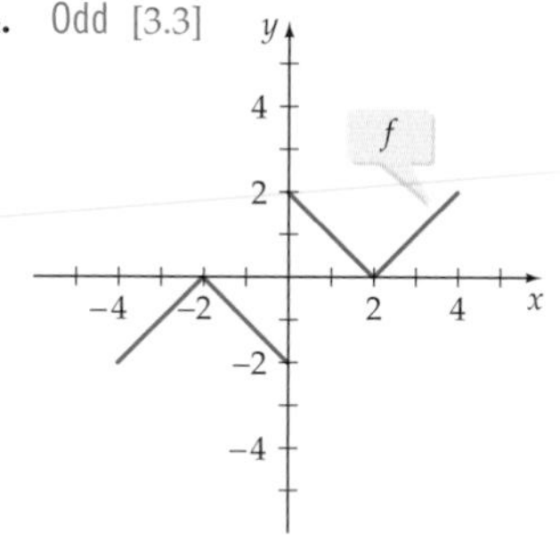

In Exercises 25 to 30, use composition of functions to determine whether the two functions are inverses of one another.

25. $f(x) = 2x + 6; g(x) = \frac{1}{2}x - 3$ Yes [3.2]

26. $f(x) = \frac{2}{3}x - 2; g(x) = \frac{3}{2}x + 1$ No [3.2]

27. $f(x) = x + 1; g(x) = \frac{1}{x + 1}$ No [3.2]

28. $f(x) = 3x - 1; g(x) = \frac{1}{3}x - 1$ No [3.2]

29. $f(x) = x^2 - 1, x \geq 0, g(x) = \sqrt{x + 1}, x \geq -1$ Yes [3.2]

30. $f(x) = \frac{1}{x + 1}; g(x) = \frac{1 - x}{x}$ Yes [3.2]

In Exercises 31 to 36, find the inverse function of the given function.

31. $f(x) = 3x + 12$ $f^{-1}(x) = \frac{1}{3}x - 4$ [3.2]

32. $f(x) = \frac{1}{2}x - 1$ $f^{-1}(x) = 2x + 2$ [3.2]

33. $f(x) = x^2 + 4, x \geq 0$ $f^{-1}(x) = \sqrt{x - 4}, x \geq 4$ [3.2]

34. $f(x) = \sqrt{x - 1}$ $f^{-1}(x) = x^2 + 1, x \geq 0$ [3.2]

35. $f(x) = \frac{2}{x - 1}$ $f^{-1}(x) = \frac{x + 2}{x}, x \neq 0$ [3.2]

36. $f(x) = \frac{x}{x + 4}$ $f^{-1}(x) = \frac{4x}{1 - x}, x \neq 1$ [3.2]

Business and Economics

37. ***Revenue*** Suppose the price per unit (in dollars) of an inkjet printer is given by $p(x) = 150 - \frac{x}{500}$. Find the revenue function for the manufacturer of this printer. Find the manufacturer's revenue for selling 20,000 printers. $R(x) = 150x - \frac{x^2}{500}$; \$2,200,000 [3.1]

38. ***Manufacturing*** The price per unit (in dollars) at which a manufacturer can sell a kitchen faucet is given by $p(x) = 50 - \frac{x}{250}$. The cost to produce x faucets is given by $C(x) = 25x + 450$. Find the profit function for the manufacturer of these faucets. $P(x) = -\frac{x^2}{250} + 25x - 450$ [3.1]

Life and Health Sciences

39. ***Pressure on a Diver*** The pressure on a diver in the ocean is directly proportional to the depth of the diver below the ocean's surface. If the pressure at a depth of 20 feet is 8.9 pounds per square inch, find the pressure at a depth of 50 feet. 22.25 lb/in.^{2} [3.4]

40. ***Loudness of a Sound*** The impact of sound on the human ear, measured in decibels, from a stereo speaker is inversely proportional to the square of the distance of the listener from the speaker. If the loudness is 35 decibels at a distance of 10 feet from the speaker, what is the loudness when the listener is 6 feet from the speaker? Round to the nearest decibel. 97 decibels [3.4]

Physical Sciences and Engineering

41. ***Construction*** The perimeter of a rectangular cement slab is 100 feet. Find the area of the rectangle as a function of the length of the rectangle. $A(L) = -L^2 + 50L$ [3.1]

42. ***Period of a Pendulum*** The period T of a pendulum (the time it takes the pendulum to make one complete oscillation) varies directly as the square root of its length L. A pendulum 5 feet long has a period of 2.5 seconds. Find the period of a pendulum that is 8 feet long. Round to the nearest tenth. 3.2 s [3.4]

Chapter 3 Test

—Answer graphs to Exercises 7 and 10–12 are on page AA9.

1. Let $f(x) = 2x - 5$ and $g(x) = 1 - x^2$. Find:

   a. $(f - g)(-3)$ -3 **b.** $(f + g)(x)$ $-x^2 + 2x - 4$

   c. $(f \cdot g)(x)$ $-2x^3 + 5x^2 + 2x - 5$ **d.** $(f \div g)(2)$ $\frac{1}{3}$ [3.1]

2. Let $f(x) = -2x + 1$ and $g(x) = x^2 - x$. Find:

   a. $(f \circ g)(-2)$ -11 **b.** $(f \circ g)(x)$ $-2x^2 + 2x + 1$

   c. $(g \circ f)(0)$ 0 **d.** $(g \circ f)(x)$ $4x^2 - 2x$ [3.1]

3. Use composition of functions to determine whether $f(x) = 3x + 15$ is the inverse function of $g(x) = \frac{1}{3}x - 5$. Yes [3.2]

4. Given $f(x) = 1 - x^2$, find the difference quotient $\dfrac{f(2 + h) - f(2)}{h}$. $-4 - h$ [3.1]

5. Given $f(x) = 3x - 6$, find $f^{-1}(x)$. $f^{-1}(x) = \frac{1}{3}x + 2$ [3.2]

6. Given $f(x) = \dfrac{x + 1}{x}$, find $f^{-1}(x)$. $f^{-1}(x) = \dfrac{1}{x - 1}$ [3.2]

7. Given the graph of f below, draw the graph of f^{-1}.

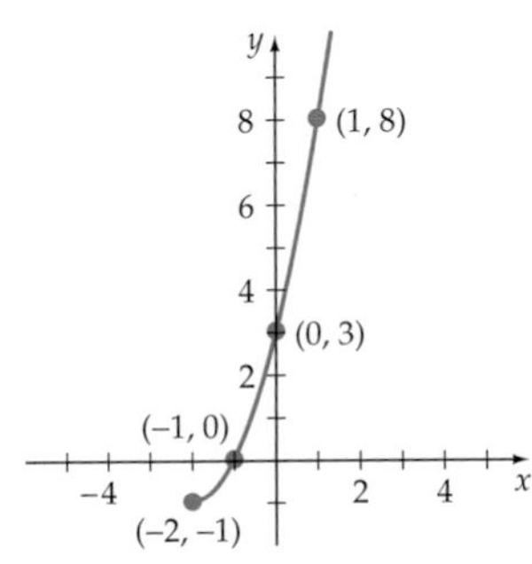

8. The domain of each of the following functions is the set of real numbers. Which of these functions does not have an inverse that is a function? $g(x) = x^2 + 1$ [3.2]

$$f(x) = -\frac{2}{3}x + 1,\ g(x) = x^2 + 1,\ h(x) = x^3 - 2$$

9. Let the domain of $f(x) = x^2 - 2x - 3$ be $\{x \mid x \leq 1\}$. Then the range of f is $\{y \mid y \geq -4\}$. What are the domain and range of f^{-1}? Domain: $\{x \mid x \geq -4\}$; range: $\{y \mid y \leq 1\}$ [3.2]

10. Use the graph below to draw a graph of **a.** $y = f(x + 1) - 2$ and **b.** $y = f(2x)$.

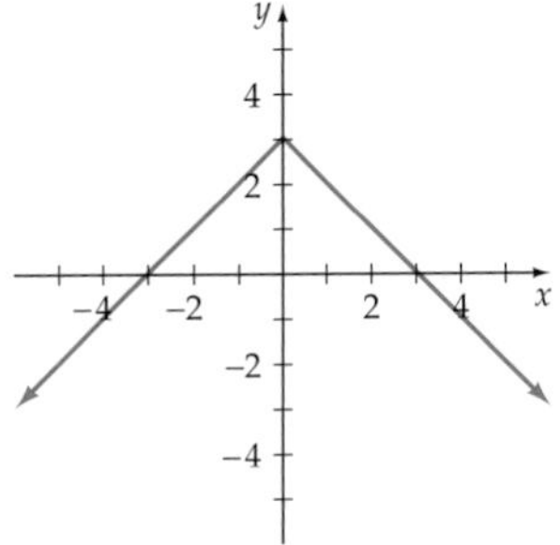

11. Complete the graph below so that **a.** the resulting graph represents an even function and **b.** the resulting graph is symmetrical to the origin.

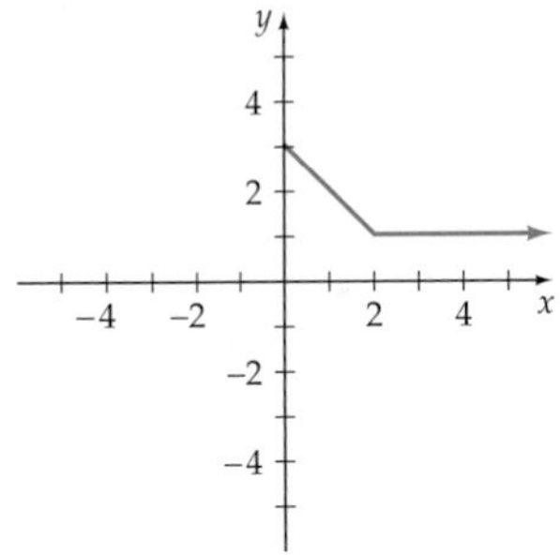

12. Use the graph below to draw the graph of $y = -f(x + 2) - 1$.

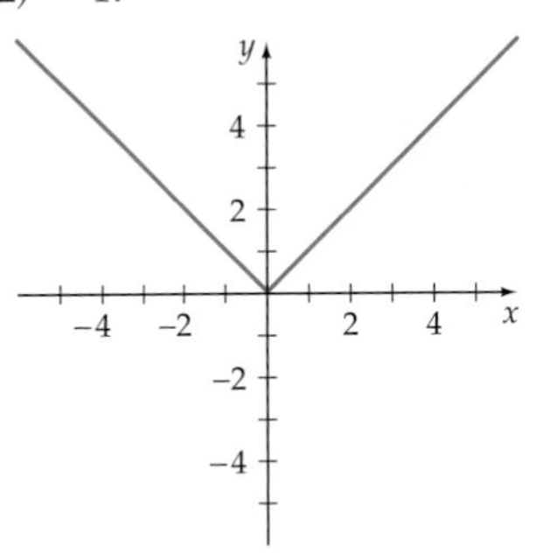

13. Classify each of the following as an even function, an odd function, or neither. State whether the graph of the function is symmetric with respect to the x-axis, the y-axis, or the origin.

a. $f(x) = x^4 - 3x^2 + 5$ Even; y-axis symmetry **b.** $f(x) = 2x^3 - 4x$ Odd; origin symmetry

c. $f(x) = x - 1$ Neither [3.3]

14. The price per unit (in dollars) at which a manufacturer can sell a car tire is given by $p(x) = 75 - \frac{x}{500}$. The cost to produce x tires is given by $C(x) = 20x + 400$. Find the manufacturer's profit for selling 15,000 tires. \$374,600 [3.1]

15. The range of a projectile is directly proportional to the square of its velocity. If a ball thrown with a velocity of 88 feet per second travels 140 feet, find the distance a ball on a similar trajectory will travel if the velocity is increased to 100 feet per second. Round to the nearest foot. 181 ft [3.4]

Cumulative Review Exercises

In Exercises 1 and 2, simplify the expression.

1. $\left(\frac{2x}{y}\right)^{-2}(x^3y)^2$ $\frac{x^4y^4}{4}$ [P.2] **2.** $\sqrt{8x^7y^2} - 2x\sqrt{2x^5y^2}$ 0 [P.6]

3. Multiply: $(2x^2 - 3x + 1)(3x - 2)$ $6x^3 - 13x^2 + 9x - 2$ [P.3]

4. Divide: $(4x^2 + 2x - 1) \div (x + 2)$ $4x - 6 + \frac{11}{x + 2}$ [P.3]

5. Simplify: $(2 + 3i)(-3 + 4i)$ $-18 - i$ [P.7]

6. Solve: $6x^2 - 5x = 6$ $\frac{3}{2}, -\frac{2}{3}$ [1.2]

7. Solve: $|2x + 5| < 1$ $\{x \mid -3 < x < -2\}$ [1.4]

8. What is the midpoint of the line segment joining the points $P_1(-1, 3)$ and $P_2(3, -5)$? $(1, -1)$ [2.1]

9. Is $(2, -3)$ an ordered pair of $g(x) = \frac{3}{x - 1}$? No [2.2]

10. Does the equation $x^2 + y^2 = 1$ define y as a function of x? No [2.2]

11. Find the slope of the line between the points $P_1(3, -2)$ and $P_2(5, 1)$. $\frac{3}{2}$ [2.3]

12. If $P(-2, 5)$ is a point on a line whose slope is $-\frac{1}{2}$, find the equation of the line. $y = -\frac{1}{2}x + 4$ [2.4]

13. Find the vertex of the parabola given by $f(x) = 2x^2 + 6x - 1$. $\left(-\frac{3}{2}, -\frac{11}{2}\right)$ [2.5]

14. The graph of $y = f(x)$ is shown in the next column. Graph $y = -f(x + 1)$.

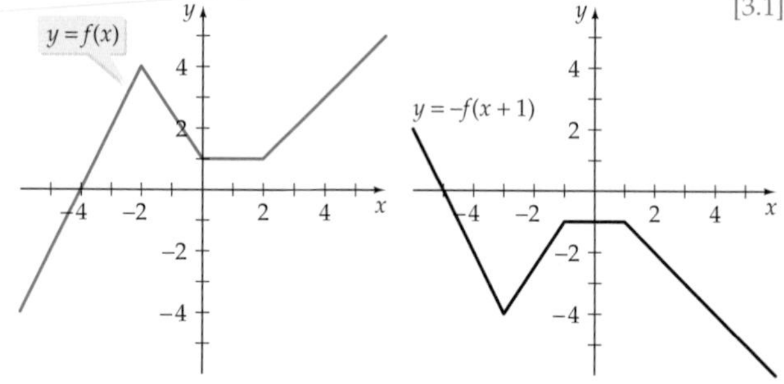

15. Discuss the symmetry of the graph of $f(x) = 2x^3 - 4x$ with respect to the x-axis, the y-axis, and the origin. Origin symmetry [3.3]

16. Given $f(x) = -2x + 3$ and $g(x) = 3x - 1$, find **a.** $f(x) - g(x)$ and **b.** $f(x) \cdot g(x)$.
a. $-5x + 4$ **b.** $-6x^2 + 11x - 3$ [3.1]

17. If $f(x) = \frac{2}{3}x + 1$, find $f^{-1}(x)$. $f^{-1}(x) = \frac{3}{2}x - \frac{3}{2}$ [3.2]

18. Suppose the value V, in dollars, of a car x years after it is purchased new is given by $V = 18{,}000 - 1800x$. At what rate, in dollars per year, is the value of the car decreasing? \$1800 per year [2.3]

19. Pure gold measures 24 karats. How much pure gold must be added to 2 ounces of 14-karat gold to produce an alloy that is 18-karat gold? $\frac{4}{3}$ oz [1.1]

20. On a certain road surface, the skidding distance a car will travel with its brakes locked is directly proportional to the square of the speed of the car when the brakes are applied. If a car is traveling 45 miles per hour when the brakes are applied and skids 50 feet, how far will the car skid when it is traveling 60 miles per hour when the brakes are applied? Round to the nearest whole number. 89 ft [3.4]

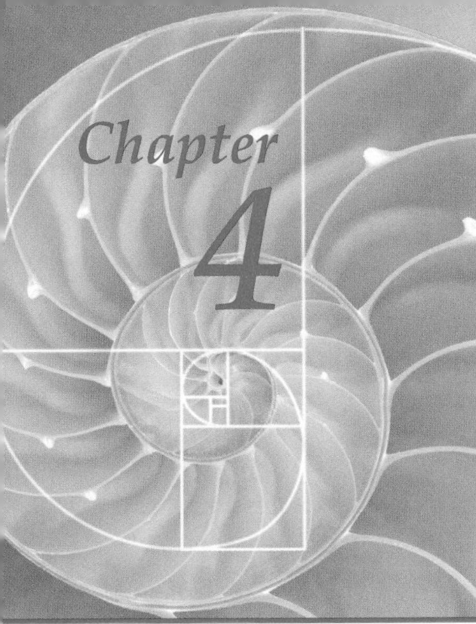

Chapter 4

Polynomial and Rational Functions

Wedding Expenses

In this chapter you will study polynomial and rational functions. These types of functions have many practical applications. They can be used to model and analyze wedding expenses, as shown below.

The average cost of a wedding was \$15,208 in 1990, \$19,104 in 1997, and about \$23,000 in 2001. The table to the right lists some of the average costs associated with a wedding in 1990 and in 1997.

The polynomial function $D(t) = 4.8t + 793$ models the average cost of a wedding dress, and the function $W(t) = 38t^2 + 291t + 15{,}208$ models the average cost of a wedding, where $t = 0$ represents the year 1990 and $0 \le t \le 12$. The rational function

$$R(t) = \frac{4.8t + 793}{38t^2 + 291t + 15{,}208}$$

represents the relative cost of a wedding dress compared to the cost of a wedding. For $t = 0$, $t = 7$, and $t = 12$, we find that $R(0) \approx 5.2\%$, $R(7) \approx 4.3\%$, and $R(12) \approx 3.5\%$.

Thus, although the average price of a wedding dress has steadily increased over the last few years, the relative cost of a wedding dress, compared to the average cost of a wedding, has decreased.

Another wedding expense application is given in Exercise 57, page 373.

Category	Average cost, 1990	Average cost, 1997
Flowers	\$478	\$756
Music	\$882	\$830
Rehearsal dinner	\$501	\$698
Wedding dress	\$794	\$823
Reception	\$5900	\$7635

Source: Bride's Magazine, 1997 Millennium Report (**EmeraldWeddings.com**)

Chapter Prep Quiz *Do these exercises to prepare for Chapter 4.*

1. Find the quotient and remainder of $(x^2 + 4x + 7) \div (x + 3)$. [P.3] Quotient $x + 1$; remainder 4
2. Find the quotient and remainder of $(x^3 - 4x^2 - 5x + 6) \div (x - 2)$. [P.3] Quotient $x^2 - 2x - 9$; remainder -12
3. Factor: $x^4 - 81$ [P.4] $(x^2 + 9)(x + 3)(x - 3)$
4. Find the x-intercepts of the graph of $y = 6x^2 + x - 12$. [2.5] $\left(\frac{4}{3}, 0\right), \left(-\frac{3}{2}, 0\right)$
5. Find the minimum of $P(x) = 3x^2 + 6x - 1$. [2.5] -4
6. Find the maximum of $P(x) = -x^2 + 8x + 1$. [2.5] 17
7. Find the product: $(x - 1)(x - 2)(x + 3)$ [P.3] $x^3 - 7x + 6$
8. Find the product: $[x - (2 + i)][x - (2 - i)]$ [P.7] $x^2 - 4x + 5$
9. Simplify: $(x^2 - 2x - 15) \div (x + 3)$ [P.3] $x - 5$
10. Write $\dfrac{2x - 3}{2x - 1} - 3x$ as a rational expression. [P.5] $\dfrac{-6x^2 + 5x - 3}{2x - 1}$

Problem Solving Strategies

Narrow the Search

An important aspect of problem solving involves narrowing the list of possible solutions to a problem and then checking to see which of these solutions are actual solutions. Here is a simple example that illustrates this technique.

Problem Find all three-digit palindromic numbers that are also perfect squares. A *palindromic number* is a natural number that reads the same from left to right as it reads from right to left. For instance, 4114 and 53235 are both palindromic numbers.

Solution At first this appears to be a difficult problem. After all, there are 900 three-digit numbers that range from 100 to 999. Let's try to narrow our search. Make a list of the three-digit perfect squares. (A number is a perfect square if it is the square of a natural number.)

$10^2 = 100$	$11^2 = 121$	$12^2 = 144$
$13^2 = 169$	$14^2 = 196$	$15^2 = 225$
$16^2 = 256$	$17^2 = 289$	$18^2 = 324$
$19^2 = 361$	$20^2 = 400$	$21^2 = 441$
$22^2 = 484$	$23^2 = 529$	$24^2 = 576$
$25^2 = 625$	$26^2 = 676$	$27^2 = 729$
$28^2 = 784$	$29^2 = 841$	$30^2 = 900$
$31^2 = 961$		

The above list of three-digit perfect squares only has 22 numbers. We have just narrowed our search from 900 numbers to 22 numbers! A close examination of the above numbers shows that 121, 484, and 676 are the only palindromic numbers that have three digits and are also perfect squares.

You will use this technique of narrowing your search in this chapter when you find the rational zeros of a polynomial function by applying the Theorem on Rational Zeros.

INSTRUCTOR NOTE

The *Chapter Prep Quiz* is a means to test your students' mastery of prerequisite material that is assumed in the coming chapter. Each question identifies the section to review (if necessary) and all answers are provided in the Answers to Selected Exercises.

SECTION 4.1 The Remainder Theorem and the Factor Theorem

- The Remainder Theorem
- The Factor Theorem
- Reduced Polynomials

The Remainder Theorem

If $P(x)$ is a polynomial function, then the values of x for which $P(x)$ is equal to 0 are called the **zeros** of $P(x)$. For instance, -1 is a zero of $P(x) = 2x^3 - x + 1$ because

$$\begin{aligned} P(-1) &= 2(-1)^3 - (-1) + 1 \\ &= -2 + 1 + 1 \\ &= 0 \end{aligned}$$

QUESTION Is 0 a zero of $P(x) = 2x^3 - x + 1$?

Much of the work in this chapter concerns finding the zeros of a polynomial function. Sometimes the zeros of a polynomial function can be determined with the aid of the following theorem.

The Remainder Theorem

If a polynomial function $P(x)$ is divided by $x - c$, then the remainder equals $P(c)$.

The following example illustrates the Remainder Theorem by showing that the remainder of $(x^2 + 9x - 16) \div (x - 3)$ is the same as $P(x) = x^2 + 9x - 16$ evaluated at $x = 3$.

Let $x = 3$ and $P(x) = x^2 + 9x - 16$.

$$\begin{aligned} \text{Then } P(3) &= (3)^2 + 9(3) - 16 \\ &= 9 + 27 - 16 \\ &= 20 \end{aligned}$$

$$\begin{array}{r} x + 12 \\ x - 3\overline{)x^2 + 9x - 16} \\ \underline{x^2 - 3x} \\ 12x - 16 \\ \underline{12x - 36} \\ 20 \end{array}$$

$P(3)$ is equal to the remainder of $P(x)$ divided by $(x - 3)$.

It is generally easier to evaluate a polynomial function by direct substitution than it is to use polynomial division, as shown above. However, a procedure called **synthetic division** can expedite the division process. To apply the synthetic division procedure, the divisor must be a polynomial of the form $x - c$, where c is a constant. In the synthetic division procedure, the variables that occur in the polynomials are not listed. To understand how synthetic division is performed,

ANSWER No. $P(0) = 2(0)^3 - 0 + 1 = 1$. Because $P(0) \neq 0$, we know 0 is not a zero of $P(x)$.

examine the following long division on the left and the related synthetic division on the right.

Long division ***Synthetic division***

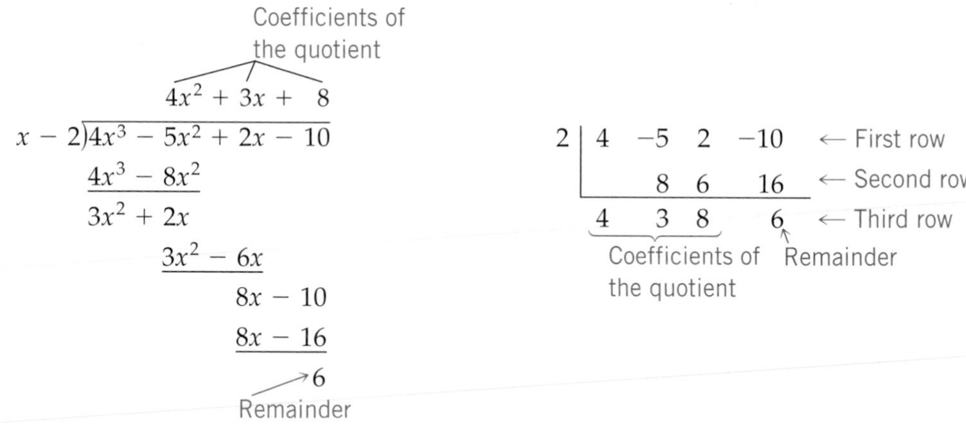

In the long division above, the dividend is $4x^3 - 5x^2 + 2x - 10$ and the divisor is $x - 2$. Because the divisor is of the form $x - c$, with $c = 2$, the division can be performed by the synthetic division procedure. Observe that in the accompanying synthetic division

1. the constant c is listed as the first number in the first row, followed by the coefficients of the dividend.
2. the first number in the third row is the leading coefficient of the dividend.
3. each number in the second row is determined by computing the product of c and the number in the third row of the preceding column.
4. each of the numbers in the third row, other than the first number, is determined by adding the numbers directly above it.

The following explanation illustrates the steps used to find the quotient and remainder of $(2x^3 - 8x + 7) \div (x + 3)$ by using synthetic division. The divisor $x + 3$ is written in $x - c$ form as $x - (-3)$, which indicates that $c = -3$. The dividend $2x^3 - 8x + 7$ is missing an x^2 term. If we insert $0x^2$ for the missing term, the dividend becomes $2x^3 + 0x^2 - 8x + 7$.

Coefficients of the dividend

$$\begin{array}{r|rrrr} -3 & 2 & 0 & -8 & 7 \\ & \downarrow & & & \\ \hline & 2 & & & \end{array}$$

Write the constant c, -3, followed by the coefficients of the dividend. Bring down the first coefficient in the first row, 2, as the first number of the third row.

$$\begin{array}{r|rrrr} -3 & 2 & 0 & -8 & 7 \\ & & -6 & & \\ \hline & 2 & -6 & & \end{array}$$

Multiply c times the first number in the third row, 2, to produce the first number of the second row, -6. Add the 0 and the -6 to produce the next number of the third row, -6.

$$\begin{array}{r|rrrr} -3 & 2 & 0 & -8 & 7 \\ & & -6 & 18 & \\ \hline & 2 & -6 & 10 & \end{array}$$

Multiply c times the second number in the third row, -6, to produce the next number of the second row, 18. Add the -8 and the 18 to produce the next number of the third row, 10.

$$\begin{array}{r|rrrr} -3 & 2 & 0 & -8 & 7 \\ & & -6 & 18 & -30 \\ \hline & 2 & -6 & 10 & -23 \end{array}$$

Coefficients of the quotient

Remainder

Multiply c times the third number in the third row, 10, to produce the next number of the second row, -30. Add the 7 and the -30 to produce the last number of the third row, -23.

The last number in the bottom row, -23, is the remainder. The other numbers in the bottom row are the coefficients of the quotient. The quotient of a synthetic division always has a degree that is *one less* than the degree of the dividend. Thus the quotient in this example is $2x^2 - 6x + 10$. The results of the above synthetic division can be expressed in **fractional form** as

TAKE NOTE

$2x^2 - 6x + 10 + \dfrac{-23}{x + 3}$ can also be written as $2x^2 - 6x + 10 - \dfrac{23}{x + 3}$

$$\frac{2x^3 - 8x + 7}{x + 3} = 2x^2 - 6x + 10 + \frac{-23}{x + 3}$$

or as

$$2x^3 - 8x + 7 = (x + 3)(2x^2 - 6x + 10) - 23$$

In Example 1 we illustrate the compact form of synthetic division, obtained by condensing the process explained above.

Alternative to Example 1
Exercise 4, page 320

EXAMPLE 1 Use Synthetic Division to Divide Polynomials

Use synthetic division to divide $x^4 - 4x^2 + 7x + 15$ by $x + 4$.

Solution Because the divisor is $x + 4$, we perform synthetic division with $c = -4$.

$$\begin{array}{r|rrrrr} -4 & 1 & 0 & -4 & 7 & 15 \\ & & -4 & 16 & -48 & 164 \\ \hline & 1 & -4 & 12 & -41 & 179 \end{array}$$

The quotient is $x^3 - 4x^2 + 12x - 41$ and the remainder is 179.

$$\frac{x^4 - 4x^2 + 7x + 15}{x + 4} = x^3 - 4x^2 + 12x - 41 + \frac{179}{x + 4}$$

CHECK YOUR PROGRESS 1 Use synthetic division to divide $5x^3 - 6x^2 - 19$ by $x - 2$.

Solution *See page S19.* $5x^2 + 4x + 8 - \dfrac{3}{x - 2}$

INTEGRATING TECHNOLOGY

A TI-82/83 synthetic-division program called SYDIV is available on the Internet at

math.college.hmco.com

The program prompts you to enter the degree of the dividend, the coefficients of the dividend, and the constant c from the divisor $x - c$. For instance, to perform the synthetic division in Example 1, enter 4 for the degree of the dividend, followed by the coefficients 1, 0, −4, 7, and 15. See Figure 4.1. Press ENTER followed by −4 to produce the display in Figure 4.2. Press ENTER to produce the display in Figure 4.3. Press ENTER again to produce the display in Figure 4.4.

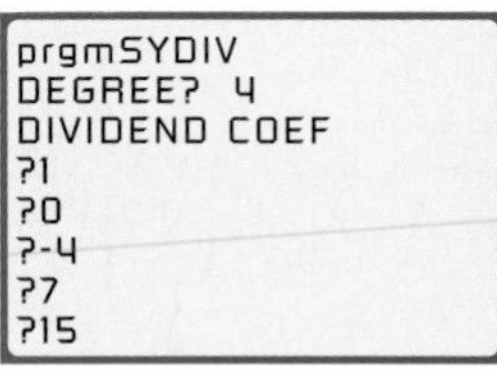

Figure 4.1

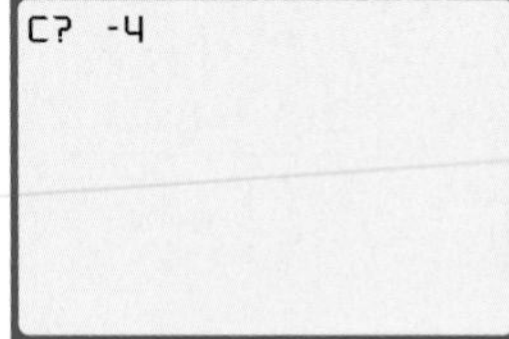

Figure 4.2

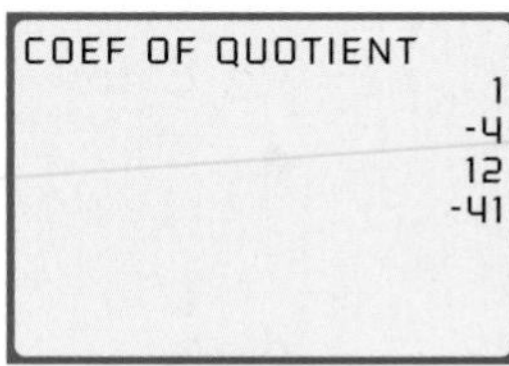

Figure 4.3

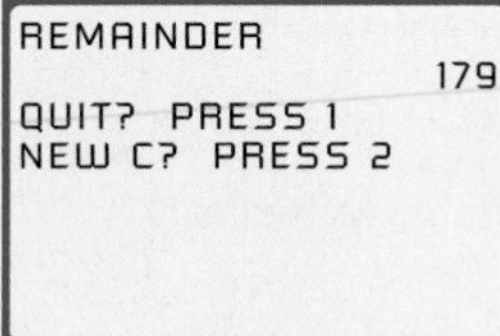

Figure 4.4

In Example 2 we use synthetic division and the Remainder Theorem to evaluate a polynomial function.

Alternative to Example 2
Exercise 16, page 320

EXAMPLE 2 Use the Remainder Theorem to Evaluate a Polynomial Function

Let $P(x) = 2x^3 + 3x^2 + 2x - 2$. Use the Remainder Theorem to find $P(c)$ for $c = -2$ and $c = \frac{1}{2}$.

Solution

Perform synthetic division with $c = -2$ and $c = \frac{1}{2}$ and examine the remainders.

$$\begin{array}{r|rrrr} -2 & 2 & 3 & 2 & -2 \\ & & -4 & 2 & -8 \\ \hline & 2 & -1 & 4 & -10 \end{array}$$

The remainder is −10. Therefore, $P(-2) = -10$.

$$\begin{array}{r|rrrr} \frac{1}{2} & 2 & 3 & 2 & -2 \\ & & 1 & 2 & 2 \\ \hline & 2 & 4 & 4 & 0 \end{array}$$

The remainder is 0. Therefore, $P\left(\frac{1}{2}\right) = 0$.

Visualize the Solution

A graph of P shows that the points $(-2, -10)$ and $\left(\frac{1}{2}, 0\right)$ are on the graph.

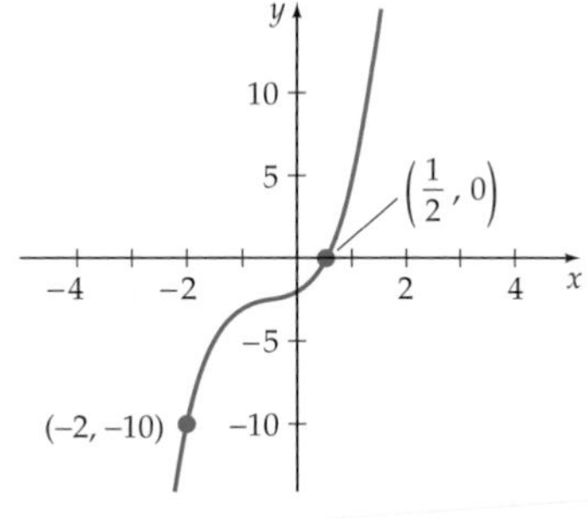

$P(x) = 2x^3 + 3x^2 + 2x - 2$

Continued ➤

CHECK YOUR PROGRESS 2 Let $P(x) = 2x^3 - x^2 + 3x - 1$. Use the Remainder Theorem to find $P(c)$ for $c = 3$.

Solution *See page S19.* $P(3) = 53$

Using the Remainder Theorem to evaluate a polynomial function is often faster than evaluating the polynomial function by direct substitution. For instance, to evaluate $P(x) = x^5 - 10x^4 + 35x^3 - 50x^2 + 24x$ by substituting 7 for x, we must do the following work.

$$\begin{aligned} P(7) &= 7^5 - 10(7)^4 + 35(7)^3 - 50(7)^2 + 24(7) \\ &= 16{,}807 - 10(2401) + 35(343) - 50(49) + 24(7) \\ &= 16{,}807 - 24{,}010 + 12{,}005 - 2450 + 168 \\ &= 2520 \end{aligned}$$

TAKE NOTE

Because $P(x)$ has a constant term of 0, we must include 0 as the last number in the first row of the synthetic division at the right.

Using the Remainder Theorem to perform the above evaluation requires only the following work.

$$\begin{array}{r|rrrrrr} 7 & 1 & -10 & 35 & -50 & 24 & 0 \\ & & 7 & -21 & 98 & 336 & 2520 \\ \hline & 1 & -3 & 14 & 48 & 360 & 2520 \leftarrow P(7) \end{array}$$

INSTRUCTOR NOTE

Point out that although the Factor Theorem provides a method of *testing* whether $(x - c)$ is a factor of a polynomial, it does not provide a procedure for *detecting* such a factor.

▪ The Factor Theorem

Note from Example 2 that $P\left(\frac{1}{2}\right) = 0$. The number $\frac{1}{2}$ is a zero of P because $P(x) = 0$ when $x = \frac{1}{2}$.

The following theorem is a direct result of the Remainder Theorem. It points out the important relationship between a zero of a given polynomial function and a factor of the polynomial function.

The Factor Theorem

A polynomial function $P(x)$ has a factor $(x - c)$ if and only if $P(c) = 0$. That is, $(x - c)$ is a factor of $P(x)$ if and only if c is a zero of P.

Alternative to Example 3
Exercise 30, page 320

EXAMPLE 3 Apply the Factor Theorem

Use synthetic division and the Factor Theorem to determine whether $(x + 5)$ or $(x - 2)$ is a factor of $P(x) = x^4 + x^3 - 21x^2 - x + 20$.

Solution

$$\begin{array}{r|rrrrr} -5 & 1 & 1 & -21 & -1 & 20 \\ & & -5 & 20 & 5 & -20 \\ \hline & 1 & -4 & -1 & 4 & 0 \end{array}$$

Continued ➤

The remainder of 0 indicates that $(x + 5)$ is a factor of $P(x)$.

2	1	1	−21	−1	20
		2	6	−30	−62
	1	3	−15	−31	−42

The remainder of −42 indicates that $(x - 2)$ is not a factor of $P(x)$.

CHECK YOUR PROGRESS 3 Use synthetic division and the Factor Theorem to determine whether $(x + 6)$ is a factor of $P(x) = x^3 + 4x^2 - 27x - 90$.

Solution *See page S19.* $(x + 6)$ is a factor of $P(x)$.

QUESTION Is −5 a zero of $P(x)$ given in Example 3?

Here is a summary of the important role played by the remainder in the division of a polynomial by $(x - c)$.

The Remainder of a Polynomial Division

In the division of the polynomial function $P(x)$ by $(x - c)$, the remainder is

- equal to $P(c)$.
- 0 if and only if $(x - c)$ is a factor of P.
- 0 if and only if c is a zero of P.

Also, if c is a real number, then the remainder of $P(x) \div (x - c)$ is 0 if and only if $(c, 0)$ is an x-intercept of the graph of P.

Reduced Polynomials

In Example 3, we determined that $(x + 5)$ is a factor of the polynomial function $P(x) = x^4 + x^3 - 21x^2 - x + 20$ and that the quotient of $x^4 + x^3 - 21x^2 - x + 20$ divided by $(x + 5)$ is $Q(x) = x^3 - 4x^2 - x + 4$. Thus

$$P(x) = (x + 5)(x^3 - 4x^2 - x + 4)$$

The quotient $Q(x) = x^3 - 4x^2 - x + 4$ is called a **reduced polynomial** or a **depressed polynomial** of $P(x)$ because it is a factor of $P(x)$ and its degree is 1 less than the degree of $P(x)$. Reduced polynomials will play an important role in Sections 4.3 and 4.4.

ANSWER Yes. Because $(x + 5)$ is a factor of $P(x)$, the Factor Theorem states that $P(-5) = 0$, and thus −5 is a zero of $P(x)$.

Alternative to Example 4
Exercise 46, page 320

EXAMPLE 4 Find a Reduced Polynomial

Verify that $(x - 3)$ is a factor of $P(x) = 2x^3 - 3x^2 - 4x - 15$ and write $P(x)$ as the product of $(x - 3)$ and the reduced polynomial $Q(x)$.

Solution

3	2	−3	−4	−15
		6	9	15
	2	3	5	0

Coefficients of the reduced polynomial $Q(x)$

Thus $(x - 3)$ and the reduced polynomial $2x^2 + 3x + 5$ are both factors of $P(x)$. That is,

$$P(x) = 2x^3 - 3x^2 - 4x - 15 = (x - 3)(2x^2 + 3x + 5)$$

CHECK YOUR PROGRESS 4 Verify that $(x + 1)$ is a factor of $P(x) = x^4 + 5x^3 + 3x^2 - 5x - 4$ and write $P(x)$ as the product of $(x + 1)$ and the reduced polynomial $Q(x)$.

Solution *See page S19.* $P(x) = (x + 1)(x^3 + 4x^2 - x - 4)$

Topics for Discussion

1. Explain the meaning of the phrase *zero of a polynomial.*

2. If $P(x)$ is a polynomial function of degree $n \geq 2$, what is the degree of the quotient $\dfrac{P(x)}{x - c}$?

3. Discuss how the Remainder Theorem can be used to determine whether a number is a zero of a polynomial function.

4. A zero of $P(x) = x^3 - x^2 - 14x + 24$ is -4. Discuss how this information and the Factor Theorem can be used to solve $x^3 - x^2 - 14x + 24 = 0$.

5. Discuss the advantages and disadvantages of using synthetic division rather than substitution to evaluate a polynomial function at $x = c$.

EXERCISES 4.1

— Suggested Assignment: Exercises 1–57, odd; and 58–63, all.

In Exercises 1 to 14, use synthetic division to divide the first polynomial by the second.

1. $4x^3 - 5x^2 + 6x - 7,\ \ x - 2$ $\quad 4x^2 + 3x + 12 + \dfrac{17}{x - 2}$

2. $4x^3 + 5x^2 - 11x + 2,\ \ x - 3$ $\quad 4x^2 + 17x + 40 + \dfrac{122}{x - 3}$

3. $4x^3 - 2x + 3,\ \ x + 1$ $\quad 4x^2 - 4x + 2 + \dfrac{1}{x + 1}$

4. $6x^3 - 4x^2 + 17, \quad x + 3$ $\quad 6x^2 - 22x + 66 - \dfrac{181}{x+3}$

5. $x^5 - 10x^3 + 5x - 1, \quad x - 4$ $\quad x^4 + 4x^3 + 6x^2 + 24x + 101 + \dfrac{403}{x-4}$

6. $6x^4 - 2x^3 - 3x^2 - x, \quad x - 5$ $\quad 6x^3 + 28x^2 + 137x + 684 + \dfrac{3420}{x-5}$

7. $x^5 - 1, \quad x - 1$ $\quad x^4 + x^3 + x^2 + x + 1$

8. $x^4 + 1, \quad x + 1$ $\quad x^3 - x^2 + x - 1 + \dfrac{2}{x+1}$

9. $8x^3 - 4x^2 + 6x - 3, \quad x - \dfrac{1}{2}$ $\quad 8x^2 + 6$

10. $12x^3 + 5x^2 + 5x + 6, \quad x + \dfrac{3}{4}$ $\quad 12x^2 - 4x + 8$

11. $x^8 + x^6 + x^4 + x^2 + 4, \quad x - 2$
$x^7 + 2x^6 + 5x^5 + 10x^4 + 21x^3 + 42x^2 + 85x + 170 + \dfrac{344}{x-2}$

12. $-x^7 - x^5 - x^3 - x - 5, \quad x + 1$
$-x^6 + x^5 - 2x^4 + 2x^3 - 3x^2 + 3x - 4 - \dfrac{1}{x+1}$

13. $x^6 + x - 10, \quad x + 3$
$x^5 - 3x^4 + 9x^3 - 27x^2 + 81x - 242 + \dfrac{716}{x+3}$

14. $2x^5 - 3x^4 - 5x^2 - 10, \quad x - 4$
$2x^4 + 5x^3 + 20x^2 + 75x + 300 + \dfrac{1190}{x-4}$

In Exercises 15 to 24, use the Remainder Theorem to find $P(c)$.

15. $P(x) = 3x^3 + x^2 + x - 5, c = 2$ $\quad 25$

16. $P(x) = 5x^3 - 4x^2 + x - 7, c = 3$ $\quad 95$

17. $P(x) = 4x^4 - 6x^2 + 5, c = -2$ $\quad 45$

18. $P(x) = 6x^3 - x^2 + 4x, c = -3$ $\quad -183$

19. $P(x) = -2x^3 - 2x^2 - x - 20, c = 10$ $\quad -2230$

20. $P(x) = -x^3 + 3x^2 + 5x + 30, c = 8$ $\quad -250$

21. $P(x) = -x^4 + 1, c = 3$ $\quad -80$

22. $P(x) = x^5 - 1, c = 1$ $\quad 0$

23. $P(x) = x^4 - 10x^3 + 2, c = 3$ $\quad -187$

24. $P(x) = x^5 + 20x^2 - 1, c = -5$ $\quad -2626$

In Exercises 25 to 34, use synthetic division and the Factor Theorem to determine whether the given binomial is a factor of $P(x)$.

25. $P(x) = x^3 + 2x^2 - 5x - 6, x - 2$ $\quad$ Yes

26. $P(x) = x^3 - 6x^2 - 30x + 40, x + 4$ $\quad$ Yes

27. $P(x) = 2x^3 + x^2 - 3x - 1, x + 1$ $\quad$ No

28. $P(x) = 3x^3 + 4x^2 - 27x - 36, x - 4$ $\quad$ No

29. $P(x) = x^4 - 25x^2 + 144, x + 3$ $\quad$ Yes

30. $P(x) = x^4 - 25x^2 + 144, x - 3$ $\quad$ Yes

31. $P(x) = x^5 + 2x^4 - 22x^3 - 50x^2 - 75x, x - 5$ $\quad$ Yes

32. $P(x) = 9x^4 - 6x^3 - 23x^2 - 4x + 4, x + 1$ $\quad$ Yes

33. $P(x) = 16x^4 - 8x^3 + 9x^2 + 14x + 4, x - \dfrac{1}{4}$ $\quad$ No

34. $P(x) = 10x^4 + 9x^3 - 4x^2 + 9x + 6, x + \dfrac{1}{2}$ $\quad$ Yes

In Exercises 35 to 44, use synthetic division to show that c is a zero of $P(x)$.

35. $P(x) = 3x^3 - 8x^2 - 10x + 28, c = 2$

36. $P(x) = 4x^3 - 10x^2 - 8x + 6, c = 3$

37. $P(x) = x^4 - 1, c = 1$

38. $P(x) = x^3 + 8, c = -2$

39. $P(x) = 3x^4 + 8x^3 + 10x^2 + 2x - 20, c = -2$

40. $P(x) = x^4 - 2x^2 - 100x - 75, c = 5$

41. $P(x) = 2x^3 - 18x^2 - 50x + 66, c = 11$

42. $P(x) = 2x^4 - 34x^3 + 70x^2 - 153x + 45, c = 15$

43. $P(x) = 3x^2 - 8x + 4, c = \dfrac{2}{3}$

44. $P(x) = 5x^2 + 12x + 4, c = -\dfrac{2}{5}$

In Exercises 45 to 48, verify that the given binomial is a factor of $P(x)$, and write $P(x)$ as the product of the binomial and its reduced polynomial $Q(x)$.

45. $P(x) = x^3 + x^2 + x - 14, x - 2$
$(x - 2)(x^2 + 3x + 7)$

46. $P(x) = x^4 + 7x^3 + 17x^2 + 21x + 18, x + 3$
$(x + 3)(x^3 + 4x^2 + 5x + 6)$

47. $P(x) = x^4 - x^3 - 9x^2 - 11x - 4, x - 4$
$(x - 4)(x^3 + 3x^2 + 3x + 1)$

48. $P(x) = 2x^5 - x^4 - 7x^3 + x^2 + 7x - 10, x - 2$
$(x - 2)(2x^4 + 3x^3 - x^2 - x + 5)$

Business and Economics

49. ***Cost of a Wedding*** The average cost of a wedding, in dollars, is modeled by

$$C(t) = 38t^2 + 291t + 15{,}208$$

where $t = 0$ represents the year 1990 and $0 \le t \le 12$. Use the Remainder Theorem to estimate the average cost of a wedding in

a. 1998. $19,968

b. 2001. $23,007

Life and Health Sciences

50. ***Selection of Bridesmaids*** A bride-to-be has several girlfriends, but she has decided to have only five bridesmaids, including the maid of honor. The number of different ways n girlfriends can be chosen and assigned a position, such as maid of honor, first matron, second matron, and so on, is given by the polynomial function

$$P(n) = n^5 - 10n^4 + 35n^3 - 50n^2 + 24n$$

a. Use the Remainder Theorem to determine the number of ways the bride can select her bridesmaids if she chooses from $n = 7$ girlfriends. 2520

b. Evaluate $P(n)$ for $n = 7$ by substituting 7 for n. How does this result compare with the result obtained in part **a.**? 2520; They are the same.

Sports and Recreation

51. ***Selection of Cards*** The number of ways you can select three cards from a stack of n cards, in which the order of selection is important, is given by

$$P(n) = n^3 - 3n^2 + 2n, n \ge 3$$

a. Use the Remainder Theorem to determine the number of ways you can select three cards from a stack of $n = 8$ cards. 336

b. Evaluate $P(n)$ for $n = 8$ by substituting 8 for n. How does this result compare with the result obtained in part **a.**? 336; They are the same.

52. ***Rocket Launch*** A model rocket is projected upward from an initial height of 4 feet with an initial velocity of 158 feet per second. The height of the rocket is given by $s = -16t^2 + 158t + 4$, where $0 \le t \le 9.9$ seconds. Use the Remainder Theorem to determine the height of the rocket at

a. $t = 5$ seconds. 394 ft

b. $t = 8$ seconds. 244 ft

53. ***House of Cards*** The number of cards C needed to build a house of cards with r rows (levels) is given by the function $C(r) = 1.5r^2 + 0.5r$.

Use the Remainder Theorem to determine the number of cards needed to build a house of cards with

a. $r = 8$ rows. 100 cards

b. $r = 20$ rows. 610 cards

Social Sciences

54. ***Election of Class Officers*** The number of ways a class of n students can elect a president, a vice-president, a secretary, and a treasurer is given by the function $P(n) = n^4 - 6n^3 + 11n^2 - 6n$, where $n \ge 4$. Use the Remainder Theorem to determine the number of ways the class can elect officers if the class consists of

a. $n = 12$ students. 11,880 ways

b. $n = 24$ students. 255,024 ways

55. ***Population Density of a City*** The population density D (in people per square mile) of a city is related to the distance x, in miles, from the center of the city by

$D = -45x^2 + 190x + 200, 0 < x < 5$. Use the Remainder Theorem to determine the population density of the city at a distance of

a. $x = 2$ miles. 400 people/mi^2 **b.** $x = 4$ miles. 240 people/mi^2

Physical Sciences and Engineering

56. ***Volume of a Solid*** The volume of the following solid is given by $V(x) = x^3 + 3x^2$.

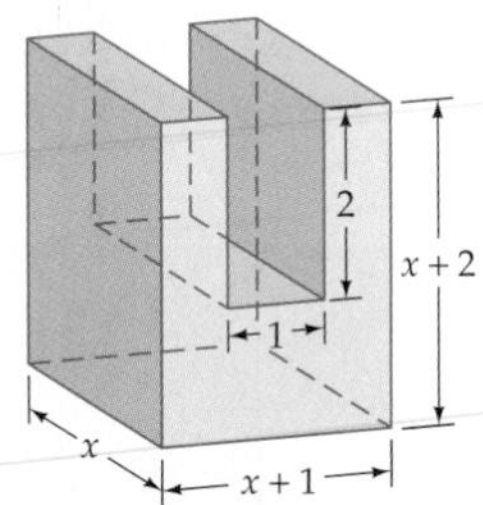

Use the Remainder Theorem to determine the volume of the solid if

a. $x = 7$ inches. 490 in^3 **b.** $x = 11$ inches. 1694 in^3

57. ***Volume of a Solid*** The volume of the following solid is given by $V(x) = x^3 + x^2 + 10x - 8$.

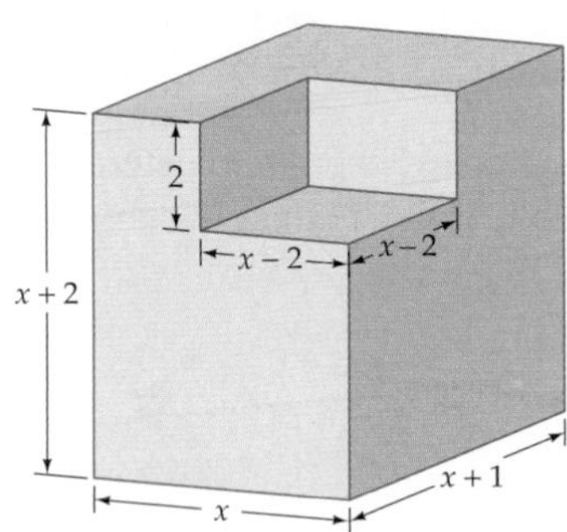

Use the Remainder Theorem to determine the volume of the solid if

a. $x = 6$ inches. 304 in^3 **b.** $x = 9$ inches. 892 in^3

Prepare for Section 4.2

58. Find the minimum value of $P(x) = x^2 - 4x + 6$. [2.5] 2

59. Find the maximum value of $P(x) = -2x^2 - x + 1$. [2.5] $\frac{9}{8}$

60. Find the interval on which $P(x) = x^2 + 2x + 7$ is increasing. [2.2] $[-1, \infty)$

61. Find the interval on which $P(x) = -2x^2 + 4x + 5$ is decreasing. [2.2] $[1, \infty)$

62. Factor: $x^4 - 5x^2 + 4$ [P.4] $(x + 1)(x - 1)(x + 2)(x - 2)$

63. Find the x-intercepts of the graph of $P(x) = 6x^2 - x - 2$. [2.5] $\left(\frac{2}{3}, 0\right), \left(-\frac{1}{2}, 0\right)$

Explorations

1. ***Horner's Polynomial Form*** William Horner (1786–1837) devised a method of writing a polynomial in a form that does not involve any exponents other than 1. For instance, $4x^4 + 2x^3 - 5x^2 + 7x - 11$ can be written in each of the following forms.

$$4x^4 + 2x^3 - 5x^2 + 7x - 11$$

$$= (4x^3 + 2x^2 - 5x + 7)x - 11$$ ▪ Factor an x from the first four terms.

$$= [(4x^2 + 2x - 5)x + 7]x - 11$$ ▪ Factor an x from the first three terms inside the innermost parentheses.

$$= \{[(4x + 2)x - 5]x + 7\}x - 11$$ ▪ Factor an x from the first two terms inside the innermost parentheses.

Horner's form $\{[(4x + 2)x - 5]x + 7\}x - 11$ is easier to evaluate than the descending exponent form $4x^4 + 2x^3 - 5x^2 + 7x - 11$. Horner's form is sometimes used by computer programmers to make their programs run faster.

a. Let $P(x) = 3x^5 - 4x^4 + 5x^3 - 2x^2 + 3x - 8$. Find $P(6)$ by direct substitution. $P(6) = 19{,}162$

b. Use Horner's method to write $P(x)$ in a form that does not involve any exponents other than 1. Now use this form to evaluate $P(6)$. Which was easier to perform, the evaluation in part **a.** or in part **b.**?

$P(x) = (\{[(3x - 4)x + 5]x - 2\}x + 3)x - 8$; 19,162; part **b.**

SECTION 4.2 Polynomial Functions of Higher Degree

- Far-Left and Far-Right Behavior
- Maximum and Minimum Values
- Real Zeros of a Polynomial Function
- Even and Odd Powers of $(x - c)$ Theorem

TAKE NOTE

The **general form of a polynomial** is given by $a_nx^n + a_{n-1}x^{n-1} + \cdots + a_0$. In this text the coefficients $a_n, a_{n-1}, \ldots, a_0$ are all real numbers unless specifically stated otherwise.

Table 4.1 summarizes information developed in Chapter 2 about graphs of polynomial functions of degree 0, 1, or 2.

Table 4.1

Polynomial Function $P(x)$	Graph
$P(x) = a$ (degree 0)	Horizontal line through $(0, a)$
$P(x) = ax + b$ (degree 1), $a \neq 0$	Line with y-intercept $(0, b)$ and slope a.
$P(x) = ax^2 + bx + c$ (degree 2), $a \neq 0$	Parabola with vertex $\left(-\frac{b}{2a}, P\left(-\frac{b}{2a}\right)\right)$

Polynomial functions of degree 3 or higher can be graphed by the technique of plotting points; however, some additional knowledge about polynomial functions will make graphing easier.

All polynomial functions have graphs that are **smooth continuous curves.** The terms *smooth* and *continuous* are defined rigorously in calculus, but for the present, a smooth curve is a curve that does not have sharp corners such as that shown in Figure 4.5a. A continuous curve does not have a break or hole such as those shown in Figure 4.5b.

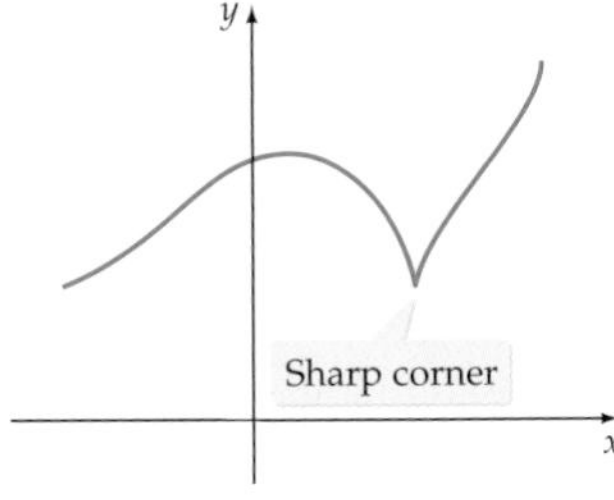

a. Continuous, but not smooth

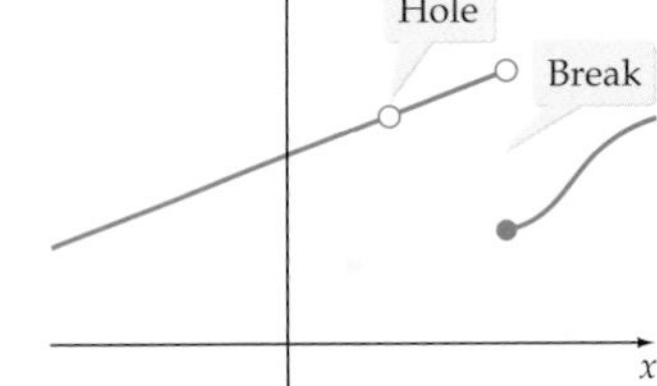

b. Not continuous

Figure 4.5

TAKE NOTE

The leading term of a polynomial function in x is the nonzero term that contains the largest power of x. The leading coefficient of a polynomial function is the coefficient of the leading term.

Far-Left and Far-Right Behavior

The graph of a polynomial function may have several up and down fluctuations; however, the graph of every polynomial function will eventually increase or decrease without bound as the graph moves far to the left or far to the right. The **leading term** a_nx^n is said to be the **dominate term** of the polynomial function $P(x) = a_nx^n + a_{n-1}x^{n-1} + \cdots + a_1x + a_0$ because as $|x|$ becomes large, the absolute value of a_nx^n will be much larger than the absolute value of any of the other terms. Because of this condition, you can determine the **far-left and far-right behavior** of the polynomial by examining the **leading coefficient** a_n and the degree n of the polynomial.

Table 4.2 indicates the far-left and far-right behavior of a polynomial function $P(x)$ with leading term $a_n x^n$.

Table 4.2 Far-Right and Far-Left Behavior of the Graph of a Polynomial Function with Leading Term $a_n x^n$

	n is even	n is odd
$a_n > 0$	Up to left and up to right	Down to left and up to right
$a_n < 0$	Down to left and down to right	Up to left and down to right

Alternative to Example 1
Exercise 2, page 332

EXAMPLE 1 Determine the Far-Left and Far-Right Behavior of a Polynomial Function

Examine the leading term to determine the far-left and far-right behavior of the graph of each polynomial function.

a. $P(x) = x^3 - x$

b. $S(x) = \frac{1}{2}x^4 - \frac{5}{2}x^2 + 2$

c. $T(x) = -2x^3 + x^2 + 7x - 6$

d. $U(x) = 9 + 8x^2 - x^4$

Solution

a. Because $a_n = 1$ is *positive* and $n = 3$ is *odd*, the graph of P goes down to its far left and up to its far right. See Figure 4.6.

b. Because $a_n = \frac{1}{2}$ is *positive* and $n = 4$ is *even*, the graph of S goes up to its far left and up to its far right. See Figure 4.7.

c. Because $a_n = -2$ is *negative* and $n = 3$ is *odd*, the graph of T goes up to its far left and down to its far right. See Figure 4.8.

d. The leading term of $U(x)$ is $-x^4$ and the leading coefficient is -1. Because $a_n = -1$ is *negative* and $n = 4$ is *even*, the graph of U goes down to its far left and down to its far right. See Figure 4.9.

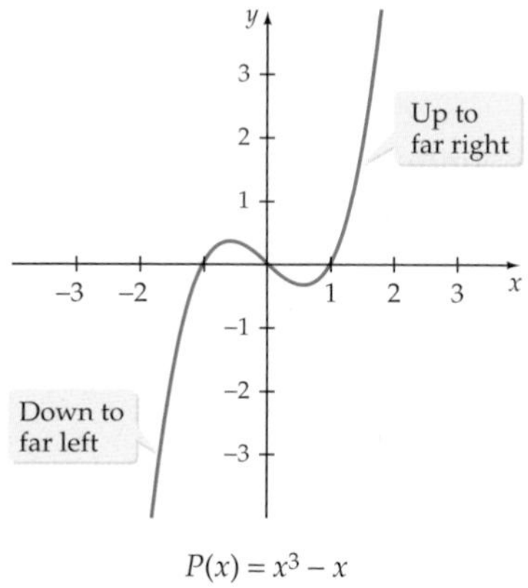

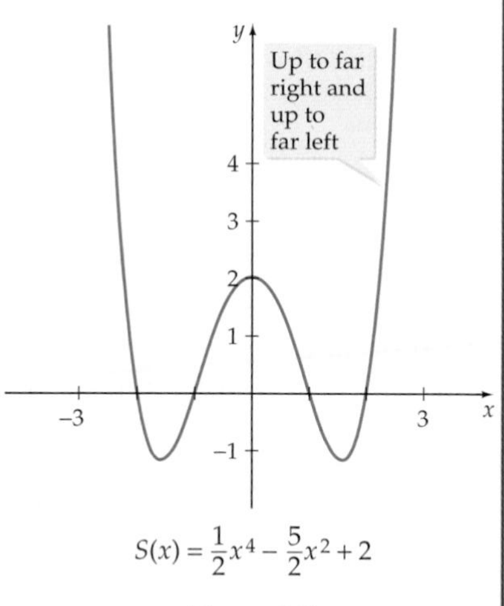

Figure 4.7

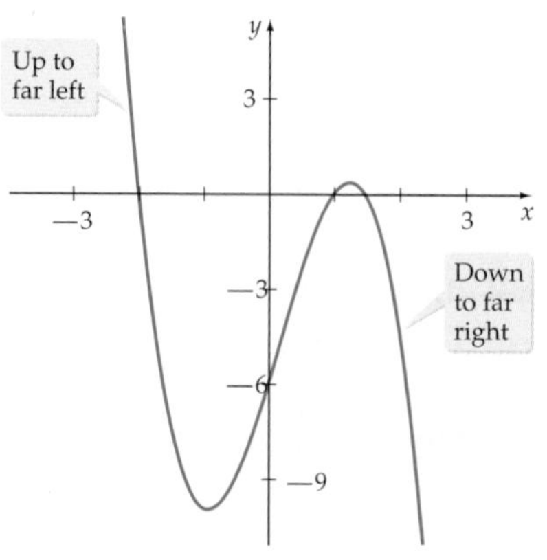

Figure 4.8

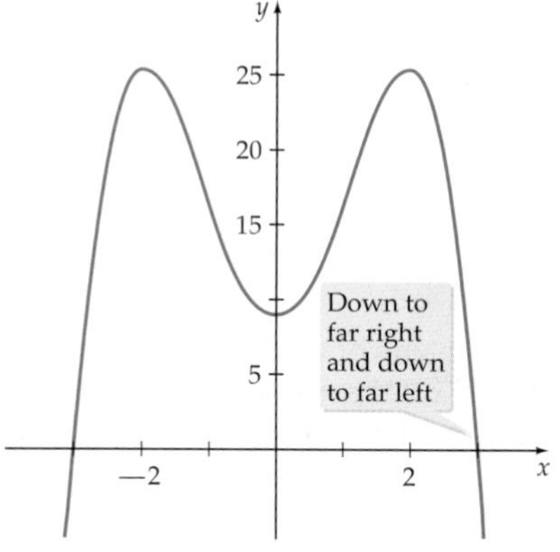

Figure 4.9

Up to far right
Down to far left
$P(x) = x^3 - x$

Figure 4.6

Continued ➤

CHECK YOUR PROGRESS 1 Examine the leading term to determine the far-left and far-right behavior of the graph of $P(x) = -2x^3 - 6x^2 + 5x - 1$.

Solution *See page S19.* Up to the far left, down to the far right

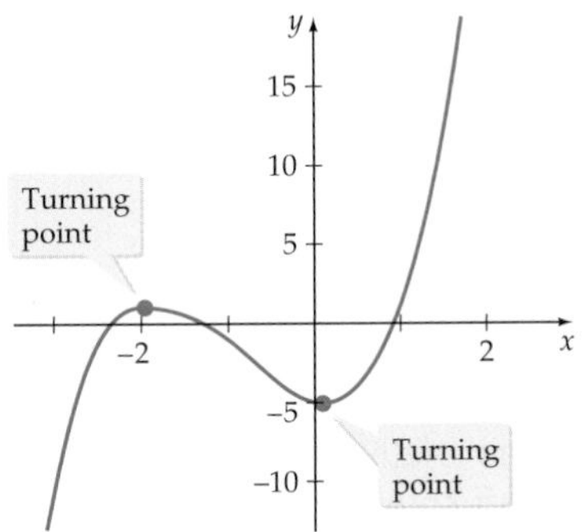

$P(x) = 2x^3 + 5x^2 - x - 5$

Figure 4.10

Maximum and Minimum Values

Figure 4.10 illustrates the graph of a polynomial function of degree 3 with two **turning points,** points at which the function changes from an increasing function to a decreasing function, or vice versa. In general, the graph of a polynomial function of degree n has at most $n - 1$ turning points.

Turning points can be related to the concepts of maximum and minimum values of a function. These concepts were introduced in the discussion of graphs of second-degree equations in two variables earlier in the text. Recall that the minimum value of a function f is the smallest range value of f. It is often called the **absolute minimum.** The maximum value of a function f is the largest range value of f. The maximum value of a function is also called the **absolute maximum.** For the function whose graph is shown in Figure 4.11, the y value of point E is the absolute minimum. There are no y values less than y_5.

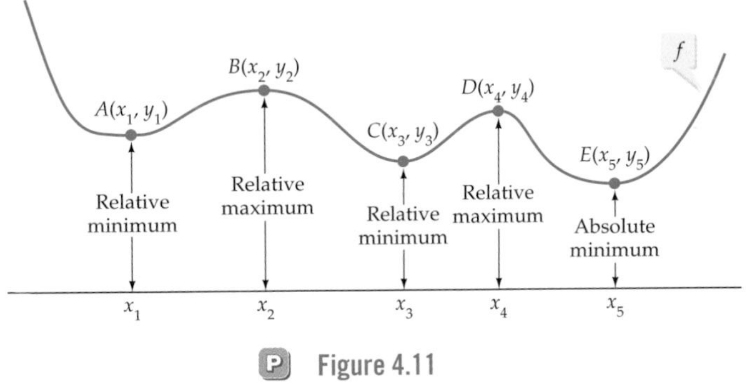

Figure 4.11

INTEGRATING TECHNOLOGY

A web applet is available to explore the relationships between the graph of a polynomial and its coefficients. This applet, Fourth Degree Polynomial, can be found on our web site at

math.college.hmco.com

Now consider y_1, the y value of turning point A in Figure 4.11. It is not the smallest y value of every point on the graph of f; however, it is the smallest y value if we *localize* our field of view to a small open interval containing x_1. It is for this reason that we refer to y_1 as a **local minimum**, or **relative minimum**, of f. The y value of point C is also a relative minimum of f.

The function does not have an absolute maximum because it goes up both to its far left and to its far right. The y value of the point B is a relative maximum, as is the y value of point D. The formal definitions of **relative maximum** and **relative minimum** are presented below.

Relative Minimum and Relative Maximum

If there is an open interval I containing c on which

- $f(c) \le f(x)$ for all x in I, then $f(c)$ is a **relative minimum** of f.
- $f(c) \ge f(x)$ for all x in I, then $f(c)$ is a **relative maximum** of f.

QUESTION Is the absolute minimum y_5 shown in Figure 4.11 also a relative minimum of f?

INTEGRATING TECHNOLOGY

A graphing utility can estimate the minimum and maximum values of a function. To use a TI-83 calculator to estimate the relative maximum of

$$P(x) = 0.3x^3 - 2.8x^2 + 6.4x + 2$$

use the following steps.

1. Enter the function in the Y = menu. Choose your window settings.
2. Select 4:maximum from the CALC menu, which is located above the TRACE key. The graph of Y1 is displayed.
3. Press ◀ or ▶ to select an x-value that is to the left of the relative maximum point. Press ENTER. A left bound is displayed in the bottom left corner.
4. Press ▶ to select an x-value that is to the right of the relative maximum point. Press ENTER. A right bound is displayed in the bottom left corner.
5. The word **Guess?** is now displayed in the bottom left corner. Press ◀ to move to a point near the maximum point. Press ENTER.
6. The cursor appears on the relative maximum point and the coordinates of the relative maximum point are displayed. In this example, the y value 6.312608 is the relative maximum.

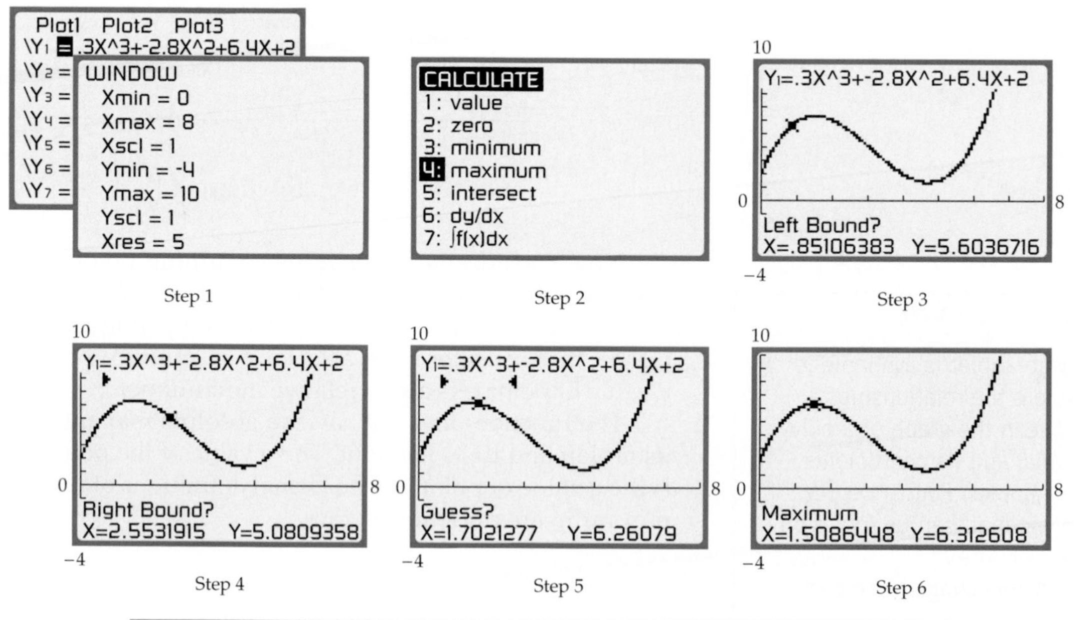

ANSWER Yes, the absolute minimum y_5 also satisfies the requirements of a relative minimum.

The following example illustrates the role a maximum may play in an application.

Alternative to Example 2
Exercise 48, page 333

EXAMPLE 2 Solve an Application

A rectangular piece of cardboard measures 12 inches by 16 inches. An open box is formed by cutting congruent squares that measure x inches by x inches from each of the corners of the cardboard and folding as shown below.

a. Express the volume V of the box as a function of x.

b. Determine (to the nearest 0.1 inch) the x value that maximizes the volume.

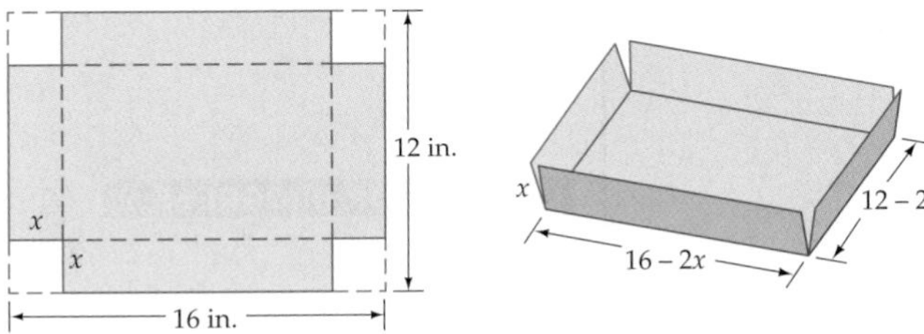

Solution

a. The height, width, and length of the open box are x, $12 - 2x$, and $16 - 2x$. The volume is given by

$$V(x) = x(12 - 2x)(16 - 2x)$$
$$= 4x^3 - 56x^2 + 192x$$

b. Use a graphing utility to graph $y = V(x)$. The graph is shown in Figure 4.12.

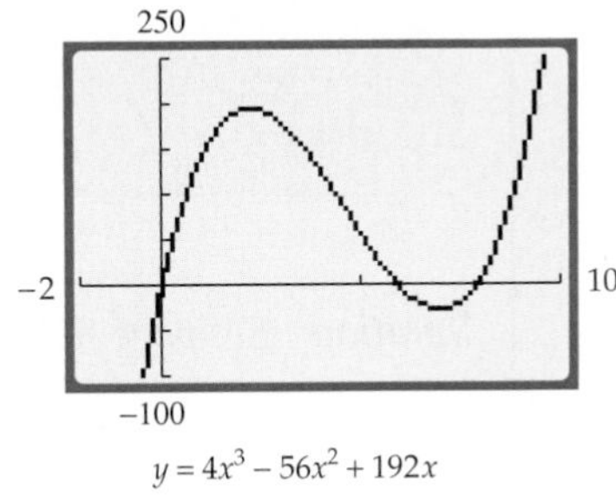

Figure 4.12

Note that we are interested only in the part of the graph for which $0 < x < 6$. This is because the length of each side of the box must be positive. In other words,

$$x > 0, \quad 12 - 2x > 0, \quad \text{and} \quad 16 - 2x > 0$$
$$x < 6 \qquad\qquad x < 8$$

Continued ➤

INTEGRATING TECHNOLOGY

A TI graphing calculator program is available that simulates the construction of a box by cutting out squares from each corner of a rectangular piece of cardboard. This program, CUTOUT, can be found on our web site at

math.college.hmco.com

The domain of V is the intersection of the solution sets of the three inequalities. Thus the domain is $\{x | 0 < x < 6\}$.

Now use a graphing utility to find that V attains its maximum of about 194.06736 when $x \approx 2.3$. See Figure 4.13.

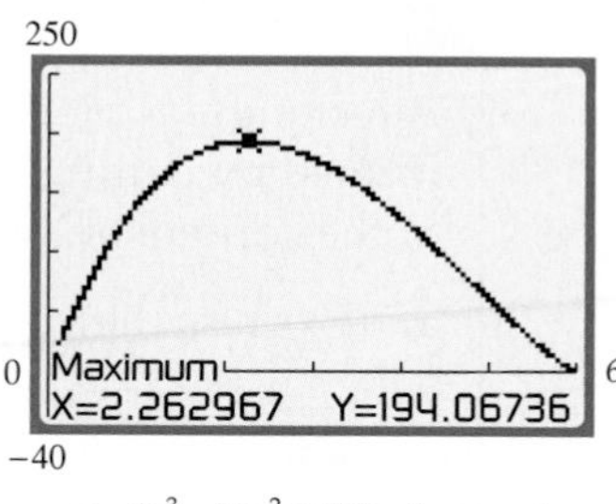

$y = 4x^3 - 56x^2 + 192x, 0 < x < 6$

Figure 4.13

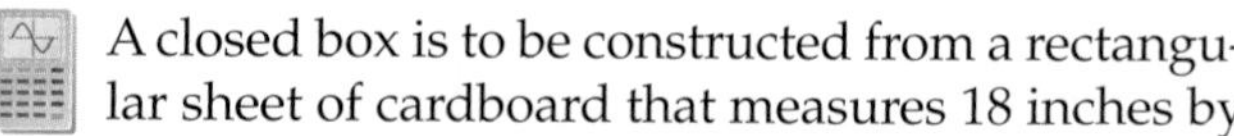

CHECK YOUR PROGRESS 2 A closed box is to be constructed from a rectangular sheet of cardboard that measures 18 inches by 42 inches. The box is made by cutting rectangles that measure x inches by $2x$ inches from two of the corners and by cutting two squares that measure x inches by x inches from the top and bottom of the rectangle, as shown in the following figure. What value of x (to the nearest 0.001 inch) will produce a box with the maximum volume? What is the maximum volume (to the nearest 0.1 cubic inch)?

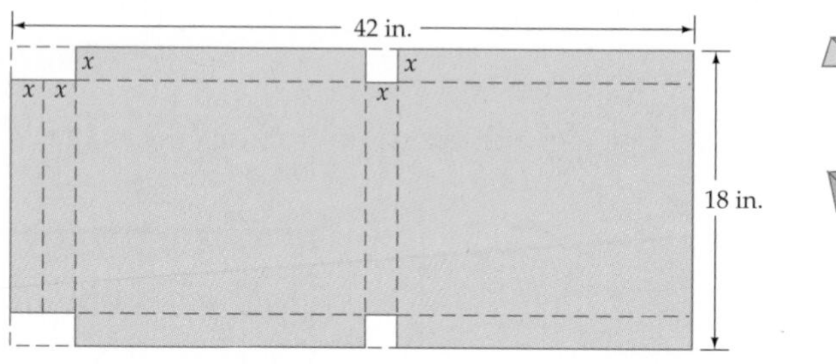

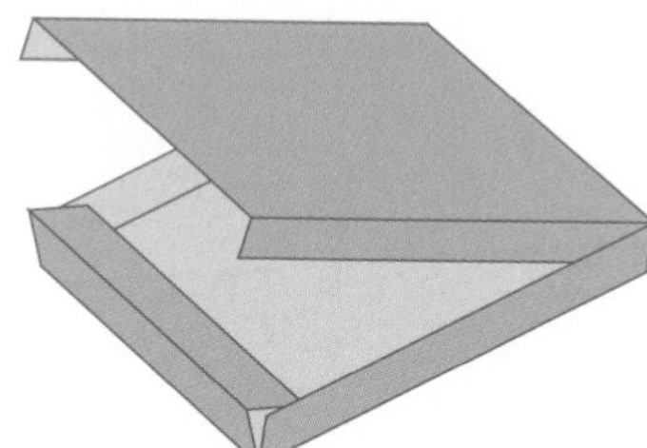

Solution *See page S20.* $x \approx 3.571$ in., $V \approx 606.6$ in^3

▪ Real Zeros of a Polynomial Function

Sometimes the real zeros of a polynomial function can be determined by using the factoring procedures developed in the previous chapters. We illustrate this concept in the next example.

Alternative to Example 3
Exercise 28, page 333

EXAMPLE 3 Factor to Find the Real Zeros of a Polynomial Function

Factor to find the three real zeros of $P(x) = x^3 + 3x^2 - 4x$.

Continued ➤

Solution

$P(x)$ can be factored as shown below.

$$P(x) = x^3 + 3x^2 - 4x$$
$$= x(x^2 + 3x - 4) \quad \text{• Factor out the common factor } x.$$
$$= x(x - 1)(x + 4) \quad \text{• Factor the trinomial } x^2 + 3x - 4.$$

The real zeros of $P(x)$ are $x = 0$, $x = 1$, and $x = -4$.

Visualize the Solution

The graph of $P(x)$ has x-intercepts at (0, 0), (1, 0), and (−4, 0).

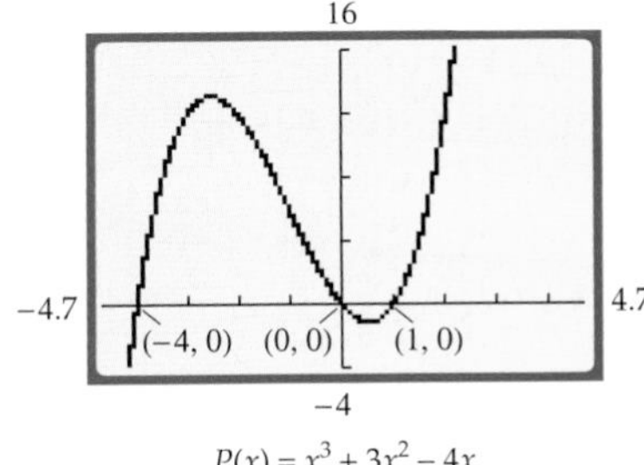

$P(x) = x^3 + 3x^2 - 4x$

CHECK YOUR PROGRESS 3 Factor to find the four real zeros of $P(x) = x^4 - 29x^2 + 100$.

Solution *See page S20.* −5, −2, 2, 5

The graph of every polynomial function P is a smooth continuous curve, and if the value of P changes sign on an interval, then $P(c)$ must equal zero for at least one real number c in the interval. This result is known as the *Zero Location Theorem.*

INSTRUCTOR NOTE

The Zero Location Theorem is a special case of the Intermediate Value Theorem (IVT), which states that if f is continuous on $[a, b]$ and $f(a) \neq f(b)$, then on $[a, b]$, f takes on every value between $f(a)$ and $f(b)$. You may wish to discuss the IVT with your students and explain that the Zero Location Theorem is a *corollary* of the IVT because it follows directly from the IVT.

The Zero Location Theorem

Let $P(x)$ be a polynomial function and let a and b be two distinct real numbers. If $P(a)$ and $P(b)$ have opposite signs, then there is at least one real number c between a and b such that $P(c) = 0$.

In other words, if the value of P is negative at $x = a$ and positive at $x = b$, then there is at least one real number c between a and b such that $P(c) = 0$. See Figure 4.14.

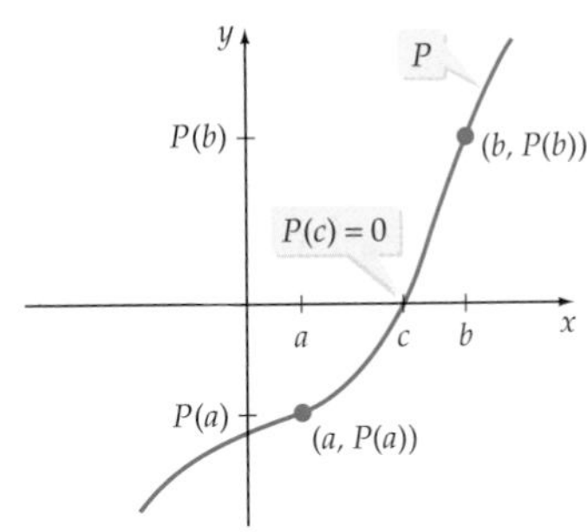

$P(a) < 0, P(b) > 0$

Figure 4.14

Alternative to Example 4
Exercise 36, page 333

EXAMPLE 4 Apply the Zero Location Theorem

Use the Zero Location Theorem to verify that $S(x) = x^3 - x - 2$ has a real zero between 1 and 2.

Solution

Use synthetic division to evaluate S for $x = 1$ and $x = 2$. If S changes sign between these two values, then S has a real zero between 1 and 2.

$$\begin{array}{r|rrrr} 1 & 1 & 0 & -1 & -2 \\ & & 1 & 1 & 0 \\ \hline & 1 & 1 & 0 & -2 \end{array}$$

■ $S(1)$ is negative.

$$\begin{array}{r|rrrr} 2 & 1 & 0 & -1 & -2 \\ & & 2 & 4 & 6 \\ \hline & 1 & 2 & 3 & 4 \end{array}$$

■ $S(2)$ is positive.

The graph of S is continuous because S is a polynomial function. Also, $S(1)$ is negative and $S(2)$ is positive. Thus the Zero Location Theorem indicates that there is a real zero between 1 and 2.

Visualize the Solution

The graph of S crosses the x-axis between $x = 1$ and $x = 2$. Thus S has a real zero between 1 and 2.

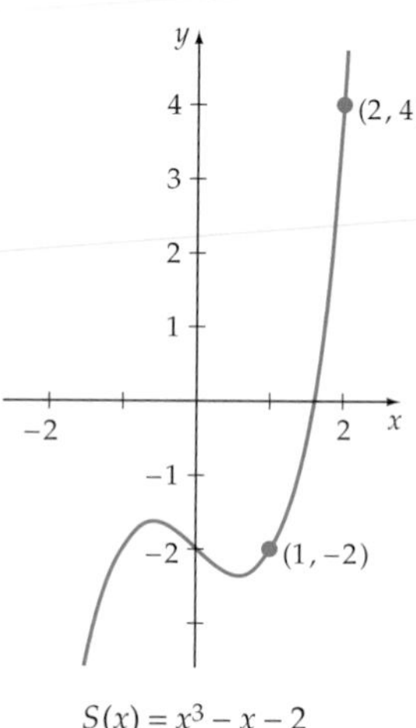

$S(x) = x^3 - x - 2$

CHECK YOUR PROGRESS 4 Use the Zero Location Theorem to verify that $P(x) = 4x^3 - x^2 - 6x + 1$ has a real zero between 0 and 1.

Solution *See page S20.* Because P is a polynomial function, the graph of P is continuous. Also, $P(0) = 1$ and $P(1) = -2$ have opposite signs. Thus by the Zero Location Theorem we know that P must have a real zero between 0 and 1.

The following theorem summarizes important relationships among the real zeros of a polynomial function, the x-intercepts of its graph, and its factors that can be written in the form $(x - c)$, where c is a real number.

INSTRUCTOR NOTE

To fully understand the concepts in this chapter, the student will need to know that the four statements marked by the bullets are equivalent.

P

Polynomial Functions, Real Zeros, Graphs, and Factors $(x - c)$

If P is a polynomial function and c is a real number, then all of the following statements are equivalent in the sense that if any one statement is true, then they are all true, and if any one statement is false, then they are all false.

- $(x - c)$ is a factor of P.
- $x = c$ is a real solution of $P(x) = 0$.
- $x = c$ is a real zero of P.
- $(c, 0)$ is an x-intercept of the graph of $y = P(x)$.

Sometimes it is possible to make use of the preceding theorem and a graph of a polynomial function to find factors of a function. For example, the graph of

$$S(x) = x^3 - 2x^2 - 5x + 6$$

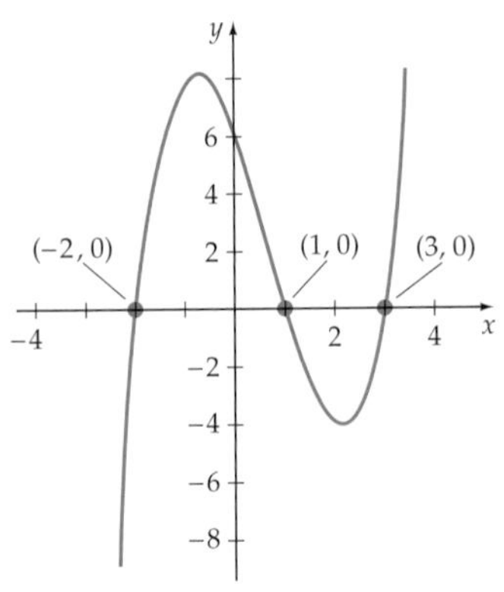

Figure 4.15

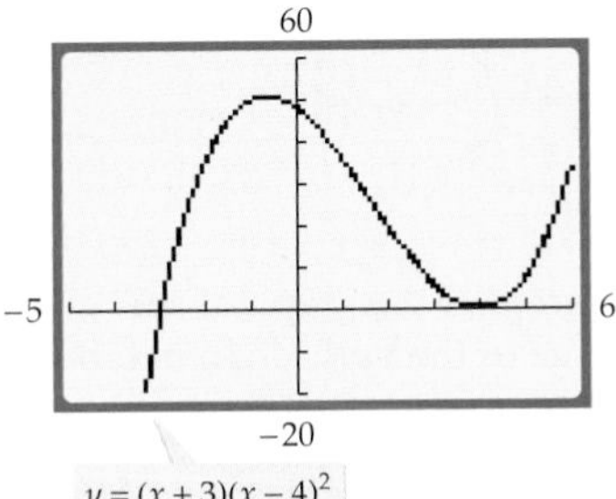

Figure 4.16

Alternative to Example 5
Exercise 42, page 333

is shown in Figure 4.15. The x-intercepts are $(-2, 0)$, $(1, 0)$, and $(3, 0)$. Hence, -2, 1, and 3 are zeros of S, and $[x - (-2)]$, $(x - 1)$, and $(x - 3)$ are all factors of S.

Even and Odd Powers of $(x - c)$ Theorem

Use a graphing utility to graph $P(x) = (x + 3)(x - 4)^2$. Compare your graph with Figure 4.16. Examine the graph near the x-intercepts $(-3, 0)$ and $(4, 0)$. Observe that the graph of P

- crosses the x-axis at $(-3, 0)$.
- intersects the x-axis but does not cross the x-axis at $(4, 0)$.

The following theorem can be used to determine at which x-intercepts the graph of a polynomial function will cross the x-axis and at which x-intercepts the graph will intersect but not cross the x-axis.

Even and Odd Powers of $(x - c)$ Theorem

If c is a real number and the polynomial function $P(x)$ has $(x - c)$ as a factor exactly k times, then the graph of P will

- intersect but not cross the x-axis at $(c, 0)$, provided k is an even positive integer.
- cross the x-axis at $(c, 0)$, provided k is an odd positive integer.

EXAMPLE 5 Apply the Even and Odd Powers of $(x - c)$ Theorem

Determine where the graph of $P(x) = (x + 3)(x - 2)^2(x - 4)^3$ crosses the x-axis and where the graph intersects but does not cross the x-axis.

Solution The exponents of the factors $(x + 3)$ and $(x - 4)$ are odd integers. Therefore, the graph of P will cross the x-axis at the x-intercepts $(-3, 0)$ and $(4, 0)$.

The exponent of the factor $(x - 2)$ is an even integer. Therefore, the graph of P will intersect but not cross the axis at $(2, 0)$.

Use a graphing utility to check these results.

CHECK YOUR PROGRESS 5 Determine where the graph of $P(x) = (x + 2)(x - 6)^2$ crosses the x-axis and where the graph intersects but does not cross the x-axis.

Solution *See page S20.* The graph crosses the x-axis at $(-2, 0)$. The graph intersects but does not cross the x-axis at $(6, 0)$.

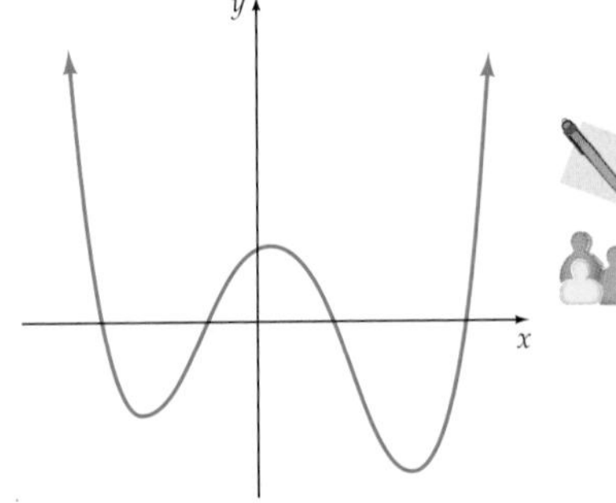

Figure 4.17

Topics for Discussion

1. Discuss the meaning of the phrase *polynomial function*. Give examples of polynomial functions and of functions that are not polynomial functions.
2. Is it possible for the graph of the polynomial function shown in Figure 4.17 to be the graph of a polynomial function that has an odd degree? If so, explain how. If not, explain why not.
3. Explain the difference between a relative minimum and an absolute minimum.

4. Discuss how the Zero Location Theorem can be used to find a real zero of a polynomial function.

5. Let $P(x)$ be a polynomial function with real coefficients. Explain the relationships among a real zero of the polynomial function, the x-coordinate of the x-intercept of the graph of the polynomial function, and the solution of the equation $P(x) = 0$.

EXERCISES 4.2

—Suggested Assignment: Exercises 1–55, odd; and 57–62, all.
—Answer graphs to Exercises 15–26 are on page AA10.

In Exercises 1 to 8, examine the leading term and determine the far-left and far-right behavior of the graph of the polynomial function.

1. $P(x) = 3x^4 - 2x^2 - 7x + 1$
   Up to the far left, up to the far right
2. $P(x) = 2x^3 + 5x^2 - 2x + 1$
   Down to the far left, up to the far right
3. $P(x) = 5x^5 - 4x^3 - 17x^2 + 2$
   Down to the far left, up to the far right
4. $P(x) = -6x^4 - 3x^3 + 5x^2 - 2x + 5$
   Down to the far left, down to the far right
5. $P(x) = 2 - 3x - 4x^2$ Down to the far left, down to the far right
6. $P(x) = -16 + x^4$ Up to the far left, up to the far right
7. $P(x) = \frac{1}{2}(x^3 + 5x^2 - 2)$ Down to the far left, up to the far right
8. $P(x) = -\frac{1}{4}(x^4 + 3x^2 - 2x + 6)$
   Down to the far left, down to the far right
9. The following graph is the graph of a third-degree (cubic) polynomial function. What do the far-left and far-right behavior of the graph say about the leading coefficient a?

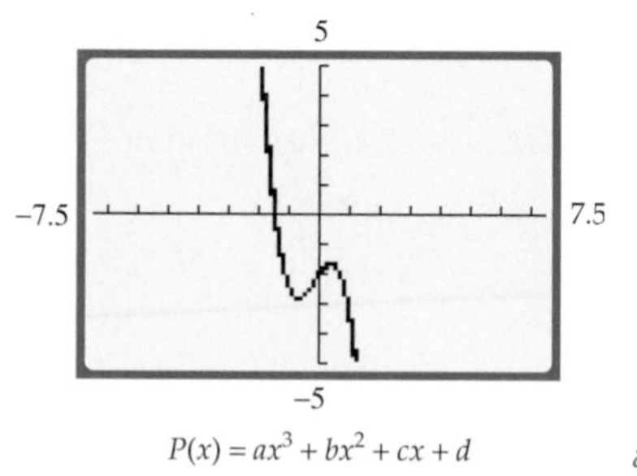

$P(x) = ax^3 + bx^2 + cx + d$ $a < 0$

10. The following graph is the graph of a fourth-degree (quartic) polynomial function. What do the far-left and far-right behavior of the graph say about the leading coefficient a?

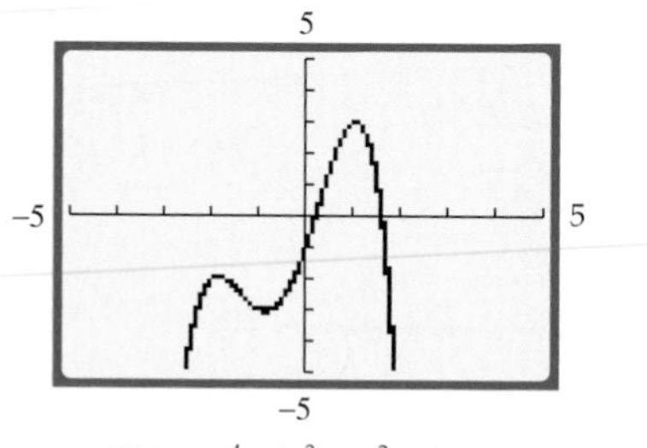

$P(x) = ax^4 + bx^3 + cx^2 + dx + e$ $a < 0$

In Exercises 11 to 14, state the vertex of the graph of the function and use your knowledge of the vertex of a parabola to find the maximum or minimum of each function.

11. $P(x) = x^2 + 4x - 1$ Vertex is $(-2, -5)$, minimum is -5.
12. $P(x) = x^2 + 6x + 1$ Vertex is $(-3, -8)$, minimum is -8.
13. $P(x) = -x^2 - 8x + 1$ Vertex is $(-4, 17)$, maximum is 17.
14. $P(x) = -2x^2 + 8x - 1$ Vertex is $(2, 7)$, maximum is 7.

In Exercises 15 to 20, sketch the graph of the polynomial function.

15. $P(x) = x^3 - x^2 - 2x$
16. $P(x) = x^3 + 2x^2 - 3x$
17. $P(x) = -x^3 - 2x^2 + 5x + 6$
18. $P(x) = -x^3 - 3x^2 + x + 3$
19. $P(x) = x^4 - 4x^3 + 2x^2 + 4x - 3$
20. $P(x) = x^4 - 6x^3 + 8x^2$

In Exercises 21 to 26, use a graphing utility to graph each polynomial. Use the maximum and minimum features of the graphing utility to estimate, to the nearest tenth, the coordinates of the points where $P(x)$ has a relative maximum or a relative minimum. For each point, indicate whether the y value is a relative maximum or a relative minimum. The number in parentheses to the right of the polynomial is the total number of relative maxima and minima.

21. $P(x) = x^3 + x^2 - 9x - 9$ (2)
    Relative maximum $y \approx 5.0$ at $x \approx -2.1$, relative minimum $y \approx -16.9$ at $x \approx 1.4$

22. $P(x) = x^3 + 4x^2 - 4x - 16$ (2)
Relative maximum $y \approx 5.0$ at $x \approx -3.1$, relative minimum $y \approx -16.9$ at $x \approx 0.4$

23. $P(x) = x^3 - 3x^2 - 24x + 3$ (2)
Relative maximum $y \approx 31.0$ at $x \approx -2.0$, relative minimum $y \approx -77.0$ at $x \approx 4.0$

24. $P(x) = -2x^3 - 3x^2 + 12x + 1$ (2)
Relative maximum $y \approx 8.0$ at $x \approx 1.0$, relative minimum $y \approx -19.0$ at $x \approx -2.0$

25. $P(x) = x^4 - 4x^3 - 2x^2 + 12x - 5$ (3) Relative maximum $y \approx 2.0$ at $x \approx 1.0$, relative minima $y \approx -14.0$ at $x \approx -1.0$ and $y \approx -14.0$ at $x \approx 3.0$

26. $P(x) = x^4 - 10x^2 + 9$ (3) Relative maximum $y \approx 9.0$ at $x \approx 0.0$, relative minima $y \approx -16.0$ at $x \approx -2.2$ and $y \approx -16.0$ at $x \approx 2.2$

In Exercises 27 to 32, find the real zeros of each polynomial function by factoring. The number in parentheses to the right of each polynomial indicates the number of real zeros of the given polynomial function.

27. $P(x) = x^3 - 2x^2 - 15x$ (3) $-3, 0, 5$

28. $P(x) = x^3 - 6x^2 + 8x$ (3) $0, 2, 4$

29. $P(x) = x^4 - 13x^2 + 36$ (4) $-3, -2, 2, 3$

30. $P(x) = 4x^4 - 37x^2 + 9$ (4) $-3, -\frac{1}{2}, \frac{1}{2}, 3$

31. $P(x) = x^5 - 5x^3 + 4x$ (5) $-2, -1, 0, 1, 2$

32. $P(x) = x^5 - 25x^3 + 144x$ (5) $-4, -3, 0, 3, 4$

In Exercises 33 to 38, use the Zero Location Theorem to verify that *P* has a zero between *a* and *b*.

33. $P(x) = 2x^3 + 3x^2 - 23x - 42$; $a = 3, b = 4$

34. $P(x) = 5x^3 - 4x^2 - 11x + 2$; $a = 1, b = 2$

35. $P(x) = 3x^3 + 7x^2 + 3x + 7$; $a = -3, b = -2$

36. $P(x) = 2x^3 - 21x^2 - 2x + 21$; $a = 10, b = 11$

37. $P(x) = 4x^4 + 7x^3 - 11x^2 + 7x - 15$; $a = 1, b = 1\frac{1}{2}$

38. $P(x) = 5x^3 - 16x^2 - 20x + 64$; $a = 3, b = 3\frac{1}{2}$

In Exercises 39 to 46, determine the *x*-intercepts of the graph of *P*. For each *x*-intercept, use the Even and Odd Powers of $(x - c)$ Theorem to determine whether the graph of *P* crosses the *x*-axis or intersects but does not cross the *x*-axis.

39. $P(x) = (x - 1)(x + 1)(x - 3)$
Crosses the x-axis at $(-1, 0)$, $(1, 0)$, and $(3, 0)$

40. $P(x) = (x - 2)(x + 3)(x + 1)$
Crosses the x-axis at $(-3, 0)$, $(-1, 0)$, and $(2, 0)$

41. $P(x) = -(x - 3)^2(x - 7)^5$
Crosses the x-axis at $(7, 0)$; intersects but does not cross at $(3, 0)$

42. $P(x) = (x + 2)^3(x - 6)^{10}$
Crosses the x-axis at $(-2, 0)$; intersects but does not cross at $(6, 0)$

43. $P(x) = (2x - 3)^4(x - 1)^{15}$
Crosses the x-axis at $(1, 0)$; intersects but does not cross at $\left(\frac{3}{2}, 0\right)$

44. $P(x) = (5x + 10)^6(x - 2.7)^5$
Crosses the x-axis at $(2.7, 0)$; intersects but does not cross at $(-2, 0)$

45. $P(x) = x^3 - 6x^2 + 9x$
Crosses the x-axis at $(0, 0)$; intersects but does not cross at $(3, 0)$

46. $P(x) = x^4 + 3x^3 + 4x^2$
Intersects but does not cross the x-axis at $(0, 0)$

Business and Economics

47. ***Construction of a Box*** A company constructs boxes from rectangular pieces of cardboard that measure 10 inches by 15 inches. An open box is formed by cutting squares that measure x inches by x inches from each corner of the cardboard and folding, as shown in the following figure.

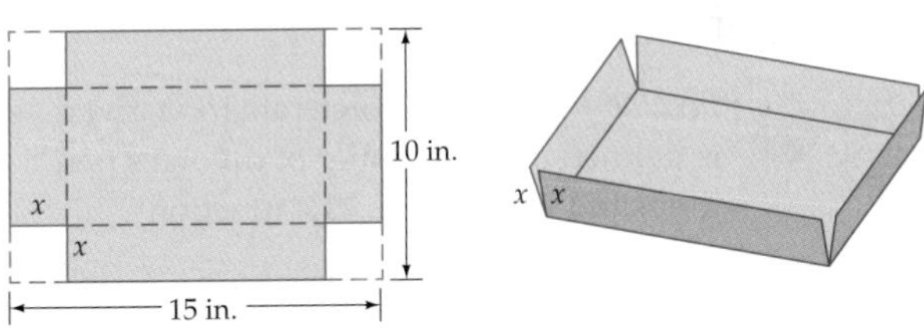

a. Express the volume V of the box as a function of x.
$V(x) = x(15 - 2x)(10 - 2x) = 4x^3 - 50x^2 + 150x$

b. Determine (to the nearest hundredth of an inch) the x value that maximizes the volume of the box. 1.96 in.

48. ***Profit*** A software company produces a computer game. The company has determined that its profit P, in dollars, from the manufacture and sale of x games is given by

$$P(x) = -0.000001x^3 + 96x - 98{,}000$$

where $0 < x \leq 9000$.

a. What is the maximum profit, to the nearest thousand dollars, the company can expect from the sale of its games? \$264,000

b. How many games, to the nearest unit, does the company need to produce and sell to obtain the maximum profit? 5657 games

49.  ***Advertising Expenses*** A company manufactures digital cameras. The company estimates that the profit from camera sales is

$$P(x) = -0.02x^3 + 0.01x^2 + 1.2x - 1.1$$

where P is the profit in millions of dollars and x is the amount, in hundred-thousands of dollars, spent on advertising.

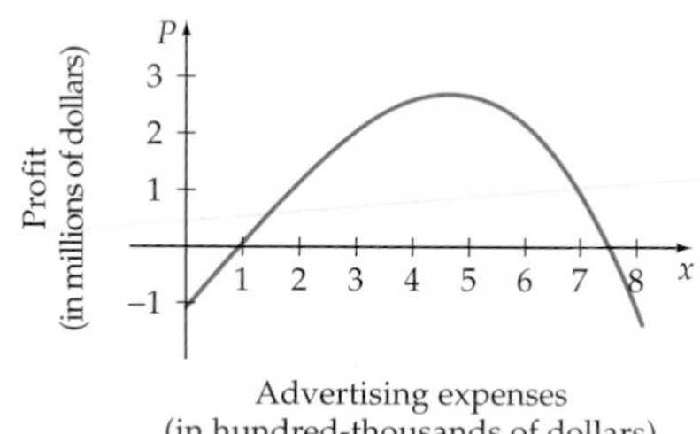

Determine the amount, rounded to the nearest thousand dollars, the company needs to spend on advertising if it is to generate the maximum profit. $464,000

Social Sciences

50.  ***Divorce Rate*** The divorce rate for a given year is defined as the number of divorces per thousand population. The function

$$D(t) = 0.00001807t^4 - 0.001406t^3 + 0.02884t^2 - 0.003466t + 2.1148$$

approximates the U.S. divorce rate for the years 1960 ($t = 0$) to 1999 ($t = 39$). Use $D(t)$ and a graphing utility to estimate

a. the year during which the U.S. divorce rate reached its absolute maximum for the period from 1960 to 1999. 1981

b. the absolute minimum divorce rate, rounded to the nearest 0.1, during the period from 1960 to 1999. 2.1 divorces per thousand population

51.  ***Marriage Rate*** The marriage rate for a given year is defined as the number of marriages per thousand population. The function

$$M(t) = -0.00000115t^4 + 0.000252t^3 - 0.01827t^2 + 0.4438t + 9.1829$$

approximates the U.S. marriage rate for the years 1900 ($t = 0$) to 1999 ($t = 99$).

Use $M(t)$ and a graphing utility to estimate

a. during what year the U.S. marriage rate reached its maximum for the period from 1900 to 1999. 1918

b. the relative minimum marriage rate, rounded to the nearest 0.1, during the period from 1950 to 1970. 9.5 marriages per thousand population

Life and Health Sciences

52. ***Gazelle Population*** A herd of 204 African gazelles is introduced into a wild animal park. The population of the gazelles, $P(t)$, after t years is given by $P(t) = -0.7t^3 + 18.7t^2 - 69.5t + 204$, where $0 < t \leq 18$.

a. Use a graph of P to determine the absolute minimum gazelle population (rounded to the nearest single gazelle) that is attained during this time period. 134 gazelles

b. Use a graph of P to determine the absolute maximum gazelle population (rounded to the nearest single gazelle) that is attained during this time period. 1013 gazelles

53. ***Medication Level*** Pseudoephedrine hydrochloride is an allergy medication. The function

$$L(t) = 0.03t^4 + 0.4t^3 - 7.3t^2 + 23.1t$$

where $0 \leq t \leq 5$, models the level of pseudoephedrine hydrochloride, in milligrams, in the bloodstream of a patient t hours after 30 milligrams of the medication have been taken.

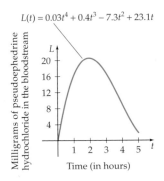

$L(t) = 0.03t^4 + 0.4t^3 - 7.3t^2 + 23.1t$

Milligrams of pseudoephedrine hydrochloride in the bloodstream

Time (in hours)

a. Use a graphing utility and the function $L(t)$ to determine the maximum level of pseudoephedrine hydrochloride in the patient's bloodstream. Round your result to the nearest 0.01 milligram. 20.69 mg

b. At what time t, to the nearest minute, is this maximum level of pseudoephedrine hydrochloride reached? 118 min

54. ***Squirrel Population*** The population of squirrels in a wilderness area is given by

$$P(t) = 0.6t^4 - 13.7t^3 + 104.5t^2 - 243.8t + 360$$

where $0 \le t \le 12$ years.

a. What is the absolute minimum number of squirrels (rounded to the nearest single squirrel) attained on the interval $0 \le t \le 12$? 185 squirrels

b. The absolute maximum of P is attained at the endpoint, where $t = 12$. What is this absolute maximum (rounded to the nearest single squirrel)? 1250 squirrels

Sports and Recreation

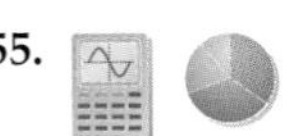

55. ***Beam Deflection*** The deflection D, in feet, of an 8-foot beam that is center loaded is given by

$$D(x) = (-0.0025)(4x^3 - 3 \cdot 8x^2), 0 < x \le 4$$

where x is the distance, in feet, from one end of the beam.

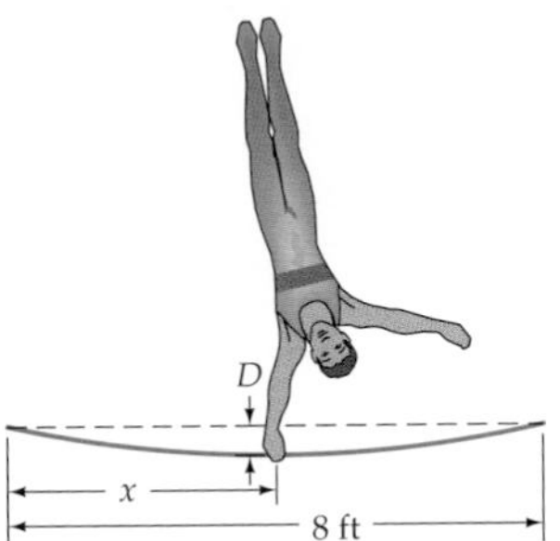

a. Determine the deflection of the beam when $x = 3$ feet. Round to the nearest hundredth of an inch. 3.24 in.

b. At what point does the beam achieve its maximum deflection? What is the maximum deflection? Round to the nearest hundredth of an inch. 4 ft from an end; 3.84 in.

c. What is the deflection at $x = 5$ feet? 3.24 in.

Physical Sciences and Engineering

56. ***Engineering*** A cylindrical log with a diameter of 22 inches is to be cut so that it will yield a beam that has a rectangular cross section of depth d and width w.

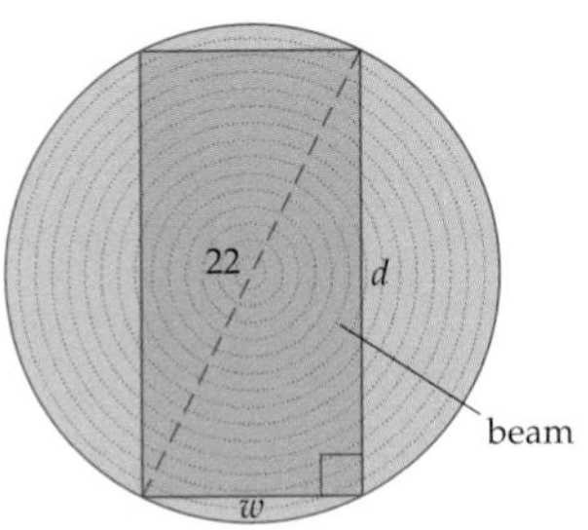

Cross section of a log

An engineer has determined that the stiffness S of the resulting beam is given by $S = 1.15wd^2$, where $0 < w < 22$ inches. Find the width and the depth that will maximize the stiffness of the beam. Round each result to the nearest hundredth of an inch. (*Hint:* Use the Pythagorean Theorem to solve for d^2 in terms of w^2.)
$w = 12.70$ in., $d = 17.96$ in.

Prepare for Section 4.3

57. Find the zeros of $P(x) = 6x^2 - 25x + 14$. [4.2] $\frac{2}{3}, \frac{7}{2}$

58. Use synthetic division to divide $2x^3 + 3x^2 + 4x - 7$ by $x + 2$. [4.1] $2x^2 - x + 6 - \frac{19}{x+2}$

59. Use synthetic division to divide $3x^4 - 21x^2 - 3x - 5$ by $x - 3$. [4.1] $3x^3 + 9x^2 + 6x + 15 + \frac{40}{x-3}$

60. List all natural numbers that are factors of 12. [P.1]
1, 2, 3, 4, 6, 12

61. List all integers that are factors of 27. [P.1]
$\pm 1, \pm 3, \pm 9, \pm 27$

62. Given $P(x) = 4x^3 - 3x^2 - 2x + 5$, find $P(-x)$. [3.1]
$P(-x) = -4x^3 - 3x^2 + 2x + 5$

Explorations 

1. **Real Zeros and the Degree of a Polynomial** This project examines the connection between the number of real zeros of a polynomial function and its degree.

   a. Graph the polynomial functions $P(x) = x^4 + 1$, $Q(x) = x^4 - x^3 - 11x^2 - x - 12$, and

   $$R(x) = x^4 - 2x^3 - 13x^2 + 14x + 24$$

   How many distinct real zeros does each of the polynomial functions $P(x)$, $Q(x)$, and $R(x)$ have? Graph some other polynomial functions of degree 4. On the basis of your graphs, make a conjecture about the number of distinct real zeros a polynomial function of degree 4 can have. $P(x)$ has no real zeros; $Q(x)$ has two real zeros; $R(x)$ has four real zeros. Conjectures will vary.

   b. Graph the polynomial functions $P(x) = x^5$, $Q(x) = x^5 - 3x^4 - 11x^3 + 27x^2 + 10x - 24$, and

   $$R(x) = x^5 - 2x^4 - 10x^3 + 10x^2 - 11x + 12$$

   How many distinct real zeros does each of the polynomial functions $P(x)$, $Q(x)$, and $R(x)$ have? Graph some other polynomial functions of degree 5. On the basis of your graphs, make a conjecture about the number of distinct real zeros a polynomial function of degree 5 can have. $P(x)$ has one real zero; $Q(x)$ has five real zeros; $R(x)$ has three real zeros. Conjectures will vary.

   c. Graph some second-degree and third-degree polynomial functions and note the number of real zeros of each function. On the basis of all your graphs and the graphs in **a.** and **b.**, make a conjecture about the number of distinct real zeros a polynomial function of degree n can have. Conjectures will vary.

2. A student thinks that $P(n) = n^3 - n$ is always a multiple of 6 for all natural numbers n. What do you think? Provide a mathematical argument to show that the student is correct or a counterexample to show that the student is wrong.
The student is correct. The polynomial function $P(n) = n^3 - n$ can be written in factored form as $P(n) = n(n - 1)(n + 1)$. In this form it is easy to see that $P(n)$ is the product of three consecutive natural numbers, one of which must be an even number and one of which must be a multiple of three. Thus $P(n)$ must be a multiple of 6 for any natural number n.

SECTION 4.3 Zeros of Polynomial Functions

- Multiple Zeros of a Polynomial Function
- The Rational Zero Theorem
- Upper and Lower Bounds for Real Zeros
- Descartes' Rule of Signs
- Zeros of a Polynomial Function
- Applications of Polynomial Functions

▪ Multiple Zeros of a Polynomial Function

Recall that if $P(x)$ is a polynomial function, then the values of x for which $P(x)$ is equal to 0 are called the *zeros* of $P(x)$ or the **roots** of the equation $P(x) = 0$. A zero of a polynomial function may be a **multiple zero.** For example, $P(x) = x^2 + 6x + 9$ can be expressed in factored form as $(x + 3)(x + 3)$. Setting each factor equal to zero yields $x = -3$ in both cases. Thus $P(x) = x^2 + 6x + 9$ has a zero of -3 that occurs twice. The following definition will be most useful when we are discussing multiple zeros.

Definition of Multiple Zeros of a Polynomial Function

If a polynomial function $P(x)$ has $(x - r)$ as a factor exactly k times, then r is a **zero of multiplicity k** of the polynomial function $P(x)$.

The graph of the polynomial function

$$P(x) = (x - 5)^2(x + 2)^3(x + 4)$$

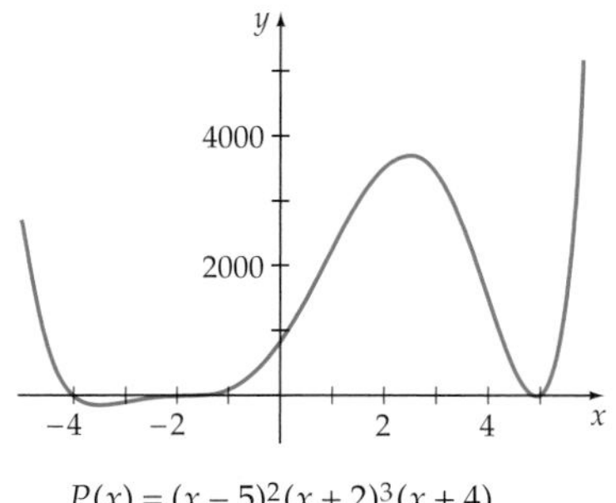

$P(x) = (x - 5)^2(x + 2)^3(x + 4)$

Figure 4.18

is shown in Figure 4.18. This polynomial function has

- 5 as a zero of multiplicity 2.
- −2 as a zero of multiplicity 3.
- −4 as a zero of multiplicity 1.

A zero of multiplicity 1 is generally referred to as a **simple zero.**

When searching for the zeros of a polynomial function, it is important that we know how many zeros to expect. This question is answered completely in Section 4.4. For the work in this section, the following result is valuable.

Number of Zeros of a Polynomial Function

A polynomial function P of degree n has at most n zeros, where each zero of multiplicity k is counted k times.

The Rational Zero Theorem

The rational zeros of polynomial functions with integer coefficients can be found with the aid of the following theorem.

TAKE NOTE

The Rational Zero Theorem is one of the most important theorems of this chapter. It enables us to narrow the search for rational zeros to a finite list.

The Rational Zero Theorem

If $P(x) = a_nx^n + a_{n-1}x^{n-1} + \cdots + a_1x + a_0$ has *integer* coefficients ($a_n \neq 0$) and $\frac{p}{q}$ is a rational zero (in lowest terms) of P, then

- p is a factor of the constant term a_0 and
- q is a factor of the leading coefficient a_n.

The Rational Zero Theorem often is used to make a list of all possible rational zeros of a polynomial function. The list consists of all rational numbers of the form $\frac{p}{q}$, where p is an integer factor of the constant term a_0 and q is an integer factor of the leading coefficient a_n.

Alternative to Example 1
Exercise 12, page 346

EXAMPLE 1 Apply the Rational Zero Theorem

Use the Rational Zero Theorem to list all possible rational zeros of

$$P(x) = 4x^4 + x^3 - 40x^2 + 38x + 12$$

Solution List all integers p that are factors of 12 and all integers q that are factors of 4.

$$p: \quad \pm1, \pm2, \pm3, \pm4, \pm6, \pm12$$
$$q: \quad \pm1, \pm2, \pm4$$

Continued ➤

TAKE NOTE

The Rational Zero Theorem gives the possible rational zeros of a polynomial function. That is, if P has a rational zero, then it must be one indicated by the theorem. However, P may not have any rational zeros. In the case of the polynomial function in Example 1, the only rational zeros are $-\frac{1}{4}$ and 2. The remaining rational numbers in the list are not zeros of P.

Form all possible rational numbers using $\pm1, \pm2, \pm3, \pm4, \pm6$, and ±12 for the numerator and $\pm1, \pm2$, and ±4 for the denominator. By the Rational Zero Theorem, the possible rational zeros are

$$\pm1, \pm\frac{1}{2}, \pm\frac{1}{4}, \pm2, \pm3, \pm\frac{3}{2}, \pm\frac{3}{4}, \pm4, \pm6, \pm12$$

It is not necessary to list a factor that is already listed in reduced form. For example, $\pm\frac{6}{4}$ is not listed because it is equal to $\pm\frac{3}{2}$.

CHECK YOUR PROGRESS 1 Use the Rational Zero Theorem to list all possible rational zeros of $P(x) = 3x^3 + 11x^2 - 6x - 8$.

Solution *See page S20.* $\pm1, \pm2, \pm4, \pm8, \pm\frac{1}{3}, \pm\frac{2}{3}, \pm\frac{4}{3}, \pm\frac{8}{3}$

QUESTION If $P(x) = a_nx^n + a_{n-1}x^{n-1} + \cdots + a_1x + a_0$ has integer coefficients and a leading coefficient of $a_n = 1$, must all of the rational zeros of P be integers?

Upper and Lower Bounds for Real Zeros

A real number b is called an **upper bound** of the zeros of the polynomial function P if no zero is greater than b. A real number b is called a **lower bound** of the zeros of P if no zero is less than b. The following theorem is often used to find positive upper bounds and negative lower bounds for the real zeros of a polynomial function.

Upper- and Lower-Bound Theorem

Let $P(x)$ be a polynomial function with real coefficients and a positive leading coefficient. Use synthetic division to divide $P(x)$ by $x - b$, where b is a nonzero real number.

Upper bound If $b > 0$ and all the numbers in the bottom row of the synthetic division of P by $x - b$ are either positive or zero, then b is an upper bound for the real zeros of P.

Lower bound If $b < 0$ and all the numbers in the bottom row of the synthetic division of P by $x - b$ alternate in sign (the number zero can be considered positive or negative), then b is a lower bound for the real zeros of P.

Upper and lower bounds are not unique. For example, if b is an upper bound for the real zeros of P, then any number greater than b is also an upper bound. Likewise, if a is a lower bound for the real zeros of P, then any number less than a is also a lower bound.

ANSWER Yes. By the Rational Zero Theorem, the rational zeros of P are of the form $\frac{p}{q}$, where p is an integer factor of a_0 and q is an integer factor of a_n. Thus $q = \pm1$ and $\frac{p}{q} = \frac{p}{\pm1} = \pm p$.

Alternative to Example 2
Exercise 18, page 346

TAKE NOTE

When you check for bounds, you do not need to limit your choices to the possible zeros given by the Rational Zero Theorem. For instance, in Example 2 the integer −4 is a lower bound; however, −4 is not one of the possible zeros of P as given by the Rational Zero Theorem.

EXAMPLE 2 Find Upper and Lower Bounds

According to the Upper- and Lower-Bound Theorem, what is the smallest positive integer that is an upper bound and the largest negative integer that is a lower bound of the real zeros of $P(x) = 2x^3 + 7x^2 - 4x - 14$?

Solution To find the smallest positive-integer upper bound, use synthetic division with 1, 2, . . . , as test values.

1	2	7	−4	−14
		2	9	5
	2	9	5	−9

2	2	7	−4	−14
		4	22	36
	2	11	18	22

▪ All positive signs

Thus 2 is the smallest positive-integer upper bound.

Now find the largest negative-integer lower bound.

−1	2	7	−4	−14
		−2	−5	9
	2	5	−9	−5

−2	2	7	−4	−14
		−4	−6	20
	2	3	−10	6

−3	2	7	−4	−14
		−6	−3	21
	2	1	−7	7

−4	2	7	−4	−14
		−8	4	0
	2	−1	0	−14

▪ Alternating signs

Thus −4 is the largest negative-integer lower bound.

CHECK YOUR PROGRESS 2 According to the Upper- and Lower-Bound Theorem, what is the smallest positive integer that is an upper bound and the largest negative integer that is a lower bound of the real zeros of $P(x) = x^3 - 19x - 28$?

Solution *See page S20.* 5 is the smallest positive integer upper bound and −5 is the largest negative integer lower bound.

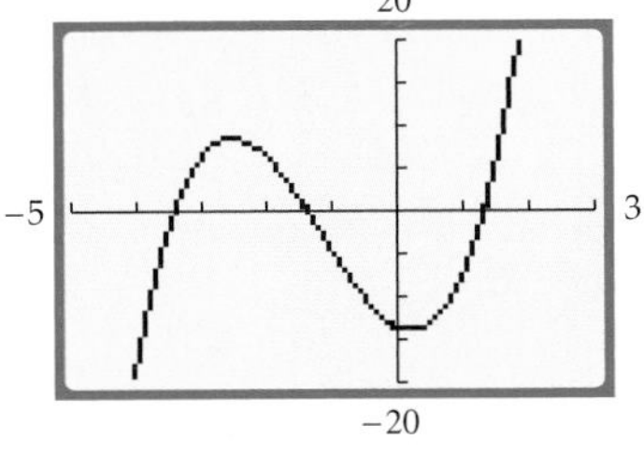

$P(x) = 2x^3 + 7x^2 - 4x - 14$

Figure 4.19

INTEGRATING TECHNOLOGY

You can use the Upper- and Lower-Bound Theorem to determine Xmin (the lower bound) and Xmax (the upper bound) for the viewing window of a graphing utility. This will ensure that all the real zeros, which are the x-coordinates of the x-intercepts of the polynomial function, will be shown. Note in Figure 4.19 that the zeros of $P(x) = 2x^3 + 7x^2 - 4x - 14$ are between −4 (a lower bound) and 2 (an upper bound).

▪ Descartes' Rule of Signs

Descartes' Rule of Signs is another theorem that is often used to obtain information about the zeros of a polynomial function. In Descartes' Rule of Signs, the number of **variations in sign** of the coefficients of $P(x)$ or $P(-x)$ refers to sign changes of the coefficients from positive to negative or from negative to positive that we find when we examine successive terms of the function. The terms are assumed to appear in order of descending powers of x. For example, the polynomial function

$$P(x) = \underbrace{+3x^4 - 5x^3}_{1} \underbrace{- 7x^2 + x}_{2} \underbrace{\; - 7}_{3}$$

has three variations in sign. The polynomial function

$$\begin{aligned} P(-x) &= +3(-x)^4 - 5(-x)^3 - 7(-x)^2 + (-x) - 7 \\ &= +\ 3x^4 \underbrace{+\ 5x^3 \ -\ 7x^2}_{1} - x - 7 \end{aligned}$$

has one variation in sign.

Terms that have a coefficient of 0 are not counted as variations in sign and may be ignored. For example,

$$P(x) = \underbrace{-x^5 + 4x^2}_{1} + 1$$

has one variation in sign.

Point of Interest

Descartes' Rule of Signs first appeared in his *La Géométrie* (1673). Although a proof of Descartes' Rule of Signs is beyond the scope of this course, we can see that a polynomial function with no variations in sign cannot have a positive zero. For instance, consider $P(x) = x^3 + x^2 + x + 1$. Each term of P is positive for any positive value of x. Thus P is never zero for $x > 0$.

Descartes' Rule of Signs

Let $P(x)$ be a polynomial function with real coefficients and with the terms arranged in order of decreasing powers of x.

1. The number of positive real zeros of $P(x)$ is equal to the number of variations in sign of $P(x)$, or to that number decreased by an even integer.
2. The number of negative real zeros of $P(x)$ is equal to the number of variations in sign of $P(-x)$, or to that number decreased by an even integer.

Alternative to Example 3
Exercise 34, page 347

EXAMPLE 3 Apply Descartes' Rule of Signs

Use Descartes' Rule of Signs to determine both the number of possible positive and the number of possible negative real zeros of each polynomial function.

a. $P(x) = x^4 - 5x^3 + 5x^2 + 5x - 6$ **b.** $P(x) = 2x^5 + 3x^3 + 5x^2 + 8x + 7$

Solution

a.

$$P(x) = \underbrace{+x^4 - 5x^3}_{1}\underbrace{\;+ 5x^2}_{2} + \underbrace{5x - 6}_{3}$$

There are three variations in sign. By Descartes' Rule of Signs, there are either three or one positive real zeros. Now examine the variations in sign of $P(-x)$.

$$P(-x) = x^4 + 5x^3 + \underbrace{5x^2 - 5x}_{1} - 6$$

There is one variation in sign of $P(-x)$. By Descartes' Rule of Signs, there is one negative real zero.

b. $P(x) = 2x^5 + 3x^3 + 5x^2 + 8x + 7$ has no variation in sign, so there are no positive real zeros.

$$P(-x) = -2x^5 \underbrace{- 3x^3 + 5x^2}_{1} \underbrace{\;}_{2}\underbrace{- 8x + 7}_{3}$$

INSTRUCTOR NOTE

A real application of some of the concepts in this chapter can be found in *A Genuine Application of Synthetic Division, Descartes' Rule of Signs, and All That Stuff* by Dwight D. Freund [*The College Mathematics Journal*, vol. 26, no. 2 (March 1995)].

Continued ➤

$P(-x)$ has three variations in sign, so there are either three or one negative real zeros.

CHECK YOUR PROGRESS 3 Use Descartes' Rule of Signs to determine the number of possible positive and negative real zeros of $P(x) = x^3 - 19x - 30$.

Solution *See page S20.* One positive and two or no negative real zeros

QUESTION If $P(x) = ax^2 + bx + c$ has two variations in sign, must $P(x)$ have two positive real zeros?

In applying Descartes' Rule of Signs, we count each zero of multiplicity k as k zeros. For instance,

$$P(x) = x^2 - 10x + 25$$

has two variations in sign. Thus, by Descartes' Rule of Signs, $P(x)$ must have either two or no positive real zeros. Factoring $P(x)$ produces $(x - 5)^2$, from which it can be observed that 5 is a positive zero of multiplicity 2.

Zeros of a Polynomial Function

Guidelines for Finding the Zeros of a Polynomial Function with Integer Coefficients

1. ***Gather general information.*** Determine the degree n of the polynomial function. The number of zeros of the polynomial function is at most n. Apply Descartes' Rule of Signs to find the possible number of positive zeros and also the possible number of negative zeros.
2. ***Check suspects.*** Apply the Rational Zero Theorem to list rational numbers that are possible zeros. Use synthetic division to test numbers in your list. If you find an upper or a lower bound, then eliminate from your list any number that is greater than the upper bound or less than the lower bound.
3. ***Work with the reduced polynomials.*** Each time a zero is found, you obtain a reduced polynomial.
   - If a reduced polynomial is of degree 2, find its zeros either by factoring or by applying the quadratic formula.
   - If the degree of a reduced polynomial is 3 or greater, repeat the above steps for this polynomial.

Example 4 illustrates the procedures discussed in the above guidelines.

ANSWER No. According to Descartes' Rule of Signs, $P(x)$ will either have two positive real zeros or no positive real zeros.

Alternative to Example 4
Exercise 40, page 347

INTEGRATING TECHNOLOGY

If you have a graphing utility, you can produce a graph similar to the one below. By looking at the x-intercepts of the graph, you can reject as possible zeros some of the values suggested by the Rational Zero Theorem. This will reduce the amount of work that is necessary to find the zeros of the polynomial function.

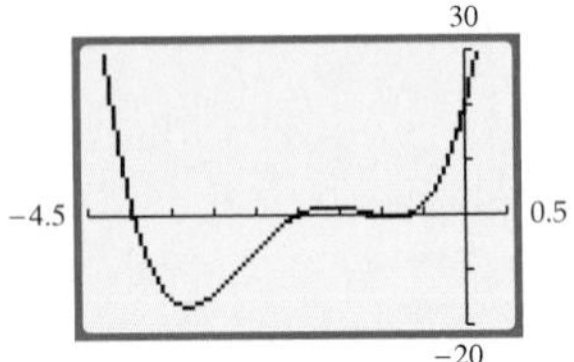

$P(x) = 3x^4 + 23x^3 + 56x^2 + 52x + 16$

EXAMPLE 4 Find the Zeros of a Polynomial Function

Find the zeros of $P(x) = 3x^4 + 23x^3 + 56x^2 + 52x + 16$.

Solution

1. *Gather general information.* The degree of P is 4. Thus the number of zeros of P is at most 4. By Descartes' Rule of Signs, there are no positive zeros, and there are either four, two, or no negative zeros.

2. *Check suspects.* By the Rational Zero Theorem, the possible negative rational zeros of P are

$$\frac{p}{q}: \quad -1, -2, -4, -8, -16, -\frac{1}{3}, -\frac{2}{3}, -\frac{4}{3}, -\frac{8}{3}, -\frac{16}{3}$$

Use synthetic division to test the possible rational zeros. The following work shows that -4 is a zero of P.

$$\begin{array}{r|rrrrr} -4 & 3 & 23 & 56 & 52 & 16 \\ & & -12 & -44 & -48 & -16 \\ \hline & 3 & 11 & 12 & 4 & 0 \end{array}$$

Coefficients of the first reduced polynomial

3. *Work with the reduced polynomials.* Because -4 is a zero, $(x + 4)$ and the first reduced polynomial $(3x^3 + 11x^2 + 12x + 4)$ are both factors of P. Thus

$$P(x) = (x + 4)(3x^3 + 11x^2 + 12x + 4)$$

All remaining zeros of P must be zeros of $3x^3 + 11x^2 + 12x + 4$. The Rational Zero Theorem indicates that the only possible negative rational zeros of $3x^3 + 11x^2 + 12x + 4$ are

$$\frac{p}{q}: \quad -1, -2, -4, -\frac{1}{3}, -\frac{2}{3}, -\frac{4}{3}$$

Synthetic division is again used to test possible zeros.

$$\begin{array}{r|rrrr} -2 & 3 & 11 & 12 & 4 \\ & & -6 & -10 & -4 \\ \hline & 3 & 5 & 2 & 0 \end{array}$$

Coefficients of the second reduced polynomial

Because -2 is a zero, $(x + 2)$ is also a factor of P. Thus

$$P(x) = (x + 4)(x + 2)(3x^2 + 5x + 2)$$

The remaining zeros of P must be zeros of $3x^2 + 5x + 2$.

$$3x^2 + 5x + 2 = 0$$
$$(3x + 2)(x + 1) = 0$$
$$x = -\frac{2}{3} \quad \text{and} \quad x = -1$$

Continued ➤

The zeros of $P(x) = 3x^4 + 23x^3 + 56x^2 + 52x + 16$ are $-4, -2, -\frac{2}{3}$, and -1.

CHECK YOUR PROGRESS 4 Find the zeros of $P(x) = x^3 - 19x - 30$.

Solution *See page S20.* $-3, -2, 5$

▪ Applications of Polynomial Functions

In the following example we make use of an upper bound to eliminate several of the possible zeros that are given by the Rational Zero Theorem.

Alternative to Example 5
Exercise 60, page 347

EXAMPLE 5 Solve an Application

Glasses can be stacked to form a triangular pyramid.

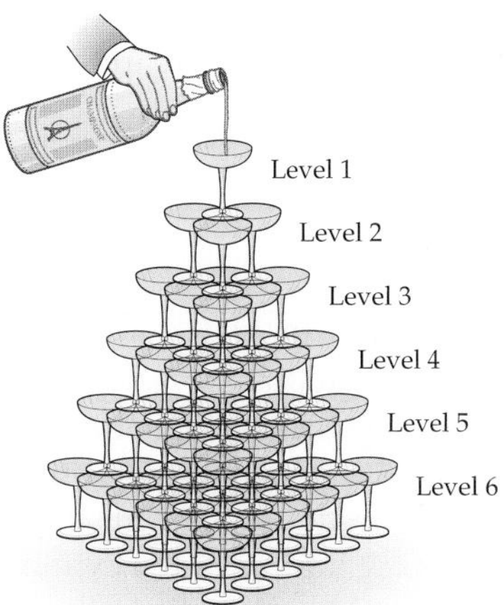

The total number of glasses in one of these pyramids is given by

$$T = \frac{1}{6}(k^3 + 3k^2 + 2k)$$

where k is the number of levels in the pyramid. If 220 glasses are used to form a triangular pyramid, how many levels are in the pyramid?

Solution We need to solve $220 = \frac{1}{6}(k^3 + 3k^2 + 2k)$ for k. Multiplying each side of the equation by 6 produces $1320 = k^3 + 3k^2 + 2k$, which can be written as $k^3 + 3k^2 + 2k - 1320 = 0$. The number 1320 has many natural number divisors, but we can eliminate many of these by showing that 12 is an upper bound.

$$\begin{array}{r|rrrr} 12 & 1 & 3 & 2 & -1320 \\ & & 12 & 180 & 2184 \\ \hline & 1 & 15 & 182 & 864 \end{array}$$

← Each number in the bottom row is positive. Thus 12 is an upper bound.

Continued ➤

TAKE NOTE

The reduced polynomial $k^2 + 13k + 132$ has zeros of $k = \frac{-13 \pm i\sqrt{359}}{2}$. These zeros are not solutions of this application because the number of levels must be a natural number.

The only natural number divisors of 1320 that are less than 12 are 1, 2, 3, 4, 5, 6, 8, 10, and 11. The following synthetic division shows that 10 is a zero of $k^3 + 3k^2 + 2k - 1320$.

$$\begin{array}{r|rrrr} 10 & 1 & 3 & 2 & -1320 \\ & & 10 & 130 & 1320 \\ \hline & 1 & 13 & 132 & 0 \end{array}$$

The pyramid has 10 levels. There is no need to seek additional solutions, because the number of levels is uniquely determined by the number of glasses.

CHECK YOUR PROGRESS 5 If 560 glasses are used to form a triangular pyramid, how many levels are in the pyramid?

Solution *See page S21.* 14 levels

The procedures developed in this section will not find all solutions of every polynomial equation. However, a graphing utility can be used to estimate the real solutions of any polynomial equation. In Example 6 we utilize a graphing utility to solve an application.

Alternative to Example 6
Exercise 68, page 349

EXAMPLE 6 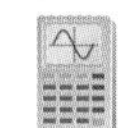 **Use a Graphing Utility to Solve an Application**

A CO_2 (carbon dioxide) cartridge for a paintball rifle has the shape of a right circular cylinder with a hemisphere at each end. The cylinder is 4 inches long and the volume of the cartridge is 2π cubic inches (approximately 6.3 cubic inches). In the figure at the right, the common interior radius of the cylinder and the hemispheres is denoted by x. Use a graphing utility to estimate, to the nearest hundredth of an inch, the length of the radius x.

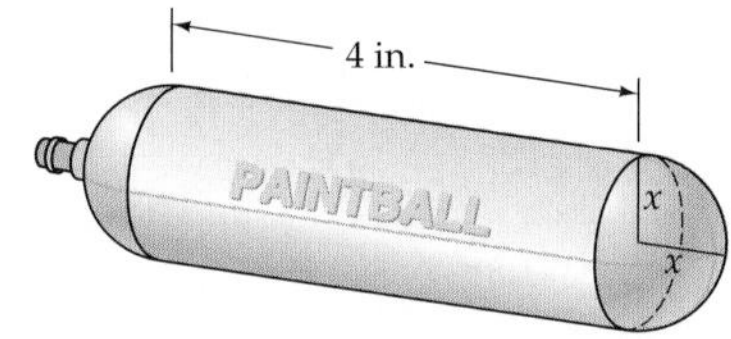

Solution The volume of the cartridge is equal to the volume of the two hemispheres plus the volume of the cylinder. Recall that the volume of a sphere of radius x is given by $\frac{4}{3}\pi x^3$. Therefore, the volume of a hemisphere is $\frac{1}{2}\left(\frac{4}{3}\pi x^3\right)$. The volume of a right circular cylinder is $\pi x^2 h$, where x is the radius of the base and h is the height of the cylinder. Thus the volume V of the cartridge is given by

$$V = \frac{1}{2}\left(\frac{4}{3}\pi x^3\right) + \frac{1}{2}\left(\frac{4}{3}\pi x^3\right) + \pi x^2 h$$

$$= \frac{4}{3}\pi x^3 + \pi x^2 h$$

Replacing V with 2π and h with 4 yields

$$2\pi = \frac{4}{3}\pi x^3 + 4\pi x^2$$

$$2 = \frac{4}{3}x^3 + 4x^2 \qquad \text{■ Divide by } \pi.$$

$$3 = 2x^3 + 6x^2 \qquad \text{■ Multiply by } \tfrac{3}{2}.$$

Continued ➤

Here are two methods that can be used to solve

$$3 = 2x^3 + 6x^2 \quad \text{(I)}$$

for x with the aid of a graphing utility.

1. **Intersection Method** Use a graphing utility to graph $y = 2x^3 + 6x^2$ and $y = 3$ on the same screen, with $x > 0$. The x-coordinate of the point of intersection of the two graphs is the desired solution. The graphs intersect at $x \approx 0.64$ inch.

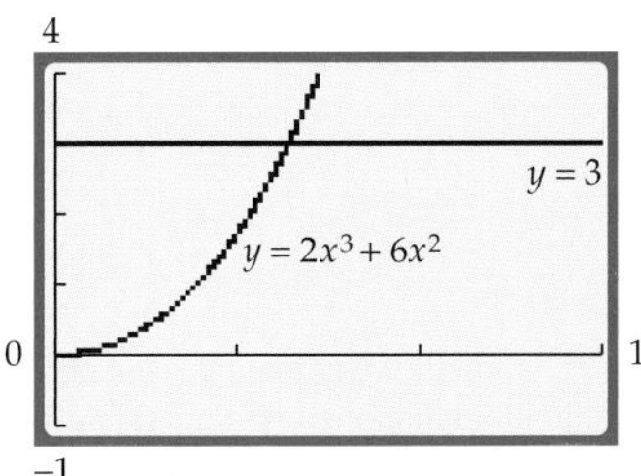

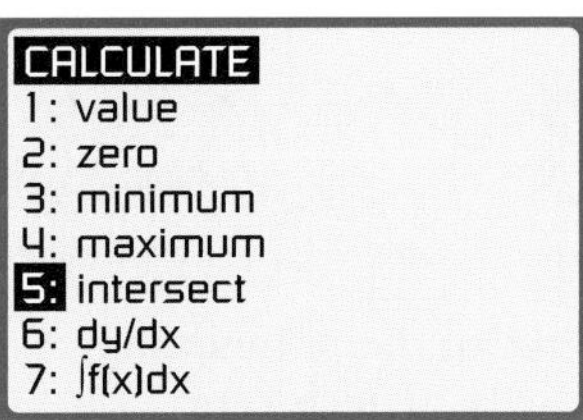

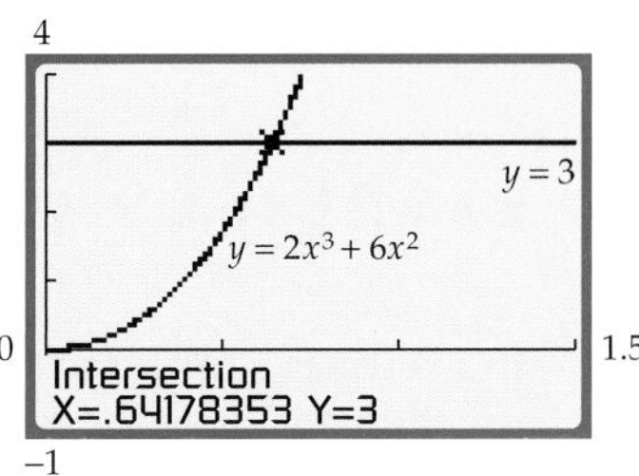

The length of the radius is approximately 0.64 inch.

2. **Intercept Method** Rewrite Equation (I) as $2x^3 + 6x^2 - 3 = 0$. Graph $y = 2x^3 + 6x^2 - 3$ with $x > 0$. Use a graphing utility to find the x-intercept of the graph. This method also shows that $x \approx 0.64$ inch.

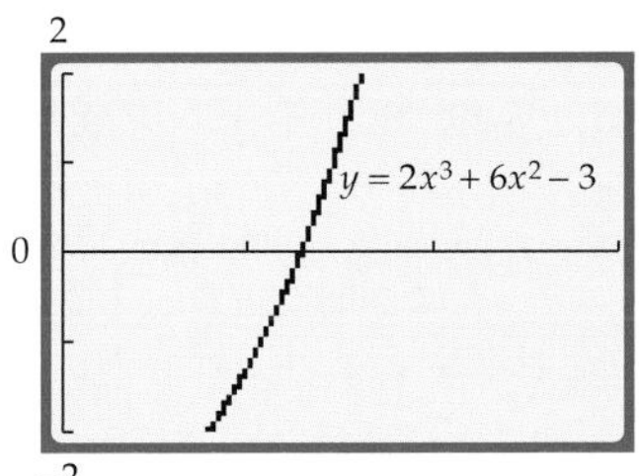

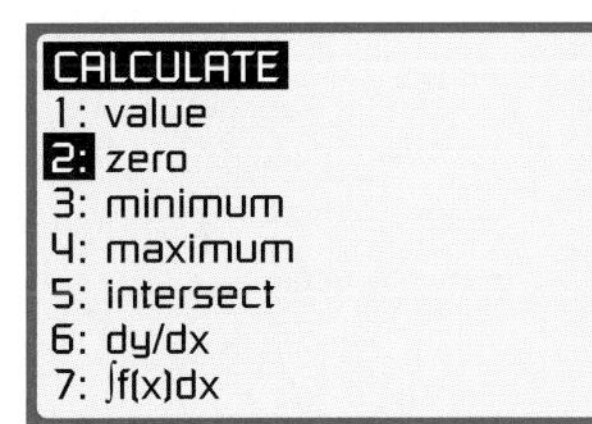

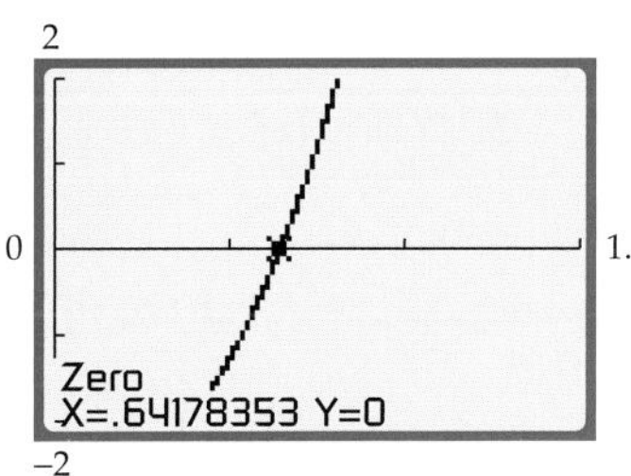

The length of the radius is approximately 0.64 inch.

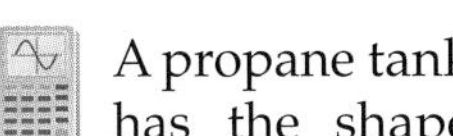

A propane tank has the shape of a circular cylinder with a hemisphere at each end. The cylinder is 6 feet long and the volume of the tank is 9π cubic feet. Find, to the nearest thousandths of a foot, the length of the radius x.

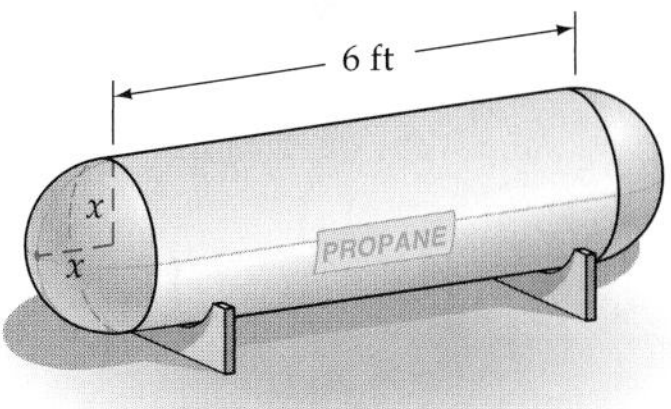

Solution *See page S21.* 1.098 ft

Topics for Discussion

1. What is a multiple zero of a polynomial function? Give an example of a polynomial function that has -2 as a multiple zero.
2. Discuss how the Rational Zero Theorem is used.
3. Let $P(x)$ be a polynomial function with real coefficients. Explain why $(a, 0)$ is an x-intercept of the graph of $P(x)$ if a is a real zero of $P(x)$.

4. Let $P(x)$ be a polynomial function with integer coefficients. Suppose that the Rational Zero Theorem is applied to $P(x)$ and that after testing each possible rational zero, it is determined that $P(x)$ has no rational zeros. Does this mean that all of the zeros of $P(x)$ are irrational numbers?

EXERCISES 4.3

— *Suggested Assignment: Exercises 1–71, odd; 75, and 76–81, all.*

In Exercises 1 to 6, find the zeros of the polynomial function and state the multiplicity of each zero.

1. $P(x) = (x - 3)^2(x + 5)$ 3 (multiplicity 2), -5 (multiplicity 1)
2. $P(x) = (x + 4)^3(x - 1)^2$ -4 (multiplicity 3), 1 (multiplicity 2)
3. $P(x) = x^2(3x + 5)^2$ 0 (multiplicity 2), $-\frac{5}{3}$ (multiplicity 2)
4. $P(x) = x^3(2x + 1)(3x - 12)^2$ 0 (multiplicity 3), $-\frac{1}{2}$ (multiplicity 1), 4 (multiplicity 2)
5. $P(x) = (x^2 - 4)(x + 3)^2$ 2 (multiplicity 1), -2 (multiplicity 1), -3 (multiplicity 2)
6. $P(x) = (x + 4)^3(x^2 - 9)^2$ -4 (multiplicity 3), 3 (multiplicity 2), -3 (multiplicity 2)

In Exercises 7 to 16, use the Rational Zero Theorem to list possible rational zeros for each polynomial function.

7. $P(x) = x^3 + 3x^2 - 6x - 8$ $\pm 1, \pm 2, \pm 4, \pm 8$
8. $P(x) = x^3 - 19x - 30$ $\pm 1, \pm 2, \pm 3, \pm 5, \pm 6, \pm 10, \pm 15, \pm 30$
9. $P(x) = 2x^3 + x^2 - 25x + 12$ $\pm 1, \pm 2, \pm 3, \pm 4, \pm 6, \pm 12, \pm\frac{1}{2}, \pm\frac{3}{2}$
10. $P(x) = 2x^3 + 3x^2 - 11x - 6$ $\pm 1, \pm 2, \pm 3, \pm 6, \pm\frac{1}{2}, \pm\frac{3}{2}$
11. $P(x) = 6x^4 + 23x^3 + 19x^2 - 8x - 4$ $\pm 1, \pm 2, \pm 4, \pm\frac{1}{2}, \pm\frac{1}{3}, \pm\frac{2}{3}, \pm\frac{4}{3}, \pm\frac{1}{6}$
12. $P(x) = 2x^3 + 9x^2 - 2x - 9$ $\pm 1, \pm 3, \pm 9, \pm\frac{1}{2}, \pm\frac{3}{2}, \pm\frac{9}{2}$
13. $P(x) = 4x^4 - 12x^3 - 3x^2 + 12x - 7$ $\pm 1, \pm 7, \pm\frac{1}{2}, \pm\frac{7}{2}, \pm\frac{1}{4}, \pm\frac{7}{4}$
14. $P(x) = x^5 - x^4 - 7x^3 + 7x^2 - 12x - 12$ $\pm 1, \pm 2, \pm 3, \pm 4, \pm 6, \pm 12$
15. $P(x) = x^5 - 32$ $\pm 1, \pm 2, \pm 4, \pm 8, \pm 16, \pm 32$
16. $P(x) = x^4 - 1$ ± 1

In Exercises 17 to 26, find the smallest positive integer and the largest negative integer that, by the Upper- and Lower-Bound Theorem, are upper and lower bounds for the real zeros of each polynomial function.

17. $P(x) = x^3 + 3x^2 - 6x - 6$ Upper bound 2, lower bound -5
18. $P(x) = 2x^3 + 3x^2 - 11x - 4$ Upper bound 2, lower bound -4
19. $P(x) = 2x^3 + x^2 - 25x + 10$ Upper bound 4, lower bound -4
20. $P(x) = 3x^3 + 11x^2 - 6x - 9$ Upper bound 2, lower bound -5
21. $P(x) = 6x^4 + 23x^3 + 19x^2 - 8x - 4$ Upper bound 1, lower bound -4
22. $P(x) = 2x^3 + 9x^2 - 2x - 9$ Upper bound 1, lower bound -5
23. $P(x) = 4x^4 - 12x^3 - 3x^2 + 12x - 7$ Upper bound 4, lower bound -2
24. $P(x) = x^5 - x^4 - 7x^3 + 7x^2 - 12x - 12$ Upper bound 4, lower bound -3
25. $P(x) = x^5 - 32$ Upper bound 2, lower bound -1
26. $P(x) = x^4 - 1$ Upper bound 1, lower bound -1

In Exercises 27 to 36, use Descartes' Rules of Signs to state the number of possible positive and negative real zeros of each polynomial function.

27. $P(x) = x^3 + 3x^2 - 6x - 8$ One positive zero, two or no negative zeros
28. $P(x) = x^3 - 3x^2 - 18x + 40$ Two or no positive zeros, one negative zero
29. $P(x) = 2x^3 + x^2 - 25x + 12$ Two or no positive zeros, one negative zero
30. $P(x) = 3x^3 + 11x^2 - 6x - 8$ One positive zero, two or no negative zeros
31. $P(x) = 6x^4 + 23x^3 + 19x^2 - 8x - 4$ One positive zero, three or one negative zeros

32. $P(x) = 2x^3 + 9x^2 - 2x - 9$
One positive zero, two or no negative zeros

33. $P(x) = 4x^4 - 12x^3 - 3x^2 + 12x - 7$
Three or one positive zeros, one negative zero

34. $P(x) = x^5 - x^4 - 7x^3 + 7x^2 - 12x - 12$
Three or one positive zeros, two or no negative zeros

35. $P(x) = x^5 - 32$ One positive zero, no negative zeros

36. $P(x) = x^4 - 1$ One positive zero, one negative zero

In Exercises 37 to 58, find the zeros of each polynomial function. If a zero is a multiple zero, state its multiplicity.

37. $P(x) = x^3 + 3x^2 - 6x - 8$ $2, -1, -4$

38. $P(x) = x^3 - 3x^2 - 18x + 40$ $5, 2, -4$

39. $P(x) = 2x^3 + x^2 - 25x + 12$ $3, -4, \frac{1}{2}$

40. $P(x) = 3x^3 + 11x^2 - 6x - 8$ $1, -4, -\frac{2}{3}$

41. $P(x) = 6x^4 + 23x^3 + 19x^2 - 8x - 4$ $\frac{1}{2}, -\frac{1}{3}, -2$ (multiplicity 2)

42. $P(x) = 2x^3 + 9x^2 - 2x - 9$ $1, -1, -\frac{9}{2}$

43. $P(x) = 2x^4 - 9x^3 - 2x^2 + 27x - 12$ $\frac{1}{2}, 4, \sqrt{3}, -\sqrt{3}$

44. $P(x) = 3x^3 - x^2 - 6x + 2$ $\frac{1}{3}, \sqrt{2}, -\sqrt{2}$

45. $P(x) = x^3 - 8x^2 + 8x + 24$ $6, 1 + \sqrt{5}, 1 - \sqrt{5}$

46. $P(x) = x^3 - 7x^2 - 7x + 69$ $-3, 5 + \sqrt{2}, 5 - \sqrt{2}$

47. $P(x) = 2x^4 - 19x^3 + 51x^2 - 31x + 5$ $5, \frac{1}{2}, 2 + \sqrt{3}, 2 - \sqrt{3}$

48. $P(x) = 4x^4 - 35x^3 + 71x^2 - 4x - 6$ $3, -\frac{1}{4}, 3 + \sqrt{7}, 3 - \sqrt{7}$

49. $P(x) = 3x^6 - 10x^5 - 29x^4 + 34x^3 + 50x^2 - 24x - 24$
$1, -1, -2, -\frac{2}{3}, 3 + \sqrt{3}, 3 - \sqrt{3}$

50. $P(x) = 2x^4 + 3x^3 - 4x^2 - 3x + 2$
$-2, -1, \frac{1}{2}, 1$

51. $P(x) = x^3 - 3x - 2$
$2, -1$ (multiplicity 2)

52. $P(x) = 3x^4 - 4x^3 - 11x^2 + 16x - 4$ $-2, \frac{1}{3}, 1, 2$

53. $P(x) = x^4 - 5x^2 - 2x$ $0, -2, 1 + \sqrt{2}, 1 - \sqrt{2}$

54. $P(x) = x^3 - 2x + 1$ $1, \frac{-1 + \sqrt{5}}{2}, \frac{-1 - \sqrt{5}}{2}$

55. $P(x) = x^4 + x^3 - 3x^2 - 5x - 2$ -1 (multiplicity 3), 2

56. $P(x) = 6x^4 - 17x^3 - 11x^2 + 42x$ $-\frac{3}{2}, 0, 2, \frac{7}{3}$

57. $P(x) = 2x^4 - 17x^3 + 4x^2 + 35x - 24$ $-\frac{3}{2}, 1$ (multiplicity 2), 8

58. $P(x) = x^5 + 5x^4 + 10x^3 + 10x^2 + 5x + 1$ -1 (multiplicity 5)

59. ***Find the Dimensions*** A cube measures n inches on each edge. If a slice 2 inches thick is cut from one face of the cube, the resulting solid has a volume of 567 cubic inches. Find n. $n = 9$ in.

60. ***Find the Dimensions*** A cube measures n units on each edge. If a slice 1 inch thick is cut from one face of the cube, and then a slice 3 inches thick is cut from another face of the cube as shown, the resulting solid has a volume of 1560 cubic inches. Find the dimensions of the original cube.

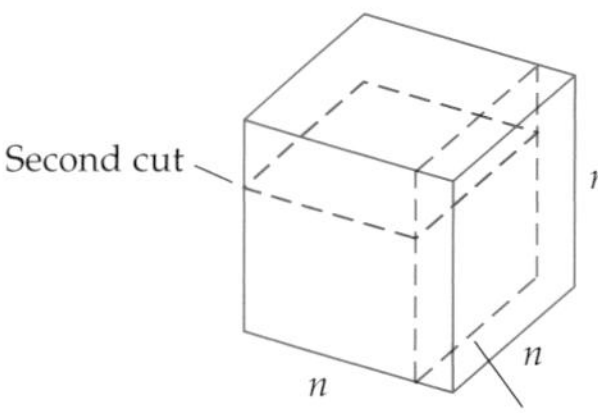

13 in. on each edge

61. ***Dimensions of a Solid*** For what value of x will the volume of the following solid be 112 cubic inches?

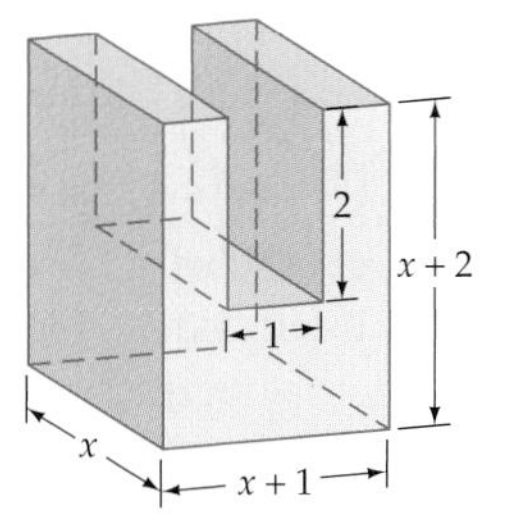

$x = 4$ in.

62. ***Dimensions of a Box*** The length of a rectangular box is 1 inch more than twice the height of the box, and the width is 3 inches more than the height. If the volume of the box is 126 cubic inches, find the dimensions of the box.

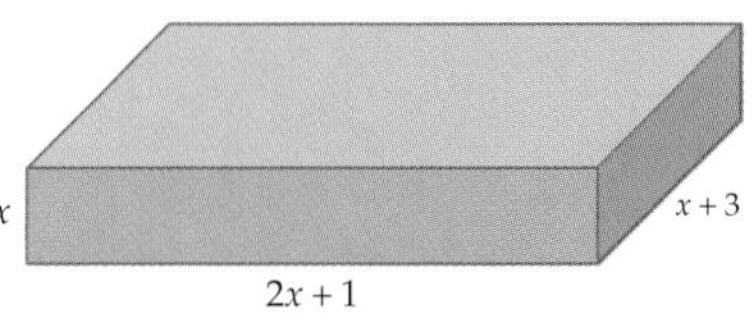

3 in. by 7 in. by 6 in.

63. ***Pieces and Cuts*** One straight cut through a thick piece of cheese produces two pieces. Two straight cuts can produce a maximum of four pieces. Three straight cuts can produce a maximum of eight pieces.

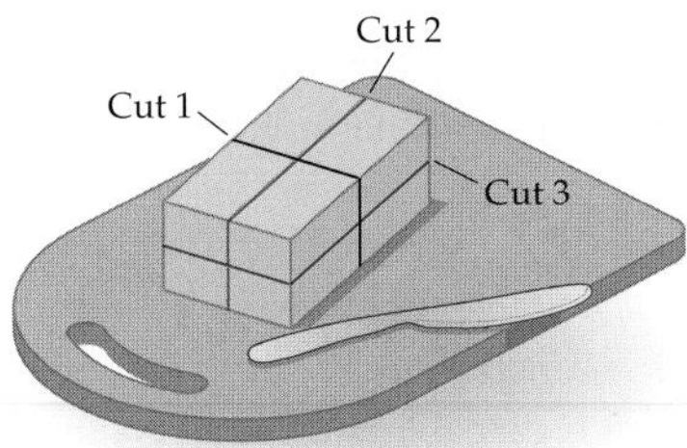

You might be inclined to think that every additional cut doubles the previous number of pieces. However, for four straight cuts, you get a maximum of 15 pieces. The maximum number of pieces P that can be produced by n straight cuts is given by

$$P(n) = \frac{n^3 + 5n + 6}{6}$$

a. Use the above function to determine the maximum number of pieces that can be produced by five straight cuts. 26 pieces

b. What is the fewest number of straight cuts that are needed to produce 64 pieces? 7 cuts

64. ***Inscribed Quadrilateral*** Isaac Newton discovered that if a quadrilateral with sides of lengths a, b, c, and x is inscribed in a semicircle with diameter x, then the lengths of the sides are related by the following equation.

$$x^3 - (a^2 + b^2 + c^2)x - 2abc = 0$$

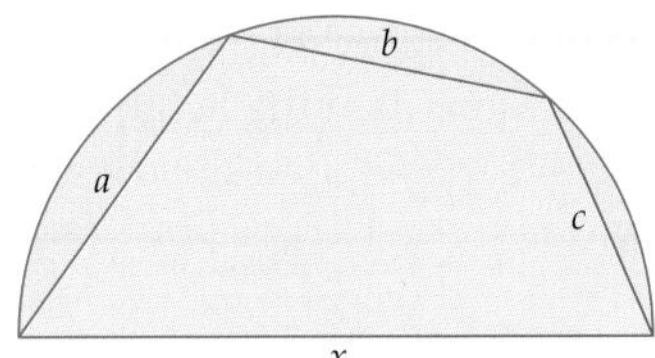

Given $a = 6$, $b = 5$, and $c = 4$, find x. Round to the nearest hundredth. 10.04

65. ***Cannonball Stacks*** Cannonballs can be stacked to form a pyramid with a square base. The total number of cannonballs T in one of these square pyramids is

$$T = \frac{1}{6}(2n^3 + 3n^2 + n)$$

where n is the number of rows (levels). If 140 cannonballs are used to form a square pyramid, how many rows are in the pyramid? 7 rows

Business and Economics

66. ***Advertising Expenses*** A company manufactures digital cameras. The company estimates that the profit from camera sales is

$$P(x) = -0.02x^3 + 0.01x^2 + 1.2x - 1.1$$

where P is the profit in millions of dollars and x is the amount, in hundred-thousands of dollars, spent on advertising.

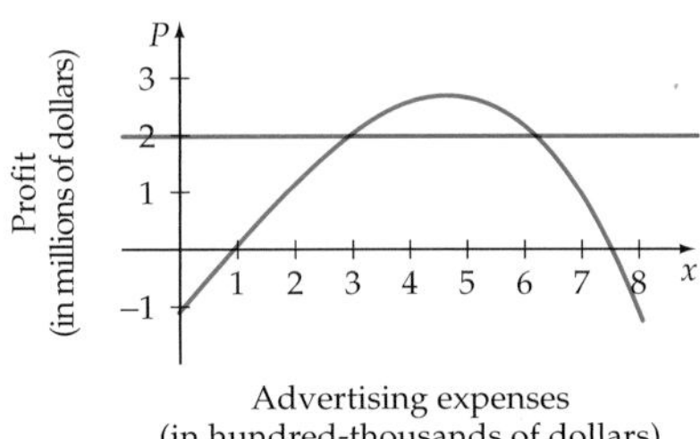

Determine the minimum amount, rounded to the nearest thousand dollars, the company needs to spend on advertising if it is to receive a profit of $2,000,000. $293,000

67. ***Cost Cutting*** At the present time a nutrition bar in the shape of a rectangular solid measures 0.75 inch by 1 inch by 5 inches.

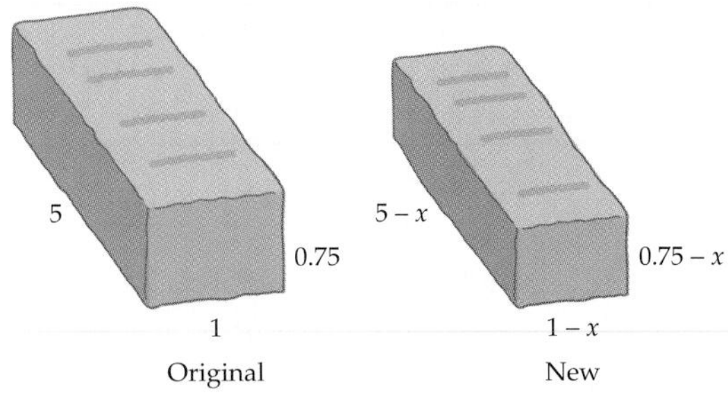

To reduce costs the manufacturer has decided to decrease each of the dimensions of the nutrition bar by x inches. What value of x, rounded to the nearest thousandth of an inch, will produce a new nutrition bar with a volume that is 0.75 cubic inch less than the present bar's volume? $x = 0.084$ in.

68. ***Silo Construction*** A farmer plans to construct a silo with a cylindrical base topped by a hemisphere. The height of the cylindrical base is to be 20 feet and the volume of the silo needs to be 864π cubic feet. Determine the length of the radius r.

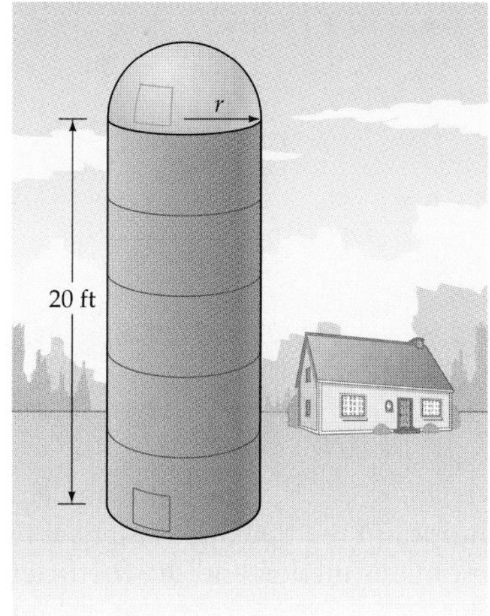

6 ft

Social Sciences

69. ***Divorce Rate*** The divorce rate for a given year is defined as the number of divorces per thousand population. The polynomial function

$$D(t) = 0.00001807t^4 - 0.001406t^3 + 0.02884t^2 - 0.003466t + 2.1148$$

approximates the U.S. divorce rate for the years 1960 ($t = 0$) to 1999 ($t = 39$). Use $D(t)$ and a graphing utility to determine during what years the U.S. divorce rate attained a level of 5.0. 1977 and 1986

Life and Health Sciences

70. ***Medication Level*** Pseudoephedrine hydrochloride is an allergy medication. The polynomial function

$$L(t) = 0.03t^4 + 0.4t^3 - 7.3t^2 + 23.1t$$

where $0 \le t \le 5$, models the level of pseudoephedrine hydrochloride, in milligrams, in the bloodstream of a patient t hours after 30 milligrams of the medication have been taken.

At what times, to the nearest minute, does the level of pseudoephedrine hydrochloride in the bloodstream reach 12 milligrams? After 39 min and after 3 h 38 min

71. ***Weight and Height of Giraffes*** A veterinarian at a wild animal park has determined that the average weight w, in pounds, of an adult male giraffe is closely approximated by the function

$$w = 8.3h^3 - 307.5h^2 + 3914h - 15230$$

where h is the giraffe's height in feet, and $15 \le h \le 18$. Use the above function to estimate the height of a giraffe that weighs 3150 pounds. Round to the nearest tenth of a foot. 16.9 ft

Sports and Recreation

72. ***Selection of Cards*** The number of ways one can select three cards from a group of n cards (the order of the selection matters), where $n \ge 3$, is given by $P(n) = n^3 - 3n^2 + 2n$. For a certain card trick a magician has determined that there are exactly 504 ways to choose three cards from a given group. How many cards are in the group? 9 cards

Physical Sciences and Engineering

73. ***Floating Sphere*** A spherical buoy with radius $r = 1$ foot is made from a type of material that has a specific gravity of 0.6. This means that a cubic foot of the material weighs 60% as much as a cubic foot of water. When placed in water, 60% of the sphere will be below the surface of the water.

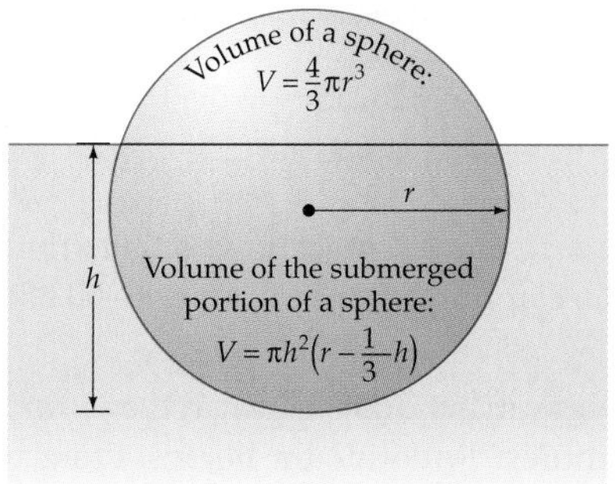

Find, to the nearest hundredth foot, the depth h to which the buoy will sink in water. (*Hint:* The volume of a sphere of radius r is $\frac{4}{3}\pi r^3$. Thus 60% of the given sphere has a volume of $\frac{4}{3}\pi(0.6)$ cubic feet.) Archimedes was able to show that if h is the depth to which a sphere sinks in water, then the volume of the submerged portion of the sphere is

$$\pi h^2\left(r - \frac{1}{3}h\right)$$

Because this is also the volume of 60% of the sphere, the value of this expression with $r = 1$ must equal $\frac{4}{3}\pi(0.6)$. This gives us the equation

$$\pi h^2\left(1 - \frac{1}{3}h\right) = \frac{4}{3}\pi(0.6)$$

which simplifies to the cubic equation $5h^3 - 15h^2 + 12 = 0$. 1.13 ft

74. ***Floating Sphere*** A sphere with a radius of 2 feet has a specific gravity of 0.85. Find, to the nearest hundredth of a foot, the depth to which the sphere will sink in water (which has a specific gravity of 1). See Exercise 73 for an explanation of specific gravity. 3.02 ft

75. ***Digits of Pi*** In 1999, Professor Yasumasa Kanada of the University of Tokyo used a supercomputer to compute 206,158,430,000 digits of pi (π). (*Source: Guinness World Records 2001,* Bantam Books, page 252) Computer scientists often try to find mathematical models that approximate the time a computer program takes to complete a calculation or mathematical procedure. Procedures for which the completion time can be closely modeled by a polynomial are called **polynomial time procedures.** Here is an example. A student finds that the time, in seconds, required to compute $n \times 10{,}000$ digits of pi on a personal computer using the mathematical program MAPLE is closely approximated by

$$T(n) = 0.23245n^3 + 0.53797n^2 + 7.88932n - 8.53299$$

a. Evaluate $T(n)$ to estimate how long, to the nearest second, the computer takes to compute 50,000 digits of pi. 73 s

b. About how many digits of pi can the computer compute in 5 minutes? Round to the nearest thousand digits. 93,000 digits

Prepare for Section 4.4

76. What is the conjugate of $3 - 2i$? [P.7] $3 + 2i$

77. What is the conjugate of $2 + i\sqrt{5}$? [P.7] $2 - i\sqrt{5}$

78. Find $(x - 1)(x - 3)(x - 4)$. [P.3] $x^3 - 8x^2 + 19x - 12$

79. Find $[x - (2 + i)][x - (2 - i)]$. [P.3/P.7] $x^2 - 4x + 5$

80. Solve: $x^2 + 9 = 0$ [1.2] $-3i, 3i$

81. Solve: $x^2 - x + 5 = 0$ [1.2] $\frac{1}{2} - \frac{1}{2}i\sqrt{19}, \frac{1}{2} + \frac{1}{2}i\sqrt{19}$

Explorations 

1. ***Relationships between Zeros and Coefficients*** Consider the polynomial function

$$P(x) = x^n + C_1x^{n-1} + C_2x^{n-2} + \cdots + C_n$$

with zeros $r_1, r_2, r_3, \ldots, r_n$. The following equations illustrate important relationships between the zeros of the polynomial function and the coefficients of the polynomial.

- **The sum of the zeros.**

$$r_1 + r_2 + r_3 + \cdots + r_{n-1} + r_n = -C_1$$

- **The sum of the product of the zeros taken two at a time.**

$$r_1r_2 + r_1r_3 + \cdots + r_{n-2}r_n + r_{n-1}r_n = C_2$$

- **The sum of the product of the zeros taken three at a time.**

$$r_1r_2r_3 + r_1r_2r_4 + \cdots + r_{n-2}r_{n-1}r_n = -C_3$$
$$\vdots$$

- **The product of the zeros.**

$$r_1r_2r_3r_4 \cdots r_{n-1}r_n = (-1)^nC_n$$

a. Show that each of the previous equations holds true for the function $P(x) = x^3 - 6x^2 + 11x - 6$, which has zeros of 1, 2, and 3.

b. Create a polynomial function of degree 4 with four real zeros. Illustrate that each of the above equations holds true for your polynomial function. *Hint*: The polynomial function

$$P(x) = (x - a)(x - b)(x - c)(x - d)$$

has $a, b, c,$ and d as zeros.
Responses will vary.

2. ***Cubic Function Applet*** The point $\left(-\frac{b}{3a}, P\left(-\frac{b}{3a}\right)\right)$ is called the **center of rotation** of the graph of the **cubic polynomial** function $P(x) = ax^3 + bx^2 + cx + d$, $a \neq 0$. A web applet is available to explore the relationships that exist between the coefficients of $P(x)$ and the position of its center of rotation. This applet, called *cubic function activity,* can be found on our web site at **math.college.hmco.com.**

Experiment with this applet to discover the following relationships.

a. What happens to the center of rotation as d is increased? The center of rotation moves straight upward.

b. If a and b have the same sign, what happens to the center of rotation as c is increased? If a and b have opposite signs, what happens to the center of rotation as c is increased?
As c is increased, the center of rotation moves straight downward when a and b have the same sign and straight upward when a and b have opposite signs.

SECTION 4.4 The Fundamental Theorem of Algebra

- The Fundamental Theorem of Algebra
- The Number of Zeros of a Polynomial Function
- The Conjugate Pair Theorem
- Find a Polynomial Function with Given Zeros

▪ The Fundamental Theorem of Algebra

The German mathematician Carl Friedrich Gauss (1777–1855) was the first to prove that every polynomial function has at least one complex zero. This concept is so basic to the study of algebra that it is called the **Fundamental Theorem of Algebra.** The proof of the Fundamental Theorem is beyond the scope of this text; however, it is important to understand the theorem and its consequences. As you consider each of the following theorems, keep in mind that the terms **complex coefficients** and **complex zeros** include real coefficients and real zeros, because the set of real numbers is a subset of the set of complex numbers.

> **The Fundamental Theorem of Algebra**
>
> If $P(x)$ is a polynomial function of degree $n \geq 1$ with complex coefficients, then $P(x)$ has at least one complex zero.

▪ The Number of Zeros of a Polynomial Function

Let $P(x)$ be a polynomial function of degree $n \geq 1$ with complex coefficients. The Fundamental Theorem implies that $P(x)$ has a complex zero—say, c_1. The Factor Theorem implies that

$$P(x) = (x - c_1)Q(x)$$

where $Q(x)$ is a polynomial of degree one less than the degree of $P(x)$. Recall that the polynomial $Q(x)$ is called a reduced polynomial. Assuming that the degree of $Q(x)$ is 1 or more, the Fundamental Theorem implies that it must also have a zero. A continuation of this reasoning process leads to the following theorem.

Point of Interest

Carl Friedrich Gauss (1777–1855) has often been referred to as the Prince of Mathematics. His work covered topics in algebra, calculus, analysis, probability, number theory, non-Euclidean geometry, astronomy, and physics, to name but a few. The following quote by Eric Temple Bell gives credence to the fact that Gauss was one of the greatest mathematicians of all time. "Archimedes, Newton, and Gauss, these three, are in a class by themselves among the great mathematicians, and it is not for ordinary mortals to attempt to range them in order of merit."*

**Men of Mathematics* by E. T. Bell. New York: Simon and Schuster, Inc., 1937.

Alternative to Example 1
Exercise 2, page 358

The Linear Factor Theorem

If $P(x)$ is a polynomial function of degree $n \geq 1$ with leading coefficient $a_n \neq 0$,

$$P(x) = a_n x^n + a_{n-1}x^{n-1} + \cdots + a_1 x^1 + a_0$$

then $P(x)$ has exactly n linear factors

$$P(x) = a_n(x - c_1)(x - c_2) \cdots (x - c_n)$$

where $c_1, c_2, \ldots, c_n$ are complex numbers.

The following theorem follows directly from the Linear Factor Theorem.

The Number of Zeros of a Polynomial Function Theorem

If $P(x)$ is a polynomial function of degree $n \geq 1$ with complex coefficients, then $P(x)$ has exactly n complex zeros, provided each zero is counted according to its multiplicity.

The Linear Factor Theorem and the Number of Zeros of a Polynomial Function Theorem are referred to as **existence theorems.** They state that an nth degree polynomial will have n linear factors and n complex zeros, but they do not provide any information on how to determine the linear factors or the zeros. In Example 1 we make use of previously developed methods to actually find the linear factors and zeros of some polynomial functions.

EXAMPLE 1 Find the Zeros and Linear Factors of a Polynomial Function

Find all the zeros of each of the following polynomial functions, and write each polynomial as a product of linear factors.

a. $P(x) = x^4 - 4x^3 + 8x^2 - 16x + 16$

b. $S(x) = x^4 - 6x^3 + 10x^2 + 2x - 15$

Solution

a. We know that $P(x)$ will have four zeros and four linear factors. The possible rational zeros are $\pm 1, \pm 2, \pm 4, \pm 8, \pm 16$. Synthetic division can be used to show that 2 is a zero of multiplicity 2.

$$\begin{array}{r|rrrrr} 2 & 1 & -4 & 8 & -16 & 16 \\ & & 2 & -4 & 8 & 16 \\ \hline & 1 & -2 & 4 & -8 & 0 \end{array}$$

$$\begin{array}{r|rrrr} 2 & 1 & -2 & 4 & -8 \\ & & 2 & 0 & 8 \\ \hline & 1 & 0 & 4 & 0 \end{array}$$

Continued ➤

The final reduced polynomial is $x^2 + 4$. The zeros of $x^2 + 4$ can be found by solving $x^2 + 4 = 0$, as shown below.

$$\begin{aligned} x^2 + 4 &= 0 \\ x^2 &= -4 \\ x &= \pm\sqrt{-4} \\ x &= \pm 2i \end{aligned}$$

TAKE NOTE

To review concepts involving complex numbers, see Section P.7.

Thus the four zeros of $P(x)$ are $2, 2, -2i$, and $2i$. The linear factored form of $P(x)$ is

$$P(x) = (x - 2)(x - 2)[x - (-2i)][x - 2i]$$

or

$$P(x) = (x - 2)^2(x + 2i)(x - 2i)$$

b. We know that $S(x)$ will have four zeros and four linear factors. The possible rational zeros are $\pm1, \pm3, \pm5, \pm15$. Synthetic division can be used to show that 3 and -1 are zeros of $S(x)$.

$$\begin{array}{r|rrrrr} 3 & 1 & -6 & 10 & 2 & -15 \\ & & 3 & -9 & 3 & 15 \\ \hline & 1 & -3 & 1 & 5 & 0 \end{array}$$

$$\begin{array}{r|rrrr} -1 & 1 & -3 & 1 & 5 \\ & & -1 & 4 & -5 \\ \hline & 1 & -4 & 5 & 0 \end{array}$$

The final reduced polynomial is $x^2 - 4x + 5$. We can find the remaining zeros by using the quadratic formula to solve $x^2 - 4x + 5 = 0$.

$$\begin{aligned} x &= \frac{-(-4) \pm \sqrt{(-4)^2 - 4(1)(5)}}{2(1)} \\ &= \frac{4 \pm \sqrt{-4}}{2} \\ &= 2 \pm i \end{aligned}$$

Thus the four zeros of $S(x)$ are $3, -1, 2 + i$, and $2 - i$. The linear factored form of $S(x)$ is

$$S(x) = (x - 3)[x - (-1)][x - (2 + i)][x - (2 - i)]$$

or

$$S(x) = (x - 3)(x + 1)(x - 2 - i)(x - 2 + i)$$

CHECK YOUR PROGRESS 1 Find all the zeros of $P(x) = x^3 - 3x^2 + 7x - 5$, and write $P(x)$ as a product of linear factors.

Solution *See page S21.* $1, 1 - 2i, 1 + 2i$; $P(x) = (x - 1)(x - 1 + 2i)(x - 1 - 2i)$

▪ The Conjugate Pair Theorem

You may have noticed that the complex zeros of the polynomial function in Example 2 were complex conjugates. The following theorem shows that this is not a coincidence.

The Conjugate Pair Theorem

If $a + bi$ $(b \neq 0)$ is a complex zero of a polynomial function *with real coefficients,* then the conjugate $a - bi$ is also a complex zero of the polynomial function.

Alternative to Example 2
Exercise 12, page 358

EXAMPLE 2 Use the Conjugate Pair Theorem to Find Zeros

Find all the zeros of $P(x) = x^4 - 4x^3 + 14x^2 - 36x + 45$ given that $2 + i$ is a zero.

Solution Because the coefficients are real numbers and $2 + i$ is a zero, the Conjugate Pair Theorem implies that $2 - i$ must also be a zero. Using synthetic division with $2 + i$ and then $2 - i$, we have

$$\begin{array}{r|rrrrr} 2+i & 1 & -4 & 14 & -36 & 45 \\ & & 2+i & -5 & 18+9i & -45 \\ \hline & 1 & -2+i & 9 & -18+9i & 0 \end{array}$$

▪ The coefficients of the reduced polynomial

$$\begin{array}{r|rrrr} 2-i & 1 & -2+i & 9 & -18+9i \\ & & 2-i & 0 & 18-9i \\ \hline & 1 & 0 & 9 & 0 \end{array}$$

▪ The coefficients of the next reduced polynomial

The resulting reduced polynomial is $x^2 + 9$, which has $3i$ and $-3i$ as zeros. Therefore, the four zeros of $x^4 - 4x^3 + 14x^2 - 36x + 45$ are $2 + i, 2 - i, 3i,$ and $-3i$.

CHECK YOUR PROGRESS 2 Find all the zeros of $P(x) = 3x^3 - 29x^2 + 92x + 34$ given that $5 + 3i$ is a zero.

Solution *See page S21.* $5 + 3i, 5 - 3i, -\frac{1}{3}$

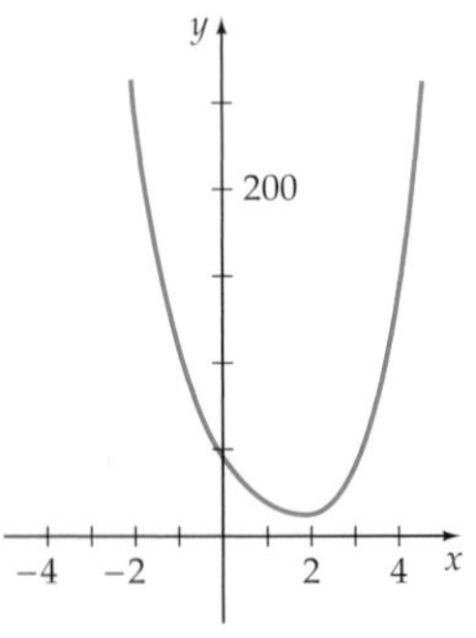

$P(x) = x^4 - 4x^3 + 14x^2 - 36x + 45$

Figure 4.20

A graph of $P(x) = x^4 - 4x^3 + 14x^2 - 36x + 45$ is shown in Figure 4.20. Because the polynomial in Example 2 is a fourth-degree polynomial and because we have verified that $P(x)$ has four imaginary solutions, it comes as no surprise that the graph does not intersect the x-axis.

When performing synthetic division with complex numbers, it is helpful to write the coefficients of the given polynomial as complex coefficients. For instance, -10 can be written as $-10 + 0i$. This technique is illustrated in the next example.

Alternative to Example 3
Exercise 14, page 358

INTEGRATING TECHNOLOGY

Many graphing calculators can be used to do computations with complex numbers. The following TI-83 screen display shows that the product of $3 - 5i$ and $-7 - 5i$ is $-46 + 20i$. The i symbol is located above the decimal point key.

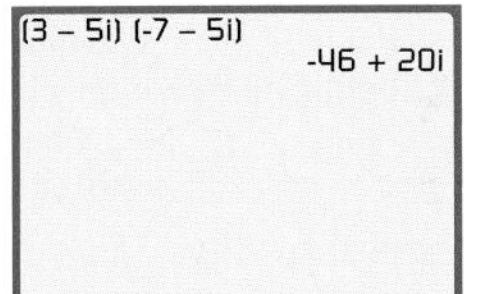

EXAMPLE 3 Apply the Conjugate Pair Theorem

Find all the zeros of $P(x) = x^5 - 10x^4 + 65x^3 - 184x^2 + 274x - 204$ given that $3 - 5i$ is a zero.

Solution Because the coefficients are real numbers and $3 - 5i$ is a zero, $3 + 5i$ must also be a zero. Use synthetic division to produce

$$\begin{array}{r|rrrrrr}
3-5i & 1 & -10+0i & 65+0i & -184+0i & 274+0i & -204 \\
 & & 3-5i & -46+20i & 157-35i & -256+30i & 204 \\
\hline
3+5i & 1 & -7-5i & 19+20i & -27-35i & 18+30i & 0 \\
 & & 3+5i & -12-20i & 21+35i & -18-30i & \\
\hline
 & 1 & -4 & 7 & -6 & 0 &
\end{array}$$

Descartes' Rule of Signs can be used to show that the reduced polynomial $x^3 - 4x^2 + 7x - 6$ has three or one positive zeros and no negative zeros.

$$\frac{p}{q} = 1, 2, 3, 6$$

$$\begin{array}{r|rrrr}
2 & 1 & -4 & 7 & -6 \\
 & & 2 & -4 & 6 \\
\hline
 & 1 & -2 & 3 & 0
\end{array}$$

Use the quadratic formula to solve $x^2 - 2x + 3 = 0$.

$$x = \frac{-(-2) \pm \sqrt{(-2)^2 - 4(1)(3)}}{2(1)} = \frac{2 \pm \sqrt{-8}}{2} = \frac{2 \pm 2\sqrt{2}i}{2} = 1 \pm \sqrt{2}i$$

The zeros of $P(x) = x^5 - 10x^4 + 65x^3 - 184x^2 + 274x - 204$ are $3 - 5i$, $3 + 5i$, 2, $1 + \sqrt{2}i$, and $1 - \sqrt{2}i$.

CHECK YOUR PROGRESS 3 Find all the zeros of $P(x) = x^5 - 6x^4 + 22x^3 - 64x^2 + 117x - 90$ given that $3i$ is a zero.

Solution *See page S21.* $3i, -3i, 2, 2 + i, 2 - i$

QUESTION Is it possible for a third-degree polynomial function with real coefficients to have two real zeros and one complex zero?

Recall that the real zeros of a polynomial function P are the x-coordinates of the x-intercepts of the graph of P. This important connection between the real zeros of a polynomial function and the x-intercepts of the graph of the polynomial function is the basis for using a graphing utility to solve equations. Careful analysis of the graph of a polynomial function and your knowledge of the properties of polynomial functions can be used to solve many polynomial equations.

ANSWER No. Because the coefficients of the polynomial are real numbers, the complex zeros of the polynomial function must occur as conjugate pairs.

Alternative to Example 4
Exercise 24, page 358

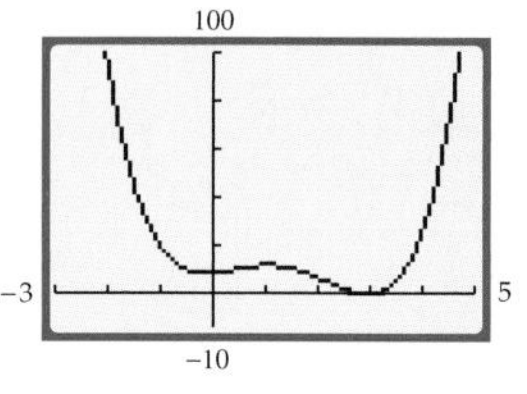

$P(x) = x^4 - 5x^3 + 4x^2 + 3x + 9$

Figure 4.21

EXAMPLE 4 Solve a Polynomial Equation

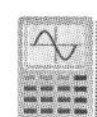
Solve: $x^4 - 5x^3 + 4x^2 + 3x + 9 = 0$

Solution Let $P(x) = x^4 - 5x^3 + 4x^2 + 3x + 9$. The x-intercepts of the graph of P are the real solutions of the equation. Use a graphing utility to graph P. See Figure 4.21.

From the graph, it appears that (3, 0) is an x-intercept and the only x-intercept. Because the graph of P intersects but does not cross the x-axis at (3, 0), we know that 3 is a multiple zero of P with an even multiplicity.

$$\begin{array}{r|rrrrr} 3 & 1 & -5 & 4 & 3 & 9 \\ & & 3 & -6 & -6 & -9 \\ \hline & 1 & -2 & -2 & -3 & 0 \end{array}$$

▪ Coefficients of P

▪ The remainder is zero. Thus 3 is a zero.

By the Number of Zeros Theorem, there are three more zeros of P. Use synthetic division to show that 3 is also a zero of the reduced polynomial $x^3 - 2x^2 - 2x - 3$.

$$\begin{array}{r|rrrr} 3 & 1 & -2 & -2 & -3 \\ & & 3 & 3 & 3 \\ \hline & 1 & 1 & 1 & 0 \end{array}$$

▪ Coefficients of reduced polynomial

▪ The remainder is zero. Thus 3 is a zero of multiplicity 2.

We now have 3 as a double root of the original equation, and from the last line of the preceding synthetic division, the remaining solutions must be solutions of $x^2 + x + 1 = 0$. Use the quadratic formula to solve this equation.

$$x = \frac{-1 \pm \sqrt{1^2 - 4(1)(1)}}{2(1)} = \frac{-1 \pm \sqrt{-3}}{2} = \frac{-1 \pm i\sqrt{3}}{2}$$

The solutions of $x^4 - 5x^3 + 4x^2 + 3x + 9 = 0$ are $3, 3, -\frac{1}{2} + \frac{\sqrt{3}}{2}i$, and $-\frac{1}{2} - \frac{\sqrt{3}}{2}i$.

CHECK YOUR PROGRESS 4 Solve: $4x^3 + 3x^2 + 16x + 12 = 0$

Solution *See page S22.* $-\frac{3}{4}, -2i, 2i$

▪ Find a Polynomial Function with Given Zeros

Many of the problems in this section and in Section 4.3 dealt with the process of finding the zeros of a given polynomial function. Example 5 considers the reverse process, finding a polynomial function when the zeros are given.

Alternative to Example 5
Exercise 36, page 359

EXAMPLE 5 Determine a Polynomial Function Given Its Zeros

Find each polynomial function.

a. A polynomial function of degree 3 that has 1, 2, and −3 as zeros

b. A polynomial function of degree 4 that has real coefficients and zeros $2i$ and $3 - 7i$

Continued ➤

Solution

a. Because 1, 2, and -3 are zeros, $(x - 1)$, $(x - 2)$, and $(x + 3)$ are factors. The product of these factors produces a polynomial function that has the indicated zeros.

$$P(x) = (x - 1)(x - 2)(x + 3) = (x^2 - 3x + 2)(x + 3) = x^3 - 7x + 6$$

b. By the Conjugate Pair Theorem, the polynomial function also must have $-2i$ and $3 + 7i$ as zeros. The product of the factors $x - 2i$, $x - (-2i)$, $x - (3 - 7i)$, and $x - (3 + 7i)$ produces the desired polynomial function.

$$\begin{aligned} P(x) &= (x - 2i)(x + 2i)[x - (3 - 7i)][x - (3 + 7i)] \\ &= (x^2 + 4)(x^2 - 6x + 58) \\ &= x^4 - 6x^3 + 62x^2 - 24x + 232 \end{aligned}$$

CHECK YOUR PROGRESS 5 Find a polynomial function of degree 3, with real coefficients, that has $3 + 2i$ and 7 as zeros.

Solution *See page S22.* $P(x) = x^3 - 13x^2 + 55x - 91$

A polynomial function that has a given set of zeros is not unique. For example, $P(x) = x^3 - 7x + 6$ has zeros 1, 2, and -3, but so does any nonzero multiple of $P(x)$, such as $S(x) = 2x^3 - 14x + 12$. This concept is illustrated in Figure 4.22. The graphs of the two polynomial functions are different, but they have the same x-intercepts.

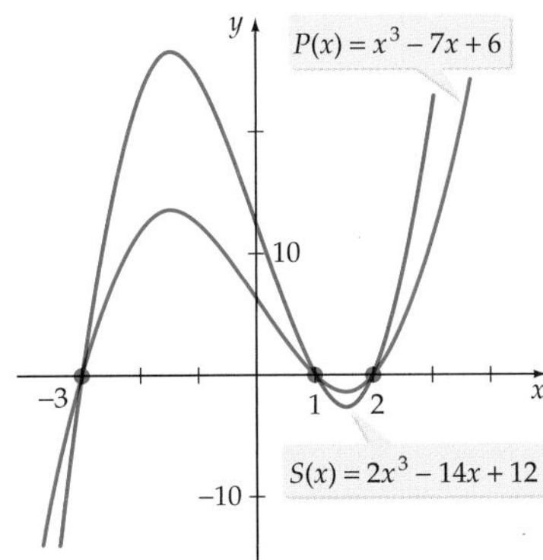

Figure 4.22

Topics for Discussion

1. What is the Fundamental Theorem of Algebra and why is this theorem so important?
2. Let $P(x)$ be a polynomial function of degree n with real coefficients. Discuss the number of *possible* real zeros of this polynomial function. Include in your discussion the cases when n is even and when n is odd.
3. Consider the graph of a polynomial function in Figure 4.23. Is it possible that the degree of the polynomial is 3? Explain.

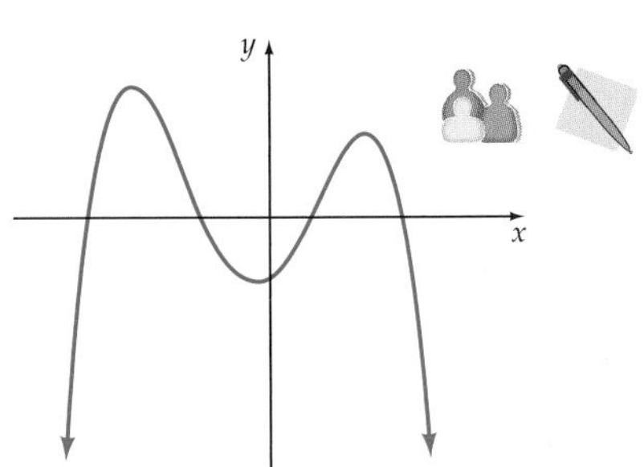

Figure 4.23

4. If two polynomial functions have exactly the same zeros, do the graphs of the polynomial functions look exactly the same?

5. Does the graph of every polynomial function have at least one x-intercept?

EXERCISES 4.4

— *Suggested Assignment: Exercises 1–49, odd; and 51–56 all.*

In Exercises 1 to 10, find all the zeros of the polynomial function and write the polynomial as a product of linear factors. (*Hint:* First determine the rational zeros.)

1. $P(x) = x^4 + x^3 - 2x^2 + 4x - 24$
   $2, -3, 2i, -2i$; $P(x) = (x-2)(x+3)(x-2i)(x+2i)$

2. $P(x) = x^4 - 3x^3 + 5x^2 - 27x - 36$
   $4, -1, 3i, -3i$; $P(x) = (x-4)(x+1)(x-3i)(x+3i)$

3. $P(x) = 2x^4 + x^3 + 39x^2 + 136x - 78$
   $\frac{1}{2}, -3, 1+5i, 1-5i$; $P(x) = \left(x - \frac{1}{2}\right)(x+3)(x-1-5i)(x-1+5i)$

4. $P(x) = x^3 - 13x^2 + 65x - 125$
   $5, 4+3i, 4-3i$; $P(x) = (x-5)(x-4-3i)(x-4+3i)$

5. $P(x) = x^5 - 9x^4 + 34x^3 - 58x^2 + 45x - 13$
   1 (multiplicity 3), $3+2i, 3-2i$; $P(x) = (x-1)^3(x-3-2i)(x-3+2i)$

6. $P(x) = x^4 - 4x^3 + 53x^2 - 196x + 196$
   2 (multiplicity 2), $7i, -7i$; $P(x) = (x-2)^2(x-7i)(x+7i)$

7. $P(x) = 2x^4 - x^3 - 15x^2 + 23x + 15$
   Answers to Exercises 7–9 are on page 359.

8. $P(x) = 3x^4 - 17x^3 - 39x^2 + 337x + 116$

9. $P(x) = 2x^4 - 14x^3 + 33x^2 - 46x + 40$

10. $P(x) = 3x^4 - 10x^3 + 15x^2 + 20x - 8$
    $-1, \frac{1}{3}, 2+2i, 2-2i$; $P(x) = (x+1)\left(x - \frac{1}{3}\right)(x-2-2i)(x-2+2i)$

In Exercises 11 to 22, use the given zero to find the remaining zeros of each polynomial function.

11. $P(x) = 2x^3 - 5x^2 + 6x - 2$; $1 + i$ $\quad 1 - i, \frac{1}{2}$

12. $P(x) = x^3 - 7x^2 + 17x - 15$; $2 - i$ $\quad 2 + i, 3$

13. $P(x) = x^3 + 3x^2 + x + 3$; $-i$ $\quad i, -3$

14. $P(x) = x^4 - 6x^3 + 71x^2 - 146x + 530$; $2 + 7i$
    $2 - 7i, 1 + 3i, 1 - 3i$

15. $P(x) = x^4 - 4x^3 + 14x^2 - 4x + 13$; $2 - 3i$ $\quad 2 + 3i, i, -i$

16. $P(x) = 8x^4 - 2x^3 + 199x^2 - 50x - 25$; $-5i$ $\quad 5i, -\frac{1}{4}, \frac{1}{2}$

17. $P(x) = x^4 - 4x^3 + 19x^2 - 30x + 50$; $1 + 3i$
    $1 - 3i, 1 + 2i, 1 - 2i$

18. $P(x) = x^5 - x^4 - 4x^3 - 4x^2 - 5x - 3$; i
    $-i, 3, -1$ (multiplicity 2)

19. $P(x) = x^5 - 3x^4 + 7x^3 - 13x^2 + 12x - 4$; $-2i$
    $2i, 1$ (multiplicity 3)

20. $P(x) = x^4 - 8x^3 + 18x^2 - 8x + 17$; i $\quad -i, 4 + i, 4 - i$

21. $P(x) = x^4 - 17x^3 + 112x^2 - 333x + 377$; $5 + 2i$

22. $P(x) = 2x^5 - 8x^4 + 61x^3 - 99x^2 + 12x + 182$; $1 - 5i$

21. $5 - 2i, \frac{7}{2} + \frac{\sqrt{3}}{2}i, \frac{7}{2} - \frac{\sqrt{3}}{2}i$ $\quad$ 22. $1 + 5i, -1, \frac{3}{2} + \frac{\sqrt{5}}{2}i, \frac{3}{2} - \frac{\sqrt{5}}{2}i$

In Exercises 23 to 30, use a graph and your knowledge of the zeros of polynomial functions to determine the *exact* values of all the solutions of each equation.

23. $2x^3 - x^2 + x - 6 = 0$ $\quad \frac{3}{2}, -\frac{1}{2} + \frac{\sqrt{7}}{2}i, -\frac{1}{2} - \frac{\sqrt{7}}{2}i$

24. $2x^3 - 5x^2 + 18x - 45 = 0$ $\quad \frac{5}{2}, 3i, -3i$

25. $24x^3 - 62x^2 - 7x + 30 = 0$ $\quad -\frac{2}{3}, \frac{3}{4}, \frac{5}{2}$

26. $12x^3 - 52x^2 + 27x + 28 = 0$ $\quad -\frac{1}{2}, \frac{4}{3}, \frac{7}{2}$

27. $x^4 - 4x^3 + 5x^2 - 4x + 4 = 0$ $\quad -i, i, 2$ (multiplicity 2)

28. $x^4 + 4x^3 + 8x^2 + 16x + 16 = 0$ $\quad -2$ (multiplicity 2), $2i, -2i$

29. $x^4 + 4x^3 - 2x^2 - 12x + 9 = 0$
    -3 (multiplicity 2), 1 (multiplicity 2)

30. $x^4 + 3x^3 - 6x^2 - 28x - 24 = 0$ $\quad 3, -2$ (multiplicity 3)

In Exercises 31 to 40, find a polynomial function of lowest degree with integer coefficients that has the given zeros.

31. $4, -3, 2$
    $P(x) = x^3 - 3x^2 - 10x + 24$

32. $-1, 1, -5$
    $P(x) = x^3 + 5x^2 - x - 5$

33. $3, 2i, -2i$
    $P(x) = x^3 - 3x^2 + 4x - 12$

34. $0, i, -i$
    $P(x) = x^3 + x$

35. $3 + i, 3 - i, 2 + 5i, 2 - 5i$
$P(x) = x^4 - 10x^3 + 63x^2 - 214x + 290$

36. $2 + 3i, 2 - 3i, -5, 2$ $P(x) = x^4 - x^3 - 9x^2 + 79x - 130$

37. $P(x) = x^5 - 22x^4 + 212x^3 - 1012x^2 + 2251x - 1830$

37. $6 + 5i, 6 - 5i, 2, 3, 5$

38. $\frac{1}{2}, 4 - i, 4 + i$
$P(x) = 2x^3 - 17x^2 + 42x - 17$

39. $\frac{3}{4}, 2 + 7i, 2 - 7i$
$P(x) = 4x^3 - 19x^2 + 224x - 159$

40. $\frac{1}{4}, -\frac{1}{5}, i, -i$
$P(x) = 20x^4 - x^3 + 19x^2 - x - 1$

In Exercises 41 to 48, find a polynomial function $P(x)$ with real coefficients that has the indicated zeros and satisfies the given conditions.

41. Zeros: $2 - 5i, -4$; degree 3 $P(x) = x^3 + 13x + 116$

42. Zeros: $1 - 3i, 4$; degree 3 $P(x) = x^3 - 6x^2 + 18x - 40$

43. Zeros: $4 + 3i, 5 - i$; degree 4
$P(x) = x^4 - 18x^3 + 131x^2 - 458x + 650$

44. Zeros: $i, 3 - 5i$; degree 4 $P(x) = x^4 - 6x^3 + 35x^2 - 6x + 34$

45. Zeros: $-1, 2, 3$; degree 3; $P(1) = 12$
$P(x) = 3x^3 - 12x^2 + 3x + 18$

46. Zeros: $3i, 2$; degree 3; $P(3) = 27$ $P(x) = \frac{3}{2}x^3 - 3x^2 + \frac{27}{2}x - 27$

47. Zeros: $3, -5, 2 + i$; degree 4; $P(1) = 48$
$P(x) = -2x^4 + 4x^3 + 36x^2 - 140x + 150$

48. Zeros: $\frac{1}{2}, 1 - i$; degree 3; $P(4) = 140$
$P(x) = 4x^3 - 10x^2 + 12x - 4$

49. Verify that $P(x) = x^3 - x^2 - ix^2 - 9x + 9 + 9i$ has $1 + i$ as a zero and that its conjugate $1 - i$ is not a zero. Explain why this does not contradict the Conjugate Pair Theorem. The Conjugate Pair Theorem does not apply because some of the coefficients of the polynomial are not real numbers.

50. Verify that $P(x) = x^3 - x^2 - ix^2 - 20x + ix + 20i$ has a zero of i and that its conjugate $-i$ is not a zero. Explain why this does not contradict the Conjugate Pair Theorem. The Conjugate Pair Theorem does not apply because some of the coefficients of the polynomial are not real numbers.

Prepare for Section 4.5

51. Simplify: $\frac{x^2 - 9}{x^2 - 2x - 15}$ [P.5] $\frac{x-3}{x-5}$

52. Evaluate $\frac{x + 4}{x^2 - 2x - 5}$ for $x = -1$. [P.1] $-\frac{3}{2}$

53. Evaluate $\frac{2x^2 + 4x - 5}{x + 6}$ for $x = -3$. [P.1] $\frac{1}{3}$

54. For what values of x does the denominator of $\frac{x^2 - x - 5}{2x^3 + x^2 - 15x}$ equal zero? [1.3] $x = 0, -3, \frac{5}{2}$

55. Determine the degree of the numerator and the degree of the denominator of $\frac{x^3 + 3x^2 - 5}{x^2 - 4}$. [P.3]
Degree of numerator: 3; degree of denominator: 2

56. Write $\frac{x^3 + 2x^2 - x - 11}{x^2 - 2x}$ in $Q(x) + \frac{R(x)}{x^2 - 2x}$ form. [4.1]
$x + 4 + \frac{7x - 11}{x^2 - 2x}$

Explorations

1. ***Investigate the Roots of a Cubic Equation*** Hieronimo Cardano, using a technique he learned from Nicolo Tartaglia, was able to solve some cubic equations.

 a. Show that the cubic equation $x^3 + bx^2 + cx + d = 0$ can be transformed into the "reduced" cubic $y^3 + my = n$, where m and n are constants, depending on b, c, and d, by using the substitution $x = y - \frac{b}{3}$.

 b. Cardano then showed that a solution of the reduced cubic is given by

$$\sqrt[3]{\frac{n}{2} + \sqrt{\frac{n^2}{4} + \frac{m^3}{27}}} - \sqrt[3]{-\frac{n}{2} + \sqrt{\frac{n^2}{4} + \frac{m^3}{27}}}$$

 Use Cardano's procedure to solve the equation $x^3 - 6x^2 + 20x - 33 = 0$. See the Instructor's Resource Manual.

7. $-3, -\frac{1}{2}, 2 + i, 2 - i$; $P(x) = (x + 3)\left(x + \frac{1}{2}\right)(x - 2 - i)(x - 2 + i)$

8. $-4, -\frac{1}{3}, 5 + 2i, 5 - 2i$; $P(x) = (x + 4)\left(x + \frac{1}{3}\right)(x - 5 - 2i)(x - 5 + 2i)$

9. $4, 2, \frac{1}{2} + \frac{3}{2}i, \frac{1}{2} - \frac{3}{2}i$; $P(x) = (x - 4)(x - 2)\left(x - \frac{1}{2} - \frac{3}{2}i\right)\left(x - \frac{1}{2} + \frac{3}{2}i\right)$

SECTION 4.5 Graphs of Rational Functions and Their Applications

- Vertical and Horizontal Asymptotes
- A Sign Property of Rational Functions
- A General Graphing Procedure
- Slant Asymptotes
- Graph Rational Functions That Have a Common Factor
- Applications of Rational Functions

Vertical and Horizontal Asymptotes

If $P(x)$ and $Q(x)$ are polynomials, then the function F given by

$$F(x) = \frac{P(x)}{Q(x)}$$

is called a **rational function.** The domain of F is the set of all real numbers except those for which $Q(x) = 0$. For example, the domain of

$$F(x) = \frac{x^2 - x - 5}{x(2x - 5)(x + 3)}$$

is the set of all real numbers except $0, \frac{5}{2}$, and -3.

The graph of $G(x) = \dfrac{x + 1}{x - 2}$ is given in Figure 4.24. The graph shows that G has the following properties:

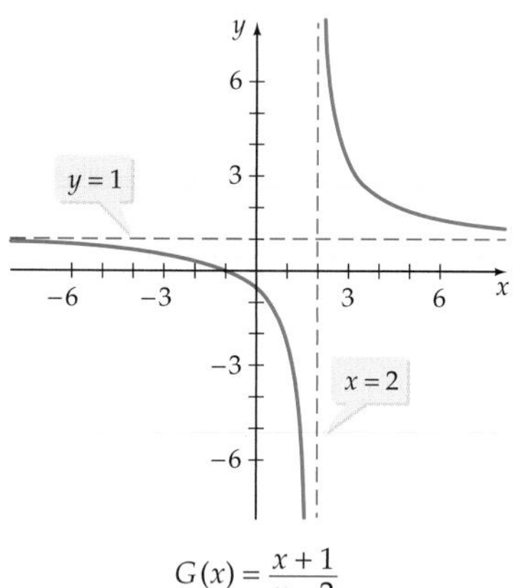

$G(x) = \dfrac{x+1}{x-2}$

Figure 4.24

- The graph has an x-intercept at $(-1, 0)$ and a y-intercept at $\left(0, -\frac{1}{2}\right)$.
- The graph does not exist when $x = 2$.

Note the behavior of the graph as x takes on values that are close to 2 but *less* than 2. Mathematically, we say that "x approaches 2 from the left."

x	1.9	1.95	1.99	1.995	1.999
$G(x)$	-29	-59	-299	-599	-2999

From this table and the graph, it appears that as x approaches 2 from the left, the functional values $G(x)$ decrease without bound.

- In this case, we say that "$G(x)$ approaches negative infinity."

Now observe the behavior of the graph as x takes on values that are close to 2 but *greater* than 2. Mathematically, we say that "x approaches 2 from the right."

x	2.1	2.05	2.01	2.005	2.001
$G(x)$	31	61	301	601	3001

From this table and the graph, it appears that as x approaches 2 from the right, the functional values $G(x)$ increase without bound.

- In this case, we say that "$G(x)$ approaches positive infinity."

Now consider the values of $G(x)$ as x *increases* without bound. The following table gives values of $G(x)$ for selected values of x.

x	1000	5000	10,000	50,000	100,000
$G(x)$	1.00301	1.00060	1.00030	1.00006	1.00003

- As x increases without bound, the values of $G(x)$ become closer to 1.

Now let the values of x *decrease* without bound. The table below gives the values of $G(x)$ for selected values of x.

x	−1000	−5000	−10,000	−50,000	−100,000
$G(x)$	0.997006	0.999400	0.999700	0.999940	0.999970

- As x decreases without bound, the values of $G(x)$ become closer to 1.

When we are discussing graphs that increase or decrease without bound, it is convenient to use mathematical notation. The notation

$$f(x) \to \infty \quad \text{as} \quad x \to a^+$$

means that the functional values $f(x)$ increase without bound as x approaches a from the right. Recall that the symbol ∞ does not represent a real number but is used merely to describe the concept of a variable taking on larger and larger values without bound. See Figure 4.25a.

The notation

$$f(x) \to \infty \quad \text{as} \quad x \to a^-$$

means that the function values $f(x)$ increase without bound as x approaches a from the left. See Figure 4.25b.

The notation

$$f(x) \to -\infty \quad \text{as} \quad x \to a^+$$

means that the functional values $f(x)$ decrease without bound as x approaches a from the right. See Figure 4.25c.

The notation

$$f(x) \to -\infty \quad \text{as} \quad x \to a^-$$

means that the functional values $f(x)$ decrease without bound as x approaches a from the left. See Figure 4.25d.

Each graph in Figure 4.25 approaches a vertical line through $(a, 0)$ as $x \to a^+$ or a^-. The line is said to be a **vertical asymptote** of the graph.

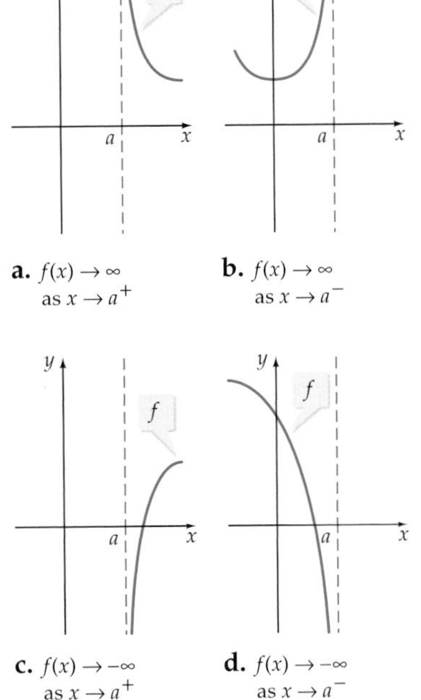

Figure 4.25

Definition of a Vertical Asymptote

The line $x = a$ is a **vertical asymptote** of the graph of a function F provided

$$F(x) \to \infty \quad \text{or} \quad F(x) \to -\infty$$

as x approaches a from either the left or right.

In Figure 4.24, the line $x = 2$ is a vertical asymptote of the graph of G. Note that the graph of G in Figure 4.24 also approaches the horizontal line $y = 1$ as $x \to \infty$ and as $x \to -\infty$. The line $y = 1$ is a **horizontal asymptote** of the graph of G.

Definition of a Horizontal Asymptote

The line $y = b$ is a **horizontal asymptote** of the graph of a function F provided

$$F(x) \to b \quad \text{as} \quad x \to \infty \quad \text{or} \quad x \to -\infty$$

Figure 4.26 illustrates some of the ways in which the graph of a rational function may approach its horizontal asymptote. It is common practice to display the asymptotes of the graph of a rational function by using dashed lines. Although a rational function may have several vertical asymptotes, it can have at most one horizontal asymptote. The graph may intersect its horizontal asymptote.

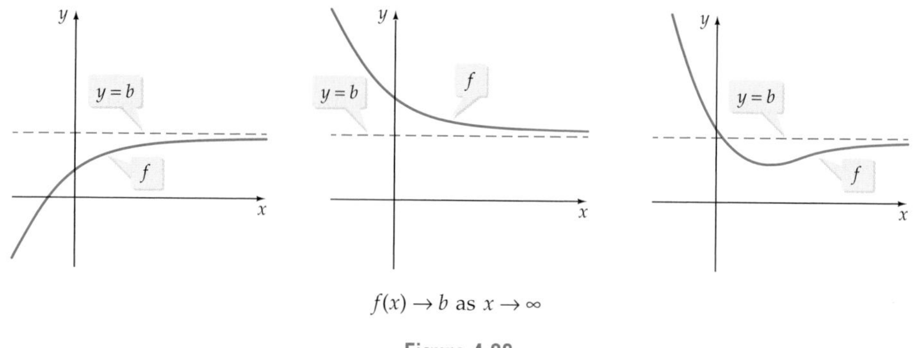

$f(x) \to b$ as $x \to \infty$

Figure 4.26

? QUESTION Can a graph of a rational function cross its vertical asymptote? Why or why not?

Geometrically, a line is an asymptote of a curve if the distance between the line and a point $P(x, y)$ on the curve approaches zero as the distance between the origin and the point P increases without bound.

Vertical asymptotes of the graph of a rational function can be found by using the following theorem.

INSTRUCTOR NOTE

Stress that the Theorem on Vertical Asymptotes is valid only for rational functions whose numerator and denominator have no common factor.

Theorem on Vertical Asymptotes

If the real number a is a zero of the denominator $Q(x)$, then the graph of $F(x) = P(x)/Q(x)$, where $P(x)$ and $Q(x)$ have no common factors, has the vertical asymptote $x = a$.

? ANSWER No. If $x = a$ is a vertical asymptote of a rational function R, then $R(a)$ is undefined.

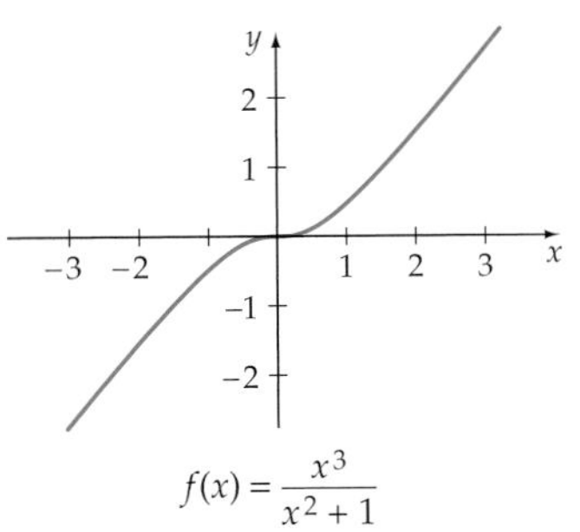

$f(x) = \dfrac{x^3}{x^2+1}$

Figure 4.27

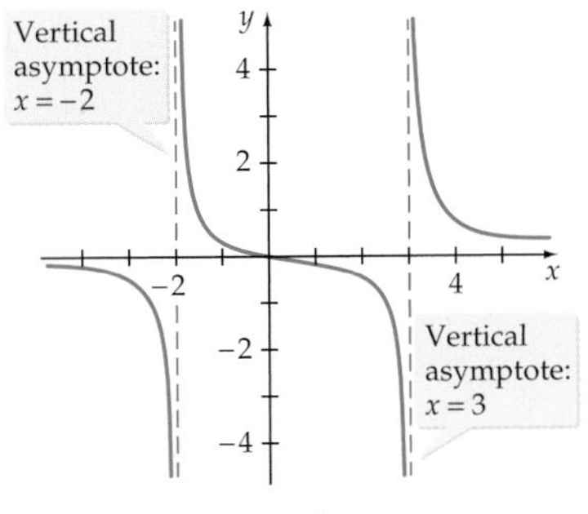

$g(x) = \dfrac{x}{x^2-x-6}$

Figure 4.28

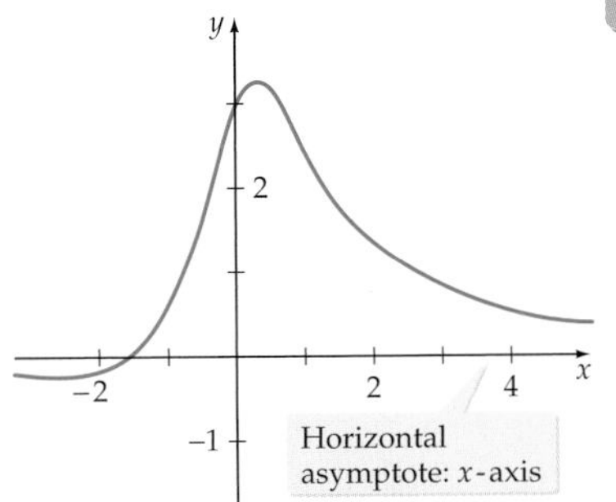

$f(x) = \dfrac{2x+3}{x^2+1}$

Figure 4.29

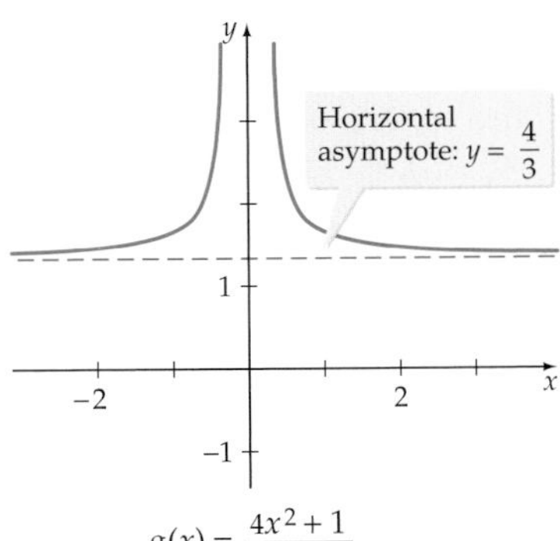

$g(x) = \dfrac{4x^2+1}{3x^2}$

Figure 4.30

EXAMPLE 1 Find the Vertical Asymptotes of a Rational Function

Find the vertical asymptotes of each rational function.

Alternative to Example 1
Exercise 2, page 372

a. $f(x) = \dfrac{x^3}{x^2 + 1}$ **b.** $g(x) = \dfrac{x}{x^2 - x - 6}$

Solution

a. To find the vertical asymptotes, determine the real zeros of the denominator. The denominator $x^2 + 1$ has no real zeros, so the graph of f has no vertical asymptotes. See Figure 4.27.

b. The denominator $x^2 - x - 6 = (x - 3)(x + 2)$ has zeros of 3 and −2. The numerator has no common factors with the denominator, so $x = 3$ and $x = -2$ are both vertical asymptotes of the graph of g, as shown in Figure 4.28.

CHECK YOUR PROGRESS 1 Find the vertical asymptotes of $F(x) = \dfrac{3x^2 + 5}{x^2 - 4}$.

Solution *See page S22.* $x = 2, x = -2$

The following theorem indicates that a horizontal asymptote can be determined by examining the leading terms of the numerator and the denominator of a rational function.

P

Theorem on Horizontal Asymptotes

Let
$$F(x) = \frac{a_n x^n + a_{n-1}x^{n-1} + \cdots + a_1 x + a_0}{b_m x^m + b_{m-1}x^{m-1} + \cdots + b_1 x + b_0}$$

be a rational function with numerator of degree n and denominator of degree m.

1. If $n < m$, then the x-axis, which is the line $y = 0$, is the horizontal asymptote of the graph of F.
2. If $n = m$, then the line $y = a_n/b_m$ is the horizontal asymptote of the graph of F.
3. If $n > m$, the graph of F has no horizontal asymptote.

EXAMPLE 2 Find the Horizontal Asymptote of a Rational Function

Find the horizontal asymptote of each rational function.

a. $f(x) = \dfrac{2x + 3}{x^2 + 1}$ **b.** $g(x) = \dfrac{4x^2 + 1}{3x^2}$ **c.** $h(x) = \dfrac{x^3 + 1}{x - 2}$

Alternative to Example 2
Exercise 6, page 372

Solution

a. The degree of the numerator $2x + 3$ is less than the degree of the denominator $x^2 + 1$. By the Theorem on Horizontal Asymptotes, the x-axis is the horizontal asymptote of f. See the graph of f in Figure 4.29.

b. The numerator $4x^2 + 1$ and the denominator $3x^2$ of g are both of degree 2. By the Theorem on Horizontal Asymptotes, the line $y = \frac{4}{3}$ is the horizontal asymptote of g. See the graph of g in Figure 4.30.

c. The degree of the numerator $x^3 + 1$ is larger than the degree of the denominator $x - 2$, so by the Theorem on Horizontal Asymptotes, the graph of h has no horizontal asymptotes.

CHECK YOUR PROGRESS 2 Find the horizontal asymptote of

$$F(x) = \frac{3x^3 - 27x^2 + 5x - 11}{x^5 - 2x^3 + 7}.$$

Solution *See page S22.* $y = 0$

The proof of the Theorem on Horizontal Asymptotes makes use of the technique employed in the following verification. To verify that

$$y = \frac{5x^2 + 4}{3x^2 + 8x + 7}$$

has a horizontal asymptote of $y = \frac{5}{3}$, divide the numerator and the denominator by the largest power of the variable x (x^2 in this case).

$$y = \frac{\dfrac{5x^2 + 4}{x^2}}{\dfrac{3x^2 + 8x + 7}{x^2}} = \frac{5 + \dfrac{4}{x^2}}{3 + \dfrac{8}{x} + \dfrac{7}{x^2}}, \quad x \neq 0$$

As x increases without bound or decreases without bound, the fractions $\frac{4}{x^2}$, $\frac{8}{x}$, and $\frac{7}{x^2}$ approach zero. Thus

$$y \to \frac{5 + 0}{3 + 0 + 0} = \frac{5}{3} \quad \text{as} \quad x \to \pm\infty$$

and hence the line $y = \frac{5}{3}$ is a horizontal asymptote of the graph.

▪ A Sign Property of Rational Functions

The zeros and vertical asymptotes of a rational function F divide the x-axis into intervals. In each interval, $F(x)$ is positive for all x in the interval or $F(x)$ is negative for all x in the interval. For example, consider the rational function

$$g(x) = \frac{x + 1}{x^2 + 2x - 3}$$

which has vertical asymptotes of $x = -3$ and $x = 1$ and a zero of -1. These three numbers divide the x-axis into the four intervals $(-\infty, -3)$, $(-3, -1)$, $(-1, 1)$, and $(1, \infty)$. Note in Figure 4.31 that the graph of g is negative for all x such that $x < -3$, positive for all x such that $-3 < x < -1$, negative for all x such that $-1 < x < 1$, positive for all x such that $x > 1$.

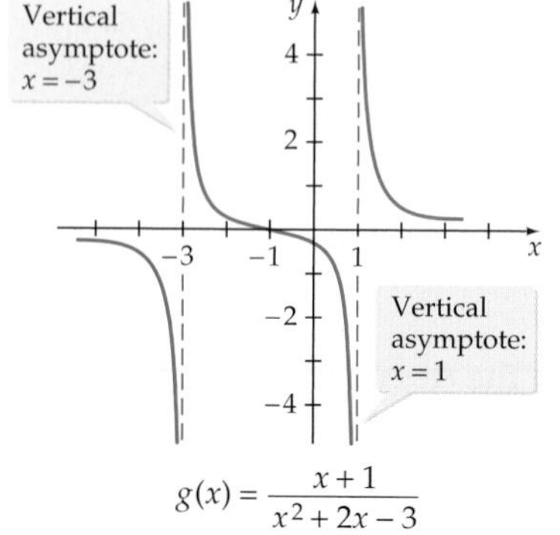

Figure 4.31

▪ A General Graphing Procedure

If $F(x) = P(x)/Q(x)$, where $P(x)$ and $Q(x)$ are polynomials that have no common factors, then the following general procedure offers useful guidelines for graphing F.

P

General Procedure for Graphing Rational Functions That Have No Common Factors

1. *Asymptotes* Find the real zeros of the denominator $Q(x)$. For each zero a, draw the dashed line $x = a$. Each line is a vertical asymptote of the graph of F. Also graph any horizontal asymptotes.
2. *Intercepts* Find the real zeros of the numerator $P(x)$. For each zero a, plot the point $(a, 0)$. Each such point is an x-intercept of the graph of F. For each x-intercept use the even and odd powers of $(x - c)$ to determine if the graph crosses the x-axis at the intercept or if the graph intersects but does not cross the x-axis. Also evaluate $F(0)$. Plot $(0, F(0))$, the y-intercept of the graph of F.
3. *Symmetry* Use the tests for symmetry to determine whether the graph of the function has symmetry with respect to the y-axis or symmetry with respect to the origin.
4. *Additional points* Plot at least two points that lie in the intervals between and beyond the vertical asymptotes and the x-intercepts.
5. *Behavior near asymptotes* If $x = a$ is a vertical asymptote, determine whether $F(x) \to \infty$ or $F(x) \to -\infty$ as $x \to a^-$ and also as $x \to a^+$.
6. *Complete the sketch* Use all the information obtained above to sketch the graph of F.

Alternative to Example 3
Exercise 16, page 372

INSTRUCTOR NOTE
Example 3 shows how graphing technology can be used in conjunction with the analytical concepts that were discussed in this section. The idea is that technology is useful for displaying graphs and that analytical concepts can be used to ensure that the graph reflects what was intended and that no errors were made when entering the function. If the intercepts, symmetry, and asymptotes that are determined analytically are not reflected in the graph, then either the expression was entered incorrectly or the analysis was performed incorrectly.

EXAMPLE 3 Graph a Rational Function

Sketch a graph of $f(x) = \dfrac{2x^2 - 18}{x^2 + 3}$.

Solution

Asymptotes The denominator $x^2 + 3$ has no real zeros, so the graph of f has no vertical asymptotes. The numerator and denominator both are of degree 2. The leading coefficients are 2 and 1, respectively. By the Theorem on Horizontal Asymptotes, the graph of f has a horizontal asymptote of $y = \frac{2}{1} = 2$.

Intercepts The zeros of the numerator occur when $2x^2 - 18 = 0$ or, solving for x, when $x = -3$ and $x = 3$. Therefore, the x-intercepts are $(-3, 0)$ and $(3, 0)$. The factored numerator is $2(x + 3)(x - 3)$. Each linear factor has an exponent of 1, an odd number. Thus the graph crosses the x-axis at its x-intercepts. To find the y-intercept, evaluate f when $x = 0$. This gives $y = -6$. Therefore, the y-intercept is $(0, -6)$.

Symmetry Below we show that $f(-x) = f(x)$, which means that f is an even function and therefore its graph is symmetric with respect to the y-axis.

$$f(-x) = \frac{2(-x)^2 - 18}{(-x)^2 + 3} = \frac{2x^2 - 18}{x^2 + 3} = f(x)$$

Additional points The intervals determined by the x-intercepts are $x < -3$, $-3 < x < 3$, and $x > 3$. Generally, it is necessary to determine points in all intervals. However, because f is an even function, its graph is symmetric with respect to the y-axis. The following table lists a few points for $x > 0$. Symmetry can be used to locate corresponding points for $x < 0$.

Continued ➤

x	1	2	6
$f(x)$	-4	$-\frac{10}{7} \approx -1.43$	$\frac{18}{13} \approx 1.38$

Behavior near asymptotes As x increases or decreases without bound, $f(x)$ approaches the horizontal asymptote $y = 2$.

To determine whether the graph of f intersects the horizontal asymptote at any point, solve the equation $f(x) = 2$.

There are no solutions of $f(x) = 2$ because

$$\frac{2x^2 - 18}{x^2 + 3} = 2 \quad \text{implies} \quad 2x^2 - 18 = 2x^2 + 6 \quad \text{implies} \quad -18 = 6$$

This is not possible. Thus the graph of f does not intersect the horizontal asymptote but approaches it from below as x increases or decreases without bound.

Complete the sketch Use the summary in Table 4.3, to the left, to finish the sketch. The completed graph is shown in Figure 4.32.

Table 4.3

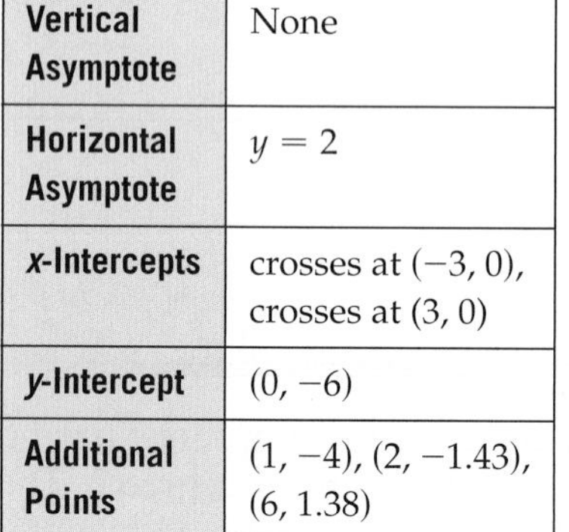

Vertical Asymptote	None
Horizontal Asymptote	$y = 2$
x-Intercepts	crosses at $(-3, 0)$, crosses at $(3, 0)$
y-Intercept	$(0, -6)$
Additional Points	$(1, -4)$, $(2, -1.43)$, $(6, 1.38)$

Horizontal asymptote: $y = 2$

$f(x) = \frac{2x^2 - 18}{x^2 + 3}$

Figure 4.32

CHECK YOUR PROGRESS 3 Sketch a graph of $F(x) = \frac{1}{x - 2}$.

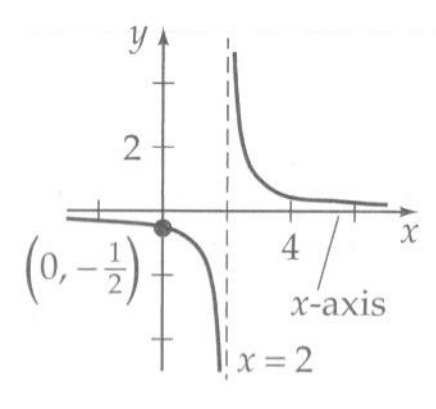

Solution *See page S22.*

Alternative to Example 4
Exercise 22, page 372

EXAMPLE 4 Graph a Rational Function

Sketch a graph of $h(x) = \frac{x^2 + 1}{x^2 + x - 2}$.

Solution

Asymptotes The denominator $x^2 + x - 2 = (x + 2)(x - 1)$ has zeros -2 and 1; because there are no common factors of the numerator and the denominator, the lines $x = -2$ and $x = 1$ are vertical asymptotes.

Continued ➤

The numerator and denominator both are of degree 2. The leading coefficients of the numerator and denominator are both 1. Thus h has the horizontal asymptote $y = \frac{1}{1} = 1$.

Intercepts The numerator $x^2 + 1$ has no real zeros, so the graph of h has no x-intercepts. Because $h(0) = -0.5$, h has the y-intercept $(0, -0.5)$.

Symmetry By applying the tests for symmetry, we can determine that the graph of h is not symmetric with respect to the origin or to the y-axis.

Additional points The intervals determined by the vertical asymptotes are $(-\infty, -2)$, $(-2, 1)$, and $(1, \infty)$. Plot a few points from each interval.

x	-5	-3	-1	0.5	2	3	4
$h(x)$	$\frac{13}{9}$	$\frac{5}{2}$	-1	-1	$\frac{5}{4}$	1	$\frac{17}{18}$

The graph of h will intersect the horizontal asymptote $y = 1$ exactly once. This can be determined by solving the equation $h(x) = 1$.

$$\frac{x^2 + 1}{x^2 + x - 2} = 1$$

$$x^2 + 1 = x^2 + x - 2 \qquad \text{■ Multiply both sides by } x^2 + x - 2.$$

$$1 = x - 2$$

$$3 = x$$

The only solution is $x = 3$. Therefore, the graph of h intersects the horizontal asymptote at $(3, 1)$.

Behavior near asymptotes As x approaches -2 from the left, the denominator $(x + 2)(x - 1)$ approaches 0 but remains positive. The numerator $x^2 + 1$ approaches 5, which is positive, so the quotient $h(x)$ increases without bound. Stated in mathematical notation,

$$h(x) \to \infty \quad \text{as} \quad x \to -2^-$$

Similarly, it can be determined that

$$h(x) \to -\infty \quad \text{as} \quad x \to -2^+$$

$$h(x) \to -\infty \quad \text{as} \quad x \to 1^-$$

$$h(x) \to \infty \quad \text{as} \quad x \to 1^+$$

Complete the sketch Use the summary in Table 4.4 to obtain the graph sketched in Figure 4.33.

Continued ➤

Table 4.4

Vertical Asymptote	$x = -2, x = 1$
Horizontal Asymptote	$y = 1$
***x*-Intercepts**	None
***y*-Intercept**	$(0, -0.5)$
Additional Points	$(-5, 1.\overline{4})$, $(-3, 2.5)$, $(-1, -1)$, $(0.5, -1)$, $(2, 1.25)$, $(3, 1)$, $(4, 0.9\overline{4})$

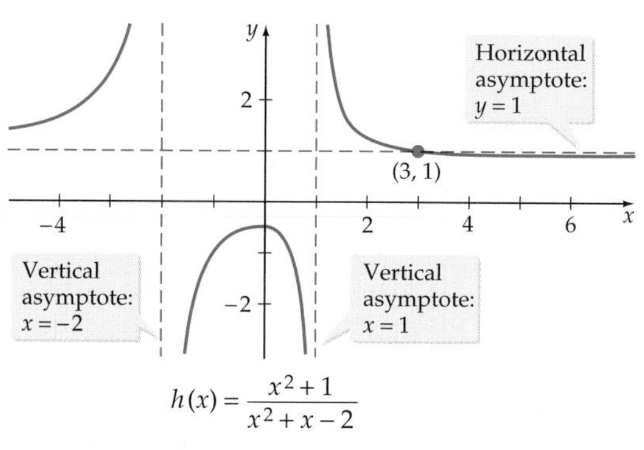

Figure 4.33

CHECK YOUR PROGRESS 4 Sketch a graph of $F(x) = \dfrac{x^2}{x^2 - 6x + 9}$.

Solution *See page S22.*

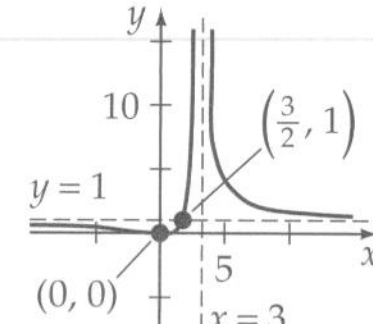

▪ Slant Asymptotes

Some rational functions have an asymptote that is neither vertical nor horizontal, but slanted.

> **Theorem on Slant Asymptotes**
>
> The rational function given by $F(x) = P(x)/Q(x)$, where $P(x)$ and $Q(x)$ have no common factors, has a **slant asymptote** if the degree of the polynomial $P(x)$ in the numerator is one greater than the degree of the polynomial $Q(x)$ in the denominator.

To find the slant asymptote, divide $P(x)$ by $Q(x)$ and write $F(x)$ in the form

$$F(x) = \frac{P(x)}{Q(x)} = (mx + b) + \frac{r(x)}{Q(x)}$$

where the degree of $r(x)$ is less than the degree of $Q(x)$. Because

$$\frac{r(x)}{Q(x)} \to 0 \quad \text{as} \quad x \to \pm\infty$$

we know that $F(x) \to mx + b$ as $x \to \pm\infty$.

The line represented by $y = mx + b$ is the slant asymptote of the graph of F.

Alternative to Example 5
Exercise 32, page 372

EXAMPLE 5 Find the Slant Asymptote of a Rational Function

Find the slant asymptote of $f(x) = \dfrac{2x^3 + 5x^2 + 1}{x^2 + x + 3}$.

Continued ➤

Solution Because the degree of the numerator $2x^3 + 5x^2 + 1$ is exactly one larger than the degree of the denominator $x^2 + x + 3$ and f is in simplest form, f has a slant asymptote. To find the asymptote, divide $2x^3 + 5x^2 + 1$ by $x^2 + x + 3$.

$$\begin{array}{r} 2x + 3 \\ x^2 + x + 3 \overline{)2x^3 + 5x^2 + 0x + 1} \\ \underline{2x^3 + 2x^2 + 6x } \\ 3x^2 - 6x + 1 \\ \underline{3x^2 + 3x + 9} \\ -9x - 8 \end{array}$$

Therefore,

$$f(x) = \frac{2x^3 + 5x^2 + 1}{x^2 + x + 3} = (2x + 3) + \frac{-9x - 8}{x^2 + x + 3}$$

and the line given by $y = 2x + 3$ is the slant asymptote for the graph of f. Figure 4.34 shows the graph of f and its slant asymptote.

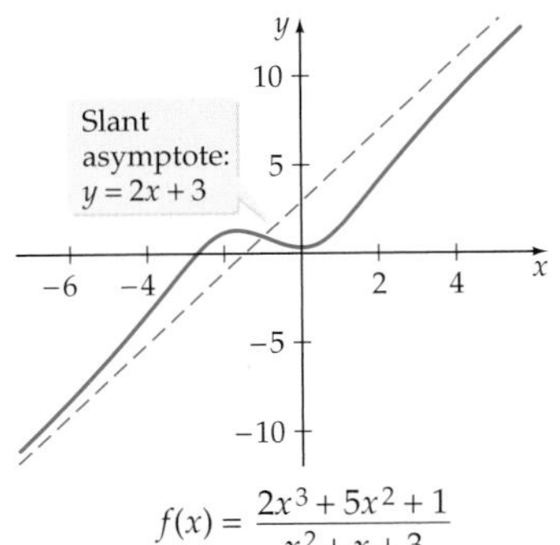

$f(x) = \dfrac{2x^3 + 5x^2 + 1}{x^2 + x + 3}$

Figure 4.34

CHECK YOUR PROGRESS 5 Find the slant asymptote of $F(x) = \dfrac{x^3 - 2x^2 + 3x + 4}{x^2 - 3x + 5}$.

Solution *See page S22.* $y = x + 1$

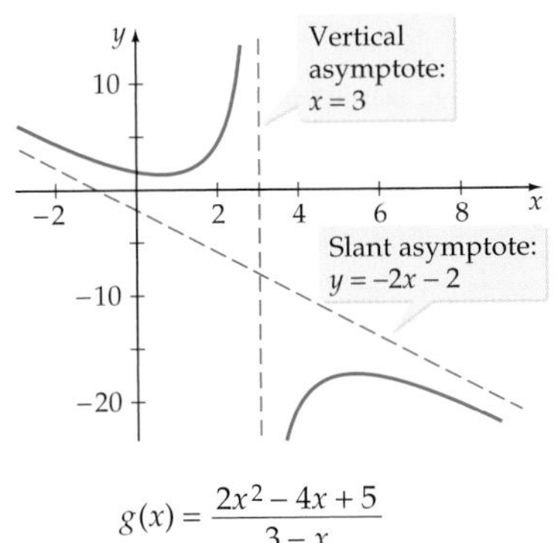

$g(x) = \dfrac{2x^2 - 4x + 5}{3 - x}$

Figure 4.35

The function f in Example 5 does not have a vertical asymptote because the denominator $x^2 + x + 3$ does not have any real zeros. However, the function

$$g(x) = \frac{2x^2 - 4x + 5}{3 - x}$$

has both a slant asymptote and a vertical asymptote. The vertical asymptote is $x = 3$, and the slant asymptote is $y = -2x - 2$. Figure 4.35 shows the graph of g and its asymptotes.

▪ Graph Rational Functions That Have a Common Factor

If a rational function has a numerator and denominator that have a common factor, then you should reduce the rational function to lowest terms before you apply the general procedure for sketching the graph of a rational function.

Alternative to Example 6
Exercise 48, page 373

EXAMPLE 6 Graph a Rational Function That Has a Common Factor

Sketch the graph of $f(x) = \dfrac{x^2 - 3x - 4}{x^2 - 6x + 8}$.

Continued ➤

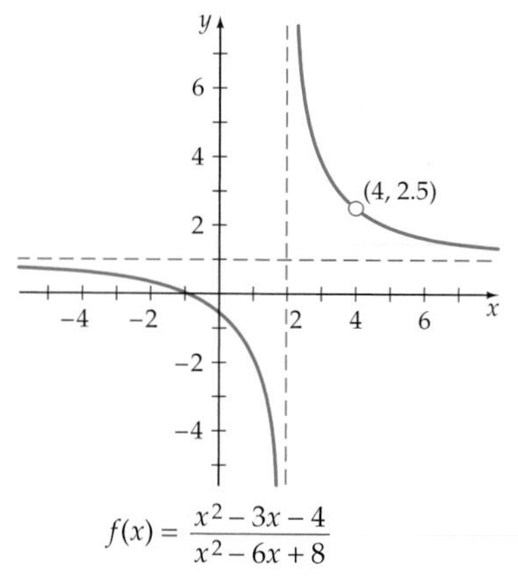

Figure 4.36

Solution Factor the numerator and denominator to obtain

$$f(x) = \frac{x^2 - 3x - 4}{x^2 - 6x + 8} = \frac{(x + 1)(x - 4)}{(x - 2)(x - 4)}, \quad x \neq 2, x \neq 4$$

Thus for all x values other than $x = 4$, the graph of f is the same as the graph of

$$G(x) = \frac{x + 1}{x - 2}$$

Figure 4.24 on page 360, shows a graph of G. The graph of f will be the same as this graph, except that it will have an open circle at (4, 2.5) to indicate that it is undefined at $x = 4$. See the graph of f in Figure 4.36.

CHECK YOUR PROGRESS 6 Sketch the graph of $F(x) = \frac{x^2 - x - 12}{x^2 - 2x - 8}$.

Solution *See page S22.*

QUESTION Does $F(x) = \frac{x^2 - x - 6}{x^2 - 9}$ have a vertical asymptote at $x = 3$?

Applications of Rational Functions

Alternative to Example 7
Exercise 60, page 374

Figure 4.37

EXAMPLE 7 Solve an Application

A cylindrical soft drink can is to be constructed so that it will have a volume of 21.6 cubic inches. See Figure 4.37.

a. Write the total surface area A of the can as a function of r, where r is the radius of the can in inches.

b. Use a graphing utility to estimate the value of r (to the nearest tenth of an inch) that produces the minimum surface area.

Continued ➤

ANSWER No. $F(x) = \frac{x^2 - x - 6}{x^2 - 9} = \frac{(x - 3)(x + 2)}{(x - 3)(x + 3)} = \frac{x + 2}{x + 3}, x \neq 3$. As $x \to 3$, $F(x) \to \frac{5}{6}$.

Solution

a. The formula for the volume of a cylinder is $V = \pi r^2 h$, where r is the radius and h is the height. Because we are given that the volume is 21.6 cubic inches, we have

$$21.6 = \pi r^2 h$$

$$\frac{21.6}{\pi r^2} = h \qquad \text{■ Solve for } h.$$

The surface area of the cylinder is given by

$$A = 2\pi r^2 + 2\pi rh$$

$$A = 2\pi r^2 + 2\pi r\left(\frac{21.6}{\pi r^2}\right) \qquad \text{■ Substitute for } h.$$

$$A = 2\pi r^2 + \frac{2(21.6)}{r} \qquad \text{■ Simplify.}$$

$$A = \frac{2\pi r^3 + 43.2}{r} \qquad (1)$$

b. Use Equation (1) with $y = A$ and $x = r$ and a graphing utility to determine that A is a minimum when $r \approx 1.5$ inches. See Figure 4.38.

INTEGRATING TECHNOLOGY

A web applet is available to explore the relationship between the radius of a cylinder with a given volume and the surface area of the cylinder. This applet, CYLINDER, can be found on our web site at **math.college.hmco.com**

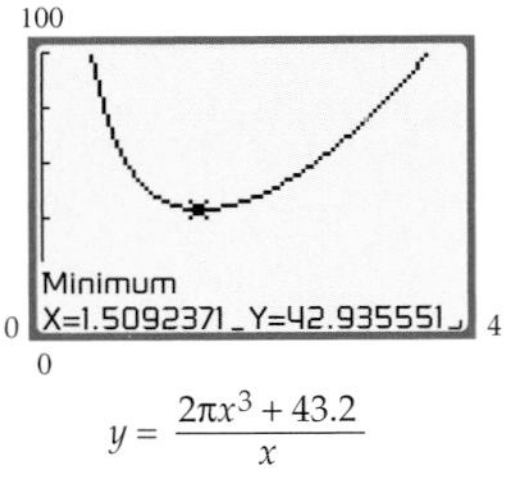

$$y = \frac{2\pi x^3 + 43.2}{x}$$

Figure 4.38

CHECK YOUR PROGRESS 7 The cost of producing x cellular telephones is given by $C(x) = 0.0006x^2 + 9x + 401{,}000$. The average cost per telephone is

$$\overline{C}(x) = \frac{C(x)}{x} = \frac{0.0006x^2 + 9x + 401{,}000}{x}$$

a. Find the average cost per telephone of producing 1000, 10,000, and 100,000 cellular telephones.

b. What is the minimum average cost per telephone? How many cellular telephones should be produced to minimize the average cost per telephone?

Solution *See page S23.* **a.** \$410.60, \$55.10, \$73.01 **b.** \$40.02; 25,852 telephones

Topics for Discussion

1. What is a rational function? Give examples of functions that are rational functions and of functions that are not rational functions.
2. Does the graph of every rational function have at least one vertical asymptote? If so, explain why. If not, give an example of a rational function without a vertical asymptote.
3. Does the graph of every rational function have a horizontal asymptote? If so, explain why. If not, give an example of a rational function without a horizontal asymptote.
4. Can the graph of a polynomial function have a vertical asymptote? a horizontal asymptote?

EXERCISES 4.5

— *Suggested Assignment: Exercises 1–61, odd.*
— *Answer graphs to Exercises 9–30, 35–52, 55c., and 61a. are on pages AA10–AA12.*

In Exercises 1 to 4, find all vertical asymptotes of each rational function.

1. $F(x) = \dfrac{2x - 1}{x^2 + 3x}$ $x = 0, x = -3$

2. $F(x) = \dfrac{4x + 1}{x^2 - 9}$ $x = 3, x = -3$

3. $F(x) = \dfrac{x^2 + 11}{6x^2 - 5x - 4}$ $x = -\dfrac{1}{2}, x = \dfrac{4}{3}$

4. $F(x) = \dfrac{3x - 5}{x^3 - 8}$ $x = 2$

In Exercises 5 to 8, find the horizontal asymptote of each rational function.

5. $F(x) = \dfrac{4x^2 + 1}{x^2 + x + 1}$ $y = 4$

6. $F(x) = \dfrac{4x - 3}{2x + 1}$ $y = 2$

7. $F(x) = \dfrac{15{,}000x^3 + 500x - 2000}{700 + 500x^3}$ $y = 30$

8. $F(x) = 6000\left(1 - \dfrac{25}{(x + 5)^2}\right)$ $y = 6000$

In Exercises 9 to 30, determine the vertical and horizontal asymptotes and sketch the graph of the rational function *F*. Label all intercepts and asymptotes.

9. $F(x) = \dfrac{1}{x + 4}$ $x = -4, y = 0$

10. $F(x) = \dfrac{3}{x^2 + 1}$ no vertical asymptote; $y = 0$

11. $F(x) = \dfrac{-4}{x - 3}$ $x = 3, y = 0$

12. $F(x) = \dfrac{-3}{x + 2}$ $x = -2, y = 0$

13. $F(x) = \dfrac{4}{x}$ $x = 0, y = 0$

14. $F(x) = \dfrac{-4}{x}$ $x = 0, y = 0$

15. $F(x) = \dfrac{x}{x + 4}$ $x = -4, y = 1$

16. $F(x) = \dfrac{x}{x - 2}$ $x = 2, y = 1$

17. $F(x) = \dfrac{x + 4}{2 - x}$ $x = 2, y = -1$

18. $F(x) = \dfrac{x + 3}{1 - x}$ $x = 1, y = -1$

19. $F(x) = \dfrac{1}{x^2 - 9}$ $x = 3, x = -3, y = 0$

20. $F(x) = \dfrac{-2}{x^2 - 4}$ $x = 2, x = -2, y = 0$

21. $F(x) = \dfrac{1}{x^2 + 2x - 3}$ $x = -3, x = 1, y = 0$

22. $F(x) = \dfrac{1}{x^2 - 2x - 8}$ $x = 4, x = -2, y = 0$

23. $F(x) = \dfrac{x^2}{x^2 + 4x + 4}$ $x = -2, y = 1$

24. $F(x) = \dfrac{2x^2}{x^2 - 1}$ $x = -1, x = 1, y = 2$

25. $F(x) = \dfrac{10}{x^2 + 2}$ no vertical asymptote; $y = 0$

26. $F(x) = \dfrac{-20}{x^2 + 4}$ no vertical asymptote; $y = 0$

27. $F(x) = \dfrac{2x^2 - 2}{x^2 - 9}$ $x = 3, x = -3, y = 2$

28. $F(x) = \dfrac{6x^2 - 5}{2x^2 + 6}$ no vertical asymptote; $y = 3$

29. $F(x) = \dfrac{x^2 + x + 4}{x^2 + 2x - 1}$ $x = -1 + \sqrt{2}, x = -1 - \sqrt{2}, y = 1$

30. $F(x) = \dfrac{2x^2 - 14}{x^2 - 6x + 5}$ $x = 5, x = 1, y = 2$

In Exercises 31 to 34, find the slant asymptote of each rational function.

31. $F(x) = \dfrac{3x^2 + 5x - 1}{x + 4}$ $y = 3x - 7$

32. $F(x) = \dfrac{2x^2 - 3x + 1}{2x + 1}$ $y = x - 2$

33. $F(x) = \dfrac{x^3 - 1}{x^2}$ $y = x$

34. $F(x) = \dfrac{4000 + 20x + 0.0001x^2}{x}$ $y = 0.0001x + 20$

In Exercises 35 to 44, determine the vertical and slant asymptotes and sketch the graph of the rational function *F*.

35. $F(x) = \dfrac{x^2 - 4}{x}$ $x = 0, y = x$

36. $F(x) = \dfrac{x^2 + 10}{2x}$ $x = 0, y = \dfrac{1}{2}x$

37. $F(x) = \dfrac{x^2 - 3x - 4}{x + 3}$ $x = -3, y = x - 6$

38. $F(x) = \dfrac{x^2 - 4x - 5}{2x + 5}$ $x = -\dfrac{5}{2}, y = \dfrac{1}{2}x - \dfrac{13}{4}$

39. $F(x) = \dfrac{2x^2 + 5x + 3}{x - 4}$ $x = 4, y = 2x + 13$

40. $F(x) = \dfrac{4x^2 - 9}{x + 3}$ $x = -3, y = 4x - 12$

41. $F(x) = \dfrac{x^2 - x}{x + 2}$ $x = -2, y = x - 3$

42. $F(x) = \dfrac{x^2 + x}{x - 1}$ $x = 1, y = x + 2$

43. $F(x) = \dfrac{x^3 + 1}{x^2 - 4}$ $x = 2, x = -2, y = x$

44. $F(x) = \dfrac{x^3 - 1}{3x^2}$ $x = 0, y = \dfrac{1}{3}x$

In Exercises 45 to 52, sketch the graph of the rational function F. (*Hint:* First examine the numerator and denominator to determine whether there are any common factors.)

45. $F(x) = \dfrac{x^2 + x}{x + 1}$

46. $F(x) = \dfrac{x^2 - 3x}{x - 3}$

47. $F(x) = \dfrac{2x^3 + 4x^2}{2x + 4}$

48. $F(x) = \dfrac{x + 3}{x^2 + x - 6}$

49. $F(x) = \dfrac{-2x^3 + 6x}{2x^2 - 6x}$

50. $F(x) = \dfrac{x^3 + 3x^2}{x(x + 3)(x - 1)}$

51. $F(x) = \dfrac{x^2 - 3x - 10}{x^2 + 4x + 4}$

52. $F(x) = \dfrac{2x^2 + x - 3}{x^2 - 2x + 1}$

Business and Economics

53. ***Average Cost of Golf Balls*** The cost, in dollars, of producing x golf balls is given by

$$C(x) = 0.43x + 76{,}000$$

The average cost per golf ball is given by

$$\overline{C}(x) = \frac{C(x)}{x} = \frac{0.43x + 76{,}000}{x}$$

a. Find the average cost of producing 1000, 10,000, and 100,000 golf balls. $76.43, $8.03, $1.19

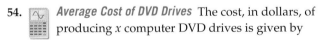

b. What is the equation of the horizontal asymptote of the graph of $\overline{C}$? Explain the significance of the horizontal asymptote as it relates to this application.

53. b. $y = 0.43$. As the number of golf balls produced increases, the average cost per golf ball approaches $.43.

54. ***Average Cost of DVD Drives*** The cost, in dollars, of producing x computer DVD drives is given by

$$C(x) = 0.001x^2 + 101x + 245{,}000$$

The average cost per DVD is given by

$$\overline{C}(x) = \frac{C(x)}{x} = \frac{0.001x^2 + 101x + 245{,}000}{x}$$

a. Find the average cost of producing 1000, 10,000, and 100,000 DVD drives. $347, $135.50, $203.45

b. What is the minimum average cost per DVD drive? How many DVD drives should be produced to minimize the average cost per DVD drive?
≈$132.30; ≈15,652 DVD drives

55. ***Desalinization*** The cost C, in dollars, to remove $p\%$ of the salt in a tank of seawater is given by

$$C(p) = \frac{2000p}{100 - p}, \quad 0 \le p < 100$$

a. Find the cost of removing 40% of the salt. $1333.33

b. Find the cost of removing 80% of the salt. $8000

c. Sketch the graph of C.

56. ***Manufacturing*** A large electronics firm finds that the number of computers it can produce per week after t weeks of production is approximated by

$$C(t) = \frac{2000t^2 + 20{,}000t}{t^2 + 10t + 25}, \quad 0 \le t \le 50$$

a. Find the number of complete computers the firm produced during the first week. 611 computers

b. Find the number of computers the firm produced during the tenth week. 1777 computers

c. What is the equation of the horizontal asymptote of the graph of C? $y = 2000$

d. Use a graphing utility to estimate how many weeks pass, to the nearest tenth of a week, until the firm can produce 1900 computers in a single week. 17.4 weeks

57. ***Wedding Expenses*** The function

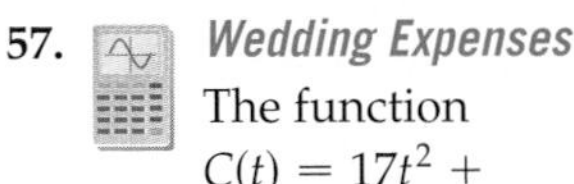

$C(t) = 17t^2 + 128t + 5900$ models the average cost of a wedding reception and the function $W(t) = 38t^2 + 291t + 15{,}208$ models the average cost of a wedding, where $t = 0$ represents the year 1990 and $0 \le t \le 12$. The rational function

$$R(t) = \frac{C(t)}{W(t)} = \frac{17t^2 + 128t + 5900}{38t^2 + 291t + 15{,}208}$$

gives the relative cost of the reception compared to the cost of a wedding.

a. Use $R(t)$ to estimate the relative cost of the reception compared to the cost of a wedding for the years $t = 0$, $t = 7$, and $t = 12$. Round your results to the nearest tenth of a percent.
$R(0) \approx 38.8\%$, $R(7) \approx 39.9\%$, $R(12) \approx 40.9\%$

b. According to the function $R(t)$, what percent of the total cost of a wedding, to the nearest tenth of a percent, will the cost of the reception approach as the years go by? ≈44.7%

58. ***Income Tax Theory*** The economist Arthur Laffer conjectured that if taxes were increased starting from very low levels, then the tax revenue received by the government would increase. But, as tax rates continued to increase, there would be a point at which the tax revenue would start to decrease. The underlying concept was that if taxes were increased too much, people would not work as hard because much of their additional income would be taken from them by the increase in taxes. Laffer illustrated his concept by drawing a curve similar to the following.

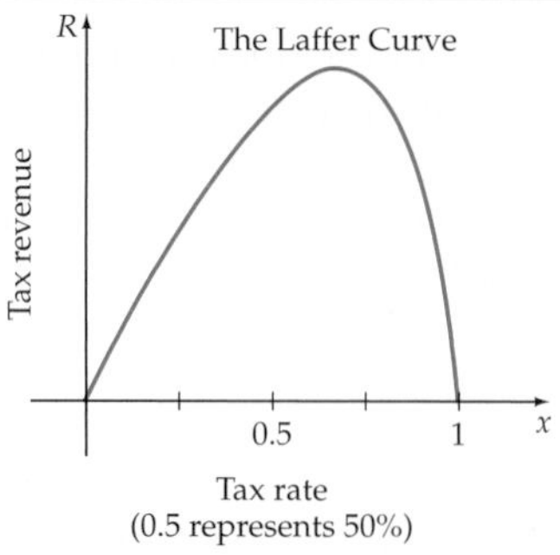

Laffer's curve shows that if the tax rate is 0%, the tax revenue will be \$0 and if the tax rate is 100%, the tax revenue will also be \$0. Laffer assumed that most people would not work if all of their income went for taxes.

Most economists agree with Laffer's basic concept, but there is much disagreement about the equation of the actual tax revenue curve R and the tax rate x that will maximize the government's tax revenues.

a. Assume that Laffer's curve is given by

$$R(x) = \frac{-6.5(x^3 + 2x^2 - 3x)}{x^2 + x + 1}$$

where R is measured in trillions of dollars. Use a graphing calculator to determine the tax rate x, to the nearest tenth of a percent, that would produce the maximum tax revenue. (*Hint:* Use a domain of $[0, 1]$ and a range of $[0, 4]$.) ≈39.7%

b. Assume that Laffer's curve is given by

$$R(x) = \frac{-1500(x^3 + 2x^2 - 3x)}{x^2 + x + 400}$$

where R is measured in trillions of dollars. Use a graphing calculator to determine the tax rate x, to the nearest tenth of a percent, that would produce the maximum tax revenue. ≈53.5%

Social Sciences

59.  ***A Population Model*** The population of a suburb, in thousands, is given by

$$P(t) = \frac{420t}{0.6t^2 + 15}$$

where t is the time in years after June 1, 1996.

a. Find the population of the suburb for $t = 1, 4$, and 10 years. 26,923, 68,293, 56,000

b. In what year will the population of the suburb reach its maximum? 2001

c. What will happen to the population as $t \to \infty$? The population will approach 0.

Life and Health Sciences

60. ***A Medication Model*** The rational function

$$M(t) = \frac{0.5t + 400}{0.04t^2 + 1}$$

models the number of milligrams of medication in the bloodstream of a patient t hours after 400 milligrams of the medication have been injected into the patient's bloodstream.

a. Find $M(5)$ and $M(10)$. Round to the nearest milligram. 201 mg, 81 mg

b. What will M approach as $t \to \infty$? 0 mg

Physics and Engineering

61. ***Minimizing Surface Area*** A cylindrical soft drink can is to be made so that it will have a volume of 354 milliliters. If r is the radius of the can in centimeters, then the total surface area A of the can is given by the rational function

$$A(r) = \frac{2\pi r^3 + 708}{r}$$

a. Graph A and use the graph to estimate (to the nearest tenth of a centimeter) the value of r that produces the minimum value of A. 3.8 cm

b. Does the graph of A have a slant asymptote? No

c. Explain the meaning of the following statement as it applies to the graph of A.

$$\text{As } r \longrightarrow \infty, A \longrightarrow 2\pi r^2.$$

As the radius r increases without bound, the surface area approaches twice the area of a circle with radius r.

62. ***Resistors in Parallel*** The following electronic circuit shows two resistors connected in parallel.

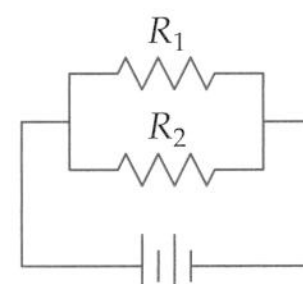

One resistor has a resistance of R_1 ohms and the other has a resistance of R_2 ohms. The total resistance for the circuit, measured in ohms, is given by the formula

$$R_T = \frac{R_1 R_2}{R_1 + R_2}$$

Assume R_1 has a fixed resistance of 10 ohms.

a. Compute R_T for $R_2 = 2$ ohms and for $R_2 = 20$ ohms. $\frac{5}{3}$ ohms, $\frac{20}{3}$ ohms

b. What happens to R_T as $R_2 \longrightarrow \infty$? $R_T \longrightarrow 10$ ohms

Explorations 

1. 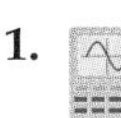  ***Parabolic Asymptotes*** It can be shown that the rational function $F(x) = R(x)/S(x)$, where $R(x)$ and $S(x)$ have no common factors, has a parabolic asymptote provided the degree of $R(x)$ is *two* greater than the degree of $S(x)$. For instance, the rational function

$$F(x) = \frac{x^3 + 2}{x + 1}$$

has a parabolic asymptote given by $y = x^2 - x + 1$.

a. Use a graphing utility to graph $F(x)$ and the parabola given by $y = x^2 - x + 1$ in the same viewing window. Does the parabola appear to be an asymptote for the graph of F? Explain. Yes. As $x \longrightarrow \infty$ and as $x \longrightarrow -\infty$, the graph of F approaches the graph of the parabola.

b. Write a paragraph that explains how to determine the equation of the parabolic asymptote for a rational function $F(x) = R(x)/S(x)$, where $R(x)$ and $S(x)$ have no common factors and the degree of $R(x)$ is two greater than the degree of $S(x)$. Divide $R(x)$ by $S(x)$ to find the quotient of $Q(x)$. The equation $y = Q(x)$ is the equation of the parabolic asymptote.

c. What is the equation of the parabolic asymptote for the rational function $G(x) = \dfrac{x^4 + x^2 + 2}{x^2 - 1}$? Use a graphing utility to graph $G(x)$ and the parabolic asymptote in the same viewing window. Does the parabola appear to be an asymptote for the graph of G? $y = x^2 + 2$; yes

d. Create a rational function that has $y = x^2 + x + 2$ as its parabolic asymptote. Explain the procedure you used to create your rational function. Answers will vary.

Chapter 4 Summary

Key Terms

absolute maximum/minimum **[p. 325]**
complex coefficients **[p. 351]**
complex zeros **[p. 351]**
cubic polynomial **[p. 351]**
depressed polynomial **[p. 318]**
Descartes' Rule of Signs **[p. 339]**
dominate term of a polynomial **[p. 323]**
existence theorems **[p. 352]**
far-left and far-right behavior **[p. 323]**
fractional form **[p. 315]**

Fundamental Theorem of Algebra **[p. 351]**
horizontal asymptote **[p. 362]**
intercept method **[p. 345]**
intersection method **[p. 345]**
leading term of a polynomial **[p. 323]**
local maximum/minimum **[p. 325]**
lower bound **[p. 338]**
multiple zero **[p. 336]**
polynomial time procedure **[p. 350]**
quartic polynomial **[p. 332]**
rational function **[p. 360]**
reduced polynomial **[p. 318]**
relative maximum/minimum **[p. 325]**
roots of a polynomial equation **[p. 336]**
simple zero **[p. 337]**
slant asymptotes **[p. 368]**
smooth continuous curve **[p. 323]**
synthetic division **[p. 313]**
turning points **[p. 325]**
upper bound **[p. 338]**
variation in sign **[p. 339]**
vertical asymptote **[p. 361]**
zero of a polynomial function **[p. 313]**
zero of multiplicity k **[p. 336]**

Essential Concepts and Formulas

- **Zeros of a Polynomial Function**
  If $P(x)$ is a polynomial function, then the values of x for which $P(x)$ is equal to 0 are called the zeros of $P(x)$. **[p. 313]**

- **The Remainder Theorem**
  If a polynomial $P(x)$ is divided by $(x - c)$, then the remainder equals $P(c)$. **[p. 313]**

- **The Factor Theorem**
  A polynomial $P(x)$ has a factor $(x - c)$ if and only if $P(c) = 0$. **[p. 317]**

- **Properties Used in Graphing Polynomial Functions**
  1. Polynomial functions are smooth continuous curves and the graph of a polynomial function of degree n has at most $n - 1$ turning points.
  2. The leading term can be used to determine the behavior of the graph of a polynomial function at the far right or the far left.
  3. The real zeros of the polynomial function determine the x-intercepts. **[pp. 323–330]**

- **Relative Minima and Maxima**
  Let f be a function defined on the open interval I, and let c be an element of I. Then
  $f(c)$ is a relative minimum of f on I if $f(c) \leq f(x)$ for all x in I.
  $f(c)$ is a relative maximum of f on I if $f(c) \geq f(x)$ for all x in I.
  [p. 325]

- **The Zero Location Theorem**
  Let $P(x)$ be a polynomial and let a and b be two distinct real numbers. If $P(a)$ and $P(b)$ have opposite signs, then there is at least one real number c between a and b such that $P(c) = 0$. **[p. 329]**

- **Even and Odd Powers of $(x - c)$ Theorem**
  If c is a real number and the polynomial function $P(x)$ has $(x - c)$ as a factor exactly k times, then the graph of P will
  - intersect but not cross the x-axis at $(c, 0)$, provided k is an even positive integer.
  - cross the x-axis at $(c, 0)$, provided k is an odd positive integer. **[p. 331]**

- **Definition of Multiple Zeros of a Polynomial**
  If a polynomial $P(x)$ has $(x - r)$ as a factor exactly k times, then r is said to be a zero of multiplicity k of the polynomial $P(x)$. **[p. 336]**

- **The Rational Zero Theorem**
  If

$$P(x) = a_n x^n + a_{n-1}x^{n-1} + \cdots + a_1 x + a_0$$

  has integer coefficients and p/q is a rational zero (in lowest terms) of $P(x)$, then p is a factor of a_0 and q is a factor of a_n. **[p. 337]**

- **Upper- and Lower-Bound Theorem**
  Let $P(x)$ be a polynomial function with real coefficients and a positive leading coefficient. Use synthetic division to divide $P(x)$ by $x - b$ where b is a nonzero real number.

Upper Bound If $b > 0$ and all the numbers in the bottom row of the synthetic division of P by $x - b$ are either positive or zero, then b is an upper bound for the real zeros of P.
Lower Bound If $b < 0$ and the numbers in the bottom row of the synthetic division of P by $x - b$ alternate in sign, then b is a lower bound for the real zeros of P. **[p. 338]**

■ Descartes' Rule of Signs

Let $P(x)$ be a polynomial function with real coefficients and with terms arranged in order of decreasing powers of x.

1. The number of positive real zeros of $P(x)$ is equal to the number of variations in sign of $P(x)$, or to that number decreased by an even integer.
2. The number of negative real zeros of $P(x)$ is equal to the number of variations in sign of $P(-x)$, or to that number decreased by an even integer. **[p. 340]**

■ Zeros of a Polynomial Function

The zeros of some polynomial functions with integer coefficients can be found by using the guidelines stated on page 341.

■ The Fundamental Theorem of Algebra

If $P(x)$ is a polynomial function of degree $n \geq 1$ with complex coefficients, then $P(x)$ has at least one complex zero. **[p. 351]**

■ The Linear Factor Theorem

If $P(x)$ is a polynomial function of degree $n \geq 1$ with leading coefficient $a_n \neq 0$,

$$P(x) = a_n x^n + a_{n-1}x^{n-1} + \cdots + a_1 x^1 + a_0$$

then $P(x)$ has exactly n linear factors

$$P(x) = a_n(x - c_1)(x - c_2) \cdots (x - c_n)$$

where $c_1, c_2, \ldots, c_n$ are complex numbers. **[p. 352]**

■ The Number of Zeros of a Polynomial Function

If $P(x)$ is a polynomial function of degree $n \geq 1$ with complex coefficients, then $P(x)$ has exactly n complex zeros, provided each zero is counted according to its multiplicity. **[p. 352]**

■ The Conjugate Pair Theorem

If $a + bi$ $(b \neq 0)$ is a complex zero of a polynomial function with real coefficients, then the conjugate $a - bi$ is also a complex zero of the function. **[p. 354]**

■ Definition of a Rational Function

If $P(x)$ and $Q(x)$ are polynomials, then the function F given by $F(x) = \dfrac{P(x)}{Q(x)}$ is called a rational function. **[p. 360]**

■ Theorem on Vertical Asymptotes

If the real number a is a zero of the denominator $Q(x)$, then the graph of $F(x) = P(x)/Q(x)$, where $P(x)$ and $Q(x)$ have no common factors, has the vertical asymptote $x = a$. **[p. 362]**

■ Theorem on Horizontal Asymptotes

Let $F(x) = \dfrac{a_n x^n + a_{n-1}x^{n-1} + \cdots + a_1 x + a_0}{b_m x^m + b_{m-1}x^{m-1} + \cdots + b_1 x + b_0}$ be a rational function with numerator of degree n and denominator of degree m.

1. If $n < m$, then the x-axis is the horizontal asymptote of the graph of F.
2. If $n = m$, then the line $y = a_n/b_m$ is the horizontal asymptote of the graph of F.
3. If $n > m$, the graph of F has no horizontal asymptote. **[p. 363]**

■ General Procedure for Graphing Rational Functions That Have No Common Factors

1. Find the real zeros of the denominator. For each zero a, the vertical line $x = a$ will be a vertical asymptote. Use the Theorem on Horizontal Asymptotes to determine the equation of any horizontal asymptote. Graph any horizontal asymptotes.
2. Find the real zeros of the numerator. For each zero a, plot $(a, 0)$. These points are the x-intercepts.
3. Use the tests for symmetry to determine whether the graph has symmetry with respect to the y-axis or to the origin.
4. Find additional points that lie in the intervals between and beyond the x-intercepts and the vertical asymptotes.
5. Determine the behavior of the graph near the asymptotes.
6. Use the information obtained in the above steps to sketch the graph. **[p. 365]**

■ Theorem on Slant Asymptotes

The rational function given by $F(x) = P(x)/Q(x)$, where $P(x)$ and $Q(x)$ have no common factors, has a slant asymptote if the degree of $P(x)$ is one greater than the degree of $Q(x)$. **[p. 368]**

Chapter 4 True/False Exercises

In Exercises 1 to 14, answer true or false. If the statement is false, explain why the statement is false or give an example to show that the statement is false.

1. The complex zeros of a polynomial function with complex coefficients always occur in conjugate pairs.
False; $P(x) = x - i$ has a zero of i, but it does not have a zero of $-i$.
2. Descartes' Rule of Signs indicates that the polynomial function $P(x) = x^3 - x^2 + x - 1$ must have three positive zeros.
3. The polynomial $2x^5 + x^4 - 7x^3 - 5x^2 + 4x + 10$ has two variations in sign. True
4. If 4 is an upper bound of the zeros of the polynomial function P, then 5 is also an upper bound of the zeros of P.
True
5. The graph of every rational function has a vertical asymptote. False; $F(x) = \dfrac{x}{x^2 + 1}$ does not have a vertical asymptote.
6. The graph of the rational function $F(x) = \dfrac{x^2 - 4x + 4}{x^2 - 5x + 6}$ has a vertical asymptote of $x = 2$.
7. If 7 is a zero of the polynomial function P, then $x - 7$ is a factor of P. True

2. False; Descartes' Rule of Signs indicates that $P(x) = x^3 - x^2 + x - 1$ has three or one positive zeros. In fact, P has only 1 positive zero.

8. According to the Zero Location Theorem, the polynomial function $P(x) = x^3 + 6x - 2$ has a real zero between 0 and 1. True
9. Synthetic division can be used to show that $3i$ is a zero of $P(x) = x^3 - 2x^2 + 9x - 18$. True
10. Every fourth-degree polynomial function with complex coefficients has exactly four complex zeros, provided each zero is counted according to its multiplicity. True
11. The graph of a rational function never intersects any of its vertical asymptotes. True
12. The graph of a rational function can have at most one horizontal asymptote. True
13. Descartes' Rule of Signs indicates that the polynomial function $P(x) = x^3 + 2x^2 + 4x - 7$ does have a positive zero. True
14. Every polynomial function has at least one real zero.
False; $P(x) = x^2 + 1$ does not have a real zero.

6. False; $F(x) = \dfrac{(x-2)^2}{(x-3)(x-2)} = \dfrac{x-2}{x-3}$, $x \neq 2$. The graph of F has a hole at $x = 2$.

Chapter 4 Review Exercises

—Answer graphs to Exercises 15–20 and 47–54 are on page AA12.

In Exercises 1 to 6, use synthetic division to divide the first polynomial by the second.

1. $4x^3 - 11x^2 + 5x - 2$, $x - 3$ $\quad 4x^2 + x + 8 + \dfrac{22}{x-3}$ [4.1]
2. $5x^3 - 18x + 2$, $x - 1$ $\quad 5x^2 + 5x - 13 - \dfrac{11}{x-1}$ [4.1]
3. $3x^3 - 5x + 1$, $x + 2$ $\quad 3x^2 - 6x + 7 - \dfrac{13}{x+2}$ [4.1]
4. $2x^3 + 7x^2 + 16x - 10$, $x - \dfrac{1}{2}$ $\quad 2x^2 + 8x + 20$ [4.1]
5. $3x^3 - 10x^2 - 36x + 55$, $x - 5$ $\quad 3x^2 + 5x - 11$ [4.1]
6. $x^4 + 9x^3 + 6x^2 - 65x - 63$, $x + 7$ $\quad x^3 + 2x^2 - 8x - 9$ [4.1]

In Exercises 7 to 10, use the Remainder Theorem to find $P(c)$.

7. $P(x) = x^3 + 2x^2 - 5x + 1$, $c = 4$ $\quad 77$ [4.1]
8. $P(x) = -4x^3 - 10x + 8$, $c = -1$ $\quad 22$ [4.1]
9. $P(x) = 6x^4 - 12x^2 + 8x + 1$, $c = -2$ $\quad 33$ [4.1]
10. $P(x) = 5x^5 - 8x^4 + 2x^3 - 6x^2 - 9$, $c = 3$ $\quad 558$ [4.1]

In Exercises 11 to 14, use synthetic division to show that c is a zero of the given polynomial function.
The verifications in Exercises 11–14 make use of the concepts from Section 4.1.

11. $P(x) = x^3 + 2x^2 - 26x + 33$, $c = 3$
12. $P(x) = 2x^4 + 8x^3 - 8x^2 - 31x + 4$, $c = -4$

13. $P(x) = x^5 - x^4 - 2x^2 + x + 1, c = 1$

14. $P(x) = 2x^3 + 3x^2 - 8x + 3, c = \frac{1}{2}$

In Exercises 15 to 20, graph the polynomial function.

15. $P(x) = x^3 - x$ [4.2]

16. $P(x) = -x^3 - x^2 + 8x + 12$ [4.2]

17. $P(x) = x^4 - 6$ [4.2]

18. $P(x) = x^5 - x$ [4.2]

19. $P(x) = x^4 - 10x^2 + 9$ [4.2]

20. $P(x) = x^5 - 5x^3$ [4.2]

In Exercises 21 to 26, use the Rational Zero Theorem to list all possible rational zeros for each polynomial function.

21. $P(x) = x^3 - 7x - 6$
$\pm 1, \pm 2, \pm 3, \pm 6$ [4.3]

22. $P(x) = 2x^3 + 3x^2 - 29x - 30$
22. $\pm 1, \pm 2, \pm 3, \pm 5, \pm 6, \pm 10, \pm 15, \pm 30, \pm\frac{1}{2}, \pm\frac{3}{2}, \pm\frac{5}{2}, \pm\frac{15}{2}$ [4.3]

23. $P(x) = 15x^3 - 91x^2 + 4x + 12$
Answer is on p. 380

24. $P(x) = x^4 - 12x^3 + 52x^2 - 96x + 64$
$\pm 1, \pm 2, \pm 4, \pm 8, \pm 16, \pm 32, \pm 64$ [4.3]

25. $P(x) = x^3 + x^2 - x - 1$
± 1 [4.3]

26. $P(x) = 6x^5 + 3x - 2$
$\pm 1, \pm 2, \pm\frac{1}{6}, \pm\frac{1}{3}, \pm\frac{1}{2}, \pm\frac{2}{3}$ [4.3]

In Exercises 27 to 30, use Descartes' Rule of Signs to state the number of possible positive and negative real zeros of each polynomial function.

27. $P(x) = x^3 + 3x^2 + x + 3$
No positive real zeros and three or one negative real zero [4.3]

28. $P(x) = x^4 - 6x^3 - 5x^2 + 74x - 120$
Three or one positive real zero, one negative real zero [4.3]

29. $P(x) = x^4 - x - 1$
One positive real zero and one negative real zero [4.3]

30. $P(x) = x^5 - 4x^4 + 2x^3 - x^2 + x - 8$
Five, three, or one positive real zero, no negative real zeros [4.3]

In Exercises 31 to 36, find the zeros of the polynomial function.

31. $P(x) = x^3 + 6x^2 + 3x - 10$ $1, -2, -5$ [4.3]

32. $P(x) = x^3 - 10x^2 + 31x - 30$ $2, 5, 3$ [4.3]

33. $P(x) = 6x^4 + 35x^3 + 72x^2 + 60x + 16$ -2 (multiplicity 2), $-\frac{1}{2}, -\frac{4}{3}$ [4.3]

34. $P(x) = 2x^4 + 7x^3 + 5x^2 + 7x + 3$ $-\frac{1}{2}, -3, i, -i$ [4.4]

35. $P(x) = x^4 - 4x^3 + 6x^2 - 4x + 1$ 1 (multiplicity 4) [4.3]

36. $P(x) = 2x^3 - 7x^2 + 22x + 13$ $-\frac{1}{2}, 2 + 3i, 2 - 3i$ [4.4]

In Exercises 37 and 38, use the given zero to find the remaining zeros of each polynomial function.

37. $P(x) = x^4 - 4x^3 + 6x^2 - 4x - 15; 1 - 2i$ $-1, 3, 1 + 2i$

38. $P(x) = x^4 - x^3 - 17x^2 + 55x - 50; 2 + i$ $-5, 2, 2 - i$

39. Find a third-degree polynomial function with integer coefficients and zeros of 4, -3, and $\frac{1}{2}$.
$P(x) = 2x^3 - 3x^2 - 23x + 12$ [4.4]

40. Find a fourth-degree polynomial function with zeros of 2, -3, i, and $-i$.
$P(x) = x^4 + x^3 - 5x^2 + x - 6$ [4.4]

41. Find a fourth-degree polynomial function with real coefficients that has zeros of 1, 2, and $5i$.
$P(x) = x^4 - 3x^3 + 27x^2 - 75x + 50$ [4.4]

42. Find a fourth-degree polynomial function with real coefficients that has -2 as a zero of multiplicity 2 and also has $1 + 3i$ as a zero.
$P(x) = x^4 + 2x^3 + 6x^2 + 32x + 40$ [4.4]

In Exercises 43 to 46, find the vertical, horizontal, and slant asymptotes for each rational function.

43. $f(x) = \frac{3x + 5}{x + 2}$
43. Vertical asymptote: $x = -2$, horizontal asymptote: $y = 3$ [4.5]

44. $f(x) = \frac{2x^2 + 12x + 2}{x^2 + 2x - 3}$
44. Vertical asymptotes: $x = -3$, $x = 1$, horizontal asymptote: $y = 2$ [4.5]

45. $f(x) = \frac{2x^2 + 5x + 11}{x + 1}$
45. Vertical asymptote: $x = -1$, slant asymptote: $y = 2x + 3$ [4.5]

46. $f(x) = \frac{6x^2 - 1}{2x^2 + x + 7}$
46. No vertical asymptote, horizontal asymptote: $y = 3$ [4.5]

In Exercises 47 to 54, graph each rational function.

47. $f(x) = \frac{3x - 2}{x}$ [4.5]

48. $f(x) = \frac{x + 4}{x - 2}$ [4.5]

49. $f(x) = \frac{6}{x^2 + 2}$ [4.5]

50. $f(x) = \frac{4x^2}{x^2 + 1}$ [4.5]

51. $f(x) = \frac{2x^3 - 4x + 6}{x^2 - 4}$ [4.5]

52. $f(x) = \frac{x}{x^3 - 1}$ [4.5]

53. $f(x) = \frac{3x^2 - 6}{x^2 - 9}$ [4.5]

54. $f(x) = \frac{-x^3 + 6}{x^2}$ [4.5]

Business and Economics

55. ***Average Cost of Skateboards*** The cost, in dollars, of producing x skateboards is given by

$$C(x) = 5.75x + 34{,}200$$

The average cost per skateboard is given by

$$\overline{C}(x) = \frac{C(x)}{x} = \frac{5.75x + 34{,}200}{x}$$

a. Find the average cost per skateboard, to the nearest cent, of producing 5000 and 50,000 skateboards.
\$12.59, \$6.43

b. What is the equation of the horizontal asymptote of the graph of $\bar{C}$? Explain the significance of the horizontal asymptote as it relates to this application.
$y = 5.75$. As the number of skateboards produced increases, the average cost per skateboard approaches \$5.75. [4.5]

Life and Health Sciences

56. *Food Temperature* The temperature F (measured in degrees Fahrenheit) of a dessert placed in a freezer for t hours is given by the rational function

$$F(t) = \frac{60}{t^2 + 2t + 1}, \quad t \geq 0$$

a. Find the temperature of the dessert after it has been in the freezer for 1 hour. 15°F

b. Find the temperature of the dessert after 4 hours.
2.4°F

c. What temperature will the dessert approach as $t \to \infty$?
0°F [4.5]

57. *Physiology* One of Poiseuille's Laws states that the resistance R encountered by blood flowing through a blood vessel is given by

$$R(r) = C\frac{L}{r^4}$$

where C is a positive constant determined by the viscosity of the blood, L is the length of the blood vessel, and r is its radius.

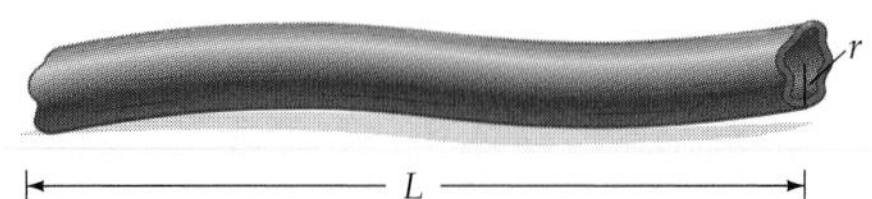

a. Explain the meaning of $R(r) \to \infty$ as $r \to 0$.
As the radius of the blood vessel approaches 0, the resistance gets larger.

b. Explain the meaning of $R(r) \to 0$ as $r \to \infty$.
As the radius of the blood vessel gets larger, the resistance approaches zero. [4.5]

23. $\pm 1, \pm 2, \pm 3, \pm 4, \pm 6, \pm 12, \pm\frac{1}{3}, \pm\frac{2}{3}, \pm\frac{4}{3}, \pm\frac{1}{5}, \pm\frac{2}{5}, \pm\frac{3}{5}, \pm\frac{4}{5}, \pm\frac{6}{5}, \pm\frac{12}{5}, \pm\frac{1}{15}, \pm\frac{2}{15}, \pm\frac{4}{15}$ [4.3]

Chapter 4 Test

—Answer graphs for Exercises 17 and 18 are on page AA12.

1. Use synthetic division to divide:

$$(3x^3 + 5x^2 + 4x - 1) \div (x + 2)$$

$3x^2 - x + 6 - \frac{13}{x + 2}$ [4.1]

2. Use the Remainder Theorem to find $P(-2)$ if

$$P(x) = -3x^3 + 7x^2 + 2x - 5$$

43 [4.1]

3. Show that $x - 1$ is a factor of $x^4 - 4x^3 + 7x^2 - 6x + 2$.
The verification for Exercise 3 makes use of the concepts from Section 4.1.

4. Examine the leading term of the function given by the equation $P(x) = -3x^3 + 2x^2 - 5x + 2$ and determine the far-left and far-right behavior of the graph of P.
Up to the far left and down to the far right [4.2]

5. Find the real solutions of $3x^3 + 7x^2 - 6x = 0$. $0, \frac{2}{3}, -3$ [4.2]

6. Use the Zero Location Theorem to verify that

$$P(x) = 2x^3 - 3x^2 - x + 1$$

has a zero between 1 and 2.

7. Find the zeros of

$$P(x) = (x^2 - 4)^2(2x - 3)(x + 1)^3$$

and state the multiplicity of each.
2 (multiplicity 2), -2 (multiplicity 2), $\frac{3}{2}$ (multiplicity 1), -1 (multiplicity 3) [4.3]

6. $P(1) < 0$, $P(2) > 0$. Therefore, by the Zero Location Theorem, the polynomial function P has a zero between 1 and 2. [4.2]

8. Use the Rational Zero Theorem to list the possible rational zeros of $P(x) = 6x^3 - 3x^2 + 2x - 3$.
$\pm 1, \pm 3, \pm\frac{1}{2}, \pm\frac{3}{2}, \pm\frac{1}{3}, \pm\frac{1}{6}$ [4.3]

9. Find, by using the Upper- and Lower-Bound Theorem, the smallest positive integer and the largest negative integer that are upper and lower bounds for the real zeros of the polynomial function

$$P(x) = 2x^4 + 5x^3 - 23x^2 - 38x + 24$$

Upper bound 4, lower bound -5 [4.3]

10. Use Descartes' Rule of Signs to state the number of possible positive and negative real zeros of

$$P(x) = x^4 - 3x^3 + 2x^2 - 5x + 1$$

4, 2, or 0 positive zeros, no negative zero [4.3]

11. Find the zeros of $P(x) = 2x^3 - 3x^2 - 11x + 6$
$\frac{1}{2}, 3, -2$ [4.3]

12. Given that $2 + 3i$ is a zero of

$$P(x) = 6x^4 - 5x^3 + 12x^2 + 207x + 130$$

find the remaining zeros. $2 - 3i, -\frac{2}{3}, -\frac{5}{2}$ [4.4]

13. Find all the zeros of

$$P(x) = x^5 - 6x^4 + 14x^3 - 14x^2 + 5x$$

0, 1 (multiplicity 2), $2 + i$, $2 - i$ [4.4]

14. Find a polynomial of lowest degree that has real coefficients and zeros $1 + i$, 3, and 0.
$P(x) = x^4 - 5x^3 + 8x^2 - 6x$ [4.4]

15. Find all vertical asymptotes of the graph of

$$f(x) = \frac{3x^2 - 2x + 1}{x^2 - 5x + 6}$$

Vertical asymptotes: $x = 3, x = 2$ [4.5]

16. Find the horizontal asymptote of the graph of

$$f(x) = \frac{3x^2 - 2x + 1}{2x^2 - 1}$$

Horizontal asymptote: $y = \frac{3}{2}$ [4.5]

17. Graph $f(x) = \dfrac{x^2 - 1}{x^2 - 2x - 3}$. Use an open circle to show where f is undefined. [4.5]

18. Graph $f(x) = \dfrac{2x^2 + 2x + 1}{x + 1}$ and label the slant asymptote with its equation. [4.5]

19. The rational function

$$w(t) = \frac{70t + 120}{t + 40}, \quad t \geq 0$$

models Rene's typing speed, in words per minute, after t hours of typing lessons.

a. Find $w(1)$, $w(10)$, and $w(20)$. Round to the nearest word per minute. 5 words/min, 16 words/min, 25 words/min

b. What will Rene's typing speed approach as $t \to \infty$?
70 words/min [4.5]

20. ***Maximizing Volume*** You are to construct an open box from a rectangular sheet of cardboard that measures 16 inches by 22 inches. To assemble the box, make the four cuts shown in the figure below and then fold on the dashed lines. What value of x (to the nearest 0.001 inch) will produce a box with maximum volume? What is the maximum volume (to the nearest 0.1 cubic inch)? 2.137 in., 337.1 in^3 [4.3]

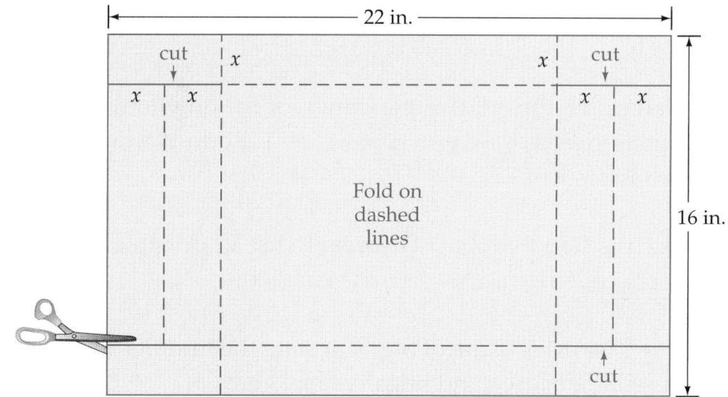

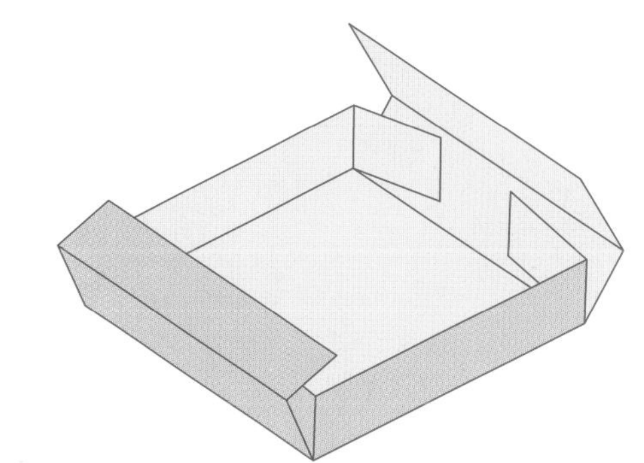

Cumulative Review Exercises

1. Write $\dfrac{3 + 4i}{1 - 2i}$ in $a + bi$ form. $-1 + 2i$ [P.7]

2. Use the quadratic formula to solve $x^2 - x - 1 = 0$. $\dfrac{1 \pm \sqrt{5}}{2}$ [1.2]

3. Solve: $\sqrt{2x + 5} - \sqrt{x - 1} = 2$ 2, 10 [1.3]

4. Solve: $|x - 3| \leq 11$ $\{x \mid -8 \leq x \leq 14\}$ [1.4]

5. Find the distance between the points (2, 5) and (7, −11).
$\sqrt{281}$ [2.1]

6. Explain how to use the graph of $y = x^2$ to produce the graph of $y = (x - 2)^2 + 4$.
Translate the graph of $y = x^2$ 2 units to the right and 4 units up. [3.3]

7. Find the difference quotient for the function $P(x) = x^2 - 2x - 3$. $2a + h - 2$ [3.1]

8. Given $f(x) = 2x^2 + 5x - 3$ and $g(x) = 4x - 7$, find $(f \circ g)(x)$. $32x^2 - 92x + 60$ [3.1]

9. Find the inverse function of $f(x) = 2x - 5$.
$f^{-1}(x) = \frac{1}{2}x + \frac{5}{2}$ [3.2]

10. Use synthetic division to divide $(4x^4 - 2x^2 - 4x - 5)$ by $(x + 2)$. $4x^3 - 8x^2 + 14x - 32 + \frac{59}{x + 2}$ [4.1]

11. Use the Remainder Theorem to find $P(3)$ for $P(x) = 2x^4 - 3x^2 + 4x - 6$. 141 [4.1]

12. Determine the far-right behavior of the graph of $P(x) = -3x^4 - x^2 + 7x - 6$. The graph goes down. [4.2]

13. Determine the relative maximum of the polynomial function $P(x) = -3x^3 - x^2 + 4x - 1$. Round to the nearest 0.0001. 0.3997 [4.2]

14. Use the Rational Zero Theorem to list all possible rational zeros of $P(x) = 3x^4 - 4x^3 - 11x^2 + 16x - 4$. $\pm 1, \pm 2, \pm 4, \pm\frac{1}{3}, \pm\frac{2}{3}, \pm\frac{4}{3}$ [4.3]

15. Use Descartes' Rule of Signs to state the number of possible positive and negative real zeros of $P(x) = x^3 + x^2 + 2x + 4$.
Zero positive real zeros, three or one negative real zero [4.3]

16. Find all zeros of $P(x) = x^3 + x + 10$. $-2, 1 + 2i, 1 - 2i$ [4.4]

17. Find a polynomial function of lowest degree that has real coefficients and -2 and $3 + i$ as zeros.
$P(x) = x^3 - 4x^2 - 2x + 20$ [4.4]

18. Write $P(x) = x^3 - 2x^2 + 9x - 18$ as a product of linear factors. $(x - 2)(x + 3i)(x - 3i)$ [4.4]

19. Determine the vertical and horizontal asymptotes of the graph of $F(x) = \frac{4x^2}{x^2 + x - 6}$.
Vertical asymptotes: $x = -3$, $x = 2$; horizontal asymptote: $y = 4$ [4.5]

20. Find the equation of the slant asymptote for the graph of $F(x) = \frac{x^3 + 4x^2 + 1}{x^2 + 4}$. $y = x + 4$ [4.5]

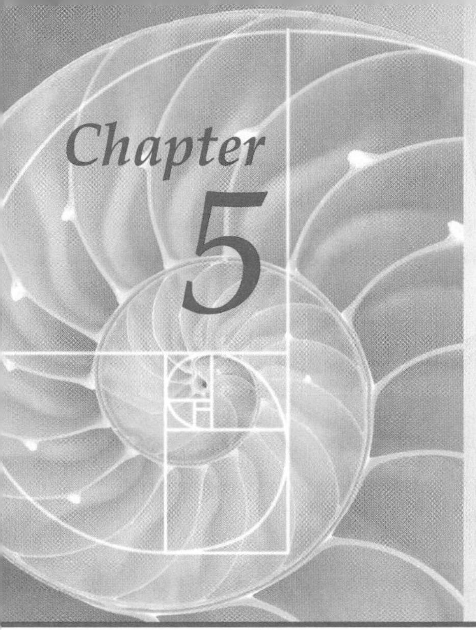

Chapter 5

Exponential and Logarithmic Functions

Pay It Forward

The movie *Pay It Forward* is based on Catherine Ryan Hyde's book by the same name. In the movie Trevor McKinney, played by Haley Joel Osment, is given a school assignment to "think of an idea to change the world—and then put it into action." In response to this assignment, Trevor develops his *pay it forward* project. In this project, anyone who benefits from another person's good deed must do a good deed for three additional people. Each of these three people are then obligated to do a good deed for another three people, and so on.

The photo above shows that after the first pay it forward round, three people will be beneficiaries of a good deed. After the second round, a total of 12 people (the original three plus nine more) will be beneficiaries. After the 10th round, the number of beneficiaries is 88,572. The number of beneficiaries after n rounds can be determined by evaluating an exponential function. See Exercise 67, page 398 for more details on this "pay it forward" function.

Chapter Prep Quiz

Do these exercises to prepare for Chapter 5.

1. Evaluate: 2^5 [P.1] 32

2. Evaluate: 3^{-4} [P.2] $\frac{1}{81}$

3. Evaluate: 2^0 [P.2] 1

4. State the domain and range of $f(x) = x^2 + 1$. [2.2/2.5] Domain: Set of all real numbers; range: $\{y \mid y \geq 1\}$

5. Determine the y-intercept of the graph of $f(x) = 3x + 2$. [2.3] (0, 2)

6. Explain how to obtain the graph of $g(x) = 2x + 3$ by using a transformation of the graph of $f(x) = 2x$. [3.3]
Shift the graph of $f(x)$ upward 3 units.

7. Explain how to produce the graph of $g(x) = 2(-x) + 1$ by using a transformation of the graph of $f(x) = 2x + 1$. [3.3]
Reflect the graph of $f(x)$ across the y-axis.

8. Explain how to produce the graph of $g(x) = 2(x - 3) + 1$ by using a transformation of the graph of $f(x) = 2x + 1$. [3.3]
Shift the graph of $f(x)$ to the right 3 units.

9. What is a necessary condition for a function to have an inverse function? [3.2]
The function must be a one-to-one function.

10. Find $f^{-1}(x)$, given $f(x) = \frac{1}{2}x - 3$. [3.2] $f^{-1}(x) = 2x + 6$

Problem Solving Strategies

Is the Solution Reasonable?

For most application problems, it is advisable to use your intuition to make an estimate of the solution to the problem and then compare your actual solution with this estimate. If your solution does not seem reasonable, you should re-examine your work to see if you can find a mistake.

Some of the applications in this chapter may provide a strong challenge to your intuitive powers. For instance, consider the classic problem in which you plan to save 1 cent on the first day of a month, 2 cents on the second day, 4 cents on the third day, 8 cents on the fourth day, and continue this procedure of saving twice the amount saved on the preceding day for 31 days. Use your intuition to estimate the total amount of money you will have saved.

Most people estimate an amount far below the actual total. The correct total is $2^{31} - 1$ cents, which is \$21,474,836.47. Do you find this to be a surprisingly large result? To better comprehend that the total amount saved can be so large, consider the number of cents saved on the first few days: 1¢, 2¢, 4¢, 8¢, 16¢, 32¢, 64¢, and 128¢. By the eighth day you need to save over 100 times more money than you did on the first day. If you take one dollar from the eighth day and double it successively over the next 7 days, then on the 15th day of the month you will need to save over \$100. Similarly, after another 7 days, you will need to save over \$10,000 on the 22nd day of the month. On the 29th day of the month you will need to save well over \$1,000,000. Using this procedure we see that the approximate accumulated total of 21 million dollars for the complete month doesn't seem quite so unreasonable.

INSTRUCTOR NOTE

The *Chapter Prep Quiz* is a means to test your students' mastery of prerequisite material that is assumed in the coming chapter. Each question identifies the section to review (if necessary) and all answers are provided in the Answers to Selected Exercises.

SECTION 5.1 Exponential Functions and Their Applications

- Exponential Functions
- Graphs of Exponential Functions
- The Natural Exponential Function
- Applications of Exponential Functions

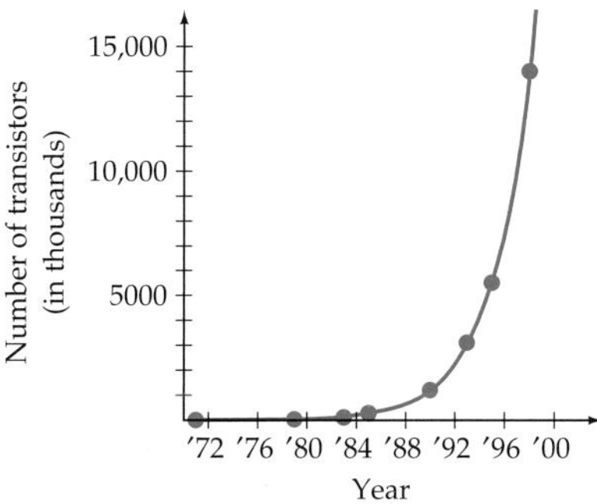

Figure 5.1 Moore's Law

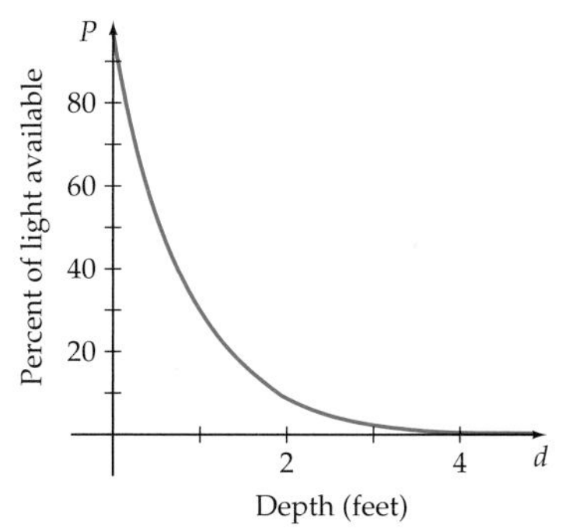

Figure 5.2

Exponential Functions

In 1965, Gordon Moore, one of the cofounders of Intel Corporation, observed that the maximum number of transistors that could be placed on a microprocessor seemed to be doubling every 18 to 24 months. Table 5.1 below shows how the maximum number of transistors on various Intel processors has changed over time. (*Source:* Intel Museum home page)

Table 5.1

Year	1971	1979	1983	1985	1990	1993	1995	1998
Number of transistors per microprocessor (in thousands)	2.3	31	110	280	1200	3100	5500	14,000

The curve that approximately passes through the points is a mathematical model of the data. See Figure 5.1. The model is based on an *exponential* function.

When light enters water, the intensity of the light decreases with the depth of the water. The graph in Figure 5.2 shows a model, for Lake Michigan, of the decrease in the percent of available light as the depth of the water increases. This model is also based on an exponential function.

Definition of an Exponential Function

The **exponential function** with base b is defined by

$$f(x) = b^x$$

where $b > 0$, $b \neq 1$, and x is any real number.

The base b of $f(x) = b^x$ is required to be positive. If the base were a negative number, the value of the function would be a complex number for some values of x. For instance, if $b = -4$ and $x = \frac{1}{2}$, then $f\left(\frac{1}{2}\right) = (-4)^{1/2} = 2i$. To avoid complex number values of a function, the base of any exponential function must be a nonnegative number. Also, b is defined such that $b \neq 1$ and $b \neq 0$ because both $f(x) = 1^x = 1$ and $f(x) = 0^x = 0$ are constant functions.

In the following examples we evaluate $f(x) = 2^x$ at $x = 3$ and $x = -2$.

$$f(3) = 2^3 = 8 \qquad f(-2) = 2^{-2} = \frac{1}{2^2} = \frac{1}{4}$$

To evaluate the exponential function $f(x) = 2^x$ at an irrational number such as $x = \sqrt{2}$, we use a rational approximation of $\sqrt{2}$, such as 1.4142, and a calculator to obtain an approximation of the function. For instance, if $f(x) = 2^x$, then $f\left(\sqrt{2}\right) = 2^{\sqrt{2}} \approx 2^{1.4142} \approx 2.6651$.

Alternative to Example 1

Evaluate $f(x) = 5^x$ at $x = 3$, $x = -2$, and $x = \sqrt{2}$.

- $f(3) = 125$, $f(-2) = \dfrac{1}{25}$, $f(\sqrt{2}) = 5^{\sqrt{2}} \approx 9.7385$

EXAMPLE 1 Evaluate an Exponential Function

Evaluate $f(x) = 3^x$ at $x = 2$, $x = -4$, and $x = \pi$.

Solution

$$f(2) = 3^2 = 9$$

$$f(-4) = 3^{-4} = \frac{1}{3^4} = \frac{1}{81}$$

$$f(\pi) = 3^{\pi} \approx 3^{3.1415927} \approx 31.54428$$

- Evaluate with the aid of a calculator.

CHECK YOUR PROGRESS 1 Evaluate $g(x) = \left(\frac{1}{2}\right)^x$ at $x = 3$, $x = -1$, and $x = \sqrt{3}$.

Solution *See page S23.* $g(3) = \frac{1}{8}$; $g(-1) = 2$; $g(\sqrt{3}) = \left(\frac{1}{2}\right)^{\sqrt{3}} \approx 0.30102$

Graphs of Exponential Functions

The graph of $f(x) = 2^x$ is shown in Figure 5.3. The coordinates of some of the points on the curve are given in Table 5.2.

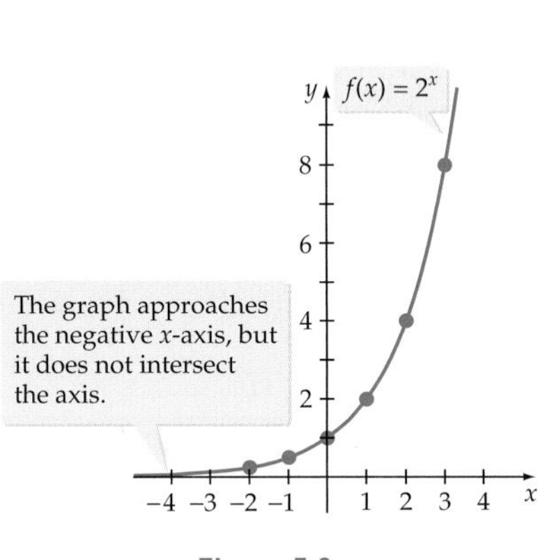

Figure 5.3

Table 5.2

x	$y = f(x) = 2^x$	(x, y)
−2	$f(-2) = 2^{-2} = \frac{1}{4}$	$\left(-2, \frac{1}{4}\right)$
−1	$f(-1) = 2^{-1} = \frac{1}{2}$	$\left(-1, \frac{1}{2}\right)$
0	$f(0) = 2^0 = 1$	$(0, 1)$
1	$f(1) = 2^1 = 2$	$(1, 2)$
2	$f(2) = 2^2 = 4$	$(2, 4)$
3	$f(3) = 2^3 = 8$	$(3, 8)$

Note the following properties of the graph of an exponential function $f(x) = b^x$ for which the base b is greater than 1. (In Figure 5.3, $b = 2 > 1$.)

- The y-intercept is $(0, 1)$.
- The graph passes through $(1, b)$.
- As x decreases without bound (that is, as $x \to -\infty$), $f(x) \to 0$.
- The graph is a smooth continuous increasing curve.

Now consider the graph of an exponential function for which the base is between 0 and 1. The graph of $f(x) = \left(\frac{1}{2}\right)^x$ is shown in Figure 5.4. The coordinates of some of the points on the curve are given in Table 5.3.

Table 5.3

x	$y = f(x) = \left(\frac{1}{2}\right)^x$	(x, y)
−3	$f(-3) = \left(\frac{1}{2}\right)^{-3} = 8$	$(-3, 8)$
−2	$f(-2) = \left(\frac{1}{2}\right)^{-2} = 4$	$(-2, 4)$
−1	$f(-1) = \left(\frac{1}{2}\right)^{-1} = 2$	$(-1, 2)$
0	$f(0) = \left(\frac{1}{2}\right)^{0} = 1$	$(0, 1)$
1	$f(1) = \left(\frac{1}{2}\right)^{1} = \frac{1}{2}$	$\left(1, \frac{1}{2}\right)$
2	$f(2) = \left(\frac{1}{2}\right)^{2} = \frac{1}{4}$	$\left(2, \frac{1}{4}\right)$

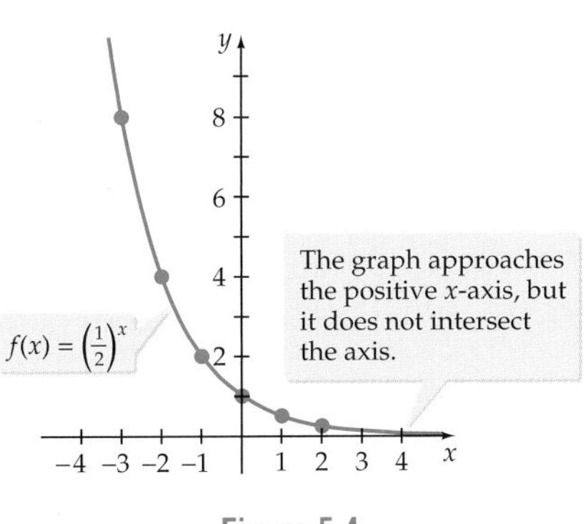

Figure 5.4

Note the following properties of the graph of $f(x) = \left(\frac{1}{2}\right)^x$ in Figure 5.4:

- The y-intercept is $(0, 1)$.
- The graph passes through $\left(1, \frac{1}{2}\right)$.
- As x decreases without bound, the y-values decrease toward 0. That is, as $x \to \infty$, $f(x) \to 0$.
- The graph is a smooth continuous decreasing curve.

P

Properties of $f(x) = b^x$

For positive real numbers b, $b \neq 1$, the exponential function defined by $f(x) = b^x$ has the following properties:

1. The function f is a one-to-one function. It has the set of real numbers as its domain and the set of positive real numbers as its range.
2. The graph of f is a smooth continuous curve with a y-intercept of $(0, 1)$, and the graph passes through $(1, b)$.
3. If $b > 1$, f is an increasing function and the graph of f is asymptotic to the negative x-axis. [As $x \to -\infty$, $f(x) \to 0$.] See Figure 5.5a.
4. If $0 < b < 1$, f is a decreasing function and the graph of f is asymptotic to the positive x-axis. [As $x \to \infty$, $f(x) \to 0$.] See Figure 5.5b.

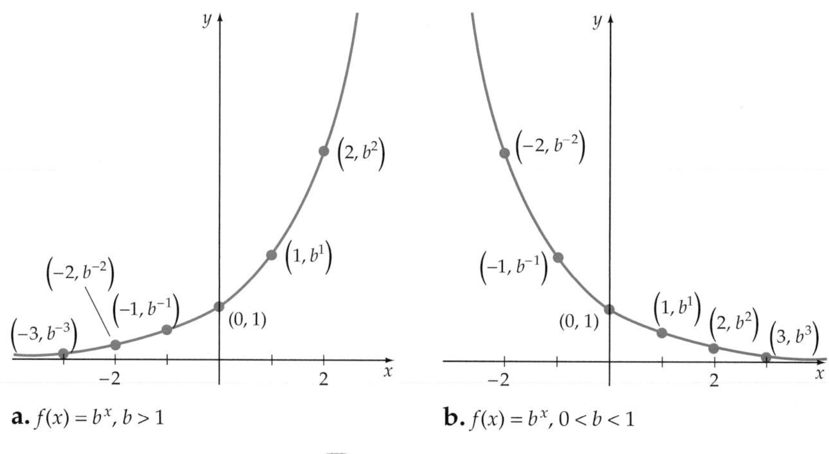

a. $f(x) = b^x, b > 1$ **b.** $f(x) = b^x, 0 < b < 1$

Figure 5.5

QUESTION What is the x-intercept of the graph of $f(x) = \left(\frac{1}{3}\right)^x$?

Alternative to Example 2
Exercise 38, page 395

EXAMPLE 2 Graph an Exponential Function

Graph $g(x) = \left(\frac{3}{4}\right)^x$.

Solution Because the base $\frac{3}{4}$ is less than 1, we know that the graph of g is a decreasing function that is asymptotic to the positive x-axis. The y-intercept of the graph is the point (0, 1), and the graph also passes through $\left(1, \frac{3}{4}\right)$. Plot a few additional points (see Table 5.4), and then draw a smooth curve through the points as in Figure 5.6.

Table 5.4

x	$y = g(x) = \left(\frac{3}{4}\right)^x$	(x, y)
−3	$\left(\frac{3}{4}\right)^{-3} = \frac{64}{27}$	$\left(-3, \frac{64}{27}\right)$
−2	$\left(\frac{3}{4}\right)^{-2} = \frac{16}{9}$	$\left(-2, \frac{16}{9}\right)$
−1	$\left(\frac{3}{4}\right)^{-1} = \frac{4}{3}$	$\left(-1, \frac{4}{3}\right)$
2	$\left(\frac{3}{4}\right)^{2} = \frac{9}{16}$	$\left(2, \frac{9}{16}\right)$
3	$\left(\frac{3}{4}\right)^{3} = \frac{27}{64}$	$\left(3, \frac{27}{64}\right)$

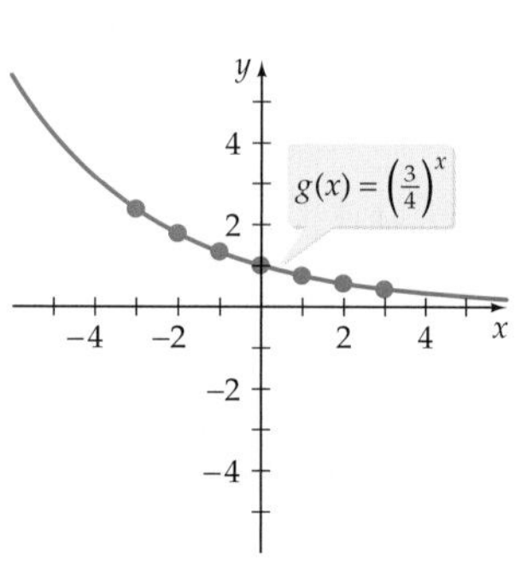

Figure 5.6

Continued ➤

ANSWER The graph does not have an x-intercept. As x increases, the graph approaches the x-axis, but it does not intersect the x-axis.

2.

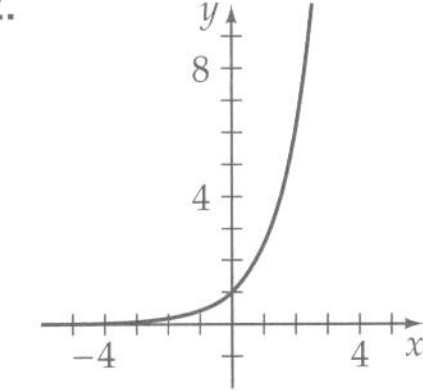

CHECK YOUR PROGRESS 2 Graph $f(x) = \left(\frac{5}{2}\right)^x$.

Solution *See page S23.* See the graph at the left.

Consider the functions $F(x) = 2^x - 3$ and $G(x) = 2^{x-3}$. You can construct the graphs of these functions by plotting points; however, it is easier to construct their graphs by using a translation of the graph of $f(x) = 2^x$, as shown in Example 3.

Alternative to Example 3
Exercise 44, page 396

EXAMPLE 3 Use a Translation to Graph a Function

Sketch the graph of each function. **a.** $F(x) = 2^x - 3$ **b.** $G(x) = 2^{x-3}$

Solution

a. Let $f(x) = 2^x$. Then $F(x) = 2^x - 3 = f(x) - 3$. The graph of F is a vertical translation of f down 3 units, as shown in Figure 5.7.

b. Let $f(x) = 2^x$. Then $G(x) = 2^{x-3} = f(x - 3)$. The graph of G is a horizontal translation of f to the right 3 units, as shown in Figure 5.8.

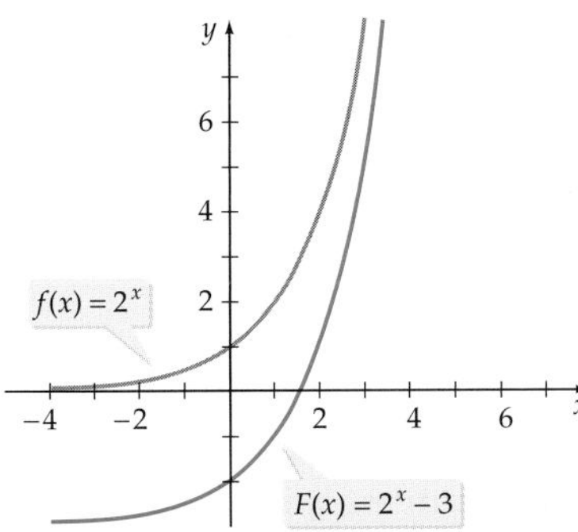

Figure 5.7

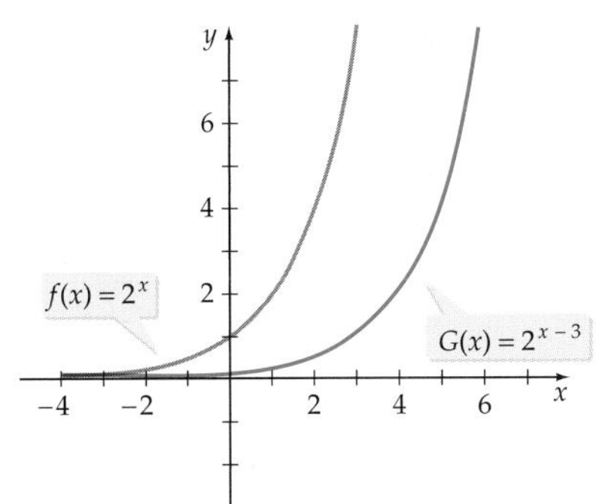

Figure 5.8

CHECK YOUR PROGRESS 3 Sketch the graph of each function.

a. $J(x) = \left(\frac{1}{2}\right)^x + 2$ **b.** $K(x) = 2^{x+2}$

Solution *See page S23.* See the graphs at the left.

3.a.

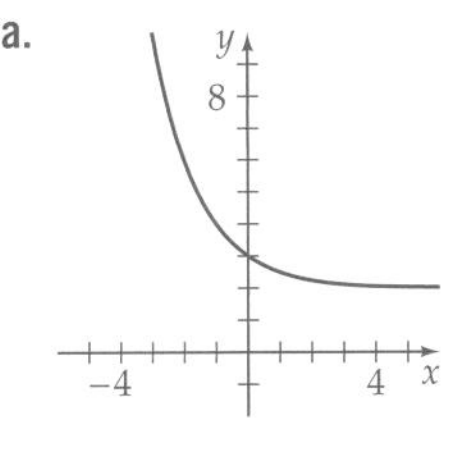

b.

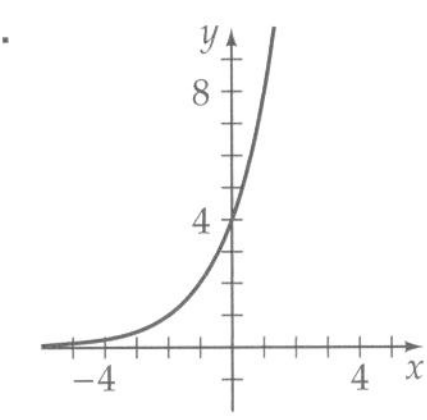

The graphs of some functions can be constructed by stretching, or compressing, or reflecting the graph of an exponential function.

Alternative to Example 4
Exercise 42, page 396

EXAMPLE 4 Use Stretching or Reflecting Procedures to Graph a Function

Sketch the graph of each function. **a.** $M(x) = 2(2^x)$ **b.** $N(x) = 2^{-x}$

Continued ➤

Solution

a. Let $f(x) = 2^x$. Then $M(x) = 2(2^x) = 2f(x)$. The graph of M is a vertical stretching of f, as shown in Figure 5.9. If (x, y) is a point on the graph of $f(x) = 2^x$, then $(x, 2y)$ is a point on the graph of M.

b. Let $f(x) = 2^x$. Then $N(x) = 2^{-x} = f(-x)$. The graph of N is the graph of f reflected across the y-axis, as shown in Figure 5.10. If (x, y) is a point on the graph of $f(x) = 2^x$, then $(-x, y)$ is a point on the graph of N.

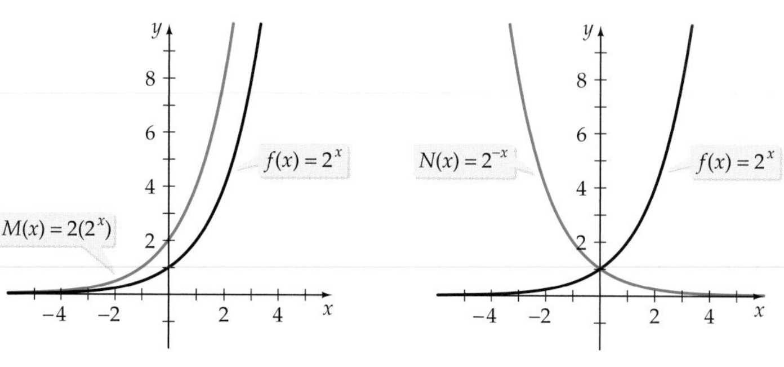

Figure 5.9

Figure 5.10

CHECK YOUR PROGRESS 4 Sketch the graph of each function.

a. $R(x) = \frac{1}{4}(2^x)$ **b.** $S(x) = -2^x$

Solution *See page S23.* See the graphs at the left.

4.a.

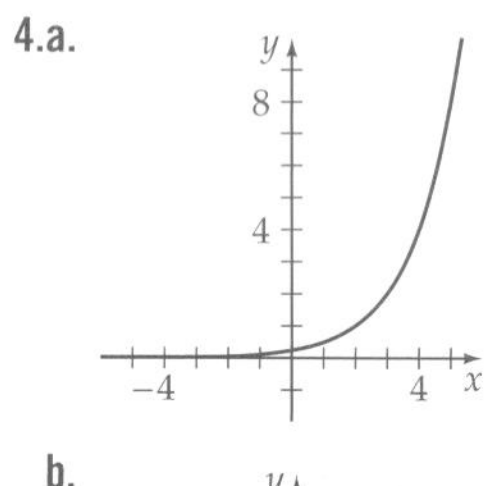

b.

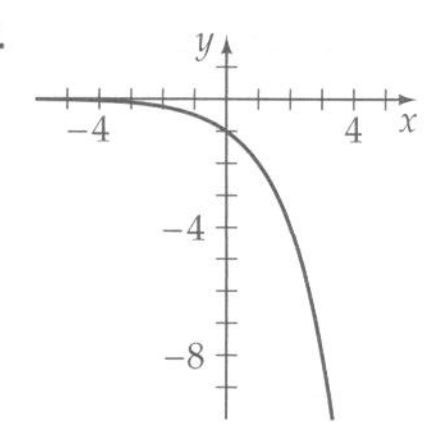

▪ The Natural Exponential Function

The irrational number π is often used in applications that involve circles. Another irrational number, denoted by the letter e, is useful in applications that involve growth or decay.

Definition of *e*

The number e is defined as the number that

$$\left(1 + \frac{1}{n}\right)^n$$

approaches as n increases without bound.

The letter e was chosen in honor of the Swiss mathematician Leonhard Euler. He was able to compute the value of e to several decimal places by evaluating $\left(1 + \frac{1}{n}\right)^n$ for large values of n, as shown in Table 5.5.

Point of Interest

Leonhard Euler (1707–1783)
Some mathematicians consider Euler to be the greatest mathematician of all time. He certainly was the most prolific writer of mathematics of all time. He made substantial contributions in the areas of number theory, geometry, calculus, differential equations, differential geometry, topology, complex variables, and analysis, to name but a few. Euler was the first to introduce many of the mathematical notations that we use today. For instance, he introduced the symbol i for the square root of -1, the symbol π for pi, the functional notation $f(x)$, and the letter e for the base of the natural exponential function. Euler's computational skills were truly amazing. The mathematician François Arago remarked, "Euler calculated without apparent effort, as men breathe, or as eagles sustain themselves in the wind."

INSTRUCTOR NOTE

Is $e^{\pi\sqrt{164}}$ an integer? For more information on this unusual power, see the following conflicting reports. Evidence suggesting that it is an integer is given in *More Joy of Mathematics* by Theoni Pappas (San Carlos, CA: Wide World Publishing/Tetra, 1991, p. 182). For evidence against this hypothesis, see Stephen Wolfram, *Mathematica: A System for Doing Mathematics* (Reading, MA: Addison-Wesley, 1988, p. 31).

Table 5.5

Value of n	Value of $\left(1 + \frac{1}{n}\right)^n$
1	2
10	2.59374246
100	2.704813829
1000	2.716923932
10,000	2.718145927
100,000	2.718268237
1,000,000	2.718280469
10,000,000	2.718281693

The value of e accurate to eight decimal places is 2.71828183.

The Natural Exponential Function

For all real numbers x, the function defined by

$$f(x) = e^x$$

is called the **natural exponential function.**

A calculator can be used to evaluate e^x for specific values of x. For instance,

$$e^2 \approx 7.389056, \quad e^{3.5} \approx 33.115452, \quad \text{and} \quad e^{-1.4} \approx 0.246597$$

On many calculators the e^x function is located above the LN key.

The graph of the natural exponential function can be constructed by plotting a few points or by using a graphing utility.

Alternative to Example 5 Exercise 50, page 396

EXAMPLE 5 Graph the Natural Exponential Function

Graph $f(x) = e^x$.

Solution Use a calculator to find the range values for a few domain values. The range values in Table 5.6 have been rounded to the nearest tenth.

Table 5.6

x	−2	−1	0	1	2
$f(x) = e^x$	0.1	0.4	1.0	2.7	7.4

Plot the points given in Table 5.6 and then connect the points with a smooth curve. Because $e > 1$, we know that the graph is an increasing function. To the far left, the graph will approach the x-axis. The y-intercept is (0, 1). See Figure 5.11.

Continued ➤

INTEGRATING TECHNOLOGY

The graph below was produced on a TI-83 graphing calculator by entering e^x in the Y= menu.

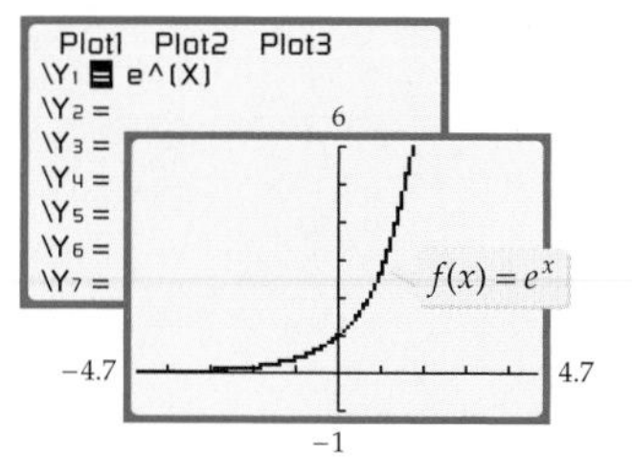

Note in Figure 5.12 how the graph of $f(x) = e^x$ compares with the graphs of $g(x) = 2^x$ and $h(x) = 3^x$. You may have anticipated that the graph of $f(x) = e^x$ would lie between the two other graphs because e is between 2 and 3.

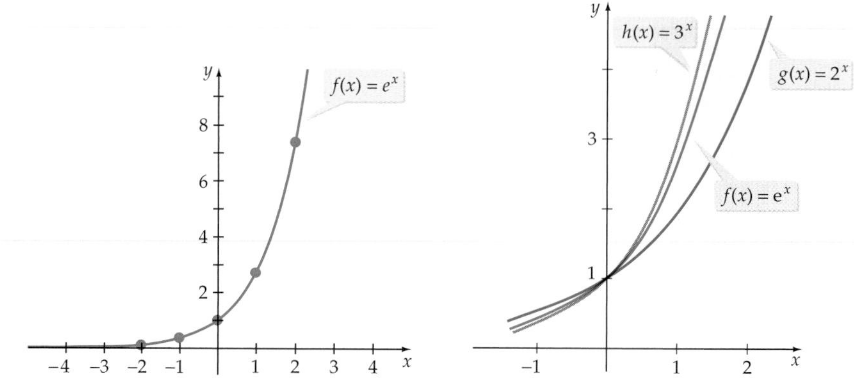

Figure 5.11

Figure 5.12

CHECK YOUR PROGRESS 5 Graph $f(x) = e^{-x} + 2$.

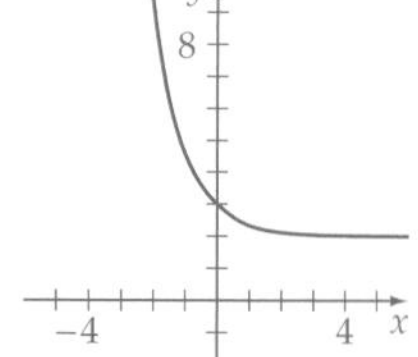

Solution *See page S23.*

INSTRUCTOR NOTE

Applications concerning compound interest have been placed in Chapter 8: The Mathematics of Finance.

Alternative to Example 6

Exercise 70, page 398

Applications of Exponential Functions

Many applications can be effectively modeled by an exponential function. For instance, in Example 6 we make use of an exponential function to model the temperature of a cup of coffee.

EXAMPLE 6 Use a Mathematical Model

A cup of coffee is heated to 160°F and placed in a room that maintains a temperature of 70°F. The temperature T of the coffee, in degrees Fahrenheit, after t minutes is given by

$$T = 70 + 90e^{-0.0485t}$$

a. Find the temperature of the coffee, to the nearest degree, 20 minutes after it is placed in the room.

b. Use a graphing utility to determine when the temperature of the coffee will reach 90°F.

Solution

a.
$$\begin{aligned} T &= 70 + 90e^{-0.0485t} \\ &= 70 + 90e^{-0.0485\cdot(20)} && \text{Substitute 20 for } t. \\ &\approx 70 + 34.1 \\ &\approx 104.1 \end{aligned}$$

After 20 minutes the temperature of the coffee is about 104°F.

Continued ➤

TAKE NOTE

In Example 6**b.** we use a graphing utility to solve the equation $90 = 70 + 90e^{-0.0485t}$. Analytical methods of solving this type of equation without the use of a graphing utility will be developed in Section 5.4.

b. Graph $T = 70 + 90e^{-0.0485t}$ and $T = 90$. See the following figure.

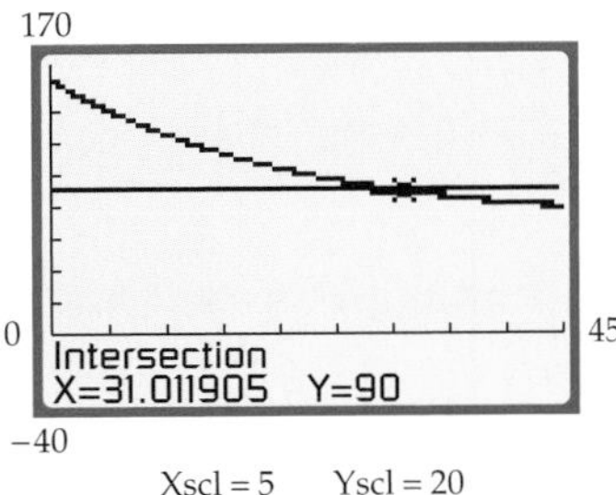

Xscl = 5 Yscl = 20

The graphs intersect at about (31.01, 90). It takes the coffee about 31 minutes to cool to 90°F.

CHECK YOUR PROGRESS 6 The function $A(t) = 200e^{-0.014t}$ gives the amount of medication, in milligrams, in a patient's bloodstream t minutes after the medication has been injected into the patient's bloodstream.

a. Find the amount of medication, to the nearest milligram, in the patient's bloodstream after 45 minutes.

b. Use a graphing utility to determine how long it will take, to the nearest minute, for the amount of medication in the patient's bloodstream to reach 50 milligrams.

Solution *See page S23.* **a.** 107 mg **b.** 99 min

Alternative to Example 7
Exercise 58, page 396

EXAMPLE 7 Use a Mathematical Model

The weekly revenue R, in dollars, from the sale of a product varies with time according to the function

$$R(x) = \frac{1760}{8 + 14e^{-0.03x}}$$

where x is the number of weeks that have passed since the product was put on the market. What will the weekly revenue approach as time goes by?

Solution

Method 1: Use a graphing utility to graph $R(x)$ and use the TRACE feature to see what happens to the revenue as the time increases. The following graph shows that as the weeks go by, the weekly revenue will increase and approach \$220.00 per week.

Continued ➤

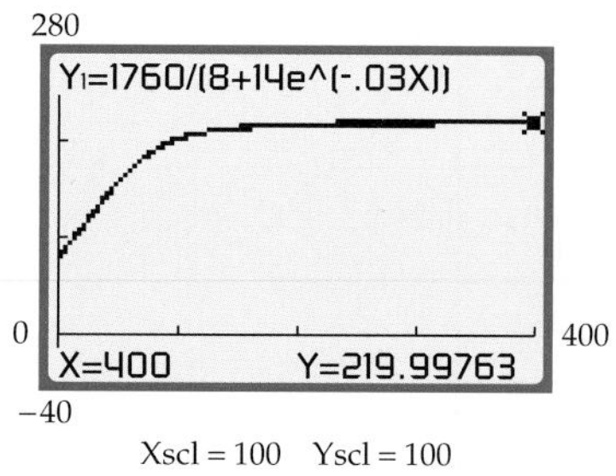

Method 2: Write the revenue function in the following form.

$$R(x) = \frac{1760}{8 + \dfrac{14}{e^{0.03x}}}$$

■ $14e^{-0.03x} = \dfrac{14}{e^{0.03x}}$

As x increases without bound, $e^{0.03x}$ increases without bound, and the fraction $\dfrac{14}{e^{0.03x}}$ approaches 0. Therefore, as $x \to \infty$, $R(x) \to \dfrac{1760}{8 + 0} = 220$. Both methods indicate that as the number of weeks increases, the revenue approaches \$220 per week.

CHECK YOUR PROGRESS 7 The monthly revenue R, in dollars, from the sale of a product varies with time according to the function $R(t) = 520 + 480e^{-0.022t}$, where t is the number of months that have passed since the product was put on the market. What does the monthly revenue approach as time goes by?

Solution *See page S24.* \$520 per month

Topics for Discussion

1. Explain how to use the graph of $f(x) = 2^x$ to produce the graph of $g(x) = 2^{(x-3)} + 4$.
2. At what point does the function $g(x) = e^{-x^2/2}$ take on its maximum value?
3. Without using a graphing utility, determine whether the revenue function $R(t) = 10 + e^{-0.05t}$ is an increasing function or a decreasing function.
4. Discuss the properties of the graph of $f(x) = b^x$ when $b > 1$.
5. What is the base of the natural exponential function? How is it calculated? What is its approximate value?

EXERCISES 5.1

— *Suggested Assignment: Exercises 1–69, odd; and 72–77, all.*
— *Answer graphs to Exercises 33–54 are on page AA13.*

In Exercises 1 to 10, evaluate each power.

1. 3^4 81

2. 5^3 125

3. 10^{-2} $\frac{1}{100}$

4. 10^0 1

5. e^0 1

6. 3^{-3} $\frac{1}{27}$

7. $64^{1/3}$ 4

8. 2.5^{-2} $\frac{4}{25}$

9. 0.4^{-2} 6.25

10. $216^{-1/3}$ $\frac{1}{6}$

In Exercises 11 to 20, evaluate each power. Round to the nearest hundredth.

11. $3^{\sqrt{2}}$ 4.73

12. $5^{\sqrt{3}}$ 16.24

13. $10^{\sqrt{7}}$ 442.34

14. $10^{\sqrt{11}}$ 2073.12

15. $e^{5.1}$ 164.02

16. $e^{-3.2}$ 0.04

17. $e^{\sqrt{3}}$ 5.65

18. $e^{\sqrt{5}}$ 9.36

19. $e^{-0.031}$ 0.97

20. $e^{-0.42}$ 0.66

In Exercises 21 to 30, evaluate each functional value, given that $f(x) = 3^x$ and $g(x) = e^x$. Round to the nearest hundredth.

21. $f(\sqrt{15})$ 70.45

22. $f(2\pi)$ 995.04

23. $f(e)$ 19.81

24. $f(-\sqrt{15})$ 0.01

25. $g(e)$ 15.15

26. $g(-3.4)$ 0.03

27. $f[g(2)]$ 3353.33

28. $f[g(-1)]$ 1.50

29. $g[f(2)]$ 8103.08

30. $g[f(-1)]$ 1.40

31. Examine the following four functions and the graphs labeled **a**, **b**, **c**, and **d**. For each graph, determine which function has been graphed.

$f(x) = 5^x$ $\qquad g(x) = 1 + 5^{-x}$

$h(x) = 5^{x+3}$ $\qquad k(x) = 5^x + 3$

a.
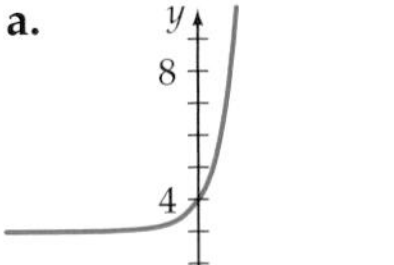

b.
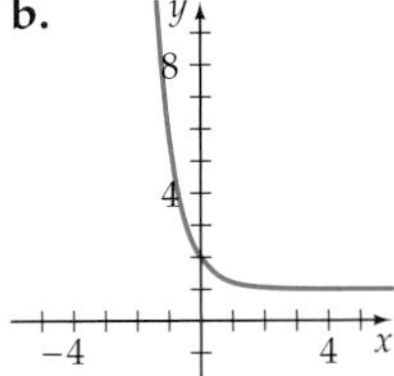

c.
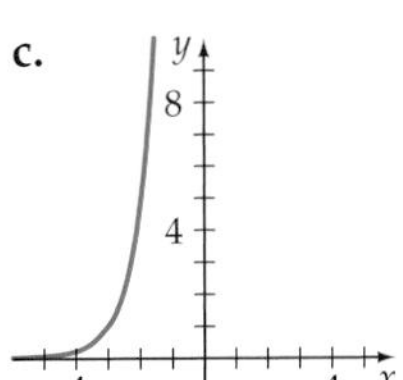

d.
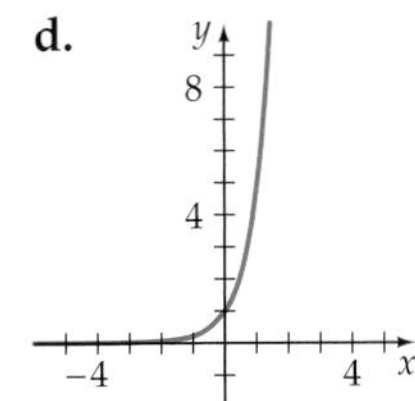

a. $k(x)$ b. $g(x)$ c. $h(x)$ d. $f(x)$

32. Examine the following four functions and the graphs labeled **a**, **b**, **c**, and **d**. For each graph, determine which function has been graphed.

$$f(x) = \left(\frac{1}{4}\right)^x \qquad g(x) = \left(\frac{1}{4}\right)^{-x}$$

$$h(x) = \left(\frac{1}{4}\right)^{x-2} \qquad k(x) = 3\left(\frac{1}{4}\right)^x$$

a.
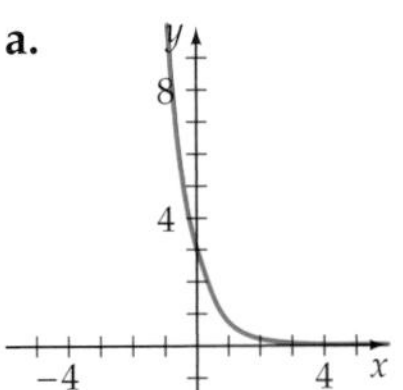

b.
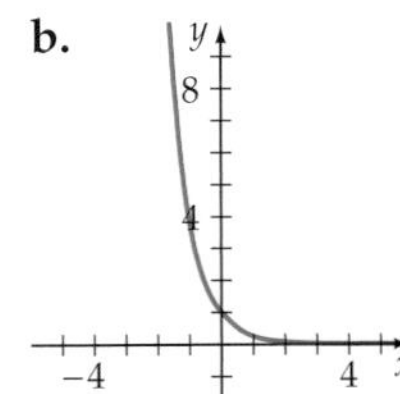

c.
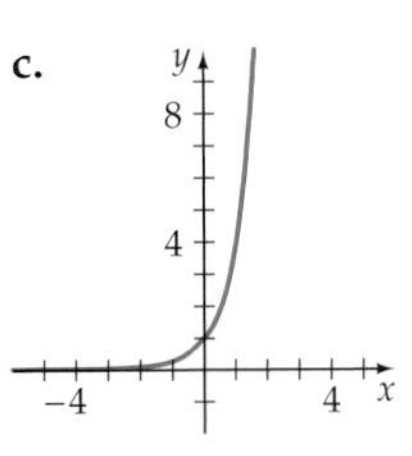

d.
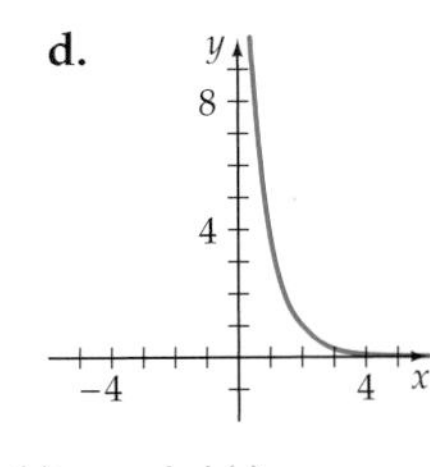

a. $k(x)$ b. $f(x)$ c. $g(x)$ d. $h(x)$

In Exercises 33 to 46, sketch the graph of each function.

33. $f(x) = 3^x$

34. $f(x) = 4^x$

35. $f(x) = \left(\frac{3}{2}\right)^x$

36. $f(x) = \left(\frac{4}{3}\right)^x$

37. $f(x) = \left(\frac{1}{3}\right)^x$

38. $f(x) = \left(\frac{2}{3}\right)^x$

39. $f(x) = \left(\frac{1}{2}\right)^{-x}$ **40.** $f(x) = \left(\frac{1}{3}\right)^{-x}$

41. $f(x) = \frac{5^x}{2}$ **42.** $f(x) = \frac{10^x}{10}$

43. $f(x) = 2^{x+2}$ **44.** $f(x) = 2^{x+3}$

45. $f(x) = 3^x - 1$ **46.** $f(x) = 3^x + 1$

In Exercises 47 to 54, use a graphing utility to graph each function. If the function has a horizontal asymptote, state the equation of the horizontal asymptote.

47. $f(x) = \frac{3^x + 3^{-x}}{2}$
No horizontal asymptote

48. $f(x) = 4 \cdot 3^{-x^2}$
Horizontal asymptote: $y = 0$

49. $f(x) = \frac{e^x - e^{-x}}{2}$
No horizontal asymptote

50. $f(x) = \frac{e^x + e^{-x}}{2}$
No horizontal asymptote

51. $f(x) = -e^{(x-4)}$
Horizontal asymptote: $y = 0$

52. $f(x) = 0.5e^{-x}$
Horizontal asymptote: $y = 0$

53. $f(x) = \frac{10}{1 + 0.4e^{-0.5x}}$, $x \geq 0$
Horizontal asymptote: $y = 10$

54. $f(x) = \frac{10}{1 + 1.5e^{-0.5x}}$, $x \geq 0$
Horizontal asymptote: $y = 10$

Business and Economics

55. ***Internet Connections*** Data from Forrester Research suggest that the number of broadband [cable and digital subscriber line (DSL)] connections to the Internet can be modeled by $f(x) = 1.353(1.9025)^x$, where x is the number of years after January 1, 1998, and $f(x)$ is the number of connections in millions.

a. How many broadband Internet connections, to the nearest million, does this model predict will exist on January 1, 2004? 64 million connections

b. According to the model, in what year will the number of broadband connections first reach 150 million? (*Hint:* Use the intersect feature of a graphing utility to determine the x-coordinate of the point of intersection of the graphs of $f(x)$ and $y = 150$.) 2005

56. ***Demand for a Product*** The price p, in dollars, you charge for a product and the customer demand d for the product are generally related. For instance, the demand, in items per week, for a specific product is given by

$$d(p) = 10 + 630e^{-0.1p}$$

a. What will be the weekly demand, to the nearest unit, when the price of the product is \$6 and when the price is \$10? 356 items per week; 242 items per week

b. What will happen to the demand as the price increases without bound? The demand will approach 10 items per week.

57. ***Demand for a Product*** The demand d for a specific product, in items per month, is given by

$$d(p) = 25 + 880e^{-0.18p}$$

where p is the price, in dollars, of the product.

a. What will be the monthly demand, to the nearest unit, when the price of the product is \$8 and when the price is \$18? 233 items per month; 59 items per month

b. What will happen to the demand as the price increases without bound? The demand will approach 25 items per month.

58. ***Sales*** The monthly income I, in dollars, from a new product is given by

$$I(t) = 24{,}000 - 22{,}000e^{-0.005t}$$

where t is the time, in months, since the product was first put on the market.

a. What was the monthly income after the 10th month and after the 100th month? \$3072.95; \$10,656.33

b. What will the monthly income from the product approach as the time increases without bound? \$24,000

59. ***A Probability Function*** The manager of a home improvement store finds that between 10 A.M. and 11 A.M., customers enter the store at the average rate of 45 customers per hour. The following function gives the probability that a customer will arrive within t minutes of 10 A.M. (*Note:* A probability of 0.6 means there is a 60% chance that a customer will arrive during a given time period.)

$$P(t) = 1 - e^{-0.75t}$$

a. Find the probability, to the nearest hundredth, that a customer will arrive within 1 minute of 10 A.M. 0.53

b. Find the probability, to the nearest hundredth, that a customer will arrive within 3 minutes of 10 A.M. 0.89

c. Use a graph of $P(t)$ to determine how many minutes, to the nearest tenth of a minute, it takes for $P(t)$ to equal 98%. 5.2 min

d. Write a sentence that explains the meaning of the answer in part **c**. There is a 98% probability that at least one customer will arrive between 10:00 A.M. and 10:05.2 A.M.

60. ***A Probability Function*** The owner of a sporting goods store finds that between 9 A.M. and 10 A.M., customers enter the store at the average rate of 12 customers per hour. The following function gives the probability that a customer will arrive within t minutes of 9 A.M.

$$P(t) = 1 - e^{-0.2t}$$

a. Find the probability, to the nearest hundredth, that a customer will arrive within 5 minutes of 9 A.M. 0.63

b. Find the probability, to the nearest hundredth, that a customer will arrive within 15 minutes of 9 A.M. 0.95

c. Use a graph of $P(t)$ to determine how many minutes, to the nearest 0.1 minute, it takes for $P(t)$ to equal 90%. 11.5 min

d. Write a sentence that explains the meaning of the answer in part **c**. There is a 90% probability that at least one customer will arrive between 9:00 A.M. and 9:11.5 A.M.

Exercises 61 and 62 involve the factorial function $x!$, which is defined for whole numbers x as

$$x! = \begin{cases} 1, & \text{if } x = 0 \\ x \cdot (x-1) \cdot (x-2) \cdot \cdots \cdot 3 \cdot 2 \cdot 1, & \text{if } x \geq 1 \end{cases}$$

For example, $3! = 3 \cdot 2 \cdot 1 = 6$ and $5! = 5 \cdot 4 \cdot 3 \cdot 2 \cdot 1 = 120$.

61. ***Queuing Theory*** During the 30-minute period before a Broadway play begins, the members of the audience arrive at the theatre at the average rate of 12 people per minute. The probability that x people will arrive during a particular minute is given by $P(x) = \dfrac{12^x e^{-12}}{x!}$. Find the probability, to the nearest 0.1%, that

a. 9 people will arrive during a given minute. 8.7%

b. 18 people will arrive during a given minute. 2.6%

62. ***Queuing Theory*** During the period from 2:00 P.M. to 3:00 P.M., a bank finds that an average of seven people enter the bank every minute. The probability that x people will enter the bank during a particular minute is given by $P(x) = \dfrac{7^x e^{-7}}{x!}$. Find the probability, to the nearest 0.1%, that

a. only two people will enter the bank during a given minute. 2.2%

b. 11 people will enter the bank during a given minute. 4.5%

Life and Health Sciences

63. ***E. Coli Infection*** *Escherichia coli* (*E. coli*) is a bacterium that can reproduce at an exponential rate. The *E. coli* reproduce by dividing. A small number of *E. coli* bacteria in the large intestine of a human can trigger a serious infection within a few hours. Consider a particular *E. coli* infection that starts with 100 *E. coli* bacteria. Each bacterium splits into two parts every half hour. Assuming none of the bacteria die, the size of the *E. coli* population after t hours is given by $P(t) = 100 \cdot 2^{2t}$, where $0 \leq t \leq 16$.

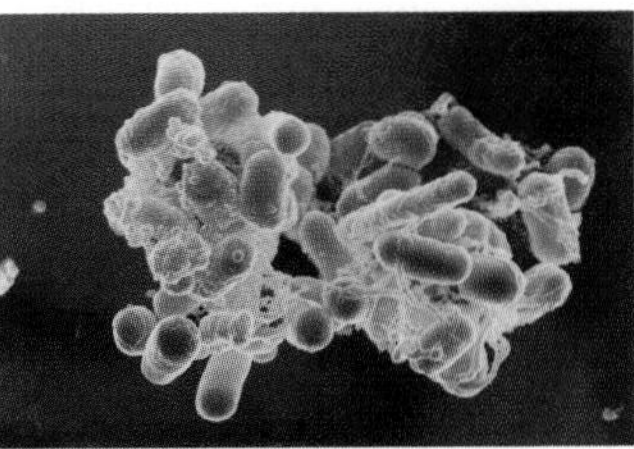

a. Find $P(3)$ and $P(6)$. 6400; 409,600

b. Use a graphing utility to find the time, to the nearest 0.1 hour, it takes for the *E. coli* population to number 1 billion. 11.6 h

64. ***Radiation*** Lead shielding is used to contain radiation. The percent of a certain radiation that can penetrate x millimeters of lead shielding is given by $I(x) = 100e^{-1.5x}$.

a. What percent of radiation, to the nearest tenth of a percent, will penetrate a lead shield that is 1 millimeter thick? 22.3%

b. How many millimeters of lead shielding are required so that less than 0.05% of the radiation penetrates the shielding? Round to the nearest millimeter. 5 mm

65. ***AIDS*** An exponential function that approximates the number of people in the U.S. who have been infected with AIDS is given by $N(t) = 138{,}000(1.39)^t$, where t is the number of years after January 1, 1990.

a. According to this function, how many people had been infected with AIDS as of January 1, 1994? Round to the nearest thousand. 515,000 people

b. Use a graph to estimate during what year the number of people in the U.S. who had been infected with AIDS first reached 1.5 million. 1997

Sports and Recreation

66. 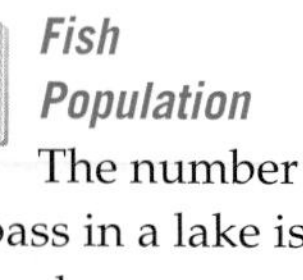 *Fish Population* The number of bass in a lake is given by

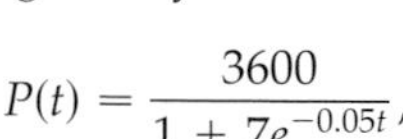

$$P(t) = \frac{3600}{1 + 7e^{-0.05t}},$$

where t is the number of months that have passed since the lake was stocked with bass.

a. How many bass were in the lake immediately after it was stocked? 450 bass

b. How many bass were in the lake 1 year after the lake was stocked? $\approx$744 bass

c. What will happen to the bass population as t increases without bound?
The bass population will increase, approaching 3600.

Social Sciences

67. 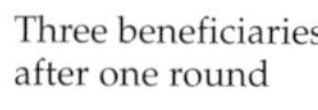 *The Pay It Forward Model* The chapter opener on page 383 described the "pay it forward" project from the movie *Pay It Forward.* The following diagram shows the number of people who have been a beneficiary of a random act of kindness after 1 round and after 2 rounds of this project.

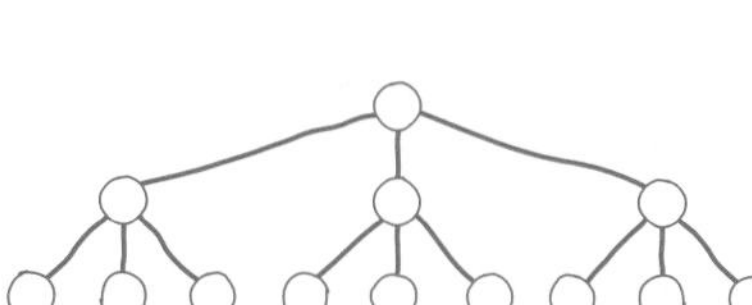

A mathematical model of the number of pay it forward beneficiaries after n rounds is given by the exponential function $B(n) = \frac{3^{n+1} - 3}{2}$. Use this model to determine

a. the number of beneficiaries after 5 rounds and after 10 rounds. Assume that no single person is a beneficiary of more than one random act of kindness.
363 beneficiaries; 88,572 beneficiaries

b. how many rounds are required to produce at least 2 million beneficiaries. 13 rounds

Physical Sciences and Engineering

68. *Intensity of Light* The percent $I(x)$ of the original intensity of light striking the surface of a lake that is available x feet below the surface of the lake is given by $I(x) = 100e^{-0.95x}$.

a. What percent of the light, to the nearest tenth of a percent, is available 2 feet below the surface of the lake?
15.0%

b. At what depth, to the nearest hundredth of a foot, is the intensity of the light one-half the intensity at the surface? 0.73 ft

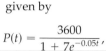

69. *A Temperature Model* A cup of coffee is heated to 180°F and placed in a room that maintains a temperature of 65°F. The temperature of the coffee after t minutes is given by $T(t) = 65 + 115e^{-0.042t}$.

a. Find the temperature, to the nearest degree, of the coffee 10 minutes after it is placed in the room.
141°F

b. Use a graphing utility to determine when, to the nearest minute, the temperature of the coffee will reach 100°F. After 28 min

70. *A Temperature Model* Soup that is at a temperature of 170°F is poured into a bowl in a room that maintains a constant temperature. The temperature of the soup decreases according to the model given by $T(t) = 75 + 95e^{-0.12t}$, where t is time in minutes after the soup is poured.

a. What is the temperature, to the nearest tenth of a degree, of the soup after 2 minutes? 149.7°F

b. A certain customer prefers soup at a temperature of 110°F. How many minutes, to the nearest 0.1 minute, after the soup is poured does the soup reach that temperature? 8.3 min

c. What is the temperature of the room? 75°F

71. *Musical Scales* Starting on the left side of a standard 88-key piano, the frequency, in vibrations per second, of the nth note is given by $f(n) = (27.5)2^{(n-1)/12}$.

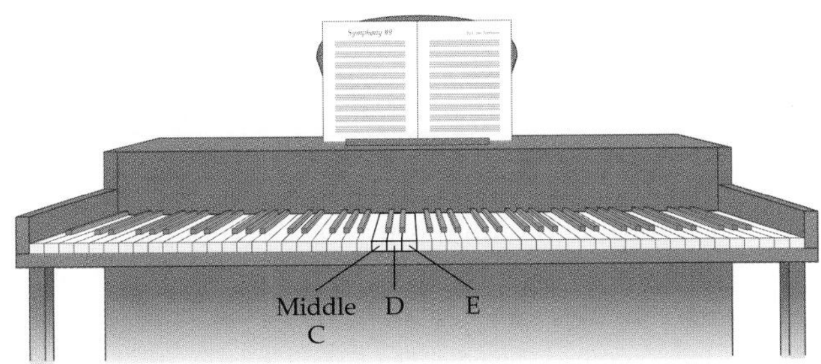

a. Using this formula, determine the frequency, to the nearest hundredth of a vibration per second, of middle C, key number 40 on an 88-key piano.
261.63 vibrations per second

b. Is the difference in frequency between middle C (key number 40) and D (key number 42) the same as the difference in frequency between D (key number 42) and E (key number 44)? Explain.
No. The function $f(n)$ is not a linear function. Therefore, the graph of $f(n)$ does not increase at a constant rate.

Prepare for Section 5.2

72. If $2^x = 16$, determine the value of x. [5.1] 4

73. If $3^{-x} = \frac{1}{27}$, determine the value of x. [5.1] 3

74. If $x^4 = 625$, determine the value of x. [5.1] 5

75. Find the inverse of $f(x) = \frac{2x}{x+3}$. [3.2] $f^{-1}(x) = \frac{3x}{2-x}$

76. State the domain of $g(x) = \sqrt{x-2}$. [2.2] $\{x \mid x \geq 2\}$

77. If the range of $h(x)$ is the set of all positive real numbers, then what is the domain of $h^{-1}(x)$? [3.2]
The set of all positive real numbers

Explorations

—Answer graph to Exercise 1**a.** is on page AA13.

1. ***The Saint Louis Gateway Arch*** The Gateway Arch in Saint Louis was designed in the shape of an inverted catenary, as shown by the red curve in the following drawing. The Gateway Arch is one of the largest optical illusions ever created. As you look at the arch (and its basic shape defined by the catenary curve), it appears to be much taller than it is wide. However, this is not the case. The height of the catenary is given by

$$h(x) = 693.8597 - 68.7672\left(\frac{e^{0.0100333x} + e^{-0.0100333x}}{2}\right)$$

where x and $h(x)$ are measured in feet and $x = 0$ represents the position at ground level that is directly below the highest point of the catenary.

a. Use a graphing utility to graph $h(x)$.

b. Use your graph to find the height of the catenary for $x = 0$, 100, 200, and 299 feet. Round each result to the nearest tenth of a foot. 625.1 ft, 587.5 ft, 433.5 ft, 1.6 ft

c. What is the width of the catenary at ground level and what is the maximum height of the catenary? Round each result to the nearest tenth of a foot.
Width $\approx$ 598.5 ft, height $\approx$ 625.1 ft

d. By how much does the maximum height of the catenary exceed its width at ground level? Round to the nearest tenth of a foot. 26.6 ft

2. ***The Tower of Hanoi*** The Tower of Hanoi is a puzzle invented by Edouard Lucas in 1883. The puzzle consists of three pegs and a number of disks of distinct diameter stacked on one of the pegs such that the largest disk is on the bottom, the next largest is placed on the largest disk, and so on, as shown in the figure.

The object of the puzzle is to transfer the tower to one of the other pegs. The rules require that *only one disk be moved at a time* and that *a larger disk may not be placed on a smaller disk.* All pegs may be used.

Determine the *minimum* number of moves required to transfer all of the disks to another peg for each of the following situations.

a. You start with only one disk. 1

b. You start with two disks. 3

c. You start with three disks. (*Note:* You can use a stack of various size coins to simulate the puzzle, or you can use one of the many sites on the web that provide a JavaScript simulation of the puzzle.) 7

d. You start with four disks. 15

e. You start with five disks. 31

f. You start with n disks. Write your answer as an exponential function of n. $f(n) = 2^n - 1$

g. Lucas included with the Tower of Hanoi puzzle a legend about a tower that had 64 gold disks on one of three diamond needles. A group of priests had the task of transferring the 64 disks to one of the other needles using the same rules as the Tower of Hanoi puzzle. When they had completed the transfer, the tower would crumble and the universe would cease to exist. Assuming that the priests could transfer one disk to another needle every second, how many years would it take for them to transfer all of the 64 disks to one of the other needles? About 5.85×10^{11} yrs

SECTION 5.2 Logarithmic Functions and Their Applications

- Logarithmic Functions
- Graphs of Logarithmic Functions
- Domains of Logarithmic Functions
- Common and Natural Logarithms
- Applications of Logarithmic Functions

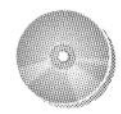

▪ Logarithmic Functions

Every exponential function of the form $g(x) = b^x$ is a one-to-one function and therefore has an inverse function. Sometimes we can determine the inverse of a function represented by an equation by interchanging the variables of its equation and then solving for the dependent variable. If we attempt to use this procedure for $g(x) = b^x$, we obtain

$$g(x) = b^x$$
$$y = b^x$$
$$x = b^y \qquad \text{▪ Interchange the variables.}$$

None of our previous methods can be used to solve the equation $x = b^y$ for the exponent y. Thus we need to develop a new procedure. One method would be to merely write

$$y = \text{the power of } b \text{ that produces } x$$

Although this would work, it is not very concise. We need a compact notation to represent "y is the power of b that produces x." This more compact notation is given in the following definition.

Point of Interest

Logarithms were developed by John Napier (1550–1617) as a means of simplifying the calculations of astronomers. One of his ideas was to devise a method by which the product of two numbers could be determined by performing an addition.

TAKE NOTE

The notation $\log_b x$ replaces the phrase "the power of b that produces x." For instance, "3 is the power of 2 that produces 8" is abbreviated $3 = \log_2 8$. In your work with logarithms, remember that a logarithm is an *exponent*.

Definition of a Logarithm and a Logarithmic Function

If $x > 0$ and b is a positive constant ($b \neq 1$), then

$$y = \log_b x \quad \text{if and only if} \quad b^y = x$$

The notation $\log_b x$ is read "the **logarithm** (or log) base b of x." The function defined by $f(x) = \log_b x$ is a **logarithmic function** with base b. This function is the inverse of the exponential function $g(x) = b^x$.

It is essential to remember that $f(x) = \log_b x$ is the inverse function of $g(x) = b^x$. Because these functions are inverses and because functions that are inverses have the property that $f(g(x)) = x$ and $g(f(x)) = x$, we have the following important relationships.

Composition of Logarithmic and Exponential Functions

Let $g(x) = b^x$ and $f(x) = \log_b x$ ($x > 0, b > 0, b \neq 1$). Then

$$g(f(x)) = b^{\log_b x} = x \quad \text{and} \quad f(g(x)) = \log_b b^x = x$$

As an example of these relationships, let $g(x) = 2^x$ and $f(x) = \log_2 x$. Then

$$2^{\log_2 x} = x \quad \text{and} \quad \log_2 2^x = x$$

The equations

$$y = \log_b x \quad \text{and} \quad b^y = x$$

are different ways of expressing the same concept.

Exponential Form and Logarithmic Form

The **exponential form** of $y = \log_b x$ is $b^y = x$.
The **logarithmic form** of $b^y = x$ is $y = \log_b x$.

These concepts are illustrated in the next two examples.

Alternative to Example 1

Write $4 = \log_3 81$ in its exponential form.

■ $3^4 = 81$

EXAMPLE 1 Change from Logarithmic to Exponential Form

Write each equation in its exponential form.

a. $3 = \log_2 8$ **b.** $2 = \log_{10}(x + 5)$ **c.** $\log_e x = 4$ **d.** $\log_b b^3 = 3$

Solution Use the definition $y = \log_b x$ if and only if $b^y = x$.

Logarithms are exponents.

a. $3 = \log_2 8$ if and only if $2^3 = 8$

Base

Continued ➤

b. $2 = \log_{10}(x + 5)$ if and only if $10^2 = x + 5$.
c. $\log_e x = 4$ if and only if $e^4 = x$.
d. $\log_b b^3 = 3$ if and only if $b^3 = b^3$.

CHECK YOUR PROGRESS 1 Write each equation in its exponential form.

a. $2 = \log_5 25$ **b.** $3 = \log_{10}(2x)$ **c.** $\log_e 3 = x$ **d.** $\log_b b^4 = 4$

Solution *See page S25.* **a.** $5^2 = 25$ **b.** $10^3 = 2x$ **c.** $e^x = 3$ **d.** $b^4 = b^4$

Alternative to Example 2
Write $5^4 = 625$ in its logarithmic form.
■ $4 = \log_5 625$

EXAMPLE 2 Change from Exponential to Logarithmic Form

Write each equation in its logarithmic form.

a. $3^2 = 9$ **b.** $5^3 = x$ **c.** $a^b = c$ **d.** $b^{\log_b 5} = 5$

Solution The logarithmic form of $b^y = x$ is $y = \log_b x$.

Exponent

a. $3^2 = 9$ if and only if $2 = \log_3 9$

Base

b. $5^3 = x$ if and only if $3 = \log_5 x$.
c. $a^b = c$ if and only if $b = \log_a c$.
d. $b^{\log_b 5} = 5$ if and only if $\log_b 5 = \log_b 5$.

CHECK YOUR PROGRESS 2 Write each equation in its logarithmic form.

a. $10^2 = 100$ **b.** $e^{1.5} = x$ **c.** $a = c^b$ **d.** $b^{\log_b 3} = 3$

Solution *See page S25.* **a.** $\log_{10} 100 = 2$ **b.** $\log_e x = 1.5$ **c.** $\log_c a = b$ **d.** $\log_b 3 = \log_b 3$

The definition of a logarithm and the definition of inverse functions can be used to establish many properties of logarithms. For instance:

- $\log_b b = 1$ because $b = b^1$.
- $\log_b 1 = 0$ because $1 = b^0$.
- $\log_b(b^x) = x$ because $b^x = b^x$.
- $b^{\log_b x} = x$ because $f(x) = \log_b x$ and $g(x) = b^x$ are inverse functions. Thus $g[f(x)] = x$.

We will refer to the above properties as the *basic logarithmic properties.*

Basic Logarithmic Properties

1. $\log_b b = 1$ **2.** $\log_b 1 = 0$ **3.** $\log_b(b^x) = x$ **4.** $b^{\log_b x} = x$

Alternative to Example 3

Evaluate each of the following.

a. $\log_2 1$ b. $\log_8 8$
c. $5^{\log_5 9}$ d. $\log_3(3^5)$

■ a. 0 b. 1 c. 9 d. 5

EXAMPLE 3 Apply the Basic Logarithmic Properties

Evaluate each of the following logarithms.

a. $\log_8 1$ **b.** $\log_5 5$ **c.** $\log_2(2^4)$ **d.** $3^{\log_3 7}$

Solution

a. By Property 2, $\log_8 1 = 0$.
b. By Property 1, $\log_5 5 = 1$.
c. By Property 3, $\log_2(2^4) = 4$.
d. By Property 4, $3^{\log_3 7} = 7$.

CHECK YOUR PROGRESS 3 Evaluate each of the following logarithms.

a. $\log_{10} 10$ **b.** $\log_e 1$ **c.** $\log_5 5^3$ **d.** $2^{\log_2 3}$

Solution *See page S25.* a. 1 b. 0 c. 3 d. 3

Some logarithms can be evaluated just by remembering that a logarithm is an exponent. For instance, $\log_5 25$ equals 2 because the base 5 raised to the second power equals 25.

- $\log_{10} 100 = 2$ because $10^2 = 100$.
- $\log_4 64 = 3$ because $4^3 = 64$.
- $\log_7 \dfrac{1}{49} = -2$ because $7^{-2} = \dfrac{1}{7^2} = \dfrac{1}{49}$.

? QUESTION What is the value of $\log_5 625$?

■ Graphs of Logarithmic Functions

Because $f(x) = \log_b x$ is the inverse function of $g(x) = b^x$, the graph of f is a reflection of the graph of g across the line $y = x$. The graph of $g(x) = 2^x$ is shown in Figure 5.13. Table 5.7 below shows some of the ordered pairs on the graph of g.

Table 5.7

x	−3	−2	−1	0	1	2	3
$g(x) = 2^x$	$\frac{1}{8}$	$\frac{1}{4}$	$\frac{1}{2}$	1	2	4	8

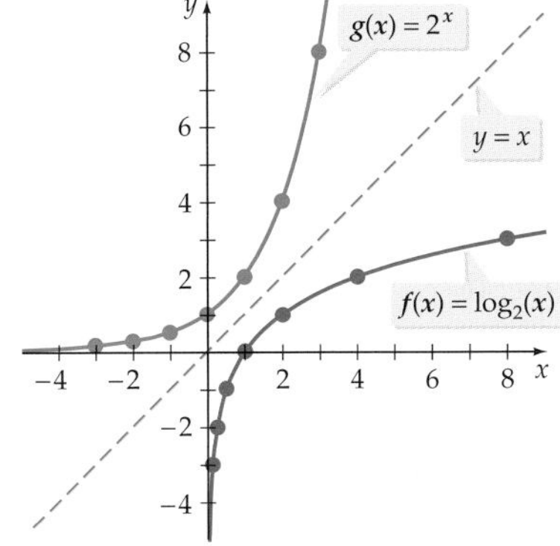

Figure 5.13

The graph of the inverse of g, which is $f(x) = \log_2 x$, is also shown in Figure 5.13. Some of the ordered pairs of f are shown in Table 5.8. Note that if (x, y) is a point on the graph of g, then (y, x) is a point on the graph of f. Also notice that the graph of f is a reflection of the graph of g across the line $y = x$.

? ANSWER $\log_5 625 = 4$ because $5^4 = 625$.

Table 5.8

x	$\frac{1}{8}$	$\frac{1}{4}$	$\frac{1}{2}$	1	2	4	8
$f(x) = \log_2 x$	-3	-2	-1	0	1	2	3

The graph of a logarithmic function can be drawn by first rewriting the function in its exponential form. This procedure is illustrated in Example 4.

Alternative to Example 4
Exercise 32, page 409

EXAMPLE 4 Graph a Logarithmic Function

Graph $f(x) = \log_3 x$.

Solution To graph $f(x) = \log_3 x$, consider the equivalent exponential equation $x = 3^y$. Because this equation is solved for x, choose values of y and calculate the corresponding values of x, as shown in Table 5.9.

Table 5.9

$x = 3^y$	$\frac{1}{9}$	$\frac{1}{3}$	1	3	9
y	-2	-1	0	1	2

Now plot the ordered pairs and connect the points with a smooth curve, as shown in Figure 5.14.

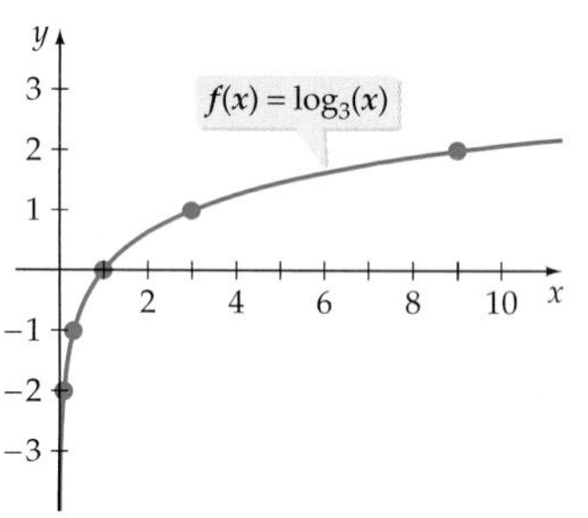

Figure 5.14

CHECK YOUR PROGRESS 4 Graph $f(x) = \log_6 x$.

Solution *See page S25.*

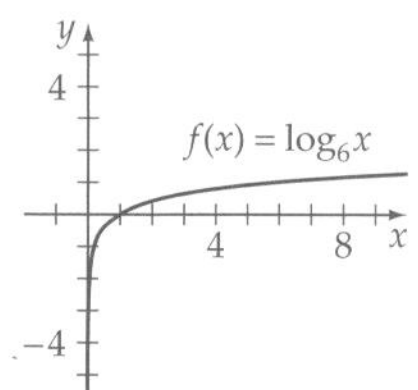

We can use a similar procedure to draw the graph of a logarithmic function with a fractional base. For instance, consider $y = \log_{2/3} x$. Rewriting this in exponential form gives us $\left(\frac{2}{3}\right)^y = x$. Choose values of y and calculate the corresponding x-values. See Table 5.10. Plot the points corresponding to the ordered pairs (x, y) and then draw a smooth curve through the points, as shown in Figure 5.15.

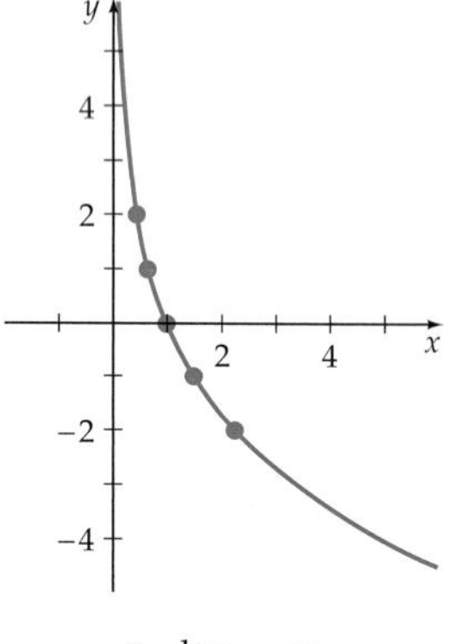

Figure 5.15

Table 5.10

$x = \left(\frac{2}{3}\right)^y$	$\left(\frac{2}{3}\right)^{-2} = \frac{9}{4}$	$\left(\frac{2}{3}\right)^{-1} = \frac{3}{2}$	$\left(\frac{2}{3}\right)^{0} = 1$	$\left(\frac{2}{3}\right)^{1} = \frac{2}{3}$	$\left(\frac{2}{3}\right)^{2} = \frac{4}{9}$
y	-2	-1	0	1	2

Properties of $f(x) = \log_b x$

For all positive real numbers b, $b \neq 1$, the function $f(x) = \log_b x$ has the following properties.

1. The domain of f consists of the set of positive real numbers and its range consists of the set of all real numbers.
2. The graph of f has an x-intercept of $(1, 0)$ and passes through $(b, 1)$.
3. If $b > 1$, f is an increasing function and its graph is asymptotic to the negative y-axis. [As $x \to 0$ from the right, $f(x) \to -\infty$.] See Figure 5.16a.
4. If $0 < b < 1$, f is a decreasing function and its graph is asymptotic to the positive y-axis. [As $x \to 0$ from the right, $f(x) \to \infty$.] See Figure 5.16b.

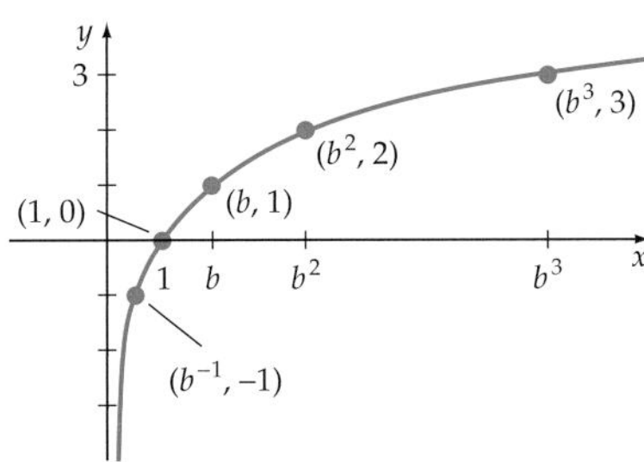

a. $f(x) = \log_b x$, $b > 1$

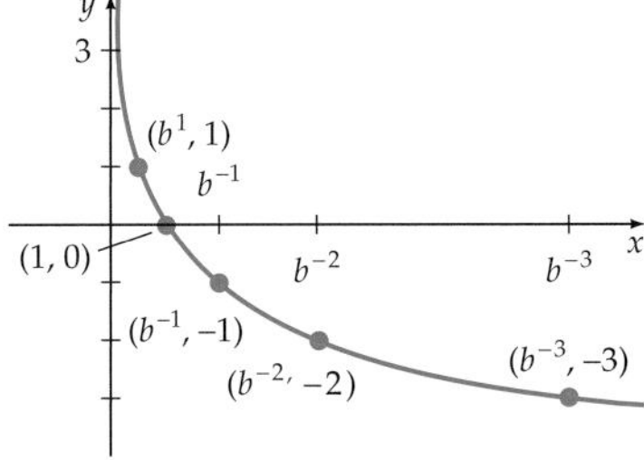

b. $f(x) = \log_b x$, $0 < b < 1$

Figure 5.16

Domains of Logarithmic Functions

The function $f(x) = \log_b x$ has the set of positive real numbers as its domain. The function $f(x) = \log_b(g(x))$ has as its domain the set of all x for which $g(x) > 0$. To determine the domain of a function such as $f(x) = \log_b(g(x))$, we must determine the values of x that make $g(x)$ positive. This process is illustrated in Example 5.

Alternative to Example 5
Exercise 38, page 409

EXAMPLE 5 Find the Domain of a Logarithmic Function

Find the domain of each of the following logarithmic functions.

a. $f(x) = \log_6(x - 3)$ **b.** $F(x) = \log_2|x + 2|$ **c.** $R(x) = \log_5\left(\dfrac{x}{8 - x}\right)$

Solution

a. Solving $(x - 3) > 0$ for x gives us $x > 3$. The domain of f consists of all real numbers greater than 3. In interval notation the domain is $(3, \infty)$.

b. The solution set of $|x + 2| > 0$ consists of all real numbers x except $x = -2$. The domain of F consists of all real numbers $x \neq -2$. In interval notation the domain is $(-\infty, -2) \cup (-2, \infty)$.

c. Solving $\left(\dfrac{x}{8 - x}\right) > 0$ yields the set of all real numbers x between 0 and 8. The domain of R is all real numbers x such that $0 < x < 8$. In interval notation the domain is $(0, 8)$.

Continued ➤

CHECK YOUR PROGRESS 5 Find the domain of each logarithmic function.

a. $k(x) = \log_4(5 - x)$ **b.** $p(x) = \log_{10}|x - 3|$ **c.** $s(x) = \log_7\left(\frac{x + 4}{x}\right)$

Solution *See page S25.* a. $(-\infty, 5)$ b. $(-\infty, 3) \cup (3, \infty)$ c. $(-\infty, -4) \cup (0, \infty)$

Some logarithmic functions can be graphed by using horizontal and/or vertical translations of a previously drawn graph.

Alternative to Example 6
Exercise 50, page 409

EXAMPLE 6 Use Translations to Graph Logarithmic Functions

Graph: **a.** $f(x) = \log_4(x + 3)$ **b.** $f(x) = \log_4 x + 3$

Solution

a. The graph of $f(x) = \log_4(x + 3)$ can be obtained by shifting the graph of $g(x) = \log_4 x$ three units to the left. See Figure 5.17. Note that the domain of f consists of all real numbers x greater than -3 because $x + 3 > 0$ for $x > -3$. The graph of f is asymptotic to the vertical line $x = -3$.

b. The graph of $f(x) = \log_4 x + 3$ can be obtained by shifting the graph of $g(x) = \log_4 x$ three units upward. See Figure 5.18.

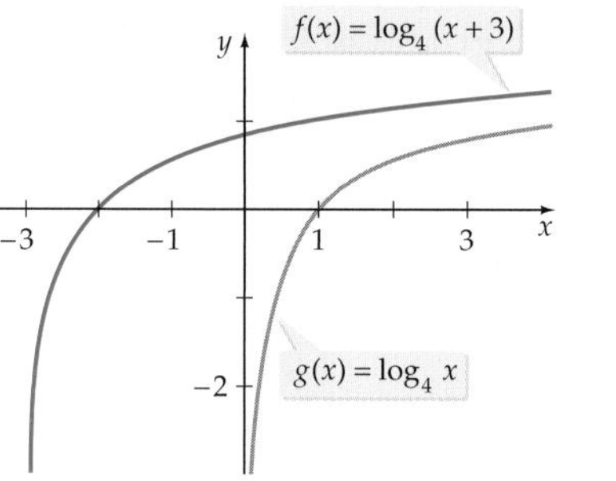

Figure 5.17

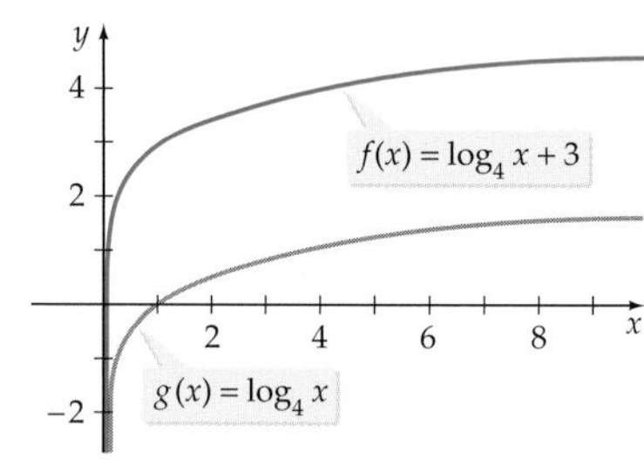

Figure 5.18

CHECK YOUR PROGRESS 6 Graph: **a.** $f(x) = \log_2(x - 4)$ **b.** $f(x) = -3 + \log_2 x$

Solution *See page S25.* See the graphs at the left.

6a.

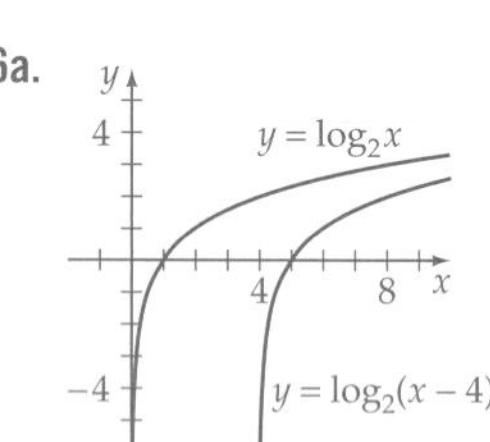

b.

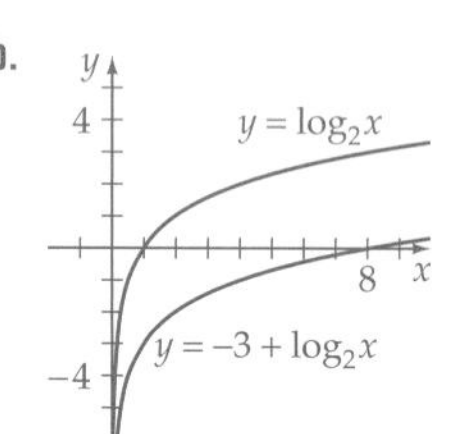

▪ Common and Natural Logarithms

Two of the most frequently used logarithmic functions are *common logarithms*, which have base 10, and *natural logarithms*, which have base e (the base of the natural exponential function).

Definition of Common and Natural Logarithms

The function defined by $f(x) = \log_{10} x$ is called the **common logarithmic function.** It is customarily written without stating the base as $f(x) = \log x$.

The function defined by $f(x) = \log_e x$ is called the **natural logarithmic function.** It is customarily written as $f(x) = \ln x$.

Most scientific or graphing calculators have a LOG key for evaluating common logarithms and a LN key to evaluate natural logarithms. For instance, using a graphing calculator,

$$\log 24 \approx 1.3802112 \quad \text{and} \quad \ln 81 \approx 4.3944492$$

The graphs of $f(x) = \log x$ and $f(x) = \ln x$ can be drawn using the same techniques we used to draw the graphs in the preceding examples. However, these graphs can also be produced with a graphing calculator by entering $\log x$ and $\ln x$ into the Y= menu. See Figure 5.19 and Figure 5.20.

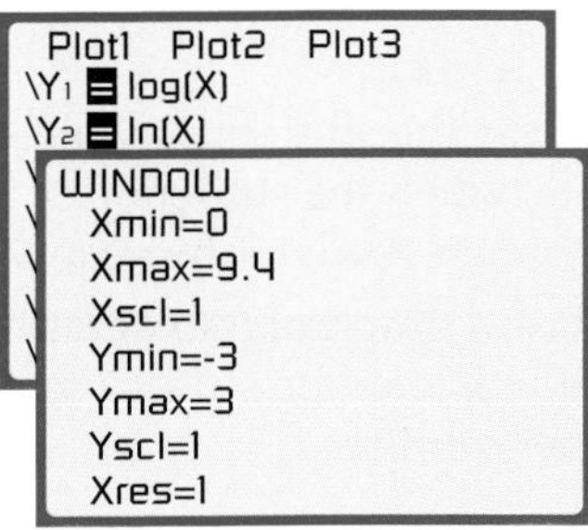

Figure 5.19

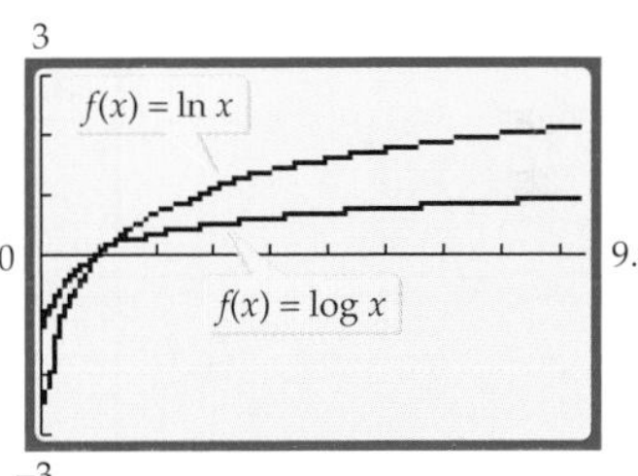

Figure 5.20

Observe that each graph passes through (1, 0). Also note that as $x \to 0$ from the right, the functional values $f(x) \to -\infty$. Thus the y-axis is a vertical asymptote for each of the graphs. The domain of both $f(x) = \log x$ and $f(x) = \ln x$ is the set of positive real numbers. Each of these functions has a range consisting of the set of real numbers.

Alternative to Example 7
Exercise 66, page 410

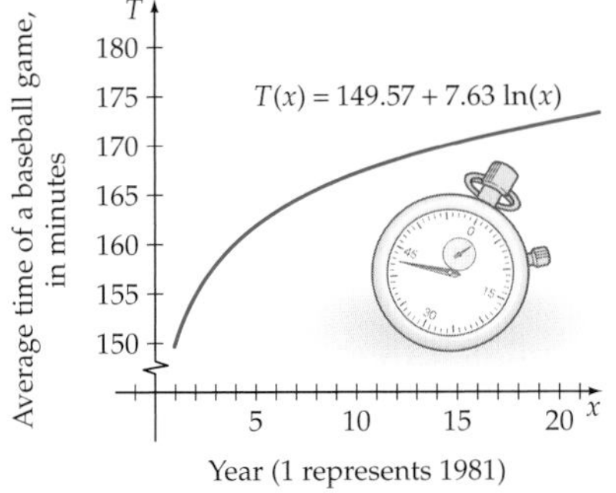

Applications of Logarithmic Functions

Many applications can be modeled by logarithmic functions.

EXAMPLE 7 Average Time of a Major League Baseball Game

From 1981 to 1999, the average time of a major league baseball game tended to increase each year. If the year 1981 is represented by $x = 1$, then the function

$$T(x) = 149.57 + 7.63 \ln x$$

approximates the average time T, in minutes, of a major league baseball game for the years 1981 to 1999—that is, $x = 1$ to $x = 19$.

Continued ➤

Point of Interest

Although logarithms were originally developed to assist with computations, logarithmic functions have a much broader use today. They are often used in such disciplines as geology, acoustics, chemistry, physics, and economics, to name a few.

a. Use the function T to determine the average time of a major league baseball game during the 1981 season and during the 1999 season.

b. By how much did the average time of a major league baseball game increase during the years 1981 to 1999?

Solution

a. The year 1981 is represented by $x = 1$ and the year 1999 by $x = 19$.

$$T(1) = 149.57 + 7.63\ln(1) = 149.57$$

In 1981 the average time of a baseball game was about 149.57 minutes.

$$T(19) = 149.57 + 7.63\ln(19) \approx 172.04$$

In 1999 the average time of a baseball game was about 172.04 minutes.

b. $T(19) - T(1) \approx 172.04 - 149.57 = 22.47$. For the years 1981 to 1999, the average time of a baseball game increased by about 22.47 minutes.

CHECK YOUR PROGRESS 7 The following function models the average typing speed S (in words per minute) of a student who has been typing for t months.

$$S(t) = 5 + 29\ln(t + 1), \qquad 0 \le t \le 16$$

a. What was the student's average typing speed, to the nearest word per minute, when the student first started to type? What was the student's average typing speed, to the nearest word per minute, after 3 months?

b. Use a graph of S to determine how long, to the nearest tenth of a month, it will take the student to achieve an average typing speed of 65 words per minute.

Solution *See page S25.* **a.** 5 words/min; about 45 words/min **b.** About 6.9 months

Topics for Discussion

1. If $m > n$, must $\log_b m > \log_b n$?

2. For what values of x is $\ln x > \log x$?

3. What is the domain of $f(x) = \log(x^2 + 1)$? Explain why the graph of f does not have a vertical asymptote.

4. The subtraction $3 - 5$ does not have an answer if we require that the answer be positive. Keep this idea in mind as you work the rest of this exercise.

   Press the MODE key of a TI-83 graphing calculator and choose "Real" from the menu. Now use the calculator to evaluate $\log(-2)$. What output is given by the calculator? Press the MODE key and choose "a + bi" from the menu. Now use the calculator to evaluate $\log(-2)$. What output is given by the calculator? Write a sentence or two that explain why the output is different for these two evaluations.

EXERCISES 5.2

— Suggested Assignment: Exercises 1–69, odd; and 73–78, all.
— Answer graphs to Exercises 31–36, 47–52, and 55–64 are on page AA14.

In Exercises 1 to 10, change each equation to its exponential form.

1. $\log 10 = 1$ $10^1 = 10$
2. $\log 10{,}000 = 4$ $10^4 = 10{,}000$
3. $\log_8 64 = 2$ $8^2 = 64$
4. $\log_4 64 = 3$ $4^3 = 64$
5. $\log_7 x = 0$ $7^0 = x$
6. $\log_3 \dfrac{1}{81} = -4$ $3^{-4} = \dfrac{1}{81}$
7. $\ln x = 4$ $e^4 = x$
8. $\ln e^2 = 2$ $e^2 = e^2$
9. $\ln 1 = 0$ $e^0 = 1$
10. $\ln x = -3$ $e^{-3} = x$

In Exercises 11 to 20, change each equation to its logarithmic form.

11. $3^2 = 9$ $\log_3 9 = 2$
12. $5^3 = 125$ $\log_5 125 = 3$
13. $4^{-2} = \dfrac{1}{16}$ $\log_4 \dfrac{1}{16} = -2$
14. $10^0 = 1$ $\log 1 = 0$
15. $b^x = y$ $\log_b y = x$
16. $2^x = y$ $\log_2 y = x$
17. $y = e^x$ $\ln y = x$
18. $5^1 = 5$ $\log_5 5 = 1$
19. $100 = 10^2$ $\log 100 = 2$
20. $2^{-4} = \dfrac{1}{16}$ $\log_2 \dfrac{1}{16} = -4$

In Exercises 21 to 30, evaluate each logarithm. Do not use a calculator.

21. $\log_4 16$ 2
22. $\log_{3/2} \dfrac{8}{27}$ -3
23. $\log_3 \dfrac{1}{243}$ -5
24. $\log_b 1$ 0
25. $\ln e^3$ 3
26. $\log_b b$ 1
27. $\log \dfrac{1}{100}$ -2
28. $\log 1{,}000{,}000$ 6
29. $\log_{0.5} 16$ -4
30. $\log_{0.3} \dfrac{100}{9}$ -2

In Exercises 31 to 36, graph each function by using its exponential form.

31. $f(x) = \log_4 x$
32. $f(x) = \log_5 x$
33. $f(x) = \log_{12} x$
34. $f(x) = \log_8 x$
35. $f(x) = \log_{1/2} x$
36. $f(x) = \log_{1/4} x$

In Exercises 37 to 46, find the domain of the function. Write the domains using interval notation.

37. $f(x) = \log_5(x - 3)$ $(3, \infty)$
38. $g(x) = \log_2(3x - 1)$ $\left(\dfrac{1}{3}, \infty\right)$
39. $k(x) = \log_{2/3}(11 - x)$ $(-\infty, 11)$
40. $H(x) = \log_{1/4}(x^2 + 1)$ $(-\infty, \infty)$
41. $P(x) = \ln(x^2 - 4)$ $(-\infty, -2) \cup (2, \infty)$
42. $J(x) = \ln\left(\dfrac{x - 3}{x}\right)$ $(-\infty, 0) \cup (3, \infty)$
43. $h(x) = \ln\left(\dfrac{x^2}{x - 4}\right)$ $(4, \infty)$
44. $R(x) = \ln(x^4 - x^2)$ $(-\infty, -1) \cup (1, \infty)$
45. $N(x) = \log_2(x^3 - x)$ $(-1, 0) \cup (1, \infty)$
46. $s(x) = \log_7(x^2 + 7x + 10)$ $(-\infty, -5) \cup (-2, \infty)$

In Exercises 47 to 52, use translations of the graphs in Exercises 31 to 36 to produce the graph of the given function.

47. $f(x) = \log_4(x - 3)$
48. $f(x) = \log_5(x + 3)$
49. $f(x) = \log_{12} x + 2$
50. $f(x) = \log_8 x - 4$
51. $f(x) = 3 + \log_{1/2} x$
52. $f(x) = 2 + \log_{1/4} x$

53. Examine the following four functions and the graphs labeled **a**, **b**, **c**, and **d**. Determine which graph is the graph of each function.

$f(x) = \log_5(x - 2)$ $g(x) = 2 + \log_5 x$
$h(x) = \log_5(-x)$ $k(x) = -\log_5(x + 3)$

a. 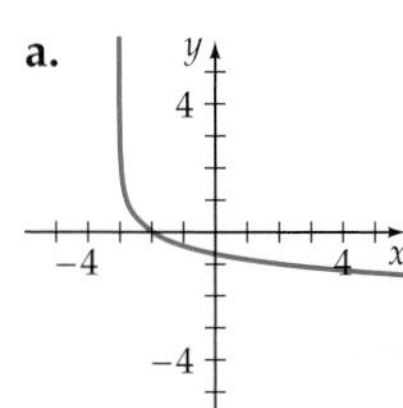
b. 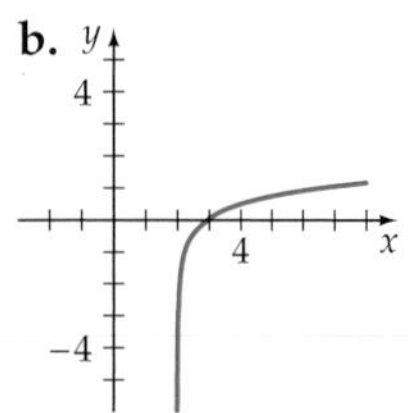
c. 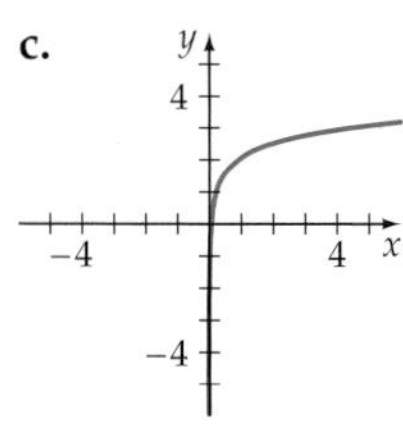
d. 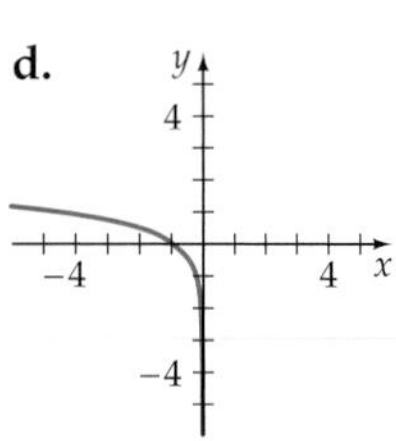

a. $k(x)$ b. $f(x)$ c. $g(x)$ d. $h(x)$

54. Examine the following four functions and the graphs labeled **a**, **b**, **c**, and **d**. Determine which graph is the graph of each function.

$f(x) = \ln x + 3$ $\qquad g(x) = \ln(x - 3)$

$h(x) = \ln(3 - x)$ $\qquad k(x) = -\ln(-x)$

a. 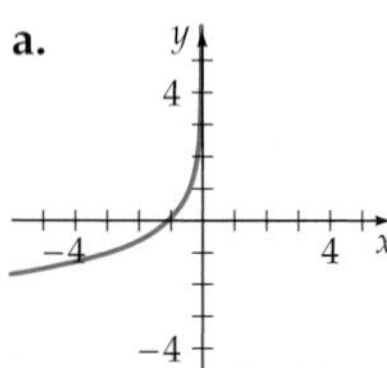
b. 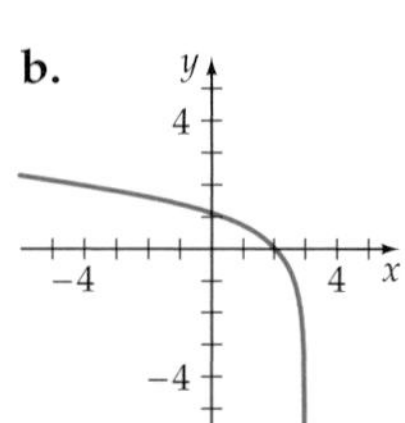
c. 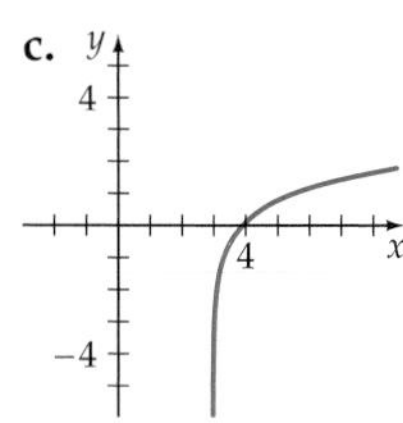
d. 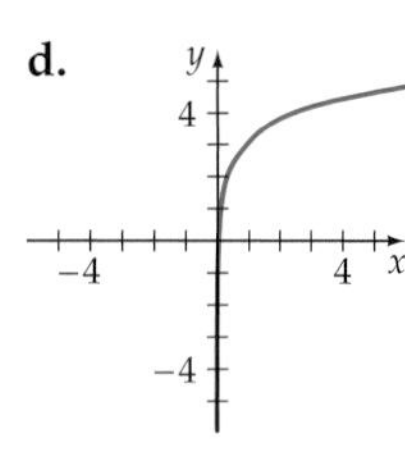

a. $k(x)$ b. $h(x)$ c. $g(x)$ d. $f(x)$

In Exercises 55 to 64, use a graphing utility to graph the function.

55. $f(x) = -2 \ln x$ **56.** $f(x) = -\log x$

57. $f(x) = |\ln x|$ **58.** $f(x) = \ln |x|$

59. $f(x) = \log \sqrt[3]{x}$ **60.** $f(x) = \ln \sqrt{x}$

61. $f(x) = \log(x + 10)$ **62.** $f(x) = \ln(x + 3)$

63. $f(x) = 3 \log |2x + 10|$ **64.** $f(x) = \frac{1}{2} \ln |x - 4|$

Business and Economics

65. ***Money Market Rates*** The function

$$r(t) = 3.56096 + 0.33525 \ln t$$

gives the annual interest rate r, as a percent, a bank will pay on its money market accounts, where t is the term (the time the money is invested) in months.

a. What interest rate, to the nearest tenth of a percent, will the bank pay on a money market account with a term of 9 months? 4.3%

b. What is the minimum number of complete months during which a person must invest to receive an interest rate of at least 5%? 74 months

66. ***Maintaining a Skill*** The function

$$S(t) = 58 - 6.8 \ln(t + 1)$$

models the typing speed S, in words per minute, of a student t months after the student completes a typing class.

a. What is the student's typing speed, to the nearest word per minute, 1 month, 3 months, and 10 months after the student completes the class? 53 wpm; 49 wpm; 42 wpm

b. How many months, to the nearest 0.1 month, will elapse before the student's average typing speed falls below 45 words per minute? 5.8 months

67. ***Advertising Costs and Sales*** The function

$$N(x) = 2750 + 180 \ln\left(\frac{x}{1000} + 1\right)$$

models the relationship between the dollar amount x spent on advertising a product and the number of units N that a company can sell.

a. Find the number of units that will be sold with advertising expenditures of \$20,000, \$40,000, and \$60,000. 3298 units; 3418 units; 3490 units

b. How many units will be sold if the company does not pay to advertise the product? 2750 units

Life and Health Sciences

68.   *Medicine* In anesthesiology it is necessary to accurately estimate the body surface area of a patient. One formula for estimating body surface area (*BSA*) was developed by Edith Boyd (University of Minnesota Press, 1935). Her formula for the *BSA* (in square meters) of a patient of height *H* (in centimeters) and weight *W* (in grams) is

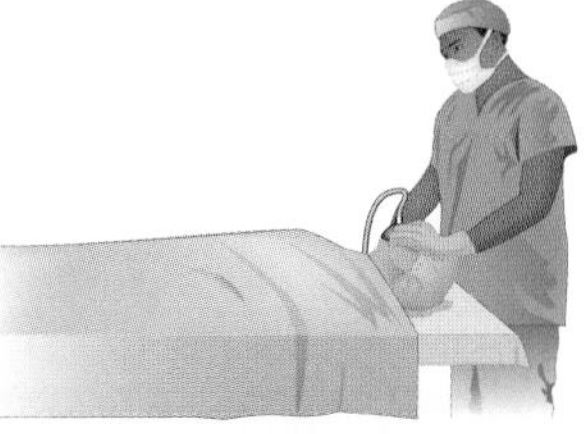

$$BSA = 0.0003207 \cdot H^{0.3} \cdot W^{(0.7285-0.0188 \log W)}$$

Use Boyd's formula to estimate the body surface area of a patient with the given weight and height. Round to the nearest hundredth of a square meter.

a. $W = 110$ pounds (49,895.2 grams), $H = 5$ feet 4 inches (162.56 centimeters) 1.50 m^2

b. $W = 180$ pounds (81,646.6 grams), $H = 6$ feet 1 inch (185.42 centimeters) 2.05 m^2

c. $W = 40$ pounds (18,143.7 grams), $H = 29$ inches (73.66 centimeters) 0.67 m^2

Sports and Recreation

69.   *World Records in the Discus Throw* The function $d(t) = -41.71 + 25.76 \ln t$ approximates the world record distance, in meters, for the men's discus throw for the years 1965 to 1985. The year 1965 is represented by $t = 65$.

a. According to the function $d(t)$, what was the world record distance in 1975 and in 1984? Round to the nearest hundredth meter. 69.51 m; 72.43 m

b. Use the Internet to check the world record distance in the men's discus throw for the current year. How does this actual result compare with the distance predicted by the function $d(t)$? Answers will vary.

Physical Sciences and Engineering

70. *Astronomy* Astronomers measure the apparent brightness of a star by a unit called the **apparent magnitude.** This unit was created in the second century B.C. when the Greek astronomer Hipparchus classified the relative brightness of several stars. In his list he assigned the number 1 to the stars that appeared to be the brightest (Sirius, Vega, and Deneb). They are first-magnitude stars. Hipparchus assigned the number 2 to all the stars in the Big Dipper. They are second-magnitude stars. The following table shows the relationship between a star's brightness relative to a first-magnitude star and the star's apparent magnitude. Notice from the table that a first-magnitude star appears in the sky to be about 2.51 times as bright as a second-magnitude star.

Brightness relative to a first-magnitude star x	Apparent magnitude $M(x)$
1	1
$\frac{1}{2.51}$	2
$\frac{1}{6.31} \approx \frac{1}{2.51^2}$	3
$\frac{1}{15.85} \approx \frac{1}{2.51^3}$	4
$\frac{1}{39.82} \approx \frac{1}{2.51^4}$	5
$\frac{1}{100} \approx \frac{1}{2.51^5}$	6

The following logarithmic function gives the apparent magnitude $M(x)$ of a star as a function of its brightness x.

$$M(x) = -2.51 \log x + 1, \quad 0 < x \le 1$$

a. Use $M(x)$ to find the apparent magnitude of a star that is $\frac{1}{10}$ as bright as a first-magnitude star. Round to the nearest hundredth. 3.51

b. Find the approximate apparent magnitude of a star that is $\frac{1}{400}$ as bright as a first-magnitude star. Round to the nearest hundredth. 7.53

c. Which star appears brighter: a star with an apparent magnitude of 12 or a star with an apparent magnitude of 15? Apparent magnitude of 12

d. Is $M(x)$ an increasing function or a decreasing function? Decreasing function

71. ***Number of Digits in b^x*** An engineer has determined that the number of digits N in the expansion of b^x, where both b and x are positive integers, is $N = \text{int}(x \log b) + 1$, where $\text{int}(x \log b)$ denotes the greatest integer of $x \log b$. (*Note:* The greatest integer of the real number x is x if x is an integer and is the largest integer less than x if x is not an integer. For example, the greatest integer of 5 is 5 and the greatest integer of 7.8 is 7.)

a. Because $2^{10} = 1024$, we know that 2^{10} has four digits. Use the equation $N = \text{int}(x \log b) + 1$ to verify this result. Answers will vary.

b. Find the number of digits in 3^{200}. 96 digits

c. Find the number of digits in 7^{4005}. 3385 digits

d. The largest known prime number in 2001 was $2^{13466917} - 1$. Find the number of digits in this prime number. (*Hint:* Because $2^{13466917}$ is not a power of 10, both $2^{13466917}$ and $2^{13466917} - 1$ have the same number of digits.) 4,053,946 digits

72. ***Number of Digits in $9^{(9^9)}$*** A science teacher has offered 10 points extra credit to any student who will write out all the digits in the expansion of $9^{(9^9)}$.

a. Use the formula from Exercise 71 to determine the number of digits in this number. 369,693,100 digits

b. Assume that you can write 1000 digits per page and that 500 pages of paper are in a ream of paper. How many reams of paper, to the nearest tenth of a ream, are required to write out the expansion of $9^{(9^9)}$? Assume that you write on only one side of each page. ≈739.4 reams

Prepare for Section 5.3

In Exercises 73 to 78, use a calculator to compare each of the given expressions.

73. $\log 3 + \log 2$; $\log 6$ [5.2] ≈0.77815 for each expression

74. $\ln 8 - \ln 3$; $\ln\left(\frac{8}{3}\right)$ [5.2] ≈0.98083 for each expression

75. $3 \log 4$; $\log(4^3)$ [5.2] ≈1.80618 for each expression

76. $2 \ln 5$; $\ln(5^2)$ [5.2] ≈3.21888 for each expression

77. $\ln 5$; $\frac{\log 5}{\log e}$ [5.2] ≈1.60944 for each expression

78. $\log 8$; $\frac{\ln 8}{\ln 10}$ [5.2] ≈0.90309 for each expression

Explorations

—Answer graph to Exploration is on page AA15.

1. ***Benford's Law*** The authors of this text know some interesting details about your finances. For instance, of the last 100 checks you have written, about 30% are for amounts that start with the number 1. Also, you have written about 3 times as many checks for amounts that start with the number 2 than you have for amounts that start with the number 7.

We are sure of these results because of a mathematical formula known as **Benford's Law.** This law was first discovered by the mathematician Simon Newcomb in 1881, and then rediscovered by the physicist Frank Benford in 1938. Benford's Law states that the probability P that the first digit of a number selected from a wide range of numbers is d is given by

$$P(d) = \log\left(1 + \frac{1}{d}\right)$$

a. Use Benford's Law to complete the table to the right and the bar graph on the following page.

d	$P(d) = \log\left(1 + \frac{1}{d}\right)$
1	0.301
2	0.176
3	0.125
4	0.097
5	0.079
6	0.067
7	0.058
8	0.051
9	0.046

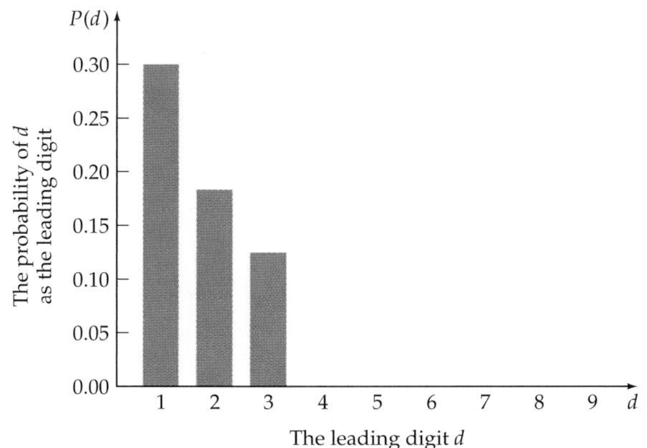

Benford's Law applies to most data with a wide range. For instance, it applies to

- the populations of the cities in the U.S.
- the numbers of dollars in the savings accounts at your local bank.
- the number of miles driven during a month by each person in a state.

b. Use the table in part **a.** to find the probability that in a U.S. city selected at random, the number of telephones in that city will be a number starting with 6.
About 0.067, or 6.7%

c. Use the table in part **a.** to estimate how many times as many purchases you have made for dollar amounts that start with a 1 than for dollar amounts that start with a 9. About 6.54 times as many

d. Explain why Benford's Law would not apply to the set of telephone numbers of the people living in a small city such as Le Mars, Iowa.
Telephone numbers in many small towns start with the same digit.

e. Explain why Benford's Law would not apply to the set of all the ages, in years, of the students at a local high school.
Most high school students are teenagers.

An Application of Benford's Law Benford's Law has been used to identify fraudulent accountants. In most cases these accountants are unaware of Benford's Law and have replaced valid numbers with numbers selected at random. Their numbers do not conform to Benford's Law. Hence, an audit is warranted.

SECTION 5.3 Working with Logarithms and Logarithmic Scales

- **Properties of Logarithms**
- **Changing the Base**
- **Logarithmic Scales**

▪ Properties of Logarithms

In Section 5.2 we introduced the following basic properties of logarithms.

$$\log_b b = 1 \qquad \text{and} \qquad \log_b 1 = 0$$

Also, because exponential functions and logarithmic functions are inverses of each other, we observed the relationships

$$b^{\log_b x} = x \qquad \text{and} \qquad \log_b(b^x) = x$$

We can use the properties of exponents to establish the following additional logarithmic properties.

TAKE NOTE

Pay close attention to these properties. Note that

$$\log_b(MN) \neq \log_b M \cdot \log_b N$$

and

$$\log_b \frac{M}{N} \neq \frac{\log_b M}{\log_b N}$$

Also,

$$\log_b(M + N) \neq \log_b M + \log_b N$$

In fact, the expression $\log_b(M + N)$ cannot be expanded at all.

INSTRUCTOR NOTE

Stress the importance of the properties of logarithms. An understanding of these properties is essential to the study and use of logarithms.

Alternative to Example 1

Use the properties of logarithms to express

$$\log_2\left(\frac{a^3b^2}{\sqrt{c}}\right)$$

in terms of logarithms of a, b, and c.

▪ $3 \log_2 a + 2 \log_2 b - \frac{1}{2} \log_2 c$

Properties of Logarithms

In the following properties, b, M, and N are positive real numbers ($b \neq 1$).

Product property	$\log_b(MN) = \log_b M + \log_b N$
Quotient property	$\log_b \frac{M}{N} = \log_b M - \log_b N$
Power property	$\log_b(M^p) = p \log_b M$
Logarithm-of-each-side property	$M = N$ implies $\log_b M = \log_b N$
One-to-one property	$\log_b M = \log_b N$ implies $M = N$

QUESTION Is it true that $\ln 5 + \ln 10 = \ln 50$?

The above properties of logarithms are often used to rewrite logarithmic expressions in an equivalent form.

EXAMPLE 1 Rewrite Logarithmic Expressions

Use the properties of logarithms to express the following logarithms in terms of logarithms of x, y, and z.

a. $\log_5(xy^2)$ **b.** $\log_b \frac{2\sqrt{y}}{z^5}$

Solution

a. $\log_5(xy^2) = \log_5 x + \log_5 y^2$ ▪ Product property

$= \log_5 x + 2 \log_5 y$ ▪ Power property

b. $\log_b \frac{2\sqrt{y}}{z^5} = \log_b(2\sqrt{y}) - \log_b z^5$ ▪ Quotient property

$= \log_b 2 + \log_b \sqrt{y} - \log_b z^5$ ▪ Product property

$= \log_b 2 + \log_b y^{1/2} - \log_b z^5$ ▪ Replace $\sqrt{y}$ with $y^{1/2}$

$= \log_b 2 + \frac{1}{2} \log_b y - 5 \log_b z$ ▪ Power property

CHECK YOUR PROGRESS 1 Use the properties of logarithms to express $\ln \frac{z^3}{\sqrt{xy}}$ in terms of logarithms of x, y, and z.

Solution *See page S25.* $3 \ln z - \frac{1}{2} \ln x - \frac{1}{2} \ln y$

The properties of logarithms are also used to rewrite expressions that involve several logarithms as a single logarithm.

ANSWER Yes. By the product property, $\ln 5 + \ln 10 = \ln(5 \cdot 10)$.

Alternative to Example 2

Use the properties of logarithms to rewrite

$5 \log_4(x - 7) - 3 \log_4 x$

as a single logarithm with a coefficient of 1.

▪ $\log_4 \frac{(x-7)}{x^3}$

EXAMPLE 2 Rewrite Logarithmic Expressions

Use the properties of logarithms to rewrite each expression as a single logarithm with a coefficient of 1.

a. $2 \log_b x + \frac{1}{2} \log_b(x + 4)$ **b.** $4 \log_3(x + 2) - 3 \log_3(x - 5)$

Solution

a. $2 \log_b x + \frac{1}{2} \log_b(x + 4)$

$= \log_b x^2 + \log_b(x + 4)^{1/2}$ ▪ Power property

$= \log_b[x^2(x + 4)^{1/2}]$ ▪ Product property

$= \log_b\left(x^2\sqrt{x + 4}\right)$

b. $4 \log_3(x + 2) - 3 \log_3(x - 5)$

$= \log_3(x + 2)^4 - \log_3(x - 5)^3$ ▪ Power property

$= \log_3 \frac{(x + 2)^4}{(x - 5)^3}$ ▪ Quotient property

CHECK YOUR PROGRESS 2 Use the properties of logarithms to rewrite the expression as a single logarithm with a coefficient of 1.

$$3 \log_2 t - \frac{1}{3} \log_2 u + 4 \log_2 v$$

Solution *See page S25.* $\log_2 \frac{t^3 v^4}{\sqrt[3]{u}}$

▪ Changing the Base

Recall that to determine the value of y in $\log_3 81 = y$, we are basically asking, "What power of 3 is 81?" Because $3^4 = 81$, we have $\log_3 81 = 4$. Now suppose that we need to determine the value of $\log_3 50$. In this case we need to find the power of 3 that produces 50. Because $3^3 = 27$ and $3^4 = 81$, the value we are seeking is somewhere between 3 and 4. But how can we achieve more accuracy? Most calculators can only evaluate logarithms of base 10 or e.

The exponential form of $\log_3 50 = y$ is $3^y = 50$. Applying logarithmic properties gives us

$\ln 3^y = \ln 50$ ▪ Logarithm-of-each-side property

$y \ln 3 = \ln 50$ ▪ Power property

$y = \frac{\ln 50}{\ln 3}$ ▪ Solve for y.

≈ 3.56088 ▪ Use a calculator.

So $\log_3 50 \approx 3.56088$. In the above procedure, we could just as well have used logarithms of any base and arrived at the same value. Thus any logarithm can be expressed in terms of logarithms of any base we wish. This general result is summarized in the following formula.

Change-of-Base Formula

If x, a, and b are positive real numbers with $a \neq 1$ and $b \neq 1$, then

$$\log_b x = \frac{\log_a x}{\log_a b}$$

Because most calculators use only common logarithms ($a = 10$) or natural logarithms ($a = e$) the change-of-base formula is most often used in the following form.

If x and b are positive real numbers and $b \neq 1$, then

$$\log_b x = \frac{\log x}{\log b} = \frac{\ln x}{\ln b}$$

Alternative to Example 3

Evaluate $\log_8 75$. Round to the nearest ten thousandth.

■ **2.0763**

EXAMPLE 3 Use the Change-of-Base Formula

Evaluate each logarithm. Round to the nearest hundred thousandth.

a. $\log_3 18$ **b.** $\log_{12} 400$

Solution To approximate these logarithms, we may use the change-of-base formula with $a = 10$ or $a = e$. For this example, we choose to use the change-of-base formula with $a = e$. That is, we will evaluate these logarithms by using the ln key on a scientific or graphing calculator.

a. $\log_3 18 = \dfrac{\ln 18}{\ln 3} \approx 2.63093$ **b.** $\log_{12} 400 = \dfrac{\ln 400}{\ln 12} \approx 2.41114$

TAKE NOTE

If common logarithms had been used for the calculation in Example **3a**, the final result would be the same.

$$\log_3 18 = \frac{\log 18}{\log 3} \approx 2.63093$$

CHECK YOUR PROGRESS 3 Evaluate each logarithm. Round your results to the nearest hundred thousandth.

a. $\log_5 50$ **b.** $\log_8 \dfrac{1}{5}$

Solution *See page S25.* **a.** 2.43068 **b.** −0.77398

The change-of-base formula and a graphing calculator can be used to graph logarithmic functions that have a base other than 10 or e. For instance, to graph $f(x) = \log_3(2x + 3)$, we rewrite the function in terms of base 10 or base e. Using base 10 logarithms, we have $f(x) = \log_3(2x + 3) = \dfrac{\log(2x + 3)}{\log 3}$. The graph is shown in Figure 5.21.

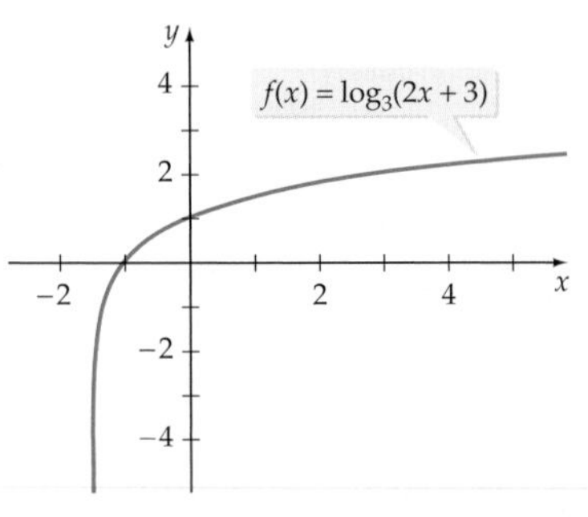

Figure 5.21

Alternative to Example 4
Exercise 30, page 423

EXAMPLE 4 Use the Change-of-Base Formula to Graph a Logarithmic Function

Graph $f(x) = \log_2|x - 3|$.

Solution Rewrite f using the change-of-base formula. We will use the natural logarithm function; however, the common logarithm function could be used instead.

$$f(x) = \log_2|x - 3| = \frac{\ln|x - 3|}{\ln 2}$$

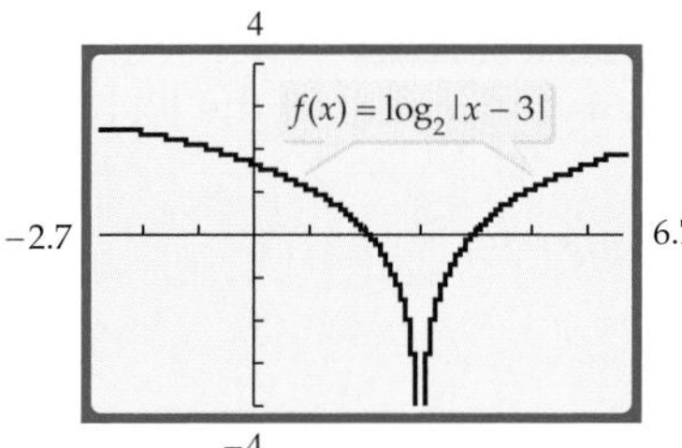

- Enter $\dfrac{\ln|x - 3|}{\ln 2}$ into Y1. Note that the domain of $f(x) = \log_2|x - 3|$ is all real numbers except 3, because $|x - 3| = 0$ when $x = 3$ and $|x - 3|$ is positive for all other values of x.

CHECK YOUR PROGRESS 4  Graph $g(x) = \log_8(5 - x)$.

Solution *See page S25.*

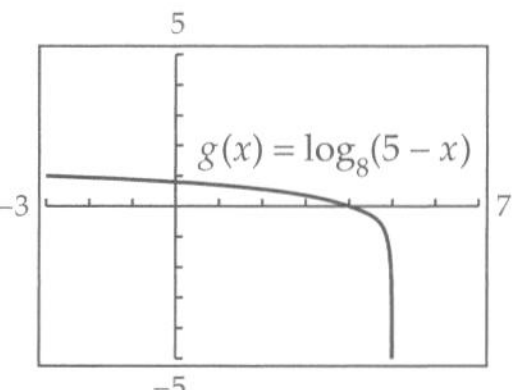

Point of Interest

The Richter scale was created by the seismologist Charles F. Richter in 1935. Notice that a tenfold increase in the intensity level of an earthquake only increases the Richter scale magnitude of the earthquake by 1.

▪ Logarithmic Scales

Logarithmic functions are often used to scale very large (or very small) numbers into numbers that are easier to comprehend. For instance, the *Richter scale* magnitude of an earthquake uses a logarithmic function to convert the intensity of the earthquake's shock waves I into a number M, which for most earthquakes is in the range of 0 to 10. The intensity I of an earthquake is often given in terms of the constant I_0, where I_0 is the intensity of the smallest earthquake (called a **zero-level earthquake**) that can be measured on a seismograph near the earthquake's epicenter. The following formula is used to compute the Richter scale magnitude of an earthquake.

The Richter Scale Magnitude of an Earthquake

An earthquake with an intensity of I has a **Richter scale magnitude** of

$$M = \log\left(\frac{I}{I_0}\right)$$

where I_0 is the measure of the intensity of a zero-level earthquake.

Alternative to Example 5

Find the Richter scale magnitude of an earthquake with intensity $I = 8{,}250{,}000I_0$

■ **6.9**

TAKE NOTE

Notice in Example 5 that we didn't need to know the value of I_0 to determine the Richter scale magnitude of the quake.

EXAMPLE 5 Determine the Magnitude of an Earthquake

Find the Richter scale magnitude (to the nearest 0.1) of the 1999 Joshua Tree, California earthquake that had an intensity of $I = 12{,}589{,}254I_0$.

Solution

$$M = \log\left(\frac{I}{I_0}\right) = \log\left(\frac{12{,}589{,}254I_0}{I_0}\right) = \log(12{,}589{,}254) \approx 7.1$$

The 1999 Joshua Tree earthquake had a Richter scale magnitude of 7.1.

CHECK YOUR PROGRESS 5 What is the Richter scale magnitude of an earthquake whose intensity is twice that of the Joshua Tree earthquake in Example 5?

Solution *See page S25.* Approx. 7.4

If you know the Richter scale magnitude of an earthquake, you can determine the intensity of the earthquake.

Alternative to Example 6

Find the intensity of an earthquake measuring 9.2 on the Richter scale.

■ $\approx$**1,584,893,000**I_0

EXAMPLE 6 Determine the Intensity of an Earthquake

Find the intensity of the 1999 Taiwan earthquake, which measured 7.6 on the Richter scale.

Solution

$$\log\left(\frac{I}{I_0}\right) = 7.6$$

$$\frac{I}{I_0} = 10^{7.6}$$ ■ Change to exponential form

$$I = 10^{7.6}I_0$$ ■ Solve for I

$$I \approx 39{,}810{,}717I_0$$

The 1999 Taiwan earthquake had an intensity that was approximately 39,811,000 times the intensity of a zero-level earthquake.

CHECK YOUR PROGRESS 6 On September 3, 2000, an earthquake measuring 5.2 struck the Napa Valley, 50 miles north of San Francisco. Find the intensity of the quake.

Solution *See page S25.* Approx. $158{,}489I_0$

In Example 7 we make use of the Richter scale magnitudes of two earthquakes to compare the intensities of the earthquakes.

Alternative to Example 7

How many times more intense is an earthquake with a Richter scale magnitude of 8.5 than a 6.8 quake?

■ **≈50 times**

TAKE NOTE

The results of Example 7 show that if an earthquake has a Richter scale magnitude of M_1 and a smaller earthquake has a Richter scale magnitude of M_2, then the larger earthquake is $10^{M_1 - M_2}$ times as intense as the smaller earthquake.

Damage from the 1999 Izmit earthquake in western Turkey

EXAMPLE 7 Compare Earthquakes

The 1960 Chile earthquake had a Richter scale magnitude of 9.5. The 1989 San Francisco earthquake had a Richter scale magnitude of 7.1. Compare the intensities of the earthquakes.

Solution Let I_1 be the intensity of the Chilean earthquake and I_2 the intensity of the San Francisco earthquake. Then

$$\log\left(\frac{I_1}{I_0}\right) = 9.5 \qquad \text{and} \qquad \log\left(\frac{I_2}{I_0}\right) = 7.1$$

$$\frac{I_1}{I_0} = 10^{9.5} \qquad\qquad \frac{I_2}{I_0} = 10^{7.1}$$

$$I_1 = 10^{9.5}I_0 \qquad\qquad I_2 = 10^{7.1}I_0$$

To compare the intensities of the earthquakes, we compute the ratio I_1/I_2.

$$\frac{I_1}{I_2} = \frac{10^{9.5}I_0}{10^{7.1}I_0} = \frac{10^{9.5}}{10^{7.1}} = 10^{9.5-7.1} = 10^{2.4} \approx 251$$

The earthquake in Chile was approximately 251 times as intense as the San Francisco earthquake.

CHECK YOUR PROGRESS 7 On August 17, 1999, a 7.4 magnitude quake hit western Turkey, followed by many aftershocks. One aftershock on August 19 measured 5.0, and a November aftershock measured 7.2. How much more intense was the November aftershock than the earlier one?

Solution *See page S25.* Approx. 158 times more intense

Seismologists generally determine the Richter scale magnitude of an earthquake by examining a *seismogram*. See Figure 5.22.

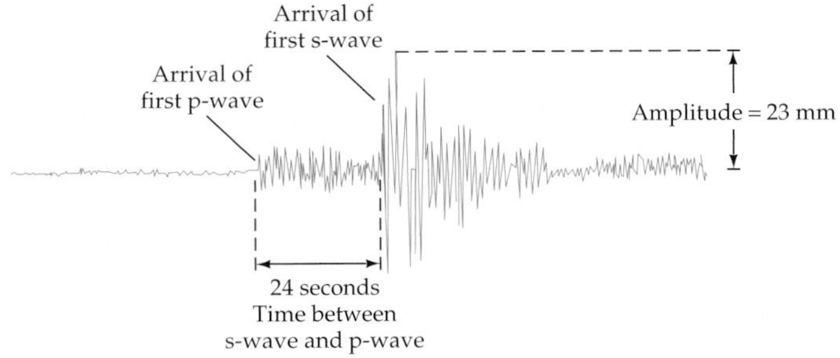

Figure 5.22

The magnitude of an earthquake cannot be determined just by examining the amplitude of a seismogram, because this amplitude decreases as the distance between the epicenter of the earthquake and the observation station increases. To account for the distance between the epicenter and the observation station, a seismologist examines a seismogram for both small waves called **p-waves** and larger waves called s-waves. The Richter scale magnitude M of the earthquake is a function of both the amplitude A of the s-waves and the difference in time t between the occurrence of the s-waves and the p-waves. In the 1950s, Charles Richter developed the following formula to determine the magnitude of an earthquake from the data in a seismogram.

Amplitude-Time-Difference Formula

The Richter scale magnitude M of an earthquake is given by

$$M = \log A + 3 \log 8t - 2.92$$

where A is the amplitude, in millimeters, of the s-waves on a seismogram and t is the difference in time, in seconds, between the s-waves and the p-waves.

Alternative to Example 8
Exercise 64, page 425

EXAMPLE 8 Determine the Magnitude of an Earthquake from Its Seismogram

Find the magnitude of the earthquake that produced the seismogram in Figure 5.22.

Solution

$$\begin{aligned} M &= \log A + 3 \log 8t - 2.92 \\ &= \log 23 + 3 \log[8 \cdot 24] - 2.92 \quad \text{Substitute 23 for } A \text{ and 24 for } t. \\ &\approx 1.36173 + 6.84990 - 2.92 \\ &\approx 5.3 \end{aligned}$$

The earthquake had a magnitude of 5.3 on the Richter scale.

TAKE NOTE
The Richter scale magnitude is usually rounded to the nearest tenth.

CHECK YOUR PROGRESS 8 Find the magnitude of the earthquake that produced the seismogram shown below.

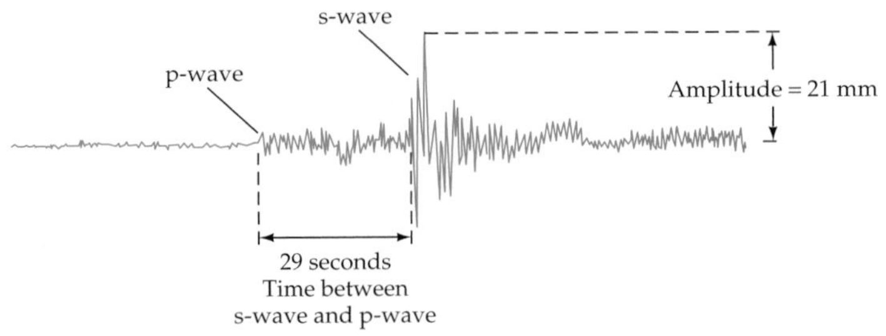

Solution *See page S25.* 5.5

Logarithmic scales are also used in chemistry. One example is when we determine the pH of a liquid, which is a measure of the liquid's acidity or alkalinity. (You may have tested the pH of a swimming pool or an aquarium.) Pure water, which is considered neutral, has a pH of 7.0. The pH scale ranges from 0 to 14, with 0 corresponding to the most acidic solutions and 14 to the most alkaline. Lemon juice has a pH of about 2, whereas household ammonia measures about 11.

Specifically, the pH of a solution is a function of the hydronium-ion concentration of the solution. Because the hydronium-ion concentration of a solution can be very small (with values such as 0.00000001), pH uses a logarithmic scale.

TAKE NOTE

One mole is equivalent to 6.022×10^{23} ions.

The pH of a Solution

The **pH of a solution** with a hydronium-ion concentration of H^+ moles per liter is given by

$$\text{pH} = -\log[\text{H}^+]$$

Alternative to Example 9

Find the pH of a solution with $H^+ = 5.77 \times 10^{-6}$ mole per liter.

■ $\approx$**5.2**

EXAMPLE 9 Find the pH of a Solution

Find the pH of each liquid. Round to the nearest tenth.

a. Orange juice with $H^+ = 2.8 \times 10^{-4}$ mole per liter

b. Milk with $H^+ = 3.97 \times 10^{-7}$ mole per liter

c. Rainwater with $H^+ = 6.31 \times 10^{-5}$ mole per liter

d. A baking soda solution with $H^+ = 3.98 \times 10^{-9}$ mole per liter

Solution

a. $\text{pH} = -\log[\text{H}^+] = -\log(2.8 \times 10^{-4}) \approx 3.6$
The orange juice has a pH of 3.6.

b. $\text{pH} = -\log[\text{H}^+] = -\log(3.97 \times 10^{-7}) \approx 6.4$
The milk has a pH of 6.4.

c. $\text{pH} = -\log[\text{H}^+] = -\log(6.31 \times 10^{-5}) \approx 4.2$
The rainwater has a pH of 4.2.

d. $\text{pH} = -\log[\text{H}^+] = -\log(3.98 \times 10^{-9}) \approx 8.4$
The baking soda solution has a pH of 8.4.

Point of Interest

The pH scale was created by the Danish biochemist Søren Sørensen in 1909 to measure the acidity of water used in the brewing of beer. pH is an abbreviation for *pondus hydrogenii*, which translates as "potential hydrogen."

CHECK YOUR PROGRESS 9 Find the pH of each liquid. Round to the nearest tenth.

a. A cleaning solution containing bleach with $H^+ = 2.41 \times 10^{-13}$ mole per liter

b. A cola soft drink containing $H^+ = 5.07 \times 10^{-4}$ mole per liter

Solution *See page S25.* **a.** 12.6 **b.** 3.3

Figure 5.23 illustrates the pH scale, along with the corresponding hydronium-ion concentrations. A solution on the left half of the scale, with a pH of less than 7, is an **acid**, and a solution on the right half of the scale is an **alkaline solution** or a **base.** Because the scale used is logarithmic, a solution with a pH of 5 is 10 times more acidic than a solution with a pH of 6. From Example 9 we see that the orange juice, rainwater, and milk are acids, whereas the baking soda solution is a base.

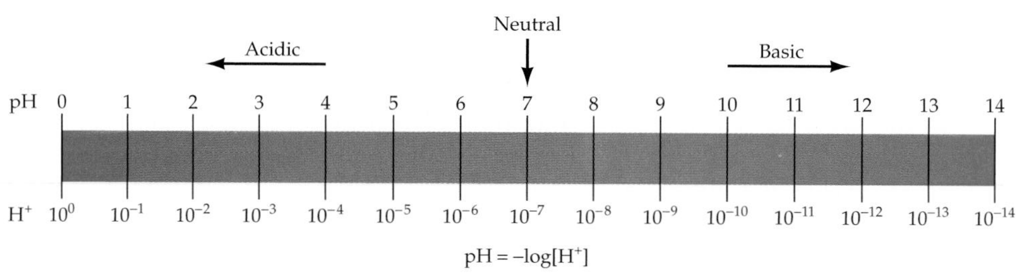

Figure 5.23

Alternative to Example 10

Find the hydronium-ion concentration of cleaning solution with pH 3.4.

■ **3.98×10^{-4} mol/L**

EXAMPLE 10 Find the Hydronium-Ion Concentration

A sample of blood has a pH of 7.3. Find the hydronium-ion concentration of the blood.

Solution

$$\mathrm{pH} = -\log[\mathrm{H}^+]$$
$$7.3 = -\log[\mathrm{H}^+]$$ ■ Substitute 7.3 for pH.
$$-7.3 = \log[\mathrm{H}^+]$$ ■ Multiply both sides by -1.
$$10^{-7.3} = \mathrm{H}^+$$ ■ Change to exponential form.
$$5.0 \times 10^{-8} \approx \mathrm{H}^+$$

The hydronium-ion concentration of the blood is about 5.0×10^{-8} mole per liter.

CHECK YOUR PROGRESS 10 The water in the Great Salt Lake in Utah has a pH of 10.0. Find the hydronium-ion concentration of the water.

Solution *See page S25.* 10^{-10} mole per liter

Topics for Discussion

1. The function $f(x) = \log_b x$ is defined only for $x > 0$. Explain why this condition is imposed.
2. If p and q are positive numbers, explain why $\ln(p + q)$ isn't normally equal to $\ln p + \ln q$.
3. If $f(x) = \log_b x$ and $f(c) = f(d)$, can we conclude that $c = d$?
4. Give examples of situations in which it is advantageous to use logarithmic scales.

EXERCISES 5.3

— *Suggested Assignment: Exercises 1–29, odd; 31–40, all; 45–57, odd; 60 and 66–71, all.*
— *Answer graphs to Exercises 23–30 are on page AA15.*

In Exercises 1 to 8, write the given logarithm in terms of logarithms of x, y, and z.

1. $\log_b(xyz)$ $\log_b x + \log_b y + \log_b z$

2. $\log_b(x^2y^3)$ $2\log_b x + 3\log_b y$

3. $\ln \dfrac{x}{z^4}$ $\ln x - 4\ln z$

4. $\log_5 \dfrac{xy^2}{z^4}$ $\log_5 x + 2\log_5 y - 4\log_5 z$

5. $\log_2 \dfrac{\sqrt{x}}{y^3}$ $\dfrac{1}{2}\log_2 x - 3\log_2 y$

6. $\log_b\left(x\sqrt[3]{y}\right)$ $\log_b x + \dfrac{1}{3}\log_b y$

7. $\log_7 \dfrac{\sqrt{xz}}{y^2}$ $\dfrac{1}{2}\log_7 x + \dfrac{1}{2}\log_7 z - 2\log_7 y$

8. $\ln \sqrt[3]{x^2\sqrt{y}}$ $\dfrac{2}{3}\ln x + \dfrac{1}{6}\ln y$

In Exercises 9 to 14, write each logarithmic expression as a single logarithm with a coefficient of 1. Simplify when possible.

9. $\log(x + 5) + 2\log x$ $\log[x^2(x + 5)]$

10. $5\log_3 x - 4\log_3 y + 2\log_3 z$ $\log_3 \dfrac{x^5z^2}{y^4}$

11. $\ln(x^2 - y^2) - \ln(x - y)$ $\ln(x + y)$

12. $\dfrac{1}{2}\log_8(x + 5) - 3\log_8 y$ $\log_8 \dfrac{\sqrt{x+5}}{y^3}$

13. $3\log x + \dfrac{1}{3}\log y + \log(x + 1)$ $\log\left[x^3 \cdot \sqrt[3]{y}(x + 1)\right]$

14. $\ln(xz) - \ln\left(x\sqrt{y}\right) + 2\ln\dfrac{y}{z}$ $\ln \dfrac{y^{3/2}}{z}$

In Exercises 15 to 22, use the change-of-base formula to approximate the logarithm accurate to the nearest ten thousandth.

15. $\log_7 20$ 1.5395

16. $\log_5 37$ 2.2436

17. $\log_{11} 8$ 0.8672

18. $\log_{50} 22$ 0.7901

19. $\log_6 \dfrac{1}{3}$ -0.6131

20. $\log_3 \dfrac{7}{8}$ -0.1215

21. $\log_9 \sqrt{17}$ 0.6447

22. $\log_4 \sqrt{7}$ 0.7018

In Exercises 23 to 30, use a graphing utility and the change-of-base formula to graph the logarithmic function.

23. $f(x) = \log_4 x$

24. $s(x) = \log_7 x$

25. $g(x) = \log_8(x - 3)$

26. $t(x) = \log_9(5 - x)$

27. $h(x) = \log_3(x - 3)^2$

28. $J(x) = \log_{12}(-x)$

29. $F(x) = -\log_5 |x - 2|$

30. $n(x) = \log_2 \sqrt{x - 8}$

In Exercises 31 to 40, determine if the statement is true or false for all $x > 0$, $y > 0$. If it is false, write an example that disproves the statement.

31. $\log_b(x + y) = \log_b x + \log_b y$
False; $\log 10 + \log 10 = 2$ but $\log(10 + 10) = \log 20 \neq 2$.

32. $\log_b(xy) = \log_b x \cdot \log_b y$
False; $\log(10 \cdot 10) = 2$ but $\log 10 \cdot \log 10 = 1$.

33. $\log_b(xy) = \log_b x + \log_b y$ True

34. $\log_b x \cdot \log_b y = \log_b x + \log_b y$
False; $\log 10 \cdot \log 10 = 1$ but $\log 10 + \log 10 = 2$.

35. $\log_b x - \log_b y = \log_b(x - y)$, $x > y$
False; $\log 100 - \log 10 = 1$ but $\log(100 - 10) = \log 90 \neq 1$.

36. $\log_b \dfrac{x}{y} = \dfrac{\log_b x}{\log_b y}$ False; $\log \dfrac{100}{10} = \log 10 = 1$ but $\dfrac{\log 100}{\log 10} = \dfrac{2}{1} = 2$.

37. $\dfrac{\log_b x}{\log_b y} = \log_b x - \log_b y$ False; $\dfrac{\log 100}{\log 10} = \dfrac{2}{1} = 2$ but $\log 100 - \log 10 = 1$.

38. $\log_b(x^n) = n\log_b x$ True

39. $(\log_b x)^n = n\log_b x$ False; $(\log 10)^2 = 1$ but $2\log 10 = 2$.

40. $\log_b \sqrt{x} = \dfrac{1}{2}\log_b x$ True

41. Evaluate the following *without* using a calculator.

$$\log_3 5 \cdot \log_5 7 \cdot \log_7 9 \quad 2$$

42. Evaluate the following *without* using a calculator.

$$\log_5 20 \cdot \log_{20} 60 \cdot \log_{60} 100 \cdot \log_{100} 125 \quad 3$$

43. Which is larger, 500^{501} or 506^{500}? These numbers are too large for most calculators to handle. (They each have

1353 digits!) (*Hint:* Let $x = 500^{501}$ and $y = 506^{500}$ and then compare $\ln x$ with $\ln y$.) 500^{501}

44. Which number is smaller, $\dfrac{1}{50^{300}}$ or $\dfrac{1}{151^{233}}$? $\dfrac{1}{50^{300}}$

Business and Economics

45. ***Animated Maps*** A software company that creates interactive maps for web sites has designed an animated zooming feature so that when a user selects the zoom-in option, the map appears to expand on a location. This is accomplished by displaying several intermediate maps to give the illusion of motion. The company has determined that zooming in on a location is more informative and pleasing to observe when the scale of each step of the animation is determined using the equation

$$S_n = S_0 \cdot 10^{\frac{n}{N}(\log S_f - \log S_0)}$$

where S_n represents the scale of the current step n ($n = 0$ corresponds to the initial scale), S_0 is the starting scale of the map, S_f is the final scale, and N is the number of steps in the animation following the initial scale. (If the initial scale of the map is 1:200, then $S_0 = 200$.) Determine the scales to be used at each intermediate step if a map is to start with a scale of 1:1,000,000 and proceed through five intermediate steps to end with a scale of 1:500,000.
1:870,551; 1:757,858; 1:659,754; 1:574,349; 1:500,000

46. ***Animated Maps*** Use the equation in Exercise 45 to determine the scales for each stage of an animated map zoom that goes from a scale of 1:250,000 to a scale of 1:100,000 in four steps (following the initial scale).
1:198,818; 1:158,114; 1:125,743; 1:100,000

Life and Health Sciences

47. ***pH*** Milk of magnesia has a hydronium-ion concentration of about 3.97×10^{-11} mole per liter. Determine the pH of milk of magnesia and state whether it is an acid or a base.
10.4; base

48. ***pH*** Vinegar has a hydronium-ion concentration of 1.26×10^{-3} mole per liter. Determine the pH of vinegar and state whether it is an acid or a base. 2.9; acid

49. ***Hydronium-ion Concentration*** A morphine solution has a pH of 9.5. Determine the hydronium-ion concentration of the morphine solution. 3.16×10^{-10} mol/L

50. ***Hydronium-ion Concentration*** A rainstorm in New York City produced rainwater with a pH of 5.6. Determine the hydronium-ion concentration of the rainwater.
2.51×10^{-6} mol/L

Sports and Recreation

51. ***Decibel Level*** The range of sound intensities that the human ear can detect is so large that a special decibel scale (named after Alexander Graham Bell) is used to measure and compare sound intensities. The **decibel level** dB of a sound is given by

$$dB(I) = 10 \log\left(\frac{I}{I_0}\right)$$

where I_0 is the intensity of sound that is barely audible to the human ear. Find the decibel level for the following sounds. Round to the nearest tenth dB.

Sound	Intensity
a. Automobile traffic 82.0 dB	$I = 1.58 \times 10^8 \cdot I_0$
b. Quiet conversation 40.3 dB	$I = 10{,}800 \cdot I_0$
c. Fender guitar 115.0 dB	$I = 3.16 \times 10^{11} \cdot I_0$
d. Jet engine 152.0 dB	$I = 1.58 \times 10^{15} \cdot I_0$

52. ***Comparison of Sound Intensities*** A team in Arizona installed a 48,000-watt sound system in a Ford Bronco that it claims can output 175-decibel sound. The human pain threshold for sound is 125 decibels. How many times more intense is the sound from the Bronco than the human pain threshold? 100,000 times more intense

53. ***Comparison of Sound Intensities*** How many times more intense is a sound that measures 120 decibels than a sound that measure 110 decibels? 10 times more intense

54. ***Decibel Level*** If the intensity of a sound is doubled, what is the increase in the decibel level? (*Hint:* Find $dB(2I) - db(I)$.) $\approx$3.0103 dB

Physical Sciences and Engineering

55. ***Earthquake Magnitude*** What is the Richter scale magnitude of an earthquake with an intensity of $I = 100{,}000I_0$? 5

56. ***Earthquake Magnitude*** The Colombia earthquake of 1906 had an intensity of $I = 398{,}107{,}000I_0$. What did it measure on the Richter scale? 8.6

57. ***Earthquake Intensity*** The Coalinga, California earthquake of 1983 had a Richter scale magnitude of 6.5. Find the intensity of this earthquake.
$10^{6.5}I_0$ or about $3{,}162{,}277.7I_0$

58. ***Earthquake Intensity*** The earthquake that occurred just south of Concepcion, Chile in 1960 had a Richter scale magnitude of 9.5. Find the intensity of this earthquake. $3{,}162{,}277{,}660 I_0$

59. ***Comparison of Earthquakes*** Compare the intensity of an earthquake that measures 5 on the Richter scale to the intensity of an earthquake that measures 3 on the Richter scale by finding the ratio of the larger intensity to the smaller intensity. 100 to 1

60. ***Comparison of Earthquakes*** How many times more intense was the 1960 earthquake in Chile, which measured 9.5 on the Richter scale, than the San Francisco earthquake of 1906, which measured 8.3 on the Richter scale? $10^{1.2} \approx 15.8$ times more intense

61. ***Comparison of Earthquakes*** On March 2, 1933, an earthquake of magnitude 8.9 on the Richter scale struck Japan. In October 1989, an earthquake of magnitude 7.1 on the Richter scale struck San Francisco, California. Compare the intensity of the larger earthquake to the intensity of the smaller earthquake by finding the ratio of the larger intensity to the smaller intensity. $10^{1.8}$ to 1 or about 63 to 1

62. ***Comparison of Earthquakes*** An earthquake that occurred in China in 1978 measured 8.2 on the Richter scale. In 1988, an earthquake in California measured 6.9 on the Richter scale. Compare the intensity of the larger earthquake to the intensity of the smaller earthquake by finding the ratio of the larger intensity to the smaller intensity. $10^{1.3}$ to 1 or about 20 to 1

63. ***Earthquake Magnitude*** Find the magnitude of the earthquake that produced the seismogram in the following figure. 5.5

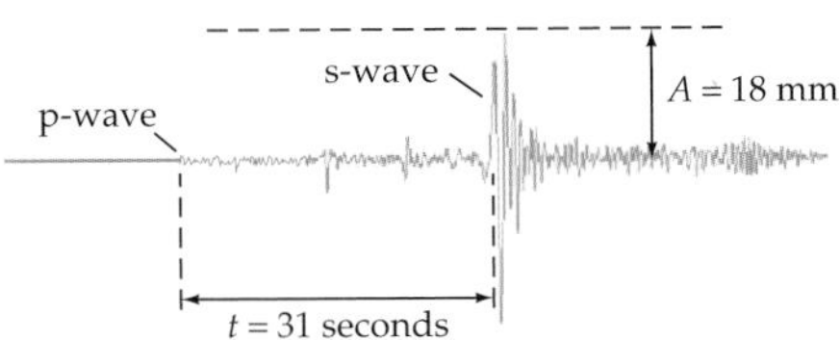

64. ***Earthquake Magnitude*** Find the magnitude of the earthquake that produced the seismogram in the following figure. 4.9

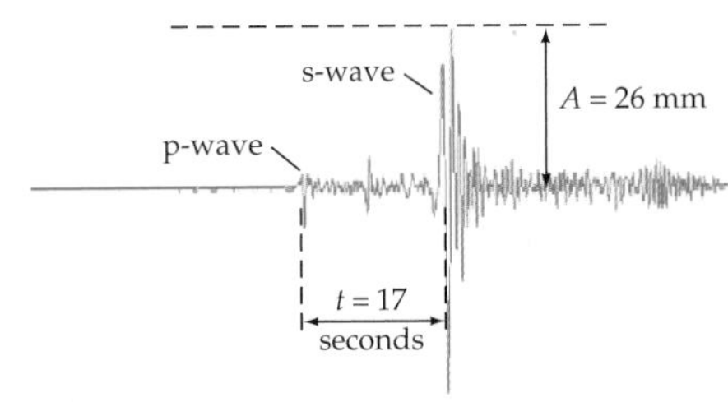

65. ***Nomograms and Logarithmic Scales*** A **nomogram** is a diagram used to determine a numerical result by drawing a line across numerical scales. The following nomogram, used by Richter, determines the magnitude of an earthquake from its seismogram. To use the nomogram, mark the amplitude of a seismogram on the amplitude scale and mark the time between the s-wave and the p-wave on the S-P scale. Draw a line between these marks. The magnitude of the earthquake that produced the seismogram is shown by the intersection of the line and the center scale. The example below shows that an earthquake with a seismogram amplitude of 23 millimeters and an S-P time of 24 seconds has a Richter scale magnitude of about 5.

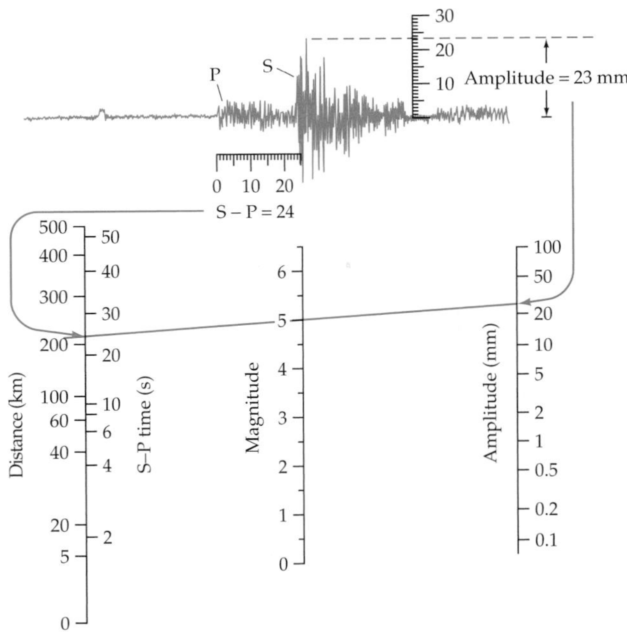

Richter's earthquake nomogram

The amplitude and the S-P time are shown on logarithmic scales. On the amplitude scale, the distance from 1 to 10 is the same as the distance from 10 to 100, because $\log 100 - \log 10 = \log 10 - \log 1$.

Use the nomogram on the preceding page to determine the Richter scale magnitude of an earthquake with a seismogram

a. amplitude of 50 millimeters and S-P time of 40 seconds. $M \approx 6$

b. amplitude of 1 millimeter and S-P time of 30 seconds. $M \approx 4$

c. How do the results in parts **a.** and **b.** compare with the Richter scale magnitude produced by using the amplitude-time-difference formula?
The results are close to the magnitudes produced by the amplitude-time-difference formula.

Prepare for Section 5.4

66. Use the definition of a logarithm to write the exponential equation $3^6 = 729$ in logarithmic form. [5.2]
$\log_3 729 = 6$

67. Use the definition of a logarithm to write the logarithmic equation $\log_5 625 = 4$ in exponential form. [5.2]
$5^4 = 625$

68. Use the definition of a logarithm to write the exponential equation $a^{x+2} = b$ in logarithmic form. [5.2]
$\log_a b = x + 2$

69. Solve for x: $4a = 7bx + 2cx$ [1.1] $x = \dfrac{4a}{7b + 2c}$

70. Solve for x: $165 = \dfrac{300}{1 + 12x}$ [1.3] $x = \dfrac{3}{44}$

71. Solve for x: $A = \dfrac{100 + x}{100 - x}$ [1.3] $x = \dfrac{100(A - 1)}{A + 1}$

Explorations 

—*Answer graphs to Exercises 1a., 2a., and 2b. are on page AA15.*

1. ***Logarithmic Scales*** Sometimes logarithmic scales are used to better view a collection of data that span a wide range of values. For instance, consider the table below, which lists the approximate masses of various marine creatures in grams. Next we have attempted to plot the masses on a number line.

Animal	Mass (g)
Rotifer	0.000000006
Dwarf goby	0.30
Lobster	15,900
Leatherback turtle	851,000
Giant squid	1,820,000
Whale shark	4,700,000
Blue whale	120,000,000

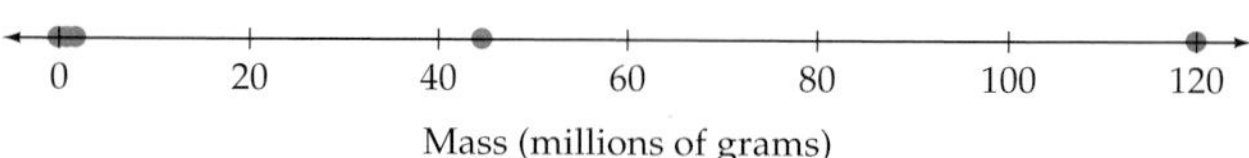

As you can see, we had to use such a large span of numbers that the data for most of the animals are bunched up at the left. Visually, this number line isn't very helpful for any comparisons.

a. Make a new number line, this time plotting the logarithm (base 10) of each of the masses.

b. Which number line is more helpful to compare the masses of the different animals?
The logarithmic number line

c. If the data points for two animals on the logarithmic number line are 1 unit apart, how do the animals' masses compare? What if the points are 2 units apart?
One is 10 times heavier; one is 100 times heavier

2. ***Logarithmic Scales*** The distances of the planets in our solar system from the sun are given in the table below.

Planet	Distance (million km)
Mercury	58
Venus	108
Earth	150
Mars	228
Jupiter	778
Saturn	1427
Uranus	2871
Neptune	4497
Pluto	5913

a. Draw a number line with an appropriate scale to plot the distances.

b. Draw a second number line, this time plotting the logarithm (base 10) of each distance.

c. Which number line do you find more helpful to compare the different distances? Answers will vary.

d. If two distances are 3 units apart on the logarithmic number line, how do the distances of the corresponding planets compare?
The distance of one planet is 1000 times greater than that of the other.

SECTION 5.4 Exponential and Logarithmic Equations

- Exponential Equations
- Logarithmic Equations

Exponential Equations

If a variable appears in an exponent of a term of an equation, such as $2^{x+1} = 32$, then the equation is called an **exponential equation.** Because the variable is in the exponent, we can't solve such equations using the same algebraic techniques we discussed in Chapter 4. However, if each side of the equation can be written as an exponential expression with the same base, then the following theorem states that we can solve the equation by equating exponents.

> **Equality of Exponents Theorem**
>
> If $b^x = b^y$, then $x = y$, provided $b > 0$ and $b \neq 1$.

Alternative to Example 1
Solve: $4^{3x-5} = 16$
- $x = \dfrac{7}{3}$

EXAMPLE 1 Solve an Exponential Equation

Use the Equality of Exponents Theorem to solve $2^{x+1} = 32$.

Solution

$$\begin{aligned} 2^{x+1} &= 32 \\ 2^{x+1} &= 2^5 && \text{• Write each side as a power of 2.} \\ x + 1 &= 5 && \text{• Equate the exponents.} \\ x &= 4 && \text{• Solve the resulting equation.} \end{aligned}$$

CHECK

$$\begin{aligned} 2^{x+1} &\stackrel{?}{=} 32 \\ 2^{4+1} &\stackrel{?}{=} 32 && \text{• Let } x = 4. \\ 2^5 &\stackrel{?}{=} 32 \\ 32 &= 32 \end{aligned}$$

CHECK YOUR PROGRESS 1 Solve the exponential equation $3^{5-2x} = \frac{1}{9}$.

Solution *See page S26.* $x = \dfrac{7}{2}$

If you have access to a graphing utility and need only an approximate solution (or want to check your algebraic work), you can find the solution to an equation such as the one in Example 1 visually. In general, a graphing utility can be used to find the solutions of any equation of the form $f(x) = g(x)$. Either of the following two methods, which were introduced in Example 6 of Section 4.3, can be employed.

Using a Graphing Utility to Find the Solutions of $f(x) = g(x)$

Intersection Method

Graph $y_1 = f(x)$ and $y_2 = g(x)$ on the same screen. The solutions of $f(x) = g(x)$ are the x-coordinates of the points of intersection of the graphs.

Intercept Method

The solutions of $f(x) = g(x)$ are the x-coordinates of the x-intercepts of the graph of $y = f(x) - g(x)$.

The following graphs illustrate the use of each of these methods for solving the equation $2^{x+1} = 32$ from Example 1.

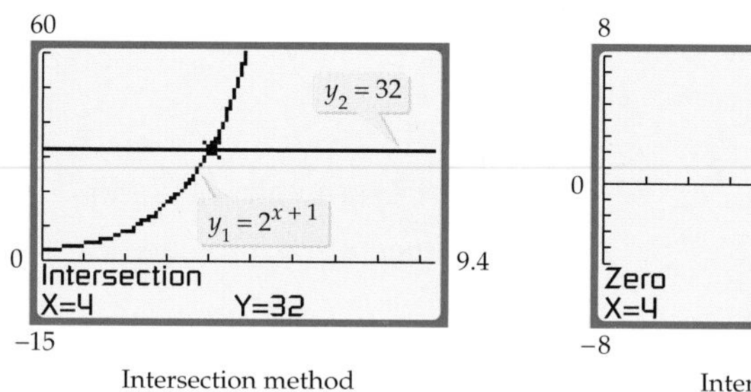

Intersection method

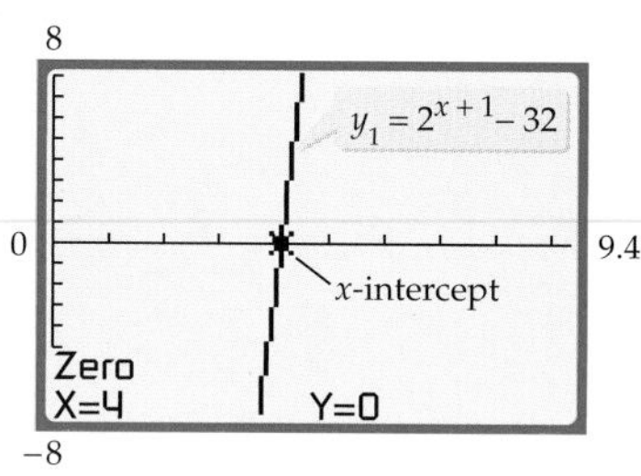

Intercept method

In Example 1, we were able to write each side of the equation as a power with the same base. If this is not possible, we can solve the equation by first taking logarithms of both sides. We can then use the properties of logarithms to solve for the variable. The procedure is illustrated in Example 2.

Alternative to Example 2

Solve: $7^x = 111$

▪ $x = \dfrac{\log 111}{\log 7} \approx 2.420$

TAKE NOTE

We could have taken ln of both sides rather than log; the final result would be the same.

EXAMPLE 2 Solve an Exponential Equation

Solve: $5^x = 40$

Solution

$$5^x = 40$$

$$\log(5^x) = \log 40 \quad \text{▪ Take the logarithm of each side.}$$

$$x \log 5 = \log 40 \quad \text{▪ Power property}$$

$$x = \frac{\log 40}{\log 5} \quad \text{▪ Exact solution}$$

$$x \approx 2.292 \quad \text{▪ Decimal approximation}$$

Visualize the Solution

We can use the intersection method. The solution of $5^x = 40$ is the x-coordinate of the point of intersection of the graphs of $y = 5^x$ and $y = 40$.

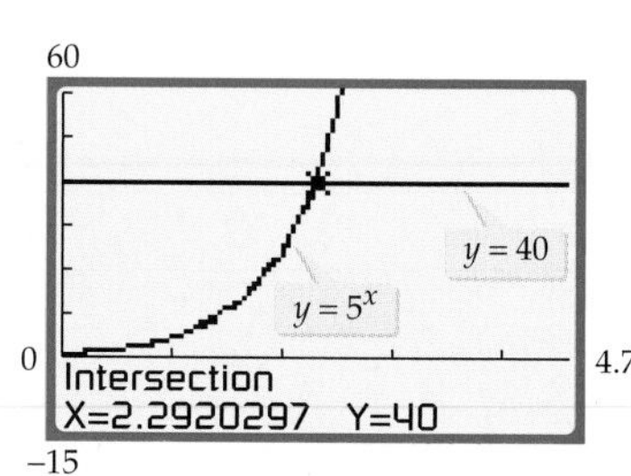

Continued ➤

CHECK YOUR PROGRESS 2 Find the exact solution of the equation $3^x = 175$.

Solution *See page S26.* $x = \dfrac{\log 175}{\log 3}$

An alternative approach to solving the equation in Example 2 is to rewrite the exponential equation in logarithmic form: $5^x = 40$ is equivalent to the logarithmic equation $\log_5 40 = x$. Using the change-of-base formula, we find that $x = \log_5 40 = \dfrac{\log 40}{\log 5}$. In the following example, however, we must take logarithms of both sides to reach a solution.

Alternative to Example 3
Solve: $4^x = 6^{1-2x}$
- $x = \dfrac{\ln 6}{\ln 4 + 2 \ln 6}$
≈ 0.361

EXAMPLE 3 Solve an Exponential Equation

Solve: $5^{3-x} = 7^x$

Solution

$$5^{3-x} = 7^x$$
$$\ln 5^{3-x} = \ln 7^x \qquad \text{Take the natural logarithm of each side.}$$
$$(3 - x)\ln 5 = x \ln 7 \qquad \text{Power property}$$
$$3 \ln 5 - x \ln 5 = x \ln 7 \qquad \text{Distributive property}$$
$$3 \ln 5 = x \ln 7 + x \ln 5 \qquad \text{Solve for } x.$$
$$3 \ln 5 = x(\ln 7 + \ln 5) \qquad \text{Factor.}$$
$$x = \frac{3 \ln 5}{\ln 7 + \ln 5} \qquad \text{Exact solution}$$
$$x \approx 1.36 \qquad \text{Decimal approximation}$$

Visualize the Solution

Here we use the intercept method. The solution of $5^{3-x} = 7^x$ is the x-coordinate of the x-intercept of the graph of $y = 5^{3-x} - 7^x$.

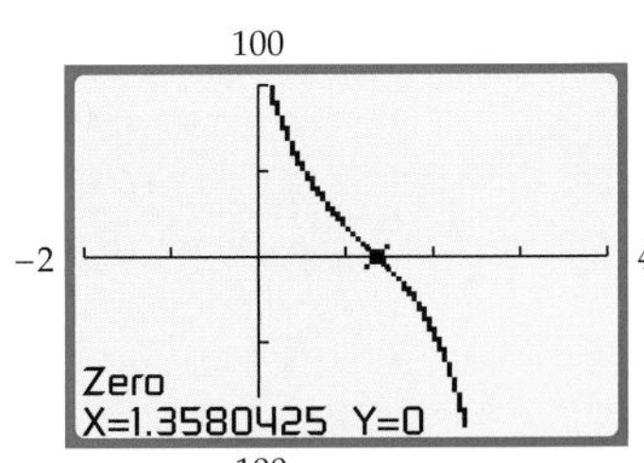

CHECK YOUR PROGRESS 3 Solve for x: $3^{2x+1} = 8^x$. Give both an exact solution and a decimal approximation rounded to the nearest thousandth.

Solution *See page S26.* $x = \dfrac{\ln 3}{\ln 8 - 2 \ln 3} \approx -9.327$

Alternative to Example 4
Exercise 52, page 437

EXAMPLE 4 Use a Mathematical Model

In Example 6 of Section 5.1, the model $T = 70 + 90e^{-0.0485t}$ was used to find the temperature of a cup of coffee after t minutes. In that example, we graphically estimated how long it would take for the coffee to reach 90°F. Determine algebraically the time required for the coffee to reach 90°F.

Continued ➤

Solution Substitute 90 for T and solve for t.

$$90 = 70 + 90e^{-0.0485t}$$
$$20 = 90e^{-0.0485t}$$
$$\frac{20}{90} = e^{-0.0485t}$$
$$\ln\frac{2}{9} = -0.0485t \qquad \text{■ Rewrite in logarithmic form.}$$
$$\frac{1}{-0.0485}\ln\frac{2}{9} = t \qquad \text{■ Solve for } t.$$
$$t \approx 31.01$$

As we saw in Section 5.1, it takes about 31 minutes for the coffee to cool to 90°F.

CHECK YOUR PROGRESS 4 In Check Your Progress 6 of Section 5.1, the amount of medication (in milligrams) in the blood t minutes after the medication was administered was given by $A(t) = 200e^{-0.014t}$. Find algebraically the time required for the amount of medication in the bloodstream to decrease to 50 milligrams. Round to the nearest tenth.

Solution *See page S26.* 99.0 min

Alternative to Example 5
Exercise 58, page 437

EXAMPLE 5 Use a Mathematical Model

Biologists stocked a lake with 500 specimens of an endangered species of fish, hoping to help the fish to establish a stable population in the lake. The number of fish in the lake after t years can be modeled by the function

$$P(t) = \frac{12{,}000}{1 + 23e^{-0.373t}}$$

a. After how many years will the fish population in the lake reach 10,000?

b. Use a graph of the function to describe the fish population over time.

Solution

a. Substitute 10,000 for $P(t)$ and solve for t.

$$10{,}000 = \frac{12{,}000}{1 + 23e^{-0.373t}}$$
$$10{,}000\,(1 + 23e^{-0.373t}) = 12{,}000$$
$$10{,}000 + 230{,}000e^{-0.373t} = 12{,}000$$
$$230{,}000e^{-0.373t} = 2000$$
$$e^{-0.373t} = \frac{2000}{230{,}000}$$
$$-0.373t = \ln\frac{2}{230} \qquad \text{■ Rewrite in logarithmic form.}$$

Continued ➤

$$t = \frac{1}{-0.373} \ln \frac{2}{230}$$

$$t \approx 12.72$$

It will take almost 12 and three-fourths years for the fish population to reach 10,000.

b.

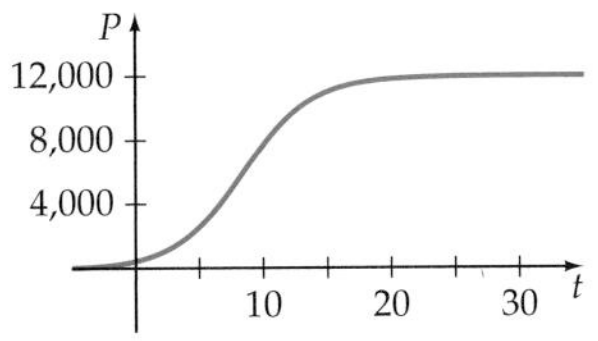

P: 12,000; 8,000; 4,000
t: 10; 20; 30

The graph appears to approach a population of 12,000 fish as t increases. We can verify this mathematically by observing that as t increases, $e^{-0.373t} = \frac{1}{e^{0.373t}}$ approaches 0. Thus as $t \to \infty$, $P(t) \to \frac{12{,}000}{1 + 23(0)} = 12{,}000$.

CHECK YOUR PROGRESS 5 A company provides training in the assembly of computer circuits to new employees. Past experience has shown that the number of correctly assembled circuits per week can be modeled by

$$N(t) = \frac{250}{1 + 249e^{-0.503t}}$$

where t is the number of weeks of training. How many weeks (to the nearest week) of training are needed before a new employee will correctly assemble 140 circuits?

Solution *See page S26.* 11 weeks

▪ Logarithmic Equations

If the variable of an equation is contained within a logarithm, such as $\log_3 x = 7$, the equation is called a **logarithmic equation.** The properties of logarithms, along with the definition of a logarithm, are often used to help find solutions to such equations.

Alternative to Example 6
Solve: $\log_5(2x + 9) = 2$
▪ $x = 8$

EXAMPLE 6 Solve a Logarithmic Equation

Solve: $\log(3x - 5) = 2$

Solution Recall that $\log(3x - 5)$ means $\log_{10}(3x - 5)$.

$$\log_{10}(3x - 5) = 2$$

$$3x - 5 = 10^2 \quad \text{▪ Definition of a logarithm}$$

$$3x = 105 \quad \text{▪ Solve for } x.$$

$$x = 35$$

▪ **CHECK** $\log[3(35) - 5] = \log 100 = 2$

Continued ➤

CHECK YOUR PROGRESS 6 Solve: $\ln(2x + 3) = 4$

Solution *See page S26.* $x = \dfrac{e^4 - 3}{2}$

QUESTION Can a negative number be a solution of a logarithmic equation?

If an equation contains more than one logarithmic expression, use the properties of logarithms to first consolidate these expressions into one logarithmic expression, as shown in the next example.

Alternative to Example 7

Solve:

$\log_4(3x + 5) - \log_4(2x) = 3$

■ $x = \dfrac{1}{25}$

EXAMPLE 7 Solve a Logarithmic Equation

Solve: $\log_3(x + 5) - \log_3 4x = 2$

Solution

$$\log_3(x + 5) - \log_3 4x = 2$$

$$\log_3 \frac{x + 5}{4x} = 2 \quad \text{■ Quotient property}$$

$$\frac{x + 5}{4x} = 3^2 \quad \text{■ Definition of logarithm}$$

$$x + 5 = 36x \quad \text{■ Solve for } x.$$

$$-35x = -5$$

$$x = \frac{1}{7}$$

We can check the solution by substituting $\frac{1}{7}$ into the original equation.

Visualize the Solution

We plot the graphs of $y = \log_3(x + 5) - \log_3 4x$ and $y = 2$. The x-value of the point of intersection is the solution of the equation $\log_3(x + 5) - \log_3 4x = 2$. The graph confirms our conclusion that the solution is $x = \frac{1}{7} \approx 0.1429$.

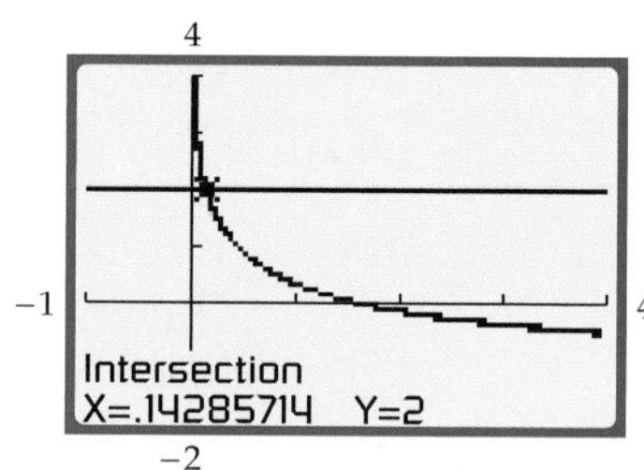

TAKE NOTE

Use the change-of-base formula to enter the function into a graphing calculator:

$$y = \log_3(x + 5) - \log_3(4x)$$
$$= \frac{\ln(x + 5)}{\ln 3} - \frac{\ln(4x)}{\ln 3}$$

Enter the above expression as Y1.

CHECK YOUR PROGRESS 7 Solve: $\log(2x) - \log(x - 3) = 1$

Solution *See page S26.* $x = \dfrac{15}{4}$

The next example shows that the process of solving a logarithmic equation by using logarithmic properties may introduce extraneous solutions.

ANSWER Yes. For instance, -10 is a solution of $\log(-x) = 1$.

Alternative to Example 8
Solve: $\log_3(x - 8) + \log_3 x = 2$
■ $x = 9$

EXAMPLE 8 Solve a Logarithmic Equation

Solve: $\log_4 x + \log_4(x - 6) = 2$

Solution

$$\log_4 x + \log_4(x - 6) = 2$$
$$\log_4[x(x - 6)] = 2 \quad \text{■ Product property}$$
$$x(x - 6) = 4^2 \quad \text{■ Definition of logarithm}$$
$$x^2 - 6x - 16 = 0 \quad \text{■ Solve for } x.$$
$$(x - 8)(x + 2) = 0$$
$$x = 8 \quad \text{or} \quad x = -2$$

TAKE NOTE

Always check your results. Some, or all, of your results may be extraneous solutions.

We can verify that $x = 8$ is a solution by substituting 8 into the original equation. However, if we substitute -2 into the original equation, we get $\log_4(-2)$, which is undefined. Thus -2 is an extraneous solution; the only valid solution is 8.

Visualize the Solution

The graphs of the equations $y = \log_4 x + \log_4(x - 6)$ and $y = 2$ intersect at only one point. Thus there is only one real solution.

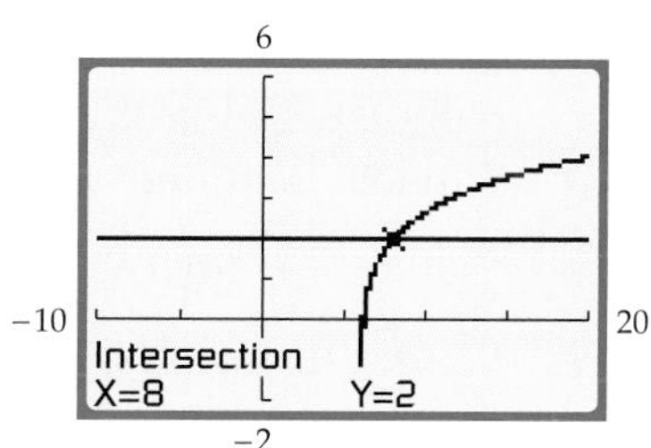

CHECK YOUR PROGRESS 8 Solve for x: $\log_5(x + 4) = 2 - \log_5(x + 4)$

Solution *See page S26.* $x = 1$

Alternative to Example 9
Exercise 64, page 438

EXAMPLE 9 Velocity of a Falling Object Experiencing Air Resistance

The time t, in seconds, required for an object that is dropped to reach a velocity of v feet per second is given by $t = \frac{v}{32}$, when the object is not being slowed down by air resistance. (This is the case if an object falls in a vacuum.) If air resistance is considered, one possible model is given by $t = 2.43 \ln \frac{150 + v}{150 - v}$, $0 \le v < 150$. Use this equation to find the velocity of an object that was dropped from a helicopter and has been falling for 5 seconds.

Continued ➤

Solution Substitute 5 for t and solve for v.

$$t = 2.43 \ln \frac{150 + v}{150 - v}$$

$$5 = 2.43 \ln \frac{150 + v}{150 - v}$$ ▪ Replace t by 5.

$$\frac{5}{2.43} = \ln \frac{150 + v}{150 - v}$$ ▪ Divide by 2.43.

$$e^{5/2.43} = \frac{150 + v}{150 - v}$$ ▪ If $a = \ln b$, then $e^a = b$.

$$e^{5/2.43}(150 - v) = 150 + v$$ ▪ Solve for v.

$$e^{5/2.43}(150) - e^{5/2.43}v = 150 + v$$

$$-v - e^{5/2.43}v = 150 - e^{5/2.43}(150)$$

$$v(-e^{5/2.43} - 1) = 150 - e^{5/2.43}(150)$$

$$v = \frac{150 - e^{5/2.43}(150)}{-e^{5/2.43} - 1}$$

$$v \approx 116.014$$

After 5 seconds, the velocity of the object will be approximately 116 feet per second.

CHECK YOUR PROGRESS 9 One model that can be used to find the time t, in seconds, required for a skydiver to reach a velocity of v feet per second is $t = -\frac{175}{32}\ln\left(1 - \frac{v}{175}\right)$. Use this equation to determine the velocity of a skydiver after 10 seconds. Round to the nearest tenth.

Solution *See page S27.* 146.9 ft/s

Alternative to Example 10

A company's advertising budget is given by

$$A(x) = 15.4 + 3.8 \ln t$$

where $A(x)$ is the budget in millions of dollars and t is the number of years after 1990. In what year will the advertising budget reach \$26 million?

▪ **2006**

EXAMPLE 10 Average Time of a Major League Baseball Game

In Example 7 of Section 5.2, the function $T(x) = 149.57 + 7.63 \ln x$ was suggested as a model for the average time T, in minutes, of a major league baseball game, where x represents the number of years after 1980. If this trend continues, in what year will a major league game be, on average, 3 hours long?

Solution Three hours is 180 minutes, so substitute 180 for $T(x)$ and solve for x.

$$180 = 149.57 + 7.63 \ln x$$

$$30.43 = 7.63 \ln x$$

$$\frac{30.43}{7.63} = \ln x$$

$$e^{30.43/7.63} = x$$ ▪ Definition of logarithm

$$x \approx 53.96$$

Continued ➤

The model predicts that it will be almost 54 years, or the year 2034, before a major league baseball game averages 3 hours in length.

CHECK YOUR PROGRESS 10 The function $d(t) = -41.71 + 25.76 \ln t$ approximates the world record distance, in meters, for the men's discus throw for the years 1965 to 1985, where $t = 0$ corresponds to the year 1900. (See Exercise 69 in Section 5.2.) If this model remains accurate, in what year will the world record distance be 80 meters?

Solution *See page S27.* 2013

Topics for Discussion

1. Discuss two different methods of solving the equation $b^x = c$ for x.
2. Discuss how to solve the equation $a = \log_b x$ for x.
3. What is the domain of $y = \log_4(2x - 5)$? Explain why this means that the equation $\log_4(x - 3) = \log_4(2x - 5)$ has no real number solution.
4. In solving the logarithmic equation $\log_2 x + \log_2(x + 2) = 3$, we arrive at two solutions.

$$\begin{aligned} \log_2 x + \log_2(x + 2) &= 3 \\ \log_2[x(x + 2)] &= 3 \\ x(x + 2) &= 2^3 \\ x^2 + 2x &= 8 \\ x^2 + 2x - 8 &= 0 \\ (x + 4)(x - 2) &= 0 \\ x = -4 \quad \text{or} \quad x &= 2 \end{aligned}$$

If you check these results, you will find that -4 is an extraneous solution. Discuss at which step in the solution the extraneous solution -4 was introduced.

EXERCISES 5.4

— *Suggested Assignment: Exercises 1–45, odd; 49, 50, 53–59, odd; 63, 65, 68, and 69–74, all.*
— *Answer graphs to Exercises 58**a**., 61**a**., 66**a**., and 67**a**. are on pages AA15–AA16.*

In Exercises 1 to 8, solve for x algebraically.

1. $2^x = 64$ 6
2. $3^x = 243$ 5
3. $3^{4-x} = 27$ 1
4. $5^{2x+7} = 125$ -2
5. $2^{5x+3} = \frac{1}{8}$ $-\frac{6}{5}$
6. $4^{3-2x} = \frac{1}{16}$ $\frac{5}{2}$
7. $49^x = \frac{1}{343}$ $-\frac{3}{2}$
8. $9^x = \frac{1}{243}$ $-\frac{5}{2}$

In Exercises 9 to 20, solve for x. Give both an exact answer and a decimal approximation rounded to three decimal places. If you have a graphing utility, check your answer graphically.

9. $4^x = 70$ $\frac{\log 70}{\log 4} \approx 3.065$
10. $6^x = 50$ $\frac{\log 50}{\log 6} \approx 2.183$
11. $3^{-x} = 120$
12. $7^{-x} = 63$ $-\frac{\log 63}{\log 7} \approx -2.129$
13. $10^{2x+3} = 315$
14. $e^{x+1} = 20$ $\ln 20 - 1 \approx 1.996$
15. $5^{x+1} = 7^x$
16. $8^{-x} = 5^{x+2}$ $-\frac{2 \ln 5}{\ln 8 + \ln 5} \approx -0.873$

11. $-\frac{\log 120}{\log 3} \approx -4.358$ 13. $\frac{\log 315 - 3}{2} \approx -0.251$ 15. $\frac{\ln 5}{\ln 7 - \ln 5} \approx 4.783$

17. $4^{x+3} = 5^x$ $\dfrac{3\ln 4}{\ln 5 - \ln 4} \approx 18.638$

18. $5^{3x} = 3^{x+4}$ $\dfrac{4\ln 3}{3\ln 5 - \ln 3} \approx 1.178$

19. $2^{1-x} = 3^{x+1}$ $\dfrac{\ln 2 - \ln 3}{\ln 2 + \ln 3} \approx -0.226$

20. $3^{x-2} = 4^{2x+1}$ $\dfrac{\ln 4 + 2\ln 3}{\ln 3 - 2\ln 4} \approx -2.141$

In Exercises 21 to 30, solve for x algebraically and check your answers.

21. $\log(4x - 18) = 1$ 7

22. $\log(x^2 + 19) = 2$ $-9, 9$

23. $\ln(3 - x) = 1$ $3 - e$

24. $\ln(2x + 5) = 3$ $\dfrac{e^3 - 5}{2}$

25. $\ln(x^2 - 12) = \ln x$ 4

26. $\log_5(x^2 + 15) = \log_5(8x)$ $3, 5$

27. $\log_2 x + \log_2(x - 4) = 2$ $2 + 2\sqrt{2}$

28. $\log_3 x + \log_3(x + 6) = 3$ 3

29. $\log_4(x + 3) = 2 - \log_4(x + 3)$ 1

30. $\log_6(x - 4) = 1 - \log_6(x + 1)$ 5

In Exercises 31 and 32, solve for x. If you have a graphing utility, check your answer graphically.

31. $\log(5x - 1) = 2 + \log(x - 2)$ $\dfrac{199}{95}$

32. $1 + \log(3x - 1) = \log(2x + 1)$ $\dfrac{11}{28}$

In Exercises 33 to 36, use a graphing utility to estimate the solutions of the equation to the nearest hundredth.

33. $2^{-x+3} = x + 1$ 1.61

34. $2e^{x+2} + 3x = 2$ -1.05

35. $\log(x + 2) = 5 - 3x$ 1.49

36. $\ln x = -x^2 + 4$ 1.84

In Exercises 37 to 42, solve for x algebraically.

37. $\log(\log x) = 1$ 10^{10}

38. $\ln(\ln x) = 2$ e^{e^2}

39. $e^{e^{x-1}} = 3$ $\ln(\ln 3) + 1$

40. $10^{10^{3-x}} = 4$ $3 - \log(\log 4)$

41. $\dfrac{10^x + 10^{-x}}{10^x - 10^{-x}} = 5$ $\dfrac{1}{2}\log\dfrac{3}{2}$

42. $\dfrac{e^x + e^{-x}}{e^x - e^{-x}} = 3$ $\dfrac{\ln 2}{2}$

In Exercises 43 to 46, solve for x by graphing. Round to the nearest thousandth. (These equations are difficult to solve algebraically.)

43. $2^x = 2 + \sqrt{x}$ 1.729

44. $\log_2 x = \dfrac{x - 2}{3^x}$ 0.612

45. $3^{-x} = \ln x$ 1.278

46. $e^x + \ln x = \sqrt[3]{x}$ 0.449

47. A common mistake that students make is to write $\log(x + y)$ as $\log x + \log y$. However, the expressions are not generally equal. For what values of x and y does $\log(x + y) = \log x + \log y$? (*Hint:* Solve for x in terms of y.) $x = \dfrac{y}{y - 1}$

Business and Economics

48. ***Retirement Planning*** The retirement account for a graphic designer contains \$250,000 on January 1, 2003 and earns interest at a rate of 0.5% per month. On February 1, 2003, the designer withdraws \$2000 and plans to continue withdrawing this amount as retirement income each month. The value V of the account after x months is given by $V(x) = 400{,}000 - 150{,}000(1.005)^x$.

If the designer wishes to leave \$100,000 to a scholarship foundation, what is the maximum number of monthly withdrawals (to the nearest month) the designer can make from this account and still have \$100,000 to donate? 138 withdrawals

49. ***Retirement Planning*** The retirement account for an assembly line shift manager contains \$300,000 on January 1, 2003 and earns interest at a rate of 0.75% per month. On February 1, 2003, the manager withdraws \$3000 and plans to continue withdrawing this amount as retirement income each month. The value V of the account after x months is given by

$$V = 700{,}000 - 400{,}000(1.0075)^x$$

What is the maximum number of monthly withdrawals (to the nearest month) the manager can make and still have a balance of more than \$3000?

74 withdrawals

Life and Health Sciences

50. ***Physical Fitness*** After a race, a runner's pulse rate R in beats per minute decreases according to the function $R(t) = 145e^{-0.092t}$, $0 \le t \le 15$, where t is measured in minutes.

a. Find the runner's pulse rate at the end of the race and 1 minute after the end of the race. Round to the nearest beat per minute. 145 beats/min; 132 beats/min

b. How long, to the nearest minute, after the end of the race will the runner's pulse rate be 80 beats per minute? 6 min

51. ***Medicine*** During surgery, a patient's circulatory system requires at least 50 milligrams of an anesthetic. The amount of anesthetic, in milligrams, present t hours after 80 milligrams of anesthetic have been administered is given by $T(t) = 80(0.727)^t$.

   a. How much of the anesthetic is present in the patient's circulatory system 30 minutes after the anesthetic is administered? 68 mg

   b. How long, to the nearest minute, can the operation last if the patient does not receive additional anesthetic? 88 min

Sports and Recreation

52. ***Rate of Cooling*** A caterer placed several bottles of soda at 79°F in a refrigerator that maintains a constant temperature of 36°F to chill them to 45°F before serving. The temperature T of the bottles t minutes after they are placed in the refrigerator is given by $T(t) = 36 + 43e^{-0.058t}$.

   a. Find the temperature of the soda 10 minutes after it is placed in the refrigerator. $\approx$60°F

   b. When, to the nearest minute, will the temperature of the soda be 45°F? 27 min

53. ***Probability*** In Exercise 59 of Section 5.1, the function $P(t) = 1 - e^{-0.75t}$ was used to give the probability that a customer will arrive at a home improvement store within t minutes of 10 A.M. Find algebraically the number of minutes required for the probability of someone entering the store to be 98%. Round to the nearest tenth of a minute. 5.2 min

54. ***Probability*** In Exercise 60 of Section 5.1, the function $P(t) = 1 - e^{-0.2t}$ was given to compute the probability that a customer will enter a sporting goods store within t minutes of 9:00 A.M.

   a. Find algebraically the smallest amount of time that gives a 90% probability of a customer entering the store. Round to the nearest tenth of a minute. 11.5 min

   b. According to this function, is there an amount of time that corresponds to a 100% probability of a customer entering the store? No

Social Sciences

55. ***Population Growth*** The population P of a city grows exponentially according to the function $P(t) = 8500(1.1)^t$, $0 \le t \le 8$, where t is measured in years.

   a. Find the population at time $t = 0$ and at time $t = 2$. 8500 people; 10,285 people

   b. When, to the nearest year, will the population reach 15,000? In 6 years

56. ***Population Growth*** The population P of India can be modeled by the equation $P(t) = 368(1.02)^t$, where t is the number of years since 1950 and P is given in millions.

   a. Use the function to estimate the population of India in the year 2010. Round to the nearest million. 1207 million

   b. Use the function to determine the year during which the population of India will reach 2 billion. 2035

57. ***Psychology*** According to a software company, users of its typing tutorial can expect to type $N(t)$ words per minute after t hours of practice with the product, according to the function $N(t) = 100(1.04 - 0.99^t)$.

   a. How many words per minute can a user expect to type after 2 hours of practice? 6 wpm

   b. How many words per minute can a user expect to type after 40 hours of practice? 37 wpm

   c. According to the function N, how many hours (to the nearest hour) of practice are required before a user can expect to type 60 words per minute? 82 h

58. ***Psychology*** Industrial psychologists study employee training programs to assess the effectiveness of the instruction. In one study, the percent score P on a test taken by a person who has completed t hours of training was given by $P = \dfrac{100}{1 + 30e^{-0.088t}}$.

   a. Use a graphing utility to graph the equation for $t \ge 0$.

   b. Use the graph to estimate (to the nearest hour) the number of hours of training necessary to achieve a score of 70% on the test. 48 h

   c. Determine the answer to part **b.** algebraically. 48 h

59. ***Learning Theory*** Suppose that historical records of employee training at a company show that the percent

score on a product information test is given by $P = \dfrac{100}{1 + 25e^{-0.095t}}$, where t is the number of hours of training. How many hours (to the nearest hour) of training are needed before a new employee will answer at least 75% of the questions correctly? $\approx$45 h

60. ***Law*** A lawyer has determined that the number of people P who have been exposed to a news item after t days is given by the function $P(t) = 1{,}200{,}000(1 - e^{-0.03t})$.

    a. How many days after a major crime is reported will 40% of the population have heard of the crime? 17 days

    b. A defense lawyer knows it will be very difficult to pick an unbiased jury after 80% of the population has heard of the crime. After how many days will 80% of the population have heard of the crime? 54 days

61. ***Predicting Adequacy of Resources*** The adequacy of a city's resources can sometimes be modeled by a gamma density function. The function enables city planners to determine the probability that certain city services can be maintained. Suppose a city has determined that the probability of its being able to provide more than x million liters of water per day is given by

$$P = \left(\frac{1}{3}x + 1\right)e^{-x/3}$$

    a. Use a graphing utility to graph $P(x)$ for $x \geq 0$. Use [0, 1] for the range.

    b. Determine the probability that the city can supply more than 5 million liters of water per day. Round to the nearest thousandth. 0.504

    c. As $x \to \infty$, $P \to 0$. Explain why this makes sense in the context of this application.
    As the number of liters of water demanded per day becomes very large ($x \to \infty$), the probability of providing the water approaches 0.

Physical Sciences and Engineering

62. ***Physics*** The electrical current $I(t)$ (measured in amperes) of a circuit is given by the function $I(t) = 6(1 - e^{-2.5t})$, where t is the number of seconds after the current is turned on. (See the following graph.)

    a. Find the current when $t = 0$. 0 amps

    b. Find the current, to the nearest hundredth ampere, when $t = 0.5$. 4.28 amps

    c. Solve the equation for t. $t = -\dfrac{2}{5}\ln\left[1 - \dfrac{I(t)}{6}\right]$

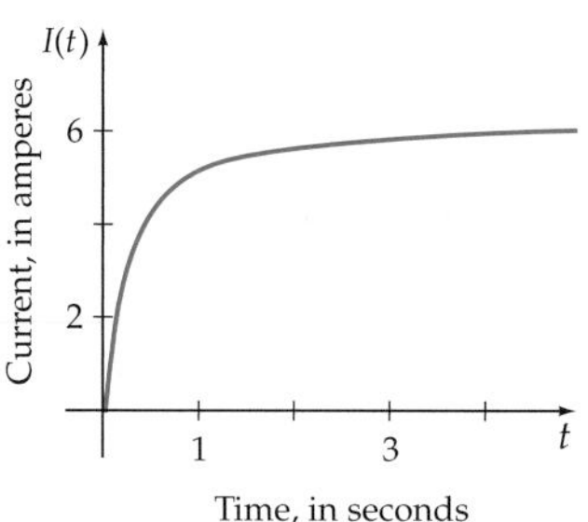

63. ***Effects of Air Resistance on Velocity*** If we assume that air resistance is proportional to the square of velocity, then the time t in seconds required for a moving object to reach a velocity of v feet per second is given by

$$t = \frac{9}{24}\ln\frac{24 + v}{24 - v}$$

    a. Determine the velocity of the object after 1.5 seconds. Round to the nearest hundredth of a foot per second. 23.14 ft/s

    b. The graph of this function has a vertical asymptote at $v = 24$. Explain the meaning of this asymptote in the context of this problem.
    The velocity of the object cannot exceed 24 ft/s.

64. ***Terminal Velocity with Air Resistance*** The velocity v of an object t seconds after it has been dropped from a height above the surface of the earth is given by the equation $v = 32t$ feet per second, assuming no air resistance. If we assume that air resistance is proportional to the square of the velocity, then the velocity, in feet per second, after t seconds is given by $v = 100\left(\dfrac{e^{0.64t} - 1}{e^{0.64t} + 1}\right)$. In how many seconds will the velocity be 50 feet per second? Round to the nearest hundredth. 1.72 s

65. ***Terminal Velocity with Air Resistance*** If we assume that air resistance is proportional to the square of the velocity, then the velocity v, in feet per second, of an object t seconds after it has been dropped is given by $v = 50\left(\dfrac{e^{1.6t} - 1}{e^{1.6t} + 1}\right)$. (See Exercise 64; the reason for the difference in the equations is that the proportionality constants are different.) In how many seconds (to the nearest hundredth) will the velocity be 20 feet per second? 0.53 s

66. ***Effects of Air Resistance on Distance*** The distance s, in feet, that the object in Exercise 64 will fall in t seconds is given by

$$s = \frac{100^2}{32}\ln\left(\frac{e^{0.32t} + e^{-0.32t}}{2}\right)$$

a. Use a graphing utility to graph this equation for $t \geq 0$.

b. Use the graph to estimate (to the nearest tenth of a second) the time it takes for the object to fall 100 feet. 2.6 s

67. ***Effects of Air Resistance on Distance*** The distance s, in feet, that the object in Exercise 65 will fall in t seconds is given by

$$s = \frac{50^2}{32} \ln\left(\frac{e^{0.8t} + e^{-0.8t}}{2}\right)$$

a. Use a graphing utility to graph this equation for $t \geq 0$.

b. Use the graph to estimate (to the nearest tenth of a second) the time it takes for the object to fall 100 feet. 2.4 s

68. ***Hanging Cable*** The height h, in feet, of any point P on the cable shown in the diagram is given by

$$h(x) = 10(e^{x/20} + e^{-x/20}), \qquad -15 \leq x \leq 15$$

where x is the horizontal distance, in feet, between P and the y-axis.

a. What is the lowest height of the cable? 20 ft

b. What is the height of the cable 10 feet to the right of the y-axis? Round to the nearest tenth of a foot. 22.6 ft

c. How far to the right of the y-axis is the cable 24 feet in height? Round to the nearest tenth of a foot. 12.4 ft

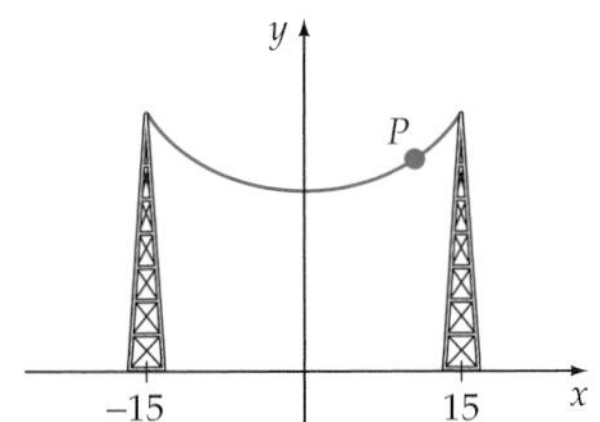

Prepare for Section 5.5

69. Use a calculator to evaluate $e^{-0.57}$. Round your result to the nearest thousandth. [5.1] 0.566

70. Use a calculator to evaluate $545e^{0.18(3)}$. Round your result to the nearest hundredth. [5.1] 935.22

71. Evaluate the expression Ae^{kt} when $A = 8500$, $k = -0.148$, and $t = 12$. Round your answer to the nearest tenth. [5.1] 1439.2

72. Determine the value of $\ln(e^5)$ without using a calculator. [5.2] 5

73. Use a property of logarithms to simplify $\ln(e^{0.862t})$. [5.2] $0.862t$

74. Find the value of x, rounded to the nearest hundredth, in the equation $M = A + Bx^3$ when $M = 86$, $A = 47$, and $B = 114$. [1.3] 0.70

Explorations 

1. ***Navigating*** The pilot of a boat is trying to cross a river to a point O that is located 2 miles due west of the boat's starting position by always pointing the nose of the boat toward O. Suppose the speed of the current is w miles per hour and the speed of the boat is v miles per hour. If point O is the origin and the boat's starting position is (2, 0) (see the diagram below), then the equation of the boat's path is given by

$$y = \left(\frac{x}{2}\right)^{1-(w/v)} - \left(\frac{x}{2}\right)^{1+(w/v)}$$

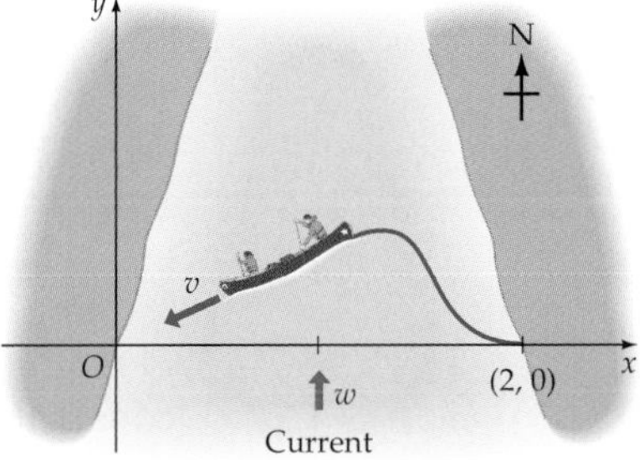

a. If the speed of the current and the speed of the boat are the same, can the pilot reach point O by always pointing the nose of the boat toward O? If not, at what point will the pilot arrive? Explain your answer.

b. If the speed of the current is greater than the speed of the boat, can the pilot reach point O by always pointing the nose of the boat toward O? If not, at what point will the pilot arrive? Explain.

c. If the speed of the current is less than the speed of the boat, can the pilot reach point O by always pointing the nose of the boat toward O? If not, at what point will the pilot arrive? Explain.

1. **a.** No. If $w = v$, the equation becomes $y = 1 - \frac{x^2}{4}$. When $x = 0$, $y = 1$, so the boat will arrive at point (0, 1).
1. **b.** No. If $w > v$, then $1 - (w/v)$ is negative and the expression $\left(\frac{x}{2}\right)^{1-(w/v)}$ is undefined for $x = 0$. The closer x is to 0, the larger y becomes. Thus the boat will follow a path that curves north and never reaches the other side.
1. **c.** Yes. If $w < v$, then $y = 0$ when $x = 0$, and the boat reaches O.

SECTION 5.5 Exponential Growth and Decay

- Exponential Growth
- Exponential Decay
- Carbon Dating
- Newton's Law of Cooling

Exponential Growth

In many applications, a quantity changes at a *constant percentage rate* rather than by a fixed amount. A common example is population. A population that increases at a constant percentage rate adds more and more members the larger it gets. In these types of situations, the quantity always can be modeled by the same type of function.

Exponential Growth Function

If a quantity N increases (or decreases) at a constant percentage rate, the quantity can be described by the function

$$N(t) = N_0 e^{kt}$$

where t is time, N_0 is the value of N at time $t = 0$, and k is a constant.
If $k > 0$, the quantity N is increasing, and N exhibits **exponential growth.** If $k < 0$, the quantity is decreasing, and the function describes **exponential decay.**

QUESTION Is the function $N(t) = 200e^{(-0.2t)}$ an example of exponential growth or exponential decay?

Populations that increase at a constant continuous percentage rate exhibit exponential growth. Thus they can be described by the function $N(t) = N_0 e^{kt}$. The constant k is called the **continuous growth rate** and represents the *continuous* percentage rate at which the population grows.

For instance, say a city currently has 20,000 residents and is growing at a continuous rate of 3% per year. Then its population is described by $N(t) = 20{,}000e^{0.03t}$, where t is the time in years from the present. The following example examines the total annual increase that results from a population growing at a continuous percentage rate.

Alternative to Example 1
Exercise 26, page 450

EXAMPLE 1 Population Increasing at a Continuous Rate

A suburban city has a population of 17,400 residents on January 1, 2003. The city's planning department predicts that the city's population will grow at a continuous rate of 5.3% per year.

a. Write a function that gives the population of the city t years after January 1, 2003.

b. What does the function predict the population will be on January 1, 2004?

c. How long will it take for the population to double?

Continued ➤

ANSWER Exponential decay, because the constant $k = -0.2$ is a negative number.

Solution

a. The population follows exponential growth, so we use $N(t) = N_0e^{kt}$ with $N_0 = 17{,}400$ and $k = 0.053$: $N(t) = 17{,}400e^{0.053t}$.

b. $N(1) = 17{,}400e^{0.053(1)} \approx 18{,}347$. Thus the function predicts that the population will reach 18,347 in 1 year's time, an increase of 947 people. Notice that 5.3% of 17,400 is about 922. Why did our function predict a larger increase? The city's population is growing at 5.3% *continuously,* and as more people are added to the population, they start contributing to the 5.3% growth rate. By the end of a year, the population increases by a little more than 5.3% of the starting population.

c. Find t such that $N(t) = 2(17{,}400) = 34{,}800$.

$$34{,}800 = 17{,}400e^{0.053t}$$

$$2 = e^{0.053t} \quad \text{■ Divide each side by 17,400.}$$

$$\ln 2 = \ln e^{0.053t} \quad \text{■ Take the natural logarithm of each side.}$$

$$\ln 2 = 0.053t \quad \text{■ Solve for } t.$$

$$\frac{\ln 2}{0.053} = t$$

$$t \approx 13.08$$

The function predicts that it should take a little more than 13 years for the population to double.

CHECK YOUR PROGRESS 1 A small town in the midwest currently has only 1370 residents, but expects to grow at a continuous rate of 7.5% per year. Write a function that gives the population t years from now, and use your function to predict the population in 5 years. How long will it take for the population to triple?

Solution *See page S27.* $N(t) = 1370e^{0.075t}$; 1993; 14.6 yr

INSTRUCTOR NOTE

You can demonstrate that if r is a total annual percentage increase and k is a continuous rate of increase, then

$$e^k = 1 + r.$$

We saw in Example 1 that a population growing continuously at 5.3% actually experiences a total increase of more than 5.3% after one year. We can find the specific relationship by rewriting the function.

$$N(t) = 17{,}400e^{0.053t}$$

$$= 17{,}400(e^{0.053})^t \quad \text{■ Properties of exponents}$$

$$\approx 17{,}400(1.0544)^t \quad \text{■ } e^{0.053} \approx 1.0544$$

So $N(t) = 17{,}400e^{0.053t}$ and $N(t) = 17{,}400(1.0544)^t$ are essentially the same function. The form $N(t) = 17{,}400(1.0544)^t$ shows that the initial population of 17,400 becomes 105.44% of 17,400 the next year, and that each year the population grows to a total of 105.44% of the previous year. Thus the total percentage increase is 5.44% per year, whereas the continuous growth rate is 5.3% per year.

Alternative to Example 2

An investment fund is increasing at a total rate of 6% per year. At what continuous rate is the fund growing?

■ ≈5.83%

EXAMPLE 2 Find an Equivalent Continuous Growth Rate

A company wants to show a total increase in its profits of 9% per year. At what continuous rate should the company aim to increase its profits?

Continued ➤

Solution To increase profits by 9% per year, each year the company must make 109% of the previous year's profits. If we let N_0 represent the profit for the year corresponding to $t = 0$, then a function describing the desired profit is $N(t) = N_0(1.09)^t$. This function is an example of an exponential growth function, but in order to determine the equivalent continuous growth rate, we must rewrite the function in the form $N(t) = N_0e^{kt}$.

$$\begin{aligned} N(t) &= N_0(1.09)^t \\ &= N_0(e^k)^t && \text{■ } e^k = 1.09, \text{ so } k = \ln 1.09 \approx 0.0862. \\ &\approx N_0(e^{0.0862})^t \\ &= N_0e^{0.0862t} \end{aligned}$$

Because $k \approx 0.0862$, the equivalent continuous growth rate is about 8.62% per year.

CHECK YOUR PROGRESS 2 An Internet service provider (ISP) estimates that it is attracting new customers at a continuous rate of 16.5% per year. Find the total annual percentage increase in membership the ISP is experiencing.

Solution *See page S27.* 17.9%

If we know that a quantity is growing exponentially, then we need only two data points to determine the exponential growth function $N(t) = N_0e^{kt}$, as the next example demonstrates.

Alternative to Example 3
Exercise 32, page 451

EXAMPLE 3 Find an Exponential Growth Function

The student enrollment at a university is growing exponentially. In 1990, 8700 students attended the school, and in 2000, the enrollment was 13,300.

a. Find an exponential growth function for the university's enrollment.

b. Use the function from part **a.** to predict the number of students attending the school in 2012.

Solution

a. We need to determine N_0 and k in $N(t) = N_0e^{kt}$. If we let $t = 0$ correspond to the year 1990, then our given information is $N(0) = 8700$ and $N(10) = 13{,}300$. N_0 is defined to be $N(0)$, so $N_0 = 8700$. To determine k, substitute 10 for t and 8700 for N_0.

$$\begin{aligned} N(10) &= 8700e^{k\cdot 10} \\ 13{,}300 &= 8700e^{10k} && \text{■ Substitute 13,300 for } N(10). \\ \frac{13{,}300}{8700} &= e^{10k} && \text{■ Solve for } e^{10k}. \\ \ln\frac{13{,}300}{8700} &= 10k && \text{■ Take the natural logarithm of each side.} \\ \frac{1}{10}\ln\frac{13{,}300}{8700} &= k && \text{■ Solve for } k. \\ k &\approx 0.0424 \end{aligned}$$

The exponential growth function is $N(t) \approx 8700e^{0.0424t}$.

TAKE NOTE

Be careful not to round off the value of k to too few digits. Because k appears in the exponent of the function, small inaccuracies can create large differences in output values.

Continued ➤

b. The year 1990 is represented by $t = 0$, so $t = 22$ corresponds to the year 2012. Therefore,

$$N(t) \approx 8700e^{0.0424t}$$
$$N(22) \approx 8700e^{0.0424 \cdot 22}$$
$$\approx 22{,}112$$

Rounding to the nearest hundred, the function predicts that the university's enrollment will be 22,100 in the year 2012.

CHECK YOUR PROGRESS 3 Find an exponential growth function to model a population of trout in a lake that numbered 4800 in 1980 and 11,500 in 2000. Then use the function to predict the number of trout in the lake, to the nearest hundred, in the year 2008.

Solution *See page S27.* $N(t) = 4800e^{0.0437t}$ ($t = 0$ corresponds to 1980); 16,300 trout

▪ Exponential Decay

If the growth constant k in the function $N(t) = N_0e^{kt}$ is negative, the amount $N(t)$ *decreases* at a continuous percentage rate over time, and the function is called an *exponential decay function.* The value of $|k|$ is the continuous decay rate. Radioactive materials exhibit this behavior; their atoms disintegrate at a continuous rate.

Alternative to Example 4

Ten ounces of a radioactive material are decaying so that after 8 days there are 7.2 ounces remaining. Find an exponential decay function that gives the amount of material remaining after t days and determine the number of days required so that only 1 ounce of material remains.

▪ $N(t) \approx 10e^{-0.0411t}$; ≈56 days

EXAMPLE 4 Find an Exponential Decay Function

Suppose we start with 12 grams of a radioactive substance that is decaying. In 5 days, there are 9.8 grams of the substance remaining.

a. Find an exponential decay function that gives the amount of the substance remaining after t days.

b. Find the number of days required for half of the material to disintegrate.

Solution

a. We need to determine N_0 and k in $N(t) = N_0e^{kt}$. N_0 is defined to be $N(0)$, so $N_0 = 12$. We are given $N(5) = 9.8$. Thus

$$N(5) = 12e^{k \cdot 5}$$
$$9.8 = 12e^{5k} \quad \text{▪ Substitute 9.8 for } N(5).$$
$$\frac{9.8}{12} = e^{5k} \quad \text{▪ Solve for } e^{5k}.$$
$$\ln\frac{9.8}{12} = 5k \quad \text{▪ Take the natural logarithm of each side.}$$
$$\frac{1}{5}\ln\frac{9.8}{12} = k \quad \text{▪ Solve for } k.$$
$$k \approx -0.0405$$

Continued ➤

The exponential decay function is $N(t) \approx 12e^{-0.0405t}$. Notice that the value of k is negative, as expected, because the quantity is decreasing. Because $k \approx -0.0405$, we know that the substance is decaying at a continuous rate of about 4.05% per day.

b. To find the number of days required for only 6 grams of the substance to remain, solve for t using $N(t) = 6$.

$$12e^{-0.0405t} = 6$$
$$e^{-0.0405t} = \frac{1}{2}$$
$$-0.0405t = \ln\frac{1}{2}$$
$$t = \frac{\ln(1/2)}{-0.0405}$$
$$t \approx 17.1$$

Half of the substance will disintegrate in approximately 17.1 days.

CHECK YOUR PROGRESS 4 A 4-gram sample of a radioactive material is decaying rapidly. It only takes 22 minutes for 3 grams of the material to disintegrate. Write an exponential decay function to describe the amount of sample remaining after t minutes, and use your function to find the amount remaining after 1 hour.

Solution *See page S27.* $N(t) \approx 4e^{-0.0630t}$; ≈ 0.091 g

Be careful how you interpret the result of Example 4**b.**; the answer does *not* tell us that if we wait another 17.1 days, the remaining 6 grams of the radioactive material will disappear. After that time, half of the 6 grams will erode, leaving 3 grams. If we wait another 17.1-day period, half of the 3 grams will disintegrate.

Because the rate of decrease is a percentage of the amount remaining, a radioactive substance never completely disappears (even though the amount is always decreasing). Consequently, the rate of decay is often referred to in terms of **half-life,** which is defined as the time required for half of the atoms in an amount of radioactive material to disintegrate. So in part **b.** of Example 4, we determined that the half-life of the material is about 17.1 days.

Some radioactive substances decay extremely quickly, while others decay very slowly. The half-lives of several radioactive isotopes are shown in Table 5.11. In the next example, we will use a known half-life to find the exponential decay function.

Alternative to Example 5

If a radioactive substance has a half-life of 242 days, find an exponential decay function for the amount of substance remaining after t days.

▪ $N(t) \approx N_0e^{-0.00286t}$

EXAMPLE 5 Use Half-Life to Find an Exponential Decay Function

Find an exponential decay function for the amount of phosphorus (^{32}P) that remains in a sample after t days.

Continued ➤

Table 5.11

Isotope	Half-Life
Carbon (^{14}C)	5730 years
Radium (^{226}Ra)	1660 years
Polonium (^{210}Po)	138 days
Phosphorus (^{32}P)	14 days
Polonium (^{214}Po)	1/10,000 of a second

Solution We don't have a specific starting amount, so we leave N_0 in the equation as a constant. Then we need to find the value of k in the function $N(t) = N_0e^{kt}$. Because the phosphorus has a half-life of 14 days (from Table 5.11), we know $N(14) = \frac{1}{2}N_0$. Then

$$N(14) = N_0e^{k\cdot 14}$$

$$\frac{1}{2}N_0 = N_0e^{14k} \qquad \text{▪ Substitute } \tfrac{1}{2}N_0 \text{ for } N(14).$$

$$\frac{1}{2} = e^{14k} \qquad \text{▪ Divide each side by } N_0.$$

$$\ln\frac{1}{2} = 14k$$

$$\frac{1}{14}\ln\frac{1}{2} = k$$

$$k \approx -0.0495$$

The exponential decay function is $N(t) \approx N_0e^{-0.0495t}$.

CHECK YOUR PROGRESS 5 Find an exponential decay function for the amount of radium (^{226}Ra) remaining after t years in a sample that initially contains 3 ounces.

Solution *See page S28.* $N(t) = 3e^{-0.000418t}$

▪ Carbon Dating

If an archeologist finds an animal skeleton, how does she know how long ago the animal died? The mystery can be solved by measuring the percentage of a radioactive material in the bones and using an exponential decay function.

The bone tissue in all living animals contains both carbon-12, which is nonradioactive, and carbon-14, which is radioactive with a half-life of approximately 5730 years. As long as the animal is alive, the ratio of carbon-14 to carbon-12 remains constant. When the animal dies ($t = 0$), the initial amount of carbon-14, N_0, begins to decay. See Figure 5.24. Thus a bone that has a smaller ratio of carbon-14 to carbon-12 is older than a bone that has a larger ratio.

The amount of carbon-14 present at time t can be determined by using the fact that the half-life of carbon-14 is 5730 years. As in Example 5, we can use the exponential decay equation to write $N_0e^{k\cdot 5730} = \frac{1}{2}N_0$. We can then determine the value of k:

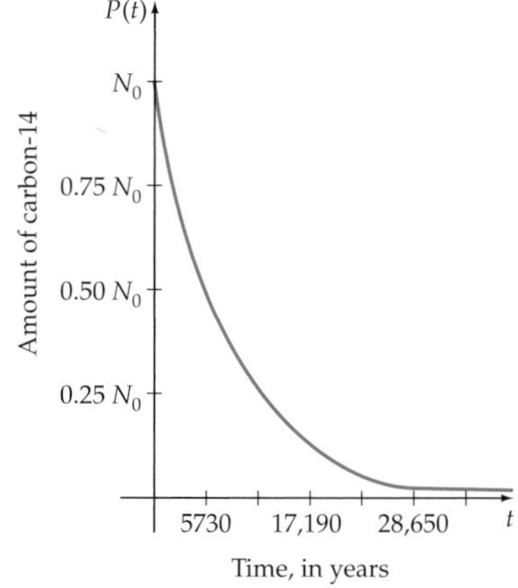

The exponential decay of carbon-14. N_0 represents the amount of carbon-14 present at time $t = 0$.

Figure 5.24

$$N_0e^{k\cdot 5730} = \frac{1}{2}N_0$$

$$e^{5730k} = \frac{1}{2} \qquad \text{▪ Divide each side by } N_0.$$

$$5730k = \ln\frac{1}{2} \qquad \text{▪ Rewrite in logarithmic form.}$$

$$k = \frac{1}{5730}\ln\frac{1}{2} \qquad \text{▪ Solve for } k.$$

$$k \approx -0.000121$$

Thus the amount of carbon-14 present at time t is $N(t) = N_0e^{-0.000121t}$. In Example 6 we use this function to estimate the age of a bone. The process of estimating the age of an item from the amount of carbon-14 present is called **carbon dating.**

Alternative to Example 6

Determine the age of an elephant tusk that now contains 74% of the carbon-14 it had when the elephant died.

■ **≈2488 years**

EXAMPLE 6 Application to Archeology

Determine the age of an animal bone if it now contains 85% of the carbon-14 it had when the animal died ($t = 0$).

Solution Let t be the time at which $N(t) = 0.85N_0$.

$$0.85N_0 = N_0e^{-0.000121t}$$

$$0.85 = e^{-0.000121t} \qquad \text{■ Divide each side by } N_0.$$

$$\ln 0.85 = -0.000121t \qquad \text{■ Rewrite in logarithmic form.}$$

$$\frac{\ln 0.85}{-0.000121} = t \qquad \text{■ Solve for } t.$$

$$t \approx 1343$$

Rounded to the nearest 10 years, the bone is about 1340 years old.

Point of Interest

The chemist Willard Frank Libby developed the carbon-14 dating technique in 1947. He was awarded the Nobel Prize in chemistry in 1960 for discovering the technique.

CHECK YOUR PROGRESS 6 An archeologist on a dig discovered an elephant bone that currently contains 56% of the carbon-14 it had when the elephant died. Estimate the age of the bone. Round to the nearest ten years.

Solution *See page S28.* 4790 years old

■ Newton's Law of Cooling

When a cup of hot coffee is poured, the temperature of the coffee decreases and approaches room temperature. You may have noticed that the coffee cools more quickly at first. Isaac Newton discovered that the rate at which objects change temperature is a percentage of the *difference* between the object's temperature and the temperature of its surroundings. This concept is now known as **Newton's Law of Cooling.**

If the temperature of an object at time t is given by T, T_0 is the initial temperature of the object, and A is the temperature of the room (or surrounding material), then the temperature T is given by a function that is very similar to the exponential decay function used with radioactive materials:

$$T(t) - A = (T_0 - A)e^{kt}$$

or

$$T(t) = A + (T_0 - A)e^{kt}$$

Because T is a decreasing function, k will be negative.

Alternative to Example 7
Exercise 16, page 449

EXAMPLE 7 Find a Cooling Function

A cup of tea at 160°F is placed in a room with a temperature of 72°F. If after 15 minutes the temperature of the tea is 125°F, write a function that gives the temperature of the tea after t minutes. Use your function to predict the temperature of the tea after 25 minutes. How long will it take the tea to reach 80°F?

Solution We are given that $T(0) = 160 = T_0$, $A = 72$, and $T(15) = 125$. Substitute these values into the equation $T(t) = A + (T_0 - A)e^{kt}$ to determine the value of k.

$$T(15) = 72 + (160 - 72)e^{k \cdot 15} \qquad \blacksquare\ t = 15, A = 72, T_0 = 160$$
$$125 = 72 + 88e^{15k} \qquad \blacksquare\ T(15) = 125$$
$$53 = 88e^{15k}$$
$$\frac{53}{88} = e^{15k}$$
$$\ln \frac{53}{88} = 15k \qquad \blacksquare\ \text{Rewrite in logarithmic form.}$$
$$\frac{1}{15} \ln \frac{53}{88} = k$$
$$k \approx -0.0338$$

Thus the function is $T(t) \approx 72 + 88e^{-0.0338t}$.

After 25 minutes, $T(25) \approx 72 + 88e^{-0.0338 \cdot 25} \approx 109.8$°F. To find the time for the tea to reach 80°F, solve for t using $T(t) = 80$.

$$80 = 72 + 88e^{-0.0338t}$$
$$8 = 88e^{-0.0338t}$$
$$\frac{8}{88} = e^{-0.0338t}$$
$$\ln \frac{8}{88} = -0.0338t \qquad \blacksquare\ \text{Rewrite in logarithmic form.}$$
$$\frac{1}{-0.0338} \ln \frac{8}{88} = t$$
$$t \approx 70.9$$

It will take almost 71 minutes for the tea to cool to 80°F.

Visualize the Solution

Plot the graphs of $y = 72 + 88e^{-0.0338x}$ and $y = 80$; the x-value of the point of intersection is the solution of the equation when $T(t) = 80$.

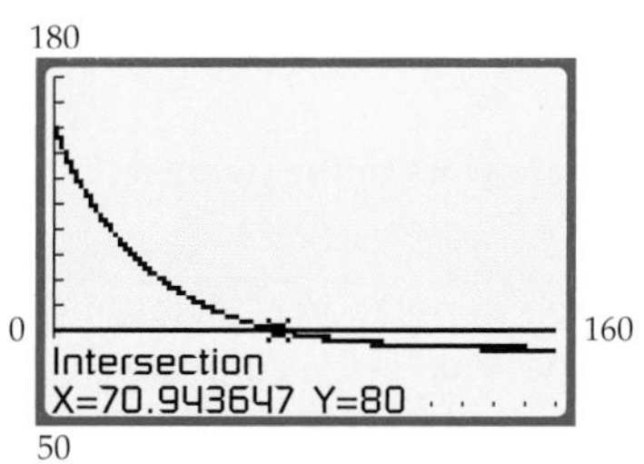

INTEGRATING TECHNOLOGY

To enter the equation into a graphing calculator, use x in place of t.

Continued ➤

Notice that the graph of $T(t)$ is steeper at the start of the cooling process. The curve starts to level out as it slowly approaches $y = 72$, the room temperature. In fact, $y = 72$ is a horizontal asymptote for the function.

CHECK YOUR PROGRESS 7 A freshly poured cappuccino has a temperature of 180°F and cools to 140°F after 12 minutes in a 75°F room. Find a cooling function that gives the temperature of the cappuccino t minutes after it was poured. Use your function to estimate how long it will take for the temperature of the cappuccino to reach 100°F. Round to the nearest tenth of a minute.

Solution *See page S28.* $T(t) = 75 + 105e^{-0.0400t}$; 35.9 min

Topics for Discussion

1. Explain the difference between a population that is growing at a continuous rate of 8% per year and a population that is growing at a total annual rate of 8%.
2. What is an exponential growth model? Give an example of an application for which an exponential growth model might be appropriate.
3. What is an exponential decay model? Give an example of an application for which an exponential decay model might be appropriate.
4. How is the function given by Newton's Law of Cooling similar to an exponential decay model? How is it different?

EXERCISES 5.5

— Suggested Assignment: Exercises 1–21, odd; 14, 24, 25, 27, 29; 35–47, odd; 44, and 49–54, all.

In Exercises 1 to 8, solve for t algebraically. Round your answer to the nearest hundredth.

1. $e^{0.63t} = 3$ 1.74
2. $e^{1.2t} = 12$ 2.07
3. $8e^{0.7t} = 59$ 2.85
4. $84e^{-0.23t} = 71$ 0.73
5. $5750e^{-0.42t} = 2600$ 1.89
6. $488e^{0.91t} = 5150$ 2.59
7. $11{,}460e^{0.31t} = 15{,}340$ 0.94
8. $22{,}825e^{-0.11t} = 18{,}330$ 1.99

In Exercises 9 to 12, round your answers to the nearest hundredth of a percent.

9. ***Equivalent Growth Rate*** A state's population is increasing at a total rate of 6.3% annually. What is the equivalent continuous rate at which the population is growing? 6.11%

10. ***Equivalent Growth Rate*** The bird population in a forest is growing at a total rate of 3.8% each year. Find the equivalent continuous rate at which the bird population is increasing. 3.73%

11. ***Equivalent Growth Rate*** An environmental organization claims that it is increasing its membership at a continuous rate of $7\frac{1}{2}\%$ per year. What is the total annual percent increase in membership? 7.79%

12. ***Equivalent Growth Rate***

    a. Find the continuous growth rate that is equivalent to a total increase of 17.2% per year. 15.87%

    b. Find the total percentage annual increase that is equivalent to a continuous growth rate of 17.2% per year. 18.77%

13. Find the constant k that will make $f(t) = 2.2^t$ and $g(t) = e^{kt}$ represent essentially the same function. Round to the nearest thousandth. $k \approx 0.788$

Business and Economics

14. ***Agriculture*** A farmer knows that planting the same crop in the same field year after year reduces the yield. If the yield on each succeeding year's crop is 90% of the preceding year's yield, then the yield at any time t can be modeled by an exponential decay function.

    a. If Y_0 represents the yield when $t = 0$, write an exponential decay function that gives the yield for year t. $Y(t) = Y_0(0.90)^t$

    b. In how many years, to the nearest whole year, will the yield be 60% of Y_0? 5 yr

15. ***Newton's Law of Cooling*** A canned soda drink is placed in a refrigerator that maintains a constant temperature of 34°F. It takes the soda 5 minutes to cool from 75°F to 65°F.

    a. Use Newton's Law of Cooling to write a function that gives the temperature of the soda t minutes after being placed in the refrigerator. $T(t) \approx 34 + 41e^{-0.0559t}$

    b. What will the temperature (to the nearest degree) of the soda be after 30 minutes? 42°F

    c. How long, to the nearest minute, will it take the soda to cool to 36°F? 54 min

16. ***Newton's Law of Cooling*** A sheet of steel is heated to 1500°F and then plunged into 60°F water in order to harden the steel. In 2 minutes the steel cools to 300°F.

    a. Use Newton's Law of Cooling to write a function that gives the temperature of the steel t minutes after being placed in the water. $T(t) \approx 60 + 1440e^{-0.896t}$

    b. What will the temperature (to the nearest degree) of the steel be after 8 minutes? 61°F

17. ***Depreciation*** If an automobile depreciates at a constant percentage rate, its value can be described by the exponential decay function $V(t) = V_0(1 - r)^t$, where $V(t)$ is the value, in dollars, after t years, V_0 is the original value, and r is the yearly depreciation rate.

    a. A car is purchased for $18,500 and has an annual depreciation rate of 14%. What is the value of the car after 4 years? $10,119.65

    b. Determine in how many years (to the nearest 0.1 year) the car will depreciate to half its original value. 4.6 yr

18. ***Depreciation*** Home computers depreciate in value rapidly; their value over time can be modeled by an exponential decay function. (See Exercise 17.)

    a. A student buys a new computer for $2800, and after 2 years it is worth only $1200. Write an exponential decay function that gives the value of the computer t years after it was purchased. $V(t) \approx 2800(0.65465)^t$

    b. Use your function to predict the value, to the nearest dollar, of the computer 5 years after it was purchased. $337

    c. At what total annual rate is the computer depreciating? Round to the nearest tenth of a percent. 34.5%

Life and Health Sciences

19. ***Medicine*** Sodium-24 is a radioactive isotope of sodium that is used to study circulatory dysfunction. Assuming that 4 micrograms of sodium-24 are injected into a person, the amount A, in micrograms, remaining in that person after t hours is given by the function $A = 4e^{-0.046t}$.

    a. What amount of the sodium-24, to the nearest hundredth microgram, remains after 5 hours? 3.18 micrograms

    b. What is the half-life of sodium-24? Round to the nearest hundredth of an hour. 15.07 h

    c. In how many hours, to the nearest hundredth, will the amount of sodium-24 be 1 microgram? 30.14 h

    d. At what continuous rate is the isotope decaying? 4.6%

20. ***Water Purification*** If an activated carbon filter removes 82% of the toxins in the water that passes through it, what is the minimum number of times that the same body of water would need to be passed (and repassed) through the filter in order to remove at least 99.9% of the toxins? 4 times

21. ***Medication Level*** Aspirin is introduced into a patient's bloodstream. The body removes the aspirin from the blood at a constant percentage rate, so the amount of aspirin in the bloodstream follows exponential decay. If the patient receives 1.5 grams of aspirin and the half-life of aspirin in the bloodstream is 2 hours, find the amount of aspirin remaining in the patient's blood after 5 hours. Round to the nearest hundredth of a gram. 0.27 g

22. ***Medication Level*** Suppose the patient in Exercise 21 is given three dosages of aspirin, with each dose of 1 gram given 3 hours after the preceding dose. The amount of aspirin A in the patient's bloodstream t hours after the first dose is administered is given by

$$A(t) = \begin{cases} (0.5)^{t/2} & 0 \le t < 3 \\ (0.5)^{t/2} + (0.5)^{(t-3)/2} & 3 \le t < 6 \\ (0.5)^{t/2} + (0.5)^{(t-3)/2} + (0.5)^{(t-6)/2} & t \ge 6 \end{cases}$$

Find the amount of aspirin (to the nearest hundredth of a gram) in the patient's body when

a. $t = 1$. 0.71 g **b.** $t = 4$. 0.96 g **c.** $t = 9$. 0.52 g

23. ***Medication Level*** Use a graphing calculator and the dosage formula in Exercise 22 to determine when the amount of aspirin in the patient's body first reaches 0.25 gram. Round to the nearest hundredth of an hour. 11.1 h

24. ***Bacteria Growth*** Two biology students wrote functions to describe the growth of bacteria they were studying. One arrived at $F(x) = 1.4^x$ and the other used $G(x) = e^{0.336x}$. Explain why both of these functions are essentially the same. $e^{0.336x} = (e^{0.336})^x \approx 1.4^x$

Social Sciences

25. ***Population Growth*** A large city currently has 640,000 residents and its population is growing at a continuous rate of 3% per year.

a. Write an exponential growth function that gives the population t years from the present. $N(t) \approx 640{,}000e^{0.03t}$

b. Use your function to find the population in 10 years. Round to the nearest thousand. 864,000

c. How long, to the nearest tenth of a year, will it take for the city's population to reach one million? 14.9 yr

26. ***Population Growth*** A metropolitan city with a population of 225,000 in the year 2000 is experiencing 4.2% continuous growth per year.

a. Write an exponential growth function that gives the population t years after 2000. $N(t) \approx 225{,}000e^{0.042t}$

b. Use your function to find the population in 2012. Round to the nearest thousand. 372,000

c. How long, to the nearest year, will it take for the city's population to reach a half-million? 19 yr

27. ***Population Growth*** A successful new company is expanding rapidly. Currently it has 1350 employees, but the founders believe they will be growing at a continuous rate of 14% per year.

a. If this growth rate continues, write an exponential growth function that gives the number of employees of the company t years from the present. $N(t) \approx 1350e^{0.14t}$

b. Use your function to predict the number of employees in 3 years. 2055

c. How many years, to the nearest tenth, will it take for the company to triple its number of employees? 7.8 yr

28. ***Population Growth*** A bacteria culture that started with 1500 bacteria is growing at a continuous rate of 22% per hour.

a. Write an exponential growth function that gives the number of bacteria t hours from the present. $N(t) \approx 1500e^{0.22t}$

b. Use your function to estimate the number of bacteria after 1 day. $\approx$294,555

c. How quickly will the number of bacteria increase to 1600? Round to the nearest minute. After 18 min

29. ***Population Growth*** The population of Los Angeles, California for the years 1992 through 1996 can be modeled by the function $P(t) = 10{,}130(1.005)^t$, where t is in years, $t = 0$ corresponds to January 1, 1992, and P is measured in thousands.

a. According to this function, what was the population of Los Angeles on January 1, 1992? 10,130,000

b. At what total annual rate is the population increasing? 0.5%

c. If the population of Los Angeles continues to be modeled by this equation, determine in what year the population will reach 13 million. 2042

30. ***Population Growth*** One model for the population of the Las Vegas, Nevada metropolitan area for the years 1990 to 2000 is $P(t) = 764e^{0.0548t}$, where $t = 0$ corresponds to 1990 and P is measured in thousands.

a. At what continuous rate is the population increasing? 5.48%

b. Find the equivalent total annual rate at which the population is growing. Round to the nearest hundredth of a percent. 5.63%

c. Use the model to predict the year the population of the Las Vegas metropolitan area will reach two million. 2007

31. *Population Growth* A town had a population of 22,600 in 1990 and a population of 24,200 in 1995.

a. If the town's population is growing exponentially, find an exponential growth function for the population of the town. Use $t = 0$ to represent the year 1990. $N(t) \approx 22{,}600e^{0.01368t}$

b. Use your function to predict the population of the town in 2010. Round to the nearest hundred. 29,700

32. *Population Growth* A city had a population of 53,700 in 1996 and a population of 58,100 in 2000.

a. If the town's population is growing exponentially, find an exponential growth function for the population of the town. Use $t = 0$ to represent the year 1996. $N(t) \approx 53{,}700e^{0.01969t}$

b. Use your function to predict the population of the town in 2008. Round to the nearest hundred. 68,000

33. *Population Growth* The population of the Tucson, Arizona metropolitan area was 667,000 in 1992 and 791,000 in 1998.

a. If the population of the region is following exponential growth, write a function that gives the population for year t. Use $t = 0$ to represent the year 1992. $N(t) \approx 667{,}000e^{0.0284t}$

b. Use your function to predict the population of the region in 2005. Round to the nearest thousand. 965,000

34. *Population Growth* The population of the Philippines was 51,092,000 in 1980 and 81,160,000 in 2000.

a. If the population of the country is growing exponentially, write an exponential growth function that gives the population for year t. Use $t = 0$ to represent the year 1980. $N(t) \approx 51{,}092{,}000e^{0.02314t}$

b. Use your function to estimate in what year the population of the Philippines will reach 100,000,000. 2009

35. *Comparing Populations* One city has 16,000 residents and is growing at a continuous rate of 5% per year, and another city currently has a population of 13,000 but is growing continuously at 6% per year. Determine when (to the nearest tenth of a year) the two cities will have the same population. After 20.8 yr

36. *Comparing Populations* Two bacteria cultures are growing exponentially. One started with 200 bacteria that numbered 1150 after 3 hours. The other started at the same time with 350 bacteria and contained 925 bacteria after 3 hours. Find (to the nearest tenth of an hour) the time when the two cultures contained the same number of bacteria. 2.2 h

37. *Comparing Decay Rates* Two radioactive materials are decaying. The first material starts with 6 grams and has a half-life of 620 years. The second material starts with 4 grams and has a half-life of 835 years. How long, to the nearest year, before the amount of each material remaining is the same? 1409 yr

38. *Comparing Populations* The town of Springfield had 15,200 residents in 1980 and grew exponentially to 16,600 residents in 2000. Shelbyville's population was 18,700 in 1980 and 21,300 in 2000, and is growing linearly. Use a graphing calculator to determine in what year the two cities' populations will be the same. 2285

39. *Oil Spills* Crude oil leaks from a tank at a rate that depends on the amount of oil that remains in the tank. Because $\frac{1}{8}$ of the oil in the tank leaks out every 2 hours, the volume of oil $V(t)$ in the tank after t hours is given by $V(t) = V_0(0.875)^{t/2}$, where $V_0 = 350{,}000$ gallons is the number of gallons in the tank when the tank starts to leak $(t = 0)$.

a. How many gallons does the tank hold after 3 hours? Round to the nearest gallon. 286,471 gal

b. How many gallons does the tank hold after 5 hours? Round to the nearest gallon. 250,662 gal

c. How long will it take until 90% of the oil has leaked from the tank? Round to the nearest hour. 34 h

Physical Sciences and Engineering

40. *Radioactive Decay* A liquid solution initially contains 8 grams of a radioactive material. After 14 days, 5.5 grams remain.

a. Find an exponential decay function that gives the amount of the material remaining in the solution after t days. $N(t) \approx 8e^{-0.02676t}$

b. Compute the amount of material remaining after 30 days. Round to the nearest hundredth of a gram. 3.58 g

c. Find the half-life of the material. Round to the nearest tenth of a day. 25.9 days

41. ***Radioactive Decay*** A 3-ounce sample of a radioactive substance is decaying such that after 5 years, 2.1 ounces remain.

 a. Find an exponential decay function that gives the amount of the substance remaining after t years. $N(t) \approx 3e^{-0.07133t}$

 b. Determine how long, to the nearest tenth of a year, it will be until only 1 ounce of the sample remains. 15.4 yr

 c. Find the half-life of the substance. Round to the nearest hundredth of a year. 9.72 yr

42. ***Radioactive Decay*** Polonium (^{210}Po) has a half-life of 138 days. Find the exponential decay function for the percent of polonium (^{210}Po) that remains in a sample after t days. $N(t) \approx N_0e^{-0.005023t}$

43. ***Radioactive Decay*** A radioactive isotope has a half-life of 280 years. If we start with a 6-gram sample, write an exponential decay function for the amount of the isotope remaining after t years. $N(t) \approx 6e^{-0.002476t}$

44. ***Radioactive Decay*** The amount, in grams, of a radioactive material remaining after t days is given by the function $N(t) = 12e^{-0.031t}$.

 a. How much of the material was initially present? 12 g

 b. What is the continuous decay rate, to the nearest tenth of a percent, of the material? 3.1%

 c. Find the half-life, to the nearest hundredth of a day, of the material. 22.36 days

45. ***Carbon Dating*** Geologists have determined that Crater Lake in Oregon was formed by a volcanic eruption. Chemical analysis of a wood chip that is assumed to be from a tree that died during the eruption has shown that it contains approximately 45% of its original carbon-14. Determine how long ago, to the nearest hundred years, the volcanic eruption occurred. 6600 years ago

46. ***Carbon Dating*** Determine the age of a bone if it now contains 65% of its original carbon-14. Round to the nearest hundred years. 3600 years old

47. ***Carbon Dating*** The Rhind papyrus, named after A. Henry Rhind, contains most of what we know today of ancient Egyptian mathematics. A chemical analysis of a sample from the papyrus has shown that it contains approximately 75% of its original carbon-14. What is the age, to the nearest year, of the Rhind papyrus? 2378 years old

48. ***Carbon Dating*** In 2000 an archeologist discovered a sandal made from yucca fiber that she believed was from A.D. 700. Estimate the percentage of the yucca fiber's original carbon-14 that remained when the sandal was found. Round to the nearest tenth of a percent. 85.4%

Prepare for Section 5.6

49. Find the value of $328.5 \cdot 1.47^{4.2}$ rounded to the nearest tenth. [5.1] 1656.8

50. For the equation $y = a \cdot b^x$, find the value of y when $a = 2418.3$, $b = 2.5912$, and $x = 8.5$. Round your answer to the nearest hundred. [5.1] 7,911,700

51. Find the value of $74.6 + 8.201 \ln 49$ rounded to the nearest hundredth. [5.2] 106.52

52. For the equation $y = a + b \ln x$, find the value of y when $a = 109.52$, $b = -21.442$, and $x = 135$. Round your answer to the nearest hundredth. [5.2] 4.34

53. Find the value of $\dfrac{670.8}{1 + 5.23e^{-0.7934(14)}}$ rounded to the nearest hundredth. [5.1] 670.75

54. For the equation $y = \dfrac{c}{1 + ae^{-bx}}$, find the value of y when $a = 9.394$, $b = 0.16342$, $c = 1438.5$, and $x = 40$. Round your answer to the nearest tenth. [5.1] 1419.2

Explorations 

1. ***Biological Diversity*** To discuss the variety of species that live in a certain environment, a biologist needs a precise definition of diversity. Let $p_1, p_2, \ldots, p_n$ be the proportions of n species that live in an environment. The **biological diversity** D of this system is

$$D = -(p_1 \log_2 p_1 + p_2 \log_2 p_2 + \cdots + p_n \log_2 p_n)$$

Suppose an ecosystem has exactly five different varieties of grass: rye (R), Bermuda (B), blue (L), fescue (F), and St. Augustine (A). The various proportions of these grasses are shown in the following tables.

Table 1

R	B	L	F	A
$\frac{1}{5}$	$\frac{1}{5}$	$\frac{1}{5}$	$\frac{1}{5}$	$\frac{1}{5}$

Table 2

R	B	L	F	A
$\frac{1}{8}$	$\frac{3}{8}$	$\frac{1}{16}$	$\frac{1}{8}$	$\frac{5}{16}$

Table 3

R	B	L	F	A
0	$\frac{1}{4}$	0	0	$\frac{3}{4}$

Table 4

R	B	L	F	A
0	0	0	0	1

a. Calculate the diversity of this ecosystem if the proportions of these grasses are as shown in Table 1. Round to the nearest hundredth. 2.32

b. Because Bermuda and St. Augustine are virulent grasses, after a time the proportions will be as shown in Table 2. Calculate the diversity of this system. Does this system have more or less diversity than the system given in Table 1? 2.06; less

c. After an even longer time period, the Bermuda and St. Augustine grasses completely overrun the environment and the proportions are as shown in Table 3. Calculate the diversity of this system. Round to the nearest hundredth. (*Note:* Although the equation is not technically correct, for purposes of the diversity definition we may say that $0 \log_2 0 = 0$. By using very small values of p_i, we can demonstrate that this definition makes sense.) Does this system have more or less diversity than the system given in Table 2? 0.81; less

d. Finally, the St. Augustine grass overruns the Bermuda and the proportions are as shown in Table 4. Calculate the diversity of this system. Write a sentence that explains the meaning of the value you obtained.
0; With only one variety of grass, the system has no diversity.

SECTION 5.6 Modeling Data with Exponential and Logarithmic Functions

- Analyze Scatter Plots
- Applications

▪ Analyze Scatter Plots

In Chapter 2 we used linear and quadratic functions to model several data sets. However, in some applications, data can be modeled more closely by using exponential or logarithmic functions. For instance, Figure 5.25 illustrates some scatter plots that can be effectively modeled by exponential and logarithmic functions.

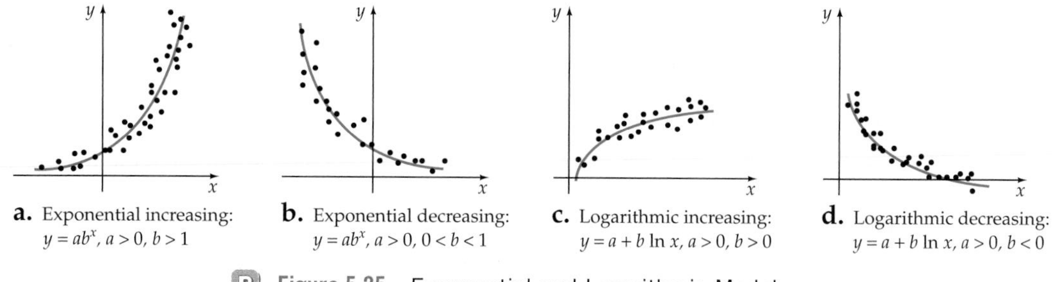

y
x
a. Exponential increasing: $y = ab^x, a > 0, b > 1$
b. Exponential decreasing: $y = ab^x, a > 0, 0 < b < 1$
c. Logarithmic increasing: $y = a + b \ln x, a > 0, b > 0$
d. Logarithmic decreasing: $y = a + b \ln x, a > 0, b < 0$

Figure 5.25 Exponential and Logarithmic Models

The terms *concave upward* and *concave downward* are often used to describe a graph. For instance, Figures 5.26a. and 5.26b. show the graphs of two increasing functions that join the points P and Q. The graphs of f and g differ in that they bend in different directions. We can distinguish between these two types of "bending" by examining the positions of *tangent lines* to the graph. In Figures 5.26c. and 5.26d., tangent lines (in red) have been drawn to the graphs of f and g. The graph of f lies above its tangent lines and the graph of g lies below its tangent lines. The function f is said to be concave upward, and g is concave downward.

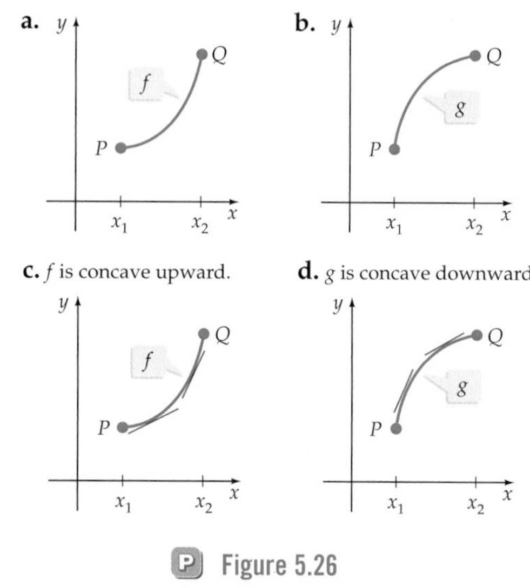

Figure 5.26

Definition of Concavity

If the graph of f lies above all of its tangents on an interval $[x_1, x_2]$, then f is **concave upward** on $[x_1, x_2]$.

If the graph of f lies below all of its tangents on an interval $[x_1, x_2]$, then f is **concave downward** on $[x_1, x_2]$.

An examination of the graphs in Figure 5.25 shows that the graphs of all exponential functions (of the form $y = ab^x, a > 0, b > 0, b \neq 1$) are concave upward. The graphs of increasing logarithmic functions are concave downward, and the graphs of decreasing logarithmic functions are concave upward.

In Example 1 we analyze scatter plots by determining whether the shape of the scatter plot can best be approximated by an increasing or a decreasing function, and by a function that is concave upward or concave downward.

QUESTION Is the graph of $y = 5 - 2\ln x$ concave upward or concave downward?

Alternative to Example 1
Exercise 6, page 461

EXAMPLE 1 Analyze Scatter Plots

For each of the following data sets, determine whether the most suitable model of the data would be an exponential function or a logarithmic function.

$$A = \{(1, 0.6), (2, 0.7), (2.8, 0.8), (4, 1.3), (6, 1.5), (6.5, 1.6), (8, 2.1), (11.2, 4.1), (12, 4.6), (15, 8.2)\}$$

$$B = \{(1.5, 2.8), (2, 3.5), (4.1, 5.1), (5, 5.5), (5.5, 5.7), (7, 6.1), (7.2, 6.4), (8, 6.6), (9, 6.9), (11.6, 7.4), (12.3, 7.5), (14.7, 7.9)\}$$

Solution For each set construct a scatter plot of the data. See Figure 5.27.

TAKE NOTE

See Section 2.6 if you need to review the steps needed to create a scatter plot on a TI-83 calculator.

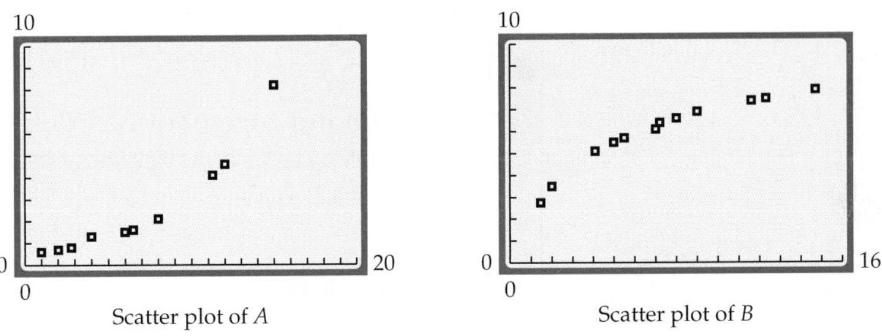

Figure 5.27

The scatter plot of A suggests that A is an increasing function that is concave upward. Thus A can be effectively modeled by an increasing exponential function.

The scatter plot of B suggests that B is an increasing function that is concave downward. Thus B can be effectively modeled by an increasing logarithmic function.

CHECK YOUR PROGRESS 1 For each of the following data sets, determine whether the most suitable model of the data would be an exponential function or a logarithmic function.

$$A = \{(5, 2.3), (7, 3.9), (9, 4.5), (12, 5.0), (16, 5.4), (21, 5.8), (26, 6.1)\}$$

$$B = \{(3, 0.50), (4, 0.56), (6, 0.70), (11, 1.24), (14, 1.74), (22, 4.31)\}$$

Solution *See page S28.* Set A, logarithmic function; set B, exponential function

---

ANSWER The graph is concave upward because the b-value, -2, is less than zero.

▪ Applications

The methods used to model data using exponential or logarithmic functions are similar to the methods used in Chapter 2 to model data using linear or quadratic functions. Here is a summary of the modeling process.

The Modeling Process

Use a graphing utility to:

1. **Construct a *scatter plot* of the data** to determine which type of function will effectively model the data.
2. **Find the *regression equation*** of the modeling function and the correlation coefficient for the regression.
3. **Examine the *correlation coefficient*** and ***view a graph*** that displays both the modeling function and the scatter plot to determine how well your function fits the data.

In the following example we use the above modeling process to find a function that closely models the value of a diamond as a function of its weight.

Alternative to Example 2
Exercise 18a., page 462

TAKE NOTE

The value of a diamond is generally determined by its color, cut, clarity, and carat weight. These characteristics of a diamond are known as the four c's. In Example 2 we have assumed that the color, cut, and clarity of all of the diamonds are similar. This assumption enables us to model the value of each diamond as a function of just its carat weight.

EXAMPLE 2 Model an Application with an Exponential Function

A diamond merchant has determined the value of several white diamonds that have different weights (measured in carats), but are *similar in quality*. See Table 5.12.

Table 5.12

4.00 ct	3.00 ct	2.00 ct	1.75 ct	1.50 ct	1.25 ct	1.00 ct	0.75 ct	0.50 ct
$14,500	$10,700	$7,900	$7,300	$6,700	$6,200	$5,800	$5,000	$4,600

Find a function that models the value of the diamonds as a function of their weights and use the function to predict the value of a 3.5-carat diamond of similar quality.

Solution

1. **Construct a scatter plot of the data.**

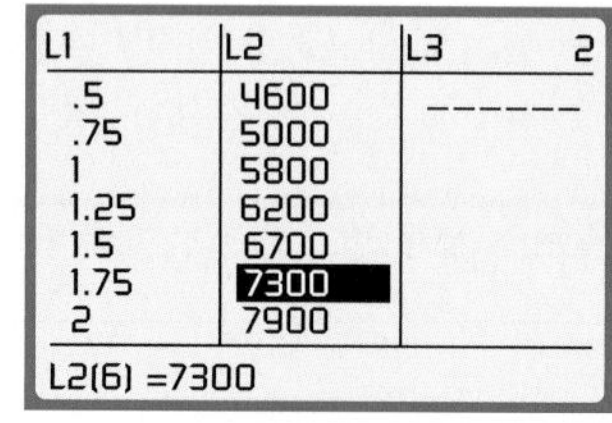

18000

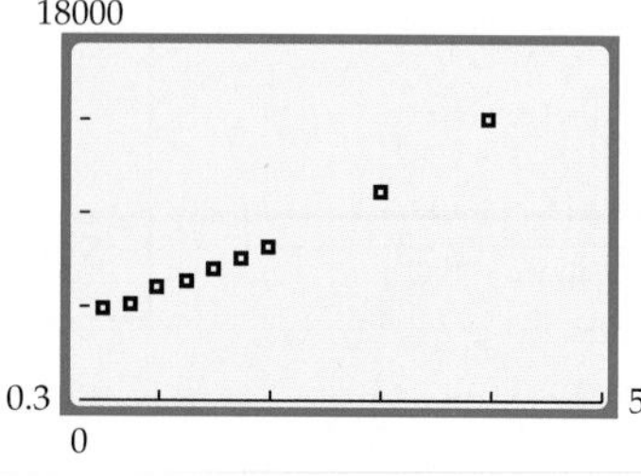

Figure 5.28

Continued ➤

INTEGRATING TECHNOLOGY

Most graphing utilities have built-in routines that can be used to determine the exponential or logarithmic regression function that best models a set of data. On a TI-83, the ExpReg instruction is used to find the exponential regression function and the LnReg instruction is used to find the logarithmic regression function. The TI-83 does not show the value of the regression coefficient *r* unless the DiagnosticOn command has been entered. The DiagnosticOn command is in the CATALOG menu.

From the scatter plot in Figure 5.28 it appears that the data can be closely modeled by an exponential function of the form $y = ab^x$, $b > 1$.

2. **Find the regression equation.** The calculator display in Figure 5.29 shows that the exponential regression equation is $y \approx 4067.6(1.3816)^x$, where x is the carat weight of the diamond and y is the value of the diamond.

```
ExpReg
 y=a*b^x
 a=4067.641145
 b=1.381644186
 r²=.994881215
 r=.9974373238
```

Figure 5.29 ExpReg display (DiagnosticOn)

3. **Examine the correlation coefficient.** The correlation coefficient $r \approx 0.9974$ is close to 1. This indicates that the exponential regression function $y \approx 4067.6(1.3816)^x$ provides a good fit for the data. The graph in Figure 5.30 also shows that the exponential regression function provides a good model for the data.

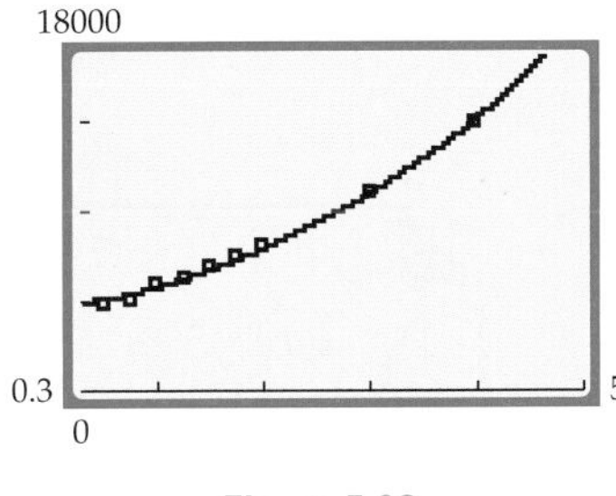

Figure 5.30

To estimate the value of a 3.5-carat diamond, replace x in the exponential regression function with 3.5.

$$y \approx 4067.6(1.3816)^{3.5} \approx \$12,610$$

According to the exponential regression function, the value of a 3.5-carat diamond of similar quality is about \$12,610.

Point of Interest

The Hope Diamond, shown below, is the world's largest deep blue diamond. It has a weight of 45.52 carats. We should not expect the function $y \approx 4067.6 \times 1.3816^x$ in Example 2 to yield an accurate value of the Hope Diamond because the Hope Diamond is not the same type of diamond as the diamonds in Table 5.12 and its weight is much larger than the weights of the diamonds in Table 5.12.

The Hope Diamond is on display at the Smithsonian Museum of Natural History in Washington, D.C.

CHECK YOUR PROGRESS 2  Table 5.13 shows the Earth's atmospheric pressure P at an altitude of a kilometers. Find a suitable function that models the atmospheric pressure as a function of the altitude. Use the function to estimate the atmospheric pressure at an altitude of 24 kilometers. Round to the nearest tenth of a newton per square centimeter.

Continued ➤

Table 5.13

Altitude a above sea level, in kilometers	Atmospheric pressure P, in newtons per square centimeter
0	10.3
2	8.0
4	6.4
6	5.1
8	4.0
10	3.2
12	2.5
14	2.0
16	1.6
18	1.3

Solution *See page S29.* $y \approx 10.147(0.89104)^x$

In the next example we consider a data set that can be effectively modeled by more than one type of function.

Alternative to Example 3
Exercise 16, page 462

EXAMPLE 3 Choosing the Best Model

Table 5.14 shows the winning times in the women's Olympic 100-meter freestyle event for the years 1968 to 1996.

Table 5.14 Women's Olympic 100-Meter Freestyle, 1968 to 1996

Year	Time (in seconds)	Year	Time (in seconds)
1968	60.0	1984	55.92
1972	58.59	1988	54.93
1976	55.65	1992	54.64
1980	54.79	1996	54.50

Source: Time Almanac 2002

a. Determine whether the data in Table 5.14 can best be modeled by an exponential function or a logarithmic function.

b. Use the function you chose in part **a.** to predict the winning time in the women's Olympic 100-meter freestyle event for the year 2004.

Continued ➤

Solution

a. Construct a scatter plot of the data. In this example we have represented the year 1968 by $x = 68$, the year 1996 by $x = 96$, and the winning time by y.

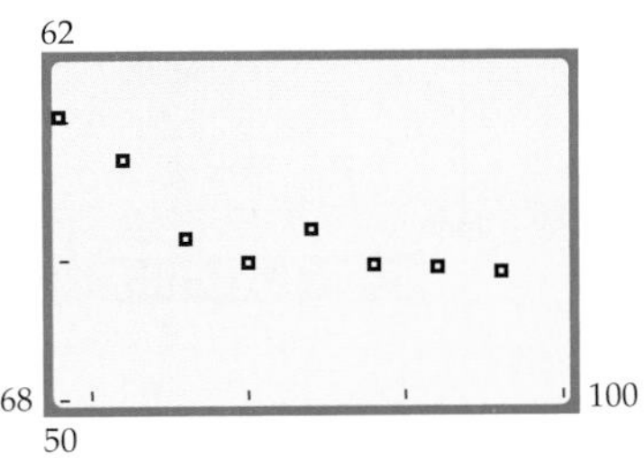

Figure 5.31

From the scatter plot in Figure 5.31, it appears that the data can be effectively modeled by a decreasing exponential function and also by a decreasing logarithmic function. Use a graphing utility to determine both an exponential regression function and a logarithmic regression function for the data. Figure 5.32 shows the exponential regression function and Figure 5.33 shows the logarithmic regression function.

```
ExpReg
 y=a*b^x
 a=72.31443696
 b=.9969076216
 r²=.7151837083
 r=-.8456853483
```

Figure 5.32

```
LnReg
 y=a+blnx
 a=120.7473143
 b=-14.68497634
 r²=.7454398922
 r=-.8633886102
```

Figure 5.33

In this example, the regression coefficients are both negative. In such cases the regression function that has a correlation coefficient closer to -1 provides the better fit for the given data. Thus the logarithmic model provides a slightly better fit for the data in this example. The logarithmic regression function is $y \approx 120.75 - 14.685 \ln x$. The graph of $y \approx 120.75 - 14.685 \ln x$, along with a scatter plot of the data, is shown in Figure 5.34.

TAKE NOTE

For the year 2000, the regression function $y \approx 120.75 - 14.685 \ln x$ yields a winning time of $120.75 - 14.685 \ln(100) \approx 53.12$ seconds. The actual winning time for the women's Olympic 100-meter freestyle event in the year 2000 was 58.83 seconds. These results illustrate that just because a regression function provides a good fit for a given data set, that does not guarantee the regression function will provide good predictions of future events.

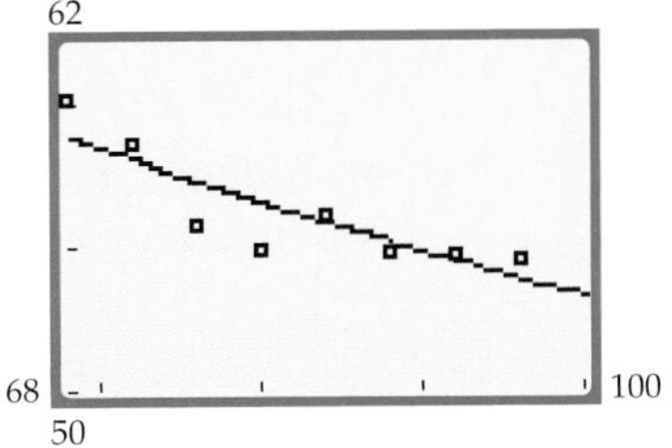

Figure 5.34

Continued ➤

b. To predict the winning time for the women's Olympic 100-meter freestyle event in the year 2004, replace x in the logarithmic regression function with 104.

$$y \approx 120.75 - 14.685 \ln(104) \approx 52.55$$

According to the logarithmic regression function, the winning time for the women's Olympic 100-meter freestyle event in the year 2004 will be about 52.55 seconds.

CHECK YOUR PROGRESS 3   Table 5.15 lists the world record times in the men's 400-meter race for the years 1948 to 2002.

Table 5.15 World Record Times in the Men's 400-Meter Race, 1948 to 2002

Year	Time (in seconds)	Year	Time (in seconds)
1948	45.9	1963	44.9
1950	45.8	1964	44.9
1955	45.4	1967	44.5
1956	45.2	1968	44.1
1960	44.9	1968	43.86
1960	44.9	1988	43.29
		1999	43.18

Source: Track and Field Statistics, **http://trackfield.brinkster.net/Main.asp**

a. Determine whether the data in Table 5.15 can be modeled better by an exponential function or a logarithmic function. Let $x = 48$ represent the year 1948.

b. Use the function you chose in part **a.** to predict the world record time in the men's 400-meter race for the year 2008. Round to the nearest hundredth of a second.

Solution *See page S29.* **a.** Logarithmic function **b.** 42.52 s

Topics for Discussion

1. A student tries to determine the exponential regression equation for the following data.

x	1	2	3	4	5
y	8	2	0	−1.5	−2

The student's calculator displays an ERROR message. Explain why the calculator was unable to determine the exponential regression equation for the data.

2. Consider the logarithmic model $h(x) = 6 - 2 \ln x$.

a. Is h an increasing or a decreasing function?

b. Is h concave up or concave down on the interval $(0, \infty)$?

c. Find, if possible, $h(0)$ and $h(e)$.

d. Does h have a horizontal asymptote? Explain.

2. Linear function; quadratic function; decreasing exponential function; decreasing logarithmic function

5. Linear function; quadratic function; decreasing logarithmic function; decreasing exponential function

EXERCISES 5.6

— Suggested Assignment: Exercises 1 to 29, odd; and 30–36, all.
— Answer graphs to Exercises 1–6 are on page AA16.

In Exercises 1 to 6, use a scatter plot to determine which of the following types of functions might provide a suitable model of the data: a linear function; a quadratic function; an exponential function; a logarithmic function. (*Note:* In some exercises the data can be closely modeled by more than one type of function.)

1. $\{(1, 3), (1.5, 4), (2, 6), (3, 13), (3.5, 19), (4, 27)\}$
Quadratic function; increasing exponential function

2. $\{(1.0, 1.12), (2.1, 0.87), (3.2, 0.68), (3.5, 0.63), (4.4, 0.52)\}$

3. $\{(-2, 11), (-1, 5), (0, 1), (1.5, -1.3), (3, 1), (5, 11), (6, 19)\}$
Quadratic function

4. $\{(0, 1.3), (0.5, 1.7), (1, 2.16), (2, 3.2), (3, 4.4), (4.6, 6.1)\}$
Linear function; quadratic function; increasing exponential function

5. $\{(1, 2.5), (1.5, 1.7), (2, 0.7), (3, -0.5), (3.5, -1.3), (4, -1.5)\}$

6. $\{(1, 3), (1.5, 3.8), (2, 4.4), (3, 5.2), (4, 5.8), (6, 6.6)\}$
Linear function; quadratic function; increasing logarithmic function

In Exercises 7 to 10, use a graphing utility to find the exponential regression function for the data. State the correlation coefficient r. Round a, b, and r to the nearest hundred thousandth.

7. $\{(10, 6.8), (12, 6.9), (14, 15.0), (16, 16.1), (18, 50.0), (19, 20.0)\}$ $y \approx 0.99628(1.20052)^x$; $r \approx 0.85705$

8. $\{(2.6, 16.2), (3.8, 48.8), (5.1, 160.1), (6.5, 590.2), (7, 911.2)\}$
$y \approx 1.48874(2.50469)^x$; $r \approx 0.99999$

9. $\{(0, 1.83), (1, 0.92), (2, 0.51), (3, 0.25), (4, 0.13), (5, 0.07)\}$
$y \approx 1.81505(0.51979)^x$; $r \approx -0.99978$

10. $\{(4.5, 1.92), (6.0, 1.48), (7.5, 1.14), (10.2, 0.71), (12.3, 0.49)\}$
$y \approx 4.23016(0.83937)^x$; $r \approx -0.99999$

In Exercises 11 to 14, use a graphing utility to find the logarithmic regression function for the data. State the correlation coefficient r. Round a, b, and r to the nearest hundred thousandth.

11. $\{(5, 2.7), (6, 2.5), (7.2, 2.2), (9.3, 1.9), (11.4, 1.6), (14.2, 1.3)\}$
$y \approx 4.89060 - 1.35073 \ln x$; $r \approx -0.99921$

12. $\{(11, 15.75), (14, 15.52), (17, 15.34), (20, 15.18), (23, 15.05)\}$
$y \approx 18.02743 - 0.94970 \ln x$; $r \approx -0.99997$

13. $\{(3, 16.0), (4, 16.5), (5, 16.9), (7, 17.5), (8, 17.7), (9.8, 18.1)\}$
$y \approx 14.05858 + 1.76393 \ln x$; $r \approx 0.99983$

14. $\{(8, 67.1), (10, 67.8), (12, 68.4), (14, 69.0), (16, 69.4)\}$
$y \approx 60.08692 + 3.36076 \ln x$; $r \approx -0.99932$

Business and Economics

15.  ***Interest Rates on Auto Loans*** The following table shows the annual interest rates for new car loans in 2002 based on the length (term) of the loan.

Term t, in months	12	24	36	48
Annual interest rate	5.72%	5.97%	6.23%	6.50%

a. Find an exponential model for the data in the table and use the model to predict the interest rate, to the nearest 0.01%, on an auto loan with a term of 60 months. Round a, b, and r to the nearest hundred thousandth.
$y \approx 5.48184(1.00356)^x$; 6.78%

b. According to your model in part **a.**, what is the term of a loan with a 7.00% interest rate? Round to the nearest month. 69 months

Life and Health Sciences

16.  ***Generation of Garbage*** According to the U.S. Environmental Protection Agency, the amount of garbage generated per person has been increasing over the last few decades. The following table shows the per capita garbage, in pounds per day, generated in the United States.

Year, t	1960	1970	1980	1990	2000
Pounds per day, p	2.66	3.27	3.61	4.00	4.30

Represent the year 1960 by $t = 60$.

a. Use a graphing utility to find a linear model and a logarithmic model for the data. Use t as the independent variable (domain) and p as the dependent variable (range).
LinReg: $p \approx 0.0401t + 0.36$, LnReg: $p \approx -10.23519 + 3.161541 \ln t$

b. Examine the correlation coefficients of the two regression models to determine which model provides a better fit for the data. Linear model: $r \approx 0.99096$; logarithmic model: $r \approx 0.99738$. The logarithmic model provides a slightly better fit.

c. Use the model you selected in part **b.** to predict the amount of garbage that will be generated per capita per day in 2005. Round to the nearest hundredth of a pound. 4.48 lb per capita per day

17. ***The Henderson-Hasselbach Function*** The scientists Henderson and Hasselbach determined that the pH of blood is a function of the ratio q of the amounts of bicarbonate and carbonic acid in the blood.

a. Use a graphing utility and the data in the following table to determine a linear model and a logarithmic model for the data. Use q as the independent variable (domain) and pH as the dependent variable (range). State the correlation coefficient for each model. Round a and b to 5 decimal places and r to 6 decimal places. Which model provides the better fit for the data?

q	7.9	12.6	31.6	50.1	79.4
pH	7.0	7.2	7.6	7.8	8.0

b. Use the model you chose in part **a.** to find the q-value associated with a pH of 8.2. Round to the nearest tenth.
126.0

18. ***World Population*** The following table lists the years in which the world's population first reached 3, 4, 5, and 6 billion.

17. a. LinReg: pH $\approx 0.01353q + 7.02852$, $r \approx 0.956627$; LnReg: pH $\approx 6.10251 + 0.43369 \ln q$, $r \approx 0.999998$. The logarithmic model provides a better fit.

World Population Milestones

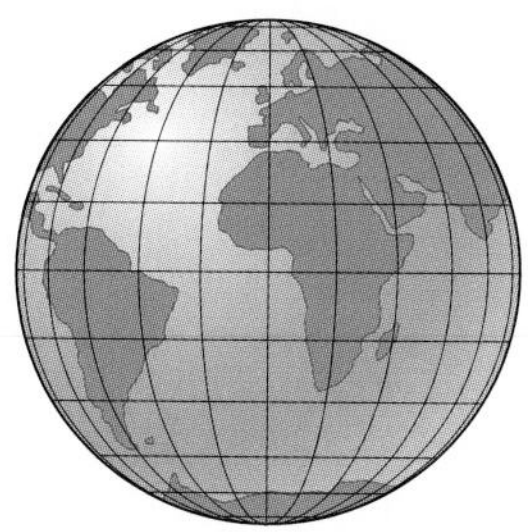

1960	3 billion
1974	4 billion
1987	5 billion
1999	6 billion

Source: Time Almanac 2002, page 708.

a. Find an exponential model for the data in the table. Let $x = 0$ represent the year 1960. $y \approx 3.05401(1.0179)^x$

b. Use the model to predict the year in which the world's population will first reach 7 billion. 2006

19.  ***Panda Population*** One estimate gives the world panda population as 3200 in 1980 and 590 in 2000.

a. Find an exponential model for the data and use the model to predict the year in which the panda population p will be reduced to 200. (Let $t = 0$ represent the year 1980.) $p \approx 3200(0.91894)^t$; 2012

b. Because the exponential model in part **a.** fits the data perfectly, does this mean that the model will accurately predict future panda populations? Explain.
No. The model fits the data perfectly because there are only two data points.

Sports and Recreation

20.  ***Olympic Records*** The following table shows the Olympic gold medal distances for the women's high jump from 1968 to 2000.

Women's Olympic High Jump, 1968 to 2000

Year	Distance	Year	Distance
1968	5 ft $11\frac{3}{4}$ in.	1984	6 ft $7\frac{1}{2}$ in.
1972	6 ft $3\frac{5}{8}$ in.	1988	6 ft 8 in.
1976	6 ft 4 in.	1992	6 ft $7\frac{1}{2}$ in.
1980	6 ft $5\frac{1}{2}$ in.	1996	6 ft $8\frac{3}{4}$ in.
		2000	6 ft 7 in.

Source: Time Almanac 2002

Represent the year 1968 by 68.

a. Use a graphing utility to determine a linear model and a logarithmic model for the data, with the distance measured in inches. State the correlation coefficient r for each model. LinReg: $y \approx 0.22448x + 58.87986$, $r \approx 0.86012$; LnReg: $y \approx -7.07160 + 19.17358 \ln x$, $r \approx 0.88386$

b. Use the correlation coefficient for each of the models in part **a.** to determine which model provides the better fit for the data. The logarithmic model provides the better fit.

c. Use the model you selected in part **b.** to predict the women's Olympic gold medal high jump distance in 2012. Round to the nearest tenth of an inch. 83.4 in. (6 ft 11.4 in.)

21.  ***Baseball Statistics*** The following table shows the average times of nine-inning major league baseball (MLB) games from 1981 to 1999.

Average Times of Nine-Inning MLB Games, 1981 to 1999

Year	Time (in hours and minutes)	Year	Time (in hours and minutes)
1981	2:33	1991	2:49
1982	2:34	1992	2:49
1983	2:36	1993	2:48
1984	2:35	1994	2:54
1985	2:40	1995	2:50
1986	2:44	1996	2:51
1987	2:48	1997	2:52
1988	2:45	1998	2:47
1989	2:46	1999	2:53
1990	2:48		

Source: Elias Sports Bureau (reported in the *San Diego Union-Tribune*)

a. Use a graphing utility to find a linear model and a logarithmic model, with the time measured in minutes, for the data. Represent 1981 by $x = 1$. State the correlation coefficient r for each model.

b. Use the correlation coefficient for each of the models in part **a.** to determine which model provides the better fit for the data. The logarithmic model provides a better fit.

c. Use the model you selected in part **b.** to predict the first year in which the average time of a nine-inning game will exceed 2 hours 56 minutes. During the 2012 season

22.  ***Baseball Statistics***

a. Use a graphing utility to determine a logarithmic model, with the time measured in minutes, for the data in Exercise 21. Represent the year 1981 by $t = 81$. State the correlation coefficient r. LnReg: Time $\approx -259.07255 + 94.36317 \ln t$, $r \approx 0.893807$

b. Does the logarithmic model from part **a.** provide a better fit for the data than the logarithmic model from part **a.** of Exercise 21? No

c. Explain why the two logarithmic models do not have the same correlation coefficients.

Social Sciences

23.  ***Number of Automobiles*** In 1900, the number of automobiles in the United States was around 8000. By 2000, the number of automobiles in the United States had reached 200 million.

a. Find an exponential model for the data and use the model to predict the number of automobiles, to the nearest 100,000, in the United States in 2010. Use $t = 0$ to represent the year 1900. ExpReg: $a \approx 8000(1.10657)^t$; 550,500,000 automobiles

b. According to the model, in what year will the number of automobiles in the United States first reach 300 million? 2004

Physical Sciences and Engineering

24.  ***Temperature of Coffee*** A cup of coffee is placed in a room that maintains a constant temperature of 70°F. The following table shows both the coffee temperature T after t minutes and the difference between the coffee temperature and the room temperature after t minutes.

Time t (minutes)	0	5	10	15	20	25
Coffee temp. T (°F)	165°	140°	121°	107°	97°	89°
$T - 70°$	95°	70°	51°	37°	27°	19°

a. Use a graphing utility to find an exponential model for the difference $T - 70°$ as a function of t. $T - 70° \approx 96.16777(0.93787)^t$

b. Use the model to predict how long it will take (to the nearest minute) for the coffee to cool to 80°F. 35 min

21. a. LinReg: Time $\approx 1.04035t + 154.96491$, $r \approx 0.88458$; LnReg: Time $\approx 149.56876 + 7.63077 \ln t$, $r \approx 0.93101$

22. c. The graph of $y = a + b \ln t$ becomes flatter and flatter as $t \to \infty$. Thus the function $y = a + b \ln t$ does not provide the same fit for small values of t as it does for larger values of t.

25. ***Desalination*** The following table shows the amount of fresh water w (in cubic yards) produced from saltwater after t hours of a desalination process.

t	1	2.5	3.5	4.0	5.1	6.5
w	18.2	46.6	57.4	61.5	68.7	76.2

a. Use a graphing utility to find a linear model and a logarithmic model for the data.
LinReg: $w \approx 10.17227t + 16.45111$, $r \approx 0.95601$;
LnReg: $w \approx 18.26750 + 31.03499 \ln t$, $r \approx 0.99996$

b. Examine the correlation coefficients of the two regression models to determine which model provides the better fit for the data. State the correlation coefficient r for each model. The logarithmic model provides a better fit.

c. Use the model you selected in part **b.** to predict the amount of freshwater that will be produced after 10 hours of the desalination process. Round to the nearest tenth of a cubic yard. 89.7 yd^3

26. ***A Correlation Coefficient of 1*** A scientist uses a graphing utility to model the data set $\{(2, 5), (4, 6)\}$ with a logarithmic function. The following display shows the results.

```
LnReg
 y=a+blnx
 a=4
 b=1.442695041
 r²=1
 r=1
```

What is the significance of the fact that the correlation coefficient for the above regression equation is $r = 1$?

27. ***Duplicate Data Points*** An engineer needs to model the data in set A with an exponential function.

$$A = \{(2, 5), (3, 10), (4, 17), (4, 17), (5, 28)\}$$

Because the ordered pair (4, 17) is listed twice, the engineer decides to eliminate one of these ordered pairs and model the data in set B.

$$B = \{(2, 5), (3, 10), (4, 17), (5, 28)\}$$

Determine whether A and B both have the same exponential regression function.
A and *B* have different exponential regression functions.

28. ***Domain Error*** A scientist needs to model the data in set A.

$$A = \{(0, 1.2), (1, 2.3), (2, 2.8), (3, 3.1), (4, 3.3), (5, 3.4)\}$$

The scientist views a scatter plot of the data and decides to model the data with a logarithmic function of the form $y = a + b \ln x$.

a. When the scientist attempts to use a graphing calculator to determine the logarithmic regression equation, the calculator displays the message

"ERR:DOMAIN"

Explain why the calculator was unable to determine the logarithmic regression equation for the data.
The x-coordinate of the first ordered pair is 0, and 0 is not in the domain of $y = \ln x$.

b. Explain what the scientist could do so that the data in set A could be modeled by a logarithmic function of the form $y = a + b \ln x$.

29. ***Power Functions*** A function that can be written in the form $y = ax^b$ is said to be a **power function.** Some data sets can best be modeled by a power function. On a TI-83, the PwrReg instruction is used to produce a power regression function for a set of data.

a. Use a graphing utility to find an exponential regression function and a power regression function for the following data. State the correlation coefficient r for each model. ExpReg: $y \approx 1.81120(1.61740)^x$, $r \approx 0.96793$;
PwrReg: $y \approx 2.0985(x)^{1.40246}$, $r \approx 0.99999$

x	1	2	3	4	5	6
y	2.1	5.5	9.8	14.6	20.1	25.8

b. Which of the two regression functions produces the better fit for the data?
The power regression function provides the better fit.

30. ***Period of a Pendulum*** The following table shows the time t (in seconds) of the period of a pendulum (the time it takes the pendulum to complete a swing to the left and back) of length l (in feet).

a. Use a graphing utility to determine the equation of the best model for the data. Your model must be a power function or an exponential function.
The power function $t \approx 1.11088(l^{0.50113})$

Length l	1	2	3	4	6	8
Time t	1.11	1.57	1.92	2.25	2.72	3.14

b. According to the model you chose in part **a.**, what is the length of a pendulum, to the nearest tenth of a foot, that has a period of 12 seconds? 115.4 ft

Prepare for Section 5.7

31. Determine the far-right behavior of the graph of $f(x) = e^{-0.2x}$. [5.1] As $x \to \infty$, $f(x) \to 0$.

32. Given $P(t) = \dfrac{600}{1 + 2.2e^{-0.05t}}$, find $P(0)$. Round to the nearest tenth. [5.1] 187.5

33. Given $P(t) = \dfrac{600}{1 + 2.2e^{-0.05t}}$, find $P(4)$. Round to the nearest hundredth. [5.1] 214.19

In Exercises 34 to 36, provide the exact solution and an approximation rounded to 5 decimal places.

34. Solve $e^{-0.12x} = 0.4$ for x. [5.4] $-\dfrac{\ln 0.4}{0.12} \approx 7.63576$

35. Solve $1 + e^{-0.04t} = 1.25$ for t. [5.4] $-\dfrac{\ln 0.25}{0.04} \approx 34.65736$

36. Solve $300 = \dfrac{400}{1 + e^{-0.2t}}$ for t. [5.4] $-5\ln\left(\dfrac{1}{3}\right) = 5\ln 3 \approx 5.49306$

Explorations

1. *A Modeling Exploration* The purpose of this Exploration is for you to find data that can be modeled by an exponential or a logarithmic function. Choose data from a *real-life* situation that you find interesting. Search for the data in a magazine, newspaper, or almanac, or on the Internet. If you wish, you can collect your data by performing an experiment. Responses will vary.

a. List the source of your data. Include the date, page number, and any other specifics about the source. If your data were collected by performing an experiment, then provide all the details of the experiment.

b. Explain what you have chosen as your variables. Which variable is the dependent variable and which variable is the independent variable?

c. Use the three-step modeling process to find a regression equation that models the data.

d. Graph the regression equation on the scatter plot of the data. What is the regression coefficient for the model? Do you think that your regression equation accurately models your data? Explain.

e. Use the regression equation to predict the value of

- the dependent variable for a specific value of the independent variable.
- the independent variable for a specific value of the dependent variable.

f. Write a few comments about what you have learned from this Exploration.

26. The graph of the logarithmic regression equation passes through both of the data points.

28. b. Use a horizontal translation. For instance, add 1 to each of the x-coordinates. Find the logarithmic regression function for this new data set. Remember that each x-value in the regression function represents $x - 1$ in the original data.

SECTION 5.7 Logistic Growth Models and Their Applications

- Logistic Growth Models
- Applications of Logistic Growth Models
- Use Regression to Find a Logistic Growth Model
- The Inflection Point Theorem

Point of Interest

Pierre Verhulst (1804–1849) worked in many areas of mathematics, but he is best known for his research on population growth. He showed that forces that tend to prevent the growth of a biological population grow in proportion to the ratio of the "excess" population to the total population.

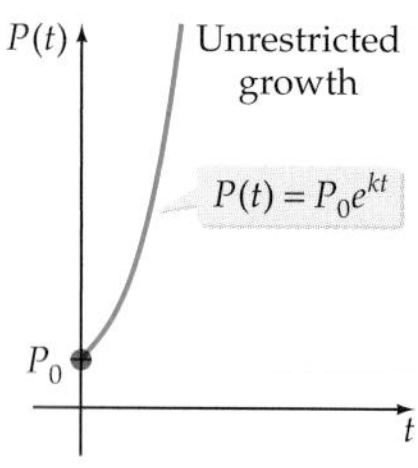

Malthusian growth model

Figure 5.35

TAKE NOTE

In logistic growth models the value $P(0)$ is often written as P_0.

Logistic Growth Models

The population growth function $P(t) = P_0e^{kt}$ is called a **Malthusian growth model,** after Robert Malthus (1766–1834), who wrote about population growth in *An Essay On the Principle of Population Growth,* which was published in 1798. The Malthusian model does not consider that there are limited resources and that these limitations will eventually curb population growth. A growth model that takes into consideration the effects of limited resources is the **logistic growth model,** which was developed by Pierre Verhulst in 1836.

The Logistic Growth Model

The magnitude of a population at time $t \geq 0$ is given by

$$P(t) = \frac{c}{1 + ae^{-bt}}$$

where b and c are positive constants.

Figure 5.35 illustrates the typical type of growth displayed by a Malthusian growth model. Figures 5.36a., 5.36b., and 5.36c. illustrate the three different types of growth that can be modeled by a logistic growth model.

a.

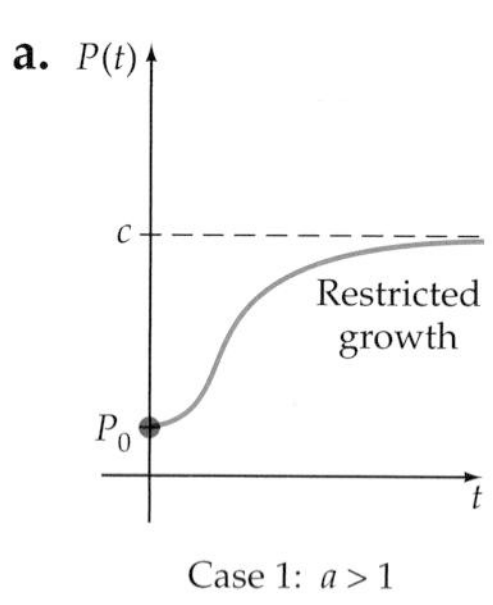

Case 1: $a > 1$

b.

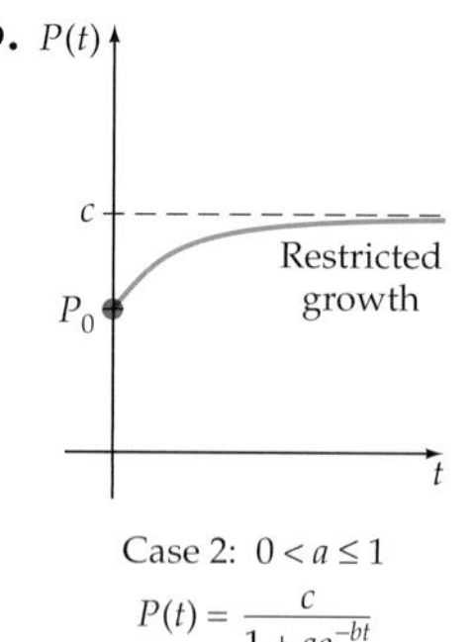

Case 2: $0 < a \leq 1$

c.

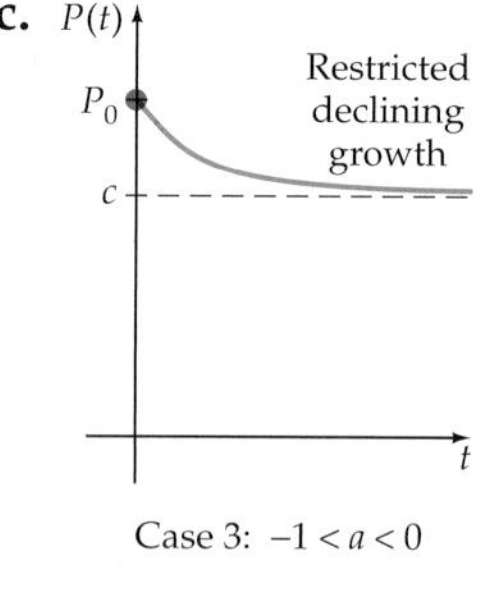

Case 3: $-1 < a < 0$

$P(t) = \frac{c}{1 + ae^{-bt}}$

Logistic growth models

Figure 5.36

Applications of Logistic Growth Models

In Example 1 we use a logistic growth model to approximate the spread of a virus.

Alternative to Example 1

The wolf population in a reserve is given by

$$P(t) = \frac{600}{1 + 1.5e^{-0.06t}}$$

where t is the time, in years, after January 1, 2003. According to the logistic growth model, in what year will the wolf population reach 300?

■ **2009**

EXAMPLE 1 Model the Spread of a Virus

In a small city, a flu virus starts by infecting five people. The logistic growth model

$$P(t) = \frac{3500}{1 + 699e^{-0.45t}}, \; t \geq 0$$

approximates the number of people in the city who have been infected by the virus t days after the initial outbreak.

a. Use $P(t)$ to find how many people are infected with the virus 10 days after the initial outbreak.

b. How many days after the initial outbreak will 3000 people have become infected?

Solution

a. When $t = 10$ we have

$$P(10) = \frac{3500}{1 + 699e^{-0.45(10)}} = \frac{3500}{1 + 699e^{-4.5}} \approx 399$$

After 10 days, about 399 people have been infected. See Figure 5.37.

b. Replace $P(t)$ with 3000 and solve for t.

$$3000 = \frac{3500}{1 + 699e^{-0.45t}} \quad \text{■ Substitute 3000 for } P(t).$$

$$1 + 699e^{-0.45t} = \frac{3500}{3000} \quad \text{■ Solve for } t.$$

$$699e^{-0.45t} = \frac{7}{6} - 1$$

$$e^{-0.45t} \approx 0.000238$$

$$-0.45t \approx \ln(0.000238)$$

$$-0.45t \approx -8.34141$$

$$t \approx 18.5$$

Therefore, 3000 people will have become infected by the 18th day of the virus outbreak. See Figure 5.37.

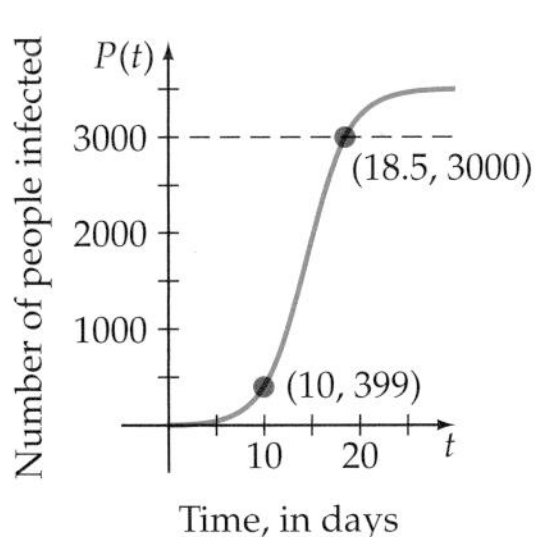

Figure 5.37

CHECK YOUR PROGRESS 1 In 1980, only 625 households in a town had a VCR. The logistic growth model

$$P(t) = \frac{6500}{1 + 9.4e^{-0.35t}}$$

approximates the number of households with a VCR t years after 1980. Use the model to estimate

a. how many households had a VCR in 1988. Round to the nearest 100.

b. in what year the number of households with a VCR first reached 5500.

Solution *See page S29.* **a.** 4100 households **b.** 1991

TAKE NOTE

The logistic growth model

$$P(t) = \frac{800}{4 + 8e^{-0.6t}}$$

does not have a carrying capacity of 800. To find the carrying capacity, multiply the numerator and denominator of $P(t)$ by $\frac{1}{4}$ to produce

$$P(t) = \frac{200}{1 + 2e^{-0.6t}}$$

which is now in the form

$$P(t) = \frac{c}{1 + ae^{-bt}}.$$

Now, by comparing constants, we see that the carrying capacity c is 200 and that $a = 2$.

Alternative to Example 2
Exercise 24, page 474

QUESTION Are all logistic growth functions increasing functions?

The constant c in a logistic growth model is referred to as the **carrying capacity** of P because $P(t)$ approaches c as t increases without bound. In Example 1, the value of c is 3500. Thus as the time t increases, the number of people who become infected approaches, but never exceeds, 3500.

The constant b in the logistic growth model is the **growth rate constant.** Reducing the size of the growth rate constant causes the population $P(t)$ to grow more slowly.

The population at time $t = 0$ is denoted by P_0. By letting $t = 0$ in the equation $P_0 = \frac{c}{1 + ae^{-bt}}$, we can see that $P_0 = \frac{c}{1 + a}$. Solving $P_0 = \frac{c}{1 + a}$ for a gives us $a = \frac{c - P_0}{P_0}$.

If we know any two of the three constants P_0, a, and c, we can use one of the above equations to find the unknown constant. In Example 2, we make use of the above relationships to determine a logistic growth model for the given data.

EXAMPLE 2 Find a Logistic Growth Model for the World's Population

In 1987, the world's population first reached 5 billion, and in 1999 it reached 6 billion. Assume that the carrying capacity of the Earth is 14 billion.

a. Use the above data to determine the growth rate constant and the logistic growth model that approximates the world's population.

b. Use the logistic model from part **a.** to predict the year in which the world's population will first reach 8 billion.

Solution

a. We represent the year 1987 by $t = 0$. The year 1999 is then represented by $t = 12$. Because $P_0 = 5$ and $c = 14$, we know

$$a = \frac{c - P_0}{P_0} = \frac{14 - 5}{5} = 1.8$$

Substitute to find the logistic growth model.

$$P(12) = \frac{14}{1 + 1.8e^{-b \cdot 12}}$$ ▪ Substitute 1.8 for a, 14 for c, and 12 for t.

$$6 = \frac{14}{1 + 1.8e^{-b \cdot 12}}$$ ▪ Replace $P(12)$ with 6.

$$6(1 + 1.8e^{-b \cdot 12}) = 14$$ ▪ Solve for b.

$$6 + 10.8e^{-b \cdot 12} = 14$$

$$10.8e^{-b \cdot 12} = 8$$

Continued ➤

ANSWER No. The logistic growth function $P(t) = \frac{c}{1 + ae^{-bt}}$, $-1 < a < 0$, is a decreasing function.

$$e^{-b \cdot 12} = \frac{8}{10.8}$$

$$-b \cdot 12 = \ln\left(\frac{8}{10.8}\right)$$

$$b = -\frac{1}{12}\ln\left(\frac{8}{10.8}\right)$$

$$b \approx 0.02501$$ ▪ **The growth rate constant**

For the given data, the logistic growth model of the world's population is $P(t) \approx \dfrac{14}{1 + 1.8e^{-0.02501t}}$

b. To determine the year in which the logistic growth model predicts that the world's population will first reach 8 billion, replace $P(t)$ with 8 and solve for t.

$$8 = \frac{14}{1 + 1.8e^{-0.02501t}}$$ ▪ **Let $P(t) = 8$.**

$$8 \cdot (1 + 1.8e^{-0.02501t}) = 14$$ ▪ **Solve for t.**

$$8 + 14.4e^{-0.02501t} = 14$$

$$14.4e^{-0.02501t} = 6$$

$$e^{-0.02501t} = \frac{6}{14.4}$$

$$-0.02501t = \ln\left(\frac{6}{14.4}\right)$$

$$t = \frac{1}{-0.02501}\ln\left(\frac{6}{14.4}\right)$$

$$t \approx 35.005$$

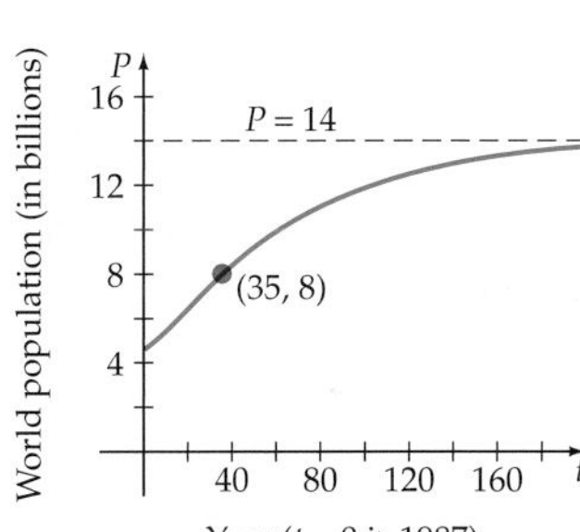

$P(t) = \dfrac{14}{1 + 1.8e^{-0.02501t}}$

Figure 5.38

Thus, according to the logistic growth model, the world's population will first reach 8 billion people in a little more than 35 years after 1987—that is, in 2022. The graph of $P(t) = 14/(1 + 1.8e^{-0.02501t})$ is shown in Figure 5.38. Note that when $t = 35$, $P \approx 8$ billion. Also note that as $t \to \infty$, the graph approaches the horizontal asymptote $P = 14$.

CHECK YOUR PROGRESS 2 In 1940, the population of San Francisco was 635,000, and in 1950 its population reached 775,000. Assume that the carrying capacity of San Francisco is 1 million.

a. Use the above data to determine the logistic growth model that can be used to approximate San Francisco's population. Let $t = 0$ represent the year 1940.

b. Use the logistic growth model from part **a.** to predict the year in which San Francisco's population will reach 995,000.

Solution *See page S29.* **a.** $P(t) \approx \dfrac{1{,}000{,}000}{1 + 0.574803e^{-0.068303t}}$ **b.** 2009

▪ Use Regression to Find a Logistic Growth Model

If a scatter plot of a set of data suggests that the data can be effectively modeled by a logistic growth model, then you can use the logistic regression feature of a graphing utility to find the logistic growth model that provides the best fit for the data. Essentially we use the same three-step modeling process that we used in Section 5.6, where we determined exponential and logarithmic regression models. This process is illustrated in Example 3.

Alternative to Example 3
Exercise 28, page 475

EXAMPLE 3 Use Logistic Regression to Find a Logistic Growth Model

Table 5.16 shows the population of deer in an animal preserve for the years 1985 to 1999.

Table 5.16 Deer Population at the Wild West Animal Preserve

Year	Population	Year	Population	Year	Population
1985	320	1990	1150	1995	2620
1986	410	1991	1410	1996	2940
1987	560	1992	1760	1997	3100
1988	730	1993	2040	1998	3300
1989	940	1994	2310	1999	3460

Use a graphing utility to find a logistic regression model that approximates the deer population as a function of the year. Use the model to predict the deer population in the year 2005.

INTEGRATING TECHNOLOGY

On a TI-83 graphing calculator, the logistic growth model is given in the form

$$y = \frac{c}{1 + ae^{-bx}}$$

Think of the variable x as the time t and the variable y as $P(t)$.

Solution

1. **Construct a scatter plot of the data.** Enter the data into a graphing utility and then use the utility to display a scatter plot of the data. In this example we represent the year 1985 by $x = 0$, the year 1999 by $x = 14$, and the deer population by y.

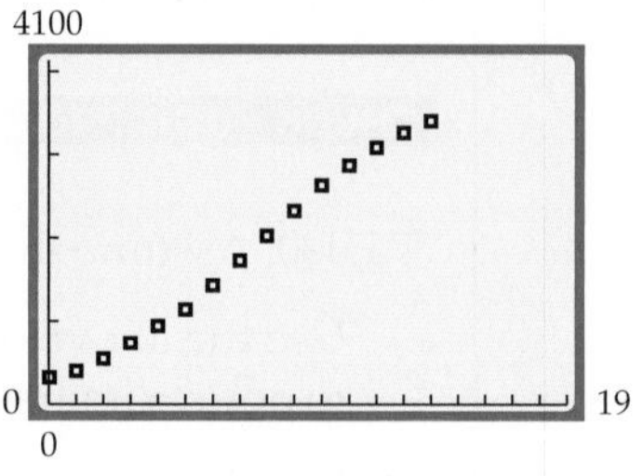

Figure 5.39

Figure 5.39 shows that the data can be closely approximated by a logistic growth model.

Continued ➤

2. **Find the regression function.** Use a graphing utility to perform a logistic regression on the data. The logistic regression function for the data is $y \approx \dfrac{3965.3}{1 + 11.445e^{-0.31152x}}$. See Figure 5.40.

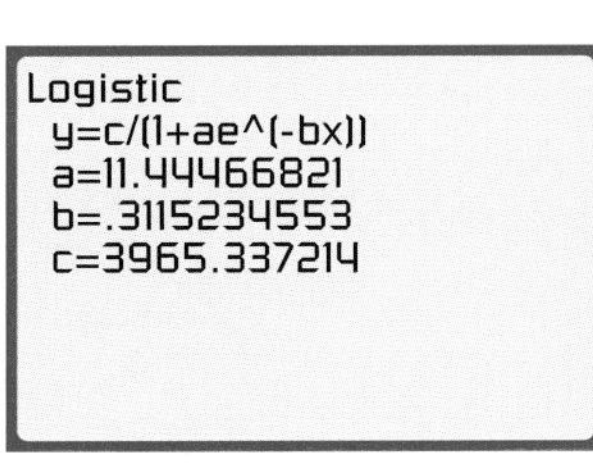

Figure 5.40

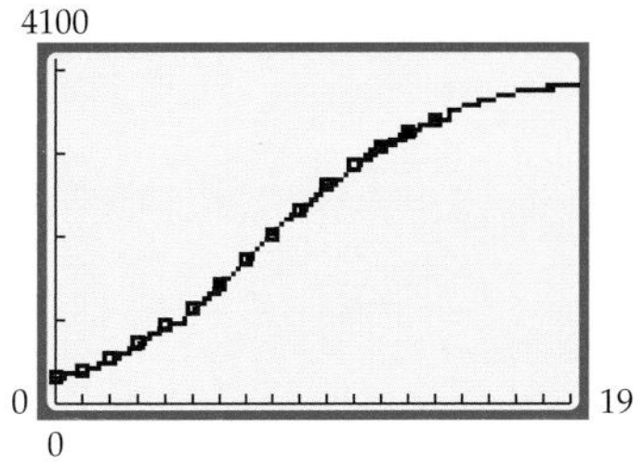

4100

0 19

0

Figure 5.41

3. **Examine the fit.** A TI-83 calculator does not give a correlation coefficient for a logistic regression. However, Figure 5.41 shows that the graph of $y \approx \dfrac{3965.3}{1 + 11.445e^{-0.31152x}}$ provides a good fit to the data. The logistic model predicts that in the year 2005 ($x = 20$), the deer population will be about 3878.

$$y \approx \frac{3965.3}{1 + 11.445e^{-0.31152(20)}} \approx 3878$$

CHECK YOUR PROGRESS 3  Table 5.17 shows the population of the state of Hawaii for selected years from 1950 to 2001.

Table 5.17 Population of the State of Hawaii

Year	Population	Year	Population	Year	Population
1950	499,000	1970	762,920	1990	1,113,491
1955	529,000	1975	875,052	1995	1,196,854
1960	642,000	1980	967,710	2000	1,212,281
1965	704,000	1985	1,039,698	2001	1,224,398

Source: economagic.com
(**http://www.economagic.com/em-cgi/data.exe/beapi/a15300**)

a. Use the logistic regression feature of a graphing utility to find a logistic growth model that approximates the population of the state of Hawaii as a function of the year. Use $t = 0$ to represent the year 1950.

b. Use the model from part **a.** to predict the population of the state of Hawaii in the year 2010. Round to the nearest ten thousand.

c. What is the carrying capacity of the model? Round to the nearest thousand.

Solution *See page S30.* **a.** $P(t) \approx \dfrac{1{,}544{,}022}{1 + 2.249350e^{-0.433335t}}$ **b.** 1,320,000 **c.** 1,544,000

The Inflection Point Theorem

A point P on a graph is called an **inflection point** if the graph changes from concave upward to concave downward or from concave downward to concave upward at P. Some logistic growth models have an inflection point and some do not. See Figures 5.42a. and 5.42b.

a.

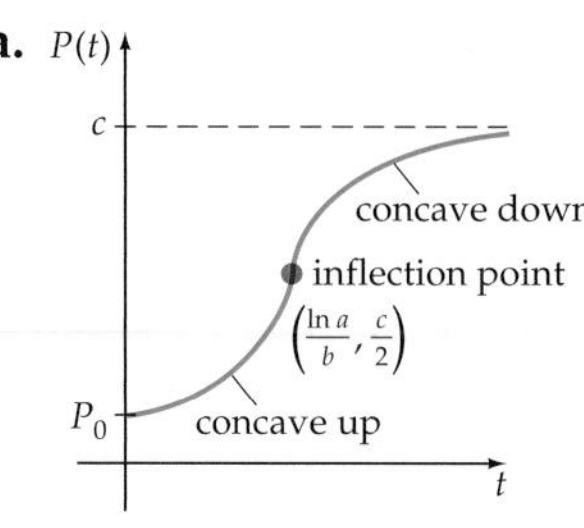

Inflection point on the graph of $P(t) = \frac{c}{1 + ae^{-bt}}$ with $a > 1$

b.

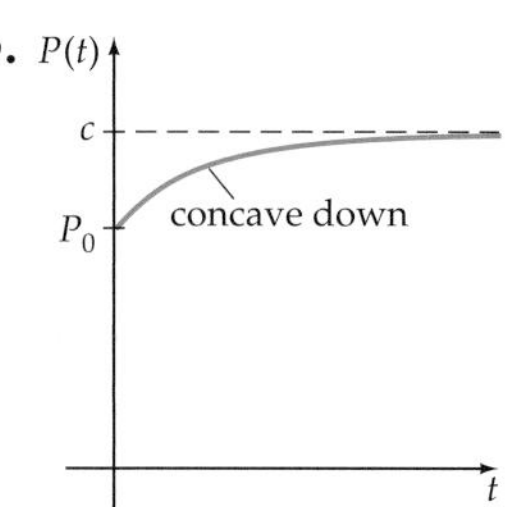

No inflection point on the graph of $P(t) = \frac{c}{1 + ae^{-bt}}$ with $0 < a \leq 1$

 Figure 5.42

The following theorem can be used to determine whether a logistic growth model has a point of inflection. For those logistic growth models that have an inflection point, it also gives the coordinates of the point.

The Inflection Point Theorem

The graph of the logistic growth model

$$P(t) = \frac{c}{1 + ae^{-bt}}, t \geq 0$$

has an inflection point at $\left(\frac{\ln a}{b}, \frac{c}{2}\right)$, provided $a > 1$. For a logistic growth model that has an inflection point, the t-value of the inflection point, $\frac{\ln a}{b}$, is the time at which the *largest growth rate occurs.*

Alternative to Example 4
Exercise 22, page 474

EXAMPLE 4 Find the Inflection Point of a Logistic Growth Model

Determine whether the graph of the logistic growth model of the world's population, as determined in Example 2, has an inflection point. If it has an inflection point, name the inflection point and explain the significance of the point.

Solution From Example 2 we know that $a = 1.8$, which is greater than 1. Thus by the Inflection Point Theorem, the graph of $P(t) = \frac{14}{1 + 1.8e^{-0.02501t}}$ has an inflection point. The inflection point is

$$\left(\frac{\ln a}{b}, \frac{c}{2}\right) = \left(\frac{\ln 1.8}{0.02501}, \frac{14}{2}\right) \approx (23.5, 7)$$

Continued ➤

Logistic growth models experience their largest growth rate at the t-value of the inflection point. Therefore, the model in Example 2 predicts that the world population will experience its largest growth rate 23.5 years after 1987—that is, in the year 2010.

CHECK YOUR PROGRESS 4 Determine whether the graph of the logistic growth model of San Francisco's population, as determined in Check Your Progress 2, has an inflection point. If it has an inflection point, name the inflection point and explain the significance of the point.

Solution *See page S30.* The logistic growth model does not have an inflection point. This indicates that the population of San Francisco reached its greatest rate of growth before 1940.

Topics for Discussion

1. Consider the logistic growth model $g(t) = \dfrac{14}{1 + 0.5e^{-0.1t}}$.
   a. Find $g(0)$.
   b. What does $g(t)$ approach as $t \to \infty$?

2. Does the graph of the logistic growth model $P(t) = \dfrac{2200}{1 + 1.5e^{-0.24t}}$, $t \geq 0$ have an inflection point? Explain.

3. The logistic growth model $P(t) = \dfrac{25}{1 + 0.2e^{-0.3t}}$ and the logistic growth model $K(t) = \dfrac{25}{1 + 0.2e^{-0.4t}}$ are the same except for their growth rate constants. Explain the major difference between their graphs.

4. Explain the difference between the procedures used in Example 2 and the procedures used in Example 3.

7. $P(t) \approx \dfrac{5500}{1 + 12.75e^{-0.37263t}}$

8. $P(t) \approx \dfrac{9500}{1 + 0.53226e^{-0.05675t}}$

9. $P(t) \approx \dfrac{100}{1 + 4.55556e^{-0.22302t}}$

EXERCISES 5.7

— *Suggested Assignment: Exercises 1–37, odd.*

In Exercises 1 to 6, determine the following constants for the given logistic growth model.

a. **The carrying capacity**

b. **The growth rate constant**

c. P_0

1. $P(t) = \dfrac{1900}{1 + 8.5e^{-0.16t}}$
   a. 1900 b. 0.16 c. 200

2. $P(t) = \dfrac{32{,}550}{1 + 0.75e^{-0.08t}}$
   a. 32,550 b. 0.08 c. 18,600

3. $P(t) = \dfrac{157{,}500}{1 + 2.5e^{-0.04t}}$
   a. 157,500 b. 0.04 c. 45,000

4. $P(t) = \dfrac{51}{1 + 1.04e^{-0.03t}}$
   a. 51 b. 0.03 c. 25

5. $P(t) = \dfrac{2400}{1 + 7e^{-0.12t}}$
   a. 2400 b. 0.12 c. 300

6. $P(t) = \dfrac{320}{1 + 15e^{-0.12t}}$
   a. 320 b. 0.12 c. 20

In Exercises 7 to 12, use algebraic procedures to find the logistic growth model for the data.

7. $P_0 = 400$, $P(2) = 780$, and the carrying capacity is 5500.

8. $P_0 = 6200$, $P(8) = 7100$, and the carrying capacity is 9500.

9. $P_0 = 18$, $P(3) = 30$, and the carrying capacity is 100.

10. $P_0 = 3200$, $P(22) \approx 5565$, and the growth rate constant is 0.056. $P(t) \approx \dfrac{8000}{1 + 1.5e^{-0.056t}}$

11. $P_0 = 500$, $P(8) \approx 3758$, and the growth rate constant is 0.45. $P(t) \approx \dfrac{4600}{1 + 8.2e^{-0.45t}}$

12. $P_0 = 1200$, $P(4) \approx 1612$, and $a = 14$. $P(t) \approx \dfrac{18{,}000}{1 + 14e^{-0.08t}}$

15. $P(t) \approx \dfrac{799.91097}{1 + 14.23484e^{-0.75065t}}$

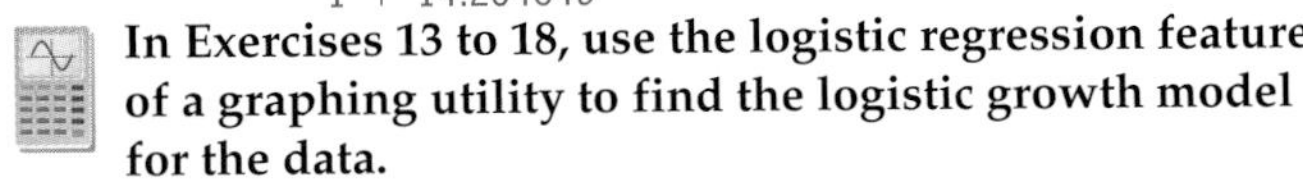

In Exercises 13 to 18, use the logistic regression feature of a graphing utility to find the logistic growth model for the data.

13. {(10, 268), (14, 380), (16, 425), (18, 490), (21, 505)} $P(t) \approx \dfrac{563.96280}{1 + 10.07473e^{-0.21933t}}$

14. {(0, 25.22), (1, 47.81), (2, 65.55), (3, 73.84), (4, 76.71)} $P(t) \approx \dfrac{78.06375}{1 + 2.09609e^{-1.19816t}}$

15. {(1.6, 151.2), (2.5, 251.8), (4, 468.6), (8, 772.7), (12, 798.6)}

16. {(0.5, 466), (2, 872), (3, 1225), (6, 2191), (10, 2626)} $P(t) \approx \dfrac{2691.43401}{1 + 6.29713e^{-0.55296t}}$

17. {(0, 42), (10, 54), (20, 59), (30, 67), (40, 72)} $P(t) \approx \dfrac{80.91413}{1 + 0.89461e^{-0.04857t}}$

18. {(0, 1.42), (2, 1.86), (3, 2.12), (4, 2.36)} $P(t) \approx \dfrac{4.94779}{1 + 2.48821e^{-0.20545t}}$

19. Find the inflection point of the graph defined by

$$P(t) = \frac{80}{1 + 5e^{-0.034t}}. \qquad \approx(47.34, 40)$$

20. Find the inflection point of the graph defined by

$$P(t) = \frac{8500}{1 + 2e^{-0.05t}}. \qquad \approx(13.86, 4250)$$

Business and Economics

21. *Revenue* The annual revenue, in dollars, of a new company can be closely modeled by the logistic growth function

$$R(t) = \frac{625{,}000}{1 + 3.1e^{-0.045t}}$$

where t is the time in years since the company was founded.

a. According to the model, what will the company's annual revenue approach in the long-term future? \$625,000

b. According to the model, how many years will it take for the company to achieve its largest growth rate? Round to the nearest hundredth of a year. 25.14 yr

22. 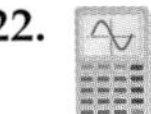 *New Car Sales* The number of cars sold annually by an automobile dealership can be closely modeled by the logistic growth function

$$A(t) = \frac{1650}{1 + 2.4e^{-0.055t}}$$

where t is the time in years since the company was founded.

a. According to the model, what will the annual number of car sales approach in the long-term future? 1650 cars

b. According to the model, how many years, to the nearest year, will it take for the dealership to achieve its largest growth rate? 16 yr

Life and Health Sciences

23. 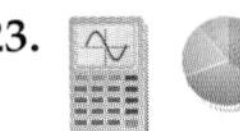 *World Population* The following table lists the years in which the world's population first reached 3, 4, 5, and 6 billion.

World Population Milestones

1960	3 billion
1974	4 billion
1987	5 billion
1999	6 billion

Source: Time Almanac 2002, page 708.

a. Find a logistic growth model, $P(t)$, for the data in the table. Let t represent the number of years after 1960 ($t = 0$ represents the year 1960). $P(t) \approx \dfrac{11.26828}{1 + 2.74965e^{-0.02924t}}$

b. According to the logistic growth model, what will the world's population approach as $t \to \infty$? Round to the nearest billion. 11 billion people

24. *Logistic Growth* The population of wolves in a reserve satisfies a logistic growth model in which $P_0 = 312$ in the year 1999, $c = 1600$, and $P(6) = 416$.

a. Determine the logistic growth model for the population, where t is the number of years after 1999.

b. Use the logistic growth model to predict the size of the wolf population in 2009. 497 wolves

25. *Logistic Growth* The population of groundhogs on a ranch satisfies a logistic growth model in which $P_0 = 240$ in the year 1998, $c = 3400$, and $P(1) = 310$.

24. a. $P(t) \approx \dfrac{1600}{1 + 4.12821e^{-0.06198t}}$

a. Determine the logistic growth model, $P(t)$, for the population, where the year 1998 is represented by $t = 0$. $P(t) \approx \dfrac{3400}{1 + 13.16667e^{-0.27833t}}$

b. Use the logistic growth model from part **a.** to predict during what year the groundhog population will experience its most rapid growth. 2007

If $-1 < a < 0$, then the logistic growth function $P(t) = \dfrac{c}{1 + ae^{-bt}}$ decreases as t increases. Biologists often use this type of logistic growth function to model populations that decrease over time. See Figure 5.36c. Exercises 26 and 27 involve declining populations.

26. ***A Declining Fish Population*** A biologist finds that the fish population in a small lake can be closely modeled by the logistic growth function

$$P(t) = \frac{1000}{1 + (-0.3333)e^{-0.05t}}$$

where t is the time, in years, since the lake was first stocked with fish.

a. What was the fish population when the lake was first stocked with fish? 1500 fish

b. According to the logistic growth function, what will the fish population approach in the long-term future? 1000 fish

27. ***A Declining Deer Population*** The deer population in a reserve is given by the logistic growth model

$$P(t) = \frac{1800}{1 + (-0.25)e^{-0.07t}}$$

where t is the time, in years, since July 1, 1995.

a. What was the deer population on July 1, 1995? 2400 deer

b. According to the logistic growth function, what will the deer population approach in the long-term future? 1800 deer

Sports and Recreation

28.  ***Olympic Distances*** The following table shows the winning Olympic distances for the men's shot put for the years 1948 to 2000.

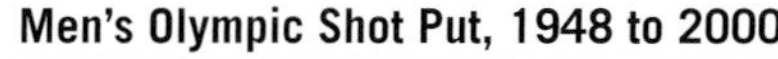

Men's Olympic Shot Put, 1948 to 2000

Year	Distance	Year	Distance
1948	56 ft 2 in.	1976	69 ft $\frac{3}{4}$ in.
1952	57 ft $1\frac{1}{2}$ in.	1980	70 ft $\frac{1}{2}$ in.
1956	60 ft 11 in.	1984	69 ft 9 in.
1960	64 ft $6\frac{3}{4}$ in.	1988	73 ft $8\frac{3}{4}$ in.
1964	66 ft $8\frac{1}{4}$ in.	1992	71 ft $2\frac{1}{2}$ in.
1968	67 ft $4\frac{3}{4}$ in.	1996	70 ft $11\frac{1}{4}$ in.
1972	69 ft 6 in.	2000	69 ft $10\frac{1}{4}$ in.

Source: Time Almanac 2002

Represent the year 1948 by $t = 48$.

a. Use the regression features of a graphing utility to determine a logistic growth model and a logarithmic model for the data.

b. Use graphs of the models in part **a.** to determine which model provides the better fit for the data. Logistic growth model

c. Use the model you selected in part **b.** to predict the men's shot put distance for the year 2008. Round to the nearest hundredth of a foot. 71.65 ft

Physical Sciences and Engineering

29. ***Relationships Between Models*** In some science courses the logistic growth model is defined as

$$P(t) = \frac{mP_0}{P_0 + (m - P_0)e^{-kt}} \qquad \text{(I)}$$

whereas we have defined the logistic growth model as

$$P(t) = \frac{c}{1 + ae^{-bt}} \qquad \text{(II)}$$

To determine a relationship between the constants used in the two definitions, divide the numerator and denominator of equation (I) by P_0. The compare your results with equation (II).

For the two definitions,

a. what is the relationship between m and c? $m = c$

b. what is the relationship between k and b? $k = b$

28. a. Logistic: distance in ft $\approx \dfrac{71.84158}{1 + 13.77825e^{-0.07915t}}$; logarithmic: distance in ft $\approx -22.58293 + 20.91655 \ln t$

c. what does a equal in terms of m and P_0? $a = \dfrac{m - P_0}{P_0}$

In Exercises 30 to 33, each logistic growth function is given in $P(t) = \dfrac{c}{1 + ae^{-bt}}$ form. For each function, determine its equivalent $P(t) = \dfrac{mP_0}{P_0 + (m - P_0)e^{-kt}}$ form.

30. $P(t) = \dfrac{500}{1 + 1.5e^{-0.06t}}$

31. $P(t) = \dfrac{180}{1 + 2e^{-0.11t}}$

32. $P(t) = \dfrac{2000}{1 + 3e^{-0.25t}}$

$P(t) = \dfrac{1{,}000{,}000}{500 + 1500e^{-0.25t}}$

33. $P(t) = \dfrac{80}{1 + \left(\dfrac{1}{3}\right)e^{-0.2t}}$

$P(t) = \dfrac{4800}{60 + 20e^{-0.2t}}$

In Exercises 34 to 37, each logistic growth function is given in $P(t) = \dfrac{mP_0}{P_0 + (m - P_0)e^{-kt}}$ form. For each function, determine its equivalent $P(t) = \dfrac{c}{1 + ae^{-bt}}$ form.

34. $P(t) = \dfrac{1200}{20 + 40e^{-0.05t}}$ $P(t) = \dfrac{60}{1 + 2e^{-0.05t}}$

35. $P(t) = \dfrac{110{,}000}{200 + 350e^{-0.17t}}$ $P(t) = \dfrac{550}{1 + 1.75e^{-0.17t}}$

36. $P(t) = \dfrac{1{,}230{,}000}{600 + 1450e^{-0.09t}}$ $P(t) = \dfrac{2050}{1 + \left(\dfrac{29}{12}\right)e^{-0.09t}}$

37. $P(t) = \dfrac{2975}{35 + 50e^{-0.34t}}$ $P(t) = \dfrac{85}{1 + \left(\dfrac{10}{7}\right)e^{-0.34t}}$

38. Some science texts state the Inflection Point Theorem as follows.

$$P(t) = \frac{c}{1 + ae^{-bt}}, t \geq 0$$

has an inflection point at $\left(\dfrac{\ln a}{b}, \dfrac{c}{2}\right)$ if and only if $P_0 < \dfrac{c}{2}$. Show that the condition $a > 1$ is equivalent to the condition $P_0 < \dfrac{c}{2}$.

30. $P(t) = \dfrac{100{,}000}{200 + 300e^{-0.06t}}$ 31. $P(t) = \dfrac{10{,}800}{60 + 120e^{-0.11t}}$

Explorations 

1. ***A Modeling Exploration*** The purpose of this Exploration is for you to find data that can be modeled by a logistic growth model. Choose data from a *real-life* situation that you find interesting. Search for the data in a magazine, newspaper, or almanac, or on the Internet. Responses will vary.

   a. List the source of your data. Include the date, page number, and any other specifics about the source.

   b. Explain what you have chosen as your variables. Which variable is the dependent variable and which variable is the independent variable?

   c. Use the three-step modeling process to find a regression equation that models the data.

   d. Graph the regression equation on the scatter plot of the data. Do you think that your regression equation accurately models your data? Explain.

   e. Use the regression equation to predict the value of

   - the dependent variable for a specific value of the independent variable.
   - the independent variable for a specific value of the dependent variable.

   f.  Write a few comments about what you have learned from this Exploration.

38. $a > 1 \Rightarrow \dfrac{m - P_0}{P_0} > 1 \Rightarrow m - P_0 > P_0 \Rightarrow m > 2P_0$ or $P_0 < \dfrac{m}{2}$. Because $m = c$ we have $P_0 < \dfrac{c}{2}$.

Chapter 5 Summary

Key Terms

acid **[p. 421]**
acidity of a solution **[p. 420]**
alkaline solution **[p. 421]**
apparent magnitude **[p. 411]**
base **[p. 421]**
Benford's Law **[p. 412]**
biological diversity **[p. 452]**
carbon dating **[p. 446]**
carrying capacity **[p. 468]**
catenary **[p. 399]**
common logarithm **[p. 407]**
concave upward (downward) **[p. 454]**
continuous growth rate **[p. 440]**
decibel level **[p. 424]**
e (base of natural exponential function) **[p. 390]**
exponential decay **[p. 440]**
exponential equation **[p. 427]**
exponential form **[p. 401]**
exponential function **[p. 385]**
exponential growth **[p. 440]**
factorial function **[p. 397]**
growth rate constant **[p. 468]**
half-life **[p. 444]**
inflection point **[p. 472]**
logarithm **[p. 401]**
logarithmic equation **[p. 431]**
logarithmic form **[p. 401]**
logarithmic function **[p. 401]**
logarithmic scale **[pp. 417/426]**
logistic growth model **[p. 466]**
Malthusian growth model **[p. 466]**
natural exponential function **[p. 391]**
natural logarithm **[p. 407]**
Newton's Law of Cooling **[p. 446]**
nomogram **[p. 425]**
pH of a solution **[p. 421]**
power function **[p. 464]**
p-wave **[p. 419]**
Richter scale magnitude of an earthquake **[p. 417]**
s-wave **[p. 420]**
zero-level earthquake **[p. 417]**

Essential Concepts and Formulas

- **Exponential Functions**
  - For $b > 0$ and $b \neq 1$, the exponential function $f(x) = b^x$ has the following properties:
    f has the set of real numbers as its domain and the set of positive real numbers as its range.
    f has a graph with a y-intercept of $(0, 1)$, and the graph passes through $(1, b)$.
    f is a one-to-one function, and the graph of f is a smooth continuous curve that is asymptotic to the x-axis.
    f is an increasing function if $b > 1$.
    f is a decreasing function if $0 < b < 1$. **[p. 387]**
  - As n increases without bound, $\left(1 + \frac{1}{n}\right)^n$ approaches an irrational number denoted by e. The value of e accurate to eight decimal places is 2.71828183. **[p. 390]**
  - The function defined by $f(x) = e^x$ is called the natural exponential function. **[p. 391]**

- **Logarithmic Functions**
  - *Definition of a Logarithm* If $x > 0$ and b is a positive constant ($b \neq 1$), then $y = \log_b x$ if and only if $b^y = x$. **[p. 401]**
  - The *exponential form* of $y = \log_b x$ is $b^y = x$. **[p. 401]**
  - The *logarithmic form* of $b^y = x$ is $y = \log_b x$. **[p. 401]**
  - *Basic Logarithmic Properties* **[p. 402]**
    $\log_b b = 1 \qquad \log_b 1 = 0 \qquad \log_b b^x = x \qquad b^{\log_b x} = x$
  - For all positive real numbers b, $b \neq 1$, the logarithmic function defined by $f(x) = \log_b x$ has the following properties:
    f has the set of positive real numbers as its domain and the set of real numbers as its range.
    f has a graph with an x-intercept of $(1, 0)$. The graph passes through $(b, 1)$.

f is a one-to-one function, and the graph of f is a smooth, continuous curve that is asymptotic to the y-axis.
f is an increasing function if $b > 1$.
f is a decreasing function if $0 < b < 1$. **[p. 405]**
- The function $f(x) = \log_{10} x$ is the *common logarithmic function.* It is customarily written as $f(x) = \log x$. **[p. 407]**
- The function $f(x) = \log_e x$ is the *natural logarithmic function.* It is customarily written as $f(x) = \ln x$. **[p. 407]**

Logarithms and Logarithmic Scales

- If b, M, and N are positive real numbers ($b \neq 1$), and p is any real number, then **[p. 414]**
  $\log_b(MN) = \log_b M + \log_b N$
  $\log_b\left(\frac{M}{N}\right) = \log_b M - \log_b N$
  $\log_b(M)^p = p \log_b M$
  If $\log_b M = \log_b N$, then $M = N$.
  If $M > 0$, $N > 0$, and $M = N$, then $\log_b M = \log_b N$.
- *Change-of-Base Formula* If x, a, and b are positive real numbers with $a \neq 1$ and $b \neq 1$, then **[p. 416]**
  $$\log_b x = \frac{\log_a x}{\log_a b}$$
- An earthquake with an intensity of I has a *Richter scale magnitude* of $M = \log\left(\frac{I}{I_0}\right)$, where I_0 is the measure of the intensity of a zero-level earthquake. **[p. 417]**
- The pH of a solution with a hydronium-ion concentration of H^+ moles per liter is given by $\text{pH} = -\log[H^+]$. **[p. 421]**

Exponential and Logarithmic Equations

- *Equality of Exponents Theorem* If b is a positive real number ($b \neq 1$) such that $b^x = b^y$, then $x = y$. **[p. 427]**
- Exponential equations of the form $b^x = b^y$ can be solved by using the Equality of Exponents Theorem. **[p. 427]**
- Exponential equations of the form $b^x = c$ can be solved by taking either the common logarithm or the natural logarithm of each side of the equation. **[p. 428]**
- Logarithmic equations can often be solved by using the properties of logarithms and the definition of a logarithm. In solving logarithmic equations, extraneous solutions may be introduced. **[p. 431]**

Exponential Growth and Decay

- The function defined by $N(t) = N_0 e^{kt}$ is called an *exponential growth function* if k is a positive constant and is called an *exponential decay function* if k is a negative constant. The constant k is the *continuous growth rate.* **[p. 440]**
- Radioactive decay is referred to in terms of *half-life,* which is defined as the time required for half of the atoms in an amount of radioactive material to disintegrate. **[p. 444]**
- *Newton's Law of Cooling* The rate at which objects change temperature is a percentage of the *difference* between the object's temperature and the temperature of its surroundings. If the temperature of an object at time t is given by T, T_0 is the initial temperature of the object, and A is the temperature of the room (or surrounding material), then $T(t) = A + (T_0 - A)e^{kt}$. **[p. 446]**

Modeling Data with Exponential and Logarithmic Functions

- If the graph of f lies above all of its tangents on an interval $[x_1, x_2]$, then f is *concave upward* on $[x_1, x_2]$. **[p. 454]**
- If the graph of f lies below all of its tangents on an interval $[x_1, x_2]$, then f is *concave downward* on $[x_1, x_2]$. **[p. 454]**
- *The Modeling Process* Use a graphing utility to:
  1. Construct a scatter plot of the data to determine which type of function will best model the data.
  2. Find the regression equation of the modeling function and the correlation coefficient for the regression.
  3. Examine the correlation coefficient, if possible, and view a graph that displays both the function and the scatter plot to determine how well the function fits the data. **[p. 456]**

Logistic Growth Models

- The logistic growth model is given by
  $$P(t) = \frac{c}{1 + ae^{-bt}}, \ t \geq 0$$
  where $P(t)$ is the population at time t, and b and c are positive constants. **[p. 466]**
- The constant c in a logistic growth model $P(t)$ is the *carrying capacity* of P. The constant b is the *growth rate constant.* **[p. 468]**
- A point on a graph is called an *inflection point* if the graph changes from concave upward to concave downward or from concave downward to concave upward at the point. **[p. 472]**
- The graph of the logistic growth model
  $$P(t) = \frac{c}{1 + ae^{-bt}}, \ t \geq 0$$
  has an inflection point at $\left(\frac{\ln a}{b}, \frac{c}{2}\right)$, provided $a > 1$. For a logistic growth model that has an inflection point, the t-value of the inflection point is the time at which the largest growth rate occurs. **[p. 472]**

Chapter 5 True/False Exercises

In Exercises 1 to 11, answer true or false. If the statement is false, explain why it is false or give an example to show the statement is false.

1. If $7^x = 40$, then $\log_7 40 = x$. True

2. If $\log_4 x = 3.1$, then $4^{3.1} = x$. True

3. If $f(x) = \log x$ and $g(x) = 10^x$, then $g[f(x)] = x$ for all real numbers x. False; because f is not defined for $x \le 0$, $g[f(x)]$ is not defined for $x \le 0$.

4. The exponential function $h(x) = b^x$ is an increasing function. False; $h(x)$ is not an increasing function for $0 < b < 1$.

5. The exponential growth function $N(t) = 560e^{1.1t}$ has a continuous growth rate of 560. False; the exponential growth function has a continuous growth rate of $1.1 = 110\%$.

6. An earthquake that measure 6.0 on the Richter scale is twice as intense as a quake measuring 3.0. False; an earthquake with a 6.0 Richter scale magnitude is 1000 times as intense as an earthquake with a 3.0 Richter scale magnitude.

7. A population that is growing exponentially at a continuous rate of 10% per year will double its population within 5 years. False; the population will double in about 6.9 years.

8. If a radioactive material has a half-life of 15 days, all of the material will decay in 30 days. False; in 30 days 75% of the material will decay.

9. If $x > 0$ and $y > 0$, then $\log(x + y) = \log x + \log y$. False; $\log x + \log y = \log(xy) \ne \log(x + y)$

10. The carrying capacity of the population given by $P(t) = \dfrac{14{,}000}{2 + 4e^{-0.5t}}$ is 14,000. False; the carrying capacity is $\dfrac{14{,}000}{2} = 7000$.

11. The graph of the logistic growth model $P(t) = \dfrac{7800}{1 + 0.4e^{-0.3t}}$, $t \ge 0$, has an inflection point. False; because $a < 1$, the given logistic growth model with $t \ge 0$ does not have an inflection point.

Chapter 5 Review Exercises

— *Answer graphs to Exercises 13–24 are on pages AA16–AA17.*

In Exercises 1 to 12, solve each equation. Do not use a calculator.

1. $\log_5 25 = x$ 2 [5.2/5.4]
2. $\log_3 81 = x$ 4 [5.2/5.4]
3. $\ln e^3 = x$ 3 [5.2/5.4]
4. $\ln e^{\pi} = x$ π [5.2/5.4]
5. $3^{2x+7} = 27$ -2 [5.4]
6. $5^{x-4} = 625$ 8 [5.4]
7. $2^x = \dfrac{1}{8}$ -3 [5.4]
8. $27(3^x) = 3^{-1}$ -4 [5.4]
9. $\log x^2 = 6$ ± 1000 [5.4]
10. $\dfrac{1}{2}\log |x| = 5$ $\pm 10^{10}$ [5.4]
11. $10^{\log 2x} = 14$ 7 [5.2/5.4]
12. $e^{\ln x^2} = 64$ ± 8 [5.2/5.4]

In Exercises 13 to 22, sketch the graph of each function.

13. $f(x) = (2.5)^x$ [5.1]
14. $f(x) = \left(\dfrac{1}{4}\right)^x$ [5.1]
15. $f(x) = 3^{|x|}$ [5.1]
16. $f(x) = 4^{-|x|}$ [5.1]
17. $f(x) = 2^x - 3$ [5.1]
18. $f(x) = 2^{(x-3)}$ [5.1]
19. $f(x) = \dfrac{1}{3}\log x$ [5.2]
20. $f(x) = 3 \log x^{1/3}$ [5.2]
21. $f(x) = -\dfrac{1}{2}\ln x$ [5.2]
22. $f(x) = -\ln |x|$ [5.2]

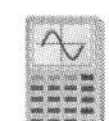

In Exercises 23 and 24, use a graphing utility to graph each function.

23. $f(x) = \dfrac{4^x + 4^{-x}}{2}$ [5.1]
24. $f(x) = \dfrac{3^x - 3^{-x}}{2}$ [5.1]

In Exercises 25 to 28, change each logarithmic equation to exponential form.

25. $\log_4 64 = 3$ $4^3 = 64$ [5.2]
26. $\log_{1/2} 8 = -3$ $\left(\dfrac{1}{2}\right)^{-3} = 8$ [5.2]
27. $\log_{\sqrt{2}} 4 = 4$ $\left(\sqrt{2}\right)^4 = 4$ [5.2]
28. $\ln 1 = 0$ $e^0 = 1$ [5.2]

In Exercises 29 to 32, change each exponential equation to logarithmic form.

29. $5^3 = 125$ $\log_5 125 = 3$ [5.2]

30. $2^{10} = 1024$ $\log_2 1024 = 10$ [5.2]

31. $10^0 = 1$ $\log 1 = 0$ [5.2]

32. $8^{1/2} = 2\sqrt{2}$ $\log_8(2\sqrt{2}) = \frac{1}{2}$ [5.2]

In Exercises 33 to 36, write the given logarithm in terms of logarithms of *x*, *y*, and *z*.

33. $\log_b \frac{x^2y^3}{z}$ $2\log_b x + 3\log_b y - \log_b z$ [5.3]

34. $\log_b \frac{\sqrt{x}}{y^2z}$ $\frac{1}{2}\log_b x - 2\log_b y - \log_b z$ [5.3]

35. $\ln xy^3$ $\ln x + 3\ln y$ [5.3]

36. $\ln \frac{\sqrt{xy}}{z^4}$ $\frac{1}{2}\ln x + \frac{1}{2}\ln y - 4\ln z$ [5.3]

In Exercises 37 to 40, write each logarithmic expression as a single logarithm with a coefficient of 1.

37. $2\log x + \frac{1}{3}\log(x+1)$ $\log\left(x^2\sqrt[3]{x+1}\right)$ [5.3]

38. $5\log x - 2\log(x+5)$ $\log\frac{x^5}{(x+2)^2}$ [5.3]

39. $\frac{1}{2}\ln(2xy) - 3\ln z$ $\ln\frac{\sqrt{2xy}}{z^3}$ [5.3]

40. $\ln x - (\ln y - \ln z)$ $\ln\frac{xz}{y}$ [5.3]

In Exercises 41 to 44, use the change-of-base formula and a calculator to approximate each logarithm. Round to the nearest ten thousandth.

41. $\log_5 101$ 2.8675 [5.3]

42. $\log_3 40$ 3.3578 [5.3]

43. $\log_4 0.85$ −0.1172 [5.3]

44. $\log_8 0.3$ −0.5790 [5.3]

In Exercises 45 to 60, solve each equation for *x*. Give exact answers. Do not use a calculator.

45. $4^x = 30$ $\frac{\log 30}{\log 4}$ or $\frac{\ln 30}{\ln 4}$ [5.4]

46. $5^{x+1} = 41$ $\frac{\ln 41}{\ln 5} - 1$ [5.4]

47. $\ln 3x - \ln(x-1) = \ln 4$ 4 [5.4]

48. $\ln(3x) + \ln 2 = 1$ $\frac{1}{6}e$ [5.4]

49. $e^{\ln(x+2)} = 6$ 4 [5.4]

50. $10^{\log(2x+1)} = 31$ 15 [5.4]

51. $\frac{4^x + 4^{-x}}{4^x - 4^{-x}} = 2$ $\frac{\ln 3}{2\ln 4}$ [5.4]

52. $\frac{5^x + 5^{-x}}{2} = 8$ $\frac{\ln\left(8 \pm 3\sqrt{7}\right)}{\ln 5}$ [5.4]

53. $\log(\log x) = 3$ 10^{1000} [5.4]

54. $\ln(\ln x) = 2$ $e^{(e^2)}$ [5.4]

55. $\log\sqrt{x-5} = 3$ 1,000,005 [5.4]

56. $\log x + \log(x-15) = 1$ $\frac{15+\sqrt{265}}{2}$ [5.4]

57. $\log_4(\log_3 x) = 1$ 81 [5.4]

58. $\log_7(\log_5 x^2) = 1$ $\pm 125\sqrt{5}$ [5.4]

59. $\log_5 x^3 = \log_5 16x$ 4 [5.4]

60. $25 = 16^{\log_4 x}$ 5 [5.4]

In Exercises 61 to 64, find the exponential growth/decay function $N(t) = N_0e^{kt}$ that satisfies the given conditions.

61. $N(0) = 1, N(2) = 5$ $N(t) \approx e^{0.80472t}$ [5.5]

62. $N(0) = 2, N(3) = 11$ $N(t) \approx 2e^{0.56825t}$ [5.5]

63. $N(1) = 4, N(5) = 5$ $N(t) \approx 3.78297e^{0.05579t}$ [5.5]

64. $N(-1) = 2, N(0) = 1$ $N(t) \approx e^{-0.69315t}$ [5.5]

65. Consider the logistic growth model

$$P(t) = \frac{126}{1 + 5e^{-0.27t}}$$

a. Find P_0. 21

b. Find the inflection point of the graph of $P(t)$. Round the coordinates to the nearest hundredth. (5.96, 63.00)

c. Explain the significance of the inflection point.

d. What does $P(t)$ approach as $t \to \infty$? 126 [5.7]

65. c. The function P attains its largest rate of growth when $t \approx 5.96$. The value 5.96 is the first coordinate of the inflection point.

Business and Economics

66. ***Counterfeit Money*** The amount of counterfeit money being created on computers has been increasing from year to year. The Federal Reserve Board reports, "According to the U.S. Secret Service, \$47.5 million in counterfeit money entered into circulation in fiscal year 2001. Of this amount, 39% was computer generated, compared with only 0.5% in 1995." (*Source:* The Federal Reserve Board, **http://www.federalreserve.gov/**) For recent years, the function $D(t) \approx 0.174(2.177)^t$ closely models the amount of computer-generated counterfeit money, in millions of dollars, in circulation where $t = 0$ represents the year 1995. Use $D(t)$ to predict the year in which the amount of computer-generated counterfeit money will first exceed \$250 million. 2004 [5.4]

67. ***Depreciation*** The scrap value S of a product with an expected life span of n years is given by $S(n) = P(1-r)^n$, where P is the original purchase price of the product and r is the annual rate of depreciation. A taxicab is purchased for \$18,400 and is expected to last 3 years. What is its scrap value if it depreciates at a rate of 41% per year? Round to the nearest hundred dollars. \$3800 [5.1]

Life and Health Sciences

68. **Medicine** A skin wound heals according to the function $N(t) = N_0 e^{-0.12t}$, where N is the number of square centimeters of unhealed skin t days after the injury, and N_0 is the number of square centimeters of skin covered by the original wound.

   a. What percent, to the nearest tenth of a percent, of the wound will be healed after 10 days? $\approx$69.9%

   b. How many days, to the nearest day, will it take for 50% of the wound to heal? 6 days

   c. How long, to the nearest day, will it take for 90% of the wound to heal? 19 days [5.5]

69. **Population Growth** **a.** Find the exponential growth function for a city whose population was 25,200 in 1998 and 26,800 in 1999. Use $t = 0$ to represent the year 1998. **b.** Use the growth function to predict, to the nearest 100, the population of the city in 2005.
a. $N(t) \approx 25{,}200e^{0.06156t}$ b. 38,800 people [5.5]

70. **Population Growth** A city currently has a population of 74,000 and its population is growing at a continuous rate of 4.6% per year.

   a. Write a function that gives the population t years from the present. $N(t) \approx 74{,}000e^{0.046t}$

   b. Use your function to find the population in 8 years. Round to the nearest hundred. 106,900 people

   c. How long will it take for the city's population to double? Round to the nearest tenth of a year. 15.1 yr [5.5]

71. **Logistic Growth** The population of coyotes in a national park satisfies a logistic growth model with $P_0 = 210$ in 1992, $c = 1400$, and $P(3) = 360$ (the population in 1995).

   a. Determine the logistic growth model for the data. $P(t) \approx \dfrac{1400}{1 + 5.66667e^{-0.22458t}}$

   b. Use the model to predict, to the nearest 10, the coyote population in 2005. 1070 coyotes

   c. Use the model to predict the year in which the coyote population will have its most rapid growth. During 1999 [5.7]

72. **Mortality Rate** The following table shows the infant mortality rate in the United States for selected years from 1950 to 1998.

U.S. Infant Mortality Rate (per 1000 live births)

Year	Rate
1950	29.2
1960	26.0
1970	20.0
1980	12.6
1990	9.2
1995	7.6
1998	7.2

Source: Time Almanac 2002

   a. Use a graphing utility to find an exponential model for the infant mortality rate R as a function of the year. Let $t = 50$ represent the year 1950. ExpReg: $R \approx 161.03059(0.96884)^t$

   b. Use the model to predict, to the nearest tenth, the infant mortality rate for 2008. 5.3 deaths per 1000 live births [5.6]

Social Sciences

73. **Active Military Duty Personnel** The following table shows the number of U.S. military personnel on active duty for each year from 1990 to 2000.

Active Military Duty Personnel, 1990–2000

1990	2,043,705	1995	1,518,224
1991	1,985,555	1996	1,471,722
1992	1,807,177	1997	1,438,562
1993	1,705,103	1998	1,406,830
1994	1,610,490	1999	1,379,756
		2000	1,370,237

Source: Time Almanac 2002

   a. Use a graphing utility to find a linear model, an exponential model, and a logarithmic model for the number of active-duty personnel P as a function of the year. Let $t = 90$ represent the year 1990. State the correlation coefficient r to the nearest ten-thousandth.
LnReg: $P \approx -69{,}667.52t + 8{,}230{,}901.59$, $r \approx -0.9547$
ExpReg: $P \approx 87{,}535{,}810.36(0.958728)^t$, $r \approx -0.9670$
LnReg: $P \approx 31{,}854{,}352.17 - 6{,}641{,}714.035 \ln t$, $r \approx -0.9588$

b. Examine the correlation coefficients of the three regression models to determine which model provides the best fit for the data. Exponential model

c. Use the model you selected in part **b.** to predict, to the nearest 10,000, the number of active-duty military personnel for the year 2005. 1,050,000 people [5.6]

Physical Sciences and Engineering

74. ***Earthquake Magnitude*** Determine, to the nearest tenth, the Richter scale magnitude of an earthquake with an intensity of $I = 51{,}782{,}000I_0$. 7.7 [5.3]

75. ***Earthquake Magnitude*** A seismogram has an amplitude of 18 millimeters and a time delay of 21 seconds. Find, to the nearest tenth, the Richter scale magnitude of the earthquake that produced the seismogram. 5.0 [5.3]

76. ***Comparison of Earthquakes*** An earthquake had a Richter scale magnitude of 7.2. Its aftershock had a Richter scale magnitude of 3.7. Compare the intensity of the earthquake to the intensity of the aftershock by finding, to the nearest unit, the ratio of the larger intensity to the smaller intensity. 3162 to 1 [5.3]

77. ***Comparison of Earthquakes*** An earthquake has an intensity of 600 times the intensity of a second earthquake. Find, to the nearest tenth, the difference between the Richter scale magnitudes of the earthquakes. 2.8 [5.3]

78. ***Chemistry*** Find the pH, to the nearest tenth, of tomatoes that have a hydronium-ion concentration of 6.28×10^{-5}. 4.2 [5.3]

79. ***Chemistry*** Find the hydronium-ion concentration of rainwater that has a pH of 5.4. $\approx 3.98 \times 10^{-6}$ mol/L [5.3]

80. ***Carbon Dating*** Determine, to the nearest 10 years, the age of a bone if it now contains 96% of its original amount of carbon-14. The half-life of carbon-14 is 5730 years. 340 years old [5.5]

81. ***Radioactive Decay*** A radioactive isotope has a half-life of 26 days. If we start with a 5-gram sample, write an exponential decay function for the amount of the isotope remaining after t days. $N(t) = 5\left(\frac{1}{2}\right)^{t/26}$ or $N(t) \approx 5e^{-0.02666t}$ [5.5]

82. ***Newton's Law of Cooling*** A cup of 140°F hot chocolate is served in a room with a temperature of 71°F. It takes the liquid 12 minutes to cool to 105°F.

a. Use Newton's Law of Cooling to write a function that gives the temperature of the hot chocolate t minutes after being served. $T(t) \approx 71 + 69e^{-0.05898t}$

b. How long will it take, to the nearest tenth of a minute, for the hot chocolate to cool to 80°F? 34.5 min [5.5]

Chapter 5 Test

— *Answer graphs to Exercises 3–5 are on page AA17.*

1. Evaluate *without* using a calculator: $\log_3 \frac{1}{27}$ −3 [5.2/5.4]

2. Use the change-of-base formula and a calculator to approximate $\log_4 12$. Round your result to the nearest ten thousandth. 1.7925 [5.3]

3. Graph: $f(x) = e^{x-2}$ [5.1]

4. Graph: $h(x) = -3^{x/2}$ [5.1]

5. Graph and state the domain: $g(x) = \log_5(x + 1)$
Domain: $\{x \mid x > -1\}$ [5.2]

6. Write $\log_b(5x - 3) = c$ in exponential form.
$b^c = 5x - 3$ [5.2]

7. Write $e^{t/4} = a$ in logarithmic form. $\ln a = \frac{t}{4}$ [5.2]

8. Write $\log_b \frac{z^2}{y^3\sqrt{x}}$ in terms of logarithms of x, y, and z.
$2\log_b z - 3\log_b y - \frac{1}{2}\log_b x$ [5.3]

9. Solve $5^x = 22$. Round to the nearest ten thousandth.
1.9206 [5.4]

10. Find the *exact* solution of $4^{5-x} = 7^x$. $\frac{5\ln 4}{\ln 4 + \ln 7}$ [5.4]

11. Solve: $\log_3(2x + 5) = 2$ 2 [5.4]

12. Find the *exact* solution of $\frac{3^x - 3^{-x}}{2} = 100$.
$\frac{\ln\left(100 + \sqrt{10{,}001}\right)}{\ln 3}$ [5.4]

13. Solve: $\log(x + 99) - \log(3x - 2) = 2$ 1 [5.4]

14. A pot of boiling water (212°F) is placed on the counter in a 71°F room. The temperature of the water after t minutes is given by $T = 71 + 141e^{-0.37t}$.

a. Find the temperature of the water after 5 minutes. Round to the nearest degree. 93°F

b. Find the time required for the water to cool to 80°F. Round to the nearest tenth of a minute. 7.4 min [5.5]

15. A city had a population of 34,600 in 1996 and 39,800 in 1999.

a. Find the exponential growth function for the population. Use $t = 0$ to represent 1996. $P(t) \approx 34{,}600e^{0.04667t}$

b. Use the growth function to predict the population of the city in 2006. Round to the nearest 1000. 55,000 people

c. In what year will the population reach 75,000? Round to the nearest year. 2013 [5.5]

16. Estimate the age of a bone, to the nearest 10 years, if it now contains 92% of its original amount of carbon-14. The half-life of carbon-14 is 5730 years. 690 years old [5.5]

17. The following table shows the annual interest rates for U.S. government bonds in August 2002.

Years t	0.25	0.5	1.0	2.0	5.0
Annual rate r	1.68%	1.69%	1.79%	2.19%	3.29%

a. Find the exponential regression function for the data and use the function to predict the annual interest rate, to the nearest 0.01%, on a bond invested for 3.0 years. $r(t) \approx 1.59035(1.15763)^t$; 2.47%

b. According to your function, how long (to the nearest 0.1 year) would you need to invest to receive a 4.0% annual interest rate? 6.3 yr [5.6]

18. Consider the following data.

$\{(2.0, 6.5), (3.5, 9.3), (5.0, 11.0), (6.5, 12.4), (7.0, 12.7)\}$

a. Find the logarithmic regression function for the data. Round constants to the nearest ten thousandth. $f(x) = 3.0610 + 4.9633 \ln x$

b. Use the function to predict the y-value, to the nearest tenth, associated with $x = 7.8$. 13.3 [5.6]

19. The population of raccoons in a state park satisfies a logistic growth model with $P_0 = 160$ in 1999 and $P(1) = 190$. A park ranger has estimated the carrying capacity of the park to be 1100 raccoons. Use this data to

a. find the logistic growth model for the raccoon population. $P(t) \approx \dfrac{1100}{1 + 5.875e^{-0.20429t}}$

b. predict the raccoon population in 2006. ≈457 raccoons

c. predict the year during which the raccoon population will have its most rapid growth. During 2007 [5.7]

20. A lake was stocked with 600 northern pike. The population of the northern pike increased according to the logistic growth model $P(t) = \dfrac{7200}{1 + 11e^{-0.04t}}$, where t is in months. According to the model,

a. what is the carrying capacity of the lake with respect to the northern pike population? 7200 northern pike

b. how many months, to the nearest tenth, will it take for the northern pike population to reach 3000? 51.5 months [5.7]

Cumulative Review Exercises

1. Solve $|x - 4| \le 2$. Write the solution set using interval notation. [2, 6] [1.4]

2. Solve $\dfrac{x}{2x - 6} \ge 1$. Write the solution set using set-builder notation. $\{x \mid 3 < x \le 6\}$ [1.5]

3. Find, to the nearest tenth, the distance between the points (5, 2) and (11, 7). 7.8 [2.1]

4. The height, in feet, of a ball released with an initial upward velocity of 44 feet per second and at an initial height of 8 feet is given by $h(t) = -16t^2 + 44t + 8$, where t is the time in seconds after the ball is released. Find the maximum height the ball will reach. 38.25 ft [2.5]

5. Given $f(x) = 2x + 1$ and $g(x) = x^2 - 5$, find $(g \circ f)$. $4x^2 + 4x - 4$ [3.2]

6. Find the inverse of $f(x) = 3x - 5$. $f^{-1}(x) = \frac{1}{3}x + \frac{5}{3}$ [3.2]

7. The load that a horizontal beam can safely support varies jointly as the width and the square of the depth of the beam. It has been determined that a beam with a width of 4 inches and a depth of 8 inches can safely support a load of 1500 pounds. How many pounds can a beam of the same material and the same length safely support if it has a width of 6 inches and a depth of 10 inches? Round to the nearest hundred pounds. 3500 lb [3.4]

8. Use Descartes' Rule of Signs to determine the number of possible real zeros of $P(x) = x^4 - 3x^3 + x^2 - x - 6$.
3 or 1 positive real zeros; 1 negative real zero [4.3]

9. Find the zeros of $P(x) = x^4 - 5x^3 + x^2 + 15x - 12$.
$1, 4, -\sqrt{3}, \sqrt{3}$ [4.3]

10. Find a polynomial function of lowest degree that has 2, $1 - i$, and $1 + i$ as zeros. $P(x) = x^3 - 4x^2 + 6x - 4$ [4.4]

11. Find the equations of the vertical and horizontal asymptotes of the graph of $r(x) = \dfrac{3x - 5}{x - 4}$.
Vertical asymptote: $x = 4$, horizontal asymptote: $y = 3$ [4.5]

12. Determine the domain and the range of the rational function $R(x) = \dfrac{4}{x^2 + 1}$.
Domain: all real numbers; range: $\{y \mid 0 < y \le 4\}$ [4.5]

13. State whether $f(x) = 0.4^x$ is an increasing function or a decreasing function. Decreasing function [5.1]

14. Write $\log_4 x = y$ in exponential form. $4^y = x$ [5.2]

15. Write $5^3 = 125$ in logarithmic form. $\log_5 125 = 3$ [5.2]

16. Find, to the nearest tenth, the Richter scale magnitude of an earthquake with an intensity of $I = 11{,}650{,}600I_0$.
7.1 [5.3]

17. Solve $2e^x = 15$. Round to the nearest ten thousandth.
2.0149 [5.4]

18. Find the age of a bone if it now has 94% of the carbon-14 it had at time $t = 0$. Round to the nearest ten years.
510 years old [5.5]

19. The following table shows the progression of the world record distances for the men's javelin throw for the years 1986 to 1996. (*Source:* **http://www.athletix.org/Statistics/stats.html)**

World Record Progression in the Men's Javelin Throw

Year	Distance (in meters)
1986	85.74
1987	87.66
1990	89.10
1992	91.46
1993	95.54
1993	95.66
1996	98.48

a. Use a graphing utility to determine a logarithmic model for the data. Use $x = 1$ to represent the year 1986 and $x = 11$ to represent the year 1996.
$y \approx 84.41319 + 4.88166 \ln x$

b. Use the model from part **a.** to predict the men's world record javelin throw distance for 2006. Round to the nearest hundredth of a meter. 99.28 m [5.6]

20. The wolf population in a national park satisfies a logistic growth model with $P_0 = 160$ in 1998 and $P(3) = 205$ (the population in 2001). It has been determined that the maximum population the park can support is 450 wolves.

a. Determine the logistic growth model for the data. $P(x) \approx \dfrac{450}{1 + 1.8125e^{-0.13882x}}$

b. Use the model to predict, to the nearest 10, the wolf population in 2008. 310 wolves [5.7]

Chapter 6 Systems of Linear Equations and Inequalities

Splines and Curves

Before the advent of computers, architects and designers used flexible strips of wood or metal called *splines* to draw smooth, curved paths. Lead weights called *ducks* were used to pull on portions of the splines until the desired shape was created. The ducks then maintained the shape of the spline while the curve was traced onto paper.

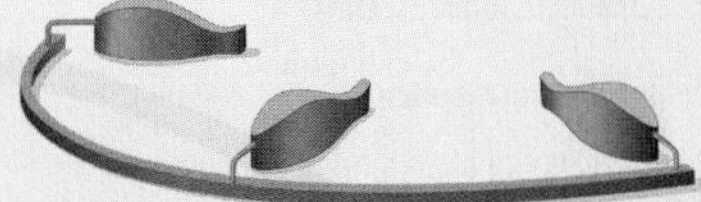

A spline held in place by ducks

Today, computer aided design (CAD) software accomplishes the same task by using mathematical equations to create smooth curves between points. Quadratic or cubic polynomial equations are often used because of the aesthetically pleasing curves that result.

A cubic spline curve requires four control points that perform a function similar to that of the ducks used with splines. Cubic *Bézier curves* (used, for example, in the Adobe PostScript printer language for fonts and graphics) are "pulled" toward the control points but generally do not pass through all of them. An *interpolating curve* or an *interpolating spline* always passes directly through all four control points. In Exercise 52 on page 530, you will compute a cubic interpolating spline curve that passes through four control points.

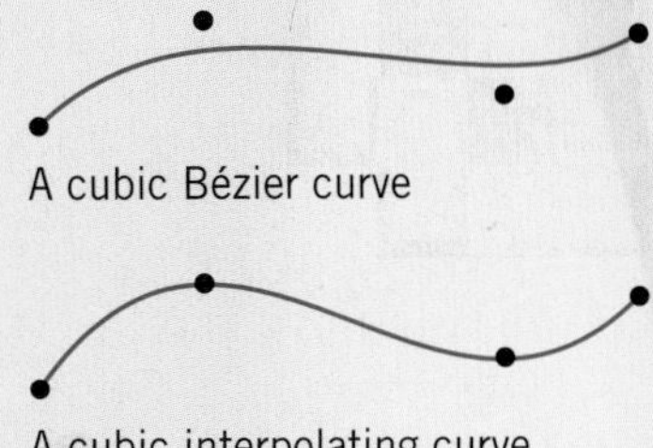

A cubic Bézier curve

A cubic interpolating curve

Chapter Prep Quiz *Do these exercises to prepare for Chapter 6.*

—Answer graphs to Exercises 4 and 5 are on page AA17.

1. What is the slope of the line given by the equation $3x - 5y = 9$? [2.3] $\frac{3}{5}$

2. Find the slope and the y-intercept of the line given by $2x - 3y = 9$. [2.3] Slope: $\frac{2}{3}$, y-intercept: $(0, -3)$

3. Find three different points on the graph of $3x + 2y = 4$. [2.3] $(0, 2)$, $(2, -1)$, $(4, -4)$ (Answers can vary.)

4. Graph $y = 2x - 4$. [2.3]

5. Graph $2x - 6y = -12$. [2.3]

6. The cost to lease a new car is \$350 per month with an initial payment of \$1800. Write a linear function that gives the cost of leasing the car after x months. [2.4] $C(x) = 350x + 1800$

7. Find $f(g(x))$ if $f(x) = 3x - 8$ and $g(x) = -2x + 5$. [3.1] $-6x + 7$

8. Find the center and radius of the circle $(x + 3)^2 + (y - 1)^2 = 7$. [2.1] Center: $(-3, 1)$, radius: $\sqrt{7}$

9. Solve the inequality $4x - 5 < 8$. [1.4] $x < \frac{13}{4}$

10. Solve the compound inequality $2x + 1 \geq 7$ or $5 - 3x > 3$. [1.4] $\left(-\infty, \frac{2}{3}\right) \cup [3, \infty)$

Problem Solving Strategies

Solve a Similar But Simpler Problem

Some problems seem at first to have too many possible solutions to be analyzed in a reasonable way. For example, consider the puzzle that starts with a checkerboard on which opposing corners have been removed. Can you place dominoes, each of which covers two squares of the checkerboard, on the board so that every square is covered and no dominoes overlap?

There are quite a few different ways the dominoes could be arranged. If your first attempt is not successful, it is difficult to predict how many tries it might take. Looking at the puzzle on a smaller scale can help expose strategies that can lead to a solution of the original puzzle. Consider a 4-by-4 checkerboard with opposite corners removed. If we start in the first square of the top row, there are only two ways a domino can be placed. These two options are shown in the checkerboards at the left. We have continued placing dominoes, in the sequences indicated, on squares on which only one placement is possible. Thus, if there were a solution, it would have to contain the dominoes indicated. You can see that neither board can be covered as prescribed.

If it is impossible to cover the smaller board with dominoes, perhaps it is impossible to cover the larger checkerboard as well. Notice the colors of the squares remaining on the smaller checkerboard. In each case we have three black squares and just one white square remaining. Because a domino will always cover one black and one white square, it is impossible to place the last two dominoes, regardless of the arrangement of the available squares. Can you use this observation to make a conclusion about the original puzzle?

INSTRUCTOR NOTE

The *Chapter Prep Quiz* is a means to test your students' mastery of prerequisite material that is assumed in the coming chapter. Each question identifies the section to review (if necessary) and all answers are provided in the Answers to Selected Exercises.

SECTION 6.1 Systems of Linear Equations in Two Variables

- Systems of Equations and Their Graphs
- Substitution Method for Solving a System of Linear Equations
- Elimination Method for Solving a System of Linear Equations
- Applications of Systems of Linear Equations

Alternative to Example 1
Exercise 44, page 499

Systems of Equations and Their Graphs

Suppose you need to rent a moving truck and you want to drop off the truck at a different location from the one at which you picked it up. A-1 Rentals will rent you a truck for \$80 plus \$1.20 per mile traveled. Then the cost of renting the truck can be described by the linear equation $C = 1.20x + 80$, where x is the number of miles traveled. Another company, EZ Movers, charges only \$50 for the rental but adds \$1.70 per mile traveled. The cost for renting the truck from this company is $C = 1.70x + 50$. To determine which company provides the better rental rate, we must know how far the truck will be driven. In the example that follows, we will determine the number of miles for which both companies charge the same price.

EXAMPLE 1 Compare Rental Prices

Two companies rent moving vans as described above. For what number of miles will the rental fee be the same for either company?

Solution If we graph both cost equations, we see that for some numbers of miles A-1 Rentals is cheaper, whereas for others EZ Movers costs less.

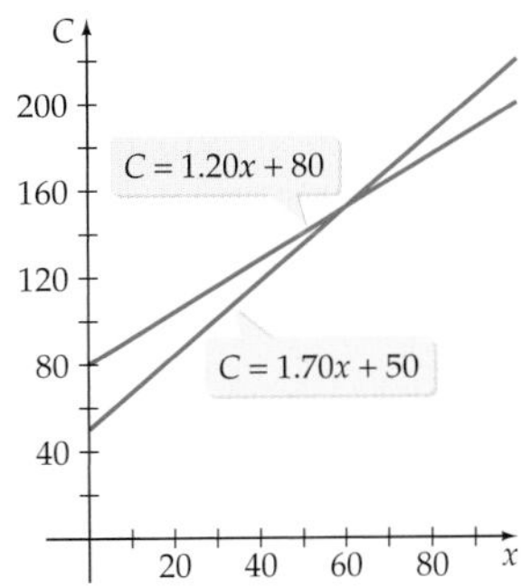

The two lines intersect at one point. It is at this point that both companies charge the same price for the same number of miles, which appears to be 60 miles. We can verify our observation using the following equations.

$$\text{A-1 Rentals: } C = 1.2(60) + 80 = 152 \qquad \blacksquare\ x = 60$$
$$\text{EZ Movers: } C = 1.7(60) + 50 = 152 \qquad \blacksquare\ x = 60$$

Thus, if the distance traveled is 60 miles, both companies charge the same price of \$152.

CHECK YOUR PROGRESS 1 A cellular phone company offers a monthly plan that costs \$15 plus 8 cents per minute of phone use. Another provider charges a \$12 monthly fee plus 12 cents for every minute of phone use. For what number of minutes will both companies charge the same total monthly cost?

Solution *See page S30.* 75 min

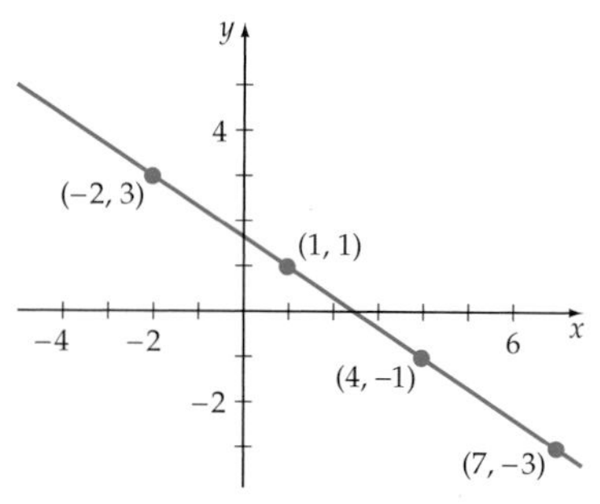

Figure 6.1

A single linear equation in two variables has many solutions. For example, $(-2, 3)$ and $(1, 1)$ are solutions of the equation $2x + 3y = 5$ because $2(-2) + 3(3) = 5$ and $2(1) + 3(1) = 5$. The graph of a linear equation, which is a straight line, shows all the possible solutions of the equation. Figure 6.1 shows the graph of $2x + 3y = 5$.

A **system of equations** is two or more equations considered together. The following system of equations is a **linear system of equations** in two variables.

$$\begin{cases} 2x + 3y = 4 \\ 3x - 2y = -7 \end{cases}$$

A **solution of a system of equations** in two variables is an ordered pair that satisfies *all* equations simultaneously.

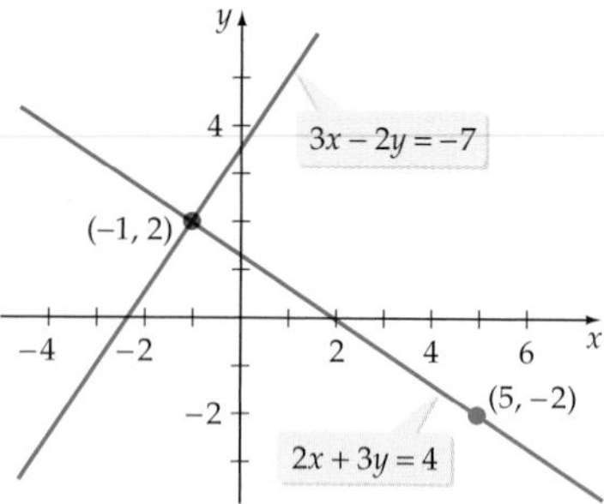

Figure 6.2

In Figure 6.2, the graphs of the two equations in the system of equations above intersect at the point $(-1, 2)$. Because this point lies on both lines, $(-1, 2)$ is a solution of both equations and thus is a solution of the system of equations. The point $(5, -2)$ is a solution of the first equation but not the second equation. Therefore, $(5, -2)$ is not a solution of the system of equations. In Example 1, the goal was to find a single value of x that gave the same value of C for both equations. The answer, $(60, 152)$, is the solution of the linear system formed by the two cost equations.

The graphs of two linear equations in two variables can intersect at a single point (see Figure 6.3a), be the same line (Figure 6.3b), or be parallel (Figure 6.3c). When the graphs intersect at a single point or are the same line, the system is called a **consistent system of equations.** The system is called an **independent system of equations** when the lines intersect at exactly one point. The system is called a **dependent system of equations** when the equations represent the same line. In this case, the system has an infinite number of solutions. When the graphs of the two equations are parallel lines, the system is called **inconsistent** and has no solution.

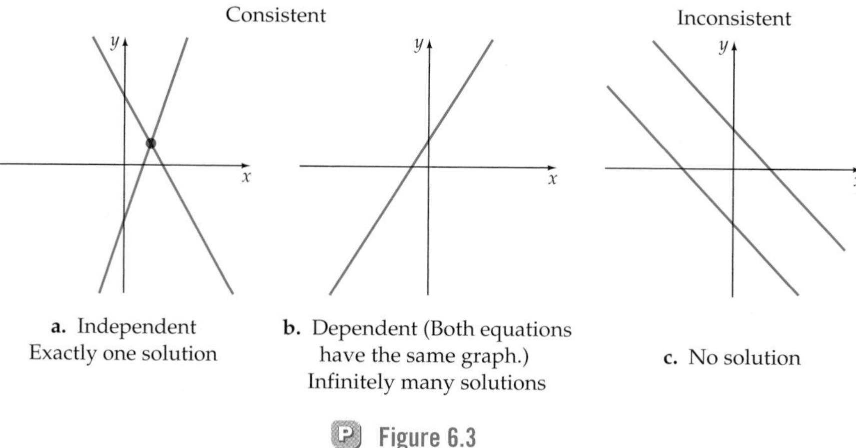

Figure 6.3

The solution of a system of equations can be found by graphing the equations and examining the intersection. It can be difficult or inconvenient, however, to find the exact value of a solution by graphing. Next we will look at two algebraic methods for solving a linear system of equations, the *substitution method* and the *elimination method*.

INSTRUCTOR NOTE

It may help some students if you explain the concept of ordered pairs as solutions of a system of equations. For instance, in Example 2, because $(x, 2x)$ is a solution of Equation (2), for the system of equations to have a solution, $(x, 2x)$ must also be a solution of Equation (1). This is the premise on which the substitution method is based.

Substitution Method for Solving a System of Linear Equations

The **substitution method** is one procedure for solving a system of equations. This method is illustrated in Example 2.

Alternative to Example 2

Solve: $\begin{cases} 5x + 2y = -4 \\ y = -3x \end{cases}$

- (4, −12)

EXAMPLE 2 Solve a System of Equations by the Substitution Method

Solve: $\begin{cases} 3x - 5y = 7 & (1) \\ y = 2x & (2) \end{cases}$

Solution

The solutions of $y = 2x$ are the ordered pairs $(x, 2x)$. For the system of equations to have a solution, ordered pairs of the form $(x, 2x)$ must also be solutions of $3x - 5y = 7$. To determine whether the ordered pairs $(x, 2x)$ are solutions of Equation (1), substitute $(x, 2x)$ into Equation (1) and solve for x. Think of this as *substituting* $2x$ for y.

$$\begin{aligned} 3x - 5y &= 7 && \text{Equation (1)} \\ 3x - 5(2x) &= 7 && \text{Substitute } 2x \text{ for } y. \\ 3x - 10x &= 7 \\ -7x &= 7 \\ x &= -1 \end{aligned}$$

$$\begin{aligned} y &= 2x && \text{Equation (2)} \\ &= 2(-1) = -2 && \text{Substitute } -1 \text{ for } x \text{ in Equation 2.} \end{aligned}$$

The only ordered-pair solution of the system of equations is $(-1, -2)$. When a system of equations has a unique solution, the system of equations is independent.

Visualize the Solution

Graphing $3x - 5y = 7$ and $y = 2x$ shows that the ordered pair $(-1, -2)$ belongs to both lines. Therefore, $(-1, -2)$ is a solution of the system of equations. See Figure 6.4.

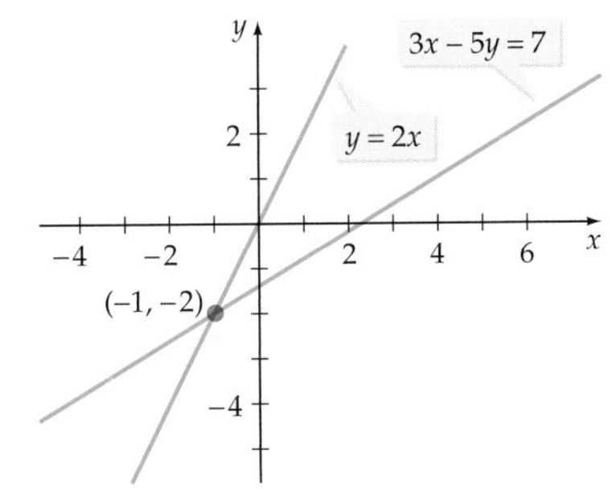

Figure 6.4

An independent system of equations

CHECK YOUR PROGRESS 2 Use the substitution method to solve the following system of equations.

$$\begin{cases} 8x + 3y = -7 \\ x = 3y + 15 \end{cases}$$

Solution *See page S30.* $\left(\frac{8}{9}, -\frac{127}{27}\right)$

EXAMPLE 3 Identify an Inconsistent System of Equations

Solve: $\begin{cases} x + 3y = 6 & (1) \\ 2x + 6y = -18 & (2) \end{cases}$

Alternative to Example 3

Solve: $\begin{cases} 3x + 2y = -4 \\ 9x + 6y = -8 \end{cases}$

- Inconsistent System

Solution

Solve Equation (1) for y:

$$x + 3y = 6$$
$$y = -\frac{1}{3}x + 2$$

The solutions of $y = -\frac{1}{3}x + 2$ are the ordered pairs $\left(x, -\frac{1}{3}x + 2\right)$. For the system of equations to have a solution, ordered pairs of this form must also be solutions of $2x + 6y = -18$. To determine whether the ordered pairs $\left(x, -\frac{1}{3}x + 2\right)$ are solutions of Equation (2), substitute $\left(x, -\frac{1}{3}x + 2\right)$ into Equation (2) and solve for x.

$$2x + 6y = -18 \quad \text{■ Equation (2)}$$
$$2x + 6\left(-\frac{1}{3}x + 2\right) = -18 \quad \text{■ Substitute } -\frac{1}{3}x + 2 \text{ for } y.$$
$$2x - 2x + 12 = -18$$
$$12 = -18 \quad \text{■ A false statement}$$

The false statement $12 = -18$ means that no ordered pair that is a solution of Equation (1) is also a solution of Equation (2). The equations have no ordered pairs in common and thus the system of equations has no solution. This is an inconsistent system of equations.

Visualize the Solution

Solving Equations (1) and (2) for y gives $y = -\frac{1}{3}x + 2$ and $y = -\frac{1}{3}x - 3$. Note that these two equations have the same slope, $-\frac{1}{3}$, and different y-intercepts. Therefore, the graphs of the two lines are parallel and never intersect. See Figure 6.5.

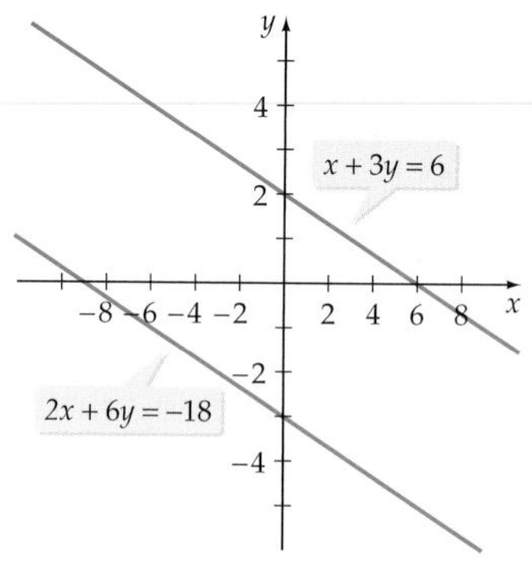

Figure 6.5

An inconsistent system of equations

CHECK YOUR PROGRESS 3 Use the substitution method to solve the following system of equations.

$$\begin{cases} 3x - 4y = 8 \\ 6x - 8y = 9 \end{cases}$$

Solution *See page S31.* Inconsistent system

EXAMPLE 4 Solve a Dependent System of Equations

Solve: $\begin{cases} 8x - 4y = 16 & (1) \\ 2x - y = 4 & (2) \end{cases}$

Continued ➤

Alternative to Example 4

Solve: $\begin{cases} 10x - 5y = 8 \\ 30x - 15y = 24 \end{cases}$

■ $\left(c, 2c - \frac{8}{5}\right)$

Solution

Solve Equation (2) for y:

$$2x - y = 4$$
$$y = 2x - 4$$

The solutions of $y = 2x - 4$ are the ordered pairs $(x, 2x - 4)$. For the system of equations to have a solution, ordered pairs of this form $(x, 2x - 4)$ must also be solutions of $8x - 4y = 16$. To determine whether the ordered pairs $(x, 2x - 4)$ are solutions of Equation (1), substitute $(x, 2x - 4)$ into Equation (1) and solve for x.

$$8x - 4y = 16 \quad \text{■ Equation (1)}$$
$$8x - 4(2x - 4) = 16 \quad \text{■ Substitute } 2x - 4 \text{ for } y.$$
$$8x - 8x + 16 = 16$$
$$16 = 16 \quad \text{■ A true statement}$$

Visualize the Solution

Solving Equations (1) and (2) for y gives $y = 2x - 4$ and $y = 2x - 4$. Note that these two equations have the same slope, 2, and the same y-intercept, $(0, -4)$. Therefore, the graphs of the two lines are exactly the same. One graph intersects the second graph infinitely often. See Figure 6.6.

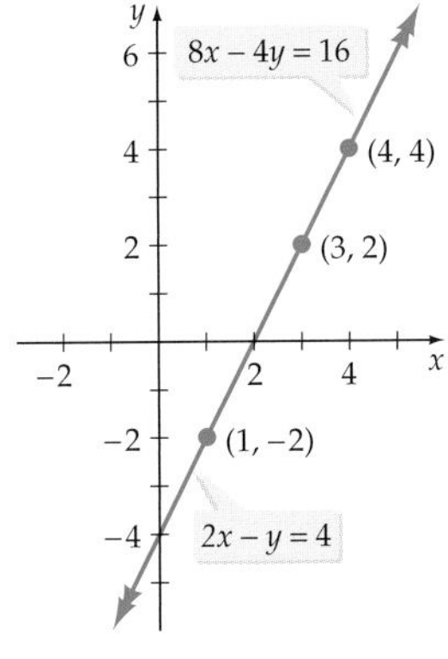

Figure 6.6

The true statement $16 = 16$ means that the ordered pairs $(x, 2x - 4)$ that are solutions of Equation (2) are also solutions of Equation (1). Because x can be replaced by any real number c, the solution of the system of equations is the set of ordered pairs $(c, 2c - 4)$. This is a dependent system of equations.

CHECK YOUR PROGRESS 4 Solve the system of equations.

$$\begin{cases} 5x + 2y = 2 \\ y = -\dfrac{5}{2}x + 1 \end{cases}$$

Solution *See page S31.* $\left(c, -\frac{5}{2}c + 1\right)$

QUESTION If a system of equations is dependent, does that mean that every ordered pair (x, y) is a solution of the system?

Some of the specific ordered-pair solutions in Example 4 can be found by choosing various values for c. The following table shows the ordered pairs that result from choosing 1, 3, and 4 as c. The ordered pairs $(1, -2)$, $(3, 2)$, and $(4, 4)$ are specific solutions of the system of equations. These points are on the graphs of Equation (1) and Equation (2), as shown in Figure 6.6.

ANSWER No. There are an infinite number of solutions, but only those ordered pairs whose coordinates satisfy the equations are solutions.

c	$(c, 2c - 4)$	(x, y)
1	$(1, 2(1) - 4)$	$(1, -2)$
2	$(3, 2(3) - 4)$	$(3, 2)$
3	$(4, 2(4) - 4)$	$(4, 4)$

TAKE NOTE

When a system of equations is dependent, there is more than one way to write the solutions of the system. The solution to Example 4 is the set of ordered pairs

$$(c, 2c - 4) \text{ or } \left(\frac{1}{2}b + 2, b\right)$$

However, there are infinitely more ways in which the ordered pairs can be expressed. For instance, let $b = 2w$. Then

$$\frac{1}{2}b + 2 = \frac{1}{2}(2w) + 2 = w + 2$$

The ordered-pair solutions, written in terms of w, are $(w + 2, 2w)$.

Before leaving Example 4, note that there is more than one way to represent the ordered-pair solutions. To illustrate this point, solve Equation (2) for x.

$$2x - y = 4 \quad \text{▪ Equation (2)}$$

$$x = \frac{1}{2}y + 2 \quad \text{▪ Solve for } x.$$

Because y can be replaced by any real number b, there are an infinite number of ordered pairs $\left(\frac{1}{2}b + 2, b\right)$ that are solutions of the system of equations. Choosing -2, 2, and 4 as b gives the same ordered pairs as above: $(1, -2)$, $(3, 2)$, and $(4, 4)$. There is always more than one way to describe the ordered pairs when writing the solution of a dependent system of equations. For Example 4, the ordered pairs $(c, 2c - 4)$ or the ordered pairs $\left(\frac{1}{2}b + 2, b\right)$ would generate all the solutions of the system of equations.

▪ Elimination Method for Solving a System of Linear Equations

Two systems of equations are **equivalent** if each system has exactly the same solutions. The systems

$$\begin{cases} 3x + 5y = 9 \\ 2x - 3y = -13 \end{cases} \quad \text{and} \quad \begin{cases} x = -2 \\ y = 3 \end{cases}$$

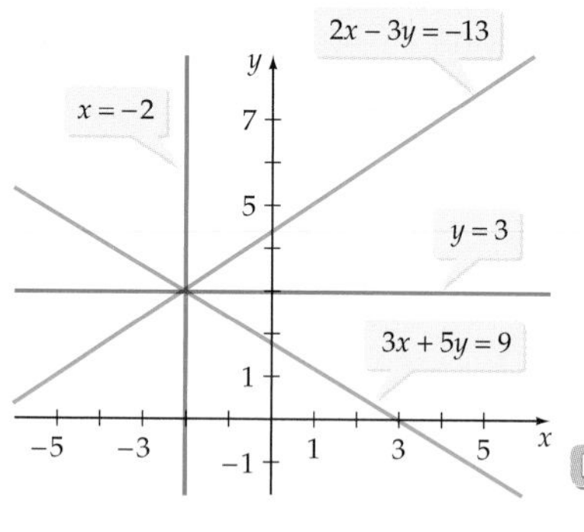

Figure 6.7

are equivalent systems of equations. Each system has the solution $(-2, 3)$, as shown in Figure 6.7.

A second technique for solving a system of equations is similar to the strategy used for solving first-degree equations in one variable. The system of equations is replaced by a series of equivalent systems until the solution becomes apparent.

Operations That Produce Equivalent Systems of Equations

1. Interchange any two equations.
2. Replace an equation with a nonzero multiple of that equation.
3. Replace an equation with the sum of that equation and a nonzero constant multiple of another equation in the system.

Because the order in which the equations are written does not affect the system of equations, interchanging the equations does not affect its solution. The second operation restates the property that says multiplying each side of an equation by the same nonzero constant does not change the solutions of the equation.

The third operation can be illustrated as follows. Consider the system of equations

$$\begin{cases} 3x + 2y = 10 & (1) \\ 2x - 3y = -2 & (2) \end{cases}$$

Multiply each side of Equation (2) by 2. (Any nonzero number would work.) Add the resulting equation to Equation (1).

$$\begin{aligned} 3x + 2y &= 10 && \text{■ Equation (1)} \\ 4x - 6y &= -4 && \text{■ 2 times Equation (2)} \\ \hline 7x - 4y &= 6 && \text{■ Add the equations.} \end{aligned}$$

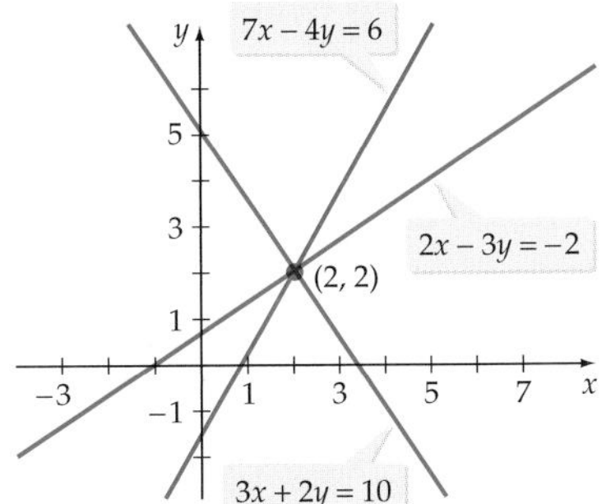

Figure 6.8

Replace Equation (1) with the new Equation (3) to produce the following equivalent system of equations.

$$\begin{cases} 7x - 4y = 6 & (3) \\ 2x - 3y = -2 & (2) \end{cases}$$

The third operation states that the above system of equations has the same solutions as the original system and is therefore equivalent to the original system of equations. Figure 6.8 shows the graph of $7x - 4y = 6$. Note that the line passes through the same point at which the lines of the original system of equations intersect, the point (2, 2).

Alternative to Example 5

Solve: $\begin{cases} 8x + 5y = 9 \\ 3x - 2y = -16 \end{cases}$

■ (–2, 5)

EXAMPLE 5 Solve a System of Equations by the Elimination Method

Solve: $\begin{cases} 3x - 4y = 10 & (1) \\ 2x + 5y = -1 & (2) \end{cases}$

Solution

Use the operations that produce equivalent systems to eliminate a variable from one of the equations. We will eliminate x from Equation (2) by multiplying each equation by a different constant to produce a new system of equations in which the coefficients of x are additive inverses.

$$\begin{aligned} 6x - 8y &= 20 && \text{■ 2 times Equation (1)} \\ -6x - 15y &= 3 && \text{■ −3 times Equation (2)} \\ \hline -23y &= 23 && \text{■ Add the equations.} \\ y &= -1 && \text{■ Solve for } y. \end{aligned}$$

Solve Equation (1) for x by substituting -1 for y.

$$\begin{aligned} 3x - 4(-1) &= 10 \\ 3x &= 6 \\ x &= 2 \end{aligned}$$

The solution of the system of equations is (2, –1).

Visualize the Solution

Graphing $3x - 4y = 10$ and $2x + 5y = -1$ shows that (2, –1) is the only point that belongs to both lines. Therefore, (2, –1) is the solution of the system of equations. See Figure 6.9.

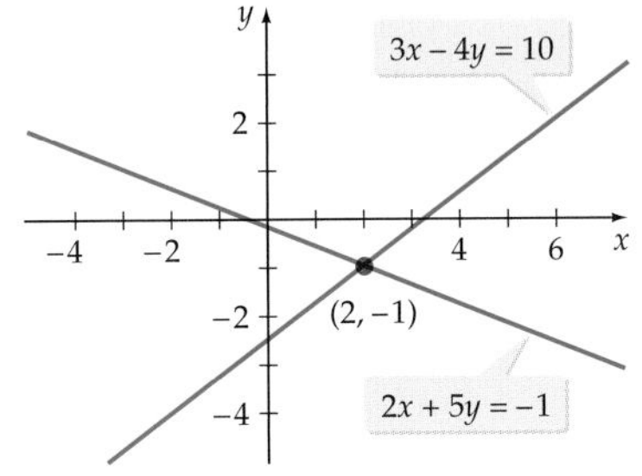

Figure 6.9

Continued ➤

CHECK YOUR PROGRESS 5 Use the elimination method to solve the following system of equations.

$$\begin{cases} 3x - 8y = -6 \\ -5x + 4y = 10 \end{cases}$$

Solution *See page S31.* $(-2, 0)$

The method just described is called the **elimination method** for solving a system of equations because it involves *eliminating* a variable from one of the equations.

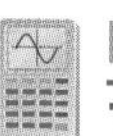

INTEGRATING TECHNOLOGY

You can use a graphing calculator to solve a system of equations in two variables. First, algebraically solve each equation for y.

$$3x - 4y = 10 \rightarrow y = 0.75x - 2.5$$
$$2x + 5y = -1 \rightarrow y = -0.4x - 0.2$$

Now graph the equations. Enter $0.75x - 2.5$ into Y1 and $-0.4x - 0.2$ into Y2 and graph the two equations in the standard viewing window. The sequence of steps shown in Figure 6.10 can be used to find the point of intersection with a TI-83 calculator.

Press 2nd CALC.
Select 5: intersect.
Press ENTER.

CALCULATE
1: value
2: zero
3: minimum
4: maximum
5: intersect
6: dy/dx
7: ∫f(x)dx

The "First curve?" shown on the bottom of the screen means to select the first of the two graphs that intersect. Just press ENTER.

10
Y1=.75X–2.5
–10 10
First curve?
X=0 Y=–2.5
–10

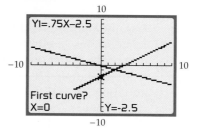

The "Second curve?" shown on the bottom of the screen means to select the second of the two graphs that intersect. Just press ENTER.

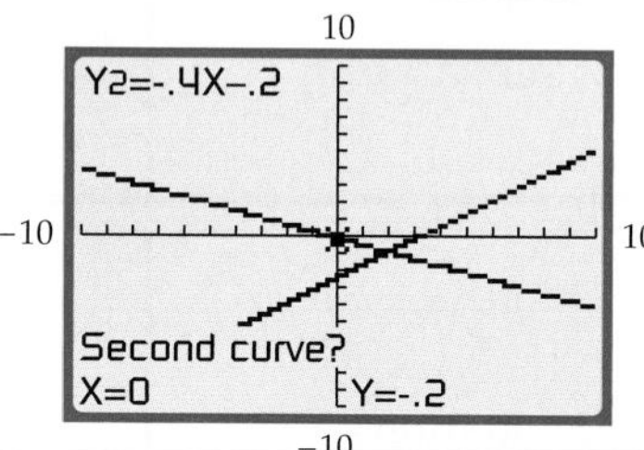

"Guess?" is shown on the bottom of the screen. Move the cursor until approximately on the point of it is intersection. Press ENTER.

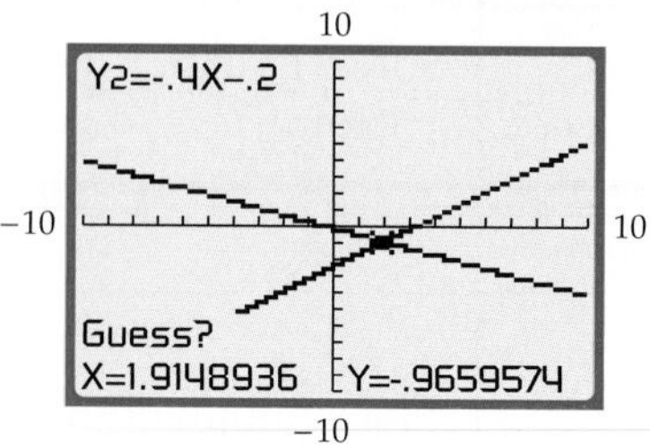

The coordinates of the point of intersection, (2, −1), are shown at the bottom of the screen.

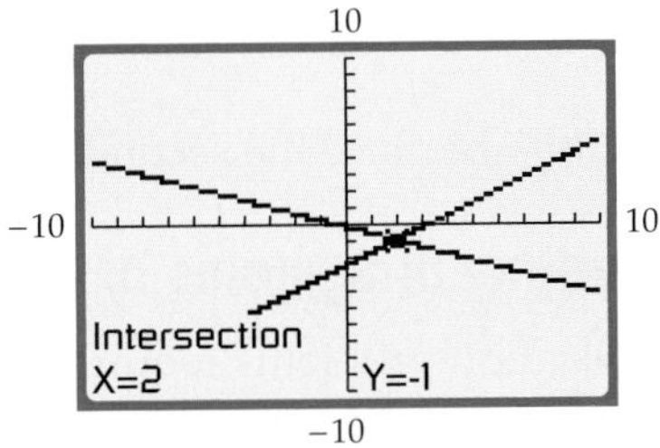

Figure 6.10

For the system of equations in Example 5, the intersection of the two graphs occurs at a point in the standard viewing window. If the point of intersection does not appear on the screen, you must adjust the viewing window so that the point of intersection is visible.

Alternative to Example 6

Solve: $\begin{cases} 2x + 6y = 4 \\ 5x + 15y = 10 \end{cases}$

■ $\left(c, -\frac{1}{3}c + \frac{2}{3}\right)$

EXAMPLE 6 Solve a Dependent System of Equations

Solve: $\begin{cases} x - 2y = 2 & (1) \\ 3x - 6y = 6 & (2) \end{cases}$

Solution Eliminate x by multiplying Equation (2) by $-\frac{1}{3}$ and then adding the result to Equation (1).

$$x - 2y = 2 \qquad \text{■ Equation (1)}$$
$$-x + 2y = -2 \qquad \text{■ } -\tfrac{1}{3} \text{ times Equation (2)}$$
$$0 = 0 \qquad \text{■ Add the two equations.}$$

Replace Equation (2) by $0 = 0$.

$$\begin{cases} x - 2y = 2 \\ 0 = 0 \end{cases} \qquad \text{■ This is an equivalent system of equations.}$$

Because the equation $0 = 0$ is a true statement, an ordered pair that is a solution of Equation (1) is also a solution of $0 = 0$. Thus the solutions are the solutions of $x - 2y = 2$. Solving for y, we find that $y = \frac{1}{2}x - 1$. Because x can be replaced by any real number c, the solutions of the system equations are the ordered pairs $\left(c, \frac{1}{2}c - 1\right)$. See Figure 6.11.

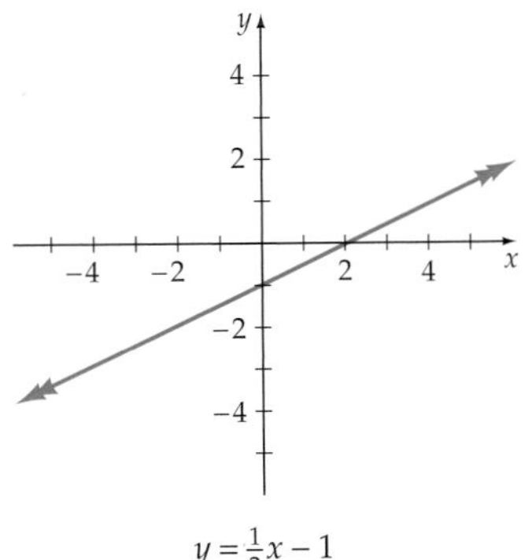

Figure 6.11

CHECK YOUR PROGRESS 6 Use the elimination method to solve the system of equations.

$$\begin{cases} 4x + 5y = 2 \\ 8x + 10y = 4 \end{cases}$$

Solution *See page S31.* $\left(c, -\frac{4}{5}c + \frac{2}{5}\right)$

TAKE NOTE

Referring again to Example 6 and solving Equation (1) for x, we have $x = 2y + 2$. Because y can be any real number b, the ordered-pair solutions of the system of equations can also be written as $(2b + 2, b)$.

If one equation of a system of equations is replaced by a false equation, the system of equations has no solution. For example, the system of equations

$$\begin{cases} x + y = 4 \\ 0 = 5 \end{cases}$$

has no solution because the second equation is false for any choice of x and y.

Applications of Systems of Linear Equations

Many application problems require the use of more than one variable. In the following examples, a system of linear equations in two variables can be used.

Alternative to Example 7

A customer at a fast-food restaurant ordered three hamburgers and three sodas for a total of \$14.10. The next customer ordered five hamburgers and two sodas for a total of \$19.15. What is the price of a hamburger and what is the price of a soda?

- **hamburger: \$3.25**
  soda: \$1.45

EXAMPLE 7 Solve an Application

A company has decided to purchase a new laptop computer for each of its 75 employees. Two models are available, a standard model and a faster (but more expensive) version. The dealer offers to sell 40 standard laptops and 35 faster models for a total of \$91,250, or 25 standard laptops and 50 faster models for \$95,000. How much is the dealer charging for each model of laptop?

Solution Let S represent the cost of a standard laptop, and let F represent the cost of the faster laptop model. Then the total cost for 40 standard laptops is $40S$, and the total cost of 35 faster models is $35F$. Thus the first offer from the dealer can be expressed as $40S + 35F = 91{,}250$. Writing the second offer in a similar manner gives the following system of equations.

$$\begin{cases} 40S + 35F = 91{,}250 & (1) \\ 25S + 50F = 95{,}000 & (2) \end{cases}$$

We can solve this system by using the elimination method.

$$\begin{aligned} 200S + 175F &= 456{,}250 && \text{5 times Equation (1)} \\ -200S - 400F &= -760{,}000 && \text{−8 times Equation (2)} \\ -225F &= -303{,}750 && \text{Add the equations.} \\ F &= 1350 && \text{Solve for } F. \end{aligned}$$

Substituting 1350 for F into Equation (1) gives

$$\begin{aligned} 40S + 35(1350) &= 91{,}250 \\ 40S &= 44{,}000 \\ S &= 1100 \end{aligned}$$

Thus the standard laptops are priced at \$1100 each, and the faster models cost \$1350 each.

CHECK YOUR PROGRESS 7 Two families are attending a play. The first family purchased two adult tickets and four child tickets for a total of \$108, and the second family spent \$120 on three adult tickets and three child tickets. What is the cost of an adult ticket and what is the cost of a child ticket?

Solution *See page S31.* Adult ticket: \$26; child ticket: \$14

Alternative to Example 8

A plane flying with the wind traveled 700 miles in 2 hours. Flying against the wind, the plane traveled only 660 miles in 2 hours. Find the rate of the plane in calm air and the rate of the wind.

- **plane: 340 mph**
  wind: 10 mph

EXAMPLE 8 Solve an Application

A rowing team rowing with the current traveled 18 miles in 2 hours. Against the current, the team rowed 10 miles in 2 hours. Find the rate of the boat in calm water and the rate of the current.

Solution Let b represent the rate of the boat in calm water, and let c represent the rate of the current.

The rate of the boat *with the current* is $b + c$.

The rate of the boat *against the current* is $b - c$.

Because the rowing team traveled 18 miles in 2 hours with the current, we use the equation $d = rt$.

$$d = r \cdot t$$

$$18 = (b + c) \cdot 2 \qquad \blacksquare\ d = 18, t = 2$$

$$9 = b + c \qquad \blacksquare\ \text{Divide each side by 2.}$$

Because the team rowed 10 miles in 2 hours against the current, we write

$$10 = (b - c) \cdot 2 \qquad \blacksquare\ d = 10, t = 2$$

$$5 = b - c \qquad \blacksquare\ \text{Divide each side by 2.}$$

Thus we have a system of two linear equations in the variables b and c.

$$\begin{cases} 9 = b + c \\ 5 = b - c \end{cases}$$

Solving the system by using the elimination method, we find that b is 7 miles per hour and c is 2 miles per hour. Thus the rate of the boat in calm water is 7 miles per hour and the rate of the current is 2 miles per hour. You should verify these solutions.

CHECK YOUR PROGRESS 8 A canoeist can travel 12 miles with the current in 2 hours. Paddling against the current, it takes the canoeist 4 hours to travel the same distance. Find the rate of the canoeist in calm water and the rate of the current.

Solution *See page S31.* Canoeist: 4.5 mph; current: 1.5 mph

Topics for Discussion

1. Explain how to find the solution of a system of linear equations in two variables graphically.
2. Explain how to use the substitution method to solve a system of linear equations in two variables.
3. Explain how to use the elimination method to solve a system of linear equations in two variables.
4. Give an example of a system of linear equations in two variables that is

   a. independent **b.** dependent **c.** inconsistent

5. How many solutions does a linear equation in two variables have? What are the possible numbers of solutions a system of linear equations in two variables can have?

6. If a system of linear equations in two variables has no solution, what does that tell us about the graphs of the equations of the system?

EXERCISES 6.1

— *Suggested Assignment: Exercises 1–39, odd; 43–49, odd; 52, 55, 57, and 60–65.*

In Exercises 1 to 4, use the graph to find the solution, if possible, of the system of equations. Then check your answer using the equations.

1. $\begin{cases} y = -x - 1 \\ y = 2x - 7 \end{cases}$ (2, −3)

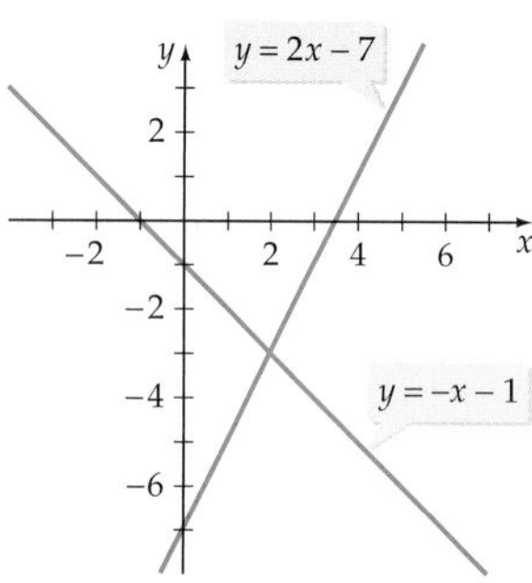

2. $\begin{cases} y = \frac{1}{3}x - 1 \\ y = -\frac{1}{2}x + 4 \end{cases}$ (6, 1)

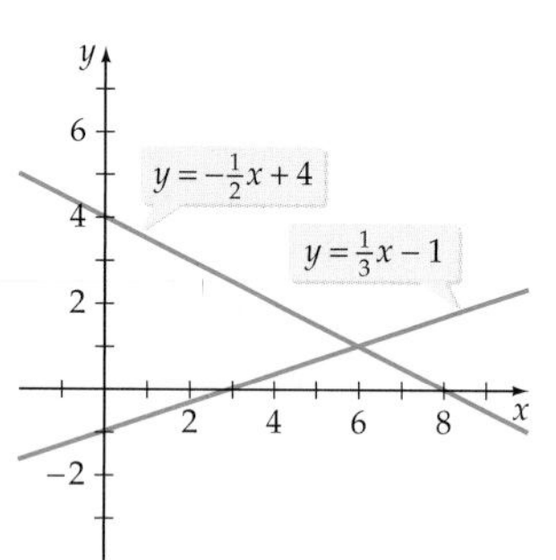

3. $\begin{cases} 2x - 3y = -5 \\ 4y - x = 5 \end{cases}$ (−1, 1)

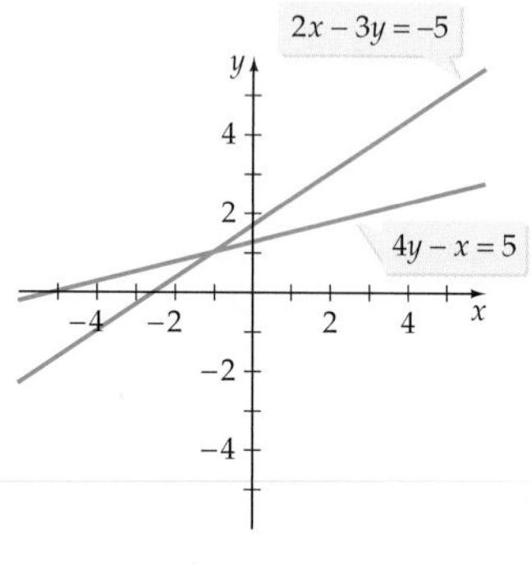

4. $\begin{cases} 2x + 3y = 5 \\ 4x + 6y = 3 \end{cases}$ No solution

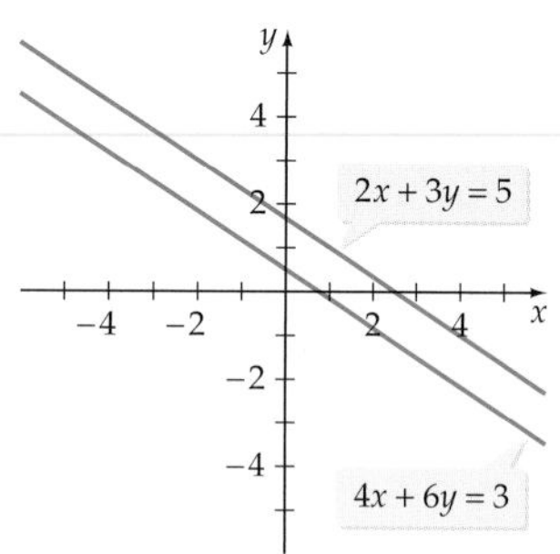

In Exercises 5 to 24, solve each system of equations by the substitution method.

5. $\begin{cases} 2x - 3y = 16 \\ x = 2 \end{cases}$ (2, −4)

6. $\begin{cases} 3x - 2y = -11 \\ y = 1 \end{cases}$ (−3, 1)

7. $\begin{cases} 3x + 4y = 18 \\ y = -2x + 3 \end{cases}$ $\left(-\frac{6}{5}, \frac{27}{5}\right)$

8. $\begin{cases} 5x - 4y = -22 \\ y = 5x - 2 \end{cases}$ (2, 8)

9. $\begin{cases} -2x + 3y = 6 \\ x = 2y - 5 \end{cases}$ (3, 4)

10. $\begin{cases} 5x + 6y = -2 \\ x = 4y + 7 \end{cases}$ $\left(\frac{17}{13}, -\frac{37}{26}\right)$

11. $\begin{cases} 6x + 5y = 1 \\ x - 3y = 4 \end{cases}$ (1, −1)

12. $\begin{cases} -3x + 7y = 14 \\ 2x - y = -13 \end{cases}$ (−7, −1)

13. $\begin{cases} 7x + 6y = -3 \\ y = \frac{2}{3}x - 6 \end{cases}$ (3, −4)

14. $\begin{cases} 9x - 4y = 3 \\ x = \frac{4}{3}y + 3 \end{cases}$ (−1, −3)

15. $\begin{cases} y = 4x - 3 \\ y = 3x - 1 \end{cases}$ (2, 5)

16. $\begin{cases} y = 5x + 1 \\ y = 4x - 2 \end{cases}$ (−3, −14)

17. $\begin{cases} y = 5x + 4 \\ x = -3y - 4 \end{cases}$ (−1, −1)

18. $\begin{cases} y = -2x - 6 \\ x = -2y - 2 \end{cases}$ $\left(-\frac{10}{3}, \frac{2}{3}\right)$

19. $\begin{cases} 3x - 4y = 2 \\ 4x + 3y = 14 \end{cases}$ $\left(\frac{62}{25}, \frac{34}{25}\right)$

20. $\begin{cases} 6x + 7y = -4 \\ 2x + 5y = 4 \end{cases}$ $(-3, 2)$

21. $\begin{cases} 3x - 3y = 5 \\ 4x - 4y = 9 \end{cases}$ No solution

22. $\begin{cases} 5x - 3y = -2 \\ 10x - 6y = 3 \end{cases}$ No solution

23. $\begin{cases} 4x + 3y = 6 \\ y = -\frac{4}{3}x + 2 \end{cases}$ $\left(c, -\frac{4}{3}c + 2\right)$

24. $\begin{cases} 3x - 7y = 5 \\ y = \frac{3}{7}x - \frac{5}{7} \end{cases}$ $\left(c, \frac{3}{7}c - \frac{5}{7}\right)$

In Exercises 25 to 40, solve each system of equations by the elimination method.

25. $\begin{cases} 3x - y = 10 \\ 4x + 3y = -4 \end{cases}$ $(2, -4)$

26. $\begin{cases} 3x + 4y = -5 \\ x - 5y = -8 \end{cases}$ $(-3, 1)$

27. $\begin{cases} 4x + 7y = 21 \\ 5x - 4y = -12 \end{cases}$ $(0, 3)$

28. $\begin{cases} 5x + 6y = -15 \\ -2x - 3y = 6 \end{cases}$ $(-3, 0)$

29. $\begin{cases} 5x - 3y = 0 \\ 10x - 6y = 0 \end{cases}$ $\left(\frac{3c}{5}, c\right)$

30. $\begin{cases} 3x + 2y = 0 \\ 2x + 3y = 0 \end{cases}$ $(0, 0)$

31. $\begin{cases} 6x + 6y = 1 \\ 4x + 9y = 4 \end{cases}$ $\left(-\frac{1}{2}, \frac{2}{3}\right)$

32. $\begin{cases} 3x + 8y = 0 \\ 6x + 4y = -3 \end{cases}$ $\left(-\frac{2}{3}, \frac{1}{4}\right)$

33. $\begin{cases} 3x + 6y = 11 \\ 2x + 4y = 9 \end{cases}$ No solution

34. $\begin{cases} 4x - 2y = 9 \\ 2x - y = 3 \end{cases}$ No solution

35. $\begin{cases} \frac{5}{6}x - \frac{1}{3}y = -6 \\ \frac{1}{6}x + \frac{2}{3}y = 1 \end{cases}$ $(-6, 3)$

36. $\begin{cases} \frac{3}{4}x + \frac{2}{5}y = 1 \\ \frac{1}{2}x - \frac{3}{5}y = -1 \end{cases}$ $\left(\frac{4}{13}, \frac{25}{13}\right)$

37. $\begin{cases} \frac{3}{4}x + \frac{1}{3}y = 1 \\ \frac{1}{2}x + \frac{2}{3}y = 0 \end{cases}$ $\left(2, -\frac{3}{2}\right)$

38. $\begin{cases} \frac{3}{5}x - \frac{2}{3}y = 7 \\ \frac{2}{5}x - \frac{5}{6}y = 7 \end{cases}$ $(5, -6)$

39. $\begin{cases} 0.5x - 1.8y = 1 \\ 2.1x + 0.6y = 0.12 \end{cases}$ $(0.2, -0.5)$

40. $\begin{cases} 5.4x + 3.6y = 39.6 \\ -4.8x + 2.4y = -7.2 \end{cases}$ $(4, 5)$

In Exercises 41 and 42, solve by using a system of equations.

41. ***Geometry*** A right triangle in the first quadrant is bounded by the lines $y = 0$, $y = \frac{1}{2}x$, and $y = -2x + 6$. Find its area. $\frac{9}{5}$ square units

42. ***Geometry*** The lines whose equations are $2x + 3y = 1$, $3x - 4y = 10$, and $4x + ky = 5$ all intersect at the same point. What is the value of k? 3

Business and Economics

43. ***Taxi Fare*** Two taxi cabs are available at the airport. One charges \$2.20 plus \$1.50 per mile, and the other charges \$3.70 plus \$1.20 per mile. For what number of miles traveled do both taxi cabs charge the same fare? 5 mi

44. ***Fax Charges*** One copy shop charges \$2.00 to send a fax, plus \$0.55 per page. Another simply charges \$0.80 per page sent. For how many pages faxed do the two businesses charge the same amount? 8 pages

45. ***Investment*** A broker invests \$25,000 of a client's money in two different municipal bonds. The annual rate of return on one bond is 6%, and the annual rate of return on the second bond is 6.5%. The investor receives a total annual interest payment from the two bonds of \$1555. Find the amount invested in each bond. \$14,000 at 6%, \$11,000 at 6.5%

46. ***Investment*** An investment of \$3000 is placed in stocks and bonds. The annual rate of return on the stocks is 4.5%, and the rate of return on the bonds is 8%. The annual return from the stocks and bonds is \$177. Find the amount invested in bonds. \$1200

Life and Health Sciences

47. ***Vitamin Supplements*** A pharmacist has two vitamin-supplement powders. The first powder is 25% vitamin B_1 and 15% vitamin B_2. The second is 15% vitamin B_1 and 20% vitamin B_2. How many milligrams of each of the two powders should the pharmacist use to make a mixture that contains 117.5 milligrams of vitamin B_1 and 120 milligrams of vitamin B_2? First powder: 200 mg, second powder: 450 mg

48. ***Veterinary Supplements*** A veterinarian is preparing a powder vitamin supplement for dogs by mixing one product that is 1.6% niacin (by weight) and 0.3% riboflavin with a second product that is 1.2% niacin and 0.25% riboflavin. How many milligrams of each product should the veterinarian use to make a mixture that contains 12 milligrams of niacin and 2.4 milligrams of riboflavin? First powder: 300 mg, second powder: 600 mg

Sports and Recreation

49. ***Rate of Wind*** Flying with the wind, a plane traveled 450 miles in 3 hours. Flying against the wind, the plane traveled the same distance in 5 hours. Find the rate of the plane in calm air and the rate of the wind. Plane: 120 mph, wind: 30 mph

50. *Rate of Wind* A plane flew 800 miles in 4 hours while flying with the wind. Against the wind, it took the plane 5 hours to travel 800 miles. Find the rate of the plane in calm air and the rate of the wind. Plane: 180 mph, wind: 20 mph

51. *Rate of Current* A motorboat traveled a distance of 120 miles in 4 hours while traveling with the current. Against the current, the same trip took 6 hours. Find the rate of the boat in calm water and the rate of the current. Boat: 25 mph, current: 5 mph

52. *Refreshment Prices* At the movie theater, a family purchased three sodas and three bags of popcorn for a total cost of \$15. Another family bought two sodas and five bags of popcorn totaling \$16.75. What is the cost of a soda and what is the cost of a bag of popcorn? Soda: \$2.75; popcorn: \$2.25

53. *Pizza Prices* A pizzeria baked five pepperoni pizzas and three vegetarian pizzas for a party and charged a total of \$94.00. Another group ordered four pepperoni and six vegetarian pizzas for a total cost of \$113. What is the price of a pepperoni pizza and the price of a vegetarian pizza? Pepperoni: \$12.50, vegetarian: \$10.50

54. *Rate of Current* A ferry can travel 45 miles with the current in 3 hours. Moving against the current, it takes the ferry 5 hours to travel the same distance. Find the rate of the ferry in calm water and the rate of the current. Ferry: 12 mph, current: 3 mph

Physical Sciences and Engineering

55. *Metallurgy* A metallurgist made two purchases. The first purchase, which cost \$1080, included 30 kilograms of an iron alloy and 45 kilograms of a lead alloy. The second purchase, at the same prices, cost \$372 and included 15 kilograms of the iron alloy and 12 kilograms of the lead alloy. Find the cost per kilogram of the iron and lead alloys. Iron: \$12/kg, lead: \$16/kg

56. *Chemistry* For \$14.10, a chemist purchased 10 liters of hydrochloric acid and 15 liters of silver nitrate. A second purchase, at the same prices, cost \$18.16 and included 12 liters of hydrochloric acid and 20 liters of silver nitrate. Find the cost per liter of each of the two chemicals. Hydrochloric acid: \$0.48/L, silver nitrate: \$0.62/L

57. *Chemistry* A goldsmith has two gold alloys. The first alloy is 40% gold; the second alloy is 60% gold. How many grams of each should be mixed to produce 20 grams of an alloy that is 52% gold? 8 gm of 40% gold, 12 gm of 60% gold

58. *Chemistry* One acetic acid solution is 70% water and another is 30% water. How many liters of each solution should be mixed to produce 20 liters of a solution that is 40% water? 70% solution: 5 L, 30% solution: 15 L

59. *Chemistry* A chemist wants to make 50 milliliters of a 16% acid solution. How many milliliters each of a 13% acid solution and an 18% acid solution should be mixed to produce the desired solution? 20 ml of 13% solution, 30 ml of 18% solution

Prepare for Section 6.2

60. Solve the equation $5a - 2b + 7c = 9$ for b. [1.1] $b = \dfrac{5a + 7c - 9}{2}$

61. If $A = 2x - 5y$ and $B = -3x - 2y$, find $4A - 6B$. [1.1] $26x - 8y$

62. If $4x + 7y - 2z = -8$, $-3y + 6z = 24$, and $z = 5$, find the values of x and y. [1.1] $x = -3, y = 2$

63. Solve the system of equations using the substitution method: $\begin{cases} 5x - 3y = 6 \\ 4x + 2y = -20 \end{cases}$ [6.1] $\left(-\dfrac{24}{11}, -\dfrac{62}{11}\right)$

64. Solve the system of equations using the elimination method: $\begin{cases} 5x + 8y = 4 \\ -3x + 2y = 20 \end{cases}$ [6.1] $\left(-\dfrac{76}{17}, \dfrac{56}{17}\right)$

65. If a system of linear equations in two variables is dependent, how many solutions does the system have? How do you express these solutions? [6.1] An infinite number; in terms of an arbitrary constant

Explorations 

1. *Independent and Dependent Conditions* Consider the following problem: "Maria and Michael drove from Los Angeles to New York in 60 hours. How long did Maria drive?" It is not possible to answer this question without additional information. She may have driven all 60 hours while Michael relaxed, or she may have relaxed while Michael drove all 60 hours. The difficulty is that there are two unknowns (how long each person drove) and only one condition (the total driving time) relating the unknowns. If we added another condition, such as Michael drove 25 hours, then we could determine how long Maria drove, which is 35 hours. In most cases, an application problem will have a single answer only when there are as many *independent* conditions as there are variables. Two conditions are independent if knowing one does *not* allow you to know the other.

   Here is an example of two conditions that are not independent: "The perimeter of a rectangle is 50 meters. The sum of the width and length is 25 meters." To see that these conditions are dependent, write the perimeter equation and divide each side by 2.

$$2w + 2l = 50$$
$$w + l = 25 \quad \text{▪ Divide each side by 2.}$$

Note that the resulting equation is the second condition: the sum of the width and length is 25. Thus knowing the first condition allows us to determine the second condition. The conditions are not independent, so there is no one solution to this problem.

For each of the problems below, determine whether the conditions are independent or dependent. For those problems that have independent conditions, find the solution (if possible). For those problems for which the conditions are dependent, find two solutions.

a. The sum of two numbers is 30. The difference between the two numbers is 10. Find the numbers.
Independent; 20 and 10

b. The area of a square is 25 square meters. Find the length of each side. Independent; 5 m

c. The area of a rectangle is 25 square meters. Find the length of each side. Dependent; 2 m by 12.5 m or 4 m by 6.25 m

d. Emily spent $1000 for carpeting and tile. Carpeting costs $20 per square yard and tile costs $30 per square yard. How many square yards of each did she purchase?
Dependent; 20 yd^2 of each or 17 yd^2 of carpet and 22 yd^2 of tile

e. The sum of two numbers is 20. Twice the smaller number is 10 minus twice the larger number. Find the two numbers. Independent; no solution

f. Make up a word problem for which there are two independent conditions. Solve the problem.
Answers will vary.

g. Make up a word problem for which there are two dependent conditions. Find at least two solutions. Answers will vary.

SECTION 6.2 Systems of Linear Equations in Three Variables

- Systems of Equations in Three Variables
- The Substitution Method
- The Elimination Method
- Dependent Systems of Equations
- Nonsquare Systems of Equations
- Applications of Systems of Linear Equations in Three Variables

▪ Systems of Equations in Three Variables

An equation of the form $Ax + By + Cz = D$, with A, B, and C not all zero, is a **linear equation in three variables.** A solution of an equation in three variables is an **ordered triple** (x, y, z).

The ordered triple $(2, -1, -3)$ is one of the solutions of the equation $2x - 3y + z = 4$. The ordered triple $(3, 1, 1)$ is another solution. In fact, an infinite number of ordered triples are solutions of the equation.

Graphing an equation in three variables requires a third coordinate axis perpendicular to the xy-plane. This third axis is commonly called the **z-axis.** The result is a three-dimensional coordinate system called the **xyz-coordinate system.** To help visualize a three-dimensional coordinate system, think of a corner of a room: the floor is the xy-plane, one wall is the yz-plane, and the other wall is the xz-plane. (See Figure 6.12 on the following page.)

Graphing an ordered triple requires three moves, the first along the x-axis, the second along the y-axis, and the third along the z-axis. Figure 6.13 is the graph of the points $(-5, -4, 3)$ and $(4, 5, -2)$.

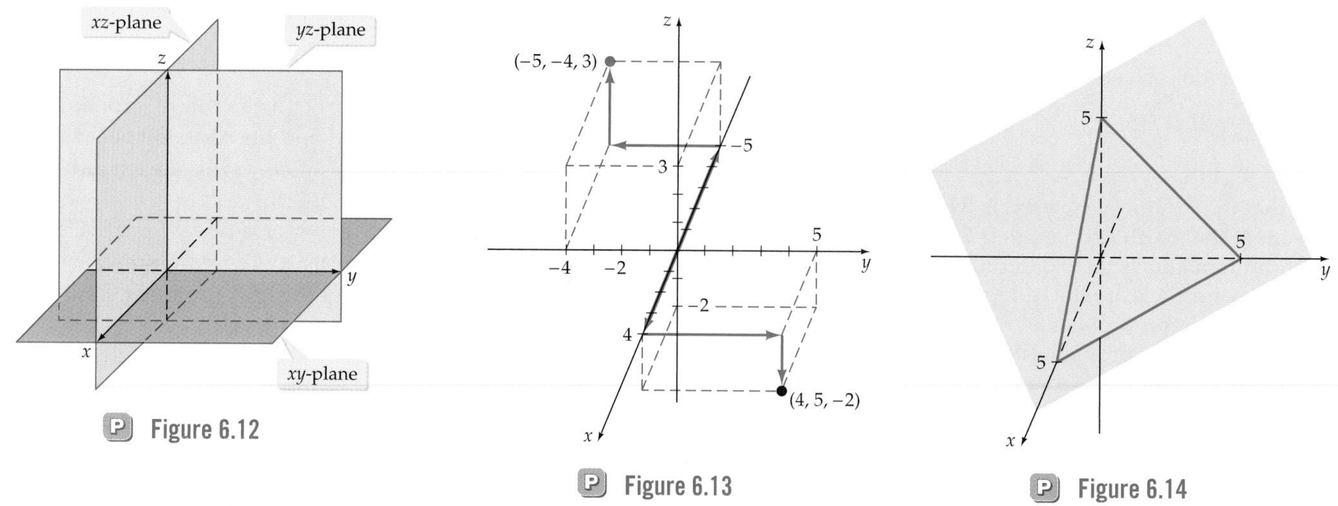

Figure 6.12

Figure 6.13

Figure 6.14

The graph of a linear equation in three variables is a plane. That is, if all the solutions of a linear equation in three variables were plotted in an xyz-coordinate system, the graph would look like a large piece of paper with infinite extent. Figure 6.14 is a portion of the graph of $x + y + z = 5$.

There are different ways in which three planes can be oriented in an xyz-coordinate system. Figure 6.15 illustrates several ways.

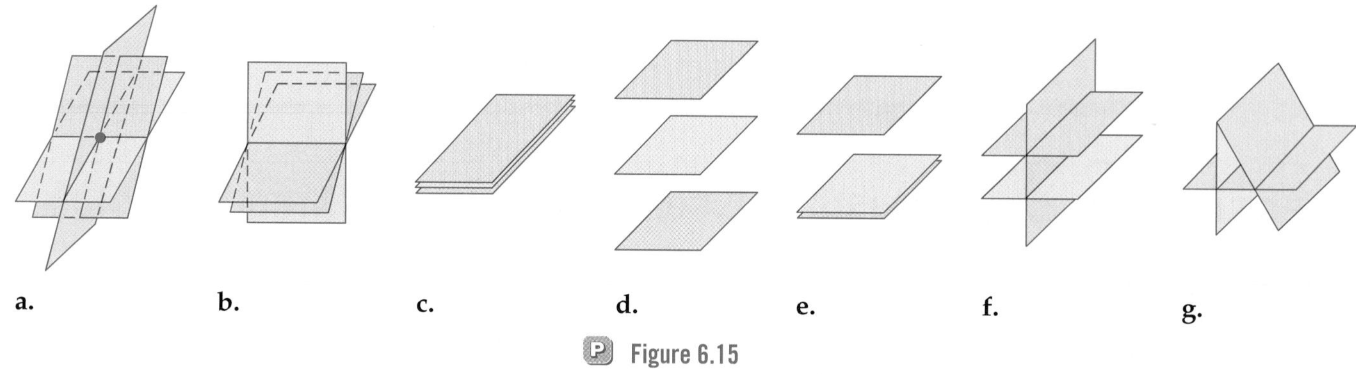

Figure 6.15

For a linear system of equations in three variables to have a solution, the graphs of the planes must intersect at a single point, they must intersect along a common line, or all equations must have a graph that is the same plane. In Figure 6.15, the graphs in **a.**, **b.**, and **c.** represent systems of equations that have a solution. The system of equations represented in Figure 6.15a is a consistent system of equations. Figures 6.15b and 6.15c are graphs of dependent systems of equations. The remaining graphs are examples of inconsistent systems of equations.

QUESTION Can a linear system in three variables have exactly two solutions?

ANSWER No. Two or more planes cannot intersect at more than one point unless the intersection is a line or a plane.

The Substitution Method

A system of equations in more than two variables can be solved by using the substitution method or the elimination method. The substitution method is illustrated in the following example.

Alternative to Example 1

Solve: $\begin{cases} 2x + 4y - z = -3 \\ x - 2y - 2z = 2 \\ -2x + y + 3z = -1 \end{cases}$

■ (4, −2, 3)

EXAMPLE 1 Solve a Linear System in Three Variables Using Substitution

Solve: $\begin{cases} x - 2y + z = 7 & (1) \\ 2x + y - z = 0 & (2) \\ 3x + 2y - 2z = -2 & (3) \end{cases}$

Solution Solve Equation (1) for x and substitute the result into Equations (2) and (3).

$$x = 2y - z + 7 \quad (4)$$

$$2(2y - z + 7) + y - z = 0$$ ■ Substitute $2y - z + 7$ for x in Equation (2).

$$4y - 2z + 14 + y - z = 0$$ ■ Simplify.

$$5y - 3z = -14 \quad (5)$$

$$3(2y - z + 7) + 2y - 2z = -2$$ ■ Substitute $2y - z + 7$ for x in Equation (3).

$$6y - 3z + 21 + 2y - 2z = -2$$ ■ Simplify.

$$8y - 5z = -23 \quad (6)$$

Now solve the system of equations formed from Equations (5) and (6).

$$\begin{cases} 5y - 3z = -14 & \text{multiply by } 8 \rightarrow & 40y - 24z = -112 \\ 8y - 5z = -23 & \text{multiply by } -5 \rightarrow & -40y + 25z = 115 \end{cases}$$
$$z = 3$$

Substitute 3 for z into Equation (5) and solve for y.

$$5y - 3z = -14$$ ■ Equation (5)

$$5y - 3(3) = -14$$
$$5y - 9 = -14$$
$$5y = -5$$
$$y = -1$$

Substitute −1 for y and 3 for z into Equation (4) and solve for x.

$$x = 2y - z + 7 = 2(-1) - (3) + 7 = 2$$

The ordered triple solution is (2, −1, 3). The graphs of the three planes intersect at a single point.

TAKE NOTE

You can check the solution to Example 1 by entering 2 for x, −1 for y, and 3 for z into all three equations of the original system:

$$2 - 2(-1) + 3 = 7$$
$$2(2) + (-1) - 3 = 0$$
$$3(2) + 2(-1) - 2(3) = -2$$

CHECK YOUR PROGRESS 1 Use the substitution method to solve the system:

$$\begin{cases} 3x + 2y - 2z = -5 \\ 2x - y - 4z = 3 \\ 2x + 3y + 3z = -3 \end{cases}$$

Solution *See page S32.* (3, −5, 2)

The Elimination Method

There are many approaches one can take to determine the solution of a system of equations by using the elimination method. For consistency, we will always follow a plan that produces an equivalent system of equations in **triangular form.** Two examples of systems of equations in triangular form are

$$\begin{cases} 2x - 3y + z = -4 \\ 2y + 3z = 9 \\ -2z = -2 \end{cases} \qquad \begin{cases} 3x - 4y + z = 1 \\ 3y + 2z = 3 \end{cases}$$

Once a system of equations is written in triangular form, the solution can be found by **back substitution**—that is, by solving the last equation of the system and substituting *back* into the previous equation. This process is continued until the value of each variable has been found.

As an example of solving a system of equations by back substitution, consider the following system of equations in triangular form.

$$\begin{cases} 2x - 4y + z = -3 & (1) \\ 3y - 2z = 9 & (2) \\ 3z = -9 & (3) \end{cases}$$

Solve Equation (3) for z. Substitute the value of z into Equation (2) and solve for y.

$$\begin{aligned} 3z &= -9 && \text{▪ Equation (3)} \\ z &= -3 \end{aligned} \qquad \begin{aligned} 3y - 2z &= 9 && \text{▪ Equation (2)} \\ 3y - 2(-3) &= 9 && \text{▪ } z = -3 \\ 3y &= 3 \\ y &= 1 \end{aligned}$$

Substitute -3 for z and 1 for y in Equation (1) and then solve for x.

$$\begin{aligned} 2x - 4y + z &= -3 && \text{▪ Equation (1)} \\ 2x - 4(1) + (-3) &= -3 \\ 2x - 7 &= -3 \\ x &= 2 \end{aligned}$$

The solution is the ordered triple $(2, 1, -3)$.

The next example illustrates a systematic approach to produce an equivalent system of equations in triangular form.

Alternative to Example 2

Solve: $\begin{cases} x - 2y - z = -5 \\ 3x + y + z = 9 \\ 2x - y - z = 1 \end{cases}$

▪ $(2, 4, -1)$

EXAMPLE 2 Solve a Linear System in Three Variables Using Elimination

Solve: $\begin{cases} x + 2y - z = 1 & (1) \\ 2x - y + z = 6 & (2) \\ 2x - y - z = 0 & (3) \end{cases}$

Continued ➤

Solution Eliminate x from Equation (2) by multiplying Equation (1) by -2 and then adding it to Equation (2). Replace Equation (2) by the new equation.

$$\begin{array}{rrl} -2x - 4y + 2z = -2 & & \text{■ } -2 \text{ times Equation (1)} \\ 2x - y + z = 6 & & \text{■ Equation (2)} \\ \hline -5y + 3z = 4 & (4) & \text{■ Add the equations.} \end{array}$$

$$\begin{cases} x + 2y - z = 1 & (1) \\ -5y + 3z = 4 & (4) \\ 2x - y - z = 0 & (3) \end{cases} \quad \text{■ Replace Equation (2) with Equation (4).}$$

Eliminate x from Equation (3) by multiplying Equation (1) by -2 and adding it to Equation (3). Replace Equation (3) by the new equation.

$$\begin{array}{rrl} -2x - 4y + 2z = -2 & & \text{■ } -2 \text{ times Equation (1)} \\ 2x - y - z = 0 & & \text{■ Equation (3)} \\ \hline -5y + z = -2 & (5) & \text{■ Add the equations.} \end{array}$$

$$\begin{cases} x + 2y - z = 1 & (1) \\ -5y + 3z = 4 & (4) \\ -5y + z = -2 & (5) \end{cases} \quad \text{■ Replace Equation (3) with Equation (5).}$$

INSTRUCTOR NOTE

We often find the solution of a system of equations by producing an equivalent system of equations in triangular form. This will help prepare students for row operations in matrices in Section 6.3. It may be helpful to point out to students that this method is not always the quickest way to the solution. However, the method always works.

Eliminate y from Equation (5) by multiplying Equation (4) by -1 and then adding it to Equation (5). Replace Equation (5) by the new equation.

$$\begin{array}{rrl} 5y - 3z = -4 & & \text{■ } -1 \text{ times Equation (4)} \\ -5y + z = -2 & & \text{■ Equation (5)} \\ \hline -2z = -6 & (6) & \text{■ Add the equations.} \end{array}$$

$$\begin{cases} x + 2y - z = 1 & (1) \\ -5y + 3z = 4 & (4) \\ -2z = -6 & (6) \end{cases} \quad \text{■ Replace Equation (5) with Equation (6).}$$

The system of equations is now in triangular form. Solve the system of equations by back substitution.

Solve Equation (6) for z. Substitute the value into Equation (4) and then solve for y.

$$\begin{aligned} -2z &= -6 && \text{■ Equation (6)} \\ z &= 3 \end{aligned}$$

$$\begin{aligned} -5y + 3z &= 4 && \text{■ Equation (4)} \\ -5y + 3(3) &= 4 && \text{■ Substitute 3 for } z. \\ -5y &= -5 && \text{■ Solve for } y. \\ y &= 1 \end{aligned}$$

Substitute 3 for z and 1 for y in Equation (1) and solve for x.

$$\begin{aligned} x + 2y - z &= 1 && \text{■ Equation (1)} \\ x + 2(1) - 3 &= 1 && \text{■ Substitute 1 for } y, \text{ 3 for } z. \\ x &= 2 \end{aligned}$$

The system of equations is consistent. The solution is the ordered triple (2, 1, 3). See Figure 6.16.

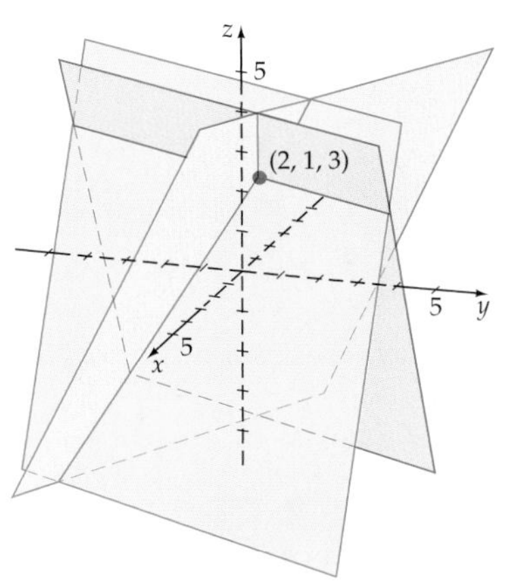

Figure 6.16

Continued ➤

CHECK YOUR PROGRESS 2 Use the elimination method to solve the system of equations.

$$\begin{cases} 3x + 2y - 5z = 6 \\ 5x - 4y + 3z = -12 \\ 4x + 5y - 2z = 15 \end{cases}$$

Solution *See page S32.* (0, 3, 0)

Alternative to Example 3

Solve: $\begin{cases} x - y - z = 6 \\ x - 3y + 2z = 2 \\ x + 5y - 5z = 3 \end{cases}$

■ Inconsistent system

EXAMPLE 3 Identify an Inconsistent System of Equations

Solve: $\begin{cases} x + 2y + 3z = 4 & (1) \\ 2x - y - z = 3 & (2) \\ 3x + y + 2z = 5 & (3) \end{cases}$

Solution Eliminate x from Equation (2) by multiplying Equation (1) by -2 and then adding it to Equation (2). Replace Equation (2). Eliminate x from Equation (3) by multiplying Equation (1) by -3 and adding it to Equation (3). Replace Equation (3). The equivalent system is

$$\begin{cases} x + 2y + 3z = 4 & (1) \\ -5y - 7z = -5 & (4) \\ -5y - 7z = -7 & (5) \end{cases}$$

Eliminate y from Equation (5) by multiplying Equation (4) by -1 and adding it to Equation (5). Replace Equation (5). The equivalent system is

$$\begin{cases} x + 2y + 3z = 4 & (1) \\ -5y - 7z = -5 & (4) \\ 0 = -2 & (6) \end{cases}$$

This system of equations contains a false equation. The system is inconsistent and has no solutions. There is no point on all three planes, as shown in Figure 6.17.

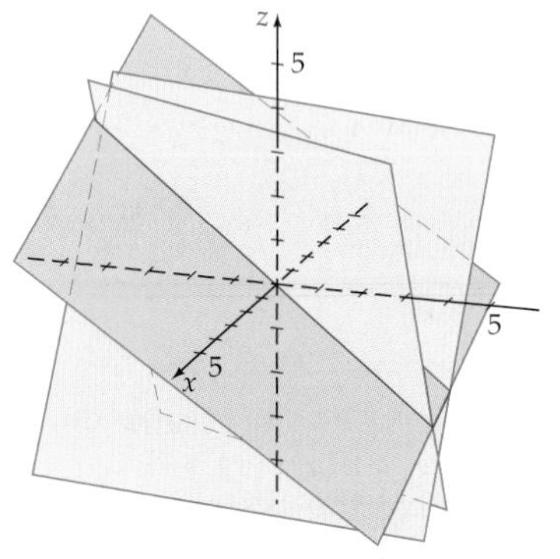

Figure 6.17

CHECK YOUR PROGRESS 3 Solve the system of equations.

$$\begin{cases} -2x + 5y + 3z = 8 \\ x - 3y - 2z = 3 \\ 2x - 4y - 2z = -5 \end{cases}$$

Solution *See page S32.* Inconsistent system

■ Dependent Systems of Equations

The next example shows how to identify a dependent system of linear equations in three variables, and how to describe the infinite number of solutions of the system.

Alternative to Example 4

Solve: $\begin{cases} 4x - 2y + 5z = 11 \\ -2x + 3y - z = -2 \\ 2x + y + 4z = 9 \end{cases}$

■ $\left(-\frac{13}{8}c + \frac{29}{8}, -\frac{3}{4}c + \frac{7}{4}, c\right)$

EXAMPLE 4 Solve a Dependent System of Equations

Solve: $\begin{cases} 2x - y - z = -1 & (1) \\ -x + 3y - z = -3 & (2) \\ -5x + 5y + z = -1 & (3) \end{cases}$

Solution Eliminate x from Equation (2) by multiplying Equation (2) by 2 and then adding it to Equation (1). Replace Equation (2) by the new equation.

$$\begin{array}{rl} 2x - y - z = -1 & \\ -2x + 6y - 2z = -6 & \\ \hline 5y - 3z = -7 & (4) \end{array}$$

■ Equation (1)
■ 2 times Equation (2)
■ Add the equations.

$$\begin{cases} 2x - y - z = -1 & (1) \\ 5y - 3z = -7 & (4) \\ -5x + 5y + z = -1 & (3) \end{cases}$$

■ Replace Equation (2) with Equation (4).

Eliminate x from Equation (3) by multiplying Equation (1) by 5 and multiplying Equation (3) by 2. Then add. Replace Equation (3) by the new equation.

$$\begin{array}{rl} 10x - 5y - 5z = -5 & \\ -10x + 10y + 2z = -2 & \\ \hline 5y - 3z = -7 & (5) \end{array}$$

■ 5 times Equation (1)
■ 2 times Equation (3)
■ Add the equations.

$$\begin{cases} 2x - y - z = -1 & (1) \\ 5y - 3z = -7 & (4) \\ 5y - 3z = -7 & (5) \end{cases}$$

■ Replace Equation (3) with Equation (5).

Eliminate y from Equation (5) by multiplying Equation (4) by -1 and then adding it to Equation (5). Replace Equation (5) by the new equation.

$$\begin{array}{rl} -5y + 3z = 7 & \\ 5y - 3z = -7 & \\ \hline 0 = 0 & (6) \end{array}$$

■ -1 times Equation (4)
■ Equation (5)
■ Add the equations.

$$\begin{cases} 2x - y - z = -1 & (1) \\ 5y - 3z = -7 & (4) \\ 0 = 0 & (6) \end{cases}$$

■ Replace Equation (5) with Equation (6).

Because any ordered triple (x, y, z) is a solution of Equation (6), the solutions of the system of equations will be the ordered triples that are solutions of Equations (1) and (4).

Solve Equation (4) for y.

$$5y - 3z = -7$$
$$5y = 3z - 7$$
$$y = \frac{3}{5}z - \frac{7}{5}$$

Continued ➤

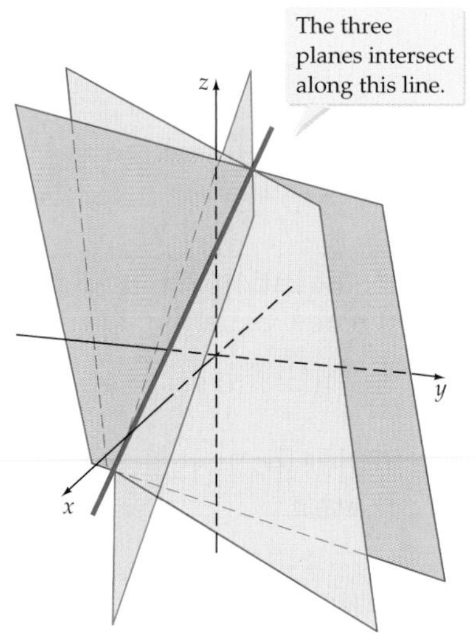
The three planes intersect along this line.
z
y
x

Figure 6.18

Substitute $\frac{3}{5}z - \frac{7}{5}$ for y in Equation (1) and solve for x.

$$2x - y - z = -1 \quad \text{▪ Equation (1)}$$

$$2x - \left(\frac{3}{5}z - \frac{7}{5}\right) - z = -1 \quad \text{▪ Replace } y \text{ by } \frac{3}{5}z - \frac{7}{5}.$$

$$2x - \frac{8}{5}z + \frac{7}{5} = -1 \quad \text{▪ Simplify and solve for } x.$$

$$2x = \frac{8}{5}z - \frac{12}{5}$$

$$x = \frac{4}{5}z - \frac{6}{5}$$

By choosing any real number c for z, we have $y = \frac{3}{5}c - \frac{7}{5}$ and $x = \frac{4}{5}c - \frac{6}{5}$. For any real number c, the ordered triple solutions of the system of equations are $\left(\frac{4}{5}c - \frac{6}{5}, \frac{3}{5}c - \frac{7}{5}, c\right)$. For instance, if we choose $c = 0$, then the corresponding solution is $\left(-\frac{6}{5}, -\frac{7}{5}, 0\right)$. The choice $c = 1$ corresponds to the solution $\left(-\frac{2}{5}, -\frac{4}{5}, 1\right)$. The solid red line shown in Figure 6.18 is a graph of the solutions.

CHECK YOUR PROGRESS 4 Solve: $\begin{cases} 2x + 3y - 6z = 4 \\ 3x - 2y - 9z = -7 \\ 2x + 5y - 6z = 8 \end{cases}$

Solution *See page S32.* $(3c - 1, 2, c)$

TAKE NOTE

Although the ordered triples

$$\left(\frac{4}{5}c - \frac{6}{5}, \frac{3}{5}c - \frac{7}{5}, c\right)$$

and

$$\left(a, \frac{3}{4}a - \frac{1}{2}, \frac{5}{4}a + \frac{3}{2}\right)$$

appear to be different, they represent exactly the same set of ordered triples. For instance, choosing $c = -1$, we have $(-2, -2, -1)$. Choosing $a = -2$ results in the same ordered triple, $(-2, -2, -1)$.

As in the case of a dependent system of equations in two variables, there is more than one way to represent the solutions of a dependent system of equations in three variables. For instance, from Example 4, let $a = \frac{4}{5}c - \frac{6}{5}$, the x-coordinate of the ordered triple $\left(\frac{4}{5}c - \frac{6}{5}, \frac{3}{5}c - \frac{7}{5}, c\right)$, and solve for c.

$$a = \frac{4}{5}c - \frac{6}{5} \rightarrow c = \frac{5}{4}a + \frac{3}{2}$$

Substitute this value of c into each component of the ordered triple.

$$\left(\frac{4}{5}\left(\frac{5}{4}a + \frac{3}{2}\right) - \frac{6}{5}, \frac{3}{5}\left(\frac{5}{4}a + \frac{3}{2}\right) - \frac{7}{5}, \frac{5}{4}a + \frac{3}{2}\right) = \left(a, \frac{3}{4}a - \frac{1}{2}, \frac{5}{4}a + \frac{3}{2}\right)$$

Thus the solutions of the system of equations can also be written as

$$\left(a, \frac{3}{4}a - \frac{1}{2}, \frac{5}{4}a + \frac{3}{2}\right)$$

▪ Nonsquare Systems of Equations

The linear systems of equations that we have solved so far contain the same number of variables as equations. These are square systems of equations. If there are fewer equations than variables—a nonsquare system of equations—the system has either no solution or an infinite number of solutions.

Alternative to Example 5

Solve: $\begin{cases} x + 2y - 3z = 5 \\ 3x + 7y - 10z = 13 \end{cases}$

■ $(c + 9, c - 2, c)$

EXAMPLE 5 Solve a Nonsquare System of Equations

Solve: $\begin{cases} x - 2y + 2z = 3 & (1) \\ 2x - y - 2z = 15 & (2) \end{cases}$

Solution Eliminate x from Equation (2) by multiplying Equation (1) by -2 and adding it to Equation (2). Replace Equation (2).

$$\begin{cases} x - 2y + 2z = 3 & (1) \\ \quad 3y - 6z = 9 & (3) \end{cases}$$

Solve Equation (3) for y.

$$3y - 6z = 9$$
$$y = 2z + 3$$

Substitute $2z + 3$ for y into Equation (1) and solve for x.

$$x - 2y + 2z = 3 \qquad \text{■ Equation (1)}$$
$$x - 2(2z + 3) + 2z = 3 \qquad \text{■ } y = 2z + 3$$
$$x = 2z + 9$$

For each value of z selected, there correspond values of x and y. If z is any real number c, then the solutions of the system are the ordered triples $(2c + 9, 2c + 3, c)$.

CHECK YOUR PROGRESS 5 Solve: $\begin{cases} x - 3y + 4z = 9 \\ 3x - 8y - 2z = 4 \end{cases}$

Solution *See page S33.* $(38c - 60, 14c - 23, c)$

■ Applications of Systems of Linear Equations in Three Variables

One application of a system of equations is "curve fitting." Given a set of points in the plane, try to find an equation whose graph passes through those points, or "fits" those points.

Alternative to Example 6

Find an equation of the form $y = ax^2 + bx + c$ whose graph passes through the points (1, 2), (2, 15), and (−2, −1).

■ $y = 3x^2 + 4x - 5$

EXAMPLE 6 Solve an Application of a System of Equations to Curve Fitting

Find an equation of the form $y = ax^2 + bx + c$ whose graph passes through the points whose coordinates are (1, 4), (−1, 6), and (2, 9).

Solution Substitute each of the given ordered pairs into the equation $y = ax^2 + bx + c$. Write the resulting system of equations.

$$\begin{cases} 4 = a(1)^2 + b(1) + c \\ 6 = a(-1)^2 + b(-1) + c \\ 9 = a(2)^2 + b(2) + c \end{cases} \quad \text{or} \quad \begin{cases} a + b + c = 4 & (1) \\ a - b + c = 6 & (2) \\ 4a + 2b + c = 9 & (3) \end{cases}$$

Solve the resulting system of equations for a, b, and c.

Continued ➤

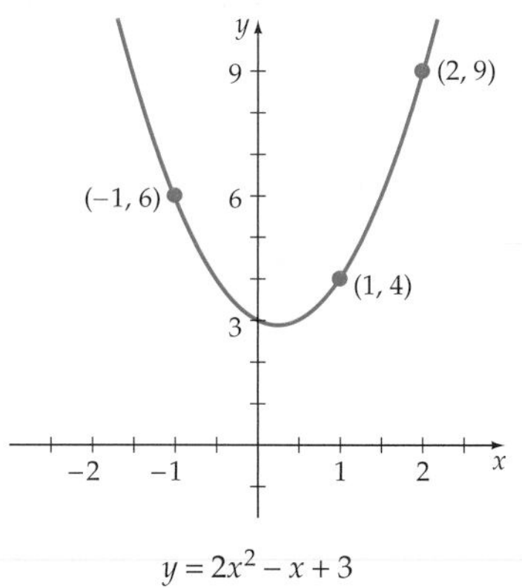

$y = 2x^2 - x + 3$

Figure 6.19

Eliminate a from Equation (2) by multiplying Equation (1) by -1 and then adding it to Equation (2). Now eliminate a from Equation (3) by multiplying Equation (1) by -4 and adding it to Equation (3). The result is

$$\begin{cases} a + b + c = 4 \\ -2b = 2 \\ -2b - 3c = -7 \end{cases}$$

Although this system of equations is not in triangular form, we can solve the second equation for b and use this value to find a and c.

Solving by substitution, we obtain $a = 2$, $b = -1$, and $c = 3$. The equation of the form $y = ax^2 + bx + c$ whose graph passes through (1, 4), (−1, 6), and (2, 9) is $y = 2x^2 - x + 3$. See Figure 6.19.

CHECK YOUR PROGRESS 6 Find an equation of the form $y = ax^2 + bx + c$ whose graph passes through the points whose coordinates are (−1, −5), (1, 5), and (3, −1).

Solution *See page S33.* $y = -2x^2 + 5x + 2$

Alternative to Example 7
Exercise 46, page 513

EXAMPLE 7 Solve an Application Using a System of Equations

An artist is creating a mobile from which three objects will be suspended from a light rod that is 18 inches long, as shown at the left. The weight, in ounces, of each object is shown in the diagram. For the mobile to balance, the objects must be positioned so that $w_1d_1 + w_2d_2 = w_3d_3$. The artist wants d_1 to be 1.5 times d_2. Find the distances d_1, d_2, and d_3 so that the mobile will balance.

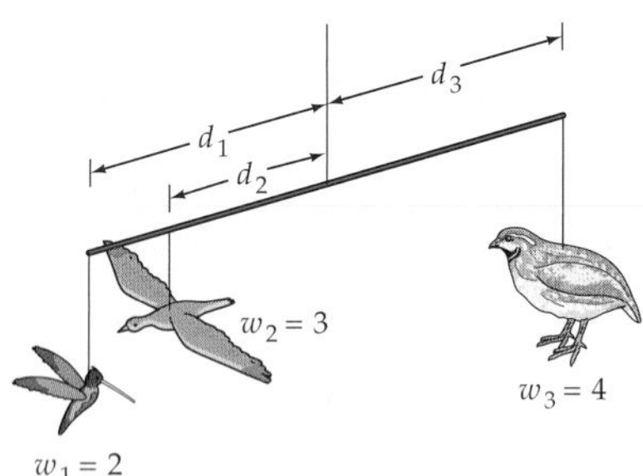

Solution The length of the rod is 18 inches. Therefore, $d_1 + d_3 = 18$. Using the diagram and the equation $w_1d_1 + w_2d_2 = w_3d_3$, we have $2d_1 + 3d_2 = 4d_3$. The artist wants d_1 to be 1.5 times d_2. Thus $d_1 = 1.5d_2$.

$$\begin{aligned} d_1 + d_3 &= 18 && (1) \\ 2d_1 + 3d_2 &= 4d_3 && (2) \\ d_1 &= 1.5d_2 && (3) \end{aligned}$$

We will solve the above system of equations by substitution. Use Equation (3) to replace d_1 in Equation (1) and in Equation (2).

$$\begin{aligned} 1.5d_2 + d_3 &= 18 \quad (4) && \text{▪ Substitute } 1.5d_2 \text{ for } d_1 \text{ in Equation (1).} \\ 2(1.5d_2) + 3d_2 &= 4d_3 && \text{▪ Substitute } 1.5d_2 \text{ for } d_1 \text{ in Equation (2).} \\ 6d_2 &= 4d_3 \quad (5) && \text{▪ Simplify.} \\ 1.5d_2 &= d_3 \quad (6) && \text{▪ Divide each side of Equation (5) by 4.} \end{aligned}$$

Replace d_3 by $1.5d_2$ in Equation (4) and solve for d_2.

$$\begin{aligned} 1.5d_2 + 1.5d_2 &= 18 \\ 3d_2 &= 18 && \text{▪ Simplify.} \\ d_2 &= 6 && \text{▪ Solve for } d_2. \end{aligned}$$

Continued ➤

From Equation (6), $d_3 = 1.5d_2 = 1.5(6) = 9$.
Substituting the value of d_3 into Equation (1), we have $d_1 = 9$. The values are $d_1 = 9$ inches, $d_2 = 6$ inches, and $d_3 = 9$ inches.

CHECK YOUR PROGRESS 7 A science museum charges \$10 for an admission ticket, but members receive a discount of \$3, and students are admitted for half price. Last Saturday, 750 tickets were sold for a total of \$5400. If 20 more student tickets than full-price tickets were sold, how many of each type of ticket were sold?

Solution *See page S33.* General admission, 190; members, 350; students, 210

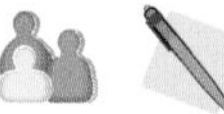

Topics for Discussion

1. Can a system of equations contain more equations than variables? If not, explain why not. If so, give an example.
2. If a system of equations contains more variables than equations, what can you say about the solution or solutions of the system?
3. If a system of three linear equations in three variables is dependent, what does that imply about the graphs of the equations of the system?
4. If a system of three linear equations in three variables is inconsistent, what does that imply about the graphs of the equations of the system?
5. Consider the plane P given by $2x + 4y - 3z = 12$. The *trace* of the graph of P is obtained by letting one of the variables equal zero. For instance, the trace in the xy-plane is the graph of $2x + 4y = 12$ that is obtained by letting $z = 0$. Determine the traces of P in the xz- and yz-planes, and discuss how the traces can be used to visualize the graph of P.

EXERCISES 6.2

— Suggested Assignment: Exercises 1–25, odd; 26, 31, 34, 40; 41–47, odd; and 50–55.

In Exercises 1 to 24, solve each system of equations.

1. $\begin{cases} 2x - y + z = 8 \\ 2y - 3z = -11 \\ 3y + 2z = 3 \end{cases}$
(2, −1, 3)

2. $\begin{cases} 3x + y + 2z = -4 \\ -3y - 2z = -5 \\ 2y + 5z = -4 \end{cases}$
(−1, 3, −2)

3. $\begin{cases} x + 3y - 2z = 8 \\ 2x - y + z = 1 \\ 3x + 2y - 3z = 15 \end{cases}$
(2, 0, −3)

4. $\begin{cases} x - 2y + 3z = 5 \\ 3x - 3y + z = 9 \\ 5x + y - 3z = 3 \end{cases}$
(1, −2, 0)

5. $\begin{cases} 3x + 4y - z = -7 \\ x - 5y + 2z = 19 \\ 5x + y - 2z = 5 \end{cases}$
(2, −3, 1)

6. $\begin{cases} 2x - 3y - 2z = 12 \\ x + 4y + z = -9 \\ 4x + 2y - 3z = 6 \end{cases}$
(1, −2, −2)

7. $\begin{cases} 2x - 5y + 3z = -18 \\ 3x + 2y - z = -12 \\ x - 3y - 4z = -4 \end{cases}$
(−5, 1, −1)

8. $\begin{cases} 4x - y + 2z = -1 \\ 2x + 3y - 3z = -13 \\ x + 5y + z = 7 \end{cases}$
(−2, 1, 4)

9. $\begin{cases} x + 2y - 3z = -7 \\ 2x - y + 4z = 11 \\ 4x + 3y - 4z = -3 \end{cases}$
(3, −5, 0)

10. $\begin{cases} x - 3y + 2z = -11 \\ 3x + y + 4z = 4 \\ 5x - 5y + 8z = -18 \end{cases}$
$\left(\frac{1 - 14c}{10}, \frac{37 + 2c}{10}, c\right)$

11. $\begin{cases} 2x - 5y + 2z = -4 \\ 3x + 2y + 3z = 13 \\ 5x - 3y - 4z = -18 \end{cases}$
(0, 2, 3)

12. $\begin{cases} 4x - 3y + 7z = 3 \\ -3x + 2y - 2z = 7 \\ 5x + 4y + 3z = 2 \end{cases}$
(−3, 2, 3)

13. $\begin{cases} 2x + y - z = -2 \\ 3x + 2y + 3z = 21 \\ 7x + 4y + z = 17 \end{cases}$
$(5c - 25, 48 - 9c, c)$

14. $\begin{cases} 3x + y + 2z = 2 \\ 4x - 2y + z = -4 \\ 11x - 3y + 4z = -6 \end{cases}$
$\left(-\frac{c}{2}, \frac{4-c}{2}, c\right)$

15. $\begin{cases} 3x - 2y + 3z = 11 \\ 2x + 3y + z = 3 \\ 5x + 14y - z = 1 \end{cases}$
$(3, -1, 0)$

16. $\begin{cases} 2x + 3y + 2z = 14 \\ x - 3y + 4z = 4 \\ -x + 12y - 6z = 2 \end{cases}$
$\left(6, \frac{2}{3}, 0\right)$

17. $\begin{cases} 2x - 3y + 6z = 3 \\ x + 2y - 4z = 5 \\ 3x + 4y - 8z = 7 \end{cases}$
No solution

18. $\begin{cases} 2x + 5y - 4z = 1 \\ x - y + 5z = 4 \\ 3x + 2y + 5z = 7 \end{cases}$
$(3 - 3c, 2c - 1, c)$

19. $\begin{cases} 2x - 3y + 5z = 14 \\ x + 4y - 3z = -2 \end{cases}$
$\left(\frac{50 - 11c}{11}, \frac{11c - 18}{11}, c\right)$

20. $\begin{cases} x + 3y - 6z = 5 \\ 4x - 6y + 3z = -7 \end{cases}$
$\left(\frac{3c + 1}{2}, \frac{3c + 3}{2}, c\right)$

21. $\begin{cases} 6x - 9y + 6z = 7 \\ 4x - 6y + 4z = 9 \end{cases}$
No solution

22. $\begin{cases} 4x - 2y + 6z = 5 \\ 2x - y + 3z = 2 \end{cases}$
No solution

23. $\begin{cases} 5x + 3y + 2z = 10 \\ 3x - 4y - 4z = -5 \end{cases}$
$\left(\frac{25 + 4c}{29}, \frac{55 - 26c}{29}, c\right)$

24. $\begin{cases} 3x - 4y - 7z = -5 \\ 2x + 3y - 5z = 2 \end{cases}$
$\left(\frac{41c - 7}{17}, \frac{16 + c}{17}, c\right)$

In Exercises 25 and 26, use the system of equations

$$\begin{cases} x - 3y - 2z = A^2 \\ 2x - 5y + Az = 9 \\ 2x - 8y + z = 18 \end{cases}$$

25. Find all values of A for which the system has no solutions. $A = -\frac{13}{2}$

26. Find all values of A for which the system has a unique solution. $2A + 13 \neq 0$ or $A \neq -\frac{13}{2}$

In Exercises 27 to 29, use the system of equations

$$\begin{cases} x + 2y + z = A^2 \\ -2x - 3y + Az = 1 \\ 7x + 12y + A^2z = 4A^2 - 3 \end{cases}$$

27. Find all values of A for which the system has a unique solution. $A \neq -3, A \neq 1$

28. Find all values of A for which the system has an infinite number of solutions. $A = 1$

29. Find all values of A for which the system has no solution. $A = -3$

30. *Curve Fitting* Find an equation of the form

$$y = ax^2 + bx + c$$

whose graph passes through the points $(1, -2)$, $(3, -4)$, and $(2, -2)$. $y = -x^2 + 3x - 4$

31. *Curve Fitting* Find an equation of the form

$$y = ax^2 + bx + c$$

whose graph passes through the points $(2, 3)$, $(-2, 7)$, and $(1, -2)$. $y = 2x^2 - x - 3$

32. *Curve Fitting* Find an equation of the form $y = ax^2 + bx + c$ whose graph passes through the points $(1, -3)$, $(2, -14)$, and $(-2, 6)$. $y = -2x^2 - 5x + 4$

33. *Curve Fitting* Find the equation of the circle whose graph passes through the points $(5, 3)$, $(-1, -5)$, and $(-2, 2)$. (*Hint:* Use the equation $x^2 + y^2 + ax + by + c = 0$.)
$x^2 + y^2 - 4x + 2y - 20 = 0$

34. *Curve Fitting* Find the equation of the circle whose graph passes through the points $(0, 6)$, $(1, 5)$, and $(-7, -1)$. (*Hint:* See Exercise 33.) $x^2 + y^2 + 6x - 4y - 12 = 0$

35. *Curve Fitting* Find the center and radius of the circle whose graph passes through the points $(-2, 10)$, $(-12, -14)$, and $(5, 3)$. (*Hint:* See Exercise 33.)
Center: $(-7, -2)$, radius: 13

36. *Curve Fitting* Find the center and radius of the circle whose graph passes through the points $(2, 5)$, $(-4, -3)$, and $(3, 4)$. (*Hint:* See Exercise 33.) Center: $(-1, 1)$, radius: 5

37. *Number Theory* The sum of the digits of a positive three-digit number is 19. The tens digit is four less than twice the hundreds digit. The number is decreased by 99 when the digits are reversed. Find the number. 685

38. *Number Theory* The sum of the digits of a positive three-digit number is 10. The hundreds digit is one less than twice the ones digit. The number is decreased by 198 when the digits are reversed. Find the number. 523

Business and Economics

39. *Exchanging Coins* Susan had been putting her change (nickels, dimes, and quarters) into a jar at the end of each day. After several months the jar was full and she decided to take the coins to the local grocery store, where there is a machine that counts change and returns cash. The machine charges 9 cents per dollar that it counts. Susan received \$60.06 for her jar of coins. If there were 660 coins in all, and three times as many nickels as dimes, how many of each coin were in the jar?
396 nickels, 132 dimes, 132 quarters

40. *Cost of Copy Machines* On Monday, a copy machine manufacturer sent out three shipments. The first order, which contained a bill for \$114,000, was for four model C400 machines, six model C600 machines, and 10 model C900 copiers. The second shipment, which contained a bill for \$72,000, was for eight C400's, three C600's, and five C900's. The third shipment, which contained a bill for \$81,000, was for two C400's, nine C600's, and five C900's. How much does the company charge for a model C600 copy machine? \$4000

41. *Investments* An investor has a total of \$18,000 deposited in three different accounts that earn annual interest of 9%, 7%, and 5%. The amount deposited in the 9% account is twice the amount in the 5% account. If the three accounts earn a total annual interest amount of \$1340, how much money is deposited in each account?
\$8000 at 9%, \$6000 at 7%, \$4000 at 5%

42. *Investments* An investor has a total of \$15,000 deposited in three different accounts that earn annual interest of 9%, 6%, and 4%. The amount deposited in the 6% account is \$2000 more than the amount in the 4% account. If the three accounts earn a total annual interest amount of \$980, how much money is deposited in each account?
\$5250 at 9%, \$5875 at 6%, \$3875 at 4%

Sports and Recreation

43. *Slot Machine Winnings* A tourist playing slot machines at a casino in Las Vegas presented a bucket full of coins—quarters, nickels, and silver dollars—to the cashier. The cashier dumped the coins into a counting machine. The machine reported a total of 740 coins with a value of \$157. If there were twice as many nickels as quarters, how many of each type of coin did the tourist have in the bucket? 220 quarters, 440 nickels, 80 silver dollars

44. *Computer Animation* An animator is designing characters and an environment with software that uses an *xyz*-coordinate system. The animator needs to draw a plane that includes the points (2, 1, 1), (−1, 2, 12), and (3, 2, 0). Find an equation for a plane that the animator could use. (*Hint:* The equation of a plane can be written as $z = ax + by + c$.) $z = -3x + 2y + 5$

45. *Design a Mobile* A sculptor is creating a mobile in which three objects will be suspended from a light rod that is 15 inches long, as shown below. The weight, in ounces, of each object is shown in the diagram. For the mobile to balance, the objects must be positioned so that $w_1d_1 = w_2d_2 + w_3d_3$. The artist wants d_3 to be three times d_2. Find the distances d_1, d_2, and d_3 so that the mobile will balance.

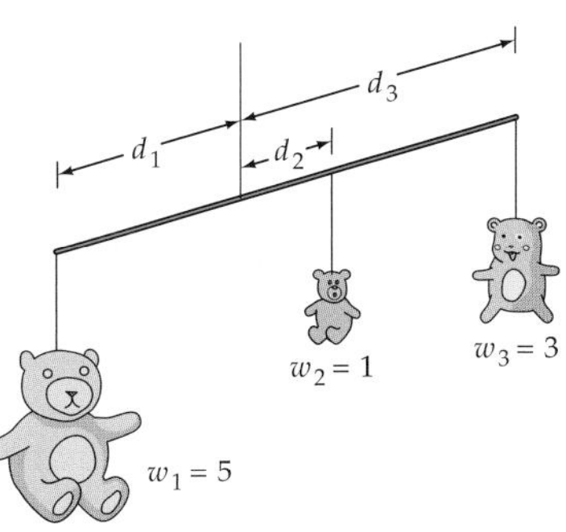

$d_1 = 6$ in., $d_2 = 3$ in., $d_3 = 9$ in.

46. *Design a Mobile* A mobile is made by suspending three objects from a light rod that is 20 inches long, as shown below. The weight, in ounces, of each object is shown in the diagram. For the mobile to balance, the objects must be positioned so that $w_1d_1 + w_2d_2 = w_3d_3$. The artist wants d_3 to be twice d_2. Find the distances d_1, d_2, and d_3 so that the mobile will balance. $d_1 = 8$ in., $d_2 = 6$ in., $d_3 = 12$ in.

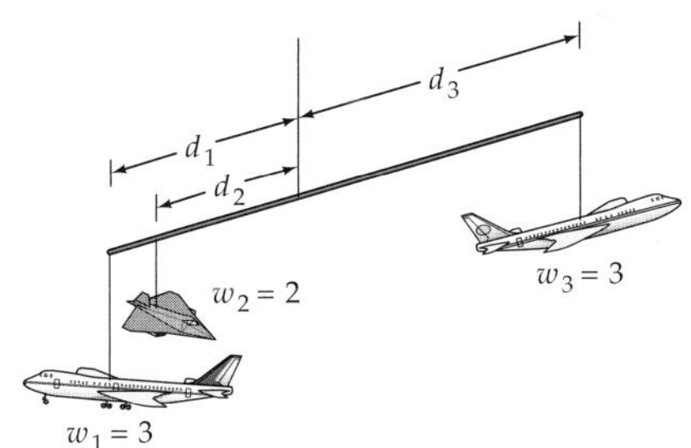

47. *Travel Industry* Smith Travel Research listed the top three hotel and motel chains in the United States according to the total number of rooms. Holiday Inn, Best Western, and Days Inn had a combined total of 558,963 rooms. Best Western and Days Inn had a combined total of 108,279 more rooms than Holiday Inn. Days Inn had 30,469 fewer rooms than Best Western. Find the total number of rooms in each of the three chains.
Holiday Inn: 255,342 rooms; Best Western: 182,045 rooms; Days Inn: 151,576 rooms

Social Sciences

48. *Labor Market* The U.S. Bureau of Labor Statistics has projected that the top three occupations with the largest numerical job decline between 1994 and 2005 are farmers, typists/word processors, and clerks (bookkeeping, accounting, and auditing). These three occupations are expected to decline by a total of

663,000 jobs. The decline in the number of jobs in farming will be 117,000 less than the combined loss in jobs for typists/word processors and clerks. The job decline in clerk positions will be 95,000 less than that for farmers. Find the job decline for each of the three occupations.
Farmers: 273,000 jobs; typists/word processors: 212,000 jobs; clerks: 178,000 jobs

49. ***Job Satisfaction*** A psychologist researched job satisfaction at a factory by surveying employees over a period of several years. In 1999 the researcher found the number of satisfied employees to be 72.2%, and in 2001 the percentage was 71.4%. In 2003 the level fell to 69%. The psychologist would like to use a quadratic equation ($y = ax^2 + bx + c$) to model the percentage of satisfied employees. Find an equation for the researcher to use that includes each of the percentages found. Use $x = 0$ to represent the year 2000. (*Hint:* The year 1999 will correspond to $x = -1$.) $y = -0.2x^2 - 0.4x + 72$

Prepare for Section 6.3

50. For the equations

$$\begin{aligned} 2x - 6y + 7z &= 8 \qquad (1) \\ -3x - 2y + 4z &= 5 \qquad (2) \end{aligned}$$

find the result of four times Equation (1) added to -7 times Equation (2). [6.2] $29x - 10y = -3$

51. For the equations

$$\begin{aligned} -4x - 3y + 3z &= -11 \qquad (1) \\ 5x + 7y - 6z &= 4 \qquad (2) \end{aligned}$$

find the result of five times Equation (1) added to four times Equation (2). [6.2] $13y - 9z = -39$

52. If $-3x - 2y + 7z = -3$, $5y - 2z = -1$, and $z = -2$, find the values of x and y. [1.1] $x = -3, y = -1$

53. If a system of linear equations has fewer equations than variables, what does this imply about the number of solutions? [6.2]

54. If a set of solutions is given as $(3a + 5, 2 - 2a, a)$, find the solution corresponding to $a = -4$. [6.2] $(-7, 10, -4)$

55. The solutions of a dependent system of equations are given by ordered triples of the form $\left(\frac{2}{7}c, \frac{1}{7} - \frac{3}{7}c, c\right)$. State three specific solutions of the system. [6.2]
Three possible solutions: $\left(0, \frac{1}{7}, 0\right), \left(\frac{2}{7}, -\frac{2}{7}, 1\right), \left(\frac{4}{7}, -\frac{5}{7}, 2\right)$

Explorations

1. ***Concept of Dimension*** In this chapter we graphed first-degree equations in three variables. If we were to attempt to graph an equation in four variables, we would need a fourth axis perpendicular to the three axes of an xyz-coordinate system. It seems impossible to imagine a fourth dimension, but incorporating it is really a quite practical matter in mathematics. In fact, there are some systems that require an infinite-dimensional coordinate system. To gain some insight into the concept of dimension, read the book *Flatland* by Edwin A. Abbott, and then write an essay explaining what this book has to do with dimension.
Answers will vary.

2. ***Abilities of a Four-Dimensional Human*** There have been a number of attempts to describe the abilities of a four-dimensional human in a three-dimensional world. Search the Internet or a library to find some of these descriptions, and then write an essay on some of the actions a four-dimensional person could perform. Answer the following question in your essay: Can a four-dimensional person remove the money from a locked safe without first opening the safe?
Answers will vary.

53. The system will either be dependent (an infinite number of solutions) or inconsistent (no solutions).

SECTION 6.3 Row Operations and Systems of Equations

- Elementary Row Operations
- Gaussian Elimination Method
- Interpolating Polynomials

TAKE NOTE

The elements of a matrix can be any type of number or variable.

Also, it is customary to use a capital letter to name a matrix and the corresponding lower-case letter to identify the elements of the matrix.

Point of Interest

The word *matrix* has the same origins as the word *mother.* James Sylvester (1814–1897), who coined the word, thought of a matrix as a place where something begins—in this case, the solution of a system of equations.

Although Sylvester helped to popularize the use of matrices, there is evidence of their existence in Chinese manuscripts that date from around 200 B.C.

▪ Elementary Row Operations

A **matrix** is a rectangular array of numbers enclosed in square brackets. The matrix below has three rows and four columns. A matrix of m rows and n columns is said to be of **order** $m \times n$ or **dimension** $m \times n$. The order of matrix A is 3×4.

$$A = \begin{bmatrix} 2 & -3 & 0 & 1 \\ 0.7 & 5 & \dfrac{1}{2} & 3 \\ \sqrt{2} & 1 & -4 & 5 \end{bmatrix}$$

If a matrix has the same number of rows and columns, it is called a **square matrix.** The numbers in a matrix are called **elements** of the matrix. The notation a_{ij} refers to the element in the ith row and the jth column of matrix A. For matrix A above, $a_{11} = 2$, $a_{31} = \sqrt{2}$, and $a_{23} = \frac{1}{2}$. The elements $a_{11}, a_{22}, a_{33}, \ldots, a_{mm}$ form the **main diagonal** of a matrix. For matrix A, the main diagonal is 2, 5, −4.

Two matrices A and B are **equal** if and only if they have the same order and their corresponding elements are equal. (Corresponding elements are elements that are in the same row and the same column.) For example, if $A = \begin{bmatrix} p & -2 & q \\ 7 & r & 1 \end{bmatrix}$ and $B = \begin{bmatrix} 3 & x & -4 \\ 7 & -1 & y \end{bmatrix}$, then $A = B$ if and only if $p = 3$, $x = -2$, $q = -4$, $r = -1$, and $y = 1$.

QUESTION If two matrices A and B are equal, do they have the same order?

Using the concept of equality of two matrices, it is possible to use matrices to represent a system of equations. For instance, by equality of matrices, the matrix equation

$$\begin{bmatrix} x - 3y + z \\ 2x + y - 3z \\ -x + 2y + 4z \end{bmatrix} = \begin{bmatrix} 8 \\ -13 \\ 9 \end{bmatrix}$$

yields the system of equations $\begin{cases} x - 3y + z = 8 \\ 2x + y - 3z = -13. \\ -x + 2y + 4z = 9 \end{cases}$

Matrices can be used to solve systems of equations. To illustrate the connection between a system of equations and matrices, consider the following system of equations.

$$\begin{cases} x - 3y - z = 8 \\ 2x \qquad - 3z = 5 \\ -4x - 5y + 2z = 4 \end{cases}$$

ANSWER Yes.

Because it is customary to write a system of equations with the constants to the right of the equal signs and the variable terms arranged in some order (alphabetical, for instance) to the left of the equal signs, it is possible to represent the above system of equations with the following matrix. A vertical bar is often used to separate the coefficients of the variables from the constants.

$$\left[\begin{array}{rrr|r} 1 & -3 & -1 & 8 \\ 2 & 0 & -3 & 5 \\ -4 & -5 & 2 & 4 \end{array}\right]$$

Note that 0 is entered as the coefficient of y in the second equation.

This matrix is called the **augmented matrix** for the system of equations. The matrix formed by the coefficients of a system of equations is called the **coefficient matrix.** The matrix formed by the constants is called the **constant matrix.** For the system of equations above,

$$\text{Coefficient matrix: } \begin{bmatrix} 1 & -3 & -1 \\ 2 & 0 & -3 \\ -4 & -5 & 2 \end{bmatrix} \qquad \text{Constant matrix: } \begin{bmatrix} 8 \\ 5 \\ 4 \end{bmatrix}$$

Alternative to Example 1

Write the augmented matrix, the coefficient matrix, and the constant matrix for the following system of equations.

$$\begin{aligned} 2x - 5y + 3z &= 8 \\ -x + 5z &= -3 \\ 4x - 7y &= 12 \end{aligned}$$

■ $\left[\begin{array}{rrr|r} 2 & -5 & 3 & 8 \\ -1 & 0 & 5 & -3 \\ 4 & 7 & 0 & 12 \end{array}\right]$, $\begin{bmatrix} 2 & -5 & 3 \\ -1 & 0 & 5 \\ 4 & 7 & 0 \end{bmatrix}$, $\begin{bmatrix} 8 \\ -3 \\ 12 \end{bmatrix}$

EXAMPLE 1 A System of Equations Represented as a Matrix

Write the augmented matrix, the coefficient matrix, and the constant matrix for the system of equations at the right.

$$\begin{cases} 3x + 4y + z = 1 \\ -2x - y + 2z = -5 \\ x - 3z = 4 \end{cases}$$

Solution

$$\text{Augmented matrix: } \left[\begin{array}{rrr|r} 3 & 4 & 1 & 1 \\ -2 & -1 & 2 & -5 \\ 1 & 0 & -3 & 4 \end{array}\right]$$

■ **Replace the missing y term in the third equation with a 0.**

$$\text{Coefficient matrix: } \begin{bmatrix} 3 & 4 & 1 \\ -2 & -1 & 2 \\ 1 & 0 & -3 \end{bmatrix} \qquad \text{Constant matrix: } \begin{bmatrix} 1 \\ -5 \\ 4 \end{bmatrix}$$

CHECK YOUR PROGRESS 1 Write the system of equations for the following augmented matrix. Use w, x, y, and z for the variables.

$$\left[\begin{array}{rrrr|r} 2 & 0 & -1 & 3 & 4 \\ 1 & -5 & 2 & 0 & -1 \\ 3 & 1 & 1 & -2 & 5 \\ 6 & -4 & 5 & 1 & 7 \end{array}\right]$$

Solution *See page S33.*

$$\begin{cases} 2w - y + 3z = 4 \\ w - 5x + 2y = -1 \\ 3w + x + y - 2z = 5 \\ 6w - 4x + 5y + z = 7 \end{cases}$$

A system of equations can be solved by writing the system as an augmented matrix and then performing operations on the matrix similar to those performed on the equations of the system. These operations are called **elementary row operations.**

Elementary Row Operations

Given the augmented matrix for a system of linear equations, each of the following row operations produces an augmented matrix for an equivalent system of equations.

1. Interchange two rows. Interchanging the ith and jth rows is shown symbolically by $R_i \leftrightarrow R_j$.
2. Multiply all the elements in a row by the same nonzero number. Multiplying the ith row by k is shown symbolically by kR_i.
3. Replace a row by the sum of that row and a nonzero multiple of another row. Replacing the ith row by the sum of that row and k times the jth row is shown symbolically by $kR_j + R_i$.

TAKE NOTE

For the third elementary row operation, the row being multiplied is *not* changed. In the example at the right, row 2 is not changed. The row being added to is replaced by the result of the operation. It is not absolutely necessary to follow this convention, but for consistency we will always use this method.

Here is an example of these operations.

$$\begin{bmatrix} 2 & 1 & -3 \\ 1 & -2 & 5 \\ -3 & 4 & -6 \end{bmatrix} \quad R_2 \leftrightarrow R_1 \quad \begin{bmatrix} 1 & -2 & 5 \\ 2 & 1 & -3 \\ -3 & 4 & -6 \end{bmatrix}$$

$$\begin{bmatrix} 2 & 1 & -3 \\ 1 & -2 & 5 \\ -3 & 4 & -6 \end{bmatrix} \xrightarrow{-2R_3} \begin{bmatrix} 2 & 1 & -3 \\ 1 & -2 & 5 \\ 6 & -8 & 12 \end{bmatrix}$$

$$\begin{bmatrix} 2 & 1 & -3 \\ 1 & -2 & 5 \\ -3 & 4 & -6 \end{bmatrix} \xrightarrow{3R_2 + R_3} \begin{bmatrix} 2 & 1 & -3 \\ 1 & -2 & 5 \\ 0 & -2 & 9 \end{bmatrix}$$

Alternative to Example 2

Perform the elementary row operations on the matrix

$$\begin{bmatrix} 1 & -4 & -1 & 2 \\ -2 & 1 & 0 & 3 \\ 3 & -1 & 1 & 5 \end{bmatrix}$$

a. $-\frac{1}{2}R_2$

b. $-3R_1 + R_3$

■ a. $\begin{bmatrix} 1 & -4 & -1 & 2 \\ 1 & -\frac{1}{2} & 0 & -\frac{3}{2} \\ 3 & -1 & 1 & 5 \end{bmatrix}$

■ b. $\begin{bmatrix} 1 & -4 & -1 & 2 \\ -2 & 1 & 0 & 3 \\ 0 & 11 & 4 & -1 \end{bmatrix}$

EXAMPLE 2 Use Elementary Row Operations

Complete the following using the given elementary row operations.

a. $\begin{bmatrix} 1 & 5 & -2 & 3 \\ -3 & 1 & 4 & -1 \\ 2 & 6 & -8 & 0 \\ 0 & -1 & 2 & 5 \end{bmatrix} \quad R_4 \leftrightarrow R_2 \quad \begin{bmatrix} & ? & \end{bmatrix}$

b. $\begin{bmatrix} 1 & 5 & -2 & 3 \\ -3 & 1 & 4 & -1 \\ 2 & 6 & -8 & 0 \\ 0 & -1 & 2 & 5 \end{bmatrix} \xrightarrow{\frac{1}{2}R_3} \begin{bmatrix} & ? & \end{bmatrix}$

c. $\begin{bmatrix} 1 & 5 & -2 & 3 \\ -3 & 1 & 4 & -1 \\ 2 & 6 & -8 & 0 \\ 0 & -1 & 2 & 5 \end{bmatrix} \xrightarrow{-2R_1 + R_3} \begin{bmatrix} & ? & \end{bmatrix}$

Continued ➤

Solution

a. $\begin{bmatrix} 1 & 5 & -2 & 3 \\ -3 & 1 & 4 & -1 \\ 2 & 6 & -8 & 0 \\ 0 & -1 & 2 & 5 \end{bmatrix} \xrightarrow{R_4 \leftrightarrow R_2} \begin{bmatrix} 1 & 5 & -2 & 3 \\ 0 & -1 & 2 & 5 \\ 2 & 6 & -8 & 0 \\ -3 & 1 & 4 & -1 \end{bmatrix}$

b. $\begin{bmatrix} 1 & 5 & -2 & 3 \\ -3 & 1 & 4 & -1 \\ 2 & 6 & -8 & 0 \\ 0 & -1 & 2 & 5 \end{bmatrix} \xrightarrow{\frac{1}{2}R_3} \begin{bmatrix} 1 & 5 & -2 & 3 \\ -3 & 1 & 4 & -1 \\ 1 & 3 & -4 & 0 \\ 0 & -1 & 2 & 5 \end{bmatrix}$

c. $\begin{bmatrix} 1 & 5 & -2 & 3 \\ -3 & 1 & 4 & -1 \\ 2 & 6 & -8 & 0 \\ 0 & -1 & 2 & 5 \end{bmatrix} \xrightarrow{-2R_1 + R_3} \begin{bmatrix} 1 & 5 & -2 & 3 \\ -3 & 1 & 4 & -1 \\ 0 & -4 & -4 & -6 \\ 0 & -1 & 2 & 5 \end{bmatrix}$

CHECK YOUR PROGRESS 2 Complete the following using the given elementary row operations.

a. $\begin{bmatrix} -3 & 6 & 9 \\ 8 & 1 & -2 \\ -2 & 3 & 5 \end{bmatrix} \xrightarrow{R_2 \leftrightarrow R_3} \begin{bmatrix} & ? & \end{bmatrix}$ $\begin{bmatrix} -3 & 6 & 9 \\ -2 & 3 & 5 \\ 8 & 1 & -2 \end{bmatrix}$

b. $\begin{bmatrix} -3 & 6 & 9 \\ 8 & 1 & -2 \\ -2 & 3 & 5 \end{bmatrix} \xrightarrow{-\frac{1}{3}R_1} \begin{bmatrix} & ? & \end{bmatrix}$ $\begin{bmatrix} 1 & -2 & -3 \\ 8 & 1 & -2 \\ -2 & 3 & 5 \end{bmatrix}$

c. $\begin{bmatrix} -3 & 6 & 9 \\ 8 & 1 & -2 \\ -2 & 3 & 5 \end{bmatrix} \xrightarrow{4R_3 + R_2} \begin{bmatrix} & ? & \end{bmatrix}$ $\begin{bmatrix} -3 & 6 & 9 \\ 0 & 13 & 18 \\ -2 & 3 & 5 \end{bmatrix}$

Solution *See page S33.*

Each of the elementary row operations has as its basis the corresponding operation that can be performed on a system of equations. These operations do not change the solution of the system of equations. Here are some examples of each row operation.

TAKE NOTE

For each example at the right, we are using the same system of equations. The solution of this system of equations is (2, 1). You should verify that (2, 1) is also a solution of each new system of equations. This demonstrates that the row operations have not changed the solution.

1. Interchange two rows.

Original system — *New system*

$$\begin{aligned} 2x - 3y &= 1 \\ 4x + 5y &= 13 \end{aligned} \quad \left[\begin{array}{cc|c} 2 & -3 & 1 \\ 4 & 5 & 13 \end{array}\right] \xrightarrow{R_1 \leftrightarrow R_2} \left[\begin{array}{cc|c} 4 & 5 & 13 \\ 2 & -3 & 1 \end{array}\right] \quad \begin{aligned} 4x + 5y &= 13 \\ 2x - 3y &= 1 \end{aligned}$$

The solution is (2, 1). The solution is (2, 1).

2. Multiply all the elements in a row by the same nonzero number.

Original system — *New system*

$$\begin{aligned} 2x - 3y &= 1 \\ 4x + 5y &= 13 \end{aligned} \quad \left[\begin{array}{cc|c} 2 & -3 & 1 \\ 4 & 5 & 13 \end{array}\right] \xrightarrow{3R_2} \left[\begin{array}{cc|c} 2 & -3 & 1 \\ 12 & 15 & 39 \end{array}\right] \quad \begin{aligned} 2x - 3y &= 1 \\ 12x + 15y &= 39 \end{aligned}$$

The solution is (2, 1). The solution is (2, 1).

3. Replace a row by the sum of that row and a multiple of any other row.

Original system | ***New system***

$$\begin{aligned} 2x - 3y &= 1 \\ 4x + 5y &= 13 \end{aligned} \quad \left[\begin{array}{cc|c} 2 & -3 & 1 \\ 4 & 5 & 13 \end{array}\right] \xrightarrow{-2R_1 + R_2} \left[\begin{array}{cc|c} 2 & -3 & 1 \\ 0 & 11 & 11 \end{array}\right] \quad \begin{aligned} 2x - 3y &= 1 \\ 11y &= 11 \end{aligned}$$

The solution is (2, 1). The solution is (2, 1).

Note that when we performed the operation $-2R_1 + R_2$ we replaced the row that follows the addition, R_2.

In certain cases, an augmented matrix represents a system of equations that can be solved by back substitution, a method we used earlier in this chapter. Consider the following augmented matrix and the corresponding system of equations.

$$\left[\begin{array}{ccc|c} 1 & 4 & -2 & 3 \\ 0 & 1 & 1 & -1 \\ 0 & 0 & 1 & -2 \end{array}\right] \rightarrow \begin{cases} a + 4b - 2c = 3 \\ b + c = -1 \\ c = -2 \end{cases}$$

Solving this system of equations by back substitution, we find that the solution is $(-5, 1, -2)$. The augmented matrix above is said to be in *row echelon form.*

P

Row Echelon Form

A matrix is in **row echelon form** if all of the following conditions are satisfied.

1. The first nonzero number in any row is a 1.
2. Rows are arranged so that the column containing the first nonzero number in any row is to the left of the column containing the first nonzero number of the next row.
3. All rows consisting entirely of zeros appear at the bottom of the matrix.

Here are some additional augmented matrices in row echelon form.

$$\left[\begin{array}{ccc|c} 1 & 4 & -3 & 2 \\ 0 & 1 & 5 & 3 \\ 0 & 0 & 1 & -4 \end{array}\right] \quad \left[\begin{array}{cc|c} 1 & -5 & 2 \\ 0 & 1 & -3 \end{array}\right] \quad \left[\begin{array}{ccc|c} 1 & -3 & 4 & 2 \\ 0 & 1 & -2 & -1 \\ 0 & 0 & 0 & 0 \end{array}\right]$$

$$\left[\begin{array}{cccc|c} 1 & -2 & 3 & 2 & -4 \\ 0 & 0 & 1 & 2 & -1 \\ 0 & 0 & 0 & 0 & 1 \\ 0 & 0 & 0 & 0 & 0 \end{array}\right]$$

QUESTION Is the augmented matrix $\left[\begin{array}{ccc|c} 1 & -2 & 3 & 2 \\ 0 & 0 & 1 & 4 \\ 0 & 1 & -1 & 3 \end{array}\right]$ in row echelon form?

ANSWER No. The matrix does not satisfy condition (2) of row echelon form.

Elementary row operations can be used to write a matrix in row echelon form.

Procedure to Write an Augmented Matrix in Row Echelon Form

1. Use row operations to change a_{11} to 1, and then change the remaining elements in the first column to 0.
2. If possible, change a_{22} to 1 and then change the remaining elements *below* a_{22} to 0.
3. Move to a_{33} and repeat the above procedure.
4. Continue moving down the main diagonal until you reach a_{nn}, or until all remaining elements on the main diagonal are zero.

As we carry out this procedure, to conserve space, we will occasionally perform more than one elementary row operation in one step. For instance, the notation

$$\xrightarrow[-5R_1 + R_3]{3R_1 + R_2}$$

means that two elementary row operations are being performed. First, multiply row 1 by 3 and add it to row 2. Replace row 2. Second, multiply row 1 by −5 and add it to row 3. Replace row 3.

TAKE NOTE

The sequence of steps used to convert a matrix to row echelon form is not unique. For instance, in Example 3, we could have started by multiplying row 1 by $\frac{1}{3}$. The sequence of steps you use may result in a row echelon form that is different from the one we show. See the Integrating Technology on the following page.

Alternative to Example 3

Write the following matrix in a row echelon form.

$$\begin{bmatrix} 0 & 3 & -5 & 1 \\ 3 & 4 & -7 & 5 \\ 1 & 2 & -1 & -3 \end{bmatrix}$$

■ $\begin{bmatrix} 1 & 2 & -1 & -3 \\ 0 & 1 & 2 & -7 \\ 0 & 0 & 1 & -2 \end{bmatrix}$

EXAMPLE 3 Write a Matrix in Row Echelon Form

Use the above procedure to write the matrix $\begin{bmatrix} -3 & 13 & -1 & -7 \\ 1 & -5 & 2 & 0 \\ 5 & -20 & -2 & 5 \end{bmatrix}$ in row echelon form.

Solution Follow the procedure to write a matrix in row echelon form. Change a_{11} to 1.

$$\begin{bmatrix} -3 & 13 & -1 & -7 \\ 1 & -5 & 2 & 0 \\ 5 & -20 & -2 & 5 \end{bmatrix} \quad R_1 \leftrightarrow R_2 \quad \begin{bmatrix} 1 & -5 & 2 & 0 \\ -3 & 13 & -1 & -7 \\ 5 & -20 & -2 & 5 \end{bmatrix}$$

Change the remaining elements in the first column to 0. This is still part of Step 1 of the procedure.

$$\begin{bmatrix} 1 & -5 & 2 & 0 \\ -3 & 13 & -1 & -7 \\ 5 & -20 & -2 & 5 \end{bmatrix} \xrightarrow[-5R_1 + R_3]{3R_1 + R_2} \begin{bmatrix} 1 & -5 & 2 & 0 \\ 0 & -2 & 5 & -7 \\ 0 & 5 & -12 & 5 \end{bmatrix}$$

Change a_{22} to 1.

$$\begin{bmatrix} 1 & -5 & 2 & 0 \\ 0 & -2 & 5 & -7 \\ 0 & 5 & -12 & 5 \end{bmatrix} \xrightarrow{-\frac{1}{2}R_2} \begin{bmatrix} 1 & -5 & 2 & 0 \\ 0 & 1 & -\frac{5}{2} & \frac{7}{2} \\ 0 & 5 & -12 & 5 \end{bmatrix}$$

Continued ➤

Change the element under a_{22} to 0.

$$\begin{bmatrix} 1 & -5 & 2 & 0 \\ 0 & 1 & -\frac{5}{2} & \frac{7}{2} \\ 0 & 5 & -12 & 5 \end{bmatrix} \xrightarrow{-5R_2 + R_3} \begin{bmatrix} 1 & -5 & 2 & 0 \\ 0 & 1 & -\frac{5}{2} & \frac{7}{2} \\ 0 & 0 & \frac{1}{2} & -\frac{25}{2} \end{bmatrix}$$

Change a_{33} to 1.

$$\begin{bmatrix} 1 & -5 & 2 & 0 \\ 0 & 1 & -\frac{5}{2} & \frac{7}{2} \\ 0 & 0 & \frac{1}{2} & -\frac{25}{2} \end{bmatrix} \xrightarrow{2R_3} \begin{bmatrix} 1 & -5 & 2 & 0 \\ 0 & 1 & -\frac{5}{2} & \frac{7}{2} \\ 0 & 0 & 1 & -25 \end{bmatrix}$$

A row echelon form for the matrix is $\begin{bmatrix} 1 & -5 & 2 & 0 \\ 0 & 1 & -\frac{5}{2} & \frac{7}{2} \\ 0 & 0 & 1 & -25 \end{bmatrix}$.

INSTRUCTOR NOTE

The algorithm that a calculator uses to express a matrix in row echelon form starts by moving the row whose first element has the greatest absolute value to row 1. This reduces rounding errors as the calculation proceeds. Note that the first row of the row echelon form in the Integrating Technology below is 1, −4, −0.4, 1. This is achieved by interchanging row 1 and row 3, then multiplying the new row 1 by $\frac{1}{5}$.

CHECK YOUR PROGRESS 3 Write the matrix $\begin{bmatrix} 3 & -6 & 12 \\ 2 & 1 & -3 \end{bmatrix}$ in row echelon form.

Solution *See page S34.* $\begin{bmatrix} 1 & -2 & 4 \\ 0 & 1 & -\frac{11}{5} \end{bmatrix}$

INTEGRATING TECHNOLOGY

ref(is the abbreviation for row echelon form. The answer is given in decimal form. An answer with fractions can be found by using the ▶Frac command.

```
Ans▶Frac
[[ 1   -4   -2/5   1  ...
 [ 0   1    -12/5  1  ...
 [ 0   0    1      -2 ...
```

INTEGRATING TECHNOLOGY

A graphing calculator can be used to find a row echelon form for a matrix. The screens below, from a TI-83 Plus calculator, show a row echelon form for the matrix in Example 3. If you need assistance with keystrokes, see the web site for this text at **math.college.hmco.com.**

```
MATRIX[A]  3 x4
[ -3    13    -1   ...
[ 1     -5    2    ...
[ 5     -20   -2   ...
```

```
NAMES MATH EDIT
0↑ cumSum(
A: ref(
B: rref(
C: rowSwap(
D: row+(
E: *row(
F: *row+(
```

```
ref([A])
[[ 1   -4   -.4    1  ...
 [ 0   1    -2.4   1  ...
 [ 0   0    1      -25...
```

Note that the row echelon form produced by the calculator is different from the one we produced in Example 3. The two forms are equivalent and either can be used for any calculation in which a row echelon form of the matrix is required.

■ Gaussian Elimination Method

The **Gaussian elimination method** is an algorithm* that uses elementary row operations to solve a system of linear equations. The goal of this method is to use elementary row operations to rewrite an augmented matrix in row echelon form, and then use back substitution to find the solution of the system of equations.

To demonstrate the Gaussian elimination method, consider the following system of equations and its augmented matrix.

$$\begin{cases} 3x + 4y = 26 \\ x - 2y = -8 \end{cases} \qquad \text{Augmented matrix: } \left[\begin{array}{cc|c} 3 & 4 & 26 \\ 1 & -2 & -8 \end{array}\right]$$

The row operations are chosen such that first there is a 1 as a_{11}; second there is a 0 as a_{21}; and third there is a 1 as a_{22}.

We begin by interchanging row 1 and row 2. The result is a 1 as a_{11}.

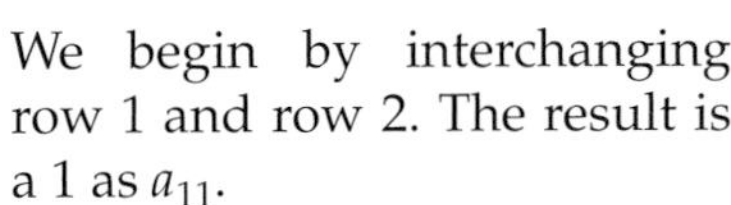

$$\left[\begin{array}{cc|c} 3 & 4 & 26 \\ 1 & -2 & -8 \end{array}\right] \quad R_1 \leftrightarrow R_2 \quad \left[\begin{array}{cc|c} 1 & -2 & -8 \\ 3 & 4 & 26 \end{array}\right]$$

Next multiply row 1 by -3 and add it to row 2. Replace row 2 with the sum. The result has a 0 as a_{21}.

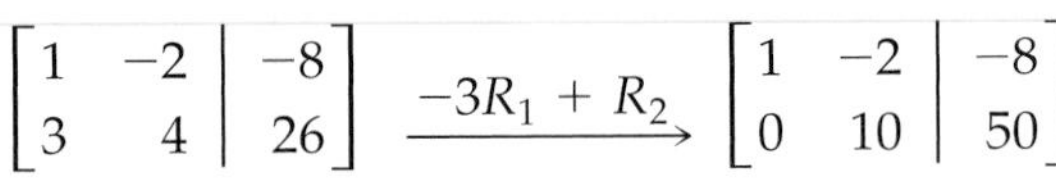

$$\left[\begin{array}{cc|c} 1 & -2 & -8 \\ 3 & 4 & 26 \end{array}\right] \xrightarrow{-3R_1 + R_2} \left[\begin{array}{cc|c} 1 & -2 & -8 \\ 0 & 10 & 50 \end{array}\right]$$

Now multiply row 2 by $\frac{1}{10}$. The result is a 1 as a_{22}.

$$\left[\begin{array}{cc|c} 1 & -2 & -8 \\ 0 & 10 & 50 \end{array}\right] \xrightarrow{\frac{1}{10}R_2} \left[\begin{array}{cc|c} 1 & -2 & -8 \\ 0 & 1 & 5 \end{array}\right]$$

The last matrix, $\left[\begin{array}{cc|c} 1 & -2 & -8 \\ 0 & 1 & 5 \end{array}\right]$, is the augmented matrix for the system of equations $\begin{cases} x - 2y = -8 \\ \quad\; y = 5 \end{cases}$.

From the last system of equations, $y = 5$. We can find x by back substitution.

$$x - 2y = -8$$
$$x - 2(5) = -8 \qquad \text{■ Replace } y \text{ by 5.}$$
$$x = 2$$

The solution of the original system of equations is (2, 5).

The next example makes use of the above procedure to solve a system of three equations in three unknowns.

Point of Interest

The Gaussian elimination method is named after Johann Carl Friedrich Gauss (1777–1855), who is considered one of the greatest mathematicians of all time. He contributed not only to mathematics, but to astronomy and physics as well. A unit of magnetism called a *gauss* was so named in his honor.

The above image is one of Gauss on a German deutsche mark note.

EXAMPLE 4 Solve a System of Equations Using the Gaussian Elimination Method

Solve by using the Gaussian elimination method: $\begin{cases} 2x - 4y + 3z = 2 \\ -3x + 4y - 2z = 5 \\ 5x - 2y + 7z = 11 \end{cases}$

Continued ➤

Alternative to Example 4

Solve by using the Gaussian elimination method:

$$\begin{aligned} 2x - 3y + 3z &= 19 \\ x - 3y - 2z &= 2 \\ -4x + 13y + 10z &= -4 \end{aligned}$$

■ (2, −2, 3)

*An *algorithm* is a procedure used in calculations. The word is derived from *Al-Khwarizmi,* the name of the author of the Arabic algebra book *Hisab al-jabr w'al-muqabala,* which was written around A.D. 825.

Solution

$$\left[\begin{array}{rrr|r} 2 & -4 & 3 & 2 \\ -3 & 4 & -2 & 5 \\ 5 & -2 & 7 & 11 \end{array}\right]$$

■ This is the augmented matrix.

$$\left[\begin{array}{rrr|r} 2 & -4 & 3 & 2 \\ -3 & 4 & -2 & 5 \\ 5 & -2 & 7 & 11 \end{array}\right] \xrightarrow{\frac{1}{2}R_1} \left[\begin{array}{rrr|r} 1 & -2 & \frac{3}{2} & 1 \\ -3 & 4 & -2 & 5 \\ 5 & -2 & 7 & 11 \end{array}\right]$$

■ Multiply row 1 by $\frac{1}{2}$ to get a 1 in a_{11}.

$$\left[\begin{array}{rrr|r} 1 & -2 & \frac{3}{2} & 1 \\ -3 & 4 & -2 & 5 \\ 5 & -2 & 7 & 11 \end{array}\right] \xrightarrow[-5R_1 + R_3]{3R_1 + R_2} \left[\begin{array}{rrr|r} 1 & -2 & \frac{3}{2} & 1 \\ 0 & -2 & \frac{5}{2} & 8 \\ 0 & 8 & -\frac{1}{2} & 6 \end{array}\right]$$

■ Get a 0 in a_{21} and a_{31}.

$$\left[\begin{array}{rrr|r} 1 & -2 & \frac{3}{2} & 1 \\ 0 & -2 & \frac{5}{2} & 8 \\ 0 & 8 & -\frac{1}{2} & 6 \end{array}\right] \xrightarrow{-\frac{1}{2}R_2} \left[\begin{array}{rrr|r} 1 & -2 & \frac{3}{2} & 1 \\ 0 & 1 & -\frac{5}{4} & -4 \\ 0 & 8 & -\frac{1}{2} & 6 \end{array}\right]$$

■ Multiply row 2 by $-\frac{1}{2}$ to get a 1 in a_{22}.

$$\left[\begin{array}{rrr|r} 1 & -2 & \frac{3}{2} & 1 \\ 0 & 1 & -\frac{5}{4} & -4 \\ 0 & 8 & -\frac{1}{2} & 6 \end{array}\right] \xrightarrow{-8R_2 + R_3} \left[\begin{array}{rrr|r} 1 & -2 & \frac{3}{2} & 1 \\ 0 & 1 & -\frac{5}{4} & -4 \\ 0 & 0 & \frac{19}{2} & 38 \end{array}\right]$$

■ Get a 0 in a_{32}.

$$\left[\begin{array}{rrr|r} 1 & -2 & \frac{3}{2} & 1 \\ 0 & 1 & -\frac{5}{4} & -4 \\ 0 & 0 & \frac{19}{2} & 38 \end{array}\right] \xrightarrow{\frac{2}{19}R_3} \left[\begin{array}{rrr|r} 1 & -2 & \frac{3}{2} & 1 \\ 0 & 1 & -\frac{5}{4} & -4 \\ 0 & 0 & 1 & 4 \end{array}\right]$$

■ Multiply row 3 by $\frac{2}{19}$ to get a 1 in a_{33}.

The last matrix is in row echelon form. The system of equations written from the matrix is

$$\begin{cases} x - 2y + \frac{3}{2}z = 1 \\ y - \frac{5}{4}z = -4 \\ z = 4 \end{cases}$$

From this system of equations, $z = 4$. Use back substitution to find that $y = 1$ and $x = -3$. Thus $(-3, 1, 4)$ is the solution of the original system of equations.

Continued ➤

CHECK YOUR PROGRESS 4 Solve by the Gaussian elimination method.

$$\begin{cases} 2x + 3y - 2z = -10 \\ 3x + y - 3z = -8 \\ -4x + 2y + z = -11 \end{cases}$$

Solution *See page S34.* (3, −2, 5)

Recall that an inconsistent system of equations is one that has no solution. If we use matrix row operations to solve an inconsistent system of equations, we will obtain a matrix that has a row with a nonzero number in the rightmost position and zeros in all the other positions of that row. For instance, if you obtain the matrix

$$\left[\begin{array}{cc|c} 3 & 4 & 6 \\ 0 & 0 & 2 \end{array}\right]$$

then you know the original system of equations is inconsistent because the bottom row of the matrix indicates that $0x + 0y = 2$, which is not true for any values of x and y.

Alternative to Example 5

Solve by using the Gaussian elimination method:

$$\begin{cases} x - 3y + 2z = 6 \\ 4x - y + 3z = 10 \\ 7x + y + 4z = -2 \end{cases}$$

▪ **Inconsistent system**

EXAMPLE 5 Use the Gaussian Elimination Method for an Inconsistent System of Equations

Show that the system of equations $\begin{cases} x - 3y + z = 5 \\ 3x - 7y + 2z = 12 \\ 2x - 4y + z = 3 \end{cases}$ is inconsistent.

Solution Use row operations to write the augmented matrix in row echelon form.

$$\left[\begin{array}{ccc|c} 1 & -3 & 1 & 5 \\ 3 & -7 & 2 & 12 \\ 2 & -4 & 1 & 3 \end{array}\right] \xrightarrow[-2R_1 + R_3]{-3R_1 + R_2} \left[\begin{array}{ccc|c} 1 & -3 & 1 & 5 \\ 0 & 2 & -1 & -3 \\ 0 & 2 & -1 & -7 \end{array}\right]$$

$$\left[\begin{array}{ccc|c} 1 & -3 & 1 & 5 \\ 0 & 2 & -1 & -3 \\ 0 & 2 & -1 & -7 \end{array}\right] \xrightarrow{\frac{1}{2}R_2} \left[\begin{array}{ccc|c} 1 & -3 & 1 & 5 \\ 0 & 1 & -\frac{1}{2} & -\frac{3}{2} \\ 0 & 2 & -1 & -7 \end{array}\right]$$

$$\left[\begin{array}{ccc|c} 1 & -3 & 1 & 5 \\ 0 & 1 & -\frac{1}{2} & -\frac{3}{2} \\ 0 & 2 & -1 & -7 \end{array}\right] \xrightarrow{-2R_2 + R_3} \left[\begin{array}{ccc|c} 1 & -3 & 1 & 5 \\ 0 & 1 & -\frac{1}{2} & -\frac{3}{2} \\ 0 & 0 & 0 & -4 \end{array}\right]$$

$$\begin{cases} x - 3y + z = 5 \\ y - \frac{1}{2}z = -\frac{3}{2} \\ 0z = -4 \end{cases}$$

▪ **Equivalent system**

Because the equation $0z = -4$ has no solution, the original system of equations has no solution. It is an inconsistent system of equations.

Continued ➤

CHECK YOUR PROGRESS 5 Show that the system of equations at the right is inconsistent.

$$\begin{cases} 2x - 3y - 3z = 19 \\ x + 4y - 7z = -18 \\ -3x + y + 8z = 11 \end{cases}$$

Solution *See page S34.*

If matrix row operations result in a row consisting entirely of zeros, the corresponding equation does not contribute to the solution of the system. If removing one or more rows of zeros leaves fewer nonzero rows than the number of variables, and the system is not inconsistent, then the system is dependent. A dependent system has an infinite number of solutions, as shown in the next example.

Alternative to Example 6

Solve by using the Gaussian elimination method:

$$\begin{cases} x + 4y + 2z = -3 \\ 4x + 5y + 4z = -4 \\ 10x + 7y + 8z = -6 \end{cases}$$

■ $\left(-\frac{6}{11}c - \frac{1}{11}, -\frac{4}{11}c - \frac{8}{11}, c\right)$

EXAMPLE 6 Use the Gaussian Elimination Method to Solve a Dependent System of Equations

Solve: $\begin{cases} x + 4y - 3z = 8 \\ 2x + 3y - z = 11 \\ -x + 2y - 3z = -2 \end{cases}$ (1)

Solution Write the augmented matrix and then use row operations to solve the system.

$$\left[\begin{array}{rrr|r} 1 & 4 & -3 & 8 \\ 2 & 3 & -1 & 11 \\ -1 & 2 & -3 & -2 \end{array}\right] \xrightarrow[R_1 + R_3]{-2R_1 + R_2} \left[\begin{array}{rrr|r} 1 & 4 & -3 & 8 \\ 0 & -5 & 5 & -5 \\ 0 & 6 & -6 & 6 \end{array}\right]$$

$$\left[\begin{array}{rrr|r} 1 & 4 & -3 & 8 \\ 0 & -5 & 5 & -5 \\ 0 & 6 & -6 & 6 \end{array}\right] \xrightarrow{-\frac{1}{5}R_2} \left[\begin{array}{rrr|r} 1 & 4 & -3 & 8 \\ 0 & 1 & -1 & 1 \\ 0 & 6 & -6 & 6 \end{array}\right]$$

$$\left[\begin{array}{rrr|r} 1 & 4 & -3 & 8 \\ 0 & 1 & -1 & 1 \\ 0 & 6 & -6 & 6 \end{array}\right] \xrightarrow{-6R_2 + R_3} \left[\begin{array}{rrr|r} 1 & 4 & -3 & 8 \\ 0 & 1 & -1 & 1 \\ 0 & 0 & 0 & 0 \end{array}\right] \quad (*)$$

The bottom row of the last matrix indicates that $0x + 0y + 0z = 0$, which is true for all values of x, y, and z, so this equation does not contribute to the solution of the system. We have 3 variables but only 2 nonzero rows in the last matrix (and the system is not inconsistent), so we know that the original system of equations is a dependent system. Its solution set consists of an infinite number of ordered triples. To determine some of these ordered triples, consider the system of equations represented by the two top rows of the above matrix.

$$\begin{cases} x + 4y - 3z = 8 \\ y - z = 1 \end{cases}$$

Solving $y - z = 1$ for y in terms of z gives $y = z + 1$.

If we substitute this value of y into $x + 4y - 3z = 8$, we can solve the resulting equation for x in terms of z.

$$x + 4y - 3z = 8$$

$$x + 4(z + 1) - 3z = 8 \qquad \text{■ } y = z + 1$$

$$x + 4z + 4 - 3z = 8 \qquad \text{■ Solve for } x.$$

$$x = -z + 4$$

Continued ➤

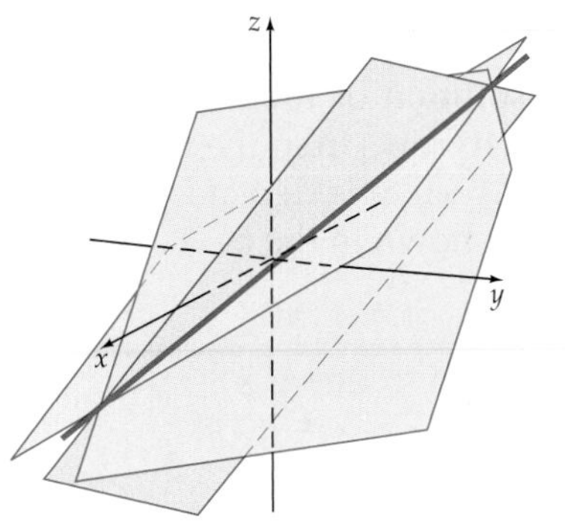

Figure 6.20

Both x and y are now expressed in terms of z. Because z can be replaced by any real number c, the solution set of the system of equations is given by $(-c + 4, c + 1, c)$. Because c can be any real number, there are an infinite number of ordered triples that satisfy the system of equations. For instance, if $c = 0$, then one such ordered triple is

$$(-(0) + 4, 0 + 1, 0) = (4, 1, 0)$$

Substituting $c = 2$ generates the ordered triple

$$(-(2) + 4, 2 + 1, 2) = (2, 3, 2)$$

A three-dimensional graph of the planes given by the equations in System (1) is shown in Figure 6.20. The line on which all the planes intersect is the graph of the ordered triples represented by $(-c + 4, c + 1, c)$. It is shown in red in Figure 6.20.

CHECK YOUR PROGRESS 6 Solve the system of equations at the right using the Gaussian elimination method.

$$\begin{cases} x - 4y + z = 3 \\ 2x + 3y - 5z = 4 \\ -x + 15y - 8z = -5 \end{cases}$$

Solution *See page S34.* $\left(\frac{17}{11}c + \frac{25}{11}, \frac{7}{11}c - \frac{2}{11}, c\right)$

INTEGRATING TECHNOLOGY

If a graphing calculator is used to find a row echelon form for the augmented matrix of System (1) of Example 6, the result is as shown below.

```
MATRIX[A] 3 x4
[ 1    4    -3   ...
[ 2    3    -1   ...
[ -1   2    -3   ...
```

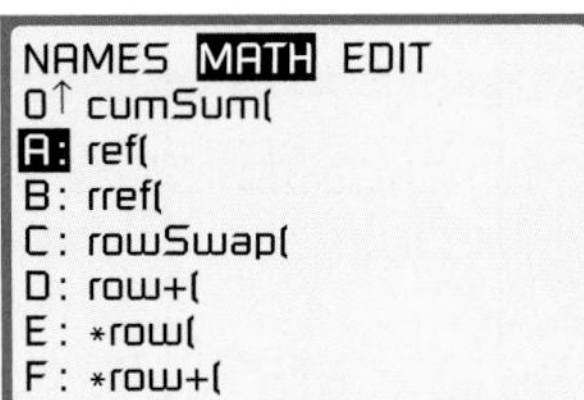


```
Ans▸Frac
[[ 1   3/2   -1/2   11  ...
 [ 0   1     -1     1   ...
 [ 0   0     0      0   ...
```

Note that the row echelon form produced by the calculator is different from the one produced in Example 6. See matrix (*) in Example 6 on page 525. The calculator row echelon form of the matrix yields the system of equations

$$\begin{cases} x + \frac{3}{2}y - \frac{1}{2}z = \frac{11}{2} \\ y - z = 1 \end{cases}$$

Solving this system of equations by back substitution gives $(-c + 4, c + 1, c)$, the same solution we obtained in Example 6.

▪ Interpolating Polynomials

Sometimes a researcher would like to know the polynomial expression whose graph passes through a given set of points. This polynomial is called the **interpolating polynomial** through the points. The degree of the interpolating polynomial is at most one less than the number of given ordered pairs.

Alternative to Example 7

Find the interpolating polynomial that passes through the points (−3, 28), (−1, 6), and (2, 3).

▪ $p(x) = 2x^2 - 3x + 1$

TAKE NOTE

Usually the degree of an interpolating polynomial through n points will be of degree $n - 1$. It is possible, however, for the degree to be less. In this case, you can still proceed by assuming the polynomial will be of degree $n - 1$, as in Example 7. The coefficient of x^{n-1} will be zero.

EXAMPLE 7 Find an Interpolating Polynomial

Find the interpolating polynomial that passes through the points whose coordinates are (−2, 13), (1, −2), and (2, 1).

Solution Because there are three given points, the degree of the interpolating polynomial will be at most 2, one less than the number of ordered pairs. The form of the polynomial will be $p(x) = a_2x^2 + a_1x + a_0$. Use this polynomial to create a system of equations.

$$p(x) = a_2x^2 + a_1x + a_0$$

$$p(-2) = a_2(-2)^2 + a_1(-2) + a_0 = 4a_2 - 2a_1 + a_0 = 13 \qquad \text{▪ } x = -2, p(-2) = 13$$

$$p(1) = a_2(1)^2 + a_1(1) + a_0 = a_2 + a_1 + a_0 = -2 \qquad \text{▪ } x = 1, p(1) = -2$$

$$p(2) = a_2(2)^2 + a_1(2) + a_0 = 4a_2 + 2a_1 + a_0 = 1 \qquad \text{▪ } x = 2, p(2) = 1$$

The system of equations and the associated augmented matrix are shown below.

$$\begin{cases} 4a_2 - 2a_1 + a_0 = 13 \\ a_2 + a_1 + a_0 = -2 \\ 4a_2 + 2a_1 + a_0 = 1 \end{cases} \qquad \left[\begin{array}{ccc|c} 4 & -2 & 1 & 13 \\ 1 & 1 & 1 & -2 \\ 4 & 2 & 1 & 1 \end{array}\right]$$

The augmented matrix in echelon form and the resulting system of equations are shown below.

$$\left[\begin{array}{ccc|c} 1 & -0.5 & 0.25 & 3.25 \\ 0 & 1 & 0 & -3 \\ 0 & 0 & 1 & -1 \end{array}\right] \qquad \begin{cases} a_2 - 0.5a_1 + 0.25a_0 = 3.25 \\ a_1 = -3 \\ a_0 = -1 \end{cases}$$

Solving the system of equations by back substitution, we have $a_0 = -1$, $a_1 = -3$, and $a_2 = 2$. The interpolating polynomial is $p(x) = 2x^2 - 3x - 1$. See Figure 6.21.

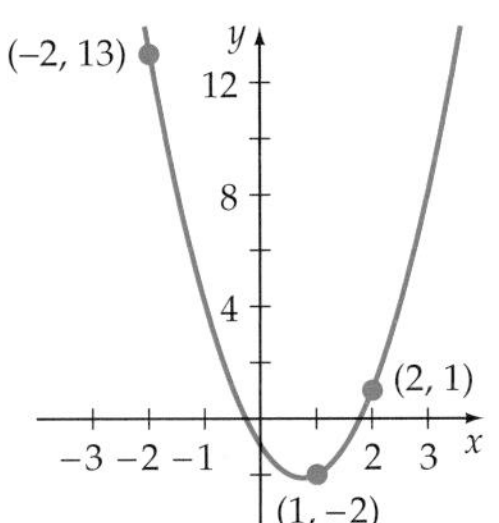

Figure 6.21

CHECK YOUR PROGRESS 7 Find the interpolating polynomial that passes through the points whose coordinates are (−1, −5), (0, 0), (1, 1), and (2, 4).

Solution *See page S35.* $p(x) = x^3 - 2x^2 + 2x$

Topics for Discussion

1. What are the elementary row operations? Give examples of each one.
2. Explain how an augmented matrix differs from the coefficient matrix for a system of equations.
3. Give examples of matrices that are in row echelon form and of matrices that are not in row echelon form.
4. After elementary row operations have been correctly performed on an augmented matrix, the result is

$$\begin{bmatrix} 1 & -2 & 3 & 0 \\ 0 & 1 & 2 & -1 \\ 0 & 0 & 0 & 3 \end{bmatrix}$$

Does this result indicate that the system of equations has a unique solution, an infinite number of solutions, or no solution?

EXERCISES 6.3

— *Suggested Assignment: Exercises 1–41, odd; 44, 45, 47, 50, 52, and 53–58.*

In Exercises 1 to 4, give the order of the matrix.

1. $\begin{bmatrix} -2 & 1 & 0 \\ 8 & -5 & 3 \\ 1 & 2 & -6 \end{bmatrix}$ 3×3

2. $\begin{bmatrix} 0 & -1 \\ 2 & 6 \\ -5 & 3 \end{bmatrix}$ 3×2

3. $[1 \quad -2 \quad 4]$ 1×3

4. $\begin{bmatrix} 1 \\ 4 \end{bmatrix}$ 2×1

In Exercises 5 to 8, write the matrix equation as a system of equations.

5. $\begin{bmatrix} 2x - 3y \\ x + 4y \end{bmatrix} = \begin{bmatrix} 1 \\ 6 \end{bmatrix}$

6. $\begin{bmatrix} 4x + 5y \\ 3x - 2y \end{bmatrix} = \begin{bmatrix} -1 \\ 5 \end{bmatrix}$

7. $\begin{bmatrix} x - 3y + 2z \\ 2x + y - 3z \\ x - 5y + z \end{bmatrix} = \begin{bmatrix} 5 \\ -4 \\ 0 \end{bmatrix}$ $\begin{cases} x - 3y + 2z = 5 \\ 2x + y - 3z = -4 \\ x - 5y + z = 0 \end{cases}$

8. $\begin{bmatrix} 3x - y + z \\ 2x + 4y - 3z \\ x - 3y + 2z \end{bmatrix} = \begin{bmatrix} 2 \\ 4 \\ 0 \end{bmatrix}$ $\begin{cases} 3x - y + z = 2 \\ 2x + 4y - 3z = 4 \\ x - 3y + 2z = 0 \end{cases}$

5. $\begin{cases} 2x - 3y = 1 \\ x + 4y = 6 \end{cases}$ 6. $\begin{cases} 4x + 5y = -1 \\ 3x - 2y = 5 \end{cases}$

In Exercises 9 to 12, write the augmented matrix, the coefficient matrix, and the constant matrix for the given system of equations.

9. $\begin{cases} 5x + 7y = 3 \\ 2x - 5y = 8 \end{cases}$ $\left[\begin{array}{cc|c} 5 & 7 & 3 \\ 2 & -5 & 8 \end{array}\right], \begin{bmatrix} 5 & 7 \\ 2 & -5 \end{bmatrix}, \begin{bmatrix} 3 \\ 8 \end{bmatrix}$

10. $\begin{cases} -4x + 5y = 21 \\ 3x - 7y = 11 \end{cases}$ $\left[\begin{array}{cc|c} -4 & 5 & 21 \\ 3 & -7 & 11 \end{array}\right], \begin{bmatrix} -4 & 5 \\ 3 & -7 \end{bmatrix}, \begin{bmatrix} 21 \\ 11 \end{bmatrix}$

11. $\begin{cases} 2x - 6y + 5z = 11 \\ 3x + 2y - 4z = 23 \\ 2x - 5y + 3z = 8 \end{cases}$ $\left[\begin{array}{ccc|c} 2 & -6 & 5 & 11 \\ 3 & 2 & -4 & 23 \\ 2 & -5 & 3 & 8 \end{array}\right], \begin{bmatrix} 2 & -6 & 5 \\ 3 & 2 & -4 \\ 2 & -5 & 3 \end{bmatrix}, \begin{bmatrix} 11 \\ 23 \\ 8 \end{bmatrix}$

12. $\begin{cases} 3w + 2x - 5y + 6z = 9 \\ 2w - 4x + 3y - z = 11 \\ -w + 7x + 4y + 3z = 2 \\ -5w - 3x - 2y - 7z = 19 \end{cases}$ Answer on page 529.

13. After row operations have been correctly performed on an augmented matrix, the result is

$$\left[\begin{array}{ccc|c} 1 & -2 & 3 & 0 \\ 0 & 1 & 2 & -1 \\ 0 & 0 & 0 & 3 \end{array}\right]$$

Does this result indicate that the system of equations has a unique solution, an infinite number of solutions, or no solution? No solution

14. After row operations have been correctly performed on an augmented matrix, the result is

$$\left[\begin{array}{ccc|c} 1 & 2 & 3 & -1 \\ 0 & 1 & 2 & -1 \\ 0 & 0 & 0 & 0 \end{array}\right]$$

Does this result indicate that the system of equations has a unique solution, an infinite number of solutions, or no solution? Infinite number of solutions

15. After row operations have been correctly performed on an augmented matrix, the result is

$$\left[\begin{array}{ccc|c} 1 & 0 & 0 & 4 \\ 0 & 1 & 0 & 2 \\ 0 & 0 & 1 & 3 \end{array}\right]$$

Does this result indicate that the system of equations has a unique solution, an infinite number of solutions, or no solution? Unique solution

16. After row operations have been correctly performed on an augmented matrix, the result is

$$\left[\begin{array}{cccc|c} 1 & -2 & 3 & 0 & 7 \\ 0 & 1 & 2 & -1 & 2 \\ 0 & 0 & 1 & 3 & 5 \\ 0 & 0 & 0 & 0 & 1 \end{array}\right]$$

Does this result indicate that the system of equations has a unique solution, an infinite number of solutions, or no solution? No solution

In Exercises 17 to 20, let $C = \left[\begin{array}{cc|c} 2 & 4 & 3 \\ 5 & 6 & 1 \end{array}\right]$ **and** $D = \left[\begin{array}{ccc|c} 1 & 2 & 6 & 4 \\ -3 & 5 & 6 & 7 \\ 1 & -4 & 3 & 4 \end{array}\right]$.

17. What row operation can be applied to C to produce $\left[\begin{array}{cc|c} 5 & 6 & 1 \\ 2 & 4 & 3 \end{array}\right]$? $R_1 \leftrightarrow R_2$

18. What row operation can be applied to D to produce
$$\left[\begin{array}{rrr|r} 1 & 2 & 6 & 4 \\ -3 & 5 & 6 & 7 \\ 0 & -6 & -3 & 0 \end{array}\right]?$$ $-R_1 + R_3$

19. What row operation can be applied to D to produce
$$\left[\begin{array}{rrr|r} 1 & 2 & 6 & 4 \\ 0 & 11 & 24 & 19 \\ 1 & -4 & 3 & 4 \end{array}\right]?$$ $3R_1 + R_2$

20. What row operation can be applied to C to produce
$$\left[\begin{array}{rr|r} 2 & 4 & 3 \\ 0 & -4 & -\frac{13}{2} \end{array}\right]?$$ $-\frac{5}{2}R_1 + R_2$

In Exercises 21 to 30, use the Gaussian elimination method to solve each system of equations.

21. $\begin{cases} x + 3y = 14 \\ 2x - 5y = -16 \end{cases}$ $(2, 4)$

22. $\begin{cases} 2x + 5y = -7 \\ 7x - 2y = 34 \end{cases}$ $(4, -3)$

23. $\begin{cases} 3x - 2y = 24 \\ 5x - 3y = 39 \end{cases}$ $(6, -3)$

24. $\begin{cases} 2x - 7y = -24 \\ 5x + 4y = -17 \end{cases}$ $(-5, 2)$

12. $\left[\begin{array}{rrrr|r} 3 & 2 & -5 & 6 & 9 \\ 2 & -4 & 3 & -1 & 11 \\ -1 & 7 & 4 & 3 & 2 \\ -5 & -3 & -2 & -7 & 19 \end{array}\right]$, $\begin{bmatrix} 3 & 2 & -5 & 6 \\ 2 & -4 & 3 & -1 \\ -1 & 7 & 4 & 3 \\ -5 & -3 & -2 & -7 \end{bmatrix}$, $\begin{bmatrix} 9 \\ 11 \\ 2 \\ 19 \end{bmatrix}$

25. $\begin{cases} x + 2y - 2z = -2 \\ 5x + 9y - 4z = -3 \\ 3x + 4y - 5z = -3 \end{cases}$ $(2, -1, 1)$

26. $\begin{cases} x - 3y + z = 8 \\ 2x - 5y - 3z = 2 \\ x + 4y + z = 1 \end{cases}$ $\left(\frac{12}{5}, -1, \frac{13}{5}\right)$

27. $\begin{cases} 3x + 7y - 7z = -4 \\ x + 2y - 3z = 0 \\ 5x + 6y + z = -8 \end{cases}$ $(1, -2, -1)$

28. $\begin{cases} 2x - 3y + 2z = 13 \\ 3x - 4y - 3z = 1 \\ 3x + y - z = 2 \end{cases}$ $(2, -1, 3)$

29. $\begin{cases} t + 2u - 3v + w = -7 \\ 3t + 5u - 8v + 5w = -8 \\ 2t + 3u - 7v + 3w = -11 \\ 4t + 8u - 10v + 7w = -10 \end{cases}$ $(2, -2, 3, 4)$

30. $\begin{cases} t + 4u + 2v - 3w = 11 \\ 2t + 10u + 3v - 5w = 17 \\ 4t + 16u + 7v - 9w = 34 \\ t + 4u + v - w = 4 \end{cases}$ $\left(2, -\frac{1}{2}, 1, -3\right)$

In Exercises 31 to 36, use the Gaussian elimination method to determine whether each system of equations is independent, dependent, or inconsistent. Solve the independent and dependent systems of equations.

31. $\begin{cases} x + 2y - 2z = 3 \\ 5x + 8y - 6z = 14 \\ 3x + 4y - 2z = 8 \end{cases}$ Dependent; $\left(-2c + 2, 2c + \frac{1}{2}, c\right)$

32. $\begin{cases} 3x - 5y + 2z = 4 \\ x - 3y + 2z = 4 \\ 5x - 11y + 6z = 12 \end{cases}$ Dependent; $(c - 2, c - 2, c)$

33. $\begin{cases} 3x + 2y - z = 1 \\ 2x + 3y - z = 1 \\ x - y + 2z = 3 \end{cases}$ Independent; $\left(\frac{1}{2}, \frac{1}{2}, \frac{3}{2}\right)$

34. $\begin{cases} 2x + 5y + 2z = -1 \\ x + 2y - 3z = 5 \\ 5x + 12y + z = 10 \end{cases}$ Inconsistent

35. $\begin{cases} x + 3y - 4z = -9 \\ 2x - y - z = 10 \\ 3x + 2y - 5z = 1 \end{cases}$ Dependent; $(c + 3, c - 4, c)$

36. $\begin{cases} x - 3y + z = -7 \\ 4x - 5y - 3z = -14 \\ -5x + y + 9z = 7 \end{cases}$ Dependent; $(2c - 1, c + 2, c)$

37. ***Interpolating Polynomial*** Find the interpolating polynomial whose graph passes through the points whose coordinates are $(-2, -7)$ and $(1, -1)$. $p(x) = 2x - 3$

38. ***Interpolating Polynomial*** Find the interpolating polynomial whose graph passes through the points whose coordinates are $(-3, -8)$ and $(1, 4)$. $p(x) = 3x + 1$

39. ***Interpolating Polynomial*** Find the interpolating polynomial whose graph passes through the points whose coordinates are $(-1, 6)$, $(1, 2)$, and $(2, 3)$. $p(x) = x^2 - 2x + 3$

40. ***Interpolating Polynomial*** Find the interpolating polynomial whose graph passes through the points whose coordinates are $(-2, -3)$, $(0, -1)$, and $(3, 17)$. $p(x) = x^2 + 3x - 1$

41. ***Interpolating Polynomial*** Find the interpolating polynomial whose graph passes through the points whose coordinates are $(-2, -12)$, $(0, 2)$, $(1, 0)$, and $(3, 8)$.
$p(x) = x^3 - 2x^2 - x + 2$

42. ***Interpolating Polynomial*** Find the interpolating polynomial whose graph passes through the points whose coordinates are $(0, -1)$, $(-1, -3)$, $(1, 1)$, and $(2, 9)$.
$p(x) = x^3 + x - 1$

By following a procedure similar to that for finding an interpolating polynomial, we can find various equations that pass through given points.

43. ***Equation of a Plane*** Find an equation of the plane that passes through the points $(-1, 0, -4)$, $(2, 1, 5)$, and $(-1, 1, -1)$. (*Suggestion:* The equation of a plane can be written as $z = ax + by + c$.) $z = 2x + 3y - 2$

44. ***Equation of a Plane*** Find an equation of the plane that passes through the points $(1, 2, -3)$, $(-2, 0, -7)$, and $(0, 1, -4)$. (*Suggestion:* The equation of a plane can be written as $z = ax + by + c$.) $z = 2x - y - 3$

45. ***Equation of a Circle*** Find an equation of the circle that passes through the points $(2, 6)$, $(-4, -2)$, and $(3, -1)$. (*Suggestion:* The equation of a circle can be written as $x^2 + y^2 + ax + by = c$.) $x^2 + y^2 + 2x - 4y = 20$

46. ***Equation of a Circle*** Find an equation of the circle that passes through the points $(2, 1)$, $(0, -7)$, and $(5, -2)$. (*Suggestion:* The equation of a circle can be written as $x^2 + y^2 + ax + by = c$.) $x^2 + y^2 - 2x + 6y = 7$

Business and Economics

47. ***Investments*** A real estate agent has set up college funds for each of her three children. The oldest child's fund earns 7% interest per year, the second child's fund earns 5%, and the youngest child's fund earns 8%. Last year the total interest earned was \$3720. If the oldest and youngest children's accounts had been switched, the total interest earned would have been \$3880. If the two younger children had traded accounts, the total interest earned would have been \$3840. How much money was in each child's fund?
Oldest child: \$28,000; second child: \$16,000; youngest child: \$12,000

48. ***Investments*** A financial planner invested \$33,000 of a client's money, part at 9%, part at 12%, and the remainder at 8%. The total annual income from these three investments was \$3290. The amount invested at 12% was \$5000 less than the combined amount invested at 9% and 8%. Find the amount invested at each rate.
\$14,000 at 12%, \$10,000 at 8%, \$9,000 at 9%

Life and Health Sciences

49. ***Emergency Supplies*** A relief organization supplies blankets, cots, and lanterns to victims of fires, floods, and other natural disasters. One week the organization purchased 15 blankets, 5 cots, and 10 lanterns at a total cost of \$1250. The next week, at the same prices, the organization purchased 20 blankets, 10 cots, and 15 lanterns at a total cost of \$2000. The following week, at the same prices, the organization purchased 10 blankets, 15 cots, and 5 lanterns at a total cost of \$1625. Find the costs of one blanket, one cot, and one lantern.
Blanket: \$25, cot: \$75, lantern: \$50

50. ***Hospital Costs*** A local hospital has three different types of rooms available for in-patient care: private rooms, semi-private rooms, and shared rooms. On Monday the hospital billed an insurance company for \$4555 for 3 private rooms, 8 semi-private rooms, and 15 shared rooms. The following day \$5250 was billed to the insurance company for 5 private, 6 semi-private, and 18 shared rooms. On the third day, the hospital billed \$3380 for 3 private, 4 semi-private, and 12 shared rooms. What is the price per day for each type of room?
Private: \$360, semi-private: \$200, shared: \$125

Physical Sciences and Engineering

51. ***Building Design*** An architect is using a CAD (computer-aided design) system to design an office building. The software uses an *xyz*-coordinate system, and the architect is trying to include a ramp that passes through the points $(1, -1, 5)$, $(2, -2, 9)$, and $(-3, -1, -1)$. Find an equation for the plane that passes through the three points. This equation could be used to design the ramp. (*Hint:* The equation of a plane can be written as $z = ax + by + c$.)
$3x - 5y - 2z = -2$

52. ***Computer Science*** A computer programmer needs to find the equation for an interpolating spline curve that goes through the points $(1, -7)$, $(2, -33)$, $(-1, 3)$, and $(-2, 11)$. Cubic polynomials are normally used in software to generate the curves. Find a cubic equation of the form $y = ax^3 + bx^2 + cx + d$ that the programmer can use for the spline curve. $y = -2x^3 - 3x^2 - 3x + 1$

Prepare for Section 6.4

53. What is the additive inverse of the real number a? [P.1] $-a$

54. What is the multiplicative identity for the real numbers? [P.1] 1

55. Are the matrices $\begin{bmatrix} 2 & 3 \\ 5 & 7 \end{bmatrix}$ and $\begin{bmatrix} 2 & 3 & 0 \\ 5 & 7 & 0 \end{bmatrix}$ equal? [6.3] No

56. Solve $\begin{bmatrix} a & 2 \\ 3 & b \end{bmatrix} = \begin{bmatrix} 5 & 2 \\ 3 & -1 \end{bmatrix}$ for a and b. [6.3] $a = 5, b = -1$

57. What is the order of the matrix $\begin{bmatrix} 2x + 3y \\ x - 4y \\ 3x + y \end{bmatrix}$? [6.3] 3×1

58. What system of equations is represented by the matrix equation at the right? [6.3]

$$\begin{bmatrix} 3x - 5y \\ 2x + 7y \end{bmatrix} = \begin{bmatrix} 16 \\ -10 \end{bmatrix}$$

$$\begin{cases} 3x - 5y = 16 \\ 2x + 7y = -10 \end{cases}$$

Explorations

1.  ***Ill-Conditioned Systems and Computers*** Solving a large system of equations (100 or more equations) by using pencil and paper is quite impractical. For these large systems of equations, using a computer is the only reasonable choice. However, computer solutions are not without their own set of problems. For instance, the decimal equivalent of $\frac{2}{7}$ is 0.285714285714285714. . . . Because the decimal representation never terminates, a computer cannot possibly store all of the digits and therefore an approximation of the number must be used. This can create quite a problem, as shown in the following example. Consider the following system of equations and its augmented matrix.

$$\begin{cases} x + \frac{1}{2}y + \frac{1}{3}z = 1 \\ \frac{1}{2}x + \frac{1}{3}y + \frac{1}{4}z = 2 \\ \frac{1}{3}x + \frac{1}{4}y + \frac{1}{5}z = 3 \end{cases} \qquad (2)$$

$$\left[\begin{array}{ccc|c} 1 & \frac{1}{2} & \frac{1}{3} & 1 \\ \frac{1}{2} & \frac{1}{3} & \frac{1}{4} & 2 \\ \frac{1}{3} & \frac{1}{4} & \frac{1}{5} & 3 \end{array}\right] \quad \textbf{Augmented matrix}$$

a. Solve System (2) by using a graphing calculator. Instead of entering the fractions in the augmented matrix, use the decimal equivalent of the fractions *rounded* to the nearest 0.01. Thus your augmented matrix is

$$\begin{bmatrix} 1.00 & 0.50 & 0.33 & 1.00 \\ 0.50 & 0.33 & 0.25 & 2.00 \\ 0.33 & 0.25 & 0.20 & 3.00 \end{bmatrix}$$

(266.67, −1433.33, 1366.67)

b. Solve System (2) by rounding the fractions in the augmented matrix to the nearest thousandth.
(30.505, −209.990, 226.697)

c. Are the answers to **a.** and **b.** approximately the same?
No

System (2) is said to be an *ill-conditioned system* of equations. In an ill-conditioned system, very small changes in the numbers in the augmented matrix can produce large changes in the solution. For systems of equations that are not ill-conditioned, approximations of the coefficients of the system produce a reasonable approximation of the solution of the system. For ill-conditioned systems of equations, this is not always the case. Ill-conditioned systems of equations can be identified by using advanced mathematical techniques; however, for the present it is important to recognize that elements of a matrix should not be rounded too drastically. The following guidelines should help reduce the size of errors due to rounding.

- Set your calculator so that it displays numbers using its maximum number of digits.
- Round only the final results.

d. Solve System (2) without rounding the fractions in the augmented matrix. (27, −192, 210)

SECTION 6.4 The Algebra of Matrices

- Addition and Subtraction of Matrices
- Scalar Multiplication of a Matrix
- Matrix Multiplication
- Applications of Matrices

Addition and Subtraction of Matrices

Spreadsheet programs such as Excel and Quattro use matrices to represent data. The rows of the spreadsheet are numbered 1, 2, 3, . . . , and the columns are identified by A, B, C, ... The partial spreadsheet below shows how a consumer's car loan is being repaid over a 5-year period. The elements in column A represent the loan amount at the beginning of the year; column B represents the amount owed at the end of the year; and column C represents the amount of interest paid during the year.

$$\begin{array}{c} \\ 1 \\ 2 \\ 3 \\ 4 \\ 5 \end{array} \begin{array}{c} \begin{array}{ccc} \quad\text{A} & \qquad\text{B} & \qquad\text{C} \end{array} \\ \begin{bmatrix} 10{,}000.00 & 8305.60 & 738.77 \\ 8305.60 & 6470.56 & 598.13 \\ 6470.56 & 4483.22 & 445.82 \\ 4483.22 & 2330.93 & 280.88 \\ 2330.93 & 0.00 & 102.24 \end{bmatrix} \end{array}$$

For instance, the element 3C means that the consumer paid \$445.82 in interest during the third year of the loan.

Matrices are effective for situations in which there are a number of items to be classified. For instance, suppose a clothing company has June sales of a certain style of T-shirt in various colors and styles as shown in the matrix below.

$$\begin{array}{r} \\ \text{Blue} \\ \text{Tan} \\ \text{White} \\ \text{Green} \end{array} \begin{array}{c} \begin{array}{ccc} \text{Small} & \text{Medium} & \text{Large} \end{array} \\ \begin{bmatrix} 45 & 56 & 81 \\ 37 & 92 & 64 \\ 74 & 12 & 46 \\ 37 & 52 & 19 \end{bmatrix} \end{array}$$

From this matrix, the store had sales of 92 medium tan T-shirts in June.

Now consider a similar matrix for July.

$$\begin{array}{r} \\ \text{Blue} \\ \text{Tan} \\ \text{White} \\ \text{Green} \end{array} \begin{array}{c} \begin{array}{ccc} \text{Small} & \text{Medium} & \text{Large} \end{array} \\ \begin{bmatrix} 56 & 71 & 37 \\ 42 & 83 & 70 \\ 55 & 46 & 38 \\ 26 & 44 & 39 \end{bmatrix} \end{array}$$

The matrices for June and July reveal that the number of large white T-shirts sold for the two months is $46 + 38 = 84$. By adding the elements in corresponding cells, we obtain the total sales for the two months of each type of T-shirt. In matrix notation, we have

$$\begin{bmatrix} 45 & 56 & 81 \\ 37 & 92 & 64 \\ 74 & 12 & 46 \\ 37 & 52 & 19 \end{bmatrix} + \begin{bmatrix} 56 & 71 & 37 \\ 42 & 83 & 70 \\ 55 & 46 & 38 \\ 26 & 44 & 39 \end{bmatrix} = \begin{bmatrix} 101 & 127 & 118 \\ 79 & 175 & 134 \\ 129 & 58 & 84 \\ 63 & 96 & 58 \end{bmatrix}$$

? QUESTION How many large green T-shirts were sold for the two months?

This example suggests that the addition of two matrices should be performed by adding the corresponding elements of the two matrices.

Addition and Subtraction of Matrices

If A and B are matrices of order $m \times n$, then:

The **sum** $A + B$ is the $m \times n$ matrix that is produced by adding each element of A to its corresponding element of B. That is,

$$A + B = [a_{ij} + b_{ij}]$$

The **difference** $A - B$ is the $m \times n$ matrix that is produced by subtracting each element of B from its corresponding element of A. That is,

$$A - B = [a_{ij} - b_{ij}]$$

Alternative to Example 1

If $A = \begin{bmatrix} -2 & 4 \\ 3 & 1 \\ 5 & 2 \end{bmatrix}$ and $B = \begin{bmatrix} -1 & 6 \\ 4 & 0 \\ -3 & 2 \end{bmatrix}$, find

a. $A + B$

b. $A - B$

■ a. $\begin{bmatrix} -3 & 10 \\ 7 & 1 \\ 2 & 4 \end{bmatrix}$

■ b. $\begin{bmatrix} -1 & -2 \\ -1 & 1 \\ 8 & 0 \end{bmatrix}$

EXAMPLE 1 Add and Subtract Matrices

Let $A = \begin{bmatrix} 4 & 1 & -3 \\ 0 & 7 & -5 \end{bmatrix}$, $B = \begin{bmatrix} -6 & 5 & 3 \\ 4 & -2 & 1 \end{bmatrix}$, and $C = \begin{bmatrix} 5 & 11 \\ 6 & 1 \end{bmatrix}$. If possible, find each of the following.

a. $A + B$ **b.** $A - B$ **c.** $A + C$

Solution

a. $A + B = \begin{bmatrix} 4 & 1 & -3 \\ 0 & 7 & -5 \end{bmatrix} + \begin{bmatrix} -6 & 5 & 3 \\ 4 & -2 & 1 \end{bmatrix}$

$$= \begin{bmatrix} 4 + (-6) & 1 + 5 & (-3) + 3 \\ 0 + 4 & 7 + (-2) & (-5) + 1 \end{bmatrix}$$

$$= \begin{bmatrix} -2 & 6 & 0 \\ 4 & 5 & -4 \end{bmatrix}$$

b. $A - B = \begin{bmatrix} 4 & 1 & -3 \\ 0 & 7 & -5 \end{bmatrix} - \begin{bmatrix} -6 & 5 & 3 \\ 4 & -2 & 1 \end{bmatrix}$

$$= \begin{bmatrix} 4 - (-6) & 1 - 5 & (-3) - 3 \\ 0 - 4 & 7 - (-2) & (-5) - 1 \end{bmatrix}$$

$$= \begin{bmatrix} 10 & -4 & -6 \\ -4 & 9 & -6 \end{bmatrix}$$

c. $A + C$ is not defined because the matrices are not of the same order.

Continued ➤

? ANSWER 58, which is the element in the fourth row, third column of the sum of the two matrices.

CHECK YOUR PROGRESS 1 Let $D = \begin{bmatrix} 2 & 9 \\ -4 & 7 \\ -3 & 8 \end{bmatrix}$ and $F = \begin{bmatrix} 3 & -2 \\ 7 & 8 \\ -2 & -3 \end{bmatrix}$. Find each of the following.

a. $D + F$ **b.** $F - D$

Solution *See page S35.* **a.** $\begin{bmatrix} 5 & 7 \\ 3 & 15 \\ -5 & 5 \end{bmatrix}$ **b.** $\begin{bmatrix} 1 & -11 \\ 11 & 1 \\ 1 & -11 \end{bmatrix}$

Just as zero is the additive identity for real numbers, the **zero matrix** is the additive identity for matrices. Because addition of two matrices requires that each matrix be of the same order, there are many identity matrices. For instance,

$$\begin{bmatrix} 0 & 0 \\ 0 & 0 \\ 0 & 0 \end{bmatrix} \quad \begin{bmatrix} 0 & 0 & 0 \\ 0 & 0 & 0 \end{bmatrix} \quad \begin{bmatrix} 0 & 0 & 0 \\ 0 & 0 & 0 \\ 0 & 0 & 0 \end{bmatrix}$$

are all examples of a zero matrix.

Scalar Multiplication of a Matrix

There are two types of products that involve matrices. The first is the product of a real number and a matrix.

Suppose VideoChips, Inc. makes three different video chips, GF1, GF2, and GFX, for home entertainment centers. Each video chip comes in various clock speeds, measured in gigahertz (ghz). The matrix below shows the wholesale cost of each chip in dollars.

	GF1	GF2	GFX
1 ghz	123	140	175
1.5 ghz	145	156	189
2.0 ghz	160	175	202
2.5 ghz	185	205	225

Because of an increase in manufacturing costs, VideoChips decides to increase the price of each chip by 2%. This can be shown in matrix form as

$$1.02\begin{bmatrix} 123 & 140 & 175 \\ 145 & 156 & 189 \\ 160 & 175 & 202 \\ 185 & 205 & 225 \end{bmatrix} = \begin{bmatrix} 1.02 \cdot 123 & 1.02 \cdot 140 & 1.02 \cdot 175 \\ 1.02 \cdot 145 & 1.02 \cdot 156 & 1.02 \cdot 189 \\ 1.02 \cdot 160 & 1.02 \cdot 175 & 1.02 \cdot 202 \\ 1.02 \cdot 185 & 1.02 \cdot 205 & 1.02 \cdot 225 \end{bmatrix}$$

$$\approx \begin{bmatrix} 125 & 143 & 179 \\ 148 & 159 & 193 \\ 163 & 179 & 206 \\ 189 & 209 & 230 \end{bmatrix}$$

where dollar amounts have been rounded to the nearest dollar. The element in row 2, column 3 of the last matrix indicates that the new wholesale cost of the GFX chip running at 1.5 ghz will be $193.

This example suggests that to multiply a matrix by a constant, we multiply each element of the matrix by the constant. When working with matrices, it is helpful to refer to any real number that multiplies a matrix as a **scalar**, to distinguish it from a matrix.

Scalar Multiplication of a Matrix

The product of a scalar c and a matrix A is the matrix obtained by multiplying the scalar times each element of the matrix. That is, $cA = [c \cdot a_{ij}]$.

As an example, consider the matrix

$$A = \begin{bmatrix} 3 & -2 & 1 \\ 4 & 0 & -5 \end{bmatrix}$$

and the scalar -3. Then

$$-3A = -3\begin{bmatrix} 3 & -2 & 1 \\ 4 & 0 & -5 \end{bmatrix} = \begin{bmatrix} -3(3) & -3(-2) & -3(1) \\ -3(4) & -3(0) & -3(-5) \end{bmatrix} = \begin{bmatrix} -9 & 6 & -3 \\ -12 & 0 & 15 \end{bmatrix}$$

The definition of scalar multiplication can also be used to factor a constant from a matrix.

$$\begin{bmatrix} 6 & -12 & 9 \\ 15 & 3 & 24 \\ -3 & 21 & 30 \end{bmatrix} = 3\begin{bmatrix} 2 & -4 & 3 \\ 5 & 1 & 8 \\ -1 & 7 & 10 \end{bmatrix}$$

Alternative to Example 2

If $A = \begin{bmatrix} 1 & -2 \\ 2 & -3 \end{bmatrix}$ and $B = \begin{bmatrix} 3 & 2 \\ 1 & 4 \end{bmatrix}$, find $3A - 2B$.

■ $\begin{bmatrix} -3 & -10 \\ -4 & -17 \end{bmatrix}$

EXAMPLE 2 Find the Sum of Two Scalar Products

Given $A = \begin{bmatrix} 3 & 1 \\ -2 & 1 \\ 0 & 4 \end{bmatrix}$ and $B = \begin{bmatrix} 9 & 4 \\ -3 & 2 \\ 1 & 4 \end{bmatrix}$, find $2A + 3B$.

Solution

$$2A + 3B = 2\begin{bmatrix} 3 & 1 \\ -2 & 1 \\ 0 & 4 \end{bmatrix} + 3\begin{bmatrix} 9 & 4 \\ -3 & 2 \\ 1 & 4 \end{bmatrix} = \begin{bmatrix} 6 & 2 \\ -4 & 2 \\ 0 & 8 \end{bmatrix} + \begin{bmatrix} 27 & 12 \\ -9 & 6 \\ 3 & 12 \end{bmatrix} = \begin{bmatrix} 33 & 14 \\ -13 & 8 \\ 3 & 20 \end{bmatrix}$$

CHECK YOUR PROGRESS 2 Given $A = \begin{bmatrix} -3 & 5 \\ 6 & -7 \end{bmatrix}$ and $B = \begin{bmatrix} 9 & -11 \\ 3 & 8 \end{bmatrix}$, find $4A - 5B$.

Solution *See page S35.* $\begin{bmatrix} -57 & 75 \\ 9 & -68 \end{bmatrix}$

■ Matrix Multiplication

Now we consider the product of two matrices. This product can be developed by considering an investor who has an investment portfolio of three stocks, IBM, Cisco Systems, and Microsoft. The table below shows the number of shares of each stock owned by the investor.

Company	Number of Shares
IBM	200
Cisco	350
Microsoft	150

Suppose that the closing prices (in dollars) for these stocks on a certain day are as in the table below.

Company	IBM	Cisco	Microsoft
Share price	104.50	17.85	62.75

The total value of the portfolio is given by

$$\text{Total value} = 104.50(200) + 17.85(350) + 62.75(150) = 36{,}560$$

The total value of the portfolio is \$36,560.

TAKE NOTE

A **row matrix** is a matrix with just one row. A **column matrix** has just one column.

In matrix terms, the closing share prices can be written as the 1×3 *row matrix* $[104.50 \quad 17.85 \quad 62.75]$. The number of shares of each stock owned can be written as the 3×1 *column matrix* $\begin{bmatrix} 200 \\ 350 \\ 150 \end{bmatrix}$. The product of the row matrix and the column matrix is

$$[104.50 \quad 17.85 \quad 62.75]\begin{bmatrix} 200 \\ 350 \\ 150 \end{bmatrix} = 104.50(200) + 17.85(350) + 62.75(150) = 36{,}560$$

In general, if A is a row matrix of order $1 \times n$ and B is a column matrix of order $n \times 1$, then the product of A and B, written AB, is

$$AB = [a_1 \; a_2 \; a_3 \; \ldots \; a_n]\begin{bmatrix} b_1 \\ b_2 \\ b_3 \\ \vdots \\ b_n \end{bmatrix} = a_1b_1 + a_2b_2 + a_3b_3 + \cdots + a_nb_n$$

TAKE NOTE

To determine the product of a row matrix and a column matrix, the number of elements in the row matrix must equal the number of elements in the column matrix. If this is not the case, the product AB is not defined. For instance, if

$$A = [2 \;\; 1 \;\; 5] \quad \text{and} \quad B = \begin{bmatrix} 3 \\ 1 \end{bmatrix},$$

then AB is not defined.

For example, if $A = [2 \;\; 3 \;\; 5]$ and $B = \begin{bmatrix} 1 \\ 4 \\ -6 \end{bmatrix}$, then

$$AB = [2 \;\; 3 \;\; 5]\begin{bmatrix} 1 \\ 4 \\ -6 \end{bmatrix} = 2(1) + 3(4) + 5(-6) = -16$$

Now consider two different investors, I_1 and I_2, whose portfolios are as shown in the following table.

Company	Shares for I_1	Shares for I_2
IBM	300	200
Cisco	150	250
Microsoft	200	100

The closing prices for these stocks on three consecutive Fridays in August, 2002 are given in the table below.

Company	IBM	Cisco	Microsoft
August 16	79.35	14.45	50.00
August 23	80.40	14.45	52.22
August 30	75.38	13.82	49.08

In terms of matrices, let the closing prices be denoted by C and the portfolios by S. Then

$$C = \begin{bmatrix} 79.35 & 14.45 & 50.00 \\ 80.40 & 14.45 & 52.22 \\ 75.38 & 13.82 & 49.08 \end{bmatrix} \quad \text{and} \quad S = \begin{bmatrix} 300 & 200 \\ 150 & 250 \\ 200 & 100 \end{bmatrix}$$

Let P denote CS. This product is calculated by extending the concept of the product of a row matrix and a column matrix. Each row of C multiplies each column of S.

$$P = \begin{bmatrix} 79.35 & 14.45 & 50.00 \\ 80.40 & 14.45 & 52.22 \\ 75.38 & 13.82 & 49.08 \end{bmatrix} \begin{bmatrix} 300 & 200 \\ 150 & 250 \\ 200 & 100 \end{bmatrix}$$

$$= \begin{bmatrix} [79.35\ 14.45\ 50.00]\begin{bmatrix} 300 \\ 150 \\ 200 \end{bmatrix} & [79.35\ 14.45\ 50.00]\begin{bmatrix} 200 \\ 250 \\ 100 \end{bmatrix} \\ [80.40\ 14.45\ 52.22]\begin{bmatrix} 300 \\ 150 \\ 200 \end{bmatrix} & [80.40\ 14.45\ 52.22]\begin{bmatrix} 200 \\ 250 \\ 100 \end{bmatrix} \\ [75.38\ 13.82\ 49.08]\begin{bmatrix} 300 \\ 150 \\ 200 \end{bmatrix} & [75.38\ 13.82\ 49.08]\begin{bmatrix} 200 \\ 250 \\ 100 \end{bmatrix} \end{bmatrix}$$

$$= \begin{bmatrix} 79.35(300) + 14.45(150) + 50.00(200) & 79.35(200) + 14.45(250) + 50.00(100) \\ 80.40(300) + 14.45(150) + 52.22(200) & 80.40(200) + 14.45(250) + 52.22(100) \\ 75.38(300) + 13.82(150) + 49.08(200) & 75.38(200) + 13.82(250) + 49.08(100) \end{bmatrix}$$

$$= \begin{bmatrix} 35{,}972.5 & 24{,}482.5 \\ 36{,}731.5 & 24{,}914.5 \\ 34{,}503 & 23{,}439 \end{bmatrix}$$

Each entry in P is the total value of an investor's portfolio on the three Fridays in August. For example, $p_{11} = 35{,}972.5$ means the value of investor I_1's portfolio on August 16 was \$35,972.50. The entry in row 3, column 2 ($p_{32} = 23{,}439$) represents the total value of investor I_2's portfolio on August 30. In each case, the subscripts on an element of p denote the date and the investor, respectively.

Using this application as a model, we now define the product of two matrices. The definition is an extension of the definition of the product of a row matrix and a column matrix.

Multiplication of Matrices

Let A be a matrix of order $m \times n$ and let B be a matrix of order $n \times p$. Then the product AB is the matrix of order $m \times p$ given by $AB = [c_{ij}]$, where each element c_{ij} is the product of the ith row matrix of A and the jth column matrix of B.

For the product of two matrices to be defined, the number of columns of the first matrix must equal the number of rows of the second matrix.

$$\underset{m \times n}{A} \cdot \underset{n \times p}{B} = \underset{m \times p}{C}$$

Must be equal (n and n)

Order of product matrix (m and p)

The product matrix has as many rows as the first matrix and as many columns as the second matrix. For example, if

$$A = \begin{bmatrix} 6 & 3 & 1 \\ 2 & 0 & 7 \\ 5 & 1 & 8 \\ 4 & 7 & 2 \end{bmatrix} \quad \text{and} \quad B = \begin{bmatrix} 3 \\ 1 \\ 8 \end{bmatrix}$$

then A is of order 4×3 and B is of order 3×1. Thus the order of AB is 4×1.

Alternative to Example 3

Let $A = \begin{bmatrix} 2 & 3 & -2 \\ 3 & 1 & 6 \end{bmatrix}$ and $B = \begin{bmatrix} -1 & -2 \\ 2 & 0 \\ 4 & 5 \end{bmatrix}$. Find AB.

■ $\begin{bmatrix} -4 & -14 \\ 23 & 24 \end{bmatrix}$

EXAMPLE 3 Find the Product of Two Matrices

Let $C = \begin{bmatrix} 3 & 2 & 1 \\ 0 & 4 & 5 \end{bmatrix}$ and $D = \begin{bmatrix} 4 & 3 \\ 7 & 1 \\ 5 & -2 \end{bmatrix}$. Find CD.

Solution Because C is of order 2×3 and D is of order 3×2, the product CD will be a matrix of order 2×2.

$$CD = \begin{bmatrix} 3 & 2 & 1 \\ 0 & 4 & 5 \end{bmatrix}\begin{bmatrix} 4 & 3 \\ 7 & 1 \\ 5 & -2 \end{bmatrix}$$

$$= \begin{bmatrix} 3(4) + 2(7) + 1(5) & 3(3) + 2(1) + 1(-2) \\ 0(4) + 4(7) + 5(5) & 0(3) + 4(1) + 5(-2) \end{bmatrix}$$

$$= \begin{bmatrix} 31 & 9 \\ 53 & -6 \end{bmatrix}$$

Continued ➤

CHECK YOUR PROGRESS 3 Let $E = \begin{bmatrix} 2 & -1 \\ 3 & 4 \\ 1 & 0 \end{bmatrix}$ and $F = \begin{bmatrix} 5 & -2 \\ 4 & 3 \end{bmatrix}$. Find EF.

Solution *See page S35.* $\begin{bmatrix} 6 & -7 \\ 31 & 6 \\ 5 & -2 \end{bmatrix}$

Generally, matrix multiplication is not commutative. That is, given two matrices A and B, $AB \neq BA$. In some cases, if we reverse the order of the matrices, the product will not be defined. For instance, if

$$A = \begin{bmatrix} -2 & 3 \\ 1 & 4 \end{bmatrix} \quad \text{and} \quad B = \begin{bmatrix} 5 & 6 & -3 \\ 4 & 1 & 2 \end{bmatrix}$$

then

$$AB = \begin{bmatrix} -2 & 3 \\ 1 & 4 \end{bmatrix}\begin{bmatrix} 5 & 6 & -3 \\ 4 & 1 & 2 \end{bmatrix} = \begin{bmatrix} 2 & -9 & 12 \\ 21 & 10 & 5 \end{bmatrix}$$

However, the product

$$BA = \begin{bmatrix} 5 & 6 & -3 \\ 4 & 1 & 2 \end{bmatrix}_{2\times 3}\begin{bmatrix} -2 & 3 \\ 1 & 4 \end{bmatrix}_{2\times 2}$$

$$\uparrow \text{columns} \neq \text{rows} \uparrow$$

is undefined.

Even in cases in which multiplication *is* defined, the products AB and BA may not be equal. For instance, if

$$A = \begin{bmatrix} 2 & -3 & 1 \\ 0 & -4 & 3 \\ 5 & 1 & 7 \end{bmatrix} \quad \text{and} \quad B = \begin{bmatrix} 4 & 6 & 1 \\ 5 & 3 & -2 \\ 1 & 0 & 5 \end{bmatrix}, \text{ then}$$

$$AB = \begin{bmatrix} 2 & -3 & 1 \\ 0 & -4 & 3 \\ 5 & 1 & 7 \end{bmatrix}\begin{bmatrix} 4 & 6 & 1 \\ 5 & 3 & -2 \\ 1 & 0 & 5 \end{bmatrix} = \begin{bmatrix} -6 & 3 & 13 \\ -17 & -12 & 23 \\ 32 & 33 & 38 \end{bmatrix} \text{ and}$$

$$BA = \begin{bmatrix} 4 & 6 & 1 \\ 5 & 3 & -2 \\ 1 & 0 & 5 \end{bmatrix}\begin{bmatrix} 2 & -3 & 1 \\ 0 & -4 & 3 \\ 5 & 1 & 7 \end{bmatrix} = \begin{bmatrix} 13 & -35 & 29 \\ 0 & -29 & 0 \\ 27 & 2 & 36 \end{bmatrix}$$

Thus $AB \neq BA$.

Although matrix multiplication is not commutative, the associative property of multiplication and the distributive property do hold for matrices.

TAKE NOTE

If A, B, and C are matrices, the associative property states that $(AB)C = A(BC)$, as long as the multiplications are defined. The distributive property holds for multiplication on the left and on the right:

$$A(B + C) = AB + AC$$

$$(B + C)A = BA + CA$$

Powers of a matrix are found by repeated multiplication, just as for real numbers. For instance, if

$$E = \begin{bmatrix} 1 & -2 & 4 \\ 2 & 0 & -3 \\ -5 & 1 & 0 \end{bmatrix}, \text{ then}$$

$$E^3 = \begin{bmatrix} 1 & -2 & 4 \\ 2 & 0 & -3 \\ -5 & 1 & 0 \end{bmatrix}^3 = \left(\begin{bmatrix} 1 & -2 & 4 \\ 2 & 0 & -3 \\ -5 & 1 & 0 \end{bmatrix}\begin{bmatrix} 1 & -2 & 4 \\ 2 & 0 & -3 \\ -5 & 1 & 0 \end{bmatrix}\right)\begin{bmatrix} 1 & -2 & 4 \\ 2 & 0 & -3 \\ -5 & 1 & 0 \end{bmatrix}$$

$$= \begin{bmatrix} -23 & 2 & 10 \\ 17 & -7 & 8 \\ -3 & 10 & -23 \end{bmatrix}\begin{bmatrix} 1 & -2 & 4 \\ 2 & 0 & -3 \\ -5 & 1 & 0 \end{bmatrix} = \begin{bmatrix} -69 & 56 & -98 \\ -37 & -26 & 89 \\ 132 & -17 & -42 \end{bmatrix}$$

? QUESTION If A is a 2×3 matrix, is it possible to calculate A^2?

INTEGRATING TECHNOLOGY

A graphing calculator can be used to perform matrix operations. Once the matrices are entered into the calculator, you can use the regular arithmetic operation keys and the variable names of the matrices to perform many operations. The screens below, from a TI-83 Plus calculator, show the operations performed in Example 2, in Example 3, and above (for E^3). For assistance with keystrokes, see the web site for this text at **math.college.hmco.com** or your calculator manual.

```
2[A] + 3[B]
          [[ 33  14 ]
           [ -13  8 ]
           [ 3   20 ]]
```

```
[C] [D]
          [[ 31   9 ]
           [ 53  -6 ]]
```

```
[E]^3
          [[ -69  56  -98 ]
           [ -37  -26  89 ]
           [ 132  -17  -42 ]]
```

Using matrix multiplication and the definition of equality of matrices, we can represent a system of equations. For instance, consider the matrix equation

$$\begin{bmatrix} 2 & -1 & 3 \\ 1 & 4 & 0 \\ 3 & 1 & -2 \end{bmatrix}\begin{bmatrix} x \\ y \\ z \end{bmatrix} = \begin{bmatrix} 2 \\ -1 \\ 4 \end{bmatrix}$$

? ANSWER No. In fact, it is only possible to find powers of square matrices.

Multiplying the matrices on the left side and using the definition of matrix equality, we have

$$\begin{bmatrix} 2 & -1 & 3 \\ 1 & 4 & 0 \\ 3 & 1 & -2 \end{bmatrix}\begin{bmatrix} x \\ y \\ z \end{bmatrix} = \begin{bmatrix} 2 \\ -1 \\ 4 \end{bmatrix}$$

$$\begin{bmatrix} 2x - y + 3z \\ x + 4y \\ 3x + y - 2z \end{bmatrix} = \begin{bmatrix} 2 \\ -1 \\ 4 \end{bmatrix} \rightarrow \begin{cases} 2x - y + 3z = 2 \\ x + 4y = -1 \\ 3x + y - 2z = 4 \end{cases}$$

Writing a system of equations as a matrix equation will be discussed further in the next section.

If a is a real number, then $1a = a$ and $a1 = a$. The number 1 is said to be the *identity element of multiplication* because multiplication of a by 1 produces the identical real number a. The **identity matrix** I_n has a property similar to that of the real number 1. The identity matrices of order 2, 3, and 4 are shown below. Notice that an identity matrix is formed by placing 1's down the main diagonal and 0's elsewhere.

$$I_2 = \begin{bmatrix} 1 & 0 \\ 0 & 1 \end{bmatrix} \qquad I_3 = \begin{bmatrix} 1 & 0 & 0 \\ 0 & 1 & 0 \\ 0 & 0 & 1 \end{bmatrix} \qquad I_4 = \begin{bmatrix} 1 & 0 & 0 & 0 \\ 0 & 1 & 0 & 0 \\ 0 & 0 & 1 & 0 \\ 0 & 0 & 0 & 1 \end{bmatrix}$$

The calculation below shows the product of I_3 and a 3×3 matrix. Observe that the resulting matrix is the original 3×3 matrix.

$$\begin{bmatrix} 1 & 0 & 0 \\ 0 & 1 & 0 \\ 0 & 0 & 1 \end{bmatrix}\begin{bmatrix} 3 & -4 & 5 \\ 1 & -2 & -1 \\ 3 & 1 & 4 \end{bmatrix} =$$

$$\begin{bmatrix} 1(3) + 0(1) + 0(3) & 1(-4) + 0(-2) + 0(1) & 1(5) + 0(-1) + 0(4) \\ 0(3) + 1(1) + 0(3) & 0(-4) + 1(-2) + 0(1) & 0(5) + 1(-1) + 0(4) \\ 0(3) + 0(1) + 1(3) & 0(-4) + 0(-2) + 1(1) & 0(5) + 0(-1) + 1(4) \end{bmatrix}$$

$$= \begin{bmatrix} 3 & -4 & 5 \\ 1 & -2 & -1 \\ 3 & 1 & 4 \end{bmatrix}$$

Identity Matrix

If A is a square matrix of order n and I_n is the identity matrix of order n, then $AI_n = I_nA = A$.

Applications of Matrices

Matrices can often be used to solve applications in which a sequence of events repeats itself over a period of time. For instance, the following application is from the field of botany.

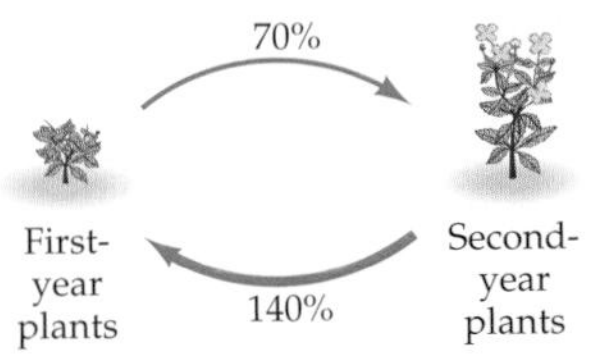

Yearly transition diagram for biennial plants

A biennial plant matures 1 year after the seed is planted. In the second year, the plant produces seeds that will become new plants in the third year. The two-year-old plant then dies. Suppose a wilderness area currently contains 500,000 of a certain biennial and that there are 225,000 plants in the first year and 275,000 in their second year. Suppose also that 70% of the one-year-old plants survive to the second year and that each 1000 second-year plants give rise to 1400 first-year plants. See the figure at the left.

With this information, a botanist can predict how many plants will be growing in the wilderness area after n years. The formula for the number of plants is given by

$$[225{,}000 \;\; 275{,}000]\begin{bmatrix} 0 & 0.7 \\ 1.4 & 0 \end{bmatrix}^n = \begin{bmatrix} \text{number of first-year plants} & \text{number of second-year plants} \end{bmatrix}$$

For instance, to find the number of each type of plant in 5 years, the botanist would calculate

$$[225{,}000 \;\; 275{,}000]\begin{bmatrix} 0 & 0.7 \\ 1.4 & 0 \end{bmatrix}^5 = [369{,}754 \;\; 151{,}263]$$

The resulting matrix indicates that after 5 years, there would be 369,754 first-year plants and 151,263 second-year plants.

Alternative to Example 4
Exercise 38, page 545

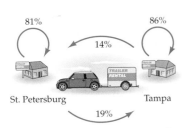

Transition diagram for the trailer rentals

EXAMPLE 4 Solve an Application

A local trailer rental agency has offices in Tampa and St. Petersburg. At the start the agency has 40% of its trailers in Tampa and the other 60% in St. Petersburg. The agency finds that each week:

- 86% of the Tampa rentals are returned to the Tampa office and the other 14% are returned to the St. Petersburg office.
- 81% of the St. Petersburg rentals are returned to the St. Petersburg office and the other 19% are returned to the Tampa office. See the accompanying figure.

The owner has determined that after n weeks, the percent T of the trailers that will be at the Tampa office and the percent S of the trailers that will be at the St. Petersburg office is given by

$$[0.4 \;\; 0.6]\begin{bmatrix} 0.86 & 0.14 \\ 0.19 & 0.81 \end{bmatrix}^n = [T \;\; S]$$

Find the percent of the trailers that will be at the Tampa office after 3 weeks and after 8 weeks.

Solution Using a calculator, we find:

$$[0.4 \;\; 0.6]\begin{bmatrix} 0.86 & 0.14 \\ 0.19 & 0.81 \end{bmatrix}^3 \approx [0.523 \;\; 0.477]$$

and

$$[0.4 \;\; 0.6]\begin{bmatrix} 0.86 & 0.14 \\ 0.19 & 0.81 \end{bmatrix}^8 \approx [0.569 \;\; 0.431]$$

Continued ➤

Thus after 3 weeks the Tampa office will have about 52.3% of the trailers, and after 8 weeks the Tampa office will have about 56.9% of the trailers.

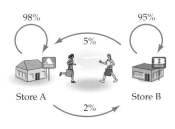

98% 95% 5% Store A Store B 2%

CHECK YOUR PROGRESS 4 A town has two grocery stores. Each month,

- Store A retains 98% of its customers and loses 2% to Store B.
- Store B retains 95% of its customers and loses 5% to Store A. See the accompanying figure.

At the start Store B has 75% of the town's customers and Store A has the other 25%. After n months, the percent of the customers who shop at Store A, denoted by a, and the percent of the customers who shop at Store B, denoted by b, is given by:

$$[0.25 \;\; 0.75]\begin{bmatrix} 0.98 & 0.02 \\ 0.05 & 0.95 \end{bmatrix}^n = [a \;\; b]$$

Find the percent, to the nearest 0.1%, of the customers who shop at Store A after 5 months.

Solution *See page S35.* Approximately 39.1%

Topics for Discussion

1. Give an example of how a spreadsheet program, such as Excel, makes use of matrices.
2. Is it always possible to add two matrices? If so, explain why. If not, discuss what conditions must be met for two matrices to be added. Is matrix addition a commutative operation?
3. Is it always possible to multiply two matrices? If so, explain why. If not, discuss what conditions must be met for two matrices to be multiplied. Is matrix multiplication a commutative operation?
4. Explain how scalar multiplication differs from the multiplication of two matrices.

EXERCISES 6.4

— *Suggested Assignment: Exercises 1–33, odd; 35, 37–40, 43, 46, and 48–53.*
— *Answers to Exercises 10–18, 23–28, 35, 44, 45, and 46 are on pages AA17–AA18.*

In Exercises 1 to 6, determine the order of each matrix. Indicate if the matrix is a square matrix, a row matrix, a column matrix, or an identity matrix.

1. $[5 \;\; -2 \;\; 7]$
1 × 3; row matrix

2. $\begin{bmatrix} 3 \\ 4 \end{bmatrix}$
2 × 1; column matrix

3. $\begin{bmatrix} 1 & 0 & 0 \\ 0 & 1 & 0 \\ 0 & 0 & 1 \end{bmatrix}$
3 × 3; square matrix and identity matrix

4. $\begin{bmatrix} 4 & -1 \\ 3 & 0 \end{bmatrix}$
2 × 2; square matrix

5. $\begin{bmatrix} 1 & 5 & 8 & 2 \\ -3 & 5 & 1 & 0 \\ 2 & 7 & -4 & 1 \end{bmatrix}$
3 × 4

6. $\begin{bmatrix} 1 & 0 & 0 & 0 \\ 0 & 1 & 0 & 0 \\ 0 & 0 & 1 & 0 \\ 0 & 0 & 0 & 1 \end{bmatrix}$
4 × 4; square matrix and identity matrix

In Exercises 7 to 12, find a. $A + B$, b. $A - B$, and c. $2A - 3B$. Do not use a calculator.

7. $A = \begin{bmatrix} 2 & -1 \\ 3 & 3 \end{bmatrix}$ $\quad B = \begin{bmatrix} -1 & 3 \\ 2 & 1 \end{bmatrix}$
a. $\begin{bmatrix} 1 & 2 \\ 5 & 4 \end{bmatrix}$ b. $\begin{bmatrix} 3 & -4 \\ 1 & 2 \end{bmatrix}$ c. $\begin{bmatrix} 7 & -11 \\ 0 & 3 \end{bmatrix}$

8. $A = \begin{bmatrix} 0 & -2 \\ 2 & 3 \end{bmatrix}$ $\quad B = \begin{bmatrix} 5 & -1 \\ 3 & 0 \end{bmatrix}$
a. $\begin{bmatrix} 5 & -3 \\ 5 & 3 \end{bmatrix}$ b. $\begin{bmatrix} -5 & -1 \\ -1 & 3 \end{bmatrix}$ c. $\begin{bmatrix} -15 & -1 \\ -5 & 6 \end{bmatrix}$

9. $A = \begin{bmatrix} 0 & -1 & 3 \\ 1 & 0 & -2 \end{bmatrix}$ $\quad B = \begin{bmatrix} -3 & 1 & 2 \\ 2 & 5 & -3 \end{bmatrix}$
a. $\begin{bmatrix} -3 & 0 & 5 \\ 3 & 5 & -5 \end{bmatrix}$ b. $\begin{bmatrix} 3 & -2 & 1 \\ -1 & -5 & 1 \end{bmatrix}$ c. $\begin{bmatrix} 9 & -5 & 0 \\ -4 & -15 & 5 \end{bmatrix}$

10. $A = \begin{bmatrix} 2 & -2 \\ 3 & 4 \\ 1 & 0 \end{bmatrix}$ $B = \begin{bmatrix} -1 & 8 \\ 2 & -2 \\ -4 & 3 \end{bmatrix}$

11. $A = \begin{bmatrix} -2 & 3 & -1 \\ 0 & -1 & 2 \\ -4 & 3 & 3 \end{bmatrix}$ $B = \begin{bmatrix} 1 & -2 & 0 \\ 2 & 3 & -1 \\ 3 & -1 & 2 \end{bmatrix}$

12. $A = \begin{bmatrix} 0 & 2 & 0 \\ 1 & -3 & 3 \\ 5 & 4 & -2 \end{bmatrix}$ $B = \begin{bmatrix} -1 & 2 & 4 \\ 3 & 3 & -2 \\ -4 & 4 & 3 \end{bmatrix}$

In Exercises 13 to 18, find AB and BA if possible. Do not use a calculator.

13. $A = \begin{bmatrix} 2 & -3 \\ 1 & 4 \end{bmatrix}$ $B = \begin{bmatrix} -2 & 4 \\ 2 & -3 \end{bmatrix}$

29. $\begin{cases} x - 2y = -1 \\ 3x + 4y = 2 \end{cases}$

14. $A = \begin{bmatrix} 3 & -2 \\ 4 & 1 \end{bmatrix}$ $B = \begin{bmatrix} -1 & -1 \\ 0 & 4 \end{bmatrix}$

30. $\begin{cases} 3x - y = 4 \\ 2x + 3y = -1 \end{cases}$

31. $\begin{cases} x = 4 \\ y = 5 \end{cases}$

15. $A = \begin{bmatrix} 2 & -1 \\ 0 & 3 \\ 1 & -2 \end{bmatrix}$ $B = \begin{bmatrix} 1 & -2 & 3 \\ 2 & 0 & 1 \end{bmatrix}$

32. $\begin{cases} x - 4y = -1 \\ y = 3 \end{cases}$

16. $A = \begin{bmatrix} -1 & 3 \\ 2 & 1 \\ -3 & -2 \end{bmatrix}$ $B = \begin{bmatrix} 0 & -1 & 2 \\ 1 & 2 & -4 \end{bmatrix}$

17. $A = \begin{bmatrix} 2 & -1 & 3 \\ 0 & 2 & -1 \\ 0 & 0 & 2 \end{bmatrix}$ $B = \begin{bmatrix} 2 & 0 & 0 \\ 1 & -1 & 0 \\ 2 & -1 & -2 \end{bmatrix}$

18. $A = \begin{bmatrix} -1 & 2 & 0 \\ 2 & -1 & 1 \\ -2 & 2 & -1 \end{bmatrix}$ $B = \begin{bmatrix} 2 & -1 & 0 \\ 1 & 5 & -1 \\ 0 & -1 & 3 \end{bmatrix}$

In Exercises 19 to 22, use the matrices $A = \begin{bmatrix} 2 & -3 \\ 1 & 1 \end{bmatrix}$ and $B = \begin{bmatrix} 3 & -1 & 0 \\ 2 & -2 & -1 \\ 1 & 0 & 2 \end{bmatrix}$ to find each of the following.

19. A^2 $\begin{bmatrix} 1 & -9 \\ 3 & -2 \end{bmatrix}$

20. A^3 $\begin{bmatrix} -7 & -12 \\ 4 & -11 \end{bmatrix}$

21. B^2 $\begin{bmatrix} 7 & -1 & 1 \\ 1 & 2 & 0 \\ 5 & -1 & 4 \end{bmatrix}$

22. B^3 $\begin{bmatrix} 20 & -5 & 3 \\ 7 & -5 & -2 \\ 17 & -3 & 9 \end{bmatrix}$

In Exercises 23 to 28, use a graphing calculator to perform the indicated operations on matrices A and B.

$$A = \begin{bmatrix} 1 & -2 & 3 & 1 \\ 0 & 1 & 2 & -2 \\ 1 & 3 & 0 & 1 \\ -1 & 2 & 1 & 3 \end{bmatrix} \quad B = \begin{bmatrix} 3 & -4 & 1 & 0 \\ 0 & 1 & 5 & -2 \\ 2 & 1 & 1 & -1 \\ 1 & -1 & 0 & 1 \end{bmatrix}$$

23. AB **24.** BA **25.** A^2

26. B^3 **27.** $2A + 5B^2$ **28.** $3A^3 + 4B$

In Exercises 29 to 34, write the system of equations that corresponds to the matrix equation.

29. $\begin{bmatrix} 1 & -2 \\ 3 & 4 \end{bmatrix}\begin{bmatrix} x \\ y \end{bmatrix} = \begin{bmatrix} -1 \\ 2 \end{bmatrix}$ **30.** $\begin{bmatrix} 3 & -1 \\ 2 & 3 \end{bmatrix}\begin{bmatrix} x \\ y \end{bmatrix} = \begin{bmatrix} 4 \\ -1 \end{bmatrix}$

31. $\begin{bmatrix} 1 & 0 \\ 0 & 1 \end{bmatrix}\begin{bmatrix} x \\ y \end{bmatrix} = \begin{bmatrix} 4 \\ 5 \end{bmatrix}$ **32.** $\begin{bmatrix} 1 & -4 \\ 0 & 1 \end{bmatrix}\begin{bmatrix} x \\ y \end{bmatrix} = \begin{bmatrix} -1 \\ 3 \end{bmatrix}$

33. $\begin{bmatrix} 2 & 1 & -3 \\ 1 & 0 & 3 \\ -2 & 1 & 4 \end{bmatrix}\begin{bmatrix} x \\ y \\ z \end{bmatrix} = \begin{bmatrix} 3 \\ 2 \\ 1 \end{bmatrix}$ $\begin{cases} 2x + y - 3z = 3 \\ x + 3z = 2 \\ -2x + y + 4z = 1 \end{cases}$

34. $\begin{bmatrix} 1 & 5 & 2 \\ 3 & 2 & 0 \\ 0 & 1 & 2 \end{bmatrix}\begin{bmatrix} x \\ y \\ z \end{bmatrix} = \begin{bmatrix} 5 \\ 1 \\ 4 \end{bmatrix}$ $\begin{cases} x + 5y + 2z = 5 \\ 3x + 2y = 1 \\ y + 2z = 4 \end{cases}$

Business and Economics

35. ***Sales Revenue*** The matrix below shows the sales revenues, in millions of dollars, that a pharmaceutical company received from various divisions in different parts of the country. The abbreviations are W = western states, N = northern states, S = southern states, and E = eastern states.

	W	N	S	E
Patented drugs	2.0	1.4	3.0	1.4
Generic drugs	0.8	1.1	2.0	0.9
Nonprescription drugs	3.6	1.2	4.5	1.5

A business report for this company anticipates a 2% decrease in sales for each division in each region of the country. Express this matrix as a scalar product and compute the anticipated sales matrix to the nearest ten thousand dollars.

36. ***Lumber Prices*** The following matrix shows the cost per foot of various grades of Douglas fir (D) and pine (P) before a 5% increase in the price per foot.

$$\begin{array}{l} \\ \text{Grade A} \\ \text{Grade B} \\ \text{Grade C} \end{array} \begin{array}{c} \begin{array}{cc} D & P \end{array} \\ \begin{bmatrix} 1.23 & 0.98 \\ 1.45 & 1.24 \\ 1.67 & 1.48 \end{bmatrix} \end{array} \quad \begin{bmatrix} 1.29 & 1.03 \\ 1.52 & 1.30 \\ 1.75 & 1.55 \end{bmatrix}$$

Find the matrix that represents the cost per foot after the price increase. Round to the nearest cent.

37. ***Manufacturing*** A manufacturer has two machines, M_1 and M_2, that are used to create CDs and DVDs. M_1 can create a CD in 2.5 minutes and a DVD in 4 minutes. M_2 can create a CD in 2.7 minutes and a DVD in 3.8 minutes. The manufacturer has three customers, C_1, C_2, and C_3, who have ordered CDs and DVDs. C_1 has ordered 500 CDs and 400 DVDs; C_2 has ordered 400 CDs and 500 DVDs; and C_3 has ordered 450 CDs and 600 DVDs. This information can be represented by the matrices below.

$$A = \begin{bmatrix} 2.5 & 4.0 \\ 2.7 & 3.8 \end{bmatrix} \begin{array}{l} M_1 \\ M_2 \end{array} \quad (\text{columns: CD, DVD}) \qquad B = \begin{bmatrix} 500 & 400 & 450 \\ 400 & 500 & 600 \end{bmatrix} \begin{array}{l} \text{CD} \\ \text{DVD} \end{array} \quad (\text{columns: } C_1, C_2, C_3)$$

a. Find AB. $\begin{bmatrix} 2850 & 3000 & 3525 \\ 2870 & 2980 & 3495 \end{bmatrix}$

b. Explain the meaning, in the context of this problem, of the element in the second row, third column of the product AB.
It takes machine M_2 3495 minutes to make 450 CDs and 600 DVDs.

38. ***Consumer Preferences*** A soft drink company has determined that every 6 months, 1.1% of its customers switch from regular soda to diet soda and 0.7% of its customers switch from diet soda to regular soda. At the present time, 55% of its customers drink regular soda and 45% drink diet soda. After n six-month periods, the percent of its customers who drink regular soda, r, and the percent of its customers who drink diet soda, d, is given by

$$[0.55 \quad 0.45]\begin{bmatrix} 0.989 & 0.011 \\ 0.007 & 0.993 \end{bmatrix}^n = [r \quad d]$$

Use a calculator and the above matrix equation to predict, to the nearest 0.1%, the percent of the customers who will be drinking diet soda

a. 1 year from now. 45.6%
b. 3 years from now. 46.7%

39. ***Consumer Preferences*** Experiment with several different values of n in the matrix equation from Exercise 38 to determine how long, to the nearest year, it will be before the number of diet soda drinkers is greater than the number of regular soda drinkers. 11 yr (10 yr 6 months)

40. ***Consumer Preferences*** Experiment with several different values of n in the matrix equation from Check Your Progress 4 on page 543 to determine how long, to the nearest month, it will be before Store A has 50% of the town's customers. 11 months

41. ***Consumer Preferences*** A video store has determined that every month, 0.8% of its customers switch from VHS to DVD and only 0.1% of its customers switch from DVD to VHS. At the present time, 88% of its customers rent VHS movies and the other 12% rent DVD movies. After n months, the percent of the customers who rent VHS movies, V, and the percent of the customers who rent DVD movies, D, is given by

$$[0.88 \quad 0.12]\begin{bmatrix} 0.992 & 0.008 \\ 0.001 & 0.999 \end{bmatrix}^n = [V \quad D]$$

Use a calculator and the above matrix equation to predict, to the nearest 0.1%, the percent of the customers who will be renting DVD movies

a. 6 months from now. 16.1%
b. 18 months from now. 23.5%

Life and Health Sciences

42. ***Fish Management*** Biologists use capture-recapture models to estimate how many animals live in a certain area. For example, a sample of fish are caught and tagged. When subsequent samples of fish are caught, a biologist can use a capture history matrix to record (with a 1) which, if any, of the fish in the original sample have been caught again. The rows of this matrix represent the particular fish (each has its own identification number), and the columns represent the number of the sample in which the fish was caught. Here is a sample capture history matrix.

	Samples 1	2	3	4
Fish A	1	0	0	1
Fish B	0	1	1	1
Fish C	0	0	1	1

a. What is the order of this matrix? Write a sentence that explains the meaning of the order in the context of the problem.
3×4; There are three different fish and four different samples.

b. What is the meaning of the 1 in row A, column 4? Fish A was caught in the fourth sample.

c. Which fish was captured the most times? Fish B

43. ***Biology*** Biologists can use a predator-prey matrix to study the relationships among animals in an ecosystem. Each row and each column represents an animal in the system. A number 1 in the matrix indicates that the animal represented by that row preys on the animal represented by that column. A 0 indicates that the animal in that row does not prey on the animal in that column. A simple predator-prey matrix is shown below. The abbreviations are H—hawk, R—rabbit, S—snake, C—coyote.

$$\begin{array}{c} \\ H \\ R \\ S \\ C \end{array} \begin{array}{c} \begin{array}{cccc} H & R & S & C \end{array} \\ \begin{bmatrix} 0 & 1 & 1 & 0 \\ 0 & 0 & 0 & 0 \\ 1 & 1 & 0 & 0 \\ 0 & 1 & 1 & 0 \end{bmatrix} \end{array}$$

a. What is the order of this matrix? Write a sentence that explains the meaning of the order in the context of the problem.
4×4; There are four categories of animals and four categories of prey.

b. What is the meaning of all zeros in column C?
No animal preys on the coyote.

c. What is the meaning of all zeros in row R?
The rabbit doesn't prey on any of the other animals.

Social Sciences

44. ***College Divisions*** The number of employees at one campus of a college is given by

	Behavioral sciences	Social sciences	Physical sciences	Life sciences	
$A =$	54	78	34	41	Faculty
	105	146	68	78	Support services
	10	14	6	7	Administration

At a second campus, the number of employees is given by

	Behavioral sciences	Social sciences	Physical sciences	Life sciences	
$B =$	65	88	46	54	Faculty
	130	165	90	91	Support services
	10	17	9	8	Administration

Find $A + B$ and write a sentence explaining the meaning of the new matrix.

45. ***Salary Schedules*** A partial current-year salary matrix for a school district is given below. College A indicates a B.A. degree, column B a B.A. degree plus 15 graduate units, column C an M.A. degree, and column D an M.A. degree plus 30 graduate units. The rows give the number of years of teaching experience. Each element of the matrix is an annual salary in thousands of dollars.

		A	B	C	D
	0 to 5	25	26.3	27.5	28.6
Years	6 to 9	30	31.7	32.9	34.2
	10 to 15	35	36.8	38.2	40.3

Express as a matrix scalar multiplication the result of the school board's approval of a 6% salary increase for all teachers in this district, and compute the scalar product to the nearest hundred dollars.

Sports and Recreation

46. ***Baseball*** The matrices for the number of wins and losses at home, H, and away, A, are shown for the top three teams of the American League East division for the 2002 season.

	W	L	
$H =$	52	28	New York
	42	39	Boston
	42	39	Toronto

	W	L	
$A =$	51	30	New York
	51	30	Boston
	36	45	Toronto

a. Find $H + A$ and write a sentence that explains its meaning in the context of the problem.

b. Find $H - A$ and write a sentence that explains its meaning in the context of the problem.

47. ***Youth Sports*** The total unit sales matrix for three soccer games is given by

	Soft drinks	Hot dogs	Candy	Popcorn	
$S =$	52	50	75	20	Game 1
	45	48	80	20	Game 2
	62	70	78	25	Game 3

The unit pricing matrix, in dollars, for the wholesale cost of each item and the retail price of each item is given by

	Wholesale	Retail	
$P =$	0.25	0.80	Soft drinks
	0.30	0.75	Hot dogs
	0.15	0.45	Candy
	0.10	0.50	Popcorn

Use matrix multiplication to find the total cost and total revenue for each concession during each game. $\begin{bmatrix} 41.25 & 122.85 \\ 39.65 & 118.00 \\ 50.70 & 149.70 \end{bmatrix}$

Prepare for Section 6.5

48. What is the multiplicative inverse of $-\frac{2}{3}$? [P.1] $-\frac{3}{2}$

49. Write the 3×3 multiplicative identity matrix. [6.4]

50. State the three elementary row operations. [6.3] See Section 6.3.

51. Complete the following:

$$\begin{bmatrix} 1 & -2 & 3 \\ 2 & -1 & 4 \\ -3 & 2 & 2 \end{bmatrix} \xrightarrow[3R_1 + R_3]{-2R_1 + R_2} \begin{bmatrix} & ? & \end{bmatrix} \quad [6.3]$$

52. Solve for x: $ax = b$. Write the answer using negative exponents. [P.2/1.1] $x = a^{-1}b$

53. What system of equations is represented by the matrix equation

$$\begin{bmatrix} 2 & 3 \\ 4 & -5 \end{bmatrix}\begin{bmatrix} x \\ y \end{bmatrix} = \begin{bmatrix} 9 \\ 7 \end{bmatrix}? \quad [6.3] \quad \begin{cases} 2x + 3y = 9 \\ 4x - 5y = 7 \end{cases}$$

Explorations

Answer graphs for Exercises 1b., 2b., 3b., 4b., and 5 are on page AA18.

Matrices in the Graphic Arts Many computer drawing programs use matrices to transform images that are stored electronically. For instance, consider the kite $ABCD$ in Figure 6.22. Let each pair of x- and y-coordinates of the vertices of the kite appear as a column of a matrix of order 2×4. This is matrix A below.

$$A = \begin{bmatrix} 8 & 10 & 8 & 6 \\ 7 & 6 & 2 & 6 \end{bmatrix}$$

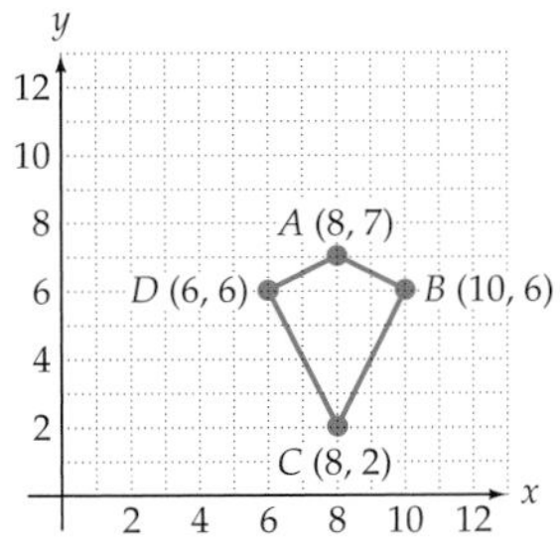

Figure 6.22

49. $\begin{bmatrix} 1 & 0 & 0 \\ 0 & 1 & 0 \\ 0 & 0 & 1 \end{bmatrix}$

51. $\begin{bmatrix} 1 & -2 & 3 \\ 0 & 3 & -2 \\ 0 & -4 & 11 \end{bmatrix}$

1. Let $R = \begin{bmatrix} 0 & -1 \\ 1 & 0 \end{bmatrix}$.

a. Find $R \cdot A$. $\begin{bmatrix} -7 & -6 & -2 & -6 \\ 8 & 10 & 8 & 6 \end{bmatrix}$

b. Using the columns of $R \cdot A$ as the x- and y-coordinates of four points, plot the points. Draw the polygon that connects the points in the order of the columns.

c. Describe how this polygon is related to the original kite. Rotation of 90° counterclockwise about the origin.

2. Let $R = \begin{bmatrix} 0 & 1 \\ 1 & 0 \end{bmatrix}$.

a. Find $R \cdot A$. $\begin{bmatrix} 7 & 6 & 2 & 6 \\ 8 & 10 & 8 & 6 \end{bmatrix}$

b. Using the columns of $R \cdot A$ as the x- and y-coordinates of four points, plot the points. Draw the polygon that connects the points in the order of the columns.

c. Describe how this polygon is related to the original kite. Reflection across the line $y = x$.

3. Let $T = \begin{bmatrix} 2 & 2 & 2 & 2 \\ -1 & -1 & -1 & -1 \end{bmatrix}$.

a. Find $A + T$. $\begin{bmatrix} 10 & 12 & 10 & 8 \\ 6 & 5 & 1 & 5 \end{bmatrix}$

b. Using the columns of $A + T$ as the x- and y-coordinates of four points, plot the points. Draw the polygon that connects the points in the order of the columns.

c. Describe how this polygon is related to the original kite. Translation 2 units to the right and one unit down.

4. a. Find $\frac{1}{2}A$. $\begin{bmatrix} 4 & 5 & 4 & 3 \\ 3.5 & 3 & 1 & 3 \end{bmatrix}$

b. Using the columns of $\frac{1}{2}A$ as the x- and y-coordinates of four points, plot the points. Draw the polygon that connects the points in the order of the columns.

c. Describe how this polygon is related to the original kite. The kite is scaled to half size and each vertex is shifted halfway to the origin.

5. Consider the vertices A, B, C, and D of the original kite and the corresponding vertices A', B', C', and D' of the polygon from Exploration Exercise 4. Graph the four lines $\overleftrightarrow{AA'}$, $\overleftrightarrow{BB'}$, $\overleftrightarrow{CC'}$, and $\overleftrightarrow{DD'}$. At what point do the four lines intersect? This point of intersection is called the *center of dilation* of the transformation. The lines intersect at the origin.

SECTION 6.5 The Inverse of a Matrix

- Inverse of a Matrix
- Singular Matrices
- Solving Systems of Equations Using Inverse Matrices
- Cryptography

Inverse of a Matrix

Recall that the multiplicative inverse of a nonzero real number c is $\frac{1}{c}$, the number whose product with c is 1. For instance, the multiplicative inverse of $\frac{2}{3}$ is $\frac{3}{2}$ because

$$\frac{2}{3} \cdot \frac{3}{2} = 1$$

For some square matrices we can define a multiplicative inverse.

> If A is a square matrix of order n, then the **inverse** of A, denoted by A^{-1}, has the property that
>
> $$A \cdot A^{-1} = A^{-1} \cdot A = I_n$$
>
> where I_n is the identity matrix of order n.

TAKE NOTE

Recall that the identity matrix of order n, denoted I_n, is the $n \times n$ matrix that has 1's for all elements on its main diagonal and 0's for all other elements.

As we will see shortly, not all square matrices have a multiplicative inverse.

QUESTION Are there any real numbers that do not have a multiplicative inverse?

Alternative to Example 1

Determine whether the following matrices are inverses of each other:

$$\begin{bmatrix} 5 & 2 \\ 8 & 3 \end{bmatrix}, \begin{bmatrix} -3 & 2 \\ 8 & -5 \end{bmatrix}$$

- Yes

EXAMPLE 1 Determine Whether Two Matrices Are Inverses of One Another

Let $A = \begin{bmatrix} 9 & 2 \\ 4 & 1 \end{bmatrix}$ and $B = \begin{bmatrix} 1 & -2 \\ -4 & 9 \end{bmatrix}$. Determine whether A and B are inverses of each other.

Solution

$$AB = \begin{bmatrix} 9 & 2 \\ 4 & 1 \end{bmatrix}\begin{bmatrix} 1 & -2 \\ -4 & 9 \end{bmatrix} = \begin{bmatrix} 9(1) + 2(-4) & 9(-2) + 2(9) \\ 4(1) + 1(-4) & 4(-2) + 1(9) \end{bmatrix} = \begin{bmatrix} 1 & 0 \\ 0 & 1 \end{bmatrix}$$

$$BA = \begin{bmatrix} 1 & -2 \\ -4 & 9 \end{bmatrix}\begin{bmatrix} 9 & 2 \\ 4 & 1 \end{bmatrix} = \begin{bmatrix} 1(9) + (-2)(4) & 1(2) + (-2)(1) \\ (-4)(9) + 9(4) & (-4)(2) + 9(1) \end{bmatrix} = \begin{bmatrix} 1 & 0 \\ 0 & 1 \end{bmatrix}$$

Because $AB = BA = I_2$, A and B are inverses of each other.

CHECK YOUR PROGRESS 1 Let $A = \begin{bmatrix} 5 & 3 \\ 7 & 4 \end{bmatrix}$ and $B = \begin{bmatrix} 4 & -3 \\ -7 & 5 \end{bmatrix}$. Determine whether A and B are inverses of each other.

Solution *See page S36.* No

ANSWER Yes. The real number 0 does not have a multiplicative inverse.

INSTRUCTOR NOTE

In the Instructor's Resource Manual there is an activity you can do with students to show how the method of finding the inverse of a matrix works.

A procedure for finding the inverse of a matrix (we will simply say *inverse* for *multiplicative inverse*) uses row operations. We will illustrate the procedure by finding the inverse of a 2×2 matrix.

Let $A = \begin{bmatrix} 2 & 7 \\ 1 & 4 \end{bmatrix}$. Inside the matrix A we include the identity matrix I_2 to the right of the elements of A. We denote this new matrix by $[A:I_2]$.

$$[A:I_2] = \left[\begin{array}{cc|cc} 2 & 7 & 1 & 0 \\ 1 & 4 & 0 & 1 \end{array}\right]$$

A ↗ ↖ I_2

Now we use row operations in a manner similar to the Gaussian elimination method. The goal is to produce

$$[I_2:A^{-1}] = \left[\begin{array}{cc|cc} 1 & 0 & b_{11} & b_{12} \\ 0 & 1 & b_{21} & b_{22} \end{array}\right]$$

I_2 ↗ ↖ A^{-1}

In this form, the inverse matrix is the matrix that is to the right of the identity matrix. That is,

$$A^{-1} = \begin{bmatrix} b_{11} & b_{12} \\ b_{21} & b_{22} \end{bmatrix}$$

To find A^{-1}, we first perform an elementary row operation on $[A:I_2]$ so that the resulting matrix has a 1 in its row 1, column 1 position.

$$\left[\begin{array}{cc|cc} 2 & 7 & 1 & 0 \\ 1 & 4 & 0 & 1 \end{array}\right] \xrightarrow{\frac{1}{2}R_1} \left[\begin{array}{cc|cc} 1 & \frac{7}{2} & \frac{1}{2} & 0 \\ 1 & 4 & 0 & 1 \end{array}\right]$$

Now continue using elementary row operations to produce the identity matrix I_2 to the left of the vertical line segment.

$$\left[\begin{array}{cc|cc} 1 & \frac{7}{2} & \frac{1}{2} & 0 \\ 1 & 4 & 0 & 1 \end{array}\right] \xrightarrow{-1R_1 + R_2} \left[\begin{array}{cc|cc} 1 & \frac{7}{2} & \frac{1}{2} & 0 \\ 0 & \frac{1}{2} & -\frac{1}{2} & 1 \end{array}\right]$$

$$\left[\begin{array}{cc|cc} 1 & \frac{7}{2} & \frac{1}{2} & 0 \\ 0 & \frac{1}{2} & -\frac{1}{2} & 1 \end{array}\right] \xrightarrow{2R_2} \left[\begin{array}{cc|cc} 1 & \frac{7}{2} & \frac{1}{2} & 0 \\ 0 & 1 & -1 & 2 \end{array}\right] \xrightarrow{-\frac{7}{2}R_2 + R_1} \left[\begin{array}{cc|cc} 1 & 0 & 4 & -7 \\ 0 & 1 & -1 & 2 \end{array}\right]$$

The inverse matrix is the matrix to the right of the identity matrix. Therefore,

$$A^{-1} = \begin{bmatrix} 4 & -7 \\ -1 & 2 \end{bmatrix}$$

Each row operation is chosen so as to advance the process of transforming the original matrix A into the identity matrix.

Alternative to Example 2

Find the inverse of the matrix

$$A = \begin{bmatrix} 1 & 1 & 4 \\ 2 & 3 & 6 \\ -1 & -1 & 2 \end{bmatrix}$$

- $A^{-1} = \begin{bmatrix} 2 & -1 & -1 \\ -\frac{5}{3} & 1 & \frac{1}{3} \\ \frac{1}{6} & 0 & \frac{1}{6} \end{bmatrix}$

EXAMPLE 2 Find the Multiplicative Inverse of a Matrix

Find the inverse of $\begin{bmatrix} 1 & -1 & 2 \\ 2 & 0 & 6 \\ 3 & -5 & 7 \end{bmatrix}$.

Solution

$$\left[\begin{array}{ccc|ccc} 1 & -1 & 2 & 1 & 0 & 0 \\ 2 & 0 & 6 & 0 & 1 & 0 \\ 3 & -5 & 7 & 0 & 0 & 1 \end{array}\right]$$

- Merge the given matrix with the identity matrix I_3.

$$\xrightarrow[-3R_1 + R_3]{-2R_1 + R_2} \left[\begin{array}{ccc|ccc} 1 & -1 & 2 & 1 & 0 & 0 \\ 0 & 2 & 2 & -2 & 1 & 0 \\ 0 & -2 & 1 & -3 & 0 & 1 \end{array}\right]$$

- Because a_{11} is already 1, we next produce zeros in a_{21} and a_{31}.

$$\xrightarrow{\frac{1}{2}R_2} \left[\begin{array}{ccc|ccc} 1 & -1 & 2 & 1 & 0 & 0 \\ 0 & 1 & 1 & -1 & \frac{1}{2} & 0 \\ 0 & -2 & 1 & -3 & 0 & 1 \end{array}\right]$$

- Produce a 1 in a_{22}.

$$\xrightarrow{2R_2 + R_3} \left[\begin{array}{ccc|ccc} 1 & -1 & 2 & 1 & 0 & 0 \\ 0 & 1 & 1 & -1 & \frac{1}{2} & 0 \\ 0 & 0 & 3 & -5 & 1 & 1 \end{array}\right]$$

- Produce a 0 in a_{32}.

$$\xrightarrow{\frac{1}{3}R_3} \left[\begin{array}{ccc|ccc} 1 & -1 & 2 & 1 & 0 & 0 \\ 0 & 1 & 1 & -1 & \frac{1}{2} & 0 \\ 0 & 0 & 1 & -\frac{5}{3} & \frac{1}{3} & \frac{1}{3} \end{array}\right]$$

- Produce a 1 in a_{33}.

$$\xrightarrow[-2R_3 + R_1]{-1R_3 + R_2} \left[\begin{array}{ccc|ccc} 1 & -1 & 0 & \frac{13}{3} & -\frac{2}{3} & -\frac{2}{3} \\ 0 & 1 & 0 & \frac{2}{3} & \frac{1}{6} & -\frac{1}{3} \\ 0 & 0 & 1 & -\frac{5}{3} & \frac{1}{3} & \frac{1}{3} \end{array}\right]$$

- Now work upward. Produce a 0 in a_{23} and a_{13}.

$$\xrightarrow{R_2 + R_1} \left[\begin{array}{ccc|ccc} 1 & 0 & 0 & 5 & -\frac{1}{2} & -1 \\ 0 & 1 & 0 & \frac{2}{3} & \frac{1}{6} & -\frac{1}{3} \\ 0 & 0 & 1 & -\frac{5}{3} & \frac{1}{3} & \frac{1}{3} \end{array}\right]$$

- Produce a 0 in a_{12}.

The inverse matrix is $A^{-1} = \begin{bmatrix} 5 & -\frac{1}{2} & -1 \\ \frac{2}{3} & \frac{1}{6} & -\frac{1}{3} \\ -\frac{5}{3} & \frac{1}{3} & \frac{1}{3} \end{bmatrix}$.

Continued ➤

You should verify that A^{-1} satisfies the condition of an inverse matrix. That is, show that $A^{-1} \cdot A = A \cdot A^{-1} = I_3$.

CHECK YOUR PROGRESS 2 Find the inverse of $\begin{bmatrix} 3 & 4 & 2 \\ 1 & 2 & 1 \\ 3 & 0 & 1 \end{bmatrix}$.

Solution *See page S36.* $\begin{bmatrix} 1 & -2 & 0 \\ 1 & -\frac{3}{2} & -\frac{1}{2} \\ -3 & 6 & 1 \end{bmatrix}$

INTEGRATING TECHNOLOGY

The inverse of a matrix can also be found by using a graphing calculator. Enter and store the matrix in some variable—say—[A]. To find the inverse of *A*, we use the x^{-1} key. For instance, here is A^{-1} for Example 2.

```
[A]
          [[1 -1  2]
           [2  0  6]
           [3 -5  7]]
```

```
[A]-1
[[5                 ...
 [.6666666667       ...
 [-1.666666667      ...
```

```
Ans▸Frac
[[5      -1/2     -1 ...
 [2/3    1/6      -1/...
 [-5/3   1/3      1/3...
```

A typical calculator display of A^{-1} is shown in the middle. Because the elements of the inverse matrix are shown as decimals, we can only see the first column of the inverse matrix. Use the arrow keys to see the remaining columns.

Another method for viewing A^{-1} is to use a command on your calculator that converts decimals to fractions. On the TI-83 calculator this command is denoted by Frac (accessed by pressing the MATH key). This command will display A^{-1} in fractional form, as shown on the right.

▪ Singular Matrices

A matrix that has a multiplicative inverse is a **nonsingular matrix.** A **singular matrix** is a matrix that does not have a multiplicative inverse. As you apply the inverse procedure described on page 549 to a singular matrix, there will come a point at which there are all zeros in a row of the *original matrix.* When that condition exists, the original matrix does not have an inverse.

Alternative to Example 3

Show that $\begin{bmatrix} 1 & -6 & 4 \\ 3 & 4 & 2 \\ 5 & 3 & 5 \end{bmatrix}$ is a singular matrix.

▪ $\left[\begin{array}{ccc|ccc} 1 & -6 & 4 & 1 & 0 & 0 \\ 6 & 1 & -\frac{5}{11} & -\frac{3}{22} & \frac{1}{22} & 0 \\ 0 & 0 & 0 & -\frac{1}{2} & -\frac{3}{2} & 1 \end{array}\right]$

EXAMPLE 3 Show That a Matrix Is a Singular Matrix

Show that $\begin{bmatrix} 1 & -1 & -1 \\ 2 & -3 & 0 \\ 1 & -2 & 1 \end{bmatrix}$ is a singular matrix.

Continued ➤

Solution

$$\left[\begin{array}{rrr|rrr} 1 & -1 & -1 & 1 & 0 & 0 \\ 2 & -3 & 0 & 0 & 1 & 0 \\ 1 & -2 & 1 & 0 & 0 & 1 \end{array}\right] \xrightarrow[-1R_1 + R_3]{-2R_1 + R_2} \left[\begin{array}{rrr|rrr} 1 & -1 & -1 & 1 & 0 & 0 \\ 0 & -1 & 2 & -2 & 1 & 0 \\ 0 & -1 & 2 & -1 & 0 & 1 \end{array}\right]$$

$$\xrightarrow{-1R_2} \left[\begin{array}{rrr|rrr} 1 & -1 & -1 & 1 & 0 & 0 \\ 0 & 1 & -2 & 2 & -1 & 0 \\ 0 & -1 & 2 & -1 & 0 & 1 \end{array}\right] \xrightarrow{R_2 + R_3} \left[\begin{array}{rrr|rrr} 1 & -1 & -1 & 1 & 0 & 0 \\ 0 & 1 & -2 & 2 & -1 & 0 \\ 0 & 0 & 0 & 1 & -1 & 1 \end{array}\right]$$

There are all zeros in a row of the original matrix. The original matrix does not have an inverse.

CHECK YOUR PROGRESS 3 Show that $\begin{bmatrix} 4 & -3 & 3 \\ 2 & 1 & -4 \\ 6 & -2 & -1 \end{bmatrix}$ is a singular matrix.

Solution *See page S36.*

▪ Solving Systems of Equations Using Inverse Matrices

Systems of linear equations can be solved by finding the inverse of the coefficient matrix of the system. The procedure is similar to solving a simple first-degree equation of the form $ax = b$, such as $2x = 6$.

$$2x = 6$$
$$(2^{-1})2x = (2^{-1})6 \qquad \text{▪ Multiply each side of the equation by the inverse of 2: } 2^{-1} = \tfrac{1}{2}.$$
$$x = 3$$

Now consider the following system.

$$\begin{cases} 3x + 7y = 5 \\ 2x + 5y = 4 \end{cases} \qquad (1)$$

Using matrix multiplication and the concept of equality of matrices, we can write System (1) as a **matrix equation.**

$$\begin{bmatrix} 3 & 7 \\ 2 & 5 \end{bmatrix}\begin{bmatrix} x \\ y \end{bmatrix} = \begin{bmatrix} 5 \\ 4 \end{bmatrix} \qquad (2)$$

This is similar to the equation $2x = 6$ in that we have a constant matrix times a variable matrix equal to a constant. If we multiply each side of Equation (2) by the inverse of the constant matrix, we have

$$\begin{bmatrix} 3 & 7 \\ 2 & 5 \end{bmatrix}^{-1}\begin{bmatrix} 3 & 7 \\ 2 & 5 \end{bmatrix}\begin{bmatrix} x \\ y \end{bmatrix} = \begin{bmatrix} 3 & 7 \\ 2 & 5 \end{bmatrix}^{-1}\begin{bmatrix} 5 \\ 4 \end{bmatrix}$$

$$\begin{bmatrix} x \\ y \end{bmatrix} = \begin{bmatrix} 5 & -7 \\ -2 & 3 \end{bmatrix}\begin{bmatrix} 5 \\ 4 \end{bmatrix} \qquad \text{▪ } \begin{bmatrix} 3 & 7 \\ 2 & 5 \end{bmatrix}^{-1} = \begin{bmatrix} 5 & -7 \\ -2 & 3 \end{bmatrix}$$

$$\begin{bmatrix} x \\ y \end{bmatrix} = \begin{bmatrix} -3 \\ 2 \end{bmatrix}$$

Thus $x = -3$ and $y = 2$.

TAKE NOTE

Because matrix multiplication is not commutative, make sure you multiply on the left on each side of the equation.

In general, to solve the matrix equation $AX = B$, multiply each side of the equation by A^{-1} on the left.

$$\begin{aligned} AX &= B \\ (A^{-1})AX &= (A^{-1})B \\ X &= A^{-1}B \end{aligned}$$

Alternative to Example 4

Use matrix algebra to solve the following system of equations.

$$\begin{cases} 3x - y + 2z = 1 \\ x + y - 4z = -9 \\ 2x + 3y + 3z = 4 \end{cases}$$

■ (−1, 0, 2)

EXAMPLE 4 Solve a System of Equations Using an Inverse Matrix

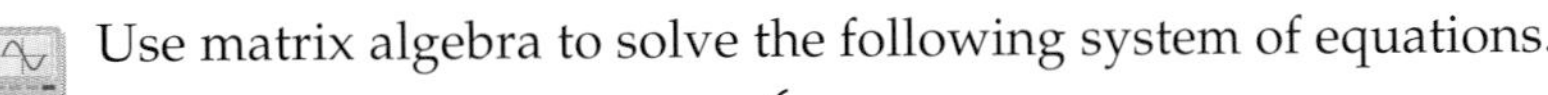

Use matrix algebra to solve the following system of equations.

$$\begin{cases} x + 7z = 20 \\ 2x + y - z = -3 \\ 7x + 3y + z = 2 \end{cases}$$

Solution Write the system of equations as a matrix equation of the form $AX = B$.

$$\begin{bmatrix} 1 & 0 & 7 \\ 2 & 1 & -1 \\ 7 & 3 & 1 \end{bmatrix} \begin{bmatrix} x \\ y \\ z \end{bmatrix} = \begin{bmatrix} 20 \\ -3 \\ 2 \end{bmatrix}$$

Find A^{-1}.

$$A^{-1} = \begin{bmatrix} -\frac{4}{3} & -7 & \frac{7}{3} \\ 3 & 16 & -5 \\ \frac{1}{3} & 1 & -\frac{1}{3} \end{bmatrix}$$

Multiply each side of the matrix equation by the inverse matrix on the left.

$$\begin{bmatrix} -\frac{4}{3} & -7 & \frac{7}{3} \\ 3 & 16 & -5 \\ \frac{1}{3} & 1 & -\frac{1}{3} \end{bmatrix} \begin{bmatrix} 1 & 0 & 7 \\ 2 & 1 & -1 \\ 7 & 3 & 1 \end{bmatrix} \begin{bmatrix} x \\ y \\ z \end{bmatrix} = \begin{bmatrix} -\frac{4}{3} & -7 & \frac{7}{3} \\ 3 & 16 & -5 \\ \frac{1}{3} & 1 & -\frac{1}{3} \end{bmatrix} \begin{bmatrix} 20 \\ -3 \\ 2 \end{bmatrix}$$

$$\begin{bmatrix} x \\ y \\ z \end{bmatrix} = \begin{bmatrix} -1 \\ 2 \\ 3 \end{bmatrix}$$

Thus $x = -1$, $y = 2$, and $z = 3$. The solution of the system of equations is the ordered triple $(-1, 2, 3)$.

CHECK YOUR PROGRESS 4 Use matrix algebra to solve the following system of equations.

$$\begin{cases} x + 2y - z = 5 \\ 2x + 3y - z = 8 \\ 3x + 6y - 2z = 14 \end{cases}$$

Solution *See page S36.* (2, 1, −1)

There is no advantage to using matrix algebra to solve a system of equations if you compute the inverse matrix by hand using row operations. The power of this method is evident when you use a graphing calculator. Then, finding the solution of a system of equations reduces to the mechanical process of entering the coefficient matrix A and the constant matrix B, and directing the calculator to compute $A^{-1}B$.

■ Cryptography

INSTRUCTOR NOTE

The applications of matrices that follow, cryptography and input-output analysis, may be easier to cover if students have access to graphing calculators.

Cryptography is the study of the techniques used to conceal the meaning of a message. The message that is to be concealed is called **plaintext** and the concealed message is called **ciphertext.** One way to change plaintext to ciphertext is to give each letter of the alphabet a numerical value. Matrices are then used to scramble the numbers so that it is difficult to determine which number is associated with each letter.

Point of Interest

When a word processor saves a file with the .txt extension, ASCII coding is used for the text of the file.

One way to assign each letter a numerical value is to use the American Standard Code for Information Interchange (ASCII). This system assigns a number to each letter and punctuation mark.

 A partial ASCII table is shown below. For a complete table, see **www.asciitable.com.**

Partial ASCII Character Table

Char	Value	Char	Value	Char	Value	Char	Value	Char	Value	Char	Value
space	32	A	65	L	76	W	87	h	104	s	115
0	48	B	66	M	77	X	88	i	105	t	116
1	49	C	67	N	78	Y	89	j	106	u	117
2	50	D	68	O	79	Z	90	k	107	v	118
3	51	E	69	P	80	a	97	l	108	w	119
4	52	F	70	Q	81	b	98	m	109	x	120
5	53	G	71	R	82	c	99	n	110	y	121
6	54	H	72	S	83	d	100	o	111	z	122
7	55	I	73	T	84	e	101	p	112	.	46
8	56	J	74	U	85	f	102	q	113	,	44
9	57	K	75	V	86	g	103	r	114	?	63

To see how this coding scheme works, consider the sentence "A picture is worth a thousand words." Group the letters of the sentence into packets. Any size packet can be used. For this example, we will use a packet of 4 and use 0 (zero) for a space. Our sentence would look like

(A0pi)(ctur)(e0is)(0wor)(th0a)(0tho)(usan)(d0wo)(rds.)

Replace each letter and punctuation mark by its numerical ASCII equivalent. For our sentence, this procedure produces

(65 48 112 105)(99 116 117 114)(101 48 105 115)(48 119 111 114)

(116 104 48 97)(48 116 104 111)(117 115 97 110)(100 48 119 111)

(114 100 115 46)

Place these numbers in a matrix, using each group of four numbers as a column. For our example, this yields

$$W = \begin{bmatrix} 65 & 99 & 101 & 48 & 116 & 48 & 117 & 100 & 114 \\ 48 & 116 & 48 & 119 & 104 & 116 & 115 & 48 & 100 \\ 112 & 117 & 105 & 111 & 48 & 104 & 97 & 119 & 115 \\ 105 & 114 & 115 & 114 & 97 & 111 & 110 & 111 & 46 \end{bmatrix}$$

Now construct a 4×4 (the packet size) matrix E, with integer elements, that has an inverse. You can use any 4×4 *nonsingular* matrix. For this example we will use

$$E = \begin{bmatrix} 1 & 2 & 3 & 1 \\ 2 & 3 & 1 & 4 \\ 2 & 1 & 1 & 0 \\ 3 & 1 & 2 & 1 \end{bmatrix}$$

Find the product $M = EW$.

$$M = \begin{bmatrix} 1 & 2 & 3 & 1 \\ 2 & 3 & 1 & 4 \\ 2 & 1 & 1 & 0 \\ 3 & 1 & 2 & 1 \end{bmatrix} \begin{bmatrix} 65 & 99 & 101 & 48 & 116 & 48 & 117 & 100 & 114 \\ 48 & 116 & 48 & 119 & 104 & 116 & 115 & 48 & 100 \\ 112 & 117 & 105 & 111 & 48 & 104 & 97 & 119 & 115 \\ 105 & 114 & 115 & 114 & 97 & 111 & 110 & 111 & 46 \end{bmatrix}$$

$$M = \begin{bmatrix} 602 & 796 & 627 & 733 & 565 & 703 & 748 & 664 & 705 \\ 806 & 1119 & 911 & 1020 & 980 & 992 & 1116 & 907 & 827 \\ 290 & 431 & 355 & 326 & 384 & 316 & 446 & 367 & 443 \\ 572 & 761 & 676 & 599 & 645 & 579 & 770 & 697 & 718 \end{bmatrix}$$

The numbers in matrix M form the ciphertext. These are the numbers that would be sent to a friend or colleague.

The person who receives your message would compute $E^{-1}M$ to restore your message to its original form.

$$W = E^{-1}M$$

$$= \begin{bmatrix} -\frac{1}{4} & 0 & \frac{1}{4} & \frac{1}{4} \\ \frac{1}{6} & \frac{1}{6} & 1 & -\frac{5}{6} \\ \frac{1}{3} & -\frac{1}{6} & -\frac{1}{2} & \frac{1}{3} \\ -\frac{1}{12} & \frac{1}{6} & -\frac{3}{4} & \frac{5}{12} \end{bmatrix} \begin{bmatrix} 602 & 796 & 627 & 733 & 565 & 703 & 748 & 664 & 705 \\ 806 & 1119 & 911 & 1020 & 980 & 992 & 1116 & 907 & 827 \\ 290 & 431 & 355 & 326 & 384 & 316 & 446 & 367 & 443 \\ 572 & 761 & 676 & 599 & 645 & 579 & 770 & 697 & 718 \end{bmatrix}$$

$$= \begin{bmatrix} 65 & 99 & 101 & 48 & 116 & 48 & 117 & 100 & 114 \\ 48 & 116 & 48 & 119 & 104 & 116 & 115 & 48 & 100 \\ 112 & 117 & 105 & 111 & 48 & 104 & 97 & 119 & 115 \\ 105 & 114 & 115 & 114 & 97 & 111 & 110 & 111 & 46 \end{bmatrix}$$

This last matrix is the original message matrix.

Topics for Discussion

1. Explain the difference between a singular matrix and a nonsingular matrix.
2. Discuss the advantages and disadvantages of solving a system of equations by using an inverse matrix.
3. Give an example of a square matrix that does not have a multiplicative inverse.
4. What is the difference between ciphertext and plaintext?

EXERCISES 6.5

— *Suggested Assignment: Exercises 1–31, odd; 34, and 35–40.*

In Exercises 1 to 4, find the product of the matrices to determine whether the matrices are inverses of each other. Do not use a calculator.

1. $\begin{bmatrix} 2 & 3 \\ -5 & -7 \end{bmatrix}$ $\begin{bmatrix} -7 & -3 \\ 5 & 2 \end{bmatrix}$ Yes

2. $\begin{bmatrix} 4 & -1 \\ 8 & -3 \end{bmatrix}$ $\begin{bmatrix} \frac{3}{4} & -\frac{1}{4} \\ 2 & -1 \end{bmatrix}$ Yes

3. $\begin{bmatrix} 2 & 2 & 1 \\ -3 & 4 & 0 \\ 1 & -2 & 2 \end{bmatrix}$ $\begin{bmatrix} \frac{4}{15} & -\frac{1}{5} & -\frac{2}{15} \\ \frac{1}{5} & \frac{1}{10} & -\frac{1}{10} \\ \frac{1}{15} & \frac{1}{5} & \frac{7}{15} \end{bmatrix}$ Yes

4. $\begin{bmatrix} 1 & 3 & -2 \\ 4 & 0 & 1 \\ -1 & 1 & 5 \end{bmatrix}$ $\begin{bmatrix} 5 & 4 & 3 \\ 1 & 0 & -1 \\ -2 & 1 & 1 \end{bmatrix}$ No

In Exercises 5 to 10, use row operations to find the inverse of the matrix.

5. $\begin{bmatrix} 1 & -3 \\ -2 & 5 \end{bmatrix}$ $\begin{bmatrix} -5 & -3 \\ -2 & -1 \end{bmatrix}$

6. $\begin{bmatrix} 1 & 2 \\ -2 & -3 \end{bmatrix}$ $\begin{bmatrix} -3 & -2 \\ 2 & 1 \end{bmatrix}$

7. $\begin{bmatrix} 1 & 4 \\ 2 & 10 \end{bmatrix}$ $\begin{bmatrix} 5 & -2 \\ -1 & \frac{1}{2} \end{bmatrix}$

8. $\begin{bmatrix} -2 & 3 \\ -6 & -8 \end{bmatrix}$ $\begin{bmatrix} -\frac{4}{17} & -\frac{3}{34} \\ \frac{3}{17} & -\frac{1}{17} \end{bmatrix}$

9. $\begin{bmatrix} 1 & 2 & -1 \\ 2 & 5 & 1 \\ 3 & 6 & -2 \end{bmatrix}$ $\begin{bmatrix} -16 & -2 & 7 \\ 7 & 1 & -3 \\ -3 & 0 & 1 \end{bmatrix}$

10. $\begin{bmatrix} 1 & 3 & -2 \\ -1 & -5 & 6 \\ 2 & 6 & -3 \end{bmatrix}$ $\begin{bmatrix} \frac{21}{2} & \frac{3}{2} & -4 \\ -\frac{9}{2} & -\frac{1}{2} & 2 \\ -2 & 0 & 1 \end{bmatrix}$

In Exercises 11 to 14, use row operations to verify that each matrix is a singular matrix.

11. $\begin{bmatrix} 2 & 3 & -9 \\ 1 & 1 & -4 \\ 3 & -1 & -8 \end{bmatrix}$ Singular matrix

12. $\begin{bmatrix} 2 & -5 & 11 \\ 3 & 4 & -18 \\ 6 & -1 & -9 \end{bmatrix}$ Singular matrix

13. $\begin{bmatrix} 2 & 3 & -11 \\ 3 & -5 & 12 \\ -2 & -1 & 5 \end{bmatrix}$ Singular matrix

14. $\begin{bmatrix} 1 & 1 & -6 \\ 3 & 5 & -20 \\ 2 & -1 & -9 \end{bmatrix}$ Singular matrix

In Exercises 15 to 20, use a graphing calculator to find the inverse of the matrix.

15. $\begin{bmatrix} 7 & 11 \\ 1 & -2 \end{bmatrix}$ $\begin{bmatrix} \frac{2}{25} & \frac{11}{25} \\ \frac{1}{25} & -\frac{7}{25} \end{bmatrix}$

16. $\begin{bmatrix} 5 & -2 \\ 7 & -3 \end{bmatrix}$ $\begin{bmatrix} 3 & -2 \\ 7 & -5 \end{bmatrix}$

17. $\begin{bmatrix} 4 & 3 & 1 \\ -2 & 1 & 0 \\ 1 & 2 & -3 \end{bmatrix}$

18. $\begin{bmatrix} 1 & -1 & 0 \\ 2 & 1 & -1 \\ 1 & 0 & -3 \end{bmatrix}$

19. $\begin{bmatrix} 1 & -1 & 2 & 1 \\ 2 & -1 & 5 & 1 \\ 3 & -3 & 7 & 5 \\ -2 & 3 & -4 & -1 \end{bmatrix}$

20. $\begin{bmatrix} 1 & 1 & -1 & 2 \\ 3 & 2 & -1 & 5 \\ 2 & 2 & -1 & 5 \\ 4 & 4 & -4 & 7 \end{bmatrix}$

Answers to 17–20 are on page 558.

In Exercises 21 to 28, solve each system of equations by using a calculator and inverse matrix methods.

21. $\begin{cases} x + 4y = 6 \\ 2x + 7y = 11 \end{cases}$ (2, 1)

22. $\begin{cases} 2x + 3y = 5 \\ x + 2y = 4 \end{cases}$ (−2, 3)

23. $\begin{cases} x + y + 2z = 4 \\ 2x + 3y + 3z = 5 \\ 3x + 3y + 7z = 14 \end{cases}$ (1, −1, 2)

24. $\begin{cases} x + 2y - z = 5 \\ 2x + 3y - z = 8 \\ 3x + 6y - 2z = 14 \end{cases}$ (2, 1, −1)

25. $\begin{cases} x + 2y + 2z = 5 \\ -2x - 5y - 2z = 8 \\ 2x + 4y + 7z = 19 \end{cases}$ (23, −12, 3)

26. $\begin{cases} x - y + 3z = 5 \\ 3x - y + 10z = 16 \\ 2x - 2y + 5z = 9 \end{cases}$ (2, 0, 1)

27. $\begin{cases} w + 2x + z = 6 \\ 2w + 5x + y + 2z = 10 \\ 2w + 4x + y + z = 8 \\ 3w + 6x + 4z = 16 \end{cases}$ (0, 4, −6, −2)

28. $\begin{cases} w - x + 2y = 5 \\ 2w - x + 6y + 2z = 16 \\ 3w - 2x + 9y + 4z = 28 \\ w - 2x - z = 2 \end{cases}$ (1, −2, 1, 3)

In Exercises 29 and 30, a message was coded using the ASCII coding table given in this section. Each message was placed in packets of 3, and the encrypting matrix *E* was used to encrypt the message *M*. Determine the original message.

29. $M = \begin{bmatrix} 640 & 691 & 592 & 650 \\ 517 & 569 & 465 & 514 \\ 438 & 465 & 258 & 426 \end{bmatrix}, E = \begin{bmatrix} 2 & 3 & 1 \\ 3 & 1 & 1 \\ 0 & 1 & 3 \end{bmatrix}$ computer bug

30. $M = \begin{bmatrix} 521 & 615 & 508 & 603 & 456 \\ 603 & 477 & 375 & 639 & 288 \\ 374 & 322 & 275 & 424 & 248 \end{bmatrix}, E = \begin{bmatrix} 3 & 1 & 2 \\ 0 & 3 & 3 \\ 1 & 2 & 1 \end{bmatrix}$ Ides of March

Business and Economics

31. ***Business Revenue*** A popular vacation spot offers helicopter tours. The price of an adult ticket is \$20; the price of a child's ticket is \$15. The tour operator's records show that 100 people took the tour on Saturday and 120 people took the tour on Sunday. The total receipts for Saturday were \$1900, and on Sunday the receipts totalled \$2275. Write and solve a system of equations, using matrices, to find the number of adults and the number of children who took the tour on Saturday and on Sunday.

On Saturday 80 adults, 20 children; On Sunday 95 adults, 25 children

32. ***Business Revenue*** A company sells a standard model and a deluxe model tape recorder. Each standard model tape recorder costs \$45 to manufacture, and each deluxe model costs \$60 to manufacture. The January manufacturing budget for 90 of these recorders was \$4650 and the February budget for 100 recorders was \$5250. Write and solve a system of equations, using matrices, to find the number of each type of recorder manufactured in January and in February.
In January 50 standard, 40 deluxe; In February 50 standard, 50 deluxe

Life and Health Sciences

33. ***Soil Science*** The following table shows the active chemical content of three different soil additives.

Additive	Grams per 100 grams		
	Ammonium Nitrate	Phosphorus	Iron
1	30	10	10
2	40	15	10
3	50	5	5

A soil chemist wants to prepare two chemical samples. The first sample requires 380 grams of ammonium nitrate, 95 grams of phosphorus, and 85 grams of iron. The second sample requires 380 grams of ammonium nitrate, 110 graphs of phosphorus, and 90 grams of iron. Write and solve a system of equations, using matrices, to determine how many grams of each additive are required for sample 1 and how many grams of each additive are required for sample 2. Sample 1: 500 g of additive 1, 200 g of additive 2, 300 g of additive 3; Sample 2: 400 g of additive 1, 400 g of additive 2, 200 g of additive 3

34. ***Nutrition*** The following table shows the carbohydrate, fat, and protein content of three food types. A nutritionist must prepare two diets from these three food types. The first diet must contain 23 grams of carbohydrate, 18 grams of fat, and 39 grams of protein. The second diet must contain 35 grams of carbohydrate, 28 grams of fat, and 42 grams of protein. Write and solve a system of equations, using matrices, to determine how many grams of each food type are required for the first diet and how many grams of each food type are required for the second diet.

Food Type	Grams per 100 grams		
	Carbohydrate	Fat	Protein
I	13	10	13
II	4	4	3
III	1	0	10

Diet 1: 100 g of food type I, 200 g of food type II, 200 g of food type III; Diet 2: 200 g of food type I, 200 g of food type II, 100 g of food type III

Prepare for Section 6.6

Answer graph to Exercise 39 is on page AA18.

35. Solve: $3x - 5 < 5x + 3$. Write the answer using interval notation. [1.4] $(-4, \infty)$

36. Solve for y: $2x - 3y \ge 6$. [1.4] $y \le \frac{2}{3}x - 2$

37. Solve: $|2x + 5| \ge 7$. Write the answer using interval notation. [1.4] $(-\infty, -6] \cup [1, \infty)$

38. Solve: $3x + 1 < 10$ and $3 - 4x < -1$. Write the answer using interval notation. [1.4] $(1, 3)$

39. Graph: $3x + 4y = -12$. [2.3]

40. Solve: $\begin{cases} x + 2y = 1 \\ 2x - 5y = 11 \end{cases}$ [6.1] $(3, -1)$

Explorations 

1. ***Digital Signatures*** One of the issues that arises when sending legal documents such as contracts over the Internet is that of signing the document. During a paper transaction, a person can sign a contract in the presence of a notary. The notary verifies that the person signing the document is, in fact, who he or she claims to be. One way to do this over the Internet is to use *digital signatures,* which use encryption and a form of a notary to guarantee a signature. Write a one-page paper on digital signatures and how they work. Answers will vary.

17. $\begin{bmatrix} \frac{3}{35} & -\frac{11}{35} & \frac{1}{35} \\ \frac{6}{35} & \frac{13}{35} & \frac{2}{35} \\ \frac{1}{7} & \frac{1}{7} & -\frac{2}{7} \end{bmatrix}$

18. $\begin{bmatrix} \frac{3}{8} & \frac{3}{8} & -\frac{1}{8} \\ -\frac{5}{8} & \frac{3}{8} & -\frac{1}{8} \\ \frac{1}{8} & \frac{1}{8} & -\frac{3}{8} \end{bmatrix}$

19. $\begin{bmatrix} \frac{19}{2} & -\frac{1}{2} & -\frac{3}{2} & \frac{3}{2} \\ \frac{7}{4} & \frac{1}{4} & -\frac{1}{4} & \frac{3}{4} \\ -\frac{7}{2} & \frac{1}{2} & \frac{1}{2} & -\frac{1}{2} \\ \frac{1}{4} & -\frac{1}{4} & \frac{1}{4} & \frac{1}{4} \end{bmatrix}$

20. $\begin{bmatrix} 0 & 1 & -1 & 0 \\ -13 & -1 & 2 & 3 \\ -6 & 0 & 1 & 1 \\ 4 & 0 & 0 & -1 \end{bmatrix}$

SECTION 6.6 Linear Inequalities and Systems of Linear Inequalities

- Graph a Linear Inequality
- Systems of Inequalities in Two Variables
- Linear Programming

Graph a Linear Inequality

An inequality of the form $ax + by > c$ or $y > mx + b$ is a **linear inequality in two variables,** where the greater than symbol can be replaced by $<$, $\leq$, or $\geq$. Examples of linear inequalities in two variables are shown below.

$$3x - 4y < 12 \qquad y > \frac{2}{3}x + 1 \qquad x \leq 1 \qquad y > -2$$

The graph of the *equation* $y = x - 1$ separates the plane into three sets: the set of points on the line, the set of points above the line, and the set of points below the line. See Figure 6.23.

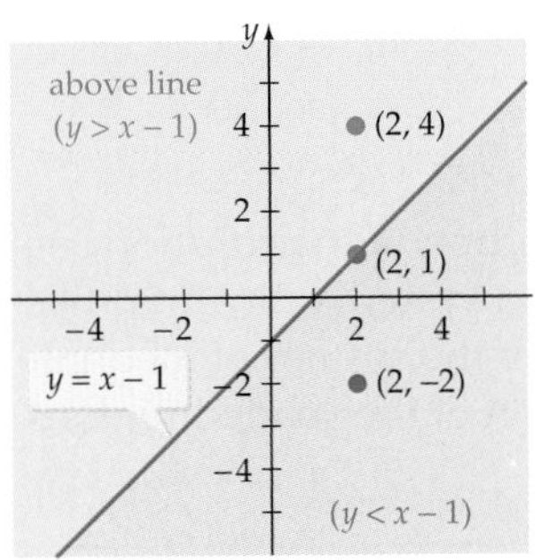

Figure 6.23

The point whose coordinates are (2, 1) is a solution of $y = x - 1$ and is a point on the line.

The point whose coordinates are (2, 4) is a solution of $y > x - 1$ and is a point above the line. The set of points above the line is the solution set of $y > x - 1$. This set of points is called a **half-plane.**

The point whose coordinates are $(2, -2)$ is a solution of $y < x - 1$ and is a point below the line. The set of points below the line is the solution set of $y < x - 1$. These points also form a half-plane.

QUESTION Is $(-3, 4)$ a solution of $y > \frac{2}{3}x + 1$?

To sketch the graph of a linear inequality, first replace the inequality symbol by an equal sign and sketch the graph of the equation. Use a dashed line for $<$ and $>$ to indicate that the line is not part of the solution set. Use a solid line for $\leq$ and $\geq$ to indicate that the line is part of the solution set. You can test that the correct half-plane has been shaded by choosing some point in the shaded region and then checking whether that point satisfies the inequality.

Alternative to Example 1
Exercise 6, page 568

EXAMPLE 1 Graph a Linear Inequality

Graph: $2x - 5y > 10$

Solution Solve $2x - 5y > 10$ for y.

$$2x - 5y > 10$$
$$-5y > -2x + 10$$
$$y < \frac{2}{5}x - 2$$

Continued ➤

ANSWER Yes. $4 > \frac{2}{3}(-3) + 1 = -1$

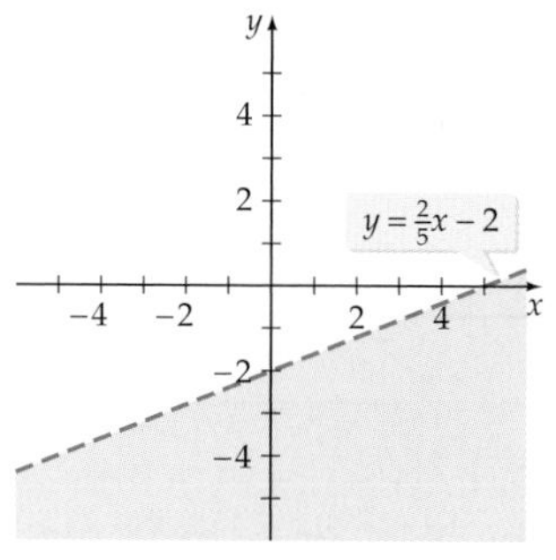

Figure 6.24

Graph $y = \frac{2}{5}x - 2$ as a dashed line. Because $y < \frac{2}{5}x - 2$, shade beneath the line. See Figure 6.24.

We can verify that the correct half-plane has been shaded by selecting any point in the shaded region—say, $(0, -4)$—and checking whether it is a solution of the inequality.

$$2x - 5y > 10$$
$$2(0) - 5(-4) \;\big|\; 10$$
$$20 > 10$$

■ A true statement

The solution checks.

CHECK YOUR PROGRESS 1 Graph $y \geq \frac{1}{2}x + 2$.

Solution *See page S37.*

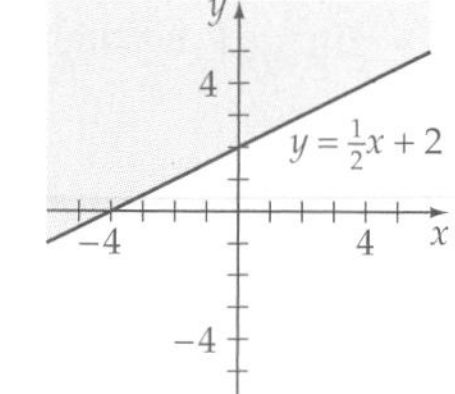

■ Systems of Inequalities in Two Variables

The **solution set of a system of inequalities** is the intersection of the solution sets of the individual inequalities. To graph the solution set of a system of inequalities, graph the solution set of each inequality. The solution set of the system of inequalities is the region of the plane represented by the intersection of the shaded regions.

Alternative to Example 2
Exercise 12, page 568

EXAMPLE 2 Graph a System of Linear Inequalities

Graph the solution set of the system of inequalities.

$$\begin{cases} 3x - 2y > 6 \\ 2x - 5y \leq 10 \end{cases}$$

Solution Solving $3x - 2y > 6$ for y, we have $y < \frac{3}{2}x - 3$. Graph $y = \frac{3}{2}x - 3$ as a dashed line and shade below it. Solving $2x - 5y \leq 10$ for y, we have $y \geq \frac{2}{5}x - 2$. Graph $y = \frac{2}{5}x - 2$ as a solid line and shade above it. The solution set of the system of inequalities is the region where the graphs of the solution sets of the two inequalities intersect. See the dark shaded region in Figure 6.25.

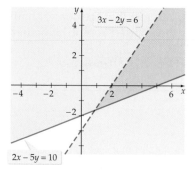

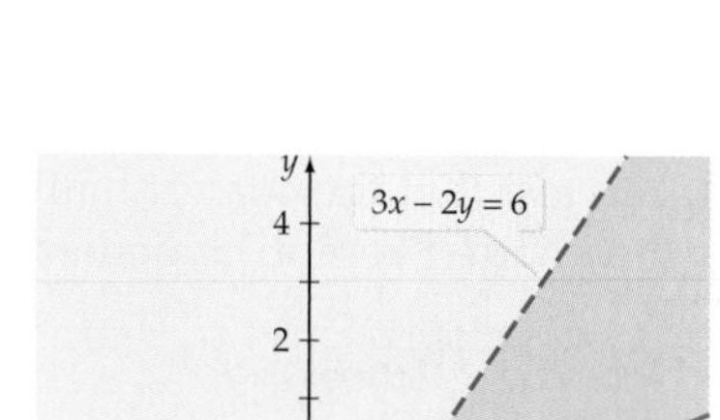

Figure 6.25

CHECK YOUR PROGRESS 2 Graph the solution set of the system of inequalities.

$$\begin{cases} 2x - 5y < -6 \\ 3x + y < 8 \end{cases}$$

Solution *See page S37.*

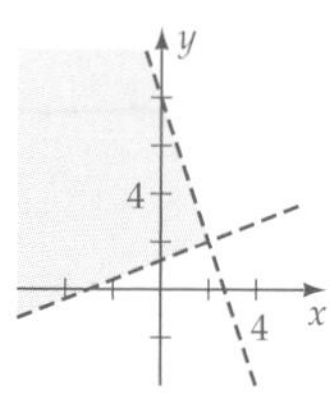

INTEGRATING TECHNOLOGY

The solution set of a linear inequality or a system of linear inequalities can be graphed with a graphing calculator. Enter the expressions to be graphed using the Y= key, and then select whether to shade above or below the line. See the web site for this text at **math.college.hmco.com** or your graphing calculator manual for assistance with this feature.

Typical screens for Example 2 are shown below.

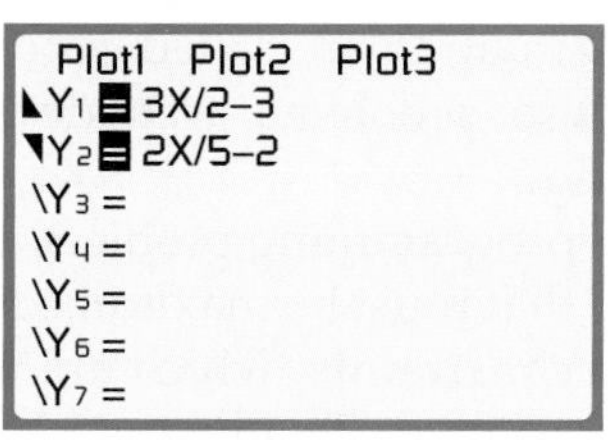

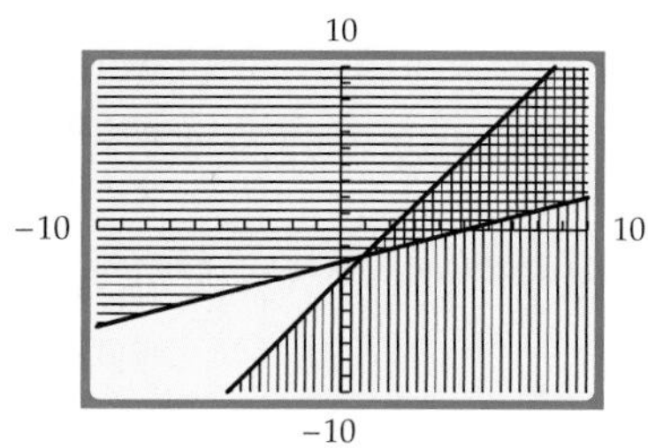

Alternative to Example 3
Exercise 20, page 568

EXAMPLE 3 Graph a System of Four Inequalities

Graph the solution set of the system of inequalities

$$\begin{cases} 2x - 3y \leq 2 \\ 3x + 4y \geq 12 \\ x \geq -1, y \geq 2 \end{cases}$$

Solution First graph the equations $x = -1$ and $y = 2$. Because $x \geq -1$ and $y \geq 2$, shade to the right of $x = -1$ and above $y = 2$. See Figure 6.26.

Solving $2x - 3y \leq 2$ for y, we have $y \geq \frac{2}{3}x - \frac{2}{3}$. Graph $y = \frac{2}{3}x - \frac{2}{3}$ as a solid line and shade above it. Solving $3x + 4y \geq 12$ for y, we have $y \geq -\frac{3}{4}x + 3$. Graph $y = -\frac{3}{4}x + 3$ as a solid line and shade above it. The solution set of the system of inequalities is the region where the graphs of the solution sets of all four inequalities intersect. This is indicated by the dark color in Figure 6.27.

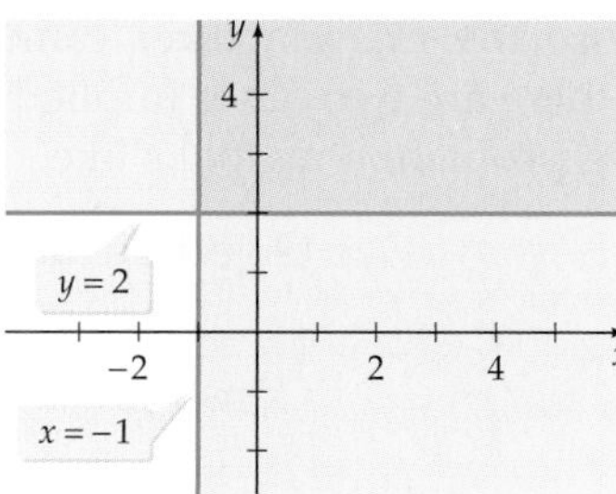

Figure 6.26

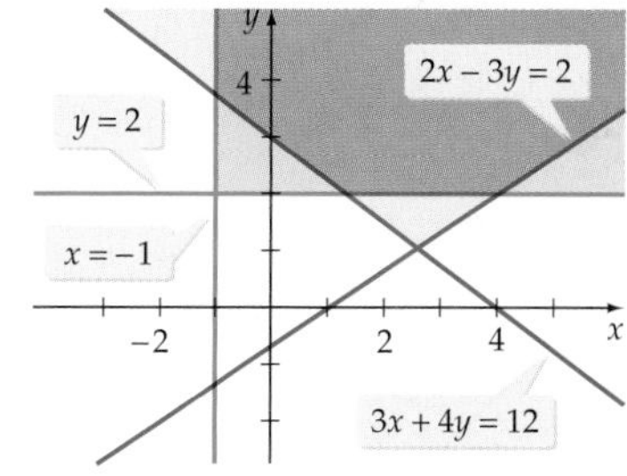

Figure 6.27

CHECK YOUR PROGRESS 3 Graph the solution set of the system of inequalities.

$$\begin{cases} 5x + y \leq 9 \\ 2x + 3y \leq 14 \\ x \geq -2, y \geq 2 \end{cases}$$

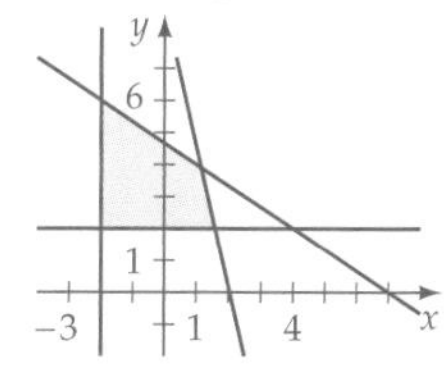

Solution *See page S37.*

▪ Linear Programming

Consider a business analyst who is trying to maximize the profit from the production of a product or an engineer who is trying to minimize the amount of energy an electrical circuit needs to operate. Generally, problems that seek to maximize or minimize a situation are called **optimization problems.** One strategy for solving some of these problems was developed in the 1940s and is called **linear programming.**

A linear programming problem involves a **linear objective function,** which is the function that must be maximized or minimized. This objective function is subject to some **constraints,** which are inequalities or equations that restrict the values of the variables. To illustrate these concepts, suppose a manufacturer produces two types of computer monitors: flat screen and CRT. Past sales experience shows that at least twice as many flat screen monitors are sold as CRT monitors. Suppose further that the manufacturing plant is capable of producing 12 monitors per day. Let x represent the number of flat screen monitors produced, and let y represent the number of CRT monitors produced. Then

$$\begin{cases} x \geq 2y \\ x + y \leq 12 \end{cases}$$ ▪ These are the constraints.

These two inequalities place constraints, or restrictions, on the manufacturer. For example, the manufacturer cannot produce 5 CRT monitors, because that would require producing at least 10 flat screen monitors, and $5 + 10 \not\leq 12$.

Suppose a profit of \$50 is earned on each flat screen monitor sold and \$75 is earned on each CRT monitor sold. Then the manufacturer's profit P, in dollars, is given by the equation

$$P = 50x + 75y$$ ▪ Objective function

The function $P = 50x + 75y$ is the objective function. The goal of this linear programming problem is to determine how many of each monitor should be produced to maximize the manufacturer's profit and at the same time satisfy the constraints.

Because the manufacturer cannot produce fewer than zero units of either monitor, there are two other implied constraints, $x \geq 0$ and $y \geq 0$. Our linear programming problem now looks like

$$\text{Objective function:} \quad P = 50x + 75y$$

$$\text{Constraints:} \quad \begin{cases} x - 2y \geq 0 \\ x + y \leq 12 \\ x \geq 0, y \geq 0 \end{cases}$$

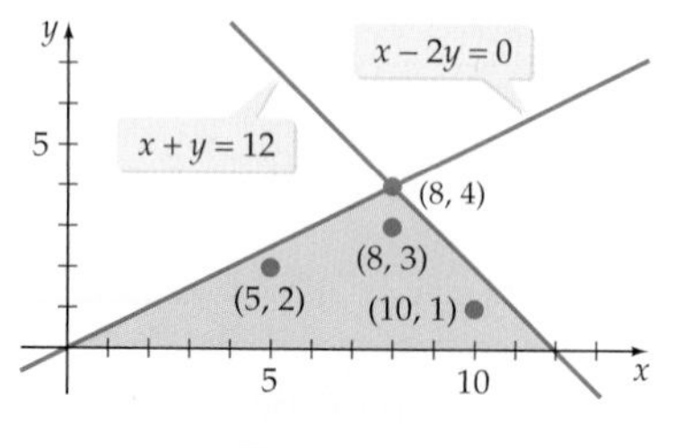

Figure 6.28

To solve this problem, graph the solution set of the constraints. The solution set of the constraints is called the **set of feasible solutions.** Ordered pairs from this set are used to evaluate the objective function to determine which ordered pair maximizes the profit. For example, (5, 2), (8, 3), and (10, 1) are three ordered pairs in the solution set. See Figure 6.28. For these ordered pairs, the profits would be

$$P = 50(5) + 75(2) = 400$$ ▪ $x = 5, y = 2$

$$P = 50(8) + 75(3) = 625$$ ▪ $x = 8, y = 3$

$$P = 50(10) + 75(1) = 575$$ ▪ $x = 10, y = 1$

TAKE NOTE

The set of feasible solutions includes ordered pairs with whole number coordinates as well as fractional coordinates. For instance, the ordered pair $\left(5, 2\frac{1}{2}\right)$ is in the set of feasible solutions. During one day, the company could produce 5 monochrome monitors and $2\frac{1}{2}$ color monitors.

It would be impossible to check every ordered pair in the set of feasible solutions to find which one maximizes profit. Fortunately, we can find that ordered pair by solving the objective function $P = 50x + 75y$ for y.

$$y = -\frac{2}{3}x + \frac{P}{75}$$

In this form, the objective function is a linear equation whose graph has slope $-\frac{2}{3}$ and y-intercept $\frac{P}{75}$. If P is as large as possible (P is a maximum), then the y-intercept will be as large as possible. Thus the maximum profit will occur on the line that has a slope of $-\frac{2}{3}$, has the largest possible y-intercept, and intersects the set of feasible solutions.

From Figure 6.29, the largest possible y-intercept occurs when the line passes through the point with coordinates (8, 4). At this point, the profit is

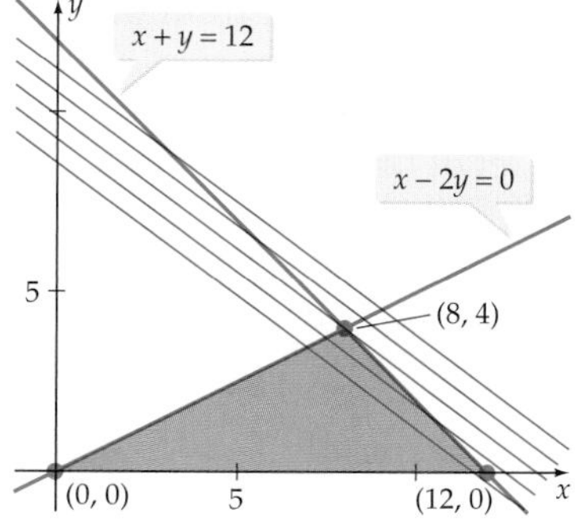

Figure 6.29

$$P = 50(8) + 75(4) = 700$$

The manufacturer will maximize profit by producing eight flat screen monitors and four CRT monitors each day. The profit will be $700 per day.

In general, the goal of any linear programming problem is to maximize or minimize the objective function, subject to the constraints. Minimization problems occur, for example, when a manufacturer wants to minimize the cost of operations.

Suppose that a cost minimization problem results in the following objective function and constraints.

Objective function: $C = 3x + 4y$

Constraints: $\begin{cases} x + y \geq 1 \\ 2x - y \leq 5 \\ x + 2y \leq 10 \\ x \geq 0, y \geq 0 \end{cases}$

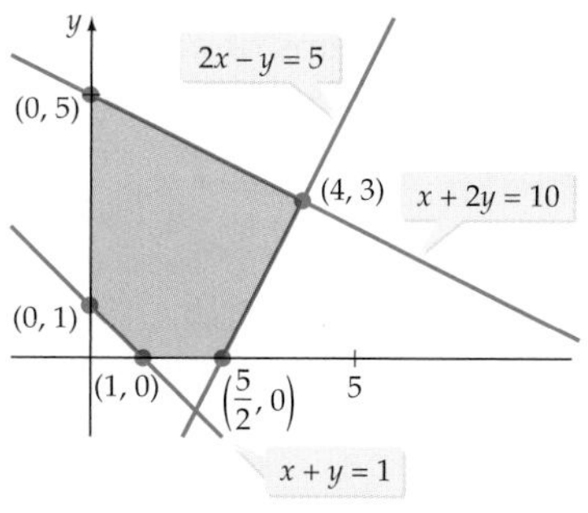

Figure 6.30

Figure 6.30 is the graph of the solution set of the constraints. The task is to find the ordered pair that satisfies all the constraints and that will give the smallest value of C. We again could solve the objective function for y and, because we want to minimize C, find the smallest y-intercept. However, a theorem from linear programming simplifies our task even more. The proof of this theorem, omitted here, is based on the techniques we used to solve our examples.

Fundamental Linear Programming Theorem

If an objective function has an optimal solution, then that solution will be at a vertex of the set of feasible solutions.

Following is a list of the values of C at the vertices. The minimum value of the objective function occurs at the point whose coordinates are (1, 0).

(x, y)	$C = 3x + 4y$	
$(1, 0)$	$C = 3(1) + 4(0) = 3$	▪ Minimum
$\left(\frac{5}{2}, 0\right)$	$C = 3\left(\frac{5}{2}\right) + 4(0) = 7.5$	
$(4, 3)$	$C = 3(4) + 4(3) = 24$	▪ Maximum
$(0, 5)$	$C = 3(0) + 4(5) = 20$	
$(0, 1)$	$C = 3(0) + 4(1) = 4$	

The maximum value of the objective function can also be determined from the list. In this case it is 24 and it occurs at the point (4, 3).

It is important to realize that the maximum or minimum value of an objective function depends on the objective function and on the set of feasible solutions. For example, using the same set of feasible solutions as in Figure 6.30 but changing the objective function to $C = 2x + 5y$ changes the maximum value of C to 25 at the ordered pair (0, 5). You should verify this result by making a list similar to the one shown above.

Alternative to Example 4
Exercise 26, page 568

EXAMPLE 4 Solve a Minimization Problem

Minimize the objective function $C = 4x + 7y$ with the constraints

$$\begin{cases} 3x + y \geq 6 \\ x + y \geq 4 \\ x + 3y \geq 6 \\ x \geq 0, y \geq 0 \end{cases}$$

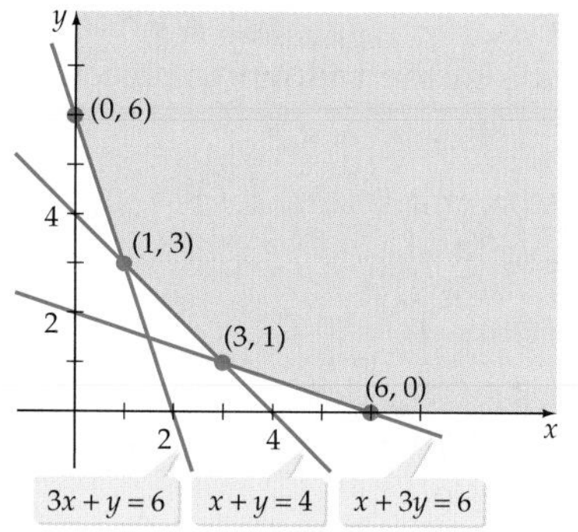

Figure 6.31

Solution Determine the set of feasible solutions by graphing the solution set of the inequalities. See Figure 6.31. Note that in this instance the set of feasible solutions is an unbounded set.

Find the vertices of the region by solving the following systems of equations. These systems are formed by the equations of the lines that intersect to form a vertex of the set of feasible solutions.

$$\begin{cases} 3x + y = 6 \\ x + y = 4 \end{cases} \qquad \begin{cases} x + 3y = 6 \\ x + y = 4 \end{cases}$$

The solutions of the two systems are (1, 3) and (3, 1), respectively. The points (0, 6) and (6, 0) are the vertices on the y- and x-axes. Evaluate the objective function at each of the four vertices of the set of feasible solutions.

(x, y)	$C = 4x + 7y$	
$(0, 6)$	$C = 4(0) + 7(6) = 42$	
$(1, 3)$	$C = 4(1) + 7(3) = 25$	
$(3, 1)$	$C = 4(3) + 7(1) = 19$	▪ Minimum
$(6, 0)$	$C = 4(6) + 7(0) = 24$	

The minimum value of the objective function is 19 at (3, 1).

Continued ➤

CHECK YOUR PROGRESS 4 Minimize $C = 4x + 3y$ with the constraints

$$\begin{cases} 2x + y \geq 8 \\ 2x + 3y \geq 16 \\ x + 3y \geq 12 \\ x \leq 20, y \leq 20 \end{cases}$$

Solution *See page S37.* Minimum is 20 at (2, 4).

Linear programming can be used to determine the best allocation of the resources available to a company. In fact, the word *programming* refers to a "program to allocate resources."

Alternative to Example 5
Exercise 42, page 569

EXAMPLE 5 Solve an Applied Minimization Problem

A manufacturer of animal food makes two grain mixtures, G_1 and G_2. Each kilogram of G_1 contains 300 grams of vitamins, 400 grams of protein, and 100 grams of carbohydrate. Each kilogram of G_2 contains 100 grams of vitamins, 300 grams of protein, and 200 grams of carbohydrate. Minimum nutritional guidelines require that a feed mixture made from these grains contain at least 900 grams of vitamins, 2200 grams of protein, and 800 grams of carbohydrate. G_1 costs \$2.00 per kilogram to produce, and G_2 costs \$1.25 per kilogram to produce. Find the number of kilograms of each grain mixture that should be produced to minimize cost.

Solution Let

$$x = \text{the number of kilograms of } G_1$$
$$y = \text{the number of kilograms of } G_2$$

The objective function is the cost function $C = 2x + 1.25y$.

Because x kilograms of G_1 contain $300x$ grams of vitamins and y kilograms of G_2 contain $100y$ grams of vitamins, the total amount of vitamins contained in x kilograms of G_1 and y kilograms of G_2 is $300x + 100y$. At least 900 grams of vitamins are necessary, so $300x + 100y \geq 900$. Following similar reasoning, we have the constraints

$$\begin{cases} 300x + 100y \geq 900 \\ 400x + 300y \geq 2200 \\ 100x + 200y \geq 800 \\ x \geq 0, y \geq 0 \end{cases}$$

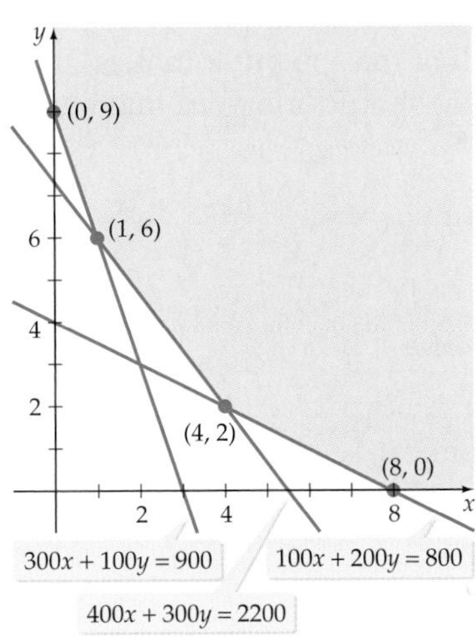

Figure 6.32

Two of the vertices of the set of feasible solutions (see Figure 6.32) can be found by solving two systems of equations. These systems are formed by the equations of the lines that intersect to form a vertex of the set of feasible solutions.

$$\begin{cases} 300x + 100y = 900 \\ 400x + 300y = 2200 \end{cases}$$
■ The vertex is (1, 6).

$$\begin{cases} 100x + 200y = 800 \\ 400x + 300y = 2200 \end{cases}$$
■ The vertex is (4, 2).

The vertices on the x- and y-axes are the x- and y-intercepts (8, 0) and (0, 9).

Continued ➤

Substitute the coordinates of the vertices into the objective function.

(x, y)	$C = 2x + 1.25y$	
$(0, 9)$	$C = 2(0) + 1.25(9) = 11.25$	
$(1, 6)$	$C = 2(1) + 1.25(6) = 9.50$	■ Minimum
$(4, 2)$	$C = 2(4) + 1.25(2) = 10.50$	
$(8, 0)$	$C = 2(8) + 1.25(0) = 16.00$	

The minimum value of the objective function is \$9.50. It occurs when the company produces a feed mixture that contains 1 kilogram of G_1 and 6 kilograms of G_2.

CHECK YOUR PROGRESS 5 An ice cream supplier has two machines that each produce vanilla and chocolate ice cream. To meet one of its contractual obligations, the company must produce at least 60 gallons of vanilla ice cream and at least 100 gallons of chocolate ice cream per day. One machine makes 4 gallons of vanilla and 5 gallons of chocolate ice cream per hour. The second machine makes 3 gallons of vanilla and 10 gallons of chocolate ice cream per hour. It costs \$28 per hour to operate machine 1 and \$25 per hour to operate machine 2. How many hours should each machine be operated to fulfill the contract at the least expense?

Solution *See page S37.* Machine 1: 12 h, machine 2: 4 h

Alternative to Example 6
Exercise 44, page 569

EXAMPLE 6 Solve an Applied Maximization Problem

A chemical firm produces two types of industrial solvents, S_1 and S_2. Each solvent is a mixture of three chemicals. Each kiloliter of S_1 requires 12 liters of chemical 1, 9 liters of chemical 2, and 30 liters of chemical 3. Each kiloliter of S_2 requires 24 liters of chemical 1, 5 liters of chemical 2, and 30 liters of chemical 3. The profit per kiloliter of S_1 is \$100, and the profit per kiloliter of S_2 is \$85. The inventory of the company shows 480 liters of chemical 1, 180 liters of chemical 2, and 720 liters of chemical 3. Assuming that the company can sell all of the solvent it makes, find the number of kiloliters of each solvent the company should make to maximize profit.

Solution Let

$$x = \text{the number of kiloliters of } S_1$$
$$y = \text{the number of kiloliters of } S_2$$

The objective function is the profit function $P = 100x + 85y$.

Because x kiloliters of S_1 require $12x$ liters of chemical 1, and y kiloliters of S_2 require $24y$ liters of chemical 1, the total amount of chemical 1 needed is $12x + 24y$. There are 480 liters of chemical 1 in inventory, so $12x + 24y \le 480$. Following similar reasoning, we have the constraints

$$\begin{cases} 12x + 24y \le 480 \\ 9x + 5y \le 180 \\ 30x + 30y \le 720 \\ x \ge 0, y \ge 0 \end{cases}$$

Continued ➤

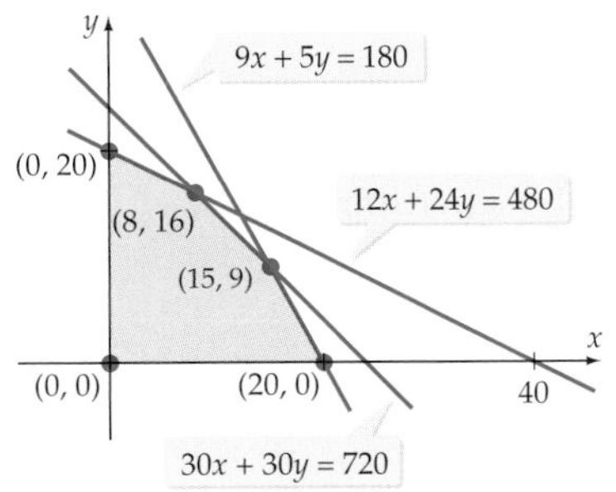

Figure 6.33

Two of the vertices of the set of feasible solutions (see Figure 6.33) can be found by solving two systems of equations. These systems are formed by the equations of the lines that intersect to form a vertex of the set of feasible solutions.

$$\begin{cases} 12x + 24y = 480 \\ 30x + 30y = 720 \end{cases}$$ ▪ The vertex is (8, 16).

$$\begin{cases} 9x + 5y = 180 \\ 30x + 30y = 720 \end{cases}$$ ▪ The vertex is (15, 9).

The vertices on the x- and y-axes are the x- and y-intercepts (20, 0) and (0, 20).

Substitute the coordinates of the vertices into the objective function.

(x, y)	$P = 100x + 85y$	
(0, 20)	$P = 100(0) + 85(20) = 1700$	
(8, 16)	$P = 100(8) + 85(16) = 2160$	
(15, 9)	$P = 100(15) + 85(9) = 2265$	▪ Maximum
(20, 0)	$P = 100(20) + 85(0) = 2000$	

The maximum value of the objective function is $2265 when the company produces 15 kiloliters of S_1 and 9 kiloliters of S_2.

CHECK YOUR PROGRESS 6 A company makes two types of telephone answering machines: the standard model and the deluxe model. Each machine passes through three processes, P_1, P_2, and P_3. One standard answering machine requires 1 hour in P_1, 1 hour in P_2, and 2 hours in P_3. One deluxe answering machine requires 3 hours in P_1, 1 hour in P_2, and 1 hour in P_3. Because of employee work schedules, P_1 is available for 24 hours, P_2 is available for 10 hours, and P_3 is available for 16 hours. If the profit is $25 for each standard model and $35 for each deluxe model, how many units of each type should the company produce to maximize profit?

Solution *See page S37.* 3 standard, 7 deluxe

Topics for Discussion

1. Does the graph of a linear inequality in two variables represent the graph of a function? Why or why not?
2. What is a half-plane?
3. Is it possible for a system of inequalities to have no solution? If so, give an example. If not, explain why not.
4. What is an optimization problem? Give an example of a situation in which optimization may be the goal.
5. What is a constraint for a linear programming problem? Explain what type of condition might be a constraint for the situation you gave in Exercise 4.
6. What is the objective function for a linear programming problem? Explain what the objective function might be for the situation you gave in Exercise 4.
7. What is the set of feasible solutions for a linear programming problem?

EXERCISES 6.6

— *Suggested Assignment: Exercises 1–31, odd; 32, 35, 38, 41, and 43.*
— *Answer graphs to Exercises 1–24 are on pages AA18–AA19.*

In Exercises 1 to 8, sketch the graph of each inequality.

1. $y \le -2$
2. $x + y > -2$
3. $y \ge 2x + 3$
4. $y < -2x + 1$
5. $2x - 3y < 6$
6. $3x + 4y \le 4$
7. $4x + 3y \le 12$
8. $5x - 2y < 8$

In Exercises 9 to 24, sketch the graph of the solution set of each system of inequalities.

9. $\begin{cases} 1 \le x < 3 \\ -2 < y \le 4 \end{cases}$
10. $\begin{cases} -2 < x < 4 \\ y \ge -1 \end{cases}$
11. $\begin{cases} 3x + 2y \ge 1 \\ x + 2y < -1 \end{cases}$
12. $\begin{cases} 2x - 5y < -6 \\ 3x + y < 8 \end{cases}$
13. $\begin{cases} 2x - y \ge -4 \\ 4x - 2y \le -17 \end{cases}$ No solution
14. $\begin{cases} 4x + 2y > 5 \\ 6x + 3y > 10 \end{cases}$
15. $\begin{cases} 4x - 3y < 14 \\ 2x + 5y \le -6 \end{cases}$
16. $\begin{cases} 3x + 5y \ge -8 \\ 2x - 3y \ge 1 \end{cases}$
17. $\begin{cases} y < 2x + 3 \\ y > 2x - 2 \end{cases}$
18. $\begin{cases} y > 3x + 1 \\ y < 3x - 2 \end{cases}$ No solution
19. $\begin{cases} 2x - 3y \ge -5 \\ x + 2y \le 7 \\ x \ge -1, y \ge 0 \end{cases}$
20. $\begin{cases} 5x + y \le 9 \\ 2x + 3y \le 14 \\ x \ge -2, y \ge 2 \end{cases}$
21. $\begin{cases} 3x + 2y \ge 14 \\ x + 3y \ge 14 \\ x \le 10, y \le 8 \end{cases}$
22. $\begin{cases} 4x + y \ge 13 \\ 3x + 2y \ge 16 \\ x \le 15, y \le 12 \end{cases}$
23. $\begin{cases} 3x + 4y \le 12 \\ 2x + 5y \le 10 \\ x \ge 0, y \ge 0 \end{cases}$
24. $\begin{cases} 5x + 3y \le 15 \\ x + 4y \le 8 \\ x \ge 0, y \ge 0 \end{cases}$

In Exercises 25 to 40, solve the linear programming problem. Assume $x \ge 0$ and $y \ge 0$.

25. Minimize $C = 4x + 2y$ with the constraints

$$\begin{cases} x + y \ge 7 \\ 4x + 3y \ge 24 \\ x \le 10, y \le 10 \end{cases}$$

The minimum is 16 at (0, 8).

26. Minimize $C = 5x + 4y$ with the constraints

$$\begin{cases} 3x + 4y \ge 32 \\ x + 4y \ge 24 \\ x \le 12, y \le 15 \end{cases}$$

The minimum is 32 at (0, 8).

27. Maximize $C = 6x + 7y$ with the constraints

$$\begin{cases} x + 2y \le 16 \\ 5x + 3y \le 45 \end{cases}$$

The maximum is 71 at (6, 5).

28. Maximize $C = 6x + 5y$ with the constraints

$$\begin{cases} 2x + 3y \le 27 \\ 7x + 3y \le 42 \end{cases}$$

The maximum is 53 at (3, 7).

29. Minimize $C = 5x + 6y$ with the constraints

$$\begin{cases} 4x - 3y \le 2 \\ 2x + 3y \ge 10 \end{cases}$$

The minimum is 20 at $\left(0, \frac{10}{3}\right)$.

30. Maximize $C = 7x + 2y$ with the constraints

$$\begin{cases} x + 3y \le 108 \\ 7x + 4y \le 280 \end{cases}$$

The maximum is 280 at (40, 0).

31. Maximize $C = 2x + 7y$ with the constraints

$$\begin{cases} x + y \le 10 \\ x + 2y \le 16 \\ 2x + y \le 16 \end{cases}$$

The maximum is 56 at (0, 8).

32. Minimize $C = 4x + 3y$ with the constraints

$$\begin{cases} 2x + y \ge 8 \\ 2x + 3y \ge 16 \\ x + 3y \ge 11 \\ x \le 20, y \le 20 \end{cases}$$

The minimum is 20 at (2, 4).

33. Minimize $C = 3x + 2y$ with the constraints

$$\begin{cases} 3x + y \ge 12 \\ 2x + 7y \ge 21 \\ x + y \ge 8 \end{cases}$$

The minimum is 18 at (2, 6).

34. Maximize $C = 2x + 6y$ with the constraints

$$\begin{cases} x + y \le 12 \\ 3x + 4y \le 40 \\ x + 2y \le 18 \end{cases}$$

The maximum is 54 at (0, 9).

35. Maximize $C = 3x + 4y$ with the constraints

$$\begin{cases} 2x + y \leq 10 \\ 2x + 3y \leq 18 \\ x - y \leq 2 \end{cases}$$

The maximum is 25 at (3, 4).

36. Minimize $C = 3x + 7y$ with the constraints

$$\begin{cases} x + y \geq 9 \\ 3x + 4y \geq 32 \\ x + 2y \geq 12 \end{cases}$$

The minimum is 36 at (12, 0).

37. Minimize $C = 3x + 2y$ with the constraints

$$\begin{cases} x + 2y \geq 8 \\ 3x + y \geq 9 \\ x + 4y \geq 12 \end{cases}$$

The minimum is 12 at (2, 3).

38. Maximize $C = 4x + 5y$ with the constraints

$$\begin{cases} 3x + 4y \leq 250 \\ x + y \leq 75 \\ 2x + 3y \leq 180 \end{cases}$$

The maximum is 325 at (50, 25).

39. Maximize $C = 6x + 7y$ with the constraints

$$\begin{cases} x + 2y \leq 900 \\ x + y \leq 500 \\ 3x + 2y \leq 1200 \end{cases}$$

The maximum is 3400 at (100, 400).

40. Minimize $C = 11x + 16y$ with the constraints

$$\begin{cases} x + 2y \geq 45 \\ x + y \geq 40 \\ 2x + y \geq 45 \end{cases}$$

The minimum is 465 at (35, 5).

Business and Economics

41. ***Maximize Profit*** A farmer is planning to raise wheat and barley. Each acre of wheat yields a profit of \$50, and each acre of barley yields a profit of \$70. To sow the crop, two machines, a tractor and a tiller, are rented. The tractor is available for 200 hours, and the tiller is available for 100 hours. Sowing an acre of barley requires 3 hours of tractor time and 2 hours of tilling. Sowing an acre of wheat requires 4 hours of tractor time and 1 hour of tilling. How many acres of each crop should be planted to maximize the farmer's profit? 20 acres of wheat; 40 acres of barley

42. ***Minimize Cost*** Nationwide Disk Group, Inc. (NDG) makes, among other things, CDs and DVDs. Existing contracts require that NDG produce at least 4200 CDs and 6000 DVDs per day. NDG has two machines available, each of which makes both CDs and DVDs. Machine 1 can make two CDs and eight DVDs per minute. Machine 2 makes eight CDs and five DVDs per minute. It costs \$40 per hour to run machine 1 and \$50 per hour to run machine 2. How many minutes should each machine be run to meet NDG's contractual obligations at the least cost?
Machine 1: 500 min; machine 2: 400 min

43. ***Maximize Profit*** A manufacturer makes two types of golf clubs: a starter model and a professional model. The starter model requires 4 hours in the assembly room and 1 hour in the finishing room. The professional model requires 6 hours in the assembly room and 1 hour in the finishing room. The total number of hours available in the assembly room is 108. There are 24 hours available in the finishing room. The profit for each starter model is \$35, and the profit for each professional model is \$55. Assuming all the clubs produced can be sold, find how many of each club should be manufactured to maximize profit.
0 starter clubs; 18 pro clubs

44. ***Maximize Profit*** A coffee producer makes medium-bodied and full-bodied coffees. Each blend is prepared by passing the coffee through a drying room, a roasting room, and a blending room. Each pound of medium-bodied coffee spends 48 minutes in the drying room, 30 minutes in the roasting room, and 50 minutes in the blending room. Each pound of full-bodied coffee spends 48 minutes in the drying room, 45 minutes in the roasting room, and 40 minutes in the blending room. On a certain day, the drying room was available for at most 480 minutes, the roasting room was available for 390 minutes, and the blending room was available for 460 minutes. The profit on 1 pound of medium-bodied coffee is \$2, and the profit on 1 pound of full-bodied coffee is \$2.50. How many pounds of each type of coffee should be produced to maximize profit for this day? What is the maximum profit?
Medium-bodied: 4 lb; full-bodied: 6 lb; \$23

Explorations 

1. ***A Parallelogram Coordinate System*** The xy-coordinate system described in this chapter consisted of two coordinate lines that intersected at right angles. It is not necessary that coordinate lines intersect at right angles for a coordinate system to exist. Draw two coordinate lines that intersect at 0 but for which the angle between the two axes is 45°. You now have a *parallelogram* coordinate system rather than a *rectangular* coordinate system. Explain the last sentence. Now experiment in this system. For example, is the graph of $3x + 4y = 12$ a straight line in a parallelogram coordinate system? In a parallelogram coordinate system, does the graph of $y = x^2$ appear to be a parabola?
Answers will vary; yes; no

Chapter 6 Summary

Key Terms

augmented matrix [p. 516]
back substitution [p. 504]
ciphertext [p. 554]
coefficient matrix [p. 516]
column matrix [p. 536]
consistent system of equations [p. 488]
constant matrix [p. 516]
constraints [p. 562]
cryptography [p. 554]
dependent system of equations [p. 488]
difference of matrices [p. 533]
dimension of a matrix [p. 515]
element of a matrix [p. 515]
elementary row operations [p. 516]
elimination method [p. 492]
equality of matrices [p. 515]
equivalent systems of equations [p. 492]
Gaussian elimination method [p. 522]
half-plane [p. 559]
identity matrix [p. 541]
inconsistent system of equations [p. 488]
independent system of equations [p. 488]
interpolating polynomial [p. 526]
inverse matrix [p. 548]
linear equation in three variables [p. 501]
linear inequality in two variables [p. 559]
linear objective function [p. 562]
linear programming [p. 562]
linear system of equations [p. 488]
main diagonal of a matrix [p. 515]
matrix [p. 515]
matrix equation [p. 552]
matrix multiplication [p. 538]
nonsingular matrix [p. 551]
nonsquare system of equations [p. 508]
optimization problem [p. 562]
order of a matrix [p. 515]
ordered triple [p. 501]
plaintext [p. 554]
row echelon form [p. 519]
row matrix [p. 536]
scalar [p. 535]
scalar multiplication of a matrix [p. 535]
set of feasible solutions [p. 562]
singular matrix [p. 551]
solution of a system of equations [p. 488]
solution set of a system of inequalities [p. 560]
square matrix [p. 515]
square system of equations [p. 508]
substitution method [p. 489]
sum of matrices [p. 533]
system of equations [p. 488]
triangular form [p. 504]
xyz-coordinate system [p. 501]
z-axis [p. 501]
zero matrix [p. 534]

Essential Concepts and Formulas

- **System of Equations**
  A system of equations is two or more equations considered together. A solution of a system of equations in two variables is an ordered pair that satisfies each equation of the system. Solutions of a system of equations correspond to points of intersection on the graph of the system. [p. 488]

- **Equivalent Systems**
  Equivalent systems of equations have the same solution set. There are two algebraic methods that can be used to solve systems of linear equations: substitution and elimination. [p. 492]

- **Consistent, Inconsistent Systems**
  A system of linear equations is consistent if it has one or more solutions. A system of linear equations is independent if it has exactly one solution. A system is dependent if it has infinitely many solutions. An inconsistent system of equations has no solution. [p. 488]

- **Operations That Produce Equivalent Systems of Equations**
  1. Interchange any two equations.
  2. Replace an equation with a nonzero multiple of that equation.

3. Replace an equation with the sum of that equation and a nonzero constant multiple of another equation in the system. **[p. 492]**

- **Solve Linear Systems of Equations in Three Variables**
To solve a system of linear equations in three variables, the substitution method or the elimination method can be used. A systematic way to implement the elimination method is to get the system into triangular form and then use back substitution. **[p. 503]**

- **Solving Systems with Matrices**
A system of linear equations can be represented by an augmented matrix. The system of equations can be solved by using elementary row operations and following the Gaussian elimination method. **[p. 516]**

- **Elementary Row Operations for a Matrix**
  1. Interchange two rows. Interchanging the ith and jth rows is shown symbolically by $R_i \leftrightarrow R_j$.
  2. Multiply all the elements in a row by the same nonzero number. Multiplying the ith row by k is shown symbolically by kR_i.
  3. Replace a row by the sum of that row and a nonzero multiple of another row. Replacing the ith row by the sum of that row and k times the jth row is shown symbolically by $kR_j + R_i$. **[p. 517]**

- **Row Echelon Form**
A matrix is in row echelon form if all of the following conditions are satisfied.
  1. The first nonzero number in any row is a 1.
  2. Rows are arranged so that the column containing the first nonzero number in any row is to the left of the column containing the first nonzero number of the next row.
  3. All rows consisting entirely of zeros appear at the bottom of the matrix. **[p. 519]**

- **Procedure to Write an Augmented Matrix in Row Echelon Form**
  1. Use row operations to change a_{11} to 1 and then change the remaining elements in the first column to 0.
  2. If possible, change a_{22} to 1 and then change the remaining elements below a_{22} to 0.
  3. Move to a_{33} and repeat the above procedure.
  4. Continue moving down the main diagonal until you reach a_{nn} or until all remaining elements on the main diagonal are zero. **[p. 520]**

- **The Gaussian Elimination Method**
The Gaussian elimination method is an algorithm used to find the solution(s) of a system of linear equations. Elementary row operations are used to rewrite an augmented matrix in row echelon form, and then back substitution is used to find the solution of the system of equations. **[p. 522]**

- **Interpolating Polynomials**
An interpolating polynomial is a polynomial whose graph passes through a given set of points, if possible. The degree of the polynomial is at most one less than the number of given points. The polynomial can be determined by solving a system of linear equations. **[p. 526]**

- **Matrix Addition and Subtraction**
Two matrices can be added or subtracted if they are of the same order. The sum (or difference) is the matrix whose elements are the sums (or differences) of the corresponding elements of the two matrices. **[p. 533]**

- **Scalar Multiplication of a Matrix**
Scalar multiplication of a matrix is accomplished by multiplying the scalar times each element of the matrix. **[p. 534]**

- **Matrix Multiplication**
If A is a matrix of order $m \times n$ and B is a matrix of order $n \times p$, then the product AB is the matrix of order $m \times p$ given by $AB = [c_{ij}]$, where each element c_{ij} is the product of the ith row matrix of A and the jth column matrix of B. **[p. 538]**

- **Identity Matrix**
The identity matrix I_n is the square matrix of order n with 1's on the main diagonal and 0's everywhere else. If A is any square matrix of order n, then $AI_n = I_nA = A$. **[p. 541]**

- **Inverse Matrix**
If A is the square matrix of order n, then the inverse matrix of A, A^{-1}, has the property that

$$A \cdot A^{-1} = A^{-1} \cdot A = I_n$$

where I_n is the identity matrix of order n. To find the inverse of a nonsingular square matrix A of order n, form an augmented matrix with A on the left and I_n on the right. Use elementary row operations to produce I_n on the left. The resulting matrix on the right is A^{-1}. **[p. 548]**

- **Matrix Equation**
A system of linear equations can be represented by a matrix equation in which the coefficient matrix times the variable matrix equals the constant matrix. The solution of the system can be found by solving the matrix equation for the variable matrix. **[p. 552]**

- **Cryptography**
Text messages can be encoded and decoded by representing letters with ASCII numbers in a matrix and using matrix multiplication. **[p. 554]**

- **Graphs of Linear Inequalities**
The graph of a linear inequality in two variables separates the xy-plane into the set of points on the line, the set of points above the line, and the set of points below the line. The solution set is a half-plane that is indicated by shading on the graph. The solution set of a system of inequalities is the intersection of the solution sets of the individual inequalities. **[p. 559]**

- **Linear Programming Problem**
  A linear programming problem is an optimization problem consisting of a linear objective function and a number of constraints, which are inequalities or equations that restrict the values of the variables. [p. 562]

- **Fundamental Linear Programming Theorem**
  If an objective function has an optimal solution, then that solution will be at a vertex of the set of feasible solutions. [p. 563]

Chapter 6 True/False Exercises

In Exercises 1 to 10, answer true or false. If the answer is false, explain why or give an example to show that the statement is false.

1. A system of equations will always have a solution as long as the number of equations is equal to the number of variables. False; $\begin{cases} x + y = 1 \\ x + y = 2 \end{cases}$ has no solution.

2. In an xyz-coordinate system, the graph of the set of points formed by the intersection of two distinct nonparallel planes is a straight line. True

3. Two systems of equations with the same solution set have the same equations in their respective systems. False; $\begin{cases} x + y = 2 \\ x + 2y = 3 \end{cases}$ and $\begin{cases} 2x + 3y = 5 \\ 2x - 2y = 0 \end{cases}$ are two systems with different equations but the same solution.

4. The systems of equations

$$\begin{cases} x = 0 \\ y = 0 \end{cases} \quad \text{and} \quad \begin{cases} y = x \\ y = -x \end{cases}$$

are equivalent systems of equations. True

5. A system of three linear equations in three variables for which the graph consists of two parallel planes and a third plane that intersects the first two is a dependent system of equations. False; inconsistent

6. The Gaussian elimination method for solving a system of linear equations can be applied only to systems of equations that have the same number of variables as equations.

6. False; if the number of equations is less than the number of variables, the Gaussian elimination method can be used to solve the system of linear equations. If the system of equations has a solution, the solution will be given in terms of an arbitrary constant.

7. If A and B are matrices, then the product AB is defined when the number of columns of A equals the number of rows of B. True

8. Every matrix has an additive inverse. (*Hint:* The matrices A and B are additive inverses if and only if $A + B = \mathbf{0}$, where $\mathbf{0}$ is the zero matrix in which every element is 0.) True

9. Every square matrix has a multiplicative inverse. False; a singular matrix does not have a multiplicative inverse.

10. If $A = \begin{bmatrix} 2 & 3 \\ 1 & 4 \end{bmatrix}$, then $A^2 = \begin{bmatrix} 4 & 9 \\ 1 & 16 \end{bmatrix}$.
False; $A^2 = A \cdot A = \begin{bmatrix} 7 & 18 \\ 6 & 19 \end{bmatrix}$.

Chapter 6 Review Exercises

Answers to Exercises 20, 45–49, 53–60, 65–68, 71 and 72, and answer graphs to Exercises 73–82 are on pages AA19–AA20.

In Exercises 1 to 18, solve each system of equations.

1. $\begin{cases} 2x - 4y = -3 \\ 3x + 8y = -12 \end{cases}$

2. $\begin{cases} 4x - 3y = 15 \\ 2x + 5y = -12 \end{cases}$ $\left(\frac{3}{2}, -3\right)$ [6.1]

3. $\begin{cases} 3x - 4y = -5 \\ y = \frac{2}{3}x + 1 \end{cases}$ $(-3, -1)$ [6.1]

4. $\begin{cases} 7x + 2y = -14 \\ y = -\frac{5}{2}x - 3 \end{cases}$ $(-4, 7)$ [6.1]

5. $\begin{cases} y = 2x - 5 \\ x = 4y - 1 \end{cases}$ $(3, 1)$ [6.1]

6. $\begin{cases} y = 3x + 4 \\ x = 4y - 5 \end{cases}$ $(-1, 1)$ [6.1]

7. $\begin{cases} 6x + 9y = 15 \\ 10x + 15y = 25 \end{cases}$ $\left(\frac{5 - 3c}{2}, c\right)$ [6.1]

8. $\begin{cases} 4x - 8y = 9 \\ 2x - 4y = 5 \end{cases}$ No solution [6.1]

1. $\left(-\frac{18}{7}, -\frac{15}{28}\right)$ [6.1]

9. $\begin{cases} 2x - 3y + z = -9 \\ 2x + 5y - 2z = 18 \\ 4x - y + 3z = -4 \end{cases}$ $\left(\frac{1}{2}, 3, -1\right)$ [6.2]

10. $\begin{cases} x - 3y + 5z = 1 \\ 2x + 3y - 5z = 15 \\ 3x + 6y + 5z = 15 \end{cases}$ $\left(\frac{16}{3}, \frac{10}{27}, -\frac{29}{45}\right)$ [6.2]

11. $\begin{cases} x + 3y - 5z = -12 \\ 3x - 2y + z = 7 \\ 5x + 4y - 9z = -17 \end{cases}$ $\left(\frac{7c - 3}{11}, \frac{16c - 43}{11}, c\right)$ [6.2]

12. $\begin{cases} 2x - y + 2z = 5 \\ x + 3y - 3z = 2 \\ 5x - 9y + 8z = 13 \end{cases}$ $\left(\frac{74}{31}, -\frac{1}{31}, \frac{3}{31}\right)$ [6.2]

13. $\begin{cases} 3x + 4y - 6z = 10 \\ 2x + 2y - 3z = 6 \\ x - 6y + 9z = -4 \end{cases}$ $\left(2, \frac{3c + 2}{2}, c\right)$ [6.2]

14. $\begin{cases} x - 6y + 4z = 6 \\ 4x + 3y - 4z = 1 \\ 5x - 9y + 8z = 13 \end{cases}$ $\left(1, -\frac{2}{3}, \frac{1}{4}\right)$ [6.2]

15. $\left(\frac{14c}{11}, -\frac{2c}{11}, c\right)$ [6.2]

15. $\begin{cases} 2x + 3y - 2z = 0 \\ 3x - y - 4z = 0 \\ 5x + 13y - 4z = 0 \end{cases}$

16. $\begin{cases} 3x - 5y + z = 0 \\ x + 4y - 3z = 0 \\ 2x + y - 2z = 0 \end{cases}$

(0, 0, 0) [6.2]

17. $\begin{cases} x - 2y + z = 1 \\ 3x + 2y - 3z = 1 \end{cases}$

$\left(\frac{c+1}{2}, \frac{3c-1}{4}, c\right)$ [6.2]

18. $\begin{cases} 2x - 3y + z = 1 \\ 4x + 2y + 3z = 21 \end{cases}$

$\left(\frac{65-11c}{16}, \frac{19-c}{8}, c\right)$ [6.2]

In Exercises 19 and 20, write the augmented matrix, the coefficient matrix, and the constant matrix for the given system of equations.

19. $\begin{cases} 5x - 7y + z = -8 \\ 3y + 4z = 11 \\ -2x - y + z = 5 \end{cases}$

$\left[\begin{array}{ccc|c} 5 & -7 & 1 & -8 \\ 0 & 3 & 4 & 11 \\ -2 & -1 & 1 & 5 \end{array}\right]$, $\begin{bmatrix} 5 & -7 & 1 \\ 0 & 3 & 4 \\ -2 & -1 & 1 \end{bmatrix}$, $\begin{bmatrix} -8 \\ 11 \\ 5 \end{bmatrix}$ [6.3]

20. $\begin{cases} -2w + x - 6y + 4z = 1 \\ 3w + 3x + 2y - 5z = -9 \\ -w - 7x + 3y - z = 4 \\ 8w + 5x + y - 3z = -12 \end{cases}$

21. After row operations have been correctly performed on an augmented matrix, the result is

$$\left[\begin{array}{ccc|c} 1 & -3 & 1 & 5 \\ 0 & 1 & -1 & 3 \\ 0 & 0 & 0 & 0 \end{array}\right]$$

Does this result indicate that the system of equations has a unique solution, an infinite number of solutions, or no solution? Infinite number of solutions [6.3]

22. After row operations have been correctly performed on an augmented matrix, the result is

$$\left[\begin{array}{ccc|c} 1 & -4 & 0 & 9 \\ 0 & 1 & 5 & -2 \\ 0 & 0 & 0 & 0 \end{array}\right]$$

Does this result indicate that the system of equations has a unique solution, an infinite number of solutions, or no solution? Infinite number of solutions [6.3]

In Exercises 23 to 38, solve the system of equations by using the Gaussian elimination method.

23. $\begin{cases} 2x - 3y = 7 \\ 3x - 4y = 10 \end{cases}$

(2, −1) [6.3]

24. $\begin{cases} 3x + 4y = -9 \\ 2x + 3y = -7 \end{cases}$

(1, −3) [6.3]

25. $\begin{cases} 4x - 5y = 12 \\ 3x + y = 9 \end{cases}$

(3, 0) [6.3]

26. $\begin{cases} 2x - 5y = 10 \\ 5x + 2y = 4 \end{cases}$

$\left(\frac{40}{29}, -\frac{42}{29}\right)$ [6.3]

27. $\begin{cases} x + 2y + 3z = 5 \\ 3x + 8y + 11z = 17 \\ 2x + 6y + 7z = 12 \end{cases}$

(3, 1, 0) [6.3]

28. $\begin{cases} x - y + 3z = 10 \\ 2x - y + 7z = 24 \\ 3x - 6y + 7z = 21 \end{cases}$

(2, 1, 3) [6.3]

29. $\begin{cases} 2x - y - z = 4 \\ x - 2y - 2z = 5 \\ 3x - 3y - 8z = 19 \end{cases}$

(1, 0, −2) [6.3]

30. $\begin{cases} 3x - 7y + 8z = 10 \\ x - 3y + 2z = 0 \\ 2x - 8y + 7z = 5 \end{cases}$

(0, 2, 3) [6.3]

31. $\begin{cases} 4x - 9y + 6z = 54 \\ 3x - 8y + 8z = 49 \\ x - 3y + 2z = 17 \end{cases}$

(3, −4, 1) [6.3]

32. $\begin{cases} 3x + 8y - 5z = 6 \\ 2x + 9y - z = -8 \\ x - 4y - 2z = 16 \end{cases}$

(4, −2, −2) [6.3]

33. $\begin{cases} x + y + 2z = -5 \\ 2x + 3y + 5z = -13 \\ 2x + 5y + 7z = -19 \end{cases}$

(−c − 2, −c − 3, c) [6.3]

34. $\begin{cases} x - 2y + 3z = 9 \\ 3x - 5y + 8z = 25 \\ x - z = 5 \end{cases}$

(5, −2, 0) [6.3]

35. $\begin{cases} w + 2x - y + 2z = 1 \\ 3w + 8x + y + 4z = 1 \\ 2w + 7x + 3y + 2z = 0 \\ w + 3x - 2y + 5z = 6 \end{cases}$

(1, −2, 2, 3) [6.3]

36. $\begin{cases} w - 3x - 2y + z = -1 \\ 2w - 5x + 3z = 1 \\ 3w - 7x + 3y = -18 \\ 2w - 3x - 5y - 2z = -8 \end{cases}$

(2, 3, −1, 4) [6.3]

37. $\begin{cases} w + 3x + y - 4z = 3 \\ w + 4x + 3y - 6z = 5 \\ 2w + 8x + 7y - 5z = 11 \\ 2w + 5x - 6z = 4 \end{cases}$

(−37c + 2, 16c, −7c + 1, c) [6.3]

38. $\begin{cases} w + 4x - 2y + 3z = 6 \\ 2w + 9x - y + 5z = 13 \\ w + 7x + 6y + 5z = 9 \\ 3w + 14x + 7z = 20 \end{cases}$

(63c + 2, −14c + 1, 5c, c) [6.3]

In Exercises 39 to 52, perform the indicated operations. Let

$A = \begin{bmatrix} 2 & -1 & 3 \\ 3 & 2 & -1 \end{bmatrix}$, $B = \begin{bmatrix} 0 & -2 \\ 4 & 2 \\ 1 & -3 \end{bmatrix}$, $C = \begin{bmatrix} 2 & 6 & 1 \\ 1 & 2 & -1 \\ 2 & 4 & -1 \end{bmatrix}$, **and**

$D = \begin{bmatrix} -3 & 4 & 2 \\ 4 & -2 & 5 \end{bmatrix}$.

39. $3A$ $\begin{bmatrix} 6 & -3 & 9 \\ 9 & 6 & -3 \end{bmatrix}$ [6.4]

40. $-2B$ $\begin{bmatrix} 0 & 4 \\ -8 & -4 \\ -2 & 6 \end{bmatrix}$ [6.4]

41. $-A + D$ $\begin{bmatrix} -5 & 5 & -1 \\ 1 & -4 & 6 \end{bmatrix}$ [6.4]

42. $2A - 3D$ $\begin{bmatrix} 13 & -14 & 0 \\ -6 & 10 & -17 \end{bmatrix}$ [6.4]

43. AB $\begin{bmatrix} -1 & -15 \\ 7 & 1 \end{bmatrix}$ [6.4]

44. DB $\begin{bmatrix} 18 & 8 \\ -3 & -27 \end{bmatrix}$ [6.4]

45. BA **46.** BD **47.** C^2

48. C^3 **49.** BAC **50.** ADB
Not possible [6.4]

51. $AB - BA$
Not possible [6.4]

52. $(A - D)C$ $\begin{bmatrix} 7 & 24 & 9 \\ -10 & -22 & 1 \end{bmatrix}$ [6.4]

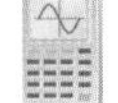

In Exercises 53 to 58, use a graphing utility to perform the indicated operations on matrices A and B.

$$A = \begin{bmatrix} 2 & 4 & -2 & 3 \\ -1 & 0 & -3 & 5 \\ -2 & 5 & 1 & -5 \\ 3 & 0 & 6 & 2 \end{bmatrix}, \quad B = \begin{bmatrix} 4 & 9 & 1 & 3 \\ 2 & -4 & 0 & 5 \\ -1 & -3 & 8 & 0 \\ -6 & 1 & 5 & 2 \end{bmatrix}$$

53. A^2 **54.** B^3 **55.** $2AB - A^3$

56. $3A - B^2$ **57.** A^{-1} **58.** B^{-1}

In Exercises 59 and 60, write the system of equations that corresponds to the matrix equation.

59. $\begin{bmatrix} 4 & 5 \\ 3 & -4 \end{bmatrix}\begin{bmatrix} x \\ y \end{bmatrix} = \begin{bmatrix} -8 \\ 10 \end{bmatrix}$ **60.** $\begin{bmatrix} -1 & 6 & 1 \\ 3 & 2 & 2 \\ 3 & -5 & 4 \end{bmatrix}\begin{bmatrix} x \\ y \\ z \end{bmatrix} = \begin{bmatrix} 4 \\ -8 \\ 1 \end{bmatrix}$

In Exercises 61 to 68, use elementary row operations to find the inverse, if it exists, of the given matrix.

61. $\begin{bmatrix} 2 & -2 \\ 3 & -2 \end{bmatrix}$ $\begin{bmatrix} -1 & 1 \\ -\frac{3}{2} & 1 \end{bmatrix}$ [6.5] **62.** $\begin{bmatrix} 3 & 4 \\ 2 & 3 \end{bmatrix}$ $\begin{bmatrix} 3 & -4 \\ -2 & 3 \end{bmatrix}$ [6.5]

63. $\begin{bmatrix} -2 & 3 \\ 2 & 4 \end{bmatrix}$ $\begin{bmatrix} -\frac{2}{7} & \frac{3}{14} \\ \frac{1}{7} & \frac{1}{7} \end{bmatrix}$ [6.5] **64.** $\begin{bmatrix} 5 & -4 \\ 3 & 2 \end{bmatrix}$ $\begin{bmatrix} \frac{1}{11} & \frac{2}{11} \\ -\frac{3}{22} & \frac{5}{22} \end{bmatrix}$ [6.5]

65. $\begin{bmatrix} 1 & 2 & 1 \\ 2 & 6 & 4 \\ 3 & 8 & 6 \end{bmatrix}$ **66.** $\begin{bmatrix} 1 & -3 & 2 \\ 3 & -8 & 7 \\ 2 & -3 & 6 \end{bmatrix}$

67. $\begin{bmatrix} 3 & -2 & 7 \\ 2 & -1 & 5 \\ 3 & 0 & 10 \end{bmatrix}$ **68.** $\begin{bmatrix} 4 & 9 & -11 \\ 3 & 7 & -8 \\ 2 & 6 & -3 \end{bmatrix}$

In Exercises 69 to 72, use a graphing utility and the inverse matrix method to solve the given system of equations for each set of constants.

69. $\begin{cases} 3x + 4y = b_1 \\ 2x + 3y = b_2 \end{cases}$

a. $b_1 = 2, b_2 = -3$
(18, −13)

b. $b_1 = -2, b_2 = 4$
(−22, 16) [6.5]

70. $\begin{cases} 2x - 5y = b_1 \\ 3x - 7y = b_2 \end{cases}$

a. $b_1 = -3, b_2 = 4$
(41, 17)

b. $b_1 = 2, b_2 = -5$
(−39, −16) [6.5]

71. $\begin{cases} 2x + y - z = b_1 \\ 4x + 4y + z = b_2 \\ 2x + 2y - 3z = b_3 \end{cases}$

a. $b_1 = -1, b_2 = 2, b_3 = 4$

b. $b_1 = -2, b_2 = 3, b_3 = 0$

72. $\begin{cases} 3x - 2y + z = b_1 \\ 3x - y + 3z = b_2 \\ 6x - 4y + z = b_3 \end{cases}$

a. $b_1 = 0, b_2 = 3, b_3 = -2$

b. $b_1 = 1, b_2 = 2, b_3 = -4$

In Exercises 73 and 74, graph the solution set of each inequality.

73. $4x - 5y < 20$ [6.6] **74.** $2x + 7y \geq -14$ [6.6]

In Exercises 75 to 82, graph the solution set of each system of inequalities.

75. $\begin{cases} 2x - 5y < 9 \\ 3x + 4y \geq 2 \end{cases}$ [6.6] **76.** $\begin{cases} 3x + y > 7 \\ 2x + 5y < 9 \end{cases}$ [6.6]

77. $\begin{cases} 2x + 3y > 6 \\ 2x - y > -2 \\ x \leq 3 \end{cases}$ [6.6] **78.** $\begin{cases} 2x + 5y > 10 \\ x - y > -2 \\ x \leq 4 \end{cases}$ [6.6]

79. $\begin{cases} 2x + 3y \leq 18 \\ x + y \leq 7 \\ x \geq 0, y \geq 0 \end{cases}$ [6.6] **80.** $\begin{cases} 3x + 5y \geq 25 \\ 2x + 3y \geq 16 \\ x \geq 0, y \geq 0 \end{cases}$ [6.6]

81. $\begin{cases} 3x + y \geq 6 \\ x + 4y \geq 14 \\ 2x + 3y \geq 16 \\ x \geq 0, y \geq 0 \end{cases}$ [6.6] **82.** $\begin{cases} 3x + 2y \geq 14 \\ x + y \geq 6 \\ 11x + 4y \leq 48 \\ x \geq 0, y \geq 0 \end{cases}$ [6.6]

In Exercises 83 to 88, solve the linear programming problem. In each problem, assume $x \geq 0$ and $y \geq 0$.

83. Objective function: $P = 2x + 2y$

Constraints: $\begin{cases} x + 2y \leq 14 \\ 5x + 2y \leq 30 \end{cases}$

Maximize the objective function.
The maximum is 18 at (4, 5). [6.6]

84. Objective function: $P = 4x + 5y$

Constraints: $\begin{cases} 2x + 3y \leq 24 \\ 4x + 3y \leq 36 \end{cases}$

Maximize the objective function.
The maximum is 44 at (6, 4). [6.6]

85. Objective function: $P = 4x + y$

Constraints: $\begin{cases} 5x + 2y \geq 16 \\ x + 2y \geq 8 \\ x \leq 20, y \leq 20 \end{cases}$

Minimize the objective function.
The minimum is 8 at (0, 8). [6.6]

86. Objective function: $P = 2x + 7y$

Constraints: $\begin{cases} 4x + 3y \geq 24 \\ 4x + 7y \geq 40 \\ x \leq 10, y \leq 10 \end{cases}$

Minimize the objective function.
The minimum is 20 at (10, 0). [6.6]

87. Objective function: $P = 6x + 3y$

Constraints: $\begin{cases} 5x + 2y \geq 20 \\ x + y \geq 7 \\ x + 2y \geq 10 \\ x \leq 15, y \leq 15 \end{cases}$

Minimize the objective function.
The minimum is 27 at (2, 5). [6.6]

88. Objective function: $P = 5x + 4y$

Constraints: $\begin{cases} x + y \leq 10 \\ 2x + y \leq 13 \\ 3x + y \leq 18 \end{cases}$

Maximize the objective function.
The maximum is 43 at (3, 7). [6.6]

89. *Interpolating Polynomial* Find an interpolating polynomial that passes through the points whose coordinates are $(-1, -4)$, $(2, 8)$, and $(3, 16)$. $y = x^2 + 3x - 2$ [6.2]

90. *Interpolating Polynomial* Find an interpolating polynomial that passes through the points whose coordinates are $(-1, 4)$, $(1, 0)$, and $(2, -5)$. $y = -x^2 - 2x + 3$ [6.2]

Business and Economics

91. *Investment* A broker invested a portion of a client's money in a bond fund with an annual interest rate of 6.5% and the remaining portion into a second bond fund with an annual interest rate of 8%. The total interest collected after 1 year was \$2070. If the investor had switched the amounts invested in the two accounts, the interest earned would have been \$2280. How much was invested in each account? \$22,000 at 6.5%; \$8000 at 8% [6.1]

92. *Investment* An investment club invested \$50,000 in three different accounts. The annual interest rates paid by the accounts were 4%, 6%, and 7%. The total interest earned in 1 year was \$2895. If 30% more money was invested at 7% than at 4%, how much was invested in each account?
\$15,000 at 4%; \$15,500 at 6%; \$19,500 at 7% [6.2/6.3]

Life and Health Sciences

93. *Dental Supplies* A few weeks ago, a dentist's office ordered three boxes of plastic bibs, eight boxes of tongue depressors, and two boxes of headrest covers at a total cost of \$136. The next week the office ordered two boxes of bibs, six boxes of tongue depressors, and three boxes of headrest covers for a total cost of \$134. Last week the office ordered five boxes of bibs, 10 boxes of tongue depressors, and four boxes of headrest covers. The total cost was \$226. What is the cost per box of each item?
Bibs, \$16; tongue depressors, \$5; headrest covers, \$24 [6.2/6.3]

94. *Nutrition* A chef wants to prepare a low-fat, low-sodium meal using lean meat, roasted potatoes, and green beans. A 1-ounce serving of meat contains 50 calories, 7 grams of protein, and 25 milligrams of sodium. A 1-ounce serving of green beans contains 20 calories, 2 grams of protein, and 12 milligrams of sodium. A 1-ounce serving of potatoes contains 24 calories, 4 grams of protein, and 40 milligrams of sodium. If the chef wants to prepare a meal that contains 350 calories, 49 grams of protein, and 323 milligrams of sodium, how many ounces of each ingredient should be prepared?
Meat, 3 oz; potatoes, 5 oz; green beans, 4 oz [6.2/6.3]

95. *Exercise Routines* A health club finds that of its regular members who exercise at least 3 hours per week, 91% will continue to exercise at least 3 hours during the following week. Of those who exercise less than 3 hours per week, only 6% will exercise at least 3 hours during the following week. During the current week, 60% of the regular members exercised at least 3 hours and 40% exercised less than 3 hours. After n weeks, the percent of regular members who exercise at least 3 hours per week, m, and the percent who exercise less than 3 hours per week, l, is given by

$$[0.60 \quad 0.40]\begin{bmatrix} 0.91 & 0.09 \\ 0.06 & 0.94 \end{bmatrix}^n = [m \quad l]$$

Use a calculator and the above matrix equation to predict, to the nearest 0.1%, the percent of the members who will be exercising at least 3 hours per week 6 weeks from now.
47.5% [6.4]

96. *Exercise Routines* Experiment with several different values of n in the matrix equation from Exercise 95 to determine how many weeks it will be before the percent of regular members who exercise at least 3 hours per week falls below 42%. 15 weeks [6.4]

Sports and Recreation

97. ***Aviation*** Flying with the wind, a small plane traveled 855 miles in 5 hours. Flying against the wind, the plane traveled 575 miles in the same amount of time. Find the rate of the wind and the rate of the plane in calm air. Plane: 143 mph; wind: 28 mph [6.1]

98. ***Lake Rentals*** A lake resort rents SeaDoos for $40 for the first hour and $25 for each additional hour. They also rent Wave Runners, which cost $28 per hour. For what number of hours does it cost the same to rent a SeaDoo as it does to rent a Wave Runner? 5 h [6.1]

99. ***Movie Theater*** A movie theater has different ticket prices for children, students, and adults. On Tuesdays, all tickets are discounted from their regular prices. The ticket prices (in dollars) are contained in the following matrices.

Regular prices (all days except Tuesdays) $A = \begin{bmatrix} 4.00 \\ 6.50 \\ 9.00 \end{bmatrix}$ children, students, adults

Discount prices (Tuesdays) $B = \begin{bmatrix} 3.00 \\ 4.50 \\ 7.00 \end{bmatrix}$ children, students, adults

The Monday, Tuesday, and Wednesday attendance for each ticket group (children, students, and adults, in that order) is given in the following matrices.

$C = [217 \quad 396 \quad 588]$ Monday

$D = [383 \quad 621 \quad 842]$ Tuesday

$E = [264 \quad 340 \quad 485]$ Wednesday

Find a matrix expression that gives the total revenue for each ticket group on Monday. CA [6.4]

100. ***Movie Theater*** For the matrices given in Exercise 101, find a matrix expression that gives the total revenue for each ticket group on Tuesday. DB [6.4]

101. ***Movie Theater*** For the matrices given in Exercise 101, find a matrix expression that gives the total number of tickets sold for each ticket group over the three-day period from Monday to Wednesday. $C + D + E$ [6.4]

Physical Science and Engineering

102. ***Chemistry*** How many liters of a 20% acid solution should be mixed with 10 liters of a 10% acid solution so that the resulting solution is a 16% acid solution? 15 L [6.1]

103. ***Building Design*** An architect is designing a slanted roof using software that uses an xyz-coordinate system. The architect wants the roof to be a plane that passes through the points (2, 1, 2), (3, 1, 0), and (−2, −3, −2). Find an equation of the form $z = ax + by + c$ for the architect to use. $z = -2x + 3y + 3$ [6.2/6.3]

104. ***Computer Science*** A programmer needs to find a cubic spline curve that passes through the points (2, −3), (4, 7), (6, 49), and (−2, −23). Find an equation of the form $y = ax^3 + bx^2 + cx + d$ for the programmer to use. $y = 0.5x^3 - 2x^2 + 3x - 5$ [6.3]

Chapter 6 Test

—Answer graphs to Exercises 17–19 are on page AA20.

In Exercises 1 to 5, solve each system of equations. If a system of equations is inconsistent, so state.

1. $\begin{cases} 3x + 2y = -5 \\ 2x - 5y = -16 \end{cases}$ $(-3, 2)$ [6.1]

2. $\begin{cases} x - \frac{1}{2}y = 3 \\ 2x - y = 6 \end{cases}$ $\left(\frac{6 + c}{2}, c\right)$ [6.1]

3. $\begin{cases} x + 3y - z = 8 \\ 2x - 7y + 2z = 1 \\ 4x - y + 3z = 13 \end{cases}$ $\left(\frac{173}{39}, \frac{29}{39}, -\frac{4}{3}\right)$ [6.2]

4. $\begin{cases} 2x - 3y + z = -1 \\ x + 5y - 2z = 5 \end{cases}$ $\left(\frac{c + 10}{13}, \frac{5c + 11}{13}, c\right)$ [6.2]

5. $\begin{cases} 4x + 2y + z = 0 \\ x - 3y - 2z = 0 \\ 3x + 5y + 3z = 0 \end{cases}$ $\left(\frac{c}{14}, -\frac{9c}{14}, c\right)$ [6.2]

6. Write the augmented matrix, the coefficient matrix, and the constant matrix for the system of equations

$$\begin{cases} 2x + 3y - 3z = 4 \\ 3x + 2z = -1 \\ 4x - 4y + 2z = 3 \end{cases}$$

6. $\left[\begin{array}{ccc|c} 2 & 3 & -3 & 4 \\ 3 & 0 & 2 & -1 \\ 4 & -4 & 2 & 3 \end{array}\right], \begin{bmatrix} 2 & 3 & -3 \\ 3 & 0 & 2 \\ 4 & -4 & 2 \end{bmatrix}, \begin{bmatrix} 4 \\ -1 \\ 3 \end{bmatrix}$ [6.3]

7. Write a system of equations that is equivalent to the augmented matrix $\left[\begin{array}{rrrr|r} 3 & -2 & 5 & -1 & 9 \\ 2 & 3 & -1 & 4 & 8 \\ 1 & 0 & 3 & 2 & -1 \end{array}\right]$.

$\begin{cases} 3w - 2x + 5y - z = 9 \\ 2w + 3x - y + 4z = 8 \\ w + 3y + 2z = -1 \end{cases}$ [6.3]

In Exercises 8 and 9, solve the system of equations by using the Gaussian elimination method.

8. $\begin{cases} x - 2y + 3z = 10 \\ 2x - 3y + 8z = 23 \\ -x + 3y - 2z = -9 \end{cases}$ $(2, -1, 2)$ [6.3]

9. $\begin{cases} w + 2x - 3y + 2z = 11 \\ 2w + 5x - 8y + 5z = 28 \\ -2w - 4x + 7y - z = -18 \end{cases}$
$(3c - 5, -7c + 14, -3c + 4, c)$ [6.3]

10. Find an equation of the circle that passes through the points $(3, 5)$, $(-3, -3)$, and $(4, 4)$. (*Hint:* Use $x^2 + y^2 + ax + by + c = 0$.) $x^2 + y^2 - 2y - 24 = 0$ [6.2/6.3]

In Exercises 11 to 14, let $A = \begin{bmatrix} -1 & 3 & 2 \\ 1 & 4 & -1 \end{bmatrix}$,

$B = \begin{bmatrix} 2 & -1 & 3 \\ 4 & -2 & -1 \\ 3 & 2 & 2 \end{bmatrix}$, **and** $C = \begin{bmatrix} 1 & -2 & 3 \\ 2 & -3 & 8 \\ -1 & 3 & -2 \end{bmatrix}$. **Perform each possible operation. If an operation is not possible, so state.**

11. a. $-3A$ b. $A + B$ c. $3B - 2C$

a. $\begin{bmatrix} 3 & -9 & -6 \\ -3 & -12 & 3 \end{bmatrix}$ b. $A + B$ is not defined. c. $\begin{bmatrix} 4 & 1 & 3 \\ 8 & 0 & -19 \\ 11 & 0 & 10 \end{bmatrix}$ [6.4]

12. a. AB b. CA

a. $\begin{bmatrix} 16 & -1 & -2 \\ 15 & -11 & -3 \end{bmatrix}$ b. CA is not defined. [6.4]

13. a. A^2 b. B^2

a. A^2 is not defined. b. $\begin{bmatrix} 9 & 6 & 13 \\ -3 & -2 & 12 \\ 20 & -3 & 11 \end{bmatrix}$ [6.4]

14. C^{-1} $\begin{bmatrix} 18 & -5 & 7 \\ 4 & -1 & 2 \\ -3 & 1 & -1 \end{bmatrix}$ [6.5]

15. Use row operations to find the inverse of the matrix $\begin{bmatrix} 2 & 4 \\ 3 & 5 \end{bmatrix}$. $\begin{bmatrix} -2.5 & 2 \\ 1.5 & -1 \end{bmatrix}$

16. Solve the system of equations using an inverse matrix and matrix algebra. (*Hint*: The answer to Problem 15 will be needed here.)

$\begin{cases} 2x + 4y = 6 \\ 3x + 5y = 5 \end{cases}$ $(-5, 4)$ [6.5]

17. Graph the inequality $3x - 4y > 8$. [6.6]

In Exercises 18 and 19, graph each system of inequalities. If the solution set is empty, so state.

18. $\begin{cases} 2x - 5y \le 16 \\ x + 3y \ge -3 \end{cases}$ [6.6] 19. $\begin{cases} x + y \ge 8 \\ 2x + y \ge 11 \\ x \ge 0, y \ge 0 \end{cases}$ [6.6]

20. A farmer has 160 acres available on which to plant oats and barley. It costs \$15 per acre for oat seed and \$13 per acre for barley seed. The labor cost is \$15 per acre for oats and \$20 per acre for barley. The farmer has \$2200 available to purchase seed and has set aside \$2600 for labor. The profit per acre for oats is \$120, and the profit per acre for barley is \$150. How many acres of oats should the farmer plant to maximize profit?
$\frac{680}{7}$ acres of oats [6.6]

Cumulative Review Exercises

1. Simplify: $\dfrac{1}{x+3} - \dfrac{5}{x-1}$ $\dfrac{-4(x+4)}{(x+3)(x-1)}$ [P.5]

2. Solve for x: $3x^2 - 2x - 4 = 0$ $x = \dfrac{1 \pm \sqrt{13}}{3}$ [1.2]

3. Solve the inequality. Use interval notation for the solution set. $5x - 11 \le 4$ and $2x + 7 > 5$ $(-1, 3]$ [1.4]

4. Find an equation of the line that passes through the two points $(-2, 5)$ and $(2, -1)$. $y = -\frac{3}{2}x + 2$ [2.4]

5. Find the vertex of the graph of the quadratic function $f(x) = 2x^2 + 8x - 5$. $(-2, -13)$ [2.5]

6. Determine whether the function $g(x) = x^3 + 2x$ is even, odd, or neither. Odd [3.3]

7. If $f(x) = x^2 - 3x$ and $g(x) = 1 - 5x$, find and simplify $(f \circ g)(x)$. $25x^2 + 5x - 2$ [3.1]

8. Find the inverse of the function $f(x) = 2 - 7x^3$. $f^{-1}(x) = \sqrt[3]{\dfrac{2-x}{7}}$ [3.2]

9. If c varies jointly as a and the square root of b, and $c = 14$ when $a = 7$ and $b = 16$, write an equation that expresses the relationship among the variables. $c = \frac{1}{2}a\sqrt{b}$ [3.4]

10. Use synthetic division and the Factor Theorem to determine whether $x - 2$ is a factor of $P(x) = x^4 - 4x^3 + 5x^2 - 7x + 4$. No [4.1]

11. Determine the far-left and far-right behavior of the graph of the polynomial function $P(x) = -3x^4 + 7x^3 - 3x^2 + 5x$.
Down to far left, down to far right [4.2]

12. Write the polynomial function $P(x) = x^3 - 3x^2 - x + 3$ as a product of linear factors.
$P(x) = (x + 1)(x - 1)(x - 3)$ [4.4]

13. Determine the vertical and horizontal asymptotes of the rational function $F(x) = \dfrac{2x^2}{x^2 - 6x + 8}$.
$x = 2, x = 4, y = 2$ [4.5]

14. Solve the equation without using a calculator: $2^{3-2x} = 128$. $x = -2$ [5.4]

15. Solve the following equation. Give an exact answer.
$\ln 6x - \ln 3 = 2$ $x = \frac{1}{2}e^2$ [5.4]

16. Write the logarithmic expression as a single logarithm with a coefficient of 1: $3 \ln x + \frac{1}{2} \ln y - 2 \ln z$ $\ln \dfrac{x^3\sqrt{y}}{z^2}$ [5.3]

17. The population of a city is growing exponentially. In 2000 the population was 36,400, and 3 years later it was 39,150. Find the exponential growth function for the city's population. Use $t = 0$ to represent the year 2000.
$N(t) \approx 36{,}400e^{0.02428t}$ [5.5]

18. Solve the system of equations algebraically.

$$\begin{cases} 2x - 4y = -10 \\ 3x + 5y = 7 \end{cases}$$
$(-1, 2)$ [6.1]

19. Use the Gaussian elimination method to solve the system of equations.

$$\begin{cases} 2x + 5y - 3z = 17 \\ x - 2y + z = -1 \\ 3x - 4y + 4z = -3 \end{cases}$$
$(3, 1, -2)$ [6.3]

20. Use row operations to find the inverse of the matrix.

$$\begin{bmatrix} 5 & -2 \\ -3 & 2 \end{bmatrix} \quad \begin{bmatrix} \frac{1}{2} & \frac{1}{2} \\ \frac{3}{4} & \frac{5}{4} \end{bmatrix}$$
[6.5]

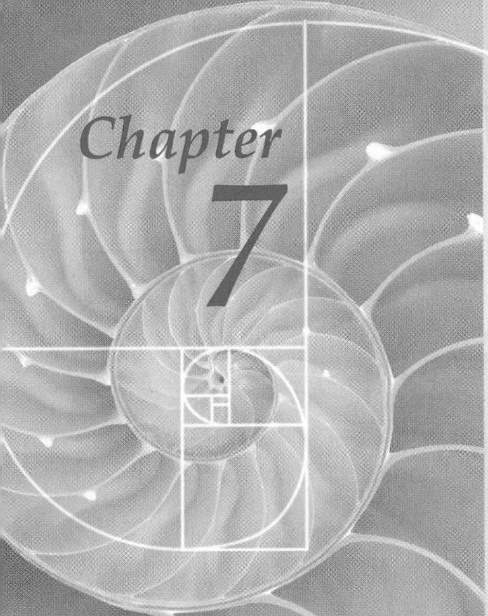

Chapter 7 Sequences and Series with Applications to the Mathematics of Finance

Interest Rates and Car Payments

The graph below shows the average annual interest rates for 48-month new car loans in the United States for the years 1995 to 2002. From the graph, the highest annual interest rate of a 48-month car loan occurred in 1995, and the lowest occurred in 2002.

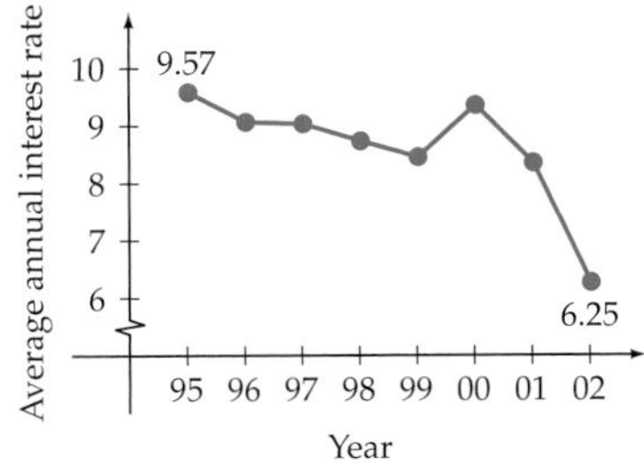

Source: Federal Reserve and Yahoo! Finance

Interest rates impact monthly car payments. For instance, the monthly payment for a car loan of $12,000 at an interest rate of 9.53% is $301.65, whereas the monthly car payment at an interest rate of 6.25% is $283.20. The savings is $18.40 per month, the approximate cost to fill the gas tank at current prices.

The calculation we used to find the monthly car payment is based on *geometric series,* one of the topics of this chapter. **Exercise 9 in Section 7.4, page 627,** asks you to find a car payment using a formula based on these series.

Chapter Prep Quiz *Do these exercises to prepare for Chapter 7.*

1. Evaluate $\frac{(-1)^n}{n^2}$ when $n = 2$ and $n = 3$. [P.1] $\frac{1}{4}, -\frac{1}{9}$

2. Evaluate $1500(1 + i)^4$ when $i = 0.005$. Round to the nearest hundredth. [P.1] 1530.23

3. Evaluate $f(n) = n^2 - n$ when $n = 1$ and $n = 4$. [P.1] 0, 12

4. Simplify: $1 - (1 + x)^{-2}$ [P.5] $\frac{x(x+2)}{(x+1)^2}$

5. Simplify: $\frac{\frac{3}{8}}{\frac{3}{4}}$ [P.5] $\frac{1}{2}$

6. Simplify: $\frac{\frac{3}{4^{n+1}}}{\frac{3}{4^n}}$ [P.5] $\frac{1}{4}$

7. Given $f(x) = 3x + 2$, find $f(n + 1) - f(n)$. [2.2] 3

8. Given $g(x) = 2^x$, find $\frac{g(n+1)}{g(n)}$. [5.1] 2

9. Solve: $(1 + x)^3 = 1.331$ [1.3] 0.1

10. Solve: $2^x = 5$ [5.4] $x = \frac{\log 5}{\log 2} \approx 2.32$

INSTRUCTOR NOTE

The *Chapter Prep Quiz* is a means to test your students' mastery of prerequisite material that is assumed in the coming chapter. Each question identifies the section to review (if necessary) and all answers are provided in the Answers to Selected Exercises.

Problem Solving Strategies

Inductive Reasoning

Consider the following sums of powers of 2.

$$2^0 + 2^1 = 1 + 2 = 3 = 2^2 - 1$$
$$2^0 + 2^1 + 2^2 = 1 + 2 + 4 = 7 = 2^3 - 1$$
$$2^0 + 2^1 + 2^2 + 2^3 = 1 + 2 + 4 + 8 = 15 = 2^4 - 1$$
$$2^0 + 2^1 + 2^2 + 2^3 + 2^4 = 1 + 2 + 4 + 8 + 16 = 31 = 2^5 - 1$$

By looking at the pattern, we can make a conjecture that

$$2^0 + 2^1 + 2^2 + 2^3 + 2^4 + 2^5 = 63 = 2^6 - 1$$

On the basis of these examples, we might make the following more general conjecture.

$$2^0 + 2^1 + 2^2 + 2^3 + \cdots + 2^n = 2^{n+1} - 1$$

Making a conjecture based on a pattern of a few examples is called **inductive reasoning,** and it is an important aspect of problem solving. However, just because it seems that a conjecture is true, we cannot assert that it is always true unless we can provide a mathematical proof. We will do this in Section 7.2.

To illustrate that making a conjecture based on a few examples does not always lead to a true conjecture, consider the statement, "For any nonnegative integer n, $n^2 + n + 41$ is a prime number." If we test the first few cases, we find

$n = 0$	$0^2 + 0 + 41 = 41$	a prime number
$n = 1$	$1^2 + 1 + 41 = 43$	a prime number
$n = 2$	$2^2 + 2 + 41 = 47$	a prime number
$n = 3$	$3^2 + 3 + 41 = 53$	a prime number

Find a counterexample that shows that the statement is false. (*Hint:* To save you some time, the value of n is greater than 35.)

SECTION 7.1 Introduction to Sequences and Series

- Introduction to Sequences
- Partial Sums and Summation Notation

Introduction to Sequences

The ordered list of numbers 1, 2, 4, 8, 16, 32, . . . is called an *infinite sequence.* The list is ordered simply because order makes a difference. The sequence 1, 2, 8, 4, 16, 32, . . . contains the same numbers, but in a different order. Therefore, it is a different infinite sequence.

An infinite sequence can be thought of as a pairing between positive integers and real numbers. For example, the sequence 2, 4, 6, 8, . . . , $2n$, . . . pairs a number with twice the number.

$$\begin{array}{cccccccc} 1 & 2 & 3 & 4 & 5 & 6 & \dots & n\dots \\ \downarrow & \downarrow & \downarrow & \downarrow & \downarrow & \downarrow & & \downarrow \\ 2 & 4 & 6 & 8 & 10 & 12 & \dots & 2n\dots \end{array}$$

This concept of a pairing of numbers leads us to define an infinite sequence as a function.

Infinite Sequence

An **infinite sequence** is a function whose domain is the set of positive integers and whose range is a subset of the real numbers.

Although the positive integers do not include zero, it is occasionally convenient to include zero in the domain of an infinite sequence. Also, we will frequently use the word *sequence* instead of the phrase *infinite sequence.*

The elements in the range of a sequence are called the **terms** of the sequence. For the above example, the terms are 2, 4, 6, 8, . . . , $2n$, ... The first term of the sequence is 2, the second term is 4, and so on. The **general term** or **nth term** is $2n$.

QUESTION What are the second term and the nth term of the sequence 1, 4, 9, . . . , n^2, . . . ?

It is customary to use subscript notation to represent a term of a sequence. We use a_n to represent the nth term of a sequence. For the sequence 2, 4, 6, 8, . . . , $2n$, . . . , the nth term is $a_n = 2n$. This gives $a_1 = 2$, $a_2 = 4$, $a_3 = 6$, and $a_4 = 8$.

There will be times when we will want to distinguish between the nth term of a sequence and the entire sequence. The sequence whose nth terms is a_n is denoted by $\{a_n\}$. Thus $a_n = 2n$ but $\{a_n\} = 2, 4, 6, 8, \dots, 2n, \dots$

A constant sequence is one for which all the terms are the same. The sequence 6, 6, 6, . . . , 6, . . . is an example of a constant sequence.

ANSWER The second term is 4. The nth term is n^2.

Alternative to Example 1

Find the first three terms and the seventh term of the sequence whose nth term is given by $a_n = n^2 - n + 1$.

▪ **1, 3, 7, 43**

EXAMPLE 1 Find the Terms of a Sequence

Find the first three terms and the eighth term of the sequence whose nth term is given by $a_n = n^2 + n$.

Solution

$$a_n = n^2 + n$$
$$a_1 = 1^2 + 1 = 2 \qquad \text{▪ Replace } n \text{ by 1.}$$
$$a_2 = 2^2 + 2 = 6 \qquad \text{▪ Replace } n \text{ by 2.}$$
$$a_3 = 3^2 + 3 = 12 \qquad \text{▪ Replace } n \text{ by 3.}$$
$$a_8 = 8^2 + 8 = 72 \qquad \text{▪ Replace } n \text{ by 8.}$$

CHECK YOUR PROGRESS 1 Find the first three terms of the sequence whose nth term is given by $2^n - 1$.

Solution *See page S38.* 1, 3, 7

Point of Interest

Fibonacci (a contraction of *filius Bonaccio*, "son of Bonaccio"), circa 1170–1250, was born in Pisa and is considered one of the greatest mathematicians of the Middle Ages. His book *Liber Abaci* (*Book of Counting*) was instrumental in bringing the Arabic numerals (the ones we use today) to the Western world.

The Fibonacci sequence can be used to model many different natural phenomena, such as pine cones, spirals on a nautilus, and flower petals. The diagram below shows a Fibonacci pattern in the growth of a plant that sprouts a new shoot from each branch once each week.

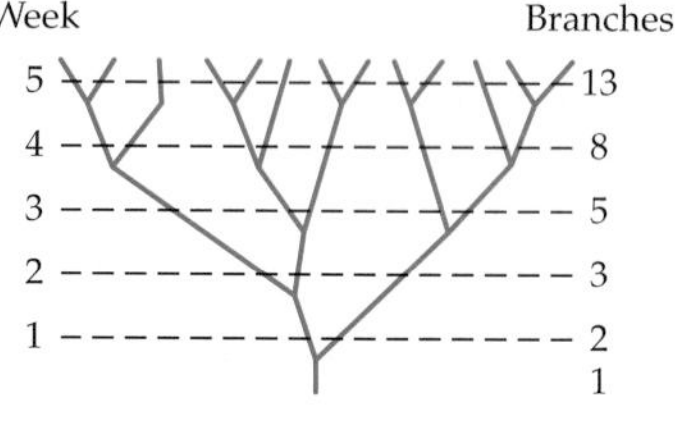

An **alternating sequence** is one in which the signs of the terms *alternate* between positive and negative values. The sequence defined by $a_n = (-1)^{n+1} \cdot \frac{1}{n^2}$ is an alternating sequence. The first three terms are shown below.

$$a_n = (-1)^{n+1} \cdot \frac{1}{n^2}$$
$$a_1 = (-1)^{1+1} \cdot \frac{1}{1^2} = 1 \qquad \text{▪ } n = 1$$
$$a_2 = (-1)^{2+1} \cdot \frac{1}{2^2} = -\frac{1}{4} \qquad \text{▪ } n = 2$$
$$a_3 = (-1)^{3+1} \cdot \frac{1}{3^2} = \frac{1}{9} \qquad \text{▪ } n = 3$$

If we continue the procedure, the first six terms of the sequence are $1, -\frac{1}{4}, \frac{1}{9}, -\frac{1}{16}, \frac{1}{25}, -\frac{1}{36}$.

A **recursive sequence** is one in which each succeeding term of the sequence is defined by using some of the preceding terms. For example, let $F_1 = 1$, $F_2 = 1$, and $F_n = F_{n-1} + F_{n-2}$, $n \geq 3$. Then

$$F_n = F_{n-1} + F_{n-2}$$
$$F_3 = F_2 + F_1 = 1 + 1 = 2 \qquad \text{▪ } n = 3$$
$$F_4 = F_3 + F_2 = 2 + 1 = 3 \qquad \text{▪ } n = 4$$
$$F_5 = F_4 + F_3 = 3 + 2 = 5 \qquad \text{▪ } n = 5$$
$$F_6 = F_5 + F_4 = 5 + 3 = 8 \qquad \text{▪ } n = 6$$

The recursive sequence 1, 1, 2, 3, 5, 8, . . . is called the *Fibonacci sequence*. It is named after Leonardo Fibonacci (c. 1170–1250), an Italian mathematician.

QUESTION What are the seventh and eighth terms of the Fibonacci sequence?

ANSWER $F_7 = F_6 + F_5 = 8 + 5 = 13$; $F_8 = F_7 + F_6 = 13 + 8 = 21$

Alternative to Example 2

Let $a_1 = 2$, $a_2 = 3$, and $a_n = a_{n-1} \cdot a_{n-2}$. Find a_3, a_4, and a_5.

■ **6, 18, 108**

EXAMPLE 2 Find Terms of a Recursive Sequence

Let $a_1 = 1$ and $a_n = na_{n-1}$. Find a_2, a_3, and a_4.

Solution

$a_2 = 2a_1 = 2 \cdot 1 = 2 \qquad a_3 = 3a_2 = 3 \cdot 2 = 6 \qquad a_4 = 4a_3 = 4 \cdot 6 = 24$

CHECK YOUR PROGRESS 2 Let $a_1 = 1$, $a_2 = 2$, and $a_{n+1} = a_n a_{n-1}$. Find a_3, a_4, and a_5.

Solution *See page S38.* 2, 4, 8

It is possible to find a formula for the nth term of the sequence defined recursively in Example 2 by $a_1 = 1$, $a_n = na_{n-1}$. Consider the term a_5 of the sequence.

$$\begin{aligned} a_5 &= 5a_4 \\ &= 5 \cdot 4a_3 \\ &= 5 \cdot 4 \cdot 3a_2 \\ &= 5 \cdot 4 \cdot 3 \cdot 2a_1 \\ &= 5 \cdot 4 \cdot 3 \cdot 2 \cdot 1 \end{aligned}$$

■ We are using $a_n = na_{n-1}$ repeatedly. For instance, $a_4 = 4a_{4-1} = 4a_3$. Therefore, we replace a_4 by $4a_3$.

Continuing in this manner for a_n, we have

$$\begin{aligned} a_n &= na_{n-1} \\ &= n(n-1)a_{n-2} \\ &= n(n-1)(n-2)a_{n-3} \\ &\;\vdots \\ &= n(n-1)(n-2) \cdot \cdots \cdot 3 \cdot 2 \cdot 1 \end{aligned}$$

This is the number n *factorial,* written $n!$.

TAKE NOTE

Here are some examples of factorials.

$$5! = 5 \cdot 4 \cdot 3 \cdot 2 \cdot 1 = 120$$

$$7! = 7 \cdot 6 \cdot 5 \cdot 4 \cdot 3 \cdot 2 \cdot 1 = 5040$$

Factorials are frequently used in statistics and probability, as we will see later in this text.

***n* Factorial**

If n is a natural number, then n **factorial** is given by

$$n! = n \cdot (n-1) \cdot (n-2) \cdot \cdots \cdot 3 \cdot 2 \cdot 1$$

We also define $0! = 1$.

■ Partial Sums and Summation Notation

Another common way of obtaining a sequence is by adding the terms of a given sequence. For example, consider the sequence whose general term is given by $a_n = \frac{1}{2^n}$. The terms of this sequence are $\frac{1}{2}, \frac{1}{4}, \frac{1}{8}, \frac{1}{16}, \frac{1}{32}, \ldots$.

From this sequence we can generate a new sequence that is the sum of the terms of $\left\{\frac{1}{2^n}\right\}$.

$$S_1 = \frac{1}{2}$$

$$S_2 = \frac{1}{2^1} + \frac{1}{2^2} = \frac{1}{2} + \frac{1}{4} = \frac{3}{4}$$

$$S_3 = \frac{1}{2^1} + \frac{1}{2^2} + \frac{1}{2^3} = \frac{1}{2} + \frac{1}{4} + \frac{1}{8} = \frac{7}{8}$$

$$S_4 = \frac{1}{2^1} + \frac{1}{2^2} + \frac{1}{2^3} + \frac{1}{2^4} = \frac{1}{2} + \frac{1}{4} + \frac{1}{8} + \frac{1}{16} = \frac{15}{16}$$

and, in general, $S_n = \frac{1}{2} + \frac{1}{4} + \frac{1}{8} + \frac{1}{16} + \cdots + \frac{1}{2^n}$.

The term S_n is called the **nth partial sum** of the sequence, and the sequence $S_1, S_2, S_3, \ldots, S_n$ is called the **sequence of partial sums.**

A convenient notation that is used for partial sums is called **summation notation.** The sum of the first n terms of a sequence $\{a_n\}$ is represented using the Greek letter Σ (sigma).

$$\sum_{i=1}^{n} a_i = a_1 + a_2 + a_3 + \cdots + a_n$$

This sum is called a **series.** The letter i is called the **index** of the summation; the letter n is called the **upper limit** of the summation; and 1 is called the **lower limit** of the summation.

Alternative to Example 3

Evaluate each series.

a. $\sum_{i=1}^{5} 3^i$ b. $\sum_{3}^{6} (-1)^{i+1} \cdot \frac{1}{i}$

- a. 363
- b. $\frac{7}{60}$

EXAMPLE 3 Evaluate a Series

Evaluate each series.

a. $\sum_{i=1}^{4} 2^i$ b. $\sum_{j=2}^{5} (-1)^j j^2$

Solution

a. $\sum_{i=1}^{4} 2^i = 2^1 + 2^2 + 2^3 + 2^4 = 2 + 4 + 8 + 16 = 30$

b. $\sum_{j=2}^{5} (-1)^j j^2 = (-1)^2 2^2 + (-1)^3 3^2 + (-1)^4 4^2 + (-1)^5 5^2$

$= 4 - 9 + 16 - 25 = -14$

TAKE NOTE

In part **b.** of Example 3, observe that the index of the summation can be a variable other than i. Note also that the lower limit of the summation can be a number other than 1.

CHECK YOUR PROGRESS 3 Evaluate $\sum_{k=1}^{5} (k^3 - k)$.

Solution *See page S38.* 210

When writing the sum of a sequence in summation notation, we use the formula for the nth term of a sequence. This is illustrated in Example 4.

Alternative to Example 4

Write the sum of the first ten terms of the sequence whose nth term is $a_n = \frac{n-1}{n^2}$ in summation notation.

■ $S_{10} = \sum_{n=1}^{10} \frac{n-1}{n^2}$

EXAMPLE 4 Use Summation Notation

Write the sum of the first eight terms of the sequence whose nth term is $a_n = \frac{n}{n+1}$ in summation notation.

Solution Use the formula for the nth term of the sequence to write the sum.

$$S_8 = \sum_{n=1}^{8} \frac{n}{n+1}$$

CHECK YOUR PROGRESS 4 Write the sum of the first five terms of the sequence whose nth term is $a_n = \frac{1}{n^3}$ in summation notation.

Solution *See page S38.* $S_5 = \sum_{n=1}^{5} \frac{1}{n^3}$

Here are some properties of summation notation.

Properties of Summation Notation

If $\{a_n\}$ and $\{b_n\}$ are sequences and c is a real number, then

1. $\sum_{i=1}^{n} (a_i \pm b_i) = \sum_{i=1}^{n} a_i \pm \sum_{i=1}^{n} b_i$
2. $\sum_{i=1}^{n} ca_1 = c\sum_{i=1}^{n} a_i$
3. $\sum_{i=1}^{n} c = nc$

Here are some examples.

$$\sum_{i=1}^{5} (i^2 + i) = \sum_{i=1}^{5} i^2 + \sum_{i=1}^{5} i = (1^2 + 2^2 + 3^2 + 4^2 + 5^2) + (1 + 2 + 3 + 4 + 5) = 70$$

$$\sum_{i=1}^{4} 3i = 3\sum_{i=1}^{4} i = 3(1 + 2 + 3 + 4) = 30$$

$$\sum_{i=1}^{5} k = \overbrace{k + k + k + k + k}^{5 \text{ times}} = 5k$$

■ Notice that the index of the summation i is not the same as the variable in the sum, k. Therefore, k does not change as i changes.

Alternative to Example 5

Given $\sum_{n=1}^{30} a_n = 50$ and $\sum_{n=1}^{30} b_n = 22$, evaluate the following series.

a. $\sum_{n=1}^{30} (4a_n - 3b_n)$ b. $\sum_{n=1}^{30} (b_n + 3)$

■ a. 134
■ b. 112

EXAMPLE 5 Evaluate a Series

Given $\sum_{n=1}^{20} a_n = 81$ and $\sum_{n=1}^{20} b_n = 50$, evaluate the following series.

a. $\sum_{n=1}^{20} 2b_n$ **b.** $\sum_{n=1}^{20} (2a_n + 3b_n)$ **c.** $\sum_{n=1}^{20} (a_n + 5)$

Continued ➤

TAKE NOTE

In part **c.**, we are summing the constant, 5, 20 times.

$$\sum_{n=1}^{20} 5 = 20(5) = 100$$

This is done using the third property of summation notation.

Also note that

$$\sum_{n=1}^{20} (a_n + 5) = 181$$

is different from

$$\sum_{n=1}^{20} a_n + 5 = 81 + 5 = 86.$$

Solution Use the properties of summation notation.

a. $\sum_{n=1}^{20} 2b_n = 2\sum_{n=1}^{20} b_n = 2(50) = 100$

b. $\sum_{n=1}^{20} (2a_n + 3b_n) = 2\sum_{n=1}^{20} a_n + 3\sum_{n=1}^{20} b_n = 2(81) + 3(50) = 312$

c. $\sum_{n=1}^{20} (a_n + 5) = \sum_{n=1}^{20} a_n + \sum_{n=1}^{20} 5 = 81 + 20(5) = 181$

CHECK YOUR PROGRESS 5 Given $\sum_{n=1}^{15} a_n = 25$ and $\sum_{n=1}^{15} b_n = 17$, find $\sum_{n=1}^{15} (3a_n - 4b_n)$.

Solution *See page S38.* 7

Topics for Discussion

1. What is an infinite sequence?
2. What is the difference between a sequence and a series?
3. What is an alternating sequence? Give an example of an alternating sequence.
4. What are the first four terms of the sequence of partial sums for $1, \frac{1}{2}, \frac{1}{3}, \ldots, \frac{1}{n}$?
5. True or false: $\left(\sum_{n=1}^{10} a_n\right)^2 = \sum_{n=1}^{10} (a_n)^2$. Explain.

EXERCISES 7.1

— Suggested Assignment: Exercises 1–57, odd; 58; and 60–65, all.

In Exercises 1 to 20, find the first three terms and the eighth term of the sequence that has the given nth term.

1. $a_n = n$ $1, 2, 3, a_8 = 8$
2. $a_n = 2n + 1$ $3, 5, 7, a_8 = 17$
3. $a_n = \frac{1}{n}$ $1, \frac{1}{2}, \frac{1}{3}, a_8 = \frac{1}{8}$
4. $a_n = 3n$ $3, 6, 9, a_8 = 24$
5. $a_n = \frac{n}{n+1}$ $\frac{1}{2}, \frac{2}{3}, \frac{3}{4}, a_8 = \frac{8}{9}$
6. $a_n = \frac{1}{n^2}$ $1, \frac{1}{4}, \frac{1}{9}, a_8 = \frac{1}{64}$
7. $a_n = \frac{(-1)^{n-1}}{2n}$ $\frac{1}{2}, -\frac{1}{4}, \frac{1}{6}, a_8 = -\frac{1}{16}$
8. $a_n = \frac{(-1)^n}{2n-1}$ $-1, \frac{1}{3}, -\frac{1}{5}, a_8 = \frac{1}{15}$
9. $a_n = \left(\frac{1}{2}\right)^n$ $\frac{1}{2}, \frac{1}{4}, \frac{1}{8}, a_8 = \frac{1}{256}$
10. $a_n = \left(-\frac{3}{4}\right)^n$ $-\frac{3}{4}, \frac{9}{16}, -\frac{27}{64}, a_8 = \frac{6561}{65{,}536}$
11. $a_n = 1 + (-1)^n$ $0, 2, 0, a_8 = 2$
12. $a_n = 1 + (-0.1)^n$ $0.9, 1.01, 0.999, a_8 = 1.00000001$
13. $a_n = (1.06)^n$ $1.06, 1.1236, 1.191016, a_8 \approx 1.5938481$
14. $a_n = (1.1)^n$ $1.1, 1.21, 1.331, a_8 = 2.14358881$
15. $a_n = \left(1 + \frac{1}{n}\right)^n$ $2, \frac{9}{4}, \frac{64}{27}, a_8 = \frac{43{,}046{,}721}{16{,}777{,}216} \approx 2.56578$
16. $a_n = \frac{n^2}{2^n}$ $\frac{1}{2}, 1, \frac{9}{8}, a_8 = \frac{1}{4}$
17. $a_n = n!$ $1, 2, 6, a_8 = 40{,}320$
18. $a_n = \frac{n!}{(n-1)!}$ $1, 2, 3, a_8 = 8$
19. $a_n = \sqrt[n]{n}$ $1, \sqrt{2}, \sqrt[3]{3}, a_8 = \sqrt[8]{8}$
20. $a_n = \frac{(-1)^{n+1}}{3n}$ $\frac{1}{3}, -\frac{1}{6}, \frac{1}{9}, a_8 = -\frac{1}{24}$

In Exercises 21 to 30, find the first three terms of each recursively defined sequence.

21. $a_1 = 3, a_n = 4a_{n-1}$ $3, 12, 48$
22. $c_1 = 1, c_n = 2c_{n-1}$ $1, 2, 4$
23. $b_1 = 2, b_n = -nb_{n-1}$ $2, -4, 12$
24. $a_1 = 1, a_n = -2na_{n-1}$ $1, -4, 24$

25. $a_1 = 2, a_n = \dfrac{n}{a_{n-1}}$ 2, 1, 3

26. $b_1 = 2, b_n = \dfrac{1}{b_{n-1}}$ $2, \frac{1}{2}, 2$

27. $c_1 = 2, c_n = 2nc_{n-1}$ 2, 8, 48

28. $a_1 = 2, a_n = (-3)n^2 a_{n-1}$ 2, −24, 648

29. $a_1 = 3, a_n = (a_{n-1})^2$ 3, 9, 81

30. $b_1 = 2, b_n = (b_{n-1})^n$ 2, 4, 64

31. ***Lucas Sequence*** A *Lucas sequence* is similar to a Fibonacci sequence except that the first two numbers are 1 and 3. Thus $a_1 = 1, a_2 = 3$, and $a_n = a_{n-1} + a_{n-2}$ for $n \geq 3$. Find the next five terms of the Lucas sequence. 4, 7, 11, 18, 29

32. ***Fibonacci Sequence*** For the Fibonacci sequence, add the first two terms and record the result. Add the first three terms and record the result. Add the first four terms and record the result. Make a conjecture as to the sum of the first five terms of the Fibonacci sequence. Based on your conjecture, what is the sum of the first 10 terms of the Fibonacci sequence? $s_n = a_{n+2} - 1$; 143

In Exercises 33 to 44, evaluate the series.

33. $\sum_{i=1}^{6} i$ 21

34. $\sum_{i=1}^{4} i^2$ 30

35. $\sum_{i=1}^{5} i(i+1)$ 70

36. $\sum_{i=1}^{6} (2i-1)$ 36

37. $\sum_{k=1}^{4} \frac{1}{k}$ $\frac{25}{12}$

38. $\sum_{k=1}^{6} \frac{1}{k(k+1)}$ $\frac{6}{7}$

39. $\sum_{j=1}^{8} 2j$ 72

40. $\sum_{i=1}^{6} (2i+1)(2i-1)$ 358

41. $\sum_{i=3}^{5} (-1)^i 2^i$ −24

42. $\sum_{i=3}^{5} \frac{(-1)^i}{2^i}$ $-\frac{3}{32}$

43. $\sum_{k=0}^{8} \frac{8!}{k!\,(8-k)!}$ 256

44. $\sum_{k=0}^{7} \frac{7!}{k!\,(7-k)!}$ 128

45. Write the sum of the first six terms of the sequence whose nth term is $a_n = n$ in summation notation. $\sum_{n=1}^{6} n$

46. Write the sum of the first eight terms of the sequence whose nth term is $a_n = 3n$ in summation notation. $\sum_{n=1}^{8} 3n$

47. Write the sum of the first 10 terms of the sequence whose nth term is $a_n = \dfrac{1}{2^n}$ in summation notation. $\sum_{n=1}^{10} \frac{1}{2^n}$

48. Write the sum of the first seven terms of the sequence whose nth term is $a_n = \dfrac{1}{n^2}$ in summation notation. $\sum_{n=1}^{7} \frac{1}{n^2}$

49. Write the sum of the first eight terms of the sequence whose nth term is $a_n = \dfrac{(-1)^{n-1}}{n}$ in summation notation. $\sum_{n=1}^{8} \frac{(-1)^{n-1}}{n}$

50. Write the sum of the first nine terms of the sequence whose nth term is $a_n = \dfrac{(-1)^{n-1}}{n!}$ in summation notation. $\sum_{n=1}^{9} \frac{(-1)^{n-1}}{n!}$

In Exercises 51 to 54, let $\sum_{n=1}^{40} a_n = 20$ and $\sum_{n=1}^{40} b_n = 40$.

51. Evaluate $\sum_{n=1}^{40} -3a_n$. −60

52. Evaluate $\sum_{n=1}^{40} (a_n - 4b_n)$. −140

53. Evaluate $\sum_{n=1}^{40} (a_n - 4)$. −140

54. Evaluate $\sum_{n=1}^{40} a_n - 4$. 16

55. ***Newton's Method*** Newton's approximation of the square root of a number N is given by the recursive sequence defined by

$$a_1 = \frac{N}{2}, \qquad a_{n+1} = \frac{1}{2}\left(a_n + \frac{N}{a_n}\right)$$

Approximate $\sqrt{7}$ by computing a_4. Compare this result with the calculator value of $\sqrt{7} \approx 2.6457513$. 2.645752048

56. ***Newton's Method*** Use the formula in Exercise 55 to approximate $\sqrt{10}$ by computing a_5. 3.16227766

57. ***Fibonacci Sequence*** Every natural number greater than 1 can be written as a sum of numbers taken from the Fibonacci sequence, where no number is used more than once. For instance, $25 = 21 + 3 + 1$. Write 81 as a sum in this way. $55 + 21 + 5$

58. ***Fibonacci and Pythagoras*** Recall that the Pythagorean Theorem states that if a and b are the lengths of the legs of a right triangle and c is the length of the hypotenuse, then $a^2 + b^2 = c^2$. Choose any four consecutive numbers of the Fibonacci sequence and call them r, s, t, and u. Now let $a = ru$, $b = 2st$, and $c = s^2 + t^2$ and show that a, b, and c satisfy the Pythagorean Theorem.

59. ***Fibonacci Sequence*** The *Binet form* for the nth term of the Fibonacci sequence is given by

$$F_n = \frac{\left(\frac{1+\sqrt{5}}{2}\right)^n - \left(\frac{1-\sqrt{5}}{2}\right)^n}{\sqrt{5}}$$

Use this formula to find F_{10} and F_{15}. $F_{10} = 55, F_{15} = 610$

Prepare for Section 7.2

60. Evaluate $\frac{n(n+1)}{2}$ for $n = 50$. [P.1] 1275

61. Evaluate $a + (n - 1)d$ when $a = -5$, $n = 20$, and $d = 3$. [P.1] 52

62. If $a = -\frac{2}{3}$ and $b = \frac{1}{2}$, write $\frac{b}{a}$ in simplest form. [P.1] $-\frac{3}{4}$

63. Solve: $10 = 5 - 2(n - 1)$ [1.1] $-\frac{3}{2}$

64. Given $f(x) = 2 - 5x$, find $f(n + 1) - f(n)$. [2.1] -5

65. Given $f(x) = \left(-\frac{1}{2}\right)^x$, find $\frac{f(n+1)}{f(n)}$ in simplest form. [2.1] $-\frac{1}{2}$

Explorations 

1. ***The Golden Rectangle and the Fibonacci Sequence*** A *golden rectangle* is a rectangle whose length and width satisfy $\frac{L}{W} = \frac{L+W}{L}$. A golden rectangle is shown at the right.

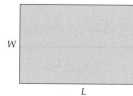

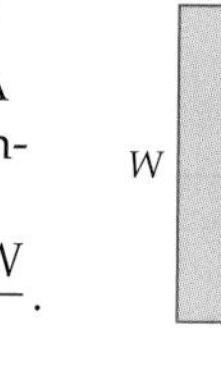

1. a. $\frac{1-\sqrt{5}}{2}, \frac{1+\sqrt{5}}{2}$

   a. We can rewrite $\frac{L}{W} = \frac{L+W}{L}$ as $\frac{L}{W} = \frac{L}{L} + \frac{W}{L} = 1 + \frac{W}{L}$. If we let $x = \frac{L}{W}$, we can substitute to get the equation $x = 1 + \frac{1}{x}$. Solve this equation for x. The positive solution of this equation is called the *golden ratio.*

   b. Consider the terms of the Fibonacci sequence 1, 1, 2, 3, 5, 8, 13, 21, ... Construct a new sequence from this one by constructing the ratios of successive terms: $\frac{1}{1}, \frac{2}{1}, \frac{3}{2}, \frac{5}{3}, \frac{8}{5}, \ldots$. Find the decimal equivalents of these terms and successive terms of this new sequence and explain how they relate to the golden ratio.
   1, 2, 1.5, 1.67, 1.6, 1.625, 1.615; The terms of the sequence approach the golden ratio.

   c. Some mathematics historians claim that the golden ratio was created by dividing a line segment into two unequal pieces such that the ratio of the longer piece to the entire line segment was equal to the ratio of the shorter piece to the longer piece. Let $x = \frac{v}{L}$. Then the equation at the right becomes $x = \frac{1}{x} - 1$. Solve this equation for x. Show that the positive solution of this equation is the reciprocal of the positive solution of the equation in part **a.** See the IRM for a complete solution.

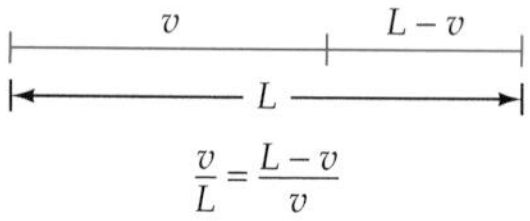

   (Figure: segment of length L divided into v and $L - v$; $\frac{v}{L} = \frac{L-v}{v}$)

   d. How are the solutions to the equation in part **a.** related to Binet's form of a Fibonacci number?
   The two solutions are part of the formula.

2. The golden rectangle is found in many works of art and architecture. Find examples of golden rectangles in Greek architecture and in the recent architecture of I. M. Pei.
   See the IRM for parts students may include in their papers.

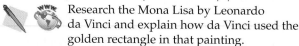

3.  Research the Mona Lisa by Leonardo da Vinci and explain how da Vinci used the golden rectangle in that painting.
   See the IRM for parts students may include in their papers.

SECTION 7.2 Arithmetic and Geometric Sequences and Series

▪ Arithmetic Sequences

Note that in the sequence

$$2, 5, 8, 11, 14, \ldots, 3n - 1, \ldots$$

the difference between successive terms is always 3. Such a sequence is called an *arithmetic sequence* or an *arithmetic progression.* These sequences have the following property: The difference between successive terms is the same constant. This constant is called the *common difference.* For the sequence above, the common difference is 3.

In general, an arithmetic sequence can be defined as follows.

> **Arithmetic Sequence**
>
> Let d be a real number. A sequence $\{a_n\}$ is an **arithmetic sequence** if $a_{i+1} - a_i = d$ for all i. The number d is the **common difference** for the sequence.

Further examples of arithmetic sequences include

$$2, 7, 12, \ldots, 5n - 3, \ldots \qquad \blacksquare\ d = 5$$

$$9, 3, -3, \ldots, -6n + 15, \ldots \qquad \blacksquare\ d = -6$$

$$1, 3, 5, \ldots, 2n - 1, \ldots \qquad \blacksquare\ d = 2$$

QUESTION What is the common difference for the arithmetic sequence $-3, 2, 7, 12, \ldots$?

Consider an arithmetic sequence in which the first term is a_1 and the common difference is d. By adding the common difference to each successive term of the arithmetic sequence, we can find a formula for the nth term.

$$a_1 = a_1$$

$$a_2 = a_1 + d$$

$$a_3 = a_2 + d = a_1 + d + d = a_1 + 2d \qquad \blacksquare\ a_2 = a_1 + d$$

$$a_4 = a_3 + d = a_1 + 2d + d = a_1 + 3d \qquad \blacksquare\ a_3 = a_1 + 2d$$

Observe the relationship between the term number and the coefficient of d. The coefficient is 1 less than the term number.

ANSWER The common difference is $2 - (-3) = 5$.

Formula for the *n*th Term of an Arithmetic Sequence

The nth term of an arithmetic sequence $\{a_n\}$ with common difference d is given by

$$a_n = a_1 + (n - 1)d$$

Alternative to Example 1

Find the thirteenth term of the arithmetic sequence whose first three terms are 10, 2, −6.

▪ −86

EXAMPLE 1 Find a Term of an Arithmetic Sequence

Find the 15th term of the arithmetic sequence whose first three terms are −5, −1, 3.

Solution Find the common difference: $d = a_2 - a_1 = -1 - (-5) = 4$. Use the formula $a_n = a_1 + (n - 1)d$ with $n = 15$.

$$a_{15} = -5 + (15 - 1)(4) = -5 + 14(4) = -5 + 56 = 51$$

The 15th term is 51.

CHECK YOUR PROGRESS 1 Find the 20th term of the arithmetic sequence whose first two terms are 5 and 2, respectively.

Solution *See page S38.* −52

Alternative to Example 2

The fifth term of an arithmetic sequence is −18 and the first term is 10. Find the ninth term.

▪ −53

EXAMPLE 2 Find a Term of an Arithmetic Sequence

The 10th term of an arithmetic sequence is 25 and the first term is 7. Find the 16th term.

Solution Solve the equation $a_n = a_1 + (n - 1)d$ for d given that $n = 10$, $a_1 = 7$, and $a_{10} = 25$.

$$\begin{aligned} a_n &= a_1 + (n - 1)d \\ 25 &= 7 + (10 - 1)d && \text{▪ } a_{10} = 25, a_1 = 7, n = 10 \\ 25 &= 7 + 9d \\ 18 &= 9d \\ 2 &= d \end{aligned}$$

Now find the 16th term.

$$\begin{aligned} a_n &= 7 + (n - 1)d \\ a_{16} &= 7 + (16 - 1)(2) = 7 + 15(2) = 7 + 30 = 37 \end{aligned}$$

The 16th term is 37.

CHECK YOUR PROGRESS 2 The fifth term of an arithmetic sequence is 24 and the 12th term is 59. Find the 20th term.

Solution *See page S38.* 99

Arithmetic Series

Consider the arithmetic sequence given by $1, 3, 5, \ldots, 2n - 1$. By adding successive terms of this sequence we can generate a sequence of partial sums. The sum of the terms of an arithmetic sequence is called an **arithmetic series.**

$$
\begin{aligned}
S_1 &= 1\\
S_2 &= 1 + 3 = 4\\
S_3 &= 1 + 3 + 5 = 9\\
S_4 &= 1 + 3 + 5 + 7 = 16\\
S_5 &= 1 + 3 + 5 + 7 + 9 = 25\\
&\vdots\\
S_n &= 1 + 3 + 5 + \cdots + (2n - 1) \stackrel{?}{=} n^2
\end{aligned}
$$

Point of Interest

Galileo (1564–1642), using the fact that the sum of the first n consecutive odd integers is n^2 (as is apparently true from the sums at the right), was able to show that objects of different weights fall at the same rate. By constructing inclines of various slopes, he measured the distances balls of different weights traveled in equal intervals of time. He concluded from his observations that the distance an object falls is proportional to the square of the time it falls and does not depend on its weight. Galileo's views were contrary to the prevailing views stated by Aristotle almost 1800 years earlier. As a result, Galileo lost his professorship at the University of Pisa.

The first five terms of the sequence of partial sums are 1, 4, 9, 16, 25. It appears from this example that the sum of the first n odd natural numbers is n^2. We can verify this conjecture using the following formula.

Formula for the nth Partial Sum of an Arithmetic Sequence

The **nth partial sum S_n of an arithmetic sequence** $\{a_n\}$ is

$$S_n = \frac{n}{2}(a_1 + a_n)$$

There is an alternative formula for the nth partial sum of an arithmetic sequence that is derived by replacing a_n by $a_n = a_1 + (n - 1)d$.

$$
\begin{aligned}
S_n &= \frac{n}{2}(a_1 + a_n)\\
&= \frac{n}{2}[a_1 + a_1 + (n - 1)d] = \frac{n}{2}[2a_1 + (n - 1)d]
\end{aligned}
$$

Alternative Formula for the nth Partial Sum of an Arithmetic Sequence

The **nth partial sum S_n of an arithmetic sequence** $\{a_n\}$ with common difference d is

$$S_n = \frac{n[2a_1 + (n - 1)d]}{2}$$

Alternative to Example 3

Find the sum of the first 30 terms of the arithmetic sequence whose first three terms are $-1, 3, 7$.

▪ 1710

EXAMPLE 3 Find the Sum of an Arithmetic Sequence

Find the sum of

a. the first 50 positive even integers.

b. the first 25 terms of the arithmetic sequence whose first three terms are 3, 7, 11.

Continued ➤

Solution Use the formula $S_n = \dfrac{n[2a_1 + (n - 1)d]}{2}$.

a. $S_{50} = \dfrac{50[2(2) + (50 - 1)2]}{2}$ ■ $n = 50, a_1 = 2, d = 2$

$= \dfrac{50[4 + 98]}{2} = 2550$

b. $S_{25} = \dfrac{25[2(3) + (25 - 1)4]}{2}$ ■ $n = 25, a_1 = 3, d = 4$

$= \dfrac{25[6 + 96]}{2} = 1275$

CHECK YOUR PROGRESS 3 Find the sum of the first 30 positive odd integers.

Solution *See page S38.* 900

The first n natural numbers 1, 2, 3, 4, . . . , n are part of an arithmetic sequence with a common difference of 1, $a_1 = 1$, and $a_n = n$. Substituting these values into $S_n = \dfrac{n(a_1 + a_n)}{2}$, we can derive a formula for the sum of the first n natural numbers.

Sum of the First *n* Natural Numbers

$$S_n = \frac{n(n + 1)}{2}$$

To find the sum of the first 85 natural numbers, use $n = 85$.

$$S_{85} = \frac{85(85 + 1)}{2} = 3655$$

Alternative to Example 4
Exercise 90, page 603

One application of arithmetic sequences and series is to the construction of a spider web. Using arithmetic sequences, a biologist can estimate the amount of "silk" the spider used to create the web.

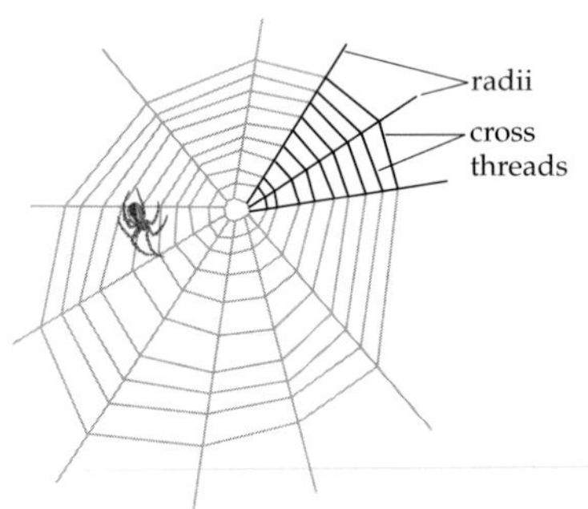

EXAMPLE 4 Solve an Application

The cross threads in a spider web are approximately equally spaced and the lengths of the cross threads form an arithmetic sequence. In a certain spider web, the length of the shortest cross thread is 4.6 millimeters. The length of the longest cross thread is 52.6 millimeters, and the difference in length between successive cross threads is 1.5 millimeters. How many cross threads are between two successive radii?

Continued ➤

Solution To find the number of cross threads, solve $a_n = a_1 + (n - 1)d$ for n.

$$a_n = a_1 + (n - 1)d$$
$$52.6 = 4.6 + (n - 1)1.5 \quad \text{■ } a_n = 52.6, a_1 = 4.6, d = 1.5$$
$$52.6 = 4.6 + 1.5n - 1.5 \quad \text{■ Solve for } n.$$
$$52.6 = 3.1 + 1.5n$$
$$49.5 = 1.5n$$
$$33 = n$$

There are 33 cross threads between two successive radii.

CHECK YOUR PROGRESS 4 Using the information in Example 4, determine the total length of all of the cross threads between two successive radii.

Solution *See page S38.* 943.8 mm

■ Geometric Sequences

An arithmetic sequence is characterized by a *common difference* between successive terms. A *geometric sequence* is characterized by a **common ratio** of successive terms.

The sequence 6, 18, 54, 162, 486, 1458, . . . is a geometric sequence. Note that the ratio of any two successive terms is 3.

$$\frac{18}{6} = 3 \qquad \frac{54}{18} = 3 \qquad \frac{162}{54} = 3 \qquad \frac{486}{162} = 3 \qquad \frac{1458}{486} = 3$$

? QUESTION What is the common ratio for the geometric sequence $2, -\frac{1}{2}, \frac{1}{8}, \ldots$?

Geometric Sequence

Let r be a nonzero constant real number. A sequence $\{a_n\}$ is a **geometric sequence** if $\frac{a_{i+1}}{a_i} = r$ for all positive integers i.

Alternative to Example 5
Which of the following are geometric sequences?
a. $1, -\frac{1}{2}, \frac{1}{4}, -\frac{1}{8}, \ldots, \left(-\frac{1}{2}\right)^{n-1}$
b. $1, 2, 6, 24 \ldots n(n - 1)$
■ a.

EXAMPLE 5 Identify a Geometric Sequence

Which of the following are geometric sequences?

a. $1, -2, 4, -8, \ldots, (-2)^{n-1}, \ldots$ **b.** $1, 4, 9, \ldots, n^2, \ldots$

Continued ➤

? ANSWER The common ratio is the ratio of two successive terms. Thus $r = \frac{-\frac{1}{2}}{2} = -\frac{1}{4}$.

TAKE NOTE

In Example 5, we must use the variable term for our calculations because the pattern of the given terms may change. For instance, there is a sequence called the RATS sequence in which the first few terms are 1, 2, 4, 8, 16. From these terms it might appear that this is a geometric sequence with a common ratio of 2. However, as shown in Exercise 75, page 602, the next two terms are 77 and 145. Thus the RATS sequence is not a geometric sequence.

Solution To determine whether the sequence is a geometric sequence, calculate the ratio of successive terms.

a. $\dfrac{a_{i+1}}{a_i} = \dfrac{(-2)^i}{(-2)^{i-1}} = (-2)^{i-(i-1)} = -2$

- $a_{i+1} = (-2)^{i+1-1} = (-2)^i$
- $a_i = (-2)^{i-1}$

The ratio of successive terms is the constant -2. The sequence is a geometric sequence.

b. $\dfrac{a_{i+1}}{a_i} = \dfrac{(i+1)^2}{i^2} = \left(1 + \dfrac{1}{i}\right)^2$

The ratio of successive terms is a variable expression. The sequence is not a geometric sequence.

CHECK YOUR PROGRESS 5 Is the sequence $1, 3, 9, \ldots, 3^{n-1}, \ldots$ a geometric sequence?

Solution *See page S39.* Yes

Consider a geometric sequence in which the first term is a_1 and the common ratio is r. By multiplying each successive term of the geometric sequence by the common ratio, we can derive a formula for the nth term.

$$a_1 = a_1$$
$$a_2 = a_1r$$
$$a_3 = a_2r = (a_1r)r = a_1r^2 \qquad \blacksquare\ a_2 = a_1r$$
$$a_4 = a_3r = (a_1r^2)r = a_1r^3 \qquad \blacksquare\ a_3 = a_1r^2$$

Observe the relationship between the number of the term and the number that is the exponent on r. The exponent on r is 1 less than the number of the term. Using this observation, we can write a formula for the nth term of a geometric sequence.

Formula for the nth Term of a Geometric Sequence

The **nth term of a geometric sequence** $\{a_n\}$ with first term a_1 and common ratio r is

$$a_n = a_1r^{n-1}$$

Alternative to Example 6

Find the formula for the nth term of the geometric sequence whose first three terms are $3, \frac{9}{4}, \frac{27}{16}$.

- $a_n = 3\left(\dfrac{3}{4}\right)^{n-1}$

EXAMPLE 6 Find the nth Term of a Geometric Sequence

Find the formula for the nth term of the geometric sequence whose first three terms are:

a. $4, \dfrac{8}{3}, \dfrac{16}{9}, \ldots$

b. $5, -10, 20, \ldots$

Continued ➤

Solution

a. $r = \dfrac{\frac{8}{3}}{4} = \dfrac{2}{3}$ and $a_1 = 4$. Thus

$$a_n = a_1 r^{n-1}$$

$$a_n = 4\left(\frac{2}{3}\right)^{n-1}$$

b. $r = \dfrac{-10}{5} = -2$ and $a_1 = 5$. Thus

$$a_n = a_1 r^{n-1}$$

$$a_n = 5(-2)^{n-1}$$

CHECK YOUR PROGRESS 6 Find the formula for the nth term of the geometric sequence whose first three terms are $1, -\frac{2}{3}, \frac{4}{9}$.

Solution *See page S39.* $a_n = \left(-\frac{2}{3}\right)^{n-1}$

Point of Interest

Suppose someone offers you a one-month job for which your salary will be 1 cent on the first day, 2 cents on the second day, 4 cents on the third day, and so on (the amount of money is doubled on each successive day of the month). Do you think this is a good salary?

Note that the amount you are making each day is given by the geometric sequence 1, 2, 4, 8, . . . , which is discussed at the right. The amount you would earn for 30 days is the sum of the first 30 terms of this geometric sequence.

Using the formula at the right, we have

$$S_{30} = \frac{1(1-2^{30})}{1-2}$$

$$\approx \$10.7 \text{ million}$$

Quite a salary for 1 month of work!

▪ Geometric Series

Consider the geometric sequence $1, 2, 4, 8, \ldots, 2^{n-1}, \ldots$. Some of the terms of the sequence of partial sums are given below.

$$S_1 = 1 = 2^1 - 1$$

$$S_2 = 1 + 2 = 3 = 2^2 - 1$$

$$S_3 = 1 + 2 + 4 = 7 = 2^3 - 1$$

$$S_4 = 1 + 2 + 4 + 8 = 15 = 2^4 - 1$$

$$\vdots$$

$$S_n = 1 + 2 + 4 + 8 + \cdots + 2^{n-1} \stackrel{?}{=} 2^n - 1$$

The first four terms of the sequence of partial sums are 1, 3, 7, 15. It appears from this example that the sum of the first n terms of $a_n = 2^{n-1}$ is $2^n - 1$. We can verify this conjecture using the following formula.

Formula for the nth Partial Sum of a Geometric Sequence

The **nth partial sum of a geometric sequence** with first term a_1 and common ratio r is

$$S_n = \frac{a_1(1-r^n)}{1-r}, \qquad r \neq 1$$

Alternative to Example 7

Evaluate: $\sum_{n=1}^{8} (-1)^n 2^{n-1}$

▪ 85

EXAMPLE 7 Find the Sum of a Geometric Sequence

a. Find S_6 for the sequence $5, 15, 45, \ldots, 5(3^{n-1}), \ldots$.

b. Evaluate: $\displaystyle\sum_{n=1}^{10} \frac{(-1)^{n-1}}{4^n}$

Continued ➤

Point of Interest

There is a type of illegal marketing plan called a *pyramid scheme* that is related to geometric sequences and series. One person, the initiator, enlists, say, seven people to invest $100 each. The initiator receives $700. These seven people (the first level) each recruit seven more people who invest $100 each. Thus there are 7^2 or 49 people on the second level. Each person on the second level receives $700 but pays 10% of the $700 to the initiator. The process is then repeated again. At the third level, the initiator would receive over $16,000. By the tenth level, 7^{10} people (more than the population of the U.S.) are required to support the pyramid.

Solution

a. $$S_n = \frac{a_1(1-r^n)}{1-r}$$

$$S_6 = \frac{5[1-3^6]}{1-3} \qquad \blacksquare\ a_1 = 5, r = 3, n = 6$$

$$= \frac{5(-728)}{-2} = 1820$$

b. When a series is given in summation form, use the expression within the summation to find the first two terms of the series. Find a_1 and a_2.

$$a_n = \frac{(-1)^{n-1}}{4^n}$$

$$a_1 = \frac{(-1)^{1-1}}{4^1} = \frac{(-1)^0}{4} = \frac{1}{4} \qquad \blacksquare\ n = 1$$

$$a_2 = \frac{(-1)^{2-1}}{4^2} = \frac{(-1)^1}{16} = -\frac{1}{16} \qquad \blacksquare\ n = 2$$

Use a_1 and a_2 to find r, the common ratio.

$$r = \frac{a_2}{a_1} = \frac{-\frac{1}{16}}{\frac{1}{4}} = -\frac{1}{4}$$

Thus

$$S_n = \frac{a_1(1-r^n)}{1-r}$$

$$S_{10} = \frac{\left(\frac{1}{4}\right)\left[1-\left(-\frac{1}{4}\right)^{10}\right]}{1-\left(-\frac{1}{4}\right)} \qquad \blacksquare\ a_1 = \frac{1}{4}, r = -\frac{1}{4}, n = 10$$

$$= 0.2$$

CHECK YOUR PROGRESS 7 Evaluate: $\sum_{n=1}^{17} 3\left(\frac{3}{4}\right)^{n-1}$

Solution *See page S39.* ≈ 11.9098

▪ Infinite Geometric Series

The following are two examples of geometric sequences for which $|r| < 1$.

$$9, 3, 1, \frac{1}{3}, \frac{1}{9}, \frac{1}{27}, \ldots, 9\left(\frac{1}{3}\right)^{n-1}, \ldots \qquad \blacksquare\ r = \frac{1}{3}$$

$$2, -1, \frac{1}{2}, -\frac{1}{4}, \frac{1}{8}, -\frac{1}{16}, \frac{1}{32}, \ldots, 2\left(-\frac{1}{2}\right)^{n-1}, \ldots \qquad \blacksquare\ r = -\frac{1}{2}$$

Observe that when the absolute value of the common ratio of a geometric sequence is less than 1, the terms of the geometric sequence become closer and closer to zero as n increases. For instance, for the geometric sequence

$$9, 3, 1, \frac{1}{3}, \frac{1}{9}, \frac{1}{27}, \ldots, 9\left(\frac{1}{3}\right)^{n-1}, \ldots$$

the first term is 9, the next term is 3 (closer to zero), then 1 (closer to zero), then $\frac{1}{3}$ (closer to zero), and so on. Each successive term of the sequence is closer to zero than the preceding term.

? QUESTION What is the value of the 20th term of the sequence given above whose nth term is $9\left(\frac{1}{3}\right)^{n-1}$?

Table 7.1

n	S_n	r^n
4	13.3333333	0.0123457
8	13.4979424	0.00015242
12	13.4999746	0.00000188
16	13.4999997	0.00000002
↓	↓	↓
∞	13.5	0.0

The partial sums

$$\sum_{i=1}^{n} 9\left(\frac{1}{3}\right)^{i-1}$$

of the above sequence for $n = 4, 8, 12$, and 16 are given in Table 7.1, along with the corresponding values of r^n. As n increases, S_n gets closer to 13.5 and r^n gets closer to zero. By finding additional values of S_n for larger values of n, we would find that S_n becomes closer and closer to 13.5 as n increases without bound. The sum of *all* the terms of the sequence is called an **infinite series.** If the sequence is a geometric sequence, the sum is called an **infinite geometric series.**

Sum of an Infinite Geometric Series

If $\{a_n\}$ is a geometric sequence with $|r| < 1$ and first term a_1, then the sum of the infinite geometric series is

$$S = \frac{a_1}{1 - r}$$

An infinite series is represented by $\sum_{n=1}^{\infty} a_n$.

Alternative to Example 8

Evaluate the infinite geometric series $\sum_{n=1}^{\infty} \left(-\frac{2}{3}\right)^{n-1}$.

■ $\frac{3}{5}$

EXAMPLE 8 Find the Sum of an Infinite Geometric Series

Evaluate the infinite geometric series $\sum_{n=1}^{\infty} \left(-\frac{3}{4}\right)^{n-1}$.

Continued ➤

? ANSWER $a_{20} = 9\left(\frac{1}{3}\right)^{20-1} \approx 7.74 \times 10^{-9}$

TAKE NOTE

The formula $S = \dfrac{a_1}{1 - r}$ can be used only for *infinite* geometric series for which $|r| < 1$.

Solution Use the formula $S = \dfrac{a_1}{1 - r}$. To find the first term a_1, let $n = 1$. Then $a_1 = \left(-\frac{3}{4}\right)^{1-1} = \left(-\frac{3}{4}\right)^0 = 1$. The common ratio is $r = -\frac{3}{4}$. Thus

$$S = \frac{a_1}{1 - r}$$
$$= \frac{1}{1 - \left(-\dfrac{3}{4}\right)} = \frac{1}{\dfrac{7}{4}} = \frac{4}{7}$$

The sum of the series is $\frac{4}{7}$.

CHECK YOUR PROGRESS 8 Evaluate the infinite geometric series $\sum_{n=1}^{\infty} 2(0.3)^{n-1}$.

Solution *See page S39.* $\dfrac{20}{7}$

Applications

One application of infinite geometric series concerns repeating decimals. Consider the repeating decimal

$$0.6666\ldots = \frac{6}{10} + \frac{6}{100} + \frac{6}{1000} + \frac{6}{10{,}000} + \cdots$$

The right-hand side of the equation is a geometric series with $a_1 = \frac{6}{10}$ and common ratio $r = \frac{1}{10}$. Thus the sum of the infinite geometric series is given by

$$S = \frac{\dfrac{6}{10}}{1 - \left(\dfrac{1}{10}\right)} = \frac{\dfrac{6}{10}}{\dfrac{9}{10}} = \frac{2}{3}$$

The repeating decimal $0.\overline{6} = \frac{2}{3}$, as you may have already known.

We can write any repeating decimal as a ratio of two integers by using the formula for the sum of an infinite geometric series.

QUESTION The repeating decimal $0.\overline{1}$ can be written as the infinite geometric series $0.\overline{1} = \frac{1}{10} + \frac{1}{100} + \frac{1}{1000} + \ldots$. What is the common ratio for this series?

ANSWER $r = \dfrac{\dfrac{1}{100}}{\dfrac{1}{10}} = \dfrac{10}{100} = \dfrac{1}{10}$

Alternative to Example 9

Write 0.23434 . . . as the ratio of two integers in lowest terms.

■ $\frac{116}{495}$

EXAMPLE 9 Write a Decimal as a Fraction

Write 0.34545 . . . as the ratio of two integers in lowest terms.

Solution

$$0.3\overline{45} = \frac{3}{10} + \left[\frac{45}{1000} + \frac{45}{100{,}000} + \frac{45}{10{,}000{,}000} + \ldots\right]$$

The terms in the brackets form an infinite geometric series. The common ratio is $r = \dfrac{\frac{45}{100{,}000}}{\frac{45}{1000}} = \dfrac{1}{100}$. Evaluate the series using $a_1 = \frac{45}{1000}$ and $r = \frac{1}{100}$, and then add the term $\frac{3}{10}$.

$$\frac{45}{1000} + \frac{45}{100{,}000} + \frac{45}{10{,}000{,}000} + \cdots = \frac{\frac{45}{1000}}{1 - \left(\frac{1}{100}\right)} = \frac{1}{22}$$

Thus $0.3\overline{45} = \frac{3}{10} + \frac{1}{22} = \frac{19}{55}$.

CHECK YOUR PROGRESS 9 Write 1.090909 . . . as the ratio of two integers in lowest terms.

Solution *See page S39.* $\frac{12}{11}$

TAKE NOTE

Remember that the sum of an infinite geometric series is $\frac{a_1}{1 - r}$ provided $|r| < 1$. For the infinite series at the right, $a_1 = \frac{D(1 + g)}{1 + i}$ and $r = \frac{1 + g}{1 + i}$. Therefore,

$$\frac{a_1}{1 - r} = \frac{\frac{D(1 + g)}{1 + i}}{1 - \left(\frac{1 + g}{1 + i}\right)} = \frac{D(1 + g)}{i - g}$$

There are many applications of infinite geometric series to the area of finance. Two such applications are **stock valuation** and the **multiplier effect.**

The Gordon model, named after Myron Gordon, is used to determine the value of a stock whose dividend is expected to increase by the same percent each year. The value of the stock is given by

$$\text{Stock value} = \sum_{n=1}^{\infty} D\left(\frac{1 + g}{1 + i}\right)^n = \frac{D(1 + g)}{i - g}, \quad g < i$$

where D is the dividend of the stock when it is purchased, g is the expected growth rate of the dividend, and i is the growth rate the investor requires. An example of stock valuation is given in Example 10.

Alternative to Example 10

Exercise 82, page 603

EXAMPLE 10 Find the Value of a Stock

Suppose a stock is paying a dividend of $1.50 and it is estimated that the dividend will increase 10% per year. An investor requires a 15% return on an investment. Using the Gordon model, determine the price the investor should pay for the stock.

Continued ➤

Solution Stock value $= \dfrac{D(1+g)}{i-g}$

$= \dfrac{1.50(1+0.10)}{0.15-0.10}$ ▪ $D = 1.50, g = 0.10, i = 0.15$

$= 33$

The investor should pay \$33 for the stock.

CHECK YOUR PROGRESS 10 If the dividend of a stock does not grow but remains constant, then the formula for the stock value is

$$\text{Stock value} = \sum_{n=1}^{\infty} \frac{D}{(1+i)^n} = \frac{D}{i}$$

where D is the dividend and i is the investor's required rate of return. Find the value of a stock whose dividend is \$2.33 and for which an investor requires a 20% rate of return.

Solution *See page S39.* \$11.65

Point of Interest

The amount that consumers will spend is referred to by economists as the *marginal propensity to consume.* For the example at the right, the marginal propensity to consume is 75%.

The *multiplier effect,* mentioned on the preceding page, can occur in various situations. Next we will examine this effect when applied to a reduction in income taxes.

Suppose that the federal government enacts a tax reduction of \$5 billion. Suppose also that an economist estimates that each person receiving a share of this reduction will spend 75% and save 25%. That means that 75% of \$5 billion, or \$3.75 billion ($0.75 \cdot 5 = 3.75$), will be spent. The amount that is spent is income for other people, who in turn spend 75% of that income, or \$2.8125 billion ($0.75 \cdot 3.75 = 2.8125$). The \$2.8125 billion then becomes income for other people, and the process is repeated. This is shown in the table below.

Amount available to spend	Multiply by 75%	New amount available to spend
\$5 billion	$0.75(5) = 3.75$	\$3.75 billion
\$3.75 billion	$0.75(3.75) = 0.75[0.75(5)]$ $= (0.75)^2 5$ $= 2.8125$	\$2.8125 billion
\$2.8125 billion	$0.75(2.8125) = 0.75[(0.75)^2 5]$ $= (0.75)^3 5$ $= 2.109375$	\$2.109375 billion
\$2.109375 billion	$0.75(2.109375) = 0.75[(0.75)^3 5]$ $= (0.75)^4 5$ $= 1.58203125$	\$1.58203125 billion

Observe that the terms in the middle column form a geometric sequence. The net effect of all the spending is found by summing an infinite geometric series.

$$5 + 5(0.75) + 5(0.75)^2 + \cdots + 5(0.75)^{n-1} + \cdots = \sum_{n=1}^{\infty} 5(0.75)^{n-1} = \frac{5}{1-0.75} = 20$$

This means that the original tax cut of $5 billion results in actual spending of $20 billion. Some economists believe this is good for economic growth; other see it as contributing to inflation and therefore undesirable.

Topics for Discussion

1. What differentiates an arithmetic sequence from any other type of sequence?
2. Give two examples of arithmetic sequences.
3. What differentiates a geometric sequence from any other type of sequence?
4. Give two examples of geometric sequences.
5. Is the sum of an infinite arithmetic series ever a finite number? If so, under what conditions?
6. Is the sum of an infinite geometric series ever a finite number? If so, under what conditions?

EXERCISES 7.2

— Suggested Assignment: Exercises 1–77, every other odd; 79–103, odd; and 105–110.

In Exercises 1 to 8, find the 10th, 25th, and *n*th terms of the arithmetic sequence.

1. $1, 3, 5, \ldots$ $a_{10} = 19$, $a_{25} = 49$, $a_n = 2n - 1$

2. $2, 4, 6, \ldots$ $a_{10} = 20$, $a_{25} = 50$, $a_n = 2n$

3. $10, 9, 8, \ldots$ $a_{10} = 1$, $a_{25} = -14$, $a_n = 11 - n$

4. $7, 5, 3, \ldots$ $a_{10} = -11$, $a_{25} = -41$, $a_n = 9 - 2n$

5. $-6, -3, 0, \ldots$ $a_{10} = 21$, $a_{25} = 66$, $a_n = 3n - 9$

6. $-9, -5, -1, \ldots$ $a_{10} = 27$, $a_{25} = 87$, $a_n = 4n - 13$

7. $1, 6, 11, \ldots$ $a_{10} = 46$, $a_{25} = 121$, $a_n = 5n - 4$

8. $3, 7, 11, \ldots$ $a_{10} = 39$, $a_{25} = 99$, $a_n = 4n - 1$

9. The second and third terms of an arithmetic sequence are 6 and 10. Find the 20th term. $a_{20} = 78$

10. The sixth and eighth terms of an arithmetic sequence are 7 and 17. Find the 15th term. $a_{15} = 52$

11. The fourth and seventh terms of an arithmetic sequence are 9 and 27. Find the 17th term. $a_{17} = 87$

12. The fourth and eighth terms of an arithmetic sequence are −1 and 19. Find the 23rd term. $a_{23} = 94$

In Exercises 13 to 20, find the *n*th partial sum of the arithmetic sequence.

13. $a_n = 2n - 1; n = 10$ 100

14. $a_n = 2n; n = 12$ 156

15. $a_n = 3n + 1; n = 15$ 375

16. $a_n = 5n - 2; n = 20$ 1010

17. $a_n = 5n; n = 12$ 390

18. $a_n = 10n; n = 14$ 1050

19. $a_n = 4 - 2n; n = 30$ −810

20. $a_n = 5 - 3n; n = 40$ −2260

In Exercises 21 to 30, determine if the sequence is arithmetic, geometric, or neither. For arithmetic sequences, find the common difference; for geometric sequences, find the common ratio.

21. $3, 9, 27, \ldots, 3^n, \ldots$ Geometric, $r = 3$

22. $1, 4, 16, \ldots, 4^{n-1}, \ldots$ Geometric, $r = 4$

23. $1, \frac{1}{2}, \frac{1}{3}, \ldots, \frac{1}{n}, \ldots$ Neither

24. $\frac{1}{2}, \frac{1}{4}, \frac{1}{8}, \ldots, \frac{1}{2^n}, \ldots$ Geometric, $r = \frac{1}{2}$

25. $1, 5, 9, \ldots, 4n - 3, \ldots$ Arithmetic, $d = 4$

26. $9, 3, -3, \ldots, 15 - 6n, \ldots$ Arithmetic, $d = -6$

27. $1, 4, 9, 16, 25, \ldots, n^2, \ldots$ Neither

28. $5, 10, 20, \ldots, 5(2^{n-1}), \ldots$ Geometric, $r = 2$

29. $2, -\frac{2}{3}, \frac{2}{9}, \ldots, 2\left(-\frac{1}{3}\right)^{n-1}, \ldots$ Geometric, $r = -\frac{1}{3}$

30. $-1, 0.4, -0.16, \ldots, (-1)^n(0.4)^{n-1}, \ldots$ Geometric, $r = -0.4$

In Exercises 31 to 44, find the *n*th term of the geometric sequence.

31. $3, 12, 48, \ldots$ $3(4)^{n-1}$

32. $2, 12, 72, \ldots$ $2(6)^{n-1}$

33. $2, -6, 18, \ldots$ $2(-3)^{n-1}$

34. $4, -8, 16, \ldots$ $4(-2)^{n-1}$

35. $1, \frac{3}{4}, \frac{9}{16}, \ldots$ $\left(\frac{3}{4}\right)^{n-1}$

36. $2, \frac{1}{2}, \frac{1}{8}, \ldots$ $2\left(\frac{1}{4}\right)^{n-1}$

37. $-9, 6, -4, \ldots$ $-9\left(-\frac{2}{3}\right)^{n-1}$

38. $-3, 5, -\frac{25}{3}, \ldots$ $-3\left(-\frac{5}{3}\right)^{n-1}$

39. $\frac{3}{10}, \frac{3}{100}, \frac{3}{1000}, \ldots$ $\frac{3}{10}\left(\frac{1}{10}\right)^{n-1}$

40. $2, \frac{1}{5}, \frac{1}{50}, \ldots$ $2\left(\frac{1}{10}\right)^{n-1}$

41. $0.6, 0.06, 0.006, \ldots$ $0.6(0.1)^{n-1}$

42. $0.12, 0.0012, 0.000012, \ldots$ $0.12(0.01)^{n-1}$

43. $0.25, 0.0025, 0.000025, \ldots$ $0.25(0.01)^{n-1}$

44. $0.5, 0.005, 0.00005, \ldots$ $0.5(0.01)^{n-1}$

45. Find the third term of the geometric sequence whose first term is 2 and whose fifth term is 162. $a_3 = 18$

46. Find the fourth term of the geometric sequence whose third term is 1 and whose eighth term is $\frac{1}{32}$. $a_4 = \frac{1}{2}$

47. Find the second term of the geometric sequence whose third term is $\frac{4}{3}$ and whose sixth term is $-\frac{32}{81}$. $a_2 = -2$

48. Find the fifth term of the geometric sequence whose fourth term is $\frac{8}{9}$ and whose seventh term is $\frac{64}{243}$. $a_5 = \frac{16}{27}$

In Exercises 49 to 56, find the sum of the geometric series.

49. $\sum_{n=1}^{5} 2^n$ 62

50. $\sum_{n=1}^{7} 3^{n-1}$ 1093

51. $\sum_{n=1}^{6} \left(\frac{2}{5}\right)^n$ $\frac{10,374}{15,625}$

52. $\sum_{n=1}^{14} \left(\frac{2}{3}\right)^n$ $\frac{9,533,170}{4,782,969}$

53. $\sum_{n=0}^{8} \left(-\frac{1}{2}\right)^n$ $\frac{171}{256}$

54. $\sum_{n=0}^{7} \left(-\frac{2}{3}\right)^n$ $\frac{1261}{2187}$

55. $\sum_{n=0}^{9} 5(3)^n$ 147,620

56. $\sum_{n=0}^{7} 2(5)^n$ 195,312

In Exercises 57 to 64, find the sum of the infinite geometric series.

57. $\sum_{n=1}^{\infty} \left(\frac{3}{4}\right)^n$ 3

58. $\sum_{n=1}^{\infty} \left(\frac{2}{5}\right)^n$ $\frac{2}{3}$

59. $\sum_{n=1}^{\infty} \left(-\frac{1}{3}\right)^n$ $-\frac{1}{4}$

60. $\sum_{n=1}^{\infty} \left(-\frac{3}{5}\right)^n$ $-\frac{3}{8}$

61. $\sum_{n=1}^{\infty} (0.2)^n$ 0.25

62. $\sum_{n=1}^{\infty} (0.4)^n$ $\frac{2}{3}$

63. $\sum_{n=0}^{\infty} (-0.3)^n$ $\frac{10}{13}$

64. $\sum_{n=0}^{\infty} (-0.7)^n$ $\frac{10}{17}$

In Exercises 65 to 74, write each rational number as the quotient of two integers in simplest form.

65. $0.\overline{2}$ $\frac{2}{9}$

66. $0.\overline{3}$ $\frac{1}{3}$

67. $0.\overline{45}$ $\frac{5}{11}$

68. $0.\overline{25}$ $\frac{25}{99}$

69. $0.1\overline{6}$ $\frac{1}{6}$

70. $0.8\overline{3}$ $\frac{5}{6}$

71. $0.\overline{123}$ $\frac{41}{333}$

72. $0.\overline{321}$ $\frac{107}{333}$

73. $0.25\overline{4}$ $\frac{229}{900}$

74. $0.37\overline{2}$ $\frac{67}{180}$

75. ***RATS Sequence*** RATS stands for Reverse, Add, Then Sort. The first few terms of this sequence are 1, 2, 4, 8, 16, 77, 145, 668.... The term 668 is derived from the preceding term, 145, by reversing the digits to get 541, adding $145 + 541 = 686$, and then sorting the digits from smallest to largest to get 668. Find the next two terms of the RATS sequence. 1345, 6677

76. If $f(x) = mx + b$, show that $f(n)$, where n is a natural number, is an arithmetic sequence.

77. If $f(x) = b^x$, show that $f(n)$, where n is a natural number, is a geometric sequence.

78. Suppose $\{a_n\}$ is an arithmetic sequence. What type of sequence is $b_n = 2^{a_n}$? Geometric

Business and Economics

79. ***Contest Prizes*** A contest offers 15 prizes. The first prize is \$5000, and each successive prize is \$250 less than the preceding prize. What is the value of the 15th prize? What is the total amount of money distributed in prizes? \$1500; \$48,750

80. ***Prosperity Club*** In 1935, the "Prosperity Club" chain letter was started. A letter was sent to five people, who were asked to send 10 cents to the name at the top of the list and then remove that person's name. Each of the five people then added his or her name to the bottom of the list and sent the letter to five friends. Assuming no one broke the chain, how much money would each of the five original letter recipients receive? \$312.50

81. ***Prosperity Club*** The population of the U.S. in 1935 was approximately 127,000,000 people. Assuming that no one broke the chain in the "Prosperity Club" chain letter (see Exercise 80) and that no one received more than one letter, how many levels would it take before the entire population received a letter? 12 levels

82. ***Gordon Model of Stock Valuation*** Use the Gordon model for the value of a stock to determine how much the manager of a mutual fund should pay for a stock whose dividend is \$1.87 and whose dividend growth rate is 15% if the manager requires a 20% rate of return on the investment. \$43.01

83. ***Gordon Model of Stock Valuation*** Suppose Myna Alton purchases a stock that pays a dividend of \$1.32 for \$67 per share. If Myna requires a 20% return on an investment, use the Gordon model of stock evaluation to determine the dividend growth rate Myna expects. 17.68%

84. ***Stock Valuation*** Suppose a stock is paying a constant dividend of \$2.94 and an investor wants to receive a 15% return on an investment. How much should the investor pay for the stock? \$19.60

85. ***Stock Valuation*** Suppose a stock is paying a constant dividend of \$3.24 and an investor pays \$16 for one share of the stock. What rate of return does the investor expect? 20.25%

86. ***Stock Valuation*** Explain why g must be less than i in the Gordon stock evaluation model. (See Example 10, page 599.) Otherwise, the common ratio would be greater than 1 and the infinite geometric series would not have a finite sum.

87. ***Multiplier Effect*** Sometimes a city will argue that having a professional sports franchise in the city will contribute to economic growth. The rationale for this statement is based on the multiplier effect. Suppose a city estimates that a professional sports franchise will bring in \$50 million of additional income and that a person receiving a portion of this money will spend 90% and save 10%. Assuming the multiplier effect model is accurate, what is the net effect of the \$50 million? (See the discussion following Check Your Progress 10 of this section.) \$500 million

88. ***Multiplier Effect*** The net effect of tourism on a city is related to the multiplier effect. Suppose a city estimates that a new convention facility will bring in an additional \$25 million of income. If a person receiving a portion of this money will spend 75% and save 25%, what is the net effect of the \$25 million in additional income? (See the discussion following Check Your Progress 10 of this section.) \$100 million

89. ***Counterfeit Money Circulation*** Suppose there is \$500,000 of counterfeit money currently in circulation and each time the counterfeit money is used, 40% of it is detected and removed from circulation. How much counterfeit money is used in transactions before it is all removed from circulation? This is another application of the multiplier effect. \$833,333.33

Life and Health Sciences

90. ***Exercise Program*** An exercise program calls for walking 15 minutes each day for a week. Each week thereafter, the amount of time spent walking increases by 5 minutes per day. In how many weeks will a person be walking 60 minutes each day? 10 weeks

91. ***Spider Web*** The cross threads in a spider web are approximately equally spaced, and the lengths of the cross threads form an arithmetic sequence. In a certain spider web, the length of the shortest cross thread is 3.5 millimeters. The length of the longest cross thread is 46.9 millimeters, and the difference in length between successive cross threads is 1.4 millimeters. How many cross threads are between two successive radii? 32 cross threads

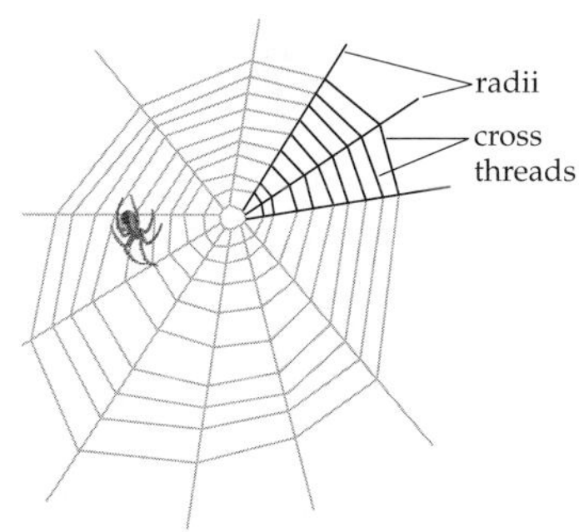

92. ***Spider Web*** What is the sum of the lengths of the cross threads between the two successive radii in Exercise 91? 806.4 mm

93. ***Medicine*** The amount of an antibiotic in the blood is given by $A + Ae^{kt} + Ae^{2kt} + \cdots + Ae^{(n-1)kt}$, where A is the amount of antibiotic in milligrams, n is the number of doses, t is the time between doses, and k is a constant that depends on how quickly the body metabolizes the antibiotic. Suppose one dose of an antibiotic increases the blood level of the antibiotic by 0.5 milligram per liter. If the antibiotic is given every 4 hours and $k = -0.867$, find the concentration of the antibiotic just before the fifth dose. Round to the nearest thousandth. 0.516 mg

94. ***Medicine*** To treat a certain disease, the sufferer must maintain a certain concentration of a medication in the blood for an extended period of time. In such cases, the amount of medication in the blood can be approximated by the infinite geometric series $A + Ae^{kt} + Ae^{2kt} + \cdots + Ae^{(n-1)kt}$. (See Exercise 93.) If the medication is given every 12 hours, $k = -0.45$, and the required amount of medication is 2 milligrams per liter, find the amount of the dosage. Round to the nearest hundredth. (*Suggestion:* Solve the infinite geometric series $A + Ae^{kt} + Ae^{2kt} + \cdots + Ae^{(n-1)kt} + \cdots = 2$ for A.) 1.99 mg

Social Sciences

95. ***Theater Seating*** The seating section in a theater has 27 seats in the first row, 29 seats in the second row, and so on, increasing by 2 seats each row for a total of 10 rows. How many seats are in the 10th row, and how many seats are there in the section? 45 seats, 360 seats

96. ***Genealogy*** Some people can trace their ancestry back 10 generations, which means two parents, four grandparents, eight great-grandparents, and so on. How many grandparents does such a family tree include? 2044 grandparents if the person tracing his or her ancestry has no children, 2046 if he or she has children

Sports and Recreation

97. ***Tennis Ties*** In some games, such as tennis, the winning player must win by at least two points. If a game is tied, play is continued until one player wins two consecutive points. It can be shown that the probability of winning two consecutive points after a game is tied is given by the infinite geometric series

$$p^2(1 + [2p(1 - p)] + [2p(1 - p)]^2 + \cdots)$$

where p is the probability that you will win a particular point. Suppose the probability that you will win a particular point is 0.55. What is the probability that you will go on to win a game that is now tied? 0.599

98. ***Winning at Craps*** A player in a game of craps can bet that when two dice are rolled, a sum of 6 will occur before a sum of 7 occurs. The probability that this will happen is given by the infinite geometric series $\frac{5}{36}\left(1 + \frac{25}{36} + \left(\frac{25}{36}\right)^2 + \left(\frac{25}{36}\right)^3 + \cdots\right)$. What is the probability of winning this bet? $\frac{5}{11}$

Physical Sciences and Engineering

99. ***Retaining Wall*** A brick retaining wall has 26 bricks on the bottom row and 14 bricks on the top row. Each row has 1 more brick than the one above it. How many bricks are required for the retaining wall? 260 bricks

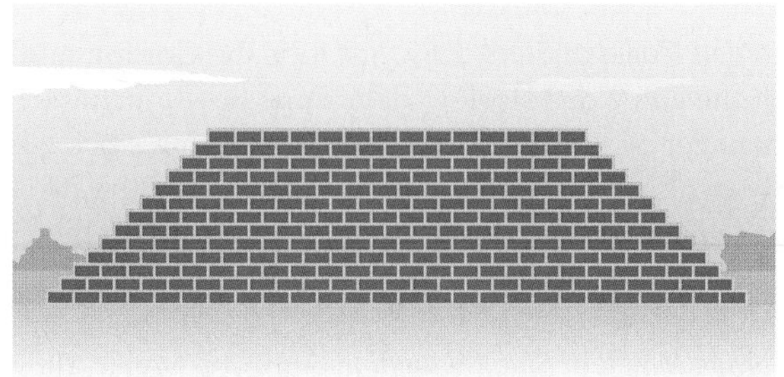

100. ***Construction*** A tower to support power lines has four identical sides. One side of the tower is shown in the diagram. What is the minimum length of steel necessary for the struts on all four sides? 217.6 ft

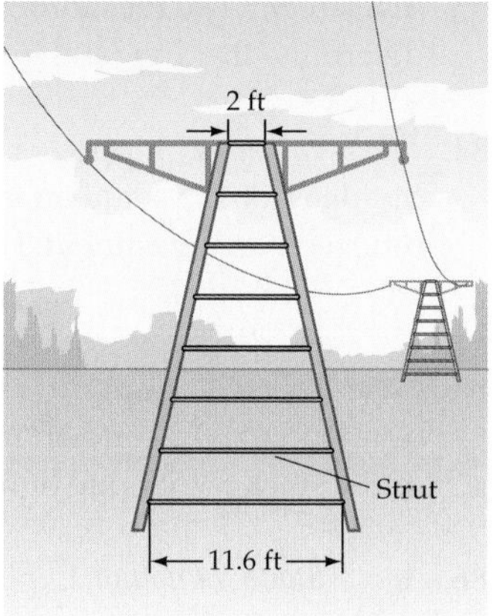

101. ***Stacking Logs*** Logs are stacked so that there are 25 logs in the bottom row, 24 logs in the second row, and so on, decreasing by 1 log each row. How many logs are stacked in the sixth row? How many logs are there in all six rows? 20 logs, 135 logs

102. ***Physics*** An object dropped from a cliff will fall 16 feet the first second, 48 feet the next second, and so on, increasing its distance by 32 feet each second. What is the total distance the object will fall in 7 seconds? 784 ft

103. ***Physics*** The distance a ball rolls down a ramp each second is given by the arithmetic sequence whose nth term is $2n - 1$ feet. Find the distance the ball rolls during the 10th second and the total distance the ball travels in 10 seconds. 19 ft, 100 ft

104. *Pendulum Motion* The bob of a pendulum swings through an arc of 30 inches on its first swing. Each successive swing is 90% of the length of the previous swing. Find the total distance the bob will travel. 300 in.

Prepare for Section 7.3

105. Evaluate $\left(1 + \frac{i}{n}\right)^n$ when $i = 0.1$ and $n = 2$. [P.1] 1.1025

106. Factor: $(1 + x)^2 + x(1 + x)^2$ [P.4] $(1 + x)^3$

107. Solve: $880 = 800 + 800x$ [1.1] 0.1

108. Solve: $1000(1 + x)^3 = 1728$ [1.4] 0.2

109. Given $P(t) = 500e^{0.05t}$, find $P(10)$. Round to the nearest hundredth. [5.1] 824.36

110. Solve: $1.05^n = 2$. Round to the nearest tenth. [5.4] 14.2

Explorations 

1. *Angles of a Triangle* The sum of the interior angles of a triangle is 180°.

a. Using this fact, what is the sum of the interior angles of a quadrilateral? 360°

b. What is the sum of the interior angles of a pentagon? 540°

c. What is the sum of the interior angles of a hexagon? 720°

d. Make a conjecture as to the sum of the interior angles of polygon with n sides. $(n - 2)180°$

2. *Perimeter of a Fractal Snowflake* A snowflake generally has a hexagonal (six-sided) shape. To create a snowflake, begin with an equilateral triangle in which each side is 1 unit long. The perimeter is 3 units. Now construct an identical but smaller triangle on the middle third of each side of the original triangle. The snowflake now consists of 12 line segments, each of length $\frac{1}{3}$ unit. The total perimeter of the figure is $12 \cdot \frac{1}{3} = 4$ units. Note that $\frac{4}{3} \cdot 3 = 4$—that is, the perimeter of the new figure is $\frac{4}{3}$ the perimeter of the original figure.

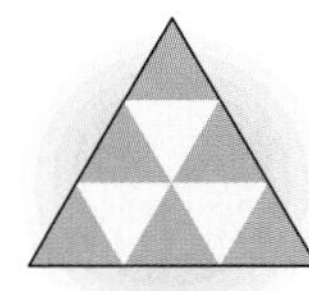

a. Repeat the procedure again and create identical but smaller triangles on the middle third of each of the 12 sides of the snowflake. What is the perimeter of the resulting figure? Show that this perimeter is $\frac{4}{3}$ the perimeter of the preceding figure. $\frac{16}{3}$ units; $\left(\frac{4}{3}\right) \cdot 4 = \frac{16}{3}$

b. Repeat the procedure again and determine the perimeter of the new figure. Show that the perimeter is $\frac{4}{3}$ the perimeter of the previous figure. $\frac{64}{9}$ units; $\left(\frac{4}{3}\right)\left(\frac{16}{3}\right) = \frac{64}{9}$

c. What type of sequence is being created by this procedure? Geometric

d. What is the perimeter of the nth figure created by this procedure? $3\left(\frac{4}{3}\right)^{n-1}$

3. *Area of a Fractal Snowflake* This Exploration uses the same scheme for creating a snowflake as Exploration 2. However, in this case, we are interested in the areas of successive snowflakes.

a. Let A be the area of the original triangle. After constructing an identical but smaller triangle on the middle third of each side of the original triangle, show that the area of the new figure is

$$A + 3\left(\frac{1}{9}\right)A = \frac{4}{3}A.$$

b. Now construct identical but smaller triangles on each of the 12 sides of the new figure. Show that the area of the new figure is $A + 3\left(\frac{1}{9}\right)A + 12\left(\frac{1}{81}\right)A = \frac{40}{27}A.$

c. Now construct identical but smaller triangles on each of the 48 sides of the new figure. Show that the area of the new figure is

$$A + 3\left(\frac{1}{9}\right)A + 12\left(\frac{1}{81}\right)A + 48\left(\frac{1}{729}\right)A = \frac{376}{243}A.$$

d. After the first term, the series in part **c.** is a geometric series. Suppose we continued this process an infinite number of times. What is the sum of the geometric series? $\frac{8A}{5}$

SECTION 7.3 Simple and Compound Interest

- Simple Interest
- Pricing Bonds: An Application of Simple Interest
- Compound Interest
- Annual Rate of Return and Effective Interest Rate
- Continuous Compounding

Simple Interest

When you deposit money into a savings account, the bank lends your money to other people to purchase a car, a home, or other items. The bank pays you for the privilege of using your money. The amount the bank pays you is called *interest*. Similarly, the person who borrowed the money must pay the bank for the privilege of using the bank's money. The amount that person pays the bank is also called interest.

The amount deposited or borrowed is called the **principal.** The interest charged is usually given as a percent of the principal, and is called the **interest rate.** Interest that is paid on the *original* principal is called **simple interest** and is the basis for all interest calculations.

Simple Interest Formula

The simple interest formula is given by $I = Prt$, where I is the simple interest, P is the principal, r is the simple interest rate, and t is the time period.

Point of Interest

The tradition of paying or receiving interest is at least 3800 years old. In Babylonia, King Hammurabi wrote the first laws regulating interest rates. The rates were set at various levels depending on what was borrowed or loaned. The interest rates on loans for silver were around 20%. The interest rates on loans for food grains were approximately $33\frac{1}{3}$%. See our web site at **math.college.hmco.com** for more information.

For the simple interest formula, the interest rate and the time period must correspond. For instance, if the interest rate is an *annual* rate, then the time period must be in years; if the interest rate is charged *monthly*, then the time period must be in months.

Alternative to Example 1
Exercise 6, page 615

EXAMPLE 1 Calculate Simple Interest

To obtain the capital to build a home, CityWide Construction secures a loan of \$300,000 at an annual simple interest rate of 8%. If the company keeps the loan for 8 months, how much interest is owed?

Solution Note that the given interest rate is in years, but the loan period is 8 months. Thus we must convert 8 months to years.

$$8 \text{ months} = \frac{8}{12} \text{ years} = \frac{2}{3} \text{ years}$$

$$\begin{aligned} I &= Prt \\ &= 300{,}000(0.08)\left(\frac{2}{3}\right) \qquad \blacksquare\ P = 300{,}000,\ r = 0.08,\ t = \tfrac{2}{3} \\ &= 16{,}000 \end{aligned}$$

The interest owed on the loan for 8 months is \$16,000.

Continued ➤

CHECK YOUR PROGRESS 1 The finance charge assessed by a credit card company is 1.8% per month on the unpaid balance of the charges on the card. What is the finance charge for a month in which the unpaid balance is $1250?

Solution *See page S39.* $22.50

When any three of the quantities in the simple interest equation $I = Prt$ are known, the fourth quantity can be computed. P, r, and t were known in Example 1 and we calculated I, the interest. The simple interest equation is also used to calculate the interest rate given I, P, and t.

For example, consider a certain type of loan, sometimes referred to as a **payday loan.** For this loan, the borrower receives an amount of money and then writes a post-dated check to the company for the amount borrowed plus a fee. The loans are typically for less than 2 weeks. For one such loan, a borrower can borrow $100 for 2 weeks for a fee of $10. Assuming the fee for the loan is equivalent to interest paid on the loan, we can calculate the equivalent simple annual interest rate for the loan by solving the simple interest equation for r. Because we are calculating the *annual* simple interest rate, we must write 2 weeks in terms of years.

Point of Interest

In Example 1, if the construction company had paid a simple annual interest rate of 260% for its loan, the interest owed would have been $520,000.

$$2 \text{ weeks} = \frac{2}{52} = \frac{1}{26} \text{ years}$$

$$I = Prt$$

$$10 = 100r\left(\frac{1}{26}\right) \qquad \text{▪ } I = 10,\ P = 100,\ t = \frac{1}{26}$$

$$10 = \frac{50}{13}r$$

$$2.6 = r \qquad \text{▪ Solve for } r.$$

To write r as a percent, we move the decimal point two places to the right. The simple annual interest rate is 260%.

▪ Pricing Bonds: An Application of Simple Interest

The purchase of municipal or corporate bonds is another application of simple interest. When a bond is offered to the investment community, it has a **face value,** which is the cost of one bond, and a **coupon rate,** which is the simple annual interest rate that an investor receives for the bond. For instance, suppose Carsey Airline corporation offers bonds with a face value of $1000 and a coupon rate of 8%. Then the amount of interest an investor would receive in 1 year on one of these bonds can be calculated by using the simple interest formula.

$$I = Prt$$

$$= 1000(0.08)(1) \qquad \text{▪ } P = 1000,\ r = 0.08,\ t = 1$$

$$= 80$$

The investor would receive $80.

Because interest rates change, the value of a bond to an investor will change. For example, if another corporation, Silver Airlines, tries to offer a bond with a face value of $1000 and an interest rate of 6%, investors will choose the Carsey

Airline bonds because the interest rate is higher. To attract buyers, Silver Airlines must *discount* their bonds. That is, they must sell their bonds for less than face value *but* pay the investor interest based on the face value. By selling the bonds for less, the investor's actual interest rate increases. The price at which the bond must be sold is the product of the ratio of the offering rate (6%) to the prevailing rate (8%) and the bond's face value ($1000). For the Silver Airlines bond we have

$$\text{Price of bond} = \frac{6\%}{8\%} \cdot 1000 = 750$$

Silver Airlines must sell its bond for $750.

Sometimes it may be necessary to pay a *premium* for a bond. In this case, the investor pays more than the face value of the bond. This occurs when the coupon rate of the bond is higher than the prevailing rate. Here is an example.

TAKE NOTE

If Silver Airlines sells its bonds for $750, then an investor who purchases one of these bonds pays $750. However, the interest rate the investor earns is based on the *face value* of the bond. Therefore, the investor receives

$$\begin{aligned} I &= Prt \\ &= 1000(0.06)(1) = 60 \end{aligned}$$

Because the investor paid only $750 for the bond, the interest rate earned by the investor can be calculated using the simple interest formula.

$$\begin{aligned} I &= Prt \\ 60 &= 750r \\ \frac{60}{750} &= r \\ 0.08 &= r \end{aligned}$$

The simple interest rate earned by the investor is 8%, the same rate the investor would have earned on the Carsey Airline bonds.

EXAMPLE 2 Find the Market Value of a Bond

Suppose the prevailing rate on a corporate bond is 8%. If a corporation has a bond with a coupon rate of 9.25% and a face value of $1000, what is the market value of this bond?

Solution

$$\begin{aligned} \text{Market value of bond} &= \frac{\text{coupon rate}}{\text{prevailing rate}} \cdot \text{face value} \\ &= \frac{9.25\%}{8\%} \cdot 1000 = 1156.25 \end{aligned}$$

The market value of the bond is $1156.25.

Alternative to Example 2
Exercise 18, page 616

CHECK YOUR PROGRESS 2 Suppose the prevailing rate on a corporate bond is 7.5%. If a corporation has a bond with a coupon rate of 7% and a face value of $1000, what is the market value of this bond?

Solution *See page S39.* $933.33

▪ Compound Interest

Although there are important applications of simple interest, the interest paid on money is usually *compound interest*. **Compound interest** is calculated not only on the original principal, but also on the interest earned.

For instance, suppose $1000 is invested in an account for 3 years at an annual interest rate of 8%, compounded annually. To find the value of the investment after 3 years, we proceed as follows.

- *End of year 1:* Calculate the simple interest earned during the first year on the original principal.

$$\begin{aligned} I &= Prt \\ &= 1000(0.08)(1) && \text{▪ } P = 1000, r = 0.08, t = 1 \\ &= 80 \end{aligned}$$

Add the interest to the original principal: $1000 + $80 = $1080.

- *End of year 2:* Calculate the simple interest earned during the second year on the new principal, \$1080, which is the original principal plus the interest earned for 1 year.

$$\begin{aligned} I &= Prt \\ &= 1080(0.08)(1) \qquad \blacksquare\ P = 1080,\ r = 0.08,\ t = 1 \\ &= 86.4 \end{aligned}$$

  Add the interest to the principal: \$1080 + \$86.40 = \$1166.40.

- *End of year 3:* Calculate the simple interest earned during the third year on the new principal, \$1166.40.

$$\begin{aligned} I &= Prt \\ &= 1166.40(0.08)(1) \qquad \blacksquare\ P = 1166.40,\ r = 0.08,\ t = 1 \\ &= 93.312 \end{aligned}$$

  Add the interest to the principal: \$1166.40 + \$93.31 = \$1259.71.

After 3 years, the value of the investment will be \$1259.71.

For the compound interest example above, the principal at the end of each year is the principal at the end of the preceding year plus 8% of that principal. This can be modeled by a geometric sequence.

$$\begin{aligned} &\text{End of year 1:} \quad & 1000(1.08) &= 1080 \\ &\text{End of year 2:} \quad & 1080(1.08) &= [1000(1.08)](1.08) \\ &&&= 1000(1.08)^2 = 1166.40 \\ &\text{End of year 3:} \quad & 1166.40(1.08) &= [1000(1.08)^2](1.08) \\ &&&= 1000(1.08)^3 = 1259.71 \end{aligned}$$

■ From year 1, replace 1080 with 1000(1.08).

■ From year 2, replace 1166.40 with $1000(1.08)^2$.

The nth term of this sequence, $1000(1.08)^n$, is the value of the investment after n years. For instance, the value of the investment after 10 years is $1000(1.08)^{10}$, which is approximately \$2158.92.

The interest earned on this investment after 3 years is the difference between the new principal and the original principal.

$$\begin{aligned} \text{Interest earned} &= \text{New principal} - \text{original principal} \\ &= 1259.71 - 1000 = 259.71 \end{aligned}$$

The interest earned is \$259.71.

If the original principal of \$1000 had been invested at an 8% *simple* interest rate, the interest earned after 3 years would have been

$$\begin{aligned} I &= Prt \\ &= 1000(0.08)(3) \qquad \blacksquare\ P = 1000,\ r = 0.08,\ t = 3 \\ &= 240 \end{aligned}$$

or \$240. By compounding the interest, the investment earned \$19.71 more than by using simple interest.

For the preceding compounding interest example, the interest was compounded annually. It is also possible to compound interest more than once a year. Interest can be compounded semiannually (twice a year), quarterly (4 times a year), monthly (12 times a year), or daily (365 times a year).

When interest is compounded more than once a year, the annual interest rate is divided equally among all compounding periods. For instance, if an investment earns 8% annual interest compounded quarterly, then the interest rate per quarter is $\frac{8\%}{4} = 2\%$. If the investment earns 8% annual interest compounded daily, the interest rate per day is $\frac{8\%}{365} \approx 0.021918\%$.

QUESTION If an investment has an annual interest rate of 6%, what is the monthly interest rate as a decimal?

Suppose \$1000 is invested in an account that earns 8% annual interest compounded *quarterly*. Because there are 4 quarters in each year, the interest rate per quarter is $\frac{8\%}{4} = 2\%$. The value of this investment after n quarters is the nth term of the geometric sequence $a_n = (1.02)^n$. For instance, after 3 years (12 quarters), the value of the investment is $(1.02)^{12}$, which is approximately \$1268.24.

Compounding quarterly, the value of \$1000 after 3 years was \$1268.24. Compounding annually, the value of \$1000 after 3 years was \$1259.71. By compounding quarterly, the investment was worth \$8.53 more. Generally, for a given interest rate and original investment, the value of the investment after a period of time will increase as the number of compounding periods per year increases.

We have seen that the value of a compound interest investment after a certain time is the nth term of a geometric sequence; it is common to state this as a formula. Before we give this formula, we will define two terms that are used when dealing with money.

The **present value** of an investment is its value today. For the example of depositing \$1000 into an account, the present value is \$1000, the amount placed in the account. The **future value** of the investment is its value after a certain time period. The future value of \$1000 in an account that earns an 8% interest rate compounded *annually* will be \$1259.71 after 3 years. The future value of \$1000 in an account that earns an 8% interest rate compounded *quarterly* will be \$1268.24 after 3 years. The following formula is used to calculate the future value of an investment.

Future Value of Compound Interest

Suppose an investment earns an annual interest rate r compounded m times per year for t years. Then the future value FV of the investment is

$$FV = PV(1 + i)^n$$

where the present value PV is the amount deposited, $i = \frac{r}{m}$, and $n = mt$.

ANSWER $\frac{6\%}{12} = \frac{0.06}{12} = 0.005$

Alternative to Example 3
Exercise 22**e.**, page 616

INTEGRATING TECHNOLOGY

If you have a scientific calculator, you can compute the future value as follows:
10000 × (1 + 0.085/365) ^(365 × 6) =

EXAMPLE 3 Find the Future Value of an Investment

A marathon runner places her $10,000 winnings into an account that earns an 8.5% annual interest rate compounded daily. Find the future value of the account after 6 years and the amount of interest earned for the 6 years.

Solution Use the compound interest formula to find the future value of the investment. The annual interest rate is 8.5% ($r = 0.085$), the interest is compounded daily ($m = 365$), and the term of the investment is 6 years ($t = 6$).

$$FV = PV(1 + i)^n$$

$$FV = 10{,}000\left(1 + \frac{0.085}{365}\right)^{2190} \qquad \blacksquare\ PV = 10{,}000,\ i = \frac{0.85}{365},\ n = 365 \times 6 = 2190$$

$$\approx 16{,}651.92322$$

The future value of the investment after 6 years is $16,651.92.

To calculate the interest earned, subtract the present value (the original investment) from its future value after 6 years.

$$\text{Interest earned} = 16{,}651.92 - 10{,}000 = 6651.92$$

The interest earned is $6651.92.

CHECK YOUR PROGRESS 3 Find the future value of a $5000 investment after 5 years placed in an account that earns a 6% annual interest rate, compounded monthly.

Solution *See page S40.* $6744.25

In Example 3, the present value of the investment was known and we calculated its future value. There will be occasions when we must calculate the present value of an investment given its future value. In this case, we solve the future value equation for present value. This results in the following formula

Present Value of Compound Interest

Suppose an investment earns an annual interest rate r compounded m times per year for t years. Then the present value PV of the investment is

$$PV = FV(1 + i)^{-n}$$

where FV is the future value of the investment, $i = \frac{r}{m}$, and $n - mt$.

Alternative to Example 4
Exercise 24**f.**, page 616

EXAMPLE 4 Find the Present Value of an Investment

The Woods family would like to have $15,000 in 3 years to use for a down payment on a house. How much must they deposit in an account today that earns a 10% interest rate, compounded quarterly, to reach their goal?

Continued ➤

Solution Because the Woods want to have \$15,000 in 3 years, the future value of their investment is \$15,000. The present value is unknown. The annual interest rate is 10% ($r = 0.10$), interest is compounded quarterly ($m = 4$), and the term of the investment is 3 years ($t = 3$). Use the compound interest formula for present value.

$$PV = FV(1 + i)^{-n}$$

$$PV = 15{,}000\left(1 + \frac{0.10}{4}\right)^{-12}$$

■ $FV = 15{,}000$, $i = \frac{0.10}{4}$, $n = 4 \times 3 = 12$

$$PV \approx 15{,}000(0.74355589)$$

$$PV \approx 11{,}153.34$$

The Woods must deposit \$11,153.34 into the account today to have \$15,000 after 3 years.

CHECK YOUR PROGRESS 4 Michael Shelley is planning on purchasing a new car in 2 years and wants to have \$5000 for a down payment. How much money must he deposit today into an account that earns an 8% interest rate, compounded daily, to reach his goal?

Solution *See page S40.* \$4260.79

■ Annual Rate of Return and Effective Interest Rate

Consider the situation of a person who wants to place \$1000 in an investment today and have the investment grow to \$2000 in 5 years. In this case, we know the present value of the investment (\$1000), its future value (\$2000), and the time (5 years). To reach the goal, the investor must determine the annual interest rate, which is called the **annual rate of return** of the investment. Assuming interest is compounded daily, $i = \frac{r}{365}$, $n = 365 \times 5 = 1825$, and we must solve the equation

$$2000 = 1000\left(1 + \frac{r}{365}\right)^{1825}$$

for r. One way to accomplish this is to use the power key on a calculator. Solving this equation is similar to solving $12 = x^5$.

$$12 = x^5$$

$$12^{1/5} = (x^5)^{1/5}$$

■ Raise each side to the $\frac{1}{5}$ power. Note that this is the reciprocal of the exponent.

$$1.64375183 \approx x$$

INTEGRATING TECHNOLOGY

To evaluate $12^{1/5}$ with a graphing calculator, key in 12^(1 ÷ 5) ENTER.

Applying this idea to our equation, we have the following.

$$2000 = 1000\left(1 + \frac{r}{365}\right)^{1825}$$

■ $FV = 2000$, $PV = 1000$, $m = 365$, $mt = 365 \times 5 = 1825$

$$2 = \left(1 + \frac{r}{365}\right)^{1825}$$

■ Divide each side of the equation by 1000.

$$2^{1/1825} = \left[\left(1 + \frac{r}{365}\right)^{1825}\right]^{1/1825}$$

■ Raise each side of the equation to the $\frac{1}{1825}$ power.

$$2^{1/1825} = 1 + \frac{r}{365} \qquad \text{▪ Simplify.}$$

$$2^{1/1825} - 1 = \frac{r}{365} \qquad \text{▪ Subtract 1 from each side of the equation.}$$

$$365(2^{1/1825} - 1) = r \qquad \text{▪ Multiply each side of the equation by 365.}$$

$$0.138656 \approx r$$

The necessary annual rate of return, to the nearest tenth of a percent, is 13.9%.

The method we used to calculate annual rate of return is also used to calculate *effective interest rate.* This is the annual simple interest rate that is equivalent to a given compound interest rate.

Effective Annual Interest Rate

Effective annual interest rate I is the annual simple interest rate that is equivalent to a given compounded interest rate. It is given by

$$I = 100\left[\left(1 + \frac{r}{m}\right)^m - 1\right]$$

where r is the annual compound interest rate and m is the number of compounding periods per year.

Alternative to Example 5
Exercise 28, page 617

TAKE NOTE

The answer to Example 5 means that the amount of interest earned in 1 year at a simple interest rate of 6.18% is the same as the amount of interest earned in 1 year at an interest rate of 6% compounded daily.

EXAMPLE 5 Find an Effective Annual Interest Rate

The interest rate on an investment is 6%, compounded daily. Find the effective annual interest rate.

Solution Use the effective annual interest rate formula.

$$I = 100\left[\left(1 + \frac{0.06}{365}\right)^{365} - 1\right] \qquad \text{▪ } r = 0.06,\ m = 356$$

$$\approx 6.1831$$

The effective annual interest rate, to the nearest hundredth of a percent, is 6.18%.

CHECK YOUR PROGRESS 5 An investment has an interest rate of 7.5%, compounded quarterly. Find, to the nearest hundredth, the effective annual interest rate.

Solution *See page S40.* 7.71%

▪ Continuous Compounding

The graphs in Figure 7.1 show the growth of a $1000 investment at an interest rate of 20% over a five-year period when the interest is compound annually, semiannually, and daily.

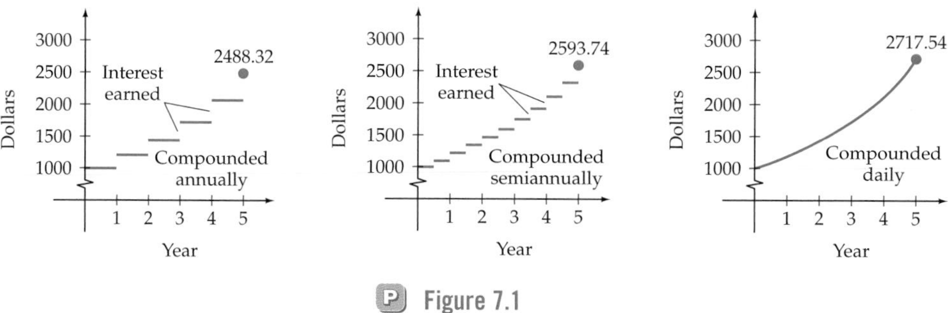

Figure 7.1

Observe that as the number of compounding periods increases, the value of the investment after 5 years increases. Also observe that as the number of compound periods per year increases, the jump in the amount of interest earned per compounding period becomes smaller. When interest is compounded daily, the jumps are so small on the scale of the graph that the curve appears smooth.

Suppose we continue to increase the number of compounding periods per year so that we are compounding interest hourly (8760 times a year), each minute (525,600 times a year), each second (31,536,000 times a year), or even more frequently. As the number of compounding periods keeps increasing, the graph of the value of the investment has smaller and smaller jumps until finally it becomes a smooth curve.

INSTRUCTOR NOTE

Exploration 2 on page 618 gives an empirical illustration of how the compound interest formula is derived. Going through this exercise with students may help them connect continuous compounding to the definition of e given in Chapter 5.

Continuous Compounding Formula

Suppose n, the number of compounding periods per year, increases without bound. In this case, interest is said to be **compounded continuously.** The future value of such an investment is given by $FV = PVe^{rt}$, where e is the base of the natural logarithm, r is the annual interest rate, and t is the number of years the investment is held.

INTEGRATING TECHNOLOGY

The calculation in Example 6 can be accomplished on a graphing calculator as follows: 1000 2nd e^x(0.2 × 5) ENTER

Alternative to Example 6

A \$2000 investment is placed in an account that earns 15% annual interest, compounded continuously. Find the future value of the investment after 10 years.

■ **\$8963.38**

EXAMPLE 6 Find the Future Value of an Investment Compounded Continuously

A \$1000 investment is placed in an account that earns 20% annual interest, compounded continuously. Find the future value of the investment after 5 years.

Solution
$$FV = PVe^{rt}$$
$$= 1000e^{0.2\times 5} \quad \text{■ } PV = 1000, r = 0.2, t = 5$$
$$\approx 2718.28$$

The future value of the investment is \$2718.28.

CHECK YOUR PROGRESS 6 Find the future value of a \$500 investment that is compounded continuously for 8 years at an annual interest rate of 7.25%.

Solution *See page S40.* \$893.02

By comparing the answer to Example 6 to the value of the same investment compounded daily (see Figure 7.1), we see that the difference between the future values is 2718.28 − 2717.54 = 0.74. This means that a $1000 investment at an interest rate of 20%, compounded continuously, is worth 74 cents more after 5 years than the same investment compounded daily.

Topics for Discussion

1. What is the difference between interest and interest rate?
2. Explain how simple interest differs from compound interest.
3. Explain the difference between present value and future value.
4. What does it mean for a bond to sell at a discount or a premium?
5. What is the meaning of effective interest rate?
6. What is the rate of return on an investment?

EXERCISES 7.3

— Suggested Assignment: Exercises 1–41, odd; and 42–47, all.

In Exercises 1 to 4, suppose $5000 is invested at the given annual simple interest rate for the given time period. Calculate the interest earned.

1. 10%, 6 months $250
2. 8%, 16 months $533.33
3. 6%, 30 months $750
4. 7%, 3 years $1050

Business and Economics

5. ***Interest on a Loan*** An airline pilot receives an 8% simple interest loan of $200,000 to complete the construction of a house. If the pilot repays the loan in 8 months, how much interest is owed? $10,666.67

6. ***Interest on a Checking Account*** Colleen Angstrom has a checking account that earns a simple annual interest rate of 2%, paid monthly, on her average monthly balance. How much interest did she earn on her account for a month in which her average monthly balance was $450.64? $.75

7. ***Interest on a Credit Card*** A credit card company charges a monthly simple interest rate of 1.8% on the unpaid balance on the credit card. How much interest is charged for a month in which the unpaid balance is $384.74? $6.93

8. ***Interest on a Credit Card*** The monthly simple interest rate on the unpaid balance on a credit card is 1.5%. What is the interest charge for a month in which the unpaid balance is $489.13? $7.34

9. ***Historical Interest Rates*** In ancient Babylonia, the interest rate charged on a loan of wheat could be as high as $33\frac{1}{3}\%$ of the amount borrowed. Suppose a farmer borrowed 60 pounds of wheat at a simple interest rate of $33\frac{1}{3}\%$. How much wheat must the farmer repay as interest 1 year later? 20 lb

10. ***Historical Interest Rates*** A typical interest rate on a loan of silver in Babylonia was 25%. Suppose a silversmith borrowed 2 pounds of silver at this interest rate. What is the total number of pounds of silver the silversmith must repay 1 year later? 2.5 lb

11. ***Monthly Payment*** Jamie Cervantes purchased a surround-sound stereo system for $1100 and financed the entire amount at an annual simple interest rate of 10% for 18 months. What are Jamie's monthly payments? $70.28

12. ***Monthly Payment*** A&W Electric purchased some equipment for its company and financed $12,000 for 2 years at an annual simple interest rate of 6.5%. What are the monthly payments for this loan? $565

13. ***Payday Loan*** Richard Brackett received a payday loan of \$250, on which he paid a fee of \$20. If Richard repaid the loan 1 week later, what was the annual simple interest rate on his loan? 416%

14. ***Payday Loan*** Lois Ngo received a payday loan of \$125, on which she paid a fee of \$15. If Lois repaid the loan 2 weeks later, what was the annual simple interest rate on her loan? 312%

15. ***Interest on a Bond*** An investor receives an annual simple interest rate of 5.5% on the face value of a bond. If the investor keeps a bond with a face value of \$10,000 for 5 years, how much interest is earned? \$2750

16. ***Interest on a Bond*** The interest earned on a corporate bond is based on 8.5% of the bond's face value. How much interest will an investor earn over 3 years on a bond from this corporation that has a face value of \$5000? \$1275

17. ***Market Value of a Bond*** The coupon rate for a \$1000 bond is 7.5%. If the prevailing interest rate is 6%, what is the market value of the bond? \$1250

18. ***Market Value of a Bond*** A \$5000 bond has a coupon rate of 8%. If the prevailing interest rate is 10%, what is the market value of the bond? \$4000

19. ***Interest Rate on a Mutual Fund*** A mutual fund manager purchased a bond with a face value of \$10,000 and a coupon rate of 9.5% for \$9000. What is the actual interest rate, to the nearest 0.1%, the fund manager received on this investment? 10.6%

20. ***Interest Rate on an Investment*** Carol Escobar purchased a bond with a face value of \$5000 and a coupon rate of 10.2% for \$5500. What is the actual interest rate, to the nearest 0.1%, she received on her investment? 9.3%

21. ***Future Value*** Suppose \$1000 is invested in an account that earns an annual interest rate of 8%. What is the future value of the investment in 5 years if interest is:

    a. simple interest? \$1400

    b. compounded annually? \$1469.33

    c. compounded semiannually? \$1480.24

    d. compounded quarterly? \$1485.95

    e. compounded monthly? \$1489.85

    f. compounded daily? \$1491.76

22. ***Future Value*** Suppose \$5000 is invested in an account that earns an annual interest rate of 6%. What is the future value of the investment in 3 years if interest is:

    a. simple interest? \$5900

    b. compounded annually? \$5955.08

    c. compounded semiannually? \$5970.26

    d. compounded quarterly? \$5978.09

    e. compounded monthly? \$5983.40

    f. compounded daily? \$5986

23. ***Present Value*** A family wishes to have \$4500 in 4 years to take a cruise. How much must the family deposit into an account today that earns an annual interest rate of 7% if interest is:

    a. simple interest? \$3515.63

    b. compounded annually? \$3433.03

    c. compounded semiannually? \$3417.35

    d. compounded quarterly? \$3409.27

    e. compounded monthly? \$3403.79

    f. compounded daily? \$3401.12

24. ***Present Value*** The owner of a cabinet shop wants to have \$10,000 in 2 years to purchase a new lathe. How much must the owner deposit into an account today that earns an annual interest rate of 8.5% if interest is:

    a. simple interest? \$8547.01

    b. compounded annually? \$8494.55

    c. compounded semiannually? \$8466.34

    d. compounded quarterly? \$8451.69

    e. compounded monthly? \$8441.71

    f. compounded daily? \$8436.82

25. ***Inflation Rate*** When an economist states an *inflation rate* for a commodity such as gasoline, food, or medical costs, the economist is stating the annual compound interest rate at which the price of that commodity is increasing. If the average price of 1 gallon of gas is \$1.78 and an economist states that the inflation rate for gasoline is 4.3%, what will be the price of 1 gallon of gas in 4 years? \$2.11

26. ***Inflation Rate*** An economist predicts that the inflation rate (see Exercise 25) for medical insurance is increasing at 9.2% per year. If the average annual cost for medical insurance is \$4600 today, what will be the average annual cost of medical insurance in 5 years? \$7142.84

27. ***Effective Annual Interest*** If \$1000 is deposited into an account that earns an annual interest rate of 7.25% compounded daily, what is the effective annual interest rate to the nearest 0.01%? 7.52%

28. ***Effective Annual Interest*** Suppose \$5000 is placed in an account that earns an annual interest rate of 6.5% compounded quarterly. What is the effective annual interest rate to the nearest 0.01%? 6.66%

29. ***Effective Annual Interest*** What quarterly compounding interest rate can an investor choose that is equivalent to an annual simple interest rate of 8%? Round to the nearest 0.01%. 7.77%

30. ***Effective Annual Interest*** What daily compounding interest rate, to the nearest 0.01%, can an investor choose that is equivalent to an annual simple interest rate of 7.5%? 7.23%

31. ***Doubling Time of an Investment*** Melissa Gwyn deposits \$2500 into an account that offers an annual interest rate of 6.25%, compounded daily. To the nearest month, how long will it take for her investment to double in value? 11 years 1 month

32. ***Doubling Time of an Investment*** Alana Lincoln selects an account that offers an annual interest rate of 7.8%, compounded daily. If Alana deposits \$2000 into the account, how long, to the nearest month, will it take for her investment to double in value? 8 years 11 months

33. ***Find an Interest Rate*** An investor wishes to double an investment of \$1000 in 5 years. What annual interest rate, compounded daily, must the investor choose to reach the goal? Round to the nearest 0.1%. 13.9%

34. ***Find an Interest Rate*** What interest rate is necessary to double the value of a \$5000 investment in 5 years if the interest is compounded quarterly? Round to the nearest 0.1%. 14.1%

35. ***Inflation*** The Consumer Price Index (CPI) is an interest rate that measures the change in the price of a commodity from one year to the next. According to data from the Bureau of Labor Statistics, since 1980, the CPI has been increasing at an annual compounding rate of approximately 3.7%. If a commodity cost \$100 in 1980, what will be the cost of the commodity in 2004? \$239.16

36. ***Inflation*** Suppose the CPI (see Exercise 35) has been increasing at an annual compounding rate of 4.7%. If a commodity cost \$100 in 1980, what will be the cost of that commodity in 2004? \$301.11

37. ***Doubling Time*** For a given compound interest rate, does the time it takes for an investment to double depend on the amount invested? For instance, will \$100 double sooner than \$1000? Give support for your answer. No. The percent increase will always be the same, no matter what the original amount.

38. ***Doubling Time*** On the basis of the results of Exercise 37, derive a formula for the time it takes an investment to double at an interest rate i, compounded annually. $t = \dfrac{\log 2}{\log(1 + i)}$, where i is the interest rate and t is the time in years.

39. ***Annual Interest Rate*** Pine Forest Management Corporation invests \$500,000 to clear a tract of land and to plant and maintain new pine trees. The corporation expects to be able to sell the land and trees in 5 years for \$1.5 million. What is the annual compounded rate of return on the corporation's investment? 24.57%

40. ***Annual Interest Rate*** A problem that occurs on some actuary exams goes like this. On January 1, 2002 and on January 1, 2003, a company has \$5000 in a bank account. On May 1, 2002 the company withdrew \$1000 and on September 1, 2002 the company deposited \$400. Assuming the account earned a constant simple interest rate for the year, find the interest rate. 13.43%

41. ***Rate of Return*** An investment advisor offers you two different opportunities. For investment A, a $1000 investment will grow to $2000 in 5 years. For investment B, a $1000 investment will grow to $3500 in 9 years. Which investment offers the greater rate of return? Give support for your answer.
The second investment. The first earns about 14.87%; the second earns about 14.93%.

Prepare for Section 7.4

42. Evaluate $\frac{1-(1+i)^{-n}}{i}$ when $i = 0.1$ and $n = 3$. Round to the nearest hundredth. [P.2] 2.49

43. Write $\frac{1-x^{-2}}{x}$ in simplest form. [P.5] $\frac{x^2-1}{x^3}$

44. Solve $a = b(1+i)^n$ for b. Write the answer using a negative exponent. [1.1] $b = a(1+i)^{-n}$

45. What is the common ratio for the geometric sequence $\frac{1}{1.1}, \frac{1}{1.1^2}, \frac{1}{1.1^3}, \frac{1}{1.1^4}, \ldots, \frac{1}{1.1^n}$? [7.2] $\frac{1}{1.1}$

46. Evaluate $\frac{2}{3} + \frac{1}{3} + \frac{1}{6} + \cdots + \frac{2}{3}\left(\frac{1}{2}\right)^{n-1}$ when $n = 8$. [7.2] $\frac{85}{64}$

47. Evaluate $\sum_{k=1}^{10} \frac{1}{1.05^k}$. Round to the nearest ten-thousandth. [7.2] 7.7217

Explorations 

1. ***Rule of 72*** The Rule of 72 states that the time t, in years, it takes an investment to double in value (called the *doubling time* of the investment) can be approximated by $t \approx \frac{72}{\text{annual interest rate}}$. For instance, the approximate doubling time of an investment at 8% is $\frac{72}{8} = 9$ years. For the accompanying table, use the Rule of 72 to calculate the approximate time it takes for an investment to double, and then calculate the actual time it takes for the investment to double (assume annual compounding). [*Hint:* The actual time $= \frac{\log 2}{\log(1 + r)}$, where r is the interest rate as a decimal.]

Interest rate r	Rule of 72	Actual time
6%	12	11.9
7%	10.3	10.2
8%	9	9.0
9%	8	8.0
10%	7.2	7.3
11%	6.5	6.6
12%	6	6.1
13%	5.5	5.7
14%	5.1	5.3

a. Based on the data in the table, complete the following sentence. "When the interest rate is below 7%, the Rule of 72 predicts a ______ doubling time than the actual doubling time." longer

b. Based on the data in the table, complete the following sentence. "When the interest rate is above 7%, the Rule of 72 predicts a ______ doubling time than the actual doubling time." shorter

2. Suppose $10,000 is placed in an account for which interest is compounded. Assuming the money is left in the account for 1 year, complete the following table by finding the future value of the investment at the given interest rate when interest is compounded daily, each hour, and each minute. (Assume a 365-day year.) Also evaluate $10{,}000e^r$ for each given interest rate.

Interest rate r	Daily	Hourly	Each minute	$10{,}000e^r$
6%	10,618.31	10,618.36	10,618.37	10,618.37
7%	10,725.01	10,725.08	10,725.08	10,725.08
8%	10,832.78	10,832.87	10,832.87	10,832.87
9%	10,941.62	10,941.74	10,941.74	10,941.74

a. On the basis of the data in the table, complete the following sentence. "As the number of compounding periods per year increases, the value of the investment approaches ______."
pe^r, where r is the annual interest rate as a decimal and p is the principal.

b. From the results in the table, do you think that a $10,000 investment at 8% could ever grow to $10,835? Give support for your answer. (*Suggestion:* If you are having trouble with this part, try compounding every second and compare the value you find to e^r.)
No. $10,835 > $10,832.87, which is the theoretical maximum amount that can be earned.

SECTION 7.4 Present Value of an Annuity and Amortization

- Present Value of an Annuity
- Amortization Schedules and Home Loans
- Annual Percentage Rate (APR)

TAKE NOTE

Recall from Section 7.3 that $PV = FV(1 + i)^{-n}$. For the loan at the right, FV is being replaced by the unknown payment PMT, $i = \frac{0.12}{12} = 0.01$, and $1 + i = 1.01$. The value of n is changing each month. For the first month, $n = 1$; for the second month, $n = 2$; for the third month, $n = 3$; and so on.

TAKE NOTE

Here are the details for calculating the sum.

$$S_n = \frac{a_1(1 - r^n)}{1 - r}$$

$$S_{12} = \frac{\frac{1}{1.01}\left(1 - \left(\frac{1}{1.01}\right)^{12}\right)}{1 - \frac{1}{1.01}}$$

$$\approx 11.25507747$$

▪ Present Value of an Annuity

An **annuity** is a payment schedule of *equal* amounts in *equal* periods of time. There are two basic types of annuities: due annuities and ordinary annuities. A due annuity is one for which the payment is made at the beginning of the time period. Paying car insurance on a monthly basis is an example of a due annuity. For an ordinary annuity, the payment is made at the end of the time period. A typical example of an ordinary annuity is a car loan. In this text, we will consider only ordinary annuities.

Calculating the amount of a car payment depends on finding the sum of a geometric series. The amount of the payment is chosen so that the sum of the present values of each payment equals the amount borrowed.

Rather than take a typical 48- or 60-month car loan, we will consider the case of a person who borrows \$500 for 1 year (12 months) at an annual interest rate of 12% compounded monthly. The loan payment is calculated so that the sum of the present values of each payment is \$500.

$$\begin{aligned} 500 &= \begin{matrix}\text{present value of}\\ \text{first payment}\end{matrix} + \begin{matrix}\text{present value of}\\ \text{second payment}\end{matrix} + \cdots + \begin{matrix}\text{present value of}\\ \text{12th payment}\end{matrix} \\ &= PMT(1.01^{-1}) + PMT(1.01^{-2}) + \cdots + PMT(1.01^{-12}) \\ &= PMT(1.01^{-1} + 1.01^{-2} + \cdots + 1.01^{-12}) \\ &= PMT\left(\frac{1}{1.01} + \frac{1}{1.01^2} + \cdots + \frac{1}{1.01^{12}}\right) \end{aligned}$$

The sum inside the parentheses is a geometric series with common ratio $\frac{1}{1.01}$. Using the formula for the sum of a geometric series from Section 7.2, we have

$$500 = PMT\left(\frac{1}{1.01} + \frac{1}{1.01^2} + \cdots + \frac{1}{1.01^{12}}\right)$$

$$500 \approx PMT(11.25507747) \qquad \text{▪ See the Take Note at the left.}$$

$$44.42439434 \approx PMT \qquad \text{▪ Divide by 11.25507747.}$$

The monthly payment is \$44.42.

Because this calculation proceeds in the same way each time, we can state a formula that relates the payment of a loan to the present value of an annuity (the amount borrowed).

Payment Formula for the Present Value of an Annuity

Suppose an annuity earns an annual interest rate of $r\%$ compounded m times a year for t years. Then the payment amount PMT for a given present value PV of the annuity is

$$PMT = PV\left[\frac{i}{1 - (1 + i)^{-n}}\right]$$

where $i = \frac{r}{m}$ and $n = mt$.

Alternative to Example 1
Exercise 10, page 627

INSTRUCTOR NOTE

One way to help students understand the present value of an annuity is by applying the concepts to a state lottery. Create this scenario. Tell students they have won a \$5,000,000 lottery that will be paid to them in annual payments of \$250,000 for 20 years. Now ask how much they would be willing to take today instead of the payments. Start with \$5 million, then \$4.5 million, and so on. The number a student chooses is his or her subjective present value. Assuming an 8% interest rate, the actual present value is approximately \$2.5 million.

EXAMPLE 1 Calculate a Monthly Payment and Total Interest

Shawn Estaban purchases a new car and secures a 9%, five-year loan for \$12,000.

a. What are Shawn's monthly payments?

b. If Shawn keeps the car for the entire five-year period, how much interest does Shawn pay on the loan?

Solution

a. Use the Payment Formula for the Present Value of an Annuity. The annual interest rate is 9% ($r = 0.09$), payments are monthly ($m = 12$), and the loan is for 5 years ($t = 5$).

$$PMT = PV\left(\frac{i}{1-(1+i)^{-n}}\right)$$

$$PMT = 12{,}000\left(\frac{0.0075}{1-(1+0.0075)^{-60}}\right)$$

$$\approx 12{,}000(0.02075836)$$

$$\approx 249.10$$

▪ $PV = 12{,}000$, $i = \frac{0.09}{12} = 0.0075$, $n = 12(5) = 60$.

The monthly payment is \$249.10.

b. The total amount repaid is the product of the number of payments (60) and the monthly payment (\$249.10). Subtracting the loan amount (12,000) from this product will give the amount of interest paid.

$$\text{Interest paid} = (249.10)60 - 12{,}000 = 2946$$

The amount of interest paid is \$2946.00.

CHECK YOUR PROGRESS 1 The Falcon Network Corporation receives a 2-year loan of \$50,000 at an annual interest rate of 11% compounded quarterly. Determine the quarterly payments.

Solution *See page S40.* \$7047.90

INTEGRATING TECHNOLOGY

Many graphing calculators have built-in programs to perform financial calculations such as those in Example 1. Here is an example using a TI-83 Plus calculator. Press APPS1.

```
CALC VARS
1:TVM Solver
2:tvm_Pmt
3:tvm_I%
4:tvm_PV
5:tvm_N
6:tvm_FV
7↓npv(
```

Press ENTER. Key in the known data. N is the number of payments (60 = 12 × 5). PN is the number of payments per year, which is 12. C/Y is the number of compounding periods per year, which is 12.

```
N=60
I%=9
PV=12000
PMT=0
FV=0
P/Y=12
C/Y=12
PMT:END BEGIN
```

In some cases, it is necessary to find the present value of an annuity given the payment. This formula is used to determine, among other things, the lump sum payment for a lottery winner and the amount due when a loan is paid off early.

Present Value of an Annuity

Suppose an annuity earns an annual interest rate of r% compounded m times a year for t years. Then the present value PV of the annuity is

$$PV = PMT\left[\frac{1-(1+i)^{-n}}{i}\right]$$

where PMT is the payment, $i = \frac{r}{m}$ and $n = mt$.

Alternative to Example 2
Exercise 2, page 627

Press the up arrow key to highlight PMT. Then press ALPHA[SOLVE]. The SOLVE function is above ENTER.

```
  N=60
  I%=9
  PV=12000
▪ PMT=-249.10026...
  FV=0
  P/Y=12
  C/Y=12
  PMT: END  BEGIN
```

The monthly payment is \$249.10. Showing the payment as a negative number is fairly typical of all financial programs, including spreadsheets. The idea is that because you are making the payments, your capital is decreasing.

TAKE NOTE

In many lotteries, the winner has the option of taking payments over a period of time or taking a lump sum amount. The lump sum amount is the present value of an annuity.

EXAMPLE 2 Determine the Present Value of an Annuity

Suppose the winner of a \$5 million lottery decides to take a lump sum amount rather than a series of annual payments of \$250,000 for 20 years. If the current annual interest rate is 6%, what lump sum amount will the winner receive?

Solution The lump sum amount is the present value of the series of payments of 250,000. The payments are once a year ($m = 1$), the annual interest rate is 6% ($r = 0.06$), and the number of years is 20 ($t = 20$).

$$PV = PMT\left[\frac{1-(1+i)^{-n}}{i}\right]$$

$$PV = 250{,}000\left[\frac{1-(1+0.06)^{-20}}{0.06}\right]$$

▪ $PMT = 250{,}000$, $i = \frac{0.06}{1} = 0.06$, $m = 1(20) = 20$

$$PV \approx 250{,}000(11.46992122)$$

$$\approx 2{,}867{,}480.305$$

The lump sum amount is \$2,867,480.31.

CHECK YOUR PROGRESS 2 The parents of a college student want to provide an allowance of \$400 per month for 4 years for their child. How much must be deposited into an account that earns 9% interest, compounded monthly, to attain that goal? (Assume the student will receive a payment every month of the year.)

Solution *See page S40.* \$16,073.91

Suppose you have a five-year car loan and decide after three years that you want to trade-in the car and get a new one. Since you have not kept the car the full length of the loan, you still owe money on the car. That amount, called the **payoff**, is the current present value of the loan and can be determined from the formula for the present value of an annuity.

Alternative to Example 3
Exercise 16, page 627

EXAMPLE 3 Determine a Loan Payoff

A real estate agent has a four-year car loan of \$9000 at an annual interest rate of 8.75% and a monthly payment of \$220.90. After making car payments for 27 months, the agent decides to trade in the car and purchase another one. What is the loan payoff?

Continued ➤

Solution To calculate the loan payoff, use the present value of an annuity formula to calculate the present value of the loan. Because the agent has made 27 payments and the original loan called for making 48 payments (4 years), there are 21 payments remaining to be made. Thus $n = 21$.

$$PV = PMT\left(\frac{1-(1+i)^{-n}}{i}\right)$$

$$PV = 220.90\left(\frac{1-\left(1+\frac{0.0875}{12}\right)^{-21}}{\frac{0.0875}{12}}\right)$$

▪ $PMT = 220.90$, $i = \frac{0.0875}{12}$, $n = 21$.

$$PV \approx 4286.74$$

The loan payoff is \$4286.74.

CHECK YOUR PROGRESS 3 What is the loan payoff after 35 monthly payments of \$177.48 have been made on a five-year loan of \$8500 at an annual interest rate of 9.25%?

Solution *See page S40.* \$4021.63

▪ Amortization Schedules and Home Loans

In 1969, the Truth in Lending Act was passed by Congress. One of the purposes of this act was to require that all lenders inform consumers regarding how much interest would be repaid on a loan and the interest rate. The interest rate must be based on the *unpaid balance* of the loan.

To illustrate, suppose that on January 1, a person gets a loan of \$1000 at an annual interest rate of 12%, compounded monthly, and makes monthly payments of \$88.85. On February 1, when the first payment is due, the borrower still owes \$1000 and must pay interest on that amount for the month of January.

$$I = Prt$$

$$\text{Interest for January} = 1000(0.12)\frac{1}{12} = 10$$

▪ Time is 1 month, which is $\frac{1}{12}$ year.

The interest for January is \$10. This means that of the monthly payment of \$88.85, the interest portion was \$10. The remaining \$78.85 went to paying off the \$1000 loan. So, after the first payment, the borrower owes \$1000 − \$78.85 = \$921.15.

On March 1, when the second payment is due, the borrower owes \$921.15 and must pay interest on that amount, not on the original \$1000.

$$I = Prt$$

$$\text{Interest for February} = 921.15(0.12)\frac{1}{12} = 9.2115$$

The interest for February is \$9.21.

Of the March 1 payment of $88.85, the interest portion was $9.21. The remaining $79.64 went to paying off the $921.15 that is still owed. After the second payment, the borrower owes $921.15 − $79.64 = $841.51.

The procedure is repeated over and over until the loan is repaid. Note from the 2 months we have shown that the amount of interest paid in the second month was less than the amount paid in the first month. As each payment is made, less principal is owed. Because the principal is declining, the amount of interest owed is declining.

Point of Interest

The term *amortize* comes from a Latin word meaning "to die." By making payments on the loan, the loan is "dying off." The word *mortgage* has the same root as the word *amortize.*

As more payments are made, the borrower owes less on the loan. The borrower is said to be **amortizing** the loan. The table below is called an **amortization schedule.** It shows how much of each payment is applied to paying off the loan (the principal) and how much of the payment is interest. The amortization schedule for the $1000 loan discussed above is shown in the table.

Amortization Schedule

Payment number	Payment amount	Amount owed	Interest portion of payment	Principal portion of payment	New principal
1	$88.85	$1000.00	$10.00	$78.85	$921.15
2	88.85	921.15	9.21	79.64	841.51
3	88.85	841.51	8.42	80.43	761.08
4	88.85	761.08	7.61	81.24	679.84
5	88.85	679.84	6.80	82.05	597.79
6	88.85	597.79	5.98	82.87	514.92
7	88.85	514.92	5.15	83.70	431.22
8	88.85	431.22	4.31	84.54	346.68
9	88.85	346.68	3.47	85.38	261.30
10	88.85	261.30	2.61	86.24	175.06
11	88.85	175.06	1.75	87.10	87.96
12	88.85	87.96	0.88	87.97	−0.01

After the 12th payment is made, the new principal is −$.01, which means the borrower has repaid the entire loan and 1 cent more.

Remember that an annuity is characterized by equal payments in equal time periods. Thus the loan payment for a home, a car, a stereo system, or any other item that is purchased by securing a loan that is paid back in equal payments at equal time intervals can be calculated using the payment formula for the present value of an annuity. When a loan for a home is secured, the loan is called a **mortgage** and the payment is referred to as the **mortgage payment.**

Alternative to Example 4

An engineer secures a 30-year mortgage of \$230,000 at an annual interest rate of 6.85%.

a. Find the monthly mortgage payment.

b. How much of the first payment is interest and how much goes toward reducing the principal?

- **a. \$1507.10**
- **b. \$194.18, \$1312.92**

Point of Interest

If the computer technician in Example 4 keeps the loan for 30 years, the technician will make 360 payments of \$859.69, for a total amount paid of 360(\$859.69) = \$309.488.40. The original loan amount was \$120,000, so the technician will pay \$189,488.40 in interest.

EXAMPLE 4 Determine a Mortgage Payment

A computer technician secures a 30-year mortgage of \$120,000 at an annual interest rate of 7.75%.

a. Find the monthly mortgage payment.

b. How much of the first payment is interest and how much goes toward reducing the principal?

Solution

a. To find the monthly payment, use the present value of an annuity formula for *PMT*.

$$PMT = PV\left[\frac{i}{1-(1+i)^{-n}}\right]$$

$$= 120{,}000\left[\frac{\frac{0.0775}{12}}{1-\left(1+\frac{0.0775}{12}\right)^{-360}}\right] \qquad \blacksquare\ PV = 120{,}000,\ i = \frac{0.0775}{12},\ n = 12(30) = 360$$

$$\approx 120{,}000(0.00716412)$$

$$\approx 859.69$$

The monthly payment is \$859.69.

b. To calculate the amount of interest, use the simple interest equation. Because this is the first payment, the amount of the mortgage (the principal) is \$120,000. The time is 1 month, or $\frac{1}{12}$ year.

$$I = Prt$$

$$= 120{,}000(0.0775)\frac{1}{12} \qquad \blacksquare\ P = 120{,}000,\ r = 0.0775,\ t = \frac{1}{12}$$

$$= 775$$

The amount of interest paid is \$775.

To find how much of the payment goes to paying off the loan, subtract the interest from the monthly mortgage payment.

$$\text{Principal paid} = 859.69 - 775.00 = 84.69$$

The amount of principal paid is \$84.69.

CHECK YOUR PROGRESS 4 A web page designer purchased a condominium and financed \$189,000 for 20 years at an annual interest rate of 8.3%.

a. Find the monthly mortgage payment.

b. How much of the first payment is interest and how much goes toward reducing the principal?

Solution *See page S40.* **a.** \$1616.34 **b.** Interest, \$1307.25; principal, \$309.09

■ Annual Percentage Rate (APR)

As we mentioned earlier, the Truth in Lending Act was designed to give consumers a standard way of comparing loans and the interest charged on them. The interest rate is called the annual percentage rate and is abbreviated APR. It is the interest rate in the present value of an annuity formula.

Example 5 shows how APR is calculated and illustrates why there was a need for uniformity. For this loan, the borrower is offered **add-on interest.** In this case, a simple interest rate is given and then the simple interest is "added on" to the loan amount. The loan amount plus the add-on interest is divided by the number of months of the loan to determine the monthly payment.

QUESTION What is the add-on interest for a loan of $2500 for 3 years if the simple interest rate is 7%?

Alternative to Example 5
Exercise 20, page 628

INSTRUCTOR NOTE
APR calculations require either a financial calculator or a graphing calculator. The solution to Example 5 uses a TI-83 graphing calculator with built-in financial functions.

EXAMPLE 5 Calculate an APR

Emily Longstreet purchases a stereo system and finances $750 for 2 years at an add-on interest rate of 12%.

a. Find the monthly payment.

b. Find the annual percentage rate. Round to the nearest tenth of a percent.

Solution

a. Find the amount of simple interest that is added to the loan.

$$\begin{aligned} I &= Prt \\ &= 750(0.12)(2) \\ &= 180 \end{aligned}$$

The simple interest added to the loan is $180. The loan amount is $750 + $180 = $930.

To find the monthly payment, divide the loan amount by 24 months.

$$\text{Monthly payment} = \frac{930}{24} = 38.75.$$ The monthly payment is $38.75.

b. To calculate the APR we must solve the present value of an annuity formula for r. This cannot be done without the aid of a graphing or financial calculator. On the following page, we use a TI-83 Plus graphing calculator.

Continued ➤

ANSWER $I = 2500(0.07)(3) = 525$. The add-on interest is $525.

Press APPS 1 ENTER and input the known values. Because the payment is money paid out, it is entered as a negative number.

```
N=24
I%=0
PV=750
PMT=-38.75
FV=0
P/Y=12
C/Y=12
PMT: END BEGIN
```

N = 24 (number of payments)
I% is unknown.
PV = 750 (amount borrowed)
PMT = −38.75 (the payment)
FV = 0 (no money is owed after all the payments)
P/Y = 12 (12 payments per year)
C/Y = 12 (interest is compounded monthly)

```
N=24
▪ I%=21.57124527
PV=750
PMT=-38.75
FV=0
P/Y=12
C/Y=12
PMT: END BEGIN
```

Use the arrow key to highlight I%. Press ALPHA [SOLVE].

The APR is 21.6%.

CHECK YOUR PROGRESS 5 Andrew McCellan purchased a HDTV and financed $1100 for 3 years at an add-on interest rate of 9%.

a. Find the monthly payment.

b. Find the APR. Round to the nearest tenth of a percent.

Solution *See page S41.* **a.** $38.81 **b.** 16.3%

Note from Example 5 that the add-on interest rate was 12%, but the APR was 21.6%. This discrepancy in values made it difficult for consumers to evaluate a loan. With the passage of the Truth in Lending Act, a standard was mandated so that the consumer could compare interest rates. That standard is APR.

Topics for Discussion

1. What is an annuity?
2. What does it mean to amortize a loan?
3. What is an amortization schedule?
4. What is APR and how does it differ from add-on interest?

EXERCISES 7.4

— *Suggested Assignment: Exercises 1–33, odd; and 35–40, all.*

1. ***Present Value*** Determine the present value of a five-year annuity for which payments of $250 are made quarterly and the annual interest rate is 8%, compounded quarterly. $4087.86
2. ***Present Value*** Determine the present value of a 10-year annuity for which payments of $100 are made monthly and the annual interest rate is 9%, compounded monthly. $7894.17
3. ***Annuity Payment*** The present value of a six-year annuity is $5000. Determine the monthly payments if the annual interest rate is 6%, compounded monthly. $82.86
4. ***Annuity Payment*** The present value of a 15-year annuity is $10,000. Determine the quarterly payments if the annual interest rate is 7.5%, compounded quarterly. $279.04
5. ***Annuity Payment*** How much money must an accountant place in an account that earns a 6.75% annual interest rate compounded quarterly to allow for quarterly withdrawals of $2500 for 6 years? $49,002.30
6. ***Annuity Payment*** How much money must a police officer place in an account that earns 7.5% annual interest rate, compounded monthly, to allow for monthly withdrawals of $250 for 10 years? $21,061.19
7. ***Annual Percentage Rate*** Find the annual percentage rate on a two-year loan of $4250 for which the monthly payments are $210 and interest is compounded monthly. Round to the nearest tenth of a percent. 16.9%
8. ***Annual Percentage Rate*** Find the annual percentage rate on a five-year loan of $25,000 for which the monthly payments are $500 and interest is compounded monthly. Round to the nearest tenth of a percent. 7.4%

Business and Economics

9. ***Car Financing*** Mary Scot purchases a car and finances $15,000 for 5 years at an annual interest rate of 8.5%, compounded monthly. Determine the monthly payment for the loan. $307.75
10. ***SUV Financing*** Leonard Sanchez purchased a sports utility vehicle and financed $17,500 for 4 years at an annual interest rate of 8.75%. Determine the monthly payment for the loan. $433.41
11. ***Sweepstakes Winner*** A person who won a supermarket sweepstakes was given the option of receiving five annual payments of $9500 or monthly payments of $425 for 10 years. If interest rates are 8%, which option has the larger present value? Assume that the number of compounding periods per year is the same as the number of payments per year. Five annual payments of $9500; present value = $37,930.75
12. ***Retirement Planning*** Allison Witherspoon has the choice of two supplemental retirement options. The first will pay her $750 per month for 10 years. The second will pay her $575 per month for 15 years. If interest rates are 7.5%, which retirement plan offers Allison the larger present value? $750/month for 10 years; present value = $63,183.56
13. ***Computer Purchase*** Minh Nguyen purchased a computer and financed $2500 for 3 years at an annual interest rate of 9.2%, compounded monthly. How much interest will Minh pay over the term of the loan? $370.28

14. ***Jet Ski Financing*** David Isaacs purchased a jet ski and financed $1700 for 2 years at an annual interest rate of 8.75%, compounded monthly. How much interest will David pay over the term of the loan? $159.28
15. ***Car Financing*** The owner of a coffee shop has a five-year car loan of $12,500 at an annual interest rate of 8.5% and a monthly payment of $256.46. After the owner makes car payments for 30 months, the car is traded in. What is the loan payoff? $6909.35
16. ***Truck Financing*** An electrician has a four-year truck loan of $15,000 at an annual interest rate of 7.9%, compounded monthly, and a monthly payment of $365.49. After the electrician makes truck payments for 36 months, the truck is traded in. What is the loan payoff? $4203.83
17. ***Mortgage Loan Payoff*** A telephone technician purchased a home and obtained a mortgage of $250,000 for 30 years at an annual interest rate of 7.5%, compounded monthly. After making mortgage payments for 5 years, the technician decides to sell the home. What is the loan payoff? (*Suggestion:* You must first calculate the mortgage payment.) $236,544.10

18. ***Mortgage Loan Payoff*** The owner of an apartment building has a mortgage of \$975,000 for 30 years at an annual interest rate of 8.25%, compounded monthly. After making monthly mortgage payments for 7 years, the owner decides to sell the building. What is the loan payoff? (*Suggestion:* You must first calculate the mortgage payment.) \$904,637.89

19. ***Mortgage Interest*** A homeowner has a 30-year mortgage of \$225,000 at an annual interest rate of 7.125%, compounded monthly. If the homeowner keeps the loan for 30 years, how much interest will be paid? \$320,713.20

20. ***Office Remodel*** Elizabeth Scallet financed \$5000 to remodel her sales office. Her loan was a 7% add-on interest loan for 3 years.

a. What are the monthly payments? \$168.06

b. What is the annual percentage rate for the loan? Round to the nearest tenth of a percent. 12.8%

21. ***Loan Repayment*** The owner of a print shop purchases a new color copier and finances \$5000. If the owner makes payments of \$300 per month and the interest rate on the loan is 7.5%, compounded monthly, how many months, to the nearest month, will it take to repay the loan? (*Suggestion:* Solve the present value of an annuity equation for t. You can do this algebraically by using logarithms or by using the TMV Solver on a TI-83 graphing calculator.) 18 months

22. ***Monthly Mortgage Interest*** The amount of interest I that is paid in any one *monthly* payment can be calculated using $I = \frac{r}{12}(PV_n)$, where r is the annual interest rate and PV_n is the present value of an annuity with n payments, including the present payment, remaining. Suppose the owner of a car has a five-year, \$7000 loan at an annual interest rate of 8.5%. How much interest is paid in the 12th payment? \$41.99

23. ***Monthly Mortgage Interest*** Suppose the owner of a home has a 30-year, \$77,000 loan at an annual interest rate of 7.25%. How much interest is paid in the 24th payment? (See Exercise 22.) \$456.29

24. ***State Lottery*** Some state lotteries offer winners of the grand prize the option of taking annual payments for 20 years or a lump sum. The lump sum is the present value of the payments. Suppose the winner of a \$3,000,000 lottery chooses the lump sum option. If interest rates are 7%, what lump sum will the lottery winner receive? \$1,589,102.14

25. ***Sweepstakes Winner*** Suppose a sweepstakes winner wins the grand prize of \$10,000,000 and the amount is paid in annual installments over 30 years or as a lump sum. If the current annual interest rate is 5%, what lump sum amount will the winner receive? \$5,124,150.29

26. ***Retirement Planning*** As a supplement to a retirement plan, a dietician deposits \$10,000 into an account that earns an interest rate of 7.5%, compounded monthly. The dietician plans on leaving the money in the account for 8 years and then making monthly withdrawals. How much money can be withdrawn each month if the dietician wants to deplete the account in 5 years? (*Suggestion:* First find the future value of the \$10,000 deposit. Then use that amount as the present value of an annuity.) \$364.43

27. ***Education Planning*** A parent deposits \$15,000 into an education account that earns an interest rate of 8.25%, compounded monthly. The plan is to leave the money in the account for 15 years and then make annual withdrawals to pay for a child's college education. How much money can be withdrawn each year if the parent wants to deplete the account in 4 years and the interest rate remains at 8.25%? (*Suggestion:* First find the future value of the \$15,000 deposit. Then use that amount as the present value of an annuity.) \$15,631.38

28. ***Corporate Bonds*** Suppose that a corporate bond pays \$250 per year for 20 years. At the end of the 20 years, the owner will also receive \$10,000. The current value of the bond is the present value of the \$250 payments plus the present value of the \$10,000. Find the value of the bond if the annual interest rate is 8%, compounded annually. \$4600.02

29. ***Corporate Bonds*** A corporate bond pays $150 semiannually for 10 years. At the end of the 10 years, the owner also receives $15,000. Find the present value of the bond if the annual interest rate is 6%, compounded semiannually. $10,536.76

Life and Health Sciences

30. ***Farming*** A farmer purchases a plot of land and obtains a 20-year loan of $150,000 at an annual interest rate of 7.25%.

    a. Find the monthly mortgage payment. $1185.56

    b. How much of the first payment is interest and how much goes to reducing the principal? $906.25, $279.31

    c. How much of the second payment is interest and how much goes to reducing the principal? Interest = $904.56, principal = $281

31. ***Veterinary Science*** A veterinarian purchased an office building and secured a 30-year loan of $250,000 at an annual interest rate of 8% compounded monthly.

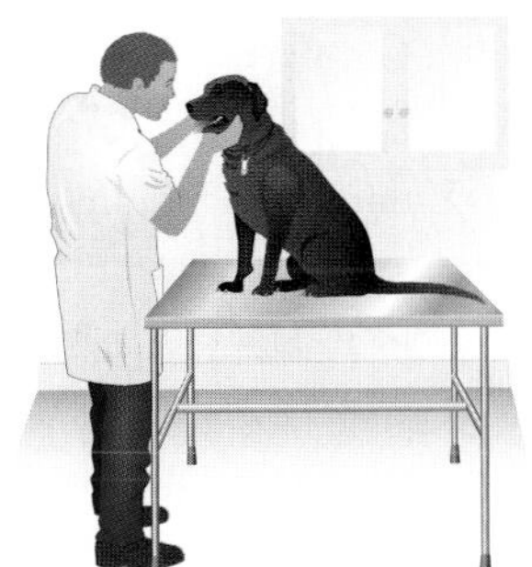

    a. Find the monthly mortgage payment. $1834.41

    b. How much of the first payment is interest and how much goes to reducing the principal? Interest = $1666.67, principal = $167.74

    c. How much of the second payment is interest and how much goes to reducing the principal? Interest = $1665.55, principal = $168.86

32. ***Dentistry*** A dentist purchases a home and obtains a 20-year mortgage of $275,000 at an annual interest rate of 8.75%, compounded monthly. If the dentist keeps the loan for 20 years, how much interest will be repaid? $308,248

33. ***Medicine*** A physician's assistant purchased a small boat and financed $2500 at 6.5% add-on interest for 2 years.

    a. What are the monthly payments? $117.71

    b. What is the annual percentage rate for the loan? Round to the nearest tenth of a percent. 12.0%

34. ***Credit Card Debt*** A dog trainer has a credit card debt of $1950 and is required to make a minimum monthly payment of $40. If the annual interest rate on the credit card is 18%, compounded monthly, how long will it take the dog trainer (assuming no new purchases) to repay the debt? (*Suggestion:* Solve the present value of an annuity equation for t. You can do this algebraically by using logarithms or by using the TMV Solver feature on a TI-83 graphing calculator.) 88 months

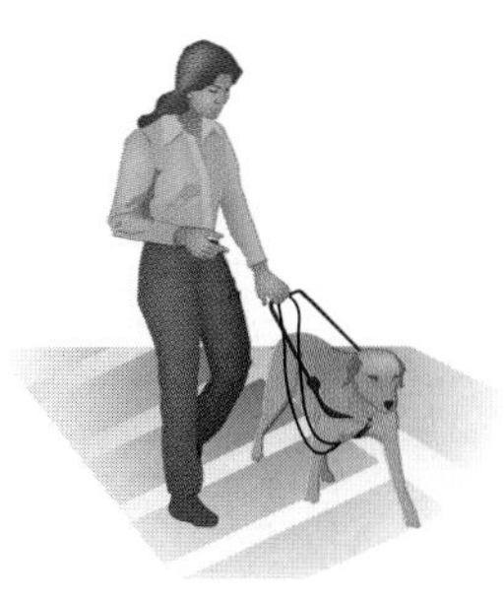

Prepare for Section 7.5

35. Evaluate $\frac{(1+i)^n - 1}{i}$ when $i = 0.05$ and $n = 4$. Round to the nearest hundredth. [P.1] 4.31

36. Solve $a = b\left[\frac{(1+i)^n - 1}{i}\right]$ for b. [1.1] $b = a\left[\frac{i}{(1+i)^n - 1}\right]$

37.  Solve $\frac{(1+x)^5 - 1}{x} = 15$. Round to the nearest hundredth. Use Xmin = 0, Xmax = 1, Ymin = 0, and Ymax = 25. [5.4] 0.57

38. Solve $2\log(1 + x) = 3$. Round to the nearest hundredth. [5.4] 30.62

39. What is the common ratio for the geometric sequence $3000, 3000(1.06)^1, 3000(1.06)^2, \ldots, 3000(1.06)^{n-1}$? [7.2] 1.06

40. Evaluate $2 + 2(1.1) + 2(1.1)^2 + \cdots + 2(1.1)^{n-1}$ when $n = 10$. Round to the nearest hundredth. [7.2] 31.87

Explorations

1. ***Mortgage Terms*** Suppose a home buyer obtains a loan of $125,000 at an annual interest rate of 7.5%, compounded monthly, and is considering a loan of 15 years, 20 years, or 30 years.

    a. Complete the table by finding the monthly mortgage payment for each loan and the total amount of interest that would be paid over the term of the loan.

Years	Monthly mortgage payment ($)	Total interest paid ($)
15	1158.77	83,578.60
20	1006.99	116,677.60
30	874.02	189,647.20

b. Are the monthly mortgage payments on a 15-year loan one-half the monthly mortgage payments on a 30-year loan? No

c. Is the total interest paid on a 15-year loan one-half the amount paid on a 30-year loan? No

d. The amount of a 20-year loan is two-thirds the amount of a 30-year loan. Is the mortgage payment on the 20-year loan two-thirds the mortgage payment on the 30-year loan? No

2. If you have a credit card that currently contains an unpaid balance, determine how many months it would take to repay the amount owed (assuming no new purchases) if you made the minimum monthly payment. Answers will vary.

3. For a given loan amount and interest rate, the payment amount will decrease as the number of years of the loan increases. However, there is a limit to how small the payment can be even if the loan period is 1000 years. Completing this Exploration can help you understand this concept. Suppose a person has a loan of $100,000 at an annual interest rate of 8% and will make monthly payments.

a. Complete the following table for the different time periods of the loan.

Years	Monthly mortgage payment
20	836.44
30	733.76
40	695.31
50	679.27
100	666.90
200	666.67
500	666.67
1000	666.67

b. From the calculations in part **a.**, what appears to be the minimum monthly payment? $666.67

c. Evaluate $\frac{r}{m}(PV)$, where r is the annual interest rate, m is the number of compounding periods per year, and PV is the amount of the mortgage. How does this result compare with the answer to part **b.**? They are the same.

SECTION 7.5 Future Value of an Annuity and Sinking Funds

- Future Value of an Annuity
- Sinking Funds

▪ Future Value of an Annuity

Recall that an annuity is a payment schedule of *equal* amounts in *equal* periods of time. In the last section, we discussed the case of a person borrowing money to buy a car and making monthly payments to pay off the loan.

An annuity can also occur when a person or company wants to save money for some future purchase. For instance, a person may make regular deposits into a savings account to save money to make a down payment on a house or to go on a vacation. A company may make regular deposits into an account so that at some future time there will be money available to replace worn-out equipment or expand the company's facilities.

Suppose a company deposits \$3000 on December 31 of each year for 5 years into an account earning 6% interest, compounded annually. The total value of all the deposits and the accrued interest is called the **future value of the annuity.** This concept is illustrated below, where we have rounded calculations to the nearest cent.

Future value of each deposit

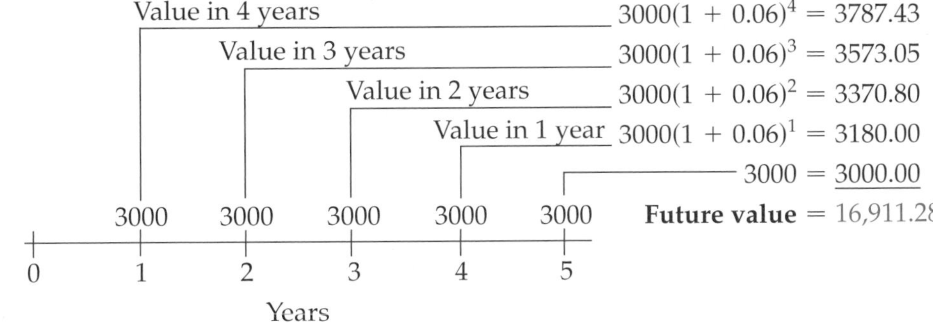

The future value of the payments is \$16,911.28.

The terms of the sum are shown below. Note that the terms form a geometric sequence with a common ratio of 1.06.

$$3000, 3000(1.06)^1, 3000(1.06)^2, 3000(1.06)^3, 3000(1.06)^4$$

Instead of calculating each term of this sequence and then adding the results (as we did above), we can use the formula for the sum of a geometric series to find the future value of this annuity.

$$S_n = \frac{a(1 - r^n)}{1 - r}$$

$$FV = \frac{3000(1 - 1.06^5)}{1 - 1.06} \approx \frac{3000(1 - 1.3382256)}{-0.06}$$

■ S_n is the future value, FV. $a = 3000$, $r = 1.06$, and n (number of terms) $= 5$

$$= \frac{3000(-0.3382256)}{-0.06}$$

$$\approx 16{,}911.28$$

Note that the future value calculated using this method is the same as the result obtained using the first method.

P

Future Value of an Ordinary Annuity

Suppose an annuity earns an annual interest rate of $r\%$ compounded m times a year for t years. Then the future value FV of the annuity is

$$FV = PMT\left[\frac{(1 + i)^n - 1}{i}\right]$$

where PMT is the amount of each deposit, $i = \dfrac{r}{m}$, and $n = mt$.

INTEGRATING TECHNOLOGY

A graphing calculator can be used to find the future value of an annuity. Typical screens for a TI-83 Plus are shown below.

Press APPS 1.

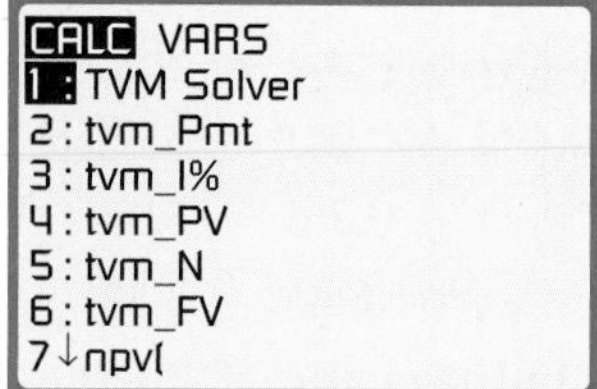

Press ENTER. Key in the known data. N is the number of payments ($40 = 2 \times 20$). The payment is entered as a negative number to indicate that money is being paid out. P/Y is the number of payments per year. C/Y is the number of compounding periods per year.

N=40
I%=6
PV=0
PMT=-1200
FV=0
P/Y=2
C/Y=2
PMT: END BEGIN

Press the up arrow key to highlight FV. Then press ALPHA [SOLVE]. The SOLVE function is above ENTER.

N=40
I%=6
PV=0
PMT=-1200
▪ FV=90481.5116
P/Y=2
C/Y=2
PMT: END BEGIN

The future value is \$90,481.51.

QUESTION Suppose an annuity earns an annual interest rate of 6%, compounded quarterly, for 5 years. Using the future value of an ordinary annuity formula, what are the values of i and n?

EXAMPLE 1 Find the Future Value of an Annuity

A biochemistry lab technician decides to set up an IRA (Individual Retirement Account) into which \$1200 will be deposited every 6 months. If the account earns 6% interest compounded semiannually, find the value of the retirement account in 20 years.

Alternative to Example 1
Exercise 2, page 636

Solution Use the future value of an annuity formula.

$$FV = PMT\left[\frac{(1+i)^n - 1}{i}\right]$$

$$= 1200\left[\frac{\left(1+\frac{0.06}{2}\right)^{40} - 1}{\frac{0.06}{2}}\right] \qquad \text{▪ } PMT = 1200,\ i = \frac{0.06}{2},\ n = 2(20) = 40$$

$$= 1200(75.40125973)$$

$$\approx 90{,}481.51$$

The value of the retirement account will be \$90,481.51.

CHECK YOUR PROGRESS 1 A customer service representative deposits \$125 per month into an account that earns 8% interest, compounded monthly. Find the value of the account in 15 years.

Solution *See page S41.* \$43,254.78

In Example 1, the lab technician was saving \$2400 per year by putting \$1200 in the account every 6 months. Suppose instead that the technician decided to put \$2400 into the account just once a year. Then

$$FV = PMT\left(\frac{(1+i)^n - 1}{i}\right)$$

$$= 2400\left(\frac{(1+0.06)^{20} - 1}{0.06}\right) \qquad \text{▪ } PMT = 2400,\ r = 0.06,\ n = 1(20) = 20$$

$$= 2400\left(\frac{1.06^{20} - 1}{0.06}\right)$$

$$\approx 88.285.42$$

This is approximately \$2200 less than the result in Example 1. Thus, although the total annual amount deposited is the same for each IRA, depositing the money semiannually results in a larger retirement value. For a given annual amount, the future value of an ordinary annuity will be greater if smaller amounts are

ANSWER $i = \frac{0.06}{4} = 0.015$; $n = 4(5) = 20$

deposited more frequently. For instance, if \$200 per month (\$2400 per year) is deposited into an account that earns 6% compounded monthly, the future value of the account in 20 years will be \$92,408.18, about \$2000 more than if the money had been deposited semiannually.

If you deposit money into an account on a regular basis, not only is the principal growing (the amount being saved), but the account is earning more and more interest.

Alternative to Example 2
Exercise 12, page 636

EXAMPLE 2 Find the Future Value of an Annuity and the Interest Earned

A veterinary assistant deposits \$75 at the end of each month into an account that earns 8% interest, compounded monthly.

a. Find the future value of the account after 10 years.

b. Find the amount of interest earned by the account after 10 years.

Solution

a. Use the future value of an annuity formula.

$$FV = PMT\left[\frac{(1+i)^n - 1}{i}\right]$$

$$= 75\left[\frac{\left(1+\frac{0.08}{12}\right)^{120} - 1}{\frac{0.08}{12}}\right]$$

▪ $PMT = 75$, $i = \frac{0.08}{12}$, $n = 12(10) = 120$

$$\approx 75(182.9460352)$$

$$\approx 13{,}720.95$$

The future value will be \$13,720.95.

b. To find the interest earned, subtract the value of the 120 payments of \$75 from the future value.

$$\text{Interest earned} = 13{,}720.95 - 120(75)$$

$$= 13{,}720.95 - 9000 = 4720.95$$

The account earned \$4720.95 interest.

CHECK YOUR PROGRESS 2 The owner of a hardware store deposits \$250 at the end of each month into an account that earns 7% interest, compounded monthly.

a. Find the future value of the account after 15 years.

b. Find the amount of interest earned by the account after 15 years.

Solution *See page S41.* **a.** \$79,240.57 **b.** \$34,240.57

TAKE NOTE

The term *sinking fund* may seem strange here. Basically, the idea is similar to someone saying, "I keep *sinking* money into this car." The person is putting money into the car to keep it going.

▪ Sinking Funds

Part of good financial management for both a company and an individual is to plan for future events. For instance, a company may know that it will need to purchase a new machine in 5 years, or a parent may have to plan for the education of a child.

Point of Interest

Recall that a bond is a type of investment sold by corporations and government agencies. The corporation or agency promises to pay the investor a certain amount of money in the future. For instance, if a corporation sells a $1000 bond that is payable in 5 years, the corporation is promising to pay an investor $1000 in 5 years. To ensure that it will have the money that will be owed to the investor, the corporation sets up a sinking fund.

Establishing a savings plan in order to have a certain amount of money available in the future is called creating a **sinking fund.** The payment is called the **sinking fund payment.** The amount of this payment can be calculated by solving the future value of an annuity formula for PMT.

Sinking Fund Payment

The amount of a payment into an annuity that earns an annual interest rate of $r\%$ compounded m times a year for t years that is necessary to attain a given future value FV is

$$PMT = FV\left[\frac{i}{(1+i)^n - 1}\right]$$

where $i = \frac{r}{m}$ and $n = mt$.

Alternative to Example 3
Exercise 6, page 636

EXAMPLE 3 Calculate a Sinking Fund Payment

Suppose a road construction company knows that it will need to purchase a new grader in 5 years. The company estimates that the cost of the graders will be $75,000. How much money should this company deposit at the end of each quarter into an account that earns 8% interest, compounded quarterly, to reach its goal of $75,000?

Solution In this case, the future value of the annuity is known ($75,000) and we must calculate the payment.

$$PMT = FV\left[\frac{i}{(1+i)^n - 1}\right]$$

$$PMT = 75{,}000\left[\frac{\frac{0.08}{4}}{\left(1+\frac{0.08}{4}\right)^{20} - 1}\right]$$

▪ $FV = 75{,}000$, $i = \frac{0.08}{4}$, $n = 4(5) = 20$

$$\approx 75{,}000(0.04115672)$$

$$\approx 3086.75$$

The company should deposit $3086.75 at the end of each quarter.

CHECK YOUR PROGRESS 3 A software sales engineer wants to take a cruise in 2 years. The engineer estimates that the cost of the cruise will be $3000. How much should the engineer deposit each month into an account that earns 6% interest, compounded monthly, to save the amount of the cruise?

Solution *See page S41.* $117.96

Using a financial calculator or a graphing calculator, it is possible to determine the interest rate an investment must pay to reach a certain goal.

Alternative to Example 4
Exercise 14, page 636

INTEGRATING TECHNOLOGY

If your graphing calculator does not support financial calculations, you can solve Example 4 by using the intersect feature. Input Y1 and Y2 as shown below.

Graph the equations. We used Xmin = 0, Xmax = 0.2, Ymin = 0, Ymax = 25000. Once the graphs are on the screen, use the intersect feature of the graphing calculator to find the point of intersection.

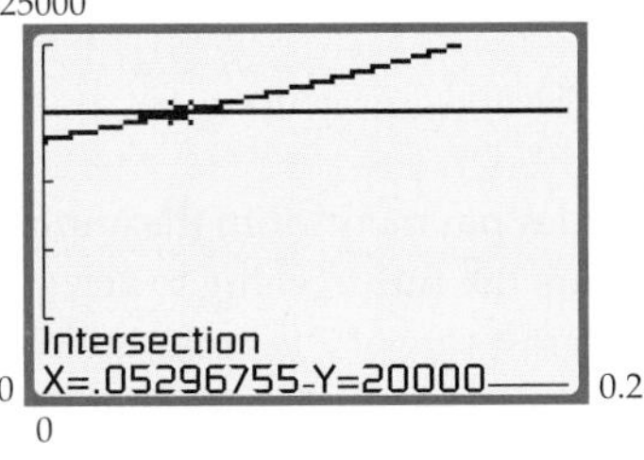

EXAMPLE 4 Find an Interest Rate

A landscape designer is planning on purchasing a house in 4 years. If the designer wants to have \$20,000 for the down payment and is willing to deposit \$375 per month into an account, what is the smallest annual interest rate, compounded monthly, the landscape designer can receive and still reach the goal?

Solution Press APPS 1 ENTER and input the known values. Because the payment is money paid out, it is entered as a negative number.

```
N=48
I%=0
PV=0
PMT=-375
FV=20000
P/Y=12
C/Y=12
PMT: END BEGIN
```

N = 48 (number of payments)
I% is unknown.
PV = 0 (No money is currently in the account.)
PMT = −375 (the payment)
FV = 20,000 (the down payment goal)
P/Y = 12 (12 payments per year)
C/Y = 12 (Interest is compounded monthly.)

```
N=48
▪ I%=5.29675505
PV=0
PMT=-375
FV=20000
P/Y=12
C/Y=12
PMT: END BEGIN
```

Use the arrow key to highlight I%. Press ALPHA [SOLVE].

The landscape designer must receive an annual interest rate of at least 5.3% (rounded to the nearest tenth).

CHECK YOUR PROGRESS 4

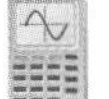

A flight attendant wants to purchase a new car in 2 years and make a down payment of \$3500. If the flight attendant deposits \$135 per month into an account, what annual interest rate, compounded monthly, must the flight attendant receive to reach the goal? Round to the nearest tenth of a percent.

Solution *See page S41.* 8.0%

Topics for Discussion

1. What is the future value of an annuity?
2. Give an example of a situation for which it is necessary to compute the future value of an annuity.
3. How does the future value of an annuity differ from the present value of an annuity?
4. What is a sinking fund?
5. Give an example of a situation in which a sinking fund might be appropriate.

EXERCISES 7.5

— *Suggested Assignment: Exercises 1–19, odd.*

1. ***Future Value*** Determine the future value of a 10-year annuity for which payments of \$500 are made at the end of each quarter and the interest rate is 8%, compounded quarterly. \$30,200.99

2. ***Future Value*** Determine the future value of a five-year annuity for which payments of \$250 are made at the end of each month and the interest rate is 9%, compounded monthly. \$18,856.03

3. ***Future Value*** Determine the future value of a 15-year annuity for which payments of \$1000 are made semiannually and the interest rate is 7%, compounded semiannually. \$51,622.68

4. ***Future Value*** Determine the future value of a 12-year annuity for which payments of \$2500 are made quarterly and the interest rate is 6%, compounded quarterly. \$173,913.05

5. ***Sinking Fund Payment*** How much money must be deposited at the end of each month into an account that has an interest rate of 8.5%, compounded monthly, to have a future value of \$10,000 in 5 years? \$134.33

6. ***Sinking Fund Payment*** How much money must be deposited at the end of each quarter into an account that has an interest rate of 6.2%, compounded quarterly, to have a future value of \$7500 in 4 years? \$416.63

Business and Economics

7. ***Value of an IRA*** Emmy Northern places \$750 at the end of every 6 months into an IRA that has an interest rate of 8.25%, compounded semiannually. Find the value of Emmy's account after 15 years. \$42,952.88

8. ***Value of a Retirement Plan*** Elias Thornberg starts a retirement plan by depositing \$125 at the end of each month into an account that has an interest rate of 7.5%, compounded monthly. What will be the value of Elias's account after 10 years? \$22,241.29

9. ***Retirement Planning*** Jonathan Wu wants to have \$50,000 in a retirement account in 15 years. How much money must Jonathan deposit at the end of each year into an account that has a 7% interest rate, compounded annually, to reach his goal? \$1989.73

10. ***Retirement Planning*** Andrea Watson wants to have \$75,000 in a retirement account in 10 years. How much money must Andrea deposit at the end of each quarter into an account that has a 7.2% interest rate, compounded quarterly, to reach her goal? \$1296.43

11. ***Equipment Replacement*** The manager of a warehouse knows that a forklift will have to be replaced in 3 years. If the estimated cost of a new forklift is \$34,000, how much money should the manager deposit at the end of each quarter into an account that earns 8% interest, compounded quarterly, to have enough money in the account to purchase the forklift in 3 years? \$2535.03

12. ***Retirement Planning*** A journalist deposits \$75 at the end of each month into an account that has an interest rate of 6.25%, compounded monthly.
    a. Find the future value of the account in 10 years. \$12,459.14
    b. How much interest will be earned on the account over the 10-year period? \$3459.14

13. ***Savings Account*** An electronics technician deposits \$90 at the end of each month into an account that earns 6.4%, compounded monthly.
    a. Find the future value of the account in 10 years. \$15,073.74
    b. The technician discontinues payments into the annuity after 10 years and transfers the future value to an account that earns an interest rate of 7%, compounded daily. Find the value of this account 10 years later. (*Suggestion:* Use the compound interest formula.) \$30,352.75

14. ***Retirement Planning*** A sales clerk for a department store can afford to deposit \$150 per month into a retirement account. If the sales clerk wants the value of the retirement account to be \$70,000 in 20 years, what is the minimum interest rate the clerk can earn on the account to reach the goal? 6.08%

15. ***Equipment Replacement*** Janice Montoya wants to replace a refrigeration unit in 3 years. If Janice deposits \$750 into an account at the end of each quarter, what annual interest rate, compounded quarterly, must the account earn so that there is \$12,000 in the account in 3 years? 20.36%

Life and Health Sciences

16. *Purchase Medical Equipment* An ophthalmologist wants to purchase a new cornea measuring device in 5 years. If the estimated cost of the device will be $110,000, how much money must the ophthalmologist deposit at the end of each month into an account that has a 6.4% interest rate, compounded monthly, to reach the goal? $1560.46

17. *Nursing* A pediatric nurse deposits $125 at the end of each month into an account that earns 6%, compounded monthly.

    a. Find the future value of the account in 15 years. $36,352.34

    b. The nurse discontinues payments into the annuity after 15 years and transfers the value to an account that earns an interest rate of 8%, compounded daily. Find the value of this account 5 years later. (*Suggestion:* This is now just compound interest, because no payments are being made into the account.) $54,228.94

Sports and Recreation

18. *Ski Vacation* Emeril Gibson wants to go on a one-week skiing vacation in 6 months. How much money should he deposit at the end of each month into an account that has an interest rate of 6%, compounded monthly, to have the estimated $2000 cost of the trip? $329.19

19. *Kayaking Vacation* Ester Manet plans on taking a kayaking trip in 10 months. How much money should she deposit at the end of each month into an account that has an interest rate of 8.1%, compounded monthly, to have the estimated $5000 cost of the trip? $485

20. *Tennis* A professional tennis player deposits $1000 at the end of each month into an account that has an interest rate of 5.5%, compounded monthly.

    a. Find the future value of the account in 15 years. $278,745.55

    b. How much interest will be earned on the account over the 15-year period? $98,745.55

Explorations

Our entire discussion up to this point has focused on ordinary annuities. Recall that these are annuities for which the payment is made at the end of a time period. For a due annuity, payments are made at the beginning of the time period. The formula for the future value of a due annuity is

$$FV = PMT(1 + i)\left[\frac{(1 + i)^n - 1}{i}\right]$$

where FV, PMT, i, and n have the same meanings as previously.

This formula differs from the formula for an ordinary annuity by the factor $(1 + i)$. This factor compensates for the fact that payments are made at the beginning of the time period.

1. Determine the future value of a 10-year due annuity for which payments of $500 are made at the beginning of each quarter and the interest rate is 8%, compounded quarterly. $30,805.01

2. Determine the future value of a 5-year due annuity for which payments of $250 are made at the beginning of each month and the interest rate is 9%, compounded monthly. $18,997.45

3. This exercise examines the difference between the *values* of two annuities that are identical except that one is an ordinary annuity and one is a due annuity.

    a. Determine the future value of a 20-year *ordinary* annuity for which payments of $500 are made at the end of each quarter and the interest rate is 9%, compounded quarterly. $109,558.78

    b. Determine the future value of a 20-year *due* annuity for which payments of $500 are made at the beginning of each quarter and the interest rate is 9%, compounded quarterly. $112,023.86

c. What is the difference in value between the two annuities? $2465.08

4. This exercise examines the difference between the *interest rates* of two annuities that are identical except that one is an ordinary annuity and one is a due annuity. You will need a graphing or financial calculator for this exercise. If you follow the procedure shown on page 632, at the bottom of the calculator screen there will be an option to select END or BEGIN. Select END for an ordinary annuity; select BEGIN for a due annuity.

a. Deposits of $100 are placed into a savings account at the end of each month. Find the interest rate necessary so that the future value of the account is $7000 in 5 years. 6.13%

b. Deposits of $100 are placed in a savings account at the beginning of each month. Find the interest rate necessary so that the future value of the account is $7000 in 5 years. 5.94%

c. What is the difference in interest rates between the two annuities? 0.19%

Chapter 7 Summary

Key Terms

alternating sequence **[p. 582]**
amortization schedule **[p. 623]**
annual percentage rate (APR) **[p. 625]**
annual rate of return **[p. 612]**
annuity **[p. 619]**
arithmetic sequence **[p. 588]**
arithmetic series **[p. 591]**
common difference **[p. 589]**
common ratio **[p. 593]**
compound continuously **[p. 614]**
compound interest **[p. 608]**
coupon rate of a bond **[p. 607]**
effective interest rate **[p. 603]**
face value of a bond **[p. 607]**
future value **[p. 631]**
future value of the annuity **[p. 632]**
geometric sequence **[p. 593]**
inductive reasoning **[p. 580]**
infinite sequence **[p. 581]**
interest rate **[p. 606]**
mortgage **[p. 623]**
n factorial **[p. 583]**
nth partial sum **[pp. 584/591]**
nth partial sum of an arithmetic sequence **[p. 591]**
nth term of a sequence **[pp. 581/589/594]**
nth term of a geometic sequence **[p. 594]**
payday loan **[p. 607]**
present value **[p. 610]**
principal **[p. 606]**
recursive sequence **[p. 582]**
sequence of partial sums **[p. 584]**
series **[p. 584]**
simple interest **[p. 606]**
sinking fund **[p. 634]**
summation notation **[p. 584]**
terms of a sequence **[pp. 581/589/594]**

Essential Concepts and Formulas

- **nth Term of an Arithmetic Sequence**
  If $\{a_n\}$ is an arithmetic sequence, then the nth term a_n of the sequence is given by $a_n = a_1 + (n - 1)d$, where a_1 is the first term of the sequence and d is the common difference. **[p. 589]**

- **nth Term of a Geometric Sequence**
  If $\{a_n\}$ is a geometric sequence, then the nth term a_n of the sequence is given by $a_n = a_1 r^{n-1}$, where a_1 is the first term of the sequence and r is the common ratio. **[p. 594]**

- **nth Partial Sum of Any Sequence**
  If $\{a_n\}$ is a sequence, then $S_n = \sum_{i=1}^{n} a_i$ is the nth partial sum of the sequence. **[p. 584]**

- **nth Partial Sum of an Arithmetic Sequence**
  If $\{a_n\}$ is an arithmetic sequence, then the nth partial sum S_n of the sequence is given by $S_n = \frac{n}{2}(a_1 + a_n)$. **[p. 591]**

- **nth Partial Sum of a Geometric Sequence**
  The nth partial sum of a geometric sequence with first term a_1 and common ratio r is $S_n = \frac{a_1(1 - r^n)}{1 - r}$, $r \neq 1$. **[p. 595]**

- **Sum of an Infinite Geometric Series**
  If $\{a_n\}$ is a geometric sequence with $|r| < 1$ and first term a_1, then the sum of the infinite geometric series is $S = \frac{a_1}{1 - r}$. **[p. 597]**

- **Simple Interest Formula**
  The simple interest formula is $I = Prt$, where I is the simple interest, P is the principal, r is the simple interest rate, and t is the time period. **[p. 606]**

- **Future Value of Compound Interest**
  Suppose an investment earns an annual interest rate of r% compounded m times a year for t years. Then the future value FV of the investment is
  $$FV = PV(1 + i)^n$$
  where the present value PV is the amount deposited, $i = \frac{r}{m}$, and $n = mt$. **[p. 610]**

- **Present Value of Compound Interest**
  Suppose an investment earns an annual interest rate of r% compounded m times a year for t years. Then the present value PV of the investment is
  $$PV = FV(1 + i)^{-n}$$
  where FV is the future value, $i = \frac{r}{m}$, and $n = mt$. **[p. 611]**

- **Effective Annual Interest Rate**
  The effective annual interest rate I is given by
  $$I = 100\left[\left(1 + \frac{r}{m}\right)^{m} - 1\right]$$
  where r is the compound interest rate and m is the number of compounding periods per year. **[p. 613]**

- **Continuous Compounding Formula**
  The future value FV of an investment that is compounded continuously is given by $FV = PVe^{rt}$, where PV is the present value, r is the annual interest rate, e is the base of the natural exponential function, and t is the time in years. **[p. 614]**

- **Payment Formula for the Present Value of an Annuity**
  Suppose an annuity earns an annual interest rate of r% compounded m times a year for t years. Then the payment amount PMT for a given present value PV is given by
  $$PMT = PV\left[\frac{i}{1 - (1 + i)^{-n}}\right]$$
  where $i = \frac{r}{m}$ and $n = mt$. **[p. 619]**

- **Present Value of an Annuity**
  The present value of an annuity that earns an annual interest rate of r% compounded m times a year for t years is given by
  $$PV = PMT\left[\frac{1 - (1 + i)^{-n}}{i}\right]$$
  where PMT is the payment, $i = \frac{r}{m}$ and $n = mt$. **[p. 620]**

- **Future Value of an Annuity**
  The future value of an annuity that earns an annual interest rate of r% compounded m times a year for t years is given by
  $$FV = PMT\left[\frac{(1 + i)^n - 1}{i}\right]$$
  where PMT is the amount of each deposit, $i = \frac{r}{m}$ and $n = mt$. **[p. 631]**

- **Sinking Fund Payment**
  The amount of a payment into an annuity that earns an annual interest rate of r% compounded m times a year for t years that is necessary to attain a given future value FV is given by
  $$PMT = FV\left[\frac{i}{(1 + i)^n - 1}\right]$$
  where PMT is the amount of each deposit, $i = \frac{r}{m}$ and $n = mt$. **[p. 634]**

Chapter 7 True/False Exercises

In Exercises 1 to 10, answer true or false. If the statement is false, give a reason or counterexample to show that it is false.

1. $\sum_{i=1}^{5} a_i b_i = \sum_{i=1}^{5} a_i \sum_{i=1}^{5} b_i$

2. No two terms of a sequence can be equal. False. In a constant sequence, all of the terms are equal.

3. $1, 8, 27, \ldots, n^3, \ldots$ is a geometric sequence. False. There is no common ratio of successive terms.

4. $a_1 = 7, a_n = a_{n-1} + 5$ defines an arithmetic sequence. True

5. Adding all the terms of an infinite sequence produces an infinite sum. False. The sum of the infinite sequence $1, \frac{1}{2}, \frac{1}{4}, \frac{1}{8}, \ldots$ is 2.

6. When referring to a loan, interest and interest rate mean the same thing. False. The interest rate is a percent. Interest is an amount of money.

7. The present value of a loan is the amount borrowed or owed. True

8. To amortize a loan means to pay it off by making a series of equal payments. True

9. If the interest rate on an investment is 12% compounded quarterly, then the interest rate earned each quarter is 3%. True

10. If two investments are made, each having the same principal and the same interest rate, the one with the greater number of compounding periods per year will have the larger value after 5 years. True

1. False. $\sum_{i=1}^{5} 2 \cdot 3 = 6 + 6 + 6 + 6 + 6 = 30$

$\sum_{i=1}^{5} 2 \sum_{i=1}^{5} 3 = (2 + 2 + 2 + 2 + 2) \cdot (3 + 3 + 3 + 3 + 3) = 150$

Chapter 7 Review Exercises

In Exercises 1 to 20, find the third and seventh terms of the sequence defined by a_n.

1. $a_n = n^2$ $a_3 = 9, a_7 = 49$ [7.1]

2. $a_n = n!$ $a_3 = 6, a_7 = 5040$ [7.1]

3. $a_n = 3n + 2$ $a_3 = 11, a_7 = 23$ [7.1]

4. $a_n = 1 - 2n$ $a_3 = -5, a_7 = -13$ [7.2]

5. $a_n = 2^{-n}$ $a_3 = \frac{1}{8}, a_7 = \frac{1}{128}$ [7.1/7.2]

6. $a_n = 3^n$ $a_3 = 27, a_7 = 2187$ [7.2]

7. $a_n = \frac{1}{n!}$ $a_3 = \frac{1}{6}, a_7 = \frac{1}{5040}$ [7.1]

8. $a_n = \frac{1}{n}$ $a_3 = \frac{1}{3}, a_7 = \frac{1}{7}$ [7.1]

9. $a_n = \left(\frac{2}{3}\right)^n$ $a_3 = \frac{8}{27}, a_7 = \frac{128}{2187}$ [7.1]

10. $a_n = \left(-\frac{4}{3}\right)^n$ $a_3 = -\frac{64}{27}, a_7 = -\frac{16{,}384}{2187}$ [7.1]

11. $a_1 = 2, a_n = 3a_{n-1}$ $a_3 = 18, a_7 = 1458$ [7.1/7.2]

12. $a_1 = -1, a_n = 2a_{n-1}$ $a_3 = -4, a_7 = -64$ [7.1/7.2]

13. $a_1 = 1, a_n = -na_{n-1}$ $a_3 = 6, a_7 = 5040$ [7.1]

14. $a_1 = 2, a_n = n^2 a_{n-1}$ $a_3 = 72, a_7 = 50{,}803{,}200$ [7.1]

15. $a_1 = 4, a_n = a_{n-1} + 2$ $a_3 = 8, a_7 = 16$ [7.1/7.2]

16. $a_1 = 3, a_n = a_{n-1} - 3$ $a_3 = -3, a_7 = -15$ [7.1/7.2]

17. $a_1 = 1, a_2 = 2, a_n = a_{n-1}a_{n-2}$ $a_3 = 2, a_7 = 256$ [7.1]

18. $a_1 = 1, a_2 = 2, a_n = \frac{a_{n-1}}{a_{n-2}}$ $a_3 = 2, a_7 = 1$ [7.1]

19. $a_1 = -1, a_n = 3na_{n-1}$ $a_3 = -54, a_7 = -3{,}674{,}160$ [7.1]

20. $a_1 = 2, a_n = -2na_{n-1}$ $a_3 = 48, a_7 = 645{,}120$ [7.1]

Classify each sequence defined in Exercises 21 to 36 as arithmetic, geometric, or neither.

21. $\sum_{n=1}^{9} (2n - 3)$ Arithmetic [7.2]

22. $\sum_{i=i}^{11} (1 - 3i)$ Arithmetic [7.2]

23. $\sum_{k=1}^{8} (4k + 1)$ Arithmetic [7.2]

24. $\sum_{i=1}^{10} (i^2 + 3)$ Neither [7.1]

25. $\sum_{n=1}^{6} 3 \cdot 2^n$ Geometric [7.2]

26. $\sum_{i=1}^{5} 2 \cdot 4^{i-1}$ Geometric [7.2]

27. $\sum_{k=1}^{9} (-1)^k 3^k$ Geometric [7.2]

28. $\sum_{i=1}^{8} (-1)^{i+1} 2^i$ Geometric [7.2]

29. $\sum_{i=1}^{10}\left(\frac{2}{3}\right)^i$ Geometric [7.2]

30. $\sum_{i=1}^{11}\left(\frac{3}{2}\right)^i$ Geometric [7.2]

31. $\sum_{n=1}^{5}\frac{(-1)^{n+1}}{n^2}$ Neither [7.1]

32. $\sum_{k=1}^{5}\frac{(-1)^{k+1}}{k!}$ Neither [7.1]

33. $\sum_{n=1}^{\infty}\left(\frac{1}{4}\right)^n$ Geometric [7.2]

34. $\sum_{i=1}^{\infty}\left(-\frac{5}{6}\right)^i$ Geometric [7.2]

35. $\sum_{k=1}^{\infty}\left(-\frac{4}{5}\right)^k$ Geometric [7.2]

36. $\sum_{j=1}^{\infty}\left(\frac{1}{5}\right)^j$ Geometric [7.2]

Business and Economics

37. ***Credit Card Debt*** A credit card company charges a monthly simple interest rate of 1.5% on the unpaid balance on that credit card. How much interest is charged for a month when the unpaid balance is \$503.56? \$7.55 [7.3]

38. ***Payday Loans*** Samuel Whitehead received a payday loan of \$100, for which he paid a fee of \$12. If Samuel repaid the loan 1 week later, what was the equivalent simple interest rate for his loan? 624% [7.3]

39. ***Interest on a Bond*** An investor receives an annual simple interest rate of 6.5% on the face value of a bond. If the investor keeps a bond with a face value of \$15,000 for 5 years, how much interest is earned? \$4875 [7.3]

40. ***Price of a Bond*** The coupon rate for a \$1000 bond is 8.25%. If the prevailing interest rate is 9%, what is the market value of the bond? \$916.67 [7.3]

41. ***Price of a Bond*** A \$5000 bond has a coupon rate of 7.5%. If the prevailing interest rate is 9.5%, what is the market value of the bond? \$3947.37 [7.3]

42. ***Interest on an Investment*** Suppose \$2500 is invested in an account that earns 6.75%. What is the future value of the investment in 10 years if interest is:

a. simple interest? \$4187.50

b. compounded annually? \$4804.18

c. compounded semiannually? \$4855.68

d. compounded quarterly? \$4882.51

e. compounded monthly? \$4900.80

f. compounded daily? \$4909.78 [7.3]

43. ***Business Planning*** The owner of a tailor shop wants to have \$12,000 in 2 years to purchase new sewing machines for the employees. How much must the owner deposit today into an account that earns 6.875% if interest is:

a. simple interest? \$10,549.45

b. compounded annually? \$10,505.80

c. compounded semiannually? \$10,482.60

d. compounded quarterly? \$10,470.64

e. compounded monthly? \$10,462.52

f. compounded daily? \$10,458.55 [7.3]

44. ***Inflation*** An economist predicts that the inflation rate for raw wheat is increasing at 2.5% per year. If the average annual cost of raw wheat is \$2.65 per bushel today, what will be the average annual cost of raw wheat in 6 years? \$3.07 [7.3]

45. ***Effective Interest Rate*** If \$3000 is deposited into an account that earns an interest rate of 10%, compounded daily, what is the effective annual interest rate? 10.52% [7.3]

46. ***Doubling Time*** What interest rate is necessary to double the value of a \$2500 investment in 10 years if the interest is compounded daily? Round to the nearest tenth of a percent. 6.9% [7.3]

47. ***Present Value*** Determine the present value of a 15-year annuity for which payments of \$125 are made monthly and the interest rate is 8%, compounded monthly. \$13,080.07 [7.4]

48. ***Car Financing*** Eugene Falstaff purchased a pickup truck and financed \$8500 for 4 years at an interest rate of 8.3%, compounded monthly. Find Eugene's monthly payments. \$208.71 [7.4]

49. ***Annuity Payments*** How much money must be placed in an account that earns an annual interest rate of 7%, compounded monthly, to allow for monthly withdrawals of \$500 for 7 years? \$33,128.64 [7.4]

50. ***Truck Financing*** A chemist has a four-year car loan of \$18,000 at an annual interest rate of 8.5% and a monthly payment of \$444.67. After the chemist makes truck payments for 36 months, the truck is traded in. What is the loan payoff? \$5098.27 [7.4]

51. ***Mortgage*** A manufacturer purchased a warehouse and secured a 25-year loan of \$500,000 at an annual interest rate of 8.25%, compounded monthly.

a. Find the monthly mortgage payment. \$3942.25

b. How much of the first payment is interest and how much goes to reducing the principal?
Interest = \$3437.50; principal = \$504.75 [7.4]

52. ***Mortgage Payoff*** An appliance repair technician purchased a home and obtained a mortgage of \$275,000 for 30 years at an annual interest rate of 8.5%. After making monthly mortgage payments for 6 years, the technician decides to sell the home. What is the loan payoff? \$259,422.88 [7.4]

53. ***Annual Percentage Rate*** A hair stylist financed \$3000 to remodel the hair salon. The loan was a 7% add-on interest loan for 3 years.

a. What are the monthly payments? \$100.83

b. What is the annual percentage rate for the loan? Round to the nearest tenth of a percent. 12.8% [7.4]

54. ***Saving for College*** A mine worker sets up an education account for a child. The miner deposits \$2500 on each birthday of the child through the child's 18th birthday. If the account earns 7% interest compounded annually, find the value of the account on the child's 18th birthday.
\$84,997.58 [7.4]

55. ***Future Value*** A chemist wants to purchase a small boat in 1 year. If the chemist estimates the cost of the boat to be \$2500, how much must the chemist deposit each month into an account that earns 6.5% annual interest, compounded monthly, to reach the goal? \$202.20 [7.5]

56. ***Stock Market*** Suppose a stock is paying a constant dividend of \$2.94 and an investor wants to receive 15% return on an investment. How much should the investor pay for the stock? \$19.60 [7.2]

57. ***Stock Market*** Use the Gordon model for the value of a stock to determine how much the manager of a mutual fund should pay for a stock whose dividend is \$2.15 and whose dividend growth rate is 16% if the manager requires a 22% rate of return on the investment. \$41.57 [7.2]

58. ***Hospitality Industry*** The net effect of tourism on a city is related to the multiplier effect. Suppose a city estimates that a new professional basketball franchise will bring an additional \$15 million of income to the city. If a person receiving a portion of this money will spend 80% and save 20%, what is the net effect of the \$15 million in additional income? \$75,000,000 [7.2]

Life and Health Sciences

59. ***Biology*** The cross threads in a spider web are approximately equally spaced and the lengths of the cross threads form an arithmetic sequence. In a certain spider web, the length of the shortest cross thread is 2.8 millimeters. The length of the longest cross thread is 44.4 millimeters, and the difference in the lengths of successive cross threads is 1.3 millimeters. How many cross threads are between two successive radii?
33 cross threads [7.2]

60. ***Biology*** What is the sum of the lengths of the cross threads between two successive radii in Exercise 59?
778.8 mm [7.2]

61. ***Medicine*** The amount of an antibiotic in the blood is given by $A + Ae^{kt} + Ae^{2kt} + \cdots + Ae^{(n-1)kt}$, where A is the number of milligrams of the antibiotic, n is the number of doses, t is the time between doses, and k is a constant that depends on how quickly the body metabolizes the antibiotic. Suppose one dose of an antibiotic increases the blood level of the antibiotic by 1.5 milligrams per liter. If the antibiotic is given every 4 hours and $k = -0.5$, find the concentration of the antibiotic in the blood just before the sixth dose. 1.73 mg [7.2]

Chapter 7 Test

In Exercises 1 and 2, find the third and fifth terms of the sequence defined by a_n.

1. $a_n = \dfrac{(-1)^{n+1}}{2n}$
$a_3 = \frac{1}{6}, a_5 = \frac{1}{10}$ [7.1]

2. $a_1 = 3, a_n = 2a_{n-1}$
$a_3 = 12, a_5 = 48$ [7.1]

In Exercises 3 and 4, classify the sequence as an arithmetic sequence, a geometric sequence, or neither.

3. $a_n = -2n + 3$
Arithmetic [7.2]

4. $a_n = 2n^2$
Neither [7.2]

5. Evaluate: $\sum_{i=1}^{6} (2i + 3)$ 60 [7.2]

6. The third term of an arithmetic sequence is 7 and the eighth term is 22. Find the 20th term. 58 [7.2]

7. Find the sum of the first 20 terms of the arithmetic sequence whose nth term is $a_n = 3n - 2$. 590 [7.2]

8. Find the sum of the first 10 terms of the geometric sequence whose nth term is $a_n = \left(\frac{3}{4}\right)^n$.
$\frac{2,968,581}{1,048,576} \approx 2.8310595$ [7.2]

9. Write $0.\overline{8}$ as the quotient of integers in simplest form. $\frac{8}{9}$ [7.2]

10. An investor receives an annual simple interest rate of 6.2% on the face value of a bond. If the investor keeps a bond with a face value of $5000 for 4 years, how much interest is earned? $1240 [7.3]

11. Suppose $3500 is invested in an account that earns 7% interest. What is the future value of the investment in 4 years if interest is:
 a. compounded annually? $4587.79
 b. compounded daily? $4630.83 [7.3]

12. A corporation invests $3 million in a new method of tinting sunglasses. The corporation expects to be able to sell the new method in 5 years for $5 million. What is the annual compounded rate of return on the corporation's investment? Round to the nearest hundredth of a percent. 10.76% [7.3]

13. Determine the present value of a 15-year annuity for which payments of $300 are made monthly and the interest rate is 8.5%, compounded monthly. $30,464.91 [7.4]

14. A dental hygienist purchased a minivan and financed $16,000 for 5 years at an annual interest rate of 9.2%. Determine the monthly payment on the loan. $333.69 [7.4]

15. A product clerk at a grocery store has a five-year car loan of $12,000 at an annual interest rate of 8.25% and a monthly payment of $246.20. After the clerk makes payments for 36 months, the car is traded in. What is the loan payoff? $5429.92 [7.4]

In Exercises 16 and 17, assume a homeowner has a 30-year mortgage of $150,000 at an annual interest rate of 7%.

16. Find the monthly mortgage payment. $997.95 [7.4]

17. How much of the first month's mortgage payment is interest? $875 [7.4]

In Exercises 18 and 19, assume a pharmacy assistant purchased a pool table and financed $1000 at an add-on interest rate of 5.5% for 2 years.

18. What are the monthly payments? $46.25 [7.4]

19. What is the annual percentage rate for the loan? Round to the nearest tenth of a percent. 10.2% [7.4]

20. An opera company wants to purchase a stage setting for a new production of *Figaro* that will be held in 5 years. If the estimated cost of the set is $50,000, how much money must the opera director place each month in an account that earns 8% interest, compounded monthly, to reach the $50,000 goal? $680.49 [7.5]

Cumulative Review Exercises

In Exercises 1 to 4, solve for x.

1. $x^2 + 4x = 6$ $-2 \pm \sqrt{10}$ [1.2]
2. $|3x - 2| > 4$ $\left\{x \mid x < -\frac{2}{3}\right\} \cup \{x \mid x > 2\}$ [1.4]
3. $\log_5 x = -2$ $\frac{1}{25}$ [5.2]
4. $e^{2x} = 2$ $\frac{1}{2}\ln 2$ [5.4]
5. Find the slope of the line and the distance between the points $P_1(-5, 1)$ and $P_2(-1, -2)$. $-\frac{3}{4}$ [2.3], 5 [2.1]
6. Find the equation of the line whose slope is $-\frac{1}{2}$ and that passes through the point $(-2, 1)$. $y = -\frac{1}{2}x$ [2.4]
7. Find the vertex of the parabola given by $f(x) = -x^2 - 2x + 1$. $(-1, 2)$ [2.5]
8. Is $f(x) = \frac{x}{x^2 + 1}$ an even function, an odd function, or neither? Odd [3.3]
9. Given the graph of f shown below, graph $g(x) = -f(x) + 2$. [3.3]

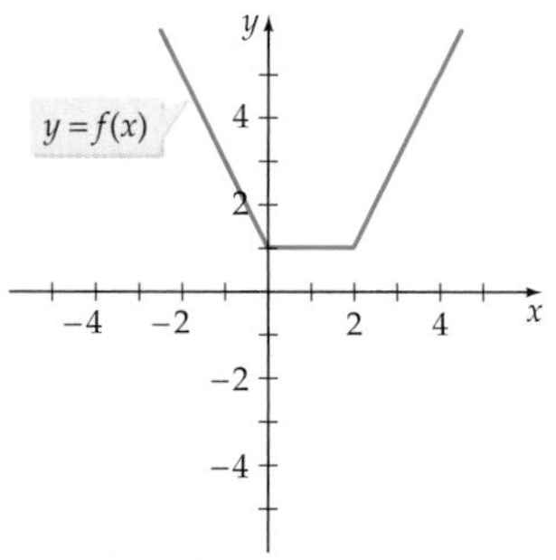

10. Given $f(x) = 1 - 3x$ and $g(x) = x^2 + 2$, find $g(f(-2))$. 51 [3.1]
11. Given $f(x) = 2x + 4$, find $f^{-1}(x)$. $f^{-1}(x) = \frac{1}{2}x - 2$ [3.2]

9.

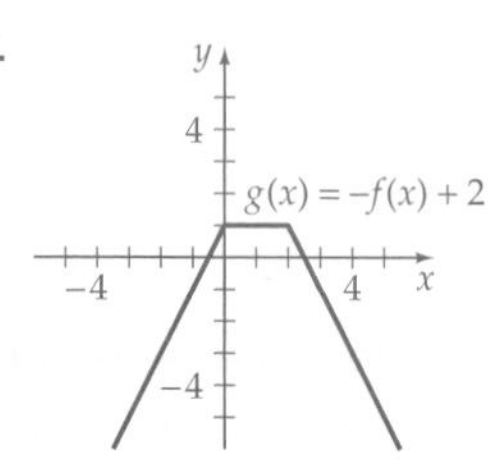

12. Is $x - 2$ a factor of $x^4 + 3x^2 - 10x - 8$? Yes [4.1]
13. Given that 3 is a zero of $P(x) = x^3 - 2x^2 - 5x + 6$, find the remaining zeros. -2 and 1 [4.3]
14. Given that $2 + i$ is a zero of $P(x) = x^2 - 4x + 5$, what is the other zero? $2 - i$ [4.4]
15. Find a polynomial of degree 3 with real coefficients that has zeros i and -1. $x^3 + x^2 + x + 1$ [4.4]
16. Find the vertical and horizontal asymptotes for $r(x) = \frac{2x}{x + 1}$. VA: $x = -1$, HA: $y = 2$ [4.5]
17. Solve: $\begin{cases} x - 3y = 7 \\ 2x + 5y = -8 \end{cases}$ $(1, -2)$ [6.1]
18. The repulsive force between the north poles of two magnets is inversely proportional to the square of the distance between them. If the repulsive force is 10 pounds when the magnets are 1 inch apart, what is the repulsive force when the magnets are 0.5 inch apart? 40 lb [3.4]
19. Carbon-11 has a half-life of 20 minutes. How much of an initial 5-milligram sample of carbon-11 will remain after 6 minutes? Round to the nearest hundredth. 4.06 mg [5.5]
20. A pharmacist has two vitamin-supplement powders. The first powder is 20% vitamin B1 and 10% vitamin B2. The second powder is 15% vitamin B1 and 20% vitamin B2. How many milligrams of each of the two powders should the pharmacist use to make a mixture that contains 130 milligrams of vitamin B1 and 80 milligrams of vitamin B2? 560 mg of B1, 120 mg of B2 [6.1]

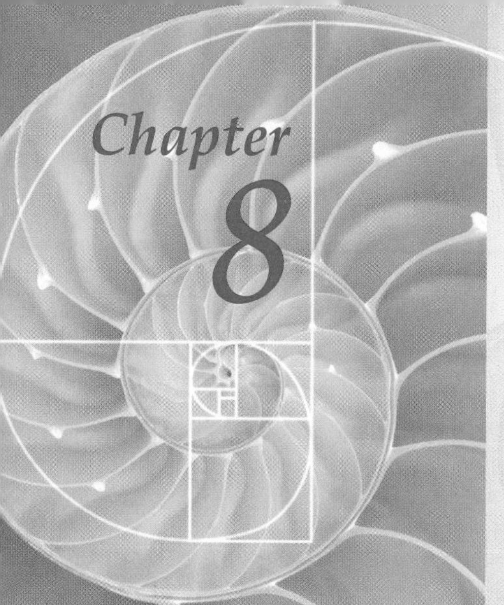

Probability and the Binomial Theorem

Winning the Lottery

In 1984, the state of California initiated a lottery game called 6-49. To play this game, a person selected six numbers from 1 to 49. If those numbers matched the six chosen by the lottery commission, the person won the jackpot. The chance of picking the correct six numbers was 1 in 13,983,816. If no one selected the correct numbers, the jackpot was increased for the next drawing. Because the chances of winning the jackpot were so small, the jackpot frequently accumulated to more than $20,000,000.

Currently, the California lottery commission is using a game called Super Lotto Plus. Five numbers are chosen, from 1 to 47, plus a sixth "MEGA number" from 1 to 27 is selected. The probability of picking the correct six numbers in this game is only 1 in 41,416,353. In February 2002 the Super Lotto Plus jackpot reached $193,000,000.

In this chapter, we focus on the mathematics needed to calculate the probability that certain events will happen, and we investigate statistical analysis of events that have already occurred. See **Exercises 64 to 67, page 681.**

Chapter Prep Quiz

Do these exercises to prepare for Chapter 8.

1. If $A = \{2, 3, 5, 11\}$ and $B = \{4, 5, 6, 7, 8\}$ find $A \cup B$. [P.1] $\{2, 3, 4, 5, 6, 7, 8, 11\}$

2. If $A = \{2, 4, 6, 8\}$ and $B = \{3, 5, 7, 9\}$ find $A \cap B$. [P.1] $\varnothing$

3. Evaluate: $10(0.2)^3(0.8)^2$ [P.2] 0.0512

4. Evaluate $1 - \left(\frac{2}{3}\right)^5$ [P.2] $\frac{211}{243}$

5. Evaluate: 8! [7.1] 40,320

6. Evaluate: 7! 4! [7.1] 120,960

7. Evaluate: $\frac{12!}{8!}$ [7.1] 11,880

8. Evaluate: $\frac{14!}{5!\,9!}$ [7.1] 2002

9. Find: $(x + 2y)^3$ [P.3] $x^3 + 6x^2y + 12xy^2 + 8y^3$

10. Find: $(2x - y)^4$ [P.3] $16x^4 - 32x^3y + 24x^2y^2 - 8xy^3 + y^4$

Problem Solving Strategies

Famous Unsolved Problems

In this text we have introduced several problem solving strategies. As you progress in your mathematical studies, you will learn even more advanced problem solving strategies that will enable you to solve many additional types of problems. Don't think, however, that for every problem there is a known problem solving strategy. One of the interesting aspects of mathematics is the fact that there are many problems in mathematics that have remained unsolved, even though some of the most talented mathematicians have given their best efforts to solve them. Here is a description of two of these famous unsolved problems. Both of these problems are easy to state and to understand, but neither problem has been solved.

1. ***Goldbach's Conjecture*** In 1742, Christian Goldbach conjectured that every even number greater than 2 can be written as the sum of two prime numbers. Many mathematicians have tried to prove or disprove this conjecture without succeeding. Show that Goldbach's conjecture is true for each of the following even numbers.

   a. 24 **b.** 50 **c.** 144 **d.** 210

2. ***The Collatz Problem*** Start with any natural number $n > 1$. Now generate a sequence of numbers using the following rules.
   - If n is even, divide n by 2.
   - If n is odd, multiply n by 3 and add 1.

   Repeat the above procedure on the new number you have just generated. Keep applying the above procedure until you obtain the number 1.

   In 1937, L. Collatz conjectured that the above procedure would always generate a sequence of numbers that would eventually reach the number 1, regardless of the starting number n. Thus far no one has been able to prove that Collatz's conjecture is true or to show that it is false.

   Show that for each of the natural numbers 2, 3, 4, . . . , 10, the Collatz procedure does generate a sequence that "returns" to 1. *Note:* The sequences generated by the Collatz procedure are sometimes called "hailstone" sequences because the numbers in the sequences tend to bounce up and down, much like a hailstone in a storm.

INSTRUCTOR NOTE

The *Chapter Prep Quiz* is a means to test your students' mastery of prerequisite material that is assumed in the coming chapter. Each question identifies the section to review (if necessary) and all answers are provided in the Answers to Selected Exercises.

SECTION 8.1 Counting Methods

- The Counting Principle
- Permutations
- Combinations

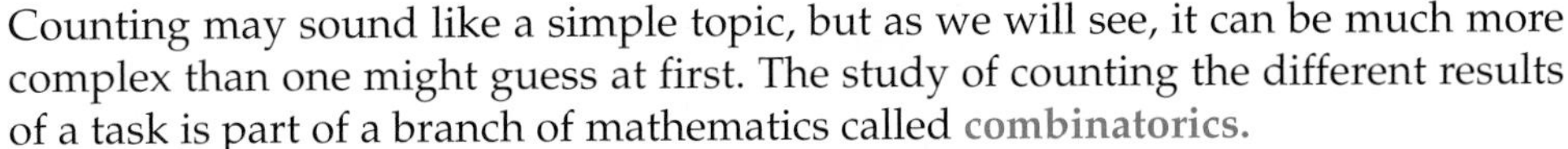

Counting may sound like a simple topic, but as we will see, it can be much more complex than one might guess at first. The study of counting the different results of a task is part of a branch of mathematics called **combinatorics.**

In combinatorics, an activity with an observable outcome is called an **experiment.** For example, if a coin is flipped, the side facing upward will be a head or a tail. The two possible outcomes can be listed as {H, T}. If a regular six-sided die is rolled, the possible outcomes are ⚀, ⚁, ⚂, ⚃, ⚄, ⚅. The outcomes can be listed as {1, 2, 3, 4, 5, 6}.

If there are not too many different outcomes, we may be able to simply list all of the possibilities.

Alternative to Example 1

Two-digit numbers are formed from the digits 5, 7, and 9. List all possible outcomes of this experiment.

- {55, 57, 59, 75, 77, 79, 95, 97, 99}

EXAMPLE 1 Outcomes of the Roll of Two Dice

Suppose we roll two standard six-sided dice, first one colored green and then a second colored red. How many different results are possible?

Solution We can organize all of the possibilities in a grid arrangement, where the first die is green and the second is red, as shown below.

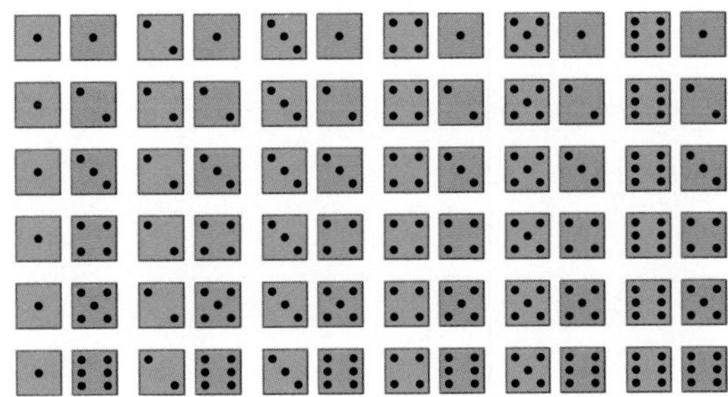

Outcomes from Rolling Two Dice

We can see that there are 36 different possibilities.

CHECK YOUR PROGRESS 1 Three coins consisting of a quarter (Q), a dime (D), and a nickel (N) are to be drawn, one at a time, from a hat. Once a coin is drawn from the hat, it is not replaced. List all possible outcomes of this experiment and determine the total number of ways these coins can be selected.

Solution *See page S41.* QDN, QND, DQN, DNQ, NQD, NDQ; 6

Rolling two dice, as shown in Example 1, is an example of a **multistage experiment.** (This experiment involved two stages: first one die was rolled, then a second.)

In multistage experiments, simply drawing or listing all the possible outcomes can be tedious and difficult to organize. To help, we can draw what is called a **tree diagram.** The technique is demonstrated in the next example.

Alternative to Example 2

A hamburger vendor offers two types of buns, two different meats, and three types of cheese. How many distinct hamburgers are available?

▪ 12

EXAMPLE 2 Draw a Tree Diagram

A computer store offers a three-component computer system. The system consists of a central processing unit (CPU), a hard drive, and a monitor. The vendor offers two different CPUs, three different hard drives, and two different monitors from which to choose. How many distinct computer systems are possible?

Solution We can organize the information by letting C_1 and C_2 represent the two CPUs; H_1, H_2, and H_3 represent the three hard drives; and M_1 and M_2 represent the two monitors.

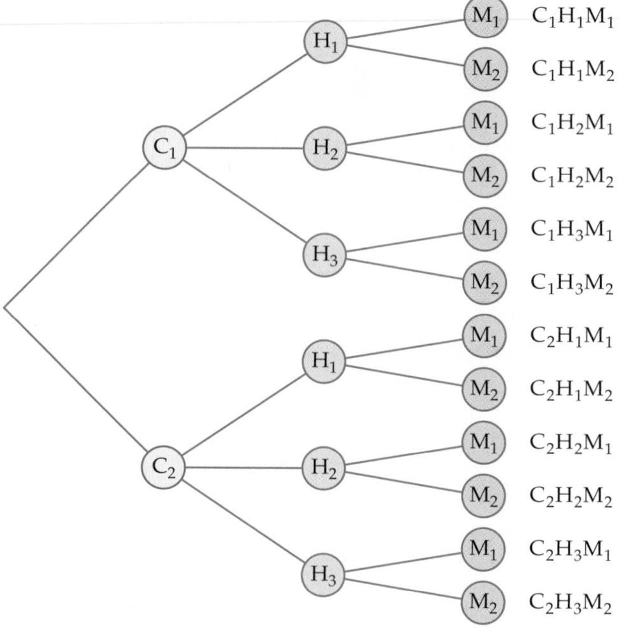

Each "branch" of the tree diagram represents a possible option; by following a path from left to right, we assemble one of the outcomes of the experiment. There are 12 possible computer systems.

CHECK YOUR PROGRESS 2 A true/false test consists of 10 questions. Draw a tree diagram to show the number of possible ways to answer the first four questions.

Solution *See page S41.* 16

T F
T F T F
T F T F T F T F
T F T F T F T F T F T F T F T F

▪ The Counting Principle

For each of the previous problems, the possible outcomes were listed and then counted to determine the number of different outcomes. However, it is not always possible or practical to list and count outcomes. For example, the number of different five-card poker hands that can be drawn from a regular deck of 52 playing cards is 2,598,960. (See Example 8 on page 654.) Trying to create a list of these hands would be quite time consuming.

In Example 1, we found that after tossing two dice, there were 36 possible outcomes. We can arrive at this result without listing the outcomes by finding the product of the number of possible outcomes of the green die and the number of possible outcomes of the red die.

$$\begin{bmatrix}\text{Outcomes}\\ \text{of green die}\end{bmatrix} \times \begin{bmatrix}\text{Outcomes}\\ \text{of red die}\end{bmatrix} = \begin{bmatrix}\text{Number of}\\ \text{outcomes}\end{bmatrix}$$
$$6 \times 6 = 36$$

Consider again the problem of selecting a computer system from three components. By using a tree diagram, we listed the 12 possible computer systems. Another way to arrive at this result is to find the product of the numbers of options available for each component.

$$\begin{bmatrix}\text{Number of}\\ \text{CPUs}\end{bmatrix} \times \begin{bmatrix}\text{Number of}\\ \text{hard drives}\end{bmatrix} \times \begin{bmatrix}\text{Number of}\\ \text{monitors}\end{bmatrix} = \begin{bmatrix}\text{Number of}\\ \text{systems}\end{bmatrix}$$
$$2 \times 3 \times 2 = 12$$

Counting all of the outcomes of a multistage experiment without listing them is accomplished by using the *counting principle.*

Counting Principle

Let E be a multistage experiment. If $n_1, n_2, n_3, \ldots, n_k$ are the number of possible outcomes of each of the k stages of E, then there are $n_1 \cdot n_2 \cdot n_3 \cdot \cdots \cdot n_k$ possible outcomes for E.

Alternative to Example 3

Eighteen people have entered a raffle where first prize is $200, second prize if $100, and third prize is $50. In how many different ways can the winners be chosen?

▪ **4896**

EXAMPLE 3 Using the Counting Principle

In horse racing, a *trifecta* consists of choosing the exact order of the first three horses across the finish line. If there are eight horses in a race, how many possible trifectas are possible, assuming there are no ties?

Solution Any one of the eight horses can be first, so $n_1 = 8$. Because a horse cannot finish both first and second, there are seven horses that can finish second; thus $n_2 = 7$. Similarly, there are six horses that can finish third; $n_3 = 6$. By the counting principle, there are $8 \cdot 7 \cdot 6 = 336$ possible trifectas.

CHECK YOUR PROGRESS 3 Nine runners are entered in a 100-meter dash for which a gold, a silver, and a bronze medal will be awarded for first, second, and third place finishes, respectively. In how many possible ways can the medals be awarded? Assume there are no ties.

Solution *See page S41.* 504

Permutations

A **permutation** is an arrangement of distinct objects in a definite order. The counting principle can be used to determine the total number of possible permutations of a group of objects.

For instance, the possible permutations of the three objects ❍ ■ ❑ are

Permutation 1 ❍ ■ ❑
Permutation 2 ❍ ❑ ■
Permutation 3 ❑ ❍ ■
Permutation 4 ❑ ■ ❍
Permutation 5 ■ ❍ ❑
Permutation 6 ■ ❑ ❍

Notice that there are three choices of symbol for first place. Once the first symbol is picked, there are two choices for second place, and finally only one choice for third place. By the counting principle, there are $3 \cdot 2 \cdot 1 = 6$ permutations of the three symbols.

Similarly, if we have four distinct objects, there are four choices for the first position, three choices for the second, two choices for the third, and only one choice for the fourth position. Thus there are $4 \cdot 3 \cdot 2 \cdot 1 = 24$ permutations of four objects.

In general, n objects can be arranged in $n \cdot (n-1) \cdot (n-2) \cdot \cdots \cdot 3 \cdot 2 \cdot 1$ different permutations. This product is given a special name.

INTEGRATING TECHNOLOGY

We discussed factorials earlier in the text. The concept of factorial is a key concept for counting, so we have restated it here.

Many calculators can directly compute factorials. (On the TI-83 graphing calculator, for example, $n!$ is accessible in the probability menu after pressing the MATH key.)

n factorial

n factorial is the product of the natural numbers 1 through n and is symbolized by $n!$.

For instance, $8! = 8 \cdot 7 \cdot 6 \cdot 5 \cdot 4 \cdot 3 \cdot 2 \cdot 1 = 40{,}320$. By convention, $0! = 1$.

Alternative to Example 4

Evaluate: **a.** $6! + 4!$ **b.** $\frac{7!}{3!}$

■ **a.** 744 **b.** 840

EXAMPLE 4 Evaluate Factorial Expressions

Evaluate: **a.** $5! - 3!$ **b.** $\frac{9!}{6!}$

Solution

a. $5! - 3! = (5 \cdot 4 \cdot 3 \cdot 2 \cdot 1) - (3 \cdot 2 \cdot 1) = 120 - 6 = 114$

b. A factorial can be written in terms of smaller factorials; this is useful for simplifying ratios of factorial expressions.

$$\frac{9!}{6!} = \frac{9 \cdot 8 \cdot 7 \cdot 6 \cdot 5 \cdot 4 \cdot 3 \cdot 2 \cdot 1}{6!} = \frac{9 \cdot 8 \cdot 7 \cdot \cancel{6!}}{\cancel{6!}} = 9 \cdot 8 \cdot 7 = 504$$

CHECK YOUR PROGRESS 4 Evaluate: **a.** $7! + 4!$ **b.** $\frac{8!}{4!}$

Solution *See page S41.* **a.** 5064 **b.** 1680

TAKE NOTE

When computing with factorials, it is sometimes convenient not to write the entire product. For instance, we can write

$$10! = 10 \cdot 9 \cdot 8!$$
$$7! = 7 \cdot 6 \cdot 5 \cdot 4!$$
$$5! = 5 \cdot 4!$$

In Example 4**b.**, we used this idea to write $9! = 9 \cdot 8 \cdot 7 \cdot 6!$.

? QUESTION Is $1! > 0!$?

There are $n!$ different ways of arranging n objects if we are arranging *all* of the objects. In many cases, however, we may use only *some* of the objects. For example, consider the five symbols ❑ ◗ ❉ ❍ ▼. When only three symbols are selected from the five, there are (by the counting principle) $5 \cdot 4 \cdot 3 = 60$ possible permutations.

The number of permutations of n distinct objects, of which k are selected, can be generalized to the following formula.

TAKE NOTE

$\frac{n!}{(n-k)!}$ is equivalent to the product

$$n \cdot (n-1) \cdot (n-2) \cdot \cdots \cdot (n-k+1)$$

Permutation Formula

The number of permutations $P(n, k)$ of n distinct objects selected k at a time is

$$P(n, k) = \frac{n!}{(n-k)!}$$

Alternative to Example 5

A recent Kentucky Derby had 18 horses entered in the race. How many different finishes of first, second, third, and fourth place were possible?

▪ 73,440

EXAMPLE 5 Choosing a Tennis Team

A university tennis team consists of six players who are ranked from 1 through 6. If a tennis coach has 10 players from which to choose, how many different tennis teams can the coach select?

Solution Because the players on the tennis team are ranked from 1 through 6, a team with player A in position 1 is different from a team with player A in position 2. Therefore, the number of different teams is the number of permutations of 10 players selected six at a time.

$$P(10, 6) = \frac{10!}{(10-6)!} = \frac{10!}{4!} = \frac{10 \cdot 9 \cdot 8 \cdot 7 \cdot 6 \cdot 5 \cdot 4!}{4!} = 151{,}200$$

There are 151,200 possible tennis teams.

CHECK YOUR PROGRESS 5 A college golf team consists of five players who are ranked from 1 through 5. If a golf coach has eight players from which to choose, how many different golf teams can the coach select?

Solution *See page S42.* 6720 teams

Point of Interest

A regular deck of playing cards consists of 52 different cards. Each shuffle of the deck results in a permutation of the cards, and there are a total of $P(52, 52) = 52! \approx 8 \times 10^{67}$ different possible arrangements. If a new deck is opened and the cards are in numerical order, how many shuffles are necessary to rearrange the deck randomly enough so that any card is equally likely to occur anywhere in the deck? Two mathematicians, Dave Bayer and Persi Diaconis, determined that seven shuffles are sufficient. Their proof has applications to other complex counting problems such as analyzing speech patterns.

▪ Combinations

In the preceding examples, the order of the arrangement of objects was important. Such an arrangement is a permutation. If a telephone extension is 2537, then the digits must be dialed in exactly that order. On the other hand, if you were to receive a one-dollar bill, a five-dollar bill, and a ten-dollar bill, you would have \$16 regardless of the order in which you received the bills. A **combination** is a collection of objects for which the order is not important. The three-letter sequences *acb* and *bca* are *different* permutations but the *same* combination.

? ANSWER No. $1! = 1$ and $0! = 1$. Thus, $1! = 0!$.

Consider the problem of finding the number of possible combinations when choosing three letters from the letters a, b, c, d, and e (without any repetitions). For each choice of three letters, there are $3! = 6$ permutations. For instance, choosing the letters a, d, and e gives the following six permutations.

ade *aed* *dea* *dae* *ead* *eda*

Because there are six permutations and each permutation is the *same* combination, the number of permutations is six times the number of combinations. This is true each time three letters are selected. Therefore, to find the number of combinations of five objects chosen three at a time, divide the number of permutations by $3! = 6$. The number of combinations of five objects chosen three at a time is

$$\frac{P(5, 3)}{3!} = \frac{5!/(5-3)!}{3!} = \frac{5!}{3! \cdot (5-3)!} = \frac{5!}{3! \cdot 2!} = \frac{5 \cdot 4 \cdot 3!}{3! \cdot 2!} = \frac{5 \cdot 4}{2 \cdot 1} = 10$$

INTEGRATING TECHNOLOGY

Some graphing calculators use nPr to represent $P(n, r)$ and nCr to represent $C(n, r)$. To find $C(11, 5)$, for instance, enter 11 nCr 5.

Combination Formula

The number of combinations of n distinct objects chosen k at a time is

$$C(n, k) = \frac{P(n, k)}{k!} = \frac{n!}{k! \cdot (n-k)!}$$

Alternative to Example 6

A guest musician on a radio program will be playing 5 tracks selected from the 12 tracks on her new CD.

a. In how many ways can she select the 5 tracks?

b. If the musician must submit a playlist, with the 5 tracks listed in order, how many playlists are possible?

■ **a. 792 b. 95,040**

EXAMPLE 6 Choosing a Basketball Team

A basketball team consists of 11 players.

a. In how many different ways can a coach choose the five starting players, assuming the position of a player is not considered?

b. In how many different ways can a coach choose the five starting players if the positions of the players are considered?

Solution

a. This is a combination problem (rather than a permutation problem) because the order in which the coach chooses the players is not important. The five starting players P_1, P_2, P_3, P_4, P_5 are the same as the five starting players P_3, P_5, P_1, P_2, P_4.

$$C(11, 5) = \frac{11!}{5! \cdot (11-5)!} = \frac{11!}{5! \cdot 6!} = \frac{11 \cdot 10 \cdot 9 \cdot 8 \cdot 7 \cdot 6!}{5! \cdot 6!}$$
$$= \frac{11 \cdot 10 \cdot 9 \cdot 8 \cdot 7}{5 \cdot 4 \cdot 3 \cdot 2 \cdot 1} = 462$$

There are 462 possible five-player starting teams.

b. This time the same five players can be chosen for different positions, and so we consider these different arrangements. Thus we are counting permutations rather than combinations.

$$P(11, 5) = \frac{11!}{(11-5)!} = \frac{11!}{6!} = \frac{11 \cdot 10 \cdot 9 \cdot 8 \cdot 7 \cdot 6!}{6!}$$
$$= 11 \cdot 10 \cdot 9 \cdot 8 \cdot 7 = 55{,}440$$

Continued ➤

There are 55,440 different ways a coach can choose the five starting players if the positions of the players are considered.

CHECK YOUR PROGRESS 6 A softball team consists of 16 players.

a. In how many ways can a coach choose the 9 starting players? Assume the position of a player is not considered.

b. In how many ways can a coach choose the batting order lineup for the first four batters from the 9 starting players?

Solution *See page S42.* **a.** 11,440 **b.** 3024

Some counting problems require the use of more than one counting technique to determine the total number of possible outcomes.

Alternative to Example 7

A magazine article is to include three color photos and two black and white photos. If there are seven color photos and five black and white photos from which to choose, in how many ways can the photographs for the article be selected?

■ 350

EXAMPLE 7 Choosing a Committee

A committee of five professors is to be chosen from five mathematicians and six economists. How many different committees are possible if the committee must include two mathematicians and three economists?

Solution Because a committee of professors A, B, C, D, and E is exactly the same as a committee of professors B, D, E, A, and C, choosing a committee is an example of a combination. There are five mathematicians from whom two are chosen, so there are $C(5, 2)$ combinations. There are six economists from whom three are chosen, so there are $C(6, 3)$ combinations. Therefore, by the counting principle, there are $C(5, 2) \cdot C(6, 3)$ ways to choose both two mathematicians and three economists.

$$C(5, 2) \cdot C(6, 3) = \frac{5!}{2! \cdot 3!} \cdot \frac{6!}{3! \cdot 3!} = 10 \cdot 20 = 200$$

There are 200 possible committees consisting of two mathematicians and three economists.

CHECK YOUR PROGRESS 7 An IRS auditor randomly chooses five tax returns to audit from a stack of 10 tax returns, four of which are from corporations and six of which are from individuals. In how many different ways can the auditor choose the tax returns if the auditor wants to include three corporate and two individual returns?

Solution *See page S42.* 60

TAKE NOTE

A standard deck of playing cards consists of 52 cards in four suits: spades (♠), hearts (♥), diamonds (♦), and clubs (♣). Each suit has 13 cards: 2 through 10, jack, queen, king, and ace. (Here we will not include the two jokers.)

Alternative to Example 8

From a standard deck of playing cards, a hand of seven cards is chosen. How many different hands consist of three aces, two kings, and two queens?

■ 144

EXAMPLE 8 Choosing Cards from a Deck

From a standard deck of playing cards, a hand of five cards is chosen.

a. How many different five-card hands are possible?

b. How many different five-card hands consist of two kings and three queens? (This hand is an example of a "full house" in the game of poker.)

c. How many different hands consist of five cards of the same suit? (This is called a "flush" in poker.)

Solution

a. Because the order in which the cards are chosen is not important, we need to determine the number of combinations of 52 cards taken five at a time.

$$C(52, 5) = \frac{52!}{5! \cdot 47!} = \frac{52 \cdot 51 \cdot 50 \cdot 49 \cdot 48}{5!} = 2{,}598{,}960$$

There are 2,598,960 possible five-card hands.

b. There are $C(4, 2)$ ways of choosing two kings from the four kings in the deck and $C(4, 3)$ ways of choosing three queens from the four queens. By the counting principle, there are $C(4, 2) \cdot C(4, 3)$ ways of choosing two kings and three queens.

$$C(4, 2) \cdot C(4, 3) = \frac{4!}{2! \cdot 2!} \cdot \frac{4!}{3! \cdot 1!} = 6 \cdot 4 = 24$$

There are 24 ways of choosing two kings and three queens.

c. First, there are four different suits from which to choose. For each suit, there are $C(13, 5)$ ways of choosing five cards from the 13 cards of that suit. By the counting principle, there are $4 \cdot C(13, 5)$ total outcomes.

$$4 \cdot C(13, 5) = 4 \cdot 1287 = 5148$$

There are 5148 ways of choosing five cards of the same suit from a regular deck of playing cards.

CHECK YOUR PROGRESS 8 From a standard deck of playing cards, a hand of five cards is chosen. How many different hands contain exactly four cards of the same suit?

Solution *See page S42.* 111,540

Topics for Discussion

1. Explain what the counting principle states and when it applies.
2. What is the difference between a permutation and a combination?
3. Give an example of an experiment that counts permutations.
4. Give an example of an experiment that counts combinations.

EXERCISES 8.1

— *Suggested Assignment: Exercises 1–21, odd; 23–26; 27–61, odd; 65, 69, 73, 76, 78, and 82–94.*
— *Answer diagrams to Exercises 11–16 are on page AA21.*

In Exercises 1 to 10, list all the possible outcomes of each experiment.

1. Select an even single-digit whole number. {0, 2, 4, 6, 8}

2. Select an odd single-digit whole number. {1, 3, 5, 7, 9}

3. Select one day from the days of the week. {Monday, Tuesday, Wednesday, Thursday, Friday, Saturday, Sunday}

4. Select one month from the months of the year. {January, February, March, April, May, June, July, August, September, October, November, December}

5. Toss a coin twice. {HH, HT, TH, TT}

In Exercises 5 to 8, H denotes a head and T denotes a tail.

6. Toss a coin three times. {HHH, HHT, HTH, HTT, THH, THT, TTH, TTT}

7. Roll a single die and then toss a coin. {1H, 1T, 2H, 2T, 3H, 3T, 4H, 4T, 5H, 5T, 6H, 6T}

8. Toss a coin and then choose a digit from the digits 1 through 4. {H1, T1, H2, T2, H3, T3, H4, T4}

9. Andy, Bob, and Cassidy need to be seated in seats A1, A2, and A3 at a concert.

10. First, second, and third prize must be awarded to the paintings of Angela, Hector, and Quinh. Assume the awards are filled in the order first, second, third. Using A for Angela, H for Hector, and Q for Quinh, the sample space is {AHQ, AQH, QAH, QHA, HAQ, HQA}.

In Exercises 11 to 16, draw a tree diagram that shows all of the possible outcomes of each experiment.

11. A dinner menu allows a customer to choose from two salads, three dinner entrees, and two desserts.

12. A new car promotion allows a buyer to choose from three body styles, two radios, and two interior color schemes.

13. Roll a four-sided die twice.

14. Toss a coin and then roll a six-sided die.

15. A two-character category code is created for employees of a company; the first character is one of the letters A through D and the second character is one of the numbers 1 through 5.

16. A three-character category code is used for customers of a store; the first character is N, S, E, or W (for North, South, East, or West), the second character is either a 0 or a 1 (for 2000 or 2001), and the third character is either Y or N (for Yes or No).

9. Assume the seats are filled in the order A1, A2, A3. Using A for Andy, B for Bob, and C for Cassidy, the sample space is {ABC, ACB, BAC, BCA, CAB, CBA}.

In Exercises 17 to 22, use the counting principle to determine the number of possible outcomes of each experiment.

17. Two digits are selected from the digits 1, 2, 3, 4, 5. Repeated digits are allowed. 25

18. Two digits are selected from the digits 1, 2, 3, 4, 5, and the same digit cannot be used twice. 20

19. A multiple choice test consisting of 15 questions is completed, and each question has four possible answers. 1,073,741,824

20. A true/false quiz consisting of 20 questions is completed. 1,048,576

21. Four-digit telephone extensions are generated, and the first digit cannot be 0, 1, 8, or 9. 6000

22. Three-letter codes are generated using only vowels (a, e, i, o, u). 125

In Exercises 23 to 26, use the following experiment. Four cards labeled A, B, C, and D are randomly placed in four boxes labeled A, B, C, and D, one card to each box.

23. In how many different ways can the cards be placed in the boxes? 24

24. In how many different ways can the cards be placed in the boxes if no card can be in a box with the same letter? 9

25. In how many different ways can the cards be placed in the boxes if *at least one* card is placed in a box with the same letter? 15

26. If you add the answers to Exercises 24 and 25, is the sum the same as the answer to Exercise 23? Why or why not? Yes, because the events account for all possible options with no overlap.

In Exercises 27 to 48, evaluate each expression.

27. $8!$ 40,320

28. $5!$ 120

29. $9! - 5!$ 362,760

30. $8! + 3!$ 40,326

31. $\frac{5!}{2!}$ 60

32. $\frac{7!}{5!}$ 42

33. $\frac{8!}{3!}$ 6720

34. $\frac{12!}{6!}$ 665,280

35. $P(7, 3)$ 210

36. $P(8, 6)$ 20,160

37. $P(9, 6)$ 60,480

38. $P(10, 5)$ 30,240

39. $P(6, 0)$ 1

40. $P(4, 4)$ 24

41. $C(9, 2)$ 36

42. $C(8, 6)$ 28

43. $C(12, 0)$ 1

44. $C(11, 11)$ 1

45. $C(6, 2) \cdot C(7, 3)$ 525

46. $C(7, 5) \cdot C(9, 4)$ 2646

47. $3! \cdot C(8, 5)$ 336

48. $\dfrac{C(10, 4) \cdot C(5, 2)}{C(15, 6)}$ $\dfrac{60}{143}$

49. If $7! = 5040$, find $8!$. 40,320

50. If $9! = 362,880$, find $10!$. 3,628,800

51. ***Geometry*** Seven points are drawn in a plane, no three of which are on the same straight line. How many different lines can be drawn that pass through two of the seven points? 21

52. ***Geometry*** A pentagon is a five-sided figure. A diagonal is a line segment connecting any two nonadjacent vertices. How many diagonals are possible? 5

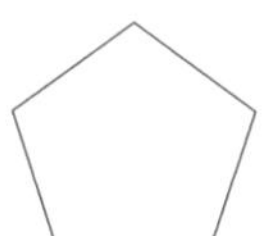

Business and Economics

53. ***Corporate Committees*** The board of directors of a corporation must select a president, a secretary, and a treasurer. In how many possible ways can this be accomplished if there are 20 members on the board of directors? Assume a member cannot hold more than one office. 6840

54. ***PIN*** The Personal Identification Number (PIN) used by a certain automatic teller machine (ATM) is a sequence of four letters.

    a. How many different PINs are possible? 456,976

    b. If no two letters in the PIN can be the same, how many different PINs are possible? 358,800

55. ***House Painting*** A house painter offers six base colors for the exterior of a home, eight different colors for the trim, and a choice of five colors for accents. How many different color combinations are possible? 240

56. ***Vehicle Purchase*** A car manufacturer offers three different body styles on its new SUV, two different engine sizes, 12 different exterior colors, and eight different interior options. How many different versions of the SUV can be purchased? 576

57. ***License Plates*** One state's automobile license plates consist of a single digit 1 through 9, followed by three letters and then three digits from 0 through 9. How many unique license plates are possible? 158,184,000

58. ***Restaurant Choices*** A restaurant offers a special pizza with five toppings. If the restaurant has 12 toppings available, how many different special pizzas are possible? Assume any topping is chosen only once per pizza. 792

59. ***Interview Scheduling*** Seven people are interviewed for a possible promotion. In how many ways can the seven candidates be scheduled for the interviews? 5040

60. ***Shift Scheduling*** One shift at a manufacturing plant requires the operation of four possible machines. If eight people are qualified to operate any of the four machines, how many different shifts are possible? 70

61. ***Shift Scheduling*** Five women and four men have volunteered to serve for 1 hour answering a phone during a 9-hour telethon.

    a. How many schedules are possible if there are no restrictions on the order of the schedule? 362,880

    b. How many schedules are possible if two people came in the same car and would like to serve during consecutive hours? 80,640

62. ***Car Rental*** A rental car agency has 12 cars and seven vans available.

    a. A group taking a field trip needs to rent six vehicles. In how many different ways is this possible? 27,132

    b. If the group needs to rent four cars and two vans, in how many different ways can it select its vehicles? 10,395

63. ***Human Resources*** The personnel director of a company must select four finalists from a group of 10 candidates for a job opening. How many different groups of four can the director select as finalists? 210

64. ***Quality Control*** A quality control inspector receives a shipment of 15 DVD players, from which the inspector

must choose three for testing. How many different groups of three players can the inspector choose? 455

65. *Panel Discussion* Twelve executives, six women and six men, are to be seated in chairs for a panel discussion.

    a. How many arrangements are possible if the men and women are to alternate seats, beginning with a woman? 518,400

    b. How many arrangements are possible if the men must all be seated together? 3,628,800

Life and Health Sciences

66. *DNA Structure* One strand of a DNA (deoxyribonucleic acid) molecule consists of sequences of elements, among which are the four bases: adenine, cytosine, guanine, and thymine. Suppose a portion of a DNA strand contains 10 of these bases.

    a. How many different sequences of these bases are possible for this strand? Assume that there are no restrictions on the bases. 1,048,576

    b. How many different sequences of these bases are possible for this strand if the strand contains three adenine, two cytosine, four guanine, and one thymine? 12,600

Social Sciences

67. *Student Committees* A committee of 16 students must select a president, a vice-president, a secretary, and a treasurer. In how many possible ways can this be accomplished? Assume a student cannot hold more than one office. 43,680

68. *Library Science* Ten volumes of an encyclopedia are arranged on a bookshelf. How many different arrangements are possible? 3,628,800

69. *Committee Selection* A committee of six students is chosen from eight juniors and eight seniors.

    a. How many different committees are possible? 8008

    b. How many different committees are possible that include three juniors and three seniors? 3136

70. *Reading List* A student must read three of seven books for an English class. How many different selections can the student make? 35

71. *True/False Quiz* A quiz consists of 10 true/false questions.

    a. In how many possible orders can a student choose to answer the 10 questions? 3,628,800

    b. In how many distinct ways can the quiz be completed if no answers are left blank? 1024

    c. In how many ways can the quiz be completed if five questions must be marked true and the other five must be marked false? 252

72. *Testing* A professor gives students 15 possible essay questions for an upcoming test. If the professor chooses three of these questions for an exam, how many different exams are possible? 455

73. *Multiple Choice Exam* An exam consists of 12 multiple choice questions, where each question has four possible answers: A, B, C, or D.

    a. In how many possible orders can a student choose to answer the 12 questions if the first six must be completed before any of the remaining six can be answered? 518,400

    b. In how many distinct ways can the exam be completed if no answers are left blank? 16,777,216

    c. In how many ways can the exam be completed if it is known that exactly three questions have "A" as their answer? 4,330,260

74. *CD Selections* A student enrolled in a CD club and can choose six free CDs from the 20 CDs offered. How many different sets of six CDs can the student order? 38,760

Sports and Recreation

75. *Olympic Events* A gold, a silver, and a bronze medal are awarded in an Olympic event. In how many possible ways can the medals be awarded for a 200-meter sprint in which there are nine runners? Assume there are no ties. 504

76. *Horseracing* Twelve horses are entered in a race. Prizes will be awarded to the owners of horses finishing first, second, third, or fourth. In how many possible ways can the owners be awarded prizes? 11,880

77. *Volleyball Teams* A volleyball coach must pick 12 players from a group of 18 volunteers. How many different teams of 12 players are possible? 18,564

78. ***Swim Team*** Twelve athletes are on a college swim team.

a. In how many ways can the coach select eight swimmers for an upcoming race? 495

b. In how many ways can the coach choose four of the swimmers to compete in a relay race, if the order of the racers is taken into account? 11,880

79. ***Softball Teams*** Eighteen people decided to play softball. In how many ways can the 18 people be divided into two teams of nine people? 48,620

80. ***Bowling Teams*** Fifteen people decide to join a bowling league. In how many ways can the 15 people be divided into three teams of five people each? 756,756

81. ***Concert Seating*** Six friends go to a concert and have tickets for six consecutive seats. How many seating arrangements are possible if each person can choose any of the six seats? 720

Card Games **In Exercises 82 to 86, use a standard deck of playing cards consisting of 52 cards divided into four suits (spades, hearts, diamonds, and clubs), where each suit consists of 13 cards: 2 through 10, jack, queen, king, and ace. Assume five cards are randomly chosen from the deck.**

82. How many hands contain four aces? 48

83. How many hands contain exactly two aces and exactly three kings? 24

84. How many hands contain exactly three jacks? 4512

85. How many hands contain exactly two 7s? 103,776

86. How many hands contain exactly two 7s and exactly two 8s? 1584

Physical Sciences and Engineering

87. ***Computer Password*** The password for a computer system consists of three letters followed by three numbers (0 through 9). If a letter or number cannot be repeated, how many different passwords are possible? 11,232,000

88. ***Internet Addresses*** Computers connected to the Internet communicate using unique *IP numbers.* Each number is of the form x.x.x.x, where x represents any one-, two-, or three-digit number from 0 through 255. (For instance, the web site for the White House, www.whitehouse.gov, corresponds to the IP address 198.137.240.91.) How many unique Internet IP addresses exist? 4,294,967,296

Prepare for Section 8.2

89. If $A = \{1, 4, 7, 10, 13\}$ and $B = \{2, 4, 6, 8, 10\}$, find $A \cup B$ and $A \cap B$. [P.1] $\{1, 2, 4, 6, 7, 8, 10, 13\}$, $\{4, 10\}$

90. If $A \cap B = \emptyset$, do A and B have any elements in common? [P.1] No

91. Let $A = \{a, b, c, d\}$, $B = \{c, d, e, f, g\}$, and $C = \{e, f, g, h, i\}$.

a. Is the number of elements in $A \cup B$ equal to the number of elements in A plus the number of elements in B? No

b. Is the number of elements in $A \cup C$ equal to the number of elements in A plus the number of elements in C? [P.1] Yes

92. From Exercise 91, when does the number of elements in the union of two sets equal the sum of the numbers of elements in each set? [P.1] When the intersection of the sets is the empty set.

93. Evaluate: $\frac{8!}{4!}$ [8.1] 1680

94. Evaluate: **a.** $P(5, 2)$ 20 **b.** $C(5, 2)$ [8.1] 10

Explorations 

1. ***Network Circuits*** The top diagram at the right can be used to represent a network, such as computers connected together or highways connecting cities. The lines represent connections between objects, which are represented by the dots (called vertices) in the figure. Some applications require finding a path through the network that goes through each vertex without visiting any vertex twice and returns to the starting vertex. Such a path is called a **Hamiltonian circuit.** A Hamiltonian circuit is shown as the zig-zag path in the middle network at the right. There are usually many Hamiltonian circuits in a network, but by comparing different paths, we can identify routes with, for example, a shortest total distance between vertices.

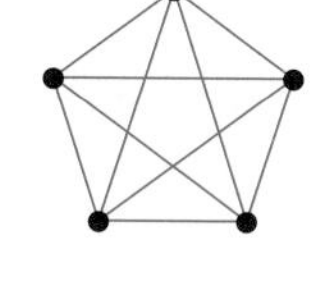

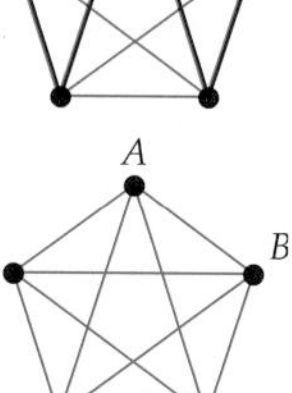

Just how many Hamiltonian circuits can there be in a particular graph? The counting principle can be used to find the number of possible paths through a network. For the network with five vertices in the bottom figure on the previous page, we begin at the home vertex, labeled A. There are four choices for the next vertex to visit, three choices for the next, then two choices for the next, and finally one choice that returns to A. By the counting principle there are $4 \cdot 3 \cdot 2 \cdot 1 = 24$ possible circuits. However, traveling the circuit *ABCDEA* is the same as traveling the circuit in reverse order, *AEDCBA*. Thus each path is actually counted twice, so there are $\frac{24}{2} = 12$ possible circuits.

In general, there are $\frac{(n-1)!}{2}$ possible Hamiltonian circuits through a network of n vertices, where $n \geq 3$, that has all possible connections drawn between vertices.

a. A network has eight vertices and all possible connections have been drawn between the vertices. How many Hamiltonian circuits are possible? 2520

b. For the network at the right, find the number of possible Hamiltonian circuits. 60

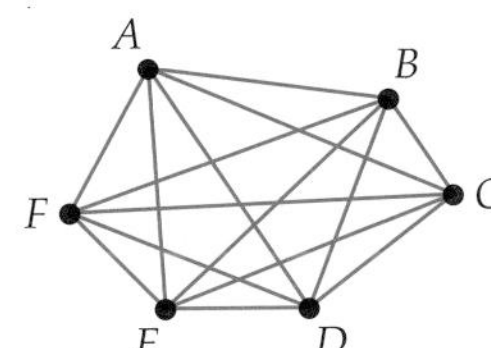

c. Suppose a network has 20 vertices (with all possible connections drawn between vertices) and a computer is available that can analyze one million paths per second. How many years would it take this computer to search all the possible Hamiltonian circuits of this network? Use 1 year = 365 days. ≈1928.7 yr

d. Suppose a network has 40 vertices (with all possible connections drawn between vertices) and a computer is available that can analyze one trillion paths per second. How many years would it take this computer to search all the possible Hamiltonian circuits? Use 1 year = 365 days. More than 3×10^{26} yr

2. ***A Secret Code*** Francis Bacon, a contemporary of William Shakespeare, invented a *cipher* (a secret code) based on permutations of the letters a and b. Research Bacon's method and the intended use of his scheme, using the Internet or a library. Write a short essay describing your findings. Answers will vary.

SECTION 8.2 Introduction to Probability

- Sample Spaces and Events
- Empirical Probability
- Applications to Genetics

In California, the likelihood of someone selecting the winning lottery numbers in the Super Lotto Plus game is approximately 1 in 41,000,000. In contrast, the likelihood of being struck by lightning is about 1 chance in 1,000,000. Comparing the likelihood of winning the Super Lotto Plus to that of being struck by lightning indicates that you are 41 times more likely to be struck by lightning than to pick the winning California lottery numbers.

The likelihood of a certain event occurring is described by a number between 0 and 1. (You can think of this as a percentage.) This number is called the **probability of the event.** An event that is not very likely has a probability close to 0; an event that is very likely has a probability close to 1 (100%). For instance, the probability of being struck by lightning is close to 0. However, if you randomly choose a basketball player from the National Basketball Association, the probability that the player is over 6 feet tall is very likely and therefore close to 1. Because any event has from 0% to 100% chance of occurring, probabilities are always between 0 and 1, inclusive.

Sample Spaces and Events

Probabilities are calculated by considering experiments, which are activities with observable outcomes. Here are some examples of experiments:

- Flip a coin and observe the outcome as head or tail.
- Select a company and observe its annual profit.
- Record the time a person spends at the checkout line in a supermarket.

The **sample space** of an experiment is the set of all possible outcomes of the experiment. For example, consider tossing a coin three times and observing the outcome as a head or a tail. Using H for head and T for tail, the sample space is

$$S = \{\text{HHH, HHT, HTH, HTT, THH, THT, TTH, TTT}\}$$

Note that the sample space consists of *every* possible outcome of tossing a coin three times.

Alternative to Example 1

Three candies are selected from a jar of red and green jelly beans. What is the sample space for this experiment?

- **{RRR, RRG, RGR, RGG, GRR, GRG, GGR, GGG}**

EXAMPLE 1 Find a Sample Space

A single die is rolled once. What is the sample space for this experiment?

Solution The sample space is the set of possible outcomes of the experiment.

$S = \{$⚀, ⚁, ⚂, ⚃, ⚄, ⚅$\}$

CHECK YOUR PROGRESS 1 A coin is tossed twice. What is the sample space for this experiment?

Solution *See page S43.* {HH, HT, TH, TT}

An **event** is a subset of a sample space. Using the sample space from Example 1, here are some possible events:

- There are an even number of pips (dots) facing up. The event is $E_1 = \{$⚁, ⚃, ⚅$\}$.
- The number of pips facing up is greater than 4. The event is $E_2 = \{$⚄, ⚅$\}$.
- The number of pips facing up is less than 20. The event is $E_3 = \{$⚀, ⚁, ⚂, ⚃, ⚄, ⚅$\}$.

  Because the number of pips facing up is always less than 20, this event will always occur. The event and the sample space are the same.
- The number of pips facing up is greater than 15. The event is $E_4 = \varnothing$, the empty set. This is an impossible event, as it is not possible for the number of pips facing up to be greater than 15.

TAKE NOTE

As an example of an experiment whose outcomes are not equally likely, consider tossing a thumbtack and recording whether it lands with the point up or on its side. There are only two possible outcomes of this experiment, but the outcomes are not equally likely.

Outcomes of some experiments are *equally likely*, which means that any one outcome is just as likely as another. For instance, if four balls of the same size but different colors—red, blue, green, and white—are placed in a box and a blindfolded person chooses one ball, the chance of choosing a blue ball is the same as the chance of choosing any other color ball.

In the case of equally likely outcomes, the probability of an event is based on the number of elements in the event and the number of elements in the sample space.

Probability of an Event

For an experiment with sample space S of *equally likely outcomes*, the probability $P(E)$ of an event E is given by

$$P(E) = \frac{n(E)}{n(S)}$$

where $n(E)$ is the number of elements in the event and $n(S)$ is the number of elements in the sample space.

Point of Interest

To win the grand lottery prize in Pennsylvania, one must correctly select 6 of 50 numbers. Once, when the prize reached $65 million, a resident suggested that one could buy a ticket with every possible combination of numbers. At $1 per combination, this is a total cost of $C(50, 6) = \$15,890,700$. Because this would guarantee a win of the grand prize, it may seem like an easy way to turn $16 million into $65 million. The strategy is flawed, however; if there are multiple winners, the prize money is split. In fact, there were 8 winners who each received approximately $8 million. So the resident's proposal is actually a gamble that no one else will choose the correct numbers!

Because each outcome of rolling a fair die is equally likely, the probability of the events E_1 through E_4 described on the preceding page can be determined from the basic probability formula.

$$P(E_1) = \frac{3}{6} \begin{array}{l}\leftarrow \text{Number of elements in } E_1 \\ \leftarrow \text{Number of elements in the sample space}\end{array}$$
$$= \frac{1}{2}$$

The probability of rolling an even number of pips on a single roll of one die is $\frac{1}{2}$ (or 50%).

$$P(E_2) = \frac{2}{6} \begin{array}{l}\leftarrow \text{Number of elements in } E_2 \\ \leftarrow \text{Number of elements in the sample space}\end{array}$$
$$= \frac{1}{3}$$

The probability of rolling a number greater than 4 on a single roll of one die is $\frac{1}{3}$.

$$P(E_3) = \frac{6}{6} \begin{array}{l}\leftarrow \text{Number of elements in } E_3 \\ \leftarrow \text{Number of elements in the sample space}\end{array}$$
$$= 1$$

The probability of rolling a number less than 20 on a single roll of one die is 1 (or 100%). The probability of any event that is *certain* to occur is 1.

$$P(E_4) = \frac{0}{6} \begin{array}{l}\leftarrow \text{Number of elements in } E_4 \\ \leftarrow \text{Number of elements in the sample space}\end{array}$$
$$= 0$$

The probability of rolling a number greater than 15 on a single roll of one die is 0.

Alternative to Example 2

A friend randomly chooses a coin, places it in your left hand, then chooses another coin for your right hand. If each coin is equally likely to be a penny, nickel, dime, or quarter, what is the probability that you are holding 26 cents?

■ $\frac{1}{8}$

EXAMPLE 2 Probability from Tossing Coins

A fair coin, one for which it is equally likely that heads or tails would result from a single toss, is tossed three times. What is the probability that two heads and one tail are tossed?

Solution Determine the number of elements in the sample space. We must include every possible toss of a head or a tail (in order) on three tosses of the coin.

$$S = \{\text{HHH, HHT, HTH, HTT, THH, THT, TTH, TTT}\}$$

The elements in the event are $E = \{\text{HHT, HTH, THH}\}$.

$$P(E) = \frac{n(E)}{n(S)} = \frac{3}{8}$$

The probability is $\frac{3}{8}$.

CHECK YOUR PROGRESS 2 If a fair die is rolled once, what is the probability that an odd number will appear on the upward face?

Solution *See page S43.* $\frac{1}{2}$

QUESTION Is it possible that the probability of some event could be 1.23?

Outcomes from Rolling Two Dice

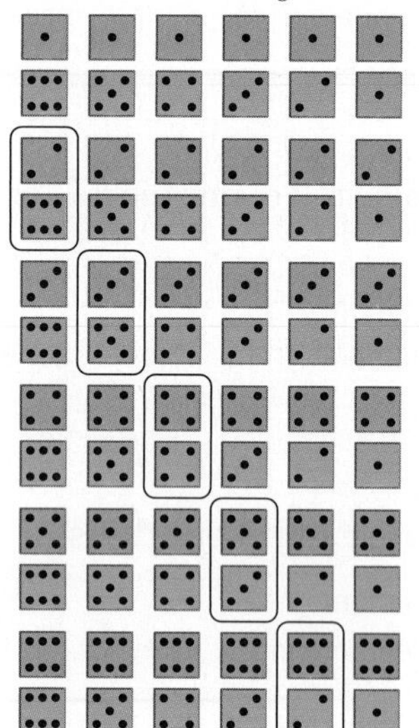

Alternative to Example 3

Two fair dice are tossed once. What is the probability that the sum of the pips on the upward faces of the two dice equals 5?

■ $\frac{1}{9}$

EXAMPLE 3 Probability from Rolling Dice

Two fair dice are tossed once. What is the probability that the sum of the pips on the upward faces of the two dice equals 8?

Solution The possible outcomes of a toss of two dice are shown in the figure at the left. From the figures, $n(S) = 36$. Let E represent the event that the sum of the pips on the upward faces is 8. The outcomes in this event are circled in the figure at the left. By counting the number of circled pairs, $n(E) = 5$.

$$P(E) = \frac{n(E)}{n(S)} = \frac{5}{36}$$

The probability that the sum of the pips is 8 on the roll of two dice is $\frac{5}{36}$.

CHECK YOUR PROGRESS 3 Two fair dice are tossed once. What is the probability that the sum of the pips on the upward faces of the two dice equals 7?

Solution *See page S43.* $\frac{1}{6}$

ANSWER No. All probabilities must be between 0 and 1, inclusive.

Alternative to Example 4

A five-card hand is dealt from a standard 52-card deck of playing cards. What is the probability that the hand consists of three red cards and two black cards?

■ $\frac{845{,}000}{2{,}598{,}960} \approx 0.325$

EXAMPLE 4 Probability from Dealing Playing Cards

A five-card hand is dealt from a standard 52-card deck of playing cards. What is the probability that the hand contains five cards of the same suit?

Solution Let E be the event of drawing five cards of the same suit. In Example 8 of Section 8.1, we determined that the total number of outcomes for E is $4 \cdot C(13, 5) = 5148$, so $n(E) = 5148$. The sample space is the total number of different five-card hands, which is given by $C(52, 5) = 2{,}598{,}960$. Thus $n(S) = 2{,}598{,}960$. The probability of E is

$$P(E) = \frac{n(E)}{n(S)} = \frac{5148}{2{,}598{,}960} \approx 0.00198$$

There is only about a 0.2% chance of being dealt five cards of the same suit.

CHECK YOUR PROGRESS 4 A committee of five professors is chosen from five mathematicians and six economists. If the members are chosen randomly, find the probability that the committee will consist of two mathematicians and three economists.

Solution *See page S43.* $\frac{100}{231} \approx 0.433$

■ Empirical Probability

A probability such as those calculated in the preceding examples is sometimes referred to as a **theoretical probability**. We assume that, in theory, we have a perfectly balanced coin in Example 2 and calculate the probability based on the fact that each outcome, heads or tails, is equally likely. Similarly, we assume that the dice in Example 3 are equally likely to land with any of the six faces upward and that the deck of cards in Example 4 is well shuffled, so that every card has an equal chance of being dealt.

When a probability is based on data gathered from an experiment, it is called an **experimental probability** or an **empirical probability**. For instance, if we tossed a thumbtack 100 times and recorded the number of times it landed "point up," the results might be as shown in the table at the left. From the table, the empirical probability of "point up" is

Point up	15
Side	85
Total	100

$$P(\text{point up}) = \frac{15}{100} = 0.15$$

Alternative to Example 5

Exercise 38, page 667

EXAMPLE 5 Empirical Probability

A survey of the registrar of voters office in a city showed the following information on the ages and party affiliations of registered voters. If one voter is chosen at random from this survey, what is the probability that the voter is a Republican?

Continued ➤

Age	Republican	Democrat	Independent	Other	Total
18–28	205	432	98	112	847
29–38	311	301	109	83	804
39–49	250	251	150	122	773
50+	272	283	142	107	804
Total	1038	1267	499	424	3228

Solution Let R be the event that a Republican is selected. Then

$$P(R) = \frac{1038}{3228} \begin{matrix} \leftarrow \text{Number of Republicans in the survey} \\ \leftarrow \text{Total number of people surveyed} \end{matrix}$$

$$\approx 0.32$$

The probability that the selected person is a Republican is about 0.32.

CHECK YOUR PROGRESS 5 Using the data from Example 5, what is the probability that a randomly selected voter is between the ages of 39 and 49?

Solution *See page S43.* $\frac{773}{3228} \approx 0.24$

Point of Interest

Gregor Mendel (1822–1884) was an Augustinian monk and teacher. His study of the transmission of traits from one generation to another was actually started to confirm the prevailing theory that environment influenced the traits of a plant. However, his research seemed to suggest that plants have certain "determiners" that dictate the characteristics of the next generation. Thus began the study of heredity. It took over 30 years for Mendel's conclusions to be accepted in the scientific community.

Applications to Genetics

The Human Genome Project (the complete set of instructions for making an organism) is a 13-year project designed to map the genetic make-up of *Homo sapiens.* Researchers hope to use this information to treat and prevent certain hereditary diseases.

The concept behind this project began with Gregor Mendel and his work on flower color and how it was transmitted from generation to generation. From his studies, Mendel concluded that flower color is predictable in future generations by making certain assumptions about a plant's color "determiner." He concluded that red was a *dominant* determiner of color and that white was a *recessive* determiner. Today, geneticists talk about the *gene* for flower color and the *allele* of the gene, which is whether the gene consists of two dominant alleles (two red), a dominant and a recessive allele (a red and a white), or two recessive alleles (two white). Because red is the dominant allele, the flower will be white only if no dominant allele is present.

Later work by Reginald Punnett (1875–1967) showed how to determine the probability of certain flower colors by using a **Punnett square.** Using a capital letter for a dominant allele (say R for red) and the corresponding lowercase letter for the recessive allele (r for white), Punnett arranged the genes of the parent in a square. A parent could be RR, Rr, or rr.

Suppose that the genotypes (genetic compositions) of two parents are as shown below (one parent is Rr and the other is rr). The genotypes of the offspring are shown in the body of the table. They are the result of combining one allele from each parent.

Parents	R	r
r	Rr	rr
r	Rr	rr

Because each of the genotypes of the offspring is equally likely, the probability that a flower is red is $\frac{1}{2}$ (two Rr genotypes of the four possible) and the probability that a flower is white is $\frac{1}{2}$ (two rr genotypes of the four possible).

Alternative to Example 6
Exercise 40, page 667

TAKE NOTE

The probability that a parent will pass a certain genetic characteristic on to a child is one-half. However, the probability that a parent has the genetic characteristic is not necessarily one-half. For instance, the probability of passing on the allele for cystic fibrosis by a parent who has the mutant allele is 0.5. However, the probability that a person randomly selected from the population has the mutant allele is less than 0.025.

EXAMPLE 6 Probability and Genetics

A child will have cystic fibrosis if the child inherits the recessive gene from both parents. Using F for the normal allele and f for the mutant allele, suppose a parent who is Ff (said to be a *carrier*) and a parent who is FF (does not have the mutant allele) decide to have a child.

a. What is the probability that the child will have cystic fibrosis? To have the disease, the child must be ff.

b. What is the probability that the child will be a carrier?

Solution Make a Punnett square.

Parents	F	F
F	FF	FF
f	Ff	Ff

a. To have the disease, the child must be ff. From the table, there is no combination of alleles that produces ff. Therefore, the child cannot have the disease, and the probability is 0.

b. For the child to be a carrier, one allele must be f. From the table, there are two cases out of four in which the child will have one f. The probability that the child will be a carrier is $\frac{2}{4} = \frac{1}{2}$.

CHECK YOUR PROGRESS 6 For a certain type of hamster, the color cinnamon, C, is dominant and the color white, c, is recessive. If both parents are Cc, what is the probability that an offspring will be white?

Solution *See page S43.* $\frac{1}{4}$

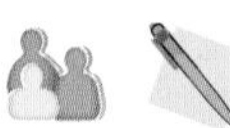

Topics for Discussion

1. If the probability of an event occurring is 0.25, how do you interpret this?
2. Explain why a probability is always between 0 and 1, inclusive.

3. What is the sample space of an experiment? What is an event?
4. How do you compute a theoretical probability?
5. What is an empirical probability?

EXERCISES 8.2

— *Suggested Assignment: Exercises 1–41, odd; 42; 45–53, odd; 57–71, odd; and 72–77.*

In Exercises 1 to 6, list the elements of the sample space for each experiment.

1. A coin is flipped three times. {HHH, HHT, HTH, THH, HTT, THT, TTH, TTT}
2. An even number between 1 and 11 is selected at random. {2, 4, 6, 8, 10}
3. One day in the first two weeks of November is selected. {11/1, 11/2, 11/3, 11/4, 11/5, 11/6, 11/7, 11/8, 11/9, 11/10, 11/11, 11/12, 11/13, 11/14}
4. A current U.S. coin is selected from a piggybank. {penny, nickel, dime, quarter, half-dollar, dollar}
5. A state is selected from the states in the United States whose name begins with the letter A. {Alaska, Alabama, Arizona, Arkansas}
6. A month is selected from the months that have exactly 30 days. {April, June, September, November}

In Exercises 7 to 15, assume that it is equally likely for a child to be born a boy or a girl, and that a family is planning on having three children.

7. List the elements of the sample space for the possible genders of the three children. {BBB, BBG, BGB, GBB, BGG, GBG, GGB, GGG}
8. List the elements of the event that the family will have two boys and one girl. {BBG, BGB, GBB}
9. List the elements of the event that the family will have at least two girls. {BGG, GBG, GGB, GGG}
10. List the elements of the event that the family will have no girls. {BBB}
11. List the elements of the event that the family will have at least one girl. {BBG, BGB, GBB, BGG, GBG, GGB, GGG}
12. Compute the probability that the family will have two boys and one girl. $\frac{3}{8}$
13. Compute the probability that the family will have at least two girls. $\frac{1}{2}$
14. Compute the probability that the family will have no girls. $\frac{1}{8}$
15. Compute the probability that the family will have at least one girl. $\frac{7}{8}$

In Exercises 16 to 22, a coin is tossed four times. Assuming that the coin is equally likely to land heads or tails, compute the probability of each event occurring.

16. Two heads and two tails $\frac{3}{8}$
17. One head and three tails $\frac{1}{4}$
18. All tails $\frac{1}{16}$
19. All four coin tosses are identical. $\frac{1}{8}$
20. At least three heads $\frac{5}{16}$
21. At least two tails $\frac{11}{16}$
22. At least one head $\frac{15}{16}$

Rolling Dice **In Exercises 23 to 32, two dice are tossed. Compute the probability that the sum of the pips on the upward faces of the two dice is each of the following. (See Example 3 for the sample space of this experiment.)**

23. 6 $\frac{5}{36}$
24. 11 $\frac{1}{18}$
25. 2 $\frac{1}{36}$
26. 12 $\frac{1}{36}$
27. 1 0
28. 14 0
29. At least 10 $\frac{1}{6}$
30. At most 5 $\frac{5}{18}$
31. An even number $\frac{1}{2}$
32. An odd number $\frac{1}{2}$
33. ***Rolling Dice*** If two dice are rolled, compute the probability of rolling doubles (both dice show the same number of pips). $\frac{1}{6}$

34. *Rolling Dice* If two dice are rolled, compute the probability of *not* rolling doubles. $\frac{5}{6}$

Business and Economics

Salary Survey **In Exercises 35 to 38, a random survey asked respondents about their current annual salaries. The results are given in the table below.**

Salary range	Number of respondents
Below \$18,000	24
\$18,000–\$27,999	41
\$28,000–\$35,999	52
\$36,000–\$45,999	58
\$45,000–\$59,999	43
\$60,000–\$79,999	39
\$80,000–\$99,999	22
\$100,000 or more	14

If a respondent of the survey is selected at random, compute the probability of each of the following. (Round to two decimal places.)

35. The respondent earns from \$36,000 to \$45,999 annually. 0.20

36. The respondent earns from \$60,000 to \$79,999 per year. 0.13

37. The respondent earns at least \$80,000 per year. 0.12

38. The respondent earns less than \$36,000 annually. $\frac{117}{293} \approx 0.40$

Life and Health Sciences

39. *Births in Alaska* During a recent year in Alaska, 5238 boys and 4984 girls were born. If a newborn had been selected at random, what is the probability that the baby would have been a girl? Round to the nearest hundredth. 0.49

40. *Genetics* The following Punnett square for flower color shows two parents, each of genotype Rr, where R corresponds to the dominant red flower allele and r represents the recessive white flower allele. (See Example 6 on page 665.)

Parents	R	r
R	RR	Rr
r	Rr	rr

What is the probability that an offspring of these parents will have white flowers? $\frac{1}{4}$

41. *Genetics* One parent plant with red flowers has genotype RR, and the other with white flowers has genotype rr, where R is the dominant allele for a red flower and r is the recessive allele for a white flower. Compute the probability of an offspring having white flowers. Hint: Draw a Punnett square. 0

42. *Genetics* The eye color of mice is determined by a dominant allele E, corresponding to black eyes, and a recessive allele e, corresponding to red eyes. If two mice, one of genotype EE and the other of type ee, have offspring, compute the probability of an offspring mouse having red eyes. Hint: Draw a Punnett square. 0

43. *Genetics* The height of a certain plant is determined by a dominant allele T, corresponding to tall plants, and a recessive allele t, corresponding to short (or dwarf) plants. If the parent plants both have genotype Tt, compute the probability that an offspring plant will be tall. Hint: Draw a Punnett square. $\frac{3}{4}$

Social Sciences

Voter Survey **In Exercises 44 to 49, use the data given in Example 5 to compute the probability that a voter randomly chosen from the survey will satisfy each of the following.**

44. The voter is a Democrat. $\frac{1267}{3228} \approx 0.39$

45. The voter is not a Republican. $\frac{2190}{3228} \approx 0.68$

46. The voter is 50 or more years old. $\frac{804}{3228} \approx 0.25$

47. The voter is under 39 years old. $\frac{1651}{3228} \approx 0.51$

48. The voter is between 39 and 49, and is registered as an Independent. $\frac{150}{3228} \approx 0.05$

49. The voter is under 29 and is registered as a Democrat. $\frac{432}{3228} \approx 0.13$

Education Level **In Exercises 50 to 53, a random survey asked 850 respondents about their highest levels of completed education. The results are given in the table below.**

Education completed	Number of respondents
No high school diploma	52
High school diploma	234
Associate's degree or two years college	274
Bachelor's degree	187
Master's degree	67
Ph.D. or professional degree	36

If a respondent of the survey is selected at random, compute the probability of each of the following.

50. The respondent did not complete high school. $\frac{26}{425} \approx 0.06$

51. The respondent has an associate's degree or two years of college (but not more). $\frac{137}{425} \approx 0.32$

52. The respondent has a Ph.D. or professional degree. $\frac{18}{425} \approx 0.04$

53. The respondent has a degree beyond a bachelor's degree. $\frac{103}{850} \approx 0.12$

Sports and Recreation

54. *Monopoly* In the game of Monopoly, a player rolls two dice. If the player rolls doubles three times in a row, the player goes to jail. What is the probability of rolling doubles three times in a row? $\frac{1}{216}$

55. *Chuck-a-luck* In the game of chuck-a-luck, a player chooses a number from 1 to 6 and then rolls three regular six-sided dice. If the number the player chose shows on at least one of the dice, the player wins. What is the probability that none of the dice will show the number chosen by the player? $\frac{125}{216}$

Playing Cards **In Exercises 56 to 61, a card is selected at random from a standard deck.**

56. Compute the probability that the card is red. $\frac{1}{2}$

57. Compute the probability that the card is a spade. $\frac{1}{4}$

58. Compute the probability that the card is a 9. $\frac{1}{13}$

59. Compute the probability that the card is a face card (jack, queen, or king). $\frac{3}{13}$

60. Compute the probability that the card is between 5 and 9. $\frac{3}{13}$

61. Compute the probability that the card is between 3 and 6. $\frac{2}{13}$

Playing Cards **In Exercises 62 to 65, a hand of five cards is dealt from a standard deck of playing cards.**

62. Find the probability that the hand will contain all four aces. $\frac{48}{2{,}598{,}960} \approx 1.85 \times 10^{-5}$

63. Find the probability that the hand will contain three jacks and two queens. $\frac{24}{2{,}598{,}960} \approx 9.23 \times 10^{-6}$

64. Find the probability that the hand will contain exactly two 7s. $\frac{103{,}776}{2{,}598{,}960} \approx 0.0399$

65. Find the probability that the hand will consist of three cards of one suit and two cards of another. $\frac{267{,}696}{2{,}598{,}960} \approx 0.103$

Roulette **Exercises 66 to 71 refer to the casino game roulette. Roulette is played by spinning a wheel with 38 numbered slots. The numbers 1 through 36 appear on the wheel, half of them colored black and half colored red, along with two slots numbered 0 and 00, colored green. A ball is placed on the spinning wheel and allowed to come to rest in one of the slots. Bets can be placed on where the ball will land.**

66. You can place a bet that the ball will stop in a black slot. If you win, the casino will pay you \$1 for each dollar you bet. What is the probability of winning this bet? $\frac{9}{19}$

67. You can bet that the ball will land on an odd number. If you win, the casino will pay you \$1 for each dollar you bet. What is the probability of winning this bet? $\frac{9}{19}$

68. You can bet that the ball will land on any number from 1 to 12. If you win, the casino will pay you \$2 for each dollar you bet. What is the probability of winning this bet? $\frac{6}{19}$

69. You can bet that the ball will land on any particular number. If you win, the casino will pay you \$35 for each dollar you bet. What is the probability of winning this bet? $\frac{1}{38}$

70. You can bet that the ball will land on one of 0 or 00. If you win, the casino will pay you \$17 for each dollar you bet. What is the probability of winning this bet? $\frac{1}{19}$

71. You can bet the ball will land on certain groups of six numbers (such as 1–6). If you win, the casino will pay you $5 for each dollar you bet. What is the probability of winning this bet? $\frac{3}{19}$

Prepare for Section 8.3

72. If $A = \{2, 3, 5, 7, 11\}$ and $B = \{1, 2, 3, 5, 8\}$, find $A \cup B$ and $A \cap B$. [P.1] $\{1, 2, 3, 5, 7, 8, 11\}; \{2, 3, 5\}$

73. If $A = \{1, 3, 5, 7\}$ and $B = \{2, 4, 6, 8\}$, how many elements are in $A \cap B$? [P.1] 0

74. Evaluate: $1 - \left(\frac{5}{6}\right)^3$ [P.1] $\frac{91}{216}$

75. Evaluate: $\frac{C(5, 2) \cdot C(8, 3)}{C(13, 5)}$ [8.1] $\frac{560}{1287} \approx 0.435$

76. A regular six-sided die is rolled once. List the elements in the event that the number on the upward face is not a 3. [8.1] $\{1, 2, 4, 5, 6\}$

77. A fair coin is tossed twice. What is the probability that neither toss results in a tail? [8.2] $\frac{3}{4}$

Explorations

1.  ***Sicherman Dice*** George Sicherman, currently a technical staff member at Lucent Technologies, proposed an alternative pair of dice that give the same sums, with the same probabilities, as a regular pair of dice. The pair is unusual in that the dice are not identical and the same number of pips can appear on more than one face. Sicherman dice have the following structure.

Die 1: $\{1, 3, 4, 5, 6, 8\}$

Die 2: $\{1, 2, 2, 3, 3, 4\}$

It has been shown that this pair of dice is the only possible pair that give the same probabilities as a regular pair of dice. (See Martin Gardner, "Mathematical Games," *Scientific American* Vol. 238, No. 2, pp. 19–32.)

a. If the two Sicherman dice are rolled, list the sample space for the experiment. (*Note:* Two numbers appear twice on the second die, but they are considered separate results.)

b. Compare the probabilities of obtaining each possible sum from rolling the two dice, and verify that these are the same probabilities that result from rolling a regular pair of dice.

c. Suppose the following pair of dice are rolled.

Die 1: $\{1, 2, 3, 4, 5, 6\}$

Die 2: $\{0, 0, 0, 6, 6, 6\}$

Show that the sum showing on the two upward faces can be any of the numbers 1 through 12, and that each sum occurs with equal probability.

2. ***The Monte Hall Problem*** A famous probability puzzle is derived from the game show *Let's Make a Deal*, of which Monte Hall was the host, and goes something like this. Suppose you appear on the show, and you are shown three closed doors. Behind one of the doors is a new car; behind the other two are goats. If you select the door hiding the car, you win the car. Clearly, the probability of randomly choosing the correct door is $\frac{1}{3}$. After you make a selection, the show's host, who knows where the car is, opens one of the other two doors and reveals a goat. He then asks if you would like to switch to another door. Should you stay with your original choice, or switch?

Most people would say at first that it makes no difference. However, computer simulations that play the game over and over have shown that you *should* switch. In fact, you will double your chances of winning the car if you give up your first choice. It can be mathematically proven that the probability of winning if you switch is $\frac{2}{3}$. Search the Internet for "Monte Hall problem" and write a short essay describing how to determine this probability, and why it is advantageous to switch doors. Answers will vary.

1. a.

1-1	3-1	4-1	5-1	6-1	8-1
1-2	3-2	4-2	5-2	6-2	8-2
1-2	3-2	4-2	5-2	6-2	8-2
1-3	3-3	4-3	5-3	6-3	8-3
1-3	3-3	4-3	5-3	6-3	8-3
1-4	3-4	4-4	5-4	6-4	8-4

1. b. $P(2) = \frac{1}{36}$, $P(3) = \frac{2}{36}$, $P(4) = \frac{3}{36}$, $P(5) = \frac{4}{36}$, $P(6) = \frac{5}{36}$, $P(7) = \frac{6}{36}$, $P(8) = \frac{5}{36}$, $P(9) = \frac{4}{36}$, $P(10) = \frac{3}{36}$, $P(11) = \frac{2}{36}$, $P(12) = \frac{1}{36}$

1. c. The sample space is

1-0	2-0	3-0	4-0	5-0	6-0
1-0	2-0	3-0	4-0	5-0	6-0
1-0	2-0	3-0	4-0	5-0	6-0
1-6	2-6	3-6	4-6	5-6	6-6
1-6	2-6	3-6	4-6	5-6	6-6
1-6	2-6	3-6	4-6	5-6	6-6

Each sum (1 through 12) occurs exactly three times, so each has probability $\frac{3}{36} = \frac{1}{12}$.

SECTION 8.3 Probability of Compound Events

- **The Addition Rule for Probabilities**
- **The Complement of an Event**
- **Independent Events**
- **Binomial Experiments**

In the previous section, we learned how to compute the probability of an event occurring. In this section, we will consider more than one event at a time. We may be interested in finding the probability that one (or more) of several events will occur, or we may wish to determine the probability that several events will all occur. These are two examples of determining the probability of a **compound event,** and as we will see, different approaches are required.

The Addition Rule for Probabilities

Suppose we draw a single card from a standard deck of playing cards. The sample space S is the 52 cards of the deck. Therefore, $n(S) = 52$. Now consider the events

E_1 = A four is drawn = {♠4, ♥4, ♦4, ♣4}
E_2 = A spade is drawn = {♠A, ♠2, ♠3, ♠4, ♠5, ♠6, ♠7, ♠8, ♠9, ♠10 ♠J, ♠Q, ♠K}

It is possible, on one draw, to satisfy the conditions of both events: The four of spades could be drawn. This card is an element of both E_1 and E_2.

Now compare the events

E_3 = A five is drawn = {♠5, ♥5, ♦5, ♣5}

and

E_4 = A king is drawn = {♠K, ♥K, ♦K, ♣K}

In this case, it is not possible to draw one card that satisfies the conditions of both events. Two events that cannot both occur at the same time are called **mutually exclusive events.** Thus events E_3 and E_4 are mutually exclusive events, whereas E_1 and E_2 are not.

Mutually Exclusive Events

Two events A and B are mutually exclusive if they cannot occur at the same time. That is, A and B are mutually exclusive when $A \cap B = \varnothing$.

? QUESTION A die is rolled once. Let E be the event that an even number is rolled and let O be the event that an odd number is rolled. Are the events E and O mutually exclusive?

The probability of either of two mutually exclusive events occurring can be determined by adding the probabilities of the individual events.

? ANSWER Yes. It is not possible to roll an even number and an odd number on a single roll of the die.

TAKE NOTE

Recall that the union of two sets *A* and *B* is the set containing the elements of *A* or *B*.

Probability of Mutually Exclusive Events

If A and B are two mutually exclusive events, then the probability of A or B occurring is

$$P(A \cup B) = P(A) + P(B) = \frac{n(A)}{n(S)} + \frac{n(B)}{n(S)}$$

Alternative to Example 1

A single card is drawn from a standard deck of playing cards. Find the probability of drawing a spade or a red card.

■ $\frac{3}{4}$

EXAMPLE 1 Probability and Playing Cards

Suppose a single card is drawn from a standard deck of playing cards. Find the probability of drawing a five or a king.

Solution Let $A = \{♠5, ♥5, ♦5, ♣5\}$ and $B = \{♠K, ♥K, ♦K, ♣K\}$. There are 52 cards in a standard deck of playing cards; thus $n(S) = 52$. Because the events are mutually exclusive, we can use the formula for the probability of mutually exclusive events.

$$P(A \text{ or } B) = P(A) + P(B)$$

■ **Formula for the probability of mutually exclusive events**

$$= \frac{1}{13} + \frac{1}{13} = \frac{2}{13}$$

■ $P(A) = \frac{4}{52} = \frac{1}{13}$, $P(B) = \frac{4}{52} = \frac{1}{13}$

The probability of drawing a five or a king is $\frac{2}{13}$.

CHECK YOUR PROGRESS 1 Two fair dice are tossed once. What is the probability of rolling a sum of 7 or 11?

Solution *See page S43.* $\frac{2}{9}$

Consider the experiment of rolling two dice. Let A be the event of rolling a sum of 8 and let B be the event of rolling a double (the same number on both dice).

$A = \{$⚁⚅, ⚅⚁, ⚂⚄, ⚄⚂, ⚃⚃$\}$

$B = \{$⚀⚀, ⚁⚁, ⚂⚂, ⚃⚃, ⚄⚄, ⚅⚅$\}$

These events are *not* mutually exclusive because it is possible to satisfy the conditions of each event on one toss of the dice—a ⚃⚃ could be rolled. Therefore $P(A \cup B)$, the probability of a sum of 8 or a double, cannot be calculated using the formula for the probability of mutually exclusive events. However, a modification of that formula can be used.

TAKE NOTE

The $P(A \cap B)$ term in the Addition Rule for Probabilities is subtracted to compensate for the overcounting of the first two terms of the formula. If two events are mutually exclusive, then $A \cap B = \varnothing$. Therefore, $n(A \cap B) = 0$ and $P(A \cap B) = \frac{n(A \cap B)}{n(S)} = 0.$ Thus, for mutually exclusive events, the Addition Rule for Probabilities is the same as the formula for the probability of mutually exclusive events.

Addition Rule for Probabilities

Let A and B be two events in a sample space S. Then

$$P(A \cup B) = P(A) + P(B) - P(A \cap B)$$

In other words, the Addition Rule states that

$$P(A \text{ or } B) = P(A) + P(B) - P(A \text{ and } B)$$

Using this formula with

$A = \{$⚁⚅, ⚅⚁, ⚂⚄, ⚄⚂, ⚃⚃$\}$

$B = \{$⚀⚀, ⚁⚁, ⚂⚂, ⚃⚃, ⚄⚄, ⚅⚅$\}$

and

$A \cap B = \{$⚃⚃$\}$

the probability of A or B can be calculated.

$$\begin{aligned} P(A \cup B) &= P(A) + P(B) - P(A \cap B) \\ &= \frac{5}{36} + \frac{6}{36} - \frac{1}{36} \\ &= \frac{10}{36} \\ &= \frac{5}{18} \end{aligned}$$

▪ $P(A) = \frac{5}{36}$, $P(B) = \frac{6}{36}$, $P(A \cap B) = \frac{1}{36}$

On a single roll of two dice, the probability of a sum of 8 or a double is $\frac{5}{18}$.

Alternative to Example 2
Exercise 50, page 680

EXAMPLE 2 Probability from Data

The table at the left shows data from an experiment to test the effectiveness of a flu vaccine. If one person is randomly selected from this population, what is the probability that the person was vaccinated or contracted the flu?

	F	No F	Total
V	21	198	219
Not V	76	195	271
Total	97	393	490

V: Vaccinated
F: Contracted the flu

Solution Let $V = \{$people who are vaccinated$\}$ and $F = \{$people who contracted the flu$\}$. These events are not mutually exclusive because there are 21 people who were vaccinated and who also contracted the flu. The sample space S consists of the 490 people who participated in the experiment. From the table, $n(V) = 219$, $n(F) = 97$, and $n(V \text{ and } F) = 21$.

$$\begin{aligned} P(V \text{ or } F) &= P(V) + P(F) - P(V \text{ and } F) \\ &= \frac{219}{490} + \frac{97}{490} - \frac{21}{490} \\ &= \frac{59}{98} \approx 0.602 \end{aligned}$$

The probability of selecting a person who was vaccinated or who contracted the flu is approximately 60.2%.

CHECK YOUR PROGRESS 2 The following table shows the starting salaries of college graduates for selected degrees. If one person is chosen at random from this population, what is the probability that the person has a degree in business or has a starting salary between \$20,000 and \$24,999?

Continued ➤

	Engineering	Business	Chemistry	Psychology
Less than \$20,000	0	4	1	12
\$20,000–\$24,999	4	16	3	16
\$25,000–\$29,999	7	21	5	15
\$30,000–\$34,999	12	35	5	7
\$35,000 or more	12	22	4	5

Solution *See page S43.* $\frac{21}{206} \approx 0.587$

▪ The Complement of an Event

Consider the experiment of tossing a single die once. The sample space is

$$S = \{⚀, ⚁, ⚂, ⚃, ⚄, ⚅\}$$

Now consider the event of tossing a ⚂, $E = \{⚂\}$. The probability of E is

$$P(E) = \frac{1}{6}$$ ← Number of elements in E / ← Number of elements in the sample space

The **complement** of E, denoted by E^c, is the event that includes all outcomes of the experiment that are not in E. For $E = \{⚂\}$,

$$E^c = \{⚀, ⚁, ⚃, ⚄, ⚅\}$$

Because E^c consists of the elements not in E, $E \cap E^c = \varnothing$. Thus E and E^c are mutually exclusive. Also, $E \cup E^c = \{⚀, ⚁, ⚂, ⚃, ⚄, ⚅\} = S$, the sample space. Using the formula for the probability of mutually exclusive events, we have

$$P(E \cup E^c) = P(S)$$

$$P(E) + P(E^c) = 1$$ ▪ $P(S) = 1$

$$P(E^c) = 1 - P(E)$$ ▪ Subtract $P(E)$ from each side of the equation.

> **Probability of the Complement of an Event**
>
> If E is an event and E^c is the complement of the event, then
>
> $$P(E^c) = 1 - P(E)$$

Continuing our example, the probability of not tossing a ⚂ is given by

$$P(\text{not a } ⚂) = 1 - P(⚂)$$

$$= 1 - \frac{1}{6} = \frac{5}{6}$$

You can also verify the probability of E^c directly:

$$P(\text{not a } ⚂) = \frac{5}{6} \begin{array}{l} \leftarrow \text{Number of elements in } E^c \\ \leftarrow \text{Number of elements in the sample space} \end{array}$$

Alternative to Example 3
The probability of winning a prize in a contest is 0.06. What is the probability of not winning a prize?
■ **0.94**

EXAMPLE 3 Using the Complement of an Event to Find a Probability

The probability of tossing a sum of 11 on the toss of two dice is $\frac{1}{18}$. What is the probability of not tossing a sum of 11?

Solution Use the formula for the probability of the complement of an event. Let $E = \{\text{toss of a sum of 11}\}$. Then $E^c = \{\text{the sum is not 11}\}$.

$$\begin{aligned} P(E^c) &= 1 - P(E) \\ &= 1 - \frac{1}{18} = \frac{17}{18} \qquad \blacksquare\ P(E) = \frac{1}{18} \end{aligned}$$

The probability of not tossing a sum of 11 is $\frac{17}{18}$.

CHECK YOUR PROGRESS 3 The probability that a person has type A blood is 34%. What is the probability that a person does not have type A blood?

Solution *See page S43.* 0.66

TAKE NOTE
The phrase "at least one" means one or more. Tossing a coin three times and asking for the probability of at least one head means to calculate the probability of one, two, or three heads.

Suppose we toss a coin three times and want to calculate the probability of having heads occur *at least once.* We could list all the possibilities of tossing three coins, as shown below, and then find all the outcomes that contain at least one head.

$$\underbrace{\{\text{HHH, HHT, HTH, HTT, THH, THT, TTH}}_{\text{At least one head}}\text{, TTT}\}$$

The probability of at least one head is $\frac{7}{8}$.

Another way to calculate this result is to use the formula for the probability of the complement of an event. Let $E = \{\text{at least one head}\}$. From the list above, note that at least one head includes every outcome except TTT (no heads). Thus $E^c = \{\text{TTT}\}$ and we have

$$\begin{aligned} P(E) &= 1 - P(E^c) \\ &= 1 - \frac{1}{8} = \frac{7}{8} \qquad \blacksquare\ P(E^c) = \frac{n(E^c)}{n(S)} = \frac{1}{8} \end{aligned}$$

This is the same result that we calculated above.

Alternative to Example 4
A pair of dice are rolled three times. What is the probability that a sum of 12 will occur at least once?
■ **≈0.081**

EXAMPLE 4 Probability from Dice

A die is tossed four times. What is the probability that a ⚅ will show on the upward face at least once?

Continued ➤

Solution Let $E = \{\text{at least one } 6\}$. Then $E^c = \{\text{no 6s}\}$. To calculate the number of elements in the sample space (all possible outcomes of tossing a die four times) and the number of items in E^c, we will use the counting principle. Because on each toss of the die there are six possible outcomes,

$$n(S) = 6 \cdot 6 \cdot 6 \cdot 6 = 1296$$

On each toss of the die, there are five numbers that are not 6s. Therefore,

$$n(E^c) = 5 \cdot 5 \cdot 5 \cdot 5 = 625$$

$$P(E) = 1 - P(E^c)$$

$$= 1 - \frac{625}{1296} = \frac{671}{1296}$$

$$\approx 0.518$$

When a die is tossed four times, the probability that a ⚅ will show on the upward face at least once is approximately 51.8%.

CHECK YOUR PROGRESS 4 A pair of dice are rolled three times. What is the probability that a sum of 7 will occur at least once?

Solution *See page S43.* 0.421

▪ Independent Events

The preceding examples demonstrate how we can compute the probability of one event *or* another event occurring (or both). We will now investigate finding the probability of two events *both* occurring.

Two events are called **independent** if the outcome of the first event does not influence the outcome of the second event. For instance, if we flip a coin twice in succession, the outcome of the first coin toss has no effect on the outcome of the second toss.

On the other hand, consider drawing two playing cards from a deck in succession without replacing the first card. The probability of drawing an ace from the deck on the second draw depends on whether or not the first card drawn was an ace. These two events are not independent.

In the case of the independent events A and B, the probability that both A *and* B will occur can be found by multiplying the individual probabilities.

TAKE NOTE

The formula for the probability of independent events can be extended to more than two events. If E_1, E_2, E_3, and E_4 are independent events, then the probability that all four events will occur is

$$P(E_1) \cdot P(E_2) \cdot P(E_3) \cdot P(E_4).$$

Probability of Independent Events

If A and B are independent events, then the probability that both A *and* B will occur is

$$P(A \text{ and } B) = P(A \cap B) = P(A) \cdot P(B)$$

EXAMPLE 5 Probability of Independent Events

A pair of dice are tossed twice. What is the probability that the first roll is a sum of 7 and the second roll is a sum of 11?

Continued ➤

Alternative to Example 5

A card is drawn from a standard deck of playing cards, then a die is rolled. What is the probability that the card will be a spade and that the die will show an even number?

■ $\frac{1}{8}$

Solution The rolls of a pair of dice are independent, since the probability of a sum of 11 on the second roll does not depend on the outcome of the first roll. Let $A = \{\text{sum of 7 on the first roll}\}$ and $B = \{\text{sum of 11 on the second roll}\}$. Then

$$P(A \text{ and } B) = P(A) \cdot P(B) = \frac{6}{36} \cdot \frac{2}{36} = \frac{1}{108}$$

CHECK YOUR PROGRESS 5 A coin is tossed and then a die is rolled. What is the probability that the coin will show a head and the die will show a 6?

Solution *See page S43.* $\frac{1}{12}$

■ Binomial Experiments

If we repeatedly roll a fair six-sided die, we can use the formula for the probability of independent events to compute the probability of, for instance, rolling five ⚅ in a row. Because each roll is an independent event, the probability is

$$P(\text{five ⚅}) = P(⚅) \cdot P(⚅) \cdot P(⚅) \cdot P(⚅) \cdot P(⚅)$$
$$= \frac{1}{6} \cdot \frac{1}{6} \cdot \frac{1}{6} \cdot \frac{1}{6} \cdot \frac{1}{6} = \frac{1}{7776}$$

What if we want to find the probability that of the first five rolls, exactly two of the rolls are 6s? Since it does not matter on which two rolls the 6s appear, we would have to account for every possible sequence of outcomes that would result in two 6s. Fortunately, we can take an alternative approach.

We can consider each roll of the die as a *trial* that has only two possible outcomes: either we get a 6 or we do not. These two outcomes are mutually exclusive, since we cannot both get a 6 and not get a 6. We will refer to each outcome as either a success or a failure. An experiment such as this is an example of a **binomial experiment.** (The *bi* in *binomial* signifies that each trial has only two outcomes.)

Characteristics of a Binomial Experiment

- The experiment consists of a fixed number of repeated trials.
- There are exactly two outcomes for each trial: success or failure.
- Each trial is independent and so has the same probability of success.

Repeatedly flipping a coin and trying to get heads is a binomial experiment because the probability of a head on any one flip is always $\frac{1}{2}$. Repeatedly drawing cards from a deck without replacing them and trying to get an ace is *not* a binomial experiment because the probability of drawing an ace depends on whether or not an ace was already drawn on previous trials.

To compute the probability of obtaining two 6s after rolling a die five times, we need to account for the probabilities of two successes and three failures, and the different sequences in which these outcomes can occur. The general result is given in the following formula.

TAKE NOTE

A failure is the complement of a success, so the two events are mutually exclusive and

$q = P(\text{failure}) = P(\text{success}^c)$
$= 1 - P(\text{success}) = 1 - p$

Binomial Probability Formula

Consider an experiment consisting of n repeated independent trials for which the probability of success on a single trial is p and the probability of failure is $q = 1 - p$. Then the probability of exactly k successes in n trials is given by

$$C(n, k)p^k q^{n-k}$$

Because the probability of rolling a die and getting a 6 is $\frac{1}{6}$, the probability of not getting a 6 is $1 - \frac{1}{6} = \frac{5}{6}$, and the probability of getting two 6s after the five rolls of the die is

$$C(5, 2)\left(\frac{1}{6}\right)^2\left(\frac{5}{6}\right)^3 = \frac{5!}{2!\,(5-2)!} \cdot \frac{1}{36} \cdot \frac{125}{216} = 10 \cdot \frac{125}{7776} \approx 0.161$$

There is about a 16% chance that two of five rolls of a die will be 6s.

Alternative to Example 6

A multiple-choice exam consists of ten questions. For each question there are four possible choices, of which only one is correct. If someone randomly guesses at the answers, what is the probability of guessing six answers correctly?

■ ≈**0.0162**

Point of Interest

In some civil cases, a judgment against a defendant will occur if at least nine members of a 12-member jury find the defendant guilty. Assuming that each juror is acting independently of the other jurors, the probability of finding a defendant guilty can be approximated by a binomial distribution. For instance, suppose that there is a 75% chance that any juror will vote guilty. Then the probability that at least nine jurors will actually vote guilty is approximately 65%.

EXAMPLE 6 Find a Binomial Probability

Approximately 10% of the world's population is left-handed. What is the probability that a small discussion group of 10 students will contain exactly two left-handed students?

Solution This experiment consists of 10 trials, so $n = 10$. Each trial has exactly two outcomes (success is a left-handed student, failure is a student who is not left-handed), and each trial is independent (that is, one student being left-handed does not influence the likelihood of any other student being left-handed). Thus this is a binomial experiment, and the probability of a success is $p = P(\text{left-handed}) = 0.1$, whereas the probability of a failure is $q = P(\text{not left-handed}) = 1 - p = 1 - 0.1 = 0.9$. Thus the probability that two of the students ($k = 2$) are left-handed is

$$C(n, k)p^k q^{n-k}$$

$$C(10, 2)(0.1)^2(0.9)^8 = 45(0.1)^2(0.9)^8 \approx 0.194$$

■ $n = 10$, $k = 2$, $p = 0.1$, $q = 0.9$.

The probability that the group will contain exactly two left-handed students is approximately 0.194, or 19.4%.

CHECK YOUR PROGRESS 6 A true-false test consists of 10 questions. If a student randomly guesses the answers to the questions, what is the probability that the student will answer seven of the questions correctly?

Solution *See page S43.* 0.117

Suppose that in Check Your Progress 6 a student must answer *at least* seven questions correctly to pass the test. This means a student will pass the test by answering seven, eight, nine, or ten of the questions correctly. Each of these events is mutually exclusive, so

$$\begin{aligned} P(\text{passing}) &= P(\text{exactly } 7) + P(\text{exactly } 8) + P(\text{exactly } 9) + P(\text{exactly } 10) \\ &\approx 0.1171875 + 0.0439453 + 0.0097656 + 0.0009766 \\ &= 0.171875 \end{aligned}$$

The probability of passing the exam by just guessing is approximately 0.172.

TAKE NOTE

Here is the calculation of $P(\text{exactly } 8)$. The calculations of the other probabilities are similar.

We have $n = 10$, $k = 8$, $p = \frac{1}{2}$, and $q = 1 - \frac{1}{2} = \frac{1}{2}$.

$P(\text{exactly } 8)$

$$= C(10, 8)\left(\frac{1}{2}\right)^8\left(\frac{1}{2}\right)^2$$

$$= 45\left(\frac{1}{2}\right)^{10} \approx 0.0439453$$

Alternative to Example 7

A manufacturer of CD-Rs (recordable CD-ROMs) has established that one of every 15 discs is defective. If a consumer purchases 12 CD-Rs, find the probability that at most two are defective.

■ **≈0.959**

EXAMPLE 7 Quality Control

Large flat-screen computer monitors (19-inch or greater) are expensive to produce because the manufacturing process for large monitors can often result in monitors that fail the manufacturer's quality assurance test. Suppose the probability that a monitor is defective is 0.30. Assuming that a monitor being defective is independent from any other monitor being defective, find the probability that of the next 12 monitors produced, at most three are defective.

Solution The phrase "at most three are defective" means zero, one, two, or three are defective. This is a binomial experiment with $n = 12$, $p = 0.3$, $q = 1 - 0.3 = 0.7$, and $k = 0, 1, 2, 3$.

$$\begin{aligned} \text{Probability(at most 3)} &= P(k = 0) + P(k = 1) + P(k = 2) + P(k = 3) \\ &= C(12, 0)(0.3)^0(0.7)^{12} + C(12, 1)(0.3)^1(0.7)^{11} \\ &\quad + C(12, 2)(0.3)^2(0.7)^{10} + C(12, 3)(0.3)^3(0.7)^9 \\ &\approx 0.0138 + 0.0712 + 0.1678 + 0.2397 = 0.4925 \end{aligned}$$

The probability that at most three are defective is approximately 0.493.

CHECK YOUR PROGRESS 7 In American roulette, a person can bet that a number between 1 and 12 (inclusive) will appear on a spin of the wheel. The probability of this event is $\frac{6}{19}$. On five spins of the wheel, what is the probability that a number between 1 and 12 (inclusive) will occur at least three times?

Solution *See page S43.* ≈0.185

Topics for Discussion

1. Discuss the difference between mutually exclusive events and events that are not mutually exclusive. Give examples of each type.
2. Discuss the Addition Rule for Probabilities and how it is used.
3. What is the complement of an event? How do you find the probability of the complement of an event?
4. What does it mean to say that two events are independent? How do you find the probability of two independent events both occurring?
5. What is necessary for an experiment to be considered a binomial experiment?
6. What is the Binomial Probability Formula and how is it used?

EXERCISES 8.3

— *Suggested Assignment: Exercises 1–51, odd; 53–58; 63–66; and 71–76.*

In Exercises 1 to 4, first verify that the compound event consists of two mutually exclusive events, and then compute the probability of the compound event occurring.

1. A single card is drawn from a deck of playing cards. Find the probability of drawing a four or an ace. $\frac{2}{13}$
2. A single card is drawn from a deck; find the probability of drawing a heart or a club. $\frac{1}{2}$
3. Two dice are rolled; find the probability of rolling a 2 or a 10. $\frac{1}{9}$
4. Two dice are rolled; find the probability of rolling a 7 or an 8. $\frac{11}{36}$

In Exercises 5 to 10, two dice are rolled. Determine the probability of each of the following.

5. Rolling a 6 or doubles $\frac{5}{18}$
6. Rolling a 7 or doubles $\frac{1}{3}$
7. Rolling an even number or doubles $\frac{1}{2}$
8. Rolling a number greater than 7 or doubles $\frac{1}{2}$
9. Rolling an odd number or a number less than 4 $\frac{19}{36}$
10. Rolling an even number or a number greater than 10 $\frac{5}{9}$

In Exercises 11 to 16, a single card is drawn from a standard deck. Find the probability of each of the following events.

11. Drawing an 8 or a spade $\frac{4}{13}$
12. Drawing an ace or a red card $\frac{7}{13}$
13. Drawing a jack or a face card (jacks, queens and kings are called face cards) $\frac{3}{13}$
14. Drawing a red card or a face card (jacks, queens, and kings are called face cards) $\frac{8}{13}$
15. Drawing a diamond or a black card $\frac{3}{4}$
16. Drawing a spade or a red card $\frac{3}{4}$

In Exercises 17 to 24, use the formula for the probability of the complement of an event.

17. Two dice are tossed. What is the probability of not tossing a sum of 7? $\frac{5}{6}$
18. Two dice are tossed. What is the probability of not getting doubles? $\frac{5}{6}$
19. Two dice are tossed. What is the probability of getting a sum of at least 4? $\frac{11}{12}$
20. Two dice are tossed. What is the probability of getting a sum of at most 11? $\frac{35}{36}$
21. A single card is drawn from a standard deck. What is the probability of not drawing an ace? $\frac{12}{13}$
22. A single card is drawn from a standard deck. What is the probability of not drawing a face card? $\frac{10}{13}$
23. A coin is flipped four times. What is the probability of getting at least one tail? $\frac{15}{16}$
24. A coin is flipped four times. What is the probability of getting at least two heads? $\frac{11}{16}$

In Exercises 25 to 30, a pair of dice are tossed twice.

25. Find the probability that both rolls give a sum of 8. $\frac{25}{1296}$
26. Find the probability that the first roll is a sum of 6, and the second is a sum of 12. $\frac{5}{1296}$
27. Find the probability that the first roll is a total of at least 10 and the second roll is a total of at least 11. $\frac{1}{72}$
28. Find the probability that both rolls result in doubles. $\frac{1}{36}$
29. Find the probability that both rolls give even sums. $\frac{1}{4}$
30. Find the probability that both rolls give a sum of at most 4. $\frac{1}{36}$
31. A die is rolled three times. What is the probability that each roll is a 5? $\frac{1}{216}$
32. A die is rolled three times. What is the probability that no roll is a 5? $\frac{125}{216}$

33. A die is rolled three times. What is the probability that at least one roll is a 5? $\frac{91}{216}$

34. A die is rolled four times. Find the probability that at least one roll is a 1. $\frac{671}{1296}$

35. A pair of dice are rolled three times. Find the probability that a sum of 5 occurs at least once. $\frac{217}{729}$

36. A pair of dice are rolled three times. Find the probability that a sum of at least 11 occurs at least once. $\frac{397}{1728}$

In Exercises 37 to 42, a card is pulled from a standard deck, replaced, and, after the deck is shuffled, another card is pulled.

37. What is the probability that both cards are aces? $\frac{1}{169}$

38. What is the probability that both cards are face cards? $\frac{9}{169}$

39. What is the probability that the first card is a spade and the second card is a diamond? $\frac{1}{16}$

40. What is the probability that the first card is red and the second card is black? $\frac{1}{4}$

41. What is the probability that the first card is an ace and the second card is not an ace? $\frac{12}{169}$

42. What is the probability that the first card is not an ace and the second card is not a face card? $\frac{120}{169}$

In Exercises 43 to 46, use the Binomial Probability Formula.

43. A die is rolled four times. What is the probability that exactly two 3s will occur? $\frac{25}{216}$

44. A die is rolled five times. Find the probability that exactly three 2s will occur. $\frac{125}{3888} \approx 0.032$

45. A coin is flipped 10 times. Find the probability of getting four heads and six tails. $\frac{105}{512}$

46. A coin is flipped eight times. What is the probability of getting three heads and five tails? $\frac{7}{32}$

Business and Economics

47. ***Manufacturing Defects*** About 1.2% of AA batteries produced by a particular manufacturer are defective. If a consumer buys a box of 12 of these batteries, what is the probability that one (but not more than one) battery is defective? ≈ 0.126

48. ***Manufacturing Defects*** Approximately 6% of a manufacturer's cellular phones are produced with defects and are discarded. Find the probability that of the next 10 phones produced, no more than one is defective. ≈ 0.882

Employment Status **In Exercises 49 to 52, use the data in the table below, which shows the employment status of individuals in a particular town by age group.**

Age	Full-time	Part-time	Unemployed
0–17	24	164	371
18–25	185	203	148
26–34	348	67	27
35–49	581	179	104
50+	443	162	173

49. If a person is randomly chosen from the town's population, what is the probability that the person is aged 26–34 or is employed part-time? $\frac{1150}{3179} \approx 0.362$

50. If a person is randomly chosen from the town's population, what is the probability that the person is at least 50 years old or is unemployed? $\frac{1428}{3179} \approx 0.449$

51. If a person is randomly chosen from the town's population, what is the probability that the person is under 18 and employed part-time? $\frac{164}{3179} \approx 0.052$

52. If a person is randomly chosen from the town's population, what is the probability that the person is 18 or older and employed full-time? $\frac{1557}{3179} \approx 0.490$

Candy Production **In Exercises 53 to 56, a snack-size bag of M&Ms candies is opened. Inside, there are 12 red candies, 12 blue, 7 green, 13 brown, 3 orange, and 10 yellow.**

53. If a candy is randomly selected, what is the probability that the candy will be red or green? $\frac{1}{3}$

54. Suppose a candy is selected, its color is noted, and the candy is returned to the bag. If this process is repeated eight times, what is the probability that a red candy will be selected exactly five times? ≈ 0.011

55. If five candies are randomly chosen, what is the probability that they will all be yellow? ≈ 0.000166

56. If six candies are randomly chosen, what is the probability that none of them will be brown? ≈ 0.21

Life and Health Sciences

57. ***Male Height*** About 5% of adult men are 6′2″ or taller. What is the probability that of a group of 20 randomly selected adult males, exactly three are 6′2″ or taller? ≈ 0.06

58. ***Medication*** A new drug causes side effects in 8% of those people who try it. Because 8% of 25 is 2, we would expect two people out of a group of 25 who use the drug to show side effects. Find the probability that of 25 people taking the drug, exactly two experience side effects. ≈ 0.28

59. ***Allergies*** About 4% of the population is allergic to bee stings. So, in a group of 50 people, one would expect two people to be allergic to bee stings. Find the probability that in a group of 50 people, exactly two are allergic to bee stings. ≈ 0.276

60. ***Colorblind*** About 4% of the population is colorblind. What is the probability that in a class of 25 students, exactly one student is colorblind? ≈ 0.375

Social Sciences

61. ***County Population*** Approximately 24% of the population of Miami-Dade county, Florida, is under age 18. If eight residents' names are randomly chosen, compute the probability that exactly half of those residents are under 18. ≈ 0.077

62. ***Taking a Quiz*** A true/false quiz consists of eight questions.

    a. If a student randomly answers the questions, what is the probability that the student will get exactly six questions correct? $\frac{7}{64} \approx 0.109$

    b. What is the probability that the student will get *at least* six questions correct? $\frac{37}{256} \approx 0.145$

63. ***Taking a Test*** A multiple choice test consists of 15 questions. Each question has four possible answers.

    a. If a student randomly answers the questions, what is the probability that the student will get exactly 12 questions correct? $\frac{12{,}285}{4^{15}} \approx 1.1441 \times 10^{-5}$

    b. What is the probability that the student will get *at least* 13 questions correct? $\frac{991}{4^{15}} \approx 9.23 \times 10^{-7}$

Sports and Recreation

California Lottery **Exercises 64 to 67 refer to the California lottery game Super Lotto Plus, in which a player selects five numbers from 1 through 47 and an additional MEGA number from 1 to 27. The jackpot, normally several million dollars, is awarded to anyone matching all six of these numbers to the numbers drawn by the lottery commission.**

64. Prize money (typically \$10,000–\$20,000) is also awarded if a player matches the first five numbers but not the MEGA number. Compute the probability that this will occur. $\frac{1}{1{,}533{,}939}$

65. Find the probability of matching all five numbers plus the MEGA number and winning the jackpot. $\frac{1}{41{,}416{,}353}$

66. A prize of \$10 is awarded for matching three of the five regular numbers. What is the probability of winning the \$10 prize? $\frac{2870}{511{,}313}$

67. Approximately \$100 is awarded if three of the five regular numbers and the MEGA number are matched. What is the probability of winning the \$100 prize? $\frac{2870}{13{,}805{,}451} \approx 0.000208$

Powerball Lottery **Exercises 68 and 69 refer to the Powerball lottery game, in which a player selects five regular numbers from 1 through 49 and an additional red number from 1 to 42. The jackpot is awarded to anyone matching all six of these numbers to the numbers drawn.**

68. Find the probability of matching all five regular numbers. $\frac{1}{1{,}906{,}884}$

69. Find the probability of matching all five regular numbers and the red number. $\frac{1}{80{,}089{,}128}$

Physical Sciences and Engineering

70. ***Computer Use*** Approximately 5% of computers in use are Apple Macintoshes. If 15 computer users enter a web contest, what is the probability that exactly three of the entrants use a Macintosh? Round to the nearest hundredth. 0.03

Prepare for Section 8.4

71. Evaluate: $\frac{9!}{5!}$ [8.1] 3024

72. Evaluate: $C(8, 3)$ [8.1] 56

73. How many distinct arrangements are possible using the letters *xxxxyyy*? [8.1] 35

74. Expand: $(a + b)^3$ [P.3] $a^3 + 3a^2b + 3ab^2 + b^3$

75. Explain the difference between a permutation and a combination. [8.1]
In a permutation, order matters. In a combination, order does not matter.

76. What is the degree of each term after $(x + y)^4$ is written in expanded form? [P.3] 4

Explorations 

1. ***Choosing Numbers in Keno*** A popular gambling game called Keno is played in many casinos. In Keno, there are 80 balls numbered from 1 to 80. The casino randomly chooses 20 balls from the 80. These are called "lucky balls" because if a gambler chooses some of the numbers on these balls, there is a possibility of winning money. The amount that is won depends on the number of lucky numbers the gambler selects. (A player can normally choose from 1 to 15 numbers.) The number of ways in which a casino can choose 20 balls from 80 is

$$C(80, 20) = \frac{80!}{20! \cdot 60!} \approx 3{,}535{,}000{,}000{,}000{,}000{,}000$$

Once the casino chooses the 20 lucky balls, the remaining 60 balls are unlucky for the gambler. A gambler who chooses five numbers will have from zero to five lucky numbers.

Let's consider the case in which two of the five numbers chosen by the gambler are lucky numbers. Because five numbers were chosen, there must also be three unlucky numbers among the five numbers. The number of ways of choosing two lucky numbers from 20 lucky numbers is $C(20, 2)$. The number of ways of choosing three unlucky numbers from 60 unlucky numbers is $C(60, 3)$. By the counting principle, there are

$$C(20, 2) \cdot C(60, 3) = 190 \cdot 34{,}220 = 6{,}501{,}800$$

ways to choose two lucky and three unlucky numbers.

In Exercises **a.**, **b.**, and **c.**, assume that a gambler playing Keno has randomly chosen five numbers.

a. In how many ways can the gambler choose exactly one lucky number? 9,752,700

b. In how many ways can the gambler choose exactly two lucky numbers? 6,501,800

c. Compute the probability of matching exactly three lucky numbers. ≈ 0.0839

d. If a gambler chooses 15 numbers and bets \$1, and matches 13 of the lucky numbers, he will be paid \$12,000. What is the probability of this occurring? $\approx 2.07 \times 10^{-8}$

e. If a gambler chooses 15 numbers and matches six of the lucky numbers, he gets his bet back but is not paid any extra. What is the probability of this occurring? ≈ 0.0863

f. Some casinos will let you choose up to 20 numbers. In this case, if you don't match any of the lucky numbers, the casino pays you! Although this may seem unusual, it is actually more difficult to *not* match any of the lucky numbers than to match a few of them. Compute the probability of not matching any of the lucky numbers at all, and compare it to the probability of matching five lucky numbers. ≈ 0.00119; ≈ 0.233

SECTION 8.4 The Binomial Theorem

- Binomial Theorem
- *i*th Term of a Binomial Expansion
- Pascal's Triangle

In certain situations in mathematics, it is necessary to write $(a + b)^n$ as the sum of its terms. Because $(a + b)$ is a binomial, this process is called **expanding the binomial.** For small values of n, it is relatively easy to write the expansion by using multiplication.

Earlier in the text, we found that

$$(a + b)^1 = a + b$$
$$(a + b)^2 = a^2 + 2ab + b^2$$
$$(a + b)^3 = a^3 + 3a^2b + 3ab^2 + b^3$$

Building on these expansions, we can write a few more.

$$(a + b)^4 = a^4 + 4a^3b + 6a^2b^2 + 4ab^3 + b^4$$
$$(a + b)^5 = a^5 + 5a^4b + 10a^3b^2 + 10a^2b^3 + 5ab^4 + b^5$$

We could continue to build on previous expansions and eventually have quite a comprehensive list of binomial expansions. Instead, however, we will look for a theorem that will enable us to expand $(a + b)^n$ directly without multiplying.

Look at the variable parts of each expansion above. Note that for each $n = 1, 2, 3, 4, 5$:

- The first term is a^n. The exponent on a decreases by 1 for each successive term.
- The exponent on b increases by 1 for each successive term. The last term is b^n.
- The degree of each term is n.

To find a pattern for the coefficients in each expansion, first note that there are $n + 1$ terms and that the coefficient of the first and last term is 1. To find the remaining coefficients, consider the expansion of $(a + b)^5$.

$$(a + b)^5 = a^5 + 5a^4b + 10a^3b^2 + 10a^2b^3 + 5ab^4 + b^5$$

$$\frac{5}{1} = 5 \qquad \frac{5 \cdot 4}{2 \cdot 1} = 10 \qquad \frac{5 \cdot 4 \cdot 3}{3 \cdot 2 \cdot 1} = 10 \qquad \frac{5 \cdot 4 \cdot 3 \cdot 2}{4 \cdot 3 \cdot 2 \cdot 1} = 5$$

Observe from these patterns that there is a strong relationship to factorials. In fact, we can express each coefficient by using factorial notation.

$$\frac{5!}{1!\,4!} = 5 \qquad \frac{5!}{2!\,3!} = 10 \qquad \frac{5!}{3!\,2!} = 10 \qquad \frac{5!}{4!\,1!} = 5$$

In each denominator, the first factorial is the same as the exponent of b and the second factorial is the same as the exponent of a.

In general, we will conjecture that the coefficient of the term $a^{n-k}b^k$ in the expansion of $(a + b)^n$ is $\dfrac{n!}{k!\,(n - k)!}$. Each coefficient of a term of a binomial expansion is called a **binomial coefficient** and is denoted by $\dbinom{n}{k}$.

TAKE NOTE

The notation $\binom{n}{k}$ and the notation $C(n, k)$ both represent

$$\frac{n!}{k!\,(n-k)!}$$

Thus, $\binom{n}{k} = C(n, k)$

Formula for a Binomial Coefficient

The coefficient of the term whose variable part is $a^{n-k}b^k$ in the expansion of $(a+b)^n$ is

$$\binom{n}{k} = \frac{n!}{k!\,(n-k)!}$$

? QUESTION Does $\binom{5}{2} = \frac{5}{2}$?

The first term of the expansion of $(a+b)^n$ can be thought of as $a^n b^0$. In that case, we can calculate the coefficient of that term as

$$\binom{n}{0} = \frac{n!}{0!\,(n-0)!} = \frac{n!}{1 \cdot n!} = 1$$

Alternative to Example 1

Evaluate: $\binom{12}{5}$

▪ 792

EXAMPLE 1 Evaluate a Binomial Coefficient

Evaluate each binomial coefficient. **a.** $\binom{9}{6}$ **b.** $\binom{10}{10}$

Solution

a. $\binom{9}{6} = \frac{9!}{6!\,(9-6)!} = \frac{9!}{6!\,3!} = \frac{9 \cdot 8 \cdot 7 \cdot 6!}{6! \cdot 3 \cdot 2 \cdot 1} = 84$

b. $\binom{10}{10} = \frac{10!}{10!\,(10-10)!} = \frac{10!}{10!\,0!} = 1$ ▪ Remember that $0! = 1$.

CHECK YOUR PROGRESS 1 Evaluate the binomial coefficient $\binom{10}{5}$.

Solution *See page S44.* 252

▪ Binomial Theorem

We are now ready to state the Binomial Theorem for positive integers.

? ANSWER No. $\binom{5}{2} = \frac{5!}{2!\,3!} = \frac{5 \cdot 4 \cdot 3!}{2! \cdot 3!} = \frac{20}{2} = 10$

Binomial Theorem for Positive Integers

If n is a positive integer, then

$$(a+b)^n = \sum_{i=0}^{n} \binom{n}{i} a^{n-i} b^i$$
$$= \binom{n}{0} a^n + \binom{n}{1} a^{n-1} b + \binom{n}{2} a^{n-2} b^2 + \cdots + \binom{n}{n} b^n$$

QUESTION How many terms occur in the expansion of $(a+b)^n$?

Alternative to Example 2
Exercise 14, page 687

EXAMPLE 2 Expand the Sum of Two Terms

Expand: $(2x^2+3)^4$

Solution

$$(2x^2+3)^4 = \binom{4}{0}(2x^2)^4 + \binom{4}{1}(2x^2)^3(3) + \binom{4}{2}(2x^2)^2(3)^2 + \binom{4}{3}(2x^2)(3)^3 + \binom{4}{4}(3)^4$$
$$= 16x^8 + 96x^6 + 216x^4 + 216x^2 + 81$$

CHECK YOUR PROGRESS 2 Expand: $(3x+2y)^4$

Solution *See page S44.* $81x^4 + 216x^3y + 216x^2y^2 + 96xy^3 + 16y^4$

Alternative to Example 3
Exercise 16, page 687

TAKE NOTE
If exactly one of the terms a or b is negative, the terms of the expansion alternate in sign.

EXAMPLE 3 Expand a Difference of Two Terms

Expand: $(\sqrt{x}-2y)^5$

Solution In this expansion $a=\sqrt{x}$ and $b=-2y$.

$$(\sqrt{x}-2y)^5 = \binom{5}{0}(\sqrt{x})^5 + \binom{5}{1}(\sqrt{x})^4(-2y) + \binom{5}{2}(\sqrt{x})^3(-2y)^2 + \binom{5}{3}(\sqrt{x})^2(-2y)^3 + \binom{5}{4}(\sqrt{x})(-2y)^4 + \binom{5}{5}(-2y)^5$$
$$= x^{5/2} - 10x^2y + 40x^{3/2}y^2 - 80xy^3 + 80x^{1/2}y^4 - 32y^5$$

CHECK YOUR PROGRESS 3 Expand: $(2x-\sqrt{y})^7$

Solution *See page S44.* $128x^7 - 448x^6\sqrt{y} + 672x^5y - 560x^4y\sqrt{y} + 280x^3y^2 - 84x^2y^2\sqrt{y} + 14xy^3 - y^3\sqrt{y}$

ANSWER $n+1$

▪ *i*th Term of a Binomial Expansion

The Binomial Theorem can also be used to find a specific term in the **expansion** of $(a + b)^n$.

TAKE NOTE

The exponent on b is 1 *less* than the term number.

P

Formula for the *i*th Term of a Binomial Expansion

The ith term in the expansion of $(a + b)^n$ is given by

$$\binom{n}{i-1} a^{n-i+1} b^{i-1}$$

Alternative to Example 4

Exercise 30, page 687

EXAMPLE 4 Find the *i*th Term of a Binomial Expansion

Find the fourth term in the expansion of $(2x^3 - 3y^2)^5$.

Solution With $a = 2x^3$ and $b = -3y^2$, and using the above theorem with $i = 4$ and $n = 5$, we have

$$\binom{5}{3}(2x^3)^2(-3y^2)^3 = -1080x^6y^6$$

The fourth term is $-1080x^6y^6$.

CHECK YOUR PROGRESS 4 Find the sixth term of $(x^{-1/2} + x^{1/2})^{10}$.

Solution *See page S44.* 252

▪ Pascal's Triangle

A pattern for the coefficients of the terms of an expanded binomial can be found by writing the coefficients in a triangular array known as **Pascal's Triangle.** See the following figure.

$(a+b)^1$:						1		1					
$(a+b)^2$:					1		2		1				
$(a+b)^3$:				1		3		3		1			
$(a+b)^4$:			1		4		6		4		1		
$(a+b)^5$:		1		5		10		10		5		1	
$(a+b)^6$:	1		6		15		20		15		6		1

Each row begins and ends with the number 1. Any other number in a row is the sum of the two closest numbers above it. For example, $4 + 6 = 10$. Thus each succeeding row can be found from the preceding row.

Pascal's Triangle can be used to expand a binomial for small values of n. For instance, the seventh row of Pascal's Triangle is

$$1 \quad 7 \quad 21 \quad 35 \quad 35 \quad 21 \quad 7 \quad 1$$

Therefore,

$$(a + b)^7 = a^7 + 7a^6b + 21a^5b^2 + 35a^4b^3 + 35a^3b^4 + 21a^2b^5 + 7ab^6 + b^7$$

Topics for Discussion

1. Can the Binomial Theorem be used to expand $(a + b)^n$, n a natural number, for any expressions a and b? Why or why not?
2. What is Pascal's Triangle and how is it related to expanding a binomial?
3. Explain how Pascal's Triangle suggests that $\binom{n-1}{k-1} + \binom{n-1}{k} = \binom{n}{k}$.

EXERCISES 8.4

— Suggested Assignment: Exercises 1–35, odd; and 37–49, every other odd.
— Answers to Exercises 9–28 are on page AA22.

In Exercises 1 to 8, evaluate the binomial coefficient.

1. $\binom{7}{4}$ 35
2. $\binom{8}{6}$ 28
3. $\binom{9}{2}$ 36
4. $\binom{11}{5}$ 462
5. $\binom{12}{9}$ 220
6. $\binom{6}{5}$ 6
7. $\binom{11}{0}$ 1
8. $\binom{14}{14}$ 1

In Exercises 9 to 28, expand the binomial.

9. $(x - y)^6$
10. $(a - b)^5$
11. $(x + 3)^5$
12. $(x - 5)^4$
13. $(2x - 1)^7$
14. $(2x + y)^6$
15. $(x + 3y)^6$
16. $(x - 4y)^5$
17. $(2x - 5y)^4$
18. $(2x + 3y)^5$
19. $\left(x + \frac{1}{x}\right)^6$
20. $\left(\sqrt{x} - 2\right)^5$
21. $(x^2 - 4)^7$
22. $(x - y^3)^6$
23. $(2x^2 + y^3)^5$
24. $(2x - y^3)^6$
25. $\left(\frac{2}{x} - \frac{x}{2}\right)^4$
26. $\left(\frac{a}{b} + \frac{b}{a}\right)^3$
27. $(s^{-2} + s^2)^6$
28. $(2r^{-1} + s^{-1})^5$

In Exercises 29 to 36, find the indicated term without expanding.

29. $(3x - y)^{10}$; eighth term $-3240x^3y^7$
30. $(x + 2y)^{12}$; fourth term $1760x^9y^3$
31. $(x + 4y)^{12}$; third term $1056x^{10}y^2$
32. $(2x - 1)^{14}$; thirteenth term $364x^2$
33. $\left(\sqrt{x} - \sqrt{y}\right)^9$; fifth term $126x^2y^2\sqrt{x}$
34. $\left(2 - \frac{1}{\sqrt{x}}\right)^8$; sixth term $-\frac{448}{x^2\sqrt{x}}$
35. $\left(\frac{a}{b} + \frac{b}{a}\right)^{11}$; ninth term $\frac{165b^5}{a^5}$
36. $\left(\frac{3}{x} - \frac{x}{3}\right)^{13}$; seventh term $\frac{5148}{x}$
37. Find the term that contains b^8 in the expansion of $(2a - b)^{10}$. $180a^2b^8$
38. Find the term that contains s^7 in the expansion of $(3r + 2s)^9$. $41472r^2s^7$

39. Find the term that contains y^8 in the expansion of $(2x + y^2)^6$. $60x^2y^8$

40. Find the term that contains b^9 in the expansion of $(a - b^3)^8$. $-56a^5b^9$

41. Find the middle term of $(3a - b)^{10}$. $-61{,}236a^5b^5$

42. Find the middle term of $(a + b^2)^8$. $70a^4b^8$

43. Find the two middle terms of $(s^{-1} + s)^9$. $126s^{-1}, 126s$

44. Find the two middle terms of $(x^{1/2} - y^{1/2})^7$. $-35x^2y^{3/2}, 35x^{3/2}y^2$

In Exercises 45 to 50, use the Binomial Theorem to simplify the power of the complex number.

45. $(2 - i)^4$ $-7 - 24i$

46. $(3 + 2i)^3$ $-9 + 46i$

47. $(1 + 2i)^5$ $41 - 38i$

48. $(1 - 3i)^5$ $316 + 12i$

49. $\left(\frac{\sqrt{2}}{2} + i\frac{\sqrt{2}}{2}\right)^8$ 1

50. $\left(\frac{1}{2} + i\frac{\sqrt{3}}{2}\right)^6$ 1

Explorations 

1. **Pascal's Triangle** Write an essay on Pascal's Triangle. Include some of the earliest known examples of the triangle and some of its applications. Answers will vary.

2. **Some Other Functions** Do some research and determine a definition of positive integers for each of the following functions. Give examples of calculations using each function.

   a. Pochhammer (m, n) b. Double factorial $(n!!)$

a. Pochhammer $(m, n) = \prod_{k=0}^{n-1} (m + k)$

b. $n!! = \begin{cases} n(n-2)\cdots 5 \cdot 3 \cdot 1, & \text{if } n \geq 1 \text{ and } n \text{ is odd} \\ n(n-2)\cdots 6 \cdot 4 \cdot 2, & \text{if } n \geq 2 \text{ and } n \text{ is even} \\ 1, & \text{if } n = -1, 0 \end{cases}$

Chapter 8 Summary

Key Terms

binomial experiment **[p. 676]**
binomial coefficient **[p. 683]**
combination **[p. 651]**
combinatorics **[p. 647]**
complement of an event **[p. 673]**
compound events **[p. 670]**
empirical probability **[p. 663]**
event **[p. 660]**
expanding a binomial **[p. 683]**
experiment **[p. 647]**
experimental probability **[p. 663]**
independent events **[p. 675]**
multistage experiment **[p. 647]**
mutually exclusive events **[p. 670]**
n factorial **[p. 650]**
Pascal's Triangle **[p. 686]**
permutation **[p. 650]**
probability of an event **[p. 659]**
Punnett square **[p. 664]**
sample space **[p. 660]**
theoretical probability **[p. 663]**
tree diagram **[p. 648]**

Essential Concepts and Formulas

- **Counting Principle**
  Let E be a multistage experiment. If $n_1, n_2, n_3, \ldots, n_k$ are the number of possible outcomes of each of the k stages of E, then there are $n_1 \cdot n_2 \cdot n_3 \cdot \cdots \cdot n_k$ possible outcomes for E. **[p. 649]**

- **Permutation Formula**
  The number of permutations $P(n, k)$ of n distinct objects selected k at a time is

$$P(n, k) = \frac{n!}{(n-k)!} \quad \textbf{[p. 651]}$$

- **Combination Formula**
  The number of combinations of n distinct objects chosen k at a time is

$$C(n, k) = \frac{P(n, k)}{k!} = \frac{n!}{k! \cdot (n - k)!} \quad \textbf{[p. 652]}$$

- **Probability of an Event**
  For an experiment with sample space S of *equally likely outcomes,* the probability of an event E is given by $P(E) = \frac{n(E)}{n(S)}$, where $n(E)$ is the number of elements in the event and $n(S)$ is the number of elements in the sample space. **[p. 661]**

- **Probability of Mutually Exclusive Events**
  If A and B are two mutually exclusive events, then the probability of A or B occurring is

$$P(A \cup B) = P(A) + P(B) \quad \textbf{[p. 671]}$$

- **Addition Rule for Probabilities**
  Let A and B be two events in a sample space S. Then $P(A \text{ or } B) = P(A \cup B) = P(A) + P(B) - P(A \cap B)$. **[p. 671]**

- **Probability of the Complement of an Event**
  If E is an event and E^c is the complement of the event, then $P(E^c) = 1 - P(E)$. **[p. 673]**

- **Probability of Independent Events**
  If A and B are independent events, then the probability that both A *and* B will occur is $P(A \text{ and } B) = P(A \cap B) = P(A) \cdot P(B)$. **[p. 675]**

- **Binomial Probability Formula**
  Consider an experiment consisting of n repeated independent trials for which the probability of success on a single trial is p and the probability of failure is $q = 1 - p$. Then the probability of exactly k successes in n trials is given by $C(n, k)p^k q^{n-k}$. **[p. 677]**

- **Formula for a Binomial Coefficient**
  The coefficient of the term whose variable part is $a^{n-k}b^k$ in the expansion of $(a + b)^n$ is

$$\binom{n}{k} = \frac{n!}{k!\,(n - k)!} \quad \textbf{[p. 684]}$$

- **Binomial Theorem for Positive Integers**
  If n is a positive integer, then

$$(a + b)^n = \sum_{i=0}^{n} \binom{n}{i} a^{n-i} b^i \quad \textbf{[p. 685]}$$

- **Formula for the *i*th Term of a Binomial Expansion**
  The ith term of the expansion of $(a + b)^n$ is

$$\binom{n}{i - 1} a^{n-i+1} b^{i-1} \quad \textbf{[p. 686]}$$

Chapter 8 True/False Exercises

In Exercises 1 to 12, answer true or false. If the statement is false, give a reason or state a counterexample to show that the statement is false.

1. The number of different ways one can choose a committee of seven people from a group of 40 is $P(40, 7)$.

1. False; in choosing the members of a committee, the order in which the members are chosen is not important. Thus the number of different ways of choosing seven people from a group of 40 is $C(40, 7)$.

2. If the probability of an event occurring is 1, then the event must occur. True

3. The probability of rolling a single die twice and getting first a 3 and then a 5 is $\frac{1}{6} + \frac{1}{6} = \frac{1}{3}$.
   False; the probability is $\frac{1}{6} \cdot \frac{1}{6} = \frac{1}{36}$.

4. A single card is pulled from a standard deck. E_1 is the event that an ace is pulled and E_2 is the event that a king is pulled. Then E_1 and E_2 are mutually exclusive events.
   True

5. Students in a class are selected at random. E_1 is the event that a female student is selected and E_2 is the event that a student taller than 6 feet is selected. Then

$$P(E_1 \text{ or } E_2) = P(E_1) + P(E_2)$$

   False; $P(E_1 \text{ or } E_2) = P(E_1) + P(E_2) - P(E_1 \cap E_2)$

6. The following are independent events: John pulls a card from a deck and places it in his pocket, then Jennifer pulls a card from the deck and places it in her pocket. False; the set of possible cards that Jennifer can pick is dependent on which card John picked.

7. The following is a binomial experiment: A jar full of different colored candies is passed around a group of children. Each child is allowed to select a piece of candy and note its color (and then eat it).

8. In the expansion of $(a + b)^8$, the exponent on a for the fifth term is 5. False; the exponent on a for the fifth term is 4.

7. False; in a binomial experiment there are exactly two outcomes for each trial, and each trial must be independent of the other trials.

Chapter 8 Review Exercises

—Answer diagram to Exercise 2 is on page AA22.

1. List all possible outcomes of the following experiment: Roll a single die and then toss a coin.
{1H, 1T, 2H, 2T, 3H, 3T, 4H, 4T, 5H, 5T, 6H, 6T} [8.1]
2. Draw a tree diagram that shows all possible outcomes of the experiment of a family planning to have three children and considering the gender of each child. [8.1]
3. A vehicle license plate consists of three digits from 0 through 9 followed by three letters, where the letter "O" isn't permitted. How many different license plates are possible? 15,625,000 [8.1]
4. Compute $C(12, 8)$. 495 [8.1]
5. Twenty people have entered a cooking contest. In how many ways can first, second, and third prize be awarded? 6840 [8.1]
6. Twenty-eight students have applied for six openings on the football team. In how many ways can six students be selected for the team? 376,740 [8.1]
7. A hand of five cards is dealt from a standard deck of playing cards. How many different hands contain three aces and two kings? 24 [8.1]
8. Two dice are tossed. Compute the probability that the sum of the pips on the upward faces is
   a. 5. $\frac{1}{9}$
   b. at least 11. $\frac{1}{12}$ [8.2]
9. A coin is tossed four times. Find the probability that at least three tails are obtained. $\frac{5}{16}$ [8.2]
10. A single card is selected from a standard deck of playing cards. What is the probability that the card is a heart? $\frac{1}{4}$ [8.2]
11. The numbers of students at a university who are currently in each class level are shown in the table.

Class level	Number of students
Freshman	642
Sophomore	549
Junior	483
Senior	445
Graduate student	376

If a student is selected at random, what is the probability that the student is an upper-division undergraduate student (junior or senior)? $\frac{928}{2495} \approx 0.37$ [8.2]

12. Compute the probability of being dealt a five-card hand that contains exactly three jacks from a standard deck of playing cards. $\frac{94}{54,145} \approx 0.0017$ [8.3]
13. A single card is pulled from a standard deck of playing cards. Find the probability of drawing a club or a face card. $\frac{11}{26}$ [8.3]
14. Two dice are rolled. What is the probability of not getting an 11? $\frac{17}{18}$ [8.3]
15. A pair of dice are rolled twice. Compute the probability that the first roll is doubles and the second roll is at least 10. $\frac{1}{36}$ [8.3]
16. A card is pulled from a deck, replaced back in the deck, and then another card is drawn. Compute the probability that the first card is a spade and the second card is red. $\frac{1}{8}$ [8.3]
17. A coin is flipped seven times. Compute the probability that at least five heads will occur. $\frac{29}{128} \approx 0.227$ [8.3]
18. At a certain college, 60% of the students are women. If a group of 20 students is selected at random, compute the probability that the group will be half women and half men. ≈ 0.117 [8.3]

Business and Economics

19. ***Quality Control*** A quality control inspector receives a shipment of 15 computer monitors, three of which are defective. If the inspector randomly chooses 5 monitors, what is the probability that the inspector will find exactly 1 defective monitor? $\frac{1485}{3003} \approx 0.495$ [8.3]
20. ***Manufacturing Defects*** Approximately 4% of a computer manufacturer's disk drives are produced with defects and must be discarded. Find the probability that, of the next 12 drives produced, no more than 1 is defective. ≈ 0.919 [8.3]

Life and Health Sciences

21. ***Genetics*** The beak color of a particular bird is determined by a dominant allele B, corresponding to a black beak, and a recessive allele b, corresponding to a gray beak. If

two parent birds are both of geontype Bb, compute the probability of an offspring bird having a black beak. $\frac{3}{4}$ [8.2]

22. *Genetics* The hair length of a particular rodent is determined by a dominant allele H, corresponding to long hair, and a recessive allele h, corresponding to short hair. If the parents are of genotypes Hh and hh, compute the probability that an offspring of the parents will have short hair. $\frac{1}{2}$ [8.2]

Social Sciences

23. *Taking a Quiz* A multiple-choice quiz consists of 8 questions. Each question has four possible answers, only one of which is correct.

    a. In how many distinct ways can the quiz be completed if no answers are left blank? 65,536

    b. If a student randomly answers the questions, what is the probability that the student will get exactly 5 questions correct? ≈ 0.023 [8.3]

In Exercises 24 to 26, use the table below, which shows the number of voters in a city who voted for or against a proposition (or abstained from voting) according to what party they are registered with.

	For	Against	Abstained
Democrat	8452	2527	894
Republican	2593	5370	1041
Independent	1225	712	686

24. *Political Science* If a voter is chosen at random, compute the probability that the person voted against the proposition. Round to the nearest thousandth. 0.366 [8.2]

25. *Political Science* If a voter is chosen at random, compute the probability that the person is a Democrat or voted for the proposition. Round to the nearest thousandth. 0.668 [8.2]

26. *Political Science* If a voter is randomly chosen, what is the probability that the person abstained from voting on the proposition and is not a Republican? Round to the nearest ten-thousandth. 0.0672 [8.2]

Sports and Recreation

27. $\frac{1}{8{,}145{,}060}$ [8.2]

27. *Lottery* A state lottery chooses 6 numbers from 1 to 45. If you choose the same 6 numbers, you win the grand prize. Compute the probability of winning the grand prize.

28. *Concert Seating* Four married couples are to be seated in a row of eight seats at a concert. How many seating arrangements are possible if no two men are to be seated next to each other? 2880 [8.1]

In Exercises 29 and 30, use the Binomial Theorem to expand each binomial.

29. $(4a - b)^5$
$1024a^5 - 1280a^4b + 640a^3b^2 - 160a^2b^3 + 20ab^4 - b^5$ [8.4]

30. $(\sqrt{a} + 2\sqrt{b})^8$ $a^4 + 16a^{7/2}b^{1/2} + 112a^3b + 448a^{5/2}b^{3/2} + 1120a^2b^2 + 1792a^{3/2}b^{5/2} + 1792ab^3 + 1024a^{1/2}b^{7/2} + 256b^4$ [8.4]

31. Find the fifth term in the expansion of $(3x - 4y)^7$.
$241{,}920x^3y^4$ [8.4]

32. Find the eighth term in the expansion of $(1 - 3x)^9$.
$-78{,}732x^7$ [8.4]

Chapter 8 Test

1. Determine the number of elements in the sample space of the following experiment: Telephone extensions at a corporation are five-digit numbers, where the first digit is a 4 or a 5 and the last digit cannot be 0. 18,000 [8.1/8.2]

2. a. How many permutations are there of nine objects selected four at a time? 3024

   b. How many combinations are there of nine objects selected four at a time? 126 [8.1]

3. A professor assigns 25 homework problems, of which 10 will be graded. How many different sets of 10 problems can the professor choose to grade? 3,268,760 [8.1]

4. A true/false quiz consists of 12 questions.

   a. In how many different ways can the quiz be completed if no questions are left blank? 4096

   b. In how many different ways can the quiz be completed if six answers must be marked true and six must be marked false? 924 [8.1]

5. If a married couple planning to have children is equally likely to have a boy or a girl and the couple plans on having four children, compute the probability that at least two children will be boys. $\frac{11}{16}$ [8.2]

6. A fair coin is tossed four times. What is the probability of getting at least three tails? $\frac{5}{16}$ [8.2]

7. A card is selected at random from a standard deck of playing cards. Compute the probability that the card
   a. is not an ace. $\frac{12}{13}$
   b. is a red card or a king. $\frac{7}{13}$
   c. is a 7 or a face card. $\frac{4}{13}$ [8.3]

8. A hand of five cards is dealt from a standard deck. Find the probability that four of the cards have the same value (four kings, four aces, etc.). $\frac{1}{4165} \approx 0.000240$ [8.3]

9. A pair of dice are rolled. Find the probability that the sum of the pips on the upward faces is less than 5. $\frac{1}{6}$ [8.2]

10. A pair of dice are rolled twice. Find the probability that the first roll is a sum of 7 and the second roll is not a sum of 11. $\frac{17}{108}$ [8.3]

11. The following table shows the number of men and the number of women that responded either positively, negatively, or neutrally to a new commercial. If one person is chosen from this group, find the probability that the person is a woman or responded positively.

	Positive	Negative	Neutral
Men	684	736	354
Women	753	642	481

$\frac{2560}{3650}$ [8.3]

12. Straight or curly hair for a hamster is determined by a dominant allele S, corresponding to straight hair, and a recessive allele s, which gives curly hair. Draw a Punnett square for parents of genotype Ss and ss, and compute the probability that offspring of the parents will have curly hair. $\frac{1}{2}$ [8.2]

13. Suppose 15% of the population is dyslexic. In a group of 18 randomly selected people, what is the probability that exactly three people are dyslexic? Round to the nearest thousandth. 0.241 [8.3]

In Exercises 14 and 15, use the Binomial Theorem to expand each binomial.

14. $(x - 3y)^4$

15. $\left(x + \frac{1}{x}\right)^5$

14. $x^4 - 12x^3y + 54x^2y^2 - 108xy^3 + 81y^4$ [8.4]

15. $x^5 + 5x^3 + 10x + \frac{10}{x} + \frac{5}{x^3} + \frac{1}{x^5}$ [8.4]

Cumulative Review Exercises

1. Determine the domain of $f(x) = \sqrt{25 - x^2}$. Write the domain using interval notation. $[-5, 5]$ [2.2]

2. Find the maximum value of $f(x) = -2x^2 + 8x - 1$. 7 [2.5]

3. Use the Rational Zero Theorem to list the possible rational zeros of $P(x) = x^3 - 4x^2 + x + 6$.
   $\pm 1, \pm 2, \pm 3, \pm 6$ [4.3]

4. Use Descartes' Rule of Signs to state the number of possible positive and possible negative real zeros of $P(x) = x^3 - 2x^2 + 3x - 5$.
   3 or 1 positive real zeros, no negative real zeros. [4.3]

5. Find, to the nearest tenth, the Richter scale magnitude of an earthquake with an intensity of $I = 8{,}750{,}600I_0$. 6.9 [5.3]

6. Solve $3e^x = 64$. Round to the nearest ten thousandth.
   3.0603 [5.4]

7. Solve $\log(x + 4) = 3 - \log(x + 94)$. 6 [5.4]

8. Find the age of a bone if it now has 98.9% of the carbon-14 it had at time $t = 0$. Round to the nearest year. The half life of carbon-14 is 5730 years. 91 years [5.5]

9. A coyote population in a state park satisfies a logistic growth model with $P_0 = 250$ in 1999 and $P(3) = 310$ (the population in 2002). It has been estimated that the maximum coyote population the park can support is 600.
   a. Determine the logistic growth model for the data. Round the growth constant b to the nearest hundred-thousandth. $P(t) = \frac{600}{1 + 1.4e^{-0.13439t}}$
   b. Use the model from part **a.** to predict, to the nearest ten, the coyote population in 2007. 410 [5.7]

10. Solve the following system of equations.
$$\begin{cases} 2x + 3y = 31 \\ 3x - y = 8 \end{cases}$$
(5, 7) [6.1]

11. Find the inverse of $\begin{bmatrix} 3 & -7 \\ 2 & 1 \end{bmatrix}$. $\begin{bmatrix} \frac{1}{17} & \frac{7}{17} \\ -\frac{2}{17} & \frac{3}{17} \end{bmatrix}$ [6.5]

12. Write the following system of equations as a matrix equation and use inverse matrix methods to solve the system.

$$\begin{cases} 3x - 2y + z = 8 \\ x + 4y - z = 2 \\ 2x - y - 3z = 5 \end{cases} \quad \begin{bmatrix} 3 & -2 & 1 \\ 1 & 4 & -1 \\ 2 & -1 & -3 \end{bmatrix}\begin{bmatrix} x \\ y \\ z \end{bmatrix} = \begin{bmatrix} 8 \\ 2 \\ 5 \end{bmatrix};$$

(2.56, −0.12, 0.08) [6.5]

13. Find the 20th term in the arithmetic sequence 2, 10, 18, 26, ... 154 [7.2]

14. Find $\sum_{n=1}^{\infty} \left(\frac{3}{10}\right)^n$. $\frac{3}{7}$ [7.2]

15. An investment of $5000 earns 4.5% annual interest, compounded monthly. Find the balance after 4 years. $5984.07 [7.3]

16. An investment of $2000 earns 4% annual interest, compounded quarterly. Find the number of years it will take for the investment to double in value. Round to the nearest tenth of a year. 17.4 years [7.3]

17. An employee savings plan allows an employee to deposit $200 at the end of each month into a savings account earning 2.75% annual interest, compounded monthly. Find the future value of this savings plan if an employee makes deposits for 5 years. $12,848.40 [7.5]

18. Find $C(8, 5)$. 56 [8.1]

19. How many different letter arrangements are possible using all of the letters of the word *bookkeeper*? 151,200 [8.1]

20. Use the Binomial Theorem to expand $(2x + y)^5$.
$32x^5 + 80x^4y + 80x^3y^2 + 40x^2y^3 + 10xy^4 + y^5$ [8.4]

APPENDIX A Conic Sections

- Parabolas
- Ellipses
- Hyperbolas

Point of Interest

Appollonius (262–200 B.C.) wrote an eight-volume treatise entitled *On Conic Sections* in which he derived the formulas for all the conic sections. He was the first to use the words *parabola, ellipse,* and *hyperbola.*

TAKE NOTE

If the intersection of a plane and a cone is a point, a line, or two intersecting lines, then the intersection is called a *degenerate conic section.*

TAKE NOTE

A web applet is available to let you experiment with parabolas by manipulating the focus, directrix, and vertex. This applet, Parabola with Horizontal Directrix, can be found on our web site at

math.college.hmco.com

Parabolas

The graph of a parabola, a circle, an ellipse, or a hyperbola can be formed by the intersection of a plane and a cone. Hence these figures are referred to as **conic sections.** See Figure A.1.

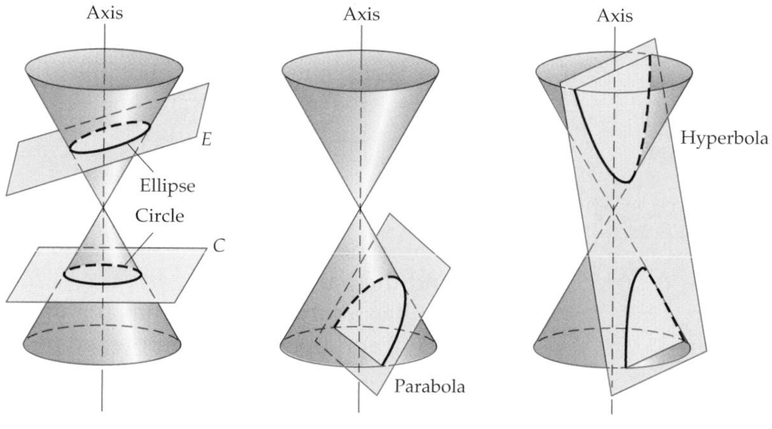

Figure A.1

Cones intersected by planes

A plane perpendicular to the axis of a cone intersects the cone in a circle (plane C). The plane *E,* tilted so that it is not perpendicular to the axis, intersects the cone in an ellipse. When the plane is parallel to a line on the surface of the cone, the plane intersects the cone in a parabola. When the plane intersects both portions of the cone, a hyperbola is formed.

Parabolas with Vertex at (0, 0)

In addition to the geometric description of a conic section just given, a conic section can be defined as a set of points. This method uses some specified conditions about the curve to determine which points in a coordinate system are points on the graph. For example, a parabola can be defined by the following set of points.

Definition of a Parabola

A **parabola** is the set of points in the plane that are equidistant from a fixed line (the **directrix**) and a fixed point (the **focus**) not on the directrix.

The line that passes through the focus and is perpendicular to the directrix is called the **axis of symmetry** of the parabola. The midpoint of the line segment

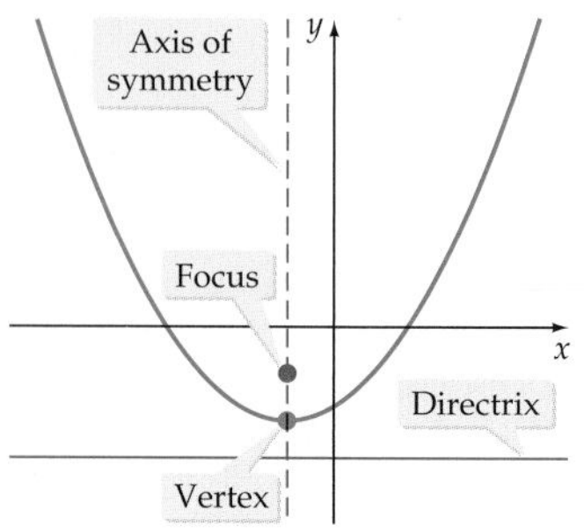

Figure A.2

between the focus and directrix on the axis of symmetry is the **vertex** of the parabola, as shown in Figure A.2.

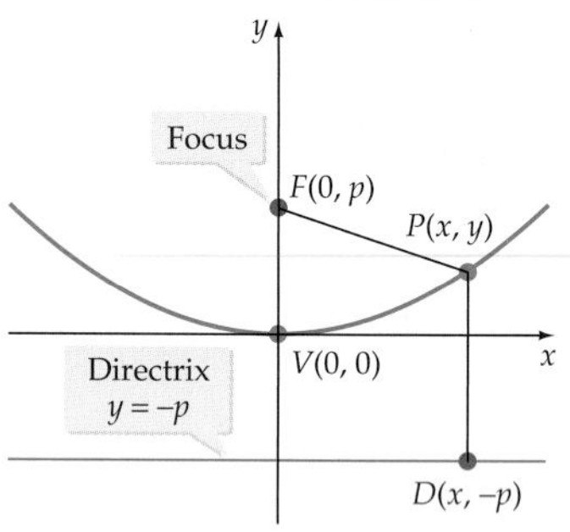

Figure A.3

Standard Forms of the Equation of a Parabola with Vertex at the Origin

Axis of Symmetry Is the y-axis

The standard form of the equation of a parabola with vertex (0, 0) and the y-axis as its axis of symmetry is $x^2 = 4py$. The focus is $(0, p)$ and the equation of the directrix is $y = -p$. See Figure A.3.

Axis of Symmetry Is the x-axis

The standard form of the equation of a parabola with vertex (0, 0) and the x-axis as its axis of symmetry is $y^2 = 4px$. The focus is $(p, 0)$ and the equation of the directrix is $x = -p$.

In the equation $x^2 = 4py$, $x^2 \geq 0$. Therefore, $4py \geq 0$. Thus if $p > 0$, then $y \geq 0$ and the parabola opens up. If $p < 0$, then $y \leq 0$ and the parabola opens down. A similar analysis shows that for $y^2 = 4px$, the parabola opens to the right when $p > 0$ and opens to the left when $p < 0$.

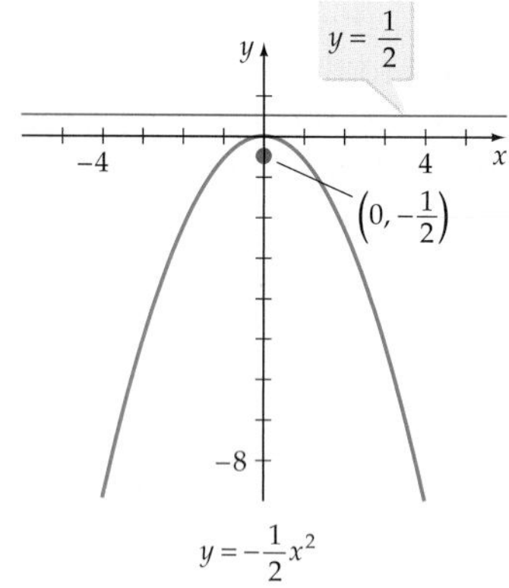

Figure A.4

EXAMPLE 1 Find the Focus and Directrix of a Parabola

Find the focus and directrix of the parabola given by the equation $y = -\frac{1}{2}x^2$.

Solution Because the x term is squared, the standard form of the equation is $x^2 = 4py$.

$$y = -\frac{1}{2}x^2$$

$$x^2 = -2y$$ ▪ Write the given equation in standard form.

Comparing this equation with $x^2 = 4py$ gives

$$4p = -2$$

$$p = -\frac{1}{2}$$

Because p is negative, the parabola opens down and the focus is below the vertex (0, 0), as shown in Figure A.4. The coordinates of the focus are $(0, -1/2)$. The equation of the directrix is $y = 1/2$.

CHECK YOUR PROGRESS 1 Find the vertex, focus, and directrix of the parabola given by $x^2 = -\frac{1}{4}y$.

Solution *See page S44.* $V(0, 0)$, $F\left(0, -\frac{1}{16}\right)$, $h = \frac{1}{16}$

EXAMPLE 2 Find the Equation of a Parabola in Standard Form

Find the equation of the parabola in standard form with the vertex at the origin and the focus at $(-2, 0)$.

Solution Because the vertex is at $(0, 0)$ and the focus is at $(-2, 0)$, $p = -2$. The graph of the parabola opens toward the focus, so in this case the parabola opens to the left. The equation in standard form of the parabola that opens to the left is $y^2 = 4px$. Substitute -2 for p in this equation and simplify.

$$y^2 = 4(-2)x = -8x$$

The equation of the parabola is $y^2 = -8x$.

CHECK YOUR PROGRESS 2 Find the equation in standard form of the parabola with vertex at the origin and focus $(5, 0)$.

Solution *See page S44.* $y^2 = 20x$

PARABOLAS WITH VERTEX AT (h, k)

The equation of a parabola with a vertical or horizontal axis of symmetry and with vertex at point (h, k) can be found by using the translations discussed previously.

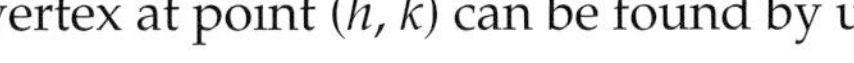

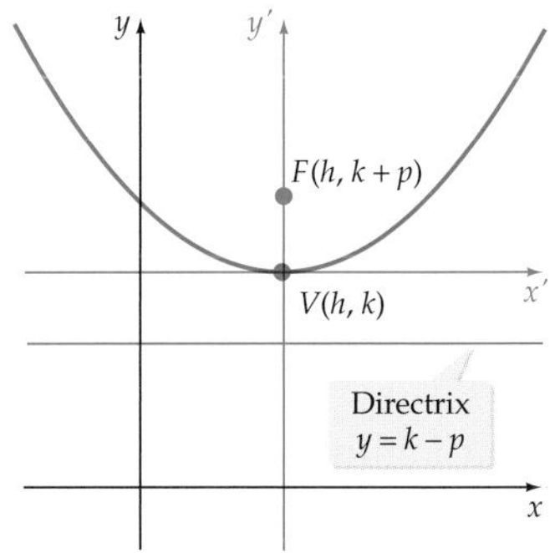

Figure A.5

Standard Forms of the Equation of a Parabola with Vertex at (h, k)

Vertical Axis of Symmetry

The standard form of the equation of the parabola with vertex (h, k) and a vertical axis of symmetry is $(x - h)^2 = 4p(y - k)$. The focus is $(h, k + p)$ and the equation of the directrix is $y = k - p$. See Figure A.5.

Horizontal Axis of Symmetry

The standard form of the equation of the parabola with vertex (h, k) and a horizontal axis of symmetry is $(y - k)^2 = 4p(x - h)$. The focus is $(h + p, k)$ and the equation of the directrix is $x = h - p$.

EXAMPLE 3 Find the Vertex, Focus, and Directrix of a Parabola

Find the equation of the directrix and the coordinates of the vertex and focus of the parabola given by the equation $3x + 2y^2 + 8y - 4 = 0$.

Solution Rewrite the equation so that the y terms are on one side of the equation, and then complete the square on y.

$$3x + 2y^2 + 8y - 4 = 0$$
$$2y^2 + 8y = -3x + 4$$
$$2(y^2 + 4y) = -3x + 4$$
$$2(y^2 + 4y + 4) = -3x + 4 + 8$$

▪ Complete the square. Note that $2 \cdot 4 = 8$ is added to each side.

Continued ➤

$$2(y+2)^2 = -3(x-4) \qquad \text{■ Simplify and then factor.}$$

$$(y+2)^2 = -\frac{3}{2}(x-4) \qquad \text{■ Write the equation in standard form.}$$

Comparing this equation to $(y-k)^2 = 4p(x-h)$, we have a parabola that opens to the left with vertex $(4, -2)$ and $4p = -3/2$. Thus $p = -3/8$.

The coordinates of the focus are

$$\left(4 + \left(-\frac{3}{8}\right), -2\right) = \left(\frac{29}{8}, -2\right)$$

The equation of the directrix is

$$x = 4 - \left(-\frac{3}{8}\right) = \frac{35}{8}$$

Choosing some values for y and finding the corresponding values of x, we plot a few points. Because the line $y = -2$ is the axis of symmetry, for each point on one side of the axis of symmetry there is a corresponding point on the other side. Two points are $(-2, 1)$ and $(-2, -5)$. See Figure A.6.

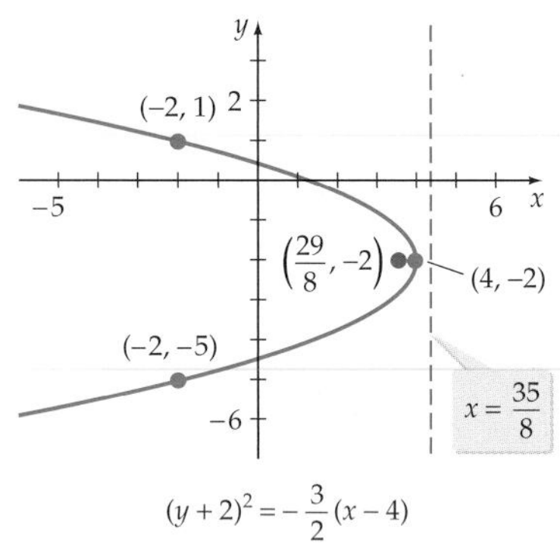

Figure A.6

CHECK YOUR PROGRESS 3 Find the vertex, focus, and directrix of the parabola given by $x^2 + 5x - 4y - 1 = 0$.

Solution *See page S44.* $V\left(-\frac{5}{2}, -\frac{29}{16}\right), F\left(-\frac{5}{2}, -\frac{13}{16}\right), y = -\frac{45}{16}$

EXAMPLE 4 Find the Equation in Standard Form of a Parabola

Find the equation in standard form of the parabola with directrix $x = -1$ and focus $(3, 2)$.

Solution The vertex is the midpoint of the line segment joining $(3, 2)$ and the point $(-1, 2)$ on the directrix.

$$(h, k) = \left(\frac{-1+3}{2}, \frac{2+2}{2}\right) = (1, 2)$$

The standard form of the equation is $(y-k)^2 = 4p(x-h)$. The distance from the vertex to the focus is 2. Thus $4p = 4(2) = 8$, and the equation of the parabola in standard form is $(y-2)^2 = 8(x-1)$. See Figure A.7.

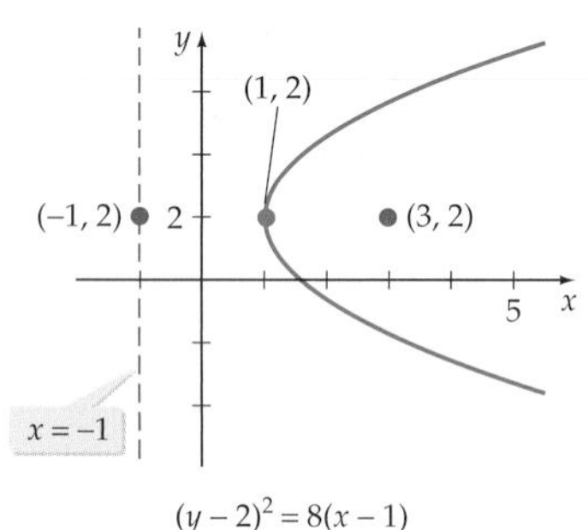

Figure A.7

CHECK YOUR PROGRESS 4 Find the equation in standard form of the parabola with vertex $(2, -3)$ and focus $(0, -3)$.

Solution *See page S44.* $(y+3)^2 = -8(x-2)$

■ Ellipses

An ellipse is another of the conic sections formed when a plane intersects a right circular cone. If β is the angle at which the plane intersects the axis of the cone and

TAKE NOTE

If a plane intersects a cone at the vertex of the cone such that the resulting figure is a point, the point is called a *degenerate ellipse.* See the accompanying figure.

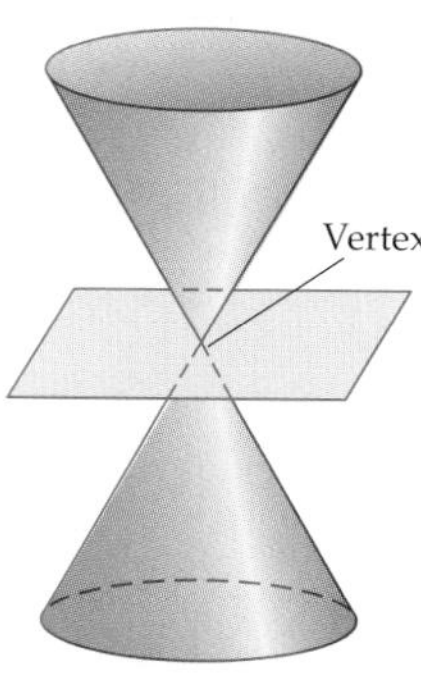

Degenerate ellipse

α is the angle shown in Figure A.8, an ellipse is formed when $\alpha < \beta < 90°$. If $\beta = 90°$, then a circle is formed.

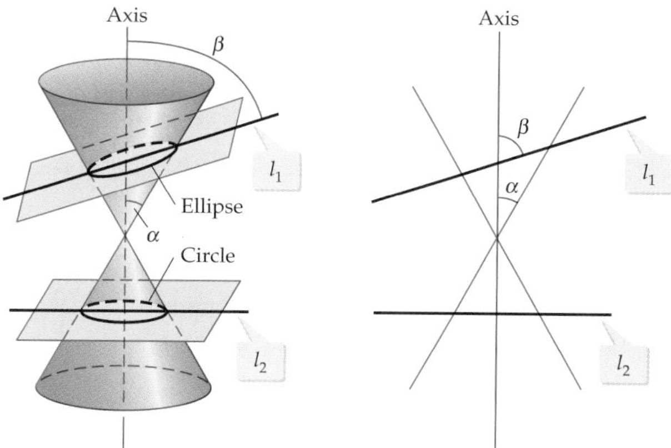

Figure A.8

As was the case for a parabola, the definition of an ellipse can be stated in terms of a certain set of points in the plane.

Definition of an Ellipse

An **ellipse** is the set of all points in the plane the sum of whose distances from two fixed points (**foci**) is a positive constant.

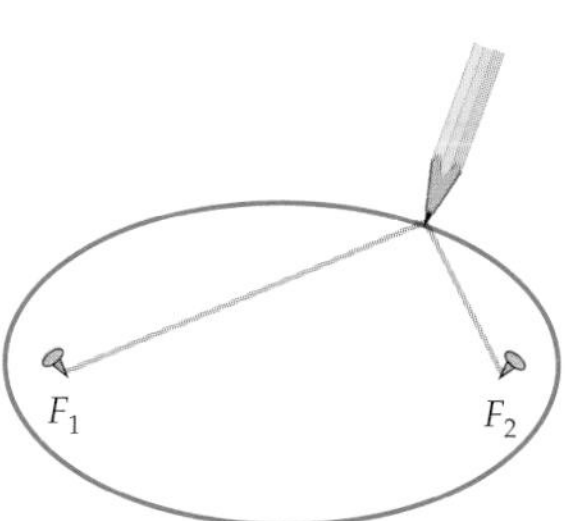

Figure A.9

We can use this definition to draw an ellipse, equipped only with a piece of string and two tacks (see Figure A.9). Tack the ends of the string to the foci, and trace a curve with a pencil held tight against the string. The resulting curve is an ellipse. The positive constant mentioned in the definition of an ellipse is the length of the string.

Ellipses with Center at (0, 0)

The graph of an ellipse has two axes of symmetry (see Figure A.10). The longer axis is called the **major axis.** The foci of the ellipse are on the major axis. The shorter axis is called the **minor axis.** It is customary to denote the length of the major axis as $2a$ and the length of the minor axis as $2b$. The **semiaxes** are one-half the axes in length. Thus the length of the semimajor axis is denoted by a and the

TAKE NOTE

A web applet is available to let you experiment with ellipses by manipulating the foci and vertices. This applet, Ellipse, can be found on our web site at **math.college.hmco.com**

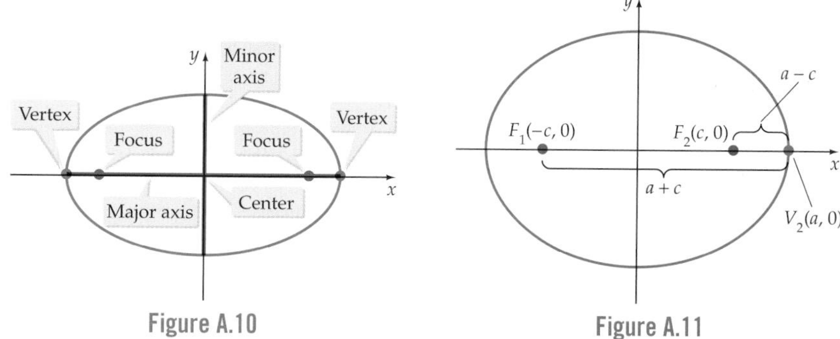

Figure A.10

Figure A.11

length of the semiminor axis is denoted by b. The **center** of the ellipse is the midpoint of the major axis. The endpoints of the major axis are the **vertices** (plural of *vertex*) of the ellipse.

Consider the point $V_2(a, 0)$, which is one vertex of an ellipse, and the points $F_2(c, 0)$ and $F_1(-c, 0)$, which are the foci of the ellipse, as shown in Figure A.11. The distance from V_2 to F_1 is $a + c$. Similarly, the distance from V_2 to F_2 is $a - c$. From the definition of an ellipse, the sum of the distances from any point on the ellipse to the foci is a positive constant. By adding the expressions $a + c$ and $a - c$, we have

$$(a + c) + (a - c) = 2a$$

Thus the positive constant referred to in the definition of an ellipse is $2a$, the length of the major axis.

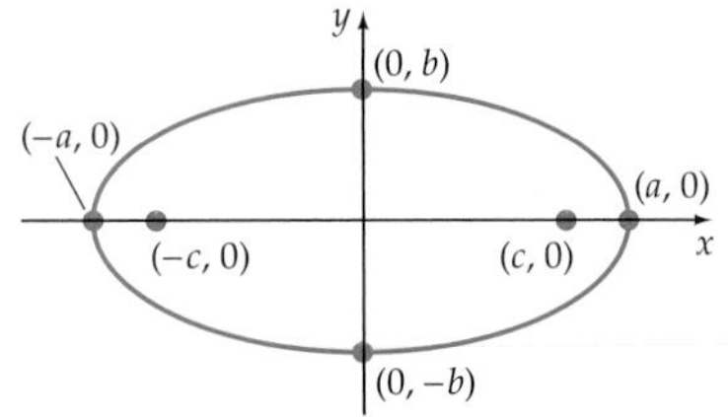

a. Major axis on x-axis

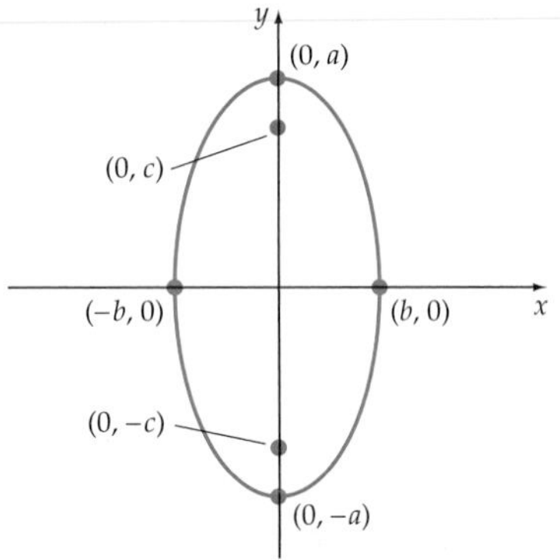

b. Major axis on y-axis

Figure A.12

Standard Forms of the Equation of an Ellipse with Center at the Origin

Major Axis on the x-axis

The standard form of the equation of an ellipse with the center at the origin and major axis on the x-axis (see Figure A.12a) is given by

$$\frac{x^2}{a^2} + \frac{y^2}{b^2} = 1, \quad a > b$$

The length of the major axis is $2a$. The length of the minor axis is $2b$. The coordinates of the vertices are $(a, 0)$ and $(-a, 0)$, and the coordinates of the foci are $(c, 0)$ and $(-c, 0)$, where $c^2 = a^2 - b^2$.

Major Axis on the y-axis

The standard form of the equation of an ellipse with the center at the origin and major axis on the y-axis (see Figure A.12b) is given by

$$\frac{x^2}{b^2} + \frac{y^2}{a^2} = 1, \quad a > b$$

The length of the major axis is $2a$. The length of the minor axis is $2b$. The coordinates of the vertices are $(0, a)$ and $(0, -a)$, and the coordinates of the foci are $(0, c)$ and $(0, -c)$, where $c^2 = a^2 - b^2$.

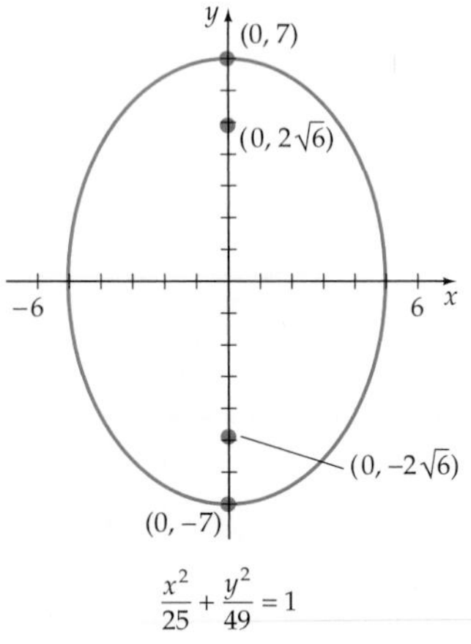

Figure A.13

EXAMPLE 5 Find the Vertices and Foci of an Ellipse

Find the vertices and foci of the ellipse given by the equation $\frac{x^2}{25} + \frac{y^2}{49} = 1$. Sketch the graph.

Solution Because the y^2 term has the larger denominator, the major axis is on the y-axis.

$$a^2 = 49 \qquad b^2 = 25 \qquad c^2 = a^2 - b^2$$
$$a = 7 \qquad b = 5 \qquad = 49 - 25 = 24$$
$$c = \sqrt{24} = 2\sqrt{6}$$

Continued ➤

The vertices are (0, 7) and (0, −7). The foci are $\left(0, 2\sqrt{6}\right)$ and $\left(0, -2\sqrt{6}\right)$. See Figure A.13.

CHECK YOUR PROGRESS 5 Find the vertices and foci of the ellipse given by $25x^2 + 12y^2 = 300$.

Solution *See page S44.* Vertices (0, 5), (0, −5); foci $\left(0, \sqrt{13}\right), \left(0, -\sqrt{13}\right)$

An ellipse with foci at (3, 0) and (−3, 0) and major axis of length 10 is shown in Figure A.14. To find the equation of the ellipse in standard form, we must find a^2 and b^2. Because the foci are on the major axis, the major axis is on the x-axis. The length of the major axis is $2a$. Thus $2a = 10$. Solving for a, we have $a = 5$ and $a^2 = 25$.

Because the foci are at (3, 0) and (−3, 0) and the center of the ellipse is the midpoint between the two foci, the distance from the center of the ellipse to a focus is 3. Therefore, $c = 3$. To find b^2, use the equation

$$c^2 = a^2 - b^2$$
$$9 = 25 - b^2$$
$$b^2 = 16$$

The equation of the ellipse in standard form is $\dfrac{x^2}{25} + \dfrac{y^2}{16} = 1$.

$\dfrac{x^2}{25} + \dfrac{y^2}{16} = 1$

Figure A.14

ELLIPSES WITH CENTER AT (*h*, *k*)

The equation of an ellipse with center (h, k) and with horizontal or vertical major axes can be found by using a translation of coordinates.

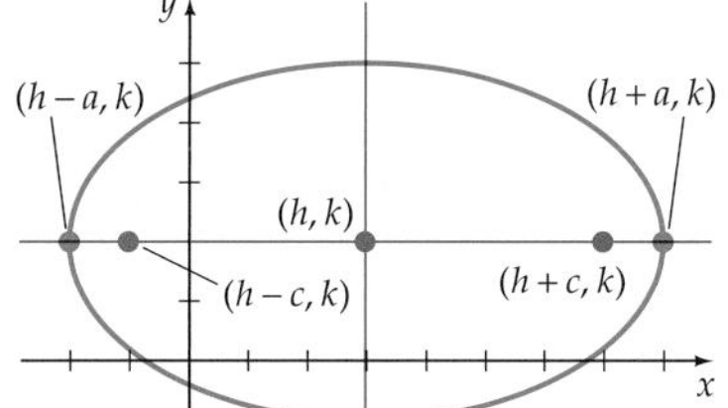

a. Major axis parallel to x-axis

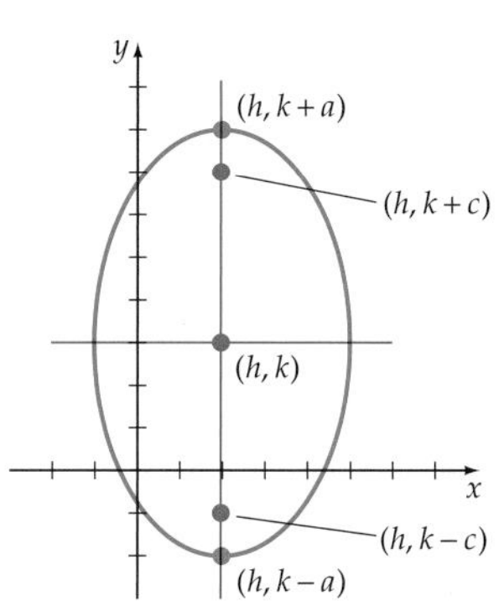

b. Major axis parallel to y-axis

Figure A.15

Standard Forms of the Equation of an Ellipse with Center at (*h*, *k*)

Major Axis Parallel to the *x*-axis

The standard form of the equation of an ellipse with the center at (h, k) and major axis parallel to the x-axis (see Figure A.15a) is given by

$$\frac{(x - h)^2}{a^2} + \frac{(y - k)^2}{b^2} = 1, \quad a > b$$

The length of the major axis is $2a$. The length of the minor axis is $2b$. The coordinates of the vertices are $(h + a, k)$ and $(h - a, k)$, and the coordinates of the foci are $(h + c, k)$ and $(h - c, k)$, where $c^2 = a^2 - b^2$.

Major Axis Parallel to the *y*-axis

The standard form of the equation of an ellipse with the center at (h, k) and major axis parallel to the y-axis (see Figure A.15b) is given by

$$\frac{(x - h)^2}{b^2} + \frac{(y - k)^2}{a^2} = 1, \quad a > b$$

The length of the major axis is $2a$. The length of the minor axis is $2b$. The coordinates of the vertices are $(h, k + a)$ and $(h, k - a)$, and the coordinates of the foci are $(h, k + c)$ and $(h, k - c)$, where $c^2 = a^2 - b^2$.

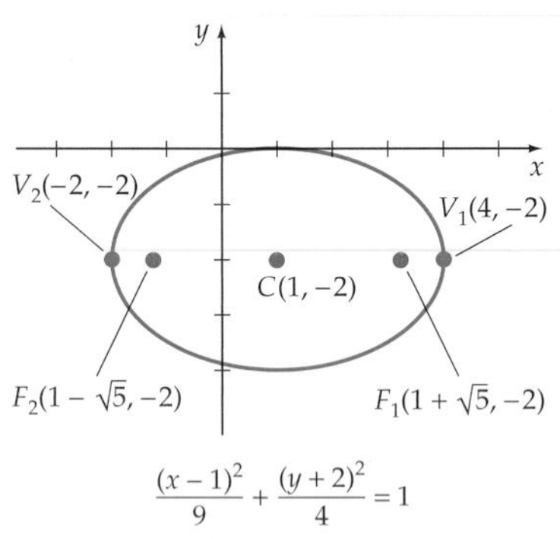

Figure A.16

EXAMPLE 6 Find the Center, Vertices, and Foci of an Ellipse

Find the vertices and foci of the ellipse given by $4x^2 + 9y^2 - 8x + 36y + 4 = 0$. Sketch the graph.

Solution Write the equation of the ellipse in standard form by completing the square.

$$
\begin{aligned}
4x^2 + 9y^2 - 8x + 36y + 4 &= 0 \\
4x^2 - 8x + 9y^2 + 36y &= -4 && \text{• Rearrange terms.} \\
4(x^2 - 2x) + 9(y^2 + 4y) &= -4 && \text{• Factor.} \\
4(x^2 - 2x + 1) + 9(y^2 + 4y + 4) &= -4 + 4 + 36 && \text{• Complete the square.} \\
4(x - 1)^2 + 9(y + 2)^2 &= 36 && \text{• Factor.} \\
\frac{(x - 1)^2}{9} + \frac{(y + 2)^2}{4} &= 1 && \text{• Divide by 36.}
\end{aligned}
$$

From the equation of the ellipse in standard form, the coordinates of the center of the ellipse are $(1, -2)$. Because the larger denominator is 9, the major axis is parallel to the x-axis and $a^2 = 9$. Thus $a = 3$. The vertices are $(4, -2)$ and $(-2, -2)$.

To find the coordinates of the foci, we find c.

$$
\begin{aligned}
c^2 &= a^2 - b^2 = 9 - 4 = 5 \\
c &= \sqrt{5}
\end{aligned}
$$

The foci are $\left(1 + \sqrt{5}, -2\right)$ and $\left(1 - \sqrt{5}, -2\right)$. See Figure A.16.

CHECK YOUR PROGRESS 6 Find the center, vertices, and foci of the ellipse given by $9x^2 + 16y^2 + 36x - 16y - 104 = 0$.

Solution *See page S44.* Vertices $\left(2, \frac{1}{2}\right), \left(-6, \frac{1}{2}\right)$; foci $\left(-2 + \sqrt{7}, \frac{1}{2}\right), \left(-2 - \sqrt{7}, \frac{1}{2}\right)$

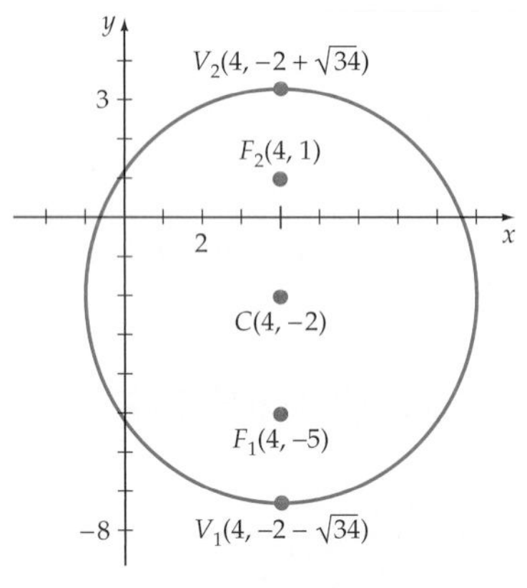

Figure A.17

EXAMPLE 7 Find the Equation of an Ellipse

Find the standard form of the equation of the ellipse with center at $(4, -2)$, $F_1(4, -5)$, and foci $F_2(4, 1)$, and minor axis of length 10, as shown in Figure A.17.

Solution Because the foci are on the major axis, the major axis is parallel to the y-axis. The distance from the center of the ellipse to a focus is c. The distance between the center $(4, -2)$ and the focus $(4, 1)$ is 3. Therefore, $c = 3$.

The length of the minor axis is $2b$. Thus $2b = 10$ and $b = 5$.

To find a^2, use the equation $c^2 = a^2 - b^2$.

$$
\begin{aligned}
9 &= a^2 - 25 \\
a^2 &= 34
\end{aligned}
$$

Thus the equation in standard form is

$$\frac{(x - 4)^2}{25} + \frac{(y + 2)^2}{34} = 1$$

Continued ➤

TAKE NOTE

If a plane intersects a cone along the axis of the cone, the resulting graph is two intersecting straight lines. This is called a *degenerate hyperbola.* See the accompanying figure.

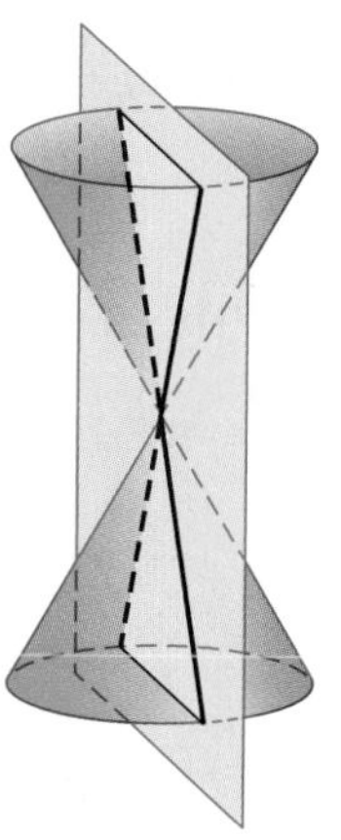

Degenerate hyperbola

CHECK YOUR PROGRESS 7 Find the equation in standard form of the ellipse with center at $(-4, 1)$, minor axis of length 8 parallel to the y-axis, and passing through the point $(0, 4)$.

Solution *See page S45.* $\frac{(x+4)^2}{256/7} + \frac{(y-1)^2}{16} = 1$

Hyperbolas

A hyperbola is the conic section formed when a plane intersects a right circular cone at a certain angle. If β is the angle at which the plane intersects the axis of the cone and α is the angle shown in Figure A.18, a hyperbola is formed when $0° < \beta < \alpha$ or when the plane is parallel to the axis of the cone.

As with the other conic sections, the definition of a hyperbola can be stated in terms of a certain set of points in the plane.

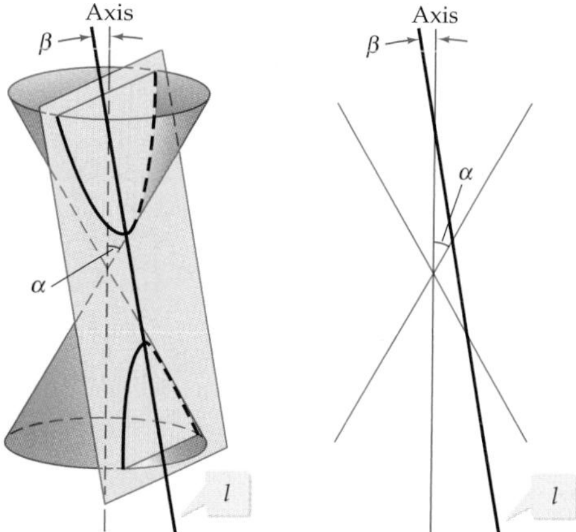

Figure A.18

TAKE NOTE

A web applet is available to let you experiment with hyperbolas by manipulating the foci and vertices. This applet, Hyperbola, can be found on our web site at **math.college.hmco.com**

Definition of a Hyperbola

A **hyperbola** is the set of all points in the plane the difference between whose distances from two fixed points (foci) is a positive constant.

This definition differs from that of an ellipse in that the ellipse was defined in terms of the *sum* of two distances, whereas the hyperbola is defined in terms of the *difference* of two distances.

Hyperbolas with Center at (0, 0)

The **transverse axis** of a hyperbola is the line segment joining the intercepts (see Figure A.19). The midpoint of the transverse axis is called the **center** of the hyperbola. The **conjugate axis** passes through the center of the hyperbola and is perpendicular to the transverse axis.

The length of the transverse axis is customarily represented as $2a$, and the distance between the two foci is represented as $2c$. The length of the conjugate axis is represented as $2b$.

The **vertices** of a hyperbola are the points at which the hyperbola intersects the transverse axis.

To determine the positive constant referred to in the definition of a hyperbola, consider the point $V_1(a, 0)$, which is one vertex of a hyperbola, and the points $F_1(c, 0)$ and $F_2(-c, 0)$, which are the foci of the hyperbola (see Figure A.20). The difference between the distance from $V_1(a, 0)$ to $F_1(c, 0)$, $c - a$, and the distance from $V_1(a, 0)$ to $F_2(-c, 0)$, $c + a$, must be a constant. By subtracting these distances, we find

$$|(c - a) - (c + a)| = |-2a| = 2a$$

Thus the constant is $2a$ and is the length of the transverse axis. The absolute value symbol is used to ensure that the distance is a positive number.

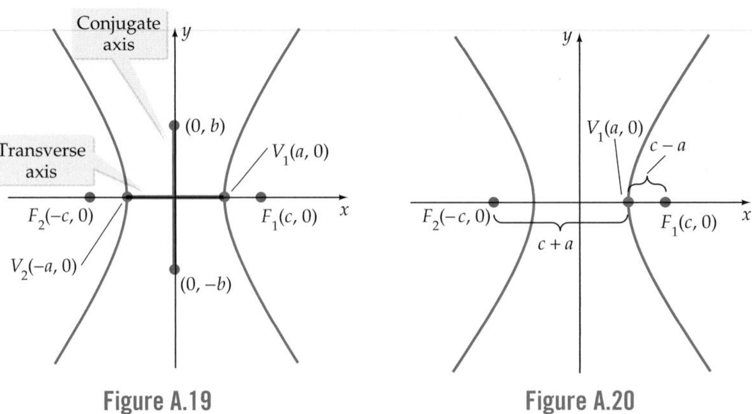

Figure A.19

Figure A.20

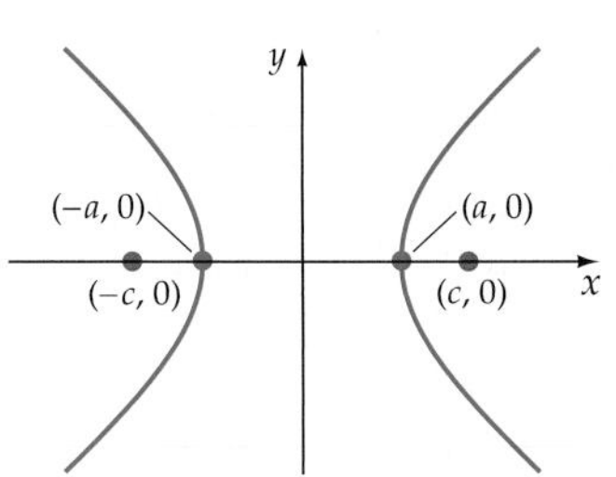

a. Transverse axis on the x-axis

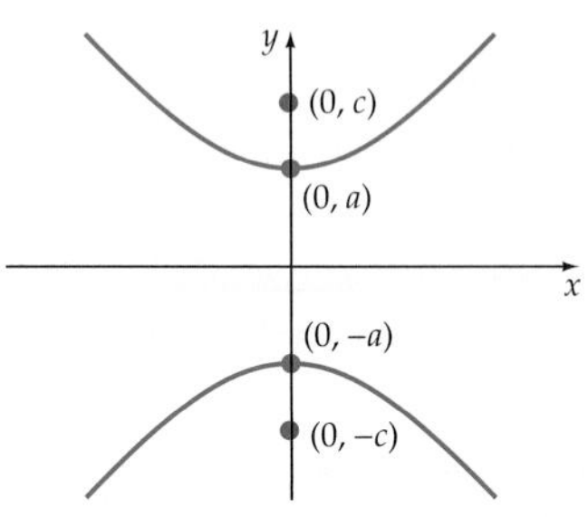

b. Transverse axis on the y-axis

Figure A.21

Standard Forms of the Equation of a Hyperbola with Center at the Origin

Transverse Axis on the x-axis

The standard form of the equation of a hyperbola with the center at the origin and transverse axis on the x-axis (see Figure A.21a) is given by

$$\frac{x^2}{a^2} - \frac{y^2}{b^2} = 1$$

The coordinates of the vertices are $(a, 0)$ and $(-a, 0)$, and the coordinates of the foci are $(c, 0)$ and $(-c, 0)$, where $c^2 = a^2 + b^2$.

Transverse Axis on the y-axis

The standard form of the equation of a hyperbola with the center at the origin and transverse axis on the y-axis (see Figure A.21b) is given by

$$\frac{y^2}{a^2} - \frac{x^2}{b^2} = 1$$

The coordinates of the vertices are $(0, a)$ and $(0, -a)$, and the coordinates of the foci are $(0, c)$ and $(0, -c)$, where $c^2 = a^2 + b^2$.

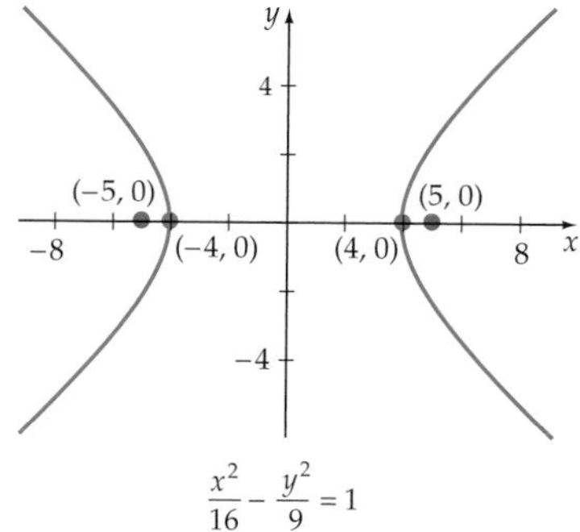

$\frac{x^2}{16} - \frac{y^2}{9} = 1$

Figure A.22

By looking at the equations, it is possible to determine the location of the transverse axis. When the x^2 term is positive, the transverse axis is on the x-axis. When the y^2 term is positive, the transverse axis is on the y-axis.

Consider the hyperbola given by the equation $\frac{x^2}{16} - \frac{y^2}{9} = 1$. Because the x^2 term is positive, the transverse axis is on the x-axis. Because $a^2 = 16$, we have $a = 4$. The vertices are $(4, 0)$ and $(-4, 0)$. To find the foci, we determine c.

$$c^2 = a^2 + b^2 = 16 + 9 = 25$$
$$c = \sqrt{25} = 5$$

The foci are $(5, 0)$ and $(-5, 0)$. The graph is shown in Figure A.22.

Each hyperbola has two asymptotes that pass through the center of the hyperbola. The asymptotes of the hyperbola are a useful guide for sketching the graph of the hyperbola.

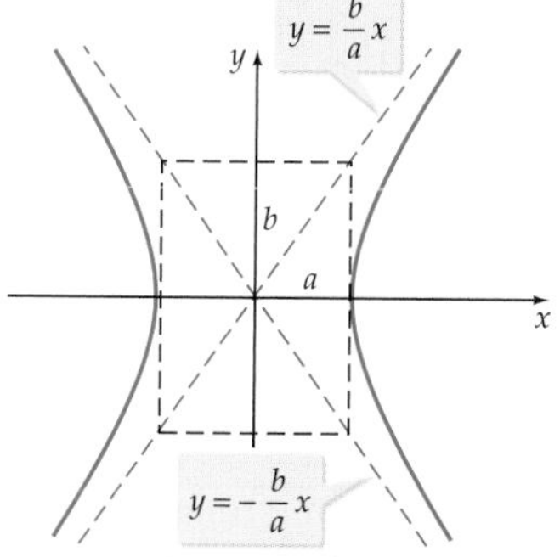

a. Asymptotes of $\frac{x^2}{a^2} - \frac{y^2}{b^2} = 1$

Asymptotes of a Hyperbola with Center at the Origin

The **asymptotes** of the hyperbola defined by $\frac{x^2}{a^2} - \frac{y^2}{b^2} = 1$ are given by the equations $y = \frac{b}{a}x$ and $y = -\frac{b}{a}x$ (see Figure A.23a).

The asymptotes of the hyperbola defined by $\frac{y^2}{a^2} - \frac{x^2}{b^2} = 1$ are given by the equations $y = \frac{a}{b}x$ and $y = -\frac{a}{b}x$ (see Figure A.23b).

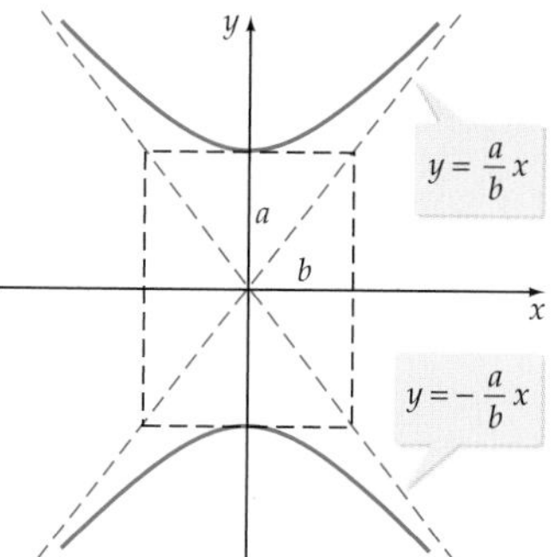

b. Asymptotes of $\frac{y^2}{a^2} - \frac{x^2}{b^2} = 1$

Figure A.23

One method for remembering the equations of the asymptotes is to write the equation of a hyperbola in standard form, but replace 1 by 0 and then solve for y.

$$\frac{x^2}{a^2} - \frac{y^2}{b^2} = 0 \quad \text{so} \quad y^2 = \frac{b^2}{a^2}x^2, \text{ or } y = \pm\frac{b}{a}x$$
$$\frac{y^2}{a^2} - \frac{x^2}{b^2} = 0 \quad \text{so} \quad y^2 = \frac{a^2}{b^2}x^2, \text{ or } y = \pm\frac{a}{b}x$$

EXAMPLE 8 Find the Vertices, Foci, and Asymptotes of a Hyperbola

Find the vertices, foci, and asymptotes of the hyperbola given by the equation $\frac{y^2}{9} - \frac{x^2}{4} = 1$. Sketch the graph.

Solution Because the y^2 term is positive, the transverse axis is on the y-axis. We know $a^2 = 9$; thus $a = 3$. The vertices are $V_1(0, 3)$ and $V_2(0, -3)$.

$$c^2 = a^2 + b^2 = 9 + 4$$
$$c = \sqrt{13}$$

The foci are $F_1(0, \sqrt{13})$ and $F_2(0, -\sqrt{13})$.

Continued ➤

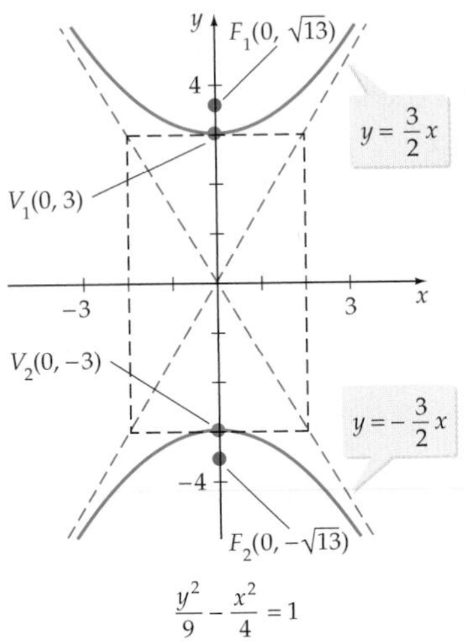

Figure A.24

Because $a = 3$ and $b = 2$ ($b^2 = 4$), the equations of the asymptotes are $y = \frac{3}{2}x$ and $y = -\frac{3}{2}x$.

To sketch the graph, we draw a rectangle that has its center at the origin and has dimensions equal to the lengths of the transverse and conjugate axes. The asymptotes are extensions of the diagonals of the rectangle. See Figure A.24.

CHECK YOUR PROGRESS 8 Find the coordinates of the center, vertices, and foci and the equations of the asymptotes of the hyperbola given by $\frac{y^2}{25} - \frac{x^2}{36} = 1$.

Solution *See page S45.* Center (0, 0); vertices (0, 5), (0, −5); foci $(0, \sqrt{61})$, $(0, -\sqrt{61})$; asymptotes $y = \pm\frac{5}{6}x$

HYPERBOLAS WITH CENTER AT (*h*, *k*)

Using a translation of coordinates similar to that used for ellipses, we can write the equation of a hyperbola with center at the point (h, k).

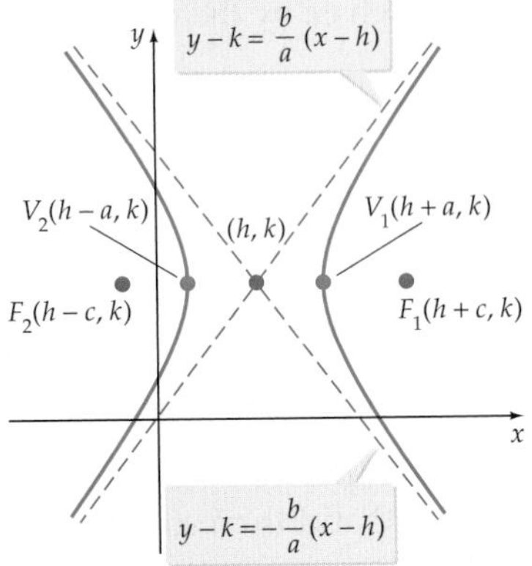

a. Transverse axis parallel to the x-axis

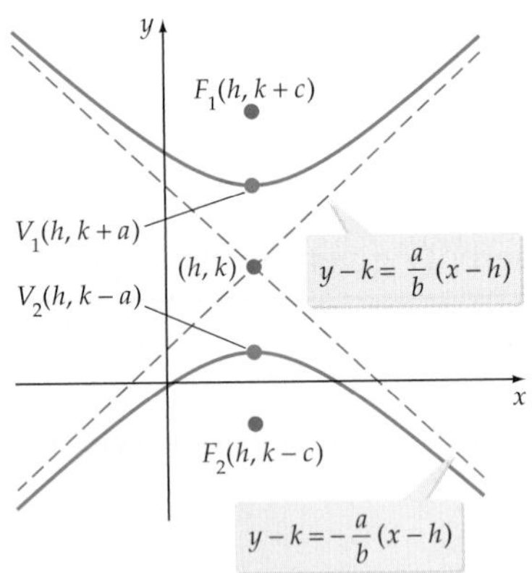

b. Transverse axis parallel to the y-axis

Figure A.25

Standard Forms of the Equation of a Hyperbola with Center at (*h*, *k*)

Transverse Axis Parallel to the x-axis

The standard form of the equation of a hyperbola with center at (h, k) and transverse axis parallel to the x-axis (see Figure A.25a) is given by

$$\frac{(x - h)^2}{a^2} - \frac{(y - k)^2}{b^2} = 1$$

The coordinates of the vertices are $V_1(h + a, k)$ and $V_2(h - a, k)$. The coordinates of the foci are $F_1(h + c, k)$ and $F_2(h - c, k)$, where $c^2 = a^2 + b^2$. The equations of the asymptotes are $y - k = \pm\frac{b}{a}(x - h)$.

Transverse Axis Parallel to the y-axis

The standard form of the equation of a hyperbola with center at (h, k) and transverse axis parallel to the y-axis (see Figure A.25b) is given by

$$\frac{(y - k)^2}{a^2} - \frac{(x - h)^2}{b^2} = 1$$

The coordinates of the vertices are $V_1(h, k + a)$ and $V_2(h, k - a)$. The coordinates of the foci are $F_1(h, k + c)$ and $F_2(h, k - c)$, where $c^2 = a^2 + b^2$. The equations of the asymptotes are $y - k = \pm\frac{a}{b}(x - h)$.

EXAMPLE 9 Find the Center, Vertices, Foci, and Asymptotes of a Hyperbola

Find the center, vertices, foci, and asymptotes of the hyperbola given by the equation $4x^2 - 9y^2 - 16x + 54y - 29 = 0$. Sketch the graph.

Continued ➤

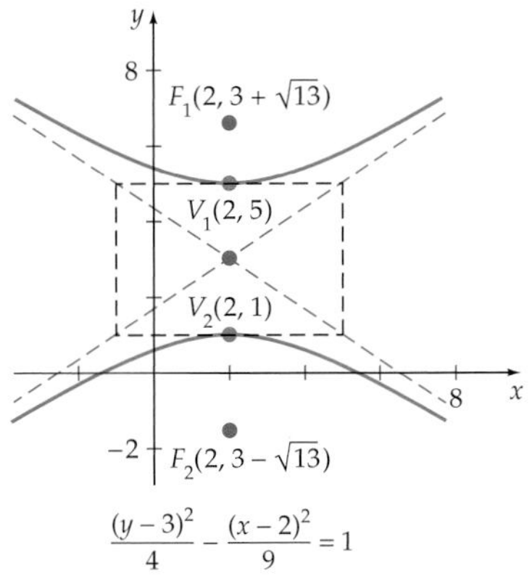

Figure A.26

Solution Write the equation of the hyperbola in standard form by completing the square.

$$\begin{aligned}
4x^2 - 9y^2 - 16x + 54y - 29 &= 0 \\
4x^2 - 16x - 9y^2 + 54y &= 29 && \text{■ Rearrange terms.} \\
4(x^2 - 4x) - 9(y^2 - 6y) &= 29 && \text{■ Factor.} \\
4(x^2 - 4x + 4) - 9(y^2 - 6y + 9) &= 29 + 16 - 81 && \text{■ Complete the square.} \\
4(x-2)^2 - 9(y-3)^2 &= -36 && \text{■ Factor.} \\
\frac{(y-3)^2}{4} - \frac{(x-2)^2}{9} &= 1 && \text{■ Divide by } -36.
\end{aligned}$$

The coordinates of the center are (2, 3). Because the term containing $(y - 3)^2$ is positive, the transverse axis is parallel to the y-axis. We know $a^2 = 4$; thus $a = 2$. The vertices are (2, 5) and (2, 1). See Figure A.26.

$$c^2 = a^2 + b^2 = 4 + 9$$
$$c = \sqrt{13}$$

The foci are $\left(2, 3 + \sqrt{13}\right)$ and $\left(2, 3 - \sqrt{13}\right)$. We know $b^2 = 9$; thus $b = 3$. The equations of the asymptotes are $y - 3 = \pm(2/3)(x - 2)$, which simplifies to

$$y = \frac{2}{3}x + \frac{5}{3} \qquad \text{and} \qquad y = -\frac{2}{3}x + \frac{13}{3}$$

CHECK YOUR PROGRESS 9 Find the coordinates of the center, vertices, and foci and the equation of the asymptotes of the hyperbola given by $16x^2 - 9y^2 - 32x - 54y + 79 = 0$.

Solution *See page S45.* Center (1, −3); vertices (1, 1), (1, −7); foci (1, 2), (1, −8); asymptotes $y = \pm\frac{4}{3}(x - 1) - 3$

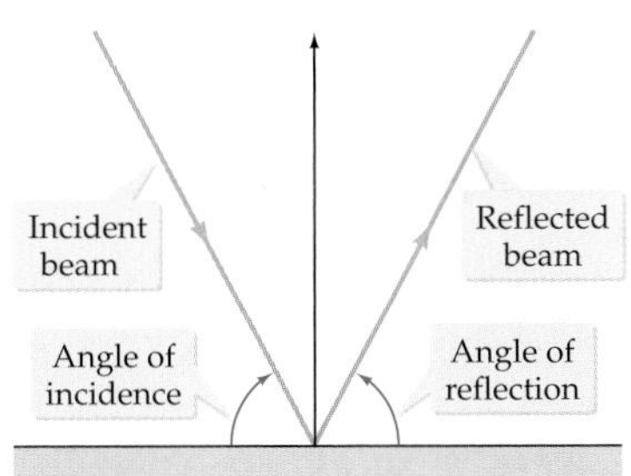

Figure A.27

Applications

A principle of physics states that when light is reflected from a point P on a surface, the angle of incidence (that of the incoming ray) equals the angle of reflection (that of the outgoing ray). See Figure A.27. This principle applied to parabolas has some useful consequences.

Optical Property of a Parabola

The line tangent to a parabola at a point P makes equal angles with the line through P parallel to the axis of symmetry and the line through P and the focus on the parabola (see Figure A.28).

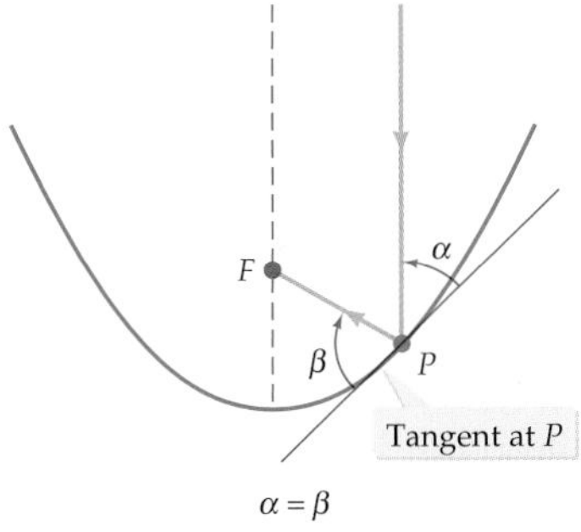

Figure A.28

A cross section of the reflecting mirror of a telescope has the shape of a parabola. The incoming parallel rays of light are reflected from the surface of the mirror to the focus. See Figure A.29.

Flashlights and car headlights also make use of this property. The light bulb is positioned at the focus of a parabolic reflector, which causes the reflected light to be reflected outward in parallel rays. See Figure A. 30.

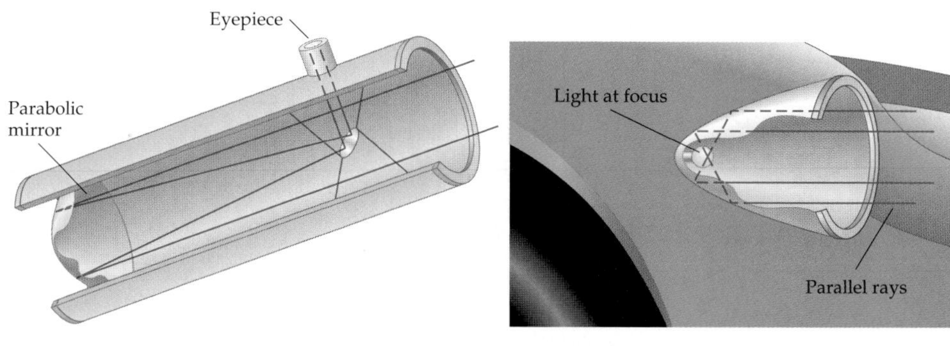

Figure A.29

Figure A.30

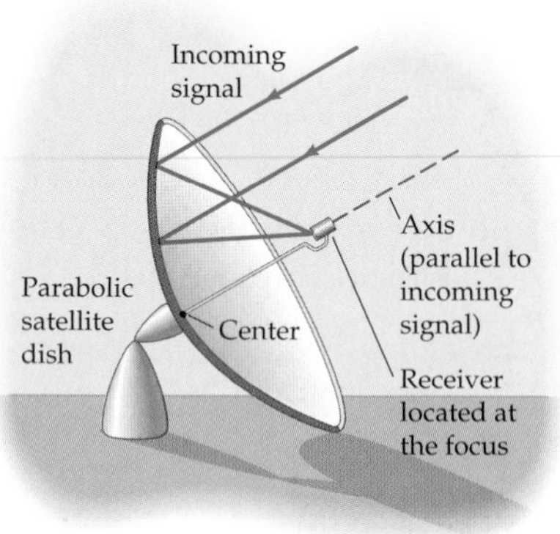

Figure A.31

EXAMPLE 10 Find the Focus of a Satellite Dish

A satellite dish has the shape of a paraboloid. The signals that it receives are reflected to a receiver that is located at the focus of the paraboloid. If the dish is 8 feet across at its opening and $1\frac{1}{4}$ feet deep at its center, determine the location of its focus.

Solution Figure A.31 shows that a cross section of the satellite dish along its axis of symmetry is a parabola. Figure A.32 shows this cross section placed in a rectangular coordinate system with the vertex of the parabola at (0, 0) and the axis of symmetry of the parabola on the y-axis. The parabola has an equation of the form

$$4py = x^2$$

Because the parabola contains the point (4, 1.25), this equation is satisfied by the substitution $x = 4$ and $y = 1.25$. Thus we have

$$\begin{aligned} 4p(1.25) &= 4^2 \\ 5p &= 16 \\ p &= \frac{16}{5} = 3.2 \end{aligned}$$

The focus of the satellite dish is on the axis of symmetry of the dish, and it is 3.2 feet above the vertex of the dish.

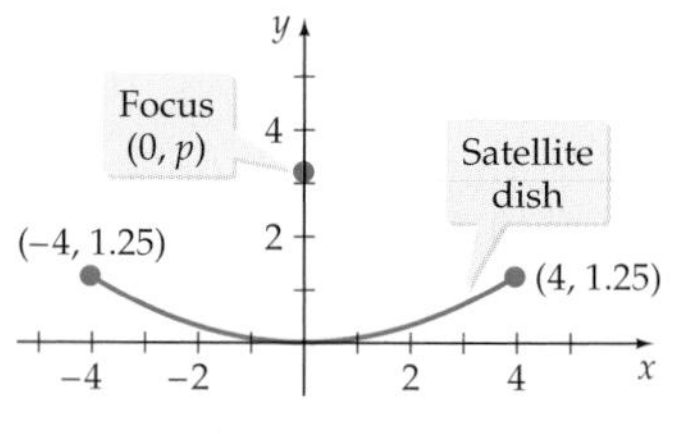

Figure A.32

CHECK YOUR PROGRESS 10 A radio telescope is in the shape of a paraboloid measuring 81 feet across, with a depth of 16 feet. Determine, to the nearest 0.1 foot, the distance from the vertex to the focus of this telescope.

Solution *See page S45.* $\approx$25.6 ft

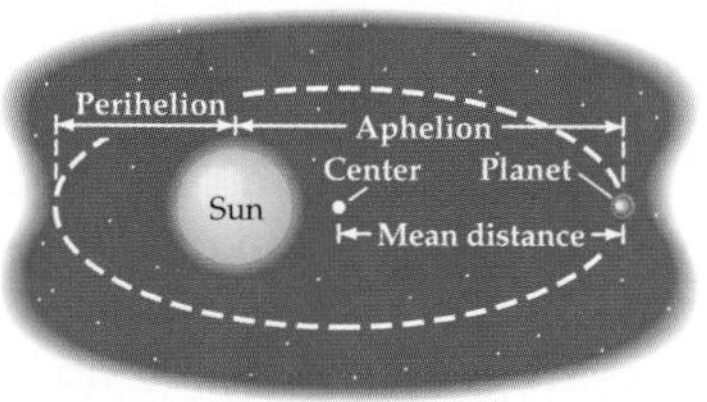

Figure A.33

The planets travel around the sun in elliptical orbits. The sun is located at a focus of each orbit. The terms *perihelion* and *aphelion* are used to denote the position of a planet in its orbit around the sun. The perihelion is the point in the orbit nearest the sun; the aphelion is the point in the orbit farthest from the sun. See Figure A.33. The length of the semimajor axis of a planet's elliptical orbit is called the *mean distance* of the planet from the sun.

EXAMPLE 11 Determine an Equation for the Orbit of Earth

Earth has a mean distance from the sun of 93 million miles and a perihelion distance of 91.5 million miles. Find an equation for Earth's orbit.

Solution A mean distance of 93 million miles implies that the length of the semimajor axis of the orbit is $a = 93$ million miles. Earth's aphelion distance is the length of the major axis less the length of the perihelion distance. Thus

$$\text{Aphelion distance} = 2(93) - 91.5 = 94.5 \text{ million miles}$$

The distance c from the sun to the center of Earth's orbit is

$$c = \text{aphelion distance} - 93 = 94.5 - 93 = 1.5 \text{ million miles}$$

The length b of the semiminor axis of the orbit is

$$b = \sqrt{a^2 - c^2} = \sqrt{93^2 - 1.5^2} = \sqrt{8646.75}$$

An equation for Earth's orbit is

$$\frac{x^2}{93^2} + \frac{y^2}{8646.75} = 1$$

CHECK YOUR PROGRESS 11 Venus has a mean distance from the sun of 67.08 million miles, and the distance from Venus to the sun at its aphelion is 67.58 million miles. Find an equation for the orbit of Venus.

Solution *See page S45.* $\frac{x^2}{67.08^2} + \frac{y^2}{67.078^2} = 1$

Sound waves, although different from light waves, have a similar reflective property. When sound is reflected from a point P on a surface, the angle of incidence equals the angle of reflection. Applying this principle to a room with an elliptical ceiling results in what are called *whispering galleries*. These galleries are based on the following theorem.

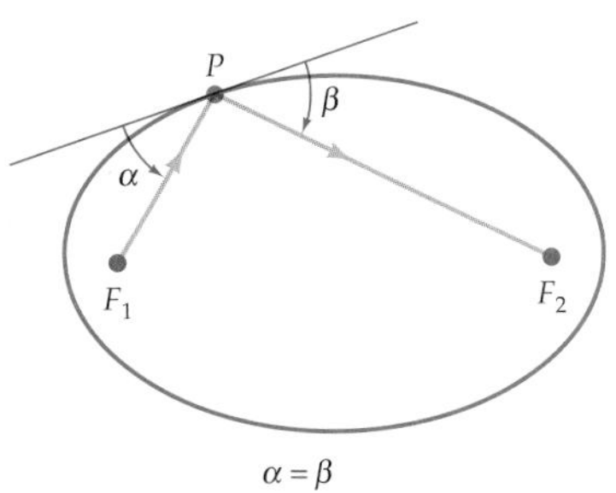

Figure A.34

The Reflective Property of an Ellipse

The lines from the foci to a point on an ellipse make equal angles with the tangent line at that point. See Figure A.34.

The Statuary Hall in the Capitol Building in Washington, D.C. is a whispering gallery. Two people standing at the foci of the elliptical ceiling can whisper and yet hear each other even though they are a considerable distance apart. The whisper from one person is reflected to the person standing at the other focus.

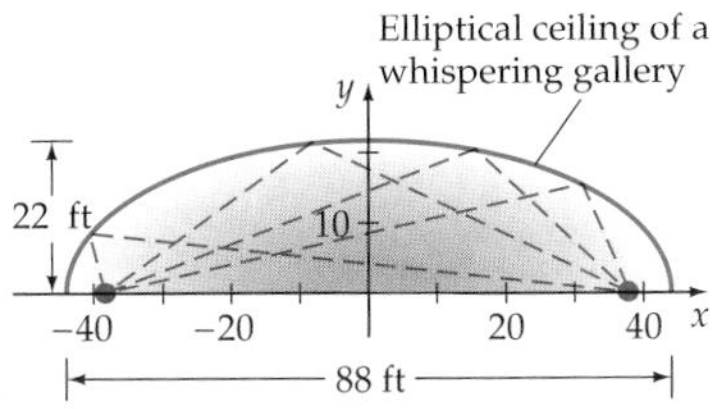

Figure A.35

EXAMPLE 12 Locate the Foci of a Whispering Gallery

A room 88 feet long is constructed to be a whispering gallery. The room has an elliptical ceiling, as shown in Figure A.35. If the maximum height of the ceiling is 22 feet, determine where the foci are located.

Continued ➤

Solution The length a of the semimajor axis of the elliptical ceiling is 44 feet. The height b of the semiminor axis is 22 feet. Thus

$$c^2 = a^2 - b^2$$
$$c^2 = 44^2 - 22^2$$
$$c = \sqrt{44^2 - 22^2} \approx 38.1 \text{ feet}$$

The foci are located about 38.1 feet from the center of the elliptical ceiling along its major axis.

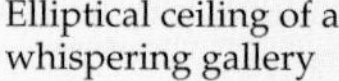

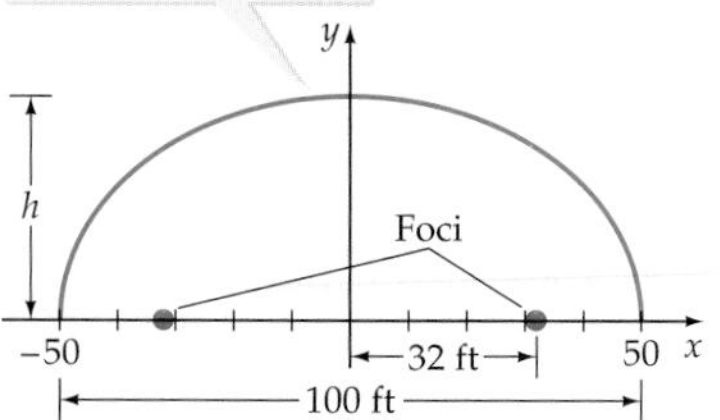

CHECK YOUR PROGRESS 12 An architect wishes to design a large room 100 feet long that will be a whispering gallery. The ceiling of the room has a cross section that is an ellipse, as shown in the figure at the left.

If the foci are to be located 32 feet to the right and left of center, find the height h of the elliptical ceiling to the nearest 0.1 foot.

Solution *See page S46.* ≈ 38.4 ft

EXERCISES *Appendix A*

— *Suggested Assignment: Exercises 1–47, odd.*
— *Answer graphs to Exercises 1–24 are on pages AA22–AA24.*

In Exercises 1 to 24, find the vertex, focus, and directrix of each parabola; find the center, vertices, and foci of each ellipse; find the center, vertices, foci, and asymptotes of each hyperbola. Graph each conic.

1. $x^2 = -4y$

2. $y^2 = \frac{1}{3}x$

3. $\frac{x^2}{16} + \frac{y^2}{25} = 1$

4. $\frac{x^2}{9} + \frac{y^2}{4} = 1$

5. $\frac{x^2}{16} - \frac{y^2}{25} = 1$

6. $\frac{y^2}{4} - \frac{x^2}{25} = 1$

7. $(x - 2)^2 = 8(y + 3)$

8. $(y + 4)^2 = -4(x - 2)$

9. $x^2 - y^2 = 9$

10. $y^2 = 16x$

11. $\frac{(x - 3)^2}{25} + \frac{(y + 2)^2}{16} = 1$

12. $\frac{(x + 2)^2}{9} + \frac{y^2}{25} = 1$

13. $\frac{(x - 3)^2}{16} - \frac{(y + 4)^2}{9} = 1$

14. $\frac{(y + 2)^2}{4} - \frac{(x - 1)^2}{16} = 1$

15. $x^2 + 4y^2 - 6x + 8y - 3 = 0$

16. $3x^2 - 4y^2 + 12x - 24y - 36 = 0$

17. $3x - 4y^2 + 8y + 2 = 0$

18. $3x + 2y^2 - 4y - 7 = 0$

19. $9x^2 + 4y^2 + 36x - 8y + 4 = 0$

20. $11x^2 - 25y^2 - 44x - 50y - 256 = 0$

21. $4x^2 - 9y^2 - 8x + 12y - 144 = 0$

22. $9x^2 - y^2 - 36x + 6y - 9 = 0$

23. $4x^2 + 28x + 32y + 81 = 0$

24. $x^2 - 6x - 9y + 27 = 0$

In Exercises 25 to 32, find the equation of the conic that satisfies the given conditions.

25. Ellipse with vertices at (7, 3) and (−3, 3); length of minor axis is 8. $\frac{(x - 2)^2}{25} + \frac{(y - 3)^2}{16} = 1$

26. Hyperbola with vertices at (4, 1) and (−2, 1); foci at (5, 1) and (−3, 1). $\frac{(x - 1)^2}{9} - \frac{(y - 1)^2}{7} = 1$

48. ***Elliptical Receivers*** Some satellite receivers are made in an elliptical shape that enables the receiver to pick up signals from two satellites. The receiver shown in the figure has a major axis of 24 inches and a minor axis of 18 inches.

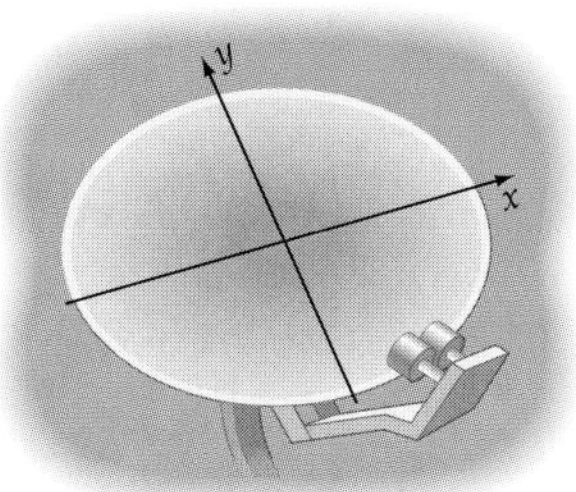

Determine, to the nearest 0.1 inch, the coordinates in the xy-plane of the foci of the ellipse. (*Note:* Because the receiver has only a slight curvature, we can estimate the location of the foci by assuming the receiver is flat.)
$(-7.9, 0), (7.9, 0)$

APPENDIX B Determinants

- **Determinants**

Determinants

Determinant of a 2 × 2 Matrix

Associated with each square matrix A is a number called the *determinant of A*. We will denote the determinant of the matrix A by $\det(A)$ or by $|A|$. For the remainder of this appendix, we assume that all matrices are square matrices.

The Determinant of a 2 × 2 Matrix

The **determinant** of the matrix $A = [a_{ij}]$ of order 2 is

$$|A| = \begin{vmatrix} a_{11} & a_{12} \\ a_{21} & a_{22} \end{vmatrix} = a_{11}a_{22} - a_{21}a_{12}$$

CAUTION Be careful not to confuse the notation for a matrix with that for a determinant. The symbol [ ] (brackets) is used for a matrix; the symbol | | (vertical bars) is used for the determinant of a matrix.

An easy way to remember the formula for the determinant of a 2×2 matrix is to recognize that the determinant is the difference between the products of the diagonal elements. That is,

$$\begin{vmatrix} a_{11} & a_{12} \\ a_{21} & a_{22} \end{vmatrix} = a_{11}a_{22} - a_{21}a_{12}$$

EXAMPLE 1 Find the Value of a Determinant

Find the value of the determinant of the matrix $A = \begin{bmatrix} 5 & 3 \\ 2 & -3 \end{bmatrix}$.

Solution

$$|A| = \begin{vmatrix} 5 & 3 \\ 2 & -3 \end{vmatrix} = 5(-3) - 2(3) = -15 - 6 = -21$$

CHECK YOUR PROGRESS 1 Evaluate: $\begin{vmatrix} 2 & 9 \\ -6 & 2 \end{vmatrix}$

Solution *See page S46.* 58

Minors and Cofactors

To define the determinant of a matrix of order greater than 2, we first need two other definitions.

> **The Minor of a Matrix**
>
> The **minor** M_{ij} of the element a_{ij} of a square matrix A of order $n \geq 3$ is the determinant of the matrix of order $n - 1$ obtained by deleting the ith row and the jth column of A.

Consider the matrix $A = \begin{bmatrix} 2 & -1 & 5 \\ 4 & 3 & -7 \\ 8 & -7 & 6 \end{bmatrix}$. The minor M_{23} is the determinant of the matrix A formed by deleting row 2 and column 3 from A.

$$M_{23} = \begin{vmatrix} 2 & -1 \\ 8 & -7 \end{vmatrix} \qquad \blacksquare \begin{vmatrix} 2 & -1 & 5 \\ 4 & 3 & -7 \\ 8 & -7 & 6 \end{vmatrix}$$

$$= 2(-7) - 8(-1) = -14 + 8 = -6$$

The minor M_{31} is the determinant of the matrix A formed by deleting row 3 and column 1 from A.

$$M_{31} = \begin{vmatrix} -1 & 5 \\ 3 & -7 \end{vmatrix} \qquad \blacksquare \begin{vmatrix} 2 & -1 & 5 \\ 4 & 3 & -7 \\ 8 & -7 & 6 \end{vmatrix}$$

$$= (-1)(-7) - 3(5) = 7 - 15 = -8$$

The second definition we need is that of the *cofactor* of a matrix.

> **Cofactor of a Matrix**
>
> The **cofactor** C_{ij} of the element a_{ij} of a square matrix A is given by $C_{ij} = (-1)^{i+j}M_{ij}$, where M_{ij} is the minor of a_{ij}.

When $i + j$ is an even integer, $(-1)^{i+j} = 1$. When $i + j$ is an odd integer, $(-1)^{i+j} = -1$. Thus

$$C_{ij} = \begin{cases} M_{ij}, & i + j \text{ is an even integer} \\ -M_{ij}, & i + j \text{ is an odd integer} \end{cases}$$

EXAMPLE 2 **Find the Minor and Cofactor of a Matrix**

Given $A = \begin{bmatrix} 4 & 3 & -2 \\ 5 & -2 & 4 \\ 3 & -2 & -6 \end{bmatrix}$, find M_{32} and C_{12}.

Solution

$$M_{32} = \begin{vmatrix} 4 & -2 \\ 5 & 4 \end{vmatrix} = 4(4) - 5(-2) = 16 + 10 = 26$$

$$C_{12} = (-1)^{1+2}M_{12} = -M_{12} = -\begin{vmatrix} 5 & 4 \\ 3 & -6 \end{vmatrix} = -(-30 - 12) = 42$$

CHECK YOUR PROGRESS 2 Evaluate M_{13} and C_{13} for $\begin{vmatrix} 3 & -2 & 3 \\ 1 & 3 & 0 \\ 6 & -2 & 3 \end{vmatrix}$.

Solution *See page S46.* −20, −20

EVALUATE A DETERMINANT USING EXPANDING BY COFACTORS

Cofactors are used to evaluate the determinant of a matrix of order 3 or greater. The technique used to evaluate a determinant by using cofactors is called *expanding by cofactors.*

Determinants by Expanding by Cofactors Theorem

Given the square matrix A of order 3 or greater, the value of the determinant of A is the sum of the products of the elements of any row or column and their cofactors. For the rth row of A, the value of the determinant of A is

$$|A| = a_{r1}C_{r1} + a_{r2}C_{r2} + a_{r3}C_{r3} + \cdots + a_{rn}C_{rn}$$

For the cth column of A, the determinant of A is

$$|A| = a_{1c}C_{1c} + a_{2c}C_{2c} + a_{3c}C_{3c} + \cdots + a_{nc}C_{nc}$$

This theorem states that the value of a determinant can be found by expanding by cofactors of *any* row or column. The value of the determinant is the same in each case. To illustrate the method, consider the matrix $A = \begin{bmatrix} 2 & 3 & -1 \\ 4 & -2 & 3 \\ 1 & -3 & 4 \end{bmatrix}$.

Expanding the determinant of A by some row, say row 2, gives

$$\begin{aligned} |A| = \begin{vmatrix} 2 & 3 & -1 \\ 4 & -2 & 3 \\ 1 & -3 & 4 \end{vmatrix} &= 4C_{21} + (-2)C_{22} + 3C_{23} \\ &= 4(-1)^{2+1}M_{21} + (-2)(-1)^{2+2}M_{22} + 3(-1)^{2+3}M_{23} \\ &= (-4)\begin{vmatrix} 3 & -1 \\ -3 & 4 \end{vmatrix} + (-2)\begin{vmatrix} 2 & -1 \\ 1 & 4 \end{vmatrix} + (-3)\begin{vmatrix} 2 & 3 \\ 1 & -3 \end{vmatrix} \\ &= (-4)9 + (-2)9 + (-3)(-9) = -27 \end{aligned}$$

Expanding the determinant of A by some column, say column 3, gives

$$\begin{aligned} |A| = \begin{vmatrix} 2 & 3 & -1 \\ 4 & -2 & 3 \\ 1 & -3 & 4 \end{vmatrix} &= (-1)C_{13} + 3C_{23} + 4C_{33} \\ &= (-1)(-1)^{1+3}M_{13} + 3(-1)^{2+3}M_{23} + 4(-1)^{3+3}M_{33} \\ &= (-1)\begin{vmatrix} 4 & -2 \\ 1 & -3 \end{vmatrix} + (-3)\begin{vmatrix} 2 & 3 \\ 1 & -3 \end{vmatrix} + 4\begin{vmatrix} 2 & 3 \\ 4 & -2 \end{vmatrix} \\ &= (-1)(-10) + (-3)(-9) + 4(-16) = -27 \end{aligned}$$

The value of the determinant of A is the same whether we expanded by cofactors of the elements of a row or by cofactors of the elements of a column. When evaluating a determinant, choose the most convenient row or column, which usually is the row or column containing the most zeros.

TAKE NOTE

Example 3 illustrates that choosing a row or column with the most zeros and then expanding about that row or column will reduce the number of calculations you must perform. For Example 3 we have $O \cdot C_{32} = O$, and it is not necessary to compute C_{32}.

EXAMPLE 3 Evaluate a Determinant by Cofactors

Evaluate the determinant of $A = \begin{bmatrix} 5 & -3 & -1 \\ -2 & 1 & -1 \\ 1 & 0 & 2 \end{bmatrix}$ by expanding by cofactors.

Solution Because $a_{32} = 0$, expand using row 3 or column 2. Row 3 will be used here.

$$\begin{aligned} |A| &= 1C_{31} + 0C_{32} + 2C_{33} = 1(-1)^{3+1}M_{31} + 0(-1)^{3+2}M_{32} + 2(-1)^{3+3}M_{33} \\ &= 1\begin{vmatrix} -3 & -1 \\ 1 & -1 \end{vmatrix} + 0 + 2\begin{vmatrix} 5 & -3 \\ -2 & 1 \end{vmatrix} = 1[3 - (-1)] + 0 + 2[5 - 6] \\ &= 4 - 2 = 2 \end{aligned}$$

CHECK YOUR PROGRESS 3 Evaluate the following determinant by expanding by cofactors.

$$\begin{vmatrix} 3 & -2 & 0 \\ 2 & -3 & 2 \\ 8 & -2 & 5 \end{vmatrix}$$

Solution *See page S46.* −45

Evaluate a Determinant Using Elementary Row Operations

TAKE NOTE

The properties of determinants stated at the right remain true when the word *row* is replaced by the word *column*. In that case, we would have elementary *column* operations.

Effects of Elementary Row Operations on the Value of a Determinant of a Matrix

If A is a square matrix of order n, then the following elementary row operations produce the indicated changes in the determinant of A.

1. Interchanging any two rows of A changes the sign of $|A|$.
2. Multiplying a row of A by a constant k multiplies the determinant of A by k.
3. Adding a multiple of a row of A to another row does not change the value of the determinant of A.

Point of Interest

There is a vast difference between computing the determinant of a matrix by expanding by cofactors and doing so by row reduction. For instance, approximately 4×10^{12} operations are necessary to evaluate a determinant of order 15 by expanding by cofactors. At 4 million operations per second, a computer would need approximately 11 days to compute the determinant. On the other hand, approximately 2000 operations are needed to find the determinant by row reduction. Using the same computer, it would take less than 0.001 second to find the determinant.

To illustrate these properties, consider the matrix $A = \begin{bmatrix} 2 & 3 \\ 1 & -2 \end{bmatrix}$. The determinant of A is $|A| = 2(-2) - 1(3) = -7$. Now consider each of the elementary row operations.

Interchange the rows of A and evaluate the determinant.

$$\begin{vmatrix} 1 & -2 \\ 2 & 3 \end{vmatrix} = 1(3) - 2(-2) = 3 + 4 = 7 = -|A|$$

Multiply row 2 of A by -3 and evaluate the determinant.

$$\begin{vmatrix} 2 & 3 \\ -3 & 6 \end{vmatrix} = 2(6) - (-3)3 = 12 + 9 = 21 = -3|A|$$

Multiply row 1 of A by -2 and add it to row 2. Then evaluate the determinant.

$$\begin{vmatrix} 2 & 3 \\ -3 & -8 \end{vmatrix} = 2(-8) - (-3)(3) = -16 + 9 = -7 = |A|$$

These elementary row operations are often used to rewrite a matrix in *triangular form.* A matrix is in **triangular form** if all elements below or above the main diagonal are zero. The matrices

$$A = \begin{bmatrix} 2 & -2 & 3 & 1 \\ 0 & -2 & 4 & 2 \\ 0 & 0 & 6 & 9 \\ 0 & 0 & 0 & -5 \end{bmatrix} \quad \text{and} \quad B = \begin{bmatrix} 3 & 0 & 0 & 0 \\ 2 & -3 & 0 & 0 \\ 6 & 4 & -2 & 0 \\ 8 & 3 & 4 & 2 \end{bmatrix}$$

are in triangular form.

Determinant of a Matrix in Triangular Form

Let A be a square matrix of order n in triangular form. The determinant of A is the product of the elements on the main diagonal.

$$|A| = a_{11}a_{22}a_{33} \cdots a_{nn}$$

For the matrices A and B given above,

$$|A| = 2(-2)(6)(-5) = 120$$
$$|B| = 3(-3)(-2)(2) = 36$$

EXAMPLE 4 Evaluate a Determinant by Elementary Row Operations

Evaluate the determinant by first rewriting it in triangular form.

$$\begin{vmatrix} 2 & 1 & -1 & 3 \\ 2 & 2 & 0 & 1 \\ 4 & 5 & 4 & -3 \\ 2 & 2 & 7 & -3 \end{vmatrix}$$

Solution Rewrite the determinant in triangular form by using elementary row operations.

$$\begin{vmatrix} 2 & 1 & -1 & 3 \\ 2 & 2 & 0 & 1 \\ 4 & 5 & 4 & -3 \\ 2 & 2 & 7 & -3 \end{vmatrix} \begin{matrix} -1R_1 + R_2 \\ -2R_1 + R_3 \\ -1R_1 + R_4 \\ = \end{matrix} \begin{vmatrix} 2 & 1 & -1 & 3 \\ 0 & 1 & 1 & -2 \\ 0 & 3 & 6 & -9 \\ 0 & 1 & 8 & -6 \end{vmatrix}$$

$$\begin{matrix} \text{Factor 3} \\ \text{from row 3.} \\ = \end{matrix} \; 3\begin{vmatrix} 2 & 1 & -1 & 3 \\ 0 & 1 & 1 & -2 \\ 0 & 1 & 2 & -3 \\ 0 & 1 & 8 & -6 \end{vmatrix} \begin{matrix} -1R_2 + R_3 \\ -1R_2 + R_4 \\ = \end{matrix} \; 3\begin{vmatrix} 2 & 1 & -1 & 3 \\ 0 & 1 & 1 & -2 \\ 0 & 0 & 1 & -1 \\ 0 & 0 & 7 & -4 \end{vmatrix}$$

$$\begin{matrix} -7R_3 + R_4 \\ = \end{matrix} \; 3\begin{vmatrix} 2 & 1 & -1 & 3 \\ 0 & 1 & 1 & -2 \\ 0 & 0 & 1 & -1 \\ 0 & 0 & 0 & 3 \end{vmatrix} = 3(2)(1)(1)(3) = 18$$

CHECK YOUR PROGRESS 4 Evaluate the following determinant by first rewriting the determinant in triangular form.

$$\begin{vmatrix} 3 & -2 & -1 \\ 1 & 2 & 4 \\ 2 & -2 & 3 \end{vmatrix}$$

Solution *See page S46.* 38

In some cases it is possible to recognize when the determinant of a matrix is zero.

Conditions for a Zero Determinant

If A is a square matrix, then $|A| = 0$ when any one of the following is true.

1. A row (column) consists entirely of zeros.
2. Two rows (columns) are identical.
3. One row (column) is a constant multiple of a second row (column).

QUESTION If I is the identity matrix of order n, what is the value of $|I|$?

ANSWER The identity matrix is in triangular form with 1s on the main diagonal. Thus $|I|$ is a product of 1s, or $|I| = 1$.

The last property of determinants that we will discuss is a product property.

Product Property of Determinants

If A and B are square matrices of order n, then

$$|AB| = |A||B|$$

A consequence of this property is the following theorem.

Existence of the Inverse of a Square Matrix

If A is a square matrix of order n, then A has a multiplicative inverse if and only if $|A| \neq 0$. Furthermore,

$$|A^{-1}| = \frac{1}{|A|}$$

EXERCISES *Appendix B*

— *Suggested Assignment: Exercises 1–55, odd.*

In Exercises 1 to 8, evaluate the determinant.

1. $\begin{vmatrix} 2 & -1 \\ 3 & 5 \end{vmatrix}$ 13

2. $\begin{vmatrix} 3 & 7 \\ -4 & -2 \end{vmatrix}$ 22

3. $\begin{vmatrix} 5 & 0 \\ 2 & -3 \end{vmatrix}$ −15

4. $\begin{vmatrix} 0 & -8 \\ 3 & 4 \end{vmatrix}$ 24

5. $\begin{vmatrix} 4 & 6 \\ 2 & 3 \end{vmatrix}$ 0

6. $\begin{vmatrix} -3 & 6 \\ 4 & -8 \end{vmatrix}$ 0

7. $\begin{vmatrix} 0 & 9 \\ 0 & -2 \end{vmatrix}$ 0

8. $\begin{vmatrix} -3 & 9 \\ 0 & 0 \end{vmatrix}$ 0

In Exercises 9 to 12, evaluate the indicated minor and cofactor for the determinant.

$$\begin{vmatrix} 5 & -2 & -3 \\ 2 & 4 & -1 \\ 4 & -5 & 6 \end{vmatrix}$$

9. M_{11}, C_{11} 19, 19

10. M_{21}, C_{21} −27, 27

11. M_{32}, C_{32} 1, −1

12. M_{33}, C_{33} 24, 24

In Exercises 13 to 16, evaluate the indicated minor and cofactor for the determinant

$$\begin{vmatrix} 3 & -2 & 3 \\ 1 & 3 & 0 \\ 6 & -2 & 3 \end{vmatrix}$$

13. M_{22}, C_{22} −9, −9

14. M_{32}, C_{32} −3, 3

15. M_{31}, C_{31} −9, −9

16. M_{23}, C_{23} 6, −6

In Exercises 17 to 26, evaluate the determinant by expanding by cofactors.

17. $\begin{vmatrix} 2 & -3 & 1 \\ 2 & 0 & 2 \\ 3 & -2 & 4 \end{vmatrix}$ 10

18. $\begin{vmatrix} 3 & 1 & -2 \\ 2 & -5 & 4 \\ 3 & 2 & 1 \end{vmatrix}$ −67

19. $\begin{vmatrix} -2 & 3 & 2 \\ 1 & 2 & -3 \\ -4 & -2 & 1 \end{vmatrix}$ 53

20. $\begin{vmatrix} 4 & -1 & 1 \\ 3 & -2 & 0 \\ 1 & 4 & -2 \end{vmatrix}$ 24

21. $\begin{vmatrix} 2 & -3 & 10 \\ 0 & 2 & -3 \\ 0 & 0 & 5 \end{vmatrix}$ 20

22. $\begin{vmatrix} 6 & 0 & 0 \\ 2 & -3 & 0 \\ 7 & -8 & 2 \end{vmatrix}$ −36

23. $\begin{vmatrix} 0 & -2 & 4 \\ 1 & 0 & -7 \\ 5 & -6 & 0 \end{vmatrix}$ 46

24. $\begin{vmatrix} 5 & -8 & 0 \\ 2 & 0 & -7 \\ 0 & -2 & -1 \end{vmatrix}$ −86

25. $\begin{vmatrix} 4 & -3 & 3 \\ 2 & 1 & -4 \\ 6 & -2 & -1 \end{vmatrix}$ 0

26. $\begin{vmatrix} -2 & 3 & 9 \\ 4 & -2 & -6 \\ 0 & -8 & -24 \end{vmatrix}$ 0

In Exercises 27 to 40, without expanding, give a reason for each equality.

27. $\begin{vmatrix} 2 & -1 & 3 \\ 0 & 0 & 0 \\ 3 & 4 & 1 \end{vmatrix} = 0$

Row 2 consists entirely of zeros, so the determinant is zero.

28. $\begin{vmatrix} 2 & 3 & 0 \\ 1 & -2 & 0 \\ 4 & 1 & 0 \end{vmatrix} = 0$

Column 3 consists entirely of zeros. Therefore, the determinant is zero.

29. $\begin{vmatrix} 1 & 4 & -1 \\ 2 & 4 & 12 \\ 3 & 1 & 4 \end{vmatrix} = 2\begin{vmatrix} 1 & 4 & -1 \\ 1 & 2 & 6 \\ 3 & 1 & 4 \end{vmatrix}$ 2 was factored from row 2.

30. $\begin{vmatrix} 1 & -3 & 4 \\ 4 & 6 & 1 \\ 0 & -9 & 3 \end{vmatrix} = -3\begin{vmatrix} 1 & 1 & 4 \\ 4 & -2 & 1 \\ 0 & 3 & 3 \end{vmatrix}$ −3 was factored from column 2.

31. $\begin{vmatrix} 1 & 5 & -2 \\ 2 & -1 & 4 \\ 3 & 0 & -2 \end{vmatrix} = \begin{vmatrix} 1 & 5 & -2 \\ 0 & -11 & 8 \\ 3 & 0 & -2 \end{vmatrix}$ Row 1 was multiplied by −2 and added to row 2.

32. $\begin{vmatrix} 1 & 1 & -3 \\ 2 & 2 & 5 \\ 1 & -2 & 4 \end{vmatrix} = \begin{vmatrix} 1 & 1 & -3 \\ 2 & 2 & 5 \\ 0 & -3 & 7 \end{vmatrix}$ Row 1 was multiplied by −1 and added to row 3.

33. $\begin{vmatrix} 4 & -3 & 2 \\ 6 & 2 & 1 \\ -2 & 2 & 4 \end{vmatrix} = 2\begin{vmatrix} 2 & -3 & 2 \\ 3 & 2 & 1 \\ -1 & 2 & 4 \end{vmatrix}$ 2 was factored from column 1.

34. $\begin{vmatrix} 2 & -1 & 3 \\ 3 & 0 & 1 \\ -4 & 2 & -6 \end{vmatrix} = 0$ Row 3 is a constant multiple of row 1. $-2R_1 = R_3$. Therefore, the determinant is zero.

35. $\begin{vmatrix} 2 & -4 & 5 \\ 0 & 3 & 4 \\ 0 & 0 & -2 \end{vmatrix} = -12$

36. $\begin{vmatrix} 3 & 0 & 0 \\ 2 & -1 & 0 \\ 3 & 4 & 5 \end{vmatrix} = -15$

35. The matrix is in triangular form. The value of the determinant is the product of the terms on the main diagonal. Therefore, the value of the determinant is −12.

37. $\begin{vmatrix} 3 & 5 & -2 \\ 2 & 1 & 0 \\ 9 & -2 & -3 \end{vmatrix} = -\begin{vmatrix} 9 & -2 & -3 \\ 2 & 1 & 0 \\ 3 & 5 & -2 \end{vmatrix}$ Row 1 and row 3 were interchanged, so the sign of the determinant was changed.

38. $\begin{vmatrix} 6 & 0 & -2 \\ 2 & -1 & -3 \\ 1 & 5 & -7 \end{vmatrix} = -\begin{vmatrix} 0 & 6 & -2 \\ -1 & 2 & -3 \\ 5 & 1 & -7 \end{vmatrix}$ Column 1 and column 2 were interchanged. Therefore, the sign of the determinant was changed.

39. $a^3\begin{vmatrix} 1 & 1 & 1 \\ a & a & a \\ a^2 & a^2 & a^2 \end{vmatrix} = \begin{vmatrix} a & a & a \\ a^2 & a^2 & a^2 \\ a^3 & a^3 & a^3 \end{vmatrix}$ Each row of the determinant was multiplied by a.

40. $\begin{vmatrix} 1 & 1 & 1 \\ 2 & 2 & 2 \\ 3 & 3 & 3 \end{vmatrix} = 0$ Columns 1, 2, and 3 are identical. Therefore, the determinant is zero.

In Exercises 41 to 50, evaluate the determinant by first rewriting the determinant in triangular form.

41. $\begin{vmatrix} 2 & 4 & 1 \\ 1 & 2 & -1 \\ 1 & 2 & 2 \end{vmatrix}$ 0

42. $\begin{vmatrix} 4 & -4 & 6 \\ 3 & -2 & -1 \\ 1 & 2 & 4 \end{vmatrix}$ 76

43. $\begin{vmatrix} 1 & 2 & -1 \\ 2 & 3 & 1 \\ 3 & 4 & 3 \end{vmatrix}$ 0

44. $\begin{vmatrix} 1 & 2 & 5 \\ -1 & 1 & -2 \\ 3 & 1 & 10 \end{vmatrix}$ 0

45. $\begin{vmatrix} 0 & -1 & 1 \\ 1 & 0 & -2 \\ 2 & 2 & 0 \end{vmatrix}$ 6

46. $\begin{vmatrix} 2 & -1 & 3 \\ 1 & 1 & 1 \\ 3 & -4 & 5 \end{vmatrix}$ −1

47. $\begin{vmatrix} 1 & 2 & -1 & 2 \\ 1 & -2 & 0 & 3 \\ 3 & 0 & 1 & 5 \\ -2 & -4 & 1 & 6 \end{vmatrix}$ −90

48. $\begin{vmatrix} 1 & -1 & -1 & 2 \\ 0 & 2 & 4 & 6 \\ 1 & 1 & 4 & 12 \\ 1 & -1 & 0 & 8 \end{vmatrix}$ 4

49. $\begin{vmatrix} 1 & 2 & 3 & -1 \\ 6 & 5 & 9 & 8 \\ 2 & 4 & 12 & -1 \\ 1 & 2 & 6 & -1 \end{vmatrix}$ 21

50. $\begin{vmatrix} 1 & 2 & 0 & -2 \\ -1 & 1 & 3 & 5 \\ 2 & 1 & 4 & 0 \\ -2 & 5 & 2 & 6 \end{vmatrix}$ 0

In Exercises 51 to 54, use a graphing calculator to find the value of the determinant of the matrix. Where necessary, round your answer to the nearest thousandth.

51. $\begin{bmatrix} 2 & -2 & 3 & 1 \\ 5 & 2 & -2 & 3 \\ 6 & -1 & 2 & 3 \\ 2 & 3 & -1 & 5 \end{bmatrix}$ 3

52. $\begin{bmatrix} 3 & -1 & 0 & 1 \\ 2 & -2 & 3 & 0 \\ -1 & -3 & 5 & 3 \\ 5 & 3 & -2 & 1 \end{bmatrix}$ 140

36. The matrix is in triangular form. The product of the elements on the main diagonal is −15. Therefore, the value of the determinant is −15.

53. $\begin{bmatrix} -\frac{2}{7} & 4 & -\frac{1}{6} \\ -2 & \sqrt{2} & -3 \\ \sqrt{3} & 3 & -\sqrt{5} \end{bmatrix}$
-38.933

54. $\begin{bmatrix} 6 & \pi & -\frac{4}{7} \\ -5 & \sqrt{7} & 2 \\ \frac{5}{6} & -\sqrt{3} & \sqrt{10} \end{bmatrix}$
122.204

55. Surveyors use a formula to find the area of a plot of land. *Surveyor's Area Formula:* If the vertices (x_1, y_1), (x_2, y_2), $(x_3, y_3), \ldots, (x_n, y_n)$ of a simple polygon are listed counterclockwise around the perimeter, the area of the polygon is

$$A = \frac{1}{2}\left\{\begin{vmatrix} x_1 & x_2 \\ y_1 & y_2 \end{vmatrix} + \begin{vmatrix} x_2 & x_3 \\ y_2 & y_3 \end{vmatrix} + \begin{vmatrix} x_3 & x_4 \\ y_3 & y_4 \end{vmatrix} + \cdots + \begin{vmatrix} x_n & x_1 \\ y_n & y_1 \end{vmatrix}\right\}$$

Use the Surveyor's Area Formula to find the area of the polygon with vertices $(8, -4)$, $(25, 5)$, $(15, 9)$, $(17, 20)$, and $(0, 10)$. 263.5

56. Show that the determinant $\begin{vmatrix} x & y & 1 \\ x_1 & y_1 & 1 \\ x_2 & y_2 & 1 \end{vmatrix} = 0$ is the equation of a line through the points (x_1, y_1) and (x_2, y_2).

Solutions to Check Your Progress

Chapter P

Section P.1

Check Your Progress 1, *page 4*

a. Integers: $-13, 29, -12$

b. Rational numbers: $-13, 4.\overline{142}, \frac{5}{2}, 29, -12, 4.32789123409$

c. Irrational numbers: $\frac{2}{\sqrt{7}}, \pi$

d. Real numbers: $-13, 4.\overline{142}, \frac{5}{2}, 29, -12, \frac{2}{\sqrt{7}}, \pi,$ 4.32789123409

e. Prime numbers: 29

Check Your Progress 2, *page 6*

a. $A \cap C = \emptyset$

b. $B \cup C = \{-4, -2, 0, 2, 3, 4, 5, 6, 7\}$

c. $A \cup (B \cap C) = \{-2, -1, 0, 1, 2, 4, 6\}$

d. $B \cap (A \cup C) = \{-2, 0, 2, 4, 6\}$

Check Your Progress 3, *page 7*

$d(x, 4) = |x - 4|$

Check Your Progress 4, *page 8*

−5 −4 −3 −2 −1 0 1 2 3 4 5

$\{x | -4 < x < 3\}$

Check Your Progress 5, *page 9*

a. −5 −4 −3 −2 −1 0 1 2 3 4 5 $[-1, \infty)$

b. −5 −4 −3 −2 −1 0 1 2 3 4 5 $\{x | -1 \le x < 1\}$

Check Your Progress 6, *page 10*

$3ab - 4(2a - 3b)$

$$\begin{aligned} 3(4)(-3) - 4[2(4) - 3(-3)] &= 3(4)(-3) - 4[8 - (-9)] \\ &= 3(4)(-3) - 4(17) \\ &= -36 - 68 = -104 \end{aligned}$$

Check Your Progress 7, *page 12*

a. $6 - 3(5a - 4) = 6 - 15a + 12 = -15a + 18$

b. $$\begin{aligned} &-3(2a - 5b + 1) + 5(3a - b + 2) \\ &= -6a + 15b - 3 + 15a - 5b + 10 \\ &= 9a + 10b + 7 \end{aligned}$$

Section P.2

Check Your Progress 1, *page 16*

$m \cdot m^2 \cdot m^3 = m^{1+2+3} = m^6$

Check Your Progress 2, *page 16*

$$\begin{aligned} (12p^4q^3)(-3p^5q^2) &= -36p^{4+5}q^{3+2} \\ &= -36p^9q^5 \end{aligned}$$

Check Your Progress 3, *page 18*

$$\begin{aligned} (-2x^3y^7)^3 &= (-2)^{1\cdot 3}x^{3\cdot 3}y^{7\cdot 3} \\ &= (-2)^3x^9y^{21} = -8x^9y^{21} \end{aligned}$$

Check Your Progress 4, *page 18*

$$\begin{aligned} (-xy^4)(-2x^3y^2)^2 &= (-xy^4)[(-2)^{1\cdot 2}x^{3\cdot 2}y^{2\cdot 2}] \\ &= (-xy^4)[4x^6y^4] = -4x^7y^8 \end{aligned}$$

Check Your Progress 5, *page 19*

$\frac{a^7b^6}{ab^3} = a^{7-1}b^{6-3} = a^6b^3$

Check Your Progress 6, *page 20*

$$\begin{aligned} \frac{-35a^6b^{-2}}{25a^{-3}b^5} &= -\frac{7a^{6-(-3)}b^{-2-5}}{5} \\ &= -\frac{7a^9b^{-7}}{5} = -\frac{7a^9}{5b^7} \end{aligned}$$

Check Your Progress 7, *page 21*

$$\begin{aligned} \left(\frac{4a^3b^2}{2a^4b^2}\right)^{-3} &= (2a^{3-4}b^{2-2})^{-3} = (2a^{-1})^{-3} \\ &= 2^{-3}a^3 = \frac{a^3}{2^3} = \frac{a^3}{8} \end{aligned}$$

■ $b^{2-2} = b^0 = 1$

Check Your Progress 8, *page 21*

$$(-3ab)(2a^3b^{-2})^{-3} = (-3ab)(2^{-3}a^{-9}b^6)$$
$$= \frac{-3a^{1+(-9)}b^{1+6}}{2^3} = -\frac{3a^{-8}b^7}{8}$$
$$= -\frac{3b^7}{8a^8}$$

Check Your Progress 9, *page 23*

a. $(2.4 \times 10^{-9})(1.6 \times 10^3) = [2.4(1.6)] \times 10^{-9+3}$
$= 3.84 \times 10^{-6}$

b. $\dfrac{5.6 \times 10^{-10}}{8.0 \times 10^{-4}} = 0.7 \times 10^{-10-(-4)} = 0.7 \times 10^{-6}$
$= 7.0 \times 10^{-7}$

Section P.3

Check Your Progress 1, *page 27*

$(-4x^2 - 3xy + 2y^2) + (3x^2 - 4y^2) = -x^2 - 3xy - 2y^2$

Check Your Progress 2, *page 28*

$(5x^2 - 3x + 4) - (-6x^3 - 2x + 8)$
$= (5x^2 - 3x + 4) + (6x^3 + 2x - 8)$
$= 6x^3 + 5x^2 - x - 4$

Check Your Progress 3, *page 28*

Profit $=$ revenue $-$ cost
$= (-0.25n^2 + 180n) - (45n + 4500)$
$= -0.25n^2 + 135n - 4500$

To find the profit for selling 310 DVD burners, evaluate the expression for profit when $n = 310$.

$-0.25n^2 + 135n - 4500$

$-0.25(310)^2 + 135(310) - 4500 = 13325$

The profit is $13,325.

Check Your Progress 4, *page 29*

a. $(5x^2 - 4x)(-3x^2) = 5x^2(-3x^2) - 4x(-3x^2)$
$= -15x^4 + 12x^3$

b. $-y^3(4y^3 - 2y - 7)$
$= (-y^3)(4y^3) - (-y^3)(2y) - (-y^3)(7)$
$= -4y^6 + 2y^4 + 7y^3$

Check Your Progress 5, *page 30*

$$\begin{array}{r} 4x^3 - 3x^2 + x - 4 \\ 2x - 5 \\ \hline -20x^3 + 15x^2 - 5x + 20 \\ 8x^4 - 6x^3 + 2x^2 - 8x \\ \hline 8x^4 - 26x^3 + 17x^2 - 13x + 20 \end{array}$$

Check Your Progress 6, *page 31*

a. $(3z - 4)(2z - 5) = 6z^2 - 15z - 8z + 20$
$= 6z^2 - 23z + 20$

b. $(4a + 5b)(4a - 5b) = 16a^2 - 20ab + 20ab - 25b^2$
$= 16a^2 - 25b^2$

Check Your Progress 7, *page 31*

Area of a triangle is given by $A = \frac{1}{2}bh$, where $b = a + 2$ and $h = a + 2$. Therefore, $A = \frac{1}{2}(a + 2)(a + 2) = \frac{1}{2}(a^2 + 4a + 4) = \frac{1}{2}a^2 + 2a + 2$. When $a = 4$

Area $= \dfrac{1}{2}(4)^2 + 2(4) + 2 = \dfrac{1}{2}(16) + 8 + 2$
$= 8 + 8 + 2 = 18$

where is 4 feet, the area is 18 square feet.

Check Your Progress 8, *page 33*

$$\begin{array}{r|l} & 2x^2 + 5x + 15 \\ x - 3 & 2x^3 - x^2 + 0x + 5 \\ & \underline{2x^3 - 6x^2} \\ & 5x^2 + 0x + 5 \\ & \underline{5x^2 - 15x} \\ & 15x + 5 \\ & \underline{15x - 45} \\ & 50 \end{array}$$

Section P.4

Check Your Progress 1, *page 38*

$6a^3b^2 - 12a^2b + 72ab^3 = 6ab(a^2b - 2a + 12b^2)$

Check Your Progress 2, *page 39*

The two factors of -28 whose sum is 12 are 14 and -2.

$b^2 + 12b - 28 = (b - 2)(b + 14)$

Check Your Progress 3, *page 40*

$57y^2 + 4y - 28 = (3y - 2)(19y + 14)$

Check Your Progress 4, *page 41*

$a = 16, b = -8, c = -35.$

$b^2 - 4ac$

$$\begin{aligned}(-8)^2 - 4(16)(-35) &= 64 - (-2240)\\ &= 64 + 2240\\ &= 2304 = 48^2\end{aligned}$$

Because 2304 is a perfect square, the trinomial is factorable over the integers.

Check Your Progress 5, *page 42*

$81b^2 - 16c^2 = (9b)^2 - (4c)^2 = (9b - 4c)(9b + 4c)$

Check Your Progress 6, *page 43*

$b^2 - 24b + 144 = (b - 12)^2$

Check Your Progress 7, *page 43*

$b^3 + 64 = b^3 + 4^3 = (b + 4)(b^2 - 4b + 16)$

Check Your Progress 8, *page 44*

$$\begin{aligned}a^2y^2 - ay^3 + ac - cy &= ay^2(a - y) + c(a - y)\\ &= (ay^2 + c)(a - y)\end{aligned}$$

Check Your Progress 9, *page 45*

$$\begin{aligned}81y^4 - 16 &= (9y^2)^2 - 4^2 = (9y^2 - 4)(9y^2 + 4)\\ &= [(3y)^2 - 2^2](9y^2 + 4)\\ &= (3y - 2)(3y + 2)(9y^2 + 4)\end{aligned}$$

Section P.5

Check Your Progress 1, *page 49*

$$\frac{2x^2 - 5x - 12}{2x^2 + 5x + 3} = \frac{(2x + 3)(x - 4)}{(2x + 3)(x + 1)} = \frac{x - 4}{x + 1}$$

Check Your Progress 2, *page 50*

$$\frac{x^2 - 16}{x^2 + 7x + 12} \cdot \frac{x^2 - 4x - 21}{x^2 - 4x} = \frac{(x - 4)(x + 4)(x + 3)(x - 7)}{(x + 3)(x + 4)x(x - 4)} = \frac{x - 7}{x}$$

Check Your Progress 3, *page 51*

$$\begin{aligned}\frac{3y - 1}{3y + 1} - \frac{2y - 5}{y - 3} &= \frac{(3y - 1)(y - 3)}{(3y + 1)(y - 3)} - \frac{(2y - 5)(3y + 1)}{(y - 3)(3y + 1)}\\ &= \frac{(3y^2 - 10y + 3) - (6y^2 - 13y - 5)}{(3y + 1)(y - 3)}\\ &= \frac{-3y^2 + 3y + 8}{(3y + 1)(y - 3)}\end{aligned}$$

Check Your Progress 4, *page 52*

$$\frac{3 - \frac{2}{a}}{5 + \frac{3}{a}} = \frac{\left(3 - \frac{2}{a}\right)a}{\left(5 + \frac{3}{a}\right)a} = \frac{3a - 2}{5a + 3}$$

Check Your Progress 5, *page 53*

$$\begin{aligned}\frac{e^{-2} - f^{-1}}{ef} &= \frac{\frac{1}{e^2} - \frac{1}{f}}{ef} = \frac{f - e^2}{e^2f} \div \frac{ef}{1}\\ &= \frac{f - e^2}{e^2f} \cdot \frac{1}{ef} = \frac{f - e^2}{e^3f^2}\end{aligned}$$

Check Your Progress 6, *page 54*

a. $$\frac{v_1 + v_2}{1 + \frac{v_1v_2}{c^2}} = \frac{1.2 \times 10^8 + 2.4 \times 10^8}{1 + \frac{(1.2 \times 10^8)(2.4 \times 10^8)}{(6.7 \times 10^8)^2}} \approx 3.4 \times 10^8 \text{ mph}$$

b. $$\frac{v_1 + v_2}{1 + \frac{v_1 \cdot v_2}{c^2}} = \frac{c^2(v_1 + v_2)}{c^2\left(1 + \frac{v_1 \cdot v_2}{c^2}\right)} = \frac{c^2(v_1 + v_2)}{c^2 + v_1 \cdot v_2}$$

Section P.6

Check Your Progress 1, *page 59*

$$\begin{aligned}\left(\frac{16x^4y^2}{2x^{-2}y^5}\right)^{-1/3} &= (8x^{4-(-2)}y^{2-5})^{-1/3} = (8x^6y^{-3})^{-1/3}\\ &= 8^{-1/3}x^{6(-1/3)}y^{-3(-1/3)} = 8^{-1/3}x^{-2}y\\ &= \frac{y}{8^{1/3}x^2} = \frac{y}{2x^2}\end{aligned}$$

Check Your Progress 2, *page 62*

$$\begin{aligned}\sqrt[3]{-81x^6y^4} &= \sqrt[3]{(-27x^6y^3)3y}\\ &= \sqrt[3]{-27x^6y^3} \cdot \sqrt[3]{3y}\\ &= -3x^2y\sqrt[3]{3y}\end{aligned}$$

Check Your Progress 3, *page 63*

$$\sqrt[3]{54xy^3} + \sqrt[3]{128xy^3} = \sqrt[3]{(27y^3)2x} + \sqrt[3]{(64y^3)2x}$$
$$= 3y\sqrt[3]{2x} + 4y\sqrt[3]{2x}$$
$$= 7y\sqrt[3]{2x}$$

Check Your Progress 4, *page 63*

$$(3\sqrt{7} - 5)(2\sqrt{7} + 3) = (3\sqrt{7} - 5)2\sqrt{7} + (3\sqrt{7} - 5)3$$
$$= (6\sqrt{49} - 10\sqrt{7}) + (9\sqrt{7} - 15)$$
$$= 6 \cdot 7 - 10\sqrt{7} + 9\sqrt{7} - 15$$
$$= 42 - \sqrt{7} - 15 = 27 - \sqrt{7}$$

Check Your Progress 5, *page 64*

a. $\dfrac{3}{\sqrt[3]{9}} = \dfrac{3}{\sqrt[3]{3^2}}\left(\dfrac{\sqrt[3]{3}}{\sqrt[3]{3}}\right) = \dfrac{3\sqrt[3]{3}}{\sqrt[3]{3^3}} = \dfrac{3\sqrt[3]{3}}{3} = \sqrt[3]{3}$

b. $\sqrt{\dfrac{5}{18x}} = \dfrac{\sqrt{5}}{\sqrt{18x}} = \dfrac{\sqrt{5}}{\sqrt{9 \cdot 2x}} = \dfrac{\sqrt{5}}{\sqrt{9}\sqrt{2x}} = \dfrac{\sqrt{5}}{3\sqrt{2x}}\left(\dfrac{\sqrt{2x}}{\sqrt{2x}}\right)$

$= \dfrac{\sqrt{10x}}{3(2x)} = \dfrac{\sqrt{10x}}{6|x|}$

Check Your Progress 6, *page 64*

$$\frac{5}{2\sqrt{3} - 3} = \frac{5}{2\sqrt{3} - 3} \cdot \frac{2\sqrt{3} + 3}{2\sqrt{3} + 3} = \frac{10\sqrt{3} + 15}{12 - 9}$$
$$= \frac{10\sqrt{3} + 15}{3}$$

Section P.7

Check Your Progress 1, *page 69*

$$4 - \sqrt{-72} = 4 - i\sqrt{72} = 4 - i\sqrt{36}\sqrt{2} = 4 - 6i\sqrt{2}$$

Check Your Progress 2, *page 70*

a. $(-3 + 5i) + (-7 - 5i) = [-3 + (-7)] + [5 + (-5)]i$

$= -10$

b. $(6 - 3i) - (6 - 4i) = (6 - 6) + [-3 - (-4)]i = i$

Check Your Progress 3, *page 71*

a. $(4 + 3i)(3 + 5i) = 12 + 20i + 9i + 15i^2$

$= 12 + 29i + 15(-1)$

$= 12 + 29i - 15$

$= -3 + 29i$

b. $(5 - \sqrt{5})(2 - 3\sqrt{-5}) = (5 - i\sqrt{5})(2 - 3i\sqrt{5})$

$= 10 - 15i\sqrt{5} - 2i\sqrt{5} + 3i^2\sqrt{25}$

$= 10 - 17i\sqrt{5} + 3(-1)(5)$

$= 10 - 17i\sqrt{5} - 15$

$= -5 - 17i\sqrt{5}$

Check Your Progress 4, *page 73*

$$\frac{3 + 2i}{5 - i} = \frac{3 + 2i}{5 - i} \cdot \frac{5 + i}{5 + i} = \frac{15 + 3i + 10i + 2i^2}{5^2 + 1}$$
$$= \frac{15 + 13i + 2(-1)}{25 + 1} = \frac{15 + 13i - 2}{26}$$
$$= \frac{13 + 13i}{26} = \frac{13}{26} + \frac{13}{26}i = \frac{1}{2} + \frac{1}{2}i$$

Check Your Progress 5, *page 73*

$i^{214} = i^2 = -1$ ▪ Remainder of $214 \div 4$ is 2.

Chapter 1

Section 1.1

Check Your Progress 1, *page 84*

Solve the equation $350 + 1.6A = 3000$.

$350 + 1.6A = 3000$

$350 + 1.6A - 350 = 3000 - 350$ ▪ Subtract 350 from each side.

$1.6A = 2650$

$\dfrac{1.6A}{1.6} = \dfrac{2650}{1.6}$ ▪ Divide each side by 1.6.

$A = 1656.25$

Thus an office measuring 1656.25 square feet can have its carpet replaced for \$3000.

Check Your Progress 2, *page 85*

$$\frac{2}{3}x + 9 = 1$$
$$\frac{2}{3}x + 9 - 9 = 1 - 9$$
$$\frac{2}{3}x = -8$$
$$\left(\frac{3}{2}\right)\left(\frac{2}{3}x\right) = \left(\frac{3}{2}\right)(-8)$$
$$x = -12$$

Check Your Progress 3, *page 85*

$\dfrac{7x}{3} - \dfrac{2}{3} = \dfrac{3}{4} - \dfrac{5x}{3}$

$12\left(\dfrac{7x}{3} - \dfrac{2}{3}\right) = 12\left(\dfrac{3}{4} - \dfrac{5x}{3}\right)$ ▪ Multiply each side of the equation by 12, the LCD of the fractions.

$28x - 8 = 9 - 20x$

$48x = 17$

$x = \dfrac{17}{48}$

Check Your Progress 4, *page 86*

$$4x(x + 2) - 1 = (2x - 3)(2x + 1)$$
$$4x^2 + 8x - 1 = 4x^2 + 2x - 6x - 3$$
$$4x^2 + 8x - 1 = 4x^2 - 4x - 3$$
$$8x - 1 = -4x - 3$$
$$12x = -2$$
$$x = -\frac{1}{6}$$

Check Your Progress 5, *page 87*

$$\text{Number of patents (in thousands)} = 5.4x + 110$$
$$150 = 5.4x + 110$$
$$40 = 5.4x$$
$$x = \frac{40}{5.4}$$
$$x \approx 7.4$$

Because x is approximately 7.4, the year is $1993 + 7.4 = 2000.4$. The number of patents first exceeded 150,000 in the year 2000.

Check Your Progress 6, *page 88*

$$ac = 2b + 3c$$
$$ac - 3c = 2b$$ ■ Subtract $3c$ from each side so that all terms containing c are on the same side.
$$c(a - 3) = 2b$$ ■ Factor c from the left side of the equation.
$$c = \frac{2b}{a - 3}$$ ■ Divide each side by $a - 3$, $a - 3 \neq 0$.

Check Your Progress 7, *page 89*

$$P = 2l + 2w, \quad w = \frac{1}{2}l + 1$$
$$110 = 2l + 2\left(\frac{1}{2}l + 1\right)$$
$$110 = 2l + l + 2$$ ■ Simplify.
$$108 = 3l$$
$$36 = l$$
$$l = 36 \text{ meters}$$
$$w = \frac{1}{2}l + 1 = \frac{1}{2}(36) + 1 = 19 \text{ meters}$$

Check Your Progress 8, *page 90*

Let t_1 = the time it takes to travel to the island.

Let t_2 = the time it takes to make the return trip.

$$t_1 + t_2 = 7.5$$
$$t_2 = 7.5 - t_1$$
$$15t_1 = 10t_2$$
$$15t_1 = 10(7.5 - t_1)$$ ■ Substitute for t_2.
$$15t_1 = 75 - 10t_1$$
$$25t_1 = 75$$
$$t_1 = 3 \text{ hours}$$
$$D = 15t_1 = 15(3) = 45 \text{ nautical miles}$$

Check Your Progress 9, *page 91*

Let x = the amount of money invested at 5%. Then $7500 - x$ represents the amount invested at 7%.

5%	x
7%	$7500 - x$

$$0.05x + 0.07(7500 - x) = 405$$
$$0.05x + 525 - 0.07x = 405$$
$$-0.02x = -120$$
$$x = 6000$$
$$7500 - x = 1500$$

$6000 was invested at 5% and $1500 was invested at 7%.

Check Your Progress 10, *page 92*

Let x = the number of liters of the 40% solution to be mixed with the 24% solution.

0.40	x
0.24	4
0.30	$4 + x$

$$0.40x + 0.24(4) = 0.30(4 + x)$$
$$0.40x + 0.96 = 1.2 + 0.30x$$
$$0.10x = 0.24$$
$$x = 2.4$$

Thus 2.4 liters of 40% sulfuric acid solution should be mixed with 4 liters of 24% sulfuric acid solution to produce the 30% solution.

Check Your Progress 11, *page 93*

Let x = the number of hours needed to print the report if both the printers are used.

Printer A prints $\frac{1}{3}$ of the report every hour.

Printer B prints $\frac{1}{4}$ of the report every hour.

$$\frac{1}{3}x + \frac{1}{4}x = 1$$
$$4x + 3x = 12 \cdot 1$$
$$7x = 12$$
$$x = \frac{12}{7} \approx 1.71$$

If both printers are used, it should take approximately 1.71 hours to print the report.

Section 1.2

Check Your Progress 1, *page 99*

a.
$$6x^2 - x - 12 = 0$$
$$(3x + 4)(2x - 3) = 0$$
$$3x + 4 = 0 \qquad 2x - 3 = 0$$
$$3x = -4 \qquad 2x = 3$$
$$x = -\frac{4}{3} \qquad x = \frac{3}{2}$$

The solutions are $-\frac{4}{3}$ and $\frac{3}{2}$.

b.
$$6x^2 + x = 15$$
$$6x^2 + x - 15 = 0$$
$$(3x + 5)(2x - 3) = 0$$
$$3x + 5 = 0 \qquad 2x - 3 = 0$$
$$3x = -5 \qquad 2x = 3$$
$$x = -\frac{5}{3} \qquad x = \frac{3}{2}$$

The solutions are $-\frac{5}{3}$ and $\frac{3}{2}$.

Check Your Progress 2, *page 100*

a.
$$2x^2 - 128 = 0$$
$$2x^2 = 128$$
$$x^2 = 64$$
$$x = \pm\sqrt{64}$$
$$x = \pm 8$$

b.
$$(x - 3)^2 = 81$$
$$x - 3 = \pm\sqrt{81}$$
$$x = 3 \pm 9$$
$$x = -6 \text{ or } 12$$

Check Your Progress 3, *page 102*

$$x^2 + 6x - 5 = 0$$
$$x^2 + 6x = 5$$
$$x^2 + 6x + 9 = 5 + 9$$
$$(x + 3)^2 = 14$$
$$x + 3 = \pm\sqrt{14}$$
$$x = -3 \pm \sqrt{14}$$

Check Your Progress 4, *page 103*

$$4x^2 + 32x - 3 = 0$$
$$4x^2 + 32x = 3$$
$$\frac{1}{4}(4x^2 + 32x) = \frac{1}{4}(3)$$
$$x^2 + 8x = \frac{3}{4}$$
$$x^2 + 8x + 16 = \frac{3}{4} + 16$$
$$(x + 4)^2 = \frac{67}{4}$$
$$x + 4 = \pm\sqrt{\frac{67}{4}}$$
$$x = -4 \pm \frac{\sqrt{67}}{2} = \frac{-8 \pm \sqrt{67}}{2}$$

Check Your Progress 5, *page 104*

a. $12x^2 - x - 6 = 0$
$$x = \frac{-b \pm \sqrt{b^2 - 4ac}}{2a}$$
$$= \frac{-(-1) \pm \sqrt{(-1)^2 - 4(12)(-6)}}{2(12)}$$
$$= \frac{1 \pm \sqrt{289}}{24} = \frac{1 \pm 17}{24} = \frac{3}{4} \text{ or } -\frac{2}{3}$$

b. $x^2 - 2x + 2 = 0$
$$x = \frac{-(-2) \pm \sqrt{(-2)^2 - 4(1)(2)}}{2(1)}$$
$$= \frac{2 \pm \sqrt{-4}}{2} = \frac{2 \pm 2i}{2} = 1 \pm i$$

Check Your Progress 6, *page 105*

a. $b^2 - 4ac = (-5)^2 - 4(4)(3) = -23$. Because the discriminant is negative, $4x^2 - 5x + 3 = 0$ has no real solutions.

b. $b^2 - 4ac = (-23)^2 - 4(6)(20) = 49$. Because the discriminant is positive, $6x^2 - 23x + 20 = 0$ has two real solutions.

c. $b^2 - 4ac = (12)^2 - 4(4)(9) = 0$. Because the discriminant is 0, $4x^2 + 12x + 9 = 0$ has one real solution.

Check Your Progress 7, *page 107*

Let $4x$ represent the width of the screen and $3x$ represent the height of the screen. Use the Pythagorean Theorem.

$$(4x)^2 + (3x)^2 = 54^2$$
$$16x^2 + 9x^2 = 2916$$
$$25x^2 = 2916$$
$$x^2 = \frac{2916}{25}$$
$$x = \sqrt{\frac{2916}{25}} = 10.8 \text{ inches}$$

The height of the screen is $3(10.8) = 32.4$ inches and the width is $4(10.8) = 43.2$ inches.

Check Your Progress 8, *page 107*

$$lwh = V$$
$$(2.5w)(w)(0.5) = 4$$
$$1.25w^2 = 4$$
$$w^2 = 3.2$$
$$w = \sqrt{3.2}$$
$$\approx 1.8$$

The width of the new candy bar should be about 1.8 inches and the length about $2.5(1.8) = 4.5$ inches.

Check Your Progress 9, *page 108*

$$0 = -16t^2 + 88t + 5$$
$$t = \frac{-(88) \pm \sqrt{(88)^2 - 4(-16)(5)}}{2(-16)}$$
$$= \frac{-88 \pm \sqrt{8064}}{-32}$$
$$\approx 5.6 \text{ or } -0.1$$

The ball will hit the ground in about 5.6 seconds. Disregard the negative solution because the time must be positive.

Section 1.3

Check Your Progress 1, *page 114*

$$|3x + 4| = 16$$
$$3x + 4 = 16 \quad \text{or} \quad 3x + 4 = -16$$
$$3x = 12 \qquad\qquad 3x = -20$$
$$x = 4 \qquad\qquad x = -\frac{20}{3}$$

Check Your Progress 2, *page 115*

$$x^4 - 36x^2 = 0$$
$$x^2(x^2 - 36) = 0$$
$$x^2(x - 6)(x + 6) = 0$$
$$x = 0, x = 6, x = -6$$

Check Your Progress 3, *page 116*

$$\frac{2x}{x-2} + 3 = \frac{4}{x-2}$$
$$(x-2)\left(\frac{2x}{x-2} + 3\right) = (x-2)\left(\frac{4}{x-2}\right)$$

■ Multiply each side by $(x - 2)$, the LCD, with $x \neq 2$.

$$2x + 3(x - 2) = 4$$
$$2x + 3x - 6 = 4$$
$$5x = 10$$
$$x = 2$$

The proposed solution is $x = 2$, but this contradicts the restriction $x \neq 2$. The equation has no solution.

Check Your Progress 4, *page 117*

$$0.52 = \frac{A}{A + 12}$$
$$(A + 12) \cdot 0.52 = (A + 12) \cdot \frac{A}{A + 12}$$

■ Multiply both sides by $(A + 12)$.

$$0.52A + 6.24 = A$$
$$0.52A + 6.24 - 0.52A = A - 0.52A$$
$$6.24 = 0.48A$$
$$\frac{6.24}{0.48} = \frac{0.48A}{0.48}$$
$$13 = A$$

A 13-year-old child would require 52% of an adult dose.

Check Your Progress 5, *page 117*

$$\sqrt{x - 5} = 6$$
$$\left(\sqrt{x - 5}\right)^2 = 6^2$$
$$x - 5 = 36$$
$$x = 41$$

Check: $\sqrt{41 - 5} = \sqrt{36} = 6$

Check Your Progress 6, *page 118*

$$x = \sqrt{5 - x} + 5$$
$$x - 5 = \sqrt{5 - x}$$
$$(x - 5)^2 = \left(\sqrt{5 - x}\right)^2$$
$$x^2 - 10x + 25 = 5 - x$$
$$x^2 - 9x + 20 = 0$$
$$(x - 5)(x - 4) = 0$$
$$x - 5 = 0 \quad \text{or} \quad x - 4 = 0$$
$$x = 5 \qquad\qquad x = 4$$

Check: $5 \stackrel{?}{=} \sqrt{5 - 5} + 5$ $\quad$ $4 \stackrel{?}{=} \sqrt{5 - 4} + 5$

$5 \stackrel{?}{=} 0 + 5$ $\quad$ $4 \stackrel{?}{=} 1 + 5$

$5 = 5$ $\quad$ $4 \neq 6$

The solution is $x = 5$.

Check Your Progress 7, *page 119*

$$5000 = \sqrt{\frac{(6.67 \times 10^{-11})(5.98 \times 10^{24})}{r}}$$
$$(5000)^2 = \left(\sqrt{\frac{(6.67 \times 10^{-11})(5.98 \times 10^{24})}{r}}\right)^2$$
$$25{,}000{,}000 = \frac{(6.67 \times 10^{-11})(5.98 \times 10^{24})}{r}$$
$$25{,}000{,}000r = (6.67 \times 10^{-11})(5.98 \times 10^{24})$$
$$r = \frac{(6.67 \times 10^{-11})(5.98 \times 10^{24})}{25{,}000{,}000}$$
$$r \approx 16{,}000{,}000$$

The radius of the satellite's orbit is approximately 16 million meters.

Check Your Progress 8, *page 120*

$$(4z + 7)^{1/3} = 2$$
$$[(4z + 7)^{1/3}]^3 = 2^3$$
$$4z + 7 = 8$$
$$4z = 1$$
$$z = \frac{1}{4}$$

Check: $\left[4\left(\frac{1}{4}\right) + 7\right]^{1/3} \stackrel{?}{=} 2$

$$8^{1/3} \stackrel{?}{=} 2$$
$$2 = 2$$

The solution is $\frac{1}{4}$.

Check Your Progress 9, *page 121*

$x^4 - 10x^2 + 9 = 0$ ■ Let $u = x^2$.

$$u^2 - 10u + 9 = 0$$
$$(u - 9)(u - 1) = 0$$

$u - 9 = 0$ or $u - 1 = 0$

$u = 9$ $\qquad$ $u = 1$

$x^2 = 9$ $\qquad$ $x^2 = 1$

$x = \pm 3$ $\qquad$ $x = \pm 1$

The solutions are 3, −3, 1, and −1.

Check Your Progress 10, *page 122*

$6x^{2/3} - 7x^{1/3} - 20 = 0$ ■ Let $u = x^{1/3}$.

$$6u^2 - 7u - 20 = 0$$
$$(3u + 4)(2u - 5) = 0$$

$3u + 4 = 0$ or $2u - 5 = 0$

$u = -\frac{4}{3}$ $\qquad$ $u = \frac{5}{2}$

$x^{1/3} = -\frac{4}{3}$ $\qquad$ $x^{1/3} = \frac{5}{2}$

$(x^{1/3})^3 = \left(-\frac{4}{3}\right)^3$ $\qquad$ $(x^{1/3})^3 = \left(\frac{5}{2}\right)^3$

$x = -\frac{64}{27}$ $\qquad$ $x = \frac{125}{8}$

The solutions are $-\frac{64}{27}$ and $\frac{125}{8}$.

Section 1.4

Check Your Progress 1, *page 129*

a. $3x - 2 > 4$

$$3x > 6$$
$$x > 2$$

The solution set is $\{x|x > 2\}$, or $(2, \infty)$.

b. $-2x + 7 \le 5$

$$-2x \le -2$$
$$x \ge 1$$

The solution set is $\{x|x \ge 1\}$, or $[1, \infty)$.

Check Your Progress 2, *page 130*

a. $x + 5 > 11$ and $2x + 3 > 7$

$x > 6$ $\qquad$ $2x > 4$

$x > 2$

The solution set is the intersection of $\{x|x > 6\}$ and $\{x|x > 2\}$. In set-builder notation the solution set is $\{x|x > 6\}$. In interval notation the solution set is $(6, \infty)$.

b. $3x < 12$ or $x + 4 > 7$

$x < 4$ $\qquad$ $x > 3$

The solution set is the union of $\{x|x < 4\}$ and $\{x|x > 3\}$. Thus the solution set is the set of all real numbers. In set-builder notation the solution set is $\{x|x \in \text{real numbers}\}$. In interval notation the solution set is $(-\infty, \infty)$.

Check Your Progress 3, *page 132*

a. $|2x - 1| \le 6$ provided $-6 \le 2x - 1 \le 6$.

$$-6 \le 2x - 1 \le 6$$
$$-5 \le 2x \le 7$$
$$-\frac{5}{2} \le x \le \frac{7}{2}$$

The solution set is $\left[-\frac{5}{2}, \frac{7}{2}\right]$.

b. $|3x + 2| > 10$ provided $3x + 2 < -10$ or $3x + 2 > 10$.

$$3x + 2 < -10 \quad \text{or} \quad 3x + 2 > 10$$
$$3x < -12 \qquad\qquad 3x > 8$$
$$x < -4 \qquad\qquad x > \frac{8}{3}$$

The solution set is $(-\infty, -4) \cup \left(\frac{8}{3}, \infty\right)$.

Check Your Progress 4, *page 133*

Let x represent the number of copies the school will produce in the next 3 years. Leasing will be less expensive provided:

$$150 \cdot 36 + 0.015x < 12{,}400 + 0.005x$$
$$0.010x < 7000$$
$$x < 700{,}000$$

Leasing will be less expensive provided the school produces fewer than 700,000 copies.

Check Your Progress 5, *page 133*

Let x represent Natasha's fourth test score.

$$\frac{85 + 64 + 92 + x}{4} \geq 84$$
$$241 + x \geq 336$$
$$x \geq 95$$

The range of test scores that Natasha can receive on her fourth test to produce an average of at least 84 is $95 \leq x \leq 100$.

Check Your Progress 6, *page 134*

Solving $C = \frac{5}{9}(F - 32)$ for F produces $F = \frac{9}{5}C + 32$.

$$41 \leq \frac{9}{5}C + 32 \leq 68$$
$$9 \leq \frac{9}{5}C \leq 36$$
$$5 \leq C \leq 20$$

The corresponding temperature range on the Celsius scale is 5°C to 20°C.

Section 1.5

Check Your Progress 1, *page 141*

a. Factor the polynomial $x^2 - 4x + 3$ to produce $(x - 1)(x - 3) \leq 0$. Draw a sign diagram for each linear factor.

$(x - 1)$	− − − − − − 0 + + +
$(x - 3)$	− − − − − − − − 0 + +
	−5 −4 −3 −2 −1 0 1 2 3 4 5

The diagram shows that the only interval on which $(x - 1)$ and $(x - 3)$ have opposite signs is $(1, 3)$. The solution set is $[1, 3]$. The critical numbers 1 and 3 are included in the solution set because they satisfy the original inequality.

b. Rewrite $2x^2 - x > 3$ as $2x^2 - x - 3 > 0$. Factor to produce $(2x - 3)(x + 1) > 0$. Draw a sign diagram for each of the linear factors. Examine the sign diagram to determine that $(2x - 3)$ and $(x + 1)$ both have the same sign on the intervals $(-\infty, -1)$ and $\left(\frac{3}{2}, \infty\right)$. The solution set is $(-\infty, -1) \cup \left(\frac{3}{2}, \infty\right)$.

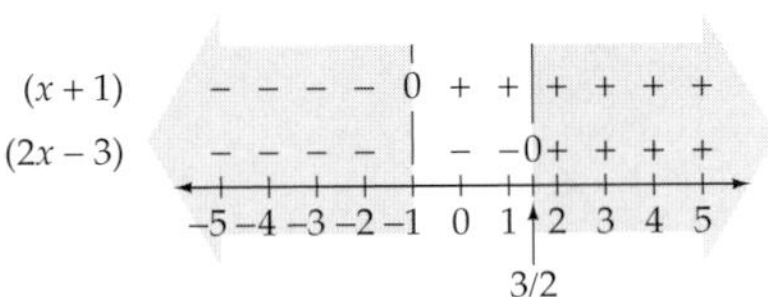

Check Your Progress 2, *page 142*

Rewrite $x^2 - 4x \leq 1$ as $x^2 - 4x - 1 \leq 0$. Use the quadratic formula to find the real zeros of $x^2 - 4x - 1$, which are $2 \pm \sqrt{5}$. Sketch a number line that shows the test intervals formed by these real zeros.

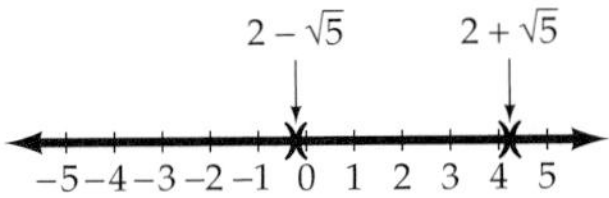

Pick a convenient test value from each of the test intervals. For instance, we picked -1, 0, and 5 as our test values. Evaluate $x^2 - 4x - 1$ for each test value.

- For $x = -1$, we have $x^2 - 4x - 1 = (-1)^2 - 4(-1) - 1 = 4$, which is positive.
- For $x = 0$, we have $x^2 - 4x - 1 = (0)^2 - 4(0) - 1 = -1$, which is negative.
- For $x = 5$, we have $x^2 - 4x - 1 = (5)^2 - 4(5) - 1 = 4$, which is positive.

These results show that $x^2 - 4x - 1$ is negative for all values of x between $2 - \sqrt{5}$ and $2 + \sqrt{5}$. Thus the solution set of $x^2 - 4x \leq 1$ is $\left[2 - \sqrt{5}, 2 + \sqrt{5}\right]$. The critical numbers $2 - \sqrt{5}$ and $2 + \sqrt{5}$ are included in the solution set because they satisfy the original inequality.

Check Your Progress 3, *page 144*

Write the inequality so that 0 appears on one side of the inequality.

$$\frac{3x+1}{x-2} \geq 4$$

$$\frac{3x+1}{x-2} - 4 \geq 0$$

Write the left side as a single rational expression.

$$\frac{3x+1}{x-2} - \frac{4(x-2)}{x-2} \geq 0$$ ▪ **Rewrite with $(x - 2)$ as the common denominator.**

$$\frac{3x+1-4x+8}{x-2} \geq 0$$ ▪ **Simplify.**

$$\frac{-x+9}{x-2} \geq 0$$

The following sign diagram shows that for all values of x on the interval $(2, 9)$ the quotient $(-x + 9)/(x - 2)$ is positive. On the other intervals the quotient $(-x + 9)/(x - 2)$ is negative.

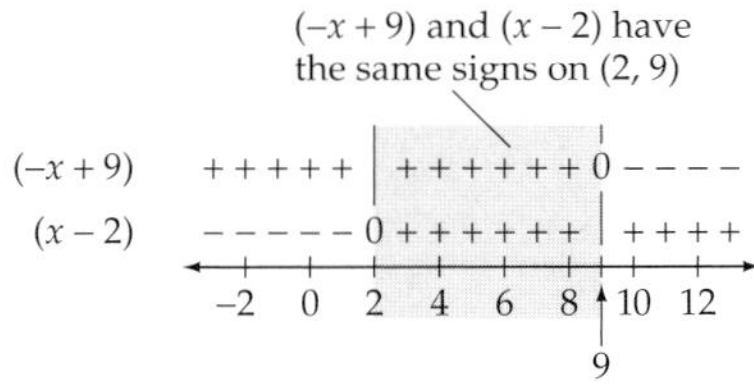

The solution set is $(2, 9]$. The number 9 is included in the solution set because $(-x + 9)/(x - 2) = 0$ for $x = 9$. However, 2 is not included in the solution set because the denominator $x - 2$ is zero for $x = 2$.

Check Your Progress 4, *page 145*

Write the inequality so that a single rational expression appears on one side and 0 appears on the other side of the inequality.

$$\frac{x^2-6x+2}{x-3} > 2$$

$$\frac{x^2-6x+2}{x-3} - 2 > 0$$

$$\frac{x^2-6x+2-2(x-3)}{x-3} > 0$$

$$\frac{x^2-8x+8}{x-3} > 0$$

The real zero of the denominator is 3. Use the quadratic formula to find the real zeros of the numerator, which are $4 \pm 2\sqrt{2}$. The critical values of the inequality are $4 - 2\sqrt{2}$, 3, and $4 + 2\sqrt{2}$. Sketch a number line that shows the test intervals formed by these critical values.

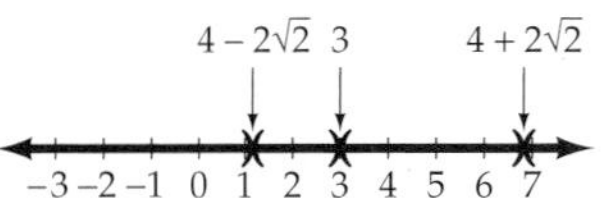

Pick a test value from each of the test intervals. For instance, we picked, 0, 2, 4, and 7 as our test values. Evaluate the rational expression $\frac{x^2-8x+8}{x-3}$ for each test value.

- For $x = 0$, the rational expression equals $-\frac{8}{3}$, which is *negative*. Thus the interval $\left(-\infty, 4 - 2\sqrt{2}\right)$ is not part of the solution set.
- For $x = 2$, the rational expression equals 4, which is *positive*. Thus the interval $\left(4 - 2\sqrt{2}, 3\right)$ is part of the solution set.
- For $x = 4$, the rational expression equals -8, which is *negative*. Thus the interval $\left(3, 4 + 2\sqrt{2}\right)$ is not part of the solution set.
- For $x = 7$, the rational expression equals $\frac{1}{4}$, which is *positive*. Thus the interval $\left(4 + 2\sqrt{2}, \infty\right)$ is part of the solution set.

The solution set is $\left(4 - 2\sqrt{2}, 3\right) \cup \left(4 + 2\sqrt{2}, \infty\right)$.

Check Your Progress 5, *page 146*

Let x be the number of additional at-bats that Sosa takes over 163. During this period his batting average will be $\frac{53+x}{163+x}$, and we wish to solve $\frac{53+x}{163+x} > 0.350$. In this application we know that $163 + x$ is positive. Multiplying each side of the inequality by $163 + x$ yields $53 + x > 57.05 + 0.350x$. Solving this inequality produces:

$$53 + x > 57.05 + 0.350x$$

$$0.650x > 4.05$$

$$x > 6.2308$$

During this hitting streak Sosa needs 7 or more at-bats to raise his average above 0.350.

Check Your Progress 6, *page 147*

Solve $\frac{2200x}{5x^2+16} > 100$ for x. Because $5x^2 + 16 > 0$, we can produce an equivalent inequality by multiplying both sides of the inequality by $5x^2 + 16$.

$$\frac{2200x}{5x^2+16} > 100$$

$$2200x > 500x^2 + 1600$$ ▪ **Multiply each side by $5x^2 + 16$.**

$$0 > 500x^2 - 2200x + 1600$$

$$0 > 5x^2 - 22x + 16$$ ▪ **Divide each term by 100.**

Use the quadratic formula to find the critical values of $0 = 5x^2 - 22x + 16$.

$$x = \frac{-(-22) \pm \sqrt{(-22)^2 - 4(5)(16)}}{2(5)}$$
$$= \frac{22 \pm \sqrt{164}}{10}$$
$$= \frac{22 \pm 2\sqrt{41}}{10}$$
$$= \frac{11 \pm \sqrt{41}}{5} \approx 3.5 \text{ or } 0.9$$

The region of the city in which the population density exceeds 100 people per square mile is the region that is more than 0.9 mile but less than 3.5 miles from the center of the city.

Check Your Progress 7, *page 149*

The profit will be at least \$550,000 provided

$$-0.01x^2 + 168x - 120{,}000 \geq 550{,}000$$
$$-0.01x^2 + 168x - 670{,}000 \geq 0$$

Use the quadratic formula to find that the approximate critical values of this inequality are 6513.2 and 10,286.8. Test values show that the inequality is positive only on the interval (6513.2, 10,286.8). The company should manufacture at least 6514 tennis racquets but not more than 10,286 tennis racquets to produce the desired profit.

Chapter 2

Section 2.1

Check Your Progress 1, *page 165*

Use the distance formula. The points are $P_1(-4, 0)$ and $P_2(-2, 5)$.

$$d = \sqrt{(x_1 - x_2)^2 + (y_1 - y_2)^2}$$
$$d = \sqrt{(-4 + 2)^2 + (0 - 5)^2} = \sqrt{(-2)^2 + (-5)^2}$$
$$= \sqrt{29} \approx 5.39$$

Check Your Progress 2, *page 166*

Use the midpoint formula with $P_1(-3, 4)$ and $P_2(4, -4)$.

$$x_m = \frac{x_1 + x_2}{2} = \frac{-3 + 4}{2} = \frac{1}{2} \qquad y_m = \frac{y_1 + y_2}{2} = \frac{4 + (-4)}{2} = \frac{0}{2} = 0$$

The coordinates of the midpoint are $\left(\frac{1}{2}, 0\right)$.

Check Your Progress 3, *page 168*

Select various values of x and calculate the corresponding values of y. Plot the ordered pairs. Then draw a smooth curve through the points.

x	$-2x + 3 = y$	(x, y)
-2	$-2(-2) + 3 = 7$	$(-2, 7)$
-1	$-2(-1) + 3 = 5$	$(-1, 5)$
0	$-2(0) + 3 = 3$	$(0, 3)$
1	$-2(1) + 3 = 1$	$(1, 1)$
2	$-2(2) + 3 = -1$	$(2, -1)$
3	$-2(3) + 3 = -3$	$(3, -3)$

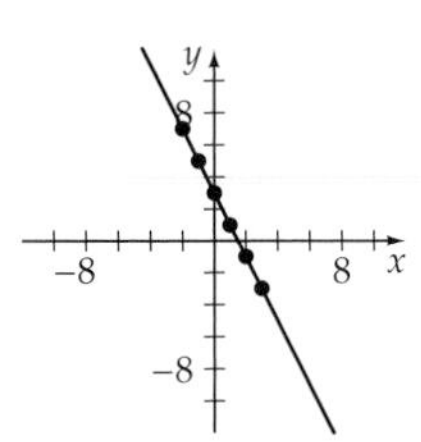

Check Your Progress 4, *page 169*

Select various values of x and calculate the corresponding values of y. Plot the ordered pairs. After the ordered pairs have been graphed, draw a smooth curve through the points.

x	$x^2 - 1 = y$	(x, y)
-3	$(-3)^2 - 1 = 8$	$(-3, 8)$
-2	$(-2)^2 - 1 = 3$	$(-2, 3)$
-1	$(-1)^2 - 1 = 0$	$(-1, 0)$
0	$(0)^2 - 1 = -1$	$(0, -1)$
1	$(1)^2 - 1 = 0$	$(1, 0)$
2	$(2)^2 - 1 = 3$	$(2, 3)$
3	$(3)^2 - 1 = 8$	$(3, 8)$

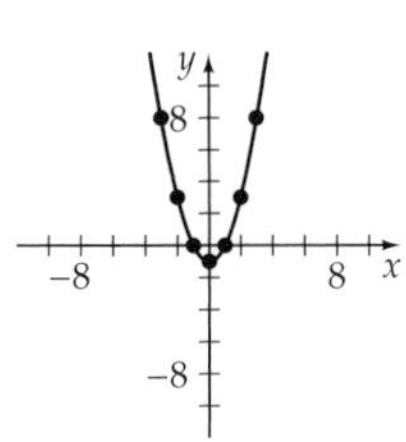

Check Your Progress 5, *page 171*

Because the point P is on the circle shown in the graph at the right, the radius r of the circle equals the distance from C to P. Use the distance formula.

$$r = \sqrt{[(-1) - (-4)]^2 + [2 - (-2)]^2}$$
$$= \sqrt{3^2 + 4^2} = \sqrt{9 + 16}$$
$$= \sqrt{25} = 5$$

Writing the standard form of the equation of a circle with $h = -4$, $k = -2$, and $r = 5$, we have

$$(x + 4)^2 + (y + 2)^2 = 25$$

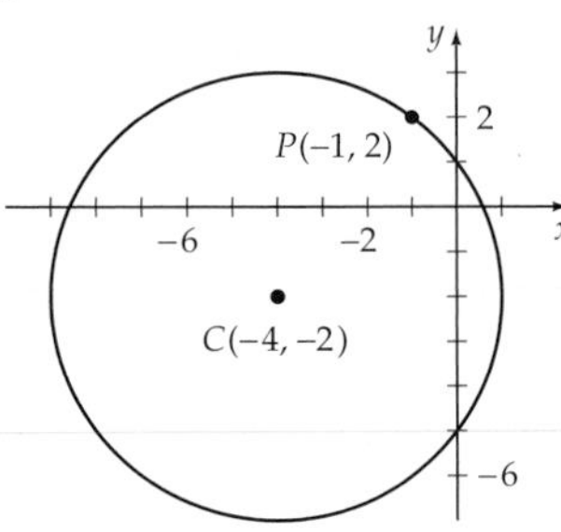

Check Your Progress 6, *page 172*

First rearrange and group the terms as shown below.

$(x^2 + 8x) + (y^2 - 5y) = -12$

Now complete the squares of $x^2 + 8x$ and $y^2 - 5y$.

$$(x^2 + 8x) + (y^2 - 5y) = -12$$
$$(x^2 + 8x + 16) + \left(y^2 - 5y + \frac{25}{4}\right) = -12 + 16 + \frac{25}{4}$$
$$(x + 4)^2 + \left(y - \frac{5}{2}\right)^2 = \frac{41}{4}$$
$$[x - (-4)]^2 + \left(y - \frac{5}{2}\right)^2 = \left(\frac{\sqrt{41}}{2}\right)^2$$

The last equation is in standard form. From this equation, the center is $C\left(-4, \frac{5}{2}\right)$ and the radius is $\frac{\sqrt{41}}{2}$.

Section 2.2

Check Your Progress 1, *page 178*

The domain is the set of first coordinates of the ordered pairs.

Domain: {1, 2, 3, 4, 5, 6, 7}

The range is the set of second coordinates of the ordered pairs.

Range: {1}

No two ordered pairs have the same first coordinate. The relation is a function.

Check Your Progress 2, *page 179*

$$f(z) = 2z^3 - 4z$$
$$f(-4) = 2(-4)^3 - 4(-4) = -128 + 16$$
$$= -112$$

Check Your Progress 3, *page 181*

Find some ordered pairs of the function by evaluating the function at various values of x.

x	$f(x) = -x^2 + 2$	(x, y)
−3	$f(-3) = -(-3)^2 + 2 = -7$	(−3, −7)
−2	$f(-2) = -(-2)^2 + 2 = -2$	(−2, −2)
−1	$f(-1) = -(-1)^2 + 2 = 1$	(−1, 1)
0	$f(0) = -(0)^2 + 2 = 2$	(0, 2)
1	$f(1) = -(1)^2 + 2 = 1$	(1, 1)
2	$f(2) = -(2)^2 + 2 = -2$	(2, −2)
3	$f(3) = -(3)^2 + 2 = -7$	(3, −7)

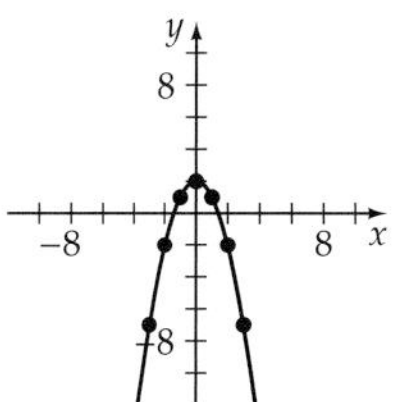

Check Your Progress 4, *page 181*

Find some ordered pairs of the function by evaluating the function at various values of x.

x	$g(x) = 2\|x\| + 4$	(x, y)
−3	$g(-3) = 2\|-3\| + 4 = 10$	(−3, 10)
−2	$g(-2) = 2\|-2\| + 4 = 8$	(−2, 8)
−1	$g(-1) = 2\|-1\| + 4 = 6$	(−1, 6)
0	$g(0) = 2\|0\| + 4 = 4$	(0, 4)
1	$g(1) = 2\|1\| + 4 = 6$	(1, 6)
2	$g(2) = 2\|2\| + 4 = 8$	(2, 8)
3	$g(3) = 2\|3\| + 4 = 10$	(3, 10)

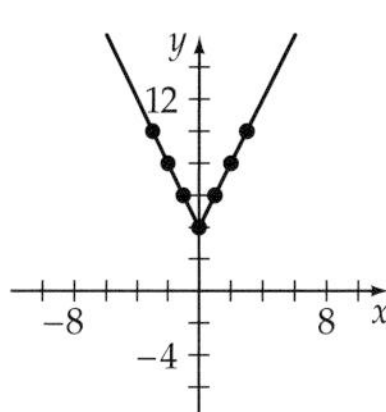

Check Your Progress 5, *page 182*

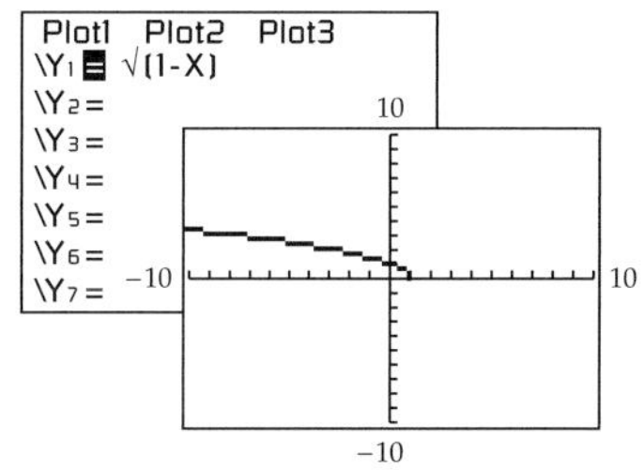

Check Your Progress 6, *page 183*

For the square root function, the radicand must be a non-negative number. Therefore, $2x - 6 \geq 0$. Solving this inequality, we have

$$2x - 6 \geq 0$$
$$2x \geq 6$$
$$x \geq 3$$

The domain of g in interval notation is $[3, \infty)$.

Check Your Progress 7, *page 184*

a. No vertical line intersects the graph at more than one point. The graph is the graph of a function.

b. There are vertical lines that intersect the graph at more than one point. The graph is not the graph of a function.

Check Your Progress 8, *page 186*

To find the total number of line segments that can be drawn between the 12 points, evaluate $N(m) = \dfrac{m(m-1)}{2}$ when $m = 12$.

$$N(m) = \frac{m(m-1)}{2}$$

$$N(12) = \frac{12(12-1)}{2} = 66$$

66 line segments can be drawn between 12 points.

Check Your Progress 9, *page 187*

Profit = revenue − cost. Thus,

$$P(x) = (13x - 0.015x^2) - (5x + 3)$$
$$= -0.015x^2 + 8x - 3$$

The profit function is $P(x) = -0.015x^2 + 8x - 3$.

Section 2.3

Check Your Progress 1, *page 196*

To find the x-intercept, let $y = 0$ and solve for x.

$$y = \frac{1}{2}x + 3$$
$$0 = \frac{1}{2}x + 3$$
$$-3 = \frac{1}{2}x$$
$$-6 = x$$

To find the y-intercept, let $x = 0$ and solve for y.

$$y = \frac{1}{2}x + 3$$
$$y = \frac{1}{2}(0) + 3$$
$$y = 3$$

The x-intercept is $(-6, 0)$; the y-intercept is $(0, 3)$.

Check Your Progress 2, *page 197*

To find the t-intercept, let $s = 0$ and solve for t.

$$s = -20t + 8000$$
$$0 = -20t + 8000$$
$$-8000 = -20t$$
$$400 = t$$

To find the s-intercept, let $t = 0$ and solve for s.

$$s = -20t + 8000$$
$$s = -20(0) + 8000$$
$$s = 8000$$

The t-intercept is $(400, 0)$. This means that the plane will land in 400 seconds. The s-intercept is $(0, 8000)$. This means that the plane was 8000 feet above the airport before starting to descend.

Check Your Progress 3, *page 199*

a. $m = \dfrac{-5 - 5}{4 - (-6)} = \dfrac{-10}{10} = -1$

b. $m = \dfrac{7 - 0}{-5 - (-5)} = \dfrac{7}{0}$ Undefined

c. $m = \dfrac{8 - (-2)}{8 - (-7)} = \dfrac{10}{15} = \dfrac{2}{3}$

d. $m = \dfrac{7 - 7}{1 - (-6)} = \dfrac{0}{7} = 0$

Check Your Progress 4, *page 200*

Place a dot at the y-intercept, $(0, -5)$. The slope is $m = \frac{3}{4}$. From the y-intercept, move 3 units up and then 4 units right. Place a dot at the new point and draw a line through the two points.

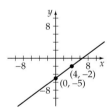

Check Your Progress 5, *page 201*

Place a dot at $P(2, 4)$. The slope is -1. $m = -1 = \dfrac{-1}{1}$.

From P, move 1 unit down and then 1 unit right. Place at dot at the new point and draw a line through the two points.

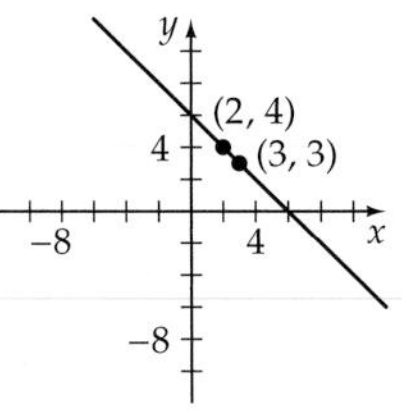

Check Your Progress 6, *page 202*

From the equation $d = 50t$, the slope is 50.

The slope means that the speed of the homing pigeon is 50 miles per hour.

Check Your Progress 7, *page 204*

To find the x-intercept, let $y = 0$ and solve for x.

$$3x + y = 6$$
$$3x + 0 = 6$$
$$3x = 6$$
$$x = 2$$

The x-intercept is (2, 0).

To find the y-intercept, let $x = 0$ and solve for y.

$$3x + y = 6$$
$$3(0) + y = 6$$
$$y = 6$$

The y-intercept is (0, 6).

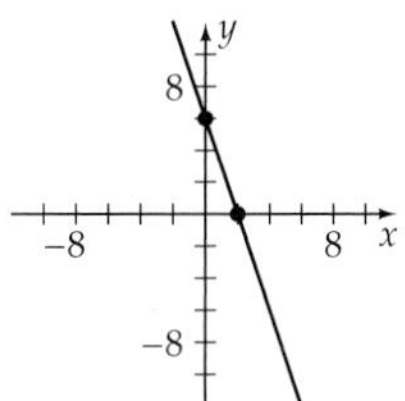

Check Your Progress 8, *page 205*

Solving $x - 1 = 0$ for x, we have $x = 1$. The graph of this equation is a vertical line through (1, 0).

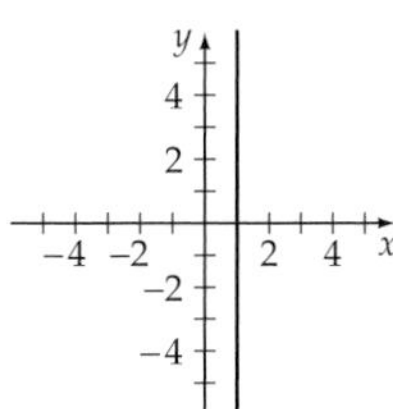

Section 2.4

Check Your Progress 1, *page 212*

The y-intercept is (0, 100) because the boiling point at sea level is 100°C. The temperature decreases 3.5°C per 1 kilometer increase in altitude. Therefore, the slope is −3.5. The equation is $y = -3.5x + 100$.

Check Your Progress 2, *page 212*

Use the point-slope formula.

$$y - y_1 = m(x - x_1)$$
$$y - 2 = -\frac{1}{2}[x - (-2)]$$
$$y - 2 = -\frac{1}{2}x - 1$$
$$y = -\frac{1}{2}x + 1$$

Check Your Progress 3, *page 213*

The given point is (1000, 3500) and $m = 2.5$. Use the point-slope formula to find the equation of the line. We will use n (number of bottles) and C (total cost) rather than x and y.

$$C - C_1 = m(n - n_1)$$
$$C - 3500 = 2.5(n - 1000)$$
$$C - 3500 = 2.5n - 2500$$
$$C = 2.5n + 1000$$

Check Your Progress 4, *page 214*

Find the slope. Then use the point-slope formula.

$$m = \frac{1 - 3}{4 - (-2)} = \frac{-2}{6} = -\frac{1}{3}$$

$$y - y_1 = m(x - x_1)$$
$$y - 3 = -\frac{1}{3}[x - (-2)]$$
$$y - 3 = -\frac{1}{3}x - \frac{2}{3}$$
$$y = -\frac{1}{3}x + \frac{7}{3}$$

Check Your Progress 5, *page 215*

a. Let $x = 0$ correspond to February. Then (0, 27,000) and (1, 26,700) are two points on the line that models the value of the car. Use these points to find the slope of the line and then use the point-slope formula.

$$m = \frac{26{,}700 - 27{,}000}{1 - 0} = -300$$

$$y - y_1 = m(x - x_1)$$
$$y - 27{,}000 = -300(x - 0)$$
$$y - 27{,}000 = -300x$$
$$y = -300x + 27{,}000$$

b. The value of the car is decreasing by \$300 each month.

c. Replace x by 11 (January of next year).

$$y = -300x + 27{,}000$$
$$= -300(11) + 27{,}000 = 23{,}700$$

Next January, the value of the car will be \$23,700.

Check Your Progress 6, *page 215*

Solve $3x + 5y = 15$ for y.

$$3x + 5y = 15$$
$$5y = -3x + 15$$
$$y = -\frac{3}{5}x + 3$$

Parallel lines have the same slope. Use the point-slope formula with $m = -\frac{3}{5}$ and $P(-2, 3)$.

$$y - y_1 = m(x - x_1)$$
$$y - 3 = -\frac{3}{5}[x - (-2)]$$
$$y - 3 = -\frac{3}{5}x - \frac{6}{5}$$
$$y = -\frac{3}{5}x + \frac{9}{5}$$

Check Your Progress 7, *page 216*

Solve $5x - 3y = 15$ for y.

$$5x - 3y = 15$$
$$-3y = -5x + 15$$
$$y = \frac{5}{3}x - 5$$

The slopes of perpendicular lines are negative reciprocals of each other. Use the point-slope formula with $m = -\frac{3}{5}$ and $P(-2, -3)$.

$$y - y_1 = m(x - x_1)$$
$$[y - (-3)] = -\frac{3}{5}[x - (-2)]$$
$$y + 3 = -\frac{3}{5}x - \frac{6}{5}$$
$$y = -\frac{3}{5}x - \frac{21}{5}$$

Check Your Progress 8, *page 217*

$m = \frac{8 - 0}{2 - 0} = \frac{8}{2} = 4$. The slope of the line through (0, 0) and (2, 8) is 4. The path of the ball is perpendicular to this line and therefore the slope is $-\frac{1}{4}$. Use the point-slope formula with $m = -\frac{1}{4}$ and (2, 8).

$$y - y_1 = m(x - x_1)$$
$$y - 8 = -\frac{1}{4}(x - 2)$$
$$y - 8 = -\frac{1}{4}x + \frac{1}{2}$$
$$y = -\frac{1}{4}x + \frac{17}{2}$$

Section 2.5

Check Your Progress 1, *page 223*

$$-\frac{b}{2a} = -\frac{0}{2(1)} = 0$$
$$f(x) = x^2 - 2$$
$$f(0) = 0^2 - 2$$
$$y = -2$$

Vertex: $(0, -2)$. Axis of symmetry: $x = 0$

Check Your Progress 2, *page 224*

To find the x-intercepts, let $y = 0$ and solve for x.

a.
$$y = 2x^2 - 5x + 2$$
$$0 = 2x^2 - 5x + 2$$
$$0 = (2x - 1)(x - 2)$$
$$2x - 1 = 0 \qquad x - 2 = 0$$
$$2x = 1 \qquad x = 2$$
$$x = \frac{1}{2}$$

The x-intercepts are $\left(\frac{1}{2}, 0\right)$ and $(2, 0)$.

$$y = 2x^2 - 5x + 2$$
$$y = 2(0)^2 - 5(0) + 2$$
$$y = 2$$

The y-intercept is $(0, 2)$.

b.
$$y = x^2 + 4x + 4$$
$$0 = x^2 + 4x + 4$$
$$0 = (x + 2)(x + 2)$$
$$x + 2 = 0 \qquad x + 2 = 0$$
$$x = -2 \qquad x = -2$$

The x-intercept is $(-2, 0)$.

$$y = x^2 + 4x + 4$$
$$y = 0^2 + 4(0) + 4$$
$$y = 4$$

The y-intercept is $(0, 4)$.

Check Your Progress 3, *page 226*

The minimum value is the y-coordinate of the vertex of the graph of $f(x) = 2x^2 - 3x + 1$.

$$-\frac{b}{2a} = -\frac{-3}{2(2)} = \frac{3}{4}$$

$$f(x) = 2x^2 - 3x + 1$$

$$f\left(\frac{3}{4}\right) = 2\left(\frac{3}{4}\right)^2 - 3\left(\frac{3}{4}\right) + 1$$

$$= -\frac{1}{8}$$

The vertex is $\left(\frac{3}{4}, -\frac{1}{8}\right)$. The minimum value of the function is $-\frac{1}{8}$.

Check Your Progress 4, *page 227*

Find the vertex of the graph of $s(t) = -16t^2 + 64t + 4$.

$$-\frac{b}{2a} = -\frac{64}{2(-16)} = 2$$

$$s(t) = -16t^2 + 64t + 4$$

$$s(2) = -16(2)^2 + 64(2) + 4$$

$$= 68$$

The vertex is (2, 68).

The ball reaches its maximum height in 2 seconds.

The maximum height of the ball is 68 feet.

Check Your Progress 5, *page 228*

Let l represent the length of the rectangle, let w represent the width of the rectangle, and let A (which is unknown) represent the area of the rectangle. Use these variables to write expressions for the perimeter and area of the rectangle.

Perimeter: $2l + 2w = 44$

Area: $A = lw$

Our goal is to maximize A. To do this, we first write A in terms of a single variable. This can be accomplished by solving $2w + 2l = 44$ for l and then substituting into $A = lw$.

$$2w + 2l = 44$$

$$l = -w + 22$$

$$A = lw$$

$$= (-w + 22)w$$

$$A = -w^2 + 22w$$

Find the w-coordinate of the vertex.

$$w = -\frac{b}{2a} = -\frac{22}{2(-1)} = 11$$

The width is 11 feet. To find l, replace w by 11 in $l = -w + 22$ and solve for l.

$$l = -w + 22$$

$$l = -11 + 22 = 11$$

The length is 11 feet. The dimensions of the rectangle are 11 feet by 11 feet.

Section 2.6

Check Your Progress 1, *page 235*

Enter the data into a calculator and then find the regression equation.

L1	L2	L3 2
3.2	4.4	
3.4	5	
3.5	5.5	
3.8	6.2	
4	7.1	
4.2	7.6	

L2(9) =

LinReg
y=ax+b
a=3.12962963
b=-5.547222222
r²=.9969979404
r=.9984978842

The regression equation is $y = 3.130x - 5.547$.

To find the speed of a camel with a stride length of 3 meters, replace x in the regression equation by 3 and solve for y.

$$y = 3.130x - 5.547$$

$$= 3.130(3) - 5.547 = 3.843$$

The predicted speed of a camel with a 3-meter stride is 3.8 meters per second.

Check Your Progress 2, *page 237*

Enter the data into your calculator. Typical graphing calculator screens are shown below. Because the screen displays only seven entries, not all data are visible.

L1	L2	L3 2
927	116	------
884	102	
804	95	
772	82	
691	75	
767	80	
771	85	

L2(7) =85

LinReg
y=ax+b
a=.1480657359
b=-30.3898439
r²=.7678050546
r=.8762448599

The correlation coefficient is approximately 0.876.

Chapter 3

Section 3.1

Check Your Progress 1, *page 253*

$f(x) = 3x - 2$, $g(x) = x^2 - 1$

a. $(f + g)(3) = f(3) + g(3)$

$$= [3(3) - 2] + (3^2 - 1) = 7 + 8 = 15$$

b. $\left(\dfrac{f}{g}\right)(4) = \dfrac{f(4)}{g(4)} = \dfrac{3(4) - 2}{4^2 - 1} = \dfrac{10}{15} = \dfrac{2}{3}$

c. $(f \cdot g)(x) = f(x) \cdot g(x) = (3x - 2)(x^2 - 1)$
$= 3x^3 - 2x^2 - 3x + 2$

d. $(f - g)(x) = f(x) - g(x) = (3x - 2) - (x^2 - 1)$
$= -x^2 + 3x - 1$

Check Your Progress 2, *page 254*

To find the profit function, find the difference between the revenue and cost functions.

Profit = revenue − cost

$$\begin{aligned} P(x) &= R(x) - C(x) \\ &= (33x - 0.05x^2) - (24x + 103) \\ &= -0.05x^2 + 9x - 103 \end{aligned}$$

The profit function is given by $P(x) = -0.05x^2 + 9x - 103$. To determine the profit for selling 50 units, evaluate the profit function when $x = 50$.

$$P(x) = -0.05x^2 + 9x - 103$$
$$P(50) = -0.05(50)^2 + 9(50) - 103 = 222$$

The profit is \$222,000.

Check Your Progress 3, *page 255*

$$\frac{f(3 + h) - f(3)}{h} = \frac{\dfrac{1}{3 + h} - \dfrac{1}{3}}{h} = \frac{\dfrac{3 - (3 + h)}{3(3 + h)}}{h} = \frac{3 - (3 + h)}{3h(3 + h)}$$
$$= \frac{-h}{3h(3 + h)} = -\frac{1}{9 + 3h}$$

Check Your Progress 4, *page 257*

Use the difference quotient with $t = 10$ and $\Delta t = 2$. (Time has changed from 10 seconds to 12 seconds, so the "change in time" is 2 seconds.)

$$\frac{s(t + \Delta t) - s(t)}{\Delta t}$$
$$\frac{s(10 + 2) - s(10)}{2} = \frac{s(12) - s(10)}{2}$$
$$= \frac{1.5(12)^2 - 1.5(10)^2}{2} = \frac{216 - 150}{2}$$
$$= \frac{66}{2} = 33$$

The average velocity for the 2-second interval is 33 feet per second.

Check Your Progress 5, *page 260*

a. $h(x) = x^2 + 1$
$h(-2) = (-2)^2 + 1 = 4 + 1 = 5$
$g(x) = 4x + 1$
$g(5) = 4(5) + 1 = 20 + 1 = 21$
$g[h(-2)] = 21$

b. $h[g(x)] = [g(x)]^2 + 1$
$= [4x + 1]^2 + 1 = [16x^2 + 8x + 1] + 1$
$= 16x^2 + 8x + 2$

Section 3.2

Check Your Progress 1, *page 264*

Reverse the coordinates of the ordered pairs of the function, plot the resulting points, and draw a graph through the points.

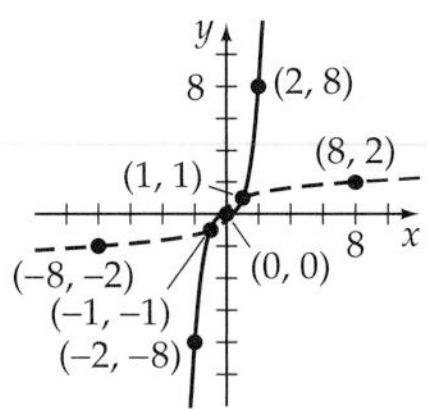

The dashed graph is the graph of the inverse function.

Check Your Progress 2, *page 265*

$$f(x) = 2x + 4$$
$$f[f^{-1}(x)] = 2\left[\frac{1}{2}x - 2\right] + 4 = [x - 4] + 4$$
$$f[f^{-1}(x)] = x$$

$$f^{-1}(x) = \frac{1}{2}x - 2$$
$$f^{-1}[f(x)] = \frac{1}{2}[2x + 4] - 2 = [x + 2] - 2$$
$$f^{-1}[f(x)] = x$$

Check Your Progress 3, *page 267*

$f(x) = 4x - 8$

$y = 4x - 8$ ▪ Replace $f(x)$ by y.

$x = 4y - 8$ ▪ Interchange x and y.

$x + 8 = 4y$ ▪ Solve for y.

$\dfrac{x + 8}{4} = y$

$\dfrac{1}{4}x + 2 = f^{-1}(x)$ ▪ Replace y by $f^{-1}(x)$.

Check Your Progress 4, *page 268*

$$f(x) = \frac{x}{x-2}$$

$$y = \frac{x}{x-2} \quad \text{▪ Replace } f(x) \text{ by } y.$$

$$x = \frac{y}{y-2} \quad \text{▪ Interchange } x \text{ and } y.$$

$$xy - 2x = y \quad \text{▪ Solve for } y.$$

$$xy - y = 2x$$

$$y(x-1) = 2x$$

$$y = \frac{2x}{x-1}$$

$$f^{-1}(x) = \frac{2x}{x-1}, x \neq 1 \quad \text{▪ Replace } y \text{ by } f^{-1}(x).$$

Check Your Progress 5, *page 269*

$$f(x) = x^2 - 6x$$

$$y = x^2 - 6x$$

$$x = y^2 - 6y$$

$$x = (y^2 - 6y + 9) - 9$$

$$x = (y-3)^2 - 9$$

$$x + 9 = (y-3)^2$$

$$\sqrt{x+9} = \sqrt{(y-3)^2}$$

$$\pm\sqrt{x+9} = y - 3$$

$$\pm\sqrt{x+9} + 3 = y$$

Because the domain of f is $\{x|x \leq 3\}$, the range of f^{-1} is $\{y|y \leq 3\}$. This means that we must choose the negative value of $\pm\sqrt{x+9}$. Thus $f^{-1}(x) = -\sqrt{x+9} + 3$.

Check Your Progress 6, *page 269*

Find the inverse of h.

$$h(x) = 3x - 10$$

$$y = 3x - 10$$

$$x = 3y - 10$$

$$x + 10 = 3y$$

$$\frac{x+10}{3} = y$$

$$\frac{1}{3}x + \frac{10}{3} = h^{-1}(x)$$

The inverse function is $h^{-1}(x) = \frac{1}{3}x + \frac{10}{3}$. Evaluate this function at 62.

$$h^{-1}(x) = \frac{1}{3}x + \frac{10}{3}$$

$$h^{-1}(62) = \frac{1}{3}(62) + \frac{10}{3}$$

$$= \frac{62}{3} + \frac{10}{3} = \frac{72}{3} = 24$$

A hat size of 24 in the U.S. is equivalent to a French hat size of 62.

Section 3.3

Check Your Progress 1, *page 276*

a. The graph of $y = f(x+3)$ is the graph of $y = f(x)$ shifted 3 units to the left.

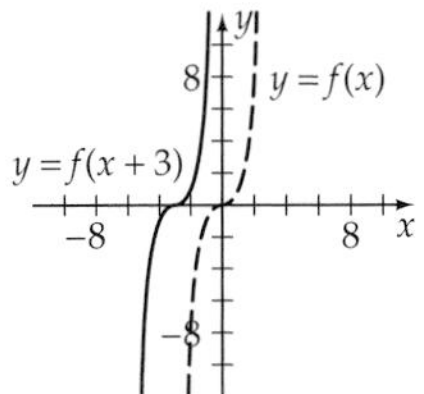

b. The graph of $y = f(x-4) + 3$ is the graph of $y = f(x)$ shifted 4 units to the right and 3 units up.

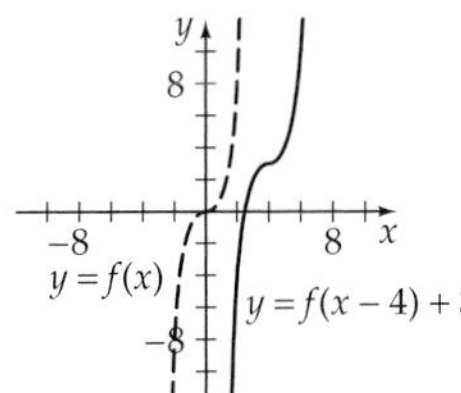

Check Your Progress 2, *page 279*

a. The graph of $y = -f(x-1) + 2$ is the graph of $y = f(x)$ moved 1 unit to the right, reflected through the x-axis, and then moved 2 units up.

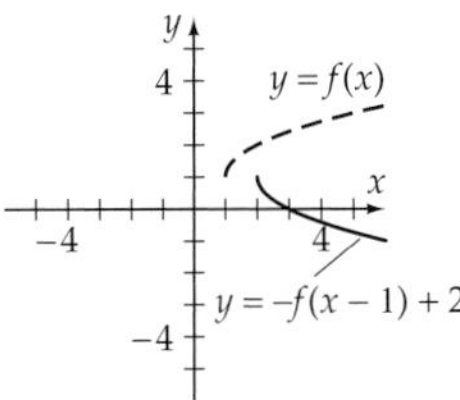

b. The graph of $y = f(-x+1) - 2$ is the graph of $y = f(x)$ shifted 1 unit to the left, reflected through the y-axis, and then shifted 2 units down. See the graph on the following page.

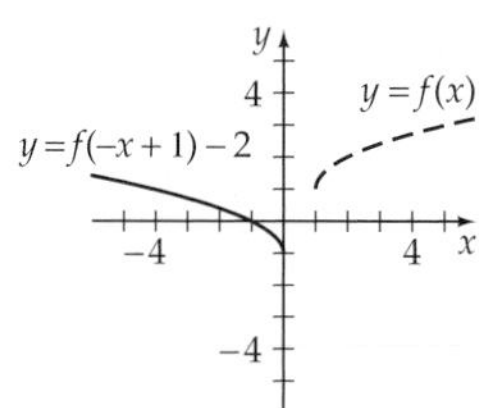

Check Your Progress 3, *page 282*

a. Each value of y on the graph of $y = \frac{1}{2}f(x)$ is $\frac{1}{2}$ the value of y on $y = f(x)$.

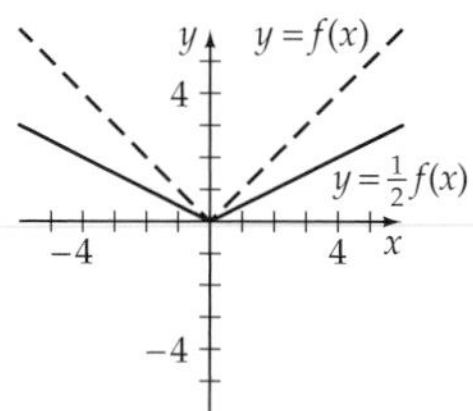

b. The graph of $y = f(3x)$ is a horizontal compression of $y = f(x)$ by a factor of $\frac{1}{3}$.

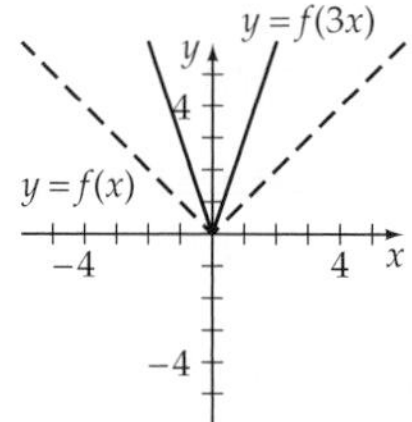

Check Your Progress 4, *page 284*

The graph has been moved 4 units to the left, reflected through the y-axis, reflected through the x-axis, and moved 2 units down. Therefore,

$$g(x) = -f(x + 4) - 2 = -\sqrt{2(x + 4) - (x + 4)^2} - 2$$
$$= -\sqrt{-x^2 - 6x - 8} - 2.$$

Check Your Progress 5, *page 286*

To determine x-axis symmetry, replace y by $-y$. If the equations are the same, the graph has x-axis symmetry.

$$y = x^2 + 2$$
$$-y = x^2 + 2$$
$$y = -x^2 - 2 \neq x^2 + 2$$

The graph does not have x-axis symmetry.

To determine y-axis symmetry, replace x by $-x$. If the equations are the same, the graph has y-axis symmetry.

$$y = x^2 + 2$$
$$y = (-x)^2 + 2$$
$$y = x^2 + 2$$

which is the same as the original equation. The graph has y-axis symmetry.

Check Your Progress 6, *page 286*

To determine origin symmetry, replace x by $-x$ and y by $-y$. If the equations are the same, the graph has origin symmetry.

$$y = x^3 + 1$$
$$-y = (-x)^3 + 1 = -x^3 + 1$$
$$y = x^3 - 1 \neq x^3 + 1$$

The graph does not have origin symmetry.

Check Your Progress 7, *page 288*

a. A function is even when $F(x) = F(-x)$.
For $F(x) = |x|$, we have
$F(-x) = |-x| = |x| = F(x)$
F is an even function. F is not an odd function.

b. For $g(x) = x^3 - x^2$, we have
$g(-x) = (-x)^3 - (-x)^2 = -x^3 - x^2$
Because $g(x) = x^3 - x^2$ and $g(-x) = -x^3 - x^2$, $g(x) \neq g(-x)$.
Therefore, g is not an even function.
$g(-x) = (-x)^3 - (-x)^2 = -x^3 - x^2$
Because $g(-x) = -x^3 - x^2$ and $-g(x) = -x^3 + x^2$, $g(-x) \neq -g(x)$.
Therefore, g is not an odd function.

Section 3.4

Check Your Progress 1, *page 298*

$d = kt$	▪ d varies directly as t.
$4020 = k(12)$	▪ $d = 4020$, $t = 12$
$335 = k$	▪ Solve for k.
$d = 335t$	▪ Replace k in $d = kt$.
$d = 335(15)$	▪ Replace t by 15.
$d = 5025$	

Sound travels 5025 meters in 15 seconds.

Check Your Progress 2, *page 299*

$s = kt^2$ ■ *s* varies directly as the square of *t*.

$144 = k(3^2)$ ■ $s = 144, t = 3$

$16 = k$ ■ Solve for *k*.

$s = 16t^2$ ■ Replace *k* in $s = kt^2$

$s = 16(7^2)$ ■ Replace *t* by 7.

$s = 784$

The object falls 784 feet in 7 seconds.

Check Your Progress 3, *page 300*

$V = \frac{k}{P}$ ■ *V* varies inversely as *P*.

$75 = \frac{k}{1.5}$ ■ $V = 75, P = 1.5$

$112.5 = k$ ■ Solve for *k*.

$V = \frac{112.5}{P}$ ■ Replace *k* in $V = \frac{k}{P}$.

$V = \frac{112.5}{2.5}$ ■ Replace *P* by 2.5.

$V = 45$

The volume is 45 milliliters.

Check Your Progress 4, *page 301*

$C = ktA$ ■ *C* varies jointly as *t* and *A*.

$175 = k(4)(2100)$ ■ $C = 175, t = 4, A = 2100$

$\frac{1}{48} = k$ ■ Solve for *k*.

$C = \frac{1}{48}tA$ ■ Replace *k* in $C = ktA$.

$C = \frac{1}{48}(6)(2400)$ ■ Replace *t* by 6 and *A* by 2400.

$C = 300$

It costs \$300 for 2400 square feet of insulation 6 inches thick.

Check Your Progress 5, *page 302*

$W = \frac{kwd^2}{L}$ ■ *W* varies jointly as *w* and d^2 and inversely as *L*.

$256 = \frac{k(4)(4^2)}{10}$ ■ $W = 256, w = 4, d = 4, L = 10$

$40 = k$ ■ Solve for *k*.

$W = \frac{40wd^2}{L}$ ■ Replace *k* in $W = \frac{kwd^2}{L}$.

$W = \frac{40(4)(6^2)}{16}$ ■ Replace *w* by 4, *d* by 6, and *L* by 16.

$W = 360$

The beam will safely support 360 pounds.

Chapter 4

Section 4.1

Check Your Progress 1, *page 315*

$$\begin{array}{r|rrrr} 2 & 5 & -6 & 0 & -19 \\ & & 10 & 8 & 16 \\ \hline & 5 & 4 & 8 & -3 \end{array}$$

Thus $\frac{5x^3 - 6x^2 - 19}{x - 2} = 5x^2 + 4x + 8 - \frac{3}{x - 2}$.

Check Your Progress 2, *page 317*

$$\begin{array}{r|rrrr} 3 & 2 & -1 & 3 & -1 \\ & & 6 & 15 & 54 \\ \hline & 2 & 5 & 18 & 53 \end{array}$$

$P(c) = P(3) = 53$

Check Your Progress 3, *page 318*

$$\begin{array}{r|rrrr} -6 & 1 & 4 & -27 & -90 \\ & & -6 & 12 & 90 \\ \hline & 1 & -2 & -15 & 0 \end{array}$$

A remainder of 0 indicates that $x + 6$ is a factor of $P(x)$.

Check Your Progress 4, *page 319*

$$\begin{array}{r|rrrrr} -1 & 1 & 5 & 3 & -5 & -4 \\ & & -1 & -4 & 1 & 4 \\ \hline & 1 & 4 & -1 & -4 & 0 \end{array}$$

The reduced polynomial is $x^3 + 4x^2 - x - 4$.

$x^4 + 5x^3 + 3x^2 - 5x - 4 = (x + 1)(x^3 + 4x^2 - x - 4)$

Section 4.2

Check Your Progress 1, *page 325*

Because $a_n = -2$ is negative and $n = 3$ is odd, the graph of *P* goes up to the far left and down to the far right.

Check Your Progress 2, *page 328*

The volume of the box is $V = lwh$, with $h = x, l = 18 - 2x$, and $w = \dfrac{42 - 3x}{2}$. Therefore, the volume is

$$V(x) = (18 - 2x)\left(\frac{42 - 3x}{2}\right)x$$
$$= 3x^3 - 69x^2 + 378x$$

Use a graphing utility to graph $V(x)$. The graph is shown below. The value of x that produces the maximum volume is 3.571 inches (to the nearest 0.001 inch). Note: Your x-value may differ slightly from 3.5705971 depending on the values you use for Xmin and Xmax. The maximum value is approximately 606.6 cubic inches.

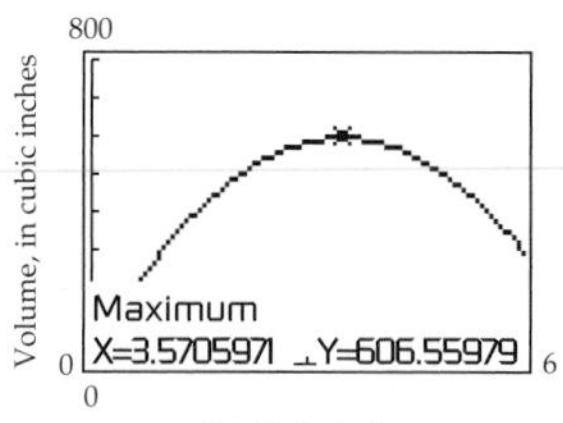

Check Your Progress 3, *page 329*

$$x^4 - 29x^2 + 100 = (x^2 - 25)(x^2 - 4)$$
$$= (x + 5)(x - 5)(x + 2)(x - 2)$$

The four real zeros of $P(x)$ are $-5, -2, 2$, and 5.

Check Your Progress 4, *page 330*

0	4	−1	−6	1	
		0	0	0	
	4	−1	−6	1	▪ $P(0) = 1$

1	4	−1	−6	1	
		4	3	−3	
	4	3	−3	−2	▪ $P(1) = -2$

Because P is a polynomial function, the graph of P is continuous. Also, $P(0)$ and $P(1)$ have opposite signs. Thus by the Zero Location Theorem we know that P must have a real zero between 0 and 1.

Check Your Progress 5, *page 331*

The exponent of $(x + 2)$ is 1, which is odd. Thus the graph of P crosses the x-axis at the x-intercept $(-2, 0)$. The exponent of $(x - 6)^2$ is even. Thus the graph of P intersects but does not cross the x-axis at $(6, 0)$.

Section 4.3

Check Your Progress 1, *page 338*

$p = \pm1, \pm2, \pm4, \pm8$

$q = \pm1, \pm3$

$\dfrac{p}{q} = \pm1, \pm2, \pm4, \pm8, \pm\frac{1}{3}, \pm\frac{2}{3}, \pm\frac{4}{3}, \pm\frac{8}{3}$

Check Your Progress 2, *page 339*

1	1	0	−19	−28
		1	1	−18
	1	1	−18	−46

2	1	0	−19	−28
		2	4	−30
	1	2	−15	−58

3	1	0	−19	−28
		3	9	−30
	1	3	−10	−58

4	1	0	−19	−28
		4	16	−12
	1	4	−3	−40

5	1	0	−19	−28
		5	25	30
	1	5	6	2

All numbers are positive, so 5 is an upper bound.

−1	1	0	−19	−28
		−1	1	18
	1	−1	−18	−10

−2	1	0	−19	−28
		−2	4	30
	1	−2	−15	2

−3	1	0	−19	−28
		−3	9	30
	1	−3	−10	2

−4	1	0	−19	−28
		−4	16	12
	1	−4	−3	−16

−5	1	0	−19	−28
		−5	25	−30
	1	−5	6	−58

These numbers alternate in sign, so −5 is a lower bound.

Check Your Progress 3, *page 341*

$P(x)$ has one positive real zero because P has one variation in sign.

$P(-x) = (-x)^3 - 19(-x) - 30 = -x^3 + 19x - 30$

$P(x)$ has two or no negative real zeros because $-x^3 + 19x - 30$ has two variations in sign.

Check Your Progress 4, *page 343*

$P(x)$ has one positive and two or no negative real zeros (see Check Your Progress 3).

5	1	0	−19	−30
		5	25	30
	1	5	6	0

The reduced polynomial is $x^2 + 5x + 6 = (x + 3)(x + 2)$, which has -3 and -2 as zeros. Thus the zeros of $P(x) = x^3 - 19x - 30$ are $-3, -2$, and 5.

Check Your Progress 5, *page 344*

We need to solve $560 = \frac{1}{6}(k^3 + 3k^2 + 2k)$ for k. Multiplying each side of the equation by 6 produces $3360 = k^3 + 3k^2 + 2k$, which can be written as $k^3 + 3k^2 + 2k - 3360 = 0$. The number 3360 has many natural number divisors, but we can eliminate many of these by showing that 15 is an upper bound.

$$\begin{array}{r|rrrr} 15 & 1 & 3 & 2 & -3360 \\ & & 15 & 270 & 4080 \\ \hline & 1 & 18 & 272 & 720 \end{array}$$

← Each number in the bottom row is positive. Thus 15 is an upper bound.

The only natural number divisors of 3360 that are less than 15 are 1, 2, 3, 4, 5, 6, 7, 8, 10, 12, and 14. The following division shows that 14 is a zero of $k^3 + 3k^2 + 2k - 3360$.

$$\begin{array}{r|rrrr} 14 & 1 & 3 & 2 & -3360 \\ & & 14 & 238 & 3360 \\ \hline & 1 & 17 & 240 & 0 \end{array}$$

Thus the pyramid has 14 levels. There is no need to seek additional solutions, because the number of levels is uniquely determined by the number of glasses.

Check Your Progress 6, *page 345*

The volume of the cartridge is equal to the volume of the two hemispheres plus the volume of the cylinder. Thus

$$\frac{4}{3}\pi x^3 + 6\pi x^2 = 9\pi$$

Dividing each term by π and multiplying by 3 produces

$$4x^3 + 18x^2 = 27$$

Intersection Method Use a graphing utility to graph $y = 4x^3 + 18x^2$ and $y = 27$ on the same screen, with $x > 0$. The x-coordinate of the point of intersection of the two graphs is the desired solution. The graphs intersect at $x \approx 1.098$ (rounded to the nearest thousandth of a foot). The length of the radius is approximately 1.098 feet.

Section 4.4

Check Your Progress 1, *page 353*

Use the Rational Zero Theorem to determine the possible rational zeros.

$$\frac{p}{q} = \pm 1, \pm 5$$

The following synthetic division shows that 1 is a zero of $P(x)$.

$$\begin{array}{r|rrrr} 1 & 1 & -3 & 7 & -5 \\ & & 1 & -2 & 5 \\ \hline & 1 & -2 & 5 & 0 \end{array}$$

Use the quadratic formula to find the zeros of the reduced polynomial $x^2 - 2x + 5$.

$$x = \frac{-(-2) \pm \sqrt{(-2)^2 - 4(1)(5)}}{2(1)} = 1 \pm 2i$$

The zeros of $P(x) = x^3 - 3x^2 + 7x - 5$ are $1, 1 - 2i$, and $1 + 2i$.

The linear factored form of $P(x)$ is

$$P(x) = (x - 1)(x - [1 - 2i])(x - [1 + 2i])$$

or

$$P(x) = (x - 1)(x - 1 + 2i)(x - 1 - 2i)$$

Check Your Progress 2, *page 354*

$$\begin{array}{r|rrrr} 5+3i & 3 & -29 & 92 & 34 \\ & & 15+9i & -97+3i & -34 \\ \hline & 3 & -14+9i & -5+3i & 0 \end{array}$$

$$\begin{array}{r|rrr} 5-3i & 3 & -14+9i & -5+3i \\ & & 15-9i & 5-3i \\ \hline & 3 & 1 & 0 \end{array}$$

The reduced polynomial $3x + 1$ has $-\frac{1}{3}$ as a zero. The zeros of $3x^3 - 29x^2 + 92x + 34$ are $5 + 3i, 5 - 3i$, and $-\frac{1}{3}$.

Check Your Progress 3, *page 355*

$$\begin{array}{r|rrrrrr} 3i & 1 & -6+0i & 22+0i & -64+0i & 117+0i & -90 \\ & & 0+3i & -9-18i & 54+39i & -117-30i & 90 \\ \hline -3i & 1 & -6+3i & 13-18i & -10+39i & 0-30i & 0 \\ & & 0-3i & 0+18i & 0-39i & 30i & \\ \hline & 1 & -6 & 13 & -10 & 0 & \end{array}$$

$$\frac{p}{q} = \pm 1, \pm 2, \pm 5, \pm 10$$

$$\begin{array}{r|rrrr} 2 & 1 & -6 & 13 & -10 \\ & & 2 & -8 & 10 \\ \hline & 1 & -4 & 5 & 0 \end{array}$$

Use the quadratic formula to solve $x^2 - 4x + 5 = 0$.

$$x = \frac{-(-4) \pm \sqrt{(-4)^2 - 4(1)(5)}}{2(1)} = \frac{4 \pm \sqrt{-4}}{2}$$

$$= \frac{4 \pm 2i}{2} = 2 \pm i$$

The zeros of $x^5 - 6x^4 + 22x^3 - 64x^2 + 117x - 90$ are $3i, -3i, 2, 2 + i$, and $2 - i$.

Check Your Progress 4, *page 356*

The graph of $P(x) = 4x^3 + 3x^2 + 16x + 12$ is shown below. Applying Descartes' Rule of Signs, we find that the real zeros are all negative numbers. From the Upper- and Lower-Bound Theorem there is no real zero less than -1, and from the Rational Zero Theorem the possible rational zeros (that are negative and greater than -1) are $\frac{p}{q} = -\frac{1}{2}, -\frac{1}{4}$, and $-\frac{3}{4}$. From the graph, it appears that $-\frac{3}{4}$ is a zero.

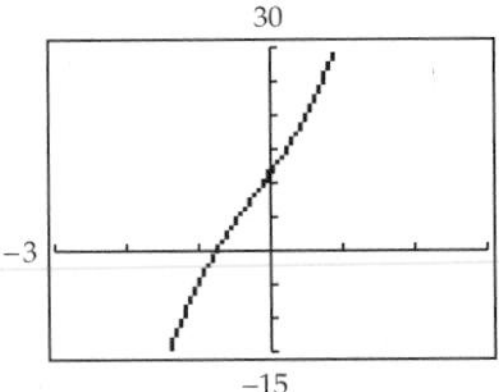

Use synthetic division with $c = -\frac{3}{4}$.

$$\begin{array}{r|rrrr} -\frac{3}{4} & 4 & 3 & 16 & 12 \\ & & -3 & 0 & -12 \\ \hline & 4 & 0 & 16 & 0 \end{array}$$

Thus $-\frac{3}{4}$ is a zero, and by the Factor Theorem,

$$4x^3 + 3x^2 + 16x + 12 = \left(x + \frac{3}{4}\right)(4x^2 + 16) = 0$$

Solve $4x^2 + 16 = 0$, to find that $x = -2i$ and $x = 2i$. The solutions of the original equation are $-\frac{3}{4}, -2i$, and $2i$.

Check Your Progress 5, *page 357*

Because P has real coefficients, use the Conjugate Pair Theorem.

$$\begin{aligned} P &= (x - [3 + 2i])(x - [3 - 2i])(x - 7) \\ &= (x - 3 - 2i)(x - 3 + 2i)(x - 7) \\ &= (x^2 - 6x + 13)(x - 7) \\ &= x^3 - 13x^2 + 55x - 91 \end{aligned}$$

Section 4.5

Check Your Progress 1, *page 363*

Set the denominator equal to zero.

$$x^2 - 4 = 0$$

$$(x - 2)(x + 2) = 0$$

$$x = 2 \quad \text{or} \quad x = -2$$

The vertical asymptotes are $x = 2$ and $x = -2$.

Check Your Progress 2, *page 364*

The horizontal asymptote is $y = 0$ (x-axis) because the degree of the denominator is larger than the degree of the numerator.

Check Your Progress 3, *page 366*

Vertical asymptote: $x - 2 = 0$

$x = 2$

Horizontal asymptote: $y = 0$

No x-intercepts.

y-intercept: $\left(0, -\frac{1}{2}\right)$

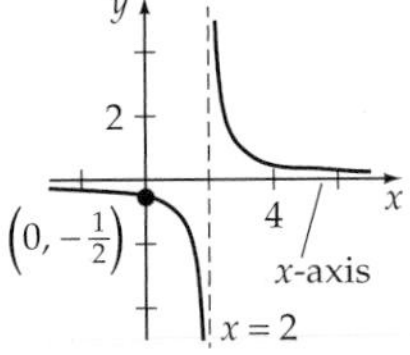

Check Your Progress 4, *page 368*

$$\begin{aligned} \text{Vertical asymptote:} \quad x^2 - 6x + 9 &= 0 \\ (x - 3)(x - 3) &= 0 \\ x &= 3 \end{aligned}$$

The horizontal asymptote is $y = \frac{1}{1} = 1$ (the Theorem on Horizontal Asymptotes) because the numerator and denominator both have degree 2. The graph crosses the horizontal asymptote at $\left(\frac{3}{2}, 1\right)$. The graph intersects, but does not cross, the x-axis at $(0, 0)$.

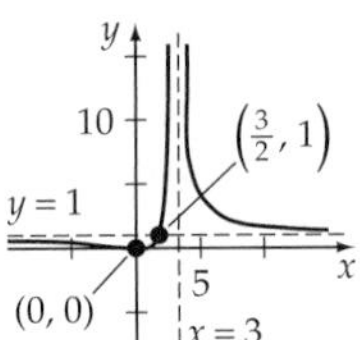

Check Your Progress 5, *page 369*

$$\begin{array}{r} x + 1 \\ x^2 - 3x + 5 \overline{) x^3 - 2x^2 + 3x + 4} \\ \underline{x^3 - 3x^2 + 5x} \\ x^2 - 2x + 4 \\ \underline{x^2 - 3x + 5} \\ x - 1 \end{array}$$

$$F(x) = x + 1 + \frac{x - 1}{x^2 - 3x + 5}$$

Slant asymptote: $y = x + 1$

Check Your Progress 6, *page 370*

$$F(x) = \frac{x^2 - x - 12}{x^2 - 2x - 8} = \frac{(x - 4)(x + 3)}{(x - 4)(x + 2)} = \frac{x + 3}{x + 2}, x \neq 4$$

The function F is undefined at $x = 4$. Thus the graph of F is the graph of $y = \frac{x + 3}{x + 2}$ with an open circle at $\left(4, \frac{7}{6}\right)$.

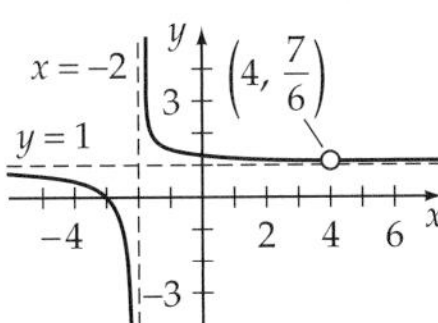

Check Your Progress 7, *page 371*

a.

$$\overline{C}(1000) = \frac{0.0006(1000)^2 + 9(1000) + 401{,}000}{1000} = \$410.60$$

$$\overline{C}(10{,}000) = \frac{0.0006(10{,}000)^2 + 9(10{,}000) + 401{,}000}{10{,}000} = \$55.10$$

$$\overline{C}(100{,}000) = \frac{0.0006(100{,}000)^2 + 9(100{,}000) + 401{,}000}{100{,}000} = \$73.01$$

b. Graph $\overline{C}$ and use the minimum feature of a graphing utility.

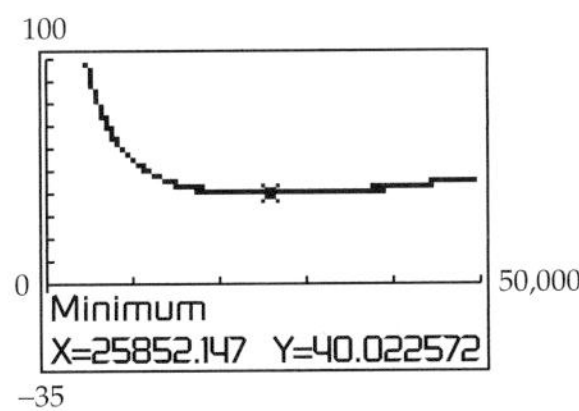

The minimum average cost per telephone is \$40.02. The minimum is achieved by producing approximately 25,852 telephones.

Chapter 5

Section 5.1

Check Your Progress 1, *page 386*

$$g(3) = \left(\frac{1}{2}\right)^3 = \left(\frac{1}{2}\right)\left(\frac{1}{2}\right)\left(\frac{1}{2}\right) = \frac{1}{8};$$

$$g(-1) = \left(\frac{1}{2}\right)^{-1} = \frac{1}{\left(\frac{1}{2}\right)} = 2;\ g(\sqrt{3}) = \left(\frac{1}{2}\right)^{\sqrt{3}} \approx 0.30102$$

Check Your Progress 2, *page 389*

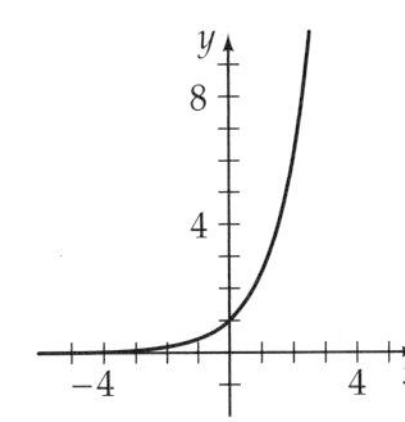

Check Your Progress 3, *page 389*

a.

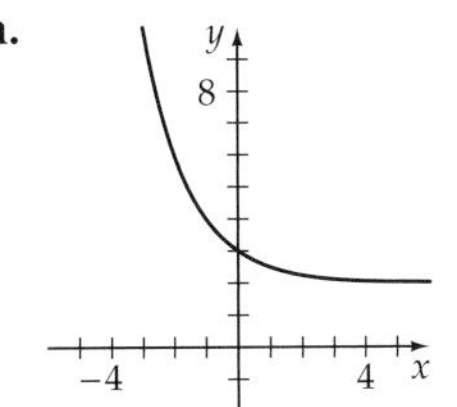

b.

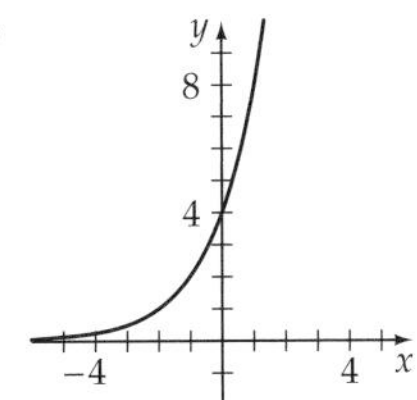

Check Your Progress 4, *page 390*

a.

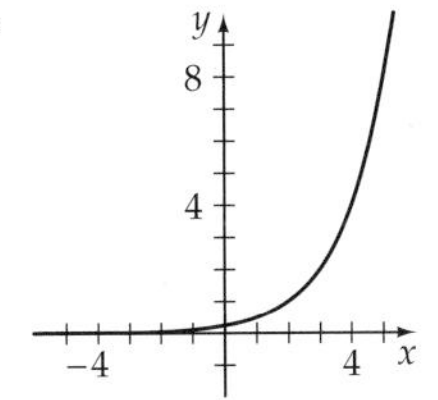

b.

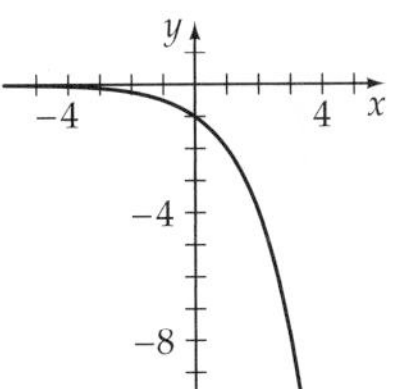

Check Your Progress 5, *page 392*

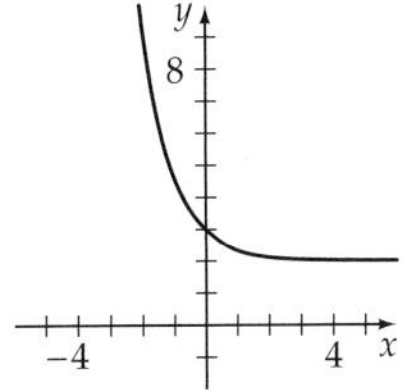

Check Your Progress 6, *page 393*

a. $A(45) = 200e^{-0.014(45)} \approx 107$ milligrams

b. Use a graphing utility to graph both $y = 200e^{-0.014x}$ and $y = 50$.

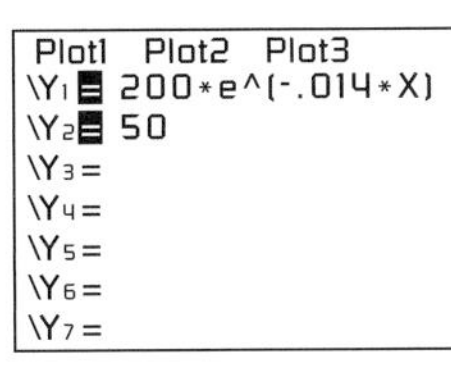

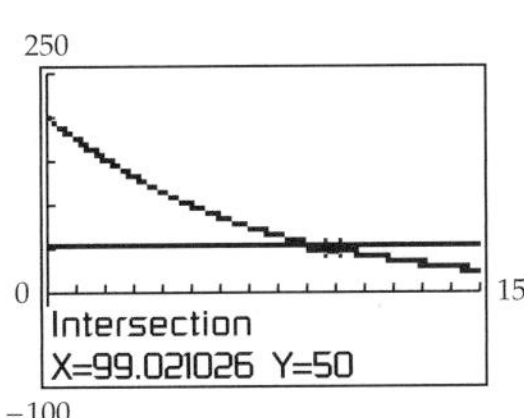

The first coordinate of the point of intersection of the graphs is $x \approx 99$. In about 99 minutes, the amount of medication in the patient's bloodstream will reach 50 milligrams.

Check Your Progress 7, *page 394*

As t increases without bound, $e^{-0.022t}$ approaches 0, as does the product $480e^{-0.022t}$. Therefore, as $t \to \infty$, $R(t) \to 520 + 0 = 520$. Thus as the number of months increases, the revenue approaches \$520 per month.

Section 5.2

Check Your Progress 1, *page 402*

a. $5^2 = 25$ **b.** $10^3 = 2x$ **c.** $e^x = 3$ **d.** $b^4 = b^4$

Check Your Progress 2, *page 402*

a. $\log_{10} 100 = 2$ **b.** $\log_e x = 1.5$ **c.** $\log_c a = b$
d. $\log_b 3 = \log_b 3$

Check Your Progress 3, *page 403*

a. Applying property 1 yields $\log_{10} 10 = 1$.
b. Applying property 2 yields $\log_e 1 = 0$.
c. Applying property 3 yields $\log_5 5^3 = 3$.
d. Applying property 4 yields $2^{\log_2 3} = 3$.

Check Your Progress 4, *page 404*

To graph $f(x) = \log_6 x$, consider the equivalent exponential equation $x = 6^y$. Because this equation is solved for x, choose values of y and calculate the corresponding values of x, as shown in the following table.

$x = 6^y$	$\frac{1}{36}$	$\frac{1}{6}$	1	6
y	−2	−1	0	1

Now plot the ordered pairs and connect the points with a smooth curve, as shown below.

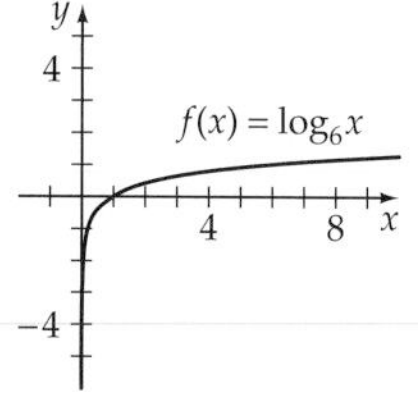

Check Your Progress 5, *page 406*

a. Solving $5 - x > 0$ for x gives us $x < 5$. Thus the domain of k consists of all real numbers less than 5. In interval notation the domain is $(-\infty, 5)$.

b. The solution set of $|x - 3| > 0$ consists of all real numbers x except $x = 3$. Thus the domain of p consists of all real numbers $x \neq 3$. In interval notation the domain is $(-\infty, 3) \cup (3, \infty)$.

c. Solving $\frac{x + 4}{x} > 0$ yields the set of all real numbers x less than -4 or greater than 0. Thus the domain of s is $\{x \mid x < -4 \text{ or } x > 0\}$. In interval notation the domain is $(-\infty, -4) \cup (0, \infty)$.

Check Your Progress 6, *page 406*

a. Translate the graph of $y = \log_2 x$ horizontally 4 units to the right.

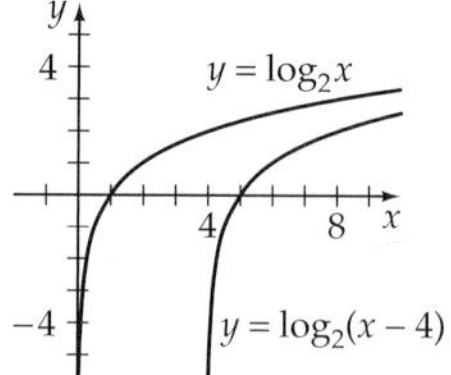

b. Translate the graph of $y = \log_2 x$ vertically 3 units down.

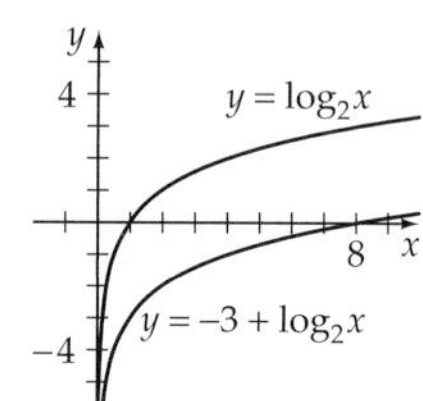

Check Your Progress 7, *page 408*

a. $S(0) = 5 + 29 \ln(0 + 1) = 5 + 0 = 5$. When starting, the student had an average typing speed of 5 words per minute. $S(3) = 5 + 29 \ln(3 + 1) \approx 45.2$. After 3 months the student's average typing speed was about 45 words per minute.

b. Use the intersection feature of a graphing utility to find the x-coordinate of the point of intersection of the graphs of $y = 5 + 29 \ln(x + 1)$ and $y = 65$.

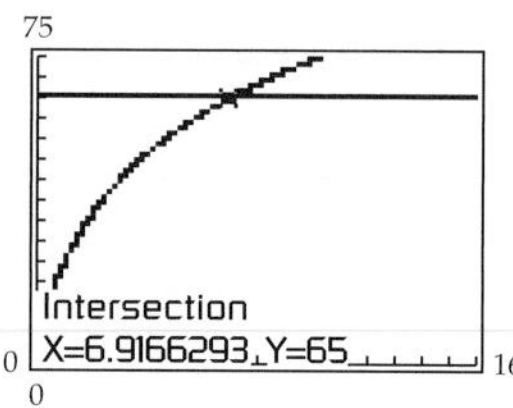

It will take the student about 6.9 months to type at a rate of 65 words per minute.

Section 5.3

Check Your Progress 1, *page 414*

$$\begin{aligned}\ln\frac{z^3}{\sqrt{xy}} &= \ln z^3 - \ln\sqrt{xy}\\ &= \ln z^3 - \ln(xy)^{1/2}\\ &= 3\ln z - \frac{1}{2}\ln(xy)\\ &= 3\ln z - \frac{1}{2}(\ln x + \ln y)\\ &= 3\ln z - \frac{1}{2}\ln x - \frac{1}{2}\ln y\end{aligned}$$

Check Your Progress 2, *page 415*

$$\begin{aligned}3\log_2 t - \frac{1}{3}\log_2 u + 4\log_2 v &= \log_2 t^3 - \log_2 u^{1/3} + \log_2 v^4\\ &= \log_2\frac{t^3}{u^{1/3}} + \log_2 v^4\\ &= \log_2\frac{t^3v^4}{u^{1/3}}\\ &= \log_2\frac{t^3v^4}{\sqrt[3]{u}}\end{aligned}$$

Check Your Progress 3, *page 416*

a. $\log_5 50 = \dfrac{\ln 50}{\ln 5} \approx 2.43068$

b. $\log_8\dfrac{1}{5} = \dfrac{\ln\frac{1}{5}}{\ln 8} = -0.77398$

Check Your Progress 4, *page 417*

$\log_8(5-x) = \dfrac{\ln(5-x)}{\ln 8}$, so enter $\dfrac{\ln(5-x)}{\ln 8}$ into Y1 on a graphing calculator.

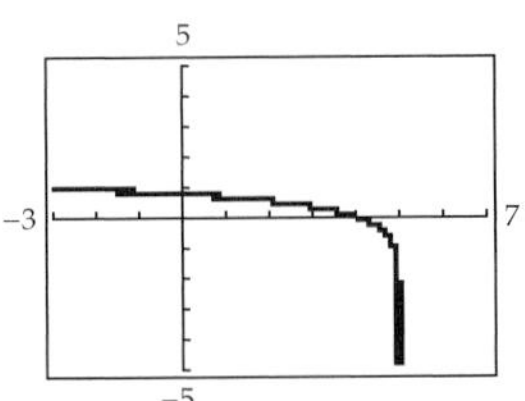

Check Your Progress 5, *page 418*

Twice the intensity of the Joshua Tree earthquake is $2\cdot 12{,}589{,}254I_0 = 25{,}178{,}508I_0$. The Richter scale magnitude is

$$M = \log\left(\frac{I}{I_0}\right) = \log\left(\frac{25{,}178{,}508I_0}{I_0}\right) = \log(25{,}178{,}508) \approx 7.4$$

Check Your Progress 6, *page 418*

$$\begin{aligned}\log\left(\frac{I}{I_0}\right) &= 5.2\\ \frac{I}{I_0} &= 10^{5.2}\\ I &= 10^{5.2}I_0\\ I &\approx 158{,}489I_0\end{aligned}$$

Check Your Progress 7, *page 419*

In Example 7 we noticed that if an earthquake has a Richter scale magnitude of M_1 and a smaller earthquake has a Richter scale magnitude of M_2, then the first earthquake is $10^{M_1-M_2}$ times as intense as the smaller earthquake. Here, $M_1 = 7.2$ and $M_2 = 5.0$. Thus $10^{M_1-M_2} = 10^{7.2-5.0} = 10^{2.2} \approx 158$. The November aftershock was approximately 158 times as intense as the earlier one.

Check Your Progress 8, *page 420*

$$\begin{aligned}M &= \log A + 3\log 8t - 2.92\\ &= \log 21 + 3\log[8\cdot 29] - 2.92\\ &\approx 1.3222 + 7.0965 - 2.92\\ &\approx 5.5\end{aligned}$$

▪ **Substitute 21 for *A* and 29 for *t*.**

Check Your Progress 9, *page 421*

a. $\text{pH} = -\log[\text{H}^+] = -\log(2.41\times 10^{-13}) \approx 12.6$; the cleaning solution has a pH of 12.6.

b. $\text{pH} = -\log[\text{H}^+] = -\log(5.07\times 10^{-4}) \approx 3.3$; the cola has a pH of 3.3.

Check Your Progress 10, *page 422*

$$\begin{aligned}\text{pH} &= -\log[\text{H}^+]\\ 10.0 &= -\log[\text{H}^+]\\ -10.0 &= \log[\text{H}^+]\\ 10^{-10.0} &= \text{H}^+\end{aligned}$$

The hydronium-ion concentration is 10^{-10} mole per liter.

Section 5.4

Check Your Progress 1, *page 427*

$$3^{5-2x} = \frac{1}{9}$$
$$3^{5-2x} = 3^{-2}$$
$$5 - 2x = -2$$
$$-2x = -7$$
$$\frac{-2x}{-2} = \frac{-7}{-2}$$
$$x = \frac{7}{2}$$

Check Your Progress 2, *page 429*

$$3^x = 175$$
$$\log(3^x) = \log 175$$
$$x \log 3 = \log 175$$
$$x = \frac{\log 175}{\log 3}$$

Note: $x = \dfrac{\ln 175}{\ln 3}$ is also a correct answer.

Check Your Progress 3, *page 429*

$$3^{2x+1} = 8^x$$
$$\ln(3^{2x+1}) = \ln(8^x)$$
$$(2x + 1) \ln 3 = x \ln 8$$
$$2x \ln 3 + \ln 3 = x \ln 8$$
$$\ln 3 = x \ln 8 - 2x \ln 3$$
$$\ln 3 = x(\ln 8 - 2 \ln 3)$$
$$\frac{\ln 3}{\ln 8 - 2 \ln 3} = x$$
$$x \approx -9.327$$

Note: $x = \dfrac{\log 3}{\log 8 - 2 \log 3}$ is also a correct answer.

Check Your Progress 4, *page 430*

Substitute 50 for $A(t)$ and solve for t.

$$50 = 200e^{-0.014t}$$
$$\frac{50}{200} = e^{-0.014t}$$
$$\ln \frac{1}{4} = -0.014t$$
$$-\frac{1}{0.014} \ln \frac{1}{4} = t$$
$$t \approx 99.0$$

Approximately 99.0 minutes will be required.

Check Your Progress 5, *page 431*

Substitute 140 for N and solve for t.

$$140 = \frac{250}{1 + 249e^{-0.503t}}$$
$$140(1 + 249e^{-0.503t}) = 250$$
$$140 + 34{,}860e^{-0.503t} = 250$$
$$34{,}860e^{-0.503t} = 110$$
$$e^{-0.503t} = \frac{110}{34{,}860}$$
$$-0.503t = \ln \frac{11}{3486}$$
$$t = -\frac{1}{0.503} \ln \frac{11}{3486}$$
$$t \approx 11$$

Approximately 11 weeks of training are needed.

Check Your Progress 6, *page 432*

$$\ln(2x + 3) = 4$$
$$2x + 3 = e^4$$
$$2x = e^4 - 3$$
$$x = \frac{e^4 - 3}{2}$$

Check Your Progress 7, *page 432*

$$\log(2x) - \log(x - 3) = 1$$
$$\log \frac{2x}{x - 3} = 1$$
$$\frac{2x}{x - 3} = 10^1$$
$$2x = 10(x - 3)$$
$$2x = 10x - 30$$
$$-8x = -30$$
$$x = \frac{-30}{-8} = \frac{15}{4}$$

Check Your Progress 8, *page 433*

$$\log_5(x + 4) = 2 - \log_5(x + 4)$$
$$\log_5(x + 4) + \log_5(x + 4) = 2$$
$$\log_5[(x + 4)^2] = 2$$
$$(x + 4)^2 = 5^2$$
$$x^2 + 8x + 16 = 25$$
$$x^2 + 8x - 9 = 0$$
$$(x + 9)(x - 1) = 0$$
$$x = -9 \quad \text{or} \quad x = 1$$

$x = -9$ is an extraneous solution because if we substitute -9 for x in the original equation, we get the expression $\log_5(-5)$, which is undefined. The solution $x = 1$ checks, so the only solution is $x = 1$.

Check Your Progress 9, *page 434*

Substitute 10 for t and solve for v.

$$10 = -\frac{175}{32}\ln\left(1 - \frac{v}{175}\right)$$

$$10\left(-\frac{32}{175}\right) = \ln\left(1 - \frac{v}{175}\right)$$

$$-\frac{64}{35} = \ln\left(1 - \frac{v}{175}\right)$$

$$e^{-64/35} = 1 - \frac{v}{175}$$

$$e^{-64/35} - 1 = -\frac{v}{175}$$

$$-175(e^{-64/35} - 1) = v$$

$$v \approx 146.9$$

The velocity is approximately 146.9 feet per second.

Check Your Progress 10, *page 435*

Substitute 80 for $d(t)$ and solve for t.

$$80 = -41.71 + 25.76 \ln t$$

$$121.71 = 25.76 \ln t$$

$$\frac{121.71}{25.76} = \ln t$$

$$e^{121.71/25.76} = t$$

$$t \approx 112.7$$

Thus the year in which record distance will be 80 meters is approximately 113 years after 1900, or 2013.

Section 5.5

Check Your Progress 1, *page 441*

$N_0 = 1370$ and $k - 0.075$, so a function that gives the population t years from now is $N(t) = N_0e^{kt} = 1370e^{0.075t}$. In 5 years the population is predicted to be $N(5) = 1370e^{0.075(5)} \approx 1993$ people. If we triple the original population, we get $1370 \cdot 3 = 4110$, so substitute 4110 for $N(t)$ and solve for t.

$$4110 = 1370e^{0.075t}$$

$$\frac{4110}{1370} = e^{0.075t}$$

$$\ln\left(\frac{411}{137}\right) = 0.075t$$

$$\frac{1}{0.075}\ln\left(\frac{411}{137}\right) = t$$

$$t \approx 14.6$$

The population will triple in approximately 14.6 years.

Check Your Progress 2, *page 442*

A function that gives the number of customers is $N(t) = N_0e^{0.165t}$.

$$N(t) = N_0e^{0.165t}$$

$$= N_0(e^{0.165})^t$$

$$\approx N_0(1.179)^t$$

This equation implies that each year, the number of customers is approximately 117.9% of the number the preceding year. Thus the total annual percentage increase is about 17.9%.

Check Your Progress 3, *page 443*

Let $t = 0$ correspond to 1980. Then $N_0 = N(0) = 4800$, so an exponential growth function is $N(t) = 4800e^{kt}$. We know that the trout population was 11,500 in 2000, so $N(20) = 11{,}500$. Substitute these values to solve for k.

$$N(20) = 11{,}500$$

$$4800e^{k\cdot 20} = 11{,}500$$

$$e^{20k} = \frac{11{,}500}{4800}$$

$$20k = \ln\frac{11{,}500}{4800}$$

$$k = \frac{1}{20}\ln\frac{11{,}500}{4800}$$

$$k \approx 0.0437$$

Thus the exponential growth function is $N(t) = 4800e^{0.0437t}$. In the year 2008, $N(28) = 4800e^{0.0437(28)} \approx 16{,}317$, so the trout population will be approximately 16,300.

Check Your Progress 4, *page 444*

We have $N_0 = N(0) = 4$ grams. If t is measured in minutes, then $N(22) = 1$ because 3 grams disintegrate after 22 minutes. Substitute these values to determine the value of k.

$$N(t) = N_0e^{kt}$$

$$N(22) = 4e^{k\cdot 22}$$

$$1 = 4e^{22k}$$

$$\frac{1}{4} = e^{22k}$$

$$\ln\frac{1}{4} = 22k$$

$$\frac{1}{22}\ln\frac{1}{4} = k$$

$$k \approx -0.0630$$

The amount of material is given by $N(t) = 4e^{-0.0630t}$. After 1 hour, the amount of material remaining is $N(60) \approx 4e^{-0.0630(60)} \approx 0.091$ gram.

Check Your Progress 5, *page 445*

We have $N_0 = N(0) = 3$ ounces. The half-life of ^{226}Ra is 1660 years, so $N(1660) = \frac{3}{2}$. Substitute these values to solve for k.

$$N(1660) = N_0 e^{k \cdot 1660}$$

$$\frac{3}{2} = 3e^{1660k}$$

$$\frac{1}{2} = e^{1660k}$$

$$\ln\frac{1}{2} = 1660k$$

$$\frac{1}{1660}\ln\frac{1}{2} = k$$

$$k \approx -0.000418$$

The amount of ^{226}Ra remaining after t years is $N(t) = 3e^{-0.000418t}$ ounces.

Check Your Progress 6, *page 446*

Let t be the time at which $N(t) = 056N_0$.

$$0.56N_0 = N_0 e^{-0.000121t}$$

$$0.56 = e^{-0.000121t}$$

$$\ln 0.56 = -0.000121t$$

$$\frac{\ln 0.56}{-0.000121} = t$$

$$t \approx 4792$$

Rounded to the nearest 10 years, the bone is about 4790 years old.

Check Your Progress 7, *page 448*

Use the equation $T(t) = A + (T_0 - A)e^{kt}$. We are given $T(0) = 180 = T_0$, $T(12) = 140$, and $A = 75$. Then

$$T(12) = 75 + (180 - 75)e^{k \cdot 12}$$

$$140 = 75 + 105e^{12k}$$

$$65 = 105e^{12k}$$

$$\frac{65}{105} = e^{12k}$$

$$\ln\frac{65}{105} = 12k$$

$$\frac{1}{12}\ln\frac{65}{105} = k$$

$$k \approx -0.0400$$

Thus the function $T(t) = 75 + 105e^{-0.0400t}$ gives the temperature of the cappuccino t minutes after it was poured. To determine how long it will take for the cappuccino to cool to 100°F, substitute 100 for $T(t)$ and solve for t.

$$100 = 75 + 105e^{-0.400t}$$

$$25 = 105e^{-0.0400t}$$

$$\frac{25}{105} = e^{-0.0400t}$$

$$\ln\frac{25}{105} = -0.0400t$$

$$-\frac{1}{0.0400}\ln\frac{25}{105} = t$$

$$t \approx 35.9$$

The cappuccino will take approximately 35.9 minutes to cool to 100°F.

Section 5.6

Check Your Progress 1, *page 455*

The scatter plot of A suggests that A is an increasing function that is concave down. The most suitable model for set A is an increasing logarithmic function.

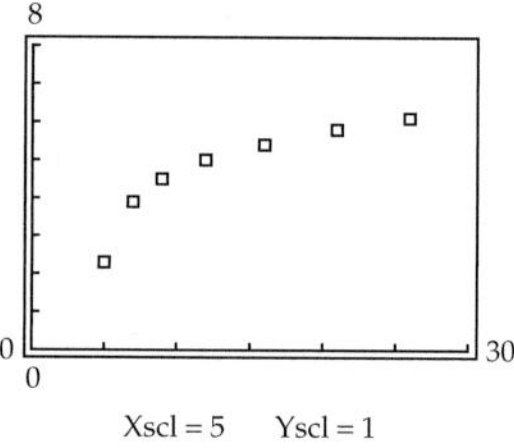

Xscl = 5 Yscl = 1

The scatter plot of B suggests that B is an increasing function that is concave upward. The most suitable model for set B is an increasing exponential function.

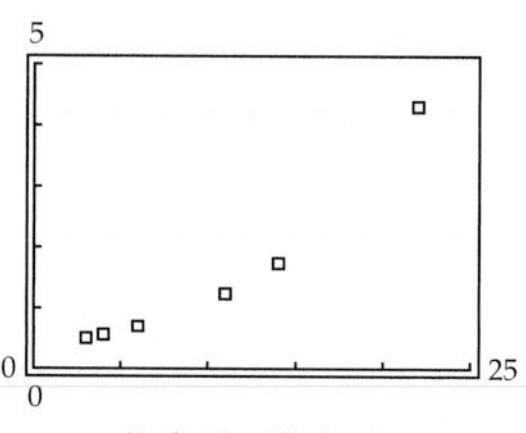

Xscl = 5 Yscl = 1

Check Your Progress 2, *page 458*

From the scatter plot in the following figure, it appears that the data can be closely modeled by an exponential function to the form $y = ab^x$, with $b < 1$.

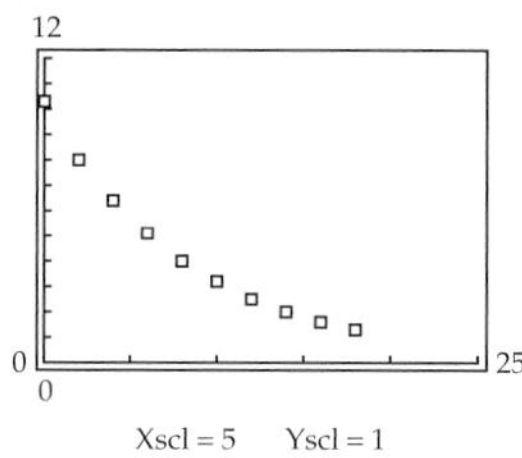

The calculator display in the following figure shows that the exponential regression equation is $y \approx 10.1468(0.89104)^x$, where x is the altitude in kilometers and y is the pressure in newtons per square centimeter.

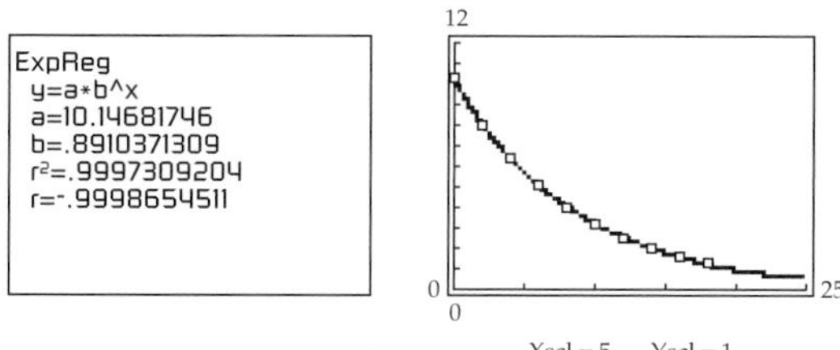

The correlation coefficient $r \approx -0.9998$ is close to -1. This indicates that the function $y \approx 10.1468(0.8910)^x$ provides a good fit for the data. The graph of y shown above also indicates that the regression function provides a good model for the data. When $x = 24$ kilometers, the atmospheric pressure is about $10.1468(0.89104)^{24} \approx 0.6$ newton per square centimeter.

Check Your Progress 3, *page 460*

a. Use a graphing utility to perform an exponential regression and a logarithm regression. For the given data, the logarithmic function $y = 61.735786 - 4.1044761 \ln x$ provides a slightly better fit than does the exponential regression function, as determined by comparing the correlation coefficients. See the calculator displays below.

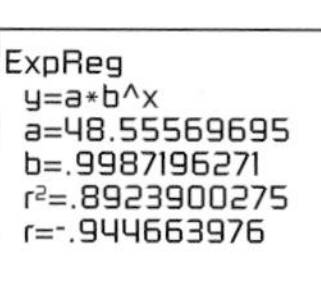

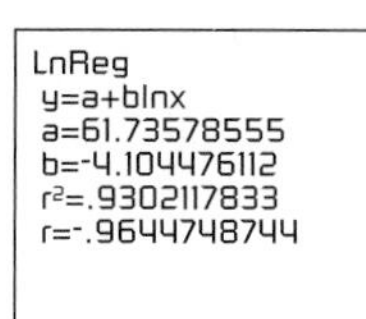

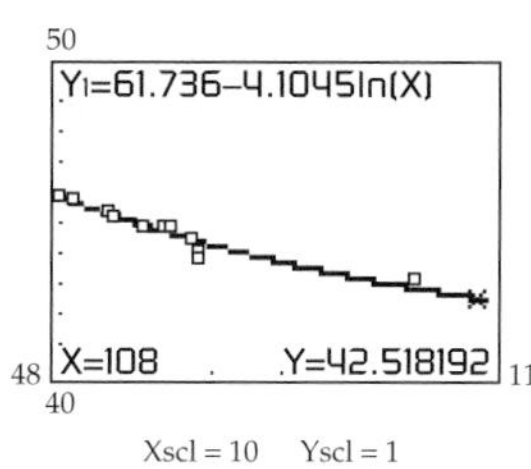

b. To predict the world record time in 2008, evaluate $y = 61.735786 - 4.104476 \ln x$ at $x = 108$. The graph on the right above shows that the predicted world record time in the men's 400-meter race for the year 2008 is about 42.52 seconds.

Section 5.7

Check Your Progress 1, *page 467*

a. $P(8) = \dfrac{6500}{1 + 9.4e^{-0.35(8)}} \approx 4136$. In 1988, the town had about 4100 households with a VCR.

b. The following graph shows that according to the logistic regression function, the number of households with a VCR reached 5500 about 11.27 years after 1980, which is the year 1991.

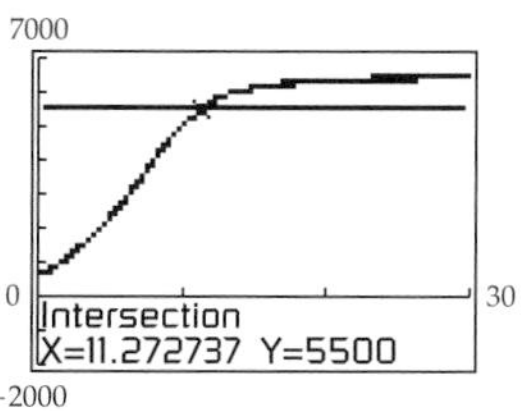

Check Your Progress 2, *page 469*

a. Represent the year 1940 by $t = 0$ and the year 1950 by $t = 10$. Because $P_0 = 635{,}000$ and $c = 1{,}000{,}000$, we have

$$a = \frac{c - P_0}{P_0} = \frac{1{,}000{,}000 - 635{,}000}{635{,}000}$$

$$= \frac{365}{635} \approx 0.574803$$

$$P(10) \approx \frac{1{,}000{,}000}{1 + 0.574803e^{-b \cdot 10}}$$ ■ Substitute 0.574803 for a, 1,000,000 for c, and 10 for t.

$$775{,}000 \approx \frac{1{,}000{,}000}{1 + 0.574803e^{-b \cdot 10}}$$ ■ Replace $P(10)$ with 775,000.

$$1 + 0.574803e^{-b \cdot 10} \approx \frac{1{,}000{,}000}{775{,}000}$$ ■ Solve for b.

$$0.574803e^{-b \cdot 10} \approx \frac{1{,}000{,}000}{775{,}000} - 1$$

$$e^{-b \cdot 10} \approx \frac{\left(\dfrac{1{,}000{,}000}{775{,}000} - 1\right)}{0.574803}$$

$$e^{-b \cdot 10} \approx 0.505082$$

$$-b \cdot 10 \approx \ln(0.505082)$$

$$b \approx -\frac{1}{10} \ln(0.505082)$$

$$b \approx 0.068303$$ ■ The growth rate constant

For the given data, the logistic growth model is

$$P(t) \approx \frac{1{,}000{,}000}{1 + 0.574803e^{-0.068303t}}$$

b. To estimate the year in which the logistic growth model predicts that San Francisco's population will first reach 995,000, replace $P(t)$ with 995,000 and solve for t.

$$995{,}000 \approx \frac{1{,}000{,}000}{1 + 0.574803e^{-0.068303t}}$$

$$995{,}000(1 + 0.574803e^{-0.068303t}) \approx 1{,}000{,}000$$

$$1 + 0.574803e^{-0.068303t} \approx \frac{1{,}000{,}000}{995{,}000} \quad \text{▪ Solve for } t.$$

$$1 + 0.574803e^{-0.068303t} \approx 1.00502513$$

$$0.574803e^{-0.068303t} \approx 0.00502513$$

$$e^{-0.068303t} \approx \frac{0.00502513}{0.574803}$$

$$-0.068303t \approx \ln\left(\frac{0.00502513}{0.574803}\right)$$

$$t \approx \frac{\ln\left(\frac{0.00502513}{0.574803}\right)}{-0.068303}$$

$$t \approx 69.39$$

Thus, according to the logistic growth model, San Francisco's population will first reach 995,000 approximately 69.39 years after 1940—that is, in 2009.

Check Your Progress 3, *page 471*

a. Use a graphing utility to perform a logistic regression on the data. The following figure shows the results obtained by using a TI-83 graphing calculator.

```
Logistic
 y=c/(1+ae^(-bx))
 a=2.249350002
 b=.0433335159
 c=1544021.968
```

The logistic regression function for the data is

$$P(t) \approx \frac{1{,}544{,}022}{1 + 2.249350e^{-0.0433335t}}.$$

b. The year 2010 is represented by $t = 60$.

$$P(60) = \frac{1{,}544{,}022}{1 + 2.249350e^{-0.0433335(60)}}$$

The logistic regression function predicts that Hawaii's population will be about 1,320,000 in 2010.

c. The carrying capacity, to the nearest thousand, of the the logistic model $P(t) = \dfrac{1{,}544{,}022}{1 + 2.249350e^{-0.0433335t}}$ is 1,544,000 people.

Check Your Progress 4, *page 473*

In Check Your Progress 2 the constant a was about 0.574803. Because $a < 1$, the graph of the logistic growth model in Check Your Progress 2 does not have an inflection point. This means that according to the logistic growth model, the population of San Francisco reached its greatest rate of growth before the year 1940.

Chapter 6

Section 6.1

Check Your Progress 1, *page 487*

The monthly cost of a cellular phone plan from the first company is $C = 15 + 0.08x$, where x is the number of minutes used. The cost of a plan from the second provider is $C = 12 + 0.12x$. Graph both equations on the same set of axes.

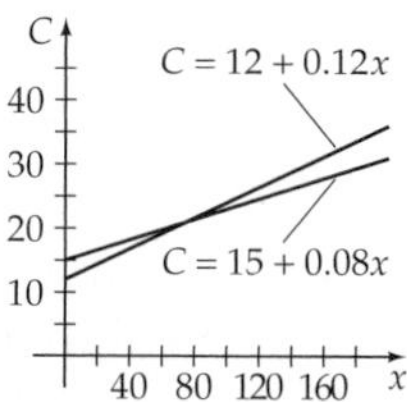

The two lines appear to intersect at $x \approx 75$. To verify that this is the case, enter 75 for x in the cost equations.

$C = 15 + 0.08(75) = 21$

$C = 12 + 0.12(75) = 21$

So both companies charge the same amount ($21.00) for 75 minutes of phone use.

Check Your Progress 2, *page 489*

$$\begin{cases} 8x + 3y = -7 & (1) \\ x = 3y + 15 & (2) \end{cases}$$

$$8(3y + 15) + 3y = -7 \quad \text{▪ Substitute } 3y + 15 \text{ for } x \text{ in Eq. (1).}$$

$$24y + 120 + 3y = -7 \quad \text{▪ Simplify.}$$

$$27y = -127$$

$$y = -\frac{127}{27}$$

$$x = 3\left(-\frac{127}{27}\right) + 15 = \frac{8}{9} \quad \text{▪ Substitute } -\frac{127}{27} \text{ for } y \text{ in Eq. (2).}$$

The solution is $\left(\frac{8}{9}, -\frac{127}{27}\right)$.

Check Your Progress 3, *page 490*

$$\begin{cases} 3x - 4y = 8 & (1) \\ 6x - 8y = 9 & (2) \end{cases}$$

$8y = 6x - 9$ ■ Solve Eq. (2) for y.

$y = \frac{3}{4}x - \frac{9}{8}$

$3x - 4\left(\frac{3}{4}x - \frac{9}{8}\right) = 8$ ■ Replace y in Eq. (1).

$3x - 3x + \frac{9}{2} = 8$ ■ Simplify.

$\frac{9}{2} = 8$

This is a false statement. Therefore, the system of equations is inconsistent and has no solution.

Check Your Progress 4, *page 491*

$$\begin{cases} 5x + 2y = 2 & (1) \\ y = -\frac{5}{2}x + 1 & (2) \end{cases}$$

$5x + 2\left(-\frac{5}{2}x + 1\right) = 2$ ■ Replace y in Eq. (1).

$5x - 5x + 2 = 2$ ■ Simplify.

$2 = 2$

This is a true statement; therefore, the system of equations is dependent. Let $x = c$. Then $y = -\frac{5}{2}c + 1$. Thus the solutions are ordered pairs of the form $\left(c, -\frac{5}{2}c + 1\right)$. Note: The solutions can also be expressed as ordered pairs of the form $\left(-\frac{2}{5}c + \frac{2}{5}, c\right)$.

Check Your Progress 5, *page 494*

$$\begin{cases} 3x - 8y = -6 & (1) \\ -5x + 4y = 10 & (2) \end{cases}$$

$3x - 8y = -6$

$-10x + 8y = 20$ ■ 2 times Eq. (2)

$-7x = 14$

$x = -2$

$3(-2) - 8y = -6$ ■ Substitute -2 for x in Eq. (1).

$-8y = 0$ ■ Solve for y.

$y = 0$

The solution is $(-2, 0)$.

Check Your Progress 6, *page 495*

$$\begin{cases} 4x + 5y = 2 & (1) \\ 8x + 10y = 4 & (2) \end{cases}$$

$-8x - 10y = -4$ ■ -2 times Eq. (1)

$8x + 10y = 4$

$0 = 0$

The system is dependent. The solutions are all ordered pairs that satisfy $4x + 5y = 2$. Solving for y gives $y = -\frac{4}{5}x + \frac{2}{5}$.

Because x can be replaced by any real number c, the solutions of the system of equations are the ordered pairs of the form $\left(c, -\frac{4}{5}c + \frac{2}{5}\right)$.

Check Your Progress 7, *page 496*

Let A represent the price of an adult ticket and C the price of a child ticket. The first family paid \$108 for two adult tickets and four child tickets, so $2A + 4C = 108$. The second family paid \$120 for three of each type of ticket, so $3A + 3C = 120$. Solve the system of equations

$$\begin{cases} 2A + 4C = 108 & (1) \\ 3A + 3C = 120 & (2) \end{cases}$$

We can solve this system by the substitution method. From Equation (1), $A = \frac{108 - 4C}{2} = 54 - 2C$. Substituting into Equation (2) gives

$3(54 - 2C) + 3C = 120$ ■ Substitute $54 - 2C$ for A.

$162 - 6C + 3C = 120$ ■ Simplify.

$-3C = -42$

$C = 14$

Then $A = 54 - 2C = 54 - 2(14) = 26$. An adult ticket costs \$26 and a child ticket costs \$14.

Check Your Progress 8, *page 497*

Let r = the rate of the canoeist.
Let w = the rate of the current.
Rate of canoeist with the current: $r + w$
Rate of canoeist against the current: $r - w$

rate · time = distance

$(r + w) \cdot 2 = 12$ (1)

$(r - w) \cdot 4 = 12$ (2)

$r + w = 6$ ■ Divide Eq. (1) by 2.

$r - w = 3$ ■ Divide Eq. (2) by 4.

$2r = 9$

$r = 4.5$

$4.5 + w = 6$

$w = 1.5$

Rate of canoeist = 4.5 mph
Rate of current = 1.5 mph

Section 6.2

Check Your Progress 1, *page 503*

$$\begin{cases} 3x + 2y - 2z = -5 & (1) \\ 2x - y - 4z = 3 & (2) \\ 2x + 3y + 3z = -3 & (3) \end{cases}$$

Solve Equation (2) for y.

$y = 2x - 4z - 3 \quad (4)$

Substitute the result into Equations (1) and (3).

$3x + 2(2x - 4z - 3) - 2z = -5$ ▪ Substitute $2x - 4z - 3$ for y in Equation (1).

$3x + 4x - 8z - 6 - 2z = -5$ ▪ Simplify.

$7x - 10z = 1 \quad (5)$

$2x + 3(2x - 4z - 3) + 3z = -3$ ▪ Substitute $2x - 4z - 3$ for y in Equation (3).

$2x + 6x - 12z - 9 + 3z = -3$ ▪ Simplify.

$8x - 9z = 6 \quad (6)$

Solve the system of equations formed from Equations (5) and (6).

$$\begin{cases} 7x - 10z = 1 & \text{Multiply by 8.} \rightarrow & 56x - 80z = 8 \\ 8x - 9z = 6 & \text{Multiply by } -7. \rightarrow & -56x + 63z = -42 \end{cases}$$

$$-17z = -34$$

$$z = 2$$

Substitute $z = 2$ into Equation (5) and solve for x.

$7x - 10(2) = 1$ ▪ Equation (5)

$x = 3$

Now substitute $x = 3$ and $z = 2$ into Equation (4) to determine the value of y.

$y = 2x - 4z - 3 = 2(3) - 4(2) - 3 = -5$

The solution is $(3, -5, 2)$.

Check Your Progress 2, *page 506*

$$\begin{cases} 3x + 2y - 5z = 6 & (1) \\ 5x - 4y + 3z = -12 & (2) \\ 4x + 5y - 2z = 15 & (3) \end{cases}$$

$15x + 10y - 25z = 30$ ▪ 5 times Eq. (1)

$-15x + 12y - 9z = 36$ ▪ -3 times Eq. (2)

$22y - 34z = 66$ ▪ Divide by 2.

$11y - 17z = 33 \quad (4)$

$12x + 8y - 20z = 24$ ▪ 4 times Eq. (1)

$-12x - 15y + 6z = -45$ ▪ -3 times Eq. (3)

$-7y - 14z = -21$ ▪ Divide by -7.

$y + 2z = 3 \quad (5)$

$11y - 17z = 33 \quad (4)$

$-11y - 22z = -33$ ▪ -11 times Eq. (5)

$-39z = 0$

$z = 0 \quad (6)$

$11y - 17(0) = 33$

$y = 3$

$3x + 2(3) - 5(0) = 6$

$x = 0$

The solution is $(0, 3, 0)$.

Check Your Progress 3, *page 506*

$$\begin{cases} -2x + 5y + 3z = 8 & (1) \\ x - 3y - 2z = 3 & (2) \\ 2x - 4y - 2z = -5 & (3) \end{cases}$$

Eliminate x from Equation (2) by multiplying Equation (2) by 2 and adding it to Equation (1). Replace Equation (2) by the new equation. Eliminate x from Equation (3) by adding Equation (1) to Equation (3). Replace Equation (3). The resulting equivalent system is

$$\begin{cases} -2x + 5y + 3z = 8 & (1) \\ -y - z = 14 & (4) \\ y + z = 3 & (5) \end{cases}$$

Eliminate y from Equation (5) by adding Equation (4) to Equation (5).

$-y - z = 14$ ▪ Equation (4)

$y + z = 3$ ▪ Equation (5)

$0 = 17 \quad (6)$ ▪ Add the equations.

The resulting equivalent system is

$$\begin{cases} -2x + 5y + 3z = 8 & (1) \\ -y - z = 14 & (4) \\ 0 = 17 & (6) \end{cases}$$

Equation (6) is a false equation, so the system is inconsistent and has no solution.

Check Your Progress 4, *page 508*

$$\begin{cases} 2x + 3y - 6z = 4 & (1) \\ 3x - 2y - 9z = -7 & (2) \\ 2x + 5y - 6z = 8 & (3) \end{cases}$$

$6x + 9y - 18z = 12$ ▪ 3 times Eq. (1)

$-6x + 4y + 18z = 14$ ▪ -2 times Eq. (2)

$13y = 26$

$y = 2 \quad (4)$

$2x + 3y - 6z = 4 \quad (1)$

$-2x - 5y + 6z = -8$ ▪ -1 times Eq. (3)

$-2y = -4$

$$\begin{aligned} y \quad &= 2 \quad (5) \\ y &= 2 \quad (4) \\ -y &= -2 \qquad \blacksquare\ -1 \text{ times Eq. (5)} \\ 0 &= 0 \quad (6) \end{aligned}$$

The equations are dependent. Let $z = c$.

$2x + 3(2) - 6c = 4$ ▪ Substitute $y = 2$ and $z = c$ in Eq. (1).

$x = 3c - 1$

The solutions are ordered triples of the form $(3c - 1, 2, c)$.

Check Your Progress 5, *page 509*

$$\begin{cases} x - 3y + 4z = 9 & (1) \\ 3x - 8y - 2z = 4 & (2) \end{cases}$$

$-3x + 9y - 12z = -27$ ▪ -3 times Eq. (1)

$3x - 8y - 2z = 4$ (2)

$y - 14z = -23$ (3)

$y = 14z - 23$ ▪ Solve Eq. (3) for y.

$x - 3(14z - 23) + 4z = 9$ ▪ Substitute $14z - 23$ for y in Eq. (1).

$x = 38z - 60$ ▪ Solve for x.

Let $z = c$. The solutions are $(38c - 60, 14c - 23, c)$.

Check Your Progress 6, *page 510*

Substitute each of the ordered pairs into the equation $y = ax^2 + bx + c$.

$$\begin{cases} -5 = a(-1)^2 + b(-1) + c \\ 5 = a(1)^2 + b(1) + c \\ -1 = a(3)^2 + b(3) + c \end{cases} \text{ or } \begin{cases} a - b + c = -5 & (1) \\ a + b + c = 5 & (2) \\ 9a + 3b + c = -1 & (3) \end{cases}$$

Solve the system of equations for a, b, and c. Eliminate a from Equation (2) by multiplying Equation (1) by -1 and adding it to Equation (2). Replace Equation (2). Then eliminate a from Equation (3) by multiplying Equation (1) by -9 and adding it to Equation (3). Replace Equation (3). The resulting equivalent system is

$$\begin{cases} a - b + c = -5 & (1) \\ 2b = 10 & (4) \\ 12b - 8c = 44 & (5) \end{cases}$$

Solving Equation (4) yields $b = 5$. Substituting this value into Equation (5) gives $c = 2$. Substituting these values into the first equation, we get $a - 5 + 2 = -5$, or $a = -2$. The polynomial whose graph passes through $(-1, -5)$, $(1, 5)$, and $(3, -1)$ is $y = -2x^2 + 5x + 2$.

Check Your Progress 7, *page 511*

$$\begin{cases} x + y + z = 750 & (1) \\ 10x + 7y + 5z = 5400 & (2) \\ z = x + 20 & (3) \end{cases}$$

Solve the system of equations by substitution. Replace z in equation (1) and equation (2) by $x + 20$.

$$\begin{aligned} x + y + (x + 20) &= 750 \\ 10x + 7y + 5(x + 20) &= 5400 \\ 2x + y &= 730 \quad (4) \\ 15x + 7y &= 5300 \quad (5) \end{aligned}$$

Multiply equation (4) by -7 and add to equation (5).

$$\begin{aligned} -14x - 7y &= -5110 \\ 15x + 7y &= 5300 \\ x &= 190 \end{aligned}$$

Substitute 190 for x into equation (4) and solve for y.

$$\begin{aligned} 2x + y &= 730 \\ 2(190) + y &= 730 \\ 380 + y &= 730 \\ y &= 350 \end{aligned}$$

Substitute the values of x and y into equation (1) and solve for z.

$$\begin{aligned} x + y + z &= 750 \\ 190 + 350 + z &= 750 \\ 540 + z &= 750 \\ z &= 210 \end{aligned}$$

The museum sold 190 full-priced tickets, 350 member tickets, and 210 student tickets.

Section 6.3

Check Your Progress 1, *page 516*

$$\begin{cases} 2w - y + 3z = 4 \\ w - 5x + 2y = -1 \\ 3w + x + y - 2z = 5 \\ 6w - 4x + 5y + z = 7 \end{cases}$$

Check Your Progress 2, *page 518*

a. $\begin{bmatrix} -3 & 6 & 9 \\ 8 & 1 & -2 \\ -2 & 3 & 5 \end{bmatrix} \quad R_2 \leftrightarrow R_3 \quad \begin{bmatrix} -3 & 6 & 9 \\ -2 & 3 & 5 \\ 8 & 1 & -2 \end{bmatrix}$

b. $\begin{bmatrix} -3 & 6 & 9 \\ 8 & 1 & -2 \\ -2 & 3 & 5 \end{bmatrix} \xrightarrow{-\frac{1}{3}R_1} \begin{bmatrix} 1 & -2 & -3 \\ 8 & 1 & -2 \\ -2 & 3 & 5 \end{bmatrix}$

c. $\begin{bmatrix} -3 & 6 & 9 \\ 8 & 1 & -2 \\ -2 & 3 & 5 \end{bmatrix} \xrightarrow{4R_3 + R_2} \begin{bmatrix} -3 & 6 & 9 \\ 0 & 13 & 18 \\ -2 & 3 & 5 \end{bmatrix}$

Check Your Progress 3, *page 521*

$$\begin{bmatrix} 3 & -6 & 12 \\ 2 & 1 & -3 \end{bmatrix} \xrightarrow{\frac{1}{3}R_1} \begin{bmatrix} 1 & -2 & 4 \\ 2 & 1 & -3 \end{bmatrix}$$

- Change a_{11} to 1 by multiplying row 1 by the reciprocal of a_{11}.

$$\begin{bmatrix} 1 & -2 & 4 \\ 2 & 1 & -3 \end{bmatrix} \xrightarrow{-2R_1 + R_2} \begin{bmatrix} 1 & -2 & 4 \\ 0 & 5 & -11 \end{bmatrix}$$

- Change a_{21} to 0 by multiplying row 1 by the opposite of a_{21} and then adding to row 2.

$$\begin{bmatrix} 1 & -2 & 4 \\ 0 & 5 & -11 \end{bmatrix} \xrightarrow{\frac{1}{5}R_2} \begin{bmatrix} 1 & -2 & 4 \\ 0 & 1 & -\frac{11}{5} \end{bmatrix}$$

- Change a_{22} to 1 by multiplying row 2 by the reciprocal of a_{22}.

Check Your Progress 4, *page 524*

$$\left[\begin{array}{ccc|c} 2 & 3 & -2 & -10 \\ 3 & 1 & -3 & -8 \\ -4 & 2 & 1 & -11 \end{array}\right]$$

- This is the augmented matrix.

$$\left[\begin{array}{ccc|c} 2 & 3 & -2 & -10 \\ 3 & 1 & -3 & -8 \\ -4 & 2 & 1 & -11 \end{array}\right] \xrightarrow{\frac{1}{2}R_1} \left[\begin{array}{ccc|c} 1 & \frac{3}{2} & -1 & -5 \\ 3 & 1 & -3 & -8 \\ -4 & 2 & 1 & -11 \end{array}\right]$$

- Multiply row 1 by $\frac{1}{2}$ to get a 1 in a_{11}.

$$\left[\begin{array}{ccc|c} 1 & \frac{3}{2} & -1 & -5 \\ 3 & 1 & -3 & -8 \\ -4 & 2 & 1 & -11 \end{array}\right] \xrightarrow[4R_1 + R_3]{-3R_1 + R_2} \left[\begin{array}{ccc|c} 1 & \frac{3}{2} & -1 & -5 \\ 0 & -\frac{7}{2} & 0 & 7 \\ 0 & 8 & -3 & -31 \end{array}\right]$$

- Get a 0 in a_{21} and a_{31}.

$$\left[\begin{array}{ccc|c} 1 & \frac{3}{2} & -1 & -5 \\ 0 & -\frac{7}{2} & 0 & 7 \\ 0 & 8 & -3 & -31 \end{array}\right] \xrightarrow{-\frac{2}{7}R_2} \left[\begin{array}{ccc|c} 1 & \frac{3}{2} & -1 & -5 \\ 0 & 1 & 0 & -2 \\ 0 & 8 & -3 & -31 \end{array}\right]$$

- Multiply row 2 by $-\frac{2}{7}$ to get a 1 in a_{22}.

$$\left[\begin{array}{ccc|c} 1 & -\frac{3}{2} & -1 & -5 \\ 0 & 1 & 0 & -2 \\ 0 & 8 & -3 & -31 \end{array}\right] \xrightarrow{-8R_2 + R_3} \left[\begin{array}{ccc|c} 1 & \frac{3}{2} & -1 & -5 \\ 0 & 1 & 0 & -2 \\ 0 & 0 & -3 & -15 \end{array}\right]$$

- Get a 0 in a_{32}.

$$\left[\begin{array}{ccc|c} 1 & \frac{3}{2} & -1 & -5 \\ 0 & 1 & 0 & -2 \\ 0 & 0 & -3 & -15 \end{array}\right] \xrightarrow{-\frac{1}{3}R_3} \left[\begin{array}{ccc|c} 1 & \frac{3}{2} & -1 & -5 \\ 0 & 1 & 0 & -2 \\ 0 & 0 & 1 & 5 \end{array}\right]$$

- Multiply row 3 by $-\frac{1}{3}$ to get a 1 in a_{33}.

The last matrix is in row echelon form. The system of equations written from the matrix is

$$\begin{cases} x + \frac{3}{2}y - z = -5 \\ \qquad y \qquad = -2 \\ \qquad\qquad z = 5 \end{cases}$$

From this system of equations, $z = 5$ and $y = -2$. Use back substitution to find $x = 3$. Thus $(3, -2, 5)$ is the solution of the system of equations.

Check Your Progress 5, *page 525*

$$\left[\begin{array}{ccc|c} 2 & -3 & -3 & 19 \\ 1 & 4 & -7 & -18 \\ -3 & 1 & 8 & 11 \end{array}\right] \quad R_1 \leftrightarrow R_2 \quad \left[\begin{array}{ccc|c} 1 & 4 & -7 & -18 \\ 2 & -3 & -3 & 19 \\ -3 & 1 & 8 & 11 \end{array}\right]$$

$$\left[\begin{array}{ccc|c} 1 & 4 & -7 & -18 \\ 2 & -3 & -3 & 19 \\ -3 & 1 & 8 & 11 \end{array}\right] \xrightarrow[3R_1 + R_3]{-2R_1 + R_2} \left[\begin{array}{ccc|c} 1 & 4 & -7 & -18 \\ 0 & -11 & 11 & 55 \\ 0 & 13 & -13 & -43 \end{array}\right]$$

$$\left[\begin{array}{ccc|c} 1 & 4 & -7 & -18 \\ 0 & -11 & 11 & 55 \\ 0 & 13 & -13 & -43 \end{array}\right] \xrightarrow{-\frac{1}{11}R_2} \left[\begin{array}{ccc|c} 1 & 4 & -7 & -18 \\ 0 & 1 & -1 & -5 \\ 0 & 13 & -13 & -43 \end{array}\right]$$

$$\left[\begin{array}{ccc|c} 1 & 4 & -7 & -18 \\ 0 & 1 & -1 & -5 \\ 0 & 13 & -13 & -43 \end{array}\right] \xrightarrow{-13R_2 + R_3} \left[\begin{array}{ccc|c} 1 & 4 & -7 & -18 \\ 0 & 1 & -1 & -5 \\ 0 & 0 & 0 & 22 \end{array}\right]$$

The equivalent system of equations is

$$\begin{cases} x + 4y - 7z = -18 \\ \qquad y - z = -5 \\ \qquad\quad 0z = 22 \end{cases}$$

Because the equation $0z = 22$ has no solution, the system of equations has no solution. It is an inconsistent system of equations.

Check Your Progress 6, *page 526*

$$\left[\begin{array}{ccc|c} 1 & -4 & 1 & 3 \\ 2 & 3 & -5 & 4 \\ -1 & 15 & -8 & -5 \end{array}\right] \xrightarrow[R_1 + R_3]{-2R_1 + R_2} \left[\begin{array}{ccc|c} 1 & -4 & 1 & 3 \\ 0 & 11 & -7 & -2 \\ 0 & 11 & -7 & -2 \end{array}\right]$$

$$\left[\begin{array}{ccc|c} 1 & -4 & 1 & 3 \\ 0 & 11 & -7 & -2 \\ 0 & 11 & -7 & -2 \end{array}\right] \xrightarrow{\frac{1}{11}R_2} \left[\begin{array}{ccc|c} 1 & -4 & 1 & 3 \\ 0 & 1 & -\frac{7}{11} & -\frac{2}{11} \\ 0 & 11 & -7 & -2 \end{array}\right]$$

$$\left[\begin{array}{ccc|c} 1 & -4 & 1 & 3 \\ 0 & 1 & -\frac{7}{11} & -\frac{2}{11} \\ 0 & 11 & -7 & -2 \end{array}\right] \xrightarrow{-11R_2 + R_3} \left[\begin{array}{ccc|c} 1 & -4 & 1 & 3 \\ 0 & 1 & -\frac{7}{11} & -\frac{2}{11} \\ 0 & 0 & 0 & 0 \end{array}\right]$$

The equivalent system of equations is

$$\begin{aligned} x - 4y + \quad z &= \ \ 3 \\ y - \frac{7}{11}z &= -\frac{2}{11} \end{aligned}$$

Solving $y - \frac{7}{11}z = -\frac{2}{11}$ for y in terms of z gives $y = \frac{7}{11}z - \frac{2}{11}$. If we substitute this value of y into $x - 4y + z = 3$, we can solve the resulting equation for x in terms of z.

$$\begin{aligned} x - 4y + z &= 3 \\ x - 4\left(\frac{7}{11}z - \frac{2}{11}\right) + z &= 3 \\ x - \frac{28}{11}z + \frac{8}{11} + z &= 3 \\ x &= \frac{17}{11}z + \frac{25}{11} \end{aligned}$$

The solutions are the ordered triples $\left(\frac{17}{11}c + \frac{25}{11}, \frac{7}{11}c - \frac{2}{11}, c\right)$ where c is any real number.

Check Your Progress 7, *page 527*

Because there are four given points, the degree of the interpolating polynomial will be at most 3, one less than the number of ordered pairs. The form of the polynomial will be $p(x) = a_3x^3 + a_2x^2 + a_1x + a_0$. Use this polynomial to create a system of equations.

$$\begin{aligned} p(x) &= a_3x^3 + a_2x^2 + a_1x + a_0 \\ p(2) &= 8a_3 + 4a_2 + 2a_1 + a_0 = 4 \\ p(-1) &= -a_3 + a_2 - a_1 + a_0 = -5 \\ p(0) &= a_0 = 0 \\ p(1) &= a_3 + a_2 + a_1 + a_0 = 1 \end{aligned}$$

The system of equations and the associated augmented matrix are shown below.

$$\begin{cases} 8a_3 + 4a_2 + 2a_1 + a_0 = 4 \\ -a_3 + a_2 - a_1 + a_0 = -5 \\ a_0 = 0 \\ a_3 + a_2 + a_1 + a_0 = 1 \end{cases} \longrightarrow \left[\begin{array}{cccc|c} 8 & 4 & 2 & 1 & 4 \\ -1 & 1 & -1 & 1 & -5 \\ 0 & 0 & 0 & 1 & 0 \\ 1 & 1 & 1 & 1 & 1 \end{array}\right]$$

The augmented matrix in echelon form and the resulting system of equations are shown below.

$$\left[\begin{array}{cccc|c} 1 & \frac{1}{2} & \frac{1}{4} & \frac{1}{8} & \frac{1}{2} \\ 0 & 1 & -\frac{1}{2} & \frac{3}{4} & -3 \\ 0 & 0 & 1 & \frac{1}{2} & 2 \\ 0 & 0 & 0 & 1 & 0 \end{array}\right] \longrightarrow \begin{cases} a_3 + \frac{1}{2}a_2 + \frac{1}{4}a_1 + \frac{1}{8}a_0 = \frac{1}{2} \\ a_2 - \frac{1}{2}a_1 + \frac{3}{4}a_0 = -3 \\ a_1 + \frac{1}{2}a_0 = 2 \\ a_0 = 0 \end{cases}$$

Solving the system of equations by back substitution, the interpolating polynomial is $p(x) = x^3 - 2x^2 + 2x$.

Section 6.4

Check Your Progress 1, *page 534*

a. $$D + F = \begin{bmatrix} 2 & 9 \\ -4 & 7 \\ -3 & 8 \end{bmatrix} + \begin{bmatrix} 3 & -2 \\ 7 & 8 \\ -2 & -3 \end{bmatrix} = \begin{bmatrix} 2+3 & 9-2 \\ -4+7 & 7+8 \\ -3-2 & 8-3 \end{bmatrix} = \begin{bmatrix} 5 & 7 \\ 3 & 15 \\ -5 & 5 \end{bmatrix}$$

b. $$F - D = \begin{bmatrix} 3 & -2 \\ 7 & 8 \\ -2 & -3 \end{bmatrix} - \begin{bmatrix} 2 & 9 \\ -4 & 7 \\ -3 & 8 \end{bmatrix} = \begin{bmatrix} 3-2 & -2-9 \\ 7+4 & 8-7 \\ -2+3 & -3-8 \end{bmatrix} = \begin{bmatrix} 1 & -11 \\ 11 & 1 \\ 1 & -11 \end{bmatrix}$$

Check Your Progress 2, *page 535*

$$\begin{aligned} 4A - 5B &= 4\begin{bmatrix} -3 & 5 \\ 6 & -7 \end{bmatrix} - 5\begin{bmatrix} 9 & -11 \\ 3 & 8 \end{bmatrix} \\ &= \begin{bmatrix} 4(-3) & 4(5) \\ 4(6) & 4(-7) \end{bmatrix} - \begin{bmatrix} 5(9) & 5(-11) \\ 5(3) & 5(8) \end{bmatrix} \\ &= \begin{bmatrix} -12 & 20 \\ 24 & -28 \end{bmatrix} - \begin{bmatrix} 45 & -55 \\ 15 & 40 \end{bmatrix} = \begin{bmatrix} -57 & 75 \\ 9 & -68 \end{bmatrix} \end{aligned}$$

Check Your Progress 3, *page 539*

$$\begin{aligned} EF &= \begin{bmatrix} 2 & -1 \\ 3 & 4 \\ 1 & 0 \end{bmatrix}\begin{bmatrix} 5 & -2 \\ 4 & 3 \end{bmatrix} \\ &= \begin{bmatrix} 2(5) + (-1)4 & 2(-2) + (-1)3 \\ 3(5) + 4(4) & 3(-2) + 4(3) \\ 1(5) + 0(4) & 1(-2) + 0(3) \end{bmatrix} = \begin{bmatrix} 6 & -7 \\ 31 & 6 \\ 5 & -2 \end{bmatrix} \end{aligned}$$

Check Your Progress 4, *page 543*

Use a graphing calculator to evaluate the expression below for $n = 5$.

$$[0.25 \quad 0.75]\begin{bmatrix} 0.98 & 0.02 \\ 0.05 & 0.95 \end{bmatrix}^n$$

$$[0.25 \quad 0.75]\begin{bmatrix} 0.98 & 0.02 \\ 0.05 & 0.95 \end{bmatrix}^5 \approx [0.391 \quad 0.609]$$

After 5 months, about 39.1% of the people shop at Store A.

Section 6.5

Check Your Progress 1, *page 548*

Find AB.

$$AB = \begin{bmatrix} 5 & 3 \\ 7 & 4 \end{bmatrix}\begin{bmatrix} 4 & -3 \\ -7 & 5 \end{bmatrix} = \begin{bmatrix} -1 & 0 \\ 0 & -1 \end{bmatrix} \neq \begin{bmatrix} 1 & 0 \\ 0 & 1 \end{bmatrix}$$

Because $AB \neq I_2$ (the 2×2 identity matrix), A and B are not inverses of one another.

Check Your Progress 2, *page 551*

$$\left[\begin{array}{ccc|ccc} 3 & 4 & 2 & 1 & 0 & 0 \\ 1 & 2 & 1 & 0 & 1 & 0 \\ 3 & 0 & 1 & 0 & 0 & 1 \end{array}\right]$$

■ Merge the given matrix with the identity matrix I_3.

$$\xrightarrow{R_1 \leftrightarrow R_2} \left[\begin{array}{ccc|ccc} 1 & 2 & 1 & 0 & 1 & 0 \\ 3 & 4 & 2 & 1 & 0 & 0 \\ 3 & 0 & 1 & 0 & 0 & 1 \end{array}\right]$$

$$\xrightarrow[-3R_1 + R_3]{-3R_1 + R_2} \left[\begin{array}{ccc|ccc} 1 & 2 & 1 & 0 & 1 & 0 \\ 0 & -2 & -1 & 1 & -3 & 0 \\ 0 & -6 & -2 & 0 & -3 & 1 \end{array}\right]$$

$$\xrightarrow{-\frac{1}{2}R_2} \left[\begin{array}{ccc|ccc} 1 & 2 & 1 & 0 & 1 & 0 \\ 0 & 1 & \frac{1}{2} & -\frac{1}{2} & \frac{3}{2} & 0 \\ 0 & -6 & -2 & 0 & -3 & 1 \end{array}\right]$$

$$\xrightarrow{6R_2 + R_3} \left[\begin{array}{ccc|ccc} 1 & 2 & 1 & 0 & 1 & 0 \\ 0 & 1 & \frac{1}{2} & -\frac{1}{2} & \frac{3}{2} & 0 \\ 0 & 0 & 1 & -3 & 6 & 1 \end{array}\right]$$

$$\xrightarrow[-R_3 + R_1]{-\frac{1}{2}R_3 + R_2} \left[\begin{array}{ccc|ccc} 1 & 2 & 0 & 3 & -5 & -1 \\ 0 & 1 & 0 & 1 & -\frac{3}{2} & -\frac{1}{2} \\ 0 & 0 & 1 & -3 & 6 & 1 \end{array}\right]$$

$$\xrightarrow{-2R_2 + R_1} \left[\begin{array}{ccc|ccc} 1 & 0 & 0 & 1 & -2 & 0 \\ 0 & 1 & 0 & 1 & -\frac{3}{2} & -\frac{1}{2} \\ 0 & 0 & 1 & -3 & 6 & 1 \end{array}\right]$$

The inverse matrix is $A^{-1} = \begin{bmatrix} 1 & -2 & 0 \\ 1 & -\frac{3}{2} & -\frac{1}{2} \\ -3 & 6 & 1 \end{bmatrix}$.

Check Your Progress 3, *page 552*

$$\left[\begin{array}{ccc|ccc} 4 & -3 & 3 & 1 & 0 & 0 \\ 2 & 1 & -4 & 0 & 1 & 0 \\ 6 & -2 & -1 & 0 & 0 & 1 \end{array}\right]$$

$$\xrightarrow{\frac{1}{4}R_1} \left[\begin{array}{ccc|ccc} 1 & -\frac{3}{4} & \frac{3}{4} & \frac{1}{4} & 0 & 0 \\ 2 & 1 & -4 & 0 & 1 & 0 \\ 6 & -2 & -1 & 0 & 0 & 1 \end{array}\right]$$

$$\xrightarrow[-6R_1 + R_3]{-2R_1 + R_2} \left[\begin{array}{ccc|ccc} 1 & -\frac{3}{4} & \frac{3}{4} & \frac{1}{4} & 0 & 0 \\ 0 & \frac{5}{2} & -\frac{11}{2} & -\frac{1}{2} & 1 & 0 \\ 0 & \frac{5}{2} & -\frac{11}{2} & -\frac{3}{2} & 0 & 1 \end{array}\right]$$

$$\xrightarrow{\frac{2}{5}R_2} \left[\begin{array}{ccc|ccc} 1 & -\frac{3}{4} & \frac{3}{4} & \frac{1}{4} & 0 & 0 \\ 0 & 1 & -\frac{11}{5} & -\frac{1}{5} & \frac{2}{5} & 0 \\ 0 & \frac{5}{2} & -\frac{11}{2} & -\frac{3}{2} & 0 & 1 \end{array}\right]$$

$$\xrightarrow{-\frac{5}{2}R_2 + R_3} \left[\begin{array}{ccc|ccc} 1 & -\frac{3}{4} & \frac{3}{4} & \frac{1}{4} & 0 & 0 \\ 0 & 1 & -\frac{11}{5} & -\frac{1}{5} & \frac{2}{5} & 0 \\ 0 & 0 & 0 & -1 & -1 & 1 \end{array}\right]$$

There are zeros in a row of the original matrix. The original matrix does not have an inverse.

Check Your Progress 4, *page 553*

Write the system of equations as a matrix equation of the form $AX = B$.

$$\begin{bmatrix} 1 & 2 & -1 \\ 2 & 3 & -1 \\ 3 & 6 & -2 \end{bmatrix}\begin{bmatrix} x \\ y \\ z \end{bmatrix} = \begin{bmatrix} 5 \\ 8 \\ 14 \end{bmatrix}$$

Find A^{-1}.

$$A^{-1} = \begin{bmatrix} 0 & 2 & -1 \\ -1 & -1 & 1 \\ -3 & 0 & 1 \end{bmatrix}$$

Multiply each side of the matrix equation by the inverse matrix.

$$\begin{bmatrix} 0 & 2 & -1 \\ -1 & -1 & 1 \\ -3 & 0 & 1 \end{bmatrix}\begin{bmatrix} 1 & 2 & -1 \\ 2 & 3 & -1 \\ 3 & 6 & -2 \end{bmatrix}\begin{bmatrix} x \\ y \\ z \end{bmatrix} = \begin{bmatrix} 0 & 2 & -1 \\ -1 & -1 & 1 \\ -3 & 0 & 1 \end{bmatrix}\begin{bmatrix} 5 \\ 8 \\ 14 \end{bmatrix}$$

$$\begin{bmatrix} x \\ y \\ z \end{bmatrix} = \begin{bmatrix} 2 \\ 1 \\ -1 \end{bmatrix}$$

Thus $x = 2$, $y = 1$, and $z = -1$. The solution of the system of equations is the ordered triple $(2, 1, -1)$.

Section 6.6

Check Your Progress 1, *page 560*

Graph $y = \frac{1}{2}x + 2$ as a solid line. Because the inequality is $y \geq \frac{1}{2}x + 2$, shade above the line.

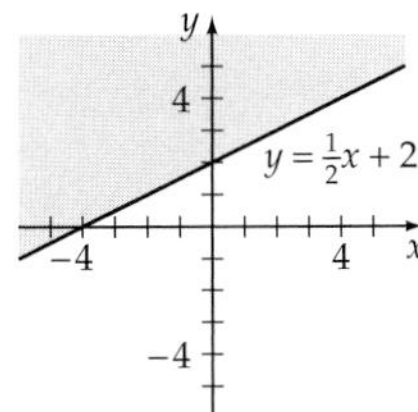

Check Your Progress 2, *page 560*

$$\begin{cases} 2x - 5y < -6 \\ 3x + \ \ y < 8 \end{cases}$$

Solving $2x - 5y < -6$ for y, we have $y > \frac{2}{5}x + \frac{6}{5}$. Graph $y = \frac{2}{5}x + \frac{6}{5}$ as a dashed line and shade above it. Solving $3x + y < 8$ for y, we have $y < -3x + 8$. Graph $y = -3x + 8$ as a dashed line and shade below it. The solution set of the system of inequalities is the region where the graphs of the solution sets of the two inequalities intersect. See the shaded region in the following figure.

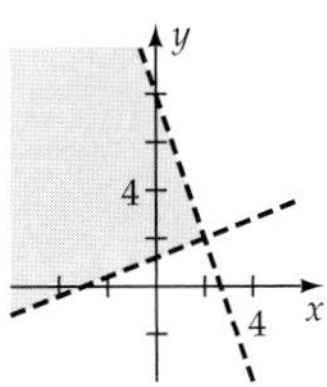

Check Your Progress 3, *page 561*

$$\begin{cases} 5x + \ \ y \leq 9 \\ 2x + 3y \leq 14 \\ x \geq -2, y \geq 2 \end{cases}$$

First graph the equations $x = -2$ and $y = 2$. Because $x \geq -2$ and $y \geq 2$, shade to the right of $x = -2$ and above $y = 2$. Solving $5x + y \leq 9$ for y, we have $y \leq -5x + 9$. Graph $y = -5x + 9$ as a solid line and shade below it. Solving $2x + 3y \leq 14$ for y, we have $y \leq -\frac{2}{3}x + \frac{14}{3}$. Graph $y = -\frac{2}{3}x + \frac{14}{3}$ as a solid line and shade below it. The solution set of the system of inequalities is the region where the graphs of the solution sets of the two inequalities intersect. See the shaded region in the following figure.

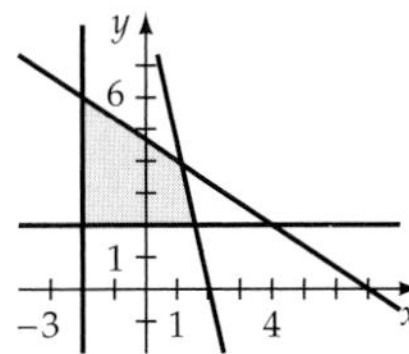

Check Your Progress 4, *page 565*

$C = 4x + 3y$

(x, y)	C
(0, 8)	24
(2, 4)	20
$\left(4, \frac{8}{3}\right)$	24
(12, 0)	48
(20, 0)	80
(20, 20)	140
(0, 20)	60

■ Minimum (at (2, 4))

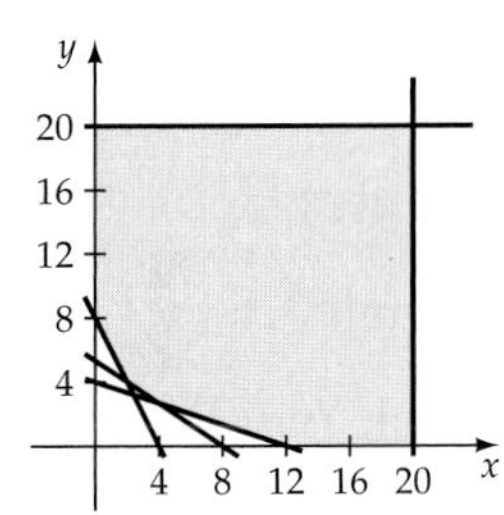

Check Your Progress 5, *page 566*

x = hours of machine 1 use
y = hours of machine 2 use
Cost = $28x + 25y$

Constraints: $\begin{cases} 4x + \ \ 3y \geq \ \ 60 \\ 5x + 10y \geq 100 \\ x \geq 0, y \geq 0 \end{cases}$

(x, y)	Cost
(0, 20)	500
(12, 4)	436
(20, 0)	560

■ Minimum (at (12, 4))

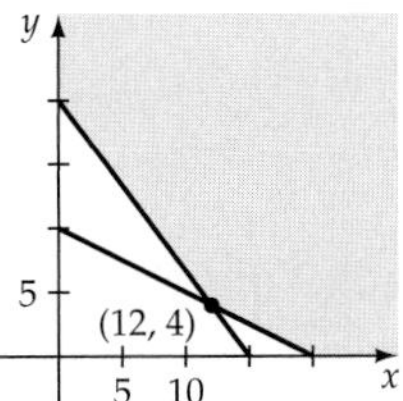

To achieve the minimum cost, use machine 1 for 12 hours and machine 2 for 4 hours.

Check Your Progress 6, *page 567*

Let x = number of standard models.
Let y = number of deluxe models.

Profit $= 25x + 35y$

Constraints: $\begin{cases} x + 3y \le 24 \\ x + \ \ y \le 10 \\ 2x + \ \ y \le 16 \\ x \ge 0, y \ge 0 \end{cases}$

(x, y)	Profit
(0, 0)	0
(0, 8)	280
(6, 4)	290
(3, 7)	320 ▪ Maximum
(8, 0)	200

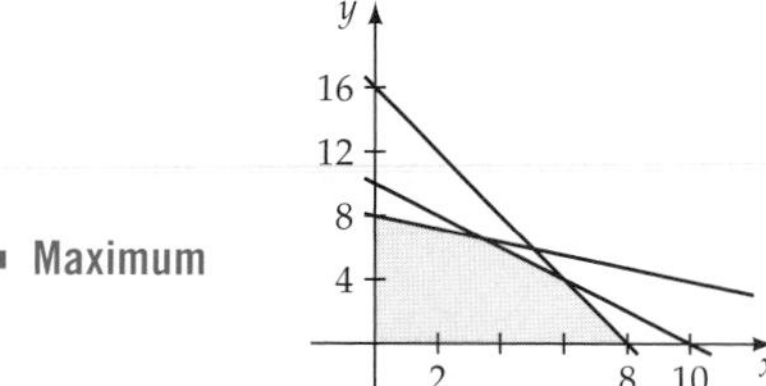

To maximize profits, produce 3 standard models and 7 deluxe models.

Chapter 7

Section 7.1

Check Your Progress 1, *page 582*

$a_n = 2^n - 1$

$a_1 = 2^1 - 1 = 1$ ▪ Replace n by 1.

$a_2 = 2^2 - 1 = 3$ ▪ Replace n by 2.

$a_3 = 2^3 - 1 = 7$ ▪ Replace n by 3.

Check Your Progress 2, *page 583*

$$a_3 = a_2 a_1 = 2 \cdot 1 = 2$$

$$a_4 = a_3 a_2 = 2 \cdot 2 = 4$$

$$a_5 = a_4 a_3 = 4 \cdot 2 = 8$$

Check Your Progress 3, *page 584*

$$\sum_{k=1}^{5} (k^3 - k) = (1^3 - 1) + (2^3 - 2) + (3^3 - 3) + (4^3 - 4) + (5^3 - 5)$$
$$= 0 + 6 + 24 + 60 + 120 = 210$$

Check Your Progress 4, *page 585*

$$\sum_{n=1}^{5} \frac{1}{n^3}$$

Check Your Progress 5, *page 586*

$$\sum_{n=1}^{15} (3a_n - 4b_n) = 3\sum_{n=1}^{15} a_n - 4\sum_{n=1}^{15} b_n$$
$$= 3(25) - 4(17)$$
$$= 75 - 68 = 7$$

Section 7.2

Check Your Progress 1, *page 590*

Use $d = a_2 - a_1$ to find the common difference.

$d = a_2 - a_1 = 2 - 5 = -3$ ▪ $a_1 = 5$, $a_2 = 2$

Now find the 20th term using the formula $a_n = a_1 + (n - 1)d$ with $n = 20$ and $d = -3$.

$a_{20} = 5 + (20 - 1)(-3) = 5 + 19(-3) = 5 - 57 = -52$

Check Your Progress 2, *page 590*

Use the equation $a_n = a_1 + (n - 1)d$ to write a system of equations given $a_5 = 24$ and $a_{12} = 59$.

$24 = a_1 + 4d$ ▪ Equation (1)

$59 = a_1 + 11d$ ▪ Equation (2)

$-35 = -7d$ ▪ Subtract the equations.

$5 = d$ ▪ Solve for d.

Substitute the value of d into Equation (1) and solve for a_1.

$24 = a_1 + 4d$

$24 = a_1 + 4(5)$

$4 = a_1$

Now use $a_n = a_1 + (n - 1)d$ to find the 20th term.

$$a_{20} = 4 + (20 - 1)5 = 4 + 19 \cdot 5 = 99$$

Check Your Progress 3, *page 592*

$$S_n = \frac{n[2a_1 + (n - 1)d]}{2}$$

$$S_{30} = \frac{30[2(1) + (30 - 1)2]}{2} \quad ▪\ n = 30, a_1 = 1, d = 2$$

$$= \frac{30(2 + 58)}{2} = 900$$

Check Your Progress 4, *page 593*

Because the first and last terms of the arithmetic series are known, use $S_n = \frac{n(a_1 + a_n)}{2}$ with $a_{33} = 52.6$, $a_1 = 4.6$, and $n = 33$. These values are taken from Example 4.

$$S_{33} = \frac{33(4.6 + 52.6)}{2} = 943.8$$

The total length is 943.8 millimeters.

Check Your Progress 5, *page 594*

Find the ratio of two successive terms using $\frac{a_{i+1}}{a_i}$.

$$\frac{a_{i+1}}{a_i} = \frac{3^{(i+1)-1}}{3^{i-1}} = \frac{3^i}{3^{i-1}} = 3^{i-(i-1)} = 3$$

The ratio of successive terms is a constant. The sequence is a geometric sequence.

Check Your Progress 6, *page 595*

$r = \frac{-\frac{2}{3}}{1} = -\frac{2}{3}$ and $a_1 = 1$. Thus

$$a_n = a_1 r^{n-1}$$
$$a_n = \left(-\frac{2}{3}\right)^{n-1}$$

Check Your Progress 7, *page 596*

Find a_1 and a_2.

$$a_n = 3\left(\frac{3}{4}\right)^{n-1}$$
$$a_1 = 3\left(\frac{3}{4}\right)^{1-1} = 3\left(\frac{3}{4}\right)^0 = 3 \qquad \blacksquare\ n = 1$$
$$a_2 = 3\left(\frac{3}{4}\right)^{2-1} = 3\left(\frac{3}{4}\right)^1 = \frac{9}{4} \qquad \blacksquare\ n = 2$$

Use a_1 and a_2 to find r, the common ratio.

$$r = \frac{a_2}{a_1} = \frac{\frac{9}{4}}{3} = \frac{9}{12} = \frac{3}{4}$$

Thus

$$S_n = \frac{a_1(1 - r^n)}{1 - r}$$
$$S_{17} = \frac{3\left[1 - \left(\frac{3}{4}\right)^{17}\right]}{1 - \frac{3}{4}} \qquad \blacksquare\ a_1 = 3, r = \frac{3}{4}, n = 17$$
$$\approx 11.9098$$

Check Your Progress 8, *page 598*

Use the formula $S = \frac{a_1}{1 - r}$. To find the first term a_1, let $n = 1$. Then $a_1 = 2(0.3)^{1-1} = 2(0.3)^0 = 2$. The common ratio is $r = 0.3$. Thus

$$S = \frac{a_1}{1 - r}$$
$$= \frac{2}{1 - 0.3} = \frac{2}{0.7} = \frac{20}{7}$$

The sum of the series is $\frac{20}{7}$.

Check Your Progress 9, *page 599*

$1.\overline{09} = 1 + \left[\frac{9}{100} + \frac{9}{10{,}000} + \frac{9}{1{,}000{,}000} + \ldots\right]$. The terms in brackets form a geometric series with $a_1 = \frac{9}{100}$ and

$r = \frac{\frac{9}{10{,}000}}{\frac{9}{100}} = \frac{1}{100}$. The sum of the infinite geometric series is

$\frac{\frac{9}{100}}{1 - \frac{1}{100}} = \frac{\frac{9}{100}}{\frac{99}{100}} = \frac{1}{11}$. Therefore,

$$1.\overline{09} = 1 + \left[\frac{9}{100} + \frac{9}{10{,}000} + \frac{9}{1{,}000{,}000} + \ldots\right] = 1 + \frac{1}{11} = \frac{12}{11}$$

Check Your Progress 10, *page 600*

Use the formula stock value $= \sum_{n=1}^{\infty} \frac{D}{(1 + i)^n} = \frac{D}{i}$ with $D = 2.33$ and $i = 0.20$.

$$\text{Stock value} = \frac{2.33}{0.20} = 11.65$$

The value of the stock is \$11.65.

Section 7.3

Check Your Progress 1, *page 607*

The finance charge is the simple interest on the unpaid balance for 1 month.

$$\text{Finance charge} = Prt$$
$$= 1250(0.018)(1) = 22.5$$

The finance charge is \$22.50.

Check Your Progress 2, *page 608*

$$\text{Market value of bond} = \frac{\text{coupon rate}}{\text{prevailing rate}} \cdot \text{face value}$$
$$= \frac{7\%}{7.5\%} \cdot 1000 \approx 933.33$$

The market value of the bond is \$933.33.

Check Your Progress 3, *page 611*

$FV = PV(1 + i)^n$

$FV = 5000\left(1 + \frac{0.06}{12}\right)^{60}$ ▪ $PV = 5000$, $i = \frac{0.06}{12}$, $n = 12 \times 5 = 60$

≈ 6744.25

The future value of the investment is \$6744.25.

Check Your Progress 4, *page 612*

$PV = FV(1 + i)^{-n}$

$= 5000\left(1 + \frac{0.08}{365}\right)^{-730}$ ▪ $FV = 5000$, $i = \frac{0.08}{365}$, $n = 365 \times 2 = 730$

$\approx 5000(0.8521587286)$ ▪ Multiply.

≈ 4260.79

Michael must deposit \$4260.79.

Check Your Progress 5, *page 613*

$I = 100\left[\left(1 + \frac{r}{m}\right)^m - 1\right]$

$I = 100\left[\left(1 + \frac{0.075}{4}\right)^4 - 1\right]$

≈ 7.71359

The effective interest rate is 7.71%.

Check Your Progress 6, *page 614*

$FV = PVe^{rt}$

$= 500e^{0.0725 \cdot 8}$ ▪ $PV = 500$, $r = 0.0725$, $t = 8$.

≈ 893.02

The future value of the investment is \$893.02.

Section 7.4

Check Your Progress 1, *page 620*

$PMT = PV\left[\frac{i}{1 - (1 + i)^{-n}}\right]$

$PMT = 50{,}000\left[\frac{0.0275}{1 - (1 + 0.0275)^{-8}}\right]$ ▪ $PV = 50{,}000$, $i = \frac{0.11}{4} = 0.0275$, $n = 4(2) = 8$

$\approx 50{,}000(0.1409579)$

≈ 7047.90

The quarterly payment is \$7047.90.

Check Your Progress 2, *page 621*

$PV = PMT\left[\frac{1 - (1 + i)^{-n}}{i}\right]$

$PV = 400\left[\frac{1 - (1 + 0.0075)^{-48}}{0.0075}\right]$ ▪ $PMT = 400$, $i = \frac{0.09}{12} = 0.0075$, $n = 12(4) = 48$

$PV \approx 400(40.184782)$

$\approx 16{,}073.91$

The parents must deposit \$16,073.91.

Check Your Progress 3, *page 622*

$PV = PMT\left[\frac{1 - (1 + i)^{-n}}{i}\right]$

$PV = 177.48\left[\frac{1 - \left(1 + \frac{0.0925}{12}\right)^{-25}}{\frac{0.0925}{12}}\right]$ ▪ $PMT = 177.48$, $i = \frac{0.0925}{12}$, $n = 60 - 35 = 25$

$PV \approx 4021.63$

The loan payoff is \$4021.63.

Check Your Progress 4, *page 624*

a. $PMT = PV\left[\frac{i}{1 - (1 + i)^{-n}}\right]$

$= 189{,}000\left[\frac{\frac{0.083}{12}}{1 - \left(1 + \frac{0.083}{12}\right)^{-240}}\right]$ ▪ $PV = 189{,}000$, $i = \frac{0.083}{12}$, $n = 12(20) = 240$

$\approx 189{,}000(0.00855207)$

≈ 1616.34

The monthly mortgage payment is \$1616.34.

b. $I = Prt$

$= 189{,}000(0.083)\frac{1}{12}$

$= 1307.25$

The amount of interest paid the first month is \$1307.25.

Principal paid = mortgage payment − interest payment

$= 1616.34 - 1307.25 = 309.09$

The amount of principal paid the first month is \$309.09.

Check Your Progress 5, *page 626*

a. Find the amount of simple interest that is added to the loan.

$$I = Prt$$
$$= 1100(0.09)(3)$$
$$= 297$$

The simple interest added to the loan is \$297. The loan amount is \$1100 + \$297 = \$1397. To find the monthly payment, divide the loan amount by 36. Monthly payment $= \frac{1397}{36} \approx 38.806$. The monthly payment is \$38.81.

b. Using a graphing or financial calculator, the APR is 16.3%.

Section 7.5

Check Your Progress 1, *page 632*

$$FV = PMT\left[\frac{(1+i)^n - 1}{i}\right]$$

$$= 125\left[\frac{\left(1+\frac{0.08}{12}\right)^{180} - 1}{\frac{0.08}{12}}\right] \quad \blacksquare\ PMT = 125,\ i = \frac{0.08}{12},\ n = 12(15) = 180$$

$$= 125(346.0382216)$$
$$\approx 43{,}254.78$$

The value of the account will be \$43,254.78.

Check Your Progress 2, *page 633*

a. $$FV = PMT\left[\frac{(1+i)^n - 1}{i}\right]$$

$$= 250\left[\frac{\left(1+\frac{0.07}{12}\right)^{180} - 1}{\frac{0.07}{12}}\right] \quad \blacksquare\ PMT = 250,\ i = \frac{0.07}{12},\ n = 12(15) = 180$$

$$= 250(316.9622967)$$
$$\approx 79{,}240.57$$

The future value is \$79,240.57.

b. The interest earned is the future value (79,240.57) minus the value of all the payments ($250 \cdot 180 = 45{,}000$).

Interest earned = 79,240.57 − 45,000 = 34,240.57

The interest earned is \$34,240.57.

Check Your Progress 3, *page 634*

$$PMT = FV\left[\frac{i}{(1+i)^n - 1}\right]$$

$$PMT = 3000\left[\frac{\frac{0.06}{12}}{\left(1+\frac{0.06}{12}\right)^{24} - 1}\right] \quad \blacksquare\ FV = 3000,\ i = \frac{0.06}{12},\ n = 12(2) = 24$$

$$\approx 3000(0.03932061)$$
$$\approx 117.96$$

The engineer must save \$117.96 per month.

Check Your Progress 4, *page 635*

Using a graphing or financial calculator, the interest rate is 8.0%.

Chapter 8

Section 8.1

Check Your Progress 1, *page 647*

The possible outcomes of the experiment:

QDN QND DQN DNQ NQD NDQ

The total number of outcomes of the experiment of selecting the three coins is six.

Check Your Progress 2, *page 648*

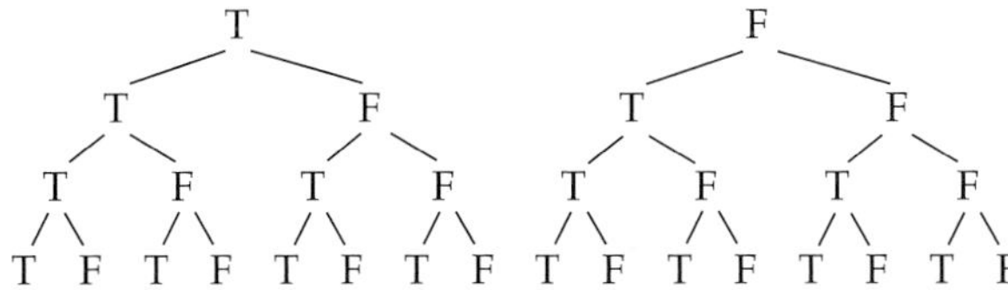

There are 16 possible ways to answer the first four questions.

Check Your Progress 3, *page 649*

Any one of the nine runners can win the gold medal, so $n_1 = 9$. Because a runner cannot finish both first and second, there are eight runners who can win the silver medal; thus $n_2 = 8$. Similarly, there are seven runners who can win the bronze medal, so $n_3 = 7$. By the counting principle, there are $9 \cdot 8 \cdot 7 = 504$ possible ways the medals can be awarded.

Check Your Progress 4, *page 650*

a. $$7! + 4! = 7 \cdot 6 \cdot 5 \cdot 4 \cdot 3 \cdot 2 \cdot 1 + 4 \cdot 3 \cdot 2 \cdot 1$$
$$= 5040 + 24 = 5064$$

b. $$\frac{8!}{4!} = \frac{8 \cdot 7 \cdot 6 \cdot 5 \cdot 4!}{4!} = 8 \cdot 7 \cdot 6 \cdot 5 = 1680$$

Check Your Progress 5, *page 651*

Because the players on the golf team are ranked from 1 through 5, a team with player A in position 1 is different from a team with player A in position 2. Therefore, the number of different teams is the number of permutations of eight players selected five at a time.

$$P(8, 5) = \frac{8!}{(8-5)!} = \frac{8!}{3!} = \frac{8 \cdot 7 \cdot 6 \cdot 5 \cdot 4 \cdot 3!}{3!} = 6720$$

There are 6720 possible golf teams.

Check Your Progress 6, *page 653*

a. This is a combination problem (rather than a permutation problem) because the order in which the coach chooses the players is not important.

$$C(16, 9) = \frac{16!}{9!\,(16-9)!} = \frac{16!}{9!\,7!} = \frac{16 \cdot 15 \cdot 14 \cdot 13 \cdot 12 \cdot 11 \cdot 10 \cdot 9!}{9! \cdot 7!}$$

$$= \frac{16 \cdot 15 \cdot 14 \cdot 13 \cdot 12 \cdot 11 \cdot 10}{7 \cdot 6 \cdot 5 \cdot 4 \cdot 3 \cdot 2 \cdot 1} = 11{,}440$$

There are 11,440 possible nine-player starting teams.

b. This time different batting orders result from choosing the same players, but in different orders. So we want to count permutations rather than combinations.

$$P(9, 4) = \frac{9!}{(9-4)!} = \frac{9!}{5!} = \frac{9 \cdot 8 \cdot 7 \cdot 6 \cdot 5!}{5!}$$

$$= 9 \cdot 8 \cdot 7 \cdot 6 = 3024$$

There are 3024 different ways a coach can choose the batting order lineup for the first four batters.

Check Your Progress 7, *page 653*

There are $C(4, 3) = 4$ ways to choose the corporate tax returns. There are $C(6, 2) = 15$ ways to choose the individual tax returns. Therefore, by the counting principle, there are $4 \cdot 15 = 60$ ways to choose the tax returns.

Check Your Progress 8, *page 654*

This problem uses the counting principle and combinations. There are four suits, so $n_1 = 4$. There are $C(13, 4)$ ways to choose four cards from one suit, so $n_2 = C(13, 4)$. Because the fifth card must be one of the 39 cards $(52 - 13 = 39)$ that are not the same suit as the four cards, $n_3 = 39$. The number of possible hands is $4 \cdot C(13, 4) \cdot 39 = 111{,}540$.

Section 8.2

Check Your Progress 1, *page 660*

{HH, HT, TH, TT}

Check Your Progress 2, *page 662*

Determine the number of elements in the sample space. This must include every possible outcome from the roll of a die.

$S = \{1, 2, 3, 4, 5, 6\}$

The elements in the event that an odd number appears on the upward face are $E = \{1, 3, 5\}$.

$$P(E) = \frac{n(E)}{n(S)} = \frac{3}{6} = \frac{1}{2}$$

The probability is $\frac{1}{2}$.

Check Your Progress 3, *page 662*

There are 36 possible outcomes as shown in the figure to the left of Example 3. This figure also shows that a sum of 7 can be obtained in 6 different ways. Thus the probability that the sum of the pips on the upward faces of the two dice equals 7 is $\frac{6}{36} = \frac{1}{6}$.

Check Your Progress 4, *page 663*

Let E be the event that two mathematicians and three economists are chosen. There are $C(5, 2)$ ways to choose two of five mathematicians. There are $C(6, 3)$ ways to choose three of six economists. Therefore, the number of possible ways to choose two mathematicians and three economists is $C(5, 2) \cdot C(6, 3)$.

The sample space consists of all possible ways five people can be chosen from 11 people. This is $C(11, 5)$. Therefore, the probability of E is

$$P(E) = \frac{n(E)}{n(S)} = \frac{C(5, 2) \cdot C(6, 3)}{C(11, 5)} = \frac{100}{231} \approx 0.433$$

Check Your Progress 5, *page 664*

Let E be the event that a voter is between 39 and 49. To find $n(E)$, add the numbers in the row representing people between 39 and 49. The sum is 773. The total number of people in the survey is 3228, so $n(S) = 3228$. Therefore,

$$P(E) = \frac{n(E)}{n(S)} = \frac{773}{3228} \approx 0.24$$

Check Your Progress 6, *page 665*

Make a Punnett square.

Parents	C	c
C	CC	Cc
c	Cc	cc

To be white, a hamster must be cc. From the table, there is one case out of four possible cases that is cc. Therefore, $P(\text{white}) = \frac{1}{4}$.

Section 8.3

Check Your Progress 1, *page 671*

By the counting principle, there are 36 possible outcomes from the roll of two dice. Thus $n(S) = 36$. Let A be the event that the sum is 7 and let B be the event that the sum is 11. Then $n(A) = 6$ and $n(B) = 2$. Because the events are mutually exclusive, we can use the formula for the probability of mutually exclusive events.

$$P(A \text{ or } B) = P(A) + P(B)$$
$$= \frac{6}{36} + \frac{2}{36} = \frac{8}{36} = \frac{2}{9}$$

The probability of rolling a sum of 7 or 11 is $\frac{2}{9}$.

Check Your Progress 2, *page 673*

Let $A = \{\text{degree in business}\}$ and $B = \{\text{salary between } \$20{,}000 \text{ and } \$24{,}999\}$. These events are not mutually exclusive because there are 16 people who received a degree in business and whose starting salaries are between \$20,000 and \$24,999. The sample space S consists of the 206 people who participated in the survey. From the table, $n(A) = 98$, $n(B) = 39$, and $n(A \text{ and } B) = 16$.

$$P(A \text{ or } B) = P(A) + P(B) - P(A \text{ and } B)$$
$$= \frac{98}{206} + \frac{39}{206} - \frac{16}{206}$$
$$= \frac{121}{206} \approx 0.587$$

The probability of selecting a person who received a degree in business or whose starting salary is between \$20,000 and \$24,999 is approximately 58.7%.

Check Your Progress 3, *page 674*

Use the formula for the probability of the complement of an event. Let $E = \{\text{person has type A blood}\}$. Then $E^c = \{\text{person does not have type A blood}\}$.

$$P(E^c) = 1 - P(E)$$
$$= 1 - 0.34 = 0.66$$

The probability that a person does not have type A blood is 66%.

Check Your Progress 4, *page 675*

Let $E = \{\text{sum of 7 at least once}\}$. Then $E^c = \{\text{sum is never 7}\}$. To calculate the number of elements in the sample space (all possible outcomes of tossing a pair of dice three times) and the number of items in E^c, we will use the counting principle. Because on each toss of two dice there are 36 possible outcomes,

$$n(S) = 36 \cdot 36 \cdot 36 = 46{,}656$$

On each toss of the dice, there are 30 ways for the sum not to be 7. Therefore,

$$n(E^c) = 30 \cdot 30 \cdot 30 = 27{,}000$$
$$P(E) = 1 - P(E^c)$$
$$= 1 - \frac{27{,}000}{46{,}656} = \frac{19{,}656}{46{,}656}$$
$$\approx 0.421$$

The probability that a sum of 7 will occur at least once is approximately 42.1%.

Check Your Progress 5, *page 676*

The toss of a coin followed by the roll of a die are independent events because the probability of a certain number on the roll of the die does not depend on the outcome of the toss of the coin. Let $A = \{\text{coin shows heads}\}$ and $B = \{\text{die shows 6}\}$. Then

$$P(A \text{ and } B) = P(A) \cdot P(B) = \frac{1}{2} \cdot \frac{1}{6} = \frac{1}{12}$$

Check Your Progress 6, *page 677*

This experiment consists of 10 trials, so $n = 10$. Each trial has exactly two outcomes (success is student answers correctly; failure is student answers incorrectly), and each trial is independent. So this is a binomial experiment for which the probability of a success is $p = \frac{1}{2}$ and the probability of a failure is $q = 1 - p = 1 - \frac{1}{2} = \frac{1}{2}$. The probability that seven questions are answered correctly is

$$C(n, k)p^k q^{n-k}$$
$$C(10, 7)\left(\frac{1}{2}\right)^7\left(\frac{1}{2}\right)^3 = 120\left(\frac{1}{2}\right)^7\left(\frac{1}{2}\right)^3$$
$$\approx 0.117$$

The probability of answering seven questions correctly is approximately 0.117, or 11.7%.

Check Your Progress 7, *page 678*

The phrase "at least three times" in five spins means three, four, or five times. This is a binomial experiment with $n = 5$, $p = \frac{6}{19}$, $q = 1 - \frac{6}{19} = \frac{13}{19}$, and $k = 3, 4, 5$.

$$\text{Probability(at least 3)} = P(k = 3) + P(k = 4) + P(k = 5)$$
$$= C(5, 3)\left(\frac{6}{19}\right)^3\left(\frac{13}{19}\right)^2 + C(5, 4)\left(\frac{6}{19}\right)^4\left(\frac{13}{19}\right)^1$$
$$+ C(5, 5)\left(\frac{6}{19}\right)^5\left(\frac{13}{19}\right)^0$$
$$\approx 0.14743 + 0.03402 + 0.00314$$
$$= 0.18459$$

The probability of at least three spins resulting in a number between 1 and 12 inclusive is approximately 0.185.

Section 8.4

Check Your Progress 1, *page 684*

$$\binom{10}{5} = \frac{10!}{5!5!} = \frac{10 \cdot 9 \cdot 8 \cdot 7 \cdot 6 \cdot 5!}{5!5!} = \frac{10 \cdot 9 \cdot 8 \cdot 7 \cdot 6}{5 \cdot 4 \cdot 3 \cdot 2 \cdot 1}$$
$$= 252$$

Check Your Progress 2, *page 685*

$$(3x + 2y)^4$$
$$= (3x)^4 + 4(3x)^3(2y) + 6(3x)^2(2y)^2 + 4(3x)(2y)^3 + (2y)^4$$
$$= 81x^4 + 216x^3y + 216x^2y^2 + 96xy^3 + 16y^4$$

Check Your Progress 3, *page 685*

$$(2x - \sqrt{y})^7 = \binom{7}{0}(2x)^7 + \binom{7}{1}(2x)^6(-\sqrt{y})$$
$$+ \binom{7}{2}(2x)^5(-\sqrt{y})^2 + \binom{7}{3}(2x)^4(-\sqrt{y})^3$$
$$+ \binom{7}{4}(2x)^3(-\sqrt{y})^4 + \binom{7}{5}(2x)^2(-\sqrt{y})^5$$
$$+ \binom{7}{6}(2x)(-\sqrt{y})^6 + \binom{7}{7}(-\sqrt{y})^7$$
$$= 128x^7 - 448x^6\sqrt{y} + 672x^5y - 560x^4y\sqrt{y}$$
$$+ 280x^3y^2 - 84x^2y^2\sqrt{y}$$
$$+ 14xy^3 - y^3\sqrt{y}$$

Check Your Progress 4, *page 686*

$$\binom{10}{6-1}(x^{-1/2})^{10-6+1}(x^{1/2})^{6-1} = \binom{10}{5}(x^{-1/2})^5(x^{1/2})^5 = 252$$

Appendix A

Check Your Progress 1, *page AP2*

Comparing $x^2 = 4py$ with $x^2 = -\frac{1}{4}y$, we have $4p = -\frac{1}{4}$ or $p = -\frac{1}{16}$.

Vertex $(0, 0)$

Focus $\left(0, -\frac{1}{16}\right)$

Directrix $y = \frac{1}{16}$

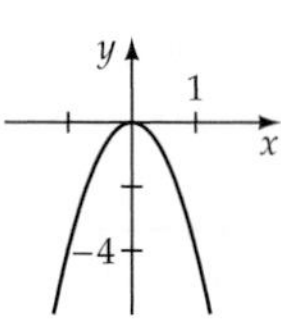

Check Your Progress 2, *page AP3*

Vertex $(0, 0)$, focus $(5, 0)$, $p = 5$ because focus is $(p, 0)$.

$y^2 = 4px$

$y^2 = 4(5)x$

$y^2 = 20x$

Check Your Progress 3, *page AP4*

$$x^2 + 5x - 4y - 1 = 0$$
$$x^2 + 5x = 4y + 1$$
$$x^2 + 5x + \frac{25}{4} = 4y + 1 + \frac{25}{4}$$ ▪ Complete the square.
$$\left(x + \frac{5}{2}\right)^2 = 4\left(y + \frac{29}{16}\right)$$ ▪ $h = -\frac{5}{2}, k = -\frac{29}{16}$
$$4p = 4$$ ▪ Compare to $(x - h)^2 = 4p(y - k)^2$.
$$p = 1$$

Vertex $\left(-\frac{5}{2}, -\frac{29}{16}\right)$

Focus $(h, k + p) = \left(-\frac{5}{2}, -\frac{13}{16}\right)$

Directrix $y = k - p = -\frac{45}{16}$

Check Your Progress 4, *page AP4*

Vertex $(2, -3)$, focus $(0, -3)$

$(h, k) = (2, -3)$, so $h = 2$ and $k = -3$.

Focus is $(h + p, k) = (2 + p, -3) = (0, -3)$.

Therefore, $2 + p = 0$ and $p = -2$.

$(y - k)^2 = 4p(x - h)$

$(y + 3)^2 = 4(-2)(x - 2)$

$(y + 3)^2 = -8(x - 2)$

Check Your Progress 5, *page AP7*

$$25x^2 + 12y^2 = 300$$
$$\frac{x^2}{12} + \frac{y^2}{25} = 1$$ ▪ $a^2 = 25, b^2 = 12, c^2 = 25 - 12$; $a = 5, b = 2\sqrt{3}, c = \sqrt{13}$

Center $(0, 0)$

Vertices $(0, 5)$ and $(0, -5)$

Foci $\left(0, \sqrt{13}\right)$ and $\left(0, -\sqrt{13}\right)$

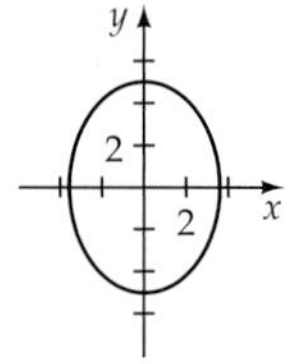

Check Your Progress 6, *page AP8*

$$9x^2 + 16y^2 + 36x - 16y - 104 = 0$$
$$9x^2 + 36x + 16y^2 - 16y - 104 = 0$$
$$9(x^2 + 4x) + 16(y^2 - y) = 104$$
$$9(x^2 + 4x + 4) + 16\left(y^2 - y + \frac{1}{4}\right) = 104 + 36 + 4$$

$$9(x+2)^2 + 16\left(y - \frac{1}{2}\right)^2 = 144$$

$$\frac{(x+2)^2}{16} + \frac{\left(y - \frac{1}{2}\right)^2}{9} = 1$$

Center $\left(-2, \frac{1}{2}\right)$

$a = 4, b = 3,$

$c = \sqrt{4^2 - 3^2} = \sqrt{7}$

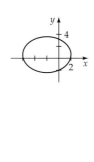

Vertices $\left(2, \frac{1}{2}\right)$ and $\left(-6, \frac{1}{2}\right)$

Foci $\left(2 + \sqrt{7}, \frac{1}{2}\right)$ and

$\left(-2 - \sqrt{7}, \frac{1}{2}\right)$

Check Your Progress 7, *page AP9*

Center $(-4, 1) = (h, k)$. Therefore, $h = -4$ and $k = 1$. Length of minor axis is 8, so $2b = 8$ or $b = 4$. The equation of the ellipse is of the form

$$\frac{(x-h)^2}{a^2} + \frac{(y-k)^2}{b^2} = 1$$

$$\frac{(x+4)^2}{a^2} + \frac{(y-1)^2}{16} = 1$$

■ $h = -4$, $k = 1$, $b = 4$

$$\frac{(0+4)^2}{a^2} + \frac{(4-1)^2}{16} = 1$$

■ The point $(0, 4)$ is on the graph. Thus $x = 0$ and $y = 4$ satisfy the equation.

$$\frac{16}{a^2} + \frac{9}{16} = 1$$

■ Solve for a^2.

$$\frac{16}{a^2} = \frac{7}{16}$$

$$a^2 = \frac{256}{7}$$

$$\frac{(x+4)^2}{256/7} + \frac{(y-1)^2}{16} = 1$$

Check Your Progress 8, *page AP12*

$$\frac{y^2}{25} - \frac{x^2}{36} = 1$$

$a^2 = 25 \qquad b^2 = 36 \qquad c^2 = a^2 + b^2 = 25 + 36 = 61$

$a = 5 \qquad b = 6 \qquad c = \sqrt{61}$

Transverse axis is on y-axis because y^2 term is positive.

Center $(0, 0)$

Vertices $(0, 5)$ and $(0, -5)$

Foci $\left(0, \sqrt{61}\right)$ and $\left(0, -\sqrt{61}\right)$

Asymptotes $y = \frac{5}{6}x$ and $y = -\frac{5}{6}x$

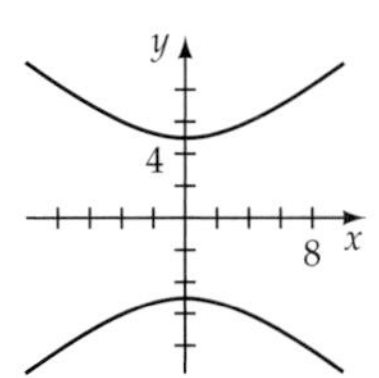

Check Your Progress 9, *page AP13*

$$16x^2 - 9y^2 - 32x - 54y + 79 = 0$$

$$16(x^2 - 2x + 1) - 9(y^2 + 6y + 9) = -79 + 16 - 81$$

$$= -144$$

$$\frac{(y+3)^2}{16} - \frac{(x-1)^2}{9} = 1$$

Transverse axis is parallel to y-axis because y^2 term is positive. Center is at $(1, -3)$; $a^2 = 16$ so $a = 4$.

Vertices $(h, k + a) = (1, 1)$

$(h, k - a) = (1, -7)$

$c^2 = a^2 + b^2 = 16 + 9 = 25$

$c = \sqrt{25} = 5$

Foci $(h, k + c) = (1, 2)$

$(h, k - c) = (1, -8)$

Because $b^2 = 9$ and $b = 3$, the asymptotes are

$y + 3 = \pm\frac{4}{3}(x - 1)$, which simplifies to $y = \frac{4}{3}x - \frac{13}{3}$ and

$y = -\frac{4}{3}x - \frac{5}{3}$

Check Your Progress 10, *page AP14*

$x^2 = 4py$

$40.5^2 = 4p(16)$

$p = \frac{40.5^2}{64}$

$p \approx 25.6$ feet

Check Your Progress 11, *page AP15*

The mean distance is $a = 67.08$ million miles.

Aphelion $= a + c = 67.58$ million miles

Thus $c = 67.58 - a = 0.50$ million miles.

$b = \sqrt{a^2 - c^2} = \sqrt{67.08^2 - 0.50^2} \approx 67.078$

An equation for the orbit of Venus is $\frac{x^2}{67.08^2} + \frac{y^2}{67.078^2} = 1$

Check Your Progress 12, *page AP16*

The length of the semimajor axis is 50 feet. Thus

$$\begin{aligned} c^2 &= a^2 - b^2 \\ 32^2 &= 50^2 - b^2 \\ b^2 &= 50^2 - 32^2 \\ b &= \sqrt{50^2 - 32^2} \\ b &\approx 38.4 \text{ feet} \end{aligned}$$

Appendix B

Check Your Progress 1, *page AP20*

$$\begin{vmatrix} 2 & 9 \\ -6 & 2 \end{vmatrix} = 2 \cdot 2 - (-6)(9) = 4 + 54 = 58$$

Check Your Progress 2, *page AP21*

$$M_{13} = \begin{vmatrix} 1 & 3 \\ 6 & -2 \end{vmatrix} = 1(-2) - 6(3) = -2 - 18 = -20$$

$$C_{13} = (-1)^{1+3} \cdot M_{13} = 1 \cdot M_{13} = 1(-20) = -20$$

Check Your Progress 3, *page AP22*

Expanding with cofactors of row 1 yields

$$\begin{aligned} \begin{vmatrix} 3 & -2 & 0 \\ 2 & -3 & 2 \\ 8 & -2 & 5 \end{vmatrix} &= 3C_{11} + (-2)C_{12} + 0 \cdot C_{13} \\ &= 3\begin{vmatrix} -3 & 2 \\ -2 & 5 \end{vmatrix} + 2\begin{vmatrix} 2 & 2 \\ 8 & 5 \end{vmatrix} + 0\begin{vmatrix} 2 & -3 \\ 8 & -2 \end{vmatrix} \\ &= 3(-15 + 4) + 2(10 - 16) + 0 \\ &= 3(-11) + 2(-6) = -33 + (-12) \\ &= -45 \end{aligned}$$

Check Your Progress 4, *page AP24*

Let $D = \begin{vmatrix} 3 & -2 & -1 \\ 1 & 2 & 4 \\ 2 & -2 & 3 \end{vmatrix}$. Then

$$D \overset{R_1 \leftrightarrow R_2}{=} -\begin{vmatrix} 1 & 2 & 4 \\ 3 & -2 & -1 \\ 2 & -2 & 3 \end{vmatrix} \overset{\substack{-3R_1+R_2 \\ -2R_1+R_3}}{=} -\begin{vmatrix} 1 & 2 & 4 \\ 0 & -8 & -13 \\ 0 & -6 & -5 \end{vmatrix}$$

$$\overset{-\frac{1}{8}R_2}{=} 8\begin{vmatrix} 1 & 2 & 4 \\ 0 & 1 & \frac{13}{8} \\ 0 & -6 & -5 \end{vmatrix} \overset{6R_2+R_3}{=} 8\begin{vmatrix} 1 & 2 & 4 \\ 0 & 1 & \frac{13}{8} \\ 0 & 0 & \frac{19}{4} \end{vmatrix}$$

$$= 8(1)(1)\left(\tfrac{19}{4}\right) = 38$$

Answers to Selected Exercises

Chapter P

Chapter Prep Quiz, *page 2*

1. -5 **2.** 60 **3.** -125 **4.** $-\frac{3}{4}$ **5.** $\frac{6}{5}$ **6.** $\frac{2}{3}$ **7.** $-\frac{1}{4}$ **8.** 24 **9.** 48 **10.** $-5x$

Exercises P.1, *page 12*

1. $-\frac{1}{5}$: rational, real; 0: integer, rational, real; -44: integer, rational, real; π: irrational, real; 3.14: rational, real; 5.050050005...: irrational, real; $\sqrt{49}$: integer, rational, real, prime; 53: integer, rational, real, prime

3. $\{-3, -2, -1, 0, 1, 2, 3, 4, 6\}$ **5.** $\{0, 1, 2, 3\}$ **7.** $\varnothing$ **9.** $\{1, 3\}$ **11.** $\{-2, 0, 1, 2, 3, 4, 6\}$ **13.** 0 **15.** -4 **17.** $a^2 + 7$

19. $|x - 3|$ **21.** $|x + 2| = 4$ **23.** $\{x \mid -2 < x < 3\}$ **25.** $\{x \mid -5 \le x \le -1\}$ **27.** $\{x \mid x \ge 2\}$ **29.** $(3, 5)$ **31.** $[-2, \infty)$ **33.** $[0, 1]$ **35.** **37.** **39.** **41.** **43.** **45.** **47.** **49.**

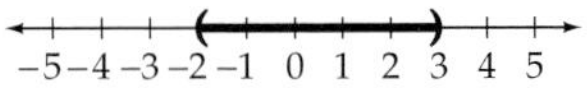

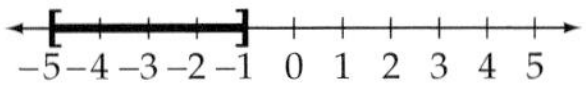

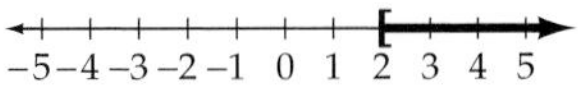

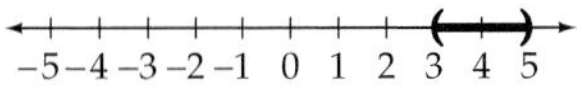

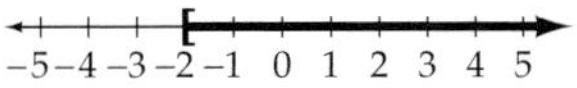

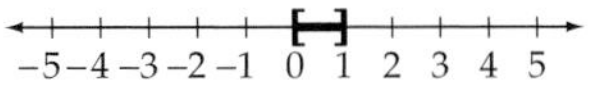

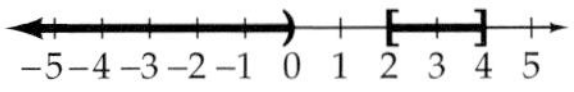

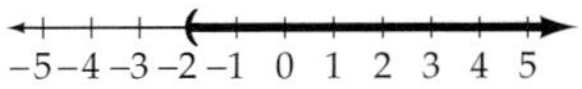

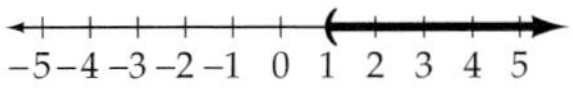

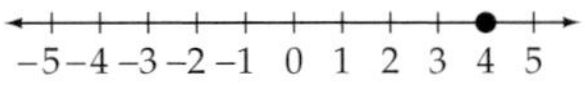

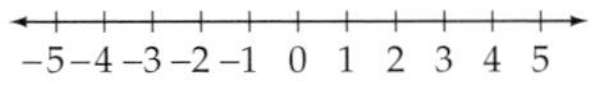

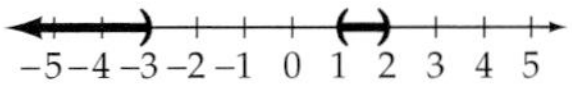

51. a. 5 **b.** $-3, 4, -1, 4$ **c.** 5 **53.** 8 **55.** 12 **57.** -72 **59.** 19 **61.** 13 **63.** -3 **65.** $6x$

67. $6 + 3x$ **69.** $\frac{3}{2}a$ **71.** $6x - 13$ **73.** $5 - 12x + 6y$ **75.** $2a$ **77.** $21a + 6$ **79.** 6 square inches **81.** \$5150

83. 66 beats per minute **85.** 100 feet **87.** Associative property of multiplication **89.** Distributive property **91.** Additive identity property **93.** Inverse property of mutliplication **95.** A **97.** $\varnothing$ **99.** All elements of B are contained in A.

101. All real numbers

Prepare for Section P.2, *page 14*

104. 32 **105.** $\frac{1}{16}$ **106.** 64 **107.** 314,000 **108.** False **109.** False

Exercises P.2, *page 23*

1. -125 **3.** 1 **5.** $\frac{1}{16}$ **7.** 32 **9.** $\frac{9}{4}$ **11.** 27 **13.** 1 **15.** $\frac{1}{x^5}$ **17.** $\frac{2}{x^4}$ **19.** $6a^3b^8$ **21.** $2a^4$ **23.** $\frac{3}{x^2}$ **25.** $-24x^6y^5$ **27.** $\frac{12}{a^3b}$

29. $-18m^5n^6$ **31.** $\frac{1}{x^6}$ **33.** $2x$ **35.** $\frac{b^{10}}{a^{10}}$ **37.** x^{7n} **39.** a^{2n^2} **41.** x^{n+1} **43.** $25a^4b^8$ **45.** 2.011×10^{12} **47.** 5.62×10^{-10}

49. 31,400,000 **51.** -0.0000023 **53.** 2.7×10^8 **55.** 1.5×10^{-11} **57.** 7.2×10^{12} **59.** 8×10^{-16} **61.** $\frac{y}{16}$ **63.** No

65. 1.01×10 cookies per person **67.** 3.125×10^7 seeds **69.** 1.5×10^{10} additions **71.** 4.836×10^8 miles

73. 1.66×10^{-24} gram **75.** $10^{(10^{10})}$

Prepare for Section P.3, *page 25*

77. $-6a + 12b$ **78.** $19 - 4x$ **79.** $3x^2 - 3x - 6$ **80.** $-x^2 - 5x - 1$ **81.** False **82.** False

Exercises P.3, *page 34*

1. a. $x^2 + 2x - 7$ **b.** 2 **c.** 1 **d.** 3 **3. a.** $x^3 - 1$ **b.** 3 **c.** 1 **d.** 2 **5. a.** $2x^4 - x^2 - 5$ **b.** 4 **c.** 2 **d.** 3 **7.** $5x^2 + 11x + 3$ **9.** $9x^3 + 8x^2 - 2x + 6$ **11.** $-2r^2 + 3r - 12$ **13.** $-3u^2 - 2u + 4$ **15.** $8x^3 + 18x^2 - 67x + 40$ **17.** $x^4 - 6x^3 + 13x^2 - 21x + 4$ **19.** $4x^4 - 6x^3 - 2x^2 - 3x + 9$ **21.** $10x^2 + 22x + 4$ **23.** $4z^2 + 13z - 12$ **25.** $4x^2 - 12x + 9$ **27.** $16z^2 - 49$ **29.** $10x^2 + 7xy - 12y^2$ **31.** $4x^2 + 20xy + 25y^2$ **33.** $12d^2 + 4d - 8$ **35.** $x^3 + y^3$ **37.** $60c^3 - 49c^2 + 4$ **39.** $x^3 + 12x^2 + 48x + 64$ **41.** $x + 1$ **43.** $x - 1$ **45.** $x + 3$ **47.** $3x - 7 + \dfrac{13}{x + 2}$ **49.** $x^2 + 4x + 4$ **51.** $x^2 + 3 + \dfrac{7}{x - 4}$ **53.** $x^2 - 4x + 20 - \dfrac{64}{x + 4}$ **55.** $2x^2 - 6x + 22 - \dfrac{71}{x + 3}$ **57.** \$11.198 trillion **59.** $-0.25x^2 + 40x$; \$1543.75 **61. a.** $-10r^3 + 6r^2$ **b.** 0.30625 meters per second **63.** $4x^3 - 140x^2 + 1200x$; 2000 cubic centimeters **65.** 97.5 feet

Prepare for Section P.4, *page 36*

67. $4x^2$ **68.** $2y$ **69.** 5 **70.** 1 **71.** $x + 5$ **72. a.** $(x^2)^3$ **b.** $(x^3)^2$

Exercises P.4, *page 46*

1. $5(x + 4)$ **3.** $-3x(5x + 4)$ **5.** $2xy(5x + 3 - 7y)$ **7.** $(x - 3)(2a + 3b)$ **9.** $(x + 3)(x + 4)$ **11.** $(a - 12)(a + 2)$ **13.** $(6x + 1)(x + 4)$ **15.** $(17x + 4)(3x - 1)$ **17.** $(3x + 8y)(2x - 5y)$ **19.** $(x^2 + 5)(x^2 + 1)$ **21.** $(6x^2 + 5)(x^2 + 3)$ **23.** Factorable over the integers **25.** Factorable over the integers **27.** Nonfactorable over the integers **29.** $(x - 3)(x + 3)$ **31.** $(2a - 7)(2a + 7)$ **33.** $(1 - 10x)(1 + 10x)$ **35.** $(x^2 + 3)(x^2 - 3)$ **37.** $(x + 3)(x + 7)$ **39.** $(x + 5)^2$ **41.** $(a - 7)^2$ **43.** $(2x + 3)^2$ **45.** $(z^2 + 2w^2)^2$ **47.** $(x - 2)(x^2 + 2x + 4)$ **49.** $(2x - 3y)(4x^2 + 6xy + 9y^2)$ **51.** $(2 - x^2)(4 + 2x^2 + x^4)$ **53.** $(x - 3)(x^2 - 3x + 3)$ **55.** $(3x + 1)(x^2 + 2)$ **57.** $(x - 1)(ax + b)$ **59.** $(3w + 2)(2w^2 - 5)$ **61.** $2(3x - 1)(3x + 1)$ **63.** $(2x - 1)(2x + 1)(4x^2 + 1)$ **65.** $a(3x - 2y)(4x - 5y)$ **67.** $b(3x + 4)(x - 1)(x + 1)$ **69.** $2b(6x + y)^2$ **71.** $(w - 3)(w^2 - 12w + 39)$ **73.** $(x + 3y - 1)(x + 3y + 1)$ **75.** Nonfactorable over the integers **77.** $(2x - 5)^2(3x + 5)$ **79.** $(2x - y)(2x + y + 1)$ **81.** 8 **83.** 64

Prepare for Section P.5, *page 47*

85. 24 **86.** $\frac{11}{12}$ **87.** $-\frac{1}{10}$ **88.** $\frac{10}{7}$ **89.** $\dfrac{xz}{wy}$ **90.** $x^2 + 2$

Exercises P.5, *page 55*

1. $\dfrac{x + 4}{3}$ **3.** $\dfrac{x - 3}{x - 2}$ **5.** $\dfrac{a^2 - 2a + 4}{a - 2}$ **7.** $-\dfrac{x + 8}{x + 2}$ **9.** $-\dfrac{4y^2 + 7}{y + 7}$ **11.** $-\dfrac{8}{a^3b}$ **13.** $\dfrac{10}{27q^2}$ **15.** $\dfrac{x(3x + 7)}{2x + 3}$ **17.** $\dfrac{x + 3}{2x + 3}$ **19.** $\dfrac{(2y + 3)(3y - 4)}{(2y - 3)(y + 1)}$ **21.** $\dfrac{1}{a - 8}$ **23.** $\dfrac{3p - 2}{r}$ **25.** $\dfrac{8x(x - 4)}{(x - 5)(x + 3)}$ **27.** $\dfrac{3y - 4}{y + 4}$ **29.** $\dfrac{7z(2z - 5)}{(2z - 3)(z - 5)}$ **31.** $\dfrac{-2x^2 + 14x - 3}{(x - 3)(x + 3)(x + 4)}$ **33.** $\dfrac{(2x - 1)(x + 5)}{x(x - 5)}$ **35.** $\dfrac{-q^2 + 12q + 5}{(q - 3)(q + 5)}$ **37.** $\dfrac{3x^2 - 7x - 13}{(x + 3)(x + 4)(x - 3)(x - 4)}$ **39.** $\dfrac{(x + 2)(3x - 1)}{x^2}$ **41.** $\dfrac{4x + 1}{x - 1}$ **43.** $\dfrac{x - 2y}{y(y - x)}$ **45.** $\dfrac{(5x + 9)(x + 3)}{(x + 2)(4x + 3)}$ **47.** $\dfrac{(b + 3)(b - 1)}{(b - 2)(b + 2)}$ **49.** $\dfrac{x - 1}{x}$ **51.** $2 - m^2$ **53.** $\dfrac{-x^2 + 5x + 1}{x^2}$ **55.** $\dfrac{-x - 7}{x^2 + 6x - 3}$ **57.** $\dfrac{2x - 3}{x + 3}$ **59.** $\dfrac{a + b}{ab(a - b)}$ **61.** $\dfrac{(b - a)(b + a)}{ab(a^2 + b^2)}$ **63.** $R\left[\dfrac{(1 + i)^n - 1}{i(1 + i)^n}\right]$ **65. a.** 136.55 miles per hour **b.** $\dfrac{2v_1v_2}{v_1 + v_2}$

Prepare for Section P.6, *page 57*

67. $-8x^5y^5$ **68.** $16x^{12}$ **69.** $4x^2 - 9y^2$ **70.** $a - 2c$ **71.** 3 **2.** $x^2 - 2x + 4$

Exercises P.6, *page 65*

1. 8 **3.** -3 **5.** -16 **7.** 16 **9.** $\frac{1}{27}$ **11.** $\frac{1}{5}$ **13.** $\frac{2}{3}$ **15.** 16 **17.** $8ab^2$ **19.** $-12x^{11/12}$ **21.** $3xy^3$ **23.** $\dfrac{4z^{1/2}}{3}$ **25.** $\dfrac{3a^{1/12}}{b}$ **27.** $\dfrac{n^2}{|m|}$ **29.** $\sqrt{3x}$ **31.** $\sqrt[3]{25x^2}$ **33.** $\sqrt{a^2+b^2}$ **35.** $(4z)^{1/3}$ **37.** $a^{3/5}$ **39.** $4x^{2/3}$ **41.** $3\sqrt{5}$ **43.** $2\sqrt[3]{3}$ **45.** $-3\sqrt[3]{5}$ **47.** $2|xy|\sqrt{6y}$ **49.** $2ay^2\sqrt[3]{2y}$ **51.** $-13\sqrt{2}$ **53.** $-10\sqrt[4]{3}$ **55.** $17y\sqrt[3]{4y}$ **57.** $-14x^2y\sqrt[3]{y}$ **59.** $17+7\sqrt{5}$ **61.** -7 **63.** $12z+\sqrt{z}-6$ **65.** $x+4\sqrt{x}+4$ **67.** $x+1+4\sqrt{x-3}$ **69.** $\sqrt{2}$ **71.** $\dfrac{\sqrt{10}}{6}$ **73.** $\dfrac{3\sqrt[3]{4}}{2}$ **75.** $\dfrac{2\sqrt[4]{x}}{x}$ **77.** $-\dfrac{3\sqrt{3}-12}{13}$ **79.** $\dfrac{3\sqrt{5}-3}{4}$ **81.** $\dfrac{1}{\sqrt{4+h}+2}$ **83.** 2 **85.** $x \le 0$ **87.** \$8167.67 **89.** 8.34 billion **91. a.** 56% **b.** 24%

Prepare for Section P.7, *page 67*

93. $2x-2$ **94.** $-15x^2-14x+8$ **95.** $-\dfrac{x+2}{x+3}$ **96.** $25x^2-20x+4$ **97.** $\dfrac{2x^2-5x-3}{x^2-1}$ **98.** b

Exercises P.7, *page 74*

1. $9i$ **3.** $7i\sqrt{2}$ **5.** $4+9i$ **7.** $5+7i$ **9.** $8-3i\sqrt{2}$ **11.** $11-5i$ **13.** $-7+4i$ **15.** $8-5i$ **17.** -10 **19.** $-2+16i$ **21.** -40 **23.** -10 **25.** $19i$ **27.** $20-10i$ **29.** $22-29i$ **31.** 41 **33.** $12-5i$ **35.** $-114+42i\sqrt{2}$ **37.** $-6i$ **39.** $3-6i$ **41.** $\frac{7}{53}-\frac{2}{53}i$ **43.** $1+i$ **45.** $\frac{15}{41}-\frac{29}{41}i$ **47.** $\frac{5}{13}+\frac{12}{13}i$ **49.** $2+5i$ **51.** $-16-30i$ **53.** $-11-2i$ **55.** $-i$ **57.** -1 **59.** $-i$ **61.** -1 **63.** $\dfrac{1}{2}+\dfrac{\sqrt{3}}{2}i$ **65.** $-\dfrac{3}{2}+\dfrac{\sqrt{3}}{2}i$ **67.** $\frac{1}{2}+\frac{1}{2}i$ **69.** $(x+4i)(x-4i)$ **71.** $(z+5i)(z-5i)$ **73.** $(2x+9i)(2x-9i)$ **79.** 0

Chapter P True/False Exercises, *page 77*

1. True **2.** True **3.** False. Let $a=\frac{1}{2}$. **4.** False. $\pi+(-\pi)=0$ **5.** False. $\sqrt{2}\cdot\sqrt{8}=\sqrt{16}=4$ **6.** True **7.** False. $a \div b$ is not a real number when $b=0$. **8.** False. $\{x \mid x \ge 2\}$ **9.** False. $\sqrt{-5}\cdot\sqrt{-5}=i\sqrt{5}\cdot i\sqrt{5}=i^2\cdot 5=-5$ **10.** False. Let $a=3$ and $b=4$.

Chapter P Review Exercises, *page 78*

1. Integer, rational, real, prime [P.1] **2.** Irrational, real [P.1] **3.** Rational, real [P.1] **4.** Rational, real [P.1] **5.** $\{1, 2, 3, 5, 7, 11\}$ [P.1] **6.** $\{5\}$ [P.1] **7.** 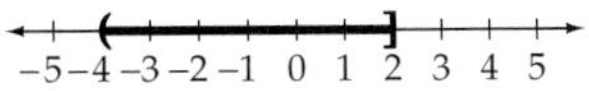

$(-4, 2]$ [P.1] **8.** 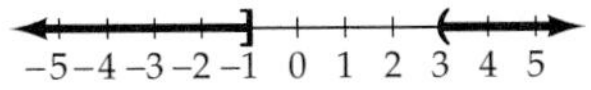

$(-\infty, -1] \cup (3, \infty)$ [P.1] **9.** 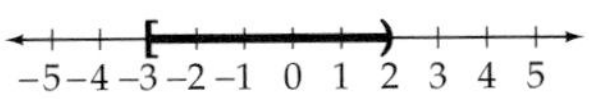

$\{x \mid -3 \le x < 2\}$ [P.1] **10.** 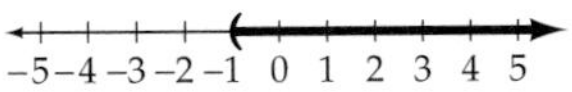

$\{x \mid x > -1\}$ [P.1] **11.** 17 [P.1] **12.** $5-\pi$ [P.1] **13.** -36 [P.1] **14.** $\frac{1}{6}$ [P.2] **15.** $12x^8y^3$ [P.2] **16.** $\dfrac{4a^2b^8}{9c^4}$ [P.2] **17.** 6.2×10^5 [P.2] **18.** 1.7×10^{-6} [P.2] **19.** 35,000 [P.2] **20.** 0.000000431 [P.2] **21.** $-a^2-2a-1$ [P.3] **22.** $2b^2+8b-8$ [P.3] **23.** $10x^2-11x-6$ [P.3] **24.** $4x^3-17x+12$ [P.3] **25.** $9y^2-30y+25$ [P.3] **26.** $3x^2+5x+16+\dfrac{44}{x-3}$ [P.3] **27.** $3(x+5)^2$ [P.4] **28.** $(5x-3y)^2$ [P.4] **29.** $4(5a+b)(5a-b)$ [P.4] **30.** $2(2a+5)(4a^2-10a+25)$ [P.4] **31.** $\dfrac{3x-2}{x+4}$ [P.5] **32.** $\dfrac{2x-5}{4x^2-10x+25}$ [P.5] **33.** $\dfrac{2x+3}{2x-5}$ [P.5] **34.** $\dfrac{2x+1}{x+3}$ [P.5] **35.** $\dfrac{x(3x+10)}{(x+3)(x-3)(x+4)}$ [P.5] **36.** $\dfrac{-5x+1}{(3x-1)(x-1)}$ [P.5] **37.** $\dfrac{2x-9}{3x-17}$ [P.5] **38.** $\dfrac{x+4}{5x+8}$ [P.5] **39.** $\frac{1}{5}$ [P.6] **40.** -9 [P.6] **41.** $x^{17/12}$ [P.6] **42.** $4x^{1/2}$ [P.6] **43.** $x^{3/4}y^2$ [P.6]

44. $x - y$ [P.6] **45.** $4|a|b^3\sqrt{3b}$ [P.6] **46.** $2a\sqrt{3ab}$ [P.6] **47.** $\frac{3y\sqrt{15y}}{5}$ [P.6] **48.** $\frac{2\sqrt{10xyz}}{5z^2}$ [P.6] **49.** $3y^2\sqrt[3]{5x^2y}$ [P.6]

50. $-5xy^2\sqrt[3]{2x}$ [P.6] **51.** 0 [P.6] **52.** $19 - 6\sqrt{3}$ [P.6] **53.** $\frac{2\sqrt{5}}{5}$ [P.6] **54.** $\frac{\sqrt{x}}{x}$ [P.6] **55.** $\frac{7\sqrt[3]{4x}}{2}$ [P.6] **56.** $\frac{5\sqrt[3]{3y^2}}{3}$ [P.6]

57. $10 + 5\sqrt{3}$ [P.6] **58.** $\frac{x - 2\sqrt{x}}{x - 4}$ [P.6] **59.** $5 - i\sqrt{7}$ [P.7] **60.** $2 + 3i\sqrt{2}$ [P.7] **61.** $6 - i$ [P.7] **62.** $-2 + 10i$ [P.7]

63. $8 + 6i$ [P.7] **64.** $29 + 22i$ [P.7] **65.** $8 + 6i$ [P.7] **66.** i [P.7] **67.** $-3 - 2i$ [P.7] **68.** $-\frac{14}{25} - \frac{23}{25}i$ [P.7]

69. 4.02 million [P.6] **70.** $343,750 [P.3] **71.** 18.6% [P.6] **72.** $118,000 [P.3] **73.** $\frac{5n + 65}{2n + 10}$ [P.5] **74.** $18,300 [P.6]

75. $20.49 [P.6] **76.** 5300 feet [P.3] **77.** 31 pounds [P.3]

Chapter P Test, *page 80*

1. $\{4, 6\}$ [P.1] **2.** 3.41×10^{-5} [P.2] **3.** 203,000,000 [P.2] **4.** $-54x^{10}y^9$ [P.2] **5.** $\frac{3y^2}{4x^3}$ [P.2] **6.** $-\frac{1}{9}$ [P.6]

7. $-x^3 + x - 8$ [P.3] **8.** $6x^3 - x^2 - 16x + 6$ [P.3] **9.** $3x + 4 + \frac{12}{x - 2}$ [P.3] **10.** $(2x + 1)(x - 3)$ [P.4]

11. $2x(2x - y)(4x^2 + 2xy + y^2)$ [P.4] **12.** $-\frac{x + 3}{x + 5}$ [P.5] **13.** $\frac{(x + 2)(x + 4)}{(x + 5)(x - 1)}$ [P.5] **14.** $\frac{x(x + 2)}{x - 3}$ [P.5] **15.** $\frac{x(2x - 1)}{2x + 1}$ [P.5]

16. $-6a|b|\sqrt{2a}$ [P.6] **17.** $\sqrt[3]{5}$ [P.6] **18.** $-\frac{19 - 17\sqrt{3}}{23}$ [P.6] **19.** $2 + 2i$ [P.7] **20.** $22 - 3i$ [P.7] **21.** $\frac{11}{26} + \frac{23}{26}i$ [P.7]

22. i [P.7] **23.** Profit $= -0.008x^2 + 40x - 3000$; $29,000 [P.3]

Chapter 1

Chapter Prep Quiz, *page 82*

1. $2\sqrt{3}$ **2.** $-6x + 5$ **3.** $(x - 8)^2$ **4.** $(x + 1)(x - 1)(x^2 + 3)$ **5.** $|x - 5| < 3$ **6.** $-2 + 2i$ **7.** $-\frac{x + 2}{x + 1}$ **8.** $\frac{2 + \sqrt{2}}{2}$
9. $(-\infty, \infty)$ **10.** $\{x | x \geq 3\}$

Exercises 1.1, *page 94*

1. 15 **3.** −4 **5.** 30 **7.** −10 **9.** $\frac{9}{2}$ **11.** $\frac{108}{23}$ **13.** $\frac{2}{9}$ **15.** 12 **17.** 16 **19.** 9 **21.** $\frac{1}{2}$ **23.** $\frac{22}{13}$ **25.** $\frac{95}{18}$ **27.** $r = \frac{3 - p}{q}$

29. $n = -\frac{A - 3m}{4}$ or $\frac{3m - A}{4}$ **31.** $x = \frac{7y}{2 + z}$ **33.** $h = \frac{3V}{\pi r^2}$ **35.** $P = \frac{A}{1 + rt}$ **37.** $v_0 = \frac{s + 16t^2}{t}$ **39.** $d = \frac{a_n - a_1}{n - 1}$ **43.** 30 feet by 57 feet **45.** 850 **47.** $7600 invested at 8%, $6400 invested at 6.5% **49.** $3750 **51.** 1200 at $14 and 1800 at $25 **53.** $48,000 **55.** 7.875 hours **57.** Maximum: 166 beats per minute; minimum: 127 beats per minute **59.** 64 liters **61.** 87 **63.** 240 meters **65.** 0.64 liter **67.** 15 minutes **69.** 2 hours **71.** $18\frac{2}{11}$ grams **73.** 11 quarts **75.** 6.25 feet **77.** 40 pounds **79.** 1384 feet

Prepare for Section 1.2, *page 97*

81. $x^2 + 8x + 16$ **82.** $(x + 6)(x - 7)$ **83.** $(2x - 3)(x + 3)$ **84.** $3i$ **85.** 1 **86.** $11 - 16i$

Exercises 1.2, *page 109*

1. −3, 5 **3.** −5, 2 **5.** $-24, \frac{3}{8}$ **7.** $0, \frac{7}{3}$ **9.** 2, 8 **11.** ±9 **13.** $\pm 2\sqrt{6}$ **15.** $\pm 2i$ **17.** −1, 11 **19.** $3 \pm 4i$ **21.** $-3 \pm 2\sqrt{2}$

23. −3, 5 **25.** $-2 \pm i$ **27.** $\frac{-3 \pm \sqrt{13}}{2}$ **29.** $\frac{-2 \pm \sqrt{6}}{2}$ **31.** $\frac{4 \pm \sqrt{13}}{3}$ **33.** −3, 5 **35.** $\frac{-1 \pm \sqrt{5}}{2}$ **37.** $\frac{-2 \pm \sqrt{2}}{2}$

39. $\frac{5}{6} \pm \frac{\sqrt{11}}{6}i$ **41.** $\frac{-3 \pm \sqrt{41}}{4}$ **43.** $-\frac{5}{6}, \frac{7}{4}$ **45.** $-2, \frac{4}{5}$ **47.** 81; two real solutions **49.** −116; no real solutions **51.** 0; one real solution **53.** 2116; two real solutions **55.** −111; no real solutions **57.** $k < 9$ **59.** Width 12 feet and length 48 feet, or width 32 feet and length 18 feet **61.** 10,000 books **63.** 1100 or 1500 items **65. a.** 45.2 square inches **b.** 10.0 inches **67.** 127.3 feet **69.** 1.8 seconds and 11.9 seconds **71.** No **73.** 9 people **75.** 2006 **77. a.** 44.8 million pounds **b.** 2007

Prepare for Section 1.3, *page 112*

79. $x(x + 4)(x - 4)$ **80.** $x^2(x + 6)(x - 6)$ **84.** 4 **82.** 64 **83.** $x + 2\sqrt{x - 5} - 4$ **84.** $x - 4\sqrt{x + 3} + 7$

Exercises 1.3, *page 123*

1. 4, −4 **3.** 7, 3 **5.** 8, −3 **7.** 2, −8 **9.** No solution **11.** 20, −12 **13.** 12, −18 **15.** 0, ±5 **17.** 0, ±3 **19.** 0, −5, 8 **21.** 0, ±4 **23.** $2, -1 \pm i\sqrt{3}$ **25.** 31 **27.** 2 **29.** No solution **31.** No solution **33.** $\frac{7}{2}$ **35.** −12 **37.** 1 **39.** 17 **41.** 40 **43.** 3 **45.** $\frac{5}{2}$ **47.** −26 **49.** 1 **51.** 1, −6 **53.** 2 **55.** 23, −31 **57.** 19 **59.** $2, -\frac{1}{8}$ **61.** $0, \frac{1}{256}$ **63.** $\pm\sqrt{7}, \pm\sqrt{2}$ **65.** $\pm 2, \pm\frac{\sqrt{6}}{2}$ **67.** $\sqrt[3]{2}, -\sqrt[3]{3}$ **69.** $-\frac{\sqrt[3]{36}}{3}, \frac{\sqrt[3]{98}}{7}$ **71.** 1, 16 **73.** $-\frac{1}{27}, 64$ **75.** $\pm\frac{\sqrt{15}}{3}$ **77.** ±1 **79.** $\frac{1}{11}, -\frac{1}{2}$ **81.** $\frac{256}{81}, 16$ **83.** ±0.62, ±1.62 **85.** $x = \frac{a \pm b}{2}$ **87.** $b = \pm\sqrt{9 - a^2}$ **89.** $f = \frac{d_0 d_1}{d_1 + d_0}$ **91.** $V_2 = \frac{P_1 V_1 T_2}{P_2 T_1}$ **93.** $x = \left(\sqrt{y} + \sqrt{z}\right)^2$ **95.** Not equivalent **97.** Equivalent **99.** $\{x \mid x \geq -4\}$ **101.** $\{x \mid x \leq -7\}$ **103.** 3, −5 **105.** 9, 36 **107. a.** $r \approx 1.63$ inches **b.** $4\sqrt{3}$ inches **109.** 35.8 feet **111.** $13\frac{1}{3}$ hours **113.** 9 years old **115.** 10 games **117.** 3 inches **119.** 13.0 feet **121.** 72 μF **123.** Approximately 19 million meters **125.** $s = \left(\frac{-275 + 5\sqrt{3025 + 176T}}{2}\right)^2$

Prepare for Section 1.4, *page 127*

127. $-16 > -18$ **128.** $\{x \mid 3 < x < 8\}$ **129.** $(-\infty, -2] \cup (5, \infty)$ **130.** $\frac{1}{5}$ **131.** 6, −1 **132.** Cost = 2.25 + 0.85m

Exercises 1.4, *page 134*

1. $\{x \mid x < 4\}$ **3.** $\{x \mid x < -6\}$ **5.** $\{x \mid x \leq -3\}$ **7.** $\left\{x \mid x \geq -\frac{13}{8}\right\}$ **9.** $\{x \mid x < 2\}$ **11.** $\left\{x \mid -\frac{3}{4} < x \leq 4\right\}$ **13.** $\left\{x \mid \frac{1}{3} \leq x \leq \frac{11}{3}\right\}$ **15.** $\left\{x \mid -\frac{3}{8} \leq x < \frac{11}{4}\right\}$ **17.** $\{x \mid x < 1\}$ **19.** $\{x \mid x > -1\}$ **21.** $\{x \mid x < 4\}$ **23.** $\{x \mid x > -28\}$ **25.** $\left(-\infty, -\frac{3}{2}\right) \cup \left(\frac{5}{2}, \infty\right)$ **27.** $(-\infty, -8] \cup [2, \infty)$ **29.** $\left[-\frac{4}{3}, 8\right]$ **31.** $(-\infty, -4] \cup \left[\frac{28}{5}, \infty\right)$ **33.** $(-\infty, \infty)$ **35.** {4} **37.** Ø **39.** $(-\infty, \infty)$ **41.** $(-\infty, -7] \cup [-1, \infty)$ **43.** $[-3, 9]$ **45.** $(-\infty, -9] \cup [15, \infty)$ **47.** $|x - 3| < 2$ **49.** $|x - 6| > 2$ **51.** $|x - 9| \leq 2$ **53.** $|x - 10| \geq 2$ **55.** $|x| \leq 4$ **57.** $|x - 6| > 3$ **59.** $|x - 33| < 2$ **61.** {21, 23, 25}, {23, 25, 27} **63.** Provided more than 100 checks per month are written **65.** ≤$1504 **67.** At least 249 telephones **69.** Minus $4.25 to plus $4.25 **71. a.** $|B - 218| > 48$ **b.** $(0, 170) \cup (266, \infty)$ **73.** Less than or equal to 26 inches **75. a.** 130.0 to 137.5 centimeters **b.** Potential stature: 168.4 to 177.5 centimeters; living stature 166.9 to 176.0 centimeters **c.** $|h - (4.74r + 54.93)| \leq 4.24$ **77.** 9.39 to 9.55 inches **79.** 2002 **81. a.** In the school year 2011–2012 **b.** In the school year 2007–2008 **83.** 4.43 to 4.48 inches

Prepare for Section 1.5, *page 139*

84. 100 **85.** 5 **86.** $4\sqrt{6}$ **87.** $(x - 7)^2$ **88.** 61 **89.** ±3

Exercises 1.5, *page 149*

1. $(-\infty, -7) \cup (0, \infty)$ **3.** $[-4, 4]$ **5.** $(-5, -2)$ **7.** $(-\infty, -4] \cup [7, \infty)$ **9.** $\left[-\frac{1}{2}, 1\right]$ **11.** $\left(-\infty, \frac{3}{2}\right) \cup \left(\frac{5}{3}, \infty\right)$ **13.** $\left[-4, \frac{1}{6}\right]$ **15.** $(-\infty, 5) \cup (8, \infty)$ **17.** $\left[-\frac{1}{2}, 6\right)$ **19.** $\left(-\infty, \frac{9}{2}\right] \cup (5, \infty)$ **21.** $(-3, 3) \cup (6, \infty)$ **23.** $\left(-\infty, \frac{5}{3}\right)$

25. $\left(-\infty, 2-\sqrt{3}\right) \cup \left(2+\sqrt{3}, \infty\right)$ **27.** $\left(-\infty, -\frac{1}{6}\right] \cup \left[\frac{9}{2}, \infty\right)$ **29.** $\left[\frac{-5-\sqrt{33}}{4}, \frac{-5+\sqrt{33}}{4}\right]$ **31.** $\left(-\infty, -\frac{3}{8}\right) \cup (6, \infty)$

33. $\left(-\infty, 3-\sqrt{5}\right] \cup \left(3, 3+\sqrt{5}\right]$ **35.** $[-1, 2) \cup [4, \infty)$ **37.** $\left(-3, \frac{1-\sqrt{21}}{2}\right) \cup \left(\frac{1+\sqrt{21}}{2}, \infty\right)$ **39.** $(-\infty, 4)$

41. At least 9791 books **43.** 100 to 200 milligrams **45.** A distance of more than 14.7 feet but less than 145.0 feet from a sideline **47.** 1900–1912 and 1984–2000 **49.** more than 1 but less than 3 seconds

Chapter 1 True/False Exercises, *page 154*

1. True **2.** False; the solution set is $\{-3, 3\}$. **3.** True **4.** False; the discriminant is $b^2 - 4ac$. **5.** False; $|-3x - 4| = 8$ is equivalent to $3x + 4 = 8$ or $3x + 4 = -8$. **6.** False. The first equation has a solution of 3, whereas the second equation has both 3 and -4 as solutions. **7.** True **8.** True **9.** True **10.** False; 3 is also a critical value.

Chapter 1 Review Exercises, *page 154*

1. $\frac{3}{2}$ [1.1] **2.** $\frac{11}{3}$ [1.1] **3.** $\frac{1}{2}$ [1.1] **4.** $\frac{11}{4}$ [1.1] **5.** $-\frac{38}{15}$ [1.3] **6.** $-\frac{1}{2}$ [1.3] **7.** -2 [1.3] **8.** $-\frac{3}{2}$ [1.3] **9.** 3, 2 [1.2] **10.** $\frac{4}{3}, -\frac{3}{2}$ [1.2] **11.** $\frac{1 \pm \sqrt{13}}{6}$ [1.2] **12.** $\frac{1}{2} \pm \frac{\sqrt{3}}{2}i$ [1.2] **13.** $0, \frac{5}{3}$ [1.3] **14.** $0, \pm 2$ [1.3] **15.** $\pm\frac{2\sqrt{3}}{3}, \pm\frac{\sqrt{10}}{2}$ [1.3]

16. $\frac{4}{9}$ [1.3] **17.** $-5, 3$ [1.3] **18.** $-4, 6$ [1.3] **19.** $7, -2$ [1.3] **20.** -79 [1.3] **21.** $-4, -2$ [1.3] **22.** $\frac{9}{4}, \frac{11}{4}$ [1.3] **23.** 1, 5 [1.3] **24.** $-9, -1$ [1.3] **25.** $-3, 2$ [1.3] **26.** $-\frac{1}{3}, 5$ [1.3] **27.** $-2, -1$ [1.3] **28.** $-4, 1, \frac{4}{3}$ [1.3] **29.** $-\frac{63}{2}, 14$ [1.3] **30.** $\frac{65}{16}$ [1.3] **31.** $m = \frac{e}{c^2}$ [1.1] **32.** $h = \frac{V}{\pi r^2}$ [1.1] **33.** $b_1 = \frac{2A - hb_2}{h}$ [1.1] **34.** $w = \frac{P - 2l}{2}$ [1.1] **35.** $t = \frac{A - P}{Pr}$ [1.1]

36. $m_1 = \frac{Fs^2}{Gm_2}$ [1.1] **37.** $(-\infty, 2]$ [1.4] **38.** $\left[\frac{6}{7}, \infty\right)$ [1.4] **39.** $(-\infty, 3)$ [1.4] **40.** $(-\infty, 9]$ [1.4] **41.** $(5, \infty)$ [1.4] **42.** $\varnothing$ [1.4] **43.** $(-\infty, 4]$ [1.4] **44.** $(-10, \infty)$ [1.4] **45.** $[-5, 2]$ [1.5] **46.** $(-\infty, -1) \cup (3, \infty)$ [1.5] **47.** $\left[\frac{145}{9}, 35\right]$ [1.4] **48.** $(86, 149)$ [1.4] **49.** $(-\infty, -3) \cup (4, \infty)$ [1.5] **50.** $(-\infty, 0] \cup [5, 7)$ [1.5] **51.** $\left(-\infty, \frac{5}{2}\right] \cup (3, \infty)$ [1.5] **52.** $\left[\frac{5}{2}, 5\right)$ [1.5] **53.** $(-\infty, 4)$ [1.5] **54.** $\left(\frac{3-\sqrt{69}}{2}, -2\right) \cup \left(\frac{3+\sqrt{69}}{2}, \infty\right)$ [1.5] **55.** $[-2, 8]$ [1.4] **56.** $(-\infty, -5) \cup (1, \infty)$ [1.4] **57.** $\left(\frac{2}{3}, 2\right)$ [1.4] **58.** $(-\infty, 1] \cup [2, \infty)$ [1.4] **59.** One possible answer: $|x - 7| < 3$ [1.4] **60.** One possible answer: $|x - 4| > 1$ [1.4] **61.** One possible answer: $\left|x - \frac{7}{2}\right| \geq \frac{5}{2}$ [1.4] **62.** One possible answer: $|x - 52| \leq 8$ [1.4] **63.** 81; two real solutions [1.2] **64.** 53; two real solutions [1.2] **65.** -116; no real solutions [1.2] **66.** -3; no real solutions [1.2] **67.** 40 [1.2] **68. a.** 35 **b.** 15 **c.** The equation $100 = \frac{1}{2}n(n - 3)$ has no natural number solutions. [1.2] **69.** \$1750 in the 4% account, \$3750 in the 6% account [1.1] **70.** 18 hours [1.3] **71.** 4024 adult tickets and 502 student tickets [1.1] **72.** \$864 [1.1] **73.** At least 14,934 watches [1.5] **74.** Between \$12 and \$24 [1.5] **75.** 1 to 4399 DVDs [1.5] **76.** 1 to 32 motorcycles [1.5] **77.** 865 to 1735 items [1.5] **78.** From 0.346 grams to 0.520 grams [1.5] **79.** 64 feet by 30 feet [1.2] **80.** 3.7 feet to 7.7 feet [1.2] **81.** Width = 12 ft, length = 15 ft [1.1] **82.** 24 nautical miles [1.1] **83.** Between 15.4% and 20.3% [1.5] **84.** 13 feet [1.3] **85.** 182 feet [1.3]

Chapter 1 Test, *page 157*

1. 1 [1.1] **2.** $-6, \frac{22}{3}$ [1.3] **3.** $-\frac{1}{2}, \frac{8}{3}$ [1.2] **4.** $\frac{4 \pm \sqrt{14}}{2}$ [1.2] **5.** $\frac{5 \pm \sqrt{37}}{6}$ [1.2] **6.** The discriminant is 9; two real solutions [1.2] **7.** $\{x \mid x \leq 8\}$ [1.4] **8.** $[-2, 5)$ [1.4] **9.** $(3, 9)$ [1.4] **10.** 23 [1.3] **11.** $\pm\sqrt{5}, \pm i\sqrt{3}$ [1.3] **12.** $[1, 2) \cup [4, \infty)$ [1.5] **13.** $\{x \mid 2 \leq x < 2.5\}$ [1.5] **14.** $w = \frac{ab}{b + 4}$ [1.1] **15.** \$28,000 at 5%, \$22,000 at 7% [1.1] **16.** 2 hours 7 minutes [1.1]

17. 9475 to 24,275 printers [1.5] **18. a.** $|d - 485| > 210$ **b.** $(0, 275) \cup (695, 900)$ [1.4] **19.** 140.23 centimeters to 147.55 centimeters [1.4] **20.** 8067 to 9076 fax machines [1.5]

Cumulative Review Exercises, *page 158*

1. -11 [P.1] **2.** 1.7×10^{-4} [P.2] **3.** $8x^2 - 30x + 41$ [P.3] **4.** $(8x - 5)(x + 3)$ [P.4] **5.** $\dfrac{2x + 17}{x - 4}$ [P.5] **6.** $a^{11/12}$ [P.6] **7.** 29 [P.7] **8.** $\frac{10}{3}$ [1.1] **9.** $\dfrac{2 \pm \sqrt{10}}{2}$ [1.2] **10.** 1, 5 [1.3] **11.** 5 [1.3] **12.** $-6, 0, 6$ [1.3] **13.** $\pm\sqrt{3}, \pm\dfrac{\sqrt{10}}{2}$ [1.3] **14.** $\{x \mid x \le -1 \text{ or } x > 1\}$ [1.4] **15.** $(-\infty, 4] \cup [8, \infty)$ [1.4] **16.** $\left\{x \middle| \frac{10}{7} \le x < \frac{3}{2}\right\}$ [1.5] **17.** Length = 58 feet, width = 42 feet [1.1] **18.** 15 hours [1.3] **19.** A score of 68 to 100 [1.4] **20. a.** Between 14.3% and 23.1% **b.** Answers will vary. [1.5]

Chapter 2

Chapter Prep Quiz, *page 162*

1. $5\sqrt{3}$ **2.** 16 **3.** 9 **4.** $y = -\frac{3}{4}x + 2$ **5.** $1 - \sqrt{3}, 1 + \sqrt{3}$ **6.** $y = 2x + 5$ **7.** 6 **8.** -3 **9.** $\sqrt{58}$ **10.** 4

Exercises 2.1, *page 173*

1.

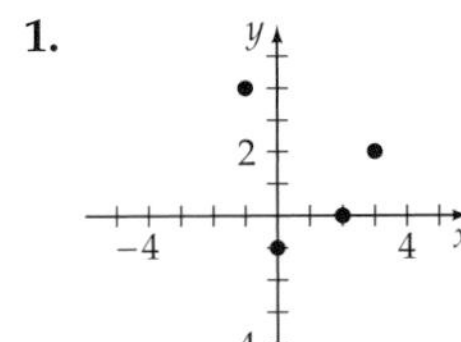

3.

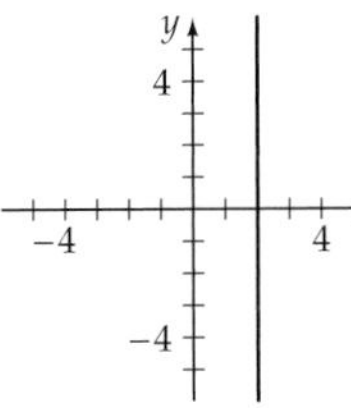

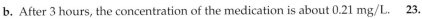

5.

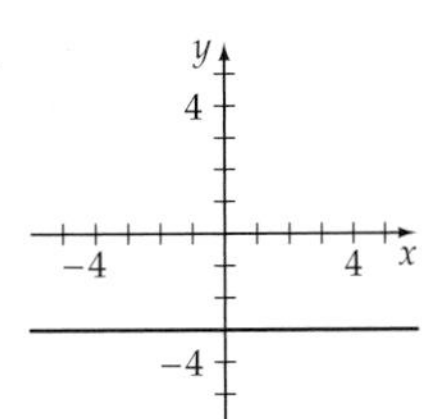

7. $\sqrt{26}$, (2.5, −1.5) **9.** $\sqrt{65}$; (−1.5, 1)

11. $2\sqrt{2}$, (−1, −2) **13.** (3.5, 3) **15.** −9, −7, −5, −3, −1, 1, 3 **17.** −1, −4, −5, −4, −1, 4, 11 **19. a.** 28, 675; 57,100; 85,275; 113,200; **b.** The revenue for customizing 30 motorhomes is $85,275. **21. a.** 0.17, 0.22, 0.21, 0.19; **b.** After 3 hours, the concentration of the medication is about 0.21 mg/L. **23.**

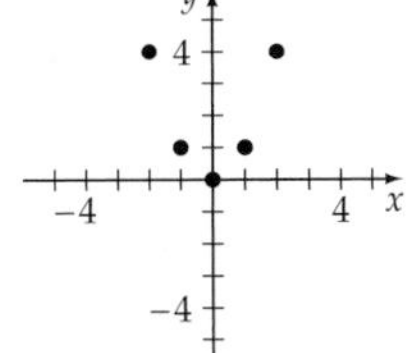

25.

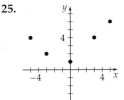

27.

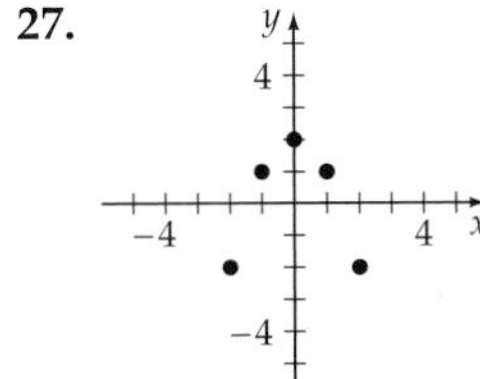

29.

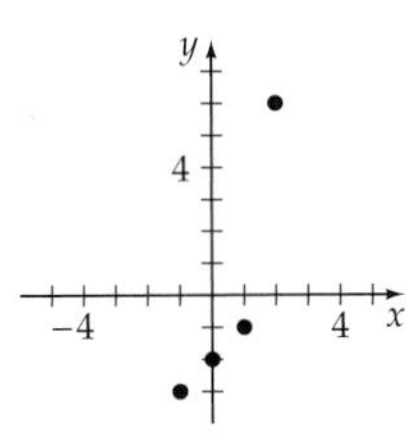

31.

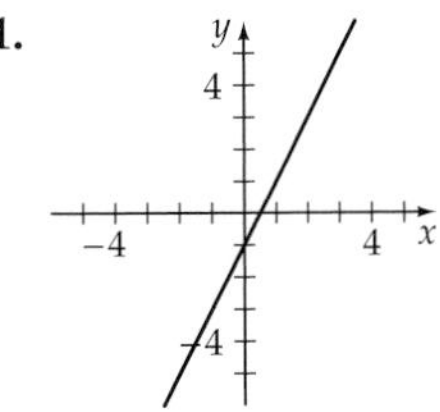

33.

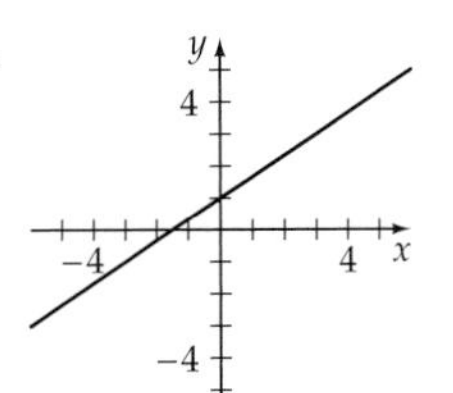

35.

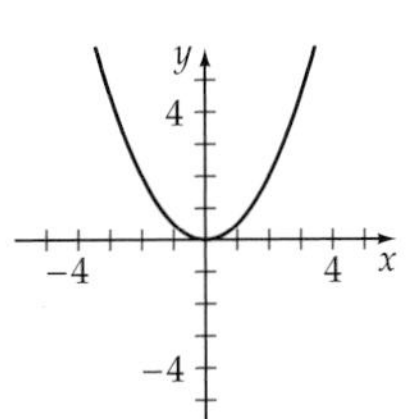

37.

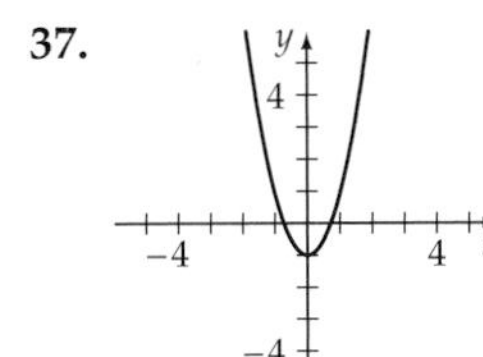

39.

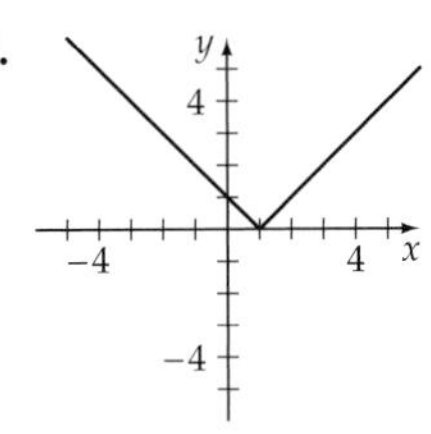

41. $(x - 1)^2 + (y + 5)^2 = 4$

43. $x^2 + (y - 1)^2 = 7$ **45.** $(x - 3)^2 + (y - 4)^2 = 20$ **47.** $x^2 + y^2 = 25$ **49.** $(x - 3)^2 + (y - 1)^2 = 13$
51. $(x + 3)^2 + \left(y - \frac{7}{2}\right)^2 = \frac{17}{4}$ **53.** $C(3, 0); r = 2$ **55.** $C(5, -2); r = 3$ **57.** $C\left(0, \frac{5}{2}\right); r = \dfrac{\sqrt{37}}{2}$

59. **61.** **63.**

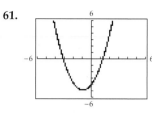

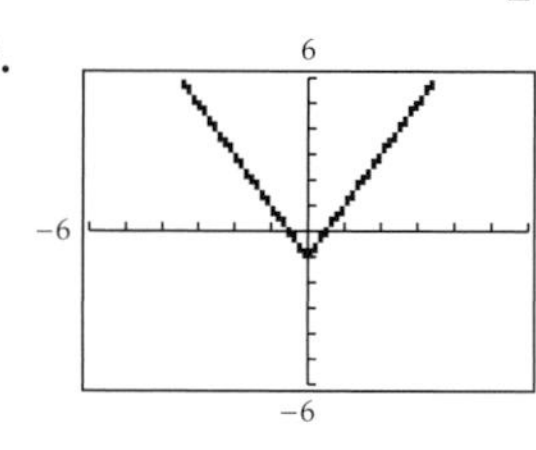

65.

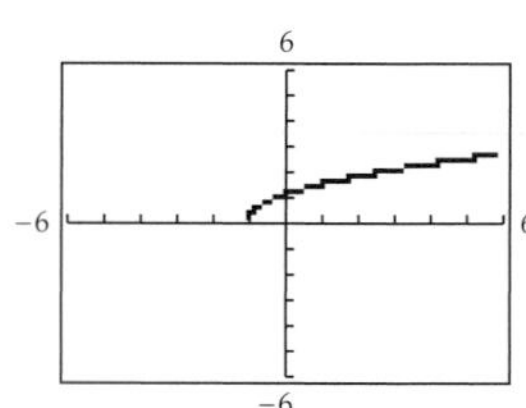

67.

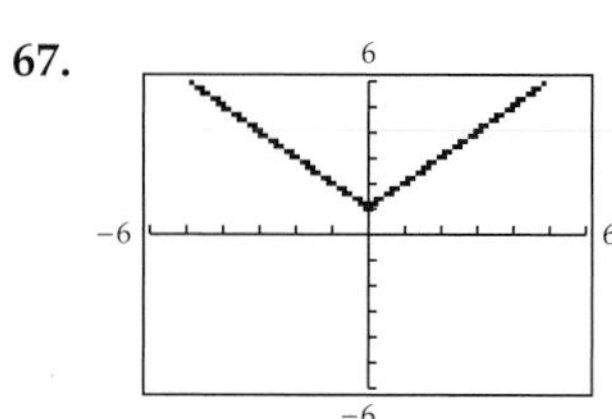

69.

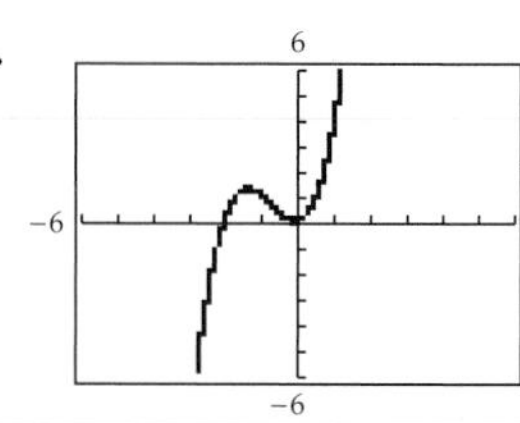

71.

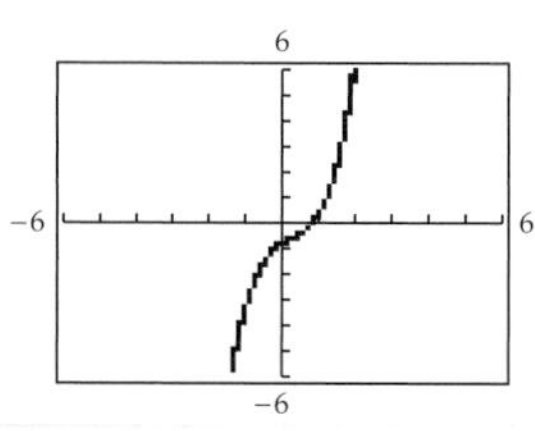

Prepare for Section 2.2, *page 175*

73. $2x^2 - x + 4$ **74.** $6x^2 + x - 2$ **75.** -1 and 1 **76.** $-1 \pm \sqrt{6}$ **77.** $<$ **78.** $>$

Exercises 2.2, *page 188*

1. Yes; domain $= \{-3, -2, 1, 4\}$, range $= \{-7, 1, 2, 5\}$ **3.** Yes; domain $= \{1, 2, 3, 4, 5\}$, range $= \{5\}$ **5.** No; domain $= \{2, 4, 6, 8\}$, range $= \{3, 5, 7, 8, 9\}$ **7.** Yes **9.** Yes **11.** Yes **13.** No **15.** No **17. a.** 13 **b.** -3 **c.** 5 **d.** 8 **e.** $4a + 5$ **f.** $8a + 5$ **19. a.** 2 **b.** 2 **c.** $a^2 - 2a - 1$ **d.** -1 **e.** $a^2 + 4a + 2$ **f.** $9c^2 + 6c - 1$ **21. a.** $\frac{8}{3}$ **b.** 1 **c.** 0 **d.** -4 **e.** 20,002 **f.** $-19{,}998$

23.

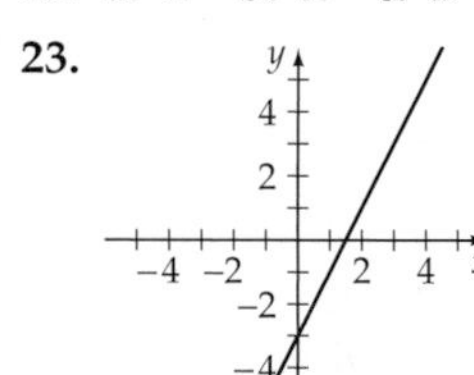

25.

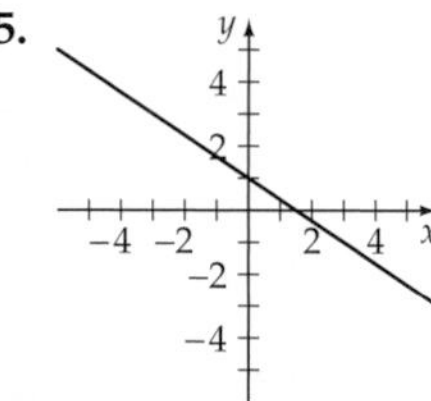

27.

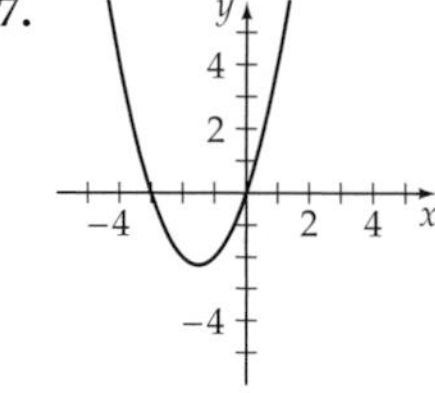

29.

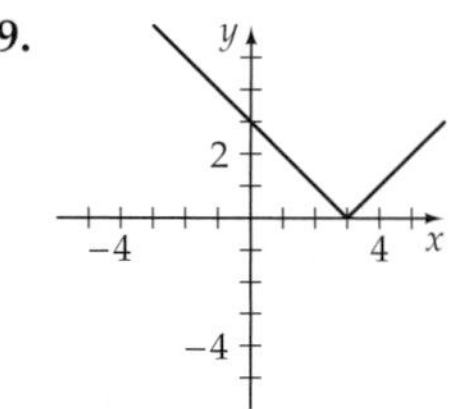

31.

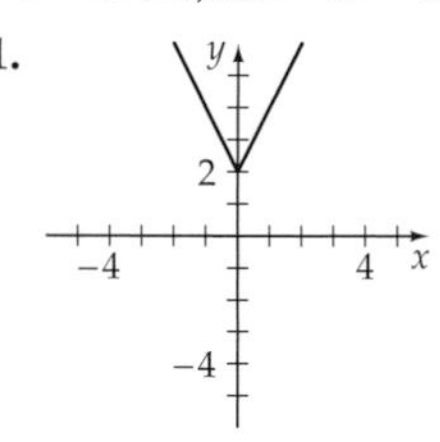

33. All real numbers **35.** All real numbers except $x = 1$ **37.** $(-\infty, 4]$ **39.** All real numbers **41. a.** Yes **b.** Yes **c.** Yes **d.** No **43.** Increasing on $(-\infty, 0]$, decreasing on $[0, \infty)$ **45.** Decreasing on $(-\infty, -2]$, constant on $[-2, 1]$, increasing on $[1, \infty)$ **47.** Constant on $(-\infty, \infty)$ **49.** Constant on $(-\infty, -2]$, $[2, \infty)$, decreasing on $[-2, 0]$, increasing on $[0, 2]$ **51.** Constant on $(-\infty, 0]$, decreasing on $[0, \infty)$ **53. a.** No **b.** Yes **c.** No **d.** Yes **e.** No **55. a.** $w(l) = 25 - l$ **b.** $A(l) = 25l - l^2$ **57. a.** $y(x) = 20 - x$ **b.** $P(x) = 20x - x^2$ **59.** 4920, 4270, 3760, 3200, 2780 **61.** $v(t) = 80{,}000 - 6500t$ **63. a.** $R(x) = 37x$ **b.** $P(x) = 9x - 400$ **c.** \$1850 **65. a.** $R(x) = 225x - \frac{1}{300}x^2$ **b.** $P(x) = \frac{13}{600}x^2 + 185x - 215$ **c.** \$1637.17 **67.** $t(x) = \dfrac{\sqrt{x^2 + 1}}{2} + \dfrac{3 - x}{8}$ **69. a.** 2.3625 tons **b.** 1.0125 tons **71. a.** 1981, 1996, 1999, 2000 **b.** 1982, 1983, 1991, 1992 **73.** $d(t) = \sqrt{9t^2 + 2500}$ **75.** $d(t) = \sqrt{100t^2 - 720t + 2025}$ **77.** Yes **79.** Yes **81.** No **83.** No **85.** Yes

87.

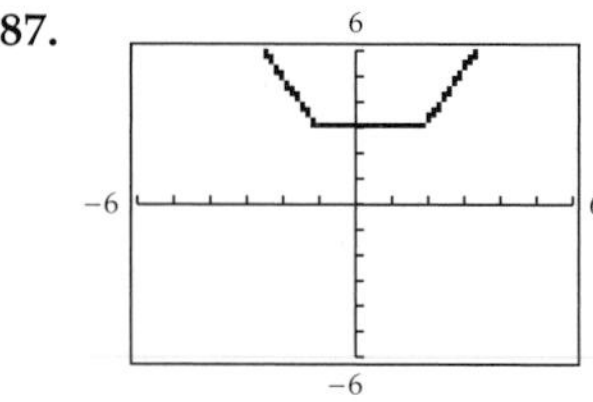

89.

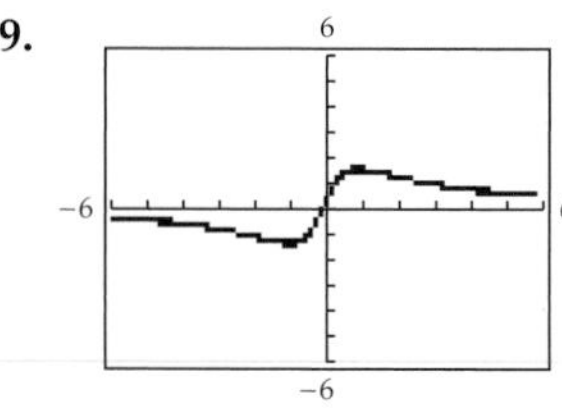

91.

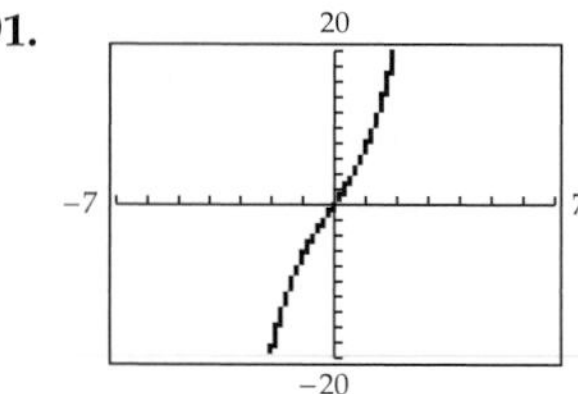

93.

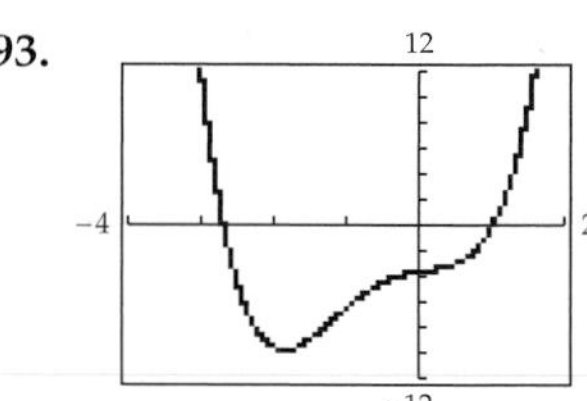

Prepare for Section 2.3, *page 193*

94. 0 **95.** 0 **96. a.** 5 **b.** 6 **97.** (1, −3) **98.** $5\sqrt{2}$ **99.** A horizontal line passing through (0, 1)

Exercises 2.3, *page 206*

1. (2, 0), (0, −6) **3.** (6, 0), (0, −4) **5.** (−4, 0), (0, −4) **7.** (4, 0), (0, 3) **9.** (4.5, 0), (0, −3) **11.** (2, 0), (0, 3) **13.** (1, 0), (0, −2) **15.** −1 **17.** $\frac{1}{3}$ **19.** $-\frac{2}{3}$ **21.** $-\frac{3}{4}$ **23.** No slope **25.** 0 **27.** Perpendicular **29.** Neither **31.** Neither **33.** Parallel **35.**

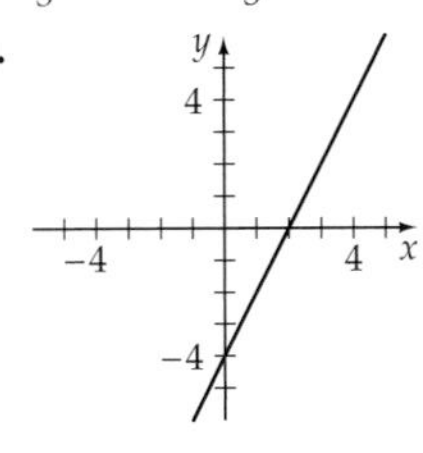

37. y, 2, −4, 2, x, −4

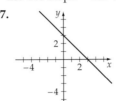

39.

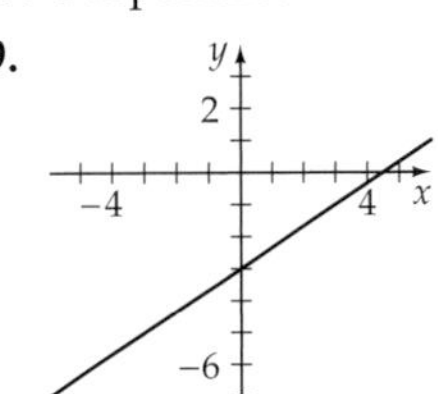

41.

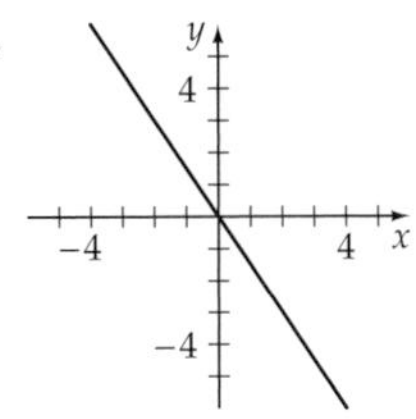

43.

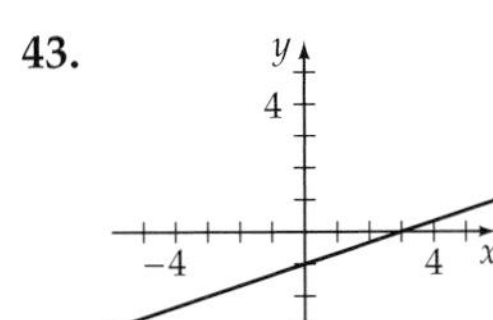

45.

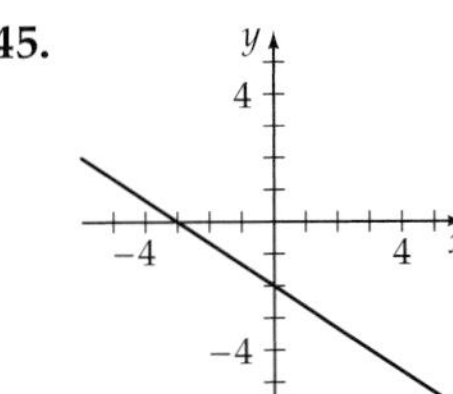

47.

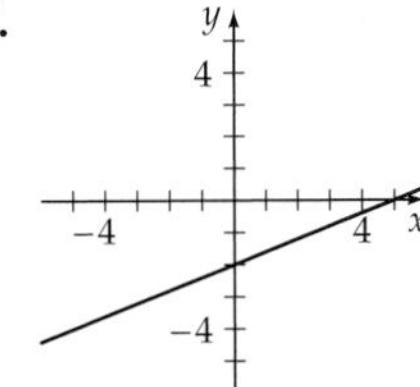

49.

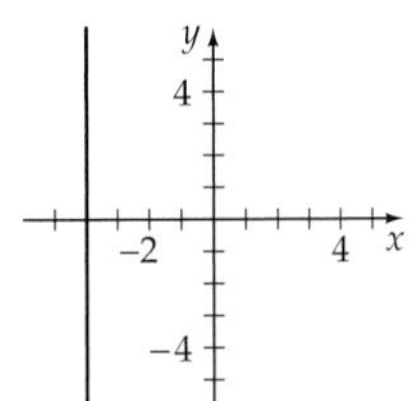

51.

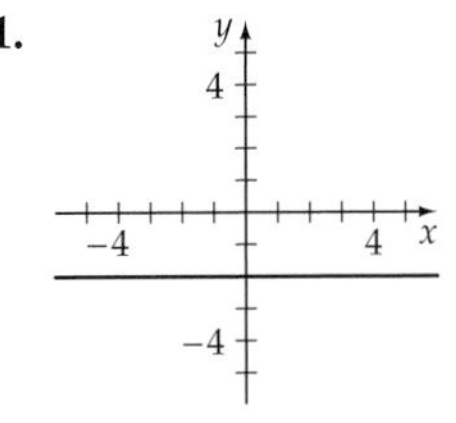

53.

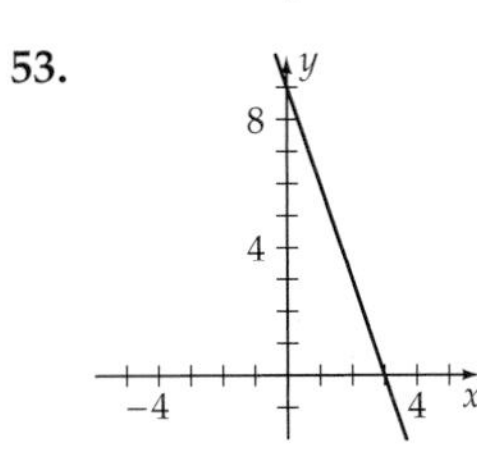

55. Vertical intercept: (0, 100,000), horizontal intercept: (40, 0). The vertical intercept represents the amount in the account ($100,000) when withdrawals begin. The horizontal intercept means that after 40 months, there is $0 remaining in the account.

57. Vertical intercept: (0, 6000), horizontal intercept: (12, 0). The vertical intercept means that Paris is6000 miles from Los Angeles. The horizontal intercept means that after 12 hours, the jet will arrive in Paris. **59.** x-intercept $= \left(\frac{30}{7}, 0\right)$. The x-intercept represents the temperature at which chirping stops. Below this temperature chirping is negative and meaningless. **61.** Slope = −5. The slope means that the oven temperature is decreasing 5°F per minute. **63.** Lois is line A. Terri is line B. The distance between them is line C. **65.** Slope = 0.04. The slope means that 0.04 megabytes are downloaded per second. **67.** 168 inches **69.** Slope = −1.08. The slope means that the water is flowing out of the pool at a rate of 1080 gallons per hour. **71.** −4 **73.** −7 **75.**

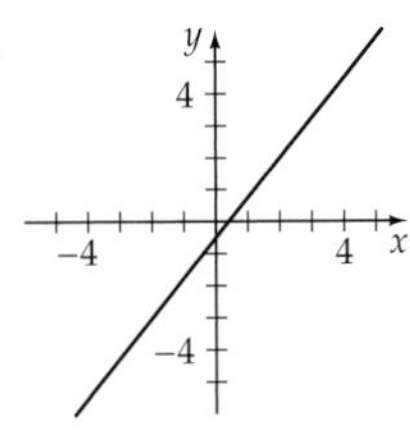

77.

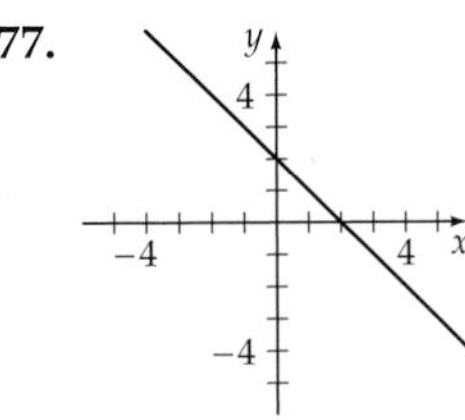

79. It decreases the slope

81. It decreases the y-intercept. **83.** Yes

Prepare for Section 2.4, *page 209*

84. −1 **85.** −2, 3 **86.** $y = -2x - 1$ **87.** $y = -\frac{1}{2}x$ **88.** A horizontal line passing through (0, −1) **89.** A diagonal line passing through the origin and quadrants I and III

Exercises 2.4, *page 218*

1. $y = 2x + 5$ **3.** $y = -3x + 4$ **5.** $y = -\frac{2}{3}x + 7$ **7.** $y = -3$ **9.** $y = \frac{3}{2}x + 3$ **11.** $y = 2x - 13$ **13.** $y = 3x + 14$ **15.** $y = -\frac{4}{5}x + 5$ **17.** $y = x + 2$ **19.** $y = -\frac{3}{2}x + 3$ **21.** $y = \frac{1}{2}x - 1$ **23.** $x = -2$ **25.** $y = x + 2$ **27.** $y = -5$ **29.** $y = -\frac{1}{2}x - \frac{1}{2}$ **31.** $x = 3$ **33.** $y = -3x + 7$ **35.** $y = \frac{2}{3}x - \frac{8}{3}$ **37.** $y = \frac{1}{3}x - \frac{1}{3}$ **39.** $y = -\frac{5}{3}x - \frac{14}{3}$ **41.** $c = \frac{1}{25}s + 1000$; \$4400 **43.** $n = -400p + 48{,}000$; 18,000 calculators **45.** $r = -\frac{3}{5}p + 545$; 485 rooms **47.** $c = 85f + 30{,}000$; \$183,000 **49. a.** $p = -3x + 215$ **b.** The slope means that blood pressure is decreasing at a rate of 3 units per milligram. **c.** 140 **51.** $d = 415t$; 1867.5 mi **53. a.** $b = 0.0052s + 100$ **b.** The slope means that the boiling point increases 0.0052°C for each additional gram of sugar added. **c.** 100.26°C **55.** $y = \frac{2}{5}x + \frac{29}{5}$

Prepare for Section 2.5, *page 220*

57. $(2x - 3)(x + 1)$ **58.** $x^2 - 6x + 9 = (x - 3)^2$ **59.** (3, 0), (0, 6) **60.** No. $x^2 + 1 \geq 0$ for all real numbers x. **61.** −4, 1 **62.** $-1 \pm \sqrt{5}$

Exercises 2.5, *page 228*

1. (2, −3) **3.** (0, −4) **5.** −5 **7.** $x = 7$ **9.** (0, −2); $x = 0$ **11.** (0, −1); $x = 0$ **13.** (0, 2); $x = 0$ **15.** (−1, −2); $x = -1$ **17.** $\left(\frac{1}{2}, -\frac{9}{4}\right)$; $x = \frac{1}{2}$ **19.** $\left(\frac{1}{4}, -\frac{41}{8}\right)$; $x = \frac{1}{4}$ **21.** (0, 0), (2, 0); (0, 0) **23.** $(-2, 0), \left(-\frac{3}{4}, 0\right)$; (0, 6) **25.** $\left(-1 - \sqrt{2}, 0\right), \left(-1 + \sqrt{2}, 0\right)$; (0, −1) **27.** None; (0, −5) **29.** $\left(\frac{4 - \sqrt{14}}{2}, 0\right), \left(\frac{4 + \sqrt{14}}{2}, 0\right)$; (0, 1) **31.** $\left(\frac{-3 - \sqrt{15}}{2}, 0\right), \left(\frac{-3 + \sqrt{15}}{2}, 0\right)$; (0, −2) **33.** 2 **35.** −3 **37.** −5 **39.** $\frac{9}{4}$ **41.** c **43.** 100 lenses **45.** \$250 **47.** 7 days **49.** 15.08 feet **51.** 29 feet **53.** 150 feet **55.** Yes **57.** 8 miles per hour **59.** $k = -8$ **61.** $k = 8, -8$

Prepare for Section 2.6, *page 230*

62. $-\frac{1}{2}$ **63.** (4, 0), (0, 3) **64.** Slope: $-\frac{1}{2}$, y-intercept: (0, 2) **65.** $y = -2x + 7$ **66.** $y = -\frac{4}{3}x - \frac{5}{3}$ **67. a.** 3 feet per minute **b.** 5 feet

Exercises 2.6, *page 239*

1. Zero **3.** Negative **5.** Figure A **7.** $y = 2.009x + 0.560$ **9.** $y = -0.723x + 9.234$ **11.** $y = 2.223x - 7.364$ **13.** $y = 4.77x - 1.27$ **15. a.** $y = 1.0268x + 6.4402$ **b.** 31 miles per gallon **17. a.** $y = 22.6029x - 21.8128$ **b.** 1199 centimeters **19. a.** $1.753x + 9.106$ **b.** 53 centimeters **21. a.** $y = -0.6747x + 69.0371$ **b.** 23.2 **23. a.** Yes **b.** $y = -0.9040x + 78.6309$ **c.** 56.0 years **25.** Yes. $r \approx 0.984$, which is very close to 1.

Chapter 2 True/False Exercises, *page 245*

1. True **2.** True **3.** False. The vertical line $x = 2$ cannot be written in the form $y = mx + b$ because a vertical line has no slope, and m in $y = mx + b$ is a real number. **4.** True **5.** False. The slope m and y-intercept are independent of each other. In $y = 2x + 4$, the slope can be changed to 3, giving $y = 3x + 4$, but the y-intercept remains the same. **6.** True **7.** True **8.** False. The product of the slopes of nonvertical perpendicular lines is −1. $y = 2x - 4$ and $y = -\frac{1}{2}x - 2$ are perpendicular because $2\left(-\frac{1}{2}\right) = -1$. **9.** False. Two lines with the same slope will never meet. $y = 2x - 4$ and $y = 2x + 4$ are parallel because they have the same slope and different y-intercepts. They will never meet. **10.** True

Chapter 2 Review Exercises, *page 246*

1. No [2.2] **2.** Yes [2.2] **3.** Yes [2.2] **4.** Yes [2.2] **5.** Yes [2.2] **6.** Yes [2.2] **7.** No [2.2] **8.** No [2.2] **9. a.** 5 **b.** −7 **c.** $c^2 - 3c - 5$ **d.** $h^2 + h - 7$ **e.** $4a^2 - 6a - 5$ **f.** $2a^2 - 6a - 10$ [2.2] **10. a.** −4 **b.** $\frac{1}{6}$ **c.** 0 **d.** $\dfrac{a+2}{a-3}$ **e.** $\dfrac{z}{z-1}$ **f.** $\dfrac{5z}{z-5}$ [2.2] **11.**

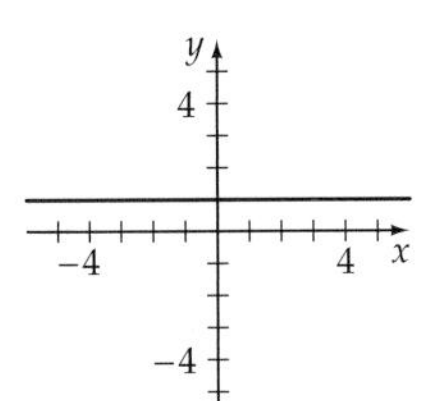

[2.1] **12.**

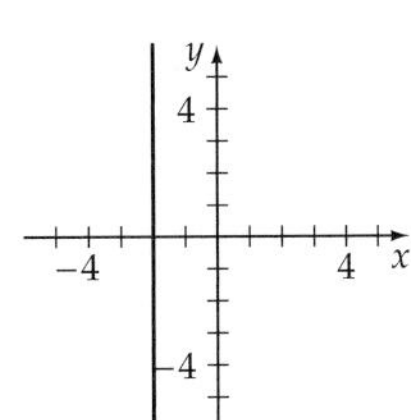

[2.1] **13.**

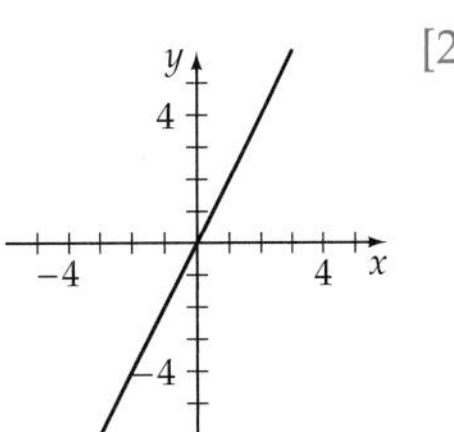

[2.1]

14.

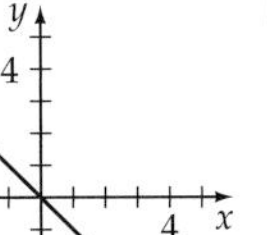

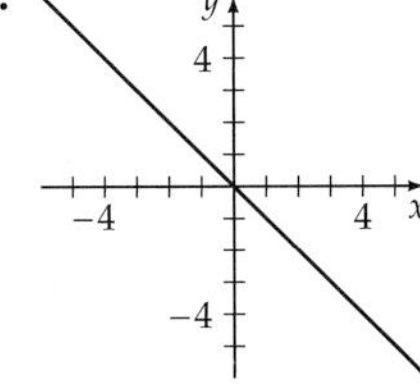

[2.1] **15.** Midpoint (−2, 3), distance 10.77 [2.1] **16.** Midpoint (0, −3), distance 5.66 [2.1]

17. Midpoint (1.5, −2), distance 7.28 [2.1] **18.** Midpoint (5.5, −2.5), distance 1.41 [2.1] **19.** $(x+3)^2 + (y-5)^2 = 9$ [2.1] **20.** $(x-1)^2 + (y-3)^2 = 8$ [2.1] **21.** $C(-3, 1); r = 4$ [2.1] **22.** $C(2, -4); r = \sqrt{19}$ [2.1] **23.**

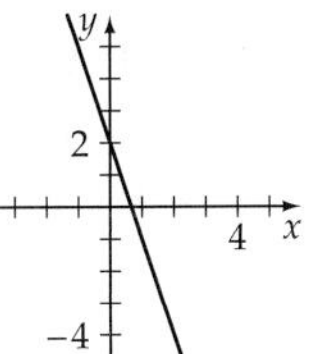

[2.1]

24.

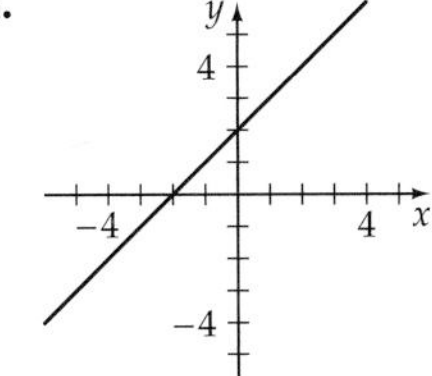

[2.1] **25.**

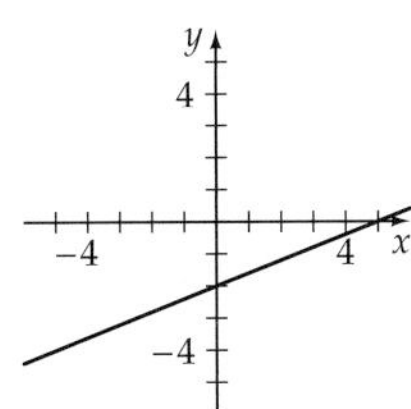

[2.3] **26.**

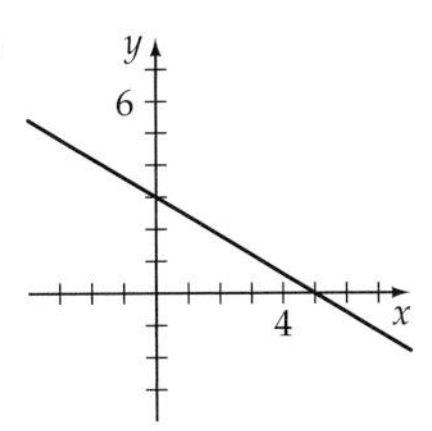

[2.3] **27.**

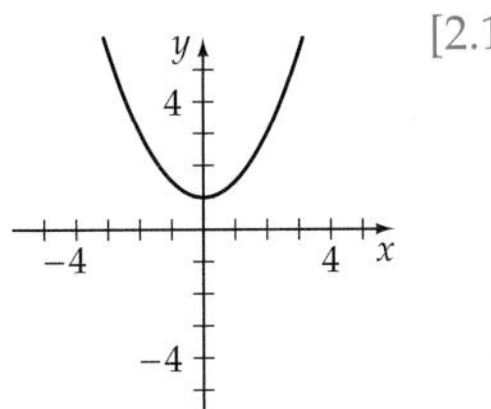

[2.1]

28.

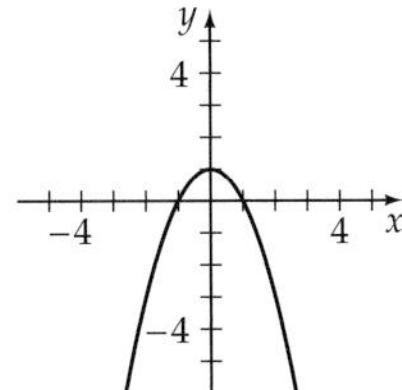

[2.1] **29.**

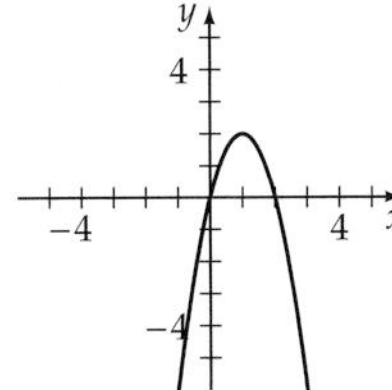

[2.2/2.5] **30.**

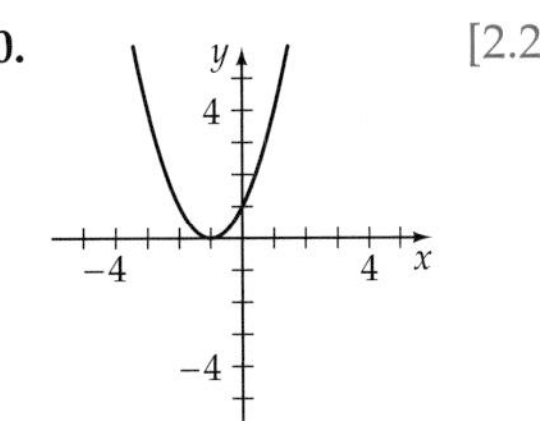

[2.2/2.5]

31.

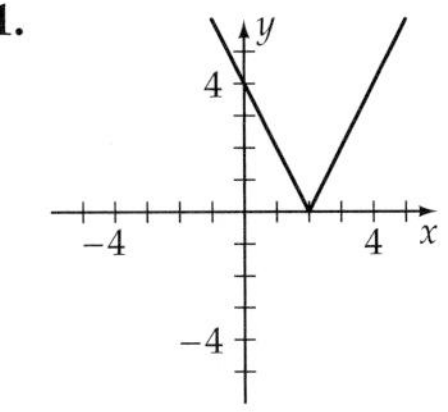

[2.2] **32.**

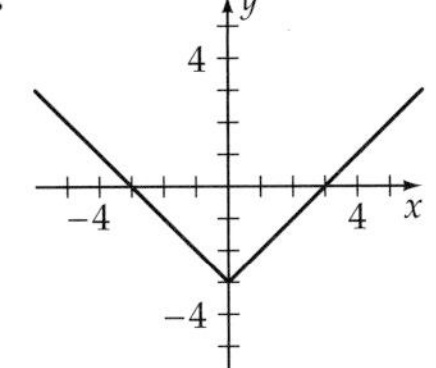

[2.2] **33.**

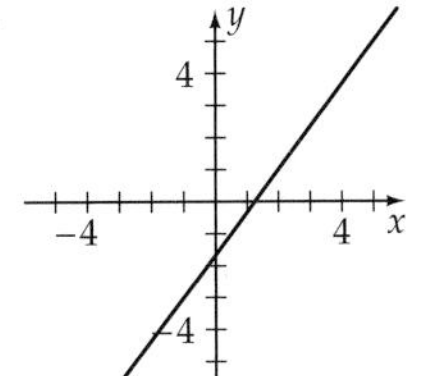

[2.3] **34.**

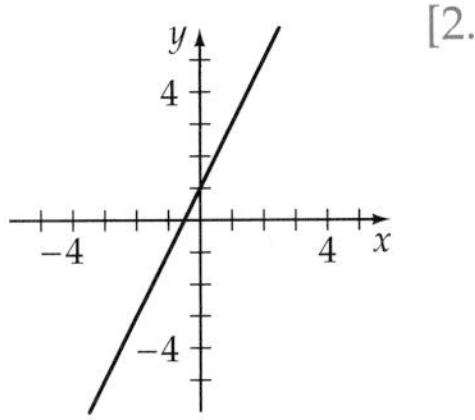

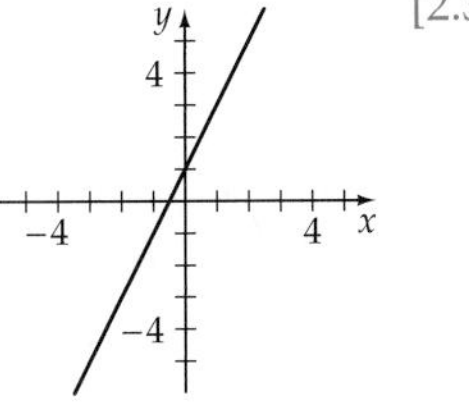

[2.3]

35. 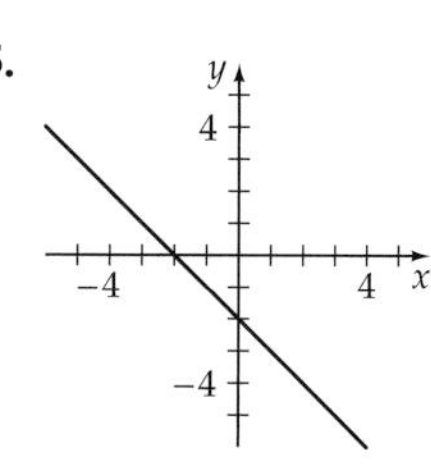

[2.3] **36.** 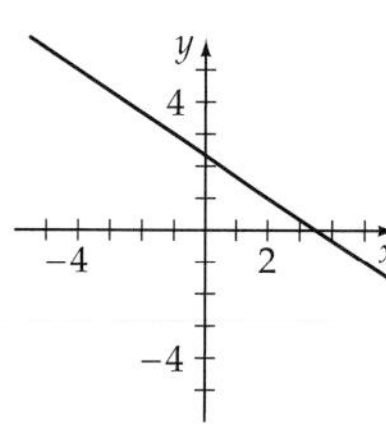

[2.3] **37.** −1 [2.3] **38.** $-\frac{4}{3}$ [2.3] **39.** No slope [2.3] **40.** 0 [2.3]

41. $-\frac{2}{3}$ [2.3] **42.** $\frac{3}{2}$ [2.3] **43.** 0 [2.3/2.4] **44.** 6 [2.4] **45.** −3 [2.4] **46.** $-\frac{1}{2}$ [2.3/2.4] **47.** Perpendicular [2.4] **48.** Parallel [2.4] **49.** Parallel [2.4] **50.** Perpendicular [2.4] **51.** $y = -2x + 10$ [2.4] **52.** $y = x + 4$ [2.4] **53.** $y = \frac{3}{4}x$ [2.4] **54.** $y = -\frac{2}{3}x$ [2.4] **55.** $y = -\frac{1}{3}x + 3$ [2.4] **56.** $y = \frac{6}{5}x - 9$ [2.4] **57.** $y = \frac{2}{3}x + \frac{14}{3}$ [2.4] **58.** $y = -2x + 6$ [2.4] **59.** $y = -\frac{2}{3}x - \frac{16}{3}$ [2.4] **60.** $y = -\frac{1}{2}x - \frac{1}{2}$ [2.4] **61.** $y = \frac{4}{3}x - 10$ [2.4] **62.** $y = -\frac{5}{3}x + 4$ [2.4] **63.** $y = -\frac{1}{2}x - 3$ [2.4] **64.** $y = -\frac{2}{5}x + \frac{4}{5}$ [2.4] **65.** (−7, 0), (3, 0); (0, −21) [2.5] **66.** $\left(-\frac{1}{2}, 0\right)$; (0, 1) [2.5] **67.** $\left(\frac{-3-\sqrt{13}}{2}, 0\right), \left(\frac{-3+\sqrt{13}}{2}, 0\right)$; (0, −1) [2.5] **68.** No x-intercepts; y-intercept is (0, 3). [2.5] **69.** Vertex: (−1.5, −5.5); axis of symmetry: $x = -1.5$ [2.5] **70.** Maximum is 2. [2.5] **71.** Minimum is −7. [2.5] **72.** Slope = 2. The slope means that the profit is increasing at a rate of $2 per game sold. [2.3] **73. a.** $y = 24.471x - 44.316$ **b.** $1277 [2.6] **74.** 250 feet by 500 feet [2.5] **75. a.** $y = -10t + 165$ **b.** The slope means that the athlete's heart rate is decreasing at a rate of 10 beats per minute. **c.** 105 beats per minute [2.3/2.6] **76.** 2000; 0.5 milligram per liter [2.5] **77. a.** $y = 0.796x + 146.646$ **b.** 584 [2.6] **78.** Slope = −10. The slope means that the speed of the car is decreasing at a rate of 10 mph per second. [2.3] **79. a.** $y = 3t$ **b.** The slope means that the height of the balloon is increasing at a rate of 3 feet per second. **c.** 600 ft [2.3/2.6]

Chapter 2 Test, *page 249*

1. 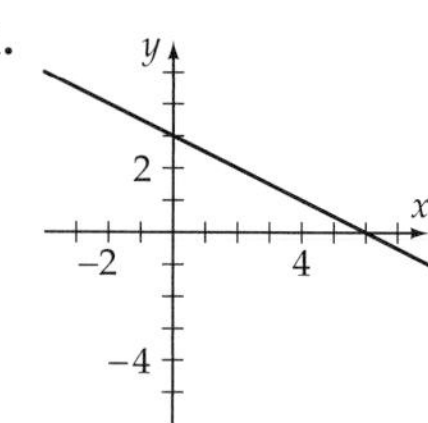

[2.2] **2.** 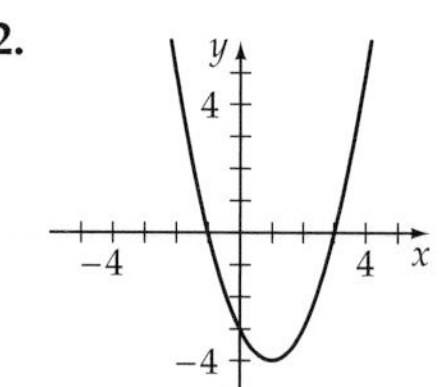

[2.5] **3.** 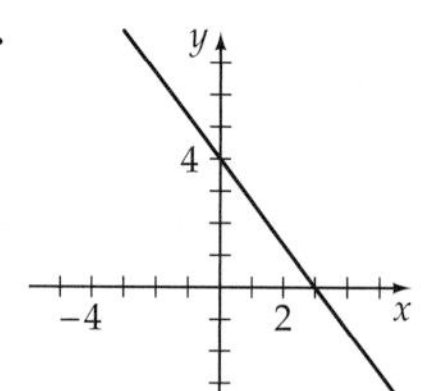

[2.3] **4.** 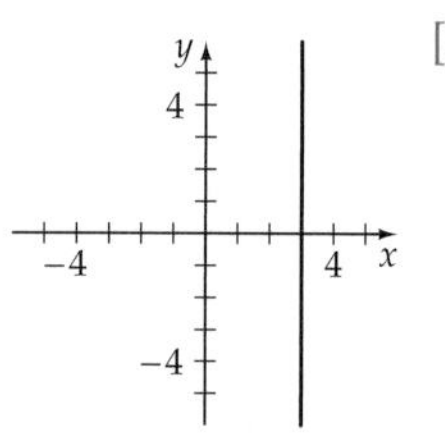

[2.3]

5. 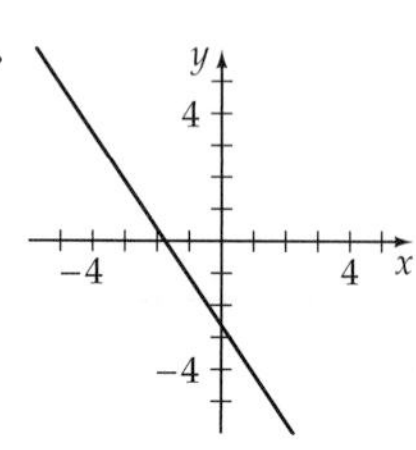

[2.3] **6.** Midpoint (4, 1), distance 6.32 [2.1] **7. a.** $-\frac{4}{3}$ **b.** No slope [2.3]

8. $(x-1)^2 + (y-2)^2 = 18$ [2.1] **9.** $y = -\frac{1}{3}x + 1$ [2.4] **10. a.** Slope = −200 **b.** The slope means that for each $1 increase in the price of a pound of strawberries, the grocer sells 200 fewer pounds. **c.** (3, 0) **d.** If the grocer charges $3 per pound, the grocer will not sell any strawberries. [2.3/2.4] **11. a.** $y = 115x + 20{,}000$ **b.** $336,250 [2.6] **12. a.** (0, −6) **b.** $\left(-\frac{3}{2}, 0\right)$ and (2, 0) **c.** $\left(\frac{1}{4}, -\frac{49}{8}\right)$ **d.** $x = \frac{1}{4}$ [2.5] **13.** 30 feet [2.5] **14. a.** $y = 0.980x - 0.723$ **b.** 6.1 ounces [2.6]

Cumulative Review Exercises, *page 250*

1. Distributive property [P.1] **2.** $-\frac{2}{3}, \sqrt{2}$ [P.1] **3.** $-10x + 8$ [P.1] **4.** $-6x^4y^6$ [P.2] **5.** $\dfrac{2a^3}{3b^2}$ [P.2] **6.** $6x^2 + 13x - 5$ [P.3] **7.** $\dfrac{x-1}{x+2}$ [P.6] **8.** $\dfrac{-2x-14}{(x+3)(x-1)}$ [P.6] **9.** 0 [1.1] **10.** $\dfrac{3 \pm \sqrt{17}}{2}$ [1.2] **11.** $-\frac{3}{2}, 2$ [1.2] **12.** $x = \dfrac{2y+14}{5}$ [1.3] **13.** $\pm\sqrt{2}, \pm i$ [1.3] **14.** $x > -4$ [1.4] **15.** $\sqrt{37}$ [2.1] **16.** 5 [2.2] **17.** $y = 2x - 5$ [2.4] **18.** 3 hours [1.1] **19.** 11:00 A.M., 2:00 P.M. [1.2] **20.** 55°F per minute [2.3]

Chapter 3

Chapter Prep Quiz, *page 252*

1. 12 **2.** $2h^2 - 13h + 20$ **3.** $2x^2 + 1$ **4.** One is the negative of the other. **5.** $-2x^3 - x^2 - 3x + 6$ **6.** $2x^3 + 4x^2 + 2x + 4$ **7.** $y = \frac{3}{5}x - 3$ **8.** $y = \dfrac{x}{1-x}$ **9.** $y = 6x - 1$ **10.** $y = x^2 + 8x + 15$

Exercises 3.1, *page 260*

1. 3 **3.** $\frac{15}{4}$ **5.** 0 **7.** 5 **9.** 10 **11.** $\frac{3}{8}$ **13.** 2 **15.** $2a + h$ **17.** $2a + h - 3$ **19.** $4a + 2h + 2$ **21.** $-\dfrac{1}{a^2 + ah}$ **23.** 12 **25.** 12 **27.** $6x + 5$ **29.** 24 **31.** 6 **33.** $x^2 + 2$ **35.** 3 **37.** 15 **39.** $4x^2 - 1$ **41.** 29 **43.** 10 **45.** $x^3 + 2$ **47. a.** $\dfrac{80x + 16{,}000}{x}$ **b.** 83.2 **c.** The selling price of a camera is $83.20 when 5000 cameras are sold.

Prepare for Section 3.2, *page 261*

49. $y = -\frac{2}{5}x + 3$ **50.** $y = \dfrac{1}{x-1}$ **51.** -1 **52.** $>$ **53.** $b \le 0$ **54.** $\{x \mid x \ge -2\}$

Exercises 3.2, *page 270*

1. 3 **3.** -3 **5.** 3 **7.** range **9.** $\{(1, -3), (2, -2), (5, 1), (-7, 4)\}$ **11.** $\{(1, 0), (2, 1), (4, 2), (8, 3), (16, 4)\}$ **13.** $f^{-1}(x) = \frac{1}{2}x - 2$ **15.** $f^{-1}(x) = \frac{1}{3}x + \frac{7}{3}$ **17.** $f^{-1}(x) = -\frac{1}{2}x + \frac{5}{2}$ **19.** $f^{-1}(x) = \dfrac{x}{x-2}, x \ne 2$ **21.** $f^{-1}(x) = \dfrac{x+1}{1-x}, x \ne 1$ **23.** $f^{-1}(x) = \sqrt{x-1}, x \ge 1$ **25.** $f^{-1}(x) = x^2 + 2, x \ge 0$ **27.** $f^{-1}(x) = \sqrt{x+4} - 2, x \ge -4$ **29.** $f^{-1}(x) = -\sqrt{x+5} - 2, x \ge -5$ **31.** Yes **33.** Yes **35.** No **37.** Yes **39.** Yes

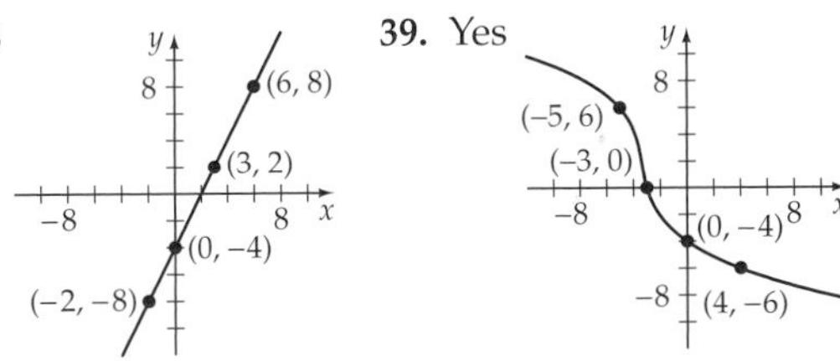

41. Yes

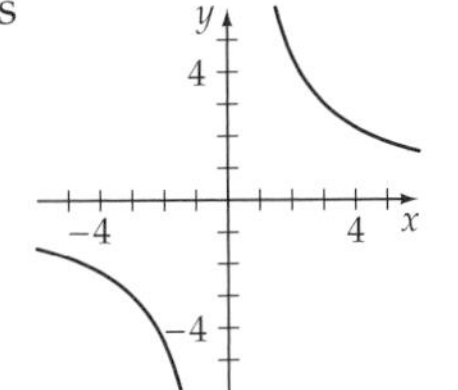

43. No

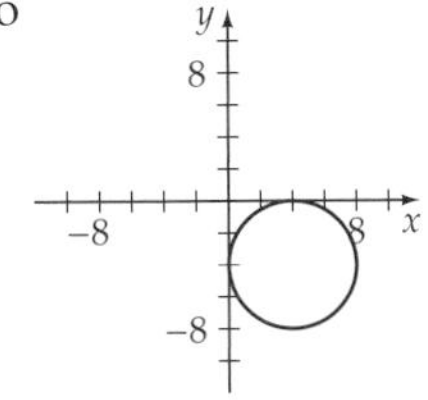

45. $V^{-1}(x) = \sqrt[3]{x}$. V^{-1} finds the length of a side of a cube given the volume.

47. Yes. Yes. A conversion function is a linear function. All linear functions have inverses that are also functions.
49. $s^{-1}(x) = \frac{1}{2}x - 12$ **51.** $E^{-1}(s) = 20s - 50{,}000$. From the monthly earnings (s), the executive can find the value of the software sold. **53.** 44205833; $f^{-1}(x) = \frac{1}{2}x + \frac{1}{2}$, $f^{-1}(44205833) = 22102917$ **55.** Because the function is increasing and 4 is between 2 and 5, c must be between 7 and 12. **57.** Between 2 and 5 **59.** Between 3 and 7 **61.** False

Prepare for Section 3.3, *page 273*

62. $x^2 - 5x + 6$ **63.** $\dfrac{x}{x+1}$ **64.** Yes **65.** Yes **66.** No **67.** No

Exercises 3.3, *page 000*

1. (2, 2) **3.** (−3, 3) **5.** (4, −4) **7.** (7, −5) **9.** (−1, 4) **11. a.**

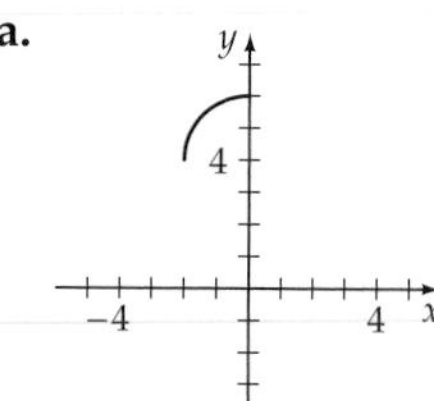

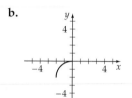

b.

c.

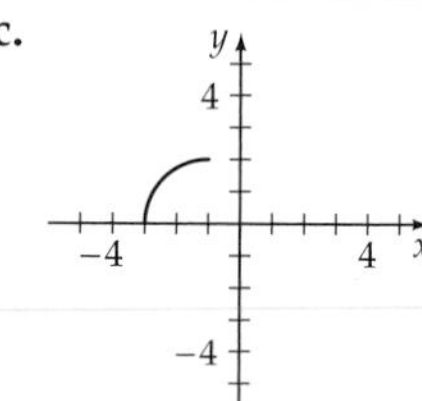

d.

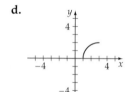

13. a.

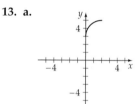

b.

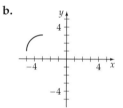

15. a.

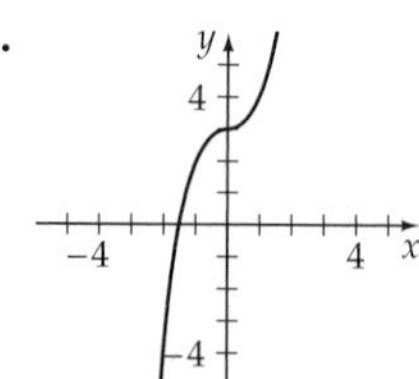

b.

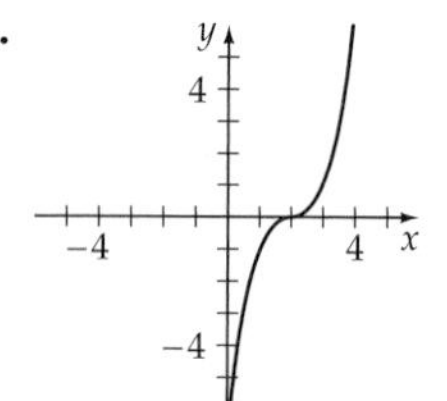

c.

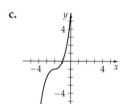

d.

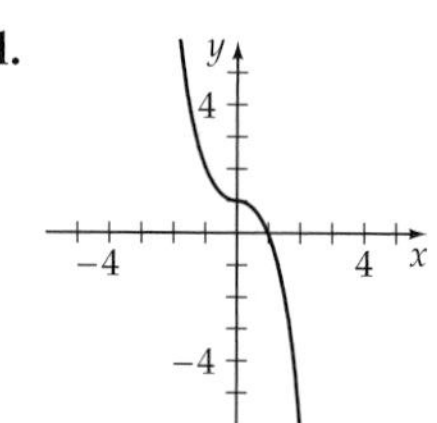

e.

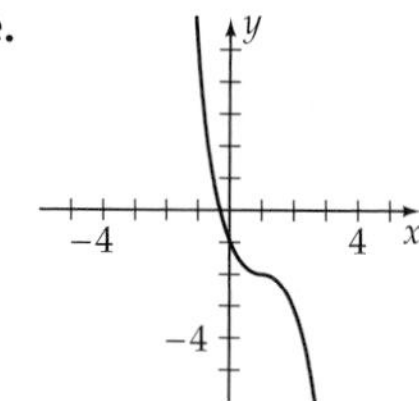

17. a.

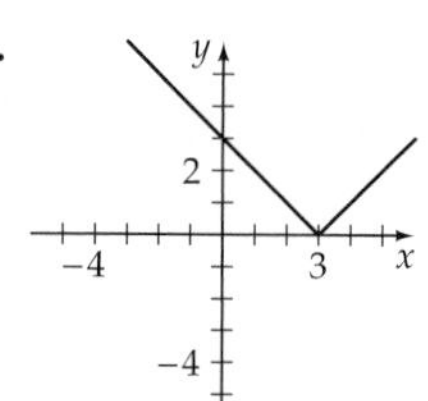

b.

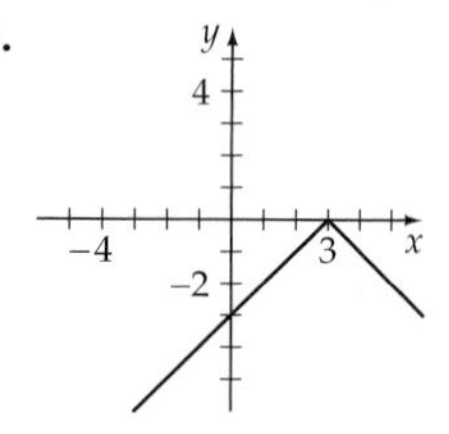

c.

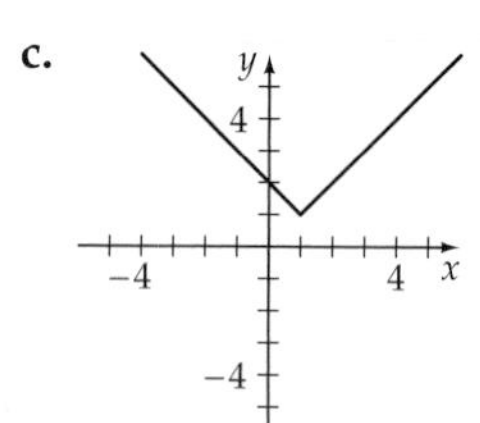

d.

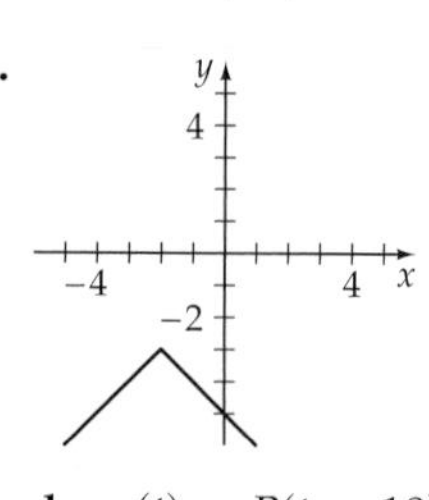

19. a. $f(t) = T(t) - 4$ **b.** $g(t) = T\left(t - \frac{1}{2}\right)$ **c.** The temperature is 2 degrees cooler and the change in temperature lags behind by 1 hour.

21. a. $f(t) = P(t) - 8000$ **b.** $g(t) = P(t - 10)$ **c.** The population is 2000 people greater than the population of Springfield 5 years ago. **23. a.**

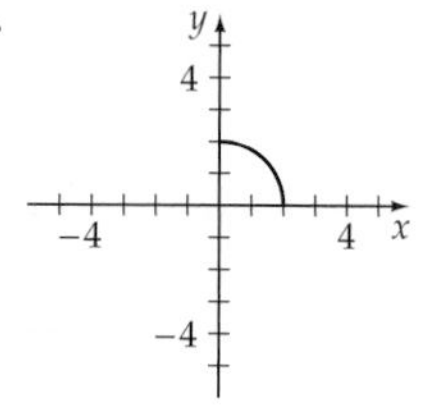

b.

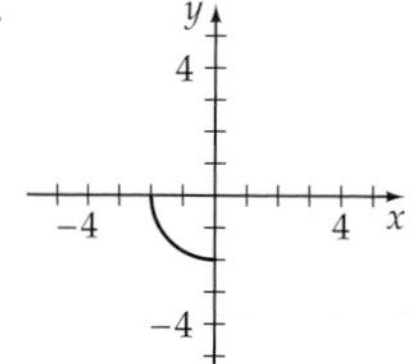

25. a.

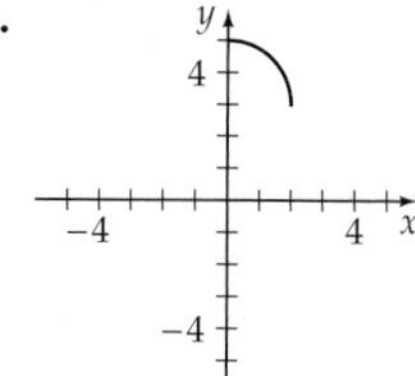

b.

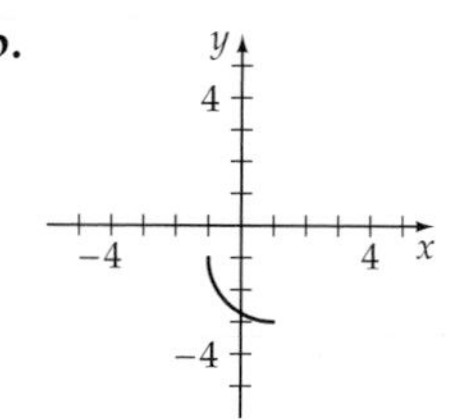

27. a. $y = (x+2)^{2/3} - 1$

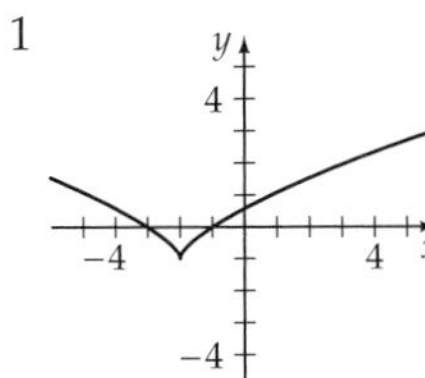

b. $y = -(x-1)^{2/3}$

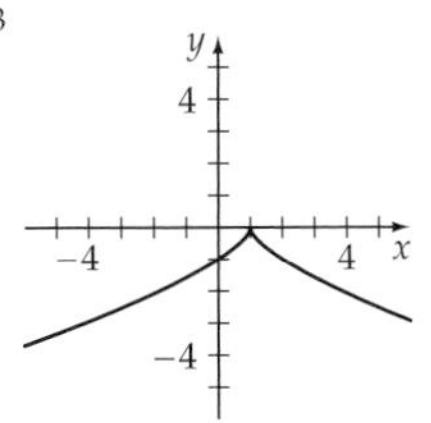

c. $y = (-x-1)^{2/3}$

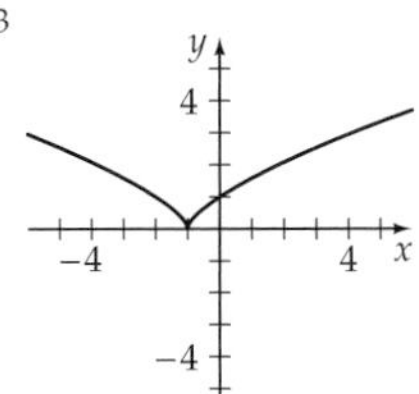

d. $y = (-x-1)^{2/3} + 2$

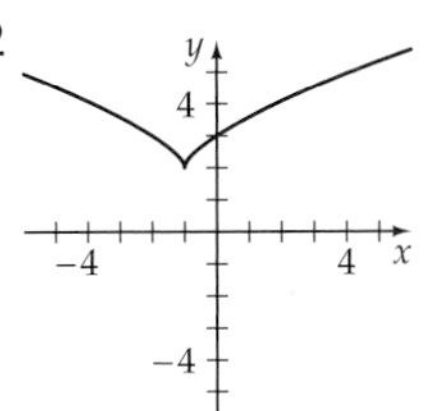

29. $f(x) = \sqrt{x-4} + 2$ **31. a.**

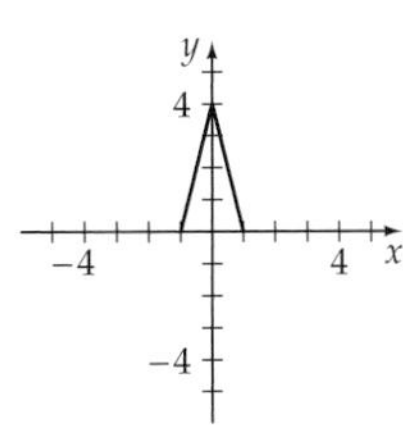

b.

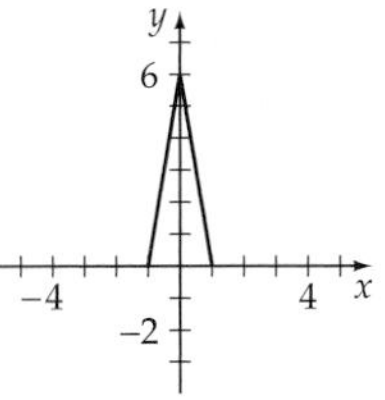

c.

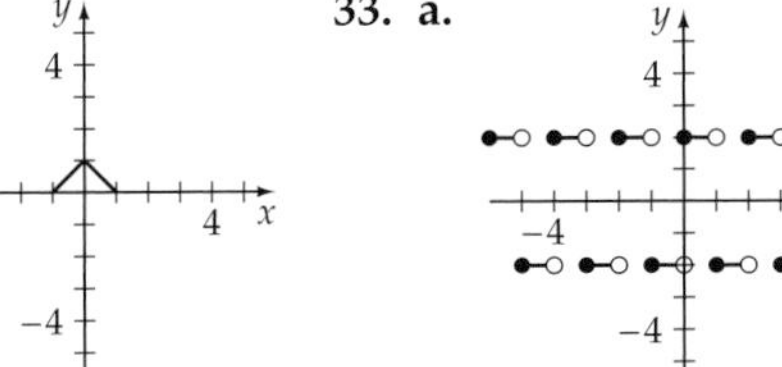

33. a.

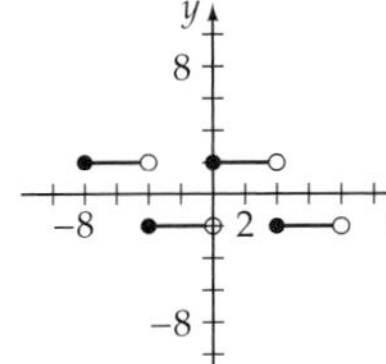

b.

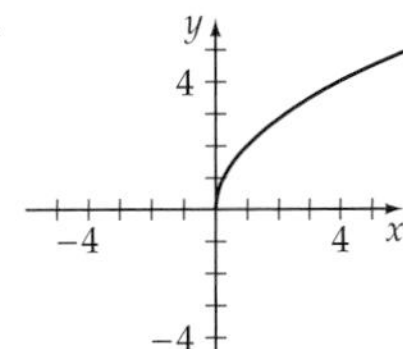

35. a.

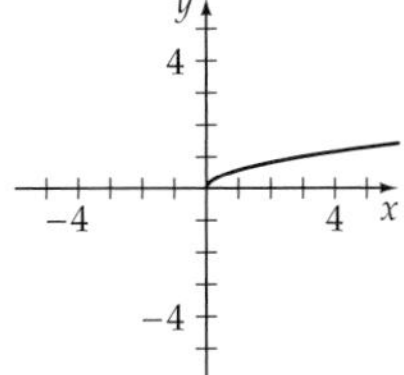

b.

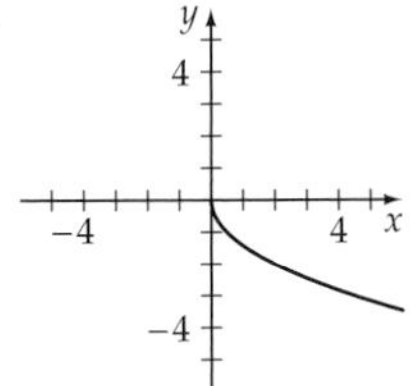

c.

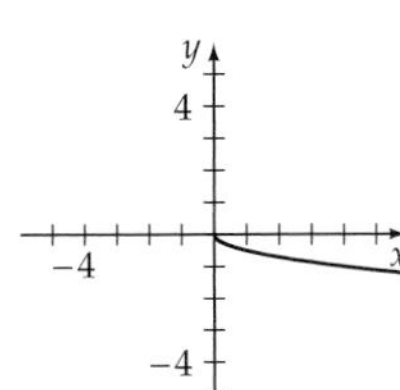

d.

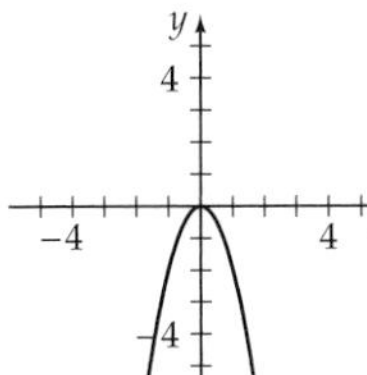

37. a.

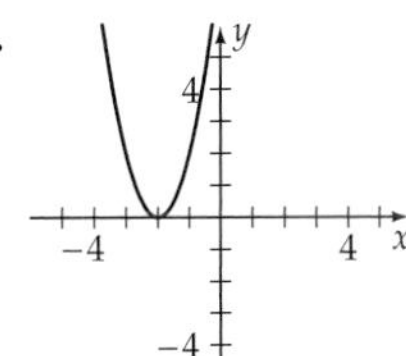

b.

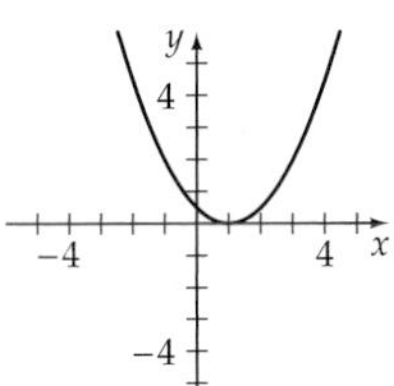

c.

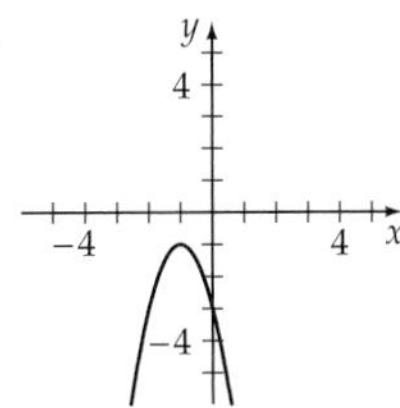

d.

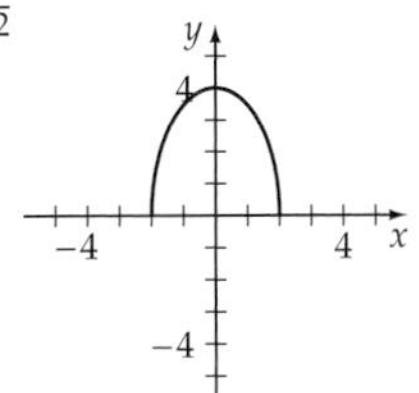

39. a. $y = 2\sqrt{4-x^2}$

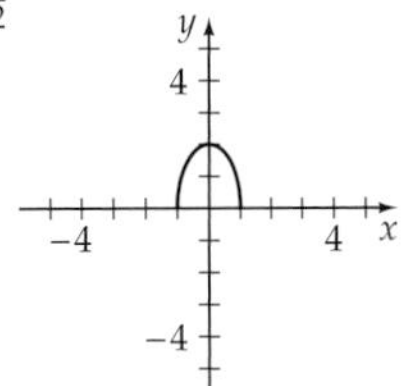

b. $y = \sqrt{4-4x^2}$

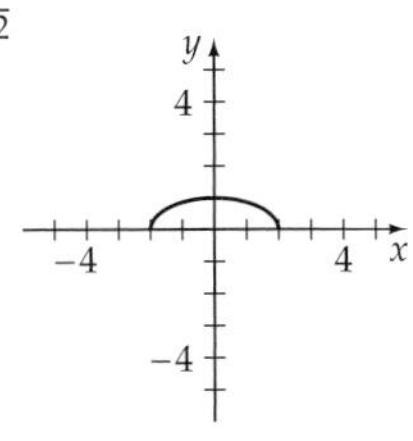

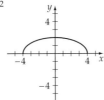

c. $y = \frac{1}{2}\sqrt{4-x^2}$

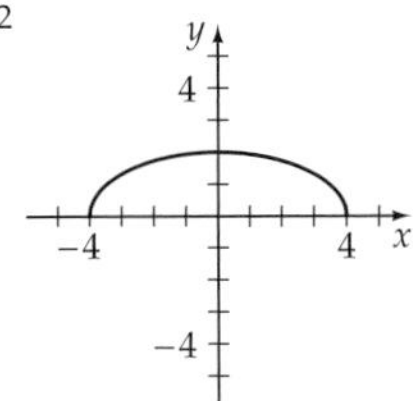

d. $y = \sqrt{4 - \frac{1}{2}x^2}$

41. VI **43.** VIII **45.** I **47.** IV

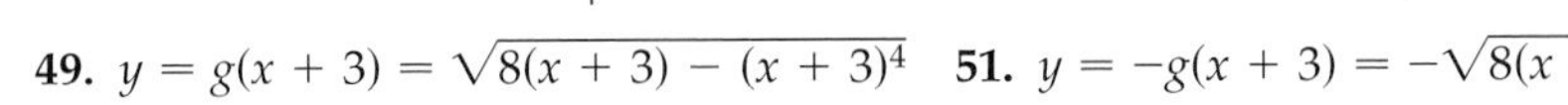

49. $y = g(x+3) = \sqrt{8(x+3) - (x+3)^4}$ **51.** $y = -g(x+3) = -\sqrt{8(x+3) - (x+3)^4}$ **53.** $y = \frac{1}{2}g(x) = \frac{1}{2}\sqrt{8x - x^4}$

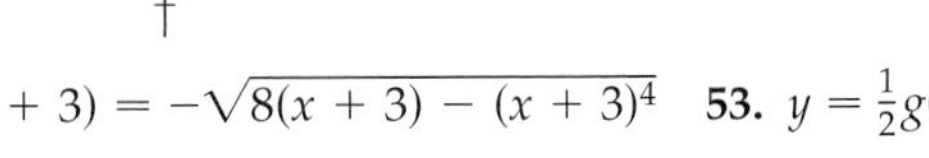

55.

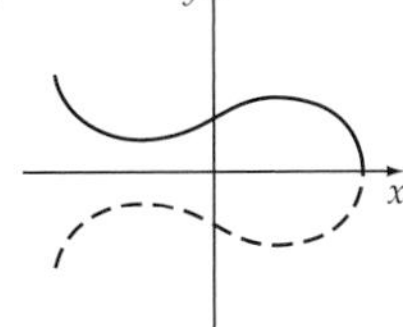

57.

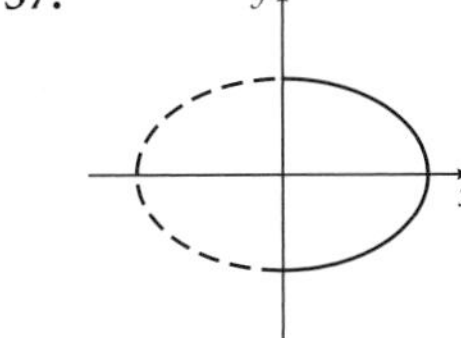

59.

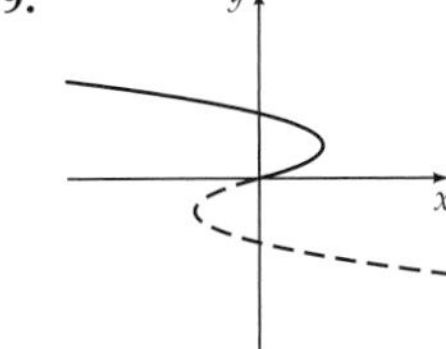

61. a. No **b.** Yes **63. a.** No **b.** No **65. a.** Yes **b.** Yes **67. a.** Yes **b.** Yes **69.** No **71.** Yes **73.** Yes **75.** No **77.** Even **79.** Odd **81.** Neither **83.** Even **85.** Odd **87.** Neither **89.** Even **91. a.**

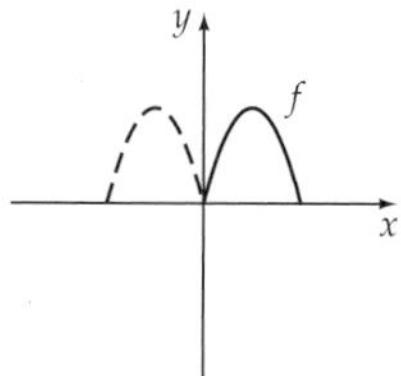

b.

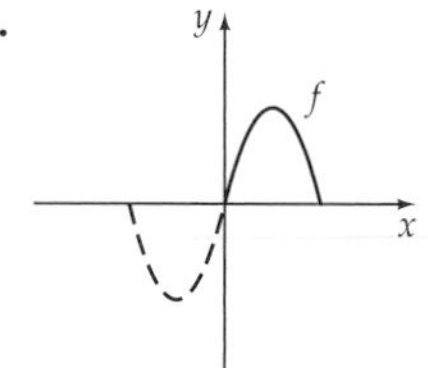

93. Odd

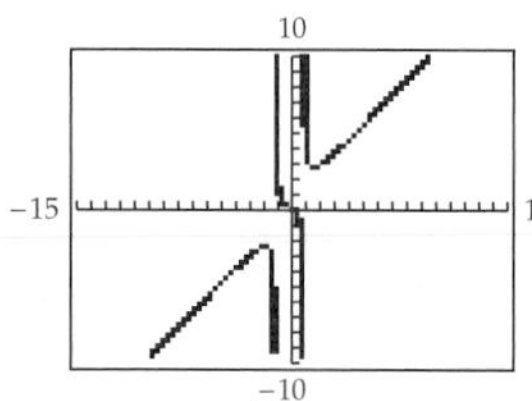

95. Even

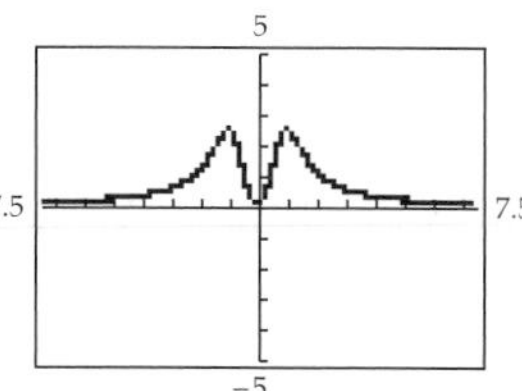

97. Even

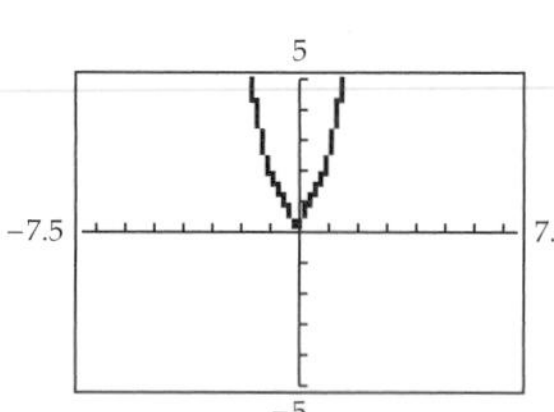

99. a. $y = \dfrac{2}{(x+1)^2+1} + 1$ **b.** $y = -\dfrac{2}{(x-2)^2+1}$

Prepare for Section 3.4, *page 296*

101. 25 **102.** 4 **103.** 2 **104.** $\frac{1}{2}$ **105.** 36 **106.** $r = \pm\sqrt{\dfrac{km}{F}}$

Exercises 3.4, *page 302*

1. $s = kt$ **3.** $y = \dfrac{k}{x}$ **5.** $S = knp$ **7.** $P = knRt$ **9.** $A = \dfrac{ks^2}{\sqrt{r}}$ **11.** $F = \dfrac{km_1m_2}{r^2}$ **13.** $s = kt, k = 2$ **15.** $v = kr^2, k = 4$ **17.** $s = kvt^2, k = \frac{1}{50}$ **19.** $V = klwh, k = 1$ **21.** \$60,000 **23.** 112 decibels **25.** 25 foot-candles **27.** 98 newton-meters **29.** 5.625 inches **31. a.** 3.5 seconds **b.** 3.3 feet **33.** 177.8 pounds **35.** 0.25 ohm **37.** 295.3 pounds **39.** 3950 pounds

Chapter 3 True/False Exercises, *page 306*

1. True. **2.** False. $\{(2, 4), (3, 4)\}$ is a function whose inverse is not a function. **3.** False. $f(x) = |x|$ is a function and $f(-2) = f(2)$, but $-2 \neq 2$. **4.** False. Let $f(x) = x^2$. Then $f(2 + 3) = f(5) = 5^2 = 25$. However, $f(2) + f(3) = 2^2 + 3^2 = 4 + 9 = 13$. **5.** True. **6.** True. **7.** False. y is halved. **8.** True. **9.** False. If $f(x) = -x$, $a = 2$, and $b = 3$, then $f(2) = -2$ and $f(3) = -3$. Thus $a < b$ but $f(a) > f(b)$. **10.** True. **11.** False. Let $f(x) = x^2$, $a = 2$, and $b = -2$. Then $a > b$, but $f(a) = f(b)$. **12.** True.

Chapter 3 Review Exercises, *page 307*

1. a. 8 **b.** $-2x^2 + x - 4$ **c.** $2x^3 - 6x^2 + x - 3$ **d.** $\frac{51}{2}$ [3.1] **2. a.** -7 **b.** $x^2 + x + 7$ **c.** $2x^3 + 5x^2 - 7x$ **d.** $\frac{2}{5}$ [3.1] **3. a.** 11 **b.** 5 **c.** 1 **d.** $4x^2 - 60x + 1$ **e.** $2x^2 - 2x - 3$; **f.** 9 [3.1] **4. a.** 74 **b.** 26 **c.** 9 **d.** $6x^2 + 3$ **e.** $12x^2 + 60x + 74$ **f.** 23 [3.1] **5.** 5 [3.1] **6.** $6a + 3h$ [3.1] **7.**

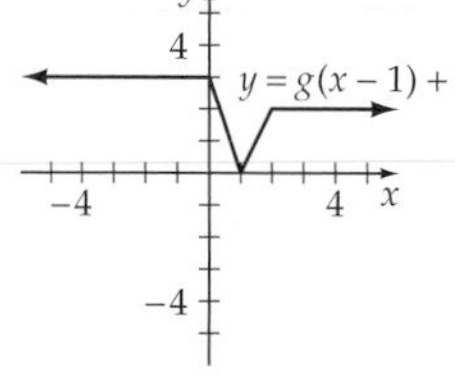

[3.3] **8.**

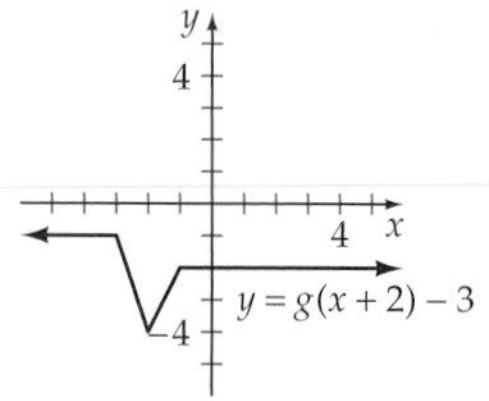

[3.3]

9. 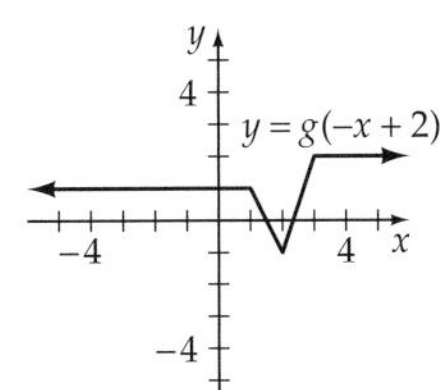

[3.3] **10.** 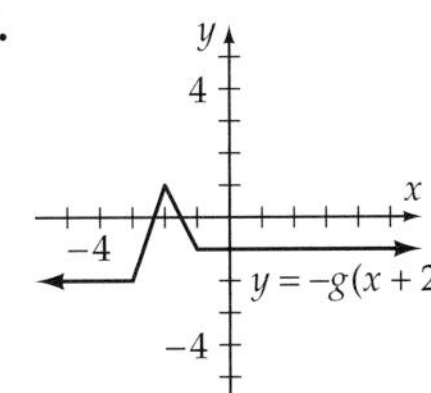

[3.3] **11.** 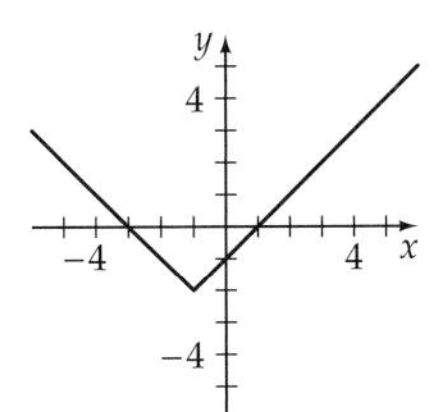

[3.3] **12.** 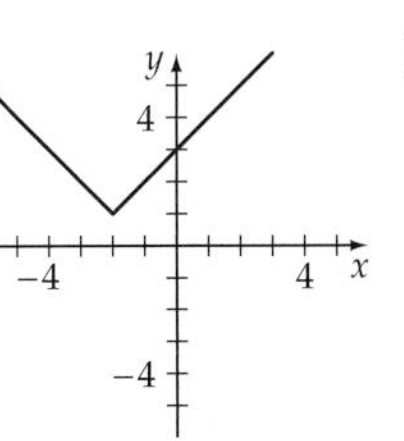

[3.3]

13. 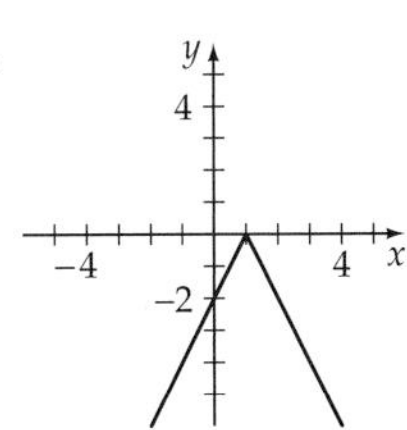

[3.3] **14.** 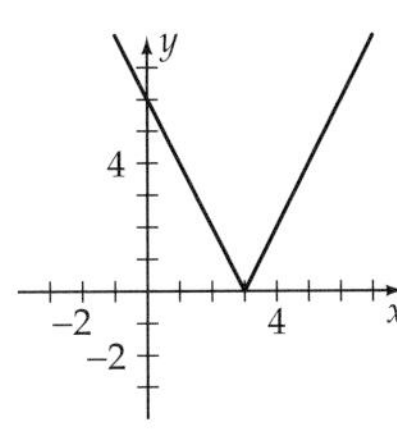

[3.3] **15.** y-axis [3.3] **16.** x-axis [3.3] **17.** Origin [3.3]

18. x-axis, y-axis, origin [3.3] **19.** x-axis, y-axis, origin [3.3] **20.** Origin [3.3] **21.** Odd [3.3] **22.** Neither [3.3] **23.** Even [3.3] **24.** Odd [3.3] **25.** Yes [3.2] **26.** No [3.2] **27.** No [3.2] **28.** No [3.2] **29.** Yes [3.2] **30.** Yes [3.2]

31. $y = \frac{1}{3}x - 4$ [3.2] **32.** $y = 2x + 2$ [3.2] **33.** $y = \sqrt{x - 4}, x \geq 4$ [3.2] **34.** $y = x^2 + 1, x \geq 0$ [3.2]

35. $y = \dfrac{x + 2}{x}, x \neq 0$ [3.2] **36.** $y = \dfrac{4x}{1 - x}, x \neq 1$ [3.2] **37.** $R(x) = 150x - \dfrac{x^2}{500}$; \$2,200,000 [3.1]

38. $P(x) = -\dfrac{x^2}{250} + 25x - 450$ [3.1] **39.** 22.25 pounds per square inch [3.4] **40.** 97 decibels [3.4] **41.** $A(L) = -L^2 + 50L$ [3.1]

42. 3.2 seconds [3.4]

Chapter 3 Test, *page 309*

1. a. -3 **b.** $-x^2 + 2x - 4$ **c.** $-2x^3 + 5x^2 + 2x - 5$ **d.** $\frac{1}{3}$ [3.1] **2. a.** -11 **b.** $-2x^2 + 2x + 1$ **c.** 0 **d.** $4x^2 - 2x$ [3.1]

3. They are inverse functions. [3.2] **4.** $-4 - h$ [3.1] **5.** $f^{-1}(x) = \frac{1}{3}x + 2$ [3.2] **6.** $f^{-1}(x) = \dfrac{1}{x - 1}$ [3.2]

7. 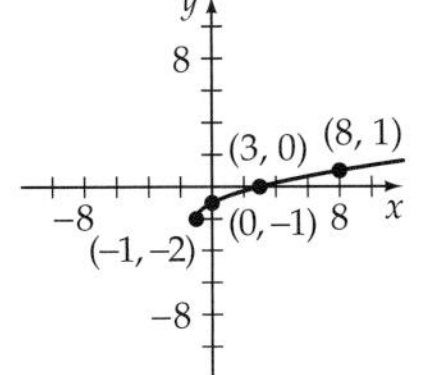

[3.2] **8.** $g(x) = x^2 + 1$ does not have an inverse function. [3.2] **9.** Domain: $\{x|x \geq -4\}$; range: $\{y|y \leq 1\}$ [3.2]

10. a. 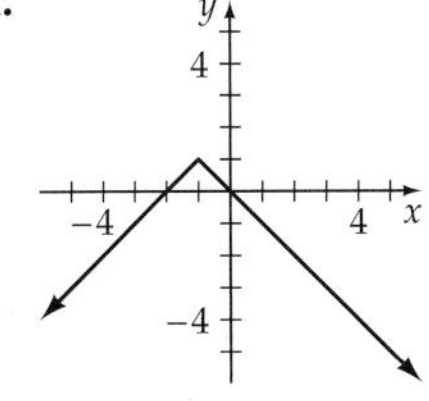

b. 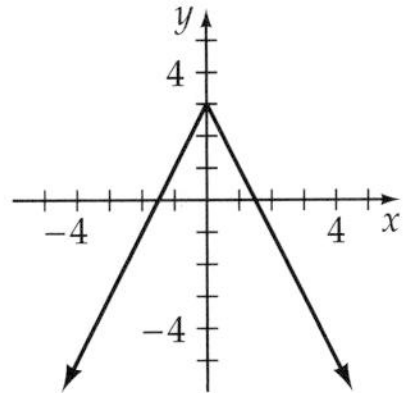

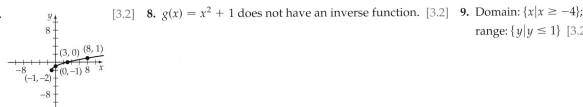

[3.3] **11. a.** 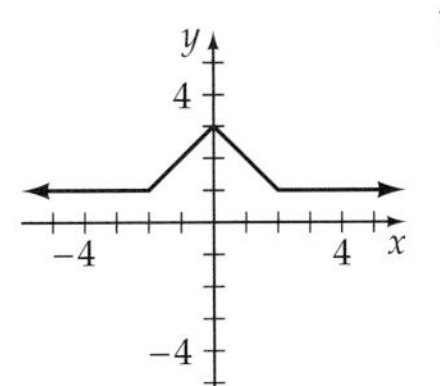

b. 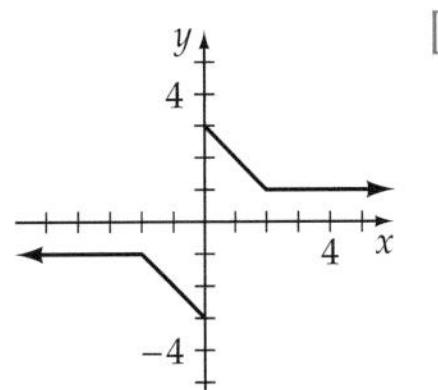

[3.3]

12. 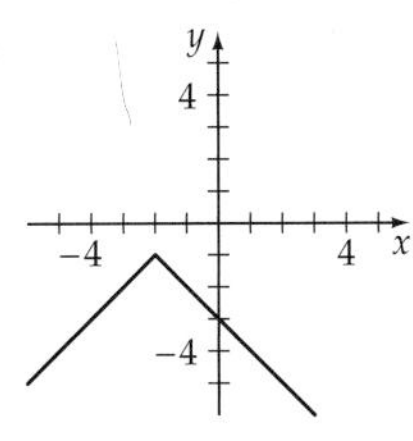

[3.3] 
13. a. Even; y-axis symmetry; **b.** Odd; origin symmetry; **c.** Neither [3.3]
14. \$374,600 [3.1] **15.** 181 feet [3.4]

Cumulative Review Exercises, *page 310*

1. $\frac{x^4y^4}{4}$ [P.2] **2.** 0 [P.6] **3.** $6x^3 - 13x^2 + 9x - 2$ [P.3] **4.** $4x - 6 + \frac{11}{x+2}$ [P.3] **5.** $-18 - i$ [P.7] **6.** $\frac{3}{2}, -\frac{2}{3}$ [1.2]
7. $\{x|-3 < x < -2\}$ [1.4] **8.** $(1, -1)$ [2.1] **9.** No [2.2] **10.** No [2.2] **11.** $\frac{3}{2}$ [2.3] **12.** $y = -\frac{1}{2}x + 4$ [2.4] **13.** $\left(-\frac{3}{2}, -\frac{11}{2}\right)$ [2.5]
14. [3.1] **15.** Origin symmetry [3.3] **16. a.** $-5x + 4$ **b.** $-6x^2 + 11x - 3$ [3.1]
17. $f^{-1}(x) = \frac{3}{2}x - \frac{3}{2}$ [3.2] **18.** \$1800 per year [2.3] **19.** $\frac{4}{3}$ ounces [1.1] **20.** 89 feet [3.4]

Chapter 4

Chapter Prep Quiz, *page 312*

1. Quotient $x + 1$; remainder 4 **2.** Quotient $x^2 - 2x - 9$; remainder -12 **3.** $(x^2 + 9)(x + 3)(x - 3)$ **4.** $\left(\frac{4}{3}, 0\right), \left(-\frac{3}{2}, 0\right)$
5. -4 **6.** 17 **7.** $x^3 - 7x + 6$ **8.** $x^2 - 4x + 5$ **9.** $x - 5$ **10.** $\frac{-6x^2 + 5x - 3}{2x - 1}$

Exercises 4.1, *page 319*

1. $4x^2 + 3x + 12 + \frac{17}{x-2}$ **3.** $4x^2 - 4x + 2 + \frac{1}{x+1}$ **5.** $x^4 + 4x^3 + 6x^2 + 24x + 101 + \frac{403}{x-4}$ **7.** $x^4 + x^3 + x^2 + x + 1$
9. $8x^2 + 6$ **11.** $x^7 + 2x^6 + 5x^5 + 10x^4 + 21x^3 + 42x^2 + 85x + 170 + \frac{344}{x-2}$ **13.** $x^5 - 3x^4 + 9x^3 - 27x^2 + 81x - 242 + \frac{716}{x+3}$
15. 25 **17.** 45 **19.** -2230 **21.** -80 **23.** -187 **25.** Yes **27.** No **29.** Yes **31.** Yes **33.** No
45. $(x - 2)(x^2 + 3x + 7)$ **47.** $(x - 4)(x^3 + 3x^2 + 3x + 1)$ **49. a.** \$19,968 **b.** \$23,007 **51. a.** 336 **b.** 336; They are the same.
53. a. 100 cards **b.** 610 cards **55. a.** 400 people per square mile **b.** 240 people per square mile **57. a.** 304 cubic inches
b. 892 cubic inches

Prepare for Section 4.2, *page 322*

58. 2 **59.** $\frac{9}{8}$ **60.** $[-1, \infty)$ **61.** $[1, \infty)$ **62.** $(x + 1)(x - 1)(x + 2)(x - 2)$ **63.** $\left(\frac{2}{3}, 0\right), \left(-\frac{1}{2}, 0\right)$

Exercises 4.2, *page 332*

1. Up to the far left, up to the far right **3.** Down to the far left, up to the far right **5.** Down to the far left, down to the far right **7.** Down to the far left, up to the far right **9.** $a < 0$ **11.** Vertex is $(-2, -5)$, minimum is -5. **13.** Vertex is $(-4, 17)$, maximum is 17. **15.** **17.** **19.**

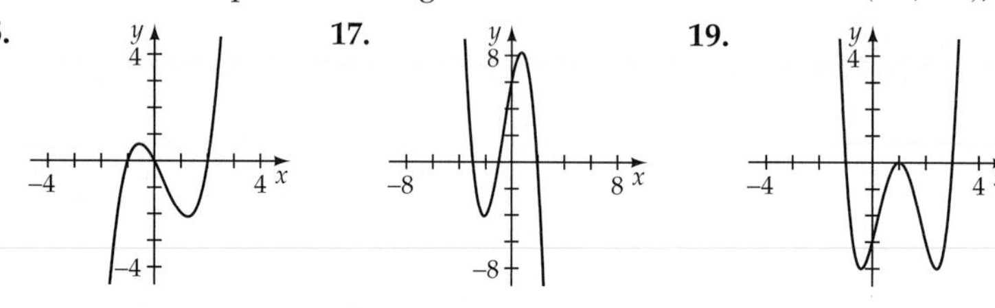

21. Relative maximum $y \approx 5.0$ at $x \approx -2.1$, relative minimum $y \approx -16.9$ at $x \approx 1.4$

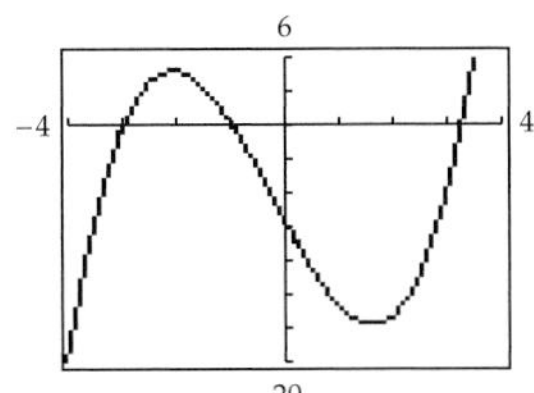

23. Relative maximum $y \approx 31.0$ at $x \approx -2.0$, relative minimum $y \approx -77.0$ at $x \approx 4.0$

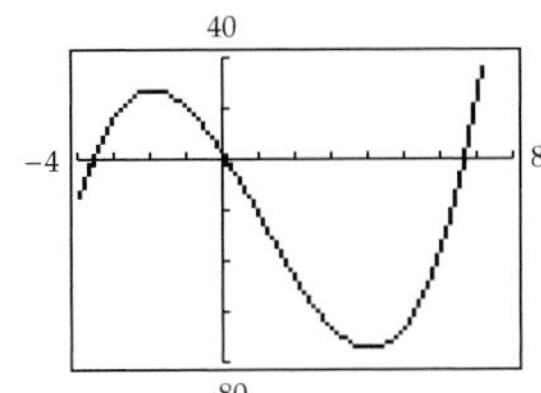

25. Relative maximum $y \approx 2.0$ at $x \approx 1.0$, relative minima $y \approx -14.0$ at $x \approx -1.0$ and $y \approx -14.0$ at $x \approx 3.0$

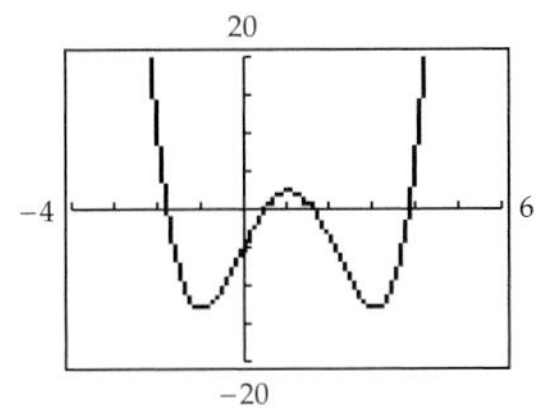

27. $-3, 0, 5$ **29.** $-3, -2, 2, 3$ **31.** $-2, -1, 0, 1, 2$ **39.** Crosses the x-axis at $(-1, 0)$, $(1, 0)$, and $(3, 0)$ **41.** Crosses the x-axis at $(7, 0)$; intersects but does not cross at $(3, 0)$ **43.** Crosses the x-axis at $(1, 0)$; intersects but does not cross at $\left(\frac{3}{2}, 0\right)$ **45.** Crosses the x-axis at $(0, 0)$; intersects but does not cross at $(3, 0)$ **47. a.** $V(x) = x(15 - 2x)(10 - 2x) = 4x^3 - 50x^2 + 150x$ **b.** 1.96 inches **49.** \$464,000 **51. a.** 1918 **b.** 9.5 marriages per thousand population **53. a.** 20.69 milligrams **b.** 118 minutes **55. a.** 3.24 inches **b.** 4 feet from either end; 3.84 inches **c.** 3.24 inches

Prepare for Section 4.3, *page 335*

57. $\frac{2}{3}, \frac{7}{2}$ **58.** $2x^2 - x + 6 - \dfrac{19}{x + 2}$ **59.** $3x^3 + 9x^2 + 6x + 15 + \dfrac{40}{x - 3}$ **60.** 1, 2, 3, 4, 6, 12 **61.** $\pm 1, \pm 3, \pm 9, \pm 27$ **62.** $P(-x) = -4x^3 - 3x^2 + 2x + 5$

Exercises 4.3, *page 346*

1. 3 (multiplicity 2), -5 (multiplicity 1) **3.** 0 (multiplicity 2), $-\frac{5}{3}$ (multiplicity 2) **5.** 2 (multiplicity 1), -2 (multiplicity 1), -3 (multiplicity 2) **7.** $\pm 1, \pm 2, \pm 4, \pm 8$ **9.** $\pm 1, \pm 2, \pm 3, \pm 4, \pm 6, \pm 12, \pm\frac{1}{2}, \pm\frac{3}{2}$ **11.** $\pm 1, \pm 2, \pm 4, \pm\frac{1}{2}, \pm\frac{1}{3}, \pm\frac{2}{3}, \pm\frac{4}{3}, \pm\frac{1}{6}$ **13.** $\pm 1, \pm 7, \pm\frac{1}{2}, \pm\frac{7}{2}, \pm\frac{1}{4}, \pm\frac{7}{4}$ **15.** $\pm 1, \pm 2, \pm 4, \pm 8, \pm 16, \pm 32$ **17.** Upper bound 2, lower bound -5 **19.** Upper bound 4, lower bound -4 **21.** Upper bound 1, lower bound -4 **23.** Upper bound 4, lower bound -2 **25.** Upper bound 2, lower bound -1 **27.** One positive zero, two or no negative zeros **29.** Two or no positive zeros, one negative zero **31.** One positive zero, three or one negative zeros **33.** Three or one positive zeros, one negative zero **35.** One positive zero, no negative zeros **37.** $2, -1, -4$ **39.** $3, -4, \frac{1}{2}$ **41.** $\frac{1}{2}, -\frac{1}{3}, -2$ (multiplicity 2) **43.** $\frac{1}{2}, 4, \sqrt{3}, -\sqrt{3}$ **45.** $6, 1 + \sqrt{5}, 1 - \sqrt{5}$ **47.** $5, \frac{1}{2}, 2 + \sqrt{3}, 2 - \sqrt{3}$ **49.** $1, -1, -2, -\frac{2}{3}, 3 + \sqrt{3}, 3 - \sqrt{3}$ **51.** $2, -1$ (multiplicity 2) **53.** $0, -2, 1 + \sqrt{2}, 1 - \sqrt{2}$ **55.** -1 (multiplicity 3), 2 **57.** $-\frac{3}{2}$, 1 (multiplicity 2), 8 **59.** $n = 9$ inches **61.** $x = 4$ inches **63. a.** 26 pieces **b.** 7 cuts **65.** 7 rows **67.** $x = 0.084$ inch **69.** 1977 and 1986 **71.** 16.9 feet **73.** 1.13 feet **75. a.** 73 seconds **b.** 93,000 digits

Prepare for Section 4.4, *page 350*

76. $3 + 2i$ **77.** $2 - i\sqrt{5}$ **78.** $x^3 - 8x^2 + 19x - 12$ **79.** $x^2 - 4x + 5$ **80.** $-3i, 3i$ **81.** $\frac{1}{2} - \frac{1}{2}i\sqrt{19}, \frac{1}{2} + \frac{1}{2}i\sqrt{19}$

Exercises 4.4, *page 358*

1. $2, -3, 2i, -2i$; $P(x) = (x - 2)(x + 3)(x - 2i)(x + 2i)$ **3.** $\frac{1}{2}, -3, 1 + 5i, 1 - 5i$; $P(x) = \left(x - \frac{1}{2}\right)(x + 3)(x - 1 - 5i)(x - 1 + 5i)$ **5.** 1 (multiplicity 3), $3 + 2i, 3 - 2i$; $P(x) = (x - 1)^3(x - 3 - 2i)(x - 3 + 2i)$ **7.** $-3, -\frac{1}{2}, 2 + i, 2 - i$; $P(x) = (x + 3)\left(x + \frac{1}{2}\right)(x - 2 - i)(x - 2 + i)$ **9.** $4, 2, \frac{1}{2} + \frac{3}{2}i, \frac{1}{2} - \frac{3}{2}i$; $P(x) = (x - 4)(x - 2)\left(x - \frac{1}{2} - \frac{3}{2}i\right)\left(x - \frac{1}{2} + \frac{3}{2}i\right)$ **11.** $1 - i, \frac{1}{2}$ **13.** $i, -3$ **15.** $2 + 3i, i, -i$ **17.** $1 - 3i, 1 + 2i, 1 - 2i$ **19.** $2i$, 1 (multiplicity 3) **21.** $5 - 2i, \dfrac{7}{2} + \dfrac{\sqrt{3}}{2}i, \dfrac{7}{2} - \dfrac{\sqrt{3}}{2}i$ **23.** $\dfrac{3}{2}, -\dfrac{1}{2} + \dfrac{\sqrt{7}}{2}i, -\dfrac{1}{2} - \dfrac{\sqrt{7}}{2}i$ **25.** $-\frac{2}{3}, \frac{3}{4}, \frac{5}{2}$ **27.** $-i, i$, 2 (multiplicity 2) **29.** -3 (multiplicity 2), 1 (multiplicity 2)

31. $P(x) = x^3 - 3x^2 - 10x + 24$ **33.** $P(x) = x^3 - 3x^2 + 4x - 12$ **35.** $P(x) = x^4 - 10x^3 + 63x^2 - 214x + 290$
37. $P(x) = x^5 - 22x^4 + 212x^3 - 1012x^2 + 2251x - 1830$ **39.** $P(x) = 4x^3 - 19x^2 + 224x - 159$ **41.** $P(x) = x^3 + 13x + 116$
43. $P(x) = x^4 - 18x^3 + 131x^2 - 458x + 650$ **45.** $P(x) = 3x^3 - 12x^2 + 3x + 18$ **47.** $P(x) = -2x^4 + 4x^3 + 36x^2 - 140x + 150$
49. The Conjugate Pair Theorem does not apply because some of the coefficients of the polynomial are not real numbers.

Prepare for Section 4.5, *page 359*

51. $\dfrac{x-3}{x-5}$ **52.** $-\frac{3}{2}$ **53.** $\frac{1}{3}$ **54.** $x = 0, -3, \frac{5}{2}$ **55.** The degree of the numerator is 3. The degree of the denominator is 2.
56. $x + 4 + \dfrac{7x - 11}{x^2 - 2x}$

Exercises 4.5, *page 372*

1 $x = 0, x = -3$ **3.** $x = -\frac{1}{2}, x = \frac{4}{3}$ **5.** $y = 4$ **7.** $y = 30$

9. $x = -4, y = 0$ **11.** $x = 3, y = 0$ **13.** $x = 0, y = 0$ **15.** $x = -4, y = 1$

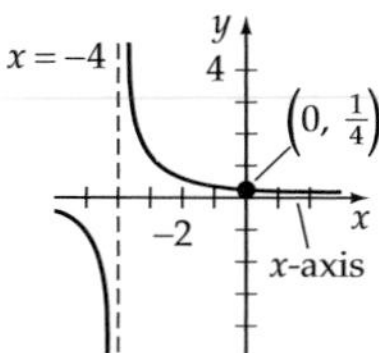

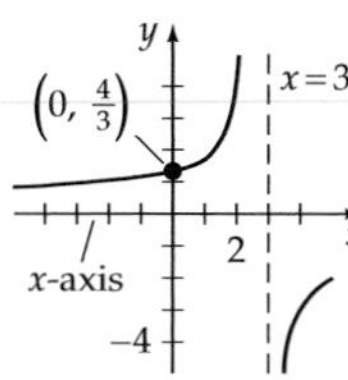

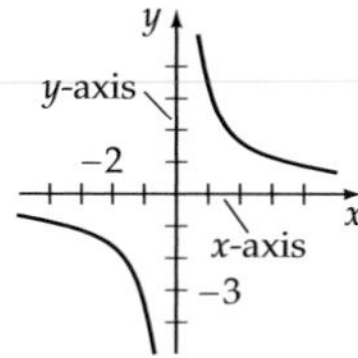

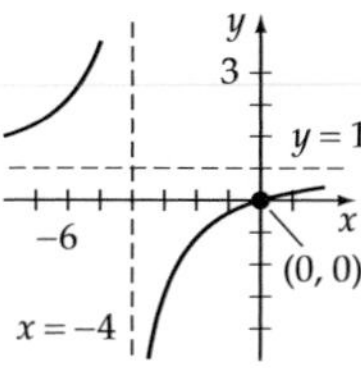

17. $x = 2, y = -1$ **19.** $x = 3, x = -3, y = 0$ **21.** $x = -3, x = 1, y = 0$ **23.** $x = -2, y = 1$

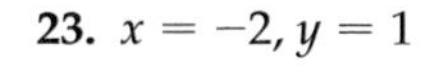

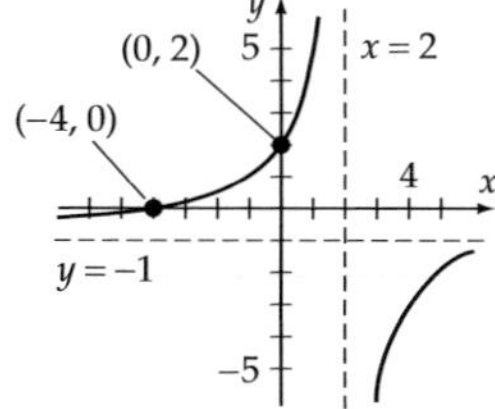

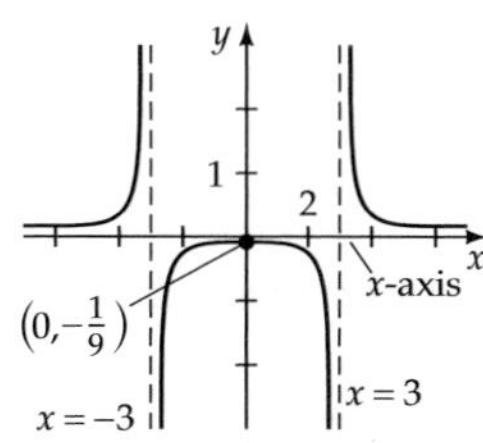

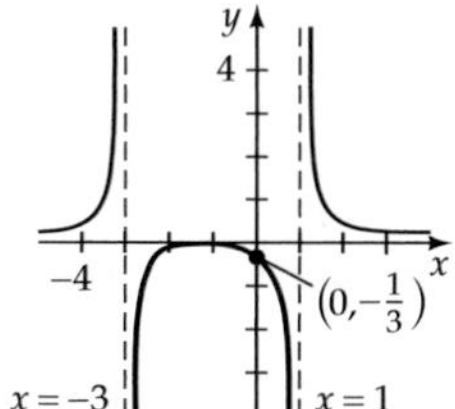

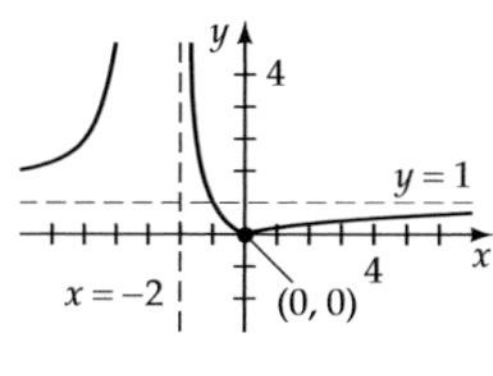

25. No vertical asymptote; horizontal asymptote: $y = 0$ **27.** $x = 3, x = -3, y = 2$ **29.** $x = -1 + \sqrt{2}, x = -1 - \sqrt{2}, y = 1$ **31.** $y = 3x - 7$

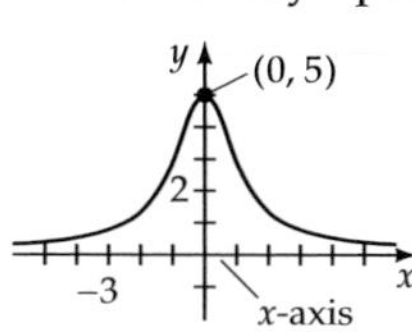

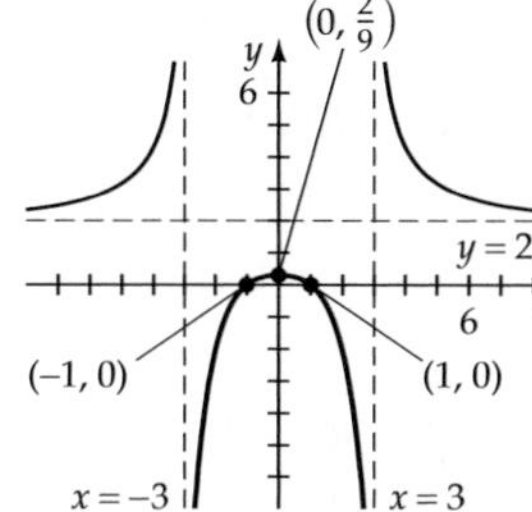

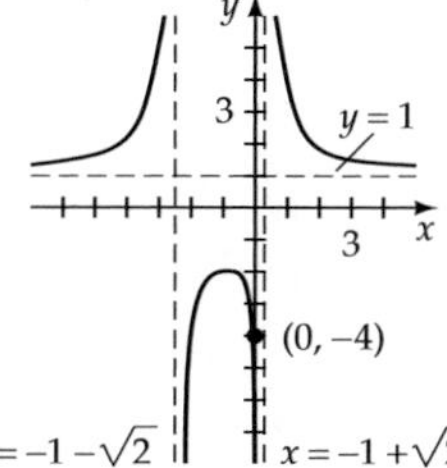

33. $y = x$ **35.** $x = 0, y = x$ **37.** $x = -3, y = x - 6$ **39.** $x = 4, y = 2x + 13$

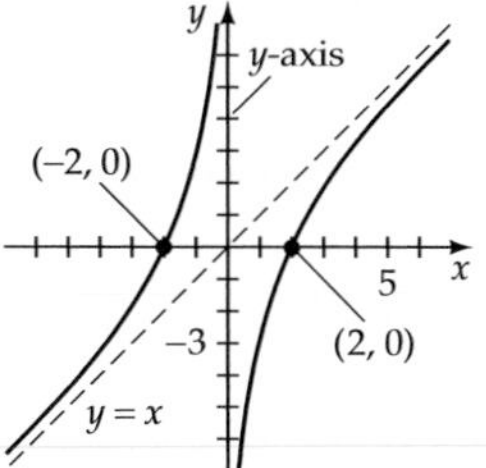

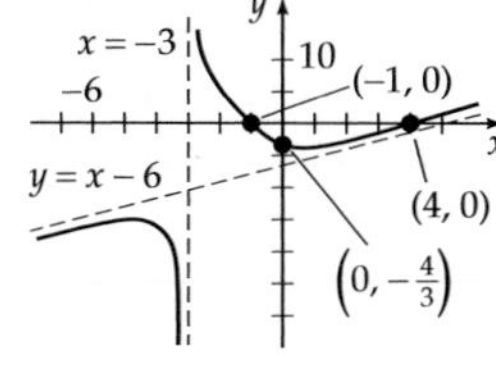

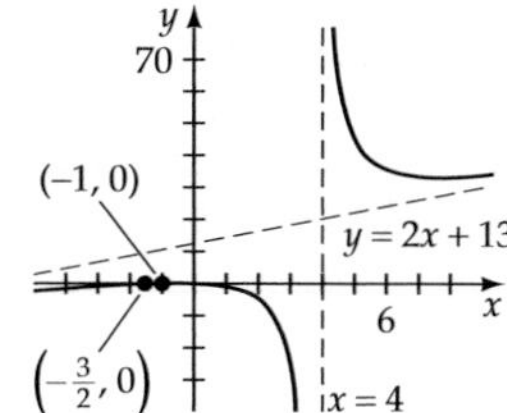

41. $x = -2, y = x - 3$

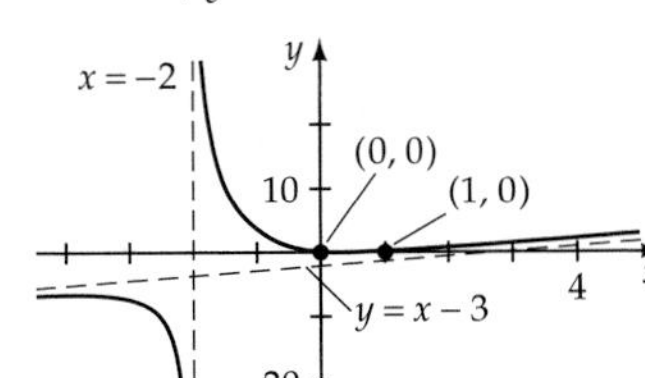

43. $x = 2, x = -2, y = x$

45.

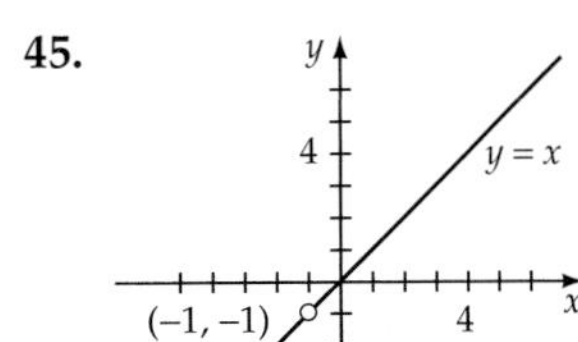

47.

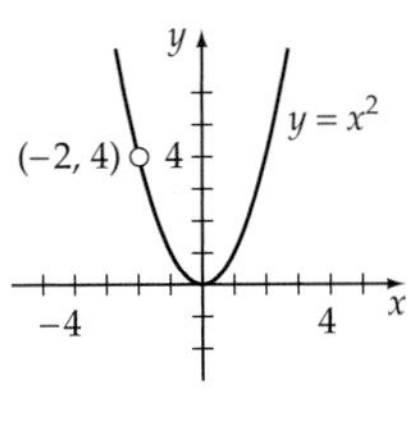

49.

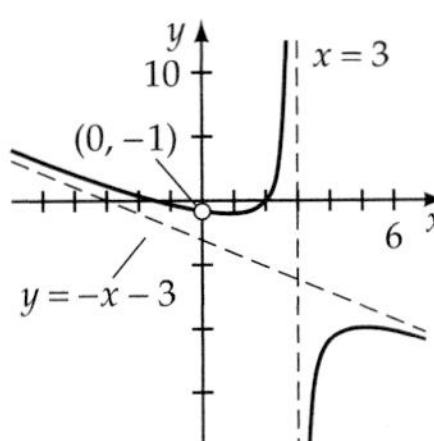

51.

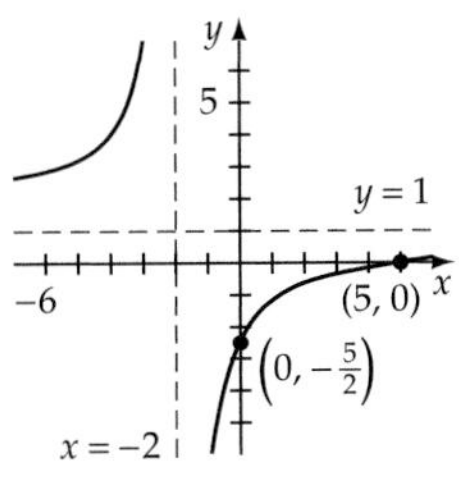

53. a. \$76.43, \$8.03, \$1.19 **b.** $y = 0.43$. As the number of golf balls produced increases, the average cost per golf ball approaches \$.43.

55. a. \$1333.33 **b.** \$8000 **c.**

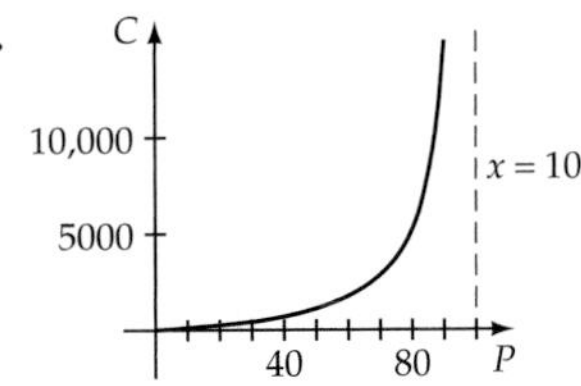

57. a. $R(0) \approx 38.8\%$, $R(7) \approx 39.9\%$, $R(12) \approx 40.9\%$ **b.** $\approx 44.7\%$

59. a. 26,923, 68,293, 56,000 **b.** 2001 **c.** The population will approach 0. **61. a.** 3.8 centimeters

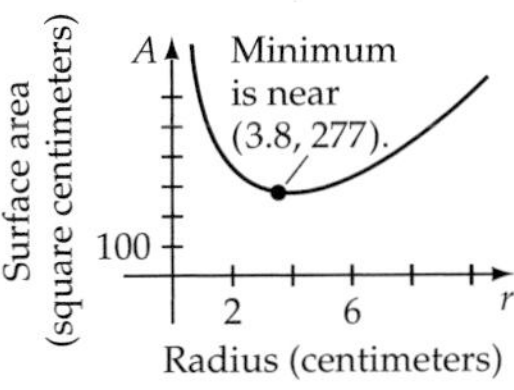

b. No **c.** As the radius r increases without bound, the surface area approaches twice the area of a circle with radius r.

Chapter 4 True/False Exercises, *page 378*

1. False; $P(x) = x - i$ has a zero of i, but it does not have a zero of $-i$. **2.** False; Descartes' Rule of Signs indicates that $P(x) = x^3 - x^2 + x - 1$ has three or one positive zeros. In fact, P has only 1 positive zero. **3.** True **4.** True **5.** False; $F(x) = \dfrac{x}{x^2 + 1}$ does not have a vertical asymptote. **6.** False; $F(x) = \dfrac{(x-2)^2}{(x-3)(x-2)} = \dfrac{x-2}{x-3}, x \neq 2$. The graph of F has a hole at $x = 2$. **7.** True **8.** True **9.** True **10.** True **11.** True **12.** True **13.** True **14.** False; $P(x) = x^2 + 1$ does not have a real zero.

Chapter 4 Review Exercises, *page 378*

1. $4x^2 + x + 8 + \dfrac{22}{x-3}$ [4.1] **2.** $5x^2 + 5x - 13 - \dfrac{11}{x-1}$ [4.1] **3.** $3x^2 - 6x + 7 - \dfrac{13}{x+2}$ [4.1] **4.** $2x^2 + 8x + 20$ [4.1]

5. $3x^2 + 5x - 11$ [4.1] **6.** $x^3 + 2x^2 - 8x - 9$ [4.1] **7.** 77 [4.1] **8.** 22 [4.1] **9.** 33 [4.1] **10.** 558 [4.1]

The verifications in Exercises 11–14 make use of the concepts from Section 4.1.

15. 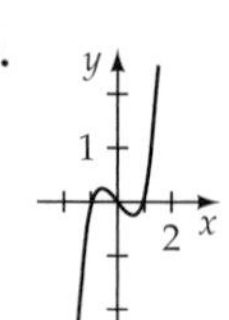

[4.2] **16.** 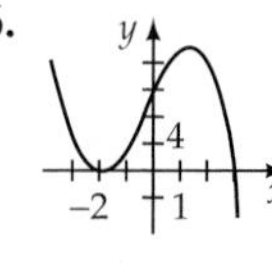

[4.2] **17.** 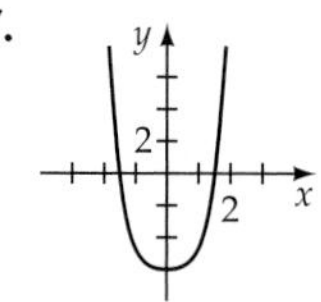

[4.2] **18.** 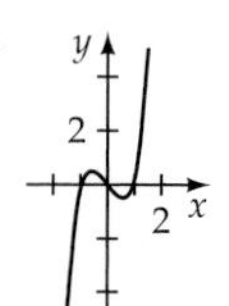

[4.2] **19.** 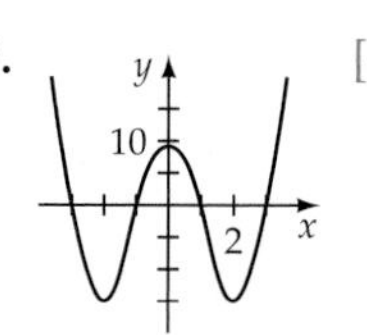

[4.2]

20. 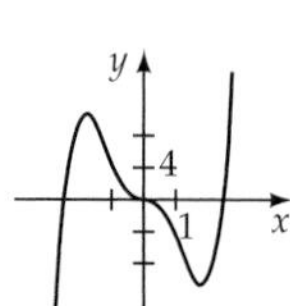

[4.2] **21.** $\pm 1, \pm 2, \pm 3, \pm 6$ [4.3] **22.** $\pm 1, \pm 2, \pm 3, \pm 5, \pm 6, \pm 10, \pm 15, \pm 30, \pm\frac{1}{2}, \pm\frac{3}{2}, \pm\frac{5}{2}, \pm\frac{15}{2}$ [4.3]

23. $\pm 1, \pm 2, \pm 3, \pm 4. \pm 6, \pm 12, \pm\frac{1}{3}, \pm\frac{2}{3}, \pm\frac{4}{3}, \pm\frac{1}{5}, \pm\frac{2}{5}, \pm\frac{3}{5}, \pm\frac{4}{5}, \pm\frac{6}{5}, \pm\frac{12}{5}, \pm\frac{1}{15}, \pm\frac{2}{15}, \pm\frac{4}{15}$ [4.3]

24. $\pm 1, \pm 2, \pm 4, \pm 8, \pm 16, \pm 32, \pm 64$ [4.3] **25.** ± 1 [4.3] **26.** $\pm 1, \pm 2, \pm\frac{1}{6}, \pm\frac{1}{3}, \pm\frac{1}{2}, \pm\frac{2}{3}$ [4.3] **27.** No positive real zeros and three or one negative real zero [4.3] **28.** Three or one positive real zero, one negative real zero [4.3] **29.** One positive real zero and one negative real zero [4.3] **30.** Five, three, or one positive real zero, no negative real zeros [4.3/4.4]

31. $1, -2, -5$ [4.3/4.4] **32.** $2, 5, 3$ [4.3/4.4] **33.** -2 (multiplicity 2), $-\frac{1}{2}, -\frac{4}{3}$ [4.3] **34.** $-\frac{1}{2}, -3, i, -i$ [4.3]

35. 1 (multiplicity 4) [4.3] **36.** $-\frac{1}{2}, 2 + 3i, 2 - 3i$ [4.3] **37.** $-1, 3, 1 + 2i$ [4.4] **38.** $-5, 2, 2 - i$ [4.4]

39. $P(x) = 2x^3 - 3x^2 - 23x + 12$ [4.4] **40.** $P(x) = x^4 + x^3 - 5x^2 + x - 6$ [4.4] **41.** $P(x) = x^4 - 3x^3 + 27x^2 - 75x + 50$ [4.4] **42.** $P(x) = x^4 + 2x^3 + 6x^2 + 32x + 40$ [4.4] **43.** Vertical asymptote: $x = -2$, horizontal asymptote: $y = 3$ [4.5] **44.** Vertical asymptotes: $x = -3, x = 1$, horizontal asymptote: $y = 2$ [4.5] **45.** Vertical asymptote: $x = -1$, slant asymptote: $y = 2x + 3$ [4.5] **46.** No vertical asymptote, horizontal asymptote: $y = 3$ [4.5]

47. 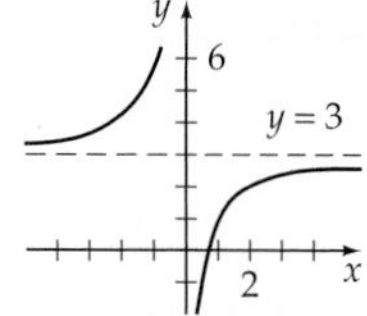

[4.5] **48.** 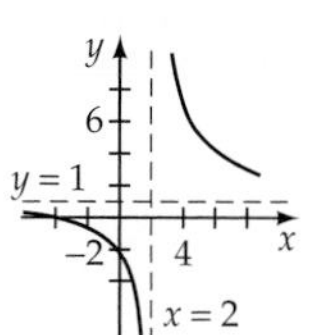

[4.5] **49.** 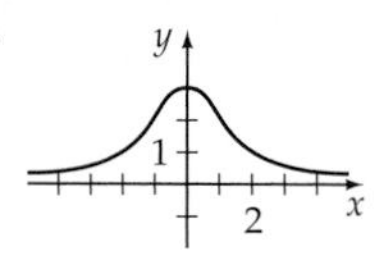

[4.5] **50.** 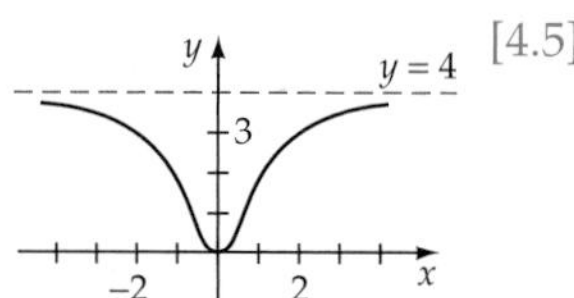

[4.5]

51. 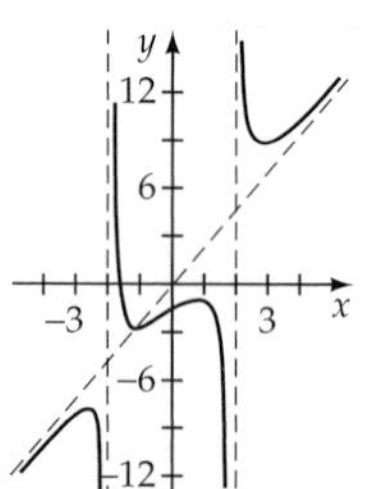

[4.5] **52.** 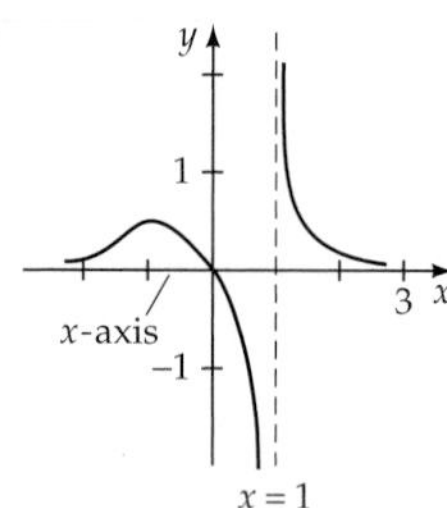

[4.5] **53.** 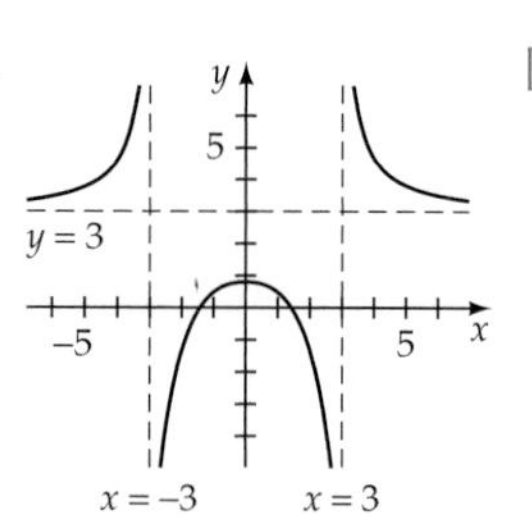

[4.5]

54. 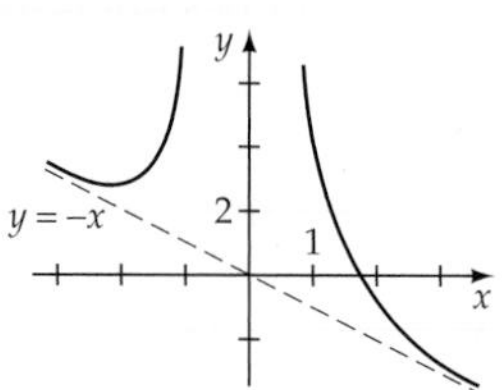

[4.5] **55. a.** \$12.59, \$6.43 **b.** $y = 5.75$. As the number of skateboards produced increases, the average cost per skateboard approaches \$5.75. [4.5]

56. a. 15°F **b.** 2.4°F **c.** 0°F [4.5] **57. a.** As the radius of the blood vessel approaches 0, the resistance gets larger. **b.** As the radius of the blood vessel gets larger, the resistance approaches zero. [4.5]

Chapter 4 Test, *page 380*

1. $3x^2 - x + 6 - \dfrac{13}{x+2}$ [4.1] **2.** 43 [4.1] **3.** The verification for Exercise 3 makes use of the concepts from Section 4.1. **4.** Up to the far left and down to the far right [4.2] **5.** $0, \dfrac{2}{3}, -3$ [4.2] **6.** $P(1) < 0, P(2) > 0$. Therefore, by the Zero Location Theorem, the polynomial function P has a zero between 1 and 2. [4.2] **7.** 2 (multiplicity 2), −2 (multiplicity 2), $\dfrac{3}{2}$ (multiplicity 1), −1 (multiplicity 3) [4.3] **8.** $\pm 1, \pm 3, \pm\dfrac{1}{2}, \pm\dfrac{3}{2}, \pm\dfrac{1}{3}, \pm\dfrac{1}{6}$ [4.3] **9.** Upper bound 4, lower bound −5 [4.3] **10.** Four, two, or zero positive zeros, no negative zero. [4.3] **11.** $\dfrac{1}{2}, 3, -2$ [4.3] **12.** $2 - 3i, -\dfrac{2}{3}, -\dfrac{5}{2}$ [4.4] **13.** 0, 1 (multiplicity 2), $2 + i, 2 - i$ [4.4] **14.** $P(x) = x^4 - 5x^3 + 8x^2 - 6x$ [4.4] **15.** Vertical asymptotes: $x = 3, x = 2$ [4.5] **16.** Horizontal asymptote: $y = \dfrac{3}{2}$ [4.5] **17.**

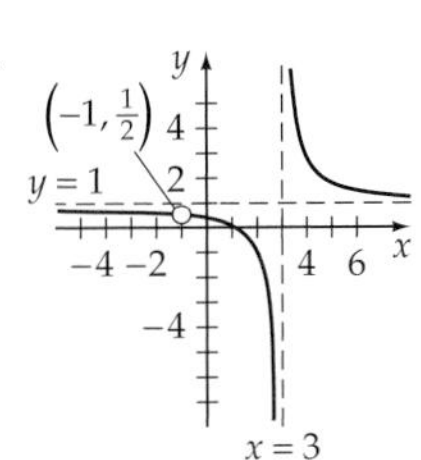

[4.5] **18.** [4.5]

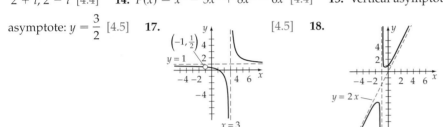

19. a. 5 words/min, 16 words/min, 25 words/min **b.** 70 words/min [4.5] **20.** 2.137 in., 337.1 in^3 [4.3]

Cumulative Review Exercises, *page 381*

1. $-1 + 2i$ [P.7] **2.** $\dfrac{1 \pm \sqrt{5}}{2}$ [1.2] **3.** 2, 10 [1.3] **4.** $\{x \mid -8 \le x \le 14\}$ [1.4] **5.** $\sqrt{281}$ [2.1] **6.** Translate the graph of $y = x^2$ 2 units to the right and 4 units up. [3.3] **7.** $2a + h - 2$ [3.1] **8.** $32x^2 - 92x + 60$ [3.1] **9.** $f^{-1}(x) = \dfrac{1}{2}x + \dfrac{5}{2}$ [3.2] **10.** $4x^3 - 8x^2 + 14x - 32 + \dfrac{59}{x+2}$ [4.1] **11.** 141 [4.1] **12.** The graph goes down. [4.2] **13.** 0.3997 [4.2] **14.** $\pm 1, \pm 2, \pm 4, \pm\dfrac{1}{3}, \pm\dfrac{2}{3}, \pm\dfrac{4}{3}$ [4.3] **15.** Zero positive real zeros, three or one negative real zero [4.3] **16.** $-2, 1 + 2i, 1 - 2i$ [4.4] **17.** $P(x) = x^3 - 4x^2 - 2x + 20$ [4.4] **18.** $(x - 2)(x + 3i)(x - 3i)$ [4.4] **19.** Vertical asymptotes: $x = -3, x = 2$; horizontal asymptote: $y = 4$ [4.5] **20.** $y = x + 4$ [4.5]

Chapter 5

Chapter Prep Quiz, *page 384*

1. 32 **2.** $\frac{1}{81}$ **3.** 1 **4.** Domain: the set of all real numbers; range: $\{y \mid y \ge 1\}$ **5.** (0, 2) **6.** Shift the graph of $f(x)$ upward 3 units. **7.** Reflect the graph of $f(x)$ across the y-axis. **8.** Shift the graph of $f(x)$ to the right 3 units. **9.** The function must be a one-to-one function. **10.** $f^{-1}(x) = 2x + 6$

Exercises 5.1, *page 395*

1. 81 **3.** $\frac{1}{100}$ **5.** 1 **7.** 4 **9.** 6.25 **11.** 4.73 **13.** 442.34 **15.** 164.02 **17.** 5.65 **19.** 0.97 **21.** 70.45 **23.** 19.81

25. 15.15 **27.** 3353.33 **29.** 8103.08 **31. a.** $k(x)$ **b.** $g(x)$ **c.** $h(x)$ **d.** $f(x)$ **33.**

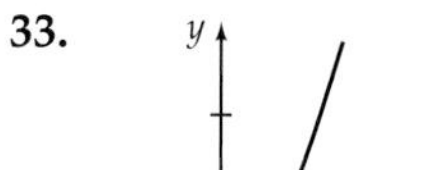

35.

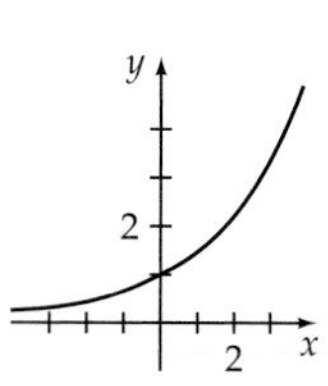

37.

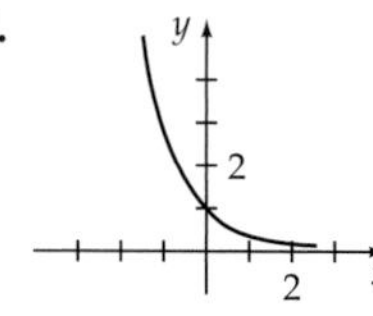

39.

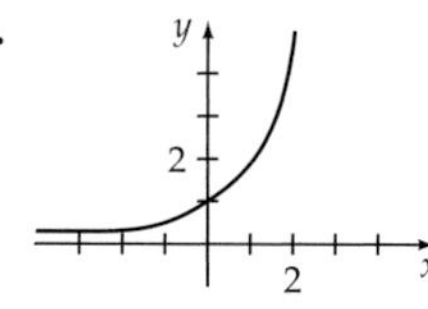

41.

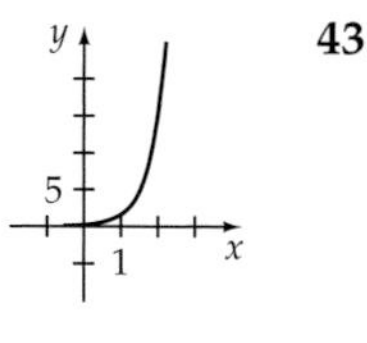

43.

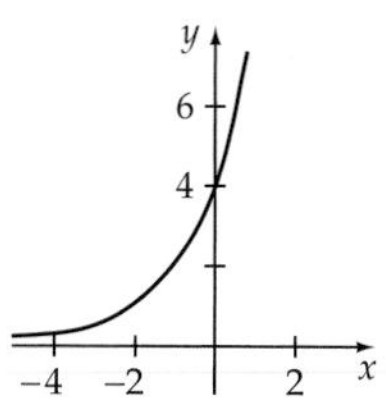

45.

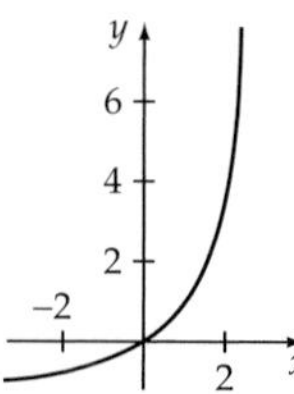

47.

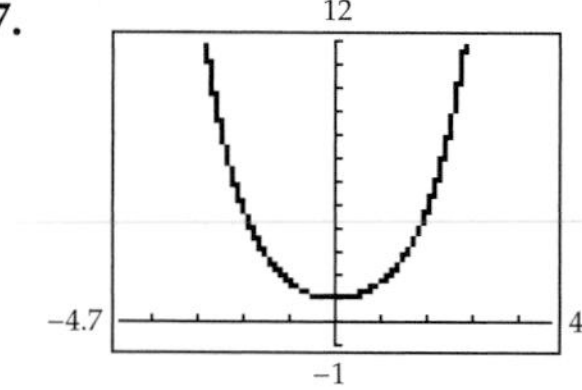

No horizontal asymptote

49.

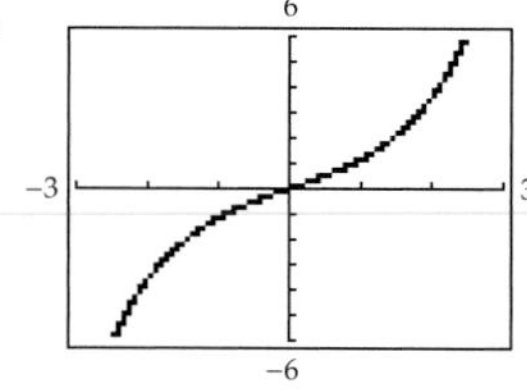

No horizontal asymptote

51.

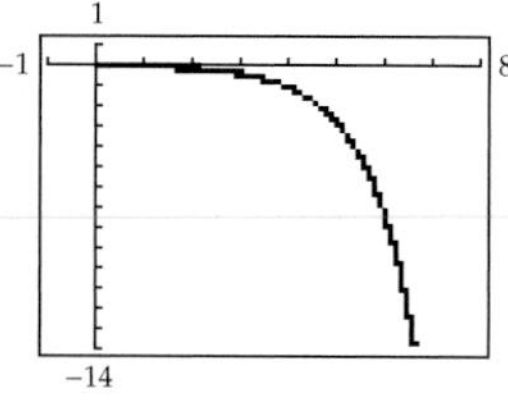

Horizontal asymptote: $y = 0$

53.

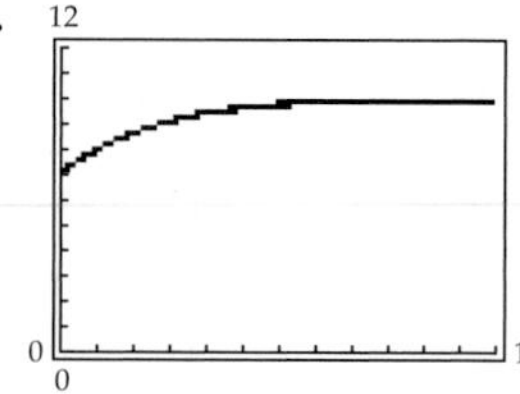

Horizontal asymptote: $y = 10$

55. a. 64 million connections **b.** 2005 **57. a.** 233 items per month; 59 items per month **b.** The demand will approach 25 items per month. **59. a.** 0.53 **b.** 0.89 **c.** 5.2 minutes **d.** There is a 98% probability that at least one customer will arrive between 10:00 A.M. and 10:05.2 A.M. **61. a.** 8.7% **b.** 2.6% **63. a.** 6400; 409,600 **b.** 11.6 h **65. a.** 515,000 people **b.** 1997 **67. a.** 363 beneficiaries; 88,572 beneficiaries **b.** 13 rounds **69. a.** 141°F **b.** After 28 minutes **71. a.** 261.63 vibrations per second **b.** No. The function $f(n)$ is not a linear function. Therefore, the graph of $f(n)$ does not increase at a constant rate.

Prepare for Section 5.2, *page 399*

72. 4 **73.** 3 **74.** 5 **75.** $f^{-1}(x) = \dfrac{3x}{2 - x}$ **76.** $\{x \mid x \geq 2\}$ **77.** The set of all positive real numbers

Exercises 5.2, *page 409*

1. $10^1 = 10$ **3.** $8^2 = 64$ **5.** $7^0 = x$ **7.** $e^4 = x$ **9.** $e^0 = 1$ **11.** $\log_3 9 = 2$ **13.** $\log_4 \frac{1}{16} = -2$ **15.** $\log_b y = x$ **17.** $\ln y = x$ **19.** $\log 100 = 2$ **21.** 2 **23.** −5 **25.** 3 **27.** −2 **29.** −4 **31.**

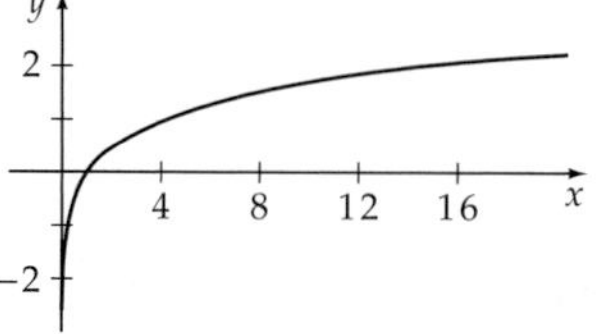

33.

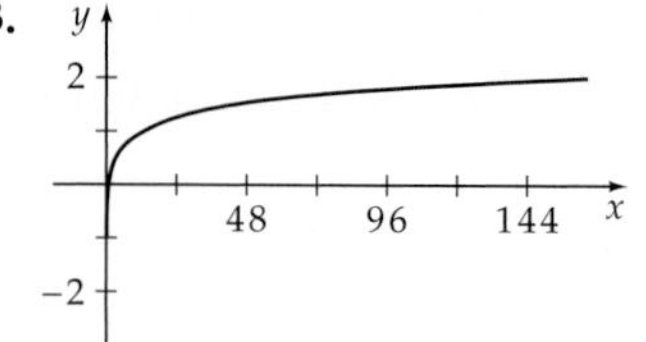

35.

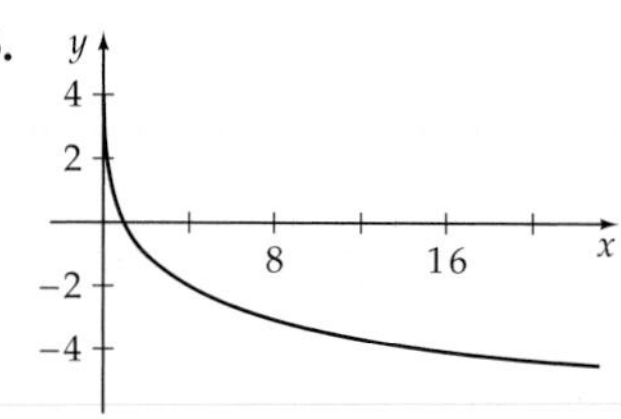

37. $(3, \infty)$ **39.** $(-\infty, 11)$ **41.** $(-\infty, -2) \cup (2, \infty)$

43. $(4, \infty)$ **45.** $(-1, 0) \cup (1, \infty)$ **47.**

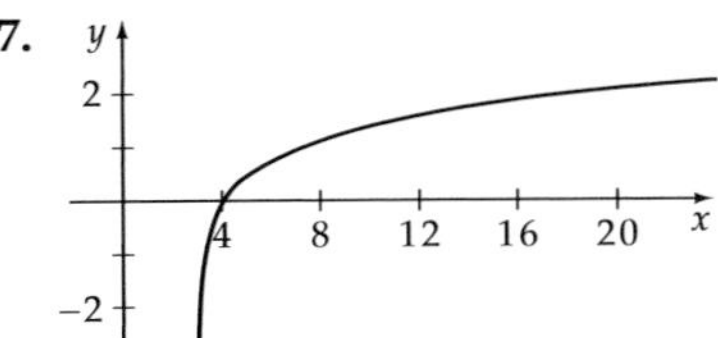

49.

51.

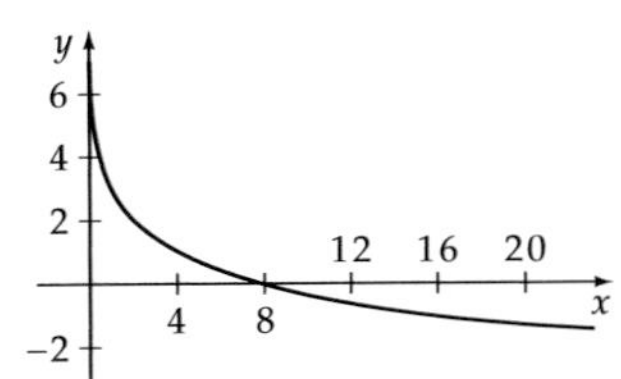

53. a. $k(x)$ **b.** $f(x)$ **c.** $g(x)$ **d.** $h(x)$ **55.**

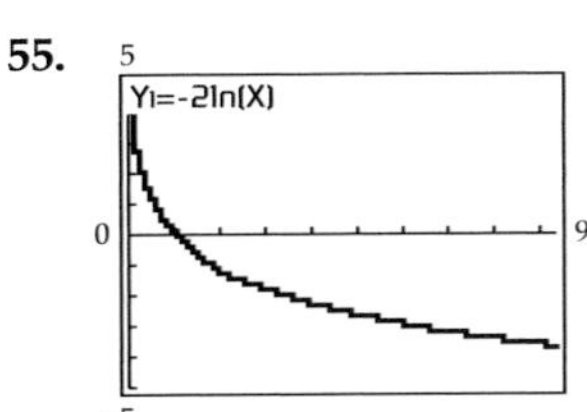

57.

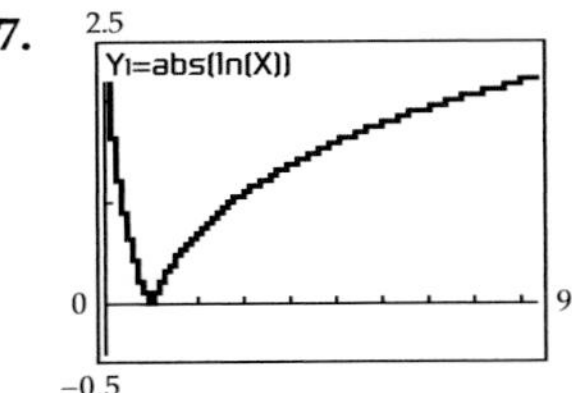

59.

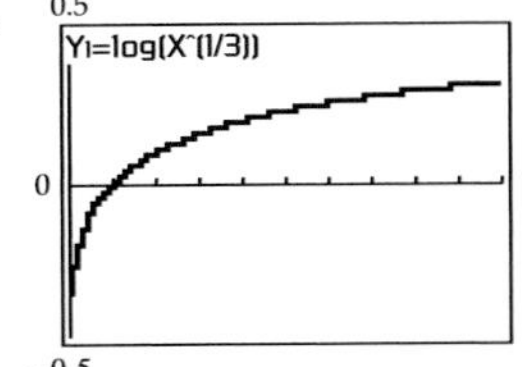

61.

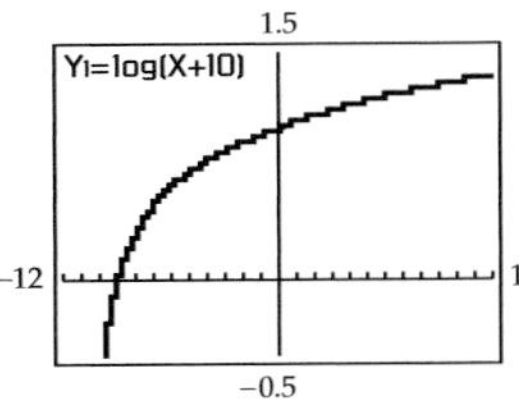

63.

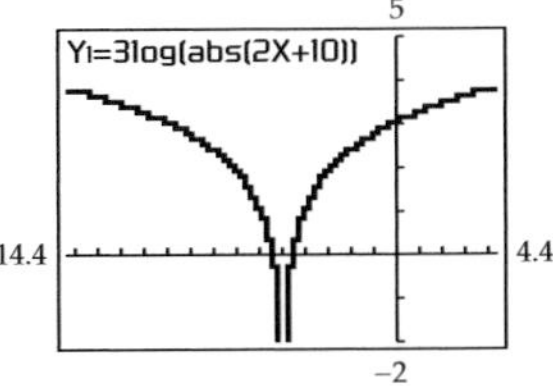

65. a. 4.3% **b.** 74 months **67. a.** 3298 units; 3418 units; 3490 units **b.** 2750 units **69. a.** 69.51 meters; 72.43 meters **b.** Answers will vary. **71. a.** Answers will vary. **b.** 96 digits **c.** 3385 digits **d.** 4,053,946 digits

Prepare for Section 5.3, *page 412*

73. ≈0.77815 for each expression **74.** ≈0.98083 for each expression **75.** ≈1.80618 for each expression **76.** ≈3.21888 for each expression **77.** ≈1.60944 for each expression **78.** ≈0.90309 for each expression

Exercises 5.3, *page 423*

1. $\log_b x + \log_b y + \log_b z$ **3.** $\ln x - 4 \ln z$ **5.** $\frac{1}{2}\log_2 x - 3\log_2 y$ **7.** $\frac{1}{2}\log_7 x + \frac{1}{2}\log_7 z - 2\log_7 y$ **9.** $\log[x^2(x + 5)]$ **11.** $\ln(x + y)$ **13.** $\log[x^3 \cdot \sqrt[3]{y}(x + 1)]$ **15.** 1.5395 **17.** 0.8672 **19.** −0.6131 **21.** 0.6447 **23.**

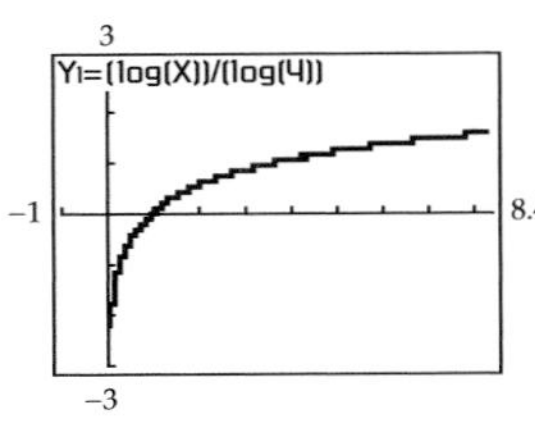

25.

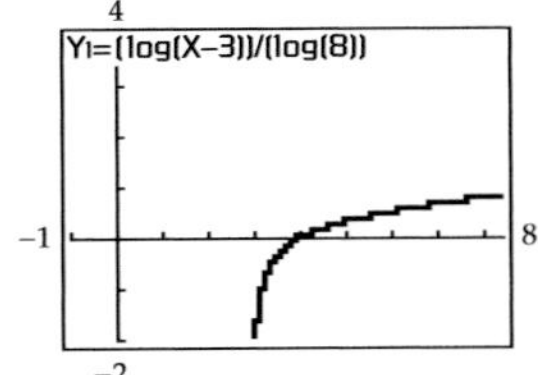

27.

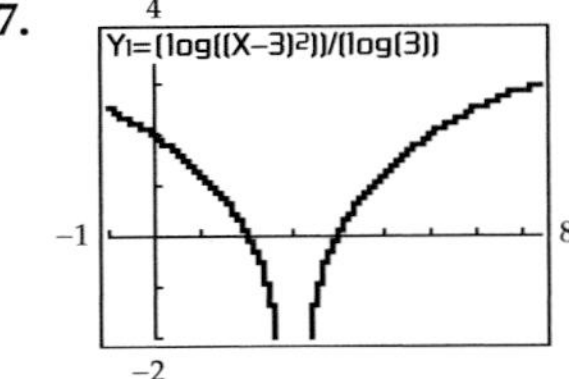

29.

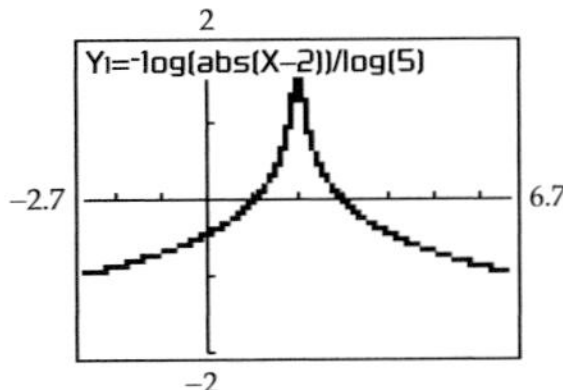

31. False; $\log 10 + \log 10 = 2$ but $\log(10 + 10) = \log 20 \neq 2$. **33.** True **35.** False; $\log 100 - \log 10 = 1$ but $\log(100 - 10) = \log 90 \neq 1$. **37.** False; $\dfrac{\log 100}{\log 10} = \dfrac{2}{1} = 2$ but $\log 100 - \log 10 = 1$. **39.** False; $(\log 10)^2 = 1$ but $2 \log 10 = 2$.

41. 2 **43.** 500^{501} **45.** 1:1870,551; 1:757,858; 1:659,754; 1:574,349; 1:500,000 **47.** 10.4; base **49.** 3.16×10^{-10} mole per liter **51. a.** 82.0 decibels **b.** 40.3 decibels **c.** 115.0 decibels **d.** 152.0 decibels **53.** 10 times more intense **55.** 5 **57.** $10^{6.5}I_0$ or about $3{,}162{,}277.7I_0$ **59.** 100 to 1 **61.** $10^{1.8}$ to 1 or about 63 to 1 **63.** 5.5 **65. a.** $M \approx 6$ **b.** $M \approx 4$ **c.** The results are close to the magnitudes produced by the amplitude-time-difference formula.

Prepare for Section 5.4, *page 426*

66. $\log_3 729 = 6$ **67.** $5^4 = 625$ **68.** $\log_a b = x + 2$ **69.** $x = \dfrac{4a}{7b + 2c}$ **70.** $x = \dfrac{3}{44}$ **71.** $x = \dfrac{100(A - 1)}{A + 1}$

Exercises 5.4, *page 435*

1. 6 **3.** 1 **5.** $-\frac{6}{5}$ **7.** $-\frac{3}{2}$ **9.** $\dfrac{\log 70}{\log 4} \approx 3.065$ **11.** $-\dfrac{\log 120}{\log 3} \approx -4.358$ **13.** $\dfrac{\log 315 - 3}{2} \approx -0.251$

15. $\dfrac{\ln 5}{\ln 7 - \ln 5} \approx 4.783$ **17.** $\dfrac{3 \ln 4}{\ln 5 - \ln 4} \approx 18.638$ **19.** $\dfrac{\ln 2 - \ln 3}{\ln 2 + \ln 3} \approx -0.226$ **21.** 7 **23.** $3 - e$ **25.** 4 **27.** $2 + 2\sqrt{2}$

29. 1 **31.** $\frac{199}{95}$ **33.** 1.61 **35.** 1.49 **37.** 10^{10} **39.** $\ln(\ln 3) + 1$ **41.** $\frac{1}{2} \log \frac{3}{2}$

43. 1.729 **45.** 1.278 **47.** $x = \dfrac{y}{y - 1}$ **49.** 74 withdrawals **51. a.** 68 milligrams **b.** 88 minutes **53.** 5.2 minutes

55. a. 8500 people; 10,285 people **b.** In 6 years **57. a.** 6 words per minute **b.** 37 words per minute **c.** 82 h

59. ≈45 hours **61. a.**

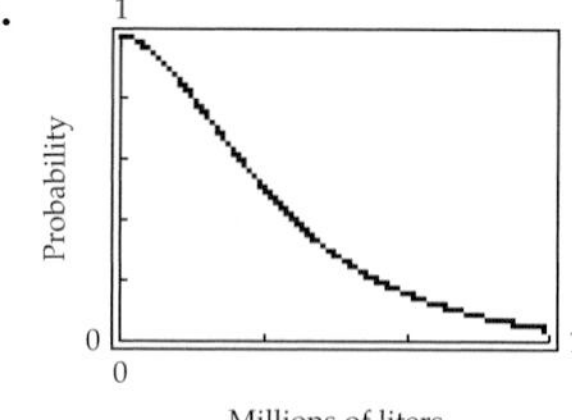

b. 0.504 **c.** As the number of liters of water demanded per day becomes very large ($x \to \infty$), the probability of providing the water approaches 0.

63. a. 23.14 feet per second **b.** The velocity of the object cannot exceed 24 feet per second. **65.** 0.53 second

67. a. 150 Distance (feet) 0 0 4 Time (seconds) **b.** 2.4 seconds

Prepare for Section 5.5, *page 439*

69. 0.566 **70.** 935.22 **71.** 1439.2 **72.** 5 **73.** $0.862t$ **74.** 0.70

Exercises 5.5, *page 448*

1. 1.74 **3.** 2.85 **5.** 1.89 **7.** 0.94 **9.** 6.11% **11.** 7.79% **13.** $k \approx 0.788$ **15. a.** $T(t) \approx 34 + 41e^{-0.0559t}$ **b.** 42°F **c.** 54 minutes **17. a.** $10,119.65 **b.** 4.6 years **19. a.** 3.18 micrograms **b.** 15.07 hours **c.** 30.14 hours **d.** 4.6% **21.** 0.27 gram **23.** 11.1 hours **25. a.** $N(t) \approx 640{,}000e^{0.03t}$ **b.** 864,000 **c.** 14.9 years **27. a.** $N(t) \approx 1350e^{0.14t}$ **b.** 2055 **c.** 7.8 years **29. a.** 10,130,000 **b.** 0.5% **c.** 2042 **31. a.** $N(t) \approx 22{,}600e^{0.01368t}$ **b.** 29,700 **33. a.** $N(t) \approx 667{,}000e^{0.0284t}$ **b.** 965,000 **35.** After 20.8 years **37.** 1409 years **39. a.** 286,471 gallons **b.** 250,662 gallons **c.** 34 hours **41. a.** $N(t) \approx 3e^{-0.07133t}$ **b.** 15.4 years **c.** 9.72 years **43.** $N(t) \approx 6e^{-0.002476t}$ **45.** 6600 years ago **47.** 2378 years old

Prepare for Section 5.6, *page 452*

49. 1656.8 **50.** 7,911,700 **51.** 106.52 **52.** 4.34 **53.** 670.75 **54.** 1419.2

Exercises 5.6, *page 461*

1. Quadratic function; increasing exponential function

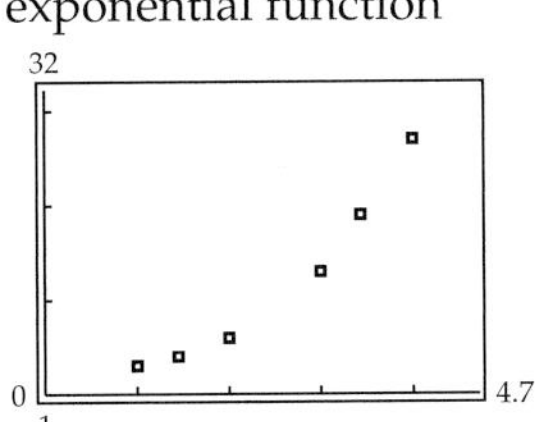

3. Quadratic function

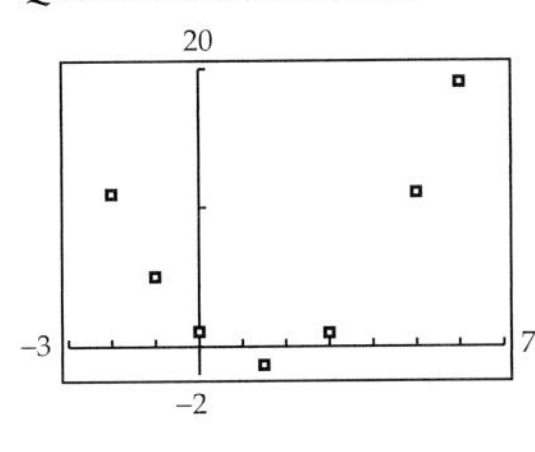

5. Linear function; quadratic function; decreasing logarithmic function; decreasing exponential function

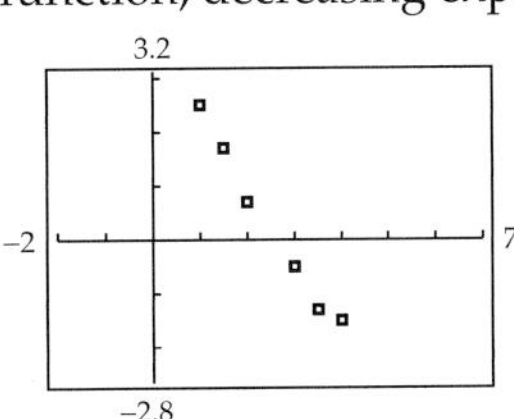

7. $y \approx 0.99628(1.20052)^x$; $r \approx 0.85705$ **9.** $y \approx 1.81505(0.51979)^x$; $r \approx -0.99978$ **11.** $y \approx 4.89060 - 1.35073 \ln x$; $r \approx -0.99921$ **13.** $y \approx 14.05858 + 1.76393 \ln x$; $r \approx 0.99983$ **15. a.** $y \approx 5.48184(1.00356)^x$; 6.78% **b.** 69 months **17. a.** LinReg: $\text{pH} \approx 0.01353q + 7.02852$, $r \approx 0.956627$; LnReg: $\text{pH} \approx 6.10251 + 0.43369 \ln q$, $r \approx 0.999998$. The logarithmic model provides a better fit. **b.** 126.0 **19.** $p \approx 3200(0.91894)^t$; 2012 **b.** No. The model fits the data perfectly because there are only two data points. **21. a.** LinReg: Time $\approx 1.04035t + 154.96491$, $r \approx 0.88458$; LnReg: Time $\approx 149.56876 + 7.63077\ln t$, $r \approx 0.93101$ **b.** The logarithmic model provides a better fit. **c.** During the 2012 season **23. a.** ExpReg: $a \approx 8000(1.10657)^t$; 550,500,000 automobiles **b.** 2004 **25. a.** LinReg: $w \approx 10.17227t + 16.45111$, $r \approx 0.95601$; LnReg: $w \approx 18.26750 + 31.03499 \ln t$, $r \approx 0.99996$ **b.** The logarithmic model provides a better fit. **c.** 89.7 cubic yards **27.** A and B have different exponential regression functions. **29. a.** ExpReg: $y \approx 1.81120(1.61740)^x$, $r \approx 0.96793$; PwrReg: $y \approx 2.0985(x)^{1.40246}$, $r \approx 0.99999$ **b.** The power regression function provides the better fit.

Prepare for Section 5.7, *page 465*

31. As $x \to \infty$, $f(x) \to 0$. **32.** 187.5 **33.** 214.19 **34.** $-\dfrac{\ln 0.4}{0.12} \approx 7.63576$ **35.** $-\dfrac{\ln 0.25}{0.04} \approx 34.65736$

36. $-5 \ln\left(\frac{1}{3}\right) = 5 \ln 3 \approx 5.49306$

Exercises 5.7, *page 473*

1. a. 1900 **b.** 0.16 **c.** 200 **3. a.** 157,500 **b.** 0.04 **c.** 45,000 **5. a.** 2400 **b.** 0.12 **c.** 300 **7.** $P(t) \approx \dfrac{5500}{1 + 12.75e^{-0.37263t}}$

9. $P(t) \approx \dfrac{100}{1 + 4.55556e^{-0.22302t}}$ **11.** $P(t) \approx \dfrac{4600}{1 + 8.2e^{-0.45t}}$ **13.** $P(t) \approx \dfrac{563.96280}{1 + 10.07473e^{-0.21933t}}$ **15.** $P(t) \approx \dfrac{799.91097}{1 + 14.23484e^{-0.75065t}}$

17. $P(t) \approx \dfrac{80.91413}{1 + 0.89461e^{-0.04857t}}$ **19.** $\approx(47.34, 40)$ **21. a.** \$625,000 **b.** 25.14 years **23. a.** $P(t) \approx \dfrac{11.26828}{1 + 2.74965e^{-0.02924t}}$

b. 11 billion people **25. a.** $P(t) \approx \dfrac{3400}{1 + 13.16667e^{-0.27833t}}$ **b.** 2007 **27. a.** 2400 deer **b.** 1800 deer **29. a.** $m = c$ **b.** $k = b$

c. $a = \dfrac{m - P_0}{P_0}$ **31.** $P(t) = \dfrac{10{,}800}{60 + 120e^{-0.11t}}$ **33.** $P(t) = \dfrac{4800}{60 + 20e^{-0.2t}}$ **35.** $P(t) = \dfrac{550}{1 + 1.75e^{-0.17t}}$ **37.** $P(t) = \dfrac{85}{1 + \left(\frac{10}{7}\right)e^{-0.34t}}$

Chapter 5 True/False Exercises, *page 479*

1. True **2.** True **3.** False; because f is not defined for $x \le 0$, $g[f(x)]$ is not defined for $x \le 0$. **4.** False; $h(x)$ is not an increasing function for $0 < b < 1$. **5.** False; the exponential growth function has a continuous growth rate of $1.1 = 110\%$. **6.** False; an earthquake with a 6.0 Richter scale magnitude is 1000 times as intense as an earthquake with a 3.0 Richter scale magnitude. **7.** False; the population will double in about 6.9 years. **8.** False; in 30 days 75% of the material will decay. **9.** False; $\log x + \log y = \log(xy) \ne \log(x + y)$ **10.** False; the carrying capacity is $\frac{14{,}000}{2} = 7000$. **11.** False; because $a < 1$, the given logistic growth model with $t \ge 0$ does not have an inflection point.

Chapter 5 Review Exercises, *page 479*

1. 2 [5.2/5.4] **2.** 4 [5.2/5.4] **3.** 3 [5.2/5.4] **4.** π [5.2/5.4] **5.** -2 [5.4] **6.** 8 [5.4] **7.** -3 [5.4] **8.** -4 [5.4] **9.** ± 1000 [5.4] **10.** $\pm 10^{10}$ [5.4] **11.** 7 [5.2/5.4] **12.** ± 8 [5.2/5.4] **13.** [5.1]

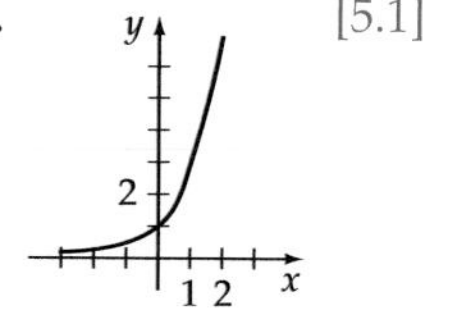

14. [5.1]

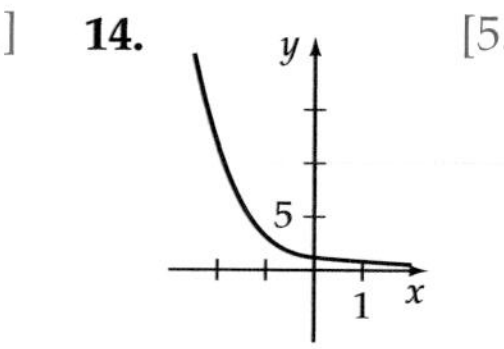

15. [5.1]

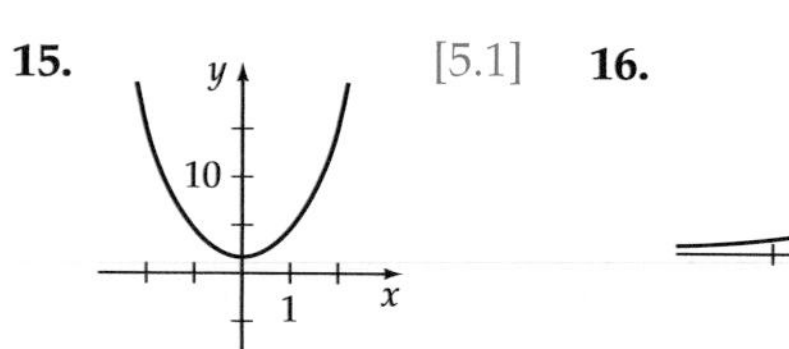

16. [5.1]

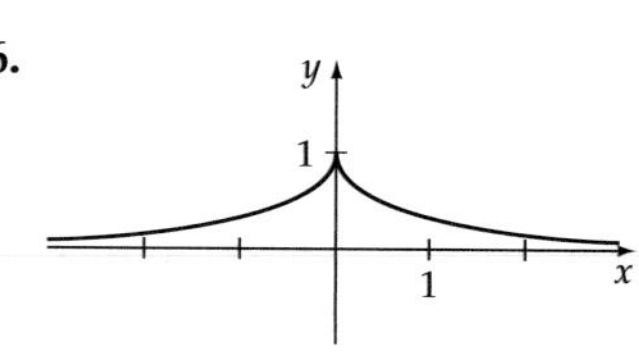

17. [5.1]

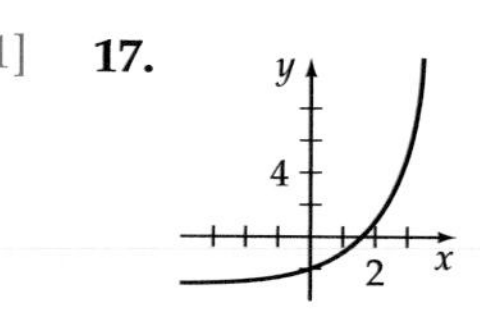

18. [5.1]

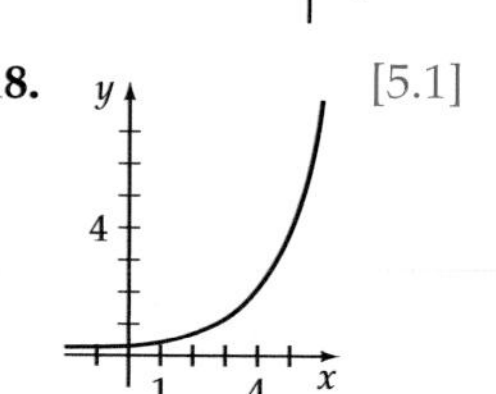

19. [5.2]

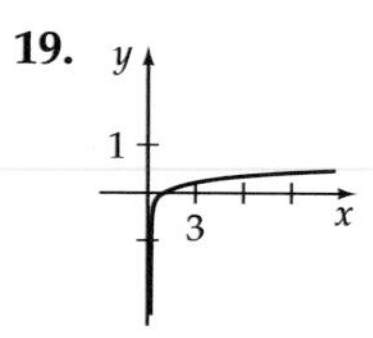

20. [5.2]

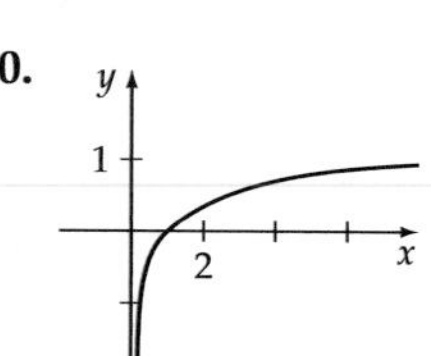

21. [5.2]

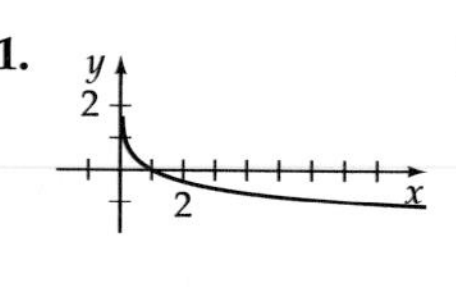

22. [5.2]

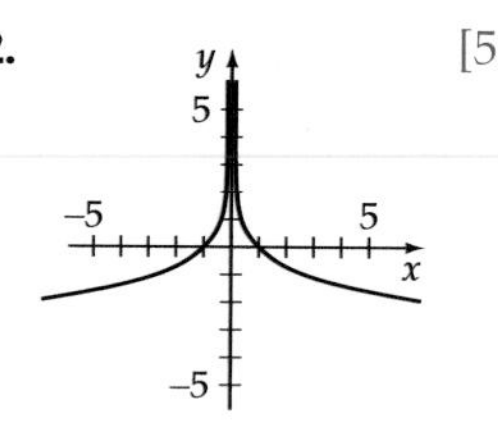

23. [5.1]

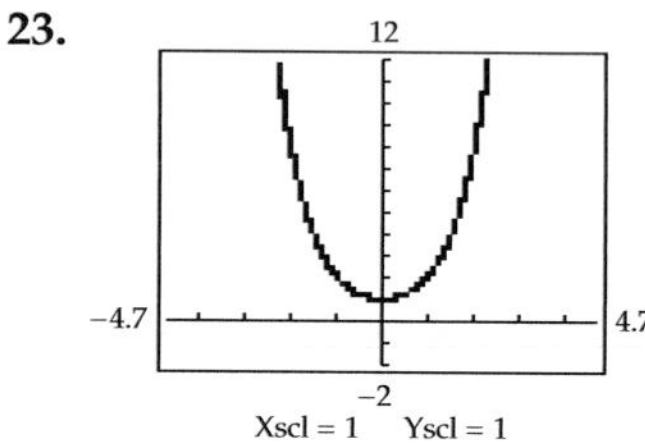

24. [5.1]

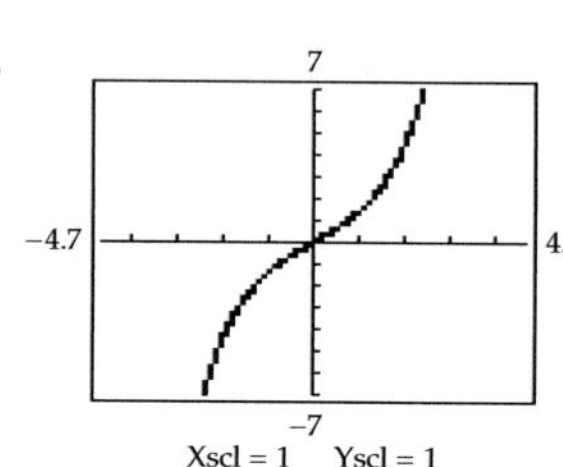

25. $4^3 = 64$ [5.2] **26.** $\left(\frac{1}{2}\right)^{-3} = 8$ [5.2]

27. $\left(\sqrt{2}\right)^4 = 4$ [5.2] **28.** $e^0 = 1$ [5.2] **29.** $\log_5 125 = 3$ [5.2] **30.** $\log_2 1024 = 10$ [5.2] **31.** $\log 1 = 0$ [5.2] **32.** $\log_8\left(2\sqrt{2}\right) = \frac{1}{2}$ [5.2] **33.** $2\log_b x + 3\log_b y - \log_b z$ [5.3] **34.** $\frac{1}{2}\log_b x - 2\log_b y - \log_b z$ [5.3] **35.** $\ln x + 3\ln y$ [5.3] **36.** $\frac{1}{2}\ln x + \frac{1}{2}\ln y - 4\ln z$ [5.3] **37.** $\log\left(x^2\sqrt[3]{x+1}\right)$ [5.3] **38.** $\log\dfrac{x^5}{(x+2)^2}$ [5.3] **39.** $\ln\dfrac{\sqrt{2xy}}{z^3}$ [5.3]

40. $\ln\dfrac{xz}{y}$ [5.3] **41.** 2.8675 [5.3] **42.** 3.3578 [5.3] **43.** -0.1172 [5.3] **44.** -0.5790 [5.3] **45.** $\dfrac{\log 30}{\log 4}$ or $\dfrac{\ln 30}{\ln 4}$ [5.4]

46. $\dfrac{\ln 41}{\ln 5} - 1$ [5.4] **47.** 4 [5.4] **48.** $\frac{1}{6}e$ [5.4] **49.** 4 [5.4] **50.** 15 [5.4] **51.** $\dfrac{\ln 3}{2\ln 4}$ [5.4] **52.** $\dfrac{\ln\left(8 \pm 3\sqrt{7}\right)}{\ln 5}$ [5.4]

53. 10^{1000} [5.4] **54.** $e^{(e^2)}$ [5.4] **55.** 1,000,005 [5.4] **56.** $\dfrac{15 + \sqrt{265}}{2}$ [5.4] **57.** 81 [5.4] **58.** $\pm 125\sqrt{5}$ [5.4] **59.** 4 [5.4]

60. 5 [5.4] **61.** $N(t) \approx e^{0.80472t}$ [5.5] **62.** $N(t) \approx 2e^{0.56825t}$ [5.5] **63.** $N(t) \approx 3.78297e^{0.05579t}$ [5.5] **64.** $N(t) \approx e^{-0.69315t}$ [5.5] **65. a.** 21 **b.** (5.96, 63.00) **c.** The function P attains its largest rate of growth when $t \approx 5.96$. The value 5.96 is the first coordinate of the inflection point. **d.** 126 [5.7] **66.** 2004 [5.4] **67.** \$3800 [5.1] **68. a.** $\approx$69.9% **b.** 6 days **c.** 19 days [5.5] **69. a.** $N(t) \approx 25{,}200e^{0.06156t}$ **b.** 38,800 people [5.5] **70. a.** $N(t) \approx 74{,}000e^{0.046t}$ **b.** 106,900 people **c.** 15.1 years [5.5] **71. a.** $P(t) \approx \dfrac{1400}{1 + 5.66667e^{-0.22458t}}$ **b.** 1070 coyotes **c.** During 1999 [5.7]

72. a. ExpReg: $R \approx 161.03059(0.96884)^t$ **b.** 5.3 deaths per 1000 live births [5.6]

73. a. LinReg: $P \approx -69{,}667.52t + 8{,}230{,}901.59$, $r \approx -0.9547$; ExpReg: $P \approx 87{,}535{,}810.36(0.958728)^t$, $r \approx -0.9670$

LnReg: $P \approx 31{,}854{,}352.17 - 6{,}641{,}714.035 \ln t$, $r \approx -0.9588$ **b.** The exponential model **c.** 1,050,000 people [5.6]
74. 7.7 [5.3] **75.** 5.0 [5.3] **76.** 3162 to 1 [5.3] **77.** 2.8 [5.3] **78.** 4.2 [5.3] **79.** $\approx 3.98 \times 10^{-6}$ [5.3]
80. 340 years old [5.5] **81.** $N(t) = 5\left(\frac{1}{2}\right)^{t/26}$ or $N(t) \approx 5e^{-0.02666t}$ [5.5] **82. a.** $T(t) \approx 71 + 69e^{-0.05898t}$ **b.** 34.5 minutes [5.5]

Chapter 5 Test, *page 482*

1. -3 [5.2/5.4] **2.** 1.7925 [5.3] **3.**

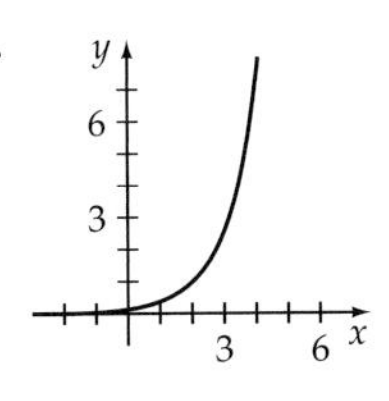

[5.1] **4.**

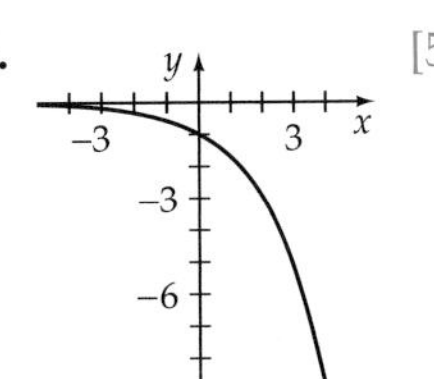

[5.1] **5.**

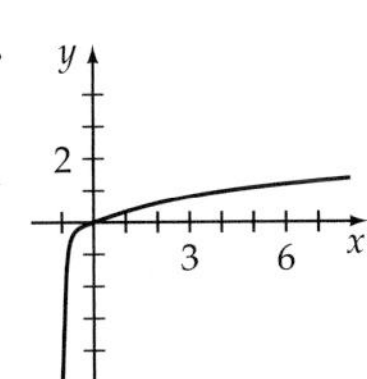

Domain: $\{x \mid x > -1\}$ [5.2]

6. $b^c = 5x - 3$ [5.2] **7.** $\ln a = \frac{t}{4}$ [5.2] **8.** $2 \log_b z - 3 \log_b y - \frac{1}{2} \log_b x$ [5.3] **9.** 1.9206 [5.4] **10.** $\frac{5 \ln 4}{\ln 4 + \ln 7}$ [5.4]

11. 2 [5.4] **12.** $\frac{\ln\left(100 + \sqrt{10{,}001}\right)}{\ln 3}$ [5.4] **13.** 1 [5.4] **14. a.** 93°F **b.** 7.4 minutes [5.5] **15. a.** $P(t) \approx 34{,}600e^{0.04667t}$

b. 55,000 people **c.** 2013 [5.5] **16.** 690 years old [5.5] **17. a.** $r(t) \approx 1.59035(1.15763)^t$; 2.47% **b.** 6.3 years [5.6]

18. a. $f(x) = 3.0610 + 4.9633 \ln x$ **b.** 13.3 [5.3] **19. a.** $P(t) \approx \frac{1100}{1 + 5.875e^{-0.20429t}}$ **b.** ≈457 raccoons **c.** During 2007 [5.7]

20. a. 7200 northern pike **b.** 51.5 months [5.7]

Cumulative Review Exercises, *page 483*

1. [2, 6] [1.4] **2.** $\{x \mid 3 < x \le 6\}$ [1.5] **3.** 7.8 [2.1] **4.** 38.25 feet [2.5] **5.** $4x^2 + 4x - 4$ [3.2] **6.** $f^{-1}(x) = \frac{1}{3}x + \frac{5}{3}$ [3.2]
7. 3500 pounds [3.4] **8.** 3 or 1 positive real zeros; 1 negative real zero [4.3] **9.** $1, 4, -\sqrt{3}, \sqrt{3}$ [4.3]
10. $P(x) = x^3 - 4x^2 + 6x - 4$ [4.4] **11.** Vertical asymptote: $x = 4$, horizontal asymptote: $y = 3$ [4.5] **12.** Domain: all real numbers; range: $\{y \mid 0 < y \le 4\}$ [4.5] **13.** Decreasing function [5.1] **14.** $4^y = x$ [5.2] **15.** $\log_5 125 = 3$ [5.2] **16.** 7.1 [5.3]
17. 2.0149 [5.4] **18.** 510 years old [5.5] **19. a.** $y \approx 84.41319 + 4.88166 \ln x$ b. 99.28 meters [5.6]

20. a. $P(x) \approx \frac{450}{1 + 1.8125e^{-0.13882x}}$ **b.** 310 wolves [5.7]

Chapter 6

Chapter Prep Quiz, *page 486*

1. $\frac{3}{5}$ **2.** Slope: $\frac{2}{3}$, y-intercept: $(0, -3)$ **3.** $(0, 2), (2, -1), (4, -4)$ (Answers can vary.) **4.**

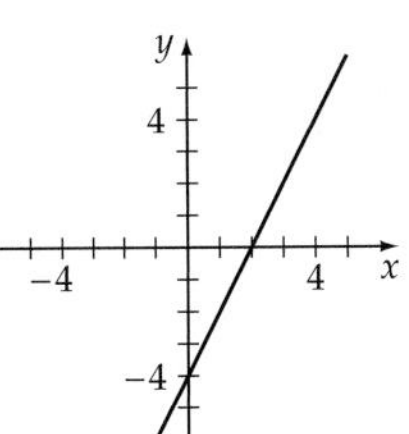

5.

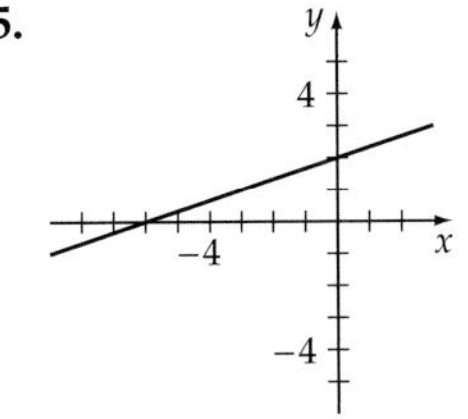

6. $C(x) = 350x + 1800$ **7.** $-6x + 7$ **8.** Center: $(-3, 1)$, radius: $\sqrt{7}$ **9.** $x < \frac{13}{4}$ **10.** $\left(-\infty, \frac{2}{3}\right) \cup [3, \infty)$

Exercises 6.1, *page 498*

1. $(2, -3)$ **3.** $(-1, 1)$ **5.** $(2, -4)$ **7.** $\left(-\frac{6}{5}, \frac{27}{5}\right)$ **9.** $(3, 4)$ **11.** $(1, -1)$ **13.** $(3, -4)$ **15.** $(2, 5)$ **17.** $(-1, -1)$ **19.** $\left(\frac{62}{25}, \frac{34}{25}\right)$ **21.** No solution **23.** $\left(c, -\frac{4}{3}c + 2\right)$ **25.** $(2, -4)$ **27.** $(0, 3)$ **29.** $\left(\frac{3c}{5}, c\right)$ **31.** $\left(-\frac{1}{2}, \frac{2}{3}\right)$ **33.** No solution **35.** $(-6, 3)$ **37.** $\left(2, -\frac{3}{2}\right)$ **39.** $(0.2, -0.5)$ **41.** $\frac{9}{5}$ square units **43.** 5 miles **45.** \$14,000 at 6%, \$11,000 at 6.5% **47.** First powder: 200 milligrams, second powder: 450 milligrams **49.** Plane: 120 miles per hour, wind: 30 miles per hour **51.** Boat: 25 miles per hour, current: 5 miles per hour **53.** Pepperoni: \$12.50, vegetarian: \$10.50 **55.** Iron: \$12 per kilogram, lead: \$16 per kilogram **57.** 8 grams of 40% gold, 12 grams of 60% gold **59.** 20 milliliters of 13% solution, 30 milliliters of 18% solution

Prepare for Section 6.2, *page 500*

60. $b = \dfrac{5a + 7c - 9}{2}$ **61.** $26x - 8y$ **62.** $x = -3, y = 2$ **63.** $\left(-\frac{24}{11}, -\frac{62}{11}\right)$ **64.** $\left(-\frac{76}{17}, \frac{56}{17}\right)$ **65.** An infinite number; in terms of an arbitrary constant

Exercises 6.2, *page 511*

1. $(2, -1, 3)$ **3.** $(2, 0, -3)$ **5.** $(2, -3, 1)$ **7.** $(-5, 1, -1)$ **9.** $(3, -5, 0)$ **11.** $(0, 2, 3)$ **13.** $(5c - 25, 48 - 9c, c)$ **15.** $(3, -1, 0)$ **17.** No solution **19.** $\left(\dfrac{50 - 11c}{11}, \dfrac{11c - 18}{11}, c\right)$ **21.** No solution **23.** $\left(\dfrac{25 + 4c}{29}, \dfrac{55 - 26c}{29}, c\right)$ **25.** $A = -\frac{13}{2}$ **27.** $A \neq -3, A \neq 1$ **29.** $A = -3$ **31.** $y = 2x^2 - x - 3$ **33.** $x^2 + y^2 - 4x + 2y - 20 = 0$ **35.** Center: $(-7, -2)$, radius: 13 **37.** 685 **39.** 396 nickels, 132 dimes, 132 quarters **41.** \$8000 at 9%, \$6000 at 7%, \$4000 at 5% **43.** 220 quarters, 440 nickels, 80 silver dollars **45.** $d_1 = 6$ inches, $d_2 = 3$ inches, $d_3 = 9$ inches **47.** Holiday Inn: 255,342 rooms; Best Western: 182,045 rooms; Days Inn: 151,576 rooms **49.** $y = -0.2x^2 - 0.4x + 72$

Prepare for Section 6.3, *page 514*

50. $29x - 10y = -3$ **51.** $13y - 9z = -39$ **52.** $x = -3, y = -1$ **53.** The system will either be dependent (an infinite number of solutions) or inconsistent (no solutions). **54.** $(-7, 10, -4)$ **55.** $\left(0, \frac{1}{7}, 0\right), \left(\frac{2}{7}, -\frac{2}{7}, 1\right), \left(\frac{4}{7}, -\frac{5}{7}, 2\right)$ (Answers can vary.)

Exercises 6.3, *page 528*

1. 3×3 **3.** 1×3 **5.** $\begin{cases} 2x - 3y = 1 \\ x + 4y = 6 \end{cases}$ **7.** $\begin{cases} x - 3y + 2z = 5 \\ 2x + y - 3z = -4 \\ x - 5y + z = 0 \end{cases}$ **9.** $\left[\begin{array}{cc|c} 5 & 7 & 3 \\ 2 & -5 & 8 \end{array}\right], \begin{bmatrix} 5 & 7 \\ 2 & -5 \end{bmatrix}, \begin{bmatrix} 3 \\ 8 \end{bmatrix}$ **11.** $\left[\begin{array}{ccc|c} 2 & -6 & 5 & 11 \\ 3 & 2 & -4 & 23 \\ 2 & -5 & 3 & 8 \end{array}\right], \begin{bmatrix} 2 & -6 & 5 \\ 3 & 2 & -4 \\ 2 & -5 & 3 \end{bmatrix}, \begin{bmatrix} 11 \\ 23 \\ 8 \end{bmatrix}$ **13.** No solution **15.** Unique solution **17.** $R_1 \leftrightarrow R_2$ **19.** $3R_1 + R_2$ **21.** $(2, 4)$ **23.** $(6, -3)$ **25.** $(2, -1, 1)$ **27.** $(1, -2, -1)$ **29.** $(2, -2, 3, 4)$ **31.** Dependent; $\left(-2c + 2, 2c + \frac{1}{2}, c\right)$ **33.** Independent: $\left(\frac{1}{2}, \frac{1}{2}, \frac{3}{2}\right)$ **35.** Dependent: $(c + 3, c - 4, c)$ **37.** $p(x) = 2x - 3$ **39.** $p(x) = x^2 - 2x + 3$ **41.** $p(x) = x^3 - 2x^2 - x + 2$ **43.** $z = 2x + 3y - 2$ **45.** $x^2 + y^2 + 2x - 4y = 20$ **47.** Oldest child: \$28,000; second child: \$16,000; youngest child: \$12,000 **49.** Blanket: \$25, cot: \$75, lantern: \$50 **51.** $3x - 5y - 2z = -2$

Prepare for Section 6.4, *page 531*

53. $-a$ (The additive inverse of 0 is 0.) **54.** 1 **55.** No **56.** $a = 5, b = -1$ **57.** 3×1 **58.** $\begin{cases} 3x - 5y = 16 \\ 2x + 7y = -10 \end{cases}$

Exercises 6.4, *page 543*

1. 1×3; row matrix **3.** 3×3; square matrix and identity matrix **5.** 3×4 **7. a.** $\begin{bmatrix} 1 & 2 \\ 5 & 4 \end{bmatrix}$ **b.** $\begin{bmatrix} 3 & -4 \\ 1 & 2 \end{bmatrix}$ **c.** $\begin{bmatrix} 7 & -11 \\ 0 & 3 \end{bmatrix}$

9. a. $\begin{bmatrix} -3 & 0 & 5 \\ 3 & 5 & -5 \end{bmatrix}$ **b.** $\begin{bmatrix} 3 & -2 & 1 \\ -1 & -5 & 1 \end{bmatrix}$ **c.** $\begin{bmatrix} 9 & -5 & 0 \\ -4 & -15 & 5 \end{bmatrix}$ **11. a.** $\begin{bmatrix} -1 & 1 & -1 \\ 2 & 2 & 1 \\ -1 & 2 & 5 \end{bmatrix}$ **b.** $\begin{bmatrix} -3 & 5 & -1 \\ -2 & -4 & 3 \\ -7 & 4 & 1 \end{bmatrix}$

c. $\begin{bmatrix} -7 & 12 & -2 \\ -6 & -11 & 7 \\ -17 & 9 & 0 \end{bmatrix}$ **13.** $AB = \begin{bmatrix} -10 & 17 \\ 6 & -8 \end{bmatrix}, BA = \begin{bmatrix} 0 & 22 \\ 1 & -18 \end{bmatrix}$ **15.** $AB = \begin{bmatrix} 0 & -4 & 5 \\ 6 & 0 & 3 \\ -3 & -2 & 1 \end{bmatrix}, BA = \begin{bmatrix} 5 & -13 \\ 5 & -4 \end{bmatrix}$

17. $AB = \begin{bmatrix} 9 & -2 & -6 \\ 0 & -1 & 2 \\ 4 & -2 & -4 \end{bmatrix}, BA = \begin{bmatrix} 4 & -2 & 6 \\ 2 & -3 & 4 \\ 4 & -4 & 3 \end{bmatrix}$ **19.** $\begin{bmatrix} 1 & -9 \\ 3 & -2 \end{bmatrix}$ **21.** $\begin{bmatrix} 7 & -1 & 1 \\ 1 & 2 & 0 \\ 5 & -1 & 4 \end{bmatrix}$ **23.** $\begin{bmatrix} 10 & -4 & -6 & 2 \\ 2 & 5 & 7 & -6 \\ 4 & -2 & 16 & -5 \\ 2 & 4 & 10 & -2 \end{bmatrix}$

25. $\begin{bmatrix} 3 & 7 & 0 & 11 \\ 4 & 3 & 0 & -6 \\ 0 & 3 & 10 & -2 \\ -3 & 13 & 4 & 5 \end{bmatrix}$ **27.** $\begin{bmatrix} 57 & -79 & -74 & 37 \\ 40 & 42 & 54 & -49 \\ 37 & -19 & 40 & -18 \\ 18 & -26 & -18 & 21 \end{bmatrix}$ **29.** $\begin{cases} x - 2y = -1 \\ 3x + 4y = 2 \end{cases}$ **31.** $\begin{cases} x = 4 \\ y = 5 \end{cases}$ **33.** $\begin{cases} 2x + y - 3z = 3 \\ x \quad\quad + 3z = 2 \\ -2x + y + 4z = 1 \end{cases}$

35. $0.98\begin{bmatrix} 2.0 & 1.4 & 3.0 & 1.4 \\ 0.8 & 1.1 & 2.0 & 0.9 \\ 3.6 & 1.2 & 4.5 & 1.5 \end{bmatrix} = \begin{bmatrix} 1.96 & 1.37 & 2.94 & 1.37 \\ 0.78 & 1.08 & 1.96 & 0.88 \\ 3.53 & 1.18 & 4.41 & 1.47 \end{bmatrix}$ **37. a.** $\begin{bmatrix} 2850 & 3000 & 3525 \\ 2870 & 2980 & 3495 \end{bmatrix}$ **b.** It takes machine M_2 3495 minutes to make 450 CDs and 600 DVDs.

39. 11 years (10 years 6 months) **41. a.** 16.1% **b.** 23.5% **43. a.** 4×4; There are four categories of animals and four categories of prey. **b.** No animal preys on the coyote. **c.** The rabbit doesn't prey on any of the other animals.

45. $1.06\begin{bmatrix} 25 & 26.3 & 27.5 & 28.6 \\ 30 & 31.7 & 32.9 & 34.2 \\ 35 & 36.8 & 38.2 & 40.3 \end{bmatrix} = \begin{bmatrix} 26.50 & 27.88 & 29.15 & 30.32 \\ 31.80 & 33.60 & 34.87 & 36.25 \\ 37.10 & 39.01 & 40.49 & 42.72 \end{bmatrix}$ **47.** $\begin{bmatrix} 41.25 & 122.85 \\ 39.65 & 118.00 \\ 50.70 & 149.70 \end{bmatrix}$

Prepare for Section 6.5, *page 547*

48. $-\frac{3}{2}$ **49.** $\begin{bmatrix} 1 & 0 & 0 \\ 0 & 1 & 0 \\ 0 & 0 & 1 \end{bmatrix}$ **50.** See Section 6.3. **51.** $\begin{bmatrix} 1 & -2 & 3 \\ 0 & 3 & -2 \\ 0 & -4 & 11 \end{bmatrix}$ **52.** $x = a^{-1}b$ **53.** $\begin{cases} 2x + 3y = 9 \\ 4x - 5y = 7 \end{cases}$

Exercises 6.5, *page 556*

1. Yes **3.** Yes **5.** $\begin{bmatrix} -5 & -3 \\ -2 & -1 \end{bmatrix}$ **7.** $\begin{bmatrix} 5 & -2 \\ -1 & \frac{1}{2} \end{bmatrix}$ **9.** $\begin{bmatrix} -16 & -2 & 7 \\ 7 & 1 & -3 \\ -3 & 0 & 1 \end{bmatrix}$ **11.** Singular matrix **13.** Singular matrix

15. $\begin{bmatrix} \frac{2}{25} & \frac{11}{25} \\ \frac{1}{25} & -\frac{7}{25} \end{bmatrix}$ **17.** $\begin{bmatrix} \frac{3}{35} & -\frac{11}{35} & \frac{1}{35} \\ \frac{6}{35} & \frac{13}{35} & \frac{2}{35} \\ \frac{1}{7} & \frac{1}{7} & -\frac{2}{7} \end{bmatrix}$ **19.** $\begin{bmatrix} \frac{19}{2} & -\frac{1}{2} & -\frac{3}{2} & \frac{3}{2} \\ \frac{7}{4} & \frac{1}{4} & -\frac{1}{4} & \frac{3}{4} \\ -\frac{7}{2} & \frac{1}{2} & \frac{1}{2} & -\frac{1}{2} \\ \frac{1}{4} & -\frac{1}{4} & \frac{1}{4} & \frac{1}{4} \end{bmatrix}$ **21.** (2, 1) **23.** (1, −1, 2) **25.** (23, −12, 3)

27. (0, 4, −6, −2) **29.** computer bug **31.** On Saturday 80 adults, 20 children; On Sunday 95 adults, 25 children **33.** Sample 1: 500 grams of additive 1, 200 grams of additive 2, 300 grams of additive 3; Sample 2: 400 grams of additive 1, 400 grams of additive 2, 200 grams of additive 3

Prepare for Section 6.6, *page 558*

35. $(-4, \infty)$ **36.** $y \le \frac{2}{3}x - 2$ **37.** $(-\infty, -6] \cup [1, \infty)$ **38.** $(1, 3)$ **39.**

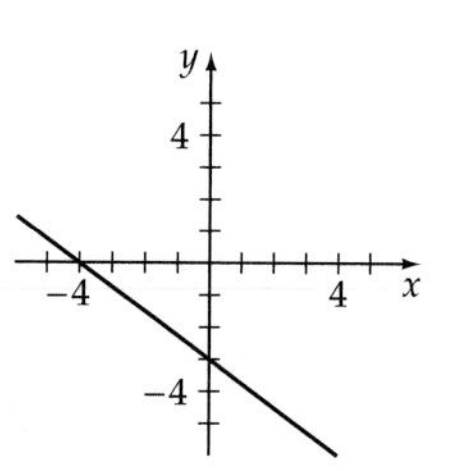

40. $(3, -1)$

Exercises 6.6, *page 568*

1.

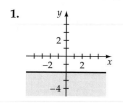

3.

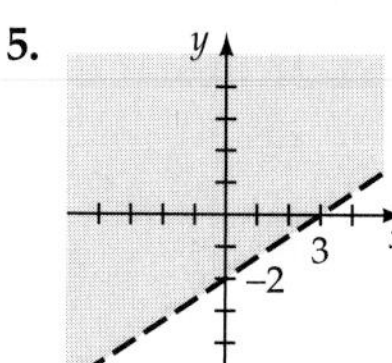

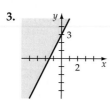

5.

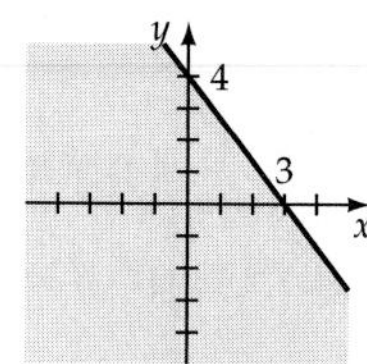

7.

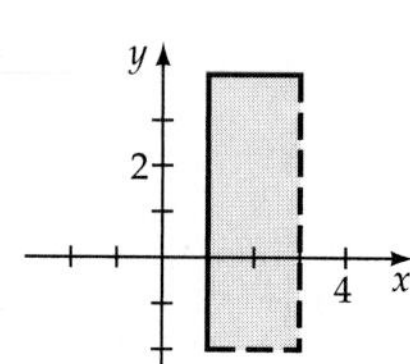

9.

11.

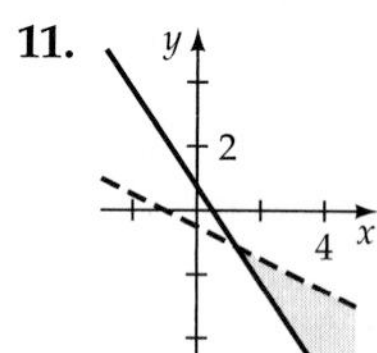

13. No solution

15.

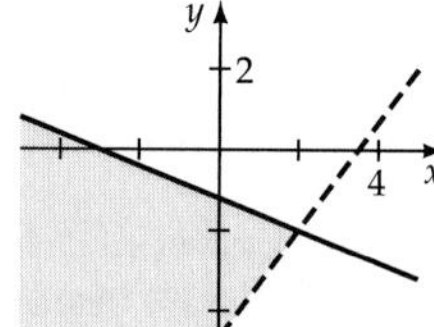

17.

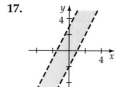

19.

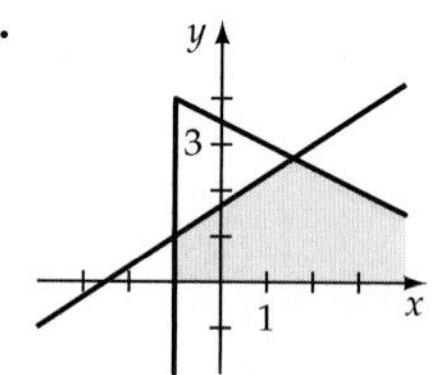

21.

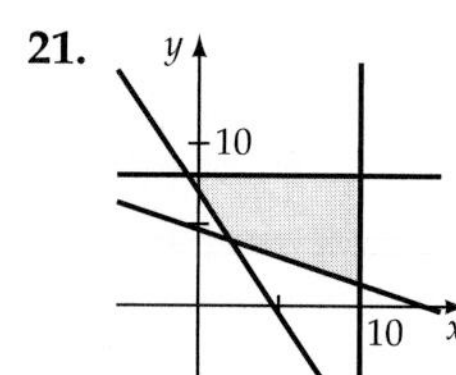

23.

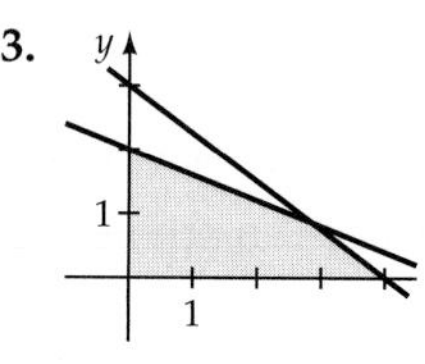

25. The minimum is 16 at (0, 8). **27.** The maximum is 71 at (6, 5).

29. The minimum is 20 at $\left(0, \frac{10}{3}\right)$. **31.** The maximum is 56 at (0, 8). **33.** The minimum is 18 at (2, 6). **35.** The maximum is 25 at (3, 4). **37.** The minimum is 12 at (2, 3). **39.** The maximum is 3400 at (100, 400). **41.** 20 acres of wheat; 40 acres of barley **43.** 0 starter clubs; 18 pro clubs

Chapter 6 True/False Exercises, *page 572*

1. False; $\begin{cases} x + y = 1 \\ x + y = 2 \end{cases}$ has no solution. **2.** True **3.** False; $\begin{cases} x + y = 2 \\ x + 2y = 3 \end{cases}$ and $\begin{cases} 2x + 3y = 5 \\ 2x - 2y = 0 \end{cases}$ are two systems with different equations but the same solution. **4.** True **5.** False; inconsistent **6.** False; if the number of equations is less than the number of variables, the Gaussian elimination method can be used to solve the system of linear equations. If the system of equations has a solution, the solution will be given in terms of an arbitrary constant. **7.** True **8.** True **9.** False; a singular matrix does not have a multiplicative inverse. **10.** False; $A^2 = A \cdot A = \begin{bmatrix} 7 & 18 \\ 6 & 19 \end{bmatrix}$.

Chapter 6 Review Exercises, *page 572*

1. $\left(-\frac{18}{7}, -\frac{15}{28}\right)$ [6.1] **2.** $\left(\frac{3}{2}, -3\right)$ [6.1] **3.** $(-3, -1)$ [6.1] **4.** $(-4, 7)$ [6.1] **5.** $(3, 1)$ [6.1] **6.** $(-1, 1)$ [6.1] **7.** $\left(\frac{5 - 3c}{2}, c\right)$ [6.1] **8.** No solution [6.1] **9.** $\left(\frac{1}{2}, 3, -1\right)$ [6.2] **10.** $\left(\frac{16}{3}, \frac{10}{27}, -\frac{29}{45}\right)$ [6.2] **11.** $\left(\frac{7c - 3}{11}, \frac{16c - 43}{11}, c\right)$ [6.2]

12. $\left(\frac{74}{31}, -\frac{1}{31}, \frac{3}{31}\right)$ [6.2] **13.** $\left(2, \dfrac{3c+2}{2}, c\right)$ [6.2] **14.** $\left(1, -\frac{2}{3}, \frac{1}{4}\right)$ [6.2] **15.** $\left(\dfrac{14c}{11}, -\dfrac{2c}{11}, c\right)$ [6.2] **16.** $(0, 0, 0)$ [6.2]

17. $\left(\dfrac{c+1}{2}, \dfrac{3c-1}{4}, c\right)$ [6.2] **18.** $\left(\dfrac{65-11c}{16}, \dfrac{19-c}{8}, c\right)$ [6.2] **19.** $\left[\begin{array}{rrr|r} 5 & -7 & 1 & -8 \\ 0 & 3 & 4 & 11 \\ -2 & -1 & 1 & 5 \end{array}\right], \begin{bmatrix} 5 & -7 & 1 \\ 0 & 3 & 4 \\ -2 & -1 & 1 \end{bmatrix}, \begin{bmatrix} -8 \\ 11 \\ 5 \end{bmatrix}$ [6.3]

20. $\left[\begin{array}{rrrr|r} -2 & 1 & -6 & 4 & 1 \\ 3 & 3 & 2 & -5 & -9 \\ -1 & -7 & 3 & -1 & 4 \\ 8 & 5 & 1 & -3 & -12 \end{array}\right], \begin{bmatrix} -2 & 1 & -6 & 4 \\ 3 & 3 & 2 & -5 \\ -1 & -7 & 3 & -1 \\ 8 & 5 & 1 & -3 \end{bmatrix}, \begin{bmatrix} 1 \\ -9 \\ 4 \\ -12 \end{bmatrix}$ [6.3] **21.** Infinite number of solutions [6.3]

22. Infinite number of solutions [6.3] **23.** $(2, -1)$ [6.3] **24.** $(1, -3)$ [6.3] **25.** $(3, 0)$ [6.3] **26.** $\left(\frac{40}{29}, -\frac{42}{29}\right)$ [6.3]

27. $(3, 1, 0)$ [6.3] **28.** $(2, 1, 3)$ [6.3] **29.** $(1, 0, -2)$ [6.3] **30.** $(0, 2, 3)$ [6.3] **31.** $(3, -4, 1)$ [6.3] **32.** $(4, -2, -2)$ [6.3]

33. $(-c-2, -c-3, c)$ [6.3] **34.** $(5, -2, 0)$ [6.3] **35.** $(1, -2, 2, 3)$ [6.3] **36.** $(2, 3, -1, 4)$ [6.3]

37. $(-37c+2, 16c, -7c+1, c)$ [6.3] **38.** $(63c+2, -14c+1, 5c, c)$ [6.3] **39.** $\begin{bmatrix} 6 & -3 & 9 \\ 9 & 6 & -3 \end{bmatrix}$ [6.4] **40.** $\begin{bmatrix} 0 & 4 \\ -8 & -4 \\ -2 & 6 \end{bmatrix}$ [6.4]

41. $\begin{bmatrix} -5 & 5 & -1 \\ 1 & -4 & 6 \end{bmatrix}$ [6.4] **42.** $\begin{bmatrix} 13 & -14 & 0 \\ -6 & 10 & -17 \end{bmatrix}$ [6.4] **43.** $\begin{bmatrix} -1 & -15 \\ 7 & 1 \end{bmatrix}$ [6.4] **44.** $\begin{bmatrix} 18 & 8 \\ -3 & -27 \end{bmatrix}$ [6.4]

45. $\begin{bmatrix} -6 & -4 & 2 \\ 14 & 0 & 10 \\ -7 & -7 & 6 \end{bmatrix}$ [6.4] **46.** $\begin{bmatrix} -8 & 4 & -10 \\ -4 & 12 & 18 \\ -15 & 10 & -13 \end{bmatrix}$ [6.4] **47.** $\begin{bmatrix} 12 & 28 & -5 \\ 2 & 6 & 0 \\ 6 & 16 & -1 \end{bmatrix}$ [6.4] **48.** $\begin{bmatrix} 42 & 108 & -11 \\ 10 & 24 & -4 \\ 26 & 64 & -9 \end{bmatrix}$ [6.4]

49. $\begin{bmatrix} -12 & -36 & -4 \\ 48 & 124 & 4 \\ -9 & -32 & -6 \end{bmatrix}$ [6.4] **50.** Not possible [6.4] **51.** Not possible [6.4] **52.** $\begin{bmatrix} 7 & 24 & 9 \\ -10 & -22 & 1 \end{bmatrix}$ [6.4]

53. $\begin{bmatrix} 13 & -2 & 0 & 42 \\ 19 & -19 & 29 & 22 \\ -26 & -3 & -40 & 4 \\ 0 & 42 & 12 & -17 \end{bmatrix}$ [6.4] **54.** $\begin{bmatrix} -345 & 117 & 546 & 171 \\ -45 & -511 & 166 & 97 \\ -69 & -285 & 396 & -195 \\ -288 & -208 & 268 & -490 \end{bmatrix}$ [6.4] **55.** $\begin{bmatrix} -154 & -30 & -230 & -49 \\ -127 & -211 & -180 & 153 \\ 19 & 212 & -83 & -97 \\ 105 & -38 & 338 & -90 \end{bmatrix}$ [6.4]

56. $\begin{bmatrix} -9 & 12 & -33 & -54 \\ 27 & -39 & -36 & 19 \\ 12 & 36 & -60 & 3 \\ 48 & 71 & -26 & 15 \end{bmatrix}$ [6.4] **57.** $\begin{bmatrix} \frac{180}{781} & -\frac{238}{781} & -\frac{144}{781} & -\frac{35}{781} \\ \frac{74}{781} & \frac{41}{781} & \frac{97}{781} & \frac{29}{781} \\ -\frac{85}{781} & \frac{69}{781} & \frac{68}{781} & \frac{125}{781} \\ -\frac{15}{781} & \frac{150}{781} & \frac{12}{781} & \frac{68}{781} \end{bmatrix}$ [6.5] **58.** $\begin{bmatrix} \frac{179}{3432} & \frac{27}{1144} & \frac{34}{429} & -\frac{157}{1144} \\ \frac{19}{264} & -\frac{5}{88} & -\frac{1}{33} & \frac{3}{88} \\ \frac{115}{3432} & -\frac{21}{1144} & \frac{53}{429} & -\frac{5}{1144} \\ \frac{21}{572} & \frac{83}{572} & -\frac{8}{143} & \frac{47}{572} \end{bmatrix}$ [6.5]

59. $\begin{cases} 4x + 5y = -8 \\ 3x - 4y = 10 \end{cases}$ [6.5] **60.** $\begin{cases} -x + 6y - z = 4 \\ 3x + 2y + 2z = -8 \\ 3x - 5y + 4z = 1 \end{cases}$ [6.5] **61.** $\begin{bmatrix} -1 & 1 \\ -\frac{3}{2} & 1 \end{bmatrix}$ [6.5] **62.** $\begin{bmatrix} 3 & -4 \\ -2 & 3 \end{bmatrix}$ [6.5]

63. $\begin{bmatrix} -\frac{2}{7} & \frac{3}{14} \\ \frac{1}{7} & \frac{1}{7} \end{bmatrix}$ [6.5] **64.** $\begin{bmatrix} \frac{1}{11} & \frac{2}{11} \\ -\frac{3}{22} & \frac{5}{22} \end{bmatrix}$ [6.5] **65.** $\begin{bmatrix} 2 & -2 & 1 \\ 0 & \frac{3}{2} & -1 \\ -1 & -1 & 1 \end{bmatrix}$ [6.5] **66.** $\begin{bmatrix} 27 & -12 & 5 \\ 4 & -2 & 1 \\ -7 & 3 & -1 \end{bmatrix}$ [6.5]

67. $\begin{bmatrix} -10 & 20 & -3 \\ -5 & 9 & -1 \\ 3 & -6 & 1 \end{bmatrix}$ [6.5] **68.** $\begin{bmatrix} 27 & -39 & 5 \\ -7 & 10 & -1 \\ 4 & -6 & 1 \end{bmatrix}$ [6.5] **69. a.** $(18, -13)$ **b.** $(-22, 16)$ [6.5]

70. a. $(41, 17)$ **b.** $(-39, -16)$ [6.5] **71. a.** $\left(-\frac{18}{7}, \frac{23}{7}, -\frac{6}{7}\right)$ **b.** $\left(-\frac{31}{14}, \frac{20}{7}, \frac{3}{7}\right)$ [6.5] **72. a.** $\left(-\frac{4}{3}, -1, 2\right)$ **b.** $(-9, -11, 6)$ [6.5]

73. 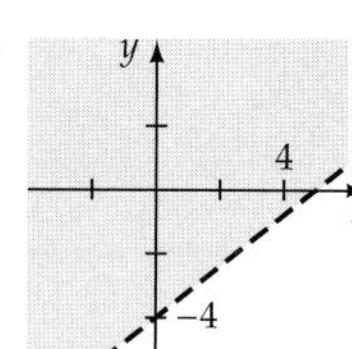

[6.6] **74.** 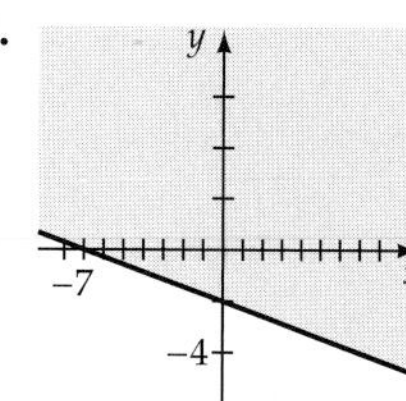

[6.6] **75.** 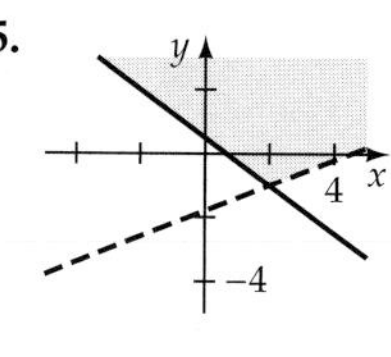

[6.6] **76.** 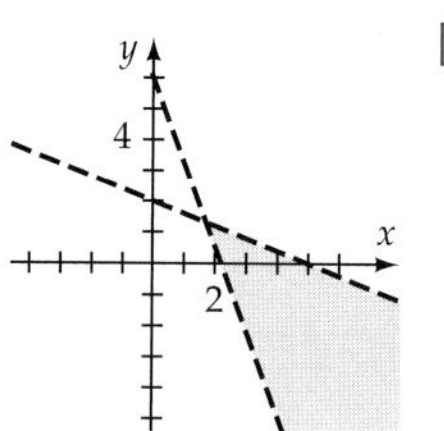

[6.6]

77. 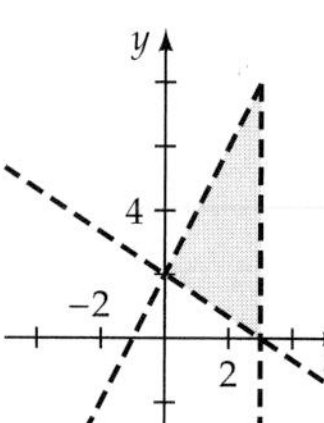

[6.6] **78.** 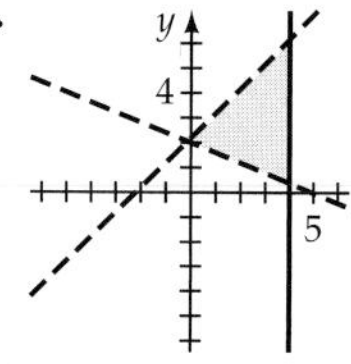

[6.6] **79.** 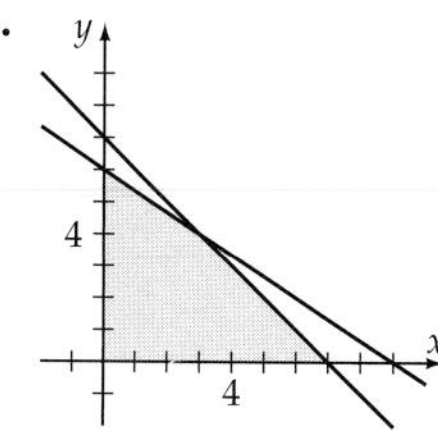

[6.6] **80.** 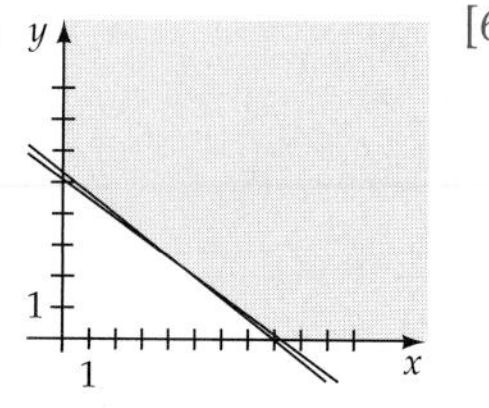

[6.6]

81. 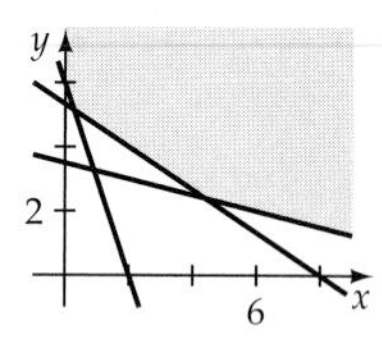

[6.6] **82.** 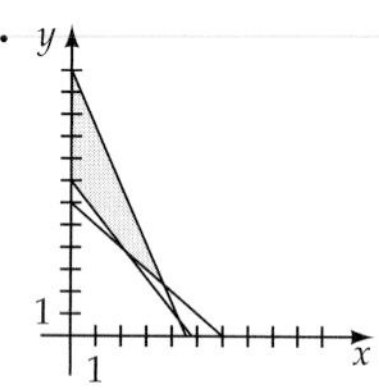

[6.6] **83.** The maximum is 18 at $(4, 5)$. [6.6]

84. The maximum is 44 at $(6, 4)$. [6.6] **85.** The minimum is 8 at $(0, 8)$. [6.6] **86.** The minimum is 20 at $(10, 0)$. [6.6] **87.** The minimum is 27 at $(2, 5)$. [6.6] **88.** The maximum is 43 at $(3, 7)$. [6.6] **89.** $y = x^2 + 3x - 2$ [6.2] **90.** $y = -x^2 - 2x + 3$ [6.2] **91.** \$22,000 at 6.5%; \$8000 at 8% [6.1] **92.** \$15,000 at 4%; \$15,500 at 6%; \$19,500 at 7% [6.2/6.3] **93.** Bibs, \$16; tongue depressors, \$5; headrest covers, \$24 [6.2/6.3] **94.** Meat, 3 ounces; potatoes, 5 ounces; green beans, 4 ounces [6.2/6.3] **95.** 47.5% [6.4] **96.** 15 weeks [6.4] **97.** Plane: 143 miles per hour; wind: 28 miles per hour [6.1] **98.** 5 hours [6.1] **99.** *CA* [6.4] **100.** *DB* [6.4] **101.** $C + D + E$ [6.4] **102.** 15 liters [6.1] **103.** $z = -2x + 3y + 3$ [6.2/6.3] **104.** $y = 0.5x^3 - 2x^2 + 3x - 5$ [6.3]

Chapter 6 Test, *page 576*

1. $(-3, 2)$ [6.1] **2.** $\left(\frac{6+c}{2}, c\right)$ [6.1] **3.** $\left(\frac{173}{39}, \frac{29}{39}, -\frac{4}{3}\right)$ [6.2] **4.** $\left(\frac{c+10}{13}, \frac{5c+11}{13}, c\right)$ [6.2] **5.** $\left(\frac{c}{14}, -\frac{9c}{14}, c\right)$ [6.2]

6. $\left[\begin{array}{rrr|r} 2 & 3 & -3 & 4 \\ 3 & 0 & 2 & -1 \\ 4 & -4 & 2 & 3 \end{array}\right], \begin{bmatrix} 2 & 3 & -3 \\ 3 & 0 & 2 \\ 4 & -4 & 2 \end{bmatrix}, \begin{bmatrix} 4 \\ -1 \\ 3 \end{bmatrix}$ [6.3] **7.** $\begin{cases} 3w - 2x + 5y - z = 9 \\ 2w + 3x - y + 4z = 8 \\ w + 3y + 2z = -1 \end{cases}$ [6.3] **8.** $(2, -1, 2)$ [6.3]

9. $(3c - 5, -7c + 14, -3c + 4, c)$ [6.3] **10.** $x^2 + y^2 - 2y - 24 = 0$ [6.2/6.3] **11. a.** $\begin{bmatrix} 3 & -9 & -6 \\ -3 & -12 & 3 \end{bmatrix}$ **b.** $A + B$ is not defined. **c.** $\begin{bmatrix} 4 & 1 & 3 \\ 8 & 0 & -19 \\ 11 & 0 & 10 \end{bmatrix}$ [6.4] **12. a.** $\begin{bmatrix} 16 & -1 & -2 \\ 15 & -11 & -3 \end{bmatrix}$ **b.** *CA* is not defined. [6.4] **13. a.** A^2 is not defined.

b. $\begin{bmatrix} 9 & 6 & 13 \\ -3 & -2 & 12 \\ 20 & -3 & 11 \end{bmatrix}$ [6.4] **14.** $\begin{bmatrix} 18 & -5 & 7 \\ 4 & -1 & 2 \\ -3 & 1 & -1 \end{bmatrix}$ [6.5] **15.** $\begin{bmatrix} -2.5 & 2 \\ 1.5 & -1 \end{bmatrix}$ [6.5] **16.** $(-5, 4)$ [6.5]

17. 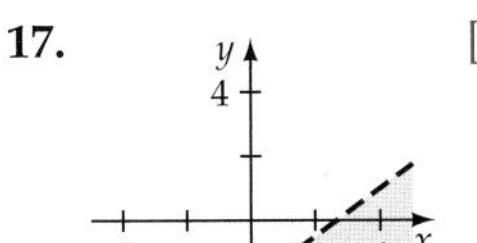 [6.6] **18.** 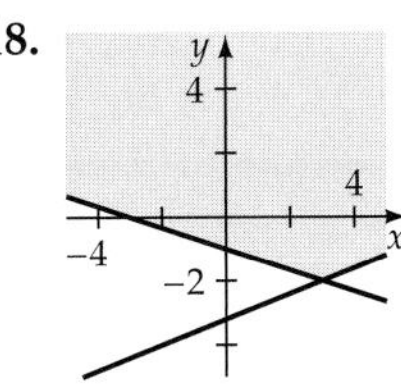 [6.6] **19.** 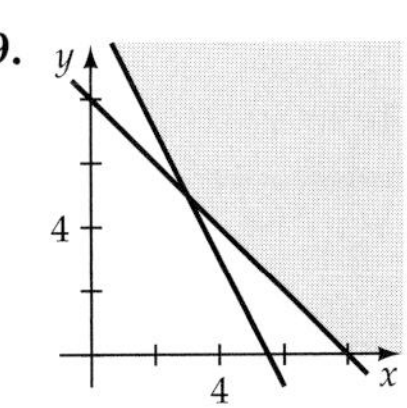 [6.6] **20.** $\frac{680}{7}$ acres of oats [6.6]

Cumulative Review Exercises, *page 577*

1. $\dfrac{-4(x+4)}{(x+3)(x-1)}$ [P.5] **2.** $x = \dfrac{1 \pm \sqrt{13}}{3}$ [1.2] **3.** $(-1, 3]$ [1.4] **4.** $y = -\frac{3}{2}x + 2$ [2.4] **5.** $(-2, -13)$ [2.5] **6.** Odd [3.1] **7.** $25x^2 + 5x - 2$ [3.2] **8.** $f^{-1}(x) = \sqrt[3]{\dfrac{2-x}{7}}$ [3.2] **9.** $c = \frac{1}{2}a\sqrt{b}$ [3.4] **10.** No [4.1] **11.** Down to far left, down to far right. [4.2] **12.** $P(x) = (x+1)(x-1)(x-3)$ [4.4] **13.** $x = 2, x = 4, y = 2$ [4.5] **14.** $x = -2$ [5.4] **15.** $x = \frac{1}{2}e^2$ [5.4] **16.** $\ln \dfrac{x^3\sqrt{y}}{z^2}$ [5.3] **17.** $N(t) = 36{,}400e^{0.02428t}$ [5.5] **18.** $(-1, 2)$ [6.1] **19.** $(3, 1, -2)$ [6.3] **20.** $\begin{bmatrix} \frac{1}{2} & \frac{1}{2} \\ \frac{3}{4} & \frac{5}{4} \end{bmatrix}$ [6.5]

Chapter 7

Chapter Prep Quiz, *page 580*

1. $\frac{1}{4}, -\frac{1}{9}$ **2.** 1530.23 **3.** 0, 12 **4.** $\dfrac{x(x+2)}{(x+1)^2}$ **5.** $\frac{1}{2}$ **6.** $\frac{1}{4}$ **7.** 3 **8.** 2 **9.** 0.1 **10.** $x = \dfrac{\log 5}{\log 2} \approx 2.32$

Exercises 7.1, *page 586*

1. $1, 2, 3, a_8 = 8$ **3.** $1, \frac{1}{2}, \frac{1}{3}, a_8 = \frac{1}{8}$ **5.** $\frac{1}{2}, \frac{2}{3}, \frac{3}{4}, a_8 = \frac{8}{9}$ **7.** $\frac{1}{2}, -\frac{1}{4}, \frac{1}{6}, a_8 = -\frac{1}{16}$ **9.** $\frac{1}{2}, \frac{1}{4}, \frac{1}{8}, a_8 = \frac{1}{256}$ **11.** $0, 2, 0, a_8 = 2$ **13.** $1.06, 1.1236, 1.191016, a_8 \approx 1.5938481$ **15.** $2, \frac{64}{27}, a_8 = \frac{43{,}046{,}721}{16{,}777{,}216} \approx 2.56578$ **17.** $1, 2, 6, a_8 = 40{,}320$ **19.** $1, \sqrt{2}, \sqrt[3]{3}, a_8 = \sqrt[8]{8}$ **21.** 3, 12, 48 **23.** 2, −4, 12 **25.** 2, 1, 3 **27.** 2, 8, 48 **29.** 3, 9, 81 **31.** 4, 7, 11, 18, 29 **33.** 21 **35.** 70 **37.** $\frac{25}{12}$ **39.** 72 **41.** −24 **43.** 256 **45.** $\displaystyle\sum_{k=1}^{6} k$ **47.** $\displaystyle\sum_{i=1}^{10} \frac{1}{2^i}$ **49.** $\displaystyle\sum_{k=1}^{8} \frac{(-1)^{k-1}}{k}$ **51.** −60 **53.** −140 **55.** 2.645752048 **57.** 55 + 21 + 5 **59.** $F_{10} = 55, F_{15} = 610$

Prepare for Section 7.2, *page 589*

60. 1275 **61.** 52 **62.** $-\frac{3}{4}$ **63.** $-\frac{3}{2}$ **64.** −5 **65.** $-\frac{1}{2}$

Exercises 7.2, *page 601*

1. $a_{10} = 19, a_{25} = 49, a_n = 2n - 1$ **3.** $a_{10} = 1, a_{25} = -14, a_n = 11 - n$ **5.** $a_{10} = 21, a_{25} = 66, a_n = 3n - 9$ **7.** $a_{10} = 46, a_{25} = 121, a_n = 5n - 4$ **9.** $a_{20} = 78$ **11.** $a_{17} = 87$ **13.** 100 **15.** 375 **17.** 390 **19.** −810 **21.** Geometric, $r = 3$ **23.** Neither **25.** Arithmetic, $d = 4$ **27.** Neither **29.** Geometric, $r = -\frac{1}{3}$ **31.** $3(4)^{n-1}$ **33.** $2(-3)^{n-1}$ **35.** $\left(\frac{3}{4}\right)^{n-1}$ **37.** $-9\left(-\frac{2}{3}\right)^{n-1}$ **39.** $\frac{3}{10}\left(\frac{1}{10}\right)^{n-1}$ **41.** $0.6(0.1)^{n-1}$ **43.** $0.25(0.01)^{n-1}$ **45.** $a_3 = 18$ **47.** $a_2 = -2$ **49.** 62 **51.** $\frac{10{,}374}{15{,}625}$ **53.** $\frac{171}{256}$ **55.** 147,620 **57.** 3 **59.** $-\frac{1}{4}$ **61.** 0.25 **63.** $\frac{10}{13}$ **65.** $\frac{2}{9}$ **67.** $\frac{5}{11}$ **69.** $\frac{1}{6}$ **71.** $\frac{41}{333}$ **73.** $\frac{229}{900}$ **75.** 1345, 6677 **79.** \$1500; \$48,750 **81.** 12 levels **83.** 17.68% **85.** 20.25% **87.** \$500 million **89.** \$833,333.33 **91.** 32 cross threads **93.** 0.516 milligrams **95.** 45 seats, 360 seats **97.** 0.599 **99.** 260 bricks **101.** 20 logs, 135 logs **103.** 19 feet, 100 feet

Prepare for Section 7.3, *page 605*

105. 1.1025 **106.** $(1 + x)^3$ **107.** 0.1 **108.** 0.2 **109.** 824.36 **110.** 14.2

Exercises 7.3, *page 615*

1. $250 **3.** $750 **5.** $10,666.67 **7.** $6.93 **9.** 20 pounds **11.** $70.28 **13.** 416% **15.** $2750 **17.** $1250 **19.** 10.6% **21. a.** $1400 **b.** $1469.33 **c.** $1480.24 **d.** $1485.95 **e.** $1489.85 **f.** $1491.76 **23. a.** $3515.63 **b.** $3433.03 **c.** $3417.35 **d.** $3409.27 **e.** $3403.79 **f.** $3401.12 **25.** $2.11 **27.** 7.52% **29.** 7.77% **31.** 11 years 1 month **33.** 13.9% **35.** $239.16 **37.** No. The percent increase will always be the same, no matter what the original amount. **39.** 24.57% **41.** The second investment. The first earns 14.87%; the second earns 14.93%.

Prepare for Section 7.4, *page 618*

42. 2.49 **43.** $\dfrac{x^2 - 1}{x^3}$ **44.** $b = a(1 + i)^{-n}$ **45.** $\dfrac{1}{1.1}$ **46.** $\dfrac{85}{64}$ **47.** 7.7217

Exercises 7.4, *page 627*

1. $4087.86 **3.** $82.86 **5.** $49,002.30 **7.** 16.9% **9.** $307.75 **11.** Five annual payments of $9500; present value = $37,930.75 **13.** $370.28 **15.** $6909.35 **17.** $236,544.10 **19.** $320,713.20 **21.** 18 months **23.** $456.29 **25.** $5,124,150.29 **27.** $15,631.38 **29.** 10,536.76 **31. a.** $1834.41 **b.** Interest = $1666.67, principal = $167.74 **c.** Interest = $1665.55, principal = $168.86 **33. a.** $117.71 **b.** 12.0%

Prepare for Section 7.5, *page 629*

35. 4.31 **36.** $b = a\left[\dfrac{i}{(1 + i)^n - 1}\right]$ **37.** 0.57 **38.** 30.62 **39.** 1.06 **40.** 31.87

Exercises 7.5, *page 636*

1. $30,200.99 **3.** $51,622.68 **5.** $134.33 **7.** $42,952.88 **9.** $1989.73 **11.** $2535.03 **13. a.** $15,073.74 **b.** $30,352.75 **15.** 20.36% **17. a.** $36,352.34 **b.** $54,228.94 **19.** $485

Chapter 7 True/False Exercises, *page 640*

1. False. $\sum_{i=1}^{5} 2 \cdot 3 = 6 + 6 + 6 + 6 + 6 = 30$ $\sum_{i=1}^{5} 2 \sum_{i=1}^{5} 3 = (2 + 2 + 2 + 2 + 2) \cdot (3 + 3 + 3 + 3 + 3) = 150$ **2.** False. In a constant sequence, all of the terms are equal. **3.** False. There is no common ratio of successive terms. **4.** True **5.** False. The sum of the infinite sequence $1, \frac{1}{2}, \frac{1}{4}, \frac{1}{8}, \ldots$ is 2. **6.** False. The interest rate is a percent that, when changed to a decimal and multiplied by the principal, yields the interest. **7.** True **8.** True **9.** True **10.** True

Chapter 7 Review Exercises, *page 640*

1. $a_3 = 9, a_7 = 49$ [7.1] **2.** $a_3 = 6, a_7 = 5040$ [7.1] **3.** $a_3 = 11, a_7 = 23$ [7.1] **4.** $a_3 = -5, a_7 = -13$ [7.2] **5.** $a_3 = \frac{1}{8}, a_7 = \frac{1}{128}$ [7.1/7.2] **6.** $a_3 = 27, a_7 = 2187$ [7.2] **7.** $a_3 = \frac{1}{6}, a_7 = \frac{1}{5040}$ [7.1] **8.** $a_3 = \frac{1}{3}, a_7 = \frac{1}{7}$ [7.1] **9.** $a_3 = \frac{8}{27}, a_7 = \frac{128}{2187}$ [7.1] **10.** $a_3 = -\frac{64}{27}, a_7 = -\frac{16,384}{2187}$ [7.1] **11.** $a_3 = 18, a_7 = 1458$ [7.1/7.2] **12.** $a_3 = -4, a_7 = -64$ [7.1/7.2] **13.** $a_3 = 6, a_7 = 5040$ [7.1] **14.** $a_3 = 72, a_7 = 50,803,200$ [7.1] **15.** $a_3 = 8, a_7 = 16$ [7.1/7.2] **16.** $a_3 = -3, a_7 = -15$ [7.1/7.2] **17.** $a_3 = 2, a_7 = 256$ [7.1] **18.** $a_3 = 2, a_7 = 1$ [7.1] **19.** $a_3 = -54, a_7 = -3,674,160$ [7.1] **20.** $a_3 = 48, a_7 = 645,120$ [7.1] **21.** Arithmetic [7.2] **22.** Arithmetic [7.2] **23.** Arithmetic [7.2] **24.** Neither [7.1] **25.** Geometric [7.2] **26.** Geometric [7.2] **27.** Geometric [7.2] **28.** Geometric [7.2] **29.** Geometric [7.2] **30.** Geometric [7.2] **31.** Neither [7.1] **32.** Neither [7.1] **33.** Geometric [7.2] **34** Geometric [7.2] **35.** Geometric [7.2] **36.** Geometric [7.2] **37.** $7.55 [7.3] **38.** 624% [7.3] **39.** $4875 [7.3] **40.** $916.67 [7.3] **41.** $3947.37 [7.3]

42. a. \$4187.50 **b.** \$4804.18 **c.** \$4855.68 **d.** \$4882.51 **e.** \$4900.80 **f.** \$4909.78 [7.3] **43. a.** \$10,549.45 **b.** \$10,505.80 **c.** \$10,482.60 **d.** \$10,470.64 **e.** \$10,462.52 **f.** \$10,458.55 [7.3] **44.** \$3.07 [7.3] **45.** 10.52% [7.3] **46.** 6.9% [7.3] **47.** \$13,080.07 [7.4] **48.** \$208.63 [7.4] **49.** \$33,128.64 [7.4] **50.** \$5098.27 [7.4] **51. a.** \$3942.25 **b.** Interest = \$3437.50; principal = \$504.75 [7.4] **52.** \$259,422.88 [7.4] **53. a.** \$100.83 **b.** 12.8% [7.4] **54.** \$84,997.58 [7.4] **55.** \$202.20 [7.5] **56.** \$19.60 [7.2] **57.** \$41.57 [7.2] **58.** \$75,000,000 [7.2] **59.** 33 cross threads [7.2] **60.** 778.8 millimeters [7.2] **61.** 1.73 milligrams [7.2]

Chapter 7 Test, *page 643*

1. $a_3 = \frac{1}{6}, a_5 = \frac{1}{10}$ [7.1] **2.** $a_3 = 12, a_5 = 48$ [7.1] **3.** Arithmetic [7.2] **4.** Neither [7.2] **5.** 60 [7.2] **6.** 58 [7.2] **7.** 590 [7.2] **8.** $\frac{2{,}968{,}581}{1{,}048{,}576} \approx 2.8310595$ [7.2] **9.** $\frac{8}{9}$ [7.2] **10.** \$1240 [7.3] **11. a.** \$4587.79 **b.** \$4630.83 [7.3] **12.** 10.76% [7.3] **13.** \$30,464.91 [7.4] **14.** \$333.69 [7.4] **15.** \$5429.92 [7.4] **16.** \$997.95 [7.4] **17.** \$875 [7.4] **18.** \$46.25 [7.4] **19.** 10.2% [7.4] **20.** \$680.49 [7.5]

Cumulative Review Exercises, *page 644*

1. $-2 \pm \sqrt{10}$ [1.2] **2.** $\left\{x \mid x < -\frac{2}{3}\right\} \cup \{x \mid x > 2\}$ [1.4] **3.** $\frac{1}{25}$ [5.2] **4.** $\frac{1}{2} \ln 2$ [5.4] **5.** $-\frac{3}{4}$ [2.3], 5 [2.1] **6.** $y = -\frac{1}{2}x$ [2.4] **7.** $(-1, 2)$ [2.5] **8.** Odd [3.3] **9.**

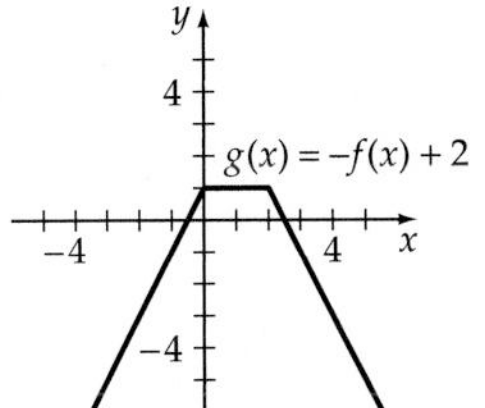

[3.3] **10.** 51 [3.1] **11.** $f^{-1}(x) = \frac{1}{2}x - 2$ [3.2] **12.** Yes [4.1] **13.** −2 and 1 [4.3] **14.** $2 - i$ [4.4] **15.** $x^3 + x^2 + x + 1$ [4.4] **16.** VA: $x = -1$, HA: $y = 2$ [4.5] **17.** $(1, -2)$ [6.1] **18.** 40 pounds [3.4] **19.** 4.06 milligrams [5.5] **20.** 560 milligrams of B1, 120 milligrams of B2 [6.1]

Chapter 8

Chapter Prep Quiz, *page 646*

1. $\{2, 3, 4, 5, 6, 7, 8, 11\}$ **2.** $\varnothing$ **3.** 0.0512 **4.** $\dfrac{211}{243}$ **5.** 40,320 **6.** 120,960 **7.** 11,880 **8.** 2002 **9.** $x^3 + 6x^2y + 12xy^2 + 8y^3$ **10.** $16x^4 - 32x^3y + 24x^2y^2 - 8xy^3 + y^4$

Exercises 8.1, *page 655*

1. $\{0, 2, 4, 6, 8\}$ **3.** {Monday, Tuesday, Wednesday, Thursday, Friday, Saturday, Sunday}

In Exercises 5 and 7, H denotes a head and T denotes a tail.

5. {HH, HT, TH, TT}

7. {1H, 1T, 2H, 2T, 3H, 3T, 4H, 4T, 5H, 5T, 6H, 6T} **9.** Assume the seats are filled in the order A1, A2, A3. Using A for Andy, B for Bob, and C for Cassidy, the sample space is {ABC, ACB, BAC, BCA, CAB, CBA}.

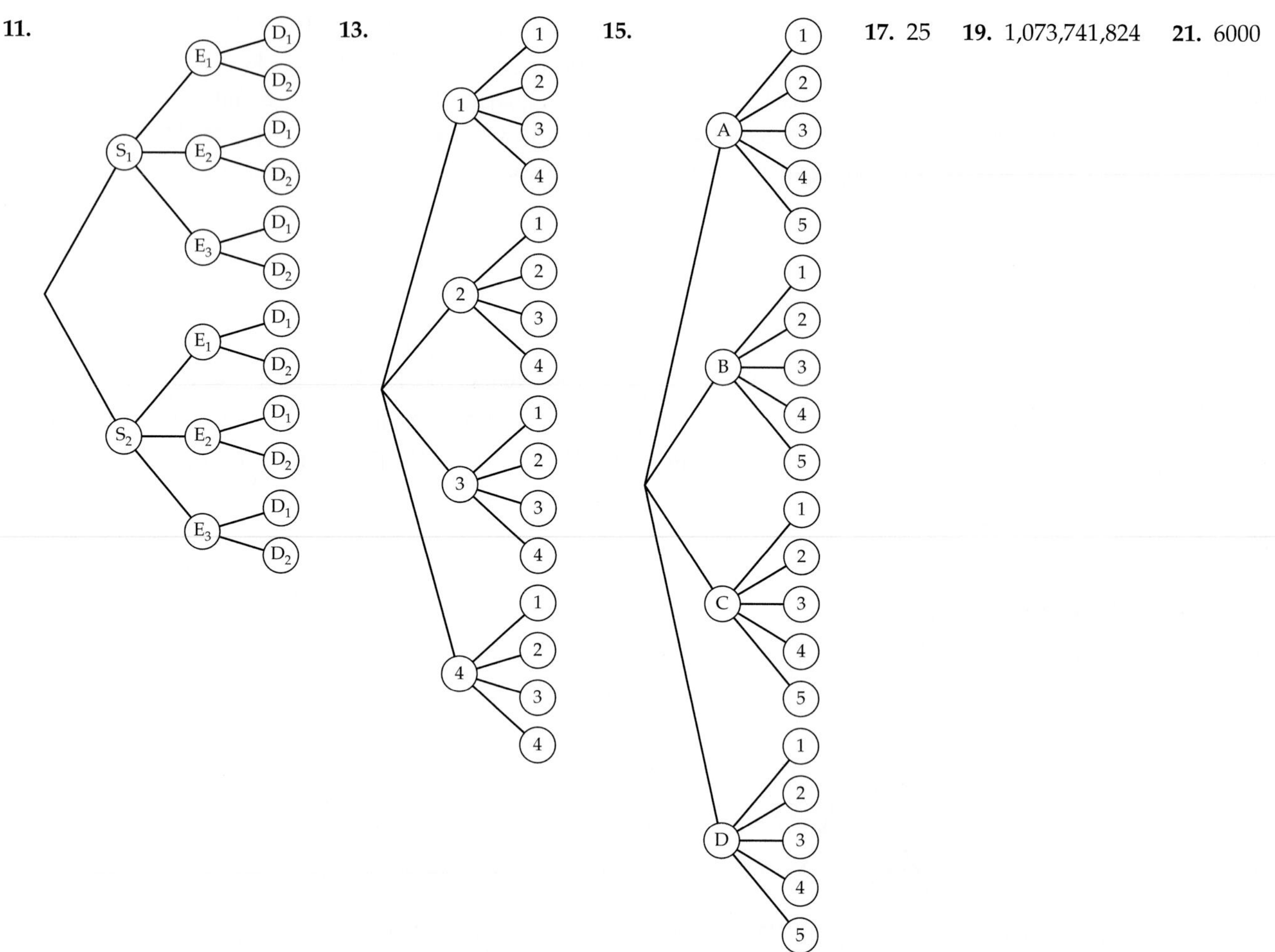

17. 25 **19.** 1,073,741,824 **21.** 6000

23. 24 **25.** 15 **27.** 40,320 **29.** 362,760 **31.** 60 **33.** 6720 **35.** 210 **37.** 60,480 **39.** 1 **41.** 36 **43.** 1 **45.** 525 **47.** 336 **49.** 40,320 **51.** 21 **53.** 6840 **55.** 240 **57.** 158,184,000 **59.** 5040 **61. a.** 362,880 **b.** 80,640 **63.** 210 **65. a.** 518,400 **b.** 3,628,800 **67.** 43,680 **69. a.** 8008 **b.** 3136 **71. a.** 3,628,800 **b.** 1024 **c.** 252 **73. a.** 518,400 **b.** 16,777,216 **c.** 4,330,260 **75.** 504 **77.** 18,564 **79.** 48,620 **81.** 720 **83.** 24 **85.** 103,776 **87.** 11,232,000

Prepare for Section 8.2, *page 658*

89. {1, 2, 4, 6, 7, 8, 10, 13}; {4, 10} **90.** No **91. a.** No **b.** Yes **92.** When the intersection of the sets is the empty set. **93.** 1680 **94. a.** 20 **b.** 10

Exercises 8.2, *page 666*

1. {HHH, HHT, HTH, THH, HTT, THT, TTH, TTT}

3. {11/1, 11/2, 11/3, 11/4, 11/5, 11/6, 11/7, 11/8, 11/9, 11/10, 11/11, 11/12, 11/13, 11/14}

5. {Alaska, Alabama, Arizona, Arkansas} **7.** {BBB, BBG, BGB, GBB, BGG, GBG, GGB, GGG} **9.** {BGG, GBG, GGB, GGG} **11.** {BBG, BGB, GBB, BGG, GBG, GGB, GGG} **13.** $\frac{1}{2}$ **15.** $\frac{7}{8}$ **17.** $\frac{1}{4}$ **19.** $\frac{1}{8}$ **21.** $\frac{11}{16}$ **23.** $\frac{5}{36}$ **25.** $\frac{1}{36}$ **27.** 0 **29.** $\frac{1}{6}$ **31.** $\frac{1}{2}$ **33.** $\frac{1}{6}$ **35.** 0.20 **37.** 0.12 **39.** 0.49 **41.** 0 **43.** $\frac{3}{4}$ **45.** $\frac{2190}{3228} \approx 0.68$ **47.** $\frac{1651}{3228} \approx 0.51$ **49.** $\frac{432}{3228} \approx 0.13$ **51.** $\frac{137}{425} \approx 0.32$

53. $\frac{103}{850} \approx 0.12$ **55.** $\frac{125}{216}$ **57.** $\frac{1}{4}$ **59.** $\frac{3}{13}$ **61.** $\frac{2}{13}$ **63.** $\frac{24}{2{,}598{,}960} \approx 9.23 \times 10^{-6}$ **65.** $\frac{267{,}696}{2{,}598{,}960} \approx 0.103$ **67.** $\frac{9}{19}$ **69.** $\frac{1}{38}$ **71.** $\frac{3}{19}$

Prepare for Section 8.3, *page 669*

72. $\{1, 2, 3, 5, 7, 8, 11\}$; $\{2, 3, 5\}$ **73.** 0 **74.** $\frac{91}{216}$ **75.** $\frac{560}{1287} \approx 0.435$ **76.** $\{1, 2, 4, 5, 6\}$ **77.** $\frac{3}{4}$

Exercises 8.3, *page 679*

1. $\frac{2}{13}$ **3.** $\frac{1}{9}$ **5.** $\frac{5}{18}$ **7.** $\frac{1}{2}$ **9.** $\frac{19}{36}$ **11.** $\frac{4}{13}$ **13.** $\frac{3}{13}$ **15.** $\frac{3}{4}$ **17.** $\frac{5}{6}$ **19.** $\frac{11}{12}$ **21.** $\frac{12}{13}$ **23.** $\frac{15}{16}$ **25.** $\frac{25}{1296}$ **27.** $\frac{1}{72}$ **29.** $\frac{1}{4}$ **31.** $\frac{1}{216}$
33. $\frac{91}{216}$ **35.** $\frac{217}{729}$ **37.** $\frac{1}{169}$ **39.** $\frac{1}{16}$ **41.** $\frac{12}{169}$ **43.** $\frac{25}{216}$ **45.** $\frac{105}{512}$ **47.** ≈ 0.126 **49.** $\frac{1150}{3179} \approx 0.362$ **51.** $\frac{164}{3179} \approx 0.052$ **53.** $\frac{1}{3}$
55. ≈ 0.000166 **57.** ≈ 0.06 **59.** ≈ 0.276 **61.** ≈ 0.077 **63. a.** $\frac{12{,}285}{4^{15}} \approx 1.14 \times 10^{-5}$ **b.** $\frac{991}{4^{15}} \approx 9.23 \times 10^{-7}$ **65.** $\frac{1}{41{,}416{,}353}$
67. $\frac{2870}{13{,}805{,}451} \approx 0.000208$ **69.** $\frac{1}{80{,}089{,}128}$

Prepare for Section 8.4, *page 681*

71. 3024 **72.** 56 **73.** 35 **74.** $a^3 + 3a^2b + 3ab^2 + b^3$
75. In a permutation, order matters. In a combination, order does not matter. **76.** 4

Exercises 8.4, *page 687*

1. 35 **3.** 36 **5.** 220 **7.** 1 **9.** $x^6 - 6x^5y + 15x^4y^2 - 20x^3y^3 + 15x^2y^4 - 6xy^5 + y^6$
11. $x^5 + 15x^4 + 90x^3 + 270x^2 + 405x + 243$ **13.** $128x^7 - 448x^6 + 672x^5 - 560x^4 + 280x^3 - 84x^2 + 14x - 1$
15. $x^6 + 18x^5y + 135x^4y^2 + 540x^3y^3 + 1215x^2y^4 + 1458xy^5 + 729y^6$ **17.** $16x^4 - 160x^3y + 600x^2y^2 - 1000xy^3 + 625y^4$
19. $x^6 + 6x^4 + 15x^2 + 20 + \frac{15}{x^2} + \frac{6}{x^4} + \frac{1}{x^6}$ **21.** $x^{14} - 28x^{12} + 336x^{10} - 2240x^8 + 8960x^6 - 21{,}504x^4 + 28{,}672x^2 - 16{,}834$
23. $32x^{10} + 80x^8y^3 + 80x^6y^6 + 40x^4y^9 + 10x^2y^{12} + y^{15}$ **25.** $\frac{16}{x^4} - \frac{16}{x^2} + 6 - x^2 + \frac{x^4}{16}$
27. $s^{-12} + 6s^{-8} + 15s^{-4} + 20 + 15s^4 + 6s^8 + s^{12}$ **29.** $-3240x^3y^7$ **31.** $1056x^{10}y^2$ **33.** $126x^2y^2\sqrt{x}$ **35.** $\frac{165b^5}{a^5}$ **37.** $180a^2b^8$
39. $60x^2y^8$ **41.** $-61{,}236a^5b^5$ **43.** $126s^{-1}$, $126s$ **45.** $-7 - 24i$ **47.** $41 - 38i$ **49.** 1

Chapter 8 True/False Exercises, *page 689*

1. False; in choosing the members of a committee, the order in which the members are chosen is not important. Thus the number of different ways of choosing seven people from a group of 40 is $C(40, 7)$. **2.** True **3.** False; the probability is $\frac{1}{6} \cdot \frac{1}{6} = \frac{1}{36}$. **4.** True **5.** False; $P(E_1 \text{ or } E_2) = P(E_1) + P(E_2) - P(E_1 \cap E_2)$. **6.** False; the set of possible cards that Jennifer can pick is dependent on which card John picked. **7.** False; in a binomial experiment there are exactly two outcomes for each trial, and each trial must be independent of the other trials. **8.** False; the exponent on a for the fifth term is 4.

Chapter 8 Review Exercises, *page 690*

1. {1H, 1T, 2H, 2T, 3H, 3T, 4H, 4T, 5H, 5T, 6H, 6T} [8.1/8.3] **2.** [8.1] **3.** 15,625,000 [8.1] **4.** 495 [8.1] **5.** 6840 [8.1] **6.** 376,740 [8.1] **7.** 24 [8.3] **8. a.** $\frac{1}{9}$ **b.** $\frac{1}{12}$ [8.2] **9.** $\frac{5}{16}$ [8.2] **10.** $\frac{1}{4}$ [8.2] **11.** $\frac{928}{2495} \approx 0.37$ [8.2] **12.** $\frac{94}{54,145} \approx 0.0017$ [8.3] **13.** $\frac{11}{26}$ [8.3] **14.** $\frac{17}{18}$ [8.3] **15.** $\frac{1}{36}$ [8.3] **16.** $\frac{1}{8}$ [8.3] **17.** $\frac{29}{128} \approx 0.227$ [8.3] **18.** ≈ 0.117 [8.3] **19.** $\frac{1485}{3003} \approx 0.495$ [8.3] **20.** ≈ 0.919 [8.3] **21.** $\frac{3}{4}$ [8.2] **22.** $\frac{1}{2}$ [8.2] **23. a.** 65,536 **b.** ≈ 0.023 [8.3] **24.** 0.366 [8.2] **25.** 0.668 [8.2] **26.** 0.0672 [8.2] **27.** $\frac{1}{8,145,060}$ [8.2] **28.** 2880 [8.1]
29. $1024a^5 - 1280a^4b + 640a^3b^2 - 160a^2b^3 + 20ab^4 - b^5$ [8.4]
30. $a^4 + 16a^{7/2}b^{1/2} + 112a^3b + 448a^{5/2}b^{3/2} + 1120a^2b^2 + 1792a^{3/2}b^{5/2} + 1792ab^3 + 1024a^{1/2}b^{7/2} + 256b^4$ [8.4]
31. $241,920x^3y^4$ [8.4] **32.** $-78,732x^7$ [8.4]

Chapter 8 Test, *page 691*

1. 18,000 [8.1/8.2] **2. a.** 3024 **b.** 126 [8.1] **3.** 3,268,760 [8.1] **4. a.** 4096 **b.** 924 [8.1] **5.** $\frac{11}{16}$ [8.2] **6.** $\frac{5}{16}$ [8.2] **7. a.** $\frac{12}{13}$ **b.** $\frac{7}{13}$ **c.** $\frac{4}{13}$ [8.3] **8.** $\frac{1}{4165} \approx 0.000240$ [8.3] **9.** $\frac{1}{6}$ [8.2] **10.** $\frac{17}{108}$ [8.3] **11.** $\frac{2560}{3650}$ [8.3] **12.** $\frac{1}{2}$ [8.2] **13.** 0.241 [8.3]
14. $x^4 - 12x^3y + 54x^2y^2 - 108xy^3 + 81y^4$ [8.4] **15.** $x^5 + 5x^3 + 10x + \frac{10}{x} + \frac{5}{x^3} + \frac{1}{x^5}$ [8.4]

Cumulative Review Exercises, *page 692*

1. $[-5, 5]$ [2.2] **2.** 7 [2.5] **3.** $\pm 1, \pm 2, \pm 3, \pm 6$ [4.3] **4.** 3 or 1 positive real zeros, no negative real zeros [4.3]
5. 6.9 [5.3] **6.** 3.0603 [5.4] **7.** 6 [5.4] **8.** 91 years [5.5] **9. a.** $P(t) = \frac{600}{1 + 1.4e^{-0.13439t}}$ **b.** 410 [5.7]
10. (5, 7) [6.1] **11.** $\begin{bmatrix} \frac{1}{17} & \frac{7}{17} \\ -\frac{2}{17} & \frac{3}{17} \end{bmatrix}$ [6.5] **12.** $\begin{bmatrix} 3 & -2 & 1 \\ 1 & 4 & -1 \\ 2 & -1 & -3 \end{bmatrix}\begin{bmatrix} x \\ y \\ z \end{bmatrix} = \begin{bmatrix} 8 \\ 2 \\ 5 \end{bmatrix}$; (2.56, −0.12, 0.08) [6.5] **13.** 154 [7.2]
14. $\frac{3}{7}$ [7.2] **15.** \$5984.07 [7.3] **16.** 17.4 years [7.3] **17.** \$12,848.40 [7.5] **18.** 56 [8.1] **19.** 151,200 [8.1]
20. $32x^5 + 80x^4y + 80x^3y^2 + 40x^2y^3 + 10xy^4 + y^5$ [8.4]

Appendix A Exercises, *page AP16*

1. Vertex: (0, 0)
Focus: (0, −1)
Directrix: $y = 1$

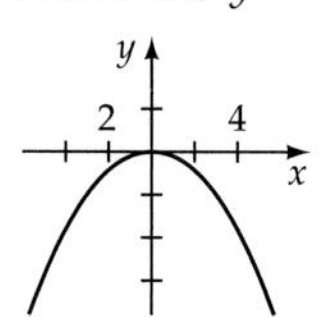

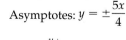

3. Center: (0, 0)
Vertices: (0, 5), (0, −5)
Foci: (0, 3), (0, −3)

5. Center: (0, 0)
Vertices: (±4, 0)
Foci: $(\pm\sqrt{41}, 0)$
Asymptotes: $y = \pm\frac{5x}{4}$

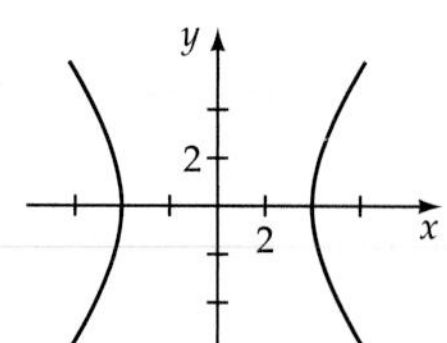

7. Vertex: (2, −3)
Focus: (2, −1)
Directrix: $y = -5$

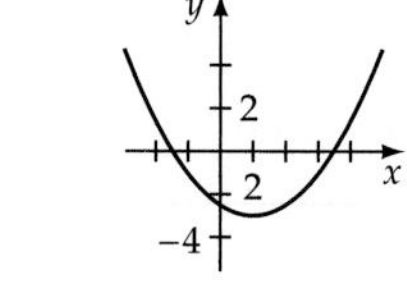

9. Center: (0, 0)
Vertices: (±3, 0)
Foci: $(\pm 3\sqrt{2}, 0)$
Asymptotes: $y = \pm x$

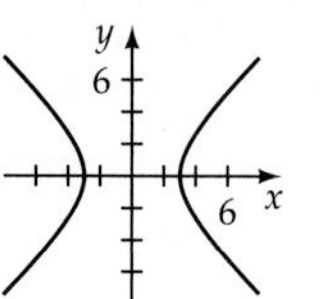

11. Center: $(3, -2)$
Vertices: $(8, -2), (-2, -2)$
Foci: $(6, -2), (0, -2)$

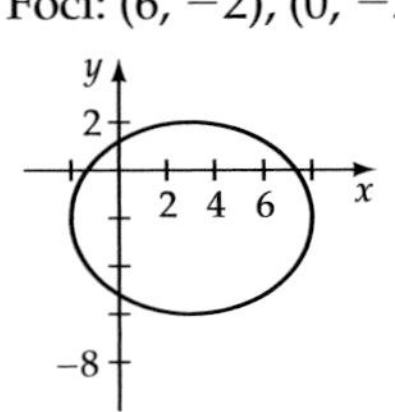

13. Center: $(3, -4)$
Vertices: $(7, -4), (-1, -4)$
Foci: $(8, -4), (-2, -4)$
Asymptotes: $y + 4 = \pm 3\frac{(x-3)}{4}$

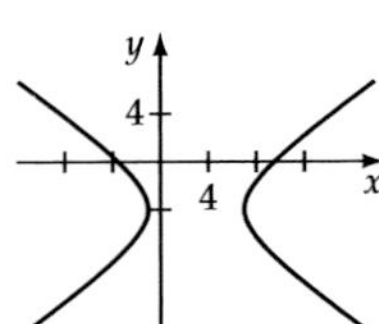

15. Center: $(3, -1)$
Vertices: $(-1, -1), (7, -1)$
Foci: $(3 \pm 2\sqrt{3}, -1)$

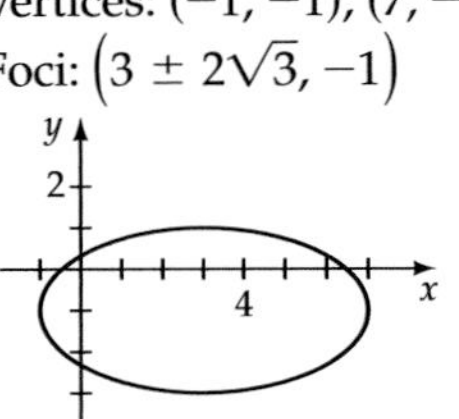

17. Vertex: $(-2, 1)$
Focus: $\left(-\frac{29}{16}, 1\right)$
Directrix: $x = -\frac{35}{16}$

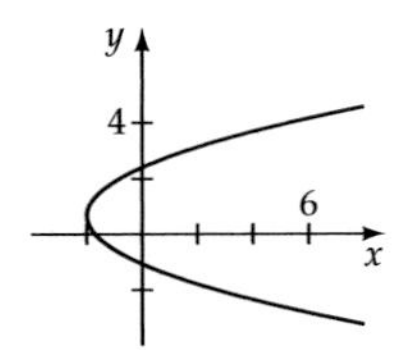

19. Center: $(-2, 1)$
Vertices: $(-2, -2), (-2, 4)$
Foci: $(-2, 1 \pm \sqrt{5})$

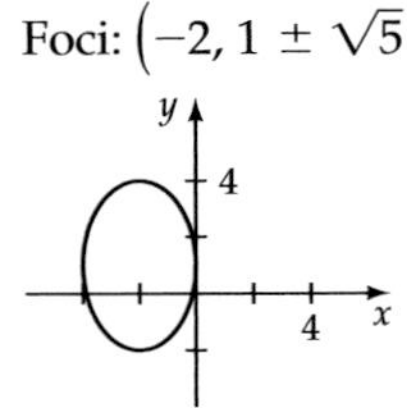

21. Center: $\left(1, \frac{2}{3}\right)$
Vertices: $\left(-5, \frac{2}{3}\right), \left(7, \frac{2}{3}\right)$
Foci: $\left(1 \pm 2\sqrt{13}, \frac{2}{3}\right)$
Asymptotes: $y - \frac{2}{3} = \pm\frac{2(x-1)}{3}$

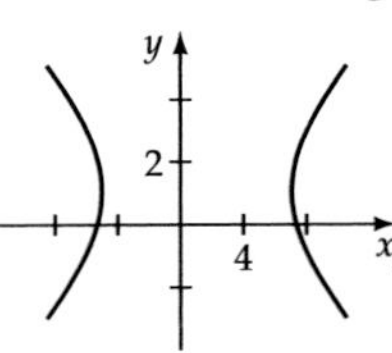

23. Vertex: $\left(-\frac{7}{2}, -1\right)$
Focus: $\left(-\frac{7}{2}, -3\right)$
Directrix: $y = 1$

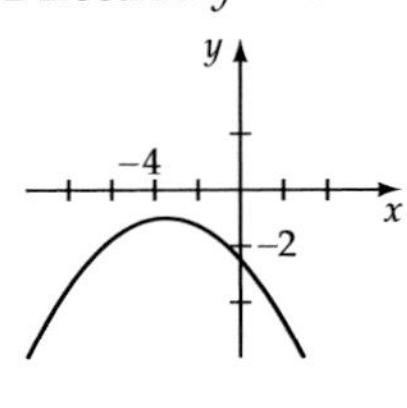

25. $\frac{(x-2)^2}{25} + \frac{(y-3)^2}{16} = 1$ **27.** $\frac{(x+2)^2}{4} - \frac{(y-2)^2}{5} = 1$ **29.** $x^2 = \frac{3(y+2)}{2}$ or $(y+2)^2 = 12x$ **31.** $\frac{x^2}{36} - \frac{y^2}{4/9} = 1$

33. $(y-3)^2 = -8x$ **35.** $\frac{(x-1)^2}{25} + \frac{(y-1)^2}{9} = 1$ **37.** On axis 4 feet above vertex **39.** 6.0 inches **41.** $a = 1.5$ inches

43. $\frac{x^2}{884.74^2} + \frac{y^2}{883.35^2} = 1$ **45.** 40 feet **47.** $\frac{\left(x - 9\sqrt{15}/2\right)^2}{324} + \frac{y^2}{81/4} = 1$

Appendix B Exercises, *page AP25*

1. 13 **3.** −15 **5.** 0 **7.** 0 **9.** 19, 19 **11.** 1, −1 **13.** −9, −9 **15.** −9, −9 **17.** 10 **19.** 53 **21.** 20 **23.** 46 **25.** 0 **27.** Row 2 consists entirely of zeros, so the determinant is zero. **29.** 2 was factored from row 2. **31.** Row 1 was multiplied by −2 and added to row 2. **33.** 2 was factored from column 1. **35.** The matrix is in triangular form. The value of the determinant is the product of the terms on the main diagonal. $2(3)(-2) = -12$. **37.** Row 1 and row 3 were interchanged, so the sign of the determinant was changed. **39.** Each row of the determinant was multipled by a. **41.** 0 **43.** 0 **45.** 6 **47.** −90 **49.** 21 **51.** 3 **53.** −38.933 **55.** 263.5

Answer Art Appendix

Chapter P

Exercises P.1, *page 13*

23. $\{x \mid -2 < x < 3\}$ 24. $\{x \mid 1 \le x \le 5\}$

25. $\{x \mid -5 \le x \le -1\}$ 26. $\{x \mid -3 < x < 3\}$

27. $\{x \mid x \ge 2\}$ 28. $\{x \mid x < 4\}$ 29. $(3, 5)$

30. $(-\infty, -1)$ 31. $[-2, \infty)$ 32. $[-1, 5)$

33. $[0, 1]$ 34. $(-4, 5]$ 35.

36. 37. 38.

39. 40. 41.

42. 43. 44.

45. 46. 47.

48. 49. 50.

Exercises P.7, *page 75*

Exploration Exercises 1–8

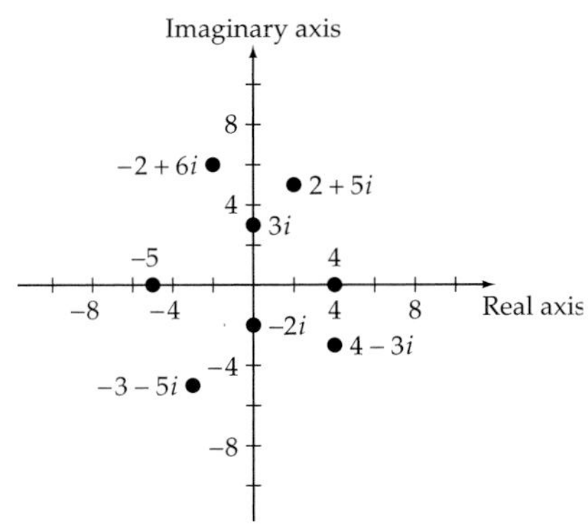

Chapter P Review Exercises, *page 78*

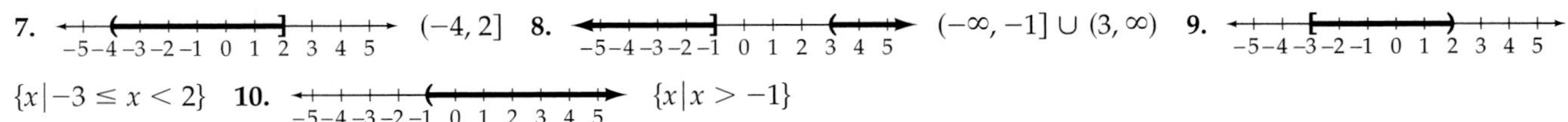

7. $(-4, 2]$ 8. $(-\infty, -1] \cup (3, \infty)$ 9. $\{x \mid -3 \le x < 2\}$ 10. $\{x \mid x > -1\}$

Chapter 2

Exercises 2.1, *page 173*

1.

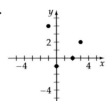

2. 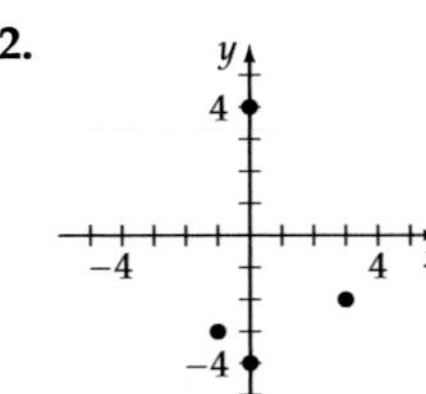

3.

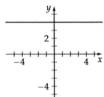

4. 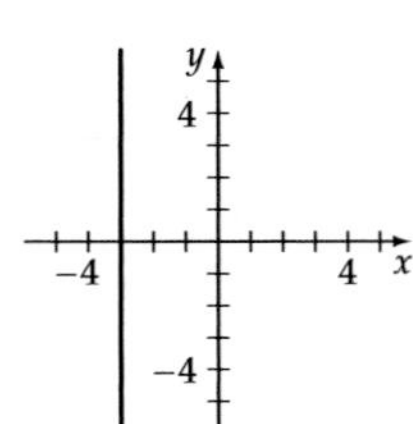

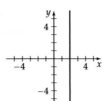

5.

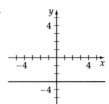

6.

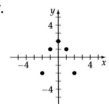

23. 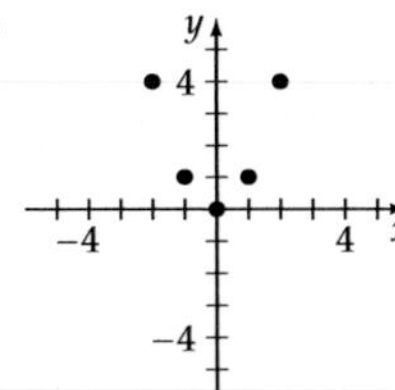

24.

25. 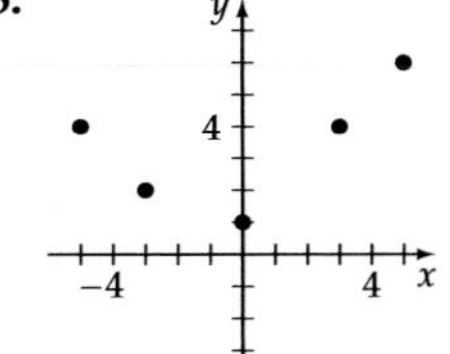

26.

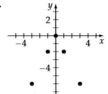

27.

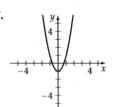

28. 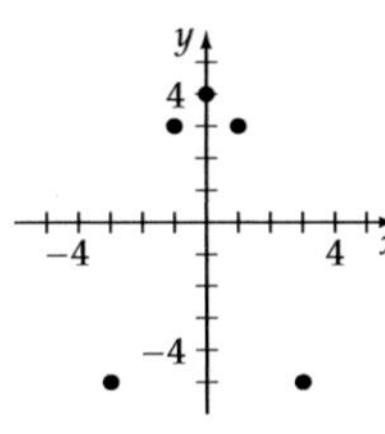

29. 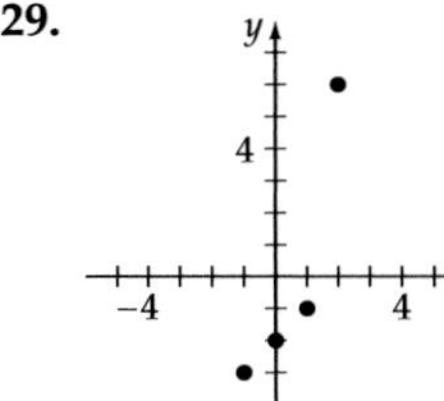

30. 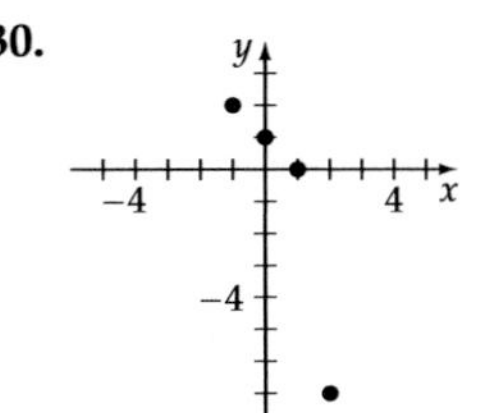

31. 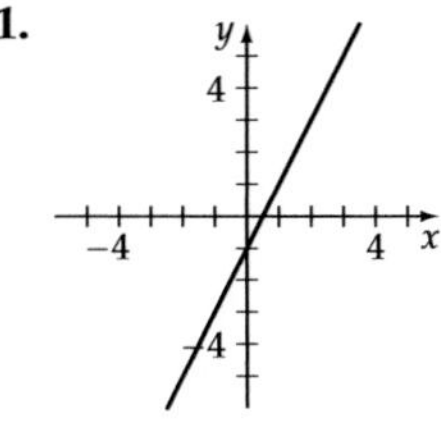

32.

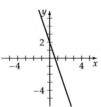

33. 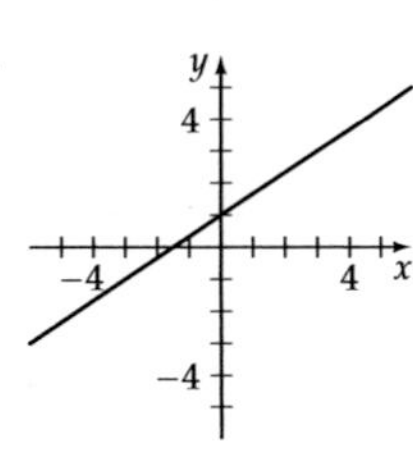

34. 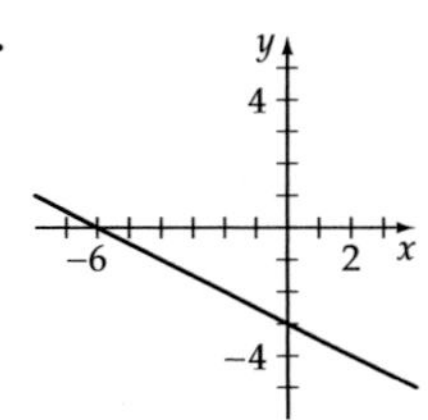

35. 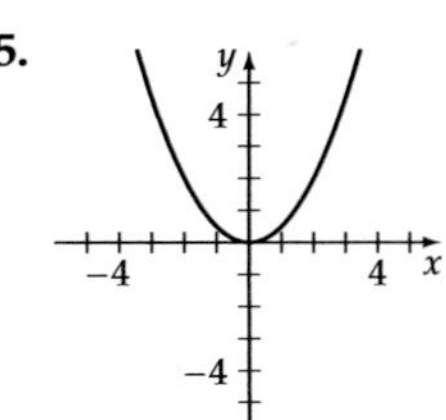

36. 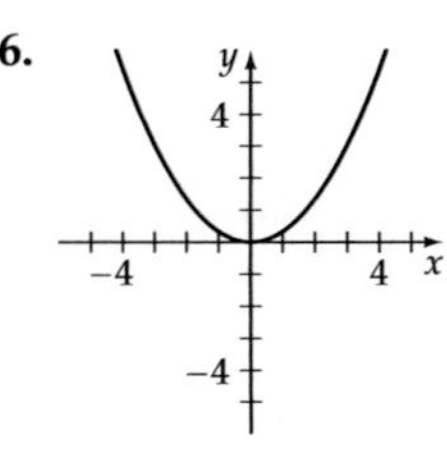

37.

38. 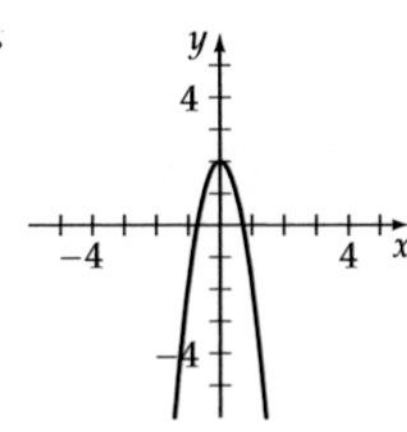

39. 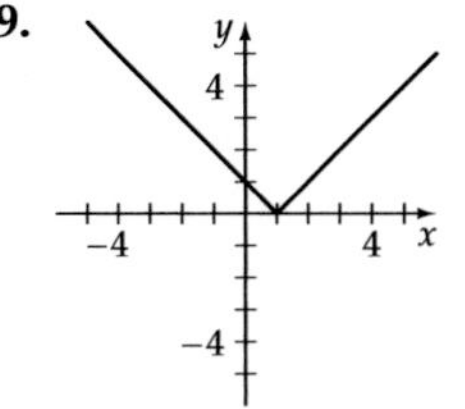

40. 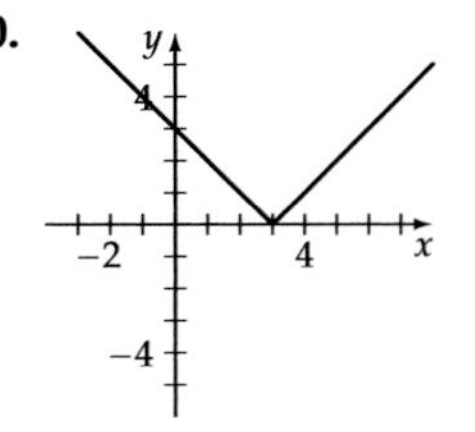

59. 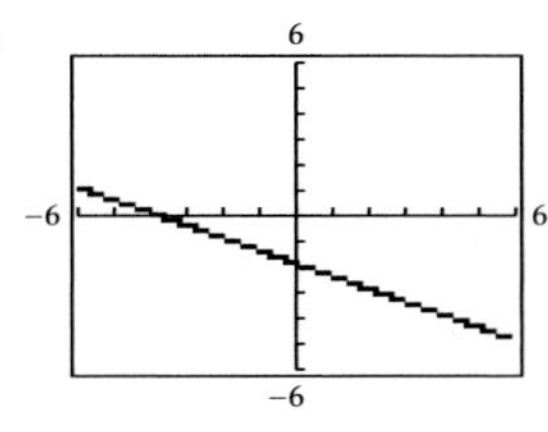

60.

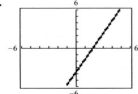

61. 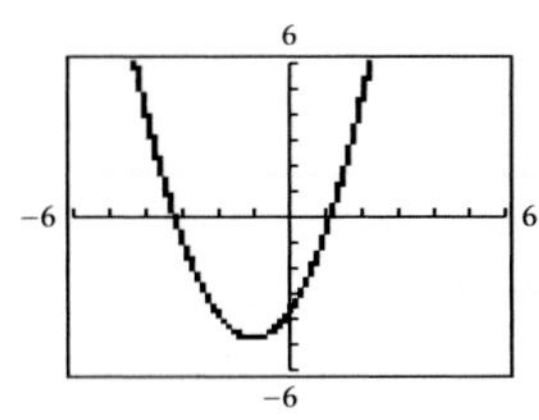

62. 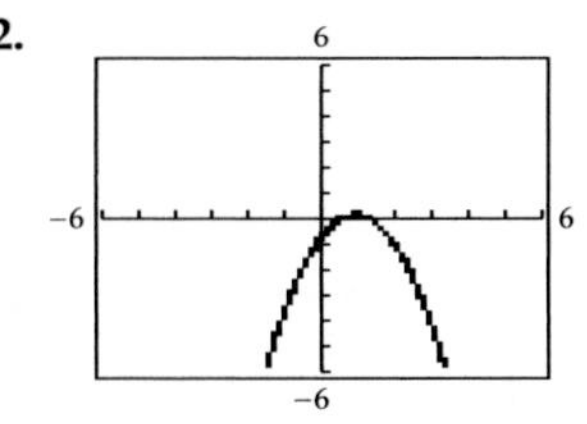

63. 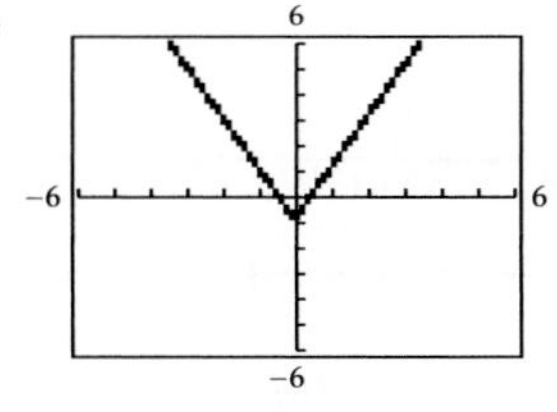

64. 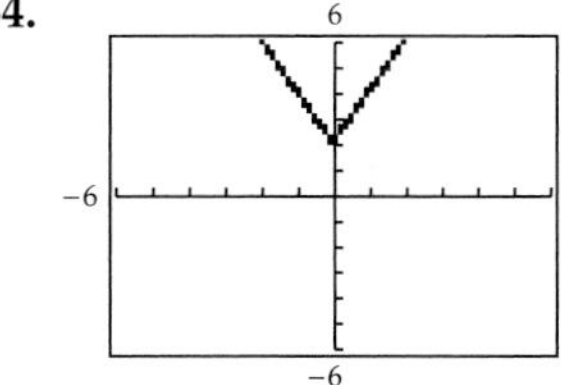

65. 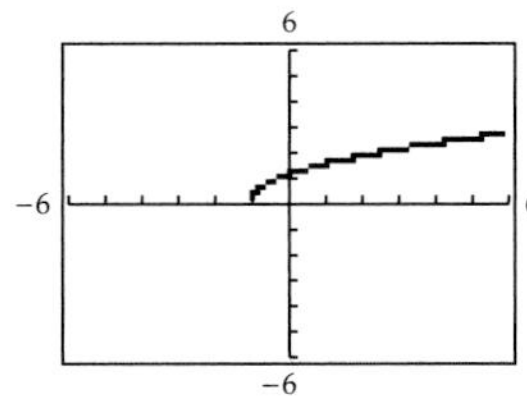

66. 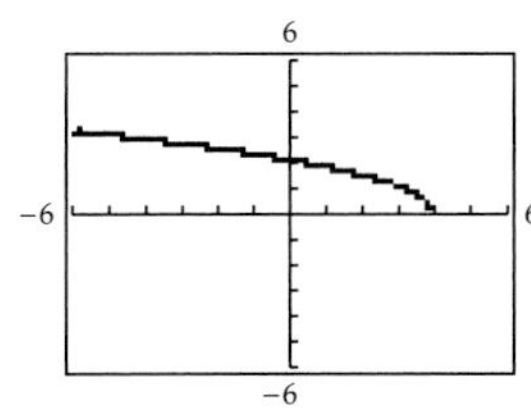

67. 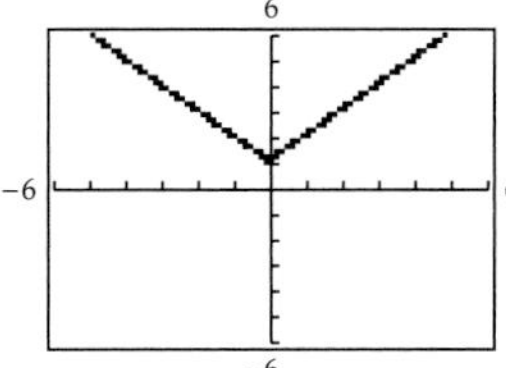

68. 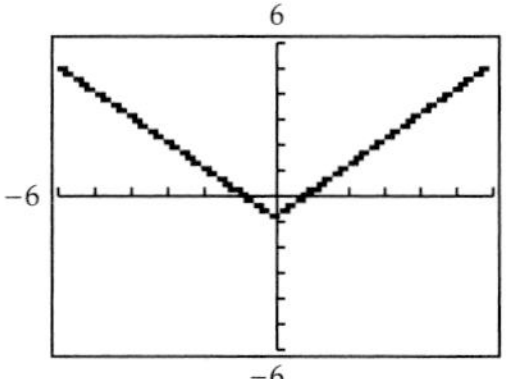

69. 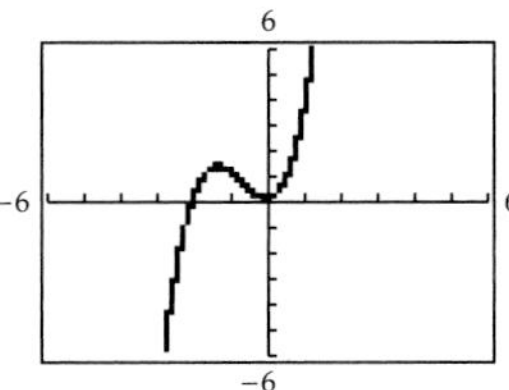

70. 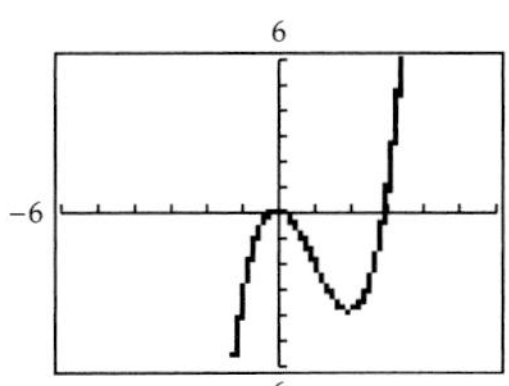

71. 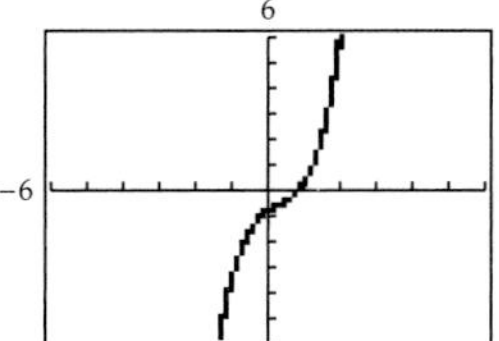

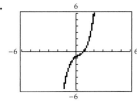

72. 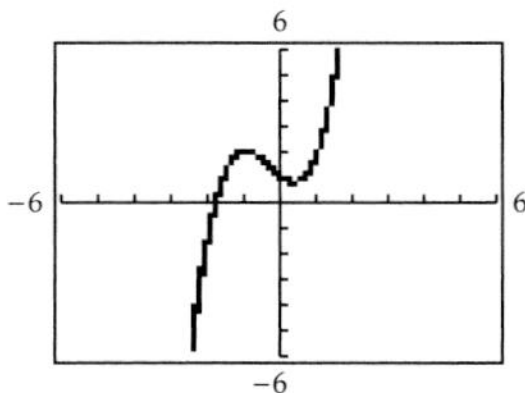

Exercises 2.2, *page 188*

23. 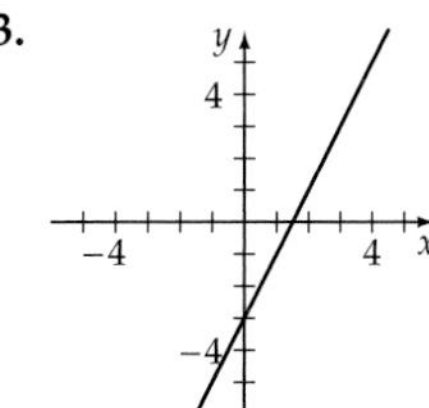

24. 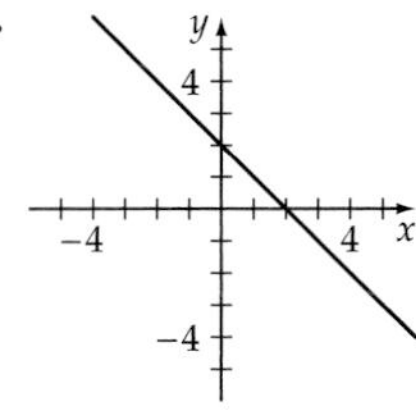

25. 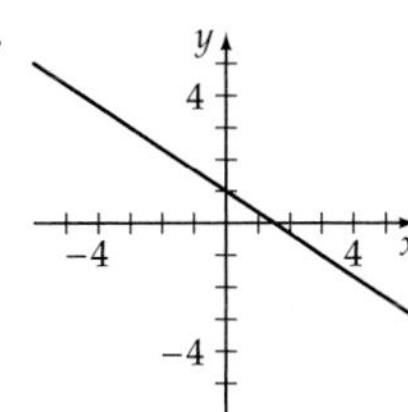

26. 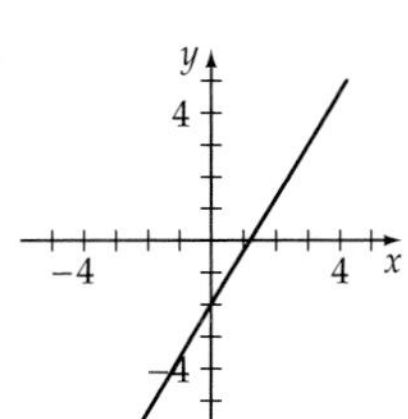

27. 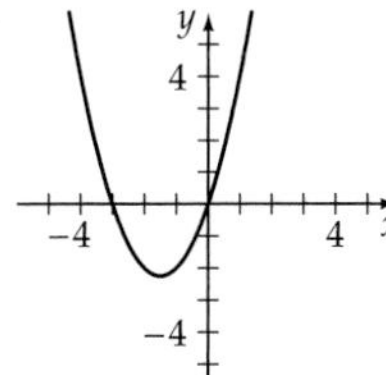

28. 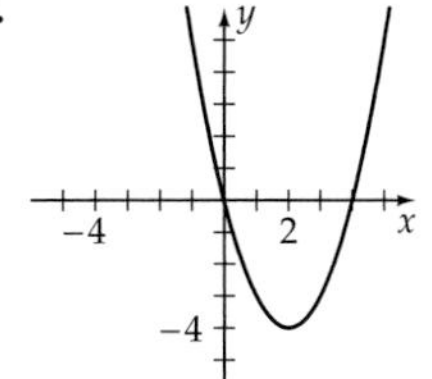

29. 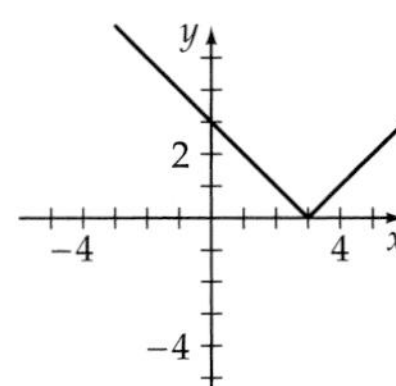

30. 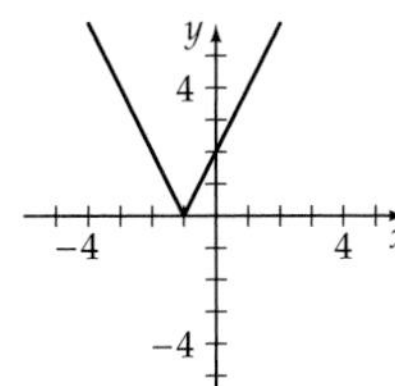

31. 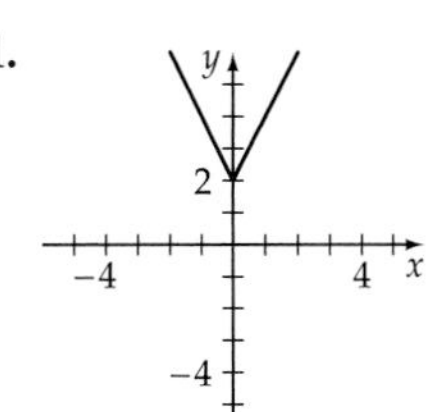

32. 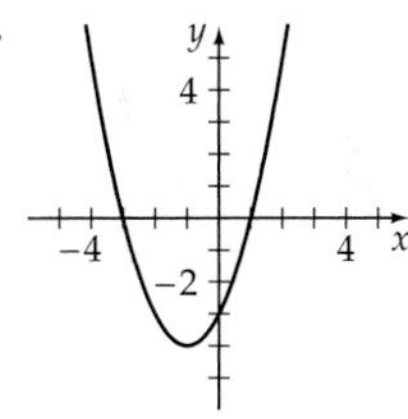

86. 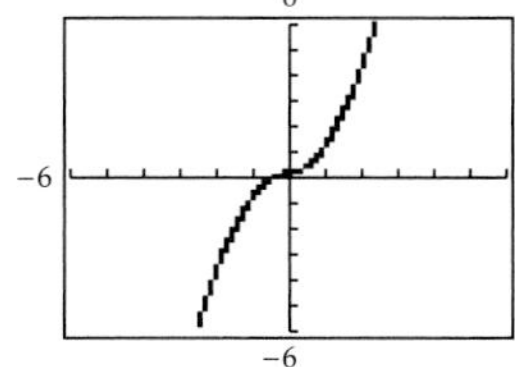

87. 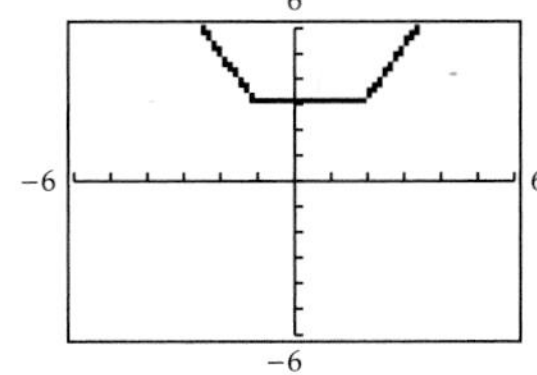

88. 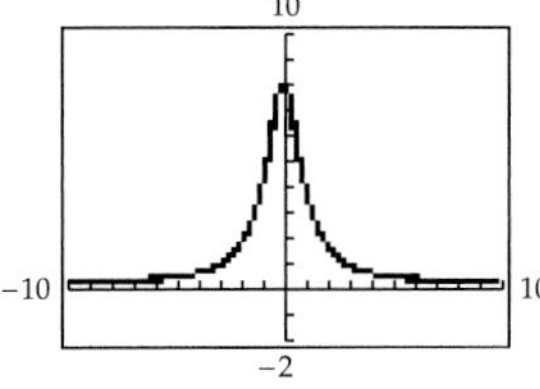

89. 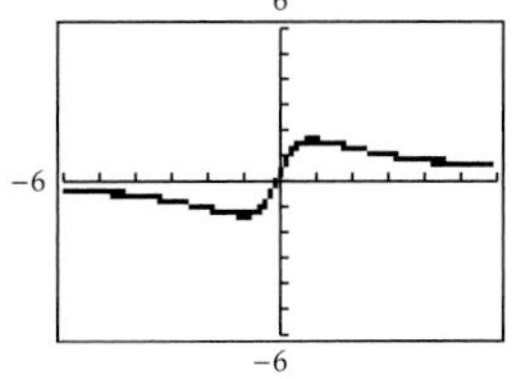

90. 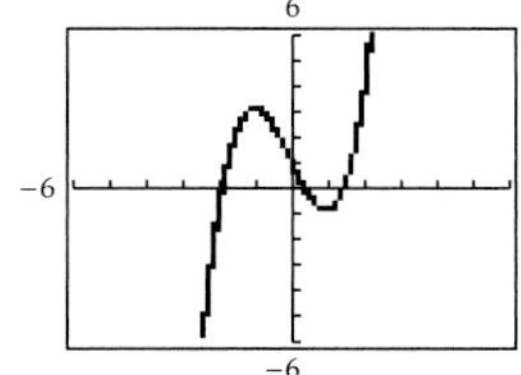

91. 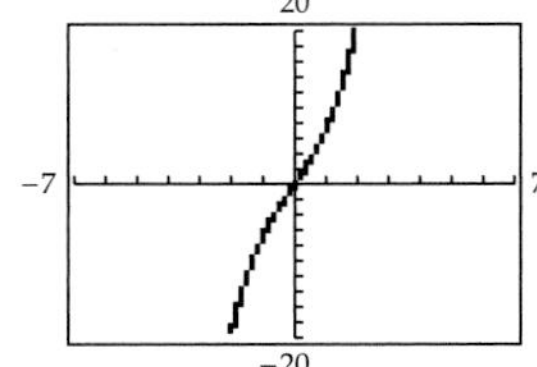

92. 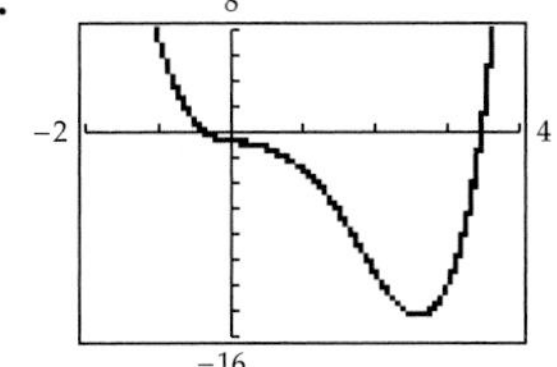

93. 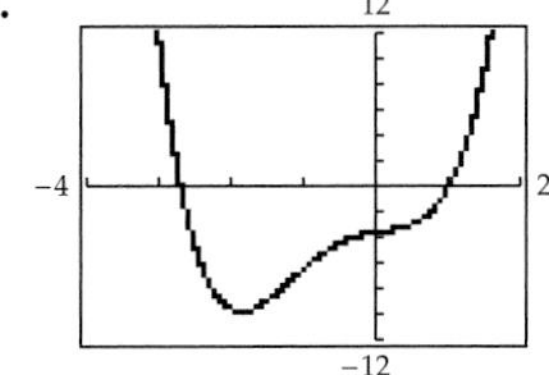

Exercises 2.3, *page 206*

35.
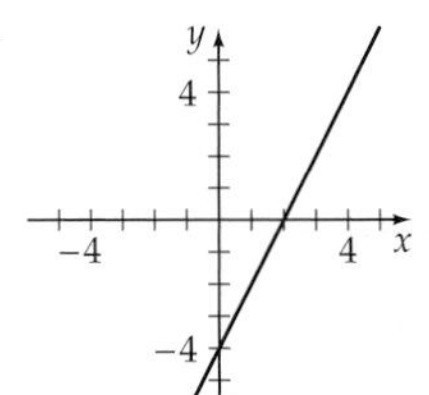

36.
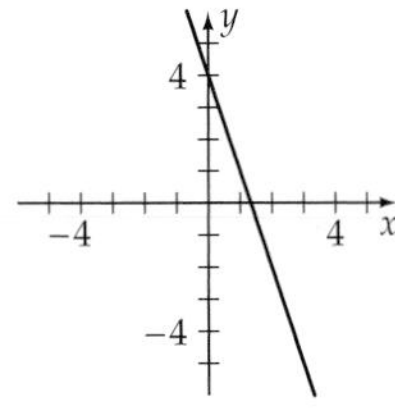

37.
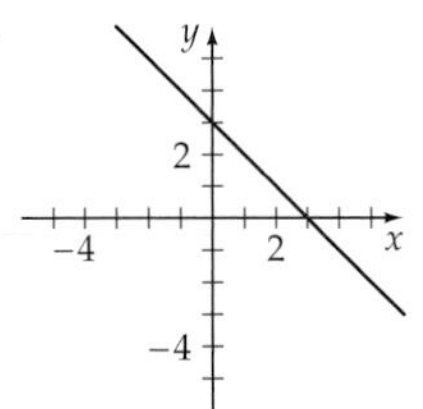

38.
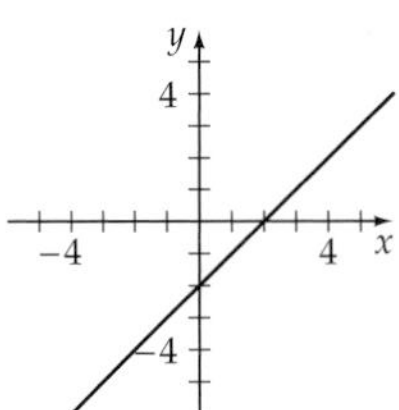

39.
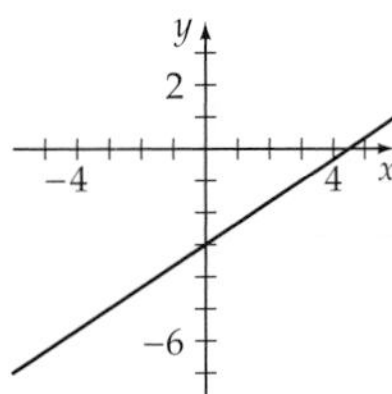

40.
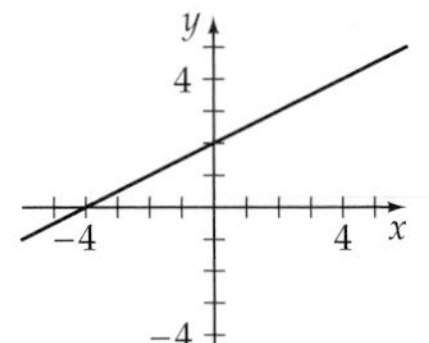

41.
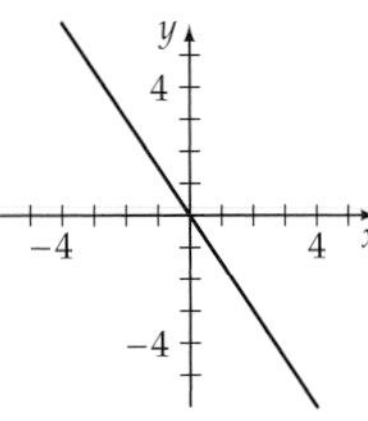

42.
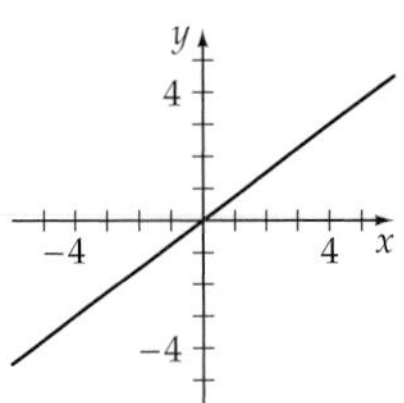

43.
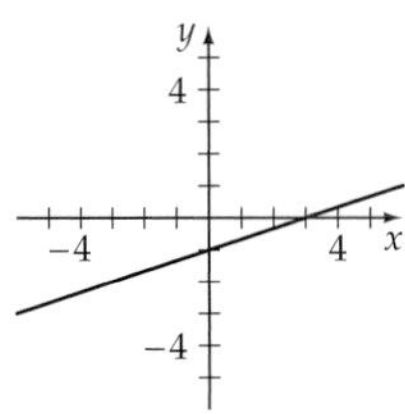

44.
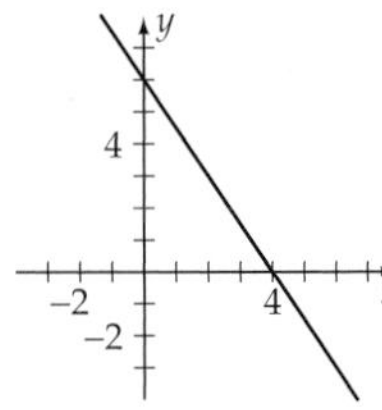

45.
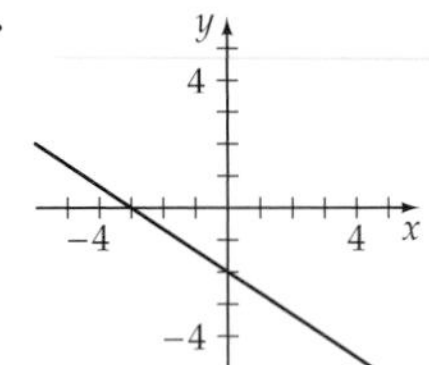

46.
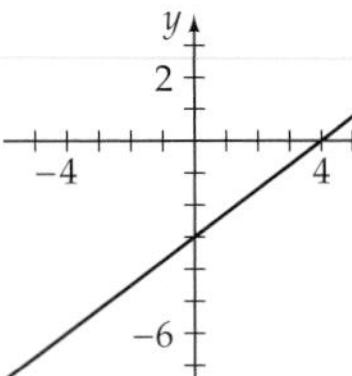

47.
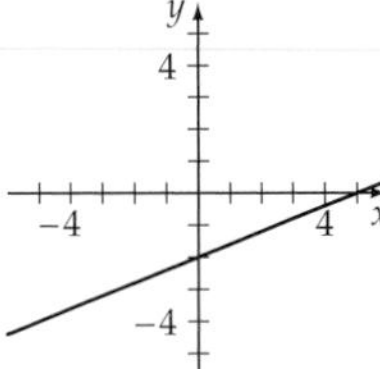

48.
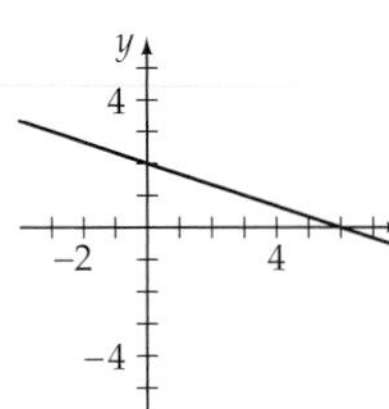

49.
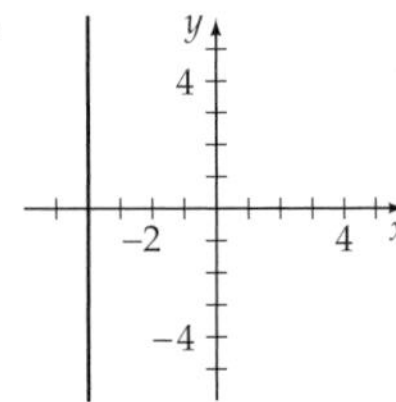

50.
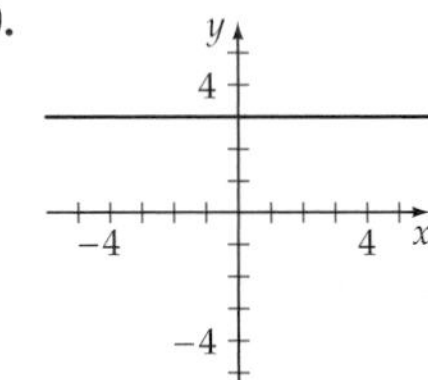

51.
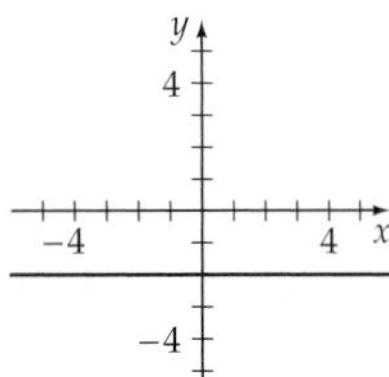

52.
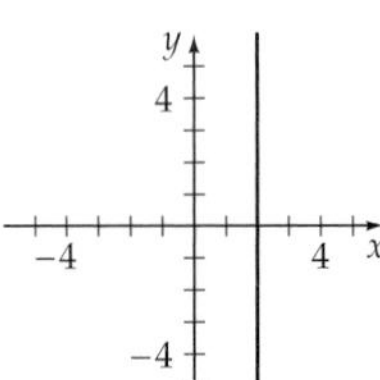

53.
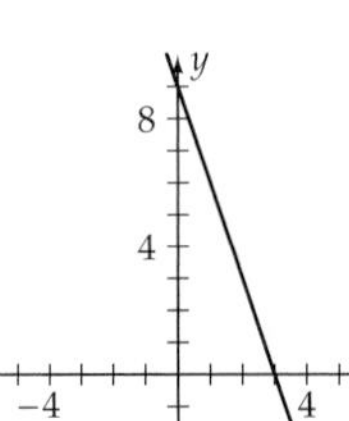

54.
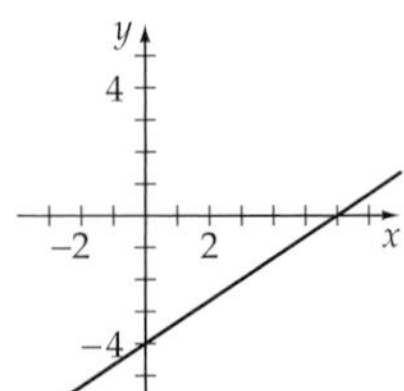

74.
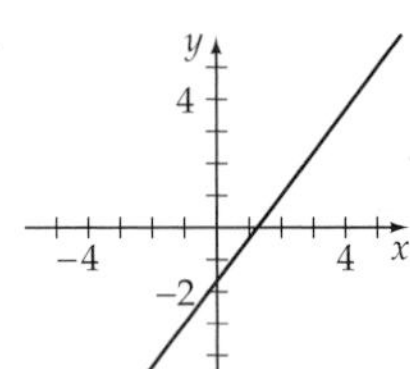

75.
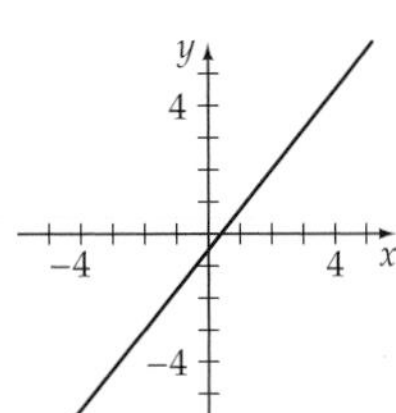

76.
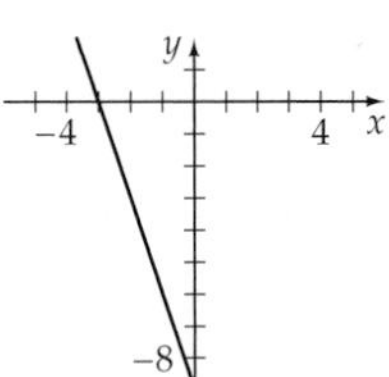

77.
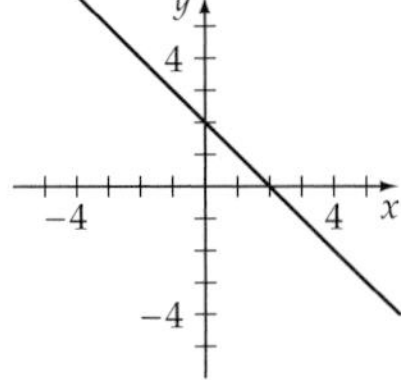

Chapter 2 Review Exercises, *page 246*

11.
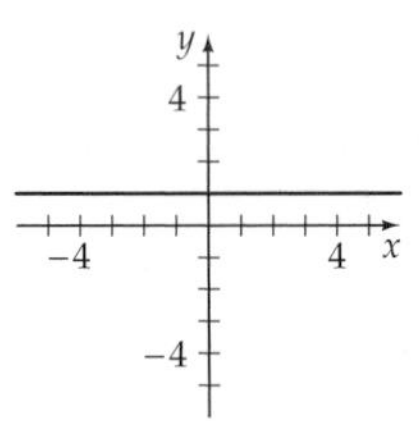

[2.1]

12.
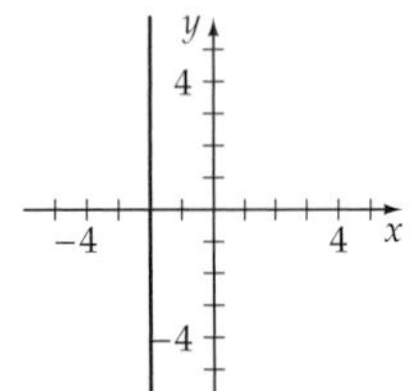

[2.1]

13.
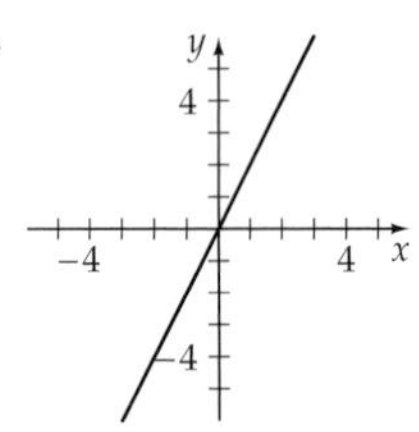

[2.1]

14.
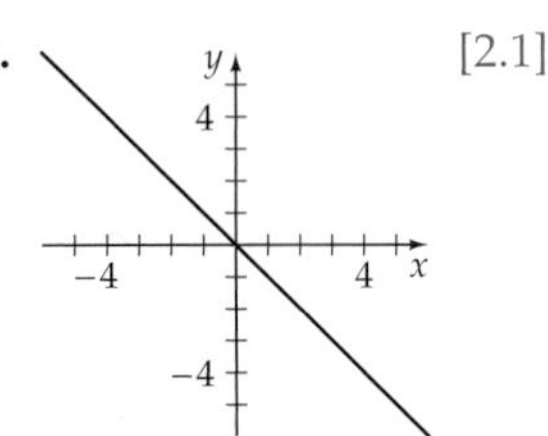

[2.1]

23. [2.1]

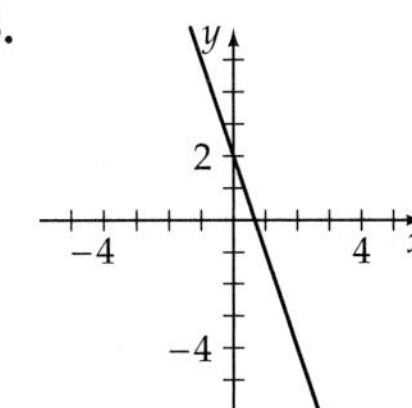

24. [2.1]

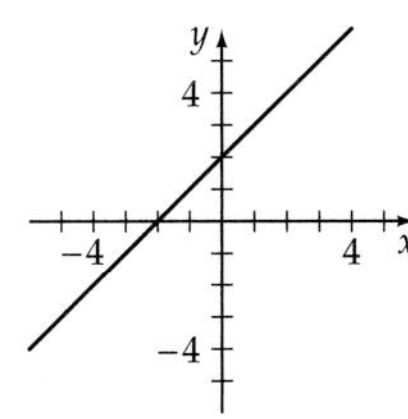

25. [2.3]

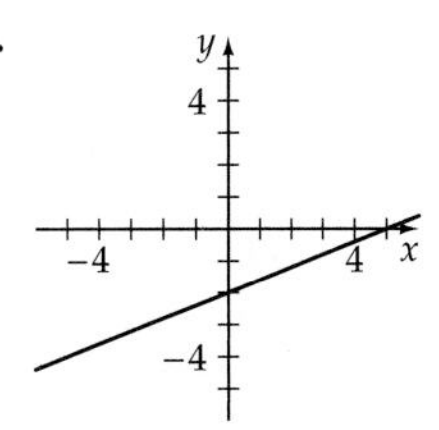

26. [2.3]

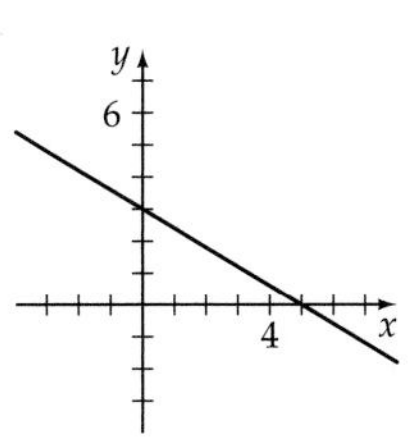

27. [2.1]

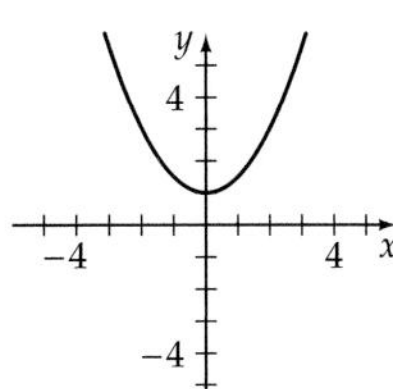

28. [2.1]

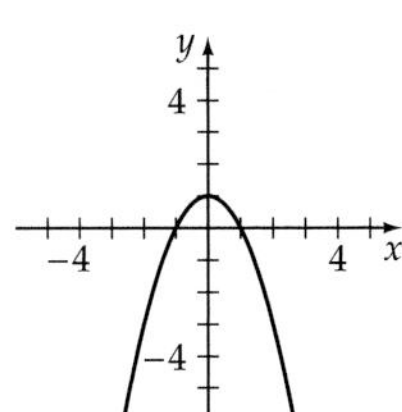

29. [2.2/2.5]

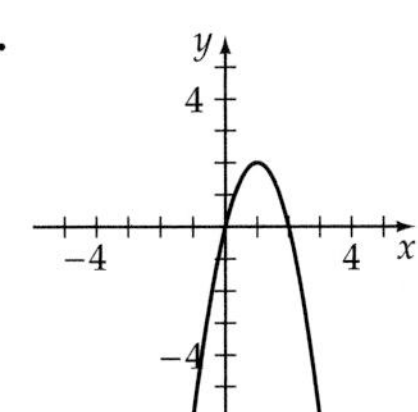

30. [2.2/2.5]

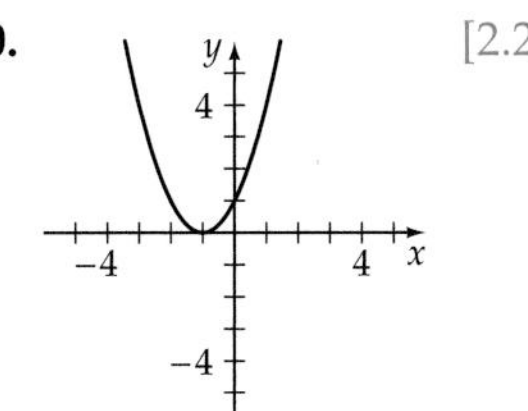

31. [2.2]

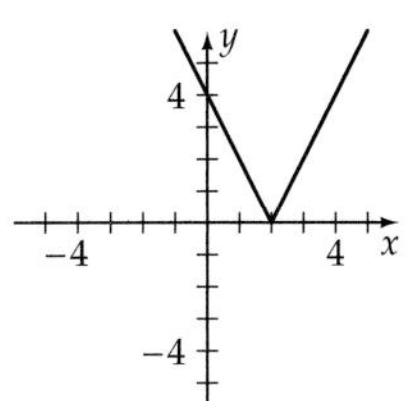

32. [2.2]

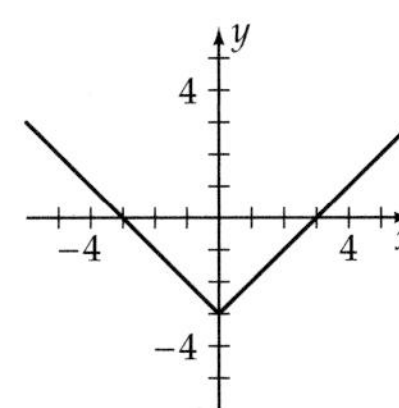

33. [2.3]

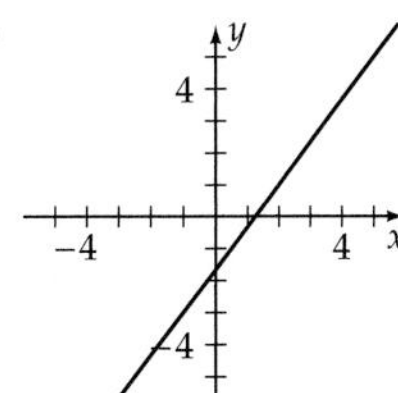

34. [2.3]

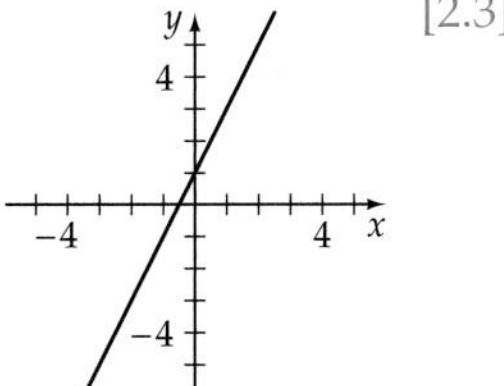

35. [2.3]

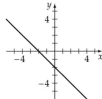

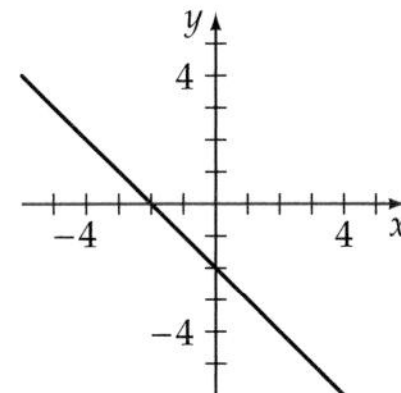

36. [2.3]

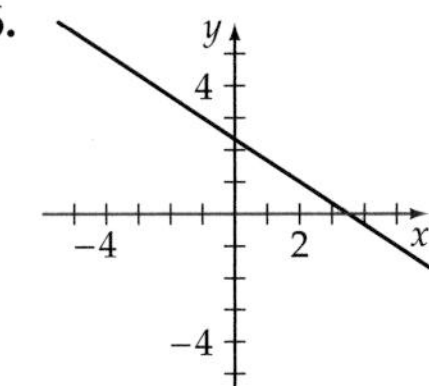

Chapter 2 Test, *page 249*

1. [2.2]

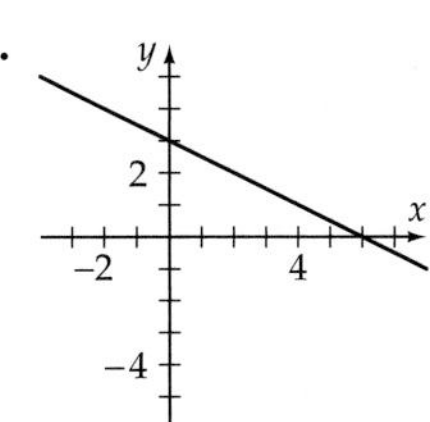

2. [2.5]

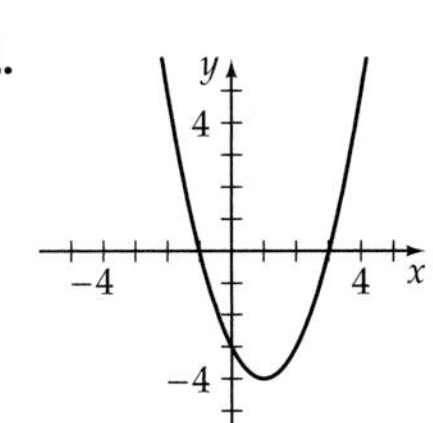

3. [2.3]

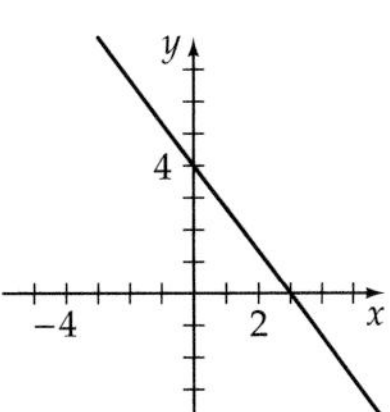

4. [2.3]

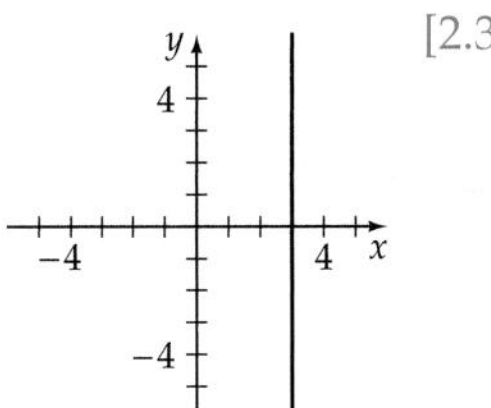

5. [2.3]

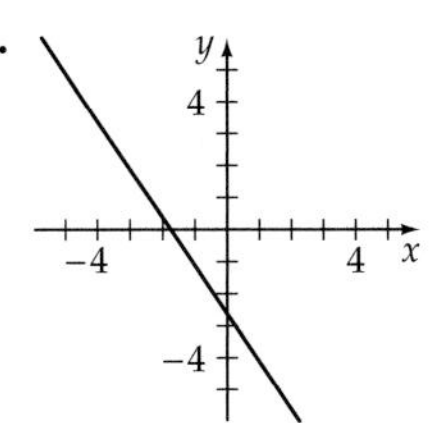

Chapter 3

Exercises 3.2, *page 270*

37.

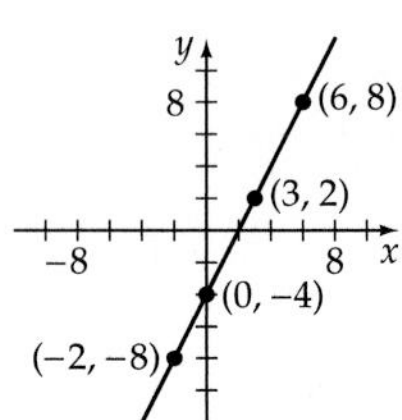

38.

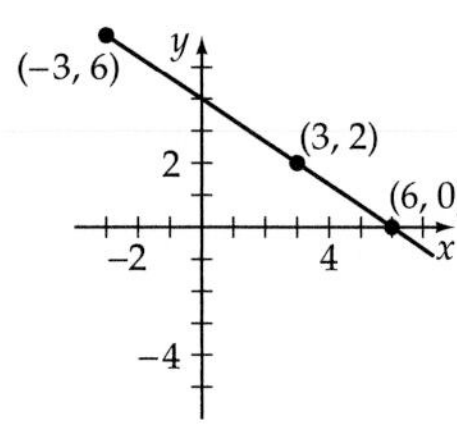

39.

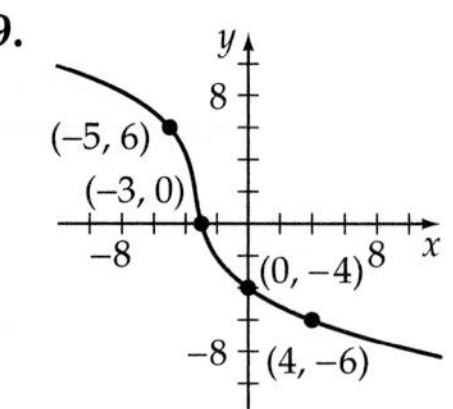

40.

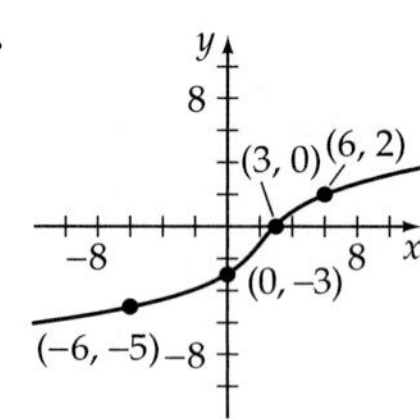

41.

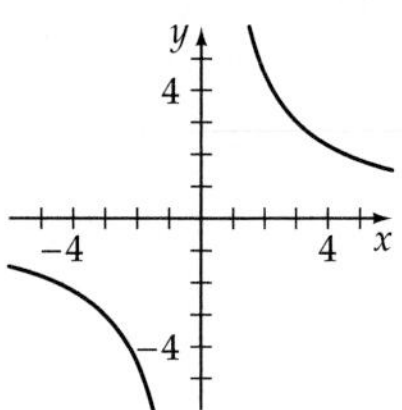

42.

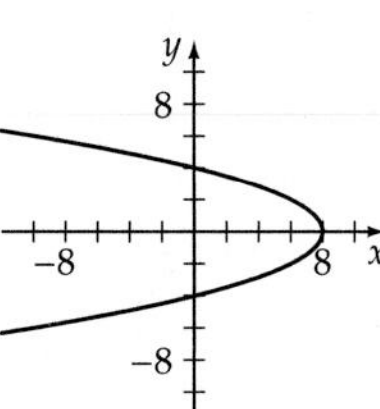

43.

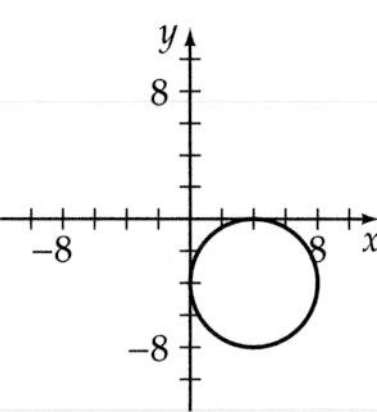

44.

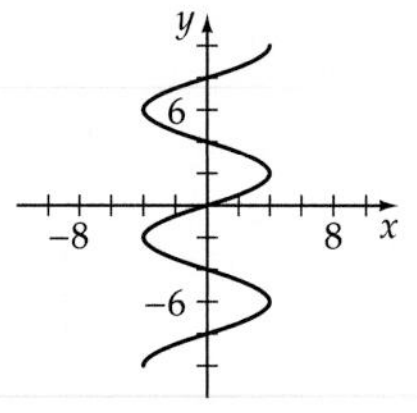

Exercises 3.3, *page 289*

11. a.

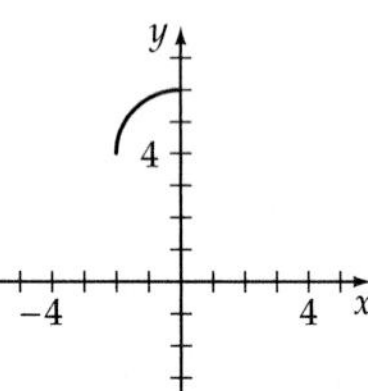

b.

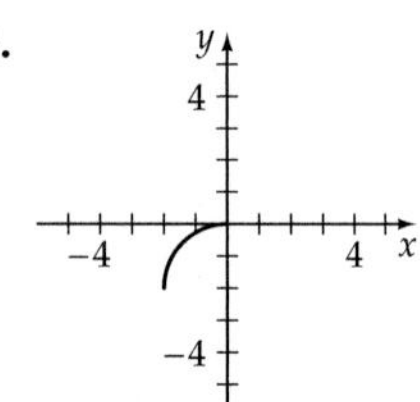

c.

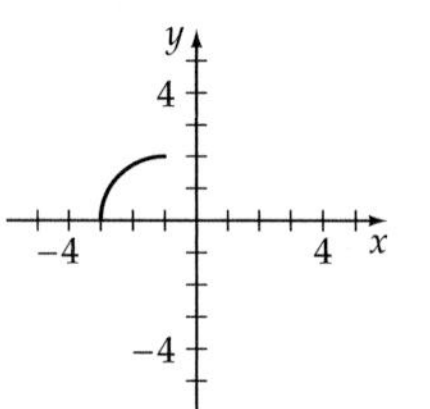

d.

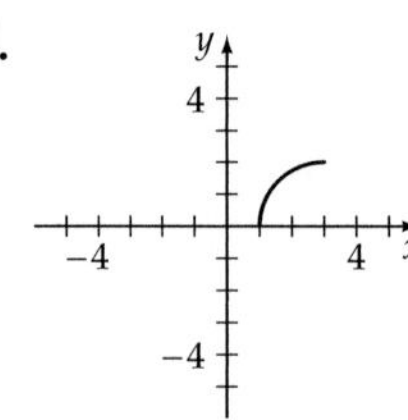

12. a.

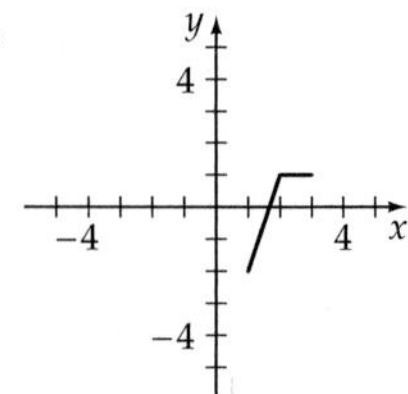

b.

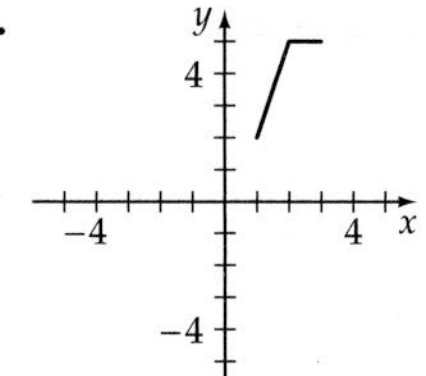

c.

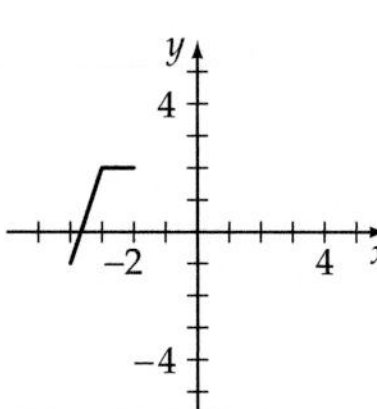

d.

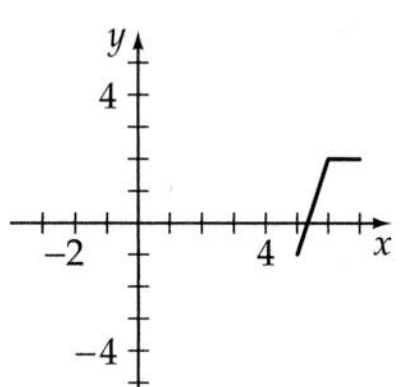

13. a.

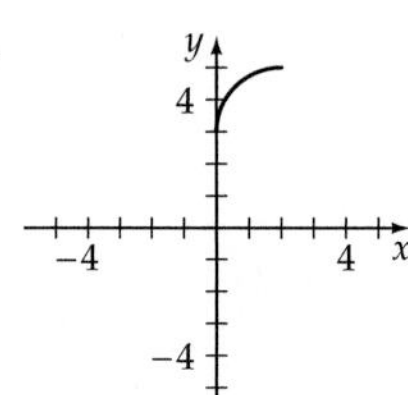

b.

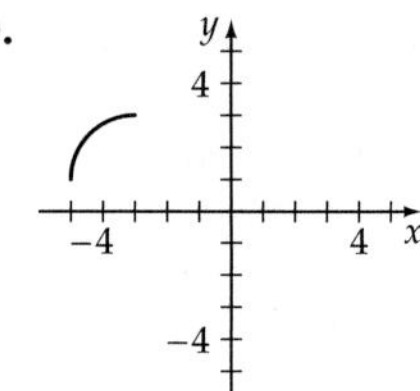

14. a.

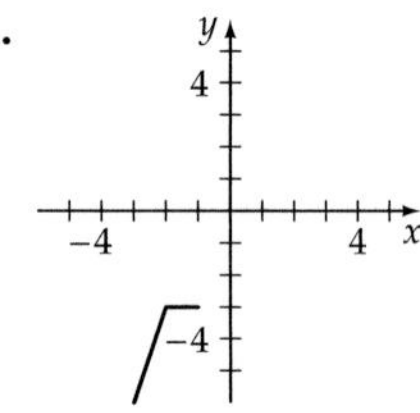

b.

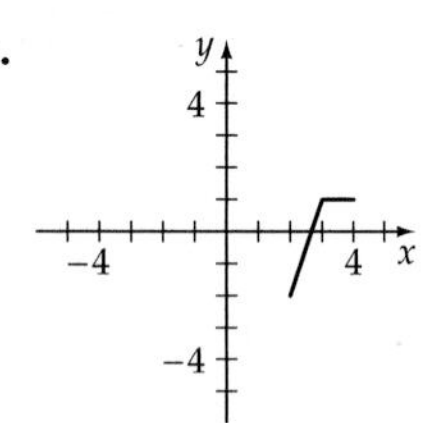

15. a.

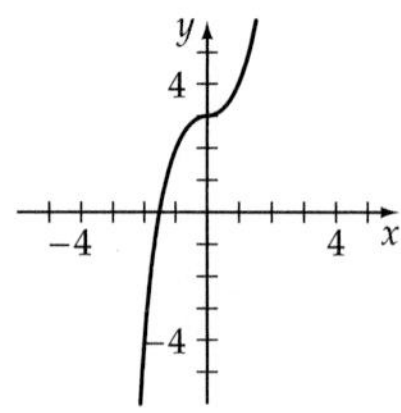

b.

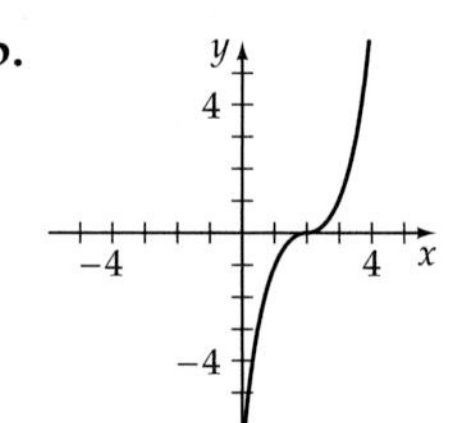

c.

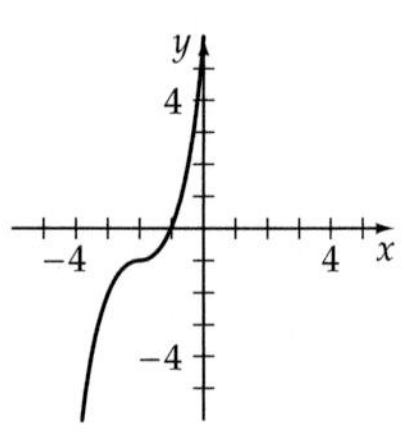

d.

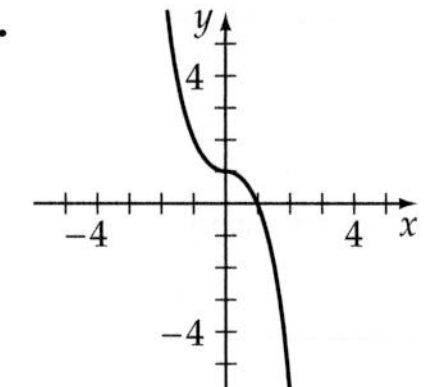

e.

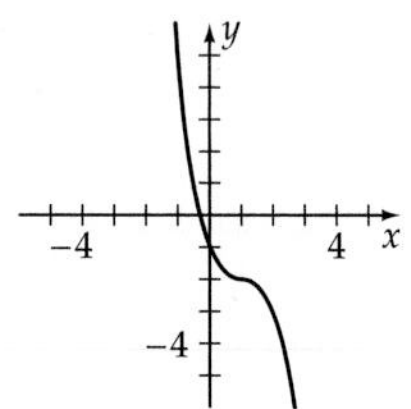

16. a.

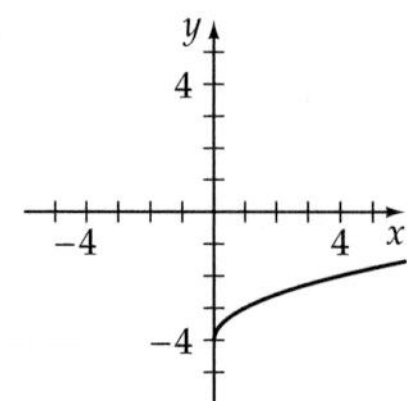

b.

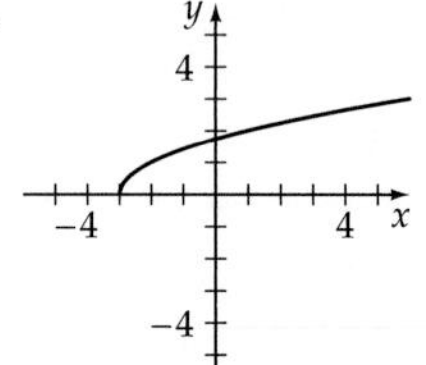

c.

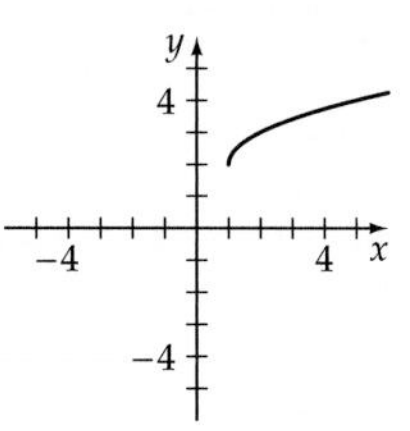

d.

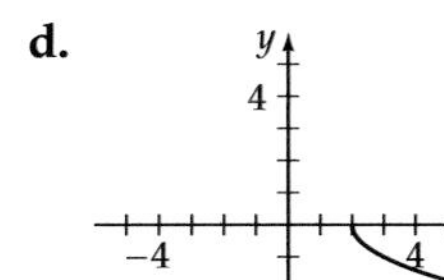

e.

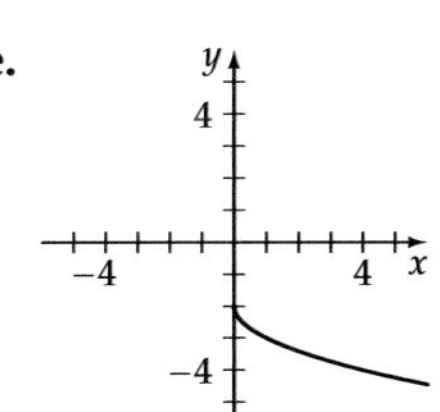

17. a.

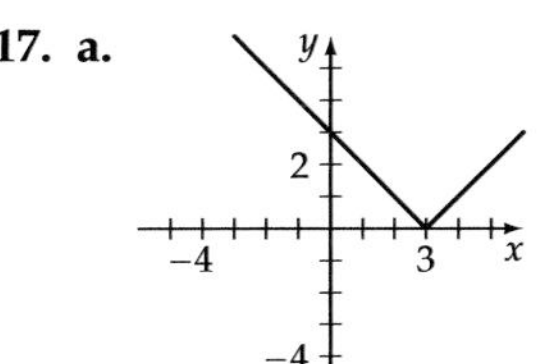

b.

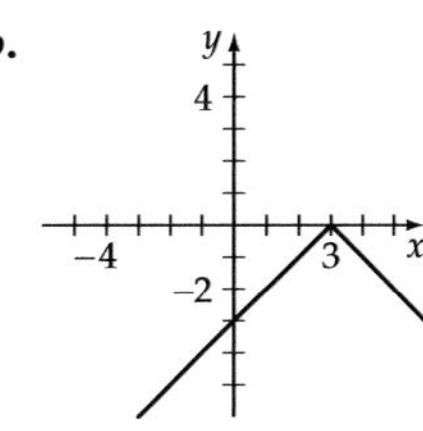

c.

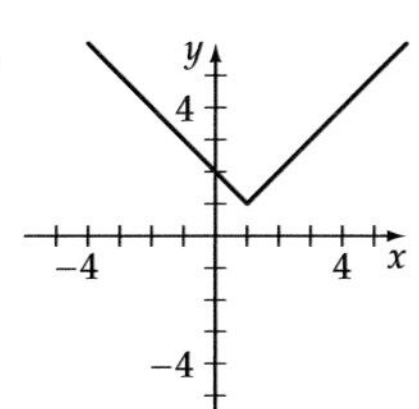

d.

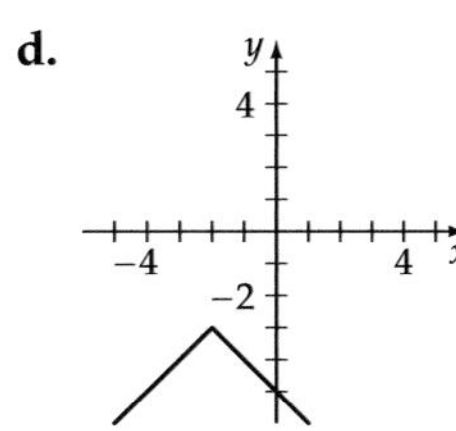

18. a.

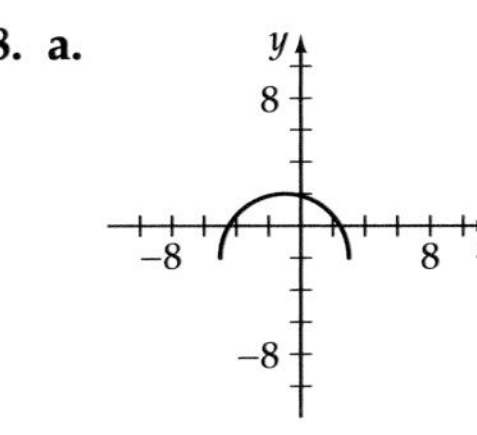

b.

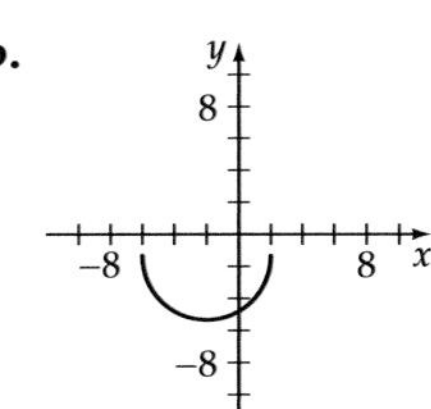

c.

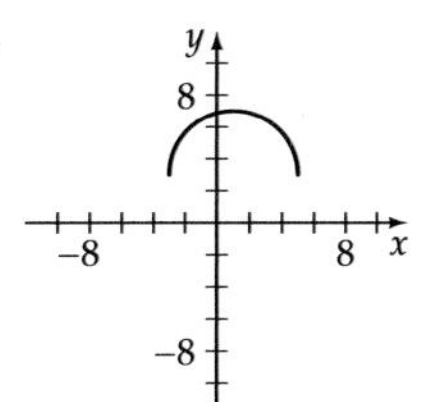

d.

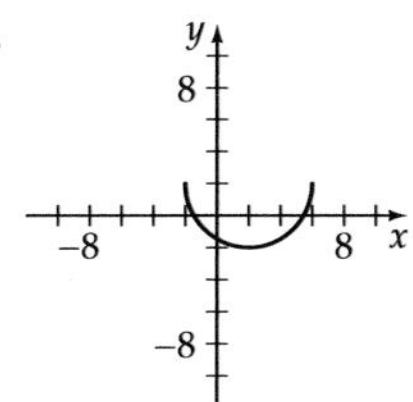

23. a.

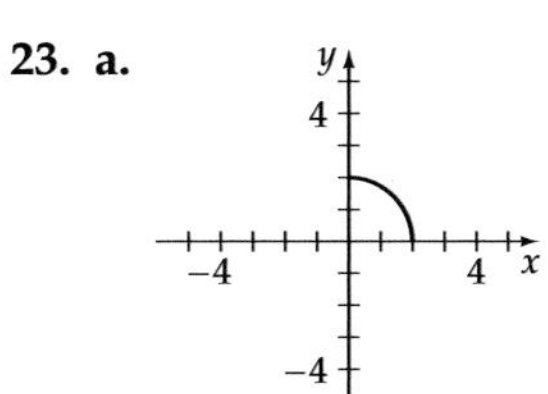

b.

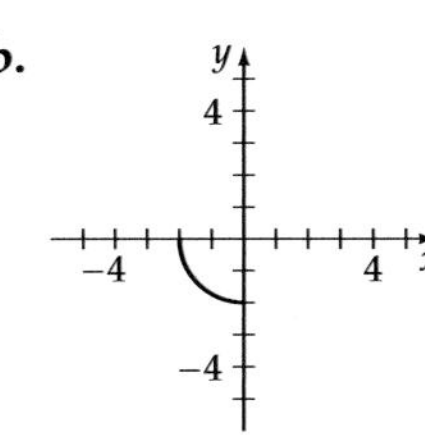

24. a.

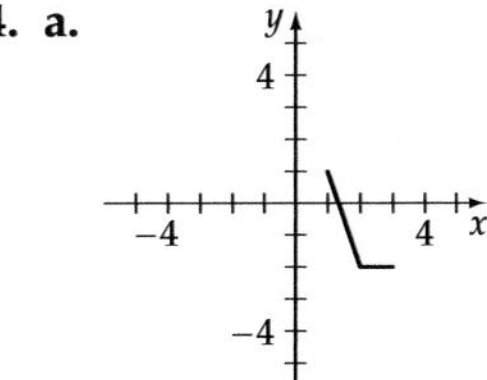

b.

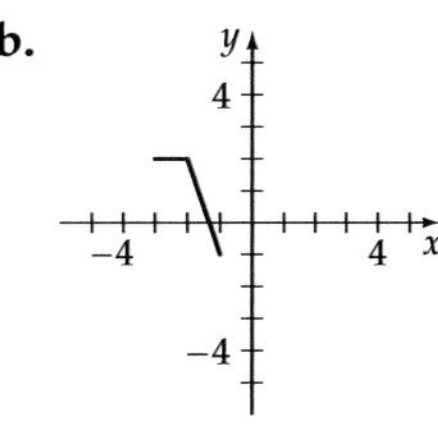

25. a

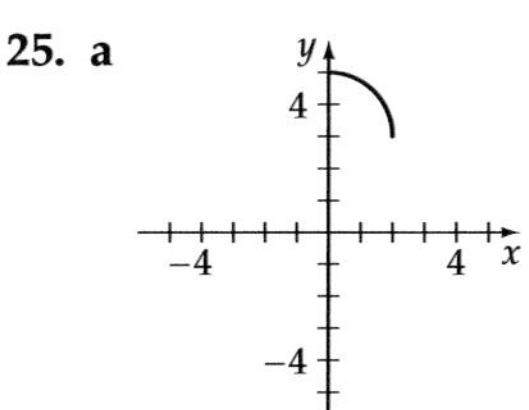

b.

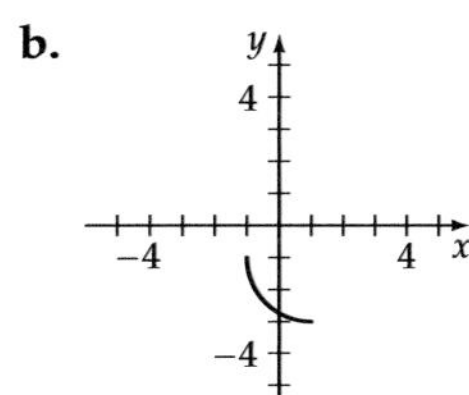

26. a.

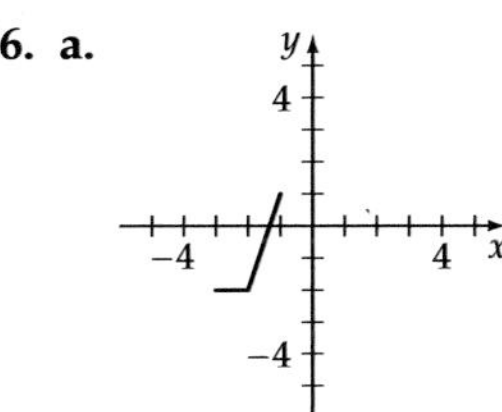

b.

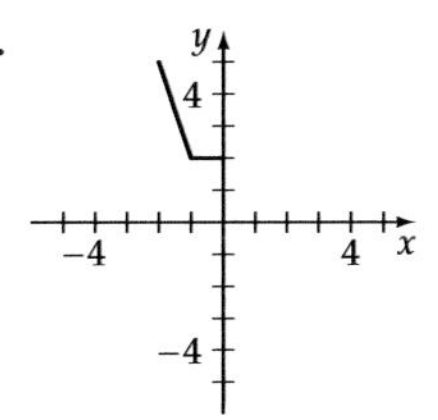

27. a.

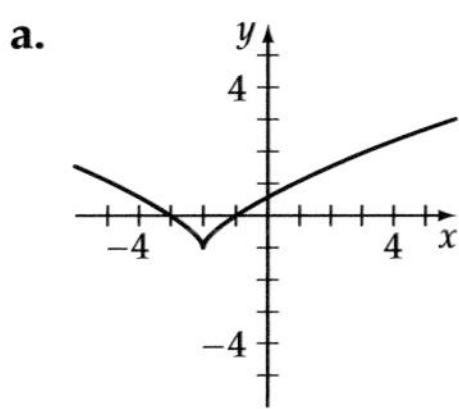

b.

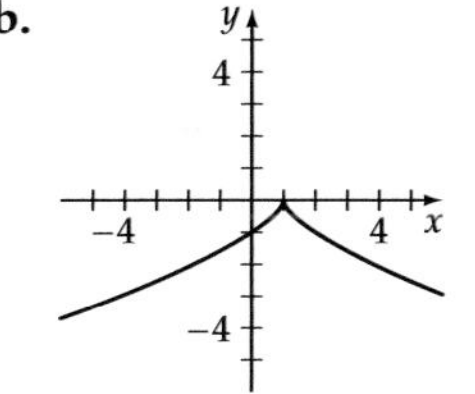

c.

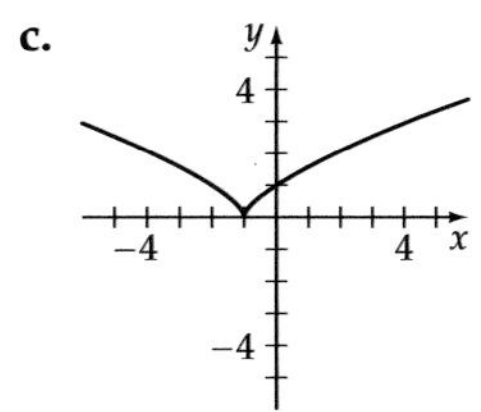

d.

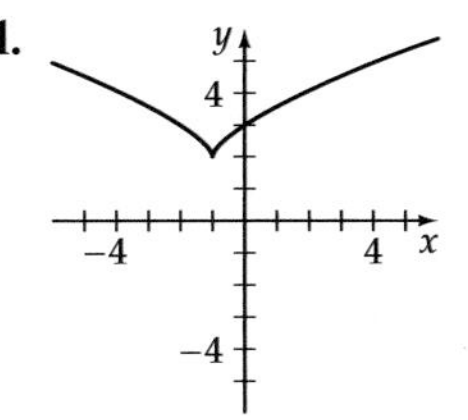

28. a.

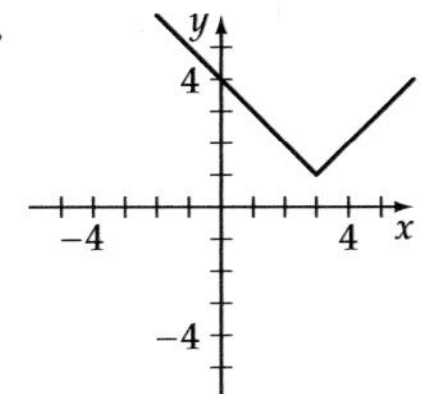

b.

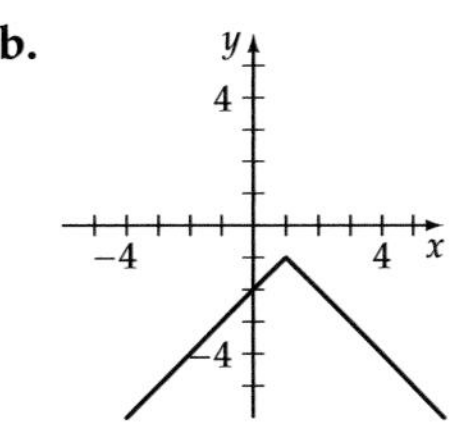

c.

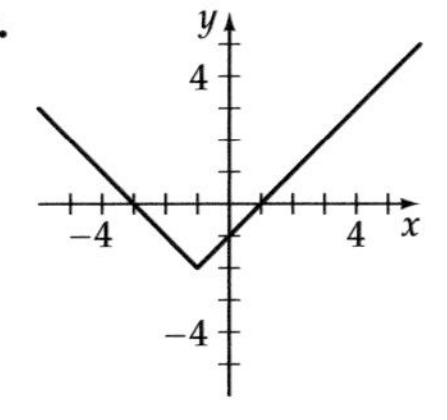

31. a.

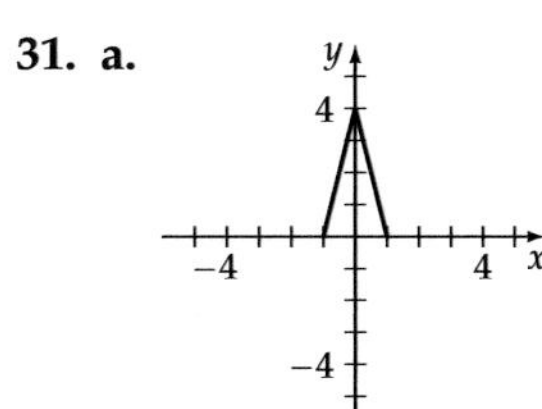

b.

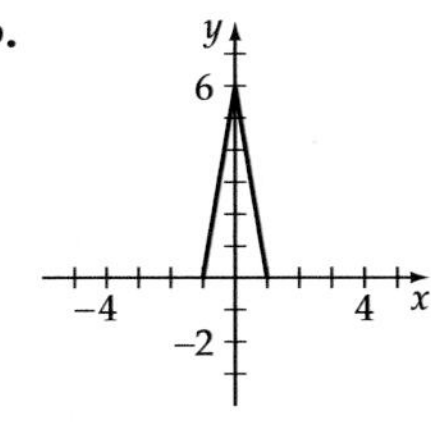

c.

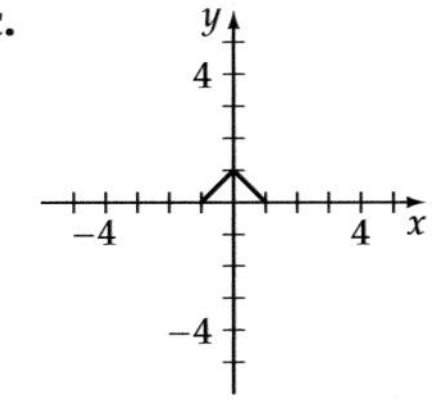

32. a.

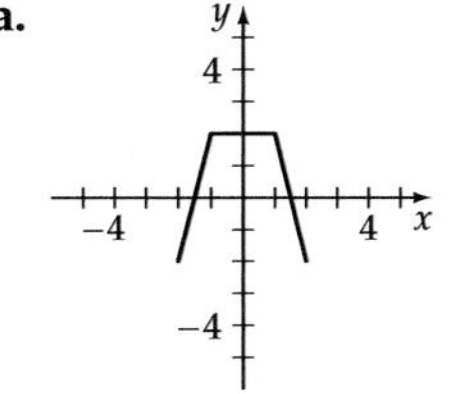

b.

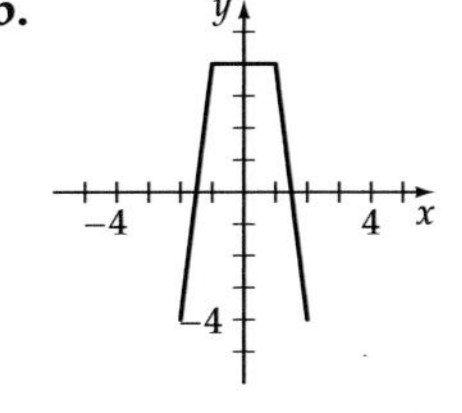

c.

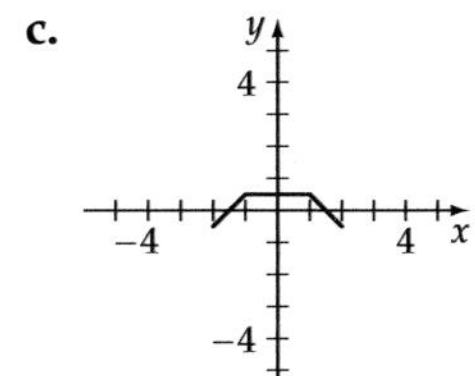

33. a.

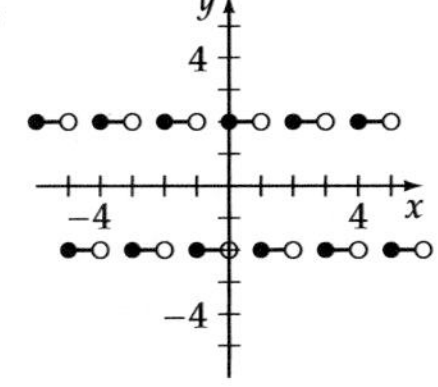

b.

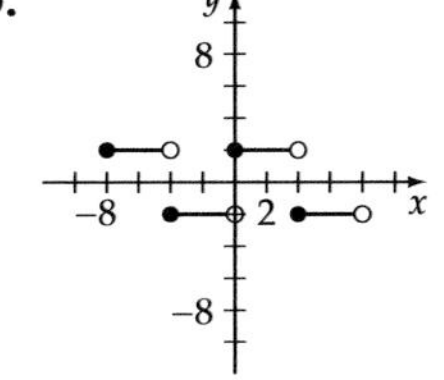

34. a.

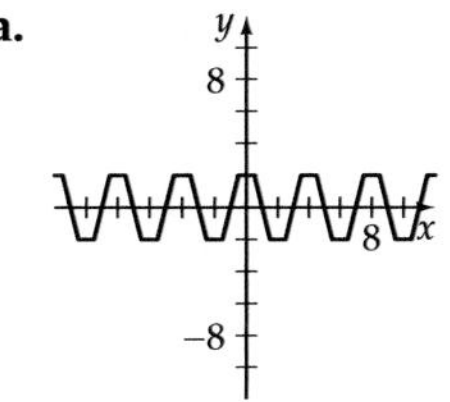

b.

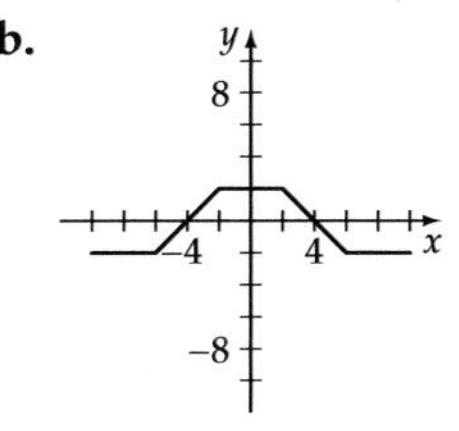

35. a.

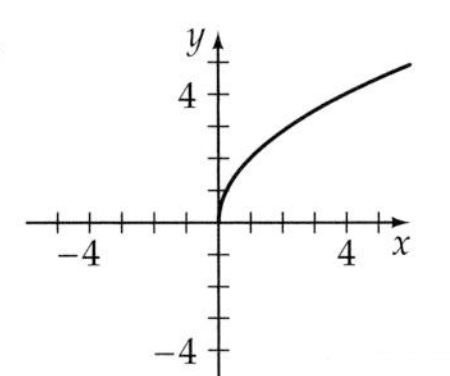

b.

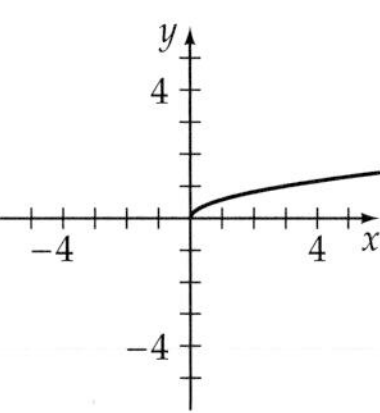

c.

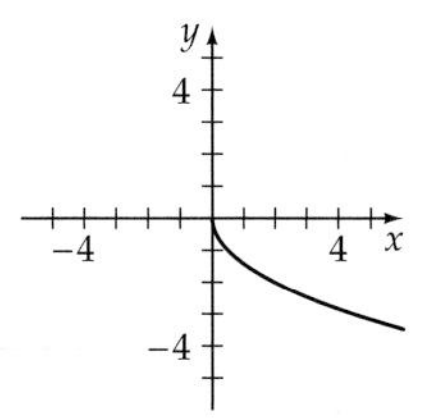

d.

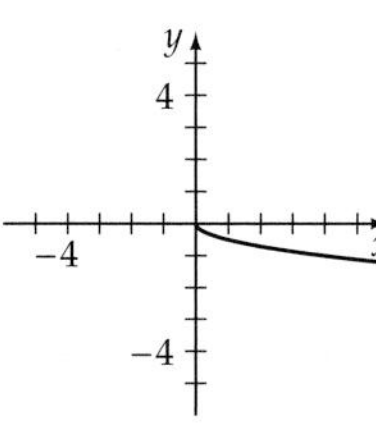

36. a.

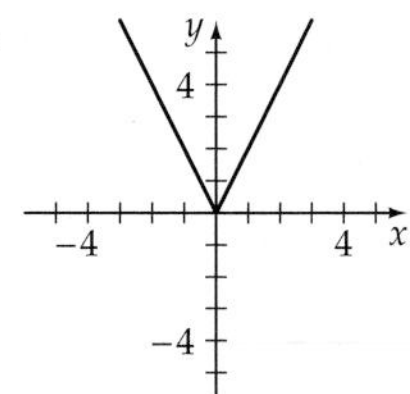

b.

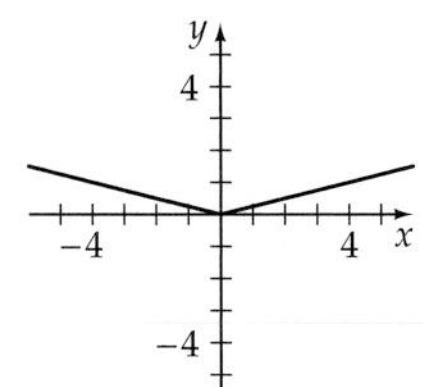

c.

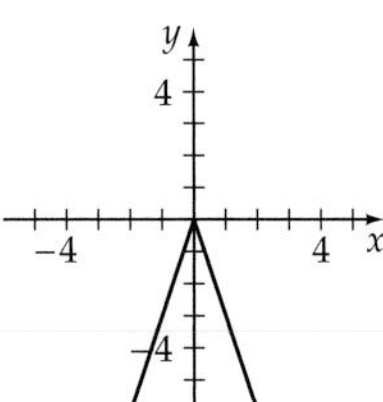

d.

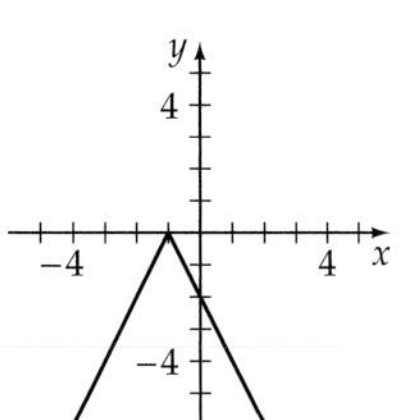

37. a.

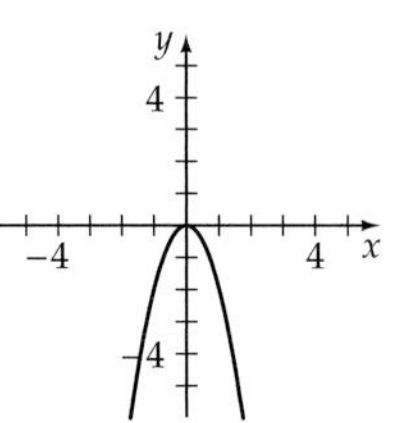

b.

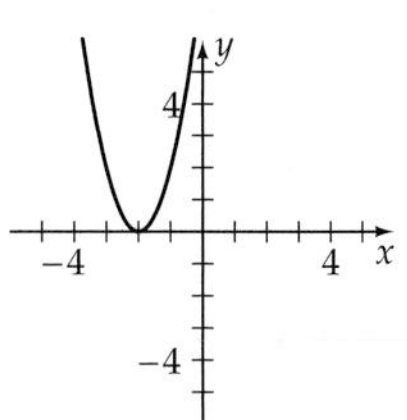

c.

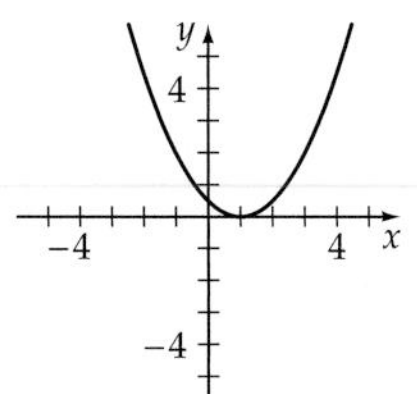

d.

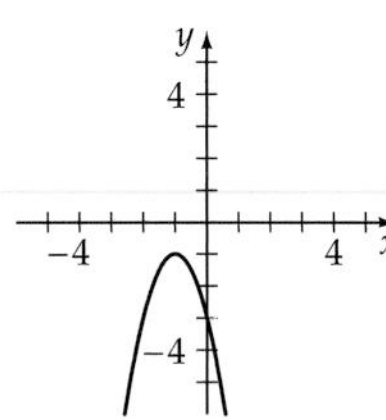

38. a.

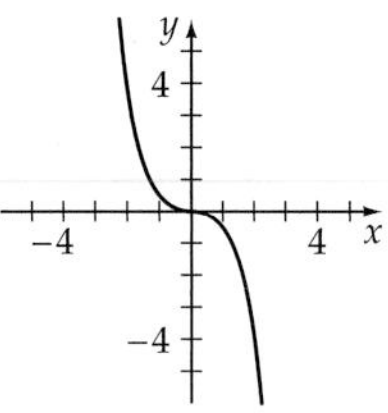

b.

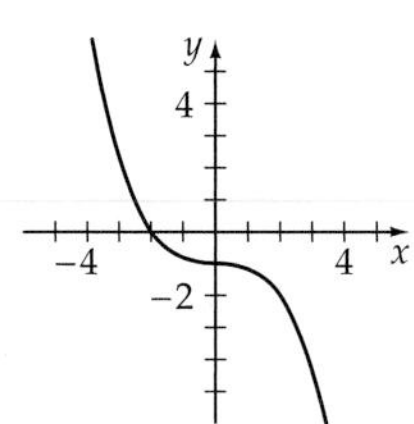

c.

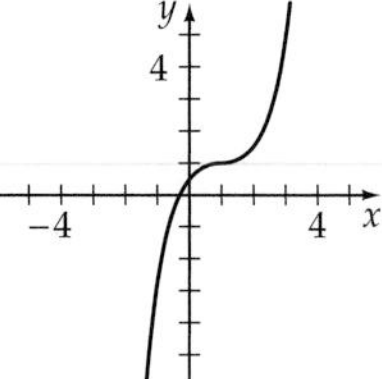

d.

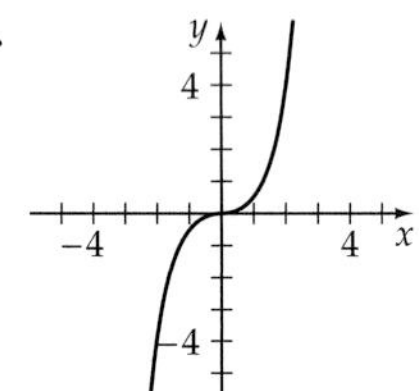

39. a.

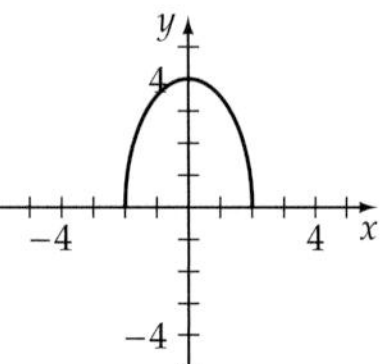

b.

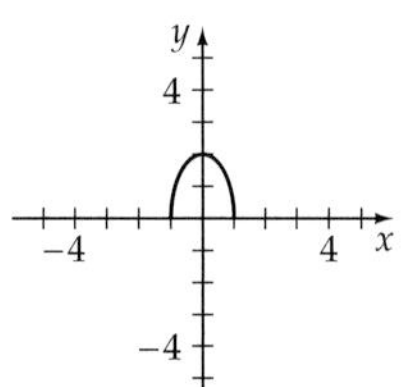

c.

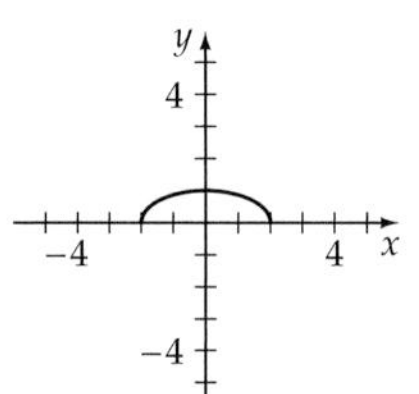

d.

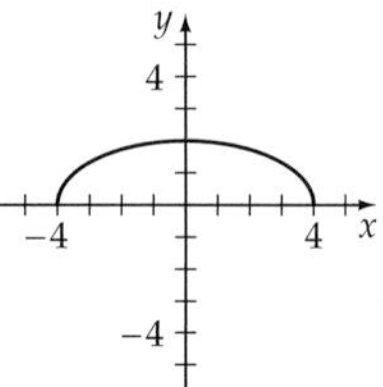

40. a.

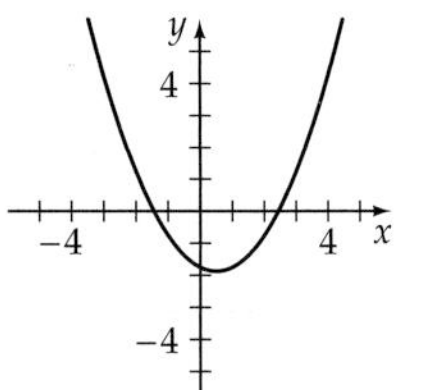

b.

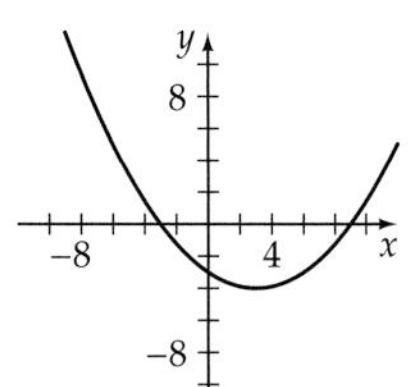

c.

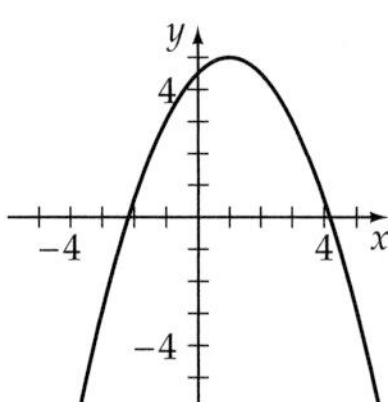

55.

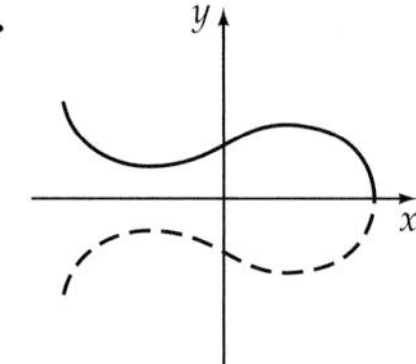

56.

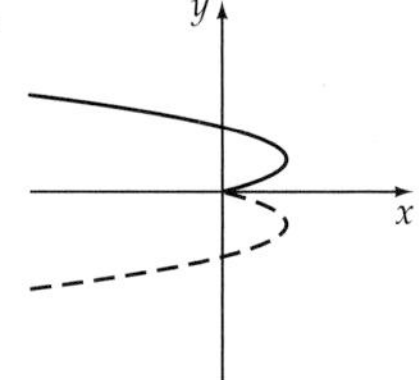

57.

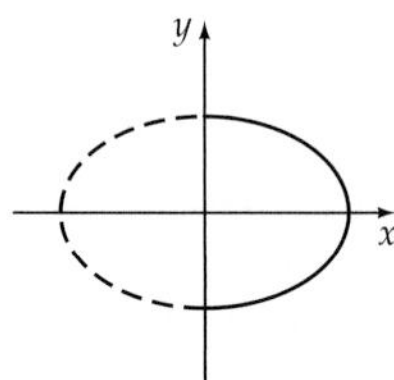

58.

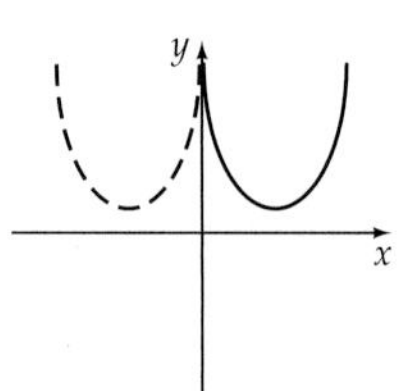

59.

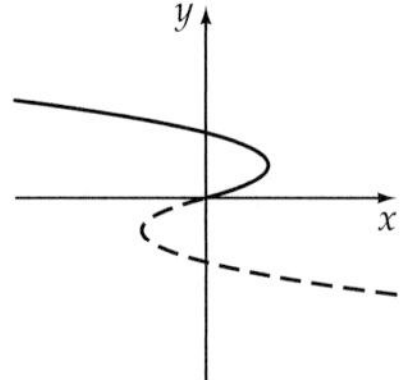

60.

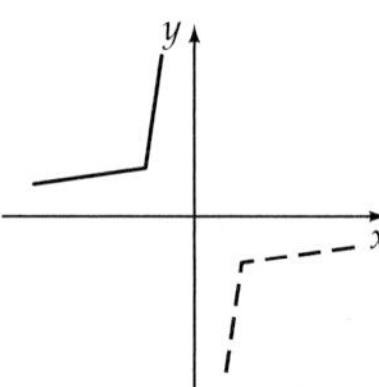

91. a.

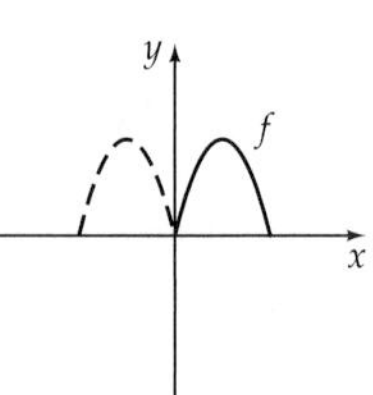

b.

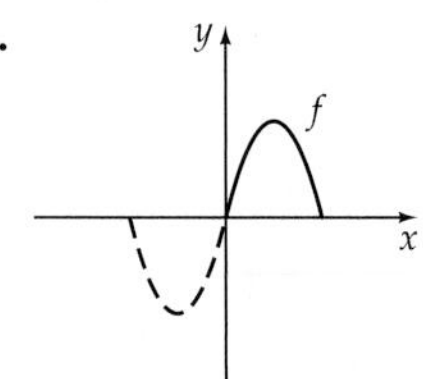

92. a.

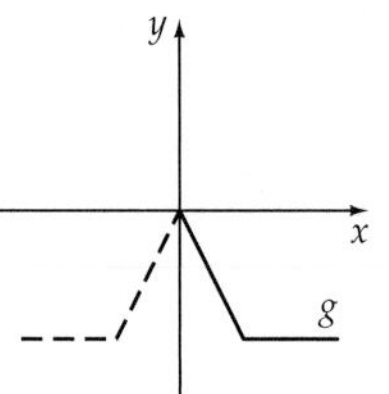

b.

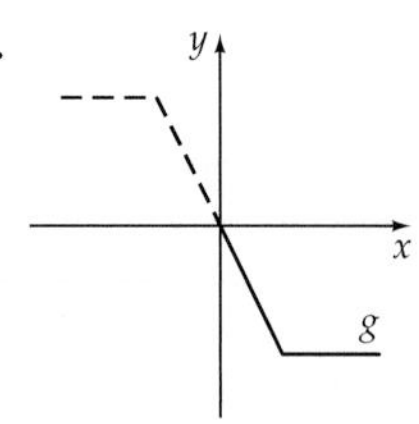

93.

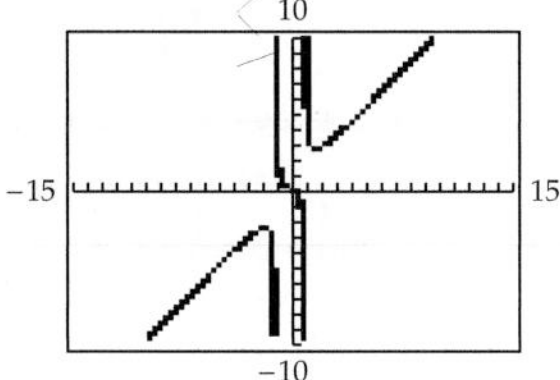

94.

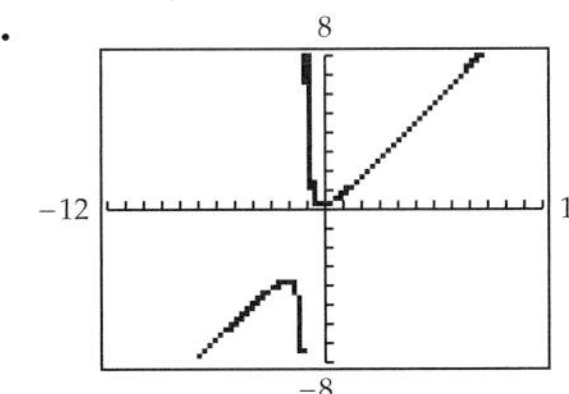

95.

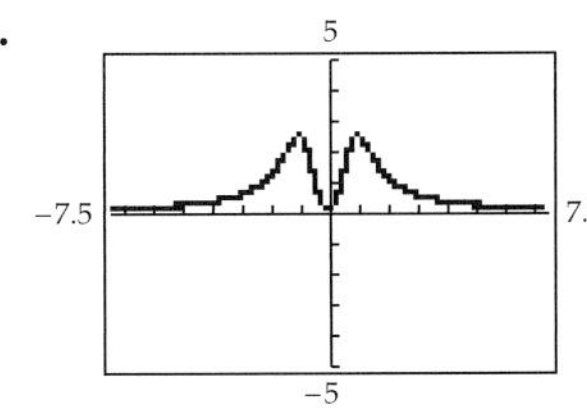

96.

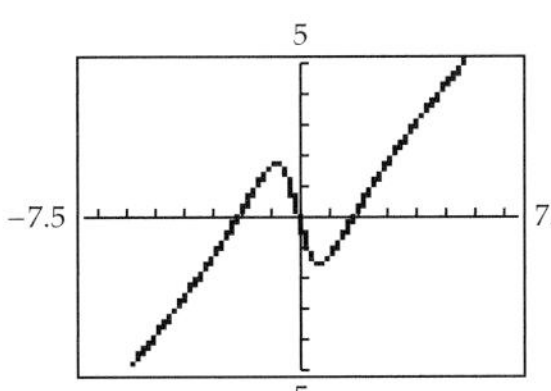

97.

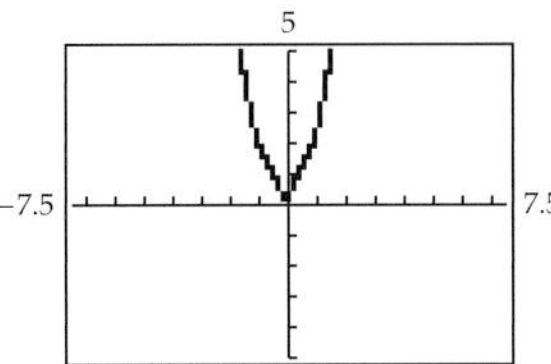

98.

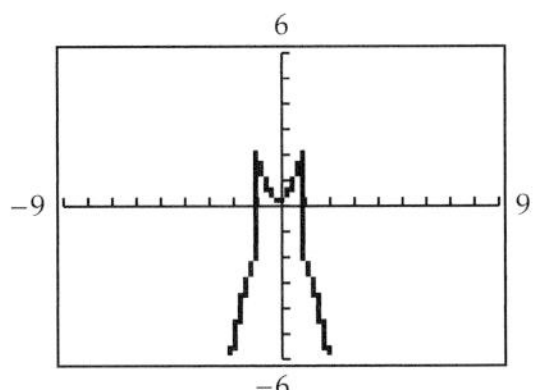

Chapter 3 Review Exercises, *page 307*

7. [3.3]

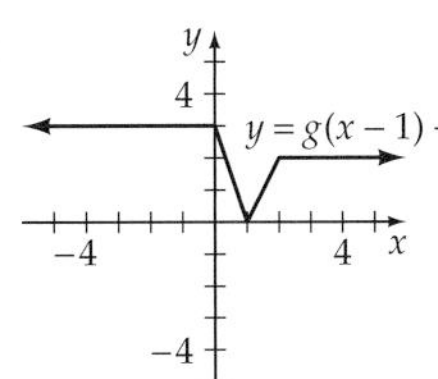

8. [3.3]

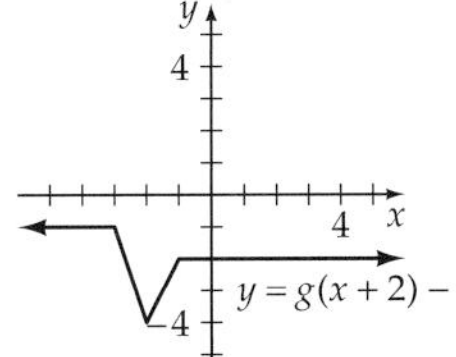

9. [3.3]

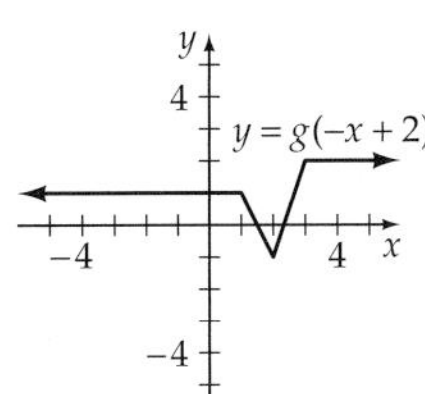

10. [3.3]

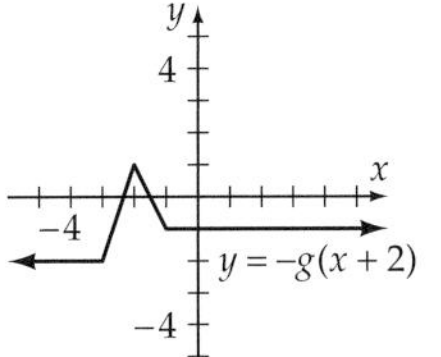

11. [3.3]

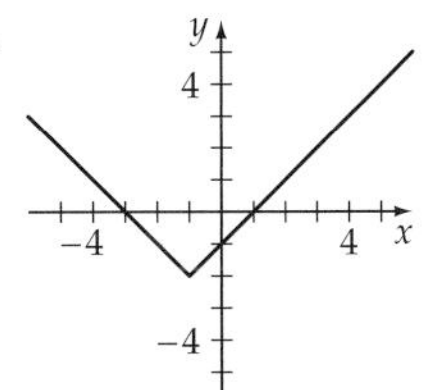

12. [3.3]

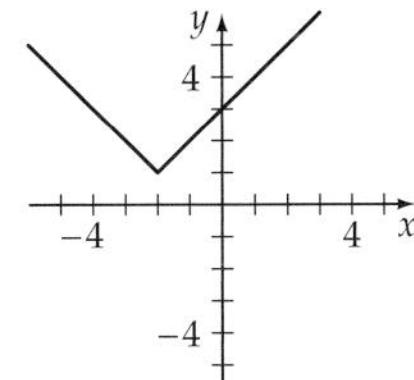

13. [3.3]

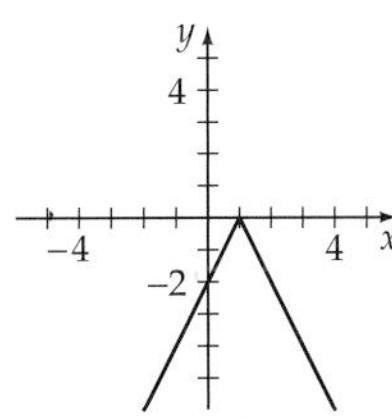

14. [3.3]

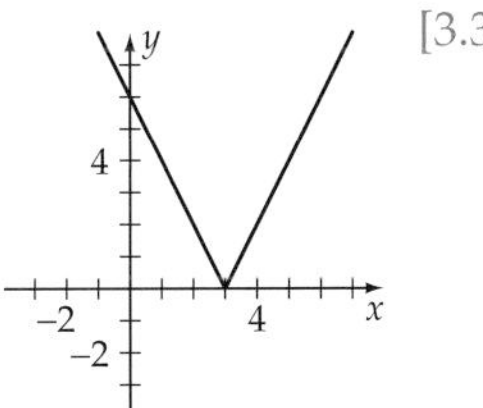

Chapter 3 Test, *page 309*

7. [3.2]

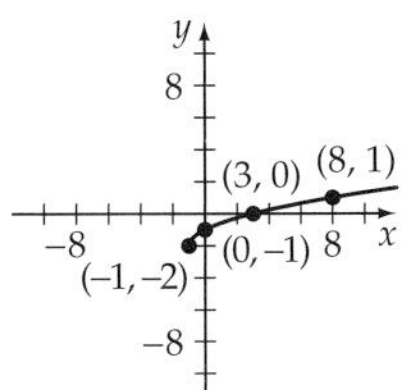

10. a.

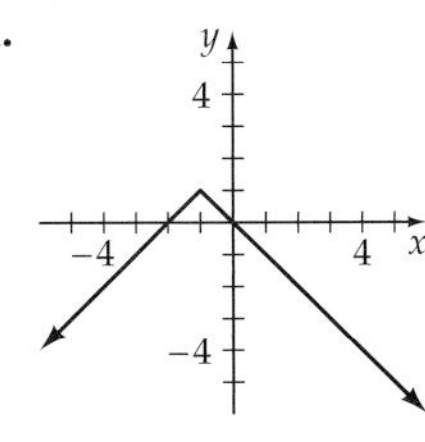

b. [3.3]

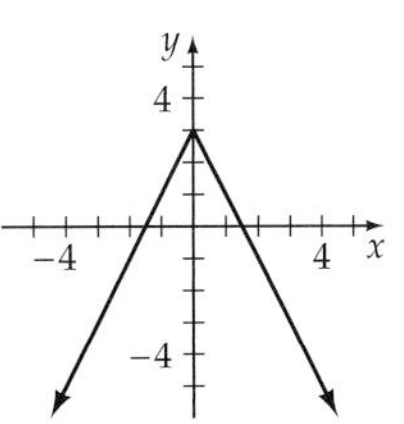

11. a.

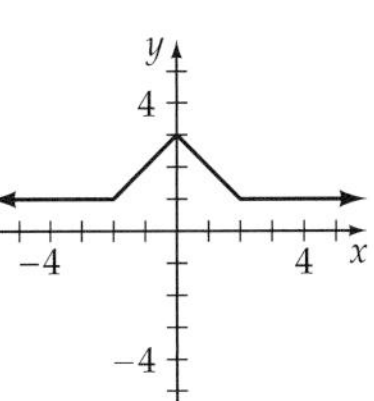
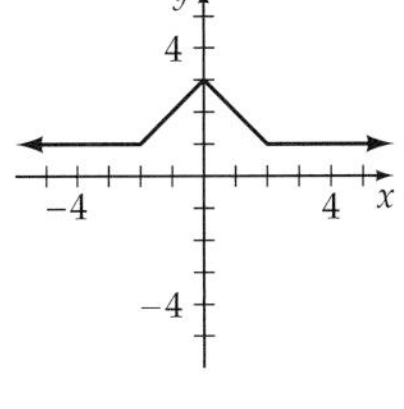

b. [3.3]

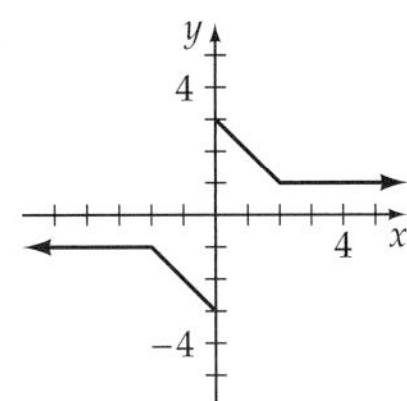

12. [3.3]

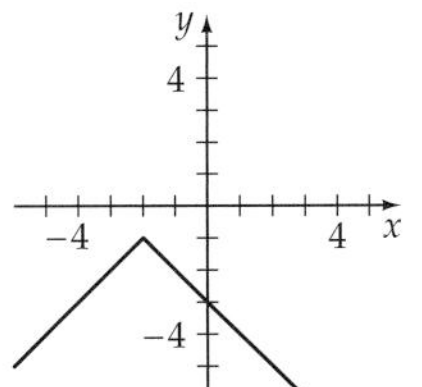

Chapter 4

Exercises 4.2, *page 332*

15. 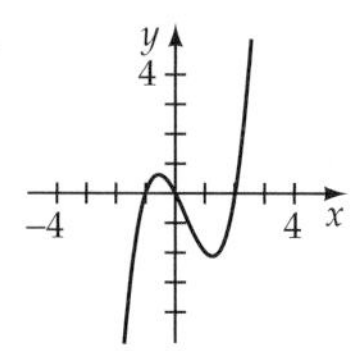

16. 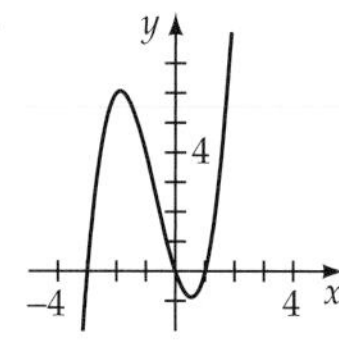

17. 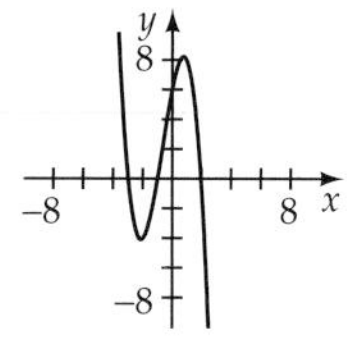

18. 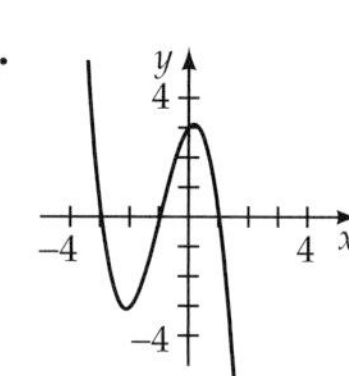

19. 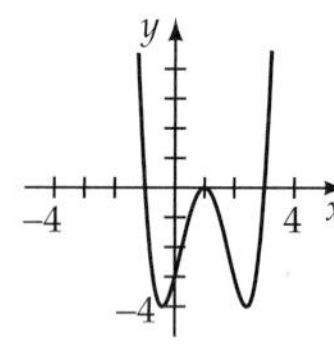

20. 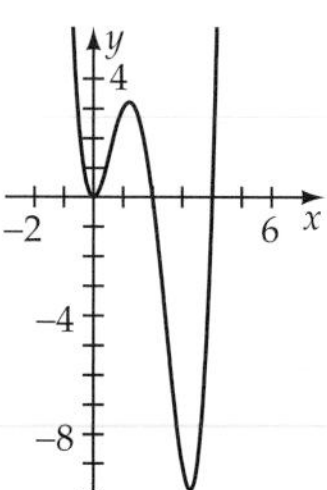

21. 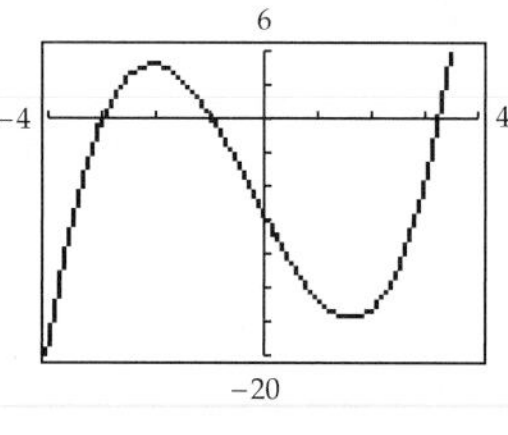

22. 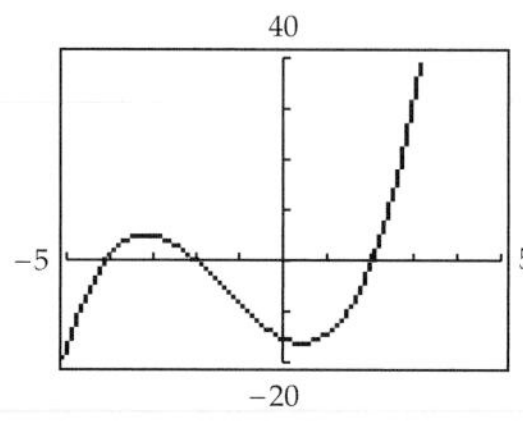

23. 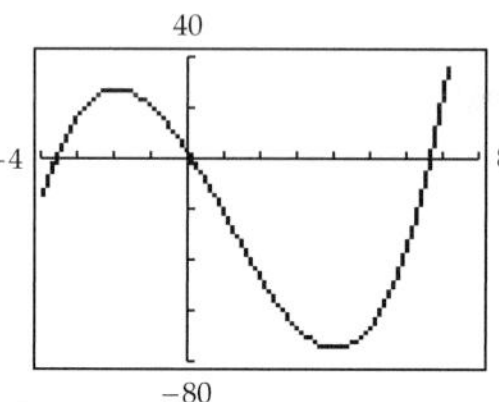

24. 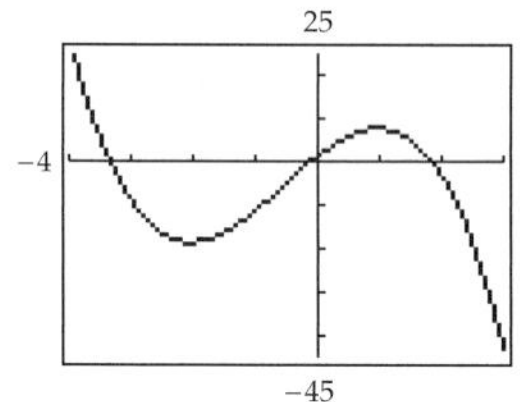

25. 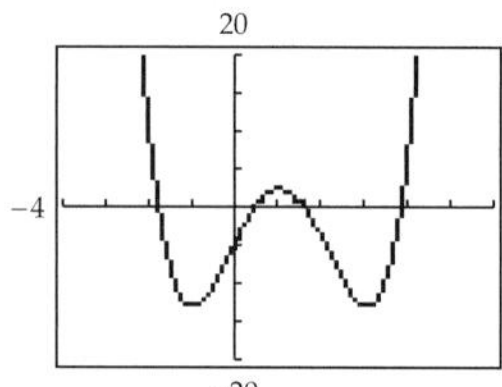

26. 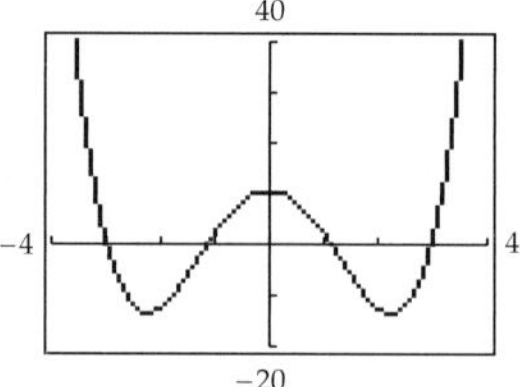

Exercises 4.5, *page 372*

9. 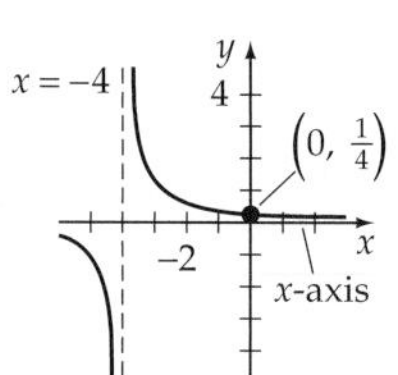

10. 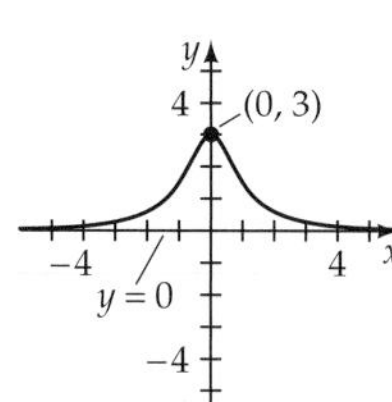

11. 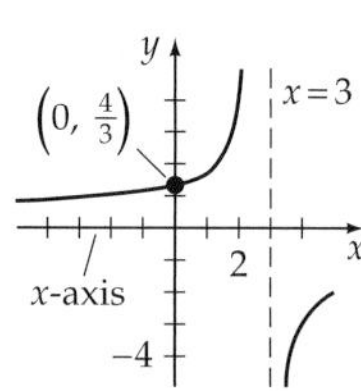

12. 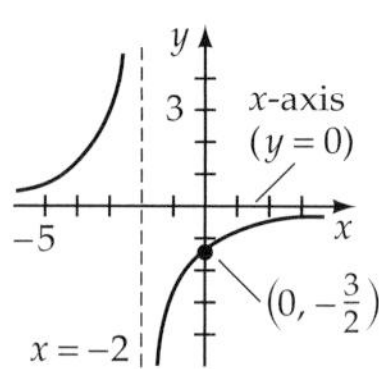

13. 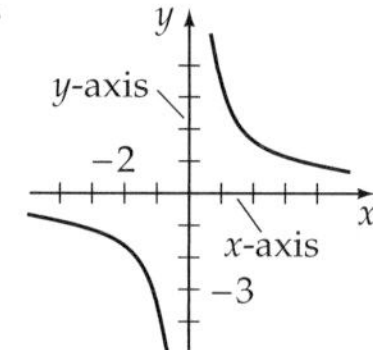

14. 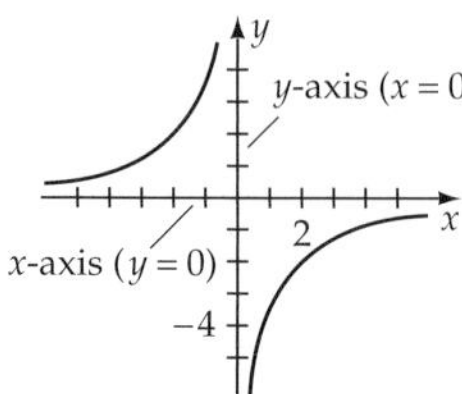

15. 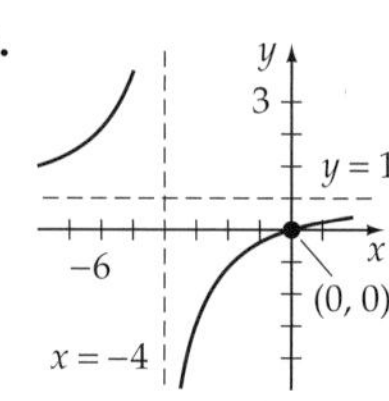

16. 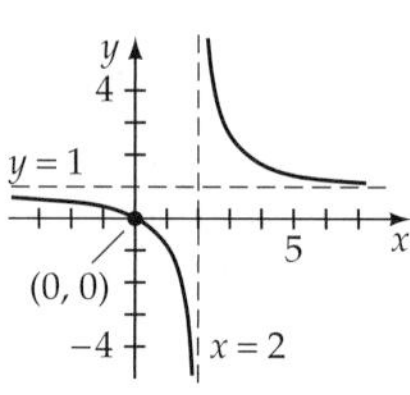

17. 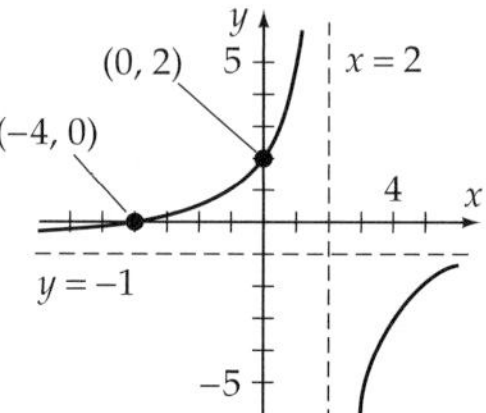

18. 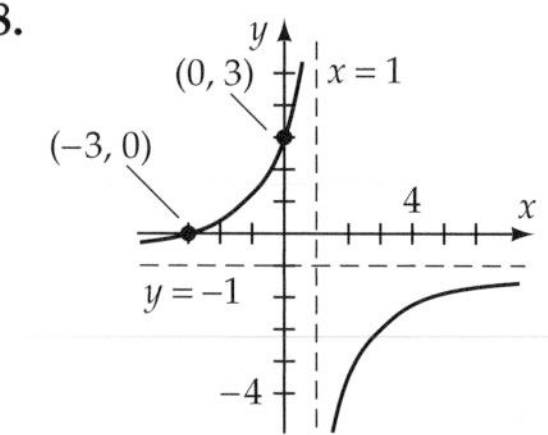

19. 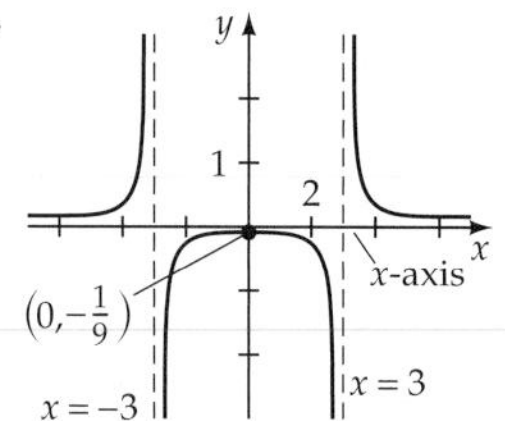

20. 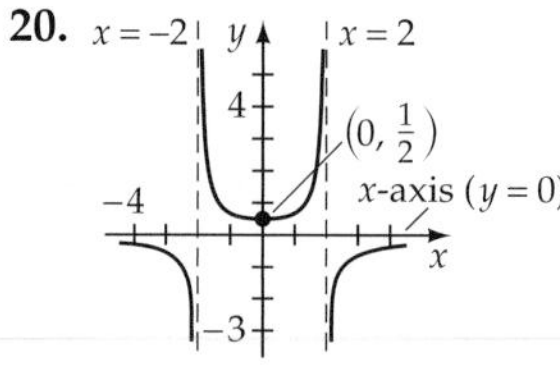

21. 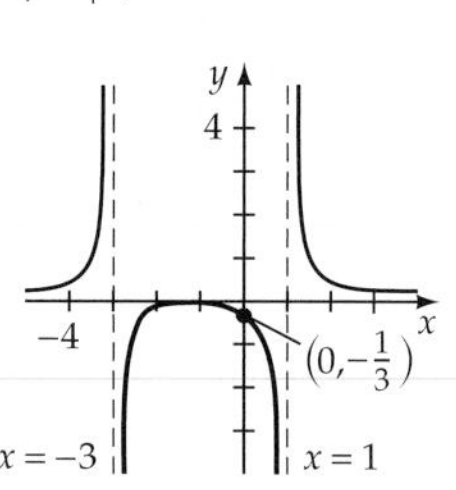

22.
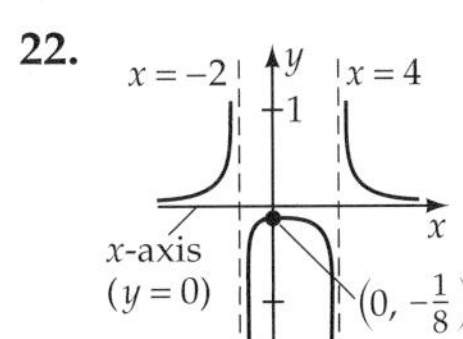

23.
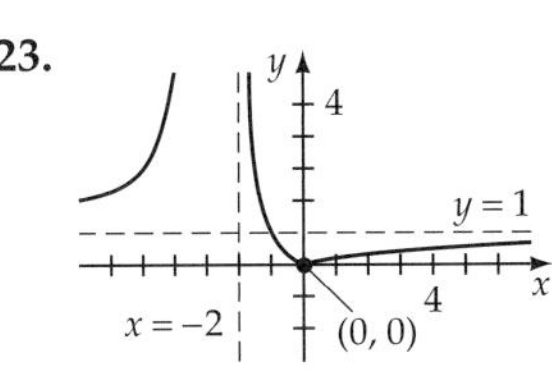

24.
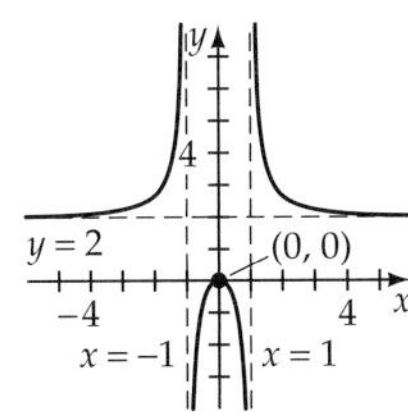

25.
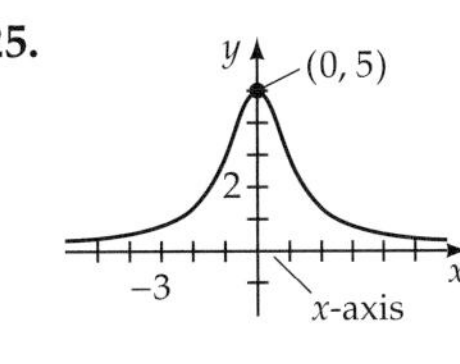

26.
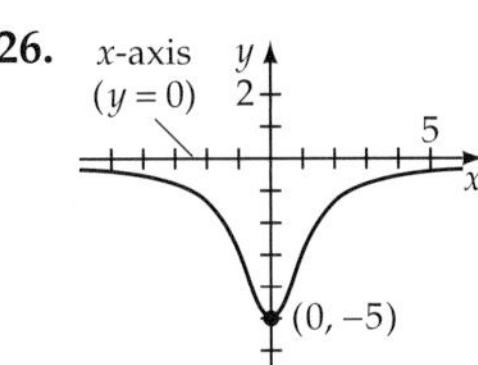

27.
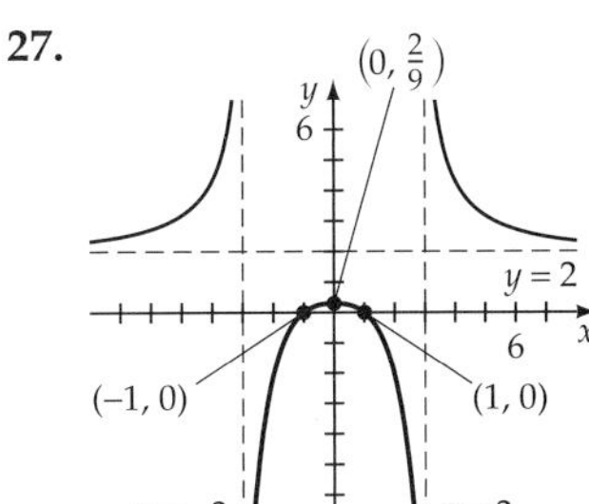

28.
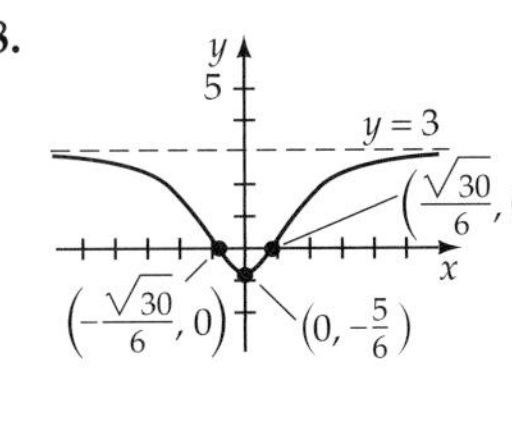

29.
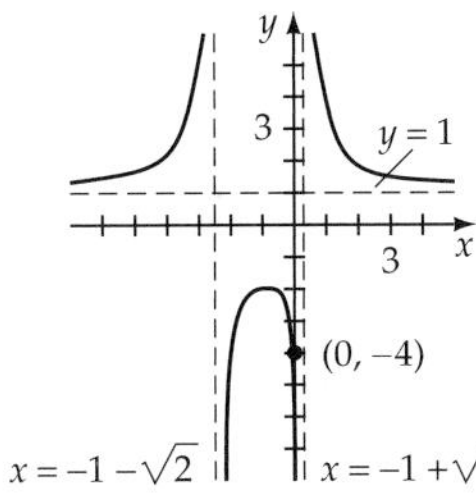

30.
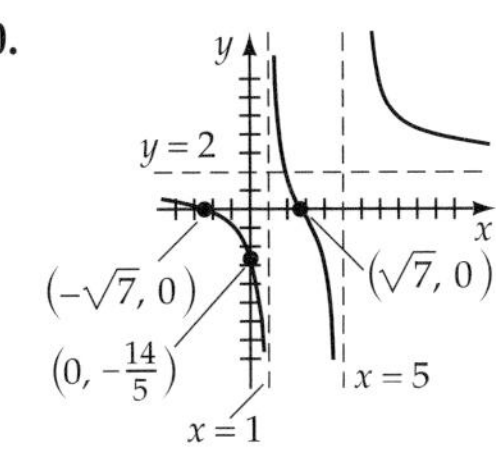

35.
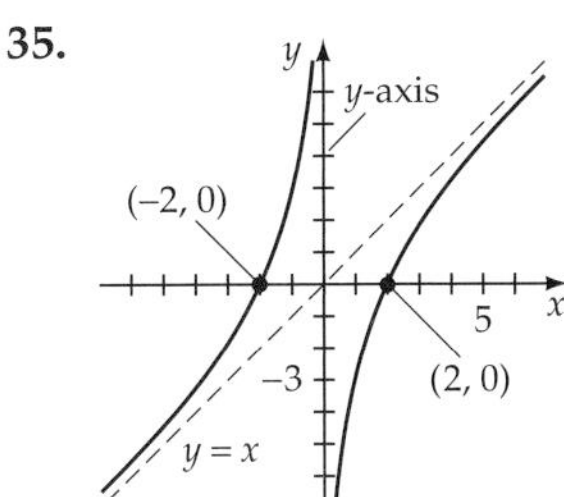

36.
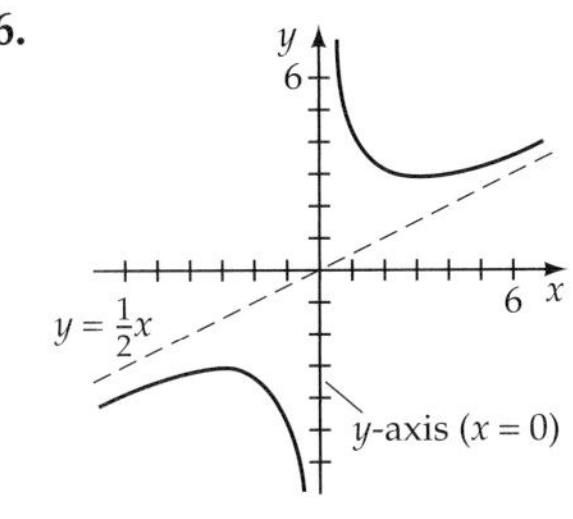

37.
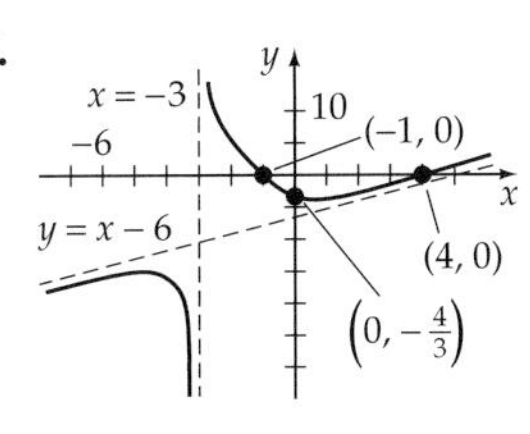

38.
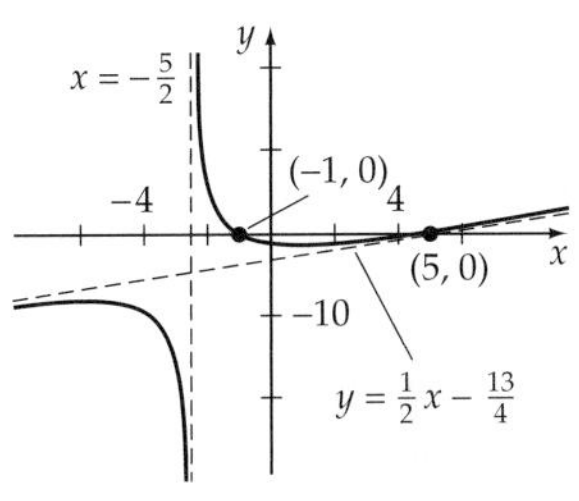

39.
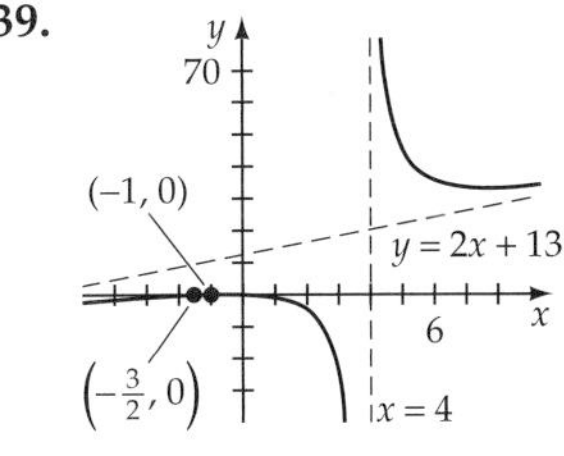

40.
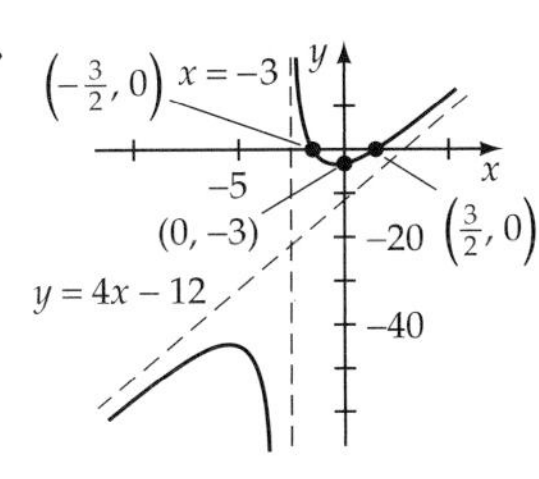

41.
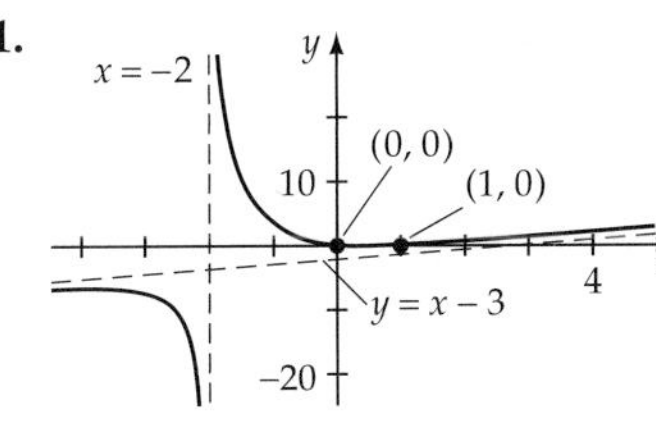

42.
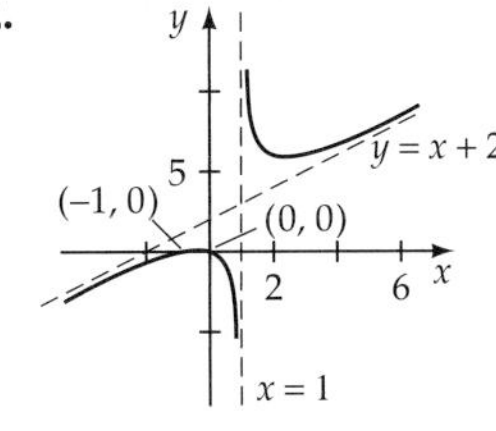

43.
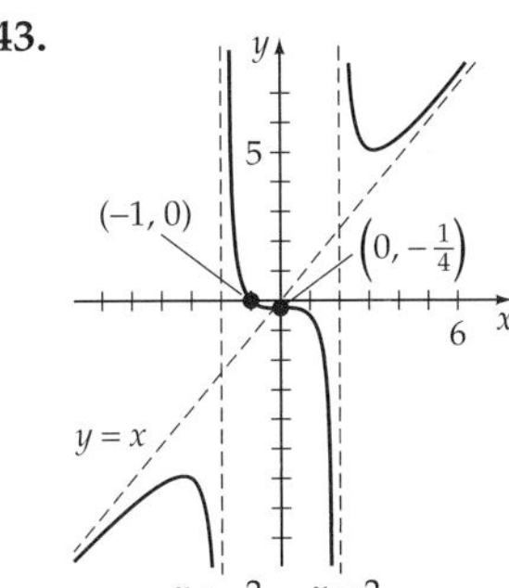

44.
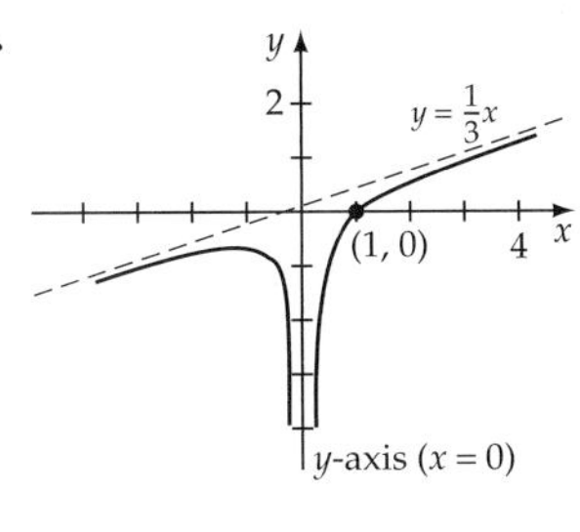

45.
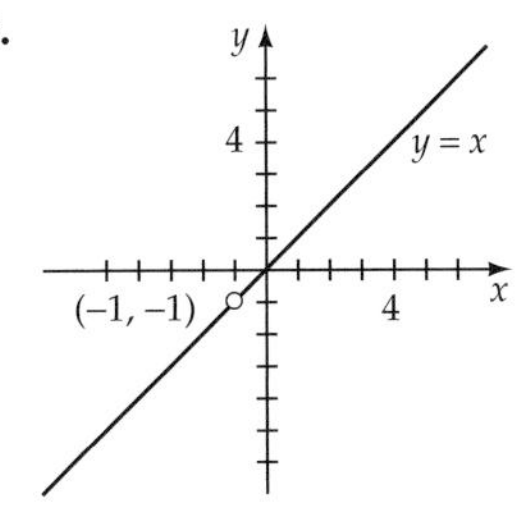

46.
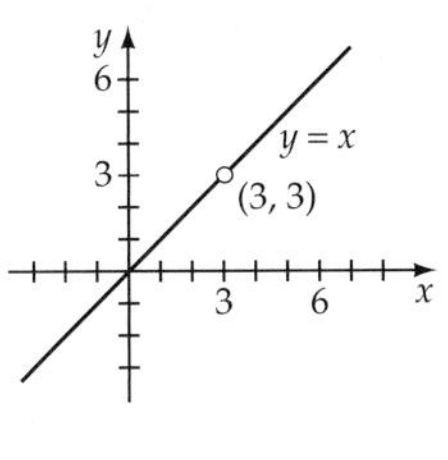

47.
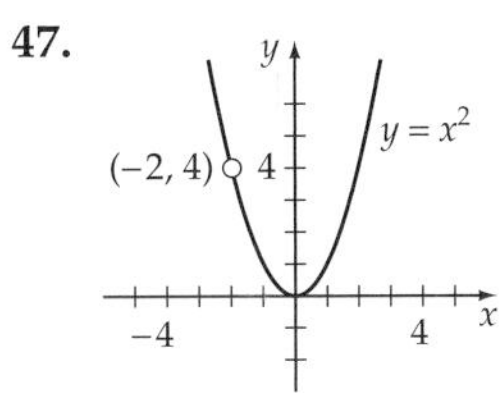

48.
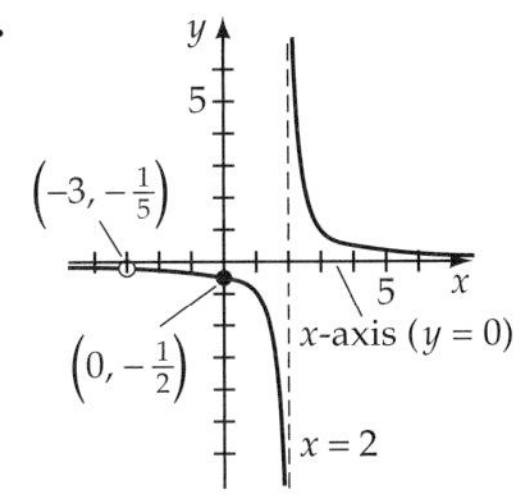

49.
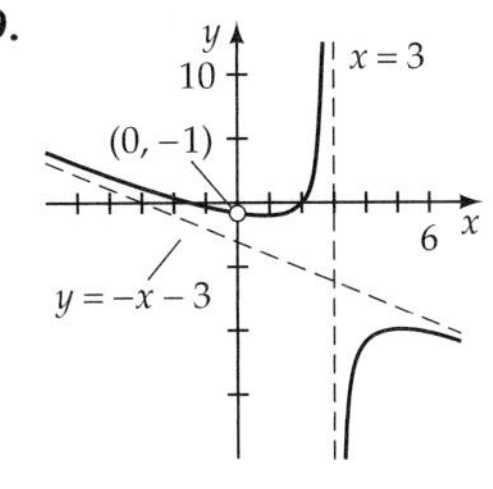

50.
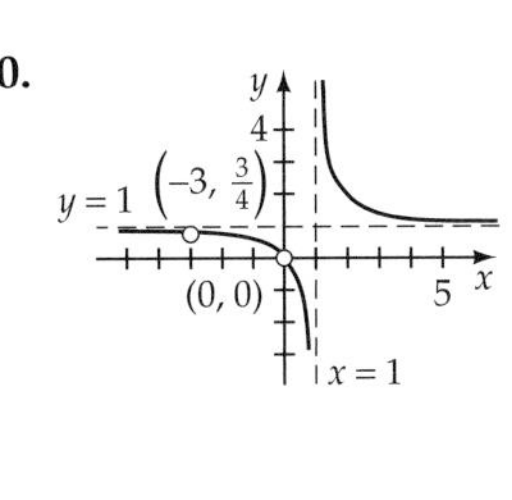

51.

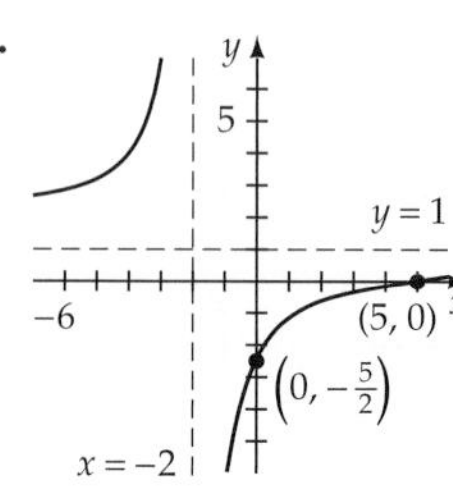

52.

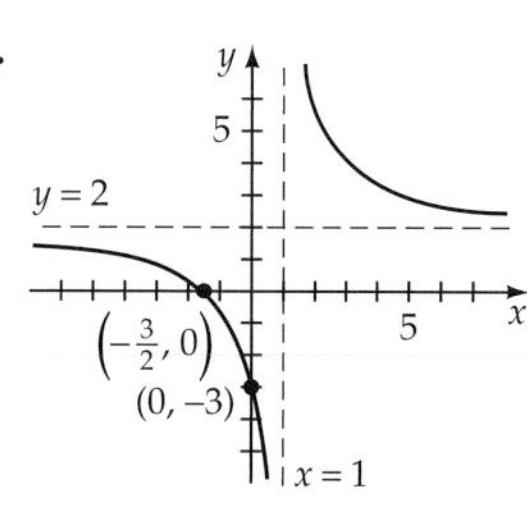

55. c.

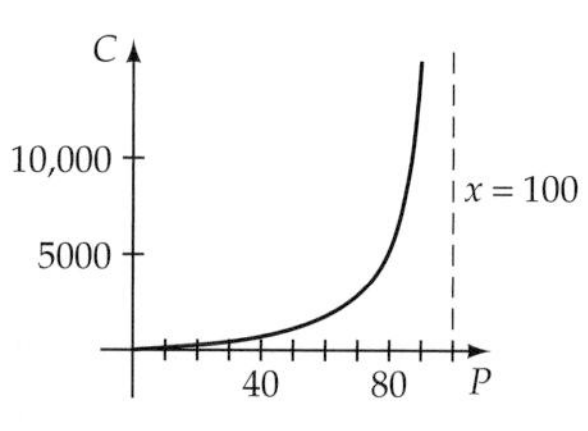

61. a.

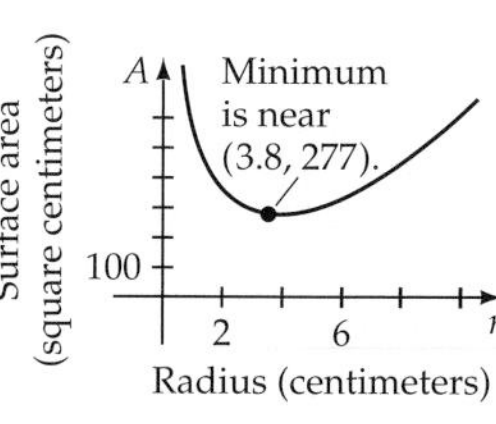

Chapter 4 Review Exercises, *page 378*

15. [4.2]

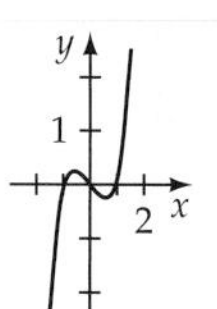

16. [4.2]

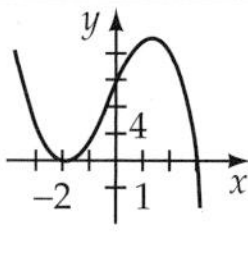

17. [4.2]

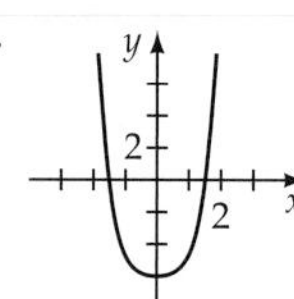

18. [4.2]

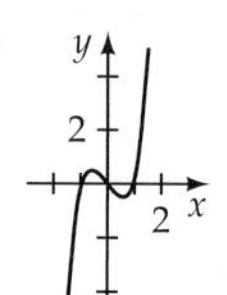

19. [4.2]

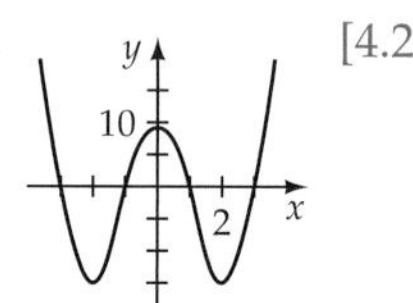

20. [4.2]

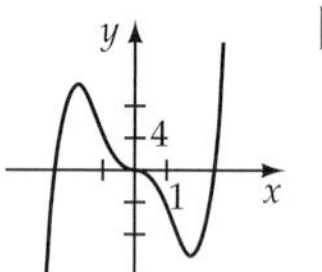

47. [4.5]

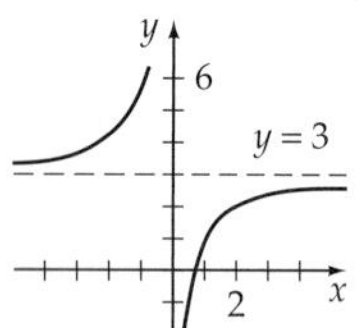

48. [4.5]

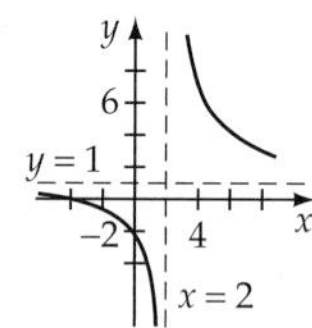

49. [4.5]

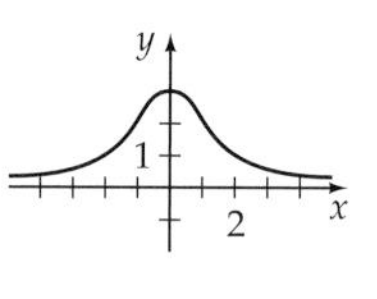

50. [4.5]

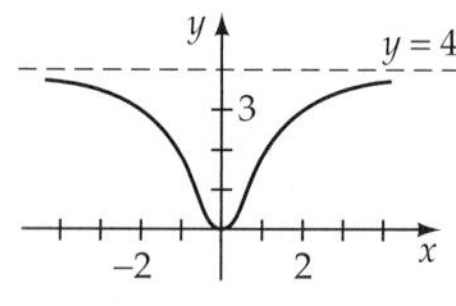

51. [4.5]

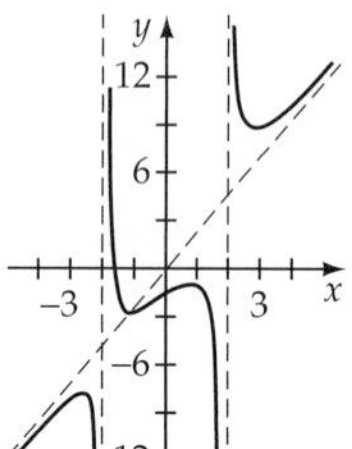

52. [4.5]

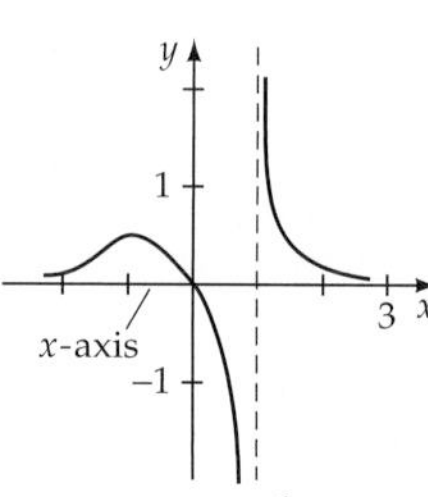

53. [4.5]

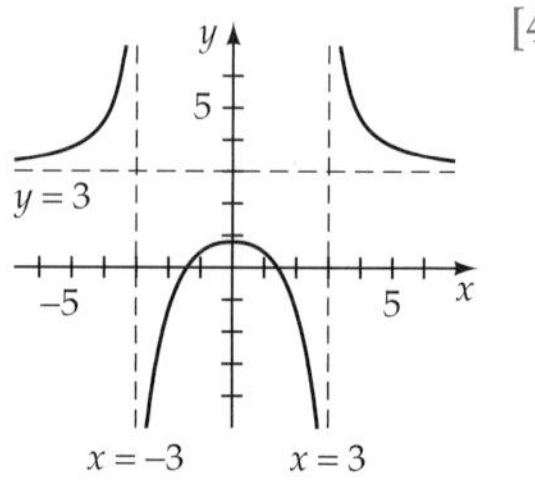

54. [4.5]

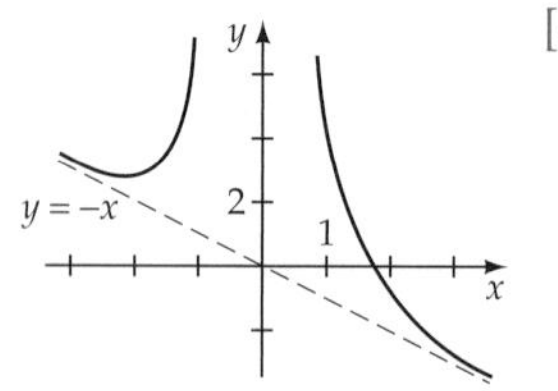

Chapter 4 Test, *page 380*

17. [4.5]

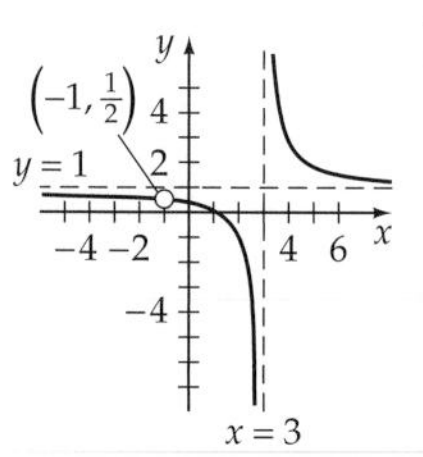

18. [4.5]

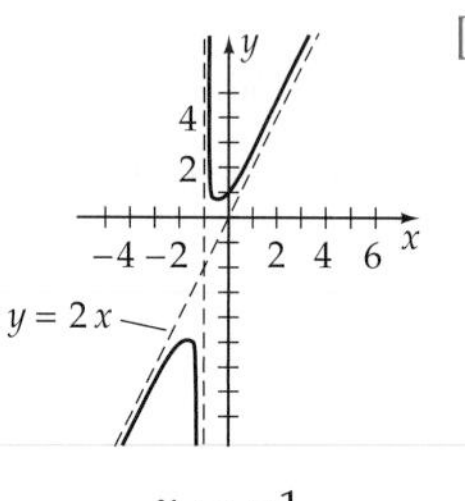

$x = -1$

Chapter 5

Exercises 5.1, *page 395*

33.
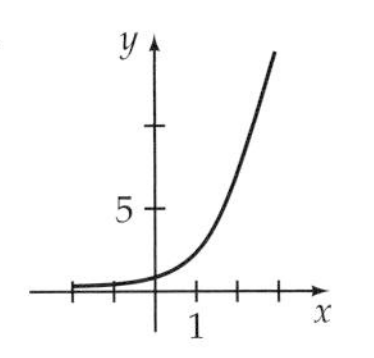

34.
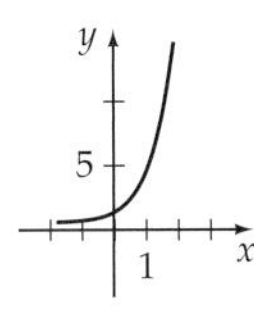

35.
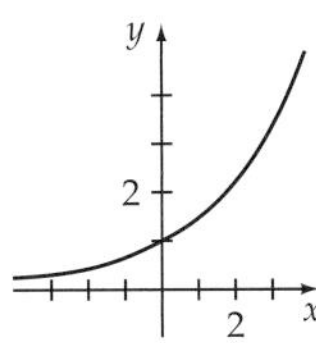

36.
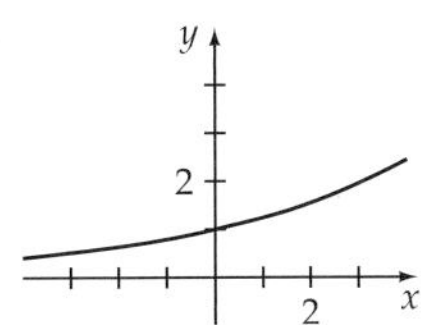

37.
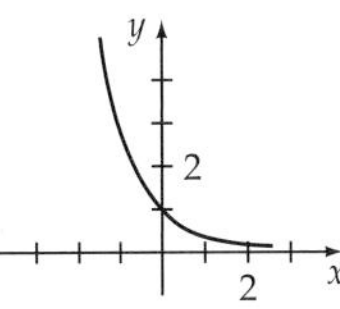

38.
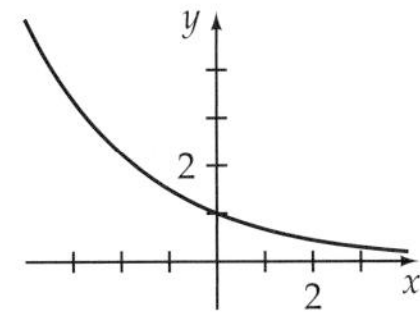

39.
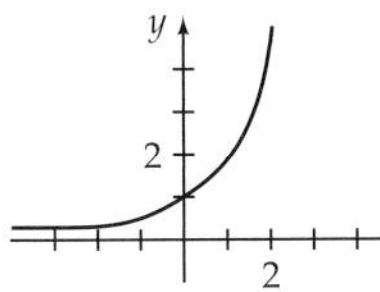

40.
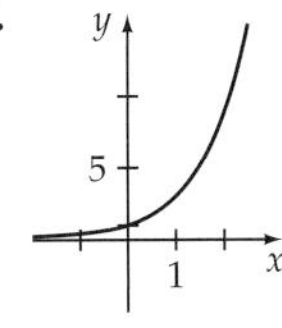

41.
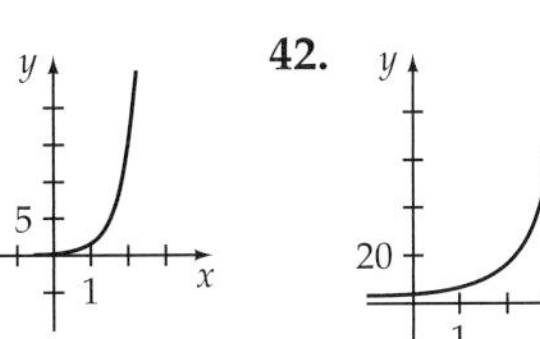

42.

43.
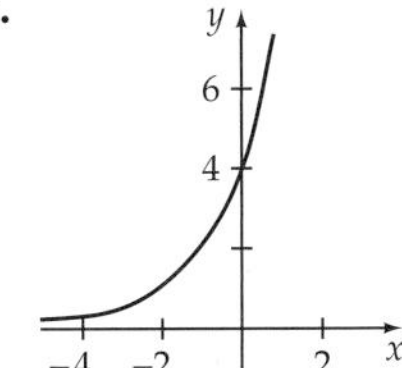

44.
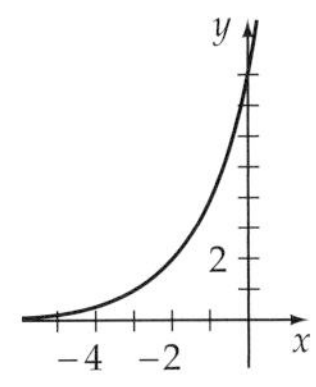

45.
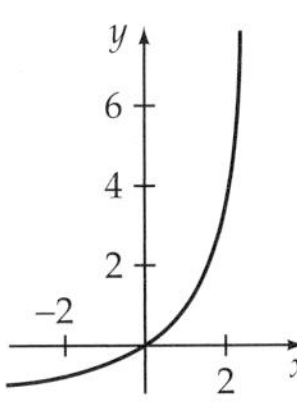

46.
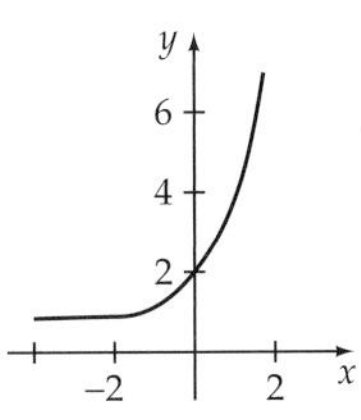

47.
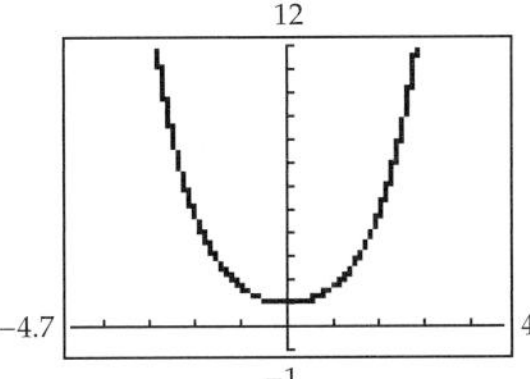

48.
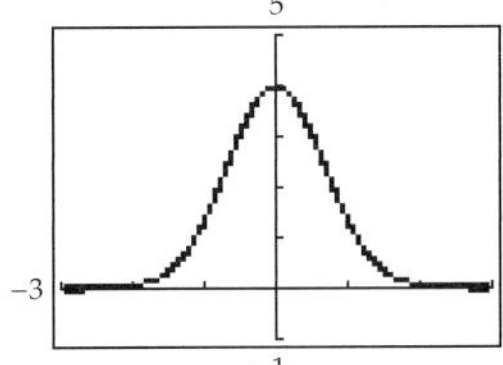

49.
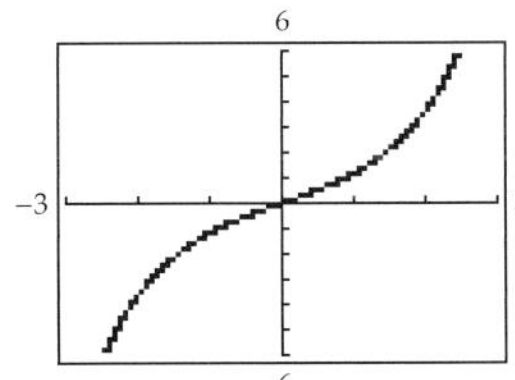

50.
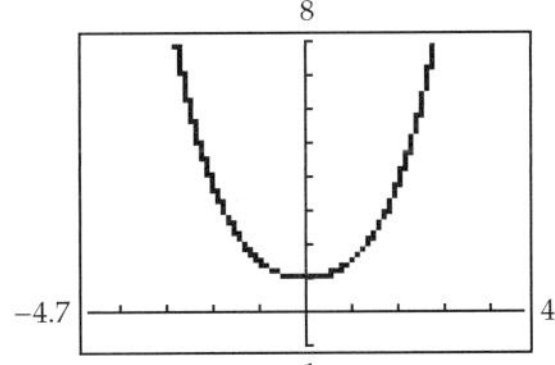

51.
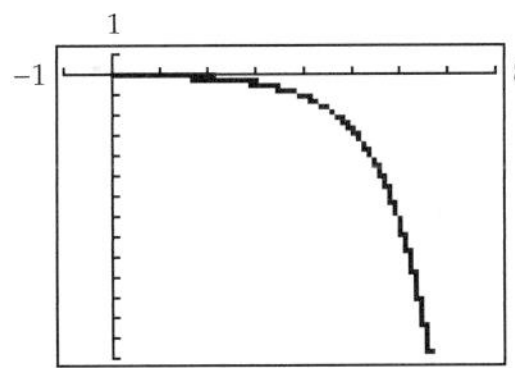

52.
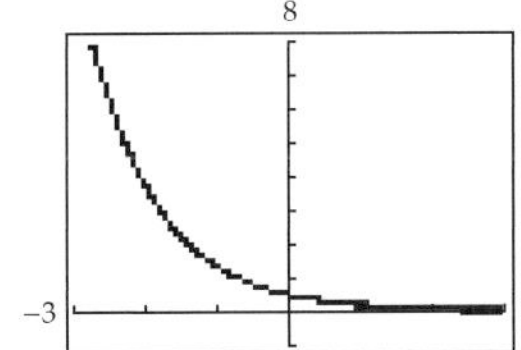

53.
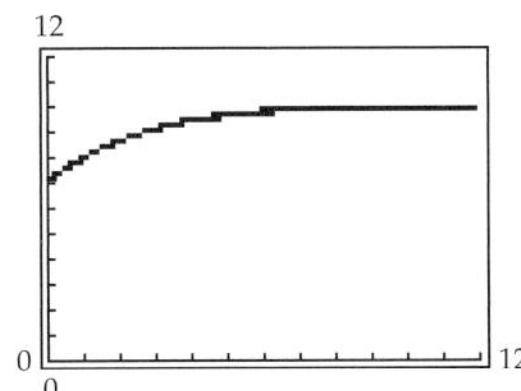

54.
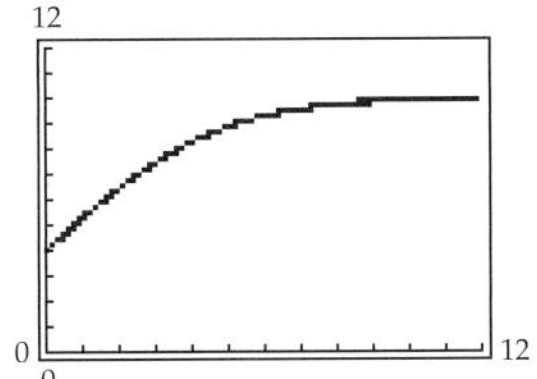

Explorations, *page 399*

1. a.
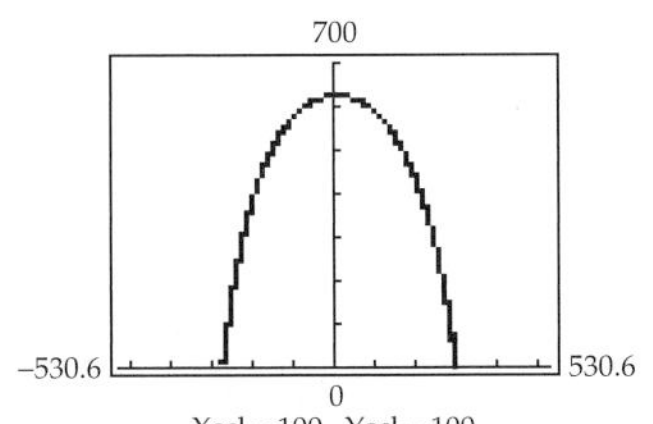

Exercises 5.2, *page 409*

31.
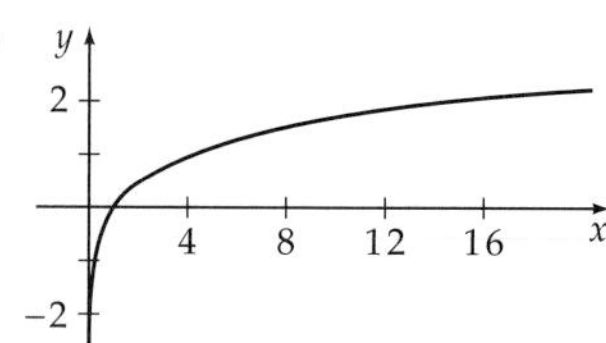

32.
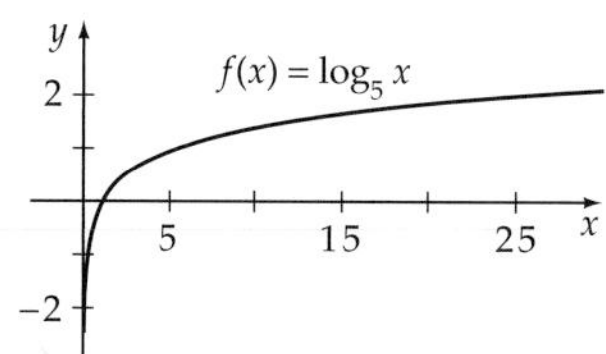

33.
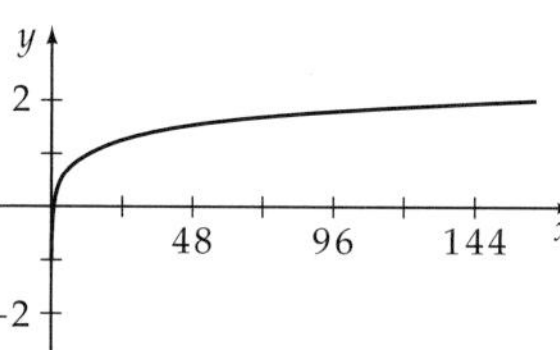

34.
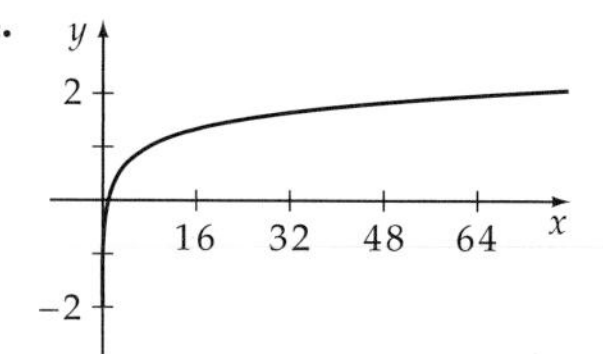

35.
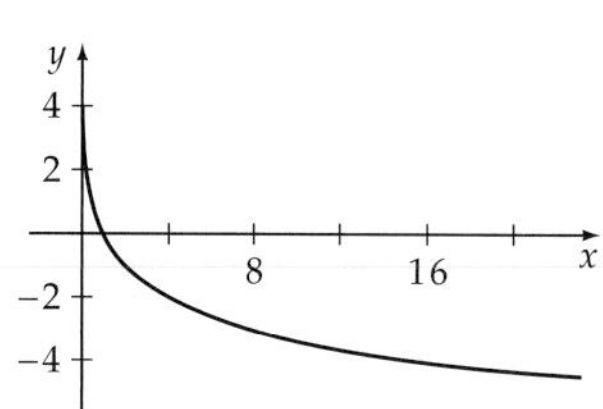

36.
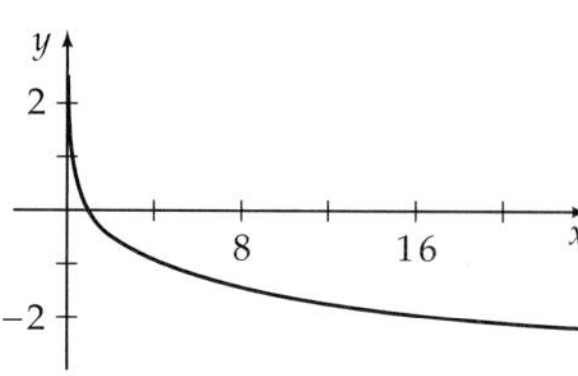

47.
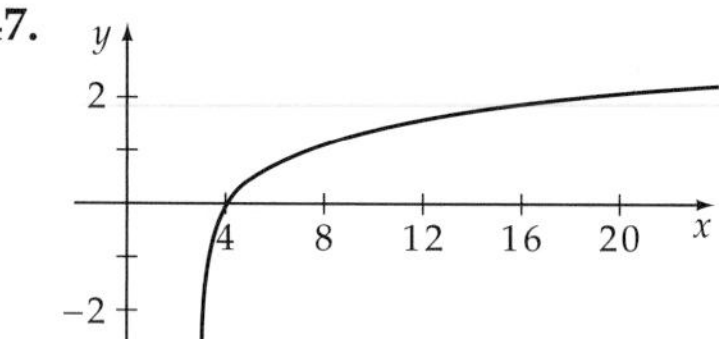

48.
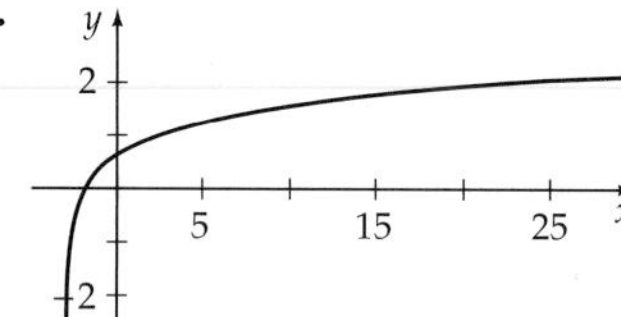

49.
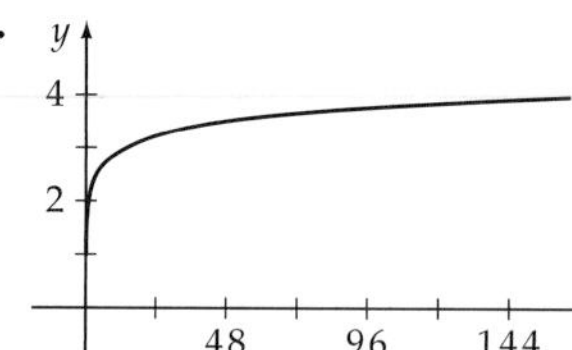

50.
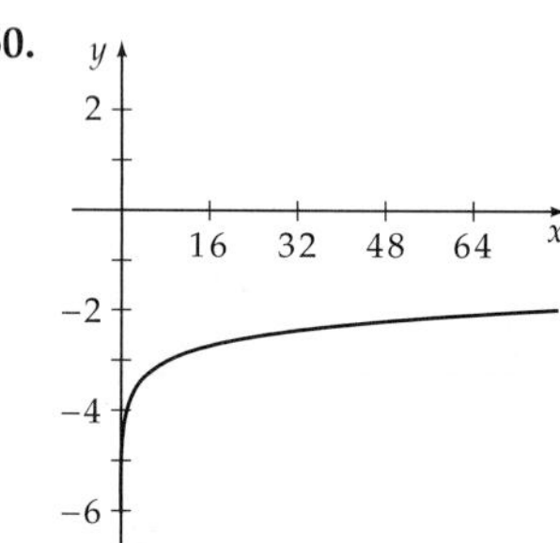

51.
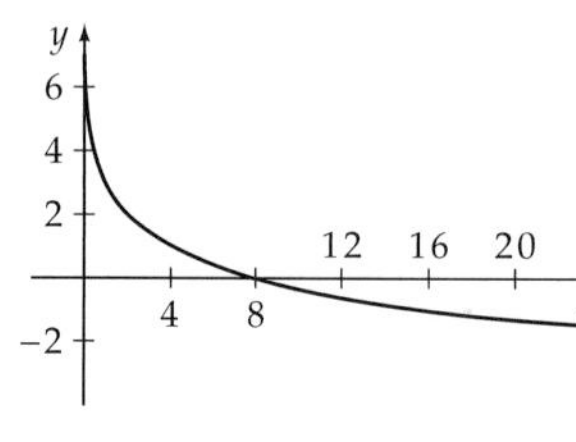

52.
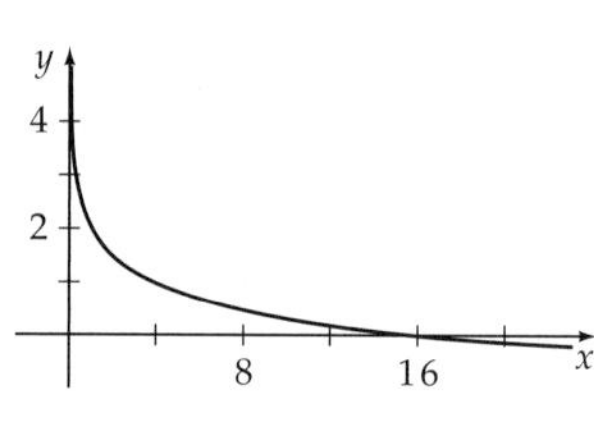

55.
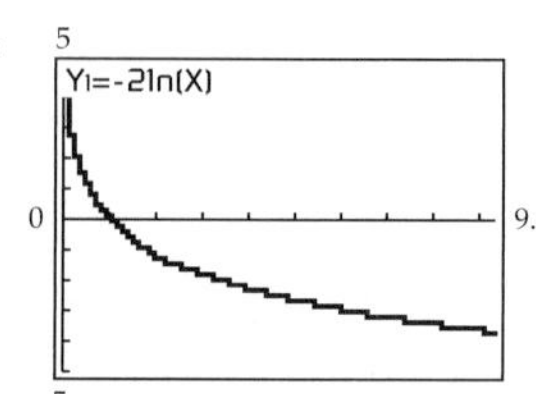

56.
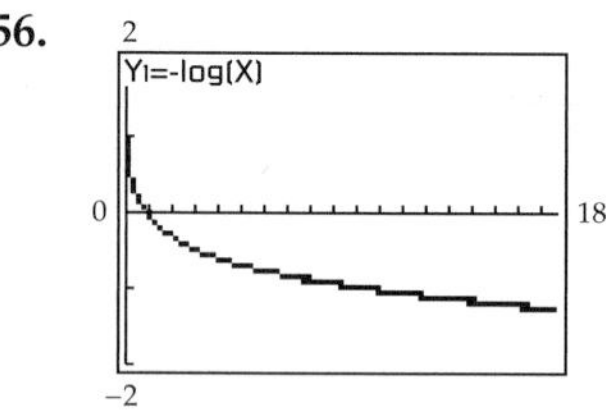

57.
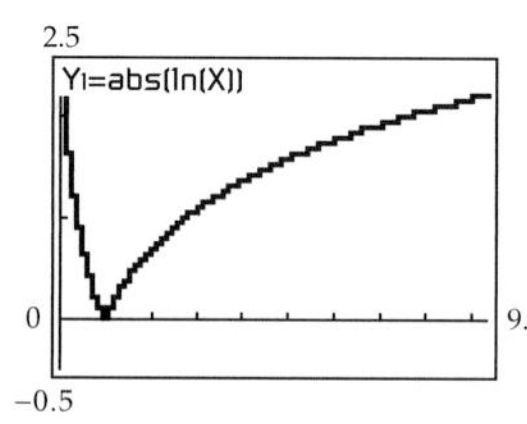

58.
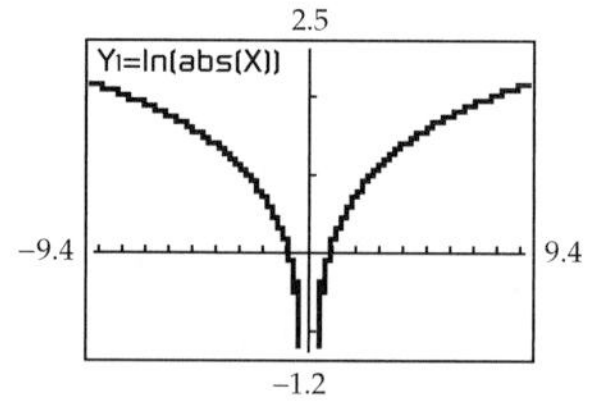

59.
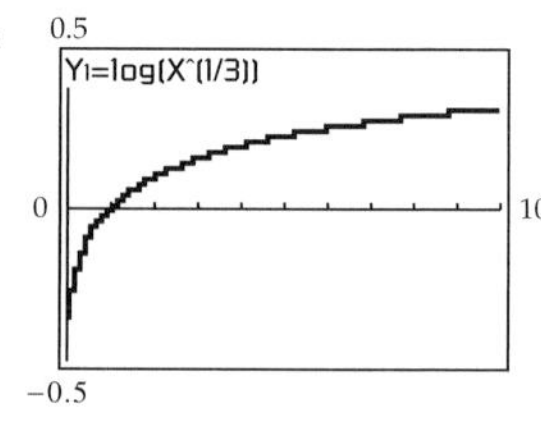

60.
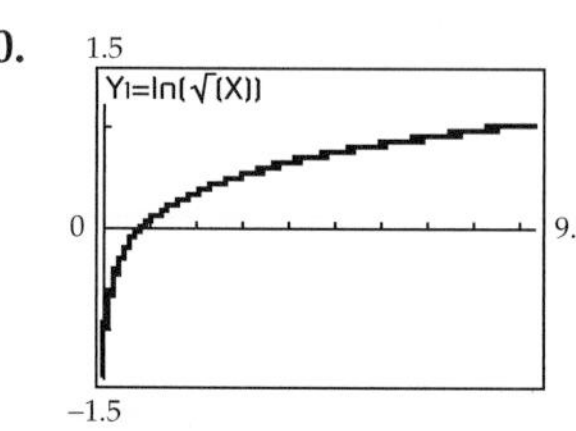

61.
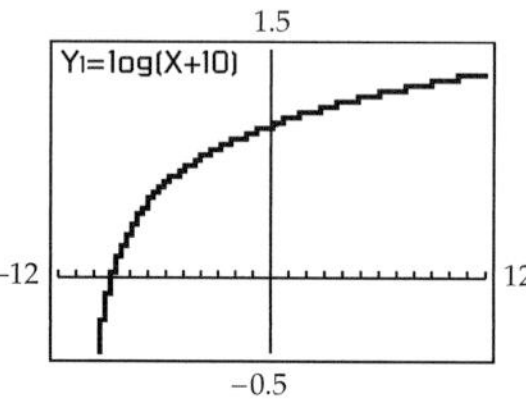

62.
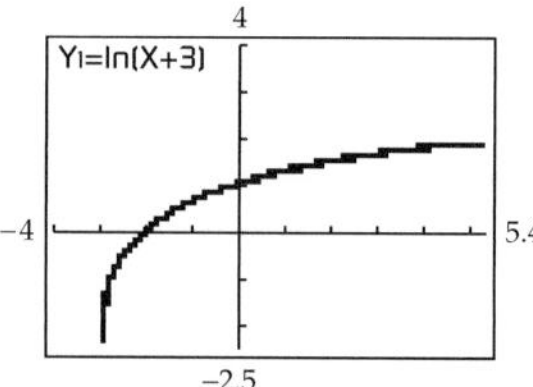

63.
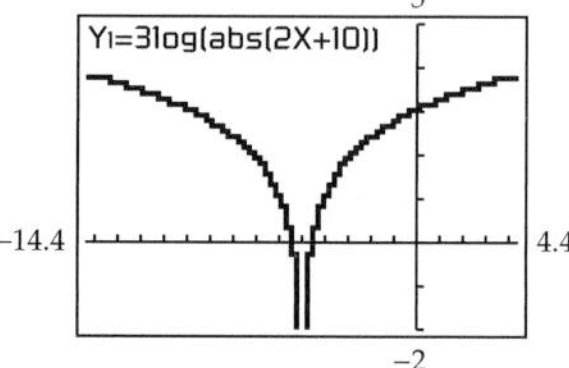

64.
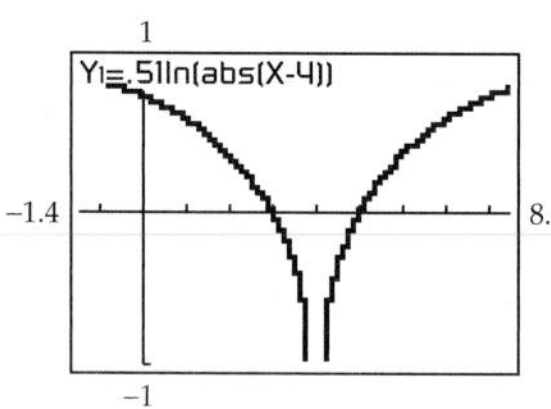

Explorations, *page 412*

1. a.

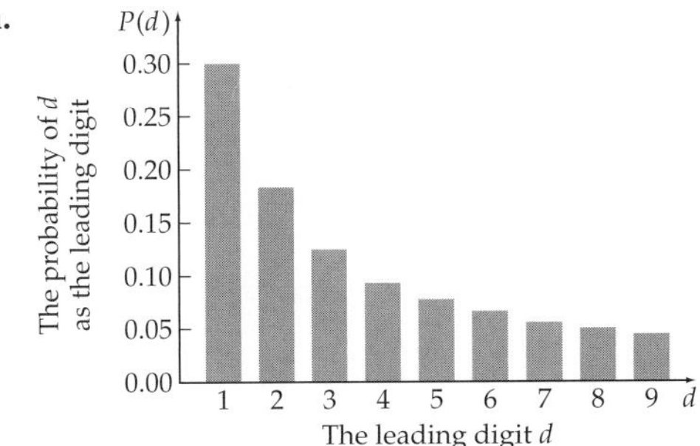

Exercises 5.3, *page 423*

23.

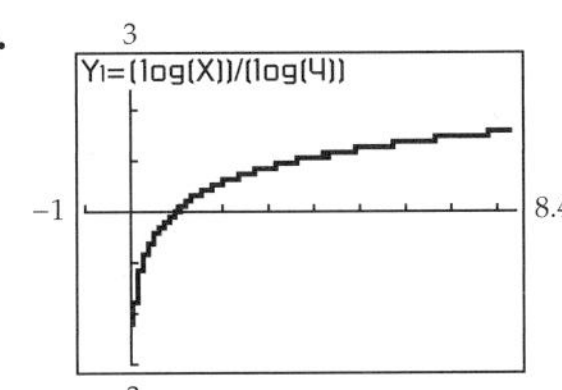

24.

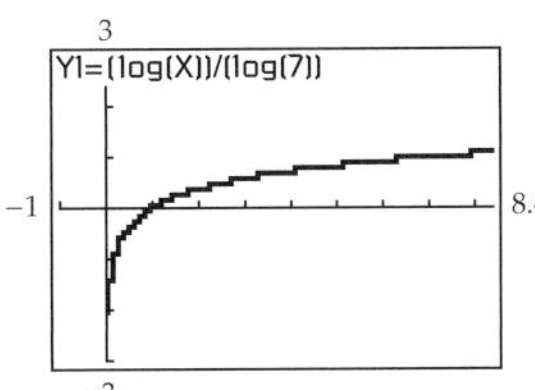

25.

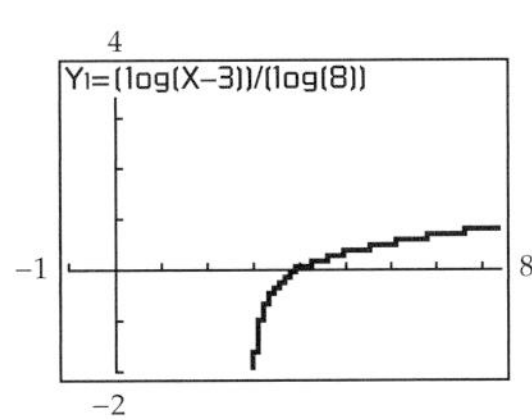

26.

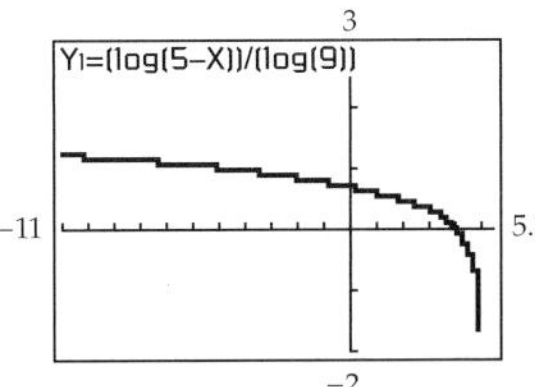

27.

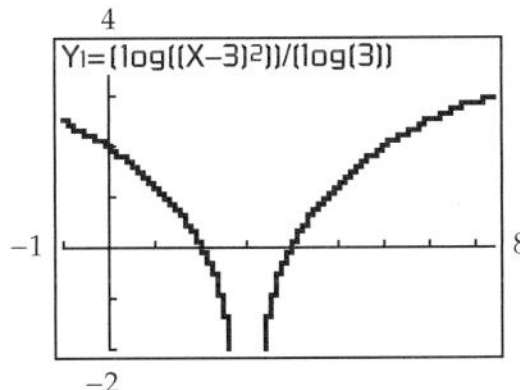

28.

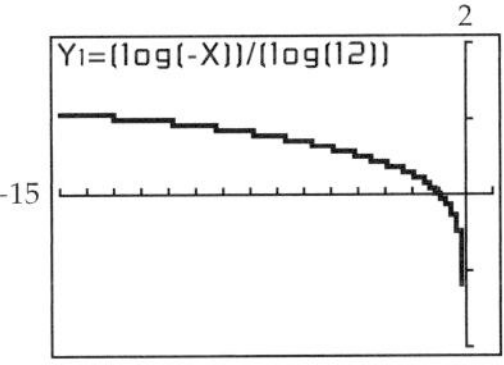

29.

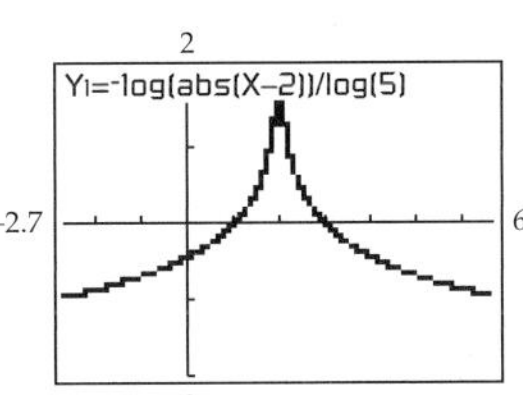

30.

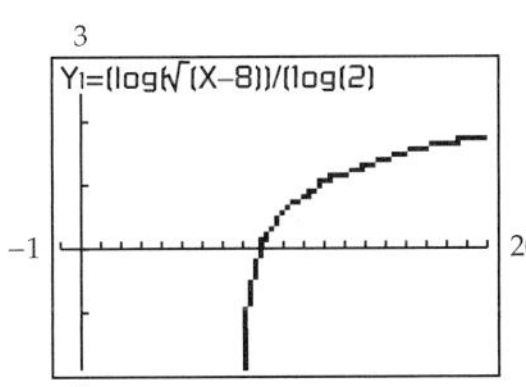

Explorations, *page 426*

1. a.

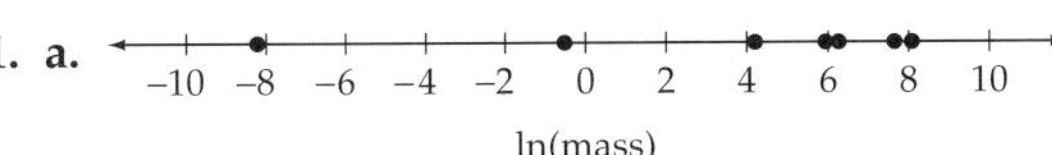

2. a.

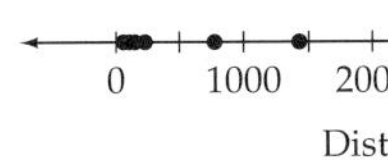

2. b.

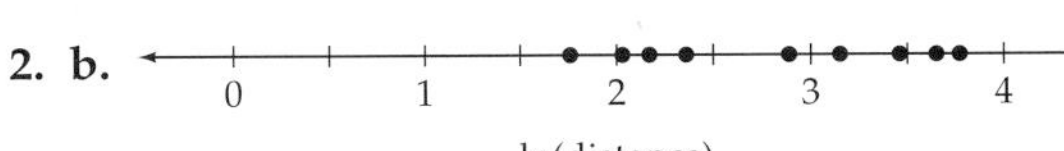

Exercises 5.4, *page 435*

58. a.

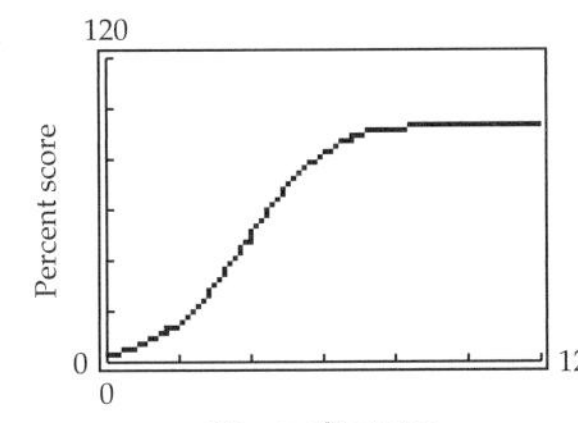

61. a.

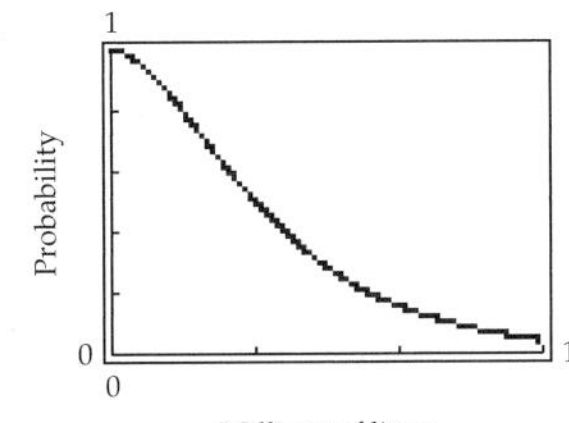

66. a.

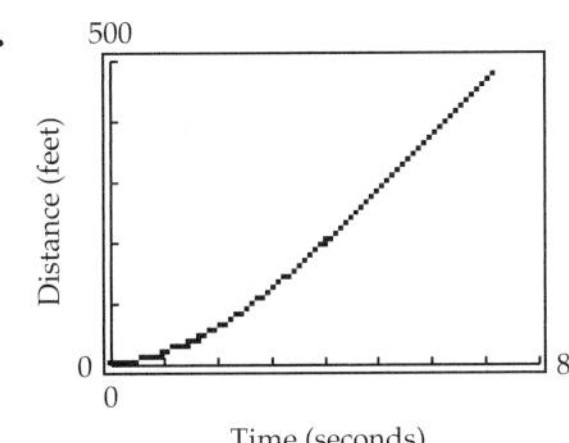

67. a.

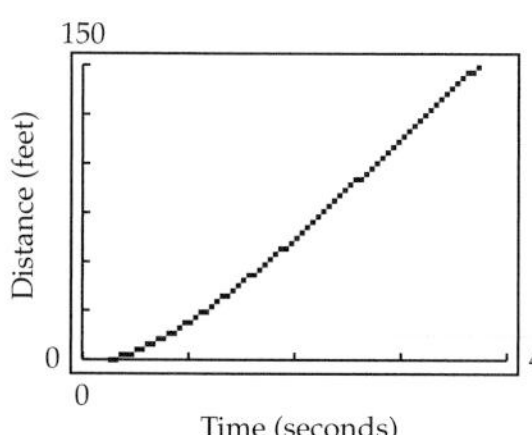

Exercises 5.6, *page 461*

1.

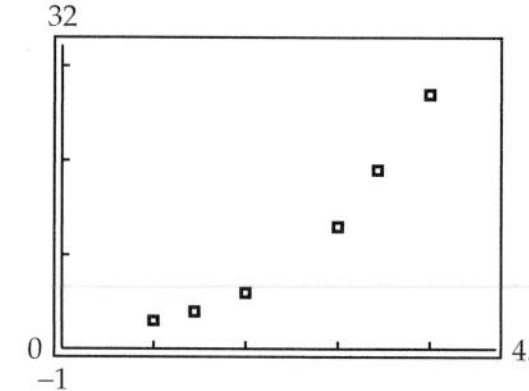

2.

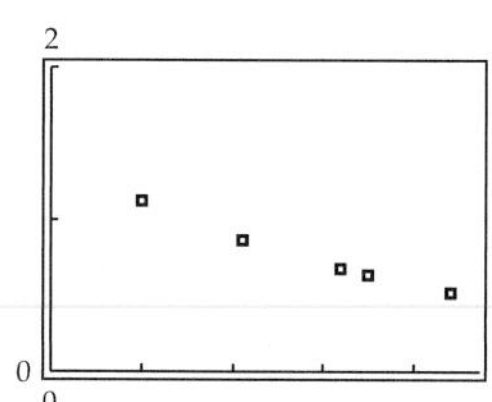

3.

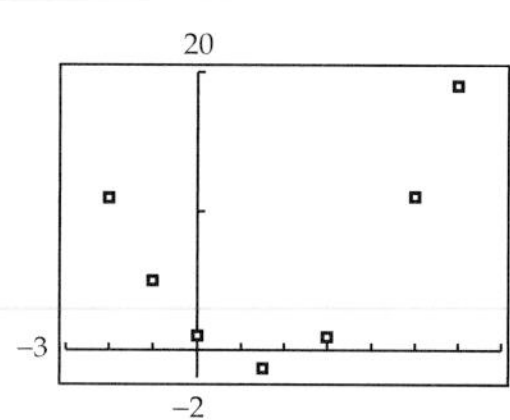

4.

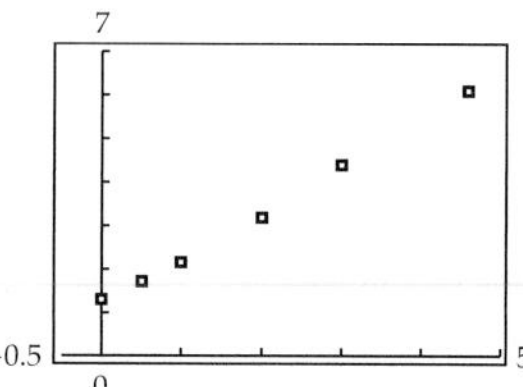

5.

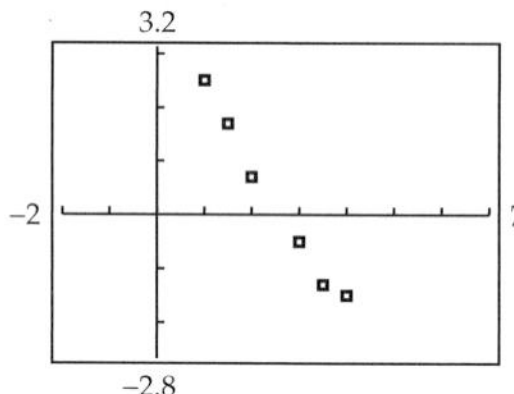

6.

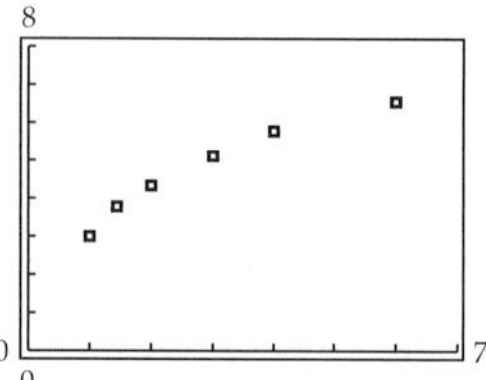

Chapter 5 Review Exercises, *page 479*

13. [5.1]

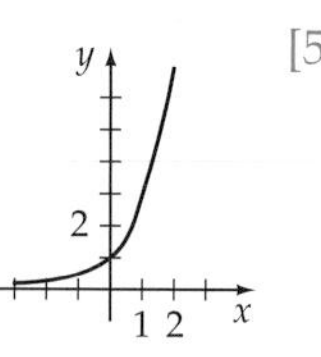

14. [5.1]

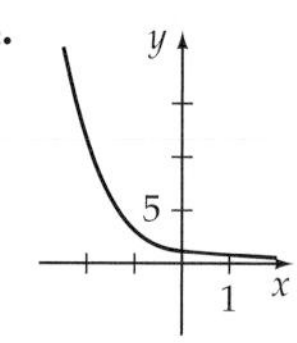

15. [5.1]

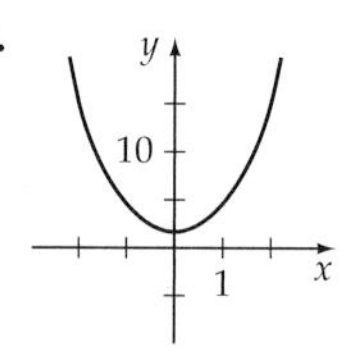

16. [5.1]

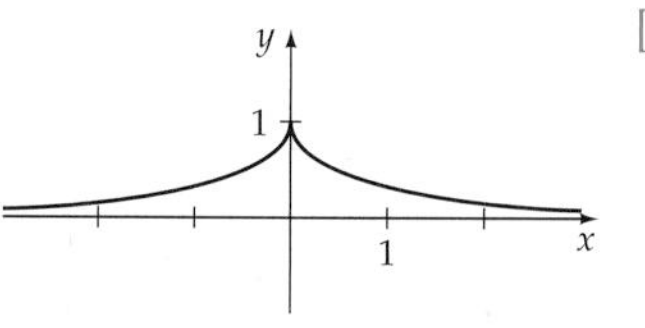

17. [5.1]

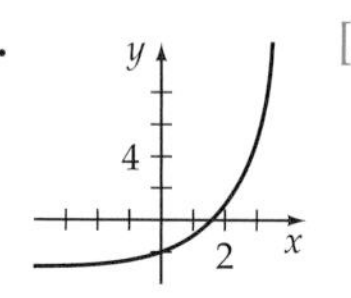

18. [5.1]

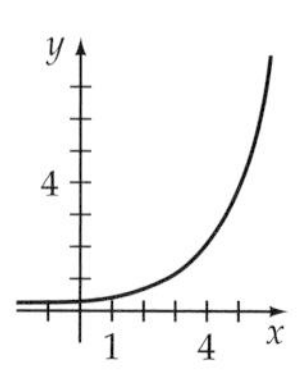

19. [5.2]

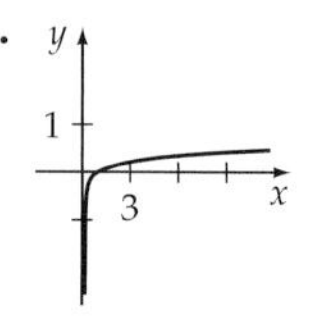

20. [5.2]

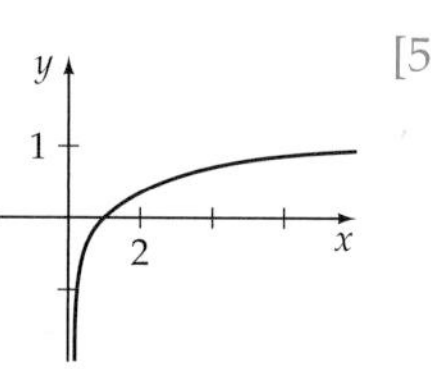

21. [5.2]

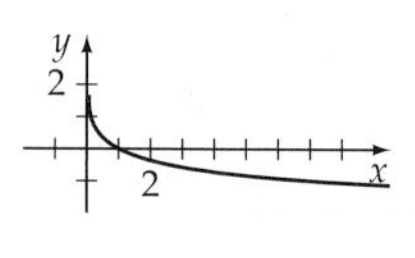

22. [5.2]

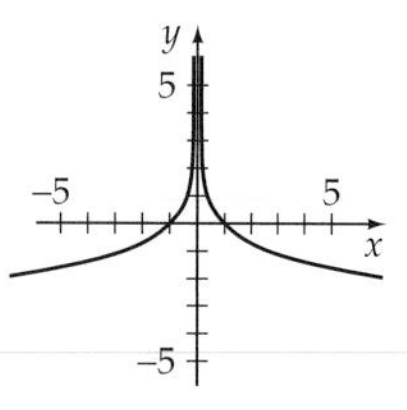

23. [5.1]

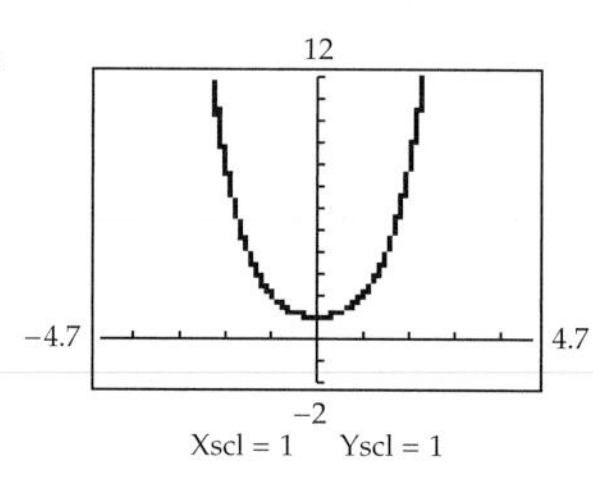
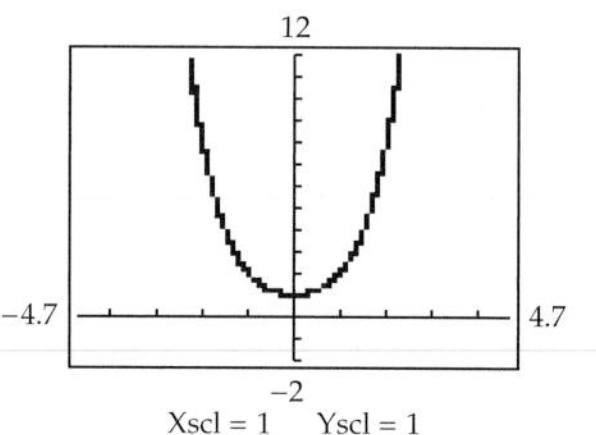

24. 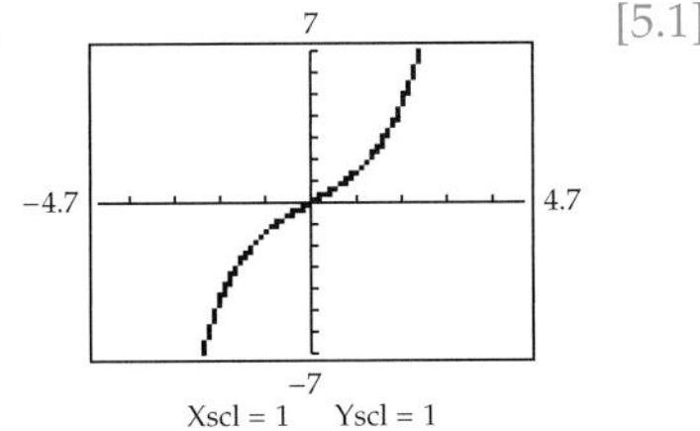

[5.1]

Chapter 5 Test, *page 482*

3. 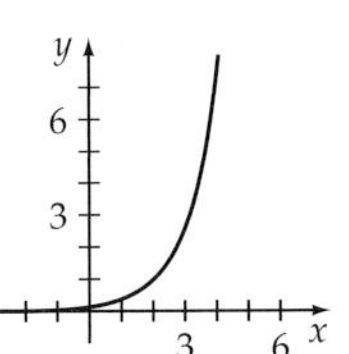 [5.1] **4.** 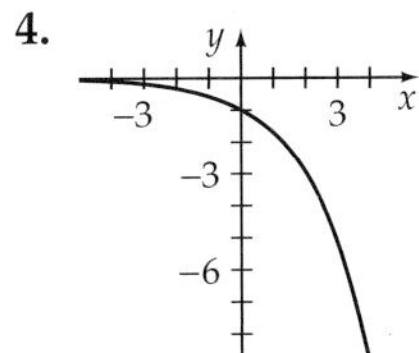 [5.1] **5.** 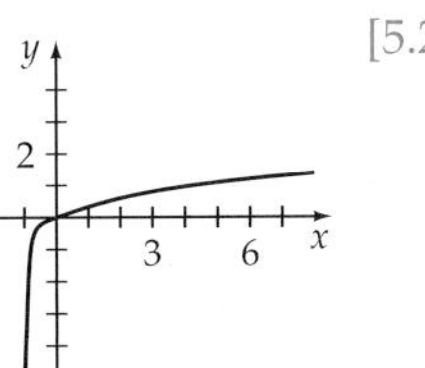 [5.2]

Chapter 6

Chapter Prep Quiz, *page 486*

4. 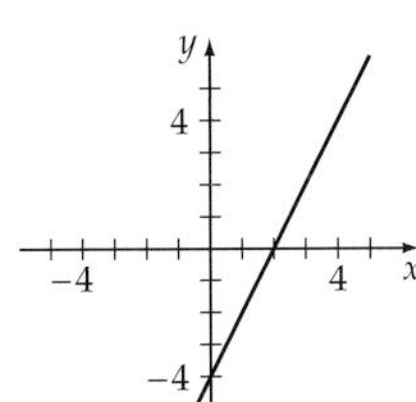 **5.** 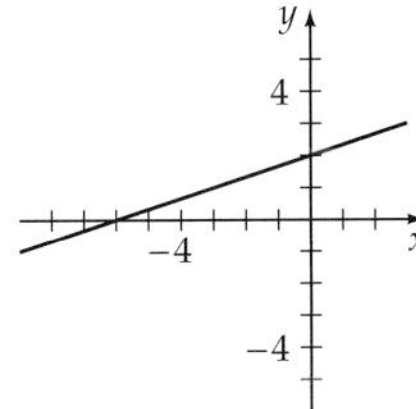

Exercises 6.4, *page 543*

10. a. $\begin{bmatrix} 1 & 6 \\ 5 & 2 \\ -3 & 3 \end{bmatrix}$ **b.** $\begin{bmatrix} 3 & -10 \\ 1 & 6 \\ 5 & -3 \end{bmatrix}$ **c.** $\begin{bmatrix} 7 & -28 \\ 0 & 14 \\ 14 & -9 \end{bmatrix}$ **11. a.** $\begin{bmatrix} -1 & 1 & -1 \\ 2 & 2 & 1 \\ -1 & 2 & 5 \end{bmatrix}$ **b.** $\begin{bmatrix} -3 & 5 & -1 \\ -2 & -4 & 3 \\ -7 & 4 & 1 \end{bmatrix}$ **c.** $\begin{bmatrix} -7 & 12 & -2 \\ -6 & -11 & 7 \\ -17 & 9 & 0 \end{bmatrix}$

12. a. $\begin{bmatrix} -1 & 4 & 4 \\ 4 & 0 & 1 \\ 1 & 8 & 1 \end{bmatrix}$ **b.** $\begin{bmatrix} 1 & 0 & -4 \\ -2 & -6 & 5 \\ 9 & 0 & -5 \end{bmatrix}$ **c.** $\begin{bmatrix} 3 & -2 & -12 \\ -7 & -15 & 12 \\ 22 & -4 & -13 \end{bmatrix}$ **13.** $AB = \begin{bmatrix} -10 & 17 \\ 6 & -8 \end{bmatrix}$, $BA = \begin{bmatrix} 0 & 22 \\ 1 & -18 \end{bmatrix}$ **14.** $AB = \begin{bmatrix} -3 & -11 \\ -4 & 0 \end{bmatrix}$, $BA = \begin{bmatrix} -7 & 1 \\ 16 & 4 \end{bmatrix}$ **15.** $AB = \begin{bmatrix} 0 & -4 & 5 \\ 6 & 0 & 3 \\ -3 & -2 & 1 \end{bmatrix}$, $BA = \begin{bmatrix} 5 & -13 \\ 5 & -4 \end{bmatrix}$ **16.** $AB = \begin{bmatrix} 3 & 7 & -14 \\ 1 & 0 & 0 \\ -2 & -1 & 2 \end{bmatrix}$, $BA = \begin{bmatrix} -8 & -5 \\ 15 & 13 \end{bmatrix}$

17. $AB = \begin{bmatrix} 9 & -2 & -6 \\ 0 & -1 & 2 \\ 4 & -2 & -4 \end{bmatrix}$, $BA = \begin{bmatrix} 4 & -2 & 6 \\ 2 & -3 & 4 \\ 4 & -4 & 3 \end{bmatrix}$ **18.** $AB = \begin{bmatrix} 0 & 11 & -2 \\ 3 & -8 & 4 \\ -2 & 13 & -5 \end{bmatrix}$, $BA = \begin{bmatrix} -4 & 5 & -1 \\ 11 & -5 & 6 \\ -8 & 7 & -4 \end{bmatrix}$ **23.** $\begin{bmatrix} 10 & -4 & -6 & 2 \\ 2 & 5 & 7 & -6 \\ 4 & -2 & 16 & -5 \\ 2 & 4 & 10 & -2 \end{bmatrix}$

24. $\begin{bmatrix} 4 & -7 & 1 & 12 \\ 7 & 12 & 0 & -3 \\ 4 & -2 & 7 & -2 \\ 0 & -1 & 2 & 6 \end{bmatrix}$ **25.** $\begin{bmatrix} 3 & 7 & 0 & 11 \\ 4 & 3 & 0 & -6 \\ 0 & 3 & 10 & -2 \\ -3 & 13 & 4 & 5 \end{bmatrix}$ **26.** $\begin{bmatrix} 8 & -82 & -80 & 53 \\ 35 & -5 & 58 & -35 \\ 33 & -21 & -10 & -2 \\ 7 & -29 & -30 & 19 \end{bmatrix}$ **27.** $\begin{bmatrix} 57 & -79 & -74 & 37 \\ 40 & 42 & 54 & -49 \\ 37 & -19 & 40 & -18 \\ 18 & -26 & -18 & 21 \end{bmatrix}$

28. $\begin{bmatrix} -12 & 53 & 106 & 66 \\ 30 & -47 & 56 & -68 \\ 44 & 91 & 16 & -10 \\ -8 & 119 & 66 & -26 \end{bmatrix}$

The matrix represents the total number of employees in each category at both campuses.

35. $0.98\begin{bmatrix} 2.0 & 1.4 & 3.0 & 1.4 \\ 0.8 & 1.1 & 2.0 & 0.9 \\ 3.6 & 1.2 & 4.5 & 1.5 \end{bmatrix} = \begin{bmatrix} 1.96 & 1.37 & 2.94 & 1.37 \\ 0.78 & 1.08 & 1.96 & 0.88 \\ 3.53 & 1.18 & 4.41 & 1.47 \end{bmatrix}$

44. $\begin{bmatrix} 119 & 166 & 80 & 95 \\ 235 & 311 & 158 & 169 \\ 20 & 31 & 15 & 15 \end{bmatrix}$;

45. $1.06\begin{bmatrix} 25 & 26.3 & 27.5 & 28.6 \\ 30 & 31.7 & 32.9 & 34.2 \\ 35 & 36.8 & 38.2 & 40.3 \end{bmatrix} = \begin{bmatrix} 26.50 & 27.88 & 29.15 & 30.32 \\ 31.80 & 33.60 & 34.87 & 36.25 \\ 37.10 & 39.01 & 40.49 & 42.72 \end{bmatrix}$

46. a. $\begin{bmatrix} 103 & 58 \\ 93 & 69 \\ 78 & 84 \end{bmatrix}$; This matrix represents the total wins and losses for each of the three teams.

46. b. $\begin{bmatrix} 1 & -2 \\ -9 & 9 \\ 6 & -6 \end{bmatrix}$; This matrix represents the difference between home wins and losses and away wins and losses.

Explorations, *page 547*

1. b.

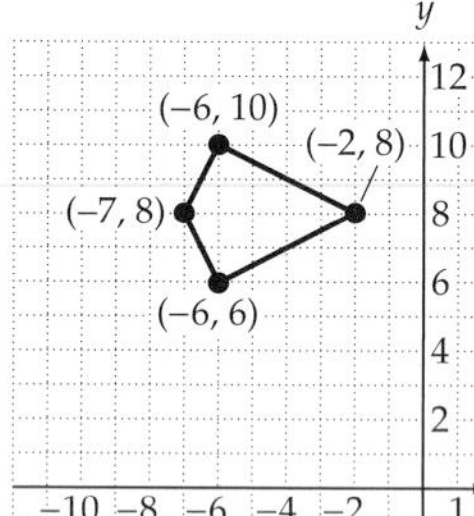

2. b.

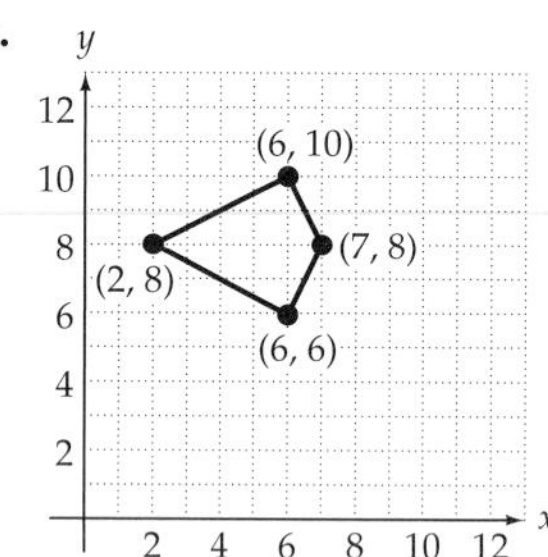

3. b.

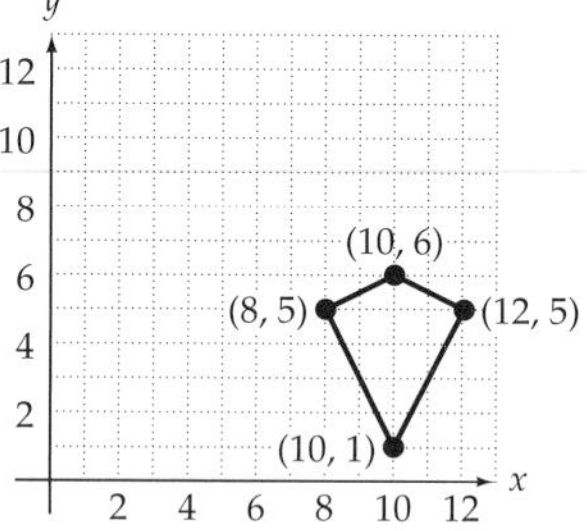

4. b.

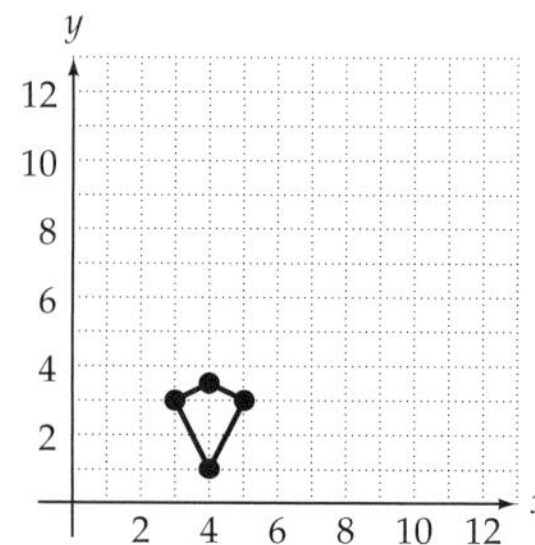

5.

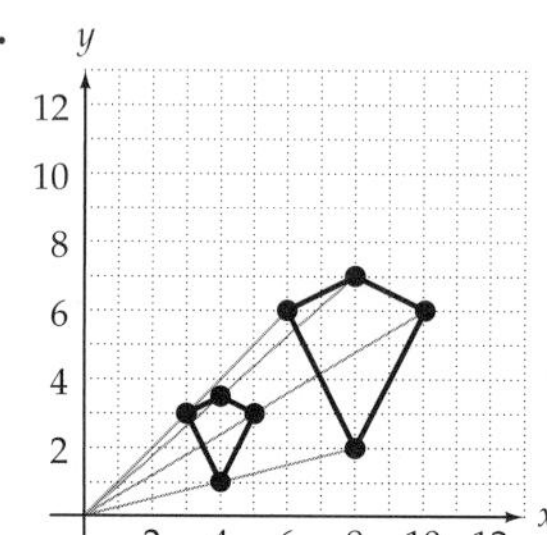

Prepare for Section 6.6, *page 558*

43.

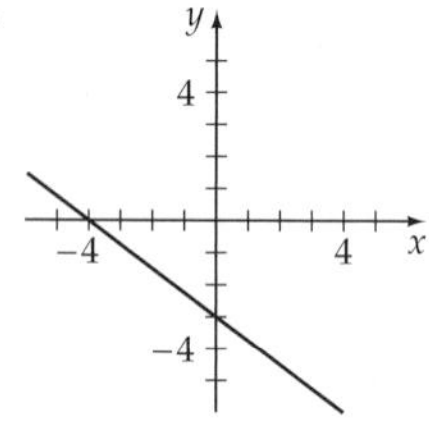

Exercises 6.6, *page 568*

1.

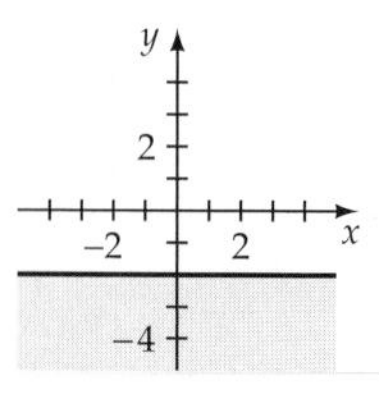

2.

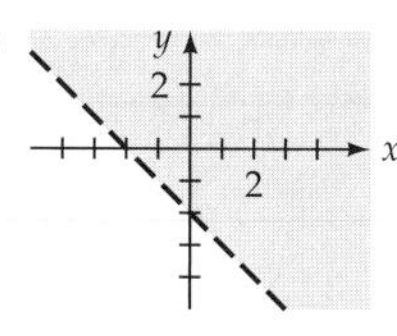

3.

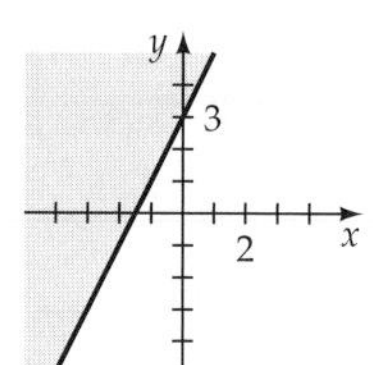

4.

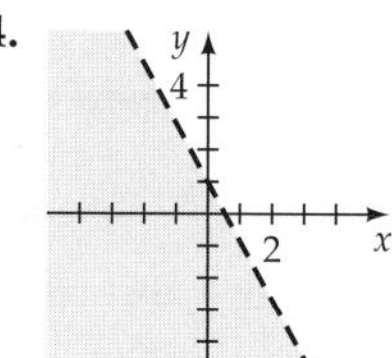

5.

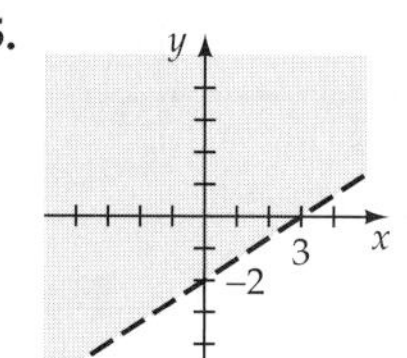

6.

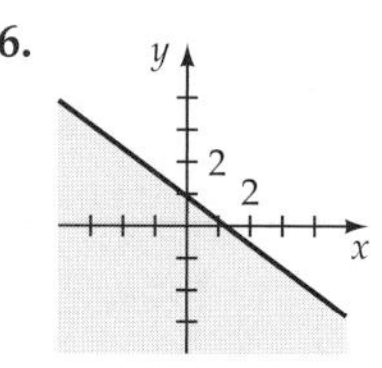

7.
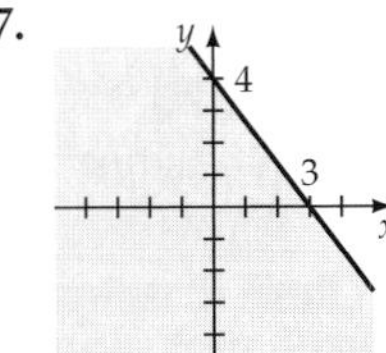

8.
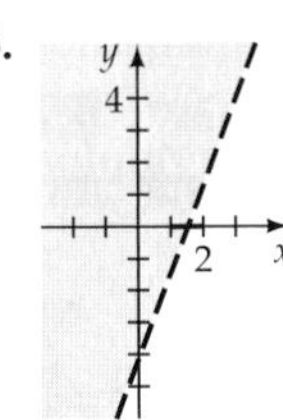

9.
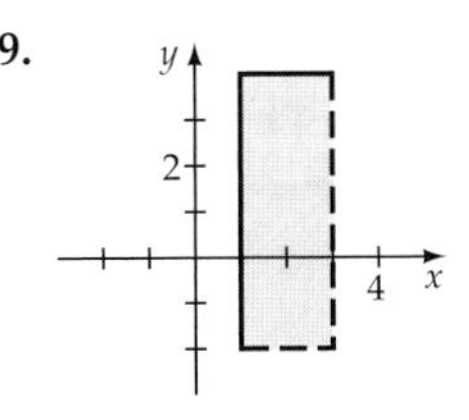

10.
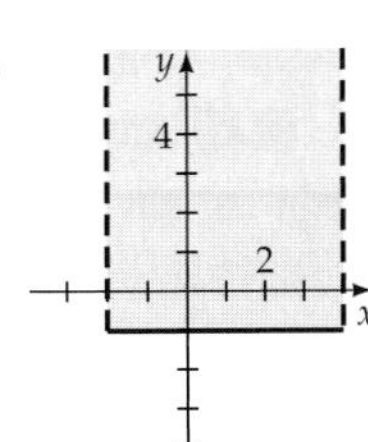

11.
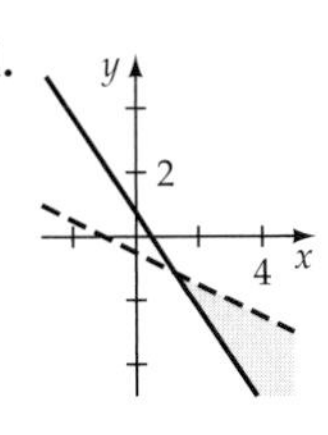

12.
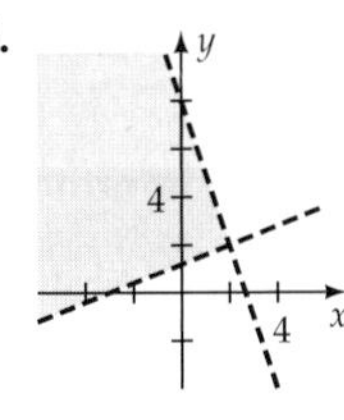

14.
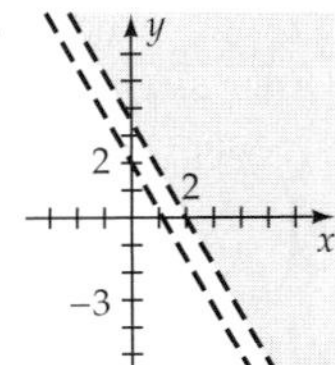

15.
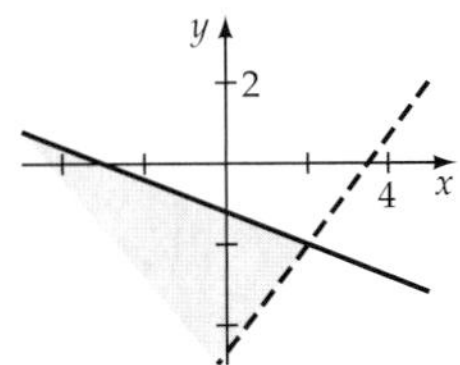

16.
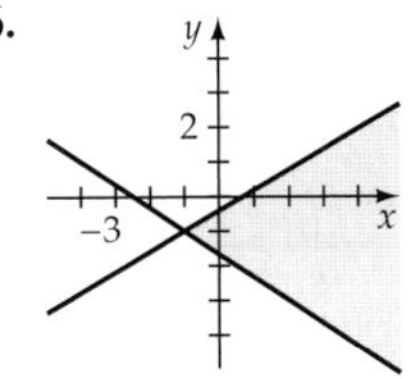

17.
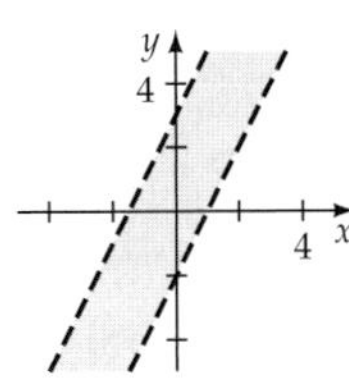

19.
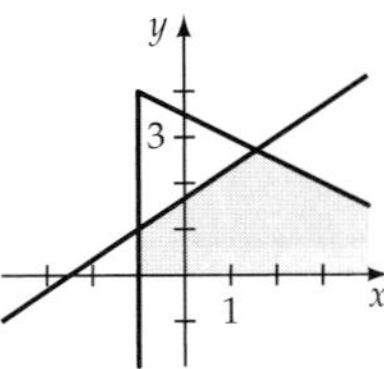

20.
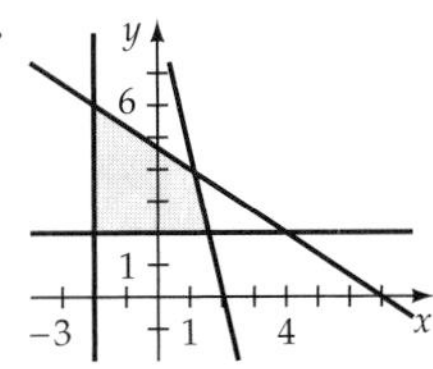

21.
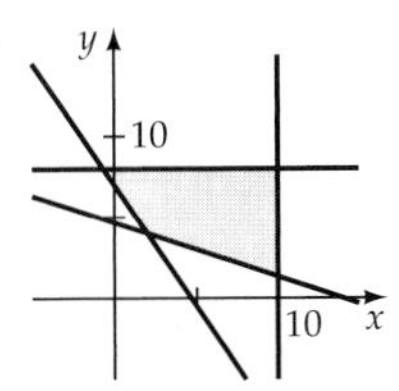

22.
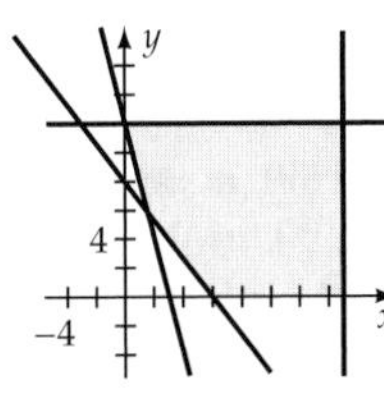

23.
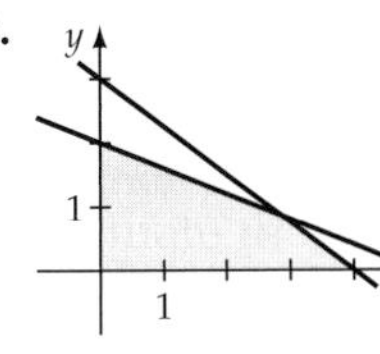

24.
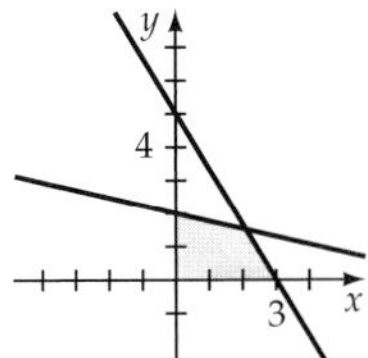

Chapter 6 Review Exercises, *page 572*

20. $\left[\begin{array}{cccc|c} -2 & 1 & -6 & 4 & 1 \\ 3 & 3 & 2 & -5 & -9 \\ -1 & -7 & 3 & -1 & 4 \\ 8 & 5 & 1 & -3 & -12 \end{array}\right], \begin{bmatrix} -2 & 1 & -6 & 4 \\ 3 & 3 & 2 & -5 \\ -1 & -7 & 3 & -1 \\ 8 & 5 & 1 & -3 \end{bmatrix}, \begin{bmatrix} 1 \\ -9 \\ 4 \\ -12 \end{bmatrix}$ [6.3] **45.** $\begin{bmatrix} -6 & -4 & 2 \\ 14 & 0 & 10 \\ -7 & -7 & 6 \end{bmatrix}$ [6.4] **46.** $\begin{bmatrix} -8 & 4 & -10 \\ -4 & 12 & 18 \\ -15 & 10 & -13 \end{bmatrix}$ [6.4]

47. $\begin{bmatrix} 12 & 28 & -5 \\ 2 & 6 & 0 \\ 6 & 16 & -1 \end{bmatrix}$ [6.4] **48.** $\begin{bmatrix} 42 & 108 & -11 \\ 10 & 24 & -4 \\ 26 & 64 & -9 \end{bmatrix}$ [6.4] **49.** $\begin{bmatrix} -12 & -36 & -4 \\ 48 & 124 & 4 \\ -9 & -32 & -6 \end{bmatrix}$ [6.4] **53.** $\begin{bmatrix} 13 & -2 & 0 & 42 \\ 19 & -19 & 29 & 22 \\ -26 & -3 & -40 & 4 \\ 0 & 42 & 12 & -17 \end{bmatrix}$ [6.4]

54. $\begin{bmatrix} -345 & 117 & 546 & 171 \\ -45 & -511 & 166 & 97 \\ -69 & -285 & 396 & -195 \\ -288 & -208 & 268 & -490 \end{bmatrix}$ [6.4] **55.** $\begin{bmatrix} -154 & -30 & -230 & -49 \\ -127 & -211 & -180 & 153 \\ 19 & 212 & -83 & -97 \\ 105 & -38 & 338 & -90 \end{bmatrix}$ [6.4] **56.** $\begin{bmatrix} -9 & 12 & -33 & -54 \\ 27 & -39 & -36 & 19 \\ 12 & 36 & -60 & 3 \\ 48 & 71 & -26 & 15 \end{bmatrix}$ [6.4]

57. $\begin{bmatrix} \frac{180}{781} & -\frac{238}{781} & -\frac{144}{781} & -\frac{35}{781} \\ \frac{74}{781} & \frac{41}{781} & \frac{97}{781} & \frac{29}{781} \\ -\frac{85}{781} & \frac{69}{781} & \frac{68}{781} & \frac{125}{781} \\ -\frac{15}{781} & \frac{150}{781} & \frac{12}{781} & \frac{68}{781} \end{bmatrix}$ [6.5] **58.** $\begin{bmatrix} \frac{179}{3432} & \frac{27}{1144} & \frac{34}{429} & -\frac{157}{1144} \\ \frac{19}{264} & -\frac{5}{88} & -\frac{1}{33} & \frac{3}{88} \\ \frac{115}{3432} & -\frac{21}{1144} & \frac{53}{429} & -\frac{5}{1144} \\ \frac{21}{572} & \frac{83}{572} & -\frac{8}{143} & \frac{47}{572} \end{bmatrix}$ [6.5] **59.** $\begin{cases} 4x + 5y = -8 \\ 3x - 4y = 10 \end{cases}$ [6.5]

60. $\begin{cases} -x + 6y - z = 4 \\ 3x + 2y + 2z = -8 \\ 3x - 5y + 4z = 1 \end{cases}$ [6.5] **65.** $\begin{bmatrix} 2 & -2 & 1 \\ 0 & \frac{3}{2} & -1 \\ -1 & -1 & 1 \end{bmatrix}$ [6.5] **66.** $\begin{bmatrix} 27 & -12 & 5 \\ 4 & -2 & 1 \\ -7 & 3 & -1 \end{bmatrix}$ [6.5] **67.** $\begin{bmatrix} -10 & 20 & -3 \\ -5 & 9 & -1 \\ 3 & -6 & 1 \end{bmatrix}$ [6.5]

68. $\begin{bmatrix} 27 & -39 & 5 \\ -7 & 10 & -1 \\ 4 & -6 & 1 \end{bmatrix}$ [6.5] **71. a.** $\left(-\frac{18}{7}, \frac{23}{7}, -\frac{6}{7}\right)$ **b.** $\left(-\frac{31}{14}, \frac{20}{7}, \frac{3}{7}\right)$ [6.5] **72. a.** $\left(-\frac{4}{3}, -1, 2\right)$ **b.** $(-9, -11, 6)$ [6.5]

73. [6.6]

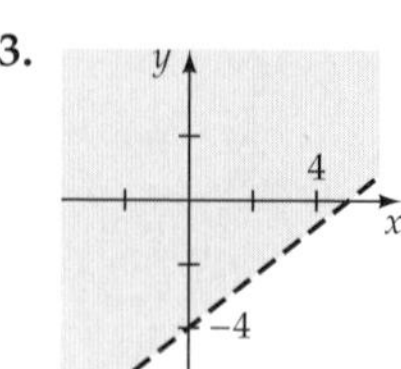

74. [6.6]

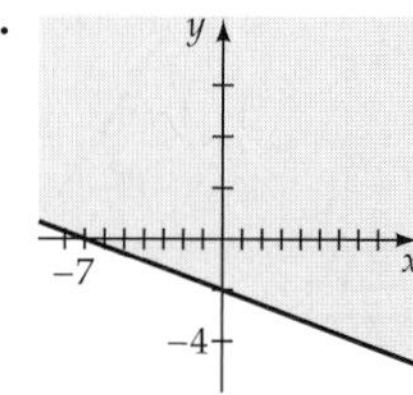

75. [6.6]

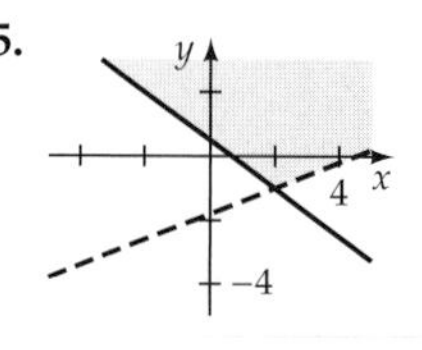

76. [6.6]

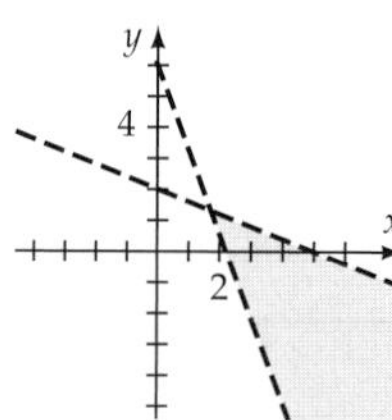

77. [6.6]

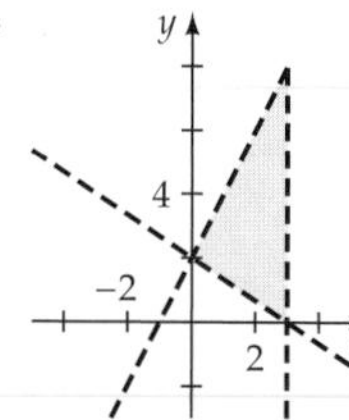

78. [6.6]

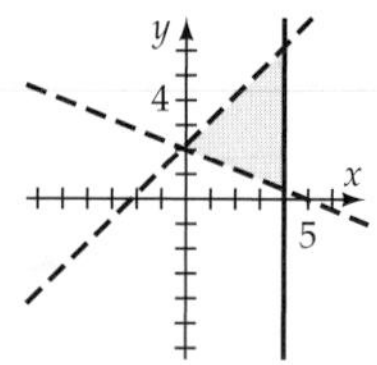

79. [6.6]

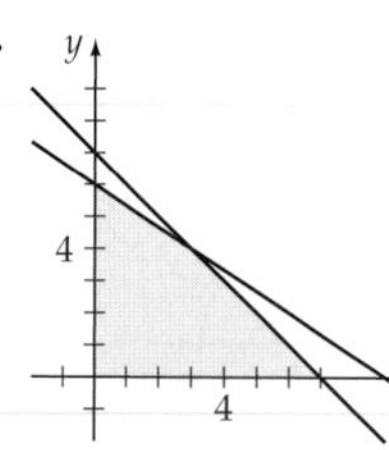

80. [6.6]

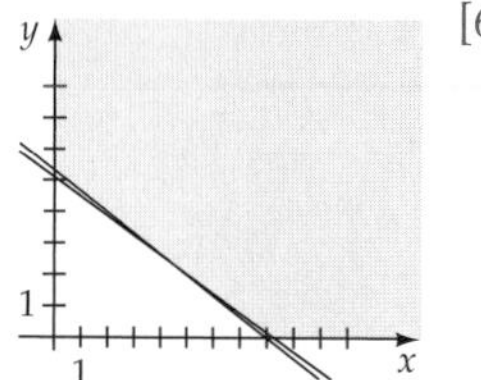

81. [6.6]

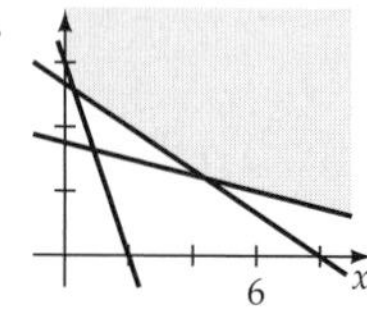

82. [6.6]

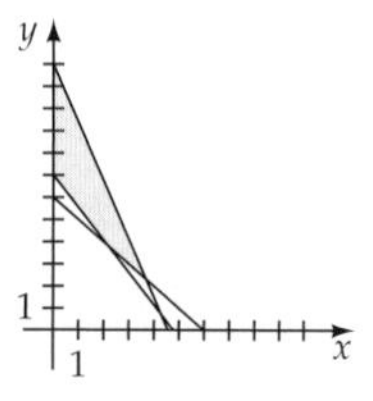

Chapter 6 Test, *page 576*

16. $\left(\begin{bmatrix} 1 & 0 & 0 \\ 0 & 1 & 0 \\ 0 & 0 & 1 \end{bmatrix} - \begin{bmatrix} 0.15 & 0.23 & 0.11 \\ 0.08 & 0.10 & 0.05 \\ 0.16 & 0.11 & 0.07 \end{bmatrix}\right)^{-1} \begin{bmatrix} 50 \\ 32 \\ 8 \end{bmatrix}$ [6.5]

17. [6.6]

18. [6.6]

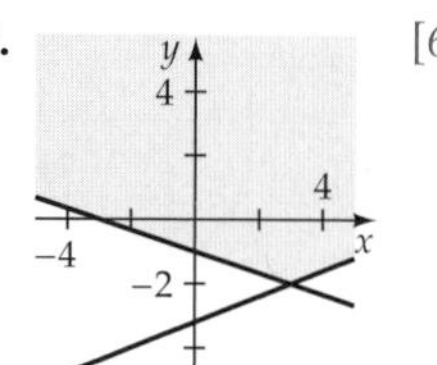

19. [6.6]

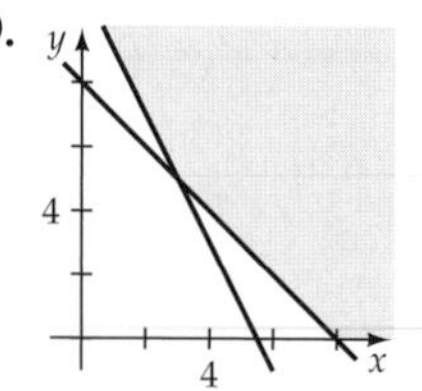

Chapter 8

Exercises 8.1, *page 655*

11.

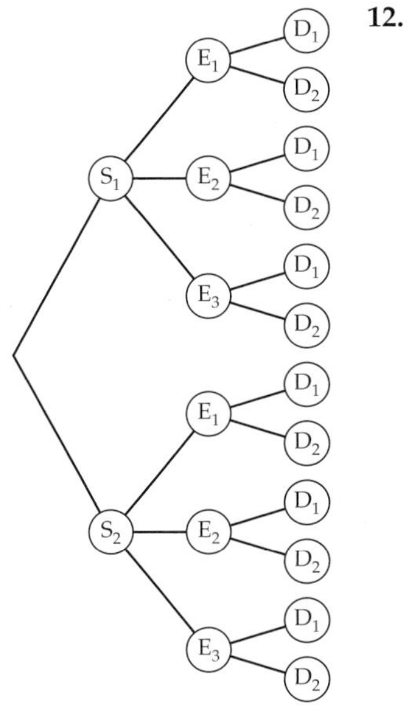

12.

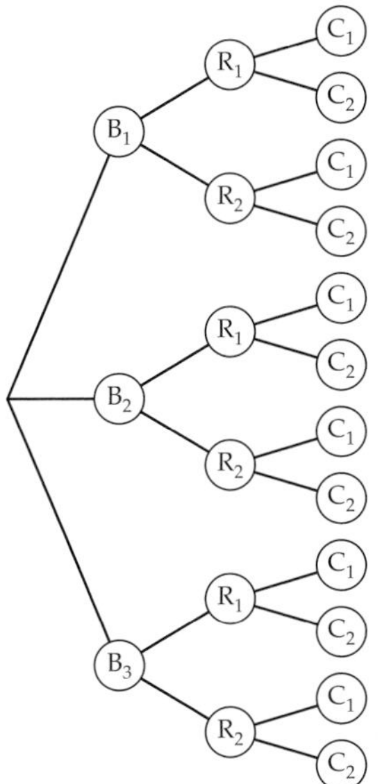

13.

14.

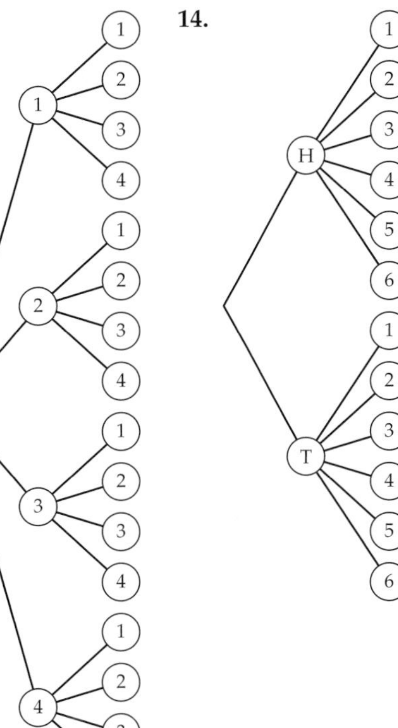

15.

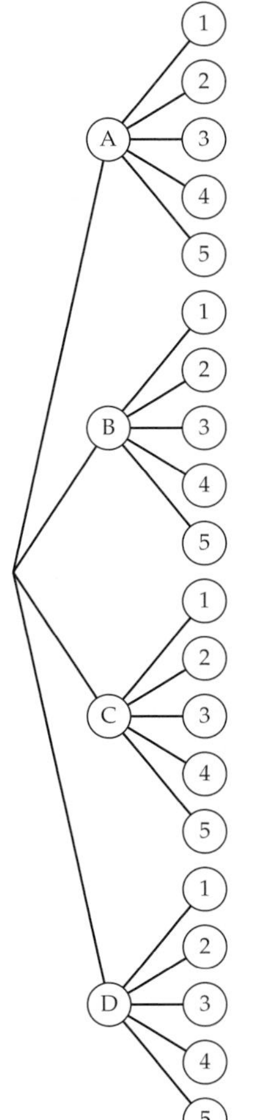

16.

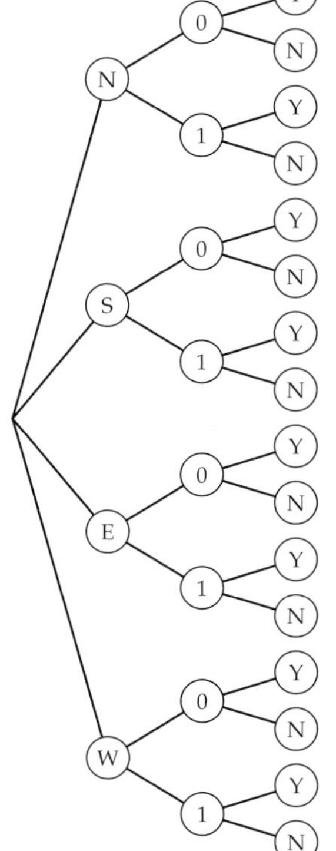

Exercises 8.4, *page 687*

9. $x^6 - 6x^5y + 15x^4y^2 - 20x^3y^3 + 15x^2y^4 - 6xy^5 + y^6$ **10.** $a^5 - 5a^4b + 10a^3b^2 - 10a^2b^3 + 5ab^4 - b^5$
11. $x^5 + 15x^4 + 90x^3 + 270x^2 + 405x + 243$ **12.** $x^4 - 20x^3 + 150x^2 - 500x + 625$
13. $128x^7 - 448x^6 + 672x^5 - 560x^4 + 280x^3 - 84x^2 + 14x - 1$
14. $64x^6 + 192x^5y + 240x^4y^2 + 160x^3y^3 + 60x^2y^4 + 12xy^5 + y^6$
15. $x^6 + 18x^5y + 135x^4y^2 + 540x^3y^3 + 1215x^2y^4 + 1458xy^5 + 729y^6$
16. $x^5 - 20x^4y + 160x^3y^2 - 640x^2y^3 + 1280xy^4 - 1024y^5$ **17.** $16x^4 - 160x^3y + 600x^2y^2 - 1000xy^3 + 625y^4$
18. $32x^5 + 240x^4y + 720x^3y^2 + 1080x^2y^3 + 810xy^4 + 243y^5$ **19.** $x^6 + 6x^4 + 15x^2 + 20 + \frac{15}{x^2} + \frac{6}{x^4} + \frac{1}{x^6}$
20. $x^2\sqrt{x} - 10x^2 + 40x\sqrt{x} - 80x + 80\sqrt{x} - 32$ **21.** $x^{14} - 28x^{12} + 336x^{10} - 2240x^8 + 8960x^6 - 21{,}504x^4 + 28{,}672x^2 - 16{,}834$
22. $x^6 - 6x^5y^3 + 15x^4y^6 - 20x^3y^9 + 15x^2y^{12} - 6xy^{15} + y^{18}$ **23.** $32x^{10} + 80x^8y^3 + 80x^6y^6 + 40x^4y^9 + 10x^2y^{12} + y^{15}$
24. $64x^6 - 192x^5y^3 + 240x^4y^6 - 160x^3y^9 + 60x^2y^{12} - 12xy^{15} + y^{18}$ **25.** $\frac{16}{x^4} - \frac{16}{x^2} + 6 - x^2 + \frac{x^4}{16}$ **26.** $\frac{a^3}{b^3} + \frac{3a}{b} + \frac{3b}{a} + \frac{b^3}{a^3}$
27. $s^{-12} + 6s^{-8} + 15s^{-4} + 20 + 15s^4 + 6s^8 + s^{12}$ **28.** $32r^{-5} + 80r^{-4}s^{-1} + 80r^{-3}s^{-2} + 40r^{-2}s^{-3} + 10r^{-1}s^{-4} + s^{-5}$

Chapter 8 Review Exercises, *page 690*

2.

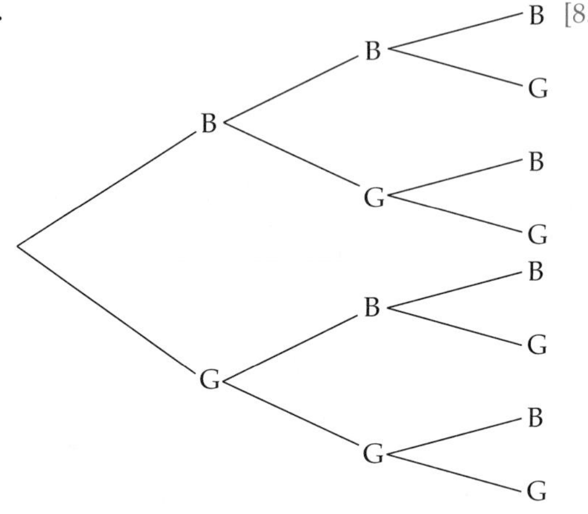

Appendix A Exercises, *page AP16*

1. Vertex: (0, 0)
Focus: (0, −1)
Directrix: $y = 1$

2. Vertex: (0, 0)
Focus: $\left(\frac{1}{12}, 0\right)$
Directrix: $x = -\frac{1}{12}$

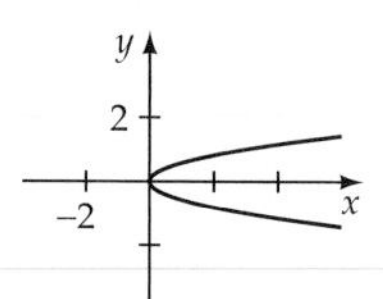

3. Center: (0, 0)
Vertices: (0, 5), (0, −5)
Foci: (0, 3), (0, −3)

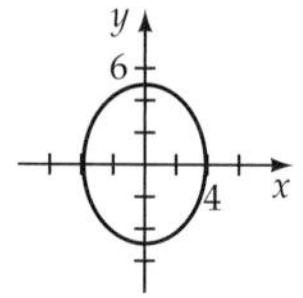

4. Center: (0, 0)
Vertices: (3, 0), (−3, 0)

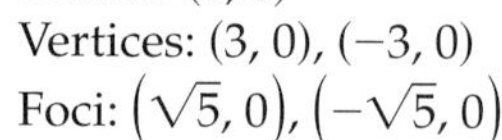
Foci: $(\sqrt{5}, 0), (-\sqrt{5}, 0)$

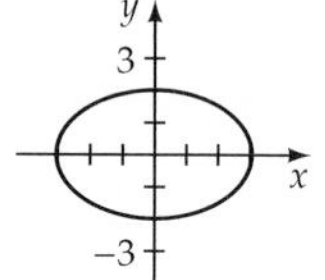

5. Center: (0, 0)
Vertices: (±4, 0)

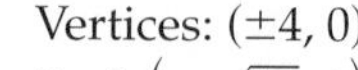
Foci: $(\pm\sqrt{41}, 0)$

Asymptotes: $y = \pm\frac{5x}{4}$

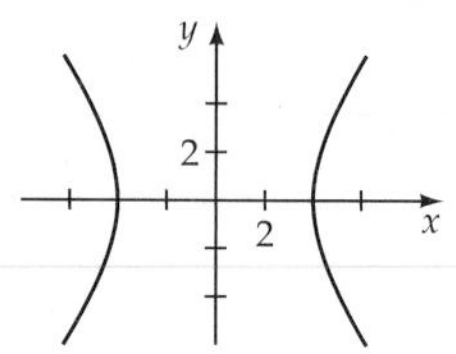

6. Center: (0, 0)
Vertices: (0, ±2)
Foci: $\left(0, \pm\sqrt{29}\right)$
Asymptotes: $y = \pm\frac{2x}{5}$

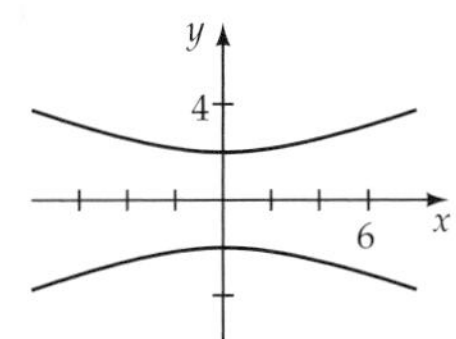

7. Vertex: (2, −3)
Focus: (2, −1)
Directrix: $y = -5$

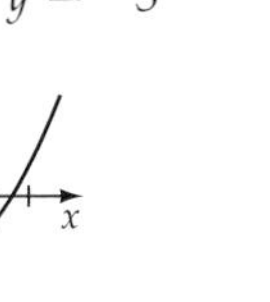

8. Vertex: (2, −4)
Focus: (1, −4)
Directrix: $x = 3$

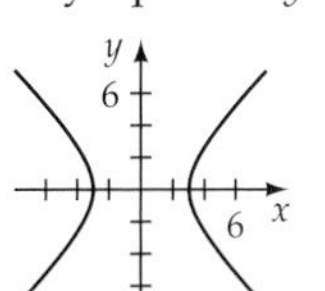

9. Center: (0, 0)
Vertices: (±3, 0)
Foci: $\left(\pm 3\sqrt{2}, 0\right)$
Asymptotes: $y = \pm x$

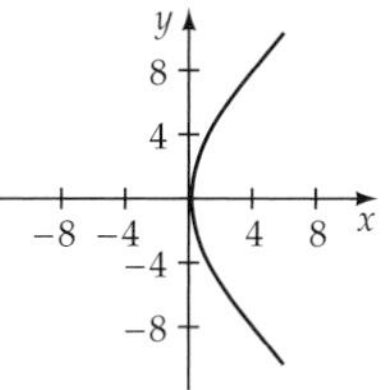

10. Vertex: (0, 0)
Focus: (4, 0)
Directrix: $x = -4$

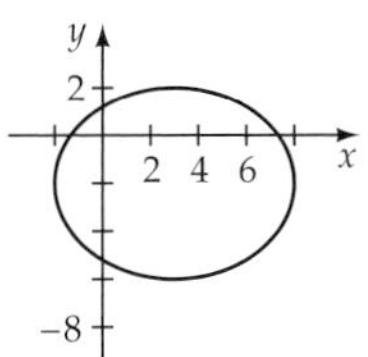

11. Center: (3, −2)
Vertices: (8, −2), (−2, −2)
Foci: (6, −2), (0, −2)

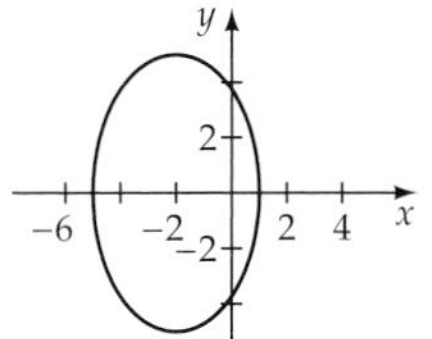

12. Center: (−2, 0)
Vertices: (−2, 5), (−2, −5)
Foci: (−2, 4), (−2, −4)

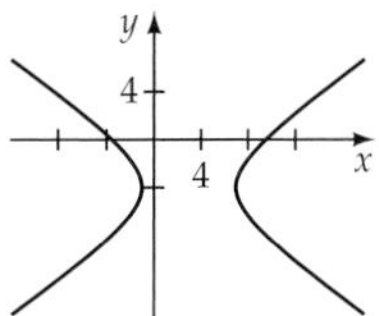

13. Center: (3, −4)
Vertices: (7, −4), (−1, −4)
Foci: (8, −4), (−2, −4)
Asymptotes: $y + 4 = \pm 3\frac{(x - 3)}{4}$

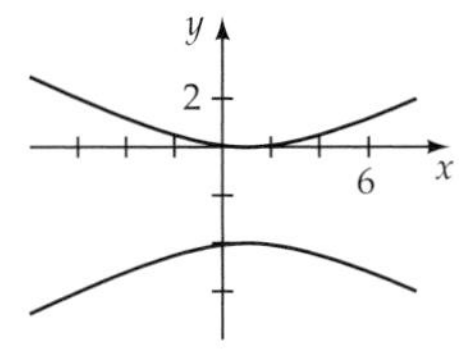

14. Center: (1, −2)
Vertices: (1, 0), (1, −4)
Foci: $\left(1, -2 \pm 2\sqrt{5}\right)$
Asymptotes: $y + 2 = \pm\frac{(x - 1)}{2}$

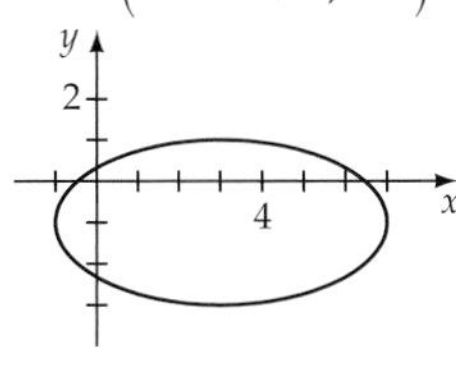

15. Center: (3, −1)
Vertices: (−1, −1), (7, −1)
Foci: $\left(3 \pm 2\sqrt{3}, -1\right)$

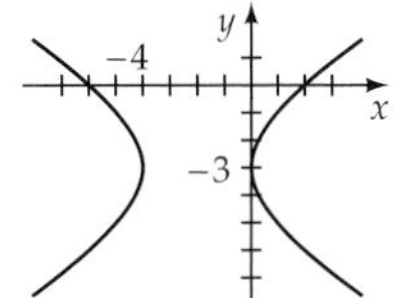

16. Vertices: (0, −3), (−4, −3)
Foci: $\left(-2 \pm \sqrt{7}, -3\right)$
Asymptotes: $(y + 3) = \pm\frac{\sqrt{3}}{2}(x + 2)$

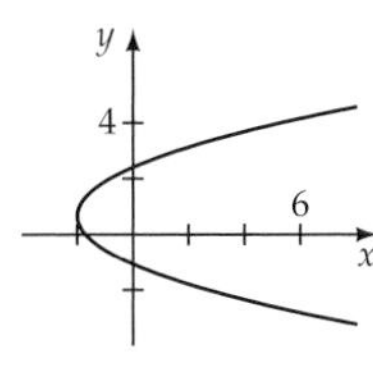

17. Vertex: (−2, 1)
Focus: $\left(-\frac{29}{16}, 1\right)$
Directrix: $x = -\frac{35}{16}$

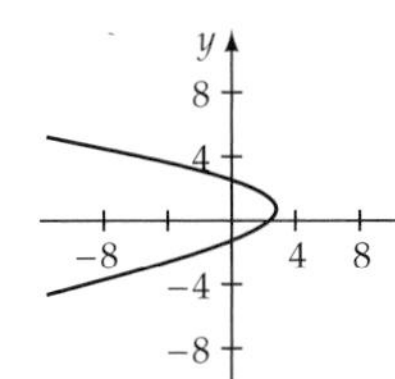

18. Vertex: (3, 1)
Focus: $\left(\frac{21}{8}, 1\right)$
Directrix: $x = \frac{27}{8}$

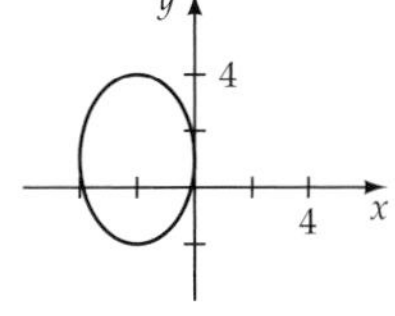

19. Center: (−2, 1)
Vertices: (−2, −2), (−2, 4)
Foci: $\left(-2, 1 \pm \sqrt{5}\right)$

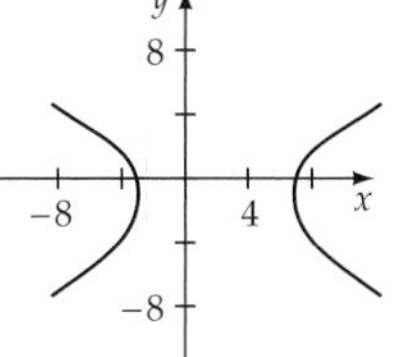

20. Center: (2, −1)
Vertices: (7, −1), (−3, −1)
Foci: (8, −1), (−4, −1)
Asymptotes: $y + 1 = \pm\frac{\sqrt{11}}{5}(x - 2)$

21. Center: $\left(1, \frac{2}{3}\right)$

Vertices: $\left(-5, \frac{2}{3}\right), \left(7, \frac{2}{3}\right)$

Foci: $\left(1 \pm 2\sqrt{13}, \frac{2}{3}\right)$

Asymptotes: $y - \frac{2}{3} = \pm 2\frac{(x-1)}{3}$

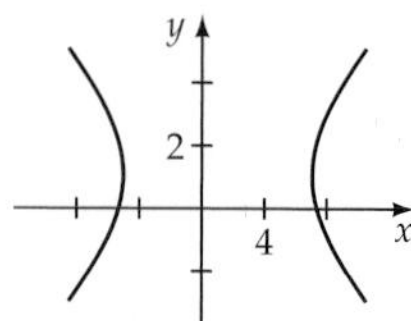

22. Center: $(2, 3)$
Vertices: $(0, 3), (4, 3)$
Foci: $\left(2 + 2\sqrt{10}, 3\right), \left(2 - 2\sqrt{10}, 3\right)$
Asymptotes: $y = 3x - 3, y = -3x + 9$

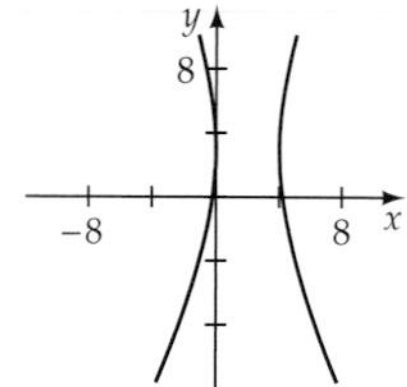

23. Vertex: $\left(-\frac{7}{2}, -1\right)$

Focus: $\left(-\frac{7}{2}, -3\right)$

Directrix: $y = 1$

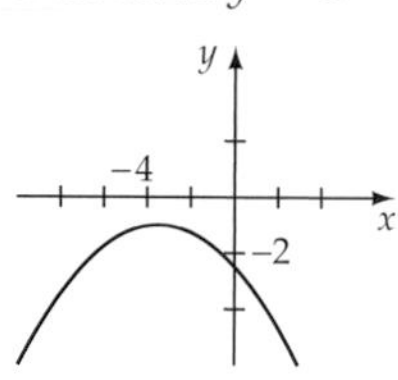

24. Vertex: $(3, 2)$

Focus: $\left(3, \frac{17}{4}\right)$

Directrix: $y = -\frac{1}{4}$

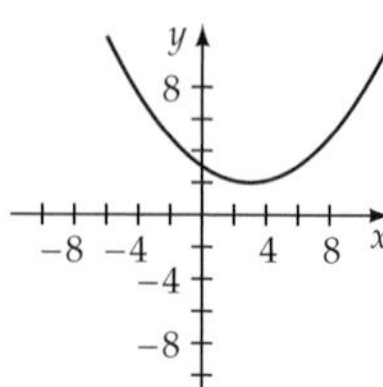

Index

Important Theorems

Pythagorean Theorem
$c^2 = a^2 + b^2$

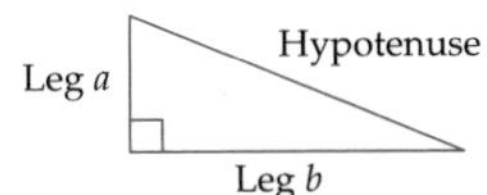

Remainder Theorem
If a polynomial function $P(x)$ is divided by $x - c$, then the remainder equals $P(c)$.

Factor Theorem
A polynomial function $P(x)$ has a factor $(x - c)$ if and only if $P(c) = 0$.

Fundamental Theorem of Algebra
If $P(x)$ is a polynomial function of degree $n \geq 1$ with complex coefficients, then $P(x)$ has at least one complex zero.

Important Formulas

Distance Formula
The *distance* between $P_1(x_1, y_1)$ and $P_2(x_2, y_2)$ is

$$d(P_1, P_2) = \sqrt{(x_1 - x_2)^2 + (y_1 - y_2)^2}$$

Midpoint Formula
The *midpoint* of the line segment from $P_1(x_1, y_1)$ to $P_2(x_2, y_2)$ is given by $\left(\frac{x_1 + x_2}{2}, \frac{y_1 + y_2}{2}\right)$.

Slope Formula
The *slope* m of a line through $P_1(x_1, y_1)$ and $P_2(x_2, y_2)$ is

$$m = \frac{y_2 - y_1}{x_2 - x_1} \quad (x_1 \neq x_2)$$

Slope-Intercept Formula
The *slope-intercept formula* for a line with slope m and y-intercept $(0, b)$ is $y = mx + b$.

Point-Slope Formula
The *point-slope formula* for a line with slope m that passes through the point $P_1(x_1, y_1)$ is $y - y_1 = m(x - x_1)$.

Quadratic Formula
If $ax^2 + bx + c = 0$, $a \neq 0$, then

$$x = \frac{-b \pm \sqrt{b^2 - 4ac}}{2a}$$

Properties of Complex Numbers

$i^2 = -1 \qquad \sqrt{-a} = i\sqrt{a}, a > 0$

$a + bi$ and $a - bi$ are complex conjugates.

$a + bi = c + di$ if and only if $a = c$ and $b = d$.

$|a + bi| = \sqrt{a^2 + b^2}$

Variation

y varies directly as x: $y = kx$

y varies inversely as x: $y = \dfrac{k}{x}$

y varies jointly as x and z: $y = kxz$

Arithmetic Sequences and Series

Common Difference $\quad a_{i+1} - a_i = d$

nth-Term Formula $\quad a_n = a_1 + (n - 1)d$

Sum of n Terms $\quad S_n = \dfrac{n}{2}(a_1 + a_n) = \dfrac{n[2a_1 + (n - 1)d]}{2}$

Geometric Sequences and Series

Common Ratio $\quad \dfrac{a_{i+1}}{a_1} = r$

nth-Term Formula $\quad a_n = a_1 r^{n-1}$

Sum of n Terms $\quad S_n = \dfrac{a_1(1 - r^n)}{1 - r} \quad (r \neq 1)$

Sum of an Infinite Series $\quad S = \dfrac{a_1}{1 - r}, |r| < 1$

Factorials, Permutations, and Combinations

In the following formulas k and n are whole numbers such that $0 \leq k \leq n$.

n Factorial $\quad n! = n \cdot (n - 1) \cdot (n - 2) \cdots 3 \cdot 2 \cdot 1$
$0! = 1 \qquad 1! = 1$

Permutations $\quad P(n, k) = \dfrac{n!}{(n - k)!}$

Combinations $\quad C(n, k) = \dfrac{n!}{k! \cdot (n - k)!}$

Binomial Expansions

$(a + b)^2 = a^2 + 2ab + b^2$

$(a + b)^3 = a^3 + 3a^2b + 3ab^2 + b^3$

$(a + b)^4 = a^4 + 4a^3b + 6a^2b^2 + 4ab^3 + b^4$

$(a + b)^5 = a^5 + 5a^4b + 10a^3b^2 + 10a^2b^3 + 5ab^4 + b^5$

Binomial Theorem

$$(a + b)^n = a^n + \binom{n}{1}a^{n-1}b + \binom{n}{2}a^{n-2}b^2 + \cdots + \binom{n}{k}a^{n-k}b^k + \cdots + b^n$$

The ith term of the expansion of $(a + b)^n$ is

$$\binom{n}{i - 1}a^{n-i+1}b^{i-1}$$

Formulas from Geometry

Formulas for Perimeter *P*, Circumference *C*, Area *A*, Surface Area *S*, and Volume *V*.

Rectangle	Square	Triangle	Circle	Parallelogram
$P = 2l + 2w$ $A = lw$	$P = 4s$ $A = s^2$	$P = a + b + c$ $A = \frac{1}{2}bh$	$C = \pi d = 2\pi r$ $A = \pi r^2$	$P = 2b + 2s$ $A = bh$
w l	s s	c h a b	r	h s b
Rectangular Solid	**Right Circular Cone**	**Sphere**	**Right Circular Cylinder**	**Frustum of a Cone**
$S = 2(wh + lw + hl)$ $V = lwh$	$S = \pi r\sqrt{r^2 + h^2} + \pi r^2$ $V = \frac{1}{3}\pi r^2 h$	$S = 4\pi r^2$ $V = \frac{4}{3}\pi r^3$	$S = 2\pi rh + 2\pi r^2$ $V = \pi r^2 h$	$S = \pi(R + r)\sqrt{h^2 + (R - r)^2} + \pi r^2 + \pi R^2$ $V = \frac{1}{3}\pi h(r^2 + rR + R^2)$
h l w	h r	r	h r	r h R

Properties of Exponents

$$x^m \cdot x^n = x^{m+n} \qquad \frac{x^m}{x^n} = x^{m-n} \qquad (x^m)^n = x^{m \cdot n}$$

$$(x^m y^n)^p = x^{m \cdot p} y^{n \cdot p} \qquad \left(\frac{x^m}{y^n}\right)^p = \frac{x^{m \cdot p}}{y^{n \cdot p}} \qquad x^{-n} = \frac{1}{x^n}$$

Properties of Radicals

$$\left(\sqrt[n]{b}\right)^m = \sqrt[n]{b^m} = b^{m/n} \qquad \sqrt[n]{a}\sqrt[n]{b} = \sqrt[n]{ab}$$

$$\frac{\sqrt[n]{a}}{\sqrt[n]{b}} = \sqrt[n]{\frac{a}{b}} \qquad \sqrt[m]{\sqrt[n]{a}} = \sqrt[mn]{a}$$

$\sqrt{x^2} = |x|$ for any real number x.

Absolute Value Equations and Inequalities

For $k > 0$:

$|E| = k$ if and only if $E = k$ or $E = -k$

$|E| \le k$ if and only if $-k \le E \le k$

$|E| \ge k$ if and only if $E \le -k$ or $E \ge k$

Properties of Logarithms

$y = \log_b x$ if and only if $b^y = x$

$$\log_b b = 1 \qquad \log_b 1 = 0 \qquad \log_b(b)^p = p$$

$$b^{\log_b x} = x \qquad \log x = \log_{10} x \qquad \ln x = \log_e x$$

$$\log_b(MN) = \log_b M + \log_b N$$

$$\log_b \frac{M}{N} = \log_b M - \log_b N$$

$$\log_b(M^p) = p \log_b M$$